ZOOLOGY

FOR
DEGREE STUDENTS – I

B.Sc. First Year

ZOOLOGY FOR DEGREE STUDENTS

B.Sc. First Year

As per UGC Model Curriculum

Unit – I : Animal Diversity–I
(Nonchordata Lower and Higher)
Unit – II : Cell Biology & Biochemistry
Unit – III : Genetics

Dr. V.K. AGARWAL
M.Sc., PhD.
Reader and Head (*Retd.*)
Department of Zoology
Meerut College, Meerut

S. Chand And Company Limited
(ISO 9001 Certified Company)

S. Chand And Company Limited

(ISO 9001 Certified Company)

Head Office: D-92, Sector–2, Noida – 201301, U.P. (India), Ph. 91-120-4682700

Registered Office: A-27, 2nd Floor, Mohan Co-operative Industrial Estate, New Delhi – 110 044, Phone: 011-49731800

www.**schandpublishing.com**; e-mail: **info@schandpublishing.com**

Marketing Offices:

Chennai : Ph: 23632120; chennai@schandpublishing.com
Guwahati : Ph: 2738811, 2735640; guwahati@schandpublishing.com
Hyderabad : Ph: 40186018; hyderabad@schandpublishing.com
Jalandhar : Ph: 4645630; jalandhar@schandpublishing.com
Kolkata : Ph: 23357458, 23353914; kolkata@schandpublishing.com
Lucknow : Ph: 4003633; lucknow@schandpublishing.com
Mumbai : Ph: 25000297; mumbai@schandpublishing.com
Patna : Ph: 2260011; patna@schandpublishing.com

First Edition 2011
Reprints 2012, 2013, 2014 (Thrice), 2015, 2018, 2019, 2020, 2021, 2022

Reprint 2023 (Twice)

ISBN: 978-81-219-3550-0 **Product Code:** H6ZOY68ZOOL10ENAA11O

PRINTED IN INDIA

By Vikas Publishing House Private Limited, Plot 20/4, Site-IV, Industrial Area Sahibabad, Ghaziabad – 201 010 and Published by S. Chand And Company Limited, A-27, 2nd Floor, Mohan Co-operative Industrial Estate, New Delhi – 110 044.

PREFACE

Author feels immense pleasure and great satisfaction in releasing this textbook. This textbook - **Zoology for Degree Students – I** contains topics of four branches of Zoology, namely, *Lower Non-chordata*, *Higher Non - chordata*, *Cell Biology* and *Genetics*. This text book is meant for Degree students of all Indian Universities. It has been developed according to UGC Model curriculum.

Author has taken a great care to present a text matter which is simple, concise, illustrative and b[illegible] on latest references. Certain keynote aspects of this textbook are as follows:

1. Carefully redrawn diagrams to explain the text.
2. Boxed information for those who always want more.
3. At the end of each chapter has been provided a wide spectrum of questions such as *Long Answer Questions*, *Short Answer Questions*, *Very Short Answer Questions*, *Fill in the Blanks*, *True and False Questions*, *Multiple Choice Questions*, etc. The answers of One Mark Questions such as VSAQs, Fill in the Blanks, True and False Questions and MCQs, have also been provided.
4. There are in all *90 chapters* in this book according to various syllabi.

Author is indebted to his wife, daughters and ever loving F_2 generation (Prafull, Muskan and Khushi) for much needed emotional warmth and time - to - time moral boosting.

Author expresses his thanks to Management and Editorial Team of S. Chand for their stipulated encouragement.

Author will like to get feedbacks from his readers regarding this endeavour.

'Neelkanth'
556, Brahmuri
Sharda Road
Meerut – 250002
(U.P.)

Author

UGC MODEL SYLLABUS

B.Sc. I Year Paper

MODEL –I

Z-101B : Animal Diversity–I

1. Principles of Classification - salient features and classification upto orders in non-chordates. Structural organization in different classes on non-chordates.
2. Protozoa - Type study and study of locomotion, osmoregulation, nutrition and reproduction in protozoa.
3. Origin of metazoa - metamerism & symmetry.
4. Porifera and coelenterata - Type study, corals and coral reefs, polymorphism in hydrozoa.
5. Ctenophora - affinities. Type study.
6. Platyhelminthes and Nemathelminthes - Type study, reproduction and parasitic adaptations.
7. Annelida - Coelome and excretory system, Type study.
8. Mollusca - Type study, torsion & detortion in gastropoda.
9. Onychopora - Type study and affinities.
10. Arthropoda - Crustacean larval forms, vision in Arthropodes, metamorphosis in insects, Type study, social life in insects, Trilobites.

Z-102B : Cell Biology

1. Diversity of cell size and shape
2. Cell theory
3. Structure of prokaryotic and eukaryotic cells
4. Isolation and growth of cells
5. Microscopic techniques for study of cells
6. Cellular organelles-separation and characterization
7. Cellular energy transactions - role of mitochondria and chloroplasts
8. Membrane transport of small molecules and the ionic basis of membrane excitability
9. Intracellular compartments and protein sorting
10. Vesicular traffic in the secretory and endocytic pathways
11. Cell signalling
12. Cytoskeleton
13. Cell-division cycle
14. The mechanics of cell division
15. Cell junctions, cell adhesion, and the extracellular matrix
16. Cell differentiation
17. Cell in development
18. Differentiated cells and the maintenance of tissues
19. Biology of cancer

MODEL –II

Z-101A : General Zoology

1. Principles of classification
2. Classification of animals, non-chrodates: Salient Geatures and classification upto classes.
3. Hemi-chordates & lower non chrodates: Salient features and classification upto sub-classes with examples for adaptive features.
4. Higher non-chordates-Salient features and clasification of extent groups upto orders with examples for adaptive features.
5. Diversity in animal from and function, like:
   i. Support and Movement
   ii. Nutrition
   iii. Gas exchange & transport
   iv. Excretory organs
   v. Sensory system
   vi. Reproductive patterns.

Z-102A : Cell Biology and Biochemistry

1. The cell
2. Methods in Cell Biology
3. Organization of Cell - extranuclear and nuclear
4. Cell Reproduction
5. An elementary idea of cell transformation and cancer
6. An elementary idea of cellular basis of immunity
7. Chemistry of living system
8. Properties of water as a biological solvent
9. Biomolecular organization
10. Chemical bonds and bond energy
11. Aminoacids, peptides and proteins
12. Enzymes
13. Biomolecules - lipids, carbohydrates, nucleic acids
14. Concepts of genetic engineering.

GENETICS

(As Per Syllabus of Several Indian Universities)

Unit–I

1. Mendelian inheritance patterns and laws of heredity
2. Co-dominance and incomplete dominance
3. Linkage & linkage maps
4. Varieties of gene expressions – multiple alleles, lethal genes, pleiotropic genes, gene interactions, epistasis.
5. Sex-chromosome systems and sex-linkage.

Unit–II

1. Non-chromosomal inheritance (Cytoplasmic inheritance)
2. Mutations and chromosomal alterations, meiotic consequences.
3. Human genetics – Chromosomal and single gene disorders; genetic counselling.

CONTENTS

UNIT – I
ANIMAL DIVERSITY –I (Nonchordata : Lower & Higher)

(PART–A)
LOWER NONCHORDATES (Invertebrates)

(PART–B)

HIGHER NONCHORDATA

UNIT – II
CELL BIOLOGY & BIOCHEMISTRY

UNIT – III
GENETICS

UNIT - I

ANIMAL DIVERSITY –I

(Nonchordata : Lower & Higher)

(PART–A)

LOWER NONCHORDATES

(Invertebrates)

(PART–A)

LOWER NONCHORDATES (Invertebrates)

1

Introduction

A living organism is primarily physico-chemical material that shows a high degree of complexity, is capable of self-regulation, possesses a metabolism and perpetuates itself through time. Many biologists believe that life is an arbitrary stage in the growing complexity of matter, with no sharp dividing line between the living and non-living worlds.

Living substance is composed of a highly structured array of macromolecules, such as proteins, lipids, nucleic acids and polysaccharides, as well as, smaller organic and inorganic molecules. A living organism has built-in regulatory mechanisms and interacts with the environment to sustain its structural and functional integrity. All reactions occurring within an individual living unit are called **metabolism**. Specific molecules (DNA or RNA) containing information in their structure are utilized both in the regulation of internal reactions and in the production of new living units.

Biology (Gr., *bios* = life + *logos* = study) is the study of life. It ranges from the study of chemicals in microscopic cells to the study of communities of organisms interacting with their physical environment. The biologist explains that how organisms and their parts look, work and behave. But this is not all. The biologist seeks to explain the mechanisms by which the present day world developed and attempts to predict where it is headed—in other words, the historical element, the evolution of the system, must be understood, for it sets the limits for life itself and shows us glimpses of our future.

Some of the important branches of biology are the following:

1. Zoology (Gr., *zoon* = animal + *logos* = study) which deals with the study of animals.

2. Botany (Gr., *botane* = herb) which deals with the study of plants.

3. Microbiology which deals with the study of micro-organisms such as viruses, bacteria, cyanobacteria (blue green algae), archaebacteria, protozoa, diatoms, unicellular algae and yeasts and other microscopic fungi.

1.1 NATURAL AND UNNATURAL GROUPS

One of the first evolutionary trees of life conceived from Darwinian (genealogical) perspective was published by **Ernst Haeckel** in 1866. Haeckel's famous **tree of life** began a tradition of depicting phylogenetic hypotheses as branching diagrams, or trees, a tradition that has persisted since that time. Since Haeckel's day, many names have been coined for the larger branches that sprout from these trees. Some of these names refer to groups of organisms that are probably **natural phylogenetic groups** (*i.e.*, groups that include an ancestor and all of its descendants), such as **metazoa** (the Animal Kingdom). Other names refer to **unnatural**, or **composite, groupings of organisms**, such as **"microbes"** (*i.e.*, any organism that is microscopic in size, such as bacteria, most protists and unicellular fungi) and **"protozoa"** (a loose assemblage of primarily unicellular heterotrophic eukaryotes).

The discovery that organisms with a cell nucleus constitute a natural group divided the living world neatly into two categories, the **prokaryotes** (those organisms lacking membrane-enclosed organelles and a nucleus, and without linear chromosomes) and the **eukaryotes** (those organisms that do possess membrane-bound organelles and a nucleus and linear chromosomes). Investigations by **Carl Woese** and others, beginning in the 1970s, led to the discovery that the prokaryotes actually comprise two distinct groups, called **Eubacteria** and **Archaea** (= Archaebacteria), both quite distinct from eukaryotes (Table 1.1).

Table 1.1 The six kingdoms of life (Source. Brusca and Brusca, 2003).

A. The Prokaryotes (the "domains" Eubacteria and Archaea)*

1. **Kingdom Eubacteria (Bacteria).** The "true" bacteria, including cyanobacteria (or blue-green algae) and spirochetes. Never with membrane-enclosed organelles or nuclei, or a cytoskeleton; none are methanogens; some use chlorophyll-based photosynthesis with peptidoglycan in cell wall; with a single known RNA polymerase.
2. **Kingdom Archaea (Archaebacteria).** Anaerobic or aerobic, largely methane-producing micro-organisms. Never with membrane enclosed organelles or nuclei, or a cytoskeleton; none use chlorophyll-based photosynthesis, without peptidoglycan in cell wall; with several RNA polymerases.

B. The Eukaryotes (the "domain" Eukaryotes, or Eukarya). Each with a variety of membrane-enclosed organelles (*e.g.*, mitochondria, lysosomes, peroxisomes) and a membrane-enclosed nucleus. Cells gain structural support from an internal network of fibrous proteins called a **cytoskeleton.**

1. **Kingdom Fungi.** The Fungi probably monophyletic group** that includes molds, mushrooms, yeasts and others. Saprobic, heterotrophic multicellular organisms. The 72,000 described species are thought to represent only 5 – 10 percent of the actual diversity.
2. **Kingdom Plantae (= Metaphyta).** The multicellular plants. Photosynthetic autotrophic, multicellular organisms that develop through embryonic tissue layering. Includes some groups of algae, the bryophytes and their kin, and the vascular plants (about 240,000 of which are thought to represent about half of Earth's actual plant diversity).
3. **Kingdom Protista.** Eukaryotic single-celled micro-organisms and certain algae. A polyphyletic grouping** of perhaps 18 phyla, including euglenids, green algae, diatoms and some other brown algae, ciliates, dinoflagellates, foraminiferans, amoebae and others. The 80,000 described species probably represent about 10 percent of the actual protist diversity on earth.
4. **Kingdom Animalia.** The multicellular animals. A monophyletic taxon, containing 34 phyla of ingestive, heterotrophic, multicellular organisms. About 1.3 million living species have been described, estimates of the number of undescribed species range from lows 10 – 30 million to high of 100 – 200 million.

* Portion of old "Kingdom Monera" are now included in the Eubacteria and the Archaea. Viruses (about 5000 described "species") and subviral organisms (Viroids, and Prions) are not included in this classification.

** See Box 1.1.

Box 1.1

1. **Monophyletic group** is a group of species that includes an ancestral species and all of its descendants-that is, a **natural group.**
2. **Paraphyletic group** is a group whose member species are descendants of a common ancestor, but that does not contain all the species descended from that ancestor.
3. **Polyphyletic group** is a group comprising species that arose from two or more different immediate ancestors. Such composite taxa have been established primarily because of insufficient knowledge concerning the species in question, *e.g.*, protozoa, radiata, etc.

It has been recently been suggested that the three main divisions of life (Eubacteria, Archaea and Eukaryota) should be recognized at a new taxonomic level, called **domains**. However, fundamental questions remain about these three "domains", including how many natural groups (kingdoms) exist in each domain, whether the domains themselves represent natural (= monophyletic) groups, and what the phylogenetic relationships are among these domains and the kingdoms, they contain. Current evidence suggests that Eukaryotes are a natural group, defined by the unique trait of a nucleus and linear chromosomes, whereas Eubacteria and Archaea may not be natural groups.

Courses and textbooks on Invertebrate Zoology often include discussion of two eukaryotic kingdoms, the **Anemalia** (= Metazoa) and certain "animal-like" (*i.e.*, heterotrophic) protist phyla loosely referred to as "**Protozoa**". Following this tradition, we intend to treat 13 out of 33 phyla of metazoa and 18 phyla of protists (many of which have traditionally been viewed as "protozoa") in this text. The vast majority of kinds (species) of living organisms that have been described are animals. The **kingdom Animalia** or **Metazoa**, is usually defined as the multicellular, integestive, heterotrophic (*Heterotrophic organisms* are those that consume other organisms or organic material as food) eukaryotes. However, its members possess other unique features as well, such as an acetylcholine-cholinsterase-based nervous system, special types of cell-cell junctions, and a unique family of connective tissue protein called **collagens**. Among the metazoa are some species that possess a backbone (or vertebral column), but most do not. Those that possess a backbone, constitute the subphylum vertebrata of the phylum **Chordata** (= animals with notochord). Those that do not possess a backbone (the remainder of the phylum chordata, plus 33 additional animal phyla) constitute **invertebrates** (see Table 1.2). Thus we can see that the division of animals (= Metazoa) into invertebrates and vertebrates is based more on tradition and convenience, reflecting a dichotomy of zoologists' interests, than it is on the recognition of natural biological groupings (see **Brusca** and **Brusca**, 2003).

Table 1.2 Approximate numbers of known extant species in various groups (Source. Brusca and Brusca 2003).

Of the 1,335,188 estimated described species of Animalia or Metazoa (excluding protists), 1,228,518 (96 per cent) are invertebrates.

A. Kingdom Protista (80,000)

B. Kingdom Animalia (Metazoa)

(*a*) *Invertebrata*

1. Porifera (5,500)	2. Cnidaria (10,000)
3. Ctenophora (100)	4. Placozoa (1)
5. Monoblastozoa (1)	6. Rhombozoa (70)
7. Orthonectids (20)	8. Platyhelminthes (20,000)
9. Nemertea (900)	10. Gnathostomulida (80)
11. Rotifera (1,800)	12. Gastrotricha (450)
13. Kinorhyncha (150)	14. Nematoda (25,000)
15. Nematomorpha (230)	16. Priapula (16)
17. Acanthocephala (700)	18. Cycliophora (1)
19. Entoprocta (150)	20. Loricifera (10)
21. Annelida (16,500)	22. Echiura (135)
23. Sipuncula (320)	24. Tardigrada (600)
25. Onychophora (110)	26. Arthropoda (1,097,631)
27. Mollusca (93,195)	28. Brachiopoda (335)
29. Ectoprocta (4,500)	30. Phornida (20)
31. Chaetognatha (100)	32. Echinodermata (7000)
33. Hemichordata (85)	

(*b*) *Chordata* (49,693)

1. Urochordata (3,000)	2. Cephalochordata (23)
3. Vertebrata (46,670)	

Major and Minor Invertebrate Phyla

The invertebrate animals have been classified into about 33 phyla which are often categorized as **major invertebrate phyla** and **minor invertebrate phyla**. Such a classification is based on two criteria:

(1) **number of animals and species present in** a phylum and (2) **participation in biotic community**, *i.e.*, representation of the phyla in great majority of biotic communities.

Major Invertebrate Phyla

These include the following ten phyla: (numerical strength of each phylum has been indicated against it (Table 1.2):

1. Porifera; 2. Cnidaria; 3. Platyhelminthes; 4. Rotifera; 5. Nematoda; 6. Annelida; 7. Arthropoda; 8. Mollusca; 9. Ectoprocta; 10. Echinodermata.

Minor Invertebrate Phyla

These are 22 in number (Table 1.2). These are 1. Ctenophora; 2. Placozoa; 3. Monoblastozoa; 4. Rhombozoa; 5. Orthonectids; 6. Nemertea; 7. Gnathostomulida; 8. Gastrotricha; 9. Kinorhyncha; 10. Nematomorpha; 11. Priapula; 12. Acanthocephala; 13. Cycliophora; 14. Entoprocta 15. Loricifera; 16. Echiura; 17. Sipuncula; 18. Tardigrada; 19. Onychophora; 20. Brachiopoda; 21. Phornida; 22. Chaetognatha.

On the basis of community-participation, two phyla—the Rotifera and the Ectoprocta are not fit to be considered as the major invertebrate phyla, because they are limited in the participation in animal communities. Thus, by substracting these two phyla from the category of 10 major phyla, the total number of major phyla falls to only 8.

Differences in Invertebrates and Vertebrates

In vertebrates, a characteristic structure the **notochord**, develops during their embryonic period and later on the same is replaced by a series of **vertebrae**, forming the vertebral column. In the invertebrate animals since the notochord is totally absent, they are also known as **non-chordates.**

The invertebrate animals differ from the vertebrate animals in many respects (see Table 1.3)

Table 1.3 Differences between invertebrates and vertebrates.

Invertebrates	*Vertebrates*
1. Body symmetry may be asymmetrical, radial or bilateral.	1. Body is bilaterally symmetrical.
2. Appendages if present are more than two pairs.	2. There are never more than two pairs of appendages.
3. May have non-living exoskeleton which is formed as a secretion of the ectoderm. Endoskeleton absent.	3. There is always an endoskeleton which is made of living tissues such as cartilage and bones.
4. Skin consists of only one layer.	4. Skin consists of two distinct layers, an outer epidermis overlying a thick dermis.
5. Alimentary canal is present or absent.	5. Alimentary canal is always present.
6. True tail is absent. Anus is terminal and opens at the posterior end of the body.	6. A true tail, which may be defined as the post-anal continuation of the body axis, may be present.
7. The breathing or respiratory organs normally develop from the ectoderm. No gill-slits are present in pharynx	7. The breathing or respiratory organs develop in relation to the wall of pharynx (an endodermal organ) and consist of gills or lungs. Pharyngeal gill-slits are present in the embryos of all vertebrates.
8. The circulatory system is mostly open type.	8. The circulatory system is closed type containing arteries, veins and capallaries.
9. The heart is always dorsal in position.	9. The heart is always ventral in position.

Invertebrates	*Vertebrates*
10. The respiratory pigment such as haemoglobin, if present, is dissolved in the plasma.	10. The haemoglobin is contained in the red blood corpuscles.
11. The central nervous system consists of solid ganglia connected by nerve fibres to form a chain.	11. The central nervous system is in the form of hollow thick-walled tube.
12. The nerve cord lies ventrally to the digestive tract.	12. The neural tube (brain + spinal cord) lies dorsally to the digestive tract.
13. Eyes develop from the skin.	13. Eyes develop as outgrowths of the brain.
14. Reproduction, sexual, asexual or both.	14. Reproduction is only sexual.

1.2 ANIMAL HABITATS

Habitat is a place where a community of organisms is found. Rocky seashores, the surface of soil and human skin are all habitats for particular communities of living organisms (**Indge**, 1997). Animals are found in all habitats. Many animals live in water. They are called **aquatic animals**. Other animals live on land. They are known as **terrestrial animals**.

1. Aquatic animals. Some aquatic animals live in the sea. They are known as **marine animals**. Other animals inhabit pools, ponds, lakes, streams and rivers. They are called **freshwater animals**. Many aquatic animals are adapted to live in **estuaries**, the regions where rivers enter the sea. The animals are classified ecologically into three categories: plankton, nekton and benthos.

(*i*) Plankton. These are small organisms that float and drifted passively in surface waters (pelagic zone; Fig. 1.1) by waves, currents and wind. They have very weak or no locomotory organs. They are confined to the photic (well lighted) zone only. Both plants (**phytoplankton**) and animals (**zooplankton**) are included among the plankton. The zooplankton includes jellyfishes, comb jellies, arrow worms, many small crustaceans, larvae and eggs.

(*ii*) Nekton. These are actively swimming animals. They have well-developed locomotory organs. They occur in surface (pelagic zones) as well as deep (benthic zone) waters and in both photic and aphotic zones. They include fishes and squids, cuttlefishes, etc.

(*iii*) Benthos. These are the bottom dwelling animals that crawl over, burrow into or attached to the bottom. Benthic animals such as most sea anemones, sponges, many snails and barnacles may live on surface of substratum, they are called **epifauna** or **epibenthic** forms. Clams, worms and sea cucumbers burrow within soft substrata. These burrowing benthos form **infauna**.

Benthic animals may also be categorized by their locomotory capabilities. Animals that are generally quite motile and are active are described as being **errant** (*e.g.*, crabs, many worms), whereas those that are firmly attached to the substratum are **sessile** (*e.g.*, sponges, corals, barnacles, etc.). Others are unattached or weakly attached, but generally do not move around much (*e.g.*, crinoids, solitary anemones, most clams); these animals are said to be **sedentary**.

2. Terrestrial animals. These animals include arthropods (isopods, insects, spiders, mites, scorpions) and some animals which are restricted to relatively moist areas, *e.g.*, snails, nematoda, and earthworms. Some terrestrial animals live underground and are called **fossorial animals**. These include earthworms, termites and other ants. Others climb walls and rocks and are termed **scansorial animals**. Some animals live on trees and are known as **arboreal animals**. Certain terrestrial animals can take to flight. These are called **volant, aerial** or **flying** animals, *e.g.*, insects. The land animals that run fast are called **cursorial animals**, *e.g.*, cockroaches. The terrestrial forms may occur in forests, grasslands, deserts and tundras.

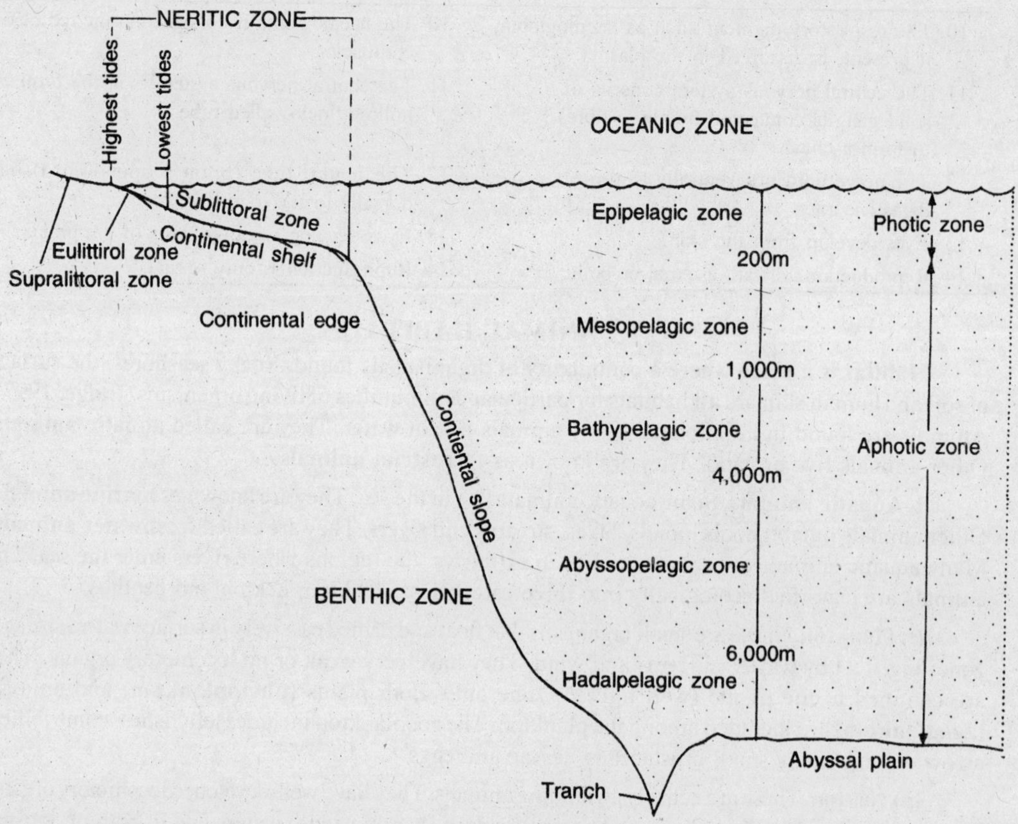

Fig. 1.1 A schematic cross section of the major habitat regions of the ocean (after Brusca and Brusca, 2003).

Box 1.2

Terrestrial animals have to pay prize for leaving the water habitat

Water provides a medium for support, for dispersing gametes, larvae and adults, and for diluting waste products, and is a source of dissolved materials needed by animals. Animals living in terrestrial environment do not enjoy these benefits of water, and must pay the prize (see **Brusca** and **Brusca**, 2003).

Mode of Existencc

Majority of animals live singly and are called **solitary animals**. Some live in groups but without helping one another, *e.g.*, sea squirts. They are known as **gregarious animals**. Certain animals live in groups and help each other. They are called **colonial animals**. They include termites, ants, bees and wasps. Many animals lead an indcpendent life and are termed **free-living animals**. Some, however, live in association with other species. An association of two different species is called **symbiosis** (Gr. *syn* = together; *bios* = life).

A Special Type of Environment: Symbiosis

As indicated earlier, many invertebrates live in an intimate association with other animals or plants. This kind of association is termed as symbiotic relationship, or simply **symbiosis**. Symbiosis was first defined in 1879 by German mycologist **H.A. DeBary** as "*unlike organisms living together*". In most symbiotic relationships, a larger organism (called the **host**) provides an environment (its body, burrow, nests, etc.) on or within which a smaller organism (the **symbiont**) lives. Some symbionts are opportunistic (**facultative**), whereas others cannot survive without their host (**obligatory**).

Symbiotic relationships can be sub-divided into following categories based on the nature of interaction between the symbiont and its host: parasitism, mutualism and commensalism.

1. Parasitism. In this form of symbiosis, the symbiont (a parasite) receives benefits at the expense of the host. Parasite may be external (**ectoparasite**) such as lice, ticks and leeches; or internal (**endoparasites**), such as liver flukes, some roundworms and tapeworms. Some parasites (*e.g.*, lice, tapeworm) live their entire adult life in association with their hosts are **permanent parasites**, whereas **temporary**, or **intermittent parasites**, such as bedbugs only feed on the host and then leave it. A **definitive host** is one in which the parasite reaches reproductive maturity. An **intermediate host** is one that is required for parasite's development, but in which the parasite does not reach reproductive maturity.

2. Mutualism. In this type of symbiosis both host and symbiont benefit. Such relationship may be extremely intimate and important for the survival of both parties; for example, the bacteria in our own large intestine are important in the production of certain vitamins and in processing material in gut. Another example is the relationship between termites and certain protists (*e.g.*, *Trichonympha*) that inhabit their digestive tracts and are responsible for the breakdown of cellulose into compounds that can be assimilated by their insect hosts (see *Animal Behaviour,* **Agarwal, V.K.,** 2009). Other mutualistic relationships may be less binding on the organism involved. Cleaner shrimps, for example, inhabit coral reef environments where they establish "cleaning stations", that are visited regularly by reef-dwelling fishes that present themselves to the shrimps for the removal of parasites. Obviously, even this rather loose association results in benefits for the shrimps (a meal) as well as for the fishes (removal of parasites). The mutualistic relationships between plants and their pollinators are essential to the survival of most flowering plants and their insect partners (and in some cases, their bird or nectar-feeding bat partners).

3. Commensalism. This type of symbiosis is usually described as an association that is advantageous to one party (the symbiont) but leaves the other (the host) unaffected. For instance, among invertebrates there are numerous examples of one species inhabiting the tubes or burrow of another (**inquilism**); the former ('symbiont–I') obtains protection, food; or both with little or no apparent effect on the latter ('symbiont–II'). A special type of commensalism is **phoresis**, where in the two symbionts "travel together", but there is no physiological or biochemical dependency on the part of either participant. Usually one **phoront** is smaller than the other and is mechanically carried about by its larger companion.

Evolutionary Role of Symbiosis—The Symbiogenesis

The Russian biologist **Konstantin Mereschkovsky** (1855 – 1921) developed the **"two plasm"** (cell within a cell) **theory**, claiming that chloroplasts originated from blue-green algae (cyanobacteria). For this process, he invented the term **symbiogenesis** (which is a macroevolutionary event). According to **serial endosymbiotic theory** (**SET**) of symbiogenesis, three classes of organelles of eukaryotic cells, namely mitochondria, cilia/flagella and plastids, are hypothesized to have once been free-living prokaryotes that were acquired endosymbiotically and in certain sequence by another (host) prokaryote (Fig. 1.2). Although symbiogenesis is an old idea, it was **Lynn Margulis** (1989) who most vigorously championed it in the twentieth century.

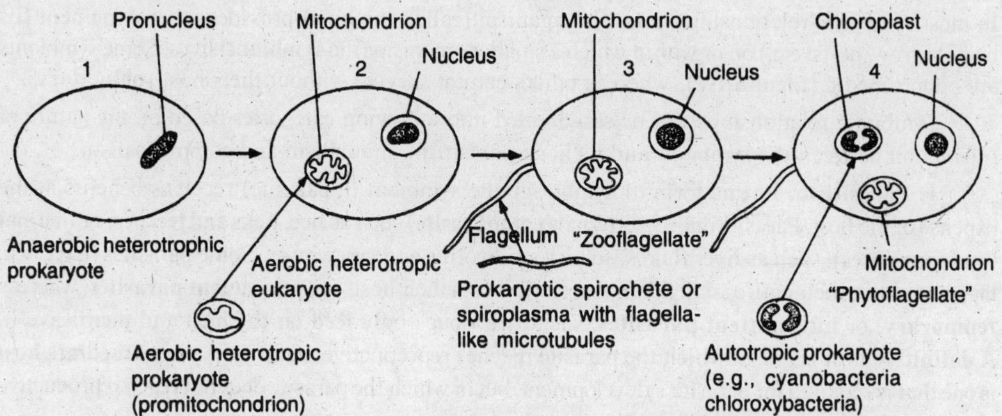

Fig. 1.2. A simple model of the origin of eukaryotic cells by symbiosis (the serial endosymbiotic theory). The three major events depicted are: acquisition of an aerobic heterotrophic prokaryote (origin of mitochondrion); acquisition of a spirochaete or spiroplasma-like prokaryote (origin of flagellum) and acquisition of an autotrophic prokaryote (origin of chloroplast) (after Brusca and Brusca, 2003).

Beyond the origin of the eukaryotic cell, symbiogenesis may be at work in many other systems. Many invertebrates are involved in such relationships, including corals and other cnidarians that serve as hosts for symbiotic dinoflagellates (called **zooxanthallae**) that live within their tissues. Various animals that harbor (and exploit) tetrodotoxin-secreting bacteria (many chaetognaths, the blue-ringed octopus, a sea star and horseshoe crab, and certain tetraodontid fishes; Box 1.3), and squids with luminous bacteria, and lichens (an intimate association between fungi and cyanobacteria or green algae) are other examples.

Box 1.3

Tetrodotoxin is a neurotoxin $C_{11}H_{17}N_3O_8$ that is found in pufferfishes and that blocks nerve conduction by suppressing permeability of nerve fiber to sodium ions.

1.3 CONCEPT OF BAUPLAN AND RELATED CONCEPTS

Stability of organismal morphology is a deep-seated notion that dates at least from the eighteenth century; the idea of limited number of plants, types or archetypes of animal forms is thus an old one. **Richard Owen** introduced the term **archetype** in 1848 to represent a model organism, or the sum of features shared by a group of related organisms. The concept of the embryological archetype, and the fact that adults are nothing more than the accumulation of features added during their development, was formalized by **Karl von Baer** and **Ernst Haeckel** in the second half of the nineteenth century. Today this concept has grown into the notion of **conserved body plans** or **bauplane** (or **Bauplan**). This concept speaks to *a stability in form that maintains itself through evolutionary time and phylogenetic divergence*.

The term **bauplan** (German for "ground plan" or "blueprint") was introduced as a technical term in Zoology in 1945 by the embryologist-turned-philosopher **Joseph Henry Woodger**. More recently **Niles Eldredge** (1989) discussed the bauplan as the common structural plan within a monophyletic taxon; **Valentine** (1986) distinguished bauplan as assemblages of homologous architectural and structural features distinguished phyla and classes; and **Gould** (1977, 1980, 1992) spoke of structural constraints leading to fundamental ground plan of anatomy. In his review of the reunion of developmental and evolutionary biology, **Atkinson** (1992) claimed that "*the single most critical concept of the reunions is that of the bauplan*".

The concept of the bauplan expresses both a notion of morphological stability and the fact that some aspects of embryonic and/or adult morphology are more free to vary than are others. That is, some stages of development are more constrained than others. The most striking evidence of developmental constraints is simple fact that despite the great variety of animals, there are relatively few basic types of animal body plans (**Brusca** and **Brusca**, 2003).

Development canalization. It is a form of constraint that channels ontogeny into restricted sets of pathways that lead to a standard phenotype in spite to genetic or environmental disturbances. The preservation of Hox gene function across phyla is a good example of developmental canalization (Box 1.4).

Box 1.4 Homeobox (Hox) Genes

Homeobox or Hox genes have played a macroevolutionary role in the origin of invertebrates. Hox genes are master regulatory genes and doing so, "select" the developmental pathways that are followed by dividing cells. Hox genes have two functions: 1. they encode short regulatory proteins that bind to particular sequences of bases of DNA and either enhance or repress gene expression, and 2. They encode proteins that are expressed in complex patterns that determine the basic geometry of the organisms. The term **Hox genes** refers specifically to those genes that *are clustered in an array on the chromosome and function primarily in establishing regional or segmental identities*. In all metazoa that have been examined, regional or segmental specialization is controlled by the spatially localized expression of these genes, which play crucial roles in determining body patterns. They underlie such fundamental attributes as anterior posterior differentiation (in both invertebrates and vertebrates) and the positioning of body wall.

Hox genes have been conserved to a remarkable degree throughout the animal kingdom, and they are now known from all animal phyla that have been examined. There is a striking corelation between the order of these genes on their chromosomes and the position of their expression in the developing animal along the main body axis. Hox proteins regulate the genes that control the cellular processes involved in morphogenesis (*i.e.*, development of the architectural features of organisms; formation and differentiation of tissues and organs). In doing so, they demarcate relative positions in animals—they do not specify the precise nature of particular structure. For example, in arthropods, Hox genes regulate where body appendages form, and they can either suppress limb development or modify it (in concert with other regulatory genes) to create unique appendage morphologies. Mutations in Hox genes, and other developmental genes, can create gross mutations (homeotic mutations or homeosis).

The characteristics of an organism's bauplan are not the same thing as its phylogenetically unique features. Instead, bauplans must be viewed as nested sets of conserved body plans, as would be predicted within an ancestor – descendant hiearchial system such as animal phylogenesis. For example, snakes possess a bauplan that differs from the bauplan of lizards, turtles or crocodiles — yet each share the reptilian bauplan. Reptiles, birds and mammals each have individual bauplan but share the vertebrate bauplan. Thus, bauplans consist of a mix of ancestral and derived characters. To understand their origin requires a knowledge of adult, larval and embryonic phases of the life cycle.

Woodger clearly asserted, as had **von Baer**, that the most basic structures defining bauplan develop early in embryonic life. Consequently, deviations early in development would have much more drastic consequences for morphology than deviations later in development. Mechanisms that establish bauplans buffer development against environmental and genetic disorders. They constrain development. **Ernst Mayr** repeatedly drew attention to the importance of such constraints, specifically in relation to bauplans, conserved morphological features and the taxonomic features used in classification.

Due to field of molecular evolutionary developmental biology, we now know that much of an animal's initial embryogeny is under maternal cytoplasmic control rather genomic control. However, at some point early in embryogenesis (*i.e.*, the origin and development of the embryo), the zygote's parental (nuclear) genome takes primary control of development. Recent work suggests that this may be one of several **pivotal points** in the control of animal ontogeny—*occasions that demarcate the fixation of bauplan*. The point at which the zygotic genome takes over control of embryogenesis has been referred to as **zootypic stage**. It may be here that *Hox* genes establish the most basic, or primary, animal body patterning (*e.g.*, anterior-posterior axis and dorsal-ventral surface).

At a later stage of embryogenesis another critical point is reached, which has been called the **phylotypic (phyletic) stage**. The *phylotype* theoretically represents the stage when the gene responsible for secondary patterning of a body plan are first fully expressed and the adult *morphogenetic fields* are positioned. This juncture is not well understood. **Anderson** (1973) identified the blastula as the phylotypic stage for the annelids and arthropods, whereas **Sander** (1976, 1983) identified the germ band stage (a 20-segmented larval stage with head, thorax and abdomen already outlined and segmented) as the phylotypic stage of insects. **Cohen** (1977, 1979), on the other hand, distinguished **phylotypic larvae** (the trochophore of annelids, for instance) from **adaptive larvae**. Phylotypic stage larvae have a simple morphology determined more by developmental (genetic) programs than by physiological requirements.

The phylotypic stage is usually thought of as the stage at which embryos within a phylum shows the greatest level of morphological similarity. Beyond this stage, the zygotic genome begins moving embryos down the individual tracks of the various lineages. In other words, early developmental stages of closely related taxa converge on a phylotype in the course of their ontogeny, only to diverge again as the adult form unfolds.

Thus, it seems likely that there are several fundamental levels of body patterning during ontogeny, and closely related taxa share critical junctures of this process in ways that recall Haeckel's "law of recapitulation".

It is already becoming clear that such developmental stages (*e.g.*, zootypic stage and the phototypic stage) canalize ontogenetic events to produce the adult bauplan. This canalization comes about in part, due to following constraints that work to maintain overall body plan:

1. **Structural constraints** (*e.g.*, constraints imposed by the limitations of patterns in early developmental stages).

2. **Genetic constraints** (*e.g.*, rate of mutation and recombination of individual alleles).

3. **Direct developmental constraints** (*e.g.*, obligatory tissue interactions).

4. **Cellular constraints** (*e.g.*, limits to rate and number of cell divisions, secretions of cell products, cell migration).

5. **Metabolic constraints** (*e.g.*, dependence on particular metabolic pathways).

6. **Functional constraints** (*e.g.*, the interconnectedness of parts of different organ systems involved in critical functions).

1.4 BODY SYMMETRY

A fundamental aspect of an animal's bauplan is its overall shape or geometry. In order to discuss invertebrate architecture and function, we must first familiarise ourselves with a basic aspect of body form: symmetry.

Symmetry. It refers to the regular arrangement of body structures relative to axis of the body. Animals that can be bisected or split along at least one plane, so that the resulting halves are similar to one another, are said to be **symmetrical**. For example, a shrimp can be bisected vertically through its midline, head to tail, to produce right and left halves that are mirror images of one another.

A few animals have no body axis and no plane of symmetry. They are said to be **asymmetrical**. Many sponges, for example, have irregular growth form and lack any clear plane of symmetry. Similarly, many protists, particularly the amoeboid forms, are asymmetrical.

Spherical symmetry. It is seen in creatures whose bodies lack an axis and have the form of a sphere, with the body parts arranged concentrically around, or radiating from, a central point (Fig. 1.3). A sphere has an infinite number of planes of symmetry that pass through its center to divide it into like halves. Spherical symmetry is rare in nature; in the strictest sense, it is found in certain protists. Organisms with spherical symmetry share an important functional attribute with asymmetrical organisms, in that both groups lack **polarity**. That is, there exists no clear differentiation along an axis. In all other forms of symmetry, some level of polarity has been achieved and with polarity comes specialization of body regions and structures.

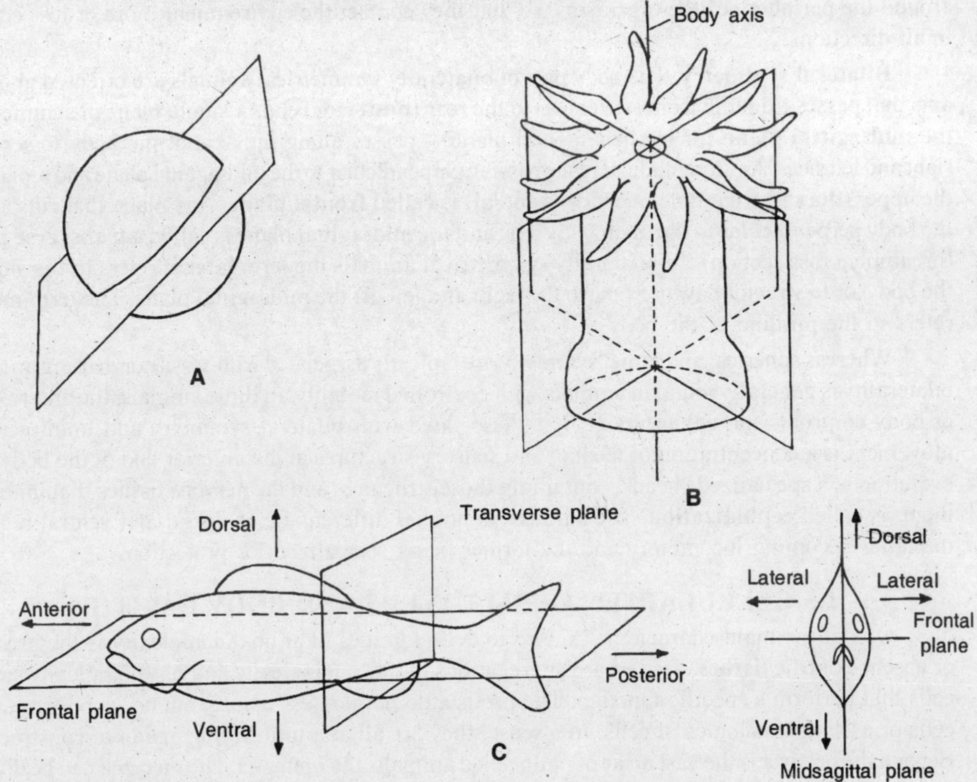

Fig. 1.3. A—Spherical symmetry; B—Radial symmetry; C—Bilateral symmetry.

Radial symmetry. A body displaying this type of symmetry has the general form of a cylinder, with one main axis around which the various parts are arranged (Fig. 1.3). In a body displaying perfect radial symmetry, the body parts are arranged equally around the axis, and any plane of sectioning that passes along that axis results is similar halves (rather like a cake being divided and sub-divided into equal halves and quarters). Nearly perfect radial symmetry occurs in some sponges (*Xetospongia*) and in many cnidarian polyps. Perfect radial symmetry is relatively rare, however, and most radially symmetrical animals have evolved modifications on this theme. **Biradial symmetry**, for example, occurs where portions of the body are specialized and only two planes of sectioning can divide the animal into perfectly similar halves. Common examples of biradial organisms are ctenophores and many sea anemones (*Epiactus*).

Further specializations of the basic radial body plan can produce nearly any combination of multiradiality. For example, many jelly fishes (*i.e.*, hydromedusa of *Scrippsia*) possess **quadriradial symmetry**. Most echinoderms (*e.g.*, sea star *Patiria* and sea bisquit *Clypeaster*) are said to display **pentaradial symmetry**.

A radially symmetrical animal has no front or back end; rather it is organized about an axis that passes through the center of its body, like an axle through a wheel. When a gut is present, this axis passed through the mouth-bearing (**oral**) surface to the opposite (**aboral**) surface. Radial symmetry is most common in sessile and sedentary animals (*e.g.*, sponges, sea stars and sea anemones) and drifting pelagic species (*e.g.*, jellyfishes and ctenophores). Given these life styles, it is clearly advantageous to be able to confront the environment equally from a variety of direction. In such creatures the feeding structures (tentacles) and sensory receptors are distributed at equal intervals around the periphery of the organisms, so that they contact the environment more or less equally in all directions.

Bilateral symmetry. The body part of bilaterally symmetrical animals are oriented about an axis that passes from the front (**anterior**) to the rear (**posterior**) end. A single plane of symmetry—the **midsagittal plane** (or median sagittal plane)—passes along the axis of the body to separate right and left sides. Any longitudinal plane passing perpendicular to the midsaggital plane and separating the upper (**dorsal**) from the underside (**ventral**) is called **frontal plane**. Any plane that cuts across the body perpendicular to the main body axis and the midsagittal plane is called a **transverse plane** (or simply cross section). In bilaterally symmetrical animals the term **lateral** refers to the sides of the body, or to structure away from (to the right and left of) the midsagittal plane. The term **medial** refers to the midline of the body.

Whereas spherical and radial symmetry are typically associated with sessile or drifting animals, bilaterality is generally found in animals with controlled mobility. In these animals, the anterior end of body confronts the environment first. Associated with bilateral symmetry and unidirectional movement is a concentration of feeding and sensory structures at the anterior end of the body. The evolution of a specialized "head', containing those structures and the nervous tissues that innervate them, is called **cephalization**. The surfaces of animal differentiate as dorsal and ventral regions, the latter becoming locomotory and the former being specialized for protection.

1.5 CELLULARITY, GERM LAYERS AND BODY CAVITIES

One of the main characteristics used to define grades of animal complexity is the presence or absence of true **tissues**. Tissues are aggregations of morphologically and physiologically similar cells that perform a specific function. The protista do not possess tissues, but occur only as single cells or as simple colonies of cells. In a sense, they are all at a **unicellular grade** of construction. Beyond the protista is the vast array of multicelled animals, the **metazoa**. The metazoa can be divided into three major levels or grades: mesozoa, parazoa and eumetazoa. These names do not represent formal taxa, but may be used to group the metazoa by their level of overall complexity. The mesozoa and parazoa are not generally considered to possess true tissues, and for this reason they are separated from the rest of the metazoa. The eumetazoa pass through distinct embryonic stages during which tissue layers form (Table 1.4).

The embryonic tissue layers of eumetazoa are called **germ layers** (from the Latin *german* = a sprout, bud, or embryonic primordium) and it is from these germ layers that all adult structures develop. The germ layers initially form as outer and inner sheets or masses of embryonic tissues, termed **ectoderm** and **endoderm** (or entoderm), respectively. In the embryogeny of the radiata phyla Cnidaria and Ctenophora, only these two germ (or germinal) layers develop (or if a middle layer does develop, it is produced by ectoderm, is largely non-cellular, and is not considered a true germ layer). These animals are regarded as **diploblastic**. In the embryogeny of most animals, however, a third cellular layer, the **mesoderm**, arises between the ectoderm and the endoderm; these metazoan groups are said to be **triploblastic.**

Table 1.4 Organization of the Protista and Animal phyla on the basis of their body construction (Brusca and Brusca, 2003).

1. **Unicellular organisms (Protista).** Phyla Euglenoida, Kinetoplastida, Ciliophora, Apicomplexa, Dinoflagellata, Stramenopiles, Rhizopoda, Actinopoda, Granuloreticulosa, Diplomonadida, Parabasilida, Chlorophyta, Cryptomonada, Microspora, Ascetospora, Myxozoa, Opalinida, Choanoflagellata.
2. **Multicellular organisms.** The **Metazoa** or **Animalia**
   A. Without tissue
   1. **The Mesozoa.** Phyla Orthonectida, Rhombozoa, Placozoa and Monoblastozoa (*Salinella*).
   2. **The Parazoa.** Phylum Porifera (sponges)

   B. With true tissues. Eumetazoa
   1. **The diploblastic eumetazoa** (without true mesoderm). The **"Radiata"**
   2. **The triploblastic eumetazoa** (with true mesoderm). The "**Bilateria**"
      (*i*) **Acoelomate.** Without a body space other than digestive tract; mesenchyme muscle fill region between gut and epidermis.
      Phyla Platyhelminthes, Gastrotricha, Entoprocta, Gnathostomulida.
      (*ii*) **Blastocoelomate** (formerly called **Pseudocoelomata**). With a persistent blastocoel between gut and body wall.
      Phyla Acanthocephala, Kinorhyncha, Loricifera, Nematoda, Nematomorpha, Rotifera
      (*iii*) **Coelomates (or eucoelomates).** With a true coelom (*i.e.*, mesodermal cavity)
      Phyla Nemertea, Phoronida, Ectoprocta, Brachiopoda, Sipuncula, Echiura, Mollusca, Priapula, Onychophora, Tardigrada, Annelida, Arthropoda, Echinodermata, Chaetognatha, Hemichordata, Chordata*

* Some of these groups (*e.g.*, Arthropoda, Mollusca) have greatly reduced coelomic spaces; often the main body cavity is a blood filled space called a **hemocoel** and is associated with an open circulatory system.

The evolution of mesoderm greatly expanded the evolutionary potential for animal complexity. Three major grades of construction are recognizable among the triploblastic metazoa: **acoelomate, blastocoelomate (pseudocoelomate)** and **eucoelomate**. An acoelomate grade occurs in several triploblastic phyla: Platyhelminthes, Entroprocta, Gnathostomulida and Gastrotricha. In these animals, the mesoderm form a more or less solid mass of tissue, sometimes with small open spaces (lacunae), between the gut and body wall (Fig. 1.4 A). In nearly all other triploblastic animals, an actual space develops as a fluid filled cavity between the body wall and the gut. In many phyla (*e.g.*, annelids and echinoderms), this cavity arises within the mesoderm itself and is completely enclosed within a thin lining called the **peritoneum**, which is derived from the mesoderm. Such a cavity is called a true **coelom (eucoelom**; Fig. 1.4C).

Several groups of triploblastic metazoa (*e.g.*, rotifers, roundworms and others) possess small or large body cavities that are neither from the mesoderm nor fully lined by peritoneum or any other form of mesodermally derived tissue. Such a cavity is used to be called a **pseudocoelom (= blastocoelom;** Fig. 1.4B). The organs of these animals actually lie free within the body cavity and bathed directly in its fluid. In most cases the space represents persistent remnants of the embryonic *blastocoel* and since there is nothing "false" about it, **Bursca** and **Bursca** (2003) used the more descriptive term **blastocoelom** instead of pseudocoelom.

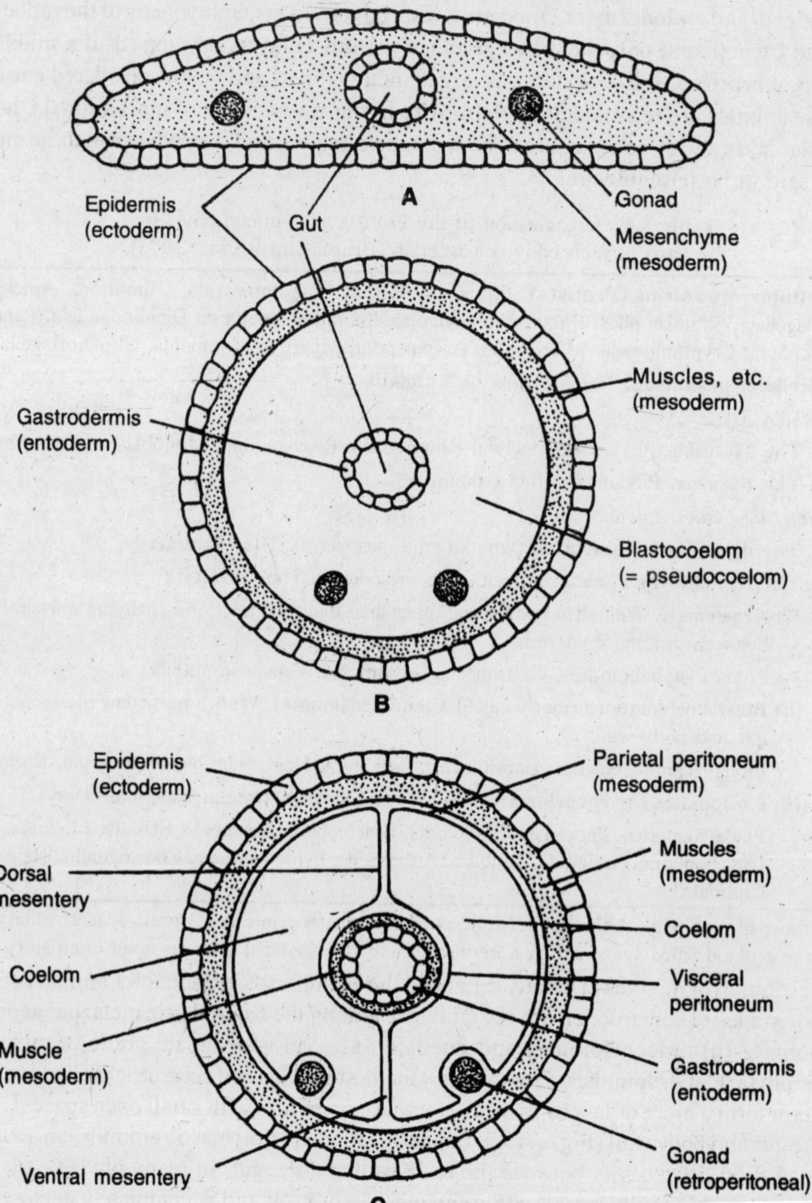

Fig. 1.4. Principal body plans of triploblastic metazoa (diagrammatic cross section). A—Acoelomate body plan; B—Blastocoelomate (= pseudocoelomate) body plan and C-Eucoelomate body plan (after Brusca and Brusca, 2003).

The evolution of this space (blastocoelom and eucoelom) created a radically new architecture, a tube-within-a tube design in which the inner tube (the gut and its associated organs) was freed from the constraint of being attached to the outer tube (the body wall), except at the very ends. The fluid-filled cavity not only served as a mechanical buffer between these two largely independent tubes, but also allowed for the development and expansion of new structures within the body, served as a storage chamber for vary body products (*e.g.*, gametes), provided a medium for circulation and was in itself an incipient hydrostatic skeleton.

1.6 GENERAL CHARACTERS OF SOME PHYLA OF INVERTEBRATA

1. Phylum Protozoa. It includes one celled, free-living, mostly aquatic animals. They are small in size, usually microscopic with distinct nucleus. Each individual consists of only one cell which has to carry on all the vital activities. They are abundantly found in water containing decaying organic matter but others live in damp soil, in fresh water or in the sea. Some are commensals, mutualists or parasites and pathogens. The examples are *Amoeba, Euglena, Paramoecium, Vorticella, Polystomella, Entamoeba histolytica* (causes amoebic dysentry in humans), *Trypanosoma gambiense* (causing sleeping sickness in humans), *Plasmodium vivax* (causing malaria in humans), *Monocystis* (a parasite of earthworms), *etc.*

2. Phylum Porifera. It includes the sponges which are found both in ocean and fresh water. These are multicellular animals having two germ layers (*i.e.*, external **pinacoderm** and internal **choanoderm**). Both layers have cells which are arranged loosely aggregated and a mostly gelatinous intermediary layer (called **mesohyl**) which carries **archaeocytes** (large amoeboid **totipotent cells** that can differentiate into all various other cell types). The body is perforated with many pores and canals through which water flows. Within sponges, water currents are generated by collared, flagellated cells called **choanocytes**, whose flagella move water out through an exhalant body opening (osculum) thereby drawing water in through external pores (ostia). This is done for filtering water for food and oxygen. The body wall of sponge is supported by a framework which is composed of lime or of silica (**spicules**) or of an organic substance known as **spongin** (collagen). The examples are *Sycon, Hyalolaema, Euplectella*, etc.

3. Phylum Cnidaria. These are lowest metazoan animals with radial symmetry of body and two definite body layers, ectoderm (epidermis) and endoderm (gastrodermis) with mesoglea in between. They are both marine and fresh water and carnivorous in habit. Their tentacular epidermis is armed with specialised cells, called **cnidocytes** that contain miniature stinging, harpoon-like organelles called **nematocysts**. Nematocysts immobilize prey and let the animal bring them to the gastric cavity (coelenteron). The examples are *Hydra, Obelia, Aurelia*, sea anemone, corals, etc.

4. Phylum Ctenophora. Ctenophora is a small group. They are all marine and planktonic. Their biradially symmetrical bodies contain eight meridionally placed ciliated plates. They resemble the cnidarians on many counts but differ from them in not having the nematocysts. Their mesoglea contain mesenchymal muscle cells. They possess a specialised aboral sense organ and the tentacles bear adhesive cells. The examples are *Beroe, Pleurobrachia, Hormiphora*, etc.

5. Phylum Platyhelminthes. It includes the free-living flatworms (Turbellaria) and parasitic flatworms (flukes and tapeworms). They have dorso-ventrally flattened body. These animals possess ectoderm and endoderm and another cell layer called mesoderm in between. For this reason they are called **triploblastic animals**. They lack a body cavity, so are called **acoelomates**. Digestive canal is incomplete with only one opening known as **mouth** and there is no anus. Excretion of waste products is done by peculiar flame cells (protonephridia). They possess extensive hermaphroditic reproductive system. The examples are leaf-like *Dugesia tigrina* (free living), leaf-like fluke (*Fasciola hepatica*), ribbon-like tape worm (*Taenia solium*), etc.

6. Phylum Nematoda. It includes round worms or thread worms having soft, tubular shape and unsegmented bodies. They contain a distinctive, fluid-filled body cavity called **pseudocoel**. They have epidermal cuticle, adhesive organs, constant cell numbers (eutely) and a digestive tract with mouth, anus and a muscular pharynx that pumps food into flaccid gut cavity. They are mostly aquatic in habit, some are endoparasitic. The examples are *Ascaris lumbricoides, Wuchereria bancrofti* (filaria), *Ancylostoma duodenale* (hook worm), *Meloidogyne* (root-knot nematode), etc.

7. Phylum Mollusca. These are unsegmented coelomates (have true body cavity). The soft parts of body are enclosed in a hard calcareous shell, as in snails and oysters. A fleshy muscular foot for locomotion is often present (they have a body plan based on creeping over shallow marine substrates) They are earliest known herbivorous mostly having a **radula**. Radula is a rasplike organ bearing chitinous teeth that unrolls from the mouth and scrapes algae from rocks. Molluscs are found on land, specially in moist places and in fresh water and salt water. The examples are freshwater mussel (*Lamellidens*), Apple snail (*Pila*), Pearl oyster (*Pinctada*), Cuttlefish (*Sepia*), etc.

8. Phylum Annelida. The soft, elongated, worm-like bilaterally symmetrical coelomate body is metamerically segmented. They can localise hydrostatic pressures to specific segments to do active sustained burrowing. Annelids include the earthworms, leeches and sandworms that live on soil, fresh water and sea water respectively.

The examples are terrestrial clitellum-bearing hermaphrodite earthworm (*Pheretima posthuma*) and fresh water leech (*Hirudinaria granulosa*) and marine *Aphrodite, Arenicola, Chaetoptera, Nereis*, etc.

9. Phylum Arthropoda. The body is bilaterally symmetrical, segmented, with definite head, thorax and abdomen. Hardened, chitinous exoskeletons with their inner projections (apodemes or endophragma) for muscle attachments, jointed appendages and development of new kinds of cephalic structures have made this phylum largest of the animal phyla (80 per cent of all known animal species today are arthropods). The examples are horseshoe crab, prawn, crab, centipede, millipede, scorpion, spider, ticks, mites, cockroach, grasshopper, locust, silkmoth, honeybee, termite, wasp, ant, house fly, mosquito, body louse, lac-insect, rice-bug, beetle, etc.

10. Phylum Echinodermata. These are spiny-skinned, marine triploblastic, unsegmented, coelomate animals. They have a mesodermal skeleton of small calcite plates or spicules lying just below the outer epidermis. They have bilateral symmetry in larvae, followed by a five-rayed (pentameral) symmetry in adults. They contain a **water vascular system** derived from coelom in which small surface **tube feet** used for locomotion and feeding connect to internal canals of circulating sea water. The examples are star fish (*Asterias*), sea-lily, sea-cucumber, sea-urchin, etc.

11. Phylum Hemichordata. They are bilateral deuterostomes, body vermiform or saccate and fundamentally trimeric, with prosome, mesosome and metasome, each with coelomic compartments. Solitary or colonial. With pharyngeal gill slits/pores. Well developed open circulatory system. Excretion by glomerulus. Gonads extra coelic. Complete gut with preoral (buccal) gut diverticulum. Deposit or suspension feeders (Box 1.5). Without a notochord. Have short, dorsal, mesosomal, occasionally hollow nerve cord. Dioecious, with external fertilization and indirect development producing tornaria larva. Strictly marine, except for *Planctosphaera pelagica*. The examples are acorn worms such as *Saccoglossus, Balanoglossus, Cephalodiscus, Rhabdopleura, Atubaria, Planctosphaera*, etc.

Box 1.5

Suspension feeding. It is the removal of suspended food particles from the surrounding medium by some sort of capture, trapping of filteration mechanism.

Deposit feeders. They are microphagous and they obtain nutrients from the sediments of soft-bottom habitats (muds and sands).

QUESTIONS

Long Answer Questions

1. Describe characteristic features of invertebrates.
2. What is the concept of major and minor phyla? Enumerate major invertebrate phyla and minor invertebrate phyla.
3. Classify invertebrate animals upto phyla. Mention main characters and one or two example of each phylum.
4. Write an essay on 'Bauplan concept'.

Short Answer Questions

1. What is a living organism?
2. Differentiate between invertebrates and vertebrates.
3. Differentiate between bilateral and radial symmetry.
4. Write a note on habitat. (*Lucknow 1997*)
5. What is symbiosis? How does it serve as a habitat of some animals?
6. Give an account of symmetry in animals.
7. What is body cavity? Describe its various types.
8. "The group Invertebrata is unnatural". Comment on this statement.
9. Write down the six constraints which maintain body plan of the animals.
10. Write short notes on the following:
    (*i*) Bauplan concept
    (*ii*) Hox gene
    (*iii*) Cephalization.

Very Short Answer Questions

1. Define the term biology and zoology.
   Ans. Biology is study of life. Biology investigates the origin, structure, functions and distribution of life. Biology includes *Botany* (study of plant life) and *Zoology* (study of animal life).
   Zoology is the scientific study of animal life.
2. What do you mean by zoology?
   Ans. Zoology is the branch of biology or life sciences that deals with the animal organisms.
3. Define Invertebrata.
   Ans. Animals which do not have a notochord and vertebral column (backbone) are called **invertebrates**. (**Note.** The division of animals into invertebrates and vertebrates is based more on tradition and convenience, reflecting a dichotomy of zoologist's interest, than it is on the recognition of natural biological groupings; see **Brusca** and **Brusca**, 2003).
4. Give examples of sessile animals.
   Ans. Sponges, corals and barnacles.
5. Give examples of sedentary animals.
   Ans. Crinoids, solitary anemones and most clams.
6. What is sagittal section.
   Ans. Median longitudinal vertical section.
7. Name the Invertebrate phyla which show metamerism.
   Ans. Annelida and Arthropoda.
8. What are triploblastic animals?
   Ans. Triploblastic animals are those which develop from three germ layers namely ectoderm, mesoderm and endoderm.
9. Name three types of body cavity found in animals.
   Ans. True coelom (Eucoelom), psuedocoelom and haemocoel.
10. Give some examples of zooplankton.
    Ans. Jelly fishes, comb jellies, arrow worms and small crustaceans and larvae.
11. Give some examples of nekton.
    Ans. Squids, cuttlefishes and fishes.

Fill in the Blanks

1. Marine animals live in the ....
2. Small organisms that passively float in surface water are referred to as ....
3. Animals which actively swim about in water are referred to as ....
4. Symmetry in *Hydra* is ....
5. Animal that runs fast on land is called ....
6. Fossorial animals live in ....
7. Protozoa form a .... group.
8. Metazoa form a .... group.
9. Kingdom Protista is a .... group.
10. Kingdom Animalia is a .... group.
11. Tardigrade is a minor ....
12. Term bauplan was coined by .... in 1945.

Multiple Choice Questions

1. Which phylum has highest number of species?
   (*a*) Protozoa (*b*) Porifera
   (*c*) Arthropoda (*d*) Mollusca
2. Platyhelminthes are commonly called
   (*a*) round worm (*b*) flat worm
   (*c*) blind worm (*d*) tubiculous worm
3. Which group includes all marine animals?
   (*a*) Mollusca (*b*) Porifera
   (*c*) Echinodermata (*d*) Cnidaria
4. A mesodermal endoskeleton among invertebrata is found in
   (*a*) Porifera (*b*) Mollusca
   (*c*) Echinodermata (*d*) Cnidaria
5. Animals living at the bottom of sea form are called as
   (*a*) plankton (*b*) nekton
   (*c*) benthos (*d*) none of these
6. Protozoans are now regarded as
   (*a*) an independent phylum (*b*) a group of monera
   (*c*) a group of protista (*d*) a group of animals

ANSWERS

Fill in the Blanks

1. sea 2. plankton 3. nekton 4. radial
5. cursorial 6. burrow 7. unnatural 8. natural
9. polyphyletic 10. monophyletic 11. phylum 12. J.H. Woodger.

Multiple Choice Questions

1. (*c*) 2. (*b*) 3. (*c*) 4. (*c*)
5. (*c*) 6. (*c*).

2

Protozoa (Characters, Classification and Types)

The **protozoa** (Gr., *protos* = first or primitive; *zoon* = animal) are microscopic, eukaryotic, unicellular, solitary or colonial, commonly motile and phagocytic heterotrophic animalcules. The term protozoa was coined by **Goldfuss** (1818) and unicellular nature of protozoa was established by **von Siebold** (1845). There are about 50,000 species of protozoa (*i.e.*, 30,000 living and 20,000 extinct) which are found in all moist places such as sea, freshwater and soil. Each protozoan functions as an independent unit and is able to perform effectively all the activities necessary for life.

2.1 GENERAL CHARACTERS

1. Protozoa are small, generally microscopic animalcules. Individual independent protozoa range in size from a micrometer (*e.g., Micromonas*) to a few millimeters in some dinoflagellates, amoebas and ciliates.
2. They are simplest and most primitive animals having **cellular level of organisation**.
3. **Unicellular** or **single-celled** animals having no tissues or organs. Specialised parts of living material (*i.e.*, protoplasm) form membrane-bound organelles or organoids.
4. The single cell performs all necessary functions. There is no **physiological division of labour**.
5. The body is bounded by the plasma membrane. The rigidity or flexibility of the body and its shape are largely dependent on the nature of the **cytoskeleton** (*e.g.*, pellicle;) and **exoskeleton** (*e.g.*, test).
6. Animalcules may be **free-living, commensal, mutualistic** or **parasitic** (both ecto-parasitic and endoparasitic). The free-living protozoans are aquatic, inhabiting fresh and salt waters. These may be **solitary** or **colonial**; in colonial forms the individuals are alike and independent.
7. Body shape variable; it may be spherical, oval, elongated or flattened.
8. Cytoplasm is differentiated into an outer **ectoplasm** (or cell cortex) and inner **endoplasm**.
9. Protozoans may have one or more **nuclei** which may be **monomorphic** or **dimorphic**. In case of dimorphic nuclei, two basic types are recognized: 1. vesicular, containing considerable nucleoplasm and 2. compact, containing little nucleoplasm.
10. Three major types of locomotor organelles namely flagella, pseudopodia and cilia occur among protozoa. The classification of phylum Protozoa mainly depends on the type and distribution of these locomotory parts. **Pseudopodia** are semipermanent to transitory extensions of protoplasm of body surface and used by sarcodina. **Flagella** are permanent or semipermanent

locomotory organelles of mastigophora. **Cilia** are many and occur in ciliata. Some protozoans are **sessile** (attached or fixed) and do not move.

11. All types of **nutrition** occur in protozoa. Thus, nutrition may be **holozoic** (animal-like), **holophytic** (plant-like, photoautotrophy), **saprozoic** (absorb soluble organic nutrients from their environment) or **parasitic**.
12. Holozoic protozoans have two general kinds of feeding: **hunting**, which involves seeking and capturing prey; and **trapping** or **suspension feeding**, which is often used by sessile protozoans.
13. Digestion occurs intracellularly within a **food vacuole** by the help of enzymes of lysosomes, and food reaches the vacuole through a cell mouth (**cytostome**) or by engulfment by pseudopodia (endocytosis and phagocytosis).
14. Respiration, excretion and egestion of waste products occur through body surface. Protozoa are mostly **ammonotelic**, *i.e.*, they excrete their nitrogenous wastes in the form of ammonia.
15. Many protozoa osmoregulate to remove excess water (called **water volume regulation**) and to adjust the concentration and proportion of their internal ions (**ionic regulation**). Osmoregulation is accomplished by **contractile vacuoles**. There is no evidence that the contractile vacuoles have any specific excretory function.
16. **Cyclosis** or cytoplasmic streaming distributes materials throughout the continuous protoplasmic mass of the organism. For example, in ciliates cyclosis occurs due to movements of food vacuoles along microtubules extending from the cytostome.
17. Protozoa are sensitive to many kinds of stimuli including touch (**thigmotaxis**), temperature changes (**thermotaxis**), light (**phototaxis**), and many chemicals (**chemotaxis**).
18. Intercellular chemical signalling (**pheromones**) in protozoans involves signal molecules such as serotonin, β-endorphin, acetylcholine and cyclic-AMP (cAMP).
19. Reproduction is asexual or sexual; **asexual reproduction** by **binary fission, multiple fission** and **budding** or **sporulation** and **sexual reproduction** by gamete formation or conjugation of adults.
20. Resistant coverings are sometimes formed for brief parts of a life cycle. This is especially true of parasites, which usually pass from one host to another as cysts or spores, covered by a resistant membrane that protects them while out of the host. Many free-living protozoa also encyst to survive unfavourable environmental conditions and be dispersed by air or water currents in a protected stage.
21. In some protozoans, the life cycles are complicated and show **alternation of generation**, *i.e.*, life cycle includes asexual and sexual phases.

2.2 CLASSIFICATION

Outline Classification of Phylum Protozoa

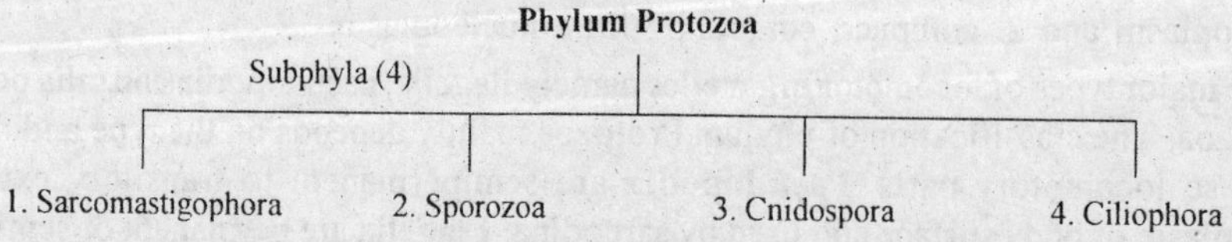

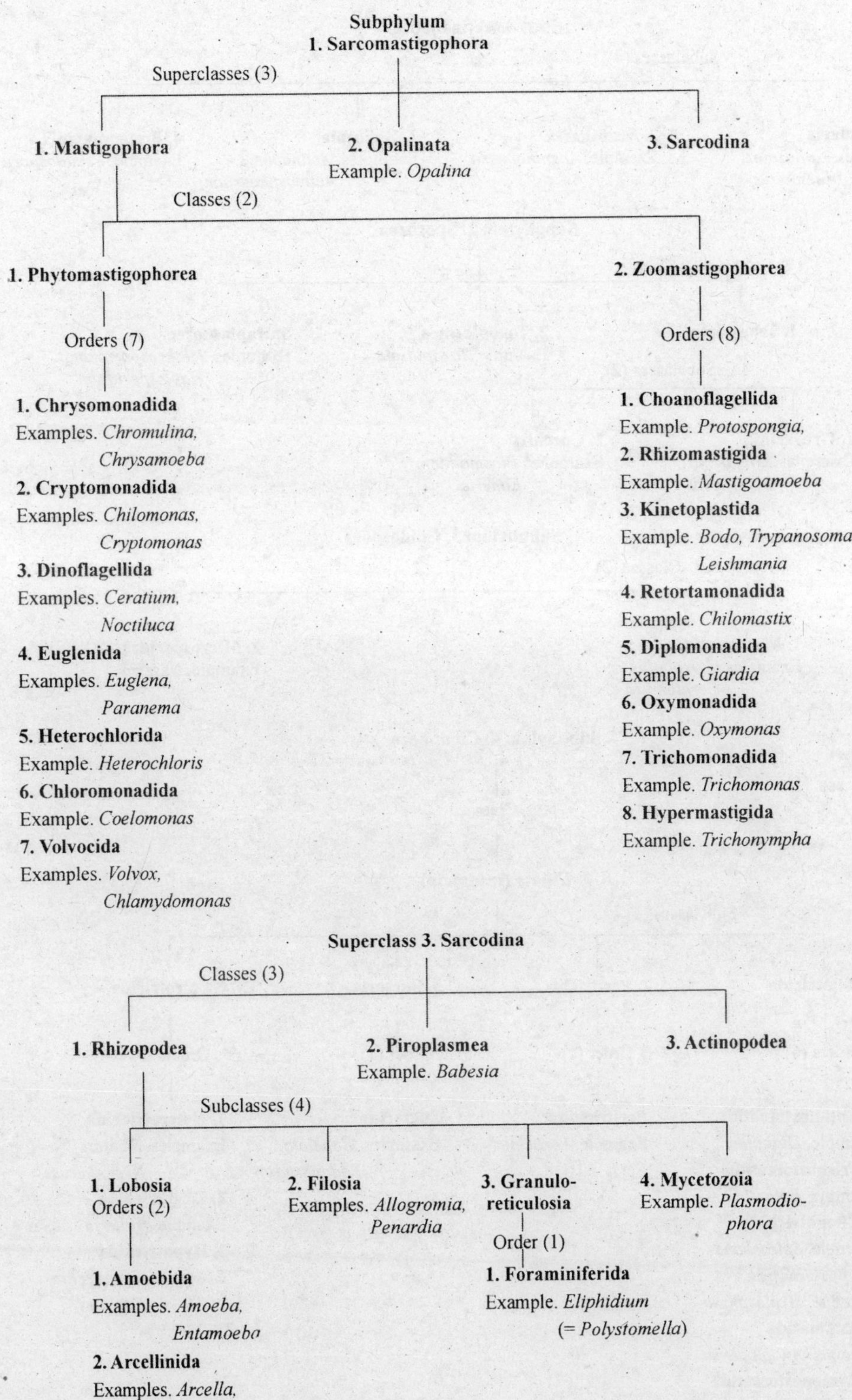
Subphylum
1. Sarcomastigophora
Superclasses (3)
1. Mastigophora
2. Opalinata
Example. Opalina
3. Sarcodina
Classes (2)
1. Phytomastigophorea
2. Zoomastigophorea
Orders (7)
Orders (8)
1. Chrysomonadida
Examples. Chromulina, Chrysamoeba
2. Cryptomonadida
Examples. Chilomonas, Cryptomonas
3. Dinoflagellida
Examples. Ceratium, Noctiluca
4. Euglenida
Examples. Euglena, Paranema
5. Heterochlorida
Example. Heterochloris
6. Chloromonadida
Example. Coelomonas
7. Volvocida
Examples. Volvox, Chlamydomonas
1. Choanoflagellida
Example. Protospongia,
2. Rhizomastigida
Example. Mastigoamoeba
3. Kinetoplastida
Example. Bodo, Trypanosoma, Leishmania
4. Retortamonadida
Example. Chilomastix
5. Diplomonadida
Example. Giardia
6. Oxymonadida
Example. Oxymonas
7. Trichomonadida
Example. Trichomonas
8. Hypermastigida
Example. Trichonympha
Superclass 3. Sarcodina
Classes (3)
1. Rhizopodea
2. Piroplasmea
Example. Babesia
3. Actinopodea
Subclasses (4)
1. Lobosia
Orders (2)
2. Filosia
Examples. Allogromia, Penardia
3. Granulo-reticulosia
Order (1)
4. Mycetozoia
Example. Plasmodiophora
1. Amoebida
Examples. Amoeba, Entamoeba
2. Arcellinida
Examples. Arcella, Difflugia, Euglypha
1. Foraminiferida
Example. Eliphidium (= Polystomella)

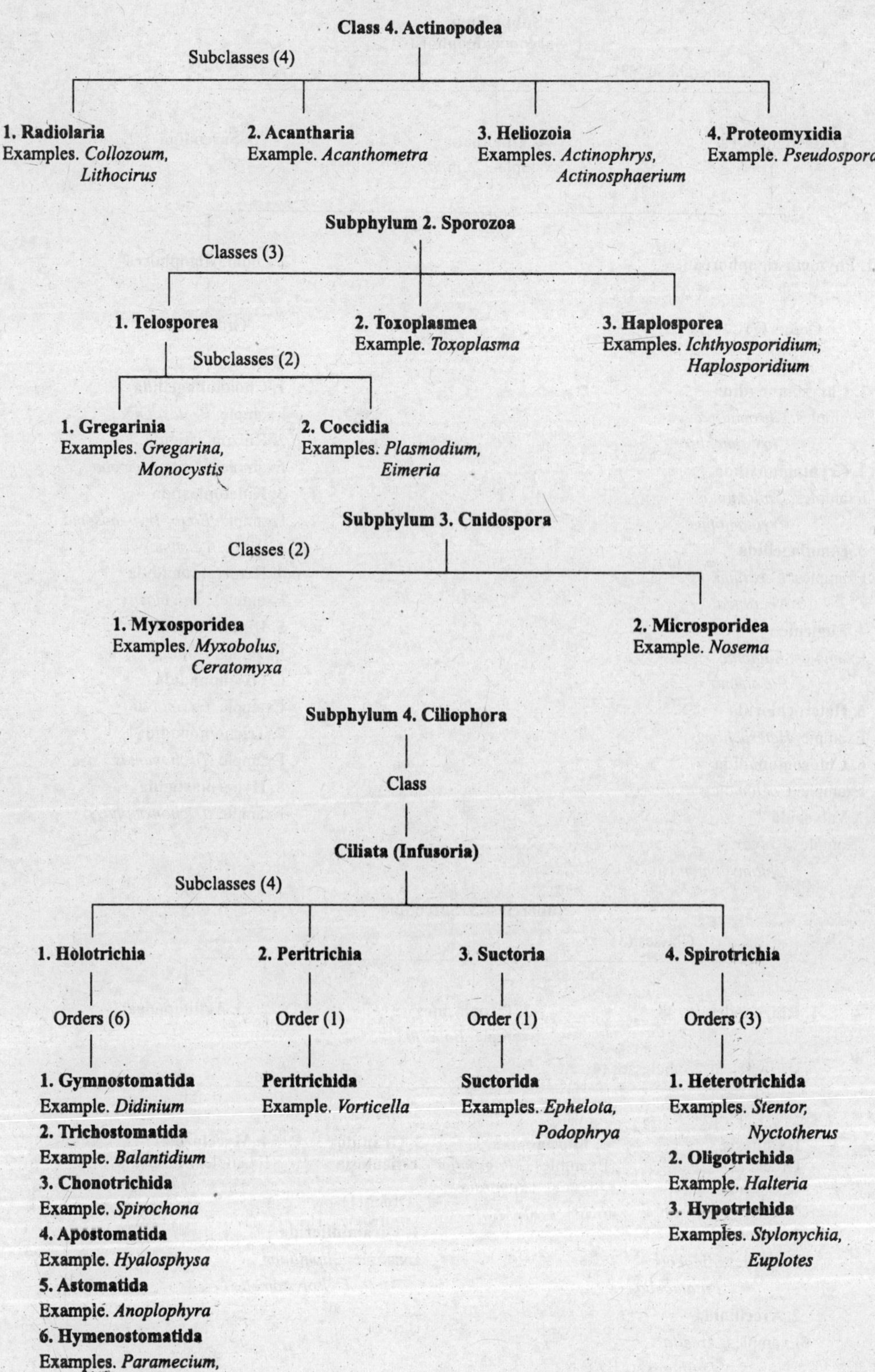

Class 4. Actinopodea
Subclasses (4)
1. Radiolaria
Examples. Collozoum, Lithocirus
2. Acantharia
Example. Acanthometra
3. Heliozoia
Examples. Actinophrys, Actinosphaerium
4. Proteomyxidia
Example. Pseudospora
Subphylum 2. Sporozoa
Classes (3)
1. Telosporea
Subclasses (2)
1. Gregarinia
Examples. Gregarina, Monocystis
2. Coccidia
Examples. Plasmodium, Eimeria
2. Toxoplasmea
Example. Toxoplasma
3. Haplosporea
Examples. Ichthyosporidium, Haplosporidium
Subphylum 3. Cnidospora
Classes (2)
1. Myxosporidea
Examples. Myxobolus, Ceratomyxa
2. Microsporidea
Example. Nosema
Subphylum 4. Ciliophora
Class
Ciliata (Infusoria)
Subclasses (4)
1. Holotrichia
Orders (6)
1. Gymnostomatida
Example. Didinium
2. Trichostomatida
Example. Balantidium
3. Chonotrichida
Example. Spirochona
4. Apostomatida
Example. Hyalosphysa
5. Astomatida
Example. Anoplophyra
6. Hymenostomatida
Examples. Paramecium, Tetrahymena
2. Peritrichia
Order (1)
Peritrichida
Example. Vorticella
3. Suctoria
Order (1)
Suctorida
Examples. Ephelota, Podophrya
4. Spirotrichia
Orders (3)
1. Heterotrichida
Examples. Stentor, Nyctotherus
2. Oligotrichida
Example. Halteria
3. Hypotrichida
Examples. Stylonychia, Euplotes

Detailed Classification

According to **Honigberg** *et al.*, (1964) phylum Protozoa has been classified into **four** subphyla: Sarcomastigophora, Sporozoa, Cnidospora and Ciliophora.

Subphylum 1. Sarcomastigophora

1. The locomotory organs are either flagella or pseudopodia or both.
2. Single type of nucleus (monomorphic).
3. Asexual reproduction by binary fission and multiple fission.
4. Sexual reproduction by syngamy.

Subphylum 2. Sporozoa

1. All are endoparasites.
2. Body covered with thick pellicle.
3. Locomotory organs are absent in adult. Flagella present only in microgametes of some group.
4. Nucleus is of the single type.
5. Saprozoic nutrition.
6. Asexual reproduction by multiple fission.
7. Sexual reproduction by syngamy followed by spore formation.
8. Spores contain one to many sporozoites which represent the infective phase.

Subphylum 3. Cnidospora

1. All are parasitic.
2. Spore formation occurs throughout life.
3. Spores have one to many cells, each with one or more **polar filaments** (which are coiled threads and can be shot out for attaching to the host) and one or more **sporoplasms** (analogous structures to sporozoites).
4. Without sporogony, zygote gives rise to one or more trophozoites, each with many nuclei.

Subphylum 4. Ciliophora

1. Most species are free-living, some are commensal or parasitic.
2. Body organization complex.
3. Presence of simple **cilia** or compound ciliary structures as locomotory or feeding organelles at some stage in the life cycle. Subpellicular cilia present even when cilia absent.
4. Nuclear dimorphism occurs, *i.e.*, they have two nuclei, a large vegetative or trophic **macronucleus** and a small generative or reproductive **micronucleus**.
5. Nutrition heterotrophic.
6. Contractile vacuole typically present.
7. Asexual reproduction by transverse binary fission; or by budding or multiple fission.
8. Sexual reproduction by conjugation.
9. Gametes are always absent.

SUBPHYLUM 1. SARCOMASTIGOPHORA

It includes **three** superclasses:

Superclass 1. Mastigophora

1. They are commonly called **flagellates.**
2. Body is covered by pellicle.

3. Locomotary organs of adults are flagella.
4. Asexual reproduction by longitudinal binary fission.
5. Free living or parasitic.

Superclass 2. Opalinata

1. Entire body surface is covered by longitudinal, oblique rows of cilia. Infraciliature characteristic of the true ciliates is lacking.
2. Cytostome absent; saprozoic nutrition.
3. Two or many monomorphic nuclei.
4. Asexual reproduction by longitudinal binary fission.
5. Sexual reproduction involves syngamy with flagellated gametes.
6. Gut (*i.e.*, rectum) parasites/commensals of anurans (frogs, toads, tadpoles) and also of fishes, salamanders and reptiles.

Examples. *Opalina, Cepedea, Zelleriella.*

(**Note.** Superclass Opalinata includes single class **Opalinatea** and single order **Opalinida**, both having characters of superclass).

Superclass 3. Sarcodina (or Rhizopoda)

1. The amoeboid form is predominant.
2. Locomotion and feeding by pseudopodia, or by locomotive protoplasmic flow without distinct pseudopodia. Flagella when present, usually restricted to developmental stages of life cycle.
3. Pseudopodia may be lobopodia, filopodia or reticulopodia.
4. Body naked or with external or internal test or skeleton.
5. Nutrition holozoic and saprozoic.
6. Asexual reproduction by fission.
7. Sexual reproduction by flagellate (rarely amoeboid) gametes.
8. Most species are solitary and free-living.

SUPERCLASS 1. MASTIGOPHORA

It is divided into **two** classes:

Class I. Phytomastigophorea

1. Chlorophyll containing chromatophores (or chloroplasts) are present.
2. There are one or two flagella.
3. Nutrition mostly holophytic (by photosynthesis).
4. Reserve food is starch or paramylon.

Class II. Zoomastigophorea

1. Chloroplasts (chromatophores) are absent.
2. There are one to many flagella; often undulating membrane is present.
3. Nutrition holozoic or saprozoic.
4. Reserve food is glycogen.
5. Parasitic, commensal or symbiotic.

CLASS I. PHYTOMASTIGOPHOREA

It is divided into following **seven** orders:

Order 1. Chrysomonadida

1. Two unequal flagella; one hairy and directed anteriorly, other trailing and smooth.
2. Chloroplasts golden brown containing chlorophyll a and c. Some species lack chloroplasts and feed holozoically.
3. Starch absent. Storage products are **chrysolaminarin** and fat.
4. Cells are either naked or with richly patterned silicified scales (cysts) or with plate-like lorica.
5. Sexual reproduction present.
6. Marine or freshwater animals.

Examples. *Chromulina, Chrysamoeba, Ochromonas.*

Order 2. Cryptomonadida

1. Cells flattened, naked, without wall or pellicle.
2. They have a gullet that reaches upto the middle of the body.
3. They have two unequal flagella.
4. Chloroplasts brown, red, olive-green, blue or yellow.
5. Reserve food includes starch and fat.
6. Sexual reproduction unknown.
7. Stigma present.
8. Marine or freshwater.

Examples. *Chilomonas, Cryptomonas.*

Order 3. Dinoflagellida

1. Cells flattened or of complex symmetry with transverse and ventral grooves and with armour of cellulose plates. Cytoplasm is vacuolated.
2. Two heterodynamic flagella, inserted apically or laterally, one ribbon-shaped with paraxial rod and single row of fine hairs, other smooth or with two rows of stiffer hairs.
3. Chloroplasts typically golden brown (xanthophyll) and green (chlorophyll a and c present).
4. Reserve food includes starch and fat.
5. Nucleus unique among eukaryotes in having permanently condensed chromosomes having less amount of histone proteins.
6. Mitosis intranuclear; asexual reproduction by fission.
7. Sexual reproduction present.
8. Stigma is present in many species.
9. Free living, pelagic, marine, freshwater; some parasitic, some mutualistic (*e.g., Symbiodinium microadriaticum* is symbiont of corals).
10. Nutrition holophytic in most species, holozoic in some.
11. Some are bioluminscent (*e.g., Noctiluca*).

Examples. *Ceratium, Noctiluca, Glenodinium, Gymnodinium.*

Order 4. Euglenida

1. Cells with helical symmetry, naked but with complex and thick pellicle of interlocking proteinaceous strips.
2. Two (rarely more) unequal flagella, one or both arising from an anterior invagination (**reservoir**). The reservoir contains the roots of flagella and receives the contents of the

contractile vacuole. It leads to so-called gullet or cytopharynx which opens to outside as mouth (cytostome). Mouth is not used in ingestion of food.

3. One flagellum bears mastigonemes distally.
4. Stigma is present.
5. Chloroplasts grass-green, but absent in many genera.
6. Reserve food is paramylon (secreted by pyrenoids of chloroplasts) in *Euglena*, fat and cyclic metaphosphates.
7. Non-spindle intranuclear mitosis.
8. Solitary or colonial, most freshwater.

Examples. *Euglena, Peranema, Phacus.*

Order 5. Heterochlorida

1. Two unequal flagella.
2. Chloroplasts yellow-green.
3. Reserve food is oil and possibly chrysolaminarin.
4. Supposedly related to xanthophyceae algae.
5. Siliceous cysts.

Examples. *Heterochloris, Myxochloris.*

Order 6. Chloromonadida

1. Dorso-ventrally flattened forms with delicate pellicle.
2. Gullet present, stigma absent.
3. Two flagella.
4. Chloroplasts are green and numerous.
5. Characteristic ring of Golgi apparatuses at anterior end.
6. Sexual reproduction by fusion of two individuals.

Examples. *Coelomonas, Vacularia.*

Order 7. Volvocida

1. Small sized cells with rigid cellulose covering.
2. Mouth and gullet absent.
3. Two or four equal, smooth, apical flagella.
4. Chloroplasts (= chromatophores) are grass-green.
5. Reserve food is starch and fat.
6. Stigma present.
7. Sexual reproduction by syngamy.
8. Mostly freshwater; some form colonies.

Examples. *Volvox, Chlamydomonas, Eudorina.*

CLASS II. ZOOMASTIGOPHOREA

It is divided into following **eight** orders:

Order 1. Choanoflagellida

1. A collar (or ring of microvilli) at the base of a single apical flagellum.
2. Cell with membraneous sheath or basket-like siliceous lorica.
3. Nutrition holozoic.
4. They are solitary or colonial and are free-living or stalked.

Example. *Proterospongia.*

Order 2. Rhizomastigida

1. Small and amoeboid.
2. Flagella one to four.
3. Locomotion by flagella and pseudopodia.
4. Freshwater and free-living.

Examples. *Mastigamoeba, Dimorpha*.

Order 3. Kinetoplastida

1. One or two flagella arising from depression.
2. Single mitochondrion extending entire length of body as single tube, hoop or network of branching tubes.
3. Gullet is absent.
4. They contain a single large **kinetoplast** which is genetically autonomous, DNA-containing, membrane-bound organelle (mitochondria) associated with the basal body at the base of flagella.
5. Most species are parasitic; some are free-living; freshwater forms are sessile.
6. Solitary or colonial.

Examples. *Oikomonas, Bodo, Trypanosoma, Leishmania*.

Order 4. Retortamonadida

1. Two or four flagella, one turned posteriorly and associated with ventrally located cytostome.
2. Mitochondria and Golgi apparatus absent.
3. Cysts present.
4. Gut parasites of insects and vertebrates.

Example. *Chilomastix*.

Order 5. Diplomonadida

1. Flagella 3 to 8, one of them directed backwards over the body and associated with cytostome or with organelles forming cell axis.
2. Mitochondria and Golgi apparatus absent.
3. One or two karyomastigont (nucleus + mastigont) are present. **Mastigont system** is a complex system formed by groups of flagella which are associated with several microtubular and fibrillar organelles.
4. Cysts present.
5. Free-living or parasitic.

Examples. *Giardia*, *Hexamita*.

Order 6. Oxymonadida

1. Uninucleate or multinucleate forms.
2. One or more karyomastigonts are present, each containing four flagella typically arranged in two pairs in motile stage.
3. Mitochondria and Golgi apparatus absent.
4. Cysts in some species.
5. Sexual reproduction in some species.
6. Parasitic.

Examples. *Foaina, Oxymonas*.

Order 7. Trichomonadida

1. Typically karyomastigonts with four to six flagella, but one genus only with one flagellum and another with no flagella at all.
2. One flagellum recurrent, free, or with proximal or entire length adherent to body surface as undulating membrane.
3. They have a longitudinal supporting structure known as an **axostyle**.
4. No cysts are formed.
5. Parasitic in humans and other animals.

Examples. *Trichomonas tenax* (mouth); *T. hominis* (intestine), *T. vaginalis* (vagina); *Chilomastix mesnili* (large intestine).

Order 8. Hypermastigida

1. Complex forms with numerous flagella.
2. Flagella bearing kinetosomes distributed in circle, in plates or in longitudinal or spiral rows meeting in a centralized structure.
3. One nucleus per cell.
4. Cysts in some forms.
5. Some species have sexual reproduction.
6. Gut parasites of termites and cockroaches.

Examples. *Trichonympha campanula* (termite gut), *Lophomonas* (cockroach intestine).

SUPERCLASS 3. SARCODINA

It includes **three** classes:

Class 1. Rhizopodea

1. Locomotory organelles are lobopodia or filopodia but never axopodia.
2. They are creeping forms.

Class 2. Piroplasmea

1. Small, round rod-shaped or amoeboid organisms.
2. Parasites in red blood cells of vertebrates.
3. Spores are not formed.

Example. *Babesia.*

(**Note.** Class Piroplasmea is not divided into orders).

Class 3. Actinopodea

1. Pseudopodia are ray-like axopodia. Each of the pseudopodia is supported by a central axial rod comprising a bundle of microtubules. Axopods are commonly used in food capturing.
2. Primarily sessile or floating (planktonic) spherical forms.
3. Test is present or absent. Test is composed of organic matter and/or silica or strontium sulphate.
4. Both asexual and sexual reproduction present.

CLASS 1. RHIZOPODEA

This class is divided into **four** subclasses:

Sub-class I. Lobosia

1. Locomotory organelles are lobopodia which are broad and blunt pseudopodia containing both ectoplasm and endoplasm.

2. Usually uninucleate forms.
3. Sporangia absent.

This sub-class is divided into **two** orders:

Order 1. Amoebida

1. Typically uninucleate amoeboid forms.
2. Test (skeleton) absent.
3. Ectoplasm and endoplasm distinct.
4. Mitochondria typically present.
5. No flagellate stage in life cycle.
6. Largely freshwater, some parasitic.

Examples. *Amoeba*, *Entamoeba*.

Order 2. Arcellinida

1. The body is encased in an one-chambered test having a single aperture through which animal extends or retracts its lobopodia.
2. Test is composed of either organic material (*e.g.*, *Arcella*) or inorganic foreign particles (*e.g.*, *Difflugia*) or both.
3. Free-living and mostly freshwater.

Examples. *Arcella*, *Difflugia*, *Euglypha*.

Sub-class II. Filosia

1. Pseudopodia are filopodia which are tapering and branching.
2. Body naked or with a shell with single aperture.
3. Ectoplasm is not distinct.
4. No flagellate or spores known.
5. Marine or freshwater.

Examples. *Allogromia*, *Penardia*, *Gromia*.

Sub-class III. Granuloreticulosia

1. Pseudopodia are reticulopodia which are delicate and finely granular.

This sub-class includes the following significant order:

Order Foraminiferida

1. Large sized protozoans with uni-or multichambered perforated calcareous test or shell.
2. Reticulopodia protruding from test's aperture, wall perforation or both.
3. Reproduction with alternation of sexual and asexual generations.
4. Gametes usually flagellate, rarely amoeboid.
5. Mostly marine.

Examples. *Globigerina*, *Elphidium* (= *Polystomella*), *Pavonina*.

Sub-class IV. Mycetozoia

1. Body large, amoeboid and multinucleate.
2. Pseudopodia numerous and blunt.
3. Form sporangia which liberate spores.
4. Nutrition phagocytic.

Examples. *Plasmodiophora* and other slime molds.

CLASS 3. ACTINOPODEA

It is divided into **four** subclasses:

Subclass I. Radiolaria

1. A perforated central spherical chitinoid capsule separates the protoplasm into inner endoplasm and outer vacuolated ectoplasm.
2. Skeleton of siliceous spicules.
3. Pseudopodia are axopodia.
4. Contractile vacuole absent.
5. Marine.

Examples. *Collozoum*, *Thalassicola*, *Lithocircus*.

Subclass II. Acantharia

1. Strontium sulphate skeleton, usually composed of 20 radial or 10 dimetrical spines.
2. Spines more or less joined in cell centre.
3. Extracellular outer (cortex) and inner envelopes usually present. Inner envelope is called **capsular membrane** closely lining cell mass.
4. Locomotion by axopodia.
5. Marine, mostly planktonic.

Example. *Acanthometra*.

Subclass III. Helizoia

1. Spherical protozoans without central capsule.
2. Skeletal structures, if present, siliceous and organic.
3. Body of most species is naked and differentiated into outer vacuolated ectoplasm and inner dense endoplasm.
4. Axopodia radiate on all sides.
5. There may be more than one nucleus.
6. Most species freshwater, some marine.

Examples. *Actinophrys* (uninucléate), *Actinosphaerium* (multinucleate), *Clathrulina* (with perforated test).

Subclass IV. Proteomyxidia

1. Pseudopodia are filopodia which are slender with pointed tips and are composed of ectoplasm only.
2. Reproduction by binary or multiple fission.
3. Mostly parasite of algae and higher plants.
4. Marine or freshwater.

Examples. *Pseudospora*, *Vampyrella*.

SUBPHYLUM 2. SPOROZOA

The subphylum sporozoa is divided into **three** classes:

Class I. Telosporea

1. Spores without polar capsules or filaments.
2. Sporozoites elongated, microgametes flagellated.
3. Trophozoites with one nucleus only.
4. Pseudopodia are absent and mature organism do locomotion by body flexion or gliding.
5. Reproduction is both asexual and sexual.

Class II. Toxoplasmea

1. Spores are not formed.
2. There are no flagella or pseudopodia at any stage.
3. Only asexual reproduction, *i.e.*, binary fission.
4. Cysts containing many naked sporozoites.
5. They infect birds and mammals causing mental retardation in humans.

Examples. *Toxoplasma*, *Sarcocystis*.

Class III. Haplosporea

1. Spores are present and have spore cases.
2. Pseudopodia may be present but flagella are absent.
3. Reproduce only asexually by schizogony (*i.e.*, multiple fission of a trophozoite).
4. Parasites of fish and invertebrates.

Examples. *Ichthyosporidium*, *Haplosporidium*, *Caelosporidium*.

Class I. Telosporea

It is subdivided into **two** subclasses:

Subclass 1. Gregarinia

1. Mature trophozoites (gamonts) are large and extracellular living as parasites in the gut and body cavity of invertebrates or lower chordates such as annelids, sipunculids, arthropods, hemichordates or ascidians.
2. Reproduction is entirely sexual with sporogony. Life cycle consisting of gametogony and sporogony.
3. Gametes similar (isogamous).
4. Gametocyte show syzygy (end-to-end union of sporonts).
5. Zygote non-motile forming oocysts within gametocytes.
6. Spores contain eight sporozoites.

Examples. *Gregarina*, *Monocystis*, *Nematocystis*, *Lankestrella*.

Subclass 2. Coccidia

1. Mature trophozoites (gamonts) are small and intracellular.
2. Gametocytes are dimorphic.
3. Sporozoites multiply by schizogony in tissue cells.
4. Life cycle characteristically consisting of merogony, gametogony and sporogony.
5. Blood or gut parasites of vertebrates.

Examples. *Eimeria*, *Isospora* and *Plasmodium*.

SUBPHYLUM 3. CNIDOSPORA

It is divided into **two** classes:

Class 1. Myxosporidea

1. Spores of multicellular origin.
2. Spores with two or more valves in their membranes.
3. Trophozoite stage well developed and is the main site of proliferation.
4. They are parasites of cold-blooded vertebrates such as fish.

Examples. *Myxobolus*, *Ceratomyxa*.

Class 2. Microsporidea

1. Spores of unicellular origin.
2. Spores with imperforate wall, having only valve.
3. There is one long tubular polar filament through which the sporoplasm emerges.
4. Mitochondria absent.
5. They are intracellular (**cytozoic**) parasites in arthropods (*e.g.*, honey bees, silk worms) and vertebrates (*e.g.*, fish).

Example. *Nosema.*

SUBPHYLUM 4. CILIOPHORA

The subphylum has the following class:

Class Ciliata

1. Most species are free-living and abundant in both marine and freshwater habitats.
2. Ciliate protozoans have a definite body form and size.
3. Body bounded externally by a firm and elastic pellicle.
4. Locomotory and feeding organelles are cilia.
5. Definite mouth (cytostome) and gullet present. Anal aperture (cytopyge) permanent.
6. Contractile vacuoles are always present at fixed positions.
7. Have two types of nuclei: larger vegetative **macronucleus** and small reproductive **micronucleus**.
8. Binary fission transverse.
9. Sexual reproduction never involves the formation of gametes.

The class Ciliata (or Infusoria) is divided into **four** subclasses:

Subclass 1. Holotrichia

1. Body cilia uniform (*i.e.*, no distinction between oral cilia and somatic cilia).
2. Cytostome often apical (or subapical) or midventral, on surface of body or at bottom of atrium or vestibulum.

Subclass 2. Peritrichia

1. Oral ciliary field prominent; body or somatic cilia absent or reduced to temporary posterior circlet of locomotor cilia.
2. Widely distributed species; many are stalked and sedentary (sessile), others are mobile.
3. Dispersal by migratory larval form called **telotroch**.
4. Conjugation total, involving fusion of microconjugants and macroconjugants.

Subclass 3. Suctoria

1. Sessile (sedentary), stalked adult ciliates with the distal end bearing few to many suctorial tentacles (each ending into a sucker-like knob).
2. Adult without cilia; migratory young larva with cilia which originate by budding from parent suctorian.

Subclass 4. Spirotrichia

1. Larged-sized free-swimming ciliates.
2. Body (somatic) cilia reduced.
3. Oral or buccal cilia well marked.
4. Cysts common in some groups.

Subclass 1. Holotrichia

This subclass is divided into following **six** orders:

Order 1. Gymnostomatida

1. Buccal (or oral) cilia absent; body or somatic ciliation uniform.
2. Cytostome opens directly to outside, vestibule (buccal cavity) absent.
3. Toxicysts (a type of trichocyst which induce paralysis or lysis of the prey) common.

Examples. *Didinium*, *Coleps*, *Nassula*.

Order 2. Trichostomatida

1. Cytostome at the bottom of a ciliated vestibule.
2. No buccal cilia; spiral rows of cilia present in the vestibule.
3. Endocommensal of vertebrate hosts such as cattle, pig, monkeys and human beings.

Examples. *Balantidium*, *Colpoda*, *Isotricha*.

Order 3. Chonotrichida

1. Vase-shaped ciliates lacking in body cilia.
2. Sessile (sedentary) forms.
3. Vestibular cilia form a spirally coiled apical funnel.
4. Marine and ectocommensal on crustaceans.

Examples. *Spirochona*, *Lobochona*, *Chilodochona*.

Order 4. Apostomatida

1. Body with spirally arranged cilia.
2. Cytostome mid-ventral, rosettle-shaped in *Foettingeria actinarium*.
3. Marine and parasites or commensals.
4. Life cycle complex, sometimes involving alternation of hosts (unique in phylum), one host is commonly a crustacean.

Examples. *Hyalosphysa*, *Polyspira*, *Foettingeria*.

Order 5. Astomatida

1. Body large or long.
2. Mouth absent.
3. Body ciliation uniform.
4. Complex infraciliary endoskeleton and often with elaborate holdfast, organelles (hooks, spines or sucker) at anterior end.
5. Intestinal endoparasites of oligochaetes (terrestrial, freshwater or marine).
6. Asexual reproduction by fission (budding).

Examples. *Anoplophrya*, *Hoplitophrya*.

Order 6. Hymenostomatida

1. Body small and with uniform somatic cilia.
2. Buccal cavity well defined; buccal ciliature consists of an undulating membrane and an adoral zone of membranelles.
3. Oral area on ventral surface, usually in anterior half of body.

Examples. *Tetrahymena*, *Paramecium*, *Colpidium*.

Subclass 2. Peritrichia

Order Peritrichida

1. Mostly attached stalked bell-shaped ciliates; the stalk is secreted by a region at the aboral end known as the **scopula**.
2. Feeding apparatus is highly specialised.

Examples. *Vorticella*, *Trichodina*.

Subclass 3. Suctoria

1. Body sessile and stalked.
2. Young with cilia, adult with tentacles.

Order Suctorida

Characters as of subclass suctoria.

Examples. *Ephelota*, *Podophrya*, *Acineta*.

Subclass 4. Spirotrichia

It is divided into **three** orders:

Order 1. Heterotrichida

1. Generally large to very large forms, often 'highly contractile, sometimes pigmented (*e.g.*, bluish *Stentor*).
2. Body encased in lorica.
3. Body cilia usually absent.
4. Macronucleus oval or often beaded.
5. Parasitic and free-living species.

Example. *Stentor*, *Bursaria*, *Blepharisma*, *Nyctotherus*.

Order 2. Oligotrichida

1. Body ovoid to elongate, sometimes with tail.
2. Pellicle thickened; body (=somatic) cilia reduced.
3. Oral (=buccal) membranelles are limited to front end.
4. Free-swimming, macrophagous, mainly pelagic.

Example. *Halteria*.

Order 3. Hypotrichida

1. Dorsoventrally flattened, highly mobile, with unique cursorial (or running) type of locomotion.
2. Body dominated by compound ciliary structures such as **cirri** which are located on the ventral side of body and used as "legs" for creeping over the substratum.

Examples. *Stylonychia*, *Euplotes*, *Oxytricha*.

2.3 RECENT CLASSIFICATION SCHEME OF PROTISTA

Brusca and **Brusca** (2003) noted in his book Invertebrates that "Ideas about protist phylogeny and classification are in such a state of flux that any scheme used at the time of this writing would be out-of-date by the time this text reached the book shelves". Thus, on the basis of modes of nutrition and locomotion, protists were divided into the *Mastigophora* (locomotion with flagella), *Ciliophora* (locomotion with cilia), *Sarcodina* (locomotion with pseudopodia), and *Sporozoa* (parasites with no obvious locomotory structures). While these divisions might accurately describe protist's roles in ecosystems, *they do not accurately reflect evolutionary relatedness*. Pseudopodia and flagella are present in many different kinds of cells (including plant and animal cells) and their presence does not indicate unique relatedness (*i.e.*, they are shared primitive features, or **symplesiomorphies**).

Photosynthetic protists contain many different types of chlorophylls and have differently constructed chloroplasts, indicating that they are not closely related. Accordingly most recent protist classification has included following seventeen phyla (see **Brusca** and **Brusca** 2003). [**Note.** Former protist group Myxozoa or Myxosporidea (*e.g.*, *Myxobolus cerebralis*) is found to include highly modified cnidarians, parasitic in certain invertebrates and vertebrates. This revelation was made possible by discovery of certain metazoans and cnidarian features (*e.g.*, collagen, nematocysts) in these animals as well as ribosomal DNA data (Box 2.1)].

Box 2.1

Myxozoan cnidarian (*e.g.*, *Myxobolus cerebralis*, Fig. 2.1) is a parasite of fish (primary host) and of *Tubifix* worm (Intermediate host). Previously it was considered to be protist. *M. cerebralis* and similar other cnidarians are now viewed as highly specialized parasitic cnidarians (**Brusca** and **Brusca**, 2003).

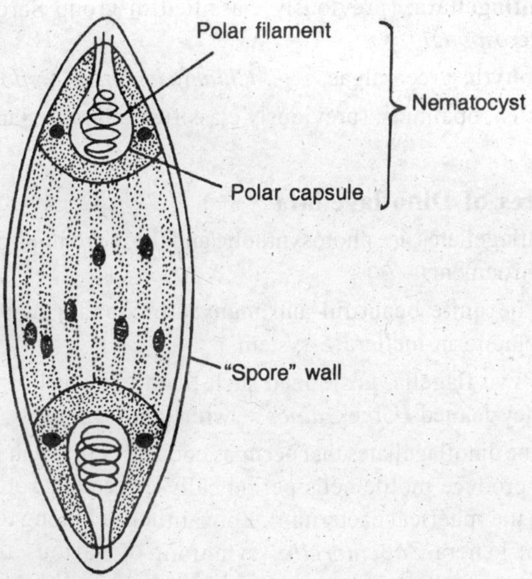

Fig. 2.1 Previously considered to be protists, myxozoans are now viewed as highly specialized, parasitic cnidarians (after Bursca and Bursca, 2003).

Current Phyla of Kingdom Protista

1. Phylum Euglenida, (previously classified in group Sarcomastigophora), *e.g.*, *Ascoglena*, *Colacium*, *Euglena*.

2. Phylum Kinetoplastida, trypanosomes (previously classified in group Sarcomastigophora) *e.g.*, *Bodo*, *Leishmania*, *Trypanosoma*.

3. Phylum Ciliophora, ciliates, *e.g.*, *Balantidium*, *Euplotes*, *Paramecium*, *Stentor*, *Vorticella*.

4. Phylum Apicomplexa (old name Sporozoa), *e.g.*, *Eimeria*, *Gregarina*, *Plasmodium*, *Toxoplasma*.

5. Phylum Dinoflagellata (previously classified in group Sarcomastigophora), *e.g.*, *Ceratium*, *Noctiluca*, *Zooxanthella*.

6. Phylum Stramenopila, *e.g.*, diatoms, brown algae, golden algae, oomycetes.

7. Phylum Rhizopoda, amoebas (previously classified in group Sarcomastigophora), *e.g.*, *Amoeba*, *Arcella*, *Entamoeba*.

8. Phylum Actinopoda, radiolarians, heliozoans, (previously classified in group Sarcomastigophora), *e.g.*, *Actinophrys*, *Actinosphaerium*, *Litocolla*.

9. Phylum Granuloreticulosa, foraminiferans, (previously classified in group Sarcomastigophora), *e.g.*, *Elphidium*, *Gromia*, *Iridia*.

10. Phylum Diplomonadida, (previously classified in group Sarcomastigophora), *e.g.*, *Giardia*, *Hexamida*.

11. Phylum Parabasilida, (previously classified in the group Sarcomastigophora), *e.g.*, *Dientamoeba*, *Trichomonas*, *Trichonympha*.

12. Phylum Cryptomonada, (previously classified in group Sarcomastigophora), *e.g.*, *Chilomonas*.

13. Phylum Microspora, microsporans, *e.g.*, *Nosema*.

14. Phylum Ascetospora, ascetosporans, *e.g.*, *Haplosporidium*.

15. Phylum Choanoflagellata, (previously classified in group Sarcomastigophora), *e.g.*, *Codosiga*, *Monosiga*, *Proterospongia*.

16. Phylum Chlorophyta, green algae, *e.g.*, *Chlamydomonas*, *Eudorina*, *Volvox*.

17. Phylum Opalinidea, opalinids, (previously classified in group Sarcomastigophora), *e.g.*, *Opalina*.

Certain Unique Features of Dinoflagellata

1. Many species of dinoflagellates are photosynthetic and these are important primary producers in many aquatic environments.
2. Dinoflagellates can be quite beautiful and many are capable of bioluminescence (*e.g.*, *Gonyaulax*) using a luciferin-luciferase system.
3. Dinoflagellates have two flagella, positioned such that they whirl or spin as they swim, the attribute to which they named (Greek *dinos* = whirling, turning).
4. **Endosymbiotic** marine dinoflagellates that occur as coccoid cells when inside their invertebrate or protist hosts, but produce motile cells periodically are called **zooxanthellae** (*coccoid* = resembling a coccus, the spherical bacterium). Zooxanthellae belong to the poorly understood but vastly important genera *Zoochlorella* (symbiont of various freshwater organisms), *Zooxanthella* (symbiont of radiolarians) and *Symbiodinium* (symbiont of cnidarians and some other metazoans). Species of *Symbiodinium* (= *Zooxanthella*) are best known as critically important mutualistic symbionts of hermatypic corals (reef-builder corals of Medreporaria and Scleractina). There are several species of *Symbiodinium* in corals. All are photosynthetic and provide nutrients to the corals and help create the internal chemical environment necessary for coral to secrete its calcium carbonate skeleton. Zooxanthellae also occur in many cnidarians other than scleractinian corals, such as milleporinids, chondrophorans, sea anemones and various medusae.
5. Some planktonic dinoflagellates occasionally undergo periodic bursts of population growth and are responsible for a phenomenon known as **red tide**. Red tides have nothing to do with actual tides, and they are only rarely red. A red tide is simply a streak or patch of ocean water discoloured (generally a pinkish orange) by the presence of billions of dinoflagellates. During a red tide, densities of these dinoflagellates may be as high as 10 to 100 million cell per liter of sea water. Exactly why the population explosions of these specific organisms occur is not entirely clear, but **organic pollutants** from terrestrial runoff may be a key culprit.

   Toxic effects of red tides. Many red tide organisms manufacture highly toxic substances. One group of toxins produced by dinoflagellate species such as *Alexandrium* spp., *Gymnodinium* *catenatum* and *Pyrodinium behamense* is called **saxitoxins**. Saxitoxins block the sodium potassium pump of nerve cells and prevent normal impulse transmission. When suspension feeders such as mussels and clams eat these protists, they store the toxins in their bodies. Extremely high concentrations of toxic dinoflagellates will even kill suspension feeders and occasionally also fish caught in the thick of the bloom. The shell fish feeding on the dinoflagellates become toxic to animals that eat them. In humans, the result is a disease known as **paralytic shellfish poisoning (PSP)**. Extreme cases of PSP result in muscular paralysis and respiratory failure. Over 300 human deaths worldwide have been documented (see **Brusca** and **Brusca**, 2003).

Gymnodinium breve releases a family of toxins called **brevetoxins** that result in **neurotoxic shellfish poisoning (NSP).** Red tides formed by *G. breve* have produced massive fish kills in Florida and along Gulf of Mexico in the recent years. Humans consuming animals that have brevetoxins accumulated in their tissues experience uncomfortable gastrointestinal side effects such as diarrhoea, vomiting and abdominal pain and also neurological problems including dizziness and reversal of temperature sensation.

A newly discovered species of dinoflagellate, *Pfiesteria piscicida* have caused in 1997 massive fish kills on the eastern shore of Maryland. The toxin in *Pfiesteria* is 1,000 times more powerful than cyanide. Outbreaks of *Pfiesteria* are thought to be linked to faecal material and urine of hogs of hog-farms which are dumped in the sea in North Carolina sea coasts (see **Brusca** and **Brusca**, 2003).

6. The shape of dinoflagellates is maintained by alveoli beneath cell's surface, and a layer of supporting microtubules. In some, the alveoli are filled with polysaccharides, typically cellulose, and these dinoflagellates are said to be **thecate** or **armored** (*e.g.*, *Ceratium*). Dinoflagellates that have empty alveoli are said to be **athecate** or **naked** (*e.g.*, *Noctiluca*).
7. In dinoflagellates, osmoregulation is performed by a unique system of double-membrane-bound tubules, called **pusules**, which open to the outside via a canal.
8. The chloroplasts of photosynthetic dinoflagellates are surrounded by three membranes, and thylakoids are arranged in stacks of three. Photosynthetic pigments include chlorophylls a and c_2, phycobilins, carotenoids (*e.g.*, β carotene) and also the xsnthophylls (*e.g.*, peridinin, found only in dinoflagellates), neoperidinin, dinoxanthin, and neodinoxanthin. These xanthophylls mask the chlorophyll pigments and account for the golden or brown colour that is commonly seen in dinoflagellates.
9. A few dinoflagellates (*e.g.*, *Noctiluca*) have a permanent cell mouth or **cytostome** supported by sheets of microtubules. The cytosome is often surrounded by **extrusomes** which are of three types: trichocysts, mucocysts and nematocysts. The most common are **trichocysts** (similar to those found in ciliates), believed to be fired in defense or to capture and bind prey. Sac-like **mucocysts** secrete mucous material on to the surface of cell. This may help in attachment to substratum (*e.g.*, *Amphidinium*) or may help to capture prey (*e.g.*, *Noctiluca*). Other dinoflagellates (*e.g.*, *Nematodinium*, *Nematopsides*, *Polykrikos*) have nematocysts that resemble, but not homologous to the stinging organelles of cnidarians of the same name.
10. The nuclei of dinoflagellates have three unusual features: (*i*) they contain five to ten times the amount of DNA that is found in most eukaryotic cells; (*ii*) the five histone proteins that are typically associated with the DNA of other eukaryotic cells are absent; and (*iii*) the chromosomes of dinoflagellates remain condensed and (*iv*) the nucleolus remains intact during interphase and mitosis. Most dinoflagellates (except *Noctiluca*) spend much of their lives as haploid cells (called **vegetative cells** to distinguish them from haploid gametes).

11. Nuclear division is by closed extranuclear pleuromitosis. No centrioles are present, and the organization center for the mitotic spindle is not obvious. Meiosis is of unusual type. Both asexual and sexual reproduction occur in the dinoflagellates.

Thus, dinoflagellates are unlike all other eukaryotes in many respects.

2.4 TYPE STUDIES

1. *Entamoeba histolytica*

Classification

Phylum	–	**Protozoa**
Subphylum	–	**Sarcomastigophora**
Superclass	–	**Sarcodina**
Class	–	**Rhizopoda**
Order	–	**Amoebida**
Genus	–	***Entamoeba***
Species	–	***histolytica***

Entamoeba histolytica is an amoeba-like microscopic disease-producing (pathogenic) endoparasite of human beings. It inhabits colon of human beings and produces the disease called **amoebiasis** or **amoebic dysentery**. **Lamble** (1859) discovered this species. Russian Zoologist **Losch** (1875) discovered its pathogenic nature.

1. Distribution. *Entamoeba histolytica* is cosmopolitan (world-wide or universal) in distribution but is more common in tropical countries such as India, China, Mexico, etc., due to poor sanitary conditions.

2. Habit and habitat. *Entamoeba histolytica* lives in the lumen of the lower portion of small intestine and the entire large intestine (colon) of humans. It may invade the lining (*i.e.*, mucosa and submucosa) of the host's large intestine. There it feeds on tissues and red blood corpuscles of intestinal wall and produces ulcers and dysentery which involves inflammation of intestine, loose stools or diarrhoea and 6 to 10 blood-stained painful mucus motions per day. In cronic cases, it may enter the blood circulation and may infect liver, lungs, kidneys, testes and even the barin. It proves fatal when it produces large abscesses in liver or perforates the intestinal wall of the host.

3. Morphology. The adult *E. histolytica* occurs in following two forms:

A. Trophozoite or magna form. It is tissue dwelling large-sized (having 20–30 μm diameter), trophic or feeding form. It lives in the mucosa of intestinal wall and feeds on fragments of tissue cells, erythrocytes (RBC), leucocytes (WBC) and bacteria. By the help of secretion of a proteolytic enzyme called **cytolysin** or **histolysin**, it destroys mucosa and produces disease. Thus, it is a **pathogenic form**.

B. Precystic or minuta form. It is lumen-dwelling, small sized (having diameter of 7–10 μm or microns), non-feeding and **non-pathogenic form**.

A. Trophozoite. Trophozoite (Fig. 2.2) represents mature parasite which is active or motile adult resembling *Amoeba* in some structural details. It is irregular or asymmetrical in form and is covered by a thin and flexible or elastic **plasmalemma** or **plasma membrane**. The cytoplasm of trophozoite is differentiated into two distinct regions: ectoplasm and endoplasm. **Ectoplasm** forms a thin, clear (hyaline), non-granular and relatively firm outer layer. It contains no organelles. The **endoplasm** is the inner, granular, fluid-like and constitutes the bulk of the body. Endoplasm contains two types of organelles: nucleus and food vacuoles. The contractile vacuole is absent in *E. histolytica* since it lives in an isotonic environment and does not face the problem of osmoregulation.

Nucleus of trophozoite is large (4–6 μm in diameter), rounded and vesicular containing six chromosomes (**Kofoid** and **Swezy**). It is bounded by a delicate **nuclear envelope** having **chromatin**

granules on its inner surface. Nucleus has a central dot-like **nucleolus** (**endosome** or **karyosome**) surrounded by a clear area called **halo**. Nucleoplasm contains distinct spoke-like striations existing between nucleolus and peripheral chromatin.

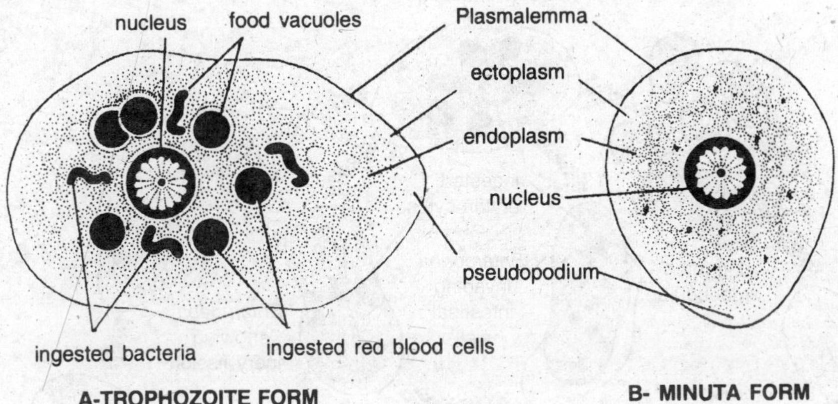

Fig. 2.2. *Entamoeba histolytica.* A—Trophozoite or magna form; B—Minuta form.

Food-vacuoles vary in number and size. They contain red blood corpuscles (a characteristic feature), white blood cells or bacteria in various stages of digestion.

4. Physiology

(*i*) Locomotion. *E. histolytica* performs locomotion by the help of a single, large, thick and blunt pseudopodium called **lobopodium**. It is called **monopodial locomotion**. Lobopodium is formed largely of ectoplasm and protrudes at the advancing end of body.

(*ii*) Nutrition. Trophozoite has **holozoic nutrition**. It dissolves the intestinal mucous membrane of the host and phagocytoses the debris, *i.e.*, blood and epithelial cells, at the level of its plasma membrane. Food is digested intracellularly inside the food vacuoles by the help of lysosomal enzymes.

(*iii*) Respiration. Trophozoite is a **facultative anaerobe**, *i.e.*, it requires very little oxygen for the breathing and can survive without oxygen. It is an adaptation for its endoparasitic mode of existence inside the colon.

(*iv*) Excretion. Trophozoite excretes its nitrogenous metabolic wastes in the form of **ammonia**. The ammonia is got rid off by diffusion through general body surface into the surrounding medium.

B. Minuta form. It represents precystic form which is small, spherical, non-motile, non-feeding and non-pathogenic. Minuta form lacks pseudopodium and food vacuoles. It contains many glycogen granules (reserve food) and black, rod-like **chromatoid bodies** (probably ribonucleoproteins). Chromatoid bodies occur singly or in the multiple of two or more but disappear in the old cysts. Minuta form undergoes encystation and helps in the transmission of parasite.

5. Reproduction. *Entamoeba histolytica* shows only asexual reproduction which involves mitosis having intranuclear karyokinesis and the cytokinesis. Asexual reproduction takes place by **binary fission**.

6. Life cycle. *E. histolytica* is a **monogenetic parasite**, *i.e.*, its life cycle involves only one host; the host being the human beings. Its life cycle includes the following steps:

(*i*) Encystment. Some of the precystic minuta forms exist in intestinal lumen and undergo **encystment** or **encystation**. However, before the encystment, they become round, eliminate food vacuoles and accumulate considerable amount of reserve food materials in the form of glycogen granules and chromatoids. Soon each parasite secretes a thin, rounded, resistant, colourless and transparent cyst wall around it. The cysts of *Entamoeba histolytica* have average size of 12 mm (or 12 microns). Each cyst has a clear cytoplasm and single nucleus and is called **mononucleate cyst**. Ultimately, the nucleus of the cyst divide twice so that each cyst becomes tetranucleate or quadranucleate. At this stage the cysts are **infective** to new hosts.

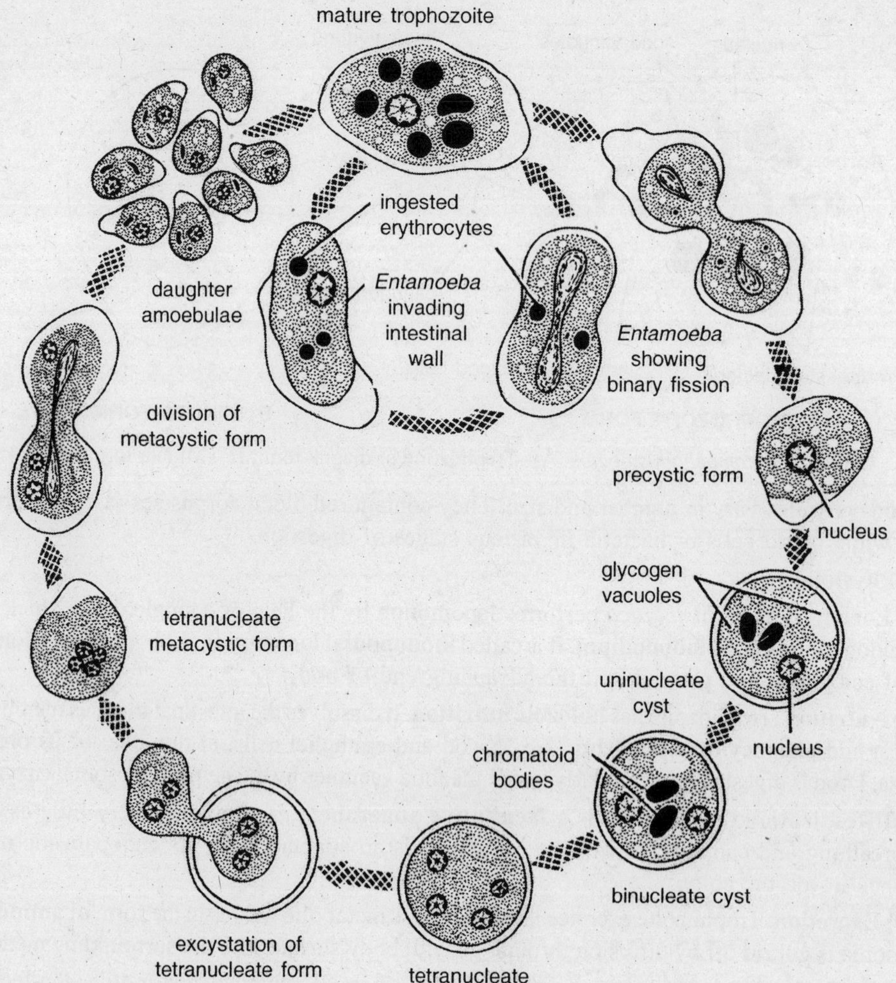

Fig. 2.3. *Entamoeba histolytica*. Life cycle.

(*ii*) Transfer to new host. The tetranucleate cysts pass out with the faecal matter of the host (These are used for diagnosis of the infection). Cysts remain alive for 10 days in the moist stool. They die if dried or if the atmospheric temperature rises to about 50°C. Cysts are also resistant to chemicals. Infection of fresh human host takes place by swallowing the infective cysts with contaminated food and drinks. Contamination of food and drinks is brought about by houseflies, cockroaches and dirty fingers of food handlers.

Through contaminated food, infective cysts pass into the lower portion of the small intestine of the host.

(*iii*) Excystment. Excystation of the parasite occurs in host's intestine according to any of two methods: 1. The cyst wall is digested by alkaline juice (trypsin enzyme) of host's small intestine. 2. In culture, a small portion of cyst is dissolved by some enzymes secreted by the pseudopodium of the parasite. Anyhow, a tetranucleate amoebae is hatched out and is called **metacystic form**.

Metacystic entamoeba produces a new generation of trophozoite by a diverse series of nuclear and cytoplasmic divisions which result in the production of eight uninucleate daughter amoebulae which represent young trophozoites. Young trophozoites feed on the contents (bacteria) of the lumen of intestine and grow in size to form the normal parasitic trophozoite.

7. Pathogenecity. *Entamoeba histolytica* causes **amoebic dysentry**. It causes ulcers and abscesses (*i.e.*, a localised collection of pus surrounded by inflammed tissue) in the intestinal wall by eroding submucosa. As the ulcers burst, the blood and mucus are passed out.

8. Symptoms. The common symptoms of infection of amoebic dysentery are abdominal pain, nausea, flatulance and bowel irregularity with fatigue and headache.

9. Treatment (therapy) and prevention (prophylaxis). Prevention from infection involves personal health and hygiene. One must keep his food-stuff, drinking water and utensils clean and well-covered, preventing their contamination by houseflies, ants, mice, cockroaches, etc. Also, one must keep his nails trimmed and wash hands with soap and water before meals.

Metronidazole, Emetin, iodine compounds (**Yatren, Diodoquin**) and arsenic compounds (**Carbarsone, Thiocarbarsone**) are used in amoebic dysentry. **Bismuth subritrate** is found beneficial in controlling amoebic dysentery. Certain **antibiotics**, such as **Fumagillin, Erythromycin, Terramycin** and **Auromycin** have proved to be more effective in case of severe infection. In case of secondary infection, *i.e.*, infection of liver, lungs, etc., **Chloroquine** is given.

2. *Entamoeba gingivalis*

Classification

Phylum	–	**Protozoa**
Subphylum	–	**Sarcomastigophora**
Superclass	–	**Sarcodina**
Class	–	**Rhizopodea**
Order	–	**Amoebida**
Genus	–	***Entamoeba***
Species	–	***gingivalis***

Entamoeba gingivalis is a nonpathogenic human parasite. It was first parasitic amoeba reported from human beings and was discovered by **Gross** in 1849.

1. Habit and habitat. *E. gingivalis* is commonly known as **mouth amoeba**. It lives in the cavities of teeth; in the tartar and plaque deposited around the base of teeth. It may also live in the abscesses of gums and pus-pockets of tonsils.

2. Morphology. The subspherical **trophozoite** of *E. gingivalis* measure 12 to 20 mm in diameter. It bears 2 or 3 small and blunt (*i.e.*, broad and rounded) pseudopodia, called **lobopodia**. The cytoplasm is divisible into a clear homogeneous peripheral **ectoplasm** and central, granular, highly vacuolated **endoplasm**. Endoplasm contains a vesicular **nucleus** with central **endosome** and several **food vacuoles**.

3. Physiology. Trophozoites has pseudopodial locomotion. It feeds mainly on bacteria present in the cellular debris around the

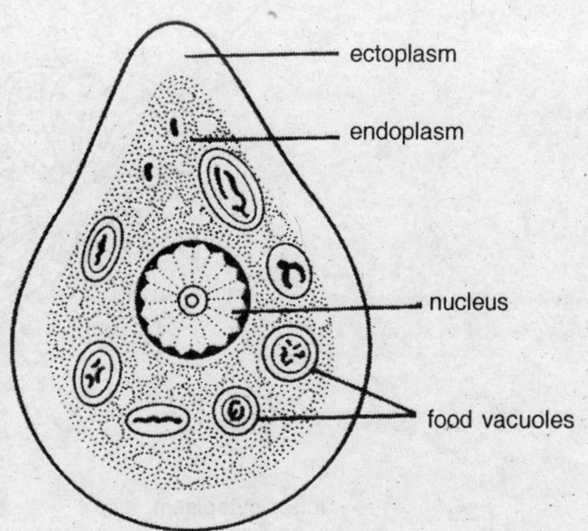

Fig. 2.4. *Entamoeba gingivalis*. Trophozoite.

roots of teeth and also on white blood cells (WBC) by phagocytosis.

4. Life cycle. Trophozoite reproduces only by asexual means, *i.e.*, by binary **fission**. It does not form cysts and is transmitted from mouth to mouth of persons either by contact during kissing or while eating or drinking by same utensils.

5. Pathogenicity. *E. gingivalis* is found in nearly 70 per cent of human population. Its occurrence is more common in the mouth of persons suffering from pyorrhoea. It was believed in the past that pyorrhoea was caused by *E. gingivalis*. However, it is now certain that pyorrhoea is caused by infection of bacteria. *E. gingivalis* is known to aggravate pyorrhoea disease by destroying the gum tissues.

3. *Elphidium crispum* (= *Polystomella crispa*)

Classification

Phylum	–	**Protozoa**
Subphylum	–	**Sarcomastigophora**
Superclass	–	**Sarcodina**
Class	–	**Rhizopodea**
Order	–	**Foraminiferida**
Genus	–	***Elphidium* (*Polystomella*)**
Species	–	***crispum* (*crispa*)**

Elphiduim crispum represents a shelled-protozoan or foraminiferan having many chambered shell and thread-like axopods. It exhibits the phenomenon of **dimorphism**, *i.e.*, *Elphidium* exists in two distinct forms—large megalospheric and small microspheric forms.

1. Habit and habitat. *Elphidium* is a marine and free-living protozoan. It is found on the bottom of sea at about the depth of 300 fathoms (one fathom = 1.8 metres). It creeps among the sea weeds.

2. Morphology

(*i*) Shell. *Elphidium* is shelled, unicellular and microscopic protozoan (having 3 to 4 mm diameter). Its shell is hard, transluscent, pale yellow in colour, and chemically is made up of calcium carbonate and small amounts of silica and magnesium sulphate. The shell is biconvex, coiled.

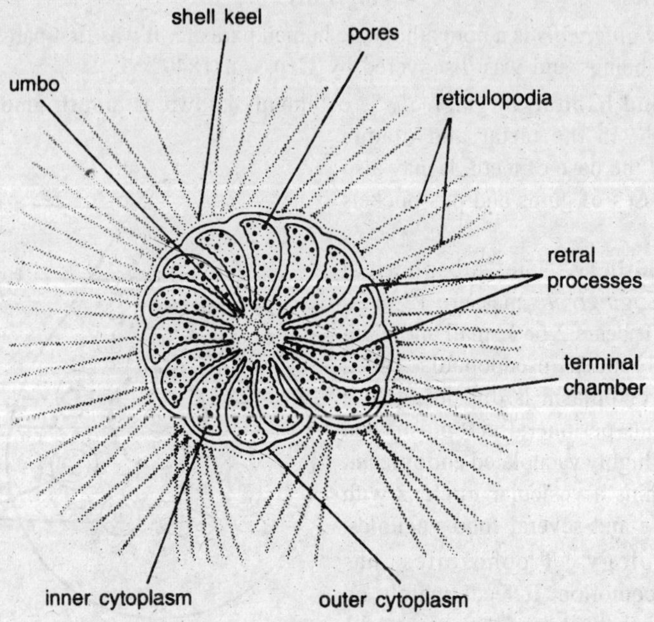

Fig. 2.5. *Elphidium crispum.*

multichambered or multilocular (= polythalamus) and perforated. Shell has a central rounded part, called **umbo** and a peripheral rigid continuous rim, called the **keel.** ✓

The shell is coiled in a **planospiral** manner, *i.e.*, all the V-shaped chambers or coils lie in the same plane or a flat spiral. All the chambers are filled with cytoplasm. Walls of each chamber bear minute pores through which the cytoplasm remains continuous from one chamber to another, and also extends to outside, to form pseudopodia. Each chamber of shell has a convex anterior surface and a concave posterior surface bearing **retral processes** filled with cytoplasm. The formation of shell begins with an initial single chamber, called **proloculum**. As the animal grows in size, successive chambers are laid down in a spiral manner, forming whorls. Each new chamber is larger than the preceding one and each new whorl partially overlaps and conceals the older one. Thus, only the last whorl is visible from outside. Overlapped portions of shell-chambers are called **alar processes.**

Shell and polymorphism. The small or large size of proloculum results in **dimorphism (polymorphism)** in *Elphidium*. There occur two forms: the large **megalospheric form (gamont)** which has a large central chamber (proloculum), a large single nucleus and sexual mode of reproduction. The small **microspheric form (agamont)** has a small central chamber, proloculum, many small nuclei and an asexual mode of reproduction.

***(ii)* Cytoplasm.** The shell's chambers are filled with inner cytoplasm or **medulla**. A thin layer of **outer cytoplasm** or **cortex** covers the shell from outside. The inner cytoplasm contains nucleus or nuclei, food particles, minute vacuoles, Golgi appartus, mitochondria, endoplasmic reticulum, ribosomes and brown granules or xanthosomes containing waste matter. Contractile vacuoles are absent.

***(iii)* Nucleus.** The medulla contains a single nucleus in megalospheric individuals and many nuclei in microspheric forms.

***(iv)* Reticulopodia.** Pseudopodia of *Elphidium* are thread-like. These threads branch and branches fuse to form a network. Such pseudo-podia are called **reticulopods** or **axopods**. Each reticulopod (= reticulopodium) is made of an inner fibrillar **axis** (compri-sing a bundle of microtubules) and outer fluid-like cortex of moving, adhesive cytoplasm (Fig. 2.6). Reticulopods are temporary extensions of cytoplasm, they can be extended out of the pores (foramina) of shell like tiny rockets and can be withdrawn with a lightining speed. They help in locomotion, nutrition and in the formation of shell.

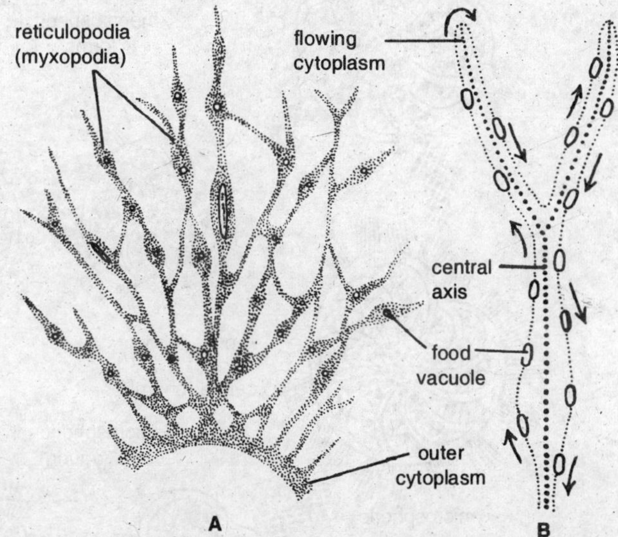

Fig. 2.6. *Elphidium*. A—Reticulopods forming a network. B—A reticulopod showing streaming circulation in the cortex.

3. Physiology

***(i)* Locomotion.** *Elphidium* creeps slowly with the help of its reticulopods on sea weeds at the bottom of the ocean. Locomotion of animal is performed by contraction and expansion (extension) of the reticulopods.

(*ii*) Nutrition. Nutrition of *Elphidium* is **holozoic**. It feeds upon minute organisms such as diatoms, algae, protozoans and microcrustaceans. The network of reticulopods secrete a layer of adhesive cytoplasm to trap the food organisms. This adhesive cytoplasm contains discharged lysosomes and their proteolytic secretions which paralyze the prey and initiate digestion even during capture (**Ruppert** and **Barnes**, 1994). The food is enclosed in food vacuoles existing in the cortex of reticulopods. Digestion of food is normally completed outside the shell and the products of digestion pass into the medulla.

4. Life cycle. The life cycle of *Elphidium* involves alternation of asexual generation (microspheric forms) with sexual generation (macrospheric forms). The macrospheric forms or gamonts produce numerous identical biflagellate gametes called **isogametes**. The gametes are released into sea water through the pores of shell. They swim for sometime and then gametes fuse in pairs (**syngamy**) to form **zygotes**. Each zygote secretes a small initial chamber (proloculum) which develops into a rnicrospheric form (Fig. 2.7).

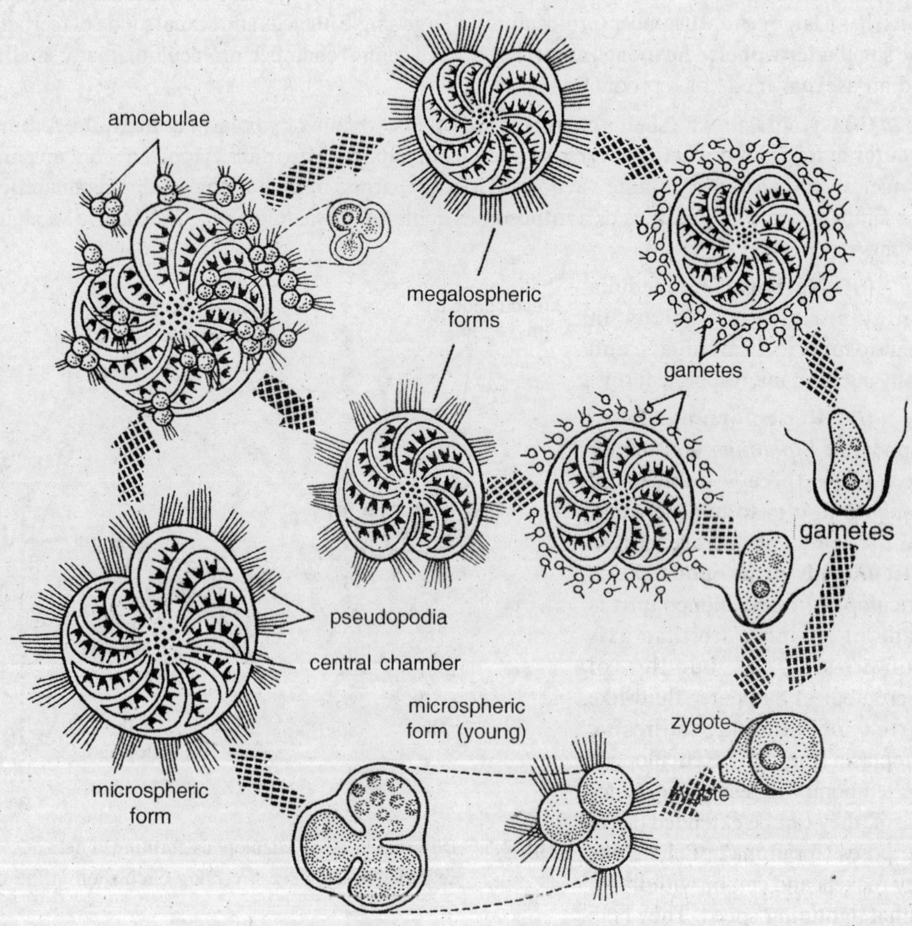

Fig. 2.7. *Elphidium crispa*. Life cycle.

4. *Plasmodium* (Malaria parasite)

Classification

Phylum	–	**Protozoa**
Subphylum	–	**Sporozoa (Apicomplexa*)**
Class	–	**Telosporea**
Subclass	–	**Coccidia**
Order	–	**Eucoccidia (Haemosporidida)**
Genus	–	***Plasmodium***

[*The term Apicomplexa is now used for Sporozoa; it is so named because of a complex of ring-like, tubular filamentous organelles at the apical end; see **Ruppert** and **Barnes**, 1994] (Fig. 2.8).

1. Geographical distribution. The tropical zone of earth is the endemic home of all malaria parasites.

2. Habit and habitat. *Plasmodium* or malaria parasite is a pathogenic **intracellular blood parasite** (= cytozoic endoparasite) of humans and other vertebrates. It lives in hepatocytes (liver cells) and red blood corpuscles (or erythrocytes). Its life cycle is **digenetic**, *i.e.*, the life cycle of the parasite is completed on the following two hosts:

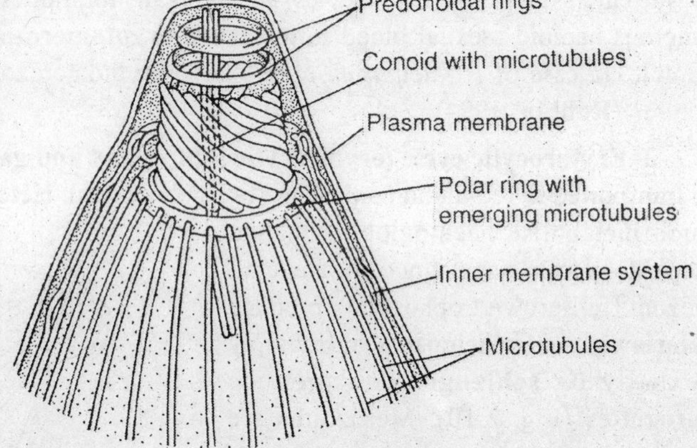

Fig. 2.8 Microstructure of the apical complex of sporozoa (after Brusca and Brusca, 2003).

(*i*) Humans. They are **primary hosts** or definitive hosts where it lives in erythrocytes (RBC) and completes its **endogenous** or **asexual cycle** (*e.g.*, schizogony).

(*ii*) Female *Anopheles* mosquito. It is the **secondary** or **intermediate host** where *Plasmodium* completes its **exogenous** or **sexual cycle** (*e.g.*, gametogony, syngamy and sporogony) inside the alimentary canal.

At present, mosquito is considered as primary host (due to presence of sexual phase of life cycle) and human being is condisered as secondary host (due to presence of asexual phase of life cycle of the parasite).

3. Morphology. *Plasmodium* exhibits typical parasitic adaptations of sporozoa, *e.g.*, feeding trophozoite of human RBC is characterized by the absence of locomotory organs, a mouth, anal opening, excretory pore and vacuoles. However, in it reproductive capacity becomes more prominent. The mode of nutrition is **saprozoic** which occurs by **osmotrophy**. Respiration is **anaerobic**.

While plasmodia are the actual infectious agents of disease (malaria), the female mosquito is the **vector** that passes the portozoan from one host to another. Following four species of *Plasmodium* are pathogenic to human beings causing different types of malaria:

1. *Plasmodium vivax*	Causes benign tertian malaria.
2. *Plasmodium falciparum*	Causes malignant tertian malaria, pernicious malaria, falciparum malaria or estivoautmnal malaria.

3. *Plasmodium malariae* Causes quartan malaria.

4. *Plasmodium ovale* Causes ovale tertian malaria.

The most virulent in human populations is *Plasmodium falciparum*. Its infection produces severe anemia, cerebral symptoms, renal failure, pulmonary edema and death.

Life Cycle

It involves the following two cycles which alternate with each other:

A. Asexual cycle in humans. It is completed in the following steps:

1. Pre-erythrocytic cycle (primary exo-erythrocytic cycle or tissue phase). Parasite (malaria sporozoite) finds its way into the blood of healthy person along with the salivary secretion (containing anticoagulant) of infected female *Anopheles* mosquito which bites for getting a blood meal. The parasite enters liver cells (hepatocytes) and asexually multiplies there by **liver schizogony** producing numerous haploid asexual blood forms called **cryptomerozoites** or **metacryptomerozoites**. For example, in case of *P. falciparum* upto 30,000 cryptomerozoites are released when a hepatocyte ruptures (**Robbin**, 1994).

2. Erythrocytic cycle (erythrocytic schizogony and gametogony). Some cryptomerozoites are transported to red blood corpuscles (or erythrocytes). Here, they become feeding **trophozoites** which metabolise haemoglobin, change morphology from ring (signet-ring stage) to schizont (full-grown trophozoite), produce **malaria pigment** or **hemozoin** and multiply asexually by **schizogony** to produce **merozoites** (Fig. 2.10). Merozoites are liberated in the blood by the rupture of RBCs, causing chills and fever. Merozoites enter fresh RBCs to repeat schizogony.

B. Sexual cycle in female *Anopheles*. During erythrocytic cycle in human blood, some merozoites develop into **gametocytes**. The male gametocytes are called **microgametocytes** and female as **macrogametocytes**. When the mosquito (female *Anopheles*) bites an infected human, it sucks up gametocytes of *Plasmodium* along with its blood meal. In the digestive tract of the mosquito the gametocytes develop into haploid male and female **gametes.** Fertilization occurs within stomach and the **zygote** (called **ookinete**) burrows into the stomach wall (Fig. 2.10). Following zygotic meiosis a series of mitosis occurs, resulting in the formation of many **sporozoites,** which migrate to the insect's salivary gland, from where they are introduced into the new host.

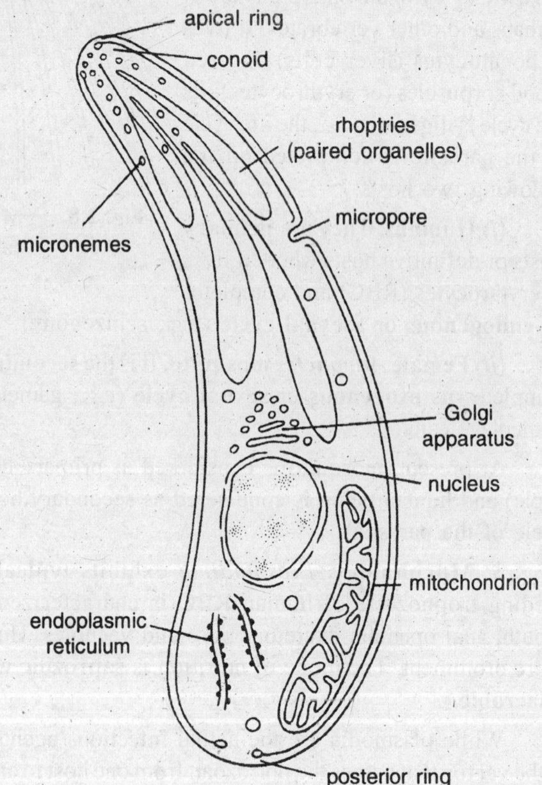

Fig. 2.9 Lateral view of a generalized sporozoan (apicomplexan); the sporozoite showing apical complex.

Thus, in the life cycle of *Plasmodium* there occurs an alternation of sexually reproducing and asexually reproducing generations or **metagenesis**.

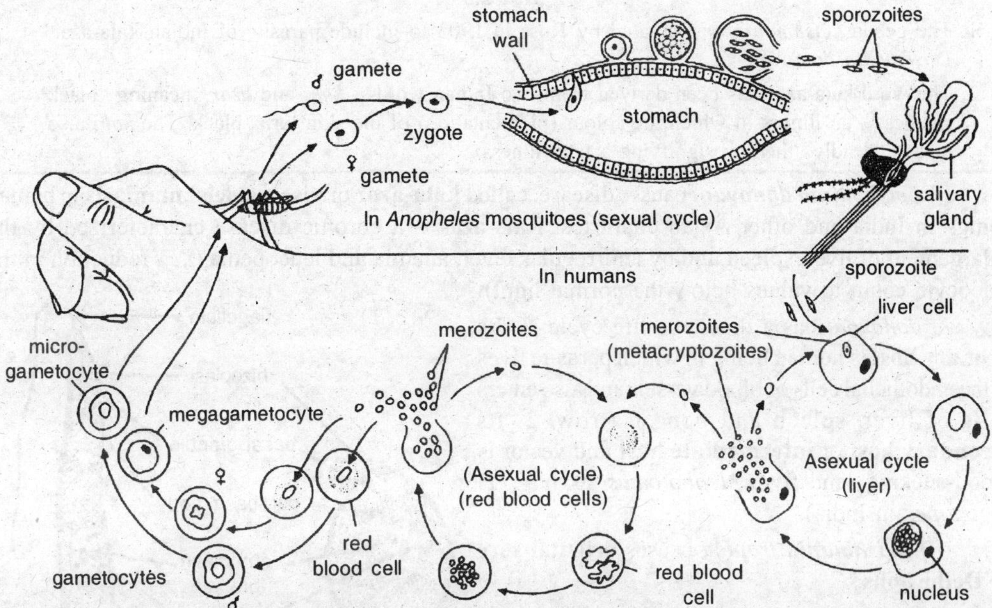

Fig. 2.10 The life cycles of *Plasmodium* in a mosquito and in a human. Reinvasion of liver cells in the tissue cycle does not occur in *Plasmodium falciparum*.

Box 2.2 Reasons for Alarming Comeback of Malaria

The development of a vaccine for the malarial parasite continues to elude researchers.

1. One of the principal causes is the dramatic rise in number of pesticide-resistant mosquitoes-*Anopheles*, the insect vector for *Plasmodium*. By 1968, 38 strains or species of *Anopheles* in India alone had been identified as largely DDT-resistant.
2. *Plasmodium falciparum*, the most deadly species of malarial parasite has become recently resistant to **chloroquine**, once a backbone of malarial drug therapy.
3. Researchers have recently discovered that *Plasmodium* species, like *Trypanosoma*, have the ability to avoid detection by the human immune system by switching among as many as 150 genes that code for different versions of the protein that coats the surface of their cells (it is this protein coat that the human immune system relies on as a recognition factor). Furthermore, because the parasite sequesters itself inside red blood cells, it is largely protected from most malarial drugs.
4. Recent evidence also suggested that the parasite may affect its mosquito host by inducing it to bite more frequently than uninfected mosquitoes (see **Brusca** and **Brusca**, 2003).

5. *Leishmania*

Classification

Phylum	–	**Protozoa**
Subphylum	–	**Sarcomastigophora (Plasmodroma)**
Superclass	–	**Mastigophora**
Class	–	**Zoomastigophorea**
Order	–	**Kinetoplastida (Protomonadina)**
Genus	–	***Leishmania***

1. Habit and habitat. *Leishmania* is a pathogenic intracellular endoparasite of humans and other mammals, causing the following three types of diseases in humans:

Box 2.3

1. The genus *Leishmania* was created by **Ross** in 1903 to include parasite of Indian kala-azar, *Leishmania donovoni.*
2. The word kala-azar has been derived from two Indian words—*kala* and *azar*, meaning "black sickness", an illness in which the colour (pigmentation) of the skin turns black. The *kala* also means "deadly" thereby signifying a fatal illness.

(*i*) *Leishmania donovani* causes disease, called **kala-azar** or **visceral leishmaniasis** in human beings in India and other Asian countries. Kala-azar is a chronic disease characterized by the enlargement of liver, spleen and by an irregular fever, anemia and leucopenia (*i.e.*, reduction in the leucocyte count to values below the normal limit).

L. donovoni has a digenetic life cycle: 1. Its **primary host** is human being in which parasite lives in the endothelial cells of blood vessels and also enters cells of liver, spleen and bone marrow. 2. Its **secondary host** or **intermediate host** and vector is blood-sucking sand fly, *Phleobotomus* sp. (*e.g.*, *P. argentipes* in India).

(*ii*) *Leishmania tropica* causes **oriental sore** or **Delhi boils**.

(*iii*) *Leishmania brassiliensis* causes mucocutaneous American leishmaniasis.

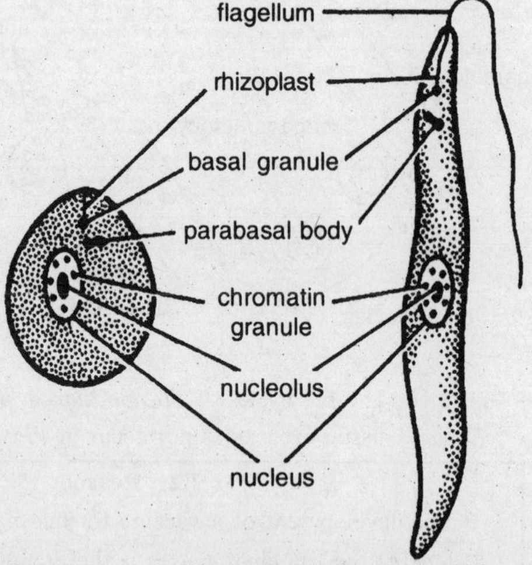

Fig. 2.11 *Leishmania.* A—Leptomonad form; B—Leishmanial form.

2. Morphology. *L. donovoni* has two phases in the life cycle, one flagellar **leishmanial form (amasti-gote stage)** in the reticulo-endothetial cells of human beings and other **leptomonad form (promastigote stage)** in the alimentary canal of sandfly. *Leishmania* is intracellular and found in the cells of reticulo-endothelial system. This form is oval leptomonad form 2–4 μm in diameter with a limiting membrane, the **pellicle**. The cytoplasm contains an oval nucleus, rod-shaped or dot-like kinetoplast and parabasal body. It reproduces asexually by **binary fission**.

Flagellar promastigote form of *L. donovoni* occurs in insect vector (sandflies). It is long slender spindle shaped measuring 15 to 20 μm in length by 1 to 2 μm in breadth. Single flagellum projects from the anterior side of body and do not form undulating membrane along the body.

3. Transmission. Amastigote forms of the pathogen are liberated in human blood stream as a result of the rupture of reticulo-endothelial cells. In the blood stream, some of free amastigote are phagocytosed by the neutrophilic granulocytes and monocytes (macrophages). In India, the vector, the blood sucking sandfly (*Phlebotomus argentipes*) draws these free amastigote forms during its blood-meal. Inside the midgut of sandfly these amastigote forms develop into **promastigote forms** which again multiply by binary fission producing numerous flagellates. Flagellates tend to spread forwards to anterior part of alimentary canal (pharynx and buccal cavity; salivary glands are not infected). The transmission of pathogen is thereby affected through bite of infected sand fly from human being to human being.

6. *Ceratium*

Classification

Phylum	–	**Protozoa**
Subphylum	–	**Sarcomastigophora (Plasmodroma)**

Class	–	**Mastigophora**
Order	–	**Dinoflagellida**
Genus	–	***Ceratium***

1. Habit and habitat. *Ceratium* (*e.g., Ceratium hirudinella*) is found in freshwater plankton. The nutrition in pigmented species is mainly holophytic (autotrophic), sometimes holozoic. Many species of *Ceratium* are known to be **luminiscent**.

2. Morphology. The body of *Ceratium* is enclosed within a thick and rigid pellicle of cellulose called **shell, theca** or **lorica**. It is made of many closely fitting platelets. In the lorica, there are three (*i.e.*, one anterior or apical and two posterior or lateral) long projections or wing-like extensions (spines) which protrude from the body, creating a bizarre shape (Fig. 2.12). The body of *Ceratium* bears two flagella. One is attached a short distance behind the middle of the body, is directed posteriorly and lies in a longitudinal groove (called **sulcus**). The other flagellum is transverse and located in a groove (called **annulus** or **cingulum**) that encircles the body. Thus, cingulum divide the body into an anterior region the **epicone** and a posterior **hypocone**. The transverse flagellum causes both rotation and forward movement; the longitudinal one drives water posteriorly and contributes to forward motion.

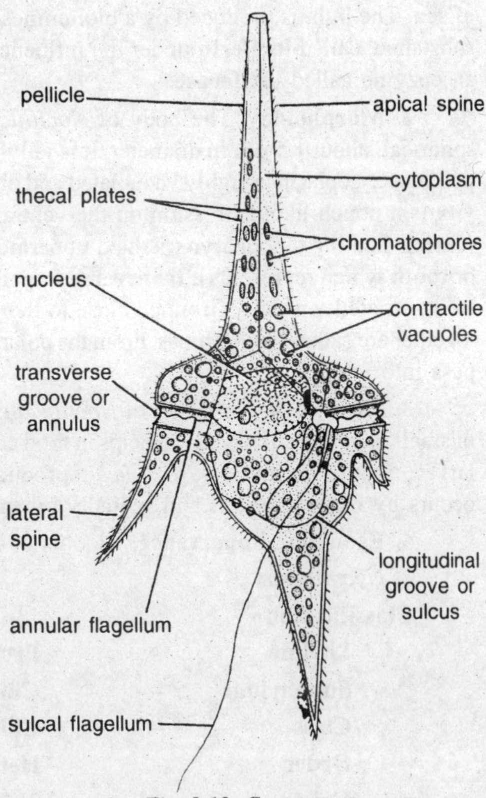

Fig. 2.12. *Ceratium.*

Cytoplasm contains a single large and central **nucleus** containing prominent **endosome** (or nucleolus) and many rod-shaped **chromatophores** which are divided into five distinct groups. Chromatophores are green (having chlorophyll) in freshwater forms, but they are yellowish-brown in marine forms. Reserve food material is starch, glycogen and fat droplets. Cytoplasm also contains non-contractile vacuole, called **pusule**, a stigma and foreign bodies such as bacteria, flagellates and diatom.

3. Physiology. Nutrition of *Ceratium* is both **holophytic** and **holozoic**. Feeding for holozoic nutrition is done by extruding large parts of cytoplasm through the pores of lorica to form an anastomosing net on the body (pseudopodia called **reticulopodia**) which entangles the prey. The partly digested food is withdrawn along with the cytoplasm.

4. Life Cycle. *Ceratium* reproduces asexually by **binary fission** and **cyst** formation.

7. *Noctiluca*

Classification

Phylum	–	**Protozoa**
Subphylum	–	**Sarcomastigophora (Plasmodroma)**
Class	–	**Phytomastigophorea**
Order	–	**Dinoflagellida**
Genus	–	***Noctiluca***

1. Habit and habitat. *Noctiluca* is a marine and pelagic phosphorescent or bioluminescent dinoflagellate. It is very common in Mardras beach and abundant in Atlantic and Pacific oceans. *Noctiluca* often becomes so numerous that it forms a thick red scum on sea surface which appears as tomato soup. It emits bluish-green light at night. Due to this character, *Noctiluca* is commonly called as "**night light**" and produces the effect of **burning of sea**. The light is produced by a bioluminescent substance, called **luciferin** under the influence of an enzyme called **luciferase**.

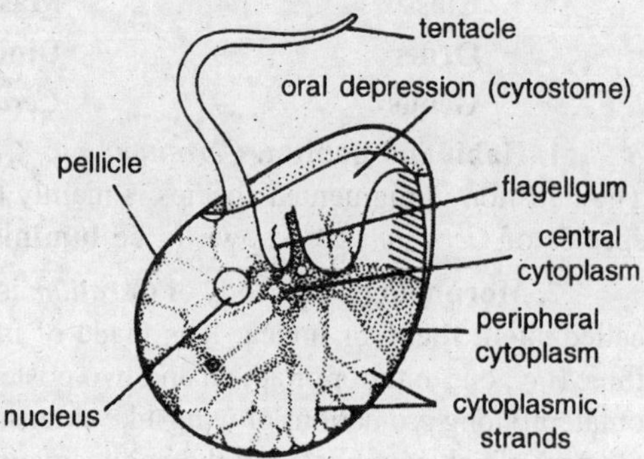

Fig. 2.13. *Noctiluca.*

2. Morphology. The body of *Noctiluca* is spherical, about 1.5 μm in diameter. It is jelly-like, pink in colour, transparent and bounded by thick **pellicle**. Cytoplasm is highly vacuolated and differentiated into endoplasm and ectoplasm. **Cytopharynx** is a pouch-like depression on the ventral surface of body and leads into an elongated mouth (during floating cytopharynx is held uppermost). Cytopharynx contains two flagella: 1. a soft flap or **tooth** which represents a **transverse flagellum**; 2. a large thick flagellum which is modified into stout, striated **tentacle**. Grouped close to the groove are a nucleus, flagella and mouth, all of them together are called a **polar mass**. From the polar mass branching and anastomosing strands of cytoplasm pass into interior.

3. Physiology. Nutrition of *Noctiluca* is **holozoic**. It being predator collects food by the help of small flagellum. Tentacle too helps in food collection. Food of *Noctiluca* includes protozoa, rotifers, larvae, minute worms, etc. *Noctiluca* reproduces both asexually and sexually. Asexual reproduction occurs by **binary fission** and sexually by **sporulation** or spore formation.

4. Economic importance. Blooms of *Noctiluca* are toxic to fish and other aquatic life.

8. *Nyctotherus*

Classification

Phylum	–	**Protozoa**
Subphylum	–	**Ciliophora**
Class	–	**Ciliata**
Order	–	**Heterotrichida (Spirotrichia)**
Genus	–	***Nyctotherus***

1. Habit and habitat. *Nyctotherus* is a common intestinal endo-commensal of invertebrates, *e.g.*, apple snail, cockroach, frog, human beings, etc. Its significant species include *N. cordiformis* (found in the rectum of frog) and *N. ovalis* (found in the intestine of cockroaches). In human beings, its infection causes dysentery.

2. Morphology. The body of *Nyctotherus* is kidney-shaped, dorso-ventrally compressed and ranges from 60 μm to 120 μm in length. Body is covered by longitudinal and slightly spiral rows of equal sized cilia. The **oral groove** or **peristome** arises from the anterior end and extend along the right side of body upto the middle of body. It is lined by a row of specially large cilia which form **adoral zone of the membranelle**. The cytopharynx is long and spirally coiled like the watch-spring (Fig. 2.14). It is lined with undulating membrane.

The cytoplasm is differentiated into a thin clear peripheral layer of cytoplasm, called **ectoplasm** and a central mass of granular cytoplasm, the **endoplasm**. The endoplasm contains a large bean-shaped **macronucleus** and a small round **micronucleus** (both in the anterior half of body), a contractile vacuole (near the posterior end) and food vacuoles. The cell anus, cytopyge or **cytoproct** is a perma-nent aperture at the posterior end of body. Glycogen granules of yellowish-brown colour are found scattered in the endoplasm. *Nyctotherus* feeds upon the intestinal debris (saprozoic).

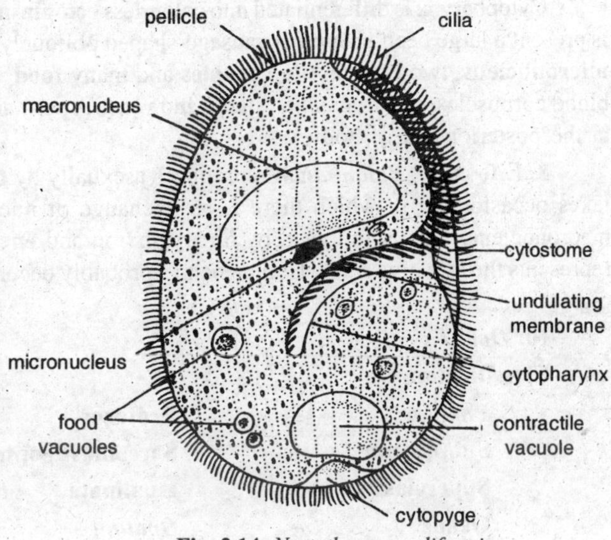

Fig. 2.14. *Nyctotherus cordiformis.*

3. Life cycle. *Nyctotherus* reproduces asexually by **binary fission**. In the rectum of tadpole these produce **exconjugant forms** which **conjugate** (to exchange the nuclear material) during metamorphosis of tadpole. In the newly metamorphosed frog, the exconjugants undergo binary fission to produce normal individuals which undergo **encystment**. The cysts pass out of the host along with its faecal matter. These cysts may be eaten up by the tadpoles.

9. *Balantidium*

Classification

Phylum	–	**Protozoa**
Subphylum	–	**Ciliophora**
Class	–	**Ciliata**
Order	–	**Holotrichia**
Genus	–	***Balantidium***

1. Habit and habitat. *Balantidium* is an intestinal endoparasite of various animals, mainly anuran amphibians and mammals. Several species are known to occur in frogs. *Balantidium coli* occurs in the colon of human beings, monkeys and pigs. It causes **balantidial dysentery** in human beings and often results in ulcers in human intestine.

2. Morphology. *Balantidium* is an active motile ciliate. It is oval in shape and measures 60-70 μm in length and 40-60 μm in width. The entire outer surface of parasite is covered by longitudinal, slightly spiral rows of small cilia. At the anterior end is a deep funnel shaped depression, the **vestibule**. It is lined by a little longer cilia called **adoral membranelle**. Cytostome is in the bottom of vestibule and leads into a cilialess **cytopharynx**.

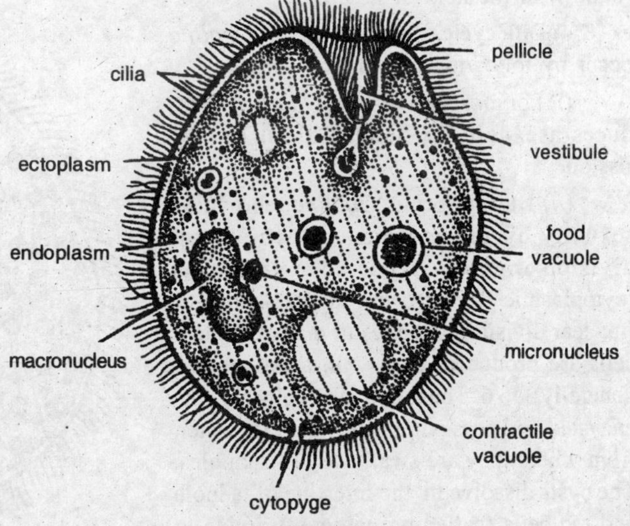

Fig. 2.15. *Balantidium coli.*

Cytopharynx is differentiated into colourless **ectoplasm** and dense **endoplasm**. Inside endoplasm is present a large bean-shaped or sausage-shaped obliquely placed **macronucleus**, a small spherical **micronucleus**, two **contractile vacuoles** and many **food vacuoles** containing ingested human red blood corpuscles, cell fragments, starch and yeast. A permanent anal opening or cytoproct is situated at the posterior end of body.

3. Life cycle. *Balantidium* reproduces asexually by **transverse binary fission. Conjugation** takes occasionally in which there is an exchange of nuclear material and reorganization of the macronucleus. This is followed by binary fission and **encystment**. The cyst is double-layered. It represents the infective stage. Human beings probably become infected by swallowing the cysts from pigs.

10. *Opalina*

Classification

Phylum	–	**Protozoa**
Subphylum	–	**Sarcomastigophora**
Superclass	–	**Opalinata**
Genus	–	***Opalina***

1. Habit and habitat. *Opalina* lives as a **commensal** in the large intestine of frogs and toads. The most common species is *Opalina ranarum* which is found in the rectum of frogs or toads.

2. Morphology. Body of *Opalina* is flat, leaf-like and oval, having body length upto 1 millimeter. Body is covered with parallel longitudinal oblique rows of many equal-sized motile fibrils which are called **cilia** or shortened **flagella**. The cytoplasm is distinguished into a clear outer **ectoplasm** and granular inner **endoplasm**. *Opalina* is **multinucleate**, *i.e.*, body contains a large number of monomorphic spherical nuclei evenly distributed throughout the cytoplasm. Each nucleus has both **trophochromatin** (of large sized chromosomes representing macronucleus) and **idiochromatin** (of small sized chromosomes representing micronucleus).

There is no cytostome or mouth, gullet, cytopyge, food vacuoles and contractile vacuole.

3. Physiology. *Opalina* feeds saprozoically, it absorbs digested food of the host by its general body surface. It usually swim but may also rotate with the help of its cilia.

4. Life cycle. Reproduction of *Opalina* occur by following methods:

(*i*) For most part of year, *Opalina* reproduces asexually by **longitudinal binary fission**.

(*ii*) During spring, *Opalina* reproduces sexually by **plasmotomy**, in which cell division involves repeated **cytokinesis** (cytoplasmic division) but no **karyokinesis** (nuclear division). As a result many daughter cells are produced, each with a few nuclei, generally 3 to 6. These daughter cells become encysted and pass out of the host into water, from where they are swallowed by tadpoles. The cysts dissolve in the intestine of tadpole and the cells (called **gametocyte**) divide to form uninucleate **microgametes** (male gamete) or **macrogametes** (female gamete).

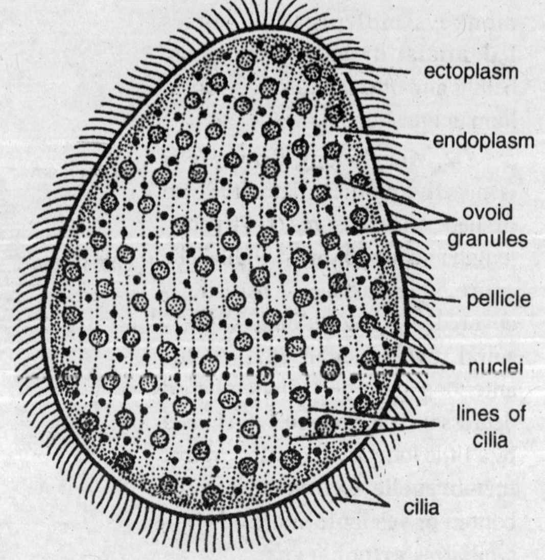

Fig. 2.16. *Opalina ranarum.*

These gametes are unequal in size and are called **anisogametes**. The male and female gametes fuse to form diploid **zygotes** which become multinucleated by the division of nucleus, and it increases in size to form an adult form.

Previously *Opalina* was placed in **Ciliophora**, then it was placed in **Flagellata**, but now visualizing its several peculiar features, it is placed in a separate superclass **Opalinata**.

11. *Vorticella*

Classification

Phylum	–	**Protozoa**
Subphylum	–	**Ciliophora**
Class	–	**Ciliata**
Subclass	–	**Peritrichia**
Order	–	**Peritrichida**
Genus	–	***Vorticella***

1. Habit and habitat. *Vorticella* is a freshwater sedentary (fixed) ciliate found attached to submerged aquatic plants, animals (**epizoic**) and stones of ponds, tanks, lakes, rivers and streams. It is a solitary animal, *i.e.*, free and independent even when occurring in gregarious form. *Vorticella* is worldwide in distribution. Its most common species is *Vorticella campanula*.

2. Morphology. Body of *Vorticella* is inverted bell-shaped with a long, contractile stalk. Stalk is secreted by a region at the aboral end of bell and is called **scopula**. The colour of the animalcule is yellowish, greenish or bluish. The rim or margin of the broad free end of the bell is thickened, called **peristomial lip** or **collar** (Fig. 2.17). Inside the peristomial collar is a narrow, shallow, circular and marginal depression called the **peristome** or **oral groove**. The peristome surrounds a broad, slightly convex circular central disc, the **peristomial disc** or **oral disc** that seems to close the opening of the bell. At one side, the peristomial disc is fused with the collar.

Between the peristome and the peristomial disc is a permanently open sunken space on the left side, it is called **vestibule** or **infundibulum**. From the vestibule a narrow **cytopharynx** leads inwards endoplasm and opens into the endoplasm. The cytopharynx lacks cilia. Between the vestibule and the cytopharynx is a **cytostome** which can open or close. **Cytoproct** is a temporary opening for expelling out the undigested food.

Cilia are present only in the peristomial region. The rest of body is without cilia. The cilia of peristo-mial groove are arranged in three concentric rows (circlets of cilia) and form 1½ spiral around the peristomial disc. Cilia of vestibule are fused in to an **undulating membrane**.

The entire body of *Vorticella* is covered by a **pellicle** which is very thick at the base of the bell. Cytoplasm is

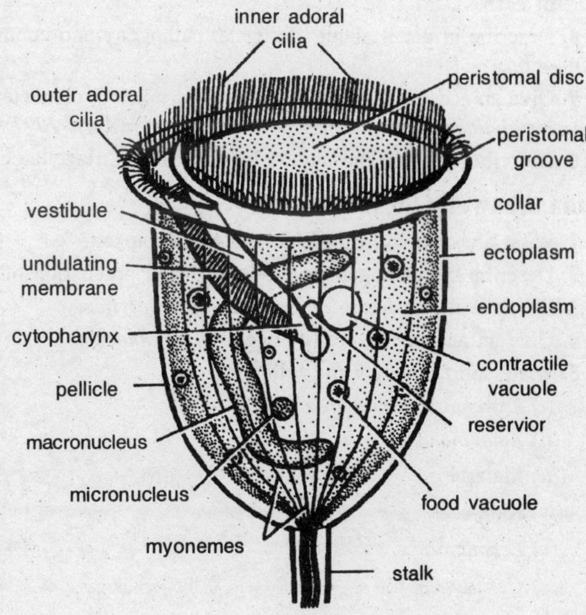

Fig. 2.17. *Vorticella campanula.*

divided into an outer layer of clear, firm, **ectoplasm** or **cortex** containing **myoneme system** and an inner fluid or granular **endoplasm** or **medulla**. The endoplasm contains a large horse-shoe-shaped **macronucleus**, small rounded **micronucleus**, a few **food vacuoles** and one **contractile vacuole**. From the bell, myonemes extend into the stalk, forming a loose spiral running through the centre of the stalk. It is called **axial filament** or **spasmoneme**.

3. Life cycle. *Vorticella* reproduces asexually by **longitudinal binary fission** and sexually by conjugation. Sometimes during unfavourable conditions, *Vorticella* develops a posterior ring of cilia to form free swimming form, called **telotroch**. It becomes detached from the stalk to swim free until a new place of attachment is found. The telotroch helps in its dispersal.

QUESTIONS

Long Answer Questions

1. Classify the phylum Protozoa upto orders giving diagnostic characters and examples of each group. *(C.C.S.U. Meerut 2005)*

2. Mention the distinguishing features of phylum Protozoa and give an outline classification of the same with characters and examples of its subclasses. *(Meerut (C.C.S.) 1989; Agra 1990, 1994; Rohilkhand 1990; Lucknow 1991, 1993; Kanpur 1993; Purvanchal 1996)*

3. Classify Protozoa upto classes, giving characters and examples of each class. *(Allahabad 1992, 1996)*

4. Give main characteristics of phylum Protozoa and classify it upto orders giving suitable examples. *(Avadh 1994; Rohilkhand 1996, 1999; CCS 1999; Agra 1999)*

5. Classify Ciliophora upto orders. Mention characters and suitable examples of each order. *(Kanpur 1994)*

6. Discuss general features of Protozoa.

7. Define Protozoa. Give an outline classification of phylum Protozoa upto orders with at least one example of each.

8. Describe in detail about life cycle, pathogeny and control measures of *Entamoeba*. *(M.S.U. Tirunelveli 2006)*

9. Give an account of habit, habitat, classification, characters and economic importance of *Entamoeba histolytica*. *(Agra 1997; Gorakhpur 1998; C.C.S.U. Meerut 2000)*

10. Describe in detail the unique features of dinoflagellates.

Short Answer Questions

1. What are distinguishing characters of Protozoa?

2. Describe the pathogenicity, treatment and prevention of *Entamoeba* species.

3. Describe the structure of *Entamoeba histolytica*.

4. Give an account of the life-cycle of *Entamoeba*. *(Purvanchal 1992; Gorakhpur 1993)*

5. Write short notes on the following:

(*i*) *Entamoeba histolytica* *(Gorakhpur 1998)*

(*ii*) *Amoebiasis* *(Gorakhpur 1998)*

(*iii*) Malaria *(Gorakhpur 1998)*

(*iv*) Kala azar *(Allahabad 1995)*

(*v*) *Leismania* *(Gorakhpur 1996)*

(*vi*) *Ceratium*

(*vii*) *Noctiluca* *(Lucknow 1999)*

(*viii*) *Nyctotherus* *(Lucknow 1996)*

(*ix*) *Vorticella* *(Lucknow 1996)*

(*x*) *Balantidium* *(Lucknow 1996)*

(*xi*) *Opalina* (*Lucknow 1996*)

(*xii*) Dinoflagellates (*Lucknow 1996*)

Very Short Answer Questions

1. Name a sedentary protozoan.

   Ans. *Vorticella.*

2. Which disease is caused by *Leishmania donovani*?

   Ans. Kala-azar (Black fever).

3. Name the enzyme secreted by *Entamoeba histolytica* to erode mucous membrane of host's intestine.

   Ans. Cytolysin.

4. Which group of protozoans has trichocysts?

   Ans. Ciliophora and *Noctiluca* (Dinoflagellata).

5. Mention one colonial protozoan.

   Ans. *Volvox.*

6. Name the group of Protozoa in which locomotory organelles are lacking.

   Ans. Sporozoa.

7. Protozoa are immortal. Is it correct?

   Ans. Yes.

8. What is the basis of classification of Protozoa?

   Ans. Locomotory organelles.

9. Whether *Myxobolus cerebralis* is a protozoan or cnidarian.

   Ans. It is a cnidarian.

Fill in the Blanks

1. *Giardia* produces the disease ....
2. Trichocyst is present in the body of ....
3. Flagellum is the main locomotory organ of the class ....
4. *Entamoeba histolytica* produces the disease ....
5. Amoebic dysentery is caused by ....
6. The chief mode of reproduction in Sporozoa is ....
7. The insect host (vector) required for the transmission of malaria to humans is ....
8. The insect vector required for the transmission of African sleeping sickness is the ....
9. *Euglena* belongs to class ....
10. Gliding movement shown by *Euglena* is called ....
11. The mode of nutrition in *Euglena* is ....
12. A basal body is derived from ....
13. Paramylum is stored strach in ....
14. *Trypanosoma gambiense* lives in human body in ....
15. *Monocystis* belongs to order ....
16. The trophozoites of *Monocystis* develop inside the ....
17. The pairing of gametocytes in *Monocystis* is called ....
18. The spores of *Monocystis* are called ....
19. Zygote formed by the fertilization of megagamete and microgamete in *Monocystis* is called ....
20. Locomotory organ of *Paramecium* is ....
21. The two types of nuclei in *Paramecium* are the .... which controls normal cellular metabolism, and the .... which involves in reproduction.
22. Intracellular digestion takes place inside ....
23. Contractile vacuole helps in ....

24. Quadrulus are specialized long cilia in the .... of *Paramecium*.
25. Response to light is called ....
26. Conjugation in *Paramecium* is a type of .... reproduction.
27. An exchange of genetic information occurs in ciliates when cells .... and exchange haploid ....
28. The unicellular algae present in the endoplasm of *Paramecium bursaria* are called ....
29. Trophozoite stage of *Entamoeba histolytica* produces .... to destroy tissue. *(C.C.S.U. Meerut 2001)*
30. The feeding process of *Entamoeba histolytica* is called as .... *(Purvanchal 1999)*
31. The causative species of sleeping sickness is .... *(Purvanchal 1999)*
32. Nutrition in *Monocystis* is .... *(Purvanchal 1999)*

Multiple Choice Questions

1. Classification of protozoa is based on *(Kanpur 1987)*
   (*a*) food vacuole (*b*) cyst structure
   (*c*) types of locomotory organelles (*d*) habit and habitat
2. *Amoeba* belongs to class
   (*a*) rhizopoda (*b*) zoomastigophorea
   (*c*) myxosporida (*d*) ciliata
3. Which of the following belongs to the class Mastigophora?
   (*a*) *Amoeba* (*b*) *Monocystis*
   (*c*) *Noctiluca* (*d*) *Paramecium*
4. To which of the following classes will you place *Plasmodium*?
   (*a*) sporozoa (*b*) mastigophora
   (*c*) rhizopoda (*d*) ciliata
5. An insect (*e.g.*, mosquito) which transmits a disease is known as
   (*a*) intermediate host (*b*) parasite
   (*c*) vector (*d*) prey
6. Which one is a communicable disease?
   (*a*) malaria (*b*) diabetes
   (*c*) hypertension (*d*) kwashiorkor
7. Malaria fever is caused by
   (*a*) female *Culex mosquito* (*b*) male *Culex* mosquito
   (*c*) female *Anopheles* mosquito (*d*) male *Anopheles* mosquito
8. Causative organism of malaria is
   (*a*) *Trypanosoma gambiense* (*b*) *Entamoeba histolytica*
   (*c*) *Leishmania donovani* (*d*) *Plasmodium vivax*
9. Syzygy is
   (*a*) union of gametes in Protozoa
   (*b*) pairing of gametocytes
   (*c*) pairing of homologous chromosomes
   (*d*) formation of spores
10. The protozoans that cause malaria in humans are
    (*a*) radiolarians (*b*) dinoflagellates
    (*c*) sporozoans (*d*) ciliates
11. Which of the following statement is true of amoeba?
    (*a*) they form pseudopodia
    (*b*) none has an outer shell
    (*c*) they have a flagellated phase in their life cycle
    (*d*) they do not have food vacuoles

12. Tongue-like projection of body of *Amoeba* is known as (*Magadh 1996*)
(*a*) rhizopodium (*b*) axopodium
(*c*) lobopodium (*d*) contractile vacuole

13. *Entamoeba histolytica* is found in man in (*Magadh 1994*)
(*a*) colon (*b*) oral cavity
(*c*) stomach (*d*) intestine

14. Which one causes amoebic dysentery in man? (*Magadh 1995*)
(*a*) *Amoeba proteus* (*b*) *Taenia solium*
(*c*) *Entamoeba histolytica* (*d*) *Ascaris*

15. *Entamoeba histolytica* lives in human body in (*Kanpur 1981*)
(*a*) stomach (*b*) duodenum
(*c*) blood (*d*) colon

16. Which of the following animals has only one host? (*Kanpur 1984*)
(*a*) *Plasmodium vivax* (*b*) *Entamoeba histolytica*
(*c*) *Trypanosoma gambiense* (*d*) None of the above.

17. Trophozoite is
(*a*) the sexual stage of parasite that produces spores
(*b*) the sexual stage of parasite that produces gametes
(*c*) the adult feeding stage of protozoan parasite
(*d*) none of the above

18. The process of cell drinking of *Amoeba* is called (*Kanpur 1984*)
(*a*) ingestion (*b*) osmoregulation
(*c*) pinocytosis (*d*) none of the above

19. Which of the following animals is able to manufacture its own food ? (*Kanpur 1984*)
(*a*) *Amoeba* (*b*) *Paramecium*
(*c*) *Euglena* (*d*) *Plasmodium*

20. Locomotion of *Euglena* takes by (*Kanpur 1981*)
(*a*) pseudopodia (*b*) flagellum
(*c*) cilia (*d*) by gliding

21. *Euglena* reproduces (*Kanpur 1980*)
(*a*) both asexually and sexually (*b*) only asexually
(*c*) only sexually (*d*) conjugation

22. *Trypanosoma gambiense* lives in human body in (*Kanpur 1980*)
(*a*) lymph (*b*) blood
(*c*) cerebrospinal fluid (*d*) blood and cerebrospinal fluid

23. Trypanosomiasis is transmitted by (*Kanpur 1986*)
(*a*) louse (*b*) may fly
(*c*) tsetse fly (*d*) fruit fly

24. *Trypanosoma gambiense* is transmitted by (*Kanpur 1983*)
(*a*) house fly (*b*) may fly
(*c*) sand fly (*d*) tsetse fly

25. Intermediate host of sleeping sickness is (*Kanpur 1982*)
(*a*) sand fly (*b*) tsetse fly
(*c*) male *Anopheles* (*d*) none of the above

26. Sleeping sickness is caused by (*Kanpur 1981*)
(*a*) *Trypanosoma* (*b*) *Entamoeba*
(*c*) *Plasmodium* (*d*) *Leishmania*

27. Polymorphism is shown by
(a) *Amoeba* (b) *Paramecium*
(c) *Trypanosoma* (d) *Sycon*

28. Slipper animalcule is the popular name of (*Kanpur 1985*)
(a) *Euglena* (b) *Amoeba*
(c) *Paramecium* (d) *Trypanosoma*

29. *Paramecium caudatum* has (*Kanpur 1985*)
(a) single nucleus (b) two micronuclei of the same size
(c) one micronucleus and one macronucleus (d) only one meganucleus

30. Paramecium and other ciliates differ from all other protozoans in
(a) having two types of nuclei (b) capturing prey with pseudopodia
(c) having contractile vacuoles (d) not using flagella for locomotion

31. Which of the following statement is not true of *Paramecium*?
(a) they are predatory ciliates
(b) they use pseudopodia for capturing prey
(c) they are preyed upon by *Didinium*
(d) they use contractile vacuoles for removing excess water

32. Animal-like nutrition is known as (*Kanpur 1987*)
(a) holophytic (b) holozoic
(c) saprophytic (d) parasitic

33. How *Paramecium* takes its food? (*Kanpur 1987*)
(a) by oral groove (b) by undulating membrane
(c) by pseudopodia (d) through body surface

34. Cyclosis is present in (*Kanpur 1987*)
(a) *Amoeba* (b) *Euglena*
(c) *Paramecium* (d) *Plasmodium*

35. The food vacuole of *Paramecium* is (*Kanpur 1984*)
(a) first alkaline then acidic (b) only alkaline
(c) only acidic (d) first acidic and then alkaline

36. Intracellular digestion occurs inside
(a) stomach (b) food vacuole
(c) contractile vacuole (d) alimentary canal

37. What is main function of contractile vacuole? (*Kanpur 1986*)
(a) food regulation (b) excretion
(c) osmoregulation (d) reproduction

38. Asexual reproduction of *Paramecium* is brought about by
(a) simple binary fission (b) multiple fission
(c) sporulation (d) transverse binary fission

39. Which of the following protozoans lives in the gut of termites and help them digest cellulose?
(a) *Plasmodium* (b) *Amoeba proteus*
(c) *Trichonympha* (d) *Trypanosoma*

40. *Glossina palpalis* transmits the infective stages of
(a) *Leishmania* (b) *Trypanosoma*
(c) *Plasmodium* (d) *Wuchereria*

41. Chagas' disease is caused by (*Lucknow 1995*)
(a) *Trypanosoma cruzi* (b) *Trypanosoma gambiense*
(c) *Trypanosoma rhodesiensi* (d) *Trypanosoma brucei*

42. Infective stage of *Entamoeba histolytica* is (*Lucknow 1996*)

(*a*) sporozoite (*b*) quadrinucleate cyst
(*c*) trophozoite (*d*) spore

43. How does locomotion takes place in *Entamoeba*? (*Lucknow 1995*)

(*a*) by lobopodia (*b*) by filopodia
(*c*) by rhizopodia (*d*) by axopodia

44. The disease kala-azar is caused by (*Lucknow 1997*)

(*a*) *Trypanosoma* (*b*) *Giardia*
(*c*) *Trichomonas* (*d*) *Leishmania*

45. The adult feeding stage in the life cycle of *Monocystis* is. (*Lucknow 1997*)

(*a*) sporocyst (*b*) gametocyst
(*c*) trophozoite (*d*) sporozoite

46. *Amoeba* has (*Rohilkhand 2000*)

(*a*) lobopodia (*b*) filopodia
(*c*) reticulopodia (*d*) axopodia

47. Sleeping sickness is caused by (*Rohilkhand 2000*)

(*a*) tse-tse fly (*b*) house fly
(*c*) dragon fly (*d*) may fly

48. In *Paramecium* nutrition is (*Rohilkhand 2000*)

(*a*) holophytic (*b*) holozoic
(*c*) mixotrophic (*d*) parasitic

49. *Paramecium caudatum* has (*Rohilkhand 2000*)

(*a*) one macro-and two micronuclei
(*b*) one macro-and one micronuclei
(*c*) two macro-and one micronuclei
(*d*) two macro-and two micronuclei

50. African sleeping sickness is caused by (*Kanpur 2001*)

(*a*) *Trypanosoma* (*b*) *Entamoeba coli*
(*c*) *Entamoeba histolytica* (*d*) *Monocystis*

51. It causes amoebiasis (*M.S.U., Tirunelveli 2006*)

(*a*) mosquito (*b*) *Trypanosoma*
(*c*) *Paramecium* (*d*) *Entamoeba*

52. Schizogony and schizont stage of *Plasmodium* occur in

(*a*) erythrocytes and liver cells of human beings
(*b*) blood of human beings
(*c*) stomach of human beings
(*d*) only liver in human beings

53. Exflagellation in malarial parasite takes place during the formation of

(*a*) ova (*b*) sperms
(*c*) gametocytes (*d*) sporozoites

54. Sporogony of malarial parasite occurs in

(*a*) liver of human beings (*b*) erythrocytes of human beings
(*c*) salivary glands of mosquito (*d*) stomach of mosquito

ANSWERS

Fill in the Blanks

1. Diarrhoea
2. *Paramecium*
3. Mastigophora
4. Amoebic dysentery
5. *Entamoeba histolytica*
6. Multiple fission
7. Female *Anopheles* mosquito
8. Tsetse fly (*Glossina sp.*)
9. Mastigophora
10. Metabody or Euglenoid movement
11. Mixotrophic
12. Centriole
13. *Euglena*
14. Cerebrospinal fluid and blood
15. Eugregarinida
16. Seminal vesicles (of earthworm)
17. Syzygy
18. Pseudonavicella
19. Sporoblast
20. Cilium
21. Macronucleus, micronucleus
22. Food vacuole
23. Osmoregulation
24. Dorsal wall of vestibule
25. Phototaxis
26. Sexual
27. Conjugate micronuclei
28. *Zoochlorella.*
29. Cytolysin
30. Holozoic
31. *Trypanosoma*
32. Saprozoic

Multiple Choice Questions

1. (*c*) **2.** (*a*) **3.** (*c*) **4.** (*a*)
5. (*c*) **6.** (*a*) **7.** (*c*) **8.** (*d*)
9. (*b*) **10.** (*c*) **11.** (*a*) **12.** (*c*)
13. (*a*) **14.** (*c*) **15.** (*d*) **16.** (*b*)
17. (*c*) **18.** (*c*) **19.** (*c*) **20.** (*b*)
21. (*b*) **22.** (*d*) **23.** (*c*) **24.** (*d*)
25. (*b*) **26.** (*a*) **27.** (*c*) **28.** (*c*)
29. (*c*) **30.** (*a*) **31.** (*b*) **32.** (*b*)
33. (*a*) **34.** (*c*) **35.** (*d*) **36.** (*b*)
37. (*c*) **38.** (*d*) **39.** (*c*) **40.** (*b*)
41. (*a*) **42.** (*b*) **43.** (*a*) **44.** (*d*)
45. (*c*) **46.** (*a*) **47.** (*a*) **48.** (*b*)
49. (*b*) **50.** (*a*) **51.** (*d*) **52.** (*a*)
53. (*b*) **54.** (*d*).

3

Euglena

3.1 CLASSIFICATION

Phylum	–	**Protozoa**
Subphylum	–	**Sarcomastigophora**
Superclass	–	**Mastigophora**
Class	–	**Phytomastigophora**
Order	–	**Euglenida**
Genus	–	***Euglena***
Species	–	***viridis***

3.2 HABIT AND HABITAT

Euglena viridis is a solitary, free-living (free-swimming) and freshwater protozoan. It is commonly found in ponds, ditches, pools, lakes, etc., especially in any other water that is rich in soluble organic matter (*i.e.*, stagnant waters containing decaying nitrogenous wastes such as dead bodies of plants and animals, and urine and faeces of animal). Under favourable environmental conditions it multiplies rapidly and forms green scum on the water surface. Other common Indian species of *Euglena* are *E. agilis*, *E. orientalis*, *E. gracilis*, *E. spirogyra* and *E. fusiformis*.

Euglena needs sunlight for photosynthesis, exhibits reactivity and rapidly multiplies by longitudinal binary fission.

3.3 MORPHOLOGY

1. Shape and size. *Euglena viridis* is green, slender, elongated and spindle-shaped in appearance. Its anterior end is bluntly rounded, the middle part is wider and posterior end is pointed.

Euglena viridis is about 4–60 μm in length and 14–20 μm in breadth at the thickest part of the body.

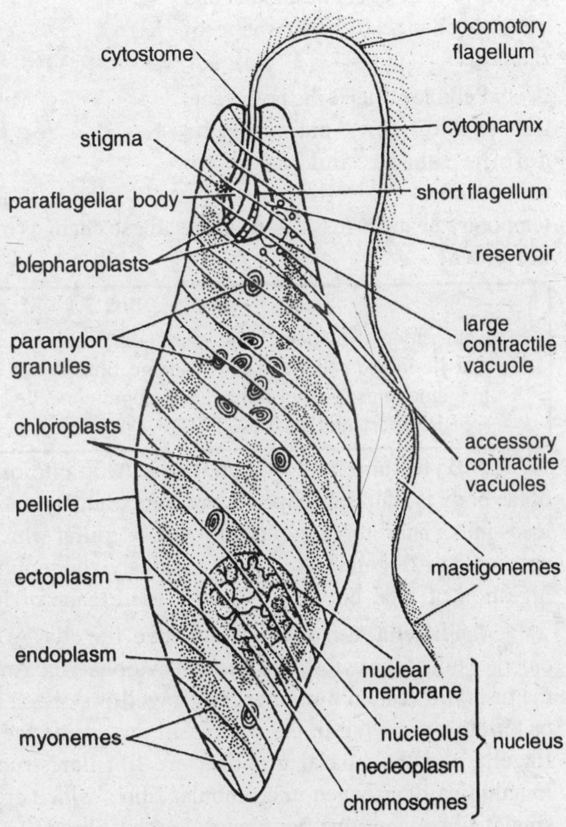

Fig. 3.1. *Euglena viridis*. Structure.

2. Pellicle. *Euglena* lacks a cell wall. The outer layer of the cell is the plasma membrane, immediately below which is the thin, flexible, strong, protei-naceous **pellicle**. Pellicle consists of a number of thickened longi-tudinal strips and microfibrils (microtubules) articulating with each other. It consists of parallel helical striations or sculpturing which can be observed by electron microscopy (Fig. 3.3). These striations extend counter-clockwise from anterior to posterior ends of body and involve longitudinal rigid strips which are interlocked with each other along their margins. When minute fibrils, called **myonemes**, occurring just beneath the pellicle in the cytoplasm contract, they cause the strips of pellicle to slide over one another and affect a change in body shape of *Euglena*.

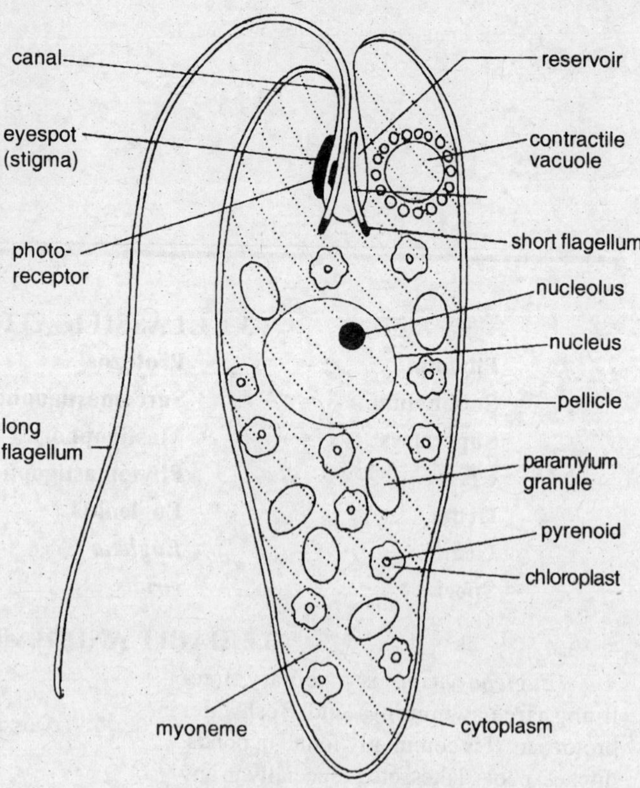

Fig. 3.2. *Euglena gracilis.* Structure.

Pellicle protects the body (acts like **exoskeleton**), maintains its definite shape and helps in movements. Pellicle of *Euglena* is flexible enough to permit temporary changes in the body shape; these changes of body shape are called **metaboly** or **euglenoid movement**.

> **Box 3.1 Myonemes**
>
> Contraction of *Euglena* body is brought about by bundles of contractile filaments, or **myonemes** that lie in the pellicle. The myoneme fibre is not composed of proteins actin and myosin, as in animal muscles, but rather another protein called **spasmin**, which requires ATP for extension (see **Ruppert** and **Barnes**, 1994).

3. Cytostome and cytopharynx. A little to one side of the centre of the anterior blunt end of the body is a funnel-shaped depression or invagination, called **cytostome** (cell mouth). Cytostome leads into a short tubular **cytopharynx** or **gullet** which in turn opens into a spherical vesicle, called **reservoir** or **flagellar sac**. Quite curiously, the mouth or cytostome and gullet are not used for the ingestion of food but as a canal for the escape of fluid from the reservoir.

4. Flagella. *Euglena* contains two flagella. A long and thin and whip-like flagellum which emerges out of the cytostome through cytopharynx. Another much shorter flagellum which lies within the reservoir (called **nonemergent flagellum**). Each flagellum arises from a minute **basal granule** or **kinetosome** lying in the cytoplasm just inside the border (plasma membrane of reservoir). The flagella and their basal granules are fibrillar structures; each flagellum contains 11 (eleven) longitudinally arranged microtubular fibres which are arranged in a specific pattern of two central singlet fibres and nine peripheral doublet fibres (9 + 2). Only the 9 peripheral fibres extend into the basal granule. The distal part of the long flagellum bears along its one margin a row of short, hair-like, contractile fibrils or processes, called **mastigonemes**. These mastigonemes are suggested to arise from two of the nine peripheral fibres (**Manton**, 1959).

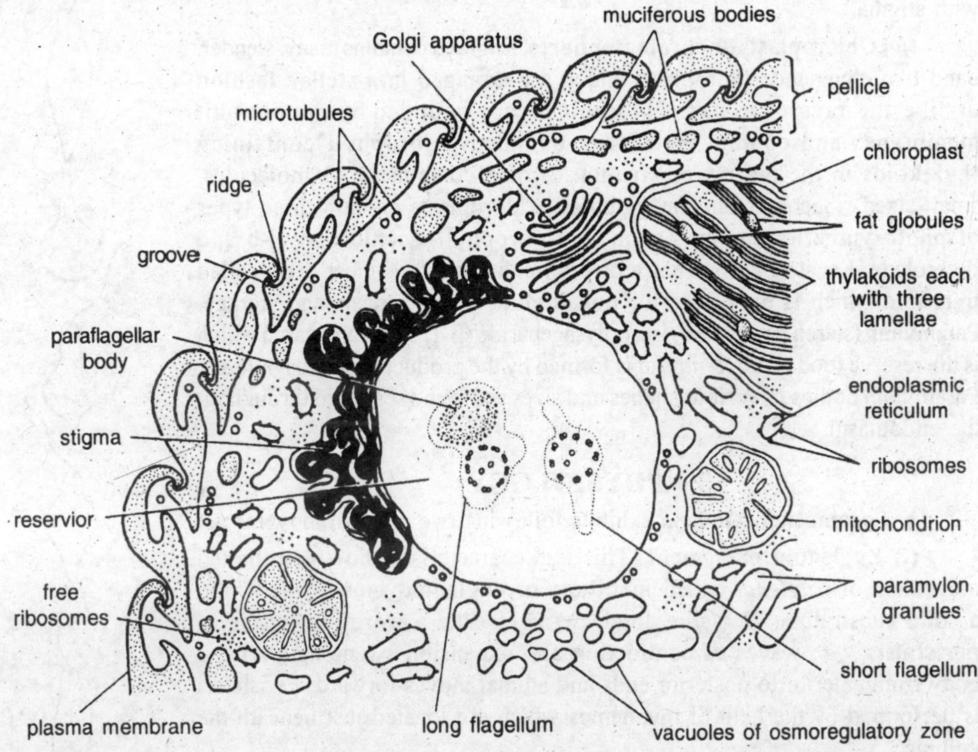

Fig. 3.3. *Euglena.* Electron microscopic structure of a part of T.S. of body passing through reservoir.

5. Cytoplasm and organelles. The cytoplasm of *Euglena* is differentiated into a thin, clear and dense outer layer of **ectoplasm** and a more fluid and granular inner (central) layer of **endoplasm**. Endoplasm contains a variety of cellular organelles such as stigma, chloroplasts (chromatophores), mitochondria, nucleus, contractile vacuole and also usual cellular organelles such as endoplasmic reticulum, Golgi apparatus, ribosomes, etc.

(i) **Nucleus.** A single, large, spherical, vesicular nucleus lies in the middle or toward posterior end of body. It contains only one chromosome (in the form of chromatin) and a few large **endosomes** or **nucleoli**. Nuclear membrane is not continuous with endoplasmic reticulum (unusual among protists; see **Brusca** and **Brusca**, 2003).

(ii) **Contractile vacuole.** A single large contractile vacuole lies near the reservoir on one side (opposite to stigma). It is surrounded by several minute **accessory** contractile vacuoles which fuse together to form the larger vacuole. The contractile vacuole discharges its watery contents into the reservoir; from there fluid escapes through cytopharynx and cytostome.

(iii) **Photoreceptor apparatus.** It includes stigma and paraflagellar body.

1. Stigma or eyespot is a small discoid red or orange-red body which is located close to the inner end of the cytopharynx adjacent to reservoir. It consists of a plate of **lipid droplet** and a granule of red **carotenoid pigment**. Stigma is a cup-shaped structure having colourless mass of oily droplets in its concavity which function as a **lens**. Stigma has a one-sided light filtering action and thus, helps in phototactic response.

2. Paraflagellar body (photoreceptor). It is a small swelling at the root of the long flagellum and is sensitive to light. The paraflagellar body acts as a photoreceptor (*i.e.*, detects light) and enables organisms to swim towards light of optimum intensity, called **phototaxis**. It also gives a directional light response in connection with stigma.

***(iv)* Chloroplasts or chromatophores.** *Euglena* contains many, slender, band-like elongated chloroplasts which are arranged in a **stellar fashion** or like the rays of a star. Chloroplasts are bounded by double unit membranes and contain membrane-bounded chlorophyll containing **thylakoids** in the **matrix** or **stroma**. Stroma contains DNA molecules, small-sized ribosomes and fat globules. Chloroplasts contain three types of photosynthetic pigments such as chlorophyll-a, chlorophyll-b and β-carotene. Located in the centre of each chloroplasts is clear area, called **pyrenoid** which is the centre for **paramylum** - biosynthesis and storage. Paramylum (starch paramylon) is a polysaccharide (β-1, 3 glycan) starch which is the reserve food of *Euglena* and is formed by the product of photosynthesis. Paramylum bodies of various shapes and sizes are found scattered throughout the endoplasm.

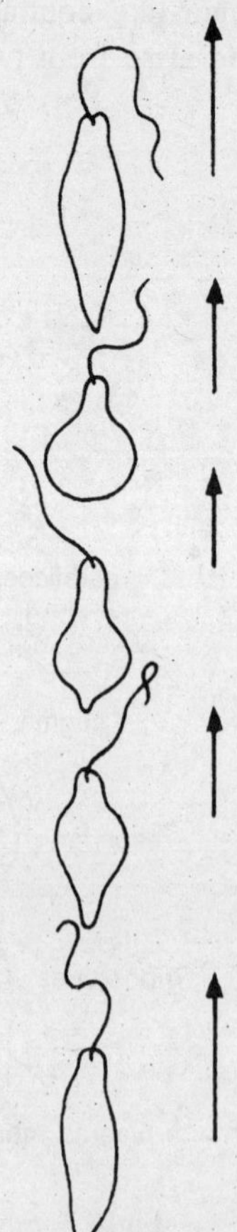

Fig. 3.4. Euglenoid movement or metaboly. It is thought to be accomplished by the sliding of microtubules against the protein strips (after Bru-sca and Brusca, 2003).

3.4 PHYSIOLOGY

1. Locomotion. *Euglena* exhibits following two types of movements:

***(i)* Euglenoid movement.** This is a characteristic slow and limited movement of *Euglena*, called **metaboly** or **euglenoid movement**. Upon a solid substratum, *Euglena* slowly wriggles like a worm by means of **peristalsis**, *i.e.*, waves of contraction and expansion sweeping over the body from anterior to posterior ends and animal moves forward. Peristalsis is performed by the help of myonemes which are located just beneath the pellicle.

***(ii)* Flagellar movement.** The locomotory flagellum is at the anterior end of the *Euglena* body and pulls the organism forward. This sort of flagellum is called **tinsel flagellum**, since it possesses minute lateral projections, called **mastigonemes** on its surface which aid this type of locomotion. The flagellum propagates an **undulatory wave** from the cell to the flagellar tip. As such waves of activity pass through the flagellum in spiral fashion from its base to its tip, they increase in amplitude and velocity. Such an activity of the flagellum causes the body of *Euglena* to rotate about its axis, at about one complete body turn per second, as well as make a corkscrew pathway through water. The organism can travel forward at a rate of 0.5 mm S^{-1} which is approximately four times its body length. *Euglena* is able to change its direction by the action of contractile myonemes.

2. Nutrition. The mode of nutrition in *Euglena* is **mixotrophic**, *i.e.*, its mode of nutrition is either holophytic or saprophytic.

***(i)* Holophytic nutrition.** *Euglena viridis* being a chlorophyll containing green organism is autotrophic. It synthesizes its own food from carbon dioxide, water, and mineral salts by the help of chlorophyll and sunlight. However, *Euglena* is dependent on an external supply of vitamins B_1 and B_{12}, which like animals, it cannot synthesize for itself. Its reserve food is **paramylum** which is stored in the form of refractile paramylum bodies.

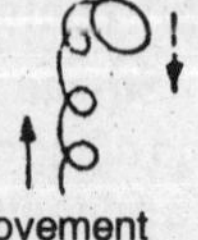

Fig. 3.5. *Euglena*. Flagellar movement by helical beat.

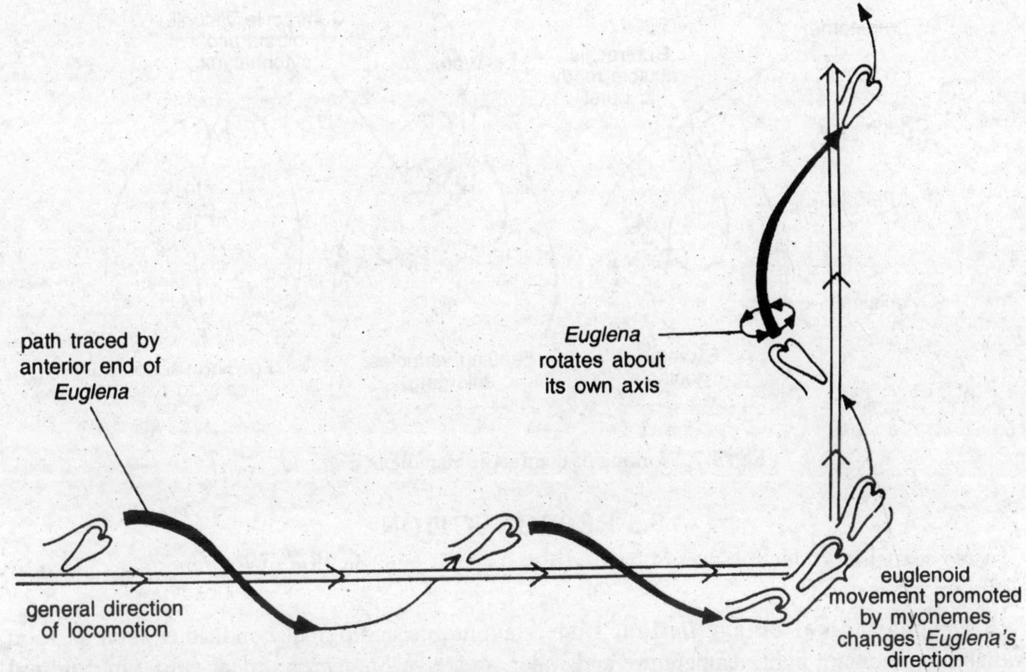

Fig. 3.6. *Euglena*. Path of flagellar movement.

(*ii*) **Saprotrophic nutrition.** In the absence of sunlight *Euglena* loses its chloroplasts, become colourless and manages its food by **saprotrophic** mode of nutrition. Certain other terms such as **saprophytic, saprozoic** or **osmotrophic** nutrition are also used for saprotrophic nutrition. In this mode of nutrition, animal absorbs through the general body surface some organic substances in solution from decaying matter. However, *Euglena* requires ammonium salts, instead of nitrates for their source of nitrogen. Chloroplasts return when the organisms are returned to light.

> **Box 3.2 Holozoic *Euglena***
>
> Some species of *Euglena* (*e.g.*, *Euglena gracilis*) are colourless and non-photosynthetic; they are heterotrophic and have following two types of nutrition:
>
> **1. Holozoic nutrition.** (Involving intracellular digestion). They ingest small food particles at the base of the gullet by pinocytosis or phagocytosis where the pellicle is absent. Food may be driven into the gullet by the action of the flagella.
>
> **2. Saprotrophic nutrition.** They carry extracellular digestion and grow most abundantly in water where putrefaction is taking place (since decaying material is rich in organic compounds).

3. Respiration. *Euglena* respires through its general body surface. Oxygen dissolved in the water diffuses in through the pellicle and carbon dioxide liberates to outside.

4. Osmoregulation. The water content of body are regulated by the **contractile vacuole** which periodically empties into the reservoir. This is called **osmoregulation**. During its **diastole**, the contractile vacuole grows in spurts because smaller accessory vacuoles, formed in its vicinity, coalesce with it one after another. After its maximum growth, the contractile vacuole undergoes **systole** by fusing with the lining of the reservoir and bursting into it. To recall, water tends to enter the animal body by **endosmosis**.

5. Excretion. *Euglena* is an ammonotelic animal; its main excretory waste is ammonia. Excretion occurs through the general body surface and contractile vacuole both.

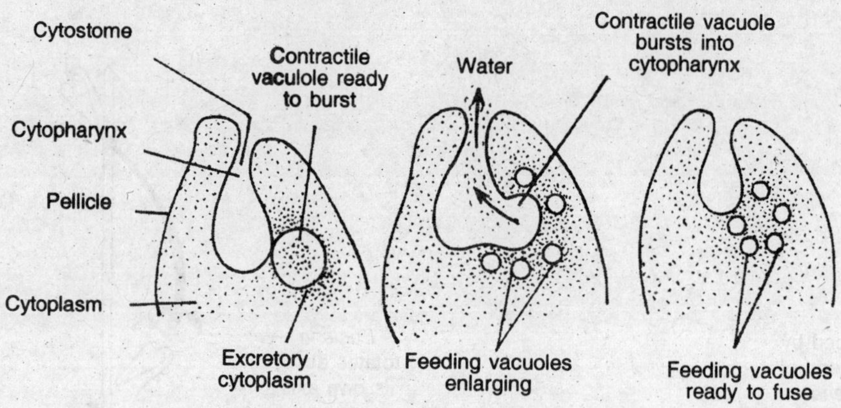

Fig. 3.7. Working of contractile vacuole in *Euglena.*

3.5 REPRODUCTION

No sexual reproduction occurs in *Euglena.* Asexual reproduction of *Euglena* takes place by the following methods:

1. Longitudinal binary fission. Under favourable ecological conditions such as food availability, optimum light, temperature and water, an active *Euglena* reproduces by **longitudinal binary fission (symmetrogenic)**. It is initiated by division of basal bodies. Nucleus is divided by peculiar **mitosis**, in which, the mitotic spindle is formed inside the intact nucleus; *i.e.*, nuclear envelope and endosome or nucleolus are not disintegrated during mitosis. Karyokinesis (nuclear division) is followed by cytokinesis (cytoplasmic division) which involves duplication of flagella, stigma, contractile vacuole, chloroplasts, mitochondria, etc. During later stages reservoir, cytopharynx and cytostome are duplicated by a longitudinal furrow. During final step of cytokinesis, a longitudinal furrow appears at anterior end of body and proceeds backward, dividing the parent body into two daughter euglenae.

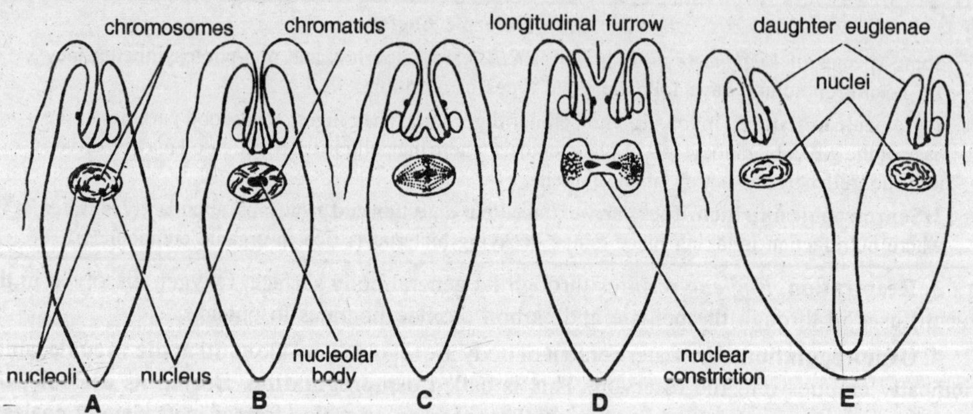

Fig. 3.8. *Euglena.* Longitudinal binary fission.

2. Multiple fission and palmella stage. Under unfavourable environmental conditions such as lack of food and oxygen, excessive heat of cold, draught, etc., *Euglena* becomes inactive and throws off its long flagellum and forms a protective thick (two layered yellowish – brown mucilaginous) **cyst** around it. A number of such encysted euglenae come together and become embedded in a common gelationous mass on the water surface. Within the cyst, each *Euglena* divides once by longitudinal fission or repeatedly forming 16 or 32 small-sized daughter forms. The latter form their own mucilaginous coats within the original cyst. A gelatinous mass containing many cysts is called **palmella stage** which resembles the palmella stage of many algae such as *Chlamydomonas*. Later on these daughter euglenae acquire normal flagella and escape out from the cyst to grow into normal euglenae.

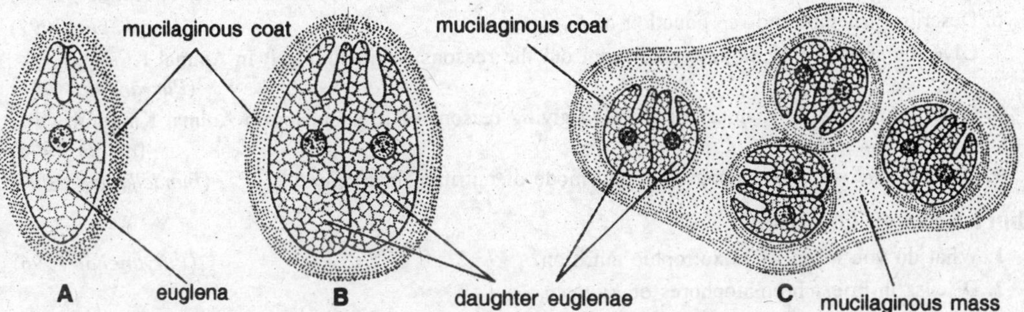

Fig. 3.9. *Euglena*. Multiple fission. A—Encysted individual; B—Fission in encysted individual; C—Palmella stage.

3. Encystment. During unfavourable conditions, *Euglena* undergoes **encystment**. For encystment, first *Euglena* becomes inactive, loses its flagella and secretes a thick rounded, red-coloured (red colour is due to **haematochrome** pigment) protective **cyst**. The cyst is secreted by the **muciferous bodies** lying below the pellicle of *Euglena*. Encysted euglenae stand a fair chance of wide dispersal. On return of favourable conditions, encysted euglenae emerge out and resume normal active life.

Economic importance. *Euglena* forms an important trophic level of an aquatic food chain. Being producer, it is consumed by herbivorous and plankton-eating fishes (primary consumers). Some species of *Euglena* have been used in experiments for waste water treatment and have been reported to extract heavy metals such as magnesium, iron and zinc from sludge (see **Brusca** and **Brusca**, 2003).

3.6 REASONS OF INCLUSION OF *EUGLENA* IN THE ANIMAL KINGDOM

Euglena shows many characters of **plants** such as chloroplasts with chlorophyll and holophytic nutrition. However, it is regarded as an **animal** due to the following facts:

1. It lacks a cellulose cell wall.
2. Its pellicle is made of proteins like other protozoan cells (*i.e.*, ciliates).
3. Presence of **stigma** and **paraflagellar** body, the 'photosensitive structure'.
4. Presence of osmoregulatory contractile vacuole.
5. Nutrition is **heterotrophic** as in other animals which may be saprotrophic or saprozoic and endocytic (pinocytic or phagocytic) in nature.
6. Presence of centrioles in the form of blepharoplasts (basal granule).
7. Presence of longitudinal binary fission.

QUESTIONS

Long Answer Questions

1. Write about the different habits of *Euglena*. Describe its morphological and anatomical features. (*Avadh 1994*)

2. Give an account of structure, nutrition and reproduction in *Euglena*. (*Gorakhpur 1994; Allahabad 1999*)
3. Describe the structure of *Euglena* and explain its movements. (*Avadh 1996; Purvanchal 1997*)
4. Draw a well labelled diagram of a transverse section of flagellum and write functions of following in *Euglena*: 1. Stigma; 2. Paramylum; 3. Contractile vacuole; and 4. Cyst. (*Avadh 1999*)
5. Give an illustrated account of the structure and life history of *Euglena viridis*. (*Allahabad 1996; Kanpur 1998*)
6. Describe structure and reproduction of *Euglena*. (*Gorakhpur 1997*)
7. Give the structure of *Euglena* and point out the reasons for including it in Animal Kingdom. (*Purvanchal 1995*)
8. Describe functional anatomy of *Euglena* giving reasons of its inclusion in Animal Kingdom. (*Avadh 1998*)
9. Describe the structure, locomotion and mode of nutrition in *Euglena*. (*Bundelkhand 2000*)

Short Answer Questions

1. What do you mean by mixotrophic nutrition? (*Purvanchal 1996*)
2. Discuss in brief chromatophores of *Euglena*.
3. Draw a well labelled diagram of *Euglena viridis*.
4. Write a detailed account of binary fission in *Euglena*.
5. Write short notes on the following:
   (*i*) Locomotion in *Euglena*, (*Allahabad 1995; Lucknow 1998*)
   (*ii*) Nutrition in *Euglena* (*Purvanchal 1991; Gorakhpur 1991, 1999; Allahabad 1997*)
   (*iii*) Functions of flagellum in *Euglena* (*Allahabad 1998*)
   (*iv*) Euglenoid movement (*Avadh 1997; Purvanchal 1998*)
   (*v*) Palmella stage (*Purvanchal 1993; Avadh 1997*)
   (*vi*) Flagellar movement (*Purvanchal 1995*)

Very Short Answer Questions

1. Name the parts of oral apparatus of *Euglena*.
   Ans. Cytosome, cytopharynx, reservoir.
2. Pellicle of *Euglena* consists of two layers. Give their names.
   Ans. Plasma membrane and epiplasm (or protein plates plus microtubules).
3. Mention the alternative terms for basal bodies.
   Ans. Blepharoplast, Kinetoplast.
4. Where is stigma is located in *Euglena*?
   Ans. On the flagellum in the cytopharyx.
5. What is endosome?
   Ans. Nucleolus in the nucleus of *Euglena*.
6. Name the species of *Euglena* that has permanently lost chlorophyll.
   Ans. *Euglena gracilis*.
7. Where does the contractile vacuole discharge its contents in *Euglena*?
   Ans. Into cytopharynx.
8. In *Euglena*, the binary fission is symmetrogenic. Is it so?
   Ans. Yes.
9. What is palmella stage?
   Ans. Multiplication of *Euglena* within gelatinous cyst.

Fill in the Blanks

1. Paramylum bodies in *Euglena* store .... as food reserve.
2. Paramylum bodies are the bioproduct of .... in *Euglena*. (*CCS 2001*)

3. The nucleoli of *Euglena* are also known as .... (*Purvanchal 1999*)

4. Sporulation in *Euglena* is found during .... (*Purvanchal 1998*)

Multiple Choice Questions

1. In *Euglena*, food is stored in the form of (*Kanpur 2001*)

(*a*) paramylum bodies (*b*) chromatophores
(*c*) mitochondria (*d*) nucleus

2. *Euglena* moves with the help of

(*a*) cilia and flagella (*b*) undulating membrane
(*c*) pseudopodia (*d*) flagellum

3. Osmoregulation in *Euglena* is done by

(*a*) flagella (*b*) nucleus
(*c*) contractile vacuole (*d*) general body surface

4. Reserve food in *Euglena* is in the form of

(*a*) starch (*b*) glycogen
(*c*) fatty acid (*d*) paramylon

5. *Euglena* lacks

(*a*) cytostome (*b*) cytopharynx
(*c*) cytopyge (*d*) contractile vacuole

6. In *Euglena* metaboly is performed by

(*a*) flagellum
(*b*) pseudopodium
(*c*) sliding of microtubules against protein strips
(*d*) none of these

7. Nutrition in *Euglena* is

(*a*) autotrophic only (*b*) saprotrophic only
(*c*) holotrophic only (*d*) mixotrophic

8. *Euglena* contains chlorophyll

(*a*) *a* and *c* (*b*) *a* and *c*
(*c*) *b* and *c* (*d*) *a*, *b* and *c*

9. In *Euglena*, the binary fission is

(*a*) longitudinal (*b*) transverse
(*c*) irregular (*d*) lacking

10. Pyrenoid of chloroplast acts as a center for

(*a*) enzyme secretion (*b*) paramylum formation
(*c*) oxygen release (*d*) none of these

ANSWERS

Fill in the Blanks

1. Starch **2.** Photosynthesis
3. Endosomes **4.** Unfavourable conditions.

Multiple Choice Questions

1. (*a*) **2.** (*d*) **3.** (*c*) **4.** (*d*)
5. (*c*) **6.** (*c*) **7.** (*d*) **8.** (*a*)
9. (*a*) **10.** (*b*).

4

Monocystis

4.1 CLASSIFICATION

Phylum	–	**Protozoa**
Subphylum	–	**Sporozoa (Apicomplexa)**
Class	–	**Telosporea**
Subclass	–	**Gregarinia**
Genus	–	***Monocystis***

4.2 HABIT AND HABITAT

Monocystis is an **extracellular endoparasite** in the seminal vesicle of earthworms (*Pheretima, Eutyphoeus* and *Lumbricus*). *Monocystis pheretimi* is parasite of *Pheritima posthuma*. Its adult feeding and growing stage is called **trophozoite** and represents vegetative phase of parasite. The stages of *Monocystis* that are usually present in the seminal vesicles are the trophozoite, cysts containing two individuals, gametes and spores in various stages of development and isolated spores. Trophozoite develops within the **sperm morula** (*i.e.*, group of developing spermatozoa) of earthworm. Developing trophozoite tends to live at the expense of the cells among which it lies. The spermatozoa of the earthworm which are deprived of nourishment by the parasite, slowly shrivel up finally becoming tiny filaments on the surface of trophozoite. As a result, young trophozoite appears ciliated because of attached tails of the dead sperms.

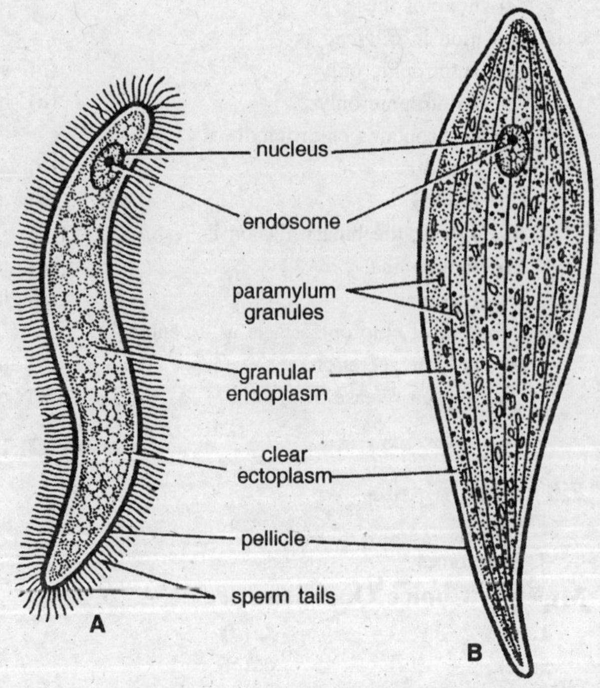

Fig. 4.1. *Monocystis*. A—Young trophozoite; B—Mature trophozoite.

4.3 MORPHOLOGY

A full grown or adult trophozoite of *Monocystis* is known as **sporadin**. The mature trophozoite is an elongated spindle or club-shaped, dorso-ventrally flattened worm-like organism. It is pointed at the both ends of body. The body of *Monocystis* is bounded externally by a thick, smooth, elastic

pellicle containing longitudinally arranged contractile microtubules. The body cytoplasm is distinctly differentiated into following two layers:

1. Cortex or ectoplasm. This is outer or peripheral, clear and firm layer of cytoplasm. It is further subdivided into three layers—the outer **epicyte**, the middle **sarcocyte** and inner **myocyte**. Myocyte consists of longitudinal and transverse contractile fibres, called **myonemes**. These cause metabolic contraction of body and characteristic wriggling or gliding movements called **gregarine movement**.

2. Medulla or endoplasm. It is inner, more fluid-like, granular part of cytoplasm. Endoplasm is rich in granules of **paraglycogen** (a special form of glycogen; reserve food of the animal), fat globules and sometimes **volutin** (a protein rich in phosphorus and nucleic acid).

Nucleus. The nucleus is single, large, vesicular and ovoid occurring in the endoplasm. It lies toward the anterior end of trophozoite's body and is bounded by a distinct **nuclear membrane** and contains a large distinctive **nucleolus**. It is haploid containing four chromosomes.

Due to its parasitic mode of life, the trophozoite of monocystis lacks parapodia, cilia, flagella, mouth, gullet and the food vacuoles and contractile vacuoles.

3. Physiology

1. Locomotion. It moves very slowly by rhythmic contraction and relaxation of myonemes, causing **gregarine movement**.

2. Nutrition. Nutrition in *Monocystis* is saprozoic; it absorbs nourishment from the fluid of the seminal vesicle by osmotrophy through the general body surface.

3. Respiration and excretion. Respiration and excretion occur by diffusion through general body surface of *Monocystis*.

4. Reproduction. *Monocystis* reproduces sexually and the process is always followed by **asexual reproduction**.

4.4 LIFE CYCLE

Monocystis is **monogenetic**, *i.e.*, its life cycle is completed in a single host, *i.e.*, earthworm. Life cycle of *Monocystis* involves **gametogony, syngamy** and **sporogony** (Fig. 4.2).

(*i*) Gametogony (or gamontogony). It is a method of sexual reproduction in which two trophozoites come together, become short and rounded and are called **gamonts** or **gametocytes**. The pair secretes a common protective two-layered cyst wall and becomes **gametocyst**. The outer layer of cyst is thick and hard and is called **ectocyst** and inner layer is thin and soft and is called **endocyst**.

The two trophozoites in the gametocyst never fuse or conjugate but remain together in pair. This stage is called **syzygy** (Fig. 4.2). Ultimately, each gametocyte of gametocyst divides by repeated mitotic cell divisions, producing a number of small cells called **gametes**.

(*ii*) Syngamy. Gametes of *Monocystis* are identical morphologically, so are called **isogametes** (**M.A. Sleigh**, 1973). The gametes unite in pairs to form diploid **zygotes** or **sporoblasts**. As a rule, two gametes uniting in a pair comes from different gametocytes of same cyst. The union and fusion of two gametes of same species is called **syngamy;** syngamy of isogametes is called **isogamy.**

(*iii*) Sporogony. Sporoblast becomes lemon-shaped, secretes a thin hard covering called **sporocyst** around it and becomes a **spore**. Spore becomes spindle-shaped or boat-shaped with a mucoid plug at each end. The spore resembles a diatom *Navicella* in shape, so it is often called a **pseudonavicella**. The nucleus and cytoplasm of the spore divide three times, first division being *meiotic* and rest two being *mitotic*, to produce eight haploid spindle-shaped **sporozoites**. The process of haploid spore formation is called **sporogony**. The spores wait for transmission to the new host.

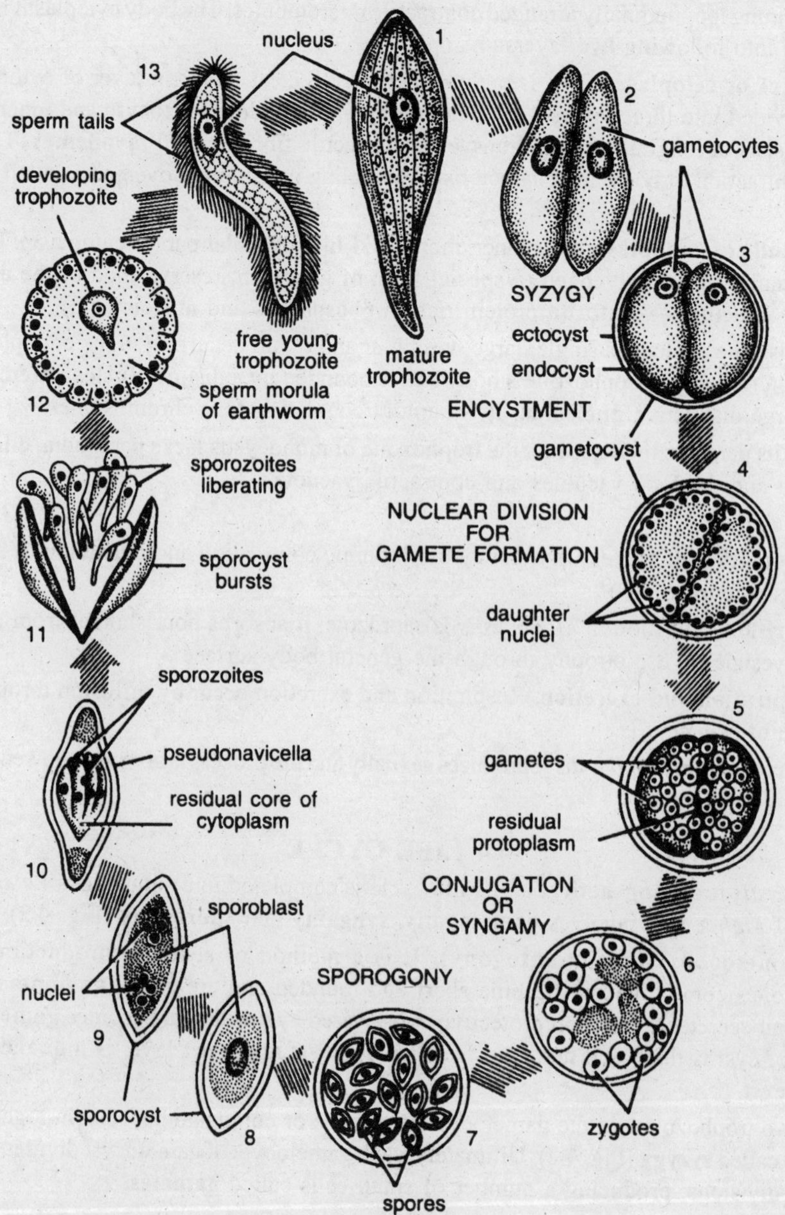

Fig. 4.2. *Monocystis.* Life cycle.

Mode of transmission. Spores of *Monocystis* may be released from the seminal vesicles by following three methods: 1. During copulation an infected host may transfer spores of monocystis to the other host along with the spermatozoa and seminal fluid. 2. When a host earthworm dies and decays, the spores become mixed with the soil. When such an infected soil is eaten by another worm, spore's reach its alimentary canal. 3. Transmission of spores also occurs by birds. When a bird swallows an infected earthworm, the monocystis spores reach, the gut of the bird. These pass out of the bird unaffected along with the faecal matter. When an earthworm eats this faecal matter, the spores reach its intestine.

In the intestine of earthworm, wall of each ingested spore is dissolved by digestive juices of the host and sporozoites are released. Sporozoites attack the intestinal wall, penetrate it and somehow reach the seminal vesicle.

4.5 PARASITIC ADAPTATIONS

The parasitic mode of life of *Monocystis* has following effects on the host and the parasite.

1. Parasitic effects on the host. Although every earthworm is infected with *Monocystis*, but its fertility is less affected. This is so because most of the seminal vesicles of earthworms are not infected. The earthworm is also able to combat some of the parasites by forming a resistant envelope around the trophozoites or by phagocytizing and killing the resistant spores.

2. On the parasite. With an abundance of food and almost no competition inside seminal vesicles of earthworms, the adult trophozoite of *Monocystis* undergoes extreme retrogression or simplification. Under the most favourable, easy and uniform conditions of life, the parasite has no need to develop the various vegetative organelles such as mouth, gullet, locomotory structures, contractile vacuoles, etc. However, the entire life cycle of *Monocystis* is passed within the same host (the earthworm), so, to continue the race, transmission to another host (worm) becomes essential. This is done by the production of sporozoites contained within spores having resistant coverings. During transmission the spores have to face ecological hazards of various sorts resulting in their large scale destruction. To compensate this high mortality, enormous number of potential individuals are produced to ensure that a sufficient number of individuals may survive to continue the race.

QUESTIONS

Long Answer Questions

1. Give an account of the structure and life cycle of *Monocystis*. (*Kanpur 2001*)
2. Describe structure of the trophozoite of *Monocystis*. Discuss morphological, physiological and reproductive adaptations of the parasite. (*Gorakhpur 1992; Agra 1992; Purvanchal 1995*)
3. Give an illustrated account of the structure and life history of *Monocystis*. (*C.C.S.U., Meerut 1991; Allahabad 1995*)
4. Describe various stages of life history of *Monocystis* giving suitable diagrams. (*Purvanchal 1995; Kanpur 1996; Gorakhpur 1998; Allahabad 1998*)
5. Describe the life history of *Monocystis*. How sporocysts are transmitted from one host to another? (*Purvanchal 1991*)
6. Describe the life cycle of *Monocystis* and comment upon the following: 1. Genetic nature of trophozoites; 2. Nutrition in Monocystis; 3. Pseudonavicella. (*Avadh 1997*)
7. Describe the life cycle of *Monocystis*. (*Purvanchal 1999*)

Short Answer Questions

1. Describe syzygy in *Monocystis*. (*Purvanchal 1996*)
2. Write down classification of *Monocystis*. (*Avadh 1993*)
3. Write short notes on the following:
   (*i*) *Monocystis* (*Avadh 1994; Gorakhpur 1998*)
   (*ii*) Gametogamy in *Monocystis* (*Avadh 1997; Purvanchal 1997*)
   (*iii*) Trophozoite of *Monocystis*. (*Kanpur 2001*)

Very Short Answer Questions

1. Define trophozoite.
   Ans. A protozoan of a vegetative form as distinguished from one of a reproductive or resting form.
2. What do you mean by vegetative?
   Ans. Having the power of growing and involving propagation by non-sexual methods.

3. What is sporogony?

Ans. Reproduction by spores; *i.e.*, formation of spores containing sporozoites that is characteristic of some sporozoans and that result, from the encystment and subsequent division of zygote.

4. What is sporozoite?

Ans. Motile, infective form of some sporozoans that is a product of sporogony and initiates an asexual cycle in the new host.

5. Define syzygy.

Ans. End-to-end union of the sporonts or gamonts of certain gregarine protozoans.

6. What is pseudonavicella?

Ans. In sporozoa, a small boat-shaped spore containing sporozoites.

Fill in the Blanks

1. *Monocystis pheretimi* is a .... of *Pheretima posthuma.*
2. A full-grown trophozoite of *Monocystis is called* ....
3. Ectoplasm of *Monocystis* is differentiated into epicyte, sarcocyte and ....
4. *Monocystis* lack contractile vacuole. Is it true? ....

Multiple Choice Questions

1. *Monocystis* is a parasite of
   (*a*) human being (*b*) cattle
   (*c*) earthworm (*d*) mosquito
2. The type of nutrition found in *Monocystis* is
   (*a*) saprozoic (*b*) mixotrophic
   (*c*) holophytic (*d*) holozoic
3. *Monocystis* is .... parasite
   (*a*) digenetic (*b*) monogenetic
   (*c*) trigenetic (*d*) none of these
4. Adult mature nutritive monocystis is called
   (*a*) sporozoite (*b*) trophozoite
   (*c*) gametocyte (*d*) none of these
5. The habitat of *Monocystis* is
   (*a*) sperm morula (*b*) colon and rectum
   (*c*) salivary glands (*d*) mucosa of small intestine
6. Life cycle of *Monocystis* includes
   (*a*) gametogony and sporogony (*b*) schizogony and gametogony
   (*c*) syzygy and gametogony (*d*) sporogony
7. Trophozoites of *Monocystis* are found in
   (*a*) seminal vesicles of earthworm (*b*) seminal vesicles of cockroach
   (*c*) both (*d*) none

ANSWERS

Fill in the Blanks

1. Parasite 2. Sporadin 3. Myocyte 4. Yes

Multiple Choice Questions

1. (*c*) 2. (*a*) 3. (*b*) 4. (*b*)
5. (*a*) 6. (*a*) 7. (*a*).

Trypanosoma

5.1 CLASSIFICATION

Phylum	–	**Protozoa**
Subphylum	–	**Sarcomastigophora**
Superclass	–	**Mastigophora (Flagellata)**
Class	–	**Zoomastigophora**
Order	–	**Kinetoplastida**
Genus	–	***Trypanosoma***

African sleeping sickness is a fatal human disease that is caused by protozoan parasite *Trypanosoma gambiense* (Fig. 5.1; Box. 5.1).

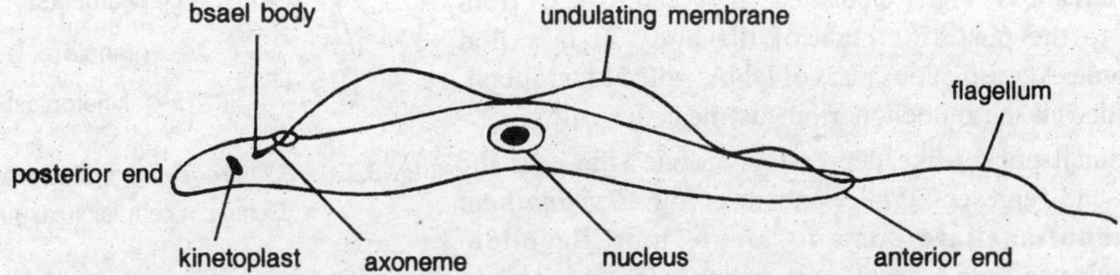

Fig. 5.1. *Trypanosoma*. Morphology of trypomastigote form.

T. gambiense is found in central Africa, West Africa (from Gambia to the Congo) and Uganda. Commonly low marshy lands near the rivers and lakes have the greatest incidence of infection because the blood sucking insect vector, the tsetse fly (*Glossina palpalis*) inhabit these areas.

Box 5.1.

The genus *Trypansoma* is parasitic in the blood of most of the vertebrates such as fishes, amphibians, reptiles, birds and mammals. Many species of *Trypanosoma* are **pathogenic** causing human disease called **trypanosomiasis**. Only three species of it are pathogenic in human beings, *viz.*, *T. gambiense, T. rhodesiense* and *T. cruzi. T. gambiense* and *T. rhodesiense* cause a fatal disease in human beings (hosts) called **sleeping sickness** in Africa, while *T. cruzi* causes **chagas disease** in children of South America. Their transmission from one vertebrate host to other takes place by invertebrate blood-sucking animals such as insects and leeches. These animals are called **vectors**.

Surra disease of horses and cattles is caused by *Trypanosoma evansi* and is transmitted by blood-sucking flies, celled *Tabanus* and *Stomoxys*.

5.2 HABIT AND HABITAT

Trypanosoma gambiense lives as a digenetic endoparasite in the blood, lymph, lymph nodes, spleen or cerebrospinal fluid of human beings and in the intestine of blood-sucking insect, *Glossina palpalis*.

5.3 MORPHOLOGY

T. gambiense is a unicellular, microscopic, slender, elongated, colourless, leaf-like protozoan. Its body measures about 10 μm to 40μm in length and 1μm to 5 μm in breadth. The anterior end of body is pointed and the posterior end is blunt.

The body is covered by tough but elastic **pellicle** which is composed of numerous microtubules. Pellicle is followed by cytoplasm which is differentiated into an outer clear layer called **ectoplasm** and an inner granular **endoplasm**. The endoplasm contains single nucleus, single mitochondrion, single Golgi apparatus, endoplasmic reticulum and ribosomes. It lacks contractile vacuole and food vacuoles, however, contains many scattered greenish refractile bodies, called **volutin granules**. These granules contain glycogen and phosphates and represent reserve food material of *Try-panosoma*.

Nucleus is single, oval, vesicular and is located in the middle of its body. It contains a large **endosome, karyosome** or **nucleolus**. As nucleus is concerned with the nutritive functions of the parasite, it is often called **trophonucleus. Mitochondrion** is single, elongated, giant and extends from anterior to the posterior ends of the body. It is called **chondriome**. A conspicuous mass of DNA, called **kinetoplast**, is embedded in the mitochondrion just near the gullet.

A small pocket-like depression on one side near the posterior end represents gullet and reservoir. *Trypanosoma* being **monoflagellate** contains single long flagellum which originates from a **basal body** which is located close to the gullet near posterior end of body. It consists of an **axoneme** comprisin microtubules in 9 + 2

free flagellum
pellicle
attached flagellum
ectoplasm
undulating membrane
endoplasm
nucleus
nucleolus
reserve food granules
attached flagellum
blepharoplast or basal granule
kinetoplast

Fig. 5.2. *Trypanosoma gambiense*: Details of cellular structure.

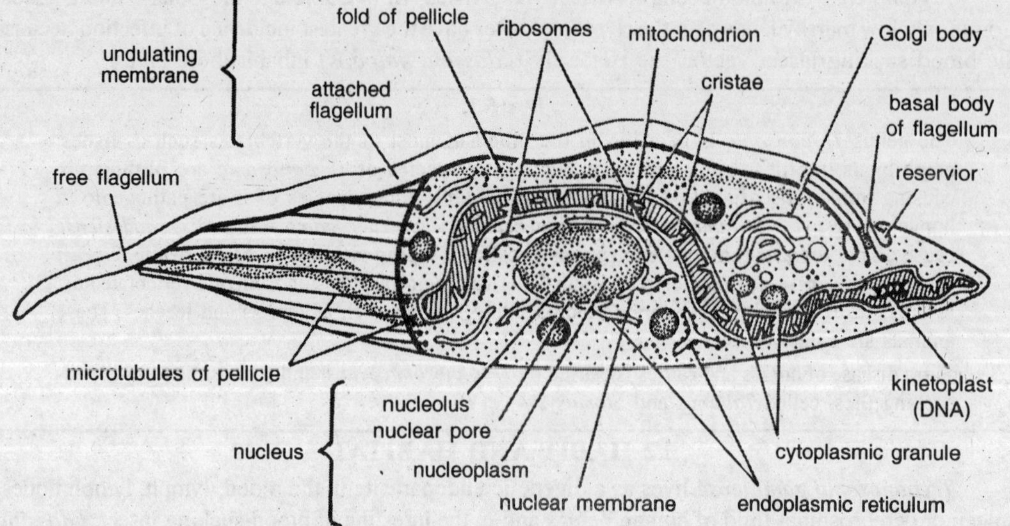

Fig. 5.3. *Trypanosoma*. An electron microscopic structure.

arrangement and bounded by the plasma membrane. Flagellum extends forward along the body to project beyond its front end, in the form of free flagellum. All along its length over the body, it is connected to the body by a fin-like **undulating membrane** formed of cytoplasm and folded pellicle. Undulating membrane is supposed to be an adaptive structure for locomotion in a viscous environment (blood, lymph) where it lives.

Polymorphism. *Trypanosoma* is a **polymorphic form**; it occurs in following four morphological forms:

1. Amastigote form (old name = Leishmanial stage). It is represented by rounded form which is without any external flagellum as in the genus *Leishmania*. It is a rounded or oval form in which reduced flagellum remain in the cytoplasm; basal body and kinetoplast are situated in the anterior part of body (Fig.5.4A).

2. Promastigote form (old name = Leptomonad stage). It is represented by an elongated form with the kinetoplast lying to the anterior to the nucleus (antenuclear kinetoplast). The flagellum arises near kinetoplast and emerges from the anterior end of body, as in the genus *Leptomonas* and others. In it, the flagellum is short, unattached and free (Fig. 5.4B).

3. Epimastigote form (old name = Crithidial stage). It is represented by an elongated form with the kinetoplast lying anterior and close to the nucleus (*i.e.*, juxtanuclear kinetoplast); the flagellum arising near it emerges from the side of the body to run along a short undulating membrane (Fig. 5.4C).

4. Trypomastigote form (old name Trypanosome stage). This term is not derived form genus *Trypanosoma*. It is represented by an elongated form with post-nuclear kinetoplast situated at the posterior end of the body. The flagellum arises near the kinetoplast and emerges from side of body to run along an undulating membrane, as in genus *Trypanosoma* (Fig. 5.4D).

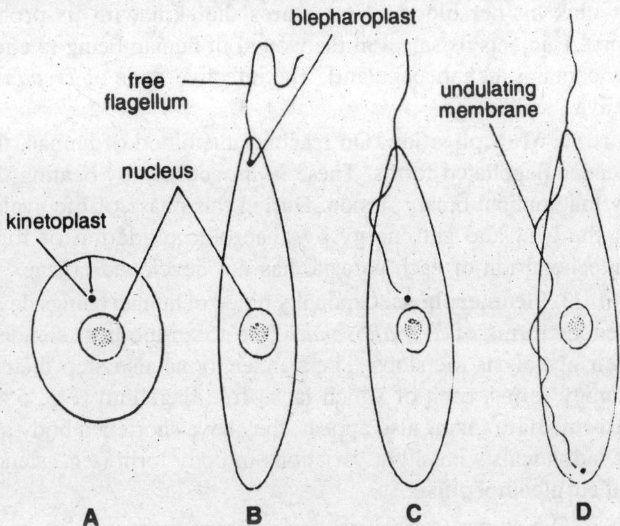

Fig. 5.4. *Trypanosoma*. Polymorphic forms. A—Amastigote or Leishmanial; B—Promastigote or Leptomonad; C—Epimastigote or Crithidial; D—Trypomastigote or Trypanosome.

Out of these four forms trypomastigote is the adult stage while other forms represent developmental stages which are formed during the part of its life cycle in the invertebrate host.

5.4 REPRODUCTION

There is no sexual production in *Trypanosoma*. It reproduces by asexual reproduction, *i.e.*, **longitudinal binary fission**. It occurs in all forms found in both hosts (*i.e.*, human beings and tsetse fly). During binary fission, the basal body (blepharoplast) divides first, then subsequently the kinetoplast and nucleus and finally the cytoplasm. Usually one daughter individual retains the old flagellum and undulating membrane, and the other daughter develops a new one, which grows out before the complete separation of the daughters.

5.5 LIFE CYCLE

Trypanosoma gambiense is a digenetic pathogen; it passes its life cycle in two different types of hosts: two vertebrate hosts and one invertebrate host. These three hosts fall into following types:

1. Primary host or definitive host in which parasite lives initially in blood outside the red blood cells (*i.e.*, extracellularly). Later on parasite lives in cerebrospinal fluid of the host, *e.g.*, human beings.

2. Reservoir host in which parasite lives in the blood but causes no harm. *e.g.*, domestic and wild animals such as pigs, antelopes, buffaloes and reed bucks.

3. Secondary host or intermediate host in which polymorphic forms of the parasite are developed. *e.g.*, blood sucking insect, tsetse fly, *Glossina palpalis* and *Glossina tachinoides*. Since tsetse fly carries the trypanosomes from person to person, it is known as **vector** or **carrier**. Both male and female tsetse fly bite by daylight, usually in the early morning and evening.

A. Life Cycle in Human Beings

1. Infection. Infective stage of *Trypanosoma* exists in the salivary glands of tsetse fly and is called **metacyclic form** (trypomastigote). When such infected fly bites a human being to suck his/her blood, it punctures the skin with its proboscis (which is a needlelike mouth part). It injects its saliva in the wound of human being to check blood clotting (the saliva of tsetse fly contains an anticoagulant). The infective stage of *Trypanosoma* enters human blood along with saliva.

2. Multiplication. On reaching the blood of human, the metacyclic forms change into long slender flagellated forms. These swim actively by beating their flagellum and multiply asexually by longitudinal binary fission. During this phase of life cycle, parasites aborb glucose from blood of the host and get energy by anaerobic oxidation of food (glycolysis). Upto this stage, the mitochondrion of each parasite has not developed cristae.

3. Pleomorphism. Gradually blood of human beings develop antibodies against the multiplying slender forms of *Trypanosoma*. Due to antibodies, slender forms stop absorbing glucose and their glycolysis too stops. The slender forms also stop their multiplication and 'change into short stumpy forms, each of which lacks free flagellum (Fig. 5.4). During this transformation certain **intermediate forms** also appear; they have shortened body and reduced free flagellum. Production of individuals with slight variations in body form (*e.g.*, slender, intermediate and stumpy form) is called **pleomorphism.**

The stumpy forms get their energy from aerobic oxidation of pyruvic acid which is the end product of glycolysis. Their mitochondria develops cristae and produces enzymes of Krebs cycle. The stumpy (ovoid leishmania forms) are sucked by the tsetse fly along with blood meal.

4. Relapse of infection. Some of the slender forms of *Trypanosoma* are not metamorphosed. They escape from the action of host's antibodies by changing the antigens which are present on their surface. Such slender forms survive and continue to multiply and cause relapse of infection.

B. Life Cycle in Tsetse Fly

1. Transfer to the tsetse fly. When the blood of an infected person or reservoir host is sucked up by a male or female tsetse fly, short and stumpy forms of *Trypanosoma* (*i.e.*, trypomastigote) enter its alimentary canal along with the blood. These exist in an unchanged form in the midgut for few hours.

2. Development in midgut. After tenth to fifteenth day short-stumpy forms develop into long slender forms. These forms pass into extra-peritrophic space (a space between the peritrophic membrane and the epithelial cells). The slender forms multiply by longitudinal binary fission. By fifteenth day they enter the lumen of the proventriculus.

3. Development in salivary glands. The long slender forms then migrate forward to the buccal cavity. They pass on the hypopharynx and eventually reach the salivary glands through the opening of salivary ducts. Here they multiply and change their morphology, first into *epimastigote* (or crithidial stage) and then into **metacyclic stage** (short stumpy form of trypomastigote) which are infective to human beings.

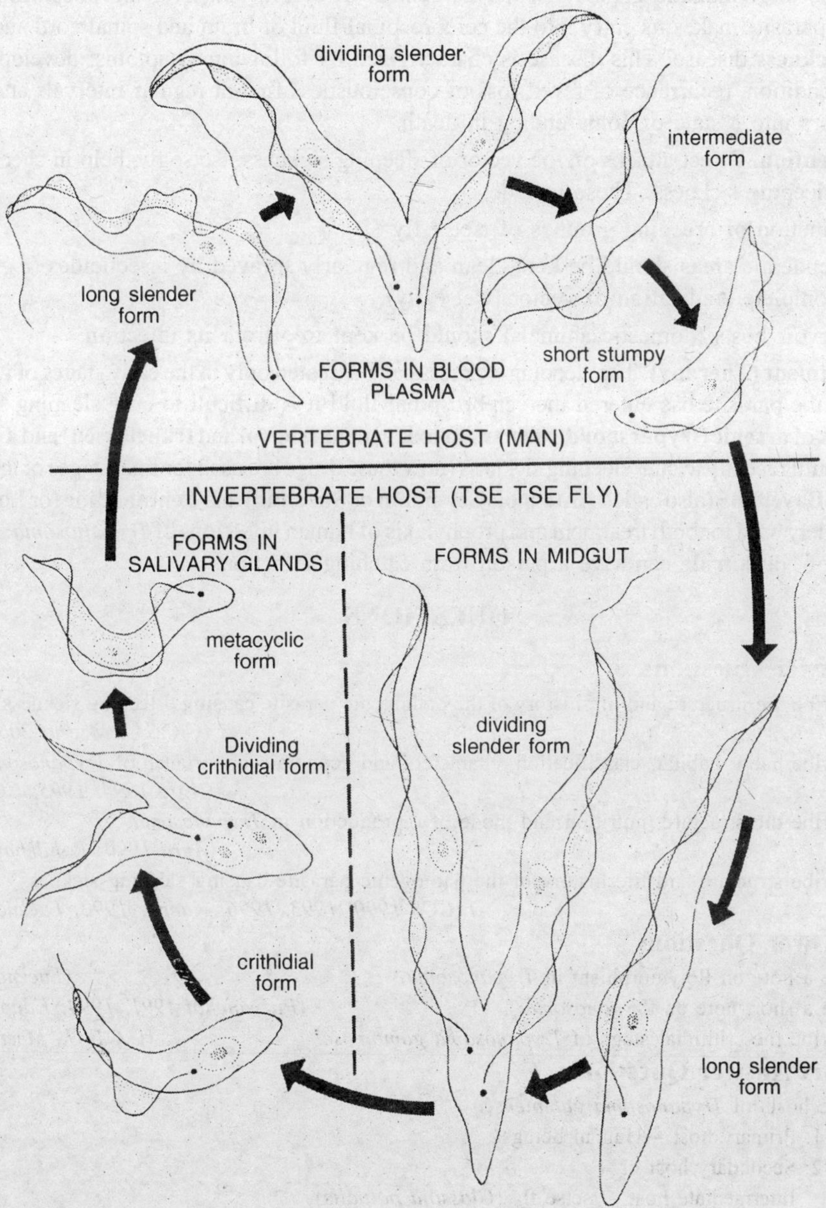

Fig. 5.5. *Trypanosoma gambiense*. Life cycle.

5.6 ECONOMIC IMPORTANCE

Presence of trypanosomes in human blood causes **gambian fever** which persists only for few days and characterized by fever and severe headache, both of which recur at regular intervals accompanied by loss of nocturnal sleep, increasing weakness, loss of weight and anaemia. Usually, the pathogenic parasite succeed in penetrating the lymphatic glands. Because of infection of trypanosome the lymphatic glands swell; enlargement of spleen and liver too becomes evident. Ultimately parasite makes its entry into the cerebrospinal fluid of brain and spinal cord and causes **sleeping sickness** disease. This disease is characterized by following symptoms: development of lethargic condition, recurrence of fever, loss of consciousness first at regular intervals and finally patient enters into a state of coma ending in death.

Prevention. Direct attacks on the vector of sleeping sickness, tsetse fly, help in checking the spread of sleeping sickness. These include:

1. Elimination of breeding grounds of tsetse fly.
2. The endemic areas should be kept clean and regularly sprayed by insecticides (*e.g.*, DDT) for complete eradication of vector (tsetse fly).
3. Reservoir hosts (domestic animals) should be kept free from its infection.

Treatment (Therapy). The sleeping sickness can be treated only in the early stages of infection, since once the parasite has entered the cerebrospinal fluid it is difficult to cure sleeping sickness. Compounds of **arsenic (tryparsamide, melarson oxide, melarsoprol** and **trimelarsen)** and **antimony** were used until recently to cure sleeping sickness (now these druge are used for final stages of infection). Two drugs, **Bayer 205** (also called **Antrypol, Germanin** or **Suramin)** and **Pentamidine** (or **Lomidine)** are now widely used for both treatment and prophylaxis of human infections of *Trypanosoma*. Periodic injections of **Naphuride** can save a person from catching infection.

QUESTIONS

Long Answer Questions

1. Describe the structure and life-history of the pathogenic parasite causing "sleeping sickness" in man. (*C.C.S.U. Meerut 2004, 2007*)
2. Describe habit, habitat, classification, characters and economic importance of *Trypanosoma*. (*Agra 1994, 1995; CCS 2000*)
3. Describe the structure, nutrition and mode of reproduction in *Trypanosoma*. (*Agra 1990; Rohilkhand 1992*)
4. Describe structure and life history of the pathogenic parasite causing sleeping sickness. (*CCS 1990, 1993, 1996; Kanpur 1992; Lucknow 1998*)

Short Answer Questions

1. Write a note on Polymorphism in *Trypanosoma*. (*Lucknow 1999*)
2. Write a short note on *Trypanosoma*. (*Purvanchal 1991, 1994; Kanpur 1998*)
3. Describe the crithidial stage of *Trypanosoma gambiense*. (*C.C.S.U., Meerut 2003*)

Very Short Answer Questions

1. Name hosts of *Trypanosoma gambiense*.

   Ans. 1. Primary host – Human beings
   2. Secondary host or
   Intermediate host – tsetse fly (*Glossina palpalis*)
   3. Reservoir host – Mammals such as pigs, buffaloes and antelopes.
2. What is reservoir host?

   Ans. Reservoir host is one which harbours the parasite but is not affected by the parasite. Reservoir host does not transmit it, *e.g.*, antelopes, pigs and buffaloes.
3. Write down zoological names of following animals.

   (*i*) Parasite causing sleeping sickness. (*Agra 1997*)

Ans. *Trypanosoma gambiense.*

(*ii*) Vector of parasite of sleeping sickness.

Ans. Tsetse fly (*Glossina palpalis*).

4. What are the infective stages in the life cycle of *Trypanosoma gambiense*?

Ans. Metacyclic forms are infective stages in the life cycle of *Trypanosoma gambiense* which are transferred to healthy human beings due to bite of tsetse fly.

Fill in the Blanks

1. African sleeping sickness is a .... human disease.

2. Insect vector of *Trypanosoma gambiense* is blood sucking .... fly.

3. Zoological name of tsetse fly is ....

4. Parasite of sleeping sickness lives in human blood outside ....

Multiple Choice Questions

1. *Trypanosoma* is a

(*a*) nonpathogenic (*b*) monogenetic parasite

(*c*) digenetic parasite (*d*) facultative parasite

2. African sleeping sickness or Gambian fever is caused by

(*a*) *Entamoeba* (*b*) *Leishmania*

(*c*) *Trichomonas* (*d*) *Trypanosoma*

3. Infective stage of *Trypanosoma gambiense* is

(*a*) crithidial form (*b*) leptomonad form

(*c*) metacyclic form (*d*) leishmania form

4. *Trypanosoma gambiense* lives in human body in

(*a*) lymph (*b*) blood

(*c*) cerebrospinal fluid (*d*) blood and cerebrospinal fluid

5. A bite of tsetse fly may transmits

(*a*) *Leishmania donovani* to human (*b*) *Plasmodium vivax* to humans

(*c*) *Entamoeba histolytica* to humans (*d*) *Trypanosoma gambiense* to humans

6. The reservoir host for *Trypanosoma gambiense* is

(*a*) rat (*b*) pig

(*c*) human being (*d*) tsetse fly

7. Nutrition of *Trypanosoma* is

(*a*) saprozoic (*b*) mixotrophic

(*c*) holophytic (*d*) none of these

8. Chagas disease of humans is caused by

(*a*) *T. gambiense* (*b*) *T. cruzi*

(*c*) *T. evansi* (*d*) *T. rhodesiense*

9. Parasite of chagas disease is transmitted through the bite of

(*a*) bug (*Triatoma*) (*b*) tsetse fly (*Glosina palpalis*)

(*c*) *Tabanus* (*d*) tabanid fly

10. *Trypanosoma evansi* causes

(*a*) surra in horse (*b*) kala azar

(*c*) sleeping sickness (*d*) chagas disease

ANSWERS

Fill in the Blanks

1. Fatal **2.** Tse-tse **3.** *Glossina palpalis* **4.** Red blood cells.

Multiple Choice Questions

1. (*c*) **2.** (*d*) **3.** (*c*) **4.** (*b*) **5.** (*d*)

6. (*b*) **7.** (*a*) **8.** (*b*) **9.** (*a*) **10.** (*a*).

6

Paramecium

6.1 CLASSIFICATION

Phylum	–	**Protozoa**
Subphylum	–	**Ciliophora**
Class	–	**Ciliata**
Subclass	–	**Holotrichia**
Genus	–	***Paramecium***
Species	–	***caudatum***

Paramecium is well-known ciliate protozoan. It demonstrates a very high degree of **cellular differentiation**, exhibiting many complex organelles which have been designed to perform specific functions for the organism. Not only its structure show specialization, but its reproductive activity is also complex. There are about 10 species of *Paramecium*. Its common species are *P. caudatum* and *P. aurelia*.

6.2 HABIT AND HABITAT

Paramecium caudatum (Gr., *paramekes* = oblong; L. *caudata* = tail) is a free-living organism having cosmopolitan (world-wide) distribution. It lives in stagnant water of freshwater, ponds, pools, ditches, and slow flowing streams rich in decaying organic matter.

6.3 STRUCTURE (MORPHOLOGY)

Size and shape. *Paramecium caudatum* is a unicellular and microscopic protozoan. It measures 170 to 290 mm. It is visible to the naked eye. Its body has a constant elongated, slipper-like shape, so the animal is commonly called **slipper animalcule**. The anterior end of animal's body is blunt and posterior end is thick and pointed. Body is **asymmetrical** exhibiting flat **ventral** or **oral surface** and convex **dorsal** or **aboral surface**. Cellular body of *Paramecium* has the following structures:

1. Pellicle. The whole body of *Paramecium* is covered by a thin, firm, flexible membrane, the **pellicle**. It maintains a definite shape of the animal and also serves as a skeletal structure. The pellicle has a uniform appearance, being composed of a lattice of hexagonally shaped pits, each centrally perforated by a single cilium. The margins of hexagonal depressions are slightly raised up as ridges. Trichocysts open in the ridges (Fig. 6.2).

Ultrastructure of pellicle. The electron microscopic study of pellicle (Fig. 6.3) by **Ehrat** and **Powers** (1957) has revealed that the hexagonal depressions correspond to regular series of cavities, the **alveoli**. All alveoli collectively form a continuous alveolar layer, which is delimited by an **outer alveolar membrane**. The outer alveolar layer is covered by **plasma membrane** which also surrounds the cilia. Thus, pellicle comprises of three membranes—1. Outer plasma membrane (outermost). 2. Outer alveolar membrane (middle membrane), and 3. Inner alveolar membrane (innermost).

2. Cilia. Entire body surface of *Paramecium* is covered with numerous (10,000 to 14,000), tiny (10–12 μm in length and 0.27 μm in diameter), hair-like projections of cytoplasm, called **cilia**.

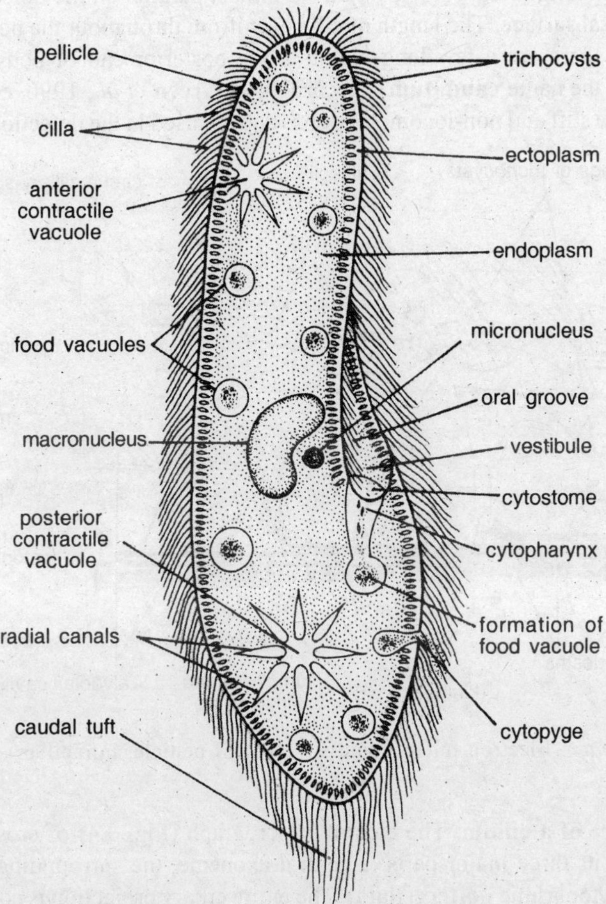

Fig 6.1. *Paramecium caudatum*. Structure.

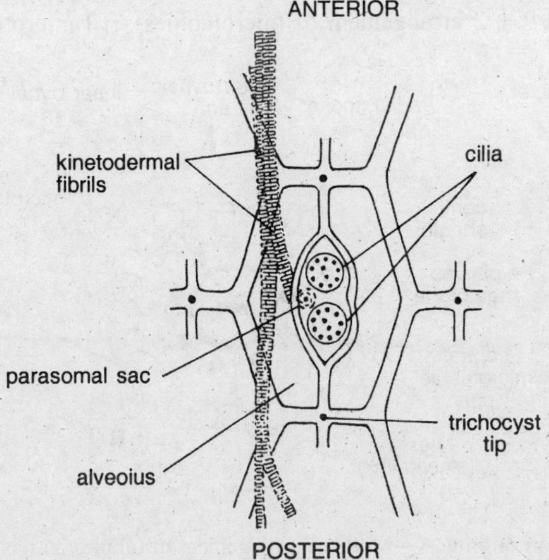

Fig. 6.2. *Paramecium aurelia*. Hexagonal area or ciliary field.

They are arranged in longitudinal rows, which are almost parallel on the dorsal surface and slightly oblique on the ventral surface. The length of cilia is unifrom throughout the body (a condition called **holotrichous**), but there are a few longer cilia at the posterior end of body. These form **caudal tuft** of cilia (hence the name **caudatum**). According to **Green** *et al.*, 1990, certain caudal cilia are **tactile cilia**, they are stiff and non-locomotory ones and are used in the detection of external stimulus.

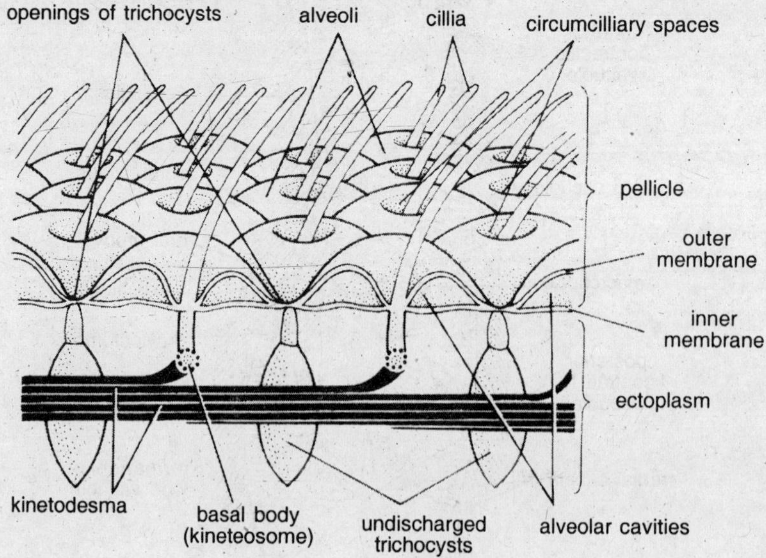

Fig. 6.3. *Paramecium*. Electron microscopic structure of pellicle, infraciliary system and associated structure.

Ultrastructure of a cilium. The electron micrograph (Fig. 6.4) of cross section of a cilium (or flagellum) exhibits three major parts: a central axoneme, the surrounding **plasma membrane** and the interposed cytoplasmic **matrix** (fluid). The entire ciliary projection is covered by a membrane that is continuous with the outer plasma membrane of pellicle. The core or shaft of the cilium, called the **axoneme**, contains an array of microtubules that run longitudinally through the entire organelle. Axoneme consists of nine peripheral doublet microtubules surrounding a central pair of single microtubules (*i.e.*, typical 9 + 2 arrangement of microtubules). All microtubules of the axoneme

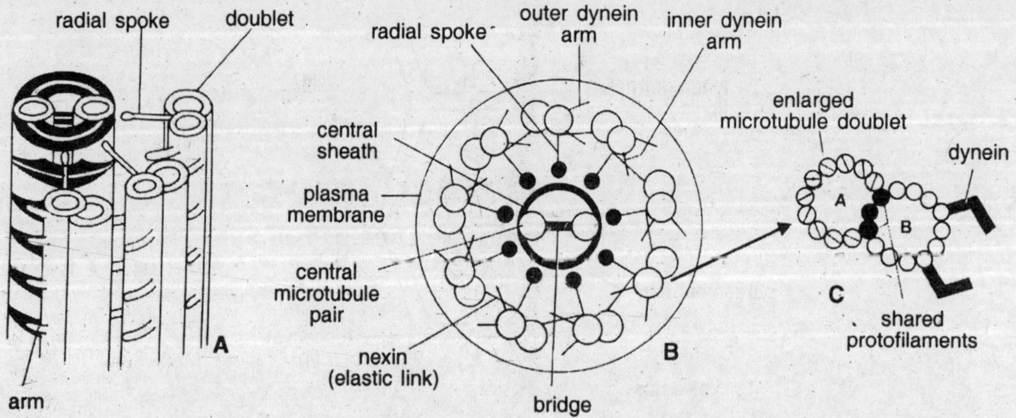

Fig. 6.4. Internal organization of cilia. A—Mode of arrangement of tubules and associated material as seen from outside. B—Cross section of cilia. C—An enlarged microtubule doublet.

have the same polarity: their plus ends are the tip of the projection and their **minus ends** at the base. Each peripheral doublet consists of one complete microtubule, the **A tubule**, and one incomplete microtubule, the **B tubule**, which contains 10 or 11 subunits (protofilaments of a and b tubulin proteins) rather than usual 13.

The central tubules are enclosed by projections that form **central sheath**, which is connected to A tubules of the peripheral doublets by a set of nine **radial spokes**. The doublets are connected to one another by an **interdoublet bridge** composed of an elastic protein, called **nexin**. Extending from each A tubule are two **dynein arms**—an **outer arm** and an **inner arm**, that are oriented in the same direction in all peripheral doublets, *i.e.*, clockwise when the axoneme is viewed from base to tip. Since dynein protein is capable of hydrolyzing ATP, it acts as an ATPase enzyme.

3. Cytoplasm. Beneath the pellicle lies the cytoplasm. The cytoplasm is differentiated into an outer narrow peripheral layer of clear and dense **plasmagel**, called **ectoplasm** and an inner large central mass of granular and semifluid **plasmasol** or **endoplasm.**

A. Ectoplasm of *Paramecium* contains the following organelles:

(*a*) Infraciliary system. It consists of basal bodies (**kinetosomes**) of cilia and **kinetodesmata** all located just beneath the alveoli of pellicle in the ectoplasm (Fig. 6.3).

(*i*) Basal body or kinetosome. The base of each cilium is connected with a tubular **basal body** or **kinetosome**, the structure from which cilium is formed. Structure of the basal body is comparable to the centriole. Like a cilium, a basal body contains nine peripheral fibres, but each consists of three microtubules rather than two. The A tubule is complete, while the B and C tubules are both incomplete. There is no central microtubules in the basal body or centriole. The A and B tubules of basal body elongate to form the doublets of the cilium (or flagellum, Fig. 6.4).

(*ii*) Kinetodesmata (kinetodesma). A single fine fibril, called **kinetodesmos** or **kinetodesmal fibril**, arises from each kinetosome and extends forwards and obliquely to its right (6.3). It joins other fibrils from adjacent kinetosomes to form a longitudinal bundle of striated fibrils, called a **kinetodesma**. The individual fibrils do not run anteriorly farther than the five kinetosomes. Thus, the number of fibrils in each kinetodesma remains five. The kinetosomes and fibrils of a particular row are collectively known as a **kinety**. All the kinety systems together form **infraciliary system** which coordinate beating of cilia and also plays an important role in the morphogenesis of *Paramecium* and other ciliates.

Motorium. There is also a dense bilobed mass or network of fibrils situated in the endoplasm near the cytostome. This is called **neuromotorium (Lund**, 1933) or **motorium.** It has branches which interconnect with the fibrils of the ectoplasm and the whole fibrillar system is believed to act as controlling centre for feeding movement of oral cilia.

M fibres. At the boundary between ectoplasm and endoplasm are large bundles of **microfilaments** called **M fibres** or **myonemes.** These are contractile in their activity and promote a change of shape in *Paramecium* to enable it to squeeze through narrow spaces.

(*b*) Trichocysts. Trichocysts are minute (4 mm in length) flask-shaped or fusiform sac-like organelles embedded in the ectoplasm. Trichocysts alternate with basal bodies and are perpendicular to surface of body. They open to outside through the permanent pores existing in the ridges of hexagonal depressions of pellicle. Trichocysts are formed by basal bodies.

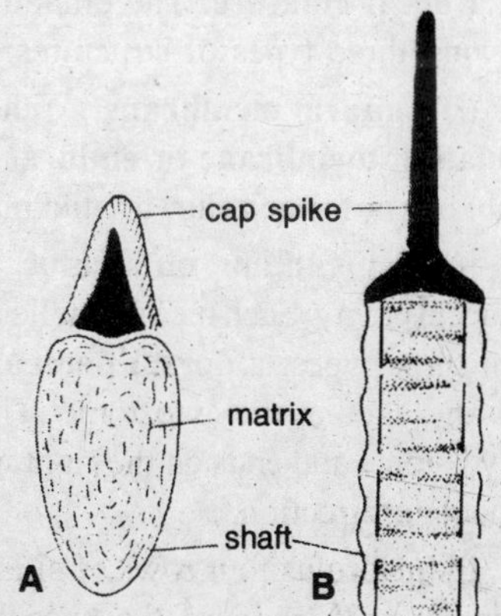

Fig. 6.5. *Paramecium*. A—An undischarged trichocyst. B—Apical portion of a discharged trichocyst.

Each trichocyst consists of an elongated **shaft** and a terminal conical **barb** or **spike** covered with a **cap** (Fig. 6.5). The shaft is filled with a dense fluid having swelling substance, an unfolded fibrous protein, called **trichinin.**

Upon stimulation, the ectoplasm suddenly contracts. This separates the cap of the trichocyst to let the water enter the sac. This results in unfolding of trichinin protein of shaft and formation of a fine 40 mm long striated filament having pointed sticky spike (Fig. 6.5). Once discharged trichocysts cannot be withdrawn, they are replaced by new ones. Trichocysts serve as a means of anchorage during feeding activity. Previously they were considered as organelle of defence, *i.e.*, the long filaments from exploded trichocysts form a sort of screen that gives some degree of safety to the organism.

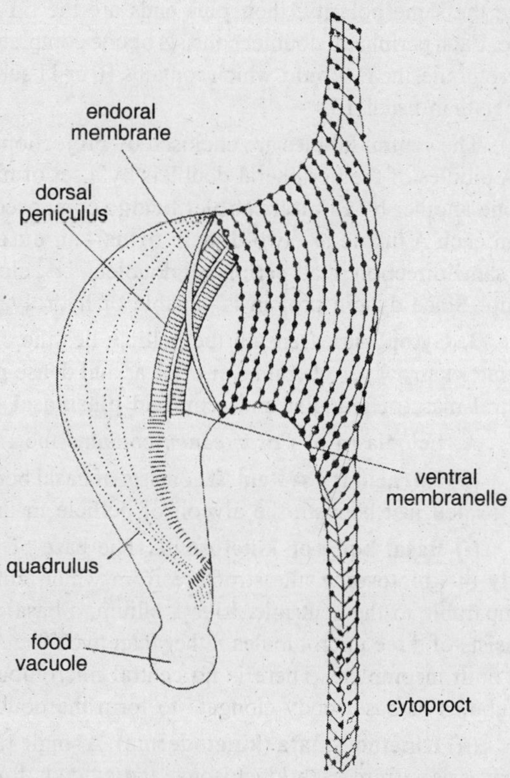

Fig. 6.6. *Paramecium.* Buccal ciliature.

(*c*) Oral apparatus. Near the anterior end of the animal and on its vental surface is a permanent ciliated broad and shallow longitudinal depression called the **oral groove** or **peristome**. It starts from the left side of the anterior end of the body and extends obliquely backwards towards the right side upto the middle of the body. There, it leads into conical funnel-like depression, called **vestibule.** The vestibule leads into a wide tubular passage, the **buccal cavity** or **gullet**. Buccal cavity extends backwards and ends in a small oval aperture, the **mouth** or **cytostome.** The mouth leads into a short tubular **oesophagus** or **cytopharynx** that opens into the endoplasm. The cytopharynx forms a food vacuole at its proximal end.

Buccal ciliature. The ciliature of buccal cavity is different and specialized. It includes the following three types of structures:

(*i*) Endoral membrane. At the junction of the vestibule and the buccal cavity is a crescentic **undulating membrane** or **endoral membrane** which runs transversely along the right wall. This membrane is formed due to sticking together of the cilia of a single row.

(*ii*) Peniculi and quadrulus. There are three membranelles (see Box 6.1) in the left wall of the buccal cavity. Each membranelle is formed by adhereing of four rows of cilia. These membranelles are of three types: a **dorsal peniculus**, a **ventral peniculus** and **quadrulus.** Ventral peniculus is short and ends at the cytostome. The dorsal peniculus is longer and crosses over to right wall at the cytostome and ends on the right wall of the cytopharynx. Cilia of two peniculi beat longitudinally in opposite direction.

In quadrulus four rows of cilia are separated and united at the ends. Quadrulus extends along the dorsal wall (roof) of the buccal cavity, crosses over to the right wall near the cytostome.

Box 6.1. Definitions

Membranelle. It is formed from two, three or four rows of cilia, all of which adhere together; actual fusion of cilia do not occur. This adherence causes all cilia of a membranelle to beat together.

Undulating membrane. It is formed by adherence of cilia of a single row.

End of the buccal cavity gives off very long **dorsal cilia** or **post buccal** cilia that project freely in the cytoplasm.

***(d)* Cytopyge. Anus, cytopyge** or **cytoproct** exists as a minute pore at a fixed point on the right side of body between the posterior end of oral groove and posterior end of the body. Undigested food is egested through it by active vacuolar activity (called **exocytosis**).

B. Endoplasm. The endoplasm includes a variety of important organelles such as nuclei, contractile vacuole, food vacuole, Golgi apparatus, mitochondria, endoplasmic reticulum, ribosomes and reserve food granules.

***(i)* Nuclei.** Dorsal to the gullet and towards the centre of body, lie two different types of nuclei (heterokaryotic). Both nuclei differ from each other not only in appearance (size and shape) but also in function. This phenomenon is called **nuclear dimorphism.** The large been-shaped **macronucleus** is **polyploid** and contains many nucleoli. It is termed **somatic** or **vegetative nucleus** since it controls metabolism and differentiation within the animals. Macronucleus divides by **amitosis** during binary fission. The old macronucleus disappears and a new one is reorganized during conjugation.

The **micronucleus** lies in a depression on the surface of the macronucleus. It is spherical and **diploid.** Micronucleus controls reproductive activities of animal and gives rise to new macronuclei. It is always active when nuclear reorganization takes place during the life history of the animal.

The macronucleus of *Paramecium* is said to contain about 500 times more DNA than that of micronucleus.

***(ii)* Contractile vacuoles.** Two fixed contractile vacuoles are present between ectoplasm and endoplasm close to dorsal surface, one at the anterior end and the other at the posterior end of animal body. Each contractile vacuole opens through a discharge canal in the pellicle on the dorsal side. Each vacuole is surrounded by 6 to 10 long and narrow **radiating** or **feeder canals.** Each feeder canal contains the following three parts:

1. Injector canal. It is the innermost part of feeder canal which opens into the vacuole.

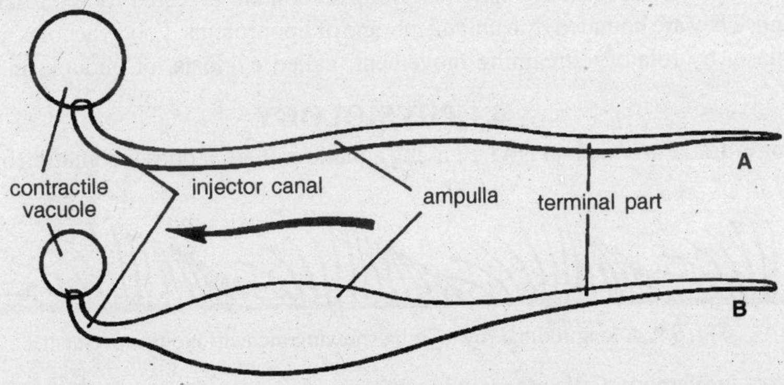

Fig. 6.7. *Paramecium.* Radial canal. A—Empty (systole); B—Full (diastole).

2. Ampulla or accessory vacuole. It is middle part of the feeder canal. It swells like a baloon when filled with water, *i.e.*, when it is in **diastole.** It collapses when it is empty, *i.e.*, when it is in **systole** (Fig. 6.7).

3. Terminal part. It is outer part of the radial or feeder canal which extends into the cytoplasm. It is also called **nephridial canal** or **collecting tubule.** The distal end of collecting tubule is surrounded by a network of fine coiled **nephridial tubules** or **spongiome tubules** (**Ruppert** and **Barnes,** 1994).

These tubules form a sponge around the collecting tubules. Spongiome or nephridial tubules collect water from the surrounding cytoplasm and transfer it to collecting tubule. Endoplasmic reticulum and mitochondria are also found around the spongiome tubules.

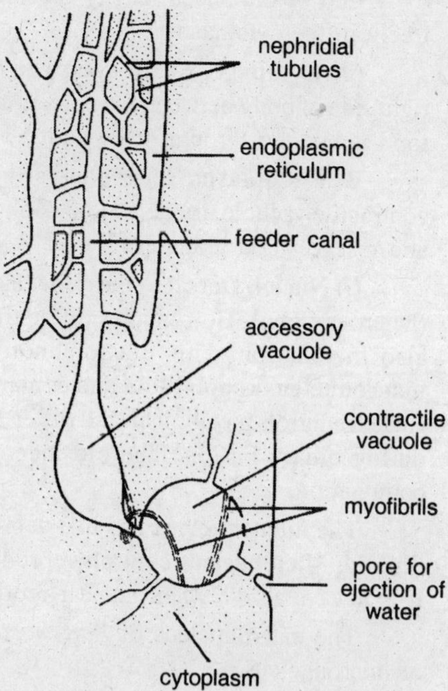

Fig. 6.8. *Paramecium.* Contractile apparatus.

Physiology of contractile vacuole. Contractile vacuoles maintain water balance (called **osmoregulation**) of the body. It pumps excess of body water out of the organism. The wall of contractile vacuoles and radiating canals contain bundles of microtubules and microfilaments (actin), called **myofibrils** and are contractile. Radiating canals become swollen on receiving water (**diastole**) and contract (**systole**) to ultimately empty into the contractile vacuole. Vacuole is reconstituted and grows in size (diastole). When the contractile vacuole reaches its maximum size it contracts suddenly (systole) and discharges its contents through the permanent pore in the pellicle. This cycle is repeated again and again to do osmoregulation.

It is reported that posterior contractile vacuole empties and fills at a faster rate than the anterior one, because of greater endosmosis in the region of gullet. *P. caudatum* can complete a cycle of vacuole filling and discharge as rapidly as every 6 seconds and expel a volume equivalent to its entire body every 15 minutes.

(*iii*) Food vacuoles (or gastrioles). Endoplasm contains a variety of food vacuoles containing food in various stages of digestion. There may be as many as 50 food vacuoles at one time in a single individual. Food vacuoles are spherical in form, contain ingested food particles and a small amount of fluid. They are bounded by a unit membrane of lipoproteins. Food vacuoles move (circulate) in the endoplasm by rotatory streaming movement, called **cyclosis**, of endoplasm.

6.4 PHYSIOLOGY

1. Locomotion. *Paramecium* is a free living aquatic animal. It does locomotion by the following methods:

Fig. 6.9. A longitudinal row of cilia showing metachronous movement.

(*i*) Ciliary movement. Cilia are main locomotory organelles in *Paramecium*. It swims in water by rhythmic beating of cilia. Each cilium beats a little in advance of the one immediately behind so that waves of ciliary activity pass over the animal (called **metachronus rhythm**). A cilium can be compared with an oar as it moves the cell in a direction perpendicular to the cilium itself. In its power stroke or effective stroke, the cilium is maintained in a rigid state (Fig. 6.10) as it pushes against the surrounding medium. In its **recovery stroke**, the cilium becomes flexible, offering relatively little resistance to the medium. The movement of cilia is controlled by the neuromotor system. Quite recently certain chemical messengers such as **Ca^{2+} ions** and **cyclic AMP (cAMP)** are reported to influence the rate and form of ciliary beat. *Paramecium* swims at the rate of about 1mm per second.

Since, cilia beat obliquely to the right, so the body of *Paramecium* while swimming forward rotates spirally slightly towards the left (Fig. 6.11). Further, effective strokes of the cilia of oral groove are more powerful than those of body cilia. This makes the anterior end of body to turn continually away from the oral side.

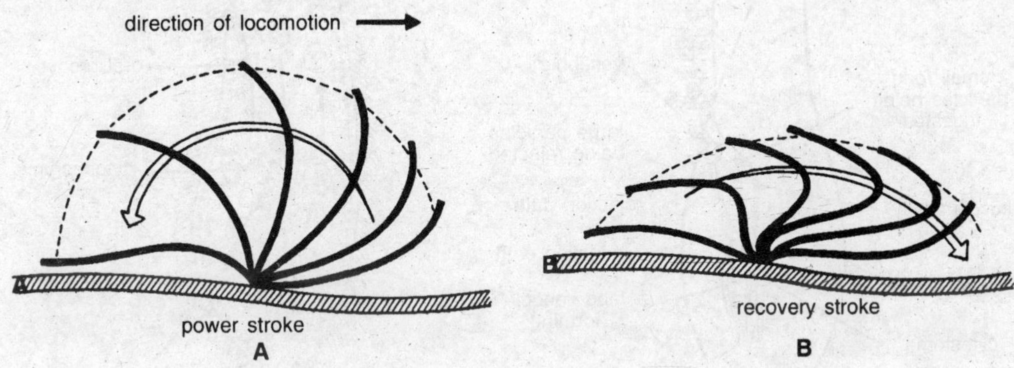

Fig. 6.10. *Paramecium*. Two stages of ciliary movement. A—Effective or power stroke; B—Recoverv stroke.

Paramecium can also swim backwards, in which case its cilia beat forwards.

***(ii)* Body contraction or metaboly.** *Paramecium* can squeeze itself through a passage narrower than its body. After this its body assumes normal shape. Such a temporary change in body shape is called **metaboly**; it is brought about by **M fibres** or **myonemes**.

2. Nutrition. *Paramecium* is **holozoic** in nutrition. The process of nutrition includes the following steps:

***(i)* Food.** *Paramecium* is an **omnivorous** and **microbivorous** animal. It feeds on a variety or microscopic biological materials such as bacteria, algae, diatoms, yeasts, small protozoans and tiny fragments of large-sized animals and plants.

***(ii)* Ingestion.** Food is ingested through **cytostome**. A current of water in the form of a vortex (whirl pool) is produced by the constant lashing movement of cilia of oral groove. Due to this the food particles are swept towards the cytostome and are carried down into the cytopharynx. These food particles are aggregated into a rounded mass (ball) by the movement of penniculi and quadrulus. Food mass along with a drop of water is enclosed in a **food vacuole** (by endocytosis) and ingested into the endoplasm.

***(iii)* Digestion and assimilation.** After their separation from the cytopharynx, the food vacuoles circulate through the endoplasm along a definite circular path (often resembling with the figure of 8). Their circulation is brought about by the rotatory streaming movements, called **cyclosis**, of endoplasm. The food vacuoles starting from cytopharynx are carried firstly behind, then forward to the dorsal surface and finally backward down tc the oral surface. During its journey, the food is digested inside the food vacuole by a variety of lysosomal enzymes such as **amylase, proteinase, dipeptidase, lipase,** etc. The food vacuole is first acidic but becomes basic later on. Digested food is absorbed and assimilated by the endoplasm during cyclosis.

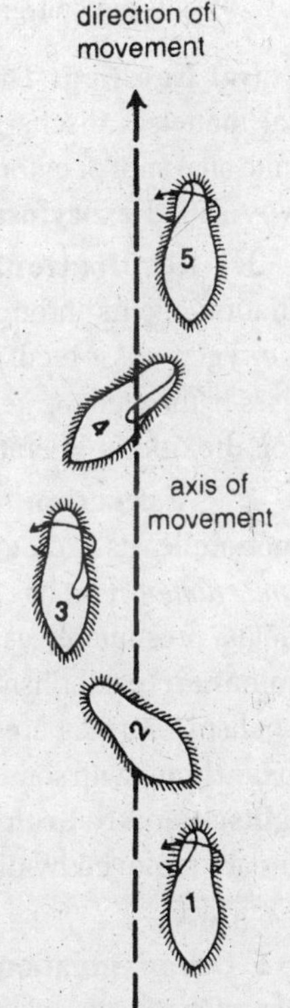

Fig. 6.11. *Paramecium*. Spiral path of movement.

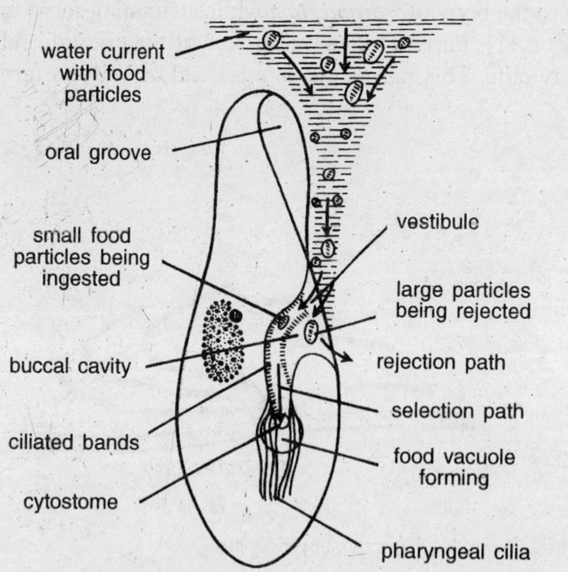

Fig. 6.12. *Paramecium.* Ciliary action creating vortex (whirl pool) and drawing food particles into the oral groove with the water current.

Fig. 6.13. *Paramecium.* Movement of food vacuoles due to cyclosis in the endoplasm.

(*iv*) Egestion. The undigested food or faecal matter is discharged outside through a definite anal spot or cytopyge by active vacuolar activity called **exocytosis**.

3. Respiration. In *Paramecium*, respiration occurs through the surface of body. The oxygen dissolved in surrounding water diffuses in through the general body surface and carbon dioxide is liberated in a similar way.

4. Excretion. *Paramecium* is **ammonotelic**; its excretory waste is **ammonia**. *Paramecium aurelia* has been shown to liberate ammonia, presumably as a result of amino acid **deamination** (catabolism), but not from purine metabolism. Purines are probably excreted as **hypoxanthine**, with some **guanine** and **adenine** (**Meglitsch** and **Schram**, 1991). Nitrogenous wastes are removed by diffusion through entire body surface.

5. Osmoregulation. The concentration of solutes of body cytoplasm of *Paramecium* is higher than that of the surrounding water (*i.e.*, cytoplasm is a **hypertonic solution** and water is a **hypotonic solution**). This results in **endosmosis** (*i.e.*, inward movement of water or

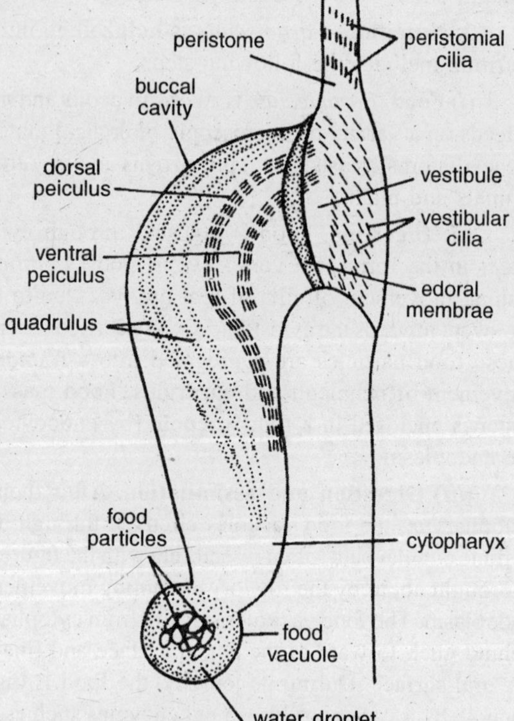

Fig. 6.14. *Paramecium.* Oral groove with food vacuole under formation.

solvent molecules through the plasma membrane (or pellicle) of *Paramecium*. Small amount of water also enters along with food (*i.e.*, via food vacuoles). This excess of water is collected and expelled out of the body by two **contractile vacuoles**. Thus, in *Paramecium* the maintenance of balance of water contents of the body or osmoregulation is performed by two contractile vacuoles. They work by contraction (systole) and expansion (diastole) at regular intervals. Posterior contractile vacuole being near the cytopharynx is able to remove far more amount of water from the body than the anterior contractile vacuole.

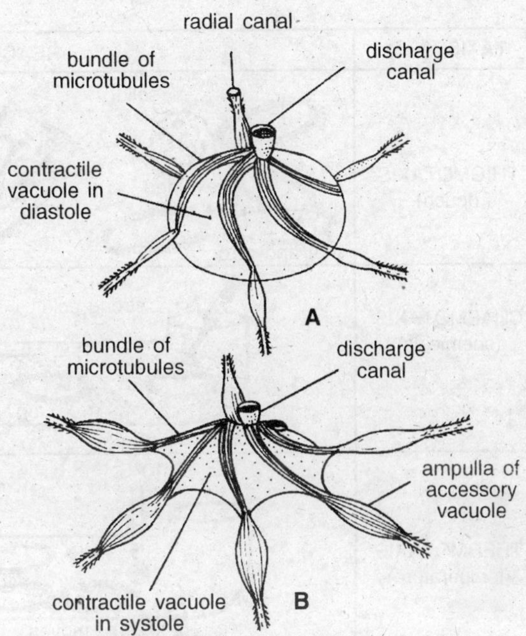

Fig. 6.15. *Paramecium.* Working of contractile vacuole. A—In diastole; B—In systole.

Excess of water from the cytoplasm is collected in the tubules of endoplasmic reticulum. From them water is passed into nephridial tubules or spongiome tubules. Collecting tubules (= nephridial canals) of radiating canals collect water from these tubules and convey it to the ampullae. When ampullae are filled with water, they ultimately discharge into contractile vacuole. The vacuole gradually expands by collecting more and more water. When grown to its maximum size, it contracts (by the help of bundles of microtubules and actin containing microfilaments in its membranes) and expels water to outside through a permanent pore in the pellicle.

6.5 BEHAVIOUR

Paramecium responds to various stimuli. It moves towards a beneficial stimulus (**positive reaction**) and move away from a harmful stimulus (**negative reaction**). Thus, *Paramecium* is sensitive to touch, different concentrations of chemicals, oxygen and carbon dioxide levels, and changes in light intensity. It avoids too hot or too cold waters and also strong light. *Paramecium* shows positive response towards a weak electric charge and weak acidic environment (This is advantageous to *Paramecium* as a slightly acidic medium contains more bacteria).

Types of Taxes in *Paramecium*

Paramecium show the following responses to different stimuli (Fig. 6.16):

1. Thigmotaxis (= Reaction to contact). Response to contact is varied in *Paramecium*. If its anterior end is lightly touched with a fine point, a strong avoiding reaction occurs.

2. Chemotaxis (= Reaction to chemicals). Generally *Paramecium* responds to a chemical stimuli by means of avoiding reaction. If a drop of weak salt solution (0.5 per cent) is introduced in a *Paramecium* population on a microslide, the animals respond with the avoiding reaction and none enters the drop. But for acid, the response of paramecia is positive.

3. Thermotaxis (= Reaction to temperature). *Paramecium* seeks an optimum temperature of 24 to 28°C. Greater heat stimulates rapid movement and avoiding reaction.

4. Phototaxis (= Reaction of light). Except green *Paramecium bursaria* which is positively phototactic, paramecia exhibit a negative phototactic reaction to bright light and to ultraviolet rays.

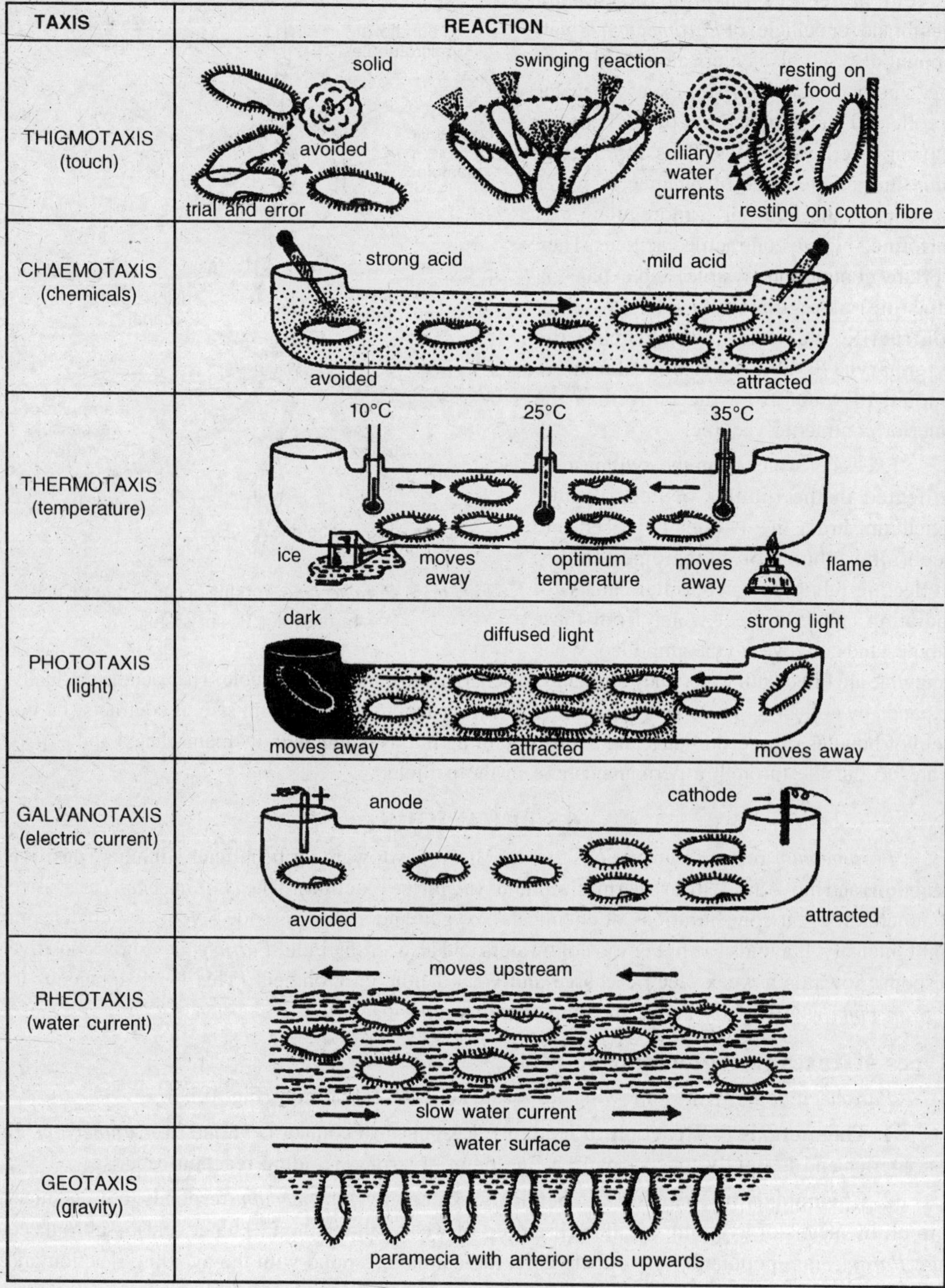

Fig. 6.16. Various taxes of *Paramecium*.

5. Galvanotaxis (= Reaction to electric current). Paramecia show positive response to negative electric pole, *i.e.*, they swim toward and concentrate at cathode (negative electrode).

6. Rheotaxis (= Reaction to water current). Paramecia show positive rheotaxis.

7. Geotaxis (= Reaction to gravity). Paramecia generally show a negative response to gravity. In a culture of *Paramecium*, many individuals gather close under the surface film with their anterior ends upper most.

Physiology of Stimuli Detection

Previously it was held that detection of external stimuli probably takes place through stiff, non-locomotory cilia at the caudal end of *Paramecium* (see **Green** *et al.*, 1990). Now it is believed that like animal receptors, *Paramecium* and other protozoans receive external stimuli as signal substances that bind to specific membrane molecules. Binding can cause a specific **ion channel** to open, allowing ions (often Na^+ and K^+) to flow down their concentration gradients (Na^+ in, K^+ out). Because the resting cell membrane is polarized with respect to the distribution of these ions, opening of the ion channels depolarizes the membrane. When the membrane is depolarized, **Ca^{++} channels** open and calcium ions enter the cell. Entry of calcium into the cell triggers other changes, such as, reversal of ciliary beat, which causes the *Paramecium* to withdraw from the disturbance.

Paramecium has at least *nine* different ion channels, some of which are localized at the front and others are at the rear of the cell. Such localized receptor fields differentiate "head" from "tail" and are, thus, analogous to the concentration of receptor organs of many animals (**Ruppert** and **Barnes**, 1994).

Trial and Error Reaction

When a *Paramecium* comes in contact with an obstacle or harmful substance, it immediately reverses its ciliary beat (*i.e.*, power stroke is exerted in the opposite direction) and the ciliate swims backward and rotates, causing it to change orientation. *Paramecium* then switches beat direction again and swims forward. This process continues until the orientation of swimming causes the ciliate to miss the object and swim pass it (For physiology of this behaviour see Box 6.2).

Box 6.2. Physiology of trial and error behaviour of *Paramecium*

1. Contact of the *Paramecium* with an object causes a depolarization of the plasma membrane and subsequent opening of voltage-gated Ca^{2+} channels.
2. Inflow of calcium ions acts (in connection with calcium binding protein, **calmodulin**) on the axoneme to cause the reversal of the direction of the beat and the backward "retreat" of ciliate.
3. Closure of calcium gates causes a drop in Ca^{2+}-levels and a return to forward swimming.
4. The rate of forward movement is regulated by cyclic AMP (cAMP) levels. If the *Paramecium* should encounter a stimulatory agent, such as chemical attractant, **K^+ – channels** open and the plasma membrane becomes **hyperpolarized** which activates the enzyme, **adenyl cyclase**, that synthesizes **cAMP**.
5. The elevated cAMP level triggers the phosphorylation of certain axonemal protein (*e.g.*, dynein) causing an increased ciliary beat frequency and an acceleration of forward directed – swimming.

6.6 REPRODUCTION

Paramecium reproduces both asexually and sexually. It multiplies asexually by **binary fission** and sexually by **conjugation**. Periodically it also undergoes several kinds of nuclear reorganization such as **endomixis, hemixis, autogamy, cytogamy** and **conjugation**.

I. Binary Fission

During favourable conditions *Paramecium* multiplies by **transverse binary fission**. It is most common method of reproduction in *Paramecium*. In this process, a full grown individual divides into two daughter individuals without leaving a parental corpse behind. Before the beginning of fission, *Paramecium* stops feeding, developing two minute pores near the middle of body for the contractile vacuoles that are to appear later for the daughter individuals. Both nuclei increase in size, elongate and pull themselves apart. The macronucleus divide **amitotically**, randomly distributing its chromosomes between the two newly formed macronuclei. The micronucleus undergoes **mitosis**. A spindle forms within its nuclear envelope and chromosomes are shared between the two micronuclei. A transverse cytoplasmic constriction furrow forms (by the help of microtubules and microfilaments) which deepens gradually dividing it into two equal sized daughter paramecia. Of the two paramecia produced, the anterior one is called **proter** and posterior one is called **opisthe**. Each half of the animal contains only those cell organelles which were originally present in it. For example, original feeding apparatus (buccal cavity, cytopharynx, etc.) remains with the anterior half while new feeding apparatus develops in the posterior half. Two new contractile vacuoles also appear, one near anterior end of opsithe and another near posterior end of proter. By the time the separation is completed, both daughters have a full complement of cell organelles. Ultimately both paramecia separate from each other and start their free-living existence. The entire process of binary fission is completed within ½ to 2 hours. All daughter paramecia produced asexually from one parent paramecium form a **clone**. All the members of a clone are hereditarily alike.

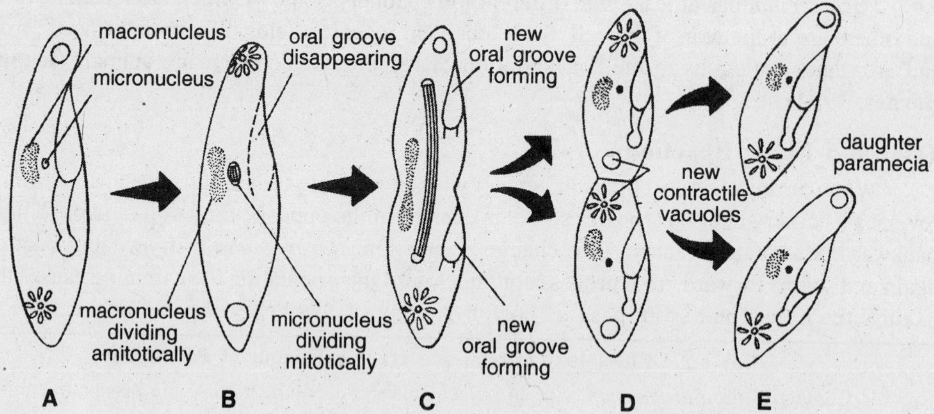

Fig. 6.17. *Paramecium*. Asexual reproduction by transverse binary fission.

II. Conjugation

Conjugation is a form of sexual reproduction. It is a temporary union of two individuals of a species for the purpose of exchange of their micronuclear material. Basically, conjugation involves **meiosis** and **nuclear exchange** and results in production of offspring with a wide variety of genotypes. Conjugation corresponds to cross-fertilization of higher animals and is essential for rejuvenation and continuity of race. It was described in *Paramecium caudatum* in 1889 by **Hertwig** and **Maupas**.

1. Mating types. According to **Sonneborn** (1947), *Paramecium caudatum* has 13 varieties (called **syngens**) which differ only in minor physiological features such as optimum temperature for growth, ability to conjugate and mating time during a 24 hour period. Each variety contains two **mating types** which differ in chemical characteristics (probably proteins) of their surface membranes. The mating types are indicated by Roman numerals, *e.g.*, variety 1 has two mating types I and II, the variety 2 has mating types III and IV and so on. Conjugation can occur between the individuals of two mating types of same variety.

2. Factors Inducing Conjugation

(*i*) Conjugation occurs usually under unfavourable living conditions such as starvation or shortage of food (*e.g.*, bacteria).

(*ii*) It occurs in individuals which have passed through desirable number (normally 300) of asexual generations. It is believed that during asexual reproduction paramecia exist in a **period of immaturity** and due to repeated binary fissions, they become **sexually mature** to conjugate.

(*iii*) The individuals entering into conjugation are generally of the same size. Both are also small sized than individuals reproducing asexually.

(*iv*) Conjugation occurs when conjugants are sticky, free of food vacuoles and slow in movement.

(*v*) Conjugation does not take place during night or darkness; it starts in early morning (dawn) and continues till afternoon.

(*vi*) A proteinous substance in the cilia of mating type individuals is said to induce conjugation.

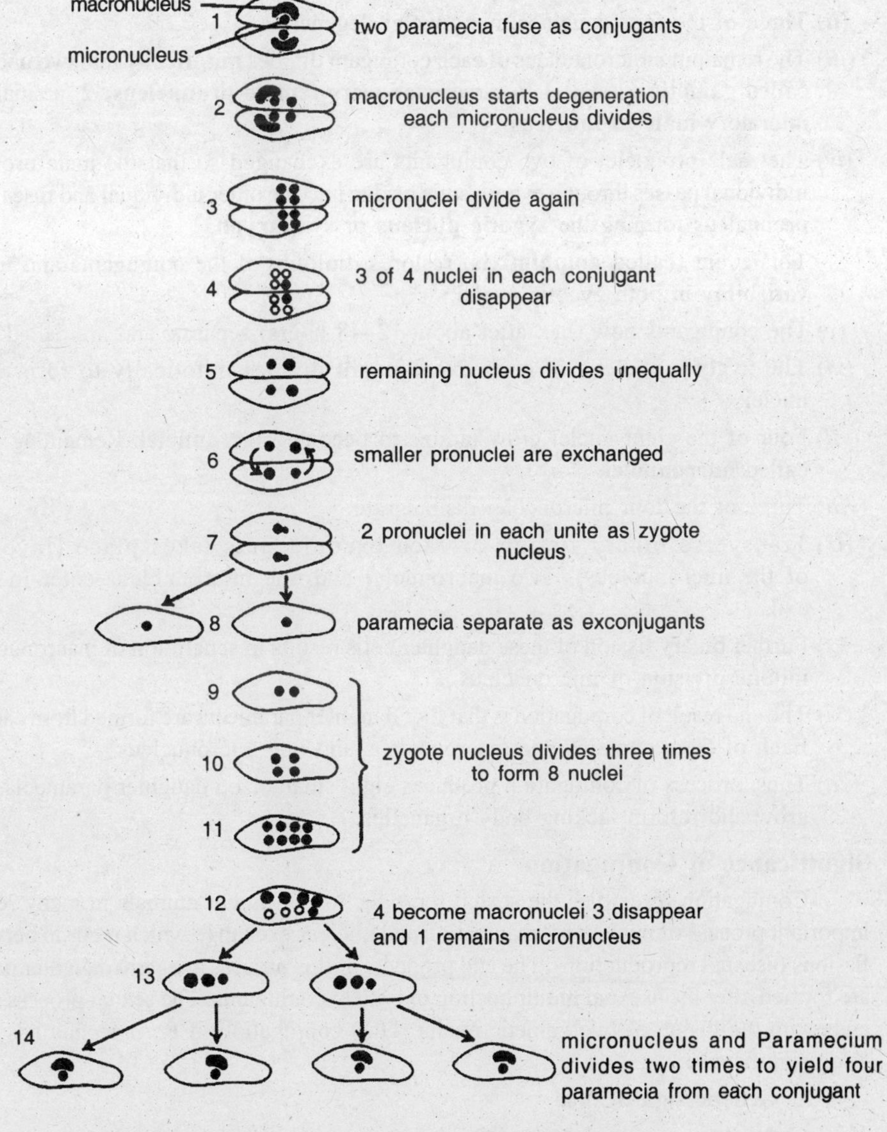

Fig. 6.18. *Paramecium caudatum*. Stages of conjugation.

3. Process of conjugation. Two different compatible mating types (**conjugants**) adhere to each other at their oral grooves. Their cilia produce a substance on the surface of body which causes adhesion of the two conjugating paramecia. At the region of contact a cytoplasmic continuity, called **cytoplasmic bridge** or **conjugation bridge** is established by the local destruction of cilia, pellicle, trichocyst, expect the basal bodies. At nuclear level, in each conjugant, following changes occur:

(A) Macronuclear changes. The macronucleus breaks up into fragments and is absorbed in the cytoplasm during the process of conjugation.

(B) Micronuclear changes.

(*i*) The diploid micronucleus divides by **meiotic cell division** and produces four haploid micronuclei.

(*ii*) Three of the four daughter micronuclei degenerate.

(*iii*) The remaining micronucleus of each conjugant divides **mitotically** into two unequal pronuclei, called **gametic nuclei:** 1. a larger stationary **female pronucleus;** 2. a smaller, active and migratory **male pronucleus**.

(*iv*) The male pronuclei of two conjugants are exchanged so that the male pronucleus of one individual passes through cytoplasmic bridge into the other individual and fuses with the female pronucleus forming the **zygotic nucleus** or **synkaryon**.

This event (called **amphimixis**) restores diploidy of the conjugants and initiates genetic variability in both zygotic nuclei.

(*v*) The conjugants now (*i.e.*, after about 12–48 hours) separate and are called **exconjugants**.

(*vi*) The zygotic nucleus of each exconjugant divides mitotically to form eight daughter nuclei.

(*vii*) Four of the eight nuclei grow in size to become **macronuclei**. Remaining four nuclei are called **micronuclei.**

(*viii*) Three of the four micronuclei disintegrate.

(*ix*) Transverse binary fission of each exconjugants takes place (involving mitosis of the micronucleus). Two macronuclei and one micronucleus enter in each daughter cell.

(*x*) Further binary fission of these daughter cells results in separation of macronuclei and another mitotic division of micronucleus.

(*xi*) The end result of conjugation is that four daughter paramecia are formed from each exconjugant. Each of them contains one macronucleus and one micronucleus.

(*xii*) Thus, process of conjugation produces eight small sized daughter paramecia, each of which grow and reform lacking body organelles.

Significance of Conjugation

Conjugation resembles the sexual reproduction of higher animals in many respects. It is an important process of nuclear reorganization and nuclear exchange which recur in between the binary fissions (asexual reproduction). The end products of this process are eight daughter paramecia which are formed after the asexual multiplication of both exconjugants. The sexual process in *Paramecium* ends with the fusion of two gametic nuclei. Thus conjugation in *Paramecium* has following two-folds significance:

1. Genetic significance

(*i*) Nuclear reorganization. The macronucleus is formed of trophocromatin which controls all the physiological activities of the individual. Continued amitotic division of the macronucleus during binary fission leads to unequal distribution of its chromosomes. This upsets the normal coordinated activity of *Paramecium* and leads to **depression**. The worn out and degenerated macronucleus is replaced by a new one during conjugation. This results in restoration of vigour and vitality of the old individuals (exconjugants).

(*ii*) **Hereditary variations.** Conjugation involves meiosis and fusion of haploid pronuclei of different individuals and this results in the formation of new combinations of genes in the descendants. This ensures their better adaptability to the new conditions of life.

2. Physiological significance. It involves **rejuvenation**, *i.e.*, restoration of vitality, of *Paramecium*. It has been found that individuals cannot continue to multiply indefinitely by asexual methods. After a definite number of asexual generations (binary fissions), individuals start losing their vigour and physiological efficiency. They appear unhealthy and ultimately die. The conjugation is the only process which infuses former vigour, restores its healthy organization and thus, saves the race from senile decay.

III. Endomixis

Endomixis is a method of nuclear reorganization followed by the division of body within single individual, without any nuclear exchange or fusion. It occurs only in the absence of conjugation and has been described by **Woodruff** and **Erdmann** (1914) in *Paramecium aurelia*, a binucleate species. Endomixis involves the following steps:

1. The macronucleus disintegrates and is absorbed in the surrounding cytoplasm.
2. The two micronuclei divide twice forming eight daughter micronuclei.
3. Six of these micronuclei disintegrate and get absorbed, leaving two micronuclei.
4. The animal with two micronuclei divides by transverse binary fission and two daughter individuals, each with one micronucleus, are formed.
5. In each daughter *Paramecium*, the micronucleus divides twice, forming four nuclei.
6. Of these four nuclei, two enlarge to become macronuclei and two remain small micronuclei.
7. Both micronuclei in each daughter individual again divide. This nuclear division is accompanied by the division of body.

Ultimately four daughter paramecia are formed from a single parent and each possesses one macronucleus and two micronuclei.

Many protozoologists (*viz.* **Diller**, 1936; **Sonneborn**, 1947; **Wichterman**, 1953; etc.) do not believe in the validity of endomixis; they consider it as a special case of autogamy or a mixture of hemixis and autogamy.

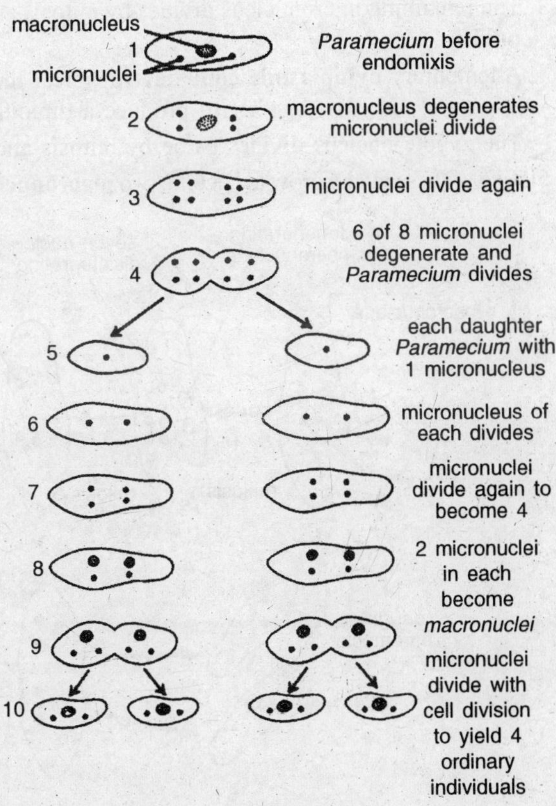

Fig. 6.19. *Paramecium aurelia*. Various stages of endomixis.

Significance of Endomixis

Like conjugation, endomixis is a process of multiplication and rejuvenation and occurs as a substitute of conjugation.

IV. Hemixis

Hemixis was described by **Diller** (1936) in *Paramecium aurelia.* It is a simple phenomenon of nuclear reorganization. In this process the micronuclei do not participate. The macronucleus gives off small fragments of chromatin, called chromatin balls which are absorbed in the surrounding and micronuclei behave normally.

Significance of Hemixis

The Hemixis is regarded as a type of nuclear reorganization in which old macronuclei are purified. The discarded chromatin balls probably contain accumulated waste substances.

V. Autogamy

Autogamy occurs within a single individual and involves fusion of two micronuclei of same individual into a synkaryon. It was described in *Paramecium aurelia* by **W.E Diller** (1936) and he called autogamy as **self-fertilization**. Similar nuclear changes occur in autogamy and conjugation, but there is no exchange of nuclei in autogamy. Autogamy involves the following steps (Fig. 6.20):

1. Macronucleus grows into a hapazard mass and break up into fragments which are absorbed by endoplasm.
2. The two micronuclei (which are diploid) divide meiotically and produce eight haploid daughter nuclei. Seven of these daughter micronuclei degenerate.
3. The remaining micronucleus divides by mitosis and produce two nuclei, now called **gamete nuclei**.
4. A temporary **cytoplasmic cone** develops near the cytostome. The two gamete nuclei enter this cone and fuse together to produce a diploid **zygote nucleus** or **synkaryon**.
5. The zygote nucleus divides twice by mitosis and produce four nuclei.
6. Two of these nuclei grow to become two **macronuclei**, and remaining two become **micronuclei**.

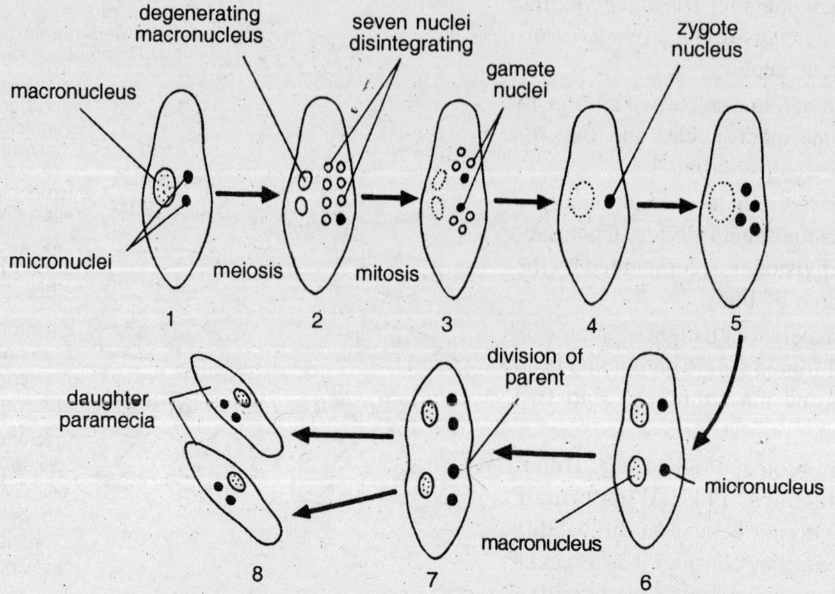

Fig. 6.20. *Paramecium aurelia.* Various steps of autogamy.

7. Each of the micronuclei and the body of *Paramecium* divide simultaneously to produce two daughter paramecia, each with a new macronucleus and two micronuclei.

Significance of Autogamy

1. Autogamy results in the production of a new macronucleus which rejuvenates *Paramecium* by increasing its vitality.
2. It is a type of sexual reproduction (by self fertilization), involving production and fusion of two haploid gamete nuclei. This may result in homozygosity of certain genes.

VI. Cytogamy

Cytogamy type reproduction was reported by **R. Witcherman** (1940) in *Paramecium caudatum*. In cytogamy, two paramecia (small, 200 mm long) form a pair and become temporarily attached at their **oral surfaces**, as in conjugation. Both of these individuals are called **cytogamonts**. Their nuclei undergo similar changes as occurring during conjugation, but there is no nuclear exchange between the cytogamonts. The two haploid gamete nuclei in each cytogamont are fused to form a synkaryon as in **autogamy**. The individuals now separate and divide as in conjugation. A new macronucleus is formed in each individual.

Significance of Cytogamy

1. During cytogamy, since a new macronucleus is formed, it rejuvenates *Paramecium*.
2. Self-fertilization also occurs during cytogamy due to which homozygosity of certain genes may occur.

Paramecium is believed never to **encyst**.

Distinction Between Conjugation and Endomixis

Conjugation and endomixis are two methods or reproduction and nuclear reorganization in *Paramecium*. Both have the following differences (see Table 6.1).

Table 6.1. Differences between conjugation and endomixis.

Conjugation	*Endomixis*
1. Conjugation is a process of sexual reproduction.	1. Endomixis is a process of nuclear reorganization and occurs when conjugation fails to occur.
2. During conjugation two paramecia of different mating types pair together.	2. Endomixis occurs in single *Paramecium*.
3. There is exchange of nuclear material between conjugants.	3. There is no exchange of material.
4. Eight daughter paramecia are formed at the end of conjugation by two parent paramecia.	4. Four daughter paramecia are formed at the end of endomixis by single *Paramecium*.

QUESTIONS

Long Answer Questions

1. Describe the structure of *Paramecium*. (*C.C.S.U., Meerut 2008*)

2. Describe the structure of *Paramecium* and explain its movements. (*Purvanchal 1998*)

3. Describe structure and function of following 1. Hypostome; 2. Pellicle. (*Avadh 1991*)

4. Describe the structure of *Paramecium* with suitable diagram. Explain process of nutrition and osmoregulation in this animal. (*Agra 1997*)

5. Describe in detail about modes of locomotion in *Paramecium*. (*M.S.U. Tirunelveli 2006*)

6. Describe the food passage in *Paramecium* showing arrangement of cilia. (*C.C.S.U. Meerut 2003*)

7. Describe the process of conjugation and its significance in *Paramecium*.
(*Lucknow 1997; Agra 1999; Gorakhpur 1999; C.C.S.U. Meerut 2006, 2009*)

8. Give an account of asexual and sexual reproduction in *Paramecium*. *(Gorakhpur 1997)*

9. Describe the behaviour of *Paramecium* under the following heads: 1. Thigmotaxis; 2. Thermotaxis; 3. Geotaxis; 4. Galvanotaxis; 5. Chemotaxis. *(Purvanchal 1994)*

10. What is cyclosis? Describe feeding, digestion and excretion of *Paramecium*. *(Avadh 1999)*

11. Describe sexual reproduction in *Paramecium*. *(Kanpur 1990, 1994)*

12. Describe conjugation, autogamy and endomixis in *Paramecium*. *(Rohilkhand 1992)*

Short Answer Questions

1. Describe the process of nutrition of *Paramecium*.
(Purvanchal, 1999; C.C.S.U. Meerut 2001; M.S.U. Tirunelveli 2006)

2. Describe locomotion in *Paramecium*.

3. Describe the digestion in *Paramecium*. *(C.C.S.U. Meerut 2003)*

4. Give an account of binary fission in *Paramecium*. *(Gorakhpur 1993)*

5. Give diagrams of process of conjugation in *Paramecium*. *(C.C.S.U. Meerut 2005)*

6. Define conjugation. Describe the process and give its significance. *(Rohilkhand 2003)*

7. How is ciliary locomotion performed in *Paramecium*? *(Purvanchal 2001)*

8. Distinguish between conjugation and autogamy in *Paramecium*. *(Avadh 2001)*

9. Write Brief/Short notes on the following:

(*i*) Structure of *Paramecium*. *(Kanpur 1994; Allahabad 1999)*
(*ii*) Infra-ciliary system. *(Avadh 1996, 1999; Purvanchal 1997)*
(*iii*) Trichocyst *(Purvanchal 1993, 1997; Avadh 1996, 1999; Lucknow 1997)*
(*iv*) Contractile vacuole. *(Avadh 1997; Allahabad 1999)*
(*v*) Osmoregulation *(Avadh 1993; Rohilkhand 1995; Lucknow 1996)*
(*vi*) Cyclosis *(Kanpur 2001)*
(*vii*) Ciliary movement *(Purvanchal 1996)*
(*viii*) Filter feeding in *Paramecium* *(Purvanchal 1992)*
(*ix*) Autogamy in *Paramecium*. *(Purvanchal 1996)*
(*x*) Binary fission *(Avadh 1994; Rohilkhand 1999)*
(*xi*) Conjugation in *Paramecium* *(Kanpur 1998, Rohilkhand 1999)*
(*xii*) Endomixis *(Allahabad 1993)*
(*xiii*) Kappa particle *(Avadh 1999)*

Very Short Answer Questions

1. Write down zoological name of following animal – slipper animalcule. *(Agra 1997)*

Ans. *Paramecium caudatum*.

2. What is axoneme?

Ans. Axial filament of a cilium or flagellum. It has 9 + 2 arrangement of microtubules.

3. What are trichocysts? *(C.C.S.U. Meerut 2004, 2006)*

Ans. Trichocysts are oval, rod or spindle-shaped minute bodies situated to right angle to the body surface of *Paramecium* in ectoplasm. They serve the purpose of defence and keeping the animal attached to substratum while it feeds.

4. Name the material found in a trichocyst.

Ans. Swelling substance.

5. What is cyclosis?

Ans. Rotary streaming movement of food vacuoles in the endoplasm (cytoplasm) is called **cyclosis.** The food vacuoles are regularly circulated by streaming endoplasm.

6. Give the location of cytostome in *Paramecium*.

Ans. Between the buccal cavity and cytopharynx.

7. Where is the endoral membrane located in *Paramecium*?

Ans. Along the right border of the junction between the vestibule and the buccal cavity.

8. What are Kappa particles? Describe. (*C.C.S.U., Meerut 2007*)

Ans. Kappa particles are cytoplasmic factors in certain paramecia (called **killers**) that mediate production of **paramecin** (a toxin) and thereby makes the medium in which such microorganisms are grown toxic to members of strains not possessing the factors (sensitives).

9. Which nucleus controls sexual reproduction in *Paramecium*?

Ans. Micronucleus.

10. Is binary fission in *Paramecium* homothetogenic?

Ans. Yes.

11. What do the term proter and opisthe indicate?

Ans. Anterior and posterior daughter paramecia formed by transverse binary fission.

12. Who discovered conjugation in *Paramecium*?

Ans. Hertwig and **Maupas**.

13. Synkaryon or Zygote. (*C.C.S.U., Meerut 2008, 2009*)

Ans. Synkaryon is a cell nucleus formed by the fusion of two pre-existing nuclei. Zygote is a cell formed by the union of two gametes.

Fill in the Blanks

1. *Paramecium* is commonly called .... animalcule/organism.
2. Pellicle of *Paramecium* is .... layered.
3. A layer of alveoli lies under the .... in *Paramecium*.
4. Pellicle of *Paramecium* has pores for .... and .... to emerge.
5. Infraciliature of *Paramecium* consists of .... and kinetodesma.
6. In *Paramecium*, the kinety comprises .... , kinetosome and ....
7. In *Paramecium*, the cytopharynx is separated from the cytoplasm by a ....
8. In *Paramecium*, the cytopharynx is lined with microtubules called ....
9. Cilia in *Paramecium* beat in a .... rhythm.
10. Macronucleus of *Paramecium* contain .... nucleoli.
11. Autogamy occurs only in *Paramecium* ....
12. In *Paramecium* the .... function is controlled by macronucleus.
13. In *Paramecium*, reproduction is controlled by ....
14. The process of self-fertilization is called .... in *Paramecium*. (*Purvanchal 1997*)
15. In *Paramecium* rotatory endoplasm carries food vacuole along a definite path, this process is called .... (*C.C.S.U., Meerut 2000*)

Multiple Choice Questions

1. Feeding in *Paramecium* takes place through

(*a*) undulating membrane (*b*) general body surface
(*c*) oral groove (*d*) none of the above

2. Which of the following one is filter feeder?

(*a*) *Amoeba* (*b*) *Trypanosoma*
(*c*) *Monocystis* (*d*) *Paramecium*

3. Method of food intake in *Paramecium* is

(*a*) autozoic (*b*) saprozoic
(*c*) holozoic (*d*) holophytic

4. Movement of food vacuole in *Paramecium* along a definite path is known as (*M.S.U. Tirunelveli 2006*)

(*a*) cytokinesis (*b*) cyclosis
(*c*) endomixis (*d*) metagenesis

5. Cytopyge is found in
   (*a*) *Paramecium* (*b*) *Amoeba*
   (*c*) *Trypanosoma* (*d*) *Entamoeba*
6. *Paramecium* moves about by
   (*a*) floating (*b*) looping
   (*c*) crawling (*d*) beating of cilia
7. Which of the following helps in anchorage and defence of *Paramecium*?
   (*a*) nematocyst (*b*) mucocyst
   (*c*) trichocyst (*d*) statocyst
8. The number of nucleus in *Paramecium* is
   (*a*) one (*b*) two
   (*c*) three (*d*) four
9. In *Paramecium*, the essential genetic information is carried by
   (*a*) micronucleus (*b*) macronucleus
   (*c*) mitochondria (*d*) both (*a*) and (*b*)
10. The main nitrogenous waste in *Paramecium* is
   (*a*) urea (*b*) uric acid
   (*c*) ammonia (*d*) amino acid
11. The main function of the contractile vacuole is (*M.S.U. Tirunelveli 2006*)
   (*a*) pumping out excess water (*b*) excretion
   (*c*) osmoregulation (*d*) respiration
12. Mention the function of the neuromotor system
   (*a*) coordination of ciliary beat (*b*) coordination of various stimuli
   (*c*) control of digestion (*d*) control of respiration movement
13. Binary fission in *Paramecium* is
   (*a*) horizontal (*b*) longitudinal
   (*c*) transverse (*d*) irregular
14. Rejuvenation of weakened *Paramecium* occurs by
   (*a*) conjugation (*b*) autogamy
   (*c*) hemixis (*d*) all of these
15. How many paramecia are produced after the conjugation?
   (*a*) two (*b*) four
   (*c*) eight (*d*) sixteen
16. Cyclosis in *Paramecium* is associated with (*Kanpur 2001*)
   (*a*) respiration (*b*) excretion
   (*c*) digestion (*d*) locomotion

ANSWERS

Fill in the Blanks

1. Slipper 2. Three 3. Plasma membran.
4. Cilia, Trichocysts 5. Kinetosome 6. Cilia, Kinetodesma
7. Membrane 8. Nematodesmata 9. Metachronous
10. Several 11. *Aurelia* 12. Vegatative
13. Reproduction 14. Autogamy 15. Cyclosis

Multiple Choice Questions

1. (*c*) 2. (*d*) 3. (*c*) 4. (*b*) 5. (*a*) 6. (*d*)
7. (*c*) 8. (*b*) 9. (*a*) 10. (*c*) 11. (*c*) 12. (*a*)
13. (*c*) 14. (*d*) 15. (*b*) 16. (*c*).

7

Locomotion in Protozoa

7.1 LOCOMOTORY ORGANELLES IN PROTOZOA

Four major types of locomotor organelles occur among protozoa; and usually each type of them is characterized by a class:

1. **Pseudopodia** are characteristics of **Sarcodina**.
2. **Flagella** are characteristics of **Flagellata (Mastigophora)**
3. **Cilia** are characteristics of **Ciliata**.
4. **Myonemes** are characteristics of **Sporozoa**.

I. Pseudopodia

Pseudopodia are temporary extensions of cytoplasm from any part of the body. They occur in those protozoa which are "naked" or have a very thin pellicle. Pseudopodia are formed from ectoplasm but also have a core of endoplasm. Among Sarcodina, the following kinds of pseudopodia are found.

1. Lobopodia. These are short finger-shaped, round-tipped pseudopodia containing both ectoplasm and endoplasm. *e.g.*, *Amoeba*, *Entamoeba*, *Arcella* and *Difflugia*.

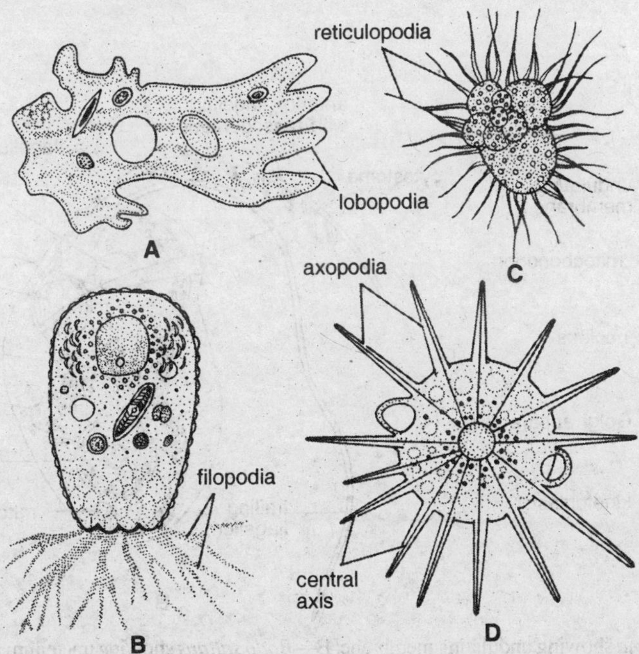

Fig. 7.1. Types of pseudopodia. A—Lobopodia of *Amoeba*; B—Filopodia of *Euglypha*; C—Reticulopodia of *Globigerina*; D—Axopodia of *Actinophrys*.

2. Filopodia. These are slender, filamentous hyaline projections, formed exclusively of ectoplasm. Filopodia have pointed tips and a tendency to branch and radiate in all directions. They never anastomose to form network, *e.g.*, *Euglypha*, *Chlamydophorus*.

3. Reticulopodia (= Rhizopodia). These are thin, long filamentous structures of endoplasm which branch extensively and fuse together to form meshworks that act as food traps, *e.g.*, *Elphidium*, *Globigerina* and *Allogromia*.

In case of *Microgromia*, the reticulopodia of several individuals unite to form a multicellular reticulum (Fig. 7.2).

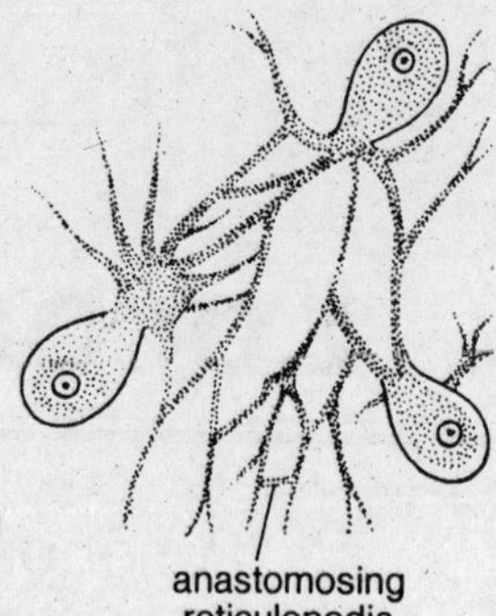

Fig. 7.2. Anastomosing reticulopodia of *Microgromia socialis*.

4. Axopodia. These are long, slender, stiff extensions of ectoplasm which radiate from circular body in all directions. Axopodia contain a slender axial filament, composed of parallel microtubules rolled into a spiral when seen in the electron microscope. The fibrils are like flagellar fibrils and cause axopodial movement, *e.g.*, *Actinophrys*, *Actinospherium*.

II. Flagella

Flagella are long, delicate, whip-like structures. They are found in flagellates (class Mastigophora) and the flagellate stages of sarcodine protozoans. Each flagellum arises from a **basal granule/body** or **kinetosome**.

Ultrastructure. The outer covering of a flagellum is continuous with the plasma membrane of the body surface. It contains a **matrix** in which a ring of nine doublet microtubules surrounds two central microtubules; a 9 + 2 arrangement also seen in cilia. Both central fibrils are connected by a bridge and are enclosed in a common central sheath. They confer bilaterality and the plane of flagellar beat may be associated with the orientation of central plane.

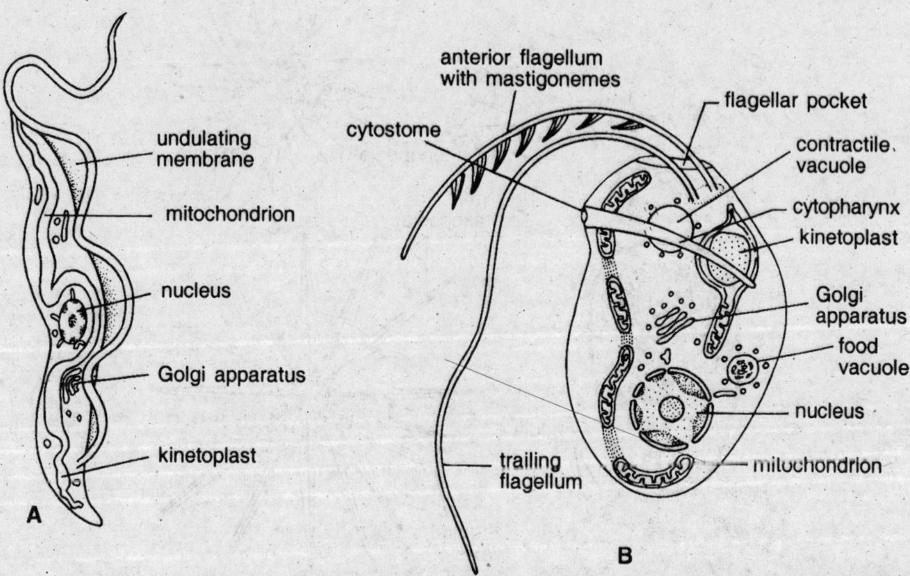

Fig. 7.3. A—*Trypanosoma* showing undulating membrane; B—*Bodo saltans* showing tractellum and trailing flagella.

At the tip of the main flagellum may be a very fine end piece or the main axis of the flagellum. Flagellum may bear fine, flexible lateral processes or **mastigonemes** on one side (Fig. 7.3B) or on both sides. Mastigoneme-bearing flagellum called **flimmer** or **ciliary flagellum**.

Number of flagella. The number and arrangement of flagella varies greatly in Mastigophora. They may be one to eight in number; several flagella occur in parasitic forms (*e.g.*, *Giardia intestinalis*). Free-living forms (*e.g.*, *Euglena*) are usually with one to two flagella. Flagella are often divided into two types: **tractella** and **pulsella**. The tractellum is situated at the end that is anterior when the organism is in motion and drags the body along; the pulsellum is generally situated posteriorly and pushes or propels the body forward. In *Bodo* (Fig. 7.3B) and *Aesonema* there are two flagella. One is directed forward, drags the body and is called **tractellum**. while the other is directed backwards along the side of the body and acts as an organ of propulsion, it is known as **trailing flagellum**.

Origin. Flagella commonly arise from the anterior end of body either directly or from a groove or depression. But in *Trypanosoma* (Fig.7.3A), flagellum originates from the posterior end. Trypanosomes have two kinetosomes, but only one has a flagellum. In many forms this flagellum lies against the side of the cell body's membrane. When the flagellum beats, the membrane of the cell is pulled up into a fold and looks like a waving or undulating membrane. This arrangement appears to be relatively efficient in moving the cell through viscous media (such as blood). Although trypanosomes can change the direction of flagellar beat in response to chemical tip of the flagellum and proceeds towards the kinetosome. This the reverse of the way flagella usually beat (from base to tip).

The kinetosomes are centriole like structures and often lie at the bottom of flagellar pockets or reservoirs. Kinetoplast (Fig. 7.3A) is the dark-staining concentration of mitochondrial DNA (mDNA). It generally is found in the part of mitochondrion lying close to the kinetosomes, although there is no known relationship between these two structures. The size, shape and position of the kinetoplast is important in taxonomy of trypanosomes and bodonids (see **Brusca** and **Brusca** 2003).

III. Cilia

The cilia are highly vibratile small ectoplasmic processes. They arise from the basal granules (or blepharoplasts or kinetosomes) in the ectoplasm. Cilia form a characteristic feature of ciliates but they are also found in the larval stages of Suctoria. In primitive forms (*e.g.*, *Opalina*, *Paramecium*, etc.) cilia cover the entire body, but in more specialized forms (*e.g.*, *Vorticella*) cilia are restricted to certain regions of body only. They are commonly arranged in longitudinal, diagonal or spiral rows. Cilia may be of equal length all over the body or may be longer at definite spots (*e.g.*, cilia of caudal tuft of *Paramecium* are longer than other body cilia).

Differences of cilia from flagella. Cilia are exactly like flagella in structure but differ from them in following points:

1. The cilia are relatively very shorter than flagella.
2. They are comparatively more numerous than flagella.
3. The basal granules of cilia are not connected with the nucleus, but, in case of flagella, the rhizoplasts join the basal granules to the nucleus. A rhizoplast is a delicate fibre running between the nucleus and the basal granule.
4. Cilia have no mastigonemes as in flagella.

Ultrastructure. The electron microscope reveals the presence of an outer elastic **membranous sheath** continuous with the plasma membrane of the cell surface and enclosing the fluid matrix. Running along the entire length of the body of cilium are **9 peripheral doublet fibrils** and **2 central singlet fibrils** collectively constituting the **axoneme**. An axoneme is centred on basal granule (= blepharoplast) which exhibits 9 peripheral subfibre **triplets**, each disposed in a twist-like fashion (Fig. 7.4).

Specialized organelles. The cilia may form the following composite motile organelles which are found in several protozoan groups. These are **membranelles** of Ciliata (*Paramecium*),

undulating membrane of Spirotricha and Peritricha and **cirri** of Hypotricha. **Cirri** are tuft-like brushes of functionally fused somatic cilia. Their kinetosomes form a basal fibre plate. **Undulating membranes** are formed of a row of oral cilia. **Membranelle** is flat plate of oral cilia with a basal fibre plate. They are arranged in a longitudinal series and beat in a coordinated fashion.

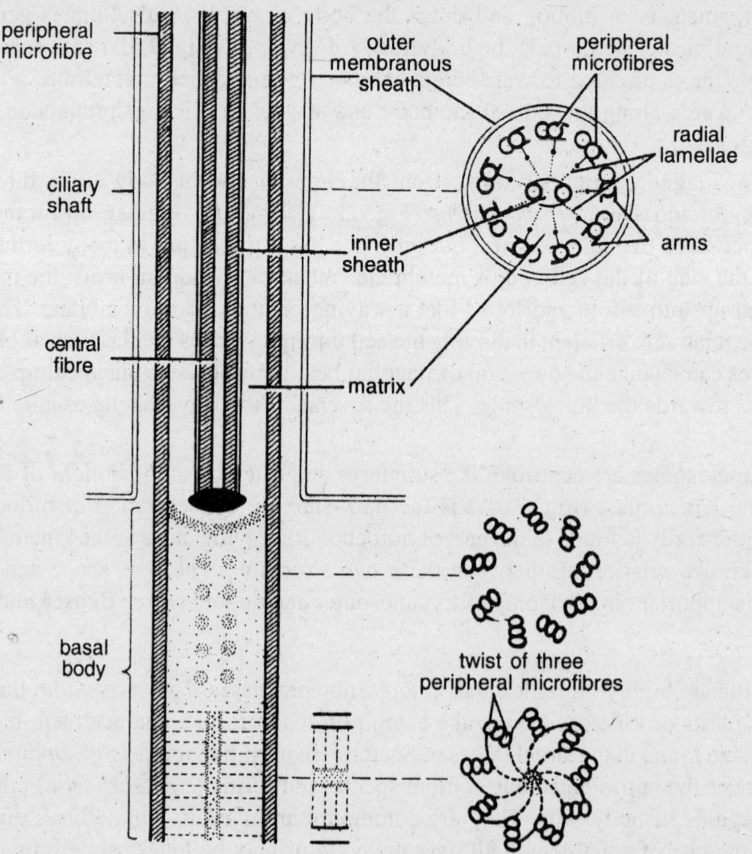

Fig. 7.4. Structure of cilium. A—L.S. of cilium; B—T.S. cilium; C—T.S. of basal granule or kinetosome.

IV. Myonemes

The myonemes are very fine cont-ractile fibrils in the pellicle of flagellates, ciliates, sporozoans and infusorians. In more complex ciliates, myonemes may extend into the endoplasm. They may be arranged longitudinally, transversely or spirally. Myonemes are primarily organelles for the **metaboly**, *e.g.*, *Euglena*, *Paramecium* and secondarily for **loco-motion** by muscle-like contraction in *Monocystis* and sporozoan such as *Plasmodium*.

7.2 MODES OF LOCOMOTION

Following four different modes of locomotion occur in protozoa:

1. Amoeboid movement

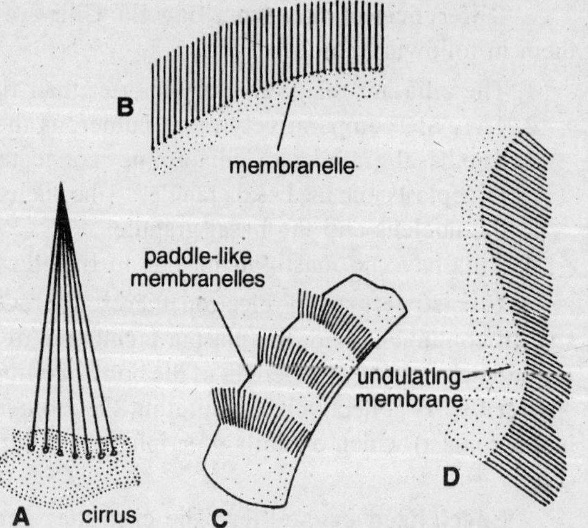

Fig. 7.5. Specialized organelles of cilia in protozoa.

2. Flagellar movement
3. Ciliary movement
4. Metabolic movement

1. Amoeboid movement (locomotion in *Amoeba*). Flowing amoeboid movement is performed by **pseudopodia** and is the characteristic of *Amoeba*, *Entamoeba* and other Sarcodina, certain mastigophores and sporozoans. The pseudopodia are finger-like temporary processes given out from any part of the body and withdrawn. They are formed by the streaming flow of the cytoplasm in the direction of movement (= locomotion). The pseudopodium is formed by the projection of ectoplasm (along with the plasma membrane) in which endoplasm flows. In fact, the cytoplasm which enters into pseudopodia, is withdrawn from the other parts of body. Various theories have been put forth to explain the way of formation of pseudopodia and resultant amoeboid movement (Box 7.1). From these theories, sol-gel theory is well accepted by zoologists.

The speed of amoeboid movement varies from 2mm to 3mm per second. This is supposed to be the most primitive type of animal movement.

Sol-gel theory. For explaining the forward flow of cytoplasm to form pseudopodia, **sol-gel theory** or **change of viscosity theory** was proposed by **Hyman** (1917) and later supported by **Pantin** (1923–26) and **Mast** (1925).

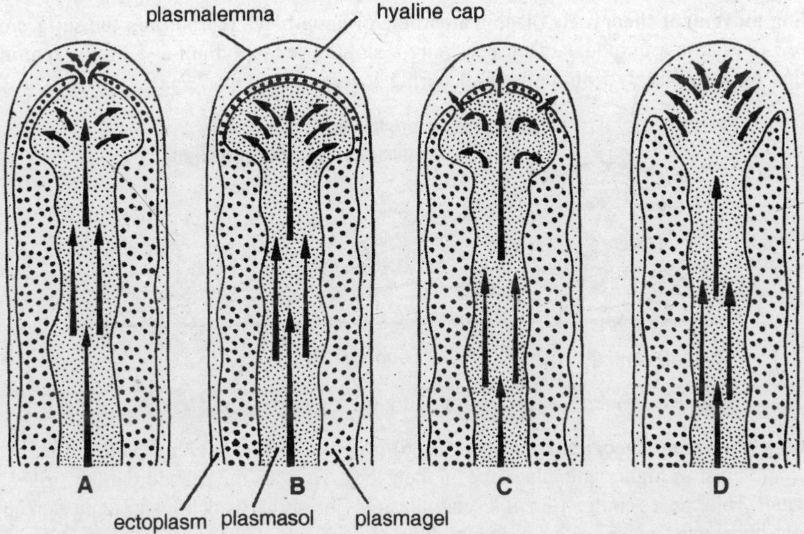

Fig. 7.6. Mechanism of amoeboid movement as explained by sol-gel theory.

According to sol-gel theroy, a delicate plasma membrane covers the surface of an amoeba. Beneath it is a clear zone free of granules, the **ectoplasm**, differentiated from the inner most granular **endoplasm**. The ectoplasm is in the gel state and makes up the **plasmagel**. The more fluid inner endoplasm is the **plasmasol**. The plasmagel exercises a constant, slight pressure on the plasmasol and can contract to increase this pressure. If for any reason the restraining influence of the plasmagel is locally insufficient, the plasmasol erupts, flowing into an extending pseudopodium (Fig. 7.6). As the plasmasol reaches the advancing tip, it spreads fountain-like, turning outward and backward to be converted into plasmagel and thus extend the plasmagel tube of the pseudopodium. If plasmagel forms too rapidly at the tip, a continuous retaining layer forms and stops further movement. If this does not happen and if new plasmasol is continuously formed at the opposite end (*i.e.*, posterior end) of the organism, movement can continue indefinitely. The change of plasmasol to plasmagel involves a loss of volume; therefore, the protoplasm may stream forward to occupy the space made available in this way. Thus, for the amoeboid movement, the following four types of changes occur in *Amoeba* (Fig. 7.12):

1. The plasma membrane of body of *Amoeba* adheres to the substratum.
2. Conversion of plasmasol to plasmagel (*i.e.*, gelation) at the "anterior" advancing tip.
3. Conversion of plasmagel into plasmasol (*i.e.*, solation) at the "posterior" end of body.
4. Contraction of the gelated protoplasm, and
5. Forward streaming of the plasmasol.

Box 7.1 Theories Regarding Amoeboid Movement

1. Adhesion theory. According to this theory, amoeba moves like a drop of water streaming out in the path of greater adhesion upon a rough-surfaced substratum.

2. Contraction hydraulic theory. Schultze (1875) postulated that due to contraction of plasmagel at the posterior end, a current is produced in the protoplasm. This current pushes the plasmosol at the anterior end and this results in the formation of new pseudopodium anteriorly and the body moves in that direction.

3. Surface tension theory. This theory was first of all proposed by **Berthold** (1886) and later supported by **Butschii** (1894) and **Rhumbler** (1898). According to this theory, constant internal tension forces the cytoplasm to shoot out from surface in the form of pseudopodium wherever surface tension suddenly lowers due to local effects.

4. Rolling movement theory. By placing a particle of lamp-black (carbon) on the surface of *Amoeba verrucosa* (which is a unipodal species forming a single pseudopodium at a time), **Jennings** (1904) noted that amoeba simply rolls like a ball during locomotion (Fig. 7.7).

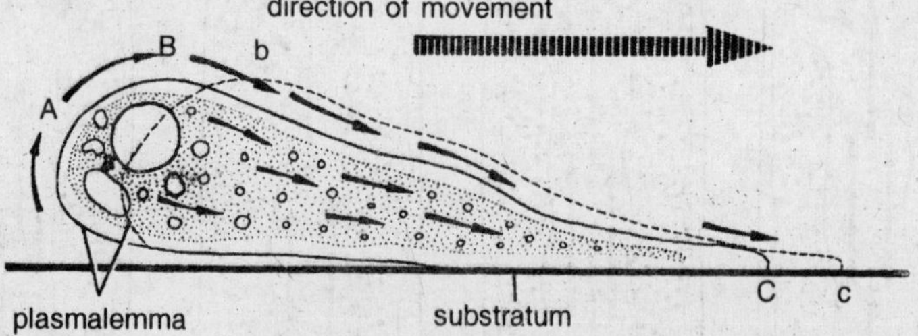

Fig. 7.7. *Amoeba verrucosa* showing rolling movement (Jennings, 1904).

5. Walking movement theories. Dellinger (1906) proposed that *Amoeba proteus* virtually "*walks on pseudopodia*" just as higher animals move on their legs. Thus, as the pseudopodium strikes an object, it sticks to it. This causes contraction of an endoplasmic fibrous network of contractile elements, pulling the whole body mass close to the pseudopodium (Fig. 7.8).

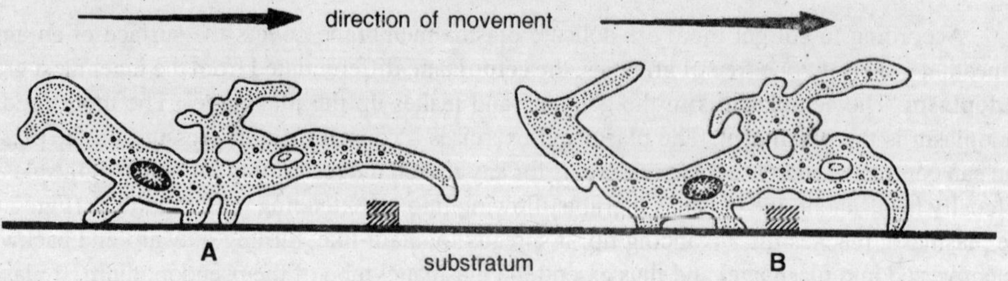

Fig. 7.8. *Amoeba proteus*. Walking movement due to contraction in the protoplasm.

6. Sol-gel theory. This theory is the most accepted theory of pseudopodium formation. It was proposed only when cytoplasm was recognized as a **colloidal** system.

7. Molecular folding-unfolding theory. Goldacre and **Lorch** (1950,52) provided a strong support to sol-gel theory by explaining the biophysical and biochemical processes involved in reversible 'sol-gel' changes and contraction of plasmagel. According to this theory, when the long-chain protein molecules contract, become tightly folded and scatter in the temporary hind region, they represent solation. However, when they become unfolded, regularly arranged and tightly bonded together at the tip of advancing pseudopodium, they represent gelatin (Fig. 7.9).

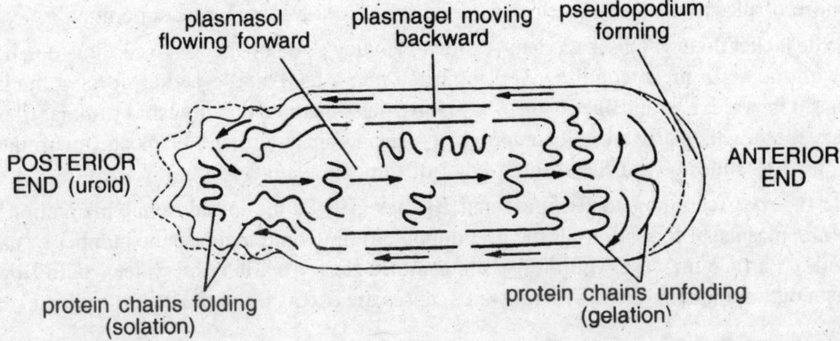

Fig. 7.9. Molecular folding-unfolding during solation and gelation of cytoplasm for amoeboid movement.

Quite recently, role of polymerization of actin subunits (G-actin) into actin microfilaments (F-actin) and vice-versa have been established for folding-unfolding or sol-gel theroy (**Karp**, 1996).

8. Sliding theroy. According to **Huxley** (1954), the proteins participating in solation and gelation are similar to **actin** and **myosin** of muscle fibres. Their folding-unfolding occurs due to sliding of actin and myosin filaments and expends great amount of ATP molecules.

9. Fountain zone contraction theory (or Front contraction theory or Pull hypothesis). Allen (1962) postulated that ectoplasm of *Amoeba* is distinguished into an outer fluid hyaline layer and an inner viscid "gel" layer, while the endoplasm has an axial core of gel surrounded by a "shear zone" of fluid sol (Fig. 7.10). The endoplasmic axial gel is pulled forward, strikes against the hyaline cap at the tip of advancing pseudopodium and fans out like a fountain to all sides, becoming continuous with fhe gel layer of ectoplasm. In the "fountain zone", the **gel contracts** by releasing some water into the hyaline cap, thus, pulling the axial core of endoplasm farther forward.

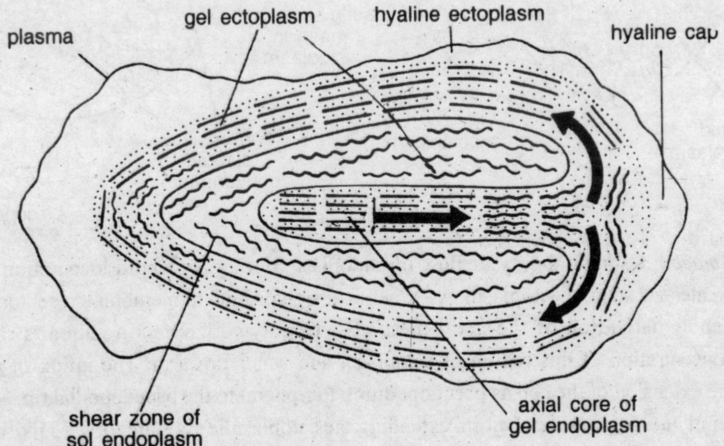

Fig. 7.10. Amoeboid movement on the basis of Allen's fountain zone contraction theory.

10. Ectoplasm contraction theory (or Push hypothesis). According to **Goldacre**, the gel portion (ectoplasm) contracts at the rear end of the body and squeezes the sol portion (endoplasm) forward, *i.e.*, force is created from behind, like the squeezing of the back of a tooth paste tube. On reaching the anterior end the sol is converted into gel.

11. Contraction-hydraulic theory. This theory was postulated by **Rinaldi** and **Jahn** (1963). It discarded Allen's fountain-zone contraction theory and supported Mast's sol-gel theoy. According to it, contraction of plasmagel layer at trailing end exerts a **hydraulic pressure** upon the plasmasol. Resistance to this pressure is minimum at the front end, highest in the middle and medium at the rear. Due to this reason, more and more of plasmasol tends to flow forward into the advancing pseudopodium.

12. Contractile jacket theory. This is modem version of Huxley's sliding theory. According to it, bonding of **myosin** with the **actin mesh** could convert the mesh into a **contractile jacket**, forcing the interior fluid endoplasm forward. **Lee** and **Jacobson** (1993) have demonstrated that **myosin I** (bipolar filaments) act as **primary motor** which drives cell locomotion. Myosin motor acts on actin filaments (microfilaments) and causes them to slide toward the leading edge of the cell.

13. Osmotic theory. According to **Ruppert** and **Barnes** (1994), the initial depolymerization of the actin mesh (*i.e.*, plasmasol formation) at the pseudopodium tip would increase the number of particles (actin subunits). Particle increase would raise the **osmotic concentration** and water would flow from the endoplasm out into the tip. The influx of water causes the extension of the cell as a pseudopodium.

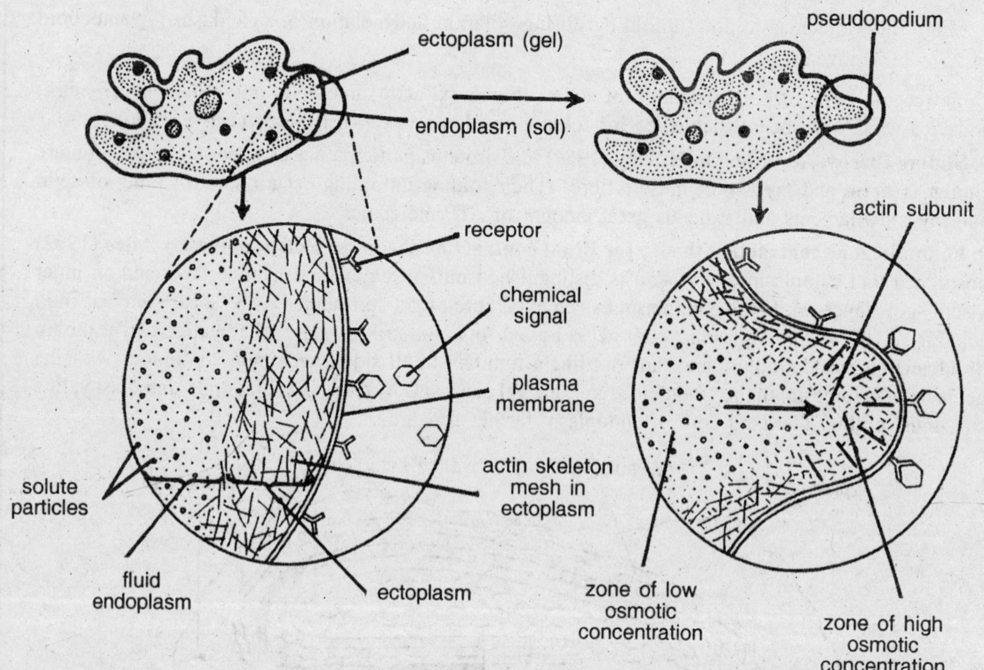

Fig. 7.11. *Amoeba.* Osmotic theory of flow of endoplasm during amoeboid locomotion. Actin mesh creates gel state of ectoplasm. A—Chemical signals bind to membrane receptors and initiate depolymerization of the actin mesh. B—Depolymerized actin subunits raise osmotic concentration of this region of ectoplasm and water flows in. The influx of water causes the extension of the cell as pseudopodium. Peripheral to the pseudopodial tip, actin subunits repolymerize as skeletal mesh, extending the ectoplasmic gel state forward like a progressive sleeve.

2. Flagellar movement (Locomotion in *Euglena*). This sort of movement is the characteristic of Mastigophora or flagellated protozoans, *e.g.*, *Euglena*. Flagellar movement is produced by continuous beating (*i.e.*, lashing movement) of long, elastic flagella. Following four types of flagellar movements have been recognized:

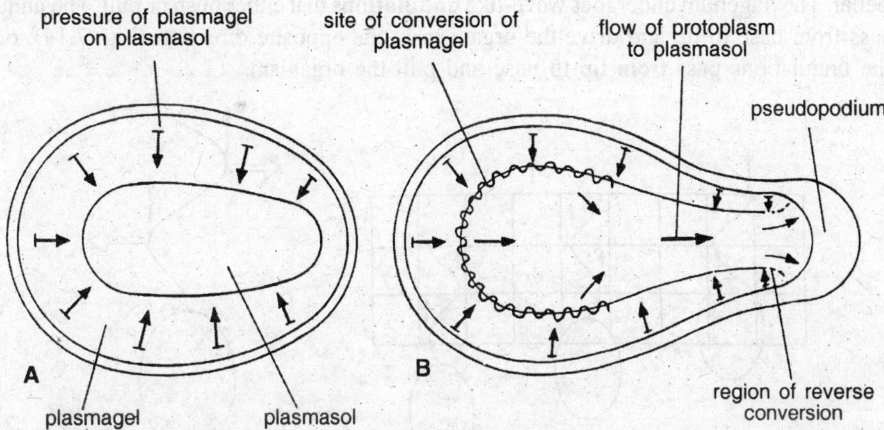

Fig. 7.12. Mode of formation of pseudopodium. Arrows with square tails show pressure of plasmagel on plasmasol; straight arrows show flow of protoplasm. Wavy line in B indicates site of conversion of plasmagel to plasmasol; broken lines at right shows region of reverse conversion where macromolecular folding reduces protoplasmic volume to facilitate its flow forward.

(*a*) **Screw propeller theory.** According to **Butschli** the spiral turning of the flagellum like a screw exerts a propelling action and pulls the animal forward. Spiral waves arise repeatedly from the base of flagellum one after the other and moving towards the tip.

(*b*) **Circular beat theory. Metzner** suggested that the flagellum beats in a circle tracing a cone and produces sufficient current to pull the animal forward.

(*c*) **Paddle stroke or sideways lashing movement.** According to **Ulehla** and **Krijsman** (1925) the flagellum beats in a side-ways lash, which consists of an **effective down stroke** or **bending** and a relaxed **recovery stroke** or **straightening** (Fig. 7.13). In the effective stroke the flagellum is held out rigidly with slight **concavity** in the direction of stroke. In the recovery stroke, however, flagellum is strongly **curved** and is brought forward again. Due to such flagellar action, the animal moves forward. Generally the flagellum beats obliquely so that during forward movement, the animal also rotates on its longitudinal axis.

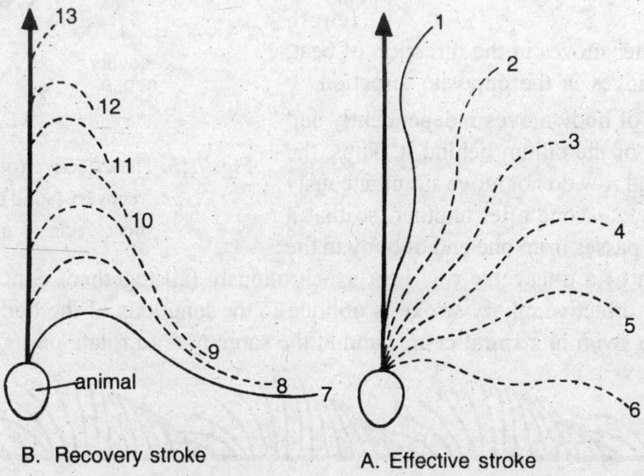

Fig. 7.13. Two stages in the lashing movement of flagellum.
A—Effective stroke; B—Recovery stroke.

(*d*) Undulating movement. Flagellar propulsion follows essentially the same principle as that of a propellar. The flagellum undergoes wave-like **undulations** that either push or pull. The undulatory waves pass from base to tip and drive the organism in the opposite direction (Fig. 7.14), or more rarely, the undulations pass from tip to base and pull the organism.

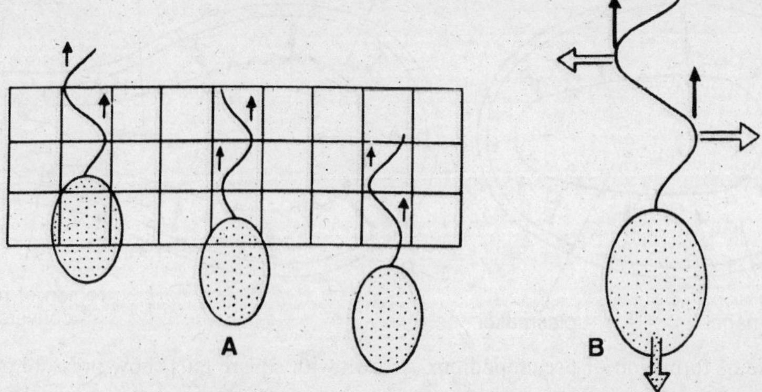

Fig. 7.14. Flagellar propulsion. A—Base to tip wave propagation. B—Forces generated by base - to - tip wave propagation. Lateral forces (outlined arrows) cancel each other. Longitudinal forces (solid arrows) combine to produce forward thrust in the opposite direction.

Many flagellates have flagella with fine lateral branches, called **mastigonemes**. The hydrodynamic effect of the branches is to cause the flagellum to **pull** rather than push even though the flagellar waves are passing from base to tip.

3. Ciliary movement (Locomotion in *Paramecium*). Ciliary movement is performed by extremely fine and very small thread-like processes of ectoplasm in cilia-tes (*e.g.*, *Paramecium*) and larval suctorians. The ciliary movement is very much similar to that of flagellar move-ment, consisting of the same effective stroke and recovery stroke, but the cilia bend throughout their length (Fig. 7.15). The water moves in the direction of beat, while the animal moves in the opposite direction.

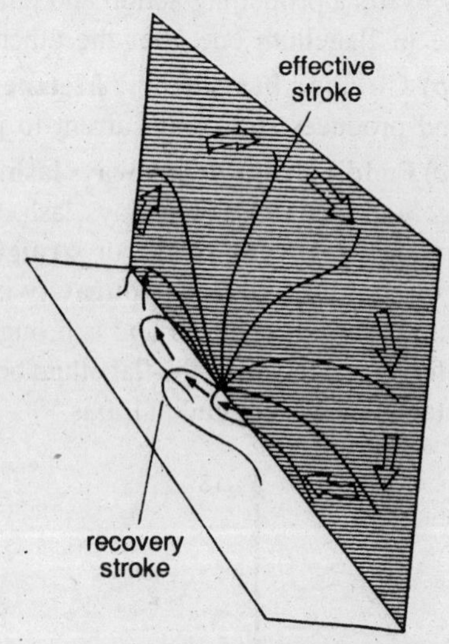

Fig. 7.15. The effective (outlined arrows) and recovery (solid arrows) stokes in the beat cycle of a single cilium.

Every cilium of body moves independently and slightly in advance of the cilium behind it. Thus, the cilia of a longitudinal row do not move simultaneously but metachronously, *i.e.*, one after another, so that a wave of propulsion passes from one end of body to the other, whereas cilia of a transverse row beat synchronously (all together). Since in *Paramecium* the direction of the effective ciliary stroke is oblique to the long axis of the body (Fig. 7.16), this causes the ciliate to swim in a spiral course and at the same time to rotate on its longitudinal axis.

Fig. 7.16. A single longitudinal row of cilia showing metachronous movement.

In fact, the movement of adjacent cilia are coupled as a result of **interference effects** of the surrounding water layers. Thus, hydrodynamic forces impose a coordination of the cilia. The beat of individual cilia is part of the metachronal waves that sweep along the length of the body. The direction and intensity of the beat of cilia are controlled by changing levels of Ca^{2+} and K^+ ions.

There is no evidence that the infraciliature functions as a conducting system in coordination of ciliary movement. Infraciliature may serve primarily in ciliary anchorage and controlling cell shape (**Ruppert** and **Barnes**, 1994).

Ciliates are the fastest moving of the protozoa. The speed of ciliary locomotion varies from 400 μm to 200 μm per second.

Fig. 7.17. Metachronal waves in *Paramecium* during forward swimming. Wave crests are shown by lines and their direction by the small solid arrows. Dotted lines indicate direction of effective stroke of cilia. Movement of ciliate is indicated by large arrow.

4. Metabolic movement (Locomotion in *Euglena* and *Monocystis*). Metabolic movement is also called **gliding, wriggling, peristaltic** or **gregarine movement**. It occurs in sporozoans (*Monocystis*) and certain flagellates (*Euglena*). The contractile **myonemes** of body contract and expand to bring about a change in shape of the body and relative displacement from the original place. Such gliding movement is performed by parasitic forms such as *Monocystis* in the body fluid of host.

Euglena shows wriggling movement by the alternative waves of contraction and expansion passing over the body from its anterior to the posterior end. Such sort of movement is called **euglenoid movement** and is performed by the help of myoneme fibrils.

Box 7.2.

Apicomplexans (= sporozoans) do not have cilia, flagella or pseudopodia. Nevertheless, they can be observed gliding and flexing. What causes these movements is not well understood, but microtubules and microfilaments underneeth alveoli may play a role. In them, microtubules originate at the apical complex and run beneath the alveoli (**Brusca** and **Brusca**, 2003).

QUESTIONS

Long Answer Questions

1. Write an essay on locomotion in protozoa.
2. Describe structure and function of locomotory organs of protozoa.

Short Answer Questions

1. Explain the mechanism of ciliary movement. (*Himachal 1995*)
2. Explain the flagellar locomotion. (*Himachal 1995*)
3. Compare the movements of *Amoeba* and *Paramecium*. (*Himachal 1993*)
4. Write a short note on amoeboid locomotion.

Very Short Answer Questions

1. What are pseudopodia?

Ans. Broad, rounded protrusions formed during amoeboid movement as portions of cell surface are pushed outward by a column of cytoplasm that flows through the interior of the cell toward the periphery.

Fill in the Blanks

1. Locomotory organelles in Protozoa are ...., pseudopodia and ....

2. Lobopodia are found in .... and ....

3. Organelles of locomotion are .... in radiolarians.

4. Locomotory organ of *Paramecium* is ....

5. Gliding movement shown by *Euglena* is called .... (*Guwahati, 1994*)

Multiple Choice Questions

1. Axopodia are locomotory organelles in
(*a*) amoebida (*b*) foraminifera
(*c*) helizoea (*d*) radiolarians

2. Lobopodia occurs in
(*a*) *Amoeba* (*b*) *Entamoeba*
(*c*) *Arcella* (*d*) all of these

3. Protozoans are mainly classified on the basis of
(*a*) locomotory organ (*b*) types of nuclei
(*c*) contractile vacuole (*d*) mode of nutrition

4. Protozoans lack which of these as locomotory organs
(*a*) pseudopodia (*b*) flagella
(*c*) cilia (*d*) parapodia

5. In *Amoeba* pseudopodia are formed due to
(*a*) contact with food
(*b*) sol-gel change
(*c*) movement towards area of higher temperature
(*d*) all the above

ANSWERS

Fill in the Blanks

1. Flagella, Cilia **2.** *Amoeba*, *Arcella* **3.** Filopodia **4.** Cilium
5. Metaboly.

Multiple Choice Questions

1. (*c*) **2.** (*d*) **3.** (*a*) **4.** (*d*)
5. (*b*).

8

Nutrition in Protozoa

The means by which living organisms obtain the substances required for growth and the maintenance of life is called **nutrition** (L., *nutrimentum* = nourishment). Thus, nutrition is the process of ingestion, digestion, assimilation and egestion of food to release energy needed for various vital activities. Protozoans acquire food in a variety of ways such as follows.

1. Holozoic or zootrophic nutrition (animal-like)
2. Holophytic or autotrophic nutrition (plant-like)
3. Saprophytic or saprozoic nutrition (fungi-like)
4. Parasitic nutrition
5. Mixotrophic or mesotrophic nutrition
6. Coprozoic.

Several types of nutrition such as holozoic, saprozoic, parasitic, mixotrophic and coprozoic nutritions are basically included in **heterotrophic nutrition.**

A. Holozoic nutrition. Holozoic nutrition is found in the majority of the free-living protozoans such as *Amoeba* and *Paramecium.* These animals utilize ready made solid food in the form of other organisms such as microorganisms, bacteria, yeast, algae, protozoans and small metazoans such as rotifers, copepod larvae and nematodes. The holozoic nutrition involves the following three steps:

1. Food capture or ingestion
2. Digestion and assimilation
3. Egestion

1. Ingestion. According to the size of food, holozoic protozoans may be **macrophagous** (feeding on large food particles, *e.g., Amoeba, Didinium*) or **microphagous** (feeding on minute food particles, *e.g., Paramecium*). Further, according to mode of food catching, the holozoic protozoans are classified into following two types:

(i) **Raptorial protozoans.** Some free living protozoans capture single organisms by direct interception, they are called **raptorial animals**. Since, raptorial protozoans are predators or hunters, so they have to seek, pursue and capture the prey. They require organelles for capturing and ingesting food. They may, move by flagella, cilia or pseudopodia and may either have a definite mouth or engulf food at any point on the cell surface. For example, an amoeba can capture the ciliate *Paramecium* (prey), which moves much more rapidly than itself, not by pursuit but by protoplasmic engulfment of its unwatchful prey by the help of its pseudopodia. On the

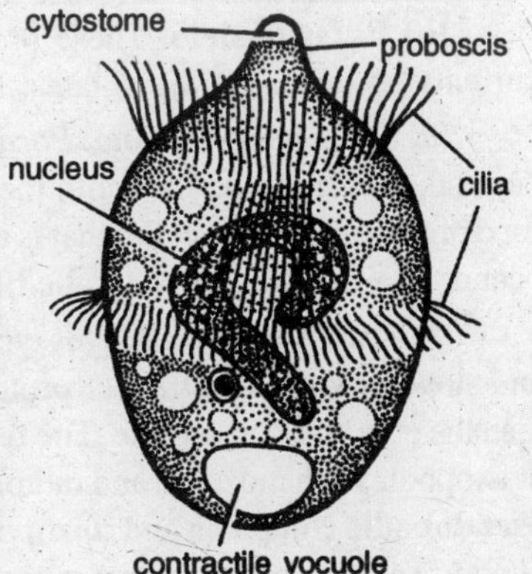

Fig. 8.1. *Didinium.*

other hand, the raptorial ciliate, *Didinium* (Fig. 8.1 and Fig. 8.2) actively pursues and captures *Paramecium* (prey). Thus, a hunter protozoan is actively motile, moving about to sample its environment and is either equipped with a large **cytostome** or **cell mouth** or can form engulfing pseudopodia.

In *Amoeba*, for example, pseudopodia extend around the prey, eventually enveloping it completely with cytoplasm (this mode of feeding may be **circumvallation** [(for the active prey) or **circumfluence** for inactive prey)] or the body surface of amoeba **invaginates** to form **food cup**. The enclosing of the captured organism by cytoplasm (phagocytosis, see Box 8.1) results in the formation of a food vacuole within the amoeba.

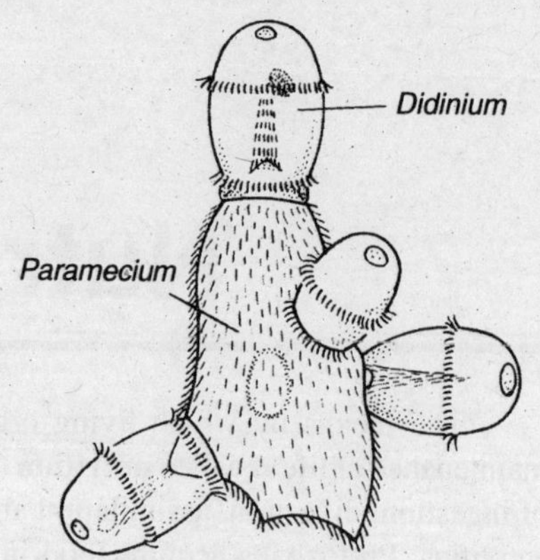

Fig. 8.2. Four small *Didinium* attacking a *Paramecium*.

Box 8.1.

How cells take up food and other substances. Substances enter protozoan cells in a variety of ways. The protein channels of the plasma membrane provide for the **passive diffusion** of water, ions and small molecules such as sugars and amino acids. Some transport proteins function as energy-requiring pumps, **actively transporting** certain molecules or moving ions in or out against their concentration gradient.

During feeding a protozoan cell has to perform **bulk transport** (a kind of active transport) by its plasma membrane; that is done by the following three methods:

1. Endocytosis. In the process of endocytosis, some extracellular material enter a cell in minute pits on the cell's plasma membrane. Later on, these pits pinch off internally. **Pinocytosis** is a nonspecific form of endocytosis in which the rate of uptake is in simple proportion to the external concentration of the material being absorbed. (Fig. 8.3A).

2. Receptor-mediated endocytosis (RME). RME is brought about by the uptake of **specific extracellular macromolecules** (called **ligands**), *e.g.*, proteins, at a rate greater than predicted by the concentration gradient. Ligands bind to and are concentrated on specific receptors of plasma membrane before these are internalized in the vesicles (Fig. 8.3B).

3. Phagocytosis. Engulfment of large particles, such as, bacteria and protozoans and their enclosure within relatively large vesicles. It requires binding of a particle to multiple membrane receptors (Fig. 8.3C).

(*ii*) Filter feeders. These protozoans collect organisms from surrounding currents or from currents produced by the ciliates. Filter feeders employ following two modes of feeding:

(*a*) Suspension feeding. *Paramecium* and *Vorticella* use a buccal ciliature in creating effective currents that bring the food into the buccal area. In case of sessile choanoflagellata, flagella sweep food organisms against collar that is then ingested by the cytoplasm at its base. In *Euglena* the ingestion occurs at a definite region, called **cytostome.**

(*b*) Trapping feeding. This type of feeding occurs in those protozoans which are sessile, floating and slow moving. The trapper protozoan increases its potential catch by **sweeping** an area greater than the effective cell surface. The **tentacles** of suctorians and **radiating pseudopodia** (reticulopodia or axopodia) of heliozoans and radiolarians, as well as, long, branching and sometimes **anastomosing pseudopodia** (filopodia and rhizopodia) of foraminiferans, are primarily used as traps in the capture of prey. Any prey organism that comes in contact with the pseudopodia become coated with an adhesive substance extruded on the pseudopod surface. This coating contains discharged lysosomes and their proteolytic secretions which paralyze the prey and initiate digestion even during capture.

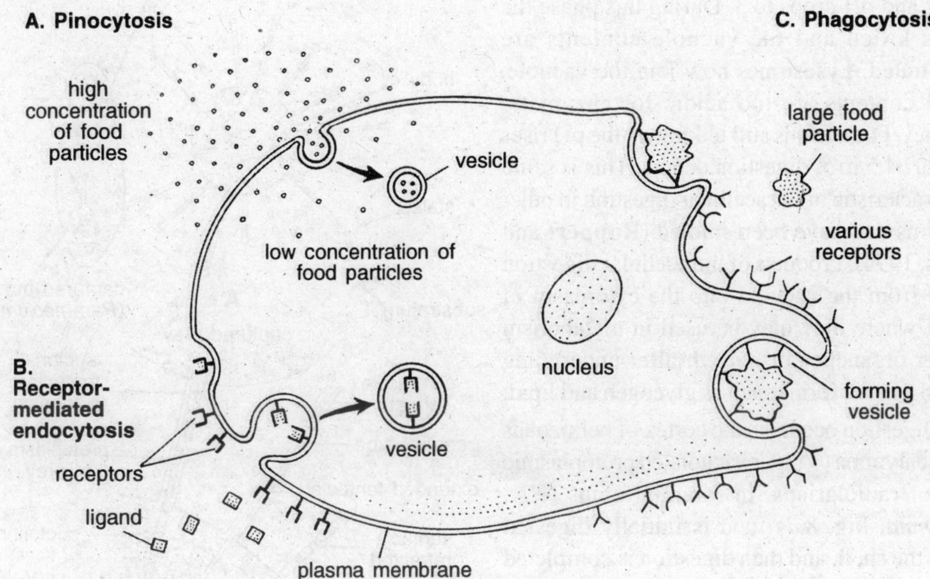

Fig. 8.3. Mode of bulk food uptake by vesicles or food-vacuoles. A—Pinocytosis; B—Receptor-mediated endocytosis; C—Phagocytosis.

The body of sessile suctoria (*e.g.*, *Ephelota*, *Podophrya*) bears tentacles which may be knobbed at the tip or shaped like long spines. Each tentacle is supported by a cylinder of microtubules and carries special organelles called **haptocysts** (Fig. 8.4). Suctorians feed on other ciliates and when prey strikes knobbed **raptorial tentacles**, the haptocysts are discharged into the prey body, anchoring it to the tentacles. The contents of the prey are then sucked through the tubular **suctorial tentacles** within a long food vacuole into the suctorian (Fig. 8.5). Thus, "suction" in suctorians provides an example of **rapid phagocytosis,** which is accelerated by microtubules within the tentacle.

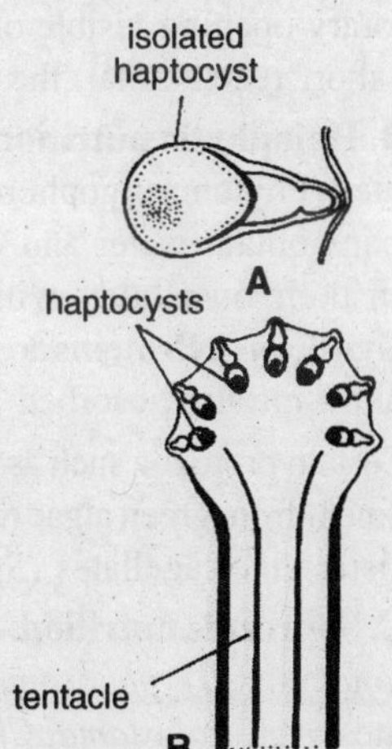

Fig. 8.4. Haptocyst of Suctoria. A—An isolated haptocyst; B—Haptocysts arranged within tip of tentacle.

Formation of food vacuole. Finally the prey is enveloped in a membrane-bounded vesicle, called **food vacuole.** In foraminiferans, heliozoans and radiolarians, food particles are enclosed in food vacuoles and drawn toward the interior of body. The axial rods of heliozoans may contract, drawing the prey into the ectoplasmic cortex, or the axopodia may liquify and surround the food, forming a vacuole at the site of capture. The vacuole, then, moves inward. In *Paramecium*, the food-vacuoles are moved along in a definite course by endoplasmic currents (cyclosis).

2. Digestion. Once food is ingested, all holozoic protozoa use food vesicles for **intracellular digestion**. In most protozoa the food vacuoles undergo definite change in hydrogen-ion concentration (pH) and in size during the process of digestion. For example, in *Paramecium*, following the formation of the food vacuole (Fig. 8.6), acidic vesicles called acidosomes, fuse with the vacuole and some unit membrane is removed from the vacuole (**Fok**

and **Allen**, 1990). As a result, the vacuole becomes smaller and pH drops to 3. During this phase the prey is killed and the vacuole contents are concentrated. **Lysosomes** now join the vacuole, but the contents are too acidic for enzymatic efficiency. For reasons still unknown, the pH rises and at pH 4.5 to 5, digestion occurs. This is same pH characteristic of intracellular digestion in other organisms that have been studied (**Ruppert** and **Barnes**, 1994). Products of intracellular digestion **diffuse** from the vacuole into the cytoplasm of the cell where they may be used in metabolism by other organelles or **stored**, after undergoing synthesis, in the forms such as glycogen and lipid.

Digestion occurs in the cortex of heliozoans and the calymma (= outer vacuolated protoplasmic layer) of radiolarians. In foraminiferans (*e.g.*, *Allogromia*, Fig. 8.7) food is initially digested outside the shell, and then digestion is completed inside small food vacuoles within the shell.

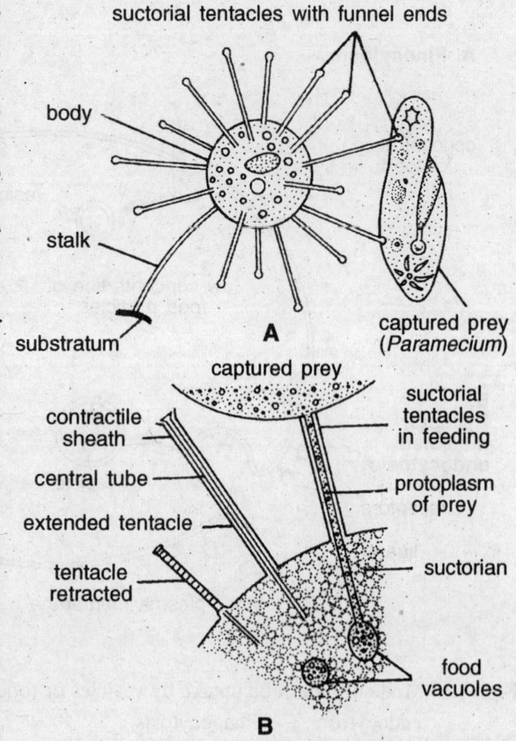

Fig. 8.5. Mode of feeding in suctoria (*Podophrya*).

3. Egestion. Indigestible material is released from the cell to the exterior by fusion of the waste-laden vacuole with the plasma membrane in a process called **exocytosis**. In *Amoeba*, egestion can take place at any point of surface of the body. Like any other naked protozoa, *Amoeba* accumulates exhausted food vacuoles and eventually leave the contents behind as they move about. Where there is a definite cell membrane covering (*e.g.*, pellicle of *Paramecium*), some exact point or **cytopyge** for egestion is required. Cytopyge may be a temporary opening visible only at the time of defaecation (*Paramecium*) or a permanent opening with a short rectal canal, the **cytoproct** (*Balantidium* and *Nyctotherus*).

B. Holophytic nutrition. Holophytic or plant-like nutrition occurs in the chlorophyll-bearing flagellates (Phytomastigophores) such as *Euglena*, *Volvox*, *Pandorina*. In this type of nutrition, the protozoans obtain water and carbon dioxide from the medium and prepare **carbohydrates** from these in their own body with the help of radiant energy of sunlight trapped by chloroplasts (**photosynthesis**). **Proteins** are then built up by combining carbohydrates with nitrogenous salts (*e.g.*, ammonium nitrate) absorbed from the medium.

Certain protozoa such as *Difflugia*, *Paramecium bursaria* and *Stentor polymorphus*, may obtain nourishment from green algae (*Zoochlorellae*) that live within them (this is called **symbiotic nutrition**). Mutualistic dinoflagellates (*Symbiodinium microadriaticum*) live as symbionts in corals.

C. Saprozoic nutrition. Saprozoic nutrition is found in the majority of the parasitic protozoans, *e.g.*, *Monocystis*, *Trypanosoma*, *Giardia*, *Opalina*, etc. It also occurs in many free-living colourless flagellates, *e.g.*, *Polystoma*, *Chilomonas*, etc. In saprozoic nutrition, the animals subsist entirely on dissolved substances in the medium. The dissolved materials are absorbed either through the entire surface of the body or through special regions. Some free-living saprozoic flagellates can flourish largely on inorganic salts, but most saprozoic forms must get organic nutrients, such as, glucose, amino acids and peptones. Some green flagellates (*e.g.*, *Euglena*) are facultative saprozoic forms. For example, *Euglena* is autotrophic in light and saprozoic in the dark.

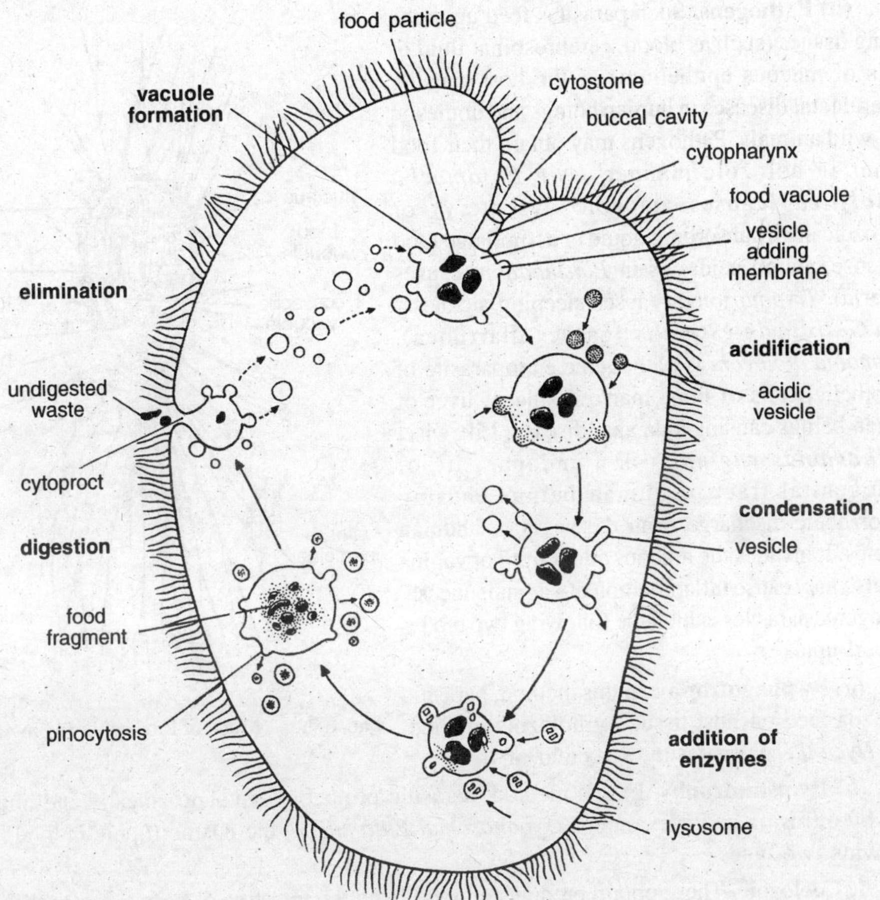

Fig. 8.6. Intracellular digestion in ciliates. In the formation of food vacuole, small discoidal vesicles add membrane to provide for its increase in size. Acidic vesicles (acidosomes) bring about a drop in pH. The food vacuole becomes smaller by the removal of membrane. Lysosomes add digestive enzymes which have their optimal activity at an acid pH.

D. Mixotrophic nutrition. Such type of nutrition occurs in many flagellates and some amoebae. In mixotrophic nutrition, protozoan feeds by more than one method at the same or different times. For example, some dinoflagellates, *e.g.*, *Ceratium* are autotrophic and holozoic simultaneously. *Entamoeba histolytica* and *Balantidium coli* which although feed holozoically, also absorb dissolved organic compounds through general body surface (*i.e.*, saprozoic nutrition). *Euglena* and *Paranema* feed holophytically but obtain certain necessary substances by passive diffusion from their surroundings.

E. Parasitic nutrition. The parasitic protozoa get their nourishment partially or wholly from the digested and decomposed substances of the food of host, either in holozoic or saprozoic manner. Parasitic protozoa may be of following two types:

(*i*) Food robbers. They feed upon the raw or digested food stuffs of the host either holozoically or by diffusion but do not harm the host. They are usually **commensals**, *e.g.*, ciliate *Balantidium coli* lives as endocommensal in the intestine of pig and is passed by means of cysts in the faeces. *Entamoeba coli* is an endocommensal of human intestine (colon) which feeds on bacteria, particles of indigested food and other debris.

(*ii*) **Pathogens.** Such parasites feed upon the living tissues (such as blood, cerebrospinal fluid or cells of mucous epithelium) of the host causing several fatal diseases in human beings and domestic and wild animals. Pathogens may obtain their food either in holozoic manner as in *Entamoeba histolytica* (causes amoebic dysentery) or saprozoically by absorbing liquid or assimilable food (glucose, amino acids) as in *Plasmodium* (causes malaria), *Trypanosoma* (causes sleeping sickness) and *Giardia intestinalis* (causes diarrhoea). *Leismania donovoni* is pathogenic endoparasite of endothelial cells of bone marrow, spleen, liver of human beings causing kala azar disease. Likewise, *Trichomonas vaginalis* is a endoparasite of urinogenital tract of human beings causing leucorrhoeic discharge from the vagina of human females. It invades the mucous epithelium of vagina and also may cause inflammation of vaginal mucosa. Pathogenic parasites exhibit the following two modes of food uptake:

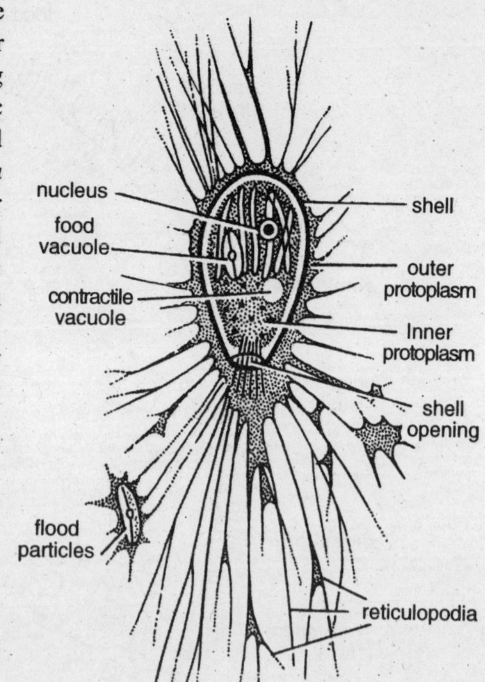

Fig. 8.7. Mode of food capturing in *Allogromia.*

(*a*) **By phagotrophy.** In this method, parasitic protozoa feed on host tissue by holozoic method, *e.g.*, *Nyctotherus* (parasite of rectum of frog).

(*b*) **By osmotrophy.** In this method, parasitic protozoa feed saprozoically and obtain liquid food by diffusion as for example, *Trypanosoma*, *Entamoeba*, etc. Osmotrophic individuals are of following two types:

1. Coelozoic. These obtain predigested food from the intestine of host, *e.g.*, *Opalina* (parasite of rectum of frog), *Entamoeba histolytica*.

2. Histozoic. Certain protozoan parasites such as *Monocystis* and *Plasmodium* feed on host tissue and are described as histozoic.

F. Coprozoic nutrition. A few free living protozoa feed upon the faecal matter of other organisms, *e.g.*, *Copromonas* (Mastigophora) is found in the faeces of frog.

QUESTIONS

Long Answer Questions

1. Describe the various modes of nutrition found in protozoa.

(*Kamraj University 1994; Guwahati 1995; Lucknow 1997*)

2. Write an essay on the modes of nutrition in protozoa studied by you.

(*Kanpur 1991, Bhopal 1993*)

Short Answer Questions

1. Mention the various types of nutrition found in protozoa. Cite one example of each.

Very Short Answer Questions

1. What is nutrition?

   Ans. The means by which living organisms obtain substances required for growth and maintenance of life is called **nutrition.**

2. Give an example of microphagus protozoan.

   Ans. *Paramecium.*

3. Give an example of macrophagus protozoan.

Ans. *Amoeba.*

4. Give an example of raptorial protozoan.

Ans. *Didinium.*

5. Which protozoan possesses cytoproct in its body?

Ans. *Balantidium.*

6. In *Paramecium*, what is the role of cytopyge?

Ans. Egestion.

7. Give an example of phagotrophous animal.

Ans. *Nyctotherus.*

8. What type of nutrition is found in parasitic protozoans?

Ans. Saprozoic nutrition.

9. Give two examples of histozoic protozoans.

Ans. *Monocystis* and *Plasmodium.*

Fill in the Blanks

1. Paramylum is a reserve food material in ....

2. Filter feeding occurs in ....

3. Prey of *Didinium* is ....

4. *Opalina* lives in the .... of ....

Multiple Choice Questions

1. Animal like nutrition is called (*Kanpur 1987*)

(*a*) holophytic (*b*) holozoic
(*c*) saprophytic (*d*) parasitic

2. Food capture in Sarcodina is the function of

(*a*) food vacuoles (*b*) eye spot
(*c*) nucleolus (*d*) pseudopodium

3. The cell drinking phenomenon is termed as

(*a*) ingestion (*b*) pinocytosis
(*c*) circumfluence (*d*) circumvallation

4. Digestion in protozoans is

(*a*) intercellular (*b*) intracellular
(*c*) extracellular (*d*) all of these

5. Intracellular digestion occurs inside

(*a*) stomach (*b*) food vacuole
(*c*) contractile vacuole (*d*) alimentary canal

6. Food capture is the function of

(*a*) food vacuole (*b*) reticulopodium
(*c*) eyespot (*d*) paramylum

7. Which of the following is a symbiont?

(*a*) *Plasmodium* (*b*) *Trichonympha*
(*c*) *Opalina* (*d*) *Nyctotherus*

8. Nutrition in *Euglena* is

(*a*) holophytic (*b*) saprotrophic
(*c*) mixotrophic (*d*) none of these

9. Contents of food vacuole are

(*a*) first acidic, then alkaline (*b*) first alkaline, then acidic
(*c*) alkaline throughout (*d*) acidic throughout

ANSWERS

Fill in the Blanks

1. *Euglena* **2.** *Paramecium*
3. *Paramecium* **4.** Rectum, amphibians.

Multiple Choice Questions

1. (*b*) **2.** (*d*) **3.** (*b*) **4.** (*b*) **5.** (*b*) **6.** (*b*)
7. (*b*) **8.** (*c*) **9.** (*a*).

9

Reproduction in Protozoa

The mode of reproduction in protozoa is variable among different groups, however, it is basically a cell division. In protozoa, reproduction takes place by **asexual** (involving **mitosis**) and **sexual** (involving **meiosis**); however, some reproduce asexually only.

9.1 ASEXUAL REPRODUCTION

Asexual reproduction involves the mitotic cell division of parent body into two or more young individuals that develop into mature organisms. It always involves a single parent and neither meiosis nor fertilization occurs. Protozoa reproduce asexually by following four different methods: 1. Binary fission; 2. Multiple fission; 3. Budding; 4. Plasmotomy.

I. Binary fission. Division of the organism into two or more progeny is called **fission**. Fission typically involves **karyokinesis** or nuclear division and **cytokinesis** or cytoplasmic division. After or during later part of karyokinesis, cytoplasmic constriction begins. The two daughters then enter a phase of **growth** and **differentiation** following cytokinesis.

When fission results in two similar and equal daughter cells, it is called **binary fission**. Where there is a definite body axis, binary fission may be of following types:

1. Simple binary fission. As in Sarcodina (*Amoeba* and *Entamoeba*) the plane of division cannot be recognized.

2. Longitudinal binary fission. In flagellates (*Euglena*) the plane of division is along the long axis of the animal's body. In this case, division is said to be **symmetrogenic**, that is, longitudinal division cuts between the rows of basal bodies and producing mirror-image of daughter cells.

3. Oblique binary fission. In dinoflagellates (*Ceratium*) the division plane is oblique. In armoured dinoflagellates, the two fission products regenerate the missing plate or form a completely new armour, the old one having been shed before division.

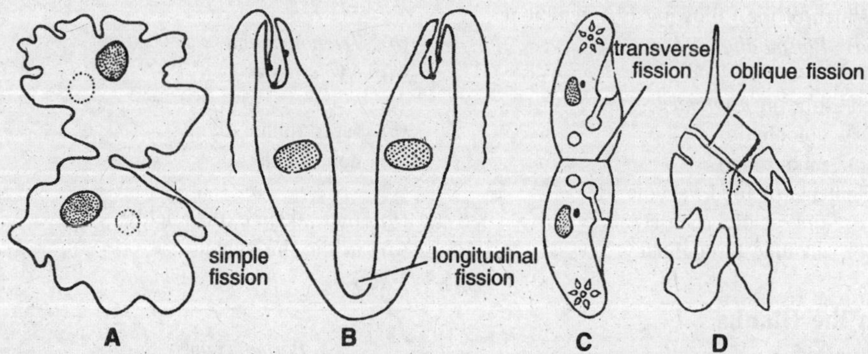

Fig. 9.1. Binary fission in protozoa. A—*Amoeba* (simple); B—*Euglena* (longitudinal); C—*Paramecium* (transverse) D—*Ceratium* (oblique).

4. Transverse binary fission. In ciliates (*Paramecium*) the division plane cuts across the kinetics *i.e.*, longitudinal rows of cilia and basal bodies. This type of division is called **homothetogenic.** Mitotic spindles are formed only in the division of the micronuclei. Division of macronucleus is usually accomplished by constriction (by amitosis).

In *Arcella* and *Euglypha*, a mass of cytoplasm extrudes from the opening of shell prior to division. This extruded mass secretes a new shell. The double-shelled animal now divides (Fig. 9.2).

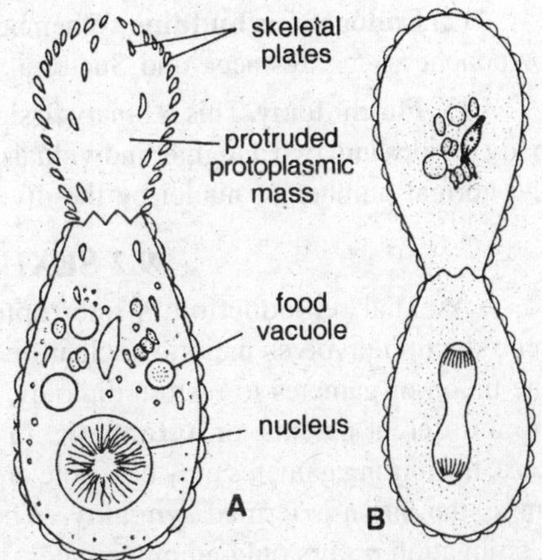

Fig. 9.2. Transverse binary fission in *Euglypha*, a shelled amoeba. A—Formation of skeletal plates on the protoplasmic mass protruding from aperture. B—Division of nucleus.

II. Multiple fission. Multiple fission is common in sporo-zoan parasites (*Monocystis*, *Plasmodium*), foraminiferans and radiolarians. Preparation for division involves repeated nuclear divisions (mitosis) to form a multinucleate organism with no cytokinesis. The cytoplasm then cleaves to form many tiny uninucleated individuals. These are set free by rupture of parent body. It is common for multiple fission to leave a mass of residual protoplasm that degenerates.

The multiple fission occurs at different phases in the life cycle of different individuals and the products are named accordingly.

(*i*) Schizogony. In schizogony, multiple fission occurs during sexual cycle in trophozoites or schizonts existing inside host cells (liver and erythrocytes) to produce primarily merozoites and later on gametocytes, *e.g.*, *Plasmodium*.

(*ii*) Gamogony. When multiple fission produces sex cells or gametes, it is termed as gamogony, *e.g.*, *Monocystis*.

(*iii*) Sporogony. The multiple fission in the zygote is known as sporogony and the offsprings as **spores**, *e.g.*, *Plasmodium*, *Eimeria*, *Monocystis*, etc.

III. Budding. Budding is a type of asexual reproduction in which one or more small daughters are produced by the parent. It may be considered a form of unequal fission. Like the fission, it involves preparatory phases before budding and a later period of growth and differentiation. Budding may be of following two types:

(1) Exogenous budding. When budding occurs at the body surface, *e.g.*, *Acanthocystis*, *Noctiluca*, *Ephelota* (Fig. 9.3) and *Entamoeba*.

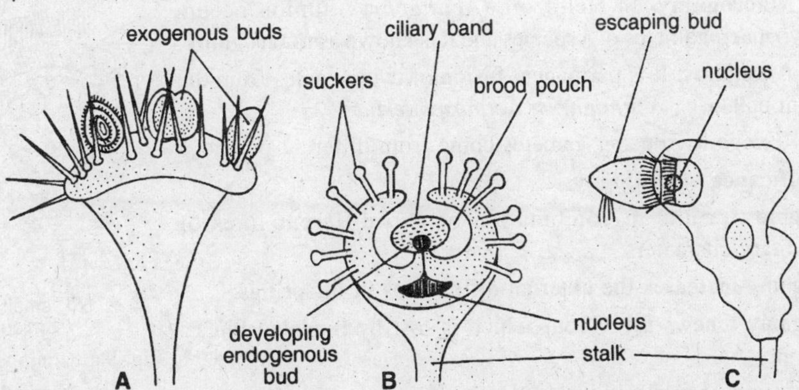

Fig. 9.3. Asexual reproduction by budding A—Formation of exogenous bud in *Ephelota*. B—Formation of endogenous bud in a Suctoria; C—Expulsion of bud from brood pouch in *Tochophyra*.

(2) Endogenous budding. When budding occurs in an internal cavity such as brood chambers or pouches, *e.g.*, Testacea and Suctoria.

IV. Plasmotomy. This is binary fission in multinucleate form such as *Opalina*. The multinucleate body divides into two daughter individuals without any nuclear division. Each of the offspring regains the normal number of nuclei by the division of parental nuclei.

9.2 SEXUAL REPRODUCTION

Asexual reproduction is interrupted at regular intervals by sexual reproduction. Sexual reproduction involves meiotic nuclear division that results in a change from diploid to haploid and the union of gametes to restore diploidy. It may be **amphimictic**, involving the union of gametes from different parents, or **automictic**, in which the gametes arise from the same parent. In either case, the uniting gametes may be whole organisms or nuclei only. Where whole organisms (gametes) unite, the union is termed **syngamy**. Where only nuclei unite, the process is called **conjugation**. Conjugation occurs only among ciliates, whereas syngamy occurs in all other groups where sexual reproduction occurs. Among protozoans the following types of sexual reproduction have been recognized: 1. Syngamy 2. Conjugation; and 3. Automixis.

1. Syngamy. It is the complete fusion of the **gametes** to form a **zygote** which gives rise to adult. The fused diploid nucleus is called **synkaryon**. Depending upon the nature of fusing gametes, a syngamy may be of the following types:

(*i*) Hologamy. It is the fusion of two mature individuals which do not produce gametes, *e.g.*, rhizopods and flagellates.

(*ii*) Paedogamy. It is the fusion of two young individuals.

(*iii*) Isogamy. The fusing gametes are similar in form and size and are known as **isogametes**. Their fusion is called **isogamy**. It is common in Foraminifera, Phytomonadida and Gregarinida.

In some isogamous species, the gametes are chemically differentiated, *i.e.*, a gamete belongs to a mating type and can unite only with a gamete from a different mating type, thus, ensuring **outbreeding**.

(*iv*) Anisogamy. The fusing gametes, dissimilar in form and size, are known as **anisogametes** and their union is called **anisogamy**. Usually the male gamete is small, active and may be flagellated and is known as **microgamete**. The female gamete is large, immotile and food-laden and is known as **macrogamete**. Anisogamy occurs in Sporozoa (*Plasmodium*) and Volvocida (= Phytomonadida; *e.g.*, *Volvox*).

(*v*) Microgamy. Certain foraminiferans and *Arcella* produce two types of **microgametes** which unite together to form synkaryon. This union is known as **microgamy**.

(*vi*) Macrogamy. In Heliozoia (*Actinophrys*), union occurs between two macrogametes of a species and it is known as **macrogamy**.

(*vii*) Autogamy. It is permanent fusion of two gametes from the same parent cell, *e.g.*, *Actinophrys*, *Actinosphaerium*.

(*viii*) Exogamy. Fusing gametes come from different parents.

Significance of syngamy.

1. Syngamy results in combination of two different lines of hereditary characters.
2. Syngamy increases the external differences in offsprings.
3. Syngamy renews the vigour which is lost by repeated binary fission.
4. The fusion of nuclei initiates the development of egg.

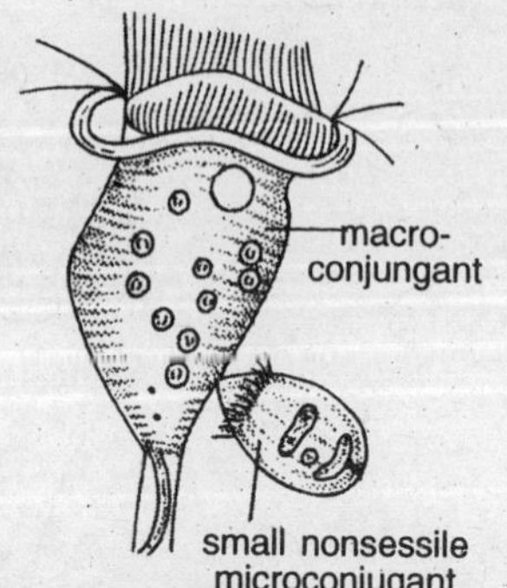

Fig. 9.4. Conjugation in *Vorticella*.

2. Conjugation. Conjugation is the temporary union of two individuals of different mating types of a species to facilitate exchange of the nuclear material. These individuals separate out after nuclear exchange. The pairing gametes (individuals) are called **conjugants** and may be either isogamous (equal) as in *Paramecium* or **anisogamous** (unequal) as in *Vorticella*. In case of *Vorticella*, the two members of a conjugating pair are of striking different sizes (Fig. 9.4). Such "dioecious" macro- and micro-conjugants represent an adaptation for conjugation in sessile species. The macroconjugant remains attached while the small bell of the microconjugant breaks free from its stalk and swims about. A synkaryon forms only in the macroconjugant from one gametic nucleus contributed by each conjugant. However, there is no separation after conjugation, and little migratory conjugant degenerates.

In the Suctoria, conjugation takes place between two attached individuals that happen to be located side by side.

Significance of conjugation. In fact, conjugation is not a mode of multiplication but instead is an important episode of reproduction which leads to:

1. New nuclear combinations and new hereditary combinations.
2. Rejuvenation and revival of the lost vigour and overcoming the period of depression resulting due to repeated binary fission. For example, it has been shown that some species of *Paramecium* can pass through only approximately 350 continuous asexual generations. If nuclear organization does not occur, the asexual line (clone) dies out, apparently because of decline in function of the macromolecules.
3. Reorganisation of a fresh macronucleus to accelerate the metabolic activities.

3. Automixis. It is a process of self-fertilization or **autogamy** that results in **homozygosis**. Automixis is the fusion of two gametic nuclei originating by the division of single nucleus of an individual. It is conjugation-like method of nuclear reorganization. The macronucleus degenerates and micronucleus divides a number of times to form eight or more nuclei. Two of these nuclei fuse to form a synkaryon; the others degenerate and disappear. The synkaryon then divides to form a new micronucleus and macronuclei as occurs in conjugation.

4. Endomixis. It is a type of nuclear reorganization which does not involve meiosis and fusion of nuclei (fertilization). Endomixis is a regular feature in the life history of *Paramecium aurelia*. It occurs in a single individual. As a result of nuclear changes a new macronucleus is formed from the micronucleus which accelerates metabolic activities and brings renewed vigour and vitality.

5. Hemixis. It is also a nuclear purification method of *Paramecium*. In hemixis, the macronucleus throws out fragments of different sizes which are absorbed in the cytoplasm. The remaining part of macronucleus functions as normal macronucleus.

9.3 REGENERATION

The capacity of regeneration (*i.e.*, replacement of lost parts) is wide-spread among free-living protozoa (*e.g.*, *Amoeba*), but is slight in parasitic protozoa. A proper part of cytoplasm and nucleus can regenerate the entire individual.

9.4 PARTHENOGENESIS

The production of young from unfertilized eggs is called **parthenogenesis**. In certain protozoa such as *Actinophrys* the gametes which fail to fertilize start their development parthenogenetically.

9.5 PLASMOGAMY

In certain rhizopods, two individuals unite partially and their nuclei do not fuse. As a result a multinucleated mass of protoplast is produced.

QUESTIONS

Long Answer Questions

1. Describe the various modes of reproduction found in protozoa studied by you.
2. Describe the various modes of reproduction found in *Euglena*, *Monocystis* and *Paramecium*.
3. Describe the different types of asexual reproduction in protozoa.

Short Answer Questions

1. Write short notes on:
   (*i*) budding
   (*ii*) binary fission
   (*iii*) syngamy
   (*iv*) automixis
   (*v*) conjugation
2. Distinguish between symmetrogenic and homothetogenic binary fission.

Very Short Answer Questions

1. Name the processes in *Paramecium* that correspond to cross-fertilization and self-fertilization in higher animals.

   Ans. Conjugation and autogamy.
2. Define syngamy.

   Ans. It is fusion of two sex cells or gametes with the formation of zygote; the zygote nucleus is called **synkaryon**. Syngamy may include hologamy, isogamy, anisogamy and autogamy depending upon the fusing gametes.

Fill in the Blanks

1. *Amoeba* reproduces by binary fission and ....
2. Homothetogenic division occurs in .... binary fission.
3. Symmetrogenic division occurs in .... binary fission.

Multiple Choice Questions

1. If two fusing nuclei are present in two different cells, then the process is known as
   (*a*) autogamy
   (*b*) paedogamy
   (*c*) hologamy
   (*d*) none of these
2. A sexual reproduction is of common occurrence in protozoa because
   (*a*) sexes have not been differentiated
   (*b*) it helps to multiply rapidly
   (*c*) majority of protozoans are parasites
   (*d*) large number can localise source of food efficiently
3. Which of the following is lacking in *Amoeba*?
   (*a*) sexual reproduction
   (*b*) multiple fission
   (*c*) binary fission
   (*d*) cyst formation

ANSWERS

Fill in the Blanks

1. Multiple fission 2. Transverse 3. Longitudinal.

Multiple Choice Questions

1. (*b*) 2. (*b*) 3. (*a*).

10

Protozoa and Disease

A large number of protozoa are parasitic to human beings and animals. Most parasites, both ectoparasites and endoparasites, are **pathogenic**, causing various types of diseases in human beings and animals. Most of these diseases are more common in the tropical and substropical regions of world.

The diseases occur as a result of the interaction between parasitic protozoans and the hosts (human beings and animals) under specific environmental conditions. This interaction is called **infection** which is always expressed in the form of **infectious disease**. All pathogenic parasites have the following common characteristics: 1. All are **obligatory parasites**, *i.e.*, they depend for their existence upon their hosts. 2. All show **host specificity**, *i.e.*, they harbour selective hosts. 3. They complete their life cycle in one or more than one host. 4. They secrete toxic substances which cause the particular disease. Parasitic species are found in all classes of protozoa and cause the following diseases in human beings and animals.

10.1 PROTOZOA AND HUMAN DISEASES

About 15 species of protozoa are pathogenic parasites of human beings causing some of the fatal diseases. They spend all or at least part of their life cycles in the body tissues, alimentary canal or blood of the host (human beings). Certain significant protozoan diseases of human beings are as following:

1. Amoebiasis (Amoebic dysentery). It is caused by *Entamoeba histolytica* which is an endoparasite and has world-wide geographical distribution but is common in tropics and subtropics. Infection of *E. histolytica* gerally takes place through drinking water. The trophozoite (= tissue-invading form) of the pathogen penetrates the walls of large intestine or colon, secretes histolytic enzymes and feeds upon its cells causing ulcers. These ulcers rupture and discharge mucus and blood along with numerous trophozoites which pass along with stools causing acute dysentery (ulcerative colitis). If the infection is allowed to continue the parasite may reach the liver, heart, lungs and brain where it causes abscesses which prove fatal.

There is no intermediate host in the life cycle of *E. histolytica*. Transmission of the parasite from person to person takes place through the tetranucleate cysts with contaminated food and drinks. These cysts are also transmitted by houseflies and cockroaches. Amoebiasis can be treated with drugs such as Emetine, Fumagillin, Metronidazole, etc.

2. Pyorrhoea (or Periodontitis). This bacterial disease is aggravated by *Entamoeba gingivalis* which lives in the tartar of teeth and abcesses of gum. Infection is caused directly by kissing or feeding in the same bowl.

Entamoeba coli is another species of *Entamoeba* which lives as a commensal in the large intestine of human beings and is non-pathogenic. It feeds on bacteria.

3. Diarrhoea. It is caused by a flagellate endoparasite, *Giardia intestinalis*. The parasite has world-wide geographical distribution and lives in small intestine (*i.e.*, duodenum and upper part of jejunum) of human beings. When conditions in the duodenum are unfavourable, encystment occurs and cysts are voided with faeces.

The parasite attaches itself with the help of its sucking disc on the convex surfaces of the epithelial cells in the intestine and may cause a disturbance of intestinal function. This leads to malabsorption of fat, persistent looseness of bowels and mild steatorrhoea (*i.e.*, passage of yellowish and greasy stools in which there is axcess of fat). The patient also suffers epigastric pain, abdominal discomfort, loss of appetite, headache and toxic effects (allergy). Transmission of *Giardia* to new host occurs via contaminated water and food. For the treatment of infection of *Giardia* the following drugs are effectively used: Atebrin, Chloroquine and Acranil.

4. Trypanosomiasis (African sleeping sickness). It is a disease of central nervous system. It is caused by the species of *Trypanosoma* (such as *T. gambiense*, *T. rhodesiense*, *T. cruzi*) which are flagellate parasites of blood plasma (in humans, the **definitive host**) and gut (in invertebrate, the **intermediate host**). The pathogen of this dreaded disease is transmitted among humans either by bite of blood-sucking insects or by bug's faeces. Following three species of *Trypanosoma* cause sleeping sickness in human beings:

(i) Trypanosoma gambiense is transmitted by tsetse fly, *Glossina palpalis*. It causes Gambian or Central African sleeping sickness.

(ii) Trypanosoma rhodesiense is transmitted by tsetse fly, *Glossina morsitans*. It causes Rhodesian or East African sleeping sickness. Infection of both of these pathogens are confined to those parts of Africa where tsetse flies (the vectors) are found. On infection by the *Trypanosoma*, initially fever is caused during which the parasite resides freely in the blood of host, where it consumes an enormous amount of glucose; then the parasite collects in the lymph glands, spleen and liver causing their enlargement; and finally they enter the cerebrospinal fluid causing **sleeping sickness**. Sleeping sickness results in coma and eventually death. In case of early and acute infection, sleeping sickness can be treated by drugs such as Suramin and Pentamidine. However, when infection of *Trypanosoma* reaches the central nervous system, arsenic containing drugs such as Tryparsamide, Melarsen, Melarsoprol (Mel B) and Trimelarsen are effectively used.

(iii) Trypanosoma cruzi is transmitted by faeces of a bug, called *Triatoma megista*. It causes **Chagas' disease** (a sleeping-sickness-like disease) or American trypanosomiasis in South and Central America. Symptoms of Chagas' disease include continuous fever, swollen lymph glands, spleen and liver. This disease can be cured by the drug Nitrofurazone.

5. Leishmaniasis. It is caused by the species of *Leishmania*, the flagellate intracellular endoparasite in the reticulo-endothelial cells of vertebrate host (*e.g.*, human beings) and the gut of an invertebrate host, the blood-sucking sandfly, *Phleobotomus*. The following three species of *Leishmania* cause different types of diseases in human beings:

(i) Leishmania donovani. It causes **kalaazar** or visceral leishmaniasis which is widespread and endemic in many places in India, China, Africa, Southern Europe, South America and Russia. In India kalaazar is common is Assam and Bengal along the coasts of the rivers Ganges and Brahmputra. In India, it is also endemic in Bihar, Orissa, Madras, and eastern U.P. In India, vector of *L. donovani* is the sandfly *Phlebotomus argentipes*.

Pathogen of kalaazar attacks the endothelial cells, bone marrow, liver, lymph glands and blood vessels of the spleen. These organs become enlarged. Symptoms of anaemia, high fever and pigmented dry and rough skin appear in the patient. Death of untreated patient occur due to profound immuno-suppressive effect inviting secondary bacterial infections. This disease has been effectively treated by following drugs: Urea stibamine, Aminostiburea, Neostibosan, etc.

(ii) Leishmania tropica. It causes **cutaneous leishmaniasis, Oriental sore (Tropical sore),** or **Delhi boil**. Its vector is sandfly, *Phlebotomus sergenti*, in India, Iraq and Persia. The pathogen is found along the shores of Mediterranean through Syria, Arabia, Mesopotamia, Iran to Central Asia and drier parts of Central and Western India. The infection is limited to a local lesion of the skin and subcutaneous tissues which turn into ulcerating wounds. Treatment of Delhi boil includes regular cleaning and dressing of the boils by pentavalent preparation of antimony.

(iii) Leishmania brasiliensis. It causes **espundia** or **muco-cutaneous leishmaniasis**. The pathogen is confined in Central and South America. It is intracellular parasite of macrophage cells of skin and mucous membrane of nose and buccal cavity. It is transmitted by the vector, the anthrophilic sandflies. Symptoms of espundia include multiple sores over large areas of the skin and oro-nasal mucosa. Ulceration in nasal cavities, mouth and pharynx is quite frequent. Treatment includes pentavalent preparation.

6. Trichomoniasis. It is caused by the species of flagellate *Trichomonas* which is pathogenic parasite of vertebrates and many invertebrates. Its life cycle includes only tropozoite phase; there is no cystic phase. Following three species of *Trichomonas* are pathogenic in human beings:

(i) Trichomonas hominis. It inhabits the ileo-caecal region of human beings and causes diarrhoea and dysentry.

(ii) Trichomonas tenax. It lives in the oral cavity, found around the tartar of the teeth and in pyorrhoeic sockets.

(iii) Trichomonas vaginalis. It lives in the urino-genital system of male and female human beings. In women, it inhabits the vagina causing **vaginitis**. The disease is characterized by inflammation of vaginal mucosa, burning sensation, annoying itch and abnormal greenish yellow discharges (called **leucorrhoea**). Transmission of pathogen is direct, taking place during coitus. Drugs containing arsenic and iodine and antibiotics such as Terramycin and Auromycin have proved effective in the treatment of this disease.

7. Malaria. This notorious and fatal disease of human beings is caused by the species of a sporozoan parasite, *Plasmodium*. It is transmitted through the bite of female *Anopheles* mosquito which is vector and secondary host. In human beings, which are the primary host, the parasite attacks and multiply asexually in liver cells and red blood cells. A toxic substance, called **hemozoin**, is released by the parasite during its asexual cycle in human beings and causes the symptoms and signs of malaria. Following four species of *Plasmodium* cause malaria in human beings:

(i) Plasmodium vivax causes benign malaria in which fever comes in every 48 hours.

(ii) Plasmodium malariae causes quartan malaria in which fever comes every 72 hours.

(iii) Plasmodium ovale causes mild tertian malaria in wich fever occurs on every 48 hours.

(iv) Plasmodium falciparum causes malignant subtertian malaria in which fever is more or less continuous (24-48 hours).

In malaria, the spleen is enlarged, erythrocytes decrease in number, pigment granules collect in the spleen and there is high temperature accompanied by chills and shivering. This fatal disease is treated by the following drugs: Quinine, Camoquine, Chloroquine, Plasmoquine, Resochin, etc.

8. Toxoplasmosis. It is caused by a sporozoan parasite, *Toxoplasma gondii* and common in Europe and United States. The pathogen is intracellular parasite of various organs and causes hydrocephalus and chorioretinitis (= inflammation of the choriod and retina of the eye).

9. Balantidial dysentery. It is caused by a ciliate parasite, *Balantidium coli*. The pathogen lives in the large intestine of human beings causing dysentery, diarrhoea and ulceration of colon. Infection is by cysts through contaminated water and food. This disease is effectively treated by drugs/ antibiotics such as Carbarsone, Diiodoquin and Oxytetracycline.

10.2 PROTOZOA AND ANIMAL DISEASES

Many parasitic protozoa of domestic and wild animals are pathogenic, causing a variety of diseases in them. Some of the pathogenic protozoa of animals are as follows:

1. *Entamoeba gingivalis*. It lives as parasite in the tartar around the teeth of dogs and cats.
2. *Trichomonas hominis*. It lives in the ileo-caecal region of monkeys, cats and dogs. It causes diarrhoea and dysentery. Infection occurs during encysted stage through contaminated food and water.
3. *Trichomonas gallinae*. It lives in the cells of epithelial lining of the oesophagus and crop of birds such as pigeons, turkeys, fowls, etc. It causes necrotic nodules in these organs.
4. *Trichomonas foetus*. It occurs in the urino-genital system of cattle, horse and sheep. It causes abortions, delayed conception and inflammation of perputial sacs. Transmission is direct during sexual intercourse.
5. *Histomonas maleagridis*. It lives in the caecum and liver of birds such as a turkeys and fowls. It causes black head disease in them.
6. *Trypanosoma*. Its following five species cause various diseases in domestic animals:
   (*i*) *Trypanosoma brucei*. It cause **nagana** disease in domestic animals. Its vector is *Glossina morsitans*.
   (*ii*) *Trypanosoma evansi*. It causes **surra** disease in horses, camels, dogs, mules and cattles. Its vectors are flies *Tabanus* and *Stomoxys*.
   (*iii*) *Trypanosoma equiperdum*. It causes **dourine** which is a disease of sex organs of horses. Disease is transmitted directly during copulation (*i.e.*, there is no vector).
   (*iv*) *Trypanosoma equinum*. It cause the disease, called **Mal de Caderas** in horses in South America. Its vector is a tabanid fly.
   (*v*) *Trypanosoma hippicum*. It causes the disease **Murina** or **Deren gadera** in horses and mules in Panama. Its vector is fly.
7. *Isospora hominis*. It lives as an endoparasite of epithelial cells of gut of dogs and cats. It causes **diarrhoea** and other abdominal discomforts.
8. *Eimeria* species. This sporozoan lives in the digestive tract of birds, sheep and cattle. It causes **dysentery** and **diarrhoea** in them. *Eimeria truncata* infests the kidneys of goose (bird) and often causes death of its host.
9. *Babesia bigemina*. It is a sporozoan parasite in erythrocytes (RBC) of cattle. It is a widespread fatal diseases of cattle in America and Australia and causes **texas fever** or **red water fever** destroying red blood cells. Its infection occurs through the bite of vector female tick, *Boophilus annulatus*.
10. *Babesia canis*. It is endoparasite of red blood corpuscles of dogs. It causes anaemia, jaundice and fever in dogs. Its infection occurs due to bite of ticks.
11. *Babesia equi*. It is an endoparasite of red blood corpuscles of horses. It causes anaemia, jaundice and paralysis of hindlimbs. Its infection occurs through the bites of ticks.
12. *Sarcocystis linedamanni*. It is a sporozoan inhabiting the striped muscle fibres of sheep's oesophagus. It causes degeneration of these muscles.
13. *Haenoproteus*. It lives in the RBC and the endothelial cells of blood vessels of birds.
14. *Plasmodium* species. Its many species live as parasite of RBC of lizards, birds and mammals and causes malaria in them.
15. *Nosema apis*. It is a sporozoan parasite which lives in the intestinal epithelium and Malpighian tubules of honey bees. It causes muscular disease in them.

QUESTIONS

Long Answer Questions

1. Write a short essay on protozoa and human diseases.
2. Describe briefly the pathogenic protozoa of human beings.
3. Write an essay on "Protozoa and disease."
4. Discuss the diseases produced in human beings by flagellate protozoans.
5. Name the diseases caused by protozoans in human beings. Give an account of any two of these.

Short Answer Questions

1. Write short notes on:
   (*i*) Sleeping sickness
   (*ii*) Giardiasis
   (*iii*) Amoebic dysentery
   (*iv*) Malaria

Very Short Answer Question

1. Mention four essential characters of a pathogen.
   Ans. 1. Obligate parasitism; 2. Host specificity; 3. Completion of life cycle of the parasite in one or two hosts; and 4. Secretion of toxin by the pathogen which results in the symptoms of the disease.
2. Where is the location of *Giardia intestinalis* in human host?
   Ans. Small intestine (*i.e.*, duodenum and jejunum).
3. Give the name of the vector of East African sleeping sickness.
   Ans. Tse-tse fly (*Glossina marsitans*).
4. Give generic name of sandfly, the vector of kalaazar.
   Ans. *Phleobotomus argentipes*.
5. What is the name of vector of pathogen of oriental sore?
   Ans. Sandfly, *Phleobotomus sergenti*.
6. Name the protozoan pathogen of honey bee.
   Ans. *Nosema apis*.
7. What is the primary host of malarial parasite?
   Ans. Female *Anopheles* mosquito.
8. The drug Resochin is used for treatment of which fatal human disease?
   Ans. Malaria.
9. Which protozoan causes leucorrhea of human females?
   Ans. Protozoan, *Trichomonas vaginalis*.

Fill in the Blanks

1. *Giardia* produces the disease .... (*Guwahati 1996*)
2. *Entamoeba histolytica* produces the disease .... (*Guwahati 1993*)
3. *Entamoeba gingivalis* may aggrevate ....
4. Amoebic dysentery is caused by .... (*Guwahati 1991*)
5. *Plasmodium* is an intracellular parasite in a .... host.
6. Vector of pathogen of sleeping sickness is ....
7. Parasitic protozoans lack .... and .... organs.
8. Two species of *Trypanosoma* cause sleeping sickness: *T.* .... and *T.* ....
9. Kalaazar is caused by ....
10. Transmission of protozoan .... occurs during coitus.

Multiple Choice Questions

1. Female *Anopheles* mosquito is a vector of a pathogen that causes
   (*a*) yellow fever
   (*b*) filariasis
   (*c*) malaria
   (*d*) dengue
2. Kalaazar is caused by
   (*a*) *Plasmodium vivax*
   (*b*) *Leishmania donovani*
   (*c*) *Trypanosoma gambiense*
   (*d*) *Wuchereria bancrofti*.

3. Kalaazar, a severe liver infection is transmitted mainly by the biting of
   (*a*) sand fly (*b*) mosquito
   (*c*) tse-tse fly (*d*) dog
4. *Leishmania tropica* causes
   (*a*) kalaazar (*b*) oriental sore
   (*c*) sleeping sickness (*d*) none of these
5. *Trypanosoma gambiensis* causes sleeping sickness in human beings. It infects the
   (*a*) blood (*b*) liver
   (*c*) brain (*d*) cerebro-spinal fluid
6. Amoebiasis is caused by
   (*a*) biting of mosquito (*b*) contaminated food and water
   (*c*) by virus (*d*) sexual transmission
7. Infective stage of *Entamoeba histolytica* for human beings is
   (*a*) uninucleate cyst (*b*) binucleate cyst
   (*c*) quadrinucleate cyst (*d*) trophozoite cyst
8. Motile elongated zygote of *Plasmodium* is found in
   (*a*) human RBCs (*b*) human liver
   (*c*) gut of mosquito (*d*) salivary gland of mosquito
9. Maurer's dots are observed in the erythrocytes of human beings if these are infected with
   (*a*) *Plasmodium vivax* (*b*) *P. ovale*
   (*c*) *P. falciparum* (*d*) *P. malariae*
10. Mepacrine and paludrine are effective drugs in treatment of
   (*a*) amoebic dysentery (*b*) malaria
   (*c*) diarrhoea (*d*) sleeping sickness
11. Bouts of malaria is due to
   (*a*) release of merozoites and their metabolites hemozoin
   (*b*) cryptozoites
   (*c*) sporozoites
   (*d*) micro-cryptomerozoites.
12. Who for the first time discovered *Plasmodium vivax* in erythrocytes of human beings?
   (*a*) Charles Laveran (*b*) Ronald Ross
   (*c*) Grass (*d*) Stephans
13. Select the matching pair
   (*a*) *Plasmodium* – Pyorrhoea (*b*) *Leishmania* – Dysentery
   (*c*) *Giardia* – Disrrhoea (*d*) *Trypanosoma* – Kalaazar
14. *Histomonas* causes this disease in poultry
   (*a*) coryza (*b*) coccidiosis
   (*c*) black head (*d*) chronic ashthma
15. Which of the following disease continues to be a major health problem in India?
   (*a*) malaria (*b*) polio
   (*c*) plague (*d*) tuber culosis

ANSWERS

Fill in the Blanks

1. Diarrhoea
2. Amoebic dysentery
3. Periodontitis
4. *Entamoeba histolytica*
5. Vertebrate
6. *Glossina palpalis*
7. Locomotory, osmoregulatory
8. *gambiense, rhodesiense*
9. *Leishmania donovani*
10. *Trichomonas vaginalis*.

Multiple Choice Questions

1. (*c*) **2.** (*b*) **3.** (*a*) **4.** (*b*) **5.** (*d*) **6.** (*b*)
7. (*c*) **8.** (*c*) **9.** (*c*) **10.** (*b*) **11.** (*a*) **12.** (*a*)
13. (*c*) **14.** (*c*) **15.** (*a*).

11

Phylum Porifera (Characters, Classification and Types)

Phylum Porifera includes animals having pores in the body wall. The name porifera (L., *porous* = a pore; *ferre* = to bear) was coined by **Robert Grant** (1836) for the plant-like sessile (permanently attached, hence nonmotile) multicellular creatures, called **sponges**. **Aristotle**, **Pliny** and other ancient naturalist considered sponges to be plants. In fact, it was not until 1765, when internal water currents and movements of the oscula were first observed by **Ellis**, that the animal nature of sponges became clearly established. The morphology and physiology of sponges were first worked out by **Grant** (1875). **Huxley** (1875) and **Solas** (1884) suggested their complete separation from other Metazoa on the basis of numerous peculiar features of sponges. **Solas** created an isolated branch of Metozoa called **Parazoa** for the sponges. In evolutionary terms, phylum Porifera is considered as "**dead-end" phylum** since it has not given rise to any group of organisms.

11.1 GENERAL CHARACTERS

Poriferans are sessile, suspension-feeding, multicellular animals that utilise flagellated cells called **choanocytes** to circulate water through a unique system of water canals. (**Note.** Suspension feeding is the removal of suspended food particles – bacteria, phytoplankton, zooplankton and some detritus – from the surrounding medium by some sort of capture, trapping or filteration mechanism). They are metazoa lacking true embryological germ layering. Not only true tissues are absent, but most of the body cells are **totipotent** – they are capable of changing form and function. Despite being multicellular, poriferans function like organisms at the unicellular grade of complexity, *i.e.*, their nutrition, cellular organization, gas exchange, and response to environmental stimuli are all protist like.

Box 11.1

Two unique organizational attributes *define* sponges and have played major roles in poriferan success: the **water current channels** or **aquiferous system** (and its **choanocytes**), and the highly **totipotent** nature of sponge cells (**Brusca** and **Brusca**, 2003). A totipotent cell is capable of developing into a complete organism or differentiation into any of its cells or tissues.

General Characters

1. Multicellular animals having **celiular grade of construction**. The cells are arranged in loose aggregations, hardly forming any tissue, organs or systems.

2. There occurs some cellular differentiation or **division of labour**, but no tissue organization. Unlike epithelium of most other animals, a basal lamina and intracellular junctions are absent.
3. All are aquatic, mostly marine, some are freshwater animals.
4. Solitary or colonial, all sessile—attached to various objects in water. (**Note.** According to **Bond** (1997, 1998) some sponges can move from one place to another).
5. Shape and size of body of sponges are variable. Size ranges between 1cm to 1m. Shape may be cylindrical, vase-shaped, cup-shaped, spherical or disc-shaped.
6. Body is either asymmetrical or superficially radially symmetrical.
7. Sponge's body is **diploblastic** with an outer ectodermal layer of flattened pinacocyte cells (called **pinacoderm**) and an inner endodermal layer of collared flagellate cells or choanocytes (called **choanoderm**). In between is a mesoglea, mesenchyme or **mesohyl** which is gelatinous and proteinaceous matrix containing collagen fibres, skeletal material and amoeboid cells (which are migratory cells of both layers). (For contradiction of these facts see Box 11.2).

Box 11.2. Sponges are monoblastic—they lack an endoderm

According to current reports body of sponges is **monoblastic**, consisting of a layer of cells which lines a non-living matrix (the mesohyl). According to them, the spongocoel and choanocyte-lining are not homologous to the gut of other metazoan animals. The choanocytes are not endodermal derivatives. **Indeed there is no endoderm in sponges**. Sponges are gutless animals and the gutless condition is primary (**Ruppert** and **Barnes**, 1994).

8. Sponge's architecture is unique, being constructed around a system of water canals, called **canal system** or **aquiferous system**, an arrangement that is correlated with sponge's sessility.
9. Body surface of sponge is perforated by numerous pores, called **ostia**, serving for the inflow of water. By collective lashing, the flagellated collar cells draw water through these pores. The water currents pass through ostia into the chambers (some of which are lined by choanocytes) and finally in the central cavity called **atrium** or **spongocoel**. From spongocoel water currents escape out through a single large exhalent opening, the **osculum**. Sponge cells obtain food and oxygen from these water current, and release carbon dioxide, excretory substance (ammonia) and reproductive products (sperm, ova) into it.

Box 11.3

The one-cell thick choanoderm may remain simple and continuous (the **asconoid** condition), or it may become folded (the **syconoid** condition), or it may become greatly subdivided into separate flagellated chambers (the **leuconoid** condition) (Fig. 11.1).

10. Sponges contain a single body cavity (*i.e.*, spongocoel). They lack gut (*i.e.*, digestive cavity) and coelom.
11. The body is strengthened by an internal skeleton of calcareous or siliceous **spicules** or horny **spongin** fibres.
12. Digestion is **intracellular** in food vacuoles as in protozoa.
13. Sponges are **ammonotelic**; their chief excretory waste is **ammonia**.
14. The sensory cells and nerve cells are absent, but each cell is directly stimulated and also transmits sensations to other cells.
15. Amoebocytes are the workhorses of sponge economy. They function in internal transport of food and for a variety of other metabolic activities.

16. Asexual reproduction occurs by **fragmentation** (regeneration), **budding**, **gemmules** and **reduction bodies**.
17. All are **hermaphrodite**, most are **protandrous**. Sexual reproduction is carried by the formation of sperm and ova. **Fertilization** is internal but cross-fertilization occurs as a rule.
18. Cleavage is **holoblastic**, development indirect through a free-swimming ciliated larva called **diploblastula, amphiblastula** or **parenchymula larva**. The larva is usually at the blastula stage of development.

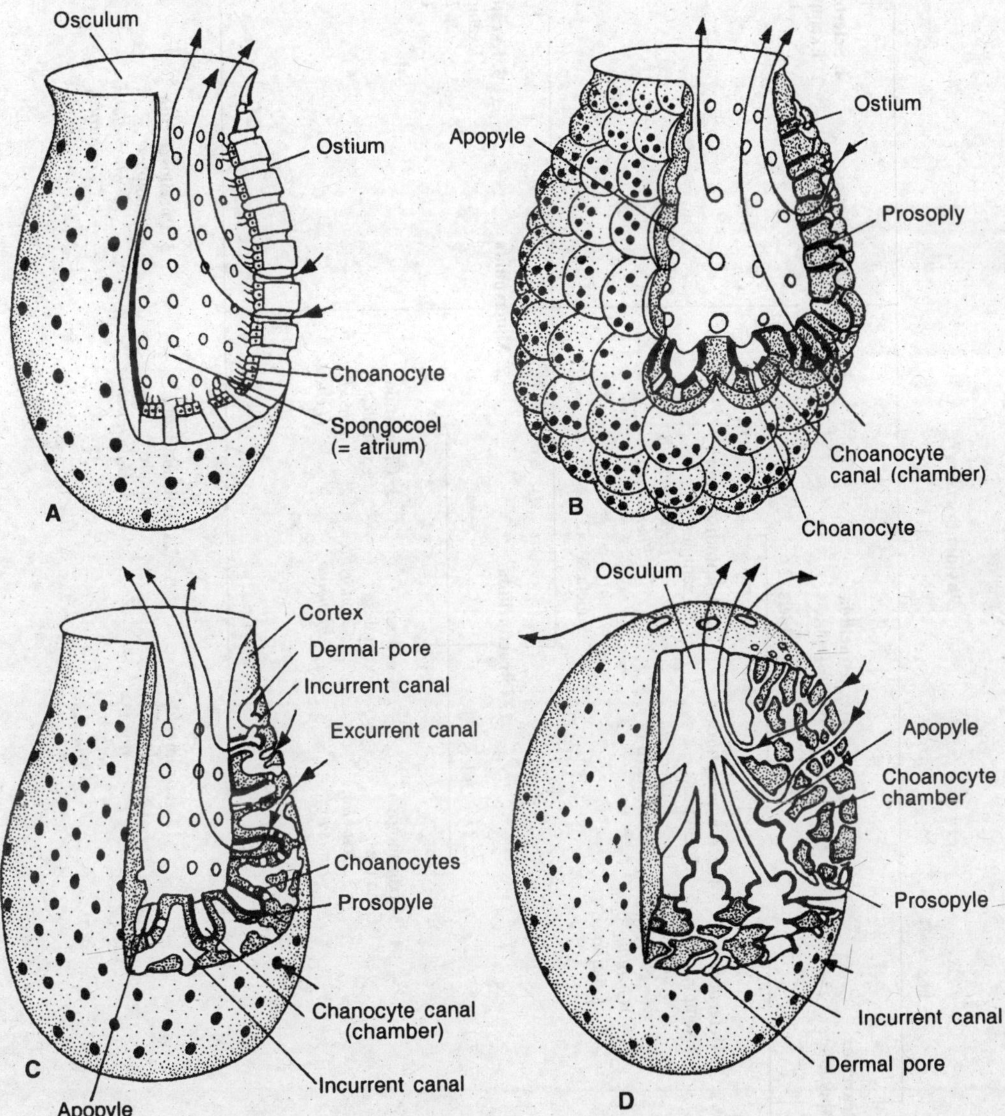

Fig. 11.1. Body complexity in sponges (arrows indicate flow of water). A—The asconoid condition; B—A simple syconoid condition; C—A complex syconoid condition with cortical growth; D—A leuconoid condition (after Brusca and Brusca, 2003).

11.2 OUTLINE CLASSIFICATION OF PHYLUM PORIFERA

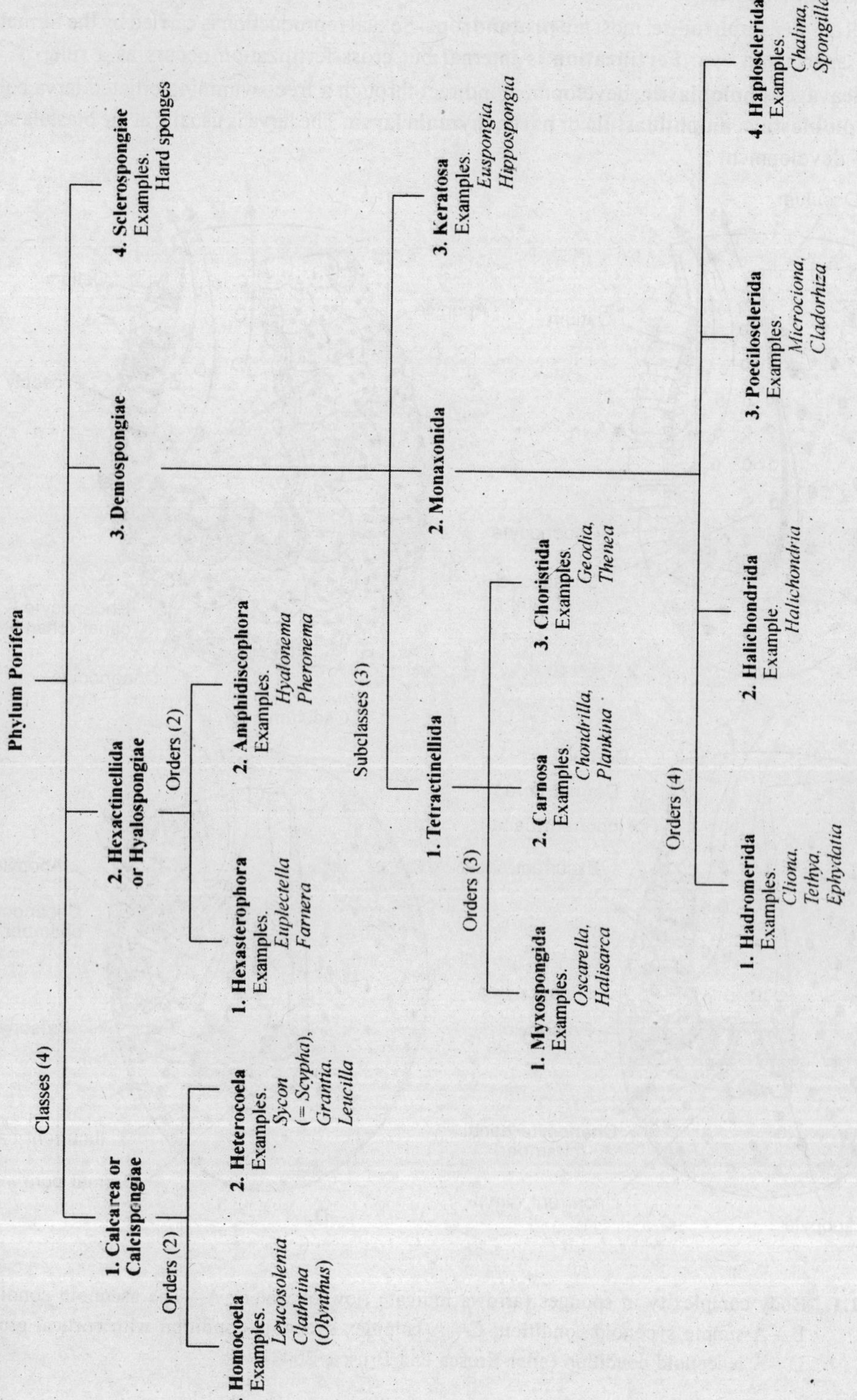

11.3 CLASSIFICATION

Sponges are the most primitive multicellular, heterotrophic, pore-bearing animals. Adult sponges are always attached (sessile) and motionless. Except for about 150 freshwater species, about 5000 described species of sponges are marine animals. They abound in all seas, whereever rocks, shells, submerged timbers, or corals provide a suitable substratum. Some species even live on soft sand or mud bottoms. Most sponges prefer relatively shallow water, but some groups, including most glass sponges, live in deep water.

On the basis of different types of spicules, sponges are classified into four classes (**Berquist**, 1978): Calcarea, Hexactinellida, Desmospongiae and Sclerospongiae. Many workers consider this classification plan as artificial one; in consequence, according to **Meglitsch** and **Schram** (1991), phylum Porifera includes two subphyla: 1. **Cellularia** (cellular organization; epidermis with pinacocytes; mesoglea of collagenous matrix, amoeboid cells and skeletal elements, various choanocyte, distinct and nucleated; myocytes present). 2. **Symplasma** (symplastic or syncytial organization; epidermis lacking pinacocytes, mesoglea as thin mesolamella, collar elements enucleate connected in reticular networks to nuclear choanoblasts, has secondary reticulum about choanoderm). Subphylum Cellularia includes 3 classes: Calcarea, Demospongie and Sclerospongiale. Subphylum Symplasma includes single class Hexactinellida.

Class I. Calcarea or Calcispongiae

1. Calcareous sponges.
2. Small sized sponges, less than 15 cm in height.
3. Skeleton composed of discrete calcareous spicules of calcium carbonate. Spongin fibre absent.
4. Spicules are **monaxon** or **tetraxon**. Tetraxon spicules lose one ray to become triradiate.
5. Spicules are differentiated into microscleres and megascleres.
6. Body organization is **asconoid**, **syconoid** and **leuconoid** type (Fig. 11.1).
7. Body is vase-like cylindrical with bristly outer surface.
8. Choanocytes are distinct, large and nucleated. They line the spongocoel in primitive forms and restricted to flagellated chambers in others.
9. They are solitary or colonial, marine occurring in shallow waters in all oceans.

Class II. Hexactinellida or Hyalospongiae

1. They are commoniy called **glass sponges**.
2. Their skeleton is composed of six-rayed (hexactine), siliceous spicules or triaxon spicules. In some hexactinellids the spicules are fused to form a lattice-like skeleton.
3. Epidermis lacks pinacocytes.
4. Mesoglea exists as thin mesolamella having collagen and spongin.
5. Spongocoel is well developed and opens to exterior by a single **osculum.** Osculum is sometimes covered by a **sieve plate**–a grate-like covering formed from spicules.
6. Glass sponges have unique structural plan. They have symplastic or **syncytial organization**. Thus, all surfaces exposed to water are covered not by pinacoderm but by a framework of syncytial strands (called **trabecular syncytium**) through which long spicules may project. Another syncytium (called **choanocyncytium**), containing flagella with collars, lines the flagellated chambers. These collar bodies lose their nuclei after being formed (Fig. 11.2).
7. Glass sponges are well-formed, vase-to-funnel-shaped individuals, attached to the substrate by a tuft of root spicules.

8. They are found in deep tropical seas (between depths of 200 to 1000 m).
9. Glass sponge may grow upto one metre.
10. Recent histological insight in glass sponges has been gained by work on *Rhabdocalyptus dawsoni* which lives in shallower waters (**Mackie, G.O.**, and **C.L. Singla**, 1983).

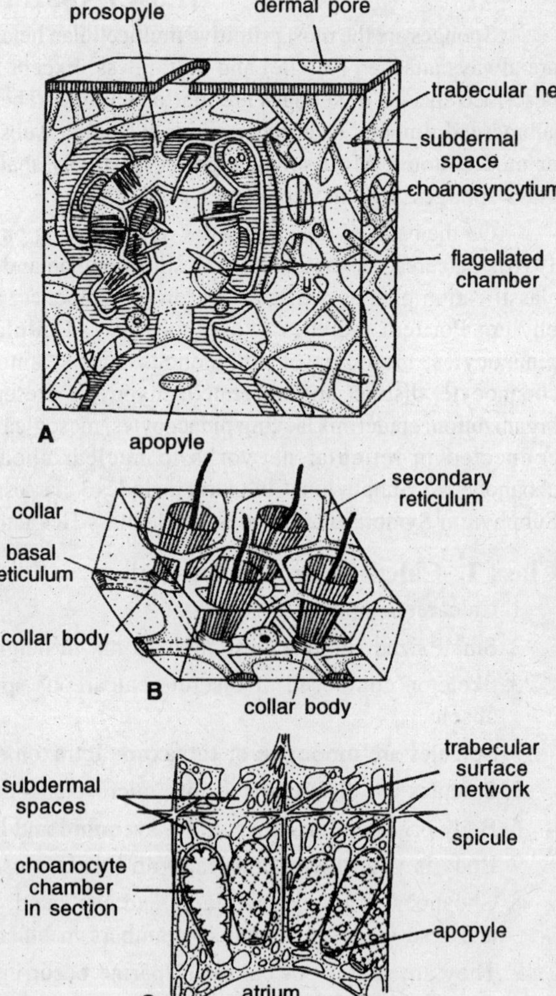

Fig. 11.2. Internal structure of glass sponge. A—Body plan of idealized hexactinellid. Water enters dermal pore, flows around trabecular net, goes through prosopyl in flagellated chambers, exits through apopyle into excurrent lacunae. B—Details of wall of flagellated chamber having secondary reticulum (internal part of trabecular system) around collars. C—Choanocyte chambers within the body wall of thc *Euplectella*.

Class III. Demospongiae

1. This large-sized class includes about 90 per cent of sponge species.
2. Generally marine, a few freshwater forms. Marine forms range in distribution from shallow water to great depth. Body shape is irregular.
3. Colouration is brilliant because of pigment granules located in the amoebocytes.
4. Skeleton may consist of siliceous spicules or spongin fibres or a combination of both. The genus, *Oscarella* is unique in lacking both spongin and spicule.
5. Siliceous skeleton of Demospongiae differ from the Hexactinellida in that their larger spicules are monaxon or tetraxon, never triaxons (hexactines).
6. Generally with small, micro-scleres, **tissue spicules** and large, megascleres, **skeletal spicules**.
7. Body organization is **leuconoid** type.

Class IV. Sclerospongiae

1. Demospongiae-like ancient sponges, called **coralline sponges.**
2. Skeleton composed of siliceous and aragonitic (calcite) spicules, and spongin fibres.
3. Leuconoid sponges.
4. Found in caves, tunnels, crevices and deep waters overhanging on coral reefs.

Examples. Hard sponges.

Box 11.4

Brusca and **Brusca** (2003) have suggested following classification of sponges:

Phylum Porifera

1. **Class Calcarea**
   (*i*) **Subclass Calcinea**, *e.g.*, *Clathrina, Dendya, Soleniscus.*
   (*ii*) **Subclass Calcaronea**, *e.g., Grantia, Leucilla, Leucosolenia, Scypha* (*Sycon*).
2. **Class Hexactinellida**
   (*i*) **Subclass Amphidiscophora**, *e.g., Hyalonema, Monorhphis, Pheronema.*
   (*ii*) **Subclass Hexasterophora**, *e.g., Euplectella, Hexactinella, Rosella, Sympagella*
3. **Class Demospongiae**
   (*i*) **Subclass Homoscleromorpha**, *e.g., Corticium, Oscarella, Plakia.*
   (*ii*) **Tetractinomorpha**, *e.g., Chondrilla, Cliona, Tethya, Tetilla, Merila*
   (*iii*) **Ceractinomorpha**, *e.g., Clathria, Spongia, Spongilla, Tedania.*

The subclass ceractinomorpha now contain some sponges previously assigned to the class sclerospongiae.

Class I. Calcarea or Calcispongiae

It is divided into following **two** orders:

Order 1. Homocoela

1. Sponges with radially symmetrical cylindrical bodies.
2. Body wall is thin and not folded. The spongocoel is lined with choanocytes.
3. Ascon type canal system.

Examples. *Leucosolenia, Clathrina.*

Order 2. Heterocoela

1. Sponges with vase-shaped body.
2. Body wall thick and folded.
3. Choanocytes are restricted to flagellated chambers or radial canals.
4. Spongocoel is lined by pinacocytes (pinacoderm).

Examples. *Scypha* (*Sycon*), *Grantia, Leucilla.*

Class II. Hexactinellida or Hyalospongiae

It is divided into following **two** orders:

Order 1. Hexasterophora

1. Spicules are six-rayed and star-shaped (hexasters).
2. Amphidiscs are absent.

Examples. *Euplectella* (Venus's flower basket).

Order 2. Amphidiscophora

1. Spicules in the form of "anchor-like" amphidiscs.
2. Hexasters absent.

Examples. *Hyalonema* (glassrope sponge), *Pheronema* (bowl sponge).

Class III. Demospongiae

This large-sized class is divided into following **three** subclasses:

Subclass 1. Tetractinellida

1. Skeleton of siliceous tetraxon (four-rayed) spicules. (These spicules are absent in order Myxospongida).
2. Spongin is absent.

Subclass 2. Monaxonida

1. Skeleton of siliceous monaxon spicules.
2. Spongin fibre may be present or absent.
3. Micro-and megascleres may be present or absent.

Subclass 3. Keratosa

1. Spicules absent.
2. Skeleton contains horny spongin fibres.
3. Includes horny sponges.

Examples. *Euspongia* (bath sponge), *Hippospongia* (horse sponge), *Hircinia* (horny sponge).

Subclass 1. Tetractinellida

It is divided into following **three** orders.

Order 1. Myxospongida

1. Simple structure.
2. Spicules absent.

Examples. *Oscarella*, *Halisarca*.

Order 2. Carnosa (or Homosclerophora)

1. Simple structure.
2. Microscleres (tissue spicules) and megascleres (skeletal spicule) are similar in shape and size.

Examples. *Alina*, *Plankina*, *Chondrilla*.

Order 3. Choristida

1. Microscleres and megascleres differ in size and shape.

Examples. *Geodia*, *Thenea*, *Stelletta*.

Subclass 2. Monaxonida

It is divided into following **four** orders.

Order 1. Hadromerida

1. Microscleres absent or aster-shaped.
2. Monaxon megascleres are in the form of tylostyles (in which both ends of the spicule are knob-like).
3. Spongin absent.

Examples. *Cliona* (boring sponge), *Tethya*, *Poterion* (Neptune's goblet), *Ephydatia* (a freshwater sponge).

Order 2. Halichondrida

1. Microscleres absent.
2. Very little spongin fibre.
3. Monaxon megascleres are often of two types, *i.e.*, **monactines and diactines**.

Example. *Halichondria* (crumb of bread sponge).

Order 3. Poecilosclerida

1. Microscleres include **sigmas** (C or S-shaped), **toxas** (bowl-shaped) and **chelas** (with recurved hooks or plates).
2. Monaxon megascleres are of two types, one type is found in the ectoderm and another type in the choanocyte layer.

Example. *Cladorhiza*.

Order 4. Haplosclerida

1. Monaxon megascleres of only one kind (diactinal or two-rayed).
2. Microscleres absent.
3. Spongin fibres present.

Examples. *Chalina* (mermaid's gloves or deadman's finger), *Spongilla*.

11.4 TYPE STUDIES

1. *EUPLECTELLA*

Classification

Phylum	–	**Porifera**
Class	–	**Hexactinellida**
Order	–	**Hexasterophora**
Genus	–	***Euplectella***
Species	–	***aspergillum***

Euplectella aspergillum (Fig. 11.3) is commonly known as **Venus's flower basket**. It is a beautiful white marine glass sponge which is found attached by its siliceous **root spicules** to the bottom of deep seas around Philippines and Japan. *Euplectella* is a solitary animal.

The size of individuals varies from 15–30 cm in length. The body is cylindrical and slightly curved (Its curved structure is an adaptation to the slow, constant water current at great depth of sea). The entire body is supported by a skeleton of six-rayed siliceous spicules which are bound together by siliceous cement so that they form a filmy lattice-like skeleton. The canal system is of syconoid type. The **parietal gaps** situated in between the lattice lead into the paragastric cavity (spongocoel). Osculum is covered with an **oscular sieve** formed of fused spicules. There is no epidermis of pinacocytes. Flagellated chambers lie radially in the sponge wall (Fig. 11.2C).

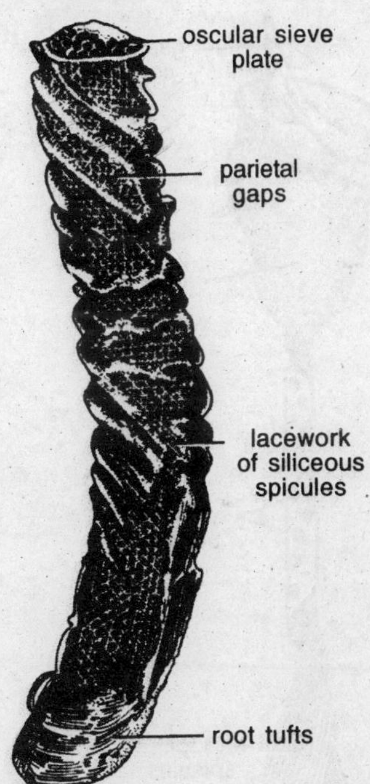

Fig. 11.3. *Euplectella.*

Venus's flower basket displays an interesting commensal relationship with certain species of shrimp (*Spongicola venusta*). A young male and a young female shrimp enter the spongocoel along with water current where they grow and cannot escape through the sieve plate covering the osculum. Thus, their entire life is spent within the sponge, where they feed on plankton brought in by the sponge's water currents. The partners remain unseparated till one of them dies. Such Venus's flower baskets were once used by Japanese as a wedding gift since each carriers felicitation of decapod couple bound in permanent marital union. The Japanese call it "*together unto old age unto same grave.*"

2. *HYALONEMA*

Classification

Phylum	–	**Porifera**
Class	–	**Hexactinellida**
Order	–	**Amphidiscophora**
Genus	–	***Hyalonema***

It is one of the most common hexactinellid sponge, with many species. *Hyalonema* is commonly called **glass rope sponge**. It is marine and found in the sea (Atlantic and Pacific near New England) at a depth of 10 to 15 metre.

Hyalonema has a round or oval body with a simple spirally-twisted rope-like root-tuft. The root-tuft is long, stalk-like and is made up of a bundle of very long glassy silicious monaxon spicules. The sponge remains attached to the substratum (sea-bottom) by the root-tuft. The root-tuft extends into the body as an axis or **columella** and often project out of the body in the form of **gastral cone**.

The middle part of the columella commonly has attached symbiotic sea anemones (of genus *Epizoanthus* or *Polythoe*). The skeleton of *Hyalonema* contains small branching five-rayed or six-rayed spicules and large and small amphidiscs. Excurrent canal opens on the gastral cone surface, spongocoel is absent.

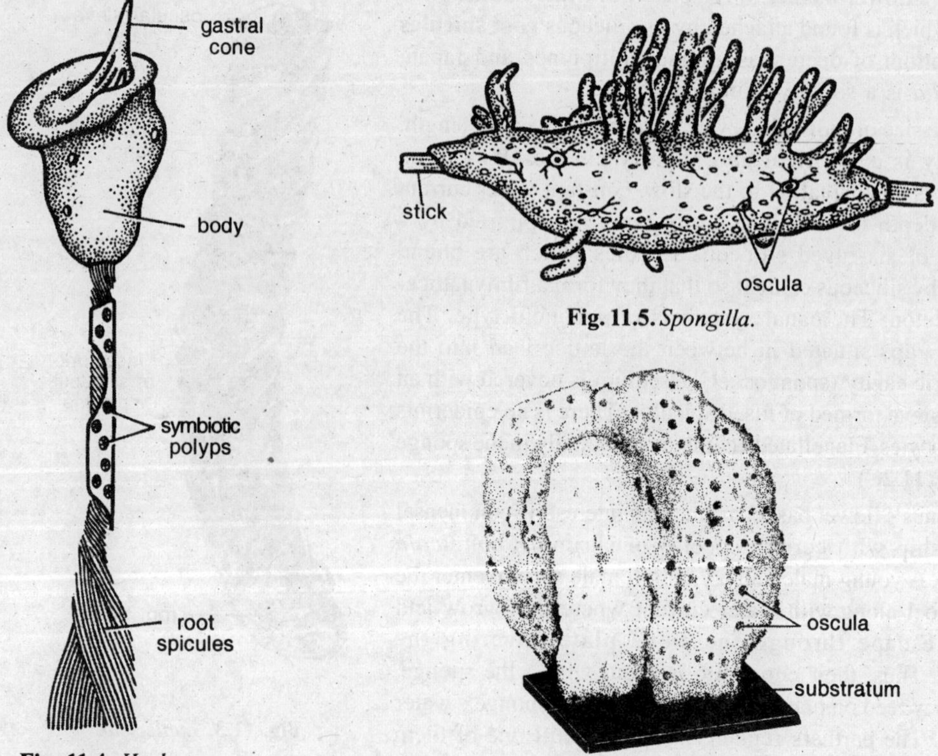

Fig. 11.4. *Hyalonema.*

Fig. 11.5. *Spongilla.*

Fig. 11.6. *Euspongia.*

3. *SPONGILLA*

Classification

Phylum	–	**Porifera**
Class	–	**Demospongiae**
Subclass	–	**Monaxonida**
Order	–	**Haplosclerina**
Genus	–	***Spongilla***

It is a well known freshwater sponge which is colonial and is abundantly found in ponds, lakes, slow streams where the water is not turbid. Profusely branched colony of *Spongilla* grows on submerged plants, sticks and stones; it shows an encrusting growth pattern. Colony of *Spongilla* is green in colour because of the presence of symbiotic *Zoochlorellae* (a green algae) in various cells of sponge such as pinacocytes, choanocytes, amoebocytes, etc.

The body wall of spongilla consists of a very thin **dermal membrane** scattered with numerous **dermal pores** and **oscula**. The canal system is **rhagon type**. The skeleton contains siliceous spicules that form a network of monaxon spicules, microscleres and megascleres embedded in spongin. Asexual reproduction takes place by gemmules. Sexual reproduction includes ciliated free-swimming larva, called **parenchymella or parenchymula**.

4. *EUSPONGIA*

Classification

Phylum	–	**Porifera**
Class	–	**Demospongiae**
Subclass	–	**Keratosa**
Genus	–	***Euspongia***

Euspongia is the common **bath sponge**. It is a horny sponge which is found attached with rocky bottom in warm shallow sea waters. It is found in plenty in the Mediterranean sea, the Gulf of Mexico and West Indies. Body is rounded and large and light brown in colour. Body surface contain many oscula.

Inside the bath sponge are numerous inhalent canals, flagellated chambers, exhalant canals, and spaces; Skeleton is made of spongin fibres only. Spongin fibres form a network, the thickest fibres run radially and each ends in small projection, called **conulus**. Body contains many conuli.

QUESTIONS

Long Answer Questions

1. Classify phylum Porifera and give diagnostic or general characters of the phylum, and examples of its various classes.
2. Enumerate the distinguishing features of phylum Porifera. Classify it giving chief characters and examples of each order.

(*Rohilkhand 1994, 2001; Kanpur 1994; Gorakhpur 1995*)

3. Give an outline classification of phylum Porifera.

Short Answer Questions

1. Classify and comment on the following:

(*i*) *Euplectella*

(*ii*) *Hyalonema*

(*iii*) *Spongilla*
(*iv*) *Euspongia*

Very Short Answer Questions

1. Mention the scientific name of Venus's flower basket.
   Ans. *Euplectella.*
2. Give the name of a sponge having commercial value.
   Ans. *Spongia* (= *Euspongia*).
3. On which main characters is the classification of sponges based.
   Ans. Skeleton and canal system.
4. What is canal system?
   Ans. It is a system of intercommunicating cavities found in a sponge body.
5. How do fresh water sponges manage to survive desiccation in summer?
   Ans. By the formation of resistant gemmules.
6. Give the order and class of *Hyalonema*.
   Ans. Amphidiscophora (order), Hexactinillida (class).
7. Cite two common examples of class Demospongiae.
   Ans. *Euspongia, Spongilla.*
8. Name the cell which secretes spongin fibres.
   Ans. Spongioblasts.
9. What is choanocyte?
   Ans. Any of the collar cells or choanate, flagellate cells lining the cavities of sponge. Choanocytes drive water through canals and chambers constituting the canal system or **aquiferous system** (**Brusca** and **Brusca**, 2003).
10. Scleroblasts are of two types. Name them.
    Ans. Calcoblasts, silicoblasts.
11. Name any freshwater sponge. Also give its order and class.
    Ans. *Spongilla*, *Haplosclerida* (order), *Demospongiae* (class).
12. What are gemmules?
    Ans. Asexual reproductive bodies.
13. Name the two pores connecting the canals of sponge internally.
    Ans. 1. **Prosopyle** which connects the incurrent canal to radial canal.
    2. **Apopyle** which connects the radial canal to excurrent canal or spongocoel.

Fill in the Blanks

1. *Spongilla* is an example of .... sponge.
2. .... is the costliest Turkish bath sponge.
3. Sponges have .... level of organisation.
4. Asexual reproductive bodies of sponges are ....
5. Life cycle of sponges includes a free-swimming larva named .... or ....
6. Phylum Porifera is divided into three classes Calcarea, Hexactinellida and ....
7. *Leucosolenia* belongs to order .... of class ....
8. A common heterocoel sponge is ....
9. The amoeboid cells that secrete silicious spicules are termed ....
10. .... secrete spongin fibres.
11. .... in sponges form the basis of classification in Porifera.

Multiple Choice Questions

1. Body organization in sponges is of
   (*a*) protoplasmic grade (*b*) tissue grade
   (c) cellular grade (*d*) organ grade

2. In sponges food is ingested by
   (*a*) scleroblasts (*b*) choanocytes
   (*c*) porocytes (*d*) pinacocytes
3. The zoological name of common bath sponge is
   (*a*) *Leucosolenia* (*b*) *Sycon*
   (*c*) *Spongia* (*d*) *Euplectella*
4. Which one of the following cells maintains current of water in the canal system of sponge?
   (*a*) porocyte (*b*) myocyte
   (*c*) chromocyte (*d*) choanocyte
5. Incurrent canals communicate with the radial canals by
   (*a*) apopyles (*b*) prosopyles
   (*c*) dermal ostia (*d*) gastral ostia
6. Glass sponges belong to the class
   (*a*) demospongiae (*b*) calcarea
   (*c*) tetractinellids (*d*) hexactinellida
7. Which of the following belongs to class Hexactinellida?
   (*a*) *Sycon* (*b*) *Euplectella*
   (*c*) *Spongilla* (*d*) *Leucosolenia*
8. Ability of sponges to replace the lost body part is
   (*a*) growth (*b*) autotomy
   (*c*) reproduction (*d*) regeneration
9. Glass rope sponge is the common name of
   (*a*) *Hyalonema* (*b*) *Sycon*
   (*c*) *Euspongia* (*d*) *Leucoslenia*
10. Gemmules play a role in
   (*a*) perennation (*b*) dispersal
   (*c*) reproduction (*d*) all of these
11. Who described the animal nature of Porifera?
   (*a*) Ellias (*b*) Ross
   (*c*) Robert Grant (*d*) Huxley
12. Parazoology is the study of
   (*a*) Protozoa (*b*) Porifera
   (*c*) Cnidaria (*d*) parasite

ANSWERS

Fill in the Blanks

1. Fresh water
2. *Spongia officianalis*
3. Cellular
4. Gemmules
5. Amphiblastula, parenchymula
6. Demospongiae
7. Homocoela, Calcarea
8. *Sycon*
9. Silicoblasts
10. Spongioblasts
11. Skeleton

Multiple Choice Questions

1. (*c*) **2.** (*b*) **3.** (*c*) **4.** (*d*) **5.** (*b*) **6.** (*d*)
7. (*b*) **8.** (*d*) **9.** (*a*) **10.** (*d*) **11.** (*a*) **12.** (*b*).

12

Leucosolenia (Phylum Porifera)

Biology of sponges can be best understood by the study of a simple asconoid type. *Leucosolenia* (Gr., *leukos* = white; *solen* = type) is one of the simplest asconoid sponges.

12.1 CLASSIFICATION AND OCCURRENCE

Phylum	–	**Porifera**
Class	–	**Calcarea**
Order	–	**Homocoela**
Genus	–	***Leucosolenia***

Leucosolenia is a small, delicate, branching, sessile, colonial marine sponge. It is found growing in shallow water below low tide mark on sea shore rocks, boulders and jetties. It does not found in calm water but is found abundantly where wave action is intense. *Leucosolenia* is very sensitive to external conditions and will die if removed from its habitat. About 100 species of this sponge occur all over the world. Some common species of *Leucosolenia* are *L. botryoides*, *L. complicata* and *L. variabilis*.

12.2 EXTERNAL FEATURES

The colony of *Leucosolenia* is whitish or yellowish in colour. It consists of radially symmetrical vase-shaped, vertical, hollow and tube-like units or individuals, united at their bases by irregular horizontal tubes and attached to the substratum through adhesive discs (Fig. 12.1). The upright cylindrical individuals may reach upto 25mm in height.

The surface of each tube-like unit or individual bears minute pores called **ostia** (singular ostium) or **incurrent pores.** The pores leads into a central spacious cavity, called **paragastric cavity**, **spongocoel** or **atrium**. Atrium of sponge opens to the outside via a single large circular opening, called **osculum**. Osculum is located at the free terminal end of the tube.

12.3 BODY WALL (MICROSCOPIC ORGANIZATION OR HISTOLOGY)

The body wall is thin and encloses the atrium or spongocoel. It consists of two thin cellular layers, the outer **pinacoderm** and inner **choanoderm**. Both layers are separated by jelly-like non-cellular layer of **mesohyl.**

1. Pinacoderm. The body of *Leucosolenia* is covered externally by **epidermis** or **pinacoderm**. This is a single-cell thick layer of thin, flat, polygonal cells called **pinacocytes**. This cellular layer ensures protection to internal organization of body. Pinacocytes are hexagonal in surface view with thin margins and bulging central part containing a nucleus. The margins of pinacocytes are contractile, so that the sponge can increase or decrease slightly in size.

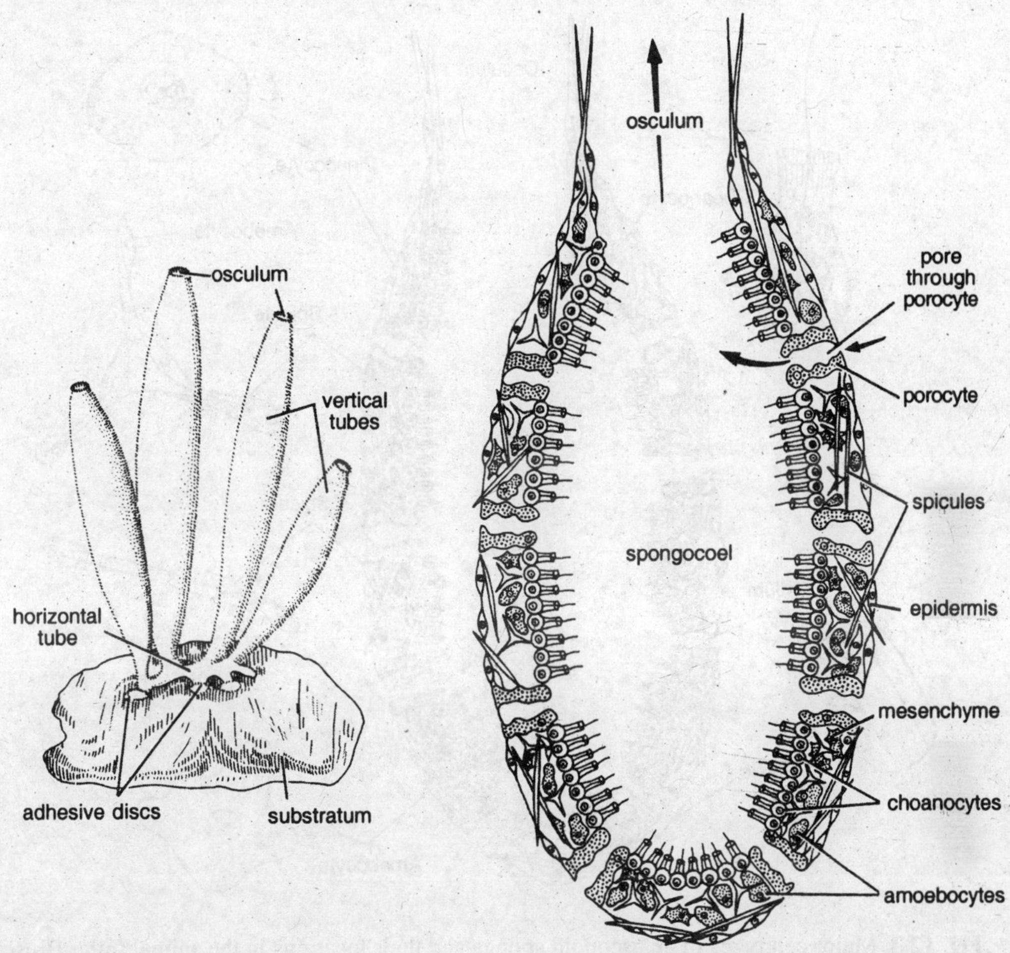

Fig. 12.1. *Leucosolenia.* Colony. **Fig. 12.2.** *Leucosolenia* Longitudinal section.

Within the body wall of *Leucosolenia* are situated special large and tubular cells, called **pore cells** or **porocytes**. Porocytes are specialized cells of pinacoderm. Each porocyte contains a central canal-like space communicating to the outside as well as the spongocoel. These spaces of porocytes are called **ostia, dermal pores** or **incurrent pores**. These permit water to flow from outside into the spongocoel.

2. Choanoderm. The atrium or spongocoel is lined by **gastrodermis** or **choanoderm**. It consists of a single layer of flagellated collar cells, the **choanocytes** (Gr., *choane* = funnel; *kytos* = cell). A choanocyte is an oval cell having a **flagellum** which arises from basal granule or kinetosome. (**Note.** According to **Brusca** and **Brusca**, 2003, the choanocyte nucleus is situated apically and the flagellum arises directly from the nucleus). The base of flagellum is surrounded by a contractile transparent protoplasmic **collar** (*i.e.*, microvilli of choanocyte). The choanocytes are basally embedded in the mesoglea. They are used in feeding and for ensuring the flow of water within the animal's water current channels or **aquiferous system** by beating of their flagella.

3. Mesohyl. It is very thin in *Leucosolenia* and is secreted by the choanocytes (some authors have suggested that mesohyl is secreted by both cellular layers-pinacoderm and choanoderm). Mesohyl

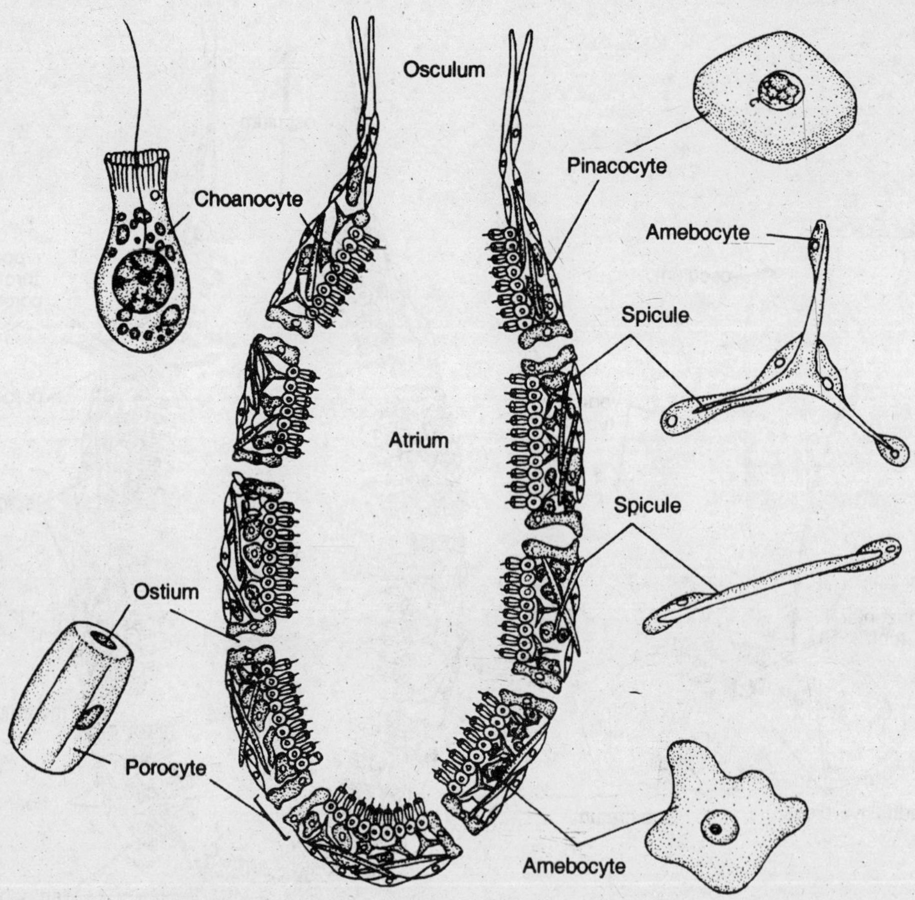

Fig. 12.3. Major cell types of an asconoid sponge and their locations in the animal (after Brusca and Brusca, 2003).

plays vital roles in digestion, gamete production, secretion of skeleton, and transport of nutrients and waste products by special amoeboid cells. The mesohyl includes the noncellular colloidal gelatinous matrix, the **mesoglea** in which are embedded **collagen** fibers, spicules (skeletal elements of $CaCO_3$) and various cells. As such, it is really a **mesenchyme**. A great number of cell types may be found in the mesohyl. Being totipotent, most of these cells are able to change from one type to another as required; but some cells are differentiated irreversibly, such as those that commit themselves to reproduction or to skeletal formation.

***(i)* Amoebocytes.** These are freely wandering amoeba-like cells of mesohyl. Amoebocytes are of many types depending upon the shapes of their pseudopodia and function. Large-sized primary amoebocytes, each with blunt pseudopodium and large nucleus, are the **archaeocytes**. These contain much RNA and they carry on all the functions essential for the life of sponge. Archaeocytes are self-replicating and also capable of giving rise to all other types of amoebocytes (*i.e.*, they are totipotent). **Scleroblasts** (or sclerocytes) are skeleton forming amoebocytes. They are **calcoblasts** as they form calcareous spicules of calcium carbonate ($CaCO_3$). **Thesocytes** are with rounded pseudopodia and reserve food material (glycogen). **Chromocytes** have colour pigments and are responsible for the colour of sponge. **Gland cells** secrete slime. Cells with thin branching pseudopodia are **collencytes**; they form connective tissues. **Myocytes** are contractile thin cells. They are distinguished by the great number of microtubules and microfilaments contained in their cytoplasm. They are regarded homologous with the smooth muscle cells of higher invertebrates. Myocytes are present around the osculum and work as **sphincter**. They are independent effectors with slow response time.

(*ii*) Spicules. They are hard skeletal elements. Spicules may be small or large and formed of crystalline calcium carbonate, so are called **calcareous spicules**. Straight needle-like spicules are **monaxons** (with single ray). Some spicules have four rays each pointing a different direction; they are called **tetraxons**. Some tetraxons secondarily become **triradiate** (triaxons) due to loss of one ray. Spicules are secreted by **calcoblasts** and remain embedded within mesenchyme. All these spicules tend to orient in the same direction. Some monaxon spicules project out of the pinacoderm. A few monaxon spicules form a scanty **fringe** around the osculum.

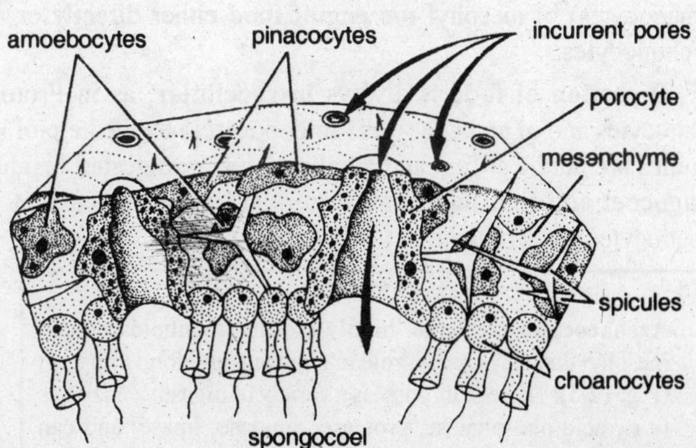

Fig. 12.4. *Leucosolenia*. Diagrammatic cross section of a portion of body wall highly magnified to its cellular organization.

(*iii*) Canal system. Ostia, spongocoel and osculum form a canal system (or aquiferous system). Canal system of *Leucosolenia* is of **ascon type**. Water moving through an asconoid sponge flows through the following route: ostia → spongocoel (over the choanoderm) → osculum. The one-cell thick choanoderm lining the spongocoel is simple and continuous.

12.4 PHYSIOLOGY

1. Locomotion and behaviour. Adult *Leucosolenia* is incapable of locomotion. This sponge is found to have local contractile powers and this power is normally restricted to region of osculum. Reactions to stimuli are very slow (Box 12.1).

> **Box 12.1 Activity and Sensitivity in Sponges**
>
> There is no conclusive evidence that sponges possess **neurons** or distinct **sensory organs**. Action potentials have never been recorded in sponges, and nothing resembling the synaptic connections of higher Metazoa are known in sponges. However, they are capable of responding to a variety of environmental stimuli by closure of ostia or oscula, canal constriction, backflow and reconstruction of flagellated chambers. The usual effect of most of these actions is to reduce or stop the flow of water through the canal system (see **Brusca** and **Brusca**, 2003).

2. Water current. Vital life processes of a sponge are dependent upon a continuous flow of water through its porous body. This is essential because sponges are sessile and have no other means of getting food and oxygen and getting rid of wastes (digestive, excretory, respiratory wastes). Water current is used by constant beating of flagella of numerous choanocytes. Water enters spongocoel through ostia and exits through osculum. Flow of water is regulated by the closing and opening of ostia. The water currents brings food and oxygen and removes digestive, respiratory, excretory and reproductive elements.

3. Digestion. Food of sponges consists of bacteria, plankton (microscopic animals and plants), detritus (organic particles) and DOM (= dissolved organic matter). DOM is taken by sponge cells of many sponges by **pinocytosis** (**Reiswig**, 1975). Other solid food is taken up by sponge cells by **phagocytosis**. In *Leucosolenia*, food particles tend to adhere to microvilli of collar of the choanocyte and at the base of collar ingested by cell body by endocytosis/phagocytosis. Amoebocytes (archaeocytes) of mesohyl too engulf food either directly or receive partially digested food from the chonocytes.

Digestion of food is always **intracellular**, as in Protozoa, occurring in food vacuoles of choanocytes and of archaeocytes (amoebocytes) by the help of lysosomal enzymes. Contents of food vacuoles are acidic at first and alkaline later. Undigested residue is eliminated by amoebocytes into spongocoel and find their way into outgoing water current. Digested food is stored in certain amoebocytes as glycogen.

Box 12.2.

Archaeocytes are large, highly motile amoeboid cells of mesohyl that play a major role in digestion and food transport (Fig. 12.5). These cells possess a variety of digestive enzymes (*e.g.*, acid phosphatase, protease, amylase, lipase) and can accept phagocytized material from the choanocytes. They also phagocytize material directly through the pinacoderm of water canals. As the principal **macrophage** of a sponge, archaeocytes carry out much of the digestive, transport and excretory activity.

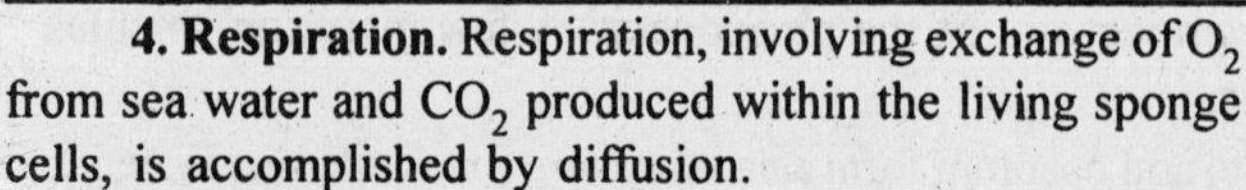

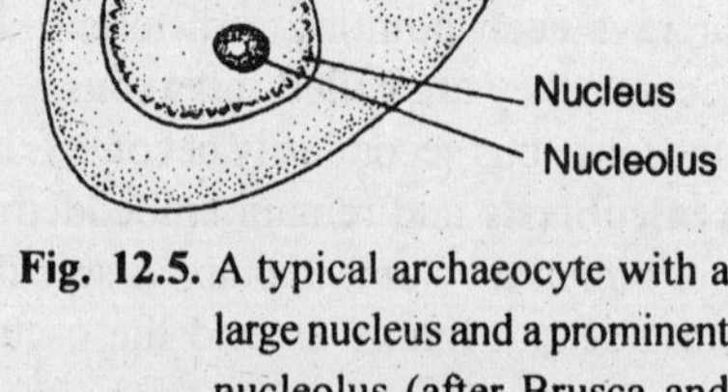

Fig. 12.5. A typical archaeocyte with a large nucleus and a prominent nucleolus (after Brusca and Brusca, 2003).

4. Respiration. Respiration, involving exchange of O_2 from sea water and CO_2 produced within the living sponge cells, is accomplished by diffusion.

5. Excretion. Elimination of metabolic nitrogenous waste (mainly ammonia) also takes place by diffusion through the general body surface or choanoderm. Archaeocytes too are involved into excretion.

12.5 REPRODUCTION AND DEVELOPMENT

Leucosolenia reproduces asexually and sexually.

1. Asexual reproduction. *Leucosolenia* reproduces asexually by **three** methods: regeneration, branching and budding.

(*i*) Regeneration. *Leucosolenia* has remarkable power of **regeneration**. Any piece of broken *Leucosolenia* is capable of growing into a complete individual. This process is slow, and months or even years may be required before normal size of young is attained.

(*ii*) Branching. New horizontal branches arise from stolon, they grow over rocks and give rise to new erect vase shaped cylinders. When an upright branch or cylinder attains sufficient size, its top breaks through as an osculum.

(*iii*) Budding. This is accomplished by evagination of body wall near the base of a vertical tube in the form of **bud** (Fig. 12.6). The bud grows in size and breaks off an osculum at its free end, thus becoming an additional individual of colony. This may again bud off new individuals.

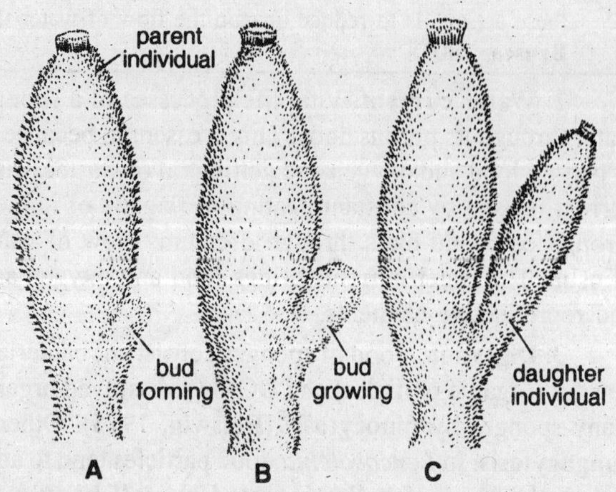

Fig. 12.6. *Leucosolenia.* Various stages of budding.

2. Sexual reproduction. *Leucosolenia* is **hermaphrodite** or **monoecious**, *i.e.*, male and female gametes (sex cells) are formed in the same individual. No special gonads are formed. Sperm and ova are derived from **archaeocytes** by gametogenesis.

Cross-fertilization occurs and it is internal fertilization. Sperms are released into sea water and they make their way into another sponge to a region of mature ovum. The development of *Leucosolenia* have been described by **Metschnikoff** (1879) and **Minchin** (1896). Mostly account of sponge embryogeny is based on **Okada** (1920) who studied embryogeny of a hexactinellid.

Development. Ovum is fertilized by a sperm within maternal body forming a diploid zygote. The zygote undergoes **equal** and **holoblastic** cleavage to form an oval hollow blastula, called **coeloblastula** (Fig. 12.7). Calcareous sponges (*e.g.*, *Leucosolenia*) often release their embryos early, as free swimming "**coeloblastula**" **larva** (see **Brusca** and **Brusca**, 2003). Wall of coeloblastula consists of a single layer of narrow, elongated, columnar and flagellated cells. At the posterior pole of larva occurs a group of large, rounded, granular, non-flagellated cells (called "**macromeres**"). These cells are believed to be archaeocytes which form all future archaeocytes of the sponge. These granular cells later migrate into cavity of coeloblastula. Some of adjacent flagellate cells also lose their flagella, become amoeboid and pass into the internal cavity. The larva now reaches a stage corresponding to the planula larva of cnidarians. It is termed as **stereogastrula**, **parenchymula** or **stereoparenchymula** and consists of an external layer of flagellated cells and an internal mass of amoeboid cells. This larval stage has no mouth opening.

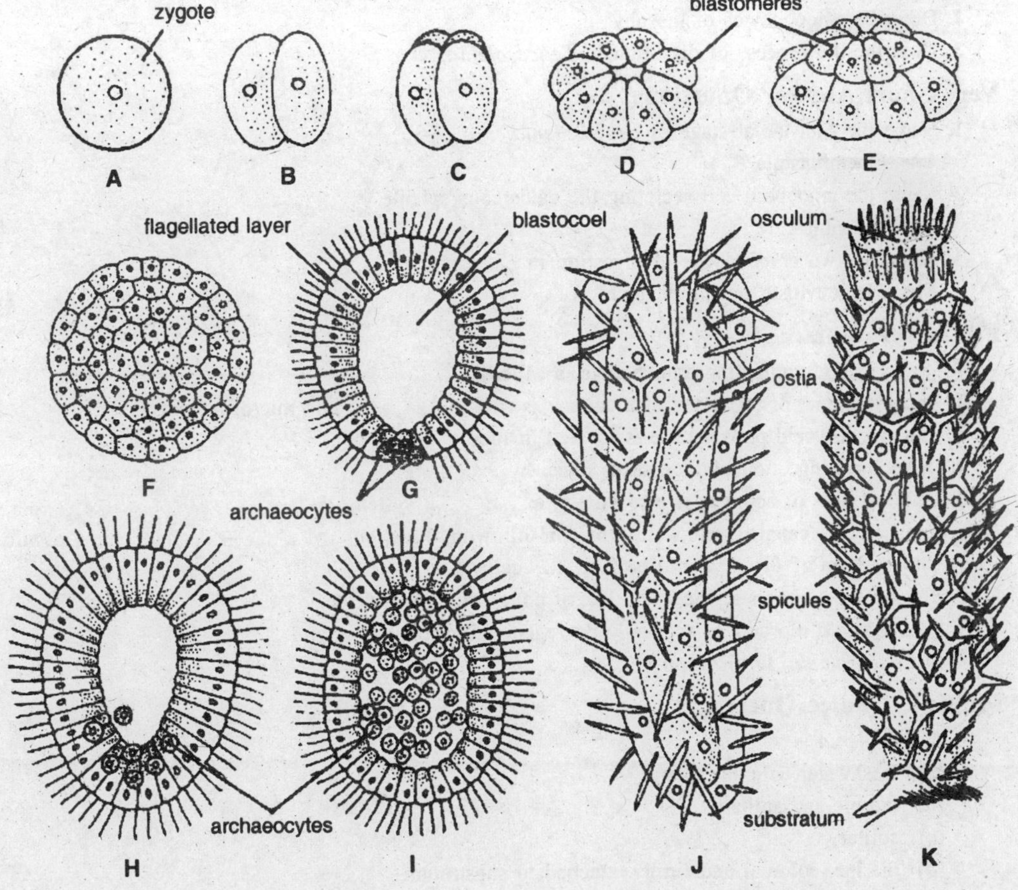

Fig. 12.7. *Leucosolenia.* Stages in development. A—Zygote, B to E—Cleavage stages, F—Early blastula, G and H—Coeloblastula, I—Parenchymula, J—Young sponge, K—Adult sponge.

The parenchymula swim freely for some hours. Then it attaches by the anterior pole and develops into a flat plate with an irregular outline. The amoeboid cells (internal cells) migrate to the external surface and form the pinacoderm and mesenchyme. The flagellated cells are, thus, enclosed and become the choanocytes. A central spongocoel appears, an osculum breaks through and spicules are secreted. After a few days of attachment, the larva is converted into the adult asconoid sponge.

QUESTIONS

Long Answer Questions

1. What is a sponge? Describe the cellular organization of *Leucosolenia*.
2. Give an account of structure of *Leucosolenia* with the help of its longitudinal section.
3. Describe structure and life cycle of a simple sponge you have studied.
4. Describe development of *Leucosolenia*.

Short Answer Questions

1. Write short notes on:
   (*i*) Coeloblastula larva (*ii*) Pinacoderm
   (*iii*) Mesohyl (*iv*) Osculum
   (*v*) Parenchymula (*vi*) Spicules of *Leucosolenia*.
2. Describe amoebocytes of mesohyl.
3. Describe the process of digestion in *Leucosolenia*.

Very Short Answer Questions

1. Name the first larval stage of *Leucosolenia*.
   Ans. Coeloblastula.
2. Name the amoeboid cell secreting the calcareous spicule.
   Ans. Calcoblast.
3. What are two chief modes of ingestion in *Leucosolenia*?
   Ans. Phagocytosis and pinocytosis.

Fill in the Blanks

1. Reserve food containing amoebocyte is called ....
2. The cytoplasm of myocytes of a sponge is rich in .... and microfilaments.
3. The enzyme acid phosphatase is located in food vacuoles of ....
4. In sponges ingestion of food takes place by ....
5. Canal system of sponges is also known as .... system.
6. In ascon type canal system water flows in following direction: .... → spongocoel → osculum.
7. Choanoderm of *Leucosolenia* is .... cell thick.
8. Phagocytosis occurs at the .... of collar of a choanocyte.
9. *Leucosolenia* reproduces asexually by regeneration, branching and ....
10. In *Leucosolenia*, fertilization is ....

Multiple Choice Questions

1. *Leucosolenia* is
   (*a*) freshwater colonial sponge
   (*b*) marine and solitary
   (*c*) solitary
   (*d*) marine, colonial and firmly attached to substratum
2. Which type of canal system is found in *Leucosolenia*?
   (*a*) ascon (*b*) rhagon
   (*c*) leucon (*d*) syconoid

3. The internal skeleton of *Leucosolenia* is formed of calcareous ....
   (*a*) monaxon spicule
   (*b*) monaxon and tri-radiate spicules
   (*c*) tri-radiate and quadri-radiate spicules
   (*d*) monaxon, tri-radiate and quadri-radiate spicules
4. The exit of *Leucosolenia* is known as
   (*a*) mesenchyme
   (*b*) osculum
   (*c*) ostia
   (*d*) spongocoel
5. In *Leucosolenia*, food is ingested by
   (*a*) scleroblasts
   (*b*) choanocytes
   (*c*) porocytes
   (*d*) pinacocytes
6. Nutrition in *Leucosolenia* is
   (*a*) intracellular
   (*b*) extracellular
   (*c*) both of these
   (*d*) none of these
7. In *Leucosolenia*, digestion takes place in
   (*a*) paragastric cavity
   (*b*) stomach
   (*c*) osculum
   (*d*) food vacuole
8. In *Leucosolenia*, asexual reproduction takes place by
   (*a*) sporulation
   (*b*) cyst formation
   (*c*) binary fission
   (*d*) external budding
9. Gametes in *Leucosolenia* are derived from
   (*a*) archaeocytes
   (*b*) choanocytes
   (*c*) porocyte
   (*d*) amoebocyte
10. The free swimming larva of *Leucosolenia* is
    (*a*) stomoblastula
    (*b*) amphiblastula
    (*c*) stereoparenchymula
    (*d*) olyanthus stage
11. First larval stage of *Leucosolenia* is
    (*a*) coeloblastula
    (*b*) parenchymula
    (*c*) olyanthus
    (*d*) stomoblastula
12. The cells surrounding the osculum in sponges is
    (*a*) archaeocytes
    (*b*) myocytes
    (*c*) choanocytes
    (*d*) gland cells
13. Fertilization in sponge is
    (*a*) internal
    (*b*) external
    (*c*) both
    (*d*) none

ANSWERS

Fill in the Blanks

1. Thesocyte 2. Microtubules 3. Archaeocytes 4. Choanocytes
5. Aqueferous 6. Ostia 7. One 8. Base
9. Budding 10. Internal.

Multiple Choice Questions

1. (*d*) 2. (*a*) 3. (*d*) 4. (*b*)
5. (*b*) 6. (*a*) 7. (*d*) 8. (*d*)
9. (*a*) 10. (*c*) 11. (*a*) 12. (*b*)
13. (*a*).

13

Scypha (Sycon) (Phylum Porifera)

13.1 CLASSIFICATION

Phylum	–	**Porifera**
Class	–	**Calcaria**
Order	–	**Heterocoela**
Type	–	***Scypha* (*Sycon*)**

Scypha or *Sycon* is commonly called **urn sponge or crown sponge**. It is a marine calcareous sponge. It is worldwide in distribution and found in abundance near North Atlantic shores. *Scypha* is found attached to submerged rocks, shells of molluscs and corals in shallow waters (upto the depth of 50 fathoms; 1 fathom = 6 feet) near the coast. It prefers places where waves provide the animal with plenty of food and well-oxygenated water. *Scypha* often occurs in the form of branching colonies.

The name of genus *Sycon* is replaced by *Scypha* by **de Laubenfels** (1936). The different species of *Scypha* are *S. ciliatum*, *S. elegans*, *S. coronata*, etc.

13.2 EXTERNAL FEATURES

1. Shape and size. *Scypha* is a small vase-shaped sponge. It measures 2.5 to 8 cm in length. Its colony has two to seven cylinders which are connected together at the base (Fig. 13.1). The base of *Scypha* is attached by a sticky secretion to some submerged solid substrata.

2. Colour. It is grey or light brown in colour.

3. Osculum. The distal free end of each cylindrical branch of *Scypha* has a single large opening, called **exhalent pore, excurrent pore** or **osculum**. The paragastric cavity or spongocoel opens to exterior through osculum. Water currents exit the sponge through this opening.

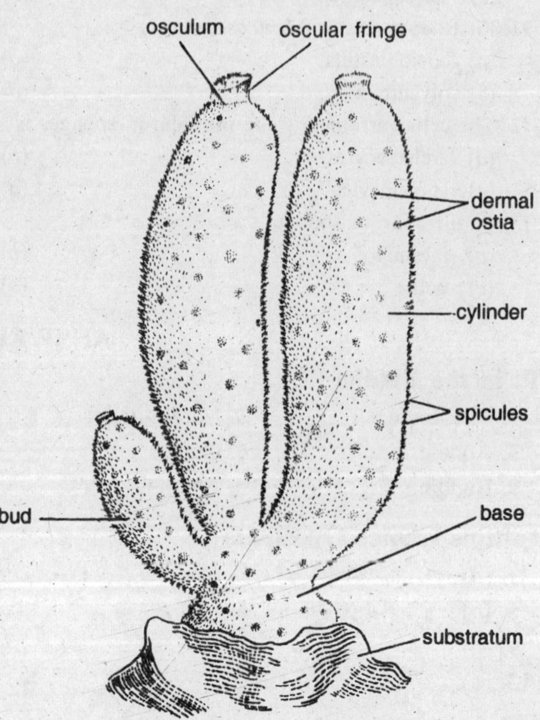

Fig. 13.1 *Scypha*. Colony

Osculum is surrounded by numerous straight needle-like calcareous monaxon spicules arranged in a circlet or crown. This is called **oscular fringe** and it prevents the entry of other animals into the sponge.

4. Ostia. The body surface of *Scypha* has numerous regularly arranged microscopic polygonal elevations. Every elevations are separated from the others by a system of depressed lines of furrows. In these furrows are located numerous microscopic apertures, the **ostia** which are **inhalent** or **incurrent pores**. Water enters the cavity of sponge through these pores. The entire body surface of *Scypha* appears bristly as numerous monaxon spicules (called **oxeote**) protrude out from it.

5. Spongocoel. Inside each cylindrical branch exists a central cavity which leads to osculum; it is called **spongocoel**, **atrium** or **paragastric cavity** (Fig. 13.2). It is not a digestive cavity and is not lined by endoderm. The wall of cylinder is thick due to increase in the amount of **mesogloea mesenchyme** or **mesohyl**. The wall of *Scypha* has folded in such a way as to form two types of canals, the **incurrent canals** and **radial canals**, both lie alternately and radially around the spongocoel. No ostium opens directly to spongocoel (so ostia are absent from the collar and basal regions of *Scypha*).

13.3 MICROSCOPIC ORGANIZATION (HISTOLOGY OR DIFFERENT TYPES OF CELLS IN *SCYPHA*)

Scypha exhibits **cellular grade of organization**, *i.e.*, its cells form loose aggregation and are not organized into tissues. It is a **monoblastic** animal having only one layer of cells lining a non-living matrix, **mesohyl** from both sides (**Note.** Previously *Scypha*, like other Porifera, was considered as a diploblastic animal having two embryonic layers, the ectoderm (dermal layer) and endoderm (gastral layer) but recent embryological data have proved that sponges lack endoderm (see **Ruppert** and **Barnes,** 1994).

According to the nature of cells the single embryonic layer of sponges forms two types of layers: **pinacocyte layer (pinacoderm)** and **choanocyte layer (choanoderm)**. Both of these layers secrete and cover the mesohyl from all sides. Pinacocyte layer and choanocyte layer lack basal laminae, inter-cellular junctions and most other ultrastructural connections that make epithelial layers of other animals (eumetazoa). Mesohyl consists of a gelatinous proteinaceous matrix containing **collagen fibres** (or spongin A), skeletal material (spicules) and amoebocytes. It is equivalent to connective tissue of other metazoans.

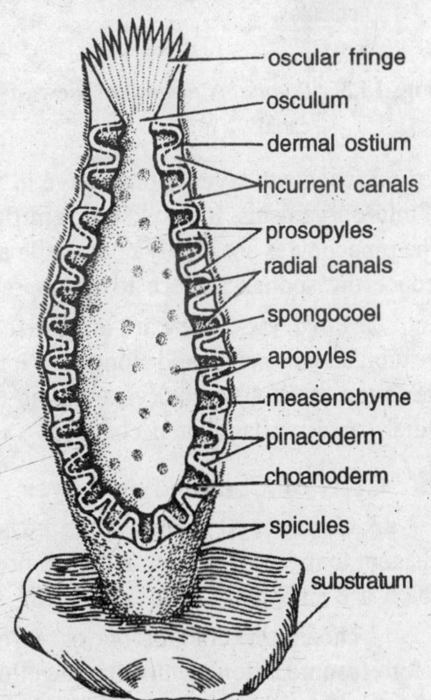

Fig. 13.2. *Scypha.* A diagrammatic L.S. of a cylinder showing gross internal structure.

The pinacoderm includes **exopinacoderm** covering the body surface except ostia and osculum and **endopinacoderm** lining the incurrent canals and spongocoel. Choanoderm lines the radial canals and consists of a single layer of choanocytes. Structure and function of various types of cells present in the sponges can be considered under the following headings:

A. Cells of Pinacocyte Layer

1. Pinacocytes. Pinacocytes are large thin polygonal scale-like flat cells. They are cemented together by their edges like the tiles in a floor. Each pinacocyte contains a nucleus at the centre which is much thicker than the periphery. Pinacocytes cover entire outer surface of the body and also lines the incurrent canal, excurrent canal and spongocoel.

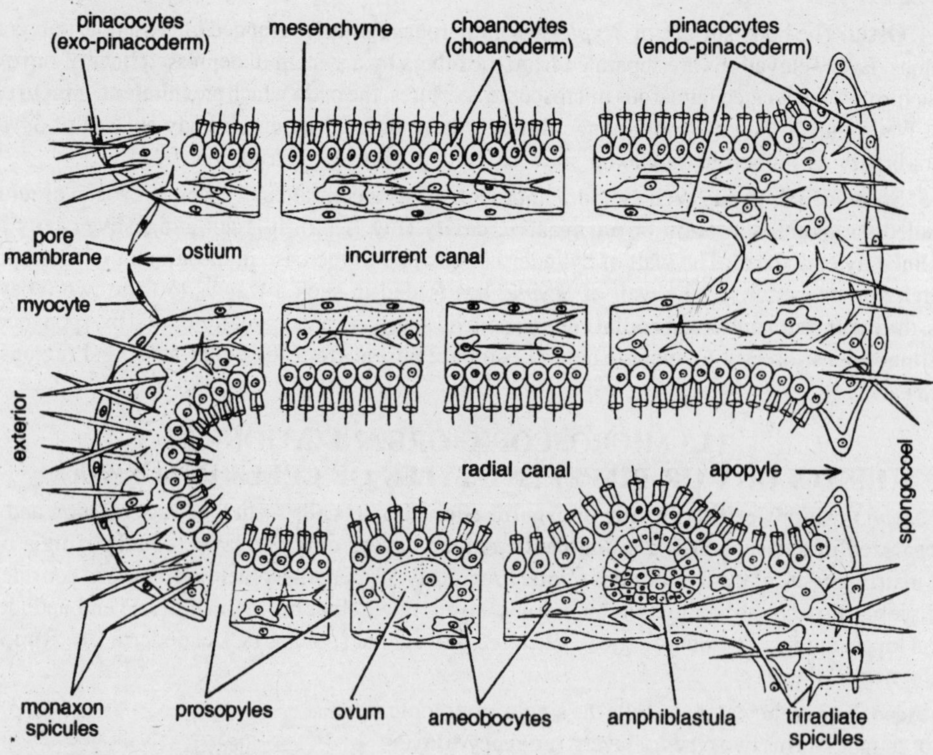

Fig. 13.3. *Scypha*. A magnified view of L.S. of a part of body wall showing one incurrent and one radial canal.

Pinacocyte layer is protective in function. It prevents mechanical injury and checks the entry of microorganisms. It also keeps semifluid mesohyl in position, preventing its dispersion into water. The pinacocytes are highly contractile and by withdrawing their margins into the central bulge, they reduce the sponge surface to an appreciable extent.

2. Porocytes. They are large-sized contractile pore cells and are modified pinacocytes. These are tubular cells perforated along their long axis by a central canal, the **prosopyle** (Fig. 13.4). Porocytes are found at frequent intervals among the pinacocytes. Prosopyle connects an incurrent canal to a radial canal or flagellated chamber.

B. Cells of Choanocyte Layer

3. Choanocytes. These cells are large, ovoid and nucleated. One end of each choanocyte exists adjacent to mesohyl, its opposite end projects into radial canal. A choanocyte has root-like processes which are embedded in mesohyl (Fig. 13.5A).

A choanocyte contains one or two contractile vacuoles, food vacuoles and a single **basal granule** or **kinetosome**. A long whip-like flagellum arises from basal granule. The basal granule is also joined to centriole and the two together forming typical **centroblepharoplast**. The flagellum contains typical 9 + 2 microtubular arrangement. It is surrounded at its base by a transparent contractile collar which is made up of a ring of 20 to 30 erect and closely set microvilli that extend from the cytoplasm (Fig. 13.5B). The microvilli or villi have microfilament cores and are connected to one another by

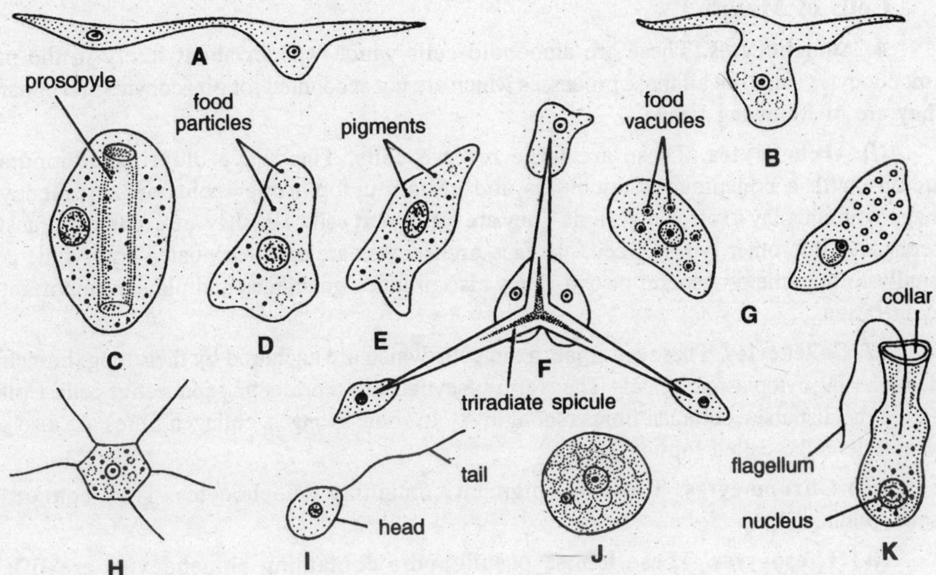

Fig. 13.4. *Scypha*. Cellular elements. A—Pinacocytes; B—Myocyte; C—Porocyte; D—Thesocyte; E—Chromocyte; F—Scleroblast; G—Archeocytes; H—Collencyte; I—Sperm; J—Ovum; K—Choanocyte.

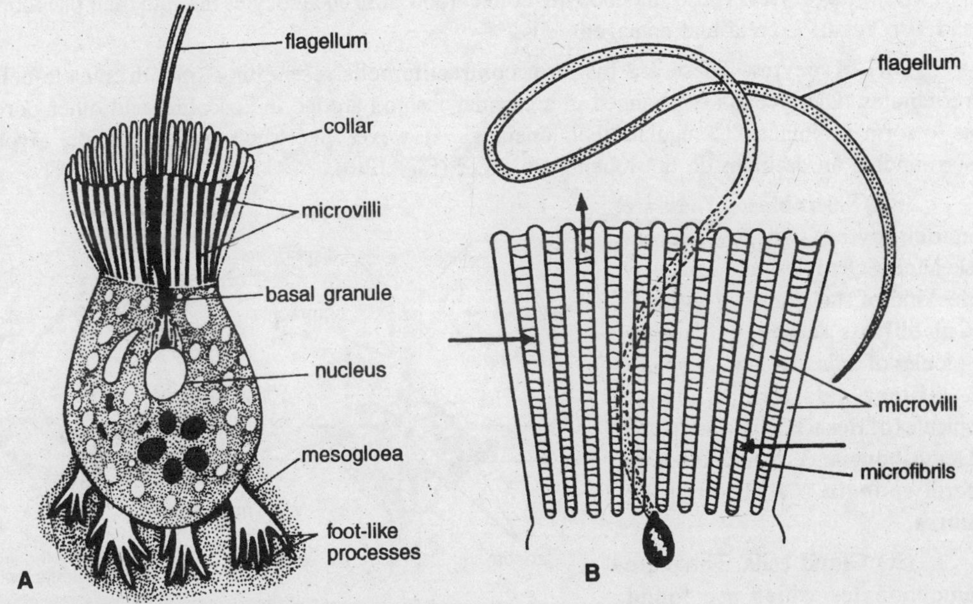

Fig. 13.5. *Scypha*. A—Ultrastructure of a choanocyte; B—Details of collar of choanocyte. Microvilli are connected by delicate microfibrils. Arrow indicates assumed flow of water through collar-sieve.

anastomosing mucous strands (a mucous reticulum). Choanocytes are held in place by inter-digitation of adjacent basal surfaces. The choanocytes are responsible for moving water through the sponge and for obtaining food. In addition, the choanocytes play an important role in digestion, excretion and reproduction (*i.e.*, fertilization).

C. Cells of Mesohyl

4. Amoebocytes. These are amoeboid cells which wander about freely in the mesohyl. Amoebocytes carry out all those processes which are not accounted for pinacocytes and choanocytes. They are of following types:

***(i)* Archeocytes.** These are large reserve cells. They have blunt pseudopodia, large nucleus with a conspicuous nucleolus and certain cytoplasmic inclusions. Archeocytes are phagocytic and play a role in digestion. They are **totipotent cells**, *i.e.*, they are capable of transforming themselves into other types of cells. In fact, archeocytes are undifferentiated embryonic cells and usually form gametes—sperm or ova. They also produce gemmules and play an important role in regeneration.

***(ii)* Collencytes.** These are small fixed cells which are anchored by their long, branching and anastomosing cytoplasmic strands. They form syncytial units representing connective cells. Collencytes secrete the dispersed collagen fibres (spongin A). In some sponges, collagen fibres are also secreted by mobile cells, called **lophocytes**.

***(iii)* Chromocytes.** These are pigment containing amoebocytes. They contain lobose pseudopodia.

***(iv)* Thesocytes.** These lobose pseudopodia containing amoebocytes are filled with reserve food. Thesocytes store food in the form of glycogen, fats and proteins (glycoproteins and lipoproteins).

***(v)* Trophocytes.** These amoebocytes are found loaded with digested food and serve to transfer it from one place (or cell) to other.

***(vi)* Phagocytes.** These amoebocytes collect food from choanocytes through their pseudopodia and also engulf excreta and damaged cells.

***(vii)* Myocytes.** These are fusiform contractile cells resembling smooth muscle cells of vertebrates. These cells are arranged in a circular fashion around the osculum and other pores so as to form a sphincter to regulate their openings. However, like true muscle cells, the myocytes surrounding an osculum do not touch each other (Fig. 13.6).

***(viii)* Scleroblasts.** These are amoebocytes which secrete skeleton. Scleroblasts are named by the kind of skeleton they form: 1. **Calcoblasts** secrete calcareous spicules of Calcarea (*e.g., Scypha*). 2. **Silicoblasts** form siliceous spicules of Hexactinellida and some Demospongia. 3. **Spongioblasts** form **spongin** (*i.e.*, spongin B) **fibres.**

***(ix)* Gland cells.** These are amoebocytes which are found attached to body surface by means of long strands. They secrete slime (mucus).

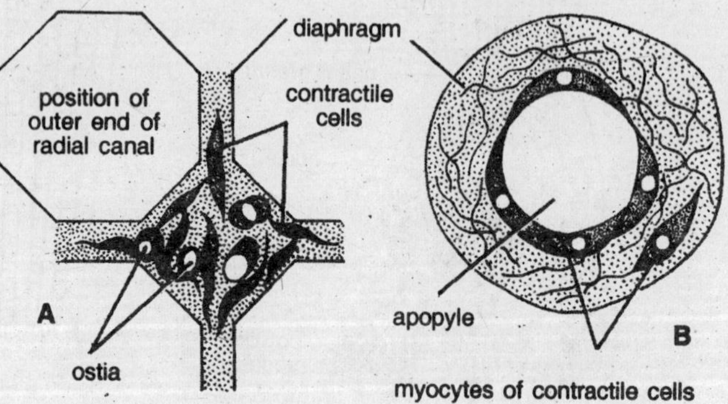

Fig. 13.6. *Scypha.* A—Pore membrane showing ostia; B—An apopyle surrounded by its diaphragm.

Box 13.1

In almost all sponges, the entire supportive matrix is built on a framework of fibrillar collagen. The cells that secrete this material are called **collencytes, lophocytes** and **spongo-cytes**. Collencytes are morphologically nearly indistinguishable from pinacocytes, whereas lophocytes are large highly motile cells that can be recognised by a collagen they typically trail behind them (Fig. 13.7). The primary func-tion of both cell types is to secrete the dispersed fibrillar **collagen** found intercellularly in virtually all sponges. Spongocytes produce the fibrous supportive collagen referred to as spongin. Spongocytes operate in groups and are always found wrapped around a spicule or spongin fibre (see **Brusca** and **Brusca**, 2003).

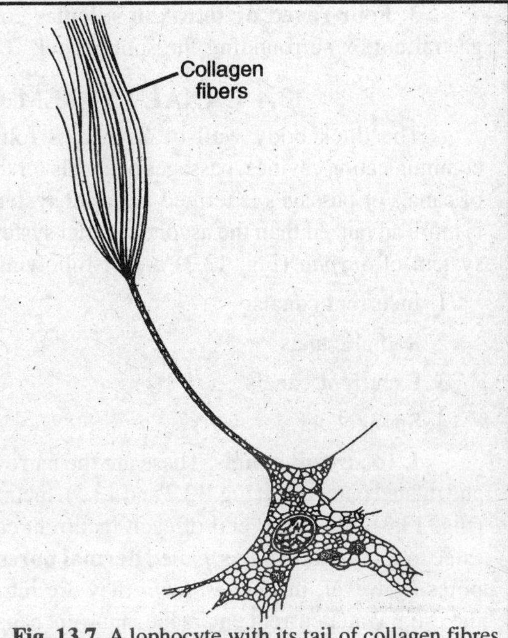

Fig. 13.7. A lophocyte with its tail of collagen fibres.

Skeleton of *Scypha*

In *Scypha* the skeleton consists of minute, crystalline structures called **sclerites** or **spicules**. These spicules are **calcareous**, *i.e.*, they are made of calcium carbonate. Spicules of *Scypha* are composed of calcium carbonate (about 87%), magnesium carbonate and other minerals (10%), water (3%) and organic materials (in traces). In *Scypha* spicules are megascleres (large) and remain separate. Depending upon the shape and size, the spicules of *Scypha* are of following types:

1. One-rayed or monaxon spicules are elongated straight needle-like spicules having only one axis. They are of following two types:

(*a*) Large one-rayed monaxon spicules are found around the osculum and are arranged in a circlet, called **oscular fringe**.

(*b*) Spear-like or club-shaped oxeote spicules occur in groups and project from the surface of *Scypha*. They give a bristly appearance to the sponge and partly conceal and protect the ostia.

2. Three-rayed or triaxon spicules consist of three axes. They occur along the radial canals with one ray pointed towards the distal end. Triaxon spicules form body supporting network of *Scypha*.

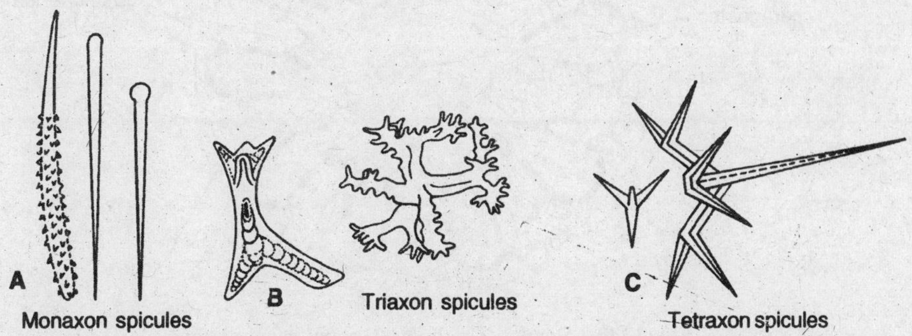

Fig. 13.8. *Scypha.* A—Monaxon spicules; B—Triaxon spicules; C—Tetraxon spicules.

3. Four-rayed or tetraxon spicules consist of four axes. They are embedded in the thick gastral cortex surrounding the spongocoel.

13.4 CANAL SYSTEM OR AQUIFEROUS SYSTEM

The thick body wall of *Scypha* is folded to produce an arrangement of complex inter-communicating cavities, passages or canals for the entrance and exit of water current. This arrangement of canals or passages is termed as **canal system**. *Scypha* has **syconoid** type of canal system which is more advanced than the asconoid canal system of *Leucosolenia*. Various pores and canals of canal system of *Scypha* (Fig. 13.3) are of following types:

1. Incurrent canals
2. Radial canals
3. Excurrent canals
4. Spongocoel

1. Incurrent canals. These are the narrow spaces between the radial canals with narrow inner and dilated outer ends (Fig. 13.2 and 13.3). Incurrent canals are lined by flattened cells, the **pinacocytes** (Fig. 13.3). The outer end of each incurrent canal is covered by a thin pore membrane containing three to four minute pores, called **dermal pores** or **ostia** (Fig. 13.6). In *Scypha*, ostia are intercellular pores (however, in *Leucosolenia* they are intracellular pores). The contractile myocytes surround the ostia. Water current enters the incurrent canal through the ostia and myocytes regulate the amount of water entering the canals by increasing or decreasing the diameter of the ostia.

The inner end of each incurrent canal is closed. However, the wall separating an incurrent canal and radial canal is pierced by many minute pores, called **prosopyles**. The prosopyles are perforations in **porocytes**. Incurrent canals open into the radial canals by prosopyles.

2. Radial canals or flagellated chambers. These are radially arranged, straight, elongated and octagonal canals of *Scypha* wall. They alternate with incurrent canals. Radial canals are lined by flagellated collar cells or **choanocytes**. They end blindly towards the outer surface but opens in the **excurrent canal** at their inner ends through the **apopyles**. The radial canals receive water current from incurrent canals through the prosopyles and convey it to spongocoel via the excurrent canals. The flagella of choanocytes beat continuously and direct the path of water current.

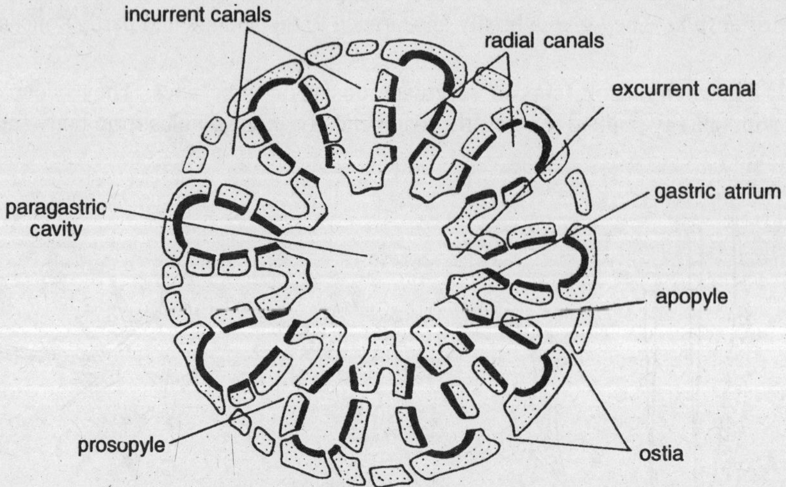

Fig. 13.9. *Scypha.* T.S. body to show canal system.

3. Excurrent canals. These are short but wide passages between the radial canals and spongocoel. Each of them is lined by flattened **pinacocytes**. The excurrent canal is separated from the radial canal by a thin elastic **diaphragm**. The diaphragm is perforated by a large central aperture, called **apopyle**. The apopyle is surrounded by myocytes and is capable of contraction and expansion, thus, regulating the diameter of apopyle. The excurrent canal opens into the spongocoel through a wide opening called **gastric ostium**.

4. Spongocoel. It is the central cavity of a cylindrical branch of colony of *Scypha*. The spongocoel of all branches remains continuous at the base of sponge. Spongocoel is lined by flattened cells or **pinacocytes**. Into it opens the excurrent canals. The spongocoel itself opens to outside through a terminal opening, the **osculum**. A group of contractile cells, called **myocytes** surround the osculum. These form an **oscular sphincter**. This sphincter regulates the diameter of the oscular opening.

Mechanism of Water Circulation

The physiology of *Scypha* and other sponges largely depends on the current of water flowing through the body. The water brings in oxygen and food and removes waste. Even sperms are moved in and out and larvae released by the water current.

The water current is produced and water is pumped into the body through ostia by the asynchronous beating of flagella of choanocytes which lines the radial canals and are oriented towards apopyle. A wave of spiral undulations passes from the base to the tip of each flagellum independently driving water from the radial canal. As a result, water is sucked into the radial canal from the incurrent canals through the prosopyles located between the bases of the choanocytes. It is then driven to the centre of the radial canal and out of the larger apopyle into an excurrent canal. Finally, the water leads into the spongocoel and is discharged to the exterior through osculum. The path of water current can be represented as follows:

$$\textbf{Water from outside} \xrightarrow[\text{ostia}]{\text{through}} \textbf{incurrent canal} \xrightarrow[\text{prosopyles}]{\text{through}} \textbf{radial canals} \xrightarrow[\text{apopyle}]{\text{through}}$$

$$\textbf{excurrent canal} \xrightarrow[\text{ostium}]{\text{through}} \textbf{spongocoel} \xrightarrow[\text{osculum}]{\text{through}} \textbf{outside}$$

Physiological importance of canal system. The water current is of great physiological importance for the sponges because it serves the following functions:

1. Nutrition. The flowing water provides food particles to nearly all the cells of *Scypha*. Silica, calcium salts and other substances are absorbed as the water passes through the sponge canal.

2. Respiration. The water entering the body of *Scypha* is rich in oxygen. Gaseous exchange occurs by simple diffusion between the flowing water and all cells of *Scypha*. Carbon dioxide is removed from sponge cells by outgoing water currents.

3. Excretion. While leaving the body, the water current removes egested waste and nitrogenous excretory products (mainly ammonia).

4. Reproduction. The spermatozoa enter other sponges along with the water current.

13.5 NUTRITION

Scypha is a **filter feeder** having **holozoic nutrition**. It depends on the stream of water passing through the body as a source of food.

Food. *Scypha* feeds on both living and dead organic food. It feeds on extremely fine particulate material which forms about 80 per cent of organic matter (organic detritus) consumed by the sponge.

The remaining 20 per cent consists of bacteria and protists such as protozoa (dinoflagellates, etc.) and other fine plankton (diatoms). There is also evidence that *Scypha* and other sponges can utilize dissolved nutrients.

Feeding. Food enters the body of *Scypha* along with water currents through ostia. It is picked up by various sponge cells by following two methods:

1. Screening of food
2. Phagocytosis

1. Screening of food. Food particles are apparently selected largely on the basis of size and are screened in the course of their passage into flagellated chamber (radial canal). Only particle smaller than a certain size can enter the ostia or pass through the prosopyles. The finest particles are removed by the choanocytes, perhaps by filteration across the microvilli of collar of the choanocyte (the microvilli are 0.03 to 0.10 mm apart).

Large particles (5–50 mm) are phagocytized by pinacocytes and porocytes lining the inhalent pathways of canal system or by archeocytes that move out of the mesohyl. Indeed, populations of archeocytes are found to clean the outer pinacoderm surface. Particles of bacterial size and below (1 mm) are removed and engulfed (phagocytized) by choanocytes on the surface of the cell and not on the microvillar collars.

2. Phagocytosis. All sponge cells can phagocytize food particles like the protozoans, *i.e.*, food particles are engulfed by the help of pseudopodia, food vacuoles are formed and **intracellular digestion** of food occurs inside food vacuoles by the help of lysosomal enzymes. Beating of flagellum of a choanocyte draws water with food particles towards the collar. Food particles adhere to the outer sticky surface of collar while water escapes through sieve-like microvilli of collar (Fig. 13.5B). These particles are then engulfed by pseudopodia arising from choanocytes (Fig. 13.10). Both archeocytes and choanocytes can transfer their engulfed particles to other cells (amoebocytes).

Digestion. Digestion of food is intracellular. It takes place inside the food vacuoles of archeocytes and digested food is absorbed across the vacuolar membrane. In *Scypha*, food digestion occurs in large sized choanocytes. The food vacuoles are first acidic and then become alkaline. The enzymes are supplied by lysosomes. Three main types of enzymes, namely proteases, carbohydrases and lipases have been detected in sponge extracts.

Assimilation. The digested food is distributed from cell to cell like the postman distributing mail by the wandering amoebocytes found in the mesohyl. The digested food is stored as glycogen, fat and glycoproteins in the storage cells called **archeocytes.** Food is also stored in **nurse cells,** called **trophocytes** that build up food reserves for use by the young oocyte or by amoebocytes preparing to form gemmules.

Egestion. The undigested food residue is egested out from the amoebocytes into the outgoing water current and ultimately expelled out of body.

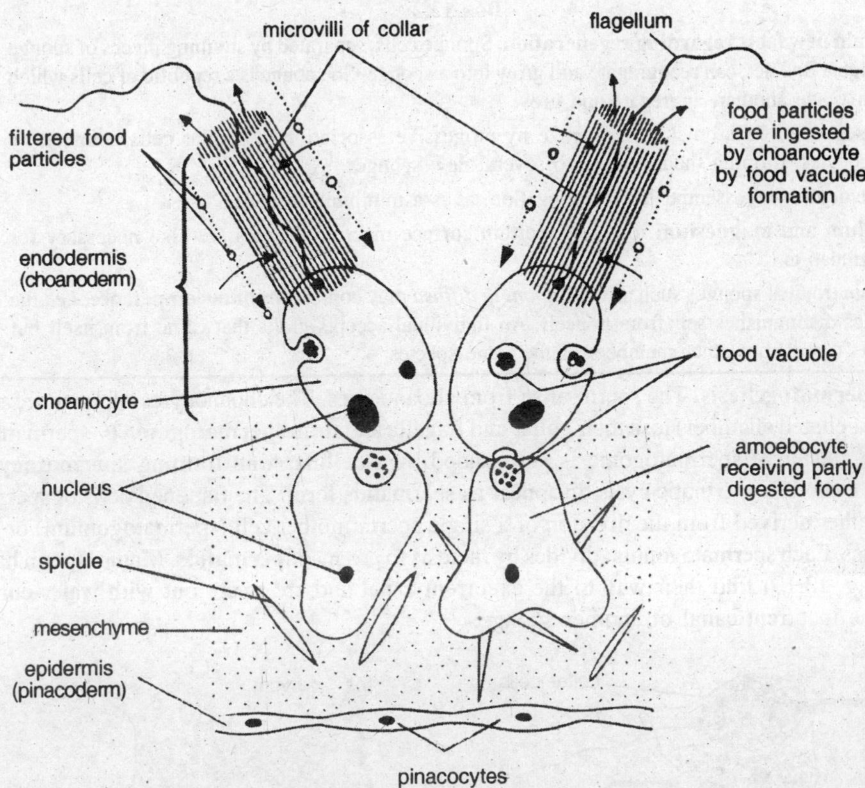

Fig. 13.10. Nutrition by choanocyte in sponge.

13.6 REPRODUCTION AND DEVELOPMENT IN *SCYPHA*

Scypha reproduces both asexually and sexually.

1. Asexual reproduction. In *Scypha* asexual reproduction occurs by

(*i*) Budding

(*ii*) Regeneration

Gemmules are not formed in *Scypha*.

(*i*) Budding. During favourable conditions *Scypha* reproduces by **budding**. A small bud appears at the base of the adult *Scypha* and grows into full size. An osculum develops at upper end of this bud. It may remain attached to the parent helping in the formation of colony or it may break free to lead an independent existence of an individual.

(*ii*) Regeneration. All sponges including *Scypha* have great power of regeneration. These sponges not only replace the parts lost during injury, but any piece of the body is capable of growing into a complete sponge. However, the process of regeneration is very slow and is completed in months or years (see Box 13.2).

2. Sexual reproduction. *Scypha* is **hermaphrodite (monoecious)**, *i.e.*, both male and female gametes are formed inside the body of same animal. It lacks definite sex organs or gonads. Both sperms and ova are produced from the **archeocytes** or **choanocytes**. Gamete production appear to be initiated by changes in water temperature, photoperiod or cellular regression depending on the species of sponge.

Box 13.2

Certain new facts regarding regeneration. Sponge cells, separated by straining pieces of sponge through a fine net, can reaggregate and grow into a sponge. So sponge is a republic of cells which identify one another, aggregate and grow.

The separated cells quickly reorganize by progessive association of similar cells bounded by pinacocytes, forming themselves into several new sponges.

Archeocytes are essential for reaggregation, as is a minimum of cells.

Calcium and **magnesium** ions plus certain surface macromolecules are also necessary for reaggregation.

Certain tropical sponges such as *Callyspongia diffusa* may contain **immunocompetence**, *i.e.*, the species distinguishes self from nonself. An individual accepts grafts that came from itself but rejects those from other members of the same species.

1. Spermatogenesis. The sperm arise from choanocytes. The choanocytes of an entire radial canal (= flagellated chamber) lose their collar and flagella and form **spermatogonia** (= sperm mother cells). The cluster of spermatogonia is surrounded by a cellular wall forming **spermatocyst** or **spermatic cyst**. The spermatic cysts disappear as spermatids form. The flattened cells of spermatic cysts are either derived from the division of a single sperm mother cell (spermatogonium) or from amoebocytes. Each spermatogonium divides by meiosis to produce **spermatids.** Mature sperm having flagella (Fig. 13.10) find their way to the excurrent canal and are swept out with water currents to reach the incurrent canal of another sponge.

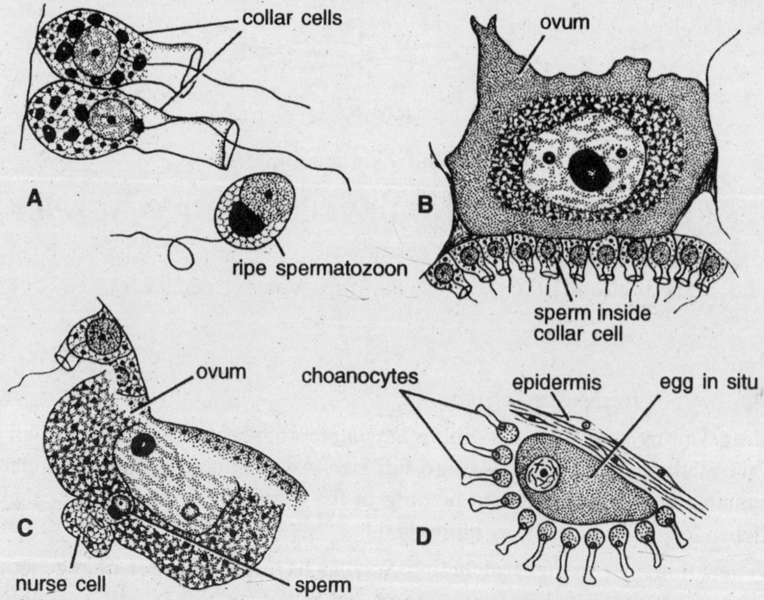

Fig. 13.11. Fertilization in *Grantia*. A—A ripe spermatozoon near two choanocytes; B—Entry of sperm in choanocyte; C—Carrier or nurse cells giving up sperm nucleus into ovum; D—Zygote.

2. Oogenesis. The egg mother cell or **ovocyte** is derived from an enlarged amoebocyte (*i.e.*, archeocyte) having a large nucleus with prominent nucleolus. It grows and acquires food for storage either by engulfing/fusing with similar amoebocytes or by engulfing nearby **nurse cells** or **trophocytes**. Meiosis occurs at the end of the growth period to produce an **ovum**. Ovum lies in the wall of radial canal (in mesohyl) ready to be fertilized by a sperm from another sponge.

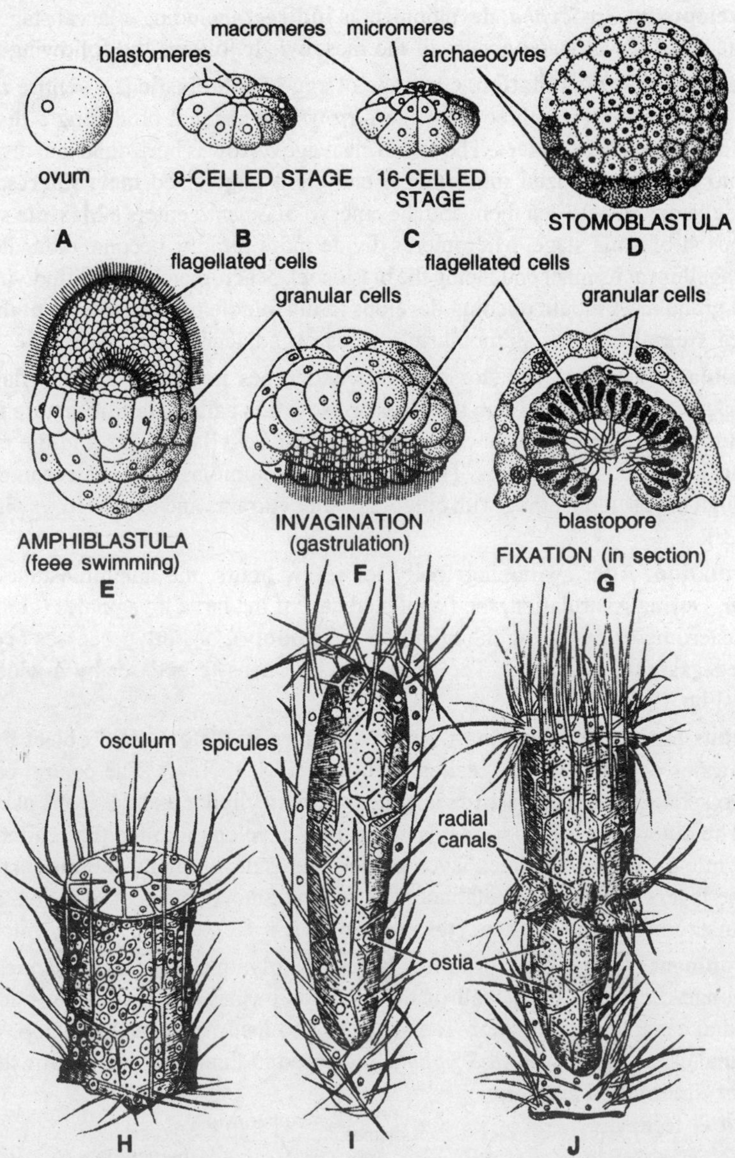

Fig. 13.12. *Scypha*. Development. A—Zygote; B—8-celled stage; C—16-cell stage; D—Stomoblastula; E—Amphiblastula; F—Gastrulation; G—Fixation of gastrula to substratum; H—Young sponge; I—Olynthus stage; J—Syconoid stage.

3. Fertilization. In *Scypha*, like the most sponges, the fertilization is **internal** and is of **cross fertilization** type. After a sperm has reached a radial canal (of another sponge), it is engulfed by a choanocyte occurring adjacent to the ripe ovum. This choanocyte has to transport the sperm to egg, it is called **nurse cell** or **carrier**. Sperm loses its flagellum. Carrier choanocyte too loses its collar and flagellum, becomes amoeboid and moves towards the ovum. After the carrier with its sperm has reached an egg, it either transfers the sperm nucleus to ovum or the carrier and sperm nuclei are engulfed by the egg. The nuclei of ovum and sperm fuse to form **zygote**.

4. Development. In *Scypha*, development is **indirect** including a larval stage in its life cycle. Development up to larval stage occurs in the mesohyl. It follows the following steps:

(*i*) Cleavage and blastulation. Cleavage of egg is **holoblastic** (*i.e.*, entire zygote is divided by the cleavage furrow). First three cleavage furrows are vertical producing a disc of eight equal sized pyramidal cells or **blastomers**. The fourth cleavage furrow is horizontal and divides blastomeres unequally into eight small-sized **micromeres** and eight large-sized **macromeres**. A cavity called **blastocoel**, develops in between them and the embryo of *Scypha* enters a **blastula stage**. In *Scypha*, larva develops at blastula stage. Micromeres divide more rapidly, become elongated and each cell develops a flagellum at its inner end facing the blastocoel. Macromeres remain undivided, they become rounded and granular. A mouth opening develops in the middle of macromeres of the embryo which is now called **stomoblastula**. Stomoblastula feeds on adjacent maternal cells.

Amphiblastula larva. The stomoblastula undergoes **inversion** and the flagellated ends of flagellated micromeres come to lie to the exterior. Now it is called **amphiblastula larva**. This larva is hollow and its one hemisphere is composed of small flagellated cells and the other hemisphere has large non-flagellated macromeres. The fully formed amphiblastula is set free into the radial canal, escapes through the osculum along with outgoing water currents and leads a free swimming existence (Fig. 13.12 E).

Gastrulation. After swimming freely for a few hours, the amphiblastula larva undergoes **gastrulation**. During gastrulation, the flagellated half of the larva invaginates (**emboly**),or the non-flagellated macromeres overgrow the micromeres (**epiboly**), or both processes occur together and a two layered **gastrula** is formed. The central cavity opens to exterior by a wide opening called **blastopore** (Fig. 13.12 G).

Olynthus larva. The gastrula of *Scypha* attaches itself to a solid object by its blastoporal end and grows as a young sponge, called **olynthus** (Fig. 13.12 I). The central cavity of gastrula develops into spongocoel. The gastrula lengthens into a cylinder and develops **osculum** at the free distal end. The outer layer of macromeres of gastrula develops into the **pinacoderm** (pinacocytes, porocytes) while the inner flagellated layer of micromeres develop into **choanoderm** (choanocytes). Both of these layers produce the gelatinous matrix and amoebocytes of mesohyl. **Scleroblasts** are formed by macromeres and **archeocytes, amoebocytes,** etc., are formed by micromeres.

Development of adult from olynthus larva. The olynthus stage of *Scypha* resembles a simple asconoid sponge. Soon the body wall of the olynthus pushes out first at the middle of the body to form a radial canal. Gradually more radial canals and incurrent canals develop. The choanocytes shift into radial canal. Body wall of *Scypha* is increased in thickness due to growth of mesohyl and formation of spicules. Thus, an adult *Scypha* is formed.

13.7 GEMMULES

The gemmules are asexual reproductive bodies of sponges. They are regarded as the **internal buds**. Gemmules are resistant packets of essential cells and are formed by all freshwater sponges (*Spongilla*, *Ephydatia*, etc.) and a few marine sponges (*Ficulina*, *Tethya*) to tide over winter and drought (*i.e.*, adverse ecological conditions).

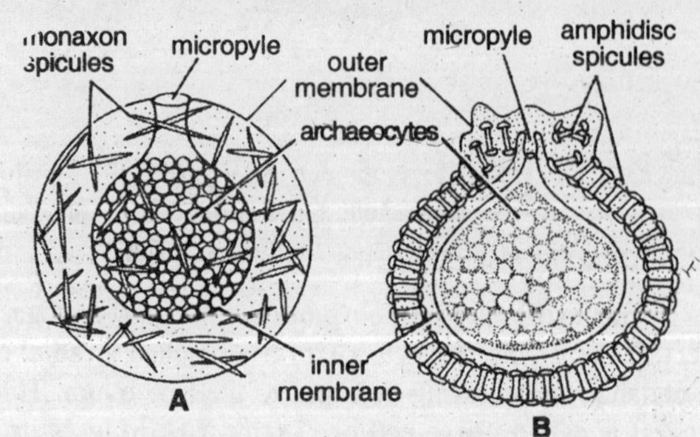

Fig. 13.13. Gemmules of freshwater sponges. A—Gemmule of *Spongilla*; B—Gemmule of *Ephydatia*.

Formation of Gemmules

Freshwater gemmules begin as masses of **archeocytes** in mesohyl (mesogloea). They are filled with food from special nurse cells, called **trophocytes.** Other amoebocytes called **spongocytes** form a covering around the central mass and differentiate into a layer of columnar cells that secrete a hard two-layered **shell** containing spongin-like material. **Scleroblasts** from the parent mesohyl form special gemmule spicule, depositing them into double-layered shell. The thick resistant double-layered shell of gemmule is broken at the **micropyle**, a pore through which the cells emerge during germination of gemmule for the formation of a new sponge.

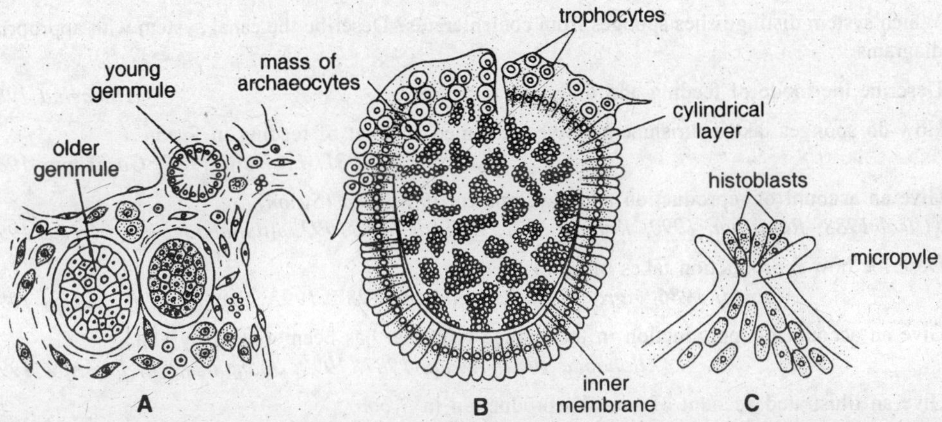

Fig. 13.14. Gemmules of marine sponges. A—Early stages in the formation of gemmule in marine sponge; B—A gemmule of marine sponge; C—Germination of gemmule.

Structure of gemmule of freshwater sponge. A gemmule is a rounded or flask-shaped structure. It is a rounded mass of food-laden archeocytes, enclosed in a double membrane capsule or shell. The inner membrane is thin and outer thick. Between the two membranes are radially arranged **spicules**. These are of monaxon type in freshwater sponges (*Spongilla*) but of amphidisc type in the marine sponges. The amphidisc spicule (of gemmule) have spiny arms and have hooks on their both ends.

Formation of marine gemmules. Marine gemmules also begin as a mass of amoebocytes (archeocytes), surrounded by other cells. The outer layer of cells develops into a flagellated surface epithelium. At maturity, gemmules have a flagellated and a non-flagellated pole. After a free-swimming period, they settle and attach at the non-flagellated end to develop into young sponges (see **Meglitsch** and **Schram**, 1991).

Germination of Gemmules

Gemmules are formed in unfavourable climatic conditions especially in winter. After the formation of gemmules, the sponge body disintegrates and its remains along with gemmules are deposited at the bottom. In spring when weather is favourable the archeocytes come out of the micropyle. Prior to their escape from the gemmule, archeocytes divides to form daughter cells, called **histoblasts**. Histoblasts differentiate into various types of cells found in sponge body. These reorganize to form body wall of the sponge.

Significance. Formation of gemmules is a life-saving mechanism. These are highly resistant and are not harmed by freezing temperature and drying while the parent sponge dies. They help in multiplication and dispersion of races. Thus, gemmules form a significant means for the continuation and multiplication of race.

QUESTIONS

Long Answer Questions

1. Describe the main histological elements found in *Sycon* and indicate their functions.
2. Draw neat and labelled diagram of L.S. of *Sycon*. Describe the structure and functions of different types of cells in it.
3. Describe different kinds of cells found in the body wall of *Sycon*/Porifera and mention their functions.
4. With the help of suitable diagrams describe the canal system in *Sycon*.
5. Which system distinguishes sponges from coelenterates? Describe the canal system with appropriate diagrams.
6. Describe the mode of feeding and digestion in *Sycon*. (*Himachal 1989*)
7. How do sponges take nourishment? Give a detailed account of feeding in *Sycon*. (*Awadh 1981; Lucknow 1988; Gorakhpur 1989*)
8. Give an account of reproduction and development in *Scypha* (*Sycon*). (*Utkal 1988; Rajasthan 1990; Kanpur 1991; Bundelkhand 1992; Allahabad 1992; Meerut 1994*)
9. Describe how reproduction takes place in *Sycon*. (*Jiwaji 1986; Agra 1988; Bundelkhand 1987, 1992; Bhopal 1990; Meerut 1994*)
10. Give an account of reproduction in the sponge type that has been studied by you. (*Allahabad 1982; Awadh 1987, 1990; Jodhpur 1990; Lucknow 1997*)
11. Give an illustrated account of sexual reproduction in *Sycon*. (*Kanpur 1987; Meerut 1987, 1991; Ravishanker 1991; Purvanchal 1991; Bhopal 1991*)
12. How does sexual and asexual reproduction takes place in *Sycon*?
13. Describe the sexual reproduction and development of *Sycon*. (*C.C.S.U. Meerut 2008*)

Short Answer Questions

1. Describe habits, habitat and external structure of *Scypha*.
2. Give the natural history and morphology of *Sycon*.
3. Give the systematic position of the *Sycon*.
4. Describe the structure of *Sycon*.
5. Describe the geographical distribution and ecology of *Sycon*.
6. Write short note on the importance of canal system of *Sycon*.
7. Give an account of canal system of *Sycon* and explain the mechanism and importance of circulation of water through it.
8. Explain the physiological importance of canal system in sponges. Describe the canal system of *Sycon*.
9. Write short notes on:

   (*i*) Gemmule (*Purvanchal 2001; C.C.S.U. Meerut 2005*)
10. Draw a neat and labelled diagram of T.S. of *Sycon* (No description is needed). (*C.C.S.U. Meerut 2007*)

Very Short Answer Questions

1. What is the common name of *Scypha*?

   Ans. Crown sponge.
2. Name the order of *Scypha*.

   Ans. Heterocoela.

3. What is the class of *Scypha*?

Ans. Calcarea.

4. Which cells maintain a current of water in *Scypha*?

Ans. Choanocytes.

5. Where are amoebocytes located in *Scypha*?

Ans. Mesohyl (Mesenchyme).

6. Name the cell that carries a sperm to an ovum during fertilization of *Scypha*.

Ans. Carrier choanocyte.

7. Which embryonic stage undergoes inversion in *Scypha*?

Ans. Stomoblastula.

8. Name the phenomenon that prevents self fertilization in *Scypha*.

Ans. Protogyny.

9. Which is the main excretory product of *Scypha*?

Ans. Ammonia.

10. Name the spicules which form the oscular fringe.

Ans. Long monaxons.

11. How many scleroblasts produce a triradiate spicule?

Ans. Three.

12. Where are oxeotes located in *Scypha*?

Ans. Oxeotes are monaxon spicules which project out from dermal cortex of mesohyl.

13. What is the name of free-swimming larval form of *Scypha*?

Ans. Amphiblastula larva.

14. What is the function of choanocytes? (*C.C.S.U., Meerut 2004*)

Ans. (*i*) Produce water current by beating of flagella and direct it into the canal system.

(*ii*) Draw food from water current toward the collar where it is captured by phagocytosis and transferred to amoebocytes for digestion.

(*iii*) Take part in reproduction by modifying into spermatogonia.

(*iv*) Help in internal fertilization by transferring sperm to oocyte.

Fill in the Blanks

1. Water current enters the *Scypha* (*Sycon*) through .... and leaves via ....

2. In *Scypha*, choanocytes line the .... canals.

3. In *Scypha*, the pinacocytes line the .... , excurrent canals and ....

4. Surface of the *Scypha* is covered by cells called ....

5. In *Scypha*, the incurrent canals communicate with the radial canals by ....

6. In *Scypha*, gastric ostia / atria lead from .... into the ....

7. In *Scypha*, the apertures between radial canal and excurrent canals are called ....

8. Spicule-secreting cells of *Scypha* are called ....

9. The food storing cells of *Scypha* are known as ....

10. The spicules of *Scypha* include monaxon, ... and tetraradiate.

Multiple Choice Questions

1. The common name of *Scypha* or *Sycon* is
(*a*) bath sponge (*b*) crown sponge
(*c*) venus flower basket (*d*) glass rope sponge

2. Canal system of *Scypha* is
(*a*) ascon type (*b*) syconoid type
(*c*) leucon type (*d*) rhagon type

3. The mesohyl is largely secreted by
(*a*) collencytes (*b*) trophocytes
(*c*) choanocytes (*d*) chromocytes

4. In *Scypha*, food is firstly ingested by choanocytes and digestion is completed in
(*a*) amoebocytes (*b*) chromocytes
(*c*) thesocytes (*d*) archaeocytes

5. The exit pore of *Scypha* is
(*a*) ostia (*b*) osculum
(*c*) spongocoel (*d*) prosopyle

6. In *Scypha*, water currents are produced by
(*a*) choanocytes (*b*) pinacocytes
(*c*) archaeocytes (*d*) thesocytes

7. What is the name of hollow blastula of *Scypha*?
(*a*) streogastrula (*b*) coeloblastula
(*c*) stomoblastula (*d*) amphiblastula

8. Name of the larva of *Scypha* is
(*a*) amphiblastula (*b*) parenchymula
(c) stomoblastula (*d*) ephyra

9. Young stage of syconoid sponge is
(*a*) coeloblastula (*b*) parenchymula
(*c*) olynthus (*d*) stomoblastula

ANSWERS

Fill in the Blanks

1. Dermal ostia, osculum **2.** Radial
3. Incurrent canal, spongocoel **4.** Pinacocytes
5. Prosopyles **6.** Excurrent canals, spongocoel
7. Apopyles **8.** Calcoblasts
9. Thesocytes **10.** Triradiate.

Multiple Choice Questions

1. (*b*) **2.** (*b*) **3.** (*c*) **4.** (*a*)
5. (*b*) **6.** (*a*) **7.** (*c*) **8.** (*a*)
9. (*c*).

14

Canal System in Sponges

The passages through which water constantly flows from outside the body to the interior of the body and then outside again, form the **canal system** in sponges. In majority of sponges the body wall is folded and is traversed by a complex system of canals (*e.g.*, incurrent canals, radial canals and excurrent canals) through which water current travels to reach the spongocoel and ultimately exit through osculum. Sponges contain the following three types of canal system:

1. Asconoid or ascon type canal system
2. Syconoid or sycon type canal system
3. Leuconoid or leucon type canal system.

14.1 ASCONOID TYPE CANAL SYSTEM

It is simplest type of canal system which is found in *Leucosolenia* and olynthus larval stage of *Scypha*. In this type of canal system, there is a radially symmetrical vase-like body consisting of a thin body wall enclosing a large central cavity, the **spongocoel**. The spongocoel opens at the free end by a narrow circular aperture, the **osculum**. An epidermis of close-fitting pinacocytes (pinacoderm) covers the surface of the ascon wall. Scattered among the pinacocytes are tubular **porocytes** with the **ostia**. Collared choanocytes (forming choanoderm) line the spongocoel. Their flagella force water toward the osculum and draw it into spongocoel through the ostia. A jelly-like **mesogloea** or **mesohyl** fills the space between the pinacoderm and choanoderm and contains wandering amoebocytes, calcareous monaxon and triaxon spicules, etc.

Thus, in asconoid type canal system, the water current enters through ostia, reaches the spongocoel and finally leaves the body through osculum.

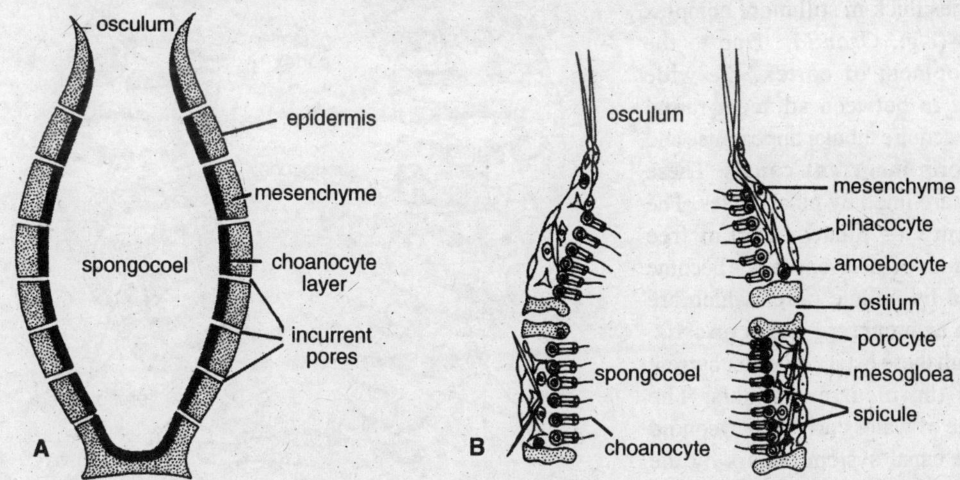

Fig. 14.1. Asconoid canal system A—Diagrammatic view; B—Schematic cellular structure of upper part of asconoid canal system.

14.2 SYCONOID TYPE CANAL SYSTEM

It is a step forward in the evolution of canal system. Syconoid canal system is formed when asconoid body wall gives out finger-like processes at regular intervals. These finger-like outgrowths are lined by choanocytes and they are called **flagellated** or **choanocyte chambers** or **radial canals**. They open to spongocoel via apopyle and excurrent canals. As choanocytes come to line the radial canals, the spongocoel is lined with flattened pinacocytes (called **endopinacocytes**). The wide opening between two adjacent radial canals forms the **ostium** and space between them is called **incurrent canal**. Incurrent canals communicate with radial canal through many **prosopyles**.

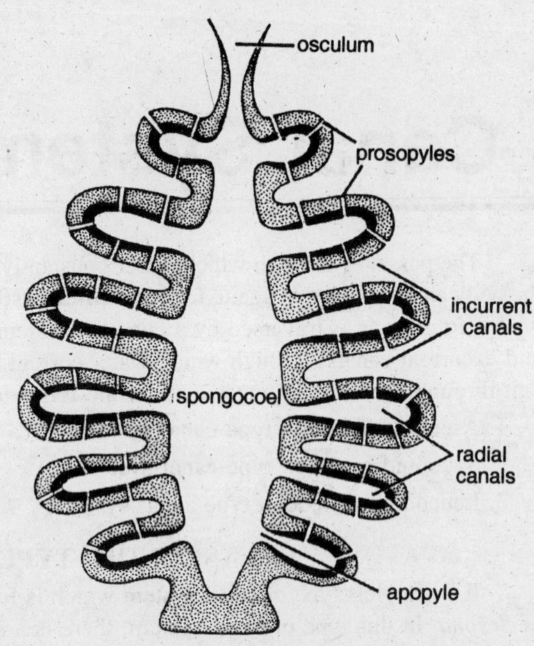

Fig. 14.2. Syconoid I type canal system.

Syconoid type canal system is of two types: **syconoid I** and **syconoid II**. Syconoid I type canal system is simple and found in *Sycetta* (*i.e.*, heterocoelous sponges). In *Sycetta*, each radial canal is a simple outpocketing of the spongocoel and is exposed directly to the surrounding water (Fig. 14.2).

In more complex forms (**syconoid II type canal system**), the outer margins of all radial canals are attached, *i.e.*, the pinacoderm and mesohyl spread over the outer surface forming **cortex** (Fig. 14.3). Cortex may be thin in simple forms (*e.g.*, *Scypha*) and becomes thick in still more complex forms (*e.g.*, *Grantia*). Due to the development of cortex, the wide spaces in between adjacent radial canals acquire tubular appearance and they form **incurrent canals**. These canals are lined by pinacocytes. The epidermis (= pinacoderm) at free distal ends of incurrent canals become pierced by minute pores which are known as **incurrent pores** or **ostia**. Through these ostia, the water current enters the incurrent canals. The passage of water current in syconoid II type canal system occurs by the following route:

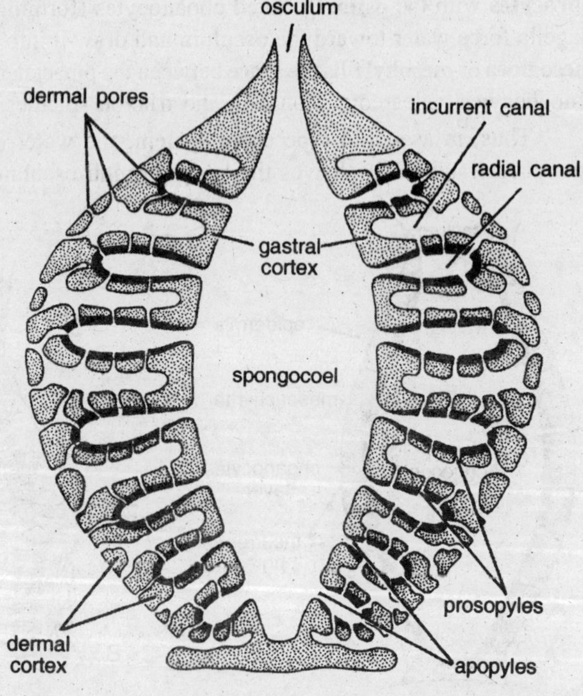

Fig. 14.3. Syconoid II type canal system having cortex.

Entry of outside water —ostia→ **Incurrent canal** —prosopyles→ **Radial canal**

—apopyle→ **Excurrent canal** —gastric ostia→ **Spongocoel** —osculum→ **Exit of water to outside**

14.3 LEUCONOID TYPE CANAL SYSTEM

Leuconoids are the most complexly organized sponges. Leuconoid structure develops by further outfolding of the radial canals of syconoid stage and thickening of mesohyl. Most leuconoid-type sponges form large masses containing many **oscula**. Each osculum marks one member of the large colonial mass, but the limits of the members are so ill-defined that it is impossible to distinguish individuals clearly.

Leuconoid organization involves folding and outpocketing of the radial canals to form clusters of round or oval **flagellated chambers** or **choanocyte chambers** (Fig. 14.4). The flagellated chambers surround an excurrent canal into which they discharge water. Smaller excurrent canals merge to form larger ones; all eventually unite to form a major excurrent canal through which water reaches the osculum. Spongocoel is, thus, obliterated in leuconoid sponges. The cortex of sponge contains a system of branching incurrent canals, delivering water in the flagellated chambers by way of small pores, the **prosopyles**. Water leaves the flagellated chambers through the **apopyles**. Ostia may lead directly into the incurrent canal or open into the **subdermal spaces** (or cavities) crossed by spicules. The leuconoid type of canal system shows the following three evolutionary grades:

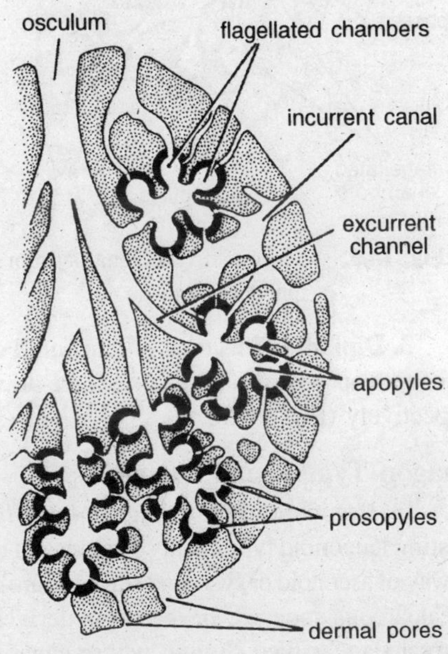

Fig. 14.4. Leuconoid type canal system showing eurypylous chambers.

1. Eurypylous
2. Aphodal
3. Diplodal

1. Eurypylous. It represents simplest type of leuconoid canal system and occurs in *Leucilla*. In eurypylous canal system, the flagellated chambers lead into excurrent canals through wide apopyles (Fig. 14.4). In it occurs the following course of water currents:

Entry of outside water —ostia→ **Subdermal cavity and incurrent canals** —prosopyles→

Flagellated chambers —apopyles→ **Excurrent canals** ——→ **Large channels** —osculum→

Exit of water to exterior.

2. Aphodal. In certain leuconoid sponges such as *Geodia* and *Stelleta*, the apopyles instead of being wide openings are drawn into narrow canals. These canals are called **aphodus** and open into excurrent canals (Fig. 14.5).

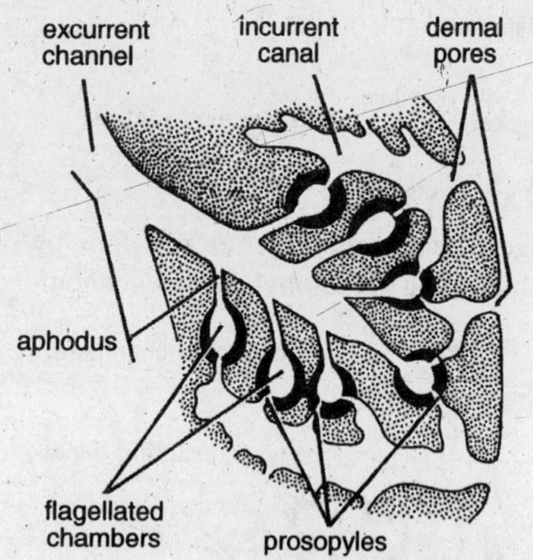

Fig. 14.5. Leuconoid type canal system with aphodal chambers.

Fig. 14.6. Leuconoid type canal system with diplodal chambers.

3. Diplodal. This type of leuconoid canal system is found in *Spongilla* and *Oscarella*. In this case, both apopyles and prosopyles are drawn out into narrow canals called **aphodus** and **prosodus**, respectively (Fig. 14.6).

Rhagon Type Canal System

In Demospongiae (*e.g.*, *Spongilla*), the existing leuconoid type canal system is not derived by way of asconoid or syconoid stages as are evident in calcareous sponges. Instead it is derived from a larval stage, called **rhagon** whose canal system is **rhagon type** (Fig. 14.7). The larval canal system includes flattened pyramid-like form with broad base, called **hypophare**, a spacious **spongocoel** and lateral walls, called **spongophares**. Flagellated chambers occurs in each spongophare. The dermal pores leading to flagellated chambers are known as **prosopyles** and their openings into spongocoel are termed as **apopyles**. The spongocoel opens to the exterior by **osculum**.

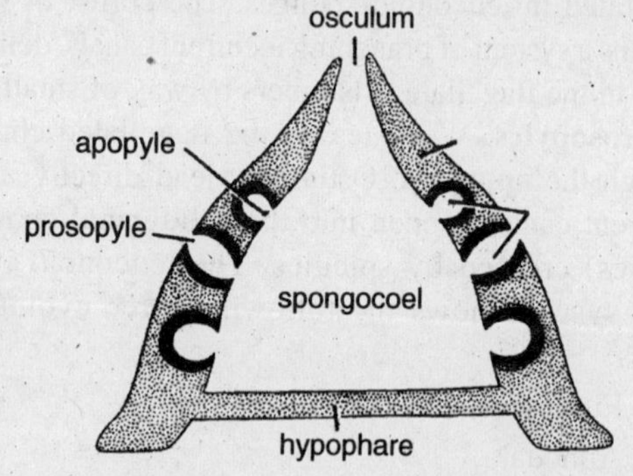

Fig. 14.7. *Rhagon* larva. V.S. of body showing rhagon type canal system.

Thus, in *Spongilla* (Fig. 14.8), between the flagellated chambers and the pinacoderm (or epidermis) lies a considerable thickness of mesohyl which is traversed by subdermal spaces and incurrent canals is as follows:

Entry of outside water —ostia→ **Subdermal spaces** ⟶ **Incurrent channels** —prosopyles→ **Flagellated chambers** —apopyles→ **Excurrent canals** ⟶ **Spongocoel** —osculum→ **Exit of water to exterior.**

Lastly, the volume of water pumped by a sponge is remarkable. For example, a specimen of *Leuconia*, a leuconoid sponge 10 cm in height and 1 cm in diameter has roughly 2,250,000 radial canals and pumps 22.5 litre of water per day through its body. The flow velocity is greatest through the osculum and slowest through the radial canals, because these two regions have, respectively, the smallest and the largest total cross sectional areas of the various passageways.

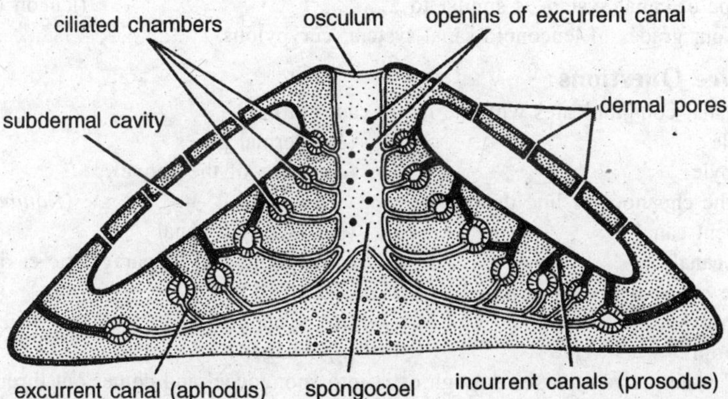

Fig. 14.8. *Spongilla*. Diagrammatic V.S. of body showing rhagon type of canal system.

QUESTIONS

Long Answer Questions

1. Describe the various types of canal systems found in sponges.
(*Magadh 1989, 1995; Allahabad 1991, 1995; Raipur 1991; BHU 1991; Rohilkhand 1992; Gorakhpur 1992, 1996; Agra 1993; Rewa 1993; Himachal 1993*)
2. Give an account of rhagon canal system. (*Himachal 1995*)
3. Define canal system. Give an account of canal system of Porifera.
(*Punjab 1990; Bhopal 1991; Jiwaji 1991; Jabalpur 1991*)
4. Why is the canal system important in the phylum Porifera? Describe the different types of canal systems found in Porifera.
5. Give an account of the canal system in Porifera and discuss its importance.
(*Allahabad 1991, 1997*)
6. Define canal system and give the details of a leuconoid type of canal system. (*Punjab 1991*)

Short Answer Questions

1. Describe syconoid canal system.
2. Discuss importance of canal system of sponges.

Very Short Answer Questions

1. Name the leuconoid canal system with normal prosopyles and apopyles.
Ans. Eurypylous.
2. Give an alternative name for leuconoid canal system in Demospongiae.
Ans. Rhagon.

Fill in the Blanks

1. Sponges acquire food by filtering water. The food-containing water current enters a sponge through .... and is then moved through the internal cavity to the .... by the flagella of .... As the water moves, food is removed by the ....
2. In sponges a constant current of water passes through ostia into spongocoel and comes out through oscula. This is known as .... (water vascular system/canal system)
3. The simplest type of canal system in sponges is ....
4. Canal system found in Demospongia is ....
5. In the leucon type of canal system when the flagellated chambers open directly by wide apopyles into the excurrent canal, it is called ....
6. The cells of a sponge that regulate the entry of water by closing pores in response to irritating stimuli are the .... cells.
7. In the leucon type of canal system when another narrow tube, called prosodus is present between the incurrent canal and the flagellated chamber, it is called .... (rhagon/diplodal/apodal)

8. Rhagon type of canal system is similar to .... (leucon type/ascon type)
9. There are four grades of leuconoid canal system: eurypylous, ... , ... , and diplodal.

Multiple Choice Questions

1. Incurrent canal communicates with the radial canal through (*Kanpur 1982*)
   (*a*) apopyle (*b*) dermal pores
   (*c*) prosopyle (*d*) none of the above
2. In *Sycon*, the choanocytes line the (*Kanpur 1982, 86, 87*)
   (*a*) excurrent canal (*b*) incurrent canal
   (*c*) radial canal (*d*) spongocoel or paragastric cavity
3. Which type of canal system is found in *Leucosolenia*? (*Lucknow 1995*)
   (*a*) asconoid (*b*) syconoid
   (*c*) leuconoid (*d*) rhagon
4. In sponges, water current passes through ostia into spongocoel and comes out through oscula. This is called
   (*a*) water vascular system (*b*) canal system
   (*c*) feeding current (*d*) circulating current
5. The canal system in the sponge *Olynthus* is of
   (*a*) asconoid type (*b*) syconoid type
   (*c*) rhagon type (*d*) leuconoid type
6. Diplodal canal system is identified by the presence of
   (*a*) aphodus only (*b*) prosodus only
   (*c*) both aphodus and prosodus (*d*) none of the above
7. Which of the following is the exhalant pore? (*Magadh 1995*)
   (*a*) ostium (*b*) prosopyle
   (*c*) apopyle (*d*) osculum
8. Flagellated cells of sponges are (*Lucknow 1997*)
   (*a*) pinacocytes (*b*) choanocytes
   (*c*) monocytes (*d*) thesocytes
9. Which one of the following cells maintains water currents in sponges?
   (*a*) porocyte (*b*) choanocyte
   (*c*) myocyte (*d*) chromocyte
10. The system through which water travels through body of a sponge is called
    (*a*) blood vascular system (*b*) canal system
    (*c*) excretory system (*d*) water vascular system
11. Spongocoel is lined by choanocytes in the canal system of
    (*a*) asconoid type (*b*) syconoid type
    (*c*) rhagon type (*d*) syconoid and leuconoid type
12. The canal system of *Euplectella* is
    (*a*) asconoid (*b*) syconoid
    (*c*) eurypylous (*d*) leuconoid

ANSWERS

Fill in the Blanks

1. Ostia, osculum, choanocytes, choanocytes
2. Canal system
3. Ascon type
4. Leucon type
5. Euryplous
6. Epidermal
7. Diplodal
8. Leucon type
9. Aphodal, prosodal

Multiple Choice Questions

1. (*c*) 2. (*c*) 3. (*a*) 4. (*b*) 5. (*a*) 6. (*c*)
7. (*d*) 8. (*b*) 9. (*b*) 10. (*b*) 11. (*a*) 12. (*c*).

15

Skeleton of Sponges

The skeleton of sponges is relatively complex and provides a supporting framework for the living cells of the animal. It is formed of spicules or organic fibre and is primarily located in the mesohyl, but spicules frequently project through the pinacoderm. The skeleton may be composed of **calcareous spicules** (of calcium carbonate), **siliceous spicules** (of silicon dioxide) or fibres of **spongin B** (a scleroprotein), or a combination of the latter two. Spicules have an axis of organic material around which is deposited the inorganic substance. The fully formed calcareous spicule is surrounded by a sheath but lacks any axial filament of an organic nature. Siliceous spicule, on the other hand, has a core in the form of an axial filament. (Box 15.1).

Box 15.1

Skeletal elements of sponges are of two types, **organic** or **inorganic**. The organic element is always **collagenous** and inorganic element is either **siliceous** (hydrated silicon dioxide) or **calcareous** (calcium carbonate in the form of calcite or aragonite). Sponges are the only animals that use hydrated silica as a skeletal material (**Bursca** and **Bursca**, 2003).

15.1 TYPES OF SPICULES

The skeletal elements of sponges exist in a variety of form and are important in the identification and classification of species. Spicules are of two general kinds—megascleres and microscleres. The microscleres are small to minute reinforcing (or packing) spicules; the megascleres are large structural spicules. The demosponges and hexactinellids have both types but calcareous sponges often have

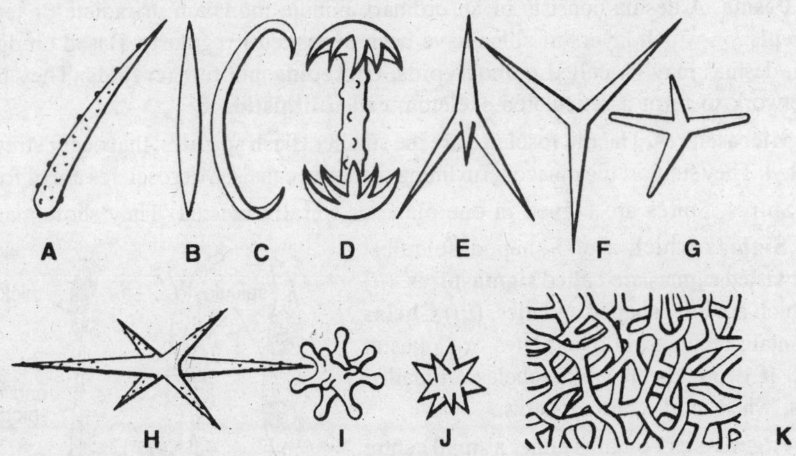

Fig. 15.1. Skeleton of sponges. A—Monactinal monaxon (style); B—Diactinal monaxon (oxea); C—Curved monaxon (sigma); D—Monaxon with hooked ends; E—Tetraxon; F—Triradiate; G—Calthrops; H—Hexactinal triaxon; I and J—Polyaxon; K—Spongin fibres.

only megascleres. The suffix—*axon* refers to the number of axes a spicule has, while—*actine* refers to the number of rays or points.

A. Megascleres. The megascleres are the larger 'skeletal spicules' that constitute the chief supporting framework of the sponge. These are of following six types:

(1) Monaxons. These spicules have single axis and are shaped like needles or rods or may be curved or straight. They may contain pointed, knobbed or hooked ends. When growth of spicule has occurred in one direction only, the spicule is called **monactinal monaxon** (Fig. 15.2) or **style.** Styles are typically rounded (called **strongylote**) at one end and pointed (called **oxeote**) at the other end. Styles in which broad end is knobbed are called **tylostyles**; those curved with thorny processes are called **acanthostyles.**

Monaxons that develop by growth in both directions from a central point are named **diactinal monaxons, diactines** or **rhabds**. Rhabds may be of following four types:

(*i*) **Oxeas** which are pointed at both ends. (*ii*) **Tornotes** which are lance-headed on both ends. (*iii*) **Strongyles** which are rounded at both ends. (*iv*) **Tylotes** which are knobed at each end, like a pin head.

2. Tetraxons. These four-rayed spicules are also called **tetractines** or **quadriradiates** spicules. Four rays of the tetraxon spicule radiate from a common point and may be equal in size; such a spicule is called **calthrops**. In **triaenes** spicule, one ray is elongated, called **rhabdome** which bears a crown of three smaller rays. When triaenes loses one smaller ray, it is called **diaene**. If the elongated ray bears a disc at both ends, it is called **amphidisc**. The **triod, triradiate** or **triactinal** spicule which is common in calcareous sponges, lacks an elongated ray.

3. Triaxons. The triaxon (three-rayed) or **hexactinal** (six-rayed) spicule consists of three axes crossing at right angles, producing six rays extending at right angles from a central point. These spicules too become modified secondarily to produce many varieties. Triaxon spicules are characteristics of class Hexactinellida.

4. Polyaxons. In polyaxon spicules several equal rays radiate from a central point.

5. Spheres. These are rounded bodies in which growth is concentric around a centre.

6. Desma. A desma consists of an ordinary minute monaxon, triradiate or tetraxon spicule called **crepis**, on which layers of silica have been deposited irregularly. Based on different shapes of crepis, desmas may be called **monocrepids, tricrepids** and **tetracrepids**. They become united into a network to form a reticulated skeleton, called **lithistid.**

B. Microscleres. The microscleres are the smaller 'flesh spicules' that occur strewn throughout the mesohyl. They support the pinacoderm lining the water canals. Microscleres are of following types:

1. Spires. Spires are curved in one plane or spirally twisted. They show many shapes:

(*i*) **Sigmas** which are C-shaped spicules. Spirally twisted sigmas are called **sigmaspires**. (*ii*) **Toxas** which are bow-shaped spicules. (*iii*) **Chelas** which contain recurved hooks, plates or flukes at each end. If two ends are alike, chelas are called **isochelas**, when unlike, **anisochelas**.

2. Asters. Aster contains either a small centre and long rays or a large centre and small rays. Among the small-centred asters are **oxyasters** with pointed rays, **strongylaster** with rounded ends and **sterrasters** with rays reduced to small projections from the spherical surface. Short spiny microscleric monaxons are known as **streptasters.**

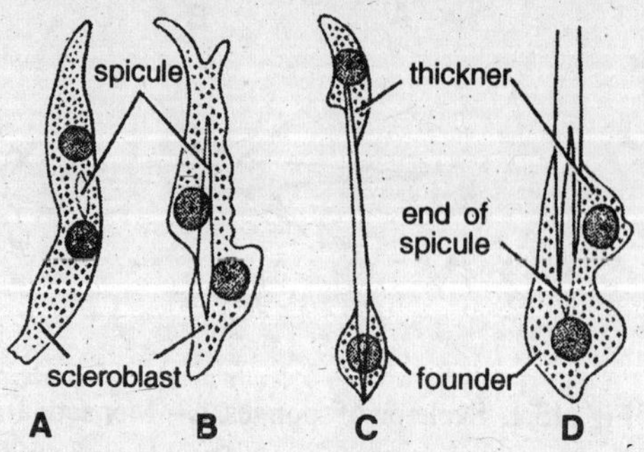

Fig. 15.2. Formation of a monaxon spicule.

15.2 DEVELOPMENT OF SPICULES

Spicules are secreted by **scleroblasts** or **sclerocytes** which are amoebocytes of mesohyl. Sclerocytes are active cells that possess abundant mitochondria, cytoplasmic microfilaments and small vacuoles. The process of development of spicules is best known only for calcareous spicules.

1. Development of monaxon spicules. In calcareous sponges monaxon spicule is secreted within a binucleate **calcoblast** which is formed due to incomplete division of an ordinary calcoblast (*i.e.*, calcareous spicule forming scleroblast). The calcium carbonate is deposited around an organic axial thread contained in the cytoplasm between two nuclei. As the spicule lengthens, the two nuclei draw apart until calcoblast divides into two cells. One cell, the **founder** is located at the inner end, the other, the **thickner** at the outer end of the spicule. The spicule is laid down chiefly by the founder cell which establishes the shape and length of spicule. The thickner cell deposits additional layers of calcium carbonate. When the formation of spicule is completed, both cells (founder and thickner) leave from the inner end of the spicule and return to mesohyl (Fig. 15.2).

The development of siliceous monaxon spicule takes place in one scleroblast called **silicoblast**.

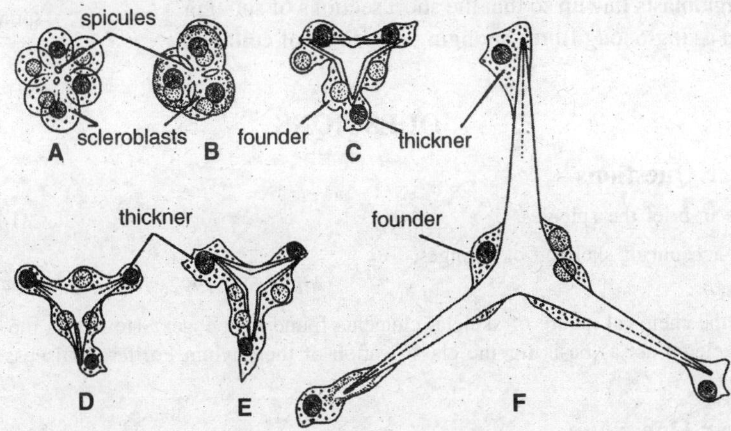

Fig. 15.3. Formation of a triaxon spicule.

2. Development of triaxon spicules. Triaxon calcareous spicules are secreted by three calcoblasts which come together in a triangle (Fig. 15.3). Each calcoblast of the trio of cells divide into two, an inner founder and outer thickner. Each pair secretes a prong or ray of the spicule; these three rays fuse at the base to form a small triradiate spicule. Each ray is then completed in the same manner as a monaxon spicule.

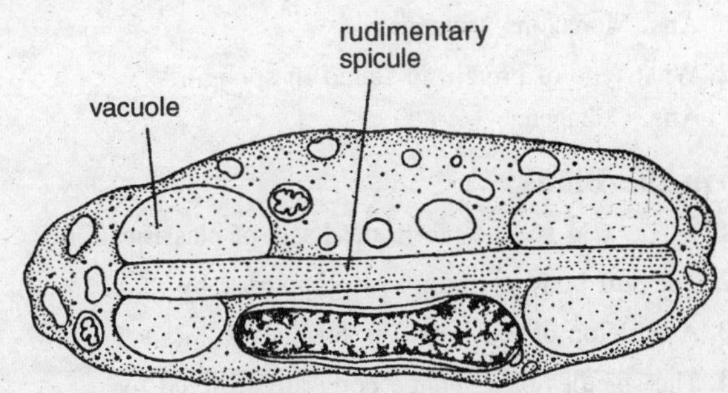

Fig. 15.4. A sclerocyte of *Mycale* (Demospongiae) with rudimentary siliceous spicule extending between two vacuoles.

3. Development of other types of spicules. In the formation of tetraxon or quadriradiate spicules, the fourth prong or ray is added to developing triradiate spicule by an additional sclero-blast. The hexactinal (siliceous) spicule arises in the centre of a multinucleate syncytial mass which is probably formed by repeated nuclear divisions of an original silicoblast.

Spongin: Structure and Secretion

Skeleton of sponges of class Demospongiae is composed of fibres of halogenated scleroprotein termed **spongin B**. Spongin exists in two forms: 1. As a cuticular secretion of a tenacious but elastic cementing substance which glues siliceous spicules together into a definite system of skeletal fibres. 2. In the form of elastic fibrillae secreted in the cells. In most sponges, spongin fibres incorporate some siliceous spicules or detritus such as sand grains, radiolarian and foraminiferan skeletons. Some sponges contain so much spongin that they are tough and rubbery.

The spongin is secreted by **spongioblasts** or **spongocytes**. Each spongioblast develops a vacuole, within which spongin material collects. Spongioblasts line up so that the short sections of spongin are united into a single long fibre. Spongin is made up of **collagen** fibrils.

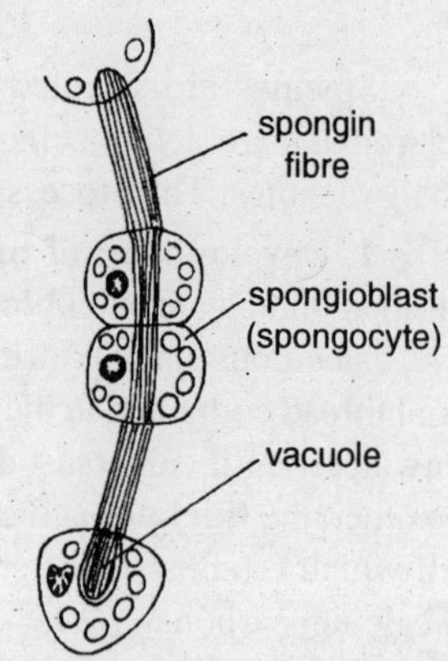

Fig. 15.5. Formation of spongin fibre by a group of spongioblasts.

QUESTIONS

Long Answer Questions

1. Describe in brief the spicules. (*Allahabad 1996*)
2. Give an account of skeleton of sponges. (*Allahabad 1992; Himachal 1994; Kanpur 1995*)
3. What is the chemical nature of skeletal elements found in sponges? How does the nature and form of the skeleton act as basis for the classification of the phylum Porifera? Discuss. (*Gorakhpur 1995*)

Short Answer Questions

1. Write a short note on spicule. (*Meerut 1991*)

Very Short Answer Questions

1. Name any two of the six major types of spicules regarding the shape.
   Ans. Monaxons, triaxons.
2. What type of protein in found in spongin.
   Ans. Collagen.

Fill in the Blanks

1. .... in sponges form the basis of classification.
2. The soft body of sponges is supported by .... (setae/spicules)
3. A skeleton of *Euspongia* is formed of .... (spongin fibre/calcareous spicules)
4. The spicules of a sponge body are secreted by .... (scleroblasts/archeocytes)
5. Spicules are of two types regarding size ....
6. Spongin fibres occur in the sponges of class ....
7. Siliceous spicules are secreted by .... (calicoblast/silicoblasts)

Multiple Choice Questions

1. Which type of spicules are found in gemmule of *Spongilla*? (*Lucknow 1995*)

(*a*) amphidisc (*b*) monaxon

(*c*) triaxon (*d*) tetraxon

2. Monaxon spicule with a knob at one end is called

(*a*) style (*b*) diactyle

(*c*) telostyle (*d*) acanthostyle

3. Most sponges have skeleton made of

(*a*) cartilage (*b*) spicules or spongin fibres

(*c*) mesogloea (*d*) pinacocytes and choanocytes

4. Skeleton in Demospongia is

(*a*) spongin fibres (*b*) calcareous spicules

(*c*) siliceous spicules (*d*) both spongin fibres and siliceous spicules

(*Kanpur 1981, 1986*)

5. Spicules are secreted by

(*a*) porocyte (*b*) choanocyte

(*c*) pinacocyte (*d*) scleroblasts

6. Hexactine spicules occur only in

(*a*) calcarea (*b*) hexactinellida

(*c*) demospongiae (*d*) calcarea and demospongiae

ANSWERS

Fill in the Blanks

1. Spicules **2.** Spicules

3. Spongin fibre **4.** Scleroblasts

5. Microscleres, macroscleres **6.** Demospongiae

7. Silicoblasts

Multiple Choice Questions

1. (*a*) **2.** (*a*) **3.** (*b*) **4.** (*a*)

5. (*d*) **6.** (*b*).

16

Affinities of Sponges (Porifera) and Origin of Metazoa

16.1 AFFINITIES OF PORIFERA

The evolutionary origin of sponges poses a number of interesting problems. The absence of organs and the low level of cellular differentiation (= division of labour) and interdependence in sponges certainly seem to be primitive characteristics. But a specialized body structure built around a canal system and lacking distinct anterior and posterior ends is found in no other groups of animals. Moreover, the cellular differentiation of sponges is unlike that of other metazoans. All these features suggest that sponges are phylogenetically remote from other metazoans (called eumetazoans).

Sponges show affinities with Protozoa as well as with Metazoa:

A. Affinities of Porifera with Protozoa

Similarities. Sponges show following similarities with *Proterospongia* (Choanoflagellida, Mastigophora; this colonial protozoa is regarded as **ancestor of Metazoa** in general since it contains a collar-ring of microvilli around a flagellum):

1. Intracellular digestion.
2. Presence of collar cells and amoeboid cells as found in *Proterospongia.*
3. Formation of skeletal spicules by independent cells.
4. The cells of sponge body are independent in their function.
5. A process known as **inversion** occurs in amphiblastula larva like those of *Volvox* (a colonial protozoan).

Differences. Sponges differ from colonial and other protozoa in following aspects:

1. Presence of canal system.
2. Presence of a characteristic skeleton.
3. The development of the multicellular body from a fertilized egg by cleavage. Due to this feature sponges become more complex than a colonial protozoan and show resemblance with Metazoa. Thus, this lone feature negates any affinity of sponges with the protozoa.

B. Affinities of Porifera with Metazoa

Similarities. Among Metazoa, coelenterates (cnidarians) show following similarities with sponges.

1. Sedentary in habit.
2. Diploblastic acoelomates, absence of cellular mesoderm.

3. Presence of non-cellular mesogloea between outer and inner layers.
4. The single body cavity, the spongocoel of sponges which opens out through a single aperture, the osculum, can be compared with the gastrovascular cavity, opening to the exterior by terminal hypostome (single opening) of coelenterates.
5. Asexual reproduction by budding to form colony.
6. The parenchymula larva of sponges is comparable with planula of Coelenterata.

Differences. The sponges differ from Metazoa in having the following features:

1. In sponges the cells are less specialized and interdependent than the cells of Metazoa.
2. Tissue formation is restricted to the formation of epithelial lining on the surface. But this epithelium lacks basal lamina and intercellular junction.
3. Sponges are diploblastic but most Metazoa are triploblastic (having three germinal layers in their body namely ectoderm, mesoderm and endoderm).
4. Osculum acts only as an exhalent aperture.
5. The sponges do not possess an anterior end or head like those of metazoa.
6. Stinging cells (cnidoblasts containing nematocysts) are not found in sponges.
7. Body surface of sponges is perforated by numerous minute pores and they possess a unique canal system in their body. All these characters are absent in Metazoa.
8. Sponges lack nervous tissue to coordinate the body activities.
9. In the development of sponges, the flagellated upper cells of the embryo situated at the animal pole become the choanoderm (= endoderm) and the granular cells of vegetal pole becomes the pinacoderm (= ectoderm). Such sort of inversion is unique and does not occur in other Metazoa. This character indicates that sponges and coelenterates have separate origin from Protozoa.

Conclusion. Visualizing the above affinities of Porifera, there can be little doubt that sponges diverged early from main line of metazoan evolution and have given rise to no other members of the Animal Kingdom. Because of their isolated phylogenetic position, the sponges often been placed in a separate subkingdom, the **Parazoa**, distinct from the other multicellular animals, the **Eumetazoa**.

16.2 ORIGIN OF METAZOA

Unraveling the origin of the multicellular animals (metazoans) has presented many problems. There is no direct proof about the ancestors of Metazoa. But all zoologists agree that the metazoans have evolved from some unicellular or acellular organisms, such as Protozoa. This is strongly evident by the facts of the embryonic development in which each metazoan passes from an unicellular (zygote) to a cellular condition (morula, blastula, gastrula, etc.) There is much disagreement over which group of unicellular organisms gave rise to them and how. The following three prevalent hypotheses (or theories) in current use are

1. That the metazoans arose from a syncytial (multinucleate) ciliated form from which cell boundaries later evolved. This is known as *syncytial ciliate hypothesis* or *syncytial theory*.
2. That the origin of metazoans was polyphyletic, or derived from more than one group of unicellular organisms. This is known as *polyphyletic origin hypothesis*.
3. That they arose from a colonial flagellated form in which the cells gradually became more specialized and interdependent. This is known as *colonial flagellate hypothesis*.

1. Syncytial Ciliate Hypothesis

Resemblance between multinucleate ciliates and acoelomate flatworms led **E.D. Hanson** (1958) and **Jovan Hadzi** (1963) to propose a ciliate origin for metazoans. Proponents of syncytial theory believed that metazoans arose from an ancestor shared with the single-celled ciliates. The common ancestor of metazoans acquired multiple nuclei with a single cell membrane and later became compartalized into the multicellular condition. It is assumed that the body form of the ancestor resembled that of modern ciliates and thus tended toward bilateral symmetry. Therefore the earliest metazoans would have been bilateral and similar to the present primitive flatworms.

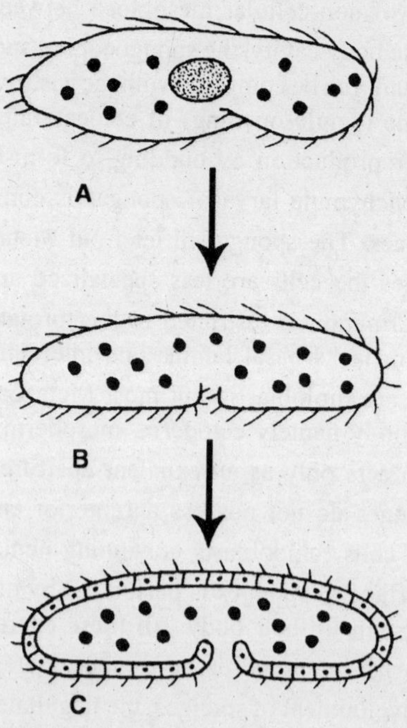

Fig. 16.1. Syncytial theory. A—Schematic representation of the hypothetical transition from a multinucleate, ciliated protist. B—The hypothetical metazoan precursor (sagittal section) as it assumed a benthic, crawling lifestyle and developed a ventral mouth and simple pharynx. C—The hypothetical metazoan precursor (sagittal section) after it achieved the acoel grade via cellularization of the epidermis, which surrounded a syncytial endodermis or entodermis (after Brusca and Brusca, 2003).

There are several objections to this hypothesis. It ignores the embryology of the flatworms in which nothing similar to cellularization occurs, it does not explain the presence of flagellated sperm in the metazoans, and perhaps more important, it implies that the radial symmetry of cnidarians is derived from a primary bilateral symmetry, for which there is no evidence.

2. Polyphyletic Origin Hypothesis.

This hypothesis was proposed by **Greenberg** (1959) and **Preston** (1967); they held that probably all Metazoa have not originated from any single ancestor. So some zoologists prefer the idea that the metazoans has a **polyphyletic origin** and suggest that the sponges, cnidarians, ctenophores and remaining eumetazoans evolved independently. Thus no single scheme might account for them al!

3. Colonial Flagellate Theory

(*i*) Haeckel's gastrea theory. The colonial theory is based upon ideas first expressed by **Ernst Haeckel** (1874), who suggested that a colonial flagellated protist gave rise to a planuloid metazoan ancestor (the planula is the basic larval type of cnidarians). The ancestral protist in this theory was a hollow sphere of flagellated cells that developed some degree of anterior-posterior locomotor orientation, and also evolved some level of specialization of cells into separate somatic and reproductive (or germinal) function. Such conditions are common in living colonial protists (*e.g.*, *Volvox*). Haeckel called this hypothetical protometazoan ancestor the **blastea** and supported its cogency by noting the widespread occurrence of **coeloblastula** among modern animals. In this projection, the first metazoan arose by invagination of the blastea; the resulting animals had a double-layered gastrula-like body (Haeckel's **gastrea**), with a blastopore-like opening to outside (Fig. 16.2 B). Similar to the gastrulae of many modern animals. Haeckel believed that these ancestral creatures (the blastea and gastrea) were recapitulated in the ontogeny of modern animals and the gastrea was viewed as the metazoan precursor to the cnidarians. In addition, the colonial theory has been supported by the argument that the body walls of many lower animals (*e.g.*, members of phyla Porifera and Cnidaria) bear monoflagellated or monociliated cell.

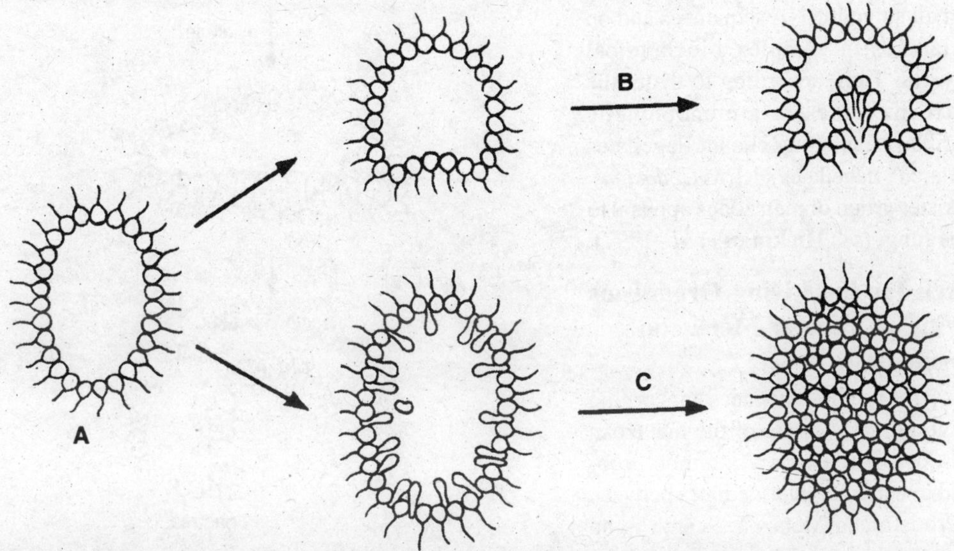

Fig. 16.2. Two versions of the colonial flagellate theory of the origin of the Metazoa. A—The hypothetical colonial flagellate ancestor, Haeckel's blastea (section). B—According to Haeckel, the transition to a multicellular condition occurred by invagination, a developmental process that resulted in hollow gastrea. C—According to Metschnikoff, the formation of a solid gastrea occurred by ingression (after Brusca and Brusca, 2003).

Modification of Haeckel's Gastrea Theory. Haeckel's original ideas have been modified over the years by various authors investigating a colonial protist ancestry to the metazoan condition (**Metschnikoff**, 1883; **Hyman, 1940**). Some have argued that the transition to a layered construction occurred by ingression rather than by invagination, and that the original Metazoa were solid not hollow (Fig. 16.2 C). This idea is based on the fact that ingression is the primitive form of gastrulation among cnidarians.

(*ii*) Butschli's plakula hypothesis. An interesting offshoot of the colonial theory was advanced by **Otto Butschli** in 1883. He suggested that the primitive metazoan was a bilaterally symmetrical flattened creature of two layers of cells; he called this hypothetical animal a ***plakula.*** According to Butschli, the plakula crawled about ingesting food through its "ventral" cell layer. Eventually this animal becomes somewhat hollow by the separation of its dorsal and ventral cell layers; this development allowed invagination of the nutrititive cells (Fig. 16.3). This formation of a "gut" chamber increased the digestive surface area and at the same time produced inner and outer cell layers, an arrangement approaching the metazoan grade of complexity.

Most evidence today points to the protist phylum protozoa's choanoflagellata as the likely ancestal group from which the Metazoa arose. Choanoflagellates possess collar cells essentially identical to those found in sponges. Genera such as *Proterospongia, Sphaeroeca* and other colonial choanoflagellates (Fig. 16.4 C, D), and are commonly cited as typifying a potential metazoan precursor (see **Brusca** and **Brusca**, 2003).

Recently, **Wainright** and coworkers (1993) have advanced molecular biological evidence in support of colonial flagellate theory. Their evidence is based on small subunit RNA sequences and on similarities in complex biochemical pathways. Thus, according to them, the metazoans represent a monophyletic assemblage including choanoflagellates ("collared" flagellates such as *Codosiga*). The sister group of metazoans appears to be the fungi (see **Hickman** *et al.*, 1997).

Search for the Living Organisms Serving as Protists-Metazoan Intermediates

Among all of the ideas concerning the evolutionary origin of the metazoan condition, there exists a common problem: the search for intermediates between the Protista and Metazoa. Search is on of a modern day metazoan creature which combine protist and metazoan traits. These organisms include multinucleate ciliates (*Paramecium*), the colonial *Volvox* and *Trichoplex*, a mesozoan.

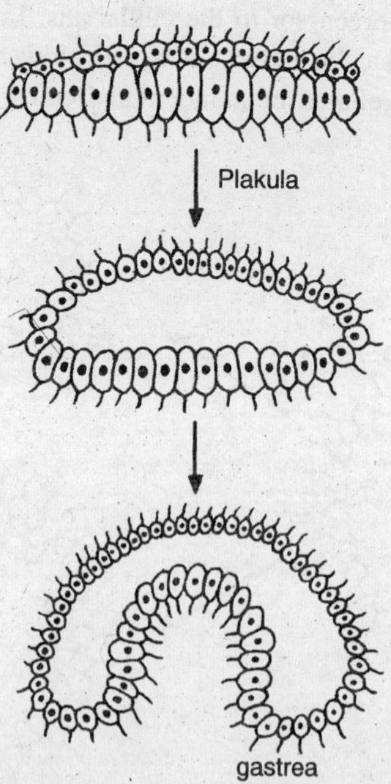

Fig. 16.3. Butschli's hypothetical plakula and its transformation to a metazoan "gastrea" by invagination of a digestive chamber (after Brusca and Brusca, 2003).

About a century ago a tiny, flagellated, but multicellular creature was discovered in marine aquarium. This animal named *Trichoplax adhaerens*, was placed in its own phylum, the Placozoa (**K.Grell**, 1972, 1982). For many years, *Trichoplax* was believed to be a larval stage of some invertebrate, but workers are now convinced that it is an adult of uncertain affinity at the mesozoan grade of organisation. *Trichoplax* has an outer, partly flagellated epithelium surrounding an inner mesenchymal cell mass. Its body margins are irregular, and animal changes shape like an amoeba. There exists some division of labour among various cells and areas of the body. When feeding, *Trichoplax* "*hunches up*" to form a temporary digestive chamber on its underside (Fig. 16.4 E) – producing a form strikingly similar to Butschli's hypothetical plakula. *Trichoplax* may represent the most primitive of all living Metazoa and perhaps even be a conservative descendant line of the ancestral metazoan type.

J. Frenzel (1892) described a tiny organism reportedly collected from salt beds in Argentina. He named this mesozoan animal *Salinella* (Fig. 16.4 F). Although *Salinella* does not possess the layered organisation of the Metazoa, it appears to display a higher level of functional organization than colonial protists. In *Salinella*, a single layer of cells forms the entire body wall and separates the digestive cavity from the outside. The digestive cavity was described as open at both ends as a mouth and anus and the animal fed on organic detritus. The phylum name **Monoblastozoa** has been proposed for this curious animal. Unfortunately, *Salinella* has not been seen since Frenzel's original report, and many zoologists now suspects that Frenzel seriously misinterpreted whatever creature he observed (see **Brusca** and **Bursca**, 2003).

Finally, animals of two mesozoan phyla, Rhombozoa and Orthonectida (Fig. 16.4 G, H), are structurally simple but display complicated life cycles; they are all endoparasites of invertebrates. Some workers believed that they are derived from another group (or groups) of Metozoa, perhaps

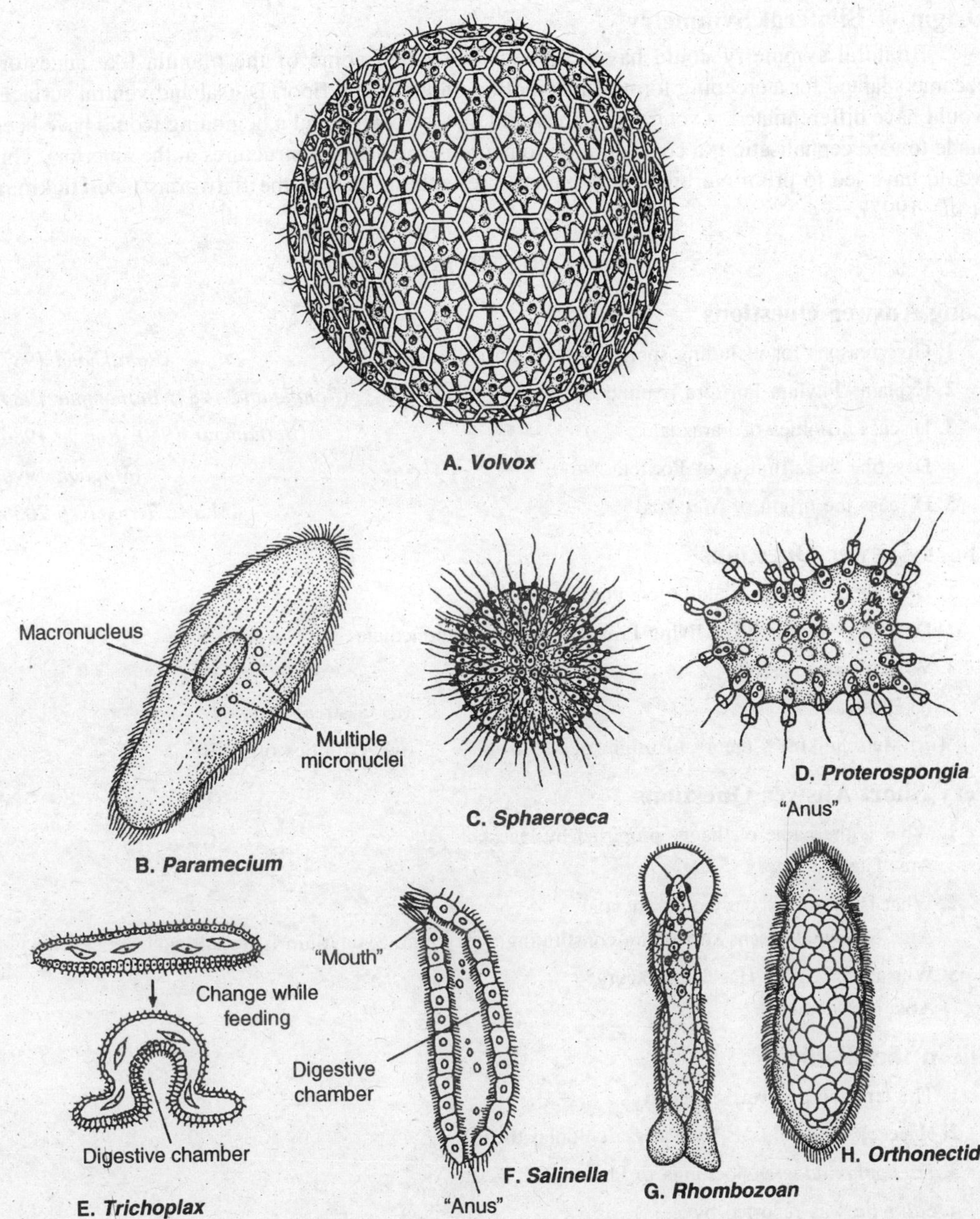

Fig. 16.4. Living organisms that have been considered as protist-metazoan intermediates or that play major roles in various hypotheses concerning the origin of the metazoan condition. A—*Volvox*, a colonial flagellate; B—A multinucleate ciliate (*Paramecium*); C, D—*Sphaeroeca volvox* and *Proterospongia*, two choanoflagellates; E—*Trichoplax*; F—*Salinella*; G—A decymid rhombozoan; H—An orthonectid (after Brusca and Brusca, 2003).

from parasitic trematodes (flukes) which undergo similar complex life cycles. Other workers have suggested that these mesozoan animals are primitively simple and thus may have been derived from early metazoan or premetazoan stock.

Origin of Bilateral Symmetry

Bilateral symmetry could have evolved later when some of the planula-like ancestors became adapted for a creeping form of locomotion on the ocean floor. Dorsal and ventral surfaces would have differentiated, a ventral mouth would have appeared and a beginning would have been made toward cephalisation (a concentration to neurons and sensory structures at the anterior). This would have led to primitive bilateral symmetry, resembling that of the flatworms (see **Hickman** *et al.*, 1997).

QUESTIONS

Long Answer Questions

1. Give reasons for including sponges in the animal kingdom. (*Rohilkhand 1987*)
2. Explain "Phylum Porifera is blind offshoot of Metazoa". (*Rohilkhand 1983; Berhampur 1984*)
3. Discuss affinities of Parazoa. (*Berhampur 1980; Ranchi 1986*)
4. Describe the affinities of Porifera. (*Magadh 1996*)
5. Discuss the origin of Metazoa. (*M.S.U. Tirunelveli 2006*)

Short Answer Questions

1. Describe Butschli's plakula hypothesis.
2. Describe the search for living Protist-Metazoan intermediate.
3. Write short notes on:
   (*i*) Blastea (*ii*) Gastrea
   (*iii*) Metschnikoff's theory of origin of Metazoa (*iv*) Affinities of Porifera.

Very Short Answer Questions

1. What is the name of theory proposed by Haeckel.
   Ans. Gastrea theory.
2. What is meant by the word syncytial?
   Ans. Syncytial means something constituting syncytium. Syncytium is a multinucleate cell.
3. Who modified the Haeckel's theory?
   Ans. Metschnikoff.

Fill in the Blanks

1. The term blastea was given by ....
2. Haeckel's .... is basically a colonial theory.
3. *Trichoplax adaerens* belongs to Phylum ....
4. *Salinella* was reported by ....
5. Butseli's hypothetical animal is ....

Multiple Choice Questions

1. Syncytial theory explains
   (*a*) formation of coelom (*b*) symmetry of animal
   (*c*) origin of life (*d*) origin of metazoa
2. Polyphyletic theory is given by
   (*a*) Greenberg and Preston (*b*) Lang and McBridge
   (*c*) Hanson and Hadzi (*d*) Lankester

3. Colonial theory was proposed by
(*a*) Butschli and Haeckel
(*b*) Hanson and Hadzi
(*c*) L.H. Hymen
(*d*) Metschnikoff

4. Gastrea theory of metazoan ancestory was presented by
(*a*) E.H. Haeckel
(*b*) L.H. Hymen
(*c*) J. Hadzi
(*d*) Lankester

5. Syncytial theory was proposed by
(*a*) Hanson and Hadzi
(*b*) L.H. Hymen
(*c*) Lankester
(*d*) E.H. Haeckel

6. Which group of protista (protozoa) support and give best explanation of colonial theory of origin of Metazoa?
(*a*) ciliates
(*b*) sarcodina
(*c*) ciliophora
(*d*) flagellate

ANSWERS

Fill in the Blanks

1. Haeckel **2.** Gastrea theory **3.** Placozoa **4.** Frenzel
5. Plakula

Multiple Choice Questions

1. (*d*) 2. (*a*) 3. (*a*) 4. (*a*)
5. (*a*) 6. (*d*).

17

Phylum Cnidaria (Coelenterata) (Characters, Classification and Types)

Cnidaria or **Coelenterata** (see Box 17.1) may be defined as '*diploblastic tentaculate metozoa with tissue grade of construction having nematocysts, a single body cavity, the coelenteron or gastrovascular cavity and a single mouth opening to the gut (= coelenteron) derived from the blastopore.*' Cnidaria includes jelly fish, sea anemones, corals and various *Hydra*-like polyps. It is a very successful phylum with about 10,000 living species and almost as many fossils.

Box 17.1

In certain old text books, the name coelenterata was used as a phylum designation to include both the **cnidaria** and **ctenophora**. Later on, it was found better to separate the groups as discrete phyletic entities. The Cnidaria closely approximate an idealised diploblastic condition and possess highly diagnostic intracellular structures, the nematocysts, while the ctenophora are triploblastic and do not possess nematocysts (**Marshall** and **Williams**, 1995).

17.1 GENERAL CHARACTERS

1. Primitive metazoans having **tissue grade of organization**. For example, epitheliomuscular tissue has **basal lamina** and intercellular junctions such as **belt junctions, septate sealing junctions** and **gap junctions** (Gap junctions are found only in class Hydrozoa).
2. They are aquatic, mostly marine except few freshwater forms such as *Hydra* and *Craspedacusta*.
3. They are sessile or free swimming and solitary or colonial.
4. Exhibit primary radial symmetry of body, so are placed in division **Radiata** of Animal Kingdom.
5. Cnidarians are **diploblastic**, that is, their bodies are constructed from only two germ layers, **ectoderm** and **endoderm**, from which two adult epithelia arise.
6. Their body wall consists of two cellular layers and one extracellular layer, *i.e.*, an outer epithelium, the **epidermis**, and an inner epithelium, the **gastrodermis**, lining the gastrovascular cavity and between these occur an extracellular gelatinous **mesogloea**.
7. Mesogloea ranges from a thin, noncellular basal lamina (called **mesolamella**) as in hydras (Hydrozoa) to thick, fibrous, jelly-like, connective tissue with or without mesenchymal cells (which have migrated from both germ layers).
8. The body wall encloses a single, central, blind sac-like body cavity called **coelenteron** or **gastrovascular cavity**. This cavity serves as gut (for digestion of food) and as circulatory system (for circulation of digested body).

9. Gastrovascular cavity lies along the polar axis of the animal and opens to outside at one end of body to form a **mouth**. Mouth serves both in ingestion and egestion. **Anus** is absent.
10. Acoelomate animals because they do not possess a second body cavity, the coelom.
11. A circle of short and slender **tentacles** surround the mouth to aid in the capture and ingestion of food. Tentacles represent evaginations of the body wall.
12. Most cnidarians show the phenomenon of **polymorphism** (see Box 17.2) but most individuals (zooids) are reducible either to hydroid (= polypoid) or medusoid types. The **medusoid** or jelly-fish form represents the sexual phase and is adapted for free-swimming mode of life. The **polypoid** or **hydroid** form represents asexual phase and is adapted for sessile mode of life. Polypoid forms may be solitary or colonial but medusoid forms are always solitary.

Box 17.2 Polymorphism

The occurrence of structurally and functionally different types of individuals, called **zooids**, within the same organism during its life history is called **polymorphism**. Cnidaria contains nine types of zooids such as **trophozooid** or **gastrozooid** (feeding), **gonophore** (sessile medusae), **gonozooid** (gonophore production), **dactylozooid** (protection, food capture), **tentaculozooid** (same as dactylozooid), **nematophore** (food capture), **pneumatophore** (float), **nectophore** (swimming) and **phyllozooid** (protective cover).

13. All cnidarians have unique stinging organs, called **nematocysts** which are a type of **cnide** (see Box 17.3). A cnide is an organ of **cnidocyte** or **cnidoblast** which is capable of eversion from cnidocyte. Cnidocytes are located in ectoderm (= epidermis) especially in tentacles. Nematocysts function in prey capture and may inject a protein toxin into the prey. About 18 types of nematocysts are found in cnidaria to serve different functions. They stun, paralyze or kill prey, entangle food and attach it to tentacles and perform other functions.

Box 17.3

Cnidarian contain following three types of **cnide** which are cellular organelles of cnidocytes and are capable of eversion from the cell:

1. Nematocysts
2. Spirocysts
3. Ptychocysts

Nematocysts are found in all cnidarians. Spirocysts and ptychocysts, however, are found only in anthozoa.

14. Cnidarians are **carnivorous**; digestion is extracellular and intracellular.
15. Skeleton either calcareous or horny and external (exoskeleton) or internal (endoskeleton).
16. Locomotion is brought about by smooth muscle fibres.
17. Respiratory, circulatory and excretory systems are absent.
18. Nervous system is a collection of multipolar and bipolar nerve cells forming an irregular net or plexus.
19. Cnidarians contain very simple sensory organs such as **dermoptic light sensitive organs, ocelli** and **statocysts**.
20. Reproduction is both by **asexual** and **sexual** methods. Asexual reproduction by **budding** or **strobilation**. Sexual reproduction produces characteristic ciliated **planula** larva.
21. Their life history shows **alternation** of a sessile and asexual **polypoid generation** and a free swimming and sexual **medusoid generation**, both being diploid phases. It is called **metagenesis** (see Box 17.4).

> **Box 17.4 What is metagenesis?**
>
> True type of alternation of generation occurs in the plants where a regular alternation between a **haploid** gametophyte and **diploid** sporophyte occurs. Such sort of alternation is not found in the cnidarians. In them occurs a somewhat modified type of alternation of generation, called **metagenesis**. Thus, in Cnidaria, alternation of two **diploid generations** occur, *i.e.*, diploid sessile asexual polypoid generation alternates with diploid free-swimming sexual generation. In them, the only haploid cells in the life cycle are the gametes. In fact, metagenesis implies postponement of sexual reproductive phases.

17.2 CLASSIFICATION

The classification of phylum Cnidaria is based on the duration of polypoid and medusoid generations. An outline classification of phylum Cnidaria upto orders and suborders including examples have been given on page 195.

Detailed Classification

Phylum Cnidaria is divided into following **four** classes:

Class 1. Hydrozoa

1. Hydrozoa (Gr., *hydra* = water; *zoios* = animal) include a polypoid, medusoid or both forms in the life cycle. However, polypoid phase is usually predominant in the life cycle.
2. Polyps are solitary or colonial.
3. Aquatic, mostly marine, a few freshwater.
4. Mesogloea is acellular or noncellular and jelly like.
5. A calcareous or cuticular skeleton, the **perisarc**, covers the body of most of the hydrozoans.
6. Gastrovascular cavity without gastric tentacles or septa (or mesentries); without gullet or stomodeum.
7. Nematocyst-containing cnidocytes are found only in epidermis.
8. Gonads develop only in epidermis. Sperm and ova are shed directly to the outside.
9. Medusae typically with **craspedon** or **velum** (called craspedote) which improves swimming efficiency of medusa.
10. Life history of some includes alternation of asexual polypoid and sexual medusoid generations.
11. Phenomenon of polymorphism is common.
12. Class Hydrozoa includes about 2700 species of most primitive cnidarians.

> **Box 17.5 Taxonomy of Cubozoa**
>
> Cubozoa forms a new class which was created by **Werner** (1975) and **Calder** and **Peters** (1975). They were separated from Scyphozoa on the following affinities. Their hydrozoan features include. 1. The presence of velum; 2. The possession of a unique type of nematocyst (stenotele); 3. Lack of a notched bell margin and rhopalia.
>
> *Chironex* (sea wasp) contain deadly poisonous nematocysts. Its sting can cause death of a human being within 3 to 20 minutes.

Class 2. Scyphozoa

1. In Scyphozoa (Gr., *skyphos* = cup + *zoios* = animal) medusa stage is predominant. Animal body is umbrella or bell-shaped.
2. All are marine, solitary, free-swimming or attached by aboral stalk. They float in cooler waters.
3. They lack velum (acraspedote).
4. Mesogloea is much thickened and cellular.

Outline Classification

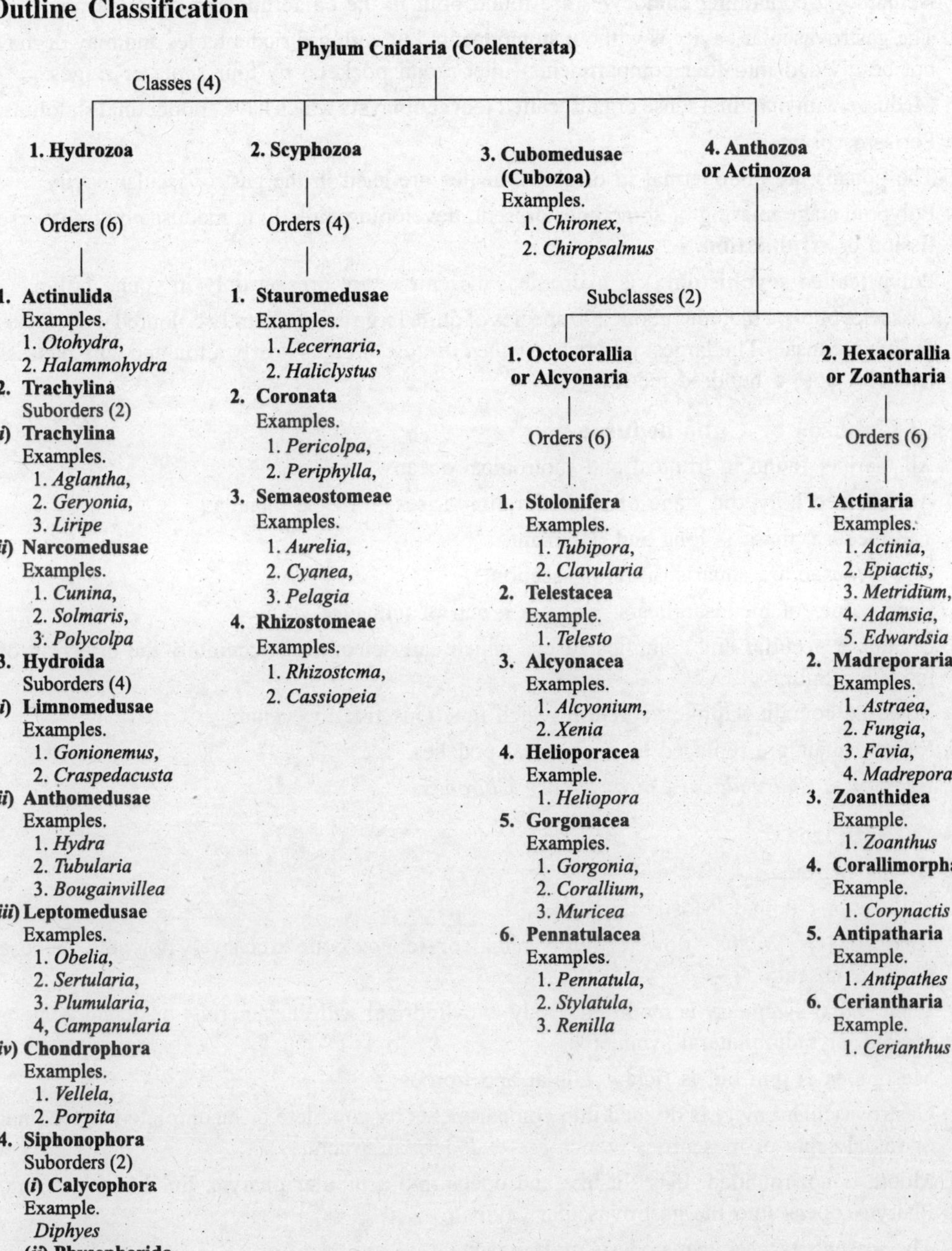

5. Nematocyst-containing cnidocytes are found both in the epidermis and gastrodermis.
6. The gastrovascular cavity is without stomodaeum but with gastric tentacles and may or may not be divided into four compartments (inter-radial pockets) by four septa or ridges.
7. Medusae with marginal sense organs, called **tentaculocysts** which have endodermal statoliths.
8. Perisarc absent.
9. The gonads are endodermal in origin. Gametes are shed in the gastrovascular cavity.
10. Polypoid stage missing or sometimes present, developing directly in medusae by transverse fission or **strobilation.**
11. Polyp (called **scyphistoma**) is hydrotuba; mesentries are present only in young polyp.
12. Class Scyphozoa includes about 200 species of quite large and brilliantly coloured jelly-fishes or true medusae. The largest jellyfish, *Cyanea arctica*, weighs nearly a ton and may possess tentacles upto a hundred feet long.

Class 3. Cubozoa or Cubomedusae

1. All marine, found in tropical and subtropical oceans.
2. An attached polypoid stage of short duration arises from the planula.
3. The medusa phase is long and dominant.
4. The medusae are squarish in cross section.
5. Each corner of medusae bears one or a group of tentacles.
6. Contain a peculiar and complicated eye which can detect very accurately the direction of incoming light.
7. Medusae contain flaplike **velarium** which functions like the velum.
8. Radial canals are replaced by four radial pouches.

Examples. *Charybdaea*, *Chiropsalmus*, *Chironex*.

Class 4. Anthozoa

1. All are marine.
2. Solitary or colonial. Mostly colonial.
3. Anthozoa (Gr., *anthos* = flower; *zoios* = animal) or Actinozoa are exclusively polypoid, medusa phase is absent.
4. Their radial symmetry is modified. Body is cylindrical with hexamerous or octamerous or biradial or radiobilateral symmetry.
5. Mesogloea is thin but is richly cellular and fibrous.
6. Gastrovascular cavity is divided into compartments by complete or incomplete longitudinal or raidal septa or mesentries (which are endodermal structures).
7. Mouth is not rounded. It is slit-like and opens into a tubular pharynx lined by ectoderm. Pharynx opens into the gastrovascular cavity.
8. The mesentries bear nematocysts and gonads.
9. Skeleton either external or internal.
10. Exoskeleton is formed from calcium carbonate. It often forms massive corals.
11. Gametes are shed in coelenteron; fertilization is external.
12. Class Anthozoa is the largest class of phylum Cnidaria. It contains over 600 species of sea anemones, corals, sea fans and sea pansies.

CLASS 1. HYDROZOA

Class Hydrozoa is divided into the following **six** orders.

Order 1. Actinulida

1. Tiny, solitary hydrozoans.
2. Body form resembles actinula larva.
3. Medusae phase absent.
4. Unlike other mature polypoid and medusoid hydrozoans, the epidermal cells are ciliated.
5. They are regarded as **neotenic** forms because they persist in the larval form (*i.e.*, actinula) and reproduce (*i.e.*, actinula does not metamorphose into medusa and becomes sexually mature).

Examples. *Otohydra*, *Halammohydra*.

Order 2. Trachylina

1. Hydroid or polypoid phase is reduced or absent.
2. Medusoid phase dominant.
3. Medusa develops from the fertilized egg. Actinula larva develops from the planula larva and the actinula develops into medusa.
4. Medusae are large, provided with marginal sense organs such as tentaculocysts and statocysts. Statoliths are endodermal.
5. This order includes most primitive members of the class Hydrozoa.

Order Trachylina is divided into the following **two** suborders.

Suborder 1. Trachymedusae

1. Margin of bell or umbrella is smooth (not scalloped).
2. With four, six, or eight radial canals.
3. Manubrium is long.
4. Gonads borne on the radial canals.

Examples. *Aglantha*, *Gonionemus*, *Geryonia*, *Liriope*.

Suborder 2. Narcomedusae

1. Margin of bell or umbrella is scalloped, *i.e.*, clefted by tentacle bases.
2. Radial canals are reduced; redial pouches are present.
3. Manubrium is absent.
4. Gonads borne in the floor of gastric cavity or stomach.
5. Tentaculocysts naked.

Examples. *Cunina*, *Solmaris*, *Polycolpa*.

Order 3. Hydroida

1. Polypoid generation is well developed. Medusa phase is present or absent.
2. Polyps are solitary or colonial.
3. Medusa phase is short lived, if present. Medusae are produced by polyp by budding.
4. Sense organs of medusae are ocelli and statocysts, all are ectodermal in origin.

Order Hydroida is divided into following **four** suborders:

Suborder 1. Limnomedusae

1. Mostly fresh water forms, some are marine.
2. Polyps are small in size and solitary.

3. Polyps lack perisarcal skeleton.
4. Some are with medusae.

Examples. Marine *Gonionemus* and freshwater *Craspedacusta* (freshwater jellyfish).

Suborder 2. Anthomedusae (or Athecata or Gymnoblastea)

1. Stalk typically covered with perisarc.
2. Perisarc does not form thecal cups for zooids, *i.e.*, hydrotheca and gonotheca are absent.
3. Tall and bell-shaped free swimming medusae.
4. Medusae bear eye spots or ocelli at the bases of tentacles.

Examples. *Hydra*, *Tubularia*, *Bougainvillea*, *Pennaria*, *Eudendrium*, *Sarsia*.

Suborder 3. Leptomedusae (Thecata or Calyptoblastea)

1. Colonial Hydrozoa.
2. Zooids are contained in skeletal cups, *i.e.*, polyps are enclosed in **hydrotheca** and blastostyles with developing medusae are enclosed in **gonotheca**.
3. Medusae, if present, are typically flattened, bowl or saucer-shaped.
4. Medusae bear gonads on the radial canals.
5. Medusae are with statocysts but without eyespots or oceli.

Examples. *Obelia*, *Sertularia*, *Plumularia*, *Campanularia*.

Suborder 4. Chondrophora

1. Free-floating polypoid colonies with high degree of polymorphism.
2. These cnidarians can also be interpreted as large, single, inverted polyps.

Examples. *Velella* (by-the-wind sailor or purple sailor), *Porpita* (blue button).

Order 4. Siphonophora

1. They form free-swimming or floating colonies of polypoid and medusoid individuals.
2. Colonies with gas-filled floats (pneumatophores) or large swimming bells which help in floating of the colony.
3. Siphonophora exhibits polymorphism of high degree. Thus, colony is composed of different types of modified polyps such as gastrozooids, gonozooids, dactylozooids and medusae. These zooids are either attached to stem or disc.
4. Polyps are without tentacles.
5. Medusae are incomplete and rarely freed.
6. Perisarc absent.
7. They are found largely in warm seas.

This order includes the following **two** suborders:

Suborder 1. Calycophora

1. Apical float or pneumatophore absent.
2. One or more swimming bells or **nectocalyces** (or nectophores) present at the upper end of the colony.

Examples. *Diphyes*, *Abyla*.

Suborder 2. Physophorida

1. Upper end of the colony bears a float or pneumatophore.

Examples. *Physalia* (Portuguese man-of-war), *Muggiae*, *Halistemma*.

Order 5. Milleporina

1. Colonial coral-like Hydrozoa without perisarc.
2. Their massive porous calcareous skeleton is secreted by ectoderm. Polyps protrude out through the pores of skeleton.
3. Colony has two kinds of zooids, the gastrozooids and the dactylozooids.
4. Gastrozooids (nutritive zooids) are short and provided with mouth and tentacles
5. Dactylozooids are elongated, hollow, slender tactile zooids. They contain tentacles but lack mouth.
6. Medusae develop in small chambers, becoming free. They lack mouth radial canals and tentacles (*i.e.*, they become degenerated).
7. Milleporina are commonly called **stinging** or **fire corals** due to the presence of powerful nematocysts in them.

   Example. *Millepora* is the only genus.

Order 6. Stylasterina

1. Colonial coral-like hydrozoa; commonly called **hydrocorals**.
2. Colony has two types of zooids, the feeding polyps or **gastrozooids** and defensive polyps or **dactylozooids**. Both types of polyps are located within star-shaped openings of the skeleton.
3. Dactylozooids are small, solid without tentacles.
4. Gastrozooids have a cup with pointed spine.
5. Medusae are permanently sessile and degenerated.

   Examples. *Stylaster*, *Allopora.*

CLASS 2. SCYPHOZOA

Class Scyphozoa is divided into following **four** orders.

Order 1. Stauromedusae

1. Adult developing directly from scyphistoma (see Box 17.6) remaining attached to seaweeds by aboral stalk.
2. Body globet or trumpet shaped.
3. Mouth cruciform (four-cornered) with small oral lobes and a short quadrangular manubrium.
4. Gastrovascular system is divided into central stomach and four preradial pouches by the four inter-radial septa.

> **Box 17.6**
>
> **Scyphistoma.** A sessile, polypoid or hydroid larva of many scyphozoa which may produce either more scyphistomae or free-swimming medusae.

5. Gonads are elongated band-like borne on the faces of septa.
6. Marginal sense organs absent or modified as tentacles.
7. Fertilization is external.
8. Larva is **planula** but lacks cilia.
9. They inhibit coastal water of colder regions. All are marine.

   Examples. *Haliclystus*, *Lucernaria.*

Order 2. Coronata

1. Free swimming Scyphozoa, found inhabiting the deep waters of ocean.
2. Medusoid body is conical, dome-shaped or flattened and grooved.

3. The umbrella is divided by a horizontal furrow into an upper cone and a lower cone.
4. The crown consists of flattened blades (pedal lobes) called **pedalia** which bear solid tentacles.
5. The bell margins are scalloped (*i.e.*, cut into curves) into **lappets** which alternate with pedalia.
6. Gastrovascular cavity is divided by septa; mouth is four cornered.
7. Tentaculocysts are four to sixteen.

Examples. *Pericolpa*, *Periphylla*, *Nausithoe*, *Atolla*.

Order 3. Semaestomeae

1. Most common free-swimming medusae found inhabiting the coastal waters of all oceans.
2. Umbrella is flat, saucer or bowl-shaped.
3. Mouth is square.
4. The corners of the mouth are prolonged into frilly **oral tentacles** or **oral arms**.
5. Without furrow on pedalia (exumbrella)
6. The margin of umbrella is fringed with hollow tentacles.
7. Eight to sixteen **rhopalia** (= little club-shaped compound sensory organs, each containing a statocyst, an ocellus and two sensory pits) are present.
8. Margin of umbrella is scalloped.
9. Gastric pouches and filaments are absent.

Examples. *Aurelia*, *Cyanea*, *Pelagia*.

Order 4. Rhizostomae

1. Free-swimming Scyphozoa found in shallow waters of tropical and subtropical oceans.
2. The umbrella is saucer or bowl-shaped or flattened. It may be concave on the top.
3. Medusae with fused oral tentacles, containing many canals and small mouths (secondary mouths). Original or primary mouth is lost due to fusion of oral arms except in *Stomolophus*.
4. Gastrovascular cavity is not divided by septa.
5. Marginal tentacles are absent but eight or more rhophalia are present.
6. Margin of umbrella is scalloped.

Examples. *Rhizostoma*, *Cassiopeia*, *Stomolophus*.

CLASS 4. ANTHOZOA

Class Anthozoa is divided into following **two** subclasses.

Subclass 1. Octocorallia (or Alcyonaria)

1. Colonial marine forms; polyps usually connected by coenenchyme (see Box 17.7).

> **Box 17.7 Coenenchyme**
>
> It is a tissue comprising mesogloea surrounding and uniting the polyps in compound anthozoans. This tissue consists of a thick mass of mesogloeá perforated by gastrodermal tubes that are continuous with the gastrovascular cavity of the polyps. The surface of entire flesh mass is covered by a layer of epidermis, which joins the epidermis of the polyp column. Only the upper portion of the polyp projects above the coenenchyme.

2. Each polyp is long or short cylinder terminating orally into a flat circular oral disc having oval or elongated mouth in the centre.
3. Each polyp bears eight feathery or pinnate, hollow tentacles.
4. Gastrovascular cavity is divided by eight single, complete septa (mesentries).
5. Single ventral **siphonoglyph** (see Box 17.8) is present.

Box 17.8 Siphonoglyph

It is a ciliated groove leading from mouth to gullet in certain anthozoans. It provides for the entry of water into gastrovascular cavity. This current of water functions to maintain an internal fluid, or hydrostatic skeleton against which the muscular system can act. Water also helps in exchange of gases through the gastrodermal surface.

6. Endoskeleton (of spicules) is the product of mesogleal cells.
7. Spicules may be calcareous or horny in nature.

Subclass 2. Hexacorallia (or Zoantharia)

1. Solitary or colonial marine forms.
2. Polyps with more than eight **tentacles** or **septa** (mesentries), typically in the cycle of 12.
3. Tentacles simple, rarely branched, hollow and cone-shaped.
4. Mesentries may be complete or incomplete.
5. Two siphonoglyphs are commonly present.
6. Endoskeleton when present is calcareous, derived from ectoderm.

Subclass 1. Octocorallia (or Alcyonaria)

Subclass Octocorallia is divided into following **six** orders.

Order 1. Stolonifera

1. They are inhabitants of shallow waters in the temperate and tropical oceans.
2. Coenenchymal mass is absent; simple polyps arise singly from creeping mats or **stolons**.
3. Skeleton of calcareous **tubes** of separate or fused calcareous spicules or horny external cuticle.
4. Spicules are formed by mesogleal cells.
5. Tubes are connected by horizontal platforms.

Examples. *Clavularia*, *Tubipora* (organ-pipe coral).

Order 2. Telestacea

1. Colony consists of simple or branched erect stems arising from a creeping base (stolons).
2. Each stem is a very elongated polyp bearing lateral polyps.
3. They contain a skeleton of calcareous spicules.

Example. *Telesto*.

Order 3. Alcyonacea

1. **Soft corals** which are mostly tropical.
2. Coenenchyme forming a rubbery or fleshy mass.
3. Colony may have a massive mushroom shape or an encrusting growth form.
4. Lower ends of their polyps fused with coenenchyme and only oral ends are protruding.
5. In some forms the polyps are dimorphic, *i.e.*, autozooids and siphonozooids.
6. Skeleton consists of separate mesogleal spicules, often consolidated, without axial core.

Examples. *Alcyonium*, *Xenia*, *Gersemia*, *Sarcophyton*.

Order 4. Helioporacea

1. They include the Indo-pacific **blue coral**.

2. They contain a massive, blue calcareous skeleton.

Example. *Heliopora* is the only genus.

Order 5. Gorgonacea

1. **Horny corals** which are found in tropical and subtropical sea shores.
2. The colony is usually upright plant-like, consists of a main stem arising from a basal plate or tuft of stolons and number of branches bearing polyps.
3. Skeleton of mesogleal spicules and axial core of horny material called **gorgonin** or calcareous material.
4. Polyps short (Fig. 17.1)

Examples. *Gorgonia* (sea fan), *Corallium* (precious red coral), *Muricea* (sea rod), *Leptogorgia* (sea whip).

Order 6. Pennatulacea

1. **Sea pens**.
2. Colony consists of a fleshy, flattened or elongated body or **rachis** (called **axial** or **primary polyp**). Secondary polyps are many small lateral dimorphic polyps attached at sides of rachis.

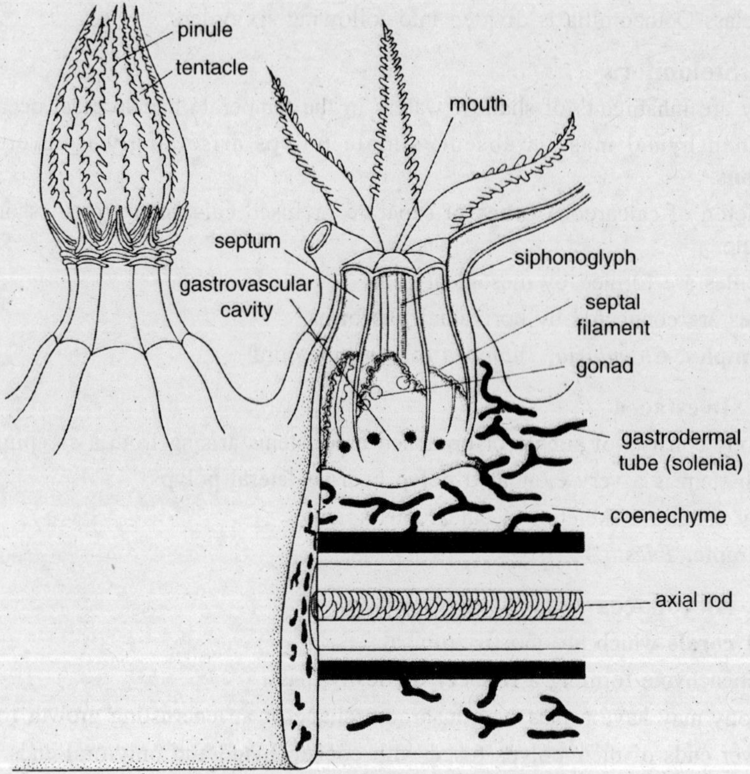

Fig. 17.1. Structure of a gorgonian coral of the anthozoan subclass Octocorallia. Note the position and structure of siphonoglyph and coenenchyme.

3. Rachis is supported by skeleton of separate calcareous spicules and axial rod.

Examples. *Pennatula* (sea pen or sea feather), *Stylatula* (sea pen); *Renilla* (sea pansy).

Subclass 2. Hexacorallia (or Zoantharia)

Subclass Hexacorallia is divided into **six** orders.

Order 1. Actiniaria

1. **Sea anemones.**
2. Solitary anthozoans with no skeleton.
3. Cylindrical polypoid body is divided into oral disc, column and base.
4. The aboral end is provided with pedal disc.
5. Tentacles and mesentries are numerous and often arranged in the multiple of six.
6. Usually with two siphonoglyphs.
7. Septa (mesentries paired, complete and incomplete).

Examples. *Actinia*, *Metridium*, *Adamsia*, *Edwardsia*, *Epiactis*.

Order 2. Madreporaria (or Scleractinia)

1. **Stony corals.**
2. Colonial forms.
3. Exoskeleton compact, hard, heavy calcareous and secreted by ectoderm (epidermis).
4. Polyps small and enclosed in cup-like depressions of the exoskeleton.
5. Siphonoglyph is absent.
6. Sclerosepta arranged in hexamerous cycles.

Examples. *Astraea* (beadlet anemones), *Fungia*, *Favia*, *Madrepora*, *Acropora* (staghorn coral).

Order 3. Zoanthidea

1. Colonial anemones.
2. Usually colonial and epizoic polyps.
3. Skeleton and pedal disc absent.
4. Polyps small with one siphonoglyph.
5. Septal pairs usually consists of one complete and one incomplete septum.

Examples. *Zoanthus*, *Palythoa*. Some such as *Epizoanthus* and *Parazoanthus* are epizoic on other invertebrates.

Order 4. Corallimorpharia

1. Coral-like animals.
2. Solitary or colonial polyps without skeleton.
3. With radially arranged tentacles.
4. They contain siphonoglyphs or basilar muscles.

Examples. *Corynactis*, *Discosoma*, *Ricordea*.

Order 5. Antipatharia

1. **Black corals** or **thorny corals** which occur mainly in deep water in tropics.
2. Colonial and upright plant-like.
3. Skeleton in the form of branched and thorny chitinous axis.
4. Polyps with two siphonoglyphs.
5. Polyps with six, ten or twelve single, complete septa and six simple or eight branched tentacles.

Examples. *Antipathes*.

Order 6. Ceriantharia

1. Long, solitary polyps adapted for burrowing. Live in self-secreted tubes which are buried in sand or mud.
2. Polyps lack skeleton and pedal disc.
3. On the oral disc, tentacles are arranged in two whorls — **marginal whorl** and **oral whorl**.
4. Only single siphonoglyph present.
5. Many single, complete septa are present.

Example. *Cerianthus*.

17.3 TYPE STUDIES

1. *PHYSALIA*

Classification

Phylum	–	**Cnidaria (Coelenterata)**
Class	–	**Hydrozoa**
Order	–	**Siphonophora**
Suborder	–	**Physophorida**
Genus	–	***Physalia***

Physalia is a marine, polymorphic, pelagic floating colony of various modified polypoid and medusoid zooids. It is one of the most famous cnidarian and is commonly known as **Portuguese man-of-war** (see Box 17.9). *Physalia* is found in tropical and sub-tropical oceans; it is seen abundantly on North Atlantic shores, both in Europe and North America. *Physalia* contains a large (may be upto 30 cm in length) bright-blue gas-filled **pneumatophore** or **float** which is formed by several **medusoids**. The float is chitinous, muscular and elongated bladder-like and is filled by a variety of gases such as carbon monoxide (CO), nitrogen (N_2), oxygen (O_2), argon (Ar), etc. Both ends of float are conical, at one end is an **apical pore**; the upper edge is produced into a median **crest** or **sail**. Float contains **gas glands** on the inner surface which secrete gases that inflates the float. The float lacks the swimming bells or nectocalyces, so is drifted to and fro by the wind. During rough weathers gas may be expelled out through the pore causing the entire colony to sink; when fresh gas is secreted, colony rises again. According to **Hyman** (1959)

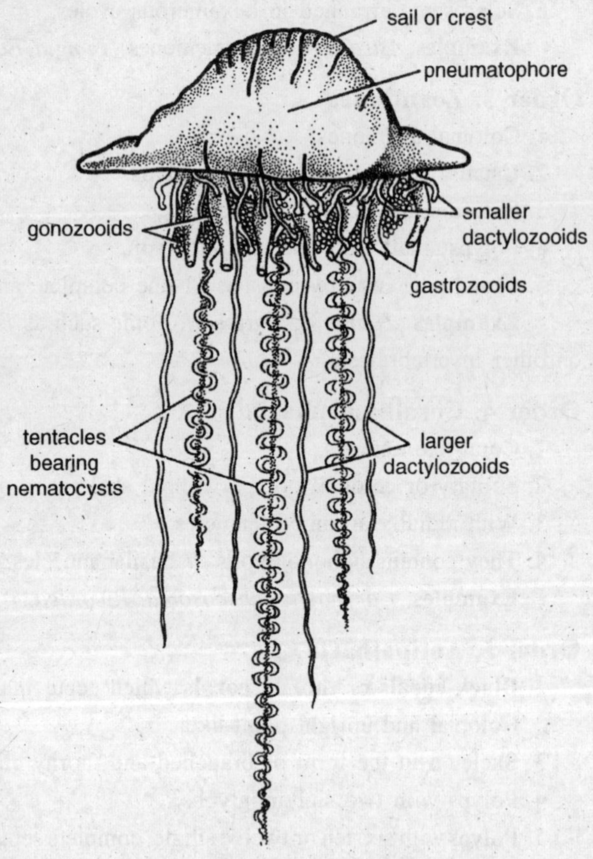

Fig. 17.2. *Physalia.*

Physalia cannot go below the surface. Thus, *Physalia* is *pleustonic*, *i.e.*, adapted entirely to surface existence, partly in water and partly outside.

Box 17.9 Star attractions of *Physalia*

Physalia's common name portuguese man-of-war may be coined because of the following reasons: (*i*) its sudden appearance and disappearance like the portuguese warships during war; (*ii*) it appears like upturned warships during war; (*iii*) brilliant peacock blue green colour of its crest resembles the cap of portuguese soldiers/or of Napoleon Bonapart.

Physalia is famous for its following three salient features:

(*i*) It forms a very good example of the phenomenon of polymorphism. It exhibits many types of zooids.

(*ii*) It has a large gas-filled, horizontal float for the colony. This float may have a brilliant peacock-blue colour or sometimes orange colour.

(*iii*) An accidental encounter with its tentacles (having batteries of nematocysts) can be a painful and even a fatal experience for the swimmers. Its nematocysts contain a proteinous **neurotoxin** which affects nervous system of victim causing severe burning pain, skin lesions, eruptions, fever, prostration and respiratory interference.

In *Physalia*, the coenosarc is not in the form of a stem or stolon, but it is flattened disc-like occurring at the under surface of the float. From the coenosarc hangs numerous **cormidia.** A cormidium is a group of polymorphic individuals which are modified polyps; they are gastrozooids or siphons, dactylozooids and gonozooids (Fig. 17.3).

(*i*) **Gastrozooids.** They are feeding zooids. They are tube-or bottle-shaped and contain only one tentacle which is also meant for sucking the food. The lips of gastrozooids are applied to the body of fish and partly digested food is sucked up in liquid form.

(*ii*) **Dactylozooids.** They are tubular, mouthless individuals, each bearing a tentacle at the base. Dactylozooids are of two types: **small** and **large**. The tentacle of each large dactylozooid is very long (upto 12 metres), brightly coloured, muscular and bears a twisted ribbon beset with **nematocysts**. The dactylozooids protect the colony and also help in capturing prey. Long tentacles, in fact, form a drift net for capturing fish for food. With the help of tentacle, *Physalia* catches, stings the prey and pulls it within the reach of gastrozooids.

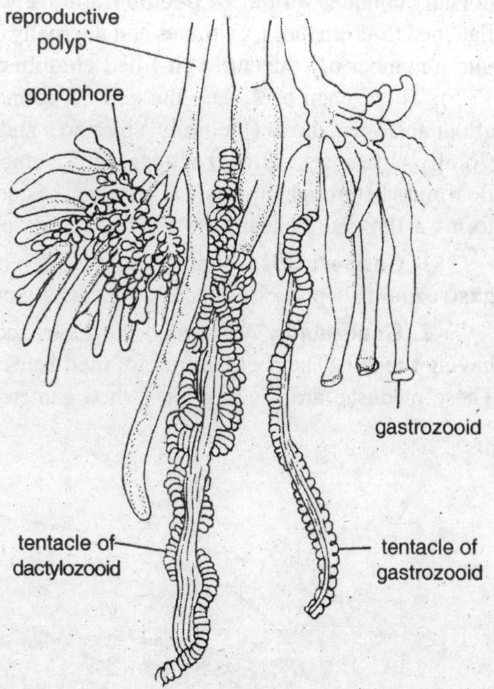

Fig. 17.3. *Physalia*. Various types of zooids.

(*iii*) **Gonozooids.** Gonozooids or **gono-dendra** are branching blastostyles having leaf-like **gonopalpons** and male and female **medusae** or **gonophores**. Each medusa bears four radial canals, the velum, sense organs but no mouth. The male gonophores are sessile, degenerate (*i.e.*, reduced to sporosacs), grape-like and developed on the proximal part of gonodendra. The female gonophores are degenerate, medusa-like, develop on the distal part of gonodendra and are set free.

The gastrozooids, dactylozooids, blastostyles and gonopalpons are modified polyps, while the gonophores and the float (pneumatophore) are modified medusae.

2. *VELELLA*

Classification

Phylum	–	**Cnidaria (Coelenterata)**
Class	–	**Hydrozoa**
Order	–	**Hydroida**
Suborder	–	**Chondrophora**
Genus	–	***Velella***

Velella is commonly called **Purple Sail or By-The-Wind Sailor**. It is a polymorphic, marine and colonial form; it looks like a single medusae (Fig. 17.4). It is found in warm seas (*i.e.*, Southern Atlantic coast). *Velella* is a pelagic hydroid which floats near the surface of water.

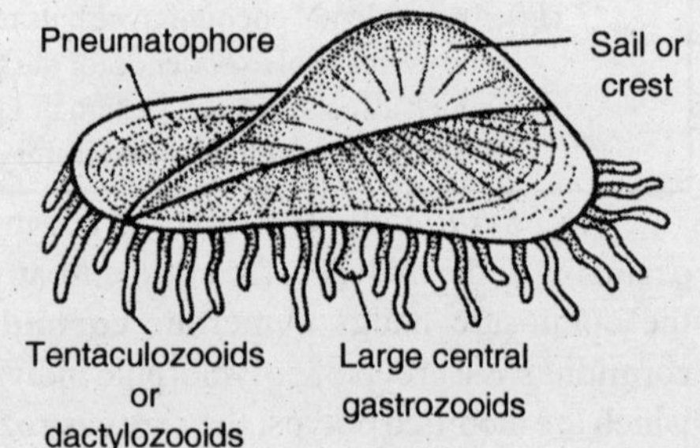

Fig. 17.4. *Velella.*

Velella has a rhomboidal body, on the upper surface of which is an oblique contractile **sail**. Body dorsally contains a **float** or **pneumatophore** which is flattened (5–8 cm long), chitinous and internally divided into a number of concentric air-filled **chambers** (Fig. 17.5). These chambers keep the colony permanently afloat and open above ('aborally') by pores and below ('orally') by canals, called **tracheae**. Air is pumped into float and sail through the tracheal system. The coenosarc forms a flat disc beneath the float and buds off the following zooids which hang downwards:

1. Gastrozooids. From the middle of ventral side of body hangs a single large polyp, called **gastrozooid**. It possesses mouth but lacks tentacles. It feeds the entire colony.

2. Gonozooids. All around the gastrozooid are numerous gonozooids or blastostyles, each having a mouth. They bear clusters of **medusae** or **gonophores** which bear gonads on the manubrium. These medusae are set free, they shed gametes and soon die as they cannot feed.

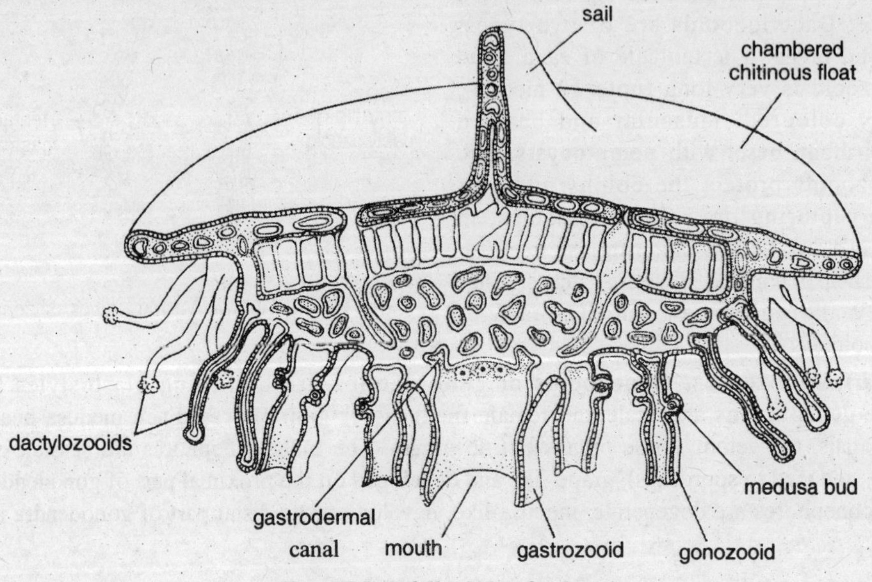

Fig. 17.5. *Velella.* V.S. of colony.

3. Dactylozooids. On the margin of colony is a circle of dactylozooids. They are long, hollow and tentacle-like and have nematocysts.

The pneumatophore (float) and its sail are a modified medusoid, while the gastrozooids, gonozooids and dactylozooids are modified polyps.

3. *PORPITA*

Classification

Phylum	–	**Cnidaria (Coelenterata)**
Class	–	**Hydrozoa**
Order	–	**Hydroida**
Suborder	–	**Chondrophora**
Genus	–	***Porpita***

Porpita is commonly called **blue button**. It has an organization similar to *Velella* and resembles a medusa (Fig.17.6). It is marine, found in all seas. *Porpita* is pelagic, floats on or near the surface of water. It represents a polymorphic floating colony.

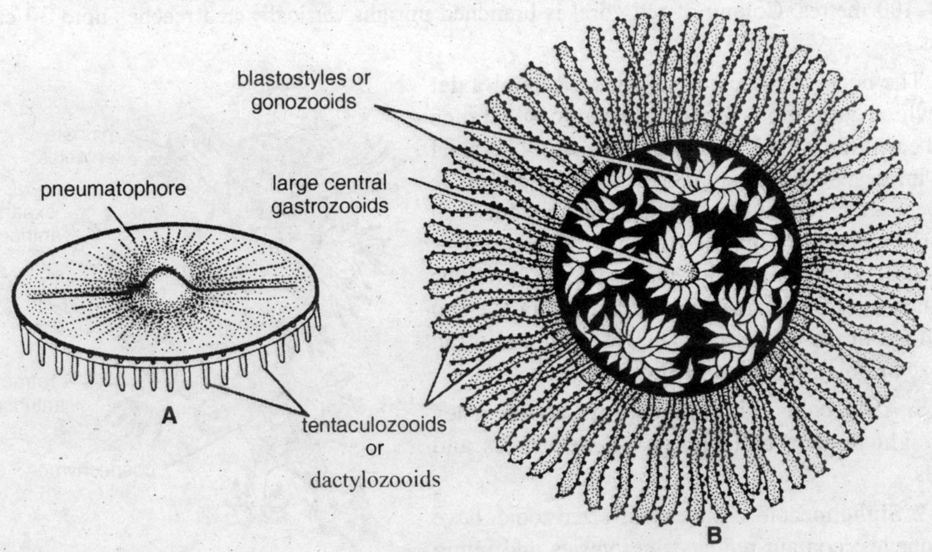

Fig. 17.6. *Porpita*. A—Colony in dorsal view; B—Colony in ventral view.

The float of *Porpita* is large (ranges from 2 to 10 cm in diameter) circular, disc-like and chitinous. In the float are several concentric air chambers, each of which communicates with the exterior by two pores. From the centre of ventral side of float hangs a single **gastrozooid**. It encloses stomach and bears mouth. The gastrozooids is surrounded by groups of closely set **gonozooids** or **blastostyles**, each having a mouth and producing **medusae** or **gonophores**. Gonophores are produced as asexual buds. These are medusa-like and contain bell or umbrella, velum. radial canals and a manubrium. They contain gonads and are set free in the sea water where they discharge gametes and die. A number of long tentacle-like **dactylozooids**, called

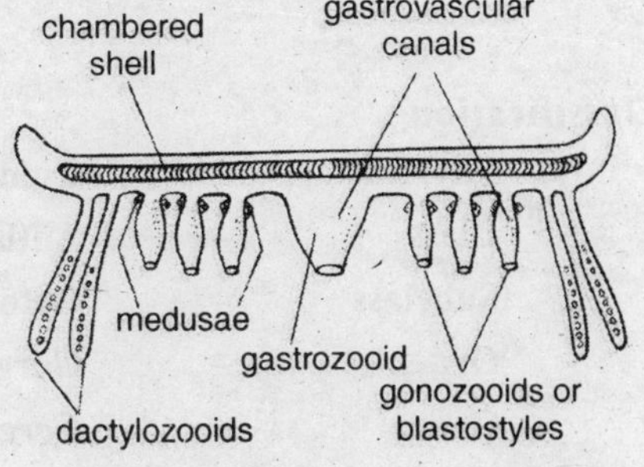

Fig. 17.7. *Porpita*. V.S. of colony.

tentaculozooids, come off from the edge of disc; they are loaded with **nematocysts**. The thick fleshy coenosarc below the float is traversed by endodermal tubes communicating with the cavities of the gastrozooid and gonozooids. Ectodermal tubes, called **tracheae**, communicate the air chambers with the outside. Each air chamber dorsally opens to outside by several pores. Contractions of the zooids are believed to circulate air in the tracheae and the air chambers, thus, helping respiration.

4. *CORALLIUM* (Red Coral)

Classification

Phylum	–	**Cnidaria (Coelenterata)**
Class	–	**Anthozoa**
Subclass	–	**Octocorallia**
Order	–	**Gorgonacea**
Genus	–	***Corallium***
Species	–	***rubrum***

Corallium rubrum is commonly called **red coral** or **red moonga**. It is a marine colonial anthozoan which inhabits rocky bottom chiefly in the Mediterranean and off coast of Japan at the depth of 30–100 metres. Colony of red coral is branched upright vertically and reaches upto 30 cm in height.

The body of red coral contains a central **axial rod** which is composed of an organic substance called **gorgonin** (which is a tanned collagen). This axial rod is impregnated with calcium carbonate. Around the axis occurs **coenenchyme**, in which are embedded red coloured calcareous **spicules**. Colony of red coral is **dimorphic**, containing following two types of polyps or anthozooids which are white in colour, retractile and embedded in the coenosarc. The free oral end of an anthozoan polyp is called **anthocodium**.

1. Autozooids. They are nutritive zooids. They bear eight pinnate tentacles, have mesentries and gonads.

2. Siphonozooids. These small sized zooids have no tentacles, contain reduced mesentries and pump water into the canals of the colony.

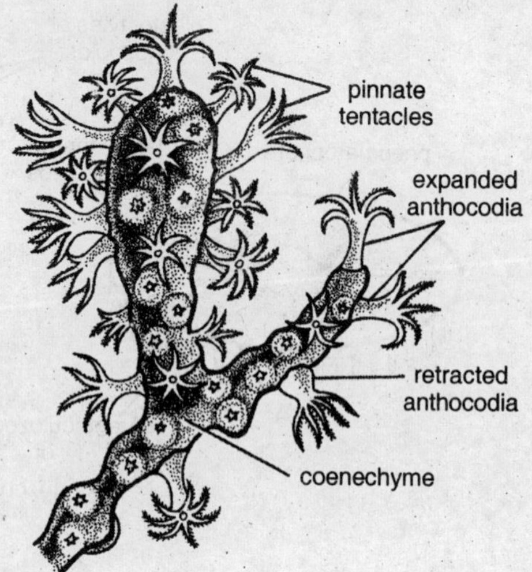

Fig. 17.8. *Corallium rubrum*. A colony.

Corallium rubrum has great economic importance. It provides precious red coral which is used extensively in jewellery.

5. *GORGONIA*

Classification

Phylum	–	**Cnidaria (Coelenterata)**
Class	–	**Anthozoa**
Subclass	–	**Octocorallia**
Order	–	**Gorgonacea**
Genus	–	***Gorgonia***

Gorgonia is commonly called **sea fan**. It is marine, colonial and found in shallow waters on the rocks of tropical and sub-tropical shores (*i.e.*, Indo-Pacific Ocean especially in Malay Archipelago).

Gorgonia produces a large (50 cm in height) upright, fan-shaped colony of yellow or red colour. In this colony, the branches unite to form a network in one plane. The colony contains a short central axial rod (the **stalk**) composed of an organic substance called gorgonin (a collagen-like protein). The stalk is attached with the substratum by a **hold fast** or **pedal/basal plate**. The colony produces branches in one plane only (like the feathers of a bird). These branches are united by numerous cross sections. This skeleton is flexible and allows bending.

Numerous small polyps, called **anthocodia** (= retractile polyps) are found in rows on two sides of branches (Fig. 17.9B). The skeleton is covered with a layer of **coenenchyme** which contains gastrovascular cavities of the polyps and loose calcareous spicules. Polyps feed on zooplankton and other smaller particles. Sexes are separate.

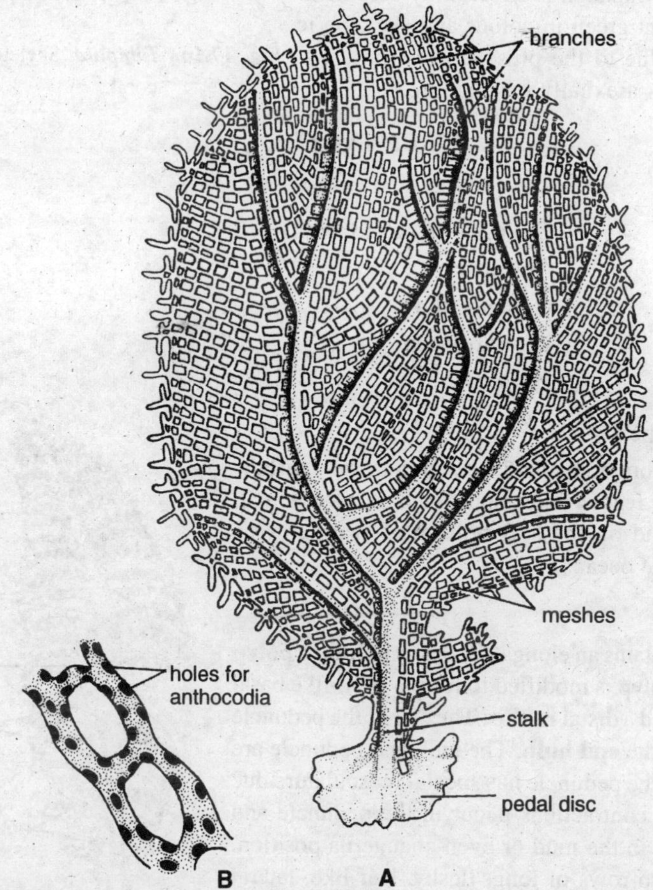

Fig. 17.9. *Gorgonia.* A—Entire colony; B—A part of colony magnified.

6. *TUBIPORA*

Classification

Phylum	–	**Cnidaria (Coelenterata)**
Class	–	**Anthozoa**
Subclass	–	**Octocorallia**
Order	–	**Stolonifera**
Genus	–	***Tubipora***
Species	–	***musica***

Tubipora is commonly called an **organ pipe coral**. It is widely distributed on coral reefs in warm waters of Atlantic, Indian and Pacific oceans. It is a colonial polypoid anthozoan. The colony consists of long parallel upright polyps enclosed in vertical (longitudinal) skeleton tubes arising from a basal plate having creeping mat of **stolons**. The mesogloeal **spicules** fuse to form a continuous vertical tube of each polyp. All the vertical tubes of different polyps are united at definite intervals by transverse calcareous **platforms** (Fig. 17.10). In the living condition this skeleton is covered by ectoderm. The polyps are bright green in colour and skeleton is dull red in colour due to the presence of iron salts. *Tubipora* reproduces asexually by budding.

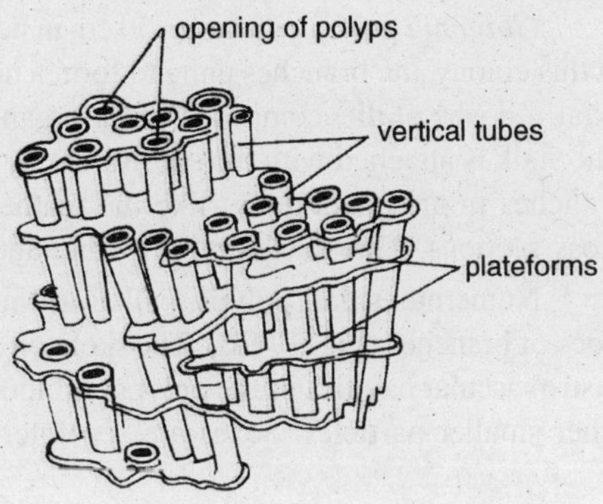

Fig. 17.10. *Tubipora.* Skeleton of organ-pipe coral.

7. *PENNATULA*

Classification

Phylum	–	**Cnidaria (Coelenterata)**
Class	–	**Anthozoa**
Subclass	–	**Octocorallia**
Order	–	**Pennatulacea**
Genus	–	***Pennatula***

Habit and Habitat

Pennatula is commonly called **sea pen** or **sea feather**, because of its pen or feather-like appearance. It is a colonial, marine animal found on the soft bottom of warmer seas (Atlantic and Pacific oceans).

Structure

Pennatula contains an elongated, fleshy **primary polyp** (Fig. 17.11). This polyp is modified to form a stem-like basal stalk or **peduncle** and a distal **rachis**. The end of the peduncle is enlarged to form the **end bulb**. The bulb and peduncle are buried in sea mud. The peduncle has special musculature due to which peristaltic contractions occur in the peduncle and sea pen can burrow in the mud or even change its position. The rachis bears two rows of long, fleshy, leaf-like, lateral branches, called **pinnules**. The colony of *Pennatula* is **dimorphic**, containing following two types of **secondary polyps:**

1. Autozooids or anthocodia. They are nutritive zooids and occur in a single row along the upper margin of pinnule. Autozooids have tentacles, mesentries and gonads.

2. Siphonozooids. These are highly modified or reduced types of zooids. They occur on the dorsal side of rachis and between the bases of the pinnules. These are without tentacles and gonads. Their mesentries are reduced, but their

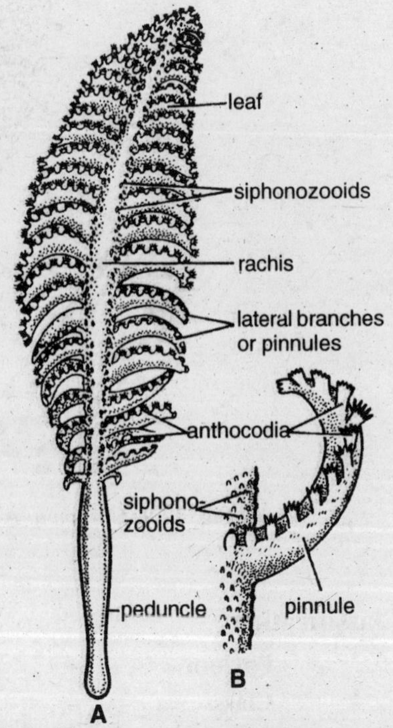

Fig. 17.11. *Pennatula.* A—Entire colony; B—A portion of colony magnified.

siphonoglyphs are enlarged. Siphonozooids pump water into the interconnected gastrovascular cavities (which are four in number) of pirmary polyp, thereby keeping the colony **turgid** and **erect**. The skeleton has a horny axis in the stem (primary polyp) and calcareous spicules are present in the mesogloea. There is no axis in the pinnules.

Pennatula looks yellow, orange, red or purple in colour due to the presence of **lipochrome** pigment in the spicules. For example, *Pennatula sulcata* has a bright orange-red colour. Its life history does not include any medusoid phase.

8. *METRIDIUM*

Classification

Phylum	–	**Cnidaria (Coelenterata)**
Class	–	**Anthozoa**
Subclass	–	**Hexacorallia**
Order	–	**Actiniaria**
Genus	–	***Metridium***

Habit and Habitat

Metridium is commonly known as **sea anemone**. It is a large sessile, brightly coloured, solitary, flower-like animal. *Metridium* is a marine animal, found attached to the rocks, piles of wharves (= landing-stages building especially along the shore for loading and unloading vessels) and solid objects from tide pools to a depth of 90 fathoms. It is found on the Atlantic Coast, Pacific Coast and Europe.

Structure

Body of *Metridium* (Fig. 17.12) is short, cylindrical and **radially symmetrical**. It is divisible into three distinct regions: oral disc, column and pedal disc.

1. Pedal disc. It is muscular, flattened base at the aboral end of the column. It is attached to the substratum.

2. Column. This is the middle cylindrical part. It is differentiated into distal or upper thin-walled short **capitulum** and the lower or proximal thickwalled **scapus**. A **collar** or **parapet** arises as a fold from scapus at the junction of these two regions.

3. Oral disc. At the oral end, the column flares slightly to form the **oral disc**. Oral disc is flat, circular and contains a slit-shaped **mouth** in its centre. Around the mouth are numerous, short, conide hollow **tentacles** in five circles, each circle having tentacles in the multiples of six. Tentacles bear numerous **spirocysts** and **nematocysts**, spirocysts are structures used in capturing of prey that have hard surfaces such as small crustaceans; each contains a single-walled capsule and a long adhesive thread.

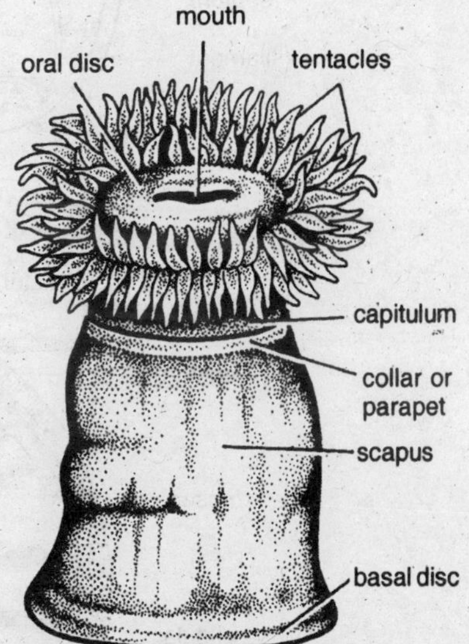

Fig. 17.12. *Metridium.*

Both exo-and endo-skeleton are absent in *Metridium*.

Internal Structure

1. Siphonoglyphs. The mouth leads downwards into a long, stout tube called **pharynx** which is lined by invaginated ciliated epidermis and hangs in the coelenteron (Fig. 17.13). In the pharynx are two longitudinal ciliated grooves called **siphonoglyphs** which are meant for the entry of water into the gastrovascular cavity. This current of water functions to maintain an internal fluid skeleton or **hydrostatic skeleton** against which the muscular system can act and also provides for the exchange of gases through the gastrodermal surface.

2. Mesentries. The gastrovascular cavity of the sea anemone is partitioned by longitudinal, radiating septa or **mesentries** which increase the internal surface area. In the sea anemone, there are two types of septa—complete and incomplete. **Complete septa** are connected to the body wall on one

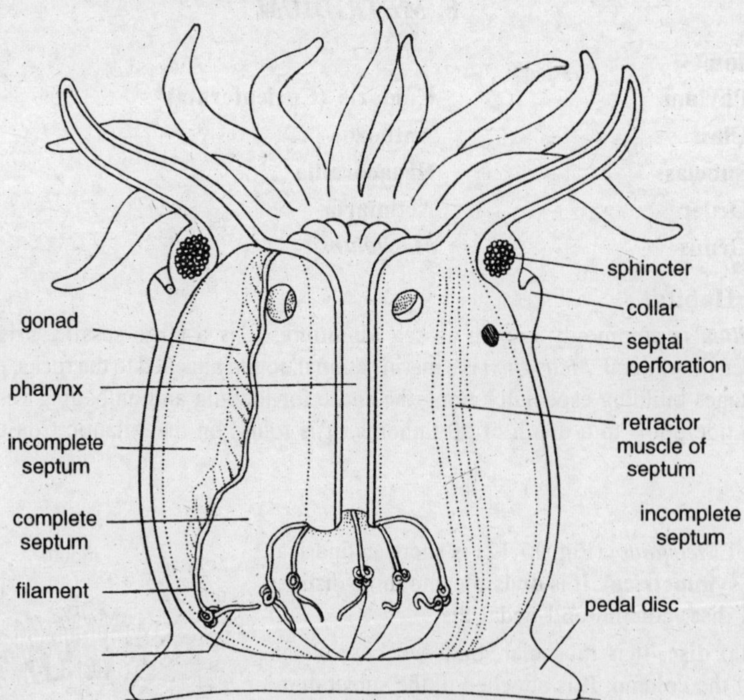

Fig. 17.13. L.S. of a sea anemone.

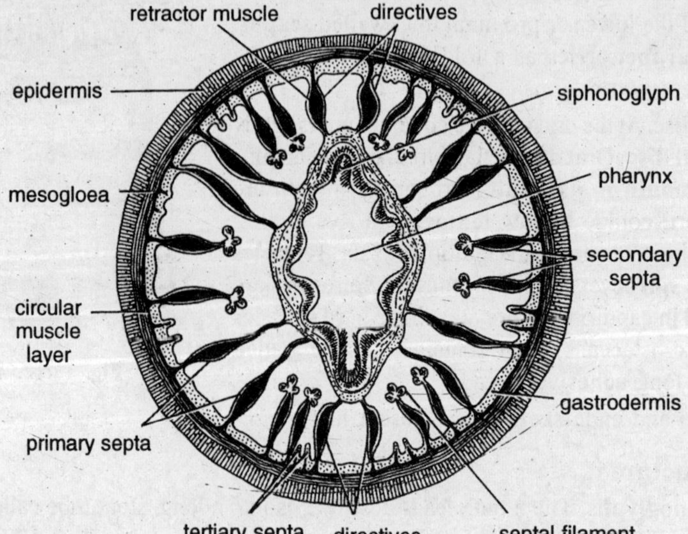

Fig. 17.14. *Metridium.* Cross section of body through the pharynx.

side and to the wall of the pharynx on the other (Fig. 17.14). **Incomplete septa** are connected only to the body wall and extend only partially in the gastrovascular cavity. Both complete and incomplete septa are arranged in adjacent pairs. In the *Metridium*, there are six pair of complete or **primary mesentries**. Two pairs of primary mesentries are joined to the siphonoglyphs and are called **directives** (Fig. 17.14). Between adjacent primary mesentry pairs are **secondary mesentries** or incomplete septa; they are six pairs. Further smaller types of incomplete septa are called **tertiaries** and **quarternaries**. Histologically septa are trilobed structures. They are composed of double layer of gastrodermis enclosing a central layer of mesogloea (Fig. 17.15). A septa contains retractor muscle and bears ciliated cells (to aid in the circulation of water), **cnidocytes, enzymatic gland cells, mucus gland cells** and **phagocytic cells** (for performing intracellular digestion inside food vacuoles). Septa also contain gonads.

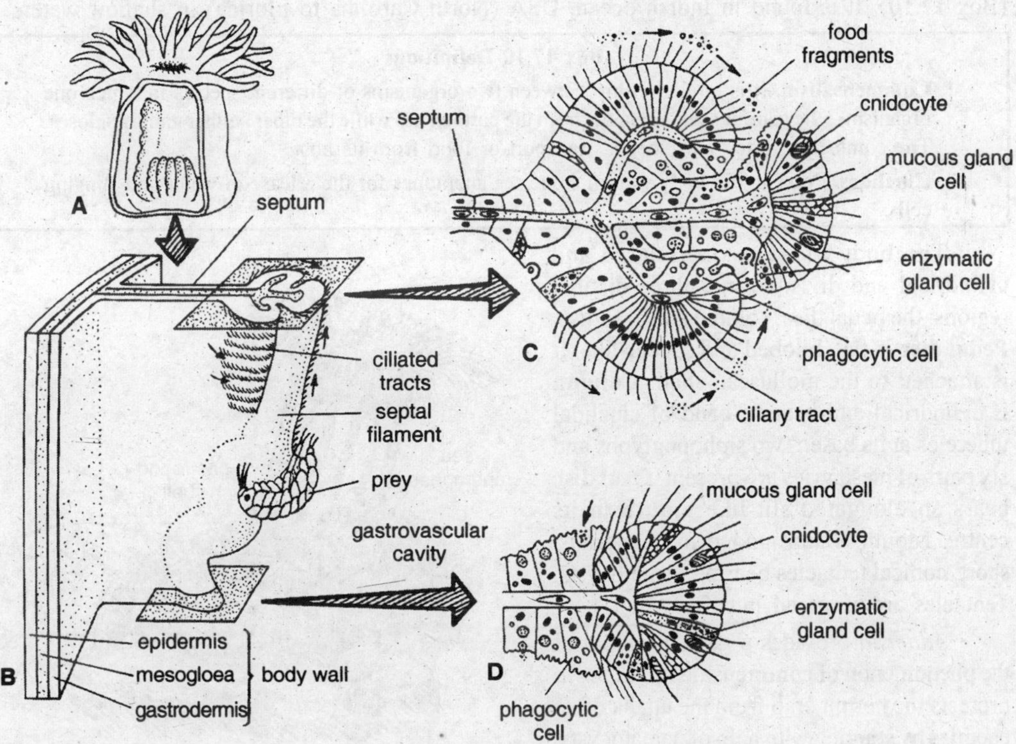

Fig. 17.15. Structure of a septal filament of sea anemone. A—Location of septum in sea anemone; B—The upper part of septal filament is trilobed and its lower part is single lobed; C—Section of trilobed filament; D—Section of single-lobed filament.

Metridium possesses curious organs in the coelenteron called **acontia** (*Gr., akontion* =dart). These are long, delicate threads springing from the inner edges of the mesentries and like the mesenterial gastric filaments are entwined around food objects during digestion. Acontia are loaded with nematocysts which can be protruded through minute apertrues in the column, called **'port holes'** or **cinclides**, when the anemone contracts sharply.

Sexes are separate. Fertilization is internal and a free-swimming ciliated **planula** originate from the zygote. It does not show alternation of generation.

9. *ADAMSIA*

Classification

Phylum	–	**Cnidaria (Coelenterata)**
Class	–	**Anthozoa**
Subclass	–	**Hexacorallia**
Order	–	**Actiniaria**
Genus	–	***Adamsia***

Adamsia is commonly called **sea anemone.** *Adamsia palliata* is a marine **commensal** (Fig. 17.16), found attached to the empty shells of gastropods inhabited by hermit crab, *Eupagurus* (Box 17.10). It is found in Indian ocean, USA (North Carolina to Florida) in shallow waters.

> **Box 17.10 Definitions**
>
> 1. **Commensalism.** It is a relationship between two organisms of different species in which one organism—the **commensal**—benefits from the partnership while the other neither gains nor loses. The commensal often gets shelter, transport or food from its host.
> 2. **Cinclides.** Pores in the body wall of some sea anemones for the release of water and stinging cells.

The body of *Adamsia* is large and cylindrical and divisible into three distinct regions: the pedal disc, column and oral disc. **Pedal disc** is flat, bilobed and sucker-like. It is attached to the molluscan shell. **Column** is cylindrical and bears a band of cinclidal tubercles at its base. Two siphonoglyphs and six pairs of mesentries are present. **Oral disc** bears an elongated slit-like **mouth** in its centre. Mouth is surrounded by numerous, short, conical tentacles bearing nematocysts. Tentacles are arranged into five whorls.

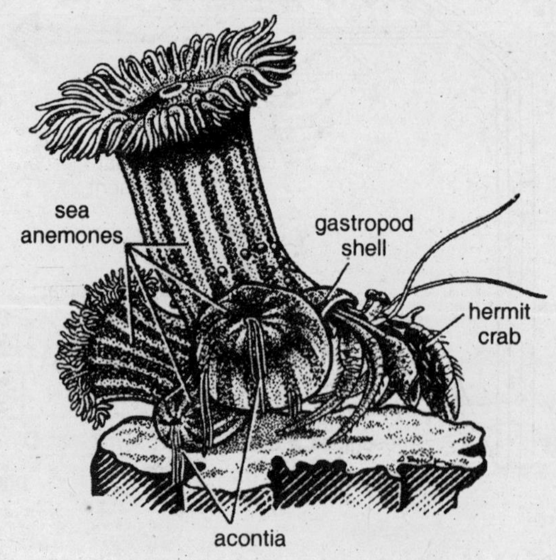

Fig. 17.16. *Adamsia*. Commensalism.

Adamsia provides a good example of the phenomenon of **commensalism**. *Adamsia* protects the hermit crab from the attack of its enemies by stinging with help of nematocysts. In return, *Adamsia* gets the advantage of transportation from one place to another by the hermit crab. Due to this action, sea anemone gets a variety of food supply.

10. *FUNGIA*

Classification

Phylum	–	**Cnidaria (Coelenterata)**
Class	–	**Anthozoa**
Subclass	–	**Hexacorallia**
Order	–	**Madreporaria**
Genus	–	***Fungia***

Fungia is commonly called **mushroom coral** (Fig. 17.17). It is solitary marine stony or scleractinian coral of coral reefs. It occurs in warm sea water of Indo-Pacific region and Gulf of California. Its skeleton is called **corallite** and it is discoidal and mushroom-like with upper surface convex and lower concave. A large number of septa (called **sclerosepta**) radiate out from the centre and are connected by calcareous rods, called **synapticulae**. Adult animal bears a single large polyp (25 cm in diameter) bearing many tentacles. Siphonoglyph is absent. It reproduces asexually (by transverse fission) and sexually (by producing a planula larva).

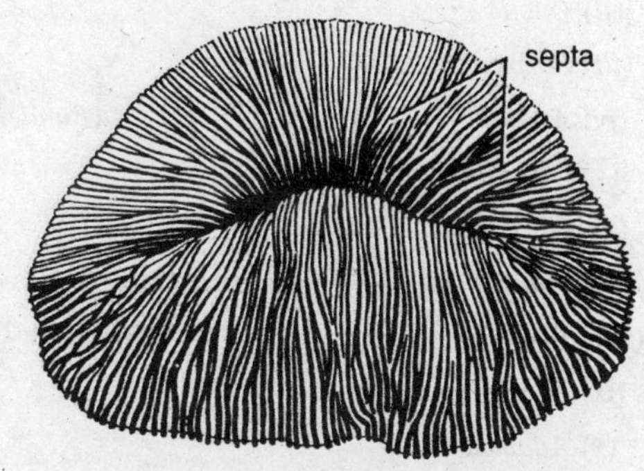

Fig. 17.17. *Fungia.*

11. *MADREPORA*

Classification

Phylum	–	**Cnidaria (Coelenterata)**
Class	–	**Anthozoa**
Subclass	–	**Hexacorallia**
Order	–	**Madreporaria**
Genus	–	***Madrepora***

Madrepora (Fig. 17.18) is commonly known as **horn coral**. It is a colonial tree-like reef-building coral, found in tropical waters, *i.e.*, Australian seas and West Indies. Small polyps are enclosed in cylindrical cups, called **corallites,** separated by perforated **coenosteum**. Terminal polyps have six **tentacles** while lateral polyps contain twelve tentacles. Corallite is made up of calcium carbonate.

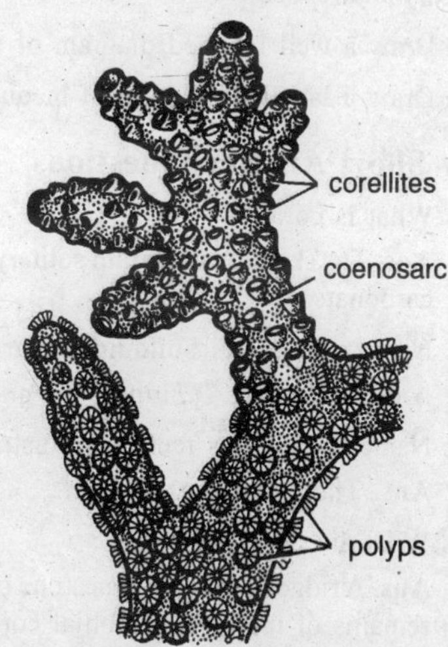

Fig. 17.18. *Madrepora.*

QUESTIONS

Long Answer Questions

1. Mention the distinguishing features of phylum Coelenterata. Give its general characters, classification and examples of each class.
   (Agra 1989, 1992; Awadh 1988; Mumbai 1983; Bundelkhand 1985, 1991; Bhopal 1988, 1991; Bilaspur 1993; Calcutta 1984; Gorakhpur 1987, 1990; Gurunanak 1986, 1990; Guwahati 1988; Indore 1990; Jiwaji 1989; Jabalpur 1992; Kanpur 1994, 1997; Meerut 1987, 1989; Osmania 1984; Purvanchal 1987, 1989; Shivaji 1989, Lucknow 1992, 1994, 1996, 1998)
2. What is the basis of classification of phylum Coelenterata. Classify it upto orders giving examples.
   (Kumaun 1990)
3. Give classification of class Hydrozoa upto orders. Mention the characteristics of each order.
   (Kanpur 1995)

Short Answer Questions

1. Give an outline classification of phylum Cnidaria or Coelenterata.
2. Classify and comment on the following:

   (*i*) *Physalia* *(Meerut or CCS 1991, 1996, 1997; Agra 1992; Bundelkhand 1990, 1994; Raipur 1990; Ravishankar 1991; Jiwaji 1993; Kanpur 1993, 1995; Gorakhpur 1992; Purvanchal 1988; Lucknow 1995, 1996)* (*ii*) *Velella*

(*iii*) *Porpita*

(*iv*) *Aurelia* (*Bundelkhand 1987; Jiwaji 1988; Agra 1989; Bhopal 1989*)

(*v*) *Corallium* (Red coral)

(*vi*) *Gorgonia*

(*vii*) *Tubipora*

(*viii*) *Pennatula* (*Meerut or CCS 1991; Indore 1994; Lucknow 1994, 1996, 1998*)

(*ix*) *Metridium* (*Lucknow 1987*)

(*x*) *Adamsia*

(*xi*) *Fungia* (*Lucknow 1993*)

(*xii*) *Madrepora* (*Lucknow 1994, 1997*)

3. Draw a well labelled diagram of subumbrella view of *Aurelia*. (*Purvanchal, 2001*)

4. Draw a labelled diagram of medusa of *Aurelia* (No description is required). (*Rohilkhand, 2003*)

Very Short Answer Questions

1. What is coral?

Ans. The skeleton of certain solitary and colonial anthozoan coelenterates: composed chiefly of calcium carbonate.

2. Name some coral building cnidarians.

Ans. *Millepora, Tubipora, Gorgonia.*

3. Name the famous reef near Australia.

Ans. The Great Barrier Reef.

4. What are coral reefs?

Ans. A ridge or mass of limestone built up of detritus material deposited around a framework of skeletal remains of molluscs, colonial corals, and massive calcareous algae.

5. Name three kinds of coral reefs.

Ans. Fringing reef, barrier reef and atoll.

6. Suggest the term for the occurrence of two or more types of individuals in an animal.

Ans. Polymorphism.

7. What is polymorphism? (*C.C.S.U. Meerut 2008*)

Ans. Polymorphism is a phenomenon of occurrence in an animal of more than one type of individuals, other than male and female, all with different form and function. The individuals of polymorphic cnidarians are known as **zooids.** Some cnidarians have only two types of zooids-polyp and medusae, while other colonial forms may contain many different types of modified polyps or zooids such as gastrozooids, gonozooids and dactylozooid.

8. What is the zoological name of Portuguese man of war?

Ans. *Physalia.*

9. Name the polymorphic hydrozoan in which the penumatophore lacks sail.

Ans. *Porpita.*

10. What is the function of dactylozooids?

Ans. Protection (They are defensive polyps).

11. Name the phylum in which nervous system appears for the first time.

Ans. Cnidaria (= Coelenterata).

12. In which class of the phylum Cnidaria is stomodaeum found?

Ans. Anthozoa.

13. Name a freshwater cnidarian.

Ans. *Hydra*.

14. Whether gonads are endodermal in *Aurelia*.

Ans. Yes.

15. What are mesentries?

Ans. The gastrovascular cavity of sea anemone is partitioned by longitudinal, radiating septa, called **mesentries**. They increase the internal surface area.

16. Define the siphonoglyph.

Ans. It is a ciliated groove leading from the mouth to the gullet in certain anthozoans.

17. Give zoological name of sea anemone studied by you.

Ans. *Adamsia palliata* (or *Metridium*).

18. What is corallite?

Ans. Exoskeleton of individual zooid of Cnidaria.

Fill in the Blanks

1. Cnidarians have .... symmetry and .... level of organisation.
2. Cnidarians have a single cavity called ....
3. Cells characteristic of cnidarians are .... and ....
4. The free-swimming larva of cnidarians is called ....
5. Cnidarians are first metazoans to have a digestive tract and ....
6. The phylum Cnidaria is divided into 4 classes .... , ..., Cubomodusae and Anthozoa.
7. The common name of *Tubipora* is ....
8. Symmetry in Anthozoa is ....
9. The true corals belong to the order ....
10. The zoological name of the red coral of commerce is ....
11. The sense organs of *Aurelia* are called ....
12. Hydrozoa have both polyp and .... form.
13. The most highly polymorphic hydrozoans form the order ....
14. Predominant body form is Scyphozoa is ....
15. .... exist in polyp form only.
16. *Vellela* forms a .... colony.
17. Zooids of *Porpita* include pneumatophore, ... , gonozooids and ....
18. Function of dactylozooids is ....
19. Members of the subclass *Octocorallia* have ....tentacles and a single .... in the stomodaeum.
20. Hexacorallina have .... tentacles and .... siphonoglyphs.

Multiple Choice Questions

1. Sea anemone is a

(*a*) anthozoan (*b*) scyphozoan

(*c*) hydrozoan (*d*) none of these

2. Sea pen is the common name of
(a) *Corallium* (b) *Pennatula*
(c) *Gorgonia* (d) *Fungia*

3. Jellyfish is a
(a) hydrozoan (b) scyphozoan
(c) anthozoa (d) fish

4. An example of soft coral is
(a) *Gorgonia* (b) *Tubipora*
(c) *Heliopora* (d) *Alcyonium*

5. The fire coral is the
(a) *Tubularia* (b) *Pennaria*
(c) *Millepora* (d) *Physalia*

6. *Gorgonia* is generally known as
(a) sea pen (b) sea fish
(c) sea fan (d) sea disc

7. Dead man's finger is
(a) *Heliopora* (b) *Fungia*
(c) *Alcyonium* (d) *Corallium*

8. Which is known as mushroom coral?
(a) *Fungia* (b) *Aurelia*
(c) *Adamsia* (d) *Tubipora*

9. Function of the polyp is
(a) reproduction (b) excretion
(c) respiration (d) nutrition

10. *Pennatula* is
(a) dimorphic (b) trimorphic
(c) tetramorphic (d) polymorphic

11. A tetramorphic colony is formed by
(a) *Gorgonia* (b) *Obelia*
(c) *Velella* (d) *Physalia*

12. Scyphistoma of *Aurelia* multiplies by
(a) budding (b) strobilation
(c) binary fission (d) none of these

13. Metagenesis is exhibited by
(a) *Aurelia* (b) *Metridium*
(c) *Obelia* (d) *Porpita*

14. The function of a dactylozooid in polymorphic colony is
(a) reproduction (b) offence and defence
(c) nutrition (d) locomotion

15. The larva of *Physalia* is
(a) planula (b) ephyra
(c) scyphistoma (d) cydippid larva

16. Polymorphism is found in
(a) *Halistemma* (b) *Leucernaria*
(c) *Zoanthus* (d) *Alcyonium*

17. Horn coral is
(a) *Madrepora* (b) *Fungia*
(c) *Astraea* (d) *Adamsia*

18. Hypnotoxin is a piosonous fluid produced by

(*a*) parasitic protozoa (*b*) nematocysts
(*c*) *Ascaris* (*d*) sponges

ANSWERS

Fill in the Blanks

1. Raidal, tissue
2. Coelenteron
3. Cnidoblast, epitheliomuscular cells
4. Planula
5. Nervous system
6. Hydrozoa, Scyphozoa
7. Organ-pipe coral
8. Biradial
9. Madreporaria
10. *Corallium rubrum*
11. Rhopalia
12. Medusa
13. Siphonophora
14. Medusa
15. Anthozoa
16. Tetramorphic
17. Gastrozooid, dactylozooids
18. Protection
19. Eight, siphonoglyph
20. Numerous, two.

Multiple Choice Questions

1. (*a*)	**2.** (*b*)	**3.** (*b*)	**4.** (*d*)	**5.** (*c*)	**6.** (*c*)
7. (*c*)	**8.** (*a*)	**9.** (*d*)	**10.** (*a*)	**11.** (*c*)	**12.** (*b*)
13. (*c*)	**14.** (*b*)	**15.** (*a*)	**16.** (*a*)	**17.** (*a*)	**18.** (*b*).

18

Obelia (Phylum Cnidaria)

18.1 CLASSIFICATION

Phylum	–	**Cnidaria (or Coelenterata)**
Class	–	**Hydrozoa**
Order	–	**Hydroida**
Suborder	–	**Leptomedusae**
Genus	–	***Obelia***
Species	–	***geniculata***

18.2 HABIT AND HABITAT

Obelia is commonly referred to as the **sea fur**. It is a marine, sedentary, colonial, delicate and filamentous organism. Its small (2 to 3 cm in height) branched colonies are attached to the rocks, stones, molluscan shells, wooden piles and seaweeds in shallow sea water. *Obelia* looks like whitish or light brown fur on the substratum. It exhibits phenomena of **polymorphism, alternation of generations** and **metagenesis**.

Geographical Distribution

Obelia has a cosmopolitan distribution. It is abundant both in the Atlantic and Pacific coastal waters. It is usually found upto the depth of 80 metres.

18.3 STRUCTURE

Obelia exists in following two distinctly different forms during its life history: 1. **Polypoid, hydroid** or **hydranth phase;** 2. **Medusoid phase.**

1. Hydroid Colony

A hydroid colony of *Obelia* contains different types of individuals, called **zooids**. Each colony of *Obelia* consists of a horizontal thread-like branched root, called **hydrorhiza** which is attached to the substratum. The hydrorhiza gives off several vertical stems called **hydrocauli.** Both hydrorhiza and hydrocauli are hollow tubes. Each hydrocaulus gives off short lateral branches in an alternate manner. Each lateral branch bears a zooid.

Microscopic structure of *Obelia* Colony

1. Coenosarc. All branches and zooids of *Obelia's* colony contain an inner tubular living portion, called **coenosarc**. Coenosarc encloses a cavity or canal, called **enteron** or **gastrovascular cavity.** Through this canal digested food is distributed throughout the hydroid colony. The wall of gastrovascular cavity is formed of outer **epidermis** (ectodermal in origin) and an inner **gastrodermis**

(endodermal in origin). A gelatinous and acellular **mesogloea** lies in between these two layers of the diploblastic animal.

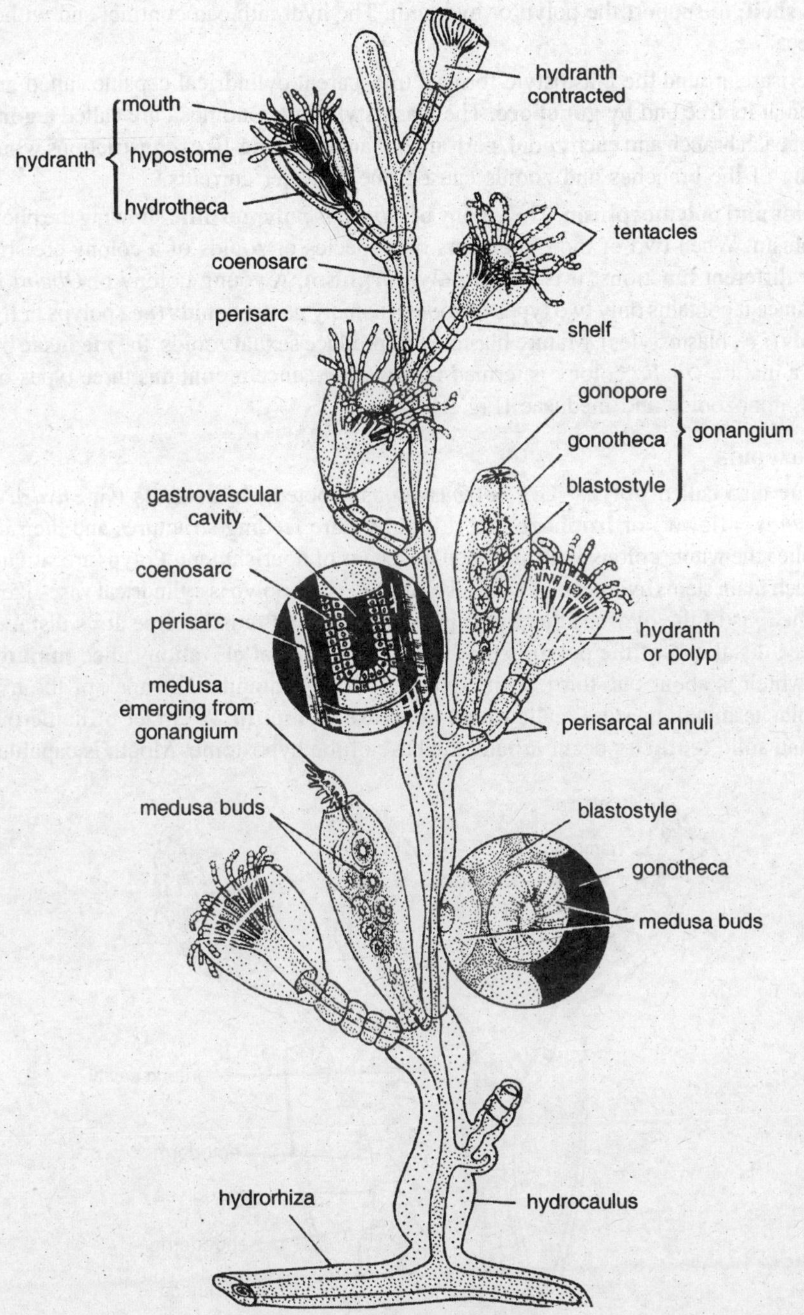

Fig. 18.1. *Obelia*. A portion of colony.

2. Perisarc. The coenosarc of entire colony is covered by a tough, yellow or brown transparent non-living protein (=chitin) layer, called **perisarc**. It is secreted by epidermis over the coenosarc and forms the **exoskeleton** of the colony. Perisarc covers hydrorhiza, hydrocauli, their branches and zooids. Around each polyp, perisarc expands to form a clear widely open wine glass-shaped structure, called **hydrotheca**. Inside the hydrotheca is present a circular, centrally perforated disc-like basal plate, called **shelf**, to support the polyp or hydranth. The hydranth can contract and withdraw into the hydrotheca.

The perisarc around the blastostyle forms a transparent cylindrical capsule called **gonotheca** which is open at its free end by **gonopore**. The blastostyle and gonotheca are called a **gonangium**. At the base of each branch and each zooid, perisarc has annular (ring-like) constrictions which permit some swaying of the branches and zooids caused due to water currents.

3. Zooids and polymorphism. The colony of *Obelia* is **polymorphic** showing the phenomenon of polymorphism. When two or more members of a species or zooids of a colony are structurally modified for different functions, it is called **polymorphism**. A young colony of *Obelia* is termed **dimorphic**, since it contains only two types of zooids, namely **gastrozooids** (*i.e.*, polyps or hydranths) and **gonozoids** (*i.e.*, blastostyles). Mature blastostyles produce sexual zooids, the **medusae** by asexual budding. So a mature *Obelia* colony is termed **trimorphic**, since it contains three types of zooids: gastrozooids, gonozooids and medusae (Fig. 18.1).

I. Gastrozooids

They are also called **polyps** (Gr., *polypus*= many-footed), **hydranths** (Gr., *hydra* or *hydor* = water + *anthos* = flower) or **trophozooids**. All polyps are feeding structures and their collective activity supplies the whole colony with an adequate means of nourishment. Polyps are situated singly at apices of each main stem (hydrocaulus) and all branches. Each polyp is cylindrical vase-like structure attached to the axis of the hydrocaulus by its proximal end and remains free at its distal end (Fig. 18.2). The free distal end of the polyp is produced into a conical elevation called **manubrium** or **hypostome** which is about one-third of the total length of hydranth. The apex of the hypostome bears a circular terminal aperture, called **oral aperture** or **mouth**. A circlet of numerous (about 24) slender and solid **tentacles** occur around the base of the hypostome. Mouth is capable of great

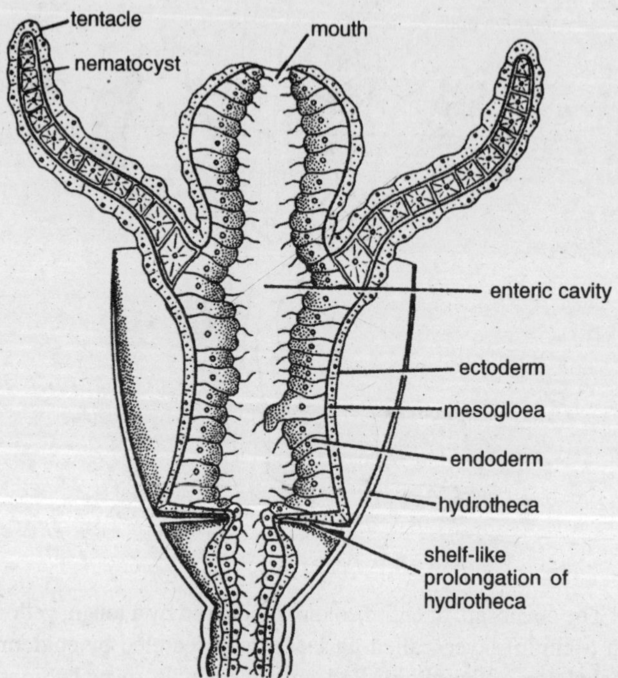

Fig. 18.2. *Obelia.* V.S. of a polyp or hydranth.

dilation and contraction. Below the hypostome is the **stomach** or **gastric region** of polyp which forms about two-third of the length of hydranth.

Mouth opens into a spacious gastrovascular cavity or **enteron** of the polypoid body, *i.e.*, cavity of hypostomal and gastric part of polyp. At the base of the polyp, the cavity is continuous with coenosarcal cavity of concerned branch (hydrocaulus). The polyp is covered by cup-shaped perisarcal hydrotheca.

Histology of polyp. The wall of hydranth is composed of an outer ectodermal **epidermis** and an inner endodermal layer of **gastrodermis**. There is an acellular, thin, transparent and gelatinous **mesogloea** in between the two layers (Fig. 18.2). The tentacles of polyp are solid with no enteron. Each tentacle consists of an outer layer of epidermal cells and a single row of thick walled cubical vacuolated gastrodermal cells.

1. Epidermis. Epidermis of polyp is a continuous layer of **epithelio-muscular cells** which are long, conical and columnar in shape (Fig. 18.3). The inner ends of these cells are produced into muscular processes which run longitudinally. Epidermis of hypostome and tentacles also consists of **sensory** and **epithelio-glandular cells** and 'batteries' of nematocyst containing **cnidocytes**. Epidermis of polyp contains very few **interstitial cells** (*i.e.*, undifferentiated cells). Branched **nerve cells** occur in deeper zones of epidermis. *Obelia* has only one kind of nematocyst called **basitrichous isorhizas** in which the capsule is oval, the butt is absent, the thread is open at the tip and has spines at its base.

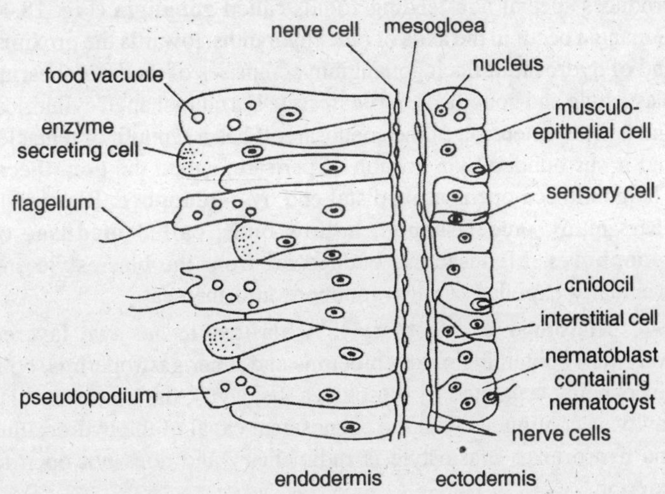

Fig. 18.3. *Obelia.* L.S. of body wall of hydranth.

2. Gastrodermis. Gastrodermis of polyp is a continuous inner layer of long, granular **endothelio-muscular cells**, their muscle processes point outwards and are circular. These gastrodermal cells have flagella towards their inner ends; flagella produce a current in the fluid of enteron. **Pseudopodial** gastrodermal cells can form pseudopodia for engulfing food. Gastrodermis of polyp also contains club-shaped **glandular cells**. Glandular cells of hypostome secrete mucus, while those of gastric region secrete the proteolytic digestive enzymes. Interstitial cells, sensory cells and nerve cells have same distribution as in epidermis. The nerve cells form an intra-gastrodermal nerve net upon the mesogloea. The two nerve nets: epidermal and gastrodermal are inter-connected.

Functions of polyp. The polyps or hydranths are nutritive zooids of the colony. They capture, ingest and digest food (see Box 18.1). *Obelia* is **carnivorous** and feeds upon aquatic crustaceans, nematodes and other worms. Tentacles help in catching and conveying the prey to the mouth. They do this job by the help of nematocysts which penetrate, hold and generally kill the prey. The highly developed longitudinal muscle tails of the tentacles holding the prey contract and succeed in carrying the prey to the mouth where it is **ingested** and passed into enteron. Digestion of food is mainly **extracellular**. It occurs in the enteron by the help of enzymes secreted by glandular gastrodermal cells. Pseudopodial gastrodermal cells ingest partly digested food (food particles) by **phagocytosis**. **Intracellular digestion** within the food vacuoles completes the digestive process. After digestion has been completed in the gastrodermis, soluble food diffuses to the epidermis through the mesogloea. Any unwanted/undigested food remains are **egested** via the oral aperture.

Box 18.1 Radial symmetry helps *Obelia* in feeding

Being sessile, *Obelia* has to rely for its food source on water currents carrying prey or prey swimming towards it. The radial symmetry of the hydranths and the branched nature of the whole colony enable the tentacles to sweep through large quantities of water and to encounter food from all directions.

Beating of flagella of gastrodermal cells and rhythmic contraction (= peristaltic movement) of hydranth circulates the digested food (along with food particles) throughout the coenosarc of the colony.

II. Gonozooid

When *Obelia* colony has reached its full development, it produces special non-feeding zooids called **gonangia** (Fig. 18.4). Gonangia occur in the axils of older hydranths, towards the proximal end of hydrocauli. Each gonangium comprises of medusae-bearing blastostyle and gonotheca. A **blastostyle** is a club-shaped cylindrical and hollow extension of the coenosarc. It has no mouth or tentacles and is surrounded by a portion of perisarc, called the **gonotheca**. The gonotheca opens at its distal end by a **gonopore**. Blastostyle bears many saucer-shaped, hollow buds, called **medusae** or **gonophores**. Medusae are budded-off from the blastostyle and eventually expelled via the gonopore into the sea.

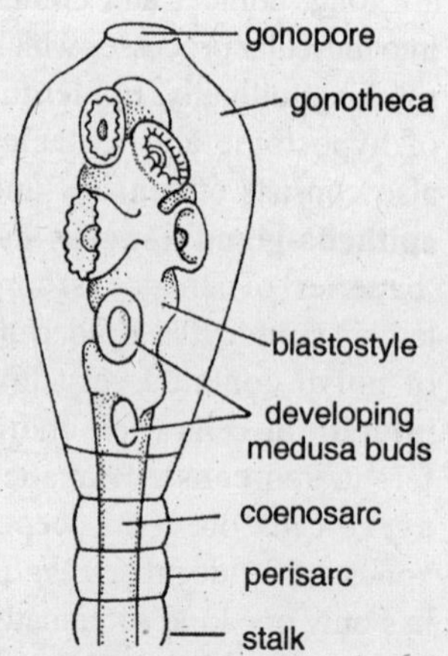

Fig. 18.4. *Obelia*. A gonangium.

Histology of blastostyle. A blastostyle has two layered body wall containing outer epidermis and inner gastrodermis, both layers being separated by mesogloea. Its cavity, the gastrovascular cavity, is continuous with the coenosarcal canal of the hydrocaulus and hydrorhiza. Blastostyle is rudimentary and does not open to exterior.

Functions of gonozooid. The main function of gonozooids is to take part in reproduction, hence are called **reproductive zooids**. Each blastostyle produces numerous lateral buds (by **asexual budding**) which develop into **medusae**. Each saucer-shaped medusa is attached by middle of its convex surface to the blastostyle, when fully formed it breaks free and escapes out in water through the gonopore. The free-swimming medusae eventually develop gonads and represents the sexual phase of the life cycle of *Obelia*.

III. Medusa

Medusae are the third type of zooids of *Obelia* colony. These are modified zooids. Medusae are produced as hollow buds from the coenosarc of blastostyle by asexual budding during spring and summer seasons. They are solitary and **pelagic**, *i.e.*, they swim freely on the surface of water. Mature madusae bear gonads and reproduce sexually. They are **unisexual (dioecious)**, *i.e.*, male gonads (testes) and female gonads (ovaries) are borne by separate individuals.

A. Structure of medusa. A medusa is radially symmetrical. It is tiny (microscopic) and circular having umbrella-like shape. The convex outer surface is known as the **exumbrella** (Fig. 18.7) and the concave inner surface is the **subumbrella**. Hanging down from the centre of the subumbrellar surface is a short, hollow **manubrium** (L.*manus* = handle) at the end of which is a four-lobed oral aperture or **mouth**. Mouth opens into a part of gastrovascular cavity, called **stomach** or **gastric cavity** situated inside the manubrium. From the stomach arise four **radial canals** which are delicate flagellated tubes, they pass outwards to the edge of umbrella and join a flagellated **circular canal**

running near the margin of umbrella. The free edge of umbrella becomes folded inward towards manubrium to form a rudimentary **velum**. The medusae with a velum are called **craspedote** (*e.g.*, *Obelia* and other hydrozoans) and those without the velum are called **acraspedote** (*e.g.*, Scyphozoa).

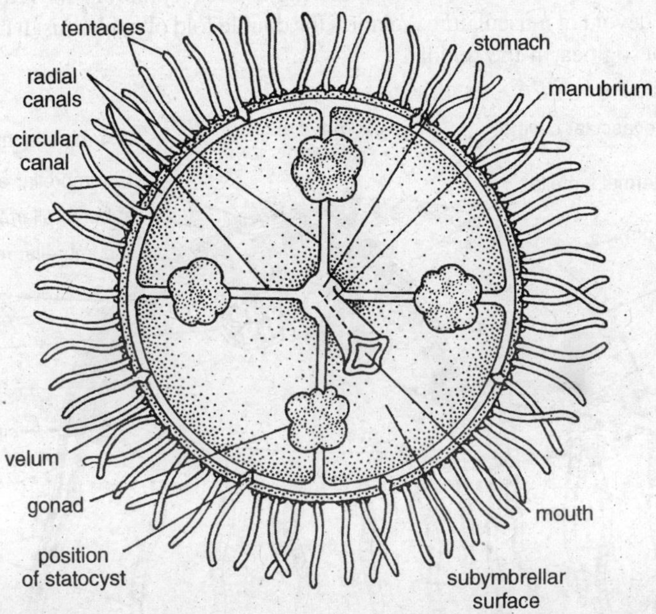

Fig. 18.5. *Obelia*. Medusa in oral view.

Tentacles. From the edge of the umbrella arise numerous small, solid and highly contractile **tentacles**. There are 16 tentacles in young medusa but this number increases in the adult stage. According to their position, 16 tentacles are classified as follows:

1. Per-radial tentacles. These are four in number and they are situated against the four radial canals.

2. Inter-radial tentacles. These are four in numbers and are located on radii, bisecting the angle between two per-radial tentacles.

3. Ad-radial tentacles. These are eight in number and found in between the per-radial and inter-radial tentacles.

4. Sub-radial tentacles. These tentacles are found in between the ad-radial and per-or inter-radial tentacles.

B. Histology of medusa. Like the hydranth, medusa is **diploblastic** having outer ectodermal **epidermis** and inner endodermal **gastrodermis**. A gelatinous **mesogloea** exists between the two epithelia; it becomes thickened towards exumbrella forming the main bulk of the body.

1. Epidermis. The epidermis covers

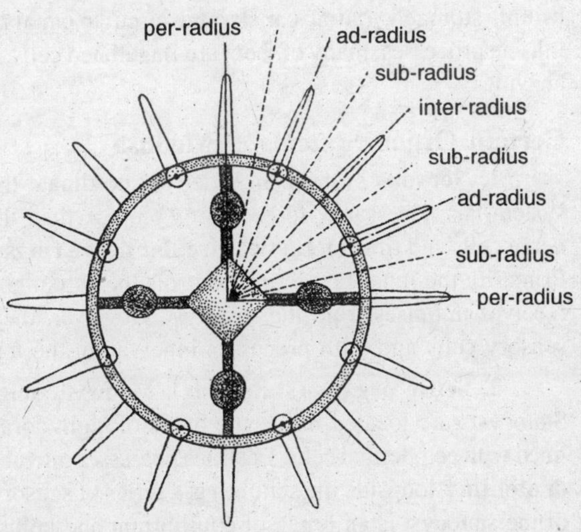

Fig. 18.6. *Obelia*. Radial symmetry and types of tentacles.

manubrium, subumbrella surface, the margins of the umbrella and entire exumbrella surface of medusa (Fig. 18.7). It consists of **epitheliomus-cular cells** which are produced into muscle processes. Muscle processes run longitudinally in the manubrium and tentacles. They form striated circular muscles and some radial muscles of sub-umbrella and bring about locomotory movements. The epidermis of exumbrella is devoid of musculature. Velum is the double fold of epidermis. It contains no endoderm but has thick mesogloea in the middle.

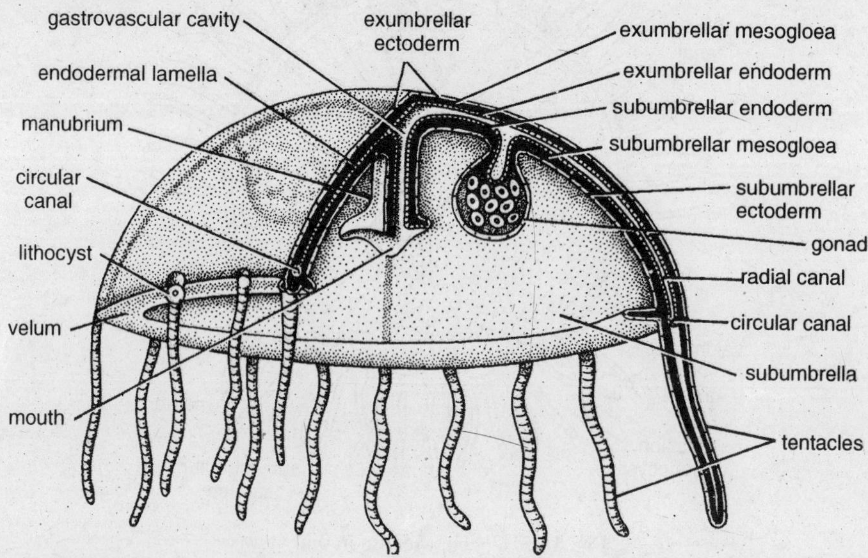

Fig. 18.7. *Obelia.* Medusa with a part cut away to show its histology and structure.

The structure of tentacles is essentially the same as in the polyp, but in addition they each possess a swelling at their base where interstitial cells have accumulated. The interstitial cells are used to replace lost or damaged tentacular **nematoblasts.**

2. Gastrodermis. The endodermis or gastrodermis lines the gastrovascular cavity of manubrium, stomach, radial canals and circular canals. The epithelial cells of endodermis have no muscle processes, many of them are flagellated cells, and some are glandular cells secreting digestive enzymes.

Certain Organ Systems of Medusa

1. Nervous system. In order to coordinate the over all activity of the medusa the nervous system has necessarily to be more complex than that of the polyp. In the medusa the epidermal nerve cells and fibres form two **circular nerve rings**, one external and one internal to circular canal. Primarily the inner nerve ring controls the subumbrellar muscle fibres, whilst the outer nerve ring receives impulses from the statocyst. However, they are also interconnected with each other, the sensory cells and with processes innervating the tentacles.

2. Sense organs (statocysts). Medusa contains eight receptor organs, called **statocysts**. Statocysts are located inside the bases of eight adradial tentacles. Each contains a fluid-filled sac lined with ectodermal cells. Each sac possesses movable particle of calcium carbonate, called **statolith** or **otolith**. Alongside the statolith is a series of sensory protoplasmic processes. The whole apparatus of the statocyst is an organ of equilibrium and helps to bring about coordination in the movement of medusa (see Box 18.2).

Box 18.2

The medusa of *Obelia* normally swims through the water with the margins of its bell horizontally aligned. If the bell is tilted, angular displacement of the statolith will cause impulses to be produced by the sensory processes of statocyst which are passed to the outer nerve ring. These impulses are then relayed to the inner nerve ring and subumbrellar musculature will be stimulated to take the appropriate **reflex corrective action** to restore equilibrium.

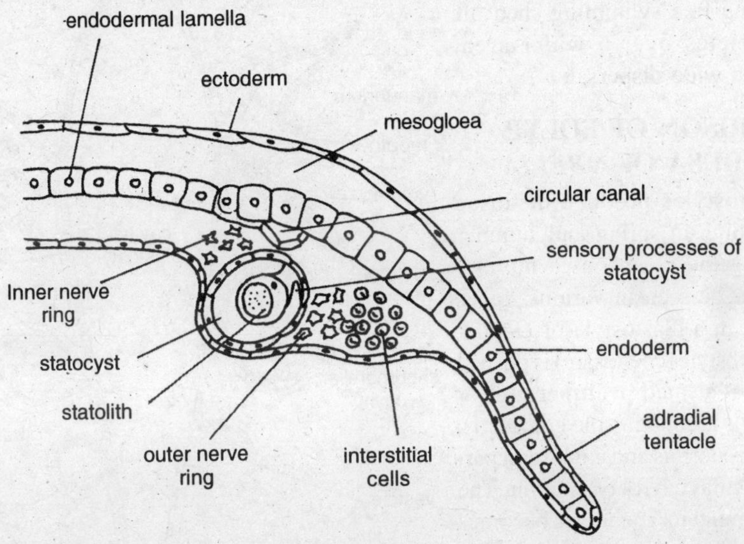

Fig. 18.8. *Obelia*. V.S. statocyst of medusa at base of adradial tentacle.

3. Reproductive organs. Each medusa of *Obelia* contains only four **gonads** which may be either testes or ovaries depending on the sex of the medusa. Halfway along each radial canal and protruding from the subumbrellar surface are the gonads. They consist of an outer layer of **ectodermal cells** and a core of **endodermal cells** derived from the radial canal itself. Germ cells originate in the ectoderm of manubrium and migrate to gonads. It is here that **meiosis** occurs. Each medusa produces either flagellate **sperms** or spherical **ova**. Liberation of the gametes occurs when the ripe gonads burst. After the release of gametes the medusa dies.

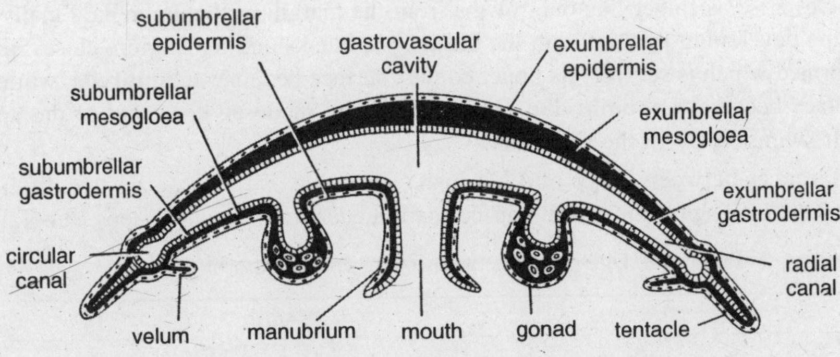

Fig. 18.9. *Obelia*. V.S. of medusa showing gonads.

Functions of medusa. Medusa is the sexual zooid. It being dioecious produces one type of gamete either sperms or ova.

Division of Labour in *Obelia*

In *Obelia* different functions are performed by different zooids:

1. The polyp or hydranth is nutritive in function. It catches and digests foods.
2. Blastostyles are reproductive zooids. They produce medusae by asexual budding.
3. Medusae are sexual zooids. They contain gonads and produce gametes. They being free swimming shed off their gametes over a wider area, ensuring a wide dispersal.

COMPARISON OF POLYP AND MEDUSA OF *OBELIA*

The blastostyles produce medusae by asexual budding in spring and summer seasons. During these seasons a number of medusae may be seen in various stages of development in a blastostyle of *Obelia*. The wall of blastostyle (coenosarc) is pushed out in the form of a small **protuberance** or **bud** (Fig. 18.10). It contains the epidermis, mesogloea, gastrodermis and encloses a part of the gastrovascular cavity or enteron. The bud grows larger and its coenosarc becomes like a **vesicle** which is attached to the blastostyle by a narrow **stalk**. The cavity of the **vesicle** is continuous with the enteron of blastostyle.

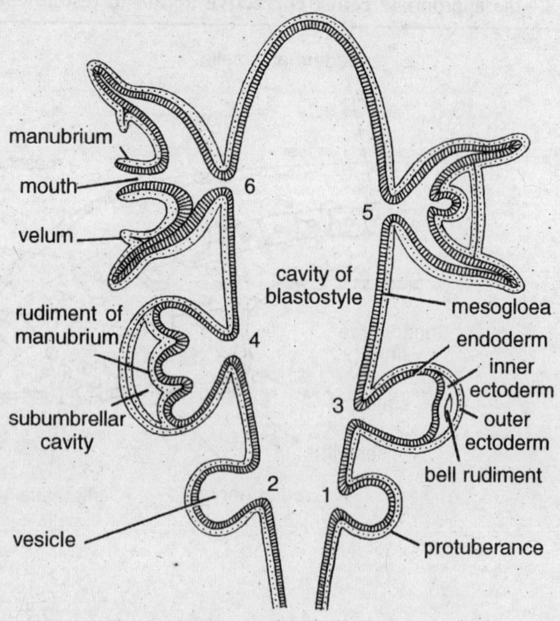

Fig. 18.10. *Obelia*. Stages of development of medusa from a blastostyle.

The epidermis at the top of the vesicle now splits into two layers. The inner of the two again splits and a small cavity, called **bell rudiment**, appears between the newly splitted epidermal layers. The innermost epidermis and the gastrodermis, now, evaginate to form **rudiment** of **manubrium**. This rudiment elongates and breaks the outer epidermis. The broken epidermis (double layered) becomes **velum** (or **craspedon**). Later the mouth opens at the free end of manubrium.

Tentacles arise as finger-like outgrowths from the margin of the umbrella. Finally the stalk connecting the developing medusa with the blastostyle breaks and its aperture closes up. Thus, a medusa is formed which is set free. Its upper convex surface becomes **exumbrella**, while its inner concave surface becomes **subumbrella**. The fully formed medusae come out of the aperture of gonotheca. It swims freely in the sea water.

3. Differences between Polyp and Medusa. The polyp and medusa are two different types of zooids of *Obelia*. Their comparison can be made under the following steps (see Table 18.1).

Table 18.1 Differences between polyp and medusa of *Obelia*.

Polyp	*Medusa*
I. Habit	
1. It is fixed or sessile.	1. It is always free-swimming and pelagic.
II. Position	
2. It is attached to the vertical branches of the colony.	2. It develops as a bud on the axial cylinder, the blastostyle, and detaches from it to live freely.

Polyp	*Medusa*
III. Morphology	
3. Body cylindrical, elongated and vase-like.	3. Body saucer-shaped or umbrella-like.
4. The hypostome or manubrium is conical in form and constitutes one third of the whole body. It is directed upward.	4. Manubrium is comparatively small and quadrangular and hangs downwards.
5. Mouth is circular.	5. Mouth is square.
6. Tentacles are normally 24 in number.	6. Tentacles are 16 in young but numerous in adult.
7. The bases of tentacles are not swollen.	7. The bases of tentacles are swollen and contain reserve interstitial cells.
8. Velum is absent.	8. Velum is present around the margin of umbrella.
9. The statocyst is absent.	9. The statocyst is present in the base of eight adradial tentacles.
IV. Histology	
10. Gastrovascular cavity simple without radial and circular canals.	10. Gastrovascular cavity is represented by stomach, four radial canals and one circular canal.
11. Endoderm (gastrodermis) does not form lamella.	11. Endoderm of the oral and aboral region constitutes endodermal lamellae.
12. Mesogloea is poorly developed being represented as a thin layer.	12. Mesogloea is enormously developed.
13. Nervous system is poorly developed; represented by few nerve cells.	13. Nervous system is well developed with a double ring of irregular nerve cells.
14. The musculature is not developed.	14. The muscle processes of epithelio-muscular cells are well developed and constitute a well developed musculature.
V. Reproduction	
15. The gonads are absent.	15. The gonads are present on the radial canals in four groups.
16. The polyp is a **nutritive zooid** and represents an **asexual** generation.	16. The medusa is a **reproductive zooid** which represents the **sexual** generation.
17. The polyp reproduces asexually by budding.	17. The medusa reproduce sexually by the formation of gametes.
18. It does not cause wide dispersal of species.	18. It causes wide dispersal of species.

4. Similarities between Polyp and Medusa. Despite their remarkable differences in structure and function, polyp and medusa are basically similar structures; they are **homologous structures** as shown by the following similarities between the two.

1. Both are **radially** symmetrical.
2. Both are **diploblastic** having outer epidermis (ectodermal) and inner gastrodermis (endodermal).
3. **Mouth** and **hypostome** (manubrium) are homologous structures in both.
4. The stomach, radial canals and circular canals of medusae are homologous to the gastrovascular cavity of polyp.
5. Both are **carnivorous**; in both the food is captured and ingested with the help of tentacles.
6. Digestion is **extracellular** as well as **intracellular** and digested food diffuses throughout the body without a circulatory system.
7. The convex outer surface of umbrella of medusa is homologous with the base of the polyp.

Advancement of Medusa over Polyp

Medusa shows many advanced features than the polyp. Few of them are as follows:

1. The epidermis of medusa resembles the epithelium of higher Metazoa. It forms a thin, protective and sensitive layer of small cells.
2. The excessive development of **mesogloea** reduces the gastrovascular cavity to a system of canals and also provides lightness which helps in buoyancy.
3. The nervous system shows differentiation into two nerve rings, which form the central nervous system, and nerve nets which form the peripheral nervous system.
4. The marginal sense organs (statocysts) present at the bases of eight tentacles are of great advantage to the free-swimming habit of medusa.
5. The mode of sexual reproduction increases the genetic variations of the species and the free swimming habit of medusa provides wide dispersal of the species.

Derivation of Medusa from Polyp or Evolutionary Significance

Despite the medusa showing slight advancement over the polyp in structural details, a medusa can be said to be a modified polyp. A medusa can be derived from polyp by assuming that the polyp is inverted and its tentacular region is pulled out. The result will be that the manubrium is now directed downwards and a disc-shaped appearance is assumed (Fig. 18.11). A saucer shaped form is developed by basal region. It, therefore, consists of two surfaces, the outer convex surface (**exumbrellar surface**) and the inner concave surface (**subumbrellar surface**). The gastrovascular cavity of the polyp can be said to form the exumbrellar surface.

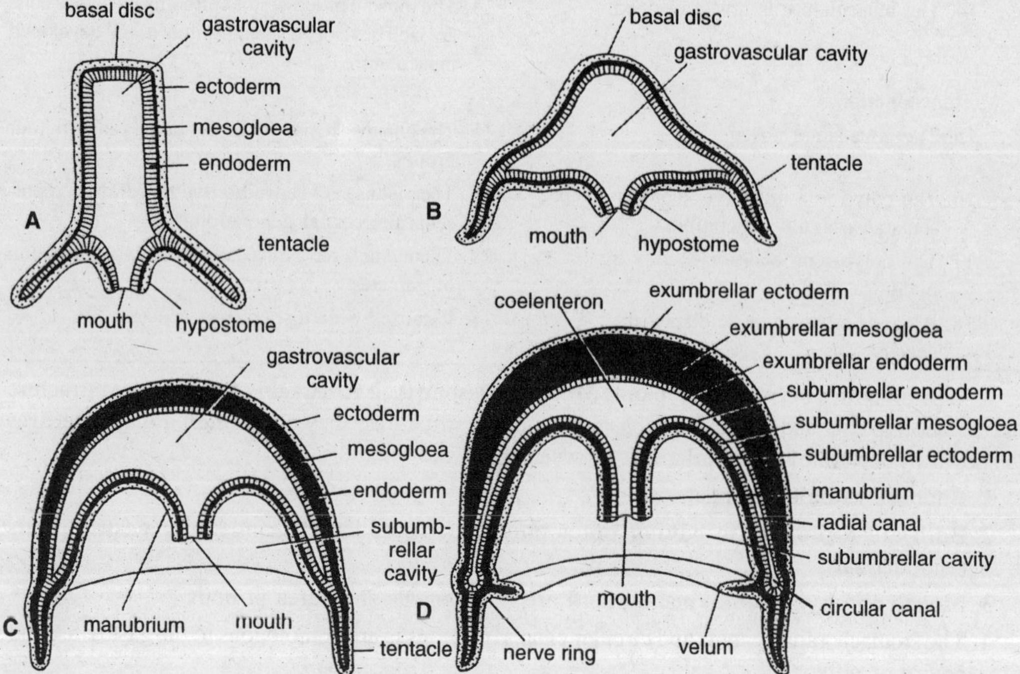

Fig. 18.11. *Obelia.* Diagram showing the derivation of the medusa from the polyp. A—Polyp in L.S., B—L.S. polyp with extended tentacular region; C—L.S. of bell; D—Vertical section of medusa.

18.4 REPRODUCTION AND LIFE CYCLE

Obelia colony propagates by the asexual method of budding producing hydranths (polyps) and blastostyles. These represent asexual generation. The blastostyles are specialized reproductive zooids forming **medusae** by budding. Medusae constitute the sexual or the reproductive individuals possessing four **gonads**. These gonads may be either male or female.

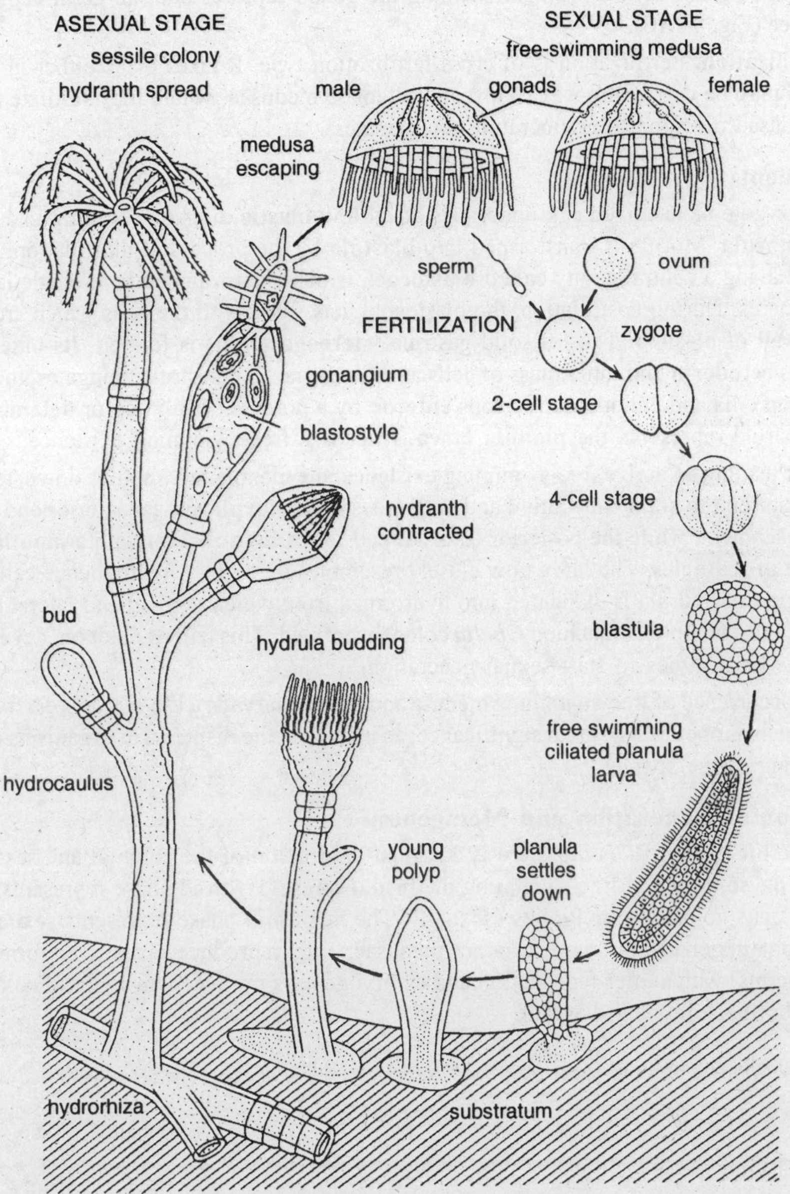

Fig. 18.12. *Obelia.* Stages in the development and life history.

Gonads

Gonads are situated on the sub-umbrellar surface of medusa, one in the middle of the course of each radial canals. They mature (*i.e.*, undergo meiosis and other steps of gametogenesis) after the medusae escape from the gonotheca. Each gonad is an oviod, knob-like body of undifferentiated interstitial cells between the epidermis (ectoderm) and gastrodermis (endoderm). Epidermis is continuous with that of the sub-umbrella and gastrodermis is continuous with that of radial canals (Fig. 18.9). The space between two layers is filled with a mass of **sex cells,** which become differentiated into haploid ova or sperms in female and male medusae respectively. On maturation of gametes, the endodermal/gastrodermal wall surrounding the gonad ruptures and the germ cells are set free in sea water (Fig. 18.12).

Fertilization. Fertilization is of cross-fertilization type. It takes place either in sea-water or the sperms may be carried by water currents to female medusae, where they fertilize the ova. The parent medusae die soon after liberating the gametes.

Development

The zygote or fertilized egg undergoes equal **holoblastic** division resulting in a solid ball of cells, the **morula**. Morula is transformed into **blastula** by the process of blastulation. The blastula is hollow, having a central cavity called **blastocoel,** which is surrounded by a single layer of cells, the blastomeres. During gastrulation, the blastocoel gets filled with the cells which are budded off from the wall of blastula. Thus, a solid gastrula (**stereogastrula**) is formed. Its outer most layer is known as **ectoderm** and inner mass of cells as **endoderm**. The gastrula elongates, the ectodermal cells develop cilia and endoderm develops **enteron** by a process of splitting or **delamination**. This ciliated gastrula represents the **planula larva**. It leads a free-swimming existence.

After leading an active free-swimming existence, the planula larva settles down, attaches itself by its anterior end to solid substratum and undergoes **metamorphosis**. Its anterior end forms **basal disc** for attachment, while the posterior or distal end forms **manubrium** with a **mouth** surrounded by a circlet of **tentacles**. The larva now closely resembles a polyp or hydra, hence called **hydrula**. The basal part of hydrula is elongated into hydrorhiza from which vertical and lateral branches are budded off and a complex branching *Obelia* colony is formed. This colony, later on, develops medusa buds on the blastostyles to start sexual generation.

The occurrence of free-swimming medusa and ciliated larva in a life-history of a fixed or sessile organism such as obelia, is of great significance; as it helps in the dispersal of organisms and prevents overcrowding of the species.

Alternation of Generation and Metagenesis

In the life-cycle of *Obelia*, there is a regular alternation of the colonial and sessile **hydroid phase** and the solitary and free-swimming **medusoid phase**. Hydroid phase represents the **asexual** phase and lacks gonads; it multiplies asexually. The medusoid phase represents **sexual** phase, the medusae contain gonads, but never produce medusae. They reproduce sexually by forming gametes (eggs or sperms) which after fertilization and embryogenesis produce new hydroid colonies, as can be shown by the following diagram:

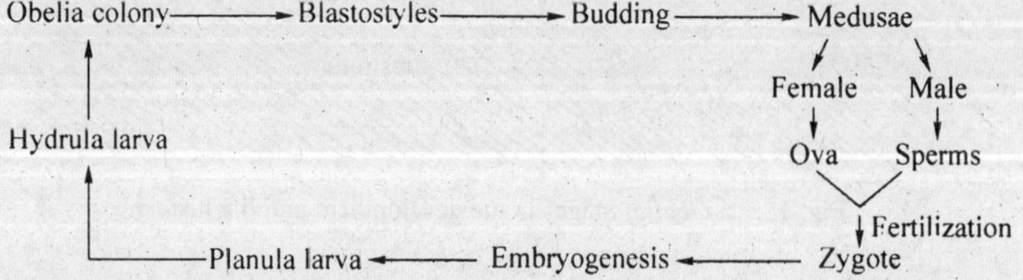

This phenomenon of alternation of sexual generation with asexual generation of life cycle of *Obelia* was termed as **alternation of generation**.

Hyman seems to be right in holding that the concept of alternation of generation is not valid in animals (*e.g.*, *Obelia*). It truely applies to plants, notably the bryophytes and pteridophytes, in which the haploid sexual generation called the **gametophyte** alternates with the diploid asexual generation, called the **sporophyte**. In contrast, in *Obelia* the apparent asexual and sexual generations, namely the polyps and medusae, respectively, are both **diploid**. In case of *Obelia*, only *the gametes are haploid* while the sexual gonad-bearing medusoid form and the hydroid form are both diploid. Further, the sexual generation of *Obelia* (*e.g.*, medusa) does not arise from a single-haploid cell. Thus, it becomes clear that there is no alternation of diploid asexual and haploid sexual generations in the life cycle of coelenterates. *Obelia*, instead shows alternation of fixed and free-swimming generations. This is advantageous for the *Obelia* as it prevents overcrowding and assures wide dispersal of the species.

It has been suggested that the hydroid colony represents **developmental** or **juvenile** stage as it does not bear gonads. Thus, the complete life cycle from egg to egg is one generation. The budding off of a medusae is merely a step in the attainment of sexual reproduction. Hence, instead of describing the phenomenon as alternation of generation, it is termed **metagenesis**, meaning postponement of the power of generation (*i.e.*, sexual reproduction).

QUESTIONS

Long Answer Questions

1. With the help of diagrams describe the structure of polyp and medusa of *Obelia*. (*Lucknow 1983; Bhopal 1984*)

2. Give an account of various types of zooids found in the *Obelia* colony. (*Gorakhpur 1991; Agra 1989, 1993; Kanpur 1983; Bundelkhand 1993; Vikram 1985; Jodhpur 1991; Rajasthan 1986; Meerut (CCS) 1992, 1996; Himachal 1992; Allahabad 1997*)

3. Why *Obelia* is called a trimorphic? Describe different types of zooids in *Obelia* colony.

4. What is polymorphism? Explain it with reference to the structure of *Obelia*. (*Awadh 1993*)

5. What is polymorphism? Describe the various types of zooids found in the *Obelia* colony. (*Magadh 1991*)

6. Describe the structure of various zooids found in the *Obelia*. List the functions of zooids of *Obelia* and *Physalia*. (*Lucknow 1998; C.C.S.U. Meerut 2008*)

7. Describe the structure and function of medusa. (*Gorakhpur 1984; Rohelkhand 1985; Kanpur 1986*)

8. Give habit and habitats of *Obelia*. What do you understand by "division of labour"? Clarify it fully with reference to *Obelia*. (*Lucknow 1986*)

9. Compare the structure of polyp and medusa and add a note on their evolutionary significance. (*Poona 1991*)

10. Describe the form and structure of a hydranth in *Obelia*. Compare it with a medusa. How can the latter be derived from the former? (*Rewa 1989, 1992; Meerut or CCS 1994, 1996*)

11. Enumerate the differences and similarities between polyp and medusa? Show how medusa can be derived from polyp? How does medusa show advance over polyp? (*Awadh 1987; Bundelkhand 1992*)

12. Give a comparative account of polyp and medusa. (*Magadh 1993; Kanpur 1997*)

(*i*) Distinguish between words: polyp and medusa. (*Magadh 1995; Himachal 1995*)

13. Justify the statement, "striking though the difference between polyp and medusa, they are strictly homologous structures." (*Rajasthan 1991; Meerut 1991, 1995; Bundelkhand 1993; Srivenkat 1991, 1994*)

14. Give a detailed account of the structure and development of medusa of *Obelia* and discuss its homologies with polyp. *(Gurunanak 1987; Bihar 1991; Meerut or CCS 1989, 1995; Jiwaji 1993; Kurukshetra 1991, 1994; Lucknow 1994, Rohilkhand 1994; Bhagalpur 1995)*

15. Give an illustrated account of the life history of *Obelia*. *(Bhagalpur 1990; Sambalpur 1990; Rajasthan 1990; Lucknow 1992, 1996; Vikram 1991; Allahabad 1992; Awadh 1992, 1994; Kanpur 1994, Gorakhpur 1997)*

16. What do you understand by "alternation of generation"? Explain the same giving the life history of *Obelia*. Mention its significance. *(Guwahati 1991, 1993, 1996; C.C.S.U., Meerut 2004, 2005)*

17. What is alternation of generation in animals? Explain it with reference to the life history of *Obelia*. *(Awadh 1986; Agra 1991; Lucknow 1990, 1992, 1995; Rajasthan 1992; Jiwaji 1992; Kanpur 1991, 1993, 1998; Jabalpur 1993; Bundelkhand 1990, 1994; Rohilkhand 1990, 1994; Meerut 1999)*

18. Explain what is meant by metagenesis by referring to the life-history of *Obelia*? *(Awadh 1986; Purvanchal 1991; Rohilkhand 1991; Utkal 1991, 1994)*

19. Describe the structure and life cycle of *Obelia*. *(Kanpur 1996)*

Short Answer Questions

1. Describe the structure of medusa of *Obelia*. Compare it with hydranth. *(C.C.S.U., Meerut 2007)*

2. Describe the structure of a polyp in *Obelia*. Compare it with a polyp. *(C.C.S.U., Meerut 2006*

3. Draw the well labelled diagrams of the following. No description is required.

(*i*) Structure of *Obelia* *(Kanpur 1992, 1995)*

(*ii*) V.S. hydranth of *Obelia* *(Meerut or CCS 1994)*

(*iii*) Medusa of *Obelia* *(Meerut or CCS, 1994; Lucknow 1995)*

(*iv*) Life cycle of *Obelia* *(Meerut or CCS 1994)*

(*v*) Life history of a hydrozoan animal *(Kanpur 1984)*

Very Short Answer Questions

1. Name the cells that contain myonemes in *Obelia*.

Ans. Epitheliomuscular cells.

2. What are the functions of polyp of *Obelia*.

Ans. Feeding and defence.

3. Tentacles of *Obelia* have a solid core of gastrodermis. Is it true?

Ans. Yes.

4. Which part of a nematocyst is discharged when cnidocil is stimulated?

Ans. Thread tube.

5. What stimulates a nematoblast?

Ans. Mechanical and chemical stimulation.

6. Which layer of cells is missing in the velum?

Ans. Gastrodermis.

7. What term is used for the alternation of generation in *Obelia*.

Ans. Metagenesis.

8. Name the various kinds of zooids found in *Obelia* colony.

Ans. Polyps, blastostyles and medusae.

9. Medusa of *Obelia* has tetramerous radial symmetry. Is it so?

Ans. Yes.

10. How many sense organs/statocysts occur in medusa of *Obelia*.

Ans. Eight.

11. What is gonangium?

Ans. Gonangium is a gonozooid and non-feeding zooid of *Obelia* colony. It comprises a medusae bearing blastostyle and gonotheca.

12. What are statocysts?

Ans. Statocysts are minute fluid-filled ectodermal sacs. They contain a movable particle of $CaCO_3$ called **statolith**. Statocysts help medusae in maintaining equilibrium (balance) while swimming.

13. What is planula larva?

Ans. It is ciliated, free-swimming larva of *Obelia* and other cnidarians.

Fill in the Blanks

1. .... forms a trimorphic colony.

2. A zooid without mouth and tentacles is called .... in *Obelia* colony.

3. Polyp of *Obelia* is protected in ....

4. Perisarc forms .... around the blastostyles.

5. Majority of the cells in epidermis of *Obelia* are ....

6. The nematoblasts contain a poisonous fluid called ....

7. The .... are the unspecialised reserve cells of *Obelia*.

8. The muscle processes of the gastrodermal cells act as the .... muscles.

9. Mesogloea of *Obelia* is secreted by ....

10. *Obelia* has both .... and .... digestion.

Multiple Choice Questions

1. *Obelia* is a hydroid of

(*a*) fresh water (*b*) marine water

(*c*) brackish water (*d*) all the above

2. The stolon of *Obelia* colony is known as

(*a*) hydranth (*b*) hydrorhiza

(*c*) hyporhiza (*d*) root

3. Medusa stage of cnidarians is advanced to polyp because

(*a*) it has great power of contractility (*b*) it has marginal sense organs

(*c*) it is motile (*d*) all of these

4. Coenosarc is

(*a*) living (*b*) non-living

(*c*) inner tubular, living (*d*) non-cellular

5. Medusa of *Obelia* is

(*a*) herbivores (*b*) carnivores

(*c*) omnivores (*d*) insectivorous

6. The number of tentacles in young medusa of *Obelia* is

(*a*) 16 (*b*) 32

(*c*) 64 (*d*) 128

7. In *Obelia*, medusae arise from

(*a*) polyps (*b*) hydrorhiza

(*c*) hydrocaulus (*d*) blastostyles

8. The nematocysts are discharged on the stimulation of
(*a*) lasso (*b*) cnidocil
(*c*) nucleus (*d*) thread tube

9. In medusae of *Obelia*, statocysts occur at the bases of
(*a*) per-radial tentacles (*b*) inter-radial tentacles
(*c*) ring canal (*d*) all these canals

10. In medusae of *Obelia* gonads developed under the
(*a*) inter-radial canals (*b*) ad-radial canals
(*c*) per-radial canals (*d*) all these canals

11. Sexual zooid in the life history of *Obelia* is
(*a*) gonozoid (*b*) hydranth
(*c*) medusa (*d*) blastostyle

12. The free-swimming larva of *Obelia* is
(*a*) planula (*b*) amphiblastula
(*c*) parenchymula (*d*) scyphistoma

13. Gonotheca of *Obelia* is a protective covering of
(*a*) polyp (*b*) medusa
(*c*) blastostyle (*d*) hydrocaulus

14. Statocyst function as the organs for
(*a*) hearing (*b*) equilibrium
(*c*) light perception (*d*) smell

ANSWERS

Fill in the Blanks

1. *Obelia* **2.** Blastostyle
3. Hydrotheca **4.** Gonotheca
5. Epitheliomuscular cells **6.** Hypnotoxin
7. Interstitial cells **8.** Circular
9. Epidermis and gastrodermis **10.** Extracellular (or intercellular), intracellular.

Multiple Choice Questions

1. (*b*) **2.** (*b*) **3.** (*d*) **4.** (*c*)
5. (*b*) **6.** (*a*) **7.** (*d*) **8.** (*b*)
9. (*d*) **10.** (*c*) **11.** (*c*) **12.** (*a*)
13. (*c*) **14.** (*b*).

19

Aurelia (Phylum Cnidaria)

19.1 CLASSIFICATION

Phylum	–	**Cnidaria (or Coelenterata)**
Class	–	**Scyphozoa**
Order	–	**Semaeostomeae**
Genus	–	***Aurelia***
Species	–	***aurita***

Aurelia is a **jellyfish**; its body is composed of jelly-like material. However, it is not a fish.

19.2 HABIT AND HABITAT

Aurelia aurita is commonly known as **moon-jelly**. It occurs in coastal waters of tropical and temperate oceans of the world. It is **pelagic**, floating or swimming freely close to the surface of water. It lives either singly or in large groups, called **shoals**. It is carnivorous and reproduces both asexually and sexually. Life cycle includes both polypoid and medusoid phases; but medusoid phase is dominant one.

19.3 STRUCTURE

The medusoid body form of *Aurelia* measures 7 to 30 cm in diameter. It is flattened, saucer-shaped structure, called **bell** or **umbrella**. It exhibits **tetramerous radial symmetry**, *i.e.*, body parts are symmetrically arranged around the oral-aboral axis in fours or some multiple of four. It is almost transparent having bluish-white body with pinkish gonads.

Umbrella-shaped body is distinctly divided into a slightly convex upper surface, called **exumbrella** and the concave lower surface, called **subumbrella**. The margin of umbrella is circular which is **scalloped**, *i.e.*, broken by eight notches into eight lobes. Each notch contains two leaf-like processes, called **marginal lappets**. Between the lappets lies a balancing sensory organ, the **statocyst** or **tentaculocyst**. The statocyst and lappets together form a **rhopalium**. The margin of umbrella in between notches is provided with numerous, small hollow marginal **tentacles**. *Aurelia* is a **acraspedote cnidarian** (Box 19.1), it lacks a true velum, however, contains a **pseudovelum** or **velarium**.

Box 19.1 Definitions

1. **Craspedote.** Cnidarians with velum or **craspedon** in their medusae, *e.g.*, hydromedusae (or medusae) of *Obelia* and other hydrozoans.
2. **Acraspedote.** Cnidarians without any velum or craspedon in their medusae, *e.g.*, *Aurelia* and other scyphozoans.

The subumbrellar surface of *Aurelia* bears in its centre a short **manubrium**. At its free end the manubrium contains a square mouth (Fig. 19.1B). All the four corners of the mouth are drawn out into four, long, mobile, delicate, frilled and tapering processes called **oral arms**. Each oral arm has a ventral ciliated groove and is provided with numerous nematocysts. Mouth leads into a short **gullet** which opens into **stomach**. The stomach gives rise to four inter-radial **gastric pouches**.

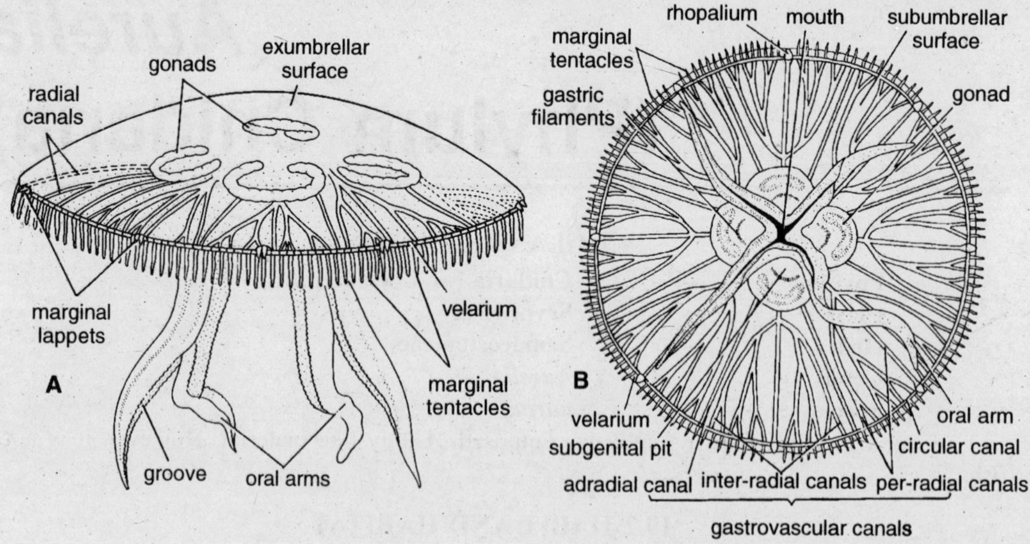

Fig. 19.1. *Aurelia aurita.* A—Lateral view; B— Ventral view.

Gonads and Subgenital Pits

Four horse-shoe-shaped pinkish gonads which later become circular are situated bet-ween oral arms in the center of the subumbrel-lar surface. Gonads are, in fact, internal lying in gastric pouches but they are visible externally. Just below each gonad in the subumbrellar surface, is a circular aperture called **subgeni-tal pit** of unknown function. The gonads, however, have no connection with subgenital pits.

Fig. 19.2. *Aurelia.* Gastrovascular system.

19.4 HISTOLOGY

The main types of cells and tissues and their arrangement in *Aurelia* are practically similar to those already described in the hydrozoan medusa, the *Obelia*.

1. Epidermis. All parts of the bell of a *Aurelia*, namely exumbrella, subumbrella, manubrium, oral arms, marginal tentacles and velarium are covered externally by epidermis (Fig. 19.3). Epidermis also lines the subgenital pits. The epidermis consists of epithelial cells on the exumbrella but epithelio-muscular cells on the subumbrella. It also contains sensory cells, nerve cells, gland cells and cnidoblasts. Epidermis forms sensory epithelium for the sense organs. The sensory epithelium consists of abundant, long, narrow sensory cells between normal epithelial cells. The muscle processes of epidermal cells are cross-striated at certain places and smooth at others.

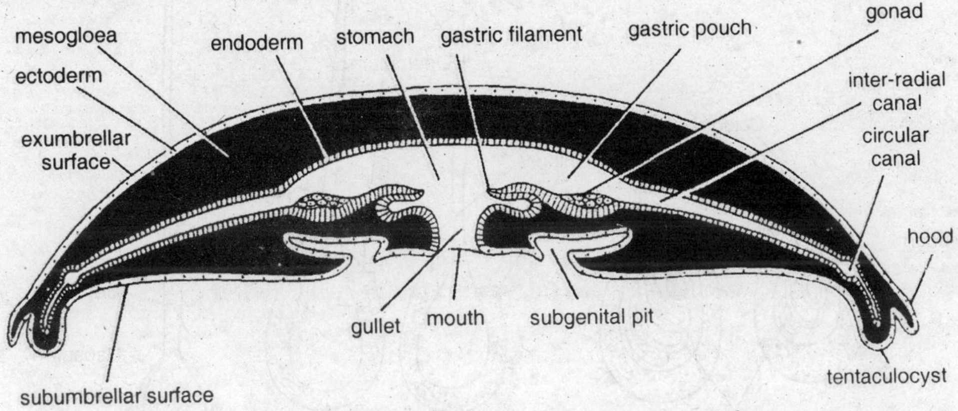

Fig. 19.3. *Aurelia aurita*. Diagrammatic vertical section along inter- radii.

2. Gastrodermis. All parts of the digestive cavity, *i.e.*, gullet, stomach, gastric pouches, radial canals and circular canals, are lined by gastrodermis. Gastrodermis consists of columnar epithelial cells which are flagellated throughout to maintain a current of water. It contains gland cells but lacks nerve cells and muscle processes. The cnidoblasts are absent except on the gastric filaments. The gastric filaments consists of a double layer of gastrodermis with a thin core of mesogloea. Gonads are also formed of the gastrodermis. Parts of the umbrella between the radial canals are solid and contain **gastrodermal lamellae** formed by fusion of the upper (exumbrellar) and lower (subumbrellar) layers of gastrodermis. The cavities of the sense organs and marginal tentacles are extensions of the gastrovascular cavity and are lined by gastrodermis. The oral arms, being prolongations of the angles of the mouth, are covered on their inner surface by gastrodermis which is flagellated here also.

3. Mesogloea. Between the epidermis and gastrodermis is a very thick layer of gelatinous and fibrous mesogloea. The latter forms the main bulk of the bell. It is not structureless like the mesogloea of *Hydra* and *Obelia*. Instead, it contains elastic fibers and free wandering amoeboid cells derived from the epidermis. It, thus, has the nature of connective tissue, so is termed **collenchyma.**

Nematocysts

In *Aurelia*, nematocysts occur on the oral arms, ex- and subumbrellar surfaces, on the marginal tentacles and also on gastric filaments. They are of three types: atrichous isorhizas or small glutinants, holotrichous isorhizas or large glutinants and heterotrichous microbasic euryteles.

The small and large glutinants have on oval capsule and a long, sticky thread-tube of uniform diameter and open at the distal end (Fig. 19.4 A, B). Both lack the butt. The thread tube is unarmed in the small glutinants and armed with a spiral row of minute spines in the large glutinants. In both cases, the thread-tube, when discharged, adheres to the body of the prey.

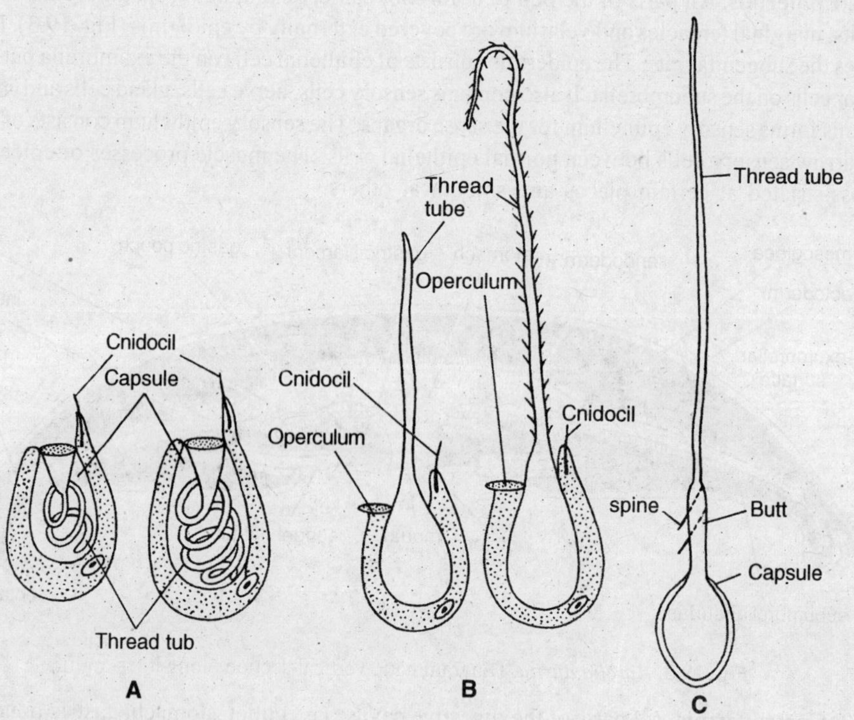

Fig. 19.4. Nematocysts of *Aurelia*. A—Small and large glutinants (undischarged); B—Small and large glutinants (discharged); C—Heterotrichous microbasic eanrytele (discharged).

The heterotrichous microbasic eurytele (Fig. 19.4 C) is characterised by the possession of an armed thread tube open at the tip and a short butt which is dilated at the distal end and bears unequal spines.

19.5 GASTROVASCULAR CANAL SYSTEM

The mouth of *Aurelia* leads into a short **gullet** within the manubrium which finally opens into a four-lobed spacious **stomach**. The stomach occupies a greater part of the center of the bell in the form of four extensions called **gastric pouches** situated inter-radially (Fig. 19.2). On the floor or subumbrellar wall of each gastric pouch lies a gonad with its concavity facing inward. On the inner surface of each gonad is row of short mobile processes, the **gastric filaments** or **phacellae**, that project freely in the gastric pouch. The phacellae are heavily armed with batteries of stinging cells and also bear gland cells.

At the four corners of the mouth are four per-radial canals situated at right angle to each other, four inter-radial canals are present in between the per-radial canals. These eight canals are much branched and they end at the marginal lappets. Eight unbranched ad-radial canals arise, two from each gastric pouch, between the inter-radial and per-radial canals. All these sixteen canals open in the **circular** or **ring canal** situated at the margin of the umbrella. Thus, the gullet, stomach with gastric pouches all the canals constitute the **gastrovascular canal system** or **enteron cavity**. The enteron cavity except gullet is lined by ciliated endodermal cells.

19.6 MUSCULATURE

The musculature of the body of *Aurelia* is well developed and confined to the subumbrellar surface only. The musculature consists of the striated and non-striated muscle processes of the epitheliomuscular cells of epidermis only. The muscle processes are arranged longitudinally in tentacles, manubrium and oral arms. The muscle processes constitute a strong, broad circular band in the peripheral part of subumbrella called **coronal muscle**. Radial muscles extend from the manubrium to the coronal muscles. The rapid and rhythmic contraction of these muscle processes brings about swimming movements in *Aurelia*.

19.7 CIRCULATION

Beating of flagella borne on the gastrodermal cells circulate a current of water through the mouth and the gullet into the stomach and thence into the gastric pouches through the inhalent channels. From here, if flows through the ad-radial canals into the circular canal. Water returns to the stomach by way of the per-radial and inter-radial canals. The per-radial canals carry the water directly into the stomach whereas the inter-radial canals first bring it to the gastric pouches where it flows outside gonads and enters the stomach through the exhalant channels. From the stomach, the water finally goes out through the manubrium and along the grooves of the oral arms. The entire circulation is completed in about 20 minutes:

Mouth → gullet → stomach → gastric pouches → eight ad-radial canals → circular canal → inter-radial and per-radial canals → gullet → exhalent grooves of oral arms.

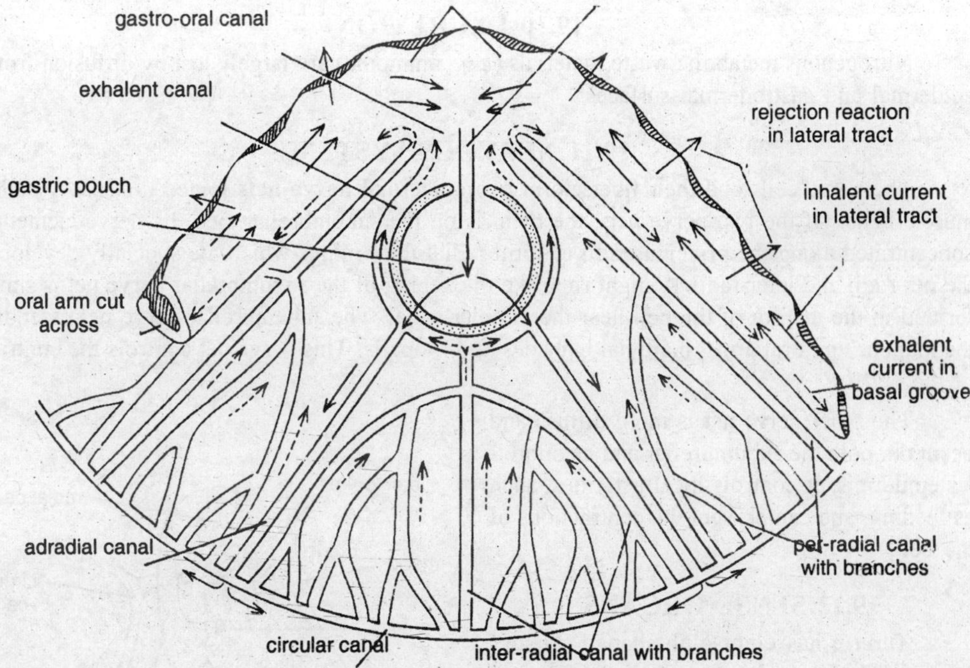

Fig. 19.5. *Aurelia aurita*. Circulation of water (only one quadrant of the body and two arms are shown).

Rhythmic contractions of the bell and movements of the oral arms also aid in the circulation of water. Water take in food, removes waste substances and helps in respiration.

19.8 NUTRITION

Aurelia is **suspension feeder** and **carnivorous** too. Its food consists of fish, crustaceans and their larvae, eggs, plankton and detritus (pieces of organic food). The food (macroplankton) is caught by oral arms and tentacles and transferred to the mouth. Nematocysts of oral arms and tentacles

In another mode of feeding flagella are involved. As aurelia sinks, plankton gets entrapped in mucus present on the subumbrellar surface, which is flagellated. Flagella sweep the food to the bell margin, from where it is scraped off by the oral arms. Flagellated grooves of the oral arms direct the food to the mouth. Here it enters the gullet along with the ingoing current of water. In the stomach and gastric pouches the alive prey is paralysed and killed by the nematocysts of gastric filaments. Here, **intercellular** or **extracellular digestion** of food is started by the enzyme containing juices from the gland cells. Partly digested food circulates in the radial and circular canals, where it is ingested by gastrodermal cells for **intracellular digestion**. The undigested particles return to stomach and escape with the outgoing water. The digested food is taken up by amoebocytes and distributed by them to ectodermal (epidermal) cells. The gastrodermis of gastric pouches stores reserve food as fat droplets and glycogen. Thus, the gastrovascular system fulfils the functions of digestion and circulation.

19.9 RESPIRATION

Respiration is aerobic. The epidermal cells are in contact with the surrounding sea water and gastrodermal cells are bathed by the water current circulating in the gastrovascular cavity. Water contains oxygen and, thus, no cell of the body of *Aurelia* is far away from it. Oxygen readily diffuses in the cells and carbon dioxide produced in them diffuses out.

19.10 EXCRETION

Nitrogenous metabolic waste materials (*e.g.*, ammonia) are largely lost by diffusion from the epidermal and gastrodermal surfaces.

19.11 NERVOUS SYSTEM

The nerve cells and their fibers form two epidermal **nerve nets** joined to sensory cells and muscle cells. Of the two nerve nets, the main is on the subumbrellar side. Its nerve elements are concentrated along the principal radii to form radial thickenings which are specially developed in the per-radii and inter-radii. A slight ring-like thickening of the subumbrellar nerve net is similarly formed in the margin of the bell near the circular canal. The subumbrellar nerve net extends into the manubrium, oral arms, marginal tentacles and rhopalia. This nerve net controls the contraction of the bell.

The other nerve net is more diffuse and occurs in both the subumbrellar and exumbrellar epidermis. It controls local activities, such as feeding, and can prevent the contractions of the bell.

19.12 SENSE ORGANS

Aurelia has eight club-shaped sensory organs, called **rhopalia**. They lie in the marginal notches between the bell between the marginal lappets (Fig. 19.6). Thus four of them are per-radial and four inter-radial in position. A rhopalium (Fig. 19.6) is a triple organ, consisting of a tentaculocyst (= statocyst) for equilibrium, a pigment spot or ocellus sensitive to light and chemosensory or olfactory pits sensitive to food and other chemicals.

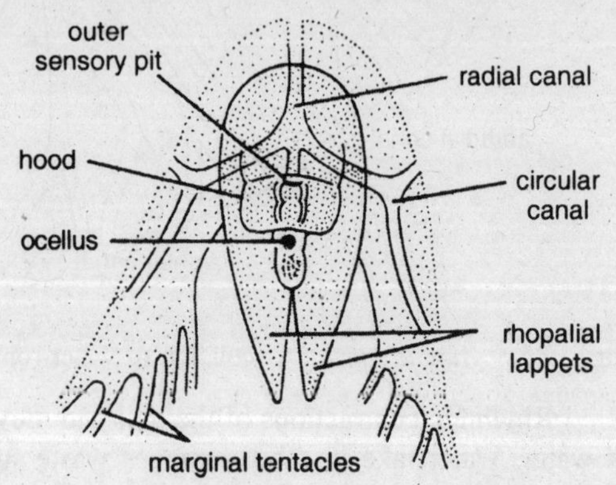

Fig. 19.6. *Aurelia aurita*. A portion of umbrella edge showing a rhopalium.

(*i*) **Tentaculocyst.** It is most prominent organ of the rhopalium. It is hollow, club-shaped-projection connected with the circular canal and hence lined by gastrodermis. The gastrodermal cells heap up in the distal part of the tentaculocyst as a mass of polygonal cells. Each cell of this mass secretes in it a calcareous particle, the **statolith**, composed of calcium sulphate mixed with a small quantity of calcium phosphate. The epidermis of the sides and base of the tentaculocyst has the nature of thick sensory epithelium. The latter contains tall sensory cell between regular epidermal cells. The subumbrellar nerve net, composed of nerve cells and nerve fibers, lie under the sensory epithelium.

The appearance of tentaculocysts mark the formation of an *organ* in cnidarians which has in general reached only a tissue grade of organisation.

(*ii*) **Ocellus.** It lies on the outer side of the tentaculocyst. It consists of an epidermal cup lined by pigmented sensory cells connected to the nerve net within.

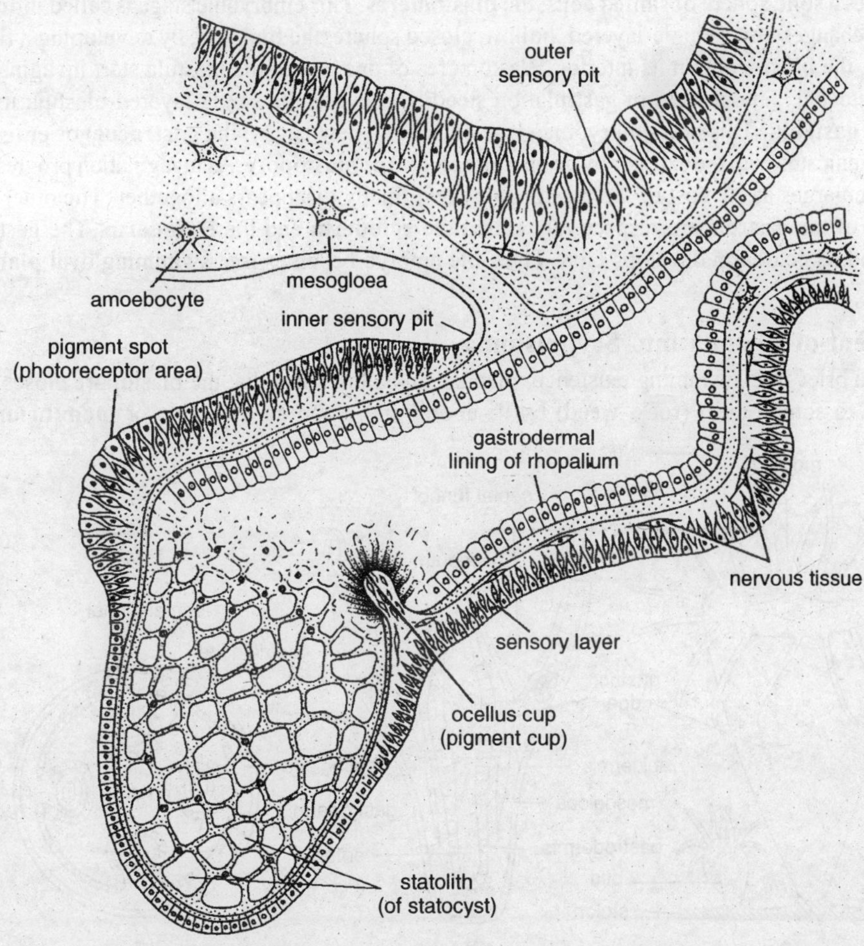

Fig. 19.7. *Aurelia.* Sectional view of rhopalium.

(*iii*) **Olfactory pits.** These are two depressions, one internal to the tentaculocyst and the other above the hood. These are respectively called the **inner** or **adoral** and **outer** or **aboral olfactory pits**. They are lined by thick sensory epithelium comprising sensory cells amongst epidermal cells.

19.13 REPRODUCTION

Aurelia is **unisexual** but there are no morphological distinctions between males and females. The gonads develop from the gastrodermis. Male gonads (testes) of mature male shed spermatozoa into the genital pouches from where they pass out through the mouth with the outgoing currents of water. Female gonads (ovaries) of mature female liberate ova in the genital pouches where they remain till fertilized. The spermatozoa from water find their way into the female with the ingoing water current and reach its genital pouches. Here, fertilization takes place. The resulting zygotes are driven out by water current and get lodged in pits in the grooves of the oral arms of the mother. In such temporary **brood chambers** they develop into ciliated planulae larvae which swim away.

19.14 DEVELOPMENT

The zygote undergoes holoblastic (*i.e.*, complete) but equal segmentation (*i.e.*, cleavage). This produces a solid sphere of similar cells, the **blastomeres**. This embryonic stage is called **morula**. The morula changes into a single-layered, hollow, closed sphere, the **blastula**, by developing a fluid-filled cavity, the **blastocoel**, in its interior. Blastomeres of one end of the blastula start invaginating into the blastocoel. The process of gastrulation steadily converts the single-layered blastula into a two-layered **gastrula**. This new embryonic stage encloses a new cavity, the **gastrocoel** or **enteron**, which communicates with the exterior by an aperture called the **blastopore**. As invagination progresses, the enteron enlarges at the cost of the blastocoel which finally disappears altogether. The outer and inner layers of the gastrula are respectively called the **ectoderm** and the **endoderm**. The gastrula elongates and its ectoderm develops cilia. Thus, the embryo becomes free-swimming oval **planula larva**.

Development of Planula into Scyphistoma

After a brief free-swimming existence, planula sinks, loses its cilia, the blastopore closes and it gets fixed to some object (rock, weed) by its aboral end. Soon an oral cone or **manubrium** is

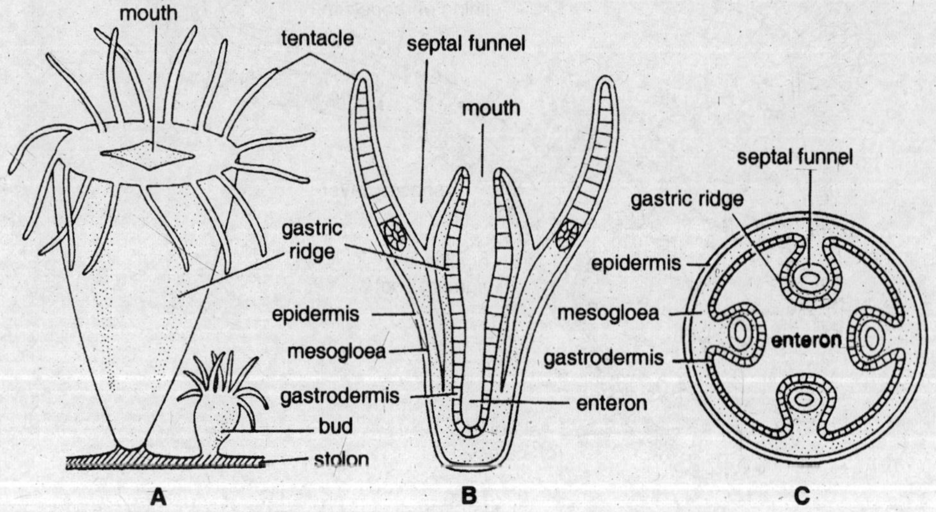

Fig. 19.8. *Aurelia*. Young scyphistoma. A—Outer view; B— V.S. through inter-radius; C—T.S. through septal tunnels.

formed at the free end of larva; a definite **mouth** appears where the blastopore closed earlier. Four hollow buds arise per-radially to become tentacles. Subsequently four inter-radial and eight ad-radial tentacles are formed. The endoderm of coelenteron forms four inter-radial longitudinal ridges called **gastric ridges** or **mesentries**. The mouth becomes square and the manubrium sinks down to form funnel-like depressions called **septal funnels** or **infundibula**. Thus, planula develops to form a trumpet-shaped polyp, called **hydratuba larva**. The root-like **stolon** arises at the base of hydrotuba, which feeds and buds off new hydrotubae throughout summer. These hydrotubae may separate from the stolon of the parent, as in *Hydra*. After summer the hydrotuba ceases to bud, it continues feeding and storing food. The hydrotuba generally winters over the first year and may bud other hydrotubae, but next winter it undergoes a process of transverse fission, called **strobilation**. The dividing hydrotuba is called **scyphistoma** or **strobila** (Fig. 19.9). It is the characteristic of the class Scyphozoa.

The transverse discs of the scyphistoma which has been produced by strobilation look like a serial pile of saucers, and each disc is an **ephyra larva**. One scyphistoma may strobilate a dozen or more ephyras in which growth begins at once in the margin. Ephyra larvae lie on each other, they are joined by muscular strands in the mesenteries which are continuous through all the ephyrae. The muscular strands contract and break so that the ephyrae are constricted off one by one and they swim away. Thus, ephyrae break away one by one from the upper surface of a scyphistoma and then they get inverted. Before the first ephyra is released from the top of its 16 tentacles are absorbed and 8 notched lobes are formed with a tentaculocyst in each lobe. The successive ephyrae released from the strobila are devoid of tentacles.

The basal part of the scyphistoma which is not segmented grows new tentacles. After a period of rest it resumes its existence as a hydrotuba, it starts strobilating again in the next winter. It may live for years feeding and budding in summer and producing ephyra in winter. Strobilation is brought about by intensive feeding followed by a drop in temperature. In *Aurelia*, if the food is plentiful and temperature is low, then many ephyrae are produced at one time; this is called **polydisc strobilation**. If there is less food and the temperature is high, then ephyrae are produced one at a time; this is called **monodisc strobilation.**

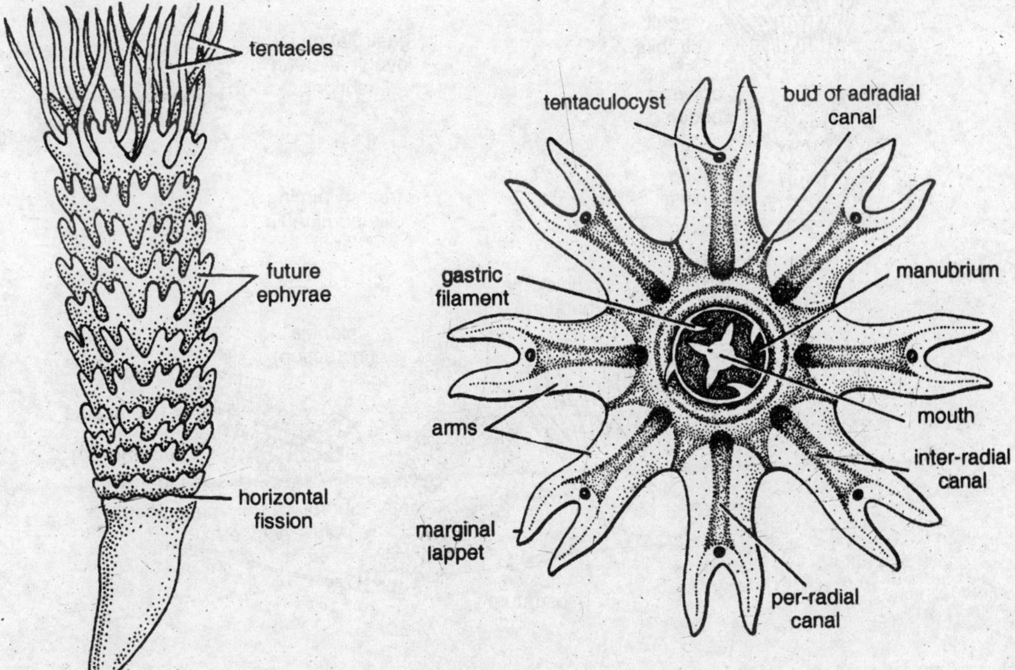

Fig. 19.9. *Aurelia.* A strobilated schyphistoma.

Fig. 19.10. *Aurelia.* An ephyra larva.

Ephyra Larva

The ephyra is a young medusoid form of *Aurelia* (Fig. 19.10). It has eight **notched lobes** or **arms** which are per-radial and inter-radial. Each lobe has two over grown **marginal lappets** with a deep groove between them having a short tentacle which becomes a **tentaculocyst**. An ephyra bears a short **stomach** or **gastric cavity** with a short **manubrium** and four-sided **mouth** on its subumbrellar surface. The gastric cavity grows into the lobes to form branched per-radial and inter-radial canals, later the ad-radial canals also appear. Four pairs of gastric filaments form in the inter-radial mesenteries.

Metamorphosis

The ephyra larva feeds mainly on protozoans which are caught by the lappets and transferred to mouth. The ephyra grows in size, the spaces between notched lobes fill up, mesogloea increases

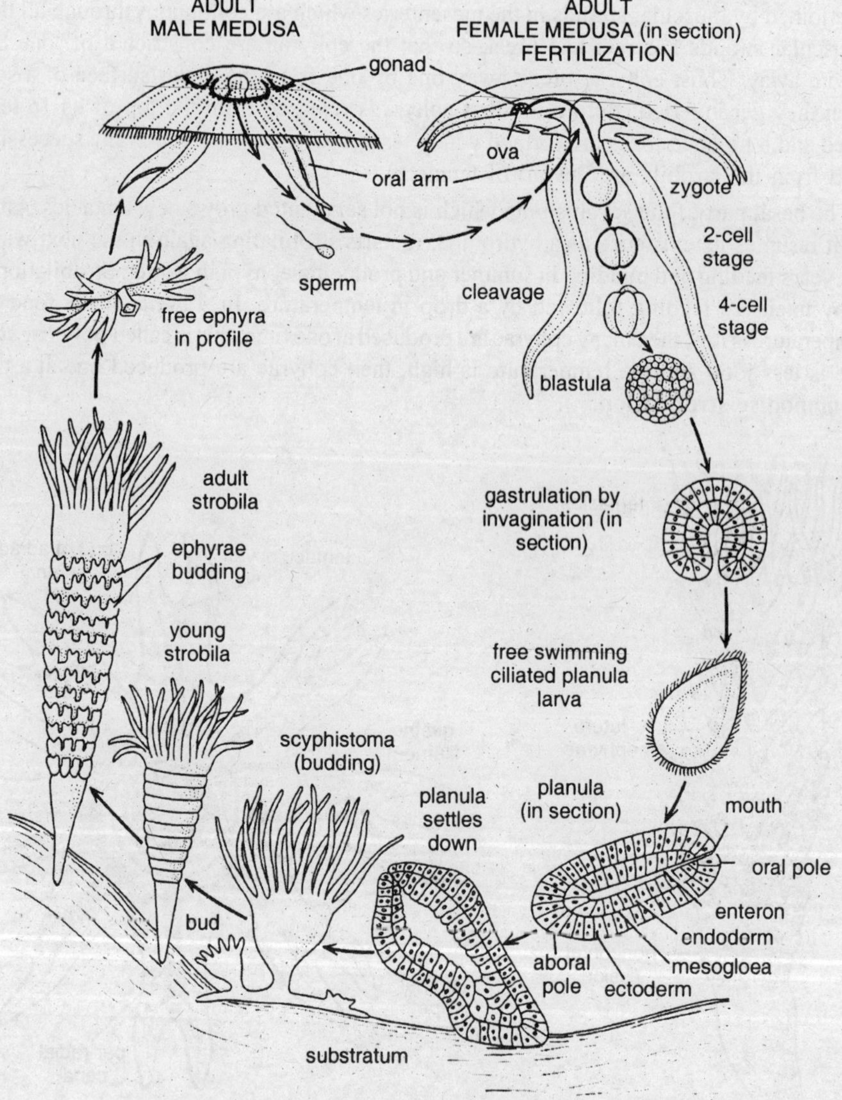

Fig. 19.11. *Aurelia aurita*. Stages in development and life history.

tremendously pushing the endoderm layers together to form an endoderm lamella except in enteron. Thus, four oral arms and marginal tentacles appear, the septal funnels become subgenital pits, and by summer the ephyra becomes an adult medusa.

The stages in the life history of *Aurelia* are: male and female medusae (adults) → gametes → zygotes → ciliated planula larvae → hydrotubae → scyphistomae → ephyrae → medusae → male and female adults.

Alternation in Generations

In the life history of *Aurelia*, the sexual *medusoid* form always alternates with the asexual *polypoid* (scyphistoma) form. Thus, in a sense, *Aurelia* exhibits the phenomenon of *alternation of generations* in its life history. The free swimming adult *Aurelia*, representing the medusoid form, is the **sexual generation** which reproduces by the formation of gametes. The zygote develops into fixed polypoid form called scyphistoma through a free swimming planula larval stage. The polypoid form scyphistoma represent **asexual generation**, which reproduces asexually by budding or fission.

DIFFERENCES BETWEEN MEDUSAE OF *AURELIA* AND *OBELIA*

Table 19.1 Differences between scyphozoan medusa (*e.g.*, *Aurelia*) and hydrozoan medusa (*e.g.*, *Obelia*).

Scyphozoan Medusa (Aurelia)	*Hydrozoan Medusa (Obelia)*
(*a*) Status	
1. Adult *Aurelia* represents large medusoid phase of life history; it is dominant stage.	1. The medusa represents one of the zooids in the life history of polymorphic colony of *Obelia*.
2. It is formed by metamorphosis of an ephyra larva.	2. It is formed from a bud arising in the colony.
(*b*) External Characters	
3. It is of large size, being 7–10 or even 30 cm in diameter.	3. It is of very small size, being only 6 mm in diameter.
4. Its body is umbrella-shaped. It is commonly known as **jelly fish** or **moon-jelly.**	4. Its body is inverted bell-like. It is commonly called **swimming bell.**
5. Manubrium is extremely short. Mouth is situated at the center of subumbrellar surface. Mouth is surrounded by four long oral arms.	5. Manubrium is fairly long. Mouth is situated at the tip of manubrium. Oral arms are not found.
6. Margin of the umbrella is very thin and flexible. It is known as the **velarium**. It contains gastrodermal canals.	6. Margin of the bell bears a narrow inwardly projecting shelf, the **velum**, which lacks gastrodermis.
7. Margin of bell bears 8 notches, each with a pair of marginal lappets.	7. Margin of bell is complete, having no notches and lappets.
8. The tentacles along the margin of the umbrella are numerous, short and hollow.	8. Marginal tentacles are solid, relatively long and highly contractile structures.
9. Gastrovascular canal system is more elaborate. It comprises gullet, stomach, gastric pouches, 16 radial canals (8 much branched and 8 unbranched) and a circular canal.	9. Gastrovascular canal system is less elaborate. It lacks gastric pouches and has only four unbranched radial canals.
10. Gastric filaments armed with stinging cells are present in the gastric pouches.	10. There are no gastric filaments.

Scyphozoan Medusa (Aurelia)	*Hydrozoan Medusa (Obelia)*
11. A current of water is circulated in the gastrovascular canal system.	11. No current of water circulates in the canai system.
(c) Histology	
12. Mesogloea contains elastic fibers and wandering cells.	12. Mesogloea lacks fibers and cells.
13. The striated and non-striated muscle fibers are endodermal in origin.	13. The muscle strands are both ecto- and endodermal in origin.
14. The marginal nerve ring is not well developed or even absent.	14. There is a double nerve ring around the margin of umbrella.
15. Sense organs are per-radial and inter-radial in position, lie in the marginal notches, are protected by lappets and hood, enclose gastrodermal canal and are called **rhopalia**. They maintain equilibrium and also have spots sensitive to light and chemicals.	15. Sense organs are ad-radial in position, lie at the bases of the tentacles, are without protective lappets and hood, and are purely ectodermal sacs and are called statocysts. They are concerned only with equilibrium.
16. The nematocysts are found all over umbrella, tentacles and oral arms. These are of three types.	16. The nematocysts are only found on manubrium and tentacles. These are of only one type.
(d) Gonads	
17. Gonads are situated on the floor of the gastric pouches and are formed of gastrodermis.	17. Gonads are on the radial canals and are formed of epidermis.
18. Gametes are shed in the gastric pouches, whence they escape into the sea by water current.	18. Gametes are shed directly in sea water.

QUESTIONS

Long Answer Questions

1. Give an illustrated account of structure and life history of *Aurelia.*
(*Lucknow 1992, 1993, 1995; Gorakhpur 1993, 1996, 1998; Kanpur 1997*)

Short Answer Questions

1. Give an account of gastro-vascular system of *Aurelia.* (*Lucknow 1994, 1997*)

2. Draw labelled diagram of oral view of *Aurelia.* (*Lucknow 1996, 1998*)

3. Write short notes on:

(*i*) Strobilation

(*ii*) Ephyra

(*iii*) Rhopalium

Very Short Answer Questions

1. Name the larvae found in the life cycle of *Aurelia.*

Ans. Plannula larvae, hydrotubae, scyphistomae and ephyrae.

2. What are the sense organs of *Aurelia.*

Ans. Rhopalia, each containing olfactory pit, statocyst and otolith.

3. What type of reproduction occurs in *Aurelia*?

Ans. Sexual reproduction.

Fill in the Blanks

1. *Aurelia* is commonly called .... or ....
2. Symmetry in *Aurelia* is ....
3. Mesogloea in *Aurelia* contains fibers and ....
4. Oral arms of *Aurelia* bear minute projections called ....
5. The .... canals are unbranched in *Aurelia*.
6. *Aurelia* has three types of nematocysts: atrichous isorhizas, ... and ....
7. *Aurelia* has a .... in each marginal notch.
8. A rhopalium consists of .... , cellus and ....
9. Marginal lappets enclose .... in *Aurelia*.
10. There are .... subgenital pits in moon jelly.
11. Phacelle project from the .... into the ....
12. *Aurelia* has a strong circular .... muscle in the peripheral part of subumbrella.
13. The sense organ for equilibrium in *Aurelia* is called ....
14. There are three larval forms in *Aurelia*; planula, ... and ephyra.
15. Scyphistoma multiplies by ....
16. Strobilation is of two types .... and polydisc.
17. *Aurelia* medusa is formed by metamorphosis of ....

Multiple Choice Questions

1. *Aurelia* is
   (*a*) herbivorous (*b*) carnivorous
   (*c*) omnivorous (*d*) all the above
2. Predominant form in *Aurelia* is
   (*a*) polypoid (*b*) medusoid
   (*c*) scyphistoma (*d*) none of these
3. Symmetry in scyphistoma is
   (*a*) bilateral (*b*) biradial
   (*c*) radial (*d*) spherical
4. Radial canals in *Aurelia* are
   (*a*) branched (*b*) unbranched
   (*c*) some branched, some unbranched (*d*) absent
5. Fimbriae are borne by
   (*a*) tentacles (*b*) lappets
   (*c*) tentaculocysts (*d*) oral arms
6. Rhopalium is a complex of
   (*a*) two sense organs (*b*) three sense organs
   (*c*) four sense organs (*d*) five sense organs
7. Ephyra is
   (*a*) a poriferan (*b*) a cnidarian
   (*c*) young *Hydra* (*d*) young *Aurelia*

8. *Aurelia* maintains the equilibrium with the help of

(*a*) elastic fibers
(*b*) nematocyst
(*c*) tentaculocysts
(*d*) radial muscles

9. In *Aurelia*, the origin of gonad is

(*a*) endodermal
(*b*) ectodermal
(*c*) mesodermal
(*d*) none of these

ANSWERS

Fill in the Blanks

1. Jelly fish, moon jelly
2. Tetramerous radial
3. Amoeboid cells
4. Fimbriae
5. Adradial
6. Holotrichous isorhizas, heterotrichous microbasic euryteles
7. Rhopalium
8. Tentaculocyst, olfactory pits
9. Rhopalia
10. Four
11. Gonads, gastric pouches
12. Coronal
13. Tentaculocysts
14. Schyphistoma
15. Strobilation
16. Monodisc
17. Ephyra.

Multiple Choice Questions

1. (*b*) **2.** (*b*) **3.** (*c*) **4.** (*c*)
5. (*d*) **6.** (*b*) **7.** (*d*) **8.** (*c*)
9. (*a*).

20

Polymorphism in Cnidaria

Polymorphism (Gr., *poly* = many + *morph* = form) is a phenomenon in which an organism has more than one type of individuals. It is more evident in class Hydrozoa and Anthozoa of Cnidaria/ Coelenterata.

Coelenterates are basically colonial forms. Individuals of a colony of coelenterates are called **zooids**. These zooids are of two basic types: polyps and medusae.

1. Polyps. Polyps are cylindrical, sessile and fixed forms. These have mouth surrounded by tentacles and a wide gastrovascular cavity. These are the **nutritive zooids** of the colony, so, feed the colony. Polyps are also called **gastrozooids** or **trophozooids**.

2. Medusae. Medusa has an umbrella-like body with marginal tentacles and a centrally located mouth on a concave side. They bear gonads and help in sexual reproduction. Therefore, these are called **sexual zooids** or **gonozooids**. They are free swimming and derived from polyps.

Hydromedusae. The medusae of class Hydrozoa are called **hydromedusae**. They are of following types:

1. Trachymedusae. Bell margin of these medusae are not scalloped. These medusae contain long manubrium or pseudomanubrium. In them gonads are present on radial canals, *e.g.*, *Aglantha* (Fig. 20.1A).

2. Narcomedusae. Bell margin of these medusae are scalloped. These medusae contain reduced radial canals and lack manubrium. In them gonads occur on the floor of gastric cavity, *e.g.*, *Cunina* (Fig. 20.1B).

3. Anthomedusae. These medusae are taller, retain the tetraradiate form, have gonads on the manubrium and usually have ocelli, *e.g.*, *Sarsia* (Fig. 20.1C), *Pennaria* (Fig. 20.1D) and other gymnoblasts.

4. Leptomedusae. These medusae are flatter, may have lost their tetraradiate symmetry by branching of the radial canals, have gonads borne on these canals. They have statocysts on the bell margin, *e.g.*, *Obelia* (Fig. 20.1G), *Tima* (Fig. 20.1F), *Aequora* (Fig. 20.1E) and other leptomedusae.

20.1 TYPES OF POLYMORPHISM

According to the types of zooids, coelenterates may be of following types:

1. Dimorphic having two types of zooids.
2. Trimorphic having three types of zooids.
3. Polymorphic having more than three types of zooids.

1. Dimorphism

The coelenterate colonies bearing two types of individuals or zooids are called **dimorphic**. Such colonies and the phenomenon is termed **dimorphism**. For example, *Bougaivillea*, *Pennatula*, etc., consist of two types of zooids: 1. **Polyps**, the gastrozooids or trophozooids and 2. **Medusae**, the sexual zooids.

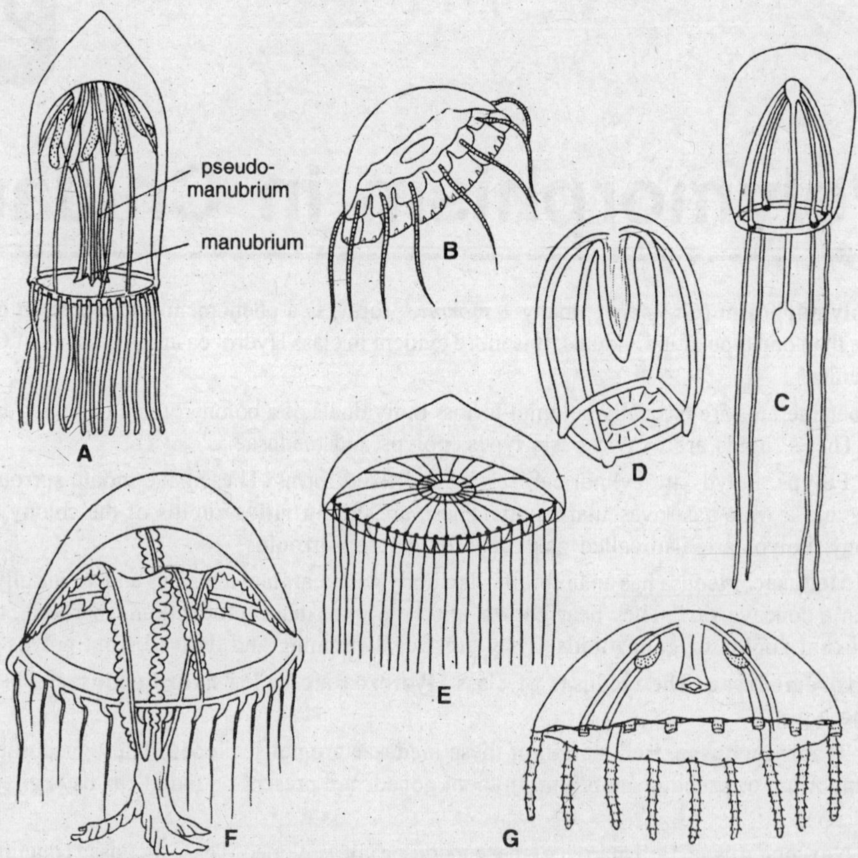

Fig. 20.1. Hydromedusae. A—Trachymedusa (*Aglantha*); B—Narcomedusa (*Cunina*); C—Anthomedusa (*Sarsia*); D—Anthomedusa (*Pennaria*); E—Leptomedusa (*Aequora*); F—Leptomedusa (*Tima*); G—Leptomedusa (*Obelia*).

Pteroids, *Corallium*, *Pennatula*, etc., of class Anthozoa show dimorphism and have two types of polyps: (1) **Autozooids** or **gastrozooids** feed the colony. (2) **Siphonozooids** circulate water current through the gastrovascular cavities of colony and bear gonads too. They lack a medusa stage in their life cycle.

2. Trimorphism

The coelenterate colonies bearing three types of zooids are called **trimorphic colonies** and the phenomenon is termed **trimorphism**. For example, *Obelia* colony has—

(*i*) Polyps or gastrozooids

(*ii*) Blastostyles or blastozooids or gonozooids which bear medusae

(*iii*) Medusae or sexual zooids.

In *Millipora*, *Plumularia*, etc., the zooids are of following three types:

(1) Gastrozooids. These are the polyps or nutritive zooids.

(2) Dactylozooids. These are the protective zooids with long and knobbed tentacles.

(3) Medusae. These are gonads-bearing sexual zooids.

3. Polymorphism

In some coelenterates both polyps and medusae are of various types and are modified to carry out different functions of the colony. For example, siphonophores (*Halistemma*, *Physalia*, *Veleila*, *Porpita*, etc.) are invariably free swimming, colonial and polymorphic. The polyps and medusae are differently modified to present a large variety of zooids. The **polypoid zooids** of siphonophores are of three types:

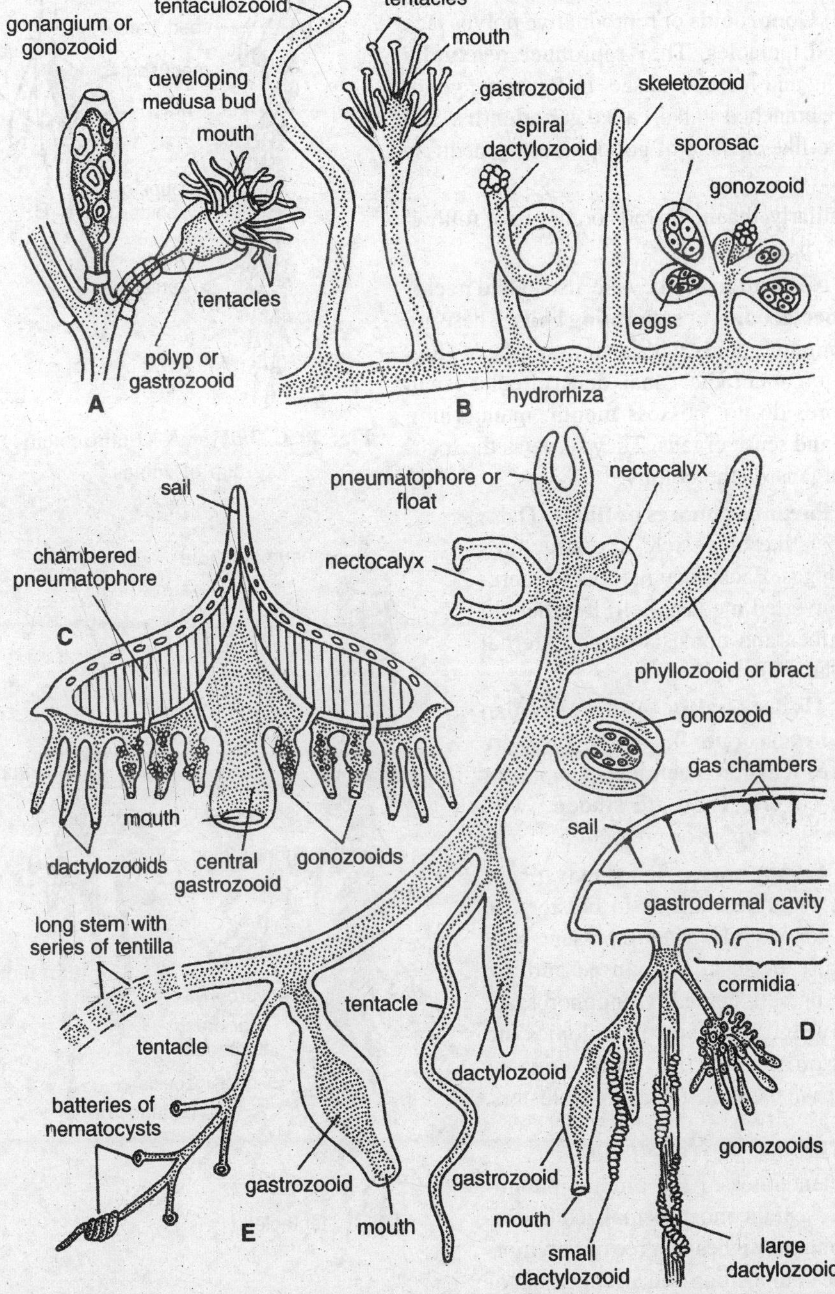

Fig. 20.2. Polymorphic colonies of cnidaria. A—*Obelia*; B—*Hydractina*; C—*Velella*; D—A single cormidium of *Physalia*; E—Generalized calycophoran Siphonophora showing different zooids.

(*i*) **Gastrozooids** or feeding polyps are each with a mouth and a long tentacle.

(*ii*) **Dactylozooids** or protective polyps are without mouth and usually with a long unbranched basal tentacles. They bear many nematocysts and are also called **palpons**, **tasters** or **feelers**.

(*iii*) **Gonozooids** or reproductive polyps lack mouth and tentacles. They reproduce **asexually** by budding and form medusae. In *Physalia*, gonozooids are branched stalks, called **gonodendra**, and bear grape-like clusters of gonophores or medusae (Fig. 20.2).

Similarly, the medusoid zooids are of following four types:

(1) Nectophores. They are also called **nectocalyces, nectozooids** or **swimming bells**. These are free-swimming zooids with swimming bells (Fig. 20.3), velum, four radial canals and a circular canal. Nectophores do not possess mouth, manubrium, tentacles and sense organs. They serve as the locomotory organs of the colony.

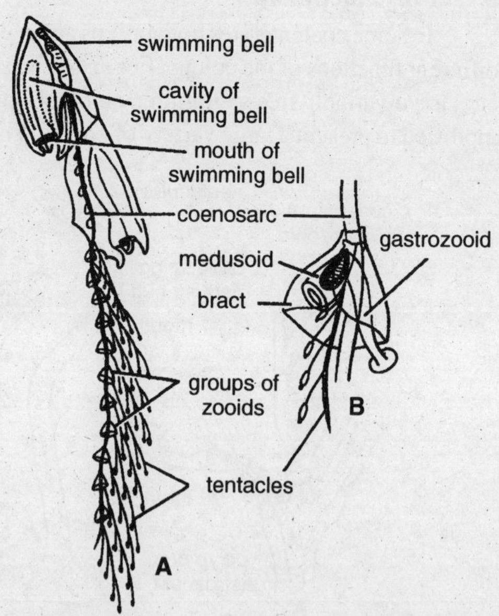

Fig. 20.3. *Diphyes*. A—Entire colony; B—Single group of zooids.

(2) Pneumatophores or floats. These are bladder-like or vesicle-like structures filled with gas. Each pneumatophore represents an inverted medusa bell; it is devoid of mesogloea and consists of an external exumbrellar wall.

(3) Hydrophyllia. These are also known as **bracts** or **phyllozooids**. These are shield-like, leaf-like, helmet - shaped or prismatic covers. They are studded with nematocysts, serving for protection.

(4) Gonophores. They may occur singly on separate stalk or in clusters on polypoid gonozooids. They represent buds of immature or sessile medusae and are plate-like or bell-shaped. Gonophores are dioecious (Fig. 20.4) but the colonies are hermaphroditic bearing both types of gonophores in the same or separate clusters.

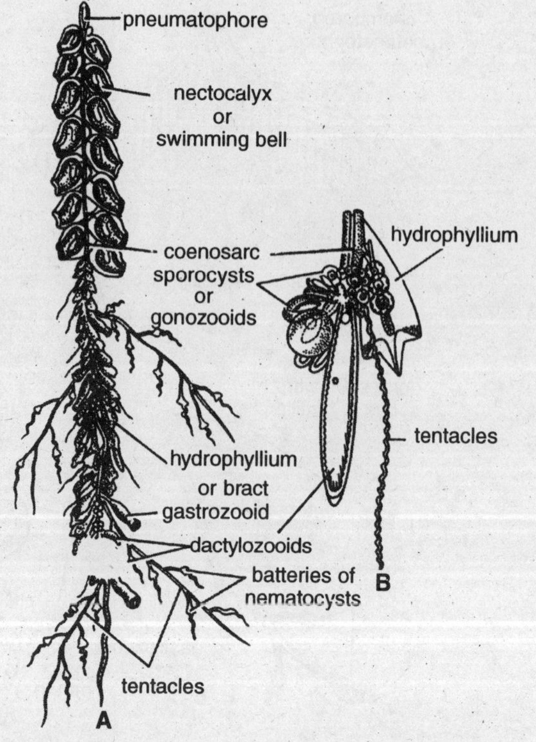

Fig. 20.4. *Halistemma*. A—Entire colony; B—One cormidium enlarged.

Polymorphism in *Halistemma*

Coelenterates of the order Siphonophora represent the most specialized hydrozoa attaining the highest degree of polymorphism. The composition and arrangement of various zooids is very different. The diversity is so great that each individual is to be studied separately. The best known polymorphic siphonophores are *Halistemma*, *Velella*, *Physalia*, *Porpita* and *Hydractinia*.

Halistemma. It possesses a long and slender **stem.** Its upper end is expanded into a bubble-like **pneumatophore** and the stem is divided into following two parts (Fig. 20.4):

1. Proximal part. It is also called **nectosome**. It bears closely set bell-shaped **swimming bells, nectocalyces** or **nectophores**. They lack mouth, tentacles and sense organs. They are modified medusae, muscular and propel the colony forward.

2. Distal part. It is also called **siphonosome**. It carries repeated groups of closely set polyps, called **cormidia**. Each cormidium consists of following types of zooids (Fig. 20.4B):

(*i*) **Bract.** It is the protective zooid and is also called **hydrophyllium**. It is in the form of a flat scale or leaf-like bract.

(*ii*) **Gastrozooid.** It is the nutritive zooid. It bears mouth and long branched basal tentacles.

(*iii*) **Dactylozooid or feeler.** It is without mouth but with a single unbranched basal tentacle.

(*iv*) **Gonozooid.** It is a reproductive zooid. It has either testes or ovaries.

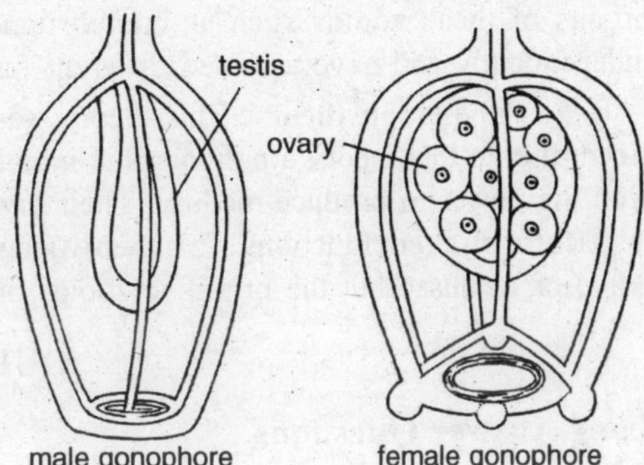

Fig. 20.5. Siphonophora. Male and female gonophores.

Forms of Polymorphism

Polymorphism has two main forms: metabolic and reproductive.

1. Metabolic form. Metabolic form of polymorphism is represented by the zooids having vegetative functions such as feeding, swimming and defence. It includes gastrozooids, dactylozooids, pneumatophores, nectocalyces, etc. All these individuals are polypoid.

2. Reproductive form. Reproductive form of polymorphism is represented by the zooids having reproductive function. It includes the asexual zooids, such as blastostyles or gonozooids and sexual zooids such as gonophores or medusae. The gonozooid is polypoid while the medusae is medusoid.

20.2 POLYMORPHISM AND ALTERNATION OF GENERATION

Polymorphism is essentially a phenomenon of "division of labour", *i.e.*, different functions are assinged to different individuals rather than to the organs of one individuals. Polymorphism is intimately associated with life history. In monomorphic forms such as *Hydra* and the class Anthozoa, the polyps reproduce both asexually and sexually, so that, life-cycle remains simple, without any larval stage and may be represented by the following, formula: **polyp-egg-polyp**. With the advent of polymorphism, the reproductive power of the organism are divided among the different individuals of the colony. In these organisms the polyps reproduce asexually to produce medusoid forms, the **gonophores**, which reproduce sexually to form polyp. The life-cycle then is represented by the following formula: **polyp-medusa-egg-planula-polyp**. Thus, the **alternation of generation** comes into existence in the life-cycle. The asexual polypoid generation alternates with the sexual medusoid generation.

20.3 ORIGIN OF POLYMORPHISM

Regarding the origin of polymorphism, there are many views. According to one view the original coelenterate was a polyp and through specialization the sexual function was relegated to secondarily developed medusoid form and this led to alternation of generation. According to another view, the ancestral coelenterate was a medusoid form, while the polypoid generation represent a persistent larval form, thus, leading to polymorphism.

Theories of Origin of Polymorphism

A number of theories have been put forward by various zoologists; some of them are as follows:

1. Polyorgan theory. This theory was put forward by **Huxley, Eschschoitz** and **Metschnikoff.** According to this theory the component zooids are organs of a single medusoid individual. Certain organs of these zooids such as manubrium, tentacles and umbrella have become multiplied independently and have assumed different forms to perform different functions.

2. Polyperson theory. This theory was put forward by **Leuckart, Vogt** and **Gegenbaur.** According to this theory a polymorphic individual is a colonial form of highly specialized polyps with the power to produce medusa. Their diversified organisms or zooids have grouped together to perform different functions. This theory maintains that the part of a polymorphic colony are either polyp or medusae but the primitive zooids of the colony are of polyp type.

QUESTIONS

Long Answer Questions

1. Write an essay on polymorphism in Coelenterata.
(*Agra 1998; Awadh 1988, 1991; Bundelkhand 1989; Jabalpur 1992, 1993; Bhopal 1993; Kurukshetera 1989; Garhwal 1989; Rohilkhand 1989; Burdwan 1986; Kanpur 1992, 1993, 1996*)
2. What is polymorphism? Describe it in the Coelenterata.
3. Select one example from Siphonophora and explain its polymorphic nature. Give a well labelled diagram of polymorphic coelenterate and comment upon its salient features. (*Meerut or C.C.S. 1990*)
4. Describe the polymorphism in *Velella* and *Physalia*. (*Delhi 1989*)

Short Answer Questions

1. Describe different types of polymorphism found in Cnidaria.
2. Give an account of polymorphism in *Halistemma*.
3. Describe the polypoid zooids of siphonophores.

Very Short Answer Questions

1. Name four types of hydromedusae.
Ans. Trachymedusae, narcomedusae, anthomedusae and Leptomedusae.
2. What is gonodendron?
Ans. A branched blastostyle (gonozooids).
3. Name the polymorphic colony with a rhomboidal float.
Ans. *Velella*.
4. Give the function of nectocalcyes.
Ans. Locomotion of colony.
5. *Hydractinia* develops four types of zooids. What are their names?
Ans. Gastrozooids, dactylozooids, gonozooids and sporasacs.
6. Mention the theories advanced to explain polymorphism in cnidarians.
Ans. Polyorgan and polyperson.
7. Which type of medusae are found in *Obelia*?
Ans. Leptomedusae.

Fill in the Blanks

1. Three grades of polymorphism are dimorphic, ... and polymorphic.
2. Examples of dimorphic colonies are .... and *Pennatula*.

3. Two types of zooids in *Pennatula are* .... and ....

4. Common trimorphic colonies are *Obelia* and ....

5. Polymorphic colonies include .... , *Physalia* and ....

6. *Physalia* colony is kept afloat by .... that contains ....

7. A group of zooids in a polymorphic colony is called ....

8. Nectocalyces are modified ....

Multiple Choice Questions

1. Polyp helps in
(*a*) reproduction (*b*) nutrition
(*c*) excretion (*d*) respiration

2. Polymorphism is shown in
(*a*) hydrozoa (*b*) anthozoa
(*c*) scyphozoa (*d*) actinozoa

3. Division of labour between the several individuals of a colony is called
(*a*) vital activity (*b*) labour division
(*c*) polymorphism (*d*) poly-function

4. Gastrozooids, gonozooids and dactylozooids are the characteristic of ...., colony
(*a*) dimorphic (*b*) trimorphic
(*c*) polymorphic (*d*) all the above

5. Modification of polyps are
(*a*) gastrozooids (*b*) dactylozooids
(*c*) gonozooids (*d*) all the above

6. Which zooid is generally motile?
(*a*) polyp (*b*) medusa
(*c*) both (*d*) none

7. Modification of medusa
(*a*) phyllozooid (*b*) nectophore
(*c*) pneumatophore (*d*) all the above

8. Cnidom is
(*a*) battery of nematoblasts (*b*) group of nematoblasts
(*c*) armed thread (*d*) unarmed thread

9. A cormidium of *Physalia* consists of
(*a*) tentacles (*b*) zooids
(*c*) both (*a*) and (*b*) (*d*) tentacles, zooids and pneumatophores

10. Pneumatophore lacks sail in
(*a*) *Physalia* (*b*) *Velella*
(*c*) *Porpita* (*d*) all the above

ANSWERS

Fill in the Blanks

1. Trimorphic **2.** *Bougainvillea*
3. Autozooids, siphonozooids **4.** *Millepora*
5. *Halistemma, Velella* **6.** Pneumatophore, gas
7. Cormidium **8.** Medusae.

Multiple Choice Questions

1. (*b*) **2.** (*a*) **3.** (*c*) **4.** (*b*)
5. (*d*) **6.** (*b*) **7.** (*d*) **8.** (*b*)
9. (*c*) **10.** (*c*).

21

Colony Formation in Cnidaria

Much of the striking diversity seen in phylum Cnidaria results from two fundamental aspects of their life-styles. First is the tendency to **form colonies** by asexual reproduction; the colony can achieve dimensions and forms unattainable by single individuals. Second, many species of cnidarians exhibit a **dimorphic life cycle** that includes two entirely different adult morphologies: a benthic **polypoid form** and a planktonic **medusoid form**. The dimorphic life cycle has major evolutionary implications touching on nearly every aspect of cnidarian biology.

Theoretically, members of Cnidaria have life-cycles that alternate between asexual sedentary **polyps** and sexual, free-swimming **medusae**. In reality, however there is vast variation within life-cycles of cnidarians. Within group Anthozoa, the medusae stage is virtually non-existent; the larva, once fusing with substratum and developing into the polyp stage, grows benthic or sessile, meaning it no longer metamorphosises into the medusae stage. Among the Scyphozoa and Cubozoa, the medusae are the dominant form in life-cycle, while the polyps are in turn are reduced or absent. Medusae are extremely varied and ranged in size from a few millimeters to over 30 meters (with tentacles). The Hydrozoa are intermediate, with significant medusoid and polyp forms.

The Siphonophora deserve special mention. These hydrozoans form **colonies** that show varying degrees of specialisation, so that in extreme cases individuals function essentially as organs of the whole.

Siphonophores are colonial, almost exclusively pelagial, freely movable Hydrozoa, composed of medusoid and polypoid zooids that are morphologically and functionally specialised. The integration of the zooids is so strong that the colony attains the character of an individual. Individual medusae are never released, they are without skeleton and individual medusoids are either irregular or bilateral in symmetry.

Types of Colonies

The colonial forms of Cnidaria may be of two types:

1. Fixed colonies, *e.g.*, *Obelia*, corals.

2. Floating colonies, *e.g.*, *Physalia*, *Rhizophysa*, *Porpita*, *Velella*, etc. (*Note*. Earlier *Porpita* and *Velella* were grouped as siphonophores, now they belong to a separate order Chondrophora; are **Brusca** and **Brusca**, 2003).

21.1 GROWTH FORMS OF COLONIES

Branched hydrozoan colonies grow in following two patterns (Fig. 21.1).

1. Monopodial growth. In this type of growth, the first polyp elongates continuously from a growth zone at the distal end of the hydrocaulus. This primary (axial) polyp may even lose its hydranth and persist merely as stalk. The primary hydrocaulus gives rise to secondary polyps by lateral budding. These secondary polyps grow and may give rise to lateral tertiary polyps in the same fashion.

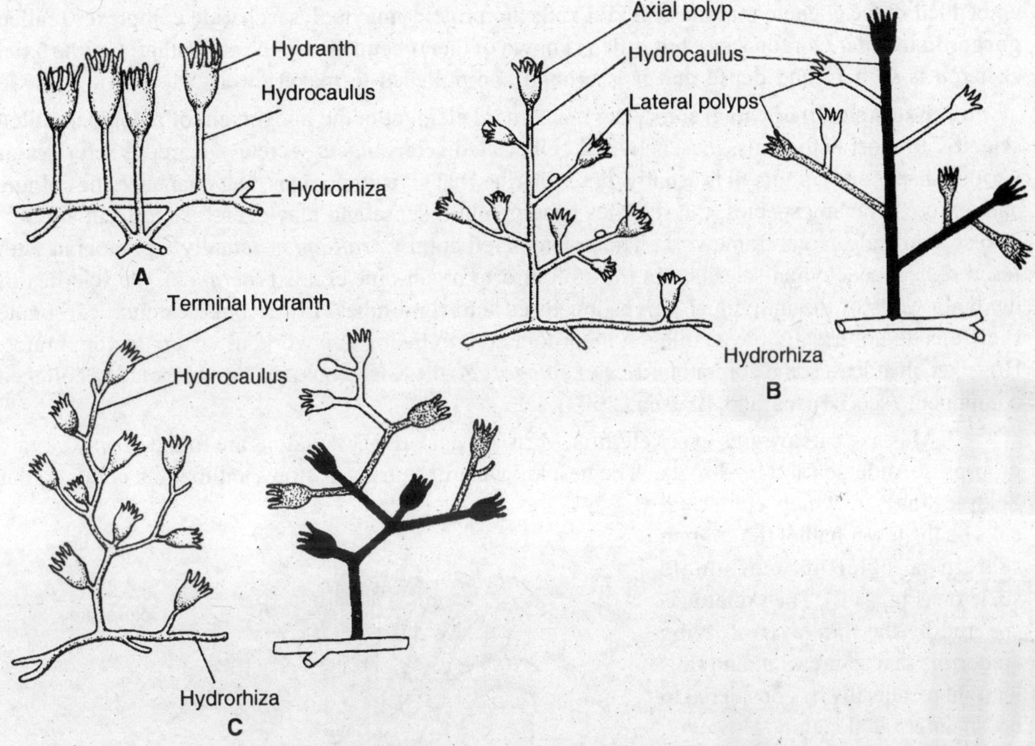

Fig. 21.1. Growth forms of hydrozoan colonies. A—Hydrorhizal colony; B—Colony displaying monopodial growth. C—Colony displaying sympodial growth. The models next to B and C illustrate age of polyps in the colony; oldest polyps are shown in black, youngest in white (after Brusca and Brusca, 2003).

2. Sympodial growth. In this type of growth of hydrozoan colonies, the primary polyp does not continue to elongate but produces one or more lateral polyps by budding and then stop growing. The new polyps extend the colony upward some distance, then stop growing and give rise to more new polyps by budding. In these colonies the main stem or axis actually represents the combined hydrocauli of many polyps and the age of these polyps decreases from base to tip along each branch.

21.2 SKELETAL SUPPORT OF COLONY

Cnidarians employ a wide range of support mechanisms. Polypoid forms rely largely on the hydrostatic qualities of the water-filled coelenteron constrained by circular and longitudinal muscles of the body wall. In addition, the mesenchyme may be stiffened with fibers, particularly in anthozoans. Colonial anthozoans may incorporate bits of sediment and shell fragments onto the column wall for further support. Many colonial hydrozoans produce a flexible, horny **perisarc**, composed largely of **chitin** secreted by the epidermis. In medusae, the principal support mechanism is the middle layer, which ranges from a fairly thin and flexible mesoglea to an extremely thick and stiffened fibrous mesenchyme, which may be almost cartilaginous in consistency.

Types of Skeleton

In addition to above soft and flexible support structure, there is a striking array of hard **skeleton** structures of three fundamental types: horny or wood-like axial skeletal structures, calcareous sclerites, and massive calcareous frameworks.

1. Horny exoskeletons. Such exoskeletons occur in several groups of colonial anthozoans such as gorgonians, sea pens and antipatharian corals (the black or horny corals of zooantherians). Amoebocytes in the coenenchyme secrete a flexible or stiff internal axial rod as a supportive base embedded in the coenchymal mass. **Axial rods** are protein-mucopolysaccharide complexes (called **gorgonin** in order Gorgonacea), but little is known of their chemistry. In the antipatharians, the axial skeleton is so hard and dense that it is ground and polished to make jewelry.

2. Exoskeleton of calcareous sclerites. A hard element in the integument of a polyp is called **sclerite.** In most octocorals, mesenchymal cells called **scleroblasts** secrete calcareous **sclerites** of various shapes and colours. It is usually these sclerites that give gorgonians their characteristic colours and texture. In many species, the sclerites become quite dense and may even fuse to form a more-or-less solid calcareous framework. The precious red coral *Corallium* is actually a gorgonian with fused red coenenchymal sclerites. In the stoloniform organ-pipe corals (*Tubipora*), the sclerites of the body walls of the individual polyps are fused into rigid tubes. Invertebrate calcium carbonate skeletons do not usually have collagen incorporated into their framework, as occurs in vertebrates. However, in at least some gorgonians (*e.g.*, *Leptogorgia*) the calcareous spicules do include a collagen component (see **Brusca** and **Brusca**, 2003).

3. Massive calcare-ous exoskeletons. Massi-ve calcareous skeletons are found in only certain grou-ps of Anthozoa and Hyd-rozoa. The best known are the true, or stony, anthozo-an corals (order Sclerac-tina), in which epidermal cells on the lower half of the column secrete a calci-um carbonate skeleton (Fig. 21.2). The skeleton is covered by the thin layer of living epidermis that se-cretes it, and thus it might technically by considered to be an internal skeleton. However, because coral colony generally sits atop a large non-living calcareous framework, most authors have regarded it as external skeleton or exoskeleton.

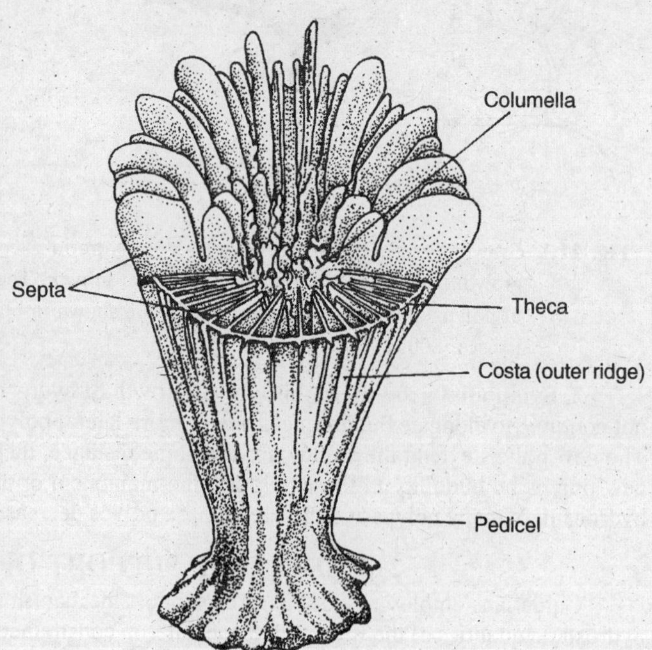

Fig. 21.2. The corallite of a solitary scleractinian coral (after Brusca and Brusca, 2003).

The entire skeleton of a scleractinian coral is termed the **corallum**, regardless of whether the animal is solitary or colonial; the skeleton of a single polyp however, is called a **corallite**. The outer wall of the corallite is the **theca**; the floor is the **basal plate** (Fig. 21.2). Rising from the center of the basal plate is often a supportive skeletal process called the **columella.** The basal plate and inner thecae walls give rise to numerous radially arranged calcareous partitions, the **septa**, which project inward and support the mesentries of the polyp. Polyps occupy only the uppermost surface of the corallum. Skeletal thickness increases as polyps grow, and the bottoms of the corallites are sealed off by the transverse calcareous partitions called **tabulae**, each of which becomes the basal support of a polyp. The corallum can assume a great variety of shapes and sizes, from simple cup-shaped structures in solitary corals to large branching forms in colonial species. The corallum of colonial forms may be upright, low and massive, or even encrust on other hard substrata.

Members of hydrozoan families Milleporidae and Stylasteridae also produce **calcareous exoskeletons**, and they are often referred to as **hydrocorals**. Like true corals, milleporid colonies may assume a variety of shapes, from erect branching forms to encrustations. The milleporid exoskeleton, termed a **coenos-teum**, is perforated by pores of two sizes that accomodates two kinds of polyps (Fig. 21.3). The gastrozooids live in large holes, or **gastropores,** and are surro-unded by a circle of smaller **dactylopores**, which house the dactylozooids. Canals lead down-ward from the pores into the coenosteum and are closed off below the transverse calcareous tubulae. As growth proceeds and the colony thickens, new tubules are formed, keeping the polyp pores at a more or less fixed depth. Hydrocoral colonies thus differ from scleractinian colonies in having the skeletal penetrated by living tissues. The stylasterine skeleton is similar to the mil-leporid skeleton, but the margins of the gastropores often bear notches that serve as dactylo-pores, and the gastrozooids and dactylozooids are supported by calcareous spine-like **gastrostyles** and low ridges called **dactylo-styles**, respectively. Stylasterine gonophores arise in chambers called **ampullae**, which connec-ted to the feeding zooids through the coenosteum.

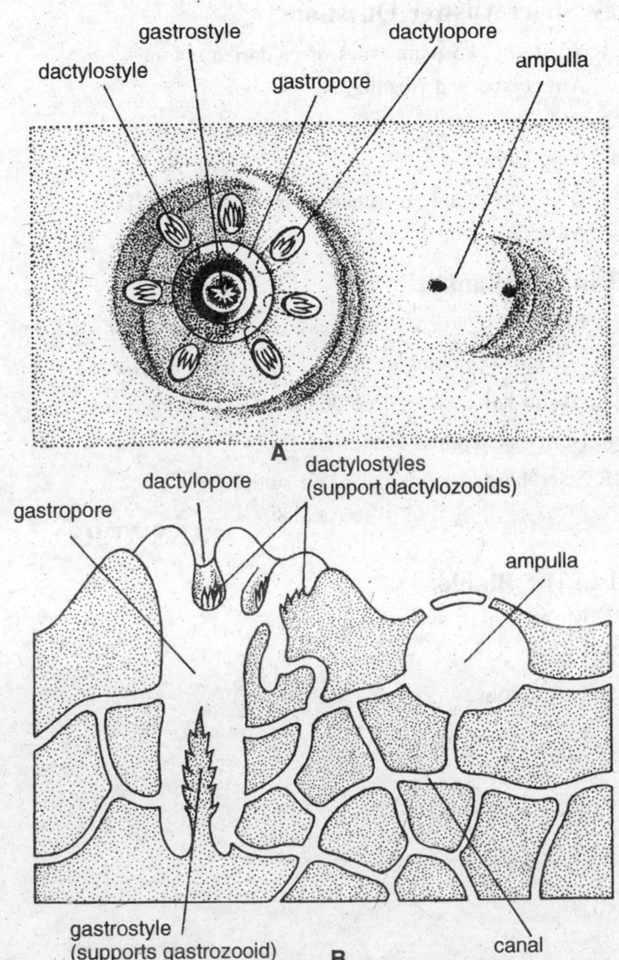

Fig. 21.3. Hydrozoan skeletons. The stylasterine hydrocoral *Allopora* has a calcareous skeleton. A—Surface view from above; B—Cross-section through the skeleton.

QUESTIONS

Long Answer Questions

1. Write an essay on "Colony formation in Cnidaria".

Short Answer Questions

1. Give an account of growth forms of hydrozoan colonies.
2. Write about skeletal support of colonial cnidarians.

Very Short Answer Questions

1. What are two main types of cnidarian colonies
   Ans. Fixed and floating.
2. Give an example of fixed hydrozoan colony.
   Ans. *Obelia.*
3. Give an example of floating colony of Cnidaria.
   Ans. *Physalia.*

Fill in the Blanks

1. Branched hydrozoan colonies have two types of growth forms: .... growth and .... growth.
2. .... colonies occur in .... , *Porpita* and *Velella.*
3. *Obelia* has .... colonies.
4. Horny skeleton is found in ....
5. Scleroblasts secrete .... sclerites.

ANSWERS

Fill in the Blanks

1. Monopodial, sympodial
2. Floating, *Physalia*
3. Fixed
4. Gorgonians
5. Calcareous.

22

Corals and Coral Reefs

22.1 CORALS

Corals are deposits of lime or $CaCO_3$ formed by coelenterates. They are formed of exoskeleton secreted by the polyps of some colonial coelenterates. They may be solid and massive or branched like the trees. Most of them belong to the class Anthozoa comprising about 2500 living 5000 fossil species.

Types of Corals

1. Hydrozoan corals. They include *Millepora* (order Milleporina) and *Stylaster* (order Stylasterina). They are colonial and secrete massive calcareous exoskeleton. The skeleton is secreted by a modified epidermis called **calicoblastic layer**. Living within the skeleton are following two types of polyps: 1. **Large gastrozooid** or **nutritive zooids**, and 2. **Small dactylozooids or protective zooids**. These corals are found with other corals, which constitute coral reefs.

2. Anthozoan corals. They are of following types:

(*i*) Horny corals. These belong to subclass Octocorallia and order Gorgonacea. They are branched and tree-like. The skeleton is formed of horny proteinaceous material, **gorgonin**, along with calcareous spicules which surround the polyp, *e.g.*, *Gorgonia*.

(*ii*) Soft corals. These belong to subclass Octocorallia and order Alcyonacea. They are colonial polyps having endoskeleton of separate spicules, embedded in a massive mesogloea, *e.g.*, *Alcyonium* (Dead man's fingers, Fig. 22.1).

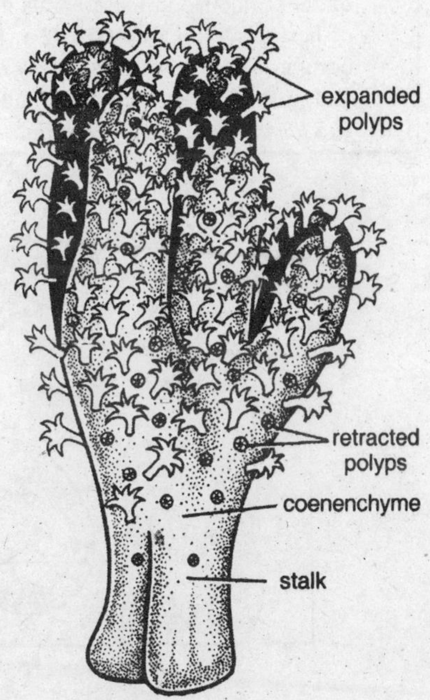

Fig. 22.1. *Alcyonium.*

(*iii*) Stony corals. These belong to subclass Octocorallia and order Stolonifera. The skeleton is made up of calcareous spicules and stained red with iron salts. The skeleton consists of vertical tubes connected by lateral platforms, *e.g.*, *Tubipora* (organ-pipe coral).

(*iv*) Blue coral. These belong to subclass Octocorallia and order Helioporacea. In its colony of polyps, the calcareous spicules form a massive skeleton. The larger cavities on the surface of skeleton contain polyps, *e.g.*, *Heliopora* (blue coral).

(*v*) True corals. These belong to subclass Hexacorallia and order Madreporaria. They form true corals. Majority of them are colonial. The colonies have different forms. These are the principal builders of the coral reef.

Organization of Corals

1. Coral skeleton. The excskeleton of a coral is known as **corallite**. It is calcareous and secreted by the epidermis. In a colonial coral, the coralite of individual polyps fuse together to form a skeletal mass which is called the **corallium**.

Each corallite is cup-shaped depression with a **basal plate** or **theca**. The cavity of the cup contains a number of vertical radiating ridges or septa, called **sclerosepta**. The inner end of sclerosepta are fused to form an irregular skeletal mass or the **columella**.

2. Coral polyp. They are small anthozoan polyp measuring about 1 cm in length. These are cylindrical without pedal disc, and oral disc bears a cycle of six tentacles. The pharynx is without siphonoglyphs. The mesenteries are restricted to the upper part of the polyp. The muscles are poorly developed. At night and on cloudy days, the polyps expand and feed on plankton.

Box 22.1 Zooxanthellate Corals

One of the most noteworthy evolutionary achievements of cnidarians is their close relationship with unicellular photosynthetic partners. The relationship is widespread and occurs in many shallow-water cnidarians. The symbionts of freshwater hydrozoans (*e.g.*, *Chlorhydra*) are single-celled species of green algae (Chlorophyta) called "zoochlorelle". In marine cnidarians, the protists are unicellular cryptomonads and dinoflagellates called "zooxanthellae" (probably many genera including *Zooxanthella*; Fig. 22.2). These algae may be capable of living free from their hosts, but very little is known about their natural history (see **Brusca** and **Brusca**, 2003). The algae typically reside in the host's gastrodermis or epidermis, although some cnidarians harbour extra-cellular zooxanthellae in mesogloea. It is usually the algal symbionts that give cnidarians their green, blue-green, or brownish colour. Corals that are reef-builders (called **hermatypic corals**) typically harbour zooxanthellae (*i.e.*, they are called **zooxanthellate corals**). Resident populations of zooxanthellae in these corals may reach a density of 30,000 algal cells per cubic millimeter of host tissue (or from 1 to 2 × 10^6 cells per centimeter of coral surface). Zooxanthellae also occur in many tropical gorgonians, anemones, and zooanthids. They also occur in some scyphozoans, and in some cases they are thought to produce much of the energy required by the host (*e.g.*, *Cassiopea*, *Linche* and *Mastigias*).

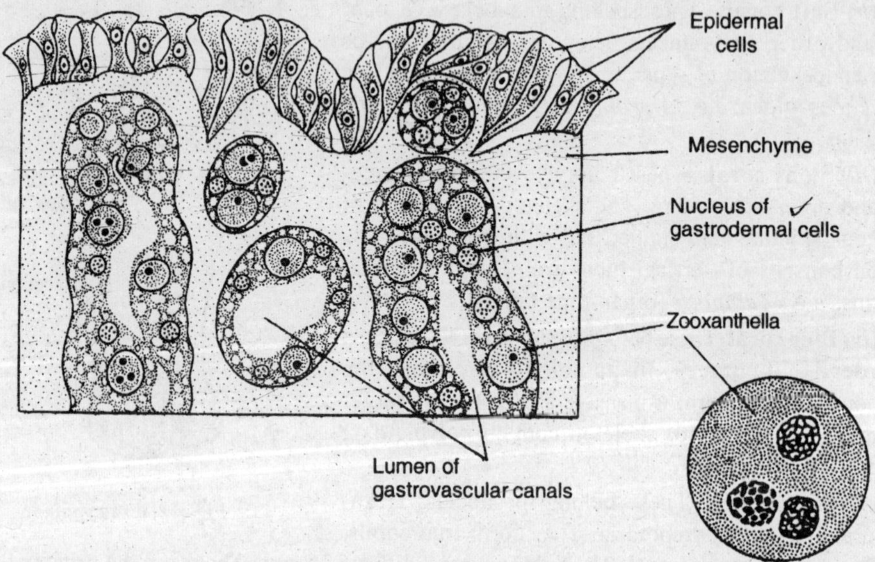

Fig. 22.2. An octocoral with zooxanthellae distributed throughout the gastrodermis (schematic section) (after Brusca and Brusca, 2003).

3. Formation of corals. The coral polyp develops from a planula which secretes a skeletal rudiment or **prototheca** with the help of ectodermal and basal plate. Radial folds of the larva secrete septa and a rim is built up around the polyp. Meanwhile, further skeletal material is added into the gaps between the septa. The septa alternate with the mesenteries of a living coelenterate.

Role of mutualistic zooxanthellae in coral formation. Corals and zooxanthellae (Fig. 22.2) exist in a **mutualistic** relationships: corals cannot live, or are handicapped without them (**Muscatine**, 1974). This restricts coral distribution to areas where zooxanthellae can thrive. Corals die after 18 days in total darkness, but they are completely carnivorous and expel zooxanthellae when starved. Zooxanthellae probably seek corals to satisfy their own nitrate and phosphorus requirements and in return help in the skeleton formation of coral. They help to dispose of H_2CO_3 released as by-product of the formation of $CaCO_3$ (**Rinkevich** and **Loya**, 1984).

22.2 CORAL REEFS

Coral reefs are limestone formations, formed of exoskeletons of stony corals (*Madreporaria*) and heavily charged with organic material, especially in upper strata where growing corals occur Calcareous algae, foraminiferans, milleporines and other calcareous Alcyonaria contribute to reef.

Coral reefs are the ridges of calcium carbonate in the sea formed by the remains of skeleton and shells of animals.

Distribution

Coral reefs grow in warm, shallow waters and limited to continental and island shores in tropical and subtropical zones. They can not survive in water having temperature less than 22°C and have luxuriant growth in shallow water having a depth less than 11 metres. Coral reefs inhabit Maldives island of Indian Ocean, Wake Island, Marshal Island, Fizi Island of Pacific Ocean, Whitesunday Island, Bahoem Island regions and the famous Great Barrier reefs of Australia.

Kinds of Coral Reefs

Coral reefs are considered to be of following three types (Fig. 22.3).

(*i*) Fringing reefs

(*ii*) Barrier reefs

(*iii*) Atolls

(1) Fringing reefs. The coral reefs are termed as **fringing reefs** which lie close to the shores of some volcanic island or part of some continent. A fringing reef extend from the shore a few metres to 1/4 mile out and consists of the **reef edges** or **front**, where the most active coral growth occurs and a slightly lower, more or less flat surface, the **reef flat**. The reef flat between the front and the shore is composed largely of coral sand and mud, dead corals and other debris usually giving a scarlet to brown green hue. The outer single beach of island may get merged into the seaward flat of the reef which already consists of **seaward slope** on which corals grow down to 20 fathoms or there may be a varying area of **boulder zone** in between. Sometime between island and boulder zone may be enclosed an **inner flat** which is hollowed out to give rise to shallow lake.

(2) Barrier reefs. Barrier reefs are same as the fringing reef but they are located some distance from the shore. They are separated from land by stretch of water or **lagoon** which may be 1/2 a miles to 10 mile or more and 60 to 1,000 feet deep.

The most notable example of barrier reef is **Great Barrier Reef** of Australia, extending over 1200 miles long and up to 90 miles from shore.

(3) Atoll or Atoll reef. They are also referred as **coral island** or **lagoon island**. They are irregular, more or less circular or horse-shoe shaped reef not enclosing any island but encircling lagoon.

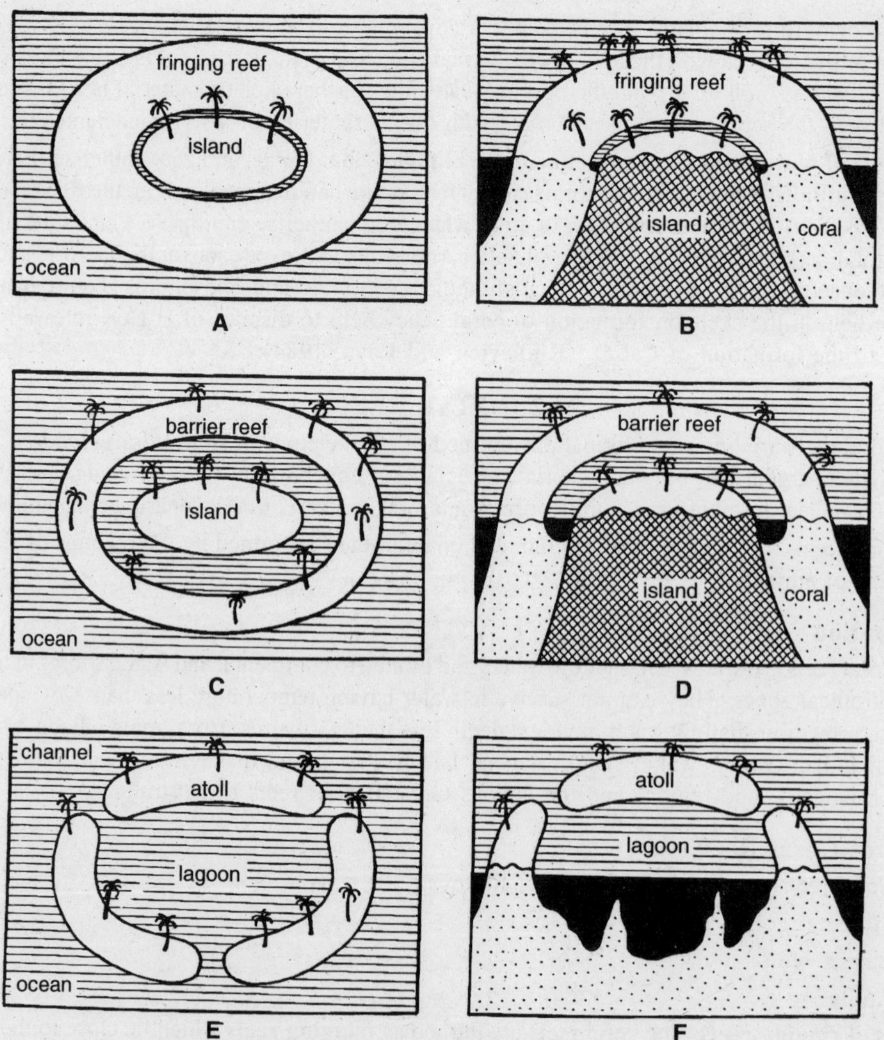

Fig. 22.3. Coral reefs. A—Fringing reef; B—Fringing reef in section; C—Barrier reef; D—Barrier reef in section; E—Atoll reef; F—Atoll reef in section.

The term atoll is derived from the Maldives language, where each government district is termed as "*atolu*" and the governor the "*atoluveri*". His province is a circular reef broken by many channels. The encircled lagoons varies from a few miles to 40 to 50 miles. The outer side of the reef slopes off rather steeply into the depth of the ocean.

Theories of Reef Formation

It is difficult to explain the great vertical thickness often attained by the coral reefs, since the reef building corals are littoral in habit and cannot grow below 150 feet. Many theories have been put forward to explain coral-reef formation. A few important ones are as follows:

1. Darwin-Dana theory of subsidence. In 1831, Darwin noticed that the place where coral reefs are formed now, a sinking of land had taken place in the past. Therefore, he assumed that the corals started their growth as fringing reefs around the sloping shores of an island in shallow tropical sea. Subsidence of the sea-floor then started in region of the reef followed by the upward and outward growth of the coral. The sinking island became smaller and smaller and finally disappeared and the barrier reef became an atoll.

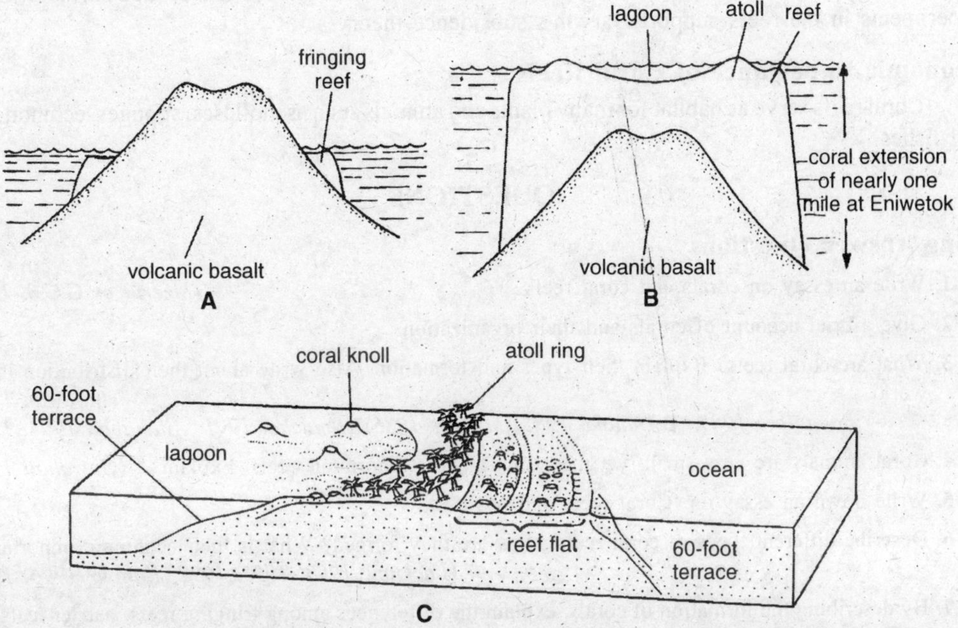

Fig. 22.4. Formation of an atoll. A—Fringing reef around an emergent volcano; B—Continuous deposition of coral as the volcanic cone subsides, leading to formation of a great coralline cap; the emergent part of the cap is the atoll; C—Section through part of the atoll.

2. Submerged bank theory. The recent scientists assumed that the coral grow to form reefs on flat pre-existing surface during or after their submergence. This is brought about by the erosion and denudation of an island both above and below water. An island is replaced completely by submerged plateau and builders will produce an atoll.

3. Daly glacial control theory. This theory is put forward by **Daly.** He assumed that during last glacial period, the widely extending polar ice caps, over a mile in thickness were formed. This lowered the sea level by at least 150 feet and exposed several flat platforms of sand and mud deposition cut out by the action of water currents during interglacial period. Due to the low temperature of island, the growth of corals are inhibited. When the ice melted, the sea level rose and more and more waters covered the platform. The sea temperature having risen to over 10°C became suitable for coral growth. This theory is based on accurate knowledge and close calculations for the very uniform depth of coral lagoon.

4. Super-Murray solution theory. According to **Sir John Murray** mounds are formed upon the sea bottom by deposition of limestone skeletons of many animals (such as shells of molluscs and starfish and deep sea corals) when they reach about 50 fathoms, the corals grow on them and reach the water surface. A barrier reef results at the outer edge and an atoll is produced by the dissolution of the inner coral rocks. The mound ultimately became flattened like plateau.

5. Struchbury's volcanic crater theory. According to **Struchbury**, the coral reefs are formed at the crater of volcano in the Pacific ocean. After formation of lagoon at the same region, the corals had continuous growth. Ultimately they came out from the water surface. But this theory is not entirely satisfactory.

The submerged bank theory and Daly's theory which supplement each other are at present most favoured although Darwin's idea still finds much support. Several lines of evidence have supported that most coral reefs grow on submerged lands, cliff's drowned valleys and eroded surfaces. Some experiments in the reets support Darwin's subsidence theory.

Economic Importance of Coral Reefs

Coral reefs serve as habitat for many plants and animals such as molluscs, sponges, echinoderms and fishes.

QUESTIONS

Long Answer Questions

1. Write an essay on corals and coral reefs. (*Meerut or C.C.S. 1991*)
2. Give a brief account of corals and their organization.
3. What are coral reefs? Explain their types and formation. Also write about their distribution in the world. (*Burdwan 1984; Bangalore 1985; Karnataka 1985; Madras 1985; Himachal 1991; Allahabad 1994, 1997*)
4. Coral animals are very small, yet they can construct enormous reefs. Explain. (*Guwahati 1995*)
5. Write down an essay on 'Coral reefs and Atoll'.
6. Describe different types of coral reefs. How are they formed? What is their economic importance? (*Lucknow 1993, 1995, 1997; Kanpur 1996, 1997*)
7. By describing the formation of corals, explain the differences among fringing reefs, barrier reefs and atoll.

Short Answer Questions

1. Write a short note on corals. (*Gorakhpur 1993; Kanpur 1993, 1995, 1997*)
2. Coral reef.

Very Short Answer Questions

1. Name two hydrozoan corals.

   Ans. *Millepora* and *Stylaster*.
2. Mention one soft coral and one horny coral.

   Ans. *Alcyonium* and *Gorgonia*.
3. Name the three types of coral reefs.

   Ans. Fringing reef, barrier reef and atoll.
4. From which coral red 'munga' is obtained.

   Ans. *Corallium rubrum*.
5. Mention an example of a solitary coral.

   Ans. *Fungia*.
6. What are corals?

   Ans. Corals are colonial, polypoid marine cnidarians living in the skeleton secreted by themselves. This skeleton is calcareous or horny and is commonly known as coral. Some coral-building cnidarians are 1. *Millepora*; 2. *Tubipora*; 3. *Gorgonia*; 4. *Alcyonium*; 5. *Fungia*.

Fill in the Blanks

1. Two main types of coral are .... corals and anthozoan corals.
2. The major reef builders are .... corals.

Multiple Choice Questions

1. The exoskeleton of the colony is

(*a*) corallum (*b*) columella

(*c*) corallite (*d*) none of these

2. Corallite is

(*a*) an animal (*b*) an individual

(*c*) a cnidarian (*d*) skeleton of a solitary coral

3. Organ-pipe coral is

(*a*) *Astrea* (*b*) *Tubipora*

(*c*) *Fungia* (*d*) *Meandrina*

4. True corals are coelenterates belonging to

(*a*) Hydroidea (*b*) Madreporaria

(*c*) Pennatulacea (*d*) Coenothecalia

5. A coral island with a central shallow lake is known as

(*a*) coral reef (*b*) lagoon

(*c*) atoll (*d*) corallite

6. Coral islands are

(*a*) Fiji (*b*) Maldives

(*c*) Laccadives (*d*) all of these

7. Coral reefs are confined to the

(*a*) tropical seas (*b*) temperate seas

(*c*) polar seas (*d*) river deltas

8. A fringing reef is usually located

(*a*) very near the shore (*b*) far away from the shore

(*c*) in deep sea (*d*) in cold sea

9. Bramuda is a

(*a*) fringing reef (*b*) coral island

(*c*) atoll reef (*d*) barrier reef

10. Most notable example of barrier reef is

(*a*) West Indies (*b*) Great Barrier reef

(*c*) Florida keys (*d*) none of these

ANSWERS

Fill in the Blanks

1. Hydrozoan **2.** Madreporarian

Multiple Choice Questions

1. (*a*) **2.** (*d*) **3.** (*b*) **4.** (*b*)

5. (*c*) **6.** (*d*) **7.** (*a*) **8.** (*a*)

9. (*b*) **10.** (*b*).

23

Phylum Platyhelminthes (Characters, Classification and Types)

Platyhelminthes, (Gr., *platys* = flat; *helmins* = worms), *i.e.*, the flatworms are dorsoventrally flattened, triploblastic, bilaterally symmetrical acoelomates with blind sac body plan and organ grade of body organization. These primitive eumetazoans are without definite anus, circulatory, skeletal or respiratory systems but with protonephridial excretory system and mesenchyma (parenchyma) filling the space between various organs of body (see Box 23.1).

The term Platyhelminthes was first proposed by **Gaugenbaur** (1859). Flatworms help biologists to understand animal evolution.

> **Box 23.1 Primitive characters of flatworms**
>
> Phylum Platyhelminthes contains about 13,000 species and has following three primitive characters:
>
> 1. Absence of circulatory system.
> 2. The use of cilia rather than muscles for locomotion.
> 3. Presence of protenephridia.
>
> These features are now believed to be the result of small body size.

23.1 GENERAL CHARACTERS

Phylum Platyhelminthes is characterized by the following characters:

1. Multicellular body shows **organ grade organization**.
2. Platyhelminthes are bilaterally symmetrical and dorso-ventrally flattened.
3. They are **triploblastic** worms, *i.e.*, their body is derived from three embryonic germ layers namely ectoderm, mesoderm and endoderm.
4. Body shape varies from moderately elongated flattened shape to long flat ribbons and leaf-like.
5. Most parasitic flatworms are white and colourless; some of them derive colour from the ingested food. The free-living forms are brown, grey, black or brilliantly coloured.
6. Animals have soft bodies and are unsegmented.
7. Anterior end of body is differentiated into the so called **head**.
8. Exoskeleton and endoskeleton are absent and hard parts of body such as spines, thorns, hooks, teeth, etc., are formed of scleroproteins.
9. Body is covered with syncytial or single layered, partly ciliated epidermis (*e.g.* Turbellaria). In parasitic forms (trematodes and cestodes) epidermis is absent and the body is covered with cuticle.

10. Great variety of adhesive secretions, organs of attachment and adhesion (*e.g.*, hooks and suckers) are present. Body surface of flatworms bears two kinds of glands: 1. **Cyanophilous glands** (*e.g.*, frontal glands) which secrete slime; 2. **Eosinophilous glands** (*e.g.*, glanduloepidermal adhesive organs) which secrete adhesive substances.
11. The **mouth** is typically the only opening to the digestive tract when the latter is present. The round or slit-like mouth appears at the site of blastopore (called **protostomia**). It may occur anywhere on the midventral surface.
12. A true body cavity or coelom is absent; so they are called **acoelomates**. Space between the organs is filled with a loose mesodermal tissue called **parenchyma** or **mesenchyma**.
13. Parenchyma helps in internal transport of materials from cell to cell distribution. Parenchymal tissue fluid is typically circulated as muscular movements squeeze and distort the parenchyma.
14. Digestive system is totally absent in Acoela and tapeworms but in other flat worms it consists of mouth, pharynx and blind and highly branched intestine. It is incomplete due to absence of **anus**.
15. No special organ for respiratory exchange or transport occurs. Respiration is **aerobic** in turbellarians but **anaerobic** in parasitic flatworms (see Box 23.2).

Box. 23.2.

In turbellarians, the aerobic respiration releases energy mainly by a **cyanide-sensitive system**, *i.e.*, after dehydrogenation, hydrogen is transferred by a cytochrome-oxidase system to the final acceptor oxygen. A small cyanide-resistant fraction is probably maintained by iron-free oxidases. Anaerobic respiration of parasitic flatworms involves glycogen degradation and formation of carbon dioxide and various organic acids. Different species release different combinations of acids, which indicate diversity among the array of respiratory enzymes.

16. Platyhelminthes are **ammonotelic**, releasing their nitrogenous waste in the form of ammonia.
17. Excretory system consists of single or paired **protonephridia** with **flame cells** or **solenocytes**. Protonephridia are absent in Acoela (see Box 23.3).

Box. 23.3.

A **flame cell** is a hollow cell that has a tuft of vibratile cilia. The protonephridium has traditionally been thought of as a primitive precursor to more complex structures of excretion seen in higher invertebrates (**Wilson** and **Webster**, 1974). Recent work indicates, however, that rather than being primitive, protonephridia are a unique solution to the filtration of extracellular fluid in a body that has no coelom or blood-vascular system (**Ruppert** and **Smith**, 1985). In this case, the solenocytes must **actively transport** selective materials out of the body fluids into the protonephridial weir (=river dam).

18. Flatworms contain a primitive ladder-like nervous system. The main nervous system consists of a pair of **cerebral ganglia** or **brain** and one to three pairs of longitudinal **nerve cords** connected to each other by transverse commissures.
19. Turbellarians have a variety of sensory organs such as **photoreceptors** (for light reception), **chemoreceptors** (for food recognition), **tangoreceptors** (for touch reception), **rheoreceptors** (for orientation to water currents) and **statocysts** (for orientation to gravity). Sense organs are greatly reduced in parasitic forms.
20. Sexes are mostly united, *i.e.*, **hermaphrodite** or **monoecious**.
21. Asexual multiplication takes place by **fission** in many freshwater Turbellaria. Endogenous or exogenous budding is common in some tapeworms.
22. Flatworms have a remarkable power of **regeneration**.
23. Reproductive system is complex and highly evolved in most of the forms. Flatworms produce two types of eggs: 1. **Endolecithal** or **entolecithal ova** with yolk reserves stored in the egg cytoplasm; 2. **Ectolecithal ova** are free of yolk but encapsulated with special **yolk cells** containing the needed food reserve (Fig. 23.1).

24. Fertilization is **internal**. It is cross-fertilization in Turbellaria and Trematoda and self-fertilization in Cestoda.

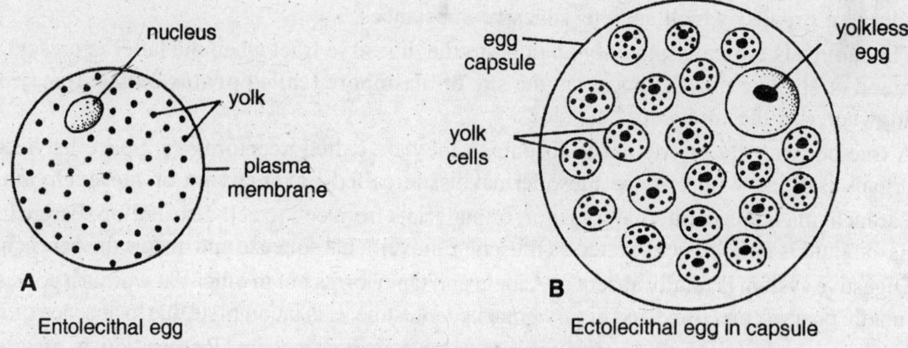

Fig. 23.1. A—Entolecithal (yolk-containing) egg; B—Ectolecithal egg.

25. Development may be direct (Turbellaria) or indirect. Life cycle is complicated and involves one or more hosts. **Digenetic life cycle** (*e.g.*, *Fasciola*) involves a **primary** or **definitive host** and an **intermediate host. Monogenetic life cycle** involves only one host.
26. Phenomena of **parthenogenesis** and **polyembryony** are common in trematodes and cestodes. Alternation of generation or **metagenesis** is also found in some cases (*e.g.*, *Fasciola*).
27. The flatworms are either **free-living** or **ecto-** or **endocommensals** or **endoparasites.**

23.2 OUTLINE CLASSIFICATION

Phylum Platyhelminthes

Classes (4)

- **1. Turbellaria**
- **2. Trematoda**
- **3. Monogenea**
  Examples: *Polystomoidella, Polystoma*
- **4. Cestoda**

Class 1. Turbellaria

Groups (2)

1. Archoophorans

Orders (6)

1. Acoela
Examples: *Convoluta, Afronta, Amphiscolops*

2. Nemertodermatida
Example: *Nemertoderma*

3. Macrostomida
Examples: *Macrostomum, Microstomum*

4. Haplopharyngida
Example: *Haplopharynx*

5. Catenulida
Examples: *Catenula, Stenostomum*

6. Polycladida
Examples: *Notoplana, Leptoplana*

2. Neophorans

Orders (5)

1. Lecithoepitheliata
Example: *Protorhyncus*

2. Prolecithophora
Example: *Hydrolimax*

3. Proseriata
Example: *Octoplana*

4. Rhabdocoela
Example: *Mesostoma*

5. Tricladida
Example: *Planaria*

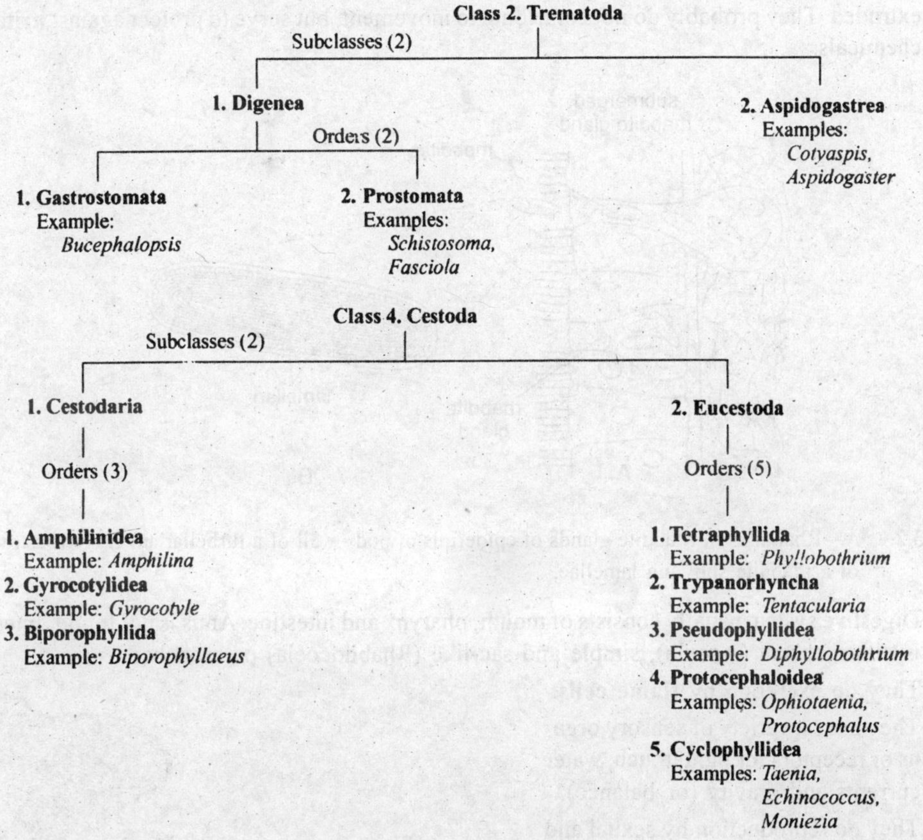

23.3 CLASSIFICATION

Phylum Platyhelminthes is divided into **four** classes of worms. Three classes are entirely parasitic and include the Trematoda, the Monogenea and the Eucestoda. The fourth class, the Turbellaria, is free-living and is certainly the group from which arose the ancestors of three parasitic classes.

Class 1. Turbellaria

(L., *turbella* = a little string)

1. Mostly free-living and carnivorous ciliated flatworms; a few are parasitic.
2. They are either aquatic, may be freshwater or marine, or terrestrial. Most aquatic forms are bottom dwellers or benthic, living in sand or mud.
3. They are mostly small but range in size from microscopic to more than 60 cm long (*e.g.*, the Lake Baikal triclad, *Rimacephalus arecepta*).
4. Bodv is unsegmented, dorsoventrally flattened and leaf-like, rounded or oval.
5. Body is covered with a cellular or syncytical **epidermis**.
6. Terrestrial species have strong epidermal ventral cilia and weak or no dorsal cilia. Epidermis of the ventral surface contains many unicellular **cyanophilous** or slime glands. Slime secretions tend to lubricate the surface; they increase water viscosity and provide resistance against which the cilia can thrust.
7. Suckers are absent, but adhesive organs are abundantly present.
8. Epidermal and subepidermal glands also form **rhabdites**. A rhabdite is a rod-shaped secretion, microscopically composed of successive lamellae (Fig. 23.2). Rhabdites dissolve rapidly when

extruded. They probably do not contribute to movement, but serve to protect against irritating chemicals.

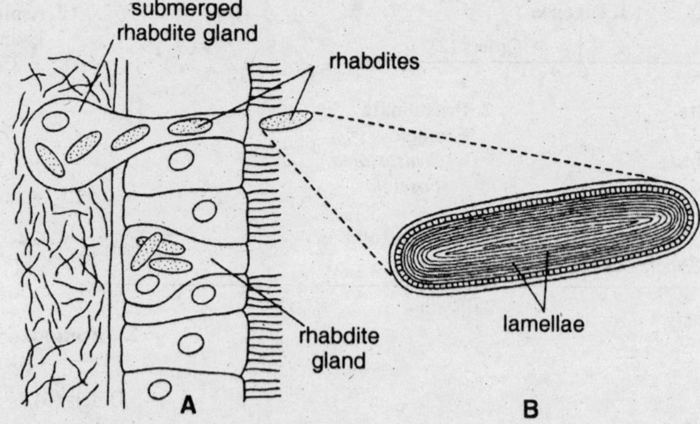

Fig. 23.2. A—Rhabdites in rhabdite glands of epidermis of body wall of a turbellarian; B—Ultrastructure of a rhabdite showing lamellae.

9. Digestive system usually consists of mouth, pharynx and intestine. Anus is not found. Intestine is either absent (Acoela), simple and sac-like (Rhabdocoela) or branched.
10. They do excretion by **flame cells**.
11. They have a variety of sensory organs or receptors for light, touch, water currents and gravity (or balance).
12. They do reproduction by sexual and asexual means. Asexual reproduction occurs by **paratomy** (or strobilation), **architomy** (= fission) and **parthenogenesis**. They have a remarkable power of regeneration.
13. Mostly hermaphrodite. Life cycle is simple. Development is **direct** in most turbellarians, but some polyclads produce free-swimming larvae that feed on plankton (**planktotrophic**). These larvae are of two types: **Goette's larvae** (having four arms or lobes) and **Mueller's larvae** (having eight arms) (Fig. 23.3). The larvae become metamorphosed into juvenile.

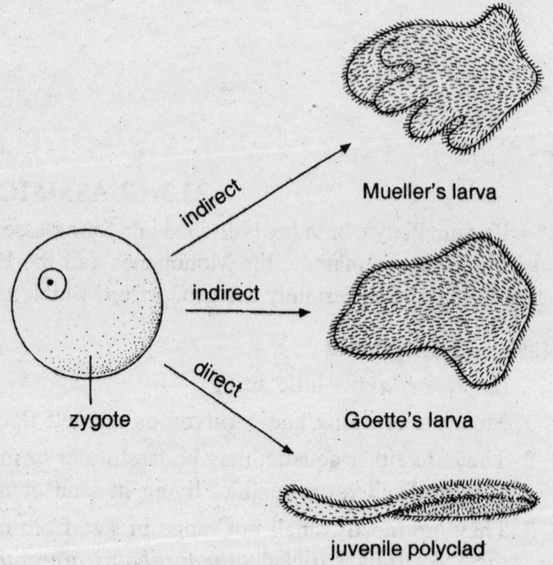

Fig. 23.3. Direct and indirect development of turbellarians.

14. About 3000 species of turbellarians have been described.

Class 2. Trematoda

(Gr., *trematodes* = having pores)

1. They are commonly called **flukes**.
2. They are **ectoparasites** or **endoparasites**.

3. Body unsegmented, dorsoventrally flattened and leaf-like.
4. Their body wall is without epidermis and cilia; instead it is covered with a thick cuticle.
5. They contain well developed suckers.
6. Their incomplete digestive system lacks anus but contains mouth, pharynx and branched intestine.
7. They are mostly **monoecious** or **hermaphrodite**.
8. They contain single ovary and two or many testes.
9. Their life history may be simple or complicated including many larval forms.

Class 3. Monogenea

1. This class includes flatworms which are mostly ecto- or endoparasites of aquatic vertebrates such as fishes, amphibians and reptiles (turtles).
2. Their body is dorsoventrally flattened and composed of a head, trunk and haptor.
3. Oral sucker is either absent or weak. Instead of it a pair of adhesive glands (head organs) are present.
4. Posterior end of body is provided with large posterior attachment organ called **opisthaptor** or **haptor**. Haptor bears **hooks** and **suckers**, allowing the parasite to cling tenaciously to the skin of fast-moving host.
5. Excretory pores are paired and situated anteriorly on the dorsolateral sides. In some species a urinary bladder is associated with each pore.
6. They have complex hermaphrodite reproductive system. Most contain a single circular or oval testis and a single ovary and extensive vitellaria.
7. Male and female gonopores are usually separate.
8. During copulation, there is a mutual exchange of sperm being passed into single or paired vagina. Males use a protrusible penis-like structure bearing hooks in copulation.
9. Their life cycle includes only **one host** (there is no intermediate host).
10. One egg (by way of a free-swimming ciliated larva called **oncomiracidium**) gives rise to only one adult worm. Hence the name **monogenea**, meaning "one generation."

    Examples. *Polystomoidella oblongum* (parasite in urinary bladder of turtles); *Benedenia melleni* (ectoparasite of fish); *Polystoma integerrinum* (endoparasite of urinary bladder of frogs and toads); *Dactylogyrus vastator* (ectoparasite on the gills of freshwater fishes)

Class 4. Cestoidea (Cestoda)

(Gr., *kestos* = girdle + *eidos* = form)

1. All are highly specialized endoparasites of vertebrates, commonly called **tapeworms**.
2. Body is flattened and ribbon-like.
3. Body is without epidermis and cilia but covered by a syncytial tegument (cuticle) having microvilli-like projections called **microtrichia**. Microtrichia plays a vital role in active transport of food molecules of various carbohydrates and amino acids.
4. Body of tapeworms is differentiated into **scolex, neck** and **body** or **strobila**.
5. Scolex is provided with adhesive structures (hooks, suckers) which are absent in subclass Cestodaria.
6. Neck is very short and narrow. It is **proliferative** giving rise to the body or strobila. Neck keeps on adding new proglottids to body by mitotic growth followed by transverse constriction.

7. Strobila consist of linearly arranged segment-like sections called **proglottids**. The youngest proglottids are at the anterior end and they increase in size and maturity as they are displaced toward the posterior end of strobila.
8. Mouth, digestive tract and sense organs are absent.
9. Excretory system consists of **protonephridia** with typical terminal **flame cells**.
10. Each mature proglottid contains one or two sets of complete hermaphroditic reproductive system.
11. Life cycle is complicated usually involving two or more hosts.
12. Embryos and larvae possess hooks.

CLASS 1. TURBELLARIA

Class Turbellaria is divided into following **two** major groups:

Group I. Archoophoran turbellarians

(*arch* = primitive + *oophoro* = ovary)

1. They have most primitive organization.
2. Vitellaria (yolk glands) are absent.
3. Eggs are entolecithal.
4. Cleavage is spiral.

Group II. Neoophoran turbellarians

(*ne* = new + *oophoro* = ovary)

1. They have an advanced level of organization.
2. Vitellaria (yolk glands) are present.
3. Eggs are ectolecithal.
4. Spiral cleavage is greatly modified.

Group I. Archoophoran turbellarians

This group includes the following **six** orders:

Order 1. Acoela

1. Minute marine turbellarians measuring less than 2 mm in length.
2. Mouth and sometimes a simple pharynx are present, but no intestine or body cavity.
3. Protonephridia are absent.
4. Gonads are often not bounded by a cellular wall.
5. Oviducts are absent.
6. Statocyst is with one statolith.
7. Usually free-living found under stones or on bottom of mud; some live in the intestine of various echinoderms.

Examples. *Convoluta, Afronta, Amphiscolops, Ectocotyla, Paratomella.*

Order 2. Nemertodermatida

1. Small marine flatworms similar to Acoela.
2. Have an epithelial digestive tract.
3. Statocyst with two statoliths.
4. They produce uniflagellate sperm.

Example. *Nemertoderma.*

Order 3. Macrostomida

1. Small (0.8 to 11 mm in length) marine and freshwater species.
2. Digestive system with a simple pharynx and a simple sac-like ciliated intestine.
3. Nervous system with one pair of ventrolateral nerve cords.
4. Testis is compact; penis armed with a stylet.
5. Yolk glands are absent.
6. Asexual reproduction with chains of zooids.

Examples. *Macrostomum, Microstomum.*

Order 4. Haplopharygida

1. Small marine species similar to macrostomids.
2. They have a proboscis.
3. A temporary anus is present.

Example. *Haplopharynx.*

Order 5. Catenulida

1. Mostly small freshwater species.
2. Digestive system contains a simple pharynx and ciliated sac-like intestine.
3. Excretory system consists of a single median protonephridia.
4. Statocyst sometimes present, usually with one statolith.
5. Gonads are unpaired. Single testis is in the form of compact mass. Penis unarmed. Male gonopore is dorsal above the pharynx.
6. Yolk glands absent. No female gonoduct.
7. Asexual fission occurs with the formation of chains of zooids.

Examples. *Catenula, Stenostomum.*

Order 6. Polycladida

1. Exclusively marine flatworms of moderate body size, averaging 3 to 20 mm in length.
2. Body usually greatly flattened and more or less oval in shape.
3. A pair of anterior marginal or dorsal tentacles may be present.
4. Many are brightly coloured.
5. Pharynx is tubular or plicate (= bell-shaped).
6. Intestine is elongated and centrally located, with many highly branched diverticula.
7. Nervous system consists of many radially arranged nerve cords.
8. Eyes are numerous.
9. Testes and ovaries are numerous and scattered.
10. They feed on coelenterates and store their undischarged nematocysts in their skin as means of self-defence.

Examples. *Notoplana, Leptoplana, Gnesioceros, Stylochus. Prostheceraeus, Pseudoceros.*

(**Note.** Order Polycladida is often divided into two suborders: 1. **Acotylea** (*e.g., Notoplana*) and 2. **Cotylea** (*e.g., Thysanozoon, Yungia*).

Group II. Neoophoran turbellarians

This group includes the following **five** orders:

Order 1. Lecithoepitheliata

1. Marine or freshwater species.
2. Mouth and pharynx are at the anterior end of body, intestine is simple.
3. Ovary produces eggs which are surrounded by follicle-like yolk cells.
4. Female ducts are simple or none.
5. Penis is with cuticular stylet.

Example. *Prorhynchus.*

Order 2. Prolecithophora

1. Usually small sized.
2. Marine and freshwater species.
3. They contain a plicate or bulbous pharynx and a simple intestine.
4. Ovary produces eggs and follicle-like yolk cells.

Examples. *Plagiostomum, Hydrolimax.*

Order 3. Proseriata

1. Small-sized marine turbellarians.
2. Statocyst with one statolith.
3. Pharynx is plicate and tubular but gut is not branched.
4. Female reproductive system consists of separate ovaries and yolk glands.

Examples. *Octoplana, Monocelis, Nemertoplana.*

Order 4. Rhabdocoela

1. It is a large group of small (less than 3 mm) marine, freshwater and terrestrial forms.
2. Digestive tract is complete having simple bulbous pharynx. Intestine unbranched sac-like without diverticula.
3. Excretory system with protonephridia.
4. One pair of nerve cords.
5. Eyes are usually present.
6. They contain one or two gonads, yolk glands present or absent.
7. Free-living, commensal or parasitic forms.

This order is divided into the following three suborders:

Suborder 1. Typhloplanoida, Example. *Mesostoma.*

Suborder 2. Dalyellioida, Examples. *Anoplodiera, Syndesmis, Kronborgia.*

Suborder 3. Kalyptorhynchia. Examples. *Gyratrix, Gnathorhynchus.*

Order 5. Tricladida

1. Relatively large (measuring from 2 to 60 cm in lenght) marine, freshwater and terrestrial turbellarians.
2. Pharynx is plicate, tubular and posteriorly directed.
3. Intestine (=gut) has three branches.
4. Eyes present in most species.
5. Protonephridia form paired lateral network.
6. Male reproductive organs consist of two to numerous testes; a penis papilla present.

7. Female reproductive organs consist of a pair of ovaries with yolk glands and a copulatory bursa.
8. Gonophore single.

This order is divided into following **three** suborders:

Suborder 1. Maricola. Examples. *Bdelloura* (It is commensal on the book gills of horseshoe crabs), *Ectoplana.*

Suborder 2. Paludicola. Examples. *Planaria, Dugesia, Dendrocoelum, Procerodes, Phagocata, Polycelis, Procotyla.*

Suborder 3. Terricola. Examples. *Bipalium, Geoplana, Orthodemus.*

CLASS 2. TREMATODA

The class Trematoda is divided into **two** subclasses: 1. Digenia and 2. Aspidogastrea.

Subclass I. Digenia

1. They include flukes which are endoparasites in vertebrates (such as fishes, amphibians, reptiles, birds and mammals) and invertebrates.
2. They contain two to four hosts in their life cycle. The **first intermediate host** is typically gastropod snail, **second intermediate host** is usually an arthropod and the **definitive host** is a vertebrate.
3. Adult digeneans range in size from approximately 0.2 mm to 6.0 cm in length.
4. They are typically dorsoventrally flattened, but some are thick and fleshy, whereas others are long and thread-like.
5. Most flukes contain two suckers which lack hooks. **Oral sucker** occurs around the mouth and aid in feeding. **Ventral sucker** or **acetabulum** is found on mid-ventral or posterior side of body and prevent dislodgement.
6. Their body is covered with a nonciliated cytoplasmic syncytium, the **tegument**. The tegument plays a vital role in the physiology of digeneans (Box 23.4).

Box 23.4.

Tegument of Digenia provides protection, especially against host's enzymes in gut-inhabiting species. Nitrogenous wastes diffuse to the exterior through the tegument and it is the site of gas exchange. In endoparasites, the tegument as well as gut absorbs **glucose** and some **amino acids.** The **protein synthesis** involved in digenean egg production and in asexual reproduction in developmental stages (polyembryony) places especially heavy demands on the amino acid supply.

7. Excretory pore (nephridiopore) single at the posterior end of body. The protonephridia excrete water and waste metabolites such as unwanted iron from haemoglobin in blood-feeding flatworms.
8. Vagina is absent: uterus is usually long having many shelled eggs.
9. The eggs are **ectolecithal,** *i.e.*, the yolk is supplied by yolk cells, which also carry the material needed in egg capsule formation.
10. Life cycle of flukes is complicated involving many larval stages such as **miracidium, sporocyst, redia** and **cercaria.**
11. Life cycle includes one or more intermediate hosts.
12. Life cycle of flukes always includes at least two infective stages which account for the name Digenea, meaning "two generations". Two or more hosts are infected before the life cycle is completed.

Subclass 2. Aspidogastrea

1. They are endoparasites in the gut of fish and reptiles and in the pericardial and renal cavities of bivalve molluscs.
2. Oral sucker is absent.
3. Anterior end of body is without paired adhesive structures.
4. The entire ventral surface of body acts as an **adhesive organ** which is either a single septate sucker or a longitudinal row of suckers. This organ lacks hooks.
5. The digestive tract contains a single intestinal caecum.
6. They contain a single nephridiopore.
7. Male reproductive system has only one testis.
8. Their life cycle is simple involving no alternation of hosts.

   Examples. *Cotyaspis, Aspidogaster.*

Subclass 1. Digenia

The subclass Digenia is divided into following **two** orders:

Order 1. Gasterostomata

1. Adults are parasite of aquatic vertebrates, mainly fishes.
2. Mouth is ventral.

   Examples. *Bucephalopsis, Bucephalus.*

Order 2. Prostomata

1. Adults are parasites on all vertebrates.
2. Mouth is subterminal or terminal in position.

   Examples. *Schistosoma, Fasciola, Opisthorchis* (or *Clonorchis*) *sinensis* (Chinese liver fluke), *Echinostoma, Paragonimus* (lung fluke).

CLASS 3. CESTOIDEA (CESTODA)

Class Cestoda is divided into following **two** subclasses:

Subclass 1. Cestodria

1. Coelomic or intestinal parasites of vertebrates (mainly fish and rarely reptiles).
2. Body is undivided, *i.e.*, no strobila (**monozoic**, *i.e.*, body consists of a single differentiated unit).
3. Scolex is absent.
4. Alimentary canal is absent.
5. Only one set of monoecious reproductive organs is present.
6. Development is indirect including **lycophore larva** bearing ten hooks.

Subclass 2. Eucestoda

1. Endoparasitic forms in gut (intestine) of vertebrate hosts.
2. Body is usually very elongated ribbon-like.
3. Body is divided into scolex, neck and strobila (**polyzoic**). Strobila contains many proglottids.
4. Scolex with hooks and adhesive suckers.
5. Each proglottid contains one set of reproductive organs.
6. Embryo (larva) with six hooks.

Subclass 1. Cestodaria

This subclass is divided into following **three** orders:

Order 1. Amphilinidea

1. Coelomic endoparasites of fishes.
2. Body flattened, oval or elongated.
3. Scolex is absent.
4. Suckers are absent.
5. Anterior end of body bears protrusible proboscis and frontal glands.
6. Male and female (= vaginal) pores are located posteriorly.
7. Uterus is very much coiled opening near the anterior end.

Example. *Amphilina.*

Order 2. Gyrocotylidea

1. Intestinal endoparasites of fishes.
2. Body is elongated and flattened.
3. An adhesive sucker and a posterior rosette-shaped adhesive organ present.
4. Anterior end of body bears eversible proboscis.
5. Male, vaginal and uterine pores are together situated in the anterior half of the body.
6. Uterus is short and straight and runs directly to pore.

Example. *Gyrocotyle* (an intestinal parasite of *Chimaera* fish).

Order 3. Biporophyllida

1. Endoparasites in the intestine of sharks.
2. Anterior end of body contains an eversible proboscis.
3. Uterus is sac-like.
4. Male and female genital pores open in a common atrium.

Example. *Biporophyllaeus.*

Subclass 2. Eucestoda

This subclass is divided into more than 12 orders; of which main **five** orders are the following:

Order 1. Tetraphyllida

1. Intestinal endoparasites of elasmobranch fishes.
2. Four ear or leaf-like outgrowths or sessile suckers (called **bothridia**) are present on the scolex. They contain hooks.
3. Testes lie anterior to ovaries.
4. Vitelline glands are scattered in two lateral bands.
5. Cirrus (or penis) is armed with spines or hooks.
6. Common genital atrium is marginal.

Example. *Phyllobothrium.*

Order 2. Trypanorhyncha

1. Parasitic in the spiral valve of digestive tract of elasmobranch fishes.
2. Body is moderate-sized.
3. Scolex with four sucking grooves or bothria and four protrusible spiny proboscides.

4. Vitellaria (yolk glands) is present in a continuous layer in cortical parenchyma.
5. Testes extend behind the ovary posteriorly.
6. Gonophores lateral; uterus opens ventrally.

Example. *Tentacularia.*

Order 3. Pseudophyllidea

1. Intestinal parasites of teleost (or bony) fishes and terrestrial vertebrates.
2. Scolex with one terminal or two lateral groove-shaped bothria.
3. Most proglottids in a strobila are at the same stage of development.
4. Testes numerous, follicular and scattered in the mesenchyma of proglottids. Ovary bilobed.
5. Vitellaria (yolk glands) are follicular and numerous.
6. Genital aperture (gonopore) is midventral.

Example. *Dipyllobothrium.*

Order 4. Protocephaloidea

1. Intestinal parasite of freshwater fishes, amphibians and reptiles.
2. Scolex is mobile and contains four cup-shaped suckers.
3. Common genital atrium is marginal.
4. Ovary is bilobed and uterus is with many lateral diverticula.

Examples. *Ophiotaenia, Proteocephalus.*

Order 5. Cyclophyllidea

1. Parasites of intestine of reptiles, birds and mammals.
2. Large-sized tapeworms.
3. Scolex bears four large cup-shaped **suckers** often with an apical **rostellum** having two rows of hooks.
4. Protonephridial excretory system consists of four longitudinal vessels.
5. Ovary is two to many lobed; uterine opening absent.
6. Gonopores on one or both margins.
7. Vitellarium (yolk gland) is single and compact.

Examples. *Taenia, Echinococcus, Moniezia.*

Remaining orders of Subclass Eucestoda their examples are as follows:

Order 6. Diphyllida.

Example. *Echinobothrium.*

Order 7. Spathebothriidea.

Example. *Spathebothrium.*

Order 8. Caryophillidea.

Example. *Caryophyllaeus.*

Order 9. Aporidea.

Example. *Gastrotaenia.*

Order 10. Nippotaeniidea.

Example. *Nippotaenia.*

Order 11. Lecanicephalidea.

Example. *Lecanicephalum.*

Order 12. Disculiceptidea.

Example. *Disculiceps.*

Box 23.5 Recent Classification of Platyhelminthes (Brusca and Brusca, 2003)

I. Class Turbellaria

1. **Order Acoela**, *e.g., Convoluta.*
2. **Order Catenulida**, *e.g., Catenula.*
3. **Order Haplopharyngida**, *e.g., Haplopharynx.*
4. **Order Lecithoepithelia**, *e.g., Prorhynchus.*
5. **Order Macrostomida**, *e.g., Macrostomum, Microstomum.*
6. **Order Nemertodermatida**, *e.g., Nemertoderma.*
7. **Order Polycladida**, *e.g., Notoplana, Planocera.*
8. **Order Prolecithophora**, *e.g., Plagiostomum.*
9. **Order Proplicastomata**, *e.g., Proplicastomata jenseni.*
10. **Order Proseriata**, *e.g., Octoplana.*
11. **Order Rabdocoela**
    (*i*) **Suborder Dalyellioidea**, *e.g., Callostoma.*
    (*ii*) **Suborder Typhloplanoida**, *e.g., Typhlorhynchus.*
    (*iii*) **Suborder Kalyptorhynchia**, *e.g., Cheliplana.*
    (*iv*) **Suborder Temnocephalida**, *e.g., Temnocephala.*
12. **Order Tricladida**, *e.g., Bipalium, Dugesia* (Formerly *Planaria*; see **Brusca** and **Brusca**, 2003), *Geoplana.*

II. Class Monogenea

1. **Subclass Monopistocotylea**, *e.g., Polystoma, Silurodescoides vistulensis* (**Rastogi** *et al.*, 2008).
2. **Subclass Polyopistocotylea**, *e.g., Diplozoon.*

III. Class Trematoda

1. **Subclass Digenea**, *e.g., Fasciola, Schistosoma, Paragonimus westermani* (lung fluke), *Fasciolopsis buski* (intestinal fluke).
2. **Subclass Aspidogastrea**, *e.g., Aspidogaster, Cotyaspis.*

IV. Class Cestoda

1. **Subclass Cestodaria**, *e.g., Amphilina, Gyrocotyl.*
2. **Subclass Eucestoda**, *e.g., Moniezia, Taenia, Echinococcus granulosus. Dipylidium caninum* (dog tapeworm), *Hymenolepis nana* (dwarf tapeworm of human beings).

23.4 TYPE STUDIES

1. *ECHINOCOCCUS GRANULOSUS*

Classification

Phylum	–	**Platyhelminthes**
Class	–	**Cestoda**
Subclass	–	**Eucestoda**
Order	–	**Cyclophyllidea**
Genus	–	***Echinococcus***
Species	–	***granulosus***

Habit and Habitat

Echinococcus granulosus is commonly called **dog tapeworm** or **hydatid worm.** It is pathogenic intestinal endoparasite of dogs, cats, foxes and wolves (**primary hosts**). It is common in Africa, Australia, New Zealand, Iceland, Holland and South America. Certain herbivores such as sheep, goats, cattle, camels, horses, kangaroos, deers, pigs and rabbits are **intermediate hosts** of *E. granulosus*. Humans are only **incidental intermediate hosts** of dog tapeworm and in them larva develops.

Size and Shape

The adult dog tapeworm measures only 3 to 6 mm having flat body.

Structure

The body of *E. granulosus* is differentiated into head or scolex, a narrow neck and three or four proglottids. **Scolex** is pear-shaped having four ovoidal **suckers** and a retractile **rostellum** which is armed with a double crown of 28 to 50 hooklets (Fig. 23.4). **Neck** is the narrow part of the scolex joining the body. **First proglottid** is small and immature containing rudiments of structures (sex organs). **Second proglottid** is mature and hermaphroditic containing a single set of male and female genital organs. Male reproductive system consists of spherical testes, vas deferens and cirrus. Female reproductive system comprises ovaries, oviduct, vitellaria, ootype, uterus and vagina. **Third** or **last segment** is gravid, elongated and biggest. It contains branched uterus filled with eggs under different stages of development (Fig. 23.4). Uterus bursts before or after the proglottid is separated from the worm.

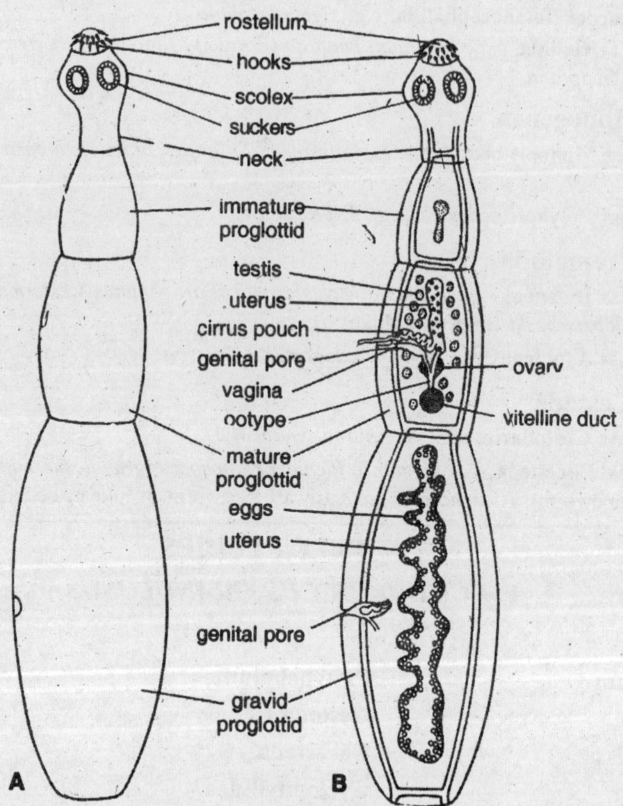

Fig. 23.4: *Echinococcus granulosus.*

Life Cycle

Eggs or gravid proglottids escape with the faeces of the primary host (dog). Inside uterus or on the ground eggs develop into **onchospheres** containing six-hooked embryos or **hexacanths**. They may be ingested by the intermediate hosts; in the intestine of which shell of onchospheres dissolves and releases hexacanths which enter mesenteric veins and make their way into various organs, specially lungs and liver. The young larva changes into a hollow **bladder** containing two layers—outer cuticular **ectocyst**, inner germinal layer or **endocyst**. The host forms an enveloping fibrous **cyst wall**, called **pericyst**. It is now called a **hydatid cyst** (Fig. 23.5).

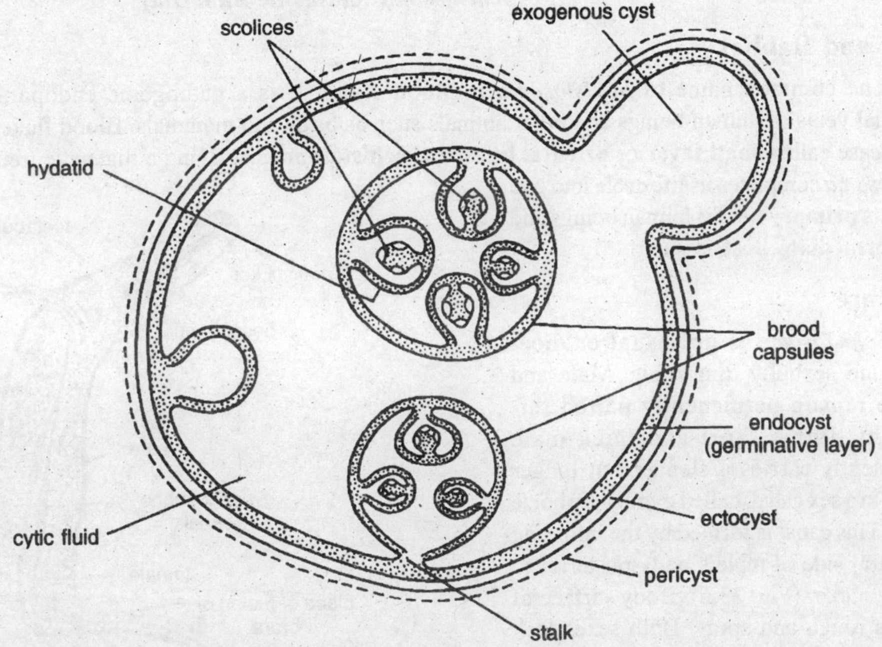

Fig. 23.5. *Echinococcus granulosus.* Hydatid cyst in cross section.

Development of the cyst is slow and after months or even years the bladder wall produces hollow **brood capsules** which remain attached by slender stalks. A cyst often throws out blind outgrowths forming **exogenous buds**. Budding is repeated within the brood capsules giving rise to the daughter buds or **hydatids (endogenous budding)**. Invagination in hydatids gives rise to small scoleces with suckers and hooks. There may be as many as two million scoleces in a single large hydatid. Brood capsules under pressure burst to liberate tiny hydatids in the cystic fluid. Hydatid upon reaching the final host is capable of giving rise to an adult worm. This happens when dogs feed raw flesh of infected sheep and goats around mutton shops.

Human beings receive infection of dog tapeworm from the pet dogs through the tongue of dogs while they are licking the hands of children or the eggs are caught under the fingernails while caressing the pets. Infected meat of sheep and goat is also the source of human infection.

Hydatid cysts (bladder worms) have a colourless toxic fluid which may be up to many litres (one litre in a cyst of 15 cm diameter). Older cysts also have a granular deposit (called **hydatid sand**) consisting of brood capsules and free scolices. They become life-threatening when growing in such places as brain, liver or eye and release their toxic fluid into the host causing severe reaction (e.g., eosilophilia). Bladder worm cysts can be removed only by surgery.

2. *SCHISTOSOMA*

Classification

Phylum	–	**Platyhelminthes**
Class	–	**Trematoda**
Subclass	–	**Digenia**
Order	–	**Prosotoma**
Genus	–	***Schistosoma* (old name *Bilharzia*)**

Habit and Habitat

The common name of *Schistosoma* is **blood fluke**. It is a pathogenic endoparasite of intestinal veins of human beings and other animals such as birds and mammals. Blood fluke causes the disease called **snail fever** or **oriental intestinal schistosomiasis**. Being a digenetic trematode, *Schistosoma* completes its life cycle into two hosts: its **primary host** is human beings and its **intermediate host** is snail.

Structure

Schistosoma is **unisexual** or **dioecious** and sexually dimorphic. Male and female remain permanently paired throughout life. A short and thick male permanently carries a slender but longer female in a sex canal, called **gynaeco-phoric canal**. This canal is formed by the infolding of ventral side of male's body posterior to ventral sucker (Fig. 23.6). Body surface of male is rough and spiny. Both sexes lack pharynx but have an anterior and a ventral sucker; the acetabulum of male is better developed than female. **Male reproductive system** contains four to seven testes which open independently into vasa efferentia, all opening into a common vas deferens. The vas deferens opens to exterior by a gonopore located below the ventral sucker. **Female reproductive system** consists of an elongated ovary (or germarium), vitelline gland (vitellarium), ootype, mehli's glands and short and coiled seminal receptacle. A tubular uterus arises from the ootype.

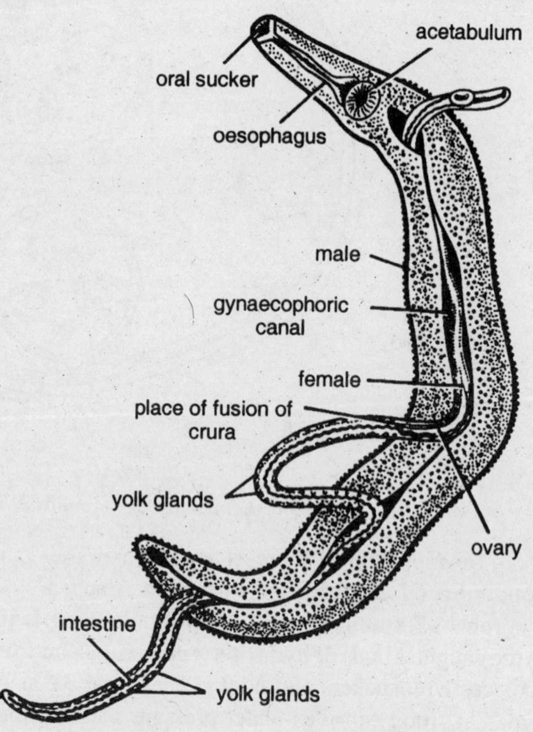

Fig. 23.6. *Schistosoma* (male and female).

> **Box 23.6**
>
> Following three species of *Schistosoma* are parasitic to human beings:
>
> 1. *S. haematobium* is found in Africa, Palestine, Iraq and Portugal in human pelvic veins; its egg capsule has a sharp terminal spine; its intermediate host is *Balinus* (snail).
> 2. *S. mansoni* is found in Africa and tropical America in the veins near the iliocaecal junction, its capsule has a lateral spine (Fig. 23.7); the intermediate host is *Planorbis*.
> 3. *S. japonicum* is found in Japan, China and Philippines in the hepatic portal and mesenteric veins; it is also parasitic in dogs, cats, cattle, horses and pigs; its egg capsule is small containing a rudimentary lateral spine; its intermediate host is *Oncomelania*.

Life Cycle

Life history of *Schistosoma* was worked out first by **Theodor Bilharz** in 1851 and later by **Wall** in 1941. When laying the eggs, the female protrudes from the groove of the male worm and deposits the eggs in small intestine venules of host. Eventually they break through by spines and enzymes into the lumen of the intestine and are released in the faeces of the host. If defecation occurs in the water, the thin walled egg capsules hatch, and the **miracidia** escape. The miracidia larva are ciliated and well supplied with sensory receptors; they seek out a particular species of freshwater snails (see Box 23.6). After penetration in the intermediate host, they transform into **sporocyst**, which eventually gives rise to a second generation of sporocysts. The sporocysts produce **cercariae** without an intermediate radia stage. Cercariae contain forked tails and are called **furocercariae** (Fig. 23.7). The cercariae leave the molluscan host, swim freely in water and on contact with human skin (when walking barefooted in water), penetrate, using enzymes of penetration glands and muscular boring movements. Cercariae develop into tailless young flukes, called **schistosomules** which are carried by the bloodstreams first to lungs, then to the liver, and finally to the veins of intestine or bladder. During this period the schistosomules gradually transform into adults. (**Note.** The eggs of *Schistosoma haematobium* occur in the vein of urinary bladder and they leave the primary host via the bladder and urine.)

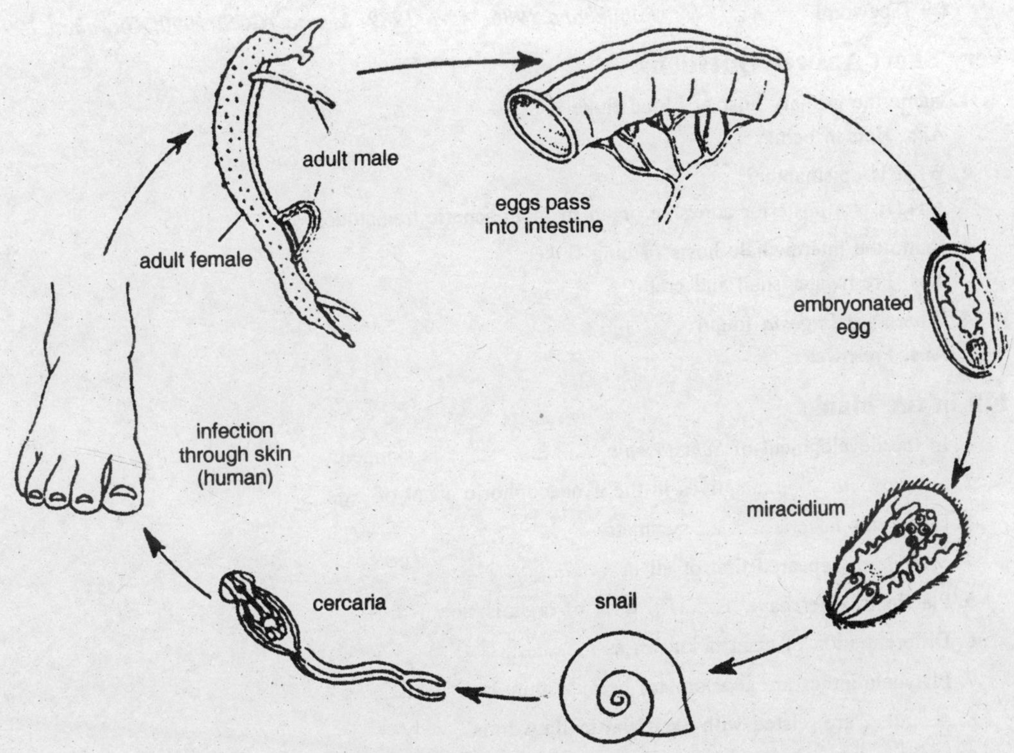

Fig. 23.7. Life cycle of *Schistosoma mansoni*.

Infection of *Schistosoma* causes headache, pain in back and legs, dysentery, daily fever, liver disorders, anaemia, eosinophilia, damage to urinogenital tract (**haematuria** which is a disorder of kidney or urinary track with discharge of blood). The disease can be prevented by sanitary control of water. It can be treated by tartar emetic or sodium antimony tartrate.

QUESTIONS

Long Answer Questions

1. Mention the distinguishing characters of phylum Platyhelminthes. Classify giving characters and examples of different classes.
(Awadh 1987; Agra 1988, 1989; Bundelkhand 1989, 1991; Gorakhpur 1993; Indore 1990, 1993; Jiwaji 1989; Kanpur 1993, 1994; Meerut 1990, 1991; Purvanchal 1988, 1991)
2. Classify phylum Platyhelminthes giving diagnostic characters and examples of each class.

Short Answer Questions

1. Classify the following animals, giving their distinguishing characters:
(*i*) *Planaria* (*Dugesia*)
(Lucknow 1988; Raipur 1990, 1991; Meerut (CCS) 1992; Punjab 1990, 1992)
(*ii*) *Echinococcus granulosus*
(Awadh 1988, 1991; Osmania 1989; Punjab 1992; Guwahati 1994; Allahabad 1994, 1996)
(*iii*) *Schistosoma* (Blood fluke) *(CCSU, Meerut, 1989, 1991; Punjab 1990, 1991; Allahabad 1991, 1993, 1994, 1997; Guwahati 1994)*
(*iv*) *Fasciola* *(Jiwaji 1988, 1989; Kanpur 1989; Agra 1989)*
(*v*) Tapeworm *(Rohilkhand 1986; Agra 1989; Meerut (CCS) 1990; Guwahati 1994)*

Very Short Answer Questions

1. Name the primary host of blood fluke.
Ans. Human being.
2. What is opisthaptor?
Ans. It is a posterior adhesive organ in monogenetic trematodes.
3. Name the intermediate hosts of lung fluke.
Ans. Freshwater snail and crab.
4. Where are *Dugesia* found?
Ans. Freshwater.

Fill in the blanks

1. In the development of *Schistosoma* .... stage is skipped.
2. *Schistosoma* .... lives in the gynaecophoric canal of ....
3. Flatworms have .... symmetry.
4. Mesoderm appeared first of all in ....
5. Platyhelminthes have .... level of organisation.
6. Differentiation of head is known as ....
7. Platyhelminthes are acoelomates because they lack ....
8. .... are related with excretion in flatworms.
9. Ovary in Platyhelminthes produces eggs but not ....
10. Epidermis is present in the class .... of phylum Platyhelminthes.
11. *Dugesia* belongs to the order .... of class ....
12. The most primitive Bilateria are free-living ....
13. The scientific name of lung fluke is ....
14. The turbellarian larva is the .... larva.

Multiple Choice Questions

1. Flatworms are
   (*a*) diploblastic (*b*) triploblastic
   (*c*) monoblastic (*d*) none of these
2. The tissue in Platyhelminthes between viscera and body wall is called
   (*a*) coelom (*b*) parenchyma
   (*c*) mesoderm (*d*) choanoderm
3. Flame cells help in
   (*a*) excretion (*b*) digestion
   (*c*) respiration (*d*) reproduction
4. Coelom in flatworms is
   (*a*) well developed (*b*) poorly developed
   (*c*) schizocoel (*d*) absent
5. Sucker is absent in
   (*a*) Turbellaria (*b*) Trematoda
   (*c*) Cestoda (*d*) all of these
6. *Schistosoma* is a parasite found in
   (*a*) lungs (*b*) intestine
   (*c*) liver (*d*) blood
7. Alimentary canal is absent in the animals of class
   (*a*) cestoda (*b*) apoda
   (*c*) gastropoda (*d*) arachnida
8. *Echinococcus* is generally known as
   (*a*) pig tapeworm (*b*) sheep tapeworm
   (*c*) human tapeworm (*d*) dog tapeworm
9. Flatworms are characterised by presence of
   (*a*) protonephridia (*b*) renette cells
   (*c*) nephridia (*d*) all of these
10. *Polystomum* is endoparasite of
    (*a*) frog (*b*) fish
    (*c*) toad (*d*) all of these
11. Opisthaptor is
    (*a*) anterior sucker (*b*) posterior sucker
    (*c*) hook (*d*) spine
12. Rhabditis are present in the epidermal cells of
    (*a*) *Dugesia* (*b*) *Taenia*
    (*c*) *Fasciola* (*d*) *Schistosoma*
13. Parasitic animals have well developed
    (*a*) respiratory system (*b*) reproductive system
    (*c*) digestive system (*d*) nervous system
14. *Paragonimum* lives in
    (*a*) intestine (*b*) blood
    (*c*) lungs (*d*) brain

15. Cysticercus larva is formed in the development of
    (a) *Dugesia* (b) *Fasciola*
    (c) *Enterobius* (d) *Taenia*

ANSWERS

Fill in the Blanks

1. Redia
2. Female, male
3. Bilateral
4. Platyhelminthes or flatworms
5. Organ-system
6. Cephalisation
7. Coelom
8. Flame cells
9. Yolk
10. *Turbellaria*
11. Tricladida, Turbellaria
12. Platyhelminthes or flatworms
13. *Paragonimus westermani*
14. Muller's

Multiple Choice Questions

1. (*b*)	**2.** (*b*)	**3.** (*a*)	**4.** (*d*)
5. (*a*)	**6.** (d)	**7.** (*a*)	**8.** (*d*)
9. (*a*)	**10.** (*d*)	**11.** (*b*)	**12.** (*a*)
13. (*b*)	**14.** (*c*)	**15.** (*d*).	

24

Dugesia tigrina (Phylum Platyhelminthes)

24.1 CLASSIFICATION

Phylum	–	**Platyhelminthes**
Class	–	**Turbellaria**
Order	–	**Tricladida**
Suborder	–	**Pauldicola**
Genus	–	***Dugesia (Euplanaria)***
Species	–	***tigrina***

24.2 HABIT AND HABITAT

Dugesia lives in damp surroundings as its body is not protected against desiccation. It is found commonly in freshwater ponds, lakes and streams and shallow rivers of cold running water. *Dugesia* prefers to keep to the bottom, where it is found resting or gliding on the undersurface of stones, logs and on aquatic leaves. It is **gregarious** (lives in groups), carnivorous and cannibalistic. *Dugesia* is worldwide in distribution.

24.3 STRUCTURE

1. Shape, Size and Colour

Dugesia has a distinct bilateral symmetry. Its body is small, thin, highly flattened, elongated, leaf-like, with a broader anterior end and pointed posterior end. It measures about 1.5 cm in length. Body wall of *Dugesia* is slimy and somewhat transparent, through which some of the internal organs are visible. Colour of the animal is dull, being greyish, brownish or blackish. Ventral surface of the animal is lighter than the dorsal side. Dorsal surface is darker, bearing spots or streaks, which accounts for the species name *Dugesia tigrina*.

2. Head End

The body of *Dugesia* is elongated with dorsal surface slightly arched and the ventral surface flat. Anterior end of body is differentiated into a broad, blunt and triangular **head**. Dorsally, the head bears a pair of dark spots, the **eyes**, and laterally on either side a small projection, the head lobe or **auricle**. Behind auricles, body is somewhat constricted to form the so-called **neck**, beyond which it gradually tapers towards the posterior end.

Behind the middle of the body on the ventral side is a **mouth** which leads into pharyngeal sheath containing a cylindrical **pharynx**. The pharynx can be everted through a mouth as a **proboscis**. Just a little behind the mouth aperture on the ventral surface, there is a **genital pore** or **gonopore** in sexually mature forms. Entire ventral surface of *Dugesia* is covered by fine, hair-like, locomotory structures, called **cilia**. However, a narrow strip along the margin of ventral surface is non-ciliated **adhesive zone**. Through it, an adhesive substance comes out from the glands. The animal clamps itself to the substratum by the adhesive zone.

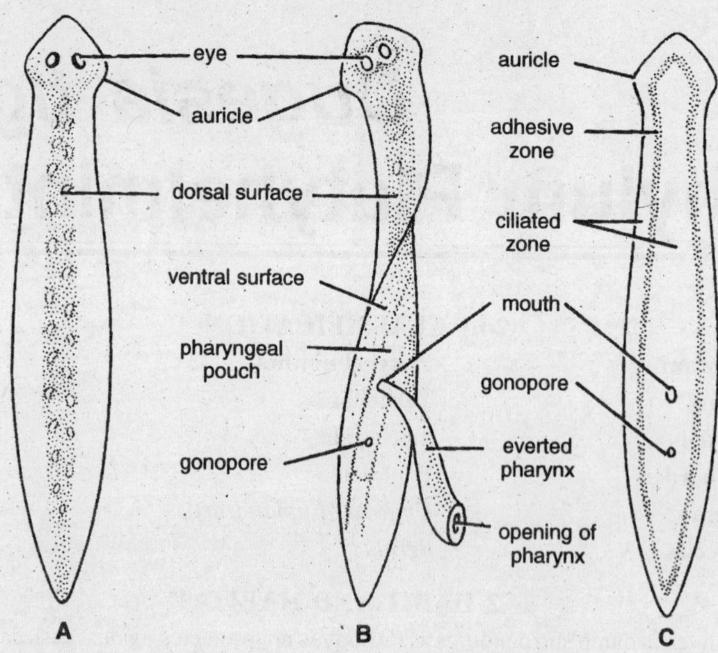

Fig. 24.1 *Dugesia*. External features. A—Dorsal surface; B—Body twisted to show a part of ventral surface; C—Ventral surface.

24.4 BODY WALL

The body wall of *Dugesia* is made up of an outer **epidermal** and inner **muscle** layers. Both of these layers are separated by a **basement membrane**. The space between muscle layer and the alimentary canal (and other internal organs) is filled with special type of tissue called **mesenchyme** or **parenchyma**. Thus, no coelom or body cavity is found in this animal.

1. Epidermis. It is single cell-layered thick and made up of cuboidal epithelial cells with rounded nuclei, situated toward the base. **Sensory cells** also occur scattered in epidermis.

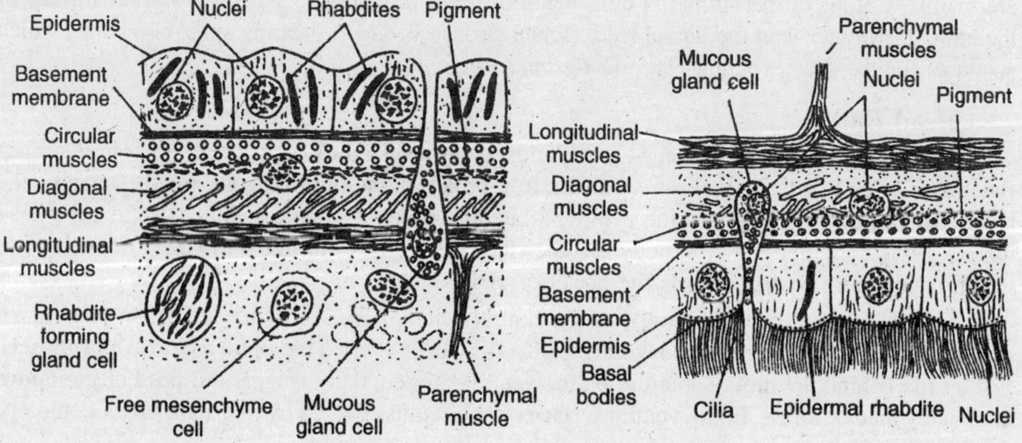

Fig. 24.2. *Dugesia.* V.L.S. dorsal body wall.

Fig. 24.3. *Dugesia.* V.L.S. ventral body wall.

(*a*) Cilia. Epidermal cells of ventral surface are richly ciliated; those of dorsal surface have no cilia. Cilia are also absent in the adhesive zone of ventral surface. Each cilium arises from a rounded body, the **basal granule**, lying below the plasma membrane of each epidermal cell.

(*b*) Rhabdites. Several rod-shaped, hyaline bodies, the **rhabdites**, occur in epidermal cells. They are more abundant on dorsal side and are completely absent in the adhesive zone and sensory cells. Presence of rhabdites is a distinctive feature of turbellarian epidermis. Rhabdites are produced by special **rhabdites forming gland cells**, situated in epidermis or mesenchyme. When epidermis is irritated, the rhabdites discharge in water to form a thick, adhesive layer around the body protecting it from desiccation and from the possible predators. Secretions of discharged rhabdites are also believed to help in obtaining living food by this planarian.

(*c*) Subepidermal glands. Numerous unicellular **mucous gland cells** and **adhesive glands** of epidermal origin lie deep in the mesenchyme. They are provided with narrow ducts or necks the opening of which exist between epidermal cells. Mucous gland cells are more abundant on the ventral surface in anterior part of body. They secrete a **slime trail** for locomotion. Adhesive gland cells produce a sticky secretion which helps in attachment of body and eggs to the substratum also in capturing the prey.

(*d*) Pigments. Below the epidermis is granules and rods of pigments.

2. Basement membrane. Beneath the epidermis is a collagen-rich basement membrane, which is often thick enough to lend some structural support to the body.

3. Muscle layer. This layer lies below the basement membrane. It consists of elongated contractile **muscle cells** produced by special cells, called **myoblasts**. The myoblasts are of mesodermal in origin. The muscle layer is differentiated into an outer layer of **circular muscles**, middle layer of **diagonal muscles** and inner layer of **longitudinal muscles**. There are present two series of diagonal muscle fibers which run at right angles to each other. The longitudinal muscle layer is more developed on the ventral surface. Extending across the body between dorsal and ventral surfaces are the **dorso-ventral muscles.**

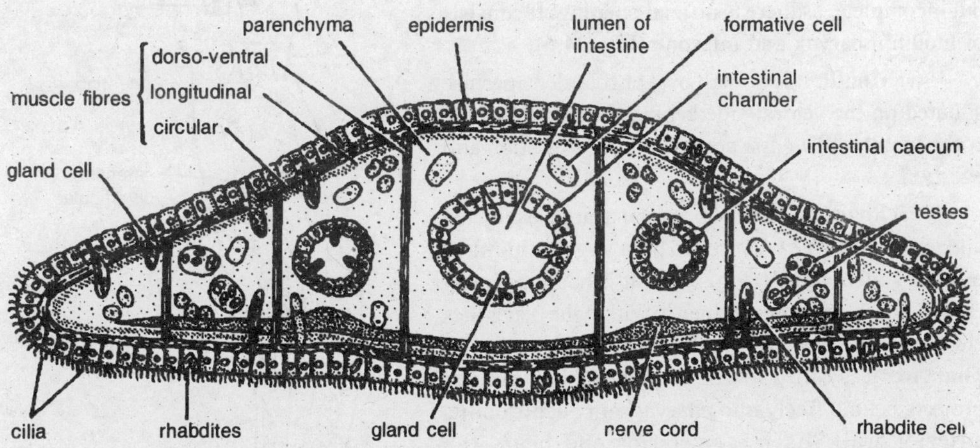

Fig. 24.4. *Dugesia.* Diagrammatic transverse section through the body.

4. Mesenchyme or parenchyma. It includes a variety of loose and fixed cells, muscle fibers, and connective tissue. Parenchyma is mesodermal in origin and is characteristic of platyhelminthes. Parenchymal cells are vacuolated, syncytial and have irregular processes, which form a loose network with fluid-filled intercellular spaces. These cells serve to transport digested food and excretory materials.

Parenchyma (mesenchyme) thus performs the role of circulatory system in which the animal lacks. In mesenchyme lie numerous free, amoeboid **formative cells** or **neoblasts**, which rush to injured areas and bring about regeneration. Various **gland cells** (which are ectodermal in origin) are also seated in the mesenchyme.

24.5 LOCOMOTION

Dugesia is aquatic but it does not swim. It always moves in contact with some substratum or on the underside of water surface film. *Dugesia* moves in two ways: 1. Gliding or ciliary movement and 2. Muscular movement.

1. Gliding or ciliary movement. This type of locomotion is performed by the help of cilia. A thin layer of mucous is first poured on the substratum. Cilia of ventral side then beat backwards in the mucous layer resulting in forward movement of the body. Head is kept slightly raised during gliding.

2. Muscular movements. These are responsible for the bending and twisting of the body in different forms and for crawling.

During crawling, the body of *Dugesia* first elongates by contraction of *circular muscles* and the head adheres to some substratum with the help of secretion of adhesive glands. This is followed by waves of contraction of *longitudinal muscles*, passing from anterior to posterior end, alternately along the right and left sides, results in the forward movement of body in a wavy way. Contraction of *diagonal* (*or oblique*) *muscles* stiffens the worm by increasing its internal turgor. By means of diagonal muscles *Dugesia* can change its direction.

24.6 DIGESTIVE SYSTEM

It includes the alimentary canal, food, feeding and digestion.

1. Alimentary Canal

Alimentary canal of *Dugesia* is distinct, extensive but incomplete, as there is no anal opening. It consists of mouth, pharynx and intestine (Fig. 24.5).

(*i*) Mouth. It is a small oval or rounded aperture situated on the ventral side behind the middle of the body. Mouth of *Dugesia* serves both for ingestion and egestion.

(*ii*) Pharynx. Mouth opens into a large, elongated and non-muscular cavity, called **pharyngeal chamber** or **pharyngeal pouch** (Fig. 24.6). It is bounded by an epithelium or **pharyngeal sheath**. Within the pouch lies elongated, thick-walled cylindrical and highly muscular **pharynx**. It is called **folded** or **plicate pharynx** and projects behind freely into pharyngeal pouch from its anterior wall. Its free posterior end bears the **pharyngeal opening**, which lies a little in front of mouth openings. The short space between these two opening is called the **buccal cavity**. While feeding, pharynx of *Dugesia* can be protruded through mouth considerably to form a **proboscis** for enveloping the prey (Fig. 24.6).

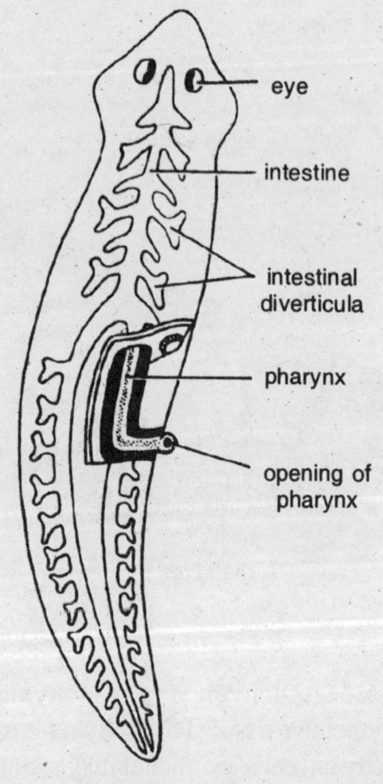

Fig. 24.5. *Dugesia*. Digestive system.

The pharyngeal wall consists of outer and inner epithelial layers, between which lie muscle layers, gland cells and nerve plexuses. The inner epithelial layer is ectodermal.

***(iii)* Intestine.** Anteriorly, pharynx leads into **intestine** which immediately forms three main branches, one *median* and two *lateral*. Median branch immediately extends to the anterior end, and each lateral branch immediately bends backwards and extends to the posterior end. This type of *three-forked intestine* is a characteristic feature of the turbellarian order Tricladida.

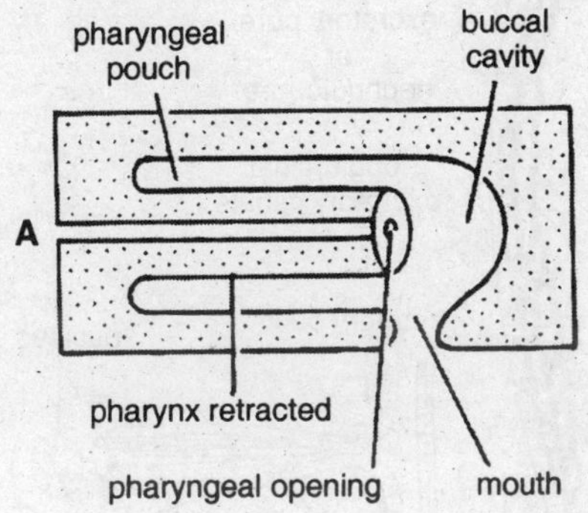

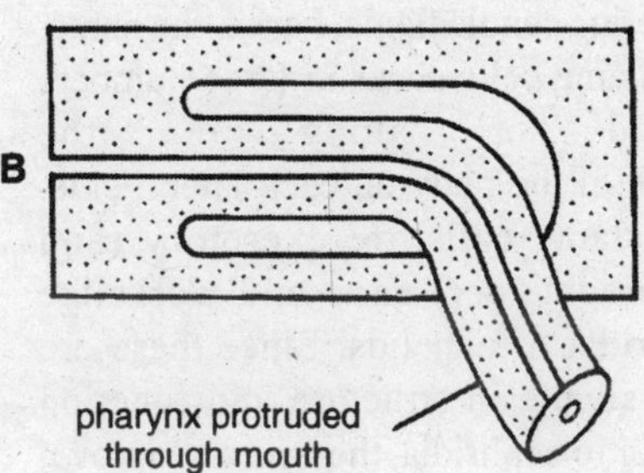

Fig. 24.6. *Dugesia.* Diagram showing retracted pharynx (A) and everted pharynx (B).

Each main branch of intestine gives out numerous lateral branches, the **diverticula** or **caeca**, which further branch and ramify. The main branches and their diverticula end blindly in solid mesenchyme (parenchyma). The much branched intestine is a means of increasing the surface area for digestion, absorption and distribution of food.

The intestine is made up of a single layer of vacuolated columnar cells containing granules (Fig. 24.4). Between the columnar cells are some gland cells of triangular shape containing reserve proteins. This layer of columnar cells is lined externally by a delicate muscle layer.

Food, feeding (ingestion) and digestion. *Dugesia* is **carnivorous**. Its food consists of small living worms, crustaceans (*e.g.*, *Daphnia*) and snails and pieces of larger dead animals. In feeding, the animal perceives the presence of food at a distance and moves towards it. Living food is often entangled in the slimy secretions of mucous glands and rhabdites, then it encloses the food in the everted pharynx and digestive juices are poured over the food, the food is broken up by pumping action of pharynx and acted upon by extruded digestive juices for extracellular digestion, after which the food is swallowed. Thus, in *Dugesia* **digestion** is both **extracellular** and **intracellular**. The mesenchyme helps to distribute digested food. Undigested food is **egested** through the mouth.

Dugesia and other planarians can live without food for long periods, they obtain nourishment by dissolving their reproductive organs, parenchyma and muscles, they get smaller in size. The missing parts (during starvation time) are regenerated when they feed again.

24.7 RESPIRATORY SYSTEM

There is no special organ for respiration. Exchange of gases takes place through the general body surface by **diffusion**.

24.8 EXCRETORY SYSTEM

Excretory system of *Dugesia* consists of numerous excretory cells, the **protonephridia** or **flame cells** and a system of **excretory tubules** (Fig. 24.7).

1. Flame cells. The flame cell is nucleated and has a number of protoplasmic branched processes reaching into the parenchyma (mesenchyme). Each flame cell has an **intracellular space** which is continued into the capillary. In the lumen of the flame cell are a number of **flagella** or **cilia** which in living animals vibrate giving the appearance of a flickering candle flame, hence, the name. The flame cells occur in large numbers placed along the length of the body on each lateral side. The flame cells are excretory (and osmoregulatory) units and work like nephridia of annelids. Since these are very simple in structure and function similar to nephridia, they are also known as **protonephridia.**

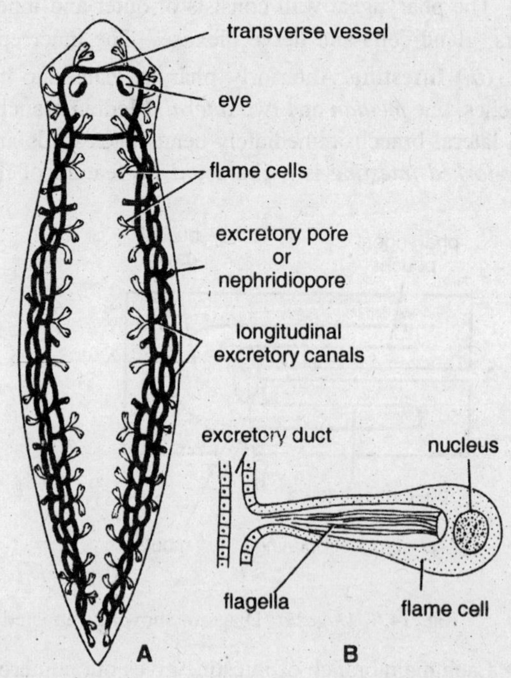

Fig. 24.7. *Dugesia.* A—Excretory system; B—A flame cell.

2. Excretory tubules. The bulbous cavity of the flame cell narrows down into a fine vessel, or **capillary duct,** which in turn opens into a lateral **longitudinal excretory duct** or **canal**. There are two such longitudinal excretory canals on each side. They extend along the entire length of body, forming a network. Longitudinal excretory canals of right and left sides are united anteriorly, in front of eyes, by a **transverse vessel.** Each longitudinal canal opens to the exterior through a number of minute **excretory pores** or **nephridiopore**, situated on the dorsal surface. Walls of lateral excretory canals are also provided with bundles of long flagella forming **lateral flames** which keep the fluids moving through them.

Physiology. Excretory substance (ammonia) along with salts and sugars, dissolved in water, is collected from the parenchyma and is transferred into the cavities of flame cells causes hydrostatic pressure by which fluid waste passes into longitudinal excretory canals. Walls of canals reabsorb useful substances, while excretory substances along with excess water, are expelled out through the nephridiopores. Much of the excretory waste is also expelled through the mouth.

Osmoregulation. Main function of flame cells or excretory system in planarians is osmoregulation. Since, it has been observed that flame cells or protonephridia are well developed in freshwater species but less developed or absent in marine species. The excretory system of *Dugesia* evidently brings about elimination of excess water, ions and salts from the animal.

24.9 NERVOUS SYSTEM

The nervous system of *Dugesia* represents the primitive type of centralised nervous system of higher animals. It consists of **brain, nerve cords** and **peripheral nerves** (Fig. 24.8).

1. Brain. At the anterior end, below epidermis and a little behind the eyes, is a prominent bilobed mass of nervous tissue, called **brain** or **cerebral ganglia**. Two lobes of brain are connected by several transverse fibers

2. Nerve cords and peripheral nerves. From the brain two **lateral longitudinal nerve cords** extend to the posterior end along the ventral side of body. Each nerve cord is in fact a strand of concentrated nerve cells.

From each lobe of brain arise numerous **periphe-ral nerves** which supply various sense organs located at the anterior end (*i.e.*, head and auricles). Each nerve cord also gives out numerous branched nerves on both the sides all along its length. Those arising from the inner side frequently anastomose with similar ones from the other nerve cord to form **transverse commissures** or **connectives**. This is also called **ladder type** of nervous system.

In addition to the centralised nervous system, *Dugesia* also possesses a **sub-epidermal** plexus (or **nerve net**) just below the epidermis like that of cnidarians. **Sub-muscular plexus** exists in the mesenchyme below the muscle layers of body wall. Both plexuses are joined to the nerve cords. Brain receives stimuli from the sense organs and conveys them to different parts of body. The nervous system of *Dugesia* acts as a coordinating center for nerve impulses.

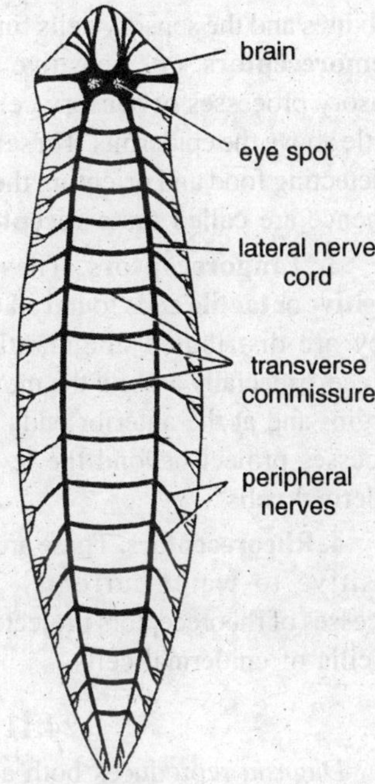

Fig. 24.8. *Dugesia.* Nervous system

24.10 SENSE ORGANS

In *Dugesia*, sense organs or receptors include **photoreceptors** (a pair of eyes or ocelli), **chemoreceptors, tangoreceptors** and **rheoreceptors**. Most of these receptors lie concentrated at the anterior end of body (head), which remains foremost during locomotion.

1. Eyes or ocelli. Eyes of *Dugesia* are a pair of conspicuous, black, rounded spots situated dorsally on head. Each eye has a cup-like form with its opening directed anterolaterally. The cup is lined by a layer of **pigment cells**. Projecting into the pigment cup are numerous bipolar neurosensory light-receptive **photosensitive** or **retinal cells**. This type of eye is called **inverse eye** and in it, the light has first to pass through photosensitive cells before reaching the pigment cells. The pigment cup serves as a shield. Expanded distal end of each retinal cell projecting into cup bears **striated border** of unknown function. Proximal end of these cells runs side by side forming the so-called nerve fibers which communicate with the brain.

Dugesia like other free-living turbellarians is negatively phototactic. The dorsal placement of the eyes and the orientation of the pigment cups facilitates the detection of light direction as well as intensity.

2. Chemoreceptors. Anterior end of body bears several grooves and pit-like depressions. They are lined by ciliated epidermal cells. A pair of whitish grooves, situated on auricles, are known as **auricular organs.** Ciliated pits, grooves and auricular organs are devoid of gland cells and

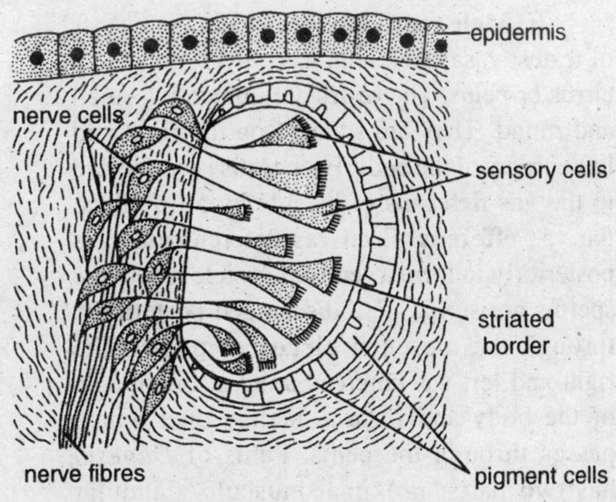

Fig. 24.9. *Dugesia.* V.S. of eye.

rhabdites and the sensory cells linking them are **chemoreceptors**, *i.e.*, sensitive to chemicals. Sensory processes of sensory cells project just a little above the epidermis. These receptors help in detecting food and orienting the body toward it, hence are called **taste receptors.**

3. Tangoreceptors. These are **touch-sensitive** or **tactile cells** found all over the body. They are distributed abundantly on ventral surface, especially around the mouth, on lateral margins and at the anterior end. Their sensory processes project beyond the level of cilia of epidermal cells.

4. Rheoreceptors. These are sensory cells sensitive to water currents. The sensory processes of rheoreceptors project much beyond the cilia of epidermal cells.

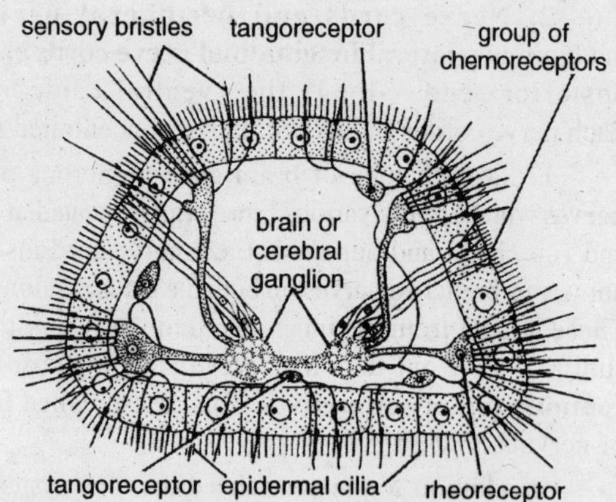

Fig. 24.10 *Dugesia.* Digrammatic T.S. of anterior end to show various receptors.

24.11 REPRODUCTIVE SYSTEM

Dugesia reproduces both asexually and sexually.

1. Asexual Reproduction

In *Dugesia*, asexual reproduction occurs by **transverse binary fission** (Fig. 24.11).

Fission occurs by the appearance of a constriction behind the pharynx. It gradually deepens and finally divides the animal into two parts, each of which regenerates the wanting structures. Thus, the separated anterior part regenerates the posterior region and rear and grows into complete planarian. Locomotion and adhesion are found essential for fission

2. Sexual Reproduction

In *Dugesia*, both male and female reproductive organs are present in the same individual. Thus, it is **hermaphrodite, monecious** or **bisexual.**

(*i*) Male reproductive organs. It consists of testes, vasa efferentia, vasa deferentia and a cirrus or penis. The **testes** are numerous, small and round. They are situated on the right and left borders of the body. Each testis is connected to the **vas deferens** of its side by a tiny duct, the **vas efferens**. Each vas deferens enlarges posteriorly to form a **seminal vesicle** where the sperms are stored until discharged posteriorly, through the muscular **cirrus** or **penis**. The right and left vas deferens unite at the middle of the body and form a median duct which passes through the penis. Penis of *Dugesia* has two parts: proximal musculo-glandular part, the **penis bulb** and distal conical parts the **penis papilla.** The penis bulb is surrounded

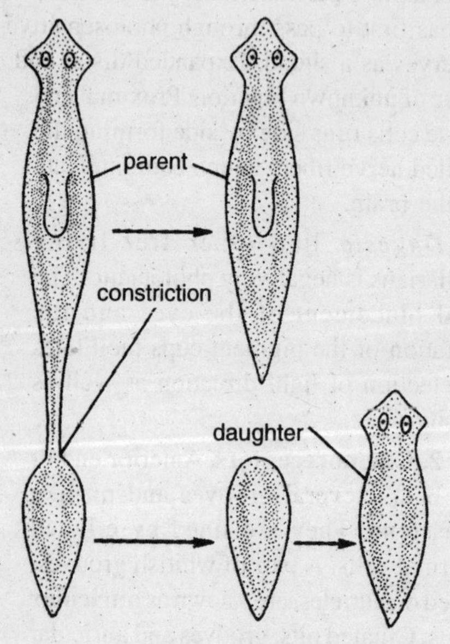

Fig. 24.11. *Dugesia.* Asexual reproduction by fission.

by numerous unicellular **prostate glands**, opening into its lumen formed by two **bulbar cavities** (Fig. 24.13). The **lumen** or **median canal** of the penis opens outside through common **gonopore**. During copulation, penis papilla extends through the gonopore. **Sperms** are long, filamentous and each provides with two long flagella.

(*ii*) **Female reproductive system.** It consists of ovaries, oviducts, vagina, uterus (or bursa copulatrix) and yolk glands (or vitelline glands). **Ovaries** are a pair of rounded bodies situated at the anterior end, one behind each eye. Posteriorly, each ovary is connected with a narrow elongated tubule, the **oviduct** running parallel and dorsal to the nerve cord of its side. Posterior ends of two oviducts unite to form a **median vagina**. Vagina opens into the genital atrium lying in front of it. Each oviduct, all along its length, is surrounded by numerous follicular **yolk glands** which open into it. It is for this reason that the oviduct is also called **ovo-vitelline duct**. Numerous minute **cement glands** surround as well as open into vagina and genital atrium.

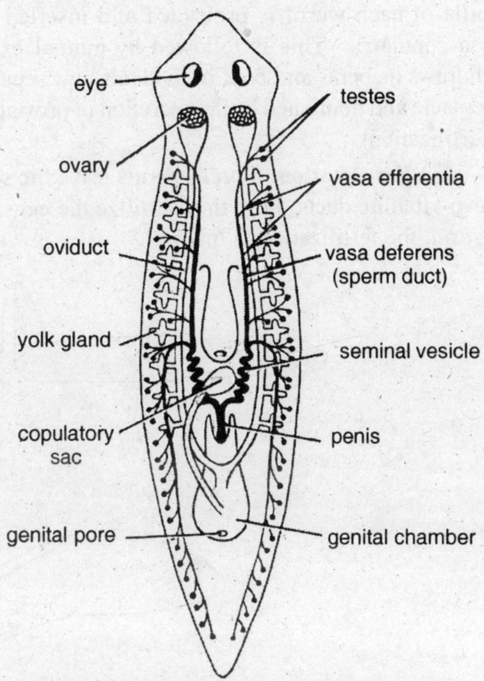

Fig. 24.12. *Dugesia*. Reproductive system.

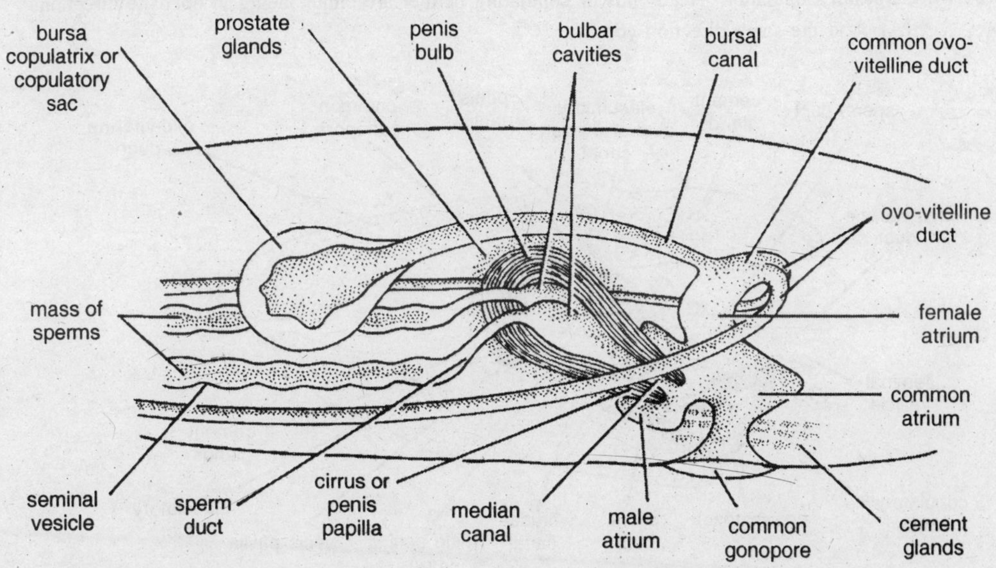

Fig. 24.13. *Dugesia*. A part of reproductive system in the lateral view.

(*iii*) **Copulation.** *Dugesia* are hermaphrodite, but self-fertilization does not occur. Instead they perform **cross-fertilization** preceded by **copulation**. During copulation, two worms come in temporary contact along their ventral surfaces with genital apertures (gonopores) opposing each other. Their head ends may face in the same or in opposite directions (Fig. 24.14). Penis papilla of each worm is protruded and inserted into the gonopore of other worm to reach its bursa copulatrix. This is followed by mutual exchange of sperms, after which each flatworm withdraws its penis and both individuals now separate. Sperms thus received are stored in seminal receptacle and nourished by the secretion of prostate glands which keeps them active for the process of fertilization.

***(iv)* Fertilization.** Stored sperms leave the seminal receptacle and migrate into the oviducts or ovo-vitelline ducts. Here they fertilize the eggs, as they are released from the ovaries. Thus, in *Dugesia*, the fertilization is internal.

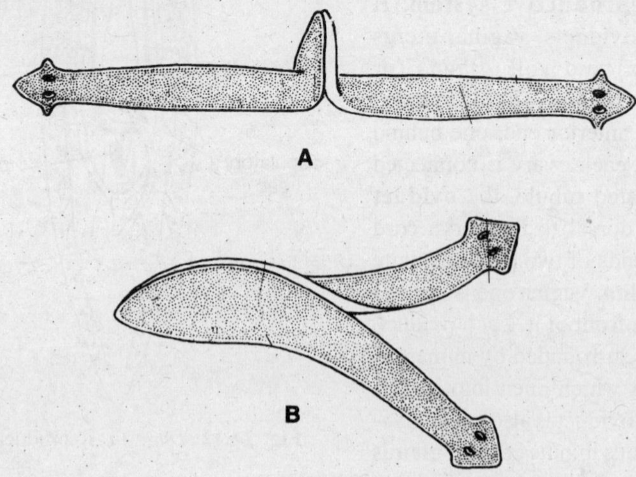

Fig. 24.14. *Dugesia*. Copulation Head ends of copulating partner are either facing in opposite directions (A) or in the same direction (B).

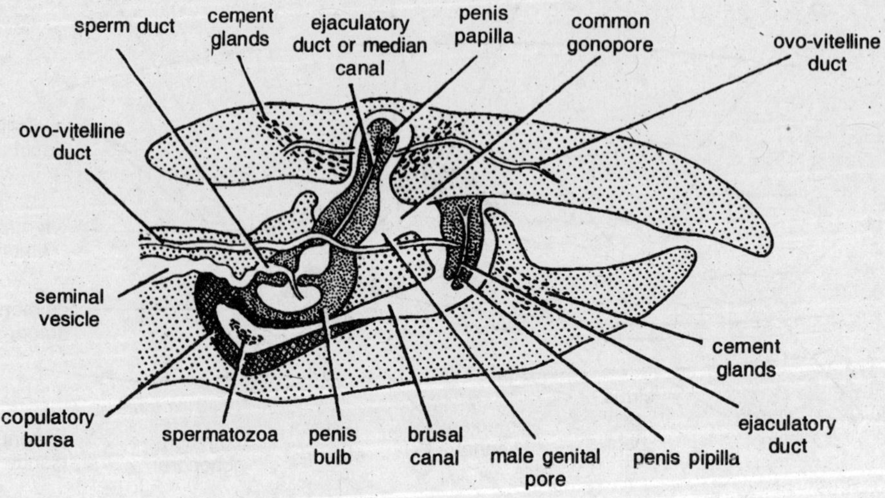

Fig. 24.15. *Dugesia*. A copulative pair in section.

***(v)* Capsule formation.** Fertilised eggs or zygotes move backwards through ovo-vitelline ducts along with yolk-laden **yolk cells** released by the yolk glands. In the genital atrium, about ten zygotes and thousands of yolk cells get enclosed in a proteinaceous shell called **cocoon** or **egg capsule**.

The capsule is spherical and is provided with a little adhesive stalk secreted by cement glands. One such egg capsule is formed after every few days in a planarian; copulation is frequent in them during the breeding season. Capsules are laid on the undersurface of some submerged object, such as aquatic weed, stone, etc., to which each egg capsule is remained attached by a short adhesive stalk.

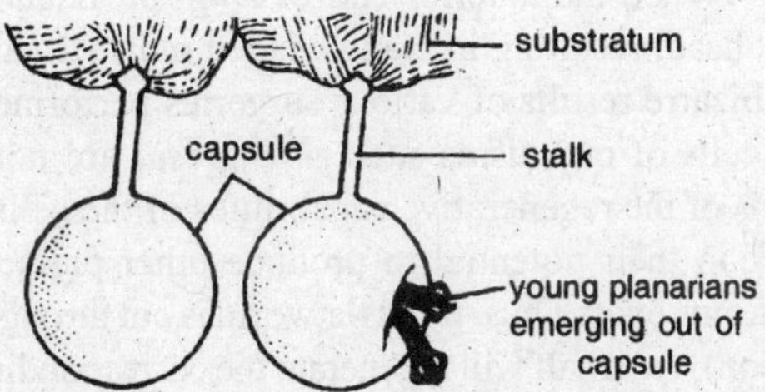

Fig. 24.16. *Dugesia*. Young planarians hatching from an egg capsule.

Development. Embryonic development of egg of *Dugesia* starts soon after the egg capsule is laid and is completed in two or three weeks. It is **direct**, *i.e.*, without the occurrence of any larval stage. Cleavage is **spiral** and **determinate**, *i.e.*, future of each blastomere is pre-determined. Each zygote develops into young one (juvenile). Yolk cells present in the egg capsule tend to provide food to the developing young ones. On completion of embryonic development, the egg capsule ruptures, probably along a definite line, and the young planarians or **juveniles** glide out of it. A newly-hatched planarian lacks reproductive organs and smaller in size than the parents.

In *Dugesia*, the reproductive organs degenerate and disappear after each mating, after which the worms reproduce asexually by fission. However, they develop again during the next breeding season.

24.12 REGENERATION

Dugesia and other planarians possess a tremendous power of regeneration. If cut across into two, three or more parts, each part regenerates into a complete and normal individual (Fig. 24.17A). Regeneration, thus, involves two complementary processes, *viz.*, **epimorphosis** in which the missing

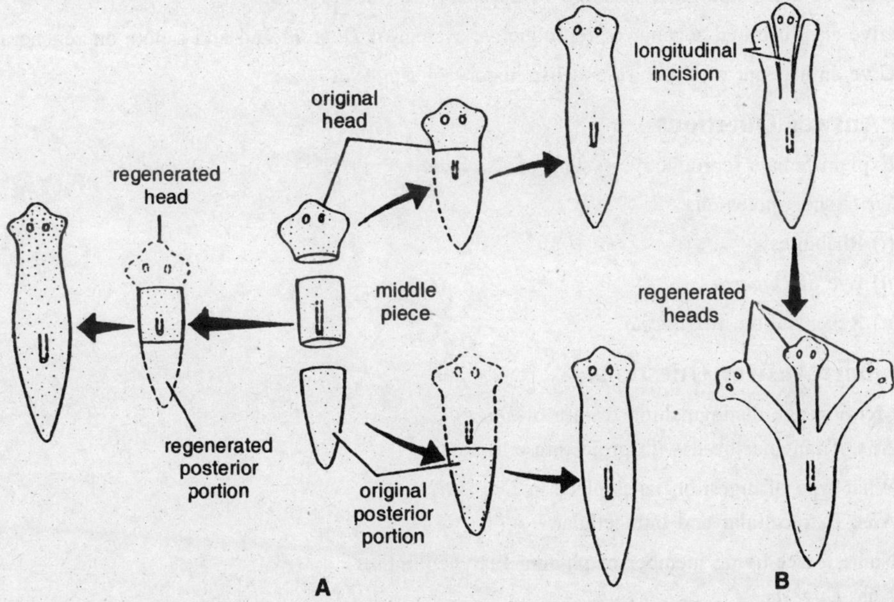

Fig. 24.17. *Dugesia*. Regeneration. A—Three individuals regenerate from an individual cut into three parts. B—Formation of a heteromorph with three heads.

parts are formed and **morphollaxis**, in which the original parts are fit to function with regenerated parts are in the new individual.

When the anterior end of *Dugesia* is cut into two or more parts, each part develops into a new head, resulting in many headed monster called **heteromorph** (Fig. 24.17B). Underlying all of the bizarre results of various surgeries performed on these animals (*i.e.*, planarians) is the fact that the cells of organisms such as *Dugesia* are not totipotent; an anterior-posterior polarity exists in terms of the regenerative capabilities of the cells. However, the cells in the midbody region are less fixed in their potential to produce other parts of the body than are those toward the anterior or posterior ends. Thus, if the flatworm is cut through the middle of the body (as it is in normal transverse fission), each half will regenerate the corresponding lost part. However, if the animal is cut transversely near one end — say, separating a smaller piece of the tail from the rest of the body — the larger piece will grow a new posterior end, but the piece of tail lacks the capability to produce an entire new anterior end. This gradient of cell potency has been of particular interest to cell biologists and medical researchers because of its relevance to healing and regeneration potential in higher animals (see **Brusca** and **Brusca**, 2003).

Recently developmental biologists have found that there is a stock of undifferentiated cells, called **neoblasts** which have remained in their embryonic state and appears to be omnipotent or totipotent cells. During the early phases of regeneration the neoblasts divide actively by mitosis and migrate toward the zones of injury to form the **regeneration blastema**. From this blastema the organs are differentiated. The neoblasts are capable of being transformed into somatic as well as germinal cells (see *Chordate Embryology* by **Verma** and **Agarwal**, 2009).

QUESTIONS

Long Answer Questions

1. Give an account of the structure, physiology and reproduction in *Dugesia*.
2. Describe the histological structure of the body wall of *Dugesia*.
3. Give an illustrated account of reproductive system of *Dugesia* and add a note on regeneration.
4. Give an account of structure and life history of *Dugesia*.

Short Answer Questions

1. Explain female reproductive system of *Dugesia*.
2. Write short notes on:
   (*i*) Rhabdites
   (*ii*) Eye of *Dugesia*
   (*iii*) Regeneration in *Dugesia*.

Very Short Answer Questions

1. Mention one distinguishing feature of *Dugesia*.
   Ans. Triangular head with projecting auricles.
2. What type of digestion takes place in *Dugesia*?
   Ans. Extracellular and intracellular.
3. Name a free-living member of phylum Platyhelminthes.
   Ans. *Dugesia*.
4. What are the structures that produce mucous in *Dugesia*?
   Ans. Mucous glands.
5. What is epimorphosis?
   Ans. Regeneration of a part or organism involving extensive cell proliferation followed by differentiation, *e.g.*, regeneration of amphibian limb.

6. What is morphalaxis?

Ans. Regeneration of a part or organism from a fragment by reorganization without cell proliferation, *e.g.*, regeneration of new individuals from body pieces in *Hydra*, *Planarians*, etc.

Fill in the Blanks

1. *Planaria* (*Dugesia*) lives in ....

2. In *Dugesia*, flame cells are units of ....

3. When a planarian is cut into two pieces, each piece develops into a separate adult. This process is called ....

Multiple Choice Questions

1. Platyhelminthes which lives in freshwaters is

(*a*) *Fasciola* (*b*) *Schistosoma*
(*c*) *Dugesia* (*d*) *Taenia*

2. In the platyhelminthes, ciliated epidermis is found in

(*a*) *Dugesia* (*b*) *Fasciola*
(c) *Taenia* (*d*) *Schistosoma*

3. Cephalisation starts from

(*a*) Porifera (*b*) Cnidaria
(*c*) Ctenophora (*d*) Planaria

4. Rhabdites occur in epidermis of

(*a*) Turbillarian (*b*) Trematoda
(*c*) Cestoda (*d*) all the above

5. *Dugesia* is

(*a*) carnivorous (*b*) omnivorous
(*c*) herbivorous (*d*) insectivorous

6. In *Dugesia*, digestion is

(*a*) intracellular only (*b*) extracellular only
(*c*) both extracellular and intracellular (*d*) intercellular

7. In *Dugesia*, respiration takes place by

(*a*) gills (*b*) general body surface
(*c*) lungs (*d*) none of these

8. The excretory organs of *Dugesia* are

(*a*) Malpighian tubules (*b*) nephridia
(*c*) green glands (*d*) flame cells

9. In *Dugesia*, tangoreceptors are sensory cells which provide stimuli for

(*a*) touch (*b*) smell
(*c*) taste (*d*) none of these

10. Cocoon formation occurs in

(*a*) *Fasciola* (*b*) *Taenia*
(*c*) *Schistosoma* (*d*) *Dugesia*

11. Asexual reproduction in *Dugesia* takes place by

(*a*) fission (*b*) fragmentation
(*c*) sporulation (*d*) budding

12. Planarians have extra power of regeneration due to the presence of
    (*a*) parenchyma (*b*) rhabdites
    (*c*) neoblasts or formative cells (*d*) interstitial cells

ANSWERS

Fiil in the Blanks

1. Freshwater 2. Excretory system 3. Regeneration

Multiple Choice Questions

1. (*c*)	2. (*a*)	3. (*d*)	4. (*a*)
5. (*a*)	6. (*c*)	7. (*b*)	8. (*d*)
9. (*a*)	10. (*d*)	11. (*a*)	12. (*c*).

25

Fasciola hepatica (Liver fluke) (Phylum Platyhelminthes)

25.1 CLASSIFICATION

Phylum	–	**Platyhelminthes**
Class	–	**Trematoda**
Order	–	**Digenea**
Genus	–	***Fasciola***
Species	–	***hepatica***

25.2 HABIT AND HABITAT

Fasciola hepatica lives as an endoparasite in the bile ducts of sheep. *Fasciola* is commonly called **liver fluke** on account of its leaf-like flattened shape and its habit to live in liver of the host. Besides sheep, it has been reported from the domestic animals such as buffaloes, goat, ass, horse, etc., and results in heavy mortality in them. *Fasciola* completes a part of its life cycle in snail. It has a worldwide distribution.

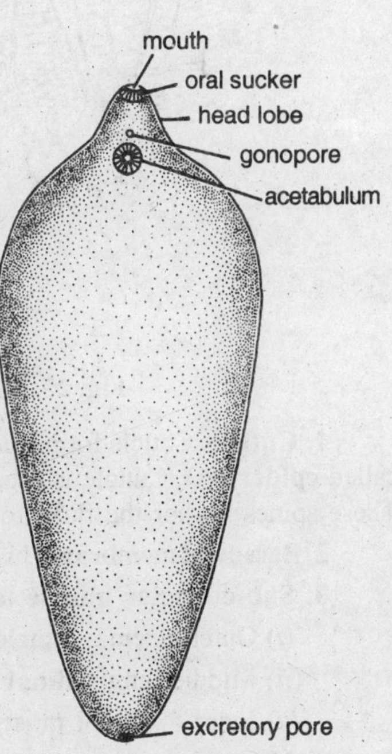

Fig. 25.1. *Fasciola hepatica.* Adult worm in ventral view.

25.3 MORPHOLOGY

Shape. Body of *Fasciola* is dorso-ventrally flattened, leaf-like and oval.

Size. The size of *Fasciola* varies from 15 to 30 mm in length and 10 to 15 mm in width. The broadest part of the body is about 1/3 distance from the anterior end.

Colouration. Generally the body of *Fasciola* is pinkish with brownish or blackish margins.

Structure

Anterior part of *Fasciola* is broad and rounded while posterior part is narrower and bluntly pointed or tapering. The anterior end of body is produced into a small conical projection, known as **oral cone, cephalic cone** or **head lobe.** The distal free end of the head lobe is perforated by a minute circular **mouth**

opening, surrounded by a muscular cup-like depression, the **oral sucker** having radially arranged muscles. On the ventral surface, a little behind the head lobe, is found a large cup-shaped sucker, the **posterior sucker, ventral sucker** or **acetabulum.** It is a glandulo-muscular organ with which the animal is attached firmly to the walls of bile passages. A sucker works primarily on the pressure vacuum principle.

The entire body of *Fasciola* is covered with scales. There are following four pores on its external surface:

1. Mouth. It lies at the anterior end of body.

2. Excretory pore. It is situated at the posterior end of the body.

3. Gonopore. Closely in front of the acetabulum on the ventral surface there is a small opening, called **gonopore.** Through it, a curved muscular organ called **cirrus** or **penis** may be found protruding in live specimens.

4. Opening of Laurer's canal. During the breeding season a temporary opening is formed on the dorsal surface of body, it is known as opening of Laurer's canal.

25.4 BODY WALL

The body wall of *Fasciola hepatica* consists of following parts (Fig 25.2).

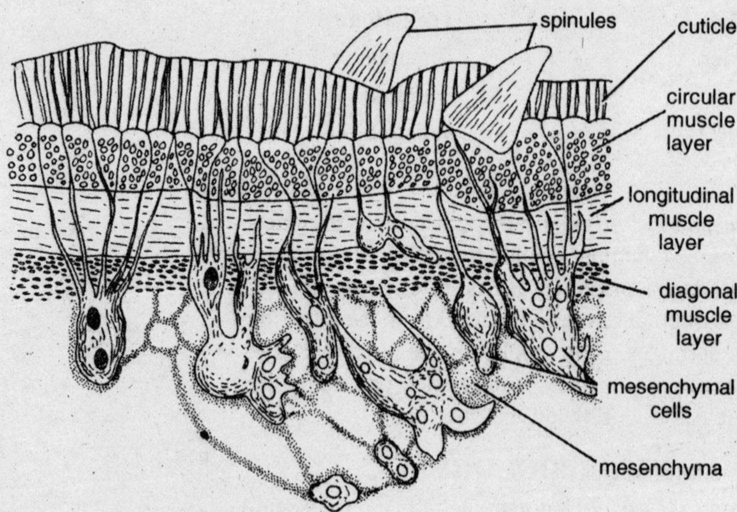

Fig. 25.2. *Fasciola hepatica*. T.S. through body wall.

1. Cuticle. Cuticle forms the outer most layer of body wall. It is of homogeneous nature (now called **epidermis**). Cuticle is rough and produced into minute spines or **spinules** or blunt **scales.** These spines anchor the fluke to the bile duct of the host.

2. Basement membrane. Just below the cuticle is present a thin and delicate basement membrane.

3. Sub-cuticular muscle layers. These layers are arranged in three groups:

(*i*) Outer **circular mucle** fibres.

(*ii*) Middle **longitudinal muscle** fibres.

(*iii*) Inner **diagonal muscle** fibres or **oblique** muscles. Muscles fibres are smooth type.

4. Mesenchyme or parenchyma. The interior of the body between the organs is occupied by large, branching fluid-filled parenchymatous cells.

5. Gland cells. Numerous unicellular gland cells are found embedded in the muscular layer. These are **ectodermal** cells which have sinked into parenchyma and are connected by protoplasmic projections.

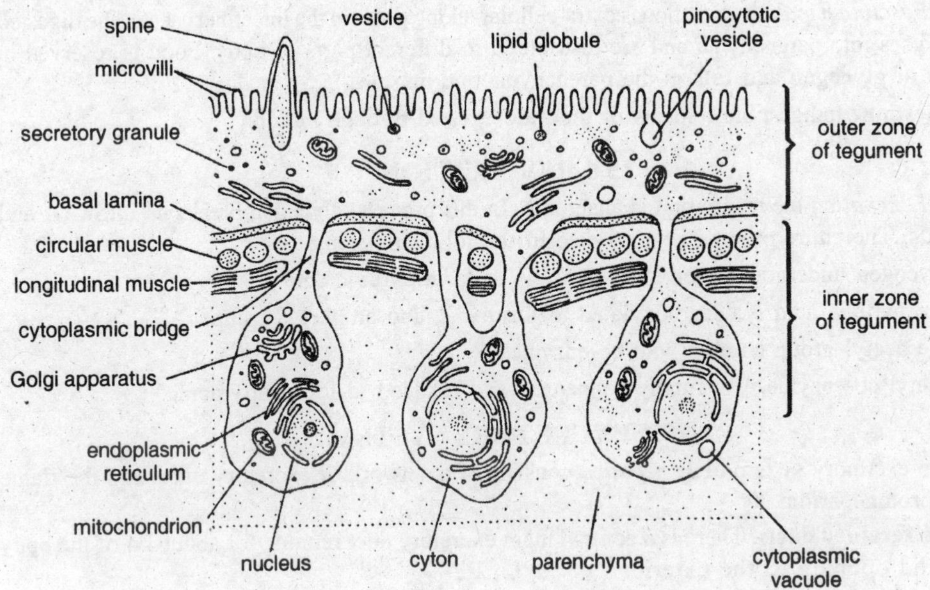

Fig. 25.3. Section through the tegument and body wall of a generalized digenean.

25.5 DIGESTIVE SYSTEM

The digestive system of *Fasciola* is complex, extensive and highly branched but without anus (*i.e.*, it is **incomplete**). It consists of following parts (Fig. 25.4):

1. Mouth. It is situated on the ventral side at the apex of head lobe and surrounded by oral sucker. Mouth leads into pharynx through the buccal cavity.

2. Oral chamber. Mouth opens into a short but funnel-shaped oral chamber.

3. Pharynx. Pharynx is short, rounded and highly muscular structure. Its walls are thick and provided with pharyngeal glands. The cavity of pharynx is small and narrow.

4. Oesophagus. The pharynx communicates with a short narrow oesophagus joining the intestine.

5. Intestine. Intestine divides into two branches or **intestinal caeca** or **crura** each running on one side to the posterior end and ending blindly. Intestinal caeca give out numerous branched **diverticula** in order to carry food to all parts of the body since there is no circulatory system. The median diverticula are short and lateral ones are long and branched.

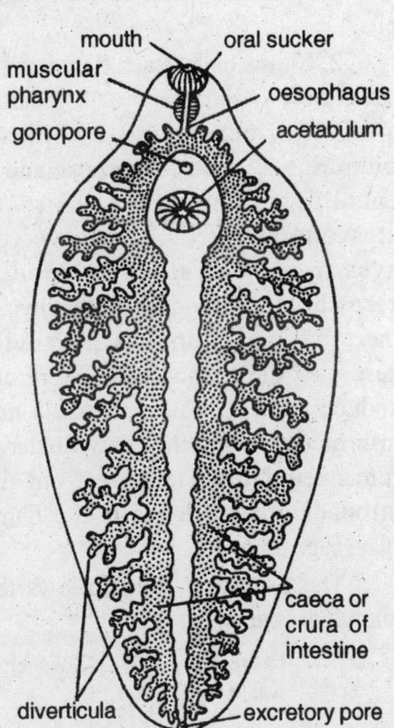

Fig. 25.4. *Fasciola hepatica*. Digestive system.

Digestion

Fasciola hepatica sucks bile and blood, lymph and cell debris from the wall of the host's bile duct by the oral sucker. The muscular pharynx acts as a suctorial apparatus. Amino acids, glucose and fructose directly diffuse into body of the fluke through general body surface from the surrounding fluid of the host.

In *Fasciola hepatica*, digestion is extra-cellular taking place in the intestinal caeca. The digested food diffuses into parenchyma and are distributed to different parts of body. Food is reserved in the form of **glycogen** and **fats** in the parenchyma and muscles.

Egestion. Indigestible remains of food are egested through mouth.

25.6 RESPIRATION

In *F. hepatica* the respiration is anaerobic. In this process, glycogen breaks up into CO_2 and fatty acids. The entire process is completed in the following steps:

1. Glycogen undergoes anaerobic glycolysis to form pyruvic acid.
2. The pyruvic acid is decarboxylated to form CO_2 and an acetyl group.
3. The acetyl group combin with co-enzymes A.
4. Acetyl co-enzyme A is finally condensed and reduced to form fatty acid.

25.7 EXCRETORY SYSTEM

The excretory system of *F. hepatica* consists of a network of excretory ducts and the flame cells or protonephridia.

1. Excretory ducts. There is a **central** main excretory duct running for about 3/4 of the body length and opening to the exterior through **excretory pore**, which is present at the posterior end of the body. The main canal through its course gives off side branches. All these branches divide and redivide into numerous minute capillaries which end into flame cells.

2. Flame cells. Each flame cell is a modified mesenchyme cell. It has a thin wall, peripheral cytoplasm containing a central nucleus and a central intracellular space, known as **intracellular duct**. Each flame cell gives out several pseudopodial processes in surrounding mesenchyme. The cilia arise from basal granules present in the flame cells. Excretory products such as water, ammonia and waste metabolites such as unwanted iron from haemoglobin diffuse from the surrounding parenchyma in the flame cells (Fig. 25.6).

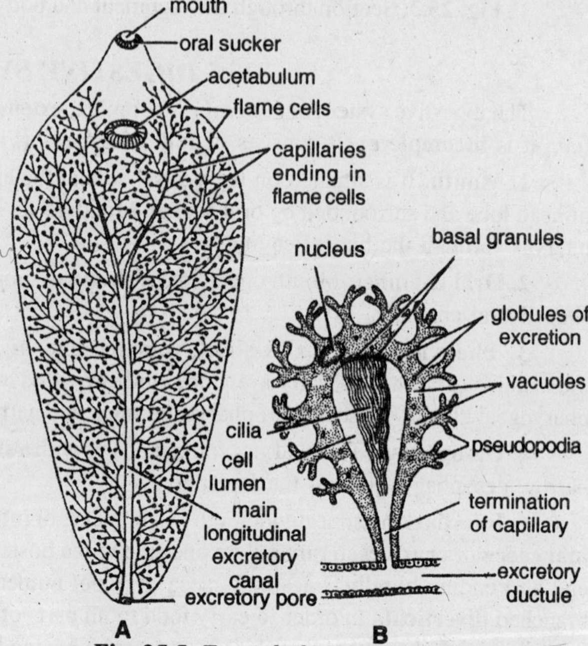

Fig. 25.5. *Fasciola hepatica*. A—Excretory system; B—A flame cell.

The excretory fluid follows the following course:

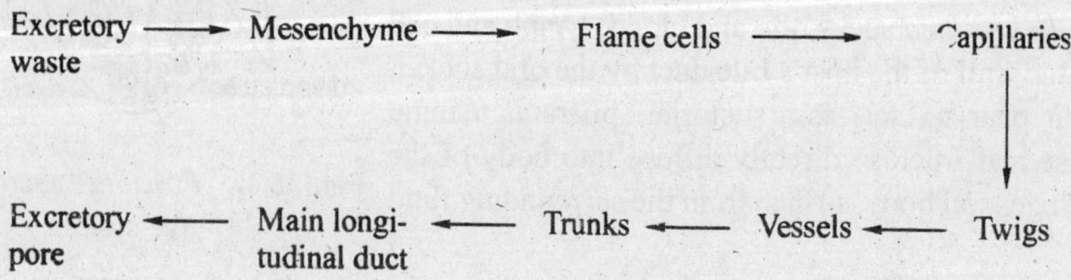

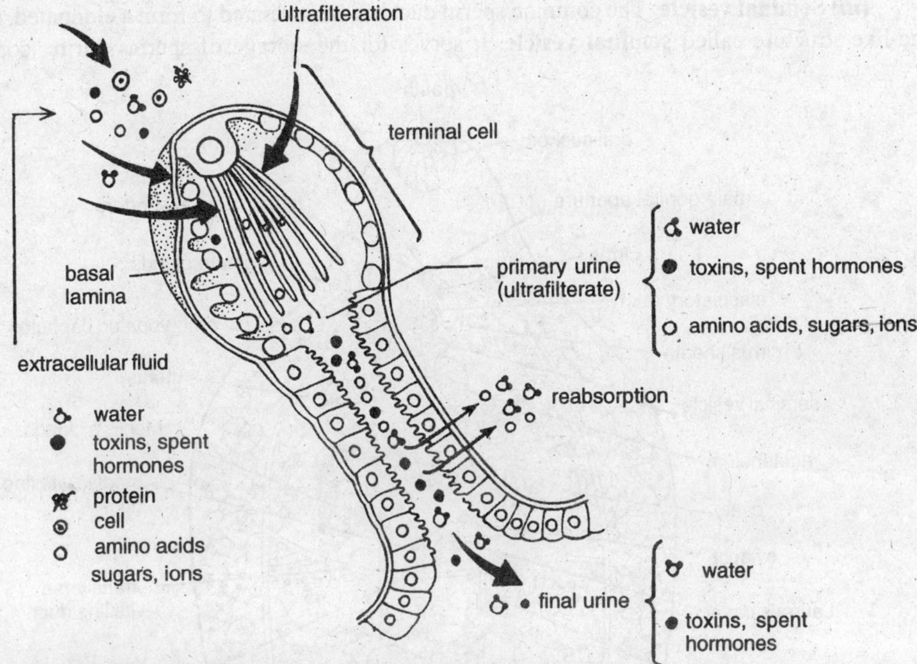

Fig. 25.6. Protonephridial excretion is a two-step process involving ultrafiltration and modification to form a final urine. Ultrafiltration occurs across the wall of a terminal cell, the flame cell. The force driving fluid across the terminal cell wall results from the action of flagella or cilia.

25.8 NERVOUS SYSTEM

The nervous system (Fig. 25.7) is well developed and consists of (*i*) A pair of cerebral ganglia. (*ii*) A pair of dorsal, a pair of ventral and a pair of lateral **nerve cords** are given off from the cerebral ganglia. (*iii*) A nerve ring exists around the oesophagus and connects the cerebral ganglia. (*iv*) Nerves are given out from the cerebral ganglia to the head lobe and to the hinder part of the body.

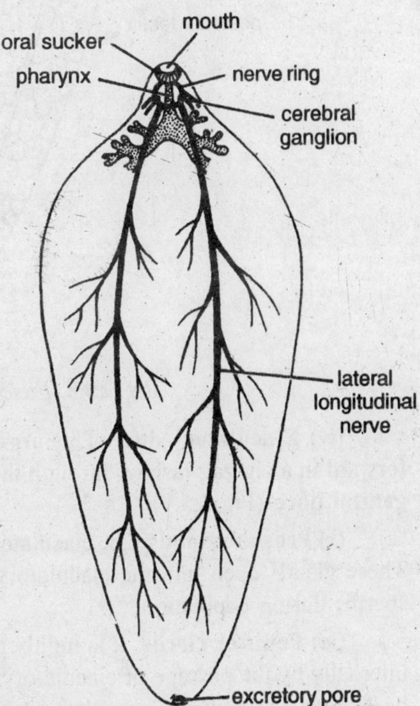

Fig. 25.7. *Fasciola hepatica.* Nervous system.

25.9 REPRODUCTIVE SYSTEM

Fasciola hepatica is **hermaphrodite** as both male and female sex organs occur in every individual. The system is studied under the following two headings:

1. Male Reproductive System

The male reproductive system consists of testes, vasa deferentia, seminal vesicle, ejaculatory duct, prostate glands, penis or cirrus and genital atrium (Fig. 25.8).

(*i*) Testes. These are two in number, greatly ramified and tubular, situated in the middle part of body, one behind the other.

(*ii*) Vasa deferentia. From each testis, a narrow delicate sperm duct or **vas deferens,** emerges out and runs forward to meet the fellow of opposite side forming the common sperm duct just below the ventral sucker.

(iii) **Seminal vesicle.** The common sperm duct becomes dilated to form a elongated, muscular, sac-like structure called **seminal vesicle**. It serves for the storage of sperms during copulation.

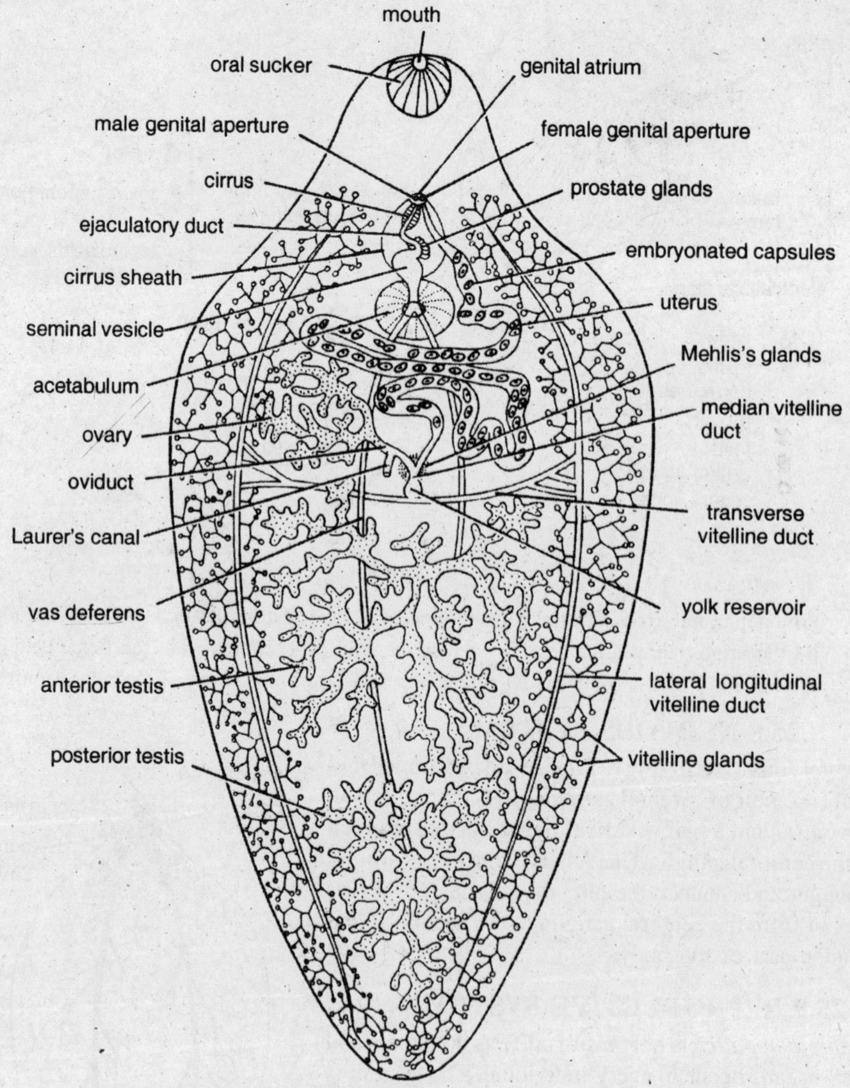

Fig. 25.8. *Fasciola hepatica*. Reproductive system.

(iv) **Ejaculatory duct.** This arises from seminal vesicle as a fine convoluted tube and runs forward in a zig-zag fashion through the cirrus to open into the **genital chamber** through the **male genital pore** (Fig. 25.9).

(v) **Prostate glands.** The ejaculatory duct is surrounded by numerous unicellular prostate glands. These glands open into the ejaculatory duct. Their alkaline secretion helps in free movement of sperms during copulation.

(vi) **Penis or cirrus.** It is highly muscular, elongated and cylindrical structure. It is traversed internally by the passage of ejaculatory duct. Cirrus opens into the **male genital aperture** and can be everted out and drawn in through the genital pore and thus helps in copulation.

The cirrus, seminal vesicle and prostate glands are enclosed in a bag-like cirrus-sheath.

2. Female Reproductive System

Female reproductive system consists of ovary, oviduct, uterus, vitelline glands, Mehlis's glands and Laurer's glands (Fig. 25.8).

***(i)* Ovary.** Ovary is single, tubular and highly branched, situated on the right side of body in front of testes. It occupies the anterior one-third of the body.

***(ii)* Oviduct.** All the branches of ovary open into short and narrow tube called **oviduct**. The oviduct travels down obliquely and opens into the median vitelline duct.

***(iii)* Uterus.** From the junction of oviduct and median vitelline duct arises a wide convoluted **uterus** having fertilized shelled **eggs** or **capsules**. Uterus runs forward to open into the genital atrium on the left side by **female genital pore**. The terminal part of uterus has muscular walls and is called **metraterm**; it helps in ejecting out the eggs. Metraterm also sometimes receives the cirrus during copulation.

***(iv)* Vitelline glands.** On each side of the body from one end to the other, there are the **yolk glands, vitelline glands** or **vitellaria** in the form of numerous minute vesicles or follicles. The yolk glands produce albuminous yolk and shell material for the eggs. A fine duct arises from each follicle or yolk gland and after uniting with similar ducts from the other follicles opens into the **lateral longitudinal vitelline ducts.** The two longitudinal ducts are connected together by a **transverse vitelline duct** placed above middle part of the body. The transverse vitelline duct is swollen in the centre to form the **yolk reservoir** or **vitelline reservoir.** From the yolk reservoir, a **median vitelline duct** starts and runs forward to join the oviduct (Fig. 25.9).

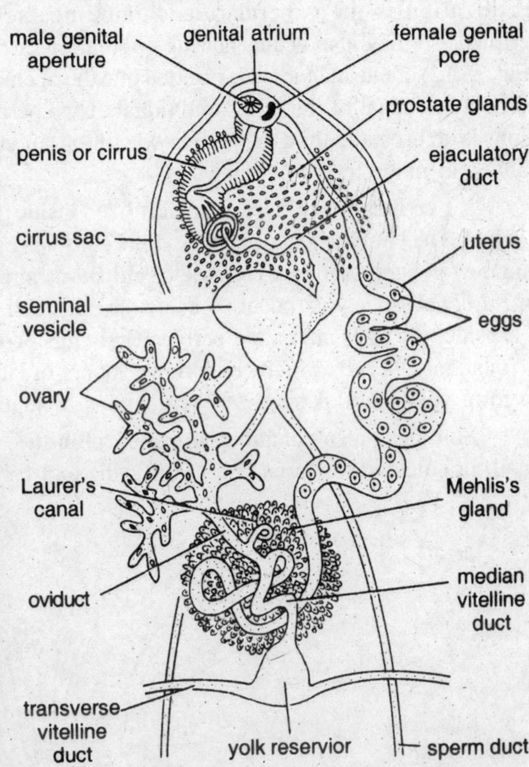

Fig. 25.9. *Fasciola hepatica.* Details of male and female reproductive organs in anterior region.

***(v)* Mehlis's glands or shell glands.** A mass of numerous unicellular shell or Mehlis's glands is found situated around the junction of median vitelline duct, oviduct and uterus (Fig. 25.10). The secretion of Mehlis's glands lubricates the passage of eggs in uterus; it probably also hardens the egg shells and activates spermatozoa (see Box 25.1).

***(vi)* Laurer's canal.** During the breeding season a temporary opening develops on the dorsal body surface. It arisers from the oviduct and acts as a vagina and is known as Laurer's canal.

Box 25.1.

The junction of oviduct and median vitelline duct is swollen to form **ootype** in certain flukes such as *Fasciola indica*. In ootype, the parts of an egg are assembled and the eggs are shaped. An ootype is not present in *Fasciola hepatica.*

25.10 LIFE HISTORY

Life cycle of *Fasciola hepatica* is completed in two hosts, so it is called a **digenetic** trematode. The **primary host** of *F. hepatica* is sheep or cattle, while the **secondary host** is a snail of the genus

Limnaea. The life cycle of sheep liver fluke is complicated by the introduction of a number of larval forms. This type of life-cycle is said to exhibit **polyembryony** and can be studied under the following headings:

1. Copulation. Copulation in *F. hepatica* occurs in the bile duct of sheep. Copulation is mutual and cross fertilization is the general rule. Sometimes self-fertilization may also occur. During copulation cirrus (Fig. 25.12) of one individual is inserted into the opening of Laurer's canal of the other individual. The sperms along with prostatic fluid make their way into Laurer's canal and move to the oviduct.

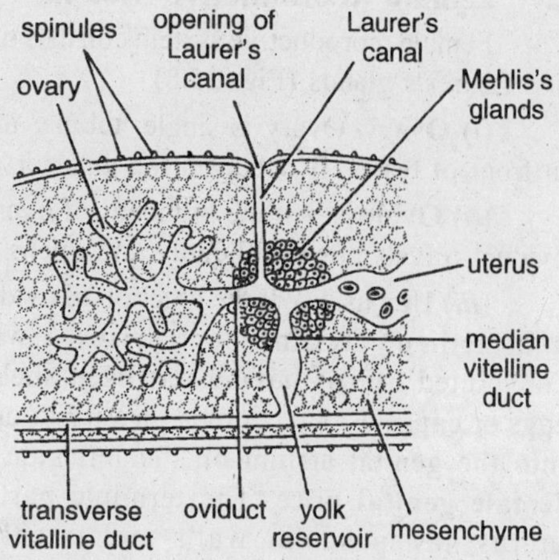

Fig. 25.10. *Fasciola hepatica*. T.S. body through Laurer's canal showing ovary and other female sex organs.

2. Fertilization. The eggs are fertilized inside the oviduct. The vitelline glands supply yolk cells to an egg and they get enclosed in a **capsule** or **chitinous shell.** In *Fasciola*, since yolk occurs outside the eggs cell in a capsule, the eggs are called **ectolecithal eggs.** Yolk is found in yolk cells which break down sooner or later to form yolk mass. A single fluke contains as many as 3,000—4,000 egg capsules in its uterus.

Yolk cells also contain many **shell globules** which are made of proteins and a phenol. The shell globules are released from yolk cells and form two-layered shell of egg capsule.

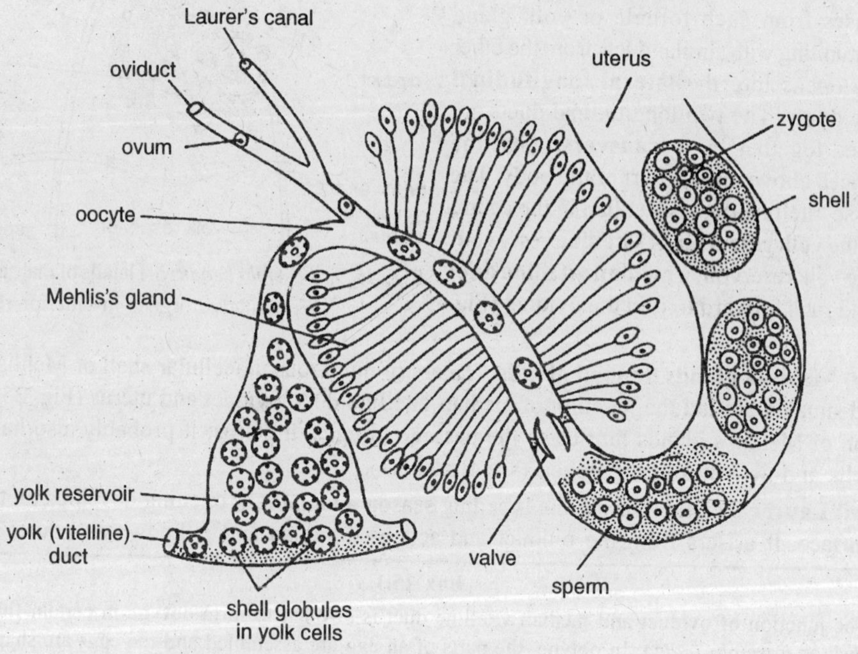

Fig. 25.11. *Fasciola hepatica*. The region of the ootype.

3. Eggs. The fertilized eggs are light brown in colour and oval in shape measuring about 130 to 150 μm in length and 63 to 90 μm in width. Each egg along with many yolk cells is surrounded by a shell or capsule which is made up of a double layer of lipoprotein and contains an **operculum** at one end. Such eggs are called **operculate.** Egg capsules come out of the gonopore into the bile duct of the sheep, they reach the intestine and are passed out with the faeces. The capsules which fall in water or damp places start development at about 75°F temperature.

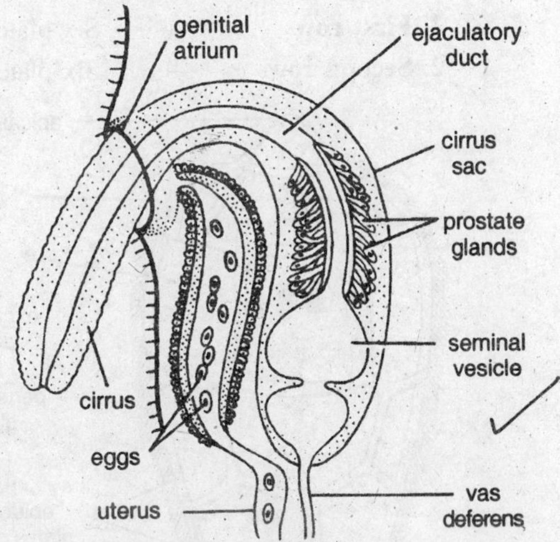

Fig. 25.12. *Fasciola hepatica.* Cirrus protrudes through gonopore.

4. Development. It includes following embryonic and larval stages (*i.e.*, it is of indirect type).

(*i*) Cleavage. Cleavage or segmentation of the fertilized egg is started when the capsule is inside the uterus and continues on the damp ground. The fertilized egg is divided into a small **propagatory cell** and a larger **somatic cell** (Fig. 25.13). The somatic cell divides and forms the **ectoderm** of the larva. The propagatory cell divides and form two cells, one cell forms the **endoderm** and **mesoderm** of the larva. The other cell forms a mass of **germ cells** at the posterior end of the larva. This method of development is repeated during formation of different larval stages during the life history

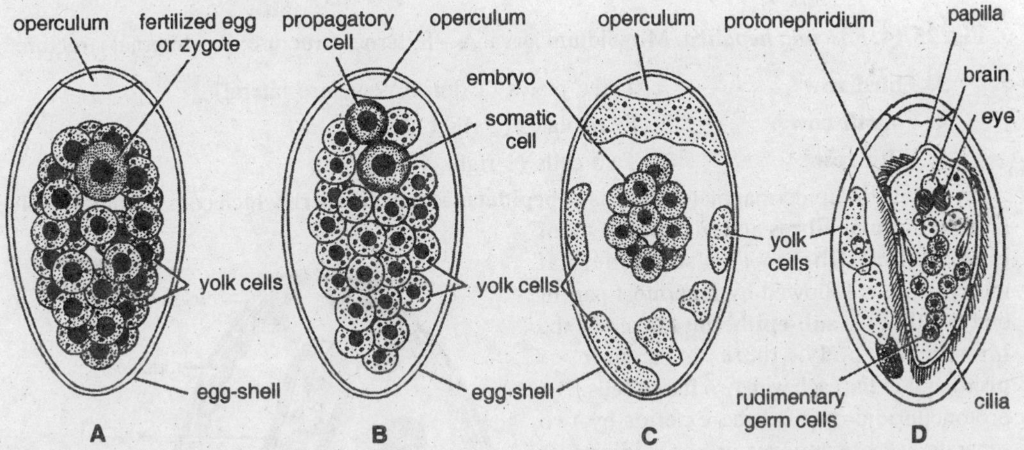

Fig. 25.13. *Fasciola hepatica.* Cleavage and larva formation. A—Fertilized egg; B—Two cell stage; C—Many cell stage (morula); D—Miracidium.

of fluke. Anyhow, within 9 to 19 days a fully formed ciliated larva, called **miracidium** is hatched out of the capsule by forcing off the operculum. The miracidium larva produces a proteolytic enzyme which erodes the lower surface of the operculum.

(*ii*) Miracidium. Miracidium is a minute (0.13 mm long), oval, elongated and ciliated free-swimming larva. Its anterior end is broader than the posterior end and is produced into an **apical lobe** or **apical papilla.** It bears openings of (*i*) a pouch-like multinucleated **apical gland** and (*ii*) a pair of unicellular **penetration glands** or **cephalic glands** (Fig. 25.14).

Except the apical lobe, entire body of miracidium larva is **ciliated** and is covered with 2 closely fitted hexagonal **epidermal plates** which are arranged in five rows or tiers. Each tier has following arrangement of plates:

1. First row	—	Six plates (2 dorsal, 2 ventral and 2 lateral)
2. Second row	—	Six plates (3 dorsal, 3 ventral)

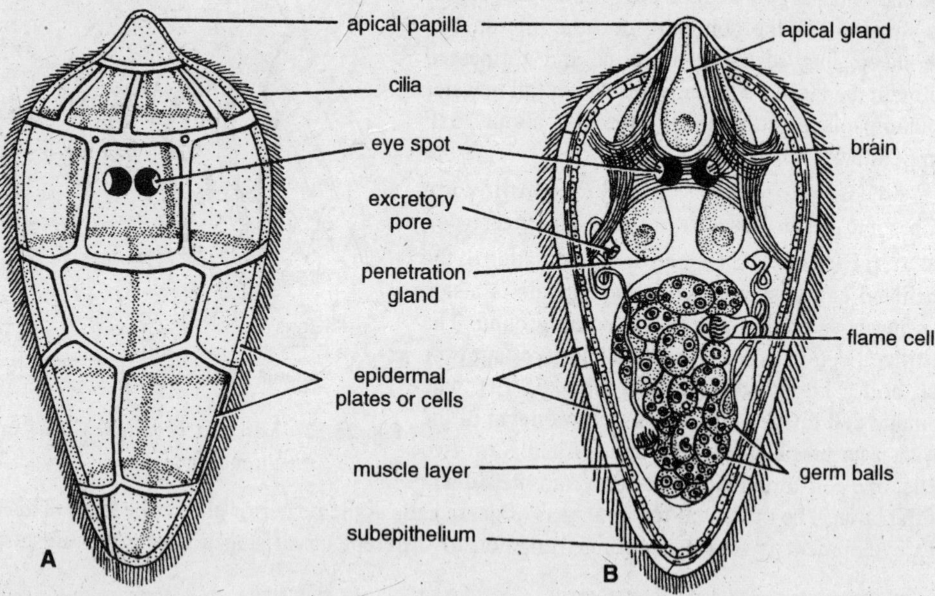

Fig. 25.14. *Fasciola hepatica*. Miracidium larva. A—External structure; B—Internal structure.

3. Third row	—	Three plates (1 dorsal, 2 ventro-lateral)
4. Fourth row	—	Four plates (2 right, 2 left)
5. Fifth row	—	Two cells (1 right, 1 left)

Below the epidermal plates there is **subepidermal musculature** which consists of (*i*) an outer layer of **circular fibres** and (*ii*) inner layer of **longitudinal fibres.** The subepidermal musculature is followed by innermost part of body wall called **sub-epithelium layer.** In the interior of body, there is a pair of **protonephridia,** each with two **flame cells.** The protonephridia open to the exterior by two separate **excretory pores** or **nephridiopores** which are located laterally in the posterior half of body. Towards the posterior side are present some **propagatory** or **germ cells,** some of which may have divided to form **germ balls** which are developing embryos.

A **brain** is present below the epidermal plates (or cells) of second row; it is associated with several nerve fibres. Above the brain is pair of X-shaped **eye spots.** The miracidium does not feed, it swims about in water. It dies in eight hours unless it can reach a suitable **intermediate host,** which is some species of amphibious snail of genus *Limnaea.* After getting a suitable host the miracidium adheres

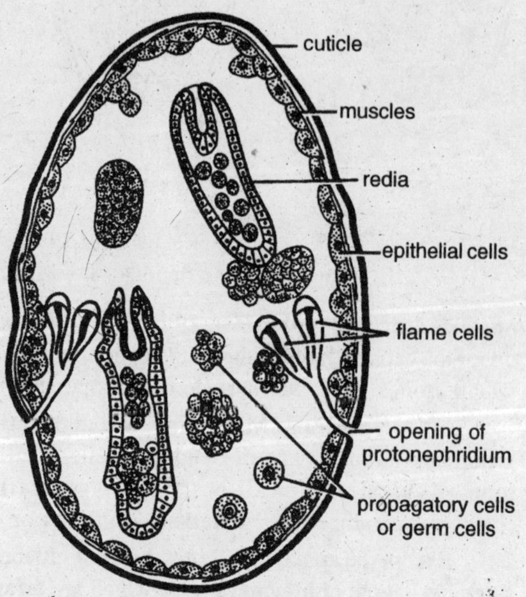

Fig. 25.15. *Fasciola hepatica.* Sporocyst.

to it by its apical papilla and enters the pulmonary sac of the snail. From the pulmonary sac, the larva penetrates into the body tissues with the aid of penetration glands and finally reaches the snail's digestive gland. In intermediate host, miracidium casts off its ciliated epidermis, loses its sense organs and the larva swells up and changes its shape to become next larval stage called **sporocyst.**

***(iii)* Sporocyst (Fig. 25.15).** This is second larval stage of *Fasciola*. It is an elongated germinal sac about 0.7 mm long and covered with a thin cuticle. Below the cuticle, there are circular and longitudinal muscles and some mesenchyme cells. The hollow interior of sporocyst has a pair of protonephridia, each with two flame cells; it also has germ cells and germ balls. Sporocyst is called as a **living cyst.** The sporocyst moves about in the host tissues and its germ cells divide and passes through embryogenesis (such as morula, blastula, gastrula, etc.) to give rise to third larval stage, called **redia larva.**

A sporocyst forms 5 to 8 radiae which are set free by rupture of the wall of sporocyst, feed on host's tissues and migrates to the liver of snail.

***(iv)* Redia larva.** It is an elongated (about 1.3 to 1.6 mm in length) cylindrical sac-like larva having two ventral processes, called **procruscula** or **lappets,** near the posterior end of body (Fig. 24.16). The body wall of redia larva contains cuticle, mesenchyme and muscles. Near the anterior end of body occurs a ring-like muscular swelling called **collar.** Collar is used in locomotion of larva. Behind the collar is present the **birth pore.** Redia has a small anterior **mouth,** a muscular **pharynx** with unicellular **pharyngeal glands** and a small sac-like **gut** or **intestine.** There is a pair of **protonephridia,** each with two pairs of **flame cells** (anterior and posterior) which open to exterior through a single **nephridiopore** on each side. The interior space of body is filled with loose parenchyma in which are found groups of **germ balls** and **germ cells.** The germ cells of redia give rise during summer months to a second generation of **daughter rediae,** but in winter they produce the fourth larval stage, the **cercaria larva.**

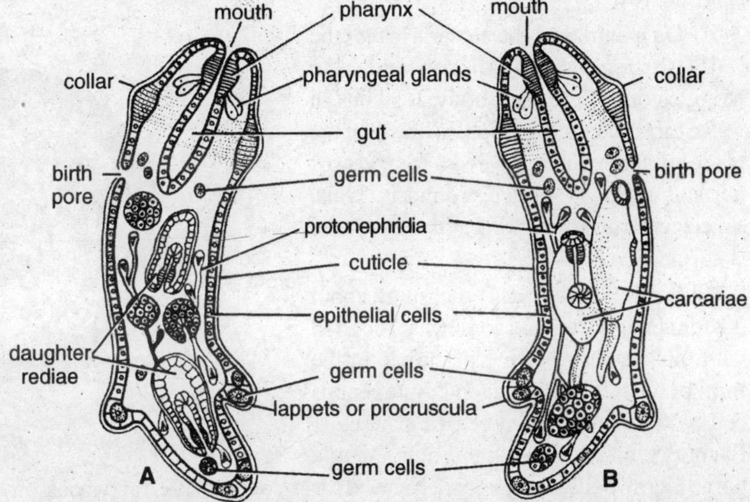

Fig. 25.16. *Fasciola hepalica.* A—Redia with daughter rediae, B—Redia with cercariae.

4. Cercaria larva (Fig. 25.17). Its body is small (0.5 mm long) rounded, oval or

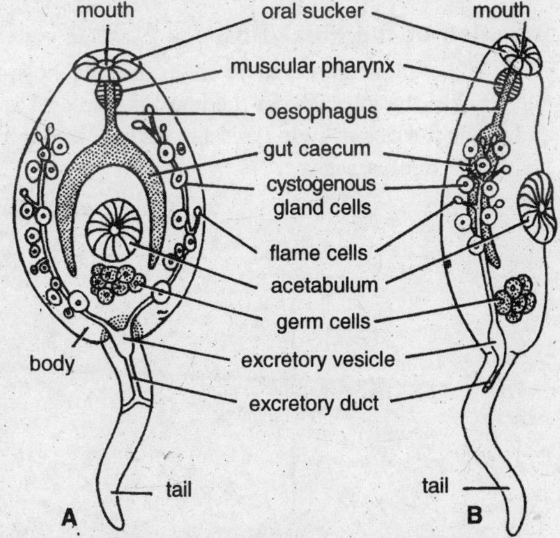

Fig. 25.17. *Fasciola hepatica.* A—Cercaria in ventral view; B—Cercaria in lateral view.

flattened with a long contractile tail. Its body surface is covered with backwardly directed **spines.** The anterior end of cercaria larva bears mouth, muscular pharynx, oesophagus and a bifid intestine. The mouth is surrounded by **oral sucker. A ventral sucker** or **acetabulum** is also present. There are numerous flame cells and the excretory ducts which unite to form the **bladder** or excretory pore situated at the base of the tail. Cercaria larva also contains germ cells and the rudiment of reproductive organs.

A number of unicellular **cystogenous glands** are situated below the larval body wall. Their secretion forms the **cyst** around the larva when it is converted into **metacercaria** (Fig. 25.18).

On maturation, the cercaria leaves the redia through the birthpore and also wriggles out of the snail body. It swims in water for some time and settles down on the blade of some aquatic weeds. It sheds off tail and a cyst is formed around it. Thus, a metacercaria representing a **juvenile fluke** is formed.

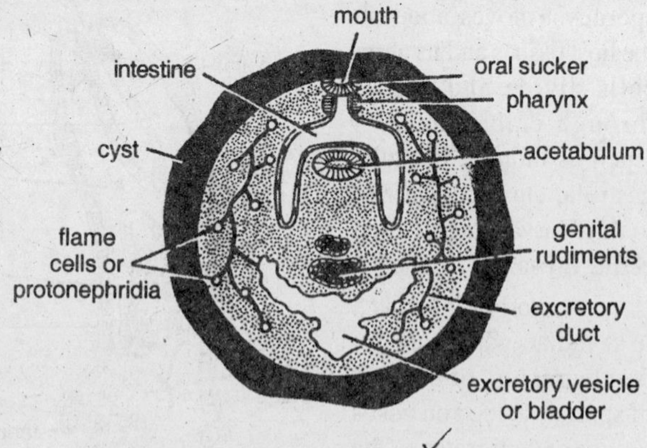

Fig. 25.18. *Fasciola hepatica.* Metacercaria.

5. Metacercaria. It measures about 0.2 mm in diameter and somewhat rounded with thick outer covering of cuticle in the form of **cyst.** It has no tail and cystogenous cells. Flame cells have increased in number. Excretory bladder opens through a single pore. Germ balls are present. Metacercaria can survive for weeks on grass and near water.

Infection of the Secondary Host

Miracidium swims freely in the water and searches for the intermediate host, the snail of genus *Limnaea* or *Planorbis*. Unable to find the host within a period of 24–30 miracidium perishes. On entering the snail, miracidium bores the pulmonary chambers, throws off cilia and makes its way into the digestive gland. Here it develops into **sporocyst larva.**

Infection of the Final Host

The cysted **metacercaria** enters the final host, the sheep, when it grazes on the aquatic weeds and reaches the intestine. In the intestine the cyst is dissolved by the action of digestive enzymes and young fluke comes out. It reaches the liver through the hepatic portal system and starts its existence inside the bile passage.

Graphic representation of the life cycle

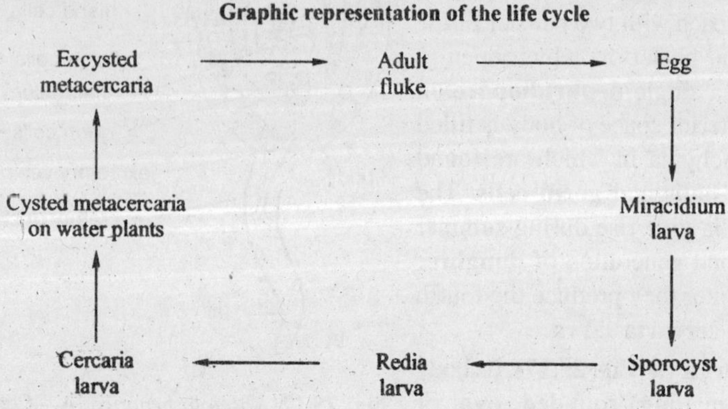

Significance of different larval stages in the life history of *Fasciola*

The different larval stages in the life cycle of *Fasciola* has following significance:

1. Alternation of generation or metagenesis. In the life cycle of *Fasciola*, a sexually reproducing adult phase alternates with asexually reproducing larval phase. Such alternation of two types of generations is called **metagenesis.**

2. Heterogamy. It was first thought (**Grobben** 1882) that the germ cells in different larva were eggs which developed into next stage by **parthenogenesis.** This type of sexual reproduction is called **heterogamy.** This hypothesis has now been totally rejected.

3. Polyembryony. According to this theory a single egg produces several embryos, it is called **polyembryony.** In the life cycle of *Fasciola* the division product of propagatory cells are called **germ cells** in the larval stage. The redia, daughter redia and cercaria are produced by **germ cells** by simple asexual multiplication. Thus, zygote gives rise to several larval stages, this shows polyembryony.

4. Extended metamorphosis. According to some, the asexual multiplication of larval forms is an extended metamorphosis due to parasitism.

5. Advanced larval stages. Body cavity, sense organs and locomotory organs are examples of advanced characters: They are absent in adult but present in larval stages. Thus, miracidium and cercaria are more advanced than the adult. The adult fluke has degenerated in many respect to suit its parasitic mode of life.

6. Complex life cycle. The presence of several larval stages has made the life-cycle of *Fasciola* quite complicated. Its life cycle is **digenetic,** *i.e.*, includes two hosts—**primary host** (sheep) and **secondary** or **intermediate host** (freshwater snail).

7. Over production. Rate of reproduction of an organism is directly proportional to the chances of death it faces at various phases of its life cycle. *Fasciola* being an endoparasite faces several threats to its survival and therefore its reproductive power is enormous.

A single fluke can produce more than 20,000 eggs in its life cycle. Each sporocyst produces 5–8 radia larvae. Each of these redia produces 8–12 daughter radia larvae and each of these radia produces 14 to 20 cercaria larva. Thus a single egg of *Fasciola* is capable of producing about 1,000 to 2,000 adults. But in nature it does not happen, only a few survive.

8. Chances of death (or mortality). The presence of a number of larval forms and a prolonged life-cycle, both increase the chances of destruction. These are:

(*i*) Dropping of eggs along with feces of sheep in dry places kills all eggs.

(*ii*) Failure of miracidium larvae to find a correct intermediate host within 24 hours kills the larvae.

(*iii*) Death of infected snail terminates the life cycle.

(*iv*) Drying of ponds kills the miracidium and cercaria.

(*v*) The sheep does not eat the plants containing metacercaria.

(*vi*) The pH and temperature of the pond water may be unsuitable for the life of larvae.

(*vii*) If the sheep swallows cercaria before encystment, the larvae are killed by the gastric juices.

To overcome the risk of disintegration of the race the parasite possesses great fertility and has evolved 'the asexual multiplication of larva'. Therefore, the extended and complicated life-history with many larval phases is an adaptation to parasitic mode of life.

25.11 ECONOMIC IMPORTANCE

1. Infection. The sheep, goat and cattle get infected with *Fasciola hepatica* by eating grass and other plants which contain **metacercaria.**

2. Pathogenesis. *Fasciola hepatica* causes serious damage to the bile duct and liver of its host. It causes the disease called **liver rot** or **fascioliasis.** The infected sheep develops the following symptoms: acute anaemia, appetite declines, loss of weight, enlargement of liver and fever. The sheep becomes weak and lazy, milk production is reduced, wool falls, eyes become yellowish and jaw muscles swell. Heavy infection of *Fasciola* causes death of the host within a few weeks.

3. Treatment. Anti-helminth drugs such as **emetin hydrochloride, phenothiazine, carbon tetrachloride, tetrachloroethane,** etc., are used to cure the sheep.

4. Preventive measures. Spread of infection of *Fasciola* can be prevented by

(*i*) Killing heavily infected sheep.

(*ii*) Introducing ducks in the ponds which eat up the intermediate host of *Fasciola*, the snail.

(*iii*) Drying the ponds.

(*iv*) By removing all vegetation around the ponds.

(*v*) By adding copper sulphate in the pond. It kills the snails.

25.12 PARASITIC ADAPTATIONS IN *FASCIOLA HEPATICA*

Adaptations can be defined as modification of an organism or its parts that make it more fit for existence under the conditions of its environment. **Parasitic adaptations** can be defined as structural, physiological and behavioural modifications developed by the organism in order to lead a parasitic mode of life.

I. Structural Adaptations

A. Morphological adaptations

1. Flat and thin body enables *Fasciola* to live in narrow spaces. It minimizes the danger of injury in narrow spaces.
2. Adhesive organs are of two types and of highly muscular nature. They occur in the form of **anterior sucker** and **ventral sucker.** By the help of these suckers, parasite attaches itself to the host's tissues.
3. In *Fasciola*, there is no distinction of body into head, trunk and tail. Body is also without any appendages, since these might prove harmful in endoparasitic mode of life.
4. Resistant tegument or the cuticle is an important parasitic adaptation. The parasite remains unaffected by alkaline pH of the bile.
5. The spines in the tegument protect the body.

B. Anatomical adaptations

1. There are no locomotory organs. As *Fasciola* is endoparasitic, it does not need to make excursion for food and safety.
2. Terminal mouth is surrounded by a muscular oral sucker which helps in feeding.
3. The pharynx is muscular and suctorial. It is adapted to suck blood, bile, etc.
4. The intestine of *Fasciola* is highly branched. Its branches supply the digested food to all parts of the body. Thus, intestine has taken over the work of circulatory system.
5. *Fasciola* feeds on already digested food so the anus is not required. Thus, alimentary canal is incomplete and intestine ends blindly.
6. Endoparasite respires **anaerobically**, as there is almost no oxygen in the bile passage of host.
7. Excretory system is well developed. It consists of extensively branched system of excretory ducts ending into flame cells. This facilitates the collection of excretory products from the interior of body.

8. Nervous system is poorly developed and sense organs are lacking, since the endoparasite does not react with the external environment, and there are least changes inside the body of host.
9. Male and female reproductive systems are well developed. The highly branched ovary and testes are capable of producing large number of gametes.
10. The eggs are covered with resistant capsule.
11. Bisexuality offers full chances of fertilization.

II. Physiological Adaptations

1. *Fasciola* is well adapted to the osmotic concentration of the bile in which it lives.
2. Respiration is anaerobic.
3. *Fasciola* can absorb certain amino acids through its body wall (tegument) as extra amino acids are needed for protein synthesis in the eggs and larval production.
4. Osmoregulation is not needed since the osmotic concentration of body fluids of *Fasciola* is almost equal to that of the host's bile.

III. Behavioural Adaptation

Fasciola hepatica is a hermaphrodite animal but in it cross-fertilization is the rule. Copulation is mutual and reciprocal. The cirrus of one partner is inserted into the vaginal canal of the other partner and sperms are ejected. After copulation both flukes lay fertilized eggs. In this way the rate of reproduction is doubled.

IV. Modifications in Life cycle

1. Very large number of fertilized eggs are produced.
2. The eggs are protected by the capsular covering.
3. The germ cells in a larva produce several new larvae asexually. This phenomenon is called **paedogenesis.**

QUESTIONS

Long Answer Questions

1. Give a detailed account of the structure of *Fasciola* and comment upon its parasitic adaptations. (*Kanpur 1989, 1992, 1994; Lucknow 1989; Bhopal 1990; Gorakhpur 1991, 1994, 1995; Rewa 1992*)
2. Where is *Fasciola* found? Give an account of its structure. (*Jiwaji 1989*)
3. Give the structure and life history of *Fasciola*. (*Punjab 1989*)
4. Give an account of morphological features of *Fasciola* and point out how these features help them to lead a parasitic mode of life?
5. Give an account of the reproductive organs of *Fasciola hepatica*. (*Bundelkhand 1988; Purvanchal 1989, 1991; Madurai 1994; Meerut (CCS) 1994*)
6. Describe the reproductive organs of *Fasciola*. (*Kanpur 1988; Lucknow 1988; Agra 1990*)
7. Describe the life-history of *Fasciola hepatica*. (*Magadh 1991, 1995; Himachal 1992, 1994; Meerut (CCS) 1992, 1993, 1996*)
8. Give a brief account of life cycle of *Fasciola* and explain the significance of different larval phases in its life history. (*Gorakhpur 1987; Bhopal 1988; Utkal 1988; Bundelkhand 1992; Meerut (CCS) 1992; Lucknow 1989; Delhi 1994; Jiwaji 1991*)
9. Give an account of the life history of *Fasciola hepatica*. Discuss its economic importance. (*Meerut (CCS) 1990; Purvanchal 1987; Bundelkhand 1987; Awadh 1993; Kanpur 1994; Raipur 1990*)
10. Name the various larval stages of liver fluke and discuss any two in detail. (*Himachal 1992; CCS 1999*)

11. What do you understand by alternation of generations or digenetic life cycle? Explain the same giving life history of *Fasciola hepatica*.
(*Agra 1987; Lucknow 1991; Meerut (CCS) 1993; Rohilkhand 1993*)

12. What do you understand by polyembryony? Explain this phenomenon with reference to the life-cycle of a digenetic parasite. (*Rohilkhand 1989; Bundelkhand 1989; Agra 1992*)

13. Name the primary and secondary hosts of *Fasciola hepatica* and describe various larval stages of this parasite. (*Magadh, 1996*)

14. Name one platyhelminth parasite of domestic cattle? State the mode of infection, pathogenic systems and control measures of parasite. (*Guwahati 1996*)

15. What is adaptation? Explain with reference to the structure and life-history of *Fasciola*.

16. Describe parasitic adaptations of *Fasciola hepatica*.
(*Awadh 1989; Jiwaji 1989; Garhwal 1989, 1992; Sagar 1990; Rewa 1994; Kanpur 1994*)

Short Answer Questions

1. Write about the parasitic adaptations of liver flukes.

2. Write short notes on the following:
(*i*) Body wall of *Fasciola hepatica*
(*ii*) Vitelline glands of liver flukes
(*iii*) Miracidium larva.

3. Draw a well labelled diagram of the following:
(*i*) Reproductive system of *Fasciola hepatica*
(*ii*) Cercaria larva
(*iii*) Metacercaria larva.

Very Short Answer Questions

1. Describe the flame cell in *Fasciola hepatica*. (*C.C.S.U. Meerut 2004*)

Ans. Flame bulbs or flame cells are protonephridia and form the osmoregulatory and excretory system of *Fasciola hepatica*. They are bulb-like bodies found distributed in the mesenchyma (= parenchyma) throughout the body of *Fasciola*. The flame cells are characteristic structures, each has a thin elastic wall with branched pseudopodia-like processes, a nucleus and an intracellular cavity having many long cilia arising from basal granules. In living condition, the cilia vibrate like a flickering flame, hence, the name flame cell. The undulating flagellar movement of cilia produces the pressure difference (force) needed to drive the **ultrafilteration** at the level of flame cell.

The cavity of each flame cell continues into the capillary-like **collecting tubule.** Removal of excess water, ions, fatty acids are some important functions of the flame cells. According to **Brusca** and **Brusca** (2003) nitrogenous waste of the flukes in the form of ammonia which is excreted, mainly across the tegument and protonephridia (*i.e.*, flame cells) are primarily osmoregulatory.

2. Name three sets of muscles found in the body wall of liver fluke?

Ans. Circular, longitudinal and diagonal.

3. What is the food of *Fasciola hepatica*?

Ans. Bile, blood and cell fragments.

4. How can you explain the lack of sense organs in liver fluke?

Ans. This is an adaptation for endoparasitic mode of life of liver fluke.

5. Name the connective tissue that fills spaces between the organs in liver fluke.

Ans. Parenchyma.

6. What is metraterm?

Ans. It is last muscular part of uterus of liver fluke.

7. What is the name of excretory cells of liver fluke?

Ans. Flame cells (*i.e.*, protonephridia).

8. What is the function of flame cells?
   Ans. Osmoregulation and excretion.
9. Miracidium larva of *Fasciola* metamorphoses into the second larval stage. Name that larva.
   Ans. Sporocyst.
10. Name the larval forms produced in a sporocyst of liver flukes.
    Ans. Rediae.
11. Which larval forms are produced in a radia (*a*) in winter and (*b*) in summer?
    Ans. Secondary rediae, Cercaria.
12. Is miracidium more advanced than the adult liver fluke?
    Ans. Yes.
13. What is the encysted stage in the life history of *Fasciola* called?
    Ans. Metacercaria.

Fill in the Blanks

1. *Fasciola hepatica* causes the .... disease in sheep and goats.
2. Liver fluke lives in the large .... of liver in sheep and goat.
3. The secondary host of liver fluke is ....
4. *Fasciola* has two suckers: oral and ....
5. Cuticle of *Fasciola hepatica* is living and called ....
6. Spaces between the viscera are filled up by .... in the liver fluke.
7. .... of liver fluke is incomplete.
8. Respiration of liver fluke is ....
9. The main function of flame cell is ....
10. In liver fluke, ammonia is mainly excreted through the ....
11. *Fasciola hepatica* has one ovary and .... testes.
12. One liver fluke may produce about .... capsule.

Multiple Choice Questions

1. Trematodes are usually described as
   (*a*) blood flukes (*b*) flukes
   (*c*) worms (*d*) flatworms
2. *Fasciola hepatica* is
   (*a*) free living (*b*) ectoparasite
   (*c*) endoparasite (*d*) ecto-and endo-parasite
3. *Fasciola hepatica* lives in the
   (*a*) intestine of sheep (*b*) liver of sheep
   (*c*) spleen of sheep (*d*) pancreas of sheep
4. The causative agent of liver rot in sheep is
   (*a*) *Dugesia* (*b*) *Fasciola*
   (*c*) *Taenia* (*d*) *Schistosoma*
5. *Fasciola hepatica* occurs in
   (*a*) bile duct of sheep (*b*) liver of sheep
   (*c*) liver of human being (*d*) human bile duct
6. Liver fluke is not affected by enzymes of host because of
   (*a*) cuticle (*b*) suckers
   (*c*) epidermis (*d*) hooks
7. The secretion of vitelline glands
   (*a*) facilitates the movement of sperms during copulation
   (*b*) forms the shell around the egg

(*c*) is deposited around the fertilized egg
(*d*) none of above

8. Laurer's canal is present in
(*a*) *Fasciola* (*b*) *Taenia*
(*c*) *Ascaris* (*d*) Cockroach

9. Secretion of Mehlis's glands serve to
(*a*) lubricate the uterus (*b*) harden the capsule
(*c*) activate the sperm (*d*) does all these

10. Liver fluke has
(*a*) self fertilization (*b*) cross fertilization
(*c*) none of these (*d*) both of these

11. Primary host of *Fasciola* is
(*a*) human being (*b*) pig
(*c*) sheep (*d*) dog

12. The intermediate host in the case of liver fluke is
(*a*) snail (*b*) human being
(*c*) pig (*d*) fly

13. Larva of *Fasciola hepatica* that bores into snail is
(*a*) cercaria (*b*) hexacanth
(*c*) miracidium (*d*) redia

14. In *Fasciola*, miracidium develops into the next stage inside
(*a*) *Bulinus* (*b*) *Limnaea truncatula*
(*c*) *Pila globosa* (*d*) *Planorbis*

15. Which of the following larval stage of *Fasciola* does not produce several larvae?
(*a*) sporocyst (*b*) secondary redia
(*c*) primary redia (*d*) cercaria

16. In *Fasciola*, germ balls of redia give rise to
(*a*) daughter sporocyst (*b*) daughter cercaria
(*c*) daughter redia (*d*) metacercaria

17. The last infective larval form of *Fasciola hepatica* is
(*a*) miracidium (*b*) sporocyst
(*c*) metacercaria (*d*) redia

18. The first larval form in liver fluke is
(*a*) redia (*b*) metacercaria
(*c*) miracidium (*d*) cercaria

ANSWERS

Fill in the Blanks

1. Liver-rot
2. Bile ducts
3. Freshwater snail, *Limnaea*
4. Ventral (acetabulum)
5. Tegument
6. Parenchyma
7. Alimentary canal
8. Anaerobic
9. Osmoregulation
10. Tegument
11. Two
12. 500,000

Multiple Choice Questions

1. (*b*)	2. (*c*)	3. (*b*)	4. (*b*)	5. (*a*)	6. (*a*)
7. (*c*)	8. (*a*)	9. (*d*)	10. (*b*)	11. (*c*)	12. (*a*)
13. (*c*)	14. (*b*)	15. (*d*)	16. (*c*)	17. (*c*)	18. (*c*).

26

Taenia solium (Phylum Platyhelminthes)

26.1 CLASSIFICATION

Phylum	–	**Platyhelminthes**
Class	–	**Cestoda**
Order	–	**Cyclophyllidea (Taenioidea)**
Genus	–	***Taenia***
Species	–	***solium***

Taenia solium is commonly called as **pork tape worm.** It is an intestinal parasite (= endoparasite) of human beings. It has a cosmopolitan distribution.

26.2 HABIT AND HABITAT

Taenia solium is a **digenetic endoparasite,** *i.e.*, its life history is completed in two hosts, human beings is the **primary host** and pig is the **secondary host.** *Taenia* is found attached to intestinal mucosa of humans by its scolex, while the rest of body lies free. It is most common in the pork eating population of tropical and subtropical regions where pork is utilized as food without being properly cooked. It may live upto 25 years in the human host.

26.3 STRUCTURE

1. Shape. The body of tapeworm is long, narrow, dorso-ventrally flattened and ribbon-like. It has two surfaces — a dorsal and a ventral.

2. Size. The size of adult tapeworm varies from 3–5 metres, but few are recorded to attain a length of about 8 metres.

3. Colouration. It is usually white in colour, but sometimes grey, yellow or cream coloured.

4. Segmentation. The body of tapeworm is differentiated into many segments, which are called **proglottids** and arranged in a chain. But this is not true segmentation, so is called **pseudosegmentation** (see Box 26.1).

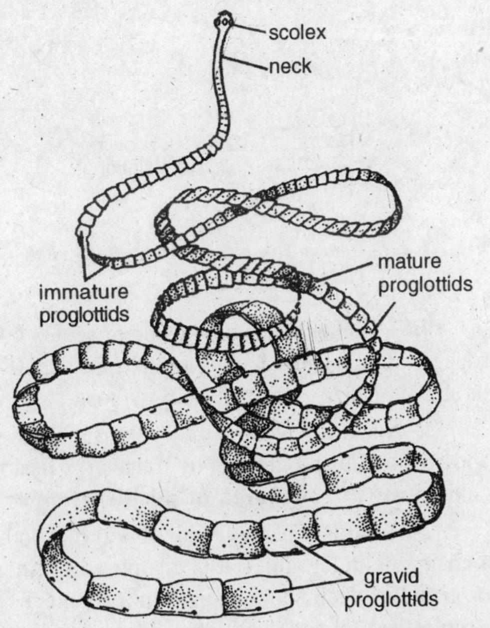

Fig. 26.1. *Taenia solium.* An adult tapeworm.

Box 26.1 Proglottids are not true segments.

The proglottids are not metameric segments like those of annelids, arthropods and other animals. In fact, these are independent, self-contained units, each with a complete set of reproductive organs both male and female and a part of excretory and nervous systems. They are formed by the process of strobilation at the zone of proliferation situated anteriorly behind the scolex. Each proglottid survives briefly after breaking away from the strobila.

5. Body divisions. The body of *Taenia* is well adapted for parasitic mode of life. It is divided into three parts: scolex, neck and body or strobila.

1. Scolex. The narrow anterior end of the body is knob-like and of the size of pin head. It is known as **scolex** (errorneously called **head**). It is a four-sided pear-shaped structure distinguished into following three parts:

(i) **Rostellum.** It is an anterior round prominence which arises from the middle of scolex. It is slightly retractile.

(ii) **Hooks.** Hooks are present at the base of rostellum. They are curved, pointed chitinous structures. Hooks are 28 in number and are of two different size. They are arranged in two circles: larger hooks in inner circle and smaller hooks in outer circle. The smaller hooks alternate with the larger ones. Each hook is made of a **base,** a **handle** directed towards the apex and a conical **blade** directed outwardly. Hooks help in the attachment of scolex with the intestinal wall.

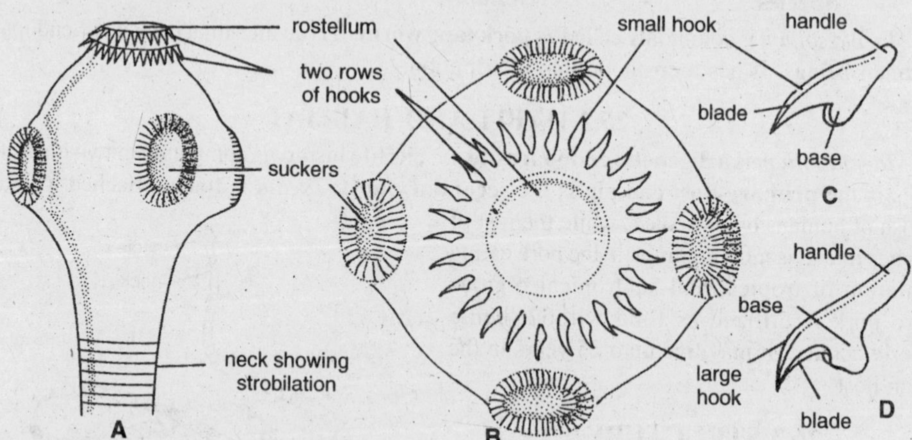

Fig. 26.2. *Taenia solium.* Scolex. A—Scolex enlarged; B—Frontal view of scolex; C—Small hook; D—Large hook.

(iii) **Suckers.** The broad side of scolex contains four cup-like pockets known as **suckers.** The suckers are muscular having radial muscles. Each sucker measures 0.5 mm in diameter. They help in attachment to intestinal wall of host.

2. Neck. It is well defined, short, narrow region behind the scolex. Neck grows continuously and proliferates proglottids by transverse fission or asexual budding. Therefore, this region is also called **growth zone, area of proliferation** or **budding zone.**

3. Strobila. The flattened, ribbon-like body after the neck is known as **strobila** (strobilus means a chain of similar units). It is composed of linear series of 800 to 1000 sets of reproductive organs or genitalia, each set contained in a segment. This linear repetition of genital organs is termed as **proglottisation** and each segment is known as **proglottid,** since the segments are budded off from the neck. The youngest segments are towards the neck and the oldest segments are posterior most. The proglottids internally remain connected together by muscles, excretory vessels and nerve cords. The proglottids are of following three types:

***(i)* Immature proglottids.** These are newly formed and youngest proglottids. Their width is more than the length. They lie next to neck region and include 200 anterior proglottids. In them the reproductive organs are either absent or in stage of development.

***(ii)* Mature or reproductive proglottids.** They are squarish 400 to 600 proglottids which are located after the immature proglottids. They are **hermaphrodite** containing fully developed male and female reproductive organs. About 100 to 150 anterior mature proglottids contain only the male reproductive system.

***(iii)* Gravid or ripe proglottids.** They are the oldest proglottids which are situated towards the posterior end of strobila. They include about 150 to 200 proglottids. They are longer than broad. Gravid proglottids contain a large, highly **branched uterus** which is packed with fertilized eggs and developing embryos. The rest of reproductive organs have already disappeared from them after performing their function.

The posterior most 5 to 6 proglottids are regularly cut off from the body and are expelled out of host body along with its feces. This process of separation of proglottids is known as **apolysis.**

Differences in a proglottid and a body segment

In a tapeworm (*Taenia solium*), each of the segment is virtually identical, differing mainly in size. Each segment, or **proglottid,** is derived from the **scolex** (head) of the tapeworm. The older proglottids are pushed back by new proglottids formed at the margin of the scolex and the most recent segment. This formation process in tapeworms constitutes a type of asexual reproduction. The proglottids are actually buds. When the gravid proglottids, swollen with self-fertilized eggs, arrive at the posterior end of the tapeworm, they break off, and surviving eggs eventually reach a new host.

In annelids, segmentation is part of a unitary body plan, and the segments are not by themselves reproductively competent. Such structures like nephridia and setae do recur, but many other structures extend through the entire body of the worm and do not exist as separate structures in each segment. Some examples are the circulatory system and the reproductive system. Also, the digestive tract is not segmented but exists as a continuous one-way tube with varying specialization occurring along its length. Thus segmentation in Annelida is an advancement that facilitates development. It continues as a **morphogenetic** (showing a hereditary pattern of shape) feature in Arthropoda as well.

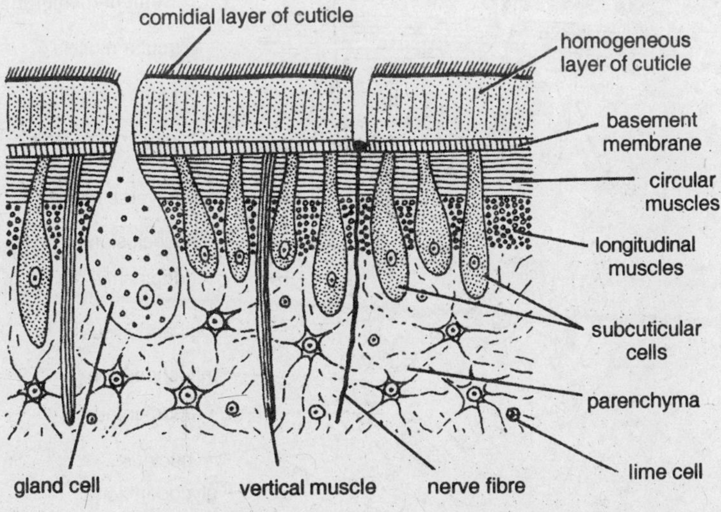

Fig. 26.3. *Taenia solium*. T.S. of body wall.

Body wall

According to classical light microscopic studies (Fig. 26.3) body wall of *Taenia* consists of following three structures: cuticle, muscular layer and parenchyma.

1. Cuticle. It forms the outermost, thick, waxy and enzyme resistant covering of body. It is composed of protein impregnated with calcium salts and is secreted by sub-cuticular cells. Cuticle is further divided into three layers: (*i*) **Comidial layer.** It is outermost layer and is produced into hair-like processes. (*ii*) **Homogeneous layer.** It is middle thick layer. (*iii*) **Basement membrane.** It is innermost thin layer.

> **Box 26.2. Basement membrane**
>
> Basement membrane or **basal lamina** is a dense membraneous layer of connective tissue that separates the layer of epithelial cells from underlying **lamina propria** (*i.e.*, highly vascular layer of connective tissue). It is secreted by epithelial cells and helps in communication between epithelium and connective tissue.

2. Subcuticular muscles. Beneath the cuticle is a layer of **circular muscles** followed by **longitudinal muscles.**

3. Parenchyma. It is fluid-filled spongy-network of mesenchyme cells. It forms a **packing** substance around the internal organs and subcuticular muscle layer. Except at the two sides in each proglottid, the parenchyma is divided into (*i*) outer **cortical region,** (*ii*) inner **medullary region.** Embedded in parenchyma are found few **gland cells, lime cells** and a large number of **cuticular cells.** Parenchyma also contains **parenchymal muscles** which help in bending movement of proglottids.

Ultrastructure of body wall. According to the electron microscopic studies of **Threadgold** (1984), there is no cuticle and no cellular epidermis in the body wall of *Taenia*. Instead of them it contains tegument and surface coat (Fig. 26.4).

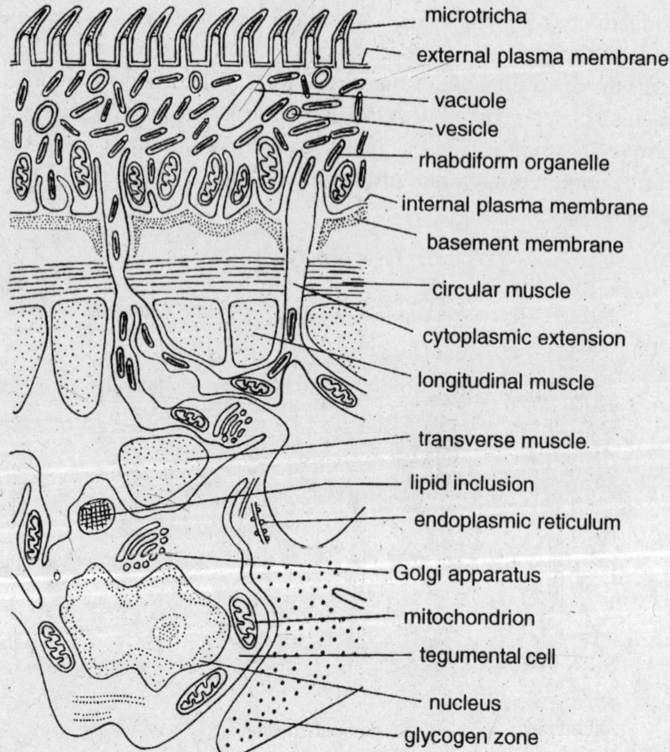

Fig. 26.4. Diagrammatic representation of electron microscopic structure of body wall of tapeworm.

Tegument and surface coat. Entire body of *Taenia* is covered with a cytoplasmic but non-cellular or **syncytium** layer. It is known as **tegument.** It varies in thickness from 1 to 30 μm. The tegument is bounded by **outer** and **inner plasma membranes.** The outer plasma membrane gives out microvilli-like folds, called **microtriches** (singular: **microthrix**). The long, smooth and flexible microtriches increase the surface area. Each of them contains an electron-dense apical tip. **Hooks** of rostellum are also formed by the tegument. The cytoplasm of tegument contains all typical cellular organelles such as mitochondria, Golgi apparatus, endoplasmic reticulum, ribosomes, glycogen granules, lipids and vacuoles. It also contains some specific structures such as **tegumental discs, dense bodies** and **multilaminar bodies.** These are required for the growth of tegument and for replacement of surface coat.

The tegument of adult tape worm does not contain **nucleus** but it is connected with nucleus-containing **subtegumental cells.** The **inner plasma membrane** of tegument is connected with the plasma membrane of finger-like extensions of parenchymal cells called **cyton** or **subtegumental cells.** Some of these cells contain stored glycogen.

The tegument is covered with a fuzzy **surface coat** which is a highly dynamic structure. It is composed of carbohydrates and mucopolysaccharides. It is renewed in every few hours. Different components of the surface coat are synthesized by the endoplasmic reticulum and Golgi apparatus of subtegumental cells and are transported by their ductules to the upper parts of the tegument.

Thus tegument is a metabolically active layer and serves the following functions:

1. Absorption of nutrients.
2. Osmoregulation and excretion.
3. Protection against the effects of digestive enzymes of host and host's immune system.

26.4 NUTRITION

Taenia solium lacks digestive organs. Digested food containing glucose, amino acids, glycerol, etc., from the host intestine is absorbed through the general body surface by diffusion. This is called **saprozoic nutrition.** Reserve food is stored as glycogen and lipids in the parenchyma.

26.5 RESPIRATION

Organs of respiration are also lacking in *T. solium*. The respiration is of **anaerobic** type. During respiration glycogen is broken into CO_2 and fatty acids/organic acids with the release of energy.

26.6 EXCRETORY SYSTEM

The excretory system is well developed and mostly helps in maintaining the **hydrostatic pressure** (water balance). Thus, it is involved in **osmoregulation.** Excretory system (Fig. 26.5) consists of following parts:

1. Longitudinal vessels. There are two **dorsal** and two **ventral vessels** extending into strobila and are linked in scolex by **nephridial plexus.** The four vessels run through proglottids (Fig. 26.5). The dorsal vessels are found in the anterior part of strobila while the ventral vessels extend its entire length. Both ventral vessels are connected except in last part of proglottid by **transverse excretory canal.** Last proglottid forms a pulsatile **caudal vesicle** opening out by an excretory pore (Fig. 26.6).

2. Secondary vessels and capillaries. Secondary vessels are formed by the union of several **capillaries.** These are ciliated while longitudinal vessels are lined by cuticle and are non-ciliated.

3. Flame cells. These are irregular bodies with granular nucleated cytoplasm and a central funnel-shaped space. Flame cells are found all over in the parenchyma. Many long flagella arise near the nucleus and flicker constantly like the flame inside the intracellular space. This space is continous with the space of excretory vessels through connecting vessels.

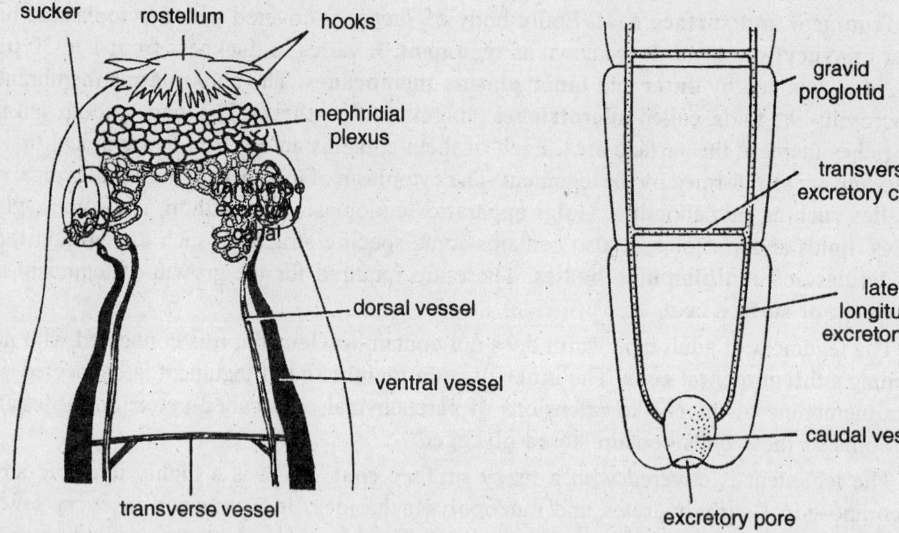

Fig. 26.5. *Taenia solium.* Excretory system of anterior part of body.

Fig. 26.6. *Taenia solium.* Excretory pore.

Function. Metabolic waste products such as ammonia, fatty acids, organic acids, etc., are removed by the excretory system of *T. solium*. Excretory system is also **osmoregulatory** in function.

26.7 NERVOUS SYSTEM

Nervous system of *Taenia* is not advanced. It consists of two **cerebral ganglia** in the scolex. These ganglia are connected by a thick transverse nerve band and by the **dorsal** and **ventral commissures** to form the **brain complex.** The brain complex is connected with another **rostellar nerve ring** having a pair of **rostellar ganglia,** in the rostellum. The suckers and rostellum are supplied by slender nerve from cerebral and rostellar ganglia. From the posterior side of brain complex, five pairs of **longitudinal nerve cords** arise and travel in the strobila. Out of these only two, the **lateral longitudinal nerves** are best developed and run along the entire length of strobila. They lie just outside the longitudinal excretory vessel and are connected in each proglottid by a **ring connective** situated below the transverse excretory canal.

26.8 REPRODUCTIVE SYSTEM

Taenia solium is **hermaphrodite.** Each mature proglottid contains a complete set of male and female reproductive organs (Fig. 26.7).

1. Male reproductive organs. The male reproductive organs are testes, vasa efferentia, vas deferens, cirrus or penis and genital atrium.

***(i)* Testes.** Testes are small, numerous, round (follicular) bodies scattered throughout in the dorsal part of the proglottid.

***(ii)* Vasa efferentia.** These are numerous minute ducts which arise from testes and collectively open into the vas deferens.

***(iii)* Vas deferens.** It is the common sperm duct. It is a long coiled tube and arises from the middle of the proglottid and transversely either to the left or right to open into the genital atrium.

***(iv)* Cirrus and cirrus sac.** The distal end of the vas deferens becomes muscular forming cirrus sac and modified into protrusible **cirrus** or **penis.** The cirrus is beset with spines, bristles or hooks and opens into the genital atrium through male genital pore.

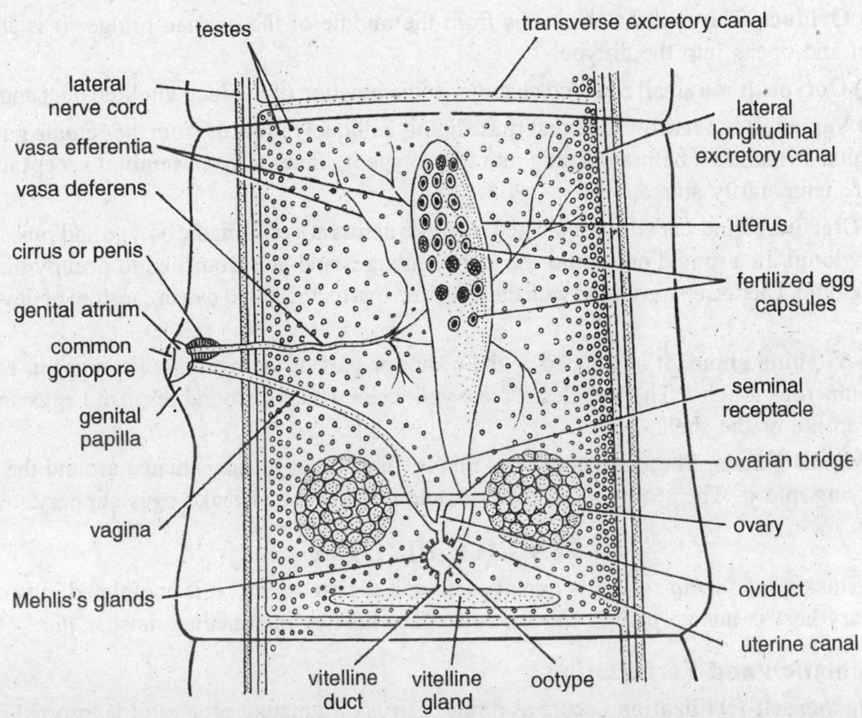

Fig. 26.7. *Taenia solium.* A mature proglottid showing male and female reproductive organs.

(*v*) **Genital atrium.** It is a cup-shaped structure, situated in the middle of proglottid at its lateral margin. Genital atrium is a common chamber and opens to the outside through a common **gonopore.** The gonpore is situated at the tip of a tiny protuberance, called the **genital papilla.** Genital papillae with gonopore occur alternately on the right and left margin of mature proglottids.

2. Female reproductive organs. The female reproductive system consists of ovary, oviduct, ootype, vagina, uterus, vitelline gland and Mehlis's glands (Fig. 26.7).

(*i*) **Ovary (Germarium).** It is a bilobed structure situated ventrally in the posterior part of the proglottid. Both the lobes of ovary are connected by a transverse ovarian bridge or **isthmus.** Each lobe of ovary is formed of radially arranged follicles. Each follicle produces yolkless ova.

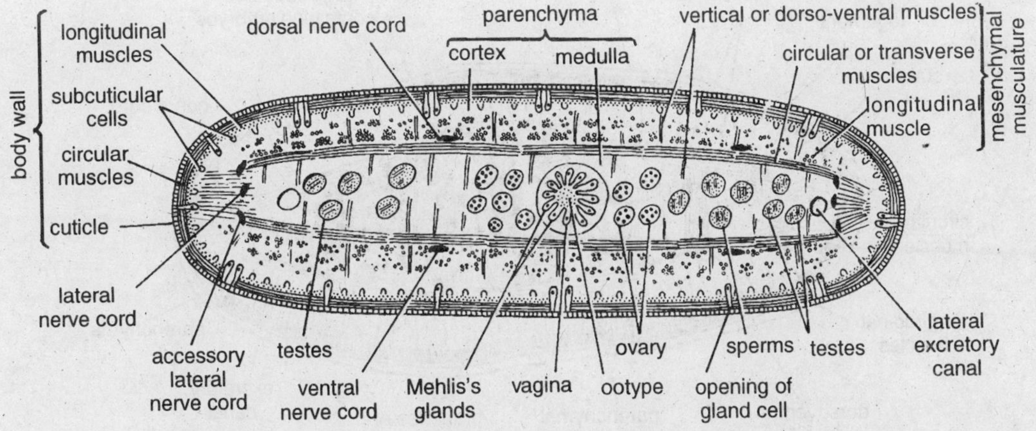

Fig. 26.8. *Taenia solium.* T.S. mature proglottid passing through ootype.

(ii) Oviduct. The oviduct originates from the middle of the ovarian bridge. It is short but wide duct and opens into the ootype.

(iii) Ootype. It is a small rounded chamber at the junction of oviduct, vitelline duct and uterus.

(iv) Vagina. It is a narrow tube that runs slightly obliquely inwards from the **female gonopore** in the genital atrium. Just before opening into the ootype it dilates to form **seminal receptacle.** The sperms are temporarily stored here.

(v) Uterus. A blind-ended, club-shaped, sac-like uterus arises from the ootype and runs forward in the proglottid. In a gravid proglottid the uterus enlarges and gets branched to occupy the whole of the proglottid. Uterus remains filled with the fertilized ova (ectolecithal ova in capsule) or developing embryos.

(vi) Vitelline gland. It is situated at the posterior part of the ovary. It is compact, elliptical mass of numerous folicles. The follicles secrete yolk cells which surround fertilized eggs and form the egg capsule or the shell.

7. Mehlis glands. These are numerous minute unicellular glands situated around the ootype and emptying into it. The secretion of these glands lubricates and make eggs slippery.

26.9 LIFE HISTORY

Life history of *Taenia solium* is complicated and **digenetic,** *i.e.*, it is completed in two hosts. The primary host is human beings and the secondary host or intermediate host is pig.

1. Copulation and Fertilization

In *Taenia* **self-fertilization** occurs as a rule. Cirrus of a mature proglottid is inserted into the vagina of the same proglottid and sperms are transferred into the vagina. From the vagina the sperms come to lie in the seminal receptacle from where they fertilize the ova in oviduct.

Cross-fertilization due to **copulation** between different proglottids of the same tapeworm is also very common. *Taenia*, in fact, is **protandrous** and its testes mature prior to ovaries (in the anterior most mature segments). Thus, anterior mature proglottids can copulate with posterior most proglottids due to folding of the strobila.

2. Capsule Formation

In the ootype, fertilized ovum, egg or **zygote** becomes associated with a yolk-cell secreted by the vitelline gland. Materials of the yolk-cell forms a hard, resistant **shell** around the zygote and

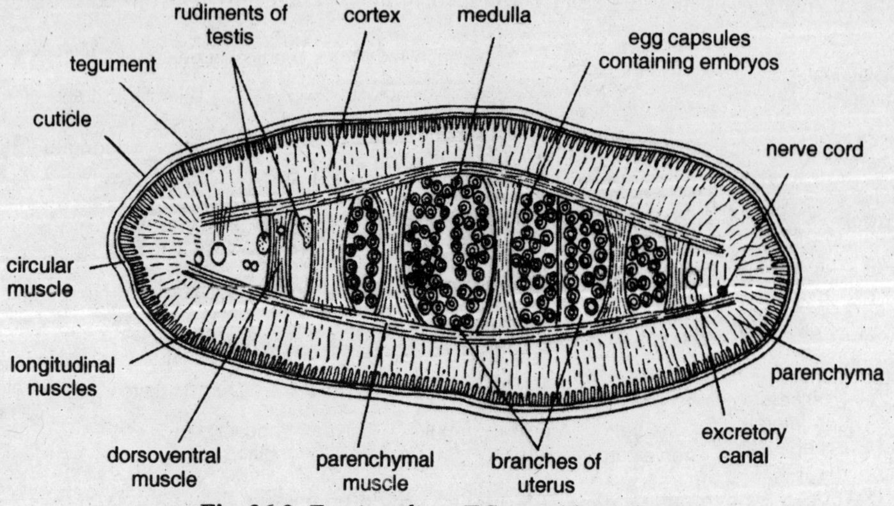

Fig. 26.9. *Taenia solium*. T.S. gravid proglottid.

yolk cell. The structure thus formed is called **capsule** which passes into the uterus. The secretion of Mehlis's gland facilitates the passage of capsule in the uterus. The uterus grows in size and becomes branched as it receives more and more (egg) capsules. Now the proglottid is known as **ripe** or **gravid** (Fig. 26.10). Soon the gravid proglottid breaks off (apolysis) and passes out of the host's intestine with his/her faeces during defaecation.

3. Fertilized Eggs

The egg capsule is very small (about 40 microns in diameter). It contains a zygote and a large yolk cell and remains surrounded by a hard and resistant egg shell (Fig. 26.11B).

4. Development

In *T. solium* development is **indirect** and includes a single larval stage.

Fig. 26.10. *T. solium.* A gravid proglottid showing branched uterus.

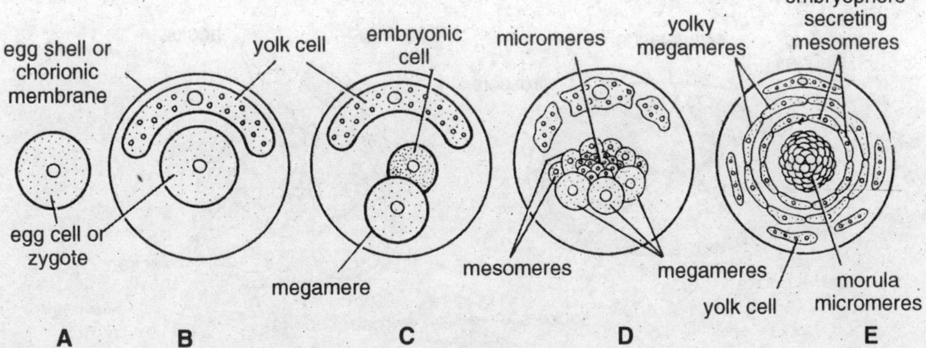

Fig. 26.11. T. solium. Early development. A—Zygote; B—Capsule; C—First cleavage; D—Embryo with micromeres, mesomeres and megameres; E—Morula.

(*i*) **Cleavage and morula formation.** Cleavage starts when the egg capsule is in the uterus (inside the intestine of host). The first division is **unequal.** It produces a large **megamere** and a small **embryonic cell.** The megamere divides repeatedly to form many megameres. The embryonic cell divides repeatedly and produces two types of cells—large **mesomeres** and small **micromeres** (Fig. 26.11D). The micromeres form a ball of cells, called **morula,** in the centre. Mesomeres form an envelope around the morula. The megameres form an outer covering around the mesomere envelope. The megameres absorb yolk from the yolk cell and supply nourishment to the developing embryo. With time the yolk cell reduces in size, breaks up and gradually disappears.

(*ii*) **Formation of hexacanth and onchosphere.** As development proceeds, the middle mesomere layer forms a thick, hard, cuticularized and striated **shell** or the **embryophore** around the morula. A basement membrane is also formed beneath the embryophore. The morula forms the **embryo** proper. Embryo develops six (= 3 pairs) chitinous **hooks.** These hooks are secreted by **onchoblast** cells of the embryo. This six-hooked embryo is called **hexacanth.** It also possesses a pair of **penetration glands** in between hooks. Hexacanth is surrounded by two **hexacanth membranes** (Fig. 26.12). The hexacanth embryo, hexacanth membranes, basement membrane, embryophore and shell together is known as **onchosphere** (Fig. 26.12B).

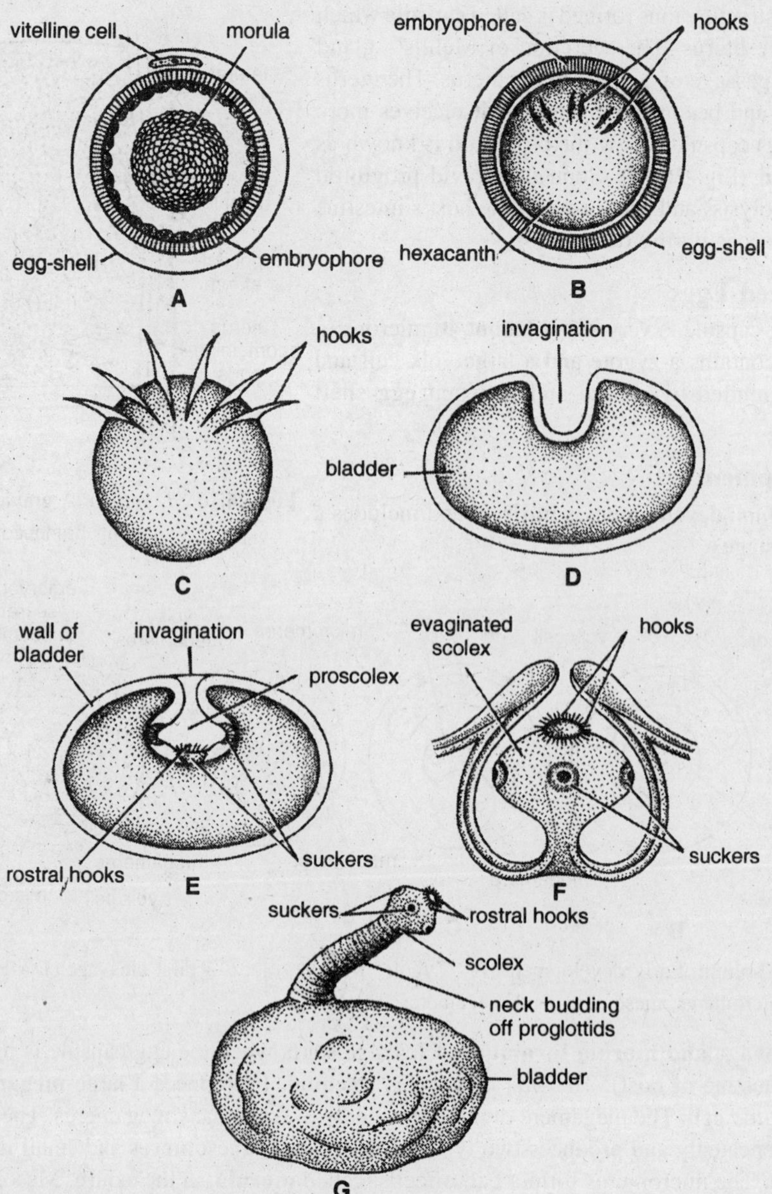

Fig. 26.12. *Taenia solium.* Stages in life cycle. A—Young onchosphere; B—Mature onchosphere; C—Free hexacanth; D—Bladderworm with invagination; E—Bladderworm with proscolex; F—Bladder-worm with evaginated scolex; G—Cysticercus with neck budding off proglottids.

By the time the onchospheres are formed, the uterus in the gravid proglottid develops 7 to 13 branches on each side and fills the entire proglottid. The uterus of a gravid proglottid contains 30,000 to 40,000 onchospheres.

5. Infection of the Secondary Host (Pig)

The oncospheres which occur either in dislocated gravid proglottids or lie freely in the host's stool are eaten by the pigs (due to pig's **coprophagous** feeding habit). On reaching the stomach of the secondary host, the capsule (shell) and embryophore are digested by the gastric juices and the hexacanth is released. The hexacanth starts boring through the intestine wall with the help of its paired penetration glands. Hooks help in anchorage of hexacanth with intestinal wall. Hexacanth loses its onchospheric membranes and hooks in the lumen of intestine and enters the submucosal blood vessels. It is carried around by the blood in the host body. Eventually, these come out of blood capillaries to settle in the striated muscles of such organs as the limbs, neck, tongue, heart, etc. Once settled in muscles, each embryo absorbs a large amount of watery fluid (plasma) of host's blood and grows to a spherical, pea-sized sac-like cyst, called **bladderworm** or **cysticercus larva.** The cysticercus of *Taenia solium* is called **cysticercus cellulosae.** The flesh of pig or pork containing many cysticerci appears white spotted resembling something like that of measles, hence, it is characteristically called **measly pork.** Thus, the pig becomes infected with larvae of *Taenia solium.*

6. Development of Bladderworm or Cysticercus Larva

The hook-less hexacanth gradually becomes encysted by the connective tissue of the host (pig). All cells of cysticercus larva fall into two layers around the fluid-filled central cavity. Cells of the outer layer coalesce to form a syncytial **tegument**. The inner layer of cells is called **mesenchyme** or germinal layer. At the future anterior end (opposite to the side where hooks were present) the wall of cysticercus thickens and then invaginates into the cavity (Fig. 26.12D). Four suckers, hooks and rostellum develop upon the knob-like invagination which thus assumes the form of an inverted scolex, called **proscolex**. Formation of cysticercus or bladderworm with proscolex is completed in host muscles in about 4 months after ingestion of eggs or onchospheres.

The cysticerci do not develop further, their further development takes place only when they are eaten up by the primary host. Thus fully formed cysticerci or bladderworms are infective to human hosts. These usually survive (in dormant state) in the flesh for several years.

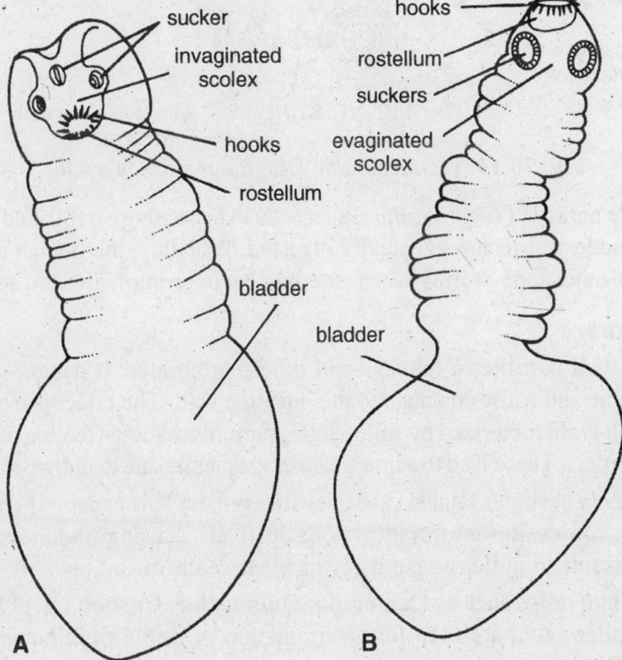

Fig. 26.13. *Taenia solium.* Cysticercus larva. A—Invaginated scolex; B—Evaginated scolex.

7. Infection of the Primary Host (Human)

Pork-eating people get the infection of *Taenia solium* by eating raw or imperfectly cooked measly pork. In their intestine, the outer cyst wall is digested off, the proscolex of a bladder worm everts out and anchors to the intestinal wall by the help of suckers and hooks. The bladder is progressively degenerated. The neck begins to produce proglottids by budding, In about 10 to 12 weeks, the parasite attains adult condition with gravid proglottids. Complete life cycle of *T. solium* has been depicted diagrammatically in Fig. 26.14.

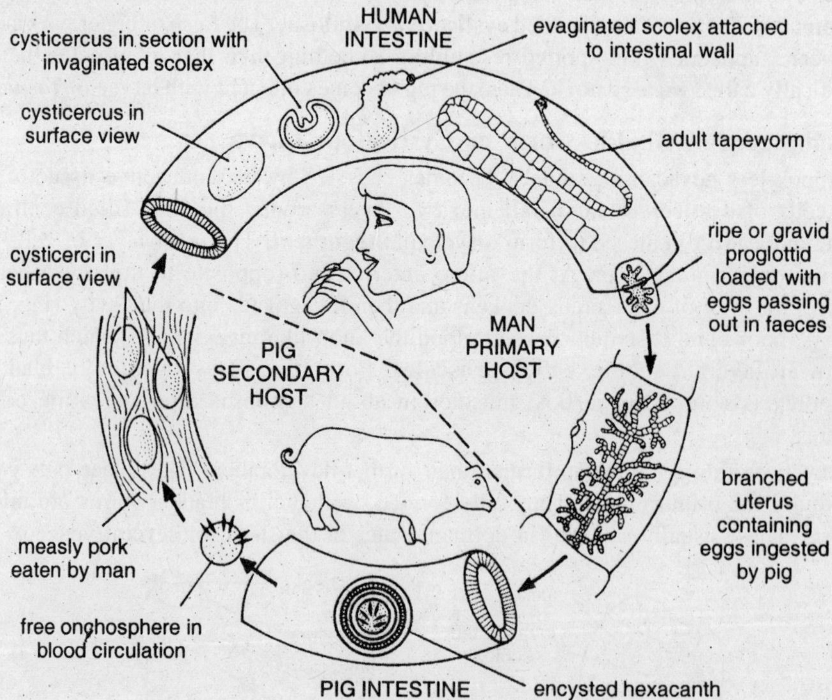

Fig. 26.14. *Taenia solium*. Diagrammatic life cycle.

Usually a single parasite (*Taenia solium*) succeeds in becoming established in human intestine; all other ingested bladderworms are eventually digested. Probably, the human host develops some kind of immunity to other tape worms when one has become implanted on its intestinal wall.

Economic Importance

Taenia is intestinal parasite of humans and other vertebrates. It depends upon the digested food from host intestine and causes damage to the intestine wall. The effect is occurrence of disease called **taeniasis** which is characterized by following symptoms: nausea (feeling to vomit), abdominal pain and nervous disorders. These lead to acute weakness, anaemia due to indigestion and eosinophilia.

Infection of *Taenia* becomes fatal in case of **cysticercosis**. This occurs when hexacanths invade eye muscles or brain to develop into cysticercs respectively causing blindness or brain necrosis. Brain necrosis may result in epilepsy, paralysis or even death of the host.

Some antihelminth drugs such as **Camoquin, Quinacrine, Carbon tetrachloride, Antiphen, Yomesun, Dichlorophen, etc.,** are used for extermination of taenia from human intestine. Scolex is detached from host's intestine by surgery.

Preventive measures against infection of *Taenia solium* include proper cooking, sterilization of nightsoil and personal hygiene.

26.10 PARASITIC ADAPTATIONS OF *TAENIA SOLIUM*

Parasitism is an association between two organisms (one is **host** on which other called **parasite** lives) for nourishment and shelter without any compensation for the host. Therefore, parasite tends to cause harm to host. Sometimes it proves so harmful that it ultimately results in the death of the host. The parasitic flatworms such as *Taenia solium* which is an intestinal endoparasite of humans have undergone profound adaptations to suit its parasitic mode of life. These adaptations are called **parasitic adaptations** and are discussed under the following two headings:

1. Morphological adaptations
2. Physiological adaptations

A. Morphological adaptations

(*i*) External form.

1. Body of *Taenia* is covered with thick tegument, provided with scales, afford suitable protection to the parasite.
2. Body is dorsoventrally flattened and ribbon-like or tape-like.
3. The anterior end of the body or the scolex is modified into a holdfast or region of attachment. It has developed suckers and hooks to ensure a firm hold of the host tissue.
4. Neck region possesses power of adding new segments to the body so that the animal grows throughout its life.
5. Body consists of large number of body segments (proglottids).

(*ii*) Internal organization.

1. Elaborate organs for nutrition are not needed because parasite can absorb the readily available digested and semi-digested food of the host.
2. In *Taenia*, respiratory organs are absent, since it has **anaerobic** type of respiration.
3. Locomotory organs and sense organs are absent. Need for quick "response to stimuli" is associated with free active life.
4. Each mature proglottid contains a complete set of reproductive organs. The posterior gravid segments contain only the fertilized eggs in the branched uterus and the developing embryo. These have power of detachment.
5. Hermaphroditism is of distinct advantage as

(*i*) It ensures copulation even when few individuals are present.

(*ii*) After copulation both the individuals lay eggs.

(*iii*) In the absence of a companion (*e.g.*, *Taenia solium*), the parasite can reproduce off-spring.

(*iv*) Germinarium (ovary) is highly branched, so that it can produce large number of eggs.

6. Eggs are covered with resistant capsule.
7. Uterus can store numerous fertilized eggs.

B. Phys.ological Adaptations

(*i*) Protective mechanisms.

1. Walls of gut of host secretes mucus, which form a protective clothing around the parasite.
2. Parasite secretes antienzymes to neutralize digestive enzymes of the host.

(*ii*) Anaerobic respiration. The worm respire anaerobically by breaking down glycogen.

***(iii)* Osmoregulation.** It maintains an equilibrium with the osmotic pressure of the host. The pressure inside the parasite and outside remains the same so that there is no difficulty in exchange of water.

***(iv)* High fertility.** In its life-time, *Taenia* produces eggs in millions.

Adaptations in the Life History

1. To cover the risk of finding individuals of different sexes in one host, the parasite becomes hermaphroditic and performs self-fertilization.
2. Each mature proglottid produces about 30 to 40 thousand eggs. Such an over-production of offsprings is necessary to maximize the chances of reaching host and ensure the perpetuation of race.
3. The complicated life-history is a remarkable feature in the parasite. Secondary or intermediate host the pig is introduced to ensure parasite's dispersal from one primary host to another.

Retrogressive and Progressive Parasitic Modifications

Parasitic adaptations of *Taenia* can also be classified into following two types:

1. Retrogressive modifications of *Taenia*. Besides the loss of surface epidermis and cilia as in flukes, the tapeworms exhibit advanced parasitic adaptation by a complete loss of digestive organs. Tapeworms obtain digested nutrients from the chyme (= partly digested food present in duodenum) of the host through body surface by absorption and pinocytosis and through surface pores. No locomotory organs and sense organs occur in adult tapeworms or their larvae, because no part of their life-cycle is passed outside host bodies.

2. Progressive modifications of *Taenia*. Besides having the protective tegument in place of surface epidermis, as in flukes, the tapeworms show advancement in parasitic adaptations by a better development of attachment organs. However, the most important advancement is shown by tapeworms over the flukes in multiplication of reproductive organ and great increase in their capacity to produce eggs. For example, an adult *Taenia* forms 7 to 8 proglottids each day throughout its life and each proglottid has a complete set of genitalia (*i.e.*, male and female reproductive systems), producing 30 to 40 thousand eggs.

26.11 COMPARISON OF IMMATURE, MATURE AND GRAVID PROGLOTTIDS

Table 26.1 Differences between immature, mature and gravid proglottid of *Taenia*.

Immature proglottids	*Mature proglottids*	*Gravid proglottids*
1. Present in the anterior part of the the body next to neck region.	1. Present in the middle part of the body.	1. Present in the posterior part of the body.
2. Their width is more than the length.	2. Length and width of each proglottid is almost the same.	2. These are long than broad.
3. The reproductive organs are either absent or in stages of development.	3. Reproductive organs are fully developed with mature male and female reproductive organs.	3. After fertilization of the eggs the male and female reproductive organs are degenerated but uterus exists.
4. No uterus.	4. Uterus is small and unbranched.	4. Uterus becomes enlarged and highly branched and fills the entire space of the proglottid.
5. —	5. Uterus is without eggs.	5. Uterus contains fertilized eggs and developing embryos.

26.12 COMPARISON OF *FASCIOLA* AND *TAENIA*

Table 26.2. Comparison of structure and life-cycle of *Fasciola hepatica* and *Taenia solium*.

Fasciola hepatica (Sheep liver fluke)	*Taenia solium* (Pork tapeworm)
I. Habit, Habitat and Distribution	
1. Endoparasitic and digenetic.	1. Endoparasitic and digenetic.
2. Adults live in bile passages of sheep and goat **(primary host).**	2. Adults live in the intestine of humans **(primary host).**
3. Larval stages live in a freshwater snail, either *Limnaea* or *Planorbis* sp. **(secondary host).**	3. Larval stages are found in pig **(intermediate** or **secondary host)**.
II. Structure	
4. Body is bilaterally symmetrical, dorsoventrally flattened oval and **leaf-like.**	4. Body is bilaterally symmetrical, dorsoventrally flattened, elongated, **tape** or **ribbon-like.**
5. Its size varies from 1.2 to 5 cm in length and 1.2 cm in width.	5. It measures 3 to 6 metres in length.
6. Pinkish or dark brown in colour.	6. Opaque white or cream (pale yellow) in colour.
7. Anterior end of body contains a conical projection, called **head lobe.** Body is **unsegmented.**	7. Body is segmented or divided into knob-like **scolex** at anterior end of body, **neck** and **strobila.** The strobila contains a chain of about 1000 proglottids (segments).
8. Adhesive organs include oral or anterior sucker around the mouth and a ventral sucker. There are no hooks.	8. Scolex acts as an adhesive organ; it bears four cup-shaped suckers (one dorsal, one ventral and two lateral) and two rows of hooks for this purpose.
9. The neck is absent.	9. Neck is present and possesses the power of proliferation. It adds new proglottids to strobila throughout life.
10. Proglottids absent.	10. Proglottids are of three types: immature, mature and gravid proglottids.
11. Body aperture consists of a mouth, genital aperture, an excretory pore and a temporary opening for Laurer's canal.	11. Body aperture consists of a common gonopore in each proglottid and one or two excretory pores.
III. Body Wall	
12. The body is covered with a cytoplasmic syncytium or the tegument. It contains backwardly directed spines or scales and numerous microvilli.	12. Body is covered by tegument like the fluke. Tegument contains numerous long microvilli-like fold or microtriches. There are no spines or scales in the tegument. The hooks of scolex are formed by the tegument.
13. Musculature consists of circular, longitudinal and diagonal muscles.	13. Musculature consists of circular and longitudinal muscles. Diagonal muscles are absent.
14. Parenchymal musculature is absent.	14. Parenchymal musculature consists of longitudinal, circular and dorsoventral muscles.
IV. Digestive System	
15. Present but incomplete (anus is absent). It consists of mouth, pharynx, oesophagus and bifurcated intestine with diverticulae.	15. Absent.

Fasciola hepatica (Sheep liver fluke)	*Taenia solium* (Pork tapeworm)
16. Mode of nutrition is holozoic; it sucks its food (bile, blood and cellular debris) by muscular and suctorial pharynx.	16. Mode of nutrition is **saprozoic;** it absorbs its food through tegument.
V. Respiratory System	
17. Lacks definite organ for respiration. Respiration is anaerobic.	17. Same as in *Fasciola.*
VI. Nervous System	
18. It contains a nerve ring bearing a pair of **cerebral ganglia** and a **ventral ganglion** surrounding the oesophagus. From the nerve ring **three** pairs of longitudinal nerve cords run posteriorly.	18. It contains a pair of **cerebral ganglia** connected by a transverse nerve band and a ganglionated **ring commissure** in the scolex. Five pairs of longitudinal nerves which travel into the strobila present.
VII. Degenerated Organs	
19. Skeletal, locomotory, circulatory organs and organs of special senses are absent. However, eyes are present in miracidium larva.	19. These organs are also absent in *Taenia.*
VIII. Excretory System	
20. It consists of numerous **flame cells** and a median **excretory duct** which opens to outside by a single **excretory pore.**	20. It consists of **flame cells** and four main **excretory ducts,** only two of these extend up to the last proglottid and open through a **common excretory pore.** When the last proglottid is lost by apolysis, the two excretory ducts open separately.
IX. Reproductive System	
21. *Fasciola* is **monoecious** or **hermaphrodite.**	21. *Taenia* is also **monoecious** or **hermaphrodite.**
22. Only one set of male and female reproductive organs is found in the body of each organism.	22. One organism has many sets of male and female reproductive organs, one set of them occurs in each mature proglottid.
(A) Male Reproductive System	
23. Male reproductive system consists of following organs: (*i*) Testes (*ii*) Vasa deferentia (*iii*) Seminal vesicle (*iv*) Cirrus and cirrus sac (*v*) Ejaculatory duct (*vi*) Prostate glands.	23. Male reproductive system consists of following organs: (*i*) Testes (*ii*) Vasa efferentia (*iii*) Vas deferens (*iv*) Cirrus and cirrus sac.
24. There is a pair of large, tubular and highly branched **testes** situated in the middle of the body one behind the other.	24. There are numerous, small follicular (rounded) **testes** found scattered throughout the proglottid.
25. There are two **vasa defferentia** arising one from each testis and running almost straight.	25. **Vas deferens** is single and convoluted. It is formed by the union of fine ducts the **vasa efferentia,** arising from the testes.
26. Both vasa differentia run forwards and join to form a pear-shaped and muscular **seminal vesicle,** from which **ejaculatory duct** arises and enters into the muscular **penis** or **cirrus.**	26. Single vas deferens runs transversely outwards and opens into the muscular penis; **seminal vesicle** and **ejaculatory duct** are absent.

Fasciola hepatica (Sheep liver fluke)	*Taenia solium* (Pork tapeworm)
27. Ejaculatory duct is surrounded by many unicellular **prostate glands.**	27. **Prostate glands** are absent.
28. A **cirrus sheath** covers the cirrus, seminal vesicle and prostate glands.	28. Cirrus sheath or sac surrounds the cirrus only.
(B) Female Reproductive System	
29. Female reproductive system consists of following organs: (*i*) Ovary (*ii*) Oviduct (*iii*) Uterus (*iv*) Laurer's canal (*v*) Vitelline glands and yolk reservoir (*vi*) Mehlis's glands (*vii*) Genital opening	29. Female reproductive system consists of following organs: (*i*) Ovary (*ii*) Oviduct (*iii*) Ootype (*iv*) Uterus (*v*) Vagina (*vi*) Seminal receptacle (*vii*) Vitelline glands (*viii*) Mehlis's glands (*ix*) Genital opening
30. **Ovary** is single, tubular and highly branched. It is situated on the right side infront of the testes.	30. **Ovary** is single but **bilobed,** both lobes are connected by a transverse ovarian bridge or **isthmus.** It is situated in the posterior part of **proglottid.**
31. An **oviduct** extends from the ovary downwards to meet the median vitelline duct.	31. An **oviduct** extends from the middle of **isthmus** and opens into ootype which receives vitelline duct.
32. **Ootype** is absent.	32. **Ootype** is present.
33. **Vagina** and **seminal receptacle** are absent.	33. A slender and tubular **vagina** extends from the genital atrium which expands near the oviduct to form seminal receptacle.
34. A blind **Laurer's canal** extends from the oviduct to the dorsal side. It opens temporarily at the time of copulation and acts as vagina.	34. Laurer's canal is absent.
35. **Vitelline glands** are numerous, rounded vesicular gland scattered on the lateral sides of the body. **Yolk reservoir** is present.	35. **Vitelline gland** is single, large, lobulated situated in posterior part of each proglottid below the ovary. Yolk reservoir is absent.
36. **Uterus** is a long, convoluted and tubular structure. It opens into the genital atrium by female gonopore.	36. **Uterus** is straight, blind and club-shaped sac.
37. Uterus enlarges to store the capsule but does not branch.	37. Uterus enlarges and develops lateral branches to store the capsules.
38. **Mehlis's glands** are numerous found around the junction of oviduct, uterus and median vitelline duct.	38. Mehlis's glands are found around the ootype.
39. Reproductive organs persist even after capsule formation.	39. All reproductive organs except uterus degenerate after capsule formation; such proglottids are called gravid.

Fasciola hepatica (Sheep liver fluke)	*Taenia solium* (Pork tapeworm)
(X) Life History	
40. Usually two different flukes copulate so that **cross fertilization** occurs but sometimes self fertilization also occurs.	40. Usually **self fertilization** occurs but interproglottid cross fertilization is also common.
41. A fertilized egg along with yolk cell is packed into a **capsule.** Many capsules are stored in uterus and are laid in bile passage of the sheep. From there these capsules finally reach to the outside with host's faeces. Capsules are with operculum.	41. Many capsules, each containing a fertilized egg and a yolk cell, are stored in the uterus. The uterus enlarges and becomes branched. Gravid proglottids packed with capsules are detached from the body of the tapeworm (apolysis) and reach to the exterior with the host's (humans) faeces. Outside the host's body, the proglottids break to release capsules. Capsules are without operculum.
42. Development starts after capsules are laid and continue to develop if dropped at suitable places of temperature, moisture, etc.	42. Development starts after capsule formation in the gravid proglottid.
43. The life-history includes following five larval stages. (*i*) Miracidium (*ii*) Sporocyst (*iii*) Redia (*iv*) Cercaria (*v*) Metacercaria	43. The life-history includes two larval stages: (*i*) Hexacanth stage (*ii*) Bladderworm stage
44. Miracidia and cercaria are free swimming stages found in water.	44. No free-living larval stage. The hexacanth larva gets freed in the alimentary canal of secondary host.
45. The larva multiply asexually or parthenogenetically (polyembryony occur).	45. No parthenogenesis and polyembryony occur.
46. Sheep gets infection by ingesting metacercaria with vegetation.	46. Human beings get infection by eating improperly cooked pork infected with cysticerci (*i.e.*, measly pork).
47. Several flukes can live in a single host.	47. Only one tapeworm can live in one host.
48. Highly pathogenic, causing **facioliasis** or **liver rot.**	48. It is also highly pathogenic causing **taeniasis** and **cysticercosis.**

In the light of above discussion it can be concluded that *Taenia solium* is better adapted and more specialized to the parasitic mode of life.

QUESTIONS

Long Answer Questions

1. Give an account of the structure of *Taenia solium.*

2. Describe the systematic position and life history of *Taenia solium.* (*Indore 1987*)

3. Give an account of the reproductive organs of *Taenia.* (*Ranchi 1990; Himachal 1991; Nagpur 1991, 1994*)

4. Give an illustrated account of the reproductive organs of *Taenia solium*. What is meant by gravid proglottid?

5. Give an illustrated account of mature proglottid of *Taenia solium*.
(*Bundelkhand 1985; Himachal 1994, 1995*)
6. Differentiate immature, mature and gravid proglottid. (*Bundelkhand 1986*)
7. Describe the life history of *Taenia*.
(*Meerut (CCS) 1987, 1991; Rajasthan 1988, 1991; Gorakhpur 1990, 1992; Allahabad 1991; Bundelkhand 1992; Awadh 1992, 1994; Jodhpur 1993; Kanpur 1995*)
8. Describe life history and economic importance of *Taenia*.
(*Meerut (CCS), 1993; Kanpur 1993; Awadh 1994; Bihar 1990, 1994*)
9. What is parasitism? Explain how an animal adapts itself to the parasitic mode of life as evidenced by the structure and life history of *Taenia*. (*Gorakhpur 1986; Bundelkhand 1986*)
10. Give an account of the parasitic adaptation of *Taenia solium*. (*Allahabad 1991*)
11. How does the segmentation of an annelid worm differ from the repeating units of the body of a tapeworm?
12. What is adaptation? Explain with reference to the structure and life-history of *Teania solium*.
13. Indicate the specialization, adaptations and degeneration of *Taenia* consequent to its parasitic mode of life. (*Utkal 1986*)
14. Describe how the flatworms are adapted to their parasitic mode of life. Write a note on how *Taenia* infection can be prevented?
15. Compare general plan of life in *Fasciola* and *Taenia* and state which in your opinion is more suited to parasitic mode of life. (*Meerut (CCS) 1993*)
16. Describe the reproductive organs of *Taenia solium* and compare it with those of *Fasciola hepatica*.
(*Rajasthan 1989; Kanpur 1991; Bundelkhand 1991; Baroda 1990, 1995; Agra 1994*)

Short Answer Questions

1. Write short notes on the following:
   (*i*) Scolex
   (*Bundelkhand 1987; Meerut (CCS) 1990; Kanpur 1994; Himachal 1990, 1994; Magadh 1996*)
   (*ii*) Strobila (*Bundelkhand 1987*)
   (*iii*) Gravid proglottid (*Magadh 1996*)
   (*iv*) Onchosphere (*Bundelkhand 1988*)
   (*v*) Cysticercus (*Bundelkhand 1987; Meerut (CCS) 1991*)
   (*vi*) Measly pork (*Magadh 1990*)
   (*vii*) Taeniasis (*Bundelkhand 1987*)
   (*viii*) Economic importance of *Taenia* (*Kanpur 1995*)
2. Draw well labelled diagrams of the following:
   (*i*) *Taenia solium* (*Meerut (CCS) 1995*)
   (*ii*) Scolex (*Punjab 1989*)
   (*iii*) A mature proglottid of *Taenia* or Reproductive organs of *Taenia*.
   (*Bundelkhand 1987; 1994; Meerut (CCS) 1988; Gorakhpur 1990; Kanpur 1990, 1993, 1995; Awadh 1991, 1994; Jiwaji 1992; Lucknow 1992*)

Very Short Answer Questions

1. What is the difference between hexacanth and onchosphere?
Ans. Hexacanth is a six-hook containing early embryonic stage of *Taenia solium*. **Oncosphere** is a tapeworm embryo that is the earliest differentiated stage of cyclophyllidean tapeworm. The oncosphere contains hexacanth embryo enclosed inside hexacanth membrane, basement membrane, embryophore and egg-shell (Fig. 26.12B).
2. Describe structure and significance of cysticercus.
Ans. Inside the secondary host of *Taenia solium*, the pig, the hexacanth grows in size and gradually becomes encysted by the connective tissue of the host. A central cavity appears in the hexacanth embryo

due to breakdown of the cells. The cavity enlarges and becomes filled with a fluid. This fluid-filled cavity is called a **bladder** and the parasite is now called **bladder-worm** or **cysticercus** (Fig. 26.12E) Its wall is contractile. The wall thickens as a knob. The knob develops **rostellum**, four **suckers** and **hooks** and thus form the **scolex** or **proscolex**. The cysticercus measures 6 to 18 mm in diameter. The cysticerci do not develop further in the secondary host. These may remain in dormant condition for years. These develop further only when eaten by the primary host (the human beings).

3. Where is Mehlis's gland found? What is its function?
Ans. Mehlis gland is one of the small unicellular glands around the ootype of tapeworm. Each gland empties into ootype and its secretion tends to lubricate and make eggs slippery.

4. Flesh is digested in the stomach of human beings but why *Taenia* is not digested?
Ans. Thick tegument of body wall of *Taenia* is enzyme resistant. Tegument also secretes antienzymes to nullify the effect of host's digestive enzymes.

5. What is scolex of *Taenia*?
Ans. Scolex is the head of *Taenia* having a muscular pad with hooks and two pairs of lateral suckers.

6. What is flame cell?
Ans. Flame cell is hollow cell that contains the terminal branches of excretory vessels in flatworms such as liver fluke and tapeworm. It is involved in osmoregulation and excretion.

7. Two species of *Taenia* commonly infect humans. Name these species.
Ans. *Taenia solium, T. saginata*.

8. Why *Taenia solium* is regarded as a digenetic parasite?
Ans. When two hosts are involved in the life cycle of an endoparasite, it is called **digenetic.** *Taenia solium* is an endoparasite and digenetic as it completes its life cycle in two hosts, *i.e.*, human being is the primary host and pig is an intermediate or secondary host.

9. What is measly pork?
Ans. The pig flesh (or the pork) infected with cysticerci of *Taenia solium* is called measly pork due to its spotted appearance.

10. Where are the proglottids formed in a tapeworm?
Ans. Neck.

11. What is the name of the organ of attachment in tapeworm?
Ans. Scolex.

12. How many suckers are there on the scolex of tapeworm?
Ans. Four

13. Name the muscles present in the tapeworm.
Ans. Longitudinal, transverse and dorsoventral.

14. Name the excretory organs of flatworms.
Ans. Protonephridia (flame cells).

15. By which process is the excretory matter formed in *Taenia*?
Ans. Ultrafilteration and reabsorption.

16. Give two functions of tegument of a tapeworm.
Ans. Intake of nutrients by active transport, evasion of host's immune response.

17. What is an ootype?
Ans. Sac at the junction of oviduct, uterus and vitelline duct.

18. How many vitelline glands occur in *Taenia*?
Ans. One.

19. Name the larval stages found in *Taenia*.
Ans. Hexacanth, Cysticercus.

Fill in the Blanks

1. Pork-eating humans are infected with a tapeworm named ....
2. *Taenia* .... infects the beef eating humans.
3. Tapeworm has a digenetic life cycle like ....

4. Nutrition in *Taenia* is ....
5. Symmetry of tapeworms scolex is ....
6. Tapeworm is attached to the intestinal wall with .... and suckers.
7. New proglottids are proliferated from the .... of tapeworm.
8. The outer layer in the tapeworm's body wall is called ....
9. *Taenia solium* has about .... proglotids of which about .... are gravid.
10. A hook of tapeworm consists of .... , guard and ....
11. The secondary host in the life history of *Taenia* is ....
12. The proglottid of *Taenia solium* which contains nothing but fertilized eggs, is called ....
13. In *Taenia solium*, the proximal portion of the oviduct that ends into the vagina is called ....
14. The first larval stage in the life history of a tapeworm is ....
15. The sac-like stage of the developing tapeworm is ....
16. The disease in human beings caused by infection of second stage larva of *Taenia solium* is known as ....

Multiple Choice Questions

1. *Taenia saginata* is commonly known as
   (*a*) pork tapeworm (*b*) beef tapeworm
   (*c*) human tapeworm (*d*) none of these
2. Sucker in tapeworm are meant for
   (*a*) attachment to intestinal wall (*b*) catching food
   (c) ingestion of food (*d*) all of these
3. In human beings, *Taenia solium* is found in
   (*a*) duodenum (*b*) large intestine
   (*c*) small intestine (*d*) rectum
4. Segments of tapeworm are called
   (*a*) scolex (*b*) cysticercus
   (*c*) proglottids (*d*) onchospheres
5. In *Taenia*, scolex bears in the middle a prominent
   (*a*) cup shaped sucker (*b*) head
   (*c*) rostellum (*d*) recurved hooks
6. The number of hooks in *Taenia solium*
   (*a*) 32 (*b*) 28
   (*c*) nil (*d*) unlimited
7. The outer layer in the body wall of tapeworm is
   (*a*) noncellular cuticle (*b*) epithelial
   (*c*) syncytial (*d*) none of above
8. *Taenia* lacks
   (*a*) excretory system (*b*) digestive system
   (*c*) reproductive system (*d*) all of these
9. The excretory product of *Taenia* is
   (*a*) ammonia (*b*) fatty acid
   (*c*) urea (*d*) ammonia and fatty acid
10. Tapeworm has
    (*a*) one testis and many ovaries (*b*) many testes and two ovaries
    (*c*) one testis and a two-lobed ovary (*d*) many testis and a two-lobed ovary
11. A proglottid is called **gravid proglottid** when it has
    (*a*) both male and female reproductive units well developed
    (*b*) only female units well developed

(*c*) only male reproductive units well developed
(*d*) branched uterus filled with fertilized eggs

12. Apolysis is a process in *Taenia* for
(*a*) regular detachment of gravid proglottids from main body
(*b*) embryonic development
(*c*) ingestion
(*d*) infection to human beings

13. Pig acquires infection of *Taenia solium* by ingestion of
(*a*) oncosphere (*b*) hexacanth
(*c*) cysticercus (*d*) none of these

14. In which form does *Taenia* pass through faecal matter in human beings?
(*a*) oncosphere (*b*) hexacanth
(*c*) cysticercus (*d*) young tapeworm.

15. Cysticercus in the larva of
(*a*) *Scypha* (*b*) *Obelia*
(*c*) *Taenia* (*d*) *Fasciola*

16. The infective stage of *Taenia* in human being is
(*a*) hexacanth (*b*) onchosphere
(*c*) bladderworm (*d*) none of the above

17. The tapeworms are found in the
(*a*) voluntary muscles of human beings (*b*) voluntary muscles of pig
(*c*) liver of pig (*d*) body of snail

18. Cysticercosis is a disease caused by accidental infection of
(*a*) hexacanth of *Taenia* (*b*) bladderworm of *Taenia*
(*c*) encysted cercaria of *Schistosoma* (*d*) eggs of *Enterobius*

ANSWERS

Fill in the Blanks

1. *Taenia solium* **2.** *saginata*
3. Liver fluke **4.** Saprozoic
5. Radial **6.** Hooks
7. Neck **8.** Tegument
9. 850, 150 to 250 **10.** Blade, handle
11. Pig **12.** Gravid proglottid
13. Uterus **14.** Oncosphere
15. Cysticercus **16.** Cysticercosis.

Multiple Choice Questions

1. (*b*)	**2.** (*a*)	**3.** (*c*)	**4.** (*c*)	**5.** (*c*)	**6.** (*b*)
7. (*c*)	**8.** (*b*)	**9.** (*d*)	**10.** (*d*)	**11.** (*d*)	**12.** (*a*)
13. (*a*)	**14.** (*a*)	**15.** (*c*)	**16.** (*c*)	**17.** (*b*)	**18.** (*b*).

27

Parasitic Adaptations in Platyhelminthes (Flatworms)

Hickman *et al*., (1996) have defined *parasitism* as the condition of an organism living in or on another organism (the host) at whose expense the parasite is maintained. It is a sort of destructive symbiosis. Also, they have defined *adaptation* as an anatomical structure, physiological process, or behavioural trait that evolved by natural selection and improves an organism's ability to survive and leave descendants.

27.1 DEFINITION

Parasitism is an association of two organisms of different sizes and species, in which smaller organism is benefited and the larger organism is harmed. The organism which is benefited is known as **parasite**, and that which suffers, is referred to as **host**. The host can live without the parasite, but the parasite cannot survive without the host.

27.2 PERKS OF PARASITES

Parasite gets nourishment from the host. It feeds on tissues or body fluids of the host or utilizes the food ingested by the host. Internal parasites, the endoparasites, may have additional advantages, such as protection, lack of competition and ensured survival.

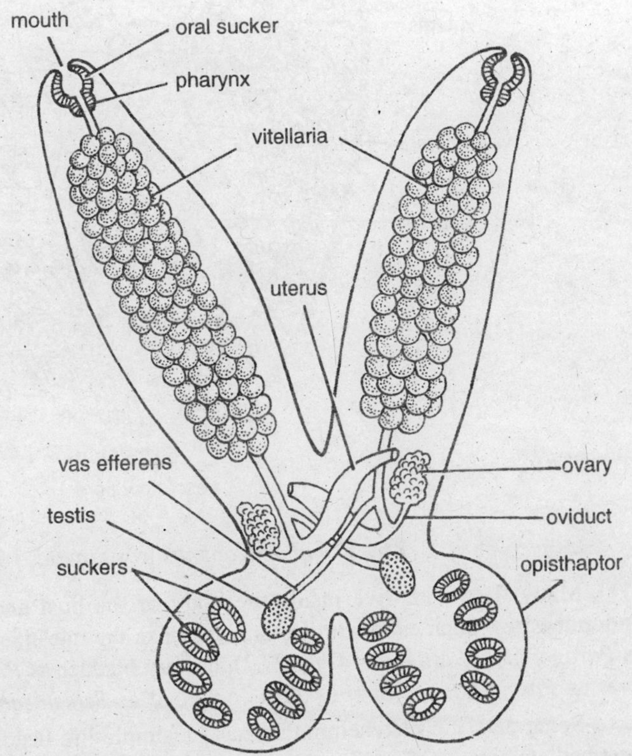

Fig. 27.1. *Diplozoon*. Two individuals are permanently united in the form of a cross.

27.3 TYPES OF PLATYHELMINTH PARASITES

The helminth parasites are parasites for the whole of their life, so are known as the **permanent parasites.** Some live on the surface of body of host and are called **external parasites** or **ectoparasites.** *Diplozoon indicum* is such an ectoparasite (Fig. 27.1). It inhibits the gills of freshwater fishes such as *Barbus* and feed on the blood of the host.

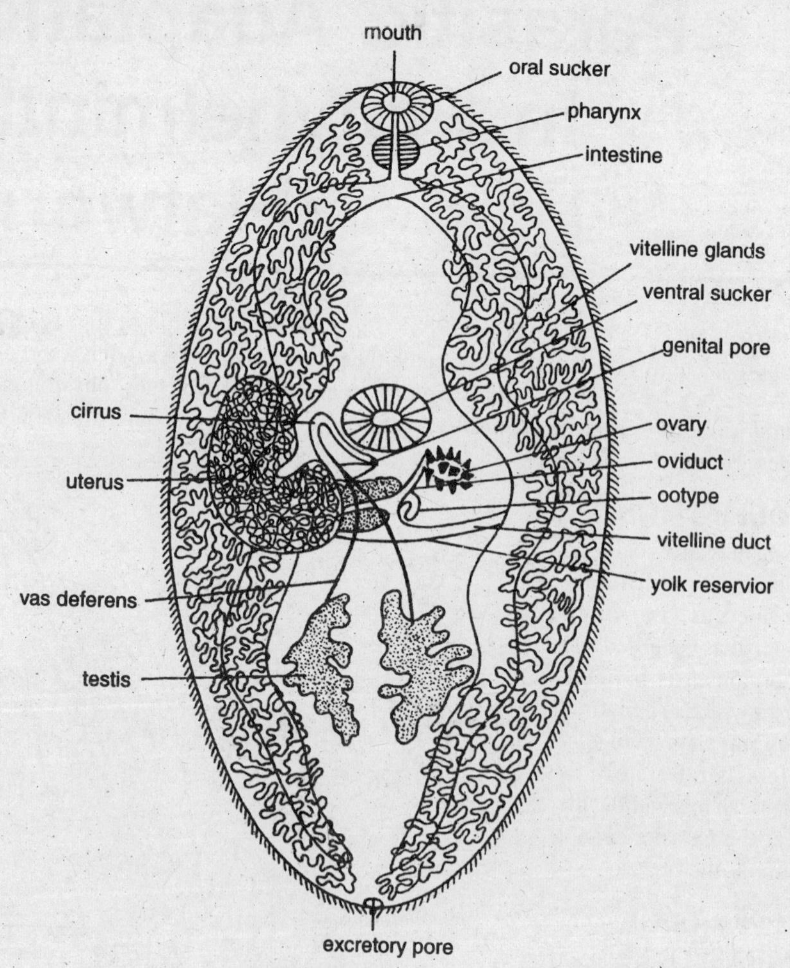

Fig. 27.2. *Paragonimus westermani*—the lung fluke.

Many flatworms live inside the body of the host and are called **internal parasites** or **endoparasites.** Such endoparasites may occur in the intestine as *Taenia solium* (Fig. 26.1), lungs as *Paragonimus westermani* (Fig. 27.2), urinary bladder as *Polystomum integerrinum* (Fig. 27.3), liver as *Fasciola hepatica* (Fig. 25.1) or blood as *Schistosoma* (Fig. 23.3).

Some parasitic platyhelminthes have a simple life history with a single larval form and no intermediate host. They are known as **monogenetic parasites**, *e.g., Polystomum, Diplozoon.* Many helminth parasites have a complex life history with several larval forms that inhabit one or more intermediate hosts, *e.g., Paragonimus, Taenia, Fasciola,* etc. The host for the adult stage of the parasite is called **definitive host** or **primary host** and that for the larval forms as the **intermediate host.**

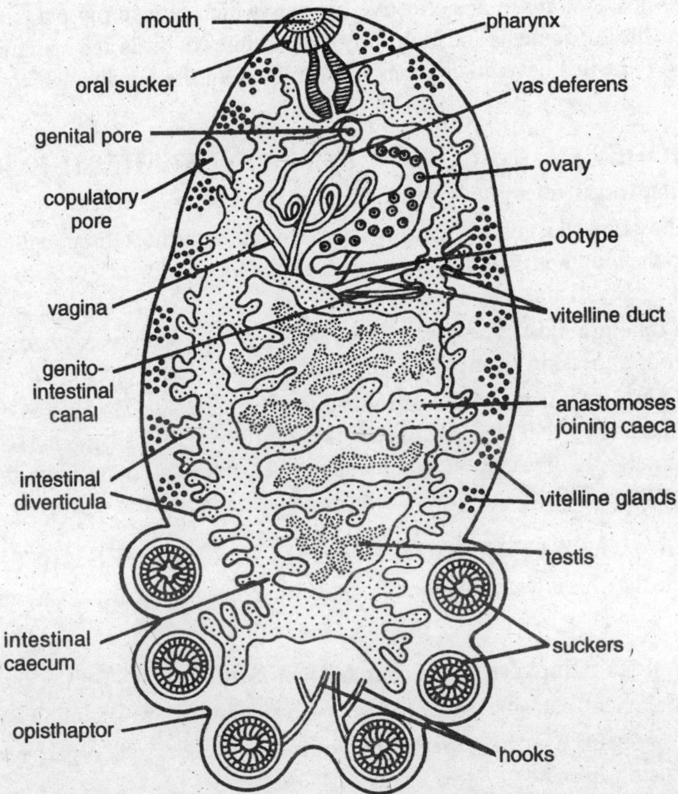

Fig. 27.3. *Polystomum*—lives in urinary bladder of frogs, toads and turtles.

Box 27.1 Kinds of Parasitism.

1. Accidental parasites. Some free-living animals are capable of leading a parasitic life for a certain period in a host, if swallowed accidentally.

2. Facultative parasites. When the animal can live both as parasite or free living, they are known as facultative parasites.

3. Obligatory parasites. These include forms which have lost power of living a completely free life, and must live on or in some other suitable organism during all or part of their lives, otherwise they die.

4. Ectoparasitism. In ectoparasitism, the animals live on the external surface of the body of host and totally depend for food on the host.

5. Endoparasitism. In endoparasitism, the animals live inside the body of host and depends entirely for the food on the host.

27.4 DIFFICULTIES OF PARASITES

Parasites face certain difficulties in the establishment and maintenance of a parasitic relationship with the host. A primary problem for the parasites is that of locating and infecting new hosts. Hosts often have defenses and respond to the invading helminth parasites as to the protozoan, bacterial and viral parasites. Any tissue damage provokes inflammation and phagocytic reactions of leucocytes (white blood cells). The host's immune system is also stimulated by the protein antigens on the parasites' surface. The parasites which overcome these defenses of the host may be walled off in the host's body. Hydrochloric acid and proteases (protein digesting enzymes) present in the vertebrate host's stomach are effective barriers which few sturdy parasites can cross. Low level of oxygen in the gut contents and high body temperatures of birds and mammals are also barriers for the parasites. Parasites have adaptations to overcome all the barriers which are posed by their hosts.

27.5 MODES OF INFECTION IN PLATYHELMINTH PARASITES

Parasitic flatworms infect new host in one of the following ways:

1. **By swallowing.** Most of the parasitic flatworms enter the host through the mouth as encysted larvae. Access to the mouth is effected with leaves, vegetables and fruits, or with improperly cooked meat (beef, pork, fish) of the intermediate hosts.

2. **Boring through skin.** The cercariae of blood flukes, such as *Schistosoma*, penetrate the human body through the skin from water of rivers, canals, tanks, etc.

3. **Direct contact.** Larvae of the gill-inhabiting ectoparasite *Diplozoon*, fix on the gills on hatching from the capsules, that are also attached to the gills by coiled filaments. Larva of *Polystomus* clings to the internal gills of a tadpole and metamorphoses into a young fluke that migrates to the urinary bladder during the metamorphic changes of the tadpole into a frog or toad.

27.6 ADAPTATIONS OF PARASITIC PLTYHELMINTHES

The parasitic flatworms have both morphological and physiological adaptations for their parasitic mode of life.

I. Morphological Adaptations of Parasitic Flatworms

The morphological parasitic adaptations of the flatworms are of following three types:

1. Loss of those organs which are useful only for free existence; such organs remain unnecessary for the parasitic life.
2. Continuation of pre-existing features found beneficial in parasitic life also.
3. Development of new structures in response to new needs of parasitic life.

1. Loss of organs. The organs lost by parasitic flatworms include locomotory, digestive and sense organs.

***(i)* Loss of locomotory organs.** The parasites have a abundant supply of readymade food and they have not to face the competition with its fellows. As a consequence, they do not need locomotory organs which have been lost. Tapeworms lack locomotory organs (cilia) throughout their life. Flukes, however, retain cilia in the free-living stage (miracidium) to move about in search of a snail.

***(ii)* Loss of digestive organs.** Endoparasitic flatworms of human intestine (tapeworm) are surrounded by digested food ready for absorption and assimilation. Therefore, they have lost the alimentary canal (the digestive organs) and have taken to saprozoic nutrition. Long and flat body of a tapeworm provides a large surface to facilitate this mode of nutrition.

***(iii)* Loss of sense organs.** The parasitic flatworms, particularly endoparasites, live in a uniform environment with practically no changes to be perceived. Their sense organs have therefore disappeared.

2. Continuation of preexisting features. The preexisting features continued in the endoparasitic flatworms include flat body and shell around the eggs.

***(i)* Thin, flat body.** The thin and flat body of the endoparasitic flatworms is beneficial in following three ways: 1. It enables the parasites to fit in narrow spaces in the host; 2. It quickens the elimination of carbon dioxide by diffusion; 3. It facilitates distribution of food digested in gut (*Fasciola*) or absorbed through the skin (*Taenia*).

(*ii*) **Shell around eggs.** A fertilized egg, alongwith one or more yolk cells, is enveloped by a resistant shell. The structure thus formed is called the **capsule**. The capsules leave the host and lie in the open, where the shell protects the egg or the developing embryo from injury, desiccation, high and low temperature, etc. This adaptation ensures survival of the parasite outside the host.

3. Development of new structures. The new organs developed by the parasitic flatworms include adhesive organs, tegument, sucking pharynx and cyst wall around the infective larvae.

(*i*) **Adhesive organs.** Adhesive organs have been developed in the endoparasitic flatworms to check dislodging of them from the host. Thus, once established within or on a host, it is advantageous for a fluke to stay more or less in one place. In that regard, nearly all of them are equipped with external organs for temporary or permanent attachment (Fig. 27.4, Fig. 27.5).

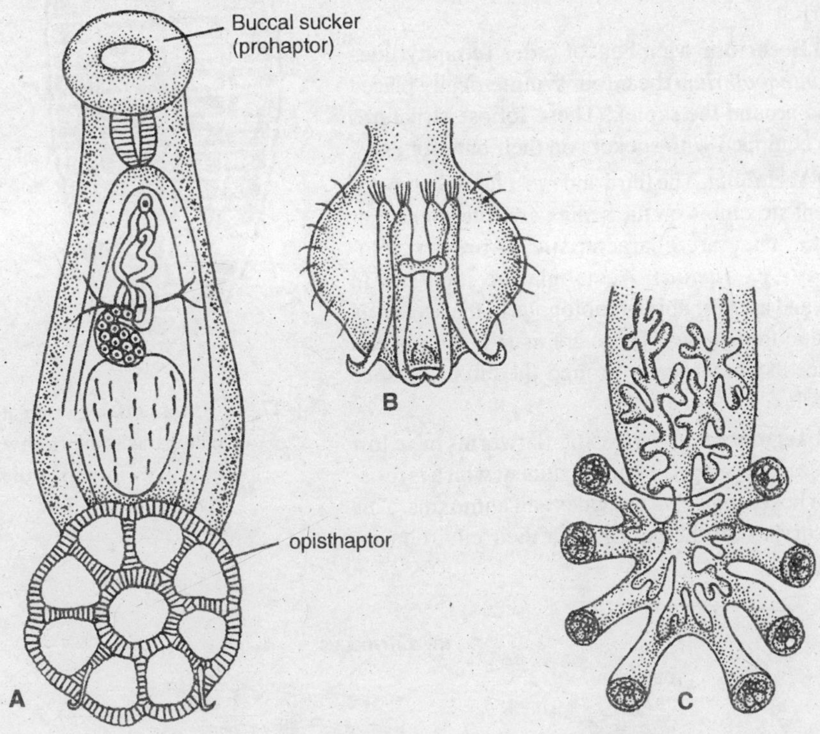

Fig. 27.4. Some attachment organs of monogenetic flukes. A—*Anoplocotyloides papillate*; B, C—Opisthaptor from monogenetic flukes.

Monogenetic flukes typically have an anterior and a posterior adhesive organs called the **prohaptor** and the **opisthaptor**, respectively. The prohaptor consists of a pair of adhesive structures, one on each side of the mouth, bearing suckers or simple adhesive pads. The opisthaptor is usually the major organ of attachment, and includes one or more well developed suckers with hooks and claws.

The digenetic fluke possess two hookless suckers. One, the **oral sucker**, surrounds the mouth, and the other, the **acetabulum** (*e.g.*, *Fasciola hepatica*). These suckers are usually supplied with adhesive gland cells, although the well developed ones operate mainly on suction produced by muscle action. The aspidogastrean flukes lack an oral sucker but have a large subdivided ventral sucker (Fig. 27.5).

Adult tapeworms do not move around much, but they are capable of muscular undulations of the body. They remain fixed to the host's intestinal wall by the scolex (in eucestoda) or adhesive organs (in cestodaria). The tip of the scolex in many cestodes (*e.g.*, *Taenia*) is equipped with a movable hook-bearing **rostellum**. In others (*e.g.*, *Cephalobothrium*) the anterior end bears a protrusible sucker, or adhesive pad, called a **myzorhynchus**. The rest of the scolex bears various suckers and sometimes hooks or spines. There are three categories of adhesive suckers upon which classification of cestodes is partially based.

(*a*) **Bothria.** They are elongate, longitudinal grooves on the scolex. They possess weak muscles but are capable of some sucking action. Bothria occur as a single pair in order Pseudophyllides (*e.g.*, *Diphyllobothrium*).

(*b*) **Bothridia.** Members of order Tetraphyllidea (*e.g.*, *Acanthobothrium*) bear four symmetrically placed **bothridia** around the scolex. These foliose structures are often equipped with suckers on their anterior ends.

(*c*) **Acetabula.** The third and most familiar type of attachment structures on the scolex are true suckers or **acetabula.** They are characteristic of order Cyclophyllidea (*e.g.*, *Taenia*). Acetabula are identical in structure and are probably homologous to the acetabula of digenetic trematodes. There are usually four acetabula, placed symmetrically around the circumference of the scolex.

(*ii*) **Tegument.** The parasitic flatworms have lost epidermis and developed a thick tegument which is resistant to the host's digestive enzymes and antitoxins. This ensures survival of the parasites in their environment (Box 27.2).

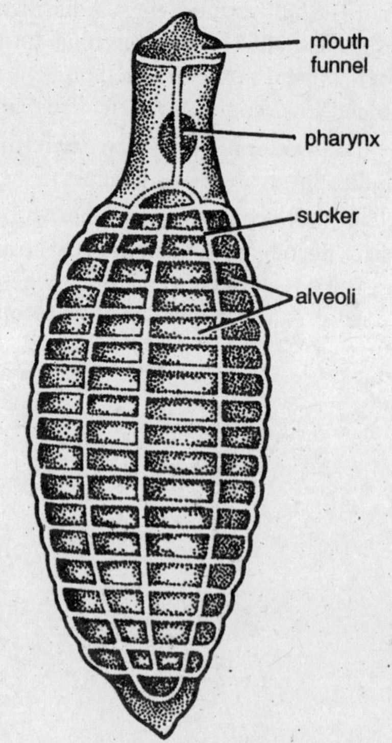

Fig. 27.5. *Aspidogaster* (a trematode). It is a endoparasite of freshwater mussel and lives in pericardial and renal cavities of the host.

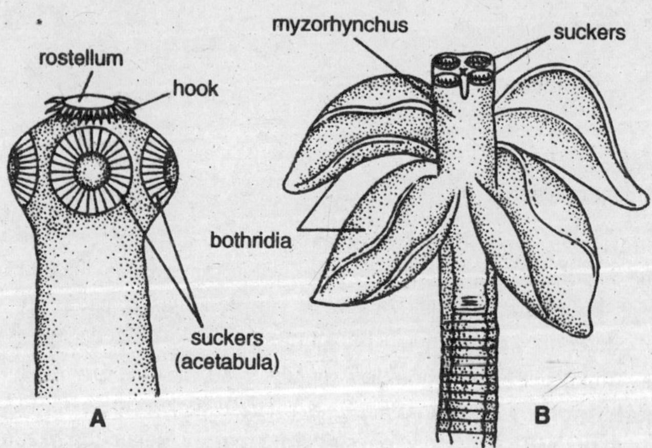

Fig. 27.6. Scoleces of two eucestodes. A—"Typical" scolex with rostellum, hooks and suckers (*Taenia solium*); B—Complex scolex with suckered myzorhynchus and leaflike bothridia (*Myzophyllobothrium*).

Box 27.2 Tegument: A key synapomorphy

The tegument of flukes and tapeworms not only provides some protection but is an important site of exchange between the body and the environment. Gases and nitrogenous wastes move across this surface by **diffusion**, and some nutrients, especially amino acids, are taken in by **pinocytosis.** In tapeworms, the uptake of nutrients occur solely across the body wall, and the surface area of the tegument is greatly increased by many tiny folds called **microtriches** (Fig. 26.4.). As one of the nature's more remarkable adaptations, these folds may interdigitate with the internal microvilli of the host's intestine and aid in absorption of nutrients.

The nature of the tegument in flukes and tapeworms is viewed by some zoologists as unique and of major phylogenetic importance (**Ehlers** 1984, 1986, 1995). The larvae of these parasitic flatworms have a "normal" ciliated epithelium over at least part of their bodies. However, this epidermis is shed, and postlarval stages develop a new, syncytial body covering, called **neodermis**, or **tegument.** This phenomenon occurs in no other animals and should be viewed as a **key synomorphy** uniting the Trematoda, Monogenea and Cestoda as a monophyletic taxon that **Ehlers** (1985) called **Neodermata** (in reference to the "new skin" of these animals) (**Brusca** and **Brusca**, 2003).

(*iii*) Cyst wall around the infective larvae. Infective larvae of many parasitic helminths get enveloped in a thick protective cyst wall that facilitates their access to the primary host. For example, the cercaria larvae of *Fasciolopsis* (Fig. 27.7) encyst on the aquatic plant *Trapa*, forming infective metacercaria larvae that are swallowed by human beings with *Trapa*. Hexacanths of *Taenia solium* encyst in muscles of pig, forming cysticerca. These encysted larvae are taken by human being with undercooked or raw pork.

II. Physiological Adaptations of Parasitic Flatworms

The physiological adaptations of parasitic flatworms include anaerobic respiration, secretion of antienzymes and mucus, osmotic adaptability, chemotaxis, excessive multiplication (high fecundity) and complex life cycle.

1. Anaerobic respiration. Many endoparasitic flatworms, such as liver flukes, intestinal flukes and tapeworms, live in media that contain little or no oxygen. They manage to live in such media by respiring in an anaerobic (anoxybiotic) manner. Without this ability, it would be difficult for them to survive.

2. Secretion of antienzymes and mucus. The intestinal flatworms secrete

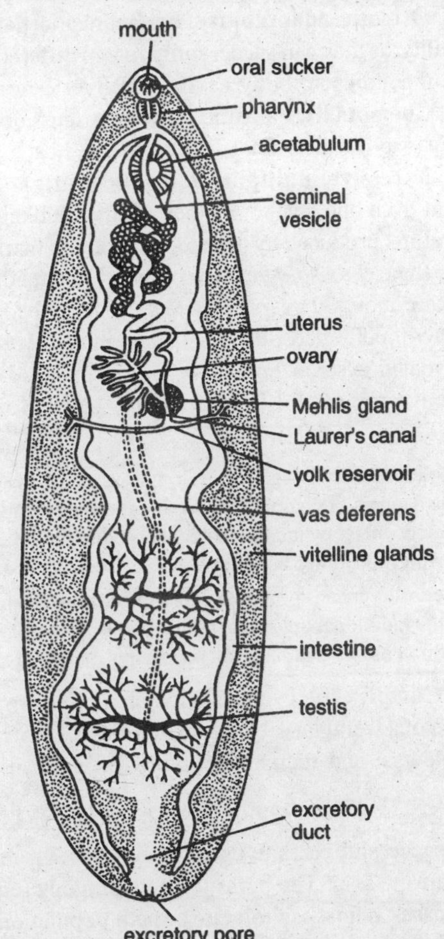

Fig. 27.7. *Fasciolopsis buski*. The intestinal fluke of human beings. Its intermediate host is snail *Planorbis* and *Segmentina*.

antienzymes which neutralize the digestive enzymes of the host. Tapeworms stimulate host's intestine to produce a large quantity of mucus, that surrounds the parasite to further reduce the action of host's enzymes. Lime cells in the body wall of the tapeworms, neutralize the acids formed in the host's gut (Box 27.3).

> **Box 27.3.**
>
> One study showed that *Hymenolepis diminuta* (a common tapeworm of rats and mice) releases proteins (*i.e.*, antitoxins or enzyme inhibitors) that appear to inhibit trypsin activity. The tapeworm can also regulate the pH of its immediate environment to about 5.0 by excreting **organic acids;** this acidic output also may inhibit the activity of trypsin (**Uglem** and **Just**, 1983).

3. Evasion of host's immune system. Some endoparasitic flatworms simply avoid host's immune system by dwelling at the sites, such as the gut, that are largely out of the reach of the immune system of the host. Blood flukes conceal themselves immunologically by absorbing the host's antigens on to their surface so that the host's immune system does not recognize them as "foreign" or "nonself".

4. Osmotic adaptability. Endoparasitic flatworms seem to possess a considerable osmotic adaptability, as they can successfully live in different media. Liver fluke of sheep, for example, passes through the stomach, body cavity and liver before reaching the bile duct.

5. Chemotaxis. Parasitic flatworms must be having marked chemotaxis that enables them to find their way to their final habitat in the host's body.

6. Excessive multiplication (Fecundity). The endoparasitic flatworms face a problem of migration from one host to another. Many are likely to perish in this adventure. Numerous capsules are, therefore, produced by the endoparasite, so that enough survive to continue the race. For example, one liver fluke (*Fasciola hepitica*) may lay about 50,000 capsules. A single tapeworm (*Taenia solium*) can live for about 30 years, and every year, it sheds about 2500 gravid proglottids, each containing 30,000 to 40,000 oncospheres. Formation of several larvae from one zygote in flukes and budding in the hydated cysts in *Echinococcus*, further add to multiplication. Bisexuality also increases the reproductive rate, as all individuals lay eggs (Box 27.4).

> **Box 27.4.**
>
> **High fecundity** is a general rule among parasites and the flukes and tapeworms are no exceptions. The dangers of complex life cycles and host location result in extremely high mortality rates that must be offset by increased zygote production or asexual processes. Flukes may produce as many as 100, 000 times as many eggs in free-living turbellarians (see **Brusca** and **Brusca**, 2003).
>
> **Fecundity** means productiveness of an individual. It is the innate potential reproductive capacity of the individual organism, as denoted by its ability to form and separate from the body the mature germ cells, *i.e.*, capable of producing many offspring.

7. Complex life cycle. The parasitic flatworms generally have complex life cycle due to the formation of several larval forms that inhabit one or more intermediate hosts. This adaptation brings about dispersal and reduces the period of exposure to the hazards of external environment.

27.7 HOST-PARASITE RELATIONSHIP

A successful parasite usually does not kill its host because the death of the host means the end of parasite also. The parasite or its progeny (eggs or larvac) must escape before the host dies. In many cases, a host can tolerate a small population of parasites without any serious consequences. If the effect of a parasitic infection is manifested in the host, the condition is called **disease.** The parasitic disease may damage host cells and tissues, and may block gut, blood vessels or bile ducts. The parasites may produce wastes or substances which have a toxic or allergic effect on the host.

The disease-producing parasites are called **pathogenic parasites** or **pathogens.** Here we can consider two serious human diseases caused by *Taenia solium*: taeniasis and cysticercosis.

1. Taeniasis. Normally only one *Taenia solium* (tapeworm) is found in the intestine of human being. It is because the presence of a tapeworm provides immunity (called **premunition**) to the host against fresh infection.

Adult tapeworm may cause mechanical injury by blocking the intestinal canal and by injuring the mucous membrane of the intestine. Symptoms of **taeniasis** include abdominal pain, increased appetite, weakness, loss of weight, nausea, difficult breathing, digestive disturbances, dizziness, insomenia, restlessness, false sensation and occasional convulsions and epileptic fits.

2. Cysticercosis. This disease is caused by cysticercus larvae. It is much more dangerous than taeniasis. It is caused by eating food contaminated with the oncospheres of tapeworm. It is also caused when the gravid proglottid is destroyed in the intestine of human beings and the released oncospheres and eggs are carried into stomach by reverse peristalsis. It is called **self-infection**. The hexacanth may reach to any part of the body including eyes, heart, spinal cord, brain and bones and form cysticercus cysts. Brain infection leads to epileptic fits, convulsions, violent headaches, giddiness, local paralysis, vomiting and brain tumours. Cysts in the bones cause spontaneous fractures. Symptoms result both from toxins produced by the parasite and mechanical pressure on adjacent tissues caused by the cysts.

QUESTIONS

Long Answer Questions

1. What are structural and functional adaptations of platyhelminthes to parasitic mode of life. (*Awadh 2001*)
2. What do you understand by parasitism and parasitic adaptations? Discuss parasitic adaptations in platyhelminthes.

Short Answer Questions

1. Write short notes on:

   (*i*) Tegument

   (*ii*) Adhesive organs of platyhelminthes.

Very Short Answer Questions

1. Where does *Schistosoma* live?

   Ans. Blood capillaries of bladder and intestine.
2. Name the intestinal fluke of human beings.

   Ans. *Fasciolopsis buski*.
3. Name a tapeworm having two intermediate hosts?

   Ans. *Diphyllobothrium latus* (*Note*. Primary host = human beings; Secondary hosts = a crustacean the cyclop and a fresh water fish).
4. Give the habitat of *Paragonimus*.

   Ans. Lungs.
5. What is a capsule?

   Ans. Egg plus one or more yolk cells enclosed in a shell.
6. Write the name of the disease caused by *Schistosoma mansoni*.

   Ans. *Schistosoma mansoni* is a blood fluke of humans which cause severe dysentery and anaemia.
7. Which organ is affected by infection of *Diphyllobothrium*.

   Ans. Human intestine.

Fill in the Blanks

1. Absorptive surface area of tapeworm is increased by the development of ....
2. Parasite gets food and .... from the host.
3. The host containing the adult stage of the parasite is called .... or .... host.
4. An intermediate host harbours the .... stages of the parasite.
5. *Diplozoon* is a .... parasite.
6. .... is a digenetic parasite.
7. *Enterobius vermicularis,* the common parasitic platyhelminth of human beings is found in ....
8. There is only single host in the life history of ....
9. Tapeworm body cannot be dissolved in the human digestive system because it has ....
10. When an intermediate host is not found, eggs and larvae of tapeworm ....

Multiple Choice Questions

1. Among the following, an ectoparasite is
   (*a*) *Fasciola* (*b*) *Paragonimus*
   (*c*) *Diplozoon* (*d*) *Taenia*
2. An unisexual fluke is
   (*a*) *Schistosoma* (*b*) *Opisthorchis*
   (c) *Fasciola* (*d*) *Paragonimus*
3. Locomotory and receptor organs are totally missing in
   (*a*) liver fluke (*b*) lung fluke
   (*c*) intestinal fluke (*d*) tapeworm
4. Flatworms have
   (*a*) ectolecithal egg (*b*) endolecithal egg
   (*c*) both a and b (*d*) neither a nor b
5. Cysticercus is the larva of
   (*a*) sponges (*b*) cnidarians
   (*c*) flukes (*d*) tapeworms
6. Tapeworm is
   (*a*) anoxybiotic (*b*) oxybiotic
   (*c*) aerobic (*d*) none of these
7. Four suckers present in *Taenia solium* is about
   (*a*) sucking food (*b*) attachment
   (*c*) crushing food (*d*) all of these
8. The parasite showing alternation of generations and alternation of hosts in its life cycle is
   (*a*) *Fasciola* (*b*) *Ascaris*
   (*c*) *Wuchereria* (*d*) *Taenia*

ANSWERS

Fill in the Blanks

1. Microtriches **2.** Shelter (protection) **3.** Definitive, primary **4.** Young (larvae)
5. Monogenetic **6.** *Fasciola* **7.** Caecum and large intestine **8.** *Polystomum*
9. Cuticle **10.** Are destroyed.

Multiple Choice Questions

1. (*c*) **2.** (*a*) **3.** (*d*) **4.** (*a*)
5. (*d*) **6.** (*a*) **7.** (*b*) **8.** (*a*).

28

Phylum Nematoda (Characters, Classification and Types)

28.1 ASCHELMINTHES

Gegenbaur (1859) created a group **Nemathelminthes** to place some pseudocoelomate animals. However, **Grobben** (1910) introduced the term **Aschelminthes** in place of Nemathelminthes to avoid confusion. The Aschelminthes (Gr., *askes* = cavity; *helmins* = worm) are a heterogenous assemblage of marine and freshwater animals, such as gastrotrichs, rotifers and roundworms (nematodes). Although previously recognized as classes within the phylum Aschelminthes, each of eight former classes now designates a distinct phylum. The informal name aschelminthes, however, is still a convenient term for reference for entire assemblage.

Definition

Aschelminthes have in common a small body size, a pseudocoelom, bilateral symmetry, a cuticle (except rotifers and acanthocephalans), adhesive tubes; a lack of complete segmentation, a digestive tract with an anterior oral and a posterior anal opening, no circulatory or respiratory systems, usually with separate sexes, and with a protonephridial system of flame cells or flame bulbs (except nematodes and tardigrades). Nuclear numbers within species are constant.

General Characters

1. Most free-living aschelminthes are tiny worm-shaped animals, ranging from microscopic size to a centimeter in length, and their small body size is correlated with several familiar design features. For example, a well developed coelom and blood vascular system are absent.
2. Cilia are used for locomotion in several groups, but in those that have a thick cuticle, muscular form of locomotion predominates.
3. Protonephridia are the typical excretory organs.
4. Body is composed of relatively few cells — some aschelminths have as few as 1000 somatic cells. The rotifer *Epiphanes senta*, for example, has 958 nuclei (or cells) and the male of the roundworm *Caenorhabditis elegans* has 1031 cells.
5. Many species have an invariant and genetically fixed number of cells—a phenomenon called **eutely**. In eutelic animals, mitosis commonly ceases following embryonic development and growth continues only through increase in the size of cells.
6. Well formed **head** is absent, but **mouth** and sense organs are present.
7. An external cuticle is present in some phyla but absent in others.
8. In general, those that have a thin flexible cuticle or that lacks a cuticle altogether have antagonistic **circular** and **longitudinal muscles,** acting on a hydrostatic skeleton. Circular

musculature is absent in aschelminthes that have thick cuticle and the action of longitudinal musculature is antagonized by the elastic cuticle.

9. Many species have many **adhesive glands** which open to outside of the body via projecting cuticular tubes.
10. The digestive tube is often a complete tube bearing a mouth and anus. A specialized pharynx is always present.
11. Most aschelminthes are **pseudocoelomate animals,** meaning that they contain a fluid-filled cavity unlined by coelomic epithelium and which arose as a persistent embryonic blastocoel. This feature however has exceptions: (1) since many aschelminths have no body cavity at all, that is, they are **acoelomate animals,** *e.g.*, gastrotrichs, most nematodes and kinorhynchs, all are acoelomate protostomes; (2) a few aschelminths, such as rotifers and large roundworms, do indeed have an expanded, fluid-filled, connective-tissue cavity, the **pseudocoel;** (3) in acanthocephalans and tardigrades, in which the body cavity is voluminous; body cavity does not develop as a persistent blastocoel (a blastocoel is absent in the development of these taxa), but rather forms anew as an enlarged space in the connective tissue, exactly as haemocoel develops in arthropods and molluscs.
12. Pseudocoelomic fluid functions as hydrostatic skeleton and as a medium for internal transport, but a heart is absent and body movement circulates the fluid irregularly.
13. Nematodes, rotifers and tardigrades that inhabit the high stress environments of water film around soil particles and mosses are capable of remarkable degree of cryptobiosis (see Box 28.1).
14. Most aschelminths are dioecious, cleavage is strongly determinate and development may be direct or indirect.

Box 28.1. Cryptobiosis

Metazons that inhabit mosses, lichens, temporary ponds and damp soils periodically face the threat of dehydration, and other physical extremes. Often, such animals have evolved reproductive adaptations, such as resistant eggs, which allow their progeny to survive the extreme conditions. This same ability to resist and survive adverse conditions occurs in the adults of some remarkable species of roundworms, rotifers and tardigrades. As their surrounding environment worsen, the animals enter a death-like state of suspended animation, called cryptobiosis (meaning "hidden life") and are revived when conditions again become favourable. Cryptobiotic animals can remain dormant for a period of months to years–some for even as long as a century before reviving. While dormant they survive dessication and temperature extremes that seem antagonistic with life. There are records of cryptobiotic animals surviving temperatures ranging from 150°C (for a few minutes) to –200°C (for days). Clearly the ability to become cryptobiotic is advantageous to animasls that live in briefly uninhabitable environments.

To successfully enter cryptobiosis, the animal must dry slowly, usually over several days, as would be typical in natural habitats. As the animals dry, they change their body shape to minimize surface area perhaps to retard the rate of desiccation. Many nematodes tightly coil their bodies, rotifers retract head and foot and adopt a spherical shape, and tardigrades contract into the shape of a barrel.

Cryptobiosis is truly a death-like state. Metabolism is almost immeasurably low, 0.01% of normal, or is entirely absent, and the water content of body decreases to less than 1%.

In the absencc of metabolism and near total dehydration, cryptobiotic animals are essentially dead but revivable. If metabolic processes actually cease, then somehow the structural order essential to life must be preserved in a sort of stable crystallized condition until water returns and life resumes. Two molecules— **glycerol** and the diasaccharide **trehalose**, appear to be important in the formation and maintenance of the crystalline state. Both are synthesized as the animals slowly enter cryptobiosis. Glycerol is believed to protect the tissues against oxidation and to replace water bound to biological macromolecules. Trehalose replaces water in membranes. In substituting for water, the glycerol and trehalose act as spacers to preserve macromolecular organization during dehydration.

Classification

The Aschelminthes include the following **eight** phyla (**Ruppert** and **Barnes,** 1994):

Phylum 1. Gastrotricha. Freshwater and marine species; glide over bottom on ventral cilia; temporarily anchor by means of adhesive tubes that contain duo-gland organs; mostly feed on bacteria which they ingest by a sucking pharynx; hermaphrodite, internal fertilization, direct development; 40 species.

Examples. *Macrodasys, Chaetonotus.*

Phylum 2. Nematoda. Marine, freshwater and terrestrial; have long, tapered and cylindrical bodies; pharynx is muscular sucking type; teeth or stylets are present in plant-feeding nematodes, mostly are animal and plant parasites; 1200 species of roundworms.

Examples. *Ascaris, Toxocara, Entrobius, Trichinella, Wuchereria, Loa, Rhabditis.*

Phylum 3. Nematomorpha. Adults are free-living occur in freshwater or damp soil; larva are parasitic on arthropods; dioecious, internal fertilization; 320 species of horse-hair worms.

Examples. *Gordius, Nectonema.*

Phylum 4. Rotifera. Cuticle absent; ciliated corona present; pharynx modified into a special grinding structure, called **mastax;** mastax bears trophi; 1500 species.

Examples. *Philodina, Notommata, Conochilus, Collotheca, Brachionus.*

Phylum 5. Acanthocephala. Parasites of gut of aquatic and terrestrial vertebrates; retractile proboscis bearing curved spines; cuticle and digestive tract are absent; food is absorbed through general body surface; 1150 species of spiny- headed worms.

Examples. *Acanthocephalus, Polymorphus.*

Phylum 6. Kinorhyncha. Chitinous cuticle and body wall are segmented; body includes 13 segments or joints; **head, neck** and eleven segmented **abdomen;** mouth is at the end of a spiny, anterior, retractile head; dioecious; development direct; molting occurs with growth; 150 species.

Examples. *Echinoderes, Pycnophyes.*

Phylum 7. Loricifera. (It is most recently discovered phylum of animals; it was described in 1983 by Danish zoologist **R. M. Kristensen**). Minute marine species with cone-shaped anterior region (introvert) that can be retracted into lorica-encased abdomen; about 9 species.

Examples. *Nanaloricus.*

Phylum 8. Tardigrade. Tiny bear-like inhabitants of beaches, soils, mosses and lichens; body bears four pairs of telescoping stubby legs, each of which is tipped with claws or adhesive discs; a pair of piercing stylets is used in combination with a sucking pharynx for feeding; dioecious, direct development; molting occurs with growth; 600 species of water bears.

Examples. *Batillipes, Macrobiotus.*

Phylogeny

The aschelminths assemblage appears to represent multiple evolutionary lines. One line includes the gastrotrichs, nematodes and perhaps, the nematomorphs; another may include the rotifers and acanthocephalans. Loriciferans may be linked to the priapulids and kinorhynchs. Tardigrades appear to be arthropod relatives.

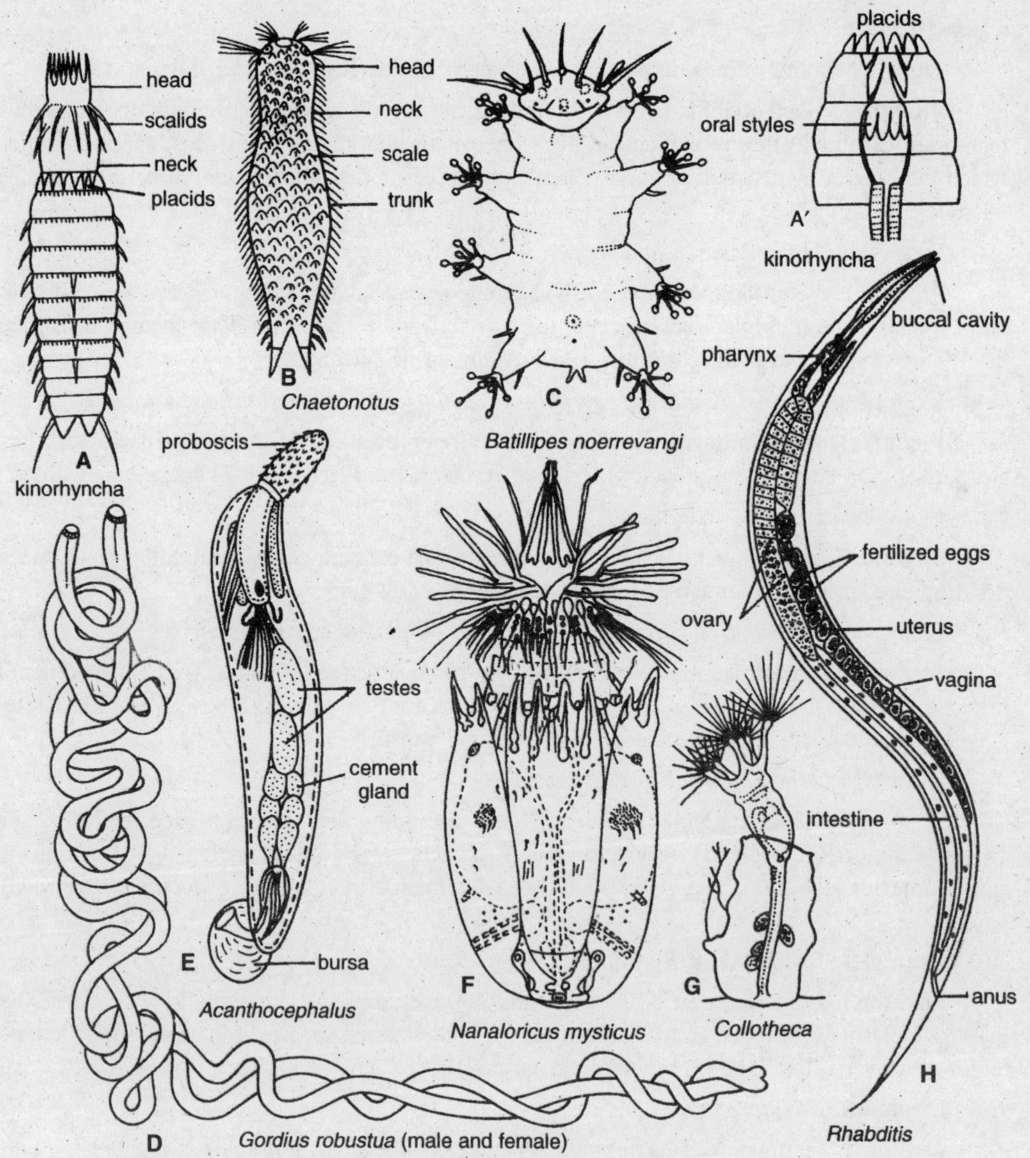

Fig. 28.1. Some Aschelminths.

28.2 PHYLUM NEMATODA

General Characters

1. Nematoda (Gr., *nema*=thread; *eidos*=form) include free-living (found in sea, freshwater and soil) or parasitic roundworms which are widely distributed.
2. The size and form of nematodes are adapted for living in interstitial spaces. Body of round worms are slender, elongated, cylindrical and tapering at both the ends. Most nematodes are microscopic in size.
3. Body is unsegmented, bilaterally symmetrical and not divisible into regions.

4. Mouth is located at somewhat rounded anterior end and is surrounded by lips and about 18 sensilla.
5. A **caudal gland** or **spinneret** is typical of many free-living nematodes.
6. Body wall consists of thick cuticle, epidermis (hypodermis) and longitudinal muscles. Circular muscles are completely absent.
7. Body is triploblastic. **Pseudocoel** is small or voluminous and filled with a fluid containing a variety of organic metabolites, including haemoglobin in some species but no circulating cells. A few **phagocytic cells** are permanently attached to the walls of the cavity and are important in internal defense.
8. Muscle fibres are arranged into four quadrants.
9. Coelom, circulatory system and respiratory system and cilia are absent.
10. Digestive system is well developed in nematodes. Nematodes are either carnivorous, phytophagous or microphagous (= feeding on diatoms, algae, fungi and bacteria).
11. Excretion of ammonium ions is performed by general body surface, excretory canal system or by excretory gland cells (called **renette**). Flame cells are absent.
12. Most nematodes move forward and rearward using sinuous eel-like undulations of the body. These undulations are in the dorsoventral plane and are produced by the alternate contraction of dorsal and ventral longitudinal muscles.
13. Sense organs are poorly developed, represented by small papillae and a pair of **amphids** on each side of head and a pair of **phasmids** at the posterior end.
14. Sexes are separate. Sexual dimorphism is well marked. In male, the tail is curved and tail bears a pair of pointed, curved, blade-like **copulatory spicules.** Sperms lack flagella and are amoeboid.
15. Fertilization is internal; parthenogenesis is common. Cleavage is strongly determinate. Development is direct.

Outline Classification

Phylum Nematoda

Classes (2)

1. Apasmida (Adenophorea)

Orders (6)

1. **Enoploidea**
   Examples.*Enoplus, Anticoma*
2. **Mermithoidea**
   Example. *Mermis, Allomermis*
3. **Dorylaimoidea**
   Examples.*Dorylaimus*
4. **Desmoscolecoidea**
   Example. *Desmoscolex, Tricoma*
5. **Monohysteroidea**
   Examples.*Monohystera, Siphonolaimus*
6. **Chromdoroidea**
   Examples.*Cromadora, Richtersia*

2. Phasmida (Secernentea)

Orders (10)

1. **Rhabditoidea**
   Examples.*Rhabditis, Heterodera*
2. **Rhabdiasoidea**
   Example. *Rabdias, Strogyloides*
3. **Oxyuroidea**
   Examples.*Oxyuris, Entrobus*
4. **Ascaroidea**
   Example. *Ascaris, Neoascaris*
5. **Strongyloidea**
   Examples.*Ancylostoma, Necator*
6. **Spiruroidea**
   Examples.*Rictularia, Thelazia*
7. **Dracunculoidea**
   Examples.*Dracunculus, Micropleura*
8. **Filarioidea**
   Examples.*Wuchereria, Loa*
9. **Tricuroidea**
   Examples.*Trichuris, Trichinella*
10. **Dioctophymoidea**
    Examples. *Dioctophyme, Hystrichis*

Detailed Classification

Phylum Nematoda is divided into **two** classes on the basis of presence or absence of phasmids.

Class 1. Aphasmida (Adenophorea)

1. Phasmids (caudal sensory organs) are absent.
2. Amphids (anterior sensory organs) are not pore like.
3. Cephalic organs, setae and hypodermal glands present.
4. Epidermal cells uninucleate. Cuticle four layered.
5. Caudal adhesive glands are present.
6. Usually two testes are present in males.
7. Most species are free-living; some are parasitic.

Class 2. Phasmida (Secernentea)

1. Phasmids are present.
2. Amphids open to outside by pores.
3. Cephalic sensory organs are present. Setae or papillae are not found in females.
4. Epidermal cells uni- or multinucleate; cuticle two or four layered.
5. Males contain single testis. Copulatory bursa is present in males.
6. Caudal adhesive glands are absent.
7. Most species are parasitic; free-living species mainly inhabit soil.

Class 1. Aphasmida

Class Aphasmida is divided into following **six** orders:

Order 1. Enoploidea

1. Mainly free-living and marine and of moderate size.
2. Anterior end with 6 labial papillae and with 10 to 12 sensory bristles.
3. Amphids are pocket-like, elongated or pore-like (called **cyanthiform**).
4. Cuticle is not ringed but with cuticular bristles.
5. Pharynx is a simple straight tube. Intestine ends in a blind pouch.

   Examples. *Enoplus*, *Anticoma*, *Synonchus*, *Triphla*, *Trilobus*.

Order 2. Mermithoidea

1. Long, slender and large sized juveniles are parasitic in invertebrates (mainly insects). Adults are free-living (in soil or water).
2. Anterior end with 6 labial papillae and no sensory bristles.
3. No buccal cavity is present, pharynx is long and intestine ends in a blind pouch. Adults do not feed, but live on food stores acquired during the parasitic existence.
4. Cuticle smooth and without bristles.
5. Amphids cyanthiform or reduced.

   Examples. *Mermis*, *Agameris*, *Paramermis*, *Allomermis*.

Order 3. Dorylaimoidea

1. Small sized, free-living nematodes found in soil and freshwater.
2. Anterior end with two circlets of labial papillae of 6 and 10 each.
3. Cuticle is smooth—without rings and bristles.

4. Buccal cavity with a hollow protrusible **spear** which is mounted on pharynx and used when sucking juices from plants and animal food organisms.
5. Rear part of pharynx is enlarged.
6. Amphids are sickle-shaped.

Examples. *Dorylamius*, *Tylencolaimus*.

Order 4. Desmoscalecoidea

1. Small sized marine nematodes.
2. Anterior end with four sensory bristles.
3. Head is armoured and distinct from body.
4. Amphids are crescent-shaped or hemispherical.
5. Cuticle is heavily ringed with prominent bristles throughout or in restricted areas.
6. Buccal cavity is reduced.

Examples. *Desmoscolex*, *Tricoma*.

Order 5. Monohysteroidea

1. Small sized, free-living, mostly marine and some are freshwater and some inhabit soil.
2. Anterior end with 4, 6, 8, or many sensory bristles.
3. Cuticle smooth or slightly ringed with scattered bristles.
4. Buccal capsule in the shape of a long tube.
5. Amphids are circular.

Examples. *Monohystera*, *Siphonolaimus*, *Cylindrolaimus*.

Order 6. Chromadoroidea

1. Small sized, free-living and mostly marine.
2. Anterior end with six lips and 10-papillae.
3. Amphids are spiral.
4. Ornamented cuticle is ringed and contains bristles.
5. Pharynx with a posterior bulb; usually with teeth in buccal cavity.

Examples. *Chromadora*, *Richtersia*, *Spilophorella*.

Class 2. Phasmida

Class Phasmida includes the following **ten** orders:

Order 1. Rhabditoidea

1. Medium sized worms. Free-living or parasitic in animals or plants.
2. Cuticle is smooth or ringed.
3. Pharynx with two bulbs.
4. Sensory bristles in the form of papillae arranged in two rings.
5. Amphids are reduced.
6. Phasmids are present in free-living forms.
7. Caudal glands are absent.
8. Males with copulatory spicules.

Examples. *Rhabditis*, *Heterodera* (lives in gall on the roots of plants), *Diplogaster*, *Tylenchus*, *Anguina*.

Order 2. Rhabdiasoidea

1. Medium sized worms. Parasite of vertebrates.
2. Cuticle is smooth.
3. Pharynx with a reduced bulb.
4. Life cycle includes both parasitic and free-living stages.
5. Parasitic stage is represented by either a protandrous hermaphrodite or a parthenogenetic female.

Examples. *Rhabdias* (lung parasite of frog), *Strongyloides* (intestinal parasite of human beings).

Order 3. Oxyuroidea

1. Small sized, pin-shaped parasitic worms.
2. Mouth is surrounded by 3–6 simple lips.
3. Pharynx contains a posterior bulb, containing valves.
4. Female contain long, slender tail.
5. Males with one spicule or two equal spicules.

Examples. *Entrobus* (Human pinworm), *Oxyuris, Heterakis, Histomonas.*

Order 4. Ascaroidea

1. Large sized, endoparasitic in digestive tract of vertebrates.
2. Mouth is surrounded by three prominent lips.
3. Pharynx is without bulb-like enlargement. Buccal capsule is absent.
4. Tail is ventrally coiled in males. Tail in female is blunt.
5. Males have two nearly equal sized copulatory spicules.

Examples. *Ascaris, Neoascaris, Toxocara.*

Order 5. Strongloidea

1. Moderate sized parasitic worms (*e.g.*, intestine of human beings).
2. Mouth without lips.
3. Pharyngeal bulb absent. Buccal capsule is well developed and has teeth.
4. Males contain a expanded **copulatory bursa** which is supported by 13 muscular rays.
5. Females contain **ovijector**.

Examples. *Necator* (commonly called **American Killer**), *Ancylostoma, Strongylus.*

Order 6. Spiruroidea

1. Moderate sized parasites.
2. Mouth with two lateral lips.
3. Pharynx has anterior muscular and posterior glandular portion; it lacks bulb.
4. The sensory papillae are present on the head. These are arranged in two circles of 6 and 4 or 8.
5. Males lack bursa but contain unequal and dissimilar copulatory spicules.
6. Males with spirally coiled tails.

Examples. *Thelazia, Rictuillria, Oxyspirura.*

Order 7. Dracunculoidea

1. Large sized parasites of vertebrates.
2. Mouth without lips and surrounded by 4 or 8 papillae.

3. Pharyngeal bulb is absent. Pharynx is muscular anteriorly and glandular posteriorly.
4. Viviparous.
5. Filiform and equal sized copulatory spicules are present; bursa is absent.
6. Females are large in size. Males are quite small.

Examples. *Dracunculus* (Guinea worm), *Philometra*, *Micropleura*.

Order 8. Filarioidea

1. Filiform, slender, moderate to large-sized worms.
2. They are parasites of body fluids and tissues of vertebrates, require a blood sucking insect as intermediate host.
3. Six labial papillae are present.
4. Mouth is without lips.
5. Pharynx is divided into an anterior muscular and posterior glandular parts; bulb is absent.
6. Males are smaller than females. Males have coiled tails.
7. In males, the bursa is absent and copulatory spicules are dissimilar and unequal.
8. In females, vulva is present near the anterior end.

Examples. *Wuchereria* (filaria), *Loa* (eye worm).

Order 9. Trichuroidea

1. Moderate sized parasitic worms of vertebrates.
2. Anterior part of body is filiform.
3. Mouth is without lips.
4. Pharynx is simple and slender and without bulb.
5. Males are without copulatory apparatus or contain an eversible spicules or cirrus.

Examples. *Trichinella*, *Trichuris*, *Capillaria*.

Order 10. Dioctophymoidea

1. Mouth is without lips but is surrounded with 6, 12, or 18 papillae.
2. Pharynx is elongated and without bulb.
3. Female has single ovary.
4. Male contains only one or no copulatory spicule.

Examples. *Dictyophyme*, *Hystrichis*, *Eustronglides*.

28.3. TYPE STUDIES

1. *WUCHERERIA BANCROFTI* (FILARIA)

Classification

Phylum	–	**Nematoda**
Class	–	**Phasmida**
Order	–	**Filarioidea**
Genus	–	***Wuchereria***
Species	–	***bancrofti***

It is commonly known as *Filaria bancrofti*.

Habit and Habitat

Wuchereria is a digenetic nematode which lives in the lymphatic vessels and lymph nodes of human. Male and female worms copulate in the lymph glands.

Structure

The body of filaria is thread-like tapering, transparent and often creamy-white in colour (Fig. 28.2). The female is usually larger than the male, measuring 8 to 13 cm in length and 0.2 to 0.3 mm in thickness. The male is only 2.5 to 4 cm long and 1 mm broad. The tail end of the male is curved and carries two spicules of unequal length. In female, tail is abruptly pointed. The male and female remain coiled together and are separated with difficulty.

The female is ovoviviparous and lays eggs which contain well developed larvae. The hatched larvae are called **microfilariae.**

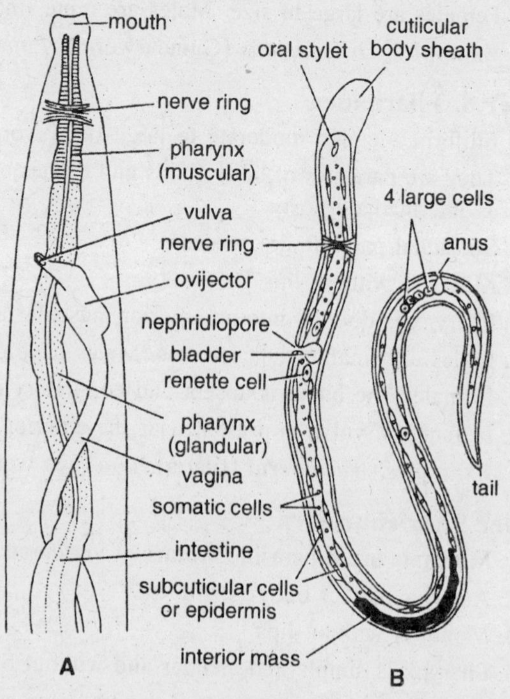

Fig. 28.2. *Wuchereria bancrofti*. A—Anterior part of female; B—Microfilaria.

Life Cycle

The life cycle of *Wuchereria bancrofti* is completed in two hosts. The primary host is human being and secondary host is mosquito—*Culex pipiens* in India.

Life cycle in human beings. A microfilaria has a colourless and transparent body with blunt anterior end and rather pointed tail end. Its body is covered in a hyaline sheath followed by **cuticula** which is lined by flattened subcuticular cells or **epidermis** and an inner column of cytoplasm containing nuclei. It contains future mouth or oral stylet, nerve ring band, nephridiopore, renette cells, a dark coloured inner mass and future anus (Fig. 28.2B). Microfilaria are active forms and can move both with and against the blood streams. These larvae are unable to develop further in the human body, unless they are taken up by the intermediate host. If these microfilariare are not sucked up by the mostquito they die within 70 days.

Periodicity of microfilariae. The microfilariae of oriental countries such as India and China exhibit **nocturnal periodicity.** They appear in peripheral circulation periodically at night, only generally between 10 pm and 4 am, but disappear during rest of day. It is believed that nocturnal periodicity of microfilariae is related with the nocturnal feeding habit of their secondary host, *Culex pipiens*.

Life cycle in mosquito. The microfilariae are sucked up by the mosquito (*Culex pipiens*) along with blood meal. These larvae are collected in the anterior part of the stomach where these cast off the hyaline sheath, penetrate the gut wall and within an hour or two migrate to the thoracic muscles and change into **first stage larvae.** They undergo moulting once or twice and form **second stage larvae.** On tenth or eleventh day the metamorphosis is completed. The tail, the digestive system, body cavity and genital organs are developed. This is the **third stage larvae** measuring 1500 to 2000 μm in length and 18 to 20 μm in diameter. These larvae are infective to humans. They become inactive and come to lie in the labium of the mosquito. When the mosquito bites the warm and moist skin of human beings, the larvae creep out of the labium to the human skin. They penetrate human skin and finally come to settle down in the lymphatics.

Pathogenicity

The disease caused by the infection of *W. bancrofti* is, in general, referred to as **wucheriasis** or **filariasis.** Accumulation of microfilaria worms blocks the human lymphatic system which results in an inflammation of lymph glands. In some cases, the obstruction of lymph vessels may result in **elephantiasis,** resulting in enormous swelling and excessive growth of the affected part such as arm, leg, scrotum or mammae.

Treatment and Prevention

Heterazan and compounds of antimony and arsenic are effective to reduce or eliminate the microfilariae from the circulatory system of human beings. The only way of prevention is to protect our bodies from mosquito bite.

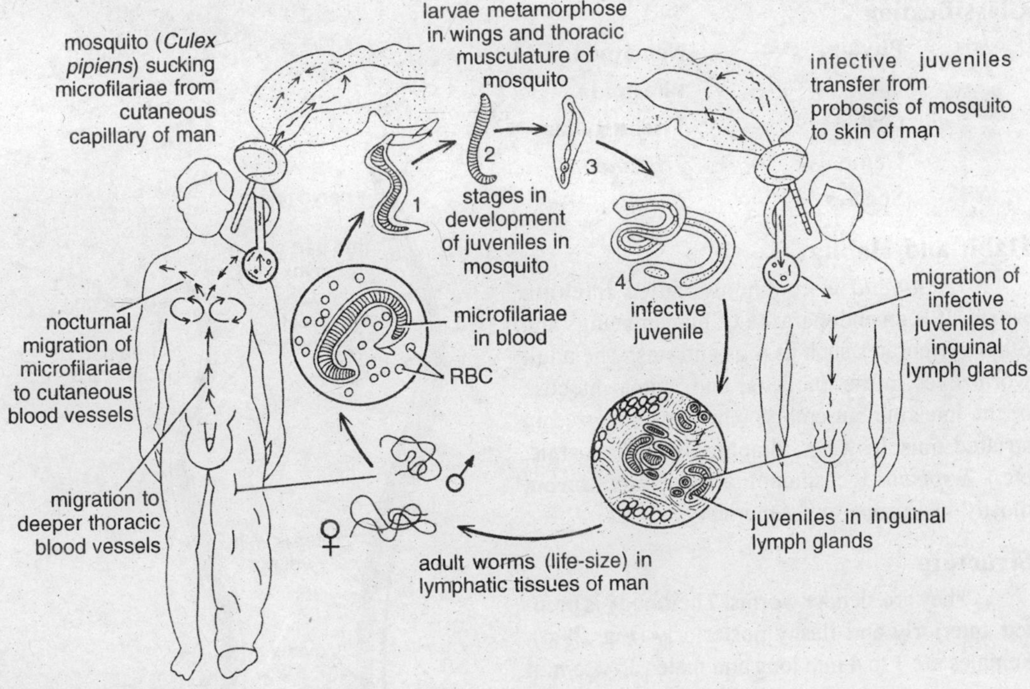

Fig. 28.3. *Wuchereria bancrofti.* Life cycle.

2. *RHABDITIS*

Classification

Phylum	–	**Nematoda**
Class	–	**Phasmida**
Order	–	**Rhabditoidea**
Genus	–	***Rhabdits***
Species	–	***elegans***

Habit and Habitat

The genus *Rhabditis* contains numerous species normally found in soil, organic matter or water and frequently in faeces of human beings and animals. *Rhabditis pellio* have, on few occasions, been found living in human vagina and larvae escaping in urine. *R. hominis* have been detected from stools of human beings and animals.

Structure

Adult male and female measures 2.0 and 1.7 mm in length respectively. Excretory system is H-shaped. Reproductive system is J-shaped. Male *Rhabditis* possesses single tubular testis and female has two ovaries. Female also contains oviducts, seminal receptacle and short uteri. Life cycle is simple. Adults breed as protandric hermaphrodites.

3. *TRICHINELIA SPIRALIS*

Classification

Phylum	–	**Nematoda**
Class	–	**Phasmida**
Order	–	**Trichuroidea**
Genus	–	***Trichinella***
Species	–	***spiralis***

Habit and Habitat

Trichinella is commonly called **trichina worm**. It is an endoparasite of human beings and other vertebrates such as pigs and rats. The adult worm lives in the duodenal and jejunal mucosa of the intestine but embryos become encysted in striated muscles (*e.g.*, diaphragm, intercostals, etc.). *T. spiralis* is cosmopolitan in distribution but mostly in Europe and America.

Structure

They are slender worms. Their body is pointed anteriorly and fleshy posteriorly (Fig. 28.4). Females are 3 to 4 mm long and males are 1.5 mm long.

The long pharynx has an anterior muscular part and posterior glandular part. Pharynx contains large granular cells, called **stichosome.** Males are with a pair of conical appendages at the posterior end. They have no copulatory spicules.

Mode of Infection

The fertilized female bores through the intestine into spiral lymph spaces and gives birth to juveniles (about 10,000 at a time). They can develop further into the adult worm only if taken up by another rat or pig. Infection normally passes from rat to rat and pig to pig. The pigs become

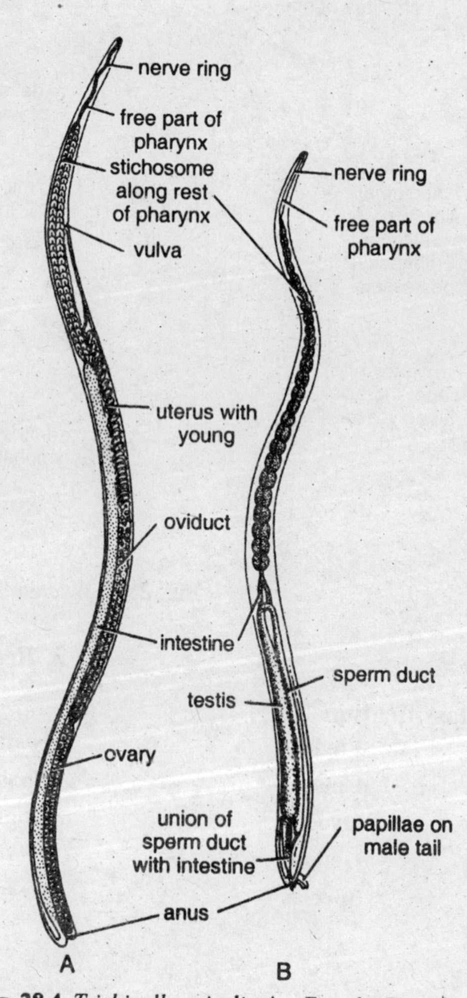

Fig. 28.4. *Trichinella spiralis*. A—Female; B—Male.

infected through the ingestion of infected pork scraps (offal) from their swill. Inside the pig these undergo the same process of development and form lemon-shaped cysts (Fig. 28.5) in the muscles. The infected pig muscles containing larvae when eaten up by human beings cause infection.

Pathogenicity

Trichinella causes a very serious and dreaded disease known as **trichinosis** which is characterized by **myocarditis** or **myositis.** It may cause paralysis of the muscles. Only 5 per cent of heavy infection end fatally and the death is usually due to the paralysis of cardiac muscles.

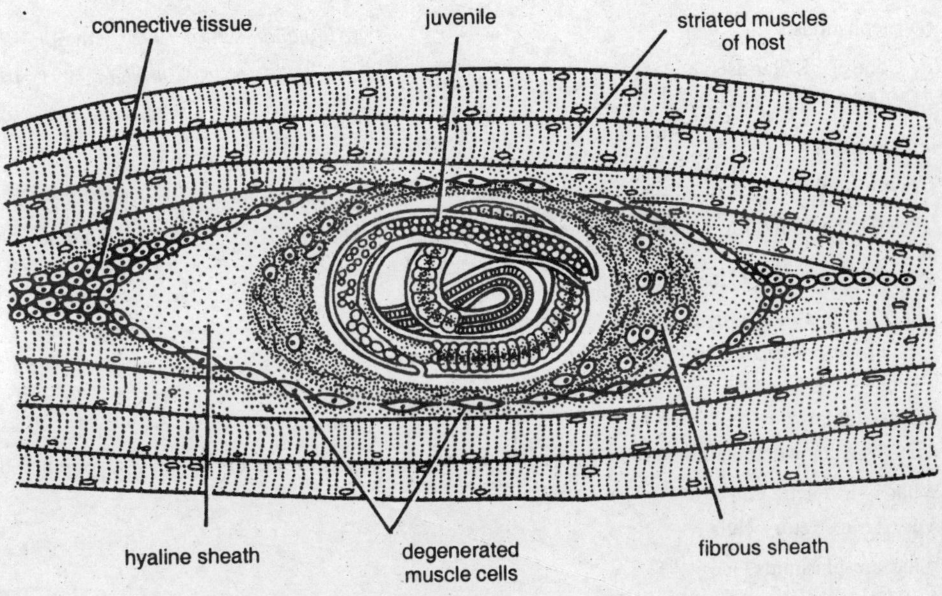

Fig. 28.5. *Trichinella spiralis.* Cyst in the striated muscle of pig.

Treatment

Daily doses of 2 to 3 gms of piperazine citrate are effective in removing the adults of *Trichinella* from human beings.

QUESTIONS

Long Answer Questions

1. Describe general characters of Aschelminthes and classify this group up to phyla/classes. *(Magadh 1989)*
2. Give an outline classification of phylum Nematoda up to orders.
3. Give distinguishing features of phylum Nematoda. Classify up to order giving character and example of each. *(Meerut or CCS, 1991)*
4. Classify Nematoda or Nemathelminthes mentioning the characters of its orders. Cite an example of each order. *(Madras 1989; Jiwaji 1994; Meerut 1994)*
5. Give zoological name of Filarial worm. In India which animal is its vector, give its zoological name. What is meant by the periodicity of microfilaria? Describe pathogenicity, treatment and prevention of disease caused by worm. *(Guwahati 1996; Allahabad 1997)*
6. Describe the morphology of the adult *Wuchereria bancrofti*. Mention the disease caused by it and the methods of its control. *(Allahabad 1989)*

7. Give an account of the structure, life history and mode of infection of *Wuchereria bancrofti*. (*Magadh 1990; Guwahati 1992, 1994; Allahabad 1993, 1995, 2001*)
8. Give an account of the life-cycle of *Wuchereria bancrofti*, adding a note on day and night periodicity. (*Magadh 1995*)
9. Describe sequential stages in life cycle of *Wuchereria bancrofti*. (*Allahabad 1996*)

Short Answer Questions

1. Write short notes on the following:
   (*i*) Elephantiasis (*Allahabad 1990, 1995; Magadh 1990*)
   (*ii*) Control of filariasis (*Allahabad 1991, 1994*)
   (*iii*) Filaria (*Punjab 1990, 1992*)
   (*iv*) *Rhabditis*
2. Draw a well labelled diagram of the life cycle of *Wuchereria bancrofti*. (*Allahabad 1992*)
3. Classify the following giving diagnostic characters of *Trichinella*. (*Allahabad 1994*)

Very Short Answer Questions

1. What is basis of classification of Nematoda into classes?
   Ans. Phylum Nematoda is classified into two classes on the basis of presence or absence of phasmids, the sense organs. Two classes of Nematoda are: Aphasmida and Phasmida.
2. What is eutely?
   Ans. Constancy of the number of cells or nuclei in the various organs and also in body, *e.g.*, nematodes.
3. What is a renette cell?
   Ans. An excretory cell.
4. What are phasmids?
   Ans. Phasmids are unicellular glandulo-sensory structures in the tail region of most nematodes.
5. Give the zoological name of the root knot eelworm.
   Ans. *Meloidogynae marioni*.
6. Name the parasite that cause elephantitiasis in human beings.
   Ans. *Wuchereria bancrofti*.
7. Does asexual reproduction occur in nematodes?
   Ans. No.
8. Name three important nematode parasites of human beings.
   Ans. *Wuchereria, Ancylostoma, Ascaris*.
9. What is elephantiasis?
   Ans. Elephantiasis or **filariasis** is a disease of human beings caused by filaria worm *Wuchereria bancrofti*. This round worm inhabits the lymphatic vessels of human beings and infection is spread through *Culex* mosquito. This disease leads to the excessive swelling of legs, scrotum and breasts as a result of blockage of lymphatic circulation by the parasite. Most effective drug for elephantiasis is Hetrazan (diethylcarbmazine).

Fill in the Blanks

1. Nematodes are the first to have a body cavity. It is called ....
2. In .... epidermis is syncytial
3. Body wall musculature of roundworms consists of ....
4. A nematode muscle cell consists of .... part and cytoplasmic part.

5. Alimentary canal is .... in nematoda.
6. Sexes are.............. in nematodes.
7. Nerve cords in nematodes are located in ....
8. The excretory tubes of nematodes lie in the lateral .... chords.
9. Cloaca is found in .... roundworms.
10. The phylum Nematoda is divided into two classes: .... and Phasmida.

Multiple Choice Questions

1. The eyeworm of human beings is .
   (a) *Wuchereria* (b) *Trichuria*
   (c) *Loa loa* (d) None
2. Nematode epidermis consists of
   (a) cuticle (b) epidermis
   (c) muscular layer (d) all of these
3. Number of cells in the various organs of the body is fixed in
   (a) flatworms (b) segmented worms
   (c) roundworms (d) walking worms
4. Gonads in male roundworms consists of
   (a) one pair of testes (b) two pairs of testes
   (c) three pairs of testes (d) unpaired testes
5. The nematode *Wuchereria bancrofti* belongs to order
   (a) Ascaroidea (b) Strongyloidea
   (c) Filarioidea (d) Oxyuroidea
6. Filarial worm is transmitted by
   (a) mosquito (b) tse-tse fly
   (c) sandfly (d) kissing bug
7. *Enterobius vermicularis* inhabits the human
   (a) large intestine (b) caecum
   (c) appendix (d) all of these
8. Trichinosis is caused by
   (a) *Trichuris* (b) *Trichinella*
   (c) *Ancylostoma* (d) None of these
9. Which of the following is a free living nematode?
   (a) *Trichuris* (b) *Trichinella*
   (c) *Rhabditis* (d) *Wuchereria*
10. Microfilaria is the larva of
    (a) tape worm (b) pin worm
    (c) round worm (d) filarial worm
11. The place where adult *Wuchereria bancrofti* lives is the
    (a) human blood vessel (b) human lymph gland
    (c) muscle of *Culex* (d) salivary gland of *Culex*
12. Elephantiasis spreads through the vector
    (a) *Anopheles* (b) *Culex*
    (c) *Musca domestica* (d) Flat worm

13. Organ system grade of body organisation is present in
   (*a*) porifera (*b*) platyhelminthes
   (*c*) aschelminthes (*d*) cnidaria
14. Intermediate host of *Wuchereria bancrofti* is
   (*a*) male *Anopheles* (*b*) female *Anopheles*
   (*c*) female *Culex* (*d*) male *Culex*
15. Which of the following is a digenetic nematode parasite?
   (*a*) *Fasciola hepatica* (*b*) *Taenia solium*
   (*c*) *Wuchereria bancrofti* (*d*) All of these
16. Transmission of *Wuchereria* to new host occurs in
   (*a*) larval stage (*b*) encyst stage
   (*c*) juvenile stage (*d*) none of the above

ANSWERS

Fill in the Blanks

1. Pseudocoel **2.** Nematoda **3.** Longitudinal **4.** Fibrillar
5. Complete **6.** Separate **7.** Epidermis **8.** Epidermal
9. Male **10.** Aphasmida.

Multiple Choice Questions

1. (*c*) **2.** (*d*) **3.** (*c*) **4.** (*d*)
5. (*c*) **6.** (*a*) **7.** (*d*) **8.** (*b*)
9. (*c*) **10.** (*d*) **11.** (*b*) **12.** (*b*)
13. (*c*) **14.** (*c*) **15.** (*c*) **16.** (*c*).

29

Ancylostoma (Phylum Nematoda)

29.1 CLASSIFICATION

Phylum	–	**Nematoda**
Class	–	**Phasmida**
Order	–	**Strongyloidea**
Genus	–	***Ancylostoma***
Species	–	***duodenale***

Ancylostoma is commonly called **hookworm** of human beings. There are two main species of hookworms — (*i*) *Ancylostoma duodenale* is the "**old world hookworm**"; (*ii*) *Necator americanus* is the "**new world hookworm**" or **American hookworm**.

Box 29.1 Homologous and Heterologous Hosts

Though primarily a human parasite, *Ancylostoma duodenale* can be induced experimentally to parasitize monkeys, cats and dogs. An animal which normally or naturally harbours a parasite is called a **homologous host,** and an animal that is artificially or experimentally infected with a parasite is termed a **heterologous host.**

29.2 HABIT AND HABITAT

A. duodenale is a dangerous nematode parasite of human beings which occur in Asia (including India), Africa, Europe, South America, etc. It lies attached to the wall of intestine (*i.e.*, jejunum of human beings and feeds on blood, lymph and mucous membrane. The infective juveniles (= larvae) find their way in the human percutaneously from the soil contaminated with faeces in which they live.

29.3 STRUCTURE

Adult *Ancylostoma duodenale* are small and cylindrical in shape. Sexes are separate; the male is about 8 mm in length and 0.4 mm in diameter, while female is generally longer about 12.5 mm in length and 0.6 mm in diameter.

The anterior end of worms of both sexes is slightly bent dorsally (hence, the name hookworm) and has a large **buccal capsule** having six **cutting plates** or **teeth** (4 hook-like on the ventral surface and 2 knob-like triangular plates called **sharp lancets**, on the dorsal surface). The posterior end of female worm tapers bluntly in a short post-anal tail, while that of male is expanded and umbrella-like (Fig. 29.1). This expanded structure is called **copulatory bursa** and surrounds the cloaca. Bursa has two large lateral lobes with six **muscular rays** in each and small median lobe. Each ray is given a name as dorsal ray, externo-dorsal ray, postero-lateral, medio-lateral, externo-lateral, latero-ventral, ventro-ventral, etc. A pair of **cephalic gland** (lie on both sides of pharynx) open in the buccal cavity.

Its secretion contains enzymes for digesting the intestinal mucosa of the host and other substances (such as **acetylcholine esterase**) which have a role in attachment of the worm to the host's intestinal wall.

The body of *Ancylostoma* is covered externally by **cuticle**, it is followed internally by the **epidermis** and **musculature** which is directed longitudinally. Its body cavity is **pseudocoel**.

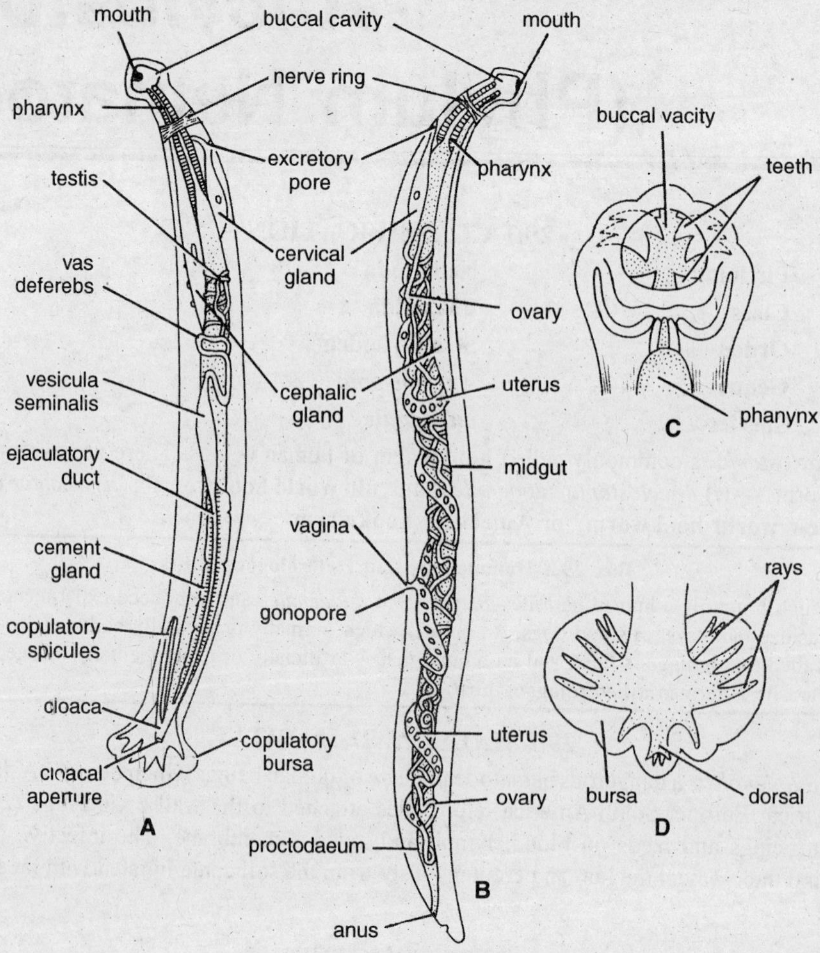

Fig. 29.1. *Ancylostoma duodenale.* A—Adult male; B—Adult female; C—Anterior end; D—Posterior end.

29.4 DIGESTIVE SYSTEM

Alimentary canal of *Ancylostoma* is tubular and very simple. It consists of the mouth, buccal capsule, muscular pharynx having a triradiate lumen lined by cuticle, oesophageal bulb, intestine, rectum and cloaca in male but anus in female. There are five glands connected with the digestive system, one of the is called **oesophageal gland** and secretes an anticoagulant secretion which prevents the clotting of blood so that the worms can suck blood from the host.

The food of this holozoic and blood-sucking nematode includes blood, lymph, tissue exudates and pieces of intestinal mucous membrane. *Ancylostoma* tends to live firmly anchored by drawing a bit of the human intestinal lining into its mouth. In the process of feeding, the tiny teeth and cutting plates of the buccal capsule make small wounds in the intestinal mucosa through which the food, blood, lymph is sucked by the pumping action of pharynx. After feeding it leaves a bleeding wound and moves to another location. An adult worm is said to suck nearly 0.8 ml of blood in a day from the host causing severe **anaemia**. Digestion of food is however, completed in the intestine.

It is held that the red blood corpuscles ingested by the *Ancylostoma* worm provide oxygen needed in the worm's metabolism, as the small intestine of the host has very little oxygen. Other systems of *Ancylostoma* such as **respiratory system, excretory system, nervous system** and **receptors** are like those of *Ascaris lumbricoides*.

29.5 REPRODUCTIVE SYSTEM

As already has been described, in *Ancylostoma* sexes are **separate** and **sexual dimorphism** is well distinct.

Male reproductive system comprises a single, tubular, thread-like **testis** twisted around the intestine in the middle of the body. Testis continues posteriorly in **vas deferens** which finally opens into an elongated, swollen, sac-like **seminal vesicle.** The seminal vesicle soon tapers to forms a narrow passage called the **ejaculatory duct** which opens into the cloaca.

Female reproductive system comprises of two much highly twisted tubules, the **ovaries**. One ovarian tubule is placed anteriorly and other posteriorly from a little behind the middle of body. Both the ovarian tubules continue into the **oviducts** and, thus, the two oviducts open into elongated and dilated **seminal receptacles**, each continues into muscular **uterus**. Thus, the two uteri (one from anterior and other from posterior) meet little behind the middle of the body to form a short tubular **vagina** which opens out by **vulva** or the **gonopore** situated at junction of the posterior and middle third of the body.

29.6 LIFE CYCLE

Ancylostoma duodenale is **monogenetic**. There is no intermediate host and human being is its main host.

1. Copulation. Copulation occurs in the intestine of the host. During copulation, male and female assume a Y-shaped figure and the copulatory bursa of male is applied on the vulva of the female and sperms are transferred. Sperms reach the seminal receptacle of female, where fertilization of eggs occur. Fertilized eggs are then pushed into uteri. About 9000 fertilized and uncleaved eggs are laid by female per day in the intestine of host and are passed out with faeces. Cleavage of eggs is started in the soil.

2. Development in soil. Under favourable environmental conditions such as moisture, oxygen and temperature (20 to 30°C), the fertilized egg divides, develops and hatch within 48 hours into the **first stage juvenile** or **rhabditiform larva**. This larva is 250 μm long and possesses the mouth, buccal capsule, elongated pharynx, bulb-like oesophagus (resembling with the nematode *Rhabditis*) and intestine. The larva feeds on bacteria and other debris in faeces or moist soil and grows rapidly. At the end of second day, it undergoes first moult (*i.e.*, shed its cuticle and grows) and changes into second stage larva, called the **strongyliform larva**. At about 5th day, it undergoes second moult to become third stage larva called **filariform larva** or **juvenile**. This is the infective, nonfeeding stage. The skin of second stage larva is retained as a protective sheath. The larva lives in the 1 cm of the top soil. It remains alive for several weeks under favourable conditions.

3. Infection of a new host. Infection of human host occurs when the filariform larvae or juveniles come in contact with human's skin. They cast off their cuticular sheath and penetrate the skin of human host. The anterior end of the larva is provided with oral spears by which it penetrates the soft skin of feet and hands, generally through the hair follicles

4. Larval migration inside host. On reaching the subcutaneous tissues the larvae enter the lymphatics and small venules. Ultimately they reach the venous circulation and are carried through the right heart into the pulmonary capillaries, where they break through the capillary walls to enter into the alveolar spaces. They then migrate in the bronchi → trachea → larynx and crawl over the epiglottis to the back of pharynx and are finally swallowed. During its migration, when larva reaches oesophagus, its **third moulting** occurs and a terminal buccal capsule is formed. The time taken in this migration is about 10 days.

Finally the growing larvae (3 to 5 mm long) settle down in the small intestine and undergo **fourth** and **final** moult to become adult. In about 3 to 4 weeks time they become sexually mature to repeat the life history again. The average **life span** of adult worm in the human intestine is about five years.

29.7 PATHOGENICITY (ECONOMIC IMPORTANCE)

The hookworms are the most dangerous parasitic nematodes because they hold on to the intestinal villi and suck blood and body fluids of the host by their muscular pharynx. They also cut holes in the intestinal mucosa and leave bleeding gaps. It results in severe **anaemia.** In children, where incidence of infection is very great, they retard the physical and mental growth. Some toxin secreted by the glands of head region of worms cause stomachache; food fermentation, diarrhoea, constipation, dyspnea (= laboured respiration), palpitation of heart, eosinophilia, ill health and the patient may finally collapse.

During penetration of larvae in the skin local irritation is caused resulting in inflammation

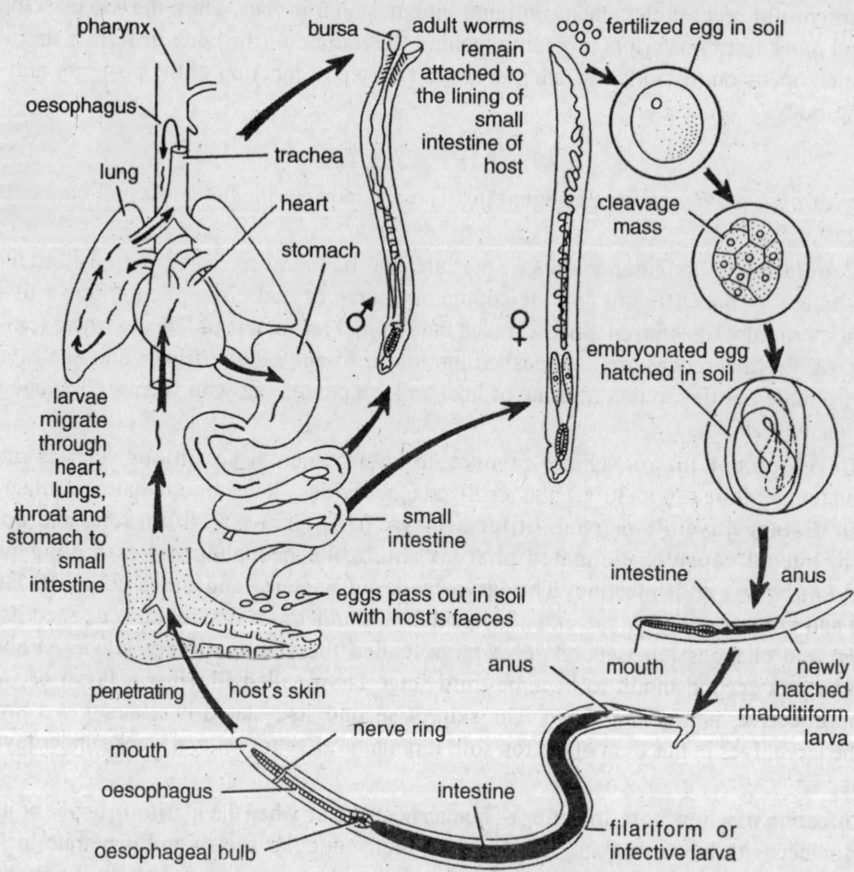

Fig. 29.2. *Ancylostoma duodenale*. Life cycle.

of the surrounding tissues; these may result into tiny sores. This condition is commonly called **miner's itch, foot itch, toe itch, dew itch** or **ground itch**. The migratory larvae in lungs may cause haemorrhage and bronchial pneumonitis (*i.e.*, coughing, sore throat and blood sputum).

29.8 TREATMENT

Drugs like **carbon tetrachloride, thymol, oil of *Chenopodium*, hexylresorcinol,** etc., are generally used effectively to control the infection of *Ancylostoma*. **Tetrachloroethylene** is an effective drug but it is dangerous to give it to alcoholic patients. **Thiabendozole, albendozole** and **pyrantel** are more effective and less toxic drugs.

Prevention and Control

Safe disposal of all human excreta is the most important prevention and control measure. Wearing of hand gloves and shoes by people working in fields of hookworm infested areas helps in avoiding catching of hookworm infection. Children should be directed to keep their hands and nails clean.

QUESTIONS

Long Answer Questions

1. Describe the structure and life history of *Ancylostoma*. (*Bundelkhand 1991; Meerut or CCS 1994, 1997, 1999, 2007*)
2. Give the habit, habitat, classification, characters and economic importance of hook worm. (*Meerut or CCS 1991*)
3. Describe the life cycle or *Ancylostoma*. (*CCSU, Meerut 2004, 2005*)

Short Answer Questions

1. What is the economic importance of hook worms? How do they infect human beings?
2. Write a note on copulatory bursa and penial spicules of *Ancylostoma*.
3. Describe the development of larva of *Ancylostoma* in soil.
4. Write short note on rhabditiform larva.
5. Draw labelled diagrams of male and female *Ancylostoma*. (*Garhwal 2001*)

Very Short Answer Questions

1. Give the zoological name of the old world hookworm.
   Ans. *Ancylostoma duodenale*.
2. What the food of hookworm?
   Ans. Blood, lymph and pieces of intestinal mucous membrane.
3. What is extracorporeal digestion?
   Ans. Digestion outside the body.
4. Give the popular name of *Necator americanus*.
   Ans. New world hookworm.
5. What is the name of a chitinous thickening on the dorsal wall of the cloaca in the male hookworm?
   Ans. Gubernaculum.
6. Name the main excretory matter of *Ancylostoma*.
   Ans. *Ammonia*.
7. Are the female reproductive organs of hookworm monadelphic or didelphic?
   Ans. Didelphic.

8. Name the larvae forms of *Ancylostoma*.
   Ans. Rhabditiform, strongyliform and filariform.
9. Are the hookworm larvae free-living or parasitic?
   Ans. Free-living.
10. How do hookworm larvae infect human beings?
    Ans. By boring through the skin.
11. What is the name of second stage larva of *Ancylostoma*?
    Ans. Strongyliform.
12. Name new world worm that causes necatorasis in human beings.
    Ans. *Necator americana.*

Fill in the Blanks

1. The popular name of *Ancylostoma duodenale* is ....
2. *Ancylostoma* belong to the order ....
3. Copulatory bursa is present in .... hookworm.
4. .... *Ancylostoma* is longer than ....
5. *Ancylostoma duodenale* or old word hookworm feeds on .... , ... , and pieces of intestinal mucosa of the host.
6. Excretory matter of the hookworm is mainly ....
7. .... helps the male hookworm in holding the female in copulation.
8. The eggs of *Ancylostoma* have a .... layered protective coat.
9. First stage larva of hookworm is called .... larva.
10. Rhabditiform larva of hookworm feeds on ....

Multiple Choice Questions

1. Strongyliform larva occurs in the life history of
   (*a*) *Ascaris* (*b*) *Fasciola*
   (*c*) *Taenia* (*d*) *Ancylostoma*
2. *Ancylostoma* is a
   (*a*) flat worm (*b*) round worm
   (c) segmented worm (*d*) none of these
3. *Ancylostoma* belongs to the order
   (*a*) Filarioidea (*b*) Strongyloidea
   (*c*) Oxyuroidea (*d*) Ascaroidea
4. Excretory organs of the hookworm are
   (*a*) flame cells (*b*) green glands
   (*c*) renette glands (*d*) nephridia
5. No secondary host is found in the life history of
   (*a*) *Fasciola* (*b*) *Taenia*
   (*c*) *Ancylostoma* (*d*) *Plasmodium*
6. Characteristic symptoms of ancylostoiasis is
   (*a*) gastro-intestinal disturbance (*b*) anaemia
   (*c*) nervous disorder (*d*) all the above

7. Which of the following penetrates through skin and enters the intestine?

(*a*) *Ascaris* (*b*) tape worm

(*c*) hook worm (*d*) *Amoeba*

8. Diagnosis of *Ancyclostoma* infection in human beings is made by examination of

(*a*) stool (*b*) urine

(*c*) blood (*d*) saliva

9. The host of *Ancyclostoma duodenale* is

(*a*) human being (*b*) mosquito

(*c*) snail (*d*) pig

10. *Ancyclostoma duodenale* is found in

(*a*) intestine of human beings (*b*) stomach of human beings

(*c*) intestine of pig (*d*) liver of human beings

11. Maximum people are infected by

(*a*) *Ascaris* (*b*) *Taenia*

(*c*) *Trypanosoma* (*d*) *Ancylostoma*

ANSWERS

Fill in the Blanks

1. Old world hookworm **2.** Strongyloidea **3.** Male

4. Female, male **5.** Blood, lymph **6.** Ammonia

7. Copulatory bursa **8.** Three **9.** Rhabditiform

10. Bacteria and organic debris.

Multiple Choice Questions

1. (*d*) **2.** (*b*) **3.** (*b*) **4.** (*c*)

5. (*c*) **6.** (*d*) **7.** (*c*) **8.** (*a*)

9. (*a*) **10.** (*a*) **11.** (*d*).

30

Ascaris (Phylum Nematoda)

30.1 CLASSIFICATION

Phylum	–	**Nematoda**
Class	–	**Phasmida**
Order	–	**Ascaroidea**
Genus	–	***Ascaris***
Species	–	***lumbricoides***

30.2 HABIT AND HABITAT

Ascaris lumbricoides is the intestinal parasite (endoparasite) of human beings. It is also found in the intestine of apes, pigs, cattle, sheep, etc. It remains free in intestine and feeds on the partly digested food (chyme) of the host.

Though it is found in all parts of the world, its infection is much more common in India, China, Korea, Phillippines and Pacific islands.

30.3 STRUCTURE

Body of the *Ascaris* is large, cylindrical and tapers at both ends. It is pinkish white or yellowish white in colour. Sexes are separate; sexual dimorphism is well marked (Table 30.1). The female is larger measuring 20–40 cm in length and 4–6 mm in diameter. Male is smaller and measures 15–30 cm in length and 2–4 mm in diameter (Fig. 30.1). The tail of male is sharply curved and contains 50 pairs of preanal and 5 pairs of post-anal genital papillae. In female, the tail is straight.

Table 30.1. External features showing sexual dimorphism in male and female Ascaris.

Male	*Female*
1. Body is smaller, 15–30 cm long and 2–4 mm thick.	1. Body is larger; 20–40 cm long and 4–6 mm thick.
2. Tail end is curved ventrally.	2. Tail end is straight.
3. Anus and genital pore are common (cloaca).	3. Anus and genital pore are separate (cloaca absent).
4. A pair of needle-like and protrusible penial setae are present in cloaca.	4. No penial setae are present.
5. Preanal and postanal papillae are present upon ventral body surface in cloacal region to help in copulation.	5. No such papillae are present.

At the anterior end three broad **lips** surround the **mouth** (Fig. 30.1C). Each lip contains fine **denticles** or **teeth** on its inner margin and **sensory papillae** on outer surface.

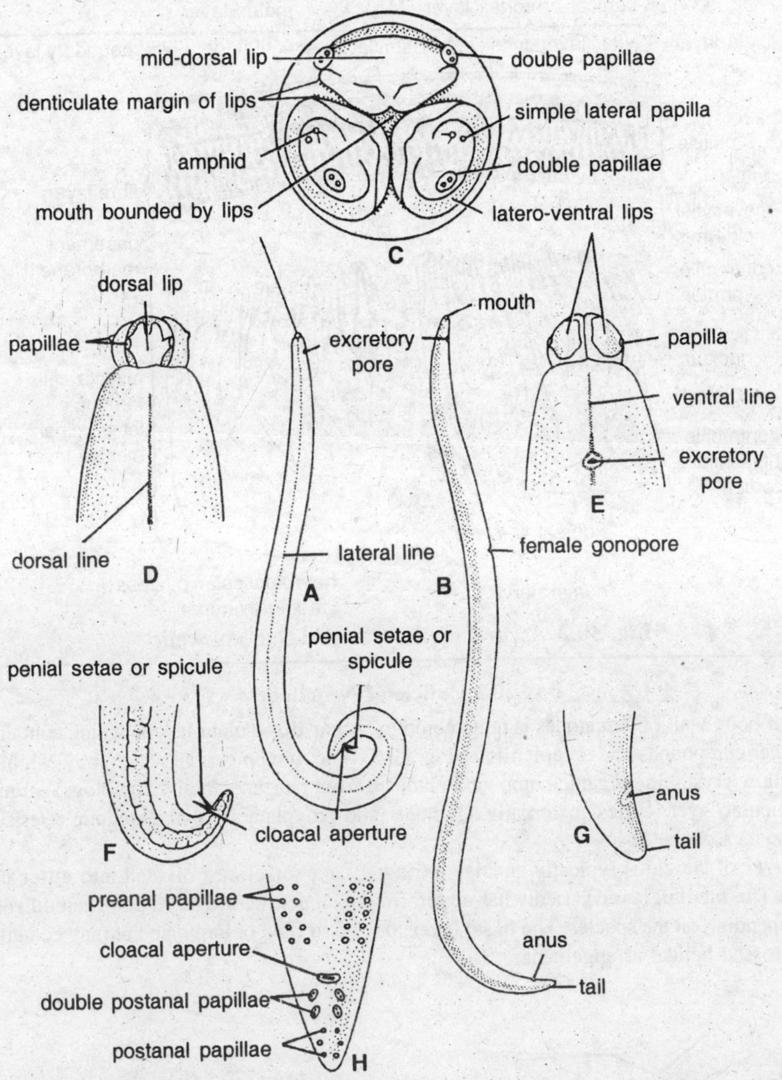

Fig. 30.1. *Ascaris lumbricoides*. A—Male; B—Female; C—Surface view of mouth and lips; D—Anterior end in dorsal view; E—Anterior end in ventral view; F—Posterior (tail) end of male, G—Tail end of female, H—Posterior end of male in ventral view showing papillae.

30.4 BODY WALL

The body wall of *Ascaris* consists of following three main layers (Fig. 30.2):

1. Cuticle;
2. Epidermis or hypodermis; and
3. Muscular layer.

1. Cuticle. The body of *Ascaris* is covered with a thick and tough non-cellular **cuticle** having transverse striations. Cuticle is a complex (Box 30.1) exoskeleton-like structure which provides protection to the body from mechanical injuries and also provides enough flexiblity and elasticity to body wall. Cuticle is followed by a thin-layer of **basement membrane.**

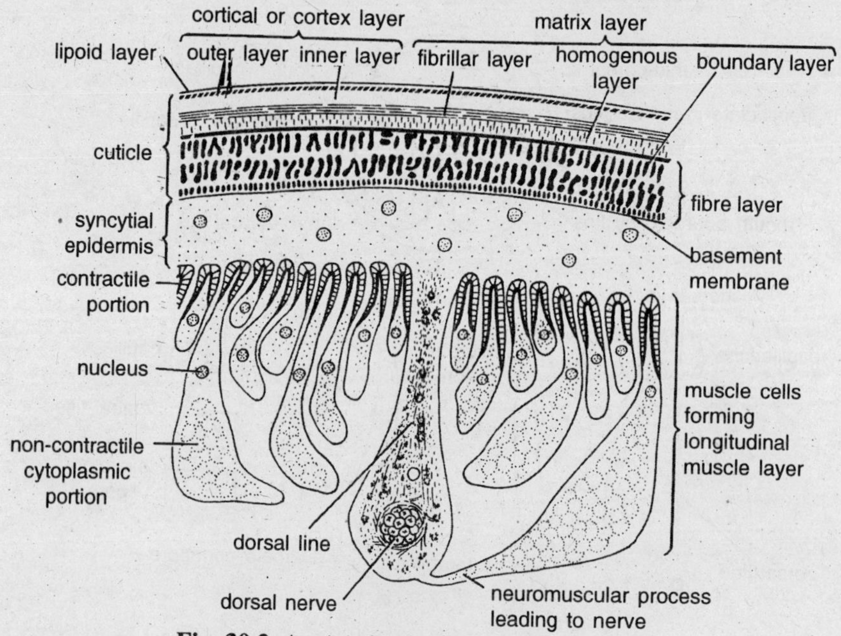

Fig. 30.2. *Ascaris lumbricoides*. T.S. body wall.

Box 30.1 Cuticle of Nematodes

The cuticle of body wall of nematodes is quite complex. It has **three** main layers, which contain **collagen** as well as other compounds and an **epicuticle** (Fig. 30.3). The thin epicuticle which may exhibit **quinone** (a golden yellow crystalline organic compound obtained from benzene and its homologs) **tanning,** cover the outer cuticular layer (cortex) externally. Quinone (and polyphenols) make epicuticle resistant to the digestive enzymes of the host.

The **outer layer** of cuticle is typically annulated (ringed) and sometimes divided into **outer cortex** and **inner cortex**.The **median layer** variedly has a uniform granular structure **os struts, skeletal rods, fibrils** or **canals,** depending on the species. The **basal layer** may be striated or laminated or may contain collagen fibres in a crossed helical arrangement.

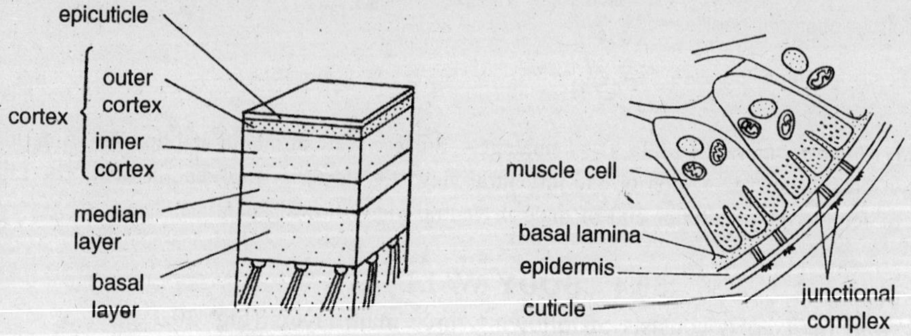

Fig. 30.3. Different parts of cuticle of nematodes.

Fig. 30.4. T.S. body wall of a nematode showing attachment of muscular layer with cuticle.

In nematodes, **growth** is accompanied by four moulting of the cuticle. Beginning at the anterior end, the old cuticle separates from the underlying epidermis and a new cuticle is secreted at least in part. The old cuticle is shed, sometimes in fragments. Moulting does not occur after the worm becomes adult, but the cuticle continues to expand as the animal grows.

2. Epidermis. The epidermis is syncytial and bulges into the pseudocoel to form mid-dorsal, mid-ventral and mid-lateral **longitudinal chords** (or lines). The epidermis secretes the cuticle, stores nutrients, bears **attachment fibres** that link the musculature to cuticle (Fig. 30.4).

3. Muscle layer. It is single-cell thick and comprises of **longitudinal muscle fibres** or **cells.** Each muscle cell has two zones: an outer fibrillar zone of longitudinally striated, spindle-shaped **muscular part** which is contractile and lies towards the epidermis; and a second **protoplasmic zone** which is club shaped, bladder-like mass of protoplasm with a nucleus and a network of supporting fibrils from a fibrous process or **muscle tail.** The muscle layer is divided into four quadrants by the four epidermal chords. Each quadrant has about 150 muscle cells. The muscle tails of all the muscle cells of two dorso-lateral groups join the dorsal nerves located in the mid-dorsal chords. Likewise, muscle tails of the two ventro-lateral muscle cell groups join the ventral-nerve cord located in the midventral longitudinal chord.

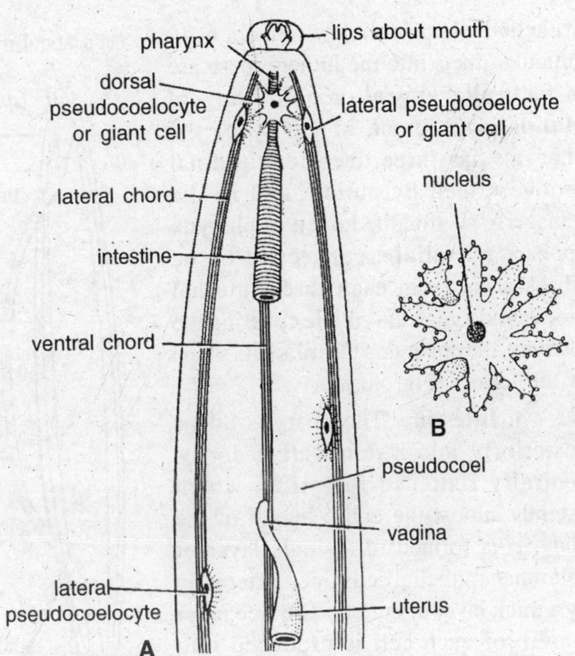

Fig. 30.5. *Ascaris lumbricoides.* A—Position of giant cells or pseudocoelomocytes (=pseudocoelocytes) in pseudocoel; B—single pseudocoelomocyte enlarged.

Pseudocoel

There is a space enclosed by the body wall, called **pseudocoel.** It is not lined by coelomic epithelium. The pseudocoel contains a clear fluid, fibrous tissue and four giant cells-called **pseudocoelomocytes** (Fig. 30.5) . Food, oxygen and waste products are transported by the pseudocoelomic fluid.

30.5 DIGESTIVE SYSTEM

The alimentary canal of *Ascaris* is complete, *i.e.,* it has mouth as well as anus. The alimentary canal is a straight tube and extends almost the entire length of body. It shows three distinct regions: 1. **Foregut** or **stomodaeum** which includes the mouth and pharynx; 2. **Midgut** which includes the intestine; and 3. **Hindgut** or **proctodaeum** which includes rectum (Fig. 30.6). The foregut and hindgut are lined by ectodermal epithelium covered with cuticle, which is continuous with the covering of body. The midgut has an endodermal lining.

1. Mouth. The mouth lies at the anterior end of the body. it is a minute, triangular aperture bounded by three **labia** or **lips**: one mid-dorsal and two latero-ventral in position (Fig. 30.1C). The lips have finely denticulate margin (*i.e.,* lips have fine teeth on their margins) and bear sensory **papillae**. The dorsal lip has two double papillae, while each latero-ventral lip has one double papilla, one simple papilla and a reduced sense organ called **amphid**. Amphid is a olfactory chemoreceptor.

2. Pharynx. The terminal mouth leads into cylindrical thick-walled **pharynx** or **oesophagus** which has a posterior swelling called end bulb which is provided with valves. The pharynx has muscular walls having radial muscle fibers which dilate its lumen (Fig. 30.6B). The pharynx has three large branched gland cells which open by cuticular ducts into the lumen; these are in fact, **pharyngeal** or **oeso-phageal glands**. The lumen or cavity of the pharynx has three deep longitudinal grooves lined by cuticle and in the transverse section the lumen of pharynx appears **triradiate**, connective tissue fibres arise from each three internal grooves and go to the cuticle covering the pharynx, they maintain the triradiate shape of the lumen (Fig. 30.7).

3. Intestine. The pharynx opens posteriorly into a thin-walled dorso-ventrally flattened **intestine**, which extends almost the entire length of the body. It is formed of a single layer of columnar epithelial cells lined externally by a thick layer of cuticle. The free inner margin of each cell is produced into several finger-like projections, the **microvilli**. They form a kind of tightly packed **brush border** (Fig. 30.6C, D). There is no muscle layer in the wall of intestine.

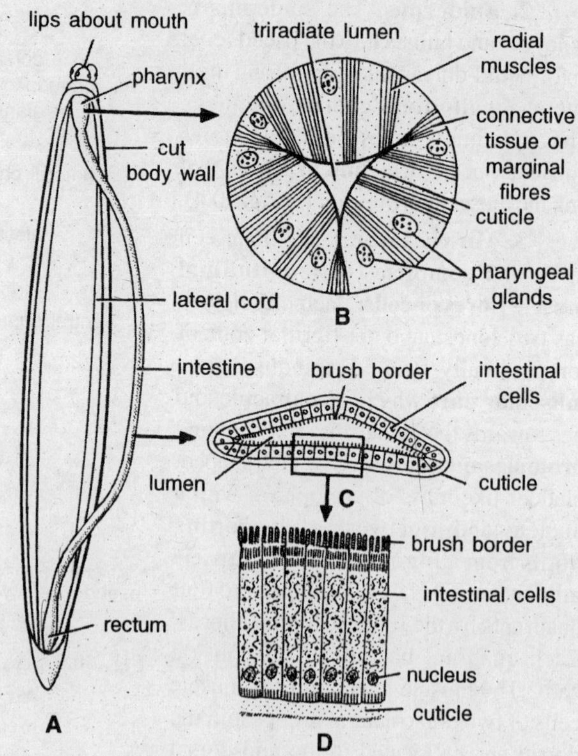

Fig. 30.6. *Ascaris*. A—Alimentary canal; B—Pharynx in T.S.; C—Intestine in T.S.; D—A portion of intestinal wall showing brush border.

4. Rectum. The intestine is followed by a short, relatively narrow chamber, called **rectum.** Its wall consists of tall columnar cells and lined internally by cuticle and externally by few muscle layers. In male, the rectum opens out by **cloaca** because it receives the ejaculatory duct but in female, the rectum opens out by a transverse aperture, the **anus** (Fig. 30.1G, H). The anal aperture is guarded by anterior and posterior lips and is provided with a few special dilator muscles running from the rectum to the body wall, called **depressor ani.** Their contraction from time to time causes the faecal matter to be discharged out. The rectum also has large unicellular **rectal glands**; three in the female and six in the male.

Food, Feeding and Digestion. Food of *Ascaris* comprises blood, tissue exudes and partly digested fluid food (chyme) from the host's intestine. Ingestion of food is accomplished by sucking action of pharynx. Sucking action of the pharynx is due to its periodic dilation by contraction of its radial muscles. Movement of the food down the pharynx and intestine is brought about partly by the pumping action of pharynx and partly by the contraction of the body wall. Absorption of food occurs in the intestine. Here the food diffuses through the intestine wall into the fluid filling the pseudocoelom. From the latter, the food is absorbed by the tissues and organs in which assimilation takes place. The unabsorbed food forms faeces in the rectum. The depressor ani muscles opens the anus (cloaca aperture in the male) periodically and the faeces is thrown out by the pressure developed in the intestine by pumping action of the pharynx.

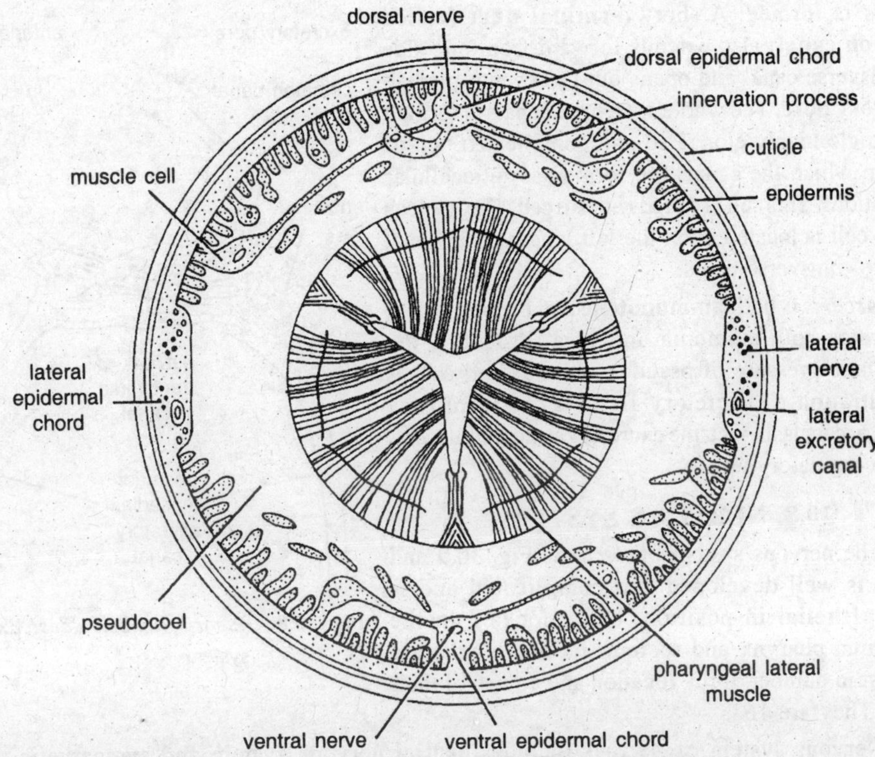

Fig. 30.7. *Ascaris lumbrircoides.* T.S. body through pharynx.

Some digestion in *Ascaris* is known to occur in the intestine. It starts in an extracellular manner in the intestine lumen and completed intracellularly. For the former, the pharyngeal gland cells and intestinal cells secrete enzymes, such as **proteases, amylase** and **lipase**. For intracellular digestion, the intestinal cells capture food particles into food vacuoles and are digested by lysosomal enzymes. Excess of food is stored mainly as glycogen and partly as lipids in the epidermis, intestinal epithelium and non-muscular parts of muscle cells. In *Ascaris*, intestine also synthesize **yolk** which is exported via pseudocoelomic fluid to the ovary for devloping oocytes.

30.6 RESPIRATION

In *Ascaris*, respiration is normally **anaerobic**, because the intestinal contents are deficient in free oxygen. In such respiration, glucose breaks down into carbon dioxide and a variety of fatty acids, mainly **valerianic acid**. It is these fatty acids, mixed perhaps with traces of other substances, which give a characteristic smell, rather like a tinned pineapple to the living *Ascaris* (see **Dhami** and **Dhami**, 2001). Energy released during respitation is utilised by the worm in its vital processes, such as locomotion, sucking of food, growth, reproduction, etc. Carbon dioxide leaves the body by diffusion through the skin.

30.7 EXCRETORY SYSTEM

Excretory system of *Ascaris* is H-shaped and consists of two slender **longitudinal excretory canals** (Fig. 30.8), one embedded in each lateral epidermal chord of body wall. A little behind the anterior end of body, both excretory canals are connected by a bridge-like network of several short and slender **transverse canals** located ventrally to oesophagus (pharynx). Both longitudinal canals extend still forwards for a short distance from the connecting bridge. Thus, a H-shaped excretory system is formed. A short **terminal duct** (called **common canal**) also extends forward from one side of transverse canal and opens out by a ventromedian **excretory pore.** The whole system of canals is formed of a single highly elongated and complicated **"giant cell"** in which the canals are formed as intracellular excavations. This cell is called **renette cell.** The nucleus of this cell is located upon the left longitudinal canal close to transverse canal.

Ascaris is both **ammonotelic** and **ureotelic,** *i.e.*, it excretes both ammonia and urea. Probably, the hydrostatic pressure of pseudocoelomic fluid causes ultrafiltration of excretory fluids from it into the excretory canals. In fact, the excretory system of *Ascaris* is osmoregulatory also.

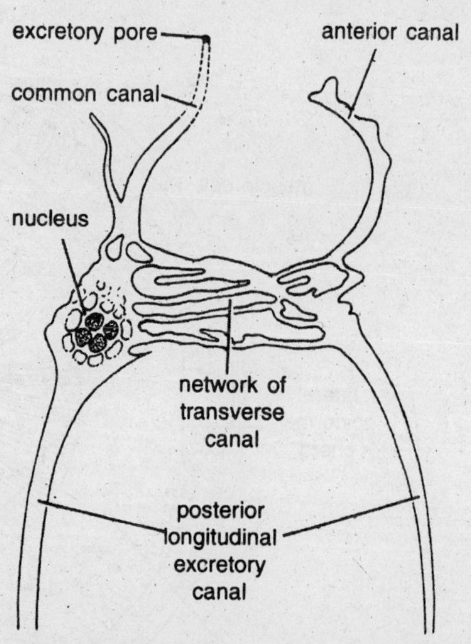

Fig. 30.8. *Ascaris lumbricoides*. Excretory system.

30.8. NERVOUS SYSTEM

The nervous system of *Ascaris* (Fig. 30.9 and 30.10) is well developed and complicated and is **intraepithelial** in position, being located in the epidermis, pharynx and rectum. Its nerve cells are constant in number, form, location and course of their fibers. They are 162.

Nervous system of *Ascaris* includes central nervous system and sympathetic nervous system.

1. Central nervous system. It includes a **circumpharyngeal nerve ring** round the pharynx. It is composed mainly of nerve fibers with very few nerve cells. From its anterior side, circumpharyngeal nerve ring gives off six **papillary** or **sensory** nerves that proceed to the sensory papillae of the lips. Two of these nerves are dorso-lateral, two lateral and two ventro-lateral in position. Each papillary nerve bears a small **papillary ganglion** just beyond its origin from the nerve ring. Associated with the nerve ring are seven ganglia. These include a median **dorsal ganglion**, a pair of **subdorsal ganglia**, a pair of **lateral ganglia**, each subdivided into six smaller ganglia; and a pair of **large ventral ganglia.** From these ganglia of the nerve ring there proceed backward eight nerves: a **mid dorsal nerve** that passes through the dorsal epidermal chord, a **midventral nerve** that passes through the ventral epidermal chord, a pair of **dorsolateral nerves** that run just above the lateral chords; a pair of **lateral nerves** that extend through the lateral epidermal chords and a pair of **ventrolateral nerves** that run just below the lateral epidermal chords. Of these the ventral nerve is most prominent and is ganglionated throughout its length. Some authors have regarded it as **ventral nerve chord** of *Ascaris.* Posteriorly the nerve chord joins a **preanal ganglion** which sends nerves to the tail. The dorsal and ventral nerves are connected at intervals by asymmetrical **dorsoventral connectives** throughout their length. the ventral and lateral nerves are also connected at intervals by **ventrolateral connectives.**

One of the six ganglia formed by subdivision of each lateral ganglion is known as **amphidial ganglion**. It sends forward an **amphidial nerve** to the amphid of its side.

The genital papillae of the male worm are innervated from special **bursal nerves** formed one on either side near the lateral nerve.

2. Sympathetic nervous system. *Ascaris* also possesses two sympathetic system, namely, the **pharyngeal** and **rectal.**

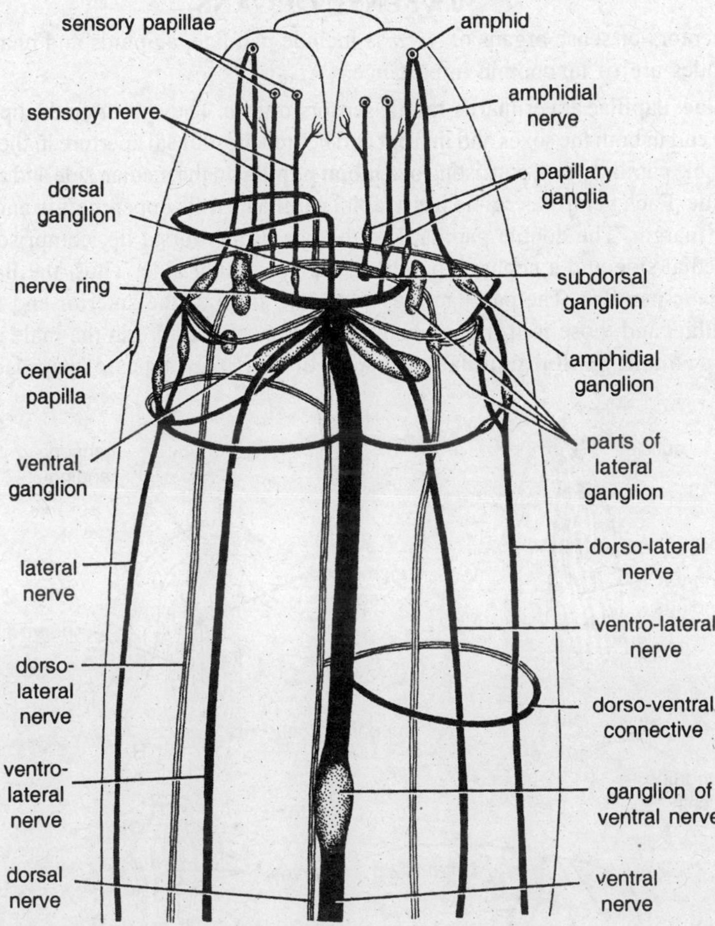

Fig. 30.9. *Ascaris lumbricoides*. Ventral view of anterior end showing central nervous system.

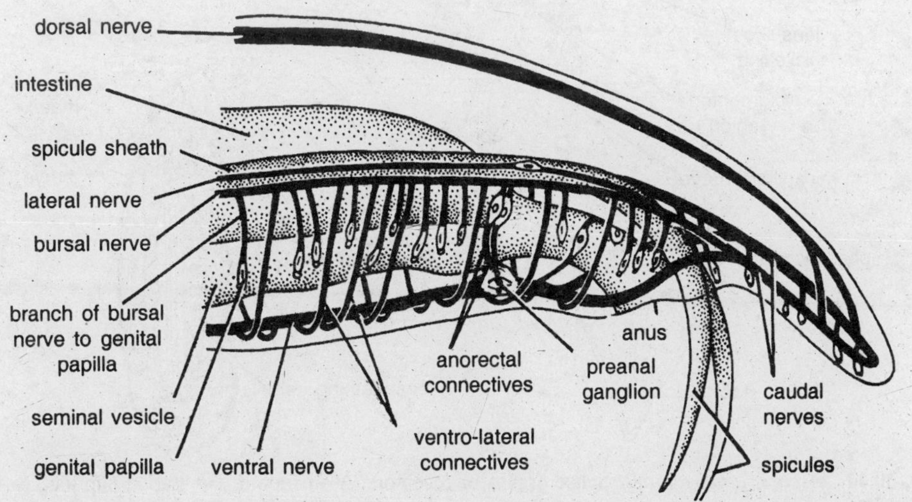

Fig. 30.10. *Ascaris lumbricoides*. Lateral view of posterior end showing nervous system.

30.9 SENSE ORGANS

The receptors or sense organs of *Ascaris* include papillae, amphids and phasmids. Receptors of the nematodes are of taxonomic importance.

I. Papillae. Papillae are primarily **tactile** sensory organs. They occur on the lips and on the sides of the anterior end in both the sexes and in front and behind the cloacal aperture in the male. The dorsal lip has two double papillae, each consisting of a **labial papilla** on the median side and a **cephalic papilla** on the outer side. Each ventrolateral lip has a labial papilla near its upper margin and a double papilla near its lower margin. The double papilla, like the one of the dorsal lip, comprises a labial papilla toward the medial side and a cephalic papilla toward the outer side. Thus, the lips bear six labial and four cephalic papillae. The papillae present on the sides of the anterior end are known as the **cervical papillae**, and those in front and behind the cloacal aperture in the male are termed as the **preanal** and **postanal genital papillae**. Structure of four main types of papillae is as follows:

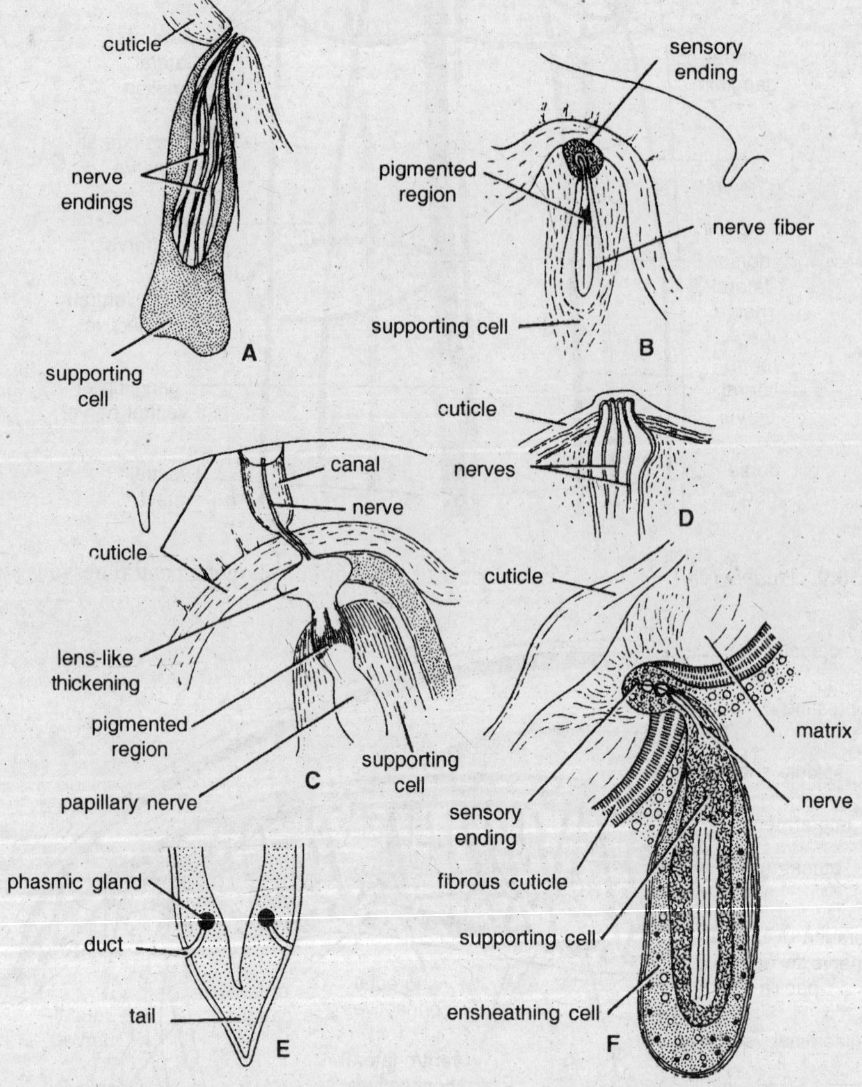

Fig. 30.11. *Ascaris lumbricoides*. Sense organs or receptors. A—Amphid; B—Labial papilla; C—Cephalic papilla; D—Genital papilla; E—Phasmid; F—Cervical papilla.

1. Labial papilla (Fig. 30.11B). It consists of a stout nerve fibre with a bulb-like ending under the cuticle and surrounded by a supporting cell. It also acts as a gustatory or taste organ.

2. Cephalic papilla (Fig. 30.11C). It also consists of a stout nerve fiber surrounded by a supporting cell, but the nerve fiber has just beneath the cuticle a lens-shaped expansion from which a canal extends to the surface.

3. Cervical papilla (Fig. 30.11F). It also consists of a nerve fiber having a bulb-like ending under the cuticle and surrounded by a supporting cell. The supporting cell is surrounded by a ensheathing cell.

4. Genital papilla (Fig. 30.11D). It consists of three nerve fibers embedded in a supporting cell and ending distally in a narrow canal that opens out by a pore.

II. Amphids (Fig. 30.11A). There are two amphids, one on each ventrolateral lip little above the labial papilla. An amphid consists of a supporting cell opening out by a pore and having a bundle of nerve fibers that directly unite into one fibre. The amphids are **olfactory chemoreceptors.**

III. Phasmids (Fig. 30.11E). They are a pair of unicellular glands that open out on the sides of tail. These are glandulo-sensory in nature (or chemoreceptors).

30.10 REPRODUCTIVE SYSTEM

In *Ascaris*, sexes are separate, *i.e.*, **dioecious.**

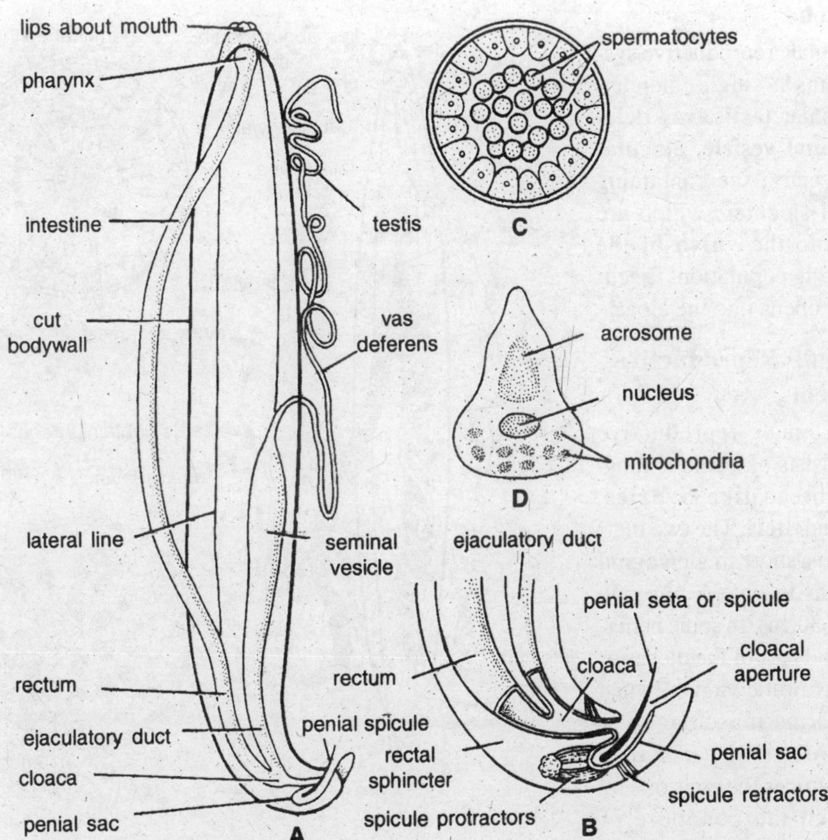

Fig. 30.12. *Ascaris lumbricoides.* A—Male reproductive system; B—Posterior end of male in lateral view showing cloaca and spicule; C—T.S. of vas deferens; D—Amoeboid sperm.

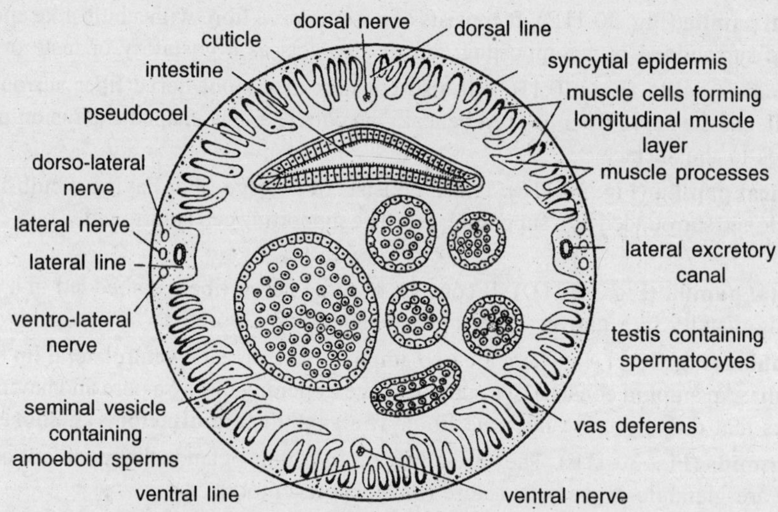

Fig. 30.13. *Ascaris lumbricoides.* T.S. of a mature male.

1. Male Reproductive System

The male reproductive system contains a single, slender, coiled, tubular **testis, vas deferens, seminal vesicle, ejaculatory duct** and a sac containing two penial spicules, which are inserted into the **vulva** of the female during copulation. Ejaculatory duct opens into the cloaca.

2. Female Reproductive System

The female reproductive system consists of a pair of long, slender, thread-like **ovaries, oviducts** and **uteri.** The two uteri join to form a short single **vagina** which opens to outside through **vulva** (Fig. 30.14). In some nematodes the end part of the uterus or the end part of the vagina, when present, forms muscular ove jectors which by peristaltic movement forces the eggs one by one through the gonopore or vulva.

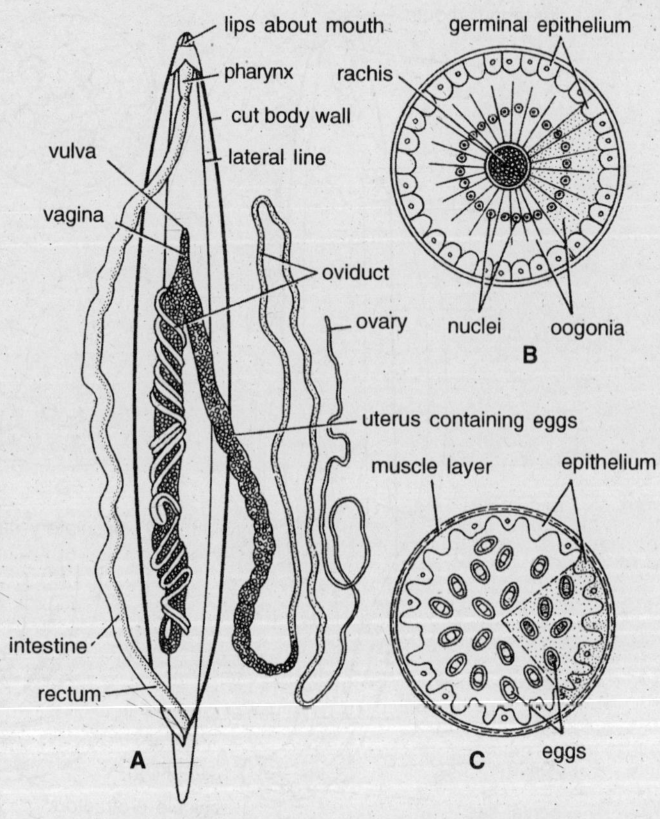

Fig. 30.14. *Ascaris lumbricoides.* A—Female reproductive system; B—T.S. of ovary; C—T.S. of uterus.

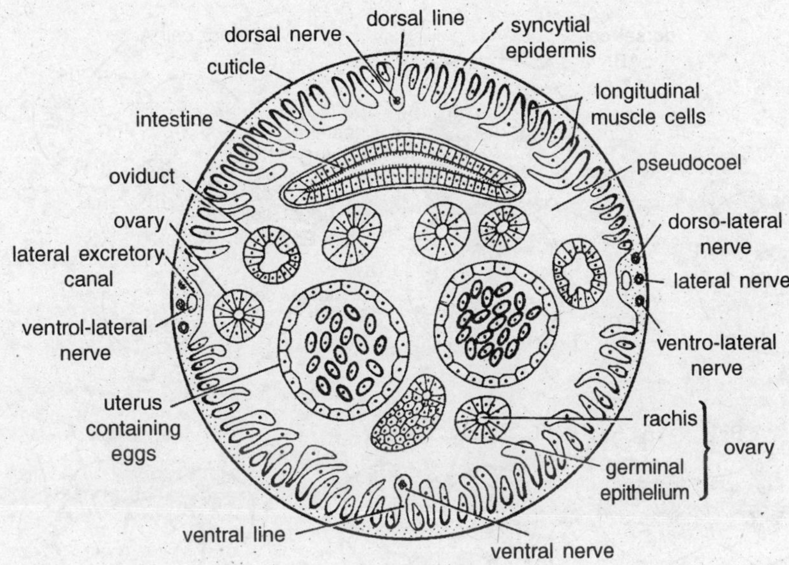

Fig. 30.15. *Ascaris lumbricoides.* T.S. of a mature female.

In *Ascaris* gonads are **telogonic,** *i.e.*, the germ cells arise at the proximal end only which is called **germinal zone** (**Note. Hologenic** means germ cells arise along entire length of gonads).

30.11 DEVELOPMENT AND LIFE HISTORY

Copulation occurs within the host's intestine. The eggs are fertilized in the oviducts of female *Ascaris*. The fertilized eggs produce a thick clear chitinous **inner shell.** As the eggs pass down, the uterine wall secretes an outer shell of albuminous protein. These eggs are called **mamillated eggs** since they contain **mamillated cysts.** A mamillated cyst is hard, rough and warty. It is resistant and impermeable to water (Fig. 30.18).

Eggs are laid in the small intestine of the host and they pass out with the faeces. The eggs remain alive in the soil for years. In a warm moist place the eggs develop in the following manner:

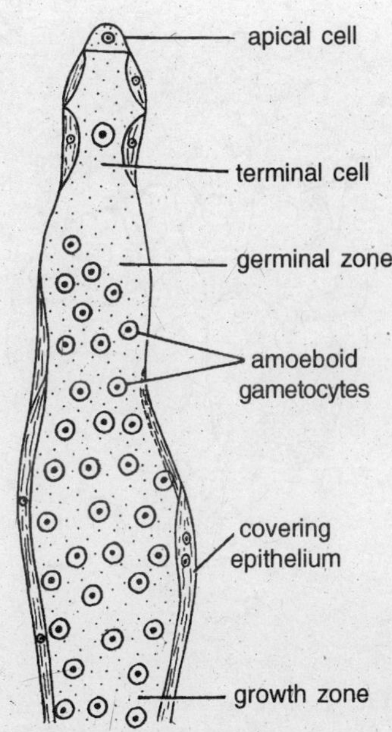

Fig. 30.16. *Ascaris lumbricoides.* Telogonic origin of sex cells in the proximal end of gonad.

1. In *Ascaris*, **cleavage** is spiral and determinate (Fig. 30.17). Within 10 to 14 days a minute larva, called **rhabditiform larva** develops in each of the eggs. It represents the infective stage.
2. If such eggs are swallowed by the human beings along with contaminated food or water, the larvae hatch and burrow into the veins or lymph vessels of the intestinal wall.

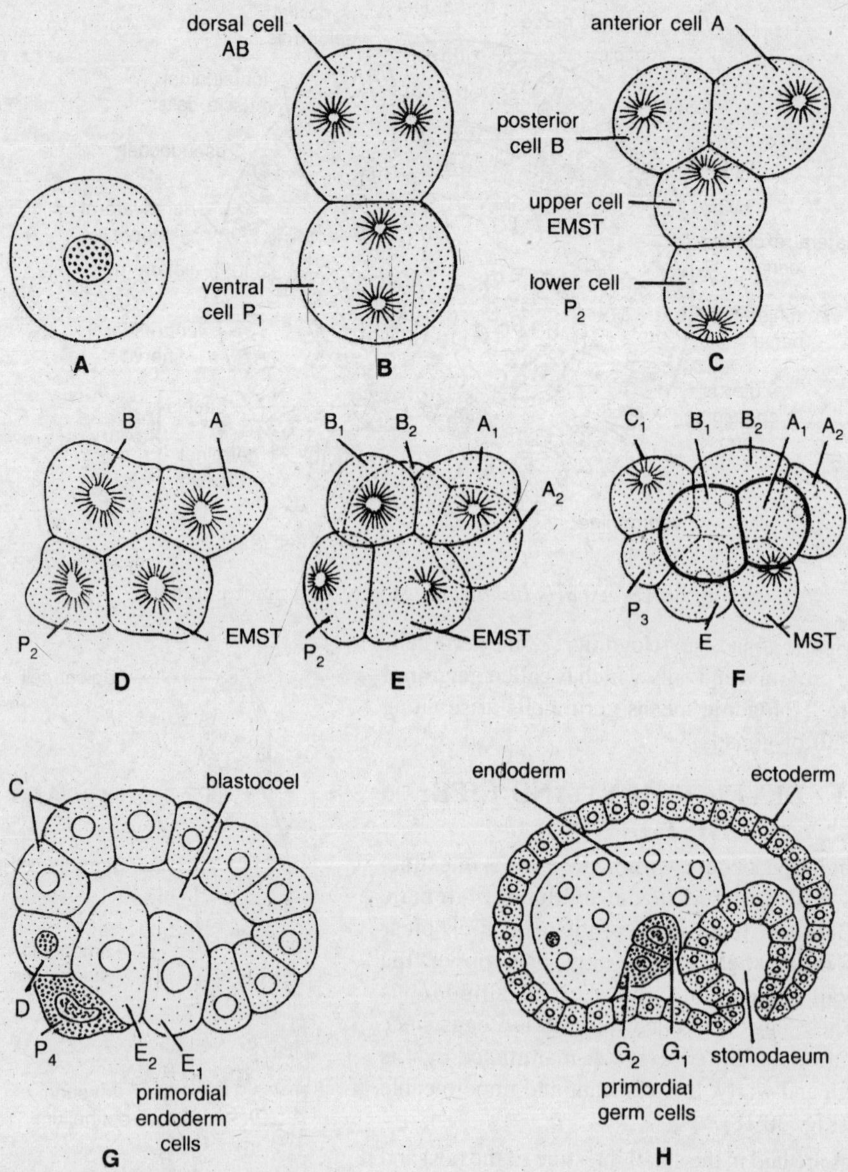

Fig. 30.17. *Ascaris.* A—Various stages of spiral and determinate cleavage. A—Fertilized egg; B—Two-cell stage; C—Four-cell stage (T-shaped); D—Four-cell stage (rhomboidal); E—Six-cell stage; F—Eight-cell stage; G—Median sagittal section through blastula; H—M.S. section of gastrula.

3. These larvae reach the capillaries of the lungs through the heart. In a few days they break into the air passages and move through the trachea, oesophagus and stomach to the intestine again, where they grow to adult stage.
4. In the 60 to 75 days they grow into adult males and females. The length of life span of the parasite in the host average only 9 months to a year.
5. During these migrations the larva moults for four times and grows.

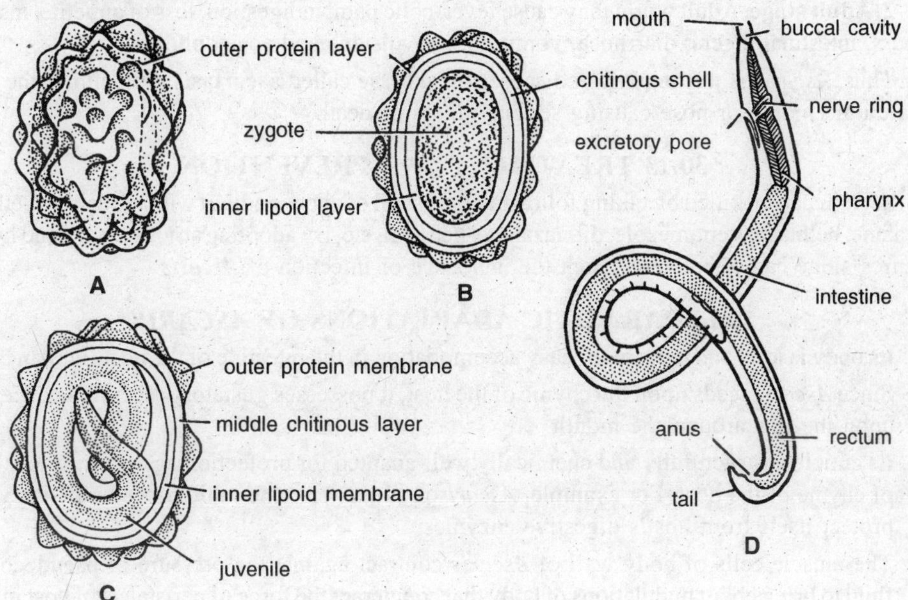

Fig. 30.18. *Ascaris.* A—An entire mammilated egg; B—T.S. of a mammilated egg; C—Embryonated egg in section; D—Rhabditiform larva.

The life-history of *Ascaris lumbricoides* can be depicted in a cyclic way as follows:

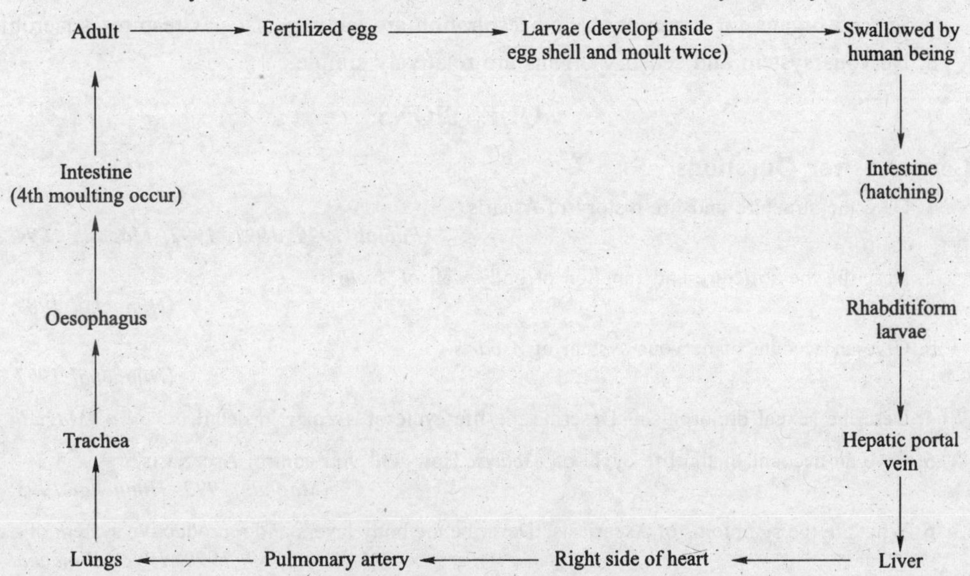

30.12 ECONOMIC IMPORTANCE

As many as 500 to 5000 *Ascaris* may be present in a single host. Both the larval stages and adults cause a variety of disorders in human beings. The larval stages are more harmful.

1. Harmful effects of larval stages. In the intestine of the host, larvae cause internal bleeding, muscular spasm, fever and anaemia. Passage of larvae through the lungs causes local inflammation and pneumonia like symptoms.

2. Adult stage. Adult worm may cause fever, colic pain, indigestion, loss of appetite, insomnia, tumours, intestinal ulcers, diarrhoea, vomiting, convulsions and eosinophilia.

Thus, *Ascaris* is **pathogenic** and causes the disease called **ascariasis.** Sometimes the worms migrate upto mouth or nose causing serious illness or death.

30.13 TREATMENT AND PREVENTION

Ascaris can be killed by taking following drugs: oil of chenopodium with tetrachloro-ethylene, piperazine, helmacid, tetramezole, dithiazanine, hetrazin, etc. By adopting strict sanitary and hygienic measures alone can effectively check the incidence of infection of *Ascaris*.

30.14 PARASITIC ADAPTATIONS OF *ASCARIS*

1. Its body is long and narrow for easy accomodation in the intestine of the host (human beings).
2. Since *Ascaris* feeds upon the chyme of the host, it possesses gustatory receptors concentrated upon the lips around the mouth.
3. Its cuticle is structurally and chemically well-adapted for protection against injurious effects of chyme of the host. For example, *Ascaris* produces enzyme inhibitors or **anti-enzymes** to protect itself from host's digestive enzymes.
4. The muscle cells of body wall of *Ascaris* contract against the pressure of pseudocoelomic fluid to bring about undulations of body that counteract the force of peristalsis of host intestine.
5. Digestive system of *Ascaris* is simple, because almost predigested nutrients are available to the worms in the chyme of the host.
6. The pharynx is adapted as a sucking and pumping organ to suck in the chyme and pump it into the intestine.
7. Special organs for locomotion and respiration are lacking. *Ascaris* respires anaerobically.
8. Nervous system and sensory organs are relatively simple.

QUESTIONS

Long Answer Questions

1. Give the structure and life-history of Ascaris.
(*Punjab 1990, 1991, 1992; Magadh 1994, 1996*)

2. Describe the structure and function of body wall of *Ascaris*.
(*Himachal 1989, 1995*)

3. Give an account of nervous system of *Ascaris*.
(*Himachal 1992, 1995*)

4. Describe sexual dimorphism. Describe the life cycle of *Ascaris* in detail. (*Magadh 1991*)

5. Give an account of the life cycle of *Ascaris*. How will you control Ascariasis?
(*Magadh 1993; Himachal 1993, 1995*)

6. What are the symptoms of Ascariasis? Describe the body layers and reproductive system of *Ascaris*.
(*Magadh 1989, 1990, 1996; Himachal 1995*)

7. Describe male and female reproductive organs of *Ascaris*. (*Himachal 1990*)

Short Answer Questions

1. Write short notes on the following:

(*i*) Excretion and osmoregulation in *Ascaris*. (*Himachal 1995*)

(*ii*) Ascariasis. (*Guwahati 1995*)

2. Draw a labelled diagram showing

(*i*) T.S. of female *Ascaris*; (*Himachal 1991, 1994*)

(*ii*) T.S. of male *Ascaris*. (*Meerut or CCS 1991*)

Very Short Answer Questions

1. How many epidermal chords occur in *Ascaris*?

Ans. Four (*i.e.*, one dorsal, one ventral and two lateral).

2. Name the animals having intracellular excretory system.

Ans. Roundworms (nematodes).

3. Name the parts of the gut lined with cuticle.

Ans. Pharynx (stomodaeum) and rectum (proctodaeum).

4. How many longitudinal nerves *Ascaris* has? Where do they lie?

Ans. Eight, epidermis.

5. Name the three layers enclosing an egg of *Ascaris*.

Ans. Proteinous (albuminous), chitinous membrane and lipoid membrane; such eggs are called **mammiliated eggs**.

6. Pseudocoel is the persistent blastocoel. Is it so?

Ans. Yes.

7. What is the function of pharynx in *Ascaris*.

Ans. Ingestion, movement of food, digestion.

8. Reproductive organs of *Ascaris* are didelphic. What does it mean?

Ans. *Ascaris* has a double set of genital organs.

9. Why is the young of *Ascaris* called juvenile and not larva.

Ans. As it resembles adult in all respects, but sexually immature.

10. What are the accessory structures that are present in penial sacs of *Ascaris*?

Ans. Penial setae.

Fill in the Blanks

1. Name of human ascarid is .... *lumbricoides* variety ....

2. .... shows distinct sexual dimorphism.

3. Hind end of .... *Ascaris* is hooked/curved.

4. Mouth of *Ascaris* is surrounded by .... lips.

5. .... of *Ascaris* is syncytial.

6. Body cavity of *Ascaris* is called ....

7. Phenomenon of having fixed number of cells in the body is termed ....

8. Alimentary canal is .... in roundworms.

9. Amphids of *Ascaris* are ....

10. The sperms of *Ascaris* are ....

Multiple Choice Questions

1. *Ascaris* inhabits the host's

(*a*) blood (*b*) lungs

(*c*) intestine (*d*) liver

2. Male *Ascaris* has
(*a*) curved/hooked hind end (*b*) anal papillae
(c) monorchic gonad (*d*) all of these

3. Body wall of *Ascaris* lacks
(*a*) muscles (*b*) longitudinal muscles
(c) circular muscles (*d*) both (*b*) and (*c*)

4. Digestion in *Ascaris* is
(*a*) intercellular (*b*) intracellular
(*c*) both (*a*) and (*b*) (*d*) neither (a) nor (b)

5. Juvenile of *Ascaris* undergoes .... molts.
(*a*) two (*b*) three
(*c*) one (*d*) four

6. In *Ascaris*, the reserve food is stored as glycogen and fats in
(*a*) intestine (*b*) muscles
(*c*) hypodermis or epidermis (*d*) all of these

7. *Ascaris* is
(*a*) host (*b*) bisexual
(*c*) dioecious (*d*) digenetic

8. In *Ascaris*, eggs are fertilized in
(*a*) oviduct (*b*) uterus
(*c*) vagina (*d*) outside human body

9. The lips of *Ascaris* are
(*a*) denticulate (*b*) smooth
(*c*) cartilaginous (*d*) bony

10. Amphids upon the lips of *Ascaris* are
(*a*) organs of smell and chemoreceptors (*b*) secrete saliva
(*c*) sense organs (*d*) adhesive organs

11. Which type of excretory canal system is present in *Ascaris*?
(*a*) C-shaped (*b*) J-shaped
(*c*) H-shaped (*d*) W-shaped

12. The higher end of male *Ascaris* is
(*a*) straight (*b*) thickened
(*c*) coiled (*d*) curved

13. In *Ascaris* penial setae arise from
(*a*) cloaca of male (*b*) anus of female
(*c*) vulva of female (*d*) none of these

14. Infective stage of *Ascaris* for new host is
(*a*) I stage juvenile (*b*) embryonated egg
(*c*) III stage juvenile (*d*) larva after II moult

15. The stage hatched from the ingested egg of *Ascaris* is called
(*a*) bladder worm (*b*) hexacanth
(*c*) maggot (*d*) rhabditiform larva

16. Embryonated egg is

(*a*) shelled egg (*b*) shelled zygote

(*c*) shelled II stage juvenile (*d*) II stage juvenile

ANSWERS

Fill in the Blanks

1. *Ascaris, humanis* **2.** *Ascaris*

3. Male **4.** Three

5. Epidermis **6.** Pseudocoel

7. Eutely **8.** Complete

9. Chemoreceptors **10.** Amoeboid.

Multiple Choice Questions

1. (*c*) **2.** (*d*) **3.** (*c*) **4.** (*c*)

5. (*d*) **6.** (*c*) **7.** (*c*) **8.** (*a*)

9. (*a*) **10.** (*a*) **11.** (*c*) **12.** (*d*)

13. (*a*) **14.** (*b*) **15.** (*d*) **16.** (*c*).

31

Parasitic Nematodes of Humans and Plants

31.1 NEMATODES AND HUMAN DISEASES

The human diseases caused due to infestation with the nematode worms constitute **nematodiasis.** Some of the common type of nematode related diseases of human beings are the following:

1. Ascariasis

Ascariasis is a very common disease of human beings caused by the large-sized nematode (roundworm) *Ascaris lumbricoides*. This roundworm is found in the intestine especially of children. In human beings the disease follows the ingestion of embryonated eggs with contaminated food (raw vegetables, fruits) and drinking water. Among children playing in the contaminated soil, there is also hand to mouth transfer of eggs by dirty fingers. Ascariasis is characterized by indigestion, protuberance of abdomen, loss of appetite and intestinal obstruction, insomnia, appendicitis, gastric ulcers, diarrhoea, liver disorders, etc. The larvae while pursuing a migratory course enter the lungs and their heavy infection affects the lungs (causing bronchitis). Ascariasis can be successfully treated by oral administration of piperazine citrate syrup, alkendazole, mebendazole, santonin and oil of chenopodium.

2. Ancylostomiasis

Ancylostomiasis is caused by two hookworms—*Ancylostoma duodenale* and *Necator americanus* (American hookworm). Both the hookworms are parasites within human intestine. They are most frequent in rural areas. Infection occurs, when human beings walk bare-foot on the faecally contaminated soil. Their filariform larvae penetrate directly through host's skin when they come in contact with it. These hookworms cause itching in skin, haemorrhage and bronchial pneumonitis in the lungs in the larval condition and the adults produce weightloss, constipation, diarrhoea, dyspnea (labored breathing), anaemia, nervous disorders, etc. Ancylostomiasis can be treated by following drugs — tetrachloroethane and carbon tetrachloride.

3. Enterobiasis

Enterobiasis (or pinworm disease) is caused by *Enterobius vermicularis* commonly called **pin-worm** or **seat-worm.** Adult pinworms live in caecum, colon and vermiform appendix of human beings (especially children). The females migrate out through colon and rectum and lay large number of eggs in the skin folds near the anus, causing intense itching. These eggs may infect the entire family through clothes, furniture, air and dust in the room. The symptoms of enterobiasis include severe itching around anus, loss of appetite, sleeplessness, bed-wetting, grinding of teeth, nausea and vomiting. Piperazine and enteric-coated capsules of gentian violet are most effective drugs in the treatment of pinworms.

4. Trichuriasis

Trichuriasis (or whipworm disease) is caused by *Trichuris trichura* and *T. ovis* (Fig. 31.1), commonly called **whip-worm.** The adult worms live in caecum and vermiform appendix of human beings. Host is infected when the embryonated eggs are swallowed with food or water. It causes abdomi-nal pain, mucous diarrhoea, often with blood, streaked stool and loss of weight. It may also show symptoms of acute appendicitis. Trichuriasis can be treated by drugs such as stibazium idodide, oeftarsone, thiabendazole and mebendazole.

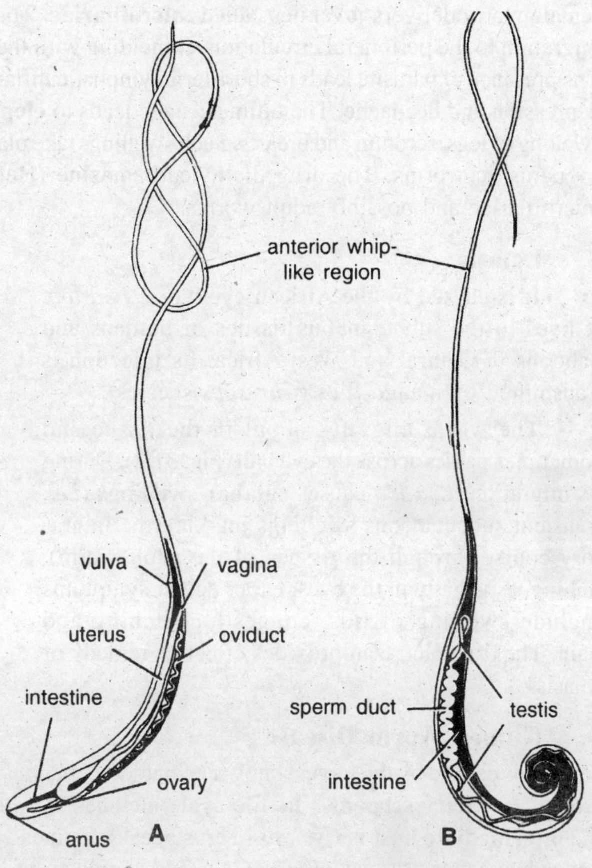

Fig. 31.1. *Trichuris ovis*. A—Female; B—Male.

5. Trichinosis

Trichinosis is caused by *Trichinella spiralis*. It is one of the smallest nematodes infecting human beings. Trichinosis is common and widespread in Europe and America. Humans become infected by *Trichinella* by eating raw or undercooked meat (pork) containing its encysted larvae. Trichinosis is characterized by eosinophilia, haemorrhage, muscular pain, thrombosis and restlessness. The drug thiabendazole is found to be effective in the treatment of trichinosis.

6. Strongyloidiasis

This disease is caused by the nematode *Strongyloides stercoralis*, commonly called **threadworm**. These worms invade the lining (mucous epithelium) of intestine of human beings. The female lays large number of eggs that pass out in faeces and develop into ***rhabditiform larvae***. These then transform into infective ***filariform larvae*** that penetrate the skin of the barefooter, enter blood vessels and pass to the lungs, where they invade alveoli. They ascend trachea, descend the oesophagus and mature to become pathenogenic females in the small intestine. Invading larvae cause dermatitis. Larvae migrating through lungs tend to produce cough, haemoptysis (= expectoration of blood in sputum) and dysponea. Severe infection of the instestine causes vomiting, diarrhoea and constipation. Its hyperinfection results in ulceration, oedoma, congestion, fibrosis and severe inflammation of the intestine. This disease can be treated by the drug thiabendazole.

7. Filariasis or Elephantiasis

Filariasis is caused by *Wuchereria bancrofti,* commonly called the filaria worm. The adult worm inhabits lymphatic vessels of human beings, most frequently those in the lymph nodes, testes and epididymis. Infection is spread through *Culex* mosquito. Following copulation the viviparous female worm delivers juveniles called **microfilariae**. These exhibit a distinct periodicity of larval migration to the peripheral circulation, coinciding with the activity of mosquito (intermediate host). The presence of parasite leads to short-term lymphatic inflammation marked by pain and fever, mental depression and headache. The ultimate stage leads to **elephantiasis** which is characterised by great swelling of legs, scrotum and breasts. Such swellings take place due to blockage of the lymph circulation by parasitic worms. The drug diethylcarbamazine (Hetrazan) is found effective in killing the microfilariae and possibly adult worms.

8. Loiasis

It is caused by the African eye worm *Loa loa*. It lives in the subcutaneous tissues of humans and baboons in Central and West Africa. Its infection is transmitted by mango flies (*Chrysops* species).

The worm migrates about in the tissue and sometimes passes across the eye ball (Fig. 31.2). During its migration, *Loa loa* causes **calabar swellings** (*i.e.*, transient subcutaneous swellings marking the migratory course through the tissues of the adult worm), itching, especially in the eyes. Other ocular symptoms include swelling of lids, congestion, itching and pain. The drug Hetrazan provides effective remedy of loiasis.

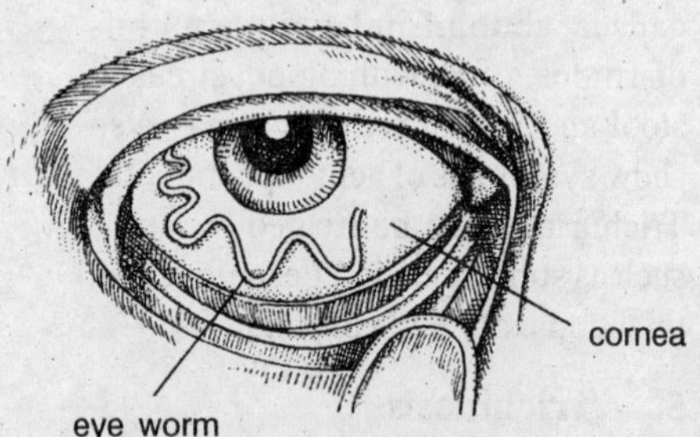

Fig. 31.2. Eye worm *Loa loa* in the cornea of human eye.

9. Guinea Worm Disease

It is caused by the largest known nematode *Dracunculus medinensis*, commonly called **"guinea worm"** or **"fiery serpent."** Its life cycle includes two hosts: **primary host** is human beings and the **intermediate host** is *Cyclops* (a crustacean). Guinea worm lives in the general viscera (coelom) and subcutaneous tissues of arms, legs and shoulders of human beings. It also infests dogs, foxes, racoons and minks. It is commonly found in dry regions of India, Arabia, Tropical Africa, South America, Persia, Turkey, Guyana and China. In India, it is found in Andhra Pradesh, Tamil Nadu, Rajasthan and foot hills of the Himalayas.

The female guinea worm is about 1 mm in diameter and up to 120 cm in length. After a period of development in the body cavity and connective tissue of the host, the gravid female migrates to the subcutaneous tissue and produces an ulcerated opening to the exterior. If the ulcerated area of the host comes in contact with water, larvae are released. After a short free living stage, the larvae are ingested by species of *Cyclops* and continue their development in the host's haemocoel. When definitive host (human being) swallows *Cyclops* in drinking water, the nematode larvae are released and penetrate the intestinal wall to reach the coelom or subcutaneous tissue.

Dracunculus secretes toxic substances that cause asthma, eosinophilia, itching, vomiting; diarrhoea, and giddiness. The worm can be surgically removed but the ancient method, still practised, is to slowly wind them out on a small stick (Fig. 31.4). A safe extraction takes 10 to 11 days. Intramuscular injection of phenothiazine emulsified in olive oil into a number of places close to worm kills the worm. Infection can be prevented by drinking filtered water.

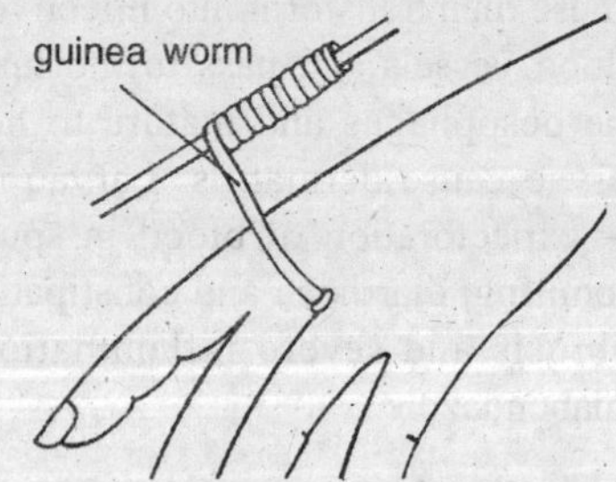

Fig. 31.3. A guinea worm being removed from the ulcerated opening of arm by slow winding on a matchstick.

Box 31.1

The dracunculid and its removal from the human is the basis of the **caduceus** (Fig. 31.3) or symbol of the medical profession. The symbol, however, has been misinterpreted as a staff (wand) with snakes instead of a nematode wrapped around a stick.

Fig. 31.4. Caduceus.

31.2. PLANT NEMATODES

Nematodes parasitic in plants attack all types of commercial crops such as grasses, flowers, vegetables, fruits and trees (see Table 31.1). Because of the damage they can cause to crop plants, many of these plant nematodes are of great economic importance.

Table 31.1 Indian examples of plant nematodes and their host plants.

Phytoparasitic nematodes	*Host plants*
1. *Heterodera avenae*	Wheat, barley, oat, maize, sorghum
2. *Heterodera cajani*	Pigeon - pea
3. *Heterodera rostochiensis*	Potato
4. *Meloidogyne incognita*	Tomato, egg-plant
5. *Meloidogyne javanica*	Okra, tomato
6. *Meloidogyne hapla*	Red clover, potato
7. *Rotylenchulus reniformis*	Castor, coffee, papaya, tomato
8. *Tylenchulus semipenetrans*	Citrus
9. *Rotylenchus robustus*	Carrot
10. *Ditylenchus dipsaci*	Onion
11. *Trichotylenchus astriatus*	Grapewine
12. *Trophurus similis*	Coconut palm
13. *Criconema punici*	Pomegranate
14. *Tylenchorhynchus phaseoli*	Leguminous crop
15. *Tylenchorhynchus vulgaris*	Maize
16. *Hirschmanniella shamimi*	Paddy
17. *Hirschmanniella indica*	Paddy
18. *Hirschmanniella oryzae*	Rice

Types of Phytoparasitic Nematodes

1. Ectoparasites of plants. Worms feed on the external cells of plants by puncturing the cell wall with oral stylets and sucking out the cell contents. *Pratylenchus*, for example, is an ectoparasite of plant roots.

2. Endoparasites of plants. Worms in juvenile stage enter the plant body and feed on the living cells; producing gall-like structures or causing tissue death. Reproduction takes place within the host, and the new generation of juvenile migrates to other plants. Cyst-forming nematode, *Heterodera,* and root-knot nematode, *Meloidogyne*, have many such species.

3. Phytoparasitic juveniles and zooparasitic female adults. The female worm produces juveniles within a plant-feeding insect host. When the insect punctures the plant tissue, the juvenile worms enter the plant and remain as endoparasites. When mature and after copulation, the female enters the larva of the insect host, which lives on the same plant. The larval host metamorphoses into an adult and deposits a new generation of juvenile worms. *Heterotylenchus aberrans*, a parasite of onion flies, illustrate this type of life cycle. Likewise, adult females of *Tylenchinema oscinellae* live in the haemocoel of fruit flies. Their larvae enter the fly. Larval flies and larval nematodes develop independently on the oat plant and as the nematode mature, the females enter the fly larvae after mating, while the male dies.

General Characters of Plant Nematodes

1. All plant nematodes belong to orders **Tylenchida** and **Aphelenchida**.
2. Phytophagous/phytoparasitic nematodes are small and the plants on which they feed are relatively huge.
3. In plant parasitic nematodes, the buccal cavity carries a long cuticular **oral stylet** which can be protruded from the mouth and used in puncturing plant cells and then the contents of plant cells are sucked out. Oral stylet is formed either by the modified lining of the buccal cavity (**stomatostyle**) or sometimes from a highly modified tooth (**odontostyle**).
4. Plant nematodes generally penetrate inside the host plants through their roots.
5. Feeding of phytoparasitic endoparasites (*e.g.*, *Nacobbus*, *Meloidogyne*) cause hyperplasia and hypoplasia resulting in a **gall**, inducing **giant nutritive cells** that allow the adult worms to be sedentary. **Hypoplasia** means the conditions of arrested development (*i.e.*, poor growth and production). **Hyperplasia** means excessive production of cells in a part.
6. Some plant nematodes show high degree of host-specialization (*e.g.*, *Heterodera*).
7. The exudations (*e.g.*, root leachates) from host plants are attractive to phytoparasitic nematodes; juveniles can find their way toward host plants by this means (*e.g.*, *Heterodera*).
8. Plant parasitic nematodes also exhibit a wide variety of **life history** patterns. In some cases, the males and females live freely in the soil and the juveniles move into and out of root on which they feed and are not restricted to one feeding site. But there are others which penetrate deeply into the plant, become sedentary and undergo sexual reproduction, the resultant juveniles move off into the soil and find new plants.

Examples of Some Plant Nematodes

1. *Meloidogyne marioni* (Root-knot Eelworm)

Root-knot is one of the oldest known diseases of plants. One of earliest references to the root galling attributable to these nematodes is that of **Berkely** (1855) who described and illustrated galls found on the roots of cucumber in English green house (Fig. 31.5). **Chitwood** (1949) reconfirmed the generic name *Meloidogyne*.

The second-stage juveniles of *Meloidogyne* enter the host plant and pierce and suck out the contents of cells. Their feeding causes hyperplasia and hypoplasia resulting in a **gall**, inducing **giant nutritive cells** that allow the adult worms to become **sedentary**. Juveniles mature in the gall. Male nematodes swell as do females, but in the fourth stage moulting they metamorphose into a vermi- form stage. They embark from the home gall into soil to seek other infected plants, where they fertilize the female and die. The females become more and more plump, losing all resemblance to the cylindrical larvae and several hundred ova form within them (Fig. 31.6B). Like *Heterodera*, females of *Meloidogyne* become semiendoparasitic, so that only the head and anterior end of their body remain embedded in host tissues and eggs are deposited outside the host.

In addition to cause direct damage to host plants, *Meloidogyne* is involved in spreading the infection of certain fungal diseases such as *Fusarium* wilt of cotton.

2. *Heterodera* (Cyst-forming Nematode)

This genus was created by **Thome** (1949). It includes all the true **cyst-forming nematodes**, *i.e.*, those nematodes in which some or all of the eggs are retained within the swollen body of the female; her body wall after her death becomes oxidized and tanned to form a tough, resistant or protective container or lemon- or pear-shaped **cyst** for the developing eggs. *Heterodera* also exhibits following two peculiar features which are not known in plant parasitic nematodes: 1. It has a high degree of host-specialisation; 2. Its hatching process depends on stimulatory substances (root leachates) emanating from the host roots. Thus, members of *Heterodera* are highly adapted to a plant parasitic existence. Species of Heterodera are best suited to relatively cool climates.

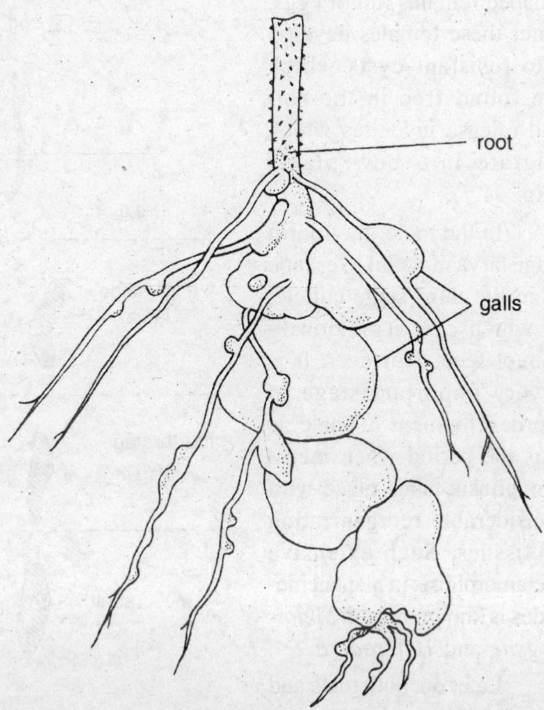

Fig. 31.5. Galls of root-knot nematode ('vibrio') on cucumber root.

Both male and female juveniles of *Heterodera* penetrate the host plant (via roots). In the female, all larval stages are separated by distinct complete moulting and the third and fourth stage larvae have well developed **oral stylets.** They continue to grow and enlarge to become sedentary pear

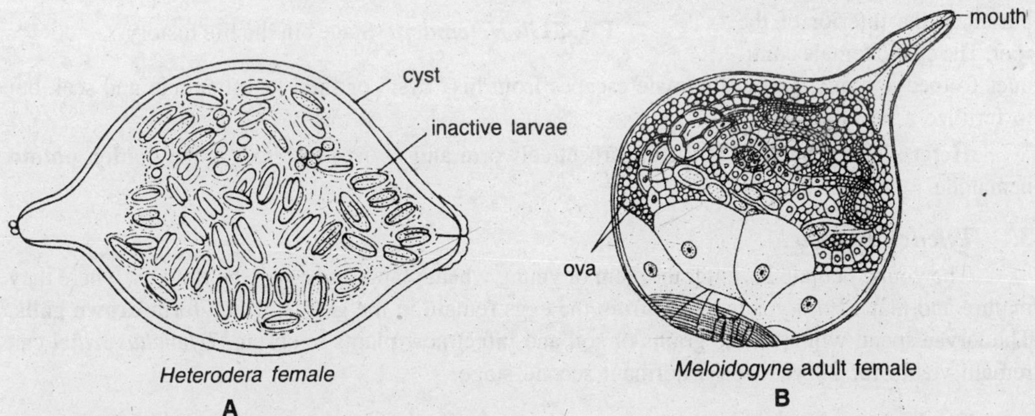

Fig. 31.6. A—*Heterodera* female after conversion into cyst filled with inactive larvae. B—*Meloidogyne*, adult female with maturing ovaries.

-shaped females full of eggs. Later these females develop into resistant cysts which are found free in the soil and release juveniles which migrate into new plants (Fig. 31.7).

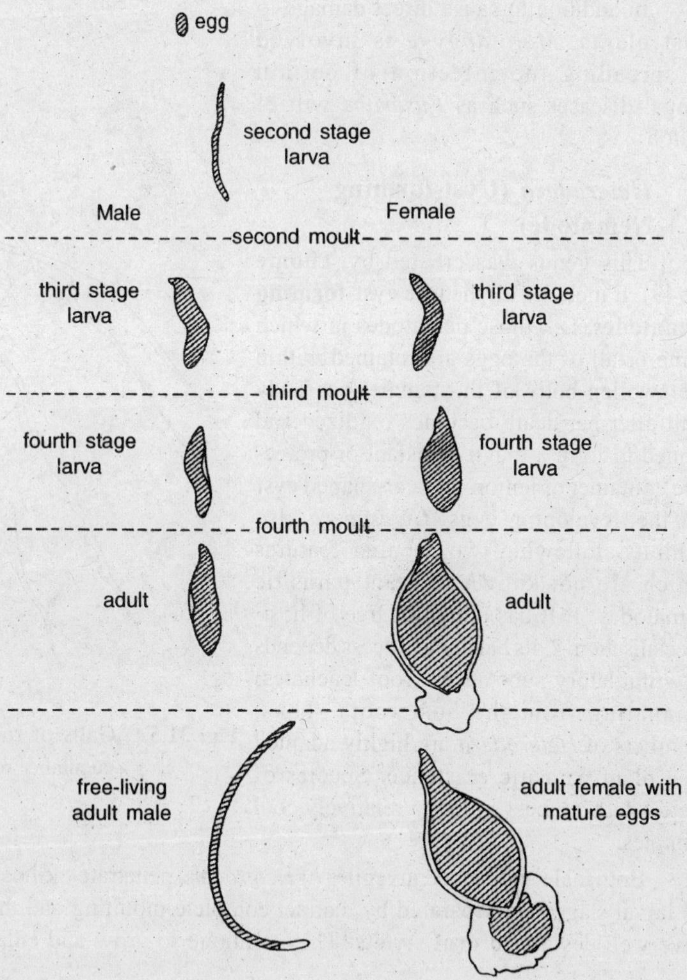

Fig. 31.7. *Heterodera.* Stages in the life history.

In the male, the fourth stage larva may fail to escape from the third stage cuticle, in which case it obviously cannot feed. However, it is a very important stage in the development of male as it is the period when **metamorphosis** takes place with considerable reorganization of tissues. Such extensive metamorphosis in plant nematodes is known only in *Meloidogyne* and *Heterodera.*

Later on, both male and female of *Heterodera* become secondarily **semi-endoparasitic**: the outer root tissues rupture under the pressure of the enlarging parasites within, so that these burst through to the surface. However, they often remain attached by their head ends which remain buried in the interior of the root. The cystic female continues to feed and enlarge but the male escapes from his "cyst" or final larval cuticle and seek out to fertilize a new female.

Heterodera schachtili is the sugarbeet eelworm and *H. rostochiensis* is the **golden potato** nematode.

3. *Tylenchus tritici*

The young nematodes enter the stem of young wheat plants and reach the flowers, where they mature and mate. Embryos hatching from the eggs remain in the grains, which form **brown galls**. The larvae spend winter in the grains or soil and infect new plants in spring. *Tylenchus tritici* can remain viable for 28 years as a dormant second stage.

4. *Trichodorus*

It is called the **stubby root nematode**. It is ectoparasite on vegetables, fodder, grain and fruit crops. Its infection checks the growth of root tip, producing stubby roots.

QUESTIONS

Long Answer Questions

1. Write an essay on nematodes and disease.

(*Sagar 1994; Vikram 1994; Jiwaji 1994; Indore 1994*)

2. Write names of human nematode parasites. Describe diseases caused by any three of them.

3. Name any three nematodes which are harmful to man and describe the life-history of one nematode studied by you. (*Magadh 1992*)

4. Discuss the economic importance of Nemathelminthes. (*Himachal 1994*)

5. Write early history of plant nematodes. Describe four phytophagus nematodes giving their common names and plants they cause damage. Add a note on their control.

(*Allahabad 1997*)

6. Name the common plant nematodes. Give a brief account of any one of them.

(*Allahabad 1993*)

Short Answer Questions

1. Suggest some measure to control phytoparasitic nematodes. (*Himachal 1995*)

2. Write short notes on the following:

(*i*) Plant parasitic nematodes (*Allahabad 1990, 1991; Himachal 1995*)

(*ii*) Root gall (*Allahabad 1990*)

(*iii*) Root-knot nematode (*Allahabad 1992*)

Very Short Answer Questions

1. How human beings get infection of *Ascaris*?

Ans. By swallowing eggs of *Ascaris* in water or food.

2. What protects the gut nematodes from digestive enzymes?

Ans. Thick cuticle of body wall.

3. Name a nematode that is parasitic in the adult stage and free-living in the young stage.

Ans. *Ancylostoma duodenale.*

4. Which parasitic nematode is introduced into the human body by a mosquito?

Ans. *Wuchereria bancrofti.*

5. Name any nematode parasite that occurs in the human muscles.

Ans. *Trichinella spiralis.*

6. Cite one example of the order Oxyuroidea.

Ans. *Enterobius* (= *Oxyuris*) *vermicularis.*

Fill in the Blanks

1. The vector of elephantiasis is ....

2. *Dracunculus medinensis* (guinea worm of humans) is a parasitic worm which invades ....

3. Nematode parasites that attack plants are called ....

4. .... are parasites which infect animal.

5. The nematode parasite that lives in the lymph is ....

6. The nematode .... attacks wheat plants.

7. The nematode parasite of horse is ....

Multiple Choice Questions

1. The intermediate host of guinea worm is
(*a*) house fly (*b*) *Cyclops*
(*c*) mosquito (*d*) none of the above

2. *Dracunculus medisensis* is a parasite of
(*a*) body tissue (*b*) body fluids
(*c*) gut (*d*) lungs

3. The common parasite of human large intestine is
(*a*) *Ascaris* (*b*) *Trichinella*
(*c*) *Enterobius* (*Oxyuris*) (*d*) *Ancylostoma*

4. *Enterobius vermicularis* inhabits the human
(*a*) large intestine (*b*) caecum
(*c*) appendix (*d*) all these

5. Trichinosis is caused by
(*a*) *Trichuris* (*b*) *Trichinella*
(*c*) *Ancylostoma* (*d*) none of these

6. *Trichuris trichura* is commonly known as
(*a*) *Ascaris* (*b*) pinworm
(*c*) whipworm (*d*) roundworm

7. The scientific name of pinworm or seatworm is
(*a*) *Trichinella* (*b*) *Enterobius*
(*c*) *Ancylostoma* (*d*) *Wuchereria*

8. The transmission of infection caused by *Loa loa* is through
(*a*) mango flies (*b*) *Anopheles*
(*c*) *Culex* (*d*) housefly

ANSWERS

Fill in the Blanks

1. *Culex* **2.** Subcutaneous tissue
3. Phytophagus **4.** Zoophagus
5. *Wuchereria bancrofti* **6.** *Tylenchus tritici*
7. *Parascaris equonium* (= *Ascaris megalocephala*).

Multiple Choice Questions

1. (*b*) **2.** (*a*) **3.** (*c*) **4.** (*d*)
5. (*b*) **6.** (*c*) **7.** (*b*) **8.** (*a*).

32

Minor Phyla

32.1 PHYLUM CTENOPHORA

1. General Characters

1. Ctenophores (Gr., *ktenes* = combs; *ophora* = carrying) are commonly called **sea walnuts, sea gooseberries** or **comb jellies.**
2. Marine, solitary, luminescent, pelagic and free-swimming animals.
3. There are only 50 species of Ctenophora. Some are abundant in coastal waters and many are oceanic.
4. Acoelomate and triploblastic animals having biradially symmetrical med-usa-like body.
5. Body delicate, transparent, gelatinous and unsegmented.
6. Body shape is usually pear-shaped, cylindrical, spherical flat or ribbon-like.
7. Body wall consists of three layers — outer **epidermis,** inner **gastrodermis** and middle thick **mesogloea** or **mesoderm** with scattered cells and muscle fibres.
8. Ctenophores perform locomotion by eight — vertical (meridional) rows of ciliary **comb plates**.
9. They are predatory (carnivorous) animals feeding on other planktonic forms (*e.g.*, copepods). Many species use in food-captur-ing their two long, branc-hed and retractile **tentacles** bearing special adhesive cells called **colloblasts** (Fig. 32.1).

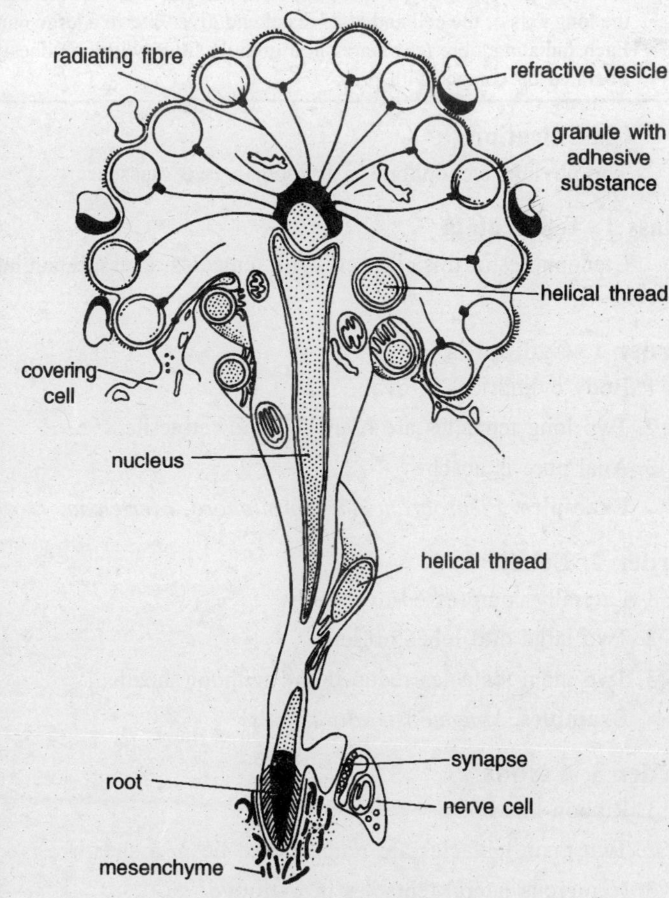

Fig. 32.1. Colloblast.

10. Nematocysts are entirely absent.
11. Digestive system consists of a **mouth**, a ciliated **pharynx, stomach** (infundibulum) and a system of gastrovascular canals.
12. Skeletal, excretory, circulatory and respiratory systems are absent.
13. Nervous system is diffuse having a subepidermal nerve network and aboral end bears a sense organ the **statocyst**.
14. They are **hermaphrodites**, *i.e.*, male and female gonads are found in the same organism.
15. Gonads are endodermal in origin.
16. Development is **indirect** type and highly **determinate**, producing a **cydippid larva**.
17. No alternation of generation occur.
18. **Regeneration** and **paedogenesis** are common.

Box 32.1.

The tentacular epidermis of a ctenophore possess peculiar adhesive cells called **colloblasts**. A colloblast cell is somewhat pear-shaped, with the narrowed end anchored in the tentacular mesogloea and having a synaptic junction with a nerve cell. An intracellular helical thread winds around the long axis of the cell and at its distal end gives rise to a large number of small radiating fibres. Each radiating fibre terminates in a granule filled with an adhesive mucoid material, which is liberated on contact with prey.

2. Classification

The phylum ctenophora is divided in **two** classes:

Class 1. Tentaculata

Ctenophores of this class are with tentacles. Class Tetaculata is divided into following **four** orders:

Order 1. Cydippida

1. Body cylindrical or oval.
2. Two long tentacles are branched and retractile.
3. Anal pore is absent.

Examples. *Pleurobranchia*, *Hormiphora*, *Mertensia*.

Order 2. Lobata

1. Laterally compressed oval body.
2. Two large oral lobes present.
3. Two main tentacles reduced and without sheaths.

Examples. *Mnemiopsis*, *Bolinopsis*.

Order 3. Cestida

1. Ribbon-like body.
2. Two main tentacles are reduced and lie in a sheath.
3. Numerous lateral tentacles in a groove.
4. Comb plates in four rows but rudimentary.

Examples. *Cestum*, *Velamen*.

Order 4. Platyctenea

1. Body greatly flattened.
2. Two well developed tentacles with sheaths are present.
3. Comb plates are reduced or absent in adult.
4. Body is adapted for creeping.

Examples. *Ctenoplana*, *Coeloplana*.

Class 2. Nuda

Ctenophores of this class are without tentacles. It includes **single** order.

Order 1. Beroidea

1. Body conical or ovoid and laterally compressed.
2. Tentacles and oral lobes absent.
3. Mouth large and muscular and stomach (stomodeum) voluminous.
4. Meridional canals are highly anastomosing.
5. Voracious feeders feeding on other ctenophores.
6. Cydippid larva is lacking.

Examples. *Beroe*.

3. Affinities of Ctenophora

The evolutionary position of ctenophores is very obscure. They have long been considered related to cnidarians (coelenterates). However there are many fundamental similarities. The triploblastic origin of their tissues and determinate cleavage are similar to some bilateral phyla. It suggests that Ctenophora exhibit following affinities:

(*i*) Resemblance with Coelenterata

1. Radial symmetry.
2. Body has tissue grade organization. General lack of organ system.
3. Tentacles and mesogloea are present.
4. Coelom absent.
5. Presence of gastrovascular cavity.
6. Diffuse nervous system.
7. Respiratory system absent.
8. Nephridia absent.
9. Presence of statocyst.
10. Endodermal origin of gonads.

Due to above similarities the ctenophores are considered by many zoologists to be a class of Coelenterata.

(*ii*) Differences with Coelenterata (Cnidarians)

1. Absence of nematocysts.
2. Tentacles oppositely placed; it suggests a biradial symmetry.
3. Presence of stomodaeum.
4. Presence of eight locomotory ciliated comb plates.
5. Presence of colloblasts over tentacles.

6. Sense organ (statocyst) is aboral in position.
7. Presence of mesenchymal muscle cells.
8. Presence of a definite organization of digestive system with anal pores.
9. Development is of determinate type.

(*iii*) Advancement over Coelentrates (Cnidarians)

1. Mesenchymal muscle cells.
2. Presence of three germ layers.
3. Apical sense organs.
4. Complete digestive organs.
5. Gonoducts are present.
6. Cleavage/development of determinate type.

Since ctenophores exhibit many striking characteristics of their own so that it appears more justified to treat them as a separate phylum rather a class of phylum Coelenterata (Cnidaria).

(*iv*) Affinities with Platyhelminthes

The *Platyctenea* is considered to be a connecting link between Ctenophora and Bilateria. Ctenophora exhibit following similarities with turbellarians (Platyhelminthes).

1. Body contain cilia (*e.g.*, comb plates, ciliated pharynx).
2. Dorso-ventrally flattened body.
3. Crawling mode of life.
4. Origin of so-called mesoderm is more or less similar.
5. Ctenophora exhibits both radial as well as biradial symmetries.
6. Similar cleavage and gastrulation.
7. Gelatinous mesenchyme with muscle fibres and cells.

The view that the primitive Bilateria have evolved through *Platyctenea* has not been accepted.

Position of Ctenophora in Animal Kingdom

Ctenophores have certain characteristics in common with the Cnidarian (coelenterates). There is no evidence about its derivation from any existing group of Cnidaria. It is believed that Ctenophora is to be an offshoot from the ancestral medusoid coelentrata (*Hydractena*), which was probably more or less spherical animal with a concentration of cilia along their meridional rows, which later develop into comb plates. Such an offshoot gave rise to not a higher form (*i.e.*, ctenophores represent a blind evolutionary offshoot).

32.2 TYPE STUDIES

1. *CESTUM*

Classification

Phylum	–	**Ctenophora**
Class	–	**Tentaculata**
Order	–	**Cestida**
Genus	–	***Cestum***

Habit and Habitat. *Cestum* (Fig. 32.2) is commonly called **venus's girdle**. It occurs in warmer seas specially the Mediterranean. It is a marine, solitary, free-swimming and bioluminescent ctenophore.

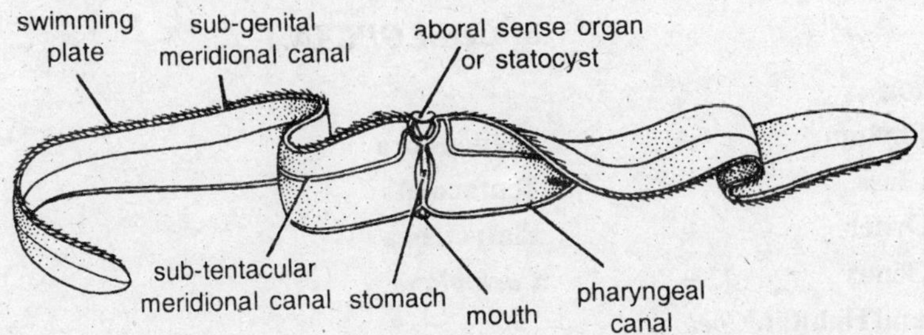

Fig. 32.2. *Cestum.*

Structure. The greatly elongated ribbon-like body is about 45 cm long and 3 cm wide and looks like a transparent celluloid belt. *Cestum* is green, blue or violet in colour. There are present eight aboral comb-plates, four of which are long and other four are rudimentary. At the middle of the oral surface of the body is the mouth. Two subtentacular canals run along the middle of each half of the body and two pharyngeal canals run along the oral surface. The base of two main tentacles are large and are enclosed in sheaths. Many small tentacles spring from grooves and extend in the whole length of oral edge. At the middle of aboral surface sensory statocyst is present. Locomotion takes place by means of long comb plates and muscular undulations.

2. *CTENOPLANA*

Classification

Phylum	–	**Ctenophora**
Class	–	**Tentaculata**
Order	–	**Platyctenea**
Genus	–	***Ctenoplana***

Habit and Habitat. *Ctenoplana* (Fig. 32.3) is a small marine solitary and planktonic ctenophore. It is found in the coastal waters of Indo-China and Japan.

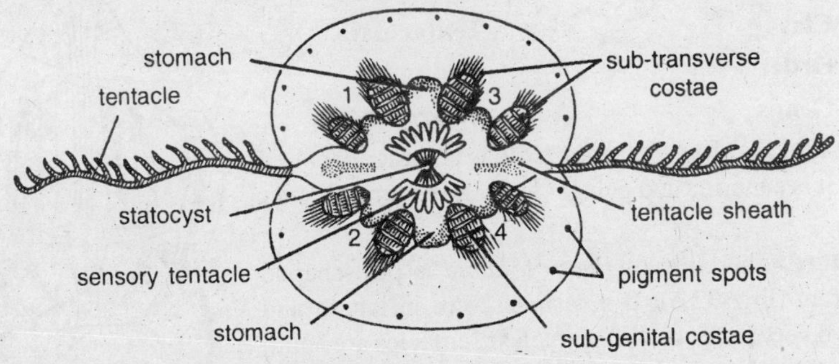

Fig. 32.3. *Ctenoplana.*

Structure. The body of *Ctenoplana* is dorso-ventrally flattened, disc-shaped and measuring about 6 mm in diameter. Colour of body on dorsal surface is mostly olive green, brown or reddish. The mouth lies in the centre of ventral or oral surface. It contains two pinnate and retractile tentacles having sheaths. In the centre of the body on the dorsal surface is a statocyst surrounded by a ring of small ciliated sensory tentacles. Eight rows of comb plates are connected to the statocyst by ciliated furrows. They help in creeping movement of *Ctenoplana*.

3. *COELOPLANA*

Classification

Phylum	–	**Ctenophora**
Class	–	**Tentaculata**
Order	–	**Platyctenea**
Genus	–	***Coeloplana***

Habit and Habitat. *Coeloplana* is marine and solitary ctenophore. It is found in water of Red Sea and the coast of Japan.

Structure. The body of *Coeloplana* (Fig. 32.4) is oval and dorso-ventrally flattened. It is elongated in the tentacular plane and measures about 60 mm in length and is transparent or pink with various shades. The comb rows are absent, though the traces of ciliated furrows exist. Mouth is ventral in position. Irregularly branched alimentary canal is present. In *Coeloplana*, locomotion takes place by creeping. Statocyst is dorsal in position. There are 12 to 60 erectile papillae found and they are the diagnostic characters of the animal. There are eight meridional canals.

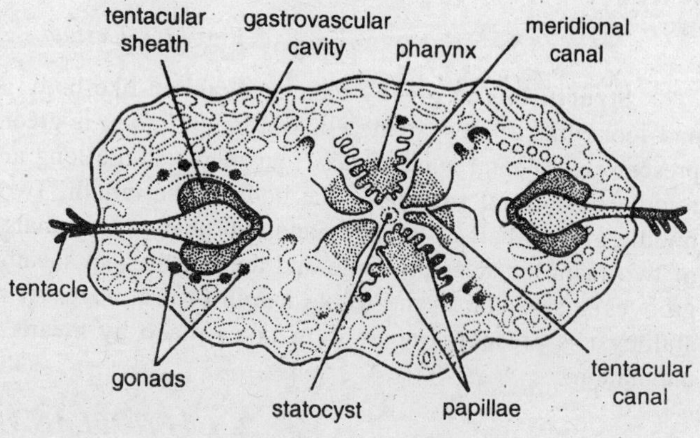

Fig. 32.4. *Coeloplana.*

4. *PLEUROBRACHIA*

Classification

Phylum	–	**Ctenophora**
Class	–	**Tentaculata**
Order	–	**Cydippida**
Genus	–	***Pleurobrachia***

Habit and Habitat. *Pleurobrachia* is commonly called **comb jelly**. It is a marine and pelagic form. It is found all over the world.

Structure. The body of *Pleurobrachia* is pear-shaped and transparent (Fig. 32.5). It measures 2 cm in length and possesses two very long (15 cm) tentacles. Tentacles are solid, highly extensile and retractile into sheaths or lateral pits, called **tentacular pouches**. There also occur eight longitudinal rows of comb plates for the locomotion. A long tubular pharynx from oral pore or **mouth** opens into stomach. The stomach extends into a system of gastrovascular canals. The sense organ, called **statocyst**, occurs at the aboral side. *Pleurobrachia* uses its colloblast cells to capture the prey. It is hermaphrodite, *i.e.*, both male and female sex organs occur in the same organism.

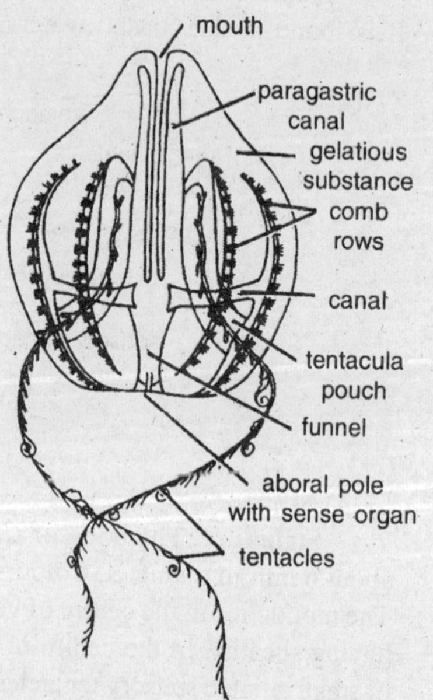

Fig. 32.5. *Pleurobrachia.*

III. General Morphology and Life History of Ctenophora

Ctenophores are commonly known as **comb jellies** or **sea walnuts**. The phylum takes its name from two Greek words: *ktenes* = combs and *ophora* = carrying.

Definition. Ctenophora is a small phylum of marine animals. These are biradially symmetrical radiates having transparent gelationous body without nematocysts but with special lasso cells (colloblasts) and with eight locomotory comb-like ciliary plates.

Structure. The generalized ctenophore is pear-shaped, roughly spherical in appearance (Fig. 32.5). Comb rows may be tinged with white orange or purple in colour. Ctenophores are noted for their **luminescence** (*i.e.*, light producing **photocytes** occur in walls of meridional digestive canal). Ctenophores range in size from few millimetres to 20 cm. Few species are as long as 1 metre or more. Their body can be divided into two hemispheres. Body can be further divided into equal sections by 8- ciliated bands. Their oral and aboral ends are distinct. The oral end lies on the lower side and bears minute oral aperture or **mouth**. The body of ctenophores contains the following distinct parts:

1. Comb plates. They are eight ciliated bands, arranged on the sides of body like meridians from aboral pole to oral pole, Each row is formed of cilia called **ctenes** or **comb plates** (Fig. 32.6A). The combs in a comb-plate are arranged in succession one behind the other. The row of plates forms roughed ridge, similar to those of walnuts and helps in locomotion.

2. Tentacles. Near the aboral end are present two long, solid and branched tentacles. These are enclosed in a pair of deep blind pouches or the **tentacular sheaths** into which tentacles can be withdrawn. Each tentacle bears a row of short branches or pinnae. It is formed of mesenchymal core covered with epidermis. Tentacles lack nematocysts but bear peculiar adhesive cells, called **colloblasts**.

3. Colloblast cell. It is modified epidermal cell (Fig. 32.6C) and contains the following parts: (*i*) **Head**. It is hemispherical and lies on the surface of epidermis. Its surface is covered with secretory granules that produce an adhesive secretion to be utilised for entangling and killing the prey. (*ii*) **Straight filament**. It connects the head to the mesenchyme core of tentacles. (*iii*) **Spiral filament**. It is spring-like and coiled around the straight filament. The colloblast lacks the mechanism of penetration.

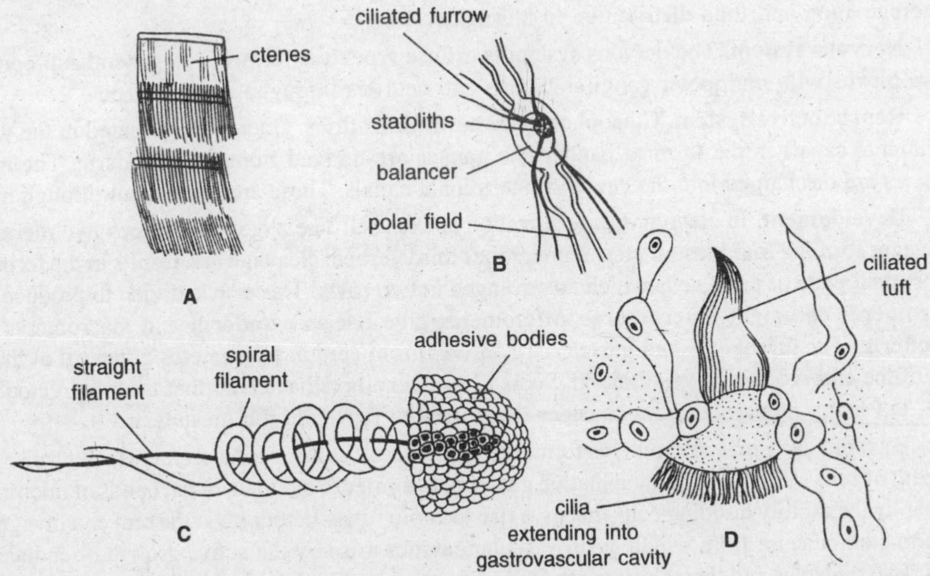

Fig. 32.6. Ctenophora. A—Comb plates or ctenes, highly magnified; B—Aboral apical sense organ; C—Colloblast; D—Rosette organ.

Body wall. The body wall is **triploblastic** and is formed of (*i*) **Epidermis**. It is outer layer, formed of syncytial cuboidal epithelium. Many types of specialized cells such as sensory, glandular, mucous and pigment cells are present in between these cells. (*ii*) **Mesogloea** or **mesoderm**. It is jelly-like layer found between epidermis and gastrodermis. It is very thick and contains amoebocytes and muscle fibres. Muscles are smooth, branched and anastomosing forming network. (*iii*) **Gastrodermis**. It is inner layer which lines the stomach and gastrovascular canals. It is formed of tall vacuolated and low ciliated cells.

Locomotion. The row of ciliated bands forms roughened ridges and helps in locomotion, *i.e.*, swimming or creeping.

Digestive system. Digestive system of ctenophores includes

(*i*) **Mouth.** It is a slit-like aperture, situated at the lower end of body.

(*ii*) **Pharynx or stomodaeum.** Mouth leads into pharynx which is a flattened chamber. Wall of pharynx is folded and is formed of long ciliated epidermal and glandular cells.

(*iii*) **Stomach.** Pharynx leads into a wide stomach or infundibulum which is connected with a system of five **gastrovascular canals**. These extend through the mesogloea in a definite arrangement up to the aboral pole.

Food, feeding and digestion. Food of ctenophores includes copepods and phytoplanktons. Food is usually caught by surface mucus and carried to mouth by cilia or entangled by colloblast on the tentacles. The tentacles are effective food traps and when they contract food is wiped off on the "lips" around mouth. In *Beroe*, which is cannibalistic, the mouth is apressed against a prey ctenophore. Then with a single gulp, the victim is swallowed.

Food is initially digested rapidly in pharynx and partly digested food enters the stomach. From there it is circulated through the gastrovascular canals where it is digested **intracellularly**.

Respiration and excretion. Ctenophores lack special respiratory organs; they exchange gases through their general body surface.

Their nitrogen is excreted in the form of **ammonia** which diffuses to the medium through body and gut surfaces. Most probably **cell rosettes** (Fig. 32.6D) are involved in excretion, osmoregulation and food distribution in mesogloea.

Nervous system. The nervous system is diffuse type. It is formed of a general sub-epithelial nerve plexus with multipolar ganglionic cells and neurites throughout the surface.

Reproductive system. Ctenophores are **hermaphrodites**. Gonads are situated in the wall of meridional canals in the form of bands. The gonads are derived from the endoderm. The mature gametes are discharged into the cavity of meridional canals. These are ejected out through mouth.

Development. In ctenophores fertilization is external. The zygote undergoes two meridional cleavages forming four blastomeres. It undergoes third vertical cleavage and results in the formation of a curved plate of eight cells which are arranged in two rows. These cells divide to produce eight **micromeres** and eight **macromeres**. Micromeres give rise to **ectoderm** and macromeres form **endoderm**. The mesogloea and muscles are derived from certain micromeres given off at the oral side of the embryo during gastrulation. Some classical authors have held that these are ectodermal cells, but many modern authors consider these cells to represent the **mesoderm**.

By rapid divisions, micromeres form a sheath of small cells over macromeres which also form a sheath of cells. As a result of invagination **gastrula** is formed. Four inter-radial bands of micromeres appear by the rapidly dividing cells that give rise to comb rows. Ectoderm at the oral end invaginates to form stomodaeum from which gastrovascular cavities arise by the active growth of endodermal cells. Two tentacles and their sheaths are formed by the ectoderm. At this stage, the embryo escapes as **cydippid larva** which is characteristic larval type for all ctenophores except the beroids.

32.3 PHYLUM ROTIFERA

Rotifera (L., *rota*=wheel; *ferre*=to bear) is also called **Rotatoria**. It represents a **pseudocoelomate** minor phylum. Rotifera includes about 1800 species of minute (or microscopic), bilaterally symmetrical, unsegmented and aquatic animals. Rotifers are commonly called "**wheel animalcules**" because their anterior end bears a **ciliated crown** or **corona** which on beating gives the appearance of a rotating wheel.

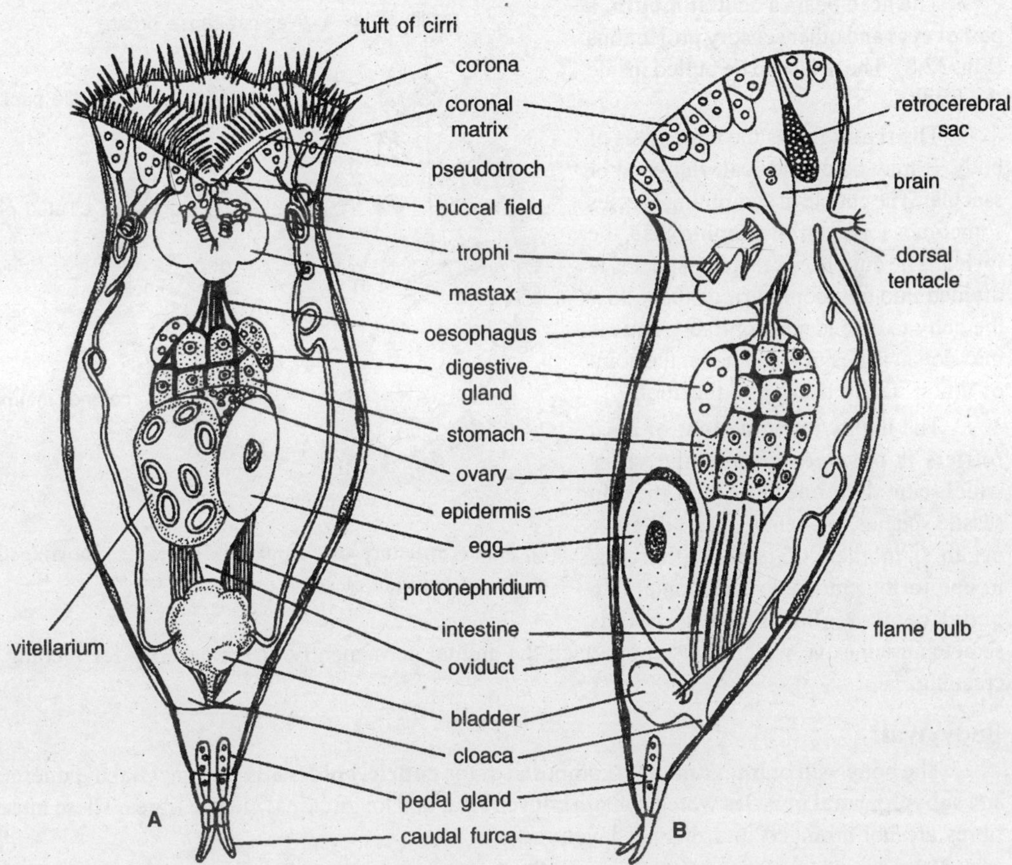

Fig. 32.7. Rotifer body plan (*Epiphanes senta*). A—Ventral view; B—Lateral view.

Habit and Habitat

Rotifers are cosmopolitan in distribution. They are found in fresh and marine water as well as in moist terrestrial habitats among mosses and lichens. Some float passively, others creep or swim actively, some are sessile and live in cuticular tubes and a few form colonies. A few rotifers are parasitic.

Structure

1. Size and shape. Rotifers are the **smallest metazoans**, approximately of the size of ciliate protozoans. Most species measure 0.1 to 1 mm in length. Their body is composed of about 1000 cells.

Shape of rotifers is highly variable. Bottom dwellers are worm-like; epizoic (attached) species are **flower-like** and floating species are usually **globular**.

2. Body plan. The body is divided into **head, trunk** and **tail** or **foot**. A **neck** may separate the head and trunk.

The head bears a retractile crown or disc, called **corona** or **trochal disc** that surrounds mouth. The corona of primitive, creeping species is composed of an evenly ciliated **buccal field** arranged around the mouth. Swimming rotifers usually have a reduced buccal field. In some rotifers, the corona bears two rings (bands or circlets) of cilia. The outer or posterior band is called **cingulum** and inner or anterior **trochus**. The cilia beats in a circular manner (metachronally) to give an impression of a revolving wheel. Corona is used in locomotion, feeding and respiration.

The head bears a central **mouth**, a pair of **eyes** and other sensory projections (Fig. 32.8). The head can be pulled inside the trunk.

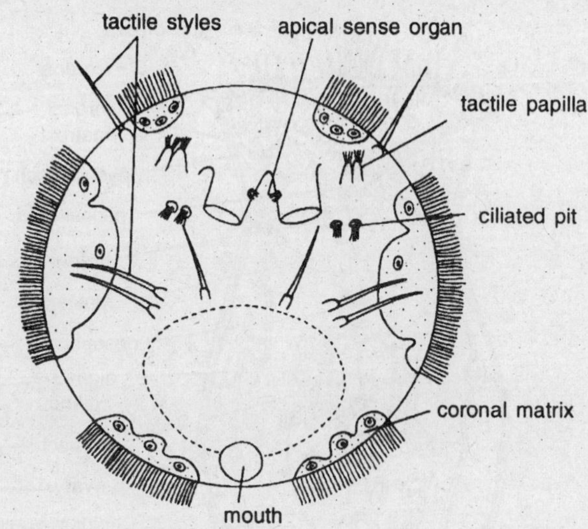

Fig. 32.8. Structure of corona of *Euchlanis*, showing the position of sense organs.

The **trunk** forms the major part of body. It may be cylindrical, flattened or saccular. The cuticle of swimming species sometimes forms an enveloping case, the **lorica**. The cuticle of creeping species is divided into telescoping rings that allow the body to extend and contract. **Anus** is mid-dorsal in position and lies at the point of union of the trunk and the foot.

The highly retractile **foot** of most rotifers is covered by cuticular rings which can slide one into the other. In sessile species, foot ends in an **adhesive organ**. In motile rotifers, foot often ends in one to four movable toes containing **pedal cement glands**. These glands secrete an adhesive which is used to attach the animal permanently or temporarily for feeding or creeping.

Body Wall

The body wall or **integument** is composed of the **cuticle, epidermis** (or syncytial hypodermis) and **subepidermal muscles** which contain both circular and longitudinal muscle fibres. These muscle fibres are not arranged into distinct layers.

The syncytial integument, with keratin -like lamina within the cytoplasm, covers the body surface and forms the lorica, spines or other surface structures. The syncytium contains a constant number of cushions that contain one or several nuclei. The modified epidermis of the corona contains a ciliary root system and large epidermal cushions, and is termed the coronal matrix. Sub cerebral glands, pedal glands and other surface glands are epidermal in origin but extend below the epidermis.

Body Cavity

It is a spacious **pseudocoel**. It is filled with perivisceral fluid containing loose syncytial branching network of amoeboid cells. It helps in internal transport. Circulatory system is absent. Respiration is mainly aerobic, sometimes anaerobic.

Locomotion

Rotifers can move about either by swimming or crawling. The beating of the coronal cilia moves the animal about in swimming. Pelagic forms maintain their buoyancy with fat globules. Leech-like crawling movement of rotifers is performed by the help of foot and its associated glands.

Digestive System

Rotifers are either **suspension feeders** or **raptorial** (=living on prey). Gut is typically developed in most rotifers. **Mouth** is ventral and is surrounded by corona. Sphincter and dilator muscles may open and close the mouth. **Pharynx** is called **mastax**. It is a muscular chamber containing a set of jaws, formed of cuticularized trophi. Jaw structure is complex (Fig. 32.9). A median **fulcrum** and pairs of **rami**, **unci** and **manubria** make up the major parts. Pharyngeal jaws of rotifers are most important distinctive structures; they are used in the identification of a species. The mastax helps in food capture and its maceration.

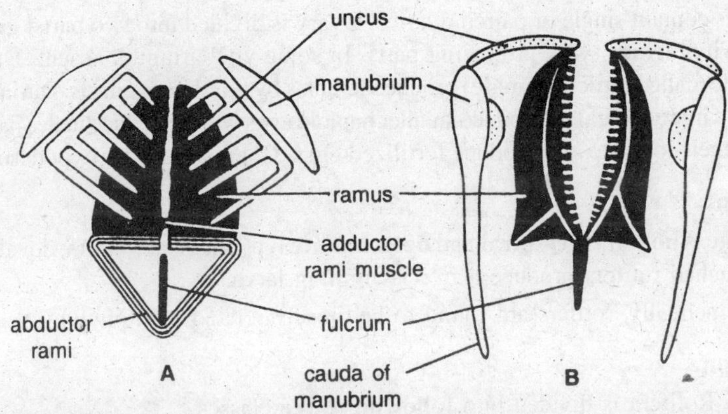

Fig. 32.9. Two types of trophi in a rotifer. A—Malleate type; B—Forcipulate type.

Pharynx opens into **oesophagus** which leads into **stomach**. Stomach is thick-walled, ciliated sac composed of constant number of cells. A delicate mesh of circular and longitudinal muscles covers its surface and a pair of syncytial **gastric glands** open into stomach. Gastric glands secrete enzymes for **extracellular digestion**. A thin-walled, syncytial and straight **intestine** is sometimes ciliated. A posterior **cloaca** receives the gut, protonephridia and oviducts.

While **feeding**, a suspension feeder remains attached to a piece of debris, etc. Rapid beating of the cilia draws a current of water towards the mouth. The water carries protozoans and algae into the mouth. The predatory rotifers shoot out their armed mastax to catch the prey. The mastax acts as pincers or forceps. **Digestion** is both extracellular and intracellular. Intracellular digestion occurs in syncytium of stomach. Carbohydrate digestion is limited. Digestion of proteins and lipids is very rapid.

Excretory System

Rotifers have two syncytial **protonephridial** tubules which bear terminal **flame bulbs** (=cells). Posteriorly the protonephridia drain into cloaca by way of common duct or a **urinary bladder**. Protonephridial system is a water-regulating device, also maintaining salt concentration of pseudocoel. Rotifers easily tolerate dehydration and freezing.

Nervous System

It is bilaterally symmetrical. A bilobed **brain** gives rise to a pair of longitudinal ventrolateral nerve trunks. Several ganglia are connected to the nerve trunks. Peculiar features of nervous system of rotifers include: (1) Clear separation of sensory and motor elements; (2) Exhibition of the phenomenon of constancy of cells (called **eutely**), *i.e.*, brain cells and nerve cells are constant in number and position.

Rotifers contain a variety of sense organs. Examples of sensory organs include photoreceptive **eyes** (*e.g.*, cerebral eyes, coronal eyes or lateral eyes), chemoreceptive **ciliated pits**, tactile **antennae** and stiff bristles (styles).

Cyclomorphosis

Rotifer phenotypes can be environmentally controlled. The form of amictic females gradually changes during the year in response to environmental conditions, a process called **cyclomorphosis**.

Reproductive System

Rotifers reproduce sexually. Sexes are separate. Males are generally small and degenerate. They lack the digestive and excretory systems. They live only for few days. Rotifers produce two types of sperms: **ciliated sperms** and **rod-shaped sperms**. The rod-shaped sperms probably help to pierce the cuticle of female integument. Some rotifers have cirrus or penis.

Females contain single or paired ovaries. Ovary is divided into two parts: **germarium** (ovary proper) and **vitellarium** (yolk producing part). In some vitellarium is absent. During most of the year the female (called **amictic female**) produce females by **parthenogenesis**. During breeding season females (called **mictic females**) produce smaller haploid eggs which produce males parthenogenetically. These males then copulate and produce fertilized eggs. Copulation is by hypodermic impregnation.

Development

Cleavage is **holoblastic, spiral** and **determinate** type. In most rotifers, development is **direct**. However, attached rotifers produce a free-swimming larva.

Phylogenetically, rotifers are found to be closely related with phylum **Acanthocephala**.

Classification

Phylum Rotifera is divided into following **three** classes:

Class 1. Seisonidea. Marine, living on gill surfaces of crustacea, body elongated, with a long neck, jaws (trophi) as forceps; small corona, paired gonads, without vitellaria, males present.

Example. *Seison.*

Class 2. Monogononta. Swimming or sessile, males with testis, females with one ovary (or germovitellarium).

Examples. *Conochilus, Hexarthra, Collotheca, Notommata, Brachionus, Keratella, Asplanchna, Gastropus.*

Class 3. Bdelloidea (Digononta). Swimming or creeping forms, with retractable anterior end, mastax adapted for grinding and contains one pair of teeth, males absent, females with two ovaries or germovitellaria).

Examples. *Philodina roseola, Rotaria, Embata.*

32.4 PHYLUM SIPUNCULA

Phylum **Sipuncula** or **Sipunculoidea** (L., *sipunculus*=a little siphon) is a **coelomate minor phylum** and includes 320 species. Sipunculans are commonly called **peanut worms** because they resemble a peanut when fully contracted (they do so when disturbed).

Habit and Habitat

Sipunculans are widespread bottom-dwelling marine animals that are mainly found in shallow water (littoral zone) of warmer seas, but also occur in polar waters and abyssal depths. They are sedentary creatures that burrow in the sand or mud (*e.g., Sipunculus*) or occupy any protected den of appropriate size, whether a rocky crevice, coral reef or an annelid tube or a molluscan shell (*e.g., Phascolion*) or even foraminiferan shell. Sipunculans are mostly detritus or deposit feeders.

Structure

1. Shape and size. Sipunculans are light brown coloured worms that measure 2 mm to more than 72 cm in length. Most are less than 10 cm.

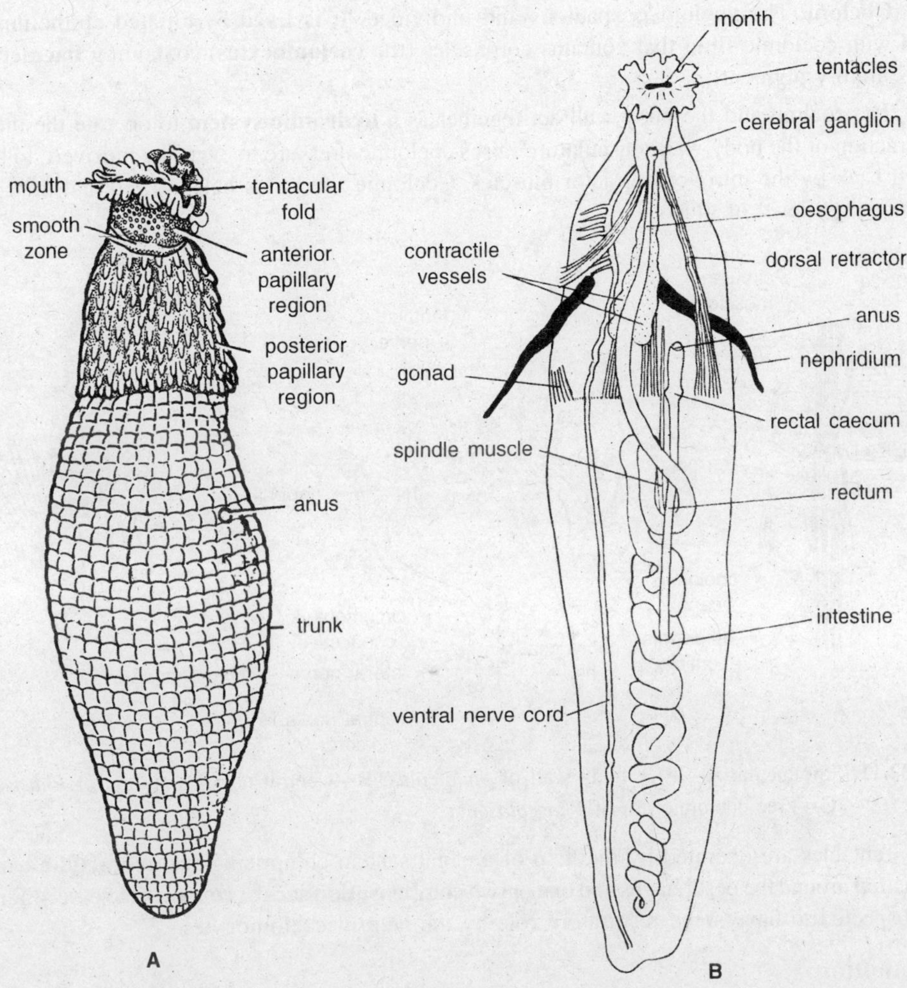

Fig. 32.10. *Sipunculus.* A—External structure; B—Internal structure.

2. Body plan. Sipunculans have a very simple external form. The body is cylindrical and vermiform without segmentation, appendages, spines or parapodia. It is divided into anterior long or short iridiscent **introvert** and posterior **trunk**.

The introvert represents head. It is narrow, funnel-shaped anterior part of body that can be involuted inside the trunk. **Mouth** lies at the tip of introvert and is surrounded by a pleated **tentacular fold** which looks like a circlet of tentacles. Tentacles are hollow, ciliated, conical, filiform or digitiform processes that bear deep **ciliated groove** on their surface. A smooth zone, called **collar**, may be found immediately behind the tentacles. Minute hooks or spines and sensory or glandular papillae often cover the surface of the introvert. The **trunk** is cylindrical bears only sensory and glandular papillae. **Anus** is mid-dorsal, at the anterior end near the base of introvert.

Body wall. Body wall of sipuncula is of annelid type in histology. It is composed of following layers: (1) **cuticle;** (2) **epidermis;** (3) **dermis** which is formed of thin connective tissue; (4) **circular muscles;** (5) **longitudinal muscles.** The circular and longitudinal muscles of *Sipunculus* can be seen through the body wall giving it a graded appearance.

Coelom. The coelom is spacious and undivided. It is lined by ciliated epithelium. It is filled with coelomic fluid that contains corpuscles (the **coelomocytes**) containing **haemerythrin** (a respiratory pigment).

The coelom and the body wall act together as a **hydraulic system** to operate the introvert. Contraction of the body wall musculature raises coelomic pressure to evert the introvert, to be later pulled back by the introvert retractor muscles. Coelomic fluid also helps in internal transport (it contains high level of glucose).

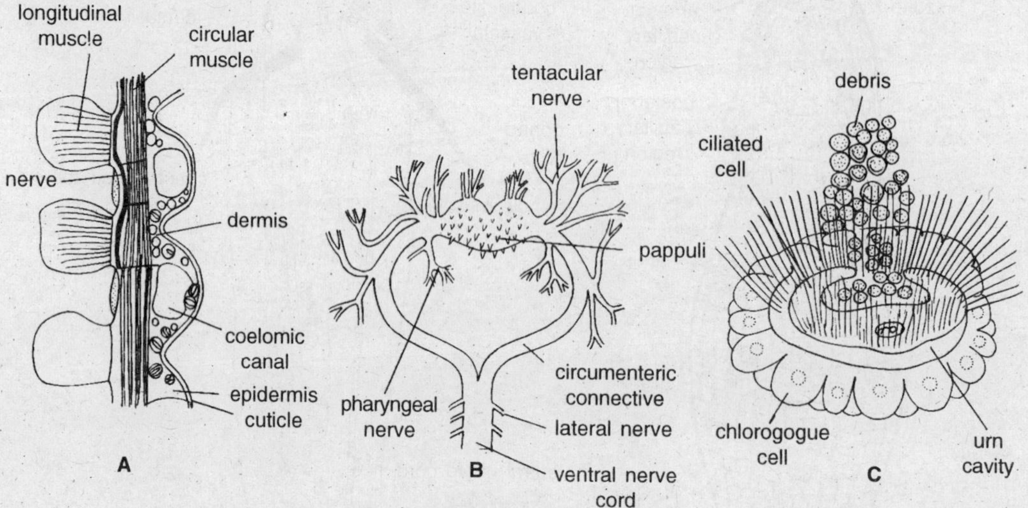

Fig. 32.11. Sipuncula. A—T. S. body wall of *Sipunculus*; B—Central nervous system of *Phascolopsis*; C—Free coelomic urn of *Phascolosoma*.

Tentacles are operated by the help of a canal system comprising two **tentacular canals**, a **ring canal** around the oesophagus and one or two **compensation sacs** or **contractile vessels**. Tentacles are suspected to have some respiratory role by the help of coelomocytes.

Locomotion

Burrowing sipunculans thrust the introvert forward into the mud, dilate the tip, and then contract the retractor muscles. This pulls the body forward.

Digestive System

Alimentary canal includes a large **mouth**, a short muscular **pharynx**, slender **oesophagus**, recurved and coiled **intestine**, **rectum** and **anus**. Most sipunculans being deposit feeders eat sand and mud and subsist on whatever digestible organic material that may yield. Diatoms, protozoa, larvae and small particles of organic detritus provide most of their nourishment. Digestion and absorption occur in the descending arm of intestine and ascending arm is involved in faeces formation.

Excretory System

The excretory organs are pair of large, saccate, brownish **metanephridia** with open nephrostomes and nephridiopores. The brownish glandular walls of the nephridia excrete nitrogenous waste (*e.g.*, ammonia) into the tubule.

Some sipunculans have curious peritoneal structures known as **coelomic urns** (Fig. 32.11C). Each urn is like a vase capped with a ciliated cell. Urns may be found fixed to the peritoneum or they may get detached and swim in the coelomic fluid. The free urns collect excretory material and form either brown bodies or trailing aggregates. The latter are discharged through the nephridia. Annelid-like chlorogogue tissue remains attached with urns.

Nervous System

It is annelidan type with the brain (**cerebral ganglion**) lying above the oesophagus, a pair of **circumoesophageal connectives** and a single **ventral nerve cord**. Ventral nerve cord lacks segmentation. Sensory organs include a pair of pigmented spots or ocelli embedded in the brain and sensory cells in **nuchal organs** or **ciliated pits** at the end of introvert.

Reproductive System

Sexes are separate (dioecious) but no sexual dimorphism. Ovaries and testes are simple masses of cells shed into the coelom. Ripe ova and sperms leave the body through nephridia. Fertilization is external.

Development

Cleavage is spiral. Development is **indirect** producing a typical **trochophore larva**. Phylogenetically sipunculans are probably related to the annelids.

Classification

Phylum Sipuncula is divided into following two classes:

Class 1. Sipunculidea. Tentacles encircling mouth; introvert hooks are simple.

Example. *Sipunculus.*

Class 2. Phascolosomatidae. Tentacles are confined to arc surrounding dorsal nuchal organ; introvert hooks recurved.

Examples. *Aspidosiphon*, *Phascolion*.

32.5 PHYLUM ECHIURA

Phylum Echiuroidea or **Echiura** (Gr., *echis* = an adder; *aura* = tail) is a small coelomate minor phylum including about 140 species. Echiurans are commonly called **spoon worms**.

Habit and Habitat

Echiurans are marine, bilaterally symmetrical, unsegmented, cylindrical and worm-like animals. They lack setae and parapodia. They are widely distributed in the shallow coastal waters of warm

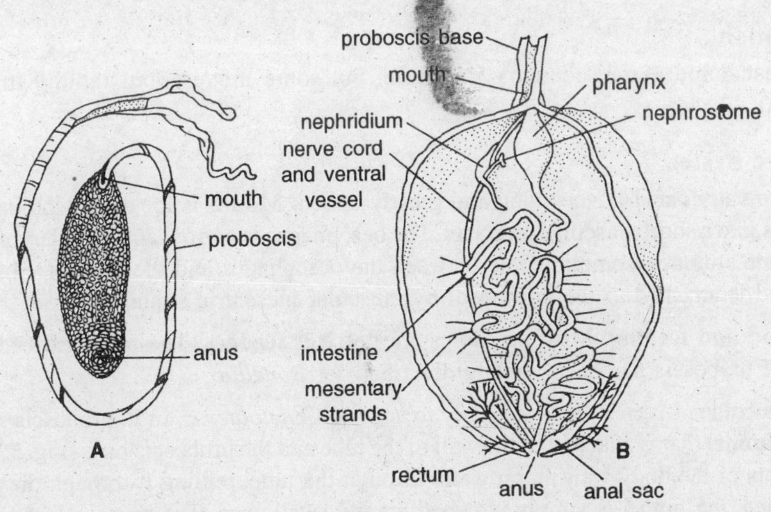

Fig. 32.12. *Bonellia viridis*. A—External structure; B—Internal structure.

and temperate seas. Echiurans are mostly bottom dwellers. Most of them dig and live in U-shaped mucus-lined tubes or burrow in sand or mud. Some live under stones, rocks and in coral crevices. Most echiurans have cryptic habits, dwelling in burrows and crevices, only thrusting out the remarkable proboscis as a food finder.

Structure

Shape, size and colour. Bilaterally symmetrical body of echiurans is cylindrical, ovoid or T-shaped. They are generally grey, red, yellow or green in colour. They are about 25 mm in length. But *Ikeda,* found in Japan, contains a trunk of 40 cm and a 1.5-m-long proboscis.

Body plan. Body of echiurans is differentiated into a cylindrical **trunk** and a prominent anterior **proboscis**. The proboscis is the centre for chemoreception and for food procurement. It is a large, flattened, spoon-like extension of the head. It contains brain, so is considered homologous to the prostomium of the annelids. Proboscis is very extensible. For example, in *Bonellia viridis* the proboscis is about 7 cm long when contracted; it can be extended fully to as much as 1.5 m almost transparent thread. However, like the Sipuncula, proboscis cannot be retracted into the trunk. Proboscis secretes a mucus cover and contains a **ciliated groove** on its ventral surface. Minute food organisms and small particles (detritus) stick to mucus and are conveyed to the mouth by the groove and muscular movements of the proboscis itself. The **mouth** lies at the base of the proboscis.

Trunk is a simple bag of guts and gonads. Its surface is either smooth or covered with **papillae** which are usually arranged in rings or are irregularly distributed. A pair of annelid-like hooked **setae** are present on the ventral surface near the anterior end of trunk. Some echiurans such as *Echiurus* possess one or two circles of **anal setae** around the posterior end of the trunk.

Body Wall and Coelom

Body wall of echiurans is highly muscular and similar to that of annelids and sipunculids. It consists of **cuticle**, **epidermis**, **dermis**, layers of **longitudinal** and **circular muscles**. **Coelom** is spacious and lined by ciliated coelomic epithelium. It contains a coelomic fluid which has amoebocytes and haemocytes containing haemoglobin. **Coelomic urns** are absent.

Locomotion

Most Echiura are sedentary in nature, but some move about rapidly in captivity (*i.e.*, aquarium).

Digestive System

Alimentry canal is very long and greatly coiled. **Mouth** is located at the base of proboscis and opens into a short muscular **pharynx**. The oesophagus is narrow and coiled. A muscular **gizzard** and saccate **stomach** sometimes lie between the oesophagus and the intestine. **Intestine** is long, coiled and is attached to the body wall by muscular mesentric strands.

Food and feeding. Most echiurans are **detritus feeders**. The mode of feeding via ciliated groove of proboscis has been already described for *Bonellia*.

Tubicolous *Urechis* feeds similarly to annelid *Chaetopterus*. In it, proboscis secretes a slimy, **mucous funnel** that is attached to the wall of the tube and the proboscis base (Fig. 32.13). Peristaltic movements of the body wall pump water through the tube, pulling it through the mucous funnel. When filled, the funnel is simply wrapped up and swallowed, The mucous funnel is an effective strainer, for its openings measure only about 40 Å.

Excretory System

Echiurans do excretion by sipunculid-like **metanephridia** which vary widely in number. For example *Bonellia* has one, whereas *Ikeda* has hundreds. Most species, however, have from one to three pairs of metanephridia. A pair of modified nephridia or **anal vesicles** open into rectum. These are contractile, simple or branched. They contain many **ciliated funnels** which open into coelom, Anal vesicles remove waste from the coelomic fluid and discharge them into the rectum.

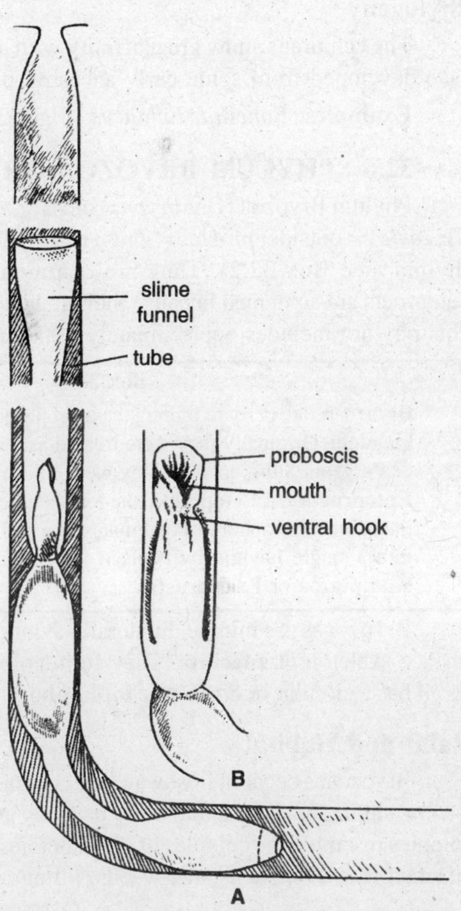

Fig. 32.13. *Urechis caupo*, in its tube with mucus funnel.

Respiration

Haemoglobin in the coelomocytes appears to serve primarily for oxygen storage during period of low oxygen availability, such as when the tide is out.

Circulatory System

It is simple and contains **dorsal vessel** and **ventral vessel** both connected by contractile **circum-intestinal** and **circum-oesophageal vessels**.

Nervous System

It is simple and built on the annelid pattern, but it contains no ganglia. The circumenteric ring is drawn out into a long loop that follows the proboscis margin and connects with ventral nerve cord. Echiurans lack special sense organs.

Reproductive System and Development

Sexes are separate and sexual dimorphism is well marked in some echiurans, *e.g.*, *Bonellia*. The trunk of *Bonellia viridis* measures 8 cm while the male is only 1 to 3 mm in length. Male *Bonellia* is ciliated and has no proboscis. Its alimentary canal is reduced and without mouth and anus.

Gonads are specialized regions of ventral mesentery. Gametes are shed into the coelom before they reach maturity. After they have ripened, they escape through the nephridia. **Fertilization** occurs externally in ocean except in *Bonellia* and *Hamingia*. In these genera, the ova are fertilized in the nephridia and remain there during development. **Cleavage** is spiral type and development is **indirect type** producing free-swimming **trochophore** larva like the annelids (this larva shows metameric segmentation).

In case of *Bonellia*, any larva that settles by itself develops into female; it becomes sexually mature in one year. A larva that settles near to or in contact with female becomes a male (**Jaccarini** *et al.*, 1983). Under the influence of hormonal secretion, the development of the male is retarded and he remains a small, ciliated organism with many juvenile characteristics. He may cling to proboscis of the female for a time, but eventually enters the female mouth and takes up residence in a fold of the nephridium. Male becomes sexually mature in one or two weeks and lives the rest of his life as an **endocommensal**; a female may contain up to 20 males.

Phylogeny

The echiurans show great affinity with annelids. It seems that they have developed from some early ancestral polychaete.

Examples. *Bonellia*, *Echiurus* (Fig. 32.14), *Urechis*, *Ikeda*.

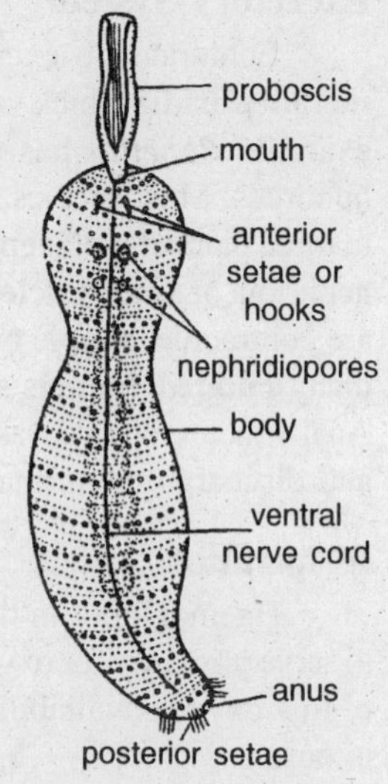

Fig. 32.14. *Echiurus*.

32.6 PHYLUM BRYOZOA OR ECTOPROCTA

Phylum Bryozoa (Gr., *bryon* = moss; *zoon* = animal) or Ectoprocta (Gr., *ektos* = outside, *proktos* = anus) is a lophophorate coelomate minor phylum (see Box 32.2). They are commonly called **moss animals**. Ectoprocts are so named because their anus lies outside the lophophore. This phylum includes approximately 4000 species.

> **Box 32.2**
>
> Bryozoans are perhaps among the most problematic of all in the animal kingdom. Originally they were treated as a single group, the **Bryozoa** or **Polyzoa**. Subsequently, they were split by **Hyman** into two phyla, **Entoprocta** and **Ectoprocta**, the former allied with pseudocoelomates, the latter with lophorate coelomates. Recently, they have been rejoined into a single phylum by **Nielsen** (1971). Entoprocta are also called **Kamptozoa** or **Endoprocta**.

Bryozoans are minute, bilaterally symmetrical sessile, colonial unsegmented coelomates with surface skeleton or **zoecium**. They form colonies of **zooids** by budding. On its anterior end, each zooid has a circular or crescentic **lophophore** bearing a series of slender ciliated post-oral tentacles.

Habit and Habitat

Bryozoans occur in freshwater and marine habitats everywhere in the world. They live in shallow waters and in depths of about 6000 m. They are also abundant in both polar and tropical seas. Their colonies resemble the colonies of coelenterates and look like sea-weeds. Bryozoan colonies are found attached to rocks, shells, sea-weeds, pilings and to other animals.

Structure

1. Size, shape and colour. An individual (zooid) measures about 0.5 mm in length. A large colony may contain as many as 2,000,000 zooids. A colony or **zoarium** of ectoproct may measure one to 50 centimetres or more in diameter. A zooid is box-like, oval or cylindrical. Mostly their colonies form a thin and flat sheet, called **sea mats**. The colonies occur in many shapes, such as encrusting, foliose, branching and globular. Colour of most bryozoan colonies is white or pale, some are red, brown, orange, blue or violet.

2. Body plan. Zooid is box-like, vase-like or tubular and is enclosed in a nonliving cuticular covering, called **zoecium**. The opening of zoecium to exterior is known as **orifice** and is covered with a lid, called **operculum**. Lophophore of zooid protrudes out of the orifice.

Each zooid contains two types of tissues: (1) **Cystid** which includes body wall tissues; (2) **Polypide** which includes business part of animal such as lophophore and the internal organs. Externally, the body of zooid is differentiated into **introvert** and **trunk**. Introvert is the anterior part of the zooid. Its free distal end bears **mouth** which is surrounded by a circlet of ciliated **tentacles**. The disc bearing mouth and tentacles is called **lophophore**. It may be circular or horse-shoe shaped. The latter bears more tentacles than the former type of lophophore. Tentacles drive the food towards mouth and also serve as tactile and respiratory organ. In some forms (*e.g.*, *Lophopus*) a ciliated lobe, called **epistome**, overhangs mouth on the anal side.

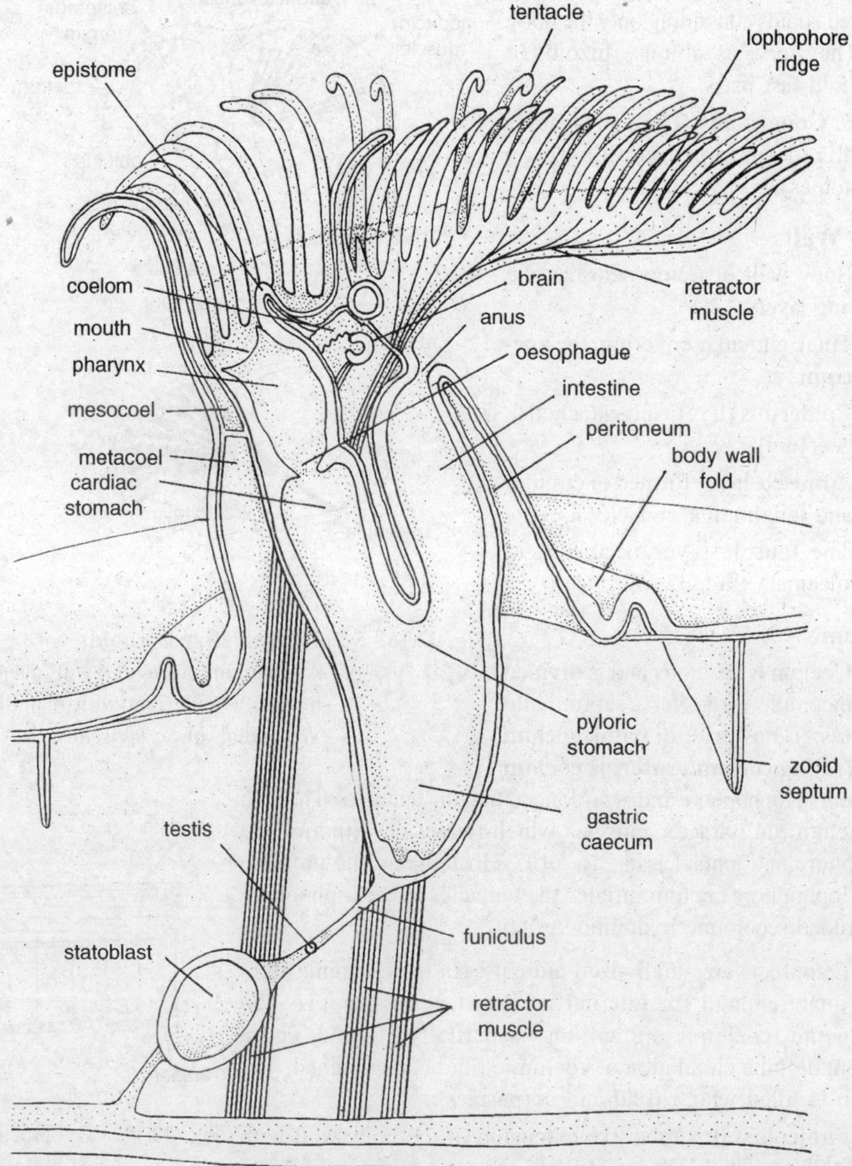

Fig. 32.15. *Cristatella.* Anatomy.

Polymorphism. Bryozoan colonies are polymorphic. They contain following types of zooids:

1. Autozooids. These are most numerous complete feeding zooids containing lophophores.

2. Vibracula. They are a type of defensive zooids in which the operculum becomes modified into a long movable bristle. These sweep away debris and settled larvae.

3. Avicularia ('little bird'). They resemble the head of a little bird. They contain flexible neck and a pair of strong beak-like pincers (Fig. 32.16). They are used to repel or crush intruders (settling larvae).

4. Kenozooids. These are highly modified zooids containing only the body wall. They serve as stolons, rhizoids or other hold-fast parts.

5. Gonozooids. These are zooids specialized for reproduction (*e.g.*, cyclostomes).

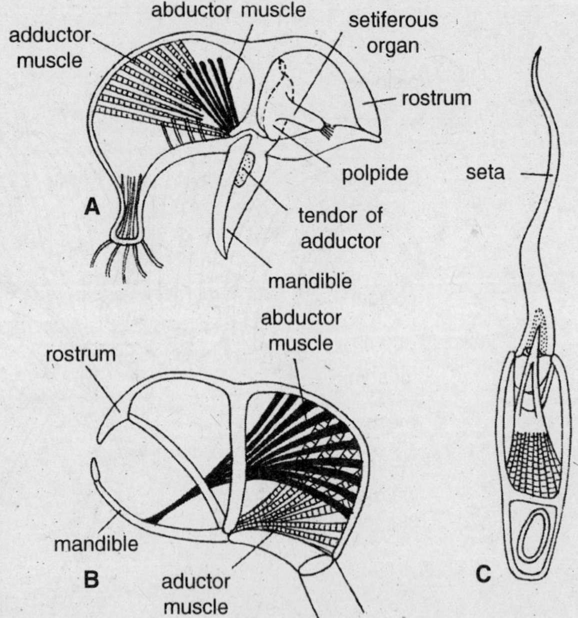

Fig. 32.16. Some specialized zooids of ectoprocta. A—Avicularium of *Bugula*, with mandible open; B—*Bugula* avicularium with mandible closed. C—Vibraculum of *Scrupocelaria.*

Body Wall

Body wall of Ectoprocta has the following layers:

(*i*) Thick cuticular covering, the **zoecium**.

(*ii*) **Epidermis** (It is firmly attached to zoecium).

(*iii*) A **muscle layer** formed of circular and longitudinal muscles.

The muscle layer is absent in Gymnolaemata (Fig. 32.17).

Coelom

Coelom is spacious and is divided by an incomplete transverse septum into two parts: (*i*) **Posterior** or **trunk coelom** and (*ii*) **Ring coelom. Anterior coelom** is present in lophophore and is prolonged into the tentacles. Through the coelom run retractor muscles which retract or withdraw the lophophore and tentacles in case of Cyclostomata. The pressure in the lophophore coelom inflates the tentacles as the lophophore is extruded (coelomic hydraulic system).

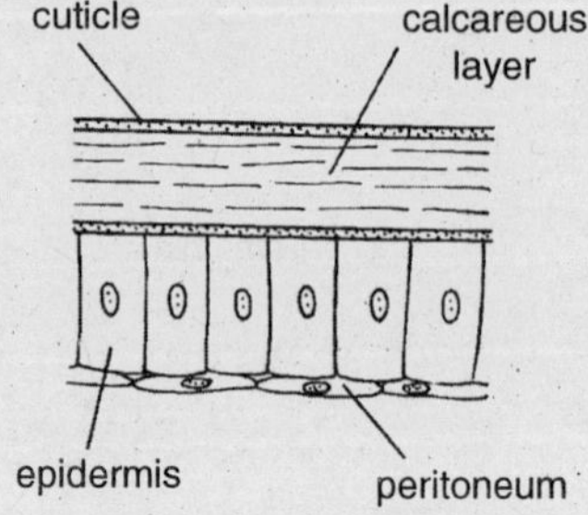

Fig. 32.17. Ectoproct. Histology of basic body wall.

Ectoprocts are small-sized animals, so, the coelomic fluid is adequate enough for internal transport within zooids. The peritoneum (coelomic epithelium) is partially ciliated, and a constant definite circulation of coelomic fluid is maintained. The coelom is filled with a fluid and corpuscles.

Funiculus. It is tube-like extension of the body wall with muscle fibres which are attached to the stomach. In phylactolaemates and cyclostomates, funiculus is found to contain fluid-filled cylinders with wall of peritoneal cells and also longitudinal muscles. Funiculus is a homologue of the circulatory system of brachiopods and phoronids.

Special **respiratory** and **excretory organs** are absent.

Digestive System

Alimentary canal of Ectoprocta is V-shaped. **Mouth** is surrounded by lophophore. **Pharynx** is short, broad, ciliated and muscular. **Oesophagus** is long, narrow and delicately muscled. **Stomach** is differentiated into three parts: **cardia, caecum** and **phylorus**. The caecum forms the bottom of U and the **rectum** is sharply set off from the pylorus. A thick-walled **gizzard** containing grinding denticles occurs at the base of oesophagus in case of some Ctenostomata.

Bryozoans filter their planktonic and detritus food particles from the surrounding water by altering the direction of beat of the impinging cilia. Ciliated currents are caused by the lophophore. The cilia on the lateral surfaces of the tentacles pull water down and out between the tentacles, directing detritus particles toward the mouth. Large particles are rejected by retraction or muscular movements of the tentacles and small particles are delivered to the mouth by ciliary currents at the lophophore base.

Digestion is very rapid and wholly extracellular. Starch is digested slowly. Proteins are hydrolyzed in stomach and absorbed there. Fat globules are phagocytized by stomach epithelial cells.

Respiration

The exchange of respiratory gases occurs at the body surface, no doubt largely through the lophophore.

Excretory System

Coelomic leucocytes collect the wastes, which are ejected out either through the lophophore or along with **brown bodies** through the alimentary canal (called **storage excretion**). Sometimes waste materials become incorporated into a body that eventually be shed or be retained in an inert condition.

Nervous System

There are no special sense organs but sensory cells are provided for reception of **tactile, chemical** and **water current stimuli.**

The nervous system is simple. In gymnolaemates, there occurs a **nerve ring** around the pharynx. A small cephalic ganglion, with peripheral neurons and central neuropile, is attached to the ring and lies within the lophophoral coelom. Two motor and two sensory fibres pass to each tentacle from the ring.

Reproductive System

Freshwater ectoprocts reproduce asexually and sexually. The **asexual reproduction** takes place by four methods: (*i*) budding; (*ii*) hyperhacula, (*iii*) brown bodies and (*iv*) by statoblast formation. Budding is also used in the growth of the colony.

Statoblasts. Buds of statoblasts appear on the funiculus. They grow and differentiate into a dark, resistant covering and a mass of yolky, undifferentiated germinal cells. Statoblasts are of three types: (1) **Floatoblasts** which contain air-filled cells in the capsule wall; (2) **Sessoblasts** which are attached to the zoecium wall by cement; (3) **Siphonoblasts** with spiny projections. Statoblasts are liberated when polypide degenerate. They are resistant and protect the species during the winter. In spring, they germinate to form a new zooid when the environmental conditions become favourable. Statoblasts are used in species identification.

Sexual reproduction. Most marine ectoprocts are protandric hermaphrodites. The **gonads** develop in the peritoneum of the coelom, on the body wall, gut or funiculus. All phylactolaemates are hermaphroditic with **ovaries** on the body wall and **testes** on the funiculus. Eggs and sperms are shed into the coelom where fertilization takes place. The fertilized eggs pass out of the coelom through a special pore. However, in many bryozoans the eggs are brooded, either in the coelom, or in a modified chamber (the **ovicell** in *Bugula*) of the zooid.

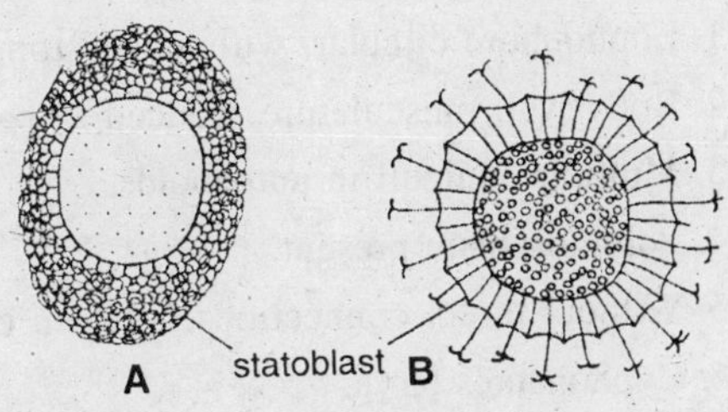

Fig. 32.18. Statoblast. A—of *Plumatella*; B—of *Cristatella*.

Development is **indirect**. They produce a modified larva, either plankton feeding **cyphonautes larva** (*e.g.*, *Frustrella*) or **coronate larva** (*e.g.*, *Bugula*) that subsists on yolk and develops in brood chamber. The larva gets attached with substratum and metamorphose to produce first zooid of new colony. The zooid undergoes budding and forms the colony.

Phylogeny

Ectoprocta show affinities with three invertebrate phyla namely—Phoronida, Brachiopoda and Endoprocta.

Classification

Phylum Ectoprocta is divided into following **three** classes:

Class 1. Phylactolaemata

1. Zooids with horse-shoe shaped lophophore.
2. Epistome present.
3. Body wall with musculature.
4. Zoecium incomplete and not calcified.
5. Coelom of adjacent zooids is continuous.
6. Polymorphism of zooids is absent.
7. Exclusively freshwater species.

Examples. *Lophopus, Cristatella, Plumatella, Pectinatella, Stolella, Fredricella (Fig. 32.19B), Plumatella.*

Class 2. Stenolaemata

Class Stenolaemata includes one living order.

Order Cyclostomata

1. Tubular, fully calcified zoecium having terminal, typically circular orifice with operculum but the orifice closed by perforated membrane.
2. Coelom divided by membranous sac into inner and outer regions, used in lophophore extrusion.
3. Polymorphism common, with formation of gonozooids, kenozoids and nannozooids.
4. Many young formed by polyembryony from single embryo in brood.
5. Exclusively marine species.

Example. *Crisia* (Fig. 32.19 C).

Class 3. Gymnolaemata

1. Lophophore circular, without epistome.
2. Body wall musculature reduced to parietal muscle.
3. Multiple funiculi in autozooids.
4 Polymorphism present.
5. Without direct connections between coeloms of adjacent zooids.
6. Cyphonautes larva.

Class Gymnolaemata is divided into following **two** orders:

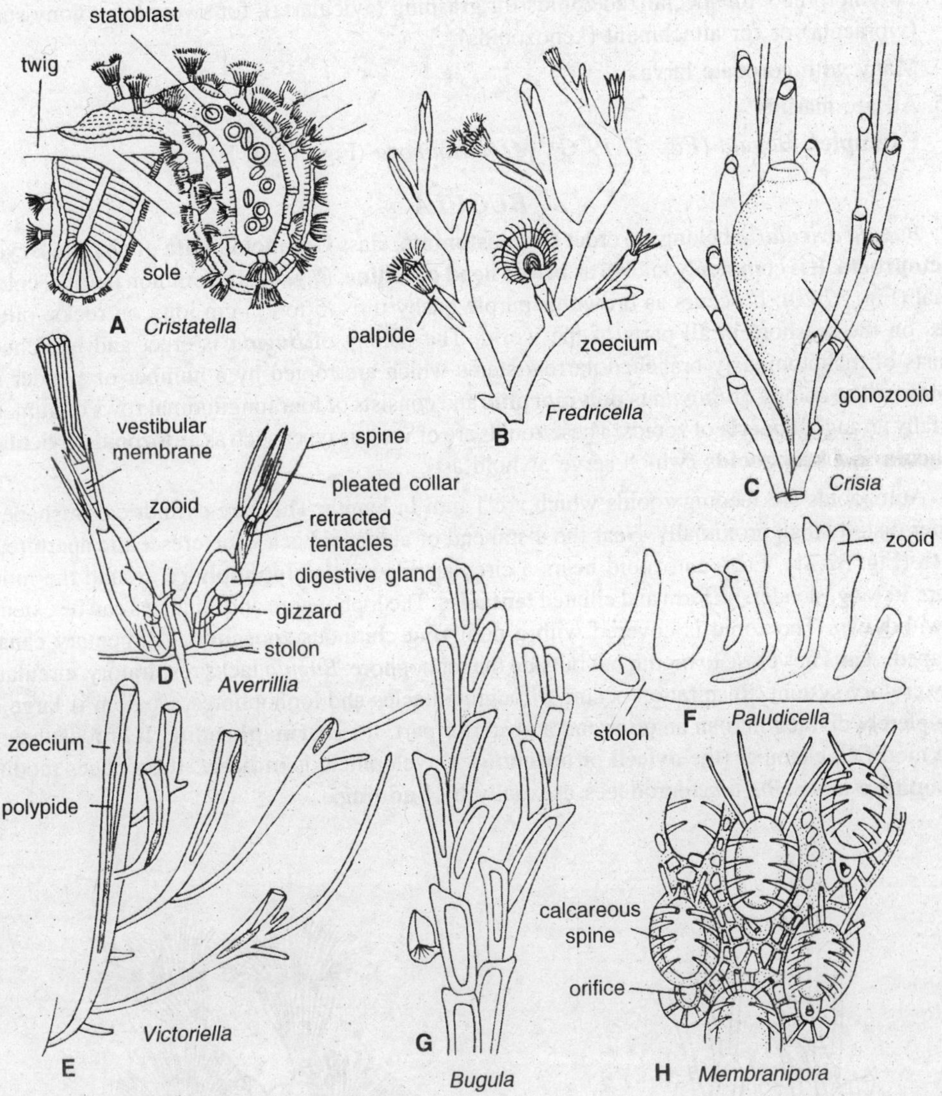

Fig. 32.19. Ectoprocta. A—*Cristatella*; B—*Fredricella*; C—*Crisia*; D—*Averrillia*; E—*Victoriella*; F—*Paludicella*; G—*Bugula*; H—*Membranipora*.

Order 1. Ctenostomata

1. With simple, flexible zoecium of chitinous material having terminal or subterminal orifice, without closing plate.
2. Only two types of zooids: feeding zooids and stolonic zooids.
3. Most are marine, some freshwater.

Examples. *Averrillia*, *Victoriella*, *Paludicella* (Fig. 32.19).

Order 2. Cheilostomata

1. With membranous or calcareous, box-like zoecium.
2. Zoecium with subterminal opening and typically, with hinged lid, operculum.

3. Polymorphic with specialized zooids for grasping (avicularia), for sweeping colony surface (vibracula) or for attachment (kenozooids).
4. Many with coronate larva.
5. All are marine.

Examples. *Bugula (Fig. 32.19 G), Membranipora* (Fig. 32.19 H).

BUGULA

Bugula avicularia belongs to order **Cheilostomata**, class **Gymnolaemata** of phylum **Bryozoa** or **Ectoprocta**. It is commonly known as **bird's head coralline**. *Bugula* is a common marine, colonial animal (Fig. 32.20). It occurs as brown or purple bushy tufts, 5 to 7.5 cm long on rocks, piles of jetties on the seashore in all parts of the world. The colony of *Bugula* is erect and branched. It consists of dichotomously branched narrow stems which are rooted by a number of slender **root filaments**. The colony of *Bugula* is **polymorphic** and consists of four longitudinal rows of numerous biserially arranged **zooecia** of zooids. These zooids are of various types such as **autozooids, avicularia, vibracula** and **kenozooids** (which serve as holdfast).

Autozooids are feeding zooids which are 1 mm in length. They are cylindrical in shape but broader distally than proximally. Near the distal end of autozooid is a wide crescentic aperture, the **mouth** (Fig. 32.21). Each autozooid bears a circular ridge called **lophophore** around the mouth; it bears 14 long, slender, filiform and ciliated **tentacles**. The lophophore and tentacles can be extended and withdrawn. The zooid is covered with a nonliving chitinous **zooecium**. Alimentary canal is U-shaped. **Anus** lies close to mouth but outside the lophophore. *Bugula* lacks respiratory, circulatory and excretory system. Respiration occurs through tentacles and lophophores. Coelom is large and incompletely divided into an anterior and a posterior part. It is **hermaphrodite**. It broods the eggs in the modified chamber (the **ovicell** or **ooecium**). Development is **indirect**. It produces modified **trochophore** larva. *Bugula* reproduces asexually by **budding**.

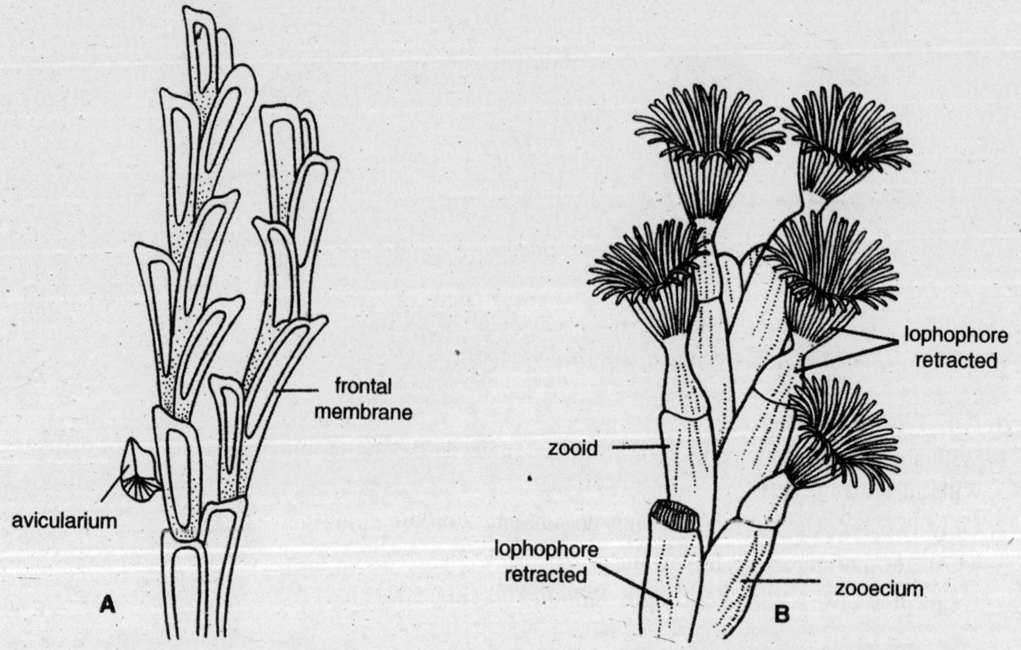

Fig. 32.20. *Bugula* colony A—A part of colony; B—Colony with extended lophophore.

Avicularia consists of a **peduncle** which is attached to the zooids of colony, a **head** which is, in fact, the modified zooecium (zoecium) and a **mandible** that is modified operculum (Fig. 32.21). The reduced parts of the polypide and cystid can be seen in the interior. The most conspicuous internal elements are the opening (**abductor**) and closing (**adduc-tor**) muscles. Avicularia snap viciously when disturbed; they grasp and hold small organisms that visit the colony. Thus, they prevent fouling of the colony by other sessile organisms.

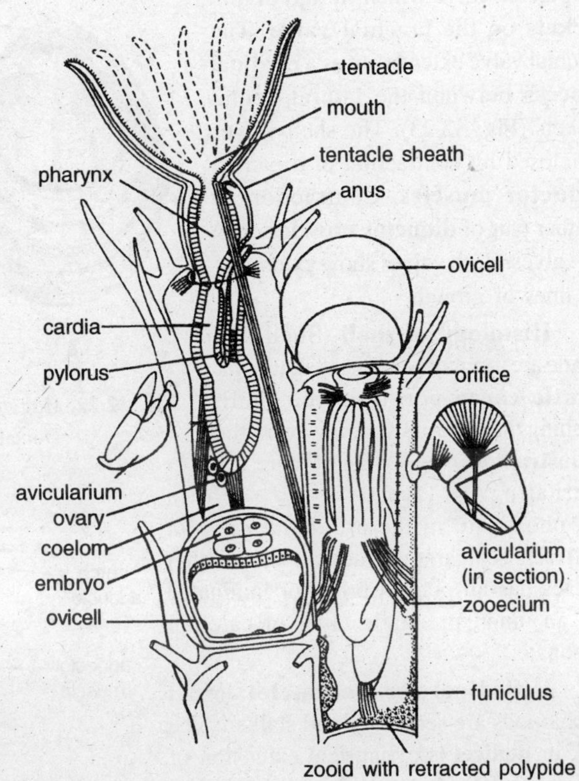

Fig. 32.21. *Bugula.* Structure of two zooids. Right zooid with retracted polypide; left zooid with extended polypide.

32.7. PHYLUM BRACHIOPODA

Phylum Brachipoda (Gr., *brachion* = arm; *podos* = foot) is a group of lophophore-bearing coelomate animals which resem-ble bivalve molluscs in possessing a mantle and a calcareous bival-ved shell. However their two shell valves are **dorsal** and **ventral;** dorsal valve being smaller than the ventral one. Brachiopods are commonly called **lamp shells**. The phylum Brachiopoda includes about 45,000 extinct species and 350 living species.

Phylum Brachiopoda made its appearance in Cambrian period (about 590–505 million years ago) and reached its peak of evolutionary development during Ordovician period.

Habit and Habitat

Brachiopods are entirely marine and found from the intertidal zone to the deep sea. Most species live attached to the rocks but some brachiopods such as *Lingula* live in burrows in sand or mud. They ate ciliary feeders.

Structure

Brachiopods range from 5 mm to 8 cm in length. Their shape may be oval, spherical, flattened or elongated. Some species contain transparent shells. Their body colour may be yellowish grey, orange, pink or red. The body includes the following parts:

1. Shell. The body is bilaterally symmetrical and dorsoventrally flattened. It is enclosed in a **shell** which is formed of two **valves**. One is **dorsal** or **brachial valve** and other is **ventral** or **pedicle valve** (Fig. 32.22). Dorsal valve is smaller and fits over the larger ventral valve. The apex of the ventral valve forms a beak which bears a hole. It looks like a Roman lamp. In inarticulate brachiopods, the two valves are held together by muscles only. In articulate brachiopods, interlocking processes form a **hinge**. Such a hinge occurs at the posterior margin of the valves and consists of hinge teeth on the pedicle valve which fit into **dental sockets** on the brachial valve. The brachial valve extends out as a **cardinal process** between the lateral dental sockets (Fig. 32.23). The shell valves are closed by contraction of a pair of **adductor muscles**. Contraction of another pair of **diductor muscles** open the valves. Both valves show externally the lines of growth.

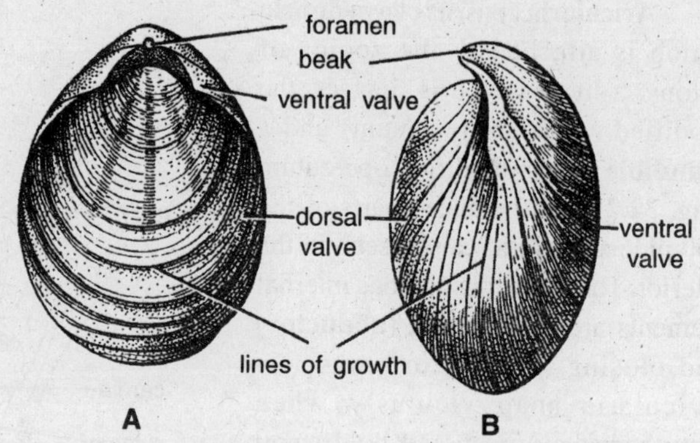

Fig. 32.22. *Magellania* (an articulate brachiopod) shell. A—Dorsal; B—Lateral.

Histology of shell. Shell and mantle are intimately associated; tubular **mantle caeca** penetrate the shell, reaching the uppermost horny layer, the **periostracum** (Fig. 32.24). There are alternating layers of chitin and phosphate in the articulate *Lingula* shell. The calcareous shell of most other species has an outer, fibrous or laminated and an inner, prismatic layer of calcium carbonate.

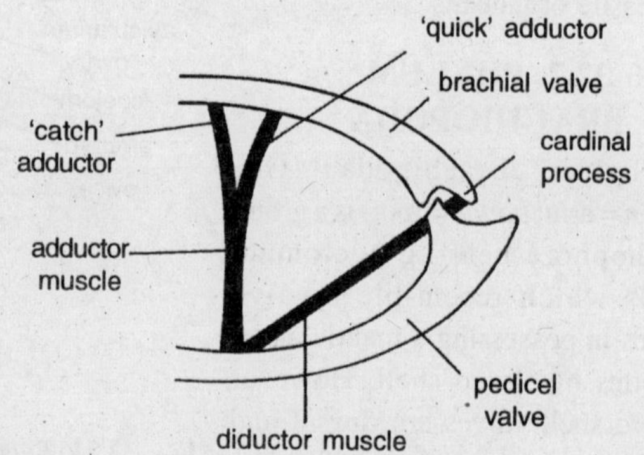

Fig. 32.23. Articulate brachiopod shell. Hinge arrangement.

2. Pedicel (or Peduncle). Most brachiopods are sedentary and stalked. The stalk or **pedicel** is cylindrical extension of the ventral body wall by which most brachiopods are attached to the substratum. Pedicel may be long, fluid-filled and muscular (in Inarticulata) or short and without muscles (in Articulata). The pedicel has been permanently lost in some brachiopods such as *Crania* (Inarticulata) and *Lacazella* (Articulata).

Inarticulates (*Lingula*) can contract the pedicel and achieve some evasive movements by moving deeper into sediments (Fig. 32.27).

3. Mantle. The two mantle lobes or folds surround the body, line the valves of the shell and enclose a large **mantle cavity**. Mantle folds are extensions of body wall. The body proper of brachiopods occupies only the posterior part of the mantle cavity. Anterior part of mantle cavity is occupied by the lophophore. The edge of the mantle bears prominent, stiff **setae**, formed by single chaetoblasts. Setae are protective and sensory in function.

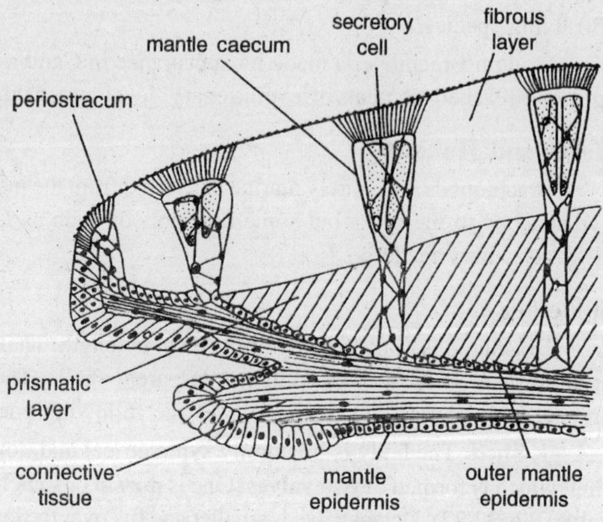

Fig. 32.24. L.S. through mantle and shell valve of an articulate brachiopod. Note caeca that extend into calcareous and fibrous parts of shell, conferring characteristic porous texture.

4. Lophophore. The body wall forms a discoid or horse-shoe shaped fold, the **lophophore**, around the mouth. Each arm of lophophore is coiled in complicated way. Each arm bears a row of numerous ciliated **tentacles**. Coiling of lophophore greatly increases the food collecting area.

The inner surface of brachial or dorsal valve is often sculptured to accommodate the lophophore, and articulates bear the **brachidium** or **lophophore** skeleton (Fig. 32.25).

Body Wall and Coelom

The body wall consists of an outer layer of epithelium, a layer of connective tissue and a muscle layer.

Coelom is large and divided incompletely into two lateral halves by a longitudinal mesentery. Each half is again divided into a **protocoel**, a **mesocoel** and a **metacoel** by two incomplete septa. Branches of mesocoel extend into tentacles, and branches of the metacoel extend into mantle.

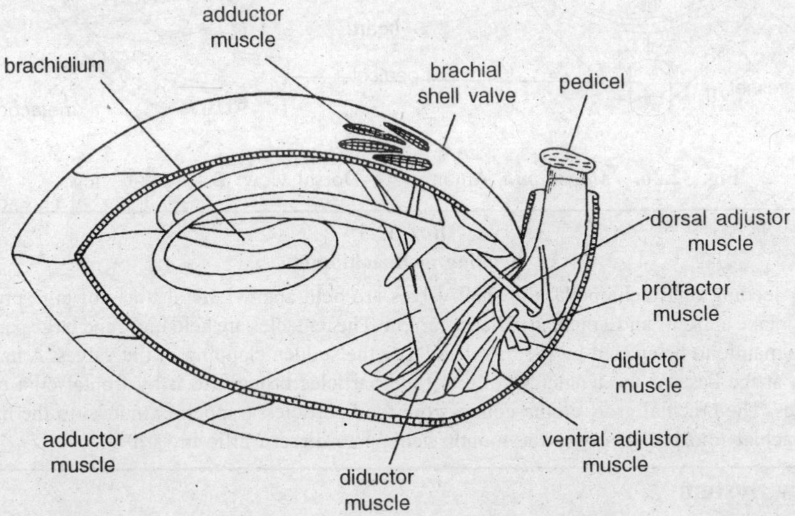

Fig. 32.25. *Magellania.* Muscles, muscle scars, brachidium and pedicel.

Coelom is filled with the coelomic fluid which contains several types of corpuscles, including haemerythrin containing corpuscles. Peritoneal cilia or flagella circulate the coelomic fluid. The fluid moves rapidly in the narrow passageways of the mantle (*i.e.*, tubular coelomic channels called **mantle canals**) and tentacles and was once thought to be blood.

Digestive System

Alimentary canal is V-shaped in Articulata (*Magellania*) having mouth, oesophagus, stomach with paired gastric glands or liver and a narrow and blind intestine (Fig. 32.26). In articulata, the intestine is looped or coiled having **anus** which opens anteriorly in the mantle cavity (metacoel). Gastric glands are probably sites of intracellular digestion and absorption (*e.g.*, *Lingula*).

The inarticulates subsist on particulate matter and are especially adept at sorting edible and inedible particles as they enter the tentacle field (see Box 32.3). They require an anus to void wastes. Articulates derive most of their nutrition from the absorption of dissolved and colloidal organic nutrients in sea water (*e.g.*, polar and subpolar areas). The gut, since is dealing with a low-waste food source, lacks an anus.

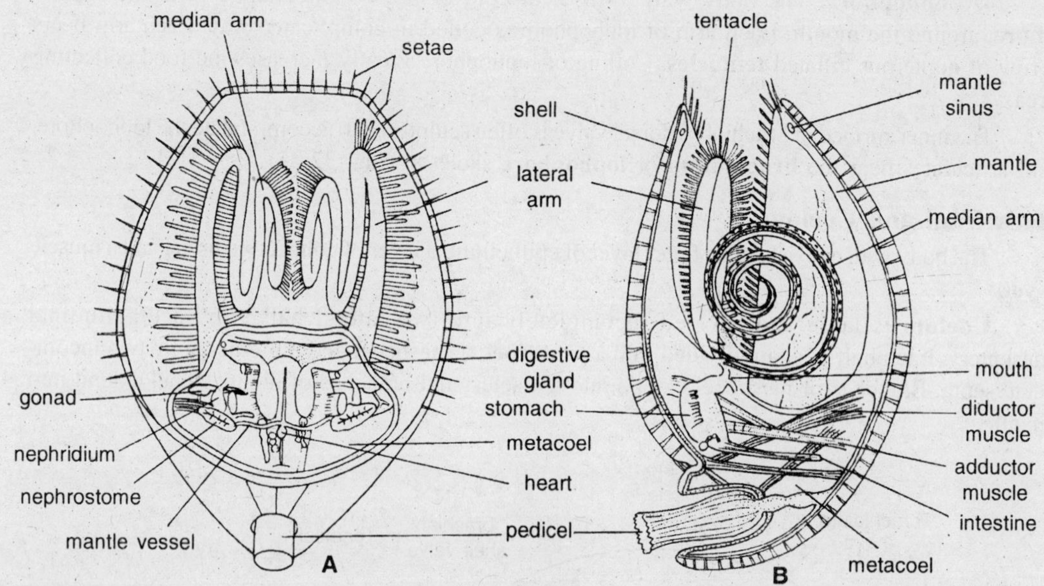

Fig. 32.26. *Magellania.* Antomy. A—Dorsal view; B—Lateral view.

Box 32.3
Feeding in Brachiopoda

During feeding of Brachiopoda, the shell valves are held apart. Lateral tracts of cilia produce lateral intake currents and a median exhaust current. The tentacles are held high, and large particles drop to mantle to be ejected by the mantle cilia or the sudden clapping of the valves. A brachial groove at the base of the tentacles collects food particles brought to it by frontal cilia on the tentacles. The brachial groove cilia conveys the food particles, trapped in mucus, to the mouth. The brachial fold projects over the mouth, helping to direct traffic inward.

Circulatory System

Colourless blood containing respiratory pigment haemerythrin in cells, circulates in an **open type** circulatory system. The contractile vesicle or **heart** lies dorsally to the stomach. It is divided into two **anterior branches**, each communicating with an extensive sinus around the gut and continuing up to the lophophore. Heart bifurcates posteriorly to form **dorsal** and **ventral mantle vessels**. Ventral mantle vessel also communicates with sinuses in the nephridia and gonads. The primary function of circulatory system of Brachiopods is circulation of nutrients.

The blood and coelomic channels communicate extensively through tissue spaces.

Respiratory System

Brachiopods respire by the help of tentacles, mantle lobes and mantle caeca of shell. They have low rates of oxygen consumption. A respiratory pigment, haemerythrin, occurs in blood and coelomic fluid cells. It does not take up oxygen as readily as haemoglobin, but gives it up very readily.

Excretory System

Excretory system includes one or two pairs of **metanephridia**. Their ruffled and flagellated **nephrostomes** open into the metacoel and the **nephridiopores** into the mantle cavity.

Lingula is mainly **ammonotelic**, although enzymes to produce urea are present (**Hammen**, 1977). Particulate nitrogenous material of coelomic fluid is taken up by **coelomic phagocytes** and by **peritoneal cells** that detach and disintegrate. Streams of particles move to nephridia through nephrostomes and are excreted out through nephridiopores. The excreted particles remain suspended in mucus strands.

Nervous System

It consists of a **nerve ring** around the oesophagus and several nerves. The nerve ring bears a **small ganglion** on its dorsal side and a **larger ganglion** on the ventral side.

Sensory Organs

Mantle margin is the most important site of sensory reception. Mantle setae transmit tactile stimuli to these sensory receptors.

Reproductive System

Sexes are separate. Four pairs of gonads are present. Sex cells or gametes are released into coelom; from there they depart through the nephridia. Fertilization is external.

Development

Cleavage is **holoblastic** and usually of **radial** type. Plankton-feeding larvae occur in Inarticulata; they develop from less yolky eggs. Yolk-subsisting larvae occur in Articulata; they develop from very yolky eggs.

Classification

Following **two** classes of brachiopods may be recognized according to whether or not the valves of the shell are joined together by a hinge.

Class 1. Inarticulata (or Ecardines)

1. Shell is without hinge held together by muscles only.
2. Lophophore is without skeletal support.
3. Digestive tract is complete with anus.

This class includes following **two** orders:

Order 1. Lingulida

1. With pedicel valve ventral but both shell valves are modified to permit pedicel passage.
2. Shell mineral is calcium phosphate.

Examples. *Lingula*, *Glottida*.

Order 2. Acrotretida

1. Without pedicel, or with pedicel emerging from notch or foramen in pedicel valve.
2. Shell minerals are calcium carbonate and/or calcium phosphate.

Examples. *Crania*, *Discinisca*.

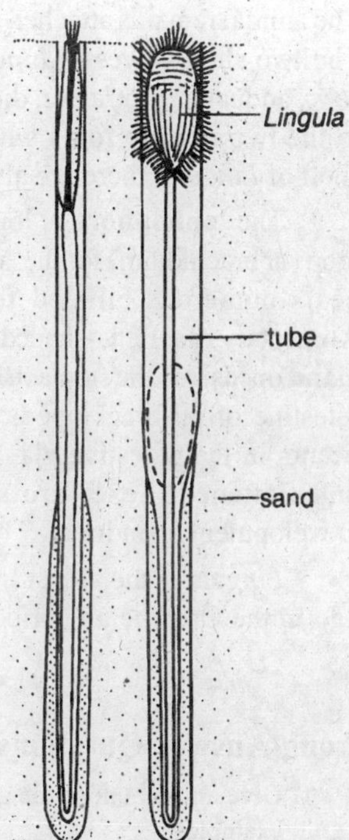

Fig. 32.27. *Lingula anatifera* in their tubes in the sand; the dotted line in the right tube indicates position of the body when retracted.

Class 2. Articulata (or Testicardines)

1. Shell valves are hinged by interlocking processes.
2. Pedicel is emerging through pedicel valve.
3. Shell mineral is calcium carbonate.
4. Lophophore is generally with skeletal support.

5. Digestive tract is incomplete (no anus is present).

Class Articulata includes following **two** orders:

Order 1. Rhynchonellida

Example. *Rhynchonella*

Order 2. Terebratulida

Examples. *Magellania*, *Terebratulina*.

LINGULA

Lingula contains about 12 species and belongs to order **Lingulida**, class **Inarticulata** and phylum **Brachiopoda**. *Lingula* is marine and lives in vertical burrows in the bottom of sand (Fig. 32.27). It is found in the tropical and subtropical waters of the Indo-West Pacific area. Japanese eat *Lingula*.

The body of *Lingula* is covered by a **shell** (Fig. 32.28). Shell consists of two almost equal and weakly convex **valves**. Shell valves are not united by a hinge. A long **pedicel** passes through the proximal end of two valves. The pedicel is extensible and can be withdrawn. The animal remains attached to the bottom of tube through the pedicel. The two shell valves of *Lingula* are connected together by muscles (*e.g.*, adductor, protractor, diductor, etc). The shell is tightly adherent to the two **mantle folds** which are extensions of body wall. In the shell of *Lingula*, there are alternating layers of chitin and phosphate.

The **lophophore** is long and spirally coiled. There is no shell loop (or brachidium) for the support of the lophophore. The lophophore bears numerous ciliated **tentacles**. Lingula is a ciliary feeder. Alimentary canal is U-shaped and complete having **anus**. The **digestive gland** opens into the stomach by four ducts. *Lingula* contains the longest intestine of all brachiopods. Digestion is chiefly **intracellular** and occurs in the digestive gland. **Metanephridia** are very long, broad and flattened. Sexes are separate and fertilization is external. Development is indirect. There is no metamorphosis in *Lingula*.

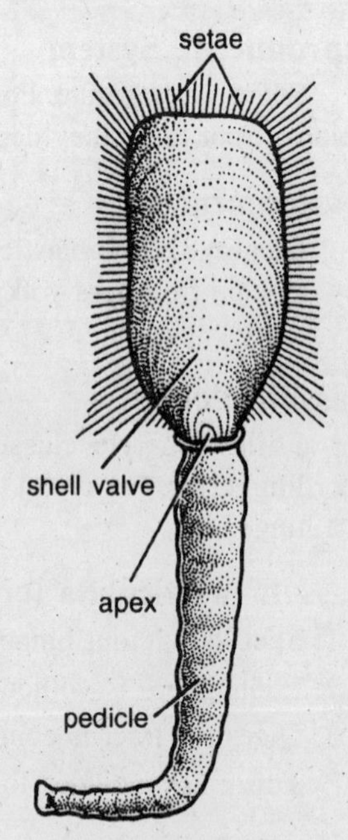

Fig. 32.28. *Lingula*. External structure.

Lingula is the oldest living multicellular animal on earth. It appeared about 400 million years ago in the Ordovician period. Since then it is surviving unchanged and is called a **"living fossil"**.

QUESTIONS

Long Answer Questions

1. Give an account of the general characteristics of phylum Ctenophora and classify it giving important examples. (*Banaras 1991*)
2. Describe the general characters and affinities of phylum Ctenophora. (*Bangalore 1990; Allahabad 1993, 1996*)
3. Give an account of morphological organization and life history of Ctenophora. (*Meerut (CCS) 1988; Berhampur 1989*)
4. Describe general characters of phylum Rotifera. (*Meerut (CCS) 1991, 1995; Sagar 1994; Indore 1994*)
5. Describe in brief the general characters of phylum Sipunculida.
6. Describe general characters of phylum Echiura or Echiuroidea. (*Merrut (CCS) 1991*)

7. Describe general characters of phylum Bryozoa.

8. Describe general characters and classification of phylum Brachiopoda. (*C.C.S.U. Meerut 1999, 2001*)

Short Answer Questions

1. How do ctenophores differ from other coelenterates? (*Bangalore 1990; Utkal 1990*)

2. Write a short note on Ctenophora. (*Allahabad 1997*)

3. Write a short note on *Sipunculus*. (*Meerut (CCS) 1990, 1992, 1996*)

4. Write a note on Bryozoa. (*Meerut (CCS) 1988; Indore 1994*)

5. Write a short note on *Bugula*. (*Meerut (CCS) 1993, 1996*)

6. Write a short note on *Lingula*.

7. Classify the following animals giving their distinguishing characters.

8. Draw a well labelled diagram of *Bonellia*.

9. Write a note on Brachiopoda.

10. Classify the following animals giving their distinguishing characters.

(*i*) *Cestum*

(*ii*) *Ctenoplana*

(*iii*) *Coeloplana*

(*iv*) *Pleurobrachia* (or *Hormiphora*)

Very Short Answer Questions

1. Name the larva of *Pleurobrachia*.
Ans. Cydippid.

2. Combs of *Hormiphora* are composed of free cilia. Is it true?
Ans. No.

3. The canals of the digestive tract are bilaterally arranged in *Pleurobrachia*. Do you agree?
Ans. Yes.

4. Where is the statocyst located in comb jellies?
Ans. At aboral pole.

5. What is the function of colloblasts?
Ans. Capturing prey.

6. Name the two classes of phylum *Ctenophora*.
Ans. *Tentaculata* and *Nuda*.

7. Give the popular name of *Cestum*.
Ans. Venus's girdle.

8. To which class does *Beroe* belong?
Ans. Nuda.

9. Give the classification of *Cestum*.
Ans. Order Cestida, Class Tentaculata, phylum Ctenophora.

10. Give the popular name of the rotifera.
Ans. Wheel bearers (wheel animalcules).

11. Cite one example of phylum Rotifera.
Ans. *Philodina roseola*.

12. What is corona of rotifers?

Ans. The head of rotifers or wheel animalcules consists of a characteristic retractile ciliary crown of disc called **corona** or **trochal disc**. It has two rings of cilia which beat in a circular manner and appear as rotating wheel.

13. Explain cyclomorphosis in context with rotifers.

Ans. In the phenomenon of cyclomorphosis, the phenotypic form of amictic female (*i.e.*, female producing the eggs parthenogenetically in case of rotifers) changes gradually in response to environmental conditions during the year.

14. How does introvert operate in *Sipunculus*?

Ans. The body of *Sipunculus* consists of **introvert** and **trunk**. The introvert may be withdrawn into the trunk. In this process the coelom and body wall together work as a **hydraulic system** to operate the introvert. The introvert is everted by the contraction of body wall musculature which raises the coelomic pressure. It is retracted by the retractor muscle of body wall.

15. What is common name of brachiopods?

Ans. Lamp shells.

16. What is the common name of animals of Phylum Ectoprocta?

Ans. Moss animals.

17. What is the popular name of sipunculans?

Ans. Peanut worms.

18. Give the common name of Echiuroidean animals?

Ans. Spoon worms.

Fill in the Blanks

1. *Pleurobrachia* is popularly called ....

2. The larva of comb jelly is named ....

3. Statocyst lies at the .... pole in the sea gooseberry.

4. *Pleurobrachia* has.............. meridional swimming plates.

5. Tentacles of comb jelly bear special cells, the ....

6. The ctenophores have .... symmetry.

7. Special adhesive cells, called .... , are the characteristics of ctenophores.

8. The ctenophores have .... level of organisation.

9. The ctenophores have .... egg, ... cleavage and .... gastrulation.

10. The larva of ctenophora is called .... larva.

Multiple Choice Questions

1. Venus's girdle is

(*a*) *Ctenoplana* (*b*) *Cestum*

(*c*) *Coeloplana* (*d*) *Beroe*

2. Mitre jelly-fish is

(*a*) *Metridium* (*b*) *Beroe*

(*c*) *Pleurobrachia* (*d*) *Ctenoplana*

3. One of the following Ctenophora is without tentacles

(*a*) *Coeloplana* (*b*) *Ctenoplana*

(c) *Beroe* (*d*) *Pleurobrachia*

4. Eight comb-like plates for locomotion are found in
(*a*) Arthropoda (*b*) Platyhelminthes
(*c*) Mollusca (*d*) Ctenophora

5. Development is indirect in Ctenophora and larva is known as
(*a*) planula (*b*) miracidium
(*c*) cydippid (*d*) ephyra

6. Select the creeping Ctenophora
(*a*) *Ctenoplana* (*b*) *Coeloplana*
(*c*) *Deiopea* (*d*) *Hormiphora*

7. *Mnemiopsis* belongs to
(*a*) Cydipidda (*b*) Platyctenea
(*c*) Lobata (*d*) Beroidea

8. Colloblasts are found in
(*a*) *Obelia* (*b*) *Aurelia*
(*c*) *Pleurobrachia* (*d*) *Metridium*

9. Digestive tract is complete in
(*a*) sea anemone (*b*) comb jelly
(*c*) *Obelia* (*d*) tapeworm

10. Symmetry in Ctenophora is
(*a*) spherical (*b*) radial
(c) biradial (*d*) bilateral

11. Ctenophores are
(*a*) monoecious (hermaphrodite) (*b*) dioecious
(c) viviparous (*d*) devoid of sex organs

12. *Sipunculus* is commonly known as
(*a*) nutworm (*b*) peanut worm
(*c*) peacock worm (*d*) clam worm

13. Colonies of zooids of Ectoprocta are called
(*a*) lamp shells (*b*) sea mats
(*c*) slipper animalcules (*d*) none of these

14. *Magellania* belongs to phylum
(*a*) Rotifera (*b*) Ectoprocta
(*c*) Brachiopoda (*d*) Acantocephala

15. Brachiopods are
(*a*) acoelomate (*b*) coelomate
(*c*) schizocoelomate (*d*) pseudocoelomate

16. Lophophore acts as a
(*a*) food catching apparatus (*b*) locomotory organ
(*c*) digestive gland (*d*) none of these

17. Brachiopods are
(*a*) filter feeders (*b*) ciliary feeders
(*c*) mucous feeders (*d*) none of these

18. Ectoprocta (Bryozoa) form colonies known as
(*a*) sea-mats (*b*) sea-colonies
(c) sea-huts (*d*) none of these

19. Complete ectoproct colony is called
(*a*) zooid (*b*) zoarium
(c) corona (*d*) none of these

20. The zoological name of Bird's Head Coralline is
(*a*) *Bugula avicularia* (*b*) *Brachionus rubens*
(*c*) *Echinorhyncus gigas* (*d*) *Oligacanthorhynchus*

21. In Ectoprocts asexual reproduction takes place by
(*a*) fragmentation (*b*) binary fission
(*c*) budding (*d*) none of these

22. The larva of *Bugula* is
(*a*) cyphonant (*b*) pilidium
(*c*) nauplius (*d*) zoa

23. Rotifers are also known as
(*a*) slipper animalcule (*b*) wheel animalcule
(*c*) ladder animalcule (*d*) none of these

24. In *Brachionus rubens*, pedal glands, secrete
(*a*) adhesive material (*b*) saliva
(*c*) both (a) and (b) (*d*) none of these

25. *Brachionus rubens* belong to phylum
(*a*) Acanthocephala (*b*) Rotifera
(*c*) Entoprocta (*d*) Ectoprocta

26. *Pedicellina* belongs to
(*a*) Entoprocta (*b*) Nemertinea
(*c*) Acanthocephala (*d*) Ectoprocta

27. Larvae form of phylum Nemertinea is
(*a*) pilidium (*b*) nauplius
(*c*) miracidium (*d*) zoea

ANSWERS

Fill in the Blanks

1. Sea gooseberry **2.** Cydippid **3.** Aboral **4.** Eight
5. Colloblasts **6.** Biradial **7.** Colloblasts **8.** Tissue
9. Lecithal, determinate, epibolic **10.** Cydippid.

Multiple Choice Questions

1. (*b*) **2.** (*b*) **3.** (*c*) **4.** (*d*)
5. (*c*) **6.** (*a*) **7.** (*c*) **8.** (*c*)
9. (*b*) **10.** (*c*) **11.** (*a*) **12.** (*b*)
13. (*b*) **14.** (*c*) **15.** (*b*) **16.** (*a*)
17. (*b*) **18.** (*a*) **19.** (*b*) **20.** (*a*)
21. (*c*) **22.** (*a*) **23.** (*b*) **24.** (*a*)
25. (*a*) **26.** (*a*) **27.** (*a*).

PART – B
HIGHER NONCHORDATA

PART B
HIGHER NONCHORDATA

33

Phylum Annelida (Characters, Classification and Types)

Phylum Annelida (L., *Annelus* = a ring) includes about 12,000 species of segmented coelomate worms. They range from a deep-sea species measuring less than 1 mm in length to giant tropical earthworms (of Australia) which measure upto 4 m in length. Most annelids are aquatic found in the sea as swimmers, crawlers, burrowers and tube dwellers or in freshwater where they usually burrow. Terrestrial annelids are mostly burrowers and some are parasites. Phylum Annelida comprises three main classes: the **Polychaeta** (marine), the **Oligochaeta** (terrestrial and freshwater) and the **Hirudinea** (terrestrial, marine and freshwater leeches).

Annelids are soft-bodied, elongated, cylindrical, bilaterally symmetrical, metamerically segmented coelomate worms having a thin covering of cuticle, a dermomuscular body wall and often chitinous setae.

33.1 GENERAL CHARACTERS

1. Annelids are triploblastic and bilaterally symmetrical coelomates.
2. Body is soft, elongated, vermiform (worm-like) and metamerically segmented (*i.e.*, body divided into similar parts or segments which are arranged in a linear series along the antero-posterior axis). The segments are marked externally by transverse grooves and internally by transverse **septa**. Each septum comprises of a layer of connective tissue sandwitched in between two layers of peritoneum.
3. The head or **acron** is represented by **prostomium** and contains the brain. It is pre-oral body part. Terminal part of body is called **pygidium** and contains **anus**. **Mouth** exists in **peristomium** (the first true segment, see Box 33.1).
4. Body wall consists of a fibrous collagenous **cuticle**, a **glandular epidermis** in which the nerve fibres are situated and a connective tissue **dermis** of varying thickness. Muscles of body wall exist in outer continuous layer of **circular muscles** (just inner to dermis) and inner layer of **longitudinal muscles**. The latter is lined by **parietal peritoneum** (coelomic epithelium). Cuticle is secreted by ectoderm (or epidermis).
5. The paired appendages if present are not jointed.
6. Locomotory organs are segmentally arranged, paired, chitinous lateral bristles called **setae** or **chaetae**. They are absent in leeches.
7. Body cavity is a true **coelom** or **perivisceral coelom** lying between the two layers of mesoderm. Coelom is a **schizocoel** (Box 33.1) and divided into segmental compartments by septa.

Box 33.1

1. Prostomium and pygidium are not true segments since they do not arise from the segmental growth zone. The formation of new segments always takes place just infront of pygidium. The oldest body segments are therefore anterior and the youngest are posterior.
2. **Schizocoel.** A coelom which appears as a space in the embryonic mesoderm.

8. **Alimentary canal** is a straight tube running from the anterior mouth to the posterior anus. Digestion is **extracellular**.
9. Respiration occurs through general body surface or gills.
10. Circulatory system or blood vascular system is well developed and **closed type**. Blood is usually red and contains **haemoglobin** (a respiratory pigment). Respiratory pigment is often dissolved in the plasma.
11. Excretory organs are segmental, ectodermal in origin, ciliated, paired and called **nephridia** (metanephridia and protonephridia).
12. The nervous system consists of an anterior, dorsal ganglionic mass or **brain**; a pair of anterior **connectives** surrounding the gut and a long double or single solid **ventral nerve cord** with ganglionic swellings and lateral nerves in each segment.

    A **giant fibre system** is used for rapid conduction and evokes fast startle contraction. *Aphrodite* and *Chaetopterus* have no giant fibre system but other annelids including *Lumbricus* have giant fibre system.
13. Gonads develop from coelomic epithelium.
14. Sexes may be united (monoecious or hermaphrodite) or separated (dioecious).
15. Development is **direct** (without any larval stage) in monoecious annelids and is **indirect** (with **trochophore larva**) in dioecious and aquatic annelids.
16. Cleavage is spiral and determinate.
17. Asexual reproduction occurs in some annelids. Power of regeneration of their lost parts occurs in many annelids.
18. Aquatic or terrestrial.

33.2 OUTLINE CLASSIFICATION

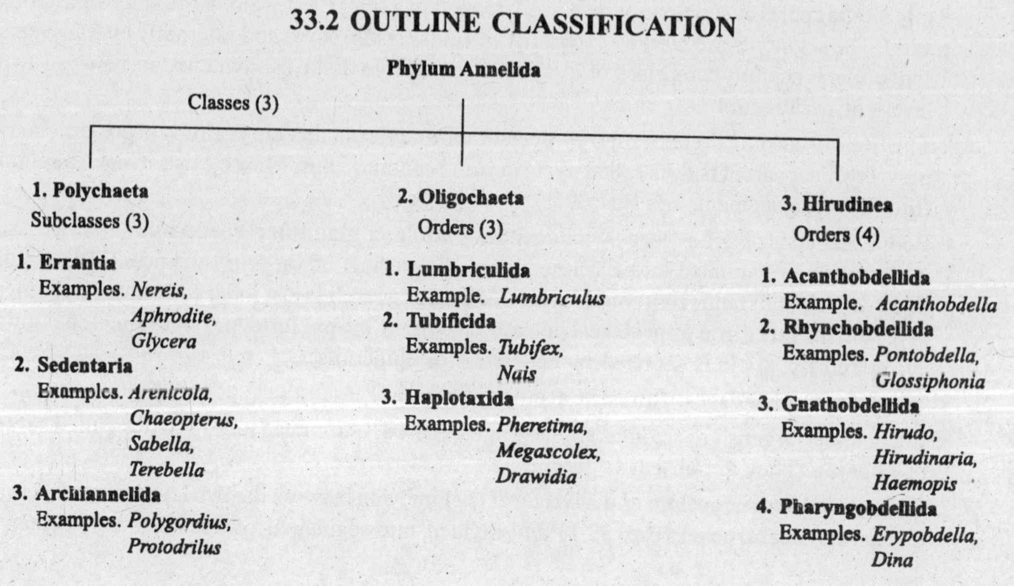

33.3 DETAILED CLASSIFICATION

Phylum Annelida is divided into **three** classes:

Class 1. Polychaeta

1. Class Polychaeta (Gr., *poly* = many; *chaete* = bristles) includes more than 8000 species having great diversity (classified into 86 families; see Box 33.4 and 33.5).
2. Exclusively marine; mostly carnivorous (predaceous or scavengers), some are herbivorous or deposit feeders.
3. Polychaetes may be errant (free-moving) or sedentary.
4. Polychaetes have varied habitats: surface-dwelling (crawling), gallery-dwelling, burrowing and tubicolous.
5. The majority are less than 10 cm long; some are as small as 1 mm and others as large as 70 cm.
6. Colour of the body may be red, pink or green.
7. Body is elongated and segmented. Each segment carries a pair of lateral, fleshy, paddle-like appendages or feet called **parapodia**. Parapodia bear many large setae or chaetae (in bundles).
8. Head is well developed (*i.e.*, cephalization well marked). It contains eyes, tentacles, cirri and palps.
9. Clitellum absent.
10. Cirri or branchiae (gills) or both may be present for respiration. Some part of parapodia is also used as a gas exchange surface.
11. Internal transport occurs through blood-vascular system or coelomic fluid or both. Gas transport uses three types of iron-containing respiratory pigments: **haemoglobin** (red in colour and most common), **chlorocruorin** (green in colour and occurs in *Serpula*, etc.) and **haemerythrin** [a nonhaem (lacking porphyrin) red protein pigment of *Magelona*]
12. Main sensory organs include **eyes, nuchal organs** or ciliated pits (chemoreceptors) and **statocysts** (*e.g.*, *Arenicola*).
13. Sexes are separate. Gonads are localized but extending throughout whole body.
14. Fertilization external (see Box 33.2); free-swimming larval stage is trochophore larva. No cocoon is formed.
15. Power of **regeneration** is well developed. For example, *Chaetopterus* can be regenerated from a single segment but this power is restricted in anterior most 14 segment.
16. Asexual reproduction occurs by lateral **budding**.

Box 33.2 Epitoky and Swarming

Copulation is rare in polychaetes. Instead synchronus emission of sperm and eggs occurs. Epitoky and swarming bring a dispersed benthic population (*e.g.*, *Neries*) together for a brief pelagic existence, when gametes are shed and the likelihood of fertilization is increased. **Epitoky** is the formation of pelagic reproductive individuals or **epitokes** (*e.g.*, *Heteroneries* and Samoan palolo worm) that are adapted for leaving bottom burrows, tubes and other habitats. **Swarming** is a synchronous behaviour in which sexually mature individuals congregates in a relatively short time to increase the likelihood of fertilization.

Class 2. Oligochaeta

1. Class Oligochaeta (Gr., *oligo* = few; *chaete* = bristles) includes about 3100 species. Oligochaetes approximate the polychaetes in size.

2. Mostly freshwater or terrestrial (*i.e.*, damp earth) forms.
3. Body segmented. Parapodia absent. Setae are usually arranged segmentally. Setae are short and embedded in the integument.
4. Head is indistinct and eyes and cephalic appendages such as tentacles and cirri are absent.
5. Clitellum is present. It secretes a cocoon for the eggs.
6. Most oligochaetes are scavengers and feed on dead organic matter, particularly vegetation. Digestive tract is straight and relatively simple containing mouth, pharynx, crop, gizzard, intestine and anus. Intestine has **typhlosole** (A dorsal longitudinal invagination of the intestine wall to increase the absorptive surface).
7. Intestine is surrounded by a layer of yellowish peritoneal cells, called **chlorogogen cells**. They have a role similar to liver of vertebrates (Box 33.3).

Box 33.3 Chlorogogen tissue

Chlorogogen tissue of earthworms plays a vital role in intermediate metabolism. It is the chief centre of glycogen and fat synthesis and storage. Storage and detoxification of toxins, haemoglobin synthesis, protein catabolism and formation of ammonia and synthesis of urea also take place in these cells. In terrestrial species silicates obtained from food material and the soil are removed from the body and deposited in the chlorogogen cells as waste concretions.

8. Oligochaetes contain two types of metanephridia—**holonephridia** and **enteronephric nephridia**–to excrete their nitrogenous wastes such as ammonia and urea.
9. Circulatory system is basically similar to that of polychaetes.
10. Oligochaetes move by peristaltic contractions. Circular muscle contraction and the consequent elongation of segments are most important in crawling and always generate a coelomic fluid pressure pulse. Longitudinal muscle contraction is more important in burrowing, dilating the burrow or anchoring the segments against the burrow wall. Setae are extended during the longitudinal muscle contraction and retracted during circular muscle contraction.
11. Oligochaetes are **hermaphrodites** with well developed reproductive systems limited to a few segments (with testes anterior to ovaries and with only one or two pairs of male and female gonads).
12. There is **copulation** and reciprocal transfer of sperm. **Fertilization** and **direct development** occur within a cocoon.

Class 3. Hirudinea

1. Class Hirudinea (L., *hirudo* = a leech) includes about 500 species of leeches.
2. Most are freshwater; some are marine or terrestrial. Some are ectoparasites.
3. Body is elongated and dorso-ventrally flattened.
4. Body consists of definite number of segments which may be further divided externally into 2 to 4 annuli. Head is indistinct.
5. **Setae** (except one genus), **parapodia** and **cephalic appendages** are absent.
6. Body is provided with a small suctorial **anterior sucker** and a large **posterior sucker** used for adhesion or locomotion. Both suckers are situated ventrally.
7. Clitellum present.
8. Coelom is greatly reduced. It is represented by a specialized circulatory system, called **haemocoelomic system.** True blood vessels are also present.
9. Coelom is filled up with characteristic **botryoidal tissue.**
10. The septa are absent. Metamerism is shown by nephridia and nervous system (*i.e.*, by gangila of the ventral nerve cord).

11. **Hermaphrodite**; fertilization is internal (cross fertilization due to copulation).
12. Development is **direct** (*i.e.*, no larval stage); eggs are laid in cocoons.
13. Vertebrates form the principal hosts of ectoparasitic, or blood-sucking leeches. Digestion of blood, which is slow, depends on exopeptidases (enzymes) produced by the leech and a symbiotic bacterial flora.

Class 1. Polychaeta

Class **Polychaeta** is divided into following **three** subclasses.

Subclass 1. Errantia

1. Subclass Errantia (L., *errans* = wandering) includes free-swimming, crawling, burrowing and tube-dwelling worms.
2. Body segments are numerous and similar except at anterior and posterior ends.
3. Each body segment carries a pair of well-developed parapodia. Parapodia contain **acicula** (stiff basal setae) and **setae.**
4. Head is well-developed, being distinguished into **prostomium** carrying eyes, tentacles and palpi. **Peristomium** with mouth.
5. Pharynx is usually protrusible and armed with chitinous jaws and teeth to capture prey.

Examples. *Nereis, Aphrodite, Glycera, Polynoe, Syllis.*

Box. 33.4 Families of Errantia

According to **Fauchald** (1977), **Parker** (1982) and **Barnes** *et al.*, (1992) subclass Errantia includes the following **thirteen** or more families:

Family 1. **Aphroditidae. Example.** *Aphrodite.*
Family 2. **Polynoidae. Example.** *Polynoe.*
Family 3. **Sigalionidae. Example.** *Sigalion.*
Family 4. **Polyodontidae. Example.** *Polyodontes.*
Family 5. **Phyllodocidae. Example.** *Phyllodoce.*
Family 6. **Amphinomidae. Example.** *Amphinome.*
Family 7. **Alciopidae. Example.** *Alciopa.*
Family 8. **Tomopteridae. Example.** *Tomopteris.*
Family 9. **Hesionidae. Example.** *Podarke.*
Family 10. **Syllide. Example.** *Syllis.*
Family 11. **Nereididae. Example.** *Nereis.*
Family 12. **Nephtyidae. Example.** *Nephtys.*
Family 13. **Glyceridae. Example.** *Glycera.*
Family 14. **Eunicea** (Considered as superfamily). **Examples.** *Eunice, Palola, Arabella.*
Family 15. **Histriobdellidae. Example.** *Histriobdella.*
Family 16. **Ichthyotomidae. Example.** *Ichthyotomus.*
Family 17. **Myzostomidae. Example.** *Myzostoma.*

Subclass 2. Sedentaria

1. Sedentary, tubicolous worms living in calcareous or chitinous tubes secreted by epidermal glands.
2. Body is divisible into two or more regions.
3. Segments are not of same type; each segment with a pair of reduced parapodia lacking acicula. Parapodia often modified to perform specific functions.

4. Head is commonly with palps, tentacles and other structures for feeding. They feed on plankton or detritus.
5. Pharynx is often non-protrusible and devoid of jaws or teeth.
6. Gills (or branchiae), when present, localized to anterior segments.

Examples. *Chaetopterus, Sabella, Arenicola, Serpula, Terebella, Spirorbis.*

Box 33.5 Families of Sedentaria

According to **Fauchald** (1977), **Parker** (1982) and **Barnes** *et al.*, (1992) subclass Sedentaria is divided into following 17 families:

Family 1. **Orbiniidae. Example.** *Orbinia.*
Family 2. **Spionidae. Example.** *Spio, Polydora.*
Family 3. **Megelonidae. Example.** *Megelona.*
Family 4. **Chaetopteridae. Example.** *Chaetopterus.*
Family 5. **Cirratulidae. Example.** *Cirratulus.*
Family 6. **Flabelligeridae. Example.** *Flabelligera.*
Family 7. **Opheliidae. Example.** *Ophelia.*
Family 8. **Capitellida. Example.** *Capitella.*
Family 9. **Arenicolidae. Example.** *Arenicola.*
Family 10. **Maldanidae. Examples.** *Maldane, Clymenella.*
Family 11. **Oweniidae. Example.** *Owenia.*
Family 12. **Sabellariidae. Example.** *Sabellaria.*
Family 13. **Pectinarridae. Example.** *Pectinaria.*
Family 14. **Ampharetidae. Example.** *Ampharete.*
Family 15. **Terebellidae. Example.** *Terebella.*
Family 16. **Sabellidae. Example.** *Sabella, Myxicola.*
Female 17. **Serpulidae. Example.** *Serpula.*

Subclass 3. Archiannelida

1. Subclass Archiannelida (Gr., *archi* = first) includes small-sized animals which are aquatic—marine or freshwater.
2. Body elongated, thread-like.
3. Segmentation chiefly internal. There is no external segmentation.
4. Parapodia and setae are reduced or absent.
5. Tentacles occur on prostomium.
6. Nephridia are either simple (protonephridia) or complex (metanephridia).
7. Nervous system is primitive, *i.e.*, it is continuous with epidermis.
8. Hermaphrodite or unisexual (sexes are separate in *Nerilla*).
9. Larva is typical trochophore.

Subclass Archiannelida includes the following **three** families:

Family 1. Nerillidae. Examples. *Nerilla, Nerillidium.*

Family 2. Polygordiidae. Examples. *Polygordius, Protodrilus.*

Family 3. Dinophilidae. Examples. *Dinophilus, Diurodrilus.*

(**Note.** Archiannelida was previously considered as a separate class of phylum Annelida. Now it is regarded as a heterogenous group of polychaetes (**Fauchald,** 1975).

Class 2. Oligochaeta

Class Oligochaeta is divided into following **three** orders on the basis of differences in the reproductive organs (**Jamieson 1978**).

Order 1. Lumbriculida

1. Male gonopores and at least one pair of testes in the same segment.
2. Each body segment consists of four pairs of setae.
3. Clitellum is one cell thick and includes male and female gonopores.
4. All are freshwater animals.

Example. *Lumbriculus.*

Order 2. Tubificida

1. Male gonopores situated in the segment immediately anterior or posterior to the segment containing testes.
2. One pair of testes followed by one pair of ovaries in adjacent segments.
3. Clitellum is one-cell thick including male and female gonopores.
4. Setae hair-like or modified. Each segment contains setae bundles-each containing two or more setae.
5. Marine, freshwater and a few terrestrial forms.

Examples. *Tubifex*, *Nais*, *Dero*, *Chaetogaster*, *Enchytraeus.*

Order 3. Haplotaxida

1. Male gonopores are situated atleast one segment posterior to last pair of testes.
2. Basically have two segments with testes followed by two segments with ovaries. One pair of testes or ovaries or often both absent. If there is one pair of testes, these are separated from the ovaries by one or two segments.
3. Clitellum one-cell or more than one cell in thickness. Male gonopore is not situated in the clitellum.

Examples. *Lumbricus*, *Megascolex*, *Pheretima*, *Drawida*, *Eutypheus.*

Class 3. Hirudinea

Class Hirudinea is divided into following **four** orders:

Order 1. Acanthobdellida

1. This order includes only one species which is parasitic on the fins of salmon fishes.
2. Body comprises thirty segments only. Each segment contains four annuli.
3. Anterior sucker is absent but posterior sucker is well developed and composed of four segments.
4. Anterior five segments are provided with two pairs of setae.
5. First five segments show internal segmentation (septa are present).
6. Proboscis is short; jaws absent; anus dorsally situated.
7. Acanthobdellida forms a connecting link between Oligochaeta and Hirudinea.

Example. *Acanthobdella peledina.*

Order 2. Rhyncobdellida

1. Blood-sucking ectoparasites of snail, fishes, frogs, tortoise and turtles. May be freshwater or marine.
2. Each typical segment consists of 3, 6, or 12 annuli.

3. Proboscis is protrusible and without jaws and teeth.
4. Setae are absent.
5. Both anterior and posterior suckers are present.
6. Coelom is reduced to sinuses but botryoidal tissue absent.
7. Circulatory system is separated from the coelomic sinuses. Blood is colourless.
8. Eyes absent.

Examples. *Pontobdella, Glossiphonia, Helobdella, Placobdella, Piscicola.*

Order 3. Gnathobdellida (or Arnchynchobdellida)

1. Most are ectoparasites on insect larva, worms, snails, amphibia, cattles and human beings; some are predators.
2. Freshwater and terrestrial forms.
3. Each typical body segment consists of five annuli.
4. Body with anterior and posterior suckers.
5. Proboscis is absent.
6. They contain three jaws with strong cuticular teeth in their buccal cavity.
7. They contain five pairs of eyes. Host finding occurs by means of olfactory sense (smell).
8. Setae are absent.
9. Coelom is reduced to sinuses and body cavity filled with botryoidal tissue.
10. Blood is red coloured.

Examples. *Hirudo, Hirudinaria, Haemopis, Macrobdella, Xerobdella.*

Order 4. Pharyngobdellida

1. Freshwater or amphibious carnivorous (predaceous) leeches.
2. Proboscis nonprotrusible. Jaws absent but they may contain one or two piercing stylets.

Examples. *Erypobdella, Dina.*

33.4 TYPE STUDIES

1. *APHRODITE*

Classification

Phylum	–	**Annelida**
Class	–	**Polychaeta**
Subclass	–	**Errantia**
Genus	–	***Aphrodite*** (New spelling *Aphrodita*; see **Ruppert** and **Barnes**, 1994)

Habit and Habitat

Aphrodite, is a type of **scale worm** which is commonly called '**sea-mouse**'. It is a small marine burrowing form found in mud in deep waters. It occurs in Europe and on the north west and north east coasts of United States. *Aphrodite* lives almost half buried in the mud; its posterior end of body is exposed at the surface. It occasionally creeps on the sea floor.

Structure

Aphrodite is about 12 cm in length. The body is short, broad and oval in shape and dorso-ventrally flattened. Body contains 30 to 35 segments. Its dorsal surface is convex and covered with

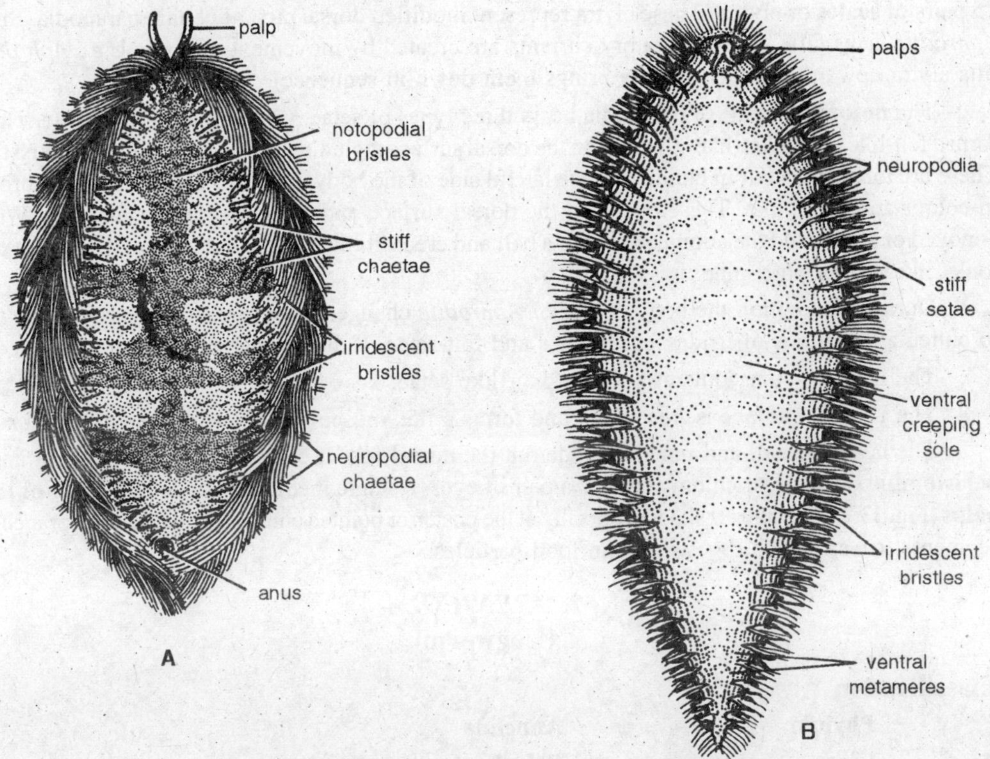

Fig. 33.1. *Aphrodite.* A—Dorsal view; B—Ventral view.

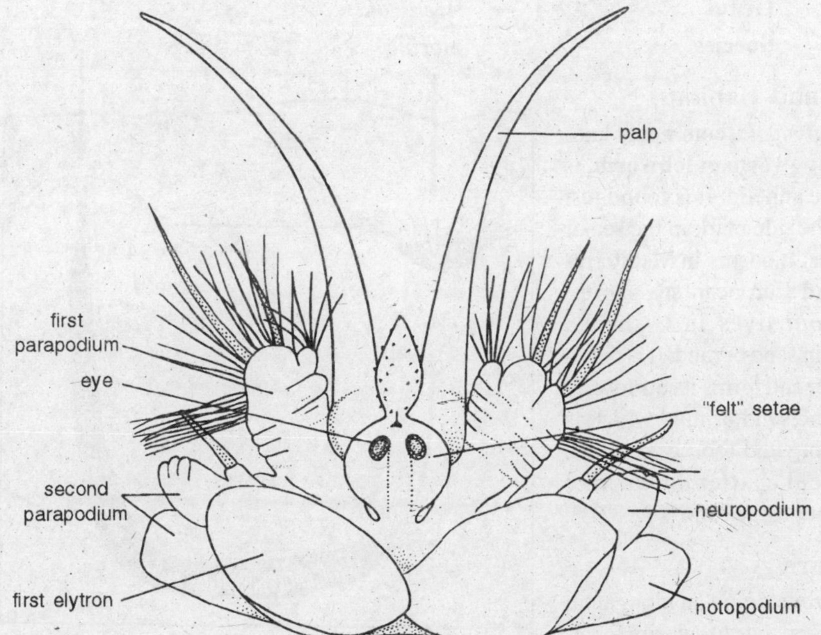

Fig. 33.2. *Aphrodite.* Anterior end, including the first pair of dorsal scales or elytra.

15 pairs of **scales** or **elytra**. These elytra represent modified dorsal cirri of certain parapodia. Since *Aphrodite* lacks cilia, in it respiratory currents are created by movement of elytra (*i.e.*, *Aphrodite* tilts elytra upward and then rapidly brings them down in sequence).

The **notopodial** part of parapodia bears three types of setae: (*i*) **Long hair-like setae.** These form a felt-like or blanket-like covering on the dorsal surface including elytra. (*ii*) **Irridescent bristles.** These are long and occur in bundles on the lateral side of the body. (***iii***) **Stiff setae.** These are brown in colour and spine-like. These occur on the dorsal surface and are protective in function. When annoyed or frightened the animal rolls into a ball and erects its stiff setae. In this condition the worm looks like a mini porcupine.

During locomotion the body colour of *Aphrodite* changes from golden to peacock blue due to cuticular striations of irridescent bristles and felt-setae. It also shines brilliantly.

The neuropodium contains soft and hair-like setae.

The **ventral surface** is free of felt and forms a flat, muscular, segmented and creeping **sole**.

The head is small and remains hidden in the dorsal felt. It consists of a **prostomium** and a **peristomium**. The prostomium contains a pair of eyes, a single median **tentacle** and a pair of long **palps** (Fig. 33.2). The anus is situated dorsally at the posterior pointed end. **Intestine** has long branching segmental **caeca** which digests minute food particles.

2. *ARENICOLA* (Lugworm)

Classification

Phylum	–	**Annelida**
Class	–	**Polychaeta**
Subclass	–	**Sedentaria**
Genus	–	***Arenicola***
Species	–	***marina***

Habit and Habitat

Arenicola, commonly known as **Lugworm** or **lobworm**, is a marine annelid. It is found just below the tide mark in the sea or estuaries. It occurs in Mediterranean and European sea shores. *Arenicola* lives in L-shaped burrow in sand or mud. It is deposit feeder and forms its burrow by eating the sand or mud which is rich in organic food or detritus. Its faecal matter looks like castings of earthworm (Fig. 33.3).

Structure

Arenicola is an elongated, cylindrical, stout, vermiform, segmented and brownish-green annelid. It measures 20 to 25 cm in length. Its body is divisible into

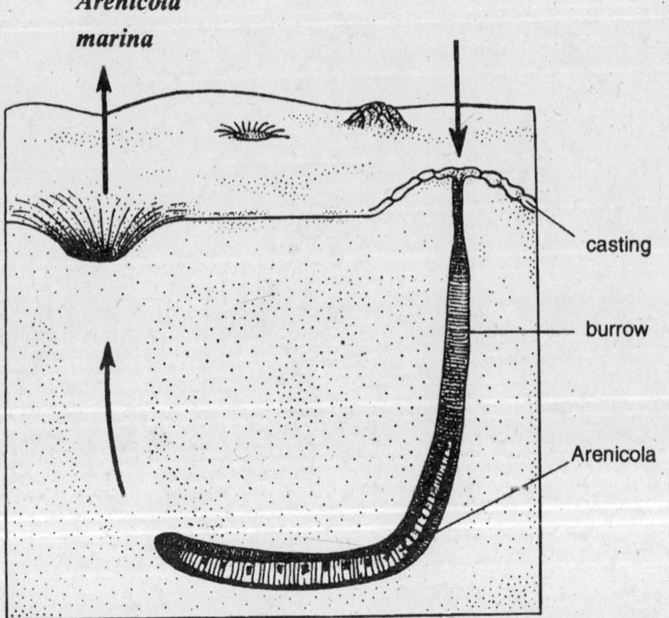

Fig. 33.3. *Arenicola.* L-shaped burrow and position of the annelid inside the burrow.

three regions (Fig. 33.4):

1. Anterior region or prebranchial region. It comprises small trilobed **prostomium** with no eyes or tentacles, a **peristomium,** an achaetous (= without setae) segment and six segments bearing setae and parapodia. The parapodia are rudimentary.

2. Middle region or branchial region. It contains thirteen segments. Each segment contains rudimentary parapodia and branched **gills** above the parapodia (Fig. 33.5). The gill are modified dorsal cirri of parapodia.

3. Posterior region or postbranchial region. It is much thinner body part than others and contains variable number of segments. It is without gills, parapodia or setae.

Each body segment bears **annuli. Mouth** lies ventral to the prostomium. The buccal region and pharynx protrude as **proboscis** which is covered by chitinized papillae (*Arenicola* uses the eversible pharynx to punch and anchor in the sediment. It pick up particles on the adhesive glandular surface). Jaws are absent. The **anus** opens through the last segment. There are six pairs of nephridia in segments 4 to 9.

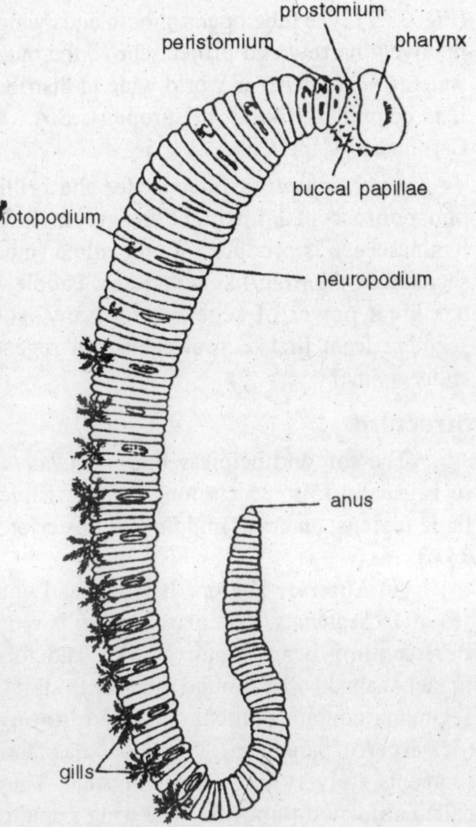

Fig. 33.4. *Arenicola*. Lateral view.

Arenicola contains two types of sensory organs— a pair of **nuchal organs** and a pair of **statocysts**. Nuchal organs consist of a pair of ciliated pits or slits, they are often eversible and situated in head region. The **statocysts** or **gravity receptors** of *Arenicola* are situated within the body wall of the head. Each statocyst is open pit or canal that opens to exterior (at lateral body surface). It contains various foreign objects such as spicules, diatom shells and sand or qurts grains all covered with chitinoid material. *Arenicola* always burrows head downward and if an aquarium containing a worm is tilted 90 degrees, the worm makes a compensating 90 degree turn in burrowing. If statocysts are destroyed this compensating ability is lost.

3. *CHAETOPTERUS*

Classification

Phylum	–	**Annelida**
Class	–	**Polychaeta**
Order	–	**Sedentaria**
Type	–	***Chaetopterus***

Habit and Habitat

Chaetopterus is commonly called **paddle worm**. It is a marine tubicolous worm. It lives permanently in U-shaped parchment-like

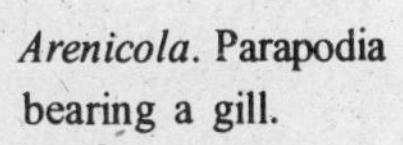

Fig. 33.5. *Arenicola*. Parapodia bearing a gill.

tube formed in the sand or mud in the intertidal zone (Fig. 33.7). The tube opens at both ends which are relatively narrow and project above the mud into water. *Chaetopterus* is world wide in distribution. It is commonly found in Europe, U.S.A., North Carolina to Cape Cod.

Chaetopterus is filter-feeder and brilliantly phosphorescent emitting blue-green light. Its luminescence is produced by the slime (mucous) secreted by the wing-like notopodia. Paddle worm has great power of regeneration; any isolated segment from first to fourteenth can regenerate entire animal.

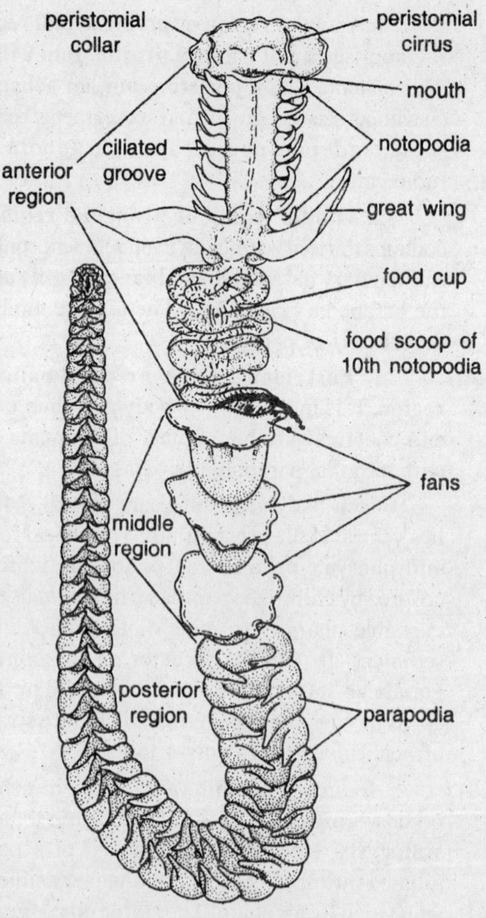

Fig. 33.6. *Chaetopterus*. Structure.

Structure

The soft and helpless body of *Chaetopterus* is usually 15 to 35 cm long. It is divisible into three regions; anterior, middle and posterior (Fig. 33.6).

(1) Anterior region. It is flat and comprises of 16 segments. The **prostomium** is reduced. **Peristomium** bears a pair of cirri and forms a funnel-shaped collar around the **mouth**. First nine segments contain conical, lobe-like **notopodia** (dorsal part of parapodia). The notopodia of the 10th segments are very large and wing-like. They are called **aliform notopo-dia.** The wing's epithelium is ciliated and con-tains numerous mucous glands. In the next three segments parapodia are reduced. Their **neuropo-dia** (ventral part of parapodia) form **anchors** or **suckers** which attach the worm within the tube. A **ciliated groove** runs on the mid-dorsal surface from the mouth to a ciliated **food-cup** or **dorsal cupule.**

(2) Middle region. It comprises of three segments only. Each segment bears a pair of biramous parapodia, each consisting of a very small **neuropodium** and a very large **notopodium.** The notopodia of both sides of each segment fuse to form semicircular **fans** that project like piston rings against the cylindrical wall of the tube.

(3) Posterior region. It consists of several segments and each segment carries typical parapodia.

Box 33.6 Feeding Mechanism of *Chaetopterus*

The beating of the fans produces a current of water that enters the chimney of the U-shaped tube near the anterior end of the form, flows through the tube, and then flows out of the opposite chimney (Fig. 33.7).

The paired, aliform (wing-like) notopodia are stretched out around the walls of the tube and a sheet of mucus is secreted between them. The mucus film is continuously secreted from each notopodium, so the sheet assumes the shape of a bag. The posterior end of the bag is grasped by a ciliated cup on the middorsal side of the worm a short distance behind the aliform notopodia. Water is brought into the tube by the rhythmic beating of the fan parapodia, passes through the mucous bag which strains suspended detritus and plankton.

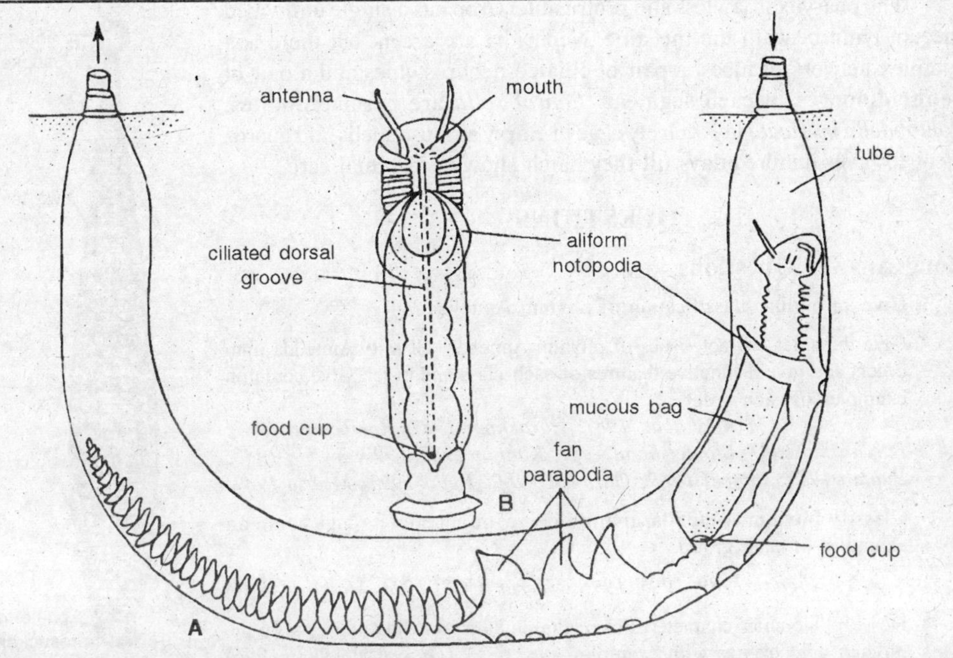

Fig. 33.7. *Chaetopterus*. Feeding mechanism. A—Anterior part of the body (dorsal view); B—Worm in tube (lateral view).

Large objects brought into the tube by the water current are detected by peristomial cilia; the aliform notopodia then are pulled back to let the large objects pass by. The food-laden mucous bag is continuously being rolled up into a ball by the dorsal cupule. When the ball reaches a certain size, the bag is cut loose from the notopodia and rolled up with the ball. The cupule then projects forward and deposits the mucous food ball into a ciliated middorsal groove which extends to the anterior of the worm and the ball is carried to the mouth. A 18 to 24 cm long specimen of *Chaetopterus* may produce mucous film for the bag at the rate approximately 1 mm/sec, with food balls averaging 3 mm in diameter.

4. *PONTOBDELLA*

Classification

Phylum	–	**Annelida**
Class	–	**Hirudinea**
Order	–	**Rhynchobdella**
Genus	–	***Pontobdella***

Habit and Habitat

Pontobdella is a marine fish leech. It is ectoparasitic on sharks, skates and rays. *Pontobdella* is found in Europe and USA.

Structure

The body is elongated, cylindrical and club-shaped measuring 20 cm in length. It consists of an anterior and a posterior **sucker.** The anterior sucker is saucer-shaped. It can be firmly applied to the tissue of the host. The posterior sucker is deep and cup-shaped. Body surface is rough and leathery. It is covered with many green tubercles or warts. Metameres are not very distinct. Eyes and gills are absent.

The pharynx is jawless and protrusible. Crop has a single undivided caecum lying beneath the intestine. Nephridia are absent but there is a complex network of tubes, a pair of ciliated nephrostomes and a pair of **nephridiopores** in each segment. *Pontobdella* are hermaphrodites. *Pontobdella muricata* lays velvety eggs in empty mollusc shells and guards them for over hundred days till they hatch showing parental care.

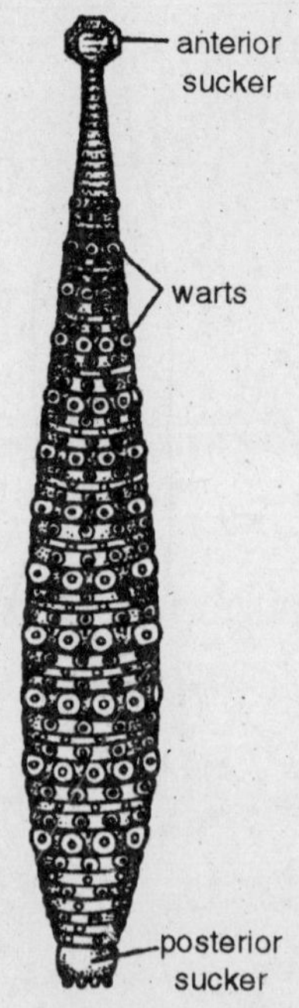

Fig. 33.8. *Pontobdella.* External structure.

QUESTIONS

Long Answer Questions

1. Give an outline classification of phylum Annelida.
2. Give the chief characteristics of phylum Annelida. Classify Annelida upto orders and give distinctive features of each class and order. Give common examples of each order.
   (*Allahabad 1987; Gorakhpur 1987; Rohilkhand 1987; Purvanchal 1987; Bundelkhand 1987; Kanpur 1987, 1990; Awadh 1988; Punjab 1989; Ranchi 1989; Garhwal 1989; Delhi 1991; Meerut 1989*)
3. Classify phylum Annelida giving diagnostic characters and common examples of each group.
   (*Magadh 1989, 1993, 1996; Meerut (CCS) 1994; Madurai 1994*)
4. Describe the chief characteristics of the phylum Annelida and classify the phylum upto classes with examples. (*Guwahati 1993, 1996*)
5. Write down the distinguishing characters of Polychaeta. (*Rewa 1989*)

Short Answer Questions

1. Assign the following animals to their systematic position giving their distinguishing features.
   (*i*) Aphrodite
   (*Allahabad 1988; Agra 1989; Purvanchal 1989, 1992; Kanpur 1989; Meerut (CCS) 1990, 1996; Awadh 1990, 1993; Kumaun 1990, 1992; Lucknow 1991; Bundelkhand 1992; Jiwaji 1993*)
   (*ii*) *Arenicola* (lug-worm)
   (*Bundelkhand 1988; Agra 1989; Awadh 1989; Raipur 1990; Ravishankar 1993*)
   (*iii*) *Chaetopterus* (*Awadh 1988; Agra 1989; Kumaun 1989; Purvanchal 1989; Srivenkat 1989; Garhwal 1990; Raipur 1991; Indore 1994*).
   (*iv*) *Pontobdella* (*Meerut (CCS) 1991; Agra 1991; Punjab 1991, 1992*)

Very Short Answer Questions

1. Name the subclass to which *Polygordius* belongs.
   Ans. Archinnelida.
2. From which germ layer is the peritoneum derived.
   Ans. Mesoderm.
3. What is pygidium?
   Ans. Terminal part of body of an annelid bearing anus.
4. Is prostominum a metamere?
   Ans. No.
5. Where are new segments formed in earthworm?
   Ans. Just infront of pigidium.
6. What is schizocoel?
   Ans. Coelom formed by split in mesodermal bands.
7. In earthworm, haemoglobin is dissolved in the plasma. Is it so?
   Ans. Yes.

8. Parapodia are absent in annelids of classes.
Ans. Oligochaeta and Hirudinea.

Fill in the Blanks

1. Segmentation in Annelida is called ....

2. The most distinguishing characteristic of the phylum Annelida is ....

3. The cuticle in Annelida .... rather than chitinoid.

4. Body cavity in annelids is ....

5. Body of an annelid may be divided into three regions: .... , trunk and ....

6. Annelids have .... circulatory system.

7. Excretion in Annelida occurs by ....

8. Larva of annelids is ....

9. Coelom of annelids is lined by peritoneum of .... origin.

10. Currently phylum Annelida is divided into .... classes.

11. Skin of annelids serves for ....

12. In annelids, respiratory pigment is present in the ....

13. In metanephridia the nephridium opens into coelom by ciliated funnel-shaped opening, called ....

14. Fans of *Chaetopterus* are modified ....

15. .... is a tubicolous polychaeta.

16. Giant fiber system is absent in ....

17. In *Arenicola*, gills are present in the .... region of body.

18. Clitellum is absent in annelids of class ....

Multiple Choice Questions

1. Members of phylum Annelida show following characters except one which is wrong. Mark it
(*a*) segmented and triploblastic (*b*) excretory organs and nephridia
(*c*) a true coelom is absent (*d*) blood vascular system is closed type

2. Segmentation in Annelida is
(*a*) homonomous (*b*) heteronomous
(*c*) both (a) and (b) (*d*) neither (a) nor (b)

3. Segmentation in Annelida affects
(*a*) external features only (*b*) internal organs only
(*c*) both (a) and (b) (*d*) none of these

4. Body cavity of annelids is
(*a*) coelom (*b*) pseudocoel
(*c*) haemocoel (*d*) none of these

5. The coelom in Annelida is
(*a*) pseudocoelic (*b*) enterocoelic
(*c*) schizocoelic (*d*) acoelous

6. *Nereis* belongs to class .... of Annelida
(*a*) Oligochaeta (*b*) Polychaeta
(*c*) Hirudinea (*d*) None of these

7. *Aphrodite* is popularly called
(*a*) scale worm (*b*) palolo worm
(*c*) lug worm (*d*) sea mouse

8. The excretory organs of annelids are
(*a*) metanephridia (*b*) malpighian tubules
(*c*) flame cells (*d*) renette cells

9. Parapodia are locomotory organs in
(*a*) Oligochaeta (*b*) Chilopoda
(*c*) Polychaeta (*d*) Diplopoda

10. Larvae form of polychaeta is known as
(*a*) tornaria (*b*) glochidium
(*c*) trochophore (*d*) bipinnaria

11. Zoological name of paddle worm is
(*a*) *Arenicola* (*b*) *Chaetopterus*
(*c*) *Polynoe* (*d*) *Pheretima*

12. *Tubifex* belongs to class .... of Annelida.
(*a*) Polychaeta (*b*) Oligochaeta
(*c*) Echiurida (*d*) Hirudinea

13. Branchiae are respiratory organs in
(*a*) *Arenicola* (*b*) *Nereis*
(*c*) *Aphrodite* (*d*) *Polynoe*

14. Locomotory structures of earthworms are
(*a*) bristles (*b*) cilia
(*c*) parapodia (*d*) setae

15. *Arenicona* is commonly known as
(*a*) lugworm (*b*) paddle worm
(*c*) sea pen (*d*) palalo worm

16. True coelom is a space that lies between
(*a*) body wall and alimentary canal (*b*) ectoderm and endoderm
(*c*) two layers of endoderm (*d*) visceral and parietal mesoderm

17. Segmentation in archiannelids is
(*a*) external only (*b*) internal only
(*c*) both external and internal (*d*) not found

18. Special cells that help in excretion in earthworm are
(*a*) chloragen cell (*b*) chlorocytes
(*c*) choanocytes (*d*) flame cells

ANSWERS

Fill in the Blanks

1. Metamerism	2. Metamerism	3. Collagenous
4. True coelom	5. Prostomium, Pygidium	6. Closed
7. Nephridia	8. Trchophore	9. Mesodermal
10. Three	11. Respiration	12. Plasma
13. Nephrostome	14. Parapodia	15. *Sabella*
16. *Chaetopterus*	17. Branchial	18. Polychaeta.

Multiple Choice Questions

1. (*c*)	2. (*c*)	3. (*c*)	4. (*a*)	5. (*c*)	6. (*b*)
7. (*d*)	8. (*a*)	9. (*c*)	10. (*c*)	11. (*b*)	12. (*b*)
13. (*a*)	14. (*d*)	15. (*a*)	16. (*d*)	17. (*b*)	18. (*a*).

34

Nereis (Neanthes) (Phylum Annelida)

34.1 CLASSIFICATION

Phylum	—	**Annelida**
Class	—	**Polychaeta**
Subclass	—	**Errantia**
Genus	—	***Nereis* (*Neanthes*)**

34.2 HABIT AND HABITAT

Nereis is commonly known as **sandworm, rag worm** or **clam worm.** It is marine and found on the sea-shore in the shallow water (*i.e.*, tide levels) in rock crevices hidden under the stones or sea weeds. Mostly it lives in temporary U-shaped burrows in sand. The mucus secreted by its body wall binds the sand particles together. Jaws are used for preparing burrows.

Nereis is **carnivorous** and **predaceous**. It feeds on small insects, molluscs and worms. It is very active during night and passive during the day, *i.e.*, *Nereis* is **nocturnal**. During night, *Nereis* come out from their hidden places and start crawling over the sand. When breeding period approaches, *Nereis* leaves the burrow and comes at the water surface to lead a pelagic life then it is called *Heteronereis*.

Distribution. *Nereis* is cosmopolitan in distribution and found in coastal waters of Pacific and North Atlantic oceans, Europe and USA.

34.3 EXTERNAL STRUCTURE

Shape and size. The body of *Nereis* is elongated, cylindrical bristle worm. It is bilaterally symmetrical, metamerically segmented and dorso-ventrally flattened (Fig. 34.1). It has convex dorsal surface and flat ventral surface. The size of body may range from 30 to 40 cm in length and 2 to 6 mm in width.

Colouration. Different species of *Nereis* have different colours. For example, *N. pelagica* is reddish brown, *N. cultrifera* is greenish, *N. limnicola* is brownish in colour while *N. virens* is bluish-green in colour, having orange or red tinge on the appendages.

Segmentation. Body of *Nereis* is metamerically segmented. It is divided upto 200 or more **segments** or **metameres** which are marked externally by narrow grooves and internally by septa. All segments, apart from those most anterior and posterior, are very similar to each other.

Division of body. The body of *Nereis* is divisible into three distinct regions, *i.e.*, head, trunk and pygidium.

1. Head. The anterior end of body has a distinct **head** which shows a great degree of cephalization (Fig. 34.2). It consists of two main parts: (*i*) **prostomium** and (*ii*) **peristomium**.

(i) Prostomium. (Gr., *pro* = before; *stoma* = mouth). It is anterior triangular fleshy outgrowth. It is situated mid-dorsally in front of mouth (Fig. 34.2). Prostomium is not the true segment of the body as it is derived from preoral lobe of larva. Prostomium carries two pairs of simple, black and rounded eyes; a pair of short, cylindrical and tactile **prostomial tentacles**; a pair of stout (fleshy) longer and tactile **prostomial palps** and a pair of **ciliary pits** or **nuchal organs**.

(ii) Peristomium. (Gr., *peri* = around; *stoma* = mouth). It is posterior ring-like portion of head. Ventrally it bears a transverse **mouth** (Fig. 34.3). On each side, peristomium bears four pairs of long slender **tactile peristomial cirri**, two are dorsal and two are ventral.

Peristomium represents fused first and second body segments. Peristomial cirri correspond to the neuropodial and notopodial cirri of parapodia of rest of the body.

2. Trunk. Posterior to the head, the rest of the body is called **trunk.** It is formed of 80 to 200 identical **somites** or **metameres** depending on species. Each metamere is broader than long and bears a pair of **parapodia** on its lateral sides.

3. Pygidium. The posterior most segment is somewhat rounded and is called **tail somite, anal segment** or **pygidium**. It bears a pair of elongated **anal cirri** (Fig. 34.4). **Anus** is terminal.

Parapodium. (Gr., *para* = side; *podus* = foot). Each body segment except the peristomium and pygidium bears on either side a flat, fleshy and hollow flap-like **parapodium**. It is **biramous** and consists of an upper **notopodium** and lower **neuropodium** (Fig. 34.5). Dark coloured chitinous rods called **acicula** support both of these processes, and two tufts of **setae** or **chaetae** emanate from each aciculum forming fan-like bodies. Two additional outgrowths of the parapodia are noticeable: a dorsal and a ventral **cirrus**.

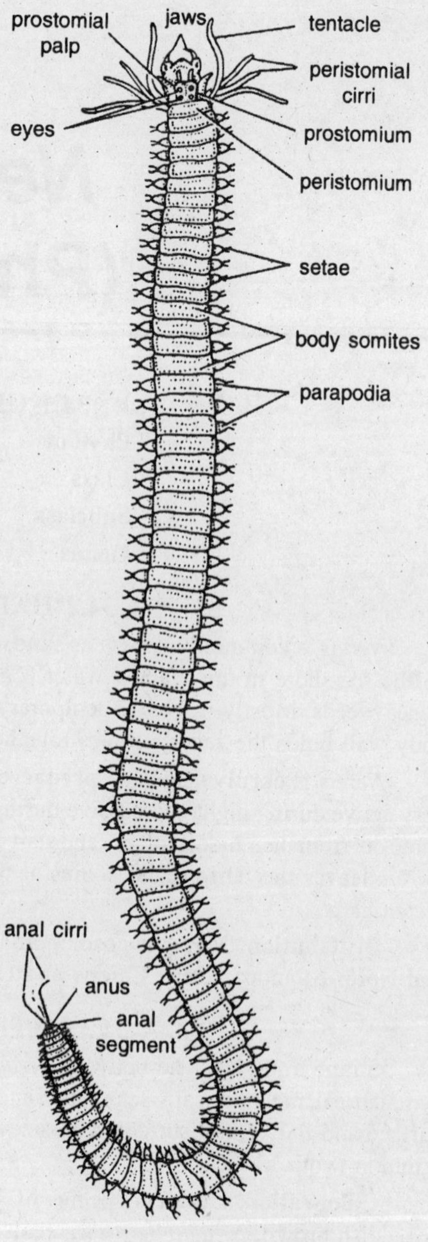

Fig. 34.1. *Nereis.* External features.

The parapodia are primarily the organs of **locomotion** used both in creeping and in swimming. Since parapodia are richly vascularized, they also serve the function of **respiration.**

Nuchal organ. These are present in the form of a pair of ciliated grooves one on either side of the prostomium. These are **chemoreceptors.** They also act as **olfactory sense organs.**

Nephridiopores. These are external openings of nephridia. They are present singly near the base of **ventral cirri** of parapodia.

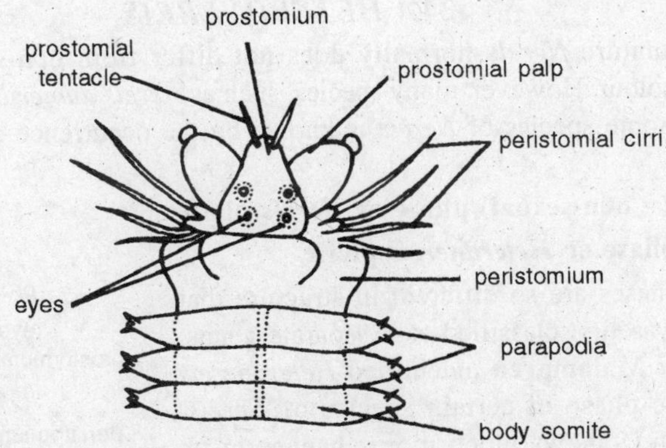

Fig. 34.2. *Nereis*. Head in dorsal view.

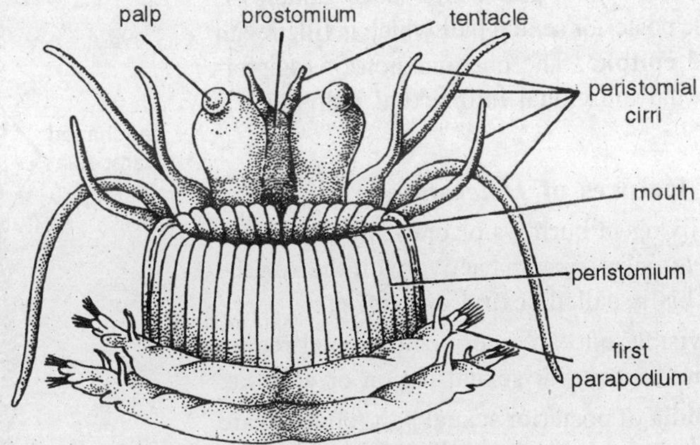

Fig. 34.3. *Nereis*. Head in ventral view.

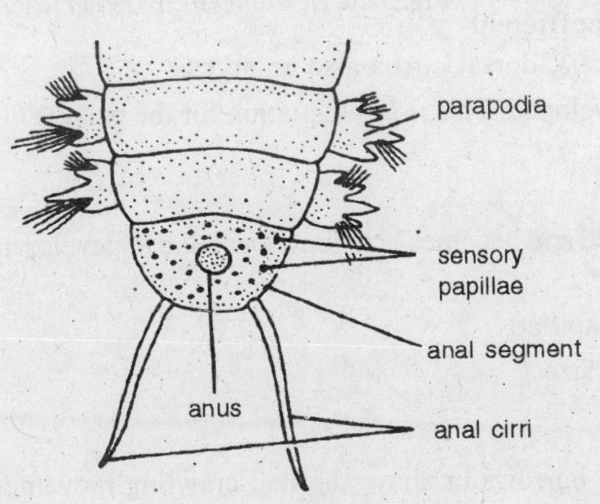

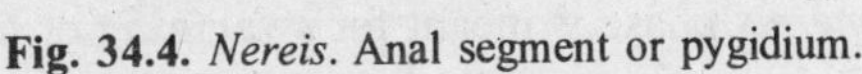

Fig. 34.4. *Nereis*. Anal segment or pygidium.

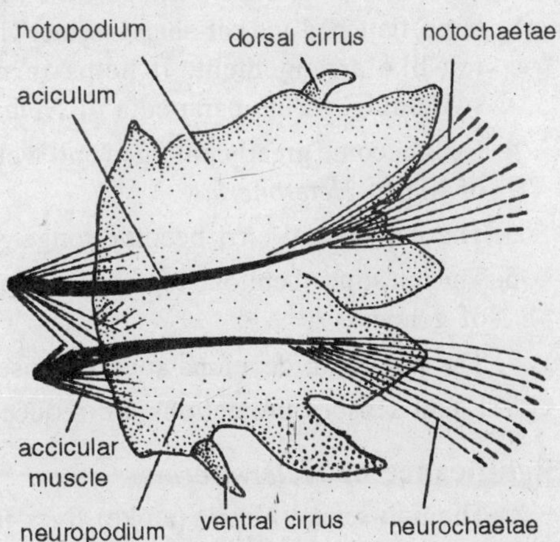

Fig. 34.5. *Nereis*. Parapodium.

34.4 *HETERONEREIS*

A sexually mature *Nereis* normally does not differ from non-sexual worm in external characters except colour. However, many species, such as *Nereis dumerilii* exhibit **polymorphism.** Polymorphism in some species of *Nereis* is known by the occurrence of two alternate forms or phases:

(*i*) Asexual (= non-sexual) phase or *Nereis* phase

(*ii*) Sexual phase or *Heteronereis* phase

These two phases are so different in structure that the sexual phase was first classified as a separate genus, *Heteronereis.* Later **Malamgren** indentified *Heteronereis* as sexually mature phase of certain species of *Nereis*. Sexually male *Nereis* shows much more changes in its body than the female. In the *Heteronereis*, gonads (*i.e.*, testes and ovaries) develop only in the posterior part of body. Its anterior non-sexual part which lacks gametes is called **atoke** while posterior sexual part which is filled with gametes is called **epitoke**. The phenomenon of transformation of nonsexual individual into sexual individual is called **epitoky**.

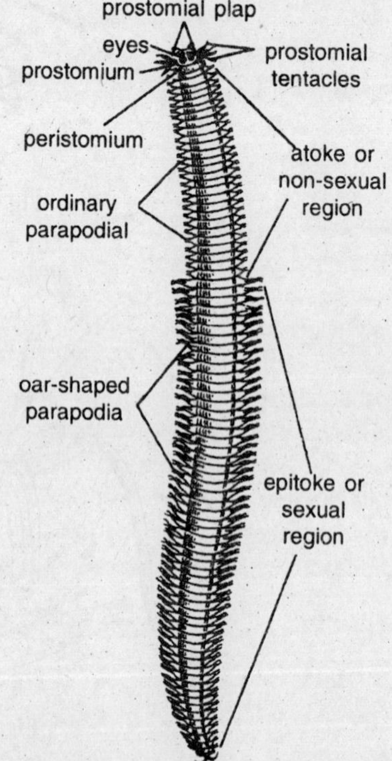

Fig. 34.6. *Heteronereis.* External features.

Characteristic features of *Heteronereis*

1. Instead of living in burrows or crawling on the sea bottom, *Heteronereis* swims actively on the surface of water. This is called **sexual swarming**.
2. Body is divisible into an anterior nonsexual region or **atoke** and a posterior sexual region or **epitoke**.
3. The **parapodia** of posterior sexual part (epitoke) are large, highly vascularized and additional leaf-like outgrowths for more rapid respiration (Fig. 34.7). The normal setae of these parapodia are replaced by new, large, flattened and oar-shaped setae. These setae show fan-like arrangement. It helps in more efficient swimming. In the parapodia of *Heteronereis*, dorsal cirrus also enlarges.
4. Eyes become greatly enlarged and well developed. This is an adaptation for the pelagic mode of life of *Heteronereis*.
5. The peristomial cirri become longer.
6. The intestine of epitoke region is compressed and becomes non-functional due to development of goands.
7. The pygidium develops special sensory papillae.
8. Usual sensory projections are reduced in size.

Significance of *Heteronereis*

The non-sexual *Nereis* (atoke) lives in the burrows or shows limited crawling movements. In contrast, the *Heteronereis* swims actively and discharges gametes to far off places in the sea and thus helps in the dispersal of species. *Heteronereis* phase is meant for **swarming** or mass **nuptial dance**.

Table 34.1 Differences between *Nereis* and *Heteronereis*.

Nereis	*Heteronereis*
1. Nereis is a non-sexual sluggish creature living in burrow or creeping at the bottom of sea.	1. *Heteronereis* is an active swimmer on the surface of water (pelagic).
2. Body is not differentiated into atoke and epitoke.	2. Body is divisible into anterior asexual **atoke** and posterior sexual **epitoke** regions.
3. The parapodia are thin, flap-like and less vascular.	3. The parapodia of posterior or sexual region are large, highly vascular and develop leaf-like outgrowths.
4. The setae on the parapodia are normal.	4. The setae are large, flattened and oar-shaped. These are arranged in fan-like shape.
5. Eyes are normal sized and the worm is less sensitive to light.	5. Eyes are large and conspicuous; worm is very sensitive to light.
6. Sensory projections are well developed.	6. Sensory projections become shrunken and anal segment develops special sensory papillae.
7. Dorsal cirri are less curved.	7. Dorsal cirri become shrunken.
8. The intestine is well formed.	8. The intestine is compressed and become functionless due to development of gonads.

Comparison of Parapodia of *Nereis* and *Heteronereis*

1. Parapodium of *Nereis*. Each trunk segment of *Nereis* bears laterally a pair of dorso-ventrally flattened, hollow flaps. For these fleshy body outgrowths **Huxley** used the term **parapodia.** The cavity of a parapodium is continuous with the body **coelom**. Each parapodium is a **biramous** structure consisting of two parts or lobes: an upper or dorsal lobe, the **notopodium** and a lower or ventral lobe the **neuropodium**. Each of these lobes is further divided into two smaller lobes, an upper and a lower. The dorsal margin of notopodium is produced into a short, cylindrical, tactile appendage, the **dorsal cirrus**. Ventral margin of neuropodium too produces such a tentacular process, called **ventral cirrus**. Of these two cirri, the dorsal one is much larger than the ventral (Fig. 34.5).

A bundle of bristle-like **setae** or **chae-tae** fans out from each lobe of parapodium (*i.e.*, notopodium and neuropodium). In each case, these form a bundle and remain embedded in a **setal sac** or **setigerous sac**. The setae can be turned in various directions by strands of muscular fibres.

In the middle of each bundle of setae and deeply embedded in the parapodium is stout, straight, thick and dark coloured chitinous rod the **aciculum**. It serves as endoskeleton of parapodium and serves to support and provide attachment to the setal muscles.

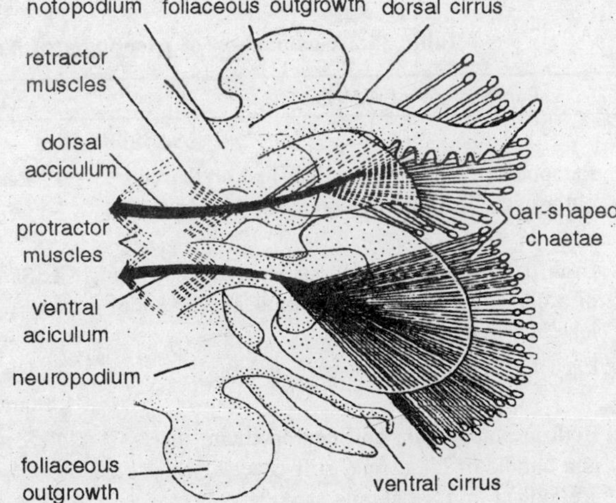

Fig. 34.7. *Heteronereis*. Parapodium.

Setae or chaetae. The setae are fine, stiff and chitinous bristles which help in locomotion by acting as minute paddles; these being protective are used for offense and defence. Setae also provide hold on the smooth surface of the burrow. Setae are lodged in setal or **satigerous sac** and each seta originates from a single large **formative cell** present at the base of setigerous sac. The setae are continuously produced by the setal sac to replace the worn out old setae. Setae can be retracted and protruded out by special sets of muscles of parapodia.

Each seta consists of two movable articulated parts—the basal **shaft** or **stalk** and the distal **blade** which can be retracted into the shaft (Fig. 34.8). The setae of *Nereis* are of three types:

(*i*) **Long blade setae** with a small slender shaft and the terminal long blade which is nearly straight, pointed and serrated at one edge.

(*ii*) **Typical setae** which have a large stout shaft, short and slightly hooked blade, and

(*iii*) **Oar-shaped setae** have a oar-shaped blade. These are found in *Heteronereis*.

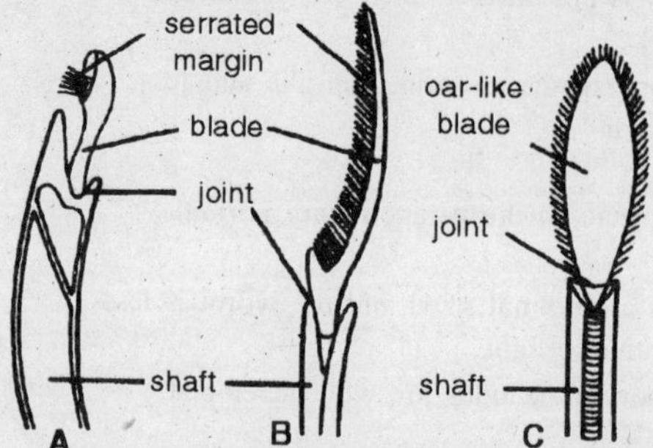

Fig. 34.8. *Nereis*. Various types of setae. A—Typical setae; B—Long bladed setae; C—Oar-shaped setae of *Heteronereis*.

The parapodia are highly muscular, well vascularized and glandular structures. The parapodia present in the middle of body are large and gradually become smaller towards two ends.

2. Parapodia of *Heteronereis*. The structure of parapodia of *Nereis* and *Heteronereis* is basically identical but parapodia of *Heteronereis* shows the following peculiar features (Fig. 34.7)

1. These are bigger and more fleshy.
2. These have foliacious outgrowths.
3. The cirri are highly enlarged and vascular.
4. The setae are oar-like and arranged in fan-shaped manner.
5. Original parapodial muscles are replaced by new muscles.

The enlarged and modified parapodia enable *Heteronereis* to swim fast and lead a pelagic life.

Table 34.2 Comparison of parapodia of *Nereis* and *Heteronereis*.

Parapodium of *Nereis*	Parapodium of *Heteronereis*
A. Similarities	
1. Parapodia are flat, fleshy, leaf-like or flap-like outgrowths from the lateral side of body (Fig. 34.5).	1. Parapodia are large, fleshy, foliaceous outgrowths (Fig. 34.7).
2. Each parapodium is biramous structure consisting of a dorsal lobe, the **notopodium** and a ventral lobe, the **neuropodium.**	2. Same.
3. Each lobe consists at its base a cirrus.	3. It also contains a dorsal cirrus and a ventral cirrus.
4. Both in notopodium and neuropodium, there is a bundle of long, fine, stiff chaetae which are lodged in a setigerous sac.	4. Same.
5. The action of setae is controlled by special sets of muscles.	5. Same.

Parapodium of *Nereis*	Parapodium of *Heteronereis*
6. The centre of each bundle of setae is occupied by a thick chitinous rod.	6. Same.
B. Differences	
1. The parapodia are small, simple, non-vascular and without any outgrowth.	1. Parapodia are greatly enlarged in the posterior part of body. These become foliacious, highly vascular and develop additional leaf-like fleshy outgrowths.
2. The setae are simple.	2. The setae are flattened and oar-shaped.
3. Setae are arranged in the form of bundles.	3. Setae are arranged in a fan-like manner.
4. Less adapted for swimming.	4. Better adapted for swimming.

34.5 DIGESTIVE SYSTEM

Alimentary canal of *Nereis* is a straight tube having two opening to the exterior, an anterior **mouth** and a posterior **anus**. It is suspended in the body cavity or coelom by the dorsal mesentery and intersegmental septa. It is differentiated into following three regions (Fig. 34.9).

1. Foregut or stomodaeum which comprises buccal cavity and pharynx;

2. Midgut or meseneteron which comprises oesophagus and stomach-intestine;

3. Hindgut or proctodaeum which comprises only rectum.

Both foregut and hindgut are internally lined by **ectoderm** and **cuticle**. Midgut is lined by endoderm.

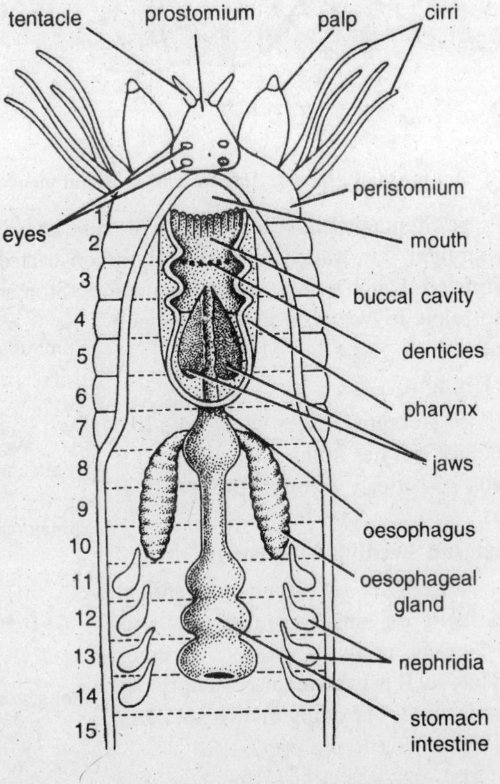

Fig. 34.9. *Nereis*. Anterior part of alimentary canal.

A. Foregut

1. Mouth. It is a transverse slit lying ventral to the prostomium. It is bordered laterally and ventrally by peristomium. The mouth opens posteriorly into buccal cavity.

2. Buccal cavity and pharynx. The buccal cavity is short but wide chamber situated inside the peristomium. The buccal cavity leads into a muscular protrusible **pharynx** which extends upto fourth segment. The buccal cavity and pharynx are bound together by a muscular sheath and internally they are lined with thick cuticle. The cuticle of buccal cavity is thickened at various points into dark brown **denticles** or **paragnaths**. In the posterior part of pharynx are found a pair of large, powerful, dark, movable, chitinous and laterally placed jaws with serrated edge. Each jaw has a rounded hollow base embedded in the muscles and an anterior pointed incurved apex, its inner edge is provided with **teeth**. Both pharynx and buccal cavity collectively form the feeding organs of *Nereis*.

The whole of buccal and pharyngeal region can be turned completely in-side out (Fig. 34.10). This is done by the help of two sets of muscles: **protractor muscles** which can evert out the pharynx and open the jaws as **introvert** or **proboscis;** and **retractor muscles** which can retract proboscis and cause closing of jaws (Fig. 34.10 and Fig. 34.11). Buccal cavity is everted due to pressure of coelomic fluid.

B. Mesenteron

3. Oesophagus. The pharynx opens posteriorly into a narrow tube, called **oesophagus** which runs upto ninth segment. A pair of large, unbranched, laterally placed glandular pouches, the **oesophageal caeca** open into oesophagus at the anterior end. The oesophagus is followed by stomach-intestine.

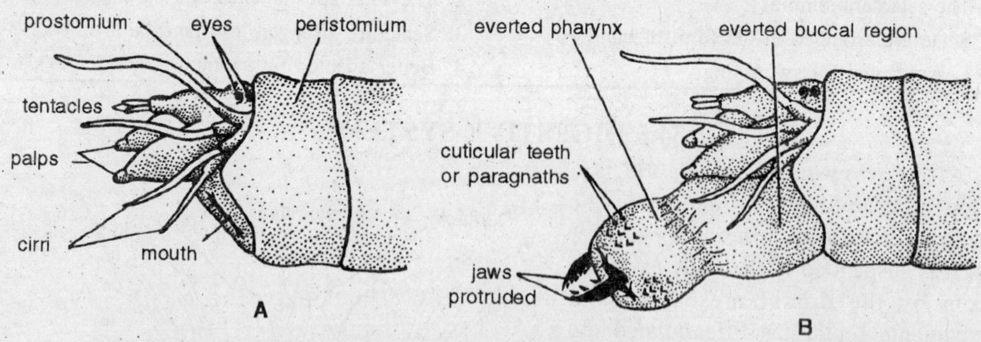

Fig. 34.10. *Nereis.* Anterior end in lateral view. A—Pharynx retracted; B—Pharynx everted.

4. Stomach-Intestine. It is a wide tube extending through the remaining length of the body. It is straight, thin-walled tube segmentally constricted by septa. Its epithelial lining contains numerous **gland cells** which secrete digestive enzymes. Stomach-intestine is the principle site of digestion and absorption. In *Nereis*, a distinct stomach is absent.

C. Hindgut

5. Rectum. It is a short tube-like structure and lies in the anal segment. It opens posteriorly through the terminal anus.

Food and Feeding Mechanism

Food. *Nereis* is a carnivorous animal and feeds on small animals such as crustaceans, molluscs (clams), sponges and larvae. It is **predaceous** or **raptorial** since its mode of feeding involves **seizing** and **devouring** living prey.

Feeding mechanism. Food is tested by the tactile **prostomial palps** and the **peristomial cirri**. The pharynx is rapidly everted, placing the jaws at the anterior end of the body and causing them to open. The food is seized by the jaws and the pharynx is retracted. The prey is, thus, dragged inside the buccal cavity during the inversion or retraction of buccopharynx. This is called **raptorial feeding** and it is performed when *Nereis* feeds on surface outside the burrow.

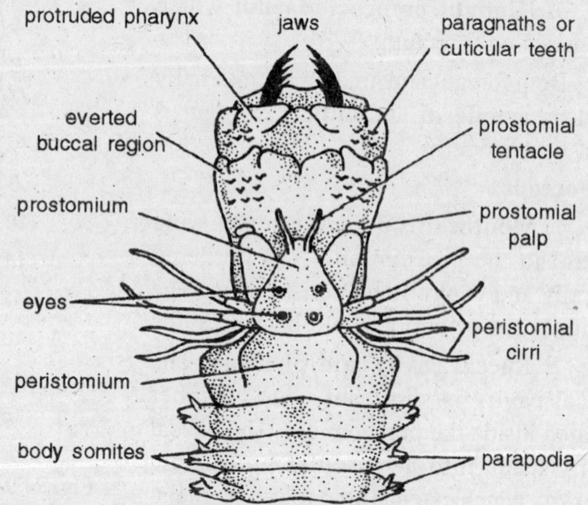

Fig. 34.11. *Nereis.* Head with everted buccal region (in dorsal view).

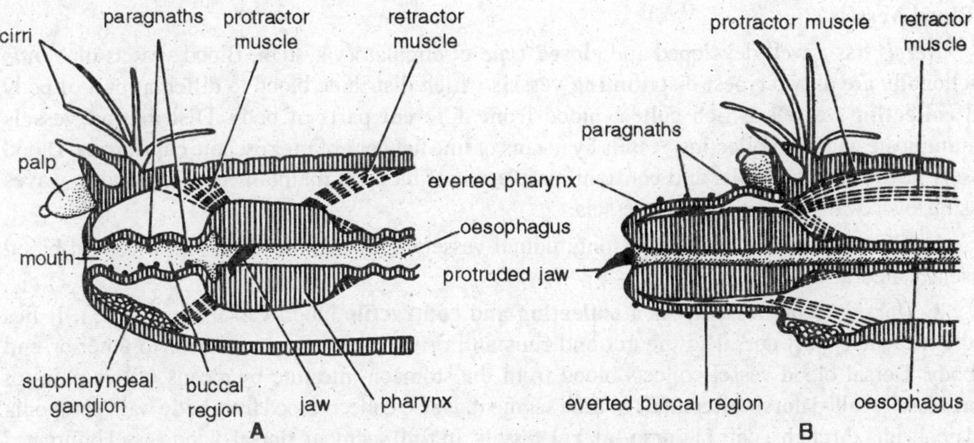

Fig. 34.12. *Nereis*. Diagrammatic representation of working of bucco-pharyngeal region. A—In inverted condition; B—In everted condition.

Nereis also feeds by many other methods (see Box 34.1). For example, in *Nereis*, another common method of feeding is **filter-feeding**. When *Nereis* lives in a U-shaped burrow, it feeds by ejecting out a cone of mucus in front of the mouth and creates a current of water by beating the parapodia. The currents of water enters the mouth end of the burrow and leave from other end. The food particles (detritus) coming along with the current of water are held in mucous cone which are later ingested by *Nereis*. This mode of feeding is called **filter-feeding**.

Box 34.1 Diversity of Feeding Habits of *Nereis*

Studies on *Nereis* by **H. Goerke** (1971) demonstrated the diversity of feeding habits that may exist even within a genus. Species of *Nereis* possess a muscular, eversible pharynx with a pair of heavy **jaws** (which are made of cross-linked tanned protein). Some species of *Nereis*, such as *Nereis pelagica*, *Nereis virens* and *Nereis diversicolor* are **omnivorous** and feed on algae, other invertebrates and even detritus. *Nereis succinea* and *Nereis longissima* feed primarily on detritus material in the substratum. *Nereis fucata*, which lives as a commensal in hermit crab shells, is **carnivorous**. *Nereis brandti*, which occurs on the Coast of USA and reaches 1.8 m in length, feeds primarily on green algae.

Digestion, absorption and egestion. The ingested food initially undergoes **mastication** in the bucco-pharyngeal region by the help of numerous denticles. The masticated food is pushed onwards inside the gut by peristaltic contraction of muscles present in the alimentary canal. Digestion is mainly **extracellular** and food is digested by enzymes secreted by the digestive glands and gland cells of the epithelial lining of stomach-intestine. Absorption of digested food occurs in stomach-intestine by the diffusion. The undigested residue is passed to rectum; from there it is ejected out through the **anus**.

34.6 BLOOD VASCULAR SYSTEM AND RESPIRATION

1. Blood

The blood of *Nereis* is bright red in colour. It consists of a fluid **plasma** and numerous amoeboid, nucleated, colourless and leucocyte-like **corpuscles**. The bright red colour of the blood is due to the presence of a respiratory pigment **erythrocruorin** which is dissolved in plasma. Blood flows through a system of blood vessels and helps in transportation of gases, food and excretory products.

2. Blood Vessels

Nereis has a well developed and **closed type** of circulatory system. Blood vessels of *Nereis* functionally are of two types: **distributing vessels** which distribute blood to different part of body and **collecting vessels** which collect blood from different parts of body. Distributing vessels communicate with the collecting vessels by means of fine network of microscopic **capillaries**. Blood vessels have muscular walls and constant circulation of blood is maintained by peristaltic waves passing over walls of large blood vessels.

In *Nereis*, there are three main longitudinal vessels: 1. dorsal blood vessel; 2. ventral blood vessel; 3. neural blood vessel.

1. Dorsal blood vessel. It is a **collecting** and **contractile** blood vessel (Fig. 34.13). It lies in the mesentery just dorsal to the gut and runs and drives blood from posterior to anterior end of body. Dorsal blood vessel collects blood from the stomach-intestine by means of two pairs of short vessels, called **dorso-intestinals** in each segment. It also collects blood from body wall, parapodia and nephridia through a pair of **dorso-lateral vessels**. In fifth segment, dorsal blood vessel bifurcates and both branches form a plexus to supply oesophageal wall.

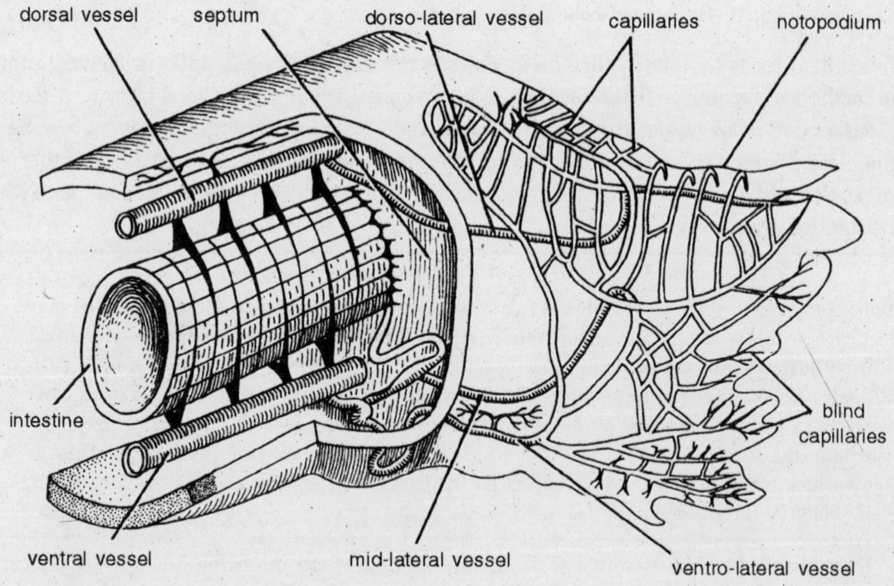

Fig. 34.13. *Nereis*. Blood vascular system.

2. Ventral blood vessel. It is mainly a **distributing** and **non-contractile** vessel. It lies mid-ventrally below the alimentary canal from fifth segment to the last. It carries blood from anterior end to the posterior end. In the fifth segment, it collects blood from the oesophageal wall by single median vessel. In the last segment, it communicates with the dorsal vessel by a **circumrectal ring**.

Blood vessels of a typical segment. In each segment except the first five segments, the dorsal and ventral blood vessels are connected on either side by a pair of loop-like **transverse vessels** or **lateral commissural vessels**. These vessels take the blood from the ventral to dorsal blood vessel but not directly. In each segment, these vessels divide into **afferent vessels** which supply blood to the body wall, nephridia and parapodia of their side. In these organs, the **afferent branches** break up into an extensive network of **capillaries**, then reunite to form the corresponding **efferent branches**. These unite to open into the dorsal vessel. In each segment, the ventral vessel also supplies blood to the stomach and intestinal wall by a pair of **afferent intestinal vessels, ventro-intestinal vessels**

These form an extensive network of capillaries in the stomach-intestinal wall (Fig. 34.14). The capillaries reunite to form a pair of **efferent intestinal vessels**, a **dorso-intestinal vessels** and carry blood to dorsal blood vessel.

3. Neural blood vessel. It is a thin blood vessel which runs below the ventral nerve cord. It collects blood from the ventral body wall and returns it to the ventral blood vessel.

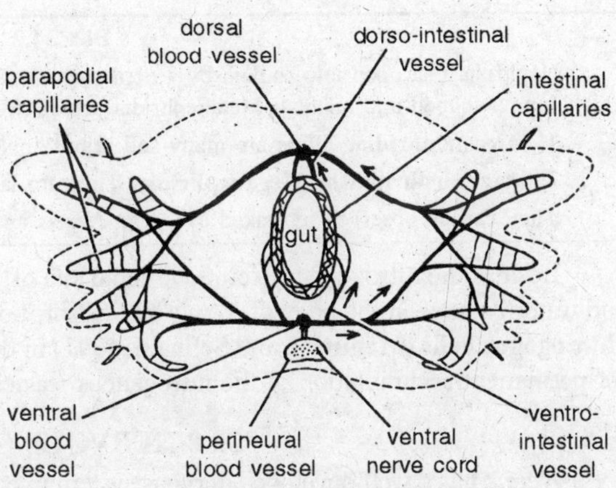

Fig. 34.14. *Nereis*. Diagrammatic representation of main blood vessels in a segment.

3. Respiration

Nereis lacks any special organs or gills for respiration. The parapodia with their blood supply and body wall (*i.e.*, general body surface) with its plexus of blood vessels serve as respiratory surfaces. Oxygen dissolved in surrounding water readily diffuses into the blood through the thin integument and parapodia. Similarly, from the blood carbon dioxide diffuses into the surrounding water. Blood plasma of *Nereis* contains the respiratory pigment, **erythrocruorin**, for transportation of respiratory gases.

34.7 EXCRETORY SYSTEM

In *Nereis*, the excretion is carried out by means of special type of coiled tube-like structures, the **nephridia**, one pair of which is found in each segment except the first and the last segment.

Structure

The nephridia of *Nereis* are macroscopic structures; they can be observed by naked eyes. Each nephridium consists of a coiled ciliated tube lying embedded in the syncytial mass of protoplasm. It is differentiated into two parts; a body and a neck. The main **body** of nephridium is an irregular, oval, compact and glandular mass which is placed transversely in the segment. It contains highly convoluted and mostly ciliated tubule. The nephridial tubule passes through the septum into the anterior segment in the form of **neck**, where it opens by a ciliated funnel or **nephrostome**. Posteriorly the tubule opens by a contractile (*i.e.*, sphinctered) **nephridiopore** located ventrally at the base of parapodium near the origin of ventral cirrus.

The nephridial tubule is an excavation in the mass of protoplasm. Such an open type of nephridium with a ciliated nephrostome is called **metanephridium**.

Function of nephridia. The outer surface of nephridium is richly supplied with blood capillaries. The gland cells present in the wall of ciliated tube remove waste materials, chiefly **ammonia** from blood. This liquid waste is expelled out of body through nephridiopores. The ciliated funnels (nephrostomes) opening into the coelom also removes dead coelomic corpuscles (coelomocytes or amoeboid corpuscles) which have eaten up bacteria, etc., and other foreign bodies.

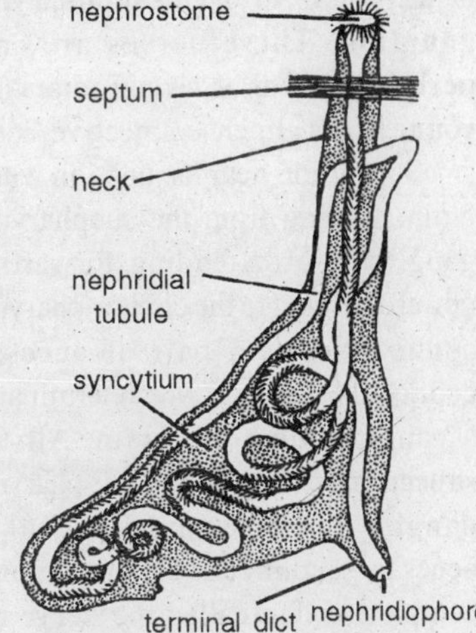

Fig. 34.15. *Nereis*. Nephridium.

Box 34.2

Nephridia that open into coelom by a **nephrostome** are called **metanephridia**. Nephrostome is funnel-shaped and ciliated. Metanephridia are of two types:

1. **Micronephridia.** These are many and minute or microscopic in size, *e.g.*, *Pheretima*.
2. **Macronephridia or Meganephridia.** They are large in size and few in number. They can be observed by naked eye, *e.g.*, *Nereis* and leech.

Besides the nitrogenous excretion, nephridia of *Nereis* do **osmoregulation;** it maintains salt and water balance in the body of *Nereis* by reabsorption of water and salts. Lastly, in *Nereis* occur **chlorogogue cells** (**Meglitsch** and **Schram**, 1991) in the visceral peritoneum; these cells are meant for permanent accumulation of the nitrogenous wastes.

34.8 NERVOUS SYSTEM

Nereis has a well developed nervous system in correlation to its active life. It is bilateral and metameric in its arrangement. It includes a central nervous system, peripheral nervous system and visceral nervous system (Fig. 34.16).

1. Central nervous sys-tem. There is a large bilobed mass of nervous tissue in the prosto-mium beneath the dorsal epider-mis. This is known as the **brain.** It is formed by the fusion of a pair of ganglia called the **suprapharyngeal ganglia** or **cerebral ganglia** (Box 34.3). A similar but smaller bilobed mass is situated in the third segment beneath the pharynx. It is called **sub-pharyngeal ganglion**. In fact, this gang-lion is formed due to fusion of two pairs of ganglia. A pair of slender cords of nervous tissue start from the sides of brain and extend downward and backward along the sides of the pharynx to join the **subpharyngeal ganglion**. These nerves are called **peripharyngeal** or **circumpharyngeal connectives**. Each connective bears a small ganglion near its junction with the brain. Arising from the subpharyngeal ganglion and extending forward and upward parallel to the circum-pharyngeal connectives is a pair of **accessory connectives**, each of which terminates in a ganglion near the brain. All these structures, *viz.*, supra- and sub-pharyngeal ganglia and circumpharyngeal and accessory connectives, thus, form a complete collar, called the **nerve ring,** round the pharynx.

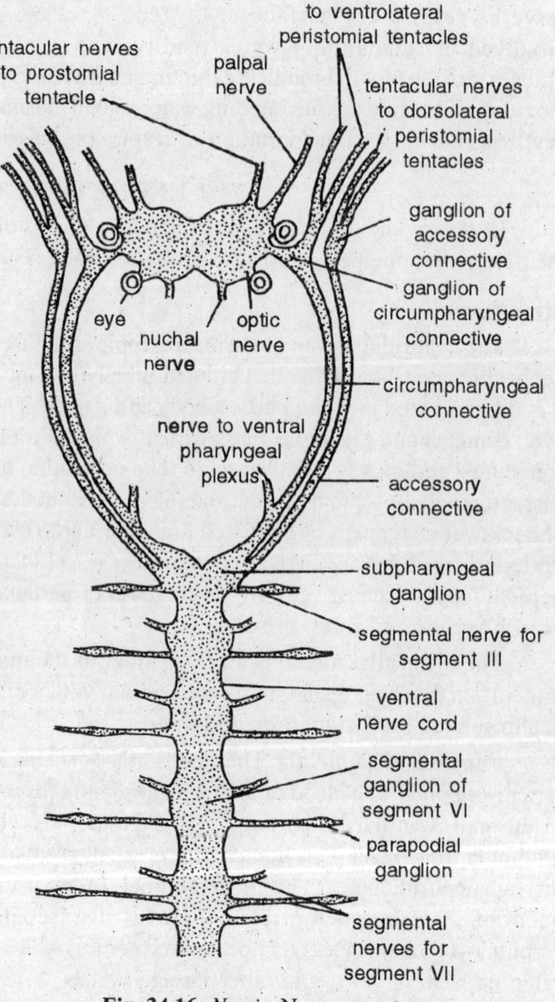

Fig. 34.16. *Nereis.* Nervous system.

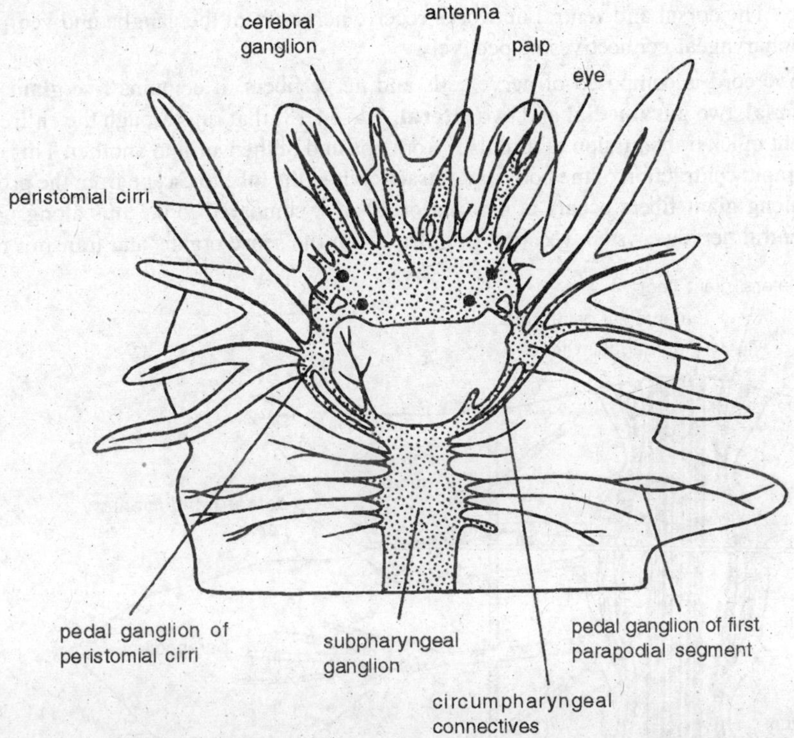

Fig. 34.17. *Nereis*. Central nervous system – nerve ring (after Brusca and Brusca, 2003).

Box 34.3

The brain of *Nereis* is often specialized into three regions, typically called the **forebrain, midbrain** and **hindbrain**. The forebrain innervates the prostomial palps, the midbrain the eyes and prostomial palps and the hindbrain the chemosensory nuchal organs. The **circumcentric connectives** (*i.e.*, circumpharyngeal plus accessory connectives) arise from forebrains and midbrains (Fig. 34.17).

A prominent cord of nervous tissue, called the **ventral nerve cord**, extends from the subpharyngeal ganglion to the end of body along the mid-ventral line of the body wall beneath the ventral blood vessel. This cord appears single but actually it consists of two solid cords fused together and surrounded by a common sheath of connective tissue. Ventral cord bears in each segment a swelling, the **segmental ganglion**, formed by the fusion of a pair of ganglia. An additional pair of **lateral inter segmental nerves** run parallel to nerve cord to connect the pedal ganglia (Fig. 34.18).

2. Peripheral nervous system. The peripheral nervous system mainly includes the nerves coming directly from the brain and ventral nerve cord to supply the different regions of the body. The brain gives out four short **optic nerves** to the eyes; two **tentacular nerves** to the prostomial tentacles; two **palpal nerves** to the palps. From the small ganglion of circumpharyngeal connectives two pairs of nerves go to the ventral peristomial cirri on each side. From the subpharyngeal ganglion a pair of nerves goes to the dorso-lateral peristomeal tentacles. The nerves that arise from the segmental ganglia innervates the body wall musculature, and parapodia (via the **pedal ganglia**) and the digestive tract.

3. Visceral nervous system. Visceral nervous system innervates the musculature of the pharynx. It consists of a network of tiny nerves with occasional small ganglia on the dorsal and ventral side of the pharynx. The dorsal and ventral networks receive nerves from the ganglia and ventral ends of the circumpharyngeal connectives respectively.

The nerve cord is composed of nerve cells and nerve fibers. It contains five **giant fibers/axons** (one **dorsal**, two **paramedial** and two **lateral**, Fig. 34.18) that run through the entire length and bring about quick transmission of impulses from one end of the worm to another. This enables the worm for rapid contraction of the body and retraction into the tube to escape from the predators. Conduction along giant fibers occurs at a speed of 20 m/s, compared to 0.5 m/s along ordinary fibers. The central nervous system receives impulses from the sense organs and transmit them to

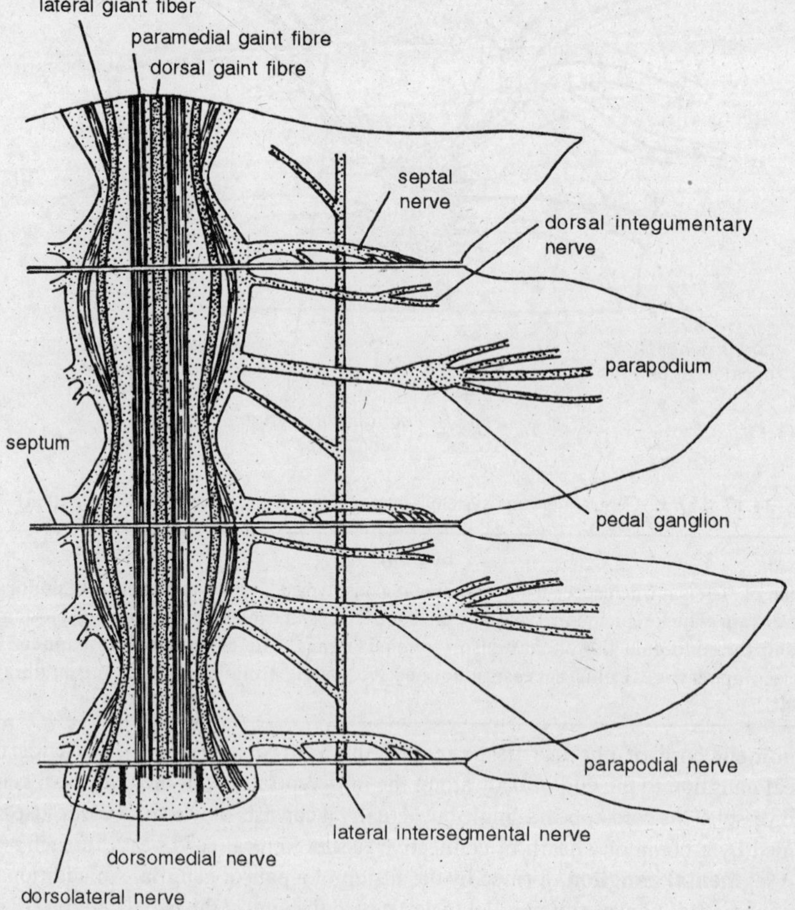

Fig. 34.18. *Nereis*. Peripheral nervous system (after Brusca and Brusca, 2003).

effectors, such as muscles of parapodia and body wall, via peripheral nervous system. The segmental ganglia act as reflex centers so that a clamworm devoid of head (*i.e.*, brain) continues normal movement.

34.9 SENSE ORGANS

In *Nereis* occur the following three types of specialized sense organs or receptors.

1. Tactile sense organs. The prostomial tentacles, prostomial palpi and peristomial cirri of head are the main **tactile** sense organs as they are sensitive to touch.

2. Chemoreceptors. Nuchal organs of *Nereis* act as chemoreceptors as well as olfactory sense organs as they are sensitive to **taste** and **smell** respectively. They are a pair of small pits in the posterior part of the prostomium on its dorsal side. They are lined by special ciliated epithelium with gland cells. They are lined by special ciliated epithelium with gland cells. They are connected with the posterior part of brain. They enable the clamworms to detect its prey present outside the burrow from the water current that enters the burrow with dissolved substances emanating from the prey.

3. Photoreceptors. There are four simple **eyes** (black spots) on the dorsal surface of the prostomium. Each eye has a cup made up of **pigmented retinal cells** produced inwards into clear rods, the opening of the cup is a **pupil** (Fig. 34.19). In fact, each retinal cell has three distinct parts: an outer nucleated part with nerve fibers, a highly pigmented main body and inner part of transparent cuticular rod. The external cuticle forms a transparent cornea. Inside the cup is a transparent, gelatinous, refractive **lens**. The retinal cells are joined to nerve fibers of the **optic nerve**. The eyes are photoreceptive; they are not related to the formation of image but they help in detecting changes in light intensity.

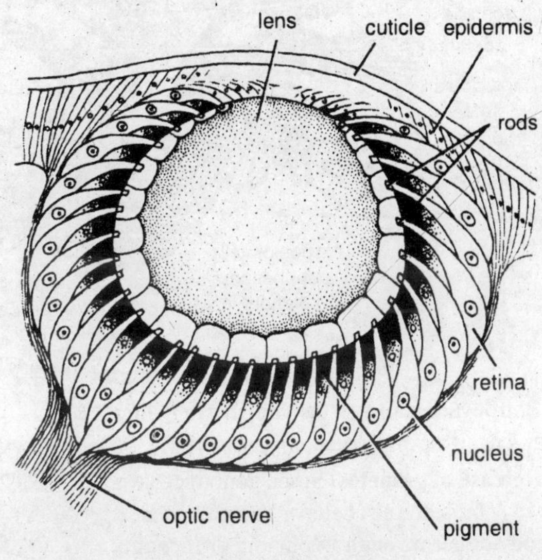

Fig. 34.19. *Nereis*. V.S. of eye.

34.10 REPRODUCTIVE SYSTEM AND DEVELOPMENT

Nereis is **dioecious**, *i.e.*, sexes are separate and **sexual dimorphism** occurs in only those species in which distinct non-sexual (*i.e.*, *Nereis*) and sexual (*i.e.*, *Heteronereis*) phases are found. The gonads are temporary or seasonal and develop only during the breeding season, *i.e.*, in summer season. They lack the **gonoducts**.

Gonads. These are formed by the proliferation of the cells of coelomic epithelium. They can be seen as groups of masses of germ cells on the ventral side in the coelomic space of each segment of body except few of anterior segments.

In male *Nereis dumerilli*, however, there is only one pair of **testes** lying in anyone segment between the 19th and 25th segments. But in other species such as *Nereis virens* and *Nereis diversicolor* the testes extend in many segments. During breeding season the germ cells which detach from the coelomic epithelium in male form the **sperm mother cells**. These cells float in the coelomic fluid, divide rapidly and undergo maturation division (meiosis) to form **spermatids** which transform into flagellated **sperms.** A mature sperm has a rod-shaped **head** and a long vibratile **flagellum** or **tail**.

In the female, the ovaries lie in many segments around blood vessels. The **oogonia** separate from coelomic epithelium, float in coelomic fluid and undergo **oogenesis** to form **ova** which are found floating in the coelomic fluid (Fig. 34.20). The ova are round in shape and filled with **yolk** granules. They are bounded by a **vitelline envelope**.

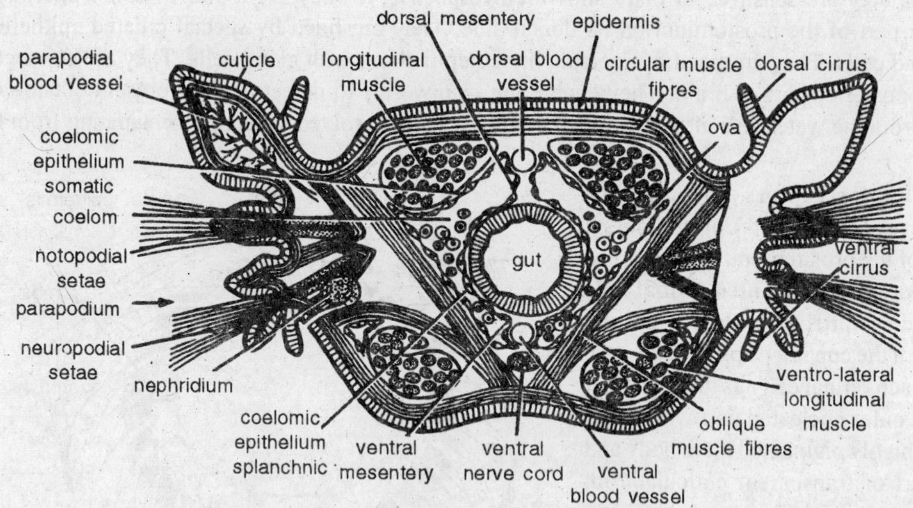

Fig. 34.20. *Nereis*. T.S. of body showing ova.

Both ovaries and testes vanish after they have given off sexual cells.

Release of gametes. Since gonoducts are absent in *Nereis*, the sex cells or gametes escape from body either through nephridia or through temporary rupture of the body wall. A pair of **dorsal ciliated organs** are found in each segment. These are in the form of short and much folded funnels of ciliated tracts opening into the body cavity by a wide opening and being closed to the exterior (Fig. 34.21). Ciliated organs resemble the coelomoducts and are said to act as **gonoducts** developing temporary openings during reproductive season.

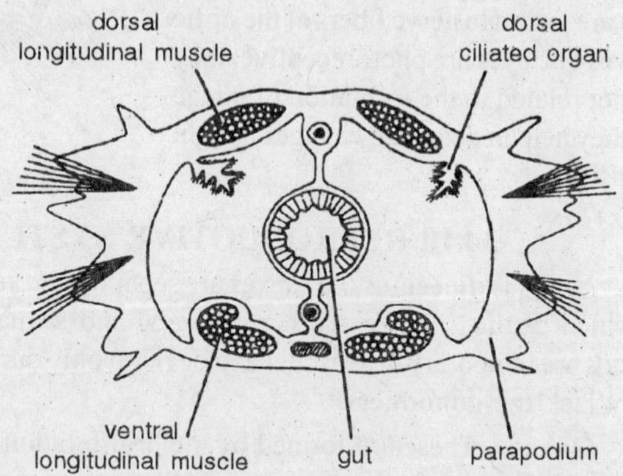

Fig. 34.21. *Nereis*. T.S. of a segment to show the position of ciliated organ.

Fertilization. The fully sexually mature clamworms leave the burrows in groups and come up at the surface of sea water to swim freely, it is called **swarming**. During swarming male and female *Nereis* (*i.e. Heteronereis*) simultaneously release gametes which fuse together during fertilization. Thus, in *Nereis* fertilization is **external** occurring in open seawater.

Development. Development of *Nereis* is **indirect** involving the formation of a trochophore larva which undergoes metamorphosis to become adult-like.

Structure of egg. The newly discharged planktonic egg of *Nereis* is spherical in shape and possesses two membranes: an outer thin **outer membrane** and a thick radially situated membrane, called **zona radiata** which encloses **ooplasm**. Ooplasm is rich in yolk having oil droplets and intact female nucleus (called **germinal vesicle**). A thick, gelatinous and transparent **vitelline envelope** surround entire ovum. However, immediately after fertilization several changes occur in egg: yolk granules aggregate at the lower end of egg to form the **vegetal pole**. The upper clear zone forms the **animal pole**. Such an egg is called **telolecithal egg**. The nucleus of ovum completes its meiosis and release **polar bodies** at the animal pole. The haploid nuclei of male and female gametes unite (during the process of **amphimixis** in fertilization) to form diploid nucleus of the **zygote**.

Cleavage and gastrulation. In *Nereis*, cleavage is **holoblastic**, **spiral** and **determinate** type. Blastula lacks a blastocoel so is called **stereoblastula.** Gastrulation involves **epiboly** and **invagination**.

The first two divisions are almost equal and vertical. The four cells produced lie in the same plane and are called A, B, C and D. Each cell gives rise to one of the quarter of the embryo. Cell D is larger than the others and forms the **dorsal** surface of the embryo; cell B forms the **ventral** surface and cell A and cell C form **lateral** side the embryo. The first quartet of cells undergoes three successive divisions (3rd, 4th and 5th). These divisions are **unequal** and at right angles to the first two. As a result three successive quartets of cells are separated; these contain comparatively smaller cells. The first four cells are called **macromeres** and the cells of three successive quartets are called **micromeres.** Since the cleavage divisions are spiral, the micromeres do not lie exactly above its macromeres, but somewhat displaced to the right in one quartet and to the left in the next quartet.

The micromeres of the first three quartets give rise to **ectoderm** of larva. The sixth cleavage separates the micromeres of fourth quartet. One of the cell (4d) of the fourth-quartet is known as **somatoblast** and gives rise to entire **mesoderm**. The macromeres form **endoderm**. The embryo at this stage is called **stereoblastula**, it is a solid blastula and lacks a blastocoel.

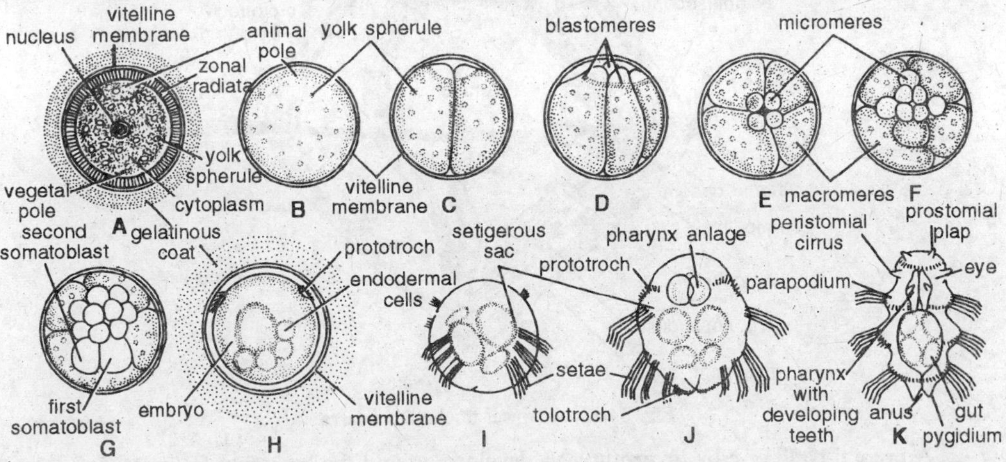

Fig. 34.22. *Nereis.* Stages in development. A—Unfertilized egg; B—Fertilized egg; C, D, E—2, 4, 8—celled stages; F—Formation of first somatoblast; G—Formation of second somatoblast; H—Young trochophore before hatching; I, J—Post-trochophore larva; K—Three segmented larva with parapodia and setae.

Gastrulation. Gastrulation of *Nereis* is of **epibolic type** and is accomplished by the spreading (epibolic growth) of micromeres over the endoderm cells (macromeres) and somatoblasts. As the somatoblasts invaginate, the **blastopore** is formed as an uncovered area at the vegetal plate. It represents the ventral surface of the larva.

Development of trochophore larva. The gastrula develops into a **trochophore** or **trochophore larva**. At the top of presumptive head end, micromeres differentiate into **apical rosette** or **apical plate**, which differentiates into **cerebral ganglia** and a tuft of cilia. Another tuft of cilia develops slightly behind the apical plate; it is known as **prototroch**. **Stomodaeum** develops as ectodermal invagination and a similar invagination at the position of blastopore form the **proctodaeum**.

The cells derived from successive divisions of somatoblasts become arranged on the presumptive ventral side of the embryo and form the **ventral plate**. The ectodermal cells present along the ventral side are thickened into **ventral nerve cord**. The deeply located derivatives of somatoblast (*i.e.*, 4d) give rise to **mesodermal bands** which develop into muscles of body wall and of alimentary canal.

Structure of Trochophore Larva

The trochophore is a minute, transparent, unsegmented and pear-shaped larva. It has broad anterior **head** and narrow posterior **anal end**. It is bounded externally by a thin ectoderm. The head end is thickened into an apical plate. It bears a tuft of **apical cilia**. A pair of **cerebral ganglia** and pigmented larval eyes are present on the **apical plate**. Larva contains a curved **gut** with a mouth, ectodermal **oesophagus**, an endodermal **stomach**, and an ectodermal **hind gut** opening by **anus**. It feeds on microorganisms.

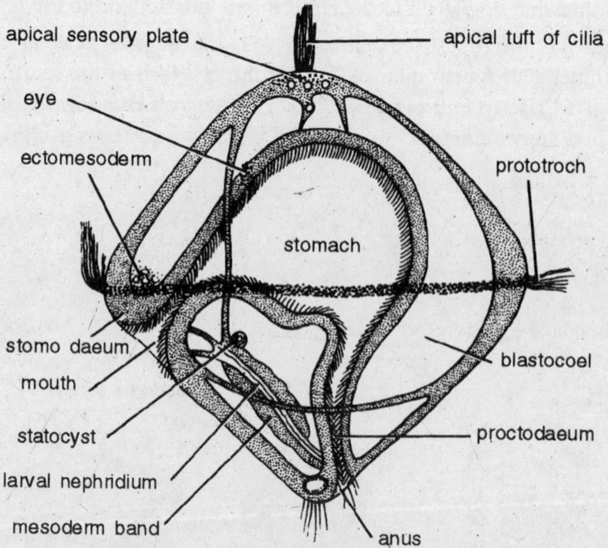

Fig. 34.23. A typical trochophore larva.

A preoral band of cilia or **prototroch** develops around the body just in front of mouth. A similar band of cilia occurs at the posterior end. This is post-oral band or **metatroch**. The larva lacks coelom but possesses a spacious **blastocoel** between the alimentary canal and ectoderm. A pair of protonephridia and larval muscles are found in the blastocoel. The mesoderm is represented by a pair of undifferentiated mass of cells. There is a statocyst near the nephridia.

A typical free-swimming trochophore larval stage occurs only in few polychaetes. In *Nereis*, larva is developed inside the egg membrane. The larva hatches out in the sea water by rupturing egg's vitelline envelope. It is called **post-trochophore** or **nectochaete**. It possesses three segments with bristles. Its larval structures are poorly developed.

Biological importance of trochophore larva. Annelid's trochophore larva is of great phylogenetic significance. It exhibits similarity with adult ctenophores and certain rotifers. It shows close affinity with the larva found in development of animals of some diverse groups such as polyclad Turbellaria (*i.e.*, Mueller's larva); Nemertinea (*i.e.*, pilibium larva), Bryozoa and Mollusca (*i.e.*, veliger's larva).

Visualizing these affinities, **Hatscheck** (1878) and others formulated the **trochophore theory**. According to which the present bilateral phyla are considered to have descended from some radially symmetrical group and trochophore represents a transitional stage in the line of emergence of group Bilateria (Rotifera) from radial group Ctenophroa.

Post-larval Development (Metamorphosis)

The trochophore larva of *Nereis* is pelagic, it drifts about in the sea swimming by its protoroch and is metamorphosed. During metamorphosis ciliated bands disappear and its pre-oral part form the prostomium having brain, tentacles and eyes. The post-oral part of larva elongates and forms segments of body by metameric segmentation (Fig. 34.24). The first segment forms the **peristomium** and last **pygidium**. The ciliary bands disappear and larva grows in length with the addition of new segments just infront of the pygidium.

Larval nephridia are replaced by permanent ones, larval setae are dropped, tentacles, palps and parapodia are formed. Young worm thus resulted settles at the bottom of the sea and starts forming its burrow and gradually attains adulthood.

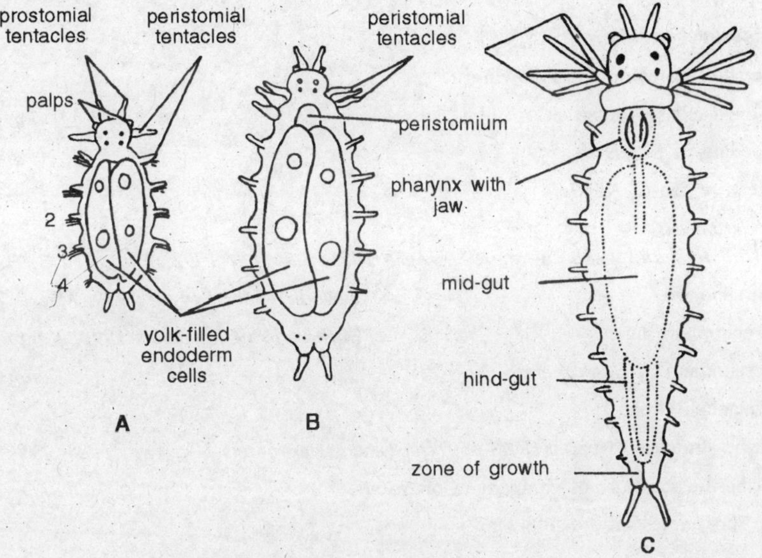

Fig. 34.24. *Nereis.* Metamorphosis and growth.

QUESTIONS

Long Answer Questions

1. Give an account of habit, habitat and external features of *Nereis*. Compare *Nereis* with *Heteronereis*. (*Bundelkhand 1988; Allahabad 1989; Jiwaji 1989; Shivaji 1989, 1992; Gorakhpur 1991; Kanpur 1992, 1995; Rohilkhand 1993*)
2. What is heteronereis phase? How do the individuals of this phase differ from *Nereis*? Compare the structure of the parapodium with that of *Heteronereis*? Give labelled diagrams. (*Magadh 1988; Awadh 1988, 1990; Agra 1989, 1992; Bundelkhand 1989, 1992; Bhopal 1990, 1992; Kanpur 1990, 1993; Purvanchal 1991; Indore 1992*)
3. Describe the alimentary canal of *Nereis*. Add a note on the mechanism of feeding. What is the food of this animal? (*Awadh 1989; Meerut or CCS 1990; Indore 1990; Lucknow 1990; Delhi 1993, 1995; Garhwal 1991; Allahabad 1991; Agra 1991, 1993; Bhopal 1991; Jabalpur 1993; Bilaspur 1993*)
4. Describe alimentary canal of *Nereis*. Discuss its adaptive features. (*Agra 1988*)
5. Describe the blood and blood vascular system of *Nereis*. How does respiration takes place in this animal? (*Vikram 1989; Kumaon 1989; Rewa 1993; Agra 1995*)

6. Give an account of the organs of excretion in *Nereis*. *(Bundelkhand 1989; Sagar 1990; Awadh 1990, 1992)*

7. Describe structure of nephridium of *Nereis*. What do you understand by enteronephric nephridial system? *(Awadh 1989, 1991; Kanpur 1992; Purvanchal 1993)*

8. Describe the reproductive system of *Nereis* and state clearly how reproduction takes place in this animal. *(Gorakhpur 1988; Bundelkhand 1989; Agra 1990)*

9. With suitable diagrams illustrate and describe the development of *Nereis*. *(Jiwaji 1989; Bhopal 1989, 1992; Kanpur 1992; Bundelkhand 1993; Indore 1993)*

10. Describe the trochophore larva of *Nereis*. Explain the trochophore formation in *Nereis*. *(Magadh 1995)*

11. How are gametes discharged in Nereis? Describe the development of this animal.

Short Answer Questions

1. Write short notes on the following:
   (*i*) Dorsal ciliated organ *(Purvanchal 1989)*
   (*ii*) Feeding in *Nereis* *(Purvanchal 1989)*
   (*iii*) Head region of *Nereis* *(Purvanchal 1989)*
   (*iv*) *Heteronereis* *(Bundelkhand 1988; Jiwaji 1988; Agra 1991; Jabalpur 1992; Indore 1993; Himachal 1995)*
   (*v*) Parapodium *(Awadh 1989; Bengal 1989; Meerut or CCS 1990)*
   (*vi*) Trochophore larva *(Awadh 1989; Lucknow 1992; Kanpur 1990, 1955)*
   (*vii*) Parapodium of *Heteronereis* *(Kanpur 1994, 1995)*
   (*viii*) Metanephridium

2. Describe structure of parapodium of *Nereis* and *Heteronereis*. *(Garhwal 1988, 1992; Sagar 1994)*

3. Describe the structure of parapodium of *Nereis*.

4. Write short notes on the following:
   (*i*) Parapodia *(Magadh 1989, 1991; Madurai 1994)*
   (*ii*) Setae *(Punjab 1990)*

5. Describe food and feeding habits of *Nereis*. *(Bundelkhand 1989)*

6. Describe structure and function of ciliated funnels or dorsal ciliated organ.

7. Explain what is epitoky?

8. What is heteronereis? Describe its characteristic features. *(Jiwaji 1993; Gwalior 1993)*

9. Describe nephridium of *Nereis*. *(Agra 1992)*

10. Describe trochophore larva of *Nereis*. Discuss its biological significance.

11. Write short note on *Heteronereis*. *(Himachal 1995)*

12. What is the significance of *Heteronereis*? Compare the external features of *Heteronereis* and *Nereis*.

13. Nuchal organs of *Nereis*. *(Gorakhpur 1991)*

14. Describe feeding in *Nereis*.

15. Describe polymorphism in *Nereis*.

16. Distinguish between *Nereis* and *Heteronereis*. *(Magadh 1995)*

17. Give difference between *Nereis* and *Heteronereis* phases. *(Sagar 1989)*

18. Compare the parapodia of *Nereis* and *Heteronereis*.

19. Draw well labelled diagrams of the following:

(*i*) Head of *Nereis* with everted buccal region

(*ii*) Parapodium of *Nereis* (*Garhwal 1988; Jiwaji 1992; Ravishankar 1993*)

(*iii*) Parapodium of *Heteronereis* (*Agra 1988; Kanpur 1990; Vikram 1993*)

(*iv*) T.S. *Nereis* through parapodia (*Agra 1988; Bundelkhand 1988; Meerut or CCS 1989;Kanpur 1989; Awadh 1989; Bhopal 1990; Osmania 1991; Jivaji 1993; Vikram 1993*)

(*v*) Nephridium of *Nereis* (*Kanpur 1995*)

(*vi*) *Heteronereis* (*Bundelkhand 1988; Jiwaji 1990; Gorakhpur 1991; Garhwal 1992; Awadh 1992; Ravishankar 1993; Kanpur 1993*)

(*vii*) Alimentary canal of *Nereis*.

(*viii*) Trochophore larva (*Garhwal 1988; Awadh 1989; Kanpur 1993, 1995; Lucknow 1995*)

Very Short Answer Questions

1. Give common name of *Nereis*?

Ans. Clam worm.

2. Why is *Nereis* called clam worm?

Ans. *Nereis* feeds mainly on clams and is found near them. For this reason *Nereis* is called **clam worm**.

3. What are paragnaths?

Ans. Paragnaths are dark brown denticles of cuticle of buccal cavity of *Nereis*. They help in mastication of food.

4. What are macronephridia?

Ans. *Nereis* have large sized and few nephridia which are called macronephridia.

5. Where are the eyes of *Nereis* located?

Ans. On prostomium.

6. What process propells the food through gut of *Nereis*?

Ans. Peristalsis.

7. Name two modes of feeding found in *Nereis*.

Ans. Raptorial and filter-feeding.

8. Name the sense organs of *Nereis*.

Ans. Tentacles, palps, nuchal organs and eyes (*i.e.*, Tactile sense organs, chemoreceptors and photoreceptors).

9. Mention the principal digestive glands of clam worm.

Ans. Oesophageal gland/caeca.

10. What type of circulatory system occur in *Nereis*?

Ans. Closed type.

11. Name the process involved in urine formation in polychaetes.

Ans. Ultrafilteration, modification of primary urine to final urine by reabsorption and secretion.

12. What is *Heteronereis*?

Ans. It is the sexually mature phase of some species of *Nereis* (*e.g., Nereis dumerilii*). *Heteronereis* (or *Heteroneanthes*) also represent free-swimming, swarming, sexual phase of *Nereis*.

13. What is epitoky?

Ans. Sexually mature phase of *Nereis* is called *Heteronereis.* It has gonads (testes or ovaries) which develop in posterior part of the body. The parapodia of posterior part of body of *Heteronereis* are also modified, for swimming mode of life. *Heteronereis's* gonad containing posterior part of body bearing modified parapodia is called **epitoke**. Its non-sexual anterior part of body which is lacking in the gonads is called **atoke**. This phenomenon of transformation of non-sexual individuals into sexual ones is called **epitoky**.

14. What is swarming in *Nereis*? Describe in brief.

Ans. Swarming means to come together, move or occur in great numbers. For breeding purpose, *Heteronereis* leave their burrows and come to the surface of the sea to swim about freely and to shed their gametes (*i.e.*, spawning). Such swarming behaviour often occur at night and during a particular phase of moon. Both sexes swarm together, sperms and eggs are released in huge numbers and this ensures fertilization. Some clamworms (*e.g.*, *Nereis limbata*) swarm on moonless nights during summer months soon after sunset. The female secretes a substance called **fertilium** to attract males and stimulate them to shed their sperms. This in turn exites the females and stimulates them to shed their eggs for fertilization. Swarming often continues for about an hour or so. The females die after liberating the eggs and the males perhaps also do not live long.

15. How many ciliary bands are found in trochophore larva? *(C.C.S.U., Meerut 2002)*

Ans. There are two ciliary bands found in trochophore larva of *Nereis*. These include a preoral ciliated band called **prototroch** and a post-oral ciliated band called **metatroch**. The former is just behind the equator and latter lies behind the mouth.

Fill in the Blanks

1. *Nereis* is commonly known as .... or ....

2. A fossorial animal such as *Nereis* lives in a .... burrow.

3. Head of *Nereis* consists of .... and peristomium.

4. A parapodium consists of two parts: dorsal notopodium and ventral ....

5. Pygidium of *Nereis* bears anal .... but lacks ....

6. Body cavity of *Nereis* is a true ....

7. Proboscis of *Nereis* consists of .... and ....

8. Digestion is .... in clam worm.

9. Wall of blood vessels in *Nereis* are made of ....

10. Larva of *Nereis* is called ....

Multiple Choice Questions

1. *Nereis* is commonly called

(*a*) ragworm (*b*) clamworm
(*c*) sandworm (*d*) all the above

2. *Nereis* is

(*a*) herbivorous (*b*) carnivorous
(*c*) omnivorous (*d*) parasite

3. Peristomium in *Nereis* is composed of

(*a*) one segment (*b*) two segments
(*c*) three segments (*d*) six segments

4. *Nereis* is

(*a*) raptorial (*b*) filter feeder
(*c*) herbivorous (*d*) both (*a*) and (*b*)

5. For excretion, clamworms have

(*a*) protonephridia (*b*) metanephridia
(*c*) malpighian tubules (*d*) podia

6. Blood of *Nereis* contains haemoglobin (= erythrocruorin) in
   (*a*) plasma (*b*) amoeboid corpuscles
   (*c*) red corpuscles (*d*) plasma as well as corpuscles
7. For Locomotion, *Nereis* uses
   (*a*) parapodia (*b*) setae
   (*c*) body wall musculature (*d*) all of these
8. Respiratory pigment of *Nereis* is
   (*a*) haemoglobin (*b*) cyanin
   (*c*) haemocyanin (*d*) erythrocruorin
9. Giant nerve fibers occur in the
   (*a*) brain (*b*) nerve cord
   (*c*) segmental nerve (*d*) ganglia
10. The chemoreceptor organs in *Nereis* is
    (*a*) prostomial palp (*b*) prostomial tentacle
    (*c*) nuchal organ (d) peristomial cirri
11. The name of the larva of *Nereis* is
    (*a*) trochophore (*b*) bipinnaria
    (*c*) zoea (*d*) cercaria
12. The clamworm having epitoke and atoke regions is called
    (*a*) heteronereis (*b*) sexually mature *Nereis*
    (*c*) heteronephrone (*d*) none

ANSWERS

Fill in the Blanks

1. Clam worm, rag worm 2. Self-made 3. Prostomium 4. Neuropodium
5. Cirri, parapodia 6. Coelom 7. Buccal cavity, pharynx
8. Extracellular 9. Basal lamina 10. Trochophore.

Multiple Choice Questions

1. (*d*) 2. (*b*) 3. (*a*) 4. (*d*) 5. (*b*) 6. (*a*)
7. (*a*) 8. (*d*) 9. (*b*) 10. (*c*) 11. (*a*) 12. (*a*).

35

Hirudinaria granulosa (Indian Cattle Leech) (Phylum Annelida)

35.1 CLASSIFICATION

Phylum	–	**Annelida**
Class	–	**Hirudinea**
Order	–	**Gnathobdellida**
Genus	–	***Hirudinaria***
Species	–	***granulosa***

Leeches are the most specialised annelids and have evolved from oligochaetes. The genus *Hirudinaria* includes following four Indian species: *H. granulosa, H. viridis, H. javanica* and *H. manillensis*. These species are also common in Burma (Myanmar), Pakistan, Bangladesh and Sri Lanka. *Haemadipsa* is a land leech of the hills of Southern India.

35.2 HABIT AND HABITAT

Hirudinaria granulosa is commonly called **'Indian cattle leech'.** It inhabits freshwater ponds and tanks, still waters near streams and rivers and water standing in rice-fields. It prefers shallow water where stones, logs, weeds, etc., provide shelter. *Hirudinaria* is an **ectoparasitic, sanguivorous** (blood sucking) animal which attaches to cattle and human beings when they visit the pond. It sucks the blood of fishes and frog also.

35.3 EXTERNAL FEATURES

Hirudinaria is an aquatic (freshwater) ectoparasitic sanguivorous (blood sucking) annelid. Its morphology and physiology is well adapted for its peculiar mode of habits and habitats.

External Structure

1. Size and Shape

Average leech measures 10 to 15 cm in length but it may grow upto a length of 35 cm. The body of *Hirudinaria* is soft, slimy, vermiform, elongated, bilaterally symmetrical and metamerically segmented. Body is dorso-ventrally flattened having flat ventral and convex dorsal surfaces. The body is broadest near the posterior end and narrowest at the anterior end. Body of leech is quite flexible whereas in contracted state it becomes cylindrical.

2. Colouration

Body of leech show protective colouration (camouflaging) being olive green dorsally, orange-yellow ventrally. Stripes of orange and black exist on the lateral sides.

3. Metamerism or Segmentation

Body of leech is divided into 33 segments, somites or **metameres** (*i.e.*, the number of segments is fixed in leeches). Except first two and last seven segments, external surface of each segment is superficially divided into variable number of **rings** or **annuli** (Fig. 35.1). All those segments which have five annuli are called **complete segments.** Segments having less than five annuli are called **incomplete segments.** Details of number of the annuli in different segments can be tabulated in the Table 35.1.

Table 35.1 Number of annuli in different body segments of *Hirudinaria.*

Segments	*Number of Annuli*	*Designation of segment*	*Nature of segment*
1, 2	One	Uniannulate	Incomplete
3	Two	Biannulate	Incomplete
4, 5, 6	Three	Triannulate	Incomplete
7 to 22	Five	Quinquannulate	Complete
23 to 26	Two	Biannulate	Incomplete
27 to 33	One	Uniannulate	Incomplete

Further, the surface of each annulus is divided by very fine longitudinal grooves (Fig. 35.3) into 18 rectangular areas on the dorsal and 18 on ventral side.

4. Division of Body

The body of leech can be divided into following six regions:

***(i)* Cephalic region.** First five segments which are incomplete form the head (Table 35.1). It contains **prostomium, anterior sucker, mouth** and **eyes.** In fact, a pair of **eyes** are situated on the dorsal side of first annulus in all the cephalic segments. These segments are, therefore, called **ocular segments.** There are no nephridiopores. On the ventral surface of cephalic region is present **anterior sucker** and **mouth.** The prostomium and first three segments form the **upper lip.**

***(ii)* Pre-clitellar region.** It comprises of three segments, *i.e.*, **6th** incomplete, **7th** and **8th** (complete). Each of these segments bear a pair of nephridiopores on its ventral surface.

***(iii)* Clitellar region.** It is composed of three complete segments, *i.e.*, **9th, 10th** and **11th.** A temporary girdle-like **clitellum** develops around this region during the breeding season. Male and female genital apertures are situated on the ventral surface of 10th and 11th segments respectively. All of these segments carry a pair of nephridiopores.

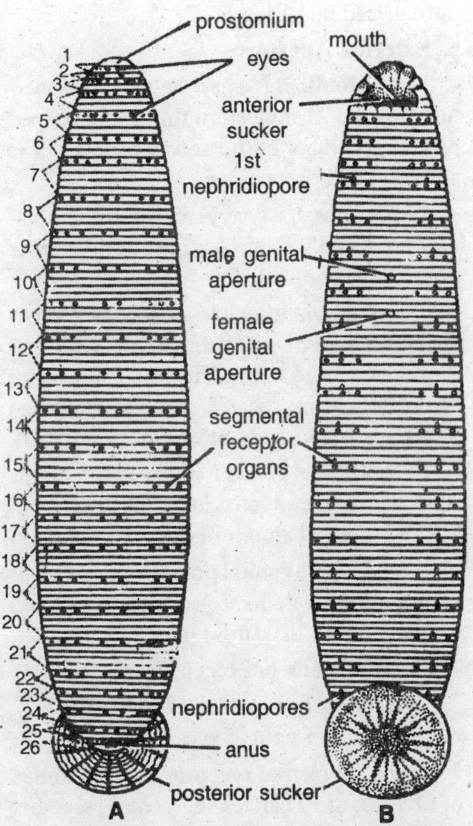

Fig. 35.1. *Hirudinaria granulosa.* External structure. A—Dorsal view; B—Ventral view

***(iv)* Middle region.** It is the largest region of the body and includes eleven complete segments (12th to 22nd). All segments contain paired nephridiopores ventrally.

***(v)* Caudal region.** It is composed of last seven incomplete segments (27th to 33rd). These segments are arranged in concentric rings.

5. Suckers

Each end of the body of leech bears a muscular adhesive structure called **sucker.** The **anterior, cep-halic** or **oral sucker** is formed by the fusion of pros-tomium and first three segments. It is oval in shape and contains a cup-like depression, called **pre-oral chamber,** on the ventral side.

The **posterior** or **anal sucker** is formed by fusion of last seven body segments. It is circular and disc-like. It is highly muscular. Posterior sucker is better developed and larger in size than the anterior sucker.

Both the suckers help the leech in adhesion (attachment) and locomotion. The anterior sucker is also used in sucking blood of host during feeding of leech.

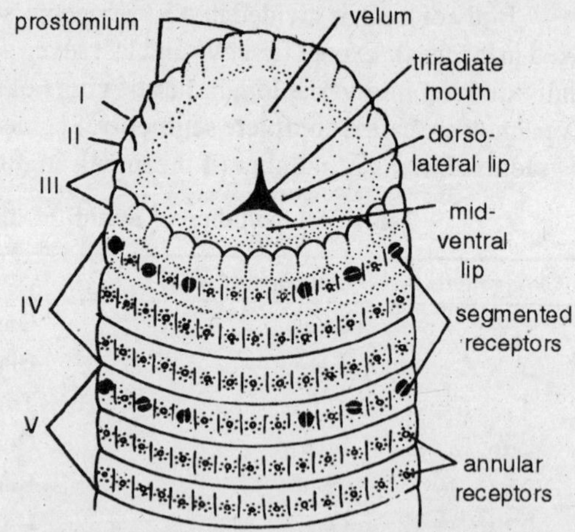

Fig. 35.2. *Hirudinaria.* Cephalic region.

6. External Apertures

***(i)* Mouth.** It is a narrow, triradiate aperture (Fig. 35.2) situated in the funnel-shaped pre-oral chamber of the anterior sucker. It is antero-ventral in position.

***(ii)* Anus.** It is a very small aperture situated mid-dorsally at the base of posterior sucker in the 26th segment.

***(iii)* Nephridiopores.** There are seventeen pairs of minute apertures, called **nephridiopores,** on the ventral surface. One pair of nephridiopores is located in the last annulus of each segment from 6th to 22nd segment.

***(iv)* Male genital pore.** It is a mid-ventral aperture situated in a groove between the 2nd and 3rd annuli of the 10th segment.

***(v)* Female genital pore.** It is a mid-ventral aperture situated in a groove between 2nd and 3rd annuli of 11th segment.

7. Sense Organs or Receptors

***(i)* Eyes.** The annulus of each of five segments carries a pair of eyes dorsally.

***(ii)* Segmental receptors.** First annulus of each segment carries segmental receptors in the form of papillae. First pairs of them occur dorsally and three pairs ventrally.

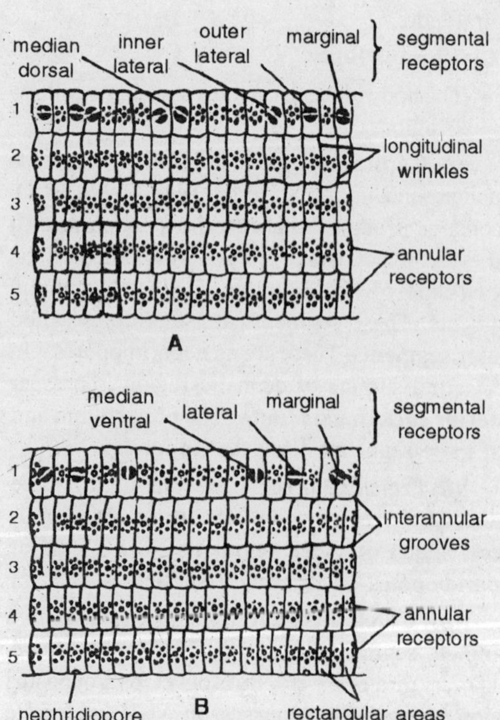

Fig. 35.3. *Hirudinaria.* A complete segment showing receptors. A—Dorsal view; B—Ventral view.

8. Annular receptors

There are thirty six **annular receptors** present in each annulus, of which eighteen are dorsal and eighteen are ventral in position. These occur in the form of minute elevated areas in 36 rectangular areas of each annulus (Fig. 35.3).

35.4 HISTOLOGY OF BODY WALL

The body wall of leech is glandular and dermo-muscular. It consists of five layers: cuticle, epidermis, dermis, muscle layers and botryoidal tissue (Fig. 35.4).

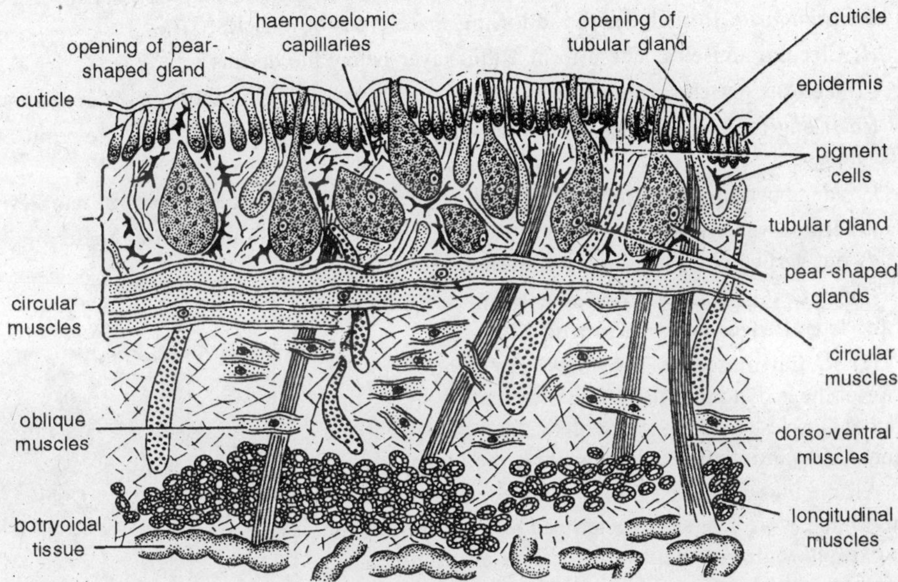

Fig. 35.4. *Hirudinaria*. T.S. of body wall.

1. Cuticle. It is the outermost, thin, transparent and elastic but firm protective layer. It is perforated by the openings of epidermal glands. Cuticle is secreted by the underlying epidermis and is cast off in patches periodically.

2. Epidermis. It consists of a single layer of hammer-shaped cells. The outer ends of these cells are closely fitted together while their narrower ends are nucleated. Some of the epidermal cells become modified to form unicellular pear-shaped glands and multicellular **receptor organs.**

Epidermal glands. These are of following types:

(i) Slime glands. These glands are pear-shaped or tubular. These are distributed all over the body and secrete a slimy mucus that covers the whole body. The main body of the glands lie in the dermis and connective tissue and their long and narrow ducts open to the surface. Each gland contains a nucleus, a thin layer of cytoplasm and numerous secretory granules.

(*ii*) Sucker glands. These glands are distributed only in the suckers and occur in clusters. These are pear-shaped in the anterior sucker and rounded in posterior sucker. The secretion of sucker glands serve to smoothen the surface for the attachment of suckers during locomotion.

(*iii*) Prostomial glands. These are pear-shaped and occur in groups in the prostomium. Their secretion plugs the ends of cocoon (=ootheca or egg case).

(*iv*) Clitelar glands. These glands are spherical in form and are found in the clitellar region (9th to 11th segment) and become active only during the breeding season. Clitellar glands are of two types: chitogenous and albumen glands. The **chitogenous glands** are located between the circular and oblique muscles and secrete a substance for the formation of wall of cocoon or ootheca. The **albumen glands** lie in the longitudinal muscles and secrete albumen which is filled up in the cocoon (it is used for the nourishment of developing embryos).

3. Dermis. It lies below the epidermis. It consists of fibrous connective tissue which is very thick and occupies much of the interior of body. Dermis consists of a network of connective tissue fibres, connective tissue cells, pigment cells, fat cells, muscle fibres and parts of haemocoelomic system. Epidermal glands also extend into this region.

4. Muscles. Musculature below the dermis is well developed forming largest part of the body wall of *Hirudinaria*. In includes the following three types of muscles:

***(i)* Circular muscles.** These form a thin layer below the dermis.

***(ii)* Oblique muscles.** These muscles are obliquely arranged below the layer of circular muscles.

***(iii)* Longitudinal muscles.** These muscles are quite powerful and form the main bulk of musculature.

Beside above muscles, there are a number of following isolated muscles in bundles:

***(a)* Dorso-ventral muscles.** They extend between the dorsal and ventral epidermis. These muscles run vertically on both sides of alimentary canal.

***(b)* Vertical muscles.** The dorso-ventral muscles situated near the lateral edges are called vertical muscles.

***(c)* Radial muscles.** These muscles extend from the interior of the body to the surface. In the pharyngeal region, the radial muscles are connected with the pharyngeal wall on the inner side and with the body wall on the outer side.

All the muscles of leech have similar basic structure. A muscle fibre is an elongated, fusiform and tubular cell having two distinct regions: the central granular **medulla** or **sarcoplasm** which contains nucleus and the peripheral, striated and contractile **cortex** or **myoplasm** (Fig. 35.5).

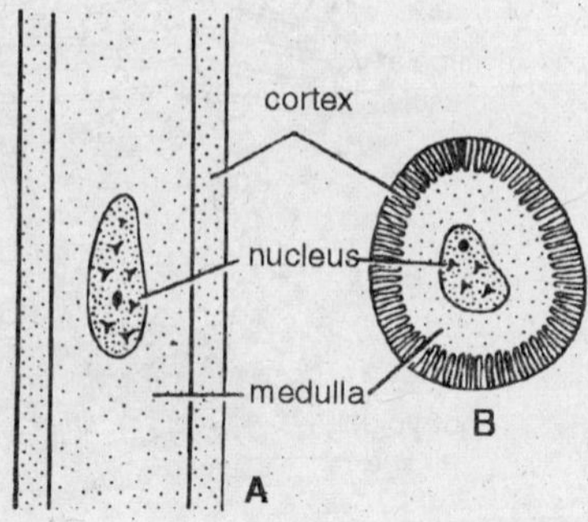

Fig. 35.5. *Hirudinaria*. A muscle fibre. A—L.S.; B—T.S.

5. Botryoidal Tissue

Botryoidal (*Gr. botryoeides* = cluster or grapes; *eidos* = shape) tissue is a characteristic tissue of jawed leeches and it resembles with bunches of grapes. It exists between the gut and muscles (longitudinal) almost completely filling the body cavity. Botryoidal tissue consists of a network of very fine capillary channels or sinuses of haemocoelomic system lined by small globular cells heavily laden with a **brown pigment** (which is a masked iron-compound probably derived from blood) partially corresponds to chloragogen tissue of oligochaetes. Its blood sinuses are roughly analogous to lymphatic system of vertebrates. Botryoidal tissue is a type of 'tissue of accumulation' which picks up particulate waste from blood and keep them forever.

Botryoidal tissue also contains another type of 'accumulation tissue' called **vaso-fibrous tissue** which contains an accumulated **green pigment** (another masked iron-compound probably derived from blood). Botryoidal tissue and vaso-fibrous tissues have excretory function.

Functions of body wall. The body wall of leech contains the following functions:

1. It protects internal organs from mechanical injury.
2. The cuticle and slimy mucous coat prevent loss of water through evaporation from the body.
3. The slimy mucous kills harmful organisms and protects the leech from infections.
4. Pigment cells of body wall produce camouflaging, protecting the leech from its predators.
5. It is richly supplied with haemocoelomic capillaries and makes gaseous exchange (respiration) possible through it.

6. It is sensory due to the presence of various types of receptors on it.
7. Clitellar glands secrete wall of cocoon and albumen.
8. Prostomial glands secrete plugs of the cocoon.
9. Body musculature helps in locomotion.

35.5 LOCOMOTION

The loss of septa, setae and coelomic compartments in leeches is correlated with a change from peristaltic burrowing to new modes of locomotion; leeches crawl or swim but never burrow using peristalsis. Leeches crawl on solid substratum by forming loops and swim in water.

Crawling

This type of locomotion is performed with the help of suckers and body musculature. During this type of locomotion:

(1) Leech fixes its posterior sucker on the substratum and contracts its circular muscles and relaxes longitudinal muscles. This causes maximum lengthening of the body. The diagonal muscles may be postural, keeping the animal rigid while it is raised and attached posteriorly.

(2) Now the leech fixes its anterior sucker on the substratum and relaxes the circular muscles and contracts the longitudinal muscles. This releases the hold of posterior sucker and shortens the body.

(3) This results in forward movement of posterior sucker. Thus, posterior end of body come closer to the anterior end and the body of leech is raised up in the middle forming a loop or D. The diagonal muscles may also enable leeches to twist their raised bodies while attached posteriorly.

(4) Once again the posterior sucker is fixed, the anterior sucker is released and the whole process is repeated again and again and leech moves forward.

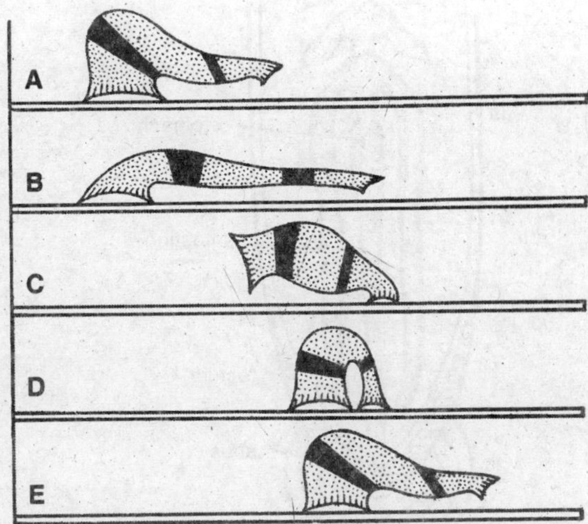

Fig. 35.6. *Hirudinaria.* Crawling.

Swimming. In the water the leech swims swiftly. During swimming the body becomes flat and the worm swims by successive undulating movements. The body is flattened by contraction of the dorsoventral musculature and waves of contraction along the longitudinal muscles produce vertical undulation.

35.6 DIGESTIVE SYSTEM

Digestive system consists of the alimentary canal and digestive glands associated with it.

1. Alimentary Canal

The alimentary canal of leech is a straight tube extending throughout the length of the body from the mouth to anus (Fig. 35.7). It is greatly modified due to its blood-sucking (sanguivorous) habit. Most part of the alimentary canal serves to store uncoagulated blood, while only a small portion serves for its digestion and absorption. The alimentary canal is differentiated into following three main parts:

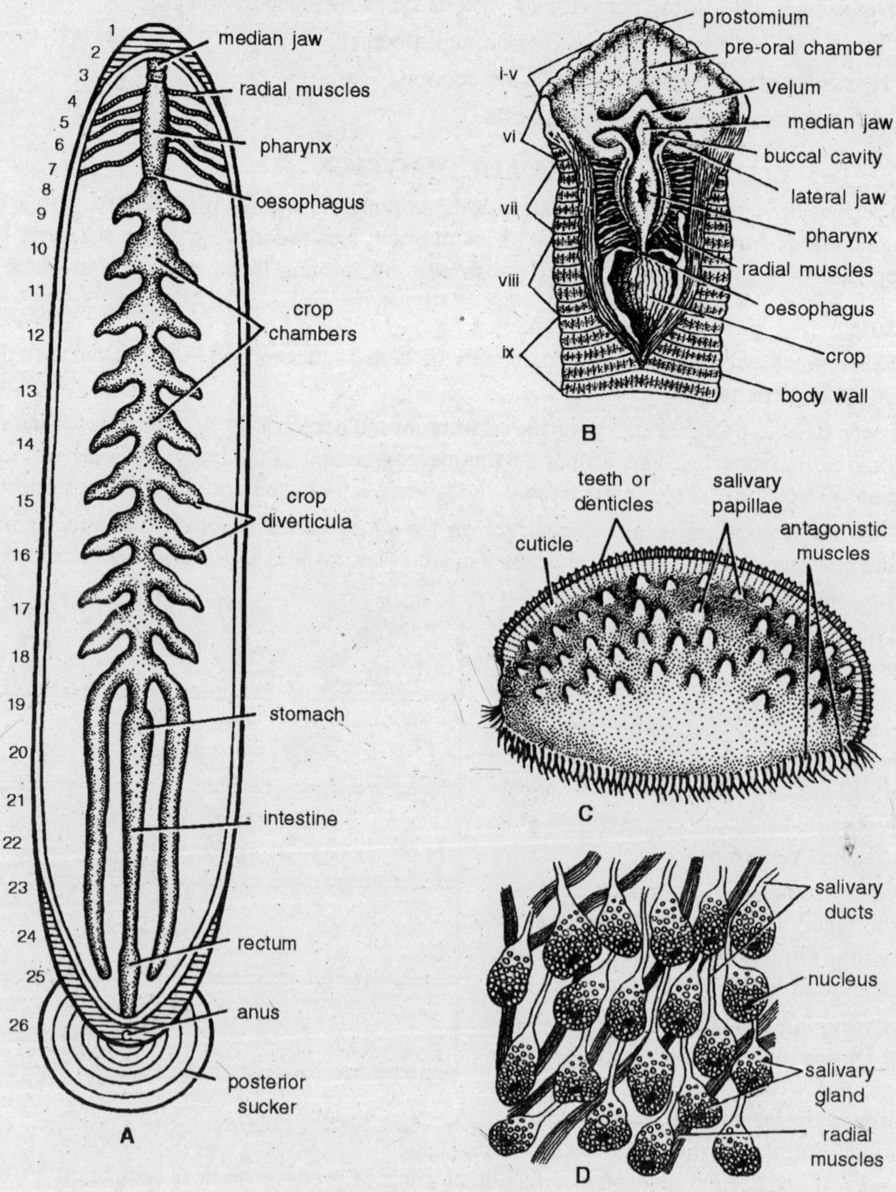

Fig. 35.7. *Hirudinaria*. A—Alimentary canal; B—Anterior part of leech dissected ventrally to show the buccal cavity and pharynx. C—A single jaw in lateral view; D—Salivary glands.

1. Stomodaeum (It is lined with ectodermal epithelium and is covered with cuticle).

2. Mesenteron (It is lined with endodermal epithelium and lacks any cuticular covering).

3. Proctodaeum (It is lined with ectodermal epithelium and covered by cuticle).

A. Stomodaeum. It includes the following three parts.

(1) Pre-oral chamber and mouth. The cup-shaped depression of the anterior sucker represents the **pre-oral chamber.** Its base is perforated by triradiate **mouth** which opens into the buccal cavity.

The pre-oral chamber and buccal cavity are separated from one another by a membranous partition, the **velum.** Velum forms three lips (two dorso-lateral and one ventral) around the mouth. Pre-oral chamber helps in blood-sucking.

(2) Buccal cavity. It is a very short chamber. In the mucous membrane of buccal cavity are embedded three crescentic **jaws,** one mid-dorsal and the other two are ventro-lateral in position. Their position corresponds to the position of three slits of mouth. Each jaw is a laterally-compressed crescentic and muscular cushion-like structure covered with cuticle. The cuticle is thickened at the free edge to form a dentigerous ridge bearing a row of minute **teeth** (denticle). As these teeth form a single series on each jaw, the jaw is termed as **monostichodont.** The median jaw bears 103 to 128 teeth, while lateral jaws bear 85 to 115 teeth each. Each lateral side of jaw carries 42–45 button-shaped salivary **papillae.** These papillae bear openings of salivary glands. The jaws are moved forward and backward by special muscles. These work simultaneously as semicircular saws producing **triradiate Y-shaped wound** in the skin of host.

(3) Pharynx. The buccal cavity leads into an oval and highly muscular sac, the **pharynx.** It extends from 5th to 8th segments. Bundles of **radial muscles** extend from the wall of pharynx to the body wall. Their contraction widen the pharyngeal cavity and creates a suction force for sucking blood. The inner surface of pharynx bears longitudinal folds.

B. Mesenteron. It includes the following four parts:

(*i*) Oesophagus. It is a short and narrow tube connecting pharynx with crop. Its internal lining is greatly folded.

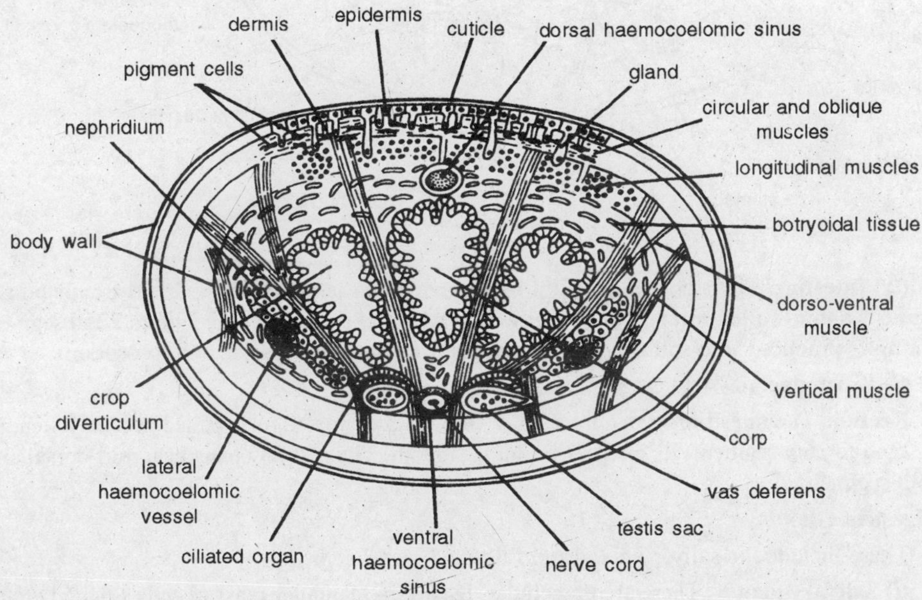

Fig. 35.8. *Hirudinaria.* T.S. through crop.

(*ii*) Crop. The crop forms the largest region of the alimentary canal extending from the ninth to the eighteenth segment. It has ten thin-walled **chambers,** one in each segment and they are connected with one another by more or less circular apertures surrounded by sphincters. Each chamber is differentiated into a small **anterior part** and a broad **posterior part.** The posterior part of each chamber is prolonged into a pair of lateral blind, backwardly directed outgrowths, the **diverticula**

or **caeca.** The size of crop chambers and caeca go on gradually increasing towards the posterior side. The last chamber of the crop is the largest and its caeca are prolonged backwards upto twenty-second segment. Crop chambers and their caeca are highly elastic and used for **storing** blood meal. One cropful of blood lasts for several months and digested slowly.

***(iii)* Stomach.** The last chamber of crop extends posteriorly into a funnel-shaped narrow tube which communicates with the heart-shaped **stomach** through a sphinctered aperture. The stomach is a short tube lying in 9th segment. It has both **secretory** and **absorptive** epithelial cells and its inner lining is produced into many anastomosing transverse folds.

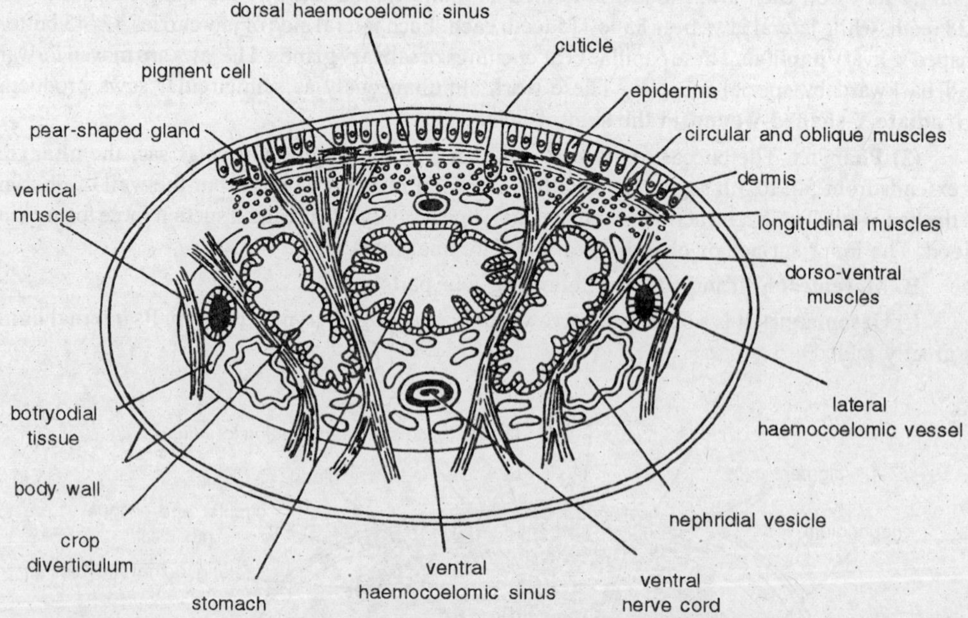

Fig. 35.9. *Hirudinaria* (leech). T.S. through stomach region.

***(iv)* Intestine.** The stomach leads into intestine which is not clearly marked from stomach. Intestine is a thin-walled, straight and narrow tube which extends from the 19th to 22nd segment. Its lining is produced into spiral villi-like process to increase the surface of absorption.

C. Proctodaeum. It consists of rectum only.

Rectum. The intestine opens into a short, thin-walled, dilated tube, called **rectum** extending from 22nd to 26th segment. Its internal lining is smooth. The rectum opens by a mid-dorsal anus in 26th segment.

2. Digestive Glands

These include the salivary glands and isolated gland cells.

***(i)* Salivary glands.** These are unicellular, pyriform, glandular mass of cells situated around and behind the pharynx (Fig. 35.7D). Each glandular cell has elongated fine ductule which finally enters the jaws to open in raised salivary papillae. The secretion of these glands is **saliva** which contains an anticoagulant substance called **anticoagulin** or **hirudin.** Hirudin prevents coagulation of blood while leech is feeding.

***(ii)* Gland cells.** These are in the form of **goblet cells,** found scattered throughout the lining (or mucous membrane) of crop. These secrete mucus in the crop.

Food and Feeding

Hirudinaria is sanguivorous in habit, feeding on blood of cattle and other mammals visiting the water source. In feeding, the leech applies its oral sucker very tightly on the skin of its prey, the jaws move towards and away from each other and they painlessly puncture the skin. The incision is anaesthetized by a substance of unknown origin. The oozing blood is sucked up by the suctorial pharynx. Pump-like action of the pharynx sucks in large quantities of blood to fill the crop. Secretion of **hirudin** prevents coagulation of blood.

At a single meal a leach usually sucks blood several times its own weight. It is sufficient for several months or even a year. After feeding, leech drops of the host and rests in concealed shelter.

Digestion

The sucked blood is stored in uncoagulated state in the crop-chambers where it is **haemolysed.** Red blood corpuscles are broken down. The haemoglobin from ruptured RBCs is separated and dissolved in plasma. Due to absorption of water from blood plasma by crop wall, the blood becomes thicker and jelly-like and dark red or blackish red in colour. From crop blood enters stomach drop by drop. This is controlled by sphincter that surrounds the opening of crop into stomach. Blood proteins (*e.g.*, globin) are digested in stomach by proteolytic enzymes.

Digestion in leech is peculiar in a number of respects. The gut secretes no amylases, lipases, or endopeptidases. The presence of only **exopeptidases** perhaps explains the fact that digestion in blood-sucking leeches is so slow. Also characteristic of the leech gut is a symbiotic bacteria flora that is important in nutrition. The bacteria (*e.g.*, *Pseudomonas hirudinicola* of *Hirudo medicinalis*) breaks down high-molecular-weight proteins, fats and carbohydrates. The bacteria may also produce vitamins and other compounds that are used by the leech host.

Rectum stores the undigested waste for sometime then egests it out through the **anus.**

35.7 RESPIRATION

Hirudinaria or leech is an aquatic animal. It lacks any special organ of respiration. The skin of leech serves as a respiratory organ. The capillaries containing haemocoelomic fluid extend in between the cells of epidermis through which exchange of respiratory gases takes place by **diffusion.** The carbon dioxide dissolved in haemocoelomic fluid passes out and the oxygen dissolved in water enters in the body. The skin is kept wet by the surrounding water, while the mucus secreted by the slime glands of the epidermis prevent it from drying.

35.8 HAEMOCOELOMIC SYSTEM

Coelom

The perivisceral coelom is absent in leech. It is obliterated and is filled with pigmented botryoidal tissue. The original coelom is replaced by a system of **longitudinal channels,** their branches and spaces or **sinuses.** It is called **haemocoelomic system** and filled with red **haemocoelomic fluid** containing colourless amoeboid corpuscles and extracellular haemoglobin dissolved in the fluid. The true coelom of leech is restricted to spaces around the testes sacs, ovisacs and gonoducts. It is filled with colourless coelomic fluid containing colourless corpuscles.

Haemocoelomic System

Leech lacks a blood vascular system but contains a haemocoelomic system (*i.e.*, modified coelomic circulatory system). Because the haemocoelomic system is evolved from the original coelom all of its vessels and channels are lined by an **endothelium,** the **peritoneum** along with connective tissue. The haemocoelomic fluid is propelled by the muscular contractions of the lateral longitudinal channels.

Haemocoelomic system of leech contains following four longitudinal channels:

1. A dorsal longitudinal haemocoelomic channel;
2. A ventral longitudinal haemocoelomic channel;
3. Two lateral longitudinal channels.

1. Dorsal Channel

The dorsal channel or dorsal sinus runs mid-dorsally below the body wall and above the alimentary canal. It is thin-walled, non-contrac-tile vessel which lacks valves. The haemocoelomic fluid flows in it from the posterior end to anterior end (from behind forward). Dorsal channel is an **distributing channel.** Anteriorly, dorsal channel ramifies in the sixth segment and supplies haemocoelomic fluid to first five segments. Posteriorly it bifurcates in the 22nd segment and two branches run downward around the rectum and join the posterior dilation of ventral sinus (Fig. 35.10).

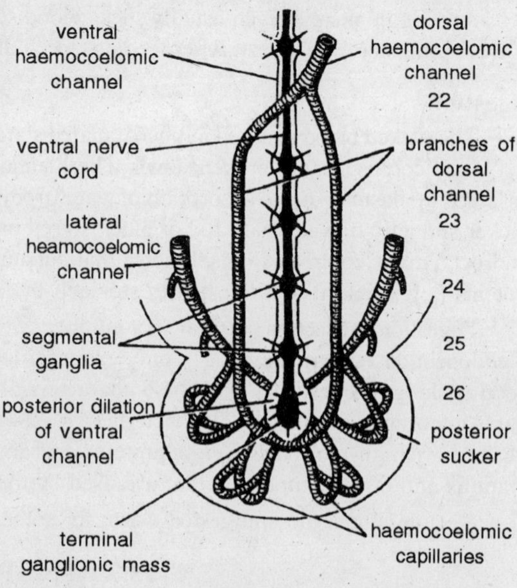

Fig. 35.10. *Hirudinaria*. Posterior end of body showing union of four haemocoelomic channels.

In each segment, the dorsal channel gives off two pairs of **dorsolaterals** taking the haemocoelomic fluid to the dorsal and dorso-lateral regions of body wall and several dorso intestinals going to the alimentary canal (Fig. 35.11).

2. Ventral Channel

The ventral channel is also a thin-walled, non-contractile distributing vessel running mid-ventrally below the alimentary canal from one end of body to the other. It is a broad vessel enclosing the entire central nervous system and lacks valves. In it, the haemocoelomic fluid flows from anterior to the posterior end. It gives off two pairs of branches in each segment.

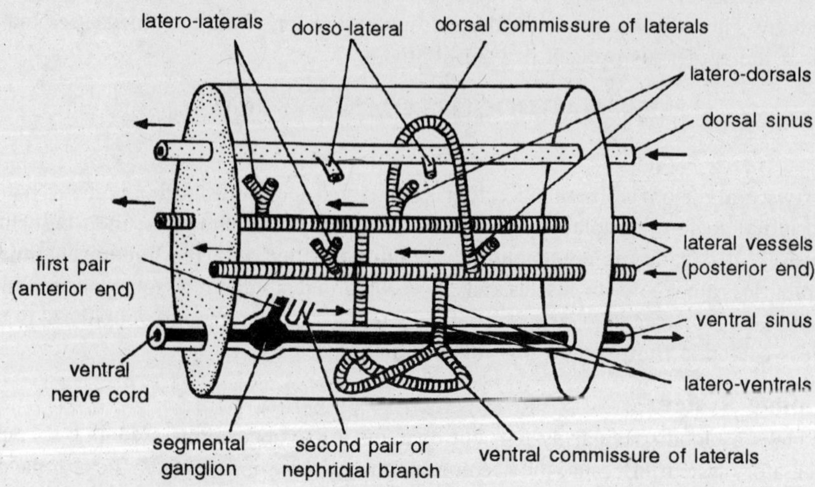

Fig. 35.11. *Hirudinaria*. Diagrammatic representation of main haemocoelomic channels of a segment.

(*i*) First pair or cutaneous branches. Two cutaneous branches are given off at the level of segmental nerve ganglion. Each branch immediately bifurcates into a **ventral branch** and an

abdomino-dorsal branch. The former forms a network in the ventro-lateral region of the body wall while the latter passes vertically upwards to dorsal body wall (Fig. 35.12).

(*ii*) Second pair or nephridial branches. These branches are issued just behind the segmental nerve ganglion. Each runs outwards to supply the testis-sac and the nephridium, where it enlarges to a **perinephrostomial ampulla** which encloses the ciliated organ. There are only eleven pairs of nephri-dial branches or one pair in each of the segments from 12th to 22nd segment.

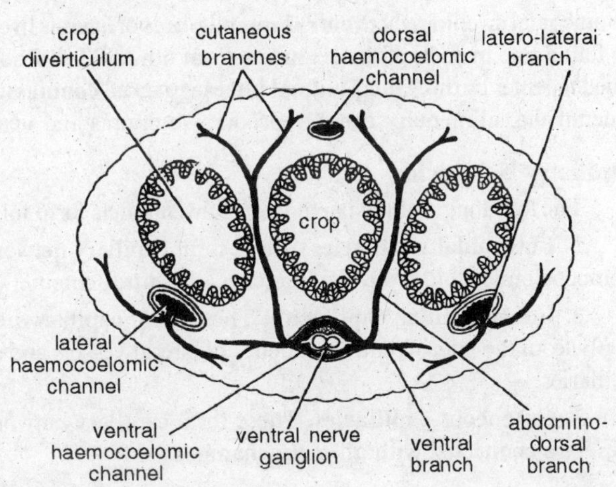

Fig. 35.12. *Hirudinaria*. T.S. body through dorso-abdominal branches.

3. Lateral Channels

A pair of lateral channels extend one on either side of the alimentary canal from one end of body to the other. Each is a wide, thick-walled, muscular, contractile vessel with seg-mentally arranged valves. In it, hae-mocoelomic fluid flows from poste-rior to anterior end (from behind forward). The lateral channels break up in the sixth segment and form capil-lary network. Posteriorly they join the dilation of ventral sinus and thus, all four channels become continuous at this level (Fig. 35.10). Both lateral channels are **collecting** as well as **distributing** channels. In each segment, the lateral channel receives on its outer side two branches a latero-lateral and a latero-dorsal and gives off one branch, the latero-ventral (Fig. 35.13).

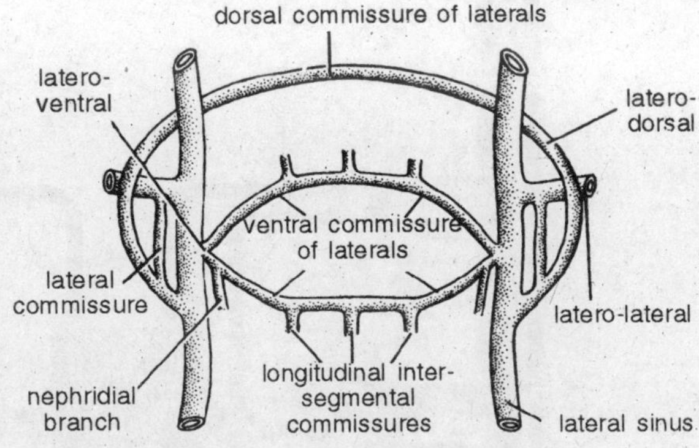

Fig. 35.13. *Hirudinaria.* Lateral vessels and their branches in a segment.

(*i*) Latero-lateral. It is a small vessel. It collects haemocoelomic fluid from the lateral region of the body wall and nephridia and opens into the lateral vessels at the level of nephridial vesicle.

(*ii*) Latero-dorsal. It is a large vessel opening in the lateral channel at the level of main lobe of the nephridium. It receives branches from the dorsal and dorso-lateral regions of the body wall, the gut wall and nephridium. In each segment the two latero-dorsals from the opposite side are connected together above the dorsal channel by a **transverse loop.** This is called **dorsal commissure** of lateral channels. There are 17 such dorsal commissures one in each segment from 6th to 22nd. A small branch, called **lateral commissure,** connects the latero-dorsal branch to the latero-lateral branch of its side.

The latero-laterals and the dorso-laterals are collecting vessels and their openings into the lateral channels are guarded with valves.

(*iii*) Latero-ventral. Each lateral channel gives out a **latero-ventral** branch which bifurcates into an **anterior** and a **posterior** branch and each join its fellow of the opposite side mid-ventrally to form a rhomboidal swelling, *i.e.*, **ventral commissures** of laterals in each segment. Eighteen such rhomboids are found one in each of the segments from 6th to 23rd. Rhomboids of adjacent segments are also joined together by three **longitudinal intersegmental commissures.** The latero-ventrals issue branches to nephridia, alimentary canal, reproductive organs and ventral side of body wall.

Capillary Networks

The four longitudinal haemocoelomic channels form following three sets of capillary network:

1. Botryoidal capillaries. These form capillary networks around the viscera. These receive haemocoelomic fluid from the branches of ventral channel and return it to the lateral vessels.

2. Intermuscular capillaries. These form capillary networks in the muscles. They receive supply from the latero-ventral branches of lateral vessels and also communicate with the botryoidal capillaries.

3. Cutaneous capillaries. These form capillary networks which extend into the epidermis. These are connected with all main channels.

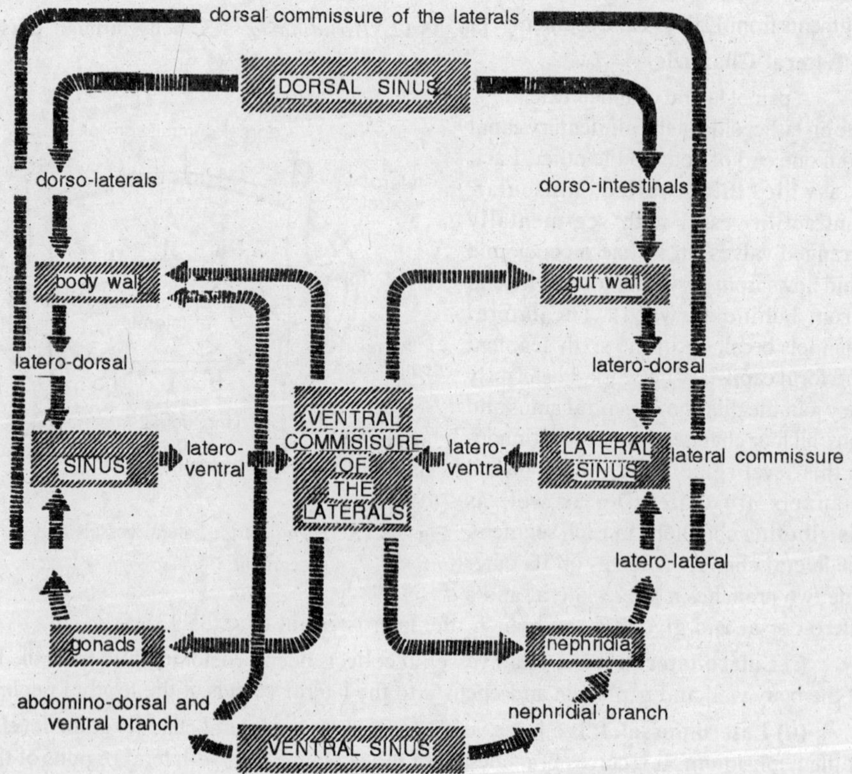

Fig. 35.14. *Hirudinaria*. Summary of course of circulation of haemocoelomic fluid.

Course of Circulation of Haemocoelomic System

The haemocoelomic fluid flows from posterior to anterior direction in the dorsal and lateral channels while it flows from anterior to posterior direction in the ventral channel. All the four longitudinal channels become continuous in the 26th segment. The dorsal and ventral channels distribute haemocoelomic fluid while the lateral channels distribute as well as collect the haemocoelomic fluid.

The **dorsal channel** supplies fluid to dorsal and dorso-lateral parts of body wall and the entire alimentary canal, from where it is collected by latero-dorsals which carry the fluid to the lateral channels.

The **ventral channel** supplies the fluid to the ventral, ventro-lateral and ventral parts of the body wall and the nephridia including ciliated organs. From all these parts, the fluid is collected by the latero-lateral branches of the lateral channels.

The **lateral channels** serve as tubular "coelomic hearts" and supply the fluid to the nephridia, reproductive organs, ventral gut-wall and the ventral body wall through latero-ventral branches. Haemocoelomic fluid from all these parts is collected by latero-lateral and latero-dorsal branches of the lateral vessels.

35.9 EXCRETORY SYSTEM

Nephridia are the excretory and osmoregulatory organs of leech. These are segmentally arranged, hollow tubes present one pair in each segment from 6th to 22nd segment. Thus, in total there are seventeen pairs of nephridia in *Hirudinaria*. These nephridia are of two types:

1. Pretesticular nephridia. First six pairs of nephridia in the segments 6–11 are without testis sacs and are called pretesticular nephridia.

2. Testicular nephridia. The remaining eleven pairs of nephridia are with testis sacs; they are called testicular nephridia. They are found in segments from 12th to 22nd.

1. Testicular Nephridia

A typical testicular nephridium is a horse-shoe shaped structure. It consists of six parts: (*i*) Main lobe; (*ii*) Vesicle and vesicle duct; (*iii*) Apical lobe; (*iv*) Inner lobe; (*v*) Initial lobe or testicular lobe; and (*vi*) Ciliated organ (Fig. 35.15).

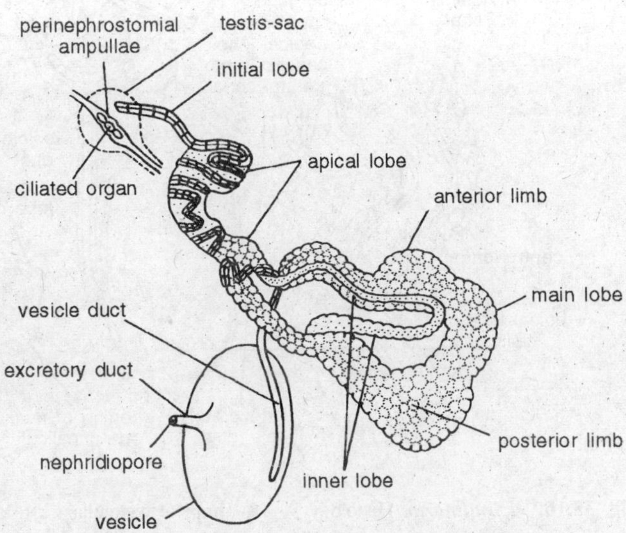

Fig. 35.15. *Hirudinaria*. A testicular nephridium.

(*i*) Main lobe. It forms horse-shoe proper of nephridium. It lies in a ventro-lateral position between the adjacent diverticula of crop. It consists of two unequal limbs: a longer **anterior limb** and a shorter **posterior limb.** Histologically main lobe comprises of large-sized polyhedral cells.

(*ii*) Vesicle and vesicle duct. From the ventral end of anterior limb of the main lobe arises a narrow **vesicle duct** which runs posteriorly to open into the vesicle. The **vesicle** or **bladder** is a large, oval, non-contractile thin-walled sac, lined internally by a non-ciliated epithelium. A short narrow **excretory duct** arises from the inner side of vesicle which opens to the exterior through **nephridiopore** surrounded by sphincter muscles.

(*iii*) Apical lobe. The posterior limb of the main lobe continues to form stout **apical lobe.** It lies in an antero-posterior position beneath the gut. The free end of apical lobe is bent on itself like the handle of a walking stick. The cells lining the apical lobe are large with regular intercellular canals.

4. Inner lobe. Running between the two ends of main lobe and situated in its concavity is found a short narrow **inner lobe.** It runs forward along the outer border of apical lobe for about half its length and then joins the apical lobe. Its cells are long and tubular. It carries a central lumen.

(v) Initial lobe. It is slender, much elongated transparent cord-like structure. It remains coiled around the apical lobe. Its posterior end joins the anterior limb of the main lobe. Anterior end of initial lobe is free and blind and extends over the testis sac to lie close to **perinephrostomial ampulla.** Histologically, initial lobe consists of a single row of elongated hollow cells placed end to end having the intracellular canal (Fig. 35.16) which gives off several diverticula in each cell.

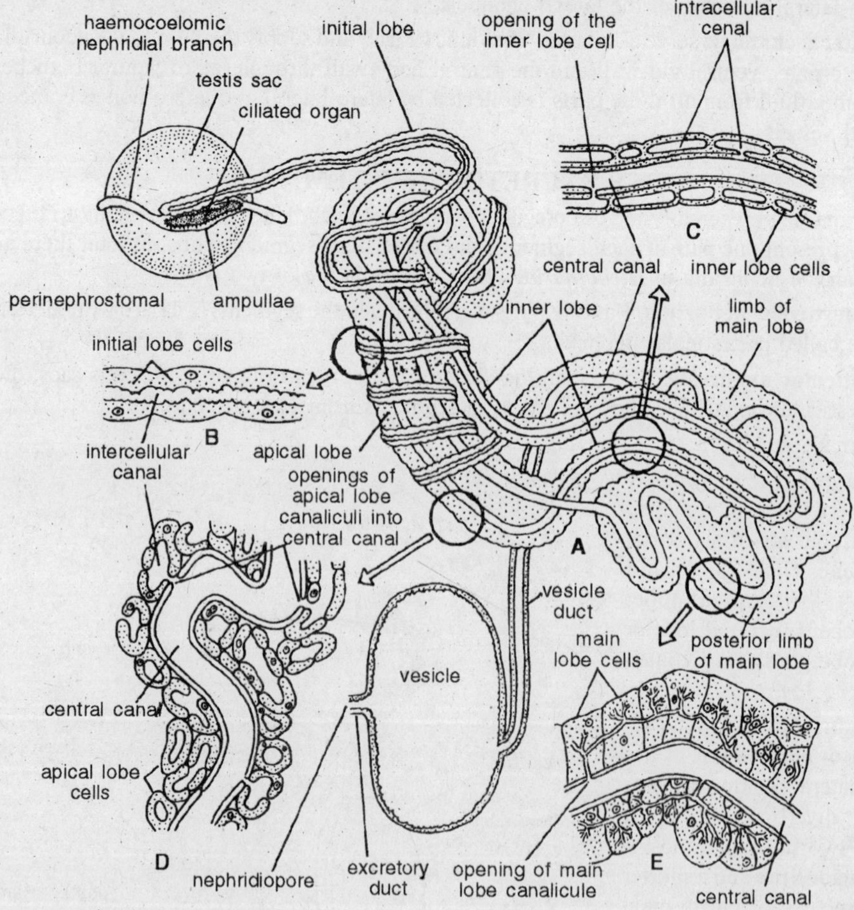

Fig. 35.16. *Hirudinaria*. Histology. A—A complete testicular nephridium showing central canal; B—A part of initial lobe in L.S.; C—A portion of inner lobe in L.S.; D—A portion of apical lobe in L.S.; E—A portion of main lobe in L.S.

(vi) Ciliated organ. A ciliated organ is completely enclosed within the perinephrostomial ampulla which is a dilation of the haemocoelomic system. A ciliated organ consists of a central **reservoir** which is perforated and a peripheral part containing closely-set **ciliated funnels.** The reservoir is spongy mass of connective tissue cells which manufacture **coelomic amoebic corpuscles.** The outer wall of ciliated funnel is perforated by numerous minute openings for the ciliated funnels. Each ciliated funnel is like an ear-lobe, with about one-fourth of its margin incomplete along one side and is covered with outwardly directed cilia on its margin and inner surface (Fig. 35.17).

Ciliated organ corresponds to coelomoduct of polychaeta and not to nephrostomes of nephridia. In the embryo, the ciliated organ remains connected with the nephridium and probably performs

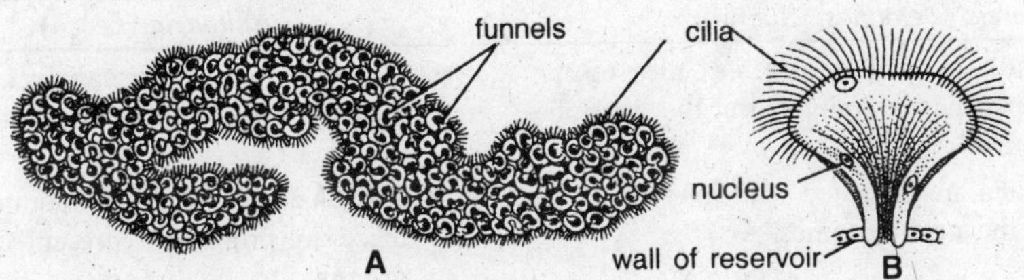

Fig. 35.17. *Hirudinaria*. A—Entire ciliated organ; B—A single funnel.

excretory function. In the adult, it loses excretory function and has no direct connection with the nephridium. Ciliated organ of adult, however, becomes associated with the haemocoelomic system and manufactures coelomic corpuscles.

There exists a central canal which originates in anterior part of apical lobe and traverses through the different lobes of nephridium and finally enters the vesicle duct. It forms a long continuous passageway across the nephridium and carries the excretory products secreted by the glandular cell masses of the nephridium and finally discharges into the vesicle through the vesicle duct.

2. Pre-testicular Nephridium

The first six pairs of nephridia are situated in segments 6th to 11th which lack testis sacs, perinephrostomial ampullae and ciliated organs. They are called pretesticular nephridia and resemble the testicular nephridia in structure. However, their initial lobes end loosely in the general connective tissue on either side of the ventral nerve cord (Fig 35.21).

Physiology

Nephridium of leech is both **osmoregulatory (Haupt 1974)** and **excretory** in function. Excess of water and nitrogenous wastes (chiefly **ammonia**) are excreted out by the nephridium. Nephridial gland cells separate the waste products from the haemocoelomic fluid. The excretory fluid is collected in the vesicle and voided through the nephridiopores. Leech excretre watery osmotic urine. The salts are reabsorbed from the urine by the nephridial tissues.

Waste particles are also picked up by botryoidal tissues (and by vasofibrous tissue) and are retained for life. Amoeboid corpuscles in the coelomic fluid are also phagocytic and play an important role in excretion.

COMPARISON OF NEPHRIDIA OF *NEREIS* AND *HIRUDINARIA*

Table 35.2 Comparison of excretory system of nephridia of *Nereis* and *Hirudinaria*.

Nereis (*Neanthes; clamworm*)	*Hirudinaria* (*Leech*)
1. Excretory system consists of **nephridia** only.	1. Excretory system consists of **nephridia** and **botryoidal tissue.**
2. **Nephridia** are segmentally arranged coiled tubes of very large size.	2. **Nephridia** are segmentally arranged coiled tubes of large size.
3. Paired nephridia are found in all segments except few anterior and posterior segments of the body.	3. Paired nephridia are found only in the segments from sixth to twenty-second.
4. All nephridia are alike.	4. Nephridia are of two types: **(*i*) Pre-testicular nephridia,** which are six pairs in number and located from segments sixth to eleventh, one pair in each segment. **(*ii*) Testicular nephridia,** which are eleven pairs in number and distributed one pair in each segment from twelth to twenty second.

Nereis (*Neanthes; clamworm*)	*Hirudinaria* (*Leech*)
5. All nephridia are **exonephric,** *i.e.*, they open out by nephridiopores situated at the base of parapodium.	5. All nephridia are **exonephric,** *i.e.*, they open directly to the exterior by nephridiopores.
6. All nephridia are provided with internal openings, the **nephrostomes.**	6. Adult's nephridia are without nephrostomes; these are probably represented by **ciliated organs.**
7. A typical nephridium is made of a small narrow neck and an oval **main body** (Fig. 34.15).	7. A typical testicular nephridium is made of a **main lobe,** a **vesicle** with duct, an **apical lobe,** an **initial lobe** and a **ciliated organ** (Fig. 35.15).
8. Each nephridium is an oval and irregular in shape.	8. Each nephridium is horse-shoe shaped.
9. Sac-like vesicle is not found.	9. Sac-like vesicle or bladder is present.
10. Nephridia remove excretory wastes from blood and coelomic fluid.	10. Nephridia remove excretory wastes from haemocoelomic fluid.
11. No other excretory organ/tissue is found.	11. The botryoidal tissue is also involved in excretion (removal and permanent storage).
12. Ammonotelic.	12. Ammonotelic.

Differences Between Testicular and Pretesticular Nephridia

Table 35.3 Differences in testicular and pretesticular nephridia of leech.

Testicular nephridia	*Pretesticular nephridia*
1. These are associated with testis-sacs.	1. These are not associated with testis-sacs.
2. These are eleven pairs in number and are present in only those segments that have testis-sacs.	2. These are six pairs and are present in those segments which do no have testis-sacs.
3. Testicular nephridium has six parts—main lobe, inner lobe, apical lobe, initial lobe, vesicle, ciliated organs.	3. Ciliated organ is absent.

35.10 SENSE ORGANS OR RECEPTORS

The receptors or sense organs of *Hirudinaria* consist of specially modified epidermal cells. These are of following types; 1. Free nerve endings; 2. Annular receptors; 3. Segmental receptors and 4. Eyes.

1. Free nerve endings. These are found all over the body between the epidermal cells. The ganglia cells of these nerve endings lie beneath the epidermis. Free nerve endings are probably **chemoreceptors.**

2. Annular receptors. About 36 annular receptors are found to be equally distributed on the dorsal and ventral sides in each annulus (*i.e.*, 18 on dorsal surface and 18 on ventral surface). Each receptor is formed of a group of flattened overlapping cells and projects out of body wall like a minute papilla (Fig. 35.18). These are **tactile organs** or **tango-receptors.** Annular receptors get their nerve supply from the anterior and posterior lateral nerves.

3. Segmental receptors. First annulus of each body segment bears four pairs of segmental receptors on the dorsal surface and three pairs on the ventral surface. These are larger than the annular receptors and occur in the form of white patches elevated on the elliptical papillae. Each segmental receptors consists of two types of cells (Fig. 35.19).

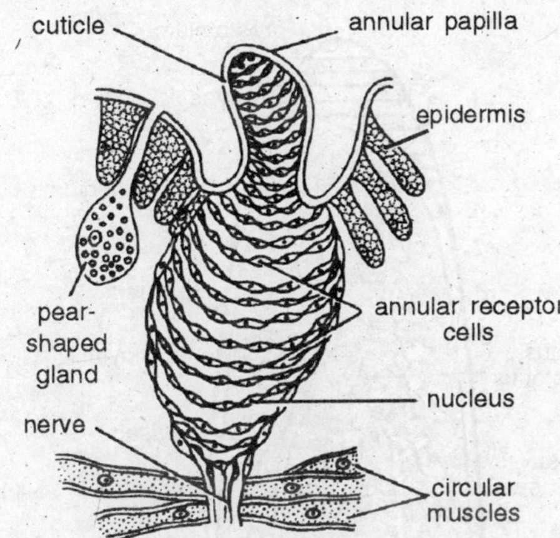

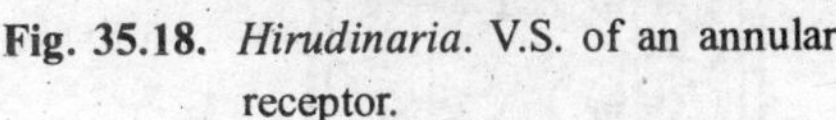

Fig. 35.18. *Hirudinaria*. V.S. of an annular receptor.

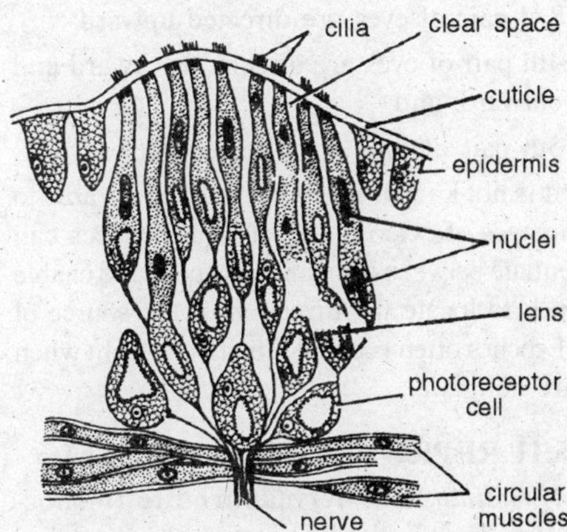

Fig. 35.19. *Hirudinaria*. V.S. of dorsal segmental receptor.

1. Tangoreceptors (or the tactile cells)
2. Photoreceptors (or the light perceiving cells)

(1) Tangoreceptors are about 5 to 10 long and slender cells which are separated from one another by clear spaces. These cells bear hair-like processes on their outer free ends. These are tactile in function (*i.e.*, sensitive to touch).

(2) Photoreceptors are found only in dorsal receptors. These consist of crescentic **optic organelle** or **lens** of hyaline substance which receives light rays.

4. Eyes. Leech possesses five pairs of simple eyes arranged as black spots in a semi-circle along the dorsal margin of the anterior sucker. A pair of eye is found in the first annulus of each of the first five segments. Each eye is cylindrical or cup-shaped with its long axis perpendicular to the body surface. Its wall is formed of black pigmented tissue and encloses numerous refractive or photoreceptor cells arranged in longitudinal rows. Each photoreceptor cell contains a large hyaline crescentic **lens** or **optic organelle** surrounded by a very thin peripheral layer of cytoplasm containing a small rounded nucleus. The distal free end of the eye is protected by a cap of transparent epidermal cells called **cornea** and a layer of cuticle. An optic nerve enters each eye at its base and runs along its median axis and gives off fine branches to all photoreceptor cells.

All the eyes are not of equal size; the first and second pairs are the largest while the remaining pairs are comparatively smaller in size, the fifth pair being the smallest. The eyes of leech are directed in different directions and each eye can receive light only from one direction:

1. 1st pair of eyes are directed forward;
2. 2nd pair of eyes are directed forward and outward;

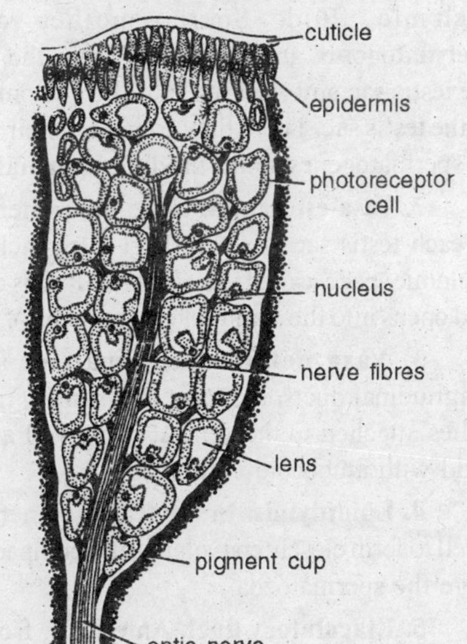

Fig. 35.20. *Hirudinaria*. V.S. of an eye.

3. 3rd pair of eyes are directed upward;
4. 4th pair of eyes are directed backward and outward; and
5. 5th pair of eyes are directed backward.

It is not known whether the eyes are able to form images of external objects but the leech can differentiate between light and darkness and enable the leech to locate the direction of the source of light. Leeches often become sensitive to light when they are hungry.

35.11 REPRODUCTIVE SYSTEM

Hirudinaria is **hermaphrodite** (monoecious), *i.e.*, each individual possesses a complete set of male and female reproductive organs (Fig. 35.21). In leech, copulation and cross-fertiliza-tion occur as a rule.

Male Reproductive Organs

The male sex organs are: 1. Testis sacs; 2. Vasa efferentia; 3. Vasa defferentia; 4. Epididy-mes; 5. Ejaculatory duct; 6. Atrium (Fig. 35.22).

1. Testis sacs. Each body segment from 12th to 22nd contains a pair of small, hollow and spherical **testis sacs** on either side of the ventral nerve cord. Thus, normally there are eleven pairs of testis sacs but sometimes 23rd and 24th segments also contain testis sacs. These testis sacs are true coelomic spaces and are filled with a colourless coelomic fluid. Sperm mother cells or **spermatogonia** are budded off from the wall of the testis sac and released into the coelomic fluid of the testis sac. There they complete their process of spermatogenesis and produce **spermatozoa.**

2. Vasa efferentia. From the posterior end of each testis sac arises a short duct enclosing a coelomic space, called **vas efferens.** It runs outward and opens into the common vas deferens of its side.

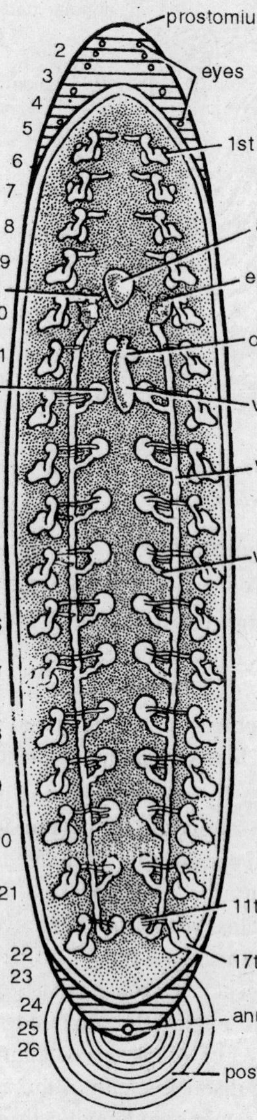

Fig. 35.21. *Hirudinaria*. Reproductive and nephridial (*i.e.*, urinogenital) system.

3. Vasa deferentia. A pair of slender longitudinal ducts, the **vasa deferentia,** run forward from twenty-second to the eleventh segment. It lies attached to the ventral body wall and is enclosed in a tubular coelomic space containing a fluid with amoeboid corpuscles.

4. Epididymis. In the tenth segment the vas deferens comes out of the coelomic space and swell to form closely convoluted and compact mass, the **sperm-vesicle** or **epididymis.** Both epididymes store the spermatozoa.

5. Ejaculatory duct. Anteriorly, from the inner side of the epididymis arises a short narrow ejaculatory duct. It runs inward horizontally and opens into a common atrium.

6. Atrium. It is a small pyriform sac situated midventrally in ninth and tenth segments (Fig. 35.22). It opens to outside in tenth segment by a mid-ventrally located **male genital pore.** The artium has following two distinct parts (Fig. 35.23):

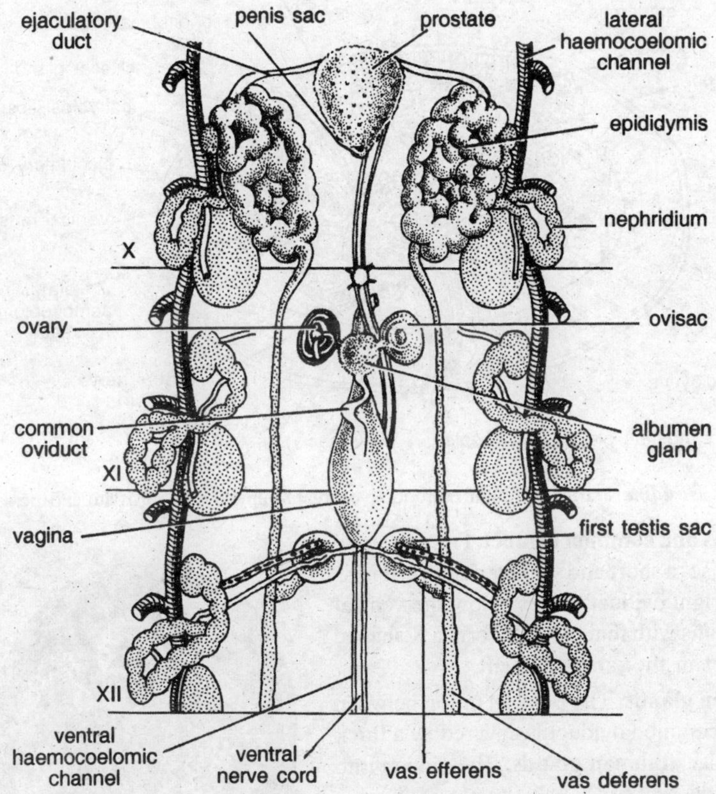

Fig. 35.22. *Hirudinaria.* Male and female reproductive organs.

(*i*) Prostate chamber. It is the anterior broad part of atrium. Its wall are thick and muscular and are surrounded by numerous unicellular **prostate glands.** Each gland opens into the prostate chamber by a slender duct. The spermatozoa are transferred into this chamber by the ejaculatory ducts. The sperms become glued together by the prostatic secretion into packets or bundles. These bundles are called **spermatophores.**

(*ii*) Penis sac. It is an elongated posterior part of atrium. It is highly muscular and contains a coiled filamentous tubular and eversible **penis.** The penis is often seen protruding through the male genital pore. It helps in copulation and transfer of spermatophores into the vagina of other leech.

Female Reproductive Organ

Female sex organs are: 1. Ovisacs; 2. Ovaries; 3. Oviducts; 4. Common oviduct; 5. Vagina; and 6. Albumen glands.

1. Ovisacs. There is a pair of ovisacs situated ventro-laterally in eleventh segment one on either side of the ventral nerve cord. Each ovisac is a hollow globular sac filled with colourless coelomic fluid. Each ovisac contains a single **ovary** (Fig. 35.23).

2. Ovaries. Leech contains a pair of ovaries, one in each ovisac. Each ovary is long, coiled, delicate, thread-like and contains a nucleated cytoplasm. The two ends of ovary are swollen and club-shaped. The ova are budded off from the ovary. The ovary lies floating in the coelomic fluid of ovisac.

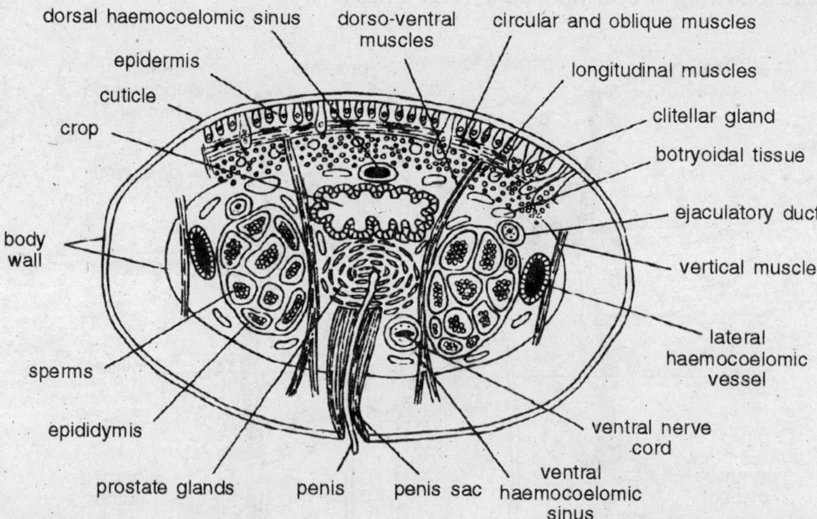

Fig. 35.23. *Hirudinaria*. T.S. through reproductive organs (epididymis, prostate and penis sac).

3. Oviducts and common oviduct. From the base of each ovisac arises a short and slender oviduct. It runs backwards, the right oviduct passes below the ventral nerve cord and joins with that of left to form a S-shaped **common oviduct** in the same segment.

4. Albumen glands. The point of union between the oviduct and common oviduct is covered by a thick layer of unicellular **albumen glands.** These open into the common oviduct.

5. Vagina. It is a large pear-shaped bag lying in the posterior part of eleventh segment (Fig. 35.22) During breeding season, vagina increases in size and becomes internally thrown into a large number of longitudinal folds. It opens to the exterior through a mid-ventral **female genital pore** in eleventh segment.

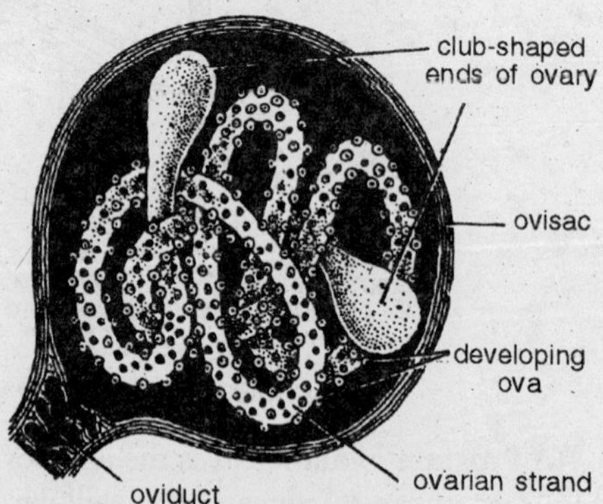

Fig. 35.24. *Hirudinaria*. An ovisac dissected to show the ovary.

The ova are budded off from the ovaries and pass through the oviducts into the vagina where they remain for some time.

Copulation and Fertilization

In leech, **copulation** takes place in the months of March and April. It occurs either on land or in water. During copulation two leeches come together by their ventral surfaces in a head-to-tail position (Fig. 35.25), so that the male genital pore of one lies against the female genital pore of the other. Then, the penis of one is inserted into the vagina of other and

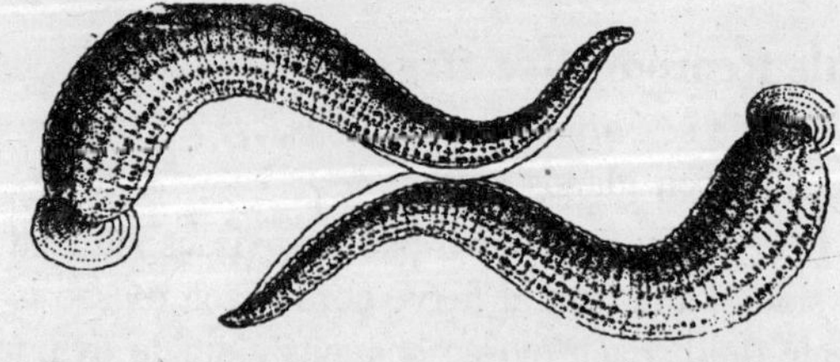

Fig. 35.25. *Hirudinaria*. Two leeches in copulation.

the spermatophores are deposited there. The **fertilization** of ova occurs in the vagina (*i.e.*, internal fertilization) and the zygotes are laid down into cocoon.

Cocoon Formation

Cocoon formation in *Hirudinaria* takes place in the months of April, May and June by the secretion of clitellum. In leech, a temporary clitellum is formed around segments ninth to eleventh during the breeding season. The clitellar glands secrete a frothy girdle which is snow white in colour. It hardens to form a **cocoon.** The albuminous gland secretes **albumen** into the cocoon which is used as nourishment by the developing embryo. The fertilized eggs are extruded into the cocoon. The leech wriggles out backwards from the cocoon (*i.e.*, cocoon is passed off over the head of each individual). The two ends of cocoon are closed by **polar plugs** (*i.e.*, an anterior plug and a posterior plug which are secreted by the prostomial glands). The complete cocoon is formed within six hours and is laid down in some moist place but never in water.

Cocoon becomes hard, barrel-shaped and light yellow in colour. It is about 25 to 30 mm long and 12 to 15 mm wide. It is filled with jelly-like albumen and contains 1 to 24 developing embryos.

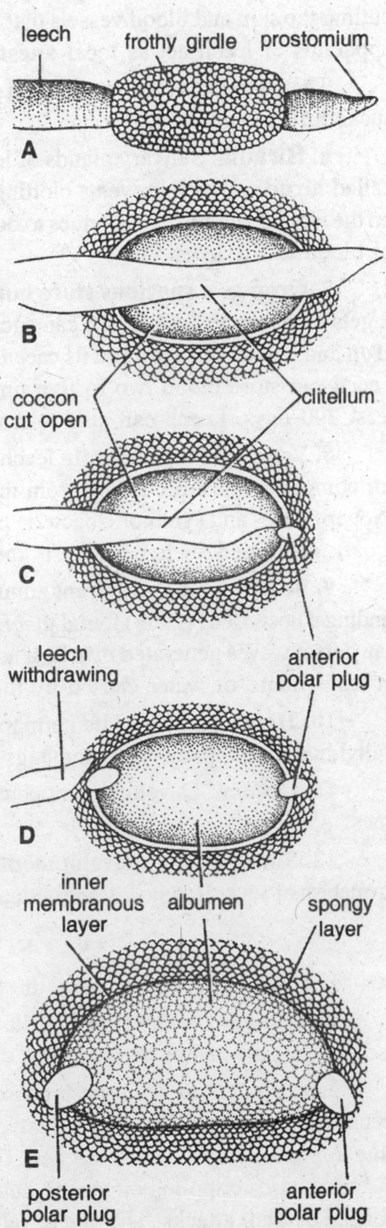

Fig. 35.26. *Hirudinaria.* Stages in cocoon formation A, B—Formation of frothy girdle. C—Formation of anterior plug; D—Formation of posterior plug; E—L.S. of a cocoon.

Development

Development in leech is **direct** involving no larva in its life cycle. It is completed in about 14 days. Young leeches after hatching out of the eggs swim in the albumen and feed on it. Soon the plugs drop off and young leeches come out of the cocoon.

35.12 PARASITIC ADAPTATIONS

Hirudinaria, the common Indian cattle leech, leads a parasitic mode of life. It is **semi-ectoparasitic** (*i.e.*, it normally leads an independent life but becomes attached to some mammal to feed) and **sanguivorous** (*i.e.*, it feeds by sucking blood of cattle and man). Its habit and habitat, morphology and physiology show the following important parasitic adaptations.

1. Habit and habitat. Leech lives in freshwater ponds, pools, slow-running streams and marshes which are regularly visited by cattles, human beings and other mammals for bathing and drinking the water. This ensures regular supply of host.

2. Body colour. The dark olive green colour helps leech in hiding undetected in weeds.

3. Shape. Its long, flattened, limbless and slimy body is well suited for its aquatic mode of life and for its swift swimming nature. Its **slimy body covering** protects it from desiccation, reduces friction during locomotion and also helps in the cutaneous respiration.

4. Suckers. Anterior and posterior **suckers** are the organs of attachment (with host as well as substratum); particularly posterior sucker is much more powerful and muscular.

5. Jaws and suctorial pharynx. The three **denticulated jaws** are so highly specialized for cutting the skin and blood vessels that host does not feel any pain. Certain secretions from the cutting apparatus of leech act as local **anesthetic.**

The pharynx serves as a suction pump. The blood flowing out of the host's wound is efficiently sucked by the pharynx.

6. Hirudin. Salivary glands of leech open on the surface of jaws and secrete an anticoagulant, called **hirudin** which prevents clotting of blood. As the jaws cut the skin of host, hirudin is poured on the wound and blood continues to ooze (without clotting). Thus, hirudin ensures continuous supply of blood to the leech.

7. Crop as a spacious store house. Getting a host is always a matter of chance. So whenever it gets a chance to feed, leech can suck and store enormous quantity of blood in its ten-chambered **crop** and **caeca.** The crop and its caeca are spacious, thin-walled, elastic and capable of great dilation. Leech can store blood two to five times its own body weight. This stored blood is enough for at least 200 days. Leech can also tolerate long period of fasting.

8. Very slow digestion. In leech, digestion is a very slow process. A full meal needs 200 days for complete digestion. Blood from the crop can enter the stomach drop by drop. It is digested by exopeptidases and symbiotic bacteria present in the intestine. After having a full meal, leech searches a dark concealed shelter and rests there for months.

9. Sense organs. Different annular and segmental receptors of leech are highly sensitive for finding a host. **Lent** (1984) found that *Hirudo*, another blood-sucking leech of warm-blooded animals, can detect waves generated by a host and is also attracted by the body secretion and very little change in temperature of water caused by the entry of a host in water.

10. Hermaphrodite. Its hermaphrodite sexuality has doubled the rate of reproduction as all individuals are capable of laying eggs in contrast to only 50 percent individuals of unisexual form.

11. Cocoon. Development occurs in cocoon which provides protection to developing young ones.

12. Short span of development. A short span of developmental period, *i.e.*, a fortnight, and formation of several (upto 24) individuals in a single cocoon ensures to maintain its regular population.

35.13 ECONOMIC IMPORTANCE

Leeches are found to have the following economic uses for human beings:

1. As food. Certain fishes, turtles, ducks and other birds, etc., eat leeches. These animals are used by human beings as food.

2. As pests. Leeches as pests cause great economic loss either directly or indirectly. For example, leeches parasitic on fishes cause great losses to fishery. Cattles and buffaloes which pass considerable time in ponds, suffer due to leeches. The wounds caused by leeches become infected with bacteria and this causes crippling effect and death of buffaloes and other cattles. Cattles and horses are often blinded by their attacks. The bite of Indian land leeches in forests is very painful. Due to their bites, human beings suffer serious loss of blood.

3. Medical Importance. Leeches owe medical importance from older time. They have been used in medicine for the following purposes.

(1) Leeches extract the blood of their victim in a painless manner. Due to this fact leeches were extensively used in **phelobotomy** (blood-letting) under the impression that these remove bad blood from the patient and cure many diseases. Tribals still use leeches for this purpose.

(2) Today, leeches are used to reduce **haematomas** (a mass of clotted blood that forms in a tissue, organ, or body space as a result of broken blood vessel) in areas of body that are difficult to treat surgically and to avoid leaving scars (see **Brusca** and **Brusca** 2003). Leeches are used as materia-medica in treatment of piles, tonsillitis and baldness.

(3) Hirudin as an anticoagulent remained in use in hospitals for a long time. However, it is not used now.

(4) In addition to the anticogulant, some leaches produce a number of other chemicals, including **anesthetics** and **vasodilators**, that are being studied for possible use in human medicine (**Conniff** 1987).

QUESTIONS

Long Answer Questions

1. Describe the external features of *Hirudinaria*. (*Lucknow 1989; Nagarjun 1994*)
2. Describe the structure and functions of body wall of leech and add a note on locomotion of leech.
3. Describe the alimentary canal of *Hirudinaria* and mention clearly the mode of sucking of the blood. (*Guwahati 1994; Meerut or CCS 1995*)
4. Describe the alimentary canal of *Hirudinaria* and mention clearly the mode of sucking blood in this animal and its peculiarities. (*Awadh 1989; Rewa 1989; Allahabad 1989; Purvanchal 1989; Rohilkhand 1989, 1992; Indore 1991; Vikram 1992; Agra 1992; Bundelkhand 1992*)
5. Give an account of the digestive organs and mechanism of feeding in leech. Explain in brief how these organs are adapted to its parasitic mode of life. (*Gorakhpur 1988; Meerut or CCS 1990, 1993; Kanpur 1991; 1993; Shivaji 1990; Jammu 1993*)
6. Give an account of the digestive system of *Hirudinaria*. (*Garhwal 1990; Bundelkhand 1990; Ravishanker 1993*)
7. Describe the reproductive organs of leech. Give neat diagram. (*Rajasthan 1988; Bundelkhand 1988, 1992; Meerut or CCS 1989, 1992, 1996; Allahabad 1989; Jiwaji 1989; Vikram 1989; Ranchi 1989; Kanpur 1989, 1994; Purvanchal 1992; Agra 1992; Rohilkhand 1992, 2001; Osmania 1993*)
8. Describe the urino-genital system of leech with labelled diagram stating its peculiarities and functions. (*Guwahati 1992, 1996*)
9. Describe the reproductive system and the process of reproduction and cocoon formation in leech. (*Tribhuwan 1988; Agra 1989, 1991; Jiwaji 1990; Rohilkhand 1990*)
10. Describe the reproductive system of leech and state clearly how reproduction takes place in the animal? (*Agra 1988; Magadh 1988; Kanpur 1989; Rohilkhand 1990*)
11. What is coelom? Give an illustrated account of the haemocoelomic system in leech. (*Awadh 1988; Bhopal 1988; Meerut or CCS 1988, 1999; Tribhuvan 1988; Agra 1989; Rajasthan 1989, 1992; Bundelkhand 1989, 1992; Purvanchal 1989, 1992; Sagar 1991; Kanpur 1992, 1994; Jodhpur 1989, 1993*)
12. Write an account of the coelom and the blood vessels of a leech and discuss the nature of the sinuses in this animal. (*Rohilkhand 1990*)
13. Describe the haemocoelomic system or circulatory system of *Hirudinaria*. (*Nagpur 1989; Mithila 1989; Bhagalpur 1989; Awadh 1990, 1992; Raipur 1990; Agra 1990; Magadh 1990, 1995; Jabalpur 1991; Jiwaji 1991; Patna 1992; Lucknow 1992; Rajasthan 1993; Indore 1994*)
14. Give an account of the lateral haemocoelomic channels of leech. (*Himachal 1989, 1991, 1995*)
15. What is nephridium? Describe the structure and function of excretory organs of *Hirudinaria*. (*Jiwaji 1988, Bhagalpur 1988; Rohilkhand 1990; Shivaji 1990; Purvanchal 1992; Osmania 1992*)
16. Describe the excretory system or nephridia of leech. (*Berhampur 1989; Meerut or CCS 1989; Sagar 1990; Allahabad 1991; Kanpur 1992; Himachal 1994*)

17. Describe lateral haemocoelomic channels and their branches in leech. *(Sagar 1988)*

18. Giving labelled diagram, describe the excretory system of leech. Add a note on the physiology of excretion. *(Garhwal 1992; Rewa 1993)*

19. Describe a typical nephridium of *Hirudinaria* and compare it with that of *Nereis*. *(Magadh 1989; Sagar 1990; Kanpur 1992; Gorakhpur 1993)*

20. Give a comparative account of the excretory organs of annelids studied by you. *(Meerut or CCS 1992)*

21. Give an account of receptor organs of leech. *(Garhwal 1988; Lucknow 1992)*

22. Describe parasitic adaptations in *Hirudinaria* and its economic importance. *(C.C.S.U. Meerut 2001)*

Short Answer Questions

1. Give the histological structure of the body wall of leech. *(Himachal 1995)*

2. Draw a labelled diagram of body wall of leech. *(Awadh 1989; 1992; Meerut or CCS 1996, 1997)*

3. Describe the locomotion of leech. *(Himachal 1994)*

4. Write a short note on the botryoidal tissue. *(Agra 1988; Rajasthan 1988; Kanpur 1988; Sagar 1989; Gorakhpur 1990; Kumaun 1991; Magadh 1991; Awadh 1992; Bhopal 1992; Lucknow 1992; Jabalpur 1992; Jiwaji 1993; Himachal 1995)*

5. Describe locomotion in leech?

6. Write a note on copulation and cocoon formation in leech. *(Kanpur 1989; Agra 1989)*

7. What is anticoagulin?

8. Discuss the role of botryoidal tissue? *(Rajasthan 1988; Bhopal 1992)*

9. What is haemocoel?

10. How leech is able to maintain a continuous flow of host blood while feeding?

11. Describe the role of radial muscles in leech.

12. Write short notes on the following:

(*i*) Cocoon

(*ii*) Crop of *Hirudinaria* *(Gorakhpur 1991)*

(*iii*) Feeding in leech

(*iv*) Haemocoel in leech *(Bundelkhand 1988; Kanpur 1992)*

(*v*) Hirudin

(*vi*) Locomotion in leech *(Bundelkhand 1988; Kanpur 1992)*

13. Write short notes on the following:

(*i*) Testicular nephridium of leech. *(Himachal 1995; Guwahati 1995)*

(*ii*) Ciliated organ. *(Garhwal 1989; Himachal 1990)*

(*iii*) Ventral haemocoelomic system of leech. *(Himachal 1990)*

(*iv*) Haemocoel. *(Himachal 1990)*

14. How cocoon formation occurs in *Hirudinaria*? *(Purvanchal 2001)*

15. Give an account of ciliated organs of leech.

16. Describe the process of respiration in leech.

17. What are basic differences between testicular and pretesticular nephridia?

18. Draw well labelled diagrams of the following:

(*i*) External features of *Hirudinaria*

or

Dorsal and ventral view of leech.

(*ii*) Alimentary canal of leech.

(*iii*) T.S. leech through stomach (*Ujjain 1993*)

(*iv*) T.S. leech through crop or through middle of body. (*Bundelkhand 1988; Agra 1989; Garhwal 1989; Awadh 1989, 1991; Jiwaji 1991; Kanpur 1995; CCS 1999*)

(*v*) Testicular nephridium of leech. (*Rohilkhand 1991; Jiwaji 1992; Agra 1991; Meerut or CCS 1993; Kanpur 1995*)

(*vi*) Reproductive organs of *Hirudinaria*. (*Purvanchal 1989; Meerut or CCS 1989; Rohilkhand 1989; Kanpur 2001*)

(*vii*) T.S. leech passing through pharynx. (*C.C.S.U. Meerut 2002*)

(*viii*) Receptors of leech

(*ix*) Epidermal glands of leech (*Sagar 1989*)

(*x*) Female reproductive organs of leech. (*Himachal 1994*)

(*xi*) Urinogenital system of leech. (*Guwahati 1993*)

(*xii*) T.S. leech through epididymis and penis sac. (*Purvanchal 1992*)

Very Short Answer Questions

1. Which segments of leech are triannulate?
Ans. Fourth, fifth and sixth.

2. What is haemolization?
Ans. Breakdown of erythrocytes to dissolve haemoglobin.

3. Name the muscle layers in the body wall of leech.
Ans. Circular, oblique and longitudinal.

4. Give the function of the chitogenous glands.
Ans. Secretion of the material for the formation of egg case.

5. What is the location of albumen glands?
Ans. In the body wall of clitellum.

6. Name the glands which secrete hirudin?
Ans. Salivary glands.

7. What is the function of crop in leech?
Ans. Storage of ingested blood.

8. Which segments of leech contain nephridia?
Ans. Segments 6 to 22.

9. Mention the position of female genital aperture.
Ans. In the groove between the second and third annuli of the segment XI on the ventral side.

10. Where does fertilization occur in leech?
Ans. In vagina.

11. What is the significant function of clitellum?
Ans. The clitellar glands secrete a snow-white frothy substance of which cocoon, ootheca or egg case is formed. Glands of clitellum also secrete an albuminous fluid to be deposited with the fertilized ova within the cocoon.

12. Why does not blood coagulate when a leech sucks it?
Ans. It is because the salivary glands of leech secrete an anticoagulant called **hirudin** besides the saliva on the wound which prevents clotting of blood while it is being sucked.

13. Which anticoagulating substance is found in leech? (*C.C.S.U. Meerut 2001*)
Ans. The anticoagulating substance, hirudin, found in the saliva of leech is secreted by salivary glands of the leech.

14. How many segments form head in leech?

Ans. The head or cephalic region of leech is formed of first five segments. Segment one and two are uniannulate segement, third segment is biannulate whereas segment fourth and fifth are triannulate.

15. What are the parts of cephalic region of leech?

Ans. The head or cephalic region of leech includes prostomium, anterior sucker, mouth and eyes.

16. Why are the segments of cephalic region of leech called ocular segments?

Ans. Since a pair of eyes are situated on the dorsal side of first annulus in all cephalic segments (*i.e.*, from segment one to five). These segments are, therefore, called **ocular segments.**

17. How many segments form the caudal region in leech?

Ans. Caudal region of leech is made up of four incomplete segments 23rd to 26th. Segment 23rd is triannulate and the rest are biannulate.

18. How will you collect leech? (*Lucknow 1989*)

Ans. Leeches are ectoparasites on vertebrates (cattles, human beings, etc.) and feed upon their blood. In nature, they are found mostly in freshwater ponds, tanks, lakes marshes and slow-moving streams. From these places leeches are collected in the old fashion. A person immerses his or her body in such muddy water where leeches are in abundance. Many leeches tend to adhere to the legs and later on are detached and collected. Sometimes, they are collected from buffaloes who love to wallow in mud and water.

19. Where is the botryoidal tissue found in leech?

Ans. Botryoidal tissue is situated beneath the longitudinal muscles surrounding the alimentary canal. It is filled in coelom reducing it into haemocoelomic channels and sinuses.

20. How many pair of pretesticular nephridia are present in leech and in which segments?

Ans. There are six pairs of pretesticular nephridia found in segments 6–11.

21. Of how many segments the posterior sucker of leech is formed?

Ans. It is formed of seven segments.

Fill in the Blanks

1. The zoological name of cattle leech is ....
2. Body of leech consists of .... segments.
3. Leech has two suckers: .... and posterior.
4. Cephalic region of the leech consists of .... segments.
5. Clitellar region of leech consists of three segments, from .... to ....
6. Posterior sucker of leech is formed of .... segments.
7. Leech has .... pairs of eyes.
8. In leech, segmental receptors occur on the .... annulus of each segment.
9. Male genital aperture of *Hirudinaria* lies in the groove between the .... and .... annuli of the segment number ....
10. The epidermal cells of the leech are .... shaped.
11. The body wall muscles are differentiated into .... layers.
12. Mouth of leech lies in a membrane called ....
13. Crop of leech consists of a row of .... chambers.
14. Triradiate wound in the skin of host is made by .... while feeding on host's blood.
15. Saliva of leech contains .... which prevents clotting of blood.
16. Coelom of leech is reduced by .... tissue.
17. Leech has .... pair of nephridia.

18. There are .... pairs of testes in *Hirudinaria*.

19. Fertilization in leech occurs in the ....

20. *Hirudinaria* belongs to the order .... of class Hirudinea.

21. Blood vascular system in leech is of .... type.

22. Body cavity of leech is filled with a network of connectives called .... tissue.

23. Coelomic corpuscles in leech are manufactured in ....

24. Ciliated organs lie in .... ampulla.

25. *Hirudinaria* is .... in nutrition.

Multiple Choice Questions

1. Common cattle leech belongs to the genus
(*a*) *Megascolex* (*b*) *Nereis*
(*c*) *Arenicola* (*d*) *Hirudinaria*

2. Locomotion in leech is brought about by
(*a*) parapodia (*b*) setae
(*c*) body musculature and suckers (*d*) suckers only

3. What is true about *Hirudinaria*?
(*a*) it has a cephalic sucker (*b*) mouth is triradiate
(*c*) segments are superficially subdivided (*d*) all of these

4. Which one of the following is an ectoparasitic annelid?
(*a*) *Hirudinaria* (*b*) *Lumbricus*
(*c*) *Pheretima* (*d*) *Ascaris*

5. Leech is a blood sucking animal nutritionally, therefore leech is
(*a*) herbivorous (*b*) carnivorous
(*c*) sanguivorous (*d*) omnivorous

6. A typical segment of leech is
(*a*) uniannulate (*b*) biannulate
(*c*) triannulate (*d*) quinquannulate

7. The anterior sucker of *Hirudinaria* is formed by the union of
(*a*) one segment (*b*) three segments
(*c*) seven segments (*d*) prostomium and three segments

8. The posterior sucker of *Hirudinaria* is formed by the union of
(*a*) one segment (*b*) seven segments
(*c*) twenty five segment (*d*) thirty three segments

9. In each segment of leech the sensillae lie on the
(*a*) first annulus (*b*) second annulus
(*c*) third annulus (*d*) fourth annulus

10. Number of annular receptors in each annulus
(*a*) 18 (*b*) 18 on dorsal and 18 on ventral
(*c*) 9 (*d*) none of the above

11. The material that fills the intervisceral spaces in leech is
(*a*) mesenchyme (*b*) mesogloea
(*c*) botryoidal (*d*) parenchyma

12. The largest organ of the alimentary canal of leech is
(*a*) buccal cavity (*b*) stomach
(*c*) intestine (*d*) crop

13. Hirudin of the saliva of leech prevents
   (*a*) digestion of blood (*b*) clotting of blood
   (*c*) putrefaction of blood (*d*) haemolysation of blood
14. Digestion of food in leech is brought about by
   (*a*) proteolytic enzymes (*b*) bacteria
   (*c*) both of these (*d*) none of these
15. Which segments possess nephridia in *Hirudinaria*?
   (*a*) 7–10 segments (*b*) 6–22 segments
   (*c*) 10–20 segments (*d*) 7–18 segments
16. Excretory matter of leech is mainly
   (*a*) ammonia (*b*) urea
   (*c*) uric acid (*d*) amino acid
17. Leech is
   (*a*) unisexual (*b*) bisexual
   (*c*) nonsexual (*d*) none of these
18. A temporary clitellum is formed in *Hirudinaria* on segments
   (*a*) 7th, 8th and 9th (*b*) 9th to 11th
   (*c*) 11th to 13th (*d*) none of these
19. In which segment do you find the ovisacs in *Hirudinaria*?
   (*a*) 8th (*b*) 11th
   (*c*) 15th (*d*) 13th
20. Number of testes in *Hirudinaria* is present in
   (*a*) 10 pairs (*b*) 11 pairs
   (*c*) 12 pairs (*d*) 13 pairs
21. Worm employed in medical practice in the past belongs to
   (*a*) *Entamoeba histolytica* (*b*) *Hirudo medicinalis*
   (*c*) *Pheretima posthuma* (*d*) *Ancylostoma duodenale*

ANSWERS

Fill in the Blanks

1. *Hirudinaria granulosa* 2. Thirty three (33)
3. Anterior 4. Five
5. Segment nine, segment eleven 6. Seven
7. Five 8. First
9. Second, third, ten 10. Hammer or Malleate
11. Three 12. Velum
13. Ten 14. Leech jaws
15. Hirudin 16. Botryoidal
17. Seventeen 18. Eleven
19. Vagina 20. Gnathobdellida
21. Haemocoelomic 22. Botryoidal
23. Ciliated organ 24. Perinephrostomial
25. Sanguivorous.

Multiple Choice Questions

1. (*d*)	2. (*c*)	3. (*d*)	4. (*a*)	5. (*c*)	6. (*d*)
7. (*c*)	8. (*b*)	9. (*a*)	10. (*b*)	11. (*c*)	12. (*d*)
13. (*b*)	14. (*b*)	15. (*b*)	16. (*a*)	17. (*b*)	18. (*b*)
19. (*b*)	20. (*b*)	21. (*b*).			

36

Metamerism and Coelom in Annelida

36.1 METAMERISM IN ANNELIDA

The annelids have successfully invaded virtually all habitats where sufficient water is available. Their success is no doubt due in part to the evolutionary plasticity of the segmented bauplan and to their exploitation of a variety of life history strategies. The basic annelid condition is characterized by a segmented body in which most internal and external parts are repeated with each segment, a situation referred to as **serial homology**. Serial homology refers to body structures with the same genetic and developmental origins that arise independently during the ontogeny of an organism. In annelids (as in arthropods) this repetition of homologous body structures results in **metamerism**, or body segmentation that arises by way of **teloblastic development** (*i.e.*, the proliferation of paired, segmental mesodermal bands from teloblast cells at a posterior growth zone in the embryo; see **Brusca** and **Brusca**, 2003).

> **Box 36.1**
>
> Annelids as a group show a **primitive metamerism** with comparatively few differences between the different somites. The introduction of metamerism by the group Annelida represents a greatest innovation seen in this phylum. An homologous but more highly specialized metamerism is seen in the arthropods.

Metamerism is encountered for the first time in the annelids, and forms the most important distinguishing character of the phylum (Box 36.1). The **prostomium**, *i.e.*, the anterior part of the body infront of mouth, and the **pygidium,** *i.e.,* the terminal part of the body carrying the anus, are not segments. New segments are formed just infront of pygidium. The oldest segments are, therefore, anterior and the youngest, posterior. Metamerism is not confined to external structures but affects the internal organs as well. Muscles, small blood vessels, nerves, excretory organs (nephridia), and gonads are repeated in each segment. Only the alimentary canal, nerve cord, and the main blood vessels are unaffected, although they too extend through each segment.

Metamerism is usually more complete or uniform throughout the body of the embryo than of the adult, in which it may become obscured due to the formation of head and other structures such as limbs. Internal organs may also become "concentrated" or restricted to certain segments. Modification of the embryonic metamerism is especially noticeable in the vertebrates.

Metamerism is of great interest, because, besides the annelids, it is also found in the most successful groups of the animal kingdoms, namely, Arthropoda and Vertebrata (Chordata). Evidently, it has originated independently in the annelids and the vertebrates. How it developed (= originated) is not clear however.

Theories Regarding Origin of Metamorphosis

1. Transverse fission theory. Some zoologists hold that the body of a metameric animal has evolved from that of an unsegmented animal by transverse fission into individuals, which remained joined end to end and gradually became integrated both morphologically and physiologically.

2. Division of ancestral organ theory. Some other zoologists maintain that the repetition of organs, such as blood vessels, nephridia and gonads, in various segments has resulted by the division of a single ancestral organ, and not by formation of new organs as the fission theory suggests.

Comparison of Pseudometamerism and Metamerism

The true metamerism seen in the annelids should not be confused with pseudometamerism found in the tapeworms (phylum Platyhelminthes). The differences between the two are given in Table 36.1.

Table 36.1 Differences between proglottids of tapeworm and true segments of Annelida.

Proglottids of tapeworm	*True segments of earthworms*
1. Number is very large and indefinite.	1. Number is small and almost definite, varying within small limits only.
2. New proglottids arise at the anterior end so that the oldest proglottid is the most posterior.	2. New segments arise from the posterior end so that the oldest segment is the most anterior.
3. Proglottids are not preformed in the embryo function.	3. Segments are preformed in the embryonic somites.
4. Proglottids are voluntarily cast off.	4. Segments are not voluntarily cast off.
5. All proglottids are identical in structure and function.	5. Though fundamentally alike, some segments are very specialized in structure as well in function.

There is thus a superficial resemblance between the proglottids of a tapeworm and segments of an earthworm.

Every mature proglottids contains a complete set of male and female reproductive organs and in this respect resembles a bisexual animal, such as liver fluke or planaria. This resemblance has led to the speculation that the tapeworm is a colony of bisexual individuals produced asexually by budding. However, the tapeworm is a single animal. The scolex contains the brain and forms its head while the strobila contains other viscera and constitutes its body. The power of increasing the number of proglottids by budding and of forming sex organs in all of them is a special adaptation to parasitic life for ensuring the continuity of the race.

Significance of Metamerism

Although a spacious, fluid-filled coelom provided an efficient hydrostatic skeleton for burrowing, precise control of body movements was not possible in the earliest coelomates. The force of muscle contraction in one area was carried throughout the body by the fluid in the undivided coelom. This defect was remedied when a series of partitions (septa) evolved in the common ancestor of annelids and arthropods. When the septa divided the coelom into series of compartments of most other body systems, such as circulatory, nervous and excretory, were repeated in each segment (*i.e.*, metamerism).

The evolutionary advent of metamerism was highly significant, because it made possible development of much greater complexity in structure and function. Metamerism not only increased the efficiency of burrowing but also made possible independent and separate movements by the separate segments. The need for fine control of movements led, in turn, to the evolution of a more sophisticated nervous system. Moreover, repetition of body parts gave the organisms a built-in redundancy, as in some human-made systems. This provided a safety factor: if one segment should fail, the others could still function. Thus an injury to one part would not necessarily be fatal.

The evolutionary potential of the metameric body plan is amply demonstrated by the large and diverse phylum Arthropoda. Metamerism also arose independently in the deuterostome line, which includes the numerous and adaptively diverse chordates (**Brusca** and **Brusca**, 2003).

36.2 COELOM IN ANNELIDA

Coelom is a cavity present between the body wall and internal organs (*i.e.*, visceral organs) and is lined with a single-layered mesodermal epithelium called the **peritoneum**. The peritoneum which lines the body wall is called the **parietal peritoneum**, and that which covers the internal or visceral organs is called the **visceral peritoneum**. The coelom contains the **coelomic fluid**, which is secreted by the peritoneum and has free mesenchyme cells, the **colomocytes** or **coelomic corpuscles**. The coelomic fluid receives materials absorbed by the alimentary canal and carries it to the tissues. Excretory substance finds its way into the coelomic fluid and thence out of body through the nephridia. Sex cells arise from the coelomic epithelium.

Status of Coelom in Bilateria

The bilateral animals can be grouped according to their body cavity type or lack of body cavity (*i.e.*, acoelomata, pseudocoelomata and eucoelomata). In the **eucoelomata bilateria** (*i.e.*, higher animals) exists a fluid-filled space that surrounds the gut (and other visceral organs). The evolution of the coelom, like bilateral symmetry, was a major advancement in animal architecture because it provides a "tube-within-tube" arrangement that allows much greater body flexibility than is possible in animals lacking an internal body cavity. The coelom also provides a space for the visceral organs and permits greater size and complexity by exposing more cells to surface exchange. The fluid-filled coelom additionally serves as a **hydrostatic skeleton** in some forms, especially many worms, aiding in such activities as movement and burrowing.

Schizocoelous and Enterocoelous Coelom

Eucoelomate bilateria possess a **true coelom** lined with mesodermal peritoneum. The true coelom arises within the mesoderm itself and may be formed by one of two methods: schizocoelous and enterocoelous formations (Fig. 36.1). Both of these terms are descriptive, for *schizo* comes from the Greek *schizein*, to split; *entero* is derived from the Greek *enteron*, meaning gut; and *coelous* comes from the Greek *koilos*, meaning hollow or cavity. In schizocoelous formation, the coelom arises, as the word implies, from the splitting of mesodermal bands that originate from cells in the blastopore region (Mesoderm is one of the three primary germ layer that appear very early in the development of bilateral animals, lying between the inner most endoderm and outermost ectoderm). In enterocoelous formation the coelom comes from pouches of the archenteron or primitive gut.

Once development is complete, the results of schizocoelous and enterocoelous formation are indistinguishable. Both processes give rise to a true coelom lined with a mesodermal peritoneum (Gr. *peritonations*, stretched around) and having mesenteries in which the visceral organs are suspended.

1. Coelom in Polychaetes and Oligochaetes. Polychaetes (*e.g., Nereis*) and oligochaetes (*e.g., Pheretima*) possess a spacious coelom extending from the anterior to the posterior end of the body. It is divided into a series of small chambers, the **coelomic compartments**, by transverse partitions, the **septa** or **coelosepta**, that extend from the annuli of the body wall to the alimentary canal. The septa are composed of a double layer of peritoneum with a thin sheet of musculo-connective tissue in between. They are usually perforated by sphinctered aperture or apertures to allow flow of coelomic fluid from one segment to another.

In polychaetes, the gut is suspended by dorsal and ventral longitudinal **mesenteries** formed of a double layer of peritoneum (Fig. 34.20). With the result, each coelomic compartment is divided into right and left halves. Actually, however, such perfect compartmentation is rarely seen because

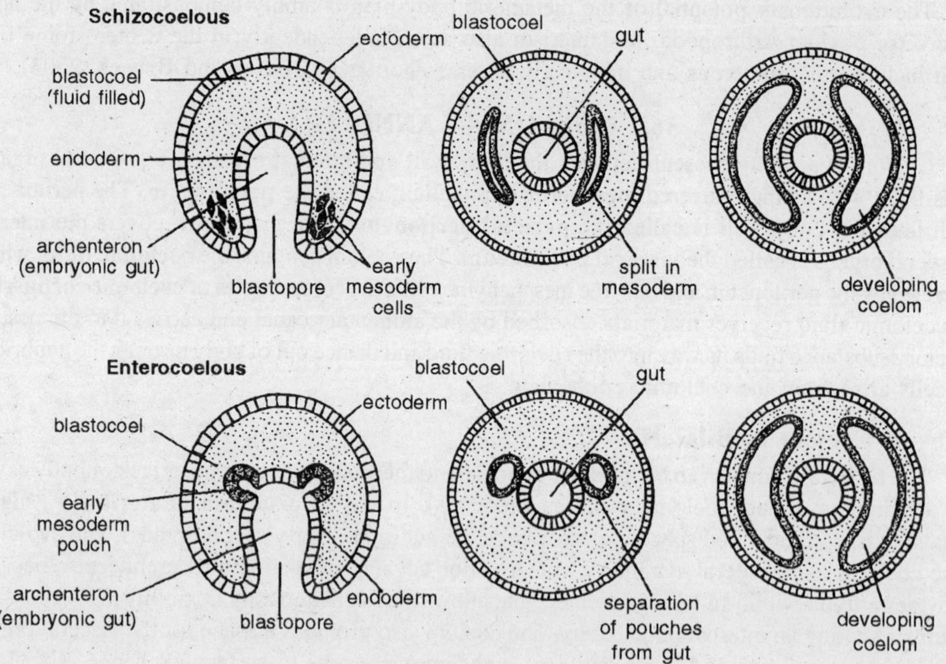

Fig. 36.1. Type of mesoderm and coelom formation. In schizocoelous formation, the mesoderm originates from the wall of archenteron near the blastopore and proliferates into a band of tissue that splits to form the coelom. In enterocoelous formation, most mesoderm originates as a series of pouches from the archenteron; these pinch off and enlarge to form the coelom. In both formations, the coeloms expand to obliterate blastocoel (after Hickman *et. al.*, 1997).

the mesenteries are almost always incomplete and the septa are weak or have partly or fully disappeared in many polychaetes. *Arenicola,* for instance, retains only three septa in the anterior region and a few in the posterior regions, those of the middle region having been lost. In *Aphrodite, Glycera* and some other annelids, the peritoneum is ciliated, and the movement of cilia brings about an active circulation of coelomic fluid in the coelom. This distributes materials to the tissues so efficiently that the blood vascular system is reduced.

In oligochaetes, the septa are relatively complete, except in the anterior and posterior parts of the body. Certain anterior septa in large earthworms are cone-shaped and very muscular (Fig. 36.3). Contraction of these septa increases the pressure of coelomic fluid,

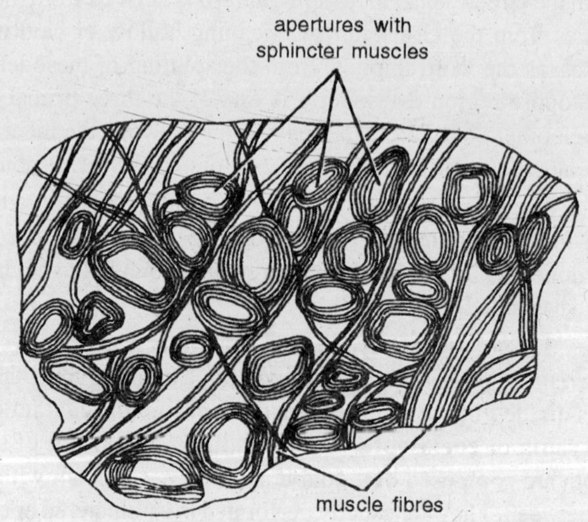

Fig. 36.2. A part of intersegmental septum of *Pheretima* (earthworm)

and this may cause eversion of the pharynx, or, if the mouth kept closed, may result in elongation and stiffening of the first segment for burrowing. In most earthworms, each coelomic compartment, except those at the ends, opens out by a sphinctered middorsal pore situated in the intersegmental groove. The dorsal pores exude coelomic fluid to keep the skin moist for respiration some giant earthworms squirt coelo-mic fluid several centimeters when disturbed.

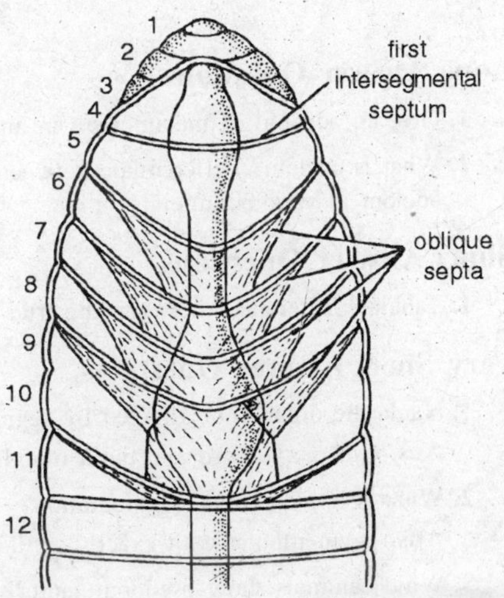

Fig. 36.3. Arrangement of septa in the anterior segments of *Pheretima*.

2. Coelom in Hirudinea. In Hirudinea, a perivisceral coelom divided into compartments by septa is found only in a single primitive leech *Acanthobdella*, the sole representative of the order Acanthobdellida. In other leeches, the septa have disappeared, and the coelom has been invaded by connective tissue that has reduced it to a system of intercommunicating coelomic sinuses. In jawed leeches, the coelom is further restricted by the proliferation of botryoidal and vasofibrous tissues from the sinus walls. The botryoidal tissue consists of grape-like masses of cells containing brown pigment, and is perhaps equivalent to the chloragogen cells of other annelids. The vasofibrous tissue consists of pigment-laden strands.

3. Coelom in Rhynchobdellida. In Rhynchobdellida (*e.g., Pontobdella*), the coelomic sinuses coexist with the blood vascular system and act as an auxillary circulatory system. There are four longitudinal sinuses: a dorsal sinus enclosing the dorsal blood vessel, a ventral sinus having the ventral blood vessel and the ventral nerve cord, and a pair of lateral sinuses. The lateral sinuses are connected with the median sinuses by irregular intermediate sinuses in each segment. A slender subepidermal sinus occurs between the epidermis and the muscles of the body wall, and communicates with the lateral and intermediate sinuses. Coelomic fluid with floating coelomocytes irregularly circulates in the coelomic sinuses as a result of body movements. Certain forms, such as *Piscicola,* have contractile vesicles in connection with the lateral sinuses for the circulation of coelomic fluid.

4. Coelom in Gnathobdellida. In Gnathobdellida (*e.g., Hirudinaria*), the original circulatory system has disappeared, and the coelomic sinuses and fluid have become modified to function as a true blood vascular system. The coelomic fluid has developed haemoglobin, which is dissolved in it, and the coelomic sinuses have acquired more distinct walls. Hence, the coelomic fluid is known as the **haemocoelomic fluid**. The latter with their branches and tributaries are said to constitute the **haemocoelomic system**. Typically there are four longitudinal haemocoelomic sinuses: a dorsal, a ventral and a pair of lateral. Dorsal sinus is not always present. The ventral sinus encloses the nerve cord and is swollen in the region of each ganglion. The lateral sinuses have muscular walls and undergo rhytmic contractions to keep the haemocoelomic fluid circulating. They are interconnected by transverse sinuses below the ventral sinus in each segment (Fig. 35.11). Nephridia have relation with the branches of the ventral sinus. The branches of the main sinuses break up into capillaries that penetrate the botryoidal tissue and body wall.

Besides the haemocoelomic sinuses, the jawed-leeches have coelomic spaces in the testis sacs and ovisacs, and around vasa differentia. These coelomic spaces have coelomic fluid.

5. Coelom in Archiannelida. The Archiannelida (*e.g., polygordius*) have a spacious and compartmented coelom.

QUESTIONS

Long Answer Questions

1. Give an account of metamerism in animals.
2. What is coelom? Differentiate between different methods of coelom formation. Give an account of coelom in various annelids.

Short Answer Questions

1. Tabulate differences between the true segments of earthworm and the proglottids of tapeworm.

Very Short Answer Questions

1. Name the organs not affected by segmentation.
   Ans. Gut, nerve cord and main blood vessels.
2. What is homonomous segmentation?
   Ans. Segmentation with exactly similar segments.
3. Which animals show pseudometamerism?
   Ans. Tapeworms.
4. In which phylum for the first time did true coelom appear?
   Ans. Annelida.
5. Define true coelom.
   Ans. The body cavity in triploblastic animals, lined with mesodermal peritoneum.

Fill in the Blanks

1. Division of the body into an anteroposterior row of similar parts is called ....
2. New segments arise just in front of the ....
3. In annelids, the oldest segments are at the .... end.
4. Metamerism is more complete in the .... than in the adult.
5. Metamerism is also found in .... and ....
6. True coelom is originated from ....
7. Coelom is always lined by mesodermal .... and has .... fluid.

Multiple Choice Questions

1. Segmentation is found in
   (*a*) Annelida (*b*) Arthropoda
   (*c*) Chordata (*d*) all of these
2. Besides Annelida and Arthropoda the metamerism is exhibited by
   (*a*) Cestoda (*b*) Mollusca
   (*c*) Chordata (*d*) Acanthocephala.

ANSWERS

Fill in the Blanks

1. Segmentation (metamerism) **2.** Pygidium **3.** Anterior **4.** Embryo
5. Arthropoda, Vertebrata **6.** Mesoderm **7.** Peritoneum, coelomic fluid

Multiple Choice Questions

1. (*d*); **2.** (*c*).

37

Trochophore Larva and its Significance

37.1 STRUCTURE OF TROCHOPHORE

Polychaetes (Annelids) have indirect development. Their life cycle includes a free-swimming trochophore larva. This larva is a characteristic of many protostomous groups (*e.g.*, molluscs, sipunculus, and echiurans). Trochophore is a minute, ciliated, unsegmented, and almost pear-shaped pelagic creature, with oral and aboral surfaces recognizable. Thus, the early, presegmental trochophore (Fig. 37.1) is characterized by a locomotor ciliary band just anterior to the region of the mouth. This ciliary band, the **prototroch,** arises from special cells, called **trochoblasts**, of the first and second quartets of micromeres. Most trochophore larvae also bear an **apical ciliary tuft** associated with an **apical sense organ** (eye and cerebral ganglia) derived from a plate of thickened ectoderm at the anterior end. In addition, there is often a perianal ciliary band called the **telotroch**. By this stage the mesentoblast has divided to form a pair of cells called **teloblasts**, which in turn proliferate a pair of mesodermal bands, one on each side of the archenteron in the region of the hindgut, an area known as the growth zone (Fig. 37.1B). Many trochophores bear larval sense organs such as ocelli and statocyst, as well as a pair of larval protonephridia. Many trochophores bear bundles of mobile chaetae, which are known to serve as a defense against predators (**Pennington** and **Chia**, 1984) and to help retard sinking.

The cilia of ciliary girdles, rings or bands, when beating give the appearance of a rotating wheel, hence the name trochophore (Gr. *Trochos* = wheel; *phoros* = bear). Digestive tract of trochophore larva is complete, and comprises **stomodaeum, stomach, intestine** and **proctodaeum** (Fig. 37.2). A spacious fluid-filled **blastocoel** surrounds the larval gut. It contains mesenchyme cells, larval muscle fibers and a pair of larval kidneys or protonephridia, one on each side of the gut. The nephridia open by a common passage. Trochophore lacks segmentation and coelom.

The fully formed trochophore may be divided into three regions: 1. **pretrochal region** comprising area about and above the mouth; 2. **pygidium** consisting of telotroch and anal area; and 3. **growth zone** lying between the mouth and telotroch. In polychaetes, the growth zone forms all the trunk segments.

Trochophore may be **planktotrophs** (*i.e.*, feeding on plankton or planktonic organisms) and have long planktonic lives, or they may be **lecithotrophs** (*i.e.*, a non-feeding larva which utilises yolk as a source of nutrition) and have a short planktonic existence. The prototroch functions as the swimming organ, and in a feeding trochophore, it also collects the suspended food particles. The feeding trochophore also has a **food groove** between the preoral prototroch and the postoral metatroch. The larger cilia of prototroch drive water and suspended food particles backward. The short metatroch cilia beat in opposite direction and drive the food particles into the food groove. The short cilia of the food groove carry the food particles to the mouth.

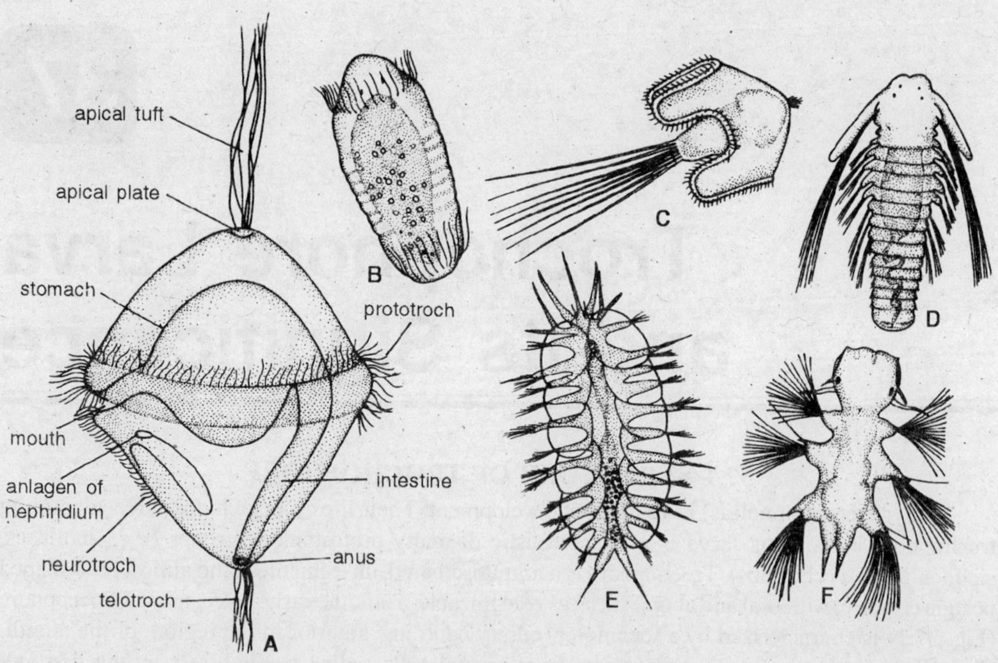

Fig. 37.1. Some polychaete larvae—trochophores and beyond. A—Anatomy of a generalised early trochophore larva; B—Young trochophore of a capillid; C—Trochophore of *Owenia*; D-Late spinoid larva; E-Late larval stage of a polynoid (Scale worm); F—Three-segmented nectochaeta "larva" of a nereid (after Brusca and Brusca, 2003).

37.2 GROWTH AND DEVELOPMENT OF TROCHOPHORE

The trochophore larva grows and elongates by proliferation of tissue in the growth zone (Fig. 37.3), while segments are produced by the anterior proliferation of mesoderm from the teloblast derivatives on either side of the gut. These packets of mesoderm hollow (schizocoely) and expand as paired coelomic spaces, which eventually obliterate the blastocoel. Thus, the production of serially arranged coelomic compartments and the formation of segments are one and the same; the anterior and posterior walls of adjacent coelomic compartments form the intersegmental septa. Proliferation of segments by this process is called **teloblastic growth.** Externally, additional ciliary bands are added at each segment. These **metatrochal bands** aid in locomotion as the animal increase in size. Such segmental larvae are sometimes called **polytroch larvae**.

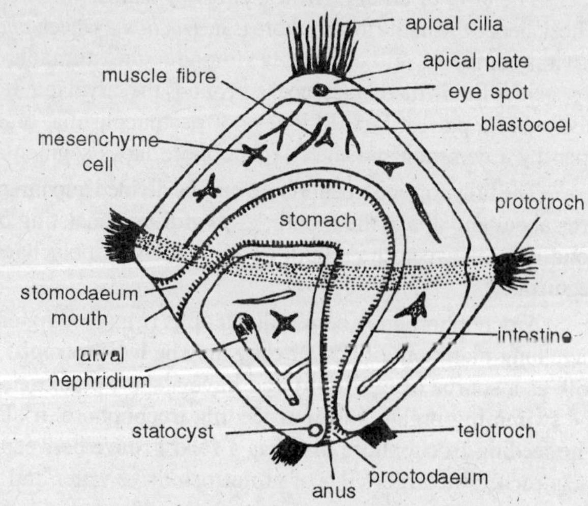

Fig. 37.2. Internal structure of a trochophore larva.

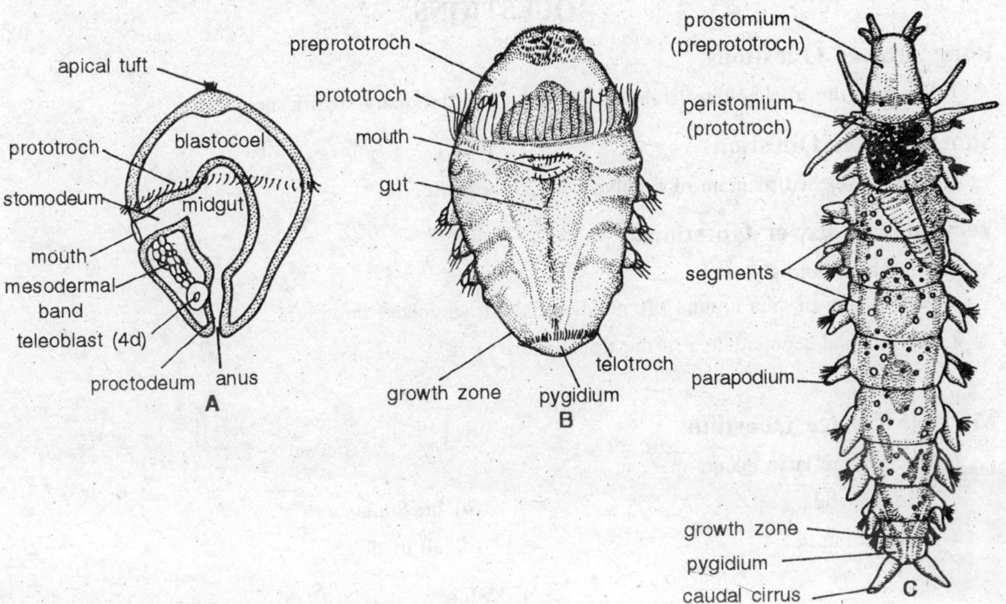

Fig. 37.3. Growth of trochophore larva. A—Trochophore larva of *Arenicola.* Note the teloblastic (4d) mesodermal bands destined to form the metameric coelomic spaces, B, C—Two stages in the development of *Eteone*; B—Early stage segmentation; C—Juvenile, showing the fates of the larval regions (after Brusca and Brusca, 2003).

The fates of the various regions of trochophore larva are now apparent (Fig 37.3). The region anterior to the protrochal ring becomes the **prostomium**, while the prototrochal area forms the **peristomium.** Note that these two parts are not involved in the proliferation of segments and are thus presegmental. The segmental, metatrochal portion of the larva forms the trunk and the growth zone and post segmental pygidium remain as the corresponding adult body parts. The apical sense organ becomes the cerebral ganglion, which is eventually joined with the developing ventral nerve cord by the formation of circumcentric connectives. The body continues to elongate as more segments form, and the juvenile worm finally drops from plankton and assumes the lifestyle of a young polychaete.

37.3 SIGNIFICANCE OF TROCHOPHORE

The trochophore larva is of great phylogenetic significance. Many animals such as polychaetes, sipunculans, echiurans, bryozoans, molluscs, brachiopods and nemertines, develop a trochophore larva, of course, with certain modifications. This has led some embryologists (*e.g.,* **Hatschek,** 1878) to suggest a theory (the so-called **trochophore theory**) that these animal groups have descended from a common hypothetical ancestor called **trochozoon** having trochophore-like features. The animal groups thought to have common ancestry may also include other phyla (Platyhelminthes and Arthropoda) which may have lost the trochophore somewhere in their evolution.

There is a striking resemblance between certain rotifers and the trochophore larvae ot some annelids, molluscs and others. This indicates that the rotifers and the animals closely related to the ancestors of the annelids, molluscs and other groups (see chapter 34).

Trochophore larva (theory) accounts satisfactorily in many ways for a common ancestor of the coelomate prostomia but not for the acoelomate groups. Many authorities, however, believe that the resemblances of trochophore larvae may be coincidental, the result of adaptive radiation and not of evolutionary significance or actual relationship.

QUESTIONS

Long Answer Questions

1. Describe the trochophore larva and discuss its evolutionary significance.

Short Answer Questions

1. Draw a labelled diagram of a polychaete trochophore.

Very Short Answer Questions

1. What is prototroch?

   Ans. A ring of cilia around the middle of the trochophore larva.

2. Do you find segmentation in the trochophore larva?

   Ans. No.

Multiple Choice Question

1. Trochophore larva occurs in

   (*a*) annelida (*b*) brachiopoda

   (*c*) mollusca (*d*) all of these.

ANSWER

Multiple Choice Questions

1. (*d*).

38

Phylum Arthropoda (Characters, Classification and Types)

The phylum **Arthropoda** (Gr., *arthros* = joint, *podos* = foot) includes nearly a million species. No other phylum of animals rival the arthropods in success: the arthropods contain the most showy group, the insects and the most variable, the crustaceans, of any other animal phyla. Much of their success can be traced to a remarkable flexibility of body plan.

Arthropods are bilaterally symmetrical and segmented animals that are equipped with jointed appendages attached to each body segment. Appendages are typically specialized for specific functions. The surface of body and appendages is covered by a continuous cuticle of complex structure, *i.e.*, **chitin** containing **exoskeleton** formed of a series of heavy skeletal plates or rings, connected by thinner, flexible articular membranes. Arthropods **lack true nephridia and cilia**. They undergo moulting periodically.

38.1 CHARACTERS

1. Body is **bilaterally symmetrical, triploblastic, metamerically segmented**. Few anterior segments (usually 5) have fused to form a distinct head.
2. They have a fixed number of segments.
3. Body is covered and protected with a thick chitinous **cuticle** forming an **exoskeleton**.
4. The exoskeleton is dead and cannot grow. Therefore, the old exoskeleton is thrown off periodically and new one is developed to take its place and to compensate the increase in size. The phenomenon is called **moulting** or **ecdysis**.
5. Body segments usually bear **paired** lateral and **jointed appendages**. Appendages are variously modified to act as organs of stimulus detection, food getting, mastication, defence, locomotion, egg carrying and copulation.
6. **Musculature** is not continuous but comprises of separate striped muscles.
7. Body cavity is a **haemocoel**. True coelom is represented by spaces within the gonads and excretory organs.
8. Digestive system is complete; **mouth** and **anus** lie at opposite ends of the body.
9. Respiration occurs through general **body surface, gills** (in aquatic forms), **trachea** or **book lungs** (in terrestrial forms).
10. Blood vascular system (or circulatory system) is of **open type**. It includes an ostia-containing **heart, arteries** but lacks capillaries. The arteries open into haemocoel.
11. True nephridia are absent. Excretion occurs through coelomoducts (*e.g.*, green glands, coxal glands) and **Malpighian tubules**.

12. Nervous system consists of a large bilobed **brain** and two ventral nerve cords.
13. Sense organs include **simple eyes, compound eyes** (with mosaic vision), chemoreceptors, tactile receptors (antennae and bristles), auditory organs and **statocysts.**
14. Cilia are entirely absent from all parts of the body.
15. Sexes are generally separate. Sexual dimorphism is common (*i.e.* male and female look different).
16. Fertilization is internal. They are either **oviparous, ovoviviparous** or **viviparous.**
17. Development is usually **indirect** involving one to several larval forms.
18. Parthenogenesis occurs in some crustaceans and insects.

38.2 OUTLINE CLASSIFICATION

An outline classification of phylum Arthropoda is as under

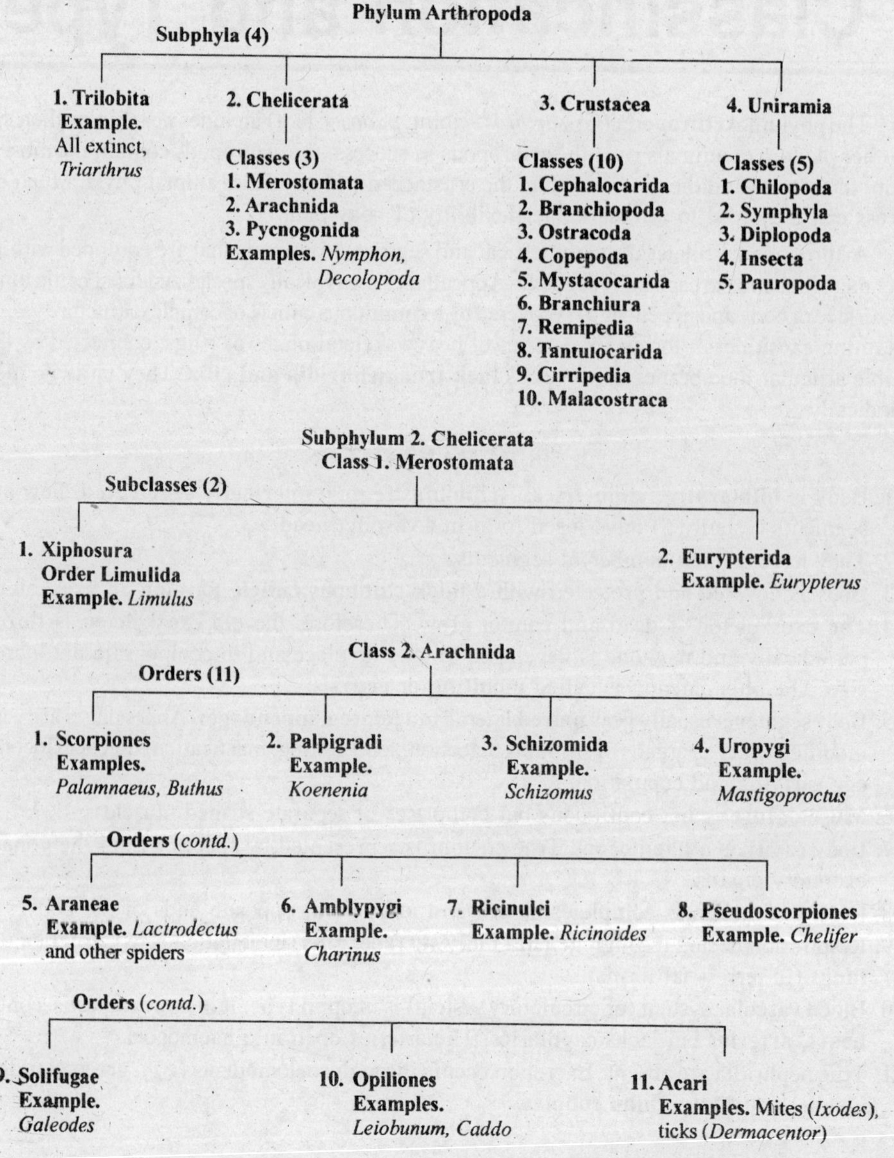

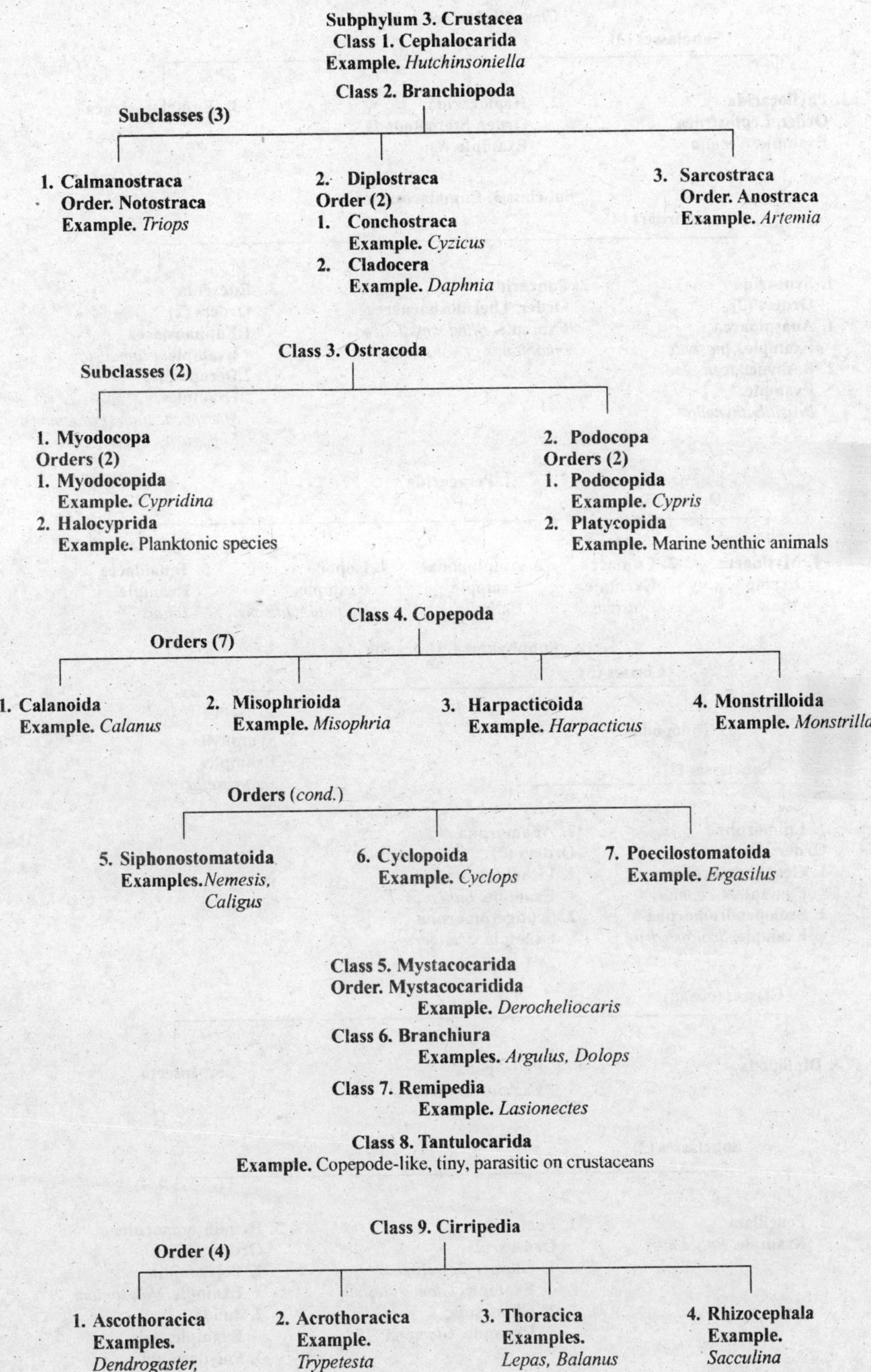
Subphylum 3. Crustacea
Class 1. Cephalocarida
Example. Hutchinsoniella
Class 2. Branchiopoda
Subclasses (3)
1. Calmanostraca
Order. Notostraca
Example. Triops
2. Diplostraca
Order (2)
1. Conchostraca
Example. Cyzicus
2. Cladocera
Example. Daphnia
3. Sarcostraca
Order. Anostraca
Example. Artemia
Class 3. Ostracoda
Subclasses (2)
1. Myodocopa
Orders (2)
1. Myodocopida
Example. Cypridina
2. Halocyprida
Example. Planktonic species
2. Podocopa
Orders (2)
1. Podocopida
Example. Cypris
2. Platycopida
Example. Marine benthic animals
Class 4. Copepoda
Orders (7)
1. Calanoida
Example. Calanus
2. Misophrioida
Example. Misophria
3. Harpacticoida
Example. Harpacticus
4. Monstrilloida
Example. Monstrilla
Orders (cond.)
5. Siphonostomatoida
Examples. Nemesis, Caligus
6. Cyclopoida
Example. Cyclops
7. Poecilostomatoida
Example. Ergasilus
Class 5. Mystacocarida
Order. Mystacocaridida
Example. Derocheliocaris
Class 6. Branchiura
Examples. Argulus, Dolops
Class 7. Remipedia
Example. Lasionectes
Class 8. Tantulocarida
Example. Copepode-like, tiny, parasitic on crustaceans
Class 9. Cirripedia
Order (4)
1. Ascothoracica
Examples. Dendrogaster, Ascothorax
2. Acrothoracica
Example. Trypetesta
3. Thoracica
Examples. Lepas, Balanus
4. Rhizocephala
Example. Sacculina

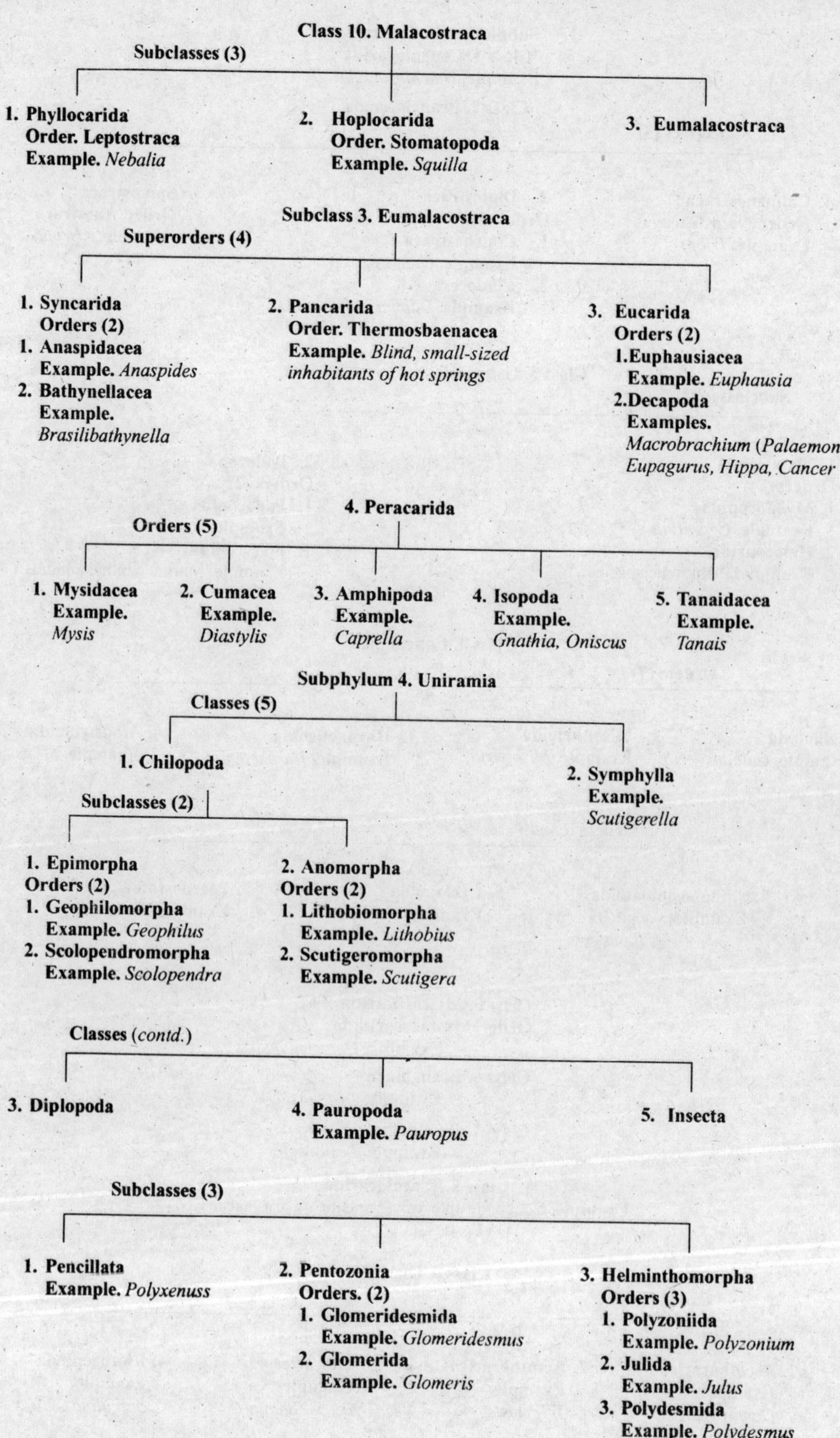
Class 10. Malacostraca
Subclasses (3)
1. Phyllocarida
Order. Leptostraca
Example. Nebalia
2. Hoplocarida
Order. Stomatopoda
Example. Squilla
3. Eumalacostraca
Subclass 3. Eumalacostraca
Superorders (4)
1. Syncarida
Orders (2)
1. Anaspidacea
Example. Anaspides
2. Bathynellacea
Example.
Brasilibathynella
2. Pancarida
Order. Thermosbaenacea
Example. Blind, small-sized
inhabitants of hot springs
3. Eucarida
Orders (2)
1.Euphausiacea
Example. Euphausia
2.Decapoda
Examples.
Macrobrachium (Palaemon),
Eupagurus, Hippa, Cancer
4. Peracarida
Orders (5)
1. Mysidacea
Example.
Mysis
2. Cumacea
Example.
Diastylis
3. Amphipoda
Example.
Caprella
4. Isopoda
Example.
Gnathia, Oniscus
5. Tanaidacea
Example.
Tanais
Subphylum 4. Uniramia
Classes (5)
1. Chilopoda
Subclasses (2)
1. Epimorpha
Orders (2)
1. Geophilomorpha
Example. Geophilus
2. Scolopendromorpha
Example. Scolopendra
2. Anomorpha
Orders (2)
1. Lithobiomorpha
Example. Lithobius
2. Scutigeromorpha
Example. Scutigera
2. Symphylla
Example.
Scutigerella
Classes (contd.)
3. Diplopoda
4. Pauropoda
Example. Pauropus
5. Insecta
Subclasses (3)
1. Pencillata
Example. Polyxenuss
2. Pentozonia
Orders. (2)
1. Glomeridesmida
Example. Glomeridesmus
2. Glomerida
Example. Glomeris
3. Helminthomorpha
Orders (3)
1. Polyzoniida
Example. Polyzonium
2. Julida
Example. Julus
3. Polydesmida
Example. Polydesmus

Class 5. Insecta

Subclasses (2)

1. Apterygota

Orders (5)

1. **Protura**
   Example. *Eosentomon*
2. **Aptera (Diplura)**
   Example. *Campodea*
3. **Collembola**
   Example. Springtails (*Entomobrya*)
4. **Microcoryphia**
   Example. Jumping springtails (*Machilis)*
5. **Thysanura**
   Example. Silver fish (*Lepisma*)

2. Pterygota

Orders (more than 17)

1. **Ephimerida**
   Example. Mayflies (*Ephemera*)
2. **Odonata**
   Example. Dragonflies (*Macromia*)
3. **Orthoptera**
   Examples. Grasshopper (*Poeciloceros*), locust (*Schistocerca*), cockroach (*Periplaneta*)
4. **Isoptera**
   Example. Termites (*Termes*)
5. **Dermaptera**
   Example. Earwig (*Forficula*)
6. **Plectoptera**
   Example. Stoneflies (*Taeniopteryx*)
7. **Mallophaga**
   Example. *Cuclogaster*
8. **Anopleura**
   Example. Human louse (*Pediculus*)
9. **Thysanoptera**
   Example. Thrips (*Taeniothrips*)
10. **Hemiptera**
    Example. Bedbug (*Cimex*)
11. **Homoptera**
    Example. *Cicada*
12. **Neuroptera**
    Example. *Mantispid*
13. **Coeloptera**
    Examples. Beetles (Lady bird beetle, *Adalia*), weevils (rice weevil, *Calandra oryzae*), fireflies (*Lamprohorous*)
14. **Lepidoptera**
    Examples. Butterflies (*Papilio*, *Kalima*), moth (*Bombyx mori*)
15. **Diptera**
    Examples. Mosquitoes (*Anopheles, Culex*), fruitflies (*Drosophila*), housefly (*Musca*)
16. **Hymenoptera**
    Examples. Wasps (*Polistes*, *Vespa*), honeybee (*Apis*)
17. **Siphonoptera**
    Example. Rat flea (*Xenopsylla*)

38.3. DETAILED CLASSIFICATION

Phylum Arthropoda is divided into **four** subphyla:

Subphylum 1. Trilobita (or Trilobitomorpha)

1. This subphylum is represented by fossil or extinct trilobites. They existed from Cambrian to Permian periods.
2. Body was divided into three lobes by two longitudinal furrows.
3. All body segments except the last contained **biramous appendages.**
4. Marine and benthozoic (bottom-dweller) arthropods.

Examples. *Triarthrus, Dalmanites.*

Subphylum 2. Chelicerata

1. Body is divided into **cephalothorax** or **prosoma** and **abdomen** or **opisthosoma.**
2. **Antennae** and **true jaws** are absent.
3. Cephalothoracic or prosomic appendages comprises of one pair of **chelicerae**, one pair of **pedipalps** and four pairs of **walking legs.**
4. Chelicerae form first pair of appendages. They are preoral and feeding appendages.
5. Second opisthosomal segment bear gonopores.
6. Excretion occurs by **coxal glands.**
7. Mostly terrestrial and predaceous.

Subphylum 3. Crustacea

1. Body is divisible into **cephalothorax** (fused head and thorax) and **abdomen.**
2. Cephalothorax is covered dorsally by **carapace.**
3. Head bears a pair of **compound eyes** and five pairs of biramous appendages: 1. two pair of antennae (*i.e.*, **antennules** and **antennae**); 2. **mandibles**; 3. two pairs of maxillae (*i.e.* **maxillules** and **maxillae**).
4. Each segment of thorax and abdomen contains a pair of branched or biramous appendages.
5. Their skeleton is hard and calcified.
6. They respire by means of gills.
7. Excretion is by green glands.
8. All are aquatic—mostly marine, some are freshwater (*e.g.*, *Daphnia*) and a few are adapted for moist land (crabs and wood lice).
9. Crustaceans are so abundant in oceans that they have been called "the insects of the sea".

Subphylum 4. Uniramia

1. Appendages are **uniramous** or unbranched.
2. The head is composed of five pairs of appendages—1. a single pair of **antennae**; 2. a pair of **labrum**; 3. a pair of **Mandibles** and 4. first and second pair of **maxillae**, with latter forming a **labium.**
3. **Trachea** are used for respiration.
4. **Malpighian tubules** are employed as excretory organs. Excretory waste is ammonia.
5. Nearly all are terrestrial

(**Note.** In old system of classification subphyla **Crustacea** and **Uniramia** were included in one subphylum **Mandibulata**. Four groups of uniramia, namely the **centipedes, millipedes, pauripods** and **symphyla**, form the common group **Myriapoda** including about 10,500 species).

Subphylum 2. Chelicerata

Subphylum Chelicerata is divided into **three** classes:

Class 1. Merostomata

1. Cephalothorax is covered by large, continuous dorsal shield.
2. Five or six pairs of abdominal appendages are modified as **gills**.
3. There is a well-developed spine or **telson** at the posterior end of body.
4. They contain a pair of compound eyes and a pair of simple eyes.
5. All are aquatic.

Class 2. Arachnida

1. This class includes about 10,000 species.
2. Body consists of two regions (or tagmata): **cephalothorax** (or **prosoma**) and **abdomen** (or **opisthosoma**).
3. The cephalothorax bears a pair of **chelicerae**, one pair of **pedipalps** and four pairs of **walking legs**. They lack **antennae** and **true jaws**.
4. Pedipalps are leg-like appendages that serve a sensory function, as in spiders, or are used for seizing the prey, as in scorpions.
5. Abdomen is without appendages.
6. Air-breathing mostly terrestrial arthropods.

Class 3. Pycnogonida (or Pentopoda)

1. This class includes about 1,000 species of sea **spiders**.
2. Arthropods with very small bodies and disproportionally **long legs**.
3. Body is narrow and comprises of **head** (cephalon), **trunk** (throax) and a short conical **abdomen.**
4. Head bears four **eyes** on a cen-tral tubercle and a cylindrical **proboscis**.
5. Trunk is formed of 4 to 6 cylindrical segments.
6. Appendages include a pair of **chelicerae** (called **chelifo-res**), a pair of **palps**, a pair of **ovigerous legs** or **ovigers** and four to six pairs of **walking legs**.
7. Ovigers are peculiar and may be used in grooming and in males to carry the eggs.
8. There are no special organs for respiration and excretion.
9. Sexes are separate; the eggs are carried on the ovigers of males and give rise to larvae, called **protonymphon**, hav-ing three pairs of legs.
10. Most pycnogonids are marine, bottom dwellers and crawl about algae, hydroids and bryozoans. They are **carnivorous** and feed on hydroids, soft corals, anemones, bryozoans, small polychaetes and sponges.

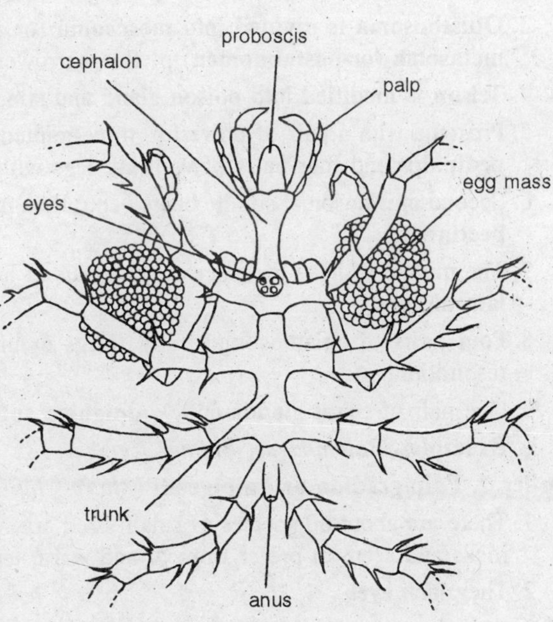

Fig. 38.1. *Nymphon rubrum*, a sea spider.

Examples. *Nymphon* (Fig. 38.1); *Pycnogonum, Pallene, Decolopoda.*

Class 1. Merostomata. Class Merostomata is divided into following **two** subclasses:

Subclass l. Xiphosura

1. Hemicircular prosoma or the cephalothorax is covered with a large horse-shoe shaped carapace.
2. Segments of abdomen (or opisthosoma) are fused and covered with a single plate.
3. Caudal spine or **telson** is as long as the body.
4. Cephalothorax with a pair of **chelicerae,** five pairs of walking legs and a pair of **chilaria.**
5. Walking legs are four to six-segmented and significant for burrowing and capturing the prey.
6. First four pairs of walking legs have projections, called **gnathobases** on the inner side of the basal coxae. They are used to fragment the prey.

   This subclass includes a **single** order.

Order Lemulida. Example. *Limulus* (king crab).

Subclass 2. Eurypterida

1. This subclass includes extinct giant arthropods that existed from the Cambrian and Permian period.
2. Cephalothorax is small and covered with a carapace.
3. Abdomen (opisthosoma) includes **mesosome** and **metasoma,** which lacks appendages.
4. Telson present.

   Examples. *Eurypterus, Pterygotus.*

Class 2. Arachnida

Class Arachnida is divided into following **eleven** orders.

Order 1. Scorpionida or Scorpiones (*Scorpions*)

1. There are about 2000 species of scorpions found in tropical and sub-tropical areas.
2. **Prosoma** is covered with a single shield or carapace.
3. **Opisthosoma** is divided into **mesosoma** (or **preabdomen**) of seven broad somites and **metasoma** (or **postabdomen**) of five narrower somites.
4. **Telson** is modified into poison gland and sting.
5. Prosoma with a pair of powerful, three-jointed chelicerae; a pair of large, distinct chelate **pedipalps** and four pairs of **walking legs** with nine podites.
6. Second opisthosomal somite bears ventrally a pair of comb-like sensory appendages, called **pectines.**
7. The middle of head carapace bears a pair of large **median eyes** and 1 to 5 pairs of small **lateral eyes.**
8. Four pairs of opisthosomal **book lungs** or unbranched **tracheal tubes** are present for respiration.
9. One pair of **coxal glands** and **Malpighian tubules** perform excretion.

   Examples. *Palamnaeus, Buthus, Centrurus.*

Order 2. Palpigradida or Palpigradi (*Micro-whip scorpions*)

1. There are about 60 species of small-sized arachnids which live in soil (in sand) and under rocks and seem to prefer tropical and warm temperate climates.
2. They lack **eyes**.
3. **Anus** is not terminal but followed by a many-segmented **post abdomen** or **flagellum** which is called **whip**.

4. Cephalothoracic carapace is formed of two pieces.
5. Chelicerae is large-sized and pedipalpi are leg-like.

Examples. *Koenenia*, *Eukoenenia*.

Order 3. Schizomida (*Schizomids*)

1. Cephalothoracic carapace is composed of three plates.
2. Pedipalps are leg-like.
3. First walking leg is antenna-like sensory limb.
4. Opisthosoma contains a large anterior portion and tiny post-abdominal region, "pygidium" of three segments with terminal flagellum.

Example. *Schizomus*.

Order 4. Uropygida or Uropygi (*Whip scorpion*)

1. This order includes about 85 species of whip scorpions.
2. Whip scorpions possess the well-developed, claw-like pedipalps of scorpions and the long, jointed-post abdomen of palpigrades.
3. **Chelicerae** is two jointed and ends in hook or fang.
4. Pedipalps are short but stout forming a pincer. They are raptorial.
5. Four pairs of eyes are usually present.
6. Abdomen (opisthosoma) is 12 segmented and bears a terminal flagellum or whip.
7. Respiration occurs by the help of two pairs of **book lungs**.
8. Excretion takes place by a pair of **coxal glands** and **Malpighian tubules**.
9. Abdomen contains a pair of large **anal glands** and when irritated the scorpion sprays the attacker with a fluid (containing acetic acid or formic acid) secreted from these glands. The secretion contains 84 per cent acetic acid and 5 per cent caprylic acid. The fluid can burn human skin. The repellent smell has given these scorpions another common name, **vinegarroon**.

Examples. *Mastigoproctus* (Fig. 38.2B), *Thelyphonellus*.

Order 5. Araneae (*Spiders*)

1. It includes about 32000 species of true spiders.
2. Prosomal segments are fused, opisthosoma is of 12 segments. Prosoma and opisthosoma are joined by a narrow **pedicel**.
3. Telson is absent.
4. Carapace of prosoma bears 8 eyes.
5. Each **chelicera** is two-jointed and it forms a hollow **fang** through which venom (poison) is injected into the prey.
6. **Pedipalps** are leg-like and are used for food handling. They are modified as copulatory organs in males.
7. The end of abdomen bears a group of modified appendages, the **spinning organs** or the **spinnerets**. These are used to make **silk webs**.
8. Spiders are extremely successful, small carnivorous animals. They play an important role in the ecological balance of their communities, serving as checks against some insect species.

Example. *Latrodectus* (black widow) and *Loxosceles reclusa* (brown recluse spider) both have extremely dangerous/fatal bites; *Lycosa* (wolf spider), *Synema* (crab spider), *Achaearanea* (house spider).

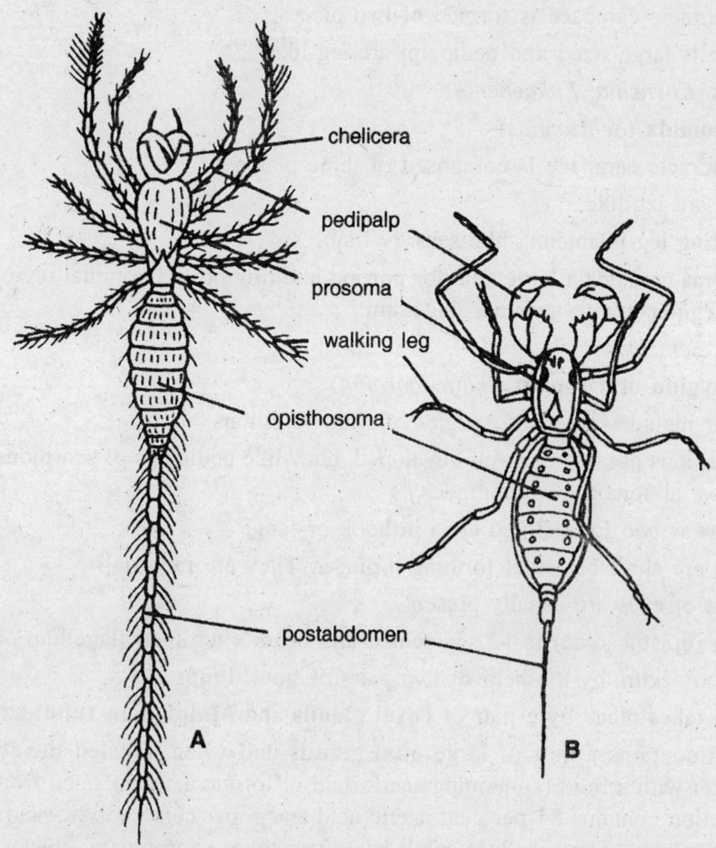

Fig. 38.2. A—*Nymphon hirsites*; B—*Mastigoproctus*.

Order 6. Amblypygida or Amblypygi (*Tail-less whip scorpion*)

1. It includes 70 species of spider-like creatures.
2. Body is flattened; carapace of prosoma bears three pairs of reduced lateral eyes, usually a pair of median ocelli.
3. Abdomen is 12-segmented and connected to cephalothorax by a narrow **pedicel** as in spiders. Abdomen is without flagellum. Telson is reduced.
4. Chelicerae are moderate.
5. Pedipalps are powerful, spiny and raptorial.
6. First pair of legs are extremely long and slender, used as tactile appendages; remaining legs are cursorial.
7. They lack spinnerets.

Examples. *Charinus, Terantula, Admetus*.

Order 7. Ricinuleida or Ricinulei (*Tick spiders*)

1. It includes 33 species of tick spiders which are found in Africa and from Brazil to southern United States.
2. Body is heavy and cuticle is very thick and often sculptured.
3. Carapace of prosoma bears a hood anteriorly, called **cucullus**, which covers chelicerae.
4. Prosoma and opisthosoma are separated by narrow pedicel; opisthosoma is wide and oval and formed of 12 united somites and ending in **anus** or short tubercle.

5. Chelicerae are two-segmented and pincer-like.
6. Pedipalps are leg-like and end in pincers.

Examples. *Ricinoides* (Africa), *Cryptocellus* (Southern USA).

Order 8. Pseudoscorpionida or Pseudoscorpiones (*False scorpions*)

1. This order includes about 2000 species of false scorpions.
2. Opisthosoma is broad and flat.
3. Carapace of prosoma is rectangular and bears one or two pairs of eyes. Eyes may be absent.
4. Chelicerae are small, two-jointed, chelate, forming pincers.
5. **Silk glands** open at the end of chelicerae. The glands are used to make a silk-lined chamber in which moulting occurs.
6. Pedipalps are powerful, chelate and contain **poison glands** that open at the tip.
7. Two pairs of spiracles lead to tracheae. Book lungs are absent.

Example. *Chelifer*.

Order 9. Solpugida or Solifugae (*Sun spiders*)

1. This order includes about 900 species of arachnids called **sun spiders** or **wind scorpions**.
2. These are not true spiders and are characteristics of hot deserts, where they hide in burrows or crevices until night, when they come out to feed. They are called 'sun spiders' because they are diurnal or 'wind scorpions' because they run very fast.
3. Prosoma is covered by shield of a single large anterior and three posterior plates. It contains a pair of closely placed eyes.
4. Abdomen (opisthosoma) is large and segmented containing 10 somites.
5. Chelicerae are two-jointed large and powerful; they form pincers.
6. Pedipalps are leg-like and contain an adhesive organ at the tip.
7. Respiration is by tracheal system.

Examples. *Galeodes* (*sun spider*).

Order 10. Phalangida or Opiliones (*Harvest* man or *Daddy longleg*)

1. This order includes about 4500 species of **daddy longlegs.** They derive their name from the fact that their legs are exceptionally long and thin.
2. Body is elliptical.
3. Undivided prosoma is broadly attached to nine-segmented opisthosoma.
4. Chelicerae are three-jointed, small and chelate.
5. Pedipalps are long and leg-like. Near the base of each pedipalp occurs a small pore that leads to **stink glands**. These give off an offensive secretion when animal is disturbed.
6. Carapace bears a pair of tubercles in the centre. Each tubercle bears an eye in centre.
7. Their reproductive system is **unique**. The male has a long tubular **penis**, and the female contains a protrusible **ovipositor**.

Examples. *Leiobunum, Caddo.*

Order 11. Acarina or Acari (*Ticks and mites*)

1. This order includes about 30,000 species of ticks (parasitic) and mites (free living).
2. Body is ovoid and compact; it is not divided into cephalothorax (prosoma) and abdomen (opisthosoma).

3. Chelicerae are typically modified for piercing.
4. Pedipalps are small but chelate.
5. Four pairs of legs with variable number of podites. They are used in creeping and climbing. In ticks each walking leg bears a pair of prehensile claws.

Examples. Mites (*Sarcoptes, Ixodes*) and tick (*Dermacentor*).

Subphylum 3. Crustacea

This subphylum includes about 38,000 aquatic species. It is divided into **ten** classes.

Class 1. Cephalocarida* (*Head shrimps*)

(*The root **carid** so widely used in the name of crustaceans taxa means "shrimp").

1. It includes about 9 species of most primitive tiny, eyeless crustaceans.
2. They are marine and found in soft bottom sediments. They feed on organic residues in bottom deposits.
3. The **head** is broadly rounded. The trunk is tapered, small and is divided into a **thoracic region** of eight segments, each bearing a pair of **biramous** appendages and **abdomen** which includes remaining 11 to 12 segments which lack appendages.
4. Antennae are short.
5. Appendages are flat phyllopods and their gnathobases-like endites help in detritus feeding.
6. They are hermaphrodites; development is **indirect** and includes metanauplius larval stage.

Examples. *Polyphemus, Hutchinsoneilla.*

Class 2. Branchiopoda (*Gill feet*)

1. It includes about 821 species which are mostly inhabitants of freshwater.
2. Small crustaceans with foliaceous trunk appendages, *i.e.*, appendages are flattened, leaf-like and act as organs of respiration and locomotion.
3. The first antennae and second maxillae are vestigial.
4. Last abdominal segment bears slender, paired processes (**cercopods**).
5. They are filter feeders.

Class Branchiopoda is divided into fullowing **three** subclasses.

Subclass 1. Calmanostraca

1. Possess carapace.

It includes **single** order.

Order 1. Notostraca (*Tadpole shrimps*)

1. The head and anterior half of trunk are covered with a large, shield-like dorsal carapace.
2. Body is composed of thorax with appendages and an abdomen without appendages.
3. They have unique external cuticular rings that do not correspond to the somites.
4. The **telson** bears a pair of annulate caudal rami (= cercopods).
5. Paired compound eyes are sessile (without eye stalk).

Tadpole shrimps prefer alkaline freshwater and are not disturbed by high turbidity.

Example. *Triops* (Fig. 38.3B).

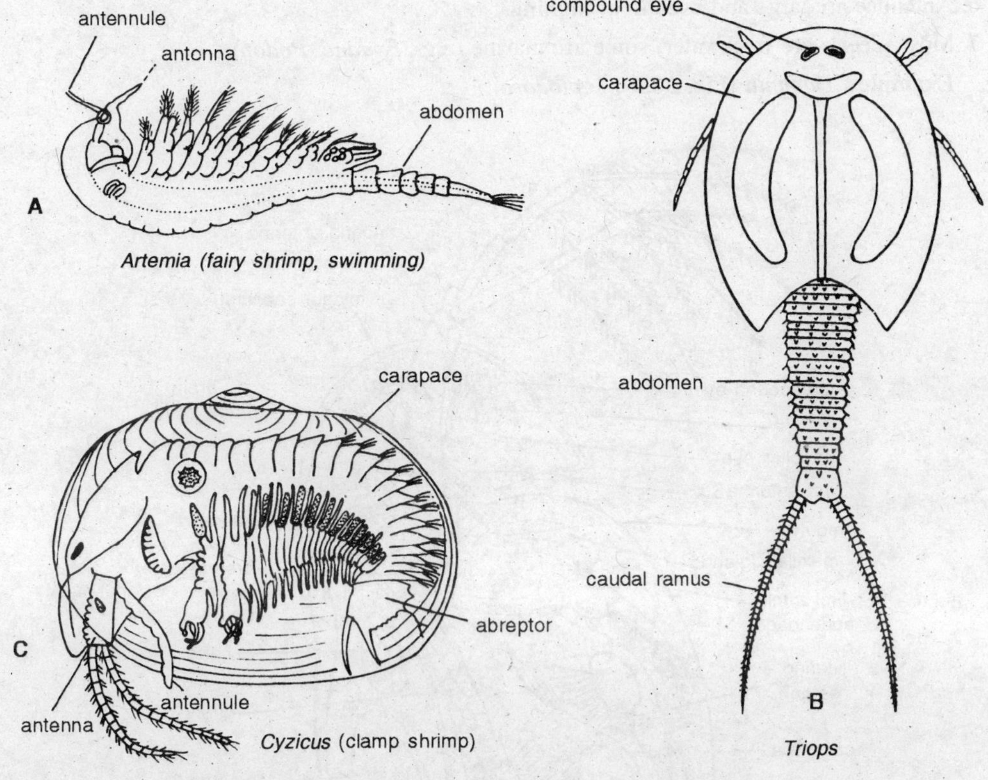

Fig. 38.3. Branchiopods A—*Artemia*; B—*Triops*; C—*Cyzicus*.

Subclass 2. Diplostraca

1. Body is enclosed within a laterally compressed bivalve carapace.

Subclass Diplostraca is divided into **two** orders:

Order 1. Conchostraca (*Clam shrimp*)

1. Entire body is completely enclosed within the carapace and animal looks strikingly like a little clam.
2. Ten to 32 trunk segments are present.
3. Second antennae are well developed, biramous and setose.
4. Compound eyes are sessile.

Examples. *Cyzicus* (Fig. 38.3C); *Lynceus*.

Order 2. Cladocera (*Water fleas*)

1. The bivalve carapace encloses the trunk but not the head and often terminates posteriorly in dorsal spine.
2. A space called **brood chamber** lies between the dorsal body surface and carapace.
3. Trunk has five or six pairs of appendages.
4. The tip end of the trunk, commonly called the **post-abdomen**, is turned ventrally and forward and bears special claws and spines for cleaning the carapace.
5. Head bears single median compound eye.

6. Antennae are large and used in swimming.
7. Most species are freshwater; some are marine (*e.g.*, *Evadne, Podon*).

Examples. *Daphnia* (Fig. 38.4), *Leptodora*.

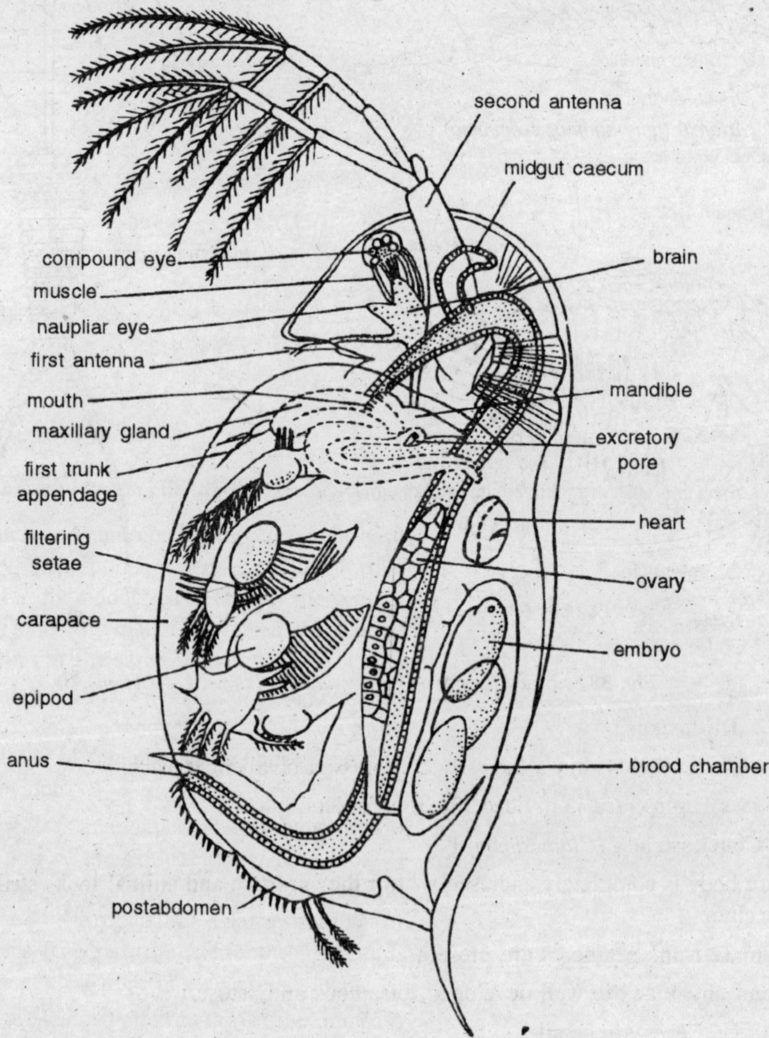

Fig. 38.4. *Daphnia pulex.* Female.

Subclass 3. Sarcostraca

1. Carapace is absent.

This subclass includes **single** order.

Order Anostraca (*Fairy shrimp*)

1. Carapace is absent.
2. Thorax contains 11 to 18 segments, each having a pair of appendages (**phyllopodia**).
3. Abdominal segments are limbless. Anal somite is with a pair of caudal rami.
4. Have stalked compound eyes.

Examples. *Branchinecta*; *Artemia* (brine shrimp, Fig. 38.3A, an inhabitant of salt lakes and ponds).

Class 3. Ostracoda

1. Include about 5,650 species of mussel or seed shrimps.
2. They are marine and freshwater minute crustaceans. Some are free living (swimmers); others live on bottom.
3. Body is highly reduced and enclosed within hinged, bivalved carapace, which is commonly impregnated with calcium carbonate.
4. Head is large and trunk is short. All external trunk segmentation have disappeared.
5. Antennules and antennae are well developed.
6. There are only two pairs of legs in trunk. Maxillae are leg-like and are modified for swimming, feeding, clasping, etc.

Class Ostracoda is divided into **two** subclasses.

Subclass 1. Myodocopa

1. All are marine.
2. Shell valves contain **antennal notch**.
3. Second antennae is adapted for swimming.
4. Two pairs of trunk appendages.

Subclass Myodocopa is divided into **two** orders.

Order 1. Myodocopida

The long worm-like second pair of trunk appendages are adapted for cleaning the interior of valves.

Example. *Cypridina*, *Gigantocypris*, *Skogsbergia*.

Order 2. Halocyprida

1. All are marine.
2. First pair of trunk appendages are present or absent.
3. Second pair of appendages are absent or short and leg-like.

Example. Planktonic species.

Subclass 2. Podocopa

1. Carapace valves are without an antennal notch.
2. One or two pairs of trunk appendages are present.

Subclass Podocopa is divided into **two** orders.

Order 1. Podocopida

1. First pair of trunk appendages are leg-like.
2. This large order includes marine as well as freshwater species.

Examples. *Cypris*, *Pontocypris*, *Candona* (See Box 38.1), *Cypridopsis*, *Mesocypris*, *Darwinula*, *Cythere*.

Box 38.1
Zenker's organ

Candona suburbana is a very common freshwater podocopid ostracod. This nonswimming bottom dweller feeds on algae and decomposing vegetation. Zenker's organ, which is a part of the male reproductive system, ejects sperm to the penis.

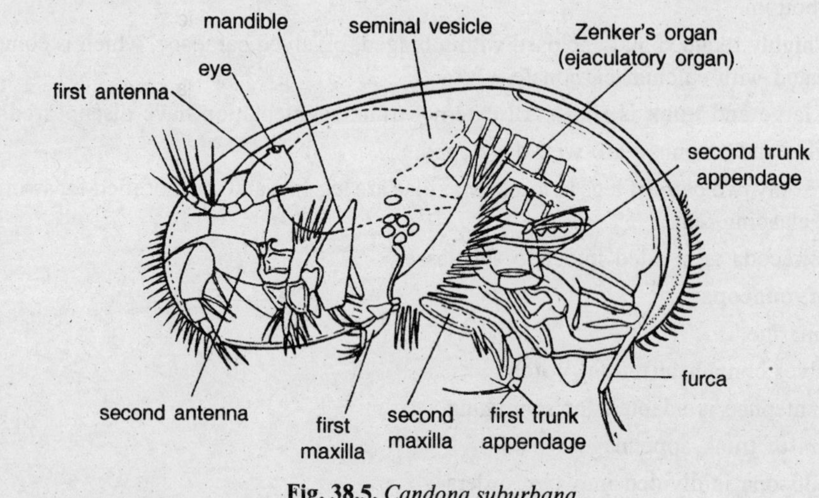

Fig. 38.5. *Candona suburbana.*

Order 2. Platycopida

1. Marine, benthic animals.
2. First pair of trunk appendages are not leg-like.

Class 4. Copepoda (*Copepods*)

1. It includes about 8405 species of small crustaceans.
2. Both marine and freshwater forms. Free living and parasitic.
3. Body is cylindrical and tapering and divided into head, thorax and abdomen.
4. Thorax bears five pairs of biramous appendages and abdomen is without appendages. Telson is with two caudal forks.
5. First pair of antennae are longer than the second pair and held outstretched.
6. Compound eyes are absent; naupliar eye is present.

Class Copepoda is divided into **seven** orders.

Order 1. Calanoida

1. Free living, planktonic forms.
2. First pair of antennae is very long having setae and 22 or more articles.

Examples. *Calanus*, *Calocalanus*, *Metridia*, *Acarita*.

Order 2. Misophrioida

1. Pelagic animals living on or above the bottom surface.
2. First pair of antennae are shorter, containing 11 to 16 articles.

Example. *Misophria*.

Order 3. Harpacticoida

1. Free living marine and freshwater forms. Bottom dwellers.
2. First pair of antennae are short having 10 articles.

Example. *Harpacticus*, *Canthocamptus*.

Order 4. Monstrilloida

1. Marine; larval stages are parasitic in polychaetes and gastropods.
2. Adults are nonfeeding, planktonic and lack second antennae and mouth parts.

Example. *Monstrilla*, *Xenocoeloma*.

Order 5. Siphonostomatoida

1. Freshwater and marine forms.
2. Adults are parasitic on fish and invertebrates.

Example. *Nemesis*, *Clevella*, *Caligus*, *Penella*.

Order 6. Cyclopoida

1. Marine, freshwater, planktonic and benthic forms.
2. Some forms are commensals and parasites.
3. First pair of antennae are short containing 10 to 16 articles.
4. Second antennae are uniramous (unbranched).

Example. *Cyclops*, *Sapphirina*, *Oncaea*.

Order 7. Poecilostomatoida

1. Marine.
2. Adults are parasites on invertebrates and fish.

Example. *Ergasilus*, *Chondracanthus*.

Class 5. Mystacocarida or Mystacarida (*Moustache shrimp*)

1. It includes 9 species of marine forms adapted for living between the intertidal sand grains.
2. Body is long and cylindrical and divided into head, thorax and abdomen.
3. Thorax contains less than six segments. Mandibular palp is present.
4. Naupliar eye is absent.
5. Both pairs of antennae are long and prominent.
6. Maxillipedes are present. Mouth appendages contain setae for collecting detritus.
7. Caudal rami work as pincers.
8. Sexes are separate. Larva is **nauplius**.

This class includes **single** order.

Order Mystacocaridida

Example. *Derocheilocaris* (Fig. 38.6A)

Class 6. Branchiura (*Gill tails*)

1. This class includes about 150 species of marine and freshwater forms.
2. They are ectoparasitic on the skin and gill cavities of fishes.
3. Head and throax are covered by a shield-like carapace.
4. Abdomen is small, bilobed and unsegmented.
5. Both pairs of antennae are reduced and modified for attachment. First pair of antennae have a large claw for attachment to the host.
6. Compound eyes are present.
7. In some (*e.g.*, *Argulus*, Fig. 38.6B), first pair of maxillae contain two large suckers for attachment. Other appendages are vestigial.

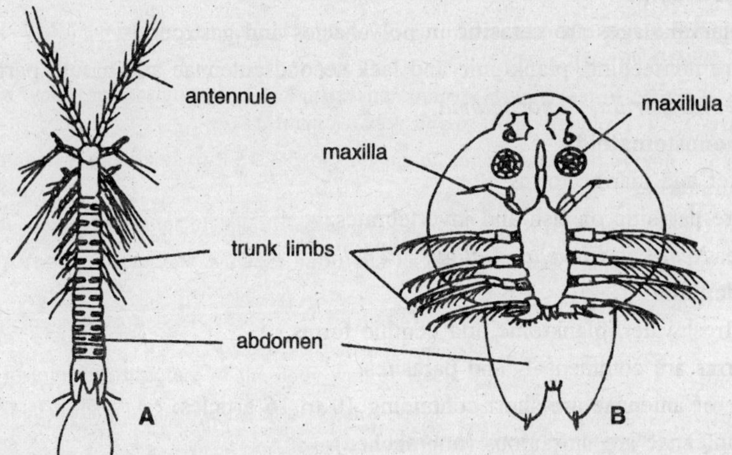

Fig. 38.6. A—*Derocheilocaris* (mystacocarid); B—*Argulus* (branchiuran).

Class 7. Remipedia

1. It includes only 2 species of crustaceans which are inhabitants of marine caves. They were discovered by **Yager** (1981).
2. Body is very long, segmented and worm-like.
3. Each body segment bears a pair of biramous appendages.

Example. *Lasionectes.*

Class 8. Tantulocarida

1. It includes only 4 species.
2. Small marine crustaceans ectoparasitic on deepwater crustaceans.
3. Body is similar to copepods but lacking trunk segments.

Example. *Copepode-like animals.*

Class 9. Cirripedia (*Curl foot barnacles*)

1. This class includes about 900 species of barnacles.
2. They are either free living or parasitic. All are marine.
3. Body is enclosed within a saccular bivalved **carapace** or mantle of lime. Carapace is divided into plates.
4. Both pairs of antennae are reduced or absent. **Mandibles** are blade-like.
5. Body is divided into head and trunk. External segmentation is very indistinct. Abdomen is absent.
6. Thorax bears six pairs of similar biramous appendages called **cirri**.
7. Most of them are **hermaphrodites**. Penis is long and tubular. Young ones hatches as **nauplius larva** that moults to produce **cypris larva**.

Class Cirripedia is divided into **four** orders:

Order 1. Ascothoracica or Ascothoracida (*Naked barnacles*)

1. They are parasitic on echinoderms and corals.
2. Bivalve or saccular carapace; without calcareous plates on the mantle.
3. First antennae are prehensile. **Mandibular palp** is present.
4. Abdomen is present.

5. Female gonopore on first thoracic segment.

Examples. *Dendrogaster, Ascothorax.*

Order 2. Acrothoracica (*Naked boring barnacles*)

1. They live attached to some calcareous substratum, especially shells of molluscs and corals. They bore into substratum.
2. They lack calcareous plates on the mantle. A chitinous attachment disc is present.
3. Three to five pairs of cirri are present.

Examples. *Trypetesa, Kochlorine, Berndita.*

Order 3. Thoracica (*Goose barnacles and Acorn barnacles*)

1. They are free living and commensal forms; with or without stalk (peduncle).
2. Carapace or mantle is covered with calcareous plates.
3. Animal lies in the carapace upside down, *i.e.*, stands on its head.
4. Six pairs of well-developed cirri are present.
5. They are monoecious.

Examples. *Lepas, Scalpellum, Verruca, Balanus, Coronula, Xenobalanus.*

Order 4. Rhizocephala

1. Parasitic on decapod crustaceans and tunicates.
2. Mantle, appendages and alimentary canal are absent.
3. Body is reduced to ramifying stolon and gonad sac.
4. Stalk or peduncle forms root-like absorptive processes.

Examples. *Sacculina, Peltogasterella, Lernaeodiscus.*

Class 10. Malacostraca

1. It includes about 22,651 species which are marine or freshwater.
2. Body is divided into head, thorax and abdomen.
3. Trunk is typically composed of 14 segments and a *telson*. Its first 8 segments form the thorax and last 6 form the abdomen.
4. Thorax may or may not be covered with carapace.
5. All body segments bear appendages.
6. **Female gonopores** are always located on the 6th thoracic segment and the **male gonopores** are on the 8th thoracic segment.

Class Malacostraca is divided into **three** subclasses:

Subclass 1. Phyllocarida

Order Leptostraca or Nebaliacea

1. It includes about 20 species of most primitive living malacostracans (appeared in Cambrian period).
2. Small sized marine animals living in bottom mud.
3. They are suspension feeders.
4. They have **eight** segments in the abdomen.
5. Thorax is enclosed in a bivalve carapace and is provided with foliaceous appendages.

Examples. *Nebalia, Paranebolia.*

Subclass 2. Hoplocarida

Order Stomatopoda (*Mantis shrimp or armed shrimps*)

1. It includes about 300 species which are marine and specialized predators of fish, crabs, shrimps and molluscs.
2. Body is dorso-ventrally flattened. Carapace is small and shield-like but does not cover the last two thoracic segments.
3. Abdomen is large, broad and segmented.
4. First antennae (*i.e.*, **antennules**) with three flagella.
5. Second pair of thoracic appendages are large and chelate; it is developed for **raptorial feeding**.
6. Thoracic limb protopod (= protopodite) is composed of two segments. Endopods (= endopodite) comprise of four segments.
7. Compound eyes are stalked.
8. Pleopods are well developed; they contain gills.
9. Telson and uropods are large.

Examples. *Squilla*, *Lysiosquilla*, *Gonodactylus*.

Subclass 3. Eumalacostraca

1. Antennae are without three flagella.
2. Endopodite is composed of five segments.
3. Gills when present, they are restricted to thoracic limbs.

Class Eumalacostraca is divided into **four** superorders:

Superorder 1. Syncarida

1. Carapace is absent.
2. Eight thoracic segments are free or the first fused with head.
3. Thoracic appendages are biramous and similar. Chelate or subchelate legs are absent.
4. Freshwater (ground water)—inhabitants or cave-dwelling.

It includes **two** orders.

Order 1. Anaspidacea

1. It includes 10 species which are suspension feeders.
2. First thoracic somite is fused with head.
3. First thoracic legs are modified as maxillipedes and have distinct gnathobases.

Example. *Anaspides*.

Order 2. Bathynellacea

1. It includes about 25 species of minute malacostracans.
2. First thoracic segment is not fused with the head.

Example. *Brasilibathynella*.

Superorder 2. Pancarida

Order Thermosbaenacea

1. They are inhabitants of hot springs and ground water.
2. Small sized and blind animals having a reduced carapace which is fused with the first thoracic segment.
3. They carry eggs in dorsal brood pouch.

Example. Hot spring inhabitants.

Superorder 3. Eucarida

1. It includes large sized malacostracans.
2. Carapace is highly developed and fused with all the thoracic segments.

3. Gills are located near the base of thoracic appendages.
4. Compound eyes are stalked.

This superorder is divided into **two** orders.

Order 1. Euphausiacea (*Krills*)

1. It includes about 90 species of brightly shining krills.
2. All are marine; pelagic and shrimp-like, upto 3 cm long.
3. Sides of carapace do not enclose the gills tightly.
4. Thoracic appendages are biramous and anterior ones are not modified as maxillipeds.
5. Many contain light producing **photophores** and are biolumnescents.

Example. *Euphausia.*

Order 2. Decapoda (*Shrimps, crabs, lobsters* and *cray fishes*)

1. It includes about 10,000 species accounting for nearly 20 per cent of the known species of crustaceans.
2. They are marine, freshwater or terrestrial.
3. Their first three pairs of thoracic appendages are modified as **maxillipedes**. The remaining five pairs (*i.e.*, 10) of thoracic appendages are **uniramous walking legs** (hence the name decapoda).
4. The carapace encloses thé gills in a branchial chamber.
5. **Statocyst** is present in the basal segment of first antennae (or antennule).

Examples. Shrimps—*Crangon* (sand shrimps), *Alpheus* (snapping shrimp), *Palaemon* or *Macrobrachium* (prawn), *Atya*; **crayfish**—*Procambarus*, *Astacus*, *Cambarus*; **lobsters**—*Nephrops*, *Homarus*; **marine burrowing shrimp**—*Thalassina*; **spiny lobsters**—*Panulirus*; **hermit crab**—*Eupagurus*, *Pagurus*; **coconut crab**—*Birgus*; **lobster**—*like Galathea*; **mole crabs**—*Hippa*, *Emerita*; *Brachyura*; **box crab**—*Calappa*, **Spider crab** *Maja*; **fiddler crab** — Uca; **mud crab**—*Xantho*; **rock crab**—*Cancer*; **blue crab**—*Callinectes*.

Superorder 4. Peracarida

1. This superorder carries 40 per cent of the known species of Crustacea.
2. They distinctly contain a **ventral brood chamber** or **marsupium** in the female. It is formed by large plate-like processes of certain thoracic coxae.
3. A carapace may be present or absent.
4. First thoracic somite is fused with the head.
5. There are eight pairs of thoracic appendages. First pair is usually a maxillipede and remaining seven are **legs**.
6. Development is direct.

Superorder Peracarida is divided into **five** orders:

Order 1. Mysidacea (*Opossum shrimps*)

1. Most are marine and some live in freshwater (*i.e*, in lakes and caves); pelagic and benthic; filter—feeder or scavengers.
2. They occur in large **swarms** forming an important food organism of economically important fish.
3. Ventral **marsupium** present *i.e.*, oostegites present from two to all of thoracic legs forming a marsupium due to which they are called **opposum shrimps**.
4. They have stalked compound eyes.
5. First and sometimes second pair of thoracic appendages are modified as maxillipedes.

6. Pleopods are present in males but reduced or absent in females.

Examples. *Mysis, Neomysis, Gnathophausia.*

> **Box 38.2.**
> *Mysis oculata* lives only in deep lakes and undergoes remarkable **diurnal movement**. During the daylight hours they stay within a metre of the bottom, but during the night they might rise to surface.

Order 2. Cumacea

1. It includes 800 species which are marine, burrowing and filter feeding creatures. They live in burrows or mucous tubes in bottom mud or sand.
2. Swarmming occurs in males.
3. Head and thorax are greatly enlarged. Carapace covers the first three to four thoracic somites forming a gill chamber.
4. Abdomen is very narrow; it is usually without pleopods in females and bear two to five pleopods in males.
5. Uropods are slender, filiform structures and used in cleaning the anterior segments.
6. Antennae are vestigial in the female but very long in males.
7. Series of filamentous gills are present on first pair of maxillipedes.

Example. *Diastylis*.

Order 3. Amphipoda (*Both foot*)

1. It includes about 5,500 species which are marine freshwater or terrestrial. They may be free living and ectoparasitic species.
2. Some are pelagic, most are benthic, a few are filter feeders.
3. Body is laterally compressed and without carapace.
4. Compound eyes are sessile lacking eyestalks.
5. There are one pair of maxillipedes and seven pairs of uniramous thoracic legs having gills.
6. The abdominal appendages include three pairs of pleopods and three pairs of uropods. These are adapted for jumping.

Example. *Caprella* (skeleton shrimp), *Hyalella*; *Paracyamus* (whale lice), *Phronima* (parasite of jelly fish or tunicates); *Orchestoidea, Rhabdosoma.*

Order 4. Isopoda (*Equal foot*)

1. It includes about 10,000 species which are freshwater, marine or terrestrial. They are free living and parasitic.
2. They are not filter feeders. Mouth parts resemble with that of insects.
3. Body is dorso-ventrally flattened. Head is shield-shaped.
4. Carapace is absent; compound eyes are sessile.
5. There is one pair of maxillipedes and seven pairs of

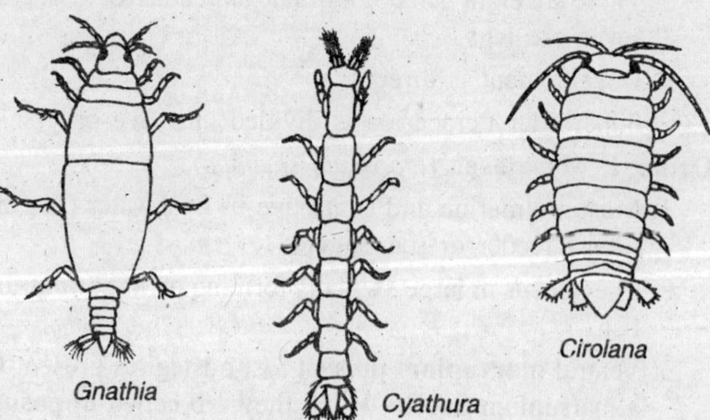

Fig. 38.7. Some isopods.

similar uniramous thoracic legs. These are adapted for crawling (they creep like insects). Some aquatic isopods can also swim.

6. Some abdominal pleopods function as gills.
7. Uropods are fan-shaped.
8. Some isopods can roll up, armadilo-like when disturbed. Land isopods, often called **pill bugs, sow bugs** or **wood lice** are frequently found under logs in wooded areas.

   Example. *Gnathia, Cyathura, Limnoria, Cirolana, Astacilla, Asellus, Phreatoicus, Bopyrus, Ligia, Oniscus.*

Order 5. Tanaidacea

1. It includes about 550 species of small, benthic marine animals.
2. They burrow in mud or live in mucous tubes.
3. A small carapace covers anterior part of body; inner surface of carapace acts as gills.
4. Eyes are lobed, present on lateral immovable processes.
5. The first thoracic appendages are maxillipedes equipped with flattened epipodite.
6. Third thoracic appendages are adapted for burrowing.
7. Females have marsupium.
8. Each of them begin life as a female. After having served in this role, it molts to a male and serves in this capacity at least once.

   Examples. *Leptochelia, Tanais.*

Subphylum 4. Uniramia

Subphylum Uniramia is divided into **five** classes:

Class 1. Chilopoda (*Centipedes*)

1. It includes about 2500 species of centipedes.
2. They live in moist places such as soil and humus and beneath stones, bark and log.

   They are nocturnal and carnivorous animals feeding upon soft insects such as cockroach, plant lice, silver fish and earthworm and slugs.
3. First pair of trunk appendages (maxillepedes) are modified into **poison claw** or **forcipules**.
4. Body is divided into head and trunk. Head is convex in scutigeromorph but flattened in other centipedes.
5. Trunk is somewhat flattened and elongated and contains 15 to 170 segments. Each segment bears one pair of legs. All the legs are similar. The last pair of legs are elongated and sensory (or defensive) in function.
6. Last segment is limbless, called **genital segment**, and it bears genital opening.
7. Centipedes run very rapidly.

   Class Chilopoda is divided into **two** subclasses:

Subclass 1. Epimorpha

1. Development is **epimorphic**, *i.e.*, young contains all segments when they hatch.
2. The female broods its eggs by winding itself around the egg mass.
3. Adults contain 21 or more pairs of legs.

   This subclass is divided into **two** orders:

Order 1. Geophilomorpha

1. Elongated, blind burrowing form with 31 to 180 pairs of legs.
2. Each trunk somite with dorsal tergite and intertergite and ventral sternite and intersternite.

3. Antennae with 14 segment.
4. Pair of spiracles are present in all but first and last somites.

   Examples. *Geophilus*, *Strigamia*, *Mecistocephalus*.

Order 2. Scolopendromorpha

1. They contain 21 to 23 pairs of legs.
2. Tergal plates correspond to sternal plates and are equally large.
3. Antennae with 17 to 31 segments.
4. Eyes are absent or composed of four ocelli.
5. They contain 9 to 11 pair of spiracles.

   Examples. *Scolopendra*, *Theatops*.

Subclass 2. Anamorpha

1. Development is **anamorphic**, *i.e.*, the youngs contain segments less than that of adults.
2. No brooding occurs.
3. Adults have 15 pairs of legs.
4. A pair of **organ of Tomosvary** is present on the head at the base of antennae. It is used in **vibration detection** and **monitoring of humidity**.

   This subclass is divided into **two** orders.

Order 1. Lithobiomorpha

1. Antennae contain 19 to 70 segments.
2. Trunk contains nine large and six small somites.
3. Large and small tergal plates alternate.
4. Spiracles are paired and lateral.
5. Eyes with many ocelli.
6. Development is anamorphic.

   Examples. *Lithobius*, *Bothropolys*.

Order 2. Scutigeromorpha

1. Legs and antennae are very long.
2. Fifteen sternal plates, but with only eight tergal plates.
3. Eyes are large and compound.
4. Spiracles are unpaired and located mid-dorsally on tergal plates.
5. Development is anamorphic.

   Example. *Scutigera*.

Class 2. Symphylla

1. It includes 160 species of myriapods which live in soil of green houses and attack plant roots and can be serious pest to vegetable and flower crops.
2. They are between 1 and 8 mm in length and superficially resemble lithobiomorph centipedes.
3. Somites of trunk are covered by 15 to 24 tergal plates.
4. Trunk contains 12 legs bearing segments.Trunk terminates in a tiny oval telson.
5. First maxilla with no palp. Second maxillae fuse to form a **labium.**
6. A single pair of spiracles opens on to the sides of the head, and the tracheae supply only the first three trunk segments.
7. **Gonopore** is present ventrally on fourth trunk segment.

8. The last (14th) segment bears a pair of spinnerets or cerci and a pair of long, sensory hairs (called **trichobothria**).
9. Eyes are absent.
10. Two organs of Tomosvary are present (Fig. 38.9).
11. Development is anamorphic.

Examples. *Scutigerella* (Fig. 38.8); *Hanseniella.*

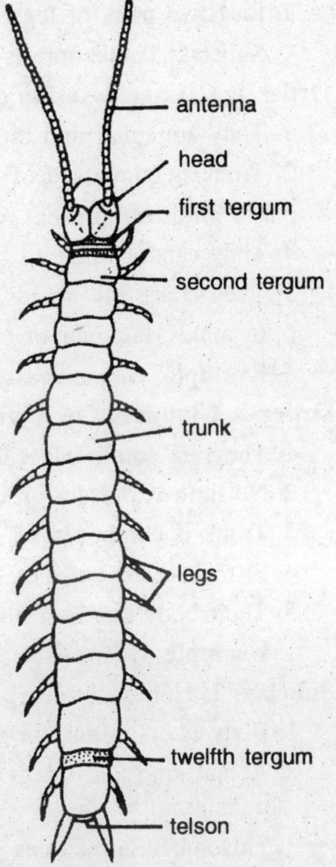

Fig. 38.8. *Scutigerella* (in dorsal view).

Class 3. Diplopoda (*'Thousand legged'* or *millipedes*)

1. It includes about 10,000 species.
2. Millipedes avoid light and live beneath leaves, stones, bark and logs and in soil. They are herbivorous and run much more slowly.
3. Diplopod (double foot) refers to the most important character of millipedes. They contain two pairs of legs in nearly each abdominal segment. This occurs because two embryonic somites fuse to form an adult somite, called **diplosegment**.
4. Body is elongated and cylindrical.
5. Body is divided into **head** (first 5 segments), **thorax** (of 3 segments) and **abdomen** of 20 to 100 segments.
6. Last head segment is called **collum** and forms a large collar behind the head.
7. Thorax contains one pair of legs in each segment.
8. Each of the diplosomites is enclosed in a continuous skeletal ring, with no articular membranes between the **tergal**, **pleural** and **sternal plates**. Each segment is covered with single tergal plate dorsally and two sternal plates ventrally. Tergites are impregnated with calcium salts like the crustacean.
9. **Stink glands** or **repugnatorial glands** are present in many millipedes (*e.g.*, *Julus*, *Apheloria*) secreting noxious substances which may be **aldehydes**, **quinones**, **phenols** and **hydrogen cyanide**.
10. Actual copulatory organs are modified trunk appendages, called **gonopods**.

Class Diplopoda is divided into following **three** subclasses:

Subclass 1. Pencillata (*Pselaphgnatha*)

1. Minute millipedes with broad and soft bodies. Integument bears tufts and rows of serrated scale-like, setae.
2. Eyes are present.
3. Gonopods and stink glands are absent.
4. Trunk bears 13 somites.

Example. *Polyxenus.*

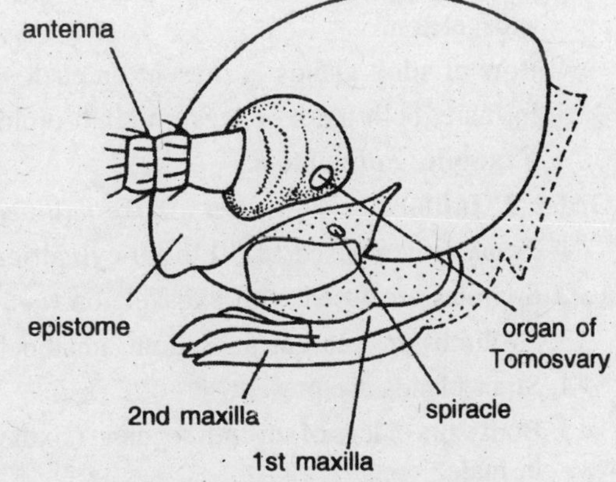

Fig. 38.9. Head of *Hanseniella.*

Subclass 2. Pentazonia

1. Tergal plates are arched.
2. Last two pairs of legs are modified for clasping.

Subclass Pentazonia is divided into **two** orders:

Order 1. Glomeridesmida or Limacomorpha

1. They are small and blind.
2. Trunk is composed of 20 to 22 arched somites.
3. No limbs are present on collum, gnathochilarium.
4. They cannot roll into ball.
5. They lack stink glands.
6. In males, last pair of legs are modified as **gonopods**.

Example. *Glomeridesmus.*

Order 2. Glomerida or Oniscomorpha

1. They are commonly called **pill millipedes.**
2. No limb is present on collum, gnathochilarium.
3. Trunk is composed of 12 to 13 somites which are flattened on the ventral surface.
4. In males, last pairs of legs are modified ; the last pair is used for sperm transfer.
5. Their body can be rolled into ball.

Example. *Glomeris.*

Subclass 3. Helminthomorpha

1. Body is cylindrical or somewhat flattened.
2. At least one pair of legs (gonopods) of the seventh segment in the male is modified for sperm transfer.
3. This subclass includes greatest number of millipedes species.

This subclass is divided into **eleven** orders, out of which following **three** are more important.

Order 1. Polyzoniida or Colobognatha (*Suctorial millipedes*)

1. Largely tropical, suctorial and elongated millipedes.
2. No limbs are present of collum; gnathochilarium.
3. Body contains 30 or more diplosomites. Sternal plates are not fused with rest of exoskeleton.
4. Row of stink glands is present on each side of body.
5. In male, both pairs of legs on sixth diplosomite is modified as gonopod.

Example. *Polyzonium.*

Order 2. Juliformia or Julida (*Snake millipedes*)

1. Trunk is composed of 30 to 90 cylindrical segments.
2. Sternites arc fused with exoskeleton (*i.e.*, pleurotergal arch).
3. No limbs are present on collum, gnathochilarium.
4. Stink glands are present.
5. Both pairs of legs of seventh segment (sixth when collum is excluded) are modified as gonopods in male.

Examples. *Julus, Nemasoma, Blaniulus, Thyroglutus.*

Order 3. Polydesmida (*Flat-backed millipedes*)

1. Relatively short body, with no limbs on collum, gnathochilarium.
2. Trunk contains 19 diplosomites, each covered by two skeletal rings, anterior **prozonite** and posterior **metazonite**.
3. Sternites are fused with exoskeleton (*i.e.*, pleurotergal arch).
4. Stink glands are present on some somites.
5. Eyes are absent.
6. In males, first pair of legs on seventh diplosomite are modified as gonopods.

Examples. *Polydesmus*, *Oxidus*, *Apheloria*.

Class 4. Pauropoda

1. It includes about 500 species.
2. They are minute, soft-bodied, grub-like animals which live in leaf mold and soil. They feed on fungi or decomposing plant tissue. Some are predatory.
3. The head bears a pair of discoidal sensory organ for moisture detection and two biramous Y-shaped antennae.
4. The trunk consists of 11 segments and a telson. Each segment, except the first, last and telson bears a pair at short legs.
5. Genital pore is present on the third trunk segment.

Example. *Pauropus*.

Class 5. Hexapoda or Insecta (L., *insectum* = being cut into)

1. This class includes more than 750,000 species of air breathing arthropods; most of which are terrestrial some are aquatic.
2. Body is divided into **head, thorax** and **abdomen**.
3. **Head** is formed of six fused segments. It bears a pair of **antennae**, a pair of **compound eyes** and mouth parts which are adapted for chewing, biting, piercing, sucking, siphoning and sponging type.
4. Heads of most insects are oriented so that the mouth parts are directed either downward (**hypognathus**), anteriorly (**prognathous**) or posteriorly (**opisthognathus**).
5. **Thorax** is formed of three segments and bears **three pairs of legs** and **one** or **two pairs** of **wings**.
6. **Abdomen** is formed of not more than 11 segments.
7. Abdomen generally lacks appendages except modified posterior gonopods.
8. Respiration takes place by the help of branched **trachea**.
9. Exoskeleton is composed of a **tergum**, a **sternum** and two **pleura**. **Sclerites** (thickened areas of cuticle) are prominent features of insect body surface. Sclerites are often separated by **sutures**.
10. Sexes are separate, fertilization is internal.

Class Insecta is divided into **two** subclasses:

Subclass 1. Apterygota (Gr., *a* = without + *ptera* = wings)

1. Wings are absent and wingless condition is primary.
2. Mouth parts either become sunk into a pouch (**entognatha**) or do not sunk into head pouch (**ectognatha**).
3. Compound eyes and Malpighian tubules are reduced or absent.
4. Metamorphosis does not occur.

Subclass Apterygota is divided into **five** orders:

Order 1..Protura

1. Small insects with a cone-shaped head without compound eyes or ocelli. They contain pseudoculi (*i.e.,* receptors for mechanical vibrations).
2. Antennae are absent and first pair of legs are used as antennae. Entognathous mouth parts.
3. They have a 12-segmented abdomen lacking cerci.
4. They contain simple or no tracheal system.
5. Development is anamorphic.
6. They live in damp humus and soil and feed on decayed organic matter.

   Examples. *Eosentomon, Acerentulus.*

Order 2. Aptera (or Diplura)

1. Small, soft, wingless insects with slender, flattened bodies.
2. They contain entognathous mouthparts.
3. They contain long antennae but lack eyes.
4. Abdomen is provided with two cerci.

   Example. *Campodea.*

Order 3. Collembola (*Spring tails*)

1. Minute, soft bodied with short antennae.
2. Eyes are either absent or represented by isolated ommatidia.
3. Simple, endognathous chewing mouthparts with sharp mandibles.
4. They contain an **abdominal jumping** or **springing organ**. Jumping organ is composed of spring **(furcula)** on fifth abdominal somite and clasp **(tentaculum)** on third abdominal somite.
5. They lack abdominal cerci as well as Malpighian tubules.
6. They feed decaying plant material and fungi.

   Examples. *Entomobrya, Achorutes.*

Order 4. Microcoryphia or Archeognatha (*Jumping bristle tails*)

1. Small to medium sized, soft, wingless insects with elongated cylindrical bodies.
2. They contain long antennae and styliform (*i.e.*, slender and elongated) abdominal cerci.
3. They have ectognathous mouthparts and feed on algae, lichens and moss.

   Example. *Machilis.*

Order 5. Thysanura (*Silverfish and bristle tails*)

1. Small insects with long, often scaly bodies.
2. They possess ectognathous biting or chewing type mouth parts.
3. Head is with long **antennae** and **compound eyes. Ocelli** may be present or absent.
4. They are fast-running insects having short cursorial legs.
5. Abdomen is with three styliform appendages—one caudal filament and two long abdominal cerci.
6. Malpighian tubules and tracheae are well developed.
7. Some species of them are found in house where they eat books and clothing.

   Example. Silver fish (*Lepisma, Thermobius*).

Subclass 2. Pterygota (Gr., *ptera* = wings)

1. Insects are with wings or they are secondarily wingless.
2. They have ectognathous mouthparts.
3. They undergo metamorphosis.

Subclass Pterygota is divided into **24** orders. Some of the important orders are as follows:

Order 1. Ephemeroptera or Ephemerida (*Mayflies*)

1. Slender body with short hair-like antennae and large compound eyes.
2. **Chewing mouthparts** are present in aquatic nymphs and vestigial mouthparts in short-lived (or ephemeral) adults.
3. Wings are net-veined, of which the first is larger than the second and both pairs are held vertically over back when at rest.
4. Abdomen contains 10 distinct somites and 11th vestigial somite.
5. They contain two very long abdominal cerci and caudal filament.
6. Metamorphosis is gradual or incomplete; with aquatic **nymphs** which have **tracheal gills** for respiration.

Example. *Ephemera.*

Order 2. Odonata (= toothed flies) (*Dragon flies and damselflies*)

1. Slender (damselflies) or stout-bodied (dragon flies), predaceous forms with strong **biting and chewing mouthparts.**
2. Head is with huge compound eyes.
3. Two pairs of similar, equal-sized, narrow, net-veined wings held straight out from body when at rest (dragon flies) or held vertically (damsel flies).
4. Long abdomen is with 10 distinct somites and two rudimentary anal somites.
5. Abdominal cerci are rudimentary.
6. Metamorphosis is gradual or incomplete; nymphs are aquatic with or without caudal gills.

Examples. Dragon flies (*Macromia, Sympetrum*).

Order 3. Orthoptera (= straight wings) (*Grasshoppers, katydids, crickets, roaches, mantids and walking sticks*)

1. Large headed insects with **strong biting and chewing mouthparts.**
2. Many species are with long, jumping (or saltatorial) hindlegs.
3. Wings are present or absent. Most winged forms have membranous hindwings folded fanlike beneath leathery forewings (called **tegmina**).
4. Largely herbivorous, sometimes causing vast crop damage (*e.g.*, Locust).
5. Metamorphosis is gradual or incomplete.

Examples. Grasshoppers (*Melanopus, Poecilocerus*); **locust** (*Schistocerca*); **cockroach** (*Periplaneta, Blatta*); **cricket** (*Gryllus*), **mole cricket** (*Gryllotalpa*); **walking sticks** (*Carausius*); **leaf insect** (*Phyllium*) and **katydids.**

(**Note.** Some classifications places the cockroaches, mantids and walking sticks in separate orders: **1. Blattaria** (cockroaches), 2. **Phasmida** (walking sticks and *Phyllium*) and 3. **Mantogea** (praying mantids). All these three orders have been included in single order **Dictyoptera**, sea **Meglitsch** and **Schram** 1991).

Order 4. Isoptera (= equal wings) (*Termites*)

1. Social insects with **caste system**; members of society are differentiated as reproductive members, workers and soldiers.

2. Colony includes both winged and wingless individuals.
3. Soft bodied, pale coloured, with abdomen broadly joined to thorax.
4. Wings are of equal size and held **horizontally** over **abdomen**.
5. Metamorphosis is gradual or incomplete.
6. **Biting mouthparts** are present.

Examples. *Termes, Zootermopsis*.

Order 5. Dermaptera (= skin wings) (*Earwig*)

1. Elongated body with tough exoskeleton; **chewing mouthparts,** compound eyes and large forceps-like **anal cerci** on the posterior end.
2. Wingless or with short forewings forming tegmina (elytera) that cover all but tip of much folded, membranous, radially-veined, fan-like hindwings.
3. They are **nocturnal** and have **omnivorous** food habits.
4. Metamorphosis is gradual or incomplete.

Example. Earwig (*Forficula*).

Order 6. Plectoptera (*Stone fly*)

1. Adults have long antennae, **chewing mouthparts** and two pairs of well-developed, membranous wings or vestigial wings.
2. Abdomen bears long cerci.
3. Metamorphosis is gradual or incomplete.
4. Nymphs are aquatic, with or without gills.

Example. *Taeniopteryx* (**stone fly**).

Order 7. Mallophaga (= *wool eater*) (*Chewing lice and bird lice*)

1. Wingless, flattened insects that live as ectoparasites on birds and mammals.
2. Eyes are reduced or absent; legs are short having stout claws and thorax is small. Head is hypognathous.
3. Abdomen with five to eight somites.
4. Metamorphosis is gradual or incomplete.
5. They contain **biting and chewing mouthparts** and feed on scales, feathers, hair, skin and sometimes dried blood around wounds.

Example. *Cuclogaster* (**chicken louse**).

Order 8. Anoplura (= unarmed tail) (*Chewing lice and bird lice*)

1. They are similar to chewing lice but mouthparts are adapted for **sucking.**
2. They are ectoparasites of birds and mammals.
3. A number of species are parasitic on domestic animals and the **head louse** or **crab louse** are parasites of humans. They cause irritation and serve as vectors of disease such as typhus fever.

Examples. *Haematopinus* (**sucking louse**); *Pediculus humanus* (**human louse**).

Order 9. Thysanoptera (*Thrips*)

1. Small, slender insects with **rasping, sucking mouthparts.**
2. They include both winged and wingless species. When winged, forewings are larger and both pairs are narrow, nearly veinless and with fringe of fine hairs on posterior margin.
3. Abdominal cerci are absent.
4. Most species feed on sap in flowers.

Example. *Taeniothrips* (**pear thrips**).

Order 10. Hemiptera (= half wings) (*True bugs*)

1. They contain **piercing** and **sucking beak** which can be folded under the body.
2. **Forewings** with a thickened basal and distal membranous section. The membranous portions of forewings overlap when at rest over the abdomen.
3. Hindwings are entirely membranous.
4. Gradual or incomplete metamorphosis.
5. They are herbivorous and predaceous.

Examples. *Cimex* (**bed bug**), *Belostoma* (**giant water bug**), *Nepa* (**water scorpion**).

Order 11. Homoptera (=same wings) (*Cicadas, leafhoppers and aphids*)

1. They include herbivorous insects related to the hemipterans, but beak arises from back of head.
2. Mouthparts adapted for **piercing** and **sap-sucking**.
3. Forewings are typically membranous. Wings are commonly held in a tent-like position over the body.
4. Metamorphosis is gradual or incomplete.

Example. Buffalo treehopper; *Cicada*.

Order 12. Neuroptera (*Lacewing, dobson flies and snake flies*)

1. They are carnivorous, with **chewing mouthparts.**
2. Antennae are usually long and slender; compound eyes are large.
3. Legs are slender. Forelegs are sometimes raptorial.
4. They contain two pairs of similar, membranous, net-veined wings which are held roof-like over the back at rest.
5. Metamorphosis is complete.
6. Larvae are predaceous and usually terrestrial.

Examples. *Mantispid*; *Chrysopa* (**lacewing**); *Sialis* (**alder fly or dobson fly**).

Order 13. Coleoptera (= shield wings) (*Beetles, weevils, fireflies*)

1. It is largest order of insects containing over 300,000 species.
2. They contain tough or hard exoskeleton, **chewing mouthparts,** antennae variable and large compound eyes.
3. Adults have two pairs of wings, of which **forewings** are modified as heavy protective covers (**elytra**), meeting in straight line at middle of back.
4. **Hindwings** are membranous and usually completely hidden by elytra when at rest.
5. Abdomen is ten-segmented lacking cerci.
6. They have complete metamorphosis.
7. Most of them are plant feeders but some are predatory too.
8. They are mostly terrestrial, a few are aquatic.

Examples. *Hippodamia*, *Melolontha* (**cockchafer beetle**).

Order 14. Lepidoptera (= scale wings) (*Butterflies, moths*)

1. Soft-bodied insects with two pairs of wings which are often large and showy. Forewings are often larger than hindwings.
2. Body and appendages (wings, etc.) are covered with overlapping pigmented scales.
3. **Mouthparts** are modified as coiled proboscis and used for **sucking** liquid food such as flower nectar.

4. Antennae are long and often feathery.
5. They contain large compound eyes.
6. Abdomen lack cerci.
7. They undergo complete metamorphosis. Larva are **caterpillars** which are usually plant feeders (herbivorous); adults feed little or not at all.

Example.

1. **Butterflies** which are diurnal with large, often brightly coloured wings which fold vertically over back at rest, have knobed antennae and soft bodies, *e.g.*, *Papilio* (**swallow-tail butterfly**); *Pieris, Kalima.*
2. **Moths** which are nocturnal and fold wings horizontally and contain pinnate antennae, stout bodies and usually dull coloured wings, *e.g.*, **royal walnut moth**; **gipsy moth**; **silk moth** (*Bombyx mori*) ; **clothes moths** which feed on textiles, clothing and fur.

Order 15. Diptera (= two wings) (*True flies, mosquitoes*)

1. **Forewings** are membranous and used for flying and **hindwings** are reduced to knoblike balancers, called **halters**.
2. Mouth parts are variable. Thus, they are either **lapping**, **sucking**, **piercing** type or vestigial.
3. Compound eyes are large; antennae are short and abdomen is without cerci.
4. They undergo complete metamorphosis.
5. Adults are often vectors of diseases; larvae are frequently damaging to vegetables and domestic animals.

Examples. Mosquitoes (*Anopheles, Culex*); **house fly** (*Musca*); **fruitfly** (*Drosophila*); **horseflies**, **midges** and **gnats**.

Order 16. Hymenoptera (= membrane wings) (*Ants, bees, wasps and sawflies*)

1. A large and varied order which includes both winged and wingless species.
2. They contain strong exoskeleton and **chewing**, **lapping** or **sucking** mouthparts.
3. Head contains slender **antennae** and large compound eyes.
4. Wings are either absent, reduced or two pairs of stiff, membranous, transparent wings are present.
5. Abdomen contains a slender pedicel.
6. They often live in **nests** in large societies with division of labour.
7. They undergo complete metamorphosis.
8. Larvae are caterpillars or are grub-like with chewing mouthparts.

Examples. Wasps (*Polistes, Vespa orientalis*); **black ant** (*Compnotus compressus*); **red ant** (*Oecophyla smaragdyna*); **mud dauber** (*Sceliphoran*), **honey bee** (*Apis indica*).

Order 17. Siphonoptera (*Fleas*)

1. They are ectoparasitic on birds and mammals.
2. Small, wingless insects with laterally flattened bodies.
3. Hind legs are long, stout and adapted for jumping.
4. Abdomen is without cerci.
5. They have **piercing** and **sucking mouthparts.**
6. They are vectors of certain epidemics, *e.g.* Indian rat flea transmits *Bacillus pestis*, a causative agent of **bubonic plague**.

7. They undergo complete metamorphosis.

Examples. Indian rat flea (*Xenopsylla cheopis*); **human flea** (*Pulex irritans*).

Box 38.3

The remaining **seven** lesser important orders of subclass **Pterygota** are the following:

1. **Grylloblatodea**, *e.g.*, *Grylloblatta*.
2. **Embioptera**, *e.g.*, **web spinners** such as *Embia*, *Oligotoma*.
3. **Psocoptera** or **Corrodentia**, *e.g.*, **Book lice** (*Ectopsocus*), **bark lice**.
4. **Zoraptera**, *e.g.*, *Zorotypus*.
5. **Strepsiptera**, *e.g.*, minute beetle-like twisted wing parasites.
6. **Mecoptera**, *e.g.*, **Scorpion flies** (*Panorpa*).
7. **Trichoptera**, *e.g.*, **Caddish flies** (*Rhyacophila*).

The animal groups which zoologist judge to have attained the "**greatest biological success**" are those which

1. Have the largest number of species and of individuals;
2. Occupying the widest stretches of territory and the greatest variety of habitats;
3. Consuming the largest amount and kinds of food; and
4. Most capable of defending themselves against their enemies.

By these standards the phylum which occupies first place among the animals (vertebrates and invertebrates) is the phylum **Arthropoda**. Horny material (**chitin**) of exoskeleton is to the arthropods what steel is to civilised human being, and it is partly to the possession of this hard cuticle that the arthropods owe their success.

38.4 TYPE STUDIES

1. *LIMULUS*
(Horse-shoe crab or King crab)

Classification

Phylum	—	**Arthropoda**
Subphylum	—	**Chelicerata**
Class	—	**Merostomata**
Subclass	—	**Xiphosura**
Order	—	**Limulida**
Genus	—	***Limulus***

Habit and Habitat

Limulus or **king crab** is a marine animal and found along the north western Atlantic Coast and the Gulf of Mexico. It lives on soft bottoms in shallow waters. During breeding season, both male and female come on land (called **nuptial visit**) and dig holes in sand at the upper limit of high tides to lay eggs. *Limulus* is a scavanger and feeds on molluscs, worms and other organisms including bottom-dwelling algae.

Limulus is **not a true crab**. It is known as a **living fossil**, because it has remained unchanged since the Triassic period nearly 200 million years ago. It is believed that *Limulus*, together with trilobites, originated from some common arthropod ancestor but remained universally isolated.

Structure

Limulus (*L. polyphemus*) reaches a length of 60 cm and is dark brown in colour. Its body is divisible into an anterior **prosoma** and a posterior **opisthosoma**. The prosoma is convex above, covered by a broad, horse-shoe shaped **carapace** bearing a pair of median simple eyes and a pair of compound eyes. *Limulus* can detect movement through eyes. On its ventral side, prosoma bears six pairs of appendages surrounding the mouth. The first pair or **chelicerae** are small, trisegmented and chelate. The next five pairs are **walking legs** which serve both the function of locomotion and food procurement. The first four pairs of walking legs are chelate. The coxa of each leg contains spines, called **gnathobases**. These macerate and move the food towards mouth. The fifth pair of legs are not chelate. The coxae of these legs bear a short process, the **flabellum**. Flabellum is used for cleaning the gills. A pair of small, oval and plate-like appendages, called the **chilaria**, are also present a little behind the mouth. Chilaria are directed vertically downwards and represent the degenerated seventh pair.

The opisthosoma is hexagonal and movably articulated with prosoma. It consists of six segmented **mesosoma**, unsegmented **metasoma** and a long **telson** or caudal spine. Mesosoma bears six pairs of appendages. The first pair is fused and forms the **genital operculum** and bears two genital pores. Remaining five pairs of appendages are flap-like and membranous. The undersurface of each flap forms leaf-like folds called **gills** or **branchial lamellae**, for respiration.

A long triangular caudal spine or **telson** articulates to the posterior end of abdomen. Telson is highly mobile and is used for correcting the body when it is accidently turned over.

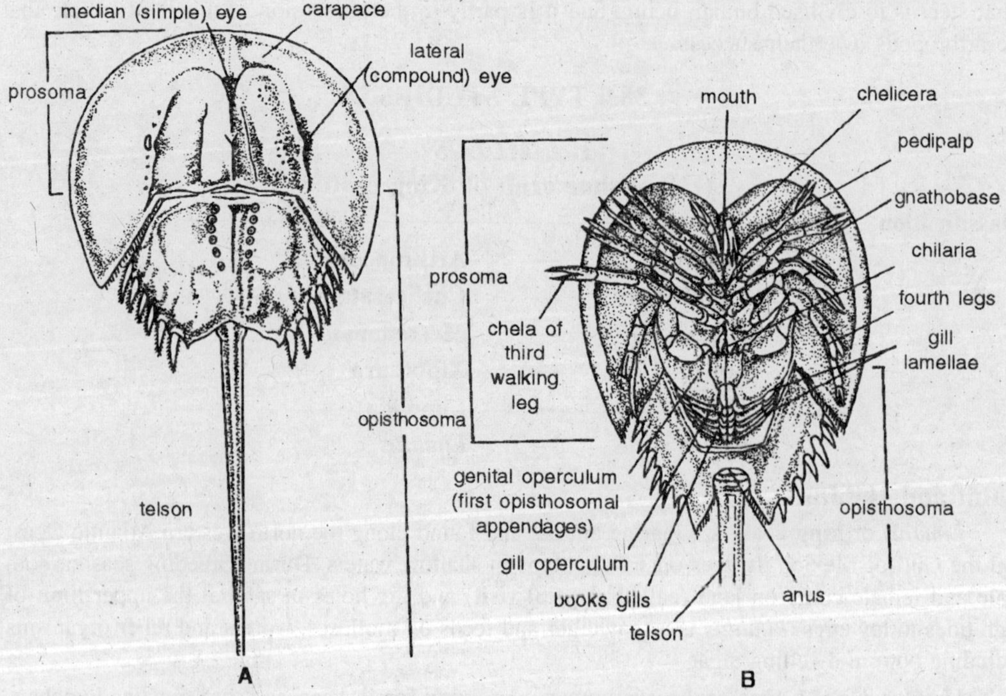

Fig. 38.10. *Limulus* (king crab). A—Dorsal view; B—Ventral view.

In *Limulus*, excretion takes place by **coxal glands** or **brick red glands**. Sexes are separate and sexual dimorphism is distinct. **Fertilization** is **external** and **cleavage** of its centrolecithal eggs is **complete** or **holoblastic**. The larva emerges from the egg as a free swimming **trilobite larva**. Average life span of *Limulus* is 19 years and it attains its sexual maturity within 9 to 12 years. *Limulus* has an economic importance too; it is fed to chickens and pigs and sometimes consumed by human beings.

2. *DAPHNIA* (Water flea)

Classification

Phylum	—	**Arthropoda**
Subphylum	—	**Crustacea**
Class	—	**Branchiopoda**
Subclass	—	**Diplostraca**
Order	—	**Cladocera**
Genus	—	***Daphnia***

Habit and Habitat

Daphnia is commonly called **water flea**. It is found in freshwater ponds, streams and ditches. *Daphnia* is widely distributed in India, Asia, Europe and U.S.A. *Daphnia* forms the food of fishes in aquatic food chain.

Structure

Daphnia is 2 mm in length. Its body is divisible into head, thorax and abdomen (Fig. 38.4). General body segmentation is, however, less distinct. **Head** is formed of six somites. Body except head is enclosed in a bivalved **shell** or carapace which terminates posteriorly as **dorsal** or **caudal spine**. The head is rounded and bent downwards and look like the beak of a bird. Head bears a pair of large biramous **antennae**, a pair of small **antennules**, a large, median, sessile **compound eye** and large **mandibles**. In adult, second pair of maxillae are absent. Head also contains a vestigial median naupliar eye and a pair of **nuchal sense organs**. Antennules contain chemoreceptor bristles. Antennae form the chief organ of locomotion of *Daphnia*. Its movement is **vertical** and **jerky**. The downstroke of the powerful antennae propels the animal upward. Then *Daphnia* uses antennae as a parachute and sinks slowly. Little **setae** present at the end of abdomen act as stabilizers.

Thorax bears five pairs of leaf-like appendages. These appendages bear fine setae which are used in food collection. *Daphnia* is suspension feeder. Each thoracic appendage also bears a thin, vesicular **epipodite** which functions as a **gill**. The **abdomen** is greatly reduced and devoid of appendages. **Postabdomen** is large and bears sensory setae.

Daphnia is **dioecious**, *i.e.*, its sexes are separate. The female carries the eggs and embryos in a large **brood chamber** on her back. It is situated between abdomen and posterior part of carapace. A pair of spine-like processes of postabdomen close the brood chamber and prevent the escape of eggs from it.

During summer when the food is in plenty and water temperature is favourable, *Daphnia* produces only female offsprings from diploid eggs **parthenogenetically**. During winter, change in water temperature decreases the food supply and induces appearance of few males. In this condition, eggs undergo meiosis and are **fertilized**. The fertilized eggs are enclosed in the brood chamber. Fertilized eggs are large and only two are produced in a batch. When female moults, brood chamber along with its fertilized eggs or embryo is cast off. The brood chamber sinks to bottom and can withstand drying and freezing. Under favourable conditions, the eggs start developing and females are produced.

3. *CYPRIS*

Classification

Phylum	—	**Arthropoda**
Subphylum	—	**Crustacea**
Class	—	**Ostracoda**
Subclass	—	**Podocopa**
Order	—	**Podocopida**
Genus	—	***Cypris***

Habit and Habitat

Cypris is commonly called **mussel shrimp**. It is a free-swimming small-sized (1 to 2 mm length) crustacean and occurs in freshwater stagnant ponds. It is **carnivorous** and feeding on small organisms.

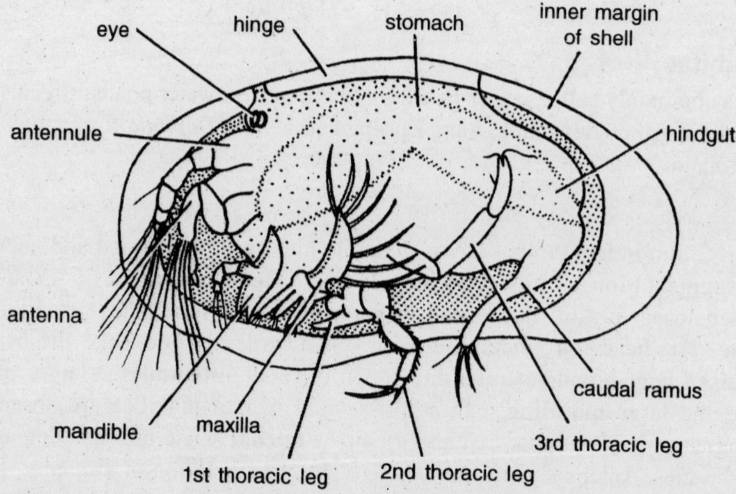

Fig. 38.11. *Cypris*. Mussel shrimp.

Structure

Body of *Cypris* is unsegmented and laterally compressed. It is completely enclosed in a bivalved **carapace**, the two halves of which are articulated in the mid-dorsal line and inter-connected by a strong adductor muscle. A large median eye is present at the anterior end of body. The compound eye and heart is absent.

Head bears four pairs of appendages such as **antennules**, **antennae**, **mandibles** and **maxillae**. Antennules and antennae are large biramous and help in swimming. *Thorax* bears three pairs of appendages, *i.e.*, legs. **Abdomen** is devoid of appendages and is terminated by a pair of small **caudal styles**. *Cypris* is **parthenogenetic** and it lacks males.

4. *CYCLOPS*

Classification

Phylum	—	**Arthropoda**
Subphylum	—	**Crustacea**
Class	—	**Copepoda**
Order	—	**Cyclopoida**
Genus	—	***Cyclops***

Habit and Habitat

Cyclops is found throughout the world in fresh and brackish water. Its movement is rapid and jerky.

Structure

Body of *Cyclops* is elongated or pear-shaped and measures 1.5 to 5 mm in length. Its body is divisible into three distinct parts: cephalothorax, thorax and abdomen. The **cephalothorax** is formed by the fusion of head and first thoracic segment. Cephalothorax is broad and has a convex dorsal surface. It is covered with a **carapace**. On the dorsal surface, there is a **median** or **naupliar eye**. *Cyclops*, in fact, derives its name after the great giant of Greek mythology, the **cyclops**, who had a single eye in his forehead.

Cephalothorax contains four pairs of appendages : **Antennules** are long, uniramous and horn-like. These are chief organs of locomotion. In the male, they are modified as **clasping organs**; these are used to hold the female during copulation. The **antennae** are short and uniramous. **Mandibles** and **maxillae** are other cephalo-thoracic appendages.

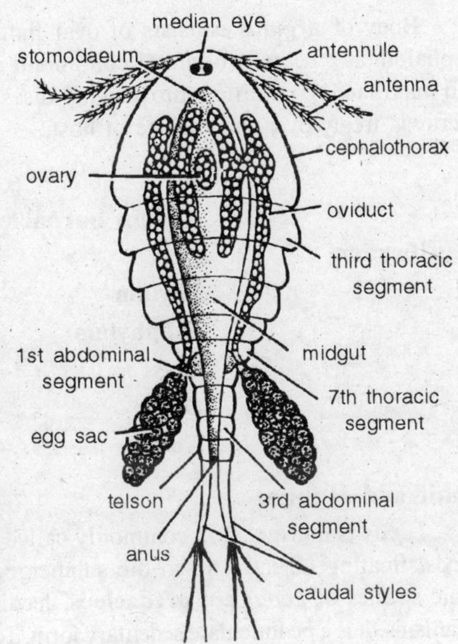

Fig. 38.12. *Cyclops*. Female.

Thorax is formed of five free thoracic segments. First four thoracic segments bear biramous appendages. The fifth thoracic segment bears a pair of vestigial limbs. The last thoracic segment bears the **genital aperture. Abdomen** is formed of four segments and bear no appendages. The last abdominal segment bears anus on its dorsal surface and a pair of branched **caudal styles** having **setae.** In the female, the first abdominal segment is fused with last thoracic segment. The fused segment carries a pair of **ovisacs** or **egg sacs** in mature females.

Cyclops is **dioecious**, *i.e.,* its sexes are separate. Fertilization is internal and ovisacs (=egg sacs) function as brood chambers. Each sac contains up to 50 or more eggs. The eggs hatch as **nauplius larvae.**

Cyclops have great economic importance. They form an important component of aquatic food chain. They are eaten as plankton by fishes. *Cyclops* serve as an intermediate host for guinea worm.

5. *ARGULUS*

Classification

Phylum	—	**Arthropoda**
Subphylum	—	**Crustacea**
Class	—	**Branchiura**
Genus	—	***Argulus***

Habit and Habitat

Argulus is commonly known as **carp lice**. It is an **ectoparasite** of the branchial chamber or skin of freshwater fishes.

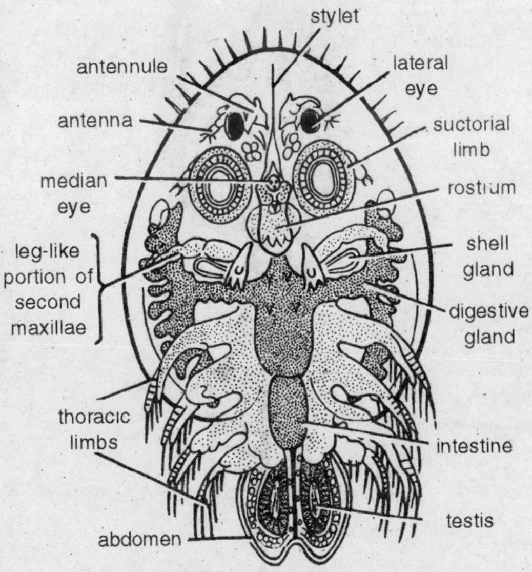

Fig. 38.13. *Argulus.*

Structure

Body of *Argulus* consists of oval flattened cephalothorax and a small bilobed abdomen. Cephalothorax bears two lateral **compound eyes** and one median simple eye. Mouth is suctorial and maxillae are modified into sucking discs. The animal is not permanently attached to host but it crawls freely over the surface of host.

6. *LEPAS*
(Ship barnacle or Goose barnacle)

Classification

Phylum	—	**Arthropoda**
Subphylum	—	**Crustacea**
Class	—	**Cirripedia**
Order	—	**Thoracica**
Genus	—	***Lepas***

Habit and Habitat

Lepas is marine. It is commonly called **ship** or **goose barnacle**. It is found attached to sea weeds, floating objects and to the submerged bottom of the ship. *Lepas* is **suspension feeder**. Some species of *Lepas* are **predacious**, feeding on copepods, isopods and other relatively large organisms. It is a pedunculate sedentary form. Too many *Lepas* remain attached at a place. A particular protein in the exoskeleton of older attached animals has been found to attract settling **cypris larvae** of *Lepas*.

Structure

Body of *Lepas* (Fig. 38.14) is divisible into a muscular, flexible **stalk** or **peduncle** and the **capitulum** or **body proper**. Peduncle is attached to substratum and bears capitulum at its distal end. It is covered with wrinkled skin and represents the preoral end of the animal. Peduncle contains

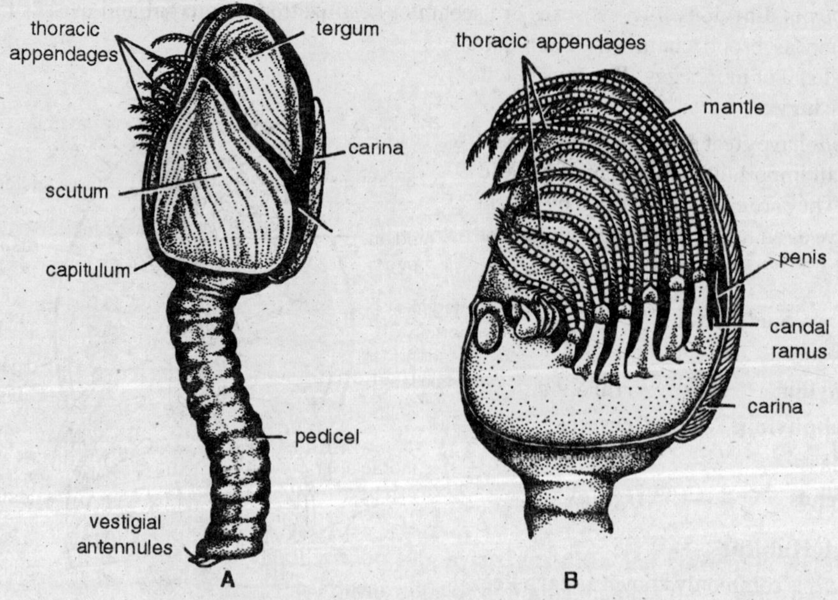

Fig. 38.14. *Lepas*. A–Entire animal; B–Carapace removed.

the vestiges of the larval **antennules** and the adhesive **cement glands** which open through the antennules.

The **body proper** or **capitulum** is enclosed in a bivalved **mantle** or **carapace**. The carapace is covered with five calcareous plates. Of these one is median and dorsal (called **carina**), two are lateral and proximal (**scuta**) and two are lateral and distal (**terga**). A large **adductor muscle** runs transversely between the two scuta. The body proper is divisible into **head** and **thorax**. The head contains vestiges of first antennae (antennules) and mouth which is surrounded by paired mandibles, maxillae and maxillulae. Second pair of antennae are absent.

Thorax bears six pairs of long, biramous appendages. These are filament-like and bear many long setae. These are called **cirri** and used in **suspension feeding**.

Lepas lacks heart, arteries, gills but contains **maxillary glands** for excretion. It is **hermaphrodite** (monoecious) but **cross-fertilization** is the rule. Fertilization is internal occurring in ovisac and sperm transfer occurs through the **penis** which is inserted into the mantle cavity of another individual. Development of *Lepas* is indirect involoving **nauplius larva** and **cypris larva**.

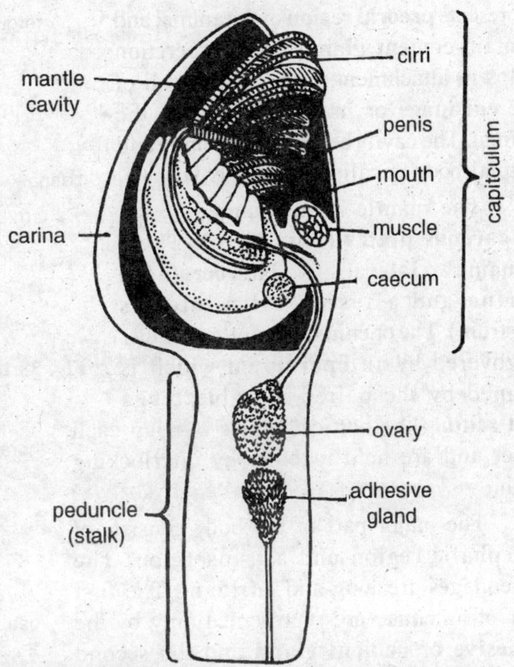

Fig. 38.15. *Lepas*. Internal structure.

Attached *Lepas* to ship bottom tend to hamper ship's speed, so much efforts and money have been expanded to develop anti-barnacle paints and other measures such as scrapping of barnacles at dry docks.

7. *BALANUS*
(Acorn-shell or rock-barnacle)

Classification

Phylum	—	**Arthropoda**
Subphylum	—	**Crustacea**
Class	—	**Cirripedia**
Order	—	**Thoracica**
Genus	—	***Balanus***

Habit and Habitat

Balanus is commonly known as **rock barnacle** or **acorn barnacle**. It is marine and found attached to rocks in great numbers between tide marks. It is also found attached to molluscan shell, floating objects (ship), whales and other animals. *Balanus* is **suspension feeder**.

Structure

Balanus is sedentary and sessile, *i.e.*, it lacks peduncle so its shell is directly attached to the rocks. Body of *Balanus* is surrounded by a calcareous shell. The broad attached under- surface of the *Balanus* is called **basis** or **disc**. It represents **preoral region** of the animal and contain **cement glands** whose secretion helps in attachment. The vertical wall of the **carapace** or **mantle** surrounds the animal. The cavity between the mantle and animal body is called **mantle cavity.**

The mantle is surrounded by six **calcareous plates** (Fig. 38.16). These contain two **laterals,** two **laterocrania**, a **carina** and a fused **rostro-laterals** (rostrum). The opening of mantle chamber is covered by an **operculum** which is formed by the paired movable **terga** and **scuta**. The vertical plates overlap each other and are held together by interlocking teeth.

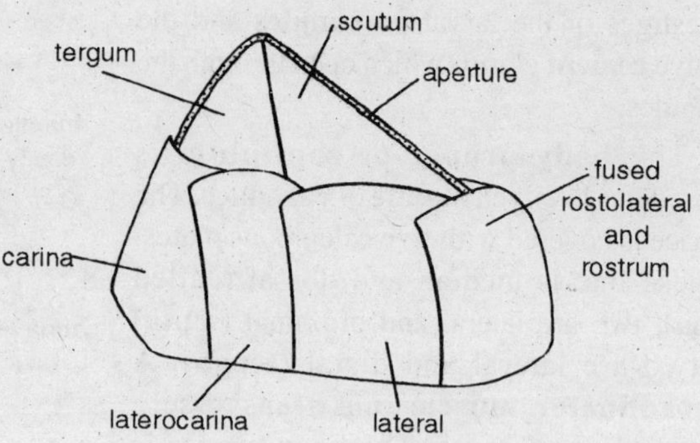

Fig. 38.16. *Balanus.* Number and position of shell plates (diagrammatic lateral view).

The major part of the body consists of a cephalic region and a trunk rgion. The appendages are long and cirriform. The first pair of antennae are represented only by the **adhesive** or **cement gland** and the second pair is present in the larva. The second pair of maxillae are united. Thoracic appendages are six pairs of delicate cirriform legs that sway in water and propel food organisms towards mouth.

Balanus is **hermaphrodite** but **cross-fertilization** is the rule. Fertilization is internal and development is indirect involving **nauplius larva** and **cypris larva**.

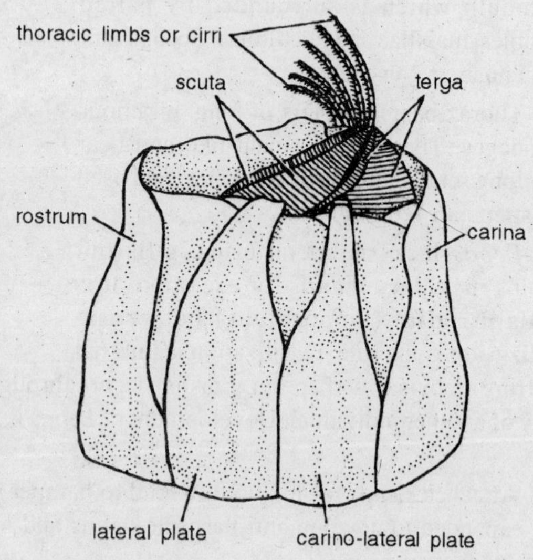

Fig. 38.17. *Balanus.*

8. *SACCULINA*

Classification

Phylum	—	**Arthropoda**
Subphylum	—	**Crustacea**
Class	—	**Cirripedia**
Order	—	**Rhizocephala**
Genus	—	***Sacculina***

Habit and Habitat

Sacculina is commonly known as **root-headed barnacle**. It is marine and found throughout the world. It is **parasitic** on decapod crustaceans such as crab.

Structure

Adult *Sacculina* (called **sacculina externa**) has the appearance of a **fleshy tumour** attached by a peduncle to the abdomen of the crab on its ventral side. From the peduncle numerous branched filaments are ramified in all parts of the body of the host except the heart and the gills. These roots absorb nourishment from the host. *Sacculina* shows great *degeneration due to parasitic mode of life*. It lacks mouth, anus, carapace, segmentation and appendages. It contains only reproductive organs such as a pair of elongated **testes** and a pair of **ovaries** with accessory **genital glands, genital atrium** and **collateral glands**. There is a single nerve ganglion. The entire body of *Sacculina* is covered with a mantle. Opening of the mantle cavity of *Sacculina* is placed at the hind end of the parasite as genital aperture or cloacal aperture.

The adult *Sacculina* is difficult to recognize as an arthropod. The study of its developmental history justifies its inclusion as a crustacean.

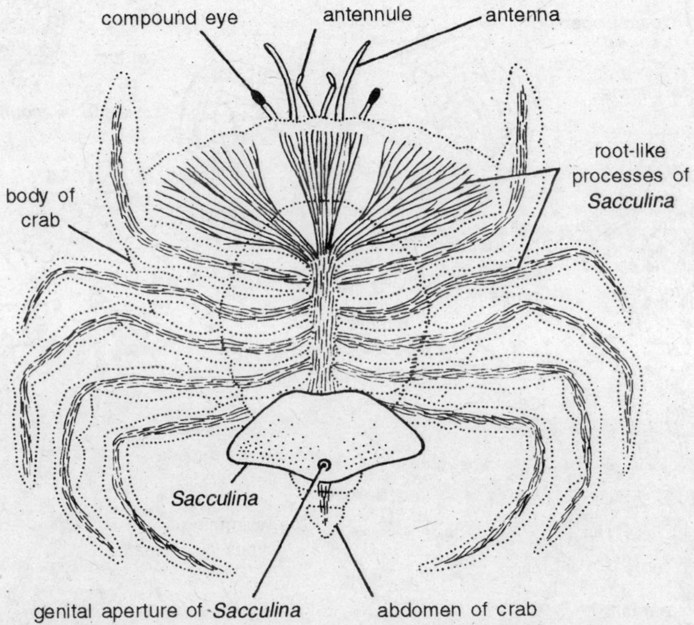

Fig. 38.18. *Sacculina*. A parasite attached to ventral surface of crab.

Life History

Sacculina is **hermaphrodite** and fertilization is internal. The youngs are hatched from the eggs as free swimming **nauplius larvae**. A nauplius larva moults into free-swimming **cypris larva**. A cypris larva on finding a crab (the host), attaches itself to the body of the crab by the help of its hook-like antennule. Now the cypris larva loses its trunk and appendages and becomes a rounded mass of cells, called **kentrogen larva** (Fig. 38.19). Kantrogen larva contains a chitinous rod known as **dart** which convey undifferentiated cells of larva into the body cavity of host. These undifferentiated cells enter the host's body and finally settles on the midgut of the host. This stage of *Sacculina* is called **sacculina interna**. It sends root-like absorptive processes throughout the body of the host except heart and gills. Such a massive parasitic attack sterilizes the host and prevents further moulting of the crab.

Finally the main body of the parasite pushes out as a swelling in the abdomen of the crab. This phase is called **sacculina externa**. *Sacculina* survives for three years on the host and then dies.

Effects of Sacculina on the Host (crab). Parasitic mode of life of *Sacculina* affects both the host and the parasite.

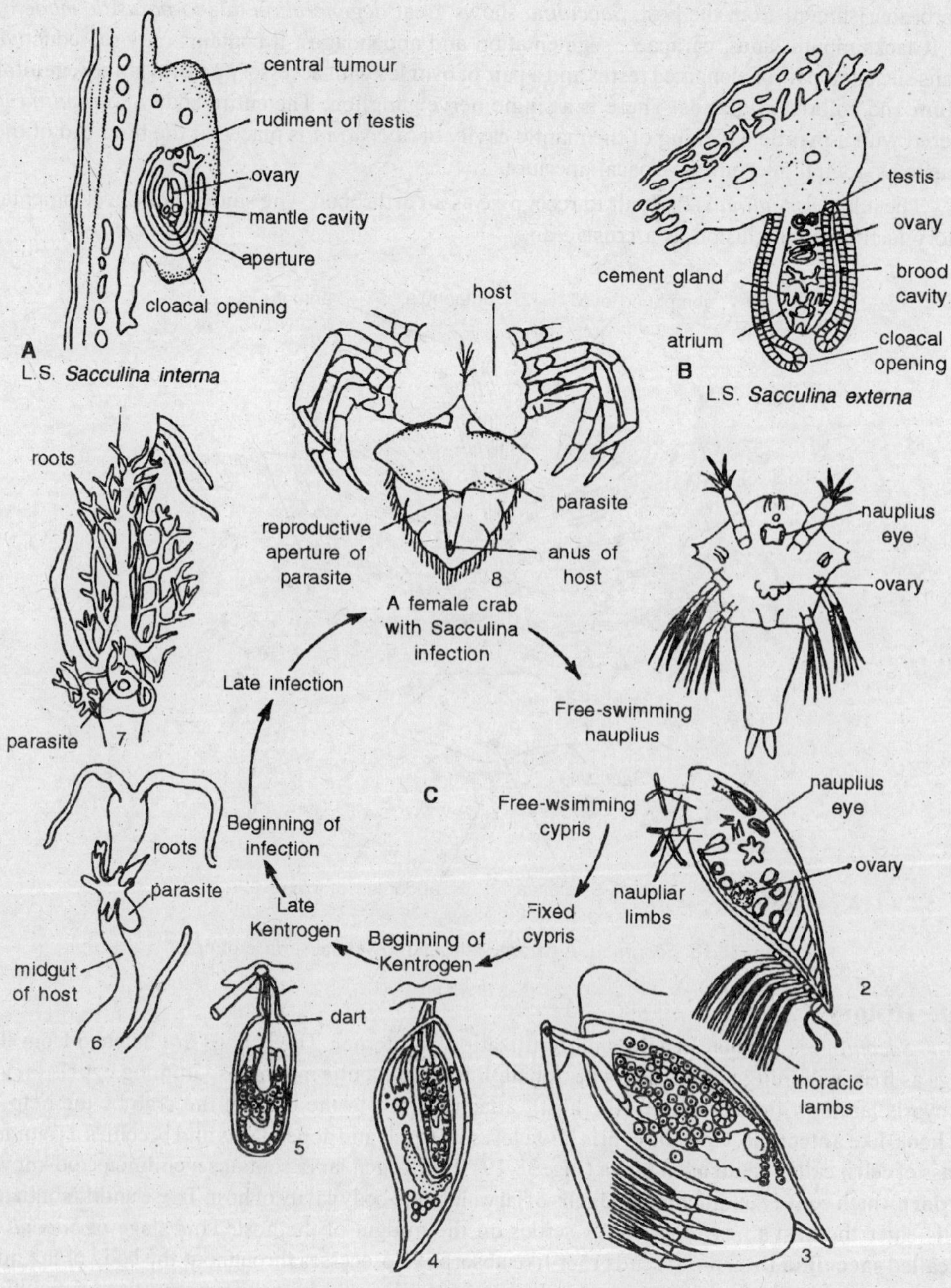

Fig. 38.19. *Sacculina*. Life cycle.

1. The crab does not moult and so does not grow in size.
2. *Sacculina* causes complete degeneration of the gonads of the host.
3. *Sacculina* changes the **secondary sexual characters** of the host. This phenomenon is called **parasitic castration**. In the female crab ovaries degenerate and ovigerous endopodites of

pleopods reduce in size. In the male crab the size of the chela reduces and they become similar to chela of female. Male's copulatory **styles** too reduce in size.

Parasite (*Sacculina*) itself undergoes several significant degenerative changes such as (i) loss of segmentation; (ii) loss of appendages; (iii) loss of different organ systems such as digestive, circulatory, respiratory and excretory systems; (iv) loss of typical crustacean features, and (v) reduction of metabolic rate.

9. *EUPAGURUS* (Hermit crab)

Classification

Phylum	—	**Arthropoda**
Subphylum	—	**Crustacea**
Class	—	**Malacostraca**
Subclass	—	**Eumalacostraca**
Superorder	—	**Eucarida**
Order	—	**Decapoda**
Genus	—	***Eupagurus***

Habit and Habitat

Eupagurus is commonly called **hermit crab**, because it leads a secluded life like a hermit (*i.e.*, hermit is a person who abandons society and lives alone, especially for religious contemplation). It is cosmopolitan in distribution and is found at Bombay and Goa beaches, from Alaska to lower California and Florida. Hermit crab lives in empty shell of a marine gastropod and often leads a commensal life with sea anemone.

Structure

Eupagurus is a peculiar crustacean having extreme modification in order to adjust in the coils of molluscan shell. Its body is soft and is divided into **cephalothorax** and **abdomen**. **Cephalothorax** is broad and flattened. It is covered with a hard carapace which is bifid posteriorly. The cephalothoracic appendages are well devoloped and help in locomotion and food capture. **Head** bears a pair of stalked elongated eyes, a pair of short **antennules** and a pair of large **antennae**. **Thorax** bears five pairs of legs. The first pair of chelate legs having unequal chela (*i.e.*, right chela is much larger than the left), the last two pairs are usually reduced. The larger first leg of left side acts as an **operculum**. It closes the mouth of the shell when the animal withdraws itself inside the shell. Normally the cephalothorax protrudes out of the shell and larger chelate leg is used to capture and crush the prey.

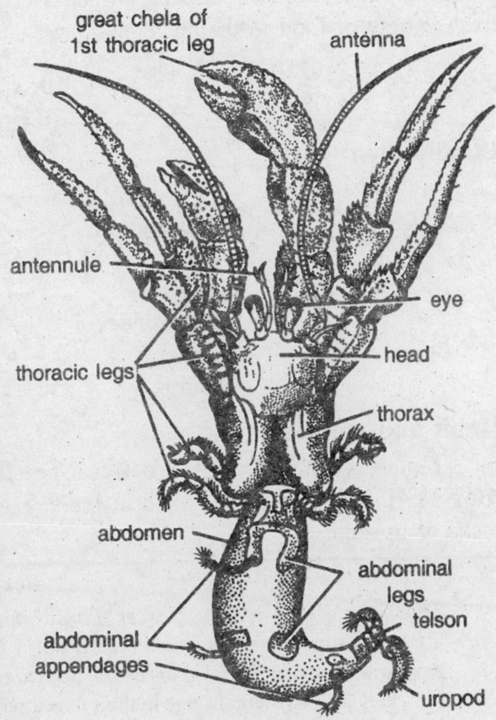

Fig. 38.20. *Eupagurus.*

The abdomen is cylindrical, soft and without external segmentation. Abdomen contains liver and gonads. It is spirally twisted to suit the shape of the shell. Abdominal appendages of the right side are absent, while those of left side are reduced (in female they are used to carry eggs). **Uropods** (last abdominal appendages) are hook-like and are adapted to cling to the shell. Surface of fourth and fifth pairs of legs and of uropods is covered with tiny tubercles that help in gripping the shell.

Sexes are separate. Spermatophores and ova are released simultaneously and fertilization is external. When the growth of the body demands, *Eupagurus* selects a new shell and deserts the old one.

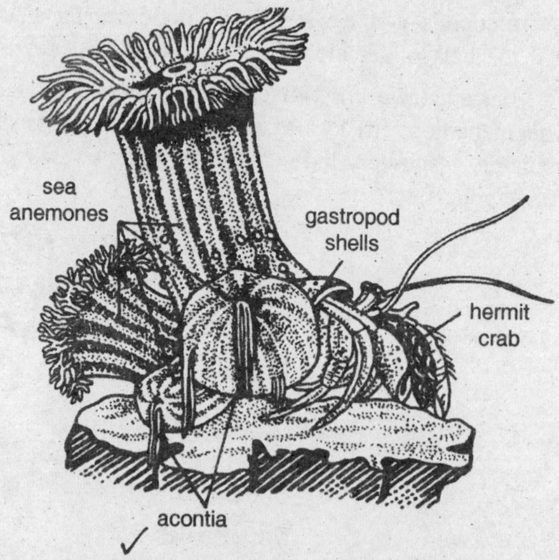

Fig. 38.21. Hermit crab and sea anemone in commensalism.

Commensalism

Hermit crab and sea anemone form a close but non-parasitic association, called **commensalism**. Sea anemone lives upon the mollusc shell while *Eupagurus* lives inside the shell. The shell and the anemone provide protection to hermit crab. Enemies of hermit crab avoid sea anemones due to their stinging cells and bad taste. In return, the hermit crab carries the sea anemone (which is a sedentry animal) from place to place and thus provides fresh variety of food to the anemone.

10. *CANCER* (True crab)

Classification

Phylum	—	**Arthropoda**
Subphylum	—	**Crustacea**
Class	—	**Malacostraca**
Subclass	—	**Eumalacostraca**
Superorder	—	**Eucarida**
Order	—	**Decapoda**
Genus	—	***Cancer***

Habit and Habitat

Cancer is commonly called **rock crab** or **true crab**. It is marine having cosmopolitan distribution (Box 38.4). It is commonly found in India, Europe and U.S.A. Cancer is found buried among rocks or in sand.

Box 38.4.

True crabs, *e.g.*, *Carcinus*, *Cancer*, *Paratelphusa*, etc., are found in all types of habitats. Most are marine or inhabit brackish water, but a number of them are freshwater, amphibious or terrestrial. Only terrestrial crabs do not need to return to water, since they have direct development (*i.e.*, they lack any larval stage in their development). *Paratelphusa masoniana* is freshwater crab of Punjab.

Structure

Body of *Cancer* is oval and flattened and is divided into **cephalothorax** and **abdomen**. Cephalothorax is quite large and is usually much broader than long. Carapace is broad, thick and hard. Eye stalks and antennae are short. **Mandibles, maxillae** and **maxillulae** are well developed. Thorax bears three pairs of **maxillipedes** and five pairs of well developed **walking legs**. Third **maxillipedes** are broad, flat and valve-like and cover the other mouth part.

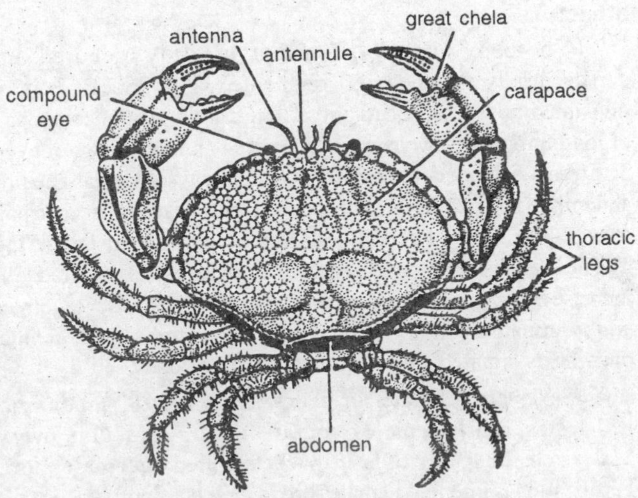

Fig. 38.22. *Cancer.*

The first pair of walking legs are **chelate** forming pinching claws. The abdomen is greatly reduced and lies permanently in a groove in the ventral or sternal surface of the thorax. Male's abdomen is narrow and V-shaped, while abdomen of female is broad and U-shaped and is adapted for carrying eggs. Abdominal appendages (**pleopods**) are much reduced. In males there are only two pairs of pleopods and it uses them as copulatory organs. Females have four pairs of pleopods which are used for carrying eggs. Uropods are absent.

Sexes are separate. Fertilization is internal. Newly laid eggs form a bright orange mass, called a **sponge**. In *Cancer*, development is indirect. Young ones hatches as **zoea** larva, which passes through **megalopa** stage before reaching maturity. Many species of crab are caught and eaten by human beings throughout the world. **Blue crabs** and **cancer crabs** are popular food item.

11. *SCOLOPENDRA*
(Centipede)

Classification

Phylum	—	**Arthropoda**
Subphylum	—	**Uniramia**
Class	—	**Chilopoda**
Subclass	—	**Epimorpha**
Order	—	**Scolopendromorpha**
Genus	—	***Scolopendra***

Habit and Habitat

Scolopendra is commonly called **centipede**. It is commonly occurs under stones, in rotten logs and in houses in damp places such as cellars and bathrooms. It is a nocturnal, active and fast moving creature. *Scolopendra* is carnivorous and voracious feeder. It feeds insects, spiders, worms and slugs. It kills its prey with the poison of the poison claws or maxillipedes which are modified first pair of trunk appendages.

Structure

The body of *Scolopendra* is brightly coloured and dorsoventrally flattened and segmented (Fig. 38.23). Body is divisible into a small head and a long trunk. Head bears a pair of long **antennae,** two pairs of **maxillae** and four pairs of **eyes.** The trunk is divided into twenty two identical segments. The first abdominal segment is closely apposed with head and bears a pair of appendages called **maxillipeds, poison claws** or **toxicognaths.** The next 21 segments bear in each a pair of legs. The legs gradually increase in size from anterior to posterior. Each leg is seven jointed and terminated with a claw having two ventral spines. Genital opening is situated in the posterior part of body.

Sexes are separate. In males, there are 20 spindle-shaped testes, while females contain a single tubular ovary. Males release sperm cells in packets called **spermatophore.** Fertilization is internal. Females brood their eggs until young ones hatch.

Scolopendra inflicts a painful bite to human beings. However, they are beneficial to human beings because they feed on household and agricultural pests such as cockroaches, crickets, ants, beetles, moths, caterpillars, flies, bedbugs, silver fish, etc.

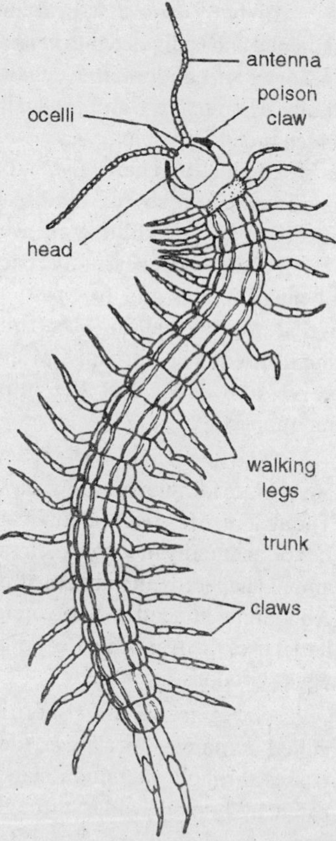

Fig. 38.23. *Scolopendra.*

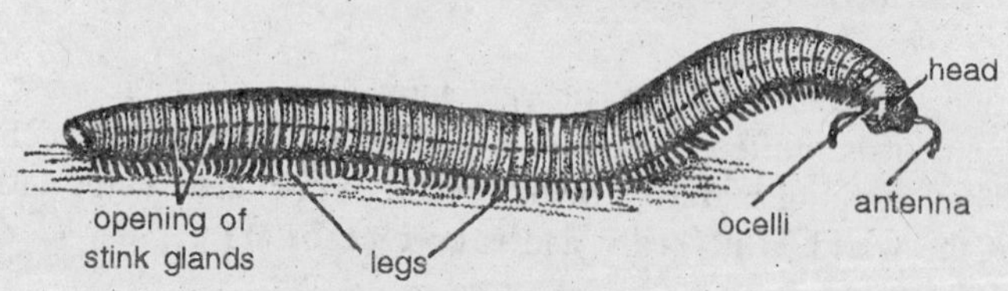

Fig. 38.24. *Julus.*

12. *JULUS*
(Millipede)

Classification

Phylum	—	**Arthropoda**
Subphylum	—	**Uniramia**
Class	—	**Diplopoda**
Subclass	—	**Helminthomorpha**
Order	—	**Julida**
Genus	—	***Julus***

Habit and Habitat

Julus or **millipede** is a sluggish and timid creature found hidden usually in dark and damp places under stones or wood or in decaying leaves. The animal is found rolled up under stones. They are **herbivorous.**

Structure

Body of *Julus* is elongated and cylindrical, consisting of large number of segments (Fig. 38.24). The body colour may be yellowish brown or reddish chestnut. The body is divisible into **head, thorax** and **abdomen**. Head bears a pair of sessile eyes, a pair of short **antennae**, a pair of **mandibles** and a pair of **maxillae** forming **gnathochilarium**, forming floor of pre-oral chamber. Poison claws are absent. Thorax contains four segments and each segment bears a pair of appendages. Abdomen has 20 to 100 segments, each abdominal segment bears two pairs of clawed legs. Body is terminated into **telson.** A pair of stink or **repugnatorial glands** occurs on side of each body segment. The secretion of these glands is toxic and repellant to other small animals.

Sexes are separate. Gonopores are located mid-ventrally on third abdominal segment. Fertilization is internal.

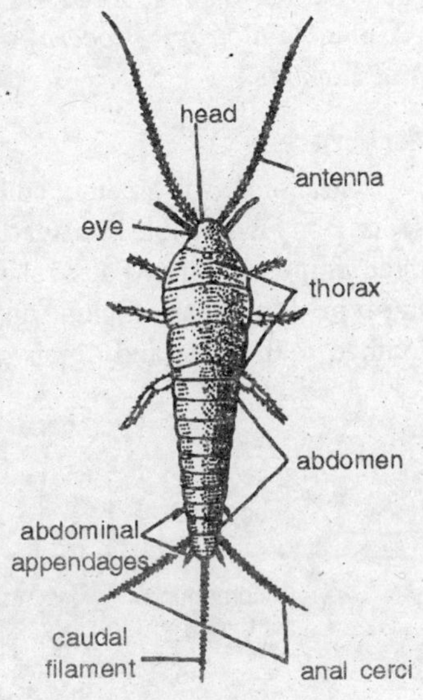

Fig. 38.25. *Lepisma.*

13. *LEPISMA* (Silver fish)

Classification

Phylum	—	**Arthropoda**
Subphylum	—	**Uniramia**
Class	—	**Insecta**
Subclass	—	**Apterygota**
Order	—	**Thysanura**
Genus	—	***Lepisma***

Habit and Habitat

Lepisma saccharina is commonly known as **silver fish**. It is. commonly found in damp, cool places and in cup-boards and book-racks where it devours starchy products.

Structure

The body of *Lepisma* is small (1.25 cm in length), flat and silver-coloured. Its body is divided into head, thorax and abdomen (Fig. 38.25). **Head** bears a pair of many segmented antennae and a pair of ocelli. Mouthparts are adapted for **biting** and **chewing**. The **thorax** is distinct and the **prothorax** forms the largest segment of body. Thorax bears three pairs of legs but lacks wings. Abdomen is eleven segmented. The last abdominal segment carries at the posterior end a pair of long, many jointed **cerci** and a long **tail** in between. The entire body of *Lepisma* is covered by shining scales. Its excrement products are in the form of dry pellets. The adults can moult and possess regenerating power. *Lepisma* cause considerable damage to the books, clothing by chewing off starch, sizing or any other articles containing glue.

14. *POICILOCERUS* (Grasshopper)

Classification

Phylum	—	**Arthropoda**
Subphylum	—	**Uniramia**
Class	—	**Insecta**
Subclass	—	**Pterygota**
Order	—	**Orthoptera**
Genus	—	***Poicilocerus***

Habit and Habitat

Poicilocerus (also *Melanopus*) is commonly known as **grasshopper**. Grasshoppers are found throughout the world in open grasslands. They feed on grass and all other leafy vegetation. They cause serious damage to the crops. They are solitary and residential. *Poicilocerous pictus* lives on Ak plant (*Calotropis procera*) in India. Spiders, frogs, lizards and birds are their enemies, they eat grasshoppers.

Structure

Melanopus is green in colour and *Poecilocerus* is yellowish or brownish in colour containing spots. Body is divisible into **head, thorax** and **abdomen**. Head bears a pair of sessile **compound eyes,** three **simple eyes** and a pair of short **antennae**. The mouthparts of a grasshopper are of **biting** and **chewing type**. Mouthparts include upper lip or **labrum**, a tongue-like **hypopharynx**, a pair of triangular, toothed **mandibles** and a pair of **maxillae**. The mandibles are used to bite and crush the food.

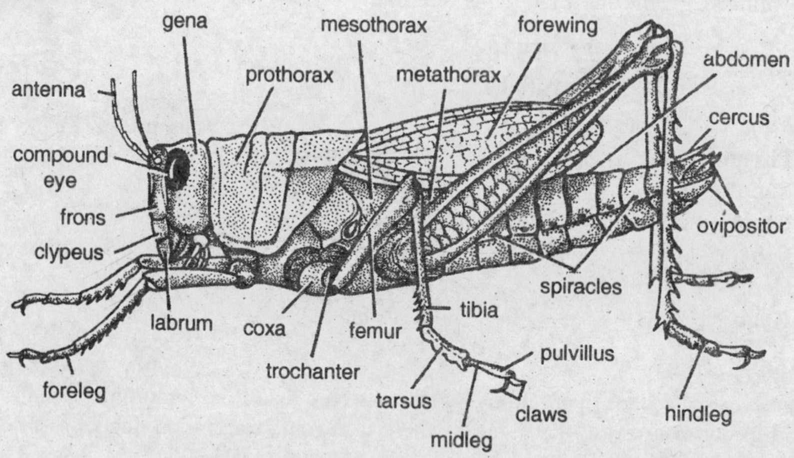

Fig. 38.26. Grasshopper (*Poicilocerus or Melanopus*).

Thorax is formed of three segments: an anterior larger **prothorax**, a middle **mesothorax** and a posterior **metathorax**. Each thoracic segment bears a pair of jointed and clawed legs. All the legs are used in walking and climbing. However hind legs are more powerful and adapted for jumping (**saltatorial legs**).

Mesothorax and metathorax bear a pair of wings each. Forewings are elongated, thick leathery and straight, called **tegmina**, while hindwings are membranous. Tegmina simply covers the hind wings and latter are used in flight. Abdomen is composed of ten segments.

Sexes are separate, sexual dimorphism is well marked. Female is provided with a long **ovipositor**. Fertilization is internal. Development is direct; nymphs are devoid of wings. Both nymph and adult grasshoppers damage crops. They are serious pests. They are eaten by human beings in several parts of world.

15. *SCHISTOCERCA* (Locust)

Classification

Phylum	—	**Arthropoda**
Subphylum	—	**Uniramia**
Class	—	**Insecta**
Subclass	—	**Pterygota**
Order	—	**Orthoptera**
Genus	—	***Schistocerca***

Habit and Habitat

Schistocerca gregaria is commonly known as **locust**. The term *locust* includes gregarious and migratory grasshopper-like forms. *Schistocerca gregaria* is found in the deserts of tropical region of world. It is most widly distributed species and commonly found in Africa, Arabia, India, Afganistan, Philippines, Japan, Mexico and U.S.A.

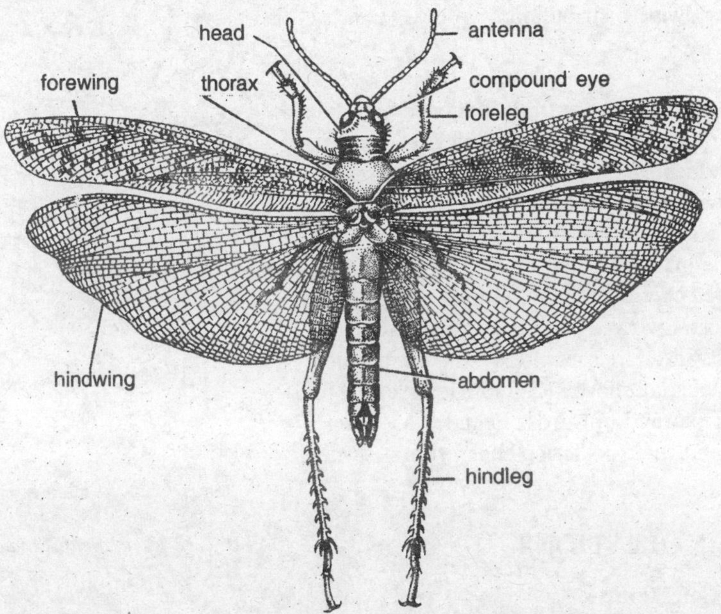

Fig. 38.27. Locust (*Schistocerca*).

Structure

Body of locust is divisible into **head**, **thorax** and **abdomen**. Head bears short **antennae** and well-developed compound eyes. Its mouth parts are **biting** and **chewing** type. Thorax is well developed and bears three pairs of legs and two pairs of wings. Forewing are elongated, thick, leathery and straight. Hindwings are large and membranous. **Abdomen** comprises ten segments with eleventh vestigial segment. The lateral sides of first abdominal segment carries a **tympanum**. Sexes are separate. Male possesses **anal styles** and female possesses **ovipositor**.

Locusts have two phases in their life: **solitary** or **non-migratory** and **gregarious** or **migratory**. In solitary phase, they have a yellow body colour and in gregarious phase, they have pinkish colour. The migratory phase appears due to some unknown reason and just before migration their population increases rapidly The swarms migrate toward green fields. The swarms of locusts move like a cloud and the direction is guided by the flow of wind. They damage crops and vegetation severely in course of their journey.

16. *PEDICULUS*
(Human head and body louse)

Classification

Phylum	—	**Arthropoda**
Subphylum	—	**Uniramia**
Class	—	**Insecta**
Subclass	—	**Pterygota**

Order	—	**Anopleura**
Genus	—	***Pediculus***

Habit and Habitat

Pediculus humanus is commonly known as **human head and body louse**. It is found as an ectoparasite of human beings. It has worldwide distribution.

Structure

Body of louse is dorsa-ventrally flattened and pale in colour with dark markings along the sides. The body is divisible into **head, thorax** and **abdomen**. Head is small and bears a pair of five-segmented antennae. Mouth parts are **piercing** and **sucking** type. Thoracic segments appear to be fused in one mass and bear clawed legs. Each leg bears a large curved claw adapted for clinging to the hairs or tissues of the host. Eyes and wings are absent. Abdomen is nine segmented. Sexes are separate; sexual dimorphism is well marked. Males are shorter (2–3 mm long) and their posterior ends are turned upwards. Females are longer (3–4 mm long) and their posterior ends are not turned upwards.

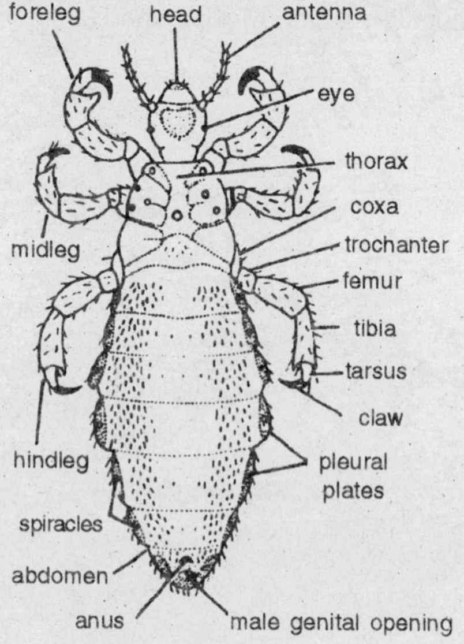

Fig. 38.28. *Pediculus humanus.*

QUESTIONS

Long Answer Questions

1. Give an outline classification of phylum Arthropoda.
2. Give chief characteristics of phylum Arthropoda mentioning the distinctive features of different classes into which the phylum is divided. Give examples of each class.
(*Agra 1988, 1990; Allahabad 1988, 1990; Kanpur 1988, 1990, 1993, 1995; Lucknow 1989; Jiwaji 1989; Delhi 1989, 1994; Awadh 1990, 1992; Bundelkhand 1990, 1993*)
3. Classify Arthropoda giving characters and examples of each class.
(*Raipur 1992; Meerut or CCS 1992, 1997; Kanpur 1993, 1995; Indore 1994; Lucknow 1997*)
4. Give the distinguishing features of main groups of Arthropoda with two examples of each.
(*Utkal 1988, 93; Vikram 1989, 1992; Gorakhpur 1990, 1995*)
5. What are distinguishing characters of class Crustacea and Arachnida?
6. State the points that indicate that the arthropods are biologically successful groups of animals.
(*Guwahati 1995*)
7. Classify Crustacea upto orders. Give important features and examples of each. (*Agra 1988*)

Short Answer Questions

1. Give classification, habits, habitat, characters and economic importance of following animals.
(*i*) *Limulus* (King crab)
(*Bundelkhand 1989; Allahabad 1989; Kanpur 1989; Meerut or CCS 1992; Lucknow 1993*)
(*ii*) *Daphnia* (*Agra 1989; Awadh 1989*)
(*iii*) *Cypris* (*Rohilkhand 1989, 1990*)
(*iv*) *Cyclops* (*Gorakhpur 1989, 1993; Awadh 1990, 1992, 1995; Lucknow 1997*)

(*v*) *Argulus* (*Agra 1989, 1990*)

(*vi*) *Lepas* (*Bhopal 1989, 1992; Agra 1990; Lucknow 1990*)

(*vii*) *Balanus* (*Lucknow, 1989, 1997; Agra 1990; Meerut or CCS 1991; Kanpur 1992*)

(*viii*) *Sacculina* (*Gorakhpur 1989, 1996; Meerut or CCS 1995; Lucknow 1995*)

(*ix*) Hermit crab (*Eupagurus*) (*Jiwaji 1989; Purvanchal 1989, 1992 ; Meerut or CCS 1990; Lucknow 1996*)

(*x*) *Cancer* (Crab)

(*xi*) *Scolopendra* (Centipede) (*Lucknow, 1992; Guwahati 1992, 1994*)

(*xii*) Millipede (*Julus*) (*Kanpur 1990; Agra 1991, 1995; Jiwaji 1993; Gwalior 1993*)

(*xiii*) *Lepisma* (Silver fish) (*Punjab 1992; Lucknow 1993*)

(*xiv*) *Poiceloceros* (Grasshopper) (*Punjab 1990; Gorakhpur 1998*)

(*xv*) Desert locust (*Schistocerca*) (*Lucknow 1993, 1997*)

(*xvi*) Head louse (*Pediculus*) (*Lucknow 1992, 1994, 1998*)

Very Short Answer Questions

1. Which animal of subclass Xiphosura is called a living fossil?

   Ans. King crab or *Limulus* of subclass Xiphosura is known as living fossil because it is living on earth since very ancient time, the Triassic period (200 million years) as primitive chelicerate, unchanged.

2. Name the two main tagmata of the arachnid body.

   Ans. Prosoma and opisthosoma.

3. With which appendages does a centipede bite?

   Ans. First pairs of legs modified into poison claws.

4. Name a crustacean parasite of crabs.

   Ans. *Sacculina.*

5. Cite two examples of order thoracica.

   Ans. *Lepas* and *Balanus*.

6. Give the zoological name and class of king crab.

   Ans. *Limulus*, Merostomata.

7. What is the common name of *Lepisma*?

   Ans. Silverfish.

8. Name two insects having complete metamorphosis.

   Ans. Butterfly and moth.

9. Which is poisonous, millipede or centipede?

   Ans. Centipede.

10. From where do locust swarms visit India periodically?

    Ans. Arabia, Persia, Afghanistan and Baluchistan.

11. Name a crustacean in which female carries eggs in ovisacs.

    Ans. *Cyclops*.

12. Which *Anopheles* (male or female) causes malaria in human beings? (*C.C.S.U. Meerut 2001*)

Ans. It is the female *Anopheles* mosquito which causes malaria in human beings. The sexual cycle of malarial parasite *Plasmodium vivax* is completed in female mosquito where as asexual cycle is passed in two phases in human beings.

13. What is the food of *Anopheles*?

Ans. Male *Anopheles* mosquito feeds on flower juices where as female *Anopheles* sucks human blood. It is because their mouth parts are modified into elongated piercing and sucking styles lying within proboscis.

14. Which part of the body of bees carry eggs?

Ans. The abdomen of queen bee (*Apis indica*) carries a large number of eggs. It lives for several years and lays about 1000 eggs per day.

Fill in the Blanks

1. The living arthropods are divided into three subphyla: .... , Crustacea and ....
2. There are two pairs of antennae in ....
3. The appendages of arthropods are ....
4. .... are ingrowths of cuticle to bear muscles.
5. Periodical replacement of exoskeleton is called .... or ....
6. Ecdysis is the phenomenon of ....
7. .... are social and polymorphic insects.
8. The eggs are .... in arthropods.
9. Stomodaeum in arthropods is lined by ....
10. Excretory organs of arthropods are .... glands and .... tubules.
11. Arthropods have .... circulatory system.
12. .... are the only invertebrates to have the power of flight.

Multiple Choice Questions

1. Members of phylum Arthropoda show following characters except one which is not correct. Mark it

(*a*) appendages are jointed (*b*) cilia are uniformly present

(*c*) exoskeleton in chitinous (d) body cavity is haemocoel.

2. Trilobites are

(*a*) extinct arthropods (*b*) connecting link between annelids and molluscs

(*c*) living fossils (*d*) insects

3. Which is the main characteristic of class Crustacea?

(*a*) cephalothorax and two pairs of antennae

(*b*) gills and green glands

(*c*) both (*a*) and (*b*)

(*d*) cephalothorax and one pair of antennae

4. Class Arachnida is characterized by

(*a*) one pair each of pedipalpi and chelicertes (*b*) four pairs of legs

(*c*) both (*a*) and (*b*) (*d*) three pairs of legs

5. Abdomen is greatly reduced in

(*a*) *Palaemon* (*b*) *Palamnaeus*

(*c*) *Cancer* (*d*) *Buthus*

6. The arthropods having three pairs of legs are members of the class
(*a*) Crustacea (*b*) Chilopoda
(*c*) Diplopoda (*d*) Insecta

7. Body cavity of arthropods is
(*a*) coelom (*b*) haemocoel
(*c*) pseudocoel (*d*) coelenteron

8. Excretory organs of crustaceans are
(*a*) nephridia (*b*) malpighian tubules
(*c*) green glands (*d*) flame cells

9. Respiratory organs in Arthropoda is
(*a*) tracheae (*b*) gills
(*c*) book lungs (*d*) all

10. Which of following arthropod respires by book lungs?
(*a*) spider (*b*) *Scolopendra*
(*c*) *Palaemon* (*d*) *Anopheles*

11. Excretory product in an insect is
(*a*) uric acid (*b*) urea
(*c*) ammonia (*d*) guanine

12. The arthropods with a sting at the hind end is
(*a*) *Scolopendra* (*b*) *Galeodes*
(*c*) *Palamnaeus* (*d*) *Lepas*

13. Which of the following has raptorial prolegs?
(*a*) locust (*b*) praying mantis
(*c*) wasp (*d*) bedbug

14. *Julus* belongs to class
(*a*) Arachnids (*b*) Diplopoda
(*c*) Chilopoda (*d*) Crustacea

15. *Galeodus* belongs to the class
(*a*) Arachnida (*b*) Crustacea
(*c*) Chilopoda (*d*) Diplopoda

16. Which of the following insect is not social?
(*a*) honeybee (*b*) ant
(*c*) termite (*d*) locust

17. Drones in honeybee are
(*a*) fertile males (*b*) fertile females
(*c*) sterile males (*d*) sterile females

18. Basic unit of compound eye is
(*a*) retina (*b*) rhabdome
(*c*) corneal facet (*d*) ommatidium

19. Which is a living fossil?
(*a*) *Limulus* (*b*) *Peripatus*
(*c*) *Lepisma* (*d*) *Pediculus*

20. The study of insects is called
(*a*) entomology (*b*) ethology
(*c*) epidemiology (*d*) none of the above

21. Termitarium is
(*a*) bunch of termites (*b*) nest of termites
(*c*) colony of termite (*d*) larvae of termites

22. In phylum Arthropoda, fertilization
(*a*) is external (*b*) is internal
(*c*) takes place in water (*d*) takes place on land

23. Complete metamorphosis is found in
(*a*) silver fish (*b*) bedbug
(*c*) grasshopper (*d*) gypsy moth

24. The moulting hormone or ecdysone is secreted from
(*a*) corpora allata (*b*) corpora cardiaca
(*c*) prothoracic glands (*d*) none of these

25. The juvenile hormone is secreted from
(*a*) corpora allata (*b*) corpora cardiaca
(*c*) corpora adiposa (*d*) none of these

26. Histolysis and histogenesis occur during
(*a*) larval stage (*b*) pupal stage
(*c*) adult stage (*d*) none of these

27. Complete metamorphosis occurs in
(*a*) locust (*b*) silver fish
(*c*) bedbug (*d*) mosquito

ANSWERS

Fill in the Blanks

1. Chelicerata, Urinamia;
2. Crustacea;
3. Jointed;
4. Apodemes:
5. Moulting, ecdysis;
6. Casting of old exoskeleton;
7. Termites;
8. Centrolecithal;
9. Cuticle;
10. Green, Malpighian;
11. Open;
12. Insects.

Multiple Choice Questions

1. (*b*)	**2.** (*a*)	**3.** (*c*)	**4.** (*c*)	**5.** (*c*)	**6.** (*d*)
7. (*b*)	**8.** (*c*)	**9.** (*d*)	**10.** (*a*)	**11.** (*a*)	**12.** (*c*)
13. (*b*)	**14.** (*b*)	**15.** (*a*)	**16.** (*d*)	**17.** (*a*)	**18.** (*d*)
19. (*a*)	**20.** (*a*)	**21.** (*b*)	**22.** (*b*)	**23.** (*d*)	**24.** (*c*)
25. (*a*)	**26.** (*b*)	**27.** (*d*)			

39

Palaemon (*Macrobrachium*) (Prawn) (Phylum Arthropoda)

The shrimp *Palaemon* (new name *Macrobrachium*) is commonly known as **prawn.** It is a freshwater crustacean arthropod and is commonly found in India. The common Indian species of *Palaemon* are *P. malcolmsonii, P. carcinus, P. idae, P. rudis* and *P. lamarrei*. *Palaemon malcolmsonii* is studied in most of the Indian Universities as a typical representative of the Crustacea.

39.1 CLASSIFICATION

Phylum	–	**Arthropoda**
Subphylum	–	**Crustacea**
Class	–	**Malacostraca**
Subclass	–	**Eumalacostraca**
Superorder	–	**Eucarida**
Order	–	**Decapoda**
Suborder	–	**Pleocyemata**
Infraorder	–	**Carida**
Genus	–	***Palaemon* (*Macrobrachium*)**

Habit and Habitat

Palaemon is found in freshwater streams, rivers, lakes and ponds. It is a nocturnal, bottom-dweller and lives within underwater crevices and aquatic vegetation. It comes out from its hides in the night in the search of food. *Palaemon* is **omnivorous** and feeds on algae, moss, other weeds, small insects and debris (*e.g.*, decaying leaves). It has dual mode of locomotion: slow walking (*i.e.*, crawling) on the bottom with the help of its walking legs and active swimming in water with the help of pleopods. Sexes are separate and sexual dimorphism is distinct. *Palaemon malcolmsonii* breeds in May, June and July and the females carry eggs attached to their pleopods. Development of *Palaemon* is direct.

39.2 EXTERNAL FEATURES

1. Shape and size. The body of *Palaemon* is elongated, laterally compressed, spindle-like and bilaterally symmetrical. Such a shape offers minimum resistance to water during swimming. The size of adult *Palaemon* varies greatly in different species. For example, *P. carcinus* from Travancore measures 90 cm and *P. malcolmsonii* from Madhya Pradesh and Madras is generally 25–38 cm long.

2. Colouration. Fresh specimen is pale-blue or greenish in colour with brown or orange red patches and bands. On exposure to air, it becomes clay-brown or yellow.

3. Division of body. The body of adult *Palaemon* consists of nineteen appendage bearing segments. It has two distinct regions: an anterior rigid (immovable) cephalothorax and a posterior flexible (movable) **abdomen** (Fig. 39.1). The entire outer surface of the body is covered by a hard exoskeleton.

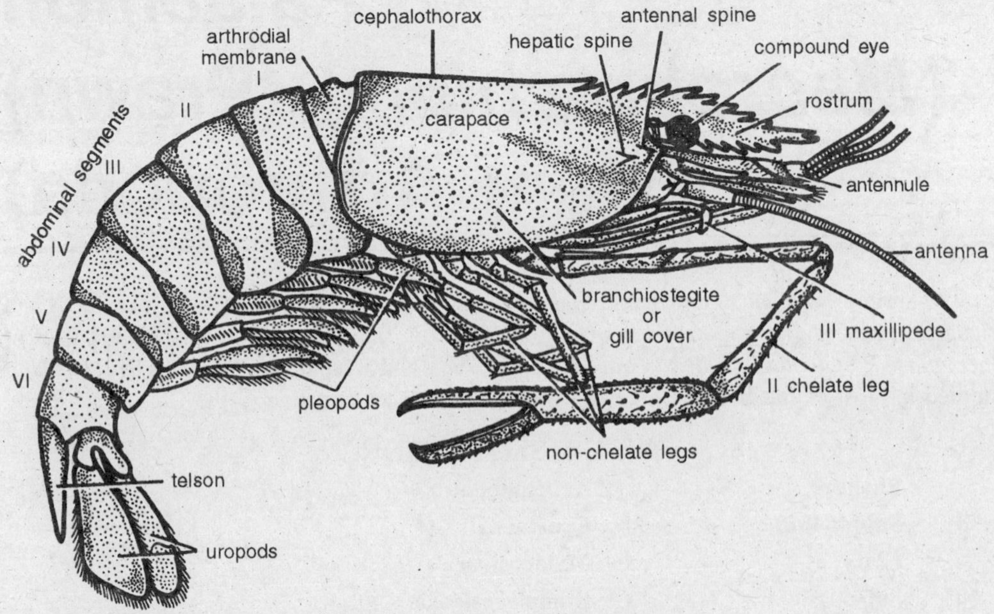

Fig. 39.1. *Palaemon*. External features.

(i) Cephalothorax. It is large, cylindrical, rigid and unjointed. It is formed by the union of 13 segments—5 of the **head** or **cephalic region** and 8 of the **thorax** or **thoracic region.** Each of these segments bears on its ventral surface a pair of appendages; thus, there are in all 13 pairs of cephalothoracic appendages. A pair of movable and stalked **compound eyes** are present in the presegmental region.

(ii) Abdomen. The abdomen is rounded dorsally but laterally compressed. It is jointed having six movable segments and a terminal conical piece, **telson.** Each abdominal segment carries a pair of jointed appendages called **pleopods** or **swimmerets.**

4. Exoskeleton. The entire body and the appendages are covered by a hard and chitinous cuticle which forms **exoskeleton.** Exoskeleton is formed by the underlying cells and becomes hard due to the deposition of calcium salts (calcium carbonate and calcium phosphate) and **scleroprotein.** The exoskeleton besides affording protection against injuries, provides space for muscle insertion. It is not continuous throughout the body but is localised to form thick **plates,** called **sclerites** which are joined by thin softer cuticle forming **arthrodial membranes.** Arthrodial membranes facilitate the movement of sclerites. Each sclerite consists of dorsal **tergal plate,** ventral **sternal plate** and two **lateral plates.**

In the cephalothorax, the tergal and the pleural plates of all sclerites are fused to form a continuous shield-like covering called **carapace.** On both ventro-lateral sides, the carapace hangs freely to form gill-chamber, gill-cover or branchiostegite. On the mid-dorsal surface the carapace is drawn into a long serrated projection called, **rostrum,** towards the anterior end. This is defensive in function.

At the base of rostrum, on either side is an **orbital notch** which accommodates the stalk of compound eye.

5. External Apertures

(1) Mouth. It is a slit-like aperture, situated mid-ventrally below the anterior end of head.

(2) Renal apertures. There are a pair of minute openings, one on a papilla on the inner surface of coxa of antennae.

(3) Female genital apertures. These are also paired crescentic apertures and lie on the inner surface of the coxae of the third pair of walking legs.

(4) Male genital apertures. These are slit-like and lie on the inner side of the arthrodial membrane connecting the fifth pair of walking legs with the thorax.

(5) Openings of statocysts. These are paired, each situated in a depression at the dorsal surface on the precoxa of antennules.

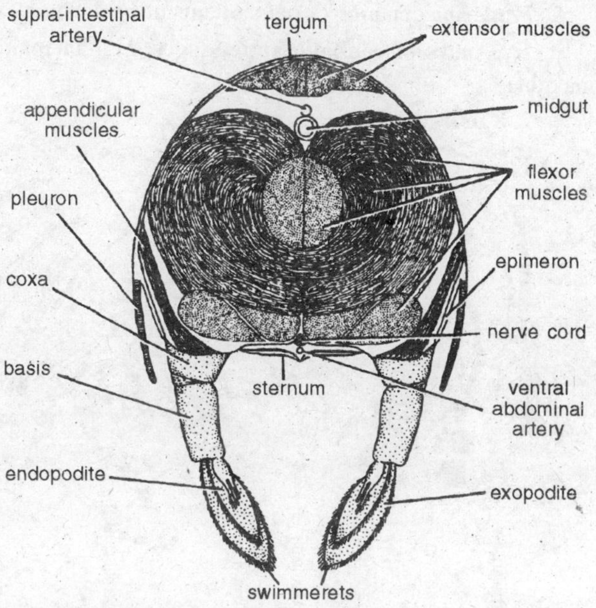

Fig. 39.2. *Palaemon*. T.S. abdominal region showing arrangement of skeleton.

(6) Anus. It is a longitudinal aperture situated mid-ventrally at the base of telson.

6. Appendages. In all there are nineteen pairs of appendages; one pair of appendages in each segment. These are of following types:

(*i*) **Cephalic appendages.** These are five pairs and include **antennules, antenna, mandibles, maxilulla** and **maxillae.**

(*ii*) **Thoracic appendages.** These are eight pairs and include first three pairs of **maxillipedes** and five pairs of **walking legs.**

(*iii*) **Abdominal appendages.** These are six pairs and include first five pairs of **pleopods** or **swimmerets** and last one pair of **uropods.**

The cephalic appendages are modified for searching the food and feeding; walking legs help in locomotion; swimmerets help in swimming and uropods help in change of direction while swimming.

39.3 APPENDAGES

Palaemon possess paired segmentally arranged jointed appendages. Each pair of appendages arises from the ventro-lateral margins of each segment. Except antennules, all appendages of *Palaemon* show a typical **biramous plan** of construction; antennules are **uniramous.** Each appendage has a basal stalk, called **base** or **protopodite** which is made up of two joints or **podomeres,** a lower coxapodite or **coxa** and an upper basopodite or **basis.** From the protopodite arise two rami or branches, an inner **endopodite** and an outer **expodite,** each of which may be composed of one to several podomeres. Appendages of *Palaemon* exhibit marked variations in form and size in response to different functions.

Types of Appendages

Palaemon contains 19 pairs of appendages which can be classified into following three types:

1. 5 pairs of cephalic appendages, *i.e.*, antennules, antennae, mandibles, maxillulae and maxillae.

2. 8 pairs of thoracic appendages, *i.e.*, 3 pairs of maxillipedes and 5 pairs of walking legs.

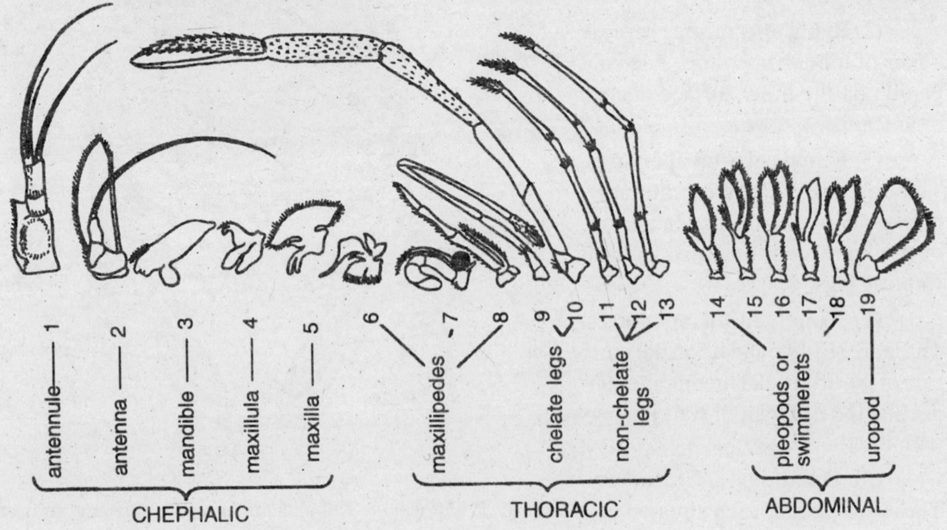

Fig. 39.3. Appendages of *Palaemon.*

3. 6 pairs of abdominal appendages, *i.e.*, 5 pairs of pleopods or swimmerets and 1 pair of uropods.

Structure and Function of Appendages

A. Cephalic Appendages

These are 5 pairs and are of following two types:

(*i*) **Pre-oral appendages,** *e.g.*, antennules and antennae;

(*ii*) **Post-oral appendages,** *e.g.*, mandibles, maxillulae and maxillae.

1. Antennules. These are uniramous pre-oral cephalic appendages on the first body segment one on either side below the eye-stalks. The **protopodite** is formed of **precoxa, coxa** and **basis.** The precoxa is large and dorso-ventrally flattened. It bears a wide depression to accommodate compound eyes and a spiny lobe, the **stylocerite,** on the outer side of the proximal end. In the depression is situated the **opening of statocyst** protected by a fold of integument. The **coxa** is short and cylindrical. **Basis** is longer than coxa and carries two whip like-elongated and many jointed **feelers.** The outer feeler is divided into an inner smaller branch and outer larger branch. The feelers are not homologous to exo- and endopodite. It is, thus, uniramous (or unbranched).

Coxa and feelers of antennule bear sensory setae. Precoxa contains a statocyst in its cavity. Antennules are **tactile** and **balancing organs.**

2. Antennae. These are also preoral cephalic appendages which are situated one on either side in 2nd segment just below and behind the antennules. The protopodite is small and is formed of **coxa** and **basis.** These two are greatly swollen due to the presence of excretory organs inside them. The excretory organ (green gland) opens to outside by a minute renal aperture on the inner side of the coxa. **Basis** bears a short spine and two rami (branches): **exo-** and **endopodite.** The endopodite is a long, many jointed **feeler.** The exopodite is broad and leaf-like plate called **squama.**

It bears setae along its inner and distal margins. Squama serves the function of a balancer during swimming. The antennae are **tactile, excretory** and **balancing** organ.

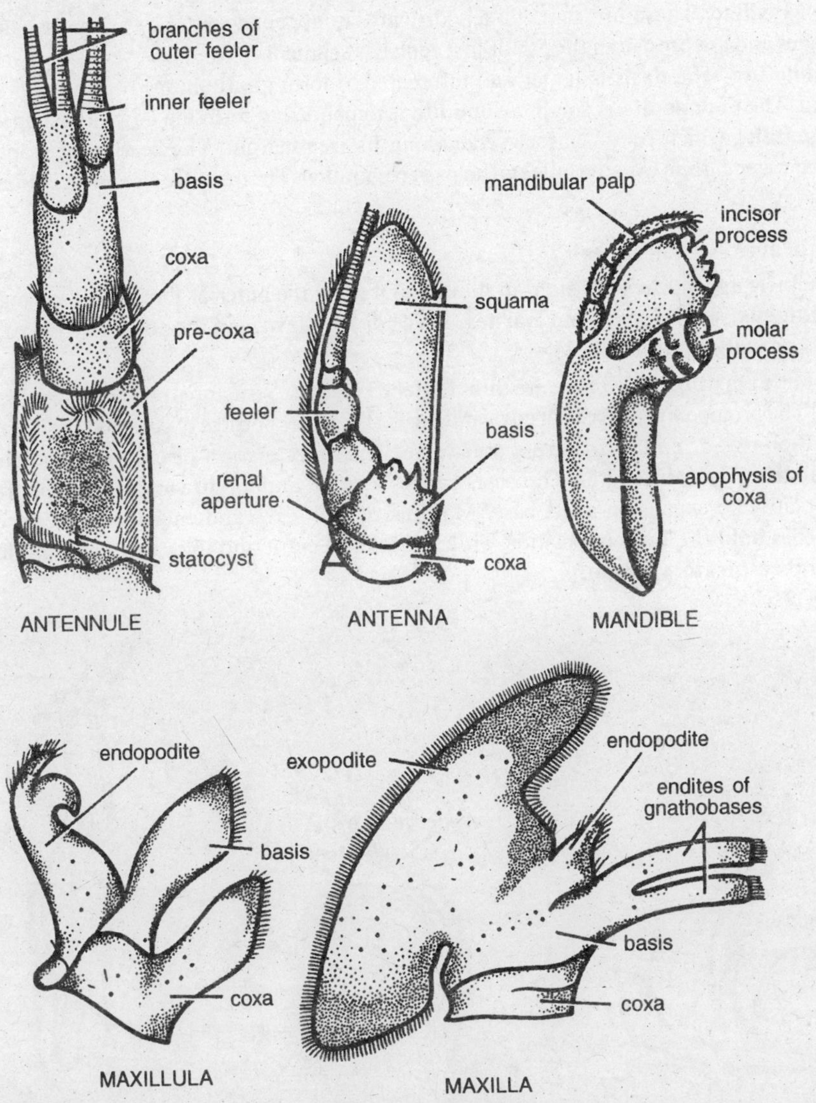

Fig. 39.4. *Palaemon*. Cephalic appendages.

3. Mandibles. These are post-oral cephalic appendages situated in third segment one on either side of mouth. Mandibles are short but stout and highly calcified structures. The greater part of each mandible consists of coxa which is differentiated into a proximal spoon-shaped **apophysis** and a solid distal **head.** The head consists of two parts: a **molar process** with five or six dental plates and a flat plate-like **incisor process** with three teeth. On the outer side, the head carries a three-segmented **mandibular palp.** It consists of a basal segment representing **basis** and the distal two segments representing **endopodite.** The expodite is absent. Mandibles are **masticatory** in function.

4. Maxillulae. These are small delicate (thin) and leaf-like post-oral appendages borne by 4th segment of the body. Each consists of **coxa, basis** and **endopodite.** The **exopodite** is absent. Coxa, and basis are broad and project inwards as jaws or **gnathobases** fringed with bristles. **Endopodite**

is curved and bifurcated terminally. It has small hooks. The maxillulae are used for passing food to mouth (or manipulation of food).

5. Maxillae. These are thin, flattened, leaf-like appendages. They are **post-oral** cephalic appendages and are present in the 5th or last cephalic segment. The **coxa** is very small and partially divided into two. The **basis** is larger and bifurcated to form **gnathobases** bearing stiff setae at its inner end. The **endopodite** is small. **Exopodite** is expanded to form the large fan-shaped **baler** or **scaphognathite** which bears sensory setae along its free margin. The beating of scaphognathite produces a water current over the gills and helps in respiration. The maxillae also assist in **manipulation of food.**

B. Thoracic Appendages

Thoracic appendages are eight in number. Of them the anterior three pairs are maxillipedes (or **maxillipeds;** see **Ruppert** and **Barnes,** 1994) or **foot-jaws** and the posterior five pairs are the **perepods** or **walking/thoracic legs.**

1. First maxillipedes. These are thin, flattened leaf-like structures present in the first thoracic segment. The protopodite consists of **coxa** and **basis.** The inner borders of coxa and basis are modified into **gnathobases** or **endities** bearing stiff setae. The outer margin of coxa bears bilobed, plate-like **epipodite** or **primitive gill.** The basis carries a small **endopodite** and long **expodite** bearing a broad plate-like expansion at its base. Margins of both exo- and endopodites bear setae. First maxillipedes **hold the food** in position while mandibles cut it and they also serve the function of **tactile** and **respiratory organ.**

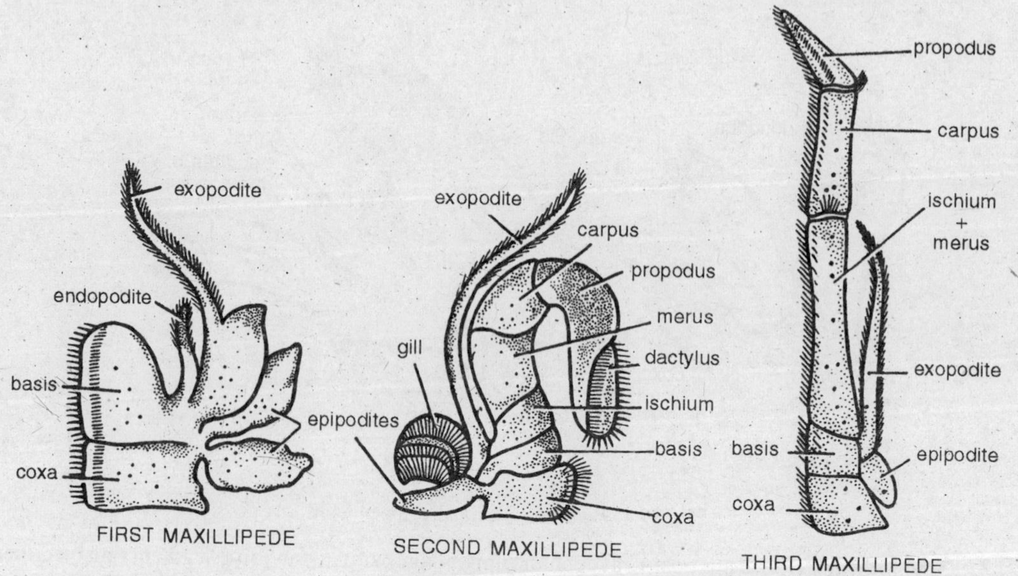

Fig. 39.5. *Palaemon*. Maxillipedes.

2. Second maxillipedes. These are not flattened. The **protopodite** consists of **coxa** and **basis.** The coxa is short and carries an **epipodite** and a **gill** (podobranch) on its outer border. The **exopodite** is long, slender, unjointed and covered with setae. The **endopodite** arises from the basis and consists of five segments or podomeres namely **ischium, merus, carpus, propodus** and **dactylus.** The distal two segments *i.e.*, propodus and dactylus are curved inwards and lie parallel to form a cutting plate with bristles. Second maxillipedes are **tactile, hold food** and **respiratory** in function.

3. Third maxillipedes. These are largest and look-like a thoracic or walking legs. The **coxa** bears a small **epipodite** on its outer side. It lacks a gill but contains setae along its inner border. The basis is small and carries an outer slender unjointed **exopodite** and an inner three-jointed **endopodite.** In the endopodite the first podomere represents fused **ischium** and **merus,** second is the **carpus** and third represents fused **propodus** and **dactylus.** The expodite is thickly covered with setae all along its length. The basis and endopodite both contain setae only along their inner margin. Third maxillipedes are **tactile, hold food** and are **respiratory** in function.

4. Thoracic or walking legs. The five pairs of walking legs of *Palaemon* differ from maxillipedes in the absence of the **exopodites** and **epipodites.** A typical walking leg such as **fourth,** consists of a two-jointed **protopodite** and five-jointed **endopodite.** The protopodite consists of a short ring-shaped **coxa** and a tringular **basis.** The endopodite consists of five podomeres, viz., **ischium, merus, carpus, propodus** and **dactylus.** All the seven podomeres are arranged in a linear series and are movably hinged together.

Walking legs are of two general types:

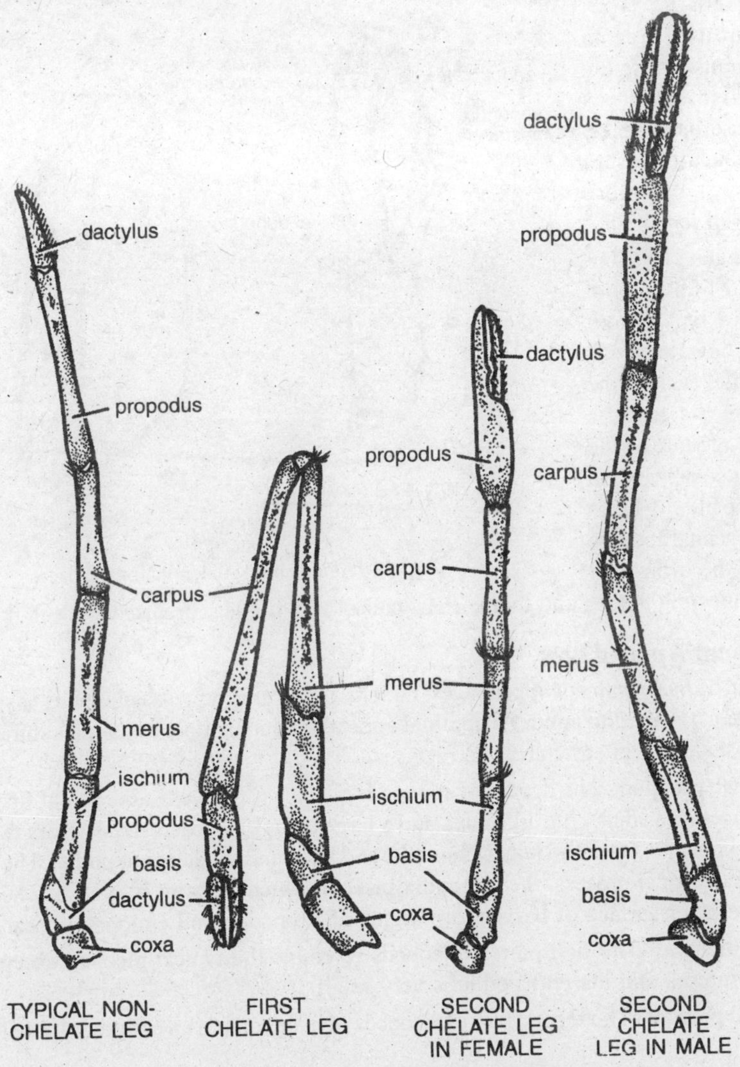

Fig. 39.6. *Palaemon.* Thoracic or walking legs.

1. Chelate legs or cephalopods. First and second pairs of walking legs bear at their distal ends a pincer-like or forceps-like structure called **chela.** In chelate legs, the propodus is very large and it extends beyond its articulation with the dactylus. Dactylus is freely movable and works against the propodus to form the chela. The biting edges of chela bear teeth and their entire surface is covered with setae. The second chelate legs are larger than the first. Chelate legs are used to catch the prey and transfer it to the mouth and also serves as organs of offense and defence. The second chelate legs in **male** are larger and more powerful than in female.

2. Nonchelate legs. Third, fourth and fifth pairs of walking legs are typical nonchelate legs, *i.e.*, they lack chela. In female, each of the third leg bears a crescentic female genital aperture on the inner side of its coxa, while in male each of the fifth leg bears a slit-like male genital aperture on its side of the arthrodial membrane connecting the leg and thorax. Last pair of walking legs are used to clean the pleopods.

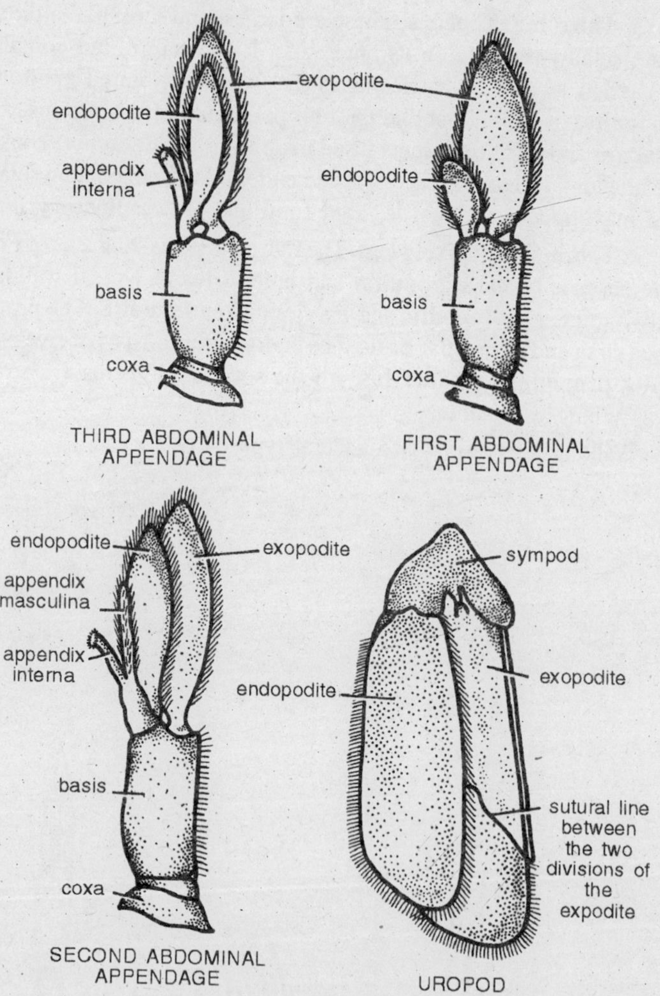

Fig. 39.7. *Palaemon.* Abdominal appendages.

C. Abdominal Appendages

Abdomen bears six pairs of appendages. First five pairs are known as **pleopods** or **swimmerets** and the sixth pair is called **uropod.** Abdominal appendages are simple and typical biramous type. They are almost similar in structure.

1. Typical pleopod. Third pair of pleopods serve as a typical abdominal apendage. Its **protopodite** contains a small ring-like **coxa** and a large cylindrical basis. Basis bears two flattened leaf-like lateral branches, outer and larger **exopodite** and inner and smaller **endopodite.** The endopodite bears a small cylindrical process, the **appendix interna** which contains a knob-like head covered with setae. The outer surface of basis and margins of exopodite and endopodite bear setae.

2. First pleopod. The first pair of pleopods resemble the typical pleopod except that they lack appendix interna and the endopodite is very small.

3. Second pleopod. The second pair of pleopods of male contain an additional rod-like process, the **appendix masculina,** in between the appendix interna and endopodite. Appendix masculina bears long spine-like setae. Female prawns do not contain this process.

Functions of pleopods include the following:

(1) These are used as paddles in swimming.

(2) During breeding season, in female, the appendix interna of opposite pleopods become interlocked to form a series of bridges on the ventral side of abdomen. The bridges support the eggs. The female carries the fertilized and developing eggs in these bridges.

(3) The second pair of pleopods of male are used in copulation.

4. Uropods. The last or the sixth pair of abdominal appendages are called **uropods** or **tail feet.** These are large, broad and flattened, stoutly-built structures. They lie on either side of telson. In the uropod, the coxa and basis are fused together to form a triangular **sympod.** The **exopodite** and **endopodite** are oval and oar-shaped. The exopodite is broad and large and is divided into two unequal parts by a oblique transverse **suture.** The endopodite is small sized and undivided. The margins of exopodite and endopodite, except the outer of the exopodite, are fringed with long sensory setae.

The uropods along with telson form a broad fan-shaped **tail fin** which acts as a balancing organ during locomotion. When endangered or disturbed, *Palaemon* suddenly jumps backwards and upwards with the help of uropods.

39.4 BODY WALL AND ECDYSIS

The body wall of *Palaemon* consists of following three parts:

1. An outer thick layer of **cuticle;**
2. A middle single layer of **epidermis** or **hypodermis;**
3. An inner **dermis.**

1. Cuticle. It is composed of two layers: an outer **epicuticle** and an inner **procuticle** (Fig. 39.8).

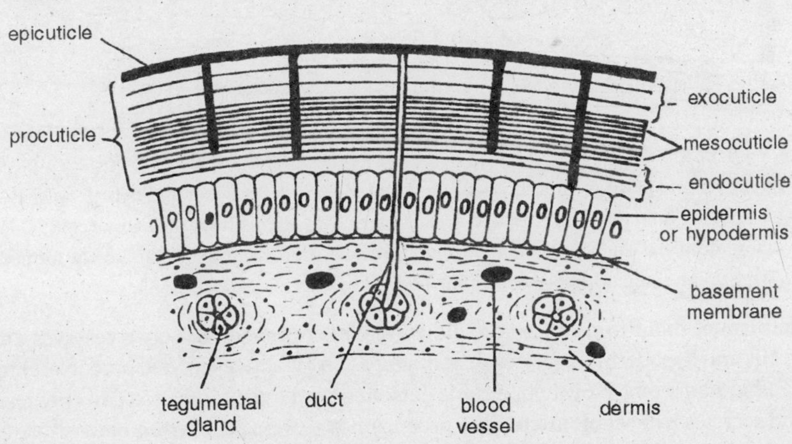

Fig. 39.8. *Palaemon*. T.S. of body wall.

***(i)* Epicuticle.** It is thin non-chitinous layer and composed of lipids, proteins, salts of calcium and pigments.

***(ii)* Procuticle.** It is a thick, laminated, chitinous but non-cellular layer. It is composed of (*i*) an **outer pigmented layer,** called **exocuticle,** of pigments, proteins, chitin and calcium salts; (*ii*) a **middle layer,** called **mesocuticle,** which is most heavily calcified and contains more chitin and less protein; and (*iii*) an **innermost thin layer,** called **endocuticle,** of protein and chitin.

Exocuticle is absent at joints and along lines where the exoskeleton tends to rupture during moulting. Since chitin is permeable to water, it is coated with a thin waxy epicuticle.

2. Epidermis or hypodermis. It lies below the cuticle and is composed of a single layer of columnar and glandular cells. This layer secretes **chitin** and the **cuticle** for the formation of **procuticle.** Hypodermis rests upon a basement membrane. It also contains trichogen cells which form the setae.

3. Dermis. Below the hypodermis is dermis. The dermis is made of connective tissue containing the blood lacunae. Dermis also contains **tegumental glands** which secrete **epicuticle.** These glands are formed by a group of cells which open into a long narrow duct opening outside on the epicuticle. Chromatophores are also present in the connective tissue.

Moulting or Ecdysis

The crustaceans (*Palaemon*) and other arthropods remain covered by a rigid, non-living exoskeleton which does not allow them to grow in size. In order to grow, the arthropods periodically shed their skeleton and secrete a new one. Periodic shedding of the exoskeleton in arthropods is called **moulting** or **ecdysis.**

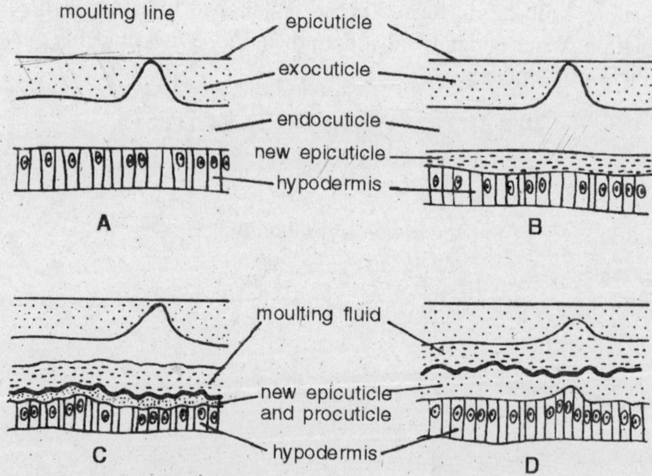

Fig. 39.9. Process of moulting in an arthropod. A—Fully formed exoskeleton and hypodermis; B—Secretion of moulting fluid (proenzymes) and separation of hypodermis from the exoskeleton; C—Enzymatic digestion of old endocuticle and secretion of new procuticle below the newly formed epicuticle (B); D—The animal enclosed within both old and new exoskeletons.

Mechanism of moulting. Before the old exoskeleton is shed, **glycogen** reserves are built up in the body cells and hypodermis secretes **proenzymes** (*i.e.*, inactive enzyme precursors) in between the **cuticle** and **hypodermis.** Now hypodermis detaches from the exoskeleton (this process is called **apolysis**) and secretes a new **epicuticle.** The proenzymes secreted earlier become activated to form **chitinase** and **protease.** The new epicuticle remains impervious for the moulting enzymes. These enzymes, however, digest the old endocuticle. The digestion products are absorbed through the new epicuticle. With the erosion of the old endocuticle, the hypodermis secretes new procuticle beneath the new (Fig. 39.9C). At this point the animal is enclosed with both an old and a new exoskeleton. The old skeleton now ruptures (splits) along certain predetermined lines and animal pulls out of old exoskeleton.

The new exoskeleton is soft and wrinkled (*i.e.*, extensible). It is stretched to accommodate the increased size of the animal. Various adaptations are used to keep the body at a maximum volume while new skeleton hardens: 1. Blood pressure is increased; 2 Insects inflate the crop with air, or 3. Crustaceans imbibe water. The tissues grow rapidly, and the new cuticle hardens during the short **postmolt period.** At this time, the stored mineral resources of crustaceans are rapidly exhausted, for they are deposited in the hardening of cuticle. Hardening of epicuticle is also helped by **tanning** of proteins which occurs by addition of phenols (Box 39.1).

Box 39.1 Moulting

1. Moulting (or ecdysis) is far more than a simple splitting of the cuticle and secretion of new one. Arthropod life consists of alternating periods of **premolt, molt, post-molt** and **intermolt.** Some arthropods (insects) escape into a permanent intermolt when they are adult; this period is known as **anecdysis.** However, others (crustaceans) never achieve stability, moulting occurs at intervals throughout adult life. The period between moults are known as **molts.**
2. Molts are recurrent crisis in the lives of arthropods. Until the new skeleton hardens, the protective case is weak, *and without hardened skeletal plates; so, skeleton is unable to support the weight of body and movement is often restricted*. To avoid danger to their life, arthropods may seek seclusion for moulting, build a protective silken retreat, retire into burrows, or have special mechanism to speed skeletal hardening. In any event, moulting remains an especially dangerous time in arthropod life cycle.
3. In *Palaemon*, the physiological processes involved in moulting are controlled by hormones. A **moult-inhibiting neurohormone** is secreted by a group of brain cells called **X-organs.** When secretion of X-organs are inhibited by **photoperiod** and **temperature,** etc., a moulting hormone or the **ecdysone** is secreted by **Y-organ** which is located beneath the muscles connected with mandibles. Ecdysone begins the process of moulting.

39.5 DIGESTIVE SYSTEM

The digestive system of *Palaemon* consists of an **alimentary canal** and a digestive gland, called **hepatopancreas.**

A. Alimentary Canal

Alimentary canal is a straight tube and is distinguished into following three parts:

1. **Stomodaeum** or **foregut** which comprises mouth, buccal cavity, oesophagus and stomach.
2. **Mesenteron** or **midgut** which comprises the intestine.
3. **Proctodaeum** or **hindgut** which comprises the rectum and the anus.

The foregut and hindgut have an internal lining of **cuticle** (intima) but the midgut has a soft lining of **endoderm.**

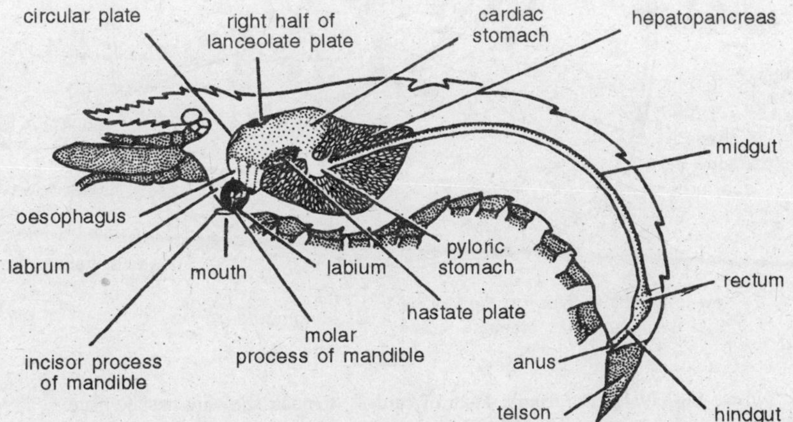

Fig. 39.10. *Palaemon*. Digestive system (alimentary canal and hepatopancreas) in lateral view.

1. Foregut

1. Mouth. It is a large longitudinal slit situated mid-ventrally in third cephalic segment. Mouth is bounded by a shield-like **labrum** in front, bilobed **labium** behind and the **incisor processes** of mandible laterally. It leads into a short buccal cavity.

2. Buccal cavity. It is an antero-posteriorly compressed small chamber. It has a thick chitinous lining which is thrown into irregular folds. The **molar processes** of **mandibles** project into the buccal cavity and lie opposite to each other. These are used to crush the food between them.

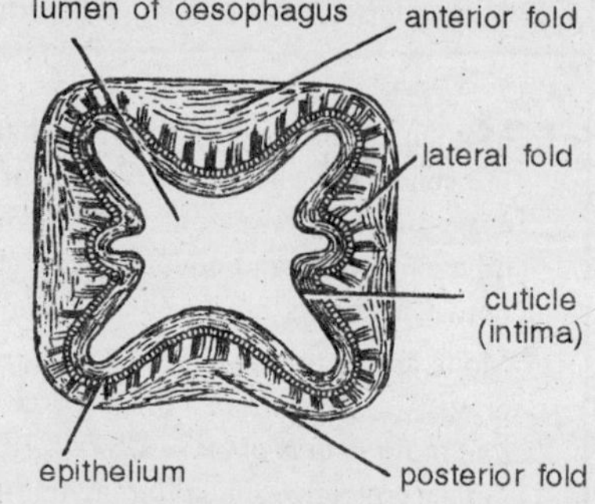

Fig. 39.11. *Palaemon.* T.S. of oesophagus.

3. Oesophagus. It is a short but broad tube which runs vertically upwards from the buccal cavity to the cardiac stomach. Internally the thick muscular walls of the oesophagus is thrown into four prominent longitudinal folds which project into its lumen. Due to these folds, the lumen of oesophagus assumes a star-shaped appearance. Of these folds, the **anterior** one is short but the **posterior** and **lateral folds** are larger and more prominent. Each of the lateral folds becomes further divided into two smaller unequal folds. The chitinous lining of oesophagus bear small bristles (Fig. 39.11).

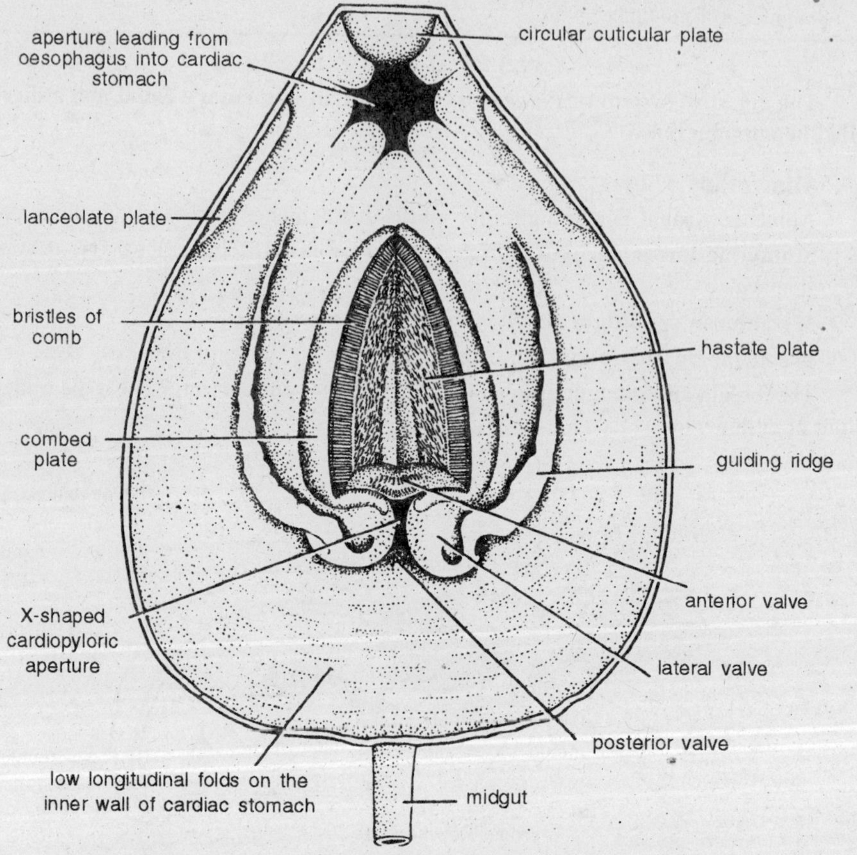

Fig. 39.12. *Palaemon.* Floor of cardiac stomach showing hastate plate.

4. Stomach. It is the largest part of alimentary canal. It occupies most of the cephalothoracic cavity. Stomach lies burried in the hepatopancreas, *i.e.*, it is covered over by the hepatopancreas laterally, posteriorly and ventrally. The stomach is divided into two unequal chambers: cardiac stomach and pyloric stomach.

(1) Cardiac stomach. It is bag-like anterior chamber of stomach. Internal chitinous lining (*i.e.*, intima) of cardiac stomach is thin and produced into numerous inconspicuous longitudinal folds covered by minute bristles (Fig. 39.12). The wall of stomach is supported by a number of calcified cuticular plates such as follows:

***(i)* Circular plate.** It is a thin plate present infront of oesophageal opening.

***(ii)* Lanceolet plate.** It is present in the roof of cardiac stomach just behind the circular plate.

***(iii)* Hastate plate.** It is a large triangular plate which is embedded in the floor of cardiac stomach. It is called **hastate plate** because it looks like the **head of a spear.** Hastate plate contains a distinct median ridge with gradually sloping sides. Its upper surface is covered with thick growth of delicate setae. Its anterior end is somewhat conical. The posterior broad part of hastate plate is depressed, fringed with setae and forms the *anterior valve* of the **cardiopyloric aperture.** Both lateral sides of hastate plate are supported by two longitudinal cuticular **supporting rods.**

***(iv)* Lateral grooves and groove plates.** A narrow lateral groove runs along the outer border of each supporting rod (*i.e.*, to each lateral border of the hastate plate). The floor of each lateral groove is covered by a cuticular plate which is like an open drain-pipe and is called the **groove plate** (Fig. 39.13).

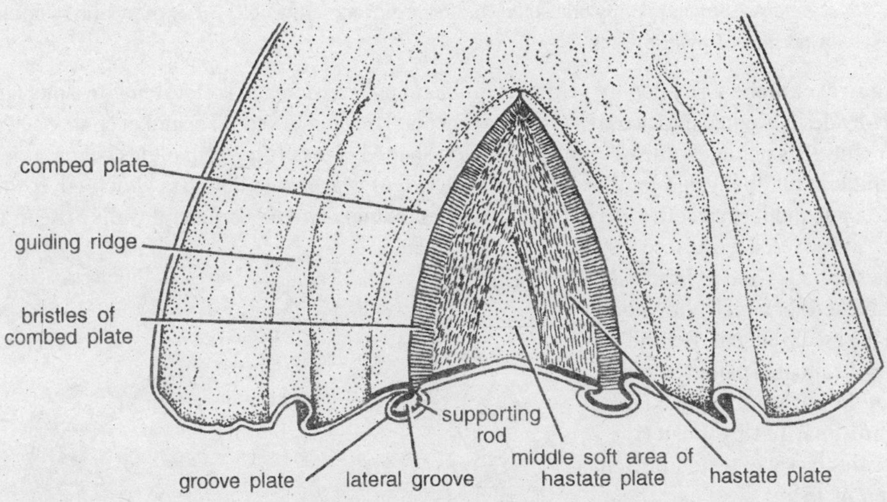

Fig. 39.13. *Palaemon.* Floor of cardiac stomach cut across.

***(v)* Combed plate.** Each lateral groove on its outer border is supported by a long **ridge plate** or **combed plate.** The two combed plates are united anteriorly but remain separated posteriorly by **cardio-pyloric aperture.** The inner border of each combed plate contains equally spaced large delicate bristles which bridge over the lateral groove and partially overlap the lateral margins of the hastate plate. The bristles of combed plate move constantly to and fro.

***(vi)* Guiding ridges.** Just outside the combed plate, the wall of cardiac stomach is folded on each lateral side to form two chitinous bluish black **lateral longitudinal folds** (Fig. 39.12). These folds are low in front but gradually increase in height behind. Posteriorly, however, these folds bend inwards to form the lateral valve of the cardiopyloric aperture. Since these lateral longitudinal folds guide the food towards the cardio-pyloric aperture, they are called **guiding ridges.**

***(vii)* Cardio-pyloric aperture.** It is a narrow X-shaped aperture and is guarded by four **valves.** The **anterior valve** is formed by the triangular posterior depressed end of hastate plate; the **lateral valves** by the large flap-like posterior ends of guiding ridges and **posterior valve** by semilunar folds of posterior wall of cardiac stomach. All these valves bear numerous setae which form a fine sieve through which only liquid food or very fine food particles can pass and enter pyloric stomach. The cardio-pyloric aperture leads into pyloric stomach.

(2) Pyloric stomach. It is very small narrow posterior chamber of the stomach. It lies below the posterior end of the cardiac stomach. Its lateral walls are thick and muscular. They divide the lumen of the pyloric stomach into a small dorsal chamber and a large ventral chamber. Both chambers are connected by a short narrow vertical slit-like aperture (Fig. 39.14).

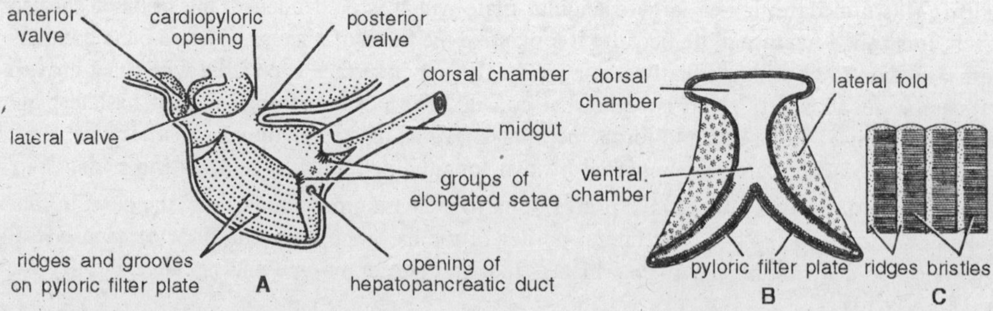

Fig. 39.14. *Palaemon*. A—Structure of pyloric stomach after removing left wall; B—T.S. of pyloric stomach; C—A portion of pyloric filter.

***(a)* Ventral chamber.** The floor of ventral chamber of pyloric stomach is elevated into a median ridge which divides the lumen into two lateral compartments. The floor of ventral chamber contains two rectangular chitinous plates. These two plates form a A-shaped **filter plate.** Both plates bear a series of alternating longitudinal ridges and grooves. All the ridges bear rows of bristles which cover the grooves like a felt and form an effective **pyloric filter** or **strainer.** The side or lateral walls of ventral chamber too are covered with closely set bristles. These bristles and pyloric filter form an efficient filter which allows only liquid food to pass through it. Behind the pyloric filter are situated paired openings of **ducts of hepatopancreas.** These openings are guarded by posteriorly directed bristles or setae which prevent regurgitation of food.

***(b)* Dorsal chamber.** The dorsal chamber bears a small, blind diverticulum or **caecum** dorsally. It leads into the midgut distally.

2. Midgut

It is a long, narrow, straight and slender tube which runs from the posterior part of cephalothorax to the sixth abdominal segment. Midgut lies above the mass of the ventral abdominal

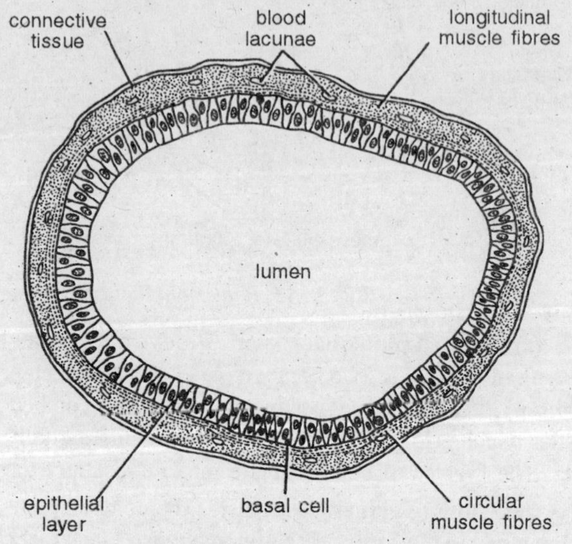

Fig. 39.15. *Palaemon*. T.S. of midgut.

muscles.

Histology. The outermost wall of the midgut is formed by a thin layer of **connective tissues.** It is followed by a thick layer of **longitudinal muscle fibres** which contains a network of blood lacunae. It is followed by a thin layer of **circular muscle fibres** and an **epithelial layer** of tall columnar cells which rest on a **basement membrane** (Fig. 39.15). Midgut is not lined by a chitinous cuticle.

The epithelial lining of posterior part of midgut is thrown into many longitudinal folds. The folds greatly reduce the cavity of the intestine (or midgut).

3. Hindgut

It extends from the posterior end of the midgut to the anus. It forms shortest part of the alimentary canal. Anteriorly, hindgut is swollen into a thick muscular sac, called the **rectum,** while the posterior part is narrow and tubular which opens to exterior through the anus. The rectum bears several thick longitudinal folds which project into the lumen of the hind gut (Fig. 39.16). Rectum is lined internally by **chitinous cuticle.**

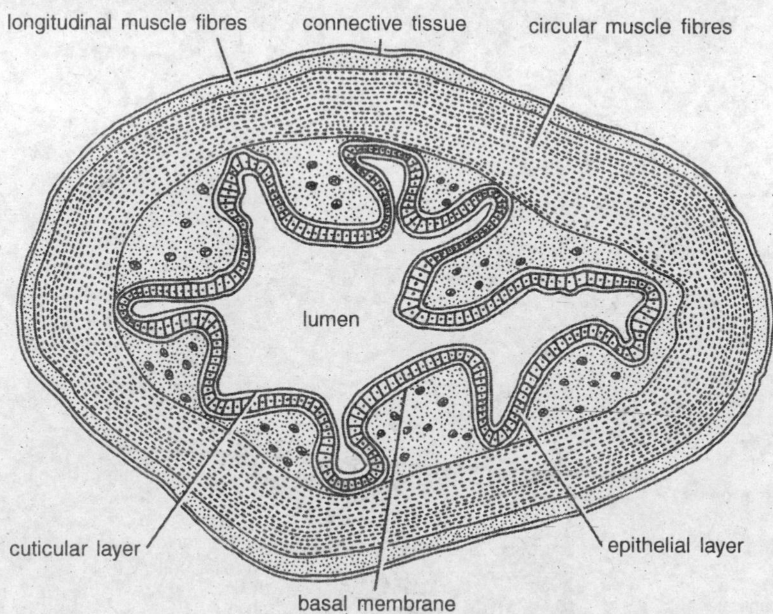

Fig. 39.16. *Palaemon.* T.S. of rectum.

4. Hepatopancreas

The **digestive gland, liver** or **hepatopancreas** is a large bilobed, compact, orange-coloured gland which fills the cephalothoracic cavity. It surrounds the stomach and anterior part of midgut. Developmentally it is formed from one pair of lateral tubular outgrowths, the hepatic caeca, of the midgut. The hepatopancreas of adult consists of numerous **glandular tubules** which are branched in a recemose manner. These branches are held together by the connective tissue as to form a compact mass. The wall of the tubules is formed of a single layer of columnar epithelium. This epithelium contains the following four types of cells: 1. **Granular cells;** 2. **Ferment cells;** 3. **Hepatic cells** and 4. **Basal** or **replacing cells.** All the epithelial cells rest on a basement membrane (Fig. 39.17).

The branching tubules join each other to form larger and larger canals which ultimately fuse to form two large **hepatopancreatic ducts.** Both ducts open into the ventral chamber of pyloric stomach. Via these ducts, digestive secretions which are produced by the hepatopancreas reach both cardiac as well as pyloric stomach.

Functions. The hepatopancreas functions like the liver, pancreas and small intestine of higher animals. It performs the following functions:

1. Like the pancreas, it secretes digestive enzymes for the digestion of carbohydrates, fats and proteins.
2. Like the small intestine it absorbs digested food.
3. Like the liver, it serves as an important storage organ. It stores glycogen, fat and calcium.
4. Some intracellular digestion also occurs in its cells.

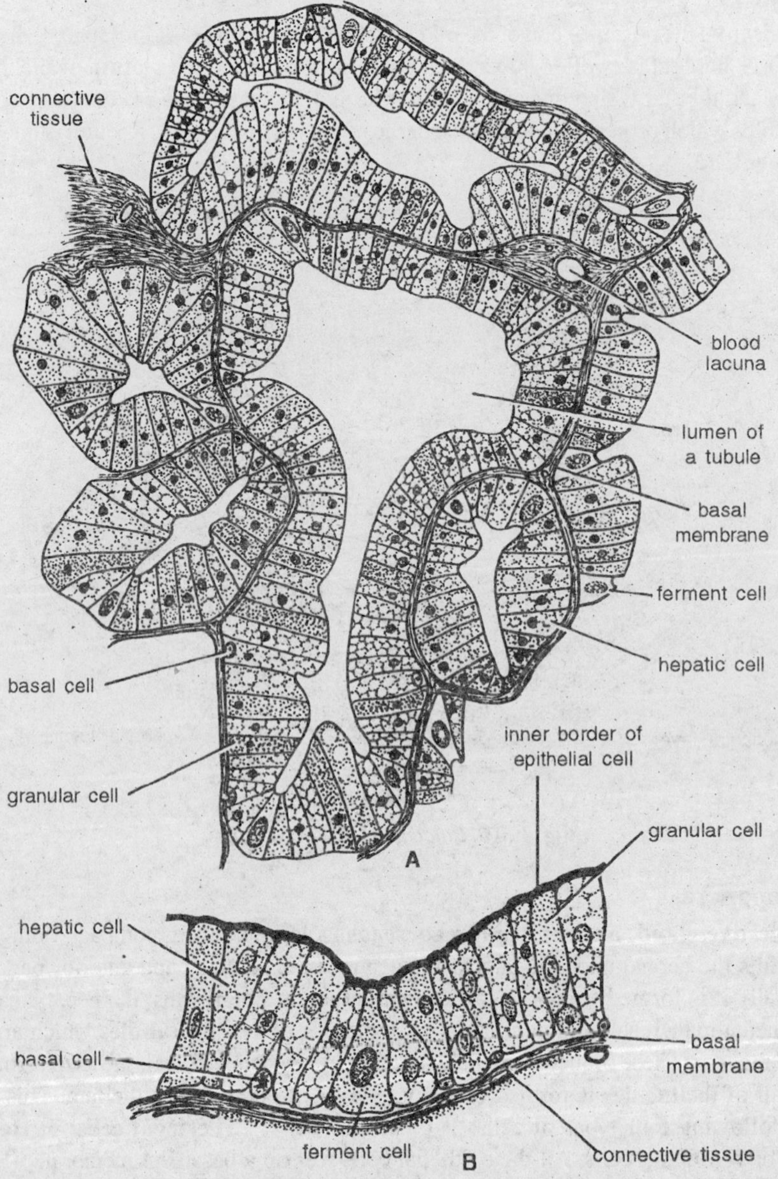

Fig. 39.17. *Palaemon.* A—T.S. through a portion of the hepatopancreas; B—T.S. of a portion of the wall of an hepatopancreatic tubule.

Food and Feeding

Palaemon is nocturnal and **omnivorous** animal. It feeds on algae, mosses and other aquatic weeds. Occasionally, it feeds on small insects, snails, tadpoles, fishes, etc.; and debris of bottom. Food is captured by the chelate legs and is passed on to the third maxillipedes which send it to other mouth parts. Coxae of second maxillipedes hold the food, while incisor processes of mandibles cut it into small pieces. These pieces are further torn by maxillae and maxillulae. The food is now swallowed into the buccal cavity with the help of 1st pair of maxillipedes, maxillae and maxillulae. Inside the buccal cavity food is masticated by molar processes of mandibles. Ultimately the food is passed on to the cardiac stomach through the oesophagus. The wall of cardiac stomach work like grinding mills.

Digestion, Absorption and Egestion

In *Palaemon*, the digestion of food is started in the cardiac stomach. The enzymatic secretions of hepatopancreas are poured into the pyloric stomach from where they reach the cardiac stomach and mix with the food. By alternate contraction and expansion of cardiac stomach the food is churned and also digested by the digestive enzymes. Guiding ridges guide the semi-digested food to pass over the hastate plate. When this food passes over the hastate plate, the moving bristles of the combed plates cut it into fine particles. The food in the semi-liquid and semi-digested form is filtered through the bristles of the combed plates into the lateral grooves below; then it is carried into the ventral chamber of pyloric stomach through the cardio-pyloric aperture. The paste-like food is again filtered through the hepatopancreatic ducts and then absorbed. As already described, the hepatopancreas serves for intracellular digestion and as well as for absorption.

The undigested food (having some unabsorbed food) ascend the dorsal chamber and passes into midgut for absorption. The remaining digested food is absorbed in the midgut and residual faecal matter is passes to the hindgut and is finally expelled out through the anus.

39.6 RESPIRATORY SYSTEM

Respiratory Organs

In *Palaemon*, the respiratory organs are well developed and they are meant for its aquatic mode of life. Its respiratory organs are

1. A pair of **branchiostegites;**
2. Three pairs of **epipodites;**
3. Eight pairs of **gills.**

These organs are contained in a pair of **gill chambers** (Fig. 39.19) which are situated one on either side of the thorax and are protected externally by the downwardly grown sides of carapace called **gill covers** or **branchiostegites.** The gill chambers open on all sides except dorsally.

1. Branchiostegites

The inner linings of branchiostegites are thin, membranous and highly vascular containing blood lacunae. These are constantly bathed in freshwater, thus, forming large respiratory surfaces. These absorbs oxygen dissolved in water and gives out carbon dioxide.

2. Epipodites

Epipodites are primitive gills. These are three pairs, simple, thin, leaf-like, highly vascular outgrowths of integument. They are attached on the outer side of coxa of the maxillipedes which are present in first three thoracic segments. Epipodites lie in the anterior part of gill chamber below the scaphognathite of maxillae. The epipodites of first pair of maxillipedes are bilobed and largest

(Fig. 39.19). Respiratory exchange of oxygen and carbon dioxide occurs through these respiratory organs.

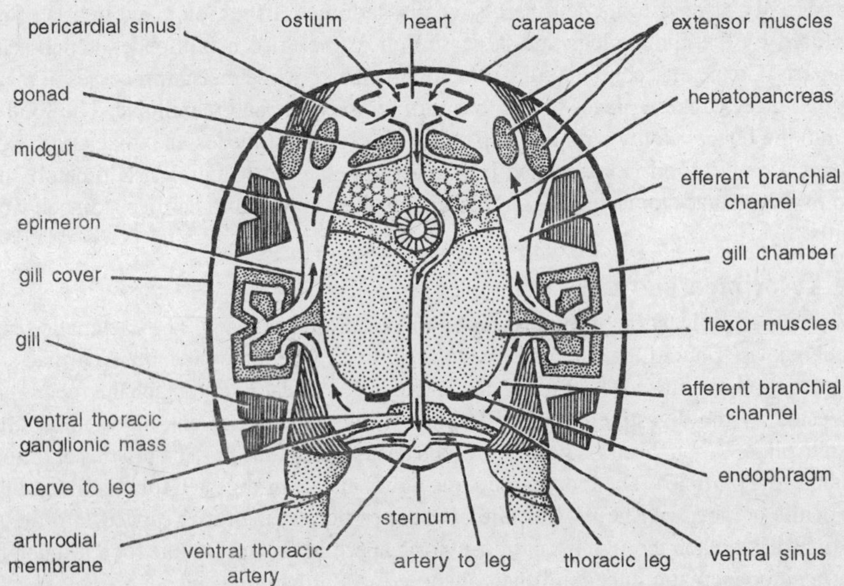

Fig. 39.18. *Palaemon.* Diagrammatic T.S. through thorax showing gill chambers and blood supply in the gills.

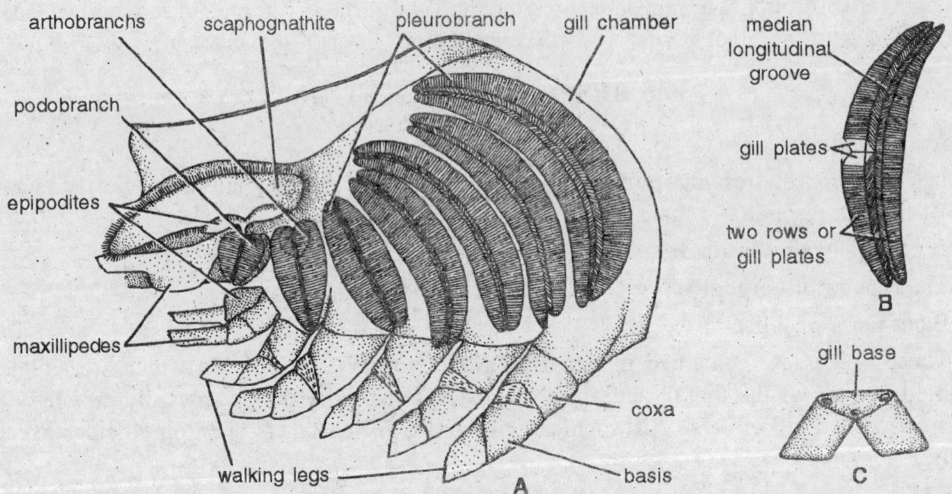

Fig. 39.19. *Palaemon.* A—Left gill-chamber is exposed to show position of epipodites and gills; B—Complete gill (phyllobranch); C—Cross-section of gill.

3. Gills

There are eight **branchiae** or **gills** in each narrow gill chamber. But only seven gills are seen on removing the gill cover as the third gill lies below the second (Fig. 39.19).

***(a)* Types of gills.** Gills are named according to their place of origin and location. There are following three types of gills in each of the gill cover:

(*i*) **Podobranchs or foot gills.** The gills attached to the coxae of appendages are called **podobranchs**. There is **one** podobranch attached to the coxa of each second maxillipede.

(*ii*) **Arthrobranchs or joint-gills.** These gills are attached to the arthrodial membranes (which connect the appendages with the body). There are **two** arthrobranchs attached to the arthrodial membrane of each third maxillipede.

(*iii*) **Pleurobranchs or side-gills.** These gills are attached to the lateral walls of the throax above the attachment of the appendages. There are **five** pleurobranchs attached to the lateral walls of the thoracic segments bearing five walking legs.

(*b*) **Branchial formula.** Branchial formula shows the number and location of gills and epipodites (primitive gills) in a single gill chamber. The branchial formula of *Palaemon* is shown in the form of a table (Table 39.1) as follows:

Table 39.1.

Appendages	*Epipodite*	*Podobranch*	*Arthrobranch*	*Pleurobranch*	*Total*
1. 1st Maxillipede	1	–	–	–	1
2. 2nd Maxillipede	1	1	–	–	2
3. 3rd Maxillipede	1	–	2	–	3
4. 1st walking leg	–	–	–	1	1
5. 2nd walking leg	–	–	–	1	1
6. 3rd walking leg	–	–	–	1	1
7. 4th walking leg	–	–	–	1	1
8. 5th walking leg	–	–	–	1	1
Total	3	1	2	5	11

(*c*) **Structure of gill.** All the eight gills are more or less crescentic or semilunar in shape. The size of gills gradually increases antero-posteriorly, *i.e.*, each gill is larger than in front of it. Each gill is attached to the thorax at its middle by a small connection known as **gill-root.** The branchial nerve and branchial channels enter and leave the gill through the gill root.

In *Palaemon*, all the gill are phyllobranchs, *i.e.*, each gill consists of two rows of leaf-like rhomboidal **gill plates** arranged parallel to one another like the leaves of a book and lie at right angles to the long axis or the base of gill (Fig. 39.19B). The two rows of gill plates are separated by a median longitudinal groove which extends lengthwise in the gill-axis. The gill plates are largest in the middle and gradually reduce at their two ends.

(*d*) **Histology.** A transverse section of a gill (Fig. 39.20) reveals that the gill base or axis is roughly triangular in outline and consists of central core of **connective tissue** enclosed by a layer of **epidermis** or **hypodermis.** Hypodermis is externally protected by a thin chitinous **cuticle.** A gill plate is made up of single layer of cells with cuticle on both sides. These cells are of two types, *viz.*, **pigmented** and **transparent,** alternating with each other.

(*e*) **Blood supply of the gills.** A **median** and two **lateral longitudinal blood channels** run in the central axis (or base) of the gill (Fig. 39.21). The two lateral channels are connected

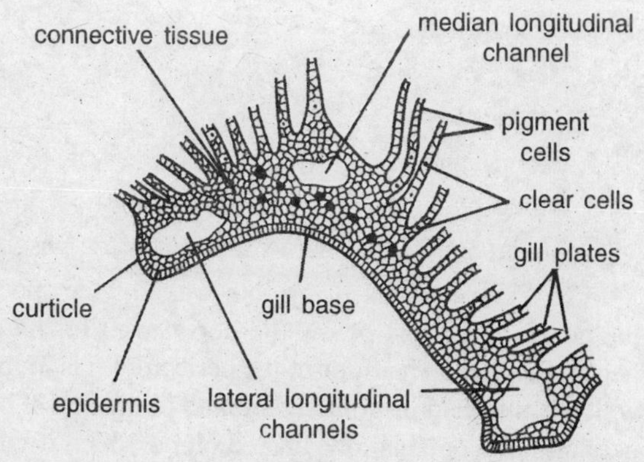

Fig. 39.20. *Palaemon*. An oblique T.S. of a gill.

by a series of **transverse channels (or transverse connectives).** These form a ladder-like structure. In each gill plate the lateral longitudinal channel gives off from its outer side a slender **marginal channel** which runs all along the free margins of the gill-plate and then joins the median longitudinal channel.

Course of blood in a gill. Deoxygenated blood to each gill is brought by the **afferent branchial channel** which empties into the transverse connective present just in front of the gill root. The blood then flows into both the **lateral longitudinal channels** and passes through the **marginal channels** and reaches the **median longitudinal channel.** During this course the blood gets oxygenated, *i.e.*, blood leaves carbon dioxide and takes up oxygen. From the median longitudinal channel of each gill all the oxygenated blood is carried by an **efferent branchial channel** to the pericardial sinus.

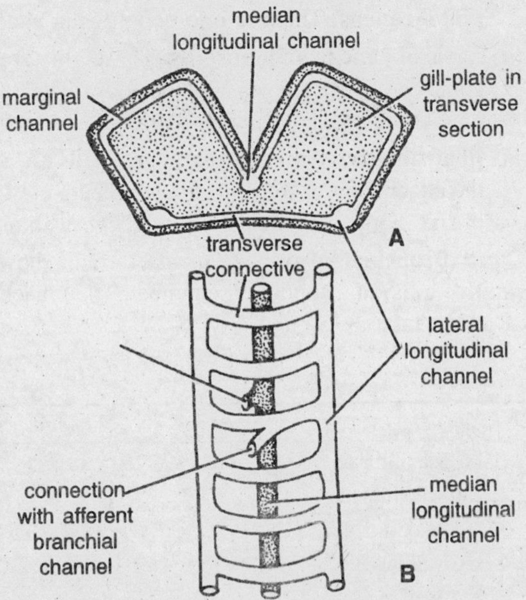

Fig. 39.21. *Palaemon.* A—Blood supply in gill-plates; B—Blood circulation in principal blood channels in a gill.

Mechanism of Respiration

The oxygen dissolved in water is utilized for respiration. A continuous current of fresh-water is maintained into the gill chamber by the constant vibrations of the scaphognathite of maxilla which bales out the used current from the anterior end of gill chamber. While the water current passes from behind forward into the gill chamber, it directly bathes the gill, epipodites and branchiostegites. The oxygen dissolved in water is taken in by blood and carbon dioxide diffuses out in the water.

39.7 BLOOD VASCULAR SYSTEM

The circulatory or blood vascular system of *Palaemon* is of ***open type,*** *i.e.*, arteries open into the cavities or **lacunae** or spaces in the body. These lacunae lacks walls. The blood vascular system of *Palaemon* consists of following parts:

1. Pericardium
2. Heart
3. Arteries
4. Blood sinuses
5. Blood channels

In *Palaemon*, capillaries and veins are absent.

1. Pericardium. It is a thin-walled wide spacious haemocoelomic chamber or **sinus** around the heart. It lies just below the dorsal wall of the thorax and above the reproductive organs and hepatopancreas. The floor of the pericardium is in the form of a thin **horizontal pericadial septum** which is attached in front and behind to the dorsal body wall and laterally to the thoracic wall. The pericardium contains oxygenated blood which it gets from the channels and passes the blood into the heart.

2. Heart. It is a triangular muscular organ with its apex in front and broad base behind. Heart is enclosed in the pericardium in the median dorsal part of thorax dorsally to the alimentary canal. It is kept in position by means of a number of muscle strands formed of muscle fibres. A median longitudinal **cardiopyloric strand** extends from the apex of the heart to the pyloric stomach and two **lateral strands,** each extending from the lateral angle of the heart to the body wall.

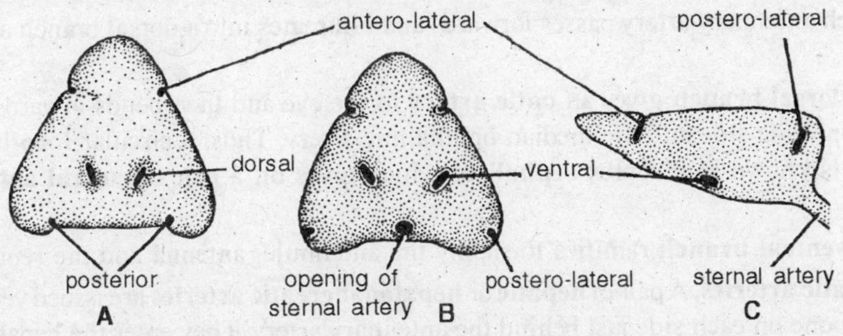

Fig. 39.22. *Palaemon*. Heart in A—Dorsal view; B—Ventral view and C—Lateral view.

The thick and muscular wall (having spongy meshwork of muscle fibres) of the heart is perforated by **five pairs** of apertures or **ostia** (*L. ostium* = a door). These ostia are located as 1. **Mid-dorsal pair,** 2. **Mid-ventral pair;** 3. **Antero-lateral pair;** 4. **Postero-lateral pair;** 5. **Posterior pair** (Fig. 39.22). Each ostium is a small, slit-like opening and its lips act as valves allowing blood to flow into the heart and not vice versa. Heart does not contain a continuous cavity.

3. Arteries

Arteries arise as thick, strong, muscular and narrow tubes from the heart to supply blood to various organs. The main arteries are as follows:

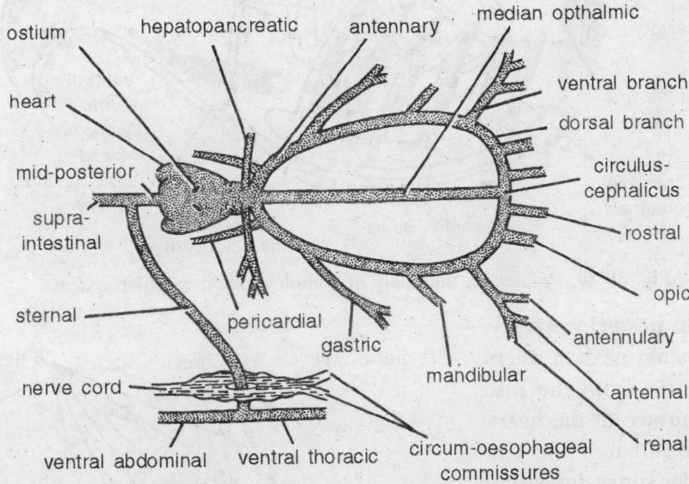

Fig. 39.23. *Palaemon*. Heart and main arteries.

1. Median ophthalmic artery. It is a single slender artery which arises from the apex of the heart. It runs forward to the head region in the mid-dorsal line above the renal sac and below the carapace. Anteriorly median ophthalmic artery reaches the roof of the oesophagus where it joins the two antennary arteries. It supplies blood to the oesophagus, cardiac stomach and the head.

***(ii)* Antennary arteries.** A pair of them originate from the apex of the heart one from either side of the median ophthalmic artery. Each runs obliquely forward along the outer border of the mandibular muscle. During its course it gives out the following three branches:

(*a*) a **pericardial branch** to the pericardium;

(*b*) a **gastric branch** to the cardiac stomach, and

(*c*) a **mandibular branch** to the mandibular muscles.

Then each antennary artery passes forwards and bifurcates into a dorsal branch and a ventral branch.

(*a*) The **dorsal branch** gives an **optic artery** to the eye and then bends inwards to meet its fellow of the opposite side and the median ophthalmic artery. Thus, a circular loop-like artery is formed which is known as **circulus cephalicus** which gives off a pair of **rostral arteries** to the rostrum.

(*b*) The **ventral branch** ramifies to supply the antennule, antenna and the renal organ.

(*iii*) Hepatic arteries. A pair of **hepatic** or **hepatopancreatic arteries** are issued ventrolaterally from the heart, one on each side just behind the antennary artery. They enter the hepatopancreatic mass and ramifies within it.

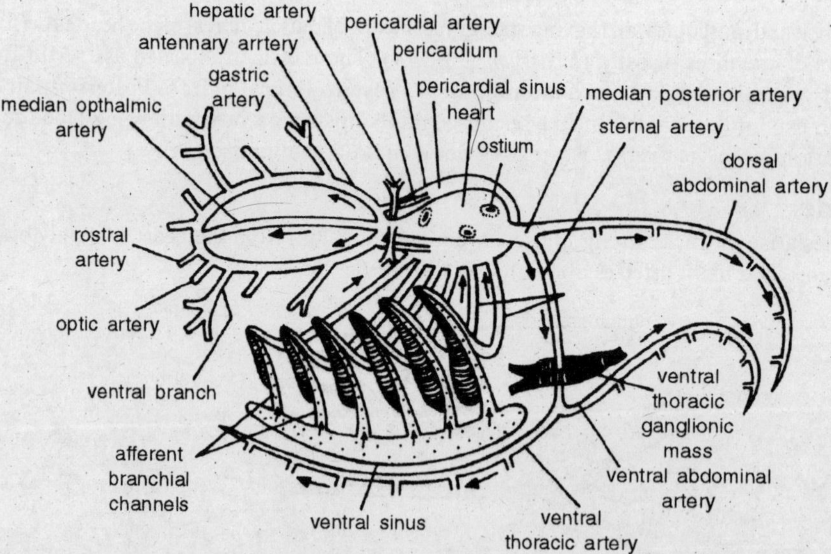

Fig. 39.24. *Palaemon.* Summary of complete blood vascular system.

(*iv*) Median posterior artery. It is a short but stout median artery running posteriorly from the mid postero-ventral surface of the heart. Immediately it bifurcates into two branches: a slender **supra-intestinal artery** and a stout **sternal artery.**

(*a*) Supra-intestinal artery. It is also called **dorsal abdominal artery.** It runs mid-dorsally above the intestine extending upto the hindgut. It supplies blood to the midgut (intestine) and the dorsal muscles of the abdomen.

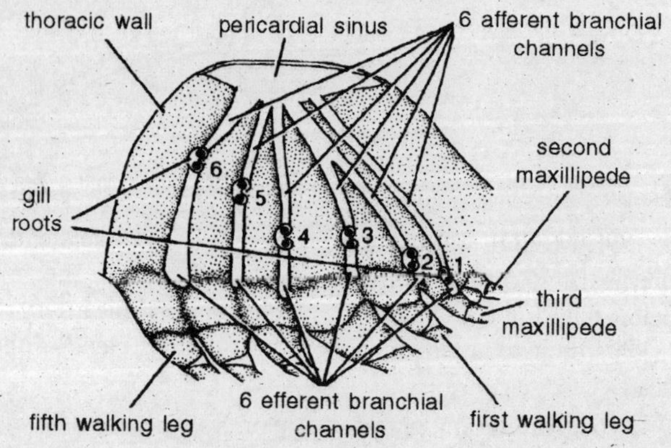

Fig. 39.25. *Palaemon.* Branchial channels and pericardial sinus.

(*b*) **Sternal artery.** It is stoutest artery in the blood vascular system of *Palaemon*. It runs downward, traverses the hepatopancreatic mass, pierces the ventral thoracic mass and finally comes to lie above the sternum (*i.e.*, below the ventral nerve cord). Here it bifurcates into the thoracic and abdominal branches. The **ventral thoracic branch** runs forward upto the mouth and supplies the sternal region, the maxillulae, maxillae and the first three pairs of walking legs. The **ventral abdominal branch** runs backwards and supplies blood to the ventral region of abdomen, the last two pairs of walking legs, the abdominal appendages and the hindgut.

4. Blood Sinuses

Various arteries branch repeatedly in vrious organs of body and finally open into **blood sinuses** or **lacunae** of the haemocoel. These lacunae lack walls and exist as large cavities among the muscles and visceral organs. All the sinuses of the body eventually meet into a pair of elongated and ill-defined **ventral sinuses** which are situated below the hepatopancreas and above the floor of the thorax. The ventral sinuses communicate with each other at various places.

5. Blood Channels

Blood channels are the lacunar tubes lacking in definite walls. From each ventral sinus blood is carried to the gill of that side by means of six **afferent branchial channels** (Fig. 39.25). As there are two ventral sinuses, the blood channels of both sides are six pairs.

The channels run upwards from the ventral sinus along the inner side of the lateral thoracic wall and enter the gill through gill roots. The first afferent branchial channel supplies the podobranch and arthrobranchs, while the remaining five channels supply blood to the pleurobranchs. Inside the gill, blood flows in following direction — transverse connective → two lateral longitudinal channels → marginal channels → median logitudinal channels (see Fig 39.26). Oxygenated blood is conveyed from the gills to the pericardium by six pairs of **efferent branchial channels.** These channels leave the gills through their gill-roots.

Blood

The blood of *Palaemon* is a clear fluid. It contains colourless amoeboid corpuscles, the **leucocytes** which are phagocytic in nature. The fluid or the **plasma** contains the respiratory pigment, the **haemocyanin.** Haemocyanin is a compound of copper and protein. It becomes bright blue when combined with oxygen, and colourless when it combines with carbon dioxide. The blood has a remarkable power of coagulation or clotting. If an appendage is injured or cut, the blood clots at once. Oxygen, carbon dioxide, digested food and waste products are transported through the blood.

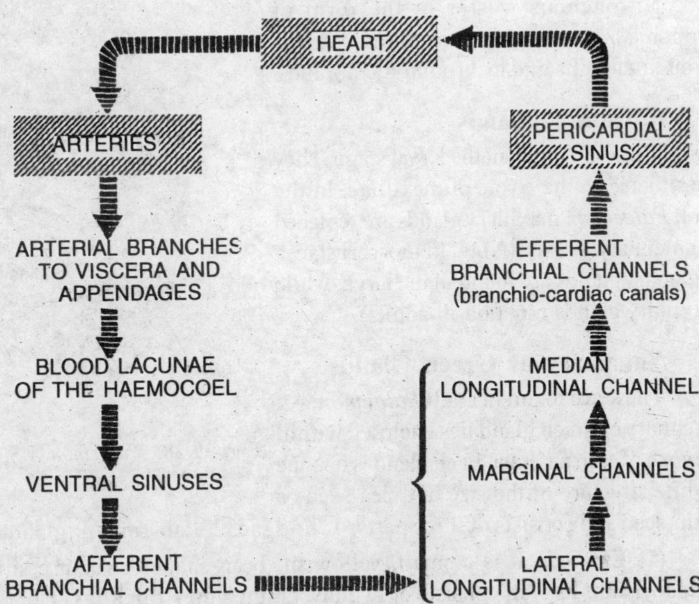

Fig. 39.26. *Palaemon.* A diagrammatic representation of the course of circulation of blood.

Course of Circulation of Blood

Due to rhythmic contraction of the heart, the blood is pumped out of the heart into the arteries which carry the oxygenated blood to all parts of the body. The course of circulation of blood in the body of *Palaemon* has been diagrammatically represented in Fig. 39.26.

39.8 EXCRETORY SYSTEM

Palaemon is an aquatic animal. Its nitrogenous wastes includes ammonia, uric acid and other nitrogen-containing compounds. Excretory system of *Palaemon* comprises the following parts:

1. Exoskeleton
2. Gills and arthrodial membrane
3. Maxillary glands
4. Antennary or green glands
5. Lateral ducts
6. Renal sac

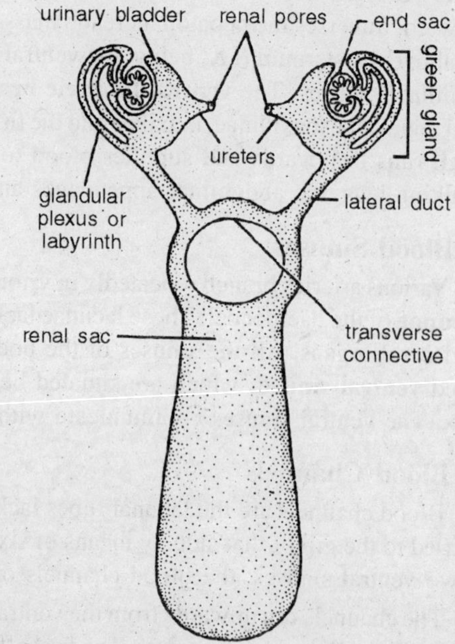

Fig. 39.27. *Palaemon*. Excretory organs: green glands, lateral ducts and renal sac.

1. Exoskeleton

The non-living exoskeleton of thick chitinous cuticle is a nitrogenous product secreted by the hypodermis. It is shed during moulting. In this way animal get rid of accumulated nitrogenous wastes in the integument.

2. Gills and Arthrodial Membrane

Nitrogenous wastes in the form of ammonia and ammonium ions are excreted through the gills and arthrodial membranes.

3. Maxillary Glands

These are found in the larval stage. They are situated in the coxae of maxilliae. In the adult *Palaemon*, maxillary glands are replaced by antennary glands. (**Note.** In most crustacea antennary glands are found in the larva, while maxillary glands function in adults).

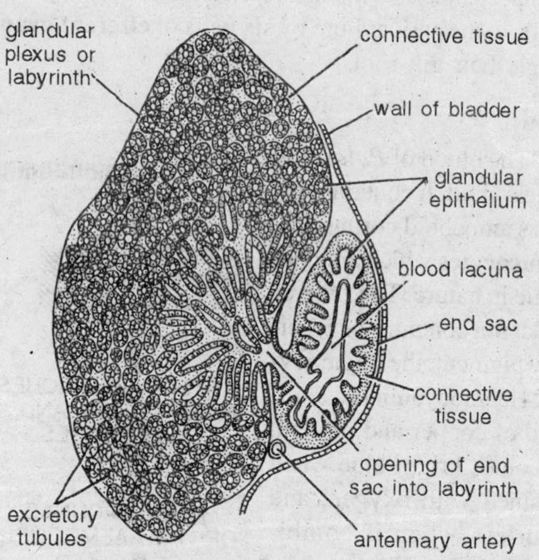

Fig. 39.28. *Palaemon*. T.S. through an antennary or green gland.

4. Antennary or Green Glands

These are found in adult *Palaemon*. An antennary or green gland lies enclosed within the coxa of each antenna. Each gland is opaque - white structure of the size of a pea-seed or gram-seed and consists of three parts: 1. End sac; 2. Labyrinth or glandular plexus, and 3. Bladder.

(1) End sac. It is a small, yellowish, bean-shaped structure situated between the bladder and labyrinth. Its wall consists of two layers, the **outer thick layer** of connective tissue and **inner thin layer** of excretory epithelium. The outer layer contains many blood lacunae. The excretory epithelium is produced into a number of **radial septa** which project into the central cavity of the end sac. The cavity of end sac contains a large **blood lacuna** which communicates with the labyrinth by a single aperture. The aperture is provided with a sphincter. The end sac serves as a filtering aparatus.

(2) Labyrinth or glandular plexus. It is relatively larger in size and green in colour. It is situated on outer side of end sac. Glandular plexus contains a mass of highly convoluted and branched excretory tubules. These tubules are embedded in the connective tissue which is richly supplied with blood lacunae.The tubules are narrow and are lined by excretory epithelium and collectively opens into the end sac by a single aperture and into bladder by many openings. Labyrinth serves as a site of reabsorption.

(3) Bladder. It is largest part of green gland. It is a thin-walled, spacious sac, lying on the inner side of the end sac. Inner wall of bladder is produced into a short **ureter** which opens to the exterior by a **renal pore.** The renal pore is located on a papilla on the inner side of coxa of the antennae. Bladder is lined by a single layer of excretory epithelium. Bladder serves as urinary bladder for temporary storage of urine.

Physiology of Antennary or Green Glands

The green glands are complex nephridia-like and perform two important functions like the vertebrate kidneys:

1. Elimination of nitrogenous waste products. The excretory epithelium of end sac is in direct contact with the blood contained in the lacuna. It collects the waste nitrogenous products from the blood. The chief excretory products are **ammonia** and **amines.** All the extracted waste is conveyed to the excretory ducts of labyrinth where selected reabsorption occurs and useful substances seprated from the filterate are reabsorbed by the blood lacunae present in the connective tissue. The urine is formed and passed on to the bladder and is expelled out of body from time to time.

2. Osmoregulation. Large quantities of water are separated from the blood by the antennary or green glands to maintain an osmotic equilibrium of body.

5. Lateral Ducts

From the bladder of each antennary gland runs posteriorly a narrow lateral duct. The lateral ducts of both sides are connected by a **transverse connective** in front of brain. These lateral ducts open into renal sac.

6. Renal Sac

Renal sac or **nephriodioperitoneal sac** is a large median, elongated, thin-walled sac situated above the cardiac stomach and just below the carapace in cephalothorax. Renal sac is lined by single cell thick excretory epithelium of flattened cells. Renal sac collects excretory matter from the posterior part of body.

39.9 NERVOUS SYSTEM

Nervous system of *Palaemon* consists of (1) Central nervous system; (2) Peripheral nervous system; and (3) Sympathetic nervous system.

A. Central Nervous System

It consists of (1) **Brain** or a pair of **supra-oesophageal ganglia;** (2) a pair of **circum-oesophageal commissures;** (3) **ventral thoracic ganglionic mass;** and (4) a ventral **nerve cord.**

1. Brain or supra-oesophageal ganglia. The brain is bilobed structure which lies at the base of rostrum anterior to the oesophagus. It is surrounded by a thick mass of fat. It is formed by the fusion of several pairs of ganglia. On each side brain gives off following nerves:

(*i*) **Antennulary nerve.** It arises from below the origin of optic nerve and enters the antennule of its side. Inside the antennule, it sends a **statocystic branch** to the statocyst.

(*ii*) **Optic nerve.** It is a stout nerve which arises from the dorsal surface of the brain and supplies to the compound eye.

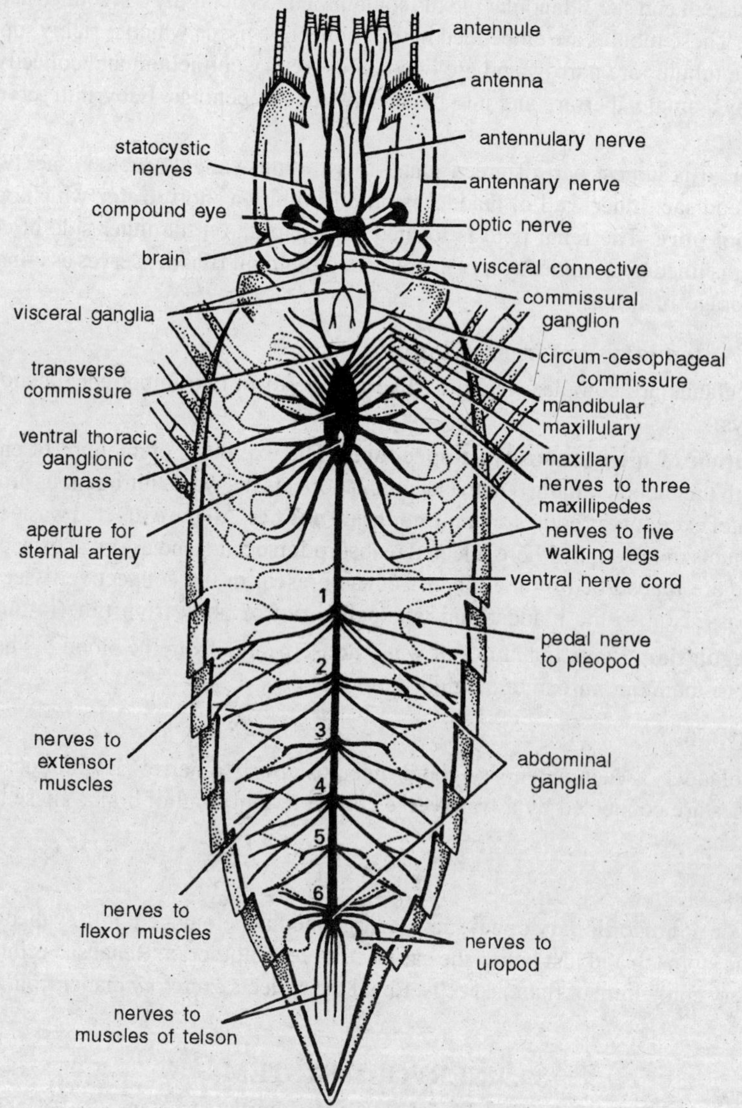

Fig. 39.29. *Palaemon.* Nervous system in dorsal view.

(*iii*) **Ophthalmic nerve.** It arises from the brain and supplies to ocular muscles in the eye-stalk.

(*iv*) **Antennary nerve.** It is a stout nerve which arises from the ventral surface of the brain. It supplies to the antenna.

(*v*) **Tegumental nerve.** It is a slender nerve which arises just behind the antennary nerve. It supplies to the labrum.

2. Circum-oesophageal commissures. These are two stout nerves, which arise from the posterior end of brain. These run backwards and downwards around the oesophagus and unite ventrally with **sub-oesophageal ganglia.** These ganglia form the indistinguishable anterior part of the **ventral thoracic ganglionic mass.** Each commissure bears a small **commissural ganglion** near its anterior end and gives off a small nerve to the mandible of its size.

The two circum-oesophageal commissures are connected with each other by a slender **transverse commissure** near the posterior end.

3. Ventral thoracic ganglionic mass. It is an elongated oval structure situated immediately above the thoracic sternal plates in the mid-ventral line. It is a composite mass which is formed as a result of fusion of **eleven pairs** of segmental nerve ganglia of the cephalothorax. It gives off eleven pairs of nerves on the lateral sides. The first three pairs are **cephalic nerves,** supplying the mandibles, maxillulae and maxillae. The last eight pairs are **thoracic nerves,** of which the first three pairs supply the three pairs of maxillipedes and the remaining five pairs supply the five pairs of walking legs. Each nerve to a leg becomes bifurcated before entering the leg.

4. Ventral nerve cord. The ventral thoracic ganglionic mass extends posteriorly to form a stout **ventral** or **abdominal nerve cord.** It runs along the mid-ventral line of the abdomen and contains an **abdominal ganglion** in each abdominal segment. The ventral nerve and abdominal ganglia are double.

Each of the first five abdominal ganglia gives off three pairs of nerves in its segment—(*i*) a pair of **pedal nerves** to the pleopods; (*ii*) a pair of nerves to the extensor muscle, and (*iii*) one pair of nerves to the flexor muscles of the succeeding segment. The last or **sixth abdominal ganglion** is comparatively large and is formed due to fusion of several ganglia. It supplies two pairs of nerves to the flexor muscles of the sixth segment, two pairs of nerves to uropods, two pairs of nerves to the telson and a single nerve to the rectum and hindgut.

B. Peripheral Nervous System

The various nerves orginating from the central nervous system, as discussed above, to innervate the different parts of the body form the peripheral nervous system.

C. Sympathetic Nervous System

The **sympathetic, visceral** or **autonomic** nervous system comprises of a few small ganglia and nerves. A small nerve arises from the mid-posterior part of the brain and bears two **visceral oesophageal ganglia,** one anterior and another posterior. The anterior viceral ganglion is joined with two **commissural ganglia** by a pair of connectives. The posterior visceral ganglion is free and gives off two pairs of nerves to the muscles of the wall of oesophagus and the cardiac stomach.

39.10 SENSE ORGANS

Palaemon has following two main sense organs.

1. Compound Eyes

Palaemon consists of a pair of black and hemispherical compound eyes. Each eye is borne on a short two-jointed movable **eye stalk** lying in an orbital notch. Eye stalk belongs to a presegmental region of cephalothorax.

Structure. Each eye of *Palaemon* is a composite structure. It is composed of a great number of independent **visual** or **optical units,** called **ommatidia;** each of which is able to form a separate image of the part of the object present in front of it. The ommatidia are all alike in structure and are arranged radially along the long axis of eye (Fig. 39.30).

The eye is protected externally by a transparent covering of cuticle known as **cornea.** It is divided into a large number of square **facets** like those of graph paper. Beneath each facet lies an ommatidium. All ommatidia are simple and are arranged radially lying side by side and separated by dark pigment cells.

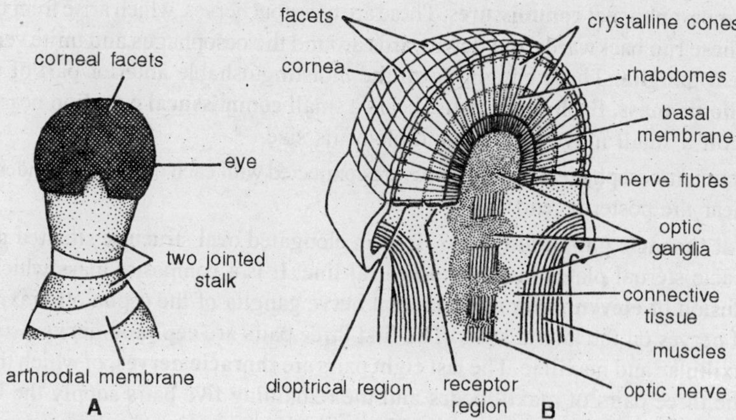

Fig. 39.30. *Palaemon.* A—Stalked compound eye; B—A diagrammatic L.S. of compound eye.

Structure of ommatidium. Each ommati-dium consists of a number of cells arranged in a linear order and comprises the following parts:

1. Cornea. It is outermost part of the ommatidium. It is the corneal facet which is thickened in the centre to form a translucent **biconvex lens.** Cornea is made of chitinous cuticle of body wall; it sheds of at the time of moulting and is again secreted by the underlying cells.

2. Corneagen cells. Below each cornea occurs a pair of **corneagen cells.** These are modified epidermal cells and secrete new cornea when the old one is cast off during moulting.

3. Cone cells or vitrelle. These are a group of four much elongated cells situated beneath the corneagen cells. Cone cells secrete and enclose a transparent or translucent **crystalline cone** which works like a **second lens.** Distal end of the cone cells are long and tapering. The **crystalline cone** serves as an optic fibre through which light is channeled to the rhabdome.

The cornea, corneagen cells and cone cells together constitute the **dioptrical region.** This region of eye is concerned with the focussing of light rays falling from the object on the inner sensitive region of eye.

4. Retinal or retinular cells. The conc cells are followed by group of seven elongated photoreceptor or **retinular cells** forming the proximal part of the axis of an ommatidium. Each retinal cell acts as a single photoreceptor unit.

5. Rhabdome. It is an elongated, spindle-

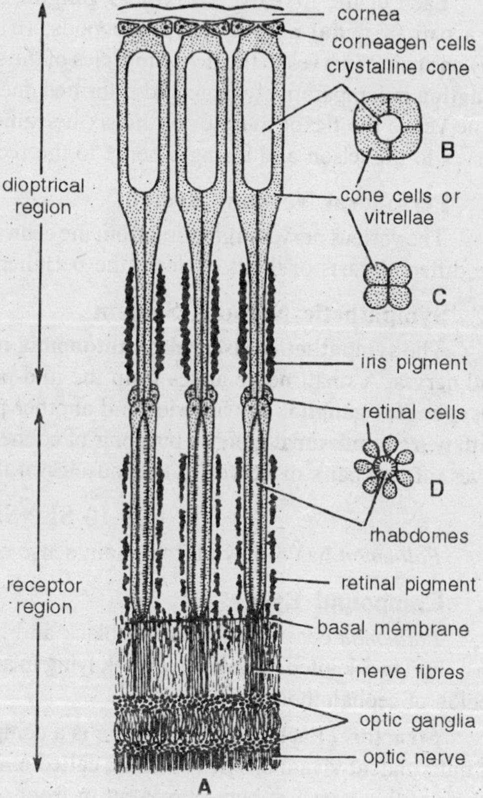

Fig. 39.31. *Palaemon.* A—Three ommatidia in L.S.; B—T.S. of an ommatidium through cone cells; C—T.S. through basal end of cone cells; D—T.S. through retinal cells and rhabdome.

shaped and transversely striated, translucent cylinder or rod which is secreted and fully enclosed by the retinal or retinular cell (Box 39.3).

6. Basement membrane and optic nerve. The retinal cells and the rhabdome rest on a **basement membrane.** A nerve fibre arises from the base of each retinal cell. Thus, a bundle of seven nerve fibres penetrates the basement membrane and makes connection with the optic ganglia within the eye stalk. A stout **optic nerve** connects the optic ganglia with the brain.

The retinal cells and rhabdome upto the basement membrane constitute the **receptor region.** Its function is to receive the light rays focussed by the dioptrical region and to form the image.

Pigment sheath. Each ommatidium remains separated from its neighbours by a sheath of movable, amoeboid, pigment cells, called **chromatophores.** The pigment cells are arranged in two series. The outer or distal series is called the **iris pigment sheath.** The inner, basal or proximal series surrounds the retinal (or retinular) cells. It is called **retinal** pigment sheath. (**Note.** According to **Ruppert** and **Barnes** (1994) the retinular cells contain black or brown pigment granules, which constitute the proximal retinal pigment). Both proximal and distal pigment cells can migrate according to variation in intensity of light.

Function of compound eye: mosaic vision. Formation of image in a compound eye is called **mosaic vision.** Each ommatidium of the compound eye produces a separate image of a small part of the object present just infront of it. The final image of the object is formed by putting several small pieces together like a mosaic painting in Roman bath or mosaic floor in which separate pieces of same size but of different or same colours are put together. In fact, the image formed by the compound eye is similar to the image produced on a television screen, on which the picture is formed by small dots of light. The apposition image formed by a compound eye is often called a **mosaic image.**

In diurnal crustaceans (and other arthropods such as butterflies, houseflies, bees and wasps), the compound eyes are adapted for bright light and it produces an **apposition** or **mosaic image.** But in nocturnal forms such as *Palaemon* (and moth, fireflies, crabs, lobsters, etc.) it is adapted for seeing in weak light and a **superposition image** is formed.

Box 39.2 Transverse striations of rhabdome

The inner surface of the retinal or retinular cells are light-senstitive and are called **rhabdomeres.** In fact, the light-sensitive surfaces are thrown into **tubular** folds or **microvilli** which are oriented at right angles to the axis of ommatidium (Fig. 39.32). The rhabdomeres fuse to form a cylindrical, translucent rhabdome. The tubular folds provide transverse striated appearance to the rhabdome.

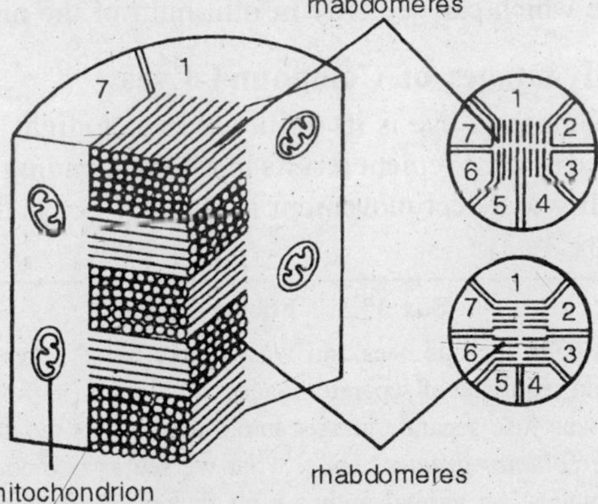

Fig. 39.32. Rhabdome of compound eye of a crayfish. Stereodiagram showing perpendicular orientation of microvilli projecting from two retinular cells. T.S. of two levels showing organisation of the rhabdomeres of seven retinular cells of one ommatidium of the compound eye.

1. Formation of apposition or mosaic image. In the **bright light,** during the day time, the pigment cells spread and completely isolate an ommatidium. As a result, light from an ommatidium cannot pass to another ommatidium. In this condition, the light rays which reach the cornea at an angle (or obliquely) are absorbed by the pigment sheath. These light rays fail to form image. The light rays which fall perpendicularly upon the cornea, travel through the crystalline cone and reach the rhabdome and the image of point of the object is formed (Fig. 39.33A).

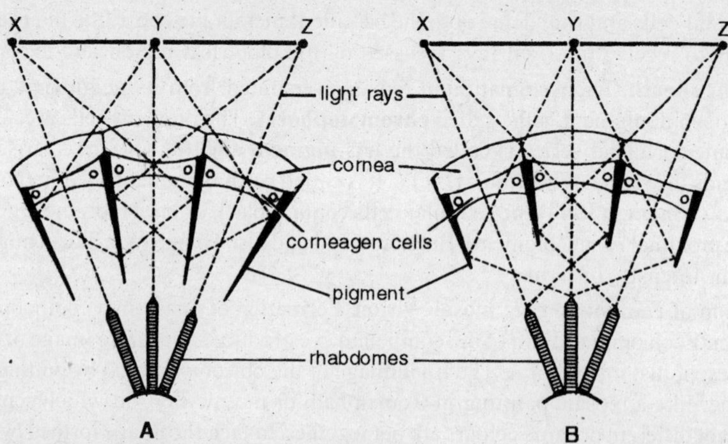

Fig. 39.33. *Palaemon*. Diagrammatic representation of image formation by a compound eye. A—Apposition or mosaic image; B—Superposition image.

2. Formation of superposition image. In the dimlight the pigment cells migrate (retract) towards the distal and basal parts of the ommatidia. In this condition, the neighbouring ommatidia work in unison. The oblique light rays can also form a point of image after passing through a number of ommatidia in their way. As a result, an overlapping of points of images takes place and a continuous image is obtained. This is called **superposition image.** It is not sharp and the animal gets only a sort of idea of the moving object in the surrounding.

Night-blind and day-blind insects. In some insects the eyes are set permanently in former or latter conditions described above. The individuals which can see only in bright light of day time are **night-blind** and those which can see only in dim light of the night are **day-blind.**

Advantages and Disadvantages of Compound Eyes

1. The advantage of compound eye is its ability to detect slight movements. A little shift in a point of light caused by a movement results in a corresponding shift in the ommatidia being stimulated. The ability to detect movement in a honey bee is three times greater than that of a human (see Box 39.3).

Box 39.3. Flicker fusion

Many active, diurnal arthropods such as bees and wasps have high flicker fusion rates. **Flicker fusion** refers to the fusing of a rapid sequence of separate images into one, as in motion picture. At approximately 16 frames per second, humans fuse separate images into a continuous moving image, but for honey bees the *flicker fusion rate is three times greater*. Thus, what we see as one continuous image a bee sees as a succession of separate images. An animal with a high flicker fusion rate can detect short increments of motion more readily than for which the fusion rate is low. A short motion in space or time might be imperceptible to us, but to a bee it might be captured in several separate images and thereby detected.

2. Another advantage of the compound eye is that the corneal surface is greatly convex, resulting in a wide visual field. This is particularly true for the stalked, compound eyes of crustacean (*Palaemon*), in which the cornea may cover an arc of 180 degrees or more.
3. The main disadvantage of compound eye is that they are not very good for distance vision compared with the large eyes of birds and mammals. So most arthropods are **short sighted.** But what matters to most arthropods is the world at close range— within 20 cm around them.
4. Images formed by compound eyes are not sharp, they are crude at best, for tne small size of the eye limits the number of photoreceptor units (ommatidia).
5. The smallest arc of the visual field subtended (*i.e.*, the side of a triangle opposite to an angle) by an ommatidium in the honey bee is 1 degree compared with 1 minute of arc (1 degree = 60 minutes) in much larger human eye. Thus, a row of closely spaced vertical bars would appear as a continuous horizontal bar to a bee.

2. Statocyst

The statocysts are a pair of small white, cuticular, bead-like or sac-like structures. Each statocyst lies inside the **precoxa** of each **antennule,** attached to its dorsal wall. The base of each statocyst is slightly depressed and the neck opens to the outside by a minute aperture situated on the concave roof of precoxa of antennule. The opening remains covered by a minute fold of body wall. The statocyst is innervated by a small **statocystic branch** of the antennulary nerve. The statocyst originates as an invagination of body wall (*i.e.*, it is ectodermal in origin).

Structure. The cavity of statocyst contains minute sand particles. On the floor of the statocyst there is an oval ring of elongated delicate **receptor setae.** Each receptor seta is attached to the inner wall of statocyst and is innervated by minute fibres of statocystic nerve. The seta consists of a swollen base and a long tapering **shaft** directed toward the sand grains. The shaft is bent in middle and its inclined part bears fine bristles (Fig. 39.34B).

Since the statocyst is formed by invagination of body wall, its lining is formed of chitinous and calcified cuticle. During moulting, the lining of statocyst and sand grains are also shed. The sand grains are replaced when the head is buried in sand or the animal actually inserts sand grains into the statocyst.

Functions. The statocysts are organs of balance. They detect the change in animal's position during swimming. In fact, they detect the force of gravity and functions as the organs of **orientation**

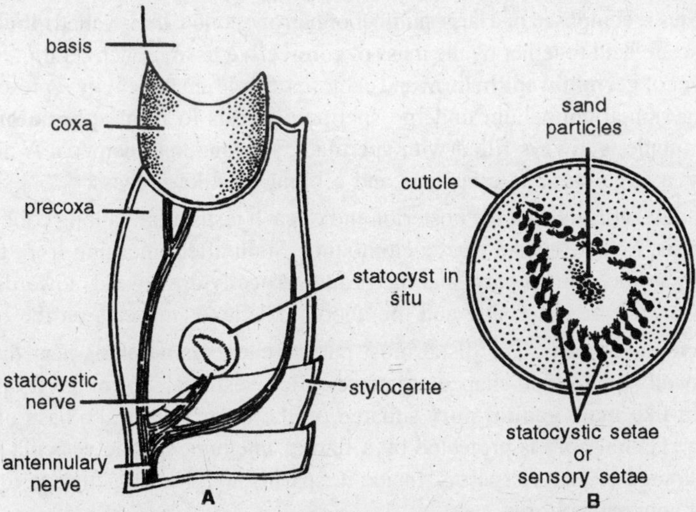

Fig. 39.34. *Palaemon*. A—Statocyst in situ; B—T.S. of statocyst.

and **equilibrium** like the semilunar canals of internal ear of vertebrates. The sand grains function as statolith.

Any change in position of swimming animal causes shift in the position of sand grains. The sand grains press some of the sensory setae and they stimulate them. This stimulus is conveyed to the brain through nerves and the animal corrects its position.

39.11 REPRODUCTIVE SYSTEM

Palaemon is **dioecious**, *i.e.*, sexes are separate and **sexual dimorphism** is well marked. Male and female individuals show the following differences:

1. The male is bigger than the female of same age.
2. Abdomen of male is narrower than that of a famale.
3. The second chelate legs of male are larger, stouter and more spiny than that of a female.
4. The second pleopod of male contains an additional process, the **appendix masculina,** between the endopodite and appendix interna. It is not found in the female.
5. The epimera of abdominal segments in the female are larger than those of male. Large epimera are adapted for carrying eggs.
6. Paired genital openings in the male lie on the coxae of 5th pair of thoracic legs. In females the oviducts open on coxae of 3rd pair of walking legs.

Male Reproductive Organs

The male reproductive organs consist of —

1. A pair of **testes**
2. A pair of **vasa deferentia**
3. A pair of **vesiculare seminales**

1. Testes. The two testes are soft, white, elongated structures lying above the posterior half of the dorsal surface of the hepatopancreas and below the pericardial sinus and the heart. The two testes remain fused at their anterior ends. There is a narrow gap in between the middle and posterior parts of the two testes. The **cardio-pyloric strand** passes through this gap to connect the heart with cardiac stomach.

Each testes is composed of a large number of narrow, coiled, thin-walled **lobules** or **seminiferous tubules.** These are held together by the mass of **connective tissue.** Each seminiferous tubule is lined by a single layer of **germinal epithelium** and encloses a large central cavity representing **true coelom.** The cells of germinal epithelium undergo spermatogenesis to form **spermatozoa.** The lumen of seminiferous tubule is always filled with spermatocytes and spermatozoa. A mature sperm has a rounded body, a large crescentic nucleus and a blunt tail-like process.

2. Vasa deferentia. From the posterior end of each testis arises a long coiled and narrow tube, the **vas deferens.** It represents a true coelomoduct. Soon after emerging from the testis, each vas deferens forms a much coiled mass and then runs vertically downwards towards fifth leg between the thoracic wall on the outer side and the abdominal flexor muscles on the inner side.

3. Vesicula seminalis. The distal most part of each vas deferens near the base of the fifth walking leg swells into a club-shaped structure called **vesicula seminalis.** It opens to exterior by means of a slit-like **male genital pore** situated on the inner side of the base of fifth thoracic leg of its side. The genital pore is protected by a flap of integument. The vesicula seminalis are used as organs of storage of mature sperms. Inside it, sperms are glued together to form white compact bodies, called **spermatophores.**

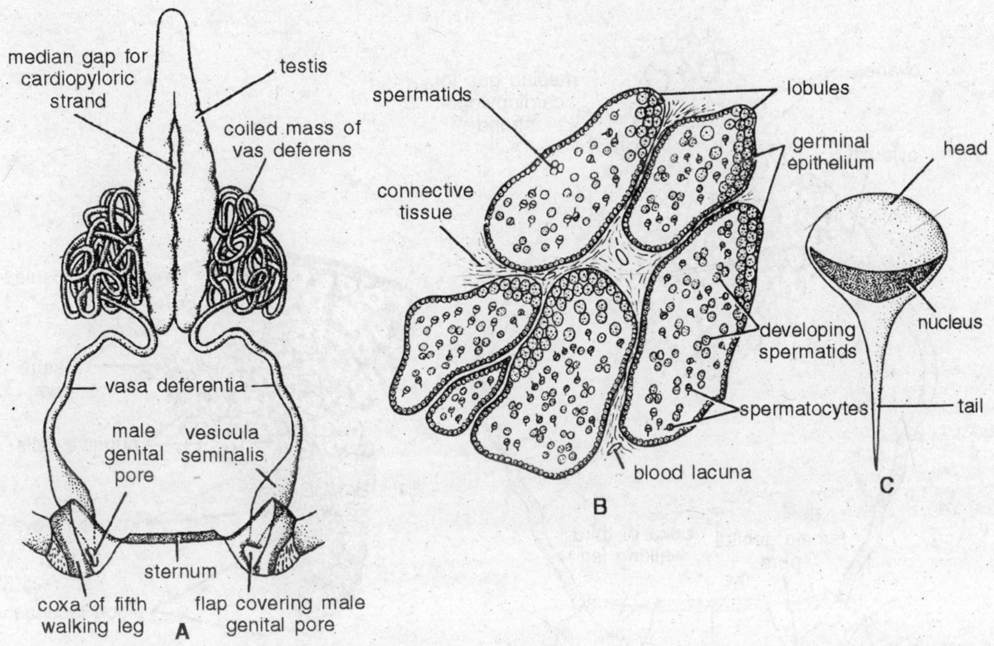

Fig. 39.35. *Palaemon* A—Male reproductive organs in dorsal view; B—T.S. of a portion of testis; C—A mature spermatozoa.

Female Reproductive Organs

The female reproductive organs consist of—

1. A pair of **ovaries**
2. A pair of **oviducts**

1. Ovaries. The ovaries in female occupy the same position as testes in male. The two ovaries are white, compact and sickle-shaped structures. They are situated above the hepatopancreas and below the pericardial sinus and heart. Both ovaries touch each other at the two ends but are widely separated in the middle by a gap for the passage of cardiopyloric muscle strand. The shape and size of ovaries vary with the age and season of the year. In the breeding season, the ovaries grow in size and extend up to the first abdominal segment.

Each ovary consists of numerous rows of ova arranged radially and enclosed in a membranous capsule. The ova are arranged in a definite sequence under different stages of development. The immature ova lie towards the centre, while the mature ova occur towards the periphery. Each developing ovum is being surrounded by a number of small nutritive cells. The mature ova are large nucleated cell with cytoplasm full of yolk granules; yolk forming a central core (centrolecithal eggs). In the centre of ovary, there is a cavity (a true coelom) which contains blood lacunae.

2. Oviducts. From the middle of the outer surface of each ovary arises a short, thin-walled tube called **oviduct.** Its anterior end is broad and forms the **oviducal funnel** which lies opposed to the surface of ovary. The oviduct runs vertically downwards to open to the exterior by **female genital pore.** It is situated on the inner side of the coxa of the third thoracic walking leg of its side. Oviduct is a true coelomoduct.

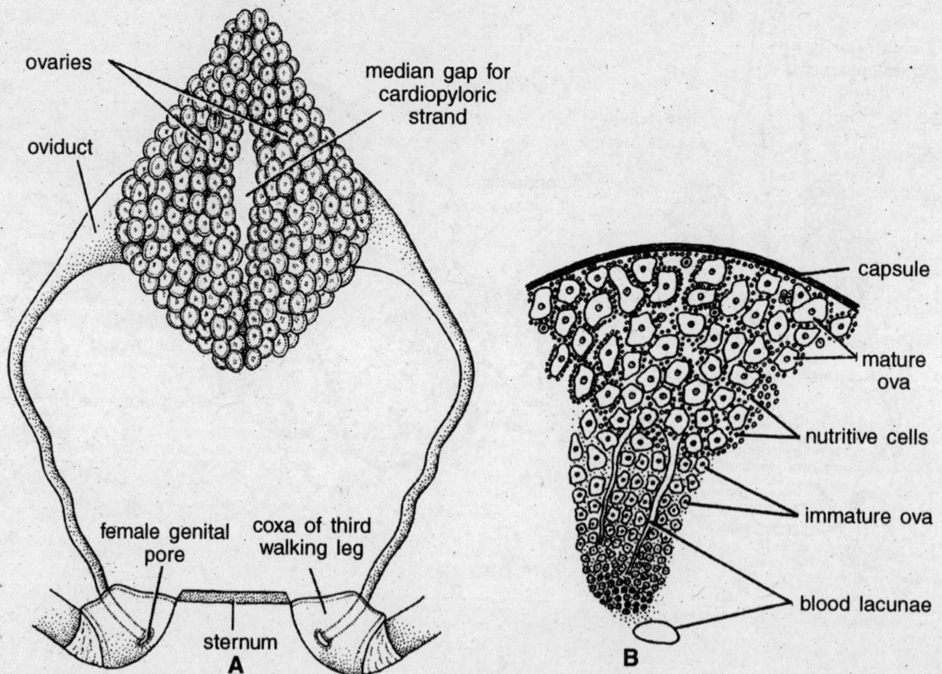

Fig. 39.36. *Palaemon* A—Female reproductive organs in dorsal view; B—T.S. of a portion of ovary.

Fertilization and Development

Fertilization. The breeding season in *Palaemon malcolmsonii* is during the months of May, June and July. Fertilization is **external.** During copulation, a male puts the female down on her back and deposits sperms (spermatophores) on the ventral surface near the female genital pores. As the eggs come out of female genital pore, they are fertilized by the sperms already present there. The female lays eggs in large numbers. The fertilized eggs are attached to the pleopods by a sticky secretion of the tegumental glands.

Development. In *Palaemon*, development is **direct** and the embryos remain attached to the pleopods until they hatch. The newly hatched young offspring looks like the adult. The adult form is reached after a series of moults.

39.12 LARVAL FORMS OF CRUSTACEA

The development of crustacea is either **direct** or **indirect.** The development is direct in case of large eggs with enough yolk and indirect in case of small eggs with a little amount of yolk. For example, some crustaceans (*e.g.*, *Palaemon*, *Astacus*, etc., develop directly from the egg and offsprings

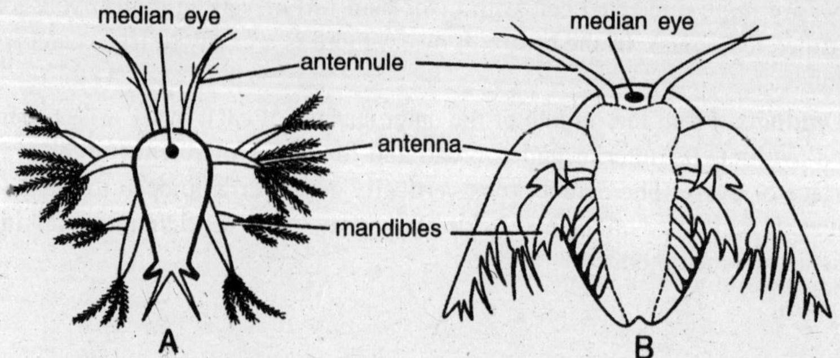

Fig. 39.37. Nauplius larva. A—Nauplius of *Penaeus*; B—Nauplius of *Apus*.

are very similar to the adult. Therefore, there is no larval stage and no metamorphosis. However, in some other crustaceans (*e.g.*, *Nebalia*, *Mysis*, etc.) development is direct but metamorphosis occurs inside the egg case. In majority of crustaceans, development is indirect being accompanied by little or more metamorphosis. In such a case, the individual hatching out of the egg is drastically different from the adult. It undergoes a series of developmental changes to acquire the adult organisation. These different young or immatue stages are known as **larvae.** In crustacea, the following larval stages have been studied:

1. Nauplius larva. Nauplius larva is the first (*i.e.*, it hatches out of the egg) and most common free-swimming larval stage in many crustacea (*i.e.*, Ostracoda, Copepoda, Branchiura, *Cirripedia*, etc). It is microscopic, oval or pear-shaped and unsegmented. Its body has a broad anterior **head** region, middle **trunk** region and bilobed **anal** region. Head bears a single sessile median eye called **nauplius eye** and a pair of **antennules** which are uniramous and unbranched. Nauplius eye is made of 3 or 4 ocelli which are pigmented cups with no lens. Antennules bear two groups of sensory cells forming the **frontal organs.** Trunk has two pairs of biramous appendages (called **feet**), *i.e.*, **antennae** and **mandibles.** Mouth opens at the anterior end, while **anus** lies on the posterior extremity. Alimentary canal is a straight tube.

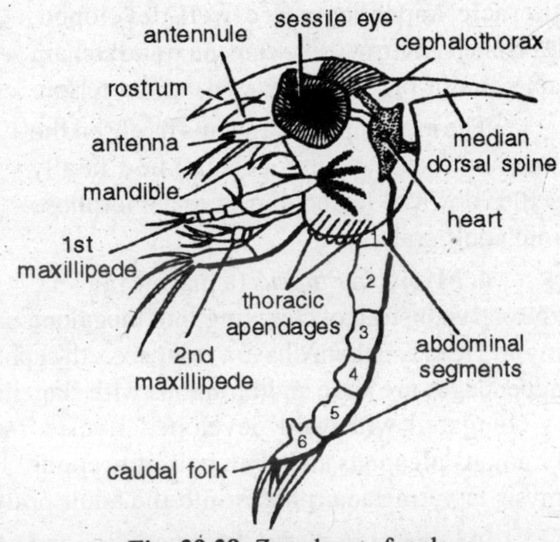

Fig. 39.38. Zoea larva of crab.

In some crustaceans (*e.g.*, most Malacostraca), nauplius exists within egg membranes and develops into the adult, but in some it passes through many interme-diate larval forms (such as **metanuplius, protozoea, zoea, cypris, mysis, megalopa, phyllosoma, alima,** etc.).

2. Zoea larva. It is fourth larval stage of the crustacea. Nauplius changes into metanauplius, metanauplius change into porotozoea and protozoea changes into zoea larva. In crabs, however, the young is hatched at the zoea stage.

Zoea is a free-swimming stage. Body of zoea comprises a large unsegmented **cephalothorax** and **long segmented abdomen.** The cephalothorax is covered with a helmet-shaped **carapace** which contains two long spines—an anterior **rostral** and a median **dorsal spine.** Two lateral spines are also present. Cephalic region contains a pair of large, **sessile compound eyes.** There are **five** pairs of well-developed head appendages (*i.e.*, antennules, antennae, mandibles, maxillulae and maxillae) and two pairs of thoracic appendages (*i.e.*, two pairs of maxillipedes). Abdomen is six-segmented and devoid of appendages. Last abdominal segment, however, bears caudal fork.

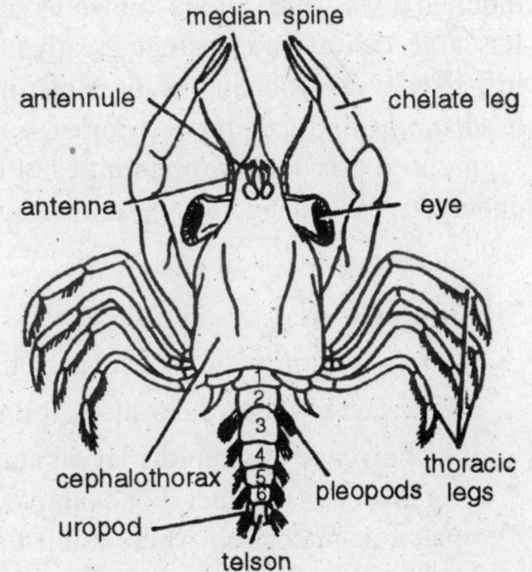

Fig. 39.39. Megalopa larva of crab.

In *Squilla*, zoea is slightly modified and is known as **alima larva.** The zoea after passing through successive molts becomes megalopa larva.

3. Megalopa larva. It is the larva of crab and it develops from zoea larva through successive moults. It has a broad crab-like cephalothorax covered by a dorsal shield of carapace. Carapace contains an anterior median **rostral spine.** The paired compound eyes are large and stalked. The cephalic and thoracic appendages are well-developed. Abdomen is narrow, six-segmented and contain six pairs of biramous pleopods and a telson.

The megalopa larva swims freely on the water surface (*i.e.*, it is pelagic) and finally settles down at the bottom to metamorphose into adult crab.

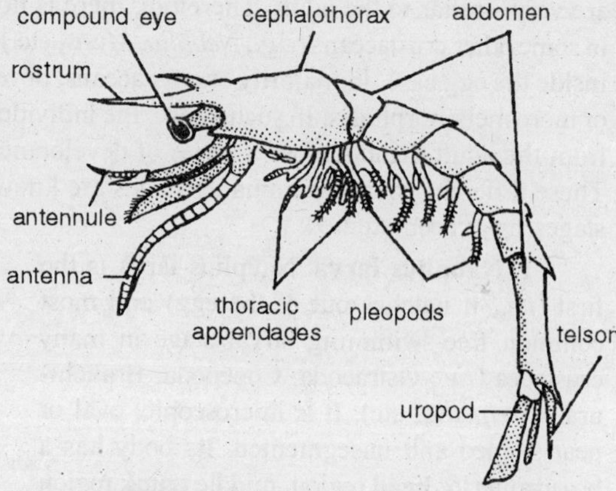

Fig. 39.40. Mysis larva of *Penaeus.*

4. Mysis. In *Pinaeus* (a marine prawn), zoea larva instead of changing into megalopa changes into **mysis** which is very similar to the adult mysis. Head and thorax have a carapace. All cephalic and thoracic appendages are present. All thoracic appendages are alike and biramous with flagellar exopodite and help in locomotion. The abdomen is elongated with well developed five pairs of biramous **pleopods** and sixth pair of **uropods.** The mysis larva metamorphoses into the adult prawn.

In lobster (*Homarus*) both nauplius and zoea are passed within the egg, it hatches as a mysis which changes into an adult.

5. Phyllosoma or glass crab. In *Palinurus* (rock lobster), the newly hatched larva is a greatly modifed mysis stage and is known as **phyllosoma.** It is large, delicate, transparent, greatly flattened and leaf-like larva. Its body has three distinct regions: head, thorax and abdomen. Abdomen is narrow and segmented but lacks appendages. Phyllosoma undergoes several moultings and transform into adult.

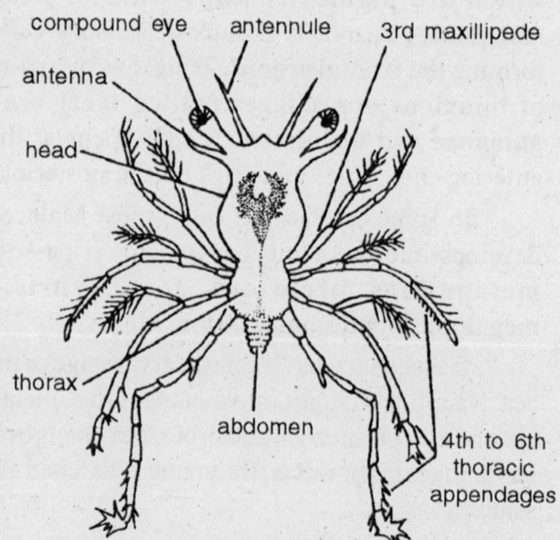

Fig. 39.41. Phyllosoma larva of *Palinurus.*

Significance of Larval Stages

1. They help in wide dispersal of the species.
2. The larval stages help in estabilishing relationship between various groups.
3. Occurrence of nauplius larval stage in all crustaceans connect the different representatives of this class together. For example, nauplius establishes relationship of some obscure animals such as *Sacculina* where adult has lost the characters of the class and even the phylum.

QUESTIONS

Long Answer Questions

1. Describe the habit and habitat and external features of *Palaemon*.
   (*W. Bengal 1989; Himachal 1989; Utkal 1992; Marathwada 1992; Kanpur 1991, 1995*)
2. Give an illustrated account of appendages of *Palaemon* and show how these structures have modified to perform different functions.
   (*Rohilkhand 1988, 1992; Bundelkhand 1989; Awadh 1989, 1992; Kanpur 1990; Rajasthan 1990; Raipur 1991; Meerut or CCS 1991; Jiwaji 1992; Lucknow 1992; Bilaspur 1993; Indore 1994*)

3. Describe a typical biramous appendage and its modifications in *Palaemon*. *(Jodhpur 1993)*

4. Enumerate the appendages of prawn and give well labelled diagrams to explain their structure (No description is required). *(Delhi 1989; Bundelkhand 1991)*

5. What is a biramous appendage? How many pairs of total appendages occur in prawn? Describe the cephalic appendages of prawn with their functions. *(Kanpur 1998)*

6. What do you understand by exopodite and endopodite? Discuss the modifications of endopodite in maxillipedes and walking legs of *Palaemon*. *(Lucknow 1994)*

7. Describe the cephalic and thoracic appendages of *Palaemon*.
(Garhwal 1988; Rajasthan 1988, 1989; Bangalore 1989; Lucknow 1989; Bundelkhand 1989; Kerala 1993; Bhopal 1993; Kanpur 1993)

8. Describe in detail the structure and function of the cephalothoracic appendages of *Palaemon*. *(Lucknow 1995)*

9. Give an account of cephalic appendages of prawn.
(Agra 1988; Sagar 1988; Bhopal 1989; Gorakhpur 1989; Rohilkhand 1989; Meerut or CCS 1989, 1992; Kumaun 1990; Punjab 1991; Bundelkhand 1991, 1993; Purvanchal 1989, 1992; Lucknow 1992)

10. Describe in detail the structure and function of the abdominal appendages of *Palaemon*. *(Punjab 1992; Lucknow 1996)*

11. Give an account of the digestive organs of prawn and the mechanism of feeding. What is the role of the hepatopancreas in prawn? *(Indore 1988; Osmania 1989; Rajasthan 1989; Sagar 1989, 1991; Bundelkhand 1989, 1993; A.P., 1989, 1993; Garhwal 1989, 1995; Rohilkhand 1990; Kanpur 1990; Purvanchal 1991; Allahabad 1991, 1992; Jiwaji 1992; Agra 1992, 1994; Lucknow 1992, 1993, 1995, 1997; Meerut or CCS 1994; Gorakhpur 1994)*

12. Describe the alimentary canal of *Palaemon*. *(Magadh 1995)*

13. Give an account of the alimentary canal in *Palaemon* and describe the physiology of digestion in this animal. *(Punjab 1990; Gorakhpur 1996)*

14. Give an account of the digestive organs of prawn. *(Rewa 1989; Garhwal 1989; Jiwaji 1989, 1991; Agra 1990; Jabalpur 1991; Awadh 1992; Allahabad 1992; Udaipur 1993)*

15. Give a detailed account of the digestive system of *Palaemon* and explain the structural peculiarities of its different parts. *(Meerut or CCS 1988; Lucknow 1989)*

16. Describe the digestive system of *Palaemon* and point out the importance of hastate plate.
(Gorakhpur 1989; Magadh 1989; Berhampur 1990; Bangalore 1994)

17. Give an account of respiratory organs of prawn (*Palaemon*).
(Kanpur 1989; Magadh 1989; Gorakhpur 1989, 1992; Bundelkhand 1989; Meerut or CCS 1989, 1991, 1993; Bhopal 1989, 1993; AP., 1990; Allahabad 1991; Vikram 1992; Nagpur 1992)

18. Describe the organs and mechanism of respiration in prawn (*Palaemon*).
(Burdwan 1988; Osmania 1988; Awadh 1988, 1990; Jiwaji 1988, 1992; Agra 1989; Bhagalpur 1990; Sagar 1990; Rohilkhand 1990; Lucknow 1991, 1993, 1996)

19. Describe in detail the respiratory system in prawn.
(Magadh 1989, 1992, 1993; Gorakhpur 1994; Himachal 1995; Kanpur 1996)

20. What is epipodite? In what respects an epipodite differs from a true gill? Describe the mechanism of respiration in *Palaemon*. *(Lucknow 1994)*

21. Describe the blood vascular system of *Palaemon*.
(Rohilkhand 1988, 1993; Allahabad 1989, 1992; Vikram 1993)

22. Describe in detail the circulatory system and the course of circulation in *Palaemon*. *(Vikram 1989)*

23. Describe the blood vascular system and course of blood circulation of prawn with proper diagram. *(Guwahati 1993, 1994; Meerut or CCS 1999)*

24. Describe the main arteries of prawn. *(Himachal 1990, 1994)*

25. Give a brief illustrated account of the excretory organs in *Palaemon*.
(Bundelkhand 1988; Purvanchal 1989; Kanpur 1992; Rohilkhand 1992)

26. Describe the anatomy and physiology of antennary and maxillary gland of *Palaemon*. How do they resemble or differ from the coxal gland of scorpion.
(Osmania 1989; Purvanchal 1991; Kanpur 1990)

27. Describe the nervous system of *Palaemon*. *(Rohilkhand 1988; Raipur 1990; Indore 1991; Himachal 1991; Lucknow 1992; Kanpur 1992; Meerut or CCS 1993; Vikram 1993)*

28. Give an account of the structure and the working of compound eye in prawn (*Palaemon*).
(Vikram 1989; Osmania 1989; Andhra 1989, 1992; Rewa 1993; Lucknow 1995; Meerut or CCS 1997)

29. What is mosaic vision? Give an account of structure and function of compound eye in *Palaemon*.
(Jodhpur 1988; Bundelkhand 1992; Purvanchal 1992; Lucknow 1992, 1998; Meerut or CCS 1995)

30. Describe the structure of the compound eye in prawn. Explain the mechanism of image formation by such an eye.
(Rajasthan 1989; Gorakhpur 1989; Agra 1988; Delhi 1989; Meerut or CCS 1990)

31. Describe the structure of compound eye and mosaic vision. *(Gorakhpur 1993; Kanpur 1997)*

32. Describe the structure and function of ommatidium. *(Himachal 1995)*

33. Give an account of the structure of the compound eyes in prawn. Explain apposition image and superposition image.

34. Describe the male and female reproductive organs of prawn. Add a note on its fertilization.
(Rohilkhand 1988; Bundelkhand 1988; Gorakhpur 1989, 1991, 1997; Agra 1991; Jiwaji 1991; Kanpur 1994; Himachal 1995)

35. Describe the male reproductive organs of *Palaemon*. *(Himachal 1995)*

36. Describe briefly the different larval forms of crustacea.
(Kanpur 1988; Himachal 1991, 1993; Guwahati 1995)

Short Answer Questions

1. Give functions of the following:
(*i*) Antenna
(*ii*) Maxilla
(*iii*) Maxillipede; and
(*iv*) Uropod.

2. Discuss role of maxillipedes.

3. What is scaphognathite and what is its function?

4. Describe structure of a typical biramous appendage of prawn.

5. What is the difference between maxilla and a maxillipede?

6. What is chelate leg?

7. Differentiate between a chelate and nonchelate walking leg. *(Himachal 1989)*

8. Define moulting or ecdysis.

9. Describe structure of stomach of prawn. *(Awadh 1992; Jiwaji 1993)*

10. Why are lateral folds in the wall of cardiac stomach called 'guiding ridges'?

11. Describe the structure of filter apparatus of *Palaemon* or write a note on pyloric filter.

12. Name the digestive gland found in *Palaemon*. Discuss its structure and function.

13. Differentiate between epipodite and gill.

14. What is mosaic vision?

15. What is the difference between superposition and apposition image? *(Himachal 1989)*

16. Write a note on antennary gland.

17. What is the function of antennary gland?

18. What is sexual dimorphism?

19. How can you identify male and female prawn?

20. Describe male reproductive organs of *Palaemon*.

21. Distinguish between.

(*i*) Second pleopod of male and female prawn. (*Himachal 1989*)

(*ii*) Cardiac stomach and pyloric stomach. (*Lucknow 1995*)

(*iii*) Podobranch and arthrobranch. (*Lucknow 1995*)

(*iv*) Antenna and antennules. (*Lucknow 1996*)

22. Describe the structure of body wall of *Palaemon*.

23. Describe the process of moulting or ecdysis. (*Meerut 1990*)

24. Write short notes on the following:

(*i*) Biramous appendage (*Purvanchal 1988*)

(*ii*) Antennule appendage (*Kanpur 1997*)

(*iii*) Mandible of prawn (*Jiwaji 1989; Bundelkhand 1990; Indore 1992*)

(*iv*) Maxillula (*Magadh 1993; Lucknow 1996*)

(*v*) Maxillipede of prawn (*Garhwal 1992*)

(*vi*) Second maxillipede of prawn

(*vii*) Pleopod (*Purvanchal 1989; Himachal 1991, 1995; Guwahati 1995*)

(*viii*) Second pleopod of prawn (*Purvanchal 1989*)

(*ix*) Uropod (*Punjab 1990; Guwahati 1991, 1993, 1996; Lucknow 1992*)

(*x*) Moulting or ecdysis (*Meerut or CCS 1990*)

(*xi*) Stomach of prawn (*Jiwaji 1993*)

(*xii*) Hastate plate (*Garhwal 1988; Allahabad 1989; Sagar 1989; Agra 1989, 1990; Purvanchal 1989, 1991; Awadh 1990; Himachal 1989, 1991, 1995; Meerut or CCS 1990; Lucknow 1993, 1996, 1998; Gorakhpur 1993, 1995, 1997*)

(*xiii*) Pyloric filter

(*xiv*) Hepatopancreas (*Himachal 1990; Madurai 1994*)

(*xv*) Respiratory gills of *Palaemon*

(*xvi*) Antennary glands (*Jiwaji 1988; Agra 1990; Jabalpur 1992*)

(*xvii*) Green gland (*Meerut or CCS 1988, 1991, 1993; Purvanchal 1988, 1990, 1992; Rohilkhand 1989; Utkal 1989; Jiwaji 1990; Gorakhpur 1990, 1996; Lucknow 1993, 1997; Himachal 1995; Kanpur 1996*)

(*xviii*) Compound eyes (*Magadh 1991*)

(*xix*) Image formation of compound eye (*Lucknow 1995*)

(*xx*) Ommatidium (*Lucknow 1995*)

25. Describe structure and function of statocyst.

26. Write a short note on statocyst of prawn. (*Magadh 1992, 1995; Gorakhpur 1993; Madurai 1994; Kanpur 1995, 1996; Meerut or CCS 1997; Lucknow 1994, 1998*)

27. Write short notes on the following.

(*i*) Eye of *Palaemon* or compound eye. (*Rajasthan 1988; Garhwal 1989; Rohilkhand 1988; Agra 1989; Delhi 1990; Meerut or CCS 1990, 1992; Gorakhpur 1991*)

(*ii*) Ommatidium (*Allahabad 1989; Lucknow 1994, 1996*)

(*iii*) Mosaic vision (*Vikram 1990*)

(*iv*) Statocyst

(*Meerut or CCS 1988; Gorakhpur 1988; Rohilkhand 1989; Garhwal 1989; Agra 198 1990, 1994, 1995, 1996; Awadh 1988, 1992; Bundelkhand 1988, 1992; Purvanchal 1989, 199 1992; Kanpur 1991, 1992, 1995, Lucknow 1991, 1994, 1995, 1996; Jabalpur 1992; Jiwaji 199.*

(*v*) Nauplius larva (*Himachal 1989, 1991; Lucknow 199*

(*vi*) Zoea larva (*Himachal 1989; Lucknow 1994, 199*

(*vii*) Megalopa larva (*BHU 199*

(*viii*) Mysis (*Himachal 199*

(*ix*) Spermatophore (*Himachal 198*

28. Draw well labelled diagrams of the following:

(*i*) External features of prawn

(*ii*) Cephalic appendages of prawn

(*Gorakhpur 1989, 1995; Awadh 1990; Agra 1992; Kanpur 199*

(*iii*) First cephalic appendage of prawn

(*iv*) Antennule of *Palaemon* (*Lucknow 1998*

(*v*) Mandible of prawn (*Himachal 1990*

(*vi*) Thoracic appendages of prawn (*Kanpur 1995*

(*vii*) Abdominal appendages of prawn (*Kanpur 1995; Gorakhpur 1997, 1998*

(*viii*) Second Maxillipede

(*ix*) Alimentary system of prawn (*Guwahati 1992*

(*x*) Inner view of cardiac stomach of prawn

(*xi*) Hastate plate (*Rohilkhand 1989; Lucknow 1995*

(*xii*) Branchial (gill) chamber of prawn with gill (*Kanpur 1988*

(*xiii*) Excretory system of prawn (*Kanpur 1998*

(*xiv*) Nervous system of prawn (*Guwahati 1994*

(*xv*) Microscopic structure of prawn's eye (*Bundelkhand 1989; Kumaon 1989; Vikram 1989*

(*xvi*) Structure of ommatidium (*Gorakhpur 1988; Kerala 1989, 1992*

(*xvii*) V.S. compound eye (*Purvanchal 1991; Kanpur 1993; Gorakhpur 1994*

(*xviii*) T.S. of body of *Palaemon* through gills (*Bundelkhand 1990, 1992; Lucknow 1992*

(*xix*) T.S. *Palaemon* through abdominal region (*Meerut or CCS 1990*

(*xx*) Male reproductive organs of prawn

(*Awadh 1989; Rohilkhand 1990; Agra 1992; Guwahati 1996*

(*xxi*) Female reproductive system of prawn (*Guwahati 1993, 1995*

(*xxii*) T.S. statocyst (*Ravishankar 1993*

(*xxiii*) T.S. green gland (*Ravishankar 1993*

(*xxiv*) T.S. of gill of *Palaemon* (*Lucknow 1997*

Very Short Answer Questions

1. How many cephalic appendages prawn has?

Ans. Five.

2. Give the names of cephalic appendages?

Ans. Antennules, atennae, mandibles, maxillulae, and maxilla.

3. What is the chemical nature of chitin?

Ans. Polyglucosamine.

4. What is ecdysis?
   Ans. Casting off old exoskeleton.
5. What is the lining of gill-cover of *Palaemon* called?
   Ans. Branchiostegite.
6. Mention the function of scaphognathite in *Palaemon*.
   Ans. Turning out water from gill chamber.
7. What are the functions of pleopods of *Palaemon*?
   Ans. Swimming, back leaping and egg carrying.
8. Mention the functions of hepatopancreas.
   Ans. Enzyme secretion, storage of glycogen, fat and calcium.
9. What types of gills prawn has?
   Ans. Phyllobranchs.
10. Where is the hastate plate found in prawn?
    Ans. Floor of cardiac stomach of *Palaemon*.
11. Where is statocyst located in prawn?
    Ans. Statocyst is located in the cavity of precoxa of antennule on either side. It opens to outside on its dorsal surface.
12. What is the shape of abdomen of *Palaemon*?
    Ans. The abdomen of *Palaemon* is rounded dorsally and little compressed laterally. It remains bent a little under the cephalothorax. Due to this bent position the animal looks like comma (,) in shape.
13. What is the location of male genital aperture in prawn?
    Ans. Male prawn has paired genital apertures present on the inner surface of coxae of 5th pair of walking legs.
14. Describe the position of female genital apertures in prawn.
    Ans. Female prawn has paired genital apertures which are situated on the inner surface of coxae of 3rd pair of walking legs.
15. What is appendix masculina? Where it is located in prawn?
    Ans. The second pair of pleopods of male prawn contains an additional rod-like process called **appendix masculina**. It lies between the appendix interna and endopodite. It bears long spine-like setae. This apparatus is used in copulation. It is lacking in the female prawn.

Fill in the Blanks

1. The zoological name of freshwater prawn is ....
2. The ability to break off a part of body in prawn is called ....
3. The exoskeleton of prawn is formed of hard plates known as ....
4. The sclerites are hardened by the deposition of ....
5. The exoskeletal sclerites articulate with one another by uncalcified .... membrane.
6. The body of *Palaemon* consists of two regions .... and abdomen.
7. The abdomen of prawn consists of .... segment and a postsegmental region, the ....
8. In prawn, the andominal appendages are called .... or ....
9. In prawn, ... is organ of orientation and balance.
10. Antennary glands are found in ....
11. The last pair of abdominal appendages are called .... These help in ....
12. First pairs of gills in *Palaemon* are called ....
13. Epipodites are present on the coxa of ....
14. Ecdysis is the phenomenon of ....

15. The last pleopod is very large and named ....

16. The cephalothorax of *Palaemon* is composed of .... segments.

17. The thoracic appendages include .... pairs of maxillipedes and .... pairs peraeopods.

18. The peraeopods of prawn are also called ....

19. In prawn, stomodaeum includes buccal cavity .... and ....

20. Stomodaeum and proctodaeum are lined by ....

21. *Palaemon* has three types of respiratory organs .... , epipodites and ....

22. *Palaemon* has .... pairs of epipodites.

23. There are .... pairs of gills in prawn.

24. Body cavity of *Palaemon* is ....

25. The excretory organs of prawn are .... and ....

Multiple Choice Questions

1. *Palaemon* belongs to the class
   (*a*) Insecta (*b*) Chilopoda
   (*c*) Diplopoda (*d*) Crustacea

2. *Palaemon* is
   (*a*) omnivorous (*b*) carnivorous
   (*c*) herbivorous (*d*) insectivorous

3. The crustacean biramous appendages have a basal part known as
   (*a*) exopodite (*b*) epipodite
   (*c*) protopodite (*d*) endopodite

4. The total number of appendages present in the prawn is
   (*a*) sixteen pairs (*b*) nineteen pairs
   (*c*) seven pairs (*d*) thirteen pairs

5. How many appendages are present in cephalothorax of prawn?
   (*a*) 8 pairs (*b*) 5 pairs
   (*c*) 13 pairs (*d*) 19 pairs

6. Which of the following walking leg are chelate?
   (*a*) first pair (*b*) third pair
   (*c*) fourth pair (*d*) fifth pair

7. In the second maxillipede of prawn which part is not present?
   (*a*) protopodite (*b*) scaphognathite
   (*c*) gnathobases (*d*) endopodite

8. Variation in appendages is based on
   (*a*) structure (*b*) function
   (*c*) adaptation (*d*) modification

9. The gill is attached on outer side of
   (*a*) I maxillipede (*b*) II maxillipede
   (*c*) III maxillipede (*d*) All maxillipedes

10. Epipodites are attached to the
    (*a*) maxillulae (*b*) maxillae
    (*c*) maxillipedes (*d*) peraeopods

11. The respiratory pigment of *Palaemon* is
    (*a*) haemoglobin (*b*) haemocyanin

(*c*) chlorocruorin (*d*) haemerythrin

12. The scientific name of "tiger prawn" is
    (*a*) *Penaeus indicus* (*b*) *Penaeus vanunamec*
    (*c*) *Penaeus monoden* (*d*) *Macrobrachium rosenbergi*
13. The monograph on *Palaemon malcolmonii* was written by
    (*a*) S.N. Dass (*b*) Ronald Ross
    (*c*) S.P. Patwardhan (*d*) P. Maheswari
14. The excretory matter of *Palaemon* consists of
    (*a*) ammonia (*b*) urea
    (*c*) amino acid (*d*) all of these
15. The number of ganglia present in the ventral thoracic ganglionic mass is
    (*a*) 10 (*b*) 12
    (*c*) 11 (*d*) 16
16. The functional unit in the compound eye of prawn is known as
    (*a*) stigma (*b*) ocellus
    (*c*) ommatidium (*d*) eye spot
17. The image formed in the compound eye in bright light is
    (*a*) superposition image (*b*) mosaic image
    (*c*) inverted image (*d*) distorted image
18. In prawn, statocysts are present in the
    (*a*) antennules (*b*) antennae
    (*c*) mandibles (*d*) rostrum
19. Blood plasma of *Daphnia* (water flea) contains
    (*a*) haemoglobin (*b*) haemocyanin
    (*c*) haemoerythrin (*d*) chlorocruorin

ANSWERS

Fill in the Blanks

1. *Palaemon malcolmsonii* 2. Autotomy
3. Sclerite 4. Calcium salts
5. Arthrodial 6. Cephalothorax
7. Six, telson 8. Pleopods, swimmerets
9. Statocyst 10. Coxa of antenae
11. Uropod, steering 12. Podobranch
13. Maxillipedes 14. Casting off old exoskeleton
15. Uropod 16. Thirteen
17. Three, five 18. Walking legs
19. Oesophagus, stomach 20. Chitin
21. Lining of branchiostegite, gills 22. Three
23. Eight 24. Haemocoel
25. Antennary (green) glands, Renal sac

Multiple Choice Questions

1. (*d*)	**2.** (*a*)	**3.** (*c*)	**4.** (*b*)	**5.** (*c*)	**6.** (*a*)
7. (*c*)	**8.** (*b*)	**9.** (*d*)	**10.** (*c*)	**11.** (*b*)	**12.** (*d*)
13. (*c*)	**14.** (*d*)	**15.** (*c*)	**16.** (*c*)	**17.** (*b*)	**18.** (*a*)
19. (*b*).					

40

Scorpion (Phylum Arthropoda)

Scorpions are the arachnids and are the oldest known terrestrial arthropods; their fossils are found from the Silurian period. Today scorpions are restricted to tropical and subtropical regions. Some common genera of scorpions are: *Palamneus, Buthus, Parabuthus, Microbuthus, Scorpio, Isometrus, etc.*

40.1 CLASSIFICATION

Classification

Phylum	—	**Arthropoda**
Subphylum	—	**Chelicerata**
Class	—	**Arachnida**
Order	—	**Scorpiones**
Genus	—	***Palamneus*** **(Scorpion)**

40.2 EXTERNAL FEATURES

Habit and Habitat

Scorpion is **solitary**, **nocturnal**, **carnivorous** and **predaceous** animal and is found in desert. It feeds on insects, spiders and other small animals. It can withstand prolonged fast and hunger but show **cannibalism**, *i.e.*, it devours its young ones. Scorpion is **viviparous** and female carries the newly-hatched young on its back.

Scorpion is fairly common in hilly regions (*i.e.*, slopes of hills) of India. *Palamneus* is an Indian scorpion found all over India. *Palamneus bengalensis* is common in North India.

Shape and Size

The body of scorpion is elongated, narrow and dorso-ventrally flattened. The size varies from 1.3 cm (*e.g., Microbuthus pusillus*) to 15 cm (*e.g., Palamneus swammerdami*).

Colouration

The colouration of body depends upon the surroundings and varies from shining black to pale yellow. The dorsal surface is much darker in colour than that of the ventral surface.

External Structure

The elongated segmented body is divisible into two major regions: **1. Prosoma** (*i.e., cephalothorax*) and **2. Opisthosoma** (*i.e., abdomen*).

***(i)* Prosoma.** It is anterior, short, broad and flat region of body. It is formed by fusion of **head** and **thorax.** Prosoma of adult is made up of six segments, each of which bears a pair of appendages on the ventral side. Dorsally, the prosoma is covered by a large thick, shield-like, square-shaped exoskeletal plate, called **cephalothoracic shield** or **dorsal carapace**. The dorsal carapace is formed by fusion of terga of this region. The anterior margin of the dorsal carapace has a deep notch forming **right** and **left frontal lobes.** The carapace bears a pair of **median eyes** in the middle and two to five pairs of smaller **lateral eyes** on the antero-lateral margins. These are simple eyes. On the ventral side of prosoma, there is a single narrow, median triangular plate, the **sternum** which is situated between the coxae of third and fourth pairs of legs. A pair of **chelicerae**, a pair of **pedipalpi** and four pairs of **walking legs** can be seen originating from the ventral side.

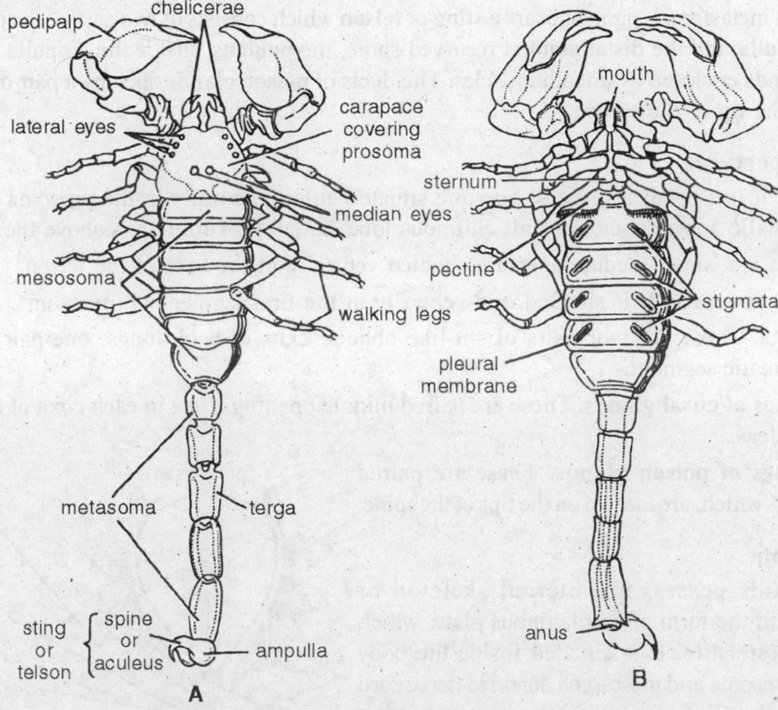

Fig. 40.1. Scorpion. A—Dorsal view; B—Ventral view.

***(ii)* Opisthosoma.** The prosoma is followed by a long, distinctly segmented posterior part of body called **opisthosoma**. It is differentiated into two parts: (*a*) an anterior broad **mesosoma** or **preabdomen** and (*b*) a posterior narrow **metasoma** or **postabdomen**.

(*a*) Mesosoma. It consists of **seven** segments. It is as broad as the prosoma anteriorly but slightly narrow posteriorly. Mesosoma lacks appendages but every segment is surrounded by a ring of exoskeleton known as **sclerite**. Each sclerite consists of a dorsal plate, the **tergum**, and a ventral plate, the **sternum**. Both of these plates are joined together by **arthrodial membrane**. On the ventral surface of mesosoma, following structures are visible:

1. Genital operculum. The sternum of first mesosomatic segment is small and bears the median genital aperture which is covered by plate-like, rounded, bifid and movable lid, called **genital operculum**.

2. Pectines (Pectenes). A pair of pectenes are present in the second mesosomatic segment. Each pectene is a comb-like structure consisting of a three-segmented **shaft** or **handle** which is attached to the median line. At its posterior margin, pectene bears a row of 4 to 36 narrow-teeth of a comb. The pectenes are said to be tactile or olfactory organs. These are comparatively larger in males and are curiously modified in females.

3. Stigmata. The sterna of third, fourth, fifth and sixth mesosomatic segments bear a pair of lateral, oblique, slit-like openings, called **stigmata.** Each stigma leads into a pulmonary sac or book-lung.

***(b)* Metasoma.** Mesosoma is followed by metasoma. Metasoma is a cylindrical, narrow (slender), five-segmented structure. It is usually held upwards arching over the back and is erroneously called **tail**. Each segment of metasoma has complete tergal and sternal plates of octagonal shape formed by the complete fusion of terga, sterna and pleurons. These segments are flexibly jointed one behind the other.

The last metastomal segment bears a **sting** or **telson** which consists of two parts—the proximal rounded **ampulla** and the distal pointed recurved spine, the **aculeus**. Inside the ampulla is a pair of **poison glands** enclosed in smooth muscles. The ducts of poison glands open by a pair of minute apertures at the tip of the spine

External Apertures

1. **Mouth.** It is a small transverse aperture situated antero-ventrally in the prosoma between the cephalic appendages. A small chitinous lobe, called **labrum** hangs above the mouth.
2. **Anus.** It is a small median aperture situated ventrally at the base of the telson.
3. **Genital apérture.** It is situated mid-ventrally in the first segment of mesosoma.
4. **Stigmata.** These are four pairs of slit-like oblique exits of book-lungs, one pair in each mesosomatic segment.
5. **Openings of coxal glands.** These are paired minute openings, one in each coxa of the third pair of legs.
6. **Openings of poison glands.** These are paired openings which, are located on the tips of the spine.

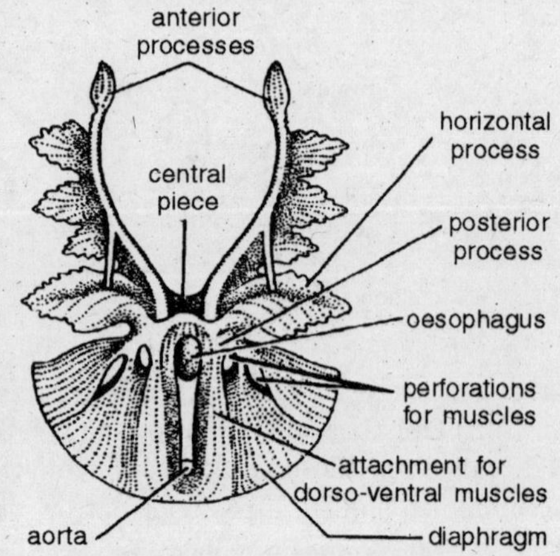

Fig. 40.2. Scorpion. Endoskeleton.

Endoskeleton

Arachnids possess an internal skeleton or endoskeleton in the form of cartilaginous plate, which is called **endosternite**. It is situated inside the body between the prosoma and mesosoma dorsal to nerve cord but ventral to the alimentary canal. In scorpion, it is a triangular struc-ture and produced into several paired processes, the anterior, posterior and horizontal which provide surface for the attachment of muscles. Posteriorly, the body of endoskeleton is fused with a chitinous plate called **diaphragm**, which is perforated by the nerve cord, alimentary canal, muscles and dorsal aorta.

40.3 APPENDAGES

In scorpion, there are six pairs of appendages in the cephalothorax. There are a pair of **chelicerae,** a pair of **pedipalpi** and four pairs of **walking legs.**

1. Chelicerae. These are the anterior-most and pre-oral appendages, lying close to each other on either side of mouth. Chelicerae are homologous to antennae of crustaceans. Each chelicera is a small, three-segmented and chelate structure (Fig. 40.3). The **first segment** is small, triangular and ring-like and lying

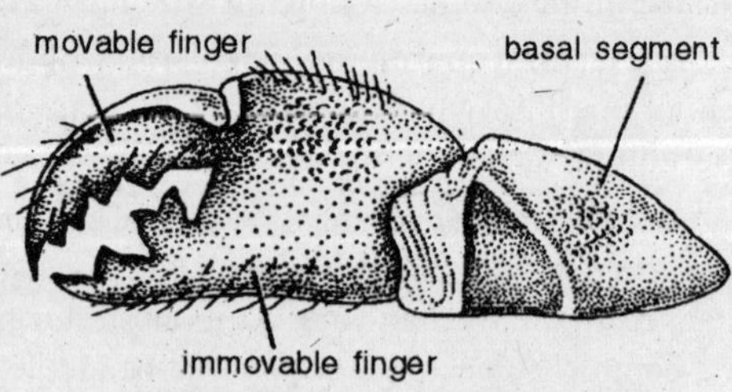

Fig. 40.3. Scorpion. Chelicera.

concealed beneath the carapace. The **second segment** is large, swollen and bears at its anterior end on the inner side a projection which is pointed and toothed. It forms the **immovable finger** of the chela. The **third segment** arises from the outside of the second segment. It is also curved and toothed and forms the **movable finger** of the chela. The chelicerae are the prehensile appendages of scorpion and are used for holding and tearing the prey.

2. Pedipalpi. The pedipalpi or large pincer claws are a second pair of large powerful and clawed appendages which are post-oral in position. Each pedipalpi is made of six segments, *viz.*, **coxa, trochanter, humerus, brachium, manus** and **movable finger**.

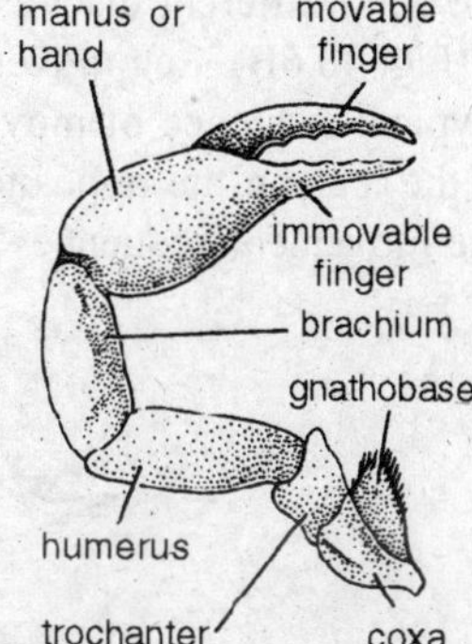

Fig. 40.4. Scorpion. Pedipalp.

***(i)* Coxa.** It is small basal segment and is situated on the side of the pre-oral cavity beneath the carapace. Coxa bears towards the mouth a blunt blade-like process, the **gnathobase**. The gnathobases of both sides serve as jaws, protruding in the pre-oral cavity and help in squeezing and crushing the body of the prey.

***(ii)* Trochanter and humerus.** Trochanter is also a small segment situated next to coxa. It bears at right angles to the longitudinal axis, a stout and powerful **humerus.**

***(iii)* Brachium.** It lies above the humerus and is directed forward.

***(iv)* Manus and movable finger.** The manus or hand is the longest segment. It forms a powerful chela with the movable finger which is movably articulated on its outer side. Teeth-like serrated edges of movable finger and the immovable finger of padipalpi help in catching the prey. The padipalpi move in the horizontal plane.

3. Walking legs. There are four pairs of identical walking legs attached to the cephalothorax. These are used in walking. Each walking leg consists of seven segments or podomeres—the **coxa, trochanter, femur, patella, tibia, protarsus** and **tarsus**. The coxae of first and second pair of legs are movable and each is provided with forwardly directed triangular **maxilliary processes** or **gnathobases**. The coxae of the first pair of legs are separate while those of second pair meet each other. The coxae of the third and fourth pairs of legs are immovable, lack maxilliary processes and are separated from those of opposite side by the sternum.

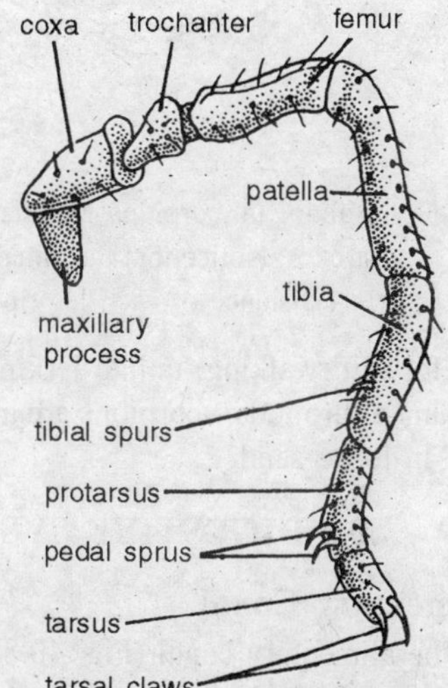

Fig. 40.5. Scorpion. First walking leg

The **trochanter** is a short and stout segment moving in all directions. The **femur** is long, straight segment moving upward and outward. The **patella** is also long and moves downwards. **Tibia** is elongated and bears a **tibial spur** distally. Protarsus bears a pair of dark tipped claw-like spurs, the **pedal spurs**. The **tarsus** bears three movable curved pointed claws, of which two are dorsal or **superior** and one is ventral or **inferior.** As the inferior claw is worn out, only two superior claws are visible.

Locomotion in Scorpions

Walking by terrestrial chelicerates demands that the body be supported off the substratum and the four pairs of legs move in sequences that maintain the animal's balance.

The details of scorpian waliking patterns have been examined by **Root** and **Bowerman** (1978). During simple forward walking, each of the eight limbs move through the usual power and recovery strokes. In scorpions the joints between the coxae and the body are virually immovable and do not contribute to the overall motion of limbs. Not all of the walking legs move in the same pattern. The tips of the anterior legs are brought quite high off the ground during their recovery strokes; they may be used to feel ahead as the animal moves (Fig. 40.6A). Also, the tips of each pair of legs extend different distances from the body, allowing for stride overlap without contact between legs (Fig. 40.6B). Leg movements in scorpions do not follow the usual metachronal model. Rather, the typical sequence of movement along one side of the body is leg 4, then 2, then 3, and finally with the legs on the opposite side being generally out of phase. The somewhat staggered overlap of the movement sequences produces a smooth gait movement sequences produces a smooth gait

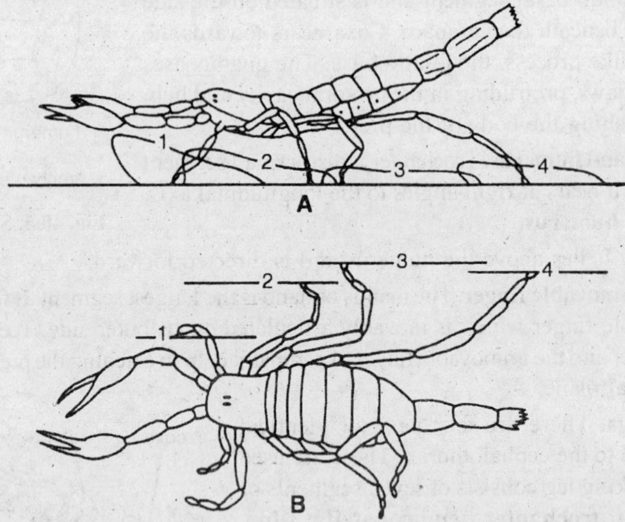

Fig. 40.6. Walking in scorpions. Numbers denote legs 1–4. A—The paths of the leg tips during their recovery strokes. Notice that the anterior legs are lifted higher off the ground than are the posterior legs. B—Distance of each leg tip from the body (after Brusca and Brusca, 2003).

during forward walking, unlike jerking motion of an insect moving by alternating tripod pattern. Like many arthropods, scorpions are also able to change speeds, turn abruptly, walk back-ward and burrow in loose sand.

40.4 DIGESTIVE SYSTEM

1. Alimentary Canal

The alimentary canal (Fig. 40.8) of scorpion is a fairly uniform and straight tube and extends from anterior mouth to posterior anus. It comprises four distinct regions: The **pre-oral cavity, foregut** (stomodaeum), **midgut** (mesenteron) and **hindgut** (proctodaeum).

1. Pre-oral cavity. It is a small cavity formed in front of mouth by the coxae of the first four pairs of appendages of prosoma. It is bounded dorsally by two **chelicerae** and a median **rostrum** which is laterally compressed cushion-like structure. Laterally, the pre-oral cavity is bounded by the **coxae** of pedipalpi and ventrally its floor is formed by two pairs of **maxillary**

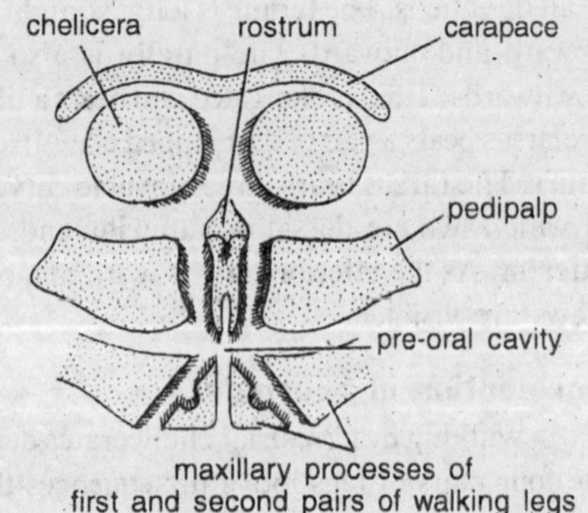

Fig. 40.7. Scorpion. T.S. through the pre-oral cavity.

processes of first two pairs of walking legs. All these structures bear bristles or hairs which protrude into the pre-oral cavity.

Below the base of the rostrum is located a narrow transverse mouth. Rostrum is also provided with a set of muscles which can roll it into a tube for sucking the fluids into the pharynx.

2. Foregut. It includes **mouth, pharynx** and **oesophagus**. The foregut is internally lined by cuticle.

***(i)* Mouth.** It is a small, narrow and transverse aperture situated below the base of rostrum and behind the pre-oral cavity. Mouth leads into the pharynx. Mouth being a narrow passage only admits juices and pulp.

***(ii)* Pharynx.** It is large, pear-shaped, bulbous and muscular structure. It is situated obliquely below the mouth. Pharynx is provided with a thick chitinous internal lining and numerous muscles outside. These muscles are of two types: sphincters and dilators. The sphincter muscles keep the pharynx contracted, while the dilator muscles dilate it considerably. Due to its muscles, pharynx serves as a powerful suctorial organ for sucking the liquid food.

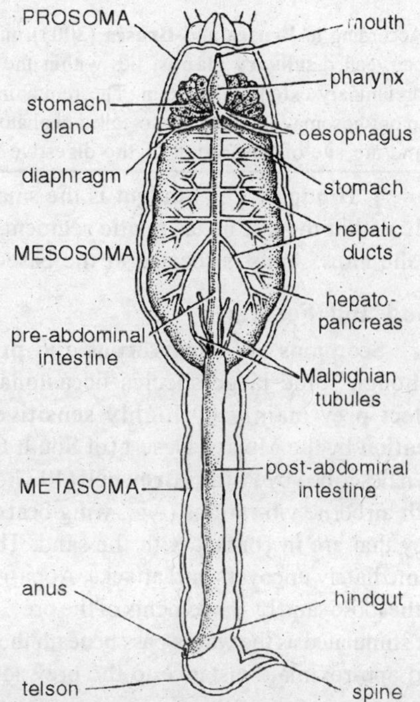

Fig. 40.8. Scorpion. Alimentary canal.

***(iii)* Oesophagus.** It is short, narrow tube arising from the postero-ventral wall of the pharynx. It extends through the nerve ring to open into the stomach. In the stomach it protrudes a short distance to form **sleeve valve** which prevents the food from coming back to the pharynx.

3. Midgut. It includes **stomach, intestine** and **digestive glands**. Midgut is internally lined by epithelium and lacks a lining of chitin.

***(i)* Stomach.** It is slightly dilated thin-walled part of alimentary canal situated in the cephalothorax and extending between the oesophagus and diaphragm. Stomach is covered above and on sides by an irregular brownish gland called **stomach gland**. This gland comprises three lobes—two lateral and one mediodorsal and opens into the stomach by a pair of short lateral ducts. **Huxley** regarded it as the **salivary gland** but **Blanchard** and **Pavlosky** called it stomach gland.

***(ii)* Intestine.** It is longest part of the alimentary canal extending from the diaphragm upto the last segment of the abdomen where it opens into the **hindgut**. The intestine is a wide tube with glandular walls and is divisible into **pre-abdominal** and **post-abdominal regions**. At the junction of these two regions, there is a narrow constriction from where are given out two pairs of narrow elongated and blind **Malpighian tubules**. These tubules are excretory in function.

***(iii)* Digestive gland.** A large, brownish and lobulated gland occupies the whole of the preabdominal cavity, it is called **digestive gland, liver, chylenteron** or **hepatopancreas**. Hepatopancreas is a racemose gland from which five pairs of lateral, narrow **hepatic ducts** open into the intestine. The wall of the lobule is composed of a layer of epithelium arranged on a thin membrane, called **tunica propria** and a peritoneal covering. The cells of the epithelial layer are of two types: 1. **absorptive cells** which are large-sized and numerous; 2. **ferment cells** which are small-sized and less in number.

Box 40.1.

According to **Brusca** and **Brusca** (2003), midgut of scorpions bear six pairs ot **digestive ceca.** The first pair, called **salivary glands,** lies within the prosoma and produces much of the digestive juice used in preliminary external digestion. The remaining five pairs of digestive ceca are highly convoluted and lie in opishosoma (*i.e.*, they are so-called hephatopancreas). These ceca produce the enzymes for final digestion and are site of absorption of the digestive products.

4. Hindgut. The hindgut is the smallest and bulbous part of the alimentary canal. It exists only within the last metasomatic segment. It is lined internally by chitin and opens to the exterior by the **anus**. Anus is located at the base of the telson.

Food and Feeding

Scorpions are **carnivorous** and **predaceous** animals. Scorpions feed mostly on insects, although some large species occasionally eat snakes and lizards. Most are nocturnal and detect prey mainly by highly sensitive mechanoreceptors. A fascinating method of **prey location** by the Mojave Desert (of South California) sand scor-pion, *Paruroctonus mesaensis,* has been described by **Phillip Brownell** (1977, 1979, 1984). Brownell noticed that *Paruroc-tonus* ignores both airborne vib-rations (*e.g.*, wing beats) and visual input, but responds imme-diately to nearby prey that are in contact with the sand. This scorpion is even able to detect buried prey, which it immediately uncovers and attacks. Appa-rently the scorpion senses slight mechanical waves set up in the loose sand by movements of the prey. Special mechanore-ceptors in the walking legs of scorpion are stimulated as the waves pass beneath them. The information is processed to determine the direction and app-roximate distance to the prey source.

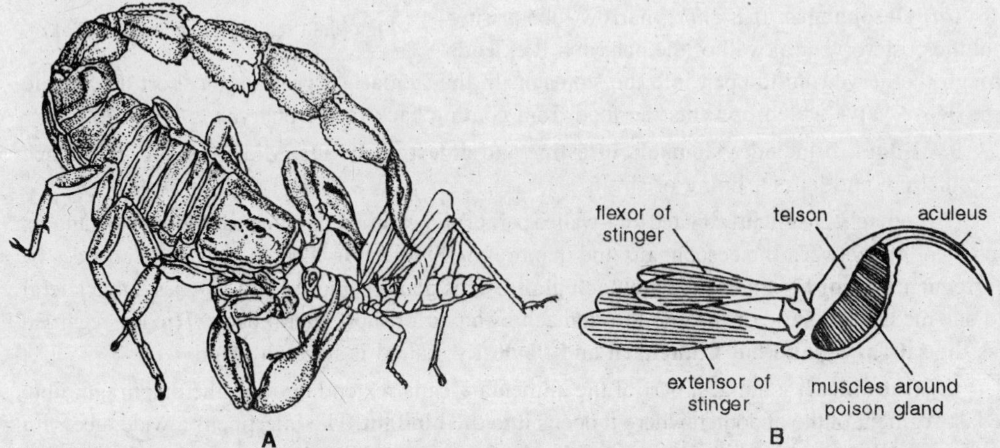

Fig. 40.9. A—The scorpion *Androctonus australis*, applying its stinging apparatus to a grasshopper while holding the prey with its pedipalps. B—The scorpion's telson and stinging apparatus. The telson is normally held flexed; extension drives the aculeus into the prey (after Brusca and Brusca, 2003).

Once a scorpion locates a victim, the prey is grasped by the chelate pedipalps. The opis-thosoma is arched over the pro-soma, bringing the telson and stinging apparatus into position to inject venom (Fig. 40.9). Mus-cular contractions force the venom through a pore in the **aculeus** and into the prey. The venom is a **neurotoxin** that can quickly paralyse and kill most prey; in fact, some scorpion venoms can kill large animals, including humans (Box 40.2).

Box 40.2.

Perhaps the most familiar of dangerous scorpions are two species that occur in the American Southwest: the bark scorpion (*Centruroides exilicauda*) and the stripe tailed scorpion (*Vaejovis spinigerus*). A north African species, *Androctonus australis,* produces a venom considered to be as potent as that of cobras. Other genera of potentially dangerous scorpions include *Buthus* and *Parabuthus* (Africa and middle east) and *Tityus* (South America) (**H.L. Keegan, 1980**).

Buthus is frequently found in our Indian houses. *Palamnaeus* is the black scorpion that grows to 10–13 cm., in length. *Scorpio, Euscorpio* and *Isometrus* are other common genera. The largest scorpion is the African *Hadogenes troglodytes* which grows to 21 cm.

The sequence of events in the ingestion of food by scorpions is typical of many arachnids Once it has been captured and stung, the prey is passed to the chelicerae, which tear it into small bits. The gnathobases grind and mash the food as digestive juices are released through the mouth; this process reduces the food to a semiliquid form. As this organic "soup" is ingested, the hard parts are discarded, and more bits of food are moved between the gnathobases for processing. One chelicera hold the food and other rips it open. The coxae of the pedipalpi and first two pairs of legs now press upon the prey so that the entire body fluid is squeezed out into pre-oral cavity. Meanwhile, the rostrum forms a groove which is applied to the prey and the liquid is sucked into pharynx and then it is driven into oesophagus and stomach.

Digestion, Absorption and Egestion

The food is reduced to pulp in the pre-oral cavity. It is partially digested by **proteolytic enzymes** present in the secretions of **alveolar glands** which are located in the maxillary processes of the first two pair of legs. Inside the stomach the partially digested food is mixed with the secretion of the **stomach gland** containing the enzymes **amylase, trypsin** and **lipase**. Most of the digestion is completed in the stomach. When the food reaches the intestine, it passes into hepatopancreas, where digestion is also performed by enzymes **amylase** and **catalase**. These enzymes are secreted by the hepatopancreas itself. Thus, absorption of digested food occurs in preabdominal intestine as well as hepatopancreas. The undigested food reaches the postabdominal intestine and hindgut to be expelled out through the anus.

40.5 BLOOD VASCULAR SYSTEM

The blood vascular system of scorpion (Fig. 40.10) is **open type**. It consists of heart enclosed in **pericardium, arteries** and **sinuses**.

1. Heart and Pericardium

The heart is an elongated and muscular tube lying in groove of the hepatopancreas. It extends back from behind the diaphragm. Heart is incompletely divided into seven distinct chambers by shallow constrictions. Each chamber bears a pair of valvular openings the **ostia** on its dorso-lateral side through which it communicates with the pericardial cavity. Valves of ostia allow flow of blood from the pericardium into the heart but not in the opposite direction. The heart is able to perform slow peristaltic movements due to its muscular walls.

Heart is enclosed in a thin-walled membranous sac, the **pericardium**. Heart is suspended into the pericardium by a pair of connective tissue bands or **ligaments** in each segment, called **ali cordi**.

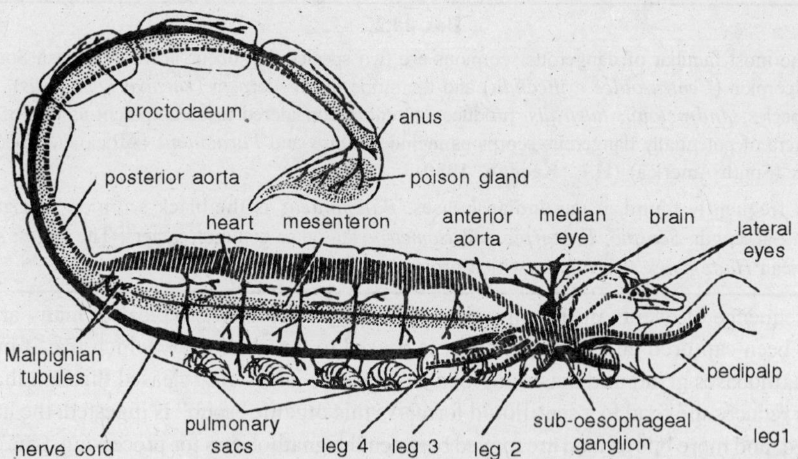

Fig. 40.10. Scorpion. Diagrammatic lateral view of the blood vascular system and other internal organs.

2. Arteries

The heart is continued anteriorly as the **anterior aorta** and posteriorly as the **posterior aorta**. A pair of lateral **systemic arteries** are also given off from each chamber of the heart.

***(i)* Anterior aorta.** It is a short and wide vessel and arises from the anterior end of the first chamber of the heart. It runs along the dorsal surface of the alimentary canal. It sends off the following branches to various organs of prosoma:

(*a*) A pair of **small visceral arteries** to the intestine and hepatopancreas.

(*b*) After piercing the diaphragm, the dorsal aorta gives off a pair of **ventro-lateral arteries** which encircle the oesophagus and unite ventrally to form a **supraneural artery**. Supraneural artery runs medially above the ventral nerve cord supplying the nerve cord and ventral muscles.

(*c*) The anterior aorta also gives of six pairs of arteries which supply the appendages of prosoma, brain and eyes.

***(ii)* Posterior aorta (or caudal artery).** It arises from the posterior end of the heart and runs backward on the dorsal side of intestine upto the last segment. It gives off a pair of arteries in each segment of the postabdomen (metasoma) and then divides into three branches: 1. A **median branch** extends ventrally to join the supraneural artery. 2. Two **lateral branches** run into the posterior lobes of the hepatopancreas and supply the muscles of the fourteenth segment. In telson, posterior aorta terminates into the poison gland.

***(iii)* Systemic arteries.** From each chamber of heart on either side arises a systemic artery. These paired arteries form a network and supply the various parts of preabdomen (mesosoma).

3. Sinuses

All the arteries break up into numerous small branches and sub-branches and pour their blood into a narrow sinus, the **lacunae**. From the lacunae, the blood is collected into larger **sinuses**. There are five sinuses in all–the **dorsal sinus** which lies just above the pericardial sinus; a large **ventral sinus** on the ventral side of the body, two **lateral sinuses** on either lateral side of the body and a **perivisceral** sinus surrounding the alimentary canal. Dorsal sinus is connected to ventral sinus through the lateral sinuses. The perivisceral sinus also communicates with ventral sinus (Fig. 40.10).

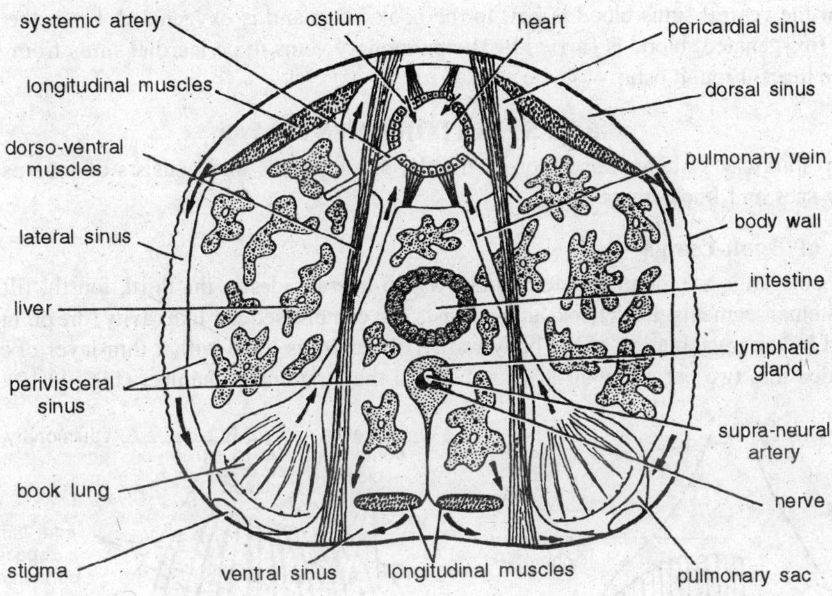

Fig. 40.11. Scorpion. T.S. passing through the middle region of body showing pulmonary and main sinuses of body.

4. Pulmonary Veins

From each pair of book-lungs the oxygenated blood is sent back to the pericardial sinus by a pair of thin-walled **pulmonary veins**. These veins are continued over the dorsal side of the book-lungs as lacunae.

5. Blood

The blood of scorpion is colourless with a bluish tinge. It contains numerous **leucocytes**. The respiratory pigment in blood is **haemocyanin** which is dissolved in the plasma. The blood becomes bluish in colour when oxygenated.

Course of Circulation

With the contraction of ligaments the cavity of the heart enlarges and the blood from the pericardial sinus rushes into the heart through ostia. With the contraction of the heart, ostia are closed and the blood is pumped to various organs of the body through the anterior and posterior aortae and the systemic arteries. The impure (deoxygenated) blood of the body is collected in the ventral

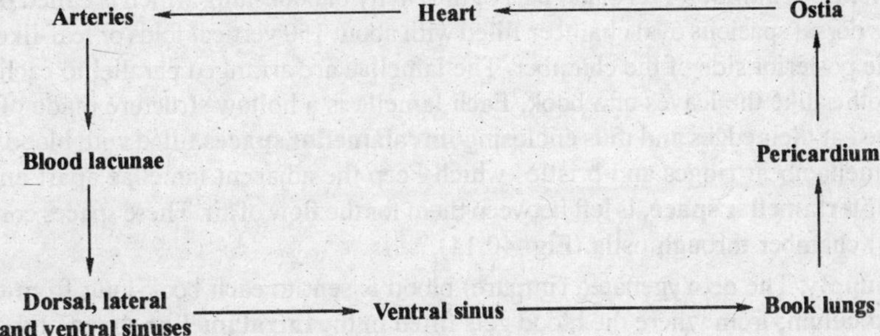

Fig. 40.12. Scorpion. Course of circulation of blood.

sinus. From the ventral sinus blood is sent to the book-lungs and is oxygenated. From the book-lungs pure (oxygenated) blood is carried by the pulmonary veins to pericardial sinus from where it enters the heart through ostia.

40.6 RESPIRATORY SYSTEM

The respiratory system of scorpion is highly characteristic and consists of four pairs of **pulmonary** sacs and **book-lungs**.

Structure of Book-Lungs

The book-lungs are found situated on the ventro-lateral sides in the third, fourth, fifth and sixth mesostomal segments. Each book-lung consists of a compressed sac-like cavity (the pulmonary sac) formed by the invagination of the body wall. This cavity is lined with a thin-layer of cuticle and is divided into two parts: the atrial chamber and the pulmonary chamber (Fig. 40.13).

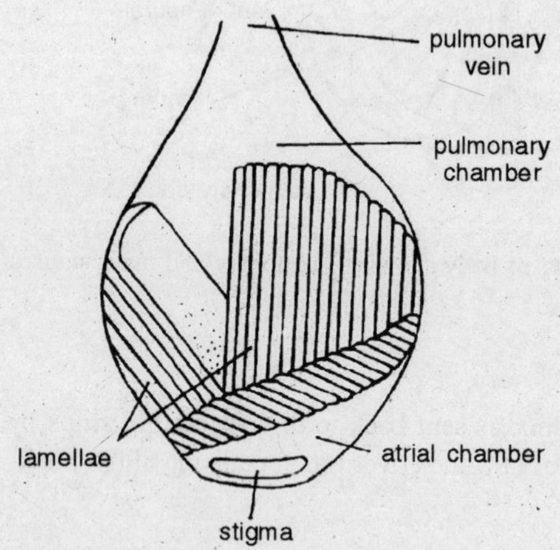

Fig. 40.13. Scorpion. A book-lung.

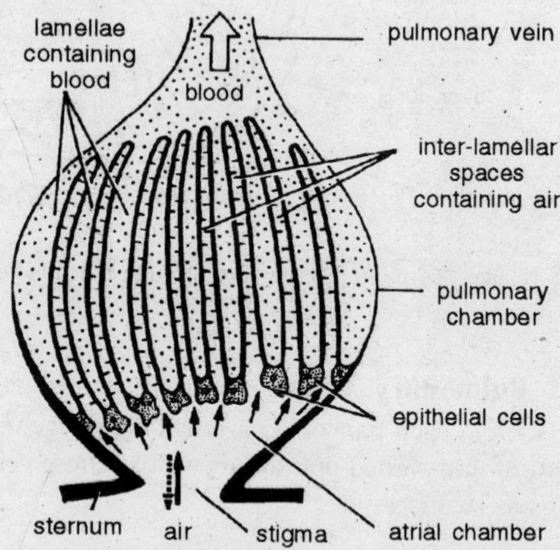

Fig. 40.14. Scorpion. Vertical section of book-lung to show the mechanism of respiration.

1. Atrial chamber. It is the proximal ventral part of the book-lung. It is in the form of small and dorsoventrally compressed air-chamber. Proximally atrial chamber opens to exterior by a narrow, oblique, slit-like opening, called **stigma**, which is placed ventrolaterally on the sternum. Distally the roof of atrial chamber is perforated by a series of minute slit-like openings or **ostia**. These ostia are placed parallel to one another. Through ostia, the atrial chamber communicates with the **interlamellar air spaces** of the pulmonary chamber.

2. Pulmonary chamber. It is distal part of the cavity of book-lung which is called **pulmonary chamber**. It is dorsal spacious oval chamber filled with about 150 vertical folds or leaf-like **lamellae**, attached to the posterior side of the chamber. The lamellae are arranged parallel to each other and one over the other like the leaves of a book. Each lamella is a hollow structure made of thin layer of cuticle united at their edges and thus enclosing **intralamellar spaces** filled with blood. The outer side of the lamellae bear ridges and bristles which keep the adjacent lamellae apart and little air space, called **interlamellar space,** is left between them for the flow of air. These spaces communicate with the atrial chamber through ostia (Fig. 40.14).

Blood supply. The deoxygenated (impure) blood is sent to each book-lung from the ventral sinus by diverticulum; from where the blood gets filled in the **intralamellar space** of the lamellae. The blood is oxygenated or purified in the lamellae of each book-lung and is collected into a **pulmonary vein** which travels dorsally and opens into the pericardium.

Mechanism of Respiration

The inspiration and expiration of air in the book-lungs is controlled by the action of the, dorso-ventral and atrial muscles. When the **dorso-ventral muscles** are contracted, the pulmonary chamber is compressed and the air filled in interlamellar air spaces is forced out into the atrial chamber. From atrial chamber this air is forced to outside through the stigma due to contraction of the **atrial muscles**. When the atrial muscles relax, the book-lungs resume their normal shape and the fresh air enters through stigma first into the atrial chamber and then into the interlamellar spaces. At this time dorso-ventral muscles relax and pulmonary chamber resumes its normal shape. The **gaseous exchange** takes place through the thin membranous walls of the lamellae and the blood becomes oxygenated and carbon dioxide is expelled out in the air.

40.7 EXCRETORY SYSTEM

Excretory system of scorpion includes the following types of excretory organs:

1. A pair of **coxal glands**
2. Two pairs of **Malpighian tubules**
3. **Hepatopancreas**
4. **Nephrocytes**

1. Coxal Glands

There is one pair of coxal glands (Fig. 40.15) which are situated near the bases of the third pair of walking legs in the fifth segment of prosoma. The coxal glands are white shining structures and are coelomoducts of coelomic origin. Each coxal gland consists of three parts: 1. saccule, 2. labyrinth 3. bladder.

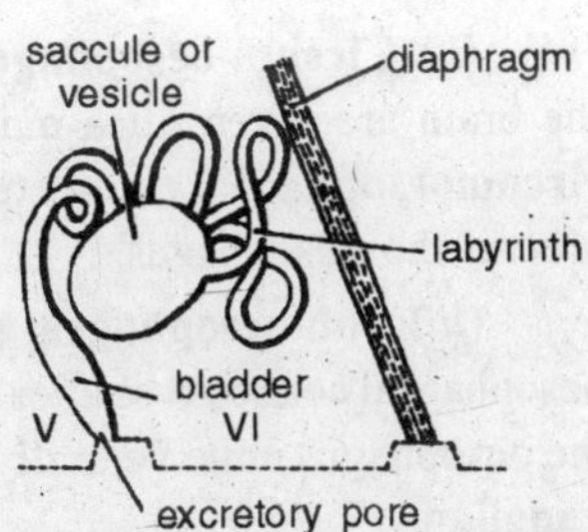

Fig. 40.15. Scorpion. Coxal glands.

1. Saccules. It is also called **end sac** or **vesicle.** It is a large sac-like structure situated in the centre of the coxal gland.

2. Labyrinth. It is a long and coiled tube arising from the saccule and opening in the bladder.

3. Bladder. It represents the swollen terminal part of labyrinth which has enlarged to take the form of a reservoir. It opens to exterior by a small opening situated on the posterior face of the coxa of fifth walking leg.

The saccule and labyrinth of each coxal gland collect excretory nitrogenous wastes from the blood and pass them outside through the excretory pore. **Urate crystals** have been found in the saccule. The coxal glands of scorpion are homologous to green glands or antennary glands of crustacea.

2. Malpighian Tubules

Two pairs of Malpighian tubules are present at the junction of preabdominal and postabdominal parts of intestine. These are endodermal in origin and lie freely in the blood from which they isolated the nitrogenous products (*i.e.*, **guanine crystals**) and discharge them into the intestine. Malpighian tubules possess thin syncytial walls which remove the waste from the blood.

3. Hepatopancreas

Hepatopancreas is regarded as an excretory organ by **Pavlvosky**. It has been found that particles of ammonical carmine injected into the body cavity of scorpion are accumulated in the form of red granules inside cells of hepatopancreas. **Pearson** believes that ferment cells of hepatopancreas pick up excretory products.

4. Nephrocytes

Large nephrocytes and lymphatic organs are specialized excretory structures. Both structures are found under the body wall in the mesosoma. These are believed to be excretory as well as phagocytic in function.

40.8 NERVOUS SYSTEM

Nervous system consists of **central nervous system, peripheral nervous system** and **visceral nervous system.**

1. Central Nervous System

The central nervous system comprises of the (*i*) brain or cerebral ganglion; (*ii*) circumoesophageal connectives; (*iii*) sub-oesophageal ganglion and (*iv*) ventral nerve cord (Fig. 40.16).

***(i)* Brain.** The **brain** or **cerebral ganglion** is a small bilobed mass situated in the prosoma just beneath the median eyes in the prosoma. Embryological studies have shown that brain develops from the fusion of eight pairs of ganglia.

***(ii)* Circum-oesophageal connectives.** From the brain are given off a pair of thick, short, stout **circumoesophageal connectives** which form a collar around the oesophagus.

***(iii)* Sub-oesophageal ganglion.** The circum-oesophageal connectives after forming a collar around the oesophagus unite ventrally in a **sub-oesophageal ganglion**.

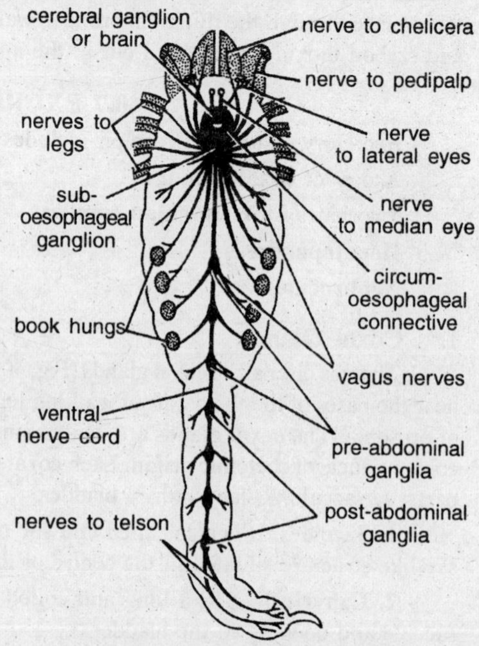

Fig. 40.16. Scorpion. Nervous system.

***(iv)* Ventral nerve cord.** The sub-oesophageal ganglion is continued backwards into the abdomen as a double **ventral nerve cord**. It runs upto the fourth segment of the post-abdomen. The nerve cord bears three ganglia in the pre-abdomen and four in the post-abdomen. Ventral nerve cord is slender and rounded in the pre-abdomen and flattened and ribbon-shaped in the post-abdomen.

2. Peripheral and Visceral Nervous System

Several nerves are given out from the brain, circumoesophageal connectives and ganglia of the ventral nerve cord, which constitute these nervous systems.

Brain gives off a pair of **optic nerves** to the *median* eyes, a pair of nerves which branch and supply the *chelicerae* and the *lateral eyes*, a pair of nerves to the *rostrum*, a pair of nerve trunks to the *pedipalpi* and four pairs of nerves to the four pairs of *walking legs*. Besides these, the brain also supplies nerves to the *pharynx, oesophagus* and *heart*. The suboesophageal ganglion may send two to four pairs of **vagus nerves** which run posteriorly into the preabdomen to supply the *genital operculum*, the *pectines* and first two pairs of *book-lungs*. The ganglia of ventral nerve cord of preabdomen send out nerves to the last two pairs of *book-lungs*, *heart*, and *muscles* of abdomen, while those from the post-abdomen supply nerves to the *segmental muscles*. The last ganglion also supplies the sting.

40.9 SENSE ORGANS

The sense organs of scorpion include **sensory setae** or **sensillae, pectines** and **eyes**.

1. Sensory sensillae. These are fine hair-like cuticular structures which are sparsed all over the body. These are much longer in legs and tail. The sensillae are **tactile organs** of scorpion and they are extremely sensitive to touch. Each sensory sensilla is connected to a sensory cell which also remains connected at its base with a nerve fibre. Sensillae are long and bristle-like on the podomeres of pedipalpi.

2. Pectines. The pectines are a pair of modified comb-like appendages lying attached to the sternum of the second segment of mesosoma joined to each other in the middle. Each pectine (Fig. 40.17) consists of a **shaft** or **handle** made of three pieces which bears a series of movable **lamellae** or **teeth** arranged like the teeth of a comb on the posterior margin of the shaft in the outer half. The number of teeth differs from species to species.

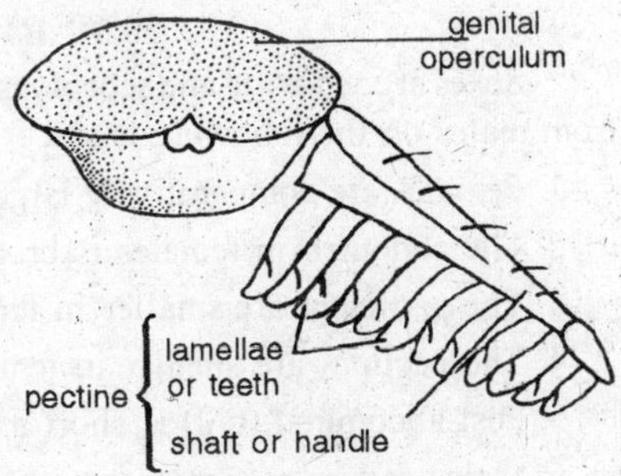

Fig. 40.17. Scorpion. Pectine.

The exact function of the pectines is still not clear. The ventral side of each tooth of the comb is provided with **chemoreceptors** and **mech-anoreceptors**. During movements of scor-pion, the pectines are held out from the sides of the body in a horizontal position so that the teeth touch the ground. They are sensitive to vibrations and appear to determine some aspects of the substratum surface, perhaps particle size. Their removal prevents sper-matophore deposition.

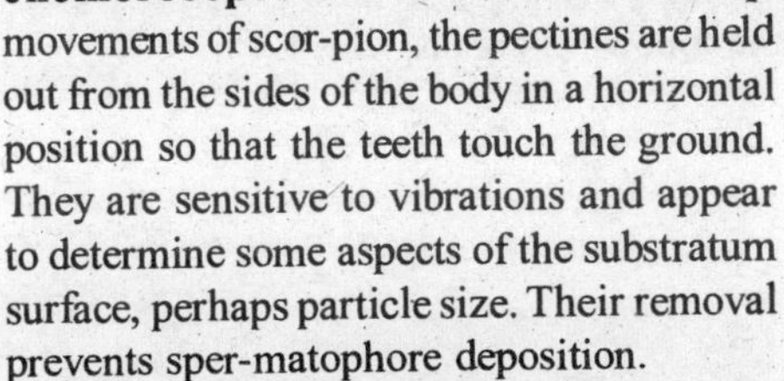

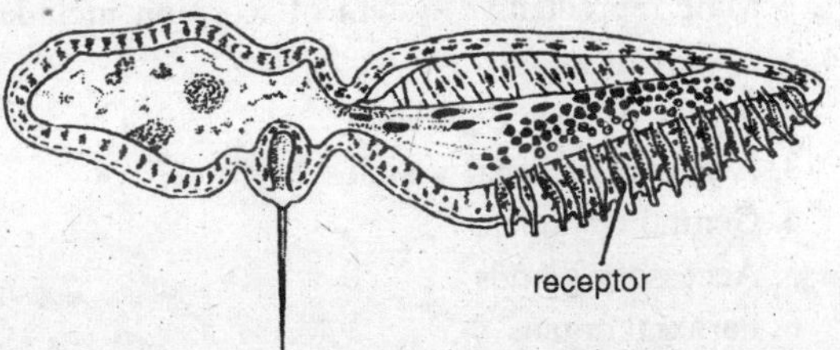

Fig. 40.18. Scorpion. T.S. through pectine.

3. Eyes. There are four pairs of eyes in scorpion. One pair of **median eyes** and three pairs of **lateral eyes.**

(*i*) Median eyes. These are cup-like in shape covered externally by a cuticular **lens** or **cornea**. Inside the pigmented cup are **rhabdomes**. Each rhabdome encloses several **retinal cells** supplied by the nerve fibres of optic nerve (Fig. 40.19).

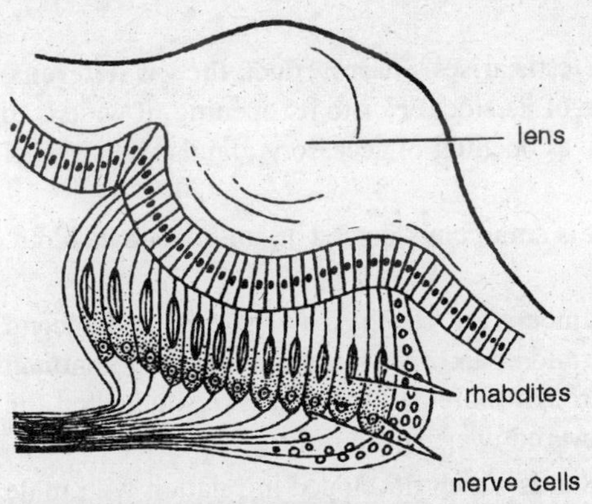

Fig. 40.19. Scorpion. V.S. of median eye.

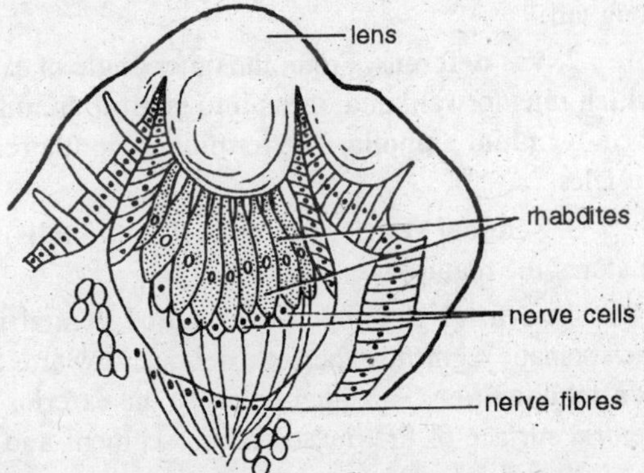

Fig. 40.20. Scorpion. V.S. of lateral eye.

The median eyes of scorpion are said to be intermediate between compound eyes and simple eyes of insects. These resemble the compound eye in having their retinal cells arranged in groups around each rhabdome as in ommatidia. But unlike the compound eyes of insects and crustaceans, the sensory retinal cells serve the function both of receiving the stimuli and transmitting the impulses.

***(ii)* Lateral eyes.** These are similar to simple eyes of insects. These are like a pigmented cup covered externally by a **biconvex lens** formed of transparent cuticle. Inside the cup are several longitudinal optic rods called **rhabdomes** associated with **retinal cells** or **retinulae**. The retinal cells receive nerve fibres (Fig. 40.20).

40.10 REPRODUCTIVE SYSTEM

Sexes are **separate** and scorpions exhibit some **sexual dimorphism**. Females can be identified from males on the following bases:

1. Females are comparatively larger in size than males.
2. The abdomen in females is broader.
3. The padipalpi are smaller in females and longer in males.
4. The pectines are smaller in females having fewer teeth. They are longer in males.
5. Post-abdominal (tail) is short and curved in female.
6. Males can be recognized easily by the hooks which are present on the opercular plates.

Male Reproductive System

Male reproductive system of scorpion includes the following organs:

1. A pair of testes
2. A pair of vasa deferentia
3. A pair of seminal vesicles
4. Genital chambers
5. Accessory glands
6. Paraxial organs

1. Testes. These are elongated ladder-like structures lying embedded in the hepatopanereatic mass in the mesosoma from third to sixth segment. Each testis consists of two slender, thread-like whitish **longitudinal tubules** lying parallel and joined to each other by four **transverse tubules** (the **cross branches**) to form three squares. Each tubule is lined internally by a single layer of germinal epithelium, the cells of which give rise to spermatozoa. The sperms are arranged in bundles, the **spermatophores**. A mature sperm of scorpion is filiform and motile having an oval body and a long tail.

2. Vas deferens. From, the outer angle of each testis arises a narrow duct, the **vas deferens**, which runs forward and opens into **genital chamber** of its side. Prior to its opening, it widens to form **terminal ampulla**. The terminal ampulla receives opening of accessory glands and seminal vesicles.

3. Seminal vesicles (or Vesicula seminalis). It is small club-shaped organ of yellow colour. It stores the mature sperms.

4. Genital chambers. The paired genital chambers are confined to the first and second mesosomatic segments. These run obliquely towards the middle and open into a small median **common genital chamber**. The latter opens to the exterior through **male genital aperture** situated on the ventral surface of first mesosomatic segment and covered over by the bifid genital operculum.

5. Accessory glands. These are four pairs of glands which developed in relation with male reproductive organs. These are **dorsal annex glands, ventral annex glands, oval glands** and **cylindrical glands**. All of them discharge their contents into the terminal ampulla of vas deferens. The fluid secreted by these glands is helpful in reproduction.

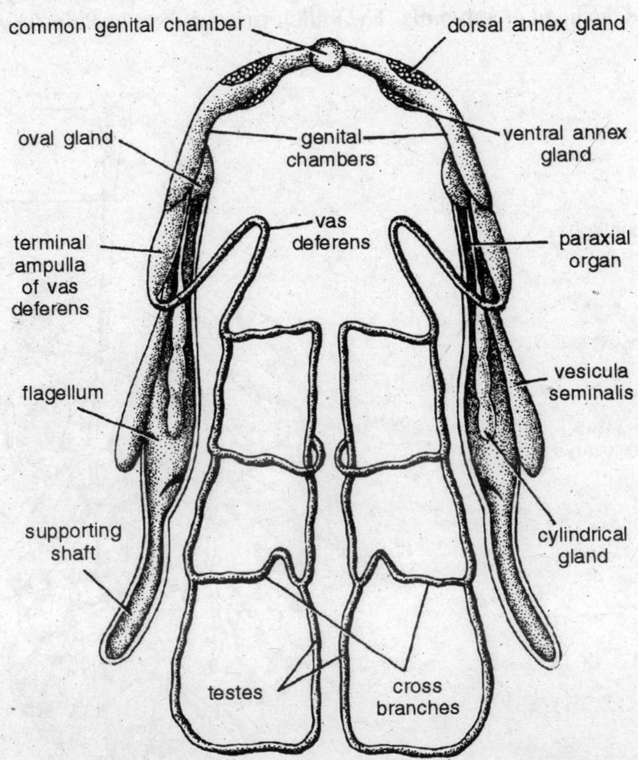

Fig. 40.21. Scorpion. Male reproductive organs.

6. Paraxial organs. Each genital chamber extends backwards on the outer side of testis into a tubular sac, the **paraxial organ**. It consists of an elongated chitinous rod, the **flagellum** which closely fits into its cavity. Flagellum consists of a supporting **shaft** with a longitudinal groove inside it. Few spines are present on the one side of the flagellum. The flagella of both sides can protrude out of the genital operculum and constitute the **double penis** of scorpion. These act as **claspers** during copulation.

Female Reproductive System

The female reproductive system of scorpion consists of following organs:

1. Ovary
2. A pair of oviducts
3. Common genital chamber

1. Ovary. The ovary is a single median structure situated in mesosoma from third to sixth segments embedded in the hepatopancreas. It consists of three elongated narrow **longitudinal tubules** known as **ovarioles**. Of them the smaller one is the median and two large ones are the lateral ovarioles. All ovarioles are connected by four hollow cross branches or the transverse **connectives**, so that the ovary assumes a ladder-like appearance. The ovarioles and their cross branches are lined by germinal epithelium, the cells of which form ova. The mature ova project into hollow bud-like processes on the surface called the **diverticula** or the **follicles** (Fig. 40.22).

In scorpions, as development of a fertilized egg takes place inside the follicle, the latter forms an elongated structure now called **uterus** (Fig. 40.23). The uterus is differentiated into a proximal swollen region or **embryonic chamber** and a distal narrow **appendix** terminating into an enlarged tip, called the **head of appendix**. The follicles are fusiform in *Palamneus* and spherical in *Buthus*.

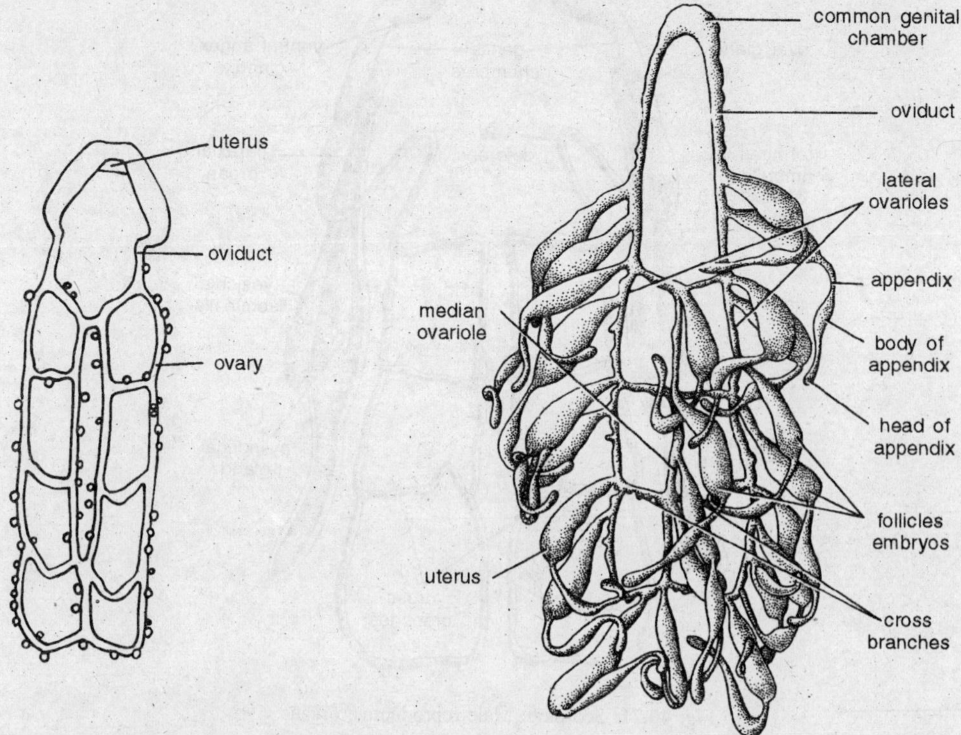

Fig. 40.22. Scorpion. Female reproductive system.

Fig. 40.23. Scorpion. Ovary with mature follicles.

2. Oviducts. The two ovarioles continue anteriorly as the **oviducts** which converge towards the mid-ventral line to open into a small median **common genital chamber**. The common genital chamber opens to outside by the **female genital aperture** situated beneath the bilobed operculum on the sternum of the first preabdominal segment.

Courtship and Development

Scorpions carry on a long courtship dance. Eventually the male persuades the female to retire with him into a burrow (see Box 40.3) In India, scorpions breed during monsoon.

In scorpions, the development is **internal** and very slow. Scorpions are viviparous. The female scorpions give birth to 2 to 95 youngs. The young ones resemble the adults. For some time the mother carries the newly hatched young ones on its back and takes care of them. The young one undergo several moults to grow into the adults.

Box 40.3.
Courtship Dance of Scorpions

During the breeding season, the male scorpion roams about and on contacting a female initiates an extended courtship. In some species the male and female face each other; each extends its abdomen high into the air and moves about in circles; in others, the male rocks. The male then seizes the female with his pedipalps, and together they walk backward and forward (called **promenade**). This behaviour may last from 10 min to hours, depending on how long it takes to locate a suitable site for **spermatophore deposition**. Eventually, the male deposits a spermatophore, which he attaches to the ground. A wing-like lever extends from the spermatophore. The male then manoeuvres the female so that her genital area is over the spermatophore (Fig. 40.24). Pressure on the spermatophore lever releases the sperm mass which is taken up into the female orifice.

All scorpions brood their eggs within the female reproductive tract and give birth to live, developed young. The **aplacental viviparous** species have large, yolky eggs; development takes place in the lumen of the ovarian tubules. The eggs of **placental viviparous species** possess little yolk. In them eggs develop within the diverticula of the ovary. Each diverticulum in turn develops a distal tubular appendage that contains a cluster of absorbing cells at the end (Fig. 40.25) These cells rest against the maternal digestive caeca, from which nutritive material is absorbed. The nutritive material passes through the tubule to the embryo at the base.

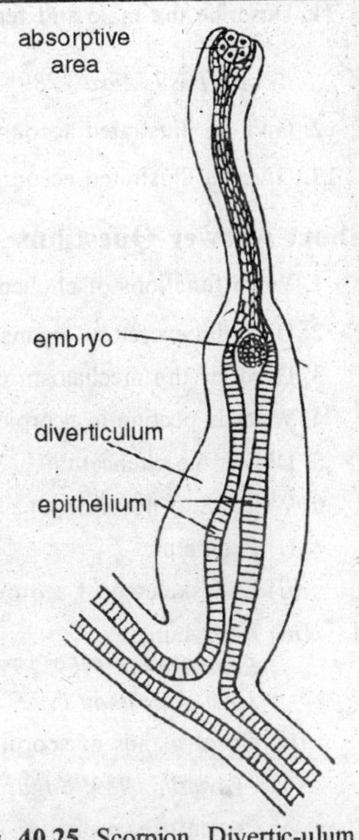

Fig. 40.25. Scorpion. Divertic-ulum of ovary with developing embryo.

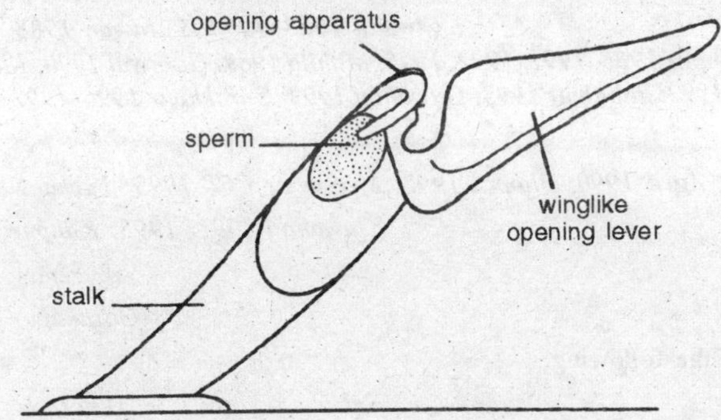

Fig. 40.24. Scorpion. Spermatophore.

QUESTIONS

Long Answer Questions

1. Describe the external features of scorpion.
(*Jiwaji 1989; Agra 1991; Kanpur 1995; Gorakhpur 1997*)

2. Describe appendages of scorpion. (*Bhopal 1992*)

3. Describe the alimentary canal, mechanism of feeding and digestion of scorpion.
(*Jammu 1989; Rewa 1993*)

4. Give an account of blood vascular system of scorpion. (*Rewa 1993; Gorakhpur 1998*)

5. Give an account of the parts that subserve respiration in scorpion.
(*Jiwaji 1988, 1990; Kanpur 1989, 1992, 1994; Agra 1990; Vikram 1993; Sagar 1993; Rewa 1993*)

6. Give an account of the respiratory organs and mechanism of respiration of scorpion.
(*Udaipur 1989; Patna 1989; Utkal 1989, 1991; Rewa 1989, 1992; Agra 1992; Kanpur 1992, 1994*)

7. Describe the excretory organs of scorpion. (*Agra 1988; Rewa 1993*)

8. Write an account of coxal glands of scorpion.

9. Describe the nervous system and sense organs of scorpion.

10. Describe sense organs of scorpion. (*Agra 1992*)

11. Describe the male and female reproductive systems of scorpion. (*Kanpur 1988, 1990, 1992, 1993, 1996; Awadh 1989; Agra 1989; Indore 1990; Jabalpur 1990; Sagar 1991; Purvanchal 1991, Jiwaji 1992*)

12. Give an illustrated account of the reproductive system of scorpion. (*Gorakhpur 1996*)

13. Give an illustrated account of female reproductive organs of scorpion. (*Gorakhpur 1995*)

Short Answer Questions

1. Write functions of chelicerae.
2. Describe excretory organs of scorpion.
3. Describe the mechanism of respiration in *Palamneus*. (*Jiwaji 1993*)
4. What is pectine in scorpion? Give its importance to the animal.
5. Define the chelicera. (*Lucknow 1994*)
6. Write short notes on the following:
   (*i*) Pedipalpi
   (*ii*) Endoskeleton of scorpion (*Agra 1989; Gorakhpur 1993*)
   (*iii*) Book-lungs (*Awadh 1988, 1990; Burdwan 1988, 1991; Bundelkhand 1989, 1991; Kanpur 1988, 1991, 1992, 1995; Mithila 1989; Garhwal 1990; Kumaun 1990; Rajasthan 1992; Agra 1992; Jabalpur 1993; Guwahati 1994; Gorakhpur 1996, 1997, 1998*)
   (*iv*) Coxal glands of scorpion (*Awadh 1989; Sagar 1989, Agra 1990; Bhopal 1992; Meerut or CCS 1993; Lucknow 1997*)
   (*v*) Pecten of scorpion (*Guwahati 1992; 1995; Kanpur 1996*)
   (*vi*) Chelicerae (*Lucknow 1997*)
   (*vii*) Classification of scorpion (*Gorakhpur 1996*)
7. Draw well labelled diagrams of the following:
   (*i*) External features of *Palamneus* (*Lucknow 1990*)
   (*ii*) Dorsal view of *Palamneus*
   (*iii*) Ventral view of *Palamneus*
   (*iv*) Appendages of *Palamneus*
   (*v*) Coxal gland of scorpion (*Awadh 1989; Agra 1991*)
   (*vi*) Male reproductive organs of scorpion (*Rohilkhand 1989; Purvanchal 1989; Jiwaji 1992; Gorakhpur 1996*)
   (*vii*) Female reproductive organs of scorpion (*Ravishankar 1993; Gorakhpur 1994, 1997*)
   (*viii*) Pectine (*Purvanchal 1992*)

Very Short Answer Questions

1. How many divisions of body occur in a scorpion?

   Ans. Two: prosoma or cephalothorax and opisthosoma or abdomen. Opisthosoma includes broad mesosoma and narrow metasoma.

2. How many appendages does a scorpion have?

   Ans. In scorpion, there are six pairs of appendages in the cephalothorax (prosoma). Appendages of scorpion include a pair of **chelicerae**, a pair of **pedipalpi** and four pairs of **walking legs**.

3. What is the function of coxal glands?

   Ans. Excretion of nitrogenous wastes.

4. What kind of vascular system is present in scorpion?

   Ans. Open type.

5. How many pairs of eyes are present in scorpion?

Ans. Four pairs of eyes (a pair of median eyes and three pairs of lateral eyes).

6. What are pectines of scorpion?

Ans. Scorpions have a pair of large, unique, comb-like structures, the pectines, on the ventral surface of the mesosoma. After detailed studies of their innervation, **Foelix** and **Schabronath** (1983) suggested that the pectines act as both mechanoreceptors and chemoreceptors. Other works has shown them to be capable of detecting slight differences in sand grain size (see **Brusca** and **Brusca**, 2003).

7. What is cannibalism?

Ans. An animal that devours members of its own species, is called **cannibal.** The phenomenon is called **cannibalism** and is found in scorpions.

Multiple Choice Questions

1. *Palamnaeus* (scorpion) belongs to class Arachnida because of
(*a*) four pairs of legs (*b*) chelate legs
(*c*) sting (*d*) none of these

2. Scorpions are
(*a*) sanguivorous (*b*) carnivorous
(*c*) herbivorous (*d*) omnivorous

3. Respiratory organs of scorpion are
(*a*) lungs (*b*) gills
(*c*) book-lungs (*d*) ctenidia

4. In how many podomeres are the walking legs of scorpions divided?
(*a*) 5 (*b*) 7
(*c*) 9 (*d*) 11

5. Solid processes of chitin in scorpion are:
(*a*) dermatida (*b*) coelodermatidia
(*c*) epiostracum (*d*) telostracum

6. Hollow processes of chitin in scorpions are
(*a*) hypostracum (*b*) coelodermatida
(*c*) coelostomata (*d*) tracheinida

7. The heart chambers in *Palamnaeus* are
(*a*) 1 (*b*) 3
(*c*) 5 (*d*) 7

8. Respiratory pigment in blood of scorpion is
(*a*) haemocyanin (*b*) erythrocyanin
(*c*) haemoglobin (*d*) none

9. The excretory system of scorpions includes
(*a*) malpighian tubules (*b*) coxal glands
(*c*) hepatopancreas (*d*) nephrocytes
(*e*) all the above

10. A characteristic feature of scorpions is
(*a*) four pairs of walking legs (*b*) vivipary
(*c*) chelicerate (*d*) pedipalpi

ANSWERS

Multiple Choice Questions

1. (*a*) **2.** (*b*) **3.** (*c*) **4.** (*b*) **5.** (*a*) **6.** (*b*)
7. (*d*) **8.** (*a*) **9.** (*e*) **10.** (*b*)

41

Zoological Importance of *Limulus*

Limulus polyphemus, the American Horse-shoe crab or king crab belongs to the subclass Xiphosura and class Merostomata of subphylum Chelicerata, Phylum Arthropoda. It is a common marine animal of the Atlantic and Gulf Coasts of North America and a favorite laboratory animal of physiologists in North America. *Limulus* inhabits shallow marine waters, generally on clean sandy bottoms. It is a burrowing, nocturnal and omnivorous animal. It can walk with legs, swim with mesosomatic appendages and leap with telson. *Limulus* often crawls about or burrows just beneath the surface, preying on other animals or scavenging. It feeds upon molluscs, worms and bottom algae. Adults attain sexual maturity in 9 to 12 years and has a life-span of 19 years.

Limulus is a large-sized arthropod, measuring upto 60 cm long (Fig. 38.10). Dorsoventrally flattened body is composed of an anterior **prosoma** (6 fused segments) and a posterior **opisthosoma** (9 segments). Prosoma is covered by a large horse shoe-shaped **carapace** marked by two lateral longitudinal grooves and bearing the median simple eyes (occlli) and two lateral compound eyes. Ventrally, prosoma bears a pair of 3-jointed preoral **chelicera** followed by five pairs of 6-jointed chelate legs. Each walking leg is formed of seven articles (coxa, trochanter, femur, patella, tibia, tarsus, pretarsus) the last two of which form the chelae. The coxal endites of the pedipalps and first three walking legs are modified as **gnathobases**. Arising from the coxae of the fourth walking legs are tiny appendage-like process called **flabella**, which function as gill cleaners. In addition, just posterior and medial to the last walking legs is a pair of reduced appendages called **chilaria**. Their function is unknown, and there is some controversy about their evolutionary significance–some specialists believe that they may reflect an additional opisthosomal segment (see **Brusca** and **Brusca**, 2003). In male *Limulus*, first pair of walking legs form clasping organs.

The broad hexagonal opisthosoma is divisible into an anterior **mesosoma** (6-segments) covered by a dorsal shield, and a posterior **metasoma** (3-segments) ending in a long pointed **telson**. Mesosoma bears 6 pairs of thin, broad appendages fused in median line. Last five pairs carry 5 pairs of **book gills**. Excretion takes place through a pair of **coxal glands**. Sexes are separate.

Mating and Development

On the onset of the breeding season, horseshoe crabs migrate into the shallow water of protected bays and estuaries. On the east coast of North America, this migration takes place in the spring and summer, and huge numbers of *Limulus* can be seen gathering near the shore in preparation for mating. **Mating** is initiated when the male climbs onto the back of female and grasps her with his modified first walking legs. The clasped pair moves to shallow water, usually at a high spring tide, and the female excavates one or more shallow depressions in the sand and deposits her eggs (2000 to 30,000 eggs per mating). The male releases sperms directly on to the eggs as they are deposited. Then the mates separate, and the females covers the fertilized eggs with sand.

Limulus has indirect development. Its early development takes place in the sand or mud "nest". Cleavage is holoblastic, producing a **stereoblastula** with most of the yolk contained in the inner cells. As development continues, the surface cells at the anterior and posterior ends of the embryo divide rapidly, forming two germinal centers. Some of these rapidly proliferating cells migrate inward as the presumptive endoderm and mesoderm. The anterior germinal center gives rise to the first four segments of the prosoma and the posterior center to the rest of the body. All of the prosomal segments fuse and are eventually covered by the developing carapace-like dorsal-shield.

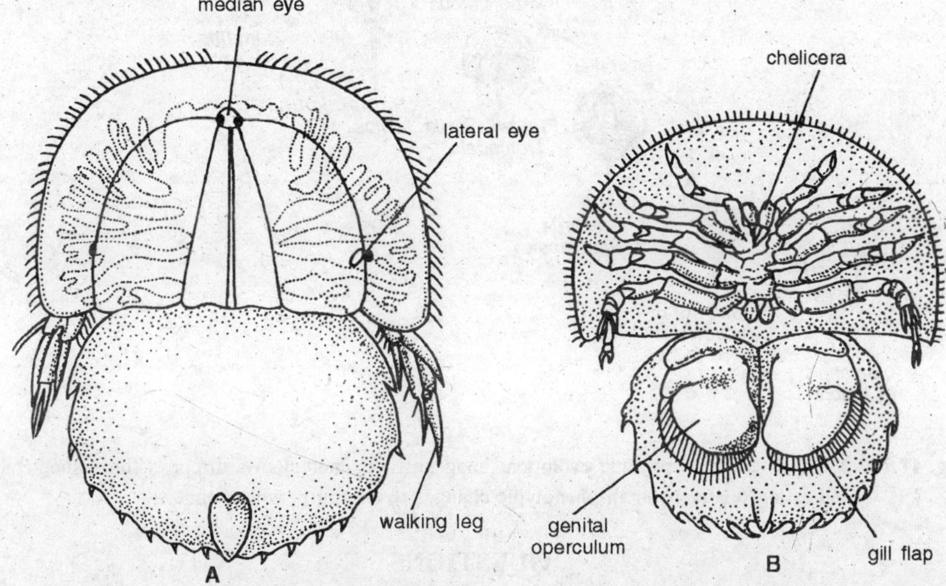

Fig. 41.1. *Limulus.* Euproops larva. A—Dorsal view; B—Ventral view (after Brusca and Brusca, 2003).

As the yolk reserves become depleted, the embryo develops into a **euproops larva** (or **"trilobite larva"**), so named because of its resemblance to the fossil carboniferous xiphosusran *Euproops* (which superficially resembled trilobites; Fig. 41.1). The larvae swim about and periodically burrow in the sand. Segments are formed and appendages added through a series of molts until the adult form is reached. Zoologists view this developmental pattern as a mixed life history. The early developmental stages are supplied with an investment of yolk by the female and are protected by the nest she constructs, but the youngs emerge as independent feeding larvae prior to maturation.

41.1 EVOLUTIONARY SIGNIFICANCE

Limulus or king crab is not a crab at all but a primitive marine arthropod resembling the scorpions and spiders. It is an evolutionary relic which has changed very little during last 200 million years. It is a **"living fossil"** whose close relatives became extinct many million years ago. *Limulus* is found to be product of **phyletic evolution** or **anagenesis** (Fig. 41.2)

Box 41.1.
Anagenesis

According to Darwin's natural selection, a single species can change through time, such successional changes which occur in a single lineage are called **phyletic evolution** or **anagenesis**. It explains evolution of sub-species level (microevolution) and is contrasted with **cladogenesis** or **phylogenetic branching** (*e.g.*, evolution of cat family).

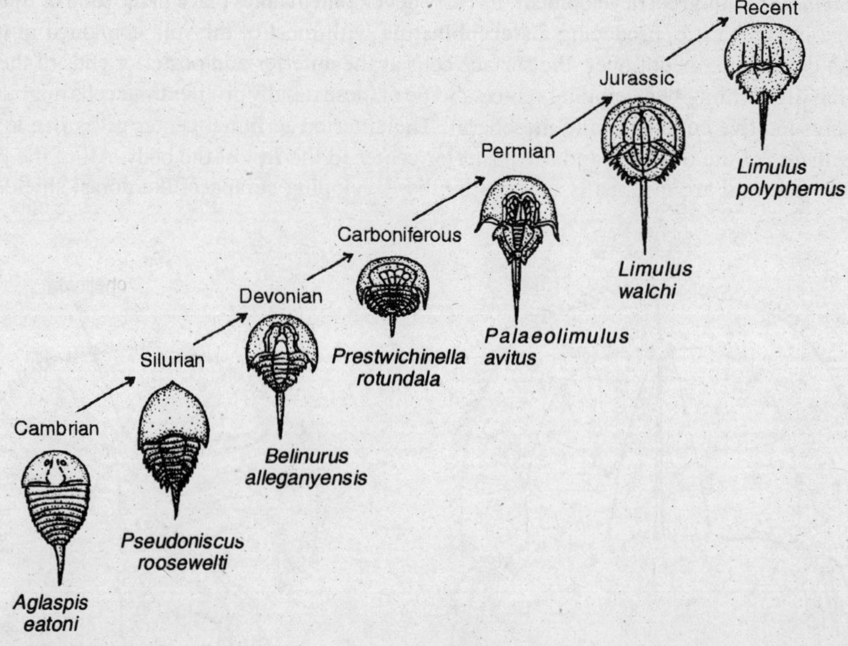

Fig. 41.2. Microevolution or phyletic evolution (anagenesis) in merostom arthropods (horseshoe-crabs), indicating relatively small phenotypic changes over long period of time.

QUESTIONS

Long Answer Questions

1. Explain the main features of *Limulus* and write about its evolutionary significance.

Short Answer Questions

1. Why *Limulus* is regarded as a living fossil?

2. Give an account of migration, mating and development of *Limulus.*

3. Write a brief ecological note on king crab.

Very Short Answer Questions

1. What is the name of larva of *Limulus*?
Ans. Euproops larva (or "trilobite larva").

2. *Limulus* is product of which type of evolution?
Ans. Anagenesis (*i.e.*, microevolution).

Fill in the Blanks

1. *Limulus* provides a good example of .... fossil.

2. King crab has .... nutrition.

3. *Limulus* is an aquatic animal. It is .... in habitat.

4. Larva of *Limulus* is called ....

ANSWERS

Fill in the Blanks

1. Living **2.** Omnivorous
3. Marine **4.** Euproops or trilobite

42

Economic Importance of Insects

Insects are most widely distributed and most powerful. They are useful as well as dangerous to humankind. They attack human beings and their domestic animals by causing disease. Insects destroy human being's property and crops. Hence, insects are of great economic importance for humans. But some insects are beneficial to human beings. Economic importance of insects to humankind can be discussed under the following headings.

A. Injurious or Harmful Insects

1. Injurious to crops. Many insects damage forest trees, growing farm crops, flowers, fruits and stored grain. They mostly belong to orders Lepidoptera, Coleoptera, Diptera and Hemiptera.

(*i*) The **brown tail moth** (*Euproctis*) and **gypsy moth** (*Lymantria*) are serious pests of shade and foliage trees; their larvae are a menace and destroy many types of forest trees.

(*ii*) The **Hessian fly** (*Myetiola*) is a small-sized midge; its larvae bore into and feed upon the pith of wheat plant.

(*iii*) The larvae of moths called **corn borer** (*Chilo* and *Pyrausta*) kill the corn and other monocotyledonous cereal plants by boring into their stems.

(*iv*) The **sugarcane leaf hopper** (*Pyrillā*) sucks the juice of sugarcane both as adult and nymph, causing great loss of sugar.

(*v*) Larvae of moth called **sugarcane stem borer** (*Argyria*), bore through the stem of sugarcane, some cereal plants and fruit trees.

(*vi*) The **Indian rice leaf-hopper** (*Nephotettix*) and **rice bug** (*Leptocorisa*) attack rice in great numbers eating the leaves and ears.

(*vii*) The nymphs and adults of *Hieroglyphus*, an orthopteran eat up the growing shoots of rice plants, thus, preventing formation of grain.

(*viii*) The **Indian cotton bug** or **Red cotton stainer** (*Dysdercus*) sucks the juice of leaves, buds and tender seeds of cotton plants. The cotton produced becomes spotted.

(*ix*) *Aphis*, a hemipteran, is a serious cotton pest of India.

(*x*) **Cabbage aphid** sucks the juice of stem and flowers of the plant of mustard, cauliflower and damage them.

(*xi*) Larvae of two Lepidoptera, *Agrotis* and *Gnorimoschema* are potato cut worms of India; the former feeds on potato leaves and cuts off stems, while the larvae of latter eat the potatoes in the field and stores. Larvae of *Agrotis* are also destructive to pea, cabbage, tobacco, groundnut, wheat and cauliflower.

(*xii*) **Red pumpkin beetle** (*Rhaphidopalpa* or *Aulacophora*) chews and destroys the young and tender plants of following vegetable plants of family Cucurbitaceae such as *Cucurbita*

(Kaddu), *Laginaria* (Lauki), *Momordica* (Karela), *Citrullus* (Tinda), *Luffa* (Torai), *Cucumis* (Kakri), etc.

(*xiii*) Larvae of *Bruchus*, a beetle, bores into pods of beans and peas kiling the seed.

(*xiv*) **Grasshoppers** (*Melanopus*, *Poecilocerus*) and locusts (*Schistocerca*) chew the tender leafy plants such as rice, wheat, maize, etc., and cause great damage. Swarms of migratory locusts cause great damage to crops and nightmares to farmers.

(*xv*) Many insects attack fruit trees, they damage roots, trunks, stems, leaves, inflorescence and fruits of these plants. *Drosicha* (**mealy bug**) causes destruction of mangoes and plums. Adults of *Ideocerus* (**mango leaf hopper**) attack the inflorescence of mango plant and suck the sap, thus, they cause considerable damage by preventing formation of mango fruit. *Psylla* (**apple bug**) lays eggs on apple and pear trees, the nymphs on hatching damage the blossoms and shoots. The larvae of *Anthonomus*, a beetle also destroys apple blossom and prevent formation of fruit.

(*xvi*) *Bagrada* (**painted bug**) destroys cruciferous plants by sucking the juice of their tender stems.

(*xvii*) Caterpillar larvae of moths such as *Earias* (**spotted bollworm**) and *Pectinophora* (**pink bollworm**), bore into flowers, young bolls and seeds of cotton, rendering these hollow and useless.

(*xviii*) **Gram caterpillar** and **corn earworm** (*Heliothis*), bore into stems of gram, corn, tomato, tobacco and some other plants.

(*xix*) Many moths (caterpillars) and beetle cause a great deal of damage to stored grains, two beetles, *Tenebrio* and *Tribolium* have similar habits and are commonly found in stores and granaries, the former is found in all stages in meal, flour and stored goods, its larvae are called **meal worms**. *Tribolium* eats stored wheat and grain. *Calandra*, a weevil bores through grain of rice and other stored grain in India.

(*xx*) **Gall insects** produce swellings on the plants to get shelter and food.

2. Injurious to Human Beings and Domestic Animals

1. Parasites. Some insects live upon (**ectoparasites**) or inside (**endoparasites**) the bodies of human beings and domestic animals. The ectoparasites suck the blood of their hosts at short intervals. Examples of such insects are the following:

(*i*) **Lice** (*Pediculus*). These insects occur between hairs of human beings and domestic animals and feathers of birds. **Human lice** are of two subspecies—head louse, *Pediculus humanus capititis* and the body louse, *Pediculus humanus corporis* (it mostly harbours the hairs of arm pits and pelvic region). Besides sucking blood, the lice cause annoyance and itching by their crawling in between hairs or upon host body.

(*ii*) **Fleas.** These are also blood-sucking ectoparasites found amongst the hairs of mammals and feathers of birds. **Indian rat flea** (*Xenopsylla*) transmits **bubonic** and other types of **plague**.

(*iii*) **Bedbugs** (*Cimex*). These are found concealed in cracks and crevices of our mats, furniture, cots, wells, etc. During night these suck human blood and cause intense itching, burning and restlessness, disturb us in our sleep. *Cimex* also transmits typhus fever.

(*iv*) The **forest fly** of cattle and horses (*Hippobosca*) sucks the blood of its hosts and often causes haemorrhage.

(*v*) The **chicken louse** (*Menopon*) sucks blood and causes destruction of fowls. Endoparasitic insects usually occur in cattle. **Botflies** (*Gasterophilus*) are most important amongst these. Larvae of botflies excavate tunnels in host skin, causing great discomfort and pain. These also render the skin useless for leather. **Sheep and horse botflies** lay eggs near the nose of their hosts. Hatched larvae may invade their brain. Larvae of **warble flies** (*Hypoderma*) bore below the skin of oxen and make holes for breathing; they thus, injure the hide but also reduce the meat and milk supply.

3. Disease transmitting insects. The insects not only cause disease by leading parasitic life, but also act as **vector** and play a dreadful role by transmitting disease causing (*i.e.*, pathogenic) viruses, bacteria, protozoans, etc., from host to host either by direct or indirect mechanical way. The various species of genus *Anopheles* (mosquito) transmit parasitic protozoa (*e.g., Plasmodium*) which causes **human malaria**. *Culex* mosquito carries *Filaria bancrofti* (Namatode) which cause **filariasis** in human beings. Deadly disease of **sleeping sickness** is transmitted by **tse-tse fly** (*Glosina*). Other dreadful diseases such as **cholera, typhoid, diarrhoea, amoebic dysentry, gangrene, gonorrhoea, leprosy** and **tuberculosis** are also spread by **housefly**; **plague** by fleas, **relasping fever** by **bedbug** and **Kala-azar** by **sandflies** (*Phelbotomous*).

Certain diseases of the domestic animals are also transmitted by certain insects. For example, **nagana** disease in horses and cattle and **surra** disease in domestic animals are conveyed by the flies *Glossina* and *Tabanus*, respectively.

4. Household pests. Human food is spoiled by cockroaches, ants, flies and weevils. Various species of moths (*Tinea, Teniola* and *Trichophaga*) are clothes moths, they lay eggs on warm clothes; the larvae on hatching eat and destroy clothes. They also feed on furs, carpets and dry fruits. *Lepisma*, the silver fish and *Liposcelis*, the book louse live in and destroy books and old manuscripts. *Termites*, the white ants cause great damage to books, carpets, wood furniture and wood-work of the buildings.

B. Beneficial Insects

1. Productive insects. Some insects are of commercial value. They manufacture various products which fetch good market value.

(*i*) **Honeybee** (*Apis*) supply to humankind millions of kilograms of honey and beeswax. **Beeswax** is obtained from its combs and is used in making candles, cream, polish, carbon ribbons, papers, cosmetics and certain ointments. **Honey** is a natural antiseptic and valuable tonic for human body. A drug prepared from the bodies of honeybees too has been used in the treatment of arthritis and snake bite. Rearing and culturing of *Apis* to obtain honey and bees wax is called **apiculture**.

(*ii*) The **lac insects**, *Tacchardia lacca*, secrete lac in the form of protective covering by the integumentary glands of female. The lac is used in the manufacture of **shellac**, varnish, polish, buttons, bangles, pots, gramophone records, sealing wax, toys and some electrical items. A dye is also obtained from dead and dried bodies of females of lac insect which is used by Indian womens for making "alta" or mahavar.

(*iii*) The dark red **cochineal dyes** in Mexico, Spain, etc., are manufactured from the dried bodies of female of the genus *Dactylopius coccus*. This dye is used in cosmetics, colouring the wine and in medicines for whooping cough.

(*iv*) **Cantharidine**, a drug used in healthy growth of hair, is obtained from **blister beetle** (*Lytta* and *Mylabris*).

(*v*) The silk moth, *Bombyx mori* and other genera, are reared the world over and especially in India, China, Japan and Europe to obtain the best quality of **silk**. The raw silk is obtained from the cocoons of these insects and the thread is reeled off to prepare the silk. The pure silk fetches a good market price and is demanded world over.

(*vi*) Gall insects are helpful in the manufacture of **tannins** which are used variously.

2. Helpful insects. Many insects are beneficial to human kind in the following ways.

(*i*) **Insect pollinators.** Insects of various groups are of great importance to agriculture, as they are the pollinating agents of various plants. Various fruit, vegetable and garden plants are mostly pollinated by insects such as **honeybee, wasps, butterflies, moths,** etc. These insects sit upon flower to flower for sucking their nectar and pollinate them in this process. Flowers of many plants are structurally adapted for insect-pollination.

(*ii*) **Scavengers.** Some insects are scavengers; they eat up dead animals and vegetable matter, thus, they prevent decay. Some ants, larvae of flies, beetles, crickets, cockroaches, etc., form common example of scavengers.

(*iii*) **Predators.** Some insects, called **entomophagous insects**, feed upon harmful insects and, thus, are useful insects. *Stagomantis*, a mantis is a voracious feeder devouring flies, grasshoppers and caterpillars which are injurious to crops. The larvae and adult of lady bird beetle (*Chilomenes*) feed upon aphids which infect the cotton plants. *Novius* (a **ladybird beetle**) destroys scale worms which are pests of orange and lemon trees. The larvae of **blister beetle** (*Epicauta*) bore through egg-capsules of locusts and destroy them. *Calasoma*, a **ground beetle**, preys upon many kinds of lepidopterous larvae which destroy cereals and cotton. Certain hymenopterous larvae and carnivorous wasps feed upon aphids in large numbers. Large dragon flies feed upon mosquitoes.

(*iv*) **Parasitoid insects.** Certain **parasitic** insects are useful for human beings in the sense that they either inhabit insects which are harmful or feed upon them. *Chaloids* and *Icheumon* flies are parasitic forms which lay eggs in cocoons and larvae of phytophagous lepidoptera. Similarly larvae of *Tachina* fly are parasites of the lepidopterous larvae.

(*v*) **Insects of food.** Insects comprise staple food for a variety of animals. Human beings are indirectly benefited by killing and using a variety of insect-eating (insectivorous) fishes, frogs and birds. Certain large sized insects such as grasshoppers and locusts are themselves used as food.

(*vi*) **Destroyer of weed plants.** Certain insects habitually attack upon and destroy weed plants in our farms and gardens.

(*vii*) **Medicinal value.** The larvae of flies such as *Lucilia* and *Phormia* are used in healing wounds of bones which do not respond to medicine

QUESTIONS

Long Answer Questions

1. Write an essay on economic importance of insects. (*Kanpur 1989, 1992; Himachal 1990, 1995; Meerut or CCS 1992; Lucknow 1997*)
2. Write an essay on useful and harmful insects.
3. Write an essay on insects in relation to man. (*Delhi 1991*)
4. Write an essay on 'insects are friends to man'. (*Meerut or CCS 1990*)
5. Give an account of beneficial insects. (*Madurai 1994*)
6. Write what you know about harmful insects? (*Himachal 1994*)

Very Short Answer Questions

1. Which insect spreads cholera?
   Ans. Housefly.
2. From which insect we get the silk?
   Ans. Cocoon formed around it by caterpillar of the silk moth (*Bombyx mori*). Cocoon is the source of silk fiber.

Fill in the Blanks

1. *Wuchereria bancrofti* is transmitted by ....
2. Rat flea spreads ....

Multiple Choice Questions

1. An insect that produces shellac is
(*a*) *Apis indica* (*b*) *Bombyx mori*
(*c*) *Laccifera lacca* (*d*) *Dictyolopius coccus*

2. Insects responsible for transmitting diseases are called
(*a*) pests (*b*) social insects
(*c*) pathogens (*d*) vectors

3. Insects are greatest competitors of human beings because most of them
(*a*) eat and destroy plants and fruits (*b*) are responsible for spreading of serious diseases
(*c*) destroy household goods and properties (*d*) are responsible for all the above

4. Insects responsible for destruction of buildings and wooden structures are
(*a*) moth (*b*) locusts
(*c*) weevils and grubs (*d*) termites

5. The insect vector for kala-azar is
(*a*) tse-tse fly (*b*) sandfly
(*c*) housefly (*d*) fleas

6. Most ticks and mites are
(*a*) free living (*b*) parasites
(*c*) predators (*d*) all of the above

7. Lac is produced as
(*a*) secretion from body (*b*) excretion from body
(*c*) excess fluid oozing out of body (*d*) faeces of lac insects

8. Cantharidin is obtained from
(*a*) ball weevils (*b*) honey bees
(*c*) beetles (*d*) white ants

9. Which of the following is a pest of paddy?
(*a*) *Dacus* (*b*) *Calandra*
(*c*) *Dysdercus* (*d*) *Leptocorisa*

10. One cocoon of silk moth gives the fiber of
(*a*) 10 feet (*b*) 100 feet
(*c*) 1000 feet (*d*) 10,000 feet

11. Bee-venom is used in the treatment of
(*a*) arthritis (*b*) malaria
(*c*) myopia (*d*) cholera

12. Bedbug is
(*a*) endoparasite (*b*) ectoparasite
(*c*) symbiotic (*d*) none of these

13. The feeding habit of bedbug is
(*a*) herbivorous (*b*) carnivorous
(*c*) omnivorous (*d*) sanguivorous

14. The pest called sugarcane leaf hopper is
(*a*) *Dysdercus* (*b*) *Pyrilla*
(*c*) *Hieroglyphus* (*d*) *Chilo*

ANSWERS

Fill in the Blanks

1. *Culex* mosquito **2.** Bubonic plague

Multiple Choice Questions

1. (*c*) **2.** (*d*) **3.** (*d*) **4.** (*d*) **5.** (*b*) **6.** (*b*)
7. (*a*) **8.** (*c*) **9.** (*d*) **10.** (*c*) **11.** (*a*) **12.** (*b*)
13. (*d*) **14.** (*b*).

43

Insect Pest Management (Insect Control)

In order to prevent damage to crops, horticulture products and to protect ourself and our domestic animals from diseases it becomes essential to control insect pests. Harmful insect pests can be controlled by following three methods:

1. Mechanical control;
2. Chemical control;
3. Biological control.

43.1 MECHANICAL CONTROL

Large-sized harmful insects can be manually picked up, or collected by nets and then destroyed. Besides this, we can also keep such insects, away from us, our cattle and our useful things by using protective meshes on doors and windows of our residences, hospitals, animal houses, storehouses, etc.

Ancient people used to collect, for killing, large numbers of such insects by using gur, sweets, etc., as baits. Modern scientists have synthesized certain sex attractant scents, called **pheromones**, for attracting insects at one place to be killed *en masses*.

Harmful insects can also be controlled through environmental changes such as follows:

1. Localized insects and bettle grubs can be destroyed by the rotation of crops.
2. Houseflies can be controlled by having proper sanitation.
3. Insects that lay eggs in stagnant water (*e.g.*, mosquitoes) can be destroyed by removing stagnant and dirty water in ponds, pools and ditches and drying them.
4. Insects can be destroyed and may be discouraged from breeding by removing heaps of garbage or by drying and burning it. This will destroy the breeding grounds of harmful insects.
5. Insects can also be killed by raising temperature of store houses, godowns or production centre. Insect pest can also be rendered inactive by lowering the temperature such as in the cold storage.

43.2 CHEMICAL CONTROL

Insect control by using various chemicals is simplest and most popular method. These chemicals are called **insecticides,** since they act as insect killers. Insecticides are toxicants and are used in various forms such as powder, sprays, solutions, emulsions or fumes. Insecticides can be divided into the following three categories:

***(i)* Stomach poisons.** The insecticides are sprayed upon food of insects. When ingested, the poisoned food kills the insects. **Lead arsenate, cryolite, sodium arsenate, sodium fluorides, sodium fluorosilicate** and **Bordeaux mixture** are common stomach poisons used for killing insects.

(*ii*) **Contact poisons.** These are strong insecticides which kill the insects instantaneously upon mere contact. Some of most popular contact insecticides are **DDT (dichloro-diphenyl-trichloro-ethane), gammexane, pyrethrum, aldrin** and **toxophene**.

(*iii*) **Fumigants.** Fumigation is a process of creating lethal concentration of a toxic gas (fumigant) for the time necessary to destroy insect infestation in a closed space. Fumigants consists of volatile chemicals which quickly vapourise and kill the insects without affecting the grain. **Ethylene dibromide, methyl bromide, aluminium phosphide, sulphur, cyanide** are some common fumigants which are used for insect pests.

Disadvantages of chemical control. Human beings have been successful in controlling insects populations by the use of insecticides. But their indiscriminate use has resulted into following problems:

1. When an insect pest is exposed to same insecticide repeatedly it tends to develop resistance to that insecticide. So that after sometime, the use of that particular insecticide becomes ineffective.
2. Ecologically DDT and other insecticides become non-biodegradable pollutant and have polluted our land, water and air. They have produced various health problems for human beings and domestic animals (they tend to result in toxicity due to their biomagnification by various food chains).

43.3 BIOLOGICAL CONTROL

This is self-perpetuating method and by it insects of a specific species can be destroyed. It involves the following methods:

(*i*) **Predation.** Biological control of insects pests can be done by introducing insectivorous animals such as lizards, snakes, frogs, toads, spiders or insectivorous birds or mammals. If the population of these insectivorous predators increases, they will devour more and more insects and will keep a check on the increase of insect population.

(*ii*) **Parasitism.** Insect pests can also be controlled by introducing those viruses, bacteria, fungus, etc., in insects which produce disease in them. These pathogens ultimately kill the insects and help in reducing their population.

(*iii*) **Sterilization.** Male insects are collected and sterilized by exposing them to X-rays or gamma rays. These are, then, released into their populations. These copulate with the females normally, but the ova remain unfertilized. Hence, there are no progenies.

QUESTIONS

Short Answer Question

1. Write a short note on insect control.

Very Short Answer Questions

1. Name two normal agricultural operations that help control insect pests.

   Ans. Ploughing and hoeing. [Hoeing means to use a hoe to turn (earth) or cut through (weeds)].

2. How does post-harvest stubble removal can control pests?

   Ans. By eliminating hiding places of pests.

3. What is a trap crop?

   Ans. An early crop sown around the main crop to attract insect pests common to both.

4. What do you mean by biocontrol?

   Ans. Control of pests by using natural enemies.

5. Name the agency in India that manages movements of useful insects from one country to another.

Ans. AICRPBC (= All India Coordinated Research Project on Biological Control of Crop Pests and Weeds).

6. Cite an example of insect pest control with a virus.

Ans. Nuclear polyhedrosis virus is used to control forest defoliating insect pest, the pine sawfly.

7. What is parasitoid?

Ans. An insect (*e.g.*, a wasp) that completes its larval development within the body of another insect which it eventually kills and that is free living as a adult.

8. Give an alternative term for pheromone.

Ans. Ectohormone.

9. Name the synthetic compound that resembles the natural pheromone dispalure of gypsymoth.

Ans. Gyplure.

10. Mention three types of allelochemicals.

Ans. Allomones, kairomones and synomones.

Fill in the Blanks

1. The two main types of insect control are .... and applied.
2. A systemic poison is taken up by .... insects from the plant sap.
3. DDT is a .... poison.
4. Calcium arsenate is a .... poison.
5. Carbon bisulphide and .... are fumigants.
6. Antifeedants protect the crops .... killing, repelling or trapping the insect pests.
7. Mimics of moulting hormone produced by plants are called ....
8. Crop rotation is a .... method of insect control.
9. .... is a predatory insect.
10. Deep and frequent tillage can kill the root-infesting .... of maize.

Multiple Choice Questions

1. Which of the following is not a stomach poison?
   (*a*) lead arsenate (*b*) paris green
   (*c*) sulphur (*d*) borax
2. Which of the following is both contact poison as well as fumigant?
   (*a*) nicotene (*b*) DDT
   (*c*) BHC (*d*) naphthalene
3. The chemicals which prevent the insect pests to feed on their host plant is called
   (*a*) repellents (*b*) trappers
   (*c*) antifeedants (*d*) fumigants
4. Ecdysterones are produced by
   (*a*) insect pests (*b*) host plants
   (*c*) all plants (*d*) all insects
5. Chlorfluazuron is an
   (*a*) insect repellent (*b*) antifeedant
   (*c*) chitin synthesis inhibitor (*d*) hormone mimic

6. One of the major difficulties in the biological control of insect pest is that
   (*a*) the method is less effective as compared with the use of insecticides
   (*b*) the practical difficulty of introducing the predator to specific areas
   (*c*) the predator develops a preference to other diets and may itself become a pest
   (*d*) the predator does not always survive when transferred to a new environment
7. Biological control of agricultural pests unlike chemical control is
   (*a*) self perpetuating (*b*) toxic
   (*c*) polluting (*d*) very expensive
8. An insecticide extracted from the leaves of neem is
   (*a*) aldicarb (*b*) malathion
   (*c*) rotenone (*d*) azardirachtin
9. Insecticides usually act upon
   (*a*) digestive system (*b*) nervous system
   (*c*) circulatory system (*d*) muscular system

ANSWERS

Fill in the Blanks

1. Natural 2. Sucking 3. Contact 4. Stomach
5. Naphthalene 6. Without 7. Phytoecdysone 8. Cultural
9. Preying mantis 10. Aphids.

Multiple Choice Questions

1. (*c*) 2. (*a*) 3. (*c*) 4. (*b*)
5. (*c*) 6. (*c*) 7. (*a*) 8. (*d*)
9. (*b*).

44

Phylum Mollusca (Characters, Classification and Types)

The phylum Mollusca (L. *molluscus* = soft bodied) includes such forms as garden slugs, clams, oyesters, squids, octopods and snail. It is the largest animal phylum after the arthropoda. Over 60,000 living species of molluscs have been described. In addition, some 35,000 fossil species are known because the phylum had a long geological history (originated in Cambrian Period). The biggest of all invertebrates (giant squid) and probably the most intelligent invertebrate (*Octopus*) are the molluscs.

Molluscan shell have always been economically important, having served as money and been fashioned into jewellery and buttons. Shells of molluscs are important components of kitchen middens of prehistoric human beings. Scallops, oysters, clams, squids and octopi are still important food items.

44.1 DEFINITION

The molluscs are soft bodied, unsegmented (except Monoplacophora), triploblastic, bilaterally symmetrical coelomates that contain a ventral muscular foot, a dorsal visceral hump and a fleshy mantle which secretes a calcareous shell.

44.2 GENERAL CHARACTERS

1. Molluscs are essentially aquatic, mostly marine, few freshwater and some terrestrial forms.
2. The body is soft, bilaterally symmetrical, unsegmented (except Monoplacophora) and without jointed appendages.
3. Body is commonly protected by an exoskeletal calcareous shell of one or more pieces.
4. Body is divided into a **head**, a **visceral mass, mantle** and a **foot**.
5. The head carries **mouth**, **eyes** and **tentacles** (Eyes and tentacles are absent in Pelecypoda and Scaphopoda).
6. The visceral mass is arranged compactly and contains vital internal organs.
7. The mantle or pallium is a thick fold of bodv wall and covers the mantle cavity like a cloak and secretes a shell.
8. The foot is ventral in position, thick, muscular and adapted for creeping, ploughing/burrowing and swimming, but is absent in oyster.
9. Body cavity is **haemocoel**. True **coelom** is confined to the lumen of pericardium, gonads and nephridia.
10. Digestive tract is simple with an anterior mouth and posterior anus. However, in gastropods, scaphopods and cephalopods the intestine becomes U-shaped bringing anus to anterior position.
11. A **radula** is present in buccal cavity of all molluscs except Pelecypoda. It is a file-like rasping organ for feeding.

12. An olfactory organ or chemoreceptor, the osphradium, is situated basally (*i.e.*, at the base of gills) to test the purity of those water current which enters mantle cavity.
13. Respiration is done by one or many **ctenidia** or **gills** enclosed in the mantle cavity.
14. Aerial respiration is done by a lung-like structure developed from the mantle in terrestrial forms.
15. Circulatory system is of open type [*i.e.,* the afferent and efferent vessels are not directly joined; instead the blood (i.e., haemolymph) coming from the heart empties into a body cavity, which is then drained by vessels leading to heart] with dorsal heart and few blood vessels. However, cephalopods show some tendency towards a closed system. Respiratory pigment is usually **haemocyanin**.
16. Excretion is performed by one or two pairs of **kidneys** (metanephridia) communicating the coelom with mantle cavity by nephridiopore.
17. Nervous system consists of paired cerebral, pedal, pleural and visceral **ganglia**. Ganglia are joined by connectives and commissures.
18. The sensory organs are eyes, tentacles, osphradium and the statocysts.
19. Sexes are usually separate (**dioecious**) but may be united (**hermaphroditic**). Gonads are usually unpaired.
20. Fertilization is either external or internal.
21. Mostly oviparous, few are viviparous.
22. Cleavage is generally spiral and determinate.
23. Development may be direct or with **veliger larva** (*i.e.*, trochophore larva).

44.3 OUTLINE CLASSIFICATION

Phylum Mollusca

Classes (7)

1. **Aplacophora** — **Example.** *Chaetoderma*
2. **Polyplacophora** — **Example.** *Chiton*
3. **Monoplacophora** — **Example.** *Neopilina*
4. **Gastropoda**
5. **Pelecypoda (Bivalvia)**
6. **Scaphopoda** — **Example.** *Dentalium*
7. **Cephalopoda**

Class 4. Gastropoda

Subclasses (3)

1. Prosobranchia
Orders (3)

1. **Archaeogastropoda**
   Examples. *Patella, Haliotis*
2. **Mesogastropoda**
   Examples. *Pila, Cypraea*
3. **Neogastropoda**
   Examples. *Murex, Buccinum*

2. Opisthobranchia
Orders (9)

1. **Anaspidea**
   Example. *Aplysia*
2. **Nudibranchia**
   Example. *Doris*
3. **Cephalaspidea**
   Example. *Bulla*
4. **Pyramidellacea**
   Example. *Turbonilla*
5. **Acochlidiacea**
   Example. *Acochlidium*
6. **Notaspidea**
   Example. *Pleurobranchus*
7. **Saccoglossa**
   Example. *Elysia*
8. **Thecosomata**
   Example. *Limacina*
9. **Gymnosomata**
   Example. *Pneumoderma*

3. Pulmonata
Orders (2)

1. **Basommatophora**
   Examples. *Planorbis, Lymnaea*
2. **Stylommatophora**
   Examples. *Helix, Limax*

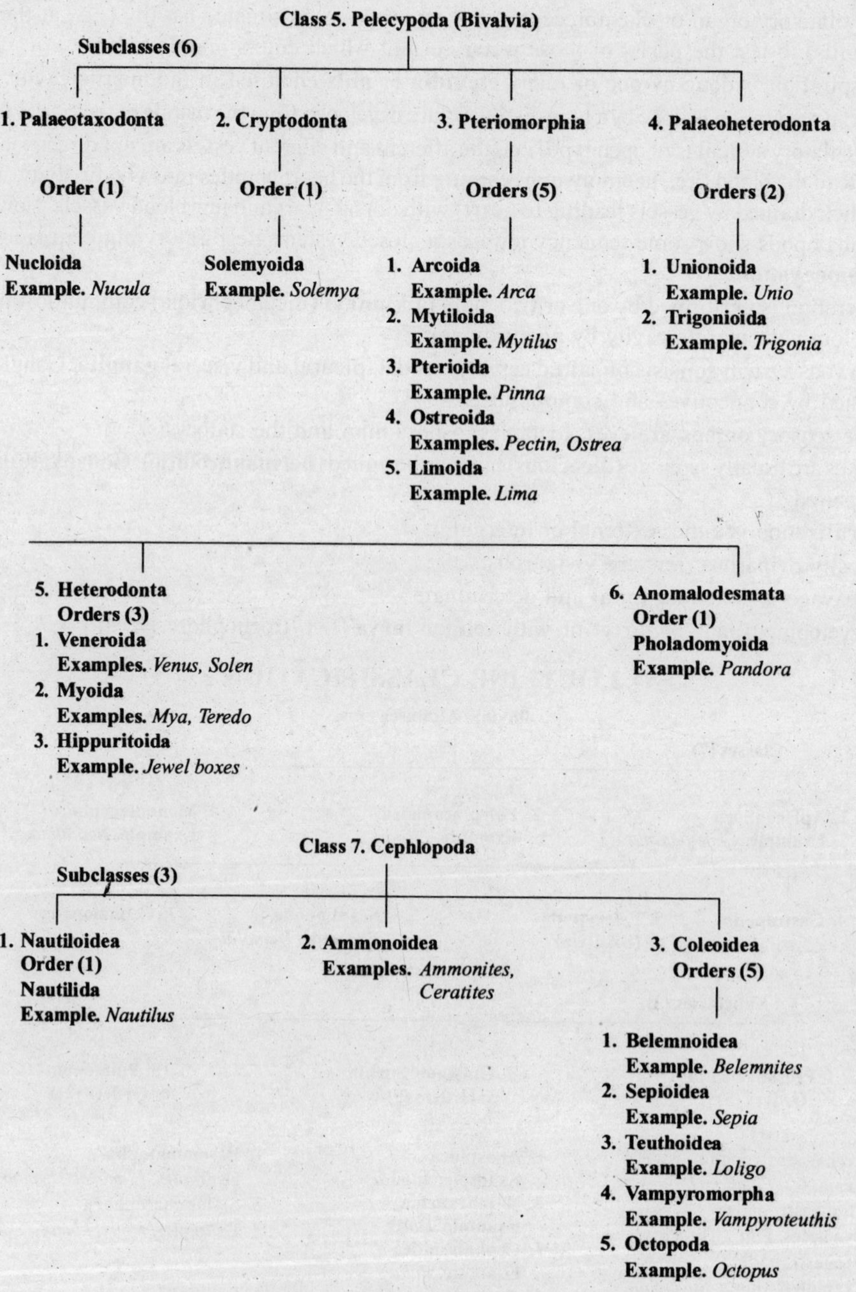

44.4 DETAILED CLASSIFICATION

Phylum Mollusca is divided into following **seven** classes:

Class 1. Aplacophora

1. Worm-like marine molluscs which are found at a depth of 9000 metres.
2. Body is less than 5 cm long, elongated, cylindrical and bilaterally symmetrical.
3. Head is poorly developed. Shell is absent.

4. Mantle is decorated with numerous calcareous spicules. It encloses the body from all sides.
5. Foot is absent or reduced to a ventral ridge.
6. Digestive tube is straight, typically with radula and with terminal mouth and anus.
7. Gills are either absent or reduced to a pair and located in posterior mantle cavity.
8. One pair of renal organs (coelomoducts) open to pericardial cavity and are used for the discharge of gametes.

Examples. *Chaetoderma*; Solenogasters such as *Neomenia*, *Paramenia*, etc.

Class 2. Polyplacophora

1. Body elliptical, convex dorsally and flattened ventrally.
2. Shell is formed by eight overlapping calcareous plates arranged in a longitudinal row.
3. Foot is ventral, broad, muscular and creeping.
4. Mantle with spicules disposed along the margins or girdles.
5. Gills are 6 to 8 pairs present in the pallial groove on the lateral sides of foot.
6. Radula is present.
7. Sexes are separate. Veliger larval stage is absent.

Examples. *Lepidopleurus, Hanleya, Chiton, Cryptocniton, Tonicella.*

Class 3. Monoplacophora

1. These are regarded as primitive molluscs and are believed to be ancestral to the gastropods, bivalves and cephalopods.
2. These are bilaterally symmetrical and metamerically segmented but segmentation is internal.
3. Shell is dorsal and consists of a single piece.
4. Head is without eyes and tentacles.
5. Mantle covers the dorsal surface of body.
6. Gills, auricles, nephridia and gonads are metamerically repeated.
7. Heart contains two auricles and two ventricles.
8. Dioecious. Gonads are four in number.

Examples. *Neopilina galatheae, Vema ewingi.*

Class 4. Gastropoda

1. It is largest class of Mollusca.
2. Gastropods are marine, freshwater and terrestrial.
3. Body is coiled and torted, *i.e.*, asymmetrical and twisted by torsion. It is also dorsoventrally elongated.
4. Foot is broad, muscular and flattened, situated ventrally below the head. It is sole-like and adapted for creeping.
5. Head is distinct bearing tentacles, eyes and mouth.
6. Visceral mass usually coiled exhibiting various degrees of torsion.
7. Mantle is a collar-like fold of body wall. Mantle cavity is restricted to posterior body.
8. Shell is univalve but spirally coiled and may be rudimentary and absent.
9. Buccal cavity contains an odontophore with a **radula** bearing rows of chitinous teeth.
10. Anus is usually situated anteriorly close to the mouth.
11. They contain a single gill or ctenidium. In absence of gill, mantle cavity functions as a lung.

12. Osphradium is present.
13. Their nervous system consists of paired ganglia such as cerebral, buccal, pleural, pedal, parietal and visceral ganglia.
14. Monoecious or dioecious.

Class 5. Pelecypoda (or Bivalvia or Lamellibranchiata)

1. They are aquatic, mostly marine, some freshwater.
2. Body is laterally compressed.
3. Shell is bilobed and consists of two lateral shell valves, both of which united or hinged mid-dorsally by hinge-ligaments.
4. There is no cephalization, *i.e.*, head, tentacles and eyes are absent. Radula is also absent.
5. Foot is highly muscular, antero-ventral, large, blade-like, adapted for burrowing. Foot is reduced in attached forms.
6. Mantle is bilobed and consists of two lateral (*i.e.*, right and left) flaps that completely enclose the body.
7. Gills or ctenidia are paired, well-developed and plate-like hence called **lamellibranchs.** They are often specialized to assist in feeding.
8. Nervous system consists of four pairs of ganglia—cerebral, pleural, pedal and visceral. Cerebral and pleural ganglia of each side usually fused into a single cerebropleural ganglion.
9. Sense organs are statocyst and osphradia.

Class 6. Scaphopoda

1. They are burrowing molluscs and live buried in sand with head downward and body steeply inclined.
2. Body is elongated and cylindrical.
3. Shell is univalved, tubular and tusk-like and opens at each end.
4. Head and sense organs are reduced. Eyes and gills are absent.
5. The large anterior opening of shell contains a cone-like foot and buccal region.
6. Adhesive knobbed tentacles, called **captacula**, surround the mouth and are used in feeding and in feeling about in mud.
7. Circulating water enters and leaves through the smaller anterior end of shell.
8. Foot small, pointed, adapted for creeping and burrowing.
9. Mantle fuses with the body ventrally and forms a tube.

   Examples. *Dentalium*, *Cadulus*.

Class 7. Cephalopoda

1. They are marine, pelagic and raptorial (or predaceous) molluscs.
2. Body is bilaterally symmetrical, somewhat cylindrical with distinct head and trunk.
3. Head is large with mouth and a pair of conspicuous eyes.
4. Mouth consists of jaws and radula.
5. Gills are bipectinate and are either one pair or two pairs.
6. Visceral mass is well developed but without torsion.
7. Mantle encloses mantle cavity posteriorly and ventrally.
8. Foot is divided into muscular, sucker-bearing arms or tentacles which are located around the buccal area.

9. Shell is either external and well formed (it may be spiral or chambered) or internal and reduced (covered by the mantle) or totally absent.
10. Water is pumped through the mantle cavity and out a ventral funnel. This jet propulsion provides the force for swimming.
11. Circulatory system is closed.
12. Nervous system is highly advanced. **Cartilage** forms a skull-like capsule around the brain for protection.
13. Sexes are separate.
14. Cleavage is meroblastic. Metamorphosis is absent.

Class 4. Gastropoda

Class Gastropoda is divided into **three** subclasses:

Subclass 1. Prosobranchia

1. Marine, freshwater and terrestrial forms.
2. Visceral mass and nervous system become twisted into figure eight due to torsion of 180°.
3. Mantle cavity opens anteriorly in front of visceral mass.
4. Head is snout-like and bears a pair of tentacles and a pair of eyes.
5. Shell is calcareous, conical and spiral. Shell opening is guarded by operculum.
6. Foot is ventral and muscular.
7. Gills are most primitive. There are two bipectinate gills in the mantle cavity.

Subclass Prosobranchia is divided into **three** orders:

Order 1. Archaeogastropoda

1. These are primitive forms.
2. Mantle cavity is primitive and is without gills, or contains a pair of ctnedia or single ctnidium.
3. Ctnedia are plume-like or bipectinate.
4. Shell has a notch or slit for anus and exhalant siphon but shell becomes imperforate when ctenidia are reduced.
5. Two auricles are present in heart.
6. They have two nephridia and often two osphradia. Each osphradium is poorly developed simple and ridge-like.
7. Proboscis, siphon, penis and prostatic gland are absent.
8. Operculum is generally absent.
9. Gonads open into the right kidney.

Examples. *Patella* (true limpet), *Acmaea* (limpet), *Haliotis* (ear shell), *Fissurella* (key-hole limpet), *Trochus* (top shell), *Astrea*, *Turbo* (cat's eye).

Order 2. Mesogastropoda (or Taenioglossa)

1. Mainly marine but with many freshwater and terrestrial genera.
2. Prosobranchs with single monopectinate gill, one auricle and one nephridium.
3. Operculum may be present or absent.
4. Shell is without slit band or hole for exhalant siphon and usually without groove for inhalant siphon.
5. Radula **taenioglossate**, *i.e.*, seven teeth in a transverse row.

6. They have a single well-developed or simple ridge-like osphradium.

   Examples. *Pila* (apple snail); *Viviparus; Valvata* (an exception with bipectinate gill); *Cypraea* (cowry); *Jonthina* (pelagic snail), *Littorina* (periwinkle); *Crepidula*.

Order 3. Neogastropoda (or Stenoglossa)

1. All are marine.
2. They have a single monopectinate (comb-like) gill, one auricle and one nephridium.
3. With an unpaired oesophageal gland which becomes poison gland in some.
4. Radula is **rachiglossate**, *i.e.*, three teeth in a transverse row.
5. Shell with more or less elongated siphon canal.
6. Have a complex osphradium with bipectinate folds.
7. They have penis.

   Examples. *Murex* (boring murex or drill); *Thais* ; *Magilus*, *Buccinum* (welk); *Conus* (cone shell), *Busycon* (conch); *Olivo* (olive shell), *Nassarius* (mud snail).

Subclass 2. Opisthobranchia

1. Mostly they are marine.
2. Many of them are secondarily, bilaterally symmetrical.
3. They show strong tendency towards reduction of shell and mantle cavity.
4. Mantle cavity is widely open and situated posteriorly.
5. They contain one or no gill, one auricle, one nephridium but display detorsion. Operculum is usually absent.
6. Head bears two pairs of cephalic tentacles.
7. Buccal cavity with a pair of jaws.
8. Gills are frequently replaced by secondary branchiae (surface gills) which lie behind the heart.
9. Nervous system is untwisted by detorsion and often become more centralized.
10. Sexes are united, *i.e.*, monoecious or hermaphrodites.

    Subclass Opisthobranchia is divided into following **nine** orders (**Note.** General students are advised to select first two orders only).

Order 1. Anaspidea (or Aplysiacea)

1. This order includes **sea hares** which are mostly found in tropical and subtropical waters.
2. Externally they are bilaterally symmetrical.
3. Shell is small, more or less is covered by mantle.
4. Visceral hump is attached along foot. Foot is with large lateral parapodia.
5. Anterior end of body bears a pair of tentacles, a pair of rhinophores and a pair of eyes. (**Rhinophores** are olfactoreceptors).
6. Jaws, gill and mantle cavity are present.
7. They contain several gastric shields and large radula with median tooth.

   Examples. *Aplysia, Akera, Creseis* (sea butterfly).

Order 2. Nudibranchia (or Acoela)

1. This order includes **sea slugs**.
2. Shell, mantle cavity, ospharidium and parapodia are absent.
3. Internal gills or ctenidium are absent. Some contain secondary gills or branchiae around anus (*e.g.*, *Doris*).

4. They contain a closed male genital groove.

Examples. *Doris* (sea lemon), *Tritonia*, *Eolis*.

Order 3. Cephalaspidea

1. This order includes **bubble snails** or **bubble shells.**
2. Dorsal surface of the head is shield-like.
3. Shell is reduced or absent.
4. They generally have gastric shield for grinding.
5. Male genital groove is open

Examples. *Bulla, Hydatina, Acteon.*

Order 4. Pyramidellacea

1. They live as ectoparasites of bivalve molluscs and of polycheates.
2. Shell and operculum are present. Shell is spirally twisted.
3. Proboscis contains a stylet instead of radula.
4. Gills and radula are absent.

Examples. *Turbonilla, Odostomia, Pyramidella.*

Order 5. Acochlidiacea

1. Shell, gill and cephalic shield are absent. They are called **naked snails.**
2. Visceral mass is sharply set off from the rest of body.
3. They are without jaws but contain radula. Gastric shield is absent.
4. They contain rhinophores.
5. Sexes united or separate.

Example. *Acochlidium.*

Order 6. Notaspidea

1. Shell is present or absent.
2. Their shell is usually internal.
3. Parapodia, mantle cavity and gastric shield are absent.
4. They contain one bipectinate gill.
5. Radula is variable but without radula sac.
6. Male genital groove is closed over.

Examples. *Pleurobranchus, Tylodina.*

Order 7. Saccoglossa

1. This order includes jawless, herbivorous forms.
2. Shell and rhinophores are present.
3. Radula bears a single row of teeth. Suctorial pharynx is adapted for feeding on algae.
4. Sperm duct is closed.

Examples. *Elysia, Alderia, Berthelinia* (bivalve).

Order 8. Thecosomata

1. This order includes the **shelled pteropods,** the **sea butterflies.**
2. They are pelagic forms with large parapodia.

3. Shell and mantle cavity are present.
4. Gill is reduced.
5. Male genital groove is open.
6. They are protandrous hermaphrodites.

Examples. *Limacina, Cavolina, Spiratella.*

Order 9. Gymnosomata

1. They include **naked pteropods** which lack shell as well as mantle cavity.
2. Foot contains parapodial fins.
3. Gills are present behind heart.
4. Male genital groove is open.

Examples. *Pneumoderma, Cliopsis.*

Subclass 3. Pulmonata

1. Mostly are terrestrial or freshwater; a few are marine.
2. Air breathing gastropods without gills (ctenidia).
3. Mantle cavity is modified into a lung (respiratory sac).
4. Shell is typically spiral or reduced or absent.
5. Operculum is absent.
6. They contain one auricle and one nephridium.
7. Nervous system is concentrated and symmetrical.
8. Hermaphrodite (monoecious); development is direct.

Subclass Pulmonata is divided into following **two** orders:

Order 1. Basommatophora

1. Primarily aquatic pulmonates found in freshwater and rarely in brackish water or marine habitats.
2. Shell is external, simple, delicate and spiral. It is with a conical spire and a large aperture.
3. Eyes are located near the base of tentacles.
4. Tentacles are one pair.
5. They usually contain separate male and female gonopores.

Examples. *Planorbis, Bulinus, Lymnaea, Helisoma, Physa, Melampus* (a salt water form).

Order 2. Stylommatophora

1. Terrestrial pulmonates.
2. Shell is rudimentary, internal or absent.
3. Tentacles are two pairs; the upper pair bears eyes at the tip.
4. Their male and female gonopores are usually united.

Examples. *Helix* (land snail); *Partule*; land slugs such as *Limax* (grey slug) and *Arion.*

Class 5. Pelecypoda

Class Pelecypoda is divided into following **six** subclasses:

Subclass 1. Palaeotaxodonta

1. Palaeotaxodonta means "primitive gills."
2. Shell valves are equal and **taxodont** *i.e.*, row of short teeth occur along hinge.
3. Shell structure is nacreous or lamellar.
4. Gills are bipectinate and **protobranchiate**, *i.e.*, gill filaments are arranged in rows on either side of a central axis and not folded into a 'W' shape.
5. Foot is ventral and flattened.
6. They have a pair of palpal tentacles.
7. Most are deposit feeders in soft bottoms.

Order 1. Nucloida

Examples. *Nucula* (nut clam), *Yoldia, Malletia.*

Subclass 2. Cryptodonta

1. Cryptodonta means "first shell".
2. Gills are protobranchiate.
3. Shell valves are thin, equal, somewhat elongated and without hinge teeth.
4. Palpal tentacles are absent.
5. Gills possess chaemosynthetic bacteria.

Order 1. Solemyoida

Example. *Solenmya.*

Subclass 3. Pteriomorphia

1. Most of them are attached by byssus threads or cemented to the substratum. Some become secondarily free living.
2. Mantle margins are not fused.
3. Foot is usually equipped with byssal gland.
4. Palps are small and lamellar.
5. Gills are **filibranchiate**, *i.e.*, gill filaments are folded into a W-shape and adjacent filaments are interconnected by cilia. Interfilamental junctions are absent.

This subclass includes following **five** orders:

Order 1. Arcoida

1. Hinge is straight and usually with vertical plications (having folded plaits such as in a fan).

Examples. Ark shells: *Arca, Barbatia, Anadara.*

Order 2. Mytiloida

1. Bivalves attached by byssal threads.
2. Hinge usually without teeth.
3. Anterior adductor muscle is reduced.

Examples. Mussels: *Mytilus* (sea mussel), *Modiolus*, *Pinctada* (pearl oyster).

Order 3. Pterioida

1. Bivalves attached by byssal threads.
2. Shell-valves are more or less equal in size.

Examples. Pen shells: *Pinna*, *Pteria*.

Order 4. Ostreoida

1. Bivalves that lie on one side; shell is either free or attached.
2. Shell valves are commonly unequal in size.

Examples. *Pecten* (scallops); *Anomia* (jingle shell or Venus' toenails); *Ostrea* (edible oyesters); *Placuna* (window pane clam).

Order 5. Limoida

1. Free or byssally attached bivalves.
2. Shell valves are equal in size.
3. Margins of mantle contain long tentacles.

Example. *Lima* (file clam).

Subclass 4. Palaeoheterodonta

1. Shell valves are of equal size and contain a few hinge teeth and many cardinal teeth.
2. Shell contains an inner nacreous layer.
3. Siphons are generally absent.
4. They contain two adductor muscles.
5. Adjacent gill filaments are joined by inter-filamental junctions.
6. Gonads contain gonopores.

This subclass is divided into **two** orders.

Order 1. Unionoida

1. Freshwater bivalves.
2. Shell valves are round to elongated in shape.

Examples. *Anodonta, Unio* (freshwater mussel), *Lamellidens, Mutela* (African freshwater oysters).

Order 2. Trigonioida

1. Marine.
2. Shell valves are triangular.

Example. *Trigonia.*

Subclass 5. Heterodonta

1. Shell valves are of equal size. Shell contains a few large cardinal teeth.
2. Shell lacks a nacreous layer.
3. Siphons are usually present.
4. Gills are eulamellibranchiate.

This subclass is divided into following **three** orders.

Order 1. Veneroida

1. Shell valves are equal in size.
2. Both adductor muscles are similar.

Examples. *Lucina*; *Venus* (hard shell clam); *Cardium* (cockle); *Corbicula* (asiatic clam); *Spisula* (giant clam); *Solecurtus* (razor clam); *Solen* (razor clam); *Petricola* (rock borer).

Order 2. Myoida

1. Thin-shelled burrowers.
2. Siphons are well developed.
3. One or no cardinal teeth are present in shell valves.

Examples. *Mya* (soft-shell clam); *Corbula*; *Teredo* (wood-boring shipworm); *Pholas* (rock borer or piddock).

Order 3. Hippuritoida

1. Shell valves are of unequal size.
2. One shell valve is cemented to substratum.

Example. *Jewel boxes* (Chamidae).

Subclass 6. Anomalodesmata

1. Shell contains equal valves.
2. Hinged teeth are absent.
3. Mantle margins are fused and siphonate (Two siphons are present).
4. They are mostly filter-feeding burrowers.
5. Gills are reduced to a horizontal muscular partition dividing the mantle cavity.
6. Foot is long, slender and lacks byssal gland.

Order 1. Pholadomyoida.

Examples. *Pandora*, *Poromya*, *Cuspidaria*.

Box 44.1.

According to old system of classification the class Pelecypoda is divided into following five orders.

Order 1. Protobranchia. Examples. *Nucula*, *Solenomya*.

Order 2. Filibranchia. Examples. *Arca*, *Mytilus*.

Order 3. Pseudolamellibranchiata. Examples. *Pecten*, *Ostrea*.

Order 4. Eulamellibranchiata. Examples. *Unio*, *Lamellidens*, *Mya*, *Teredo*.

Order 5. Septobranchiata. Examples. *Pandora*, *Cuspidaria*.

Class Cephalopoda

Class Cephalopoda is divided into **three** subclasses:

Subclass 1. Nautiloidea (Tetrabranchiata)

1. Shell is external, coiled or straight and its cavity is divided into chambers by variously shaped septa that meet the shell in simple sutures.
2. Mouth is surrounded by numerous slender suckerless tentacles.
3. Two pairs of gills and two pairs of nephridia are present.
4. Chromatophores and salivary glands are absent.
5. Pericardium communicates to the exterior directly.
6. This subclass has been in existence since the Cambrian period but all members are extinct except *Nautilus* which forms an example of **living fossil**.

Subclass Nautiloidea is divided into 14 orders, of which only Nautilida is extant, rest are extinct.

Order Nautilida. Example. *Nautilus*.

Subclass 2. Ammonoidea

1. Shell is external, coiled and with complex septa and sutures.
2. Extinct molluscs (Existed from Silurian to Cretaceous).
3. Siphon is external.

Examples. *Ammonites*, *Ceratites*.

Subclass 3. Coleoidea or Dibranchiata

1. Shell is frequently internal and reduced or absent if external. The cavity of shell is not divided by septa.
2. Mouth is surrounded by a circle of eight to ten arms bearing suckers.
3. Funnel forms a complete tube.
4. Pericardium (coelom) communicates with exterior through kidney.
5. Ctenidia (gills), auricles and nephridia are paired.
6. Osphradium is absent.
7. Chromatophores, salivary glands and ink gland are present.

Subclass Coleoidea is divided into **five** orders.

Order 1. Belemnoidea

1. All are extinct.
2. Shell is internal, chambered and with a posterior solid rostrum and a dorsal, shield-like extension.

Example. *Belemnites*.

Order 2. Sepioidea

1. This order includes cuttlefish and sepiolas.
2. They contain eight arms and two tentacles.
3. Shell is with septa but with reduced rostrum, or shell is greatly reduced or absent.
4. Body mostly short and broad or sac-like.
5. Ink gland is present.

Examples. *Sepia*, *Spirula*, *Sepiola*.

Order 3. Teuthoidea

1. This order includes squids.
2. Body is elongated.
3. They contain eight arms and two long tentacles.
4. Shell is flattened or blade-like.
5. Rostrum is absent.

Examples. *Loligo*, *Abralia*, *Architeuthis* (giant squid).

Order 4. Vampyromorpha

1. This order includes vampire squids.
2. Small, deep-water octopus-like forms.
3. They contain eight arms which are united by a web.
4. Two small filaments are also present.

Example. *Vampyroteuthis*.

Order 5. Octopoda

1. Body is globular (*i.e.*, short and round).
2. Any vestige of a shell is absent.
3. They contain eight arms.

Examples. *Octopus*, *Argonauta*.

44.5 TYPE STUDIES

1. *NEOPILINA*

Classification

Phylum	—	**Mollusca**
Class	—	**Monoplacophora**
Subclass	—	**Tergomya**
Genus	—	***Neopilina***
Species	—	***galathea***

In 1952, ten living specimens of *Neopilina*, a group of molluscs previously considered as extinct, were dredged by the Danish oceanographic research vessel, **Galathea** from a deep ocean trench off the Pacific coast of Costa Rica. The morphology of *Neopilina* provides a clue to the origin of phylum Mollusca from segmented annelid ancestor. It is regarded as a **living fossil.**

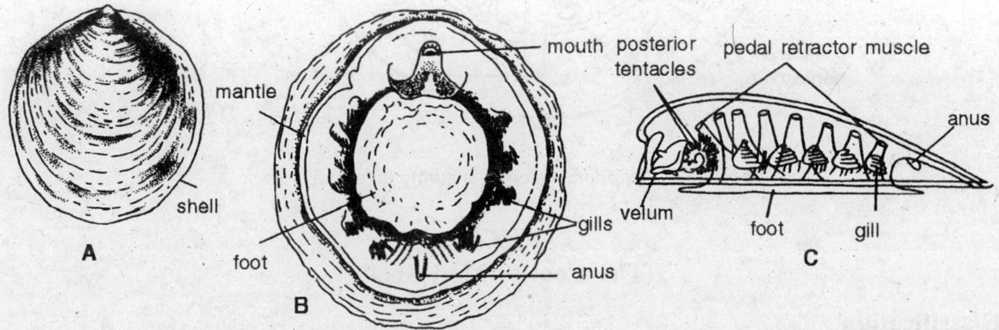

Fig. 44.1. *Neopilina*. A—Dorsal view of shell; B—Ventral view; C—Side view.

The largest specimen known is 37 mm long, 35 mm wide and 13 mm high. The body of *Neopilina* is covered by a dorsal dome-shaped shell (Fig. 44.1A). The apex of the shell is directed anteriorly. A **pallial groove** (the mantle cavity) separates the edge of ventral broad circular and flat **foot** from the **mantle** on each side (Fig. 44.1B). The **mouth** is located in front of the foot and the **anus** is located in the pallial groove at the posterior end of the body. In front of the mouth is a preoral fold, called **velum** which extends laterally on each side as a large ciliated palp-like structure.

The pallial groove contains five pairs of monopectinate **gills**. Internally, eight pairs of **retractor muscles** are present. There are six pairs of **kidneys** (or nephridia) located on each side of the body. They empty through **nephridiopores** into the pallial groove near the gill bases (Fig. 44.2). Two pairs of auricles open into two ventricles, one on each side of rectum and heart is surrounded by a paired pericardial **coelom.**

The digestive system includes a long and coiled **radula sac** and a **radula** within the buccal cavity (fore gut). The stomach contains a **style sac** and a **crystalline style** and is greatly coiled. The stomach contents consist of diatoms and sponge spicules. The sexes are separate and two pairs of **gonads** are located in the middle of the body. Each gonad is provided with a separate gonoduct that connects to one of the pairs of nephridia in middle of the body. Fertilization is external. Development is still unknown.

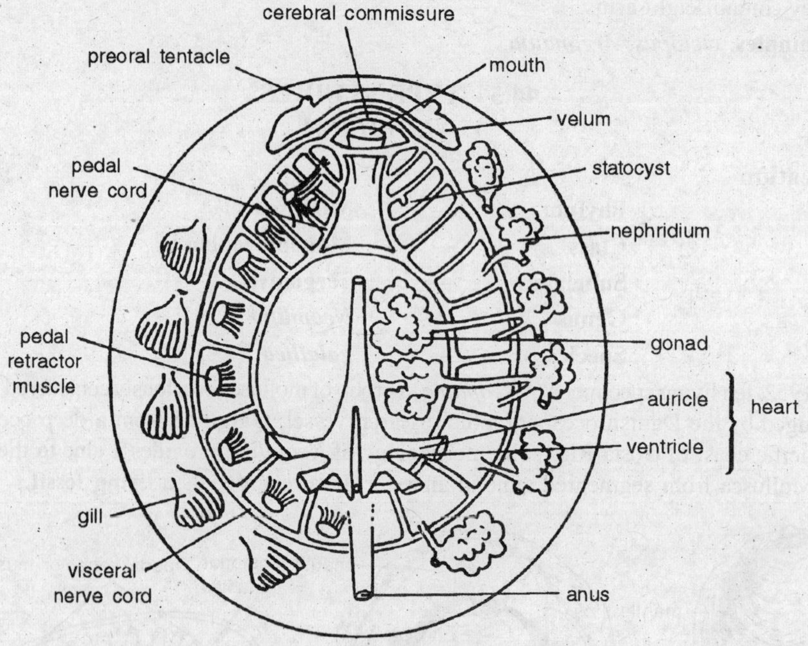

Fig. 44.2. *Neopilina*. Internal anatomy in ventral view.

2. *CHITON*
(The Coat-of-Mail Shell)

Classification

Phylum	—	**Mollusca**
Class	—	**Polyplacophora**
Genus	—	***Chiton***

Habit and Habitat

Chitons are marine. They are found in all seas except polar seas. They are very sluggish and occur in shallow water between the tide marks. They remain firmly attached to rocks, empty shells, corals and other solid objects. They are nocturnal and herbivorous.

Chiton is found all over the world. In India, they are found in Bombay, Malabar and Rameshwaram coasts.

Structure

Body of *Chiton* (Fig. 44.3) is elliptical, bilaterally symmetrical and dorso-ventrally flattened. They measure 2 to 8 cm in length. Colour of body varies from blue to grey and white. Body consists of **shell, mantle, foot, head** and **visceral mass**.

Shell is dorsal, calcareous and formed of eight hinged overlapping pieces or plates.

These enable the animal to roll like a ball. Histologically, each shell plate is composed of an upper layer or **tegmentum**, which consists of organic **conchiolin matrix** and **calcium carbonate**; and thick inner layer or **articulamentum** which consists of **calcium carbonate**.

Mantle lies beneath the shell. It covers dorsal and ventral surface of the body and secretes the shell. It covers all shell plates. The peripheral area of the mantle is thick and extends beyond the lateral margins of shell and forms girdle. The girdle contains scales, bristles or calcareous spicules.

The foot is broad, flat and muscular. It occupies most of the ventral surface of body. It provides creeping and adhesive surface to animal along with girdle. **Head** is not distinct. Eyes and tentacles are absent. On each lateral side of head there is an angular labial palp. The **mantle cavity** is present in the form of a groove on each side of the body between foot and the mantle edge. Six to eighty-eight bipectinate **gills** or ctenidia lie in a linear series on either side of the body in the mantle groove.

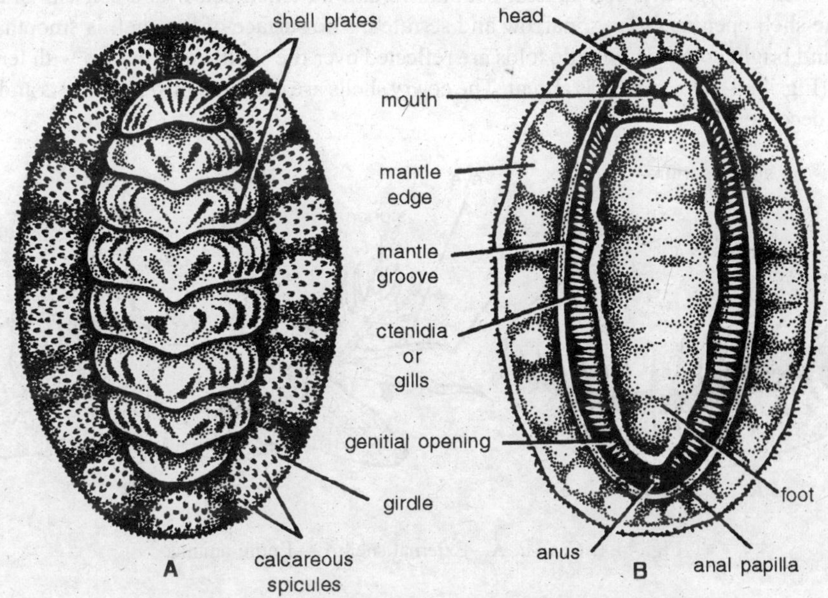

Fig. 44.3. *Chiton*. A—Dorsal view; B—Ventral view.

Chiton feeds on algae and other organisms. It scrapes the food with the help of teeth present in **radula**. Its alimentary canal consists of mouth, buccal cavity, oesophagus, a pair of pharyngeal glands, stomach, intestine and anus. **Heart** of *Chiton* consists of a pair of **auricles** and lies within the **pericardial cavity** (coelom). Respiratory pigment is **haemocyanin.** Body cavity is **haemocoel.** In *Chiton*, excretion takes place by two slender **kidneys** or metanephridia and Y-shaped excretory canal. Its **nervous system** is primitive. Ganglia are poorly developed or absent. **Sense organs** of *Chiton* include **labial palps, gustatory pits, sub-radular organs** and tactile organs **aesthetes**.

Chiton is **dioecious**. Fertilization is external. A free swimming **trochophore larva** hatches out of the egg. It metamorphoses to become adult. Veliger stage is absent.

Economic Importance

Chitons are used as food and are called as **"sea beef"** and also used for decoration.

3. *CYPREA*

Classification

Phylum	—	**Mollusca**
Class	—	**Gastropoda**
Subclass	—	**Prosobranchia**
Order	—	**Mesogastropoda**
Genus	—	***Cyprea***

Habit and Habitat

Cyprea is commonly called **cowry**. It is a marine and found among coral reefs and rocky ground. *Cyprea* is mostly found in the Indian and Pacific Oceans.

Structure

The shell of *Cyprea* is convoluted, oval above and flattened below and consists of a single whorl. The shell opening is long, narrow and serrated. The surface of the shell is smooth, highly polished and bright coloured. Mantle folds are reflected over the shell and provided with tentacular filaments (Fig. 44.4). Operculum is absent. The cowry shells are used as coins, ornaments and mantle piece for decoration.

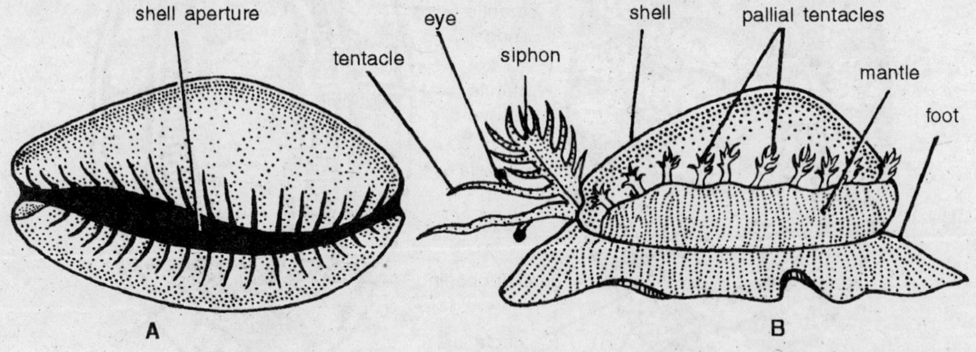

Fig. 44.4. *Cyprea*. A—External shell; B—Entire animal.

4. *APLYSIA*
(Sea hare)

Classification

Phylum	—	**Mollusca**
Class	—	**Gastropoda**
Subclass	—	**Opisthobranchia**
Order	—	**Anaspidea**
Genus	—	***Aplysia***

Habit and Habitat

Aplysia is commonly called **sea hare**. It is a marine gastropod found crawling in sea weeds. *Aplysia* is **herbivorous**, feeding mainly on sea weeds. It is able to change its body colour according to the colour of sea weeds on which it lives.

Aplysia is found in most parts of world from the Arctic to the Antarctic. It is found in West Indies, Florida coast, Asia and India.

Structure

Aplysia lacks an external shell. Its body is soft and lumpy with a thin, flexible plate-like **shell** which is almost completely covered by mantle (Fig. 44.5). The shell is internal. **Head** is indistinct and bears two pairs of **tentacles**, the anterior being larger and ear-like while the posterior pair is olfactory **rhinophores**, each bearing an eye on its base. **Mantle cavity** opens on right with single gill or ctenidium pointing backwards. **Visceral mass** is raised into a prominent hump. **Foot** is broad, muscular and ventral and bears a pair of lateral fleshy outgrowths, the **parapodia**. The parapodia help in swimming. **Anus** lies at the posterior end. Mantle possesses unicellular **ink glands**. These secrete purple ink used for defence. *Aplysia* is **hermaphrodite** with a single generative duct and with single genital pore.

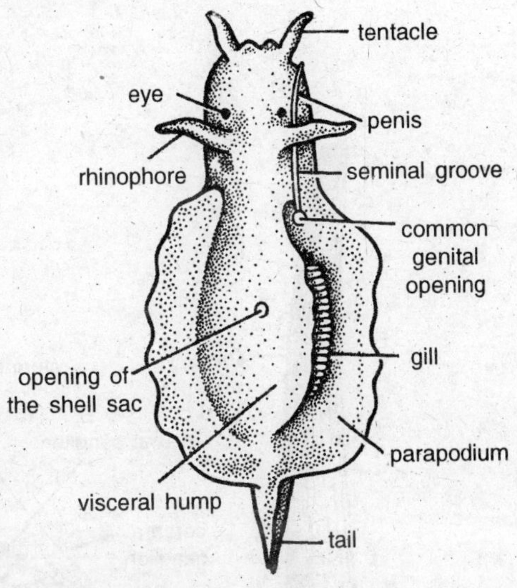

Fig. 44.5. *Aplysia*. External features.

5. *DENTALIUM*

Classification

Phylum	—	**Mollusca**
Class	—	**Scaphopoda**
Genus	—	***Dentalium***

Dentalium is commonly known as **tusk** or **tooth shell**. This name is derived from the shape of its shell, which is an elongated cylindrical tube shaped like an elephant's tusk (Fig. 44.6). Its most common species *Dentalium octogonum* is found in India.

Habit and Habitat

Dentalium is a marine and burrowing mollusc. It burrows in sand in water depths greater than 6 m. *Dentalium* feeds on minute organisms such as foraminifera and diatoms. It lives buried in sand, head downward with the body steeply inclined; only the small posterior aperture projects above the surface. *Dentalium* is found in all seas except polar.

Structure

Dentalium generally measures 2 to 5 cm in length. However, *Dentalium vernedei* which is found in Japan, is the largest living species reaching a length of 15 cm.

Shell is cylindrical slightly curved and tapering. Its both ends are open. Shell of Indian species is brilliant green in colour and bears eight longitudinal grooves.

The body of *Dentalium* (Fig. 44.6) is elongated and bilaterally symmetrical. The head and foot project out form the larger anterior aperture of the shell. Foot is long, conical, spade-like and highly extensible. It is adapted for digging and burrowing. **Mantle** unites along the ventral margin to form a tube around the body. **Mantle cavity** is large and extends the entire length of the ventral surface. The posterior aperture of mantle cavity serves for both inhalant and exhalent water currents. Water slowly enters the mantle cavity as a result of mantle ciliary action and perhaps foot protraction. After 10 to 20 minutes of inhalation, a violent muscular contraction (probably foot retraction) expels the water from the same opening it entered.

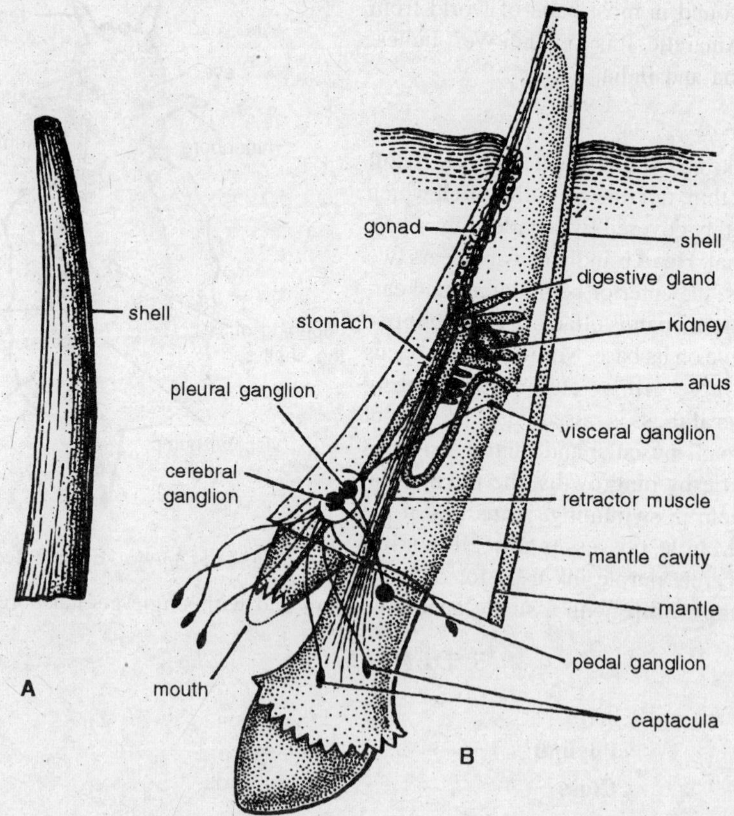

Fig. 44.6. *Dentalium.* A—Shell. B—Internal structure.

Head is reduced, bearing the **mouth** which is surrounded by a circle of long, slender, prehensile or retractile tentacles, called **captacula**. Each tentacle has a ciliated and adhesive **knob** at the tip and extends by ciliary creeping into the sand to capture food. A selected particles adhere to the knob by adhesive secretions that empty into a pit on one side of the knob. Captacula function as food gatherers, feeling about in the mud/sand and being retracted to carry food to the mouth. The buccal cavity contains a median jaw and a well developed **radula** with five teeth in each transverse row. **Digestion** is extracellular in stomach. **Anus** opens into mantle cavity.

There are no **gills** in the mantle cavity, respiratory exchange of gases takes place through the mantle surface. **Circulatory system** is reduced to a system of blood sinuses. **Heart** is absent. Excretory system consists of a pair of saccate kidneys or metanephridia having nephridiopores which are located near the anus.

The nervous system shows typical molluscan plan. Eyes and osphradium are absent. The captacula are important **tactile** organs and probably receive stimuli that mediate burrowing movements. The **subradular organ** in the buccal cavity is thought to be **chemoreceptor**. **Statocysts** are located in the foot near the pedal ganglia.

Sexes are separate, *i.e., Dentalium* is **dioecious.** Gonad is unpaired. Sperm and ova reach the outside by way of right nephridium. **Fertilization** is external. There is a free swimming **trochophore** larva which develops into a bilaterally symmetrical **veliger larva**. This larva is metamorphosed into adult.

Economic Importance

Shells of *Dentalium* are used for decoration. These were once used as currency by Red Indian Tribes of America.

6. *MYTILUS*

Classification

Phylum	—	**Mollusca**
Class	—	**Pelecypoda**
Subclass	—	**Pteriomorphia**
Order	—	**Mytiloida**
Genus	—	***Mytilus***

Habit and Habitat

Mytilus is commonly called **sea mussel**. It is marine, sedentary, and found attached to rocks between tide marks. It feeds upon the protozoans, diatoms and other organic detritus filtered from the sea water.

Mytilus is cosmopolitan in distribution. It is commonly found in India, Europe and U.S.A.

Structure

Body of *Mytilus* is enclosed in a wedge-shaped (*i.e.*, V-shaped) **shell** of two equal valves. Shell has a pointed anterior end and broadly rounded posterior end. **Mantle** is bilobed and forms an exhalent siphon posteriorly; inhalant siphon is not well marked. **Head** is rudimentary with a pair of **palps** around the **mouth**. **Foot** is small, cylindrical and tongue shaped with a ventral groove continuous with a byssus pit. **Byssus gland** secretes byssus threads which are used as organs of adhesion (Fig. 44.7). *Mytilus* remains attached to stones or rocks by the help of byssus threads. Anterior adductor muscle is greatly reduced but posterior adductor muscle is greatly enlarged. **Gills** are paired plate-like and filiform with ciliary interfilamentary junctions. Interlamellar junctions are absent. A pair of simple eyes are situated anterior to the gill lamellae. Sexes are separate. Gonads extend into mantle cavity.

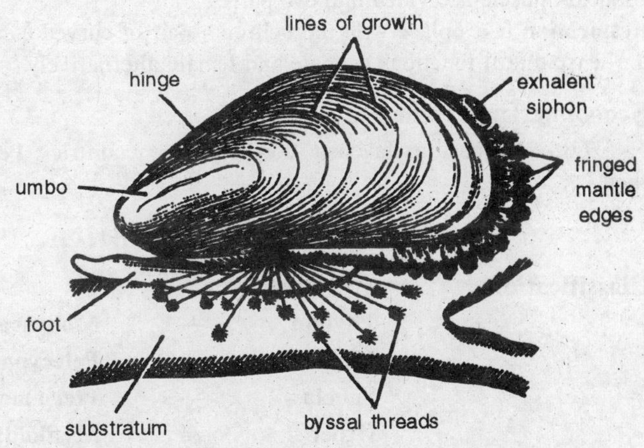

Fig. 44.7. *Mytilus*. External features.

Economic Importance

Mytilus is relished as food by poors living at sea coasts.

7. *OSTREA*

Classification

Phylum	—	**Mollusca**
Class	—	**Pelecypoda**
Subclass	—	**Pteriomorphia**
Order	—	**Ostreoida**
Genus	—	***Ostrea***

Habit and Habitat

Ostrea is commonly known as **Indian edible oyster**. It is marine and sedentary animal found attached to rocks and stones. *Ostrea* is found in all seas except the colder ones. It is found in Atlantic and Pacific coasts, India and Gulf of Mexico. In Indian Ocean, *Ostrea* is found along the Malabar coast and Cochin.

Structure

The shell is rough, irregular, thick, white and bivalved (Fig. 44.8). Two valves of the shell are unequal. The left valve is largest, thick, convex and permanently attached to the substratum. The right valve is smaller, thinner and flattened forming a short of lid. **Foot** is entirely absent in the adult. **Byssus** is also absent. The **mantle** are quite free from each other and do not form siphon. Single adductor muscle is divided into two parts.

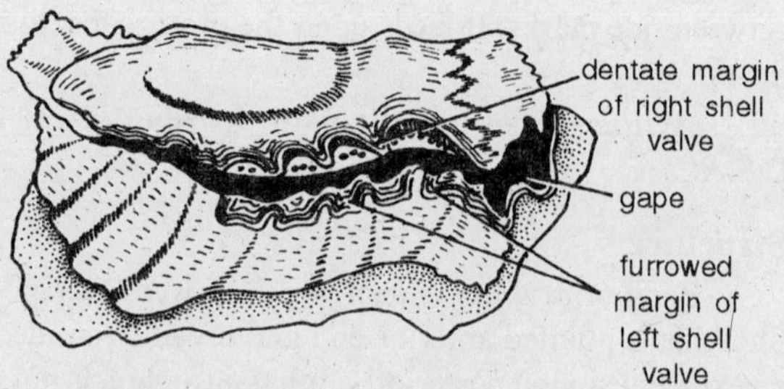

Fig. 44.8. *Ostrea*. External structure.

Respiration takes place with the help of a pair of curved ctenidia or gills. *Ostrea* is peculiar in the sense that it functions as male and female alternatively.

Economic Importance

Ostrea is used as favourite food in many countries. Pearly buttons are obtained from its shell.

8. *PINCTADA*

Classification

Phylum	—	**Mollusca**
Class	—	**Pelecypoda**
Subclass	—	**Pteriomorphia**
Order	—	**Mytiloida**
Genus	—	***Pinctada***

Habit and Habitat

Pinctada is commonly known as **Indian pearl oyster**. It is marine, sedentary animal found attached to the substratum. *Pinctada* is widely distributed except in cold zones. It is very common in the Gulf of Kutch, Tropical and Temperate seas, Pacific and Atlantic coasts.

Box 44.2.

Good quality pearls are obtained from Indian pearl oyster, *Pinctada vulgaris*, which lives in Gulf of Kutch. In India, six more species of pearl oysters are found in the Gulf of Mannar and Kutch. These Indian species are *P. fucata*, *P. margaritifera*, *P. cheminitzii*, *P. sugillata*, *P. anomioides* and *P. atropurpura*.

Pearl oyster beds are found among dead corals and rock pieces. They inhabit offshore region about 14 km away from coasts and at the depths of about 11 fathoms.

Structure

The **shell** of *Pinctada* is brown or yellowish brown in colour. Surface of the shell is coarse, irregular and ruffled. It bears radiating bands which end into finger-like projections at the margin. The shell valves are unequal and the hingeline is straight producing at both ends ear or wing-like processes. The left shell valve is large, convex and permanently attached to the rocks by a strong cable of **byssus threads**. The right shell valve is smaller and thinner and covers the viscera. Right and left **mantle lobes** are quite free. The **foot** is muscular and elongated. Just behind the foot, there are byssus threads. There occurs a single adductor muscle.

Pearl oyster is a filter-feeder. Its life span is 4 to 5 years and it grows to a size of 68 mm. The sexes are separate. Breeding occurs round the year. Fertilization is external. Its larva is **trochophore** which metamorphoses to become adult.

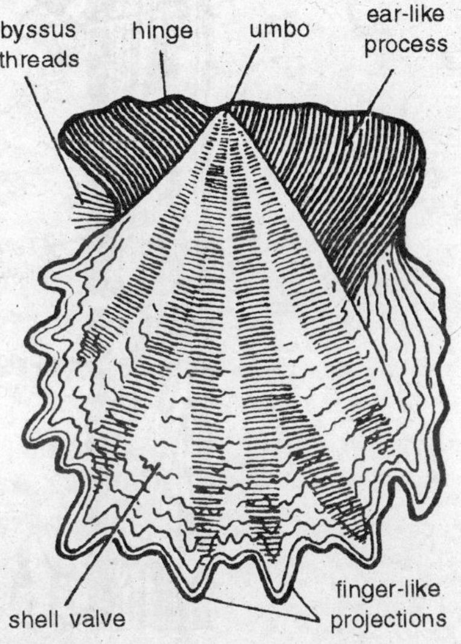

Fig. 44.9. *Pinctada*. External features.

Economic Importance

Pinctada is cultured and collected for getting **pearls**.

9. *TEREDO*

Classification

Phylum	—	**Mollusca**
Class	—	**Pelecypoda**
Subclass	—	**Heterodonta**
Order	—	**Myoida**
Genus	—	***Teredo***

Habit and Habitat

Teredo is commonly called **shipworm**. It is not a worm in the real sense but a highly specialized boring pelecypod which lives in burrows made in wooden boats, ships, other submerged woods and mangrove roots. *Teredo* eats cellulose, lignin and living microorganisms.

Teredo is cosmopolitan in distribution. It is commonly found in India, Europe and U.S.A.

Structure

The body of *Teredo* (Fig. 44.10) is elongated and vermiform with tubular **mantle** enclosing the entire body and producing a calcareous lining within the tunnel. Mantle opens anteriorly. **Foot** is very much reduced. **Shell** is reduced and limited to a very small portion at the anterior end. It consists of two-shell valves which have ridged and roughened surface and help in boring tunnels in the wood piling and floating timber. The **mouth** lies between two shell-valves. *Teredo* bores into wood reducing it into saw-dust by the rotatory action of the two shell-valves. Saw dust is swallowed by the shipworm and stored in a caecum of stomach. It is digested by the enzymes secreted by digestive gland and symbiotic bacteria which are housed within a special organ that opens into the oesophagus. These bacteria digest **cellulose** and do nitrogen fixation to compensate for the low-protein diet.

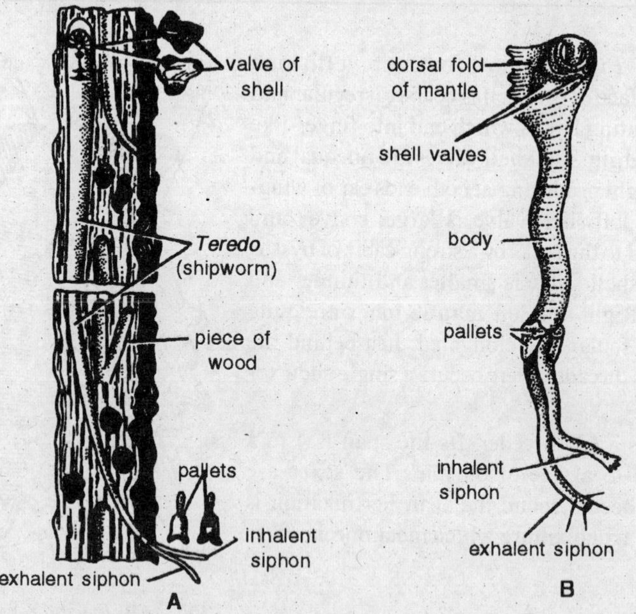

Fig. 44.10. *Teredo*. A—*Teredo* in a piece of wood; B—External features.

Mantle cavity is long and contain a pair of elongated **gills**. **Siphons** are two, elongated and retractile, both opening at the surface of wood. Both siphons are united most of their lengths and provided with **calcareous valves** or **pellets** which close the tunnel when the siphons are retracted.

Economic Importance

Shipworms form tunnels in the wood, causing great damage to ships and boat.

10. *SEPIA*

Classification

Phylum	—	**Mollusca**
Class	—	**Cephalopoda**
Subclass	—	**Coleoidea**
Order	—	**Sepioidea**
Genus	—	***Sepia***

Habit and Habitat

Sepia is commonly called cuttle-fish. It is a marine form, found in the shallow waters. *Sepia* is a good swimmer. It usually swims at night and rests flat on the bottom during day time. It is **carnivorous** and **predaceous** animal. *Sepia* feeds on small fishes, crustaceans and other animals.

Sepia is cosmopolitan in distribution. It is commonly found in India, Europe and Mediterranean region.

Structure

Body of *Sepia* (Fig. 44.11) is bilaterally symmetrical, dorsoventrally flattened and is divisible into **head**, short constricted **neck** or **collar** and **trunk**.

Head bears a pair of large highly developed eyes and five pairs of arms surrounding the **mouth**. Of the five pairs of arms, four pairs are short and stout bearing four longitudinal rows of **suckers** on the inner flat surface. The fifth pair of arms is known as **tentacles** which are long and narrow having suckers only on their free ends. The suckers are firmly applied on the body of the prey by creating a partial vacuum. Bases of ventral arms become united by a web of integument. The fourth ventral arm in male *Sepia* is modified into a copulatory organ. It is called **hectocotylised arm**.

Collar connects the head with the **trunk** which is elongated and shield-shaped, bordered by narrow **lateral fins** or **parapodia** which are used in leisure swimming. Trunk is covered with thick muscular **mantle**. The mantle encloses a large **mantle cavity** which contains viscera.

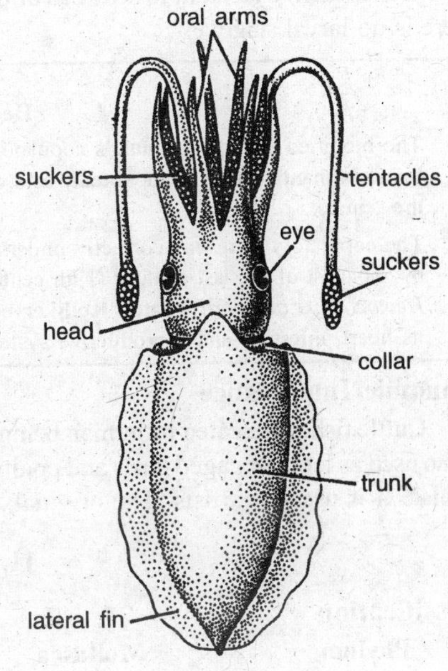

Fig. 44.11. *Sepia*. External structure in dorsal view.

The **shell** of *Sepia* is calcareous and internal. It lies enclosed in a dorsal sac of mantle and is secreted by the mantle (See Box 44.3).

> **Box. 44.3.**
> **Shell of *Sepia***
>
> The calcareous internal shell of *Sepia* is divided by thin calcareous septa. The septa enclose spaces which contain fluid and gas (largely nitrogen). The gas provides buoyancy to the cuttlefish. By regulating the amount of fluid and gas, the degree of buoyancy can be changed. Light is an important buoyancy regulating factor. Buoyancy decreases during day time and the animal lies buried in the bottom. At night buoyancy increases and the animal swims actively and hunts for food.

A **siphon** or muscular **funnel** lies below the head. Mantle cavity opens to outside through the funnel. *Sepia* moves rapidly by forcefully expelling water out of mantle cavity through the funnel. It is called **jet propulsion**.

A pair of plume-shaped **ctenidia** or **gills**, one on each side of the mantle cavity performs the respiratory function. **Excretion** is performed by a single kidney. *Sepia* contains a large **ink gland** which opens into the rectum. The gland secretes a brown or black fluid which is stored in a reservoir. When the animal is attacked, it releases the ink through the anus and resultant clouds of brown or black water confuses the attacker and *Sepia* slips unhurt.

Sepia being a predator, locates its prey with eyes and captures it by tentacles. The prey is paralysed by the poisonous secretion of the salivary glands, broken into pieces by beak-like jaws of buccal cavity and swallowed with the help of **radula**.

Sepia lacks **osphradium** but consists of **statocysts** which are large and lie on each side of brain. **Sexes** are separate. The males are usually smaller. During copulation, the male inserts its modified fourth ventral arm, called **hectocotylus arm** (See Box 44.4), into its own mantle cavity to pick up **spermatophores**. Now this arm is inserted into the mantle cavity of the female and spermatophores are deposited there. Thus **fertilization** is internal. *Sepia* deposits its eggs singly and attaches them by a stalk to seaweed or other objects. The eggs are large. Development is direct. There is no larval stage.

> **Box 44.4.**
> **Hectocotylus arm**
>
> The modified end of the male's copulatory arm is detached during mating and remains in the female's mantle cavity, undoubtedly as a device to foil the attempt of other males to inseminate the female.
>
> The copulatory arm was correctly understood by **Aristotle** but was completely misinterpreted by zoologist of the last century (19th century). **Cuvier** described it as a parasite under the name *Hectocotylus octopodis*. Later, **Kolliker** believed it to be a complete male animal and described its heart, intestine and reproductive system. It is still called a **hectocotylus** arm.

Economic Importance

Cuttle fishes are eaten by human beings. Its shell is used to make medicine. The shell of *Sepia* is also used as food for caged birds and poultry. The ink of ink glands provides brow pigment, called '**Sepia**'. It is used by artists. The original "**Indian Ink**" was obtained from *Sepia*.

11. *OCTOPUS*

Classification

Phylum	—	**Mollusca**
Class	—	**Cephalopoda**
Subclass	—	**Coleoidea**
Order	—	**Octopoda**
Genus	—	***Octopus***

Habit and Habitat

Octopus is commonly known as **devil-fish**. It is marine, bottom-dweller and nocturnal. During day time, it is found in the rocky crevice, shady underwater caverns or in coral reefs. *Octopus* is a predaceous animal.

Octopus is cosmopolitan in distribution. It is commonly found in Europe, Italy, China and India.

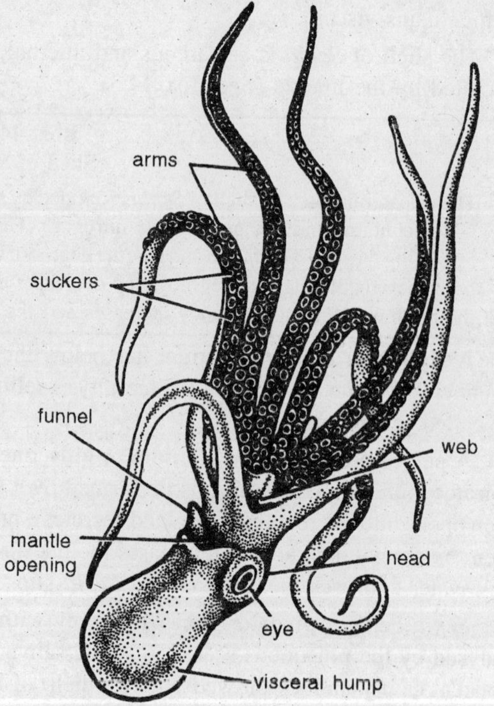

Fig. 44.12. *Octopus*. External structure.

Structure

The largest *Octopus* (*O. dofleini* of Pacific Coast) measures 36 cm and its extended arms covers an area of 9.6 metre. It lacks shell and the body is enclosed in the mantle. The body is globular and bag like.

The **head** is large and bears a pair of highly developed eyes and eight long **arms**. All thc arms are equal, similar and webbed. They carry two rows of sessile **suckers**. The third right arm in male is modified to transfer its spermatophore into female. The terminal part of this arm becomes spoon-shaped.

The **mantle edges** are fused with head dorsally and anteriorly. The mantle cavity opens to outside through **funnel**.

Usually *Octopus* crawls on the ground and is able to change its colour according to the background. It is a predaceous animal feeding on crabs, bivalves, snails and fishes. The food is captured by snake-like arms and broken into pieces by **radula** and a pair of sharp beak-like **jaws**. It tears its prey in a devilish manner. Secretions of salivary glands of *Octopus* contain a **poison** (a glycoprotein) and proteolytic enzymes.

Like *Sepia*, on the approach of its enemy, *Octopus* ejects ink from the ink gland into the surrounding water. This produces a smoky cloud to either confuse the enemy (*e.g.*, moray eel) or to make it blind. Thus, ink apparatus is a protective device.

Sexes are separate. **Fertilization** is internal. Female lays eggs and attaches them with some solid objects within the rocky crevices. Development is direct. There is no larval stage in *Octopus*.

12. *NAUTILUS*

Classification

Phylum	—	**Mollusca**
Class	—	**Cephalopoda**
Subclass	—	**Nautiloidea**
Order	—	**Nautilida**
Genus	—	***Nautilus***

Nautilus is called **living fossil** because all other members (2500 species) of its subclass are now extinct. Subclass *Nautiloidea* has been in existence since the Cambrian period.

Habit and Habitat

Nautilus is marine, found living in deep waters (at a depth of about 500 metres) near the shores and coral reefs. It is a gregarious, nocturnal and carnivorous (scavenger and predator) cephalopod. It crawls over the bottom in troops at night time in search of animal food. Food of *Nautilus* includes shrimps, crabs and shell fishes. While resting, it remains attached to the rubbles or walls of crevices with their arms. Except when feeding, it swims backwards through **jet propulsion**. While feeding, *Nautilus* swims forward.

Nautilus is found in Tropical, Indian and Pacific Oceans.

Structure

Nautilus (Fig. 44.13) is commonly called **pearly nautilus.** Body is enclosed in a calcareous spirally coiled many chambered shell (*Nautilus* is the only living cephalopod that contains an external shell). Shell is about 25 cm in diameter. It has porcelain-like white texture having outer surface brown with irregular bands of dark and inner surface of pearly white. Histologically shell of *Nautilus* is composed of outer **porcelaneous layer** and inner shining **nacreous layer.** The interior or space of the shell is divided into a series of chambers by means of transverse **septa**. The distal most chamber, called **living chamber**, is largest and encloses entire body of the animal. It is lined by **mantle**. The remaining chambers are filled with gas and are in continuation through a series of narrow openings, each of which is situated in the middle of the septum. These pores are traversed by spiral and tubular prolongations of visceral mass, the **siphon** or **sipuncle**. The sipuncle secretes gas into the chambers. The gas gives buoyancy to the shell and helps in swimming.

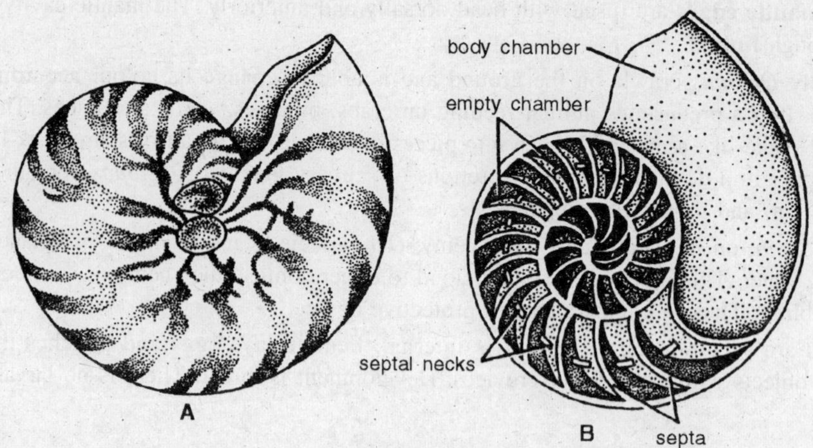

Fig. 44.13. *Nautilus.* A—Complete shell; B—Shell in section.

Head is well developed and bears a pair of simple eyes and 38 to 90 **tentacles** around it. The tentacles are suckerless but bears transverse ridges. An unusual feature of the head of *Nautilus* is the large muscular leathery and protective **dorsal hood** which serves as an operculum of gastropod, *i.e.*, it covers the opening of shell when the animal withdraws into the shell (Fig. 44.14).

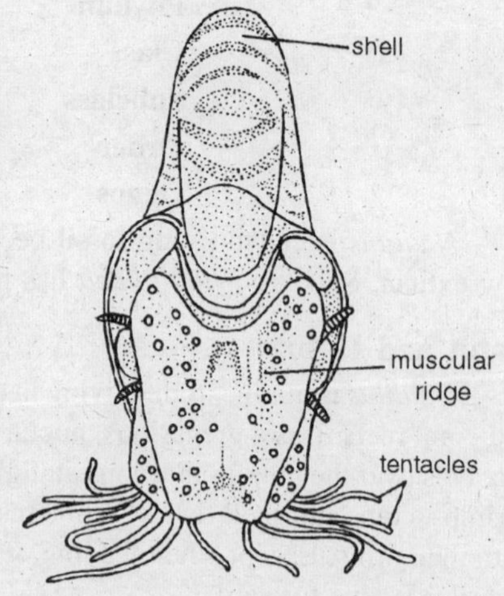

Fig. 44.14. *Nautilus.* Living animal.

In contrast to other cephalopods, in *Nautilus*, food absorption occurs mainly in the digestive gland. *Nautilus* has four **auricles** in the heart, while other cephalopods have only two auricles. Likewise, *Nautilus* has four **nephridia** while the other living cephalopods contain only two nephridia. A pair of **ospharidia** occurs near the gills only in *Nautilus*. **Statocysts** of *Nautilus* lie near the pedal ganglia. Ink glands are absent.

Nautilus is dioecious. In male, four arms are permanently modified for copulation. Female *Nautilus* has a vestigial left and functional right **oviduct**. In male the left **sperm duct** is rudimentary, while in other cephalopods the right sperm duct is missing.

Economic Importance

Nautilus is considered as an excellent food by the pacific Islanders. Its brilliantly coloured shell is extensively used for decoration.

QUESTIONS

Long Answer Questions

1. Classify Mollusca giving characters and examples of each group.
(*Awadh 1988; Gorakhpur 1988, 1996; Allahabad 1989; Kerala 1989; Kanpur 1989, 1998; Agra 1989, 1994; Himachal 1990; Rohilkhand 1990; Rajasthan 1990; Bhopal 1991; Raipur 1991; Bundelkhand 1991, 1995; Jabalpur 1992; Ravishankar 1993; Lucknow 1996; C.C.S.U; Meerut 2004*)
2. What are the distinguishing features of class Gastropoda?
(*Rewa 1989; Agra 1993; Kanpur 1993*)

3. What is the basis of the classification of phylum Mollusca? Give salient features and classification of phylum Mollusca with suitable examples. *(Gorakhpur 1998)*

4. Give an outline classification of phylum Mollusca.

Short Answer Questions

1. Write general characters of phylum Mollusca and give five examples. *(Kanpur 2001)*

2. Give five characters of phylum Mollusca. *(Rohilkhand 2001)*

3. Classify the following animals giving their distinguishing features and economic importance.

(i) *Neopilina* *(Gorakhpur 1989)*

(ii) *Chiton* *(Calcutta 1988; Utkal 1988; Bundelkhand 1988, 1991; Bhopal 1989; Jammu 1989, 1992; Agra 1989, 1993; Purvanchal 1989, 1994; Garhwal 1990; Allahabad 1992; Meerut or CCS 1992, 1997; Lucknow 1998)*

(iii) *Cyprea*

(iv) *Aplysia* (sea hare) *(Awadh 1990; Raipur 1990; Lucknow 1993, 1995, 1996; Jiwaji 1993)*

(v) *Dentalium* *(Meerut or CCS 1995)*

(vi) *Mytilus* *(Agra 1988; Bundelkhand 1989)*

(vii) *Ostrea* *(Agra 1990; Awadh 1990)*

(viii) *Pinctada*

(ix) *Teredo* *(Kanpur 1989; Agra 1989, 1991; Bundelkhand 1991, 1994; Lucknow 1992; Gorakhpur 1992)*

(x) *Sepia* *(Gorakhpur 1988; Delhi 1989, 1993; Raipur 91; Agra 1992; Ravishanker 1993)*

(xi) *Octopus* (devil fish) *(Kanpur 1988, Awadh 1989; Gorakhpur 90; Meerut or CCS 1990, 1996; Lucknow 1995, 1997)*

(xii) *Nautilus* *(Agra 1988; Lucknow 1989; Awadh 1989; Jiwaji 1989; Bundelkhand 1990; Meerut or CCS 1993)*

Very Short Answer Questions

1. Define the Mollusca.

Ans. The molluscs are nonchordates or invertebrates. They are soft-bodied, unsegmented, triploblastic, bilaterally symmetrical coelomate animals. They contain a muscular foot, a visceral hump and a mantle. The body of molluscs is usually covered with a hard shell.

2. Which invertebrate phylum is ranked as largest after Arthropoda?

Ans. The invertebrate animal phylum ranked to be largest after Arthropoda is Mollusca. More than 60,000 living species have been studied from this phylum. In addition about 35,000 fossil species provide the phylum to have its long geological history.

3. Mention a mollusc having a shell formed of eight plates.

Ans. *Chiton.*

4. Name a mollusc with external, coiled shell divided internally into many chambers by septa.

Ans. *Nautilus.*

5. Mention the three main body parts of a mollusc.

Ans. Head, visceral mass and foot.

6. Name the class of phylum Mollusca which include asymmetrical animals and define torsion.

Ans. The class of phylum Mollusca which contain asymmetrical animals is called **Gastropoda**. The body symmetry of these molluscs is lost due to two reasons: coiling and torsion. Coiling is a normal feature of gastropods in which their body is spirally coiled. Also during embryonic development their visceral mass gets twisted or torted suddenly causing the shifting of posterior end (anus) to the anterior end (mouth). Thus torsion is the phenomenon in which animals body (*i.e.*, viscero-pallium) gets rotated at an angle of 180° during larval development.

7. Differentiate between monopectinate and bipectinate gills.
   Ans. Molluscs have two types of gills, the aquatic breathing organs. When the gill or ctinedium is provided with a single row of lamellae attached to the axis, it is called **monopectinate gill,** *e.g., Pila*. The gill or ctinedium having two rows of gill lamellae (gill filaments) attached to either side of a central axis is called **bipectinate gills,** *e.g., Patella*.
8. Which respiratory pigment some molluscs have in their blood?
   Ans. Haemocyanin which is dissolved in the plasma of blood.
9. Name the order of *Helix*.
   Ans. Pulmonata.
10. Name a mollusc which makes burrows in the wood used in boats and ships.
   Ans. *Teredo*.
11. Give three examples of class Cephalopoda.
   Ans. *Sepia*, *Loligo* and *Octopus*.
12. Give the order and class of cuttle fish.
   Ans. Decapoda (order), Cephalopoda (class) (Phylum Mollusca).
13. Give the order and class of *Pinctada*.
   Ans. Pseudolamellibranchia (order), Pelecypoda (class) (Phylum Mollusca).
14. What is the common name of *Octopus*?
   Ans. Devil fish.
15. What is the common name of *Loligo*?
   Ans. Squid.
16. What is the common name of *Sepia*?
   Ans. Cuttlefish.
17. What is bysses?
   Ans. Byssus is a bunch of strong thread found in some molluscs, such as *Mytilus,* for attachment to the substratum.
18. What is the use of inksac and in which it is found?
   Ans. An **inksac** is present in *Sepia* and serves for defence. To escape from the enemies, it discharges dark clouds of ink behind it so that the enemy loses sight of it. Sepia ink for the artists is also obtained from *Sepia* (Box 44.5).

Box 44.5.

Inksac of Mollusca

In most non-nautiloid cephalopodan molluscs (*e.g., Sepia, Loligo* and *Octopus*), a large inksac is located near the intestine. An ink-producing gland lies in the wall of sac, and a duct runs from the sac to pore into the rectum. The gland secretes a brown or black fluid that contains a high concentration of melanin pigment and mucus; the fluid is stored in the inksac. When alarmed, the animal releases the ink through the anus and mantle cavity and out in the surrounding water. The cloud of inky material hangs together in the water, forming a "dummy" image that serve to confuse predators. The alkaloid nature of the ink may also act to deter predators, particularly fishes, and may interfere with their chemoreception (see **Brusca** and **Brusca**, 2003).

Fill in the Blanks

1. .... is a segmented mollusc.
2. *Neopilina* belongs to the class.................. .
3. A mollusc shell consists of 1, 2 or .... pieces.
4. *Chiton* is an example of the class ....
5. Shell in molluscs is secreted by ....
6. Foot in molluscs is mainly a .... organ.

7. Asymmetrical molluscs belong to the class ....
8. Number of arms in *Sepia* is ....
9. Some molluscs having a rasping organ, the .... in the buccal cavity.
10. Circulatory system of molluscs is of .... type.
11. Body cavity of molluscs is ....
12. Excretory organs of molluscs are ....
13. Organs of Bojanus are .... organs.
14. The eggs of molluscs undergo .... and .... cleavage.
15. .... is a living fossil among molluscs.

Multiple Choice Questions

1. Which one is a connecting link between Annelida and Mollusca?
   (*a*) *Nautilus* (*b*) *Chaetoderma*
   (*c*) *Neopilina* (*d*) *Lymnaea*
2. Pila belongs to the class............. of phylum Mollusca.
   (*a*) Monoplacophora (*b*) Pelecypoda
   (*c*) Gastropoda (*d*) Cephalopoda
3. The harmful mollusc is
   (*a*) *Chiton* (*b*) *Teredo*
   (*c*) *Pinctada* (*d*) *Unio*
4. Superior quality pearls are formed by
   (*a*) *Pinctada* (*b*) *Unio*
   (*c*) *Ostrea* (*d*) *Pecten*
5. Arms are webbed at the base in
   (*a*) *Loligo* (*b*) *Sepia*
   (*c*) *Octopus* (*d*) *all of above*
6. Shell is internal and reduced in
   (*a*) *Pila* (*b*) *Helix*
   (*c*) *Teredo* (*d*) *Limax*
7. One of the following is a living fossil
   (*a*) *Unio* (*b*) *Dentalium*
   (*c*) *Nautilus* (*d*) *Pecten*
8. Pearl oyster (*Pinctada*) belongs to the class
   (*a*) Cephalopoda (*b*) Scaphopoda
   (*c*) Gastropoda (*d*) Pelecypoda (or Bivalvia)
9. *Mytilus* belongs to molluscan class
   (*a*) Pelecypoda (*b*) Gastropoda
   (*c*) Amphineura (*d*) Cephalopoda
10. Sea hare is
    (*a*) *Teredo* (*b*) *Pila*
    (*c*) *Sepia* (*d*) *Aplysia*
11. The elephant tusk shell is
    (*a*) *Cypraea* (*b*) *Limax*
    (*c*) *Dentalium* (*d*) *Nautilus*
12. The foot of molluscs is used for
    (*a*) burrowing (*b*) swimming
    (*c*) creeping (*d*) all of these
13. Locomotion by jet propulsion mechanism is found in class
    (*a*) Scaphopoda (*b*) Cephalopoda
    (*c*) Gastropoda (*d*) Pelecypoda

14. Worm-like and without shell mollusc is included in class
(*a*) Monoplacophora (*b*) Polyplacophora
(*c*) Aplacophora (*d*) Scaphopoda

15. Torsion is the characteristic of
(*a*) Gastropoda (*b*) Scaphopoda
(*c*) Pelecypoda (*d*) Aplacophora

16. Digestive gland in mollusca is
(*a*) salivary gland (*b*) pancreas
(*c*) liver (*d*) hepatopancreas

17. The larvae of mollusca are
(*a*) veliger (*b*) trochophore
(*c*) glochidium (*d*) all of above

18. The study of Mollusca is generally known as
(*a*) malacology (*b*) concology
(*c*) helminthology (*d*) herpetology

19. Filter feeder molluscs are
(*a*) bivalves (*b*) gastropods
(*c*) *Octopus* (*d*) devil fish

20. Visceral mass of molluscan body contains
(*a*) central nervous system (*b*) gill
(*c*) main body organs (*d*) haemocoel

21. Radula is found in all molluscs except
(*a*) pelecypodes (*b*) cephalopods
(*c*) scaphopods (*d*) aplacophorans

22. The zoological name of razor shell is
(*a*) *Chiton* (*b*) *Loligo*
(*c*) *Dentalium* (*d*) *Solen*

23. The study of molluscan shell is called
(*a*) entomology (*b*) biology
(*c*) conchology (*d*) zoology

24. Shell is internal in
(*a*) Monoplacophora (*b*) Cephalopoda
(*c*) Gastropoda (*d*) Pelecypoda

25. Molluscs have
(*a*) one chambered heart (*b*) two chambered heart
(*c*) three chambered heart (*d*) none of these

ANSWERS

Fill in the Blanks

1. *Neopilina*	**2.** Monoplacophora	**3.** Eight
4. Polyplacophora	**5.** Mantle (pallium)	**6.** Locomotory
7. Gastropoda	**8.** Ten	**9.** Radula
10. Open	**11.** Haemocoel	**12.** Kidneys
13. Excretory	**14.** Spiral, determinate	**15.** *Neopilina galathea.*

Multiple Choice Questions

1. (*c*)	**2.** (*c*)	**3.** (*b*)	**4.** (*a*)	**5.** (*c*)	**6.** (*d*)
7. (*c*)	**8.** (*d*)	**9.** (*a*)	**10.** (*d*)	**11.** (*c*)	**12.** (*d*)
13. (*b*)	**14.** (*c*)	**15.** (*a*)	**16.** (*d*)	**17.** (*d*)	**18.** (*a*)
19. (*a*)	**20.** (*c*)	**21.** (*a*)	**22.** (*d*)	**23.** (*c*)	**24.** (*b*)
25. (*c*).					

45

Pila globosa (The Apple Snail) (Phylum Mollusca)

45.1 CLASSIFICATION

Phylum	—	**Mollusca**
Class	—	**Gastropoda**
Subclass	—	**Prosobranchia**
Order	—	**Mesogastropoda**
Genus	—	***Pila***
Species	—	***globosa***

45.2 HABIT AND HABITAT

Pila globosa is one of the largest freshwater snails. It is found in abundance in fresh-water ponds, pools, tanks, rice fields, streams and rivers. It is **herbivorous** and prefer clean water containing succulent aquatic plants such as *Vallisneria, Pistia*, etc. It is **amphibious** being adapted for life in water and on land. During rains it leaves ponds and make long tours on land. During drought conditions it becomes dormant and can remain buried in mud for years. This is called **aestivation.**

Pila globosa is mostly found in Northern India. Other species of *Pila* such as *P. virens, P. nevililana* and *P. theobaldi* are very common in South India.

45.3 EXTERNAL FEATURES

1. Shell

Pila lies enclosed in its shell (Fig. 45.1). The shell is single piece (**univalve**) which encloses the whole body. It is thick globose and lemon-yellow, brownish or blackish coloured. It is hollow, spirally coiled cone and contains six and half coils. A complete coil is called a **whorl.**

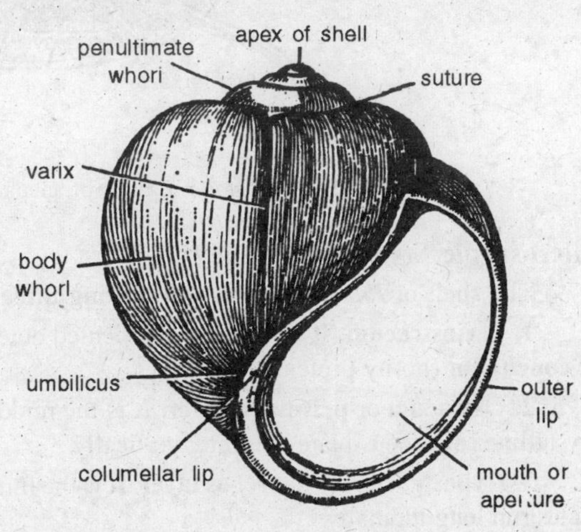

Fig. 45.1. *Pila globosa*. Shell seen from ventral surface.

The shell contains a central axis or **columella.** The topmost or the apical whorl (**apex of shell**) is smallest and oldest. Later whorls gradually increase in size and the last whorl is the largest. It is called **body whorl.** Most of the body of *Pila* lies in this whorl. The whorls above the body whorl is called **spire** (last of which is called **penultimate whorl**). There is a **line** or **suture** between every two whorls. The shell surface is marked by numerous **lines of growth** called **varices.** The body whorl opens to outside by a wide **aperture** or **mouth.** Through this mouth, the head and foot protrude. Smooth continuous margin of the mouth of the shell is called **peristome.** It contains outer lip and inner lip or **columellar lip.** Columella opens to outside by a minute opening, the **umbilicus.**

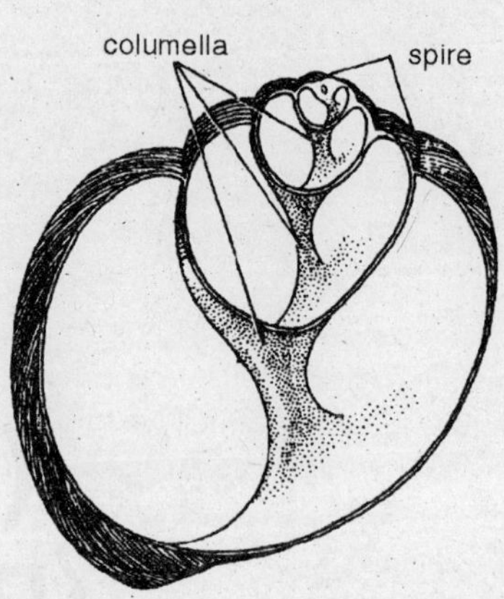

Fig. 45.2. *Pila globosa.* Shell in section showing columella.

The **shell coiling** in *Pila* is right handed, anti-clockwise or **dextral.** When the mouth (of shell) lies on the right-side of columella, the shell is called **dextrally coiled.** If the mouth facing the observer and shell is held with spire up then it is called **sinistrally coiled,** clockwise or left-handed.

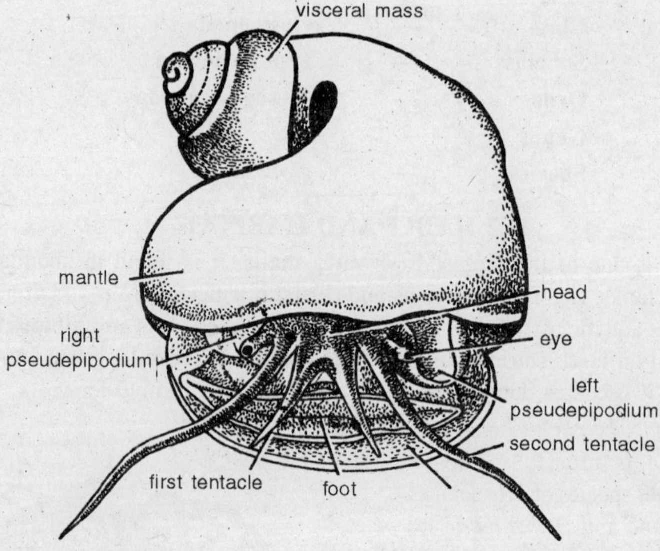

Fig. 45.3. *Pila globosa.* Body of animal after the removal of the shell.

Microscopic Structure of Shell

The shell of *Pila* is formed of following three layers:

1. Periostracum. It is thin and pigmented outermost layer. It is not calcified but composed of **conchiolin** (horny protein).

2. Ostracum or **prismatic layer.** It is the middle layer. It is thickest and calcareous having crystalline calcareous plates running vertically.

3. Hypostracum or **nacreous layer.** It is the inner most calcareous layer. In it, the calcareous plates run longitudinally.

2. Body

After removing the shell soft body of animal becomes visible. It is divisible into **head, foot, mantle** and **visceral mass.**

***(i)* Head.** It is anterior most part prolonged into a contractile snout. The snout bears a pair of **labial palps** or **first tentacles** and a pair of longer filamentous **second tentacles.** Head also bears a pair of short eye stalks or **ommatophores.** Each bears a small bead-like eye at its tip. Mouth lies between the first pair of tentacles. Tentacles are highly contractile structures.

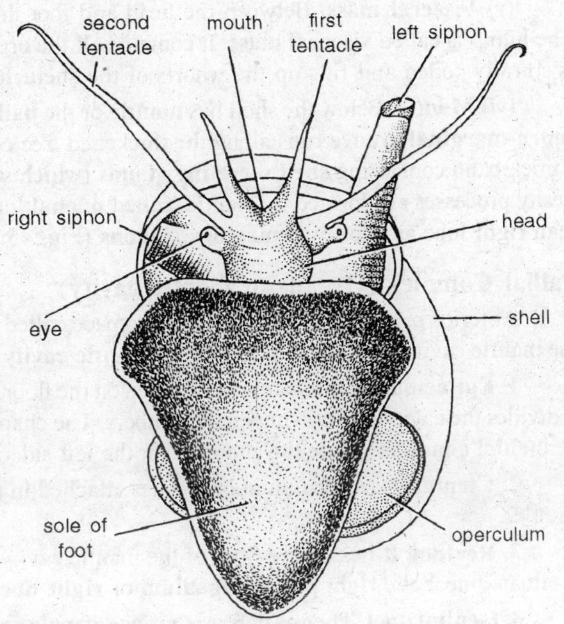

Fig. 45.4. *Pila globosa.* Animal in ventral view showing formation of elongated siphon formed by the left pseudopodium; this is done for aerial respiration.

***(ii)* Foot.** Foot is large triangular, strongly muscular and ventral in position (Fig. 45.4). The ventral surface of foot is broad, flat and smooth. It is called **sole** which is adapted for creeping movement.

***(iii)* Operculum.** Foot bears a horny disc, called **operculum,** on the posterior dorsal surface (Fig. 45.5). It is a calcareous plate and formed by the secretions of glandular cells of the foot. Outer surface of the operculum shows many concentric **rings of growth** around a small central axis or **nucleus.** The inner surface of operculum contains an elliptical area the **boss,** to which are attached the opercular muscles. Boss is surrounded by a shallow groove. When the foot is withdrawn into the shell, the operculum fits into the mouth of the shell and closes it.

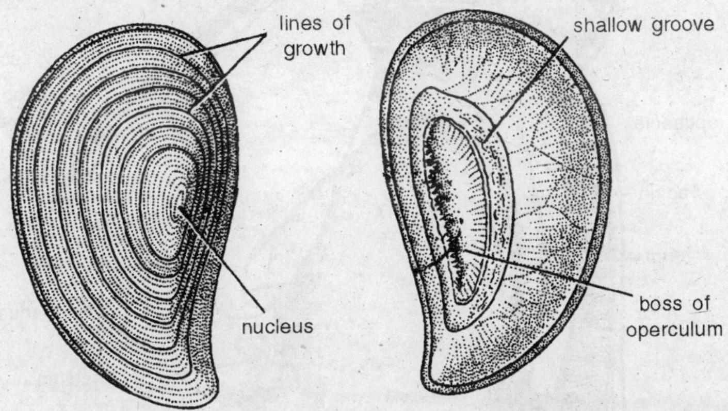

Fig. 45.5. *Pila globosa.* Operculum; A—Outer view; B—Inner view.

***(iv)* Columellar muscles.** A columellar muscle arises from the foot and is inserted in the columella, it attaches the body with the shell and it withdraws the animal inside and closes the operculum. In an expanded animal, the head and foot come out of the shell mouth but viceral mass lies inside the shell.

(v) **Visceral mass.** Between the head and foot the body forms a hump on the dorsal side. The hump is called **visceral mass.** It contains all the organ systems of the body. The visceral mass is spirally coiled and fills up the whorls of the shell. It shows the phenomenon of torsion.

(vi) **Mantle.** Below the shell lies **mantle** or the **pallium.** It is thin and delicate. A long narrow **supra-marginal groove** runs along the thickened free edge of mantle. Behind the groove there is a thick band containing **shell secreting glands** (which secrete the shell). On each side of head the fleshy processes are formed by mantle, called **nuchal lobes** or **pseudepipodia.** Left lobe is longer than right lobe and form **respiratory siphons** (Fig. 45.4).

Pallial Complex (Organs of mantle cavity)

Anterior part of body encloses a large space called **mantle cavity.** The organs present inside the mantle cavity are known as **organs of mantle cavity** or **pallial complex.** These are as follows:

1. Epitaenia. A prominent ridge is seen on the floor of the mantle cavity; it is called **epitaenia.** It divides the cavity into two unequal chambers. The chamber of the right side is small and is called **branchial chamber** while the chamber on the left side is large and called **pulmonary chamber.**

2. Ctenidium. A ctenidium or gill lies attached in the branchial chamber to the right side of mantle.

3. Rectum. It lies on the floor of the mantle cavity and opens to the exterior by **anus** which is situated near the right **pseudepipodium** or **right nuchal lobe.**

4. Genital duct. The vas deferens in case of male or the oviduct in female lies in the branchial chamber in between the rectum and epitaenia. Its opening is situated close to the **anus.** A hook-like **penis** or **copulatory organ** in male arises from the edge of mantle in front of male genital aperture. At the base of penis there is found a **hypobranchial gland.**

5. Anterior renal chamber. It projects as a reddish sac at the posterior end of the epitaenia and opens into the mantle cavity by an oblique slit.

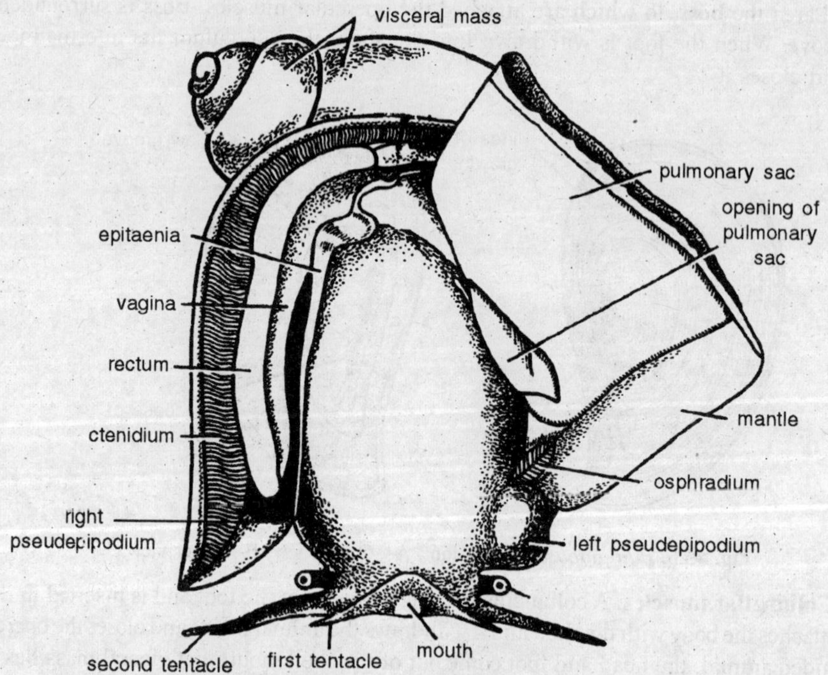

Fig. 45.6. *Pila globosa.* A female individual dissected to show the organs of pallial complex.

6. Pulmonary sac. This is so-called "lung" of *Pila* and found in the pulmonary chamber. It hangs from the roof of mantle cavity and opens into it by a large obliquely placed **pulmonary aperture.**

7. Osphradium. It is a chemo-sensory structure to test the purity of incoming water currents. It is attached to the roof of mantle cavity just above the left nuchal lobe. Course of water current through mantle cavity has been shown in Fig. 45.7.

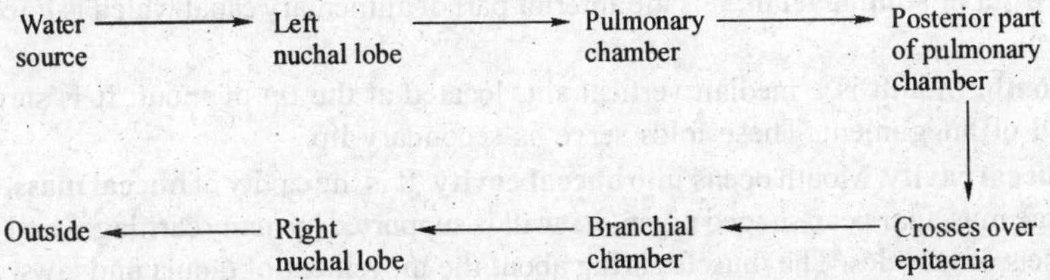

Fig. 45.7. *Pila globosa.* Course of water current through mantle cavity.

Histology of body wall. Body wall of *Pila* is composed of **epidermis** and **dermis.** The epidermis consists of a single layer of epithelial cells. The cells bear cilia in parts not protected by the shell. The epidermis also contains a large number of unicellular glands which secrete mucus, pigments and lime.

The dermis is composed of connective tissues and muscle fibres. It also contains pigment cells.

Coelom

The coelom of *Pila globosa* is represented by the cavities of kidney, gonads and pericardial cavity. The space between the visceral organs is filled with blood. This is called **haemocoel.**

45.4 DIGESTIVE SYSTEM

Digestive system of *Pila* consists of (1) Alimentary canal; (2) Digestive glands (Fig. 45.8).

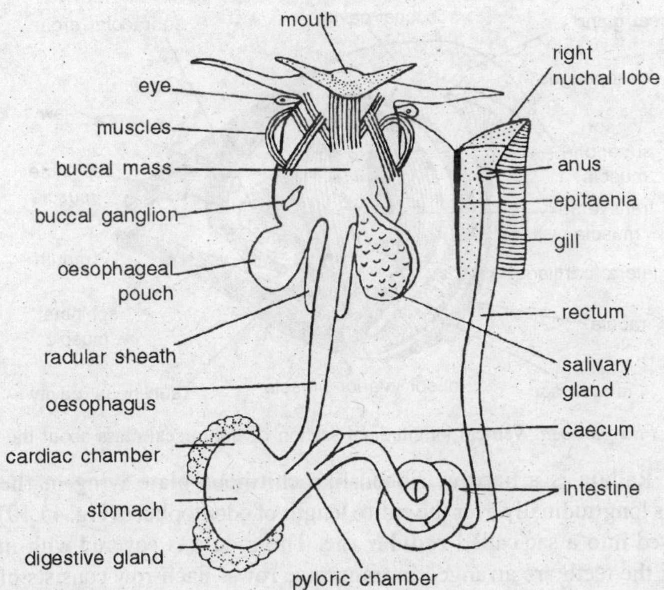

Fig. 45.8. *Pila globosa.* Digestive system: alimentary canal and digestive glands.

A. Alimentary Canal

It is curved and tubular structure and divided into following three parts:

(*i*) **Fore gut** which includes mouth, buccal cavity and oesophagus.

(*ii*) **Mid gut** which includes stomach and intestine.

(*iii*) **Hind gut** which includes rectum and anus.

1. Fore gut or stomodaeum. It is the anterior part of alimentary canal which is lined internally by **ectoderm.**

(*i*) **Mouth.** Mouth is a median vertical slit, located at the tip of snout. It is surrounded by a frilled fold of integument. These folds serve as secondary lip.

(*ii*) **Buccal cavity.** Mouth opens into **buccal cavity.** It is the cavity of **buccal mass.** The buccal mass is a large muscular pear-shaped organ. Its wall is supported by many cartilages and is provided with seven sets of muscles. The muscles bring about the movement of radula and jaws. The buccal cavity consists of following parts:

(*a*) **Jaws.** There are two jaws (Fig. 45.9). Each jaw is a flexible, somewhat elliptical chitinous structure placed dorso-laterally. The jaws form posterior margin of the vestibule. Both jaws are connected together by a thin cuticular membrane. The jaws bear numerous small and two or three large teeth.

cutting edge

ridge of attachment of muscles

Fig. 45.9. *Pila globosa.* Jaws.

(*b*) **Odontophore.** The floor of posterior part of buccal cavity is raised up as a fleshy outgrowth called **tongue mass** or **odontophore** (Fig. 45.10). The roof of buccal cavity carries a cartilage and well developed **protractor** and **retractor muscles.** It is supported by a large **'S' shaped lateral cartilage** and a pair of triangular **superior cartilage** (Fig. 45.11). The odontophore bears at its anterior end a rounded process called **sub-radular organ.** Beneath it the floor of buccal cavity has a small pouch-like **sub-lingual cavity.** The odontophore acts as a cartilaginous base of radula.

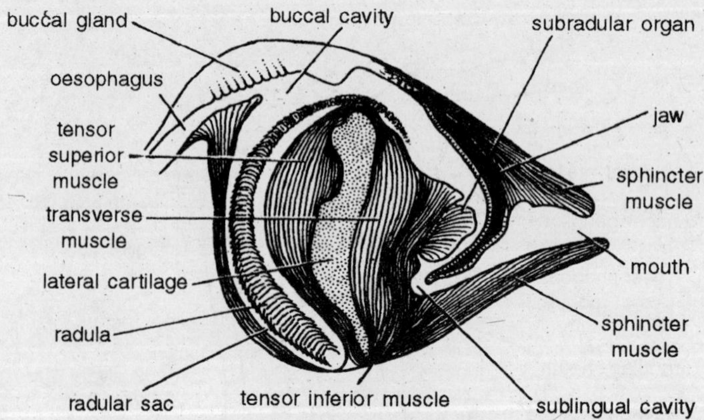

Fig. 45.10. *Pila globosa.* Vertical longitudinal section of the buccal mass about the middle.

(*c*) **Radula.** Radula is a narrow, ribbon like chitinous plate lying in the buccal cavity (Fig. 45.12). It runs longitudinally over the entire length of odontophore (Fig. 45.10). Posterior part of radula is enclosed into a sac called **radular sac.** The radula is covered with numerous horny recurved teeth. All the teeth are arranged in transverse rows. Each row consists of seven teeth in all (Fig. 45.13): the **central** or **median rachidian** (spine-like), followed by a smaller **lateral** and two needle-like **marginal teeth** on each side. The **radular formula** of *Pila* is thus, 2, 1, 1, 1, 2 (*i.e.*, Two marginal teeth + one lateral tooth + one median or central tooth + one lateral tooth + two marginal teeth). The cutting edge of central tooth bears a median and 2-3 lateral cusps on each side. The radular teeth are secreted by the **odontoblast cells** at the bottom of the radular sac.

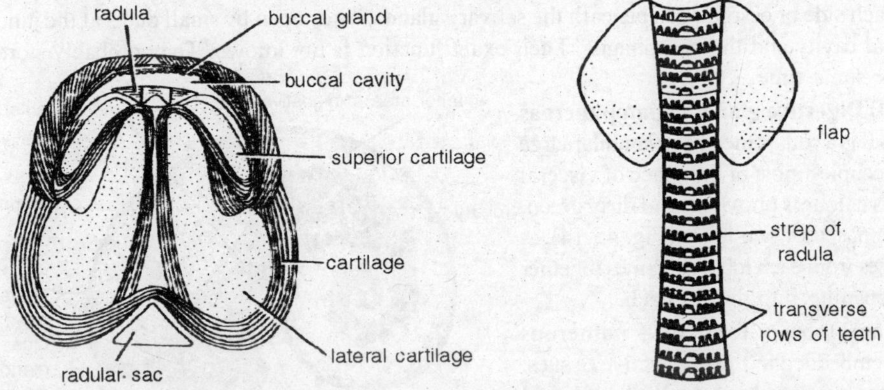

Fig. 45.11. *Pila globosa.* Buccal mass in T.S., showing position of radula on odontophore.

Fig. 45.12. *Pila globosa.* The radula.

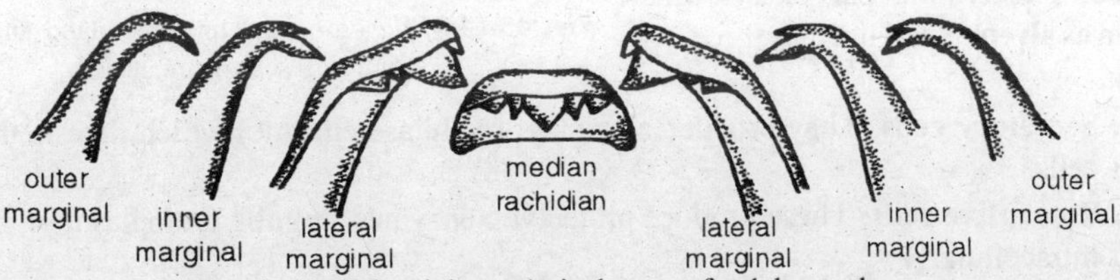

Fig. 45.13. *Pila globosa.* A single row of radular teeth.

The radula moves to and fro over the odontophore. It was called as **chain saw-like movement** by **Huxley.** These teeth act as a grater or saw and cut tissues of food by rasping.

***(iii)* Oesophagus.** The buccal cavity leads into oesophagus. It is a long narrow tube. It runs straight for a short distance and then turns to the left, enters the visceral mass and opens into the stomach.

2. Mid gut or mesenteron. It is lined by **endoderm.**

***(i)* Stomach.** It lies embedded in liver on the left side of visceral mass. It is roughly rectangular and dark red in colour. Its cavity is 'U' shaped and its posterior limb is called **cardiac chamber.** Oesophagus opens into the cardiac chamber. The anterior limb of 'U' is tubular called **pyloric chamber.** Intestine arises from this part of stomach. From the outer wall of the **pyloric chamber** a small, rounded pouch or **caecum** arise (Fig. 45.8).

***(ii)* Intestine.** The pyloric chamber of stomach opens into a long coiled tube called **intestine.** It makes 2 to 3 coils between the gonad and liver and then enters the mantle cavity to join the rectum.

3. Hind gut or proctodaeum. It is lined by ectoderm. It forms the last part of alimentary canal.

Rectum. It lies in the mantle cavity along the left side of the ctenidium. It is thick-walled tube and is lined by cuticle. It opens to outside through **anus** near the edge of right nuchal lobe.

B. Digestive Glands

There are following types of digestive glands:

(1) Buccal glands. It is a pair of glands which lies in the roof of buccal cavity. They secrete accessory digestive juices.

(2) Salivary glands. A pair of white, branched and glandular salivary glands lie on either side of oesophagus just behind the buccal mass. The ducts of salivary glands run anteriorly. The salivary juice contains mucus and digestive fluid. The mucus tends to lubricate the food and radula.

(3) Oesophageal pouches. A pair of simple, rounded, cream-coloured oesophageal pouches lie on each side of oesophagus beneath the salivary glands. They open by small ducts at the junction of buccal cavity and the oesophagus. Their exact function is not known. They probably store the food for some time.

(4) Digestive gland, hepatopancreas or liver. It is a soft, somewhat triangular large gland occupies most of the space of visceral mass. Its colour is brownish and dirty green. It is composed of two lobes (Fig. 45.14). A duct arises from each lobe and unites together before opening into the stomach.

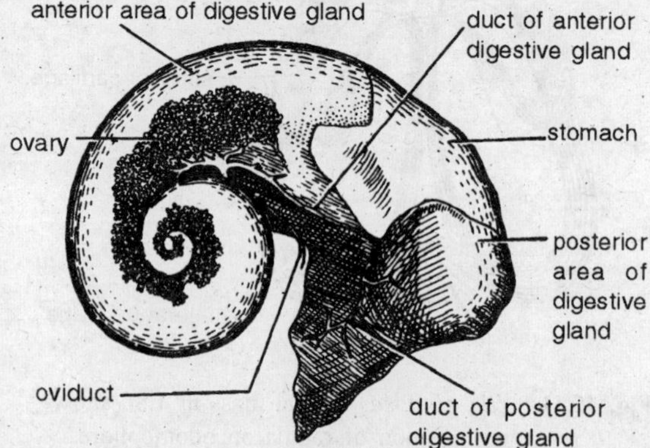

Fig. 45.14. *Pila globosa.* Digestive gland and associated structures seen from inside.

Each lobe is formed of numerous tubules embedded in the connective tissues. The tubes are lined internally by ciliated epithelium. The finer tubules join to form bigger tubules which finally combine into main duct. The terminal part of each tubule is known as **alveolus** and lined by three types of cells:

1. Secretory cells. They produce a ferment (cellulase enzyme) which digests the cellulose of plant cells.

2. Resorptive cells. These produce proteolytic enzymes, engulfs protoplasm of the food and digest it intracellularly.

3. Lime containing cells. These store phosphate of lime.

Food of *Pila*

Pila is mainly herbivorous, rarely it becomes a scavenger. Food of *Pila* includes soft and succulent leaves of aquatic plants such as *Valisneria* and *Pistia*. Sometimes they feed on dead animal tissue also.

Mode of feeding. *Pila* grasps a leaf between the sensory labial palps. By the action of sphincter and protractor muscles of buccal mass, the jaws cut the aquatic plants into pieces. The food is taken into the buccal cavity by the forward and backward movements of radula. In this way food is cut and masticated inside the buccal cavity.

Digestion

Salivary glands pour their secretion containing mucus and starch-digesting enzymes into the buccal cavity where it mixes up with the food. Therefore the digestion of starch begins in the buccal cavity. Mucus lubricates the food for its easy passage to the stomach. In stomach, digestive glands secrete digestive juices containing carbohydrate, protein and fat-digesting enzymes. Thus, the food is digested **extracellularly** in the stomach. Partly digested food particles pass into the resorptive duct of digestive glands, where the digestion of cellulose occur (This is being **intracellular digestion**).

Absorption. Absorption of digested food takes place in digestive glands and intestine.

Egestion. Undigested food passes out of rectum through anus into the branchial chamber from where it is thrown outside along with outgoing water current.

Adaptive features in the digestive system of *Pila*

1. *Pila* is herbivorous having well developed and highly efficient radula and jaws to cut off the plant leaves.

2. Alimentary canal is long with a coiled intestine.
3. For rasping of food from the ground, recurved teeth of radula move to and fro over the odontophore and help in scraping the food.
4. Teeth of radula act like a chain-saw.
5. Stomach is large and here extracellular digestion occurs.
6. Intestine is long and coiled.
7. The radula moves backward and forward and up and down by protractor and retractor muscles.
8. In digestive gland intracellular digestion occurs.
9. Digestion of cellulose is a remarkable feature, because it is a rare process in the animal world.
10. The digestive gland not only secretes the digestive enzymes but helps in the absorption of digested food.

45.5 RESPIRATORY SYSTEM

Pila is amphibious in nature therefore it possesses both **aquatic** and **terrestrial** or aerial mode of respiration. Respiratory system of *Pila* includes the following organs:

1. Ctenidium or gill for aquatic respiration
2. Pulmonary sac for aerial respiration
3. Nuchal lobes for the formation of siphons.

1. Ctenidium

Gill of *Pila* is single and monopectinate (*i.e.*, lamellae are arranged in a single row). It lies on the extreme right side of branchial chamber of mantle cavity.

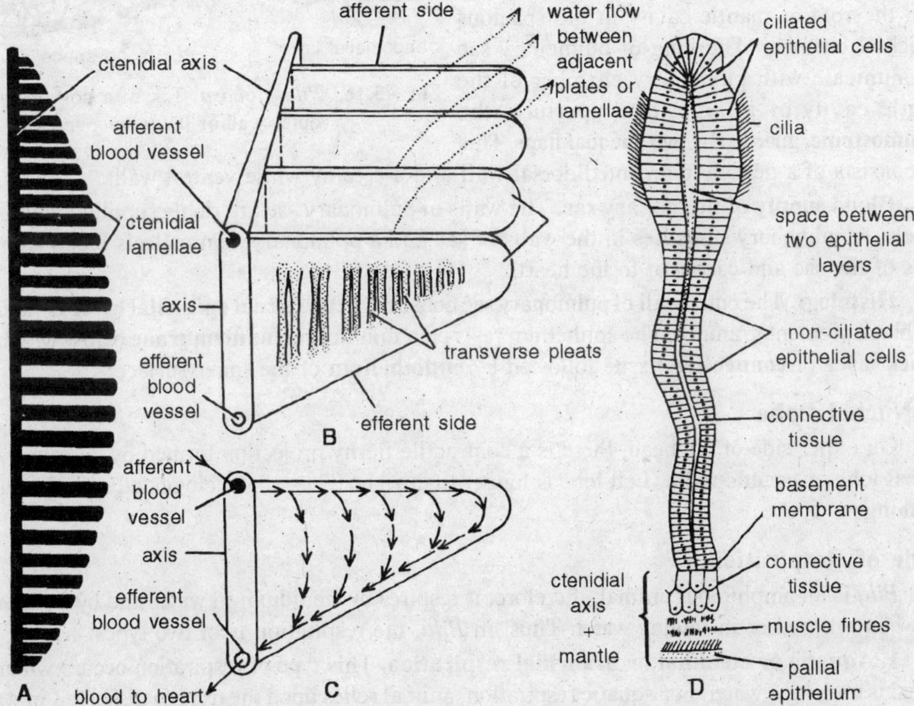

Fig. 45.15. *Pila globosa*. Respiratory organs. A—A monopectinate ctenidium; B—Movement of water currents through gill-lamellae; C—A single lamella showing flow of blood within it; D—A lamella in T.S.

The ctenidium consists of a large number of thin **leaflets** or **lamellae** (Fig. 45.15). Each lamella has a broad base and narrow apex. Lamellae are attached to the mantle through the **ctenidial axis.** The lamellae lie parallel to each other at the right angle to the ctenidial axis. In the middle they are largest and gradually become reduced in size towards the both ends. Each lamella has a smaller right side which receives blood from the afferent vessel and is called the **afferent side** and a longer left side from which blood goes into the efferent vessel called the **efferent side.** Both flat surfaces of each lamella contain transverse **ridges** or **pleats** which become smaller from the base towards the apex. These are richly supplied with blood vessels.

Blood supply of ctenidium. The afferent vein supplies deoxygenated blood to each gill lamella for oxygenation. The oxygenated blood is collected from all lamellae by the efferent vein.

Histology of ctenidium. Each lamella is formed of two layers of branchial epithelium separated by a narrow space in between. The epithelium consists of three types of cells: 1. Non-ciliated columnar cells; 2. Ciliated columnar cells and 3. Glandular cells. The epithelium rests upon basement membrane followed by a layer of connective tissues and a layer of oblique muscles. The outer most lining is of pallial epithelium or mantle epithelium.

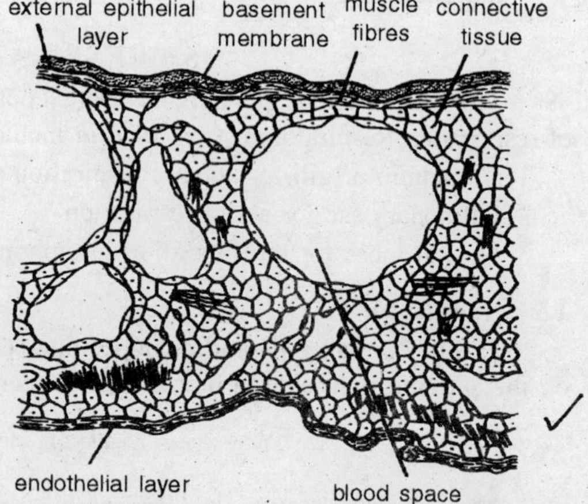

Fig. 45.16. *Pila globosa*. T.S. of a portion of the outer wall of the pulmonary sac.

2. Pulmonary Sac

It is a large bag-like structure which hangs from the roof of mantle cavity in the spacious branchial chamber. The lung or pulmonary sac communicate with pulmonary chamber of the mantle cavity by a large, oval aperture, the **pneumostome,** guarded by two unequal flaps. This sac consists of a densely pigmented dorsal wall and a creamy white ventral wall.

Blood supply of pulmonary sac. The walls of pulmonary sac are richly supplied with blood vessels. A pulmonary sinus lies in the walls of sac and a pulmonary vein collects blood from the walls of the sac and carries it to the heart.

Histology. The outer wall of pulmonary sac consists of an external **epithelial layer** containing fine black pigment granules. The epithelium rests on a thin **basement membrane** below which lies a thick layer of **connective tissue** followed by **endothelium** of the inner surface.

3. Nuchal Lobe

On either side of the head, there is a contractile fleshy projection formed by mantle, called **nuchal lobe** or **pseudopodia**. Left lobe is longer than right. These form elongated siphons during breathing.

Mode of Respiration

Pila is an amphibious animal, therefore, it respires by ctenidium in water and by pulmonary sac when on land or in muddy water. Thus, in *Pila*, the respiration is of two types:

1. Aquatic or ctenidial or branchial respiration. This type of respiration occurs when the animal is under the water. For aquatic respiration, animal relies upon the oxygen dissolved in water. The head and foot are extended out of the shell, while left nuchal lobe increases in size and forms a funnel-like channel. Water enters through this funnel, comes in contact with **osphradium.** The water reaches the posterior part of pulmonary chamber and enters the branchial chamber after crossing the epithelia and exchange of gases occur (Fig. 45.17A).

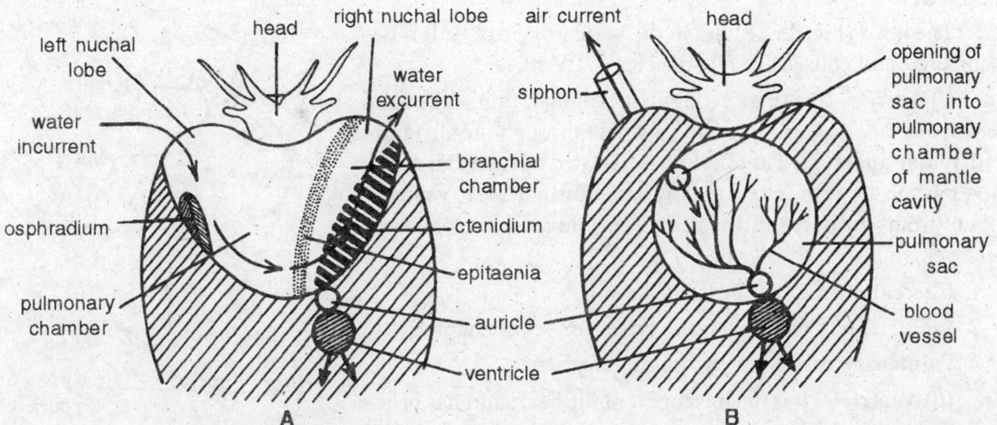

Fig. 45.17. *Pila globosa*. Diagrams showing mechanism of respiration. A—During aquatic respiration; B—During aerial respiration.

A continuous flow of water through the branchial chamber is maintained by cilia of ctenidial lamellae and by alternate rise and fall of the floor of mantle cavity. The latter is brought about by the retraction and protrusion of head.

2. Aerial, pulmonary or lung respiration. *Pila* respires through lung when the water is polluted or poor in oxygen or when it is on land. The animal after coming near the surface of water extends its left nuchal lobe out of the water which becomes elongated and tubular to form a siphon which projects above the surface of water (Fig. 45.4). The pulmonary chamber becomes greatly enlarged and is completely cut off from the branchial chamber. The air enters the pulmonary sac through siphon and the gaseous exchange occurs by the alternate contraction and expansion of the pulmonary sac (Fig. 45.17B).

While *Pila* is on land, the siphon is not formed and air enters pulmonary chamber directly through the expanded nuchal lobe.

During **hibernation** and **aestivation,** the animal lies buried in the mud for long periods. Its shell remains tightly closed by the operculum. Under these conditions, *Pila* respires through the air closed in pulmonary sac.

45.6 BLOOD VASCULAR SYSTEM

The blood vascular system of *Pila* is modified according to its amphibious mode of life. It consists of following parts:

1. Pericardium; 2. Heart; 3. Arteries; 4. Sinuses; 5. Veins.

1. Pericardium

Pericardium is situated between the pulmonary chamber and the posterior renal sac on the left side of the body whorl. It is thin-walled oval chamber enclosing the heart. It is posteriorly surrounded by the stomach and digestive glands, dorsally by the renal sacs (or chambers) and ventrally by the oesophagus. The pericardial cavity represents the true coelom and it communicates with the posterior renal chamber by **reno-pericardial aperture.** The heart, aorta and ampulla lie in the pericardial cavity.

2. Heart

The heart (Fig. 45.18) lies in the pericardial cavity. It is two chambered and consists of (*i*) Auricle; (*ii*) Ventricle.

***(i)* Auricle.** It is single, roughly triangular, thin walled and contractile sac. It opens into the ventricle through an **auriculo-ventricular aperture.** The aperture is guarded by two **semi-lunar valves** which allow the blood to flow only from auricle to ventricle but not from ventricle to auricle. The auricle receive blood from three veins:

1. Efferent ctenidial vein from the gill.
2. Efferent renal vein from the posterior renal chamber.
3. Pulmonary vein from the pulmonary sac.

***(ii)* Ventricle.** It is the lower part of the heart and lies below the auricle. It is thick walled, ovoid, muscular and spongy. It is provided with muscular strands and has a narrow cavity.

The lower end of the ventricle gives rise a short, muscular tube called **aorta.** The opening between ventricle and aorta is guarded by two **semi-lunar valves** which allow the blood flow only from the ventricle to aorta but not from aorta to ventricle.

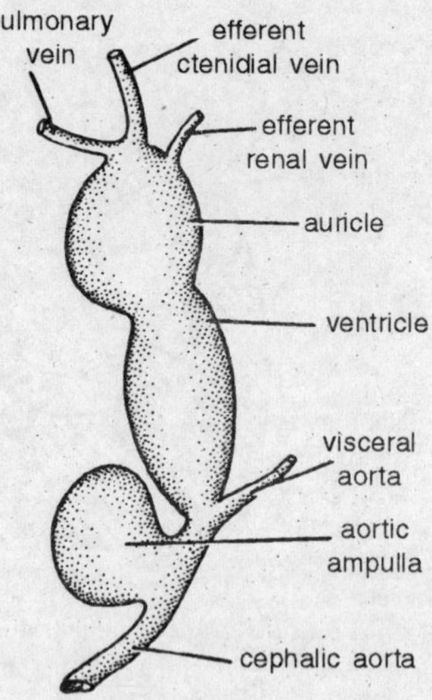

Fig. 45.18. *Pila globosa*. Heart.

3. Arteries

The aorta is divided into two sub-branches:

(*i*) The cephalic aorta; (*ii*) The visceral aorta.

***(i)* The cephalic aorta.** It supplies blood to the head region through a number of branches. Its base bears a bulb-like, thick-walled, contractile **aortic ampulla,** which aids in the flow of blood. From the outer side of aorta, three branches are given out: (*i*) a thin **cutaneous artery** which supplies blood to the body wall or skin (*ii*) a thick **oesophageal artery** which supplies blood to oesophagus; (*iii*) a thick **pallial artery** which supplies blood to the left side of the mantle, the left nuchal lobe and the osphradium. From the inner side of aorta only one branch is given out: the **pericardial artery** which supplies blood to the wall of pericardium.

After dividing into above branches cephalic aorta runs along the left side of the oesophagus and cross over to its right side. Meanwhile it sends many fine branches to the wall of oesophagus and the floor of mantle cavity. From the right side of aorta, a large branch arises which supplies the following three branches:

(*i*) A **right pallial artery** which supplies blood to the right part of mantle
(*ii*) A **right siphonal artery** to the right nuchal lobe
(*iii*) A **penial artery** to copulatory organ or penis.

Finally cephalic aorta divides into following four branches:

(*i*) **Radular sac artery** to the radular sac
(*ii*) **Optic arteries** to the eye stalk and eyes
(*iii*) **Tentacular arteries** to the tentacles
(*iv*) **Pedal arteries** to the foot after forming a network.

(2) Visceral aorta. The visceral aorta runs posteriorly into visceral mass and supplies blood through the following branches:

(*i*) A **pericardial artery** to the wall of pericardium
(*ii*) A **gastric artery** to the stomach

(*iii*) An **intestinal artery** to the intestine

(*iv*) A **hepatic artery** to the digestive gland and gonads

(*v*) Many branches to the roof of anterior renal chamber, intestine and rectum, etc.

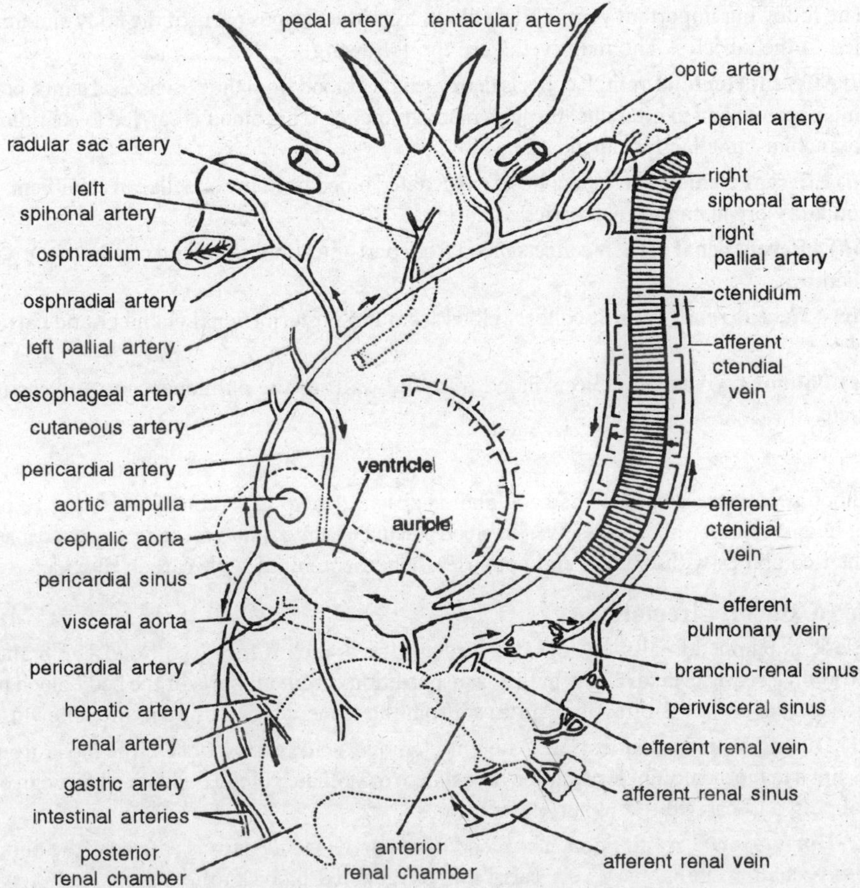

Fig. 45.19. *Pila globosa.* Heart and open type blood vascular system.

4. Sinuses

The ultimate branches of arteries widen inside the tissues or organs into blood spaces which are without definite walls. These spaces are called **blood lacunae.** Generally the blood lacunae connect together to form **sinuses.** They connect arteries to veins and serve as capillaries. The sinuses and lacunae together form the general **haemocoel.** Thus, sinuses are the space between the visceral organs and the body wall. They lack definite walls. There are following four main sinuses:

(*i*) Perivisceral sinus. It lies in the floor of mantle cavity and occupies the anterior part of the body above the foot and below the floor of mantle cavity. It surrounds the anterior part of alimentary canal. Three channels carry blood from this sinus to the renal chamber and the pulmonary sac. It contains deoxygenated blood.

(*ii*) Peri-intestinal sinus. It lies within the visceral mass along the coils of intestine. It collects blood from the digestive gland and reproductive organs.

(*iii*) Branchio-renal sinus. It lies on the right side of anterior renal chamber. It receives blood from the roof of both renal chambers.

***(iv)* Pulmonary sinus.** It lies in the wall of pulmonary sac. It receives blood from the perivisceral sinus.

5. Veins

The following important veins collect blood from the various parts of the body and this blood is carried to the auricles. The main veins are the following:

***(i)* Afferent ctenidial vein.** It collects deoxygenated blood from the perivisceral sinus, branchio-renal sinus, rectum and genital duct through many branches. This blood is carried to ctenidium then to the branchial lamellae for purification.

***(ii)* Efferent ctenidial vein.** It collects oxygenated blood from the lamellae of ctenidium, mantle and copulatory organs and then carries to auricle.

***(iii)* Afferent renal vein.** It collects blood from peri-intestinal sinus and carries to the posterior renal chamber.

***(iv)* Efferent renal vein.** It collects blood from the posterior renal chamber and carries it to the auricle.

***(v)* Pulmonary vein.** It collects blood from the walls of the pulmonary sac and carries it to the auricle.

Blood

Blood is colourless because haemoglobin is absent. However, it contains another respiratory pigment **haemocyanin** which is dissolved in blood plasma. Haemocyanin contain copper, so it becomes blue when combines with oxygen and colourless when combines with carbon dioxide.

Course of Blood Circulation

Blood is pumped to all parts of body through the branches from ventricle. The branches of cephalic and visceral aorta take part in this transportation. From all parts of the body blood reaches the **peri-visceral** and **peri-intestinal sinuses.** From the sinuses blood flows in the following ways:

(1) During aquatic respiration. Afferent-ctenidial vein carries most of the blood from peri-visceral sinus to the ctenidium for aeration. From here oxygenated blood is collected through efferent ctenidial vein and carried to the auricle.

(2) During aerial respiration. The blood from peri-visceral sinus is carried to the pulmonary sac for oxygenation. Pulmonary vein carry this oxygenated blood to the auricle.

(3) Blood from peri-intestinal sinus flows into the renal chamber to get rid of nitrogenous wastes. From here efferent renal vein carry this oxygenated blood to the auricle.

Auricle receives both aerated and non-aerated blood. Mixed blood then enters the ventricle.

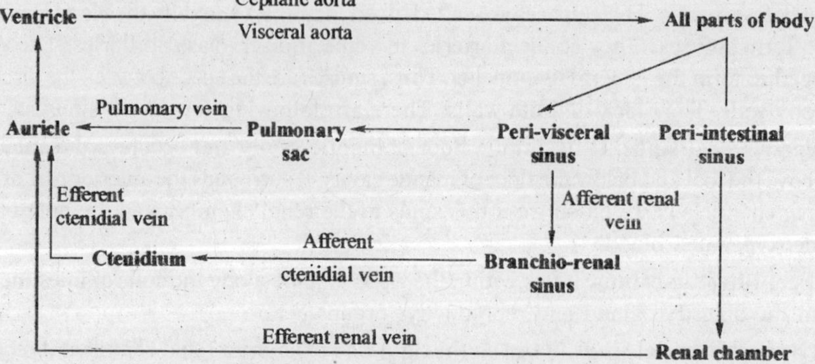

Fig. 45.20. Course of blood circulation in *Pila*.

45.7 EXCRETORY SYSTEM

All the nitrogenous wastes of *Pila* are excreted by a large **kidney, renal organ, organ of Bojanus** or **metanephridia.** It lies behind the pericardium and is a thick-walled sac which is much folded within. The renal organ contains a small anterior chamber and a large posterior chamber (Fig. 45.21 and Fig. 45.22).

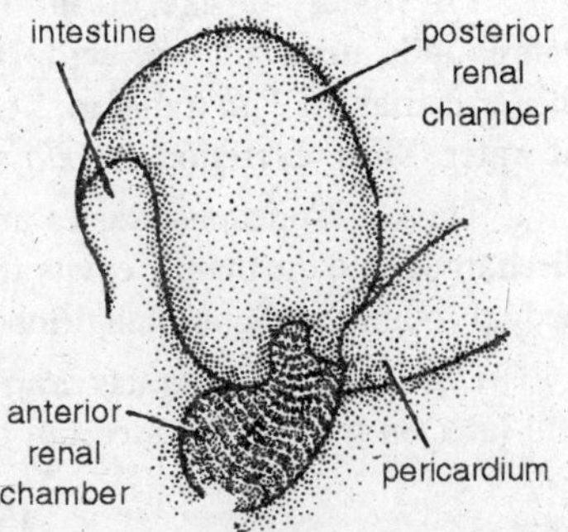

Fig. 45.21. *Pila globosa*. Renal chamber.

1. Anterior renal chamber. It is oval and reddish in colour. Anterior renal chamber is situated anterior to the pericardium. It projects into the mantle cavity and opens into the branchial chamber through a slit like opening near the epitaenia. Its other end opens into the posterior renal chamber through an internal opening. The internal cavity of the anterior chamber is reduced due to presence of triangular leaf-like process or **lamellae.**

The lamellae on the roof are arranged on either side of a median longitudinal axis, the **efferent renal sinus.** On the floor of the chamber, the lamellae are arranged on either side of a similar median axis, the **afferent renal sinus** which is the right branch of the **peri-intestinal sinus.** It branches profusely to supply lamellae on both sides. The roof or floor folds alternate to each other.

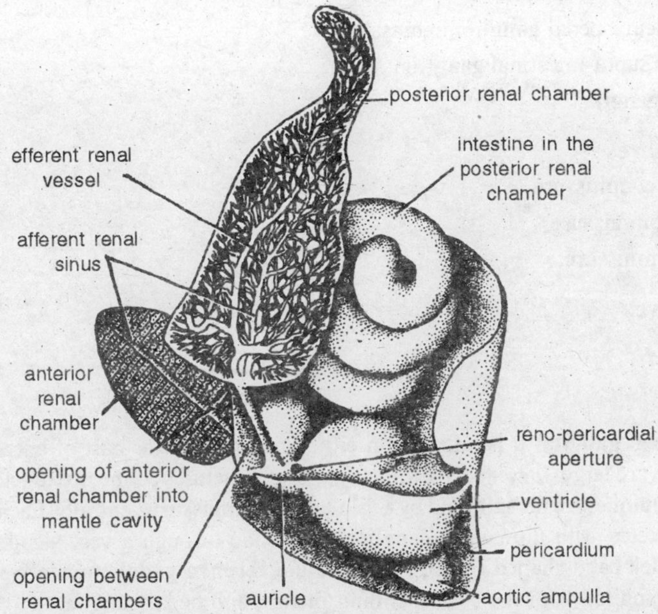

Fig. 45.22. *Pila globosa*. Pericardium and two chambers of renal organ.

2. Posterior renal chamber. This chamber is a broad, grey or brown in colour and hook-shaped. It is situated behind the anterior renal chamber in between the rectum on right side and pericardium and digestive gland on the left side. At one end it communicates with the anterior renal chamber through an aperture and the other end communicates with the pericardial cavity through an elongated slit-like **reno-pericardial aperture.** The septum between this chamber and pericardial cavity is called **reno-pericardial septum.** The roof of the chamber is richly supplied by branches of **afferent** and **efferent renal vessels.** The floor of the chamber consists of a thin lining of epithelium which separates its cavity from the organs projecting in its cavity.

Physiology of excretion. The excretory waste of *Pila* contains **ammonia, ammonium compounds, urea** and **uric acid.** During aquatic life ammonia is excreted in solution while during the terrestrial life *Pila* transforms ammonia into soluble **uric acid;** this is done for the conservation of water. Thus, excretion of *Pila* shows an adaptation for its amphibious mode of life.

These nitrogenous wastes are separated from blood by both chambers of renal organ and discharged into the mantle cavity through the renal aperture. These are ultimately passed out of the body through the right nuchal lobe along with outgoing water current.

Digestive gland also acts as an excretory organ. The cells of digestive gland engulf waste products and later on these are passed out through the stomach and intestine.

45.8 NERVOUS SYSTEM

The nervous system of *Pila* is well developed possessing a concentrated ring of most of the ganglia around the buccal mass and twisting of visceral cords in the figure of '8'. Each ganglion consists of commissures, connectives and nerves. All these structures can be summarised as follows:

1. Ganglia

(*i*) A pair of cerebral or labial ganglia

(*ii*) A pair of buccal ganglia

(*iii*) Paired pleuro-pedal ganglionic mass

(*iv*) Unpaired supra-intestinal ganglion

(*v*) Visceral ganglia.

2. Commissures

(*i*) Cerebral commissure

(*ii*) Buccal commissure

(*iii*) Pedal commissure.

3. Connectives

4. Nerves

A. Ganglia

1. Cerebral ganglia. It is situated on either side of anterior part of buccal mass. A pair of cerebral ganglia are large, flattened, roughly triangular structures connected together by thick, band-like **cerebral commissure** dorsally and by a thin labial **commissure** ventrally. Each cerebral ganglion is further connected with the buccal ganglion of its side through a very slender **cerebro-buccal connective.** Thick band-shaped **cerebro-pleural** and **cerebro-pedal connectives** serve to connect each cerebral ganglion with the corresponding pleural and pedal ganglia.

Nerves. Following nerves are given out in head region:

(*i*) Anteriorly to the skin of snout, tentacles and the buccal mass.

(*ii*) Posteriorly to the tentacles, eyes and the statocyst.

2. Buccal ganglia. A pair of small, triangular buccal ganglia situated dorso-laterally on either side of the junction of buccal mass and oesophagus. The two buccal ganglia are connected together ventrally by **buccal commissure.** A pair of **cerebro-buccal connectives** connect the buccal ganglia with cerebral ganglia.

Nerves. Each buccal ganglia sends nerves to buccal mass, radular sac, salivary glands, oesophagus and oesophageal pouches.

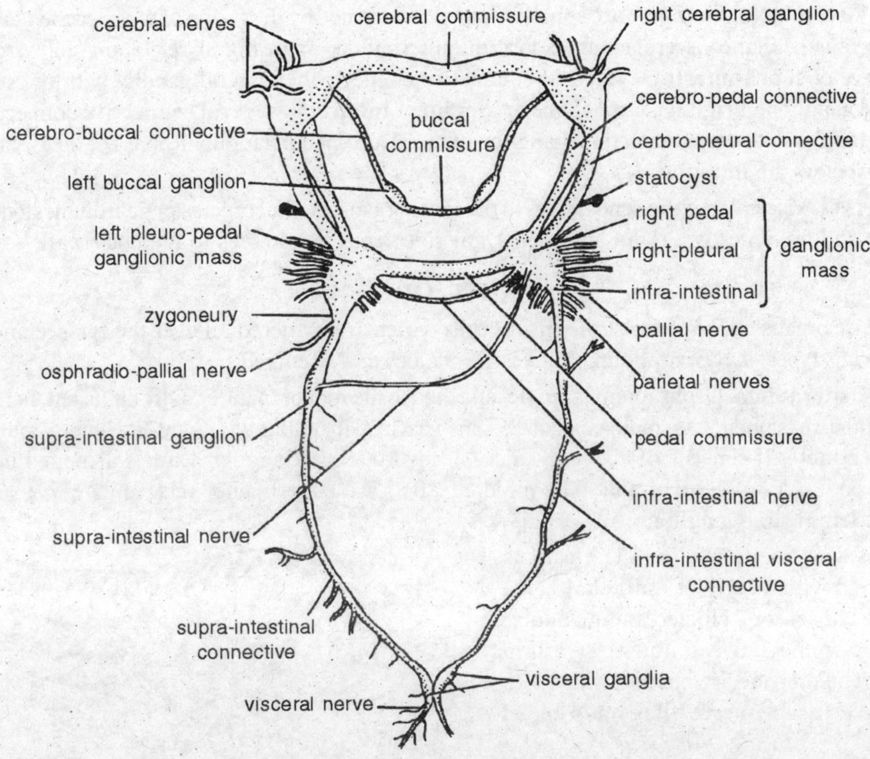

Fig. 45.23. *Pila globosa*. Nervous system.

3. Pleuro-pedal ganglionic mass. It is large, triangular and situated ventro-laterally one on either side of the buccal mass. Each ganglionic mass is formed by the fusion of an outer **pleural** and inner **pedal** ganglion. The line of fusion is shown as slight notch. Pleuro-pedal ganglionic mass of right side remains fused with **sub-intestinal ganglion.** Pedal ganglia are connected by two-thick strap-like **pedal commissures** which run on the ventral side of buccal mass. Two pleural ganglia are connected by a slender loop-like **infra-intestinal nerve.** Paired **cerebro-pleural** and **cerebro-pedal** connectives connect the cerebral ganglia with the pleural and pedal ganglia respectively.

Nerves

(*i*) Left pleural ganglion sends nerves to the body wall, mantle, osphradium, left nuchal lobe, columellar muscles and anterior part of gills.

(*ii*) Right pleural ganglion issues nerves to the body wall, epitaenia, right nuchal lobe, copulatory organs, columellar muscles and rectum.

(*iii*) Pedal ganglia sends nerves to the foot.

4. Supra-intestinal ganglion. It is unpaired, fusiform ganglion which is situated within a haemocoelomic sinus. It is connected with the pleuro-pedal ganglionic mass by a stout connection. A thin, fine **supra-intestinal nerve** connects it with the pleural ganglion of right side. It crosses over oesophagus.

Nerves

(*i*) Supra-intestinal ganglion gives off nerves to mantle, anterior part of ctenidium and pulmonary sac.

(*ii*) Supra-intestinal nerve sends a few branches to the body wall.

5. Visceral ganglia. These are spindle-shaped and formed by the fusion of two visceral ganglia. They are situated in the visceral complex near the anterior lobe of the digestive gland. A left **visceral connective** or **supra-intestinal visceral connective** connects the visceral ganglia with the supra-intestinal ganglion. A right visceral connective or **infra-intestinal visceral connective** connects the visceral ganglia with the subintestinal ganglion of the pleuro-pedal ganglionic mass of the right side and runs below the intestine.

Nerves. Visceral ganglia send nerves to the renal organs, genital organs, pericardium, stomach, intestine and the digestive gland and their right side sends few nerves to the body wall.

45.9 SENSE ORGANS

Sense organs of *Pila* consist of sensory cells which are scattered all over the surface and are of following types: 1. Osphradium, 2. Eyes, 3. Statocyst; 4. Tentacles.

1. Osphradium. Osphradium is single, attached to the roof of mantle cavity on its left, hanging over the inhalant siphon. It is small, elongated, somewhat oval in outline with right end bluntly rounded. It is **bipectinate.** It consists of about 28 thick, fleshy, roughly triangular leaflets arranged in two rows on both sides of **central axis.** The middle leaflets are largest. They receive its nerve supply from the left pleural ganglion.

Histology of osphradium. The leaf-lets are covered with a layer of **epithelial cells.** It consists of sensory, ciliated and glandular cells. The epithelium is followed by a thin **basement membrane.** Its interior is filled with connective tissues, nerve fibres and blood spaces.

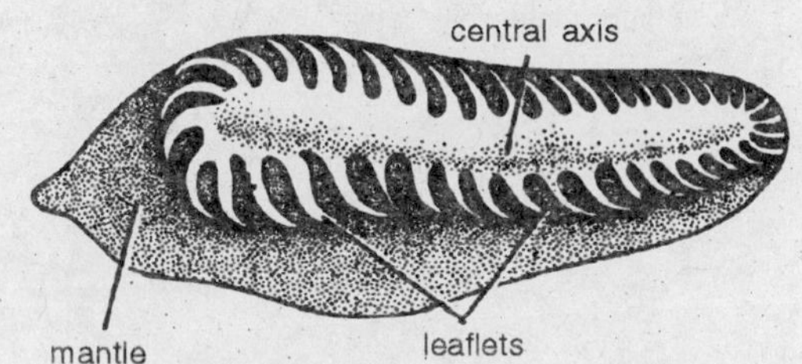

Fig. 45.24. *Pila globosa.* L.S. of osphradium.

Function. The osphradium serves as **olfactory** or **chemoreceptor organ.** It tests the chemical and physical properties of water entering the mantle cavity. If the water is not suitable for breathing, its entrance is checked by the closure of left nuchal lobe. It is also said to exercise selection over the food material.

2. Eyes. *Pila* has a pair of eyes which are short, black and stalked, lies on either side of the head outside the second pair of tentacles. The stalk of eye is called **ommatophore.**

The eye is a cup-like invagination of general epidermis. This invagination is called **optic vesicle** and is surrounded by connective tissues. The eye of *Pila* consists of the following parts:

(1) Inner cornea or pellucida internal. It lies just below the connective tissues.

(2) Lens. The central space of eye is occupied by a hyaline or structureless body called **lens.**

(3) Retina. Retina is made up of large cells. It is a sensitive part of eye. It contains two types of cells: (*i*) **Visual cells.** They are large broad cells having brush-like hair on their outer surface. (*ii*) **Supporting cells** or **packing cells.** They are scattered between the visual cells.

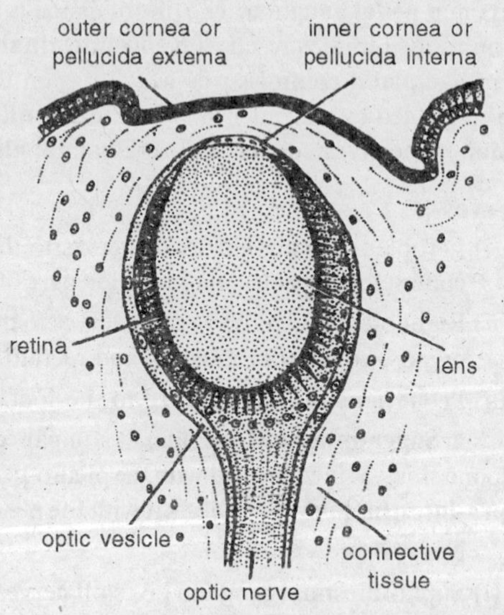

Fig. 45.25. *Pila globosa.* Horizontal longitudinal section of the eye.

4. Optic nerve. Optic nerve enters the eye at its posterior end.

Function. The eyes only respond to change in the intensity of light because sense of light is limited in *Pila*.

3. Statocyst. They are small, paired, pyriform, cream-coloured structures which are located in a depression near the pedal ganglion. Each statocyst is attached to the pedal ganglion of its side by a band of connective tissue. The statocyst is a hollow capsule surrounded by a thick layer of connective tissue. Its cavity is lined by a single layer of ciliated sensory epithelium and filled with a fluid and many minute, oval calcareous particles called as **statoconia.**

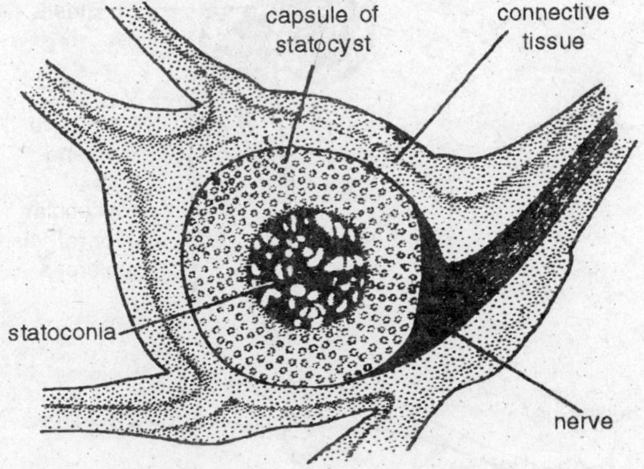

Fig. 45.26. *Pila globosa.* Statocyst.

Function. Statocyst receives its nerve supply from the cerebral ganglion and acts as a organ of equilibrium. The particles of statoconia help the animal in maintaining its position in water and on land.

4. Rhinophore or tentacular olfactory organs. In *Pila*, there are two pairs of tentacles. The surface of tentacles are covered with small ciliated papillae. These provide them a silky look. The tentacles are supplied with nerves from the cerebral ganglia.

Function. Tentacles act as chemoreceptors. They are tactile and chemosensory. They are supposed to be gustatory also.

45.10 REPRODUCTIVE SYSTEM

In *Pila* sexes are separate, *i.e.*, dioecious and **sexual dimorphism** is distinct: the shell of male animal is smaller than the female and male has a copulatory organ or penis.

A. Male Reproductive Organs

There are (*i*) Single or unpaired testis (*ii*) Vasa efferentia; (*iii*) Unpaired vasa deferens (*iv*) Vascula seminalis; (*v*) Penis; (*vi*) Hypobranchial gland (Fig. 45.27).

1. Testis. Testis of *Pila* is single, flat, roughly triangular and whitish-cream coloured. Testis is attached to the upper part of the digestive gland and extending in first or two whorls of the shell. It is enclosed in a thin cutaneous membrane. The **germinal epithelium** of testis produces two types of sperms: **(*i*) Eupyrene sperms.** These are small, thread-like containing head and tail and middle piece. They are motile and can fertilize the eggs. **(*ii*) Oligopyrene sperm.** These are large sickle-shaped or spindle-shaped, do not have head, but have tail. The tail is formed of 4 to 5 cilia. These are non-motile and unable to fertilize the eggs.

2. Vasa efferentia. They are numerous, minute ducts, originating at different levels from the testis and join to form a common large vas deferens.

3. Vas deferens. It arises from the posterior part of the testis and passes under the skin and is divided into two parts: **(*i*) Proximal tubular part.** It is distal thick-walled glandular part, runs along inner side of the digestive gland upto the posterior renal chamber and finally turns to the left and reach the pericardium and after turning to the right opens into the ventral side of vesicula seminalis. **(*ii*) Terminal glandular part.** It is thick-walled glandular part, runs along the left side of the rectum into the branchial chamber and opens into it by male genital pore. The genital pore is situated a little behind the anus to a claw-like projection of **genital papilla.**

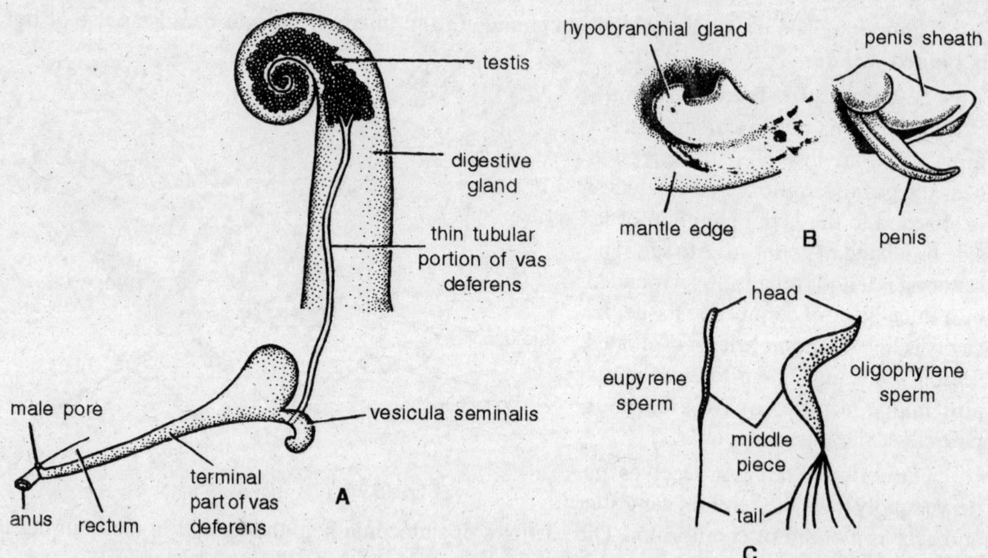

Fig. 45.27. *Pila globosa.* A—Male reproductive organs; B—Male copulatory organ in surface and ventral view; C—Sperms.

4. Vesicula seminalis or Seminal vesicle. It is situated on the right side of pericardium below the junction of anterior and posterior renal chambers. It is swollen flask-shaped sac having a blind rounded prolongation. It serves to store the sperms. The narrow tubular end of seminal vesicle is continuous with the glandular part of vas-deferens.

5. Copulatory organ (penis). Penis arise from the edge of mantle in front of anus. It is long about 1.5 cm, stout, slightly curved, flagellar with a swollen base and narrow tip. It bears a longitudinal groove. The penis is enclosed in a flap-like **penis-sheath** of yellow colour. The sheath is sac-like outgrowth of the mantle at the base of penis.

6. Hypobranchial gland. It is an oval glandular thickening at the base of the penis sheath. It consists of tall cells containing small basal nuclei. The surface of the glandular area is somewhat pleated but there is no duct and the secretions of the gland cells are poured directly on the surface.

B. Female Reproductive Organs

The female reproductive organs consist of following parts: 1. Ovary; 2. Oviduct; 3. Uterus; 4. Vagina; 5. Recepticulum seminis; 6. Copulatory organ; 7. Hypobranchial gland (Fig. 45.28).

1. Ovary. In female, the ovary occupies the same position as the testis in the male, *i.e.*, into first, second and third whorls of the shell. Ovary is a highly branched structure of a light orange colour which becomes darker in fully mature individuals. The branches of ovary, called **acini** are blind, flask-shaped structures having single layer of cells and open into the oviduct.

2. Oviduct. It is semi-transparent, narrow tube arising from the middle of ovary. It runs anteriorly just below the skin along the inner margin of the digestive gland. Near the renal organ it turns downwards.

3. Receptaculum seminis. It is a bean-shaped structure. It is situated at the junction of oviduct with the uterus and is enclosed in the cavity of posterior renal chamber. It stores the sperms received from male during copulation.

4. Uterus. It is a large, deep-yellow coloured and pear shaped structure. It lies inside the body whorl below the intestine. Uterus is differentiated into broad, vesicle-like rounded **basal part** and narrow tubular **apical part.** The former receives the opening of receptaculum seminis and latter opens into the vagina.

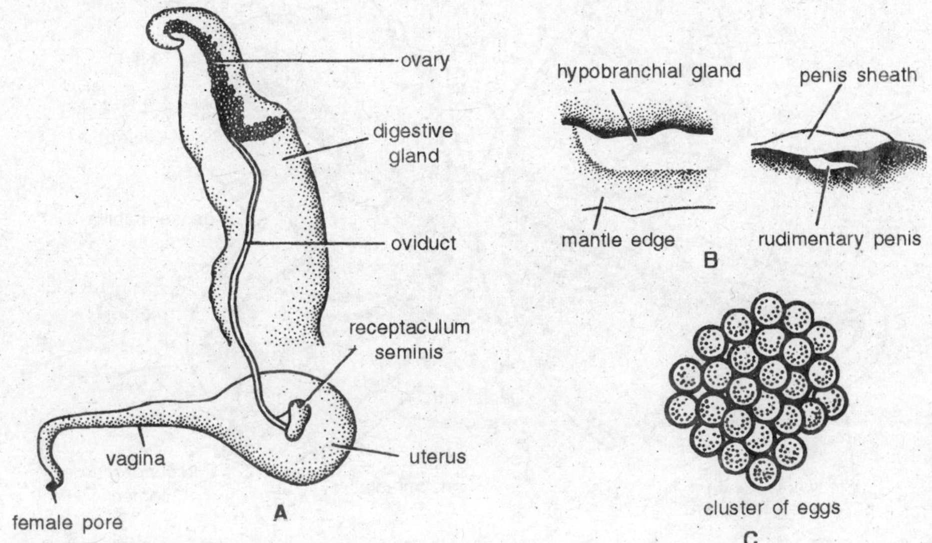

Fig. 45.28. *Pila globosa.* A—Female reproductive organs; B—Female copulatory organs in surface and ventral view; C—Eggs.

5. Vagina. The terminal narrow part of the oviduct is known as **vagina.** It is a cream-coloured tube which runs forward on the right side of rectum. It opens into the branchial chamber by **female genital aperture** which is situated behind the anus on a small papilla.

6. Copulatory organ. The female has a **rudimentary penis.** It is a small, thin flagellar structure of uniform thickness, being enclosed within the penis sheath.

7. Hypobranchial gland. It is a rudimentary glandular thickening.

Copulation

In *Pila globosa*, copulation takes place either in water or on land (Fig. 45.29). It lasts for 3 hours. Male and female snails come together facing each other. The penis of the male is expanded and gets attached to the genital papilla by its base. Then the penis and its sheath are inserted into the mantle cavity of the female. The tip of the penis is put into the female genital aperture and spermatozoa are transferred through the vagina into the receptaculum seminis.

Fertilization. Fertilization is **internal;** eggs are fertilized in the uterus. *Pila* is **oviparous** and oviposition or egg laying is started after a day or two of fertilization. The fertilized eggs are laid in masses of 200 to 800 in moist earth near ponds and lakes by the help of two temporary folds of foot.

Development

Eggs of *Pila* are round and as big as the pea seeds. Development is **direct,** *i.e.,* there is no free swimming larval stage. Young *Pila* hatches out of the egg. During development, due to torsion, the visceral mass and the shell of the embryo become spirally coiled.

Economic Importance

Pila causes great economic loss by damaging the rice crops. However, it is eaten by the poor people.

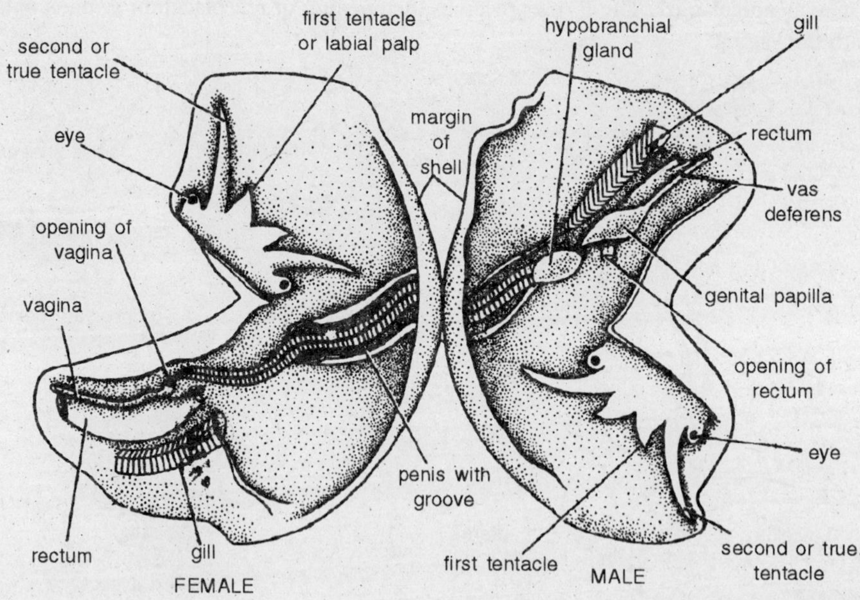

Fig. 45.29. *Pila globosa*. Copulation.

QUESTIONS

Long Answer Questions

1. Describe in detail the shell of *Pila* and give an account of contents of mantle cavity and course of water into it. (*Kanpur 1988; Jiwaji 1988; Bundelkhand 1989; Gorakhpur 1989, 1991, 1992; Awadh 1990; Meerut or CCS 1992*)
2. Describe organs of pallial complex in snail. (*Awadh 1988, 1990; Bundelkhand 1989; Lucknow 1988; Sagar 1990; Jabalpur 1993*)
3. Describe the digestive system, food, feeding mechanism and digestion in *Pila*. Why is *Pila* called a scavenger? (*Kanpur 1998*)
4. Describe the digestive system of *Pila* and discuss various adaptive features found in its digestive system. (*Agra 1989; Purvanchal 1989; Delhi 1989; Meerut or CCS 1989; Kumaun 1990; Raipur 1990; Garhwal 1990; Rohilkhand 1990, 1992; Gorakhpur 1991; Allahabad 1992; Bundelkhand 1992; Lucknow 1992; Kanpur 1992; Vikram 1993; Indore 1994*)
5. What is radula? Give its structure and function. (*Himachal 1993; Lucknow 1994*)
6. Give an account of the respiratory organs and mode of respiration in *Pila*. (*Burdwan 1988, 1996; Jiwaji 1989; Rohilkhand 1989; Marathwada 1989; Awadh 1989; Bundelkhand 1988, 1989; Meerut 1989, 1990, 1992; Gorakhpur 1989, 1994; Agra 1990, 1992, 1993; Vikram 1990; Rajasthan 1990; Kanpur 1990, 1993, 1994; Delhi 1991; Utkal 1991; Raipur 1991; Purvanchal 1992; Lucknow 1997, 1998*)
7. Describe the respiratory organs of *Pila* and explain how they are adapted for its mode of life. (*Jiwaji 1988; Bhopal 1989; Andhra Pradesh 1989; Ravi Shankar 1993*)
8. Give an account of the two types of respiration in *Pila*. (*Agru 1988; Bundelkhand 1989; Osmania 1989; Rohilkhand 1991*)
9. Describe the respiratory system of *Pila*. (*Kanpur 1997; Gorakhpur 1998*)
10. Describe the respiratory organs of *Pila*. Discuss the mechanism of aerial respiration.
11. Give an account of respiratory system and aquatic respiration in *Pila*. (*C.C.S.U. Meerut 2004*)
12. Describe the blood vascular system or circulatory system of *Pila*. (*Indore 1988; Lucknow 1989, 1996; Bundelkhand 1989; Gorakhpur 1990, 1995; Kanpur 1991; Awadh 1992; Jiwaji 1992; Rewa 1993; Vikram 1993; Agra 1997*)

13. Describe the excretory system of *Pila*. (*Kanpur 1996; Agra 1997*)
14. Give an account of the organ of Bojanus.
15. Describe the renal organ and physiology of excretion in *Pila*.(*Jiwaji 1989, 1991; Jabalpur 1991*)
16. Describe the nervous system of *Pila*. (*Gorakhpur 1997*)
17. Describe the nervous system of *Pila* and compare it with that of freshwater mussel. (*Gorakhpur 1988; Lucknow 1988; Kanpur 1988, 1989; Allahabad 1989; Rajasthan 1989; Rewa 1989; Ravishankar 1989; Meerut or CCS 1989, 1991, 1993, 1996; Agra 1990; Bhopal 1990; Bilaspur 1993*)
18. Describe the nervous system of *Pila* related to torsion. (*C.C.S.U., Meerut 2002*)
19. Give an account of sense organs in *Pila*. (*Utkal 1989; Rewa 1993*)
20. Give an account of the reproductive system of male and female reproductive organs in *Pila*. (*Agra 1988, 1991, 1992; Kanpur 1988, 1990, 1994, 1995, Shivaji 1989; Punjab 1989; Bundelkhand 1989; Gorakhpur 1989, 1993; Rohilkhand 1989, 1992; Awadh 1990*)
21. Give an account of reproductive system of *Pila*. How and where does the copulation takes place? (*Bhopal 1988; Rajasthan 1988, 1990*)
22. Describe male and female reproductive system in *Pila*. (*Himachal 1995*)
23. Write a short note on reproductive system of *Pila*. (*Lucknow 1994*)

Short Answer Questions

1. Write a note on feeding and digestion in *Pila*. (*Kanpur 1989, 1991; Bhopal 1992*)
2. Give a short account of aerial respiration in *Pila*. (*Osmania 1990; Utkal 1993*)
3. Describe renal organs of *Pila*. (*Sagar 1988*)
4. Differentiate between eupyrine and oligopyrene sperm. (*Kumaun 1989*)
5. Why do majority of organs on right side of *Pila* disappear? (*Kumaun 1988*)
6. Differentiate between male and female *Pila*. (*Bhopal 1992*)
7. Write short notes on the following:
   (*i*) Mantle cavity (*Kanpur 1995, 1997*)
   (*ii*) Shell of *Pila* (*Lucknow 1998*)
   (*iii*) Pulmonary sac (*Bhopal 1990*)
   (*iv*) Gill lamella of *Pila* (*Sagar1988; Jiwaji 1992*)
   (*v*) Digestive gland of *Pila* (*Sagar 1989*)
   (*vi*) Aquatic respiration in *Pila* (*Sagar 1988*)
   (*vii*) Organ of Bojanus (*Gorakhpur 1992; Jiwaji 1993; Kanpur 1997; Lucknow 1996*)
   (*viii*) Buccal mass (*Kanpur 1996*)
   (*ix*) Radula (*Punjab 1991; Guwahati 1995; Himachal 1995; Kanpur 1995; Lucknow 1996*)
   (*x*) Osphradium (*Punjab 1990, 1992; Magadh 1991, 1993; Kanpur 1994, 1997; Lucknow 1996, 1997; Gorakhpur 1997*)
   (*xi*) Eye in *Pila* (*Himachal 1993*)
8. Draw labelled diagrams of any two of the following (No description is required).
   (*i*) Organs visible in mantle cavity.(*Awadh 1991; Meerut or CCS 1992; Bhopal 1993; Jiwaji 1993*)
   (*ii*) Organs of pallial complex of *Pila* (*Lucknow 1993, 1998*)
   (*iii*) Molluscan shell (*Bihar 1991*)
   (*iv*) V.S. buccal cavity of *Pila* (*Awadh 1989, 1992; Rohilkhand 1992; Jiwaji 1993; Agra 1997*)
   (*v*) Heart and pericardium of *Pila* (*Sagar 1988*)
   (*vi*) Nervous system of *Pila* (*Sagar 1989; Lucknow 1989; Bundelkhand 1989; Garhwal 1989, 1992; Agra 1990; Meerut or CCS 1993, 1999*)
   (*vii*) V.S. of eye of *Pila*
   (*viii*) Reproductive organs/system of *Pila* (*Gorakhpur 1998*)

Very Short Answer Questions

1. What is the common name of *Pila*?
Ans. Apple snail.

2. Name the two parts of foot of *Pila*.
Ans. Propodium and metapodium.

3. Name the structure that divides the mantle cavity of *Pila* into right and left parts.
Ans. Epitaenia.

4. What type of digestion occur in the *Pila*?
Ans. Extracellular as well as intracellular.

5. What is radula? Describe its movement in word.
Ans. Radula is a rasping organ of *Pila*. It is file-like ribbon studded with horny or chitinous teeth, found in the mouth of *Pila* and other molluscs except Bivalvia. Radula runs longitudinally over the entire length of odontophore within the buccal mass of *Pila*. It moves to and fro over the odontophore. **Huxley** described such a movement as radula as **chain-saw** like movement.

6. Give the radular formula of *Pila*.
Ans. 2, 1, 1, 1, 2.

7. Name the digestive glands of *Pila*.
Ans. Oesophageal pouches, salivary glands and liver or hepatopancreas.

8. What type of respiration is found in *Pila*? (*C.C.S.U. Meerut, 2004*)
Ans. *Pila* is an amphibious molluscan animal, *i.e.*, it can live both on land as well as in water. To survive under both the situation it has both aquatic as well as aerial mode of respiration. Pila's aquatic respiration occurs by the help of single monopectinate gill or ctenidium and its aerial respiration takes place by the help of pulmonary sac.

9. Which nuchal lobe in *Pila* function as the inhalent channel in respiration?
Ans. Left.

10. What is the role of pneumostome?
Ans. Passage for the air breathing.

11. Where does the nephrostome opens in *Pila*?
Ans. Pericardial cavity.

12. Why do majority of organs of right side of *Pila* disappear?
Ans. Besides torsion, the reason of gastropods being asymmetrical is the loss, reduction or atrophy of paired organs from one or other side. Due to this reason, when the pallial/mantle cavity comes to lie on right, the organs present on the left side get degenerated or lost gradually leaving the organs only on the right side of body.

13. Define detortion.
Ans. It is phenomenon just opposite to torsion. Some gastropods such as *Aplysia* undergo detortion. When the shell is lost or even reduced. In this process ctenidia (gills) occupy the posterior position and visceral hump gets completely untwisted.

14. Development is direct in *Pila*. Is it true?
Ans. Yes.

Fill in the Blanks

1. Zoological name of apple snail is ....

2. *Pila* feeds on .... vegetation.

3. Period of inactivity of *Pila* in the drought period is called ....

4. *Pila* does locomotion by the help of ....

5. .... refers to the shell whorls except the body whorls.

6. The central hollow axis in *Pila* is known as ....

7. Columella opens out by an aperture, the ....

8. Head of *Pila* bears .... pairs of tentacles.

9. Mantle is also known as ....

10. The operculum of *Pila* is secreted by ....
11. Water testing organ in *Pila* is ....
12. There are .... teeth in each row of radula.
13. Cellulase is present in the secretion of ....
14. Blood of *Pila* contains a respiratory pigment name .... in the ....
15. Ctenidium of *Pila* is ....

Multiple Choice Questions

1. *Pila* shell is dextral, if the coiling is
   (*a*) right handed (*b*) left handed
   (*c*) anticlock wise (*d*) none of these
2. *Pila* is mainly
   (*a*) carnivorous (*b*) herbivorous
   (*c*) sanguivorous (*d*) omnivorous
3. How many teeth are found in a row of radula?
   (*a*) 7 (*b*) 5
   (*c*) 9 (*d*) 11
4. Body cavity in the apple snail is
   (*a*) coelom (*b*) pseudocoel
   (*c*) haemocoel (*d*) absent
5. Torsion is shown in
   (*a*) *Unio* (*b*) *Pila*
   (*c*) *Loligo* (*d*) *Mytilus*
6. Which of the following is a chemoreceptor?
   (*a*) omnatophore (*b*) umbilicus
   (*c*) nuchal lobe (*d*) osphradium
7. Mode of respiration in *Pila* is
   (*a*) aquatic (*b*) aerial
   (*c*) none (*d*) both
8. Aerial respiration in *Pila* occurs from
   (*a*) mantle (*b*) ctenidium
   (*c*) pulmonary sac (*d*) branchiae
9. Ctenidium of *Pila* performs respiration in
   (*a*) air (*b*) water
   (*c*) both the places (*d*) terrestrial
10. Pneumostome is an opening of
    (*a*) buccal cavity (*b*) pulmonary sac
    (*c*) kidney (*d*) shell
11. Blood vascular system in *Pila* is
    (*a*) open (*b*) closed
    (*c*) lacking (*d*) greatly reduced
12. Blood of *Pila* contains
    (*a*) RBC (*b*) WBC
    (*c*) ameoebocyte (*d*) monocyte

13. *Pila* has .... for excretion
    (*a*) flame cells (*b*) metanephridial system
    (*c*) malpighian tubes (*d*) gland cells
14. Statocyst of *Pila* is an organ of
    (*a*) equilibrium (*b*) defence
    (*c*) excretion (*d*) none
15. *Pila* is
    (*a*) monoecious (*b*) dioecious
    (*c*) hermaphrodite (*d*) asexual
16. *Pila* is
    (*a*) oviparous (*b*) viviparous
    (*c*) ovoviviparous (*d*) none of these

ANSWERS

Fill in the Blanks

1. *Pila globosa*
2. Succulent
3. Aestivation
4. Foot
5. Spire
6. Columella
7. Umbilicus
8. Two
9. Pallium
10. Foot glands
11. Osphradium
12. Seven
13. Hepatopancreas (liver)
14. Haemocyanin, Plasma
15. Monopectinate.

Multiple Choice Questions

1. (*a*)	2. (*b*)	3. (*a*)	4. (*c*)	5. (*b*)	6. (*d*)
7. (*d*)	8. (*c*)	9. (*b*)	10. (*b*)	11. (*a*)	12. (*c*)
13. (*b*)	14. (*a*)	15. (*b*)	16. (*a*).		

46

Unio (Lamellidens) (Phylum Mollusca)

Freshwater mussels occur all over the world. Their typical genera include *Anodonta* common in Englant, *Lamellidens* found in India and *Unio* with universal distribution. The description given here will apply almost fully to these three and other freshwater mussels.

46.1 CLASSIFICATION

Phylum	—	**Mollusca**
Class	—	**Pelecypoda**
Subclass	—	**Palaeoheterodonta**
Order	—	**Unionoida**
Genus	—	***Unio***

46.2 HABIT AND HABITAT AND EXTERNAL FEATURES

Habit and Habitat

Unio is found in freshwater ponds, lakes, rivers and streams. It is especially abundant in water containing lime. *Unio* remains buried in the mud or sand at the bottom in such a way that posterior tip of shell valves remain exposed for the ingress (entry) and egress (exit) of water. During night *Unio* lives in shallow water while in the day it migrates in deeper water. The food of *Unio* is microscopic organisms–plants (diatoms) and animals, which are taken by filter-feeding mechanism.

The flow of water through the mantle cavity by beating of cilia is one of most critical factor in the survival of *Unio* and other pelecypods. Water currents enter and leave the mantle cavity through siphons made by left and right mantle folds; they bring food and oxygen and waft away faeces, carbon dioxide, excretory wastes and genital products.

Siphons open and close in response to light, touch and other stimuli. A slight disturbance causes the withdrawal of foot and siphons and closing of shell valves. Starfishes are the natural enemy of *Unio*; they open the shell and feed on their soft viscera.

Geographical Distribution

Unio is found all over the world. *Lamellidens marginalis* is found in Indian rivers and other parts of the world.

External Features

1. Shape and size. *Unio* measures 5 to 10 cm in length about 5 cm in width. The body is soft, unsegmented, bilaterally symmetrical and laterally compressed. The anterior end of body is roughly oval (broad and blunt) and posterior end is somewhat narrower (Fig. 46.1).

2. Shell. Body of *Unio* is completely enclosed in hard, calcareous, brownish and bivalved shell (hence called Bivalvia). Both of the valves remain connected together dorsally.

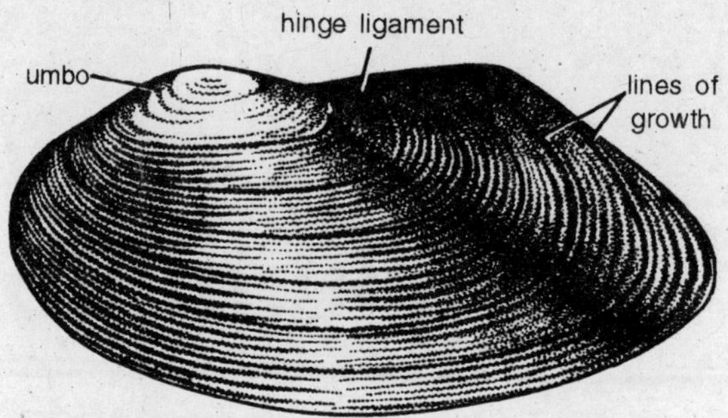

Fig. 46.1. *Unio*. Entire animal with shell (Lateral view).

***(i)* Shell valves.** The shell consists of two similar, equal sized (equivalved) valves which are known as **right** and **left valves**. Each valve has concave inner and convex outer surface.

***(ii)* Umbo.** Each shell has whitish, slightly elevated, dome-shaped structure called **umbo**. It is the oldest and thickest part of the shell.

***(iii)* Hinge ligament.** The two shell valves are joined along dorsal side by hinge ligament. It is an external, brown, tough, elastic and horny structure. It is composed of the protein **conchiolin.** They help in opening and closing of shell valves.

***(iv)* Lines of growth.** There are many crescentic lines around the umbo which run parallel to the margin of the shell. These lines are called **lines of growth.** The shell increases in size by the deposition of lime along outer rim of shell each year.

***(v)* Hinge teeth.** The inner surface of each shell valve along its hinge line at its dorsal margin bears two sharp, blade-like ridges or teeth. Teeth are separated by grooves or sockets. The teeth of one valve fit into corresponding sockets of other valve. These control the gaping of shell valves and hold them from slipping out of position.

***(vi)* Muscle scars.** The inner surface of each shell valve bears scars or impressions of muscles attached to it. There are three scars at the anterior end for **anterior adductor, anterior retractor** and **anterior protractor muscles**. The two scars at the posterior end are those of **posterior adductor** and **posterior retractor muscles.**

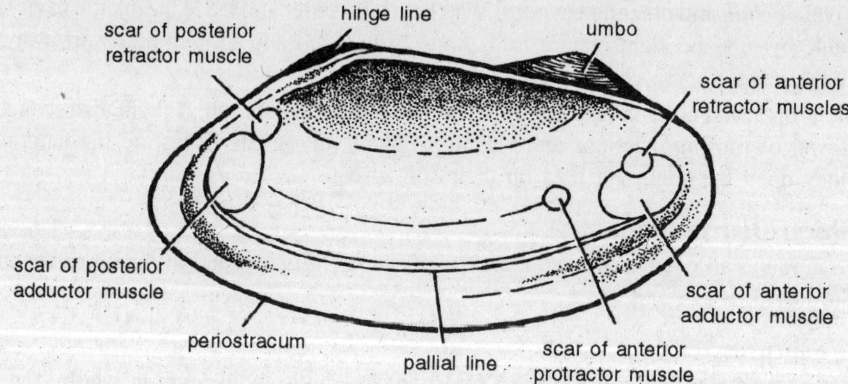

Fig. 46.2. *Unio*. Inner view of shell valve showing muscle scars.

A line joining the scars for adductor muscles of two ends is called **pallial line.** It marks the attachment of mantle to shell valve.

3. Siphons. A delicate **mantle** lies beneath the shell that envelops the whole visceral mass. At the posterior or aboral end, the mantle lobes of both sides form two short siphons: (*i*) A small, smooth edged **exhalent siphon** for the exit of water current. (*ii*) A large, fringed **inhalent siphon** for the entry of water current.

4. Foot. At times the triangular tongue-shaped foot protrudes between the shell valves at the anterior end.

46.3 MICROSCOPIC STRUCTURE OF SHELL

The microscopic structure of the shell of *Unio* reveals following three layers (Fig. 46.3):

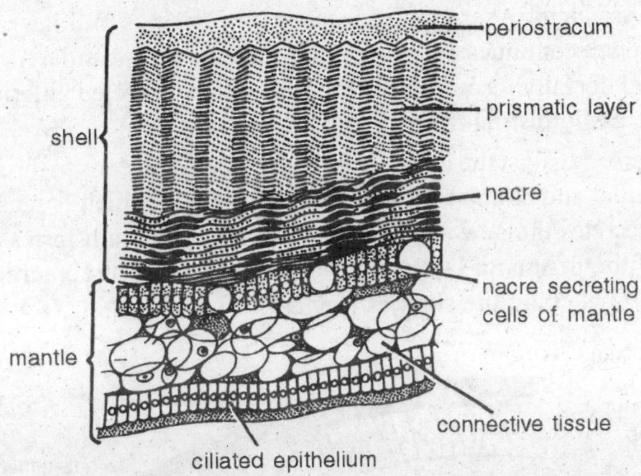

Fig. 46.3. *Unio*. Cross section of shell and mantle.

1. Periostracum

It is outermost, thin, brownish, translucent, horny layer. Periostracum is greenish in colour and formed of a special chitinous substance the **chonchiolin**. It is secreted by the edge of mantle. It is a protective layer for the under-lying calcareous prismatic layer which otherwise can be eroded out by the action of carbolic acid of the surrounding medium. The periostracum is worn away from the older part of the shell such as umbo, leaving the prismatic layer exposed to the exterior.

2. Prismatic Layer

It is formed by the minute prisms or crystals of calcium carbonate (*e.g.*, **aragonite** and **calcite**) secreted by the edge of outer fold of mantle. The prism of crystals form sheets and are always within a framework of an organic substance. This layer gives strength to the shell.

3. Nacreous Layer or Pearly Layer

It is the innermost shining layer of the shell known as the **mother of pearl** or **nacre**, since it presents a smooth irridescent surface. It is secreted by the whole outer surface of mantle and is deposited in the form of alternative transverse layer of calcium carbonate and concholin. The layers run parallel to the shell.

Constitution of Shell

The shell is mainly composed of pure carbonate of lime with little proportion of phosphate of lime and an organic base allied to chitin which is known as **conchiolin**. The proportion of carbonate of lime varies from 88 to 99% and of phosphate of lime 1 to 10%. Conchiolin forms a sort of membranous framework for the shell.

Formation of Shell

Regarding the formation of shell there exist following two views:

1. Bowerbank and Carpenter's view. The shell is an organic formation growing by interstitial deposits in the same manner as teeth and bones in higher animals.

2. Reaumer-Eising view. The shell is of nature of an excretory substance deposited like cuticle outside of the skin. It is formed simply of a number of calcareous particles which are held together by a kind of animal glue.

Mantle begins to form shell during the veliger larval stage. The two mantle folds secrete **conchiolin** to form two tiny valves connected by a narrow ligament. New material is added at the margins of the valves, producing a series of concentric growth rings (or lines). As the veliger settles to the bottom, shell growth continues but more rapidly on the ventral side. As a result, the original larval shell is displaced dorsally forming an elevation (**umbo**) directed anteriorly so that the right and left valves can be easily distinguished.

The mantle is pressed against the inner surface of the valves. It ends in three folds: **1. Innermost fold** which contains radial and circular muscles; **2. Middle fold** which is sensory in function; and **3. Outer fold** which secretes the new shell (Fig. 46.4). **Periostracum** forms at the inner margin, calcareous material of the prismatic layer at the outer border and the **nacreous material** on the whole mantle surface, thickening the shell as it ages.

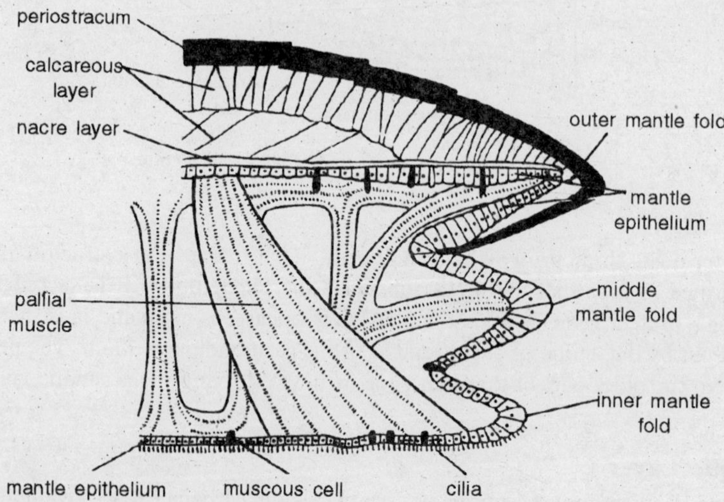

Fig. 46.4. T.S. through margin of shell and mantle of a bivalve showing mantle lobes and places of shell secretion.

There is a minute, **extrapallial space** between mantle and shell, except at points of muscle attachment. The carbonates and phosphates of calcium (lime) are separated from the blood by the cells of mantle and deposited in the extrapallial space. It is in this space that other shell materials (*e.g.*, conchiolin) are first secreted. From the extrapallial fluid of this space both the calcareous elements and the surrounding organic framework are then deposited.

46.4 PEARL FORMATION

Occasionally some foreign object or particle such as sand grain, debris, a parasite, an egg or any microscopic organism comes to lie in between the shell and mantle. The particle causes irritation to the animal. This induces epithelial cells to secrete concentric layers of nacre around it. This results in the formation of a **pearl** (Fig. 46.5). The foreign particle inside the pearl is known as **nucleus** and concentric layers of nacre are known as **mother of pearl**.

If the foreign body is enfolded within the mantle and moved about during secretion, the pearl becomes spherical or ovoid. Commonly, however, the developing pearl adheres to or even becomes completely embedded in the shell. The pearl of *Unio* is not economically precious (see Box 46.1).

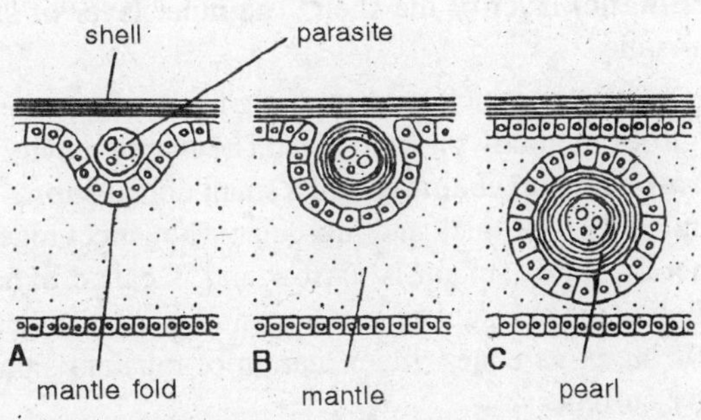

Fig. 46.5. *Unio*. Pearl formation.

Box 46.1 Cultured Pearls

Finest natural pearls are produced by the pearl oyesters, *Pinctada margaritifera*, *Pinctada vulgaris*, *Pinctada mertensi* which inhabit most of the warmer seas. There are two methods of producing **cultured pearls**. Most cultured pearls are started with a microscopic globule of liquid or ground "seeds" of unionid shells placed into a pearl oyster. The resulting year-old seed pearl is then transplanted into another oyster. A pearl of marketable size is obtained three years after transplantation.

Bead pearls are produced more quickly by starting with a spherical shell fragment only smaller than the finished pearl. This fragment is enclosed in a bag of mantle tissue and transplanted into the gonadal tissue of another pearl oyster. A calcareous veneer approximately 1 mm thick is then down laid around the bead.

46.5 INTERNAL STRUCTURE

After removing the shell of *Unio* following internal structures are observed (Fig. 46.6)

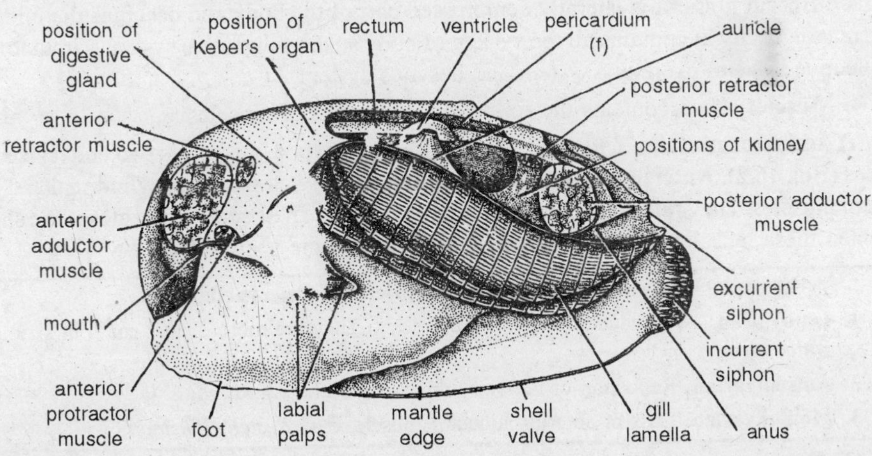

Fig. 46.6. *Unio*. Internal organs seen after the removal of the left shell valve and left mantle lobe.

1. Mantle or Pallium

Just beneath each shell valve lies a thin, creamy white, semi-transparent and soft covering of skin. It encloses the soft body parts and is called **mantle** or **pallium**. Pallium overhangs the body like a skirt. It consists of two lateral halves or **lobes**, one below left and other below right shell valves. The two mantle folds join each other and the body dorsally. The ventral edges of the mantle lobes are thick, folded and muscular. The edge contains three folds (Fig. 46.4) — an **inner**, a **middle** and an **outer fold**. The outer fold secretes the outer (**periostracum**) and middle calcareous

(prismatic) layer of the shell. The inner layer of shell or **nacre** is secreted by the entire surface of mantle.

The space enclosed by mantle folds around the body is called **mantle cavity**. It extends entire length of the body on each side. The mantle cavity is divided into two chambers, a large ventral **infra-branchial chamber** and a small dorsal **supra-branchial chamber**. At the posterior end both mantle lobes are thick and muscular. They meet together and form two short tubes or **siphons**. The upper or dorsal siphon is narrow and is called **exhalant siphon**. Water leaves the body through it. Below the exhalant siphon lies another tube, called **inhalant siphon**. It is wider and has delicate **fimbriae** at its edges. Inner margin of inhalant siphon contains papillae which test the incoming water currents.

Histology. The mantle consists of (*i*) a thin **outer layer** of columnar epithelial cells. It contains many unicellular **nacre secreting gland cells** and branches of blood vessels. It remains in contact with shell; (*ii*) a thick **middle layer** which is formed of connective tissue, muscle fibres (called **pallial muscles**), branches of blood vessels and nerves; and (*iii*) a thin inner layer of **ciliated epithelial cells**. It contains many unicellular **mucous glands**.

Function. The mantle of *Unio* protects the body, secretes shell valves and mucus and forms siphons for the flow of water.

Organs of Mantle Cavity

Mantle cavity of *Unio* contains the following structures:

1. Head. *Unio* lacks a distinct **head**. Eyes and tentacles are absent.

2. Mouth. It is a large and transverse slit. It lies immediately below the **anterior protractor muscle** and is surrounded with two pairs of **labial palps**. Labial palps are large, flattened and fleshy structures forming upper and lower lips of mouth.

3. Visceral mass. It is laterally compressed part of the body and occupies the middorsal part of the mantle cavity. It contains all the system of body such as digestive system, circulatory system, reproductive system, excretory system and nervous system.

4. Muscles. *Unio* contains the following four types of muscles:

(*i*) Adductor muscles. *Unio* is an **isomyarian** pelecypod, *i.e.*, it has two equal sized adductor muscles (Box 46.2). **Anterior** and **posterior adductor muscles** are large, cylindrical and transverse bands of muscles. They run from one shell valve to another. Their contraction closes the shell valves, and when these muscles relax the hinge ligament opens the shell valves.

Box 46.2 Adductor Muscles in Pelecypoda

1. **Isomyarian.** Two adductor muscles are present. They are more or less equal in size; *e.g.*, *Unio*, *Mya*.
2. **Anisomyarian.** Reduction of anterior adductor muscle, *e.g.*, *Mytilus*.
3. **Monomyarian.** Loss of anterior adductor muscle, *e.g.*, *Pecten*, *Pinctada*.

(*ii*) Retractor muscles. Anterior and **posterior retractor muscles** run from the foot to the shell. Their contraction withdraws the foot into the shell.

(*iii*) Protractor muscles. It is a small band of muscles. It lies close to the anterior adductor muscle. Its action compresses the visceral mass and causes extension of foot out of the shell.

(*iv*) Pallial muscles. These extend from the edge of the mantle lobes and becomes attached with the shell valves all along the pallial line.

5. Foot or podium. It is antero-ventral, hatchet-like or wedge-shaped extension of visceral mass. It is a large, muscular, laterally compressed structure. Its thick basal part contains a portion of alimentary canal, digestive gland and gonad. Its free ventral part forms a longest **keel**.

The foot has a large sinus which gets filled with blood when the foot is extended by blood pressure and muscular action of a pair of **pedal protractor muscles.** The foot becomes swollen and turgid; it is then used for crawling and for ploughing through mud in burrowing. Foot is withdrawn into the shell by the action of **anterior** and **posterior retractor muscles.**

6. Gills. A pair of large, plate-like gills or ctenidia hang in the mantle cavity. One gill lie on each side of the foot. Each gill or ctenidium is folded to form two **laminae** and each lamina contains two **lamellae**. The outer gill lamina is a little smaller than the inner one. All lamina bear numerous pores and cilia.

In addition to respiratory gaseous exchange, the gills are also used to collect food and as brooding chambers.

7. Anus. The rectum opens into the exhalant siphon through anus.

46.6 DIGESTIVE SYSTEM

Digestive system of *Unio* is distinguished into: (1) Alimentary canal, and (2) Digestive gland (Fig. 46.7).

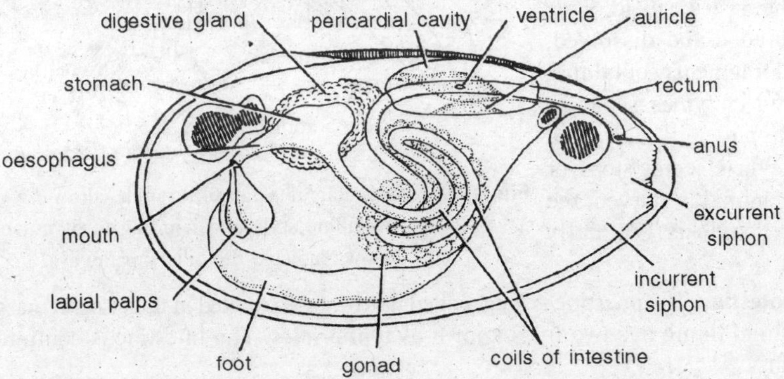

Fig. 46.7. *Unio*. Alimentary canal and digestive gland.

1. Alimentary Canal

It is a long curved tube starting from mouth at one end of the body and ending into the anus situated at the rear end. It is divided into the following parts:

1. Mouth. Mouth is a transverse slit-like aperture, lies behind the anterior adductor muscle or below anterior protractor muscle. The mouth is surrounded by two pairs of triangular ciliated flap-like **labial palps.** The labial palps of anterior-outer pair are joined together over the mouth forming upper lip, while the labial palps of posterior-inner pair are united below the mouth forming the lower lip. The lips form a ciliated oral groove, which leads into the mouth. Jaws and radula are absent. Mouth leads into the oesophagus.

2. Oesophagus. The oesophagus is a short, wide, ciliated tube which leads into the stomach.

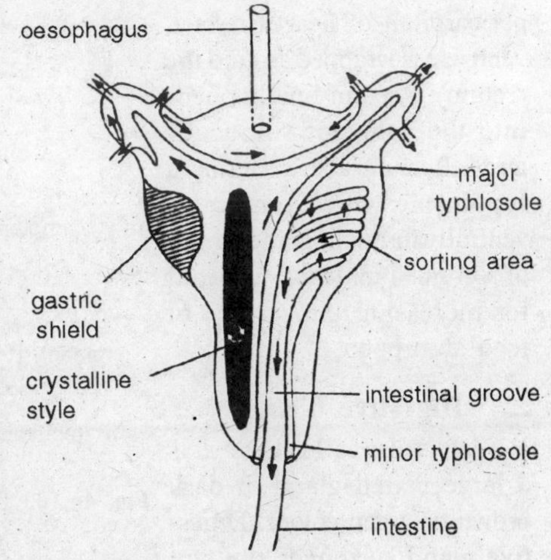

Fig. 46.8. Stomach of a typical unionoid showing typhlosoles and sorting of food particles.

3. Stomach. It is a rounded sac which is wide and thick-walled, situated dorsally in the visceral mass and surrounded by a pair of digestive glands. The stomach is divided into dorsal part (proximal globular region) and ventral part (narrowed distal region). The dorsal part of stomach contains a **sorting area**, prominent **gastric shield** and **pores of digestive gland** (Fig. 46.8). The ventral part of stomach contains a **style sac** (or **pyloric caecum**) with **crystalline style** and **intestinal grooves** with **major** and **minor typhlosoles** (Fig. 46.8).

The internal epithelial lining of the style sac bears cilia. The style sac contains a gelatinous rod called the **crystalline style**. It is secreted by the internal lining of the style sac. It is composed of mucoproteins (= consolidated mucus) and enzymes such as **carbohydrases** (*e.g.*, **amylase, cellulase, glucogenase**) and **lipase**. The cilia of the style sac rotate (clockwise) the style and move it forward (like a screw) due to which its anterior end is rubbed against a chitinous plate (*i.e.*, gastric shield). As the style strikes against the gastric shield, style end is abraded and dissolved, *i.e.*, small fragments containing mucus and enzymes separate from style and mix with food particles. When feeding stops, or when the animal is starved, the crystalline style is gradually dissolved.

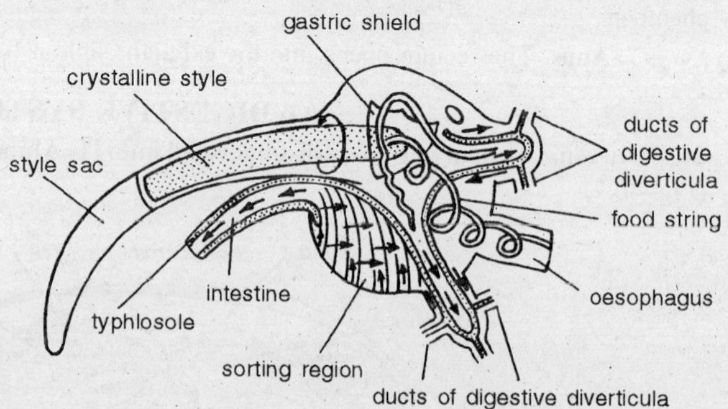

Fig. 46.9. Stomach of a lamellibranch, showing rotation of the crystalline style and winding in of mucous food string. Arrows indicate the ciliary pathways.

4. Intestine. The intestine is long, coiled narrow tube situated in the visceral mass. In its anterior part intestinal lining has two folds known as **typhlosoles**. The intestine is continued behind into the rectum.

5. Rectum. The rectum is a narrow tube running horizontally up to the posterior end in the dorsal part of the body. It passes through the ventricle and pericardium of heart (in fact, ventricle is wrapped around the rectum). Rectum finally opens into the exhalant siphon by **anus**. Rectum also contains a **typhlosole** which runs mid-ventrally throughout the length of rectum. Typhlosole is meant for increasing the surface of food absorption.

2. Digestive Gland

It is also called **liver**. It is a large paired gland of dark brown or green colour. Digestive gland surrounds the stomach and is composed of highly branched tubules. It opens into

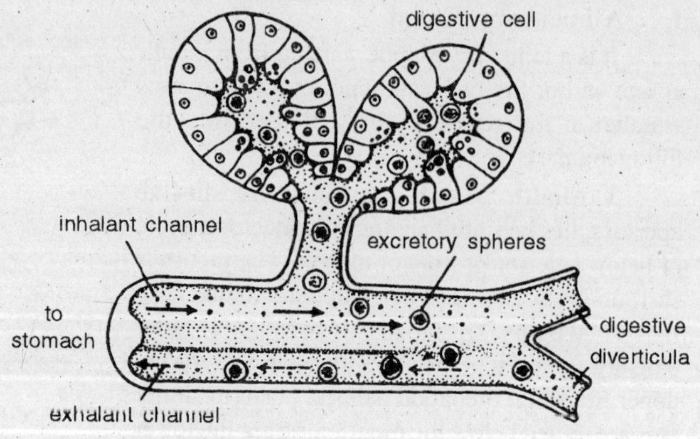

Fig. 46.10. Diagram of a section of digestive diverticulum showing the absorption and intracellular digestion of material passed inward from the stomach (solid arrow) and the outward passage of wastes (dashed arrows).

dorsal part of stomach by number of ducts to discharge the digestive enzymes and to receive partially digested food from the stomach. In fact, there is a continuous two way flow in the main duct of digestive gland. Materials for intracellular digestion and absorption enter the ducts and undigested cell fragments and other wastes (excretory waste) are returned to the stomach through the same tube (Fig. 46.10). Digestive gland of *Unio* performs the following functions:

1. Phagocytosis;
2. Intracellular digestion of proteins and lipids;
3. Secretion of enzymes for extracellular digestion in stomach;
4. Absorption of already digested food (*e.g.*, starch, cellulose);
5. Aids in excretion.

Food

Unio feeds on organic detritus and microorganisms (diatoms, protozoa, etc.) which remain suspended in water. These are brought into the mantle cavity by the respiratory water currents.

Feeding

Unio, *Lamellidens* and *Anodonta* are typical **filter feeders** or **ciliary feeders**. Their gills form the filter and cilia of the gills transport minute food particles to mouth. Beating of large **lateral cilia** of the gill filaments draw water into infra-branchial chamber of the mantle cavity through inhalant siphon. The water current brings in lot of microorganisms, detritus and silt. Water enters the gills laminae through the ostia. During this movement the water is filtered. Coarser and heavier sand particles, etc., are simply dropped from the surface of the gills to the edge of mantle (having ventral ciliated tract). These are driven towards the inhalant siphon by the cilia of the mantle edge. These rejected particles form the **pseudofaeces** and leave the mantle cavity through the inhalant siphon.

The lighter food particles are passed on to the outer gill surfaces by **latero-frontal cilia**. These particles stick to the mucus secreted by the gills. The food laden mucus mass in the form of **cord** or **string** is moved ventrally by the beating of frontal cilia of the food groove which is located at the lower edge of 'V' of each gill lamina. Cilia of the food groove move the mucus food string towards the mouth. Here the food is again sorted and conveyed by the **cilia of labial palps.** Large, heavy and indigestible particles dropped down (at the edge of the mantle and foot). The small and digestible particles are directed to the mouth (by cilia of the labial palps) through the ciliated groove formed by the lips.

Digestion

The food laden mucus mass (cord/string) reaches the stomach. Lower pH of the stomach separates food particles from mucus string. Rotation of crystalline style adds enzymes to food, churns the food and act as a windlass and pulls the food-laden mucus string from the oesophagus into stomach. Partial digestion of starch, other carbohydrates (cellulose) and lipids occur **extracellularly** in the stomach. Protein digestion occurs **intracellularly** in the digestive glands.

The rotating crystalline style continuously throws partially digested food into the ciliated sorting area of the dorsal portion of stomach. Here fine particles and fluid containing digested food are directed towards the apertures of the ducts of the digestive glands by the cilia (Fig. 46.9). Coarse and heavy particles are separated from the string and sent to the intestine. Digestion is completed intracellularly in the cells of ductules of digestive glands.

Absorption of digested food occurs chiefly through the walls of the intestine and rectum. The typhlosole increases the absorptive surface. Digestive gland also absorbs digested food.

The phagocytotic amoeboid cells found in the wall of entire alimentary canal also engulf, digest and distribute food particles.

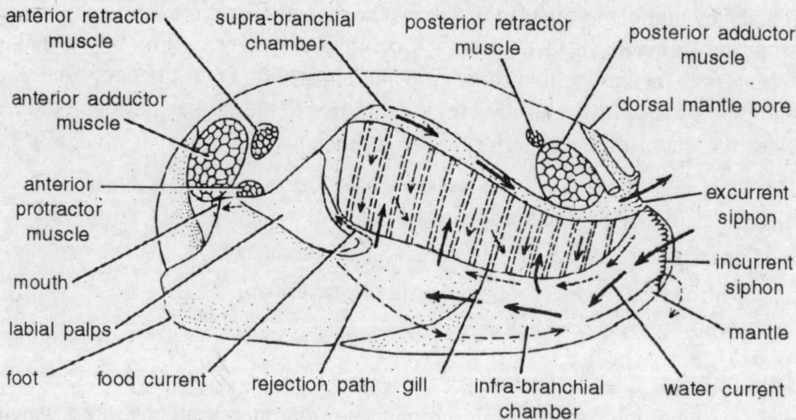

Fig. 46.11. *Unio*. Feeding mechanism and course of water current. Rejected particles and their path of removal is shown by dotted arrows.

The undigested waste is ejected through the **anus** in the exhalant siphon from where it goes out along with outgoing water currents.

46.7 RESPIRATORY SYSTEM

Ctenidia or Gills

The ctenidia or gills are highly modified derivative of mantle. A pair of large plate-like gills hang in the mantle cavity. One gill lies on each side of the foot. The gills are attached with the visceral mass and mantle lobes (Fig. 46.6).

Structure of gills. Each gill contains two **demibranchs** or **gill plates** or **laminae** — an outer and inner. The inner lamina which lies towards the visceral mass is larger than the outer lamina which lies towards the mantle lobe. These gills divide the mantle cavity into a smaller dorsal **supra-branchial cham-ber** and a larger ventral **infra-branchial chamber**. The former opens into the exhalant siphon and latter communicates with the inhalant siphon.

Each gill plate or lamina is formed of similar flaps or **gill lamellae**, an outer and an inner. Both lamellae remain connected at their ventral, anterior and posterior ends but they remain separated dorsally. They enclose a narrow space or cavity called **inter-lamellar space**. Each

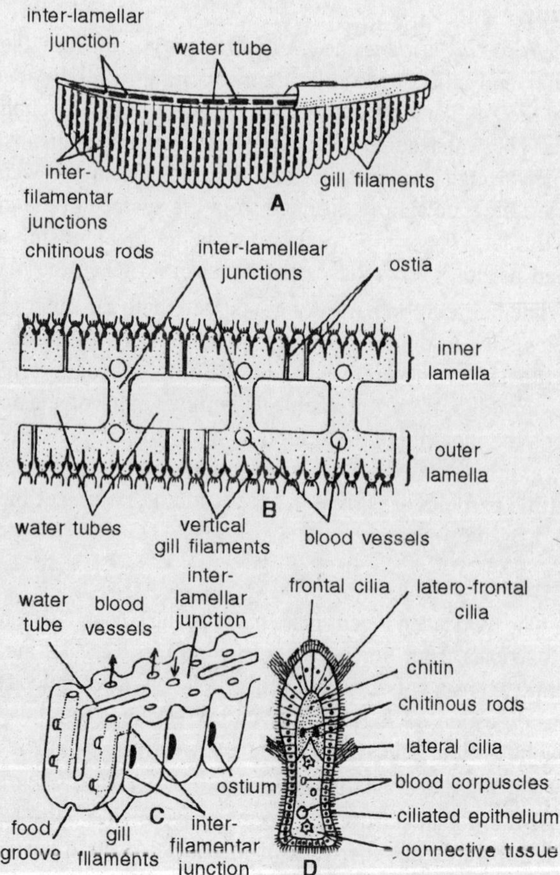

Fig. 46.12. *Unio*. Structure of gill. A—A demibranch or gill lamina seen from outside; B—A portion of gill lamina in T.S; C—A portion of gill lamina showing water tube; D—Cross section of one gill filament.

gill lamina thus forms an elongated, narrow, V-shaped bag which opens dorsally into the supra-branchial chamber (Fig. 46.12A).

Each gill lamella is formed of numerous thin, vertical and parallel **gill bars** or **gill filaments**. These produce vertical striations on the outer surface of lamella. Adjacent gill filaments of a lamella remain connected by small horizontal bars called **inter-filamentous junctions** (Fig. 46.12A). These bars produce horizontal striations on the lamellar surface. The gill filaments of both lamellae of a lamina are continuous at their ventral edge; due to this each gill lamina looks a 'V' or 'U' in a transverse section (Fig. 46.12C). Both laminae of a gill together form a 'W' like structure (Fig. 46.13).

The two lamellae of a lamina are connected by vertical cross partitions or **inter-lamellar junctions** (Fig. 46.12B). Due to these junctions the inter-lamellar space is divided into a series of compartments or **water tubes**. These tubes open dorsally into supra-branchial chamber. Each gill lamella laterally is perforated by numerous minute pores or **ostia**. Through these ostia water of infra-branchial part of mantle cavity enters into the water tubes.

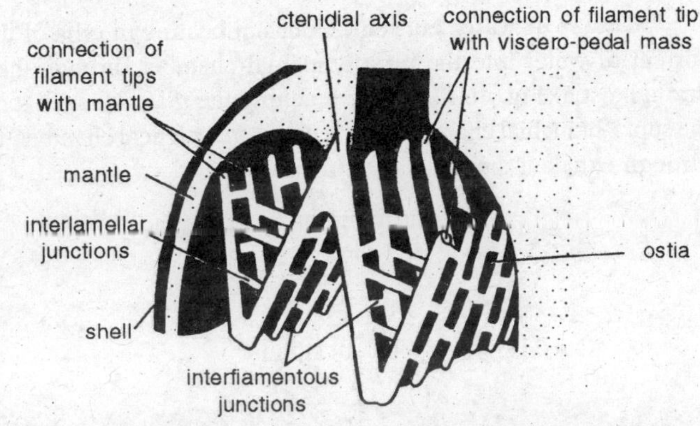

Fig. 46.13. A part of gill showing attachment.

At the ventral edge, where two arms of V meet together, each V-shaped filament bears a **notch**. The notches of all filaments collectively form a continuous **food groove** which extends throughout the length of each gill lamina.

Histology of gill filaments. A gill filament consists of a central core of **chitinous rod** and connective tissue, both are covered by a **ciliated epithelium** (Fig. 46.12D). Gill filaments are covered by following three kinds of cilia;

1. Frontal cilia are small and located on the outer face of the gill filament.

2. Lateral cilia are large and located on the lateral sides of gill filament.

3. Latero-frontal cilia are largest and located in between frontal and lateral cilia.

The gills of *Unio* are of **eulamellibranch type**, *i.e.*; the gill filaments are folded into a W-shape and adjacent filaments are connected through fleshy connections. These gills are **bipectinate**, *i.e.*, there are two rows of gill filaments—one on each side of a central **cteni-dial axis**.

Attachment of gills. Each gill has a cent-ral **ctenidial axis**. It remains fused to the

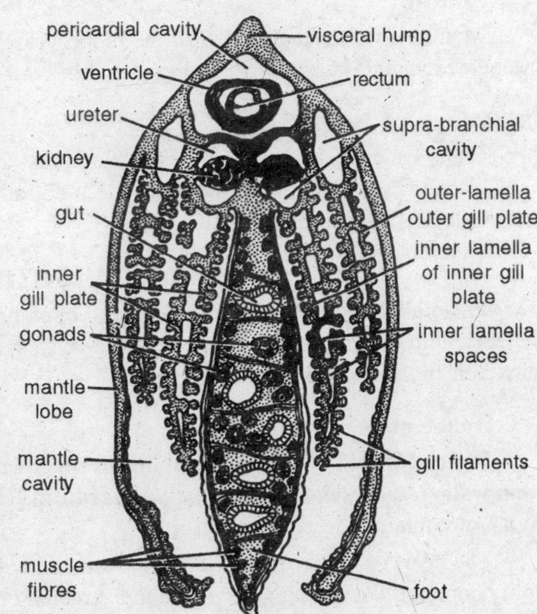

Fig. 46.14. *Unio*. T.S. of body through the middle region of gills and foot.

dorsal wall of the mantle cavity throughout but remains free near the posterior end of body. The arms of 'W' that are attached to the ctenidial axis are called **descending limbs** and the arms close to the mantle or visceral mass are called **ascending limbs**. Tips of ascending lamellae are fused with upper surface of mantle (on the outer side) and with viscro-pedal mass (on the inner side). Behind the foot the tips of ascending lamellae of inner side, fuse with that of other gill.

Blood supply of gills. The ctenidia or gills are supplied by **afferent branchial vessel** carrying deoxygenated blood from the kidneys and divides into branches at the interlamellar junctions. These branches unite to open into the **efferent branchial vessel** which carries away oxygenated blood to the heart.

Course of water current. Constant beating of cilia of the gill filaments causes a continuous current of water into the infra-branchial chamber through the inhalent siphon and enters the gill through ostia (Fig. 46.11). After reaching the gills water fills in the water tubes and flows out into the supra-branchial chamber which opens into cloacal chamber, from here water discharges to exterior through exhalant siphon (Fig. 46.15).

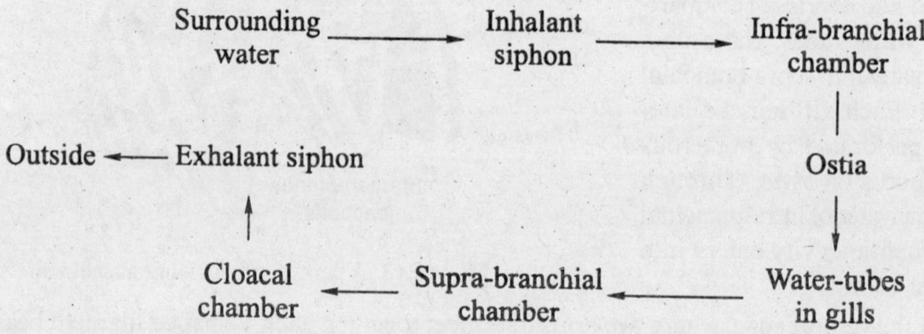

Fig. 46.15. *Unio.* Diagrammatic representation of course of water current.

2. Mantle

Mantle is richly supplied with the blood vessels and it remains in contact with water, hence, gaseous exchange takes place through its thin wall.

Physiology of Respiration

When water passes through the mantle cavity and the water tubes in the gill, gaseous exchange takes place. Thus, carbon dioxide from the blood diffuses out in the water and dissolved oxygen from water diffuses in the blood.

46.8 CIRCULATORY SYSTEM

In *Unio* the blood vascular system is of open type, *i.e.*, the blood vessels open into the spaces present within visceral organs. The circulatory system consists of heart, pericardium, arteries, sinuses, veins and blood. Capillaries are absent.

1. Heart and Pericardium

Heart lies along the mid-dorsal line infront of the posterior adductor muscle, enclosed within a thin walled triangular chamber, the **pericardium**. The cavity of pericardium is called **pericardial cavity**, it contains the **coelomic fluid**.

The heart consists of two auricles or atria and a ventricle. **Auricles** are thin walled, refractile, highly contractile and triangular chambers situated one on either side of the ventricle. The auricles are attached to the pericardial wall by thin broad bases and opens into the ventricle by an **auriculo-ventricular aperture**. The auricle receive blood from the gills, kidney and mantle. The **ventricle** is a large, thick walled and muscular chamber. It is wrapped around the rectum.

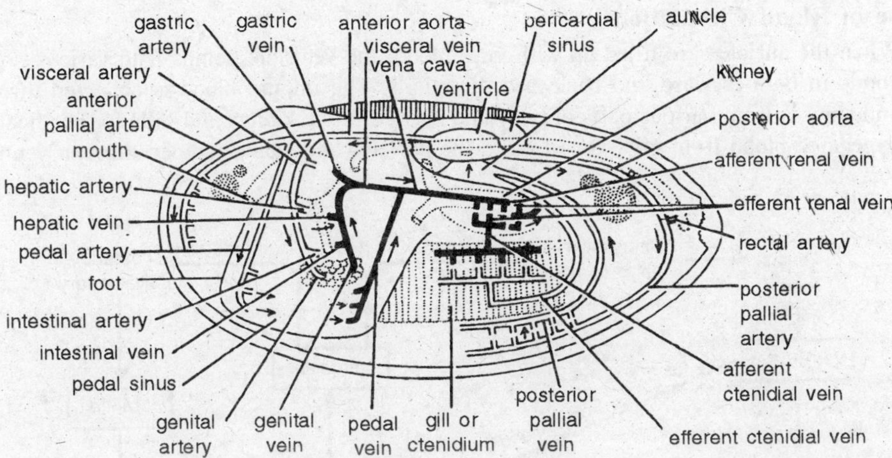

Fig. 46.16. *Unio*. Blood vascular system.

2. Arteries

Arteries arise from each end of the ventricle and are of two types: anterior and posterior aorta.

1. Anterior aorta. It runs anteriorly, dorsal to the intestine and gives off three main arteries: ***(i)* Anterior pallial artery** which supplies blood to mantle; ***(ii)* pedal artery** which supplies blood to the foot; ***(iii)* visceral artery** which sends a **gastric artery** to stomach, a **hepatic artery** to liver, an **intestinal artery** to intestine and a **gonadal artery** to gonad.

2. Posterior aorta. It runs posteriorly, ventral to the intestine. It supplies blood to the pericardium and kidneys through its fine branches. It sends a prominent **rectal artery** to the rectum and a large **posterior pallial** artery to mantle.

3. Sinuses

Sinuses are spaces in the connective tissues of organs. They lack definite walls, *i.e.*, epithelial lining is not present. Arteries open into these sinuses. From them the blood is collected by veins.

4. Veins

Veins collect blood from the different parts of the body such as

(i) **Visceral vein** collects blood from the visceral organs.

(ii) **Pedal vein** collects blood from the pedal sinus.

(iii) **Efferent renal vein** collects blood from the kidney and unite to form two **afferent ctenidial veins.** An afferent ctenidial vein runs through the central axis of the gill and sends a branch to each gill filament. The blood in the gill is oxygenated.

(iv) **Efferent ctenidial vein.** It collects oxygenated blood from the gill filaments and carry the blood to auricles.

(v) **Vena cava.** It directly sends deoxygenated blood to the auricles.

5. Blood

The blood consists of **plasma** and **leucocytes**. The plasma contains a copper-containing protein (respiratory pigment) called **haemocyanin**. Blood distributes nutrients and oxygen to all parts of body. It collects carbon dioxide and other wastes of metabolism from different parts of the body.

Course of Blood Circulation

When the auricles are filled up with pure blood, the ventricle pumps it to various organs of the body in both forward and backward directions. The impure blood is collected through veins and sinuses from various parts of body and carried to the kidney and gills for purification. The oxygenated blood from gills is carried to the auricles and again pumped through ventricle (Fig. 46.17).

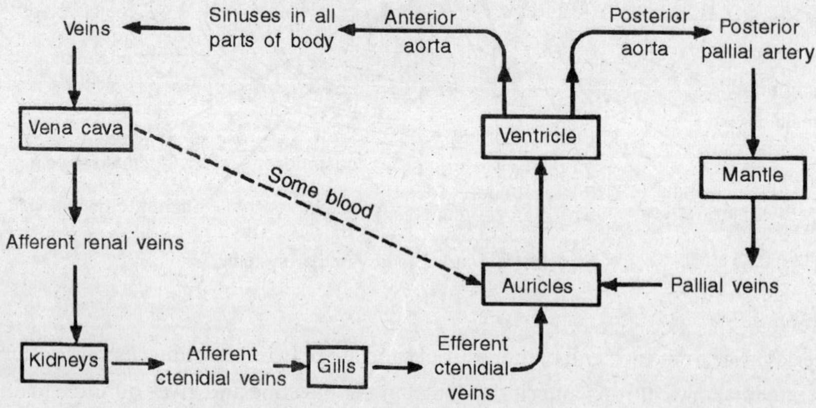

Fig. 46.17. *Unio*. Diagrammatic representation of course of blood circulation.

46.9 EXCRETORY SYSTEM

In *Unio*, excretion takes place through kidney or organs of Bojanus and the Keber's organ or the pericardial glands.

1. Kidneys or Organs of Bojanus

Kidneys or nephridia are also called **organs of Bojanus** after the name of their discoverer. There are two kidneys situated on both sides of vena cava and between the pericardium and gills. Kidneys are like a hair pin and represent true coelom. Its one arm opens into the pericardial cavity through a ciliated aperture, called the **nephrostome** or **renopericardial aperture**. The other arm opens into the mantle cavity through an **excretory** or **renal pore**. The arm of kidney opening into the pericardial cavity is brown, thick-walled, glandular and spongy. It forms kidney proper. The

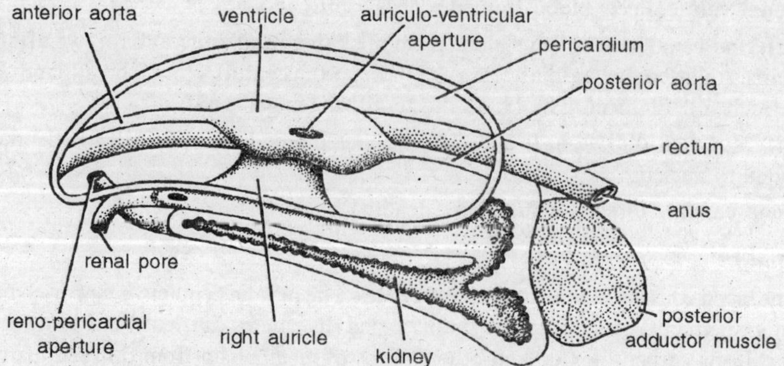

Fig. 46.18. *Unio*. Excretory system.

other arm of kidney is non-glandular, thin-walled having ciliated internal epithelium. It forms the ureter or urinary bladder and opens into the supra-branchial chamber of the mantle cavity. The bladders of both the kidneys communicate with each other through an oval aperture.

Physiology of excretion. Excretory products of *Unio* includes nitrogenous waste (*e.g.*, **ammonia** and traces of **urea** and **uric acid** and **guanine**). These are separated by glandular part of kidney from the coelomic fluid of the pericardial cavity and also from its blood that has been supplied by the vena cava.

Excretory fluid is driven out into the supra-branchial chamber by the beating of cilia of the ciliated epithelium of the bladder. From the supra-branchial it is excreted out through the outgoing water current. The kidneys reabsorb salts from the excretory fluid and also remove excess of water. In this way it also performs **osmoregulatory function.**

2. Keber's Organ or Pericardial Gland

It is a large, glandular, reddish brown mass. It is situated in the anterior part of the body just infront of pericardium. It discharges its excretory products into the pericardial cavity.

46.10 NERVOUS SYSTEM

The nervous system of *Unio* is greatly reduced due to its sedentary mode of life. The nervous system is bilaterally symmetrical and consists of three pairs of ganglia, their connectives and commissures and nerves (Fig. 46.19).

1. Cerebro-pleural ganglia. These are paired small somewhat triangular and yellowish in colour. One of the ganglia is situated on each side of the oesophagus at the base of labial palps. These are connected together by a short **cerebral commissure.** Each ganglion sends nerves to the anterior adductor muscle, labial palp and anterior part of mantle lobes.

2. Pedal ganglia. A pair of pedal ganglia lie in the foot at the junction of foot and visceral mass. The two pedal ganglia are joined to form a bilobed mass which gives off nerves to the foot,

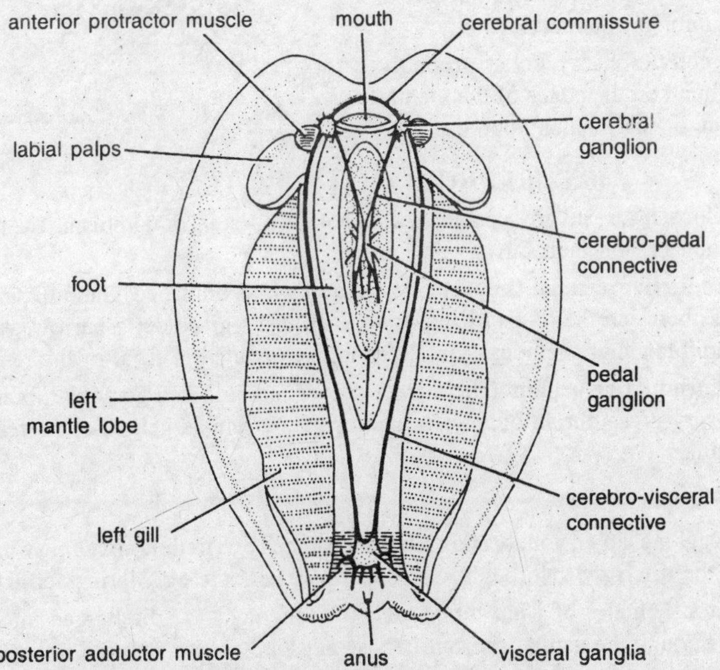

Fig. 46. 19. *Unio*. Nervous system in ventral view.

foot muscles and statocysts. Each pedal ganglia is connected with the cerebropleural ganglia through **cerebro-pedal connective.** It passes through the foot.

3. Visceral ganglia. The two visceral ganglia lie mid-ventrally on the posterior adductor muscles. These ganglia are fused together to form a flattened somewhat rectangular mass. On each side it is connected with cerebro-pleural ganglion of each side by long, delicate **cerebro-visceral connectives.**

On each side, the visceral ganglia send (*i*) **dorsal pallial nerve,** (*ii*) a **posterior pallial nerve** to the posterior part of the mantle; (*iii*) a **posterior renal nerve** to the kidney; (*iv*) a **branchial nerve** to the gill; and (*v*) a **posterior adductor nerve.**

Functions. Pedal and cerebropleural ganglia control the foot movement and the anterior adductor muscle. The posterior adductor muscle and siphons are controlled by visceral ganglia. The cerebro-pleural ganglia coordinate the movements of foot and shell valves.

46.11 SENSORY ORGANS

In *Unio* sense organs are not well developed. Eyes and tentacles are absent. Its sense organs are of following types:

1. Statocysts. Each statocyst is a minute, hollow vesicle and lies very close to a pedal ganglia. It is lined by sensory cells and contains a mass of lime called **statolith.** These help the animal in maintaining balance. Statocysts are supplied with nerve from cerebro-pleural connectives.

2. Osphradium. Beneath the posterior adductor muscles in close contact of visceral ganglia there is a path of sensory epithelium called **osphradium.** It is probably **chemosensory.** It checks the water passing through the mantle cavity.

3. Sensory cells. Sensory cells, probably **tactile** in nature, found on the edges of the mantle and on the **papillae** of the inhalent siphon.

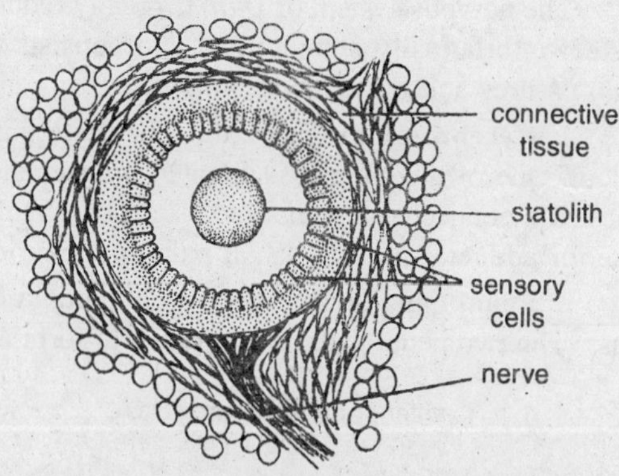

Fig. 46.20. *Unio*. T.S. of statocyst.

46.12 REPRODUCTIVE SYSTEM

Unio is **dioecious,** *i.e.*, sexes are separate. It does not show sexual dimorphism. The reproductive organs consist of gonads and their ducts.

1. Male reproductive organs. They include a pair of white, branched, glandular mass situated in the visceral mass between the loops of intestine. Each testis gives out a narrow, vas deferens, opens into the supra-branchial chamber by male gonopore situated in front of the renal aperture.

2. Female reproductive organs. There are a pair of red-coloured glandular, branched mass of tubules. Each ovary gives out an oviduct, which opens by female genital aperture, situated in front of renal opening.

Life History

1. Eggs. The eggs are larger, yolkier and each is surrounded by **vitelline envelope** which contains a minute opening, the **micropyle,** through which sperms enters the egg during fertilization.

2. Fertilization. Females of *Unio* brood their ova; the eggs are discharged into the supra-branchial chamber through the genital aperture. These eggs are carried into the water tubes of the outer gill laminae and are held by the mucus secreted by the gill epithelium. The sperm from the male escape the body through the outgoing water current. Some of them goes into the females body through inhalant siphon. On entering the gill laminae, they fuse with eggs and bring about **'internal' fertilization.**

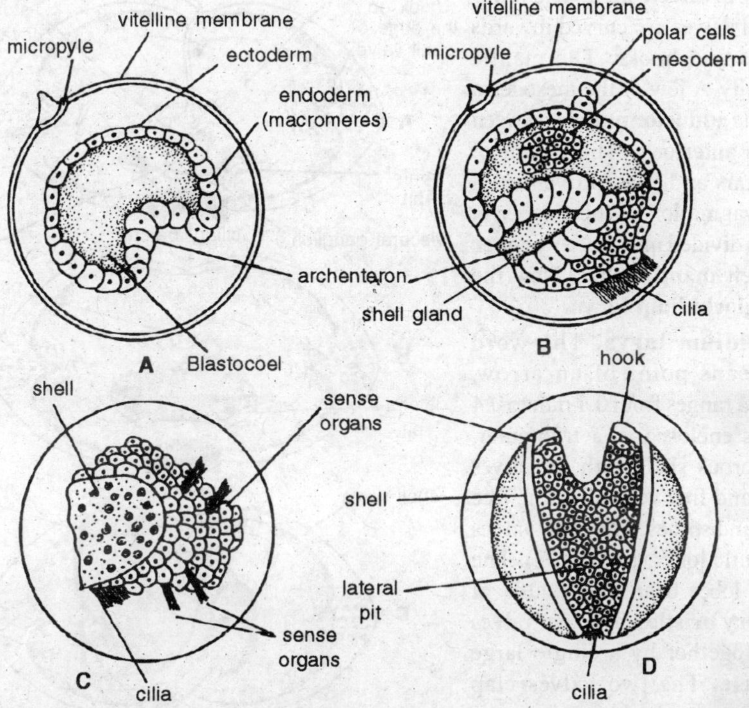

Fig. 46.21. Early four embryonic stages of freshwater mussel.

Development

Development of *Unio* is **indirect** and includes a free-swimming (and ectoparasitic) **glochidium larva** (a modified veliger larva). It occurs inside the water tubes of gills which act as brood pouches. Development of *Unio* resembles greatly with that of *Anodonta*.

1. Cleavage and blastulation. The fertilized egg or zygote undergo **complete (holoblastic)** but **unequal** and **spiral cleavage.** It produces a hollow ball of cells, the **blastula.** It contains small, yolkless and multiplying micromeres on the upper side (animal pole) and large, yolk-filled **macromeres** on the lower side.

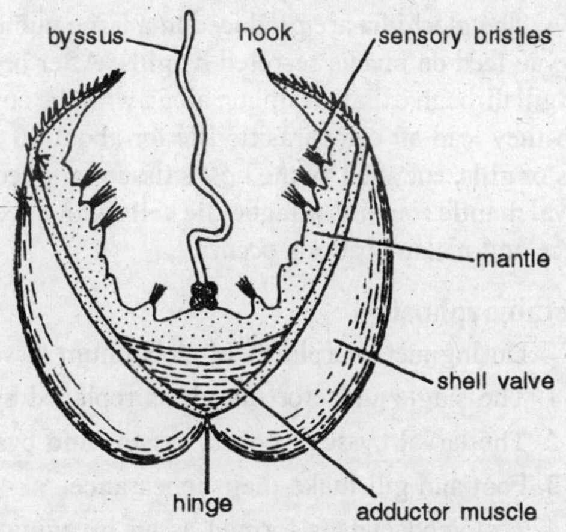

Fig. 46.22. *Unio*. A glochidium.

2. Gastrula. Gastrula is formed by invagination of macromeres (Fig. 46.21). It bears a tuft of long cilia posteriorly and a small **archenteron.** Large cells of gastrula are budded off into the blastocoel where they form **mesoderm**. At the same time a deep invagination marks the dorsal surface of the embryo. This invaginated part forms the characteristic **shell glands.**

These shell glands secrete a **univalve embryonic shell** first and later on, replace it by a triangular **bivalved shell.** The lower ends of two shell valves are curved inwards and bear spines and hooks. The mantle develops gradually. A few of the mesoderm cells constitute the **adductor muscle** between the shell valves anteriorly. The mouth of gastrula disappears and a definitive mouth is formed by invagination in the ectoderm. Thus the larva is divided into a **body proper** and right and left **mantle lobes.** Now the larva is called **glochidium larva.**

3. Glochidium larva. The word glochidium means point of an arrow. Glochidium larva ranges from 0.1 mm to 0.4 mm. Its body is enclosed in a triangular, bivalved and porous shell with its valves united dorsally and free ventrally. The free end bears hook and spines. The shell valves are lined by mantle lobes (Fig. 46.22). The edge of mantle lobes bear 3 to 4 tufts of brush like **sensory bristles.** The two valves are connected together by a single large **adductor muscle.** The two valves clap together by the action of adductor muscle. Larva lacks a mouth and an anus. Larva is without foot but contains a **glandular pouch** on the ventral surface to secrete a long, sticky thread called **provisional byssus.**

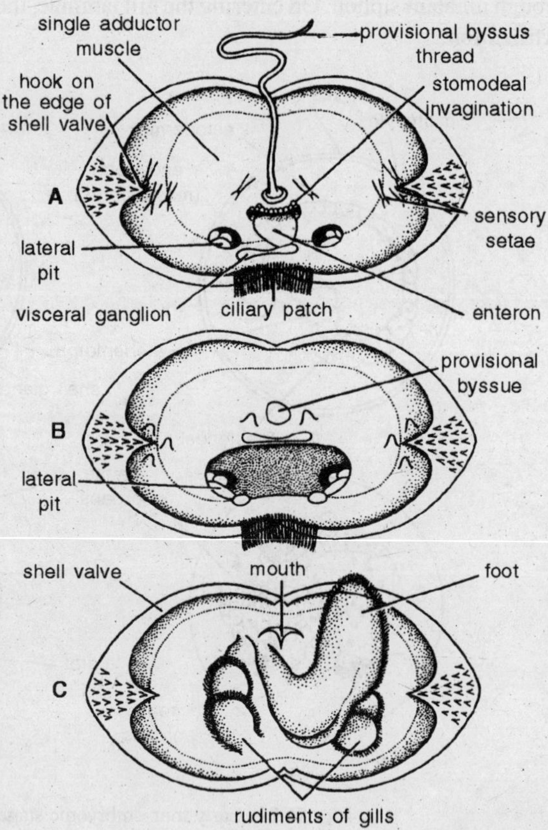

Fig. 46.23. *Unio*. Three stages of metamorphosis.

The glochidia are produced in a large number in the outer laminae and look as stuffed mass. Larvae feed on mucus secreted by gills. After being fully mature the larvae are discharged out of the gill through exhalant siphon along with the outgoing water current. When they find a fresh water fish they lead an ectoparasitic life for about 10 weeks. They become attached to the host's skin, fins or gills, encysted by the host's tissues and feed on nutrients by absorbing from body fluid. The larval mantle contains **phagocytic cells** which feed on the tissues of the host. Growth of larva takes place and metamorphosis occurs.

Metamorphosis

During metamorphosis of glochidium larva the following changes occur (Fig. 46.23):

1. The single adductor muscle is replaced by two adductors.
2. The larval byssus, sensory organs and byssus glands disappear.
3. Foot and gill make their appearance.
4. A stomodaeum is formed as an invagination and opens into archenteron. The tip of the archenteron and overlying ectoderm rupture to form **anus.**
5. The embryonic mantle folds elongate with the increase in size of shell.
6. Rudiments of visceral organs develop.
7. Two papillae or the rudiments of gills appear on each side of the foot.

After metamorphosis, the cyst ruptures and miniature clams are liberated in the water. These sinks to the bottom to assume the adult form.

Significance of Glochidium Larva

1. The developing glochidia remain protected within the cyst. Thus chance of survival increases.
2. Ectoparasitic larval stage ensures rapid dispersal of the species, carried out by the fish to far off places.
3. Mussels are sluggish and burrowing animals. Glochidium is the only way of achieving for and wide dispersal of the species.

QUESTIONS

Long Answer Questions

1. Describe habit and habitat and external features of *Lamellidens*.
2. Give an account of microscopic structure of the shell of *Unio*. How is it formed? (*Jiwaji 1988; Agra 1989; Osmania 1989; Rohilkhand 1991, 1993; Lucknow 1995*)
3. Describe microscopic structure of shell of freshwater mussel. How is pearl formed? (*Himachal 1995*)
4. Give an account of mantle, mantle cavity and its organs of *Unio*.
5. Describe the digestive organs and mode of feeding in the *Lamellidens* or *Unio*. (*Jiwaji 1989; Kanpur 1989; Indore 1989; Rohilkhand 1989; Awadh 1989; Sagar 1990; Purvanchal 1991; Gorakhpur 1991; Himachal 1992; Vikram 1993; CCS 1999*)
6. Describe the respiratory system of *Unio* (*Lamellidens*). (*Allahabad 1988; Bhopal 1988; Rajasthan 1988; Banaras 1989; Raipur 1990; Jiwaji 1991; Bundelkhand 1992; Bilaspur 1993; Vikram 1993; Indore 1993, Gorakhpur 1996*)
7. Describe the structure and mode of attachment of ctenidia of a freshwater mussel you have studied. What function do they perform? (*Jiwaji 1993, Madurai 1994*)
8. Give a clear account of the studies of respiratory organs in freshwater mussel (*Unio*). Describe the mechanism of respiration and indicate its physiological importance to the animal. (*Agra 1988; Magadh 1989; Lucknow 1989; Jabalpur 1989*)
9. Describe the respiratory organs and mode of respiration in *Lamellidens* or a bivalve mollusc. (*Purvanchal 1989; Awadh 1989, 1991; Meerut or CCS 1990, 1993, 1997; Raipur 1990; Kanpur 1990, 1992, 1998; Jabalpur 1991; Jiwaji 1991; Himachal 1991, 1995*)
10. Respiratory organs of *Unio* include gills or ctenidia and mantle.
11. Give an account of the vascular system of *Unio*. (*Bundelkhand 1988; Jiwaji 1988; Patna 1989; Mithila 1989; Awadh 1989; Rajasthan 1989, 1990; Bhopal 1989, 1990; Lucknow 1989, 1993; Kanpur 1989, 1993, 1996; Agra 1991; Ravishankar 1993; Himachal 1993; Gorakhpur 1994*)
12. Describe the heart and the circulatory system of the lamellibranch that you have studied. (*Punjab 1988; Behrampur 1989, 1992; Allahabad 1991*)
13. Give an account of excretory organs of *Unio* (*Lamellidens*). (*Gorakhpur 1996*)
14. Give an account of nervous system of *Unio* and compare it with that of *Pila*. (*Vikram 1988; Jabalpur 1989; Purvanchal 1989; Indore 1990; Raipur 1991; Meerut or CCS 1993, 1996*)
15. Describe various sense organs of *Unio*.
16. Give the structure and function of statocyst in that of *Unio*. (*Himachal 1995*)
17. Describe the reproductive organs and life history of *Lamellidense*. (*Jiwaji 1988; Purvanchal 1989; Rohilkhand 1989, 1991*)
18. Give an account of the development upto the adult stage of a bivalve mollusc. (*Agra 1989, 1990; Purvanchal 1989; Rohilkhand 1991; Osmania 1992, Kanpur 1992; Gorakhpur 1993*)

19. Give in detail the various development stages, found in the life-history of *Lamellidense*. (*Agra 1989*)

20. Describe the life history of fresh-water mussel and show how dispersal is brought about. (*Jiwaji 1988, 1992; Bhopal 1989, 1991; Osmania 1990; Rohilkhand 1990; Agra 1990, 1992; Kanpur 1992, 1997; Gorakhpur 1993*)

Short Answer Questions

1. Decribe the structure and metamorphosis of the glochidium larva and explain how it is beneficial to the animal. (*Rewa 1989*)

2. How is pearl formed in *Unio*? (*Sagar 1989; Agra 1991*)

3. Write a short account of digestion in *Unio*.

4. Describe structure of gill of *Unio*.

5. Decribe the valves of *Unio*. (*Sagar 1988*)

6. Why *Unio* is called a filter feeder? (*Magadh 1990*)

7. Describe position of gonads in *Unio*. (*Kumaun 1991*)

8. When and how fertilization occur in *Unio*? (*Kumaun 1990*)

9. Give a short account of pearl formation and pearl industry.

10. Write short notes on the following:

(*i*) Crystalline style.

(*ii*) Gills of *Unio*.

(*iii*) Glochidium larva
(*Awadh 1988; Rajasthan 1988; Allahabad 1988; Meerut or CCS 1989, 1990; Purvanchal 1989, 1992; Jabalpur 1989, 1992; Utkal 1990, 1992; Vikram 1990; Gorakhpur 1991, 1995; Kanpur 1992, 1995, 1998; Indore 1993; Himachal 1993, 1995; Garhwal 1994; Lucknow, 1995, 1998*)

(*iv*) Shell of *Unio*. (*Agra 1989; Garhwal 1992; Jiwaji 1992*)

(*v*) Nacreous layer. (*Himachal 1995*)

(*vi*) Keber's organ.
(*Bhopal 1988; Allahabad 1989; Magadh 1990; Purvanchal 1991; Himachal 1995*)

(*vii*) Organ of Bojanus. (*Himachal 1992*)

11. Draw neat and labelled diagram of the following:

(*i*) V. S. shell of Mollusca. (*Bihar 1990*)

(*ii*) T. S. of gill lamina of *Unio*. (*Rohilkhand 1989; Sagar 1989, Meerut or CCS 1997*)

(*iii*) T. S. of *Unio* through heart. (*Bundelkhand 1989; Agra 1990; Rohilkhand 1990; Awadh 1990, 1992; Lucknow 1992; Gorakhpur 1993*)

(*iv*) T. S. *Unio* through middle of body or through gill region.
(*Sagar 1988; Rajasthan 1990; Meerut or CCS 1990; Himachal 1990*)

(*v*) Blood vascular system of *Lamellidens*.

(*vi*) Nervous system of *Unio*.
(*Bundelkhand 1988; Gorakhpur 1991; Gwalior 1993; Jiwaji 1993; Meerut or CCS 1996*)

(*vii*) Glochidium larva.
(*Rajasthan 1988; Rohilkhand 1988, 1989; Meerut or CCS 1989; Kanpur 1992, 1995; Sagar 1994*)

(*viii*) T. S. *Unio* passing through branchial or gill or middle region.
(*Meerut or CCS 1994, 1996; Kanpur 1996; Lucknow 1996*)

(*ix*) T. S. of *Unio* through heart and foot. (*Gorakhpur 1994, 1995, 1997*)

12. Differentiate between gill-lamella and gill-lamina. (*Gorakhpur 1995; Lucknow 1995, 1996, 1998*)

Very Short Answer Questions

1. Give common name of *Unio* or *Lamellidens*.

Ans. Freshwater mussel.

2. What is chemical composition of shell of *Lamellidens*?

Ans. The chemical composition of shell of *Lamellidens* includes

(*i*) Calcium carbonate – 89 – 99%

(*ii*) Phosphate of lime – 1 – 2%

(*iii*) Conchiolin – Only traces

3. How are the two valves of shell of *Unio* united?

Ans. The two valves of shell of *Unio* are united by a dorsal elastic band called a **hinge ligament.** It is continuous with the two shell valves and is made of uncalcified conchiolin.

4. What is the function of adductor muscles?

Ans. Closing of shell.

5. Name the muscles which draw the foot into the shell.

Ans. Retractors.

6. What controls the opening and closing of shell valves of *Lamellidens*?

Ans. In *Lamellidens*, there are two adductor muscles for opening and closing of shell valves. These muscles are **anterior** and **posterior adductor muscles**. They are large, cylindrical and transverse bands of muscles running from one shell valve to the other. On contraction of these muscles the shell valves are closed and on their relaxation, shell valves are opened.

7. In *Unio*, which is larger, exhalant siphon or inhalant siphon.

Ans. Inhalent siphon is larger than exhalent siphon.

8. What is typhlosole?

Ans. Infold of the wall of gut (*i.e.*, rectum) into the lumen.

9. Which kind of ctenidia *Unio* has?

Ans. Lamelliform.

10. Name the respiratory pigment found in the blood of some species of *Unio*.

Ans. Haemocyanin.

11. What is the function of Keber's organs in *Unio*?

Ans. Excretion.

12. Mention the host of the larva of *Unio*.

Ans. Freshwater fish.

13. What are marsupia?

Ans. Water tubes of ctenidia that function as brood pouches.

14. Which type of development occurs in *Unio*?

Ans. Indirect development since egg of *Unio* develops into a glochidium larva which undergoes metamorphosis to become an adult-like form.

15. Select the correct alternative.

(*i*) Shell valves of *Lamellidens* are lateral/dorsal and ventral. (*Lucknow 1998*)

Ans. Lateral.

(*ii*) Water current of *Unio* is not purely/purely respiratory. (*Lucknow 1997*)

Ans. Not purely.

Fill in the Blanks

1. Shell of *Unio* consists of .... valves.
2. .... is the outer layer of the shell of *Unio*.
3. Foot of *Unio* is .... shaped.
4. Shell of *Unio* is secreted by ....
5. Mouth of *Unio* is bound on either side by .... and .... labial ....
6. The gills of *Unio* are called ....
7. The gills of *Unio*, besides carrying out gas exchange, also act as ....
8. Heart of *Unio* is three chambered, having .... atria and .... ventricle.
9. Nitrogenous waste material excreted by kidneys in *Unio* is chiefly ....
10. The osphradium of *Unio* functions as ....

Multiple Choice Questions

1. *Unio* is a member of the class
(*a*) Gastropoda (*b*) Cephalopoda
(*c*) Pelecypoda (*d*) Scaphopoda

2. The body of *Unio* is
(*a*) bilaterally symmetrical (*b*) radially symmetrical
(*c*) asymmetrical (*d*) biradially symmetrical

3. The outer layer of shell of *Unio* contains
(*a*) chitin (*b*) conchiolin
(*c*) $CaCO_3$ (*d*) Si_2O_3

4. The middle layer in the shell of *Unio* is
(*a*) prismatic layer (*b*) periostracum
(*c*) nacre (*d*) pallium

5. The muscles which are responsible to open and close the shell valve in *Unio*, are
(*a*) pallial muscles (*b*) protractor muscles
(*c*) adductor muscles (*d*) retractor muscles

6. Body cavity of *Unio* is
(*a*) coelom (*b*) haemocoel
(*c*) none of these (*d*) both of these

7. Pearl producing species of mollusc is
(*a*) *Tridacna maxima* (*b*) *Solen kempi*
(*c*) *Pinctada vulgaris* (*d*) *Mytilus viridus*

8. Typhlosole is present in *Unio* in
(*a*) stomach (*b*) intestine
(*c*) rectum (*d*) oesophagus

9. Gills in *Unio* are
(*a*) a part of visceral mass (*b*) derivatives of mantle
(*c*) derivatives of foot (*d*) modification of tentacles

10. Heart beat of *Unio* per minutes is
(*a*) 10 – 20 (*b*) 20 – 40
(*c*) 40 – 80 (*d*) 20 – 100

11. Keber's organ is the excretory organ of

(*a*) *Pila* (*b*) *Unio*
(*c*) *Neopilina* (*d*) *Doris*

12. "Organs of Bojanus" are derivatives of

(*a*) spleen (*b*) kidney
(*c*) gall bladder (*d*) liver

13. The chemoreceptor osphradium is present in

(*a*) *Pila* (*b*) *Unio*
(*c*) both (*d*) none of them

14. Statocyst in *Unio* is meant for

(*a*) olfaction (*b*) gustation
(*c*) equilibrium (*d*) photoreception

15. Larva of *Unio* is called

(*a*) trochophore (*b*) glochidium
(*c*) veligar (*d*) zoaea

16. Glochidia larva is

(*a*) endoparasite of pig (*b*) ectoparasite of pig
(*c*) ectoparasite of fish (*d*) endoparasite of sheep

ANSWERS

Fill in the Blanks

1. Two **2.** Periostracum **3.** Hatchet **4.** Mantle (pallium)
5. Outer, inner, palps **6.** Ctenidia **7.** Brood pouches **8.** Two, one
9. Ammonia **10.** Chemoreceptor

Multiple Choice Questions

1. (*c*) **2.** (*a*) **3.** (*b*) **4.** (*a*) **5.** (*c*) **6.** (*d*)
7. (*c*) **8.** (*c*) **9.** (*b*) **10.** (*c*) **11.** (*b*) **12.** (*b*)
13. (*c*) **14.** (*c*) **15.** (*b*) **16.** (*c*).

47

Torsion in Gastropods

Molluscs are typically bilaterally symmetrical animals but this symmetry is lost in Gastropoda due to two processes: **coiling** and **torsion.** Most gastropods live with the body coiled up in a spiral. In addition, while they are embryonic, the visceral mass is suddenly **torted** or **twisted** about, bringing the back end to front over the head. Most gastropods remain permanently torted while few partially untort during later growth.

47.1 DEFINITION OF TORSION

Torsion is the anticlockwise rotation of viscero-pallium at 180° from its initial position during embryonic/larval development of Gastropoda. As a result of this rotation, the posterior part of mantle cavity with its pallial complex, anus and rectum, etc., are brought to the front of body and anus comes to lie close to mouth. Entire process of torsion generally takes only a few minutes.

47.2. EFFECTS OF TORSION ON BODY STRUCTURE

The torsion introduces the following changes in the organization of *Pila* and other gastropods:

1. Displacement of mantle cavity. Before torsion the mantle cavity opens posteriorly. After torsion the mantle cavity opens in front just behind the head (Fig. 47.1).

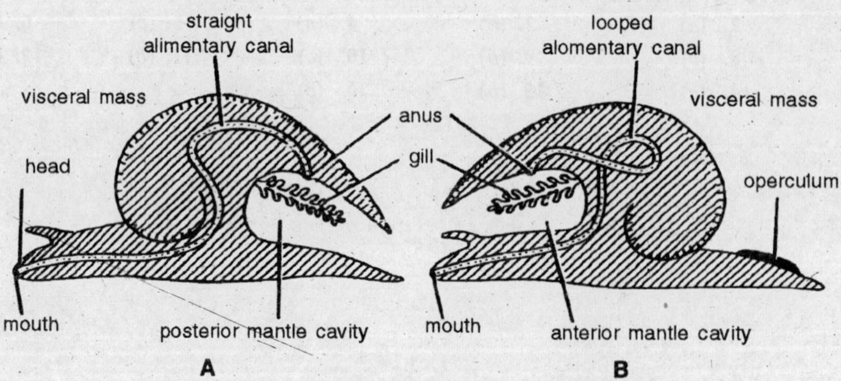

Fig. 47.1. Gastropod showing the phenomenon of torsion. A—Before torsion; B—After torsion.

2. Changes in relative position of organs. Before torsion, ctenidia point backwards, the auricles are behind the ventricle and mouth and the anus are at the opposite ends. After torsion, ctenidia come to lie in front and point anteriorly, the ctenidium of right side comes to lie on the left and that of left on the right; the auricles become anterior to the ventricle, the auricle of right side comes to lie on the left and vice versa.

3. Coiling of shell and visceral mass. The shell and visceral sac which are originally dorsal or **exogastric** become ventral or **endogastric** as a result of torsion.

4. Looping of alimentary canal. Primarily straight alimentary canal of larva becomes looped or 'U' shaped and posteriorly placed larval anus comes to occupy an anterior position.

5. Loss of symmetry. As a result of torsion original bilateral symmetry is lost due to displacement of pallial cavity and visceral sac.

6. Loss of organs or atrophy. Asymmetry of gastropods is also produced due to loss, reduction or atrophy of paired parts from one side or other. If pallial cavity comes on the right, the organs on the left side are degenerated leaving only the organ of right side.

7. Twisting of nervous system (Chiastoneury). Torsion introduces major changes in nervous system. As a result of torsion, except the visceral ganglia, all the ganglia and nerves come to occupy anterior position in body. Because of this pleural, pedal and buccal ganglia lie at the anterior end to form a nerve ring.

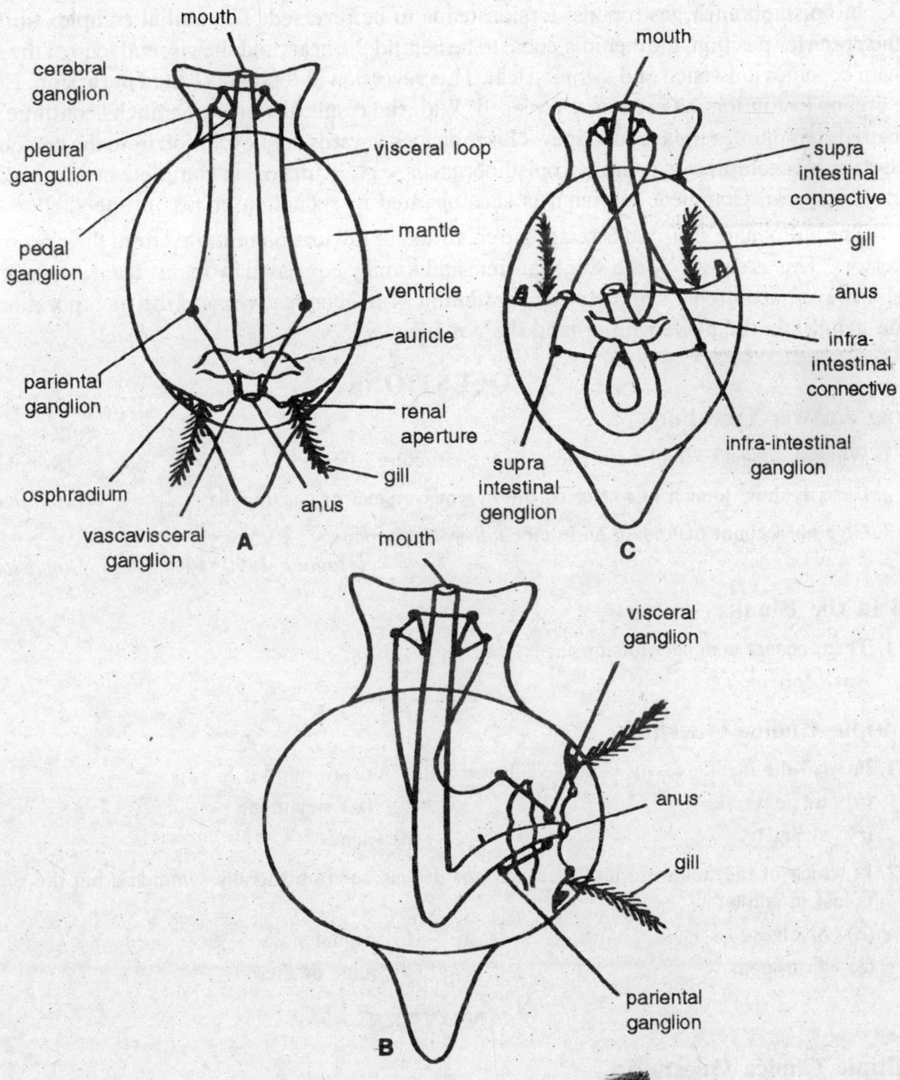

Fig. 47.2. Diagram to show position of different organs in body (A) before and after torsion(C).

The long pleuro-visceral connectives become twisted into figure of '8'. **Right connective** forms the **supra-intestinal nerve** and left connective forms **infra-intestinal nerve.** These nerves are present above and below (dorsal and ventral) to the oesophagus and join. Finally these are removed.

In some forms after the coiling of the visceral hump, torsion occurs by rotation only through 90°, so that the ctenidia and anus point laterally (Fig. 47.1B).

Significance. The anteriorly placed mantle cavity houses the head with sense organs, respiratory structures, etc., in adults due to occurrence of torsion; there is advantage. Thus, it causes an intimate contact for gaseous exchange with incoming water and also helps to test this water (**Morton** 1958)

47.3 DETORSION

Torsion occurs and persists as such only in prosobranch gastropods, *e.g.*, *Acmaea* (marine limpet), *Pomatias*, *Patella*, *Haliotis*, etc.

In opisthobranch gastropods, torsion tends to be reversed. The pallial complex shifts back to the posterior position, the ctenidia come to lie behind the heart and the visceral loop of the nervous system becomes untwisted and symmetrical. This reversion of torsion is called **detorsion.** However, the organs lost in torsion are never replaced. With the result, the opisthobranchs continue to have unpaired ctenidium, auricle and kidney. This indicates towards their evolution from the prosobranchs. Detorsion is **incomplete** in primitive opisthobranchs, *e.g.*, *Actaeon*, but **complete** in advanced forms, such as *Aplysia.* Complete detorsion is accompanied by reduction or loss of the shell.

The pulmonate gastropods retain the condition of full torsion inherited from their prosobranch ancestors. This is shown by the single auricle and kidney possessed by them. Their nervous system has, however, acquired a symmetry by shortening of the connectives and the incorporation of the main ganglia in the nerve ring around the oesophagus.

QUESTIONS

Long Answer Questions

1. What is torsion? What is its effect on the structure of *Pila*? (*Kanpur 1995*)
2. Describe how torsion has affected the nervous system of Gastropoda. (*Madurai 1994*)
3. Give an account of torsion and detorsion in Gastropoda. (*Magadh 1990, 1994, 1996, Himachal 1995*)

Fill in the Blanks

1. There occurs complete torsion in ....

   Ans. *Aplysia*.

Multiple Choice Questions

1. In snail the mantle cavity is pushed forward and outward due to

   (*a*) torsion (*b*) fast swimming
   (*c*) symmetry (*d*) none

2. In which of the following molluscan groups the larvae are bilaterally symmetric but the symmetry is lost in adults?

   (*a*) bivalves (*b*) cephalopods
   (*c*) gastropods (*d*) none of these

ANSWERS

Multiple Choice Questions

1. (*a*) 2. (*c*).

48

Phylum Onychophora and Peripatus

Onychophora (Gr., *onychus* = claws; *phorus* = bearing) is an ancient phylum and has not changed greatly since its origin in Cambrian period. It includes those invertebrates which are included into a single genus *Peripatus* having about 110 species. The genus *Peripatus* was established in 1826 by **Guilding.** He regarded it as a mollusc. In 1874, **Moseley** classified it as an primitive arthropoda discovering the tracheae.

48.1 GENERAL CHARACTERS OF ONYCHOPHORA

1. Onychophora are specialized for life in the leaf litter and plant debris and are ideally adapted for squeezing through narrow spaces and cavities. They are found in crevices of rocks, under bark, in stones and in moist places in tropical region.
2. They have a worm-like and indistinctly segmented body.
3. Their **head** is composed of three limb-bearing somites. Thus, head bears three pairs of appendages, *i.e.*, a pair of **antennae,** a pair of **peri-buccal lobes** hiding **jaws** and a pair of **oral** or **slime papillae.** At the base of each antenna is found a simple eye.
4. Each **trunk** segment bears a pair of clawed lobopodial (= lobe-like, unjointed and fleshy) limbs.
5. Body is covered by a thin, flexible and chitinous **cuticle.**
6. Paired **slime** or **adhesive** glands are present, slime is used in food (prey) capturing and self defence.
7. Respiration occurs through unbranched **tracheae.**
8. **Coelom** is reduced to gonadal cavities and nephridial sacs. Principal body cavity is haemocoel.
9. Circulatory system includes heart and haemocoelomic sinus like the artropods.
10. Excretion occurs by segmentally arranged paired **nephridia.** Excretory product most probably is **uric acid.**
11. Sexes are separate. Male sex produces **spermatophores.** Most are **viviparous.** Development is direct and resembles with arthropods.
12. Onychophora forms a connecting link between **Annelida** and **Arthropoda.**

48.2 CLASSIFICATION

Phylum Onychophora contains two families: **Peripatidae** and **Peripatopsidae;** both of whioh have peculiar geographical distribution. Each family has a wide but discontinuous distribution around the world. The members of two families are never found in the same area. The family **Peripatidae** are equatorial in distribution (*i.e.*, circum-tropical-subtropical distribution), while the family **Peripatopsidae** are limited to the Southern Hemisphere (*i.e.*, circum-south-temperate distribution).

There are several genera (such as *Peripatus*, *Peripatopsis*, *Eoperipatus*, *Opisthoperipatus*, *Paraperipatus*, etc.) but the term *Peripatus* is commonly used as the generic name for all species.

48.3 *PERIPATUS*

1. Distribution

Peripatus exhibits discontinuous geographical distribution and is found in warmer parts of the world such as tropical regions (East Indies, West Indies, Himalayas, Congo and Northern part of South America). It is also found in the temperate regions of Southern Hemisphere (Australia, New Zealand, South Africa and Andes). No species of *Peripatus* occur in the north of the Tropic of Cancer. Common name of *Peripatus* is **velvet worm.**

2. Habit and Habitat

Peripatus is **terrestrial, nocturnal** and **predaceous** animal. It is confined to dark, warm, and humid habitats, *i.e.*, it lives in crevices of rocks, under bark, stones and beneath fallen leaves (= leaf litter). It remains active during night. During cold and dry periods, *Peripatus* becomes inactive and remains hidden in burrows. *Peripatus* is predaceous and feeds on small invertebrates such as snails, insects and worm.

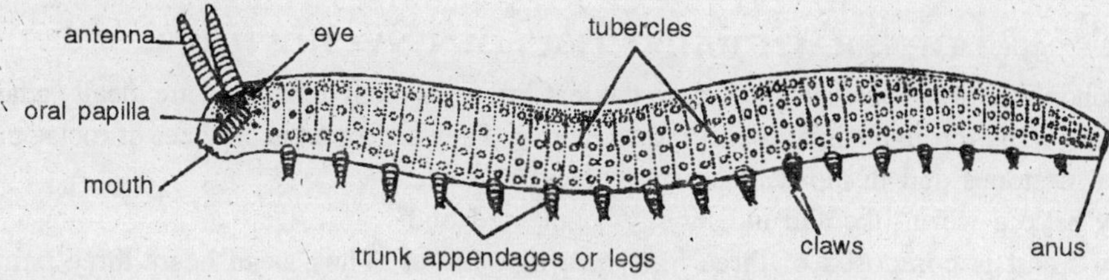

Fig. 48.1. *Peripatus*. External features.

3. Anatomical Peculiarities: External Morphology

1. *Peripatus* is elongated, cylindrical and catterpillar-like measuring 1.5 to 15.0 cm in length.
2. Body colour is blue, green, orange and black. Skin (= outer body covering) is velvety and iridescent in appearance. Skin is thrown into transverse wrinkles bearing numerous small papillae (or tubercles) armed with chitinous spines.

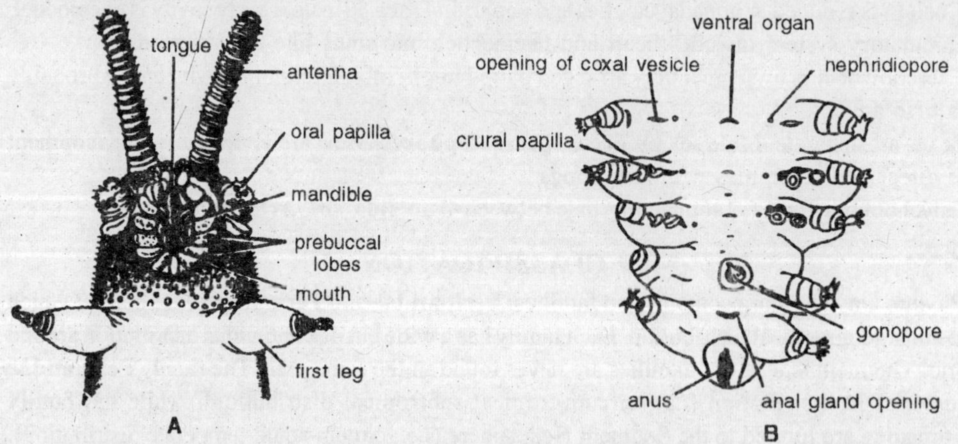

Fig. 48.2. *Peripatus*. A—Anterior end in ventral view; B—Posterior end in ventral view.

3. External segmentation is indistinct. Number of segments are represented by the number of appendages, *i.e.*, each body segment bears a pair of appendage.
4. Body is divided into **head** and **trunk.**
5. First three segments form the **head** but it is not clearly marked off from the rest of the body. First cephalic segment bears a pair of short, ringed and unjointed **antennae.** Each ring of antenna bears spines. Second cephalic segment bears a pair of short and conical **oral papillae.** Each slime gland opens on the tip of oral papilla. Third cephalic segment bears a pair of small horny **jaws.** A pair of simple **eyes** are present on the dorsal surface of head, at the base of antennae.

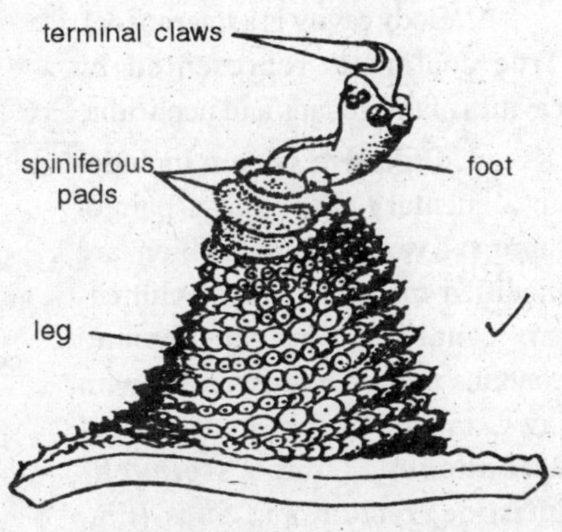

Fig. 48.3. *Peripatus.* A leg.

6. Mouth is small and mid ventral.
7. **Trunk** bears a series of paired short stumpy legs which vary in number 14 to 43 pairs depending on species. Each leg (Fig. 48.3) is a large, conical, unjointed, lobe-like and bears a pair of **claws.**
8. At the base of each leg are small **nephridiopores,** and prominences with the pores for the **crural glands** at their apex.
9. The **anus** is posterior and ventral and the **gonopore** lies directly in front of it.
10. Entire body surface is studded with minute scaled **tubercles,** some of which are equipped with sensory bristles. Tubercles are arranged in rings or bands on the body or legs.

Internal Structure (Anatomy)

1. **Body wall** is dermo-muscular with **cuticle** (which is chitinous, thin, flexible, permeable), **epidermis** (single layered), **dermis** and **muscle layers** (which are three in number—circular, diagonal and longitudinal).

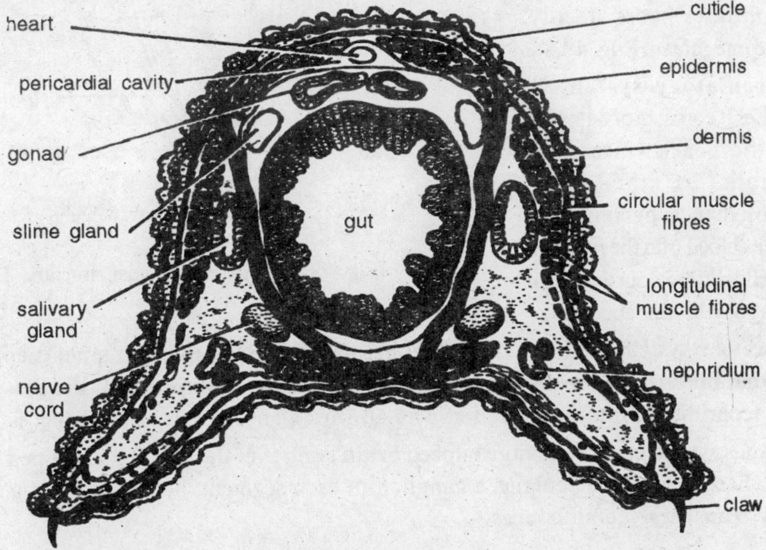

Fig. 48.4. *Peripatus.* T.S. body.

2. Body cavity is a **haemocoel.** True coelom is represented by cavities of the gonads and nephridia.

3. **Digestive system** includes an alimentary canal and a pair of large **salivary glands** (which are modified metanephridia). Alimentary canal is a straight tube and contains **mouth, buccal cavity** with jaws, muscular **pharynx,** short **oesophagus,** long **stomach-intestine, rectum** and **anus** (Fig. 48.5).

4. *Peripatus* is **champion spitters** of the animal kingdom, for it can project its secretions distances of up to 0.5 m. It can capture its prey and defend itself by ejecting a stream of nontoxic, sticky material from the paired **slime glands.**

5. *Peripatus* is adapted to **air breathing** with a **tracheal system** resembling that of terrestrial arthropods (Myriapoda). Respiratory organs are **tracheae** which are unbranched and go directly to the organ they serve. They communi-cate with the exterior by **stigma spiracles** which are minute pores deeply situated in the integument (Fig. 48.6).

6. **Circulatory system** is arthropod-like. It consists of a dorsal and contractile **heart** which runs entire length of the body and is enclosed in a **pericardial sinus.** It pumps blood into the general haemocoel. Blood is colourless and contains phagocytic amoebocytes.

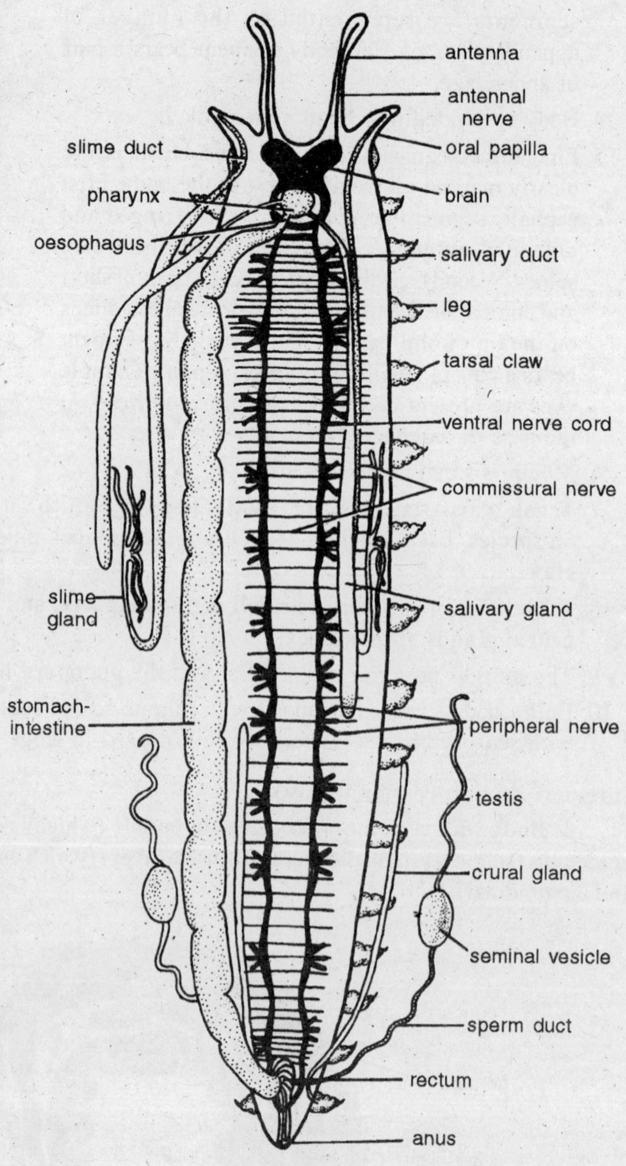

Fig. 48.5. *Peripatus.* Internal structure.

7. **Excretory system** inclu-des paired **nephridia** containing ciliated funnel **(nephrostome)** and **ectodermal bladder** and **nephridioporc** (Fig. 48.7). Each somite, except the first two, has a pair of metanephridia but the first pair serve as salivary glands. Excretory product is uric acid.

8. Nervous system contains a large bilobed **brain** lying over the pharynx and a pair of **ventral nerve cords.** Each nerve cord contains a ganglion in each segment; both ganglia of a somite are connected by **transverse commissures.**

9. Sense organs include a pair of **eyes** (ocelli), a pair of **antennae** bearing tactile endings and **taste spines** on the lips and preoral cavity. Hydroreceptors occur on anten-nae and on body surface.

10. Sexes are separate; females being larger than males. Reproductive organs are paired.

11. **Fertilization** is internal and development is direct.

12. Most species of *Peripatus* are **viviparous,** eggs develop in **uterus** of female.

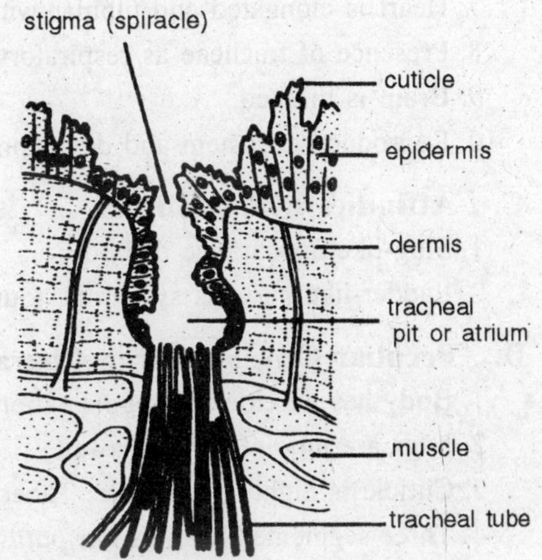

Fig. 48.6. *Peripatus*. Section through stigma (= spiracle).

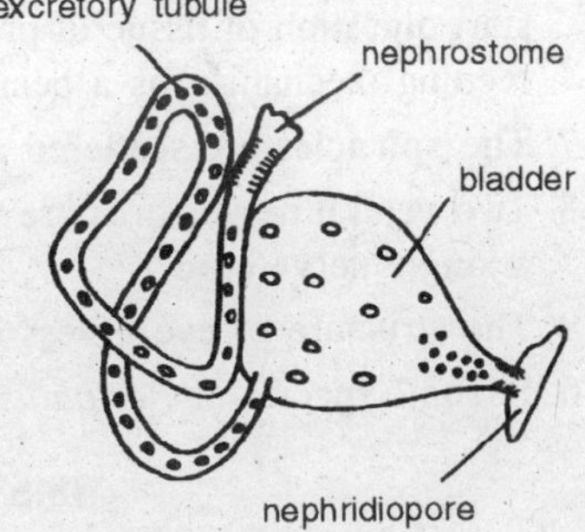

Fig. 48.7. *Peripatus*. A metanephridium.

48.4 AFFINITIES OF *PERIPATUS* (OR BIOLOGICAL IMPORTANCE OF *PERIPATUS*)

Peripatus is an animal of great biological importance. It exhibits some characteristics of Annelida and some characteristics of Arthropoda. It contains some specialized (unique) features of its own. *Peripatus* is regarded as **connecting link** between annelids and arthropods. It is also regarded as **living fossil** (*i.e.*, it is very ancient and has not changed since their origin). It has a common ancestor with uniramous arthropods **(Uniramia** like myriapods and insects) but is not itself ancestral to any group. Its affinities (similarities) with annelids, arthropods and molluscs are as follows:

A. Affinities with Annelida

1. Vermiform (worm-like) body.
2. Body wall is soft, flexible and dermo-muscular; cuticle is thin.
3. True head is absent.
4. Appendages are hollow and unjointed resembling to parapodia of Annelida.
5. Locomotion occurs by contraction and expansion of body muscles (peristaltic movements as in earthworm).
6. Simple straight alimentary canal.
7. Presence of cilia in excretory organ and in reproductive organs as in annelids.
8. The slime and coxal glands are similar to Polychaeta.
9. The structure of its **eyes** is similar to that of some polychaetes in having large **chitinous lens** and well developed retina.

B. Affinities with Arthropoda

1. Presence of antennae.
2. The cuticle is chitinous like the arthropods. It is moulted periodically like that of arthropods.
3. Presence of spines and claws.
4. Body is not divided internally by segmental septa.
5. True coelom is reduced and the body cavity is **haemocoel.**
6. Jaws are modified appendages. The pharynx and oesophagus are lined by chitin as in centipede arthropods.

7. Heart is elongated and tubular with lateral ostia.
8. Presence of tracheae as respiratory organs.
9. Brain is bilobed.
10. Reproductive system and development is similar to arthropoda.

C. Affinities with Mollusca

1. Slug-like appearance of body.
2. Ladder-like nervous system as found in Polyplacophora and Prosobranchiata.

D. Peculiar Onychophoran Characteristics

1. Body has no external segmentation.
2. Legs are unjointed but clawed.
3. Cuticle is highly permeable. Tubercles are arranged in rings on the body.
4. Three segmented head of *Peripatus* exhibits a condition mid-way between that of Annelida and Arthropoda.
5. A single pair of jaws.
6. Its claw-like jaws cut the prey. This action allows saliva to enter the body of the prey and start digestion of tissue of prey. The digested tissues are sucked by *Peripatus*. This type of feeding mechanism is a peculiar feature of *Peripatus*.
7. The **spiracles** are scattered all over the body of *Peripatus*. **Tracheae** are not branched.
8. Two ventral nerve cords are widely separated and have indistinct ganglia. Annelids contain a single nerve cord.
9. The structure of eye is less complicated.
10. In some species of *Peripatus*, the male deposits the spermatophores on the skin of female.

48.5 TAXONOMIC POSITION

Earlier *Peripatus* and related genera were classified with phylum **Arthropoda** as one of the class under **Onychophora.** According to modern view, however, this group was more wide spread and diversified. The view is supported by its discontinuous distribution and discovery of several fossils from middle Cambrain. It is therefore believed that Onychophora is an ancient independent group of segmented animal contemporary to Arthropoda and as such has been placed under separate phylum Onychophora.

QUESTIONS

Long Answer Questions

1. Give the general characters of Onychophora and discuss the importance of *Peripatus*. *(Meerut or CCS 1995)*

2. Discuss affinities and systematic position of *Peripatus*. *(Nagarjun 1994)*

3. What do you mean by living fossil? State the anatomical peculiarities and affinities of Onychophora. *(Guwahati 1994)*

4. Enumerate three annelidan, three arthropodan and one peculiar character of *Peripatus*. *(Meerut or CCS 1993)*

5. Discuss the zoological importance of *Peripatus*. *(Allahabad 1988, 1989)*

6. State the anatomical peculiarities and discuss the importance/systematics of *Peripatus*. *(Burdhwan 1990; Magadh 1990; Agra 1992; Kanpur 1994; Kurukshetra 1990, 1993)*

7. Discuss structure and affinities of *Peripatus*.
(*Osmania 1988; Meerut or CCS 1989; Bundelkhand 1989; Magadh 1996*)

8. *Peripatus* is considered to be connecting link in between Annelida and Arthropoda. Discuss.
(*Garhwal 2001*)

9. Give general characters of *Peripatus*. Discuss its affinities. (*C.C.S.U. Meerut 2005*)

Short Answer Questions

1. Write short notes on the following:

(*i*) Onychophora (*Guwahati 1993*)

(*ii*) *Peripatus* (*Bundelkhand 1989; Karanataka 1989; Garhwal 1990; Jiwaji 1990; Allahabad 1990; Awadh 1990; Rohilkhand 1991; Lucknow 1990, 1992; Kanpur 1993, CCS 1999*)

Very Short Answer Questions

1. Who established genus *Peripatus*? How many species of *Peripatus* are existing today?

Ans. R.L. Guilding (1826) established genus *Peripatus*. He described it as a leg-bearing "slug" (a mollusc). The only genus *Peripatus* of phylum onychophora has about 110 living species.

2. Describe the legs of *Peripatus*.

Ans. *Peripatus* bears about 14 to 23 pairs of short, stumpy legs in trunk region. Each leg is large, conical, unjointed, lobe-like bearing a pair of claws. At the base of each leg are found small nephridiopores and pores for crural glands.

3. Which body segment of *Peripatus* bears a pair of oral papillae?

Ans. A pair of oral papillae are present in second cephalic segment. They are short and conical with slime glands opening at their tip.

4. Why *Peripatus* is called the champion spitter of the animal kingdom?

Ans. It is because *Peripatus* can throw its secretions upto a distance of about 0.5 metres. It ejects a stream of nontoxic sticky substance to capture its prey and defend itself, from the paired slime glands.

5. Why *Peripatus* is said to be a connecting link?

Ans. *Peripatus* is known as the connecting link (or missing link) between annelids and arthropods as it exhibits some characters of Phylum Annelida and some of the Phylum Arthropoda.

6. Why is *Peripatus* regarded as a living fossil.

Ans. *Peripatus* is a very ancient animal. It remained unchanged from the time of its origin Cambrian (530 million years ago). Hence, it is called the **living fossil.**

Multiple Choice Questions

1. *Peripatus* belongs to

(*a*) Onychophora (*b*) Thysanura

(*c*) Mollusca (*d*) Alteroidea

2. Onychophora has following characters, except one which is wrong. Mark the wrong statement

(*a*) has segmentally arranged nephridia

(*b*) has chitinous exoskeleton

(*c*) has lateral segmental appendages comparable to parapodia of *Nereis*

(*d*) has haemocoel

3. *Peripatus* is connecting link between

(*a*) Annelida and Arthropoda (*b*) Annelida and Platyhelminthes

(*c*) Annelida and Mollusca (*d*) Mollusca and Arthropoda

4. One of the following shows discontinuous distribution
(*a*) *Schistocerca* (*b*) *Peripatus*
(*c*) *Periplaneta* (*d*) *Musca*

5. *Peripatus* is
(*a*) aquatic (*b*) terrestrial
(*c*) parasitic (*d*) none

6. *Peripatus* is
(*a*) herbivorous (*b*) carnivorous
(*c*) omnivorous (*d*) parasitic

7. *Peripatus* shows
(*a*) metameric segmentation (*b*) scant metamerism
(*c*) pseudometamerism (*d*) none of these

8. The number of legs found in *Peripatus* is
(*a*) 10–20 pairs (*b*) 15–40 pairs
(*c*) 14–43 pairs (*d*) none of these

9. The body wall of *Peripatus* is constructed on the typical
(*a*) arthropodan plan (*b*) annelidan plan
(*c*) molluscan plan (d) avian plan

10. In *Peripatus*, body cavity is
(*a*) schizocoel (*b*) haemocoel
(*c*) acoelomate (*d*) none of these

11. *Peripatus* breathes by
(*a*) trachea (*b*) gills
(*c*) book lungs (*d*) branchiostegite

12. The excretory structures in *Peripatus* are
(*a*) malpighian tubules (*b*) coxal glands
(*c*) metanephridia (*d*) solenocytes

13. *Peripatus* is
(*a*) oviparous (*b*) viviparous
(*c*) ovo-viviparous (*d*) none

ANSWERS

Multiple Choice Questions

1. (*a*) 2. (*b*) 3. (*a*) 4. (*b*) 5. (*b*) 6. (*b*)
7. (*b*) 8. (*c*) 9. (*c*) 10. (*b*) 11. (*a*) 12. (*c*)
13. (*b*).

49

Phylum Echinodermata (Characters, Classification and Types)

Phylum **Echinodermata** (Gr., *echinos* = hedgehog; *derma* = skin) includes about 6000 species of marine **spiny skinned animals** such as starfishes, brittle stars, sea urchins, sea cucumbers, sand dollars and sea lilies. They distinctly have a unique **water vascular system** equipped with **podia** or **tube feet** which are used in locomotion and feeding. Echinodermata forms the only major group of **deuterostome** (Box 49.1) invertebrata.

> **Box 49.1**
>
> **Protostomia *vs.* Deuterostomia**
>
> **1. Protostomes** (Gr., *proto* = first; *stoma* = mouth). Animals in which **blastopore** (which is a groove of early gastrula through which cells move inside to form endoderm and mesoderm) develops into the mouth, *e.g.*, all metazoan invertebrates except echinoderms.
>
> **2. Deuterostomes** (Gr., *deutero* = second; stoma = mouth). Animals in which blastopore becomes anus and mouth formation is secondary, *e.g.*, Echinodermata and Chordata.

49.1 DEFINITION

The echinoderms are triploblastic, unsegmented, enterocoelous coelomate animals having pentamerous radial symmetry (*i.e.*, biradial symmetry with an oral-aboral axis). Their larvae are bilaterally symmetrical. They contain an endoskeleton of calcareous plates or ossicles and a peculiar hydraulic ambulacral or water vascular system of coelomic origin. They have a complete alimentary canal, diffuse nervous system and no excretory system.

49.2 GENERAL CHARACTERS

1. The echinoderms are exclusively **marine,** free-living and mostly bottom-dwellers.
2. They are **triploblastic, coelomate** and **unsegmented** animals.
3. **Head** is absent but body is differentiated into **oral** and **aboral surfaces.**
4. Adults show **radial symmetry** (usually **pentamerous,** *i.e.*, body parts occurring in fives) but their larvae are bilaterally symmetrical.
5. They are of moderate to considerable size but none are microscopic.
6. The body wall consists of an outer layer of **epidermis,** a middle **dermis** and an inner lining of peritoneum.
7. **Endoskeleton** consists of closely fitted calcareous plates of mesodermal origin forming a shell, many called **theca** or **test** or may be composed of separate small ossicles. Skeleton bears projecting **spines** or tubercles that give the body surface a warty or spiny appearance Some echinoderms such as sea urchins have long movable spines.

8. A part of embryonic coelom is modified into a unique or distinctive **water vascular system** which consists of an array of radiating canals (ambulacral grooves) and tube-like appendages called **tube feet.**
9. Due to presence of water vascular system, the oral surface remains divided into **ambulacra** and **inter-ambulacra (inter-radii).**
10. **Coelom** is spacious.
11. **Digestive** system is well developed. The coiled alimentary canal extends from the **mouth** on oral surface to the **anus** on the aboral surface.
12. Blood vascular system (also called **haemal** or **blood lacunar system**) is present but poorly developed.
13. Respiration occurs through **body wall, podia, respiratory tree** or **gills** (or **papulae**).
14. Excretory organs are absent.
15. Nervous system is simple and consists of networks concentrated into **circum-oral ring** and **radial ganglionated nerve cords.**
16. Sense organs are poorly developed.
17. Sexes are usually separate; sexual dimorphism is not well marked. Fertilization is external.
18. Development is indeterminate and indirect including characteristic larvae which undergo metamorphosis into the radially symmetrical adults.

49.3 OUTLINE CLASSIFICATION

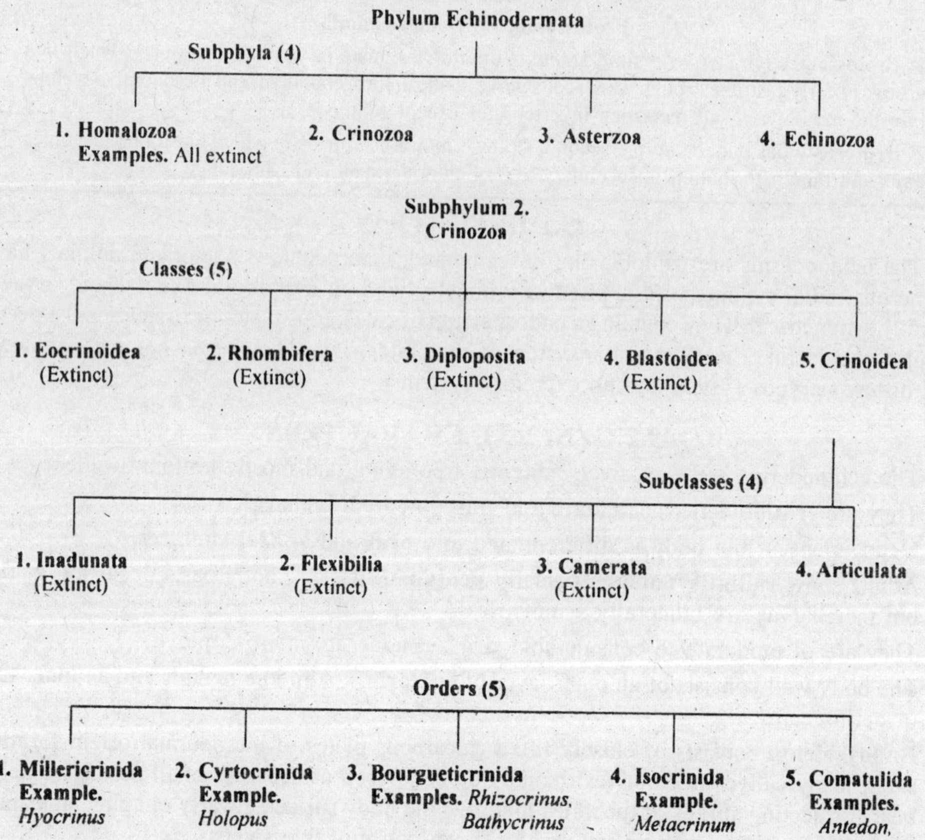

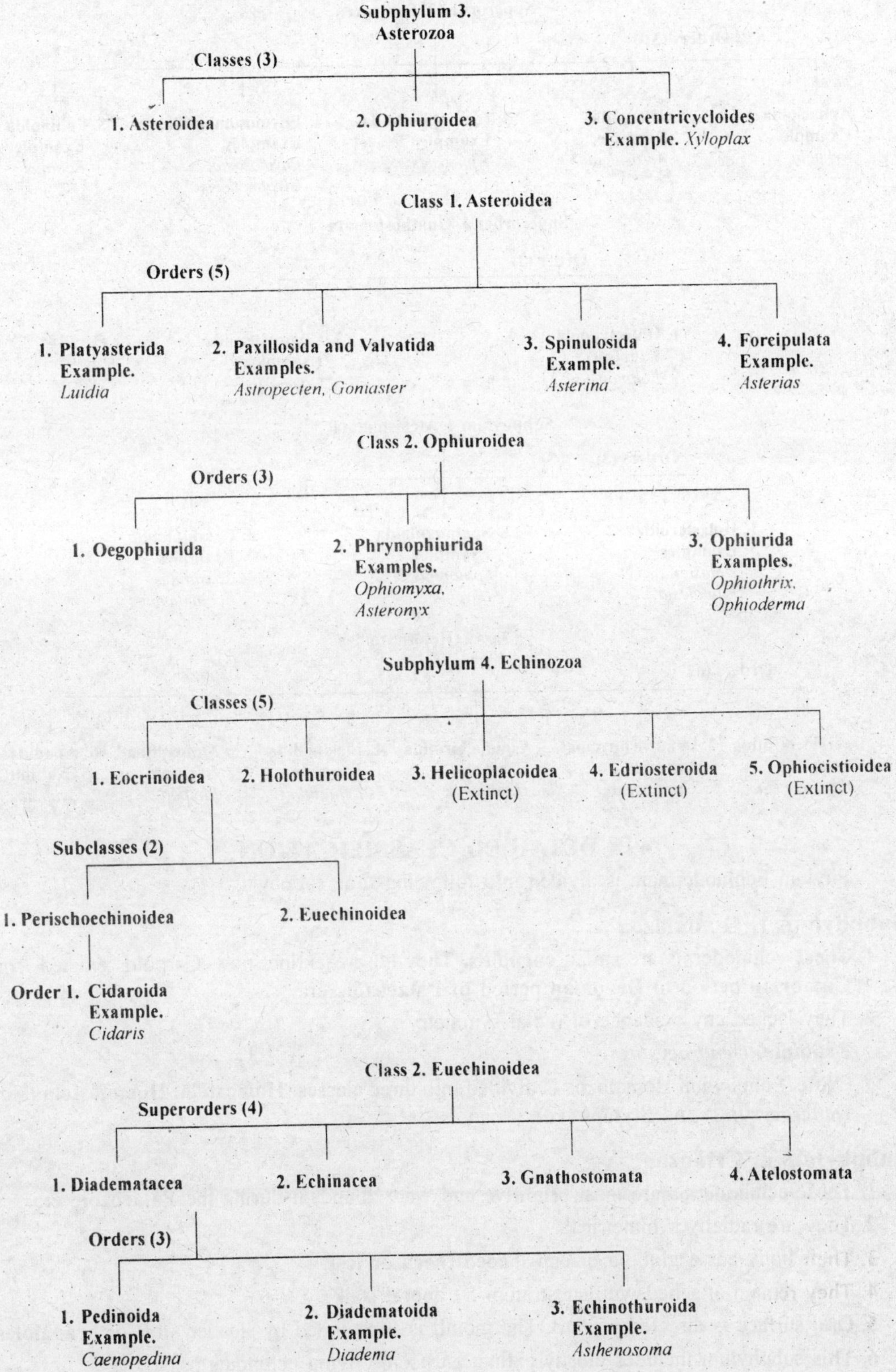
Subphylum 3.
Asterozoa
Classes (3)
1. Asteroidea
2. Ophiuroidea
3. Concentricycloides
Example. Xyloplax
Class 1. Asteroidea
Orders (5)
1. Platyasterida
Example.
Luidia
2. Paxillosida and Valvatida
Examples.
Astropecten, Goniaster
3. Spinulosida
Example.
Asterina
4. Forcipulata
Example.
Asterias
Class 2. Ophiuroidea
Orders (3)
1. Oegophiurida
2. Phrynophiurida
Examples.
Ophiomyxa,
Asteronyx
3. Ophiurida
Examples.
Ophiothrix,
Ophioderma
Subphylum 4. Echinozoa
Classes (5)
1. Eocrinoidea
2. Holothuroidea
3. Helicoplacoidea
(Extinct)
4. Edriosteroida
(Extinct)
5. Ophiocistioidea
(Extinct)
Subclasses (2)
1. Perischoechinoidea
2. Euechinoidea
Order 1. Cidaroida
Example.
Cidaris
Class 2. Euechinoidea
Superorders (4)
1. Diadematacea
2. Echinacea
3. Gnathostomata
4. Atelostomata
Orders (3)
1. Pedinoida
Example.
Caenopedina
2. Diadematoida
Example.
Diadema
3. Echinothuroida
Example.
Asthenosoma

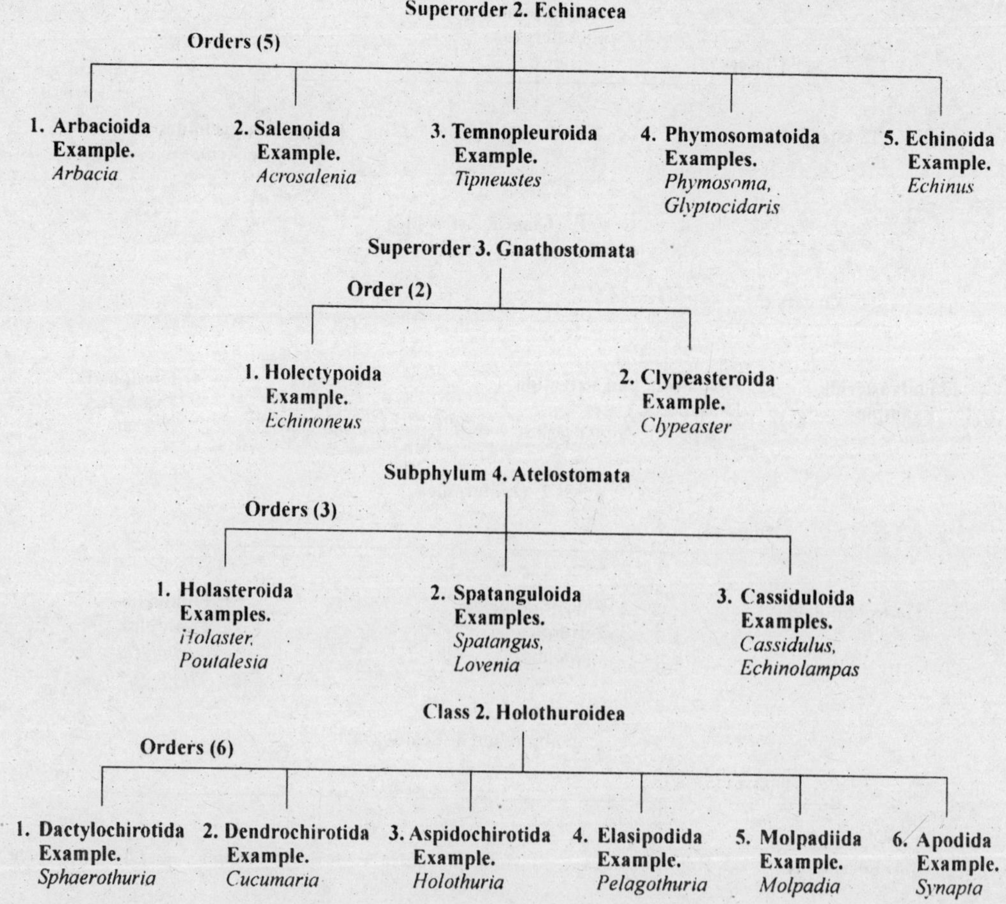

49.4 DETAILED CLASSIFICATION

Phylum Echinodermata is divided into following **four** subphyla.

Subphylum 1. Homalozoa

1. These echinoderms are called **carpoids.** They all are extinct now. Carpoids existed from **Cambrian** period to **Devonian** period of **Palaeozoic** era.
2. They lacked any evidence of radial symmetry.

   Example. *Dendrocystites.*

   (**Note.** Subphylum Homalozoa is divided into three classes: **Homostela, Homoiostela** (*e.g.*, *Dendrocystites*) and **Stylophora**).

Subphylum 2. Crinozoa

1. These echinoderms are most primitive and were abundant during the **Palaeozoic** era.
2. They are radially symmetrical.
3. Their body has a globoid or cup-shaped **theca** or **test.**
4. They remain attached with substratum by aboral stalk.
5. Oral surface is directed upward. The mouth is surrounded by slender arms or **brachioles.**
6. This subphylum includes mostly extinct and some living echinoderms.

Subphylum 3. Asterozoa

1. Unattached (free moving) radially symmetrical, star-shaped echinoderms.
2. Body is composed of a flattened central disc and radially arranged arms or rays.

Subphylum 4. Echinozoa

1. Radially symmetrical globoid or discoid echinoderms.
2. Arms or brachioles are absent.
3. Mostly unattached (*i.e.*, free moving).

SUBPHYLUM 2. CRINOZOA

Subphylum Crinozoa contains the fossil eocrinoids (class **Eocrinoidea**), cystoids (class **Cystoidea**) and the fossil and living crinoids (class **Crinoidea**). Only the classification of the class Crinoidea is covered here.

Class Crinoidea

1. It includes both extinct and living forms.
2. Living members are without stalk and free-moving but extinct forms were attached by a stalk.
3. They are commonly called **sea lilies** and **feather stars.**
4. Sea lilies are attached animals while feather stars are unattached.
5. Pentamerous body of crinoids consists of an aboral cup, the **calyx** and oral cover or roof, the **tegmen.** Oral side or surface of body is directed upward.
6. Oral surface contains a central **mouth** and eccentric **anus.**
7. They contain five movable arms. Each arm is divided into two branches at the base so that there appears to be ten arms.
8. Arms are with or without **pinnules.** Pinnate arms bear a row of pinnules on each side of the ambulacral groove.
9. They contain five ambulacral grooves which are open and extend along each of the five arm and pinnules up to tip.
10. Spines and pedicellariae are absent.
11. Water vascular system is peculiar. There are about 1500 minute ciliated funnels which perforate the membrane that covers the oral surface. **Stone canal** is present but **madreporite** is absent.
12. Sexes are separate. Gonads are indefinite masses of gametes located in the arms or pinnules. **Genital pinnules** generally lack an ambulacral groove. Males shed sperm first; this then stimulates the female to shed the eggs.
13. Larva of crinoids is non-feeding barrel-shaped **vitellaria** or **doliolaria.**

The class Crinoidea is divided into **four** subclasses.

Subclasses 1 to 3. Inadunata, Flexibilia and Camerata

1. These subclasses include extinct crinoids.
2. All were stalked Palaeozoic echinoderms.
3. They were with or without cirri.
4. Arms were branched, with or without pinnules.
5. Calyx was either (*i*) rigid, monocyclic or dicyclic and with lower arms free from it (Inadunata); (*ii*) flexible, dicyclic and with arms flexibly incorporated in it (Flexibilia); or (*iii*) rigid, mono— or dicyclic, lower arms incorporated into it (Camerata).

Examples. *Hybocystites* (Inadunata); *Forbesiocrinus* (Flexibilia), *Xenocrinus* (Camerata).

Subclass 4. Articulata

1. Extinct as well as all the living crinoids belong to this subclass.
2. Calyx flexible, pentamerous and dicyclic. Lower arms incorporated into it.
3. Tegmen leathery containing calcareous particles (ossicles) or small plates.
4. Mouth and ambulacral grooves exposed (*i.e.*, they are not covered with tegminal plates). Living members of subclass Articulata are included in the following **five** orders:

Order 1. Millericrinida

1. This order includes **sea lilies** which lack cirri.
2. They are attached crinoids. Stalk extends upto 1 metre in length.
   Examples. *Hyocrinus*, *Calamocrinus*.

Order 2. Cyrtocrinida

1. Includes **sea lilies** without stalk.
2. Aboral end of crown is attached directly to the substratum.
3. There are only two living species found in Caribbean and mid-Atlantic seas.
   Example. *Holopus*.

Order 3. Bourgueticrinida

1. Includes small **sea lilies.**
2. Stalk is slender and without cirri.
   Examples. *Rhizocrinus*, *Cenocrinus*, *Bathycrinus*.

Order 4. Isocrinida

1. Includes **sea lilies** with cirri.
2. Cirri are short, tendril-like and present aborally.
   Examples. *Metacrinum*, *Cenocrinum*.

Order 5. Comatulida

1. Includes **feather stars.**
2. Unattached stalkless crinoids.
   Examples. *Antedon*, *Florometra*, *Neometra*, *Capillaster*.

SUBPHYLUM 3. ASTEROZOA

Subphylum Asterozoa is divided into following **three** classes.

Class 1. Asteroidea

1. Body flattened, pentagonal or star-shaped.
2. They are commonly called **starfish** or **sea stars;** there are 1500 species of these.
3. Oral and aboral surface are distinct. The oral surface is directed downward and aboral surface upward.
4. Arms are wide and not sharply set-off from the central disc.
5. Ambulacral grooves are open and have podia or tube feet.
6. A large coelomic cavity is present in the selectively wide arms.
7. Endoskeleton is flexible, made of separate ossicles.
8. Development is indirect and includes **bipinnaria** or **brachiolaria larva.**

Class Asteroidea includes following **five** orders.

Order 1. Platyasterida

1. They are primitive and have tube feet without suckers.
2. Mostly extinct. Only two living genera.

Examples. *Platyasterias*, *Luidia*.

Order 2 and 3. Paxillosida and Valvatida (= Order Phanerozonia)

These two orders were formerly united within the order **Phanerozonia.**

1. Tube feet either with suckers (Paxillosida) or without suckers (Valvatida).
2. Both orders contain arms with two rows of marginal plates.
3. Pedicellariae of the sessile type (without stalk).
4. Podia or tube feet are arranged in two rows.

Examples. *Astropecten*, *Pentaceros*, *Culcita*, *Ctenodiscus*, *Goniaster* (cushion star).

Order 4. Spinulosida

1. Arms with small marginal plates.
2. Pedicellariae are absent.
3. Aboral surface is covered with short spines.
4. Tube feet are in two rows and contain suckers.

Examples. *Asterina*, *Echinaster*, *Henricia*, *Acanthaster*.

Order 5. Forcipulata

1. Marginal plates inconspicuous or absent.
2. Pedicellariae pedunculate type having a short stalk and three ossicles.
3. Tube feet are arranged in four rows and provided with suckers.

Examples. *Asterias*, *Heliaster*, *Zoroaster*, *Pycnopodia* (many-rayed star), *Leptasterias*.

Class 2. Ophiuroidea

1. This class includes **brittle stars,** of which there are about 2000 species.
2. They contain extremely long five or more arms which are sharply set off from the central axis.
3. Arms largely filled with vertebral ossicles.
4. They lack ambulacral grooves, intestine and anus.
5. Madreporite is present on oral surface.
6. Tube feet play no role in locomotion except in *Ophionereis annulata*.
7. Podia help in food collection and transport.
8. Development is indirect and includes a freely swimming pluteus larva.

Class Ophiuroidea includes following **three** orders.

Order 1. Oegophiurida

1. Mostly extinct; only one living species is present.
2. Dorsal and ventral shields or **bursae** are absent.
3. Madreporite exists at the edge of the central disc.

Example. An Indonesian species.

Order 2. Phrynophiurida

1. It includes **brittle stars** and **basket stars.**
2. Dorsal arm shields are absent.

3. In primitive brittle stars (suborder **Ophiomyxina**), disc and arm plates are covered by a thick and soft skin.
4. In basket stars, arms are simple, branched and capable of coiling vertically (suborder **Euryalina**).

Examples. Brittle stars such as *Ophiocanops, Ophiomyxa*; and **Basket stars** such as *Asteronyx, Gorgonocephalus.*

Order 3. Ophiurida

1. It includes **brittle stars** and **serpent stars.**
2. Mostly small ophiuroids, usually with five simple arms.
3. Arms are capable of transverse movement only.
4. Dorsal arm shields are present.

Examples. *Ophioderma, Ophiothrix, Ophioscotex, Ophiomaza, Ophionereis.*

Class 3. Concentricycloides

1. Minute deep-water echinoderms (which were collected in 1983 and 1984 from coast of New Zealand and described in 1986 by **Baker** *et al.*).
2. Body is disc-shaped and covered aborally with plate-like ossicles.
3. Gut is present or absent.
4. Water vascular system have two ring canal with podia arising from the outer one.

Example. *Xyloplax* (Fig. 49.1).

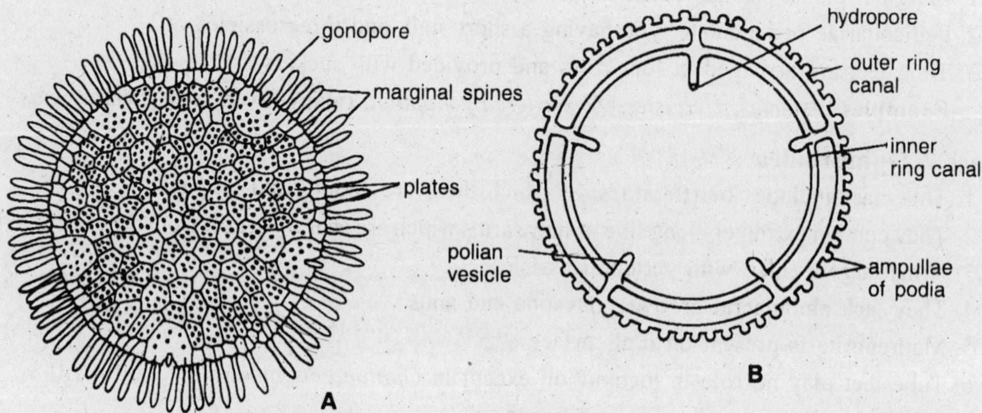

Fig. 49.1. *Xyloplax* (Class Concentricycloides). A—Dorsal view; B—Water vascular system.

SUBPHYLUM 4. ECHINOZOA

Subphylum Echinozoa is divided into **five** classes; three of which namely class **Helicoplacoidea** (*e.g., Helicoplacus*), Class **Edrioasteroidea** (*e.g., Lepidodiscus*) and class **Ophiocistoidea** (*e.g., Volchovia*) include all extinct Echinozoa. The rest two classes namely **Echinoidea** and **Holothuroidea** include living as well as extinct members.

Class 1. Echinoidea

1. It includes about 950 species of **sea urchins, heart urchins** and **sand dollars.**
2. The name Echinoidea means "like a hedgehog". It is called so because it contains movable spines (like the hedgehog) that covers the body.

3. Body is spherical, disc-shaped, oval or heart-shaped and lacks arms.
4. Body is enclosed in an endoskeletal shell or test of closely fitted calcareous plates covered with movable spines.
5. Podia or tube feet come out from the pores of ambulacral plates and are locomotory in function.
6. **Mouth** is centrally placed on oral surface. **Anus** is located at the aboral pole.
7. Mouth is generally provided with an elaborate jaw apparatus (Aristotle's lantern; Box 49.2).

Box 49.2

Aristotle's Lantern

Aristotle's lantern is a unique and highly developed scraping apparatus of sea urchins. It is associated with the digestive system (*i.e.,* mouth). It has been named so after its discoverer and its resemblance to the ancient Greek ship-lantern. The majority of sea urchins are grazers, scraping the substratum surface on which they live with their teeth which are operated by Aristotle's lantern.

Lantern is pentamerous inverted pyra-mid-like structure lying close to the peristome (Fig. 49.2). It is composed of five large, calcareous plates called **alveoli** or **pyramids,** each of which is shaped somewhat like an arrowhead with the point projecting toward mouth. The pyramids are arranged radially, with each side connected to that of adjacent pyramid by means of transverse muscle fibres (called **comminator muscles**). Passing down the midline along the inner side of each pyramid is a long, calcareous band. The oral end of the band projects beyond the tip of the pyramid as an extremely hard, pointed **tooth.** Because there is one such tooth band for each pyramid, there are five teeth projecting from the oral end of the lantern. The curled upper end of the band is enclosed within a dental sac and is the area of new tooth formation.

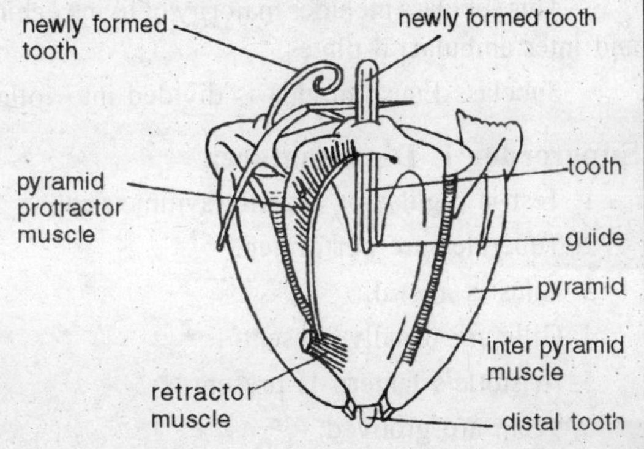

Fig. 49.2. Aristotle's lantern of echinus in lateral view.

In addition to the teeth and pyramids, Aristotle's lantern is composed of a number of smaller, rod-like pieces at the aboral end. The upper end of each pyramid bears a bow-shaped bar, the **epiphysis.** It is formed by the fusion of two ossicles. Five slender bars of radii, called **compasses**, extend radially inwards from the point of union of adjacent epiphysis. Five more pieces, the **rotules,** are also present beneath the compasses. By mean of special muscles (*e.g.,* protractor muscles and retractor muscles) the lantern can be partially protruded and retracted through the mouth. Other muscles control the teeth, which can be opened and closed. Aristotle's lantern is used in scraping and also in pulling and tearing a wide range of plant and animal materials.

8. They contain peripharyngeal coelom.
9. Ambulacral grooves are absent.
10 Pedicellariae (pinching organs) are stalked and three jawed.
11. They contain five pairs of gills for respiration.
12. Sexes are separate. Gonads are pentamerous.
13. Development is indirect and includes free swimming **echinopluteus larva.**

Class Echinoidea is divided into **two** subclasses.

Subclass 1. Perischoechinoidea

Subclass Perischoechinoidea includes mainly primitive fossil urchins of the Palaeozoic seas. It is divided into four orders. Members of three orders are completely extinct now. **Cidaroida** is the only one (order) that survived the Palaeozoic era and became the ancestor of the remaining echinoids, most of which belong to the next subclass.

Order Cidaroida

1. Test is rigid having two rows of narrow and simple plates for each ambulacrum and interambulacrum.
2. Spines are widely separated and each is encircled by small secondary spines.
3. Gills and balancing organs are absent.
4. Tridentate and globiferous pedicellariae are present.

Examples. *Eucidaris*, *Cidaris*, *Notocidaris*.

Subclass 2. Euechinoidea

This subclass includes majority of living echinods. Its members have two columns of ambulacral and inter ambulacral plates.

Subclass Euechinoidea is divided into following **four** superorders.

Superorder 1. Diadematacea

1. Test is regular or radially symmetrical.
2. Tubercles are perforated.
3. Anus is aboral.
4. Gills are usually present.
5. Aristotle's lantern is present.
6. Teeth are grooved.

This superorder is divided into following **three** orders.

Order 1. Pedinoida

1. Test is rigid and with spines.
2. Ambulacral plates ending at peristome margin.
3. Gill notches are slight.
4. Ten buccal plates exist on the peristomial membrane.

Examples. *Caenopedina* (only living genus of order Pedinoida).

Order 2. Diadematoida

1. Rigid or flexible test is globose having two rows of intrambulacral plates and two rows of compound ambulacral plates, both ending at peristome margin.
2. Spines are hollow and bear fine teeth.
3. All types of pedicellariae such as dentate, foliate and ophiocephalous (except globiferous pedicellariae) are present.
4. Ten buccal plates are present on peristomial membrane.

Examples. *Diadema*, *Plesiodiadema*.

Order 3. Echinothuroida

1. Flexible test of separated plates; ambulacral plates with peristome extending to mouth rim.
2. Simple plates are present on peristomial membrane.
3. Poisonous secondary spines are present.
4. Gills are absent.

Examples. *Asthenosoma*, *Echinothuria*.

Superorder 2. Echinacea

1. It includes sea urchins having radial symmetry, rigid test and solid spines.
2. Gills are present.
3. Peristomimal membrane with ten buccal plates.
4. Aristotle's lantern is well developed.

This superorder is divided into following **five** orders.

Order 1. Arbacioidia

1. Preriproct is covered by four or five anal plates.
2. Primary tubercles are without pores.
3. Compound ambulacral plates are composed of a median primary plate and a short plate on either side.

Example. *Arbacia.*

Order 2. Salenoida

1. Stirodont with epiphyses separate and teeth crested on inner surface.
2. Interambulacral plates with single spine.
3. Anus is not centrally located. It is situated within periproct due to the presence of a large plate (called **suranal plate**).

Example. *Acrosalenia.*

Order 3. Temnopleuroida

1. Lantern is camarodont (*i.e.*, large epiphyses are joined across pyramid top).
2. Test is sculptured in some.
3. Ambulacral plates are compound (*i.e.*, each plate is composed of two primary plates and one narrow plate between their narrow ends.)

Examples. *Tripneustes*, *Toxopneustes*, *Temnopleurus*.

Order 4. Phymosomatoida

1. Compound ambulacral plates are three; interambulacral plates with more than one tubercle.
2. Primary tubercles are without pores.
3. Lantern is camarodent.

Examples. *Phymosoma*, *Glyptocidaris*.

Order 5. Echinoida

1. Test is not sculptured.
2. Lantern is camarodent.
3. Compound ambulacral plates are composed of two primary plates and one narrow plate between their outer ends.
4. Three to sixteen pores per plate occur.
5. Gill notches are shallow.

Examples. *Echinus*, *Echinometra*, *Paracentrotus*, *Strongylocentrotus*.

Superorder 3. Gnathostomata

1. It includes irregular (bilaterally symmetrical) urchins.
2. Lantern is present.

3. Ambulacral plates are simple or weakly compound.
4. Teeth are keeled.
5. Mouth is in the centre of oral surface but anus has shifted out of centre.

This superorder is divided into following **two** orders.

Order 1. Holectypoida

1. Test is regular with apical plates centred at the aboral pole.
2. Peristome central.
3. Anus is somewhat separate from apical plates.
4. Ambulacra are not petaloid; they are narrow on oral surface.
5. Lantern teeth with lateral flanges.
6. Mostly extinct.

Examples. *Echinoneus* and *Micropetalon* are only living members.

Order 2. Clypeasteroida

1. It includes **cake urchins** or **sand dollars.**
2. Test is greatly flattened and oval or circular. Test is usually densely covered by small spines.
3. Peristome and apical system are usually central.
4. Petaloid ambulacra, *i.e.*, ambulacral areas are shaped like a petal radiating from the centre, are present.
5. Teeth are without lateral flanges.
6. Phyllodes (= oral ambulacral areas) and gills are absent. Respiration is performed by podia of petaloids.
7. Some sand dollars (*e.g.*, keyhole sand dollars or *Mellita*) contain large elongated notches or openings in testa, called **lunules.**

Examples. *Clypeaster* (sea biscuits), *Fibularia*, *Rotula*, *Mellita.*

Superorder 4. Atelostomata

1. It includes regular (bilaterally symmetrical) urchins.
2. Lantern is absent.
3. Ambulacral plates are simple.
4. Test is rigid.

This suborder is divided into following **three** orders.

Order 1. Holasteroida

1. Oval or bottle-shaped urchins with thin delicate test.
2. Petaloids and phyllodes are not developed.
3. Ambulacra of aboral surface are petal-shaped; petals are not similar.

Examples. *Holaster*, *Pourtalesia* (a deep-water species).

Order 2. Spatanguloida

1. It includes **heart urchins.**
2. Test is oval or cordiform, bilaterally symmetrical about long axis. It contains short spines.
3. Ambulacra of aboral surface is petal-shaped; petals are not similar. Petaloids are shunken.
4. Oral centre is shifted anteriorly and anus is shifted out of aboral apical centre.

5. Phylodes are present.
6. Gills are absent.

Examples. *Spatangus, Echinocardium, Moira, Meoma, Lovenia.*

Order 3. Cassiduloida

1. Somewhat similar to heart urchins.
2. Body is round or oval.
3. Ambulacra of aboral surface are petal-shaped; all petals are similar.

Examples. *Cassidulus, Echinolampas.*

Class 2. Holothuroidea

1. It includes 900 species of **sea cucumbers.**
2. Warty body is elongated along the oral-aboral axis. As a result animal lies with the side of body on the substratum.
3. Endoskeleton is very much reduced. It is represented by microscopic **ossicles.**
4. Oral or buccal podia form a circle of **tentacles** (10 to 30) around the mouth (See Box 49.3).

> **Box 49.3**
> **Tentacles in Holothuroidea**
>
> Oral tentacles of Holothuroidea may be of following four types:
>
> 1. **Digitate** or finger-like, *e.g., Molpadia, Deima.*
> 2. **Dendritic** or arborally branched, *e.g., Thyone.*
> 3. **Pinnate** or with side branches on a central axis, *e.g., Leptosynapta.*
> 4. **Peltate** or flattened, with a short stalk and terminal disc, often with horizontal branches, *e.g., Actinopyga, Peniagone.*

5. Mouth and anus are situated at the opposite ends. Arms are absent. Cloaca contains respiratory tree.
6. Ambulacra are reduced.
7. No ossicle form the jaw apparatus.
8. Tiedman's bodies are absent.
9. Some move on the bottom; others live beneath stones or in crevices.
10. They are deposit–and suspension feeders.
11. Madreporite is located in the coelom.
12. Development is indirect including **doliolaria** larva.

Class Holothuroidea is divided into following **six** orders.

Order 1. Dactylochirotida

1. It includes primitive sea cucumbers.
2. Eight to thirty digitiform (finger-like) tentacles are present.
3. Body is U-shaped and enclosed within a flexible test of imbricating plates.

Examples. *Sphaeothuria, Echinocucumis.*

Order 2. Dendrochirotida

1. Buccal podia or tentacles are 10 to 30 in number. They are branched like a tree (dendritic) and not provided with ampullae.
2. Podia (tube feet) occur on the sole, on all the ambulacra, or over the entire surface.

Examples. *Cucumaria, Thyone, Psolus.*

Order 3. Aspidochirotida

1. Oral tentacles are peltate or shield-like.
2. Many podia are present. They sometimes form a sole.
3. Respiratory trees are present.

Examples. *Holothuria*, *Actinopyga*, *Stichopus*.

Order 4. Elasipodida

1. It includes deep-sea cucumbers which contain large conical papillae and podia.
2. Tentacles are peltate.
3. They lack oral retractor muscles and respiratory trees.

Examples. *Pelagothuria*, *Peniagone*.

Order 5. Molpadiida

1. Posterior end of body tapers to form a tail.
2. Fifteen digitiform tentacles are present.
3. Podia are absent or reduced to anal papillae.
4. Respiratory tree is absent.

Examples. *Molpadia*, *Caudina*.

Order 6. Apodida

1. Worm-like sea cucumbers.
2. They contain only buccal podia or tentacles. Tentacles are digitiform or pinnate.
3. Water vascular system is reduced, lacking tube feet or radial canals.
4. Respiratory tree is absent.

Examples. *Synapta*, *Leptosynapta*, *Euapta*.

49.5 TYPE STUDIES

1. *ANTEDON*

Classification

Phylum	—	**Echinodermata**
Subphylum	—	**Crinozoa**
Class	—	**Crinoidea**
Subclass	—	**Articulata**
Order	—	**Comatulida**
Genus	—	***Antedon***

Habit and Habitat

Antedon is commonly called **feather star.** It is marine and free living crinoid that lives in littoral zones of the sea. Its seven species are found at moderate depths in all seas such as seas of Western Europe, Western Africa, Mediterranean, and West of Tropical America.

Antedon is a gregarious form and clings to the sea bottom for long periods by its cirri. However, it can creep slowly through its cirri and swim by undulating movements (*i.e.*, bending and straightening) of the arms. *Antedon* swims and crawls only for short distances. Swimming is mainly an escape response of *Antedon*.

Antedon is **nocturnal.** It hides in the crevices in coral reefs during day and keeps its arms tightly rolled. At sunset, it crawls upwards and extends it arms to feed. It is a **suspension feeder,** feeding on microscopic living organisms such as diatoms, algae, protozoa and small plankton (see Box 49.4). *Antedon bifida* feeds on the detritus.

Box 49.4
Mode of Feeding in *Antedon*

A hungry feather star (*Antedon*) clings to a rock with its arms spread wide and its pinnules at attention, forming an interlocking net to filter out small food particles carried in passing food currents.

The epidermis is thin over most of body surface but its cells are tall and ciliated in the ambulacral grooves. Food is trapped in mucus and the mucous strings move along the ambulacral groove like conveyer belts (Fig. 49.3). Interambulacral cilia on the tegmen surface cause cleansing currents that sweep towards the grooves, turn and run peripherally just outside the grooves.

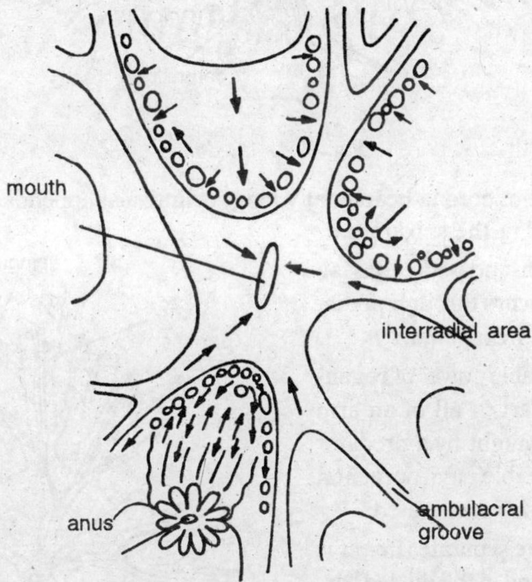

Fig. 49.3. *Antedon.* Ciliary currents on oral surface. Note the cleansing currents and food-collecting currents in food grooves.

Structure

Antedon shows a pentamerous **radial symmetry.** It consists of a **central disc** and a series of five radiating **arms.** The main part of body is called **corona** which is differentiated into an upper **oral surface** and lower **aboral surface.** It contains a cup-shaped aboral **calyx** which is formed of loose-fitting skeletal ossicles. On the oral side of disc occurs a membrane containing small skeletal ossicles, called **tegmen.** Tegmen bears the central **mouth** and the **anus** on a papilla (called **anal cone**) in an inter radius. Five long and slender flexible **arms** arise at the junction of calyx and tegmen. Each arm immediately bifurcate to form in total ten arms. Diminutive branches called **pinnules** are attached to the arm margins to give them a feathery appearance.

The skeleton of arms and pinnules is jointed; **brachial ossicles** support the arms, and **pinnulars** support the pinnules. Five ciliated **ambulacral grooves** radiate from the mouth towards the arms (Fig. 49.4) where each divides into two and the branches extend along the oral surface of the arms and pinnules. The margins of the grooves are bordered by movable flaps, called **lappets** which can expose or cover the groove. Two rows of tentacle-like **tube-feet** or **podia** lacking suckers (= ampulla) are present along the margins of ambulacral grooves. Podia are used to secrete mucus and to push food into the ambulacral groove.

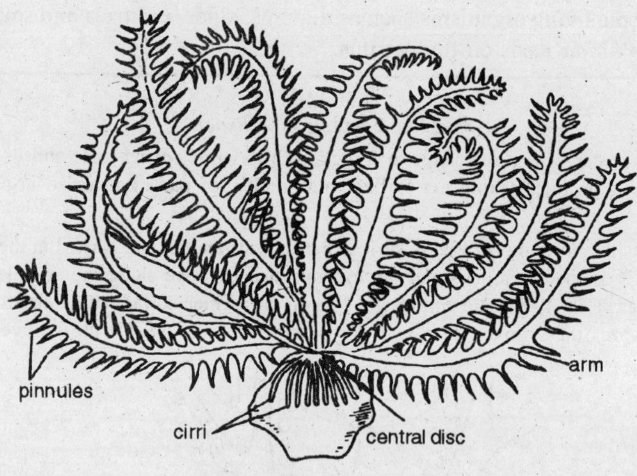

Fig. 49.4. *Antedon.*

The **aboral surface** of corona bears many long, cylindrical appendages or the cirri. Cirri are used by animal for gripping the substratum.

Spines, madreporite and **pedicellariae** are absent. Respiration occurs through podia. Ring canal of *Antedon* is hexagonal.

Antedon has remarkable power of **regeneration** and **autotomy.** Part or all of an arm can be shed if animal is caught by a predator or if subjected to unfavourable environmental conditions. The lost arm is regenerated.

In *Antedon*, sexes are separate (*i.e.*, it is **dioecious**). Gonads are found in the dilated bases of pinnules. Such **genital pinnules** lack an ambulacral groove. **Fertilization** is external and development is **indirect.** In *Antedon*, the eggs are cemented to the outer surface of the pinnules by the secretion of epidermal **cement gland** cells. The larva of *Antedon* is a non-feeding barrel-shaped **vitellaria** larva which develops into a stalked and sessile **pentacrinoid.**

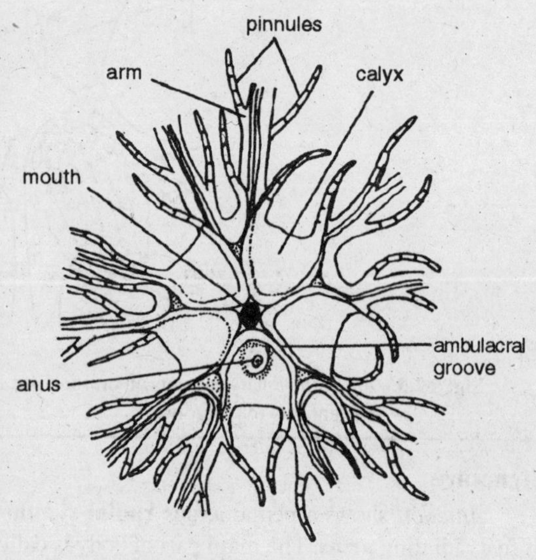

Fig. 49.5. *Antedon.* Oral view of central disc.

2. *ECHINUS* (Sea Urchin)

Classification

Phylum	—	**Echinodermata**
Subphylum	—	**Echinozoa**
Class	—	**Echinoidea**
Subclass	—	**Euechinoidea**
Superorder	—	**Echinacea**
Order	—	**Echinoda**
Genus	—	***Echinus***

Habit and Habitat

Echinus is commonly called **sea urchin** (other common sea urchins are *Arbacia, Strongylocentrotus*, etc.) It is marine and found in the sea in the rocky places. *Echinus* occurs from intertidal zone to a depth of 5000 metres. It is widely distributed in the Atlantic, Mediterranean and Pacific Oceans. In India, sea urchin (*Salmacis bicolor*) is found along Madras Coast.

Echinus moves slowly by means of highly movable spines and tube feet. It sometimes excavates holes in the rocks for hiding. It cleans its body by movement of spines and pedicellariae. *Echinus* feeds on sea weeds, algae, dead animal matter and small organisms. It itself is food for starfishes, crabs, fishes, birds, mammals (including human beings).

Structure

Body of *Echinus* is globular, somewhat flattened at oral pole. It exhibits **radial symmetry.** Body is divided into an **aboral** and an **oral hemisphere.** The oral end bears a centrally located **mouth** and is directed against the substratum. Through the mouth project five teeth of Aristotle's lantern (see Box 49.5). Mouth is surrounded by a soft membrane known as **peristome.** Peristomal membrane bears five pairs of short, heavy, modified **tube feet** or **podia,** called **oral tentacles.** It also bears five pairs of bushy **gills** or **branchiae. Anus** lies in the centre of **aboral surface.** The membrane surrounding the anus is called **periproct.** It is a small, circular membrane having a number of plates.

Box 49.5

Aristotle's lantern of *Echinus*

Sea urchins feed with a highly developed scraping apparatus called Aristotle's lantern (Fig. 49.6B). It consists of (*i*) five arrow head-like calcareous plates called **pyramids;** (*ii*) five radial **teeth;** (*iii*) five **epiphyses;** (*iv*) five **rotulae;** and (*v*) five **compasses.** These 25 plates surround the oesophagus and enclose a **lantern coelom.** Various plates are moved by intricate system of muscles. Muscles such as **retractor muscle** and **protractor muscle** also extend between the lantern and five arches arising from the corona, called **auricles.**

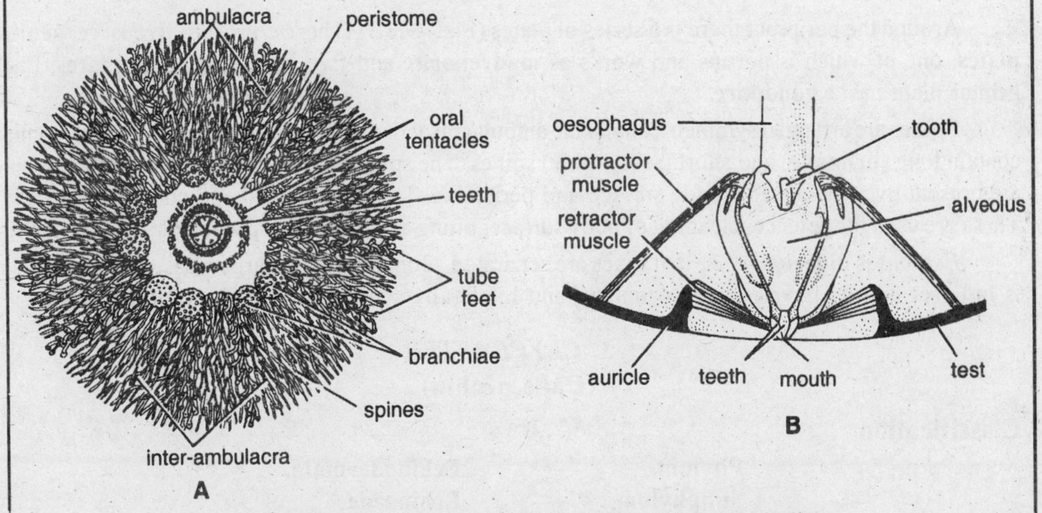

Fig. 49.6. *Echinus*. A—Oral view; B—Aristotle's lantern in situ.

The body of *Echinus* is enclosed in a rigid globular shell or **corona.** Corona is formed of closely fitting endoskeletal calcareous plates bearing round tubercles. The surface of corona is divided into five **ambulacral regions** alternating with five **interambulacral regions.** In each ambulacral region, there are two rows of tube feet.

In corona, **skeletal plates** or **ossicles** are arranged in rows. These rows run from oral pole to aboral pole. Each ambulacral area is composed of two rows of **ambulacral plates** and each interambulacral area is composed of two rows of **interambulacral plates.** There are thus 20 rows of plates. The edges of plates fit accurately and firmly together and form a rigid **shell, test** or **corona** around the body. The surface of plates contain **tubercles** for the articulation of movable **spines.** Each ambulacral plate bears pores through which tube feet are extended.

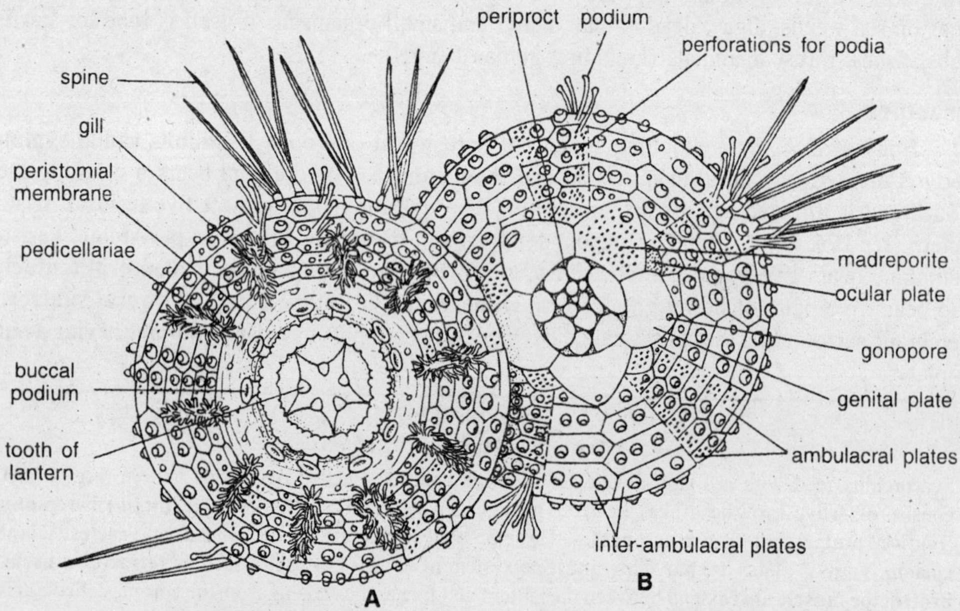

Fig. 49.7. Sea urchin (*Arbacia*). A—Oral view; B—Aboral view.

Around the periproct there is a series of plates (Fig. 49.7B). These consist of five large, **genital plates,** one of which is porous and works as **madreporite** and five smaller **occular plates.** Each genital plate has a **gonopore.**

Spines are arranged symmetrically in the ambulacral and interambulacral areas. Most sea urchins contain long (**primary**) and short (**secondary**) spines. The spines can be regenerated. **Pedicellariae** are present over the general body surface and peristome. They are stalked and contain three jaws. These are used for defence, cleaning of body surface, biting and breaking up small particles of debris.

Echinus is **dioecious,** *i.e.*, in it sexes are separated. Gonads are five large masses. Development is **indirect** and its larva is free-swimming and bilaterally symmetrical **echinopluteus larva.**

3. *CLYPEASTER*
(Cake urchin)

Classification

Phylum	—	**Echinodermata**
Subphylum	—	**Echinozoa**
Class	—	**Echinoidea**
Subclass	—	**Euechinoidea**
Superorder	—	**Gnathostomata**
Order	—	**Cypeasteroida**
Genus	—	***Clypeaster***

Habit and Habitat

Clypeaster is commonly known as **cake urchin.** It is a marine form, found in tropical and subtropical waters—Red sea and North Carolina to Brazil. *Clypeaster* is found buried in sand at the bottom in the sea. The sea biscuit, *Clypeaster rosaceus* of Florida and the West Indies, does not burrow but sits on the sea bottom. It is a selective **deposit feeder** feeding on organic debris.

Structure

Body of *Clypeaster* is flattened and oval or disc-shaped. Test is densely covered with small dark brown **spines** which help in locomotion. Body is differentiated into **oral** and **aboral surfaces.**

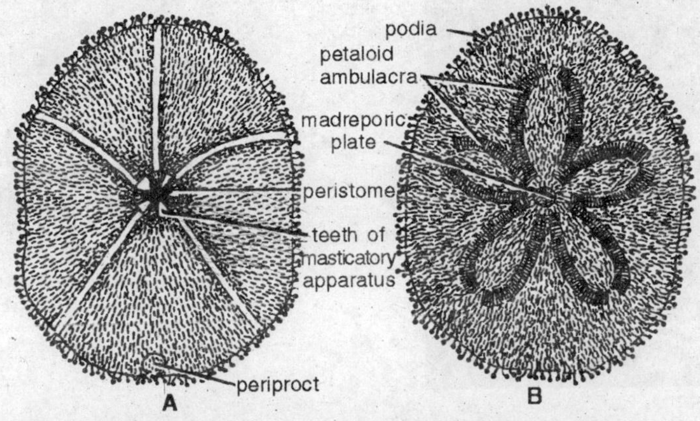

Fig. 49.8. *Clypeaster.* A—Oral view; B—Aboral view.

Mouth and anus both occur on oral surface. Mouth is central in position and anus is eccentric. Tube feet or podia are found on both oral and aboral surface.

The madreporite is central on aboral surface. From madreporite arise five wide and well marked **petaloid ambulacral areas** separated by **interambulacral areas.** In each ambulacral region, there are two rows of **tube-feet.** The podia of petaloids are modified for gas exchange.

The **oral ambulacral areas** or **phyllodes** contain modified podia for obtaining food particles. **Pedicellariae** and **sphaeridia** (ovoid sensory bodies) are present. Aristotle's lantern is present. Peristomial gills are absent.

Clypeaster is **dioeceous.** Development is indirect and includes **echinopleuteus larva.**

4. *OPHIODERMA*
(Brittle star)

Classification

Phylum	—	**Echinodermata**
Subphylum	—	**Asterozoa**
Class	—	**Stelleroidea (Ophiuroidea)**
Order	—	**Ophiurida**
Genus	—	**1. *Ophioderma;***
		2. *Ophiothrix.*

I. Ophioderma

Ophioderma is commonly called **brittle star.** It is marine and found in shallow water and at a depth of 280 metres. *Ophioderma* is littoral, tropical and mainly found in West Indies. It is free swimming, nocturnal and carnivorous echinoderm. It contains great power of **autotomy** and

regeneration, *i.e.*, it can shed one or more arms when it is disturbed or caught by a predator. The lost arm is regenerated.

Structure

Body of *Ophioderma* is olive green in colour. It comprises a **pentagonal disc** and five slender, solid, cylindrical, muscular and flexible **arms** bearing short spines. The pentagonal disc measures 10 to 20 mm in diameter and arms are 40 to 60 mm in length.

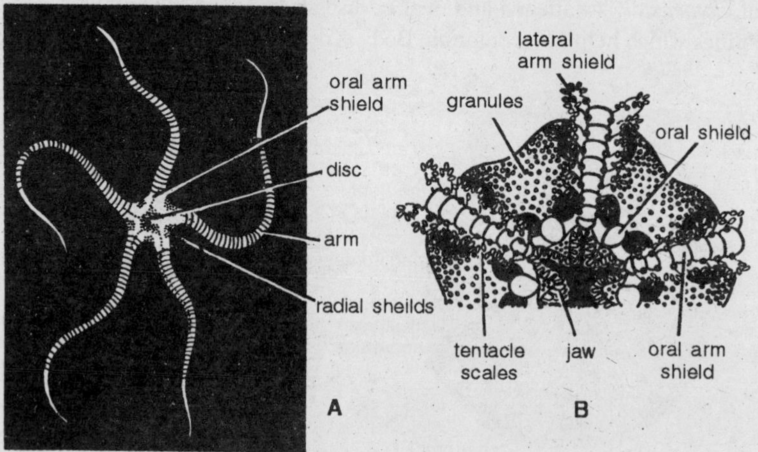

Fig. 49.9. *Ophioderma*. A—Entire brittle star in aboral view;; B—Disc magnified (oral view).

Oral surface bears pentagonal **mouth** having several papillae and is covered by numerous small closely set granules. Teeth are arranged in a single row. Four **genital bursae** occur in each inter-radius. Ambulacral grooves are covered by ossicles.

Aboral surface bears five pairs of **radial shields** and a **madreporite** but lacks **anus.** Tube feet are present on the ventrolateral side of arms in two rows.

The sexes are separate.

II. Ophiothrix

Habit and Habitat

Ophiothrix is commonly called **spiny brittle star.** It is marine and co-mmonly found along Atlantic Coast. It lives in shallow to deep waters, hid-ing beneath stones and weeds during day time. It is most active echinoderm. It swims by jerking the arms quickly. It creeps and swims by snake-like movements of the arm (*i.e.*, due to this reason *Ophiothrix* is also called **serpent star**). *Ophiothrix* quickly breaks off an arm when seized or inju-red. Lost arm is soon regenerated. It feeds on bottom debris containing microscopic life and on small crustac-eans and molluses. It is eaten by fishes.

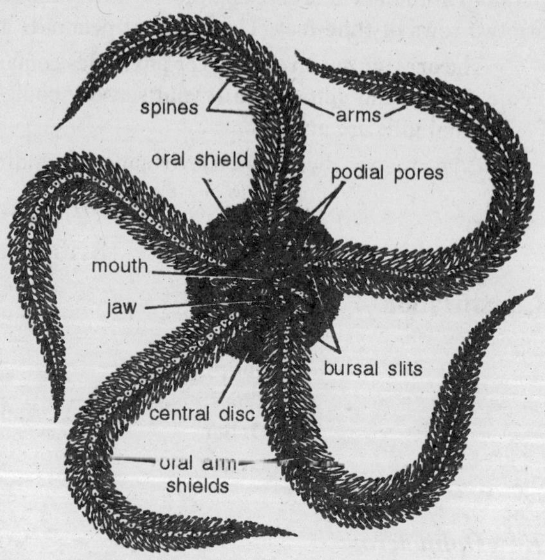

Fig. 49.10. *Ophiothrix*. Oral view.

Structure

Ophiothrix has a small rounded central **disc** and five slender **arms** which arise from the lower surface of the disc. Central disc on its **oral side** bears a central, large star-shaped **mouth**. In one of the inter-radii is located the **madreporite** which is a modified oral shield/plate. Near the base of each arm are paired ciliated grooves called **genital bursae** serving as gonopore and also helping in respiration and excretion. In the inter-radial positions, there are five **buccal plates** or **oral shields**. Internal and lateral to them are ten **angular shields**, bearing spines projecting into the mouth. Aboral surface of the disc has a series of **radial shields** around the bases of arms and a large number of tiny calcareous plates covered by the vestigial epidermis and cuticle.

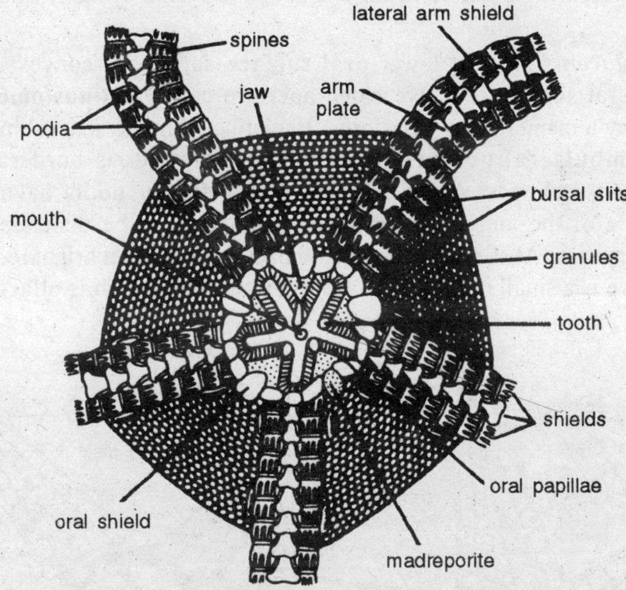

Fig. 49.11. *Ophiothrix*. Aboral view of central disc.

Each arm appears jointed since it is bounded by four longitudinal rows of calcareous plates/shields. There are two rows of **lateral shields**, one row of **aboral shields** and one row of **oral shields** (oral shields are also called **vertebrae**). Lateral shields bear 2 to 15 large spines in a vertical row. There is no **ambulacral groove** on the oral surface of arm.

The **tube feet** or **podia** are small, lack suckers and arise from tiny pores left in the sutures between ventral (aboral) and lateral plates. The pair of tube feet in each arm is directed towards the mouth and serves to taste and push the food into the mouth. There is a sensory tube foot at the tip of each arm. Pedicellariae are absent.

Ophiothrix is **dioecious**. Gonads are small. Fertilization is external. Development is **indirect**. The larva is called **ophiopluteus.**

5. *PENTACEROS*

Classification

Phylum	—	**Echinodermata**
Subphylum	—	**Asterozoa**
Class	—	**Stelleroidea (Asteroidea)**
Order	—	**Phanerozonia (Valvatida)**
Genus	—	***Pentaceros***

Habit and Habitat

Pentaceros is commonly called **sea pentagon**. It is now included in genus *Oreaster.* It is marine and found commonly in the Indo–Pacific Ocean and around the West Indies, in the Bay of Bengal and Arabian Sea. *Pentaceros* feeds on oysters, so is very harmful to the pearl industry.

Structure

The body of *Pentaceros* is very thick, regular, star-shaped and has a diameter of 25 cm. It has a large **central disc** and five short tapering **arms**. Arms are short with broad base, called **stellate arms**. They are not sharply demarcated from the central disc. The axes of arms are called **radii** and the spaces between them are **inter-radii**. The body is covered by many calcareous endoskeletal plates or **ossicles**.

Body of *Pentaceros* has flat lower **oral surface** and upper convex **aboral surface**. In the centre of the oral surface is a five sided aperture called **actinostome** which encloses a mouth surrounded by a membranous **peristome**. Radiating from the actinosome into the five arms are five narrow **ambulacral grooves**. Each of these grooves is bordered by two rows of ambulacral **spines** on each side. A row of slender **tube feet** or **podia** having sucker-like ends occurs on each side of the ambulacral groove. Tube feet can be distended and fixed to the substratum for locomotion. At the end of each ambulacral groove is a bright red **eye** made of many **ocelli**. Above the eye is a small non-retractile **sensory tentacle** which is **olfactory** and a modified tube foot.

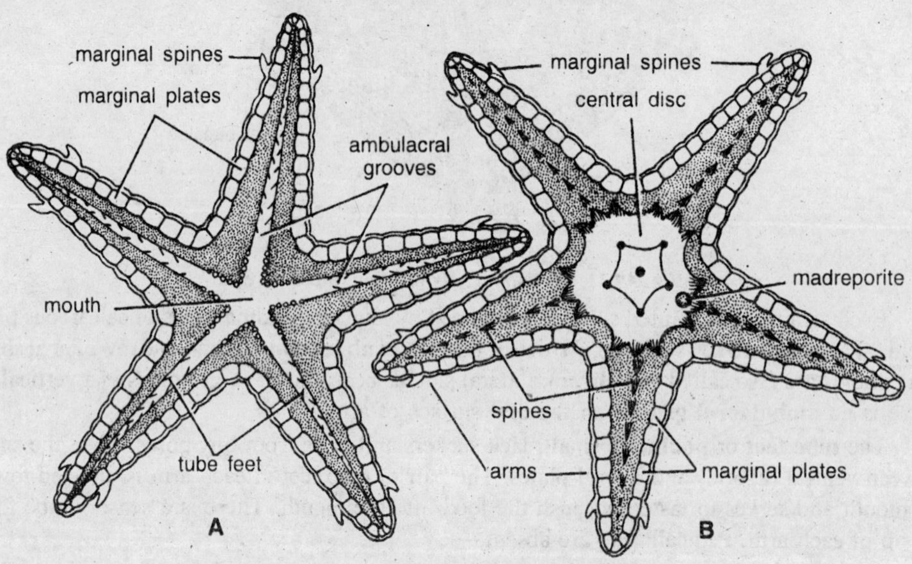

Fig. 49.12: *Pentaceros.* A—Oral view; B—Aboral view.

Aboral surface of *Pentaceros* is yellow reddish brown in colour and has many stout spines in irregular rows; all spines rest on ossicles. Between the spines, **papullae** (dermal branchiae or gills) project singly. Papullae are retractile and coelom-containing thin-walled extension of body wall meant for respiration and excretion. In the centre of aboral surface of disc is located a minute **anus**. Near the anus between two arms is a flat light coloured plate called **madreporite**. Microscopic stalked or sessile **pedicellariae** are located on both oral and aboral sides around the spines.

6. *HOLOTHURIA*
(Sea cucumber)

Classification

Phylum	—	**Echinodermata**
Subphylum	—	**Echinozoa**
Class	—	**Holothuroidea**
Order	—	**Aspidochirotida**
Genus	—	***Holothuria***

Habit and Habitat

Holothuria is commonly called **sea cucumber**. It is found in thousands on the sandy bottom of shallow tropical and subtropical waters of Indo-Pacific area. *Holothuria* is found distributed in India, West Indies and Florida. It is sluggish and sedentary. However, *Holothuria* creeps slowly engulfing microorganisms alongwith mud, *i.e.*, it is **suspension feeder** and **detritivore.**

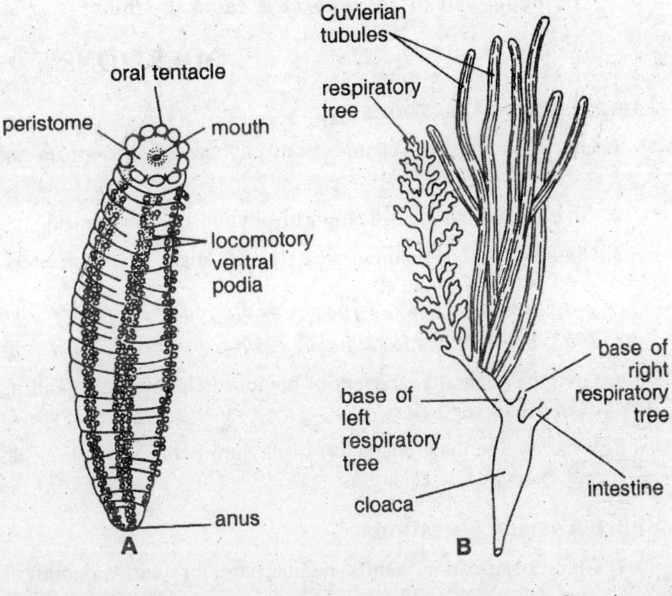

Fig. 49.13. *Holothuria*. A—Entire animal; B—Relation of respiratory tree and Cuvierian tubules with cloaca.

Structure

The body of *Holothuria* is elongated, bilaterally symmetrical animal. It lacks arms and bears mouth and anus on the opposite ends. The body wall is leathery having an endoskeleton of minute calcareous **ossicles**. Spines and pedicellariae are absent.

Five double rows of **tube feet** extend from mouth to anus. These rows represent five **ambulacral areas** and they are separated by five **interambulacral areas.** Ventral surface contains three ambulacral regions forming the **sole.** Dorsal surface contains two ambulacral area. The tube feet of ventral surface contains sucker and are used for creeping and attachment.

The mouth is surrounded by a circle of about 20–30 shield-like **tentacles** which are sensitive and contractile. The tentacles are modified tube feet and remain covered with mucus. Tentacles are extended to trap small organisms. The food-ladden tentacles are inserted into the pharynx one by one. The tentacles and mouth (introvert) can be completely withdrawn within the anterior part of body when the animal is disturbed.

The cloaca contains a pair of large, much branched **respiratory trees.** These lie in coelom. Cloaca pumps water in the tubules of respiratory trees. The cloaca dilates due to which sea water is rushed into it through the anus. When the anal sphincter closes, the cloaca contracts and water is forced into the respiratory tree. Pumping is slow. *Holothuria* requires six to ten cloacal dilations and contractions to fill the trees. All the water is expelled in one action. During expulsion whole body of *Holothuria* contracts. Gaseous exchange occurs between the water and coelomic fluid through the respiratory tree. **Haemal system** and **water-vascular system** are well developed. **Madreporite** is internal.

Holothuria exhibits a peculiar way of self-defence. It has a large mass of blind tubules (**tubules of Cuvier**) attached to the base of left tree. When the animal is attacked, the anus is directed towards the attacker, the body wall contracts, cloaca ruptures and Cuvierian tubules are shot out of the anus. During their expulsion, the tubules are greatly elongated and they break free from the respiratory tree. The tubules are sticky. They trap the attacker in a net of adhesive threads and *Holothuria* escapes. After the discharge, the Cuvierian tubules are regenerated.

Sexes are separate, gonad is single and attached to left side of the dorsal mesentery. Development is **indirect** and includes **auricularia larva.**

Economic Importance

The body wall of *Holothuria* is eaten in China.

QUESTIONS

Long Answer Questions

1. Give distinguishing characters of phylum Echinodermata and classify upto orders (*Awadh 2001; Kanpur 2001*)
2. Give an outline classification of phylum Echinodermata.
3. Classify phylum Echinodermata giving diagnostic characters and examples of each order. (*Purvanchal 1988; Agra 1988; Rohilkhand 1992; Kanpur 1996; Allahabad 1989; Lucknow 1989; Awadh 1992; Jiwaji 1993; Bundelkhand 1993; Indore 1990; Gorakhpur 1992; Jabalpur 1993; Gwalior 1993; Magadh 1994*)
4. Give the general characters of phylum Echinodermata and its various classes. Mention at least one example of each class. (*Magadh 1990; Lucknow 1995, 1998, Allahabad 2001*)
5. What are the main characters of phylum Echinodermata? Classify it by giving diagnostic characters and examples of each class. (*Gorakhpur 1995*)

Short Answer Questions

1. Give classification, habits, habitat, structure and economic importance of the following:

   (*i*) *Antedon* (*Kanpur 1988; Raipur 1991; Lucknow 1992; Meerut or CCS 1992; Ravishankar 1993; Jiwaji 1993; Gwalior 1993; Indore 1994; Himachal 1995*)

   (*ii*) *Echinus* (*Jiwaji 1989; Punjab 1992*)

   (*iii*) *Clypeaster* (Cake urchin) (*Agra 1988; Jiwaji 1989; Awadh 1992; Rohilkhand 1994*)

   (*iv*) *Ophioderma* (Brittle star) (*Meerut or CCS 1990; Rohilkhand 1991*)

   (*v*) *Pentaceros* (*Himachal 1994*)

   (*vi*) *Holothuria* (*Kanpur 1988; Gorakhpur 1989; Awadh 1990; Meerut or CCS 1992; Ravishankar 1993*)

2. Write short note on

   (*i*) Aristotle's lantern (*Kerala 1992; Berhampur 1990; BHU 1990; Gorakhpur 1991; Osmania 1992; Kanpur 1994*)

Very Short Answer Questions

1. Which system is found in Echinodermata only? What is its utility.

   Ans. The water-vascular system is the system represented in echinoderms only. It is a modified part of coelom consisting of a system of canals containing sea water and amoeboid corpuscles. It serves to bring about locomotion, feeding and aherence to the substratum.

2. Name the body symmetry in which body parts occur in fives.

   Ans. This type of body symmetry is called **pentamerous symmetry.** Adult echinoderms show radial (usually pentamerous) symmetry but their larvae are bilaterally symmetrical.

3. Body of echinoderms can be distinguished into how many surfaces? Name them.

Ans. The body of echinoderms can be distinguished into **oral** and **aboral** surfaces. Oral surface bears mouth and water-vascular system where as aboral surface has madreporite and anus on it.

4. Give an degenerate character of echinoderms.

Ans. Lack of head.

5. Mention one character in which echinoderms resemble the chordates.

Ans. Enterocoelous formation of coelom and mesoderm.

6. What is the function of pedicellariae?

Ans. To rid the body surface of debris and microorganisms.

7. Name the cells found in the coelomic fluid.

Ans. Coelomocytes (amoebocytes).

8. Radial symmetry of adult echinoderms is a secondarily developed trait. Is it so?

Ans. Yes.

9. Name three classes of subphylum Asterozoa.

Ans. Asteroidea, Ophiuroidea and Concentricycloides.

10. Name classes of living species of subphylum Echinozoa.

Ans. Two class — Echinoidea and Holothuroidea.

11. Give two examples of class Asteroidea.

Ans. *Asterias* and *Oreaster*.

12. Mention the order and class of *Clypeaster*.

Ans. Clypeastroida (order), Echinoidea (class).

13. Name one cylindrical echinoderm.

Ans. *Cucumaria*.

14. Name one disc-like echinoderm.

Ans. *Clypeaster*.

15. Mention a fixed echinoderm.

Ans. *Antedon* or feather star.

16. To which class *Echinus* belongs?

Ans. *Echinus* belongs to the class Echinoidea of the phylum Echinodermata. This animal is commonly known as **sea urchin.**

Fill in the Blanks

1. Subphylum Crinozoa has a single living class called ....
2. Adult echinoderms have .... symmetry.
3. Tube feet are mainly useful in ....
4. Sea water enters into ambulacral system through a calcareous plate called ....
5. Tube feet protrude through .... areas.
6. Dermis contains skeletal plates called ....
7. Ambulacral system is also called ....
8. Respiratory organs of echinoderms are ....
9. Circulatory system of echinoderms is called ....

10. Phylum Echinodermata is divided into four sybphyla .... , ... , ... and ....

11. Fertilization in echinoderms is ....

12. Echinoderm larvae have .... symmetry.

13. Common name of .... is cake urchin.

Multiple Choice Questions

1. Water vascular system is peculiarity of
(*a*) Porifera (*b*) Coelenterata
(*c*) Mollusca (*d*) Echinodermata

2. Which phylum is exclusively marine
(*a*) Echinodermata (*b*) Porifera
(*c*) Protozoa (*d*) Arthropoda

3. *Ophiothrix* belongs to class
(*a*) Asteroidea (*b*) Echinoidea
(*c*) Ophiuroidea (*d*) Holothuroidea

4. An echinoderm with melon-shaped body is
(*a*) sea-star (*b*) brittle star
(*c*) sea cucumber (*d*) sea urchin

5. Common name of *Antedon* is
(*a*) sea star (*b*) feather star
(*c*) star fish (*d*) cake urchin

6. Sea cucumber belongs to the class
(*a*) Holothuroidea (*b*) Asteroidea
(*c*) Echinoidea (*d*) Ophiuroidea

7. Which of the following belongs to the class Echinoidea?
(*a*) *Asterias* (*b*) *Ophiothrix*
(*c*) *Echinus* (*d*) *Cucumaria*

8. Cloacal respiration occurs in
(*a*) *Clypeaster* (*b*) *Antedon*
(*c*) *Echinus* (*d*) *Cucumaria*

9. Arms, spines and pedicellaria are absent in
(*a*) sea urchin (*b*) sea star
(*c*) sea lily (d) sea cucumber

10. The term ambulacral has to do with
(*a*) inter filament bridges (*b*) marine animals
(*c*) myosin filaments (*d*) tube feet

11. The regeneration abilities of echinoderms are
(*a*) slight (*b*) moderate
(*c*) phenomenal (*d*) occasional

12. Autotomy is common amongst
(*a*) sea lilies (*b*) sea stars
(*c*) brittle star (*d*) all of these

13. In echinoderms, skeleton is
(*a*) calcareous (*b*) chitinous
(*c*) siliceous (*d*) tunicin made

14. The animal which is commonly known as 'sand dollar' is
(*a*) *Antedon* (*b*) *Clypeaster*
(*c*) *Echinarachinus* (*d*) *Ophiothrix*

15. *Thyone* is commonly known as
(*a*) brittle star (*b*) sea cucumber
(*c*) sea lily (*d*) cake urchin

16. Sea pentagon is common name of
(*a*) star fish (*b*) sea urchin
(*c*) *Pentaceros* (*d*) *Cucumaria*

17. Aristotle's lantern is found in
(*a*) *Cucumaria* (*b*) *Antedon*
(*c*) *Echinus* (*d*) *Ophiothrix*

18. Brachiolaria larva is characteristic of the class
(*a*) Crinoidea (*b*) Asteroidea
(*c*) Ophiuroidea (*d*) Echinoidea

19. The larva that occurs in the development of holothurians
(*a*) auricularia (*b*) brachiolaria
(*c*) doliolaria (*d*) bipinnaria

20. In which class of phylum Echinodermata, pluteus larva is found?
(*a*) Asteroidea (*b*) Holothuroidea
(*c*) Ophiuroidea (*d*) Echinoidea

21. Aristotle's lantern is useful in
(*a*) photo reception (*b*) chewing
(*c*) excretion (*d*) respiration

22. Excretory products of echinoderms are
(*a*) urea (*b*) urea and creatine
(*c*) ammonia (d) uric acid

23. Respiratory organs in Holothuroidea are
(*a*) papulae (*b*) dermal branchiae
(*c*) respiratory tree (*d*) bursae

24. Ophiopluteus is
(*a*) an echinoderm (*b*) a poriferan
(*c*) an annelid (*d*) a larva of echinodermata

ANSWERS

Fill in the Blanks

1. Crinoidea **2.** Radial **3.** Locomotion **4.** Madreporite
5. Ambulacral **6.** Ossicles **7.** Water vascular system **8.** Tube feet
9. Haemal system **10.** Homalozoa, Crinozoa, Asterozoa and Echinozoa
11. External **12.** Bilateral **13.** *Clypeaster*.

Multiple Choice Questions

1. (*d*) **2.** (*a*) **3.** (*c*) **4.** (*d*)
5. (*b*) **6.** (*a*) **7.** (*c*) **8.** (*d*)
9. (*d*) **10.** (*d*) **11.** (*c*) **12.** (*c*)
13. (*a*) **14.** (*c*) **15.** (*b*) **16.** (*c*)
17. (*c*) **18.** (*b*) **19.** (*a*) **20.** (*d*)
21. (*b*) **22.** (*b*) **23.** (*c*) **24.** (*d*).

50

Asterias (Starfish or Sea star) (Phylum Echinodermata)

50.1 CLASSIFICATION

Phylum	—	**Echinodermata**
Subphylum	—	**Asterozoa**
Class	—	**Asteroidea**
Order	—	**Forcipulata**
Genus	—	***Asterias***

Asterias is commonly known as **starfish**. But this is not fish. Hence its name is misleading and now its more suitable common name is **sea star**. *Asterias* contains about 150 species, all of which have different geographical distribution. Some other common sea stars are *Pentaceros* (or *Oreaster;* it is an Indian species which is found in the Bay of Bengal and Arabian sea), *Astropecten*, *Heliaster*, *Solaster*, *Luidia*, etc.

50.2 HABIT, HABITAT AND EXTERNAL FEATURES

Habit and Habitat

Asterias is a free-living exclusively marine, bottom dwelling (or **benthonic**) animal. *Asterias* is found in shallow water in North Temperate seas and found abundantly on North-Atlantic Coast. It is found in abundance in India and U.S.A. *Asterias forbesi* occurs on the eastern sea shore from the Maine to the Gulf of Mexico and is found equally abundant on hard, rocky and sandy or soft bottom.

Most species of *Asterias* are **solitary** but under certain ecological constraints (*e.g.*, to avoid direct sunlight) many individuals may gather at some place for the purpose of protection. Most of them are **nocturnal**, remain quiet in day time but become active during night. They remain attached to rocks, shells, piles and piers, etc. Their oral surface faces the substratum. They slowly move by crawling with the help of arms on the rocky bottom. All sea stars are **carnivorous** and feed voraciously on almost any slow moving or sessile animals, chiefly on polychaetes, crustaceans, molluscs (snails, bivalves) and other echinoderms. They also feed on dead animals, so act as scavangers of sea. *Asterias* shows remarkable power of **autotomy** and **regeneration**.

External Features

1. Shape, size and colour. *Asterias* has a **radially symmetrical** star shaped body (Box 50.1). The diameter of body reaches upto 24 cm. Colour of *Asterias* varies from orange to purplish, the upper (aboral) surface being darker than the lower (oral) surface.

2. External structure. The body of *Asterias* consists of a central, pentagonal (5 sided) **central disc** from which radiate out five elongated, tapering, symmetrically spaced projections, the **rays** or **arms.** Each arm has the shape of an isosceles triangle (*i.e.*, triangle having two sides of equal length). The body lacks a well defined head and has two distinct surfaces. The lower surface, the one nor-mally kept towards the substra-tum, is flat and is known as the **oral** or **actinal surface** because it bears the mouth (Fig. 50.1). The upper surface is convex and is called the **aboral** or **abactinal surface** (Fig. 50.6). The oral and aboral surfaces are not the ventral and dorsal surfaces, but corres-pond to the left and right sides of the bilaterally symmetrical larva. The axes occupied by the arms are known as the **radii** and the regions of the central disc between the arms is the **inter-radii**.

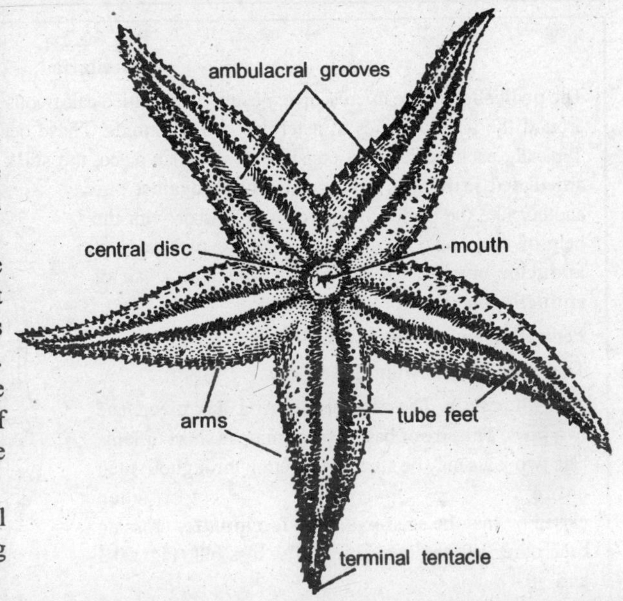

Fig. 50.1. *Asterias.* External features (oral view).

A. Oral Surface. The oral surface of *Asterias* contains following structures.

1. Mouth. Mouth is a circu-lar aperture at the top of 5-rayed (pentagonal) depression at the centre of the oral surface of the central disc. The depression is known as **actinostome**. The mouth is surrounded by a soft membrane, the **peristomial membrane** or **peristome**. Peristome is provided with a sphincter and is guarded by five groups of **oral spines** or **mouth papillae**.

Box 50.1
Symmetry of Echinodermata

Echinoderms are quite distinct in their morphology and differ from other coelomate animals in having a characteristic **pentamerous radial symmetry**. The radial symmetry is seen in majority of echinoderms except for some holothurians. The radial symmetry is not restricted to only the external form of an echinoderm, it is shown even by the arrangements, of internal organs.

In all echinoderms, the external surface is differentiated into five radial and five interradial areas. This gives pentamerous radial symmetry. The radial regions bear tube feet. These are called **ambulacral regions**. The areas between ambulacral regions are called **interambulacral regions** and they contain spines (Fig. 50.2). But in some holothurians, the tube feet are scattered all over the body surface. In Asteroidea and Ophiuroidea, the body is prolonged into arms in the direction of radii.

The radical symmetry of echinoderms is exhibited only in the organisation of the adults and is acquired at the time of metamorphosis. The larval stages of different classes of Echinodermata exhibit a **bilateral symmetry**. In fact, the radial symmetry is not total or perfect, since, most radially symmetrical forms present indication, of bilateral symmetry, *e.g.*, by the presence of a single madreporite, presence of single gonad and gonoduct in Holothruoids. Because of these reasons, it is generally believed that the radial symmetry

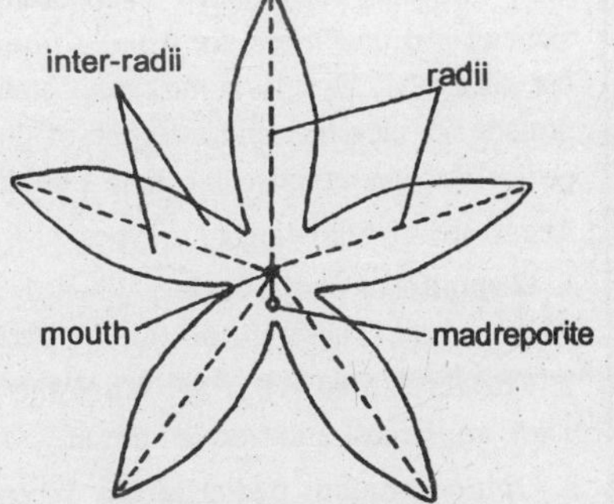

Fig. 50.2. Sea Star showing pentamerous radial symmetry.

of echinoderms has been secondarily acquired. It is also believed that the radial symmetry in ancestral echinoderms has evolved due to sedentary habit and that echinoderms have evolved from free-swimming bilaterally symmetrical, triploastic, coelomate ancestor.

Box 50.2
Pedicellaria

The **pedicellariae** are microscopic, peculiarly modified calcareous spines. They are found among the papulae around the simple spines in asteroids and echinoids. These occur on oral as well as on aboral surface. Typically each pedicellaria consists of a basilar piece, the **stalk** or **peduncle** and two or three movably articulated **jaws** (Fig. 50.3). Jaws work against one another like the blades of a forceps or scissor with the help of three paired muscles. There are two pairs of **adductor muscles** for closing the jaws and a pair of **abductor muscles** for opening them.

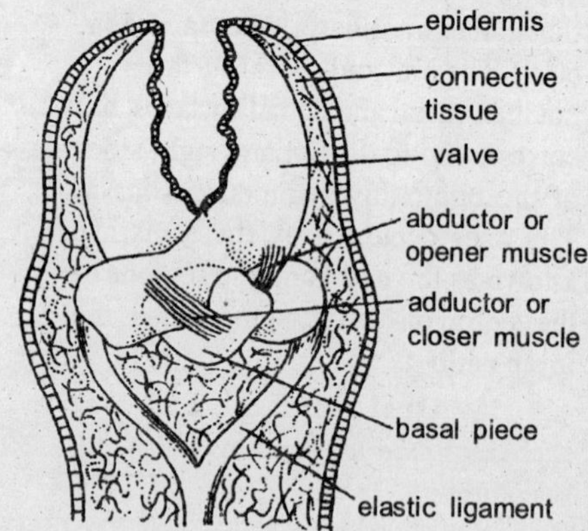

Fig. 50.3. Distal end of a scissor-type pedicellaria from *Asterias*.

Pedicellariae in Asteroidea

These are of following three types:

1. Pedunculate. These consist of a **basilar piece**, and two **jaws**. The size of basilar piece may be short or long. The two jaws may be **straight** meeting throughout their entire length when closed or may be **scissor-like** or **forcipulate**, crossing each other like the mandibles of a cross bill (Fig. 50.3 and 50.4).

2. Sessile. These pedicellariae lack the basilar piece or pendunclе and two jaws are attached directly to the ossicles by the muscles. Sessile pedicellariae may be **spiniform**, when present in clusters on the adjacent ossicles or **fascicu-late**, when the cluster of spines is present on the same ossicles.

3. Alveolar. These are similar to sessile pedicellariae but differ from them in being partly embedded in the endoskeletal depression, called **alveolus.**

Pedicellariae in Echinoidea

Pedicellariae are characteristic of all echinoids. They are located over the general body surface as well as on the peristome. The echinoid pedicellaria is composed of a long stalk surmounted by jaws. The stalk may contain a supporting skeletal rod and there are usually three opposing jaws. Muscles at the base of stalk provide for elevation and direction of the pedicellariae in response to certain stimuli.

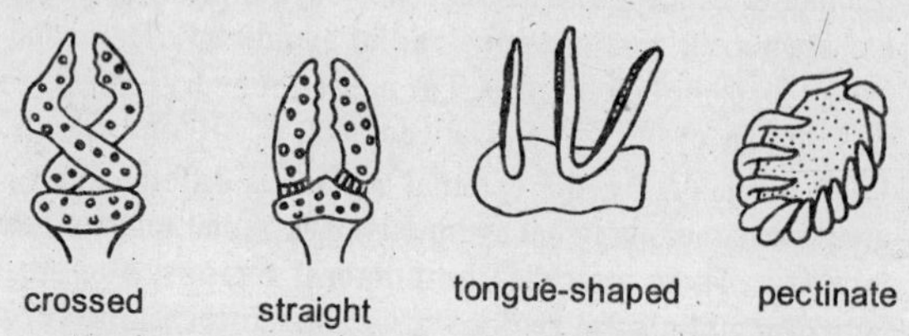

Fig. 50.4. Different type of pedicellariae of Asteroidea.

These are of following four types:

1. Gemmiform pedicellariae. When the stalk is very stiff and head is round. Each jaw is provided with poison gland. These are **poisonous pedicellariae.**

2. Tridactyle pedicellariae. When the stalk is very flexible and jaws are long tappering and serrated. They are largest and very common.

3. Ophiocephalous pedicellariae. When the stalk is very flexible and jaws are short, broad and toothed.

4. Trifoliate pedicellariae. They are smallest. Each consists of a very flexible stalk and short broad toothed jaws. Jaws do not meet at their tips.

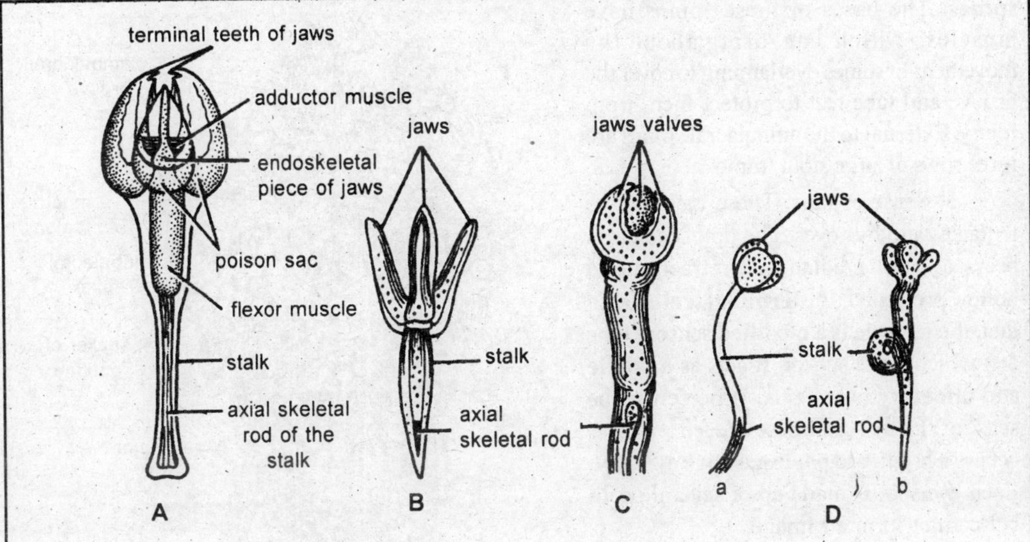

Fig. 50.5. Different types of pedicellariae in sea urchin A—Gemmiform; B—Tridactyl; C—Ophiocephalous; D—Trifoliate (*a*) of *Tripneustes* and (*b*) of *Echinus*.

Function. Pedicellariae are used for defence or for cleaning the body surface, biting and breaking up small particles of debris, which are then removed by the surface cilia. When the pedicellariae are touched on outside, they snap open; when touched on the inside, the snap shut. They also respond to chemical stimuli.

2. Ambulacral groove. There are five narrow grooves that radiate out from the angles of actionostome, one extending along the middle of the oral surface of each arm upto the tip.

3. Tube feet or podia. These are soft, thin-walled, tubular, retra-ctile structures which are arranged in four rows in each ambulacral groove. Each podium is provided with terminal sucker. Sucker func-tions as a suction pump to afford firm attachment on the surface to which it is applied. The tube feet are multipurpose organs. They mainly act as organs of locomotion and capturing food. They also help in respiration and adherence to substratum. Tube feet also act as sensory organs.

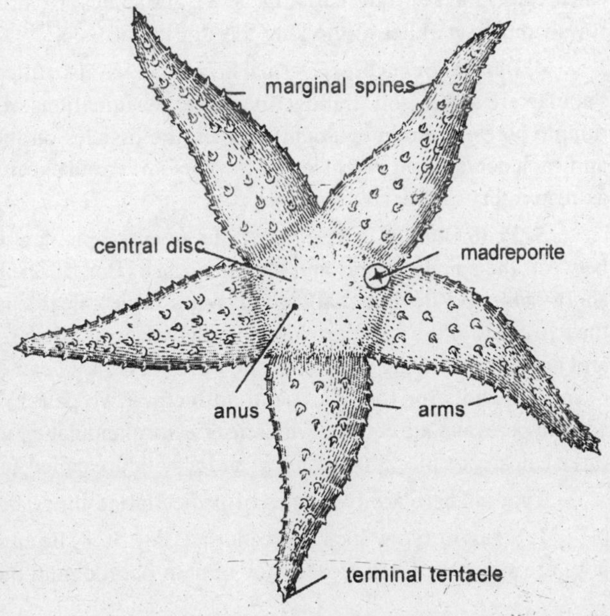

Fig. 50.6. *Asterias*. External features (aboral view).

4. Ambulacral spines. The spines are short, stout outgrowths from the calcareous plates, the **ossicles,** embedded in the body wall and covered by epidermis. On the oral surface, there are two rows of spines along either border of each ambulacral groove. These are called the **ambulacral**

spines. The bases of these spines have muscles, which can bring about the movement of spines overlapping to cover the groove and tube feet to protect them from injury. External to the ambulacral spines are three rows of large stout immovable spines.

5. Sense organs. These include five tentacles and five eyes. The tip of each arm bears a small median non-retractile and hollow projection, the **terminal tentacle**. In fact, the tentacle is a modified sensory tube feet which lacks sucker. It acts as a **tactile** and **olfactory** organ (*i.e.*, it perceives the sense of smell). At the base of each tentacles occurs a bright red photosensitive **eye spot**. Each eye spot is made up of several ocelli but cannot form an image.

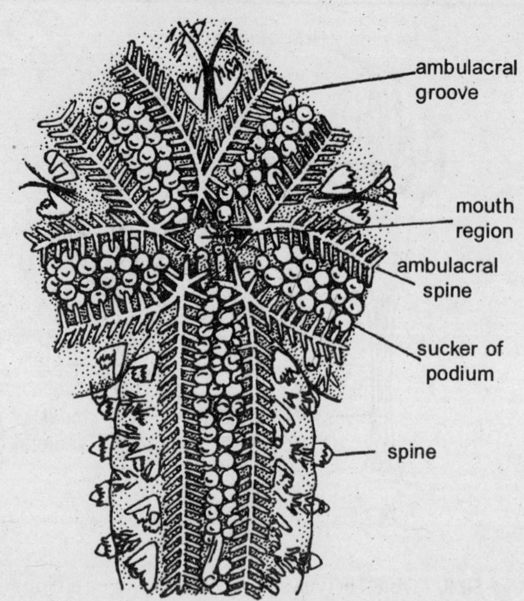

Fig. 50.7. *Asterias*. Oral surface of the disc and arm.

B. Aboral surface. The aboral surface of *Asterias* contains the following structures:

1. Anus. It is a small circular aperture. It is situated close to the centre of the central disc of aboral surface.

2. Madreporite. It is a large circular button-like porous plate. It is situated towards one side of the disc between two arms. These two arms are called **bivium** and remaining three arms—the **trivium**. Thus, madreporite introduces a bilateral symmetry in the radial symmetry of *Asterias*.

3. Spines. The entire aboral surface is covered with numerous short, strong, blunt, immovable, calcareous **spines** or **tubercles** (Fig. 50.8). The spines are variable in size and are arranged in irregular rows running parallel to the long axes of the arms.

4. Dermal branchiae. Dermal branchiae are also called **dermal gills** or **papulae** (Fig. 50.8 A). Papulae are small, soft, hollow, finger-like evaginations of the body wall. They protrude through minute pores in the integument between the ossicles on the aboral surface. They are thin-walled and each contains an extension of the coelom. Papulae can be extended and withdrawn and serve as respiratory and excretory organs.

5. Pedicellariae. These are white, microscopic, jaw-like or pincer-like structures occurring between the spines of oral and aboral surface (Box 50.2). Each pedicellaria consists of a long or short, stout and flexible **stalk**. Stalk bears three calcareous ossicles or plates: a **basilar plate** at the extremity of the stalk bearing at it top two jaws. Jaws are movably articulated with basilar plate and are serrated along their opposite edges. The jaws can be opened or closed by the contraction of minute **abductor** (one pair) and **adductor** (two pairs) muscles respectively. All three ossicles contain pores and are covered with sensory and glandular epithelium. The pedicellariae which contain three calcareous pieces (= ossicles) and a stalk are called **forcipulate pedunculate pedicellariae**.

Types. There are two types of pedicellariae in *Asterias*.

1. Straight type. Such pedicellariae (Fig. 50.8) lie among the dermal branchiae. In them two jaws are straight. When closed they remain parallel and meet throughout their length.

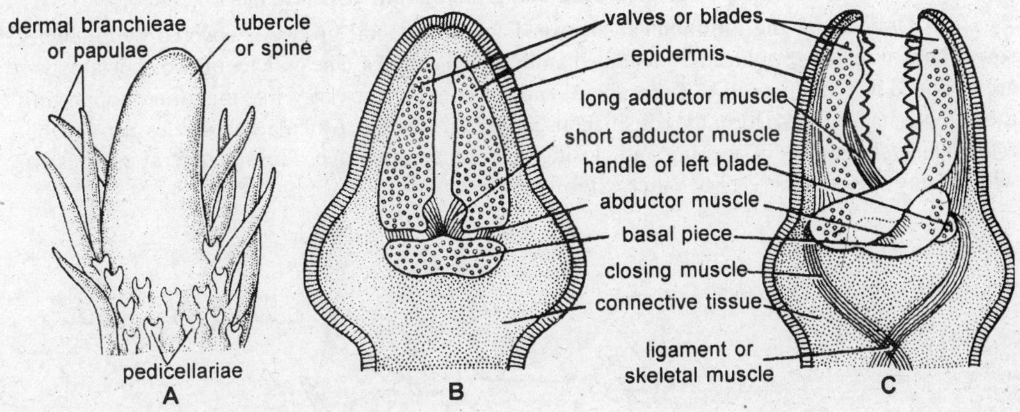

Fig. 50.8. *Asterias*. A—A cluster of pedicellariae, papulae and spine; B—Straight type pedicellaria; C—Crossed type pedicellaria.

The two jaws work against each other like the blades of forceps.

2. Crossed type. These pedicellariae are small and are arranged in rings round the white spines on the aboral surface. In them two jaws cross each other like a pair of scissors.

Function. Pedicellariae are sensitive to contact. They serve as defensive and offensive organs. They help in keeping the body surface free from debris and minute organisms (*e.g.*, larvae of sponges and coelenterates) which may settle on the body surface. In this way, they provide protection to gills. In some starfishes, they may also help in the capture of small prey.

Lastly, entire surface of the body, including that of its outgrowths such as tube feet, spines, pedicellariae and dermal branchiae is ciliated. The body of *Asterias* is enclosed in a hard but flexible integument. The hardness of the integument is due to the presence in it of calcareous ossicles.

50.3 ENDOSKELETON

Skeleton of echinoderms is derived from **mesoderm** or from **dermis** of body wall. Therefore it is called **endoskeleton**. It is formed of crystals of calcium carbonate ($CaCO_3$), *i.e.*, it is calcareous. Endoskeleton occurs in two forms:

1. In the form of **calcareous plates** or **ossicles** found in the soft dermis of bodywall.
2. In the form of **warts** or **tubercles**, attached to the calcareous plates of dermis. These projects out of the skin.

Skeleton of *Asterias* and other echinoderms differ from the skeleton of other invertebrates mainly in one character. The skeleton of invertebrates is developed from epidermis, but in echinoderms it develops from dermis or mesodermal layer. The body of starfish, thus, is enclosed in a tough, flexible integument containing numerous calcareous ossicles.

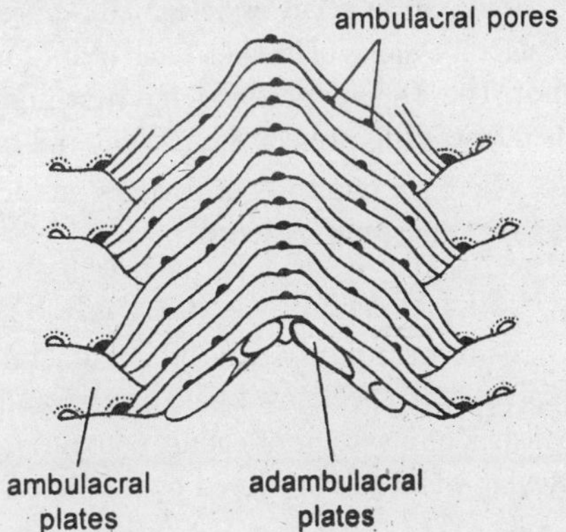

Fig. 50.9. Arrangement of ambulacral ossicles.

Arrangement of ossicles. Dermal ossicles have different but definite shapes and patterns. These are interconnected by connective tissue. In starfish, skeletal ossicles are arranged in a specific orders as follows:

(*i*) Ossicles around the mouth. Five plate-like ossicles, called **oral ossicles,** are arranged around the mouth and form a complete ring along the margin of peristome.

(ii) **Ossicles on the ambulacral groove.** Each ambulacral groove is covered by a double-row of large transversely placed, rod-shaped **ambulacral ossicles.** The ossicles of two opposite rows are arranged like an inverted V. Their aboral ends meet at the apex of V like the rafters supporting the roof of a house. These form the conspicuous **ambulacral ridge.** Ambulacral ossicles are movably articulated and permit the opening and closing of ambulacral groove. The ambulacral ossicles are without any spines, tubercles or other external appendages.

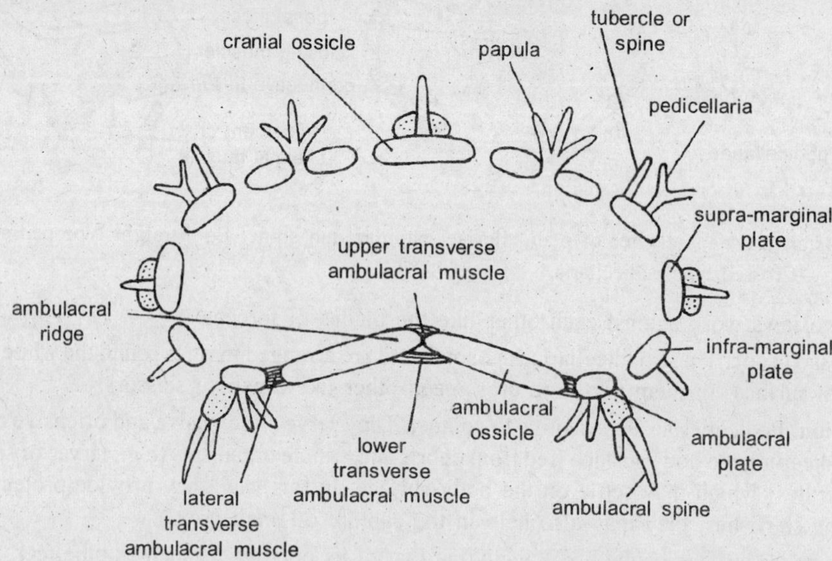

Fig. 50.10. Starfish. C.S. through the arm to show endoskeletal ossicles.

Each ambulacral ossicle has a notch on its outer as well as inner margin. The two notches of the adjacent ossicles together form an oval aperture, the **ambulacral pore** for the passage of tube-feet. The ambulacral pores are so arranged that they form two rows on each side of the ambulacral groove.

(iii) **Ambulacral ossicles.** These are present along the edges and are articulated with the ambulacral ossicles of its side. Each bears 2 to 3 movable spines on small tubercles, called **ambulacral spines.** These spines are long and bear groups of large straight **pedicellariae.** These spines can be turned inward to protect the groove and podia.

There are two rows of ossicles on the outside of ambulacral ossicles. These are called **infra** and **supramarginal ossicles.**

50.4 WATER VASCULAR SYSTEM (OR AMBULACRAL SYSTEM)

Water vascular system (or ambulacral system) is a typical and unique feature of echinoderms. It represents a specialized part of coelom, being developed from the left hydrocoel of the larva. It consists of a system of canals containing sea water and amoeboid corpuscles. The canals have muscular walls and are lined by a ciliated epithelium. It helps in locomotion.

Structure

Water vascular system of *Asterias* consists of the following parts (Fig. 50.11):

1. Madreporite
2. Stone canal

3. Ring canal
4. Tiedmann's bodies
5. Radial canals
6. Lateral or transverse canals
7. Tube feet

1. Madreporite. It is a light-coloured, sieve-like, button-shaped or circular calcareous skeletal plate. It is located on the aboral surface of the central disc in inter-radial position. The surface of madreporite is marked by a number of radiating narrow, straight or wavy grooves or **furrows** (Fig. 50.12). The furrows are covered with ciliated epithelium and at their bottom they contain about 250 minute pores. Each pore opens into **pore canals.** The pore canals unite to form larger **collecting canals** within the substance of madreporite. Collecting canals lead into a sac-like **ampulla** below the madreporite. Ampulla is continued as a canal inside the stone canal.

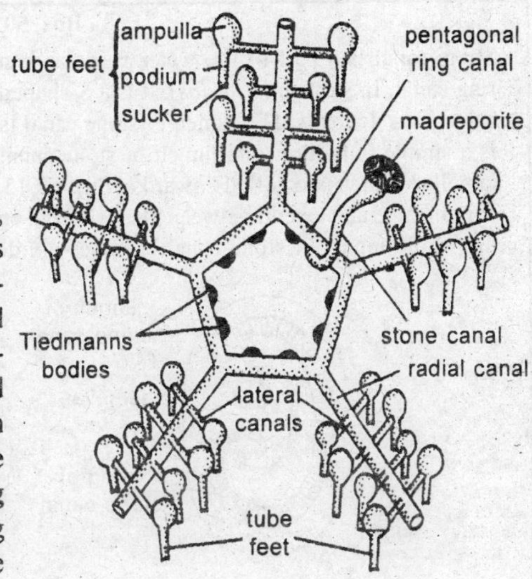

Fig. 50.11. *Asterias*. Water vascular system.

2. Stone canal. It is S-shaped verti-cal canal which extends downwards (ora-lly) and opens into a ring canal. Its wall is supported by a series of calcified rings and is lined internally with tall ciliated cells. The beating of cilia draw water into the canal. The stone canal is a simple tube in a young starfish, but in adult, its wall produces a prominent **ridge** in the lumen of stone canal. The ridge is bifurcated to form two spirally rolled **lamellae** (Fig. 50.15). They ensure water circulation.

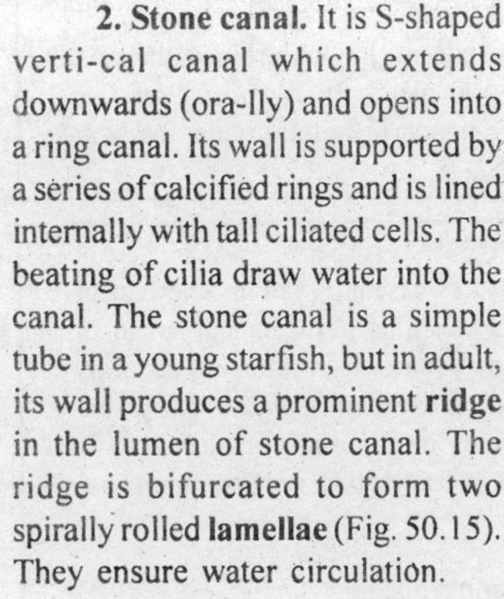

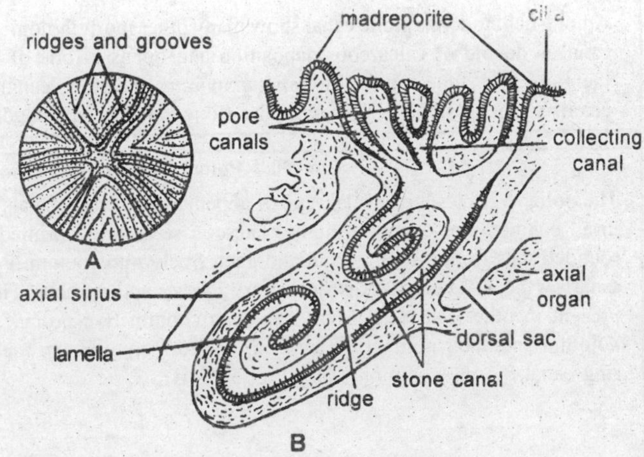

Fig. 50.12. *Asterias*. Madreporite. A—Surface view; B—Vertical section.

During its course, the stone canal along with an **axial gland** ensheathed by a wide, thin-walled tubular coelomic-sac, called **axial sinus.** This entire structure is called **axial complex.**

3. Ring canal. It is also called **ring vessel** or **water ring.** It is a wide five-sided (pentagonal) canal in the oral region around the oesophagus.

In most starfishes, but not in *Asterias*, the ring canal gives off from the inner side in each inter-radius a large, thin-walled, contractile, pear-shaped sac, the **polian vesicle.** The polian vesicles store water and help in regulating pressure in the water vascular system. They also manufacture amoeboid cells of water vascular system (Box 50.4).

4. Tiedmann's bodies. From its inner side, the ring canal produces five pairs of greatly folded pouches, called **racemose vesicles** or **Tiedmann's bodies.** Each pair of these bodies has an inter-radial position. The cavity of each Tiedmann's bodies is divided into a number of chambers which communicate with the ring canal. In *Asterias*, there are only **nine** Tiedmann's bodies. The 10th body is absent and its place is occupied by the stone canal (Fig. 50.11).

Box 50.3 Stone canal

Stone canal is a part of the water vascular system of echinoderms. It connects madreporite with the ring canal. In starfish (*Asterias*), it is a S-shaped tube. It is called **stone canal** because its wall contains calcareous deposits. The lumen of stone canal is divided by a longitudinal ridge. In *Henricia*, the ridge is a simple bulging in the lumen of stone canal. In *Asterias*, this ridge is bifurcated into two vertical lamellae which are rolled inwards (Fig. 50.13A). In *Astropecten*, the coiled lamellae become very complicated and extend between the wall from one side to another of the lumen (Fig. 50.13B). In *Culcita*, the whole lumen of stone canal becomes divided into a number of irregular chambers (Fig. 50.13C).

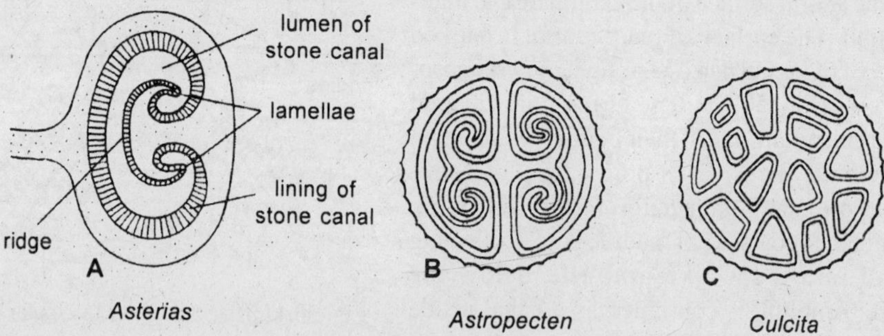

Fig. 50.13. Sectional view of stone canal in various Asteroids. A—*Asterias*; B—*Astropecten*; C—*Culcita*.

Among echinoderms stone canal show many other morpohological variations. In Ophiuroidea, the stone canal is devoid of calcareous deposition and opens in one of the oral plates. In *Trichaster elegans*, there are five stone canals. In *Ophiactis virens*, the stone canals are many. In holothurians, the stone canal is mostly single. In *Thyone*, the stone canal is branched.

Box 50.4 Polian vesicles in Echinodermata

The polian vesicles are special type of appendages of ring canals of water vascular system. They occur in all echinoderms except crinoids. These arise as thin - walled, bladder-like, pear shaped, ovoid or rounded sacs from the ring canal and hang freely into coelom. Polian vesicles open into ring canal by a narrow neck. The polian vesicles vary in size and number. They are absent in *Asterias rubens* but present in other sea stars. Many sea stars contain five polian vesicles inter-radially in the disc. In holothurians, their number varies from one to many. In *Astropecten*, a few polian vesicles open into ring canal by one common stalk (Fig. 50.14B).

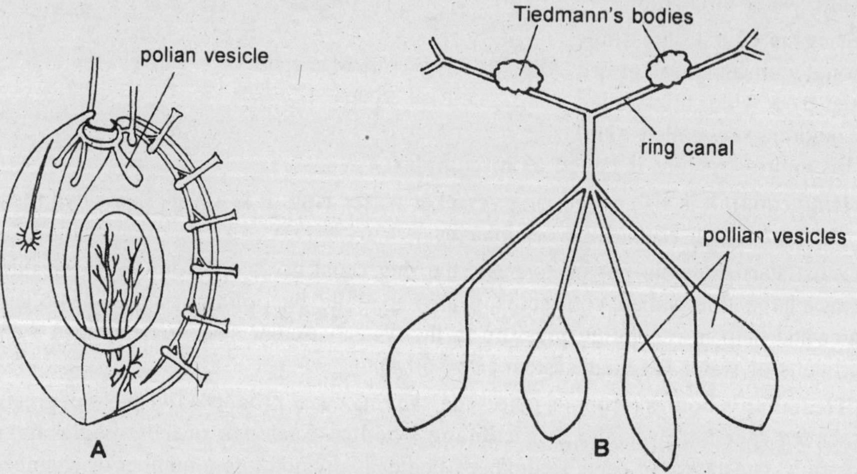

Fig. 50.14. A—Polian vesicle in *Holothuria*; B—Polian vesicles in *Astropecten*.

The actual function of Tiedmann's bodies is still unknown. They are supposed to be lymphatic glands to manufacture the **amoebocytes** of water vascular system. **Furguson** (1984) demonstrated that some sea water passes through these bodies into the perivisceral coelom. According to him these bodies participate in the production of coelomic fluid.

5. Radial canals. From the outer surface of the ring canal are given out five long, ciliated **radial canals.** One radial canal extends into each arm upto its extremity through the ambulacral groove of the arm. Radial canals run on the oral side of the ossicles and terminate as the lumen of terminal median tentacles.

6. Lateral or transverse canals. In each arm, the radial canal gives out two series of short, narrow, transverse branches called **lateral, transverse** or **podial canals.** Each lateral canal contains a **valve** to prevent backward flow of fluid into the radial canal and it opens into ampulla at the base of a tube foot.

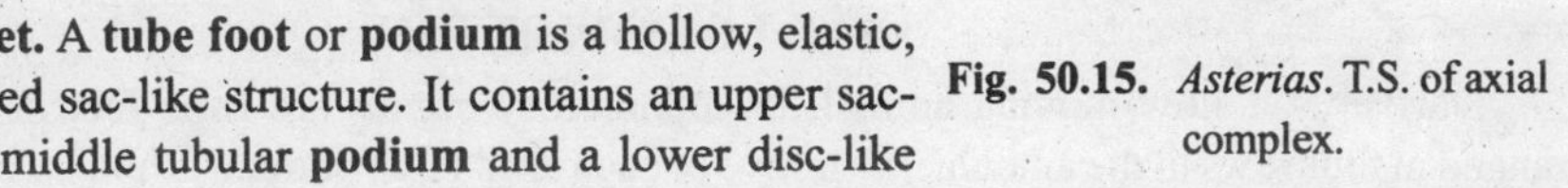

Fig. 50.15. *Asterias*. T.S. of axial complex.

7. Tube feet. A **tube foot** or **podium** is a hollow, elastic, thin-walled, closed sac-like structure. It contains an upper sac-like **ampulla,** a middle tubular **podium** and a lower disc-like flattened **sucker.** The ampulla lies within the arm, projecting into the coelom above the **ambulacral pore** which is a gap between the adjacent ambulacral ossicles for the passage of podium (Fig. 50.16). Tube feet, in fact, are external projections of the body wall; each of them has a lining of **ciliated epithelium** and a covering of **coelomic epithelium** or **peritoneum.** Between these two epithelia lie **connective tissue** and **longitudinal muscle fibres.** Contraction of the muscles of a side causes bending of a podium.

The podia are arranged in two double rows along the length of the ambulacral groove. They are chief locomotory and respiratory organs of *Asterias*.

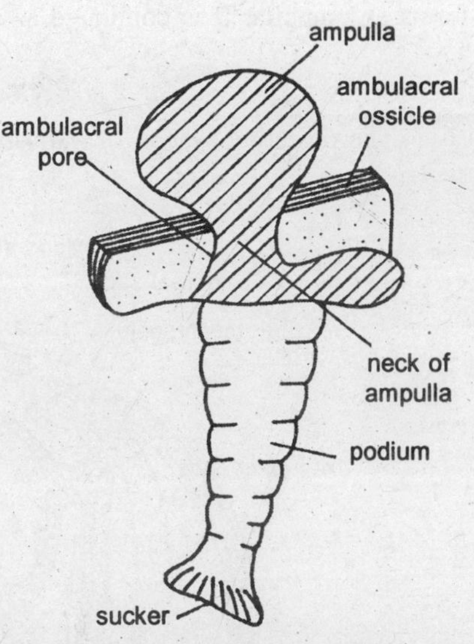

Fig. 50.16. *Asterias*. A tube foot.

Function

The water vascular system is lined with cilia and is filled with a fluid very similar to sea water. This fluid, however, contains **coelomocytes,** a little **protein** and a high concentration of **potassium ions.** The system operates as **hydraulic system** and brings about locomotion, adhesion to the substratum and the capture and handling of food. All these functions are performed by the help of tube feet.

(1) Locomotion. The beating of cilia causes water to enter through madreporite. It passes into the tube feet (= podia) and their ampullae. When the ampulla contracts, the valve in the lateral canal closes, and water (fluid) is pressed into tube feet which get distended (upto 5 cm) due to their muscular walls. The podium comes in contact with the substratum and the sucker adheres to it. Adhesion is also has a chemical base. The podium secretes a substance that makes bonds with the surface of substratum. Another secretion breaks the bonds and sucker becomes free (**Thomas** and **Hermans** 1985).

After their fixation with substratum, the tube feet contract in unison and push the body forward. Due to contraction of muscles, the fluid from each tube foot, returned to the ampulla and tube foot is shortened. This lessens the pressure at the tip so that the suckers detach. The same process is repeated again and again to bring about locomotion. In a section of arm most podia perform the same step and the animal moves forward. The action of podia is highly coordinated. During locomotion one or two arms act as leading arms and podia in all the arms move in same direction.

Other parts of the water vascular system—the madreporite, stone canal, ring canal and radial canals maintain proper fluid pressure necessary for the working of ampulae and podia. Some fluid is always lost by leakage through podial walls when fluid pressure increases in podia. To maintain constant fluid pressure some sea water is drawn through the madreporite. Star fish moves very slowly—15 cm per minute.

2. Climbing. Due to turgidity and terminal suckers, the tube feet are able to take a firm grip of the substratum and this brings about adhesion. Combined action of podial suckers enables starfish to climb vertically over rocks.

3. If a star fish is turned over, it can right itself by folding.

4. Food capture. The suckers of the tube feet are also used to capture food.

50.5 LOCOMOTION

Starfish is a slow moving animal. Its locomotory organs are **tube feet** or **podia** which are arranged in four rows in the ambulacral groove of each arm. Each podium is hollow external projection form the body wall with a terminal sucker (Fig. 50.16). Each tube foot receives a lateral, transverse or podial canal from the radial canal of water vascular system. At the base of tube feet, podial canal forms an ampulla. It is continued as a canal inside the podium.

50.6 DIGESTIVE SYSTEM

The digestive system of *Asterias* comprises the **alimentary canal** and **digestive glands.**

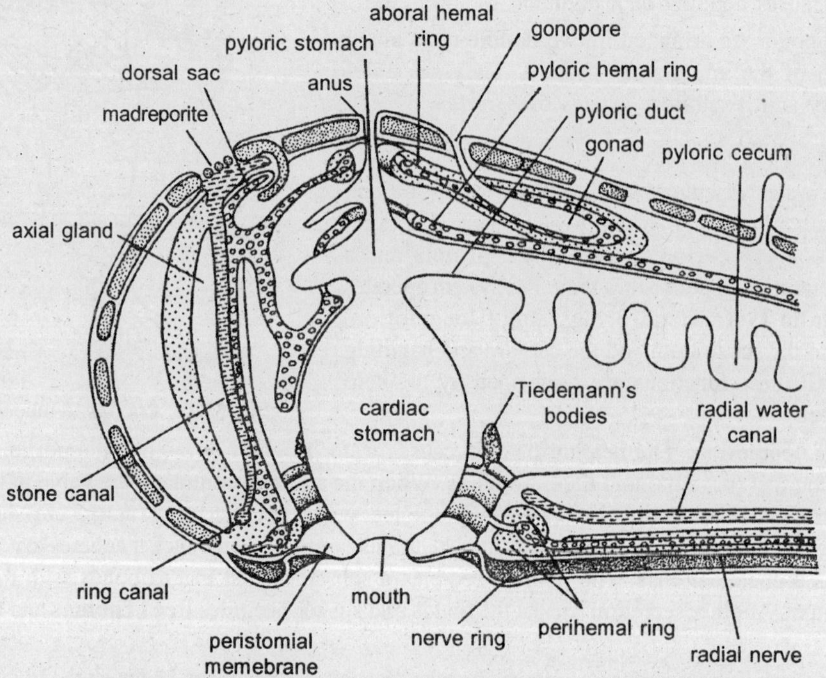

Fig. 50.17. Sea star. The central disc and base of one arm (vertical section).

Alimentary Canal

In *Asterias*, the alimentary canal is tubular, short but spacious canal which extends vertically along the oral-aboral axis in the central disc. It comprises of the following parts:

1. Mouth. The pentagonal actinostome leads into mouth which is situated on the oral surface in the centre of peristomial membrane. The mouth is surrounded by a sphincter muscle. The mouth leads upwards into the oesophagus.

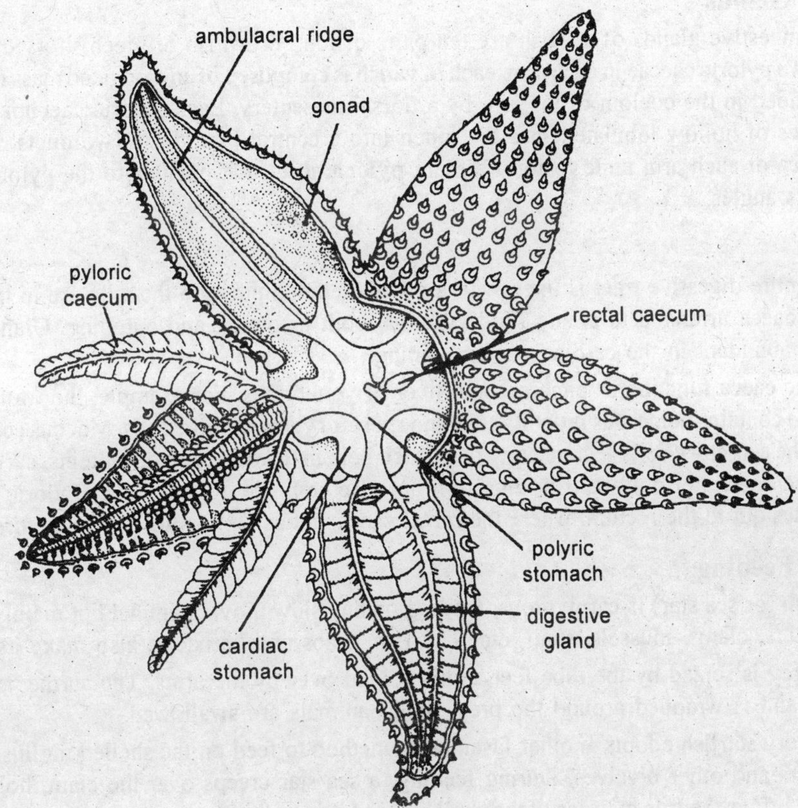

Fig. 50.18. *Asterias*. Digestive system.

2. Oesophagus. It is a very short, wide and vertical tube. It opens aborally in the stomach.

3. Stomach. The stomach is a large, thin-walled spacious five-lobed sac which fill the interior of the disc. It is typically divided by a horizontal constriction into a voluminous oral part, the **cardiac stomach** and a flattened aboral part, the **pyloric stomach.**

***(i)* Cardiac stomach.** The cardiac stomach is connected with oesophagus. It has a muscular, highly folded wall which bulges out to form five lobes, one opposite, each arm. The cardiac stomach is held in position by five pairs of triangular **mesenteries** or **gastric ligaments** which are formed of connective tissue and muscles. A pair of gastric ligaments connect the cardiac stomach to the ambulacral ridge in each arm. During the feeding process, the cardiac stomach can be everted through the mouth by the contraction of muscles of body wall. The retraction of cardiac stomach is brought about by five pairs of **retractor muscles** which arise from the lateral sides of the ambulacral ridge.

***(ii)* Pyloric stomach.** Cardiac stomach leads upwards into a small aboral pyloric stomach which is thin-walled flattened star-shaped sac. It is simply the reception chamber of the ducts from the pyloric caeca. It communicates with the intestine.

4. Intestine. It is a short, tubular, five-sided tube extending vertically from aboral side of pyloric stomach to open through a minute anus in the middle of aboral surface of disc. It gives out two or three small branched and brownish outpocketings called **rectal caecae** which are placed inter-radially. Rectal caecae are excretory glands; they secrete a brownish fluid.

5. Anus. The intestine open on the aboral surface of the central disc by a small rounded aperture, the **anus.**

Digestive Glands

The digestive glands of starfish are ten pairs of long brownish or greenish **pyloric caecae.** There are two pyloric caecae in each arm, each of which is composed of an elongated mass of glandular cells suspended in the coelom of the arm by a dorsal mesentery. Each pyloric caecum consists of double series of hollow lobulated sacs that open into a central tube duct. Two ducts of a pair of pyloric caeca of each arm unite to form a main **pyloric duct** that opens into the pyloric stomach at one of its angles.

Histology

The entire digestive tract is lined with a ciliated epithelium, and the cilia are in the ducts of the pyloric caeca arranged to create fluid currents, both incoming and outgoing. Gland cells are particularly abundant in the cardiac stomach lining.

Pyloric caeca function as pancreas and liver of vertebrates. For example, the wall of sacs of pyloric caeca contains numerous **mucous cells** and **secretory** or **granular cells.** Mucous cells produce mucus and the secretory cells secrete enzymes which help in the digestion of proteins, carbohydrates and fats. The caeca not only help in digestion but also store the reserve food. Pyloric ducts also convey wastes out to the rectum, where the rectal caeca, act as pumps for expulsion through anus.

Food and Feeding

Starfish (or sea star) is carnivorous, feeding on any slow moving animal but mainly molluscs such as oysters, clams, mussels, snail, etc. The fish, crabs and barnacles also make its food.

The prey is seized by the tube feet and is folded over by the arms. The cardiac stomach is everted out and is wrapped around the prey. Small animals are swallowed.

Predatory starfish adopts another fascinating method to feed on the shelled molluscs such as clams, oysters and other bivalves. During feeding, a sea star creeps over the clam, holds it with tube feet and brings the free margins of the shell close to its mouth. Sea star arches its body assuming a umbrella-like posture. It firmly attaches its tube feet to both shell valves and tries to pull apart them (these two shell valves are held tightly together by the powerful adductor muscles). The tube feet gripping the shell valves exert a steady pull till the adductor muscles of the clam are exhausted and give way. As the valves gape, the sea star inserts its everted cardiac stomach into the mantle cavity of clam to devour it. The everted cardiac stomach of some sea stars can squeeze through a space as slight as 0.1 mm. The gape increases as digestion ensues and the clams adductor muscles are attacked.

Digestion, Absorption and Egestion

Digestion in sea star is primarily extracellular. After everting the cardiac stomach over the captured prey, sea star pours out digestive enzymes secreted by the cardiac stomach and pyloric caeca over the prey. The enzymes digest the visceral organs of clam; they thus, converts **proteins** into **peptones** and **amino acids; polysaccharides** (carbohydrates) into **glucose** and **fats** (lipids) into **fatty acids** and **glycerol.** When the digestion is partially completed, the sea star withdraws its stomach along with the digested food by means of its retractor muscles and moves on leaving behind the empty shell of the prey. Remaining extracellular digestion of food occurs in cardiac stomach.

Products of stomach digestion are carried by ciliary tracts up the to stomach wall and through pyloric ducts into the pyloric caeca, where digestion is both **intracellular** and **extracellular** and **absorption** occurs. Products of digestion may be stored in the cells of pyloric caeca or passed through the caeca into the coelom for distribution. The undigested food is egested out through the anus.

50.7 RESPIRATION

Respiration in *Asterias* is brought about by the **dermal branchiae** or **papulae** (Fig. 50.19) and **tube feet.** Papulae are scattered over the aboral surface and tube feet project in the ambulacral grooves. Both these structures are thin-walled and enclose a part of coelom. They bring the coelomic fluid very close to the surrounding sea water for exchange of respiratory gases.

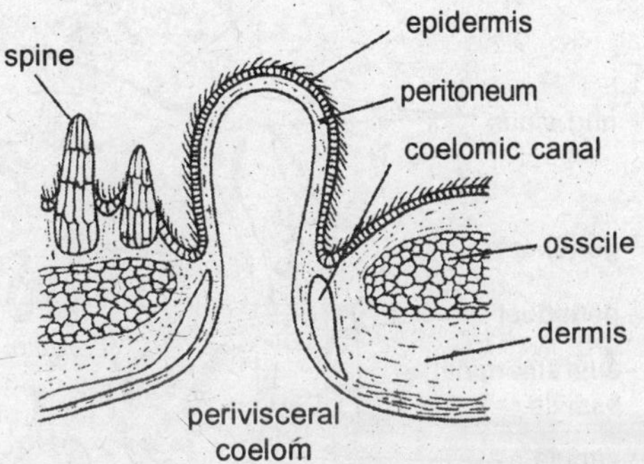

Fig. 50.19. Starfish. Section through papula.

50.8 CIRCULATORY SYSTEM

Asterias lacks a true-blood vascular system. The system which is responsible for circulation of digested food to various body organs is often known as **circulatory system.** The so-called circulatory system of *Asterias* includes following two systems (*i*) **perihaemal system;** (*ii*) **haemal system** (Fig. 50.20).

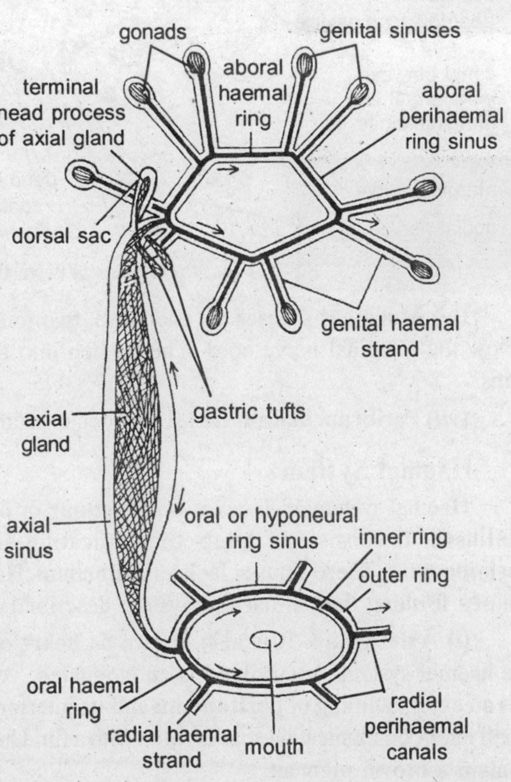

Fig. 50.20. *Asterias*. Haemal and perihaemal system.

1. Perihaemal System

It is derived from the coelom and consists of a series of tubular sinuses. These enclose the sinuses of the haemal system except the gastric haemal tufts. The perihaemal system comprises of the following perihaemal sinuses:

(*i*) Axial sinus. It is thin walled, vertical, wide tubular coelomic cavity enclosing the axial gland and stone canal. The axial sinus along with axial gland and stone canal form the **axial complex.**

(*ii*) Aboral perihaemal ring sinus. It is a tubular pentagonal sinus around the intestine, lying just inside the aboral wall of the central disc. It communicates with the axial sinus.

(*iii*) Genital sinuses. Aboral perihaemal ring sinus gives off five pairs of **genital branches,** one pair in each arm. These enclose the gonads and the genital haemal strands.

(*iv*) Oral hyponeural ring sinuses. At its oral end, the axial sinus opens into the inner division of a circular channel, called the **oral hyponeural ring sinus** which runs around the mouth and above the nerve-ring. It is a large tubular sinus and is divisible into an inner narrow and an outer wide ring by an oblique circular septum called **the haemal strand.**

(v) Radial perihaemal sinuses. From the outer part of the oral ring sinus arise five **radial hyponeural** or **perihaemal sinuses,** each extending through an arm enclosing the radial haemal strands. These sinuses also give out fine channels into the tube feet.

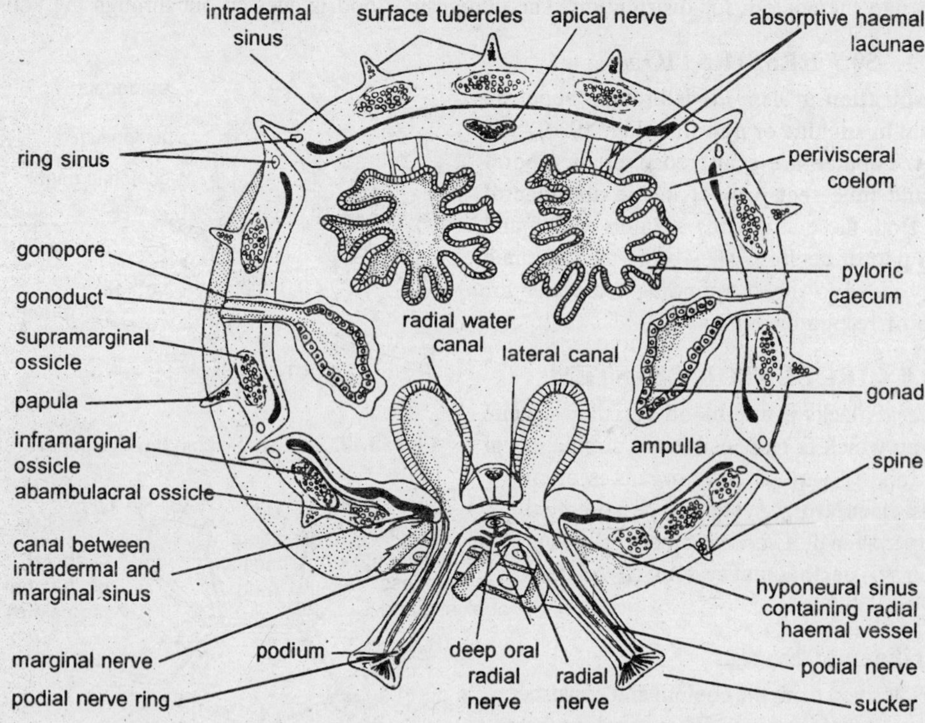

Fig. 50.21. Sea star (=starfish; *Asterias*). T.S. Arm.

(vi) Marginal sinuses. In each arm, two longitudinal marginal sinuses run one on each side below the marginal nerve cord. These open into the radial perihaemal sinuses of their respective arms.

(vii) Peribranchial sinuses. These occur as circular sinuses around the basal part of the papulae.

2. Haemal System

Haemal system of *Asterias* is of **lacunar** or **open type** like the haemocoel of Arthropoda and Mollusca. It consists of inter-communicating sinuses filled with **coelomic fluid** containing **coelomocytes.** These sinuses lack an epithelium. Haemal system is largely enclosed in the coelomic sinuses forming the perihaemal system described above. It comprises the following parts.

(i) Axial gland. It is also known as **heart** or brown **gland** and forms the principal part of the haemal system. Axial gland is an elongated, vertical, fusiform, brownish and spongy gland. It has an external lining of **peritoneum** and its interior is filled with connective tissue containing many small inter-communicating spaces filled with a fluid having amoeboid **coelomocytes.** The coelomocytes contain a brown pigment.

The axial gland is connected with the oral haemal sinus at its oral end and with the aboral haemal sinus at its aboral end. A small terminal **head process** arises from the aboral end of the axial gland. Head process lodged in a separate, closed contractile coelomic sac called **dorsal sac.** The dorsal sac is situated below the madreporite, close to the ampulla of the stone canal. A pair of **gastric tufts** arise from the haemal sinuses in the wall of cardiac stomach and open into the axial gland near its aboral end. Digested food from the stomach passes into the haemal circulation through the gastric tufts.

***(ii)* Aboral haemal ring.** It is a pentagonal ring canal situated beneath the aboral surface of the central disc. It gives off five pairs of **genital haemal strands** to the gonads.

***(iii)* Oral haemal ring.** It is the circular haemal sinus, located around the mouth just below the ring canal of the water vascular system.

***(iv)* Radial haemal sinuses.** These arise radially from the oral haemal ring. In each arm extends one radial haemal sinus along the floor of the ambulacral groove. Each radial haemal sinus gives off branches to the tube feet or podia.

Function. The haemal system acts as a pathway for the distribution of food substances carried by the coelomocytes. The flow of fluid (*i.e.* colourless blood) within it is maintained by the contractile activity of the dorsal sac (or **heart;** see **Rupert** and **Barnes,** 1994). For example, in *Asterias forbesi* dorsal sac beats rhythmically—6 beats per minute. However, the pattern of circulation of fluid is unknown.

The axial gland acts as a genital stolon, producing sex cells which reaches the gonads through the aboral haemal ring and its branches.

50.9 EXCRETION AND OSMOREGULATION

In most echinoderms dissolved nitrogenous wastes (ammonia) diffuse across body surfaces to the outside. This type of excretion occurs across the podia and papulae in sea star and other asteroids. Precipitated material and other particulate wastes are phagocytosed by certain coelomocytes in the body fluids and then discharged by various methods. In sea star, waste-laden coelomocytes accumulate in the papulae, which then pinch off their distal ends, expelling the cells and waste material. Some studies indicate that the rectal glands or rectal caecae may also be involved in excretion (**Brusca** and **Brusca**, 2003).

Echinoderms are generally considered to be strictly marine, **stenohaline** creatures (Box 50.5). Consequently, they do not have problem of osmotic and ionic regulation. The evidence to date suggests that echinoderms are **osmo-conformers.** Both water and ions pass relatively freely across thin body surfaces and the tonicity of the body fluids varies with environmental fluctuations. There appear to be some ionic regulation through active transport, but it is minimal.

Box 50.5.

Animal response to osmotic conditions of the medium

There are two extreme patterns of response to osmotic conditions of the environment. Animals may be osmotically labile (dependent) and their body fluid concentration may change with the medium. These are **osmoconformers** or **poikilosmotic animals.** Other animals are osmotically stable (independent) and when the medium changes, the internal concentrations of the body fluid remains constant or unchanged. These are **osmoregulators** or **homoisomotic animals.**

Poikilosmotic animals which can tolerate wide ranges in salinity are known as **euryhaline,** *e.g., Mytilus, Aplysia.* On the other hand poikilosmotic animals which cannot tolerate wide ranges in salinity are known as **stenohaline animals,** *e.g.,* most marine invertebrates (**Singh** and **Kumar** 2005).

5.10 REPRODUCTIVE SYSTEM

Autotomy and Regeneration

Echinoderms possess great power of **regeneration.** *Asterias* is capable to regenerate its any lost part of body at any time. Moreover, if an arm is injured or held up, *Asterias* usually cast it off near the base at the fourth or fifth ambulacral ossicle. This is called **autotomy.** The opening left in the central disc by broken off arm is immediately closed by the contraction of the adjacent body wall musculature for the protection of internal body organs and regeneration of new arm starts at that place. Autonomy is seen in most **ophiuroids,** some **asteroids** and some **holothuroids** but does not occur in **echinoids.** For example, many brittle stars, when are taken out of water, break off portions of their arms into pieces till the central disc completely devoid of arms is left.

A disc deprived of all of its arms regenerates. In *Asternia vulgaris*, a single arm with a portion of disc regenerates an entire animal. But in *Linckia*, an arm totally devoid of disc can regenerate complete animal (Fig. 50.22). Specimens with regenerating arms at the base of the large original arm are called **comets.**

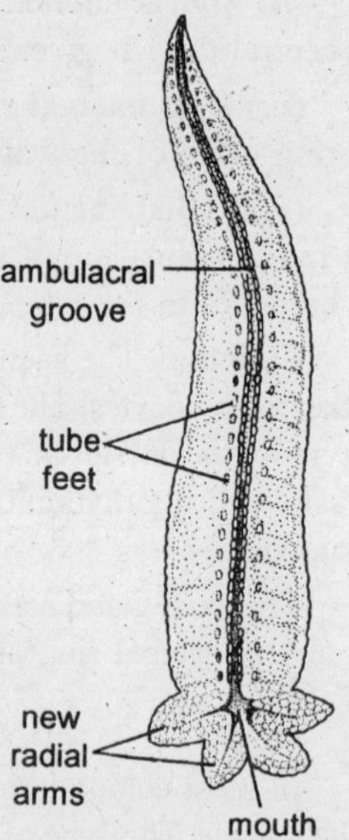

Fig. 50.22. Comet stage of *Linckia*.

In **crinoids,** the arms as well as the visceral sac of the central disc are regenerated when these become accidentally removed. In **holothuroids,** the oesophagus or cloaca with Cuvierian organs is ejected on slightest danger (autotomy) and is regenerated again.

Reproduction in *Asterias* is mainly sexual. Most species of *Asterias* are **unisexual** or **dioecious,** *i.e.*, sexes are separate. Sexual dimorphism is absent. Reproductive organs of *Asterias* are of primitive type. They lack copulatory organs, accessory glands, receptacles for storing ova and reservoirs for storing mature spermatozoa.

Gonads

The **testes** (male gonads) and **ovaries** (female gonads) are morphologically similar but they show colour variation. The testes are pale grey and the ovaries vary from pink to orange. There are five pairs of gonads in either sex, a pair in each arm. The proximal end of gonads are attached to the aboral body wall and they lie laterally freely between pyloric caeca and ampullae of the tube feet.

Fig. 50.23. *Asterias*. Gonads. A—An arm is cut open to show gonads. B—T.S. arm showing position of gonads.

The size of gonads varies with season of the year, being largest during the breeding period (late spring). At maturity gonads occupy a considerable portion of the perivisceral space. Each feathery tuft or bunch of grapes consisting of membranous and rounded follicles. It is enclosed in a genital sinus of the perihaemal system. Microscopic examination reveals that the gonad is lined by a germinal epithelium with a connective tissue matrix containing germ cells. From the proximal end of each gonad arises a short ciliated **gonoduct** which opens out laterally on the aboral surface by a minute **gonopore.**

Spermatozoa and ova are discharged by the male and female sea stars respectively into the sea water. Release of gametes is brought about by substances produced in the radial nerve.

In *Asterias*, there is a single breeding season in a year. It is the late spring and appears to be induced by the rising spring temperature. A single female lays about 200 million eggs in a season.

Fertilization

In *Asterias*, fertilization is external; it takes place in sea water.

Development

The embryological development of *Asterias* is **indirect** and includes various larval stages. The fertilized egg or zygote is spherical, 0.5 mm in diameter and contains little amount of yolk. The **cleavage** is rapid, **holoblastic,** equal, indeterminate and radial type (see Box 50.6). As a result, by the second day, a single layered, hollow, spherical, ciliated embryo, called **blastula** or **coeloblastula** is formed. Blastula contains a fluid-filled central cavity, the **blastocoel** and swims about in water freely. The blastula undergoes embolic **invagination** at its vegetal pole and becomes a two-layered, ciliated, cup-like **gastrula.** Its outer layer is formed of **ectoderm** and inner layer of **endoderm** (Fig. 50.24). The cavity of the gastrula which is lined by endoderm is known as **gastrocoel** or **archenteron.** It opens to the exterior by a wide aperture, called **blastopore.** On the ventral side of the embryo, a tubular ingrowth of ectoderm forms the **mouth.**

Box 50.6.

1. **Holoblastic cleavage.** When entire egg is divided by each cleavage furrow. When it produces equal-sized blastomeres, it is called **equal holoblastic cleavage.**
2. **Indeterminate cleavage.** When the fate of blastomere is not fixed.
3. **Radial cleavage.** When axes of the cleavage furrows are parallel or at right angles to the axis extending from animal to vegetal pole.

During gastrulation, the advancing tip of archenteron buds off **mesenchyme cells** into blastocoel, which form the **mesenchyme** or **mesoderm.** The advancing blind end of archenteron evaginates as two pouches, which finally get pinched off as the right and left sacs. These sacs enclose anterior part of archenteron, the **enterocoel,** and produce **coelom,** its mesodermal lining and water vascular system. Thus, in *Asterias*, mesoderm has a dual origin and coelom is **enterocoelous.** The embryo at this stage becomes a free-swimming larva.

Larval Development

The development of sea star includes the following larval stages:

1. Dipleurula larva. This is first larval stage and is commonly found in all echinoderms. It is egg - shaped having a bilaterally symmetrical body. The cilia which earlier cover the surface of embryo uniformly, become restricted to two ciliary wavy bands — a **peri-oral band** surrounding a concave area around the mouth and an **adoral band** lying inside the mouth. **Stomodaeum** develops as an ectodermal invagination on the ventral side of embryo and becomes continuous with the archenteron. The opening, thus, developed forms the **mouth** of larva. The archenteron gets differentiated into oesophagus, stomach and intestine. The blastopore becomes the **anus.** With these changes embryo develops into **dipleurula larva** which is capable of independent existence. This larva actively feeds on diatoms, etc. The adoral band of cilia helps in collecting the food particles. Dipleurula larva swims near the surface and rotates clock-wise with the help of cilia of peri-oral band.

2. Bipinnaria larva. In *Asterias*, the free swimming larva which is hatches out of the egg is called **bipinnaria larva.** It develops from zygote in about one week. It is a bilaterally symmetrical larva and possesses two ciliated bands: 1. **Preoral ciliated band** which surrounds the preoral lobe of larva; 2. **Postoral ciliated band** which appears to be longitudinally placed and form a complete

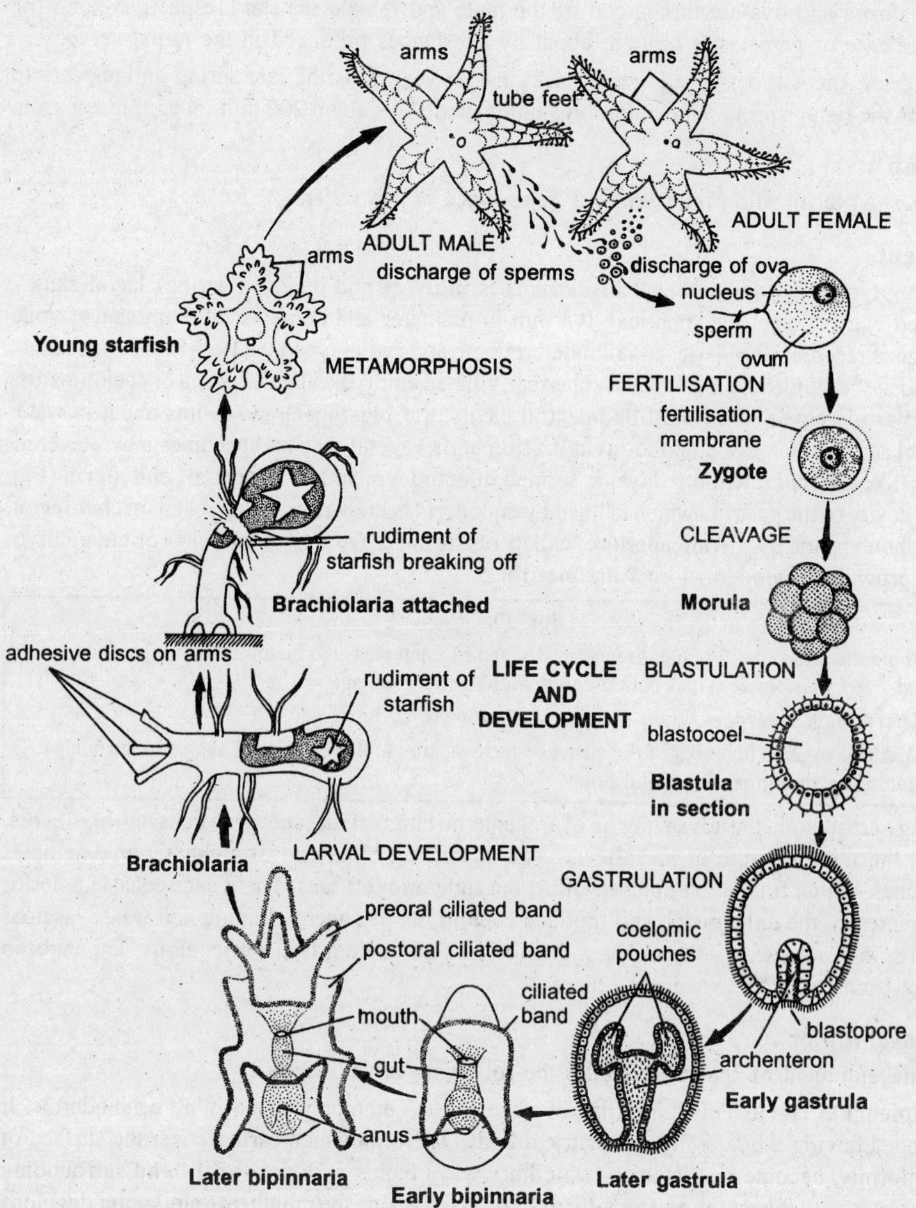

Fig. 50.24. *Asterias*. Development and life history.

ring between the mouth and anus. Both of these ciliated bands are continued over a series of prolongations or projections of the body, called **lobes** or **larval arms** (Fig. 50.25). The name and number of arms developing from preoral and postoral ciliated bands are as follows:

Name of lobe or arm	Number
1. Dorso-median	One
2. Ventro-median	One
3. Pre-oral	Two

4. Antero-dorsal	Two
5. Postero-dorsal	Two
6. Post-oral	Two
7. Postero-lateral	Two

The pre-oral and ventro-median arms develop from the pre-oral ciliated band and rest of the arms develop from the post-oral ciliated band. The arms are provided with muscles and are contractile in a nature. Inside the body occur the coelomic apparatus and the alimentary canal. The bipinnaria larva feeds, on diatoms, etc., by creating food-bearing currents by ciliary tracts in the stomodeal wall. It swims freely by forwarding its anterior end, with a clockwise rotation. After sometime bipinnaria larva develops into the next larval stage called **brachiolaria larva.**

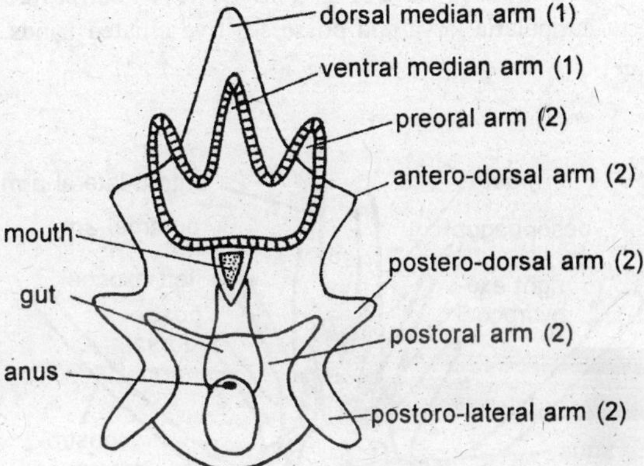

Fig. 50.25. *Asterias*. Bipinnaria larva.

3. Brachiolaria larva. The lobes of bipinnaria larva become modified into long, slender and ciliated **larval arms.** From the preoral lobes arise three short and non-ciliated processes ending into adhesive **fixation discs or suckers.** These are called **brachiolar arms.** These arms are devoid of calcareous rods and have prolongations from the coelomic cavity. Its alimentary canal consists of mouth, stomach, intestine and anus. Brachiolaria larva swims and feeds like bipinnaria larva. The appearance of sucker marks the beginning of metamorphosis.

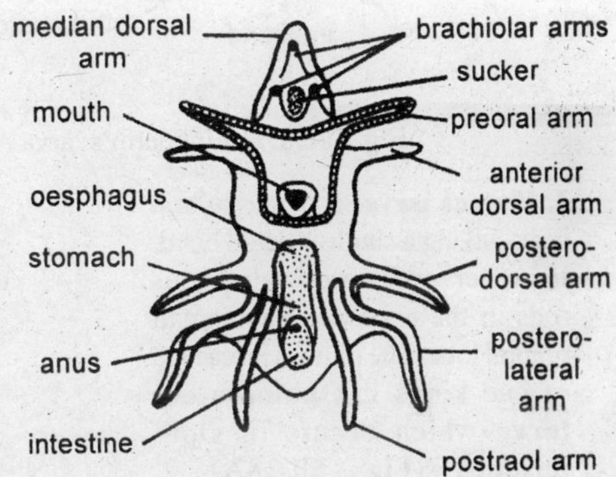

Fig. 50.26. Brachiolaria larva.

Other echinoderms produce a variety of larval forms such as **auricularia, doliolaria, ophiopluteus, echino-pluteus,** etc. (see Box. 50.7).

Box 50.7.

Other larval forms of echinoderms

1. Auricularia larva. In most holothuroids, the eggs are rich in yolk and deve-lopment is direct, but in some cases the egg hatches into a free - swimming bila-terally symmetrical larva. It is known as **auricularia larva** (Fig. 50.27A). It is transparent pelagic orga-nism of about 1 mm size. The body is provided with single longitudinal ciliated band which forms a **pre-oral loop** around the mouth and **anal loop** encircling anus. These help in swimming. The **pre-oral lobe** is very well - formed. There are no calcareous rods, being replaced by spheroids, star - shaped or wheel - like bodies.

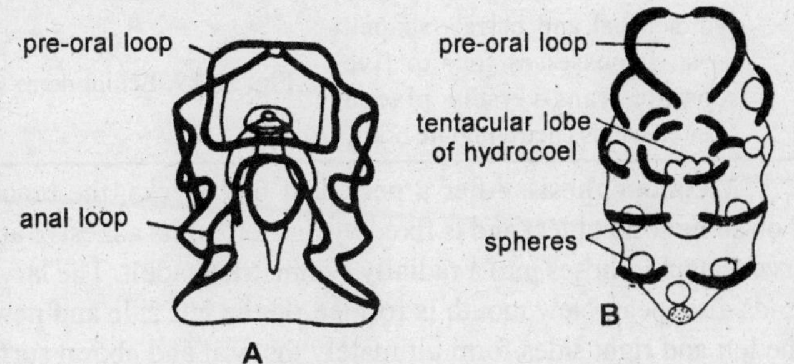

Fig. 50.27. Enhinoderm's larva A—Auricularia; B—Doliolaria.

The interior of larva is occupied by the curved gut with saccular stomach, hydrocoel and right and left stomocoels. The hydrocoel is produced into primary tentacles and communicates with the hydropore by canal.

The auricularia larva metamorphoses into second larval stage known as doliolaria larva (Fig. 50.27B).

2. Doliolaria larva (Fig. 50.27B). It is a barrel-like larva of Holothuroidea which develops from auricularia larva and possesses five ciliated bands. It is also called **pupa stage.**

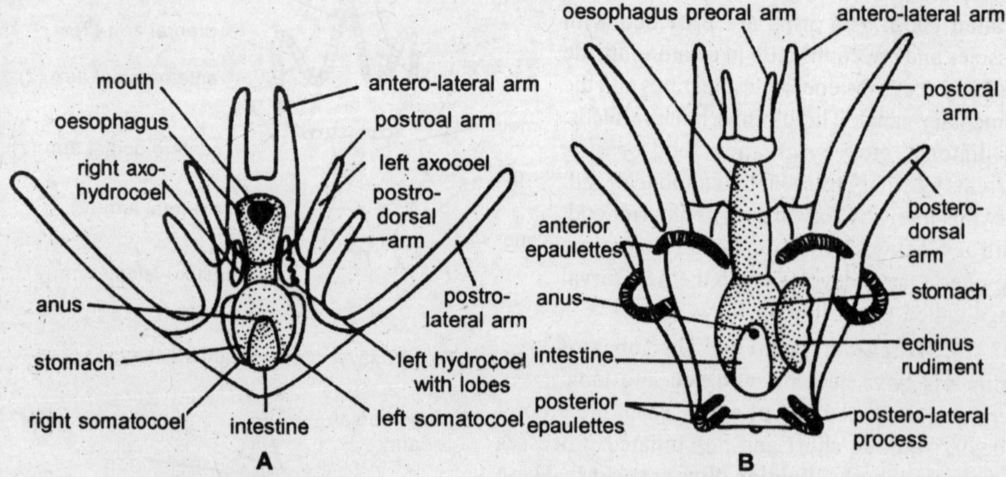

Fig. 50.28. Echinoderm's larva A—Ophiopluteus; B—Echinopluteus.

3. Pluteus larva. It is auricularia larva having a single ciliated band. However it contains calcareous rods in the arms and have small preoral lobe. The pluteus larvae are of two kinds 1. **Ophiopluteus larva** which occurs in Ophiuroidea (Fig. 50.28A) 2. **Echinopluteus larva** which occurs in Echinoidea (Fig. 50.28B).

4. Vitellaria or Yolk Larva. This larval stage is present in *Antedon*. It is free-swimming, bilaterally symmetrical and barrel- shaped larva. It possesses four to five separate trans-versely placed ciliated bands encircling the body.

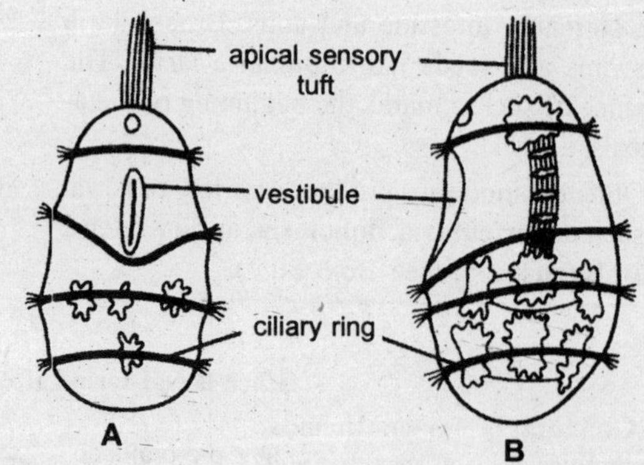

Fig. 50.29. Echinoderm's larva A—Auricularia; B—Doliolaria.

Metamorphosis. After a period of 6–7 weeks, the brachiolaria larva settles on the bottom or on some solid object and is fixed by the help of its adhesive arms. Now the bilaterally symmetrical larva metamorphoses into a radially symmetrical adult. The larval mouth and anus close. The ciliary bands disappear. New mouth is formed on the left side and new anus on the right side of the larva. The left and right sides form ultimately the oral and aboral surfaces respectively. Five lobes, called **arm rudiments,** develop around oral-aboral axis. Skeletal elements develop on the arm rudiments and the radial canals grow into the arms. The coelom of the adult develops from the right and left coelomic pouches of the larva. In each arm two pairs of outgrowths from the coelom form the first

tube feet and serve for attachment. Further complex reorganisational changes result in the formation of adult *Asterias*. The newly detached rudiment of the body of sea star is about 1 mm in diameter and contains short stubby arms.

QUESTIONS

Long Answer Questions

1. Give an account of habit and habitat of starfish. Describe its external features. (*Bundelkhand 1988; Kumaun 1992; Kanpur 1993; Purvanchal 1989; Raipur 1990; Meerut or CCS 1990, 1996; Bhopal 1993; Jabalpur 1991; Gorakhpur 1992; Rohilkhand 1993, 1994, 1996*)
2. Describe structures present on the oral and aboral surfaces of starfish. (*Agra 1988; Jiwaji 1992*)
3. Describe the external features of *Asterias*. Why the oral-aboral axis of its body is not the antero-posterior axis. (*Kanpur 1998*)
4. Give an account of skeletal system of starfish. (*Allahabad 1990; Kanpur 1990*)
5. Describe in detail the water vascular system or ambulacral system of starfish. (*Allahabad 1988; Jiwaji 1988; Rajasthan 1988, 1991; Rohilkhand 1989; Bundelkhand 1989; Madras 1989; Sagar 1989; Osmania 1989; Bhopal 1989, 1991; Awadh 1989, 1991, 1992; Meerut or CCS 1989, 1993, 2009; Guwahati 1989, 1992, 1994; Gorakhpur 1989, 1997, 1998; Himachal 1990, 1991, 1993, 1995; Kanpur 1990, 1995, 1997; Agra 1991; Magadh 1991, 1994, 1995; Lucknow 1992, 1995, 1998; Jabalpur 1992; Bilaspur 1993; Vikram 1993; Rewa 1993*)
6. Give detailed account of the structure and function of water vascular system of *Pentaceros*. (*Gorakhpur 1988, 1992, 1993; Bhopal 1990; Jabalpur 1990, 1993; Lucknow 1991; Kanpur 1992*)
7. Give an account of ambulacral system of *Asterias*. How does it help in the locomotion of the animal. (*Punjab 1991*)
8. Describe the locomotory organs and mode of locomotion in starfish. (*Bangalore 1990; Lucknow 1993*)
9. Write the structure of tube feet and how are they extended and retracted. (*Guwahati 1995*)
10. Give an account of the digestive system and mechanism of feeding in a starfish (= sea star). (*Indore 1989; Himachal 1989; Andhra 1989, 1992; Mysore 1990, 1992; Lucknow 1990, 1996, 1997, 1998; Kanpur 1992; Burdwan 1992*)
11. Write a short note on physiology of digestion. (*Lucknow 1994*)
12. Describe the circulatory system of *Asterias*. (*Jammu 1991; Madras 1992*)
13. Describe the reproductive system of *Asterias*. Where does fertilization occur in this animal? Add a note on its development. Describe various larval forms of echinoderms. (*Kerala 1989; Awadh 1990*)
14. Describe life-history of *Asterias*. (*Osmania 1989, 1992; Madurai 1990*)
15. Why is larva of starfish of particular significance? (*Guwahati 1995*)
16. Write an essay about the larval forms of Echinodermata. (*Madurai 1995*)
17. Describe locomotion, feeding, respiration and habitat in starfish. (*Himachal 1994*)
18. Describe symmetry of Echinodermata. (*Kanpur 1989; Madras 1990, 1992*)
19. Describe water vascular system in Echinodermata. (*Magadh 1992; Nagarjuna 1994*)

Short Answer Questions

1. Write short notes on the following:
   (*i*) Symmetry of Echinodermata. (*Rewa 1994*)
   (*ii*) Endoskeleton of *Echinus*. (*BHU 1991*)

(*iii*) Pedicellaria (*Sagar 1988, 1990; Jiwaji 1988, 1992; Lucknow 1988, 1991, 1996, 1998; Agra 1989; Rohilkhand 1989; Bundelkhand 1988, 1990; Purvanchal 1989, 1991; Awadh 1990; Garhwal 1990; Punjab 1990, 1992; Meerut or CCS 1990, 1994, 1997; Delhi 1991, Kanpur 1992, 1995, 1997; Himachal 1991, 1992, 1993, 1995; Gorakhpur 1992*)

(*iv*) Water vascular system (*A.P., 1989; Meerut or CCS 1999*)

(*v*) Madreporite (*Punjab 1990, 1991, 1992; Himachal 1990, 1995; Magadh 1991, 1993; Meerut or CCS 1994; Madurai 1994; Lucknow 1997*)

(*vi*) Stone canal (*Sagar 1990; Indore 1992; Meerut or CCS 1994*)

(*vii*) Polian vesicle (*Kanpur 1990; Bundelkhand 1992; Lucknow 1996*)

(*viii*) Autotomy and Regeneration (*Sagar 1989*)

(*ix*) Ambulacral groove (*Awadh 1988*)

(*x*) Tube feet (*Burdwan 1988; Awadh 1988; Agra 1988, 1990; Bundelkhand 1989; Gorakhpur 1989; Kanpur 1990, 1996; Punjab 1990; Magadh 1991, 1995; Garhwal 1992; Guwahati 1992, 1996; Meerut or CCS 1994, 1995; Himachal 1995; Lucknow 1998*)

(*xi*) Tiedmannn's bodies (*Punjab 1990; Guwahati 1991, 1993; Lucknow 1994, 1996*)

(*xii*) Locomotion in *Asterias* (*Gorakhpur 1991*)

(*xiii*) Axial organ (*Meerut or CCS 1993*)

(*xiv*) Axial gland (*Himachal 1995*)

(*xv*) Bipinnaria larva (*Awadh 1988; Sagar 1988; Burdwan 1988, 1992; Allahabad 1989; Agra 1989; Gorakhpur 1989; Bundelkhand 1989; Magadh 1989; Purvanchal 1989; Kanpur 1989, 1993; Himachal 1989, 1991, 1995; Delhi 1991; Garhwal 1992; Lucknow 1993; Rohilkhand 1993; Meerut or CCS 1994*)

(*xvi*) Auricularia larva

2. Draw well labelled diagrams of the following:

(*i*) External features of starfish. (*Rohilkhand 1993*)

(*ii*) Aboral view of starfish. (*Lucknow 1994, 1996*)

(*iii*) Oral view of starfish. (*Kanpur 1988; Rohilkhand 1989; Rajasthan 1989, 1992*)

(*iv*) Position of pyloric caeca in the arm of *Asterias*. (*Himachal 1994*)

(*v*) T.S. arm of starfish. (*Meerut or CCS 1988, 1992, 1997, 1999; A.P., 1988, 1992; Gorakhpur 1989; Agra 1989; Lucknow 1993, 1997, 1998; Rajasthan 1990; Jammu 1990; Garhwal 1990; Utkal 1992; Himachal 1995; Kanpur 1995, 1996; Awadh 1992; Purvanchal 1992; Jiwaji 1993*)

(*vi*) Water vascular system or ambulacral system of starfish. (*Sagar 1988; Purvanchal 1989; Rohilkhand 1990; Agra 1990; Jiwaji 1990, 1993; Guwahati 1994; Lucknow 1994*)

(*vii*) Pedicellaria of starfish. (*Garhwal 1990*)

(*viii*) Haemal and perihaemal system of *Pentaceros*.

(*ix*) Bipinnaria larva (*Garhwal 1989; Lucknow 1995*)

(*x*) Brachiolaria larva. (*Himachal 1991, 1993*)

(*xi*) Echinoplutcus larva

3. Differentiate between the following:

(*i*) Polian vesicles and podium (*Lucknow 1997*)

(*ii*) Sea lily and brittle star. (*Lucknow 1996*)

(*iii*) Sea urchin and sea cucumber (*Lucknow 1994*)

Very Short Answer Questions

1. Name two types of pedicellariae present in *Asterias*.

Ans. Pedicellariae with parallel jaws and with cross jaws.

2. Where is madreporite located in *Asterias*.

Ans. On the aboral surface of the central disc in one of the inter-radii.

3. What is bivium?

Ans. Two arms of *Asterias* having madreporite between them.

4. What is actinostome?

Ans. Depression in the middle of the central disc in one of the inter radii.

5. Name the type of digestion in *Asterias*.

Ans. Extracellular as well as intracellular.

6. Mention the three parts of a tube foot.

Ans. Ampulla, podium and sucker.

7. What is the function of Tiedmann's body.

Ans. Formation of phagocytic coelomocytes present in the water vascular system.

8. What is chief excretory matter of *Asterias*.

Ans. Ammonium compounds.

9. What type of cleavage does an egg of sea star (*Asterias*) undergo.

Ans. Holoblastic, radial and indeterminate (or regulative).

10. How many Teidmann's bodies are found in sea star (or star fish). (*C.C.S.U. Meerut 2002*)

Ans. There are five pairs of greatly folded pouches called **recemose vesicles** or **Teidmann's bodies** produced from the inner side of the ring canal of water vascular system of sea star (*Asterias*).

11. What do you call the property of regrowing of lost parts?

Ans. The property of regrowth of lost parts in animals is called **regeneration**. Sea star (*Asterias*) can regrow any lost part of body any time.

12. What is autotomy?

Ans. Automatic breaking of any body part by an animal is called **autotomy.** Sea stars are capable of casting off its any arm, if injured or held up, from near its base.

13. What is a comet?

Ans. A disc of *Asterias* which has lost all of its arms may regenerate all of them. A specimen which has regenerating arms at the base of the large original arm is called a **comet.**

14. Where are eyes of *Asterias* located?

Ans. The tip of each arm bears a small red pigment spot, the eye. Each eye contains several ocelli. They lie below the sensory tentacles and are sensitive to light.

15. Describe Tiedmann's bodies. (*C.C.S.U., Meerut 2009*)

Ans. These are four or five pairs of pouch-like bodies attached to the ring canal of sea stars functioning in production of coelomocytes.

Fill in the Blanks

1. *Asterias* is commonly called ....

2. Sea star belongs to the class .... and phylum ....

3. *Asterias* has .... radial symmetry.

4. Tube feet help the sea star in .... , feeding and ....

5. The respiratory organs of starfish are .... or ....

6. Pedicellariae help the sea star in ....

7. In *Asterias*, tube feet are located in the .... grooves.

8. The coelomic fluid of *Asterias* contains phagocytic cells termed ....

9. The pyloric caeca function as the ....

10. In sea star, fertilization takes place in ....

11. Sea star (*Asterias*) moves with the help of ....

12. The coelom of sea star is ....

13. Dermal papulae of sea star are useful in .... and in respiration.

Multiple Choice Questions

1. Locomotory organs of sea star are
(*a*) podia (*b*) polian vesicles
(*c*) vacuole (*d*) none

2. Epidermis of *Asterias* contains
(*a*) epithelial cells (*b*) gland cell
(*c*) neurosensory cells (*d*) all the above

3. The skeletal ossicles of sea star arise from
(*a*) ectoderm (*b*) mesoderm
(*c*) endoderm (*d*) none of these

4. Which is enterocoelic animal
(*a*) *Taenia* (*b*) *Pheretima*
(*c*) *Palaemon* (*d*) *Asterias*

5. An animal which eats its prey after digesting it is
(*a*) *Physalia* (*b*) Leech
(*c*) sea star (*d*) *Sepia*

6. Ambulacral system of *Asterias* helps in
(*a*) locomotion (*b*) attachment
(*c*) respiration (*d*) all of these

7. Ambulacral system of sea star contains
(*a*) blood (*b*) coelomic fluid
(*c*) water (*d*) mucus

8. In sea star stone canal is associated with
(*a*) blood vascular system (*b*) canal system
(*c*) digestive system (*d*) water vascular system

9. Madreporite and Tiedmann's body is associated with
(*a*) ambulacral system (*b*) nervous system
(*c*) digestion (*d*) all of these

10. Stone canal connects the
(*a*) ring canal and polian vesicle (*b*) ring canal and radial canal
(*c*) madreporite and ring canal (*d*) madreporite and polian vesicle

11. Respiration in *Asterias* occurs in the
(*a*) pedicellariae (*b*) ossicles
(*c*) papulae (*d*) madreporite

12. *Asterias* is
    (*a*) ammonotelic (*b*) uricotelic
    (*c*) ureotelic (*d*) none of these
13. The eggs of *Asterias* are
    (*a*) microlecithal (*b*) macrolecithal
    (*c*) megalecithal (*d*) oligolecithal
14. Larval stage in life history of sea star is
    (*a*) bipinnaria (*b*) brachiolaria
    (*c*) both (*d*) none

ANSWERS

Fill in the Blanks

1. Sea star (starfish)
2. Asteroidea, echinodermata
3. Pentamerous
4. Locomotion, respiration
5. Dermal branchiae, papulae
6. Clearing the body surface of debris and organisms
7. Ambulacral
8. Coelomocytes (amoebocytes)
9. Digestive gland
10. Sea water
11. Tube feet
12. Enterocoelous
13. Excretion

Multiple Choice Questions

1. (*a*) **2.** (*d*) **3.** (*b*) **4.** (*d*) **5.** (*c*) **6.** (*d*)
7. (*c*) **8.** (*d*) **9.** (*a*) **10.** (*c*) **11.** (*c*) **12.** (*a*)
13. (*a*) **14.** (*c*).

51

Phylum Hemichordata (Characters, Classification and Types)

The **hemichordata** (Gr., *hemi* = half; *chorde* = cord) is a small group of worm-like marine invertebrates. They are called hemichordata since notochord in them occurs in the anterior end of the body. Recently, notochord of hemichordates is considered as **buccal diverticulum.** Hemichordata is a group of **deuterostomes** like the Echinodermata and Chordata. It was once considered a subphylum of the chordates (as non-vertebrate chordates). Hemichordates form an evolutionary link between the echinoderms and chordates.

51.1 DEFINITION

Hemichordates are marine, soft-bodied, bilaterally symmetrical, coelomate, worm-like animals that contain diffused nervous system in the epidermis, a buccal diverticulum and paired gill-slits and no definite nephridia. Their body is divided into proboscis (protosome), collar (mesosome) and trunk (metasome).

51.2 OUTLINE CLASSIFICATION

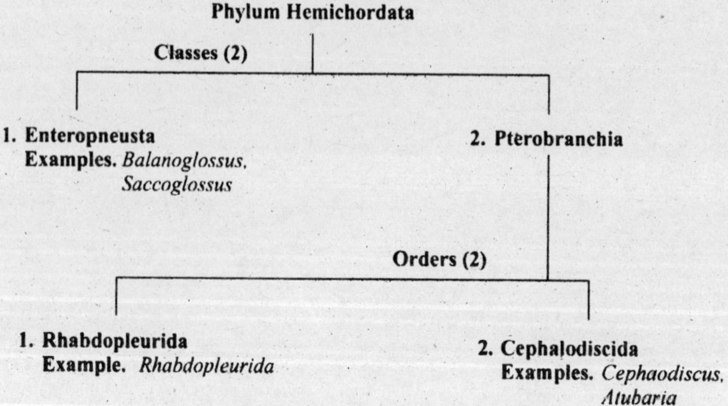

51.3 GENERAL CHARACTERS

1. Simple, solitary or colonial sedentary or free-living and marine animals, found exclusively in sea-water.
2. They are deposit or suspension feeders.

3. They are soft-bodied, bilaterally symmetrical, coelomate and worm-like.
4. Most are burrowers and some live beneath the stones.
5. The body is divided into an anterior muscular **proboscis (protosome),** a short **collar (mesosome)** and a long **trunk (metasome).** Proboscis is separated by a narrow stalk from the collar.
6. Mouth opens infront of collar.
7. Epidermis is ciliated and richly glandular. Cuticle is absent.
8. Coelom is **enterocoelous** and divisible into protocoel, mesocoel and metacoel.
9. A buccal diverticulum, formerly regarded as notochord, present in the proboscis.
10. Alimentary canal is a straight tube having a terminal anus.
11. Gill slits are numerous, paired, present in the anterior part of the trunk.
12. Blood vascular system is simple; blood flows forward in dorsal blood vessel.
13. Excretion through a single **glomerulus** situated in the proboscis.
14. Nervous system consists of an inter-epidermal nerve plexus and a dorsal hollow neural tube.
15. Movement occurs by peristaltic contraction of the proboscis.
16. Sexes are separate, fertilization is external and development may be **indirect** (having **tornaria larva)** or direct.

51.4 DETAILED CLASSIFICATION

The phylum Hemichordata is divided into **two** classes:

Class 1. Enteropneusta

1. Enteropneusta, which means **gut-breather,** includes about 70 species which are known as **acorn worm,** because of the typical shape of their proboscis.
2. They live in shallow sea-water. Most are burrowers (Fig. 51.1) Acorn worms are very **sluggish,** most live in tubes and eat sand or mud; others are free living and ciliary mucus-feeders.

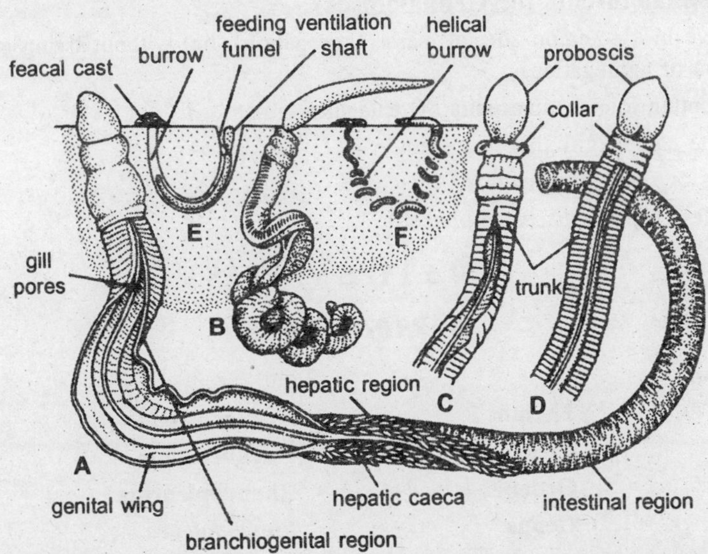

Fig. 51.1. Enteropneust diversity and burrow structure. A—*Balanoglossus*; B—*Saccoglossus*; C—*Ptychodera*; D—*Schizocardium*; E—Burrow arrangement and feeding position of *Balanoglossus*; F—Burrow and feeding position of *Saccoglossus*. Its burrow is helical and trunk is thrown into a permanent twist.

3. Most acorn worms are moderate in size, upto 30 cm or so long, but *Balanoglossus gigas* may reach a length of 2.5 m.
4. They are slimy and very soft animals, *i.e.*, they are fragile. Many smell like iodoform.
5. Many pharyngeal gill slits are present in row on trunk.
6. Alimentary canal is straight.
7. Dioecious. Gonads are located in the trunk coelom just behind the gill slits.

Examples. *Balanoglossus, Saccoglossus, Protoglossus, Planetosphaera, Ptychodera.*

Class 2. Pterobranchia

1. Pterobranchia, which means **wing gills,** includes three genera and fifteen species.
2. They are colonial, sedentary and tubicolous (*i.e.*, living in tubular **zoecium**) hemichordates.
3. Collar bears two curved hollow **arms** which bear ciliated and hollow **tentacles.**
4. Proboscis is flattened like a shield and secretes the tube.
5. Alimentary canal is U-shaped, anus is dorsal, close to the mouth.
6. They are with or without pharyngeal slits.
7. Dioecious. Development includes a ciliated larva which is different from tornaria.

Class Pterobranchia is divided into **two** orders:

Order 1. Rhabdopleurida

1. Colonial hemichordata, each living in tubular zoecium secreted by the epidermis and connected to other members by living stolon.
2. They contain two tentacular arms.
3. No gill slits are present.
4. Single gonad is present.

Example. *Rhabdopleura* (It is the only genus).

Order 2. Cephalodiscida (or Cephalodisca)

1. They live in a common zoecial mass (aggregation), but without living stolon connecting members of aggregation.
2. They contain four to nine pairs of tentacular arms.
3. Only one pair of gill slits is present.
4. There is one pair of gonads.

Examples. *Cephalodiscus, Atubaria.*

51.5 TYPE STUDIES

1. *RABDOPLEURA*

Classification

Phylum	—	**Hemichordata**
Class	—	**Pterobranchia**
Order	—	**Rhabdopleurida**
Genus	—	***Rhabdopleura***

Habit and Habitat

Rhabdopleura is sedentary and marine, found at great depths in the seas of Southern Hemisphere, Burmuda and European Coasts. It forms colonies at bottom. The individual is attached by a stolon.

The colony of *Rhabodopleura* measures about 10 cm and is enclosed in a membranous branched tubes, called **coenoecium** which gives off upright tubes for the zooids (Fig. 51.2). The colonies are attached to stones, corals and shells. The creeping tube is somewhat irregular (Fig. 51.2A) and contains the black **stolon** (Fig. 51.2B), formed of a firm tube containing a cellular column from which new buds arise. The upright tubes are built of regular, successive rings (Fig. 51.2D).

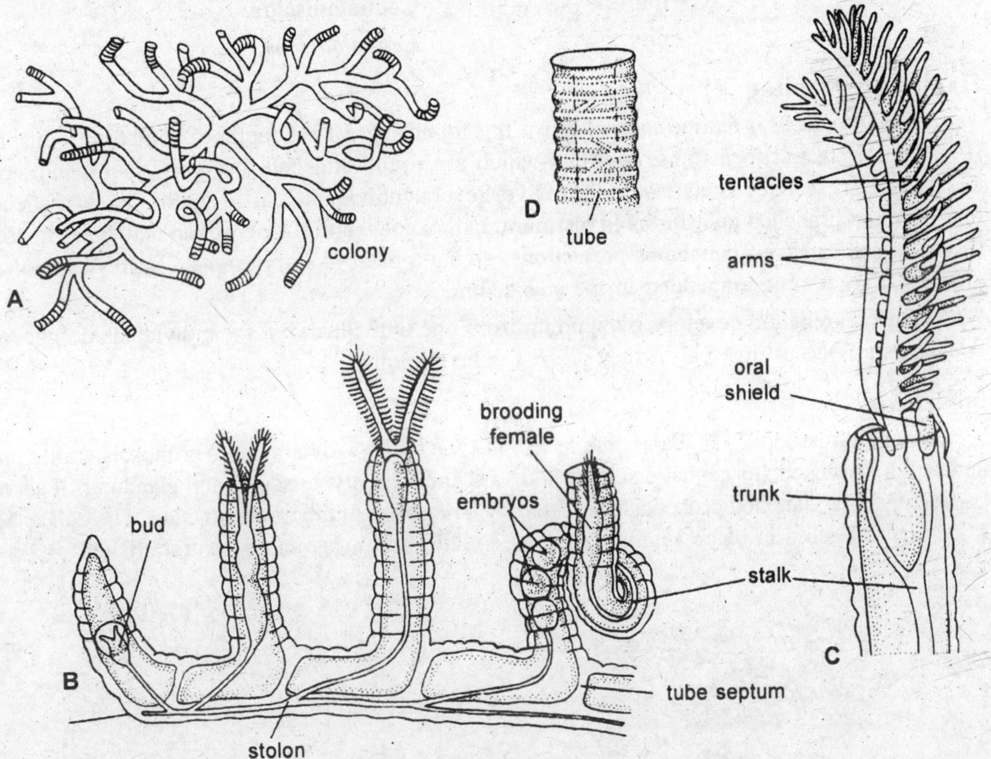

Fig. 51.2. *Rhabdopleura.* A, B—Colonial organization; C—Feeding position of a zooid with cephalic shield folded over the rim of tube; D—Various sections of a tube.

Structure

A zooid measures 1 mm in length. Body colour is brown. The body is divisible into **proboscis (oral shield** or **cephalic shield), collar, trunk** and **stalk. Proboscis** is shield-like, contains a red pigment band and secrete tubes in which animal lives. Peristaltic movement of proboscis helps in locomotion. The **collar** bears two lophophore arms, *i.e.*, they bear ciliated **tentacles.** Both the arms and tentacles are hollow. They help in capturing the food. Collar also bears **mouth** on the ventral side. *Rhabdopleura* lacks gill slits, so it deflects excess water by means of the fold-like oral lamella near the mouth.

The **trunk** is elongated and oval. From its lower surface arises a **stalk** which remains attached to the stolon.

Alimentary canal is U-shaped and anus is located on the dorsal side. **Nervous system** consists of collar-cord with ganglionic cells and nerve fibres.

Sexes are separate but **sexual dimorphism** is not well marked. The saccate gonad lies infront of the stomach. Short gonoducts open behind the collar-trunk boundary. **Fertilization** of large yolky eggs is external. **Development** is **indirect.** The larva is ciliated and becomes adult-like by the metamorphosis. A colony is formed by budding of the stalk (or common stolon).

2. *CEPHALODISCUS*

Classification

Phylum	—	**Hemichordata**
Class	—	**Pterobranchia**
Order	—	**Cephalodiscida**
Genus	–	***Cephalodiscus***

Habit and Habitat

Cephalodiscus is marine and sedentary. It is found at great sea depths of Southern Hemisphere. It is tubicolous and forms **aggregations** which are found attached to corals, shells and stones. *Cephalodiscus* secretes a gelatinous tube to live. Several individuals, called **zooids,** live in association in a common branched **gelatinous investment,** called **coenecium.** Coenecium may be stiffened by firmer spines or have filamentous projections, giving a plant-like appearance. Each zooid inhabits a firmer, darker **tube** embedded in the coenecium.

This aggregation develops by budding from one individual, but the individuals do not retain connections. Due to this reason they are not called colonial.

Structure

A zooid measures less than 5 mm in length. The body is divisible into proboscis, collar, trunk and stalk. **Proboscis** (or cephalic shield) is disc shaped, highly muscular and glandular. It secretes coenecium. The flattened proboscis is crossed by a red band of unknown function. The **collar** bears 8–16 **arms.** Each arm bears small, numerous, ciliated **tentacles** and a ventral ciliated groove or

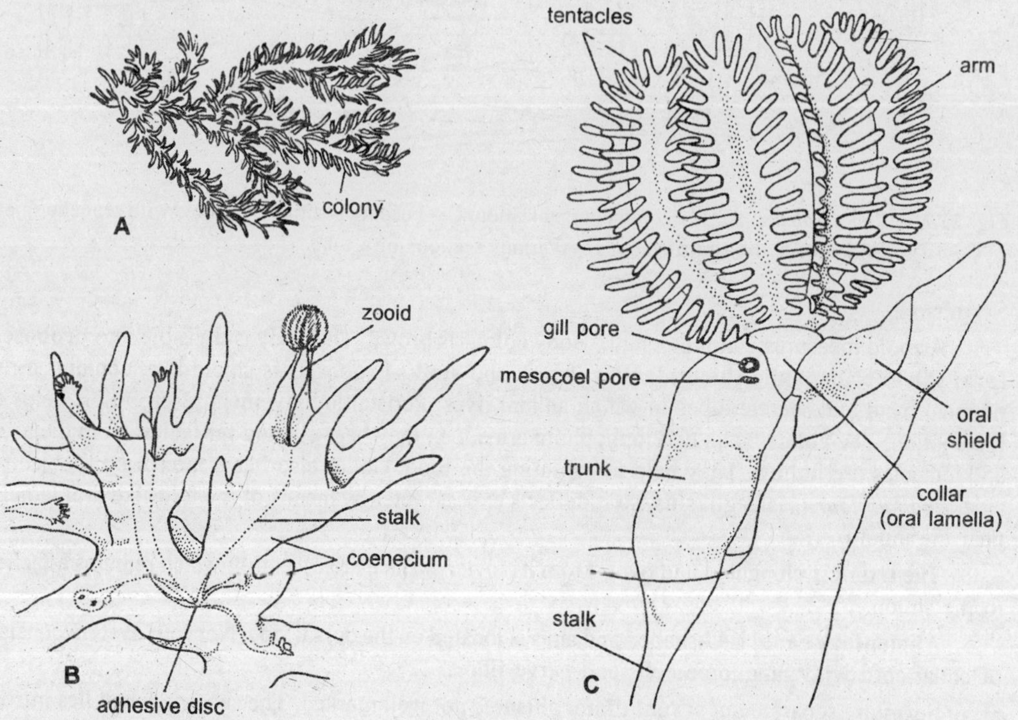

Fig. 51.3. *Cephalodiscus.* A, B—Colonial organization; zooids emerge from openings in the coenecium and perch on branch tips while feeding (B); C—Feeding posture.

gutter. Mucus glands on the dorsal side of the arms help to trap food. Collar also bears **mouth** ventrally and **anus** on dorsal side. Mouth can be closed by proboscis. Flanges of body wall **(oral lamellae)** guide food into the mouth. The **trunk** is oval and elongated, contains a pair of **gill slits.** Posterior part of the trunk continues as a long, contractile **stalk.** Stalk terminates in an adhesive region where the zooid is attached to its tube. Zooids are freely movable. They leave the tubes and cling to coenecium surface while feeding.

Nervous system is epidermal. This consists of a solid collar **ganglion** that join mid-dorsal and latero-dorsal thickenings of the nerve plexus in the cephalic shield.

Gonads are found in the trunk cavities and the development is direct. It also reproduces asexually by budding.

QUESTIONS

Long Answer Questions

1. Give the characters and classification of the phylum Hemichordata.
   (*Rajasthan 1988, 1990; Nagpur 1989; Osmania 1989; Gujarat 1990; Agra 1991; Punjab 1991; Maratha 1992; Kanpur 1992, 1995, 1997; Gwalior 1993; Bundelkhand 1993; Jiwaji 1993*)
2. Give classification, habits and habitat and structures of following:
   1. *Rabdopleura*. 2. *Cephalodiscus*.

Short Answer Questions

1. Write short note on hemichordata. (*Meerut or CCS 1999*)

Very Short Answer Questions

1. Name the excretory organ of hemichordates.
   Ans. Glomerulus.
2. Cite two examples of hemichordates bearing tentacles on the collar.
   Ans. *Rhabdopleura* and *Cephalodiscus*.
3. Name the three coeloms found in hemichordates.
   Ans. Protocoel, a pair of mesocoels, a pair of metacoels.
4. Which are primitive, enteropneusts or pterobranchs?
   Ans. Pterobranchs.
5. In which groups of hemichordates, the epidermal cells are monociliated?
   Ans. Pterobranchia.
6. Name a hemichordate that lacks tentacles on the collar.
   Ans. *Balanoglossus*.
7. Which hemichordates have a dorsal hollow nerve cord?
   Ans. Enteropneusts.
8. Name the coeloms that do not open out in *Balanoglossus*.
   Ans. Metacoels.

Fill in the Blanks

1. Phylum Hemichordata is divided into two classes: .... and ....
2. Acorn worms belongs to the class ....
3. .... are colonial hemichordates.

4. .... is a pterobranch.
5. .... are tricoelomate animals.
6. There is no peritoneum in ....
7. U-shaped digestive tract is found in ....
8. A hemichordate structure considered earlier homologous to the notochord of chordates is named ....
9. Nervous system is largely intraepidermal in ....
10. Collar with tentaculate arms is characteristic of ....

Multiple Choice Questions

1. The acorn worms belong to phylum
   (*a*) Chordata (*b*) Hemichordata
   (*c*) Urochordata (*d*) Cephalochordata
2. A non-contractile heart occurs in
   (*a*) earth worm (*b*) sea star
   (*c*) acorn worm (*d*) apple snail
3. Gastrulation in hemichordates takes place by
   (*a*) invagination (*b*) ingression
   (*c*) delamination (*d*) all these means
4. Stomochord is present in
   (*a*) *Amphioxus* (*b*) *Herdmania*
   (*c*) *Balanoglossus* (*d*) none of the above
5. *Atubaria* belongs to
   (*a*) Rhabdopleurida (*b*) Grapholita
   (*c*) Cephalodiscida (*d*) none of these
6. Larva of Hemichordata is
   (*a*) tornaria (*b*) glochidium
   (*c*) bipinnaria (*d*) nauplius

ANSWERS

Fill in the Blanks

1. Enteropneusta, Pterobranchia 2. Enteropneusta 3. Pterobranch
4. *Rhabdopleura* 5. Hemichordates 6. Hemichordates
7. Pterobranchia 8. Stomochord 9. Hemichordata
10. Pterobranchs

Multiple Choice Questions

1. (*b*) 2. (*c*) 3. (*d*) 4. (*c*) 5. (*c*) 6. (*a*).

52

Balanoglossus (Phylum Hemichordata)

52.1 HABIT, HABITAT AND EXTERNAL FEATURES

Classification

Phylum	—	**Hemichordata**
Class	—	**Enteropneusta**
Genus	—	***Balanoglossus***

Habit and Habitat

Balanoglossus (Gr., *balanus* = acorn, fruit of oak + *glossa* = tongue) is a marine, soft bodied, sluggish, burrowing worm-like animal. It is commonly called **acorn** or **tongue worm.** It lives in U-shaped burrow in shallow water or sand (Fig. 52.1). Burrow is 50–75 cm deep, with a funnel-shaped opening at the anterior end and a round posterior opening. Anterior opening of burrow includes a **feeding funnel** and **ventilation shaft.** The burrow is plastered by mucus. The faecal matter (piles of coiled rope-like castings) lies close to round opening of burrow.

Balanoglossus produces an offensive smell resembling to that of decomposing faecal matter. It is a suspension feeder. It exhibits bioluminescence, emitting bright greenish light.

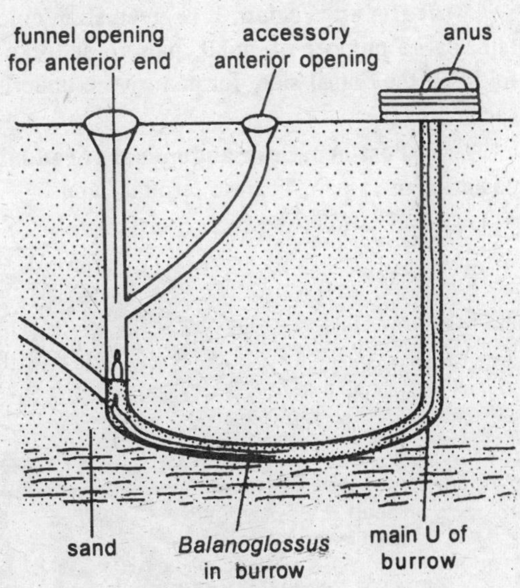

Fig. 52.1. *Balanoglossus*. U-shaped burrow and faecal cast.

External Features

Size, Shape and Colour

Most of species of *Balanoglossus* measures 2–3 cm to 2 metres or more. Body is cylindrical and worm-like. Entire body is covered with cilia. Colour of body is bright with reddish or orange tint. The various body parts are differently coloured.

Division of the Body

The body is unsegmented, bilaterally symmetrical and divided by two constrictions into three parts—(1) proboscis (**protosome**), (2) collar (**mesosome**) and (3) trunk (**metasome**) (Fig. 52.2).

1. Proboscis. It is an anterior club-shaped, highly muscular, tongue-like part. Inside the proboscis, the space is called proboscis-coelom or **proctocoel** which opens by **proboscis pore,** situated near its base (Fig. 52.3). The posterior narrow part of proboscis is called **stalk** which connect it to the collar. Below the stalk is a U-shaped ciliated **preoral ciliary organ** as a depression, to test the quality of food.

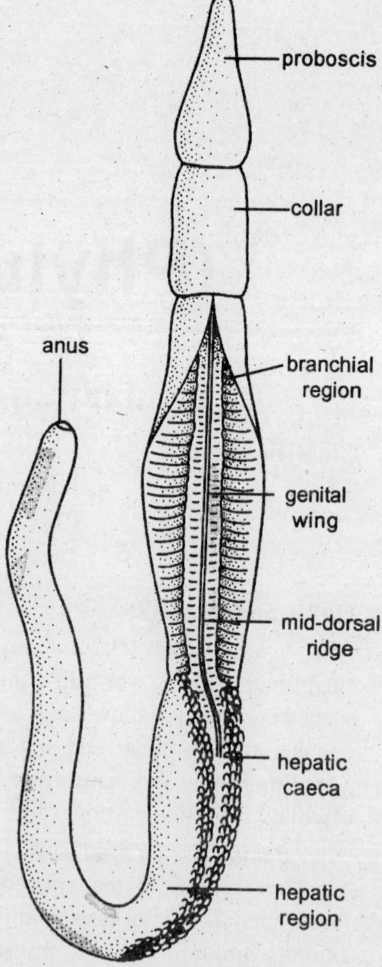

Fig. 52.2. *Balanoglossus.* External features in dorsal view.

2. Collar. It is short, cylindrical middle part of the body. Its anterior part is flap-like or funnel-like called **collarette.** It bears a wide mouth on the ventral surface. The posterior end projects backward and surrounds a part of trunk. The thick collar wall encloses buccal cavity and paired cavities of **collar coelom (mesocoel;** Fig. 52.4). Mesocoel opens into the first pair of gill pouch through **collar pores.**

3. Trunk. It is long, cylindrical with flattened posterior part and forms major part of the body. It bears mid-dorsal and mid-ventral longitudinal ridge. These ridges extend upto posterior end and the nerves and blood vessels run through these ridges. Trunk contains the **trunk coelom** or **metacoel.** Trunk is divided into following three parts:

(*i*) Branchiogenital region. It is the anterior part of trunk. It bears a row of **gill pores** on the dorsal ridge. The number of the gill pores range from a few to 100 or more pairs. Its side wall is produced into a pair of thin, flat, lateral longitudinal flaps or genital wings (Fig. 52.5). These enclose gonads which open out by microscopic gonopores.

(*ii*) Hepatic region. It is greenish in colour and contains large number of small, paired, sac-like **transverse folds** on the dorsal side, formed by the underlying hepatic caeca.

(*iii*) Post-hepatic or abdominal or caudal region. It is posterior most part, bears a terminal anus.

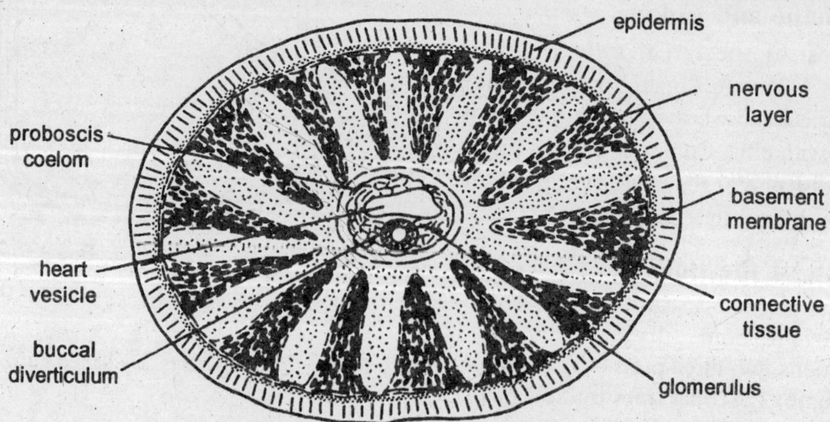

Fig. 52.3. *Balanoglossus.* T.S. of proboscis.

52.2 INTERNAL STRUCTURE

Body Wall

Body wall of *Balanoglossus* is soft, slippery and formed of (*i*) epidermis; (*ii*) muscular layer; (*iii*) peritoneum (Fig. 52.6). It lacks a cuticle.

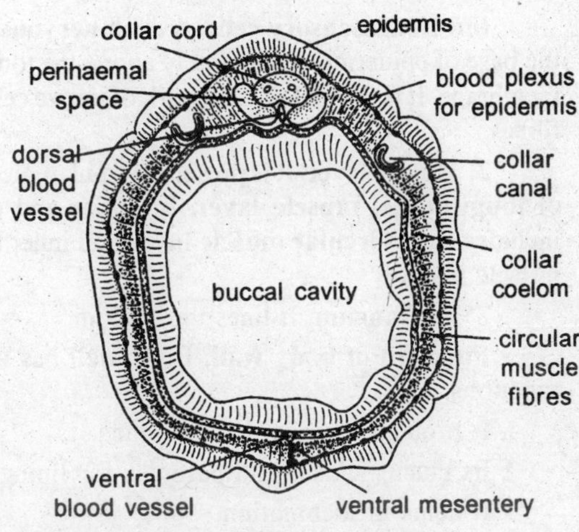

Fig. 52.4. *Balanoglossus*. T.S. of collar.

1. Epidermis. Epidermis is ciliated and richly glandular. It consists of a single layer of tall, slender, columnar cells. They are of three different types:

(*i*) Ciliated columnar cells. They are multiciliated and form bulk of epidermis.

(*ii*) Gland cells. They are most abund-ant in collar region. They secrete mucus and are of three types:

(*a*) Globlet cells. They are flask shaped and have granular cytoplasm.

(*b*) Coarsely granular gland cells.

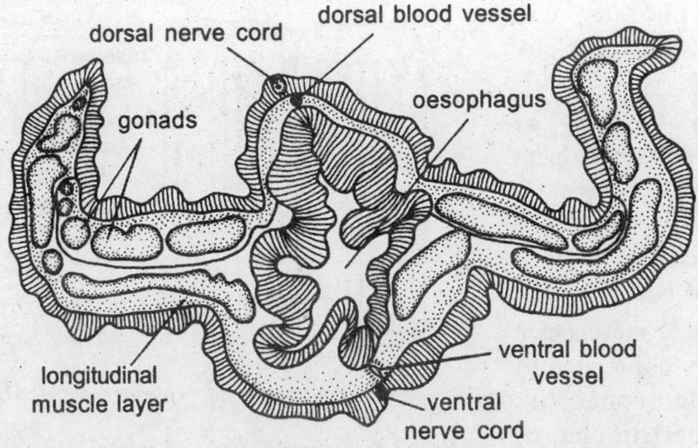

Fig. 52.5. *Balanoglossus*. C.S. through genital wings

(*c*) Reticulated mulberry-like gland cells. Some gland cells produce bromine-containing compounds that impart strong medicinal odour to acorn worms and may protect them from bacterial infection or animal predation.

A **basal lamina** and a connective tissue **dermis** lie beneath the epidermis, but the dermal fibres are well developed. This fact perhaps accounts for the fragility of body.

The collar is marked by five transverse bands of alternating glandular and non-glandular epithelium.

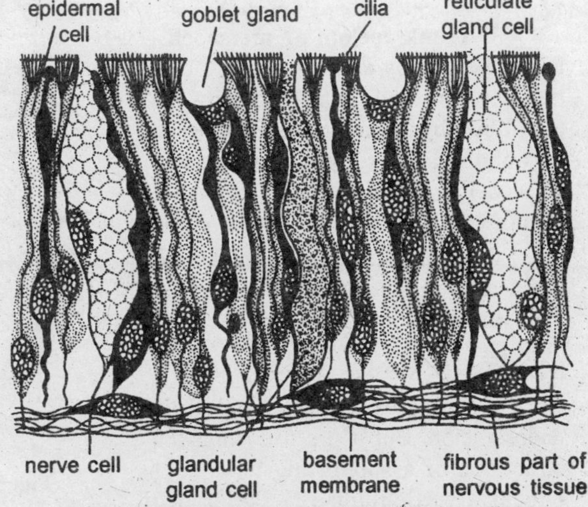

Fig. 52.6. *Balanoglossus*. V.S. of body wall.

(*iii*) **Neurosensory cells.** A thick **nervous layer** lies at the base of epidermis immediately above the tough basement membrane. It is formed of a network of **nerve cells** and **nerve fibres.**

2. Muscle layer. It is present in trunk region in the form of **longitudinal muscle layer.** In collar and proboscis, it includes outer **circular muscle layer** and inner **longitudinal muscle layer.**

3. Peritoneum. It lines the coelom.

Function of body wall. Body wall has the following functions:

1. It protects internal body organs.
2. Its mucus sticks sand particles for lining the burrow.
3. It helps in locomotion.
4. It receives stimuli.

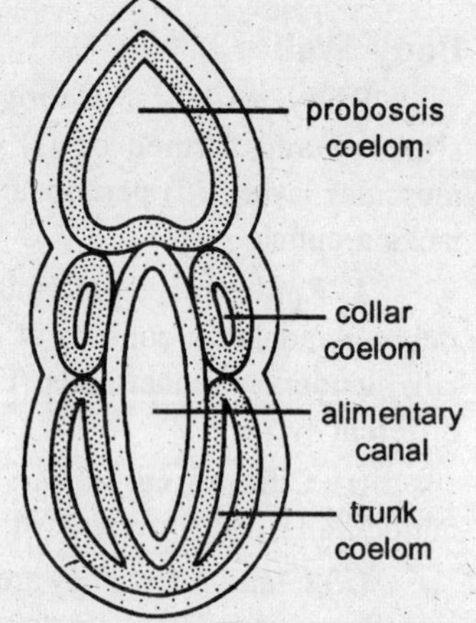

Fig. 52.7. *Balanoglossus*. Diagram of tripartite embryo showing coelomic cavities.

Coelom

Balanoglossus is **tricoelomate** like most lopho-phorate and echinoderms. Coelom is enterocoelous, having five cavities:

1. Proboscis coelom or protocoel. It is unpaired cavity, opens by **proboscis pore.** It is greatly obliterated by connective tissue and muscle strands. The central space is occupied by **proboscis complex.**

2. Collar coelom or mesocoel. It is paired and is greatly obliterated by collar musculature and connective tissue. It is also partitioned by incomplete mid-dorsal and mid-ventral mesenteries. It opens into first gill sac by a collar canal and collar pore.

3. Trunk coelom or metacoel. It is represented by a pair of closed cavities between body wall and gut. It is partitioned by incomplete **dorsal mesentery** and a complete **ventral mesentery.** This coelom is obliterated by trunk musculature.

Coelomic fluid. It is a fluid which is filled up in the collar and pro-boscis. This maintains their turgidity.

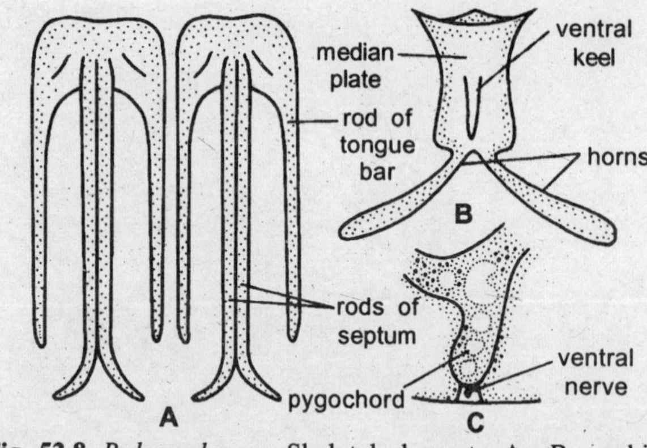

Fig. 52.8. *Balanoglossus*. Skeletal elements. A—Bra-nchial skeleton; B—Proboscis skeleton; C—Pygochord.

Skeletal System

Balanoglossus lacks a definite endoskeleton. But it has following four structures of a supporting nature: (1) buccal diverticulum; (2) skeleton of proboscis; (3) branchial skeleton; (4) pygochord (Fig. 52.8).

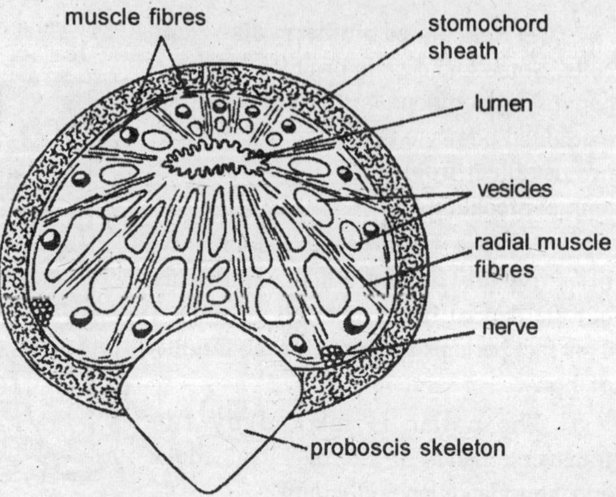

Fig. 52.9. *Balanoglossus*. T.S. stomochord.

1. Buccal diverticulum or stomochord. It is a hollow structure which arises from the roof of buccal cavity and extends upto proboscis coelom. Due to its endodermal origin and being restricted to the anterior part of the body, it is named as **sto-mochord** (Fig. 52.9).

2. Proboscis skeleton or nuchal skeleton. It lies ventrally close to the buccal diverticulum. It is a Y-shaped chitinous structure. Its median plate has a median-ventral keel and lies beneath the buccal diverticulum and two arms extend along the margin of the buccal cavity in the collar region.

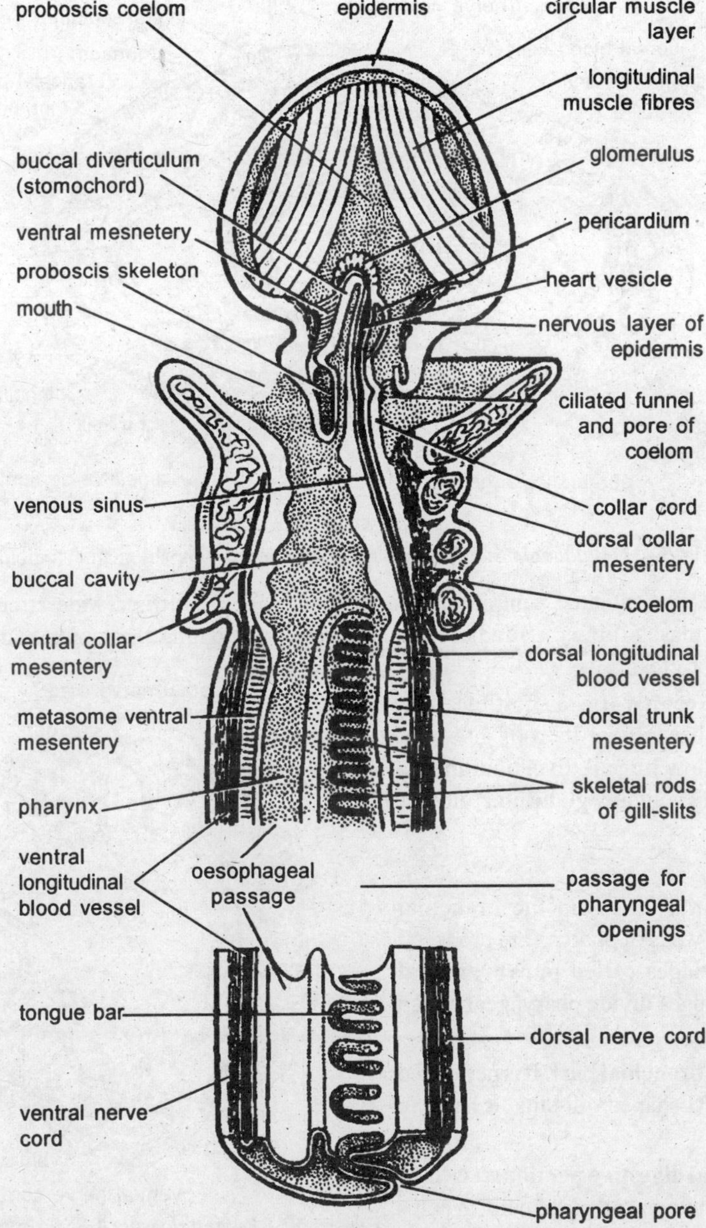

Fig. 52.10. *Balanoglossus*. M.L.S. of anterior region to show the alimentary canal.

3. Gill bars. They are supported by two types of gill-bars—primary and secondary gill bars. They are formed due to thickening of the basement membrane.

4. Pygochord. It is hard longitudinal rod-like structure which extends from the ventral side of the intestine to the body wall. Its cells are vacuolated. Pygochord may support the soft abdominal region.

52.3 DIGESTIVE SYSTEM

Alimentary canal is a straight tube supported by dorsal and ventral mesenteries and consists of the following structures: mouth, buccal cavity, pharynx, oesophagus, intestine and anus.

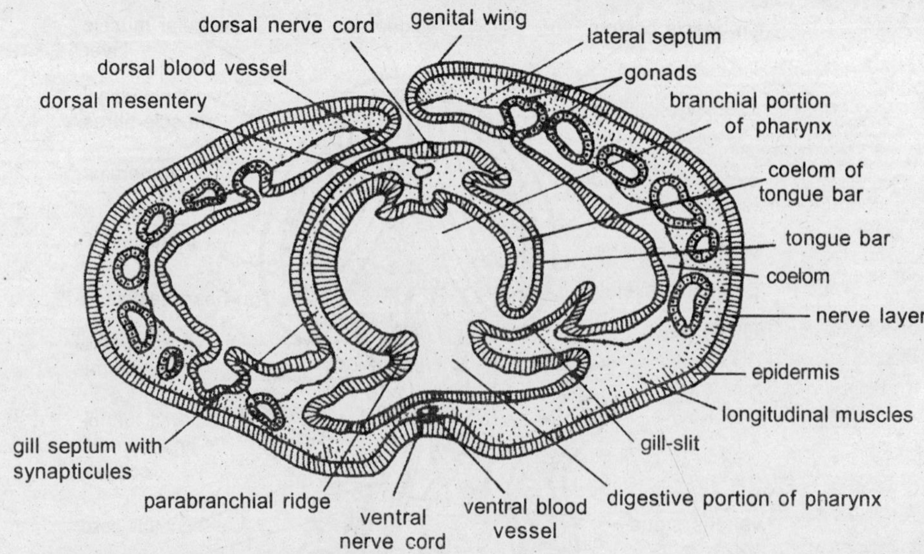

Fig. 52.11. *Balanoglossus*. T.S. through pharyngeal or branchiogenital region.

1. Mouth. It is situated ventrally at the base of proboscis. It is a wide circular aperture. It has two sets of muscle fibres: (*i*) radial muscle fibres for widening the mouth and (*ii*) concentric muscle fibres for closing it.

2. Buccal cavity. It is a short tube that occupies the collar region. Its roof gives out a short, stiff hollow **buccal diverticulum.** Its epithelium lining contains **glandular globlet cells.**

3. Pharynx. It lies in the branchial region. The side walls of pharynx are produced into a pair of ridges called **parabranchial ridges.** These ridges divide pharyngeal lumen into two parts:

(*i*) Dorsal branchial part. It is perforated by two rows of U-shaped gill-slits. It helps in respiration.

(*ii*) Ventral digestive portion. It is lined by cilia-ted epithelium and gland cells. It helps in the collection and concentration of food particles.

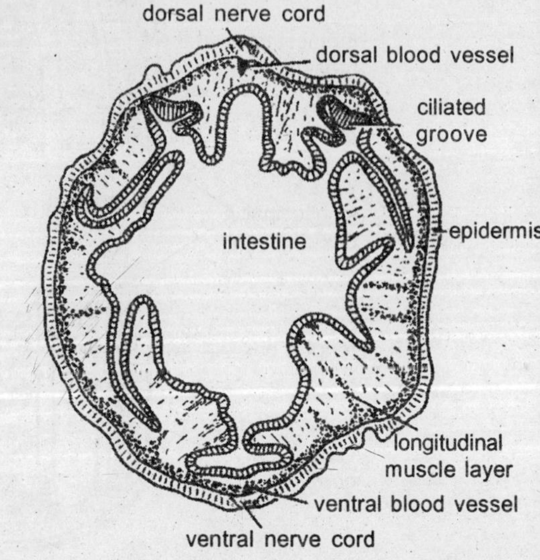

Fig. 52.12. *Balanoglossus*. T.S. through post-hepatic intestine.

4. Oesophagus. This short tube connects pharynx with intestine. Dorsal region of pharynx forms the **post-branchial canal.** It has thick, folded and glandular epithelium. The posterior part of oesophagus is narrow and its epithelium has deep furrows.

5. Intestine. It is situated in the trunk region and has two parts:

(*i*) **Hepatic region.** It is highly vascular. Numerous hepatic caeca are produced from its dorsal wall. Their cells are dark green or dark brown.

(*ii*) **Post hepatic region.** It is straight, thin walled tube and bears a pair of dorsolateral grooves. It is lined with ciliated epithelial cells.

6. Anus. It is situated at the tip of trunk. It is a circular aperture, surrounded by sphincter muscles.

Food and feeding. *Balanoglossus* is a **ciliary feeder.** It feeds on microscopic, dead and decaying organic matter present in water. Due to movement of lateral cilia of gill slits food laden water currents are directed backward which enters mouth, buccal tube, pharynx, gill slits and branchial sacs and leaves through gill-slits (Fig. 52.13). Food particles are entangled in mucus, forming mucous string. Mucus is secreted by gland cells of proboscis and contains **amylase enzyme.**

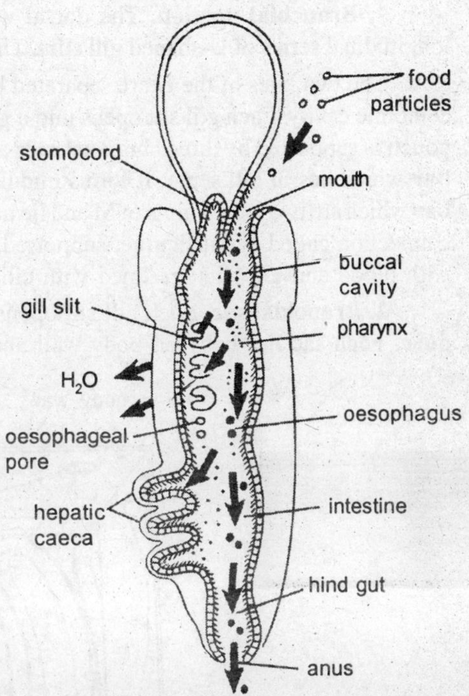

Fig. 52.13. *Balanoglossus.* Diagrammatic path of food and removal of water via gill-slits, in the pharynx.

Digestion. Digestion of food is brought about in the alimentary canal by the help of enzymes such as **amylase, maltase, lipase** and **protease.** They are secreted by pharynx, oesophagus and hepatic caeca of intestine.

52.4 RESPIRATORY SYSTEM

In *Balanoglossus* respiratory organs include:

1. Branchial portion of pharynx and gill pouches.
2. Branchial sacs.

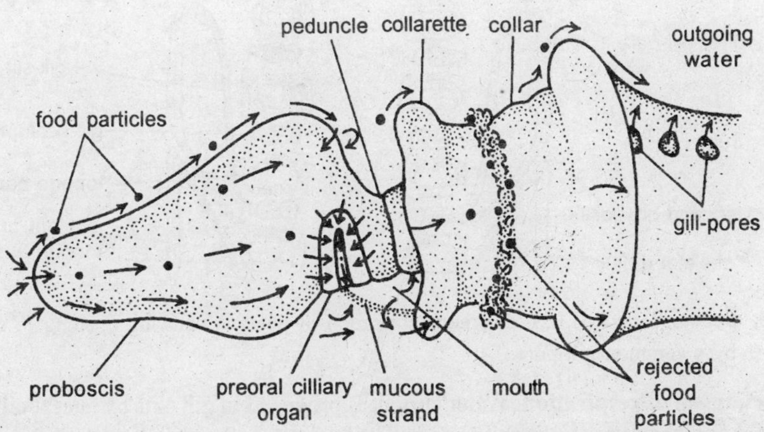

Fig. 52.14. *Balanoglossus.* Feeding current and mode of food testing by preoral ciliary organ.

1. Branchial portion. The dorsal wall of branchial portion of pharynx is perforated by a longitudinal series of U-shaped **gill slits.** Their number increases with increase in the size of animal.

The two lines of the U are separated by a **tongue bar.** The tongue bar is **hollow** and encloses coelomic cavity. Each gill slit opens into a **gill pouch** or **branchial sac** (Fig. 52.15). The wall of gill pouch is supported by three chitinous rods and form M-shaped rods. These are (*i*) **Median primary bar** which lies in gill septa. It forms middle arms of M and its lower end. (*ii*) **Lateral secondary bar** which are two lateral arms of M and lie in tongue bars. Both of the bars are connected by transverse connection called **synapticulae,** supported by chitinous rod. The gill pouches are richly supplied with blood and gill slits are lined with cilia.

2. Branchial sac. Each gill slit opens into a **branchial sac** which opens to exterior by a **gill pore.** Each sac lies between body wall and pharynx.

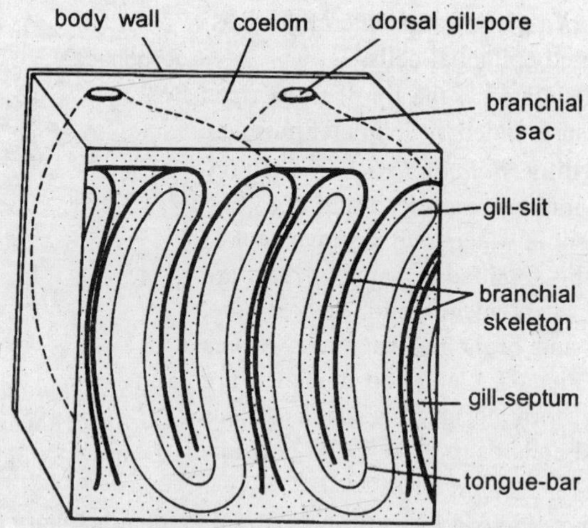

Fig. 52.15. *Balanoglossus*. Diagrammatic three-dimensional view of two gill-slits and two branchial sacs.

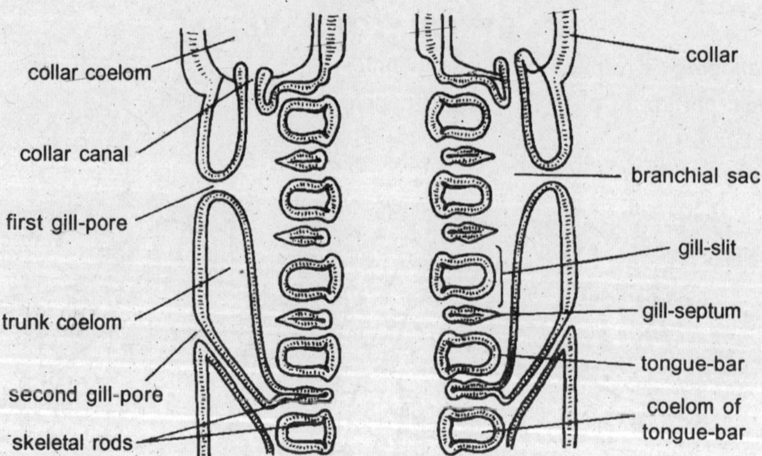

Fig. 52.16. *Balanoglossus*. H.L.S. of branchial portion of pharynx showing opening of first four gill-slits by a common gill-pore.

Mechanism of respiration. Water current is produced in gill slits by lateral cilia and enters the pharynx and finally comes out from gill pore. The exchange of gases takes place through tongue bars.

52.5 BLOOD VASCULAR SYSTEM AND EXCRETORY SYSTEM

Blood vascular system of *Balanoglossus* is of **open type** and consists of—(*i*) blood; (*ii*) dorsal and ventral vessels; (*iii*) central sinus; (*iv*) cardiac vesicle or heart and (*v*) blood vessels (Fig. 52.17).

(1) Blood. Blood is colourless. It contains a few white corpuscles. Respiratory pigment is absent.

(2) Dorsal and ventral vessels. There are two main longitudinal and highly contractile vessels, the **dorsal** and **ventral vessels,** running along the length of the body. The dorsal vessel is situated just below the dorsal nerve cord and above the alimentary canal and runs through the dorsal mesentery. The blood flows anteriorly through the dorsal vessel. The **ventral vessel** is located in the ventral mesentery and blood flows posteriorly in it.

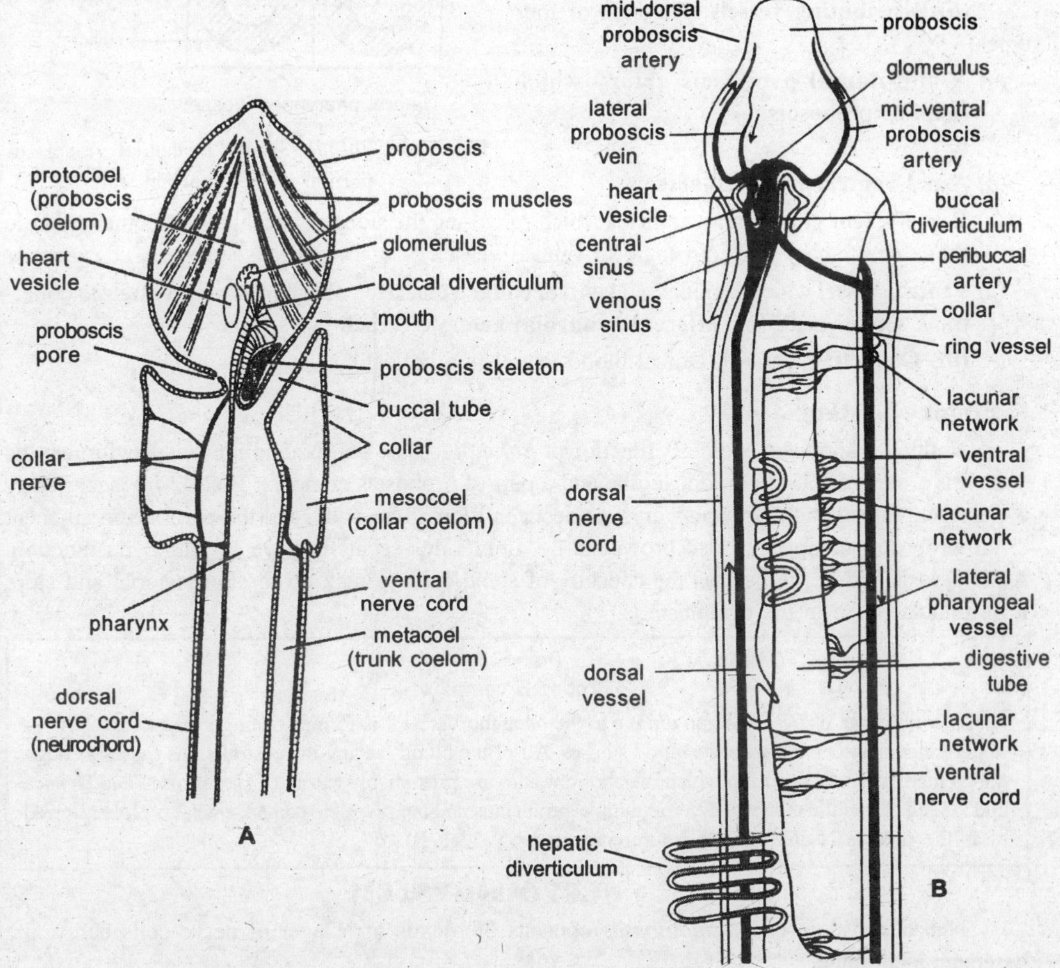

Fig. 52.17. *Balanoglossus*. A. Front end in sagittal section; B. Blood vascular system in anterior end in lateral view.

3. Central sinus. The dorsal vessel is dilated just infront of collar forming a **venous sinus** which passes anteriorly into a **central sinus.** Central sinus receives blood from the collecting vessels at its posterior end. Anteriorly it pumps blood into afferent vessels.

4. Cardiac vesicle or heart. It is situated above the central sinus. It is a closed triangular sac with muscular wall. It produces contraction in central sinus to pump the blood.

5. Blood vessels. These include (*i*) afferent vessels; (*ii*) distributing vessels and (*iii*) collecting vessels.

(*i*) Afferent vessels. They arise from the anterior end of heart (= central glomerulus sinus). These form plexus in **glomerulus.** Here the nitrogenous wastes are removed from the blood.

(*ii*) Distributing vessels. They are of four types:

(*a*) A **mid-dorsal proboscis artery** which supplies proboscis.

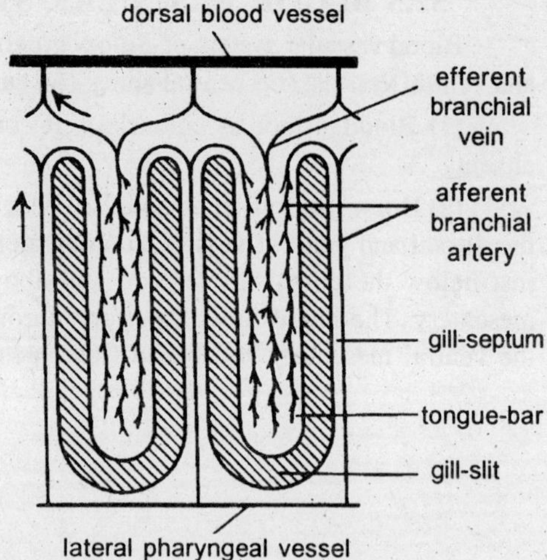

Fig. 52.18. *Balanoglossus*. Branchial vessels in portion of the branchial wall.

(*b*) A **mid-ventral proboscis artery.**

(*c*) Two **efferent glomerular arteries** which run along the sides of buccal diverticulum encircle the buccal cavity and join to form ventral vessel.

(*d*) **Ventral vessel** which includes, (*i*) **ventral collar vessel** (to collar), (*ii*) a **ring vessel** (to collar-trunk septum) and (*iii*) **afferent branchial artery** (to each gill septum).

(*iii*) Collecting vessels. Dorsal blood vessel is a collecting vessel.

Excretory System

Nothing is known of excretory function of *Balanoglossus.* The assumption that the **glomerulus** (which is a mass of blind peritoneal tube and a part of proboscis complex; Box 52.1) is excretory is purely circumstantial. Brownish crystals occur in glomerular cells, and the peritoneum adjacent to the proboscis complex is also brownish. So glomerulus seems to have an athrocytic function. **Athrocytosis** can be defined as the capacity of some cells to pick up foreign material and store it in granular form in the cytoplasm.

Box 52.1
Proboscis complex

Gloomerulus is an excretory organ unique to the hemichordates; it is formed of finger-like outpocketings of peritoneum associated with the blood sinuses. All of the blood leaving the central sinus passes through the glomerular sinuses, within which metabolic wastes are presumably extracted (see **Brusca** and **Brusca** 2003). The glomerulus, buccal diverticulum, central sinus and heart vesicle compose what is often referred to as the **proboscis complex** of enternopneusts (Fig. 52.17B).

52.6 NERVOUS SYSTEM

Nervous system of *Balanoglossus* consists of plexus or a layer of nerve cells below the basement membrane.

Nervous layer is thickened strands forming a **mid-dorsal** and a **mid-ventral nerve cord.** They run the entire length of trunk. In collar region they are connected by a circular **nerve ring.** The dorsal nerve cord is connected by another circular strand called **anterior nerve ring.**

The dorsal cord traverse collar coelom and forms nervous centre. They transport impulses over the body.

Sense Organs

Balanoglossus has simple sense organs which are of following two types:

1. Neurosensory cells. These are present in the epidermis of proboscis and in anterior part of collar.

2. Pre-oral ciliary organ. At the base of the proboscis, on the ventral side, there is a U-shaped depression bordered by long epidermal cells bearing long cilia. This structure is called the **pre-oral ciliary organ.** It is a **chemoreceptor** and it tests the quality of food and water entering the mouth.

52.7 REPRODUCTIVE SYSTEM

Reproduction in *Balanoglossus* is of following three methods:

1. Asexual reproduction. In *Balanoglossus capensis*, during summer young ones if cut into small pieces from tail, each part regenerates into a complete adult.

2. Regeneration. Proboscis, collar and trunk pieces can regenerate into new part.

3. Sexual reproduction. Sexes are separate; sexual dimorphism is not well marked. Gonads may or may not be branched, found in longitudinal rows in the genital wings. The testes and ovaries are similar.

Fertilization. Fertilization is external. Mature ova and sperm are released in surrounding sea water to fertilize each other.

52.8 DEVELOPMENT

In *Balanoglossus*, development is **indirect,** *i.e.*, embryonic development of fertilized egg forms a free swimming, ciliated and microphagous **tornaria larva** which undergoes **metamorphosis** to form an adult-like individual (Fig. 52.19).

1. Cleavage and gastrulation. Cleavage of zygote is **holoblastic** (equal and radial) and results in a sphere of blastomere called **morula.** By developing a central fluid-filled cavity, called **blastocoel,** morula develops into a single-layered **coeloblastula.**

Soon blastula undergoes an **invagination** of cells of vegetal pole to form an **archenteron** which opens outside through the **blastopore.** When ectoderm and endoderm separates from each other and blastopore is closed, the embryo is called **gastrula.**

2. Formation of larva and hatching. The gastrula elongates along the antero-posterior axis. The anterior tip of archenteron is differentiated as a coelomic vesicle, called **protocoel.** Thus, origin of coelom is **enterocoelic.** The remaining posterior part of the archenteron develops into alimentary canal. The protocoel becomes triangular in shape and opens outside on dorsal side by **hydropore.** Later on, protocoel develops into **proboscis coelom** and hydropore becomes **proboscis pore.** Collar and trunk coeloms (*i.e.*, mesocoel and metacoel) develop much later as solid invaginations of the midgut.

After the formation of protocoel, the inner end of the early gut moves towards the ventral surface and opens to outside through a **mouth.** The gut becomes differentiated into oesophagus, stomach, intestine and **anus.** Anus is formed at the place of closed blastopore. After a day or so, the entire embryo is covered uniformly by small cilia. The ciliated embryo escapes from the egg's fertilization membrane; this is called **hatching.** Free-swimming, ciliated larva is called **tornaria larva.**

3. Structure of tornaria larva. Tornaria larva was first described by **J. Muller** in 1850. It is somewhat conical, transparent, glossy and varying in size from 1–4 mm. Its body is having a number of **ciliary bands.** The **preoral ciliary band** encircling the preoral lobe. A **postoral ciliary band** is present around the mouth. It is also called **circumoral ciliary band.** It collects food from water.

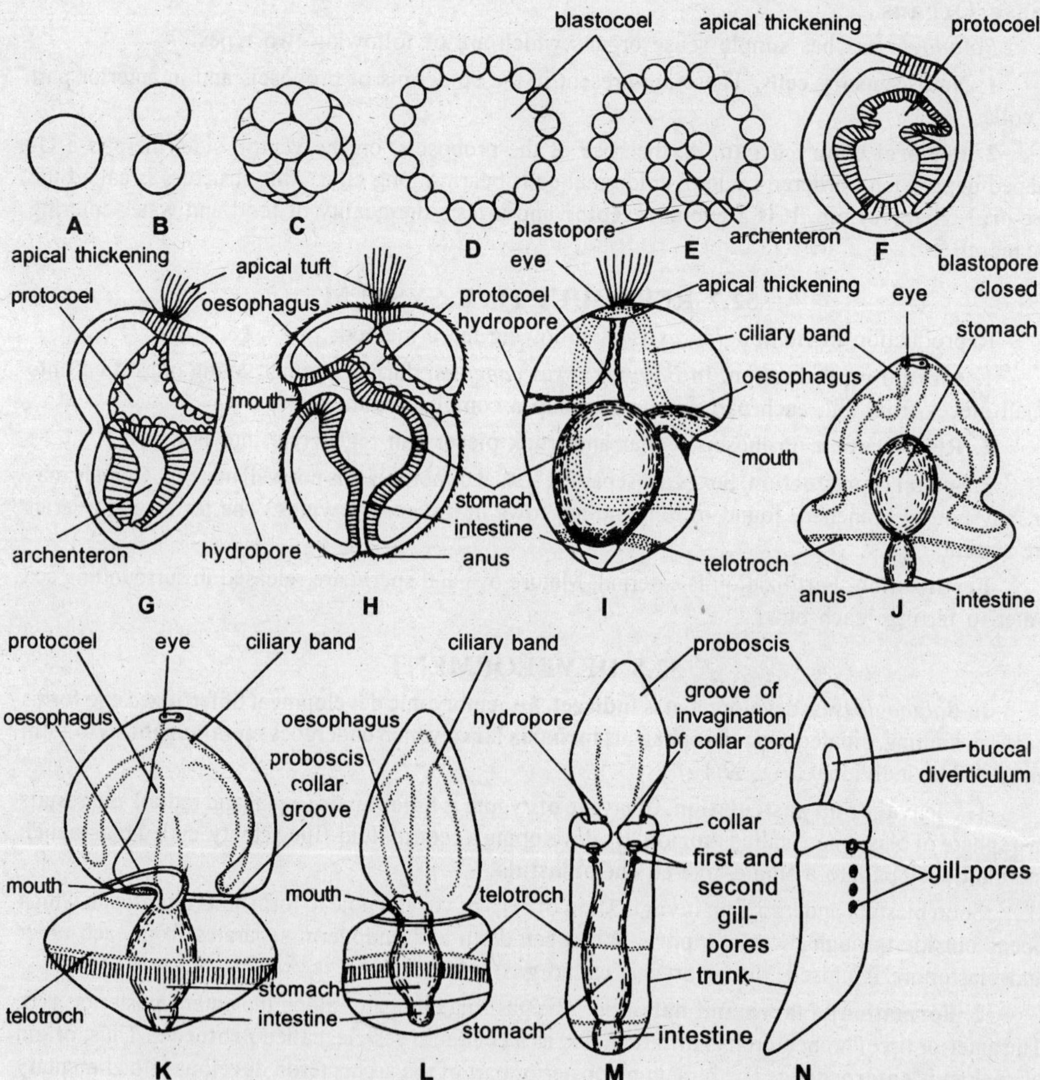

Fig. 52.19. *Balanoglossus*. Stages of development. A—Zygote; B—Two-cell stage; C—Morula; D—Coeloblastula; E—Early gastrula in L.S.; F—Gastrula; G—Early tornaria; H—Young tornaria larva in section; I—Young tornaria; J—Fully formed tornaria; K to N—Metamorphosis of tornaria into a young *Balanoglossus*.

There is another ciliary band, called **telotroch** or circum-anal ring, near the posterior end. It acts as a locomotory organ.

Anterior end of the body bears and ectodermal thickening, the **apical plate,** consisting of the nerve centre and carries nerve cells, eye spot and a band of apical cilia.

Alimentary canal is short but wide. Mouth is ventral but anus is terminal. Internally body is having **protocoel** or **proboscis cavity** which opens outside through the **proboscis pore.**

4. Metamorphosis. Tornaria larva sinks to the bottom and undergoes the following metamorphic changes:

(*i*) It reduces its size due to loss of water.

(*ii*) The ciliary bands, sensory cilia and eye spot are lost.

(*iii*) The preoral parts become uniformly ciliated to form proboscis.
(*iv*) A constriction develops at the base of proboscis to separate it from the collar.
(*v*) The trunk elongates.
(*vi*) The hydropore persists as the proboscis pore.
(*vii*) Buccal diverticulum and gill slits appear as outgrowths of the alimentary canal.

52.9 AFFINITIES

Gegenbaur established group of **Enteropneusta** for *Balanoglossus*. **Sedgwick** and **Huxley** (1814) suggested its chordate nature because of its peculiar anatomy and embryology. **Bateson** (1885) considered hemichordata as a subphylum of the phylum **Chordata.** However, **Dawydoff** (1948), **Marcus** (1958) and **Hyman** (1959) made Hemichordata an independent invertebrate phylum. *Balanoglossus* and other hemichordates show affinities with annelids, nemertines, echinoderms and chordates.

1. Affinities with Annelida

Similarities

1. Both have cylindrical worm-like body.
2. Both are of burrowing habit and tubicolous nature.
3. Both have similar mode of feeding and casting.
4. Circulatory system is similar. Heart is dorsal to gut as in Annelida.
5. Blood vessels are arranged in same manner in both.
6. In hemichordates nervous system is epidermal (partially) but in annelids, it is totally epidermal.
7. Toranaria larva resembles trochophore larva of annelids.
8. The pharyngeal pockets of hemichordates are comparable to the oesophageal pockets present in *Polygordius*.

Differences

1. Annelids do not have pharyngeal gill slits.
2. Presence of dorsal nerve in *Balanoglossus*, but it is absent in annelids.
3. Nervous system is primitive and intra-epidermal in Hemichordata.
4. The nephridium is absent in tornaria larva but it is present in **trochopore larva** of annelids.
5. Preoral proboscis coelom is present in tornaria but absent in trochophore.
6. The **blastopore** becomes **mouth** in trochophore while it forms **anus** in tornaria.
7. Process of embryonic development is entirely different in both groups.

2. Affinities with Echinodermata

Similarities in adults

1. In both coelom has enterocoelic origin.
2. Heart vesicle and glomerulus of hemichordates are homologous to dorsal sac and axial gland of Echinodermata.
3. In both nervous system is poorly developed and form epidermal nerve plexus.
4. Both are similar in habitat and ecological niches.
5. In both similar types of proteins and phosphogens are present.

Similarities in larva

1. Both are pelagic, small, transparent and oval in shape.
2. Longitudinal ciliary band is present in tornaria and in bipinnaria larva.
3. Both contain tripartite division of body.
4. In both development of coelom is similar (*i.e.*, enterocoelic origin).
5. Opening of proboscis by proboscis pore is similar to the hydrocoel of echinoderm dipleurula and shows the beginning of water vascular system.

Differences

1. Apical plate, eye-spot and sensory hair are present in tornaria larva but absent in echinoderms.
2. Transverse ciliary band is present in tornaria while absent in echinoderms.
3. Protocoel is single in tornaria while paired in echinoderm larva.

3. Affinities with Nemertines (Rhynchocephalia)

Similarities

1. Feeding and burrowing habit of both are similar.
2. Both are elongated, vermiform without external segmentation.
3. Anus is terminal in both.
4 In both, the skin is smooth containing unicellular glands.
5. Metamerically arranged simple gonads.

Differences

1. Proboscis is protrusible in nemertines and non-protrusible in *Balanoglossus*.
2. The dorsal nerve is absent in nemertines and present in *Balanoglossus*.
3. The lateral nerve is present in nemertines and absent in *Balanoglossus*.

4. Affinities with Phoronida

Similarities

1. Tornaria larva resembles the actinotrocha larva of Phoronida.
2. In *Phoronis*, the paired gastric diverticula are present but in *Balanoglossus*, single oesophageal diverticulum is present.
3. The nervous system is same in both.
4. Both possess power of regeneration.

Differences

1. The gill-slits and stomochord (= buccal diverticulum) is present in *Balanoglossus* absent in *Phoronis*.

5. Affinities with Pogonophora

Similarities

1. Body and coelom is divided into three regions.
2 Enterocoelous mode of origin of coelom.
3. In both, nervous system intraepidermal.
4. Mesosoma and metasoma are divided by a septum.
5. Pericardial sac is present in some pogonophores.
6. Gonads found in trunk.

Differences

1. Protocoelic nephridial coelomoducts present in pogonophores but not in Hemichordata.
2. Alimentary canal is absent in pogonophores but well developed in hemichordates.
3. Type of nervous system is protosoma in pogonophores but mesosoma in hemichordates.

6. Affinities with Chordates

Similarities

1. Presence of a notochord (*i.e.*, stomochord).
2. Presence of pharyngeal gill slits.
3. Presence of a dorsal tubular nerve cord.
4. Structure and function of pharynx and branchial apparatus are similar to those of Urochordata and Cephalochordata.
5. In both, origin of coelom is enterocoelous.

Differences

(*i*) Notochord

1. Notochord extends upto proboscis, while in chordates notochord extends upto full length.
2. In *Balanoglossus* notochord lies below dorsal blood vessels.
3. In *Balanoglossus* notochord is without any sheath.
4. Notochord is hollow and lined with epithelial cells in Hemichordata but solid rod-like in chordates.
5. Notochord is supporting in chordates and not in *Balanoglossus*.

(*ii*) Dorsal nerve cord

1. Nervous system of hemichordates is distinctly of invertebrate type.
2. It is intra-epidermal and consists of dorsal and a ventral nerve cord and a circumenteric nerve ring.

(*iii*) Gill slits

1. Gill slits are dorsal in *Balanoglossus* but lateral in *Amphioxus* and other chordates.
2. Differences exist in the arrangement of skeletal gill-bars.
3. Differences exist in the mode of blood supply.

Conclusion

By considering the affinities, it can be concluded that echinoderms, hemichordates and chordates have arisen from a common ancestral stock, the dipleura larva. Further, the echinoderms deviated greatly from the ancestral stock and formed blind branch in main line of evolution. The main line of evolution continued to give rise to hemichordates and chordates.

Thus, it is most logical to place hemichordates in the invertebrates as an independent phylum. Hemichordata has arisen from an ancestral stock that has given rise, on the one hand, to echinoderms and on tne other hand, to hemichordates and chordates.

QUESTIONS

Long Answer Questions

1. Describe habit, habitat and external features of *Balanoglossus*. (*Bundelkhand 1987; Jiwaji 1988, 1994; Allahabad 1989; Raipur 1991; Rewa 1993; Vikram 1994; Magadh 1995; Kanpur 1996, 1997*)

2. Describe the morphology of *Balanoglossus*.

3. Describe the habitat and habit of *Balanoglossus*. Draw only a neat and labelled diagram of the blood vascular system of this animal and show the course of blood circulation only by a graphic representation.
(*Kanpur 1998*)

4. Draw a neat and labelled diagram of median longitudinal section (M.L.S.) of anterior region of *Balanoglossus*. Describe its internal structure.
(*Lucknow 1988; Burdwan 1988, 1990; Agra 1989; Meerut or CCS 1989, 1990; Sagar 1989, 1990; Kanpur 1989, 1990; Gorakhpur 1990; Bundelkhand 1992; Rewa 1989, 1993*)

5. Describe the development of *Balanoglossus*.
(*Agra 1990; Kanpur 1992; Meerut or CCS 1994, 1997*)

6. Describe the development of *Balanoglossus* upto the formation of tornaria larva.
(*Indore 1988; Raipur 1990; Agra 1990; Kanpur 1992; Jabalpur 1992; Bhopal 1993*)

7. Give an account of development of *Balanoglossus*.
(*Agra 1988; Bundelkhand 1989; Meerut or CCS 1992, 1997*)

8. Give an account of structure of tornaria larva of *Balanoglossus* and describe its metamorphosis.
(*Garhwal 1989; Bhopal 1990; Jiwaji 1992*)

9. Discuss the affinities of *Balanoglossus* or Hemichordata. What is its latest position in the scheme of classification? (*Allahabad 1988; Kumaun 1988; Gorakhpur 1988, 1991; Bundelkhand 1989; Garhwal 1989; Agra 1989, 1991, 1992; Jiwaji 1989, 1991; Meerut or CCS 1989, 1990, 1993, 1994, 1996; Osmania 1990, 1993; Indore 1990, 1994; Jabalpur 1991, Magadh 1991, 1994; Bhopal 1993; Kanpur 1992, 1994, 1998; Nagarjuna 1994; NEHU 1996*)

10. Give general characters of Hemichordata. Discuss affinities of *Balanoglossus*.
(*Rewa 1989; Jiwaji 1989; Vikram 1990; Kanpur 1992; Magadh 1996*)

11. Give an account of external morphology of *Balanoglossus* in brief. Discuss its affinities in detail.
(*C.C.S.U., Meerut 2009*)

Short Answer Questions

1. Discuss coelom in *Balanoglossus*.

2. Write a note on tornaria larva. (*Meerut or CCS 1988; Allahabad 1988; Agra 1989*)

3. Write short notes on the following:

   (*i*) *Balanoglossus* (*Allahabad 1989*)

   (*ii*) Pharynx of *Balanoglossus*.

   (*iii*) Tongue bar (*Sagar 1990*)

   (*iv*) Tornaria larva (*Jiwaji 1988, 1992; Sagar 1989, 1990; Magadh 1989, 1990, 1993; Rajasthan 1991; Agra 1991; Kanpur 1994, 1996*)

   (*v*) Chordate affinities of *Balanoglossus* (*Sagar 1988*)

   (*vi*) Stomochord (*Magadh 1993*)

4. Draw neat and labelled diagram of following (No description is needed).

   (*i*) L.S. anterior part of *Balanoglossus* or V.S. or M.L.S. *Balanoglossus* through proboscis, collar and trunk part of body. (*Bundelkhand 1992; Bhopal 1992; Jiwaji 1993; Kanpur 1994*)

   (*ii*) T.S. *Balanoglossus* through proboscis. (*Rohilkhand 1988; Marathawada 1988, 1990; Meerut or CCS 1988, 1990; Bundelkhand 1989; Allahabad 1989; Nagpur 1990; Kanpur 1993, 1995*)

   (*iii*) T.S. *Balanoglossus* through collar (*Kanpur 1996; Meerut or CCS 1996*)

   (*iv*) T.S. *Balanoglossus* through branchial region

   (*v*) T.S. *Balanoglossus* passing through branchio-genital region (*Rajasthan 1988; Rohilkhand 1991*) (*vi*) V.S. body wall of *Balanoglossus* (*Himachal 1988, 1990; Udaipur 1992*)

(*vii*) Tornaria larva of *Balanoglossus*
(*Bundelkhand 1989; Agra 1990, 1992; Raipur 1992; Jiwaji 1993; Jabalpur 1993*)

(*viii*) *Balanoglossus* (*Meerut or CCS 1995*)

Very Short Answer Questions

1. Where is *Balanoglossus* found.

Ans. It is tubiculous and found in shallow waters between the tide marks along the coast of warm and temperate oceans.

2. What is a collarette?

Ans. Anterior part of collor around the neck.

3. Name the five coeloms of acorn worm.

Ans. Protocoel, a pair of mesocoels, a pair of metacoels.

4. Which coelom of *Balanoglossus* do not open out?

Ans. Metacoels.

5. Give the role of oesophagus of acorn worm.

Ans. Moulding of food particles and mucus into a cord.

6. Which part of the gut of *Balanoglossus* has gill slits?

Ans. Respiratory chamber of pharynx.

7. Name the site of digestion in acorn worm.

Ans. Hepatic region of intestine.

8. What is the function of preoral ciliary organ?

Ans. To test the water and food particles entering the mouth.

9. What are synapticula?

Ans. Connections between gill septa and tongue bars.

10. Heart of *Balanoglossus* is non-contractile. Is it so?

Ans. Yes.

11. What type of cleavage does the egg of acorn worm undergo?

Ans. Holoblastic, complete and radial.

12. Name the larva of *Balanoglossus*.

Ans. Tornaria.

Fill in the Blanks

1. *Balanoglossus* is the familiar example of the phylum ....

2. *Balanoglossus* lives in .... shaped, burrows.

3. Three regions of body of *Balanoglossus* are: .... , mesosome and ....

4. .... is tricoelomate.

5. *Balanoglossus* has a pair of mesocoels in the .... region.

6. Stomochord is a forward diverticulum from the ....

7. Gill slits of acorn worm are .... shaped.

8. Blood flows .... in the dorsal vessel in *Balanoglossus*.

9. Glomerulus of *Balanoglossus* functions as the .... organ in the acorn worms.

10. Sexes are .... in *Balanoglossus*.

Multiple Choice Questions

1. *Balanoglossus* belongs to class ....
(*a*) Enteropneusta (*b*) Plerobranchia
(*c*) Graplita (*d*) Planctosphaeroidea

2. The body of *Balanoglossus* is
(*a*) radially symmetrical (*b*) biradially symmetrical
(*c*) bilaterally symmetrical (*d*) laterally symmetrical

3. The coelom of *Balanoglossus* is
(*a*) enterocoelous (*b*) haemocoelous
(*c*) schizocoelous (*d*) homocoelous

4. *Balanoglossus* is
(*a*) ciliary feeder (*b*) active feeder
(*c*) filter feeder (*d*) all of the above

5. The excretory organ of *Balanoglossus*
(*a*) malpighian tubules (*b*) flame cells
(*c*) nephridia (*d*) proboscis gland

6. In *Balanoglossus*, the function of preoral ciliary organ is
(*a*) photoreceptor (*b*) chemoreceptor
(*c*) reoreceptor (*d*) tangoreceptor

7. Respiration in *Balanoglossus* is
(*a*) cutaneous (*b*) branchial
(*c*) tracheal (*d*) pulmonary

8. Acorn worm has a plexus of blood sinuses in
(*a*) gill septa (*b*) tongue bars
(*c*) both (a) and (b) (*d*) highly pressurized

9. Circulatory system of *Balanoglossus* is
(*a*) open (*b*) closed
(*c*) lacking (*d*) highly pressurized

10. Hollow nerve cord occurs in acorn worms in the
(*a*) proboscis (*b*) collar
(*c*) trunk (*d*) entire body

11. The larva of *Balanoglossus* is
(*a*) tornaria (*b*) miracidium
(*c*) brachiolaria (*d*) none of these

ANSWERS

Fill in the Blanks

1. Hemichordata	**2.** U	**3.** Protosome, metasome	**4.** metasome
5. collar	**6.** Buccal cavity	**7.** U	**8.** Forward
9. Excretory	**10.** Separate.		

Multiple Choice Questions

1. (*a*)	**2.** (*c*)	**3.** (*a*)	**4.** (*a*)	**5.** (*d*)	**6.** (*b*)
7. (*b*)	**8.** (*c*)	**9.** (*a*)	**10.** (*b*)	**11.** (*a*).	

UNIT - II
CELL BIOLOGY & BIOCHEMISTRY

UNIT – II

CELL BIOLOGY & BIOCHEMISTRY

53

Introduction

All living organisms (except viruses) consist of one or more cells. The cells are structural and functional units of life.

The biological science which deals with the study of structure, function, molecular organization, growth, reproduction and genetics of the cells, is called **cytology** (Gr., *kytos* = hollow vessel or cell; *logous* = to discourse) or **cell biology.** Much of the cell biology is devoted to the study of structures and functions of specialized cells. The results of these studies are used to formulate the generalization applied to almost all cells as well as to provide the basic understanding of how a particular cell type carries out its specific functions. The cell biologist, without losing sight of the cell as a morphologic and functional unit within the organism, has to study biological phenomena at all levels of organization and to use all the methods, techniques and concepts of other sciences.

Cytology Versus Cell Biology. The cell biology has been studied by the following three avenues: **classical cytology** dealt with only light microscopically visible structure of the cell; **cell physiology** studied biochemistry, biophysics, and functions of the cell and **cell biology** interpreted the cell in terms of molecules (macromolecules such as nucleic acids and proteins). In recent years distinction between classical cytology, cell physiology and cell biology has become blurred and outmoded and now two term—cytology and cell biology are used as the synonyms.

Discovery of Cell and its Organelles: Cells were discovered in 1665 by an English scientist, **Robert Hooke.** He saw cells for the first time in a thin slice of cork with his crude microscope. Hooke described the cork as made up of hundreds of little hexagonal boxes, giving a kind of honey-comb appearance. He called these little boxes **cells.** In 1833, a Scottish botanist, **Robert Brown** discovered and named the **nucleus** in a cell. **J.E. Purkinje** in 1839, used the term **protoplasm** for a living substance present inside the cell. For coinage of other terms in cell biology see Box 53.1.

53.1 CELL THEORY

In 1838, a German botanist **Mathias Jacob Schleiden** (1804–1881) put forth the idea that cells were the units of structure in the plants. In 1839, his co-worker, a German zoologist, **Theador Schwann** (1810–1882) applied Schleiden's thesis to the animals. Both of them, thus, postulated that the cell is the basic unit of structure and function in all life. This simple, basic and formal biological generalization is known as **cell theory** or **cell doctrine.** In fact, both **Schleiden** and **Schwann** are incorrectly credited for the formulation of the cell theory; they merely made the generalizations whieh were based on the works of their predecessors such as **Oken** (1805), **Mirbel** (1807), **Lamarck** (1809), **Dutrochet** (1824), **Turpin** (1826), etc. However, **Schleiden** was the first to describe the nucleoli and to appreciate the fact that each cell leads a double life—one independent, pertaining to its own development, and another, as integral part of a multicellular plant. **Schwann** studied both plant and animal tissues and his work with the connective tissues such as bone and cartilage led him to modify the evolving cell theory to include the idea that living things are composed of both cells and the products or secretions of the cells. **Schwann** also introduced the term **metabolism** to describe the activities of the cells.

Box 53.1
The Naming of Various Cell Components

1. Cell	*Robert Hooke* (1665)
2. Cell membrane	*C. Nageli* and *C. Cramer* (1855)
3. Plastids	*Haeckel* (1866)
4. Karyokinesis	*Schleicher* (1878)
5. Chromatin	*W. Flemming* (1879)
6. Mitosis	*W. Flemming* (1882)
7. Chloroplast	*Schimper* (1883)
8. Phagocytosis	*Metchinikoff* (1883)
9. Cytoplasm and nucleoplasm	*E. Strasburger* (1884)
10. Centrosome	*T. Boveri* (1888)
11. Chromosome	*Waldeyer* (1888)
12. Nucleic acid	*Richard Altmann* (1889)
13. Mitochondrion	*C. Benda* (1897)
14. Golgi complex	*Camillo Golgi* (1898)
15. Plasmodesmata	*E. Strasburger* (1901)
16. Meiosis	*J.B. Farmer* and *J.B. Moore* (1905)
17. Plasmalemma	*J.Q. Plowe* (1931)
18. Endoplasmic reticulum	*K.R. Porter* (1945)
19. Lysosome	*C.de. Duve* (1948)
20. Microbody	*J. Rhodin* (1954)
21. Prokaryote and Eukaryote	*Hans Ris* (1960)
22. Peroxisomes	*De Duve* (1965)
23. Glyoxisomes	*R.W. Breidenbach* (1967)
24. Viroids	*T.O. Diener* (1971)
25. Prion	*Prusiner* (1982)

In the coming years, the cell theory was to be extended and refined further. **K. Negeli** (1817–1891) showed in 1846 that plant cells arise from the division of pre-existing cells. In 1855, a German pathologist **Rudolf Virchow** (1821–1902) confirmed the Nageli's principle of the cellular basis of life's continuity. He stated in Latin that the cells arise only from the pre-existing cells (*viz.,* his actual aphorism was "*omnis cellula e cellula*"–every cell from a cell). **Virchow,** thus, established the significance of cell division in the reproduction of organisms. In 1858, **Virchow** published his classical textbook *Cellular Pathology* and in it he correctly asserted that as functional units of life, the cells were the primary sites of disease and cancer. Later, in 1865, **Louis Pasteur** (1822–1895) in France gave experimental evidence to support Virchow's extension of the cell theory.

The modem version of cell theory states that (1) All living organisms (animals, plants and microorganism) are made up of one or more cells and cell products. (2) All metabolic reactions in unicellular and multiclelluar organisms take place in cells. (3) Cells originate only from other cells, *i.e.,* no cell can originate spontaneously or *de novo,* but comes into being only by division and duplication of already existing cells. (4) The smallest clearly defined unit of life is the cell.

The cell theory had its wide biological applications. With the progress of biochemistry, it was shown that there were fundamental similarities in the chemical composition and metabolic activities of all cells. **Kolliker** applied the cell theory to embryology—after it was demonstrated that the organisms developed from the fusion of two cells—the spermatozoon and the ovum. However, in the recent years, large number of sub-cellular structures such as ribosomes, lysosomes, mitochondria, chloroplasts, etc., have been discovered and studied in detail. Consequently, it may appear that cell is no longer a basic unit of life, because life may exist without cells also. Even then, the cell theory remains a useful concept.

Exception to cell theory: Cell theory does not have universal application, *i.e.,* there are certain living organisms which do not have true cells. All kinds of true cells share the following three basic characteristics: 1. **A set of genes** which constitute the blueprints for regulating cellular activities and making new cells. 2 A limiting **plasma membrane** that permits controlled exchange of matter and energy with the external world. 3. **A metabolic machinery** for sustaining life activities such as growth, reproduction and repair of parts. **Viruses** do not easily fit in these parameters of a true cell. Thus, they lack a plasma membrane and a metabolic machinery for energy production and for the synthesis of proteins. However, like any other cellular organism, viruses have (1) a definite genetically determined macromolecular organization; (2) a genetic or hereditary material in the form of either DNA or RNA; (3) a capacity of auto-reproduction; and (4) a capacity of mutation in their genetic substance. In consequence, viruses can only reproduce inside the host cells which may belong to animals, plants or bacteria. They use their own genetic programme for reproduction but rely on the raw materials (*i.e.,* amino acids, nucleotides) and biosynthetic machinery of the host cells (*i.e.,* ribosomes, tRNA, enzymes) for their multiplication. Thus, a virus may be defined as an infectious, subcellular and ultramicroscopic particle representing an obligate cellular parasite and a potential pathogen whose reproduction (replication) in the host cell and transmission by infection cause characteristic reaction in the host cells. Outside the host cells, viruses are just like non-living inert particles and like the salt or sugar, they can be purified, crystallized and placed into jars on a shelf for years. Due to this fact, viruses have been variously described such as "*naked genes that had somehow acquired the ability to move from one cell to another*" or as "*cellular forms that have degenerated through parasitism*" or as "*primitive organisms that have not reached a cellular state.*"

There are certain other organisms such as the protozoan *Paramecium*, the fungus *Rhizopus* and the alga *Vaucheria* (Fig. 53.1B) which do not fit into the purview of the cell theory. All of these organisms have bodies containing undivided mass of protoplasm which lacks cell-like organization and has more than one nucleus. They tend to raise the question that whether cell is a basic unit of structure in them.

53.2. PROTOPLASM THEORY

Up to middle of the 19th century, greater emphasis was given to the cell wall and less to the cellular content. But soon cell biologists started to recognize the importance of "juicy" or "slimy" contents of the cells. In 1835, **Felix Dujardin** termed the jelly-like material within protozoans as **sarcode.** In 1835, **H.von Mohl** (1805–1875) described cell division. In 1839, the Czech biologist **J.E. Purkinje** (1787–1869) coined the term protoplasm to describe the contents of cells (animal embryos). **Von Mohl,** in 1846, applied the name protoplasm to the contents of embryonic cells of the plants. **Max Schultze,** in 1861, established similarity between sarcode and protoplasm of animal and plant cells and, thus, offering a theory which later on was improved and called protoplasm theory by **O. Hertwig** (1849–1922) in 1892.

Protoplasm theory holds that all living matter, out of which animals and plants are formed, is the protoplasm. The cell is an accumulation of living substance or protoplasm which is limited in space by outer membrane and possesses a nucleus. The protoplasm which is filled in the nucleus is called **nucleoplasm** and that exists between the nucleus and the plasma membrane is called **cytoplasm.**

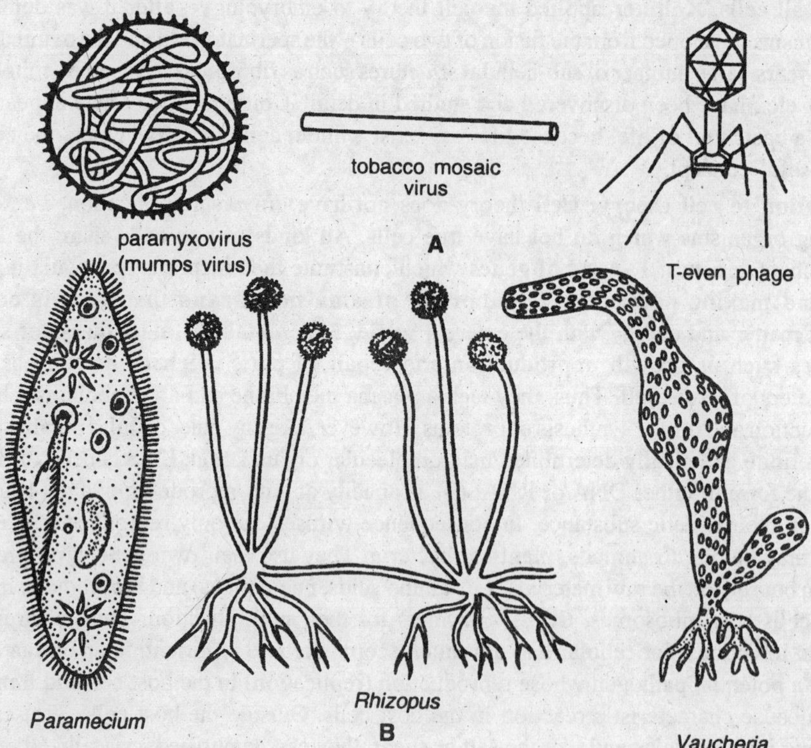

Fig. 53.1. Organisms forming exceptions to the cell theory: A—Three type of viruses; B—Three cases of cellular organization.

53.3 ORGANISMAL THEORY

The organismal theory holds that the body of all multicellular organisms is a continuous mass of protoplasm which remains divided incompletely into small centres, the cells, for the various biological activities. Thus, a multicellular organism is a highly differentiated protoplasmic individual, differing with a unicellular Protozoa only in size and degree of differentiation of the protoplasm. The differentiation involves separation of the protoplasm into subordinate semi-independent compartments, the so-called cells. Even the embryological development of a multicellular individual includes only growth and progressive internal differentiation of a small single protoplasmic individual (egg). Organismal theory too fails to ascertain the position of viruses.

QUESTIONS

Long Answer Questions

1. Give a detailed account of cell theory.

Short Answer Questions

1. What is cell biology? Who had discovered the cell?
2. Write short notes on the following:
   (*i*) Cell theory
   (*ii*) Protoplasm theory
   (*iii*) Organismal theory.

(*Osmania 1989; Bihar 1990; Guwahati (G) 1991, 1994, 1995, 1996; Rajasthan 1992; Kanpur 1995*)

Very Short Answer Question

1. Explain the contribution of any three of the following in the discipline of Cell Biology. (*Garhwal 1988*)

(*i*) Robert Hooke
(*ii*) C. Benda
(*iii*) R. Altman
(*iv*) M. Knoll and Ruska
(*v*) C. Golgi.

Fill in the Blanks

1. .... are an exception to cell theory.

2. Cell theory was first given by .... and ....

Matching of Columns

Column A	Column B
(*i*) Robert Hooke	(*a*) Protoplasm
(*ii*) Robert Brown	(*b*) Cell theory
(*iii*) Schleiden and Schwann	(*c*) The term "cell"
(*iv*) Von Mohl	(*d*) Nucleus

Multiple Choice Questions

1. Which of the following is an exception to cell theory?
(*a*) viruses
(*b*) mycoplasma
(*c*) bacteria
(*d*) RBC

2. The cell was seen for the first time under the microscope by
(*a*) Robert Hooke
(*b*) Malpighi
(*c*) Leeuwenhoek
(*d*) Grew

3. Cell theory that all plants and animals are made up of small cells was proposed by
(*a*) Robert Hooke
(*b*) Knoll and Ruska
(*c*) Schleiden and Schwann
(*d*) Watson and Crick

4. The cell theory states that
(*a*) cells are fundamental structural units of plants and animals
(*b*) cells reproduce by mitosis and meiosis
(*c*) all cells are living
(*d*) all cells have nucleus

ANSWERS

Fill in the Blanks

1. Viruses
2. Schleiden, Schwann.

Matching of the Columns

(*i*) *c*; (*ii*) *d*;
(*iii*) *b*; (*iv*) *a*.

Multiple Choice Questions

1. (*a*); **2.** (*a*); **3.** (*c*); **4.** (*a*).

54

Cell Biological Techniques

54.1. CELL FRACTIONATION AND CENTRIFUGATION

Sometimes it becomes necessary to break up tissues and cells and to isolate various parts of the cell for structural or biochemical analysis. For this purpose, the technique of cell fractionation is employed. Cell fractionation method involves, essentially the homogenisation or destruction of cell boundaries by different mechanical or chemical procedures, followed by the separation of the subcellular fractions according to mass, surface and specific gravity by centrifuges.

The centrifuge is the most important tool available to cell biologists for the separation and analysis of subcellular organelles and their constituent macromolecules. Centrifugal force is generated by placing test tubes containing samples to be fractionated in a holder called a **rotor** and then spinning the rotor at high speeds by the help of a **motor** in a centrifuge. The ultra-centrifuge, developed by the Swedish chemist, **Theodor Svedberg** between 1920 to 1940, is a special high--speed instrument capable of spinning rotors at speeds of 100,000 revolutions per minute or more. The large forces generated by these high speeds make it possible to isolate not just organelles but macromolecules such as proteins and nucleic acids as well.

Stokes formula. The rate of movement of any particle subjected to a centrifugal force is given by **Stokes formula:**

$$\frac{dx}{dt} = \frac{2r^2(\rho_P - \rho_M)}{9^\eta} \cdot g$$

where dx/dt is the velocity with which the particle moves towards the bottom of the tube, r is the particle radius, ρ_p and ρ_M are the respective densities of the particle and the suspending medium and g is the centrifugal force exerted on the particle. This formula tells us that when the density of a particle exceeds that of the medium, its velocity will be positive and the particle will migrate in the direction of the force field (towards the bottom of tube). If the particle is less dense than the medium, its velocity will be negative, meaning that it will move towards the top of the tube. Finally, if the density of particle is exactly equal to the density of the suspending medium, the particle will not move because the term, $\rho_p - \rho_M = 0$.

Types of Centrifugation

***(a)* Velocity centrifugation** (It separates organelles and molecules based on differences in size)

***(i)* Differential centrifugation**

***(ii)* Moving zone centrifugation**

***(b)* Isodensity centrifugation** (It separates organelles and molecules on the basis of differences in density).

For these centrifugations, the following three centrifugation processes are employed:

1. High speed centrifugation. In high speed centrifugation rotor runs at a speed of about 20,000 rpm and generates a gravitational pull of about 50,000 g. At this speed most microorganisms, cell debris and cell organelles form sediments. It is used in **differential centrifugation.**

2. Ultracentrifugation: Ultracentrifuge (Fig. 54.1) of **Svedburg** operates at extremely high speed (about 75,000 rpm) and produces a gravitational pull of about 500,000 g. It is used for **differential centrifugation, zonal centrifugation, density gradient centrifugation** and **isopycnic centrifugation**.

3. Analytical centrifugation: This is the latest and most sophisticated instrument. Its sample container has a quartz window and the rotor has an optical system. It is possible to observe the material sedimenting and measure the sedimentation rate during the run.

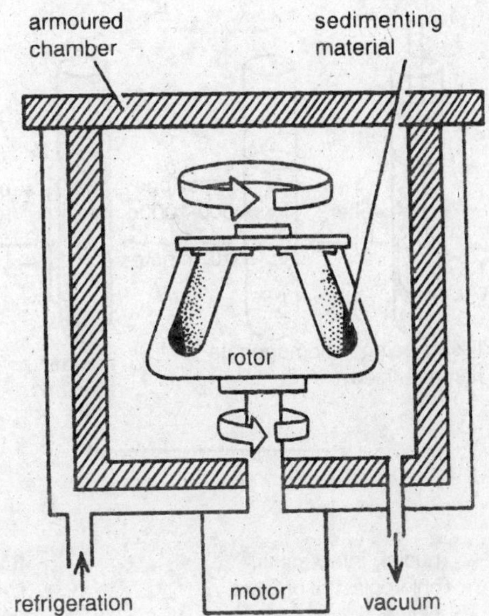

Fig. 54.1. The ultracentrifuge. The sample (*i.e.*, homogenate) is contained in tubes that are inserted into a ring of cylindrical holes in a metal rotor. Rapid rotation of the rotor generates immense centrifugal forces, which cause particles in sample to sediment. The vacuum tends to reduce friction, preventing heating of the rotor and allowing the refrigeration system to maintain the sample at 4°C.

Mechanism of Centrifugation (Cell Fractionation)

1. In the cell fractionation, the cells are gently broken by grinding a small piece of tissue in a homogeniser having a moving close-fitting glass or plastic pestle within a tube that contains a medium such as sucrose solution to preserve the cellular organelles (Fig. 54.3). The solution containing homogenised or disrupted cells, is called **homogenate**. In a homogenate the particles of different sizes are distributed homogeneously. The homogenate is subjected differential centrifugation of increasing velocity.

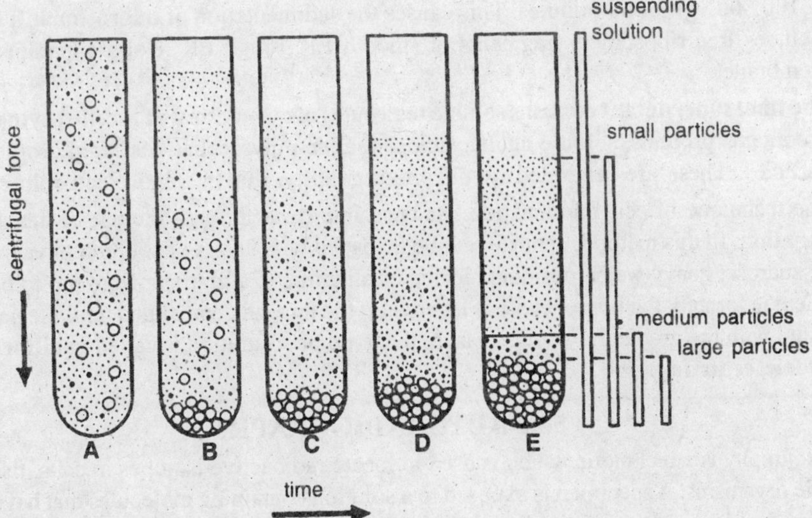

Fig. 54.2. Diagrammatic representation of the process of centrifugation and sedimentation of particles of different sizes.

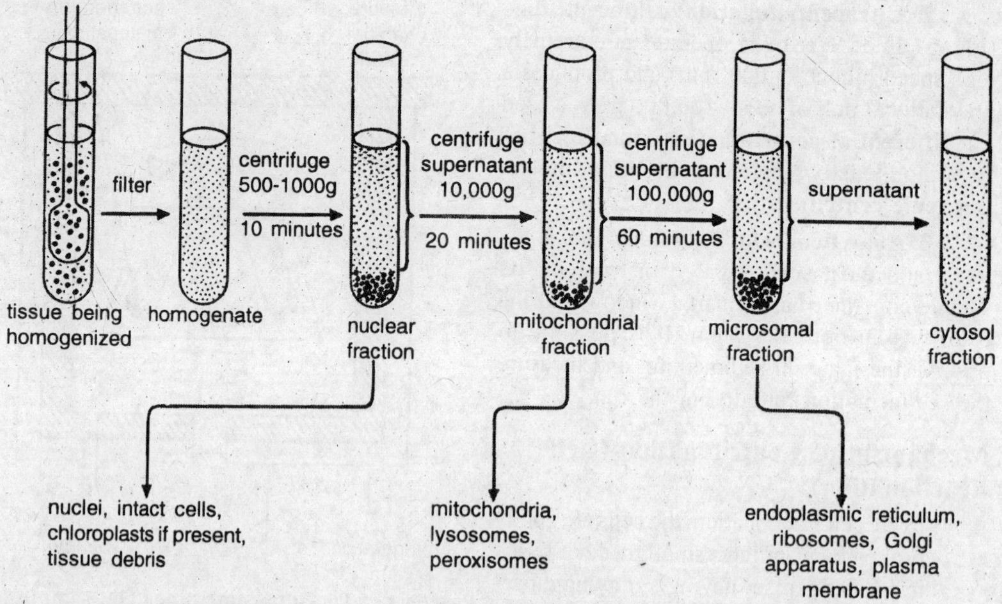

Fig. 54.3. Subcellular fractionation by differential centrifugation.

2. The tissue/liver homgenate is subjected first to low speed centrifugation at 700X g for 10 minutes. This seaiments cell nuclei to the bottom of centrifuge tube. This **sediment-I** is celled **nuclear fraction**. Nuclear fraction includes nuclues, intact cells, tissue debris and chloroplasts (in case of plant cells).
3. The overlying **supernatant-I** is transferred to another tube for second centrifugation at 10,000X g for 20 minutes. The **sediment-II** is called **mitochondrial fraction**. It contains mitochondria and also lysosomes and peroxisomes.
4. Again the **supernatant-II** is removed and subjected to third centrifugation at (~ 100,000Xg) for 60 minutes. This causes the sedimentation of **microsomal fraction** which includes free ribosomes, fragments of smooth ER, rough ER, Golgi apparatus and plasma membrane.
5. The **final supernatant** contains soluble molecules of cytoplasm and is called **cytosol fractions.** It contains proteins, soluble nucleic acids, soluble polysaccharides, lipid droplets and other particles. These are separated by **chromatography, dialysis** and **electrophoresis.**

The technique of cell fractionation has been improved through the use of **density gradient centrifugation.** In this method, centrifuge tube is loaded with layers of solution of varying densities of either sucrose, heavy water, cesium chloride or albumin, in a gradient from top to bottom. Once the gradient is formed, the homogenate is layered on the top and centrifuged until the particles reach equilibrium with the gradient. For this reason, this type of separation is called **equilibrium density** or **isopycnic centrifugation.**

54.2 AUTORADIOGRAPHY

Autoradiography is a technique which is used to locate radioactive isotopes in cells, tissues, organs and whole organisms. A specimen is exposed to a solution containing molecules that have been made radioactive by the incorporation of radioactive isotopes, such as tritium (^{3}H), carbon 14 (^{14}C), phosphorus (^{32}P) and sulphur (^{35}S). The tagged molecules are often precursor molecules used by the cell in the synthesis of other needed molecules. At intervals, samples are removed from the solution;

in case of smaller tissues, the samples are sectioned and mounted on glass slides or grids. The sections are then coated with a photographic emulsion and stored in the dark for periods ranging up to several months.

When a radioactive atom emits a beta particle (*i.e.*, electron) the photographic emulsion is affected in a manner similar to the exposure of a photographic emulsion to light. Over a period of time sufficient radioactive emissions occur to affect the silver grains of the emulsion. Black spots will appear at those sites when the emulsion is developed. Such spots will mark sites in the tissues where the radioactive atoms have accumulated. These sites can be identified by examining the stained tissue sections under the light microscope.

In the technique of autoradiography, for the study of DNA metabolism of cell ^{3}H-thymidine is used; for RNA metabolism ^{3}H-uridine is used; for protein synthesis various tritiated (^{3}H-tagged) amino acids are used; and for polysaccharides and glycoproteins tritiated monosaccharides such as ^{3}H-mannose and ^{3}H-fucose are employed.

Pulse-labelling technique. This technique is used for those cases where biological molecules undergo considerable modifications after their synthesis (*e.g.*, ribosomal RNA). Here, actively growing cells are exposed to a radioactive precursor for a short period. The labelled precursor is then removed and replaced by 'cold' (unlabelled) precursor molecules. The unlabelled precursor is then removed and replaced by 'cold' (unlabelled) precursor molecules. The unlabelled precursors are incorporated into the newly synthesized molecules and have the effect of 'chasing' the previously synthesized molecules containing the radioactive precursor through any maturation process. If the molecular species under investigation is sampled shortly after the start of the experiment, only the primary synthetic product will contain radioctivity. After longer time intervals, the original radioactive molecules will have been replaced with non-radioactive molecules. In this way the flow of radioactivity through a maturation process can be followed, together with any movement of molecular species within the cell.

54.3 OPTICAL AND ELECTRON MICROSCOPY

A **microscope** is an optical instrument consisting of a single lens or a combination of lenses for magnifying objects too small to be seen or clearly observed by the naked eye. Thus, it is used to obtain magnified images of small objects such as cells of living organisms. Investigations or studies by means of the microscope is called **microscopy**. In cell biology, various types of microscopes are used to study the structure of cell; some important ones are the following:

1. Optical or light microscopes (simple, monocular, binocular and compound microscopes)
2. Electron microscope
3. Phase contrast microscope
4. X-ray diffraction microscope
5. Interference microscope
6. Polarized light microscope
7. Ultra-violet microscope
8. Ultra-microscope or Dark field microscope

A **simple microscope** (also called dissecting microscope) has a single lens system. It is composed of a single convex lens. **A compound microscope** consists of two or more **lens systems** or **elements.** A **monocular microscope** contains one ocular or eyepiece while a **binocular microscope** has two oculars.

Here, only two types of microscopes, namely light microscope and electron microscope will be described. (For some historical microscopes see Box 54.1.

Box 54.1. Some Historical Microscopes

1. First simplest compound microscope was designed by **Francis** and **Janssen** in 1590.
2. **Robert Hooke** (1665) constructed the compound microscope (Fig. 54.4)
3. **Anton van Leeuwenhoek** (1675) was the first to use microscope for biological studies. He was first to observe living free cells. His microscope (Fig. 54.5) consisted of a single lens with a high power of magnification and was called **simple microscope.** He made 247 microscopes.

Resolving Power of Microscope

In the search for information about the structure and composition of cells, the cell biologists immediately face two limitations: the exceedingly small dimensions of cells and their component parts and the transparent nature of cells. The diameters of the majority of cells fall within a range of 0.2 and 50 μm. The human eyes have limited distinguishing or resolving power. The ability of an observational instrument such as a human eye or a microscope to reveal details of structure is expressed in terms of **limit of resolution** (l) which is defined as *the smallest distance that may separate two points on an object and still permit their observation as distinct separate points.* The unaided human eye under optimal conditions in green light (to which it is most sensitive) cannot distinguish between points less than about 0.1 μm or 100 μm apart. Structural details smaller than this, *e.g.*, cell, is unresolved unless some instrument capable of higher resolution is used. **Magnification,** *the increase in size of optical image over the size of the object being viewed,* is of no use unless the observational system can resolve the various parts of the structure being examined. Increased magnification without improved resolution results only in a large blurred image. The human eye has no power of magnification, so magnifying glasses may be used to magnify images up to about 10 times. A light compound microscope in which many lenses are combined together has a useful magnification of about 1,500 times.

The limit of resolution (l) of any optical instrument (*i.e.*, eye or microscope) is given approximately by the Abbe's relationships:

$$\text{Resolution } (l) = \frac{\text{wavelength } (\lambda)}{\text{numerical aperture } (n \sin \alpha)}$$

where λ (lambda) is the wavelength ("colour") of the illumination or radiation used to form the image, n is refractive index (a function of density) of the material (*i.e.*, mostly air or water) between the specimen and the first lens (or objective lens), and sin α is sine of the semi-angle of aperture of the first lens as viewed from the specimen. The quantity "n sin α" is often called the **numerical** aperture (NA).

Abbe's relationships make it clear that high resolution in a microscope can only be achieved by manipulating a small number of variables: the wavelength of the illuminating radiation, the refractive index and the aperture. The **aperture** is limited to something less than 90° since that would have the lens and specimen in contact with one another. In fact, 85° is about the limit in good optical microscopes. Such angles require an excellent lens. In most cases, the aperture is less because the edges of the lens introduce distortions and so cannot be used. **Refractive index** is easy to alter, but only within narrow limits. It can be increased by using oils to fill the space between the specimen and the objective lens. Transparent immersion oils used in today's microscopes (*i.e.*, **oil immersion lens**) have n up to about 1.6. Still 1.6 is big improvement over air or water ($n = 1$). In a microscope, the smallest detectable detail is equal to about one-half the **wavelength** of light with which it is observed. The smaller the object, the shorter the wavelength of light required. Hence, the wavelength of light is the area which has great chances of improvement. One can, for example, use ultraviolet light instead of visible light, thus, improving resolution as much as two-fold. In order to do that however, special lenses (*e.g.*, of quartz) must be used since ordinary glass blocks much ultraviolet light. In such a microscope, called ultraviolet microscope, the eyes cannot be used to view the image directly, for they are insensitive to ultraviolet light. Lastly, a specimen cannot absorb light of wavelength below 0.3 μm.

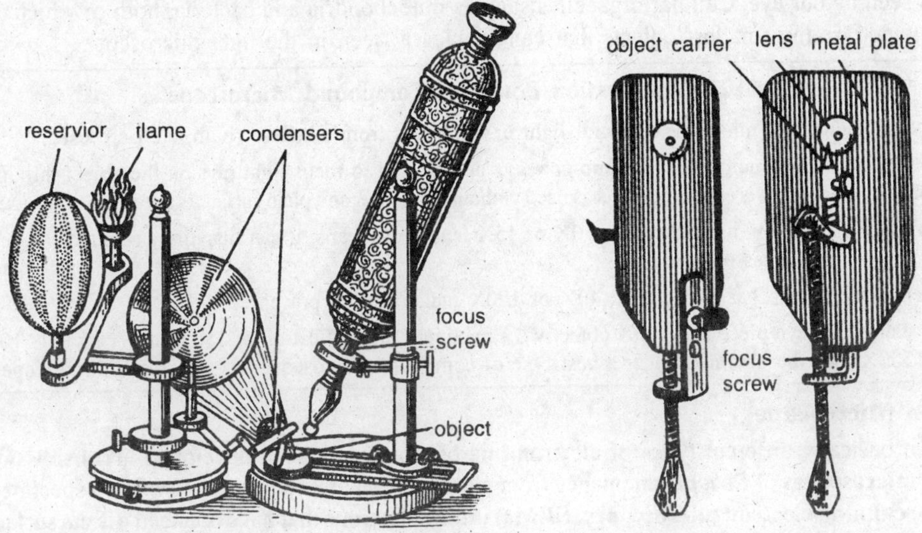

Fig. 54.4. Hooke's compound microscope. **Fig. 54.5.** Leeuwenhoek's microscope.

Thus, a good light microscope, with a numerical aperture of 1.4 and using light of short wavelength (0.4 μm) will resolve two points at about 0.17 μm separations. By such a microscope though, one can see considerable details in most cells, there is also a great deal that cannot be seen. For instance, ribosomes and chromatin threads of nucleus are about 0.02 mm in diameter and quite invisible to the light microscope. For them electron microscope is used. In cell biological studies, the following two types of microscopes are most extensively used:

Light Microscopes (Optical Microscopes)

The compound light microscope is the simplest and most commonly used microscope. It consists of the following three lens system.

(i) **Condenser lens system** occurs beneath the specimen and its function is the collection and focussing of the light rays on the object or specimen which is placed on the stage of the microscope.

(ii) **Objective lens system** remains near and above the specimen. It produces and magnifies the image of the specimen.

(iii) **Eye piece lens system** or ocular lens system remains near the eyes of the observer and it magnifies and forms the image (secondary) of the (primary) image previously produced by the objective.

Lenses. A simple lens can magnify an object only three times (3X). For getting a magnification of more than 3X, a combination of several lenses is used. Such a combination of several lenses (called **elements**) function as a single convex lens and a magnification of about 20X can be obtained. For example, an **eye piece lens** contains 2 to 3 lenses or elements. Likewise, the objective lens is made by placing 8 to 10 lenses or elements. Individual lenses are placed so close to each other that they act as a single lens. Only achromatic lenses are used in a microscope. (For certain other facts see Box 54.2).

Working. Light rays from the light source (sunlight or bulb-light) are focussed on the object by the condenser lens. The objective lens produces the enlarged image, called **primary image,** of the object or material. The eye piece lens further magnifies the primary image into secondary image which is seen by our eye. Cellular organelles such as mitochondria and bacteria, both of which are 0.5 μm wide, are the smallest objects that can be clearly seen in the light microscope.

Box 54.2 Magnification Power of Compound Microscope

1. In compound light microscope natural light or light rays from a source (bulb, etc.) are used.
2. The mirror of the microscope is **plano-concave** and is used to focus light rays on the object through substage condenser. Its concave surface is used without condenser and plain surface is used with condenser.
3. An eye piece usually has either 5X, 10X or 15X lenses. It means, it can magnify an image 5 times, 10 times or 15 times respectively.
4. Each objective too has either 10X, 40X or 100X magnification power.
5. By using 15X eye piece and 100X objective, a maximum magnification of 1500 time (*i.e.,* 15 × 100 = 1500X) may be obtained under a best type of compound microscope (*e.g.,* research microscope).

Electron Microscope

Two basically different types of electron microscopes have been developed. **Transmission electron microscopes (TEMs)** form images from electrons that are transmitted through a specimen. Whereas **scanning electron microscopes (SEMs)** utilize electrons that have bounced off the surface of the specimen. A transmission electron microscope has a very high magnification (500,000 times). The resolution TEM was designed by **Knoll** and **Ruska** of Germany in 1932. It permits direct study of biological ultrastructures (*e.g.,* structure of cell organelles). The practical limit of resolution of electron microscope is about 3 to 5 Å.

In TEM, the illuminating agent is not light but the **electrons** of short wavelength (0.500 Å), The wavelength of electrons is determined by the voltage at which these are generated. For example, at 50,000 volts, the electrons have 0.50 Å wavelength and resolution power of electron microscope can be 0.50 Å/2 = 0.25 Å. Similarly, in an electron microscope with a voltage of 100,000 volts, the wavelength of an electron is 0.004 nm (or .04 Å). Theoretically the resolution of such a microscope should be 0.02 Å. But practical resolving power of most modern electron microscope is 0.5 μm or 5 Å. This is 400 times greater than that of a light microscope.

Structure

The transmission electron microscope comprises the following parts:

1. Electron gun and anode. In TEM, the source of illumination is electrons which are produced and concentrated into an electron beam by **electron gun.** Electron gun consists of an electrically heated **tungsten filament or cathode** that emits electrons. Outside the gun is an anode which attracts electrons. Thus, between the filament and anode the electrons attain a high velocity with wavelength of .005 to .003 nm.

2. Microscope column. Electrons can travel in straight line in vacuum only, because in the air, these collide with oxygen or nitrogen atoms. The electron microscope, therefore, is enclosed in a evacuated metal tube.

3. Condenser lens. It is the **electromagnetic coil** which focusses or condenses the electron beam in the plane of object.

4. Objective lens. It is the **electromagnetic coil** which produces first magnified image of the object. It focusses the electrons which are reflected by the object and forms the first image.

5. Projector lens. It is also the electromagnetic coil which magnifies the first image formed by objective lens. It produces the final image.

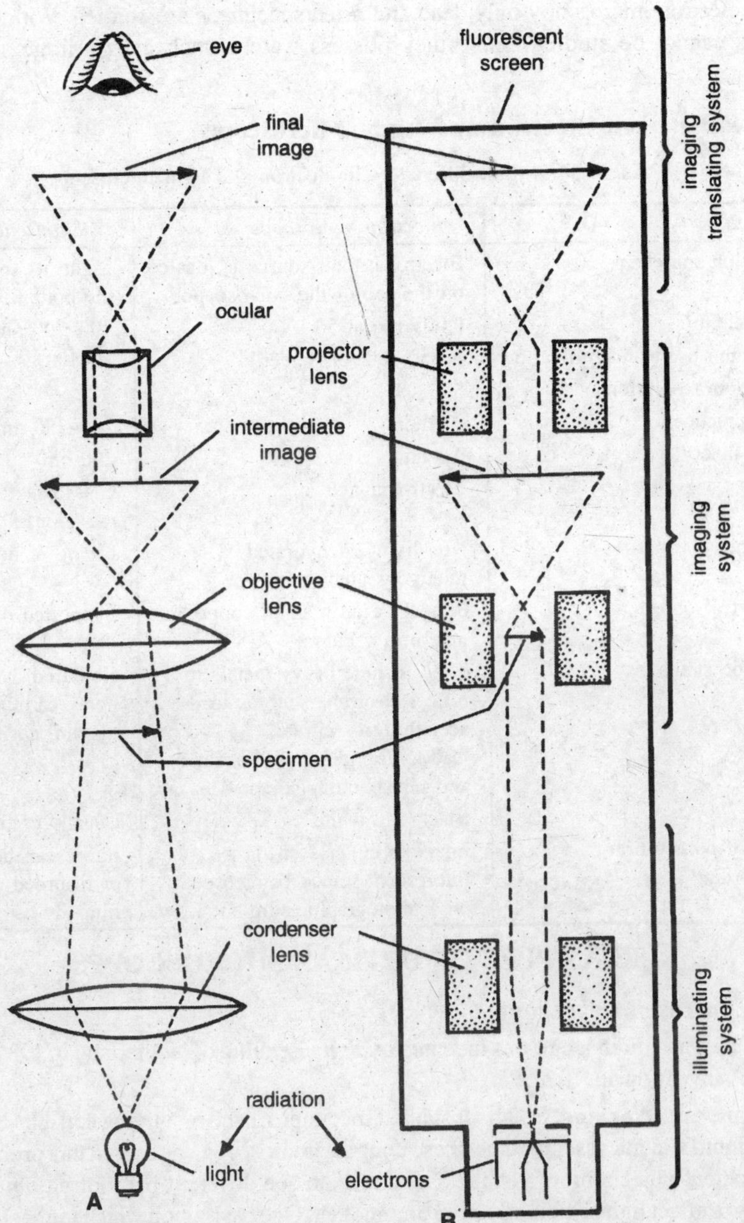

Fig. 54.6. A comparison of the lens systems of light (A) and electron microscope (B).

6. Fluorescent screen or photographic plate. Fluorescent screen is used for observing the magnified image of the object. It remains coated with a chemical (*e.g.,* zinc sulphide) which on being excited form the image as on the screen of television. The final image can also be captured on photographic film; such photographs are known as **electron micrographs.**

In the light microscope, the focussing is done by moving the lens tube up and down. In TEM, focussing is done by changing the flow of current through the coils of the electromagnetic lenses.

Further, in electron microscopy only dead and dried specimens are studied. With this microscope, living cells cannot be studied, since they possess water which causes large scattering of the electrons.

Differences Between Electron and Light Microscope

Table 54.1 Main differences in electron and light microscope

Feature	*Electron microscope*	*Light microscope*
1. Radiation source	Electrons; their source is located on the top of the microscope	Light, its source is located at the bottom of microscope
2. Wavelength	0.005 nm at 50 kV	400–700 nm
3. Maximum useful magnification	× 250,000 (on screen)	× 1500
4. Maximum resolution		
(*i*) in practice	0.5 nm	200–250 nm
(*ii*) in theory	0.2 nm	200 nm
5. Lenses	Electromagnets	Glass (quartz for ultraviolet radiation)
6. Specimen	Non-living, dehydrated, relatively small or thin.	Living or non-living
7. Support of object	Supported on a small copper grid in a vacuum	Supported on a glass slide
8. Common stains	They include heavy metals to reflect electrons such as lead acetate, lead citrate, lead hydroxide, uranyl acetate and phosphotungistic acid	Coloured dyes such as borax carmine, haema-toxylin, eosin
9. Image	Black and white	Usually coloured
10. Mode of observation or recording	Image is either seen on a fluorescent screen or recorded on a photographic film	Image can be seen with eye or recorded on photographic film by camera

EXAMPLES OF OPTICAL MICROSCOPES

1. Phase Contrast Microscope

Phase contrast microscope was invented by **Frits Zernike** of Netherland in 1932. He got Nobel Prize for this invention in 1953.

Principle and Working. Small, unstained specimens, such as a living cell, can be very difficult to see in a bright-field microscope. The phase contrast microscope has solved this problem by making highly transparent objects more visible. The ability to see different parts of an object depends on their capacity to affect light differently from one another. One basis upon which intracellular organelles differ is their refractive index. Cell organelles are made up of different proportions of various molecules: DNA, RNA, protein, lipid, carbohydrate, salts and water. Regions of different composition are likely to have different refractive indexes. Normally, however, such differences cannot be detected by our eyes. The phase contrast microscope converts differences in a refractive index into differences in intensity (relative brightness and darkness) which are then visible to the eye. The basis for this conversion centres on the ability to light waves to interact with one another, a property termed **interference**.

In contrast to bright-field microscope, phase contrast microscope performs two functions:

1. It separates the direct light (the background light of the field) from the light diffracted by the object, and,

2. It causes these two types of waves to be approximately one-half wavelength out of phase with one another so that they can destructively interact and cause changes in intensity.

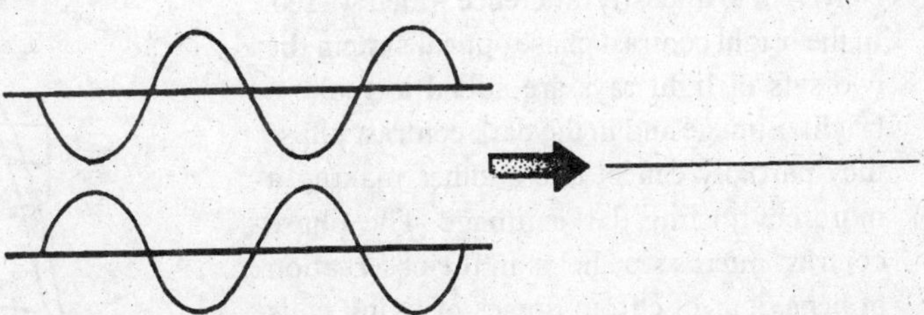

Fig. 54.7. Destructive interference. The two waves of light are one-half wavelength out of phase. When the waves meet, they combine to cancel each other out.

Suppose that all parts of the object being examined have exactly the same refractive index as the medium in which they are suspended. If this were the case, all the light from the object and the direct light would be one-half wave-length out of phase with one another, and **interference** would result in a uniform decrease in intensity (Fig. 54.7). Now consider an object whose various parts have different refractive indexes, all of which are greater than the surrounding medium. Differences in refractive index result in **phase shifts** of light rays passing through different parts of the specimen (Fig. 54.8). The relative brightness **(negative contrast)** or darkness **(positive contrast)** of each part of the image reflects the way in which the light from that part of the specimen interferes with the direct light. Since parts of the specimen have different refractive indexes, the various parts of the specimen appear in the image with varying intensity, thereby providing the necessary contrast to be visualized.

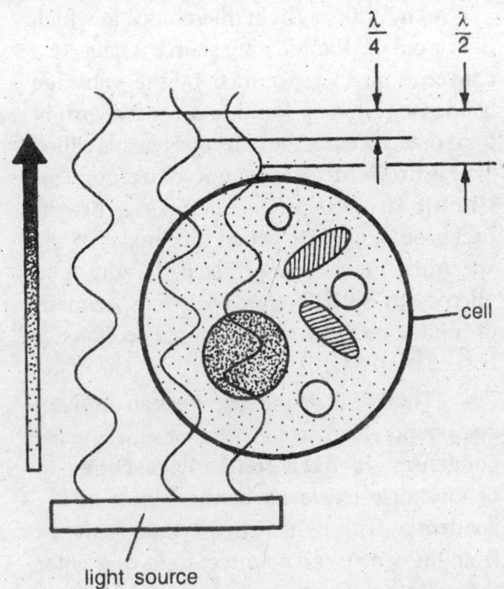

Fig. 54.8. Schematic diagram of retardation of light rays to varying degrees as they pass through different parts of a cell. Because the light rays that emerge from different parts of the cell are in *different phases* of their cycle, they interfere with the direct light to different degrees, producing different degrees of lightness and darkness in the final image.

Structure. The phase contrast microscope contains an annular phase plate in the objective and an annular diaphragm in the condenser. The annular diaphragm of phase contrast microscope permits light to pass through the condenser as a hollow cone, the remaining light being absorbed. This cone is focussed on the object. The phase plate which is placed at the back focal plane of the objective lens, is a transparent disc containing a groove (or elevation) of such a size and shape as to coincide with the direct image of the substage annular diaphragm which is formed at the back focal plane when no object is viewed. If an object is placed between the condenser and the objective, in addition to direct image a number of overlapping diffracted images of the diaphragm then appear at the back focal plane of the objective. The depth of the groove (or the elevation) in the phase plate of the objective is so made that the two sets of rays forming the direct image and the diffracted image differ in optical path by a quarter wavelength of the illuminating beam of light. Under these conditions the phase difference, which is not seen by the eye, is converted to intensify difference which we see. In the bright contrast phase optical system the two sets of light rays are added to make a brighter image and in the dark contrast phase they partially cancel one another making a more contrasting darker image. The phase contrast microscope helps in the observation of behaviour of chromosomes of living cells during mitosis or meiosis and various other cellular components such as mitochondria and vacuoles. Observations of living cells on a television screen using a video camera provide especially dramatic portraits of the dynamic activities of a cell (Table 54.2).

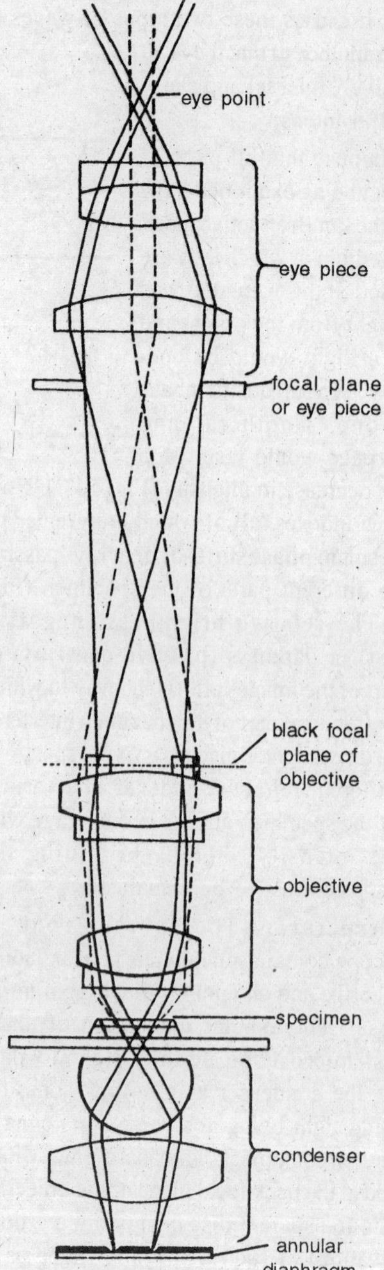

Fig. 54.9. Light path in phase contrast microscope.

2. Dark Field Microsopy

Any ordinary light microscope in which light from the illuminating source is caused to converge on the specimen by the substage condenser, thereby forming a cone of bright light that can enter the objective lens, is called a **bright field microscope.** The cone of illuminating light is seen as a bright background against which the image of the specimen must be contrasted. Bright field microscopy is ideally suited only for specimens of high contrast, such as stained sections of tissues (Fig. 54.10).

The dark field microscope is an ordinary compound microscope which has a special condenser. In **dark field microscope,** an opaque disc is placed in the centre of the condenser. The light rays are thus removed from the centre and no direct light rays enter the objective. Due to opaque disc, light rays are directed from the sides and only scattered (oblique) rays tend to enter the objective lens. As a result, the object appears illuminated (bright) against black background. Dark field microscopy improves visibility and resolution.

3. Fluorescence Microscopy

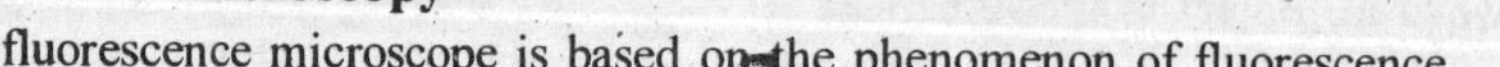

The fluorescence microscope is based on the phenomenon of fluorescence.

Table 54.2 The differences between light microscope and phase contrast microscope.

Light microscope	*Phase contrast microscope*
1. The refracted light rays are delayed by about one-quarter wavelength (0.25 λ).	1. The refracted light rays are delayed by one-half wavelength (0.50 λ).
2. All points of the image look equally bright.	2. The brightness of different points of image differs. It corresponds to differences in the refractive indexes of various cellular organelles.
3. Difference in the wavelengths of refracted and unrefracted light rays is not changed.	3. This increases the difference to produce maximum contrast (*i.e.*, interference) in the image.
4. Both living and dead (stained) specimens may be seen by this instrument.	4. Living (unstained) structures can be seen by this instrument.
5. There is no phase plate.	5. In this microscope, a glass phase plate is placed above the objective lens to increase the phase difference between the refracted and unrefracted light rays.

Fluorescence. It is found that when certain chemical compounds are illuminated by ultraviolet rays they become excited at molecular level and start to emit light of visible range. This phenomenon is known as **fluorescence**. The fluorescence may be of two kinds: **autofluorescence** and **secondary fluorescence**. Some substance, *e.g.*, chlorophyll becomes readily fluoresces or illuminates and produces red light, is an example of autofluorescence. When the fluorescence is induced in a substance by certain specific dyes or stains, the fluorescence is known as secondary fluorescence, *e.g.*, protein and carbohydrates, etc.

The source of illumination of fluorescence microscope is ultraviolet rays of higher wavelengths (3500 to 4000 Å). The different chemical compounds of the object to be seen when fluoresce by these ultraviolet rays, they emit rays of different wavelength. The rays of fluorescence can be detected and the chemical nature of the given object can be tested.

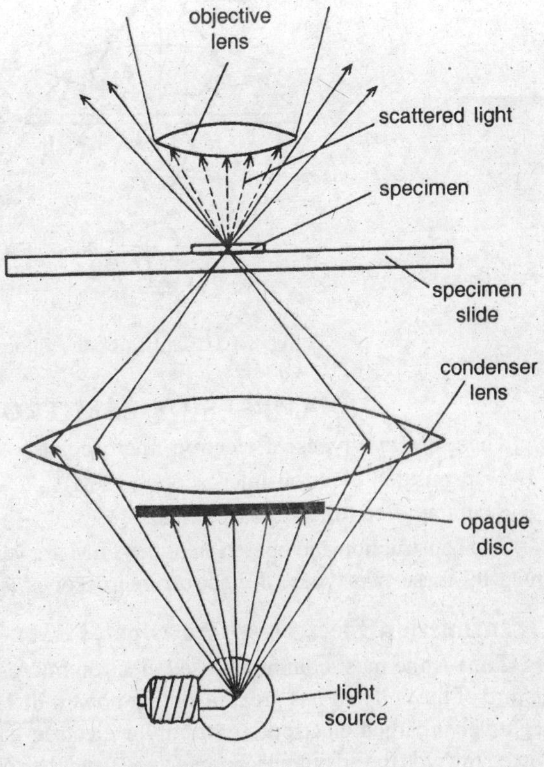

Fig. 54.10. Dark field microscope. Light path from the condenser to the objective lens.

The ordinary compound microscope can be easily converted into the fluorescence microscope by placing a special filter in between the source of illumination and eye piece.

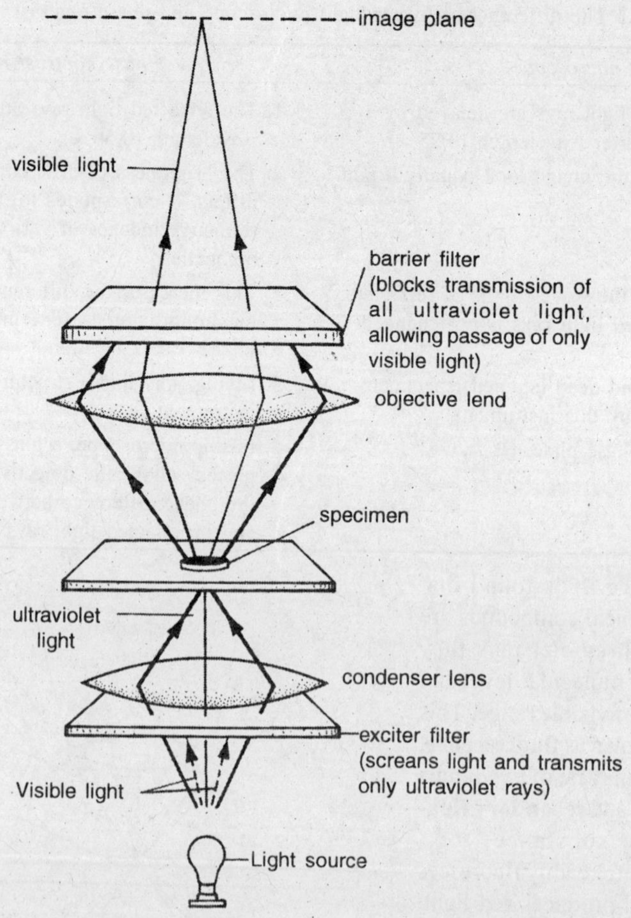

Fig. 54.11. Optics of the fluorescence microscope.

EXAMPLES OF ELECTRON MICROSCOPES

There are two types of electron microscopes:

1. Transmission electron microscope (TEM)
2. Scanning electron microscope (SEM)

The construction and operation of the TEM are very different from that of the SEM. The image formed by these two types of electron microscopes is also quite different.

1. Transmission Electron Microscope (TEM)

TEM is the most commonly used electron microscope. It is widely used to study the internal structure of the cell and has great resolving power. In TEM, an electric field propels electrons from a negatively charged electrode (**cathode** or electron gun). The electrons emitted from the gun are attracted towards the positively charged anode and then are focussed into beam by the electromagnetic **condenser**. This beam of electrons illuminates entire stained specimen. Many of these electrons pass right through the specimen and reach the electromagnetic objective. But some are absorbed or scattered by atoms of metallic stains that have combined with certain atoms of specimen. The **objective** magnifies the image of the specimen. This image is received by the third electromagnetic lens which works as **projection lens**. It further magnifies the image and projects it on a zinc sulphide or fluorescent screen at the bottom of TEM. This screen is similar to screen of a television tube. It is coated with a layer of crystals that respond to electrons by emitting visible light. The final image can also be captured on the photographic film.

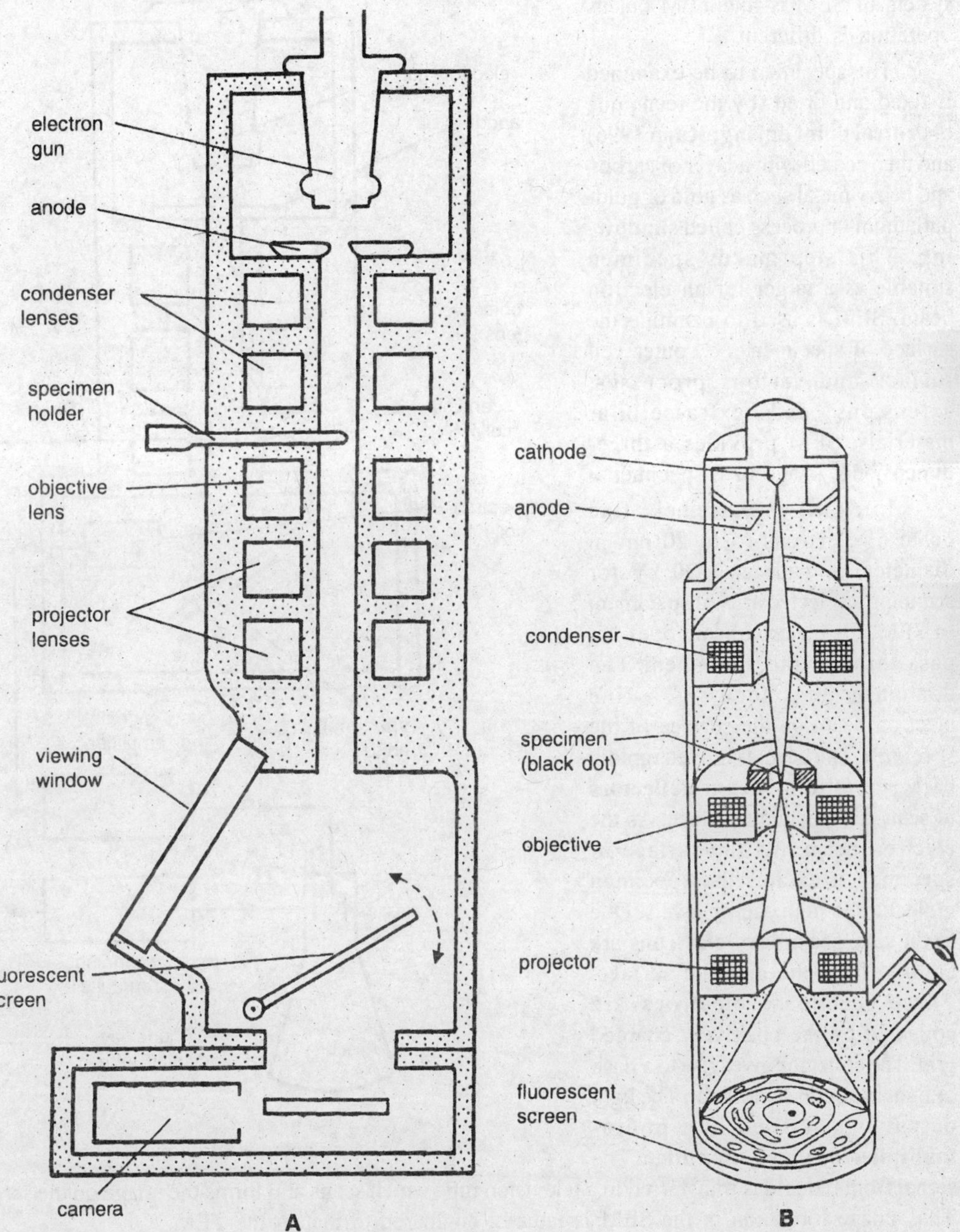

Fig. 54.12. Transmission electron microscope A–Diagrammatic view; B–Pathway of electron beam.

Electrons that are not transmitted by the specimen leave correspondingly dark regions on the viewing plate. Thus, image from a TEM is a pattern of bright and dark areas corresponding to the areas of greater or lesser electron density.

2. Scanning Electron Microscope (SEM)

The electron gun and lens system of SEM is like TEM but its operation is different.

The specimen to be examined is fixed and dried (by the technique of **critical point drying; Karp** 1996) and then coated with a layer of carbon and heavy metal such as gold or gold-palladium–a process called **shadowing.** This step makes specimen suitable as a target for an electron beam. SEM is used to examine the surface of specimen, *i.e.,* outer cell surface and various processes, extensions and extra-cellular materials. SEM provides a three-dimensional image of a specimen.

In SEM, an extremely fine beam of electrons (5 to 20 nm in diameter) is made at 3–30 kV for scanning a selected area of specimen. In SEM, the electron beam does not pass through the specimen. The condenser lens focusses a fine electron beam on the surface of the specimen. The beam is moved rapidly back and forth by **beam deflectors** to scan the specimen surface. As the electron beam hits the surface of specimen it excites the specimen molecules to high energy levels. Due to the this, **secondary electrons** are emitted from the metallic surface. These secondary electrons are collected by the positively charged grid. The **collector** gives rise to a flash of light in a solid scintillator. The light output is amplified in a **photo-multiplier** or **video amplifier.** The signal from the grid is transferred to a television tube which scans and forms the image on the screen. Thus image formation in the SEM is indirect compared to that in the TEM.

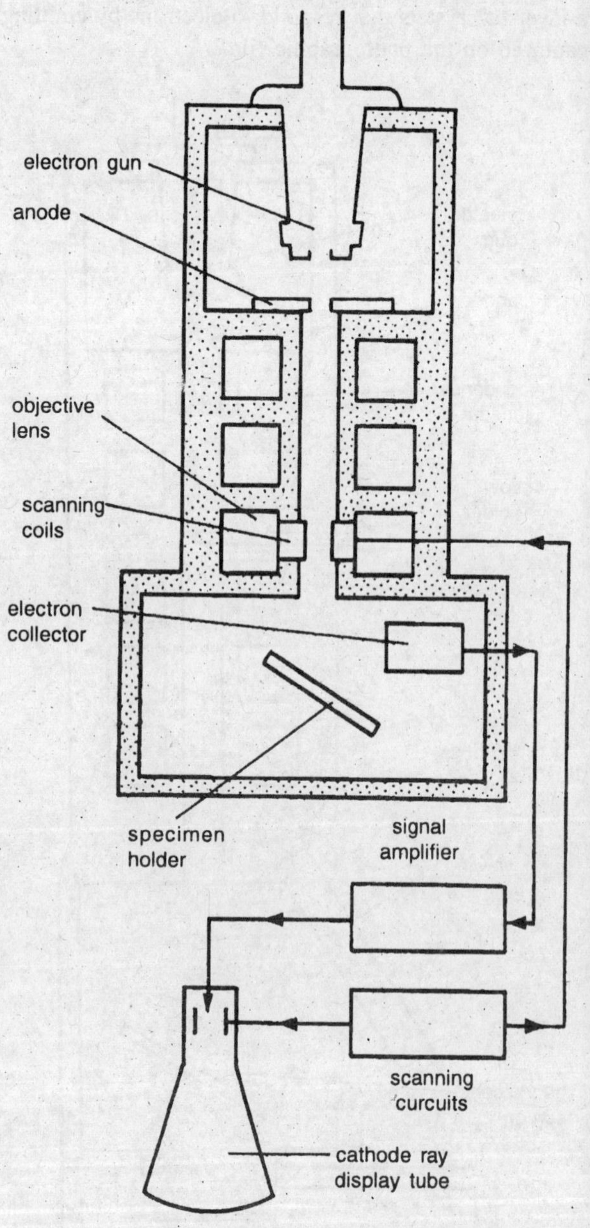

Fig. 54.13. A scanning electron microscope.

The amount of secondary electrons produced depends on the angle of specimen points with the scanning beam. Surface perpendicular to the beam produces maximum electrons while surfaces at greater angles release less electrons. Thus, the number of electrons produced depends on the three-dimensional shape of the specimen surface (or surface topology, *i.e.,* the crevices, hills and pits of the specimen). Accordingly the image contains bright (which correspond the elevations or ridges in specimen surface) and darker regions (which correspond to the valleys). These shadows give the image a three-dimensional appearance.

The resolving power of SEM is comparatively less than that of transmission electron microscope. It has an effective magnification upto 20,000 times.

54.4 CHROMATOGRAPHY

Chromatography is a technique for separating mixtures of gases, liquids or dissolved substances or the components of **proteins** (*i.e.,* amino acids), of **nucleic acids** (*i.e.,* nucleotides) or of **chlorophyll pigment.** By this technique, a mixture of dissolved components is fractionated as it moves through some type of porous matrix.

Principle of Chromatography

Chromatography is based on the phenomenon of adsorption and differential migration of components of a mixture or a compound. This needs the following two basic things:

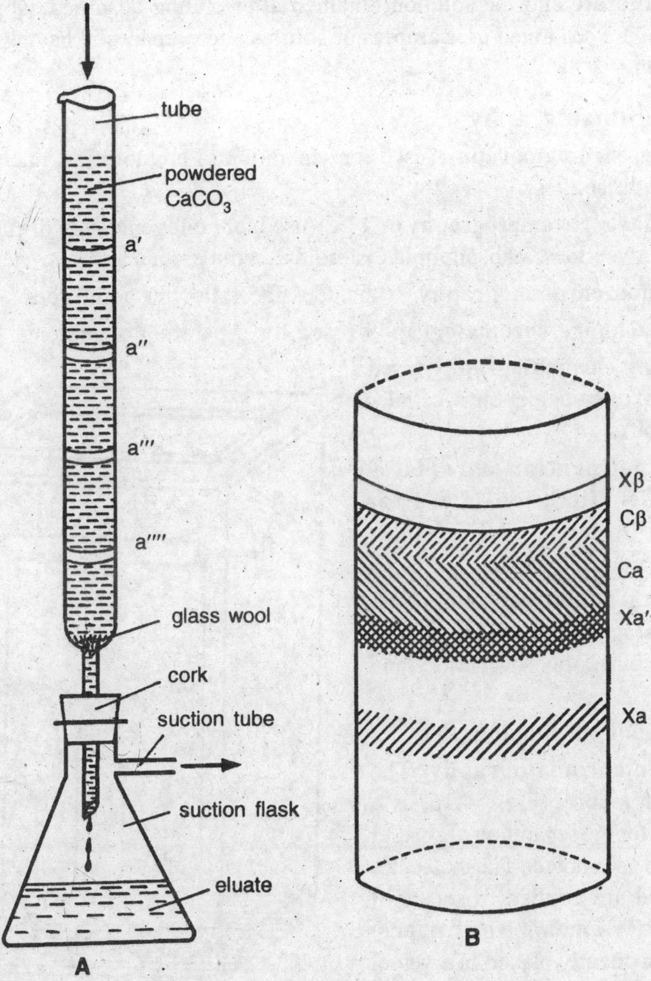

Fig. 54.14. A–An apparatus for column chromatography: B–Chromatogram of chlorophyll.

1. A force for propelling the molecules of any solution, mixture or cytosol. It is called **mobile phase** and it involves moving solvent.
2. A selective impedance causing differential movement of various molecules (*i.e.*, called **impeding medium** or **adsorbent**). It is also **immobile phase** and it involves the matrix (filter paper, $CaCO_3$, etc.) through which the solvent is moving.

During chromatography the mixture or solution is allowed to percolate in an insoluble medium or adsorbent or matrix which has different affinity for molecules of different substances. The molecules migrate through the medium matrix) at different rates and are separated. For example, if a solution of leaf pigment, chlorophyll in petroleum ether is allowed to percolate through a column of powdered calcium carbonate, the pigments are initially absorbed by pure solvent (*i.e.*, petroleum ether), the individual pigments—carotene, chlorophyll and xanthophyll separate and appear as the coloured bands or zones on the calcium carbonate column like the light rays in spectrum. Such a preparation is called **chromatogram**.

The impeding medium (or immobile phase) is called **adsorbent**. The liquid that passes through adsorbent is **filterate** and the solution obtained after eluting adsorbed substances is **eluate**. The substances that are separated are **samples** or **solutes** and preparation having these solutes is called **chromatogram**.

Types of Chromatography

A. Paper chromatography (used for separation of amino acids, nucleotides, and other low molecular weight solutes).

B. Thin layer chromatography or TLC (used for rapid separation of unsaturated and saturated fatty acids, triglycerides, phospholipids, steroids, peptides, nucleotides, etc.).

C. Column chromatography. It includes the following four types:

1. Ion-exchange chromatography (used for separation of proteins, RNA and DNA).

2. Affinity chromatography (used for separation of immunoglobulins, cellular enzymes, mRNAs).

3. Gel permeation chromatography or Gel filtration (Used for separation of proteins, nucleic acids, polysaccharides and lipids).

4. Gas chromatography or Gas-liquid chromatography or **GLC** (for the separation of lipids, oligosaccharides and amino acids).

D. Partition chromatography

1. Paper chromatography. The paper chromatography (Fig. 54.15) is a simple method for the separation of smaller molecules from one another. The molecules to be separated are applied to sheets of suitable paper **(Watman's No. 1 paper)** which are subsequently placed in a vessel which contains a suitable solvent. A distinction is made between ascending and descending paper chromatography

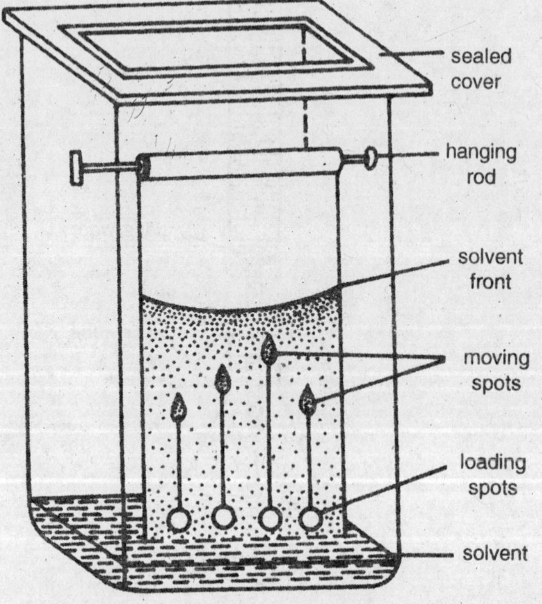

Fig. 54.15. Ascending paper chromatography.

according to whether the solvent migrates on the paper from below or above. Highly soluble components of the sample mixture will migrate at the solvent front; other substances move more slowly according to their solubility. From this it follows that the choice of solvent determines the speed of migration of the individual molecules and is crucial for a successful separation. The paper chromatography method is used for the separation of amino acids, nucleotides and other lower molecular-weight metabolic products.

If the solvent flows towards upper side on the paper, it forms the **ascending paper chromatography** (Fig. 54.15), and if it flows towards lower side then it forms the **descending paper chromatography**.

2. Column chromatography. In column chromatography (Fig. 54.16), an insoluble medium is packed into a glass tube; the length and width of this tube influence the separation of the molecules. The molecules to be separated are applied to the top of the column and their migration is started by adding a solvent. The characteristic separation which results depends on the choice of solvent and carrier material. A positively charged carrier binds negatively charged molecules; other carriers contain pores which are penetrated by the smaller molecules, which are, therefore, slowed down. The solution which flows from the column is collected in small fractions (Fig. 54.16), which contain the separated classes of molecules. Column chromatography is important for the separations of mixtures of proteins, that is, for the isolation of enzymes such as cytochrome C or RNA polymerase.

Adsorbents and solvents. A few of the adsorbents used in column chromatography are carbonates of calcium, magnesium and sodium; various forms of charcoal activated in special ways; Fuller's earth, bentonite, Lloyd's reagent, talcum, clays of different types; alumina, silica gel, cellulose, resin, starch, sucrose, inulin and benzoic acid, etc. The adsorbent column neither dissolves in the developing solvent nor reacts chemically with the solvent or the substance to be separated.

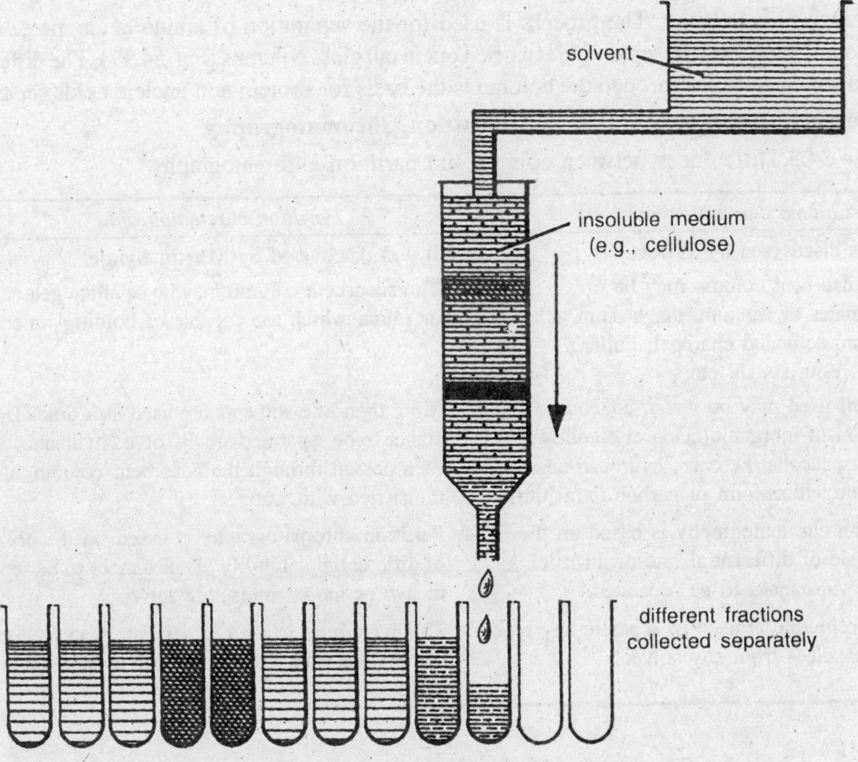

Fig. 54.16. Column chromatography.

The **solvents** commonly used are water, aqueous solutions of various acids, alkalis and salts; alcohols, ketones and cyclohexanone, hydrocarbons such as benzene, chloroform and carbon tetrachloride, etc.

3. Ion exchange chromatography. In ion exchange chromatography molecules are separated by their difference in charge. The ion-exchange resin is used as chromatographic adsorbent consists of porous polystyrene beads (polymers) to which ionizable groups are added chemically. The resins bearing negative charge are called **cation exchanger** (*e.g.,* **DEAE-** or **diethylaminoethyl-cellulose)** and positively charged resins are **anion exchangers** (*e.g.,* **CM- or carboxy methyl-cellulose**).

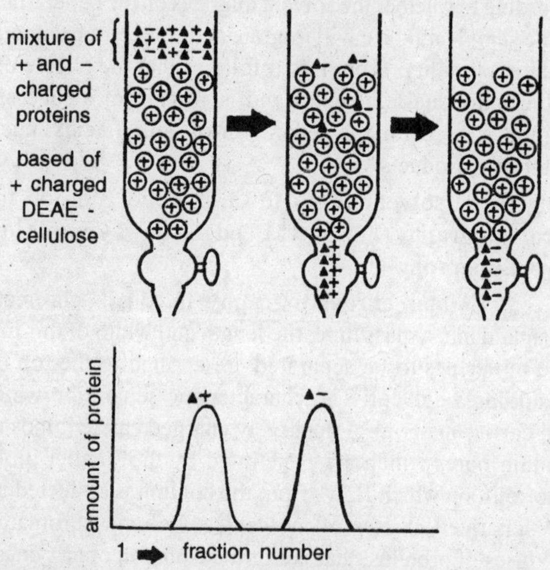

Fig. 54.17. Ion-exchange chromatography. The separation of two species of protein by DEAE-cellulose. In this case, a positively charged ion-exchange resin is used to bind the more negatively charged protein.

When a mixture of molecules possessing different charges passes through a charged column, the molecules possessing opposite charge get associated with the column, while those with same charge or with no charge pass through. The process is used for the separation of amino acids, proteins and other macromolecules and the separation is carried out in tall glass columns (Fig. 54.17). The differential rates of movements of ions through the column is the basis for protein and nucleic acids separation.

1. Differences Between Column and Partition Chromatography

Table 54.3 Differences between column and partition chromatography.

Column chromatography	*Partition chromatography*
1. It was discovered by **Tsweet**.	1. It was discovered by **Martin Synge**.
2. The adsorbent column may be of carbonates of calcium, magnesium or sodium; activated charcoal; Fuller's earth, resin, starch, etc.	2. The adsorbent column may be of silica gel, cellulose or starch which are capable of holding water.
3. Solvent used may be water, aqueous solution of inorganic acids or alkalies or salts, alcohal ketones, hydrocarbons, benzene, chloroform or carbon tetrachloride.	3. More than one solvents are used at a time. The substance to be separated are dissolved in alcohol and are then passed through the adsorbent column, already moistened with water.
4. Column chromatography is based on the principle of differential rate of diffusion of the substances to be separated.	4. Partition chromatography is based on the principle of differential solubility of substances to be separated in two or more immiscible solvents.
5. This technique is used to separate any type of substances from any solvent.	5. This technique is used to separate water soluble compounds from a mixture of two or more immiscible solvents.

2. Differences Between TLC and GLC

Table 54.4 Some differences between TLC and GLC

TLC or Thin-layer chromatography	*GLC or Gas-liquid chromatography*
1. Adsorbent may be cellulose powder, alumina or cellulose.	1. Adsorbent or the immobile phase is represented by liquid.
2. The mobile phase is the solvent (liquid).	2. The mobile phase is inert gas such as nitrogen, carbon dioxide, helium or argon.

54.5 ELECTROPHORESIS

Molecules or macromolecules may be separated in an electric-field if they are charged to different extents. The mixture of compound is applied to supporting films which dip into two containers filled with a salt solution. One of the containers holds a cathode, the other an anode. On passing an electric current, the negatively charged molecules migrate to the anode and the positively charged molecules to the cathode. Paper, agar and starch are examples of substances which may be used as supporting films.

The rate of migration of a molecule in an electric field is determined by its size and the number of charged groups per molecule. The electrophoresis method is used in the separation of proteins, nucleic acids and their building blocks.

Some of the common types of techniques of electrophoresis, which are currently used in cell biology, are the following:

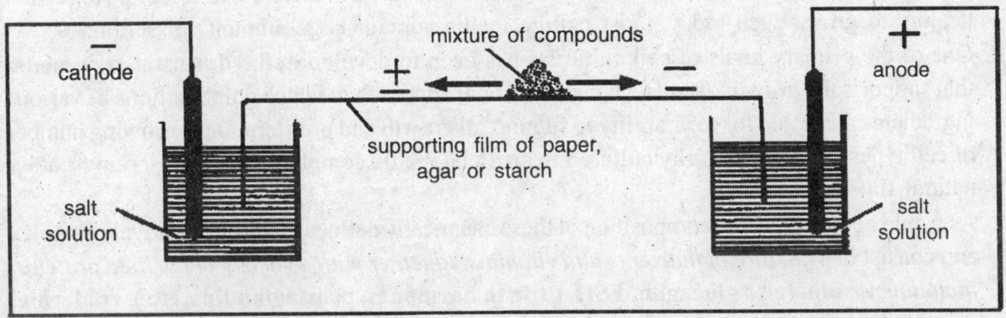

Fig. 54.18. Electrophoresis.

1. **Moving-boundary electrophoresis** (used for proteins).
2. **Gel or zone electrophoresis** (used for proteins).
3. **Discontinuous electrophoresis** (used for isolation of proteins of plasma membrane).
4. **SDS-PAGE or Sodium dodecyl sulphate-polyacrylamide gel electrophoresis** (used for separating and sizing macromolecules such as proteins, *e.g.,* membrane proteins, protein component of cytoskeleton, etc.).
5. **Maxam-Gilbert technique** (used for separation of polynuclotide fragments of RNA and DNA).
6. **Immunoelectrophoresis** (used for antigens and antibodies).

54.6 DIALYSIS

It is a sensitive method for separating lower molecular-weight component from macromolecules. A thin membrane in the form of a tube is filled with the solution containing the molecules to be separated. The pore size of the membrane allows the diffusion of small molecules such as salt or amino acids; larger molecules such as proteins or nucleic acids cannot pass through the pores and so remain inside the dialysis tube.

54.7 CELL CULTURE

The growth of cells or tissues of living organisms outside the body in a suitable sterile culture (or nutrient) medium under controlled environmental conditions is called **cell culture** or **tissue culture**. Depending on the nature of the explanted material, the techniques are called **cell, tissue, organ** or even **whole embryo culture**. The culture medium may be natural or artificial and either in solid form (*e.g.*, agar) or in liquid form (*e.g.*, physiological saline).

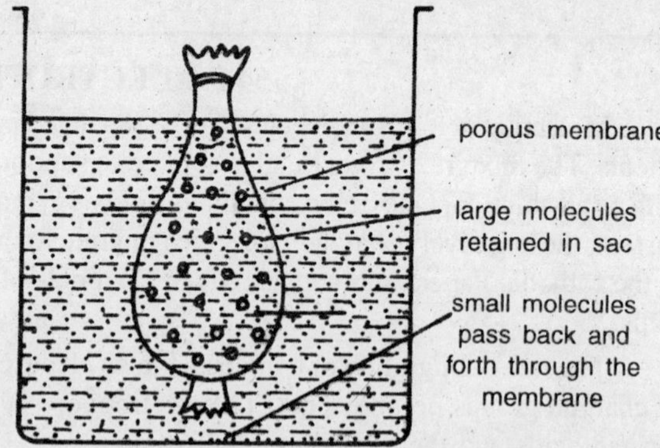

Fig. 54.19. Dialysis.

Certain Basic Requirements of Tissue Culture

1. The early tissue culture studies used media containing a great variety of *unknown* substances. Cell growth was accomplished by adding fluids obtained from living systems, such as **lymph, blood serum** or **embryo homogenate**. It was found that cells required a considerable variety of nutrients, hormones, growth factors and cofactors to remain healthy to grow. Even today, most culture media contain large amounts of serum.
2. One of the primary goals of cell culturists has been to develop **defined, serum-free media** that supports the growth of cells. Using a practical approach in which combinations of various ingredients are tested for their ability to support all growth and proliferation, a growing number of cell types have successfully cultured in artificial media completely lacking serum or other natural fluids.

   As would be expected, the composition of these chemically defined media are relatively complex and consist of a *mixture of nutrients and vitamins, together with a variety of purified proteins, including* **insulin** (also **glucagon, FSH, growth hormones, prostaglandins,** etc.), **epidermal growth factors** and **transferrin** (which provides the cells with usable iron).
3. In addition to a complex culture medium, most cells need an appropriate surface on which to grow. Unlike most cancer cells, which are capable of growing in suspension, normal cells require a solid substratum, such as glass or plastic.

Box 54.3 Historical

1. **Ross Harrison** (1907) of Yale University performed one of the classic experiments in biology. He removed a small piece of tissue from the developing nervous system of a frog embryo and placed the fragment into a tiny drop of lymphatic fluid. Harrison observed that nerve cells remained alive and healthy for many days and some of the cells sprouted processes that grew out into the surrounding medium.
2. **Alexis Carrel** showed that extracts of chicken embryo stimulate cells to grow and divide.
3. **Harry Eagle** spent many years in defining the nutritional needs of mammalian cells and formulating media to optimize their growth.
4. Once the cells settle to the bottom of the culture dish, they secrete extracellular materials such as **collagen, fibronectin** and **laminin,** which help in keeping them attached to their substratum.

Primary and Secondary Culture

The first step in cell culture is to obtain the cells. In most cases, one need only to remove a vial of frozen, previously cultured cells from a tank of liquid nitrogen, thaw the vial and transfer the cells to the waiting medium. A culture of this type is referred to as a **secondary culture** because the cells are derived from a previous culture. A **primary culture,** on the other hand, is one in which the cells are obtained from the organism itself. Most primary cultures of animal cells are obtained from embryos, whose tissues are more readily dissociated into single cells than those of adults.

To accomplish dissociation, the tissue is removed from the embryo and is either mechanically agitated or treated with a proteolytic enzyme, such as **trypsin** or **collagenase,** which is capable of digesting proteinaceous material that help hold the cells together. The tissue is then washed free of the enzyme and usually suspended in a saline solution that lacks Ca^{2+} ions and contains a substance, such as **EDTA (ethylene diamine tetra acetate)** that binds (*i.e.*, chelates) calcium ions. Calcium ions play a key role in cell-cell adhesion and their removal from the tissues greatly facilitates the separation of cells. Cell suspensions obtained by such treatments usually consist of a heterogeneous mixture of cell type.

Once the cells have been dissociated into a single-cell suspension, either they can be cultured directly, or they can be separated according to cell type and then placed in culture. Cells can be separated by a number of techniques, including differential centrifugation or the use of a **fluorescence-activated cell sorter.** In this latter approach, the cell suspension is treated with **fluorescent antibody** that binds specifically to the surface of the cell type to be cultured, and the suspension is passed through an electronic instrument that is able to deflect the fluorescently labeled cells into a separate tube from those cells lacking the label.

Once the cells have been separated, they can be grown either **suspended** in a liquid medium or **attached** to a solid surface. There are basically two types of primary cultures that can be started:

1. In a **mass culture**, a large number of cells are added to a culture dish; they settle and attach to bottom and form a relatively uniform layer of cells. Those cells that survive will grow and divide and after a number of generations, form a **monolayer** of cells that covers the bottom of dish.
2. In the other type of culture, called a **clonal culture**, a relatively small number of cells are added to a dish, each of which, after settling and attaching to surface, is at some distance from its neighbours. In this case, the proliferation of the cells generates individual colonies, or **clones**, of cells, whose members are all derived from the same original cell.

Cell Line

Normal (*i.e.*, nonmalignant) cells are capable of a limited number of cell divisions (*e.g.*, 50 cell divisions in human cells) before they undergo a process of **senescence** and **death.** Because of this, many of the cells that are commonly used in tissue culture studies have undergone **genetic modifications** (due to virus infection, chemical mutagens or radiations) that allow them to be grown indefinitely. Cells of this type (*i.e.*, transformed cells) are referred to as a **cell line** and they are typically capable of growing into malignant tumours when injected into susceptible laboratory animals. Such behaviour is typically shown by mouse cells. Among the best known cell lines are **HeLa cells** (obtained from a human carcinoma), the **L** and **$3T_3$** cells (from mouse embryo), the **BHK cells** (from baby hamster kidney) and the **CHO cells** (from Chinese hamster ovary).

Types of Tissue Culture

1. Culture of bacterial cells. The culture medium of bacteria contains an **energy source** (some simple carbohydrates or some amino acids); **mineral salts** of calcium, potassium, sulphur and phosphorus; **traces of metals** (zinc, copper, magnesium, manganese, etc.) and **growth factors**

(biochemical intermediates). The culture medium is maintained between pH 6 to 8 and at a temperature varying from 15–20°C or 30–37°C according to the species of bacteria.

The culture medium is prepared in *agar* in a petridish. Agar is a carbohydrate material derived from seaweed that is dissolved in a nutritive medium to make a semisolid gel on which bacteria can grow. When it sets, it forms a sheet on the bottom of the petridish. A very dilute suspension of bacteria is cultured in the petridish. As a result of growth and multiplication several colonies of bacteria are formed on the surface of agar. Each colony is derived from a single cell.

2. Culture of mammalian cells. Since mammalian cells are not able to synthesize essential biochemical intermediates from the raw materials, their culture media are very complex. For the culture of mammalian cells, several culture media have been developed; some of them are following:

(*a*) A mixture of serum and chick embryo extract rich in various factors required for growth.

(*b*) **Parker's 199 synthetic medium** contains different metabolites. It is supplemented with some foetal calf serum or horse serum.

Culture techniques. Mammalian cells have been cultured by adopting the following culture techniques:

(*i*) Maximov double coverslip method. A small piece of tissue to be cultured is placed on a coverslip and is covered with a drop of chick plasma and growth medium (Fig. 54.20). The plasma forms a clot over the tissue cells. The coverslip is then mounted in a cavity slide with the clot attached to the undersurface. It lies above the cavity of slide. The cavity of slide provides oxygen and allows exchange of gases. The growing cells form a sheet within the plasma clot.

(*ii*) Monolayer culture method. For the preparation of cell suspension, the chopped tissue is treated with dilute trypsin. The trypsin is removed from the cell suspension by washing in a centrifuge. The cell suspension settles down on the glass and attach to the surface of bottle. The cultured cells may be isolated by the dilute solution of trypsin and the process may be repeated again to get large number of cells.

(*iii*) Trowell method. The small fragments of an organ are placed on a raft or permeable sheet of Trowell chamber and are covered with culture medium. (Trowell is a small hand-held tool with a curved scoop for lifting plants or earth.)

3. Culture of plant cells. Many different types of plant cells such as anther, endosperm, nucellus, embryo, root, shoot tips, leaf, flower buds, etc., are capable of growing in culture. The simple undifferentiated tissues such as **primary meristems** show an organized growth in basic culture medium. But the fully differentiated tissues such as pith, secondary phloem, mesophyll and endosperm, *etc.,* in basic culture medium grow into an unorganized and undifferentiated mass of tissues, called **callus. Gauherst** *et al.,* (1939) succeeded in growing callus in vitro from isolated plant parts. By adding certain hormones, it is now possible to obtain callus culture from most plant tissues.

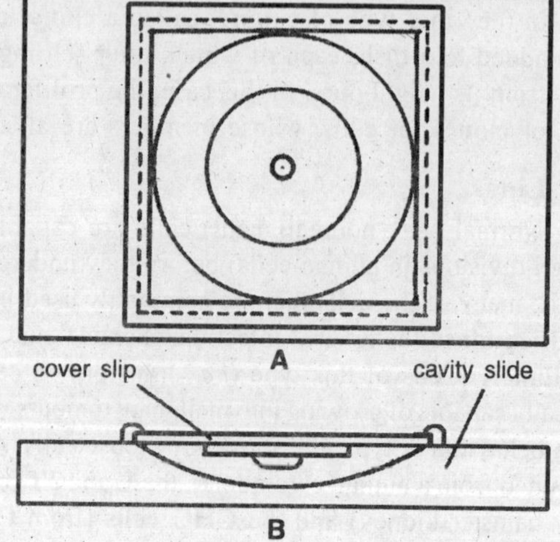

Fig. 54.20. Maximov double coverslip method of cell culture.

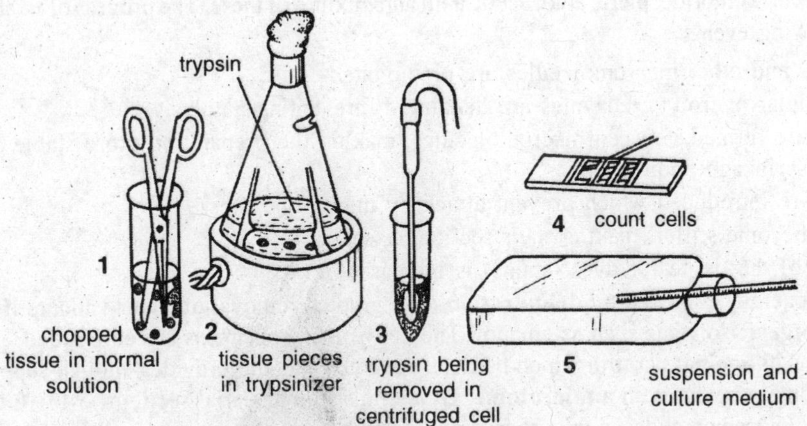

Fig. 54.21. Different steps of monolayer technique of tissue culture.

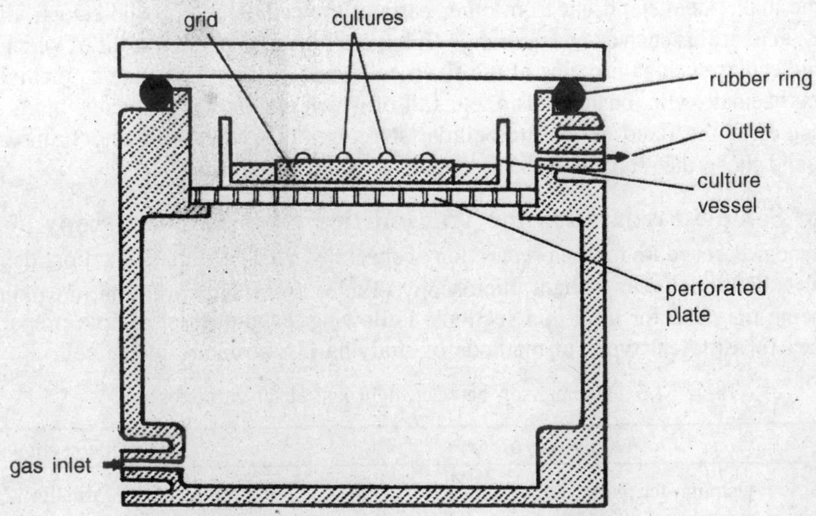

Fig. 54.22. Chamber of organ culture by Trowell method.

Culture technique. The first step in the formation of these cultures is to treat cells with the enzyme **cellulase**, which is capable of digesting away the surrounding cell wall, releasing the naked cell or **protoplast**. The protoplasts can then be grown in a chemically defined medium that promote their growth and division. Under suitable conditions, the cells can be grown into an undifferentiated clump of cells, the **callus**, which can be induced to develop shoots from which a new plant can regenerate.

54.8 SAMPLE PREPARATION FOR MICROSCOPY

Cells are transparent and optically homogeneous: so either they are viewed as such by instruments such as phase contrast microscope or to produce necessary contrast, the cells are passed through various steps of slide preparation such as killing, fixation, dehydration, embedding, sectioning, staining and mounting.

Thus, superior specimens for microscopic examination can be obtained by **killing** the cells and coagulating or **fixing** the protoplasm by preservatives, called **fixatives** such as alcohols, formaldehyde, mercuric chloride, picric acid, acetic acid and mixture of these. The process of fixation involves the following events:

1. The proteins and other macromolecules are precipitated.
2. The intracellular hydrolytic enzymes are denatured, preventing autolysis.
3. Cross links are formed between macromolecules, making the preparation more stable and minimizing shrinkage upon drying.
4. Substances are introduced which prevent attack by microorganisms.
5. The tissues become stiffer, making their sectioning easier.
6. The affinity of the tissue for dyes (stains) is increased.

Fixation is generally followed by **dehydration** (*i.e.*, gradual removal of water vapours from the tissue) by the organic solvents such as ethanol. The dehydrated specimens are **embedded**, *i.e.*, they are infiltrated with molten paraffin which hardens upon cooling and provides enough support to allow thin sections to be cut with a **microtome**. By the microtome, serial sections, 5 to 10 μm thick can be cut and placed on slides in the order of cutting and permitting a sequence of specimens for observation. These sections are stained with a non-vital-stain to increase the contrast.

Stains are the chemicals that can selectively attach to particular molecules of specific cellular structures and make them stand out from other parts of the cell. The non-vital stains fall into two main classes: **acid stains** such as eosin, orange *G*, aniline blue and fast green, all of which combine with basic molecules such as proteins of the fixed cells; and **basic stains** such as methylene blue, crystal violet, haematoxylin, basic fuchsin, etc., all of which combine with nucleic acids and other acidic molecules of the fixed cells. The cellular structures that stain with acid stains are called **acidophilic** and those that stain with basic dyes are called **basophilic**.

Methods of Sample Preparation for Transmission Electron Microscopy

The standard procedure for the preparation of specimen for TEM entails fixation, dehydration staining and sectioning similar to light microscopy (Tables 54.5). However, the most significant difference being the need for ultra-thin sections. Following techniques of sample preparation are generally used for different types of methods of studying ultrastructure of the cell:

Table 54.5 Comparison between light and electron microscopy.

Steps	*Electron microscopy*	*Light microscopy*
Fixation	Osmium tetroxide, postassium permanganate, formalin, glutaraldehyde.	Bouin's solution; formalin; Zenker's fluid.
Dehydration	Increasing concentration of ethanol (or acetone) followed by propylene oxide.	Increasing concentration of ethanol followed by benzene.
Embedding	Araldite; Vestoplaw, Epan 812; Maraglas; Durcopan.	Paraffin.
Sectioning	Usually 10–100 nm thick sections cut with a glass or diamond knife on an ultramicrotome.	Usually 6 μm thick sections cut with a razor blade on a microtome.
Mounting	On a perforated metal disc (grid) usually covered with formvar or paralodian.	On a glass slide with an egg albumin adhesive. Deparaffinized in xylol for staining.
Staining	With salts of heavy metals such as lead acetate, lead citrate, lead hydroxide, urynyl acetate; phosphotungstic acid.	Selective chromatic stains (as haematoxylin and eosin), dehydrated in ethanol series, cleared in xylol and mounted for viewing in Canada balsam or Permount.
Viewing	Grid is placed between the condenser and objective lenses in vacuum and the image is viewed on phosphorescent screen.	Slide is placed between the condenser and objective lenses and the image viewed in the ocular lens.

1. Monolayer technique. Macromolecules such as DNA and RNA are studied by monolayer technique in which the macromolecules are extended on the air-water interface before being collected on a film.

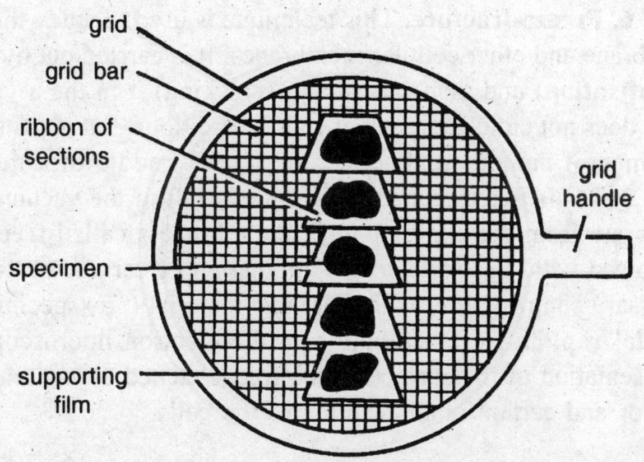

Fig. 54.23. Appearance of a ribbon of ultra-thin sections on the grid.

2. Thin sectioning. This method uses a cutting device known as **ultramicrotome** to remove ultrathin (*i.e.*, 10 nm to 100 nm thick) sections from the specimen. To withstand the passage of ultrafine diamond or glass knife without tearing, the specimen is first embedded in a hard plastic such as, epoxy resin (Table 54.5). The resin is allowed to penetrate the sample before it is polymerized. Sections are floated from the knife of ultramicrotome onto the surface of water and picked up by touching them with a fine wire mesh or small circular copper grid (*i.e.*, small discs perforated with numerous openings). Prior to its use mesh or grid is coated with a thin monolayer film (7.5 to 15 nm thick) of plastic (such as formvar or collodion) or carbon to provide a support to the sections (or sample) (Fig. 54.23). The specimen is visualized through the holes of screen.

Sections to be examined with the electron microscope are generally not stained (since no colours are seen with the electron microscope). However, contrast may be improved by "post-staining" with **electron stains** or electron-dense materials such as urynyl acetate, urynyl citrate, lead citrate, osmium tetroxide, etc. The method of thin sectioning is used to study morphology of cell.

3. Negative staining: This technique is used to study small particles such as viruses or macromolecules. Here, the specimen is embedded in a droplet of electron dense material, such as, phosphotungstic acid ($H_3PW_{12}O_{40}$). The electron stain penetrates into all the empty spaces (*i.e.*, openings and crevices) between the macromolecules. The spaces appear well defined in negative contrast. The portions of sepeimen that exclude stain transmit electrons readily, so their images can be seen.

4. Shadow casting or heavy metal shadowing. This technique is used to study three-dimensional appearance of viruses and certain macromolecules such as DNA molecules and collagen fibres. It involves placing of specimen in an evacuated chamber and evaporating at an angle, a heavy metal such as chromium, palladium, platinum or uranium from a filament of incandescent tungsten. The vapour of heavy metal is deposited on one side of the surface of the elevated particles; on the other side a shadow forms, the length of which permits determination of the height of the particle. In such a specimen, during electron microscopy, the electrons pass readily through the area of light metal content, less readily through the plane on which the particle sits, and are scattered more severely by the side of the particle on which metal has accumulated. Thus, by shadow casting, shape and profile of a particle can be observed.

5. Tracers. Several biological processes such as pinocytosis, phagocytosis and transport of molecules across plasma membrane can be studied by the use of appropriate tracers (*e.g.*, gold, mercuric sulphide, iron oxide, etc.). These tracers are detected by their electron opacity. An ideal tracer should be non-toxic, physiologically inert, composed of small-sized particles of uniform and known size and preserved *in situ* during the processing of the tissue.

6. Freeze-fracture. This technique is used to study the molecular arrangement in the plasma membrane and other cellular membranes. It is carried out by rapidly cooling or freezing the sample **(cryofixation)** and then **fracturing** (cracking) it in the a vacuum while it is still at –100°C. The knife does not cut cleanly under those conditions, but tends to **fracture** (crack) the specimen along the lines of natural weakness, such as the middle of a membrane that runs parallel to the cut (Fig. 54.24). After fracture, the sample is left in the vacuum long enough to allow some water to evaporate from the exposed surfaces, a process called **freeze etching.** The exposed face is then shadowed with electron-dense combination of carbon and metal such as platinum to provide the necessary contrast, after which organic material (*i.e.,* specimen itself) is removed by acids to leave a metal **replica** for examination in the electron microscope. Replica reveals a natural-looking representation of the surface of the freeze-etched object and is the only way of seeing membrane interior and certain other features of the cells.

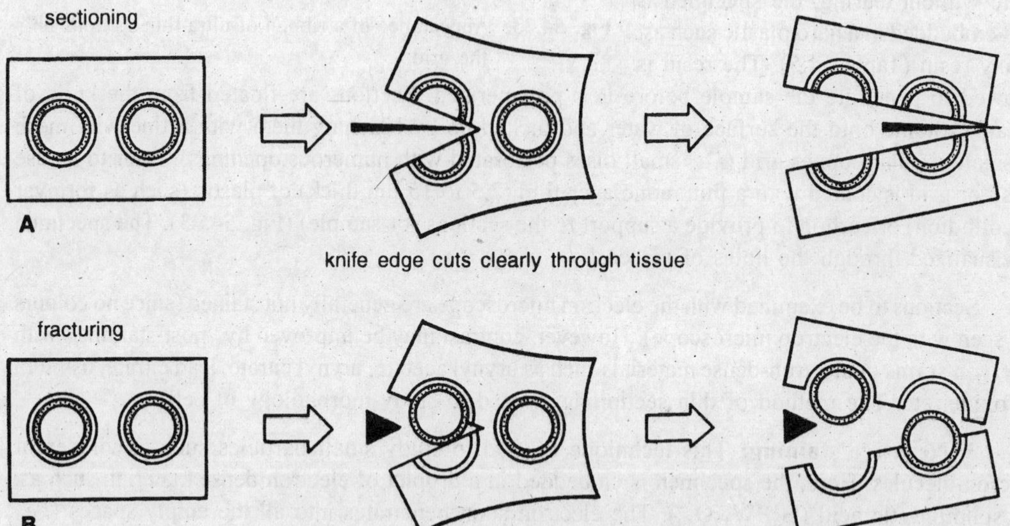

Fig. 54.24. A comparison of sectioning and fracturing tissues.

7. Whole mounts. They are often used to examine chromosomes and other relatively thick objects that can be isolated free of debris. In these methods, the specimen is neither sectioned nor stained. Thick areas will scatter electrons more strongly than thin areas, providing enough contrast to form an image.

Fixation

For microscopical study of dead cells, the process of killing and **fixation** brings about the sudden death of cells or tissues in such a manner that morphological organization and chemical composition of the cell remain unaltered up to greatest extent. The fixative or fixing solution performs following functions: it prevents bacterial decay and autolysis (self digestion) of the cells; renders the components of the cell insoluble, reduces the cell distortion and shrinkage, increases the visibility of different cellular components and prepares the cells for the staining. The fixation of the cells can be achieved either by chemicals or cooling which is as follows:

(*a*) Chemical Fixation. The choice of suitable fixative generally depends on the type of analysis and the chemical nature of different cells. For instance, for the study of the nucleus and chromosomes the **acid fixative** such as **Carnoy's solution** or **Bouin's fluid** are used. The commonly used fixatives are as follows:

***(i)* Acetic acid.** The acetic acid is used from 0.3 to 50 per cent, alone or in mixtures. It does not fix cytoplasmic proteins, but precipitates nucleoproteins, destroys Golgi apparatus and mitochonodria of the cells. Morever, the cells become swell up and soft. The acetic acid is used with basic dyes for the fixation of **nucleus** and **chromosomes**.

***(ii)* Potassium dichromate.** Potassium dichromate is generally used in mixtures and in the percentage of 2–7. It renders proteins insoluble in water and fixes lipids. Potassium dichromate is used for the fixation of chromosomes and some cytoplasmic structures.

***(iii)* Ethanol.** 70 to 100 per cent ethanol is used either alone or in mixture for the fixation of nucleus along with certain basic dyes. It does not precipitate proteins but dissolves lipids. Ethanol makes the cells to become hard but causes shrinkage.

***(iv)* Formaldehyde.** The formaldehyde (4–10 per cent) is used in mixtures for the fixation of Golgi apparatus, mitochondria and enzymes along with some basic stains.

***(v)* Osmium tetroxide.** Osmium tetroxide (0.5–2 per cent fixes cytoplasm, Golgi apparatus, mitochondria and fats of the cell but causes blackening of cell due to its oxidation which renders difficulty in staining. This fixative is widely used for electron microscopy.

***(vi)* Bouin's solution.** The Bouin's solution has 5 part picric acid, 5 parts 40% formaldehyde and one part glacial acetic acid. It is used as the fixative for chromosomes. It precipitates all proteins and causes shrinkage and softening of the cells or tissue.

***(vii)* Carnoy's fluid.** The fixative Carnoy's fluid is most commonly used for the fixation of the nucleoproteins and chromosomes because it has combined properties of ethanol and acetic acid. Carnoy's fluid contains three parts ethanol and one part glacial acetic acid.

(*b*) Fixation by Freezing. For histochemical and electron microscopic observations sometimes it is required that different chemical substances and organelles remain as such or minimum alteration occur in them. In this method the cells or tissues are killed and fixed by rapid cooling of them. Following methods of freezing fixation are used in recent cytological studies.

***(i)* Fixation by freezing-drying method.** For the electron and X-ray microscopy, the tissues or cells are fixed by rapid freezing and are dehydrated in the vacuum at a low temperature. The rapid freezing of tissues is achieved by placing the living tissue in isopentane or propane cooled upto 160°C or –190°C by liquid nitrogen or helium. After freezing, the material is dried in vaccum at 30°C to –40°C.

By freezing-drying fixation method, the tissue does not shrink, homogeneous fixation of cell takes place, soluble substances are not extracted, the chemical composition of the cell is maintained without any alteration and the fixation takes place very quickly.

***(ii)* Fixation by freezing-substitution method.** In the freezing-substitution fixation method, the living tissue is rapidly frozen and then kept frozen at a low temperature (–20 to –60°C) in a reagent that dissolves the ice-crystals. The reagent may be ethanol, methanol or acetone. This method of fixation is also used in electron microscopy and X-ray microscopy.

***(iii)* Fixation by freezing-etching method.** In the freezing-etching method of fixation the specimen is placed in 20 per cent glycerol and is frozen at –100°C. It is mounted on a chilled holder and splintered with a knife along natural cleavage planes, usually along surfaces of membranes. Occasionally the structure is cross-fractured, giving cross sections of organelles. The splintered preparation is freeze-dried and covered with a platinum and carban coating in the high, vacuum of a freezing ultramicrotame. When the preparation is dried the vacuum is broken and the preparation is placed in water to float the replica off the carrier. The replica is then washed in basic solution to remove the cellular material and the replica is mounted on a grid and dried for elctron microscopy. Such replica gives the outlines of various cell structures and verify the structures seen in thin fixed and stained sections of various cellular organelles in various types of microscopy.

Staining

The process of colouring the cells, tissues of animal or plant bodies by certain inorganic or organic dyes is known as **staining**. The selection of dye or stain for a particular material usually depends on its chemical nature, the pH value of the fixative used and the chemical reactivity of stain to the material.

Stains. Most cytological stains are solutions of dyes of aromatic organic compounds which have two kinds of active chemical groups such as, **chromophoric** and **auxochromic** groups. The **chromophoric groups** gives the colour to the dye, *e.g.*, carboxyl (–COOH), azo (= N–), nitro ($-NO_2$) quinoid (O =⬭= O) and indamin (– N=) groups. The **auxochromic group** gives to the dye the ability to attach to the tissue or the material and to dissolve and dissociate, in the water, *e.g.*, *Hydroxy* (–OH) group.

The organic stains are classified into three groups:

1. acidic stains,
2. basic stains, and
3. neutral stains.

1. Acidic stains. The acidic stains are usually used for the cytoplasm and proteins. These have great capacity for combining with the tissue at low pH than basic dyes. The most common cytological acidic stains are picric acid, acid fuchsin, Congo red, Janus green *B*, orange *G*, methyl blue, eosin, aniline blue, Bismark brown and fluorescein.

2. Basic stains. The basic stains are used to stain the nucleus, chromosomes and particularly the nucleic acids. The most common cytological basic stains are basic fuchsin, crystal violet, methyl green, safranine, acridine red, azures, methylene blue, thionine and haematoxylin.

3. Neutral stains. The neutral stains have both the properties of acidic and basic stains.

Acidiphilic and basiphilic tissues. The tissue, cell or cellular component taking acidic stains is known as **acidiphilic** tissue, cell or cellular component, *e.g.*, the cytoplasm. The nucleus, chromosomes and DNA have affinity for the basic stains and known as **basiphilic** organelles.

Metachromasia. Some basic dyes stain certain cell components with a different colour than their original colour. This property of stain is known as **metachromasia** and is very useful for **histochemical** and **physiochemical** tests. The property of metachromasia is displayed by the basic stains such as thionine, azure A and toluidine blue which react with mucopolysaccharides, nucleic acids and acidic lipids.

Mordant and lake. Certain dyes stain the proteins and cytoplasm in the presence of some metal or metallic compound which is known as **mordant.** Usually chemically a mordant is a double salt of potassium or ammonium or ferric sulphate. The mordant and the stain are collectively known as the **lake**. The most important mordant is ferric ammonium sulphate (iron alum), which is used along with the stain haematoxylin and carmine.

Staining for Light Microscopy

For light microscopy both acidic and basic stains are used for the staining of cells or tissue. The stains are specific for different types of cells and also for different organelles of cell. The cytoplasmic proteins and carbohydrates are stained by acidic stains, while the nucleus, chromosomes, etc., are stained by basic stains.

Staining for Electron Microscopy

Electron microscopy usually requires no stain because beside observing visually, the image is photographed and due to the contrast between cellular components a black and white photograph is received. But certain inorganic stains such as lead acetate, lead citrate, lead hydroxide, urynyl acetate, phosphotungstic acid, osmium tetroxide and potassium permanganate are used to increase contrast between cellular components. Sometimes certain organic stains as azure II, leucofuchsin, orcein, etc., are also used to increase the contrast.

Cytochemical Staining

In addition, there are certain specific stains, called **cytochemical** stains that bind selectively to some specific groups of cellular macromolecules such as proteins, nucleic acids, polysaccharides and lipids. For example, Millon reaction, diazonium reaction and Napthol Yellow 5 stain are used for the proteins; alkaline fast green is used for histone (basic protein); Feulgen reaction (using Schiff's reagent) is used for DNA; methyl green-pyronine stain (Unna-Pappenheim stain) is used in distinguishing between DNA and RNA and it stains DNA green and RNA red; acetocarmine and acetoorcein stains are used to stain chromosomes of dividing cells; periodic acid-Schiff (PAS) reaction is used for the demonstration of polysaccharide materials such as starch, cellulose, hemicellulose, and pectin in the plant cells and mucoproteins (glycoproteins), hyaluronic acid and chitin in animal cells; and fat soluble dyes such as Sudan Red and Sudan Black B are used for the lipids. The Sudan Black B is a specific stain for phospholipids and is used to stain Golgi apparatus.

Vital Staining

Vital stains selectively stain the intracellular structures of living cells without serious alteration of cellular metabolism and function. For example, Janus green B selectively stains mitochondria; neutral red stains plant vacuoles and methylene blue stains Golgi apparatus and also nuclear chromatin of dividing cells.

Immunostaining or usage of antibodies in Cell Biology. To use fluorescence microscopy for locating specific molecules or ions within cells, researchers must employ special indicator molecules called **fluorescent probe**. A fluorescent probe is a molecule capable of emitting fluorescent light that can be used to indicate the presence of a specific molecule or ion.

One of the most common application of fluorescent probes is **immunostaining,** a technique based on ability of **antibodies** to recognize and bind to specific molecules (the molecules to which antibodies bind are called **antigens**). Antibodies are proteins produced naturally by the immune system in response to invading microorganisms, but they can also be generated in the laboratory by injecting a foreign protein or macromolecule into an animal such as a rabbit or mouse. In this way, it is possible to produce antibodies that will bind selectively to virtually any protein that a scientist wishes to study. Antibodies are not directly visible using light microscopy, however, if they are linked to a fluorescent dye such as **fluorescein**, they emit a green fluorescence, or to **rhodamine**, they emit a red fluorescence. More recently, antibodies have been linked to **"quantum dots"**, which are tiny light-emitting crystals that are chemically more stable than traditional dyes. To identify the subcellular location of a specific protein, cell are simply stained with a fluorescent antibody directed against that protein and the location of fluorescence is then detected by viewing the cells with light of the appropriate wave-length.

Immunofluorescence micro-scopy can be performed using antibodies that are **directly labeled with fluorescent dye** (Fig. 54.25A). However, immuno-fluorescence microscopy is more commonly performed using **indirect immunofluorescence** (Fig 54.25B). In indirect immuno-fluorescence, a tissue or cell is treated with an antibody that is not labeled with dye. This antibody is called **primary antibody** and it attaches to specific antigenic sites within the tissue or cell. A second type of antibody, called **secondary antibody**, is then added. The secondary antibody is labelled with a fluorescent

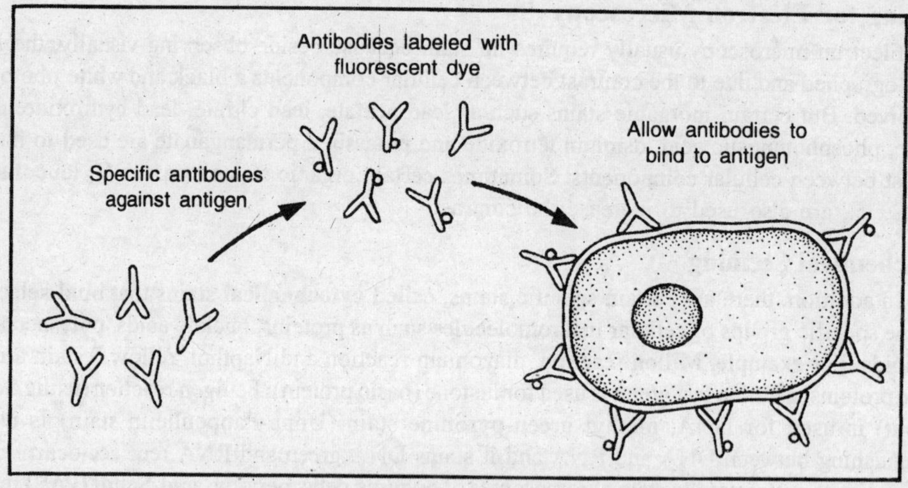

A. Immunofluorescence

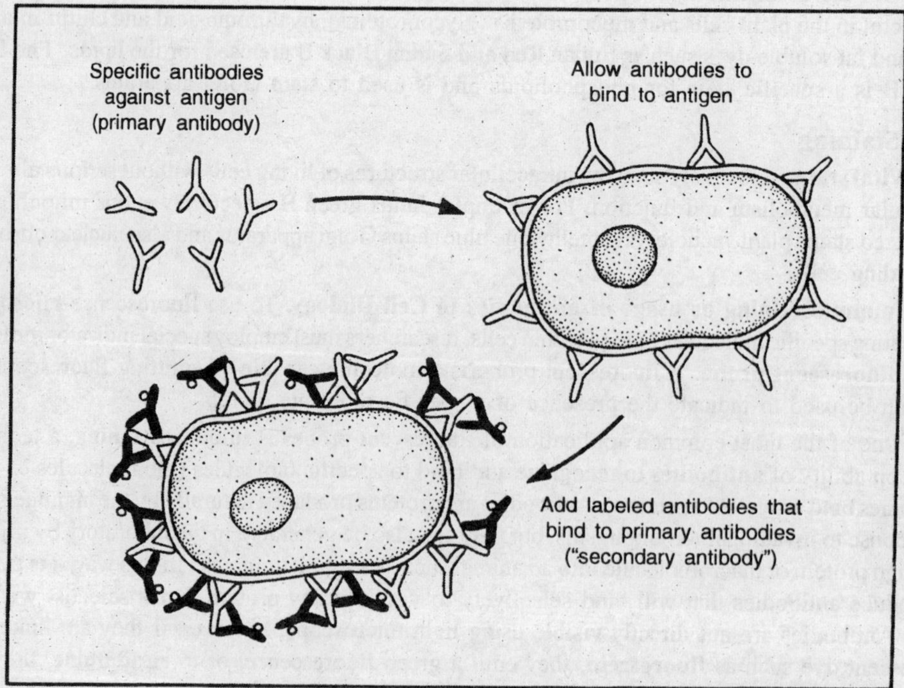

B. Indirect immunofluorescence

Fig. 54.25. Immunostaining using fluorescent antibodies (after Becker *et al.* 2007).

dye, and it attaches to the primary antibody. Because more than one primary antibody molecule can attach to an antigen, and more than one secondary antibody molecules can attach to each primary antibody, more fluorescent molecules are concentrated near each molecule that we seek to detect. As a result, indirect immunofluorescence results in signal amplification, and it is much more sensitive than the use of a primary antibody alone.

QUESTIONS

Long Answer Questions

1. Describe the method of cell fractionation.
2. What is centrifugation? Describe in brief the mechanism of centrifugation.
3. Write down the different types of centrifuges. *(B.U. Coimbatore 1997)*
4. Describe the structure and working of a light microscope. Explain as how does it differ from phase contrast microscope. Enumermate the importance of phase contrast microscope in cell biology.
5. Explain the principle, working and uses of phase contrast microscope. *(Guwahati (G) 1994)*
6. Name different kinds of microscopes used in studying the cell and cell constituents.
7. How does an electron microscope differ from a light microscope? *(Delhi 1989)*
8. Describe composition, principle of working of an electron microscope. Explain how it differs from light microscope. *(Kerala 1990; Allahabad 1992; CCS 1994; Guwahati (M) 1994, 1995)*
9. Briefly explain the principles involved in microscopy. Describe the mechanism of transmission electron microscope. *(Guwahati (M) 1995; Guwahati (G) 1993)*
10. Describe the working of transmission electron microscope and scanning electron microscope. What is the difference between the image formed by two types of electron microscopes.
11. What is resolving power of an optical system? State the working principle of phase contrast and dark field microscopy. *(Guwahati (Major) 1994)*
12. Explain the principles and working mechanism of fluorescence microscope. *(Guwahati 1992, B.U. Coimbatore 1997)*
13. What is a dark field microscope? *(CCS 1991, 1992, 1996; Guwahati (G) 1996)*
14. Name various kinds of chromatographic procedures. Describe the procedure of paper chromatography and column chromatography.
15. What do you mean by fixation? Describe the different types of fixatives used in histological study. *(Guwahati (G) 1992, 1994)*
16. What is fixation? Describe the principles of staining with reference to different types of stains used in the study of cell. *(Guwahati (G) 1993, 1995)*
17. Write an essay on 'Tissue culture'.

Short Answer Questions

1. Write a short note on the principle of autoradiography. *(Allahabad 1994, 1996)*
2. Discuss the principle of autoradiography and its application. *(Allahabad 1991)*
3. What is the limitation of a microscope? Describe the difference between light microscopy and electron microscopy with the help of schematic diagrams. *(Guwahati 1996)*
4. Write down the principle of electron microscope. *(B.U. Coimbatore 1997)*
5. Describe the principle of working of transmission electron microscope. *(Allahabad 1996)*
6. What is chromatography? Write down the principle. *(B.U. Coimbatore 1997; H.N.B. Garwal 2008)*
7. Give differences between column chromatography and partition chromatography.
8. What is the difference between TLC and GLC?
9. Write short note on the following:
   (*i*) Electron microscopy *(CCS 1996)*
   (*ii*) Resolving power of microscope (*iii*) Compound microscope *(CCS 1999)*
   (*iv*) Fundamentals of TEM *(Allahabad 1991)*
   (*v*) Chromatography *(Guwahati 1992; CCS 1994)*
   (*vi*) Ion exchange chromatography (*vii*) Electrophoresis
   (*viii*) Tissue culture (*ix*) Dialysis

(*x*) Phase contrast microscope (*H.N.B., Garwal 2008*)

(*xi*) Radioisotopes (*H.N.B., Garwal 2008*)

Very Short Answer Questions

1. What is autoradiography?

Ans. Autoradiography is a procedure for detecting the location of radioactive molecules by overlaying a sample with photographic film, which becomes darkened upon exposure to radioactivity.

2. Define chromatography.

Ans. It is a group of related techniques that utilize the flow of a fluid phase over a non-mobile absorbing phase to separate molecules based on their relative affinities for the two phases, which in turn reflect differences in size, charge, hydrophobicity or affinity for a particular chemical group.

3. What is electrophoresis?

Ans. It is a group of related techniques that utilize an electrical field to separate electrically charged molecules.

4. Define centrifugation.

Ans. It is a process of rapidly spinning a tube containing a fluid to subject its contents to a centrifugal force.

5. What is cryofixation?

Ans. It is rapid freezing of small samples so that cellular structures can be immobilized in milliseconds. It is often followed by freeze substitution, in which an organic solvent replaces the frozen water in the sample.

6. What is electron microscope?

Ans. It is an instrument that uses a beam of electrons to visualize cellular structures and thereby examine cellular architecture; the resolution is much greater than that of the light microscope, allowing detailed ultrastructural examination.

7. Define the staining.

Ans. It is incubation of tissue specimens in a solution of dye, heavy metal, or other substance that binds specifically to selected cellular constituents a distinctive colour or electron density.

Fill in the Blanks

1. Ultrastructure of cell organelles can be studied by ....

Multiple Choice Questions

1. The physical measurement of a molecule in chromatography is (*B.U. Coimbatore 1997*)

(*a*) Rg value (*b*) Rf value

(*c*) Half life (*d*) Rs value

2. In electrophoresis, the buffer establishes contact between (*B.U. Combatore 1997*)

(*a*) Two electrodes (*b*) The sample and electrodes

(*c*) The sample and supporting medium (*d*) The supporting medium and electrode

3. Separation by electrophoresis in based on (*H.N.B. Garwal 2008*)

(*a*) Charge (*b*) Density

(*c*) Solubility (*d*) Refraction

ANSWERS

Fill in the Blanks

1. Electron microscopes.

Multiple Choice Questions

1. (*b*) **2.** (*d*) **3.** (*a*).

55

Prokaryotic Cell and Eukaryotic Cell

The body of all living organisms (bacteria, blue green algae, plants and animals) except viruses has **cellular organization** and may contain one or many cells (Box 55.1). **A cell** is the smallest and complete expression of the fundamental structure and function of all living organisms. It is a unit of structure, function and heredity. The cell is the basic unit of organization or structure, of all living matter. Within a selective and retentive semipermeable membrane, it contains a complete set of different kinds of units necessary to permit its own growth and reproduction from simple nutrients. It has always been quite difficult to define a cell. Different cell biologists have defined the cell differently as follows: **A.G. Loewy** and **P. Siekevitz** (1963) have defined a cell as *a unit of biological activity delimited by a semipermeable membrane and capable of self-reproduction in a medium free of other living systems*. **Wilson** and **Morrison** (1966) have defined the cell as *an integrated and continuously changing system*.**John Paul** (1970) has defined the cell as *the simplest integrated orgainization in living systems, capable of independent survival*. The organisms with only one cell in their body are called **unicellular organisms** (e.g., bacteria, blue green algae, some algae, Protozoa, etc.). The organisms having many cells in their body are called **multicellular organisms** (*e.g.,* most plants and animals). Any cellular orgainsm may contain only one type of cell from the following types of cells:

A. Prokaryotic cell; B. Eukaryotic cell.

Box 55.1. Why viruses are not true cells?

A virus is neither an organism nor a cell, yet it consists of a core of nucleic acid (DNA or RNA) enclosed in an external mantle of protein. In the free state viruses are quite inert. They become activated only when they infect a living host cell and in the process only the nucleic acid core enter the host's cell. The nucleic acid which is the genetic substance, takes over the metabolic activity of the host cell and utilises the cell machinery for the formation of more viruses, ultimately killing the host cell. In a way, thus, viruses are cellular parasites that cannot reproduce by itself. They are primitive and simpler units of life.

55.1 THREE CELLULAR DOMAINS

The body of all living organisms except viruses has cellular organization. With the advent of electron microscopy, biologists came to recognize two fundamentally different plans of cellular organizations: the simpler one characteristic of bacteria and the more complex one found in all other kinds of cells. Based on structural differences of their cells, organisms were initially divided into two broad groups, the **prokaryotes** (bacteria) and the **eukaryotes** (all other forms of life such as protozoa, plants and animals). The most fundamental distinction between the two groups is that eukaryotic cells have a true, membrane-bounded nucleus (*eu*-is a Greek word for "true" or "genuine", *karyon*-means "nucleus") where as prokarystic cell do not (*pro*-means "before" suggesting an evolutionarily earlier form of life). The terms prokaryotic and eukaryotic were suggested by **Hans Ris** in the 1960's. On the basis of molecular and biochemical criteria, especially sequence analysis of ribosomal RNAs (rRNAs), cell biologists now recognize that the prokaryotes can be further differentiated as either **eubacteria** and **archaea**. The eubacteria ("true bacteria") include most present-day **bacteria** and **cyanobacteria** (also known as blue-green algale)—most of the commonly encountered bacteria, in other words. The archaea (also called **archaebacteria**) are similar to eubacteria in cellular structure, but are as different from eubacteria as they are from eukaryotes in terms of molecular and biochemical distinctions. Archaea are regarded as modern descendants of an evolutionarily ancient form of prokaryote that differed fundamentally from the ancestors of present eubacteria (***Archae***-is a Greek prefix meaning "ancient" or "original"). Present day archaea can be subdivided into theree main groups: the **methanobacteria**, which obtain energy by converting carbon dioxide and hydrogen into methane; the **halobacteria**, which can grow in salty environments (upto 5.5 M NaCl) and the **sulphobacteria**, which obtain energy from sulphur-containing compounds. Some of these archaea are also **thermacidophiles**, which thrive in acidic hot springs with pH as low as 2 and temperatures that can exceed 100°C.

Based on the RNA sequencing work of **Carl Woese**, **C. Fred Fox** and others (1977, 1990), cell biologists now recognize **eukaryotes, eubacteria** and **archaea** as three **fundamentally different groups** or **domains,** of organisms. In addition to the characteristics of their tRNAs, the cells in each group have other distinctive molecular, biochemical and cellular entities and properties in common, including RNA polymerases (the enzymes that synthesize RNA), sensitivity to specific inhibitors of protein and nucleic acid synthesis, and the presence or absence of memberane-bound intracellular compartments (organelles).

Origin of Eukaryotes

The crucial step in the evolution of eukaryotic cells was the acquirement of memberane-enclosed subcellular organelles allowing the development of the **complexity characteristic** of these cells. The organelles of eukaryotes are thought to have arisen by **endosymbiosis**—one cell living inside another. In particular eukaryotic organelles are thought to have evolved from prokaryotic cells living inside the ancestors of eukaryotes.

The hypothesis that eukaryotic cells evolved by endosymbiosis is particularly well supported by studies of mitochondria and chloroplasts, which are thought to have evolved from eubacteria living in larger cells (**Margulis** 1970, 1981). Both mitochondria and chloroplasts are similar to bacteria in size, and like bacteria, they reproduce by dividing in two. Most important, both mitochondria and chloroplasts contain their own DNA, which encodes some of their components. The mitochondrial and chloroplast DNAs are replicated each time the organelle divides, and the genes they encode are transcribed within the organelle and translated on organelle ribosomes. Mitochondria and chloroplasts thus contain their own genetic systems, which are distinct from the nuclear genome of the cells. Furthermore, the ribosomes and ribosomal RNAs of these organelles are more closely related to those of bacteria than to those encoded by the nuclear genomes of eukaryotes.

An endosymbiotic origin of these organelles is now generally accepted with mitochondria thought to have evolved from aerobic eubacteria and chloroplasts from photosynthetic eubacteria such as the cyanobacteria (blue-green algae). The acquisition of aerobic bacteria would have provided an anaerobic cell with the ability to carry out oxidative metabolism. The acquisition of photosynthetic bacteria would have provided the nutritional independence offorded by the ability to perform photosynthesis. Thus, these endosymbiotic associations were highly advantageous to their partners and were selected for in the course of evolution. Through time, most of the genes originally present in these bacteria evidently became incorporated into the nuclear genome of the cell, so only a few components of mitochondria and chloroplasts are still encoded by the organelle genomes (see **Cooper** and **Hausman**, 2007).

A recent hypothesis (**Rivera** and **Lake**, 2004) explained the mosaic nature of eukaryotic genomes by proposing that the genome of eukaryotes arose from a fusion of archaebacterial and eubacterial genomes (Fig. 55.1). According to this model, an endosymbiotic association between an eubacterium and an archaebacterium was followed by fusion of the two prokaryotic genomes, giving rise to an ancestral eukaryotic genome with contributions from both eubacteria and archaebacteria. The simplest version of this hypothesis is that an initial endosymbiotic relationship of an eubacterium living inside an archabacterium gave rise not only to mitochondria but also to the genome of eukaryotic cells, containing genes derived from both prokaryotic ancestors.

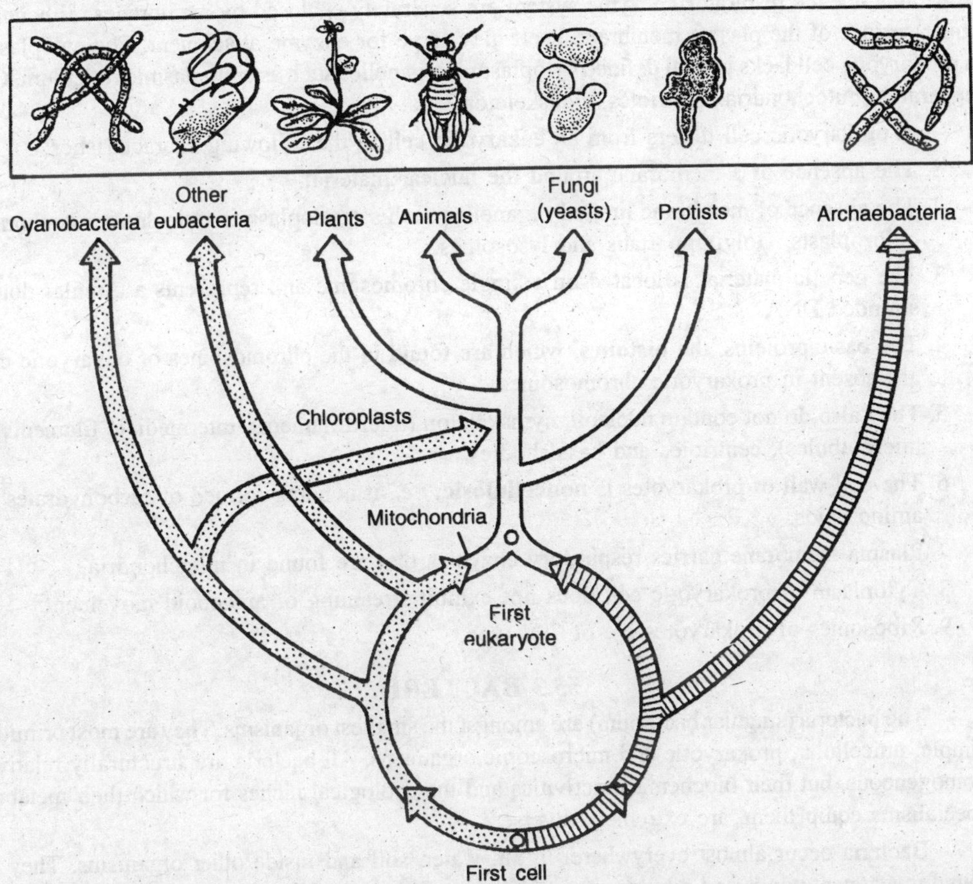

Fig. 55.1. Evolution of cells. Present day cells evolved from a common prokaryotic ancestor which diverged along two lines of descent, giving rise to archaebacteria and eubacteria. Eukaryotic cell may have arisen by endosymbiotic association of an aerobic eubacterium with an archaebacterium, leading to the development of mitochondria as well as formation of a eukaryotic genome with genes derived from both eubacteria and archaebacteria. Chloroplasts subsequently evolved as a result of the endosymbiotic association of a cyanobacterium with the ancestor of plants (after Cooper and Hausman, 2007).

55.2 PROKARYOTIC CELL

The prokaryotic (Gr., *pro* = primitive or before; *karyon* = nucleus) cells are small, simple and most primitive. They are probably the first to come into existence perhaps 3.5 billion years ago. For example, the **stromatolites** (*i.e.*, giant colonies of extinct cyanobacteria or blue green algae) of Western Australia are known to be at least 3.5 billion years old. The **eukaryotic** (Gr.,*eu* = well; *karyon* = nucleus) cells have evolved from the prokaryotic cells and the first eukaryotic (nucleated) cells may have arisen 1.4 billion years ago (**Vidal,** 1983).

The prokaryotic cells are the most primitive cells from the morphological point of view. They occur in the eubacteria (*i.e.*, mycoplasma, bacteria and cyanobacteria or blue-green algae) and archaea. A prokaryotic cell is essentially a **one-envelope system** organized in depth. It consists of central nuclear components (viz., DNA molecules, RNA molecules and nuclear proteins) surrounded by cytoplasmic ground substance, with the whole enveloped by a plasma membrane. Neither the nuclear apparatus nor the respiratory enzyme system are separately enclosed by membranes, although the inner surface of the plasma membrane itself may serve for enzyme attachment. The cytoplasm of a prokaryotic cell lacks in well defined cytoplasmic organelles such as endoplasmic reticulum, Golgi apparatus, mitochondria. centrioles, cytoskeleton, etc.

A prokaryotic cell differs from an eukaryotic cell in the following characteristics:

1. The absence of a membrane around the nuclear material.
2. The absence of membrane limited organelles such as endoplasmic reticulum, mitochondria, chloroplasts, Golgi appartaus and lysosomes.
3. The genetic material is located on a single chromosome and represents a circular double-stranded DNA.
4. The basic proteins, the **histones**, which are found in the chromosomes of eukaryotic cells, are absent in prokaryotic chromosomes.
5. They also do not contain nucleoli, cytoskeleton (microfilaments, intermediate filaments and microtubules), centrioles and basal bodies.
6. The cell wall of prokaryotes is **noncellulosic,** *i.e.*, it is being formed of carbohydrates and amino acids.
7. Plasma membrane carries respiratory enzymes that are found in mitochondria.
8. Cytoplasm of prokaryotic cell does not exhibit streaming or amoeboid movement.
9. Ribosomes of prokaryotes are of 70S type.

55.3 BACTERIA

The bacteria (singular bacterium) are amongst the smallest organisms. They are most primitive, simple, unicellular, prokaryotic and microscopic organisms. All bacteria are structurally relatively homogeneous, but their biochemical activities and the ecological niches for which their metabolic specialisms equip them, are extremely diverse.

Bacteria occur almost everywhere: in air, water, soil and inside other organisms. They are found in stagnant ponds and ditches, running streams and rivers, lakes, sea water, foods, petroleum oils from deeper regions, rubbish and manure heaps, sewage, decaying organic matter of all types, on the body surface, in body cavities and in the internal tracts of humans and animals. Bacteria thrive well in warmth, but some can survive at very cold tops of high mountains such as Alps or even in almost boiling hot springs. They occur in vast numbers. A teaspoonful of soil may contain several hundred million bacteria. They lead either an autotrophic (photoautotrophic or chaemoautotrophic), or heterotrophic (saprotrophic or parasitic) mode of existence. The saprophytic or saprotrophic species of bacteria are of great economic significance for humans. Some parasitic species of bacteria are pathogenic (disease producing) to plants, animals and humans.

Bacteria have a high ratio of surface area of volume because of their small size. They show high metabolic rate because they absorb their nutrients directly through cell membranes. They multiply at a rapid rate. In consequence, due to their high metabolic rate and fast rate of multiplication, bacteria produce marked changes in the environment in a short period.

1. Size of bacteria. Typically bacteria range between 1 µm (one micrometre) to 3 µm so they are barely visible under the light microscope. The smallest bacterium is *Dialister pneumosintes* (0.15 to 0.3 µm in length). The largest bacterium is *Spirillum volutans* (13 to 15 µm in length).

2. Forms of bacteria. Bacteria vary in their shapes. Based on their shape, bacteria are classified into the following groups:

(*i*) Cocci (singular coccus). These bacteria are spherical or round in shape. These bacterial cells may occur singly **(micrococci)**; in pairs **(diplococci**, *e.g.*, pneumonia causing bacterium, *Diplococcus pneumoniae*); in groups of four **(tetracocci)**; in a cubical arrangement of eight or more **(sarcinae)**; in irregular clumps resembling bunches of grapes **(staphylococci**,*e.g.*, boil causing bacterium, *Staphylococcus aureus*) or in a bead-like chain **(streptococci**, *e.g.*, sore throat causing bacterium, *Streptococcus pyogenes*; Fig. 55.2).

(*ii*) Bacilli (singular, bacillus). These are rod-like bacteria. They generally occur singly, but may occasionally be found in pairs **(diplobacilli)** or chains **(streptobacilli)**. Bacilli cause certain most notorious diseases of humans such as tuberculosis (*Mycobacterium* or *Bacillus tuberculosis*), tetanus (*Clostridium tetani*), typhoid (*Salmonella* or *Bacillus typhosus*), diphtheria (*Corynebacterium diphtheriae*), leprosy (*Mycobacterium leprae*), dysentery and food poisoning (*Clostridium botylinum*). Certain well known diseases of the animals are also caused by bacilli, e.g., anthrax (*Bacillus anthracis*) and black leg (*Clostridium chauvei*).

(*iii*) Spirilla (singular, spirillum). These are also called **spirochaetes**. These are spiral-shaped and motile bacteria (Fig. 55.2). Spirilla cause human disease such as syphilis (*Treponema pallidum*).

(*iv*) Vibrios (singular vibrio). These are comma-shaped or bent-rod like bacteria (Fig. 55.2). Vibrios cause human disease such as cholera (*Vibrio cholerae*).

3. Gram negative and Gram positive bacteria. On the basis of structure of cell wall and its stainability with the Gram stain, the following two types of bacteria have been recognized: Gram positive and Gram negative bacteria. The Gram staining method is named after **Christian Gram** who developed it in Denmark in 1884. In this technique, when heat-fixed bacteria are treated with the basic dye, crystal violet they become blue or purple. Such blue stained cells are treated with a mordant (*i.e.*, agent that fixes stains to tissues) such as iodine (*i.e.*, potassium iodide or KI solution) and ultimately washed with some organic solvent such as alcohol. Some bacteria retain the blue colour, while others lose it and stay colourless. The former are **Gram positive bacteria** (*e.g.*,*Bacillus subtilis, Staphylococcus*, etc.) and the latter are **Gram negative bacteria** (e.g., *Escherichia coli*, *Simonsiella*, cyanobacteria, etc.). Colourless Gram negative bacteria may thereafter be stained pink with safranin stain for their better microscopic visibility.

The long search for the chemical basis of this differentiating staining reaction ended in 1950's when it was detected that cell wall of Gram negative bacteria has high lipid content which tends to be dissolved away by alcohol. The alcohol then can enter the cell and leach out the stain, whereas the cell walls of Gram positive bacteria form a barrier (*i.e.*, peptidoglycan layers) that prevent the penetration of the solvent inside the cell.

4. Structure of bacteria. Structural details of a bacterial cell can only be seen with an electron microscope in very thin sections. A typical bacterial cell has the following components:

A. Outer covering: The outer covering of bacterial cell comprises the following three layers:

I. Plasma membrane. The bacterial protoplast is bound by a living, ultrathin (6 to 8 nm thick) and dynamic plasma membrane. The plasma membrane chemically comprises molecules of lipids and proteins which are arranged in a **fluid mosaic pattern**. That is, it is composed of a bilayer sheet of **phospholipid** molecules with their polar heads on the surfaces and their fatty-acyl chains (tails) forming the interior. The **protein** molecules are embedded within this lipid bilayer, some spanning it, some exist on its inner side and some are located on its external or outer side. These membrane proteins serve many important functions of the cell. For example, the transmembrane proteins act as **carriers** or **permeases** to carry on selective transportation of nutrients (molecules and ions) from the environment to the cell or vice versa. Certain proteins of the membrane are involved in oxidative metabolism, *i.e.*, they act as enzymes and carriers for electron flow in respiration and photosynthesis leading to phosphorylation (*i.e.*, conversion of ADP to ATP). The bacterial plasma membrane also provides a specific site at which the single circular chromosome (DNA) remains attached. It is the point from where DNA replication starts. The first stage in nuclear division involves duplication of this attachment, followed by a progressive bidirectional replication of DNA by two replication forks.

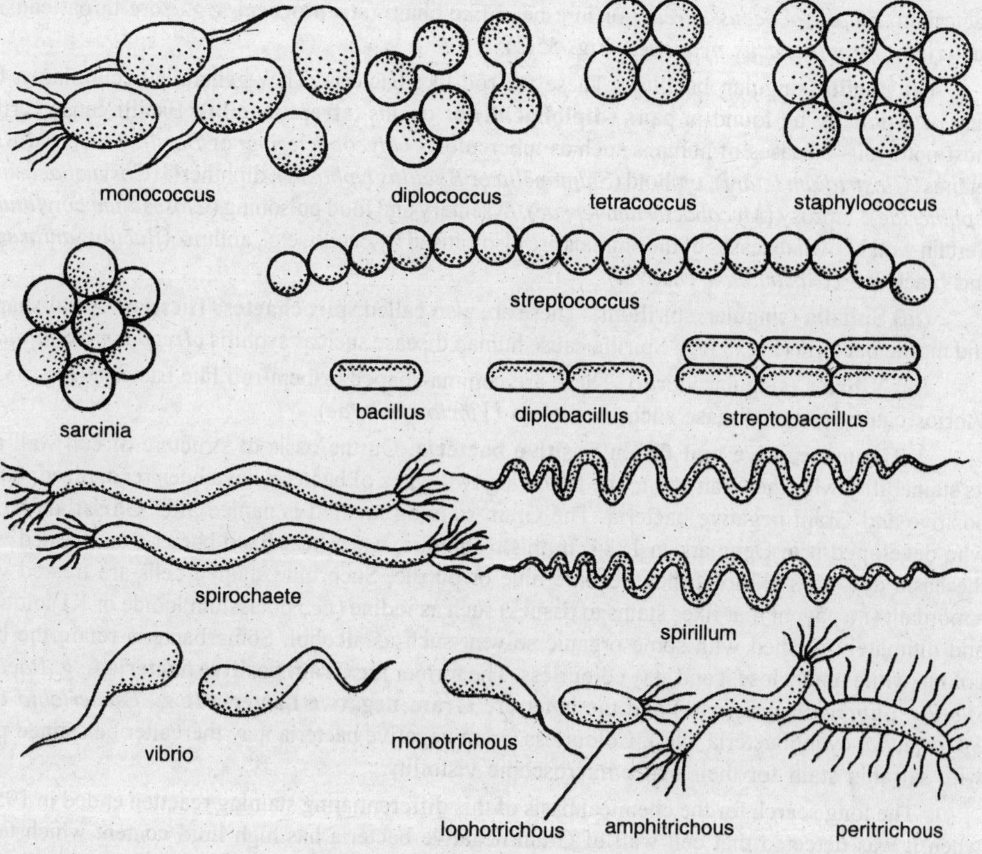

Fig. 55.2. Different forms of bacteria.

Plasma membrane intrusions. Infoldings of the plasma membrane of all Gram-positive bacteria and some Gram-negative bacteria give rise to the following two main types of structures:

1. Mesosomes (or chondrioids). They are extensions of the plasma membrane within the bacterial cell (*i.e.*, cytoplasm) involving complex whorls of convoluted membranes (Fig. 55.3). Mesosomes tend to increase the plasma membrane's surface and in turn also increase their enzymatic contents. They are seen in chemoautotrophic bacteria with high rates of aerobic respiration such as *Nitrosomonas*, and in photosynthetic bacteria such as *Rhodopseudomonas* where they are the site of photosynthetic pigments. Mesosomes are involved in crosswall (septum) formation during the division of cell.

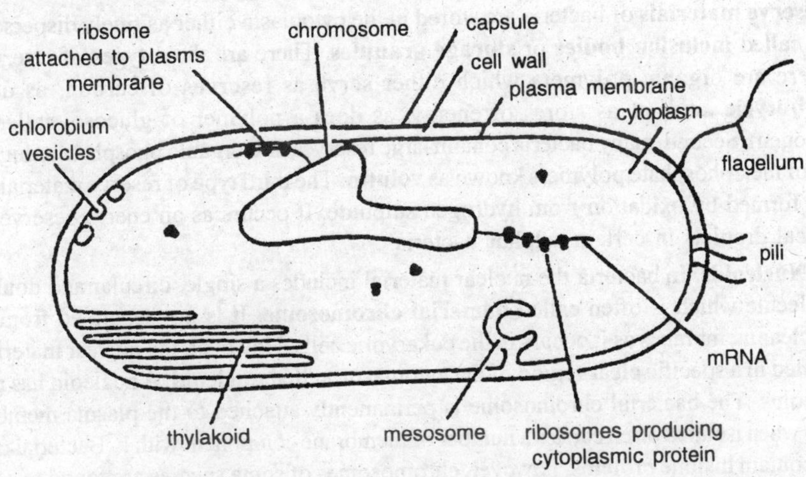

Fig. 55.3. A diagram of structures seen in the prokaryotic (bacterial) cells.

2. Chromatophores. These are photosynthetic pigment-bearing membranous structures of photosynthetic bacteria. Chromatophores vary in form as vesicles, tubes, bundled tubes, stacks, or thylakoids (as in cyanobacteria).

II. Cell wall. The plasma membrane is covered with a strong and rigid cell wall that renders mechanical protection and provides the bacteria their characteristic shapes (the cell wall is absent in *Mycoplasma*). The cell wall of bacteria differs chemically from the cell wall of plants in that it contains proteins, lipids and polysaccharides. It may also contain chitin but rarely any cellulose.

Electron microscopy has revealed the fact that the cell wall of Gram- negative bacteria comprises the following two layers: 1. Gel, proteoglycan or peptidoglycan (*e.g.,* murein or muramic acid) containing **periplasmatic space** around the plasma membrane and 2. The outer **membrane** which consists of a lipid bilayer traversed by channels of **porin** polypeptide. These channels allow diffusion of solutes. The lipids of lipid bilayer are phospholipids and lipopolysaccharides (LPS). LPS have antigenic property and anchor the proteins and polysaccharides of the surrounding capsule. The cell wall of Gram positive bacteria is thicker, amorphous homogeneous and single layered. Chemically it contains many layers of peptidoglycans and proteins, neutral polysaccharides and polyphosphate polymers such as teichoic acids and teichuronic acids.

III. Capsule. In some bacteria, the cell wall is surrounded by an additional slime or gel layer called **capsule**. It is thick, gummy mucilaginous and is secreted by the plasma membrane. The capsule serves mailnly as a protective layer against attack by phagocytes and by viruses. It also helps in regulating the concentration, and uptake of essential ions and water.

B. Cytoplasm. The plasma membrane encloses a space consisting of **hyaloplam, matrix** or **cytosol** which is the ground substance and the seat of all metabolic activities. The cytosol consists of water, proteins (including multifunctional enzymes), lipids, carbohydrates, different types of RNA molecules, and various smaller molecules. The cytosol of bacteria is often differentiated into two distinct areas: a less electron dense nuclear area and a very dense area (or dark region). In the dense cytoplasm occur thousands of particles, about 25 nm in diameter, called **ribosomes**. Ribosomes are composed of ribonucleic acid (RNA) and proteins and they are the sites of protein synthesis. Ribosomes of bacteria are 70S type and consist of two subunits (*i.e.*, a larger 50S ribosomal subunit and a small 30S ribosomal subunit). Non-functional ribosomes exist in the form of separated subunits which are suspended freely in the cytoplasm. During protein synthesis many ribosomes read the codes of single mRNA (messenger RNA) molecules and form **polyribosomes** or **polysomes**.

Reserve materials of bacteria are stored in the cytoplasm either as finely dispersed or distinct granules called **inclusion bodies** or **storage granules**. There are three types of reserve materials. First, there are organic polymers which either serve as reserves of carbon, as does **poly-β-hydroxybutyric acid**, or as stores of energy, as does a polymer of glucose, called **granulose** (*i.e.*, glycogen). Second, many bacteria contain large reserves of inorganic phosphate as highly refractile granules of metaphosphate polymers known as **volutin**. The third type of reserve material is elemental **sulphur**, formed by oxidation from hydrogen sulphide. It occurs as an energy reserve in the form of spherical droplets in certain sulphur bacteria.

C. Nucleoids. In bacteria the nuclear material includes a single, circular and double stranded DNA molecule which is often called **bacterial chromosome**. It is not separated from the cytosol by the nuclear membranes as it occurs in the eukaryotic cells. However, the nuclear material is usually concentrated in a specific clear region of the cytoplasm, called **nucleoid**. A nucleoid has no ribosome and nucleolus. The bacterial chromosome is permanently attached to the plasma membrane at one point, and when isolated often carries a number of membrane component with it. Bacterial chromosome does not contain histone proteins, however, chromosomes of some species are found to contain small quantities of a small heat-stable (HU) proteins that may be analogous to eukaryotic histones.

All three classes of RNA (*i.e.*, mRNA, tRNA, and rRNA) are formed (transcribed) by the activity of the single RNA polymerase (RNAP) species in prokaryotes. The messenger RNA formed at the chromosome is directly available for translation without processing, and so ribosomes may attach to the beginning of the mRNA strand and commence translation, while the other end of the mRN A is still being formed by transcription from DNA. Proteins for use within the cell are synthesized at cytoplasmic ribosomes; but ribosomes responsible for the synthesis of membrane proteins or proteins destined for export from the cell to form either the cell wall or secretory products, are attached to the plasma membrane. The resulting exportable polypeptides are ejected directly into or through the membrane as they are formed.

Plasmids. Many species of bacteria may also carry extra chromosomal genetic elements in the form of small, circular and closed DNA molecules, called **plasmids**. Some plasmids are merely **bacteriophage (viral) DNA** which may alternatively be incorporated within the chromosome. Other plasmids may be separated parts of the normal genome from the same or a foreign cell, and may recombine with the main chromosome. One function of some of these plasmids (called **colcinogenic factors**) is the production of antibiotically active proteins or **colicins** which inhibit the growth of other strains of bacteria in their vicinity. Some plasmids may act as **sex** or **fertility factors (F factors)** Which stimulate bacterial conjugation. **R factors** are also plasmids which carry genes for the resistance to one or more drugs such as chloramphenicol, neomycin, penicillin, streptomycin, sulphonamides and tetracyclines.

D. Flagella and other structures. Many bacteria (*e.g., E. coli*) are motile and contain one or more **flagella** for the cellular locomotion (swimming). Bacterial flagella are smaller than the eukaryotic flagella (*i.e.,* they are 15 to 20 nm in diameter and up to 20 μm long) and are also simpler in organization. A bacterial flagellum consists of a helical tube containing a single type of protein subunit, called **flagellin**. The flagellum is attached at its base, by a short flexible **hook** that is rotated, like a propeller of ship, by the flagellar rotatory **motor** *i.e.,* basal body; Fig. 55.4). The flagellar motor comprises four distinct parts: rotor (M ring), stator, bearing (S ring) and rod. The **'rotor'** is a protein disc integrated into the plasma membrane. It is driven by energy stored in the transmembrane proton H^+ gradient (not by ATP breakdown; see **Jones**, 1986) and rotates rapidly (~ 100 revolutions/second) in the lipid bilayer against another protein disc, called the **'stator'**. A **rod** links the 'rotor' to a hook and flagellum, thereby causing them to rotate. The protein **"bearing"** serves to seal the outer membrane of the cell wall as the rotating rod passes through it. The 'stator' and 'bearing' remains stationary.

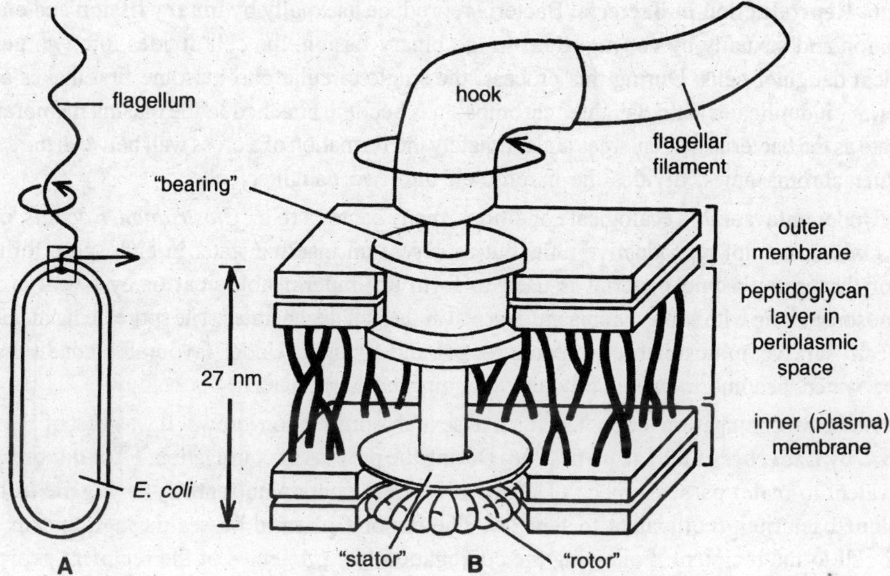

Fig. 55.4. Schematic drawing of the flagellar rotatory 'motor' of *E. coli.*

According to the number and arrangement of the flagella in a bacterial cell, following four types of flagellation patterns have been recognized: **(1) Monotrichous.** There is a single flagellum at one pole of the cell. **(2) Lophotrichous**. There are several flagella at one pole. **(3) Amphitrichous**. The cell bears at least one flagellum at each pole. **(4) Peritrichous**. 'There are flagella all over the surface of cell (Fig. 55.2). Flagella-like **axial filaments** are the characteristics of some spirochetes which move like snakes through the environment. The axial filaments do not project away from the cell but are wrapped around the cell surface.

Fimbriae or pili. Some bacteria (mostly Gram negative bacilli) contain non-flagellar, extremely fine, appendages called **fimbriae** or **pili** (singular **pilus**). Pili are non- motile but adhesive structures. They enable the bacteria to stick firmly to other bacteria, to a surface or to some eukaryote such as mold, plant and animal cells including red blood cells and epithelial cells of alimentary, respiratory and urinary tracts. Pili help in conjugation (*e.g.*, long F-pili or **sex pili** of male bacteria); in the attachment of pathogenic bacteria to their host cells (*e.g.*, attachment of gonorrhea-causing coccus, *Neisseria gonorrhoeae*, to the epithelial cells of the human urinary tract) and in acting as specific sites of attachment for the bacteriophages. Pili are known to be coded by the genes of the plasmid.

Spinae. Same Gram positive bacteria have tubular, pericellular and rigid appendages of single protein moiety, called **spinin**. They are called **spinae** and are known to help the bacterial cells to tolerate same environmental conditions such as salinity, pH, temperature, etc.

5. Nutrition in bacteria. Bacteria show wide diversity in their nutrition. Some are chemosynthetic, some are photosynthetic, but most of them are heterotrophic. Heterotrophic bacteria are mostly either saprophytic or parasitic; Parasitic bacteria live on the body of plants and animals and with few exceptions, most bacteria are pathogenic.

Modes of respiration of bacteria are both aerobic and anaerobic. Some of the end products of bacterial anaerobic respiration are useful to humans, so, they are used in the manufacture of various foods such as butter, cheese and vinegar. *Pseudomonas* is a gram negative heterotrophic aerobic form which can decompose (biodegrade) a wide variety of organic compounds such as hydrocarbons. So it is used in reducing water pollution due to petroleum spillage.

6. Reproduction in bacteria. Bacteria reproduce asexually by **binary fission** and **endospore** formation and sexually by **conjugation.** In the binary fission, the cell divides into two genetically identical daughter cells. During this process, the single circular chromasome first makes a copy of itself (*i.e.*, it duplicates) and daughter chromosomes become attached to the plasma membrane. They separate as the bacterial cell enlarges and ultimately the formation of a cross wall between the separating daughter chromasames, divides the parent cell into two daughter cells.

Under unfavaurable ecalogical conditions, many bacteria (e.g., *Clostridium, Bacillus,* etc.) form spores which are not reproductive units but represent an inactive state. In endospore formation, a part of the protoplasmic material is used to form an impermeable coat or cyst wall around the chromosome along with some cytoplasm. The rest af the cell degenerates. The spore being metabolically inert can survive an unsuitable temperature, pH and drought. Under favourable conditions spores imbibe water, become metabolically active again and germinate.

Bacterial conjugation is simplest form of sexual reproduction known. It was first of all observed in *E.coli* by **Laderberg** and **Tatum** in 1946. During the process of conjugation, $\mathbf{F^+}$ or **donor** bacterium (equivalent to male) passes a piece of DNA or plasmid containing **fertility** or **F gene** to the F^- or **recipient** bacterium (equivalent to female). The donor's plasmid passes through the sex pilus of donor cell to the recipient. Following the conjugation, the progenies of the recipient express some of the characteristics of the donor. Thus, bacterial conjugation is a means of makhig new genetic combinations or **recombinations** which are expressed in the progeny.

Examples of Prokaryotic Cells

The following three types of prokaryotic cells are well studied ones:

1. Mycoplasma (PPLO)

Among living organisms that have the smallest mass, are small bacteria called **mycoplasmas** which produce infectious diseases in animals including humans. Mycoplasmas are unicellular, prokaryotic, containing a plasma membrane, DNA, RNA and a metabolic machinery to grow and multiply in the absence of other cells (*i.e.,* they are capable of autonomous growth). They can be cultured *in vitro* like any bacteria, forming **pleomorphic** (Gr., *pleo* = many; *morphe* = forms) **colonies,** *i.e.*, depending on the type of culture medium, mycoplasmas tend to form different shaped colonies such as spheroid (fried-egg-shaped), thin, branching filaments, stellate, asteroid or irregular. They differ from the bacteria in the following respects:

1. Mycoplasmas are filterable through the bacterial filters (this fact was first demonstrated by **Iwanowsky** in 1892).
2. They do not contain cell wall and mesosomes.
3. Like the viruses and animal cells, they are resistant to antibiotics such as penicillin which kills bacteria by interfering with cell wall synthesis.
4. Their growth is inhibited by tetracyclines and similar antibiotics that act on metabolic pathways.

Mycoplasmas were discovered by French scientists, **E. Nocard** and **E. R. Roux** in 1898 while studying pleural fluids of cattle suffering from the disease **pleuropneumonia** (*i.e.*, an infectious disease of warm blooded animals producing pleural and lung inflammation). Similar organisms were later isolated from other animals such as sheep, goats, dogs, rats, mice and human beings and were named as **pleuropneumonia-like organisms (PPLO)**. PPLO were later on included under the genus *Mycoplasma* by **Nowak** (1929) and these organisms are now commonly called **mycoplasmas. W.V. Iterson** (1969) has placed PPLO in the group **My-coplasmatceae** of bacteria. Currently mycoplasmas are considered as the simplest bacteria. However, some cell biologists still prefer to place PPLO in between the viruses and bacteria.

Mycoplasmas are mostly free-living, saprophytes or parasites. For example, *Mycoplasma laidlawii* (0.1 μm in diameter) is saprophytic and is found in sewage, compost, soil, etc. *Mycoplasma gallisepticum* (0.25 μm in diameter) is parasitic and pathogenic; it is the parasite of cells and cell exudates of respiratory organs of warm-blooded animals causing in them various chronic respiratory diseases.

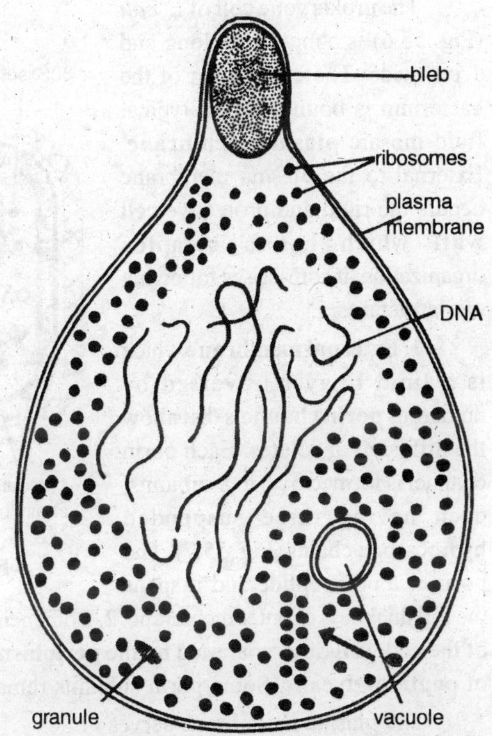

Fig. 55.5. A schemiatic diagram of typical PPLO cell.

Mycoplasmas range in size (diameter) from 0.25 to 0.1 μm. They correspond in size to some of the large viruses. The spherical cell of a mycoplasma is bounded at its surface by a 75 Å thick plasma membrane which is composed of molecules of proteins and lipids, but there is no cell wall. Internally the cell's composition is more or less diffuse. The only micro-scopically discernible features within the cell are its genetic component, the DNA and the ribosomes. The DNA molecule is contained in a membraneless and clear-nucleus-like region and it is a double helix which may exit either as the linear strands or a single circular molecule. The nuclear region is surrounded by numerous (50 to 100) 70S type ribosomes existing either freely or in the polysomes (Fig. 55.5). A variety of other cytoplasmic inclusions, such as vacuoles and granules, have also been detected, but their functions are not known. At one side of the cell occurs **bleb** (localised collection of fluid) of ill understood function. As in other prokaryotes, there is no intracellular membranous structure.

The PPLO cells contain many enzymes which may be required for DNA replication, the transcription of different kinds of RNA molecules and translation involved in protein synthesis, and also in the biosynthesis of adenosine triphosphate (ATP) by anaerobic breakdown of sugars. Unlike viruses, they are free living and do not require host cells for their duplication. PPLO reproduce by binary fission, budding, formation of small spore-like bodies and by growth of large branched filaments that ultimately fragment.

2. *Escherichia Coli* (*E. Coli*)

E. coli is a Gram negative, monotrichous, symbiotic bacillus of colon of human beings and other vertebrates. It is heterotrophic and non-pathogenic bacteria producing some vitamins (*e.g.*, vitamin K) for human use. Some strains of *E. coli* are known to recognise and bind specifically to sugar-containing target cells on the surface of gut lining of mammals (*e.g.,* D-mannose residues of epithelial cells of human gut or colon). *E. coli* is one of the best studied bacteria. It has served well in the field of molecular biology, since this bacterium is particularly easy to grow in an artificial medium where it divides every 20 minutes at 37°C under optimal conditions. Thus, a single cell become 10^9 bacteria in about 20 hours.

The prokaryotic cell of *E. coli* (Fig. 55.6) is about 2 μm long and 1 μm wide. The cyto-plasm of the bacterium is bounded by a typical fluid mosaic **plasma membrane**. External to the plasma membrane occurs the rigid and protective **cell wall** which has a complex organization; it com-prises following two structures:

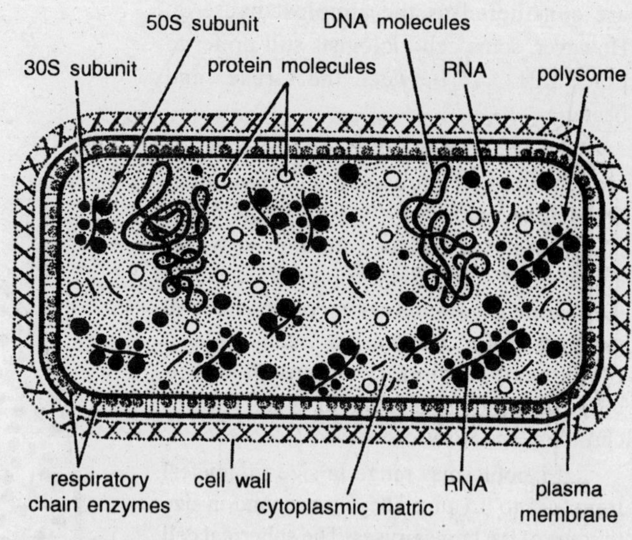

Fig. 55.6. A prokaryotic cell of *Escherichia coli.*

1. External membrane which is a lipid bilayer traversed by numerous **porin channels** that allow the diffusion of solutes. Each porin channel is formed by 6 to 8 subunits, each having three suspended hydrocarbon chains (Fig. 55.7). The porin is a polypeptide and it spans the full thickness of outer membrane. 2. Both membranes-the plasma membrane and external membrane of the cell wall- are separa-ted by the **periplasma-tic space**. This space contains a grid or reticulum of **peptido-glycans**. Some porin subunits remain attach-ed to the **peptidoglycan** grid (Fig. 55.7).

The plasma mem-brane serves as a mole-cular barrier with the surrounding medium. It comprises a variety of transport proteins, called **permeases** which control the entrance and exit of small molecules and ions. It contributes to the establishment of bacterial protoplasm. *E. coli* has both oxygen-requiring (aerobic) and non-oxygen-requiring (anaerobic) respiratory machinery for the breakdown of sugar and contains a special group of proteins called the **electron transport chain** for the generation of stored energy in the form of ATP molecules. *E. coli* lacks mitochondria, and respiratory chain enzymes such as cytochromes, enzymes of Krebs cycle, NADH, acid phosphatase, etc., are attached to inner face of the plasma membrane.

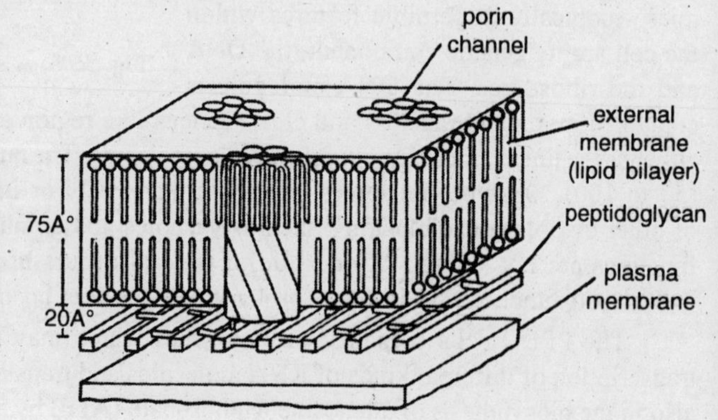

Fig. 55.7. Ultrastructure of the cell wall of a Gram-negative bacterium.

All genes of *E. coli* are contained on a single supercoiled, double-stranded, circular **DNA molecule**, which occurs in a clear zone of cytoplasm, called **nucleoid**, and is attached to the plasma membrane at one point. The total length of the DNA circle is about 1300 μm, comprising about 4.7×10^6 nucleotide pairs; this is enough DNA to code for about 4000 different proteins. The DNA of *E.coli* is naked, lacking histones, but certain polyamines may be bound to some of its phosphates. Electron microscopy of isolated chromosome of *E. coli* has shown that DNA is folded into a series of **looped domins**, *i.e.*, about 45 loops radiate out from a dense proteinaceous **scaffold**

(Fig. 55.8). The DNA of loops is in the so-called supercoiled conformation in which the double helix is itself twisted. The enzyme **DNA gyrase** is responsible for the DNA supercoiling (it is inhibited by the drug called **coumermycin**).

The colloidal cytoplasmic matrix of *E.coli* contains about 5000 distinguishable components, ranging from water to DNA (*i.e.*, three types of RNA, enzymes, glycogen, amino acids, monosaccharides and various other small molecules). Surrounding the DNA is dark dense region of matrix containing 20,000 to 30,000 70S type **ribosomes**, each existing in the form of their two subunits. During protein synthesis numerous complete ribosomes read the codes of mRNA molecules to form the polysomes (Fig. 55.6).

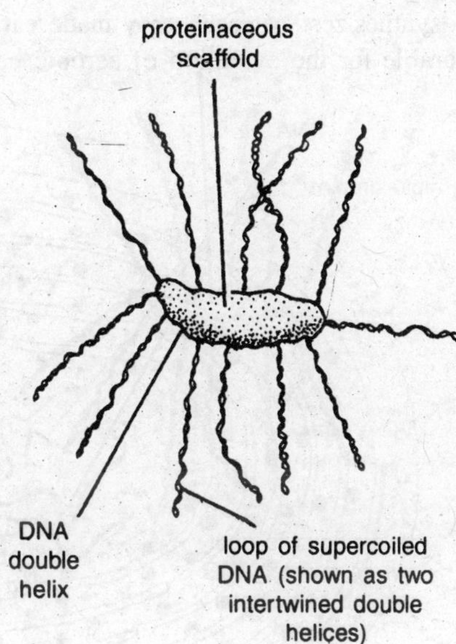

Fig. 55.8 Schematic representation of the chromosome of *E. coli*, showing only 12 of the 45 supercoiled loops.

3. Cyanobacteria (Blue green Algae)

The Gram-negative cyanobacteria or oxyphotobacteria (*i.e.*, oxygen yielding photosynthetic blue green algae) are one of the most successful and primitive (3.5 billion year old) groups of organisms on earth. They even inhabit the steaming hot springs and the undersides of icebergs. Cyanobacteria form another group of prokaryotes which include about 1500 species (85 genera and 750 species are found in India).

Cyanobacteria occur as individual cells, as small clusters or colonies of cells, or as long, filamentous chains. They lack flagella but are able to perform movement by rotatory motion or gliding over a gelatinous layer secreted through the cell surface.

A typical cell of a blue green algae is composed of outer cellular coverings and cytoplasm. The **outer cellular coverings** include an outer most **gelatinous** or **slimy layer**, **the capsule**, a middle **cell wall** and an innermost lipoproteinous **plasma membrane**. The cell wall of blue green algae resembles the cell wall of bacteria and contains an **outer bimolecular membrane** of phospholipids, lipoproteins and lipopolysaccharides, and a grid of peptidoglycans (muramic acid) in the **periplasmatic space** existing in between cell wall and plasma membrane. The **cytoplasm** of cyanobacteria appears more organized than that of other bacteria. The matrix extends throughout the cell. The cytoplasm (or protoplast) is differentiated in two regions:

1. Outer or peripheral pigmented region, the **chromoplasm** having photosynthetic lamellae or **thylakoids**.
2. Inner or central colourless region called **centroplasm** or **DNA plasm** having DNA and crystalline granules (Fig. 55.9).

Because the metabolism of the blue green algae is based on photosynthesis, therefore, the cells of them contain the photosynthetic pigment, viz., the **chlorophyll** and **carotenoid**. In addition to these pigments, these algae contain certain unique pigments collectively called **phycobilin**; one of the phycobilin is blue and called **phycocyanin**, while other type of phycobilin is red and called **phycoerythrin**. The photosynthetic pigments (chlorophylls and carotenoids) occur in flattened sacs called **lamellae** which remain arranged in parallel array. In between the lamellae occur certain granules of 400A° diameter. These granules contain phycobilin pigments and are called **cyanosomes** or **phycobilisomes**. They are attached to the outer lamellar membrane surface. Being earliest oxygenic photosynthesizers of earth, they made early earth's atmosphere aerobic providing the conditions favourable for the evolution of aerobic bacteria and eukaryotes.

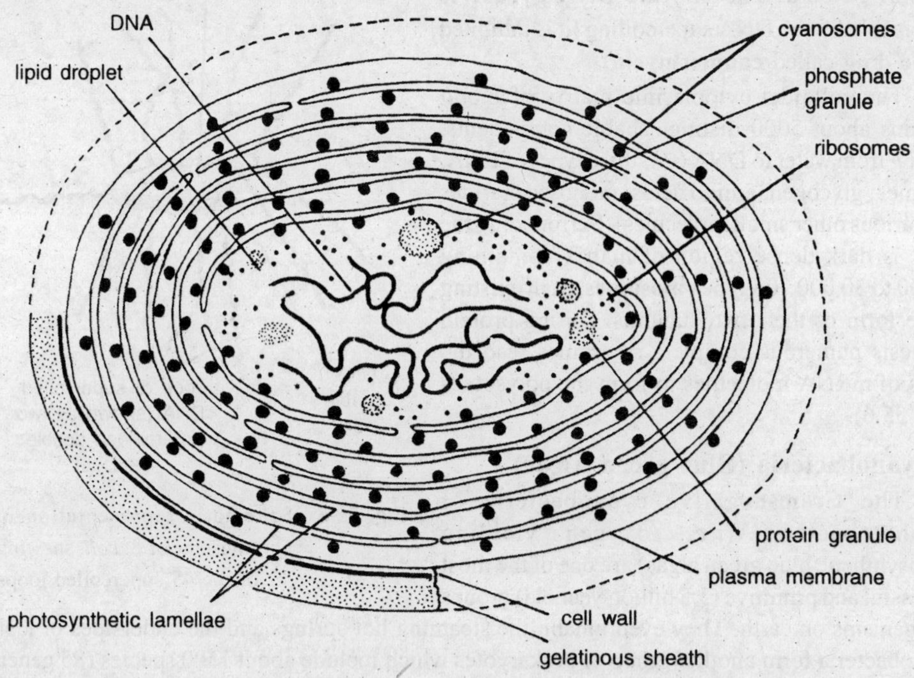

Fig. 55.9. A prokaryotic cell of cyanobacteria (electron microscopic view).

The two subunits of 70S ribosomes of cyanobacteria are freely distributed in the cytoplasm and form polyribosomes during protein synthesis. As in all prokaryotes, the DNA molecule of blue green algae is circular, double-stranded helix and occurs in the centroplasm. This area (nucleoid) is not bound by the nuclear membrane and it does not contain a nucleolus.

Cyanobacteria also contain a variety of **inclusions** in its cytoplasm. Membrane-bound inclusions are the gas vacuoles and the carboxysomes. **Gas vacuoles** are gas-filled cavities which are located in the inner part of chromatoplasm. They occur commonly in planktonic species such as *Nostoc*, *Anabaena*, *Phormidium*, *Calothrix*, *Galaeotrichia*, etc. Gas vacuoles serve the function of flotation or buoyancy. **Carboxysomes** contain enzymes involved in carbon dioxide fixation.

The cytoplasm of blue green algae also contains a variety of membrane-free inclusions such as (1) **cyanophycin granules** which are located in chromatoplasm and are protein storage products, containing large amount of arginine amino acid or copolymers of alanine and aspartic acid; (2) **myxophycean starch** which is the main food storage compound; (3) **polyglucon granules,** polyhedral bodies, lipid droplets, polyphosphate bodies, etc., are some other cytoplasmic inclusions of cyanobacteria.

Lastly, many cyanobacteria (about 20 species) tend to fix atmospheric nitrogen as ammonia, *e.g.*, *Anabaena, Nostoc, Mastigocladus*, etc. Under aerobic condition nitrogen fixation is done principally in special type of cells called **heterocysts**, as in *Nostoc*.

55.4 EUKARYOTIC CELL

The eukaryotic cells (Gr., *eu* = good, *karyotic* = nucleated) are essentially **two envelope systems** and they are very much larger than prokaryotic cells. Secondary membranes envelop the nucleus and other internal organelles and to a great extent they pervade the cytoplasm as the endoplasmic reticulum. The eukaryotic cells are the true cells which occur in the plants (from algae to angiosperms) and the animals (from Protozoa to mammals). Though the eukaryotic cells have different shape, size, and physiology; all the cells are typically composed of plasma membrane, cytoplasm and its organelles, viz., mitochondria, endoplasmic reticulum, ribosomes, Golgi apparatus, etc., and a true nucleus. Here the nuclear contents, such as DNA, RNA, nucleoproteins and nucleolus remain separated from the cytoplasm by the thin, perfotated nuclear membranes.

Cell Shape

The basic shape of the eukaryotic cell is **spherical**, however, the shape is ultimately determined by the specific function of the cell. Thus, the shape of the cell may be **variable** (*i.e.*, frequently changing the shape) or **fixed**. Variable or irregular shape occurs in *Amoeba* and white blood cells or leucocytes (In fact, leucocytes are spherical in the circulating blood, but in other conditions they may produce pseudopodia and become irregular in shape). Fixed shape of the cell occurs in almost all protists (*e.g., Euglena, Paramecium*), plants and animals. In unicellular organisms the cell shape is maintained by tough plasma membrane and exoskeleton. In a multicellular organism, the shape of the cell depends mainly on its functional adaptations and partly on the surface tension, viscosity of the protoplasm, cytoskeleton of microtubules, microfilaments and intermediate filaments, the mechanical action exerted by adjoining cells and rigidity of the plasma membrane (*i.e.*, presence of rigid cell wall in plant cells). The shape of the cell may vary from animal to animal and from organ to organ. Even the cells of the same organ may display variations in the shape. Thus, cells may have diverse shapes such as **polyhedral** (with 8, 12 or 14 sides; *e.g.*, squamous epithelium); **flattened** (*e.g.*, squamous epithelium, endothelium and the upper layers of the epidermis); **cuboidal** (*e.g.*, in thyroid gland follicles); **columnar** (*e.g.*, the cells lining the intestine); **discoidal** (*e.g.*, red blood cells or erythrocytes); **spherical** (*e.g.*, eggs of many animals); **spindle shaped** (*e.g.* smooth-muscle fibres); **elongated** (*e.g.*, nerve cells or neurons); or **branched** (*e.g.*, chromatophores or pigment cells of skin). Among plants, the cell shape also depends upon the function of the cell. For example, cells such as glandular hairs on a leaf, the guard cells of stomata and root hair cells have their special shape.

Cell Size

The eukaryotic cells are typically larger (mostly ranging between 10 to 100 µm) than the prokaryotic cells (mostly ranging between 1 to 10 µm). Size of the cells of the unicellular organisms is larger than a typical multicellular organism's cells. For example, *Amoeba proteus* is biggest among the unicellular organisms; its length being 1 µm (1000 µm). One species of *Euglena* is found up to 500 µm (0.5mm) in length. *Euplotes* (a freshwater ciliate) is 120 µm in length. Another ciliate, *Paramecium caudatum* is from 150 to 300 µm (0.15 to 0.3 mm) in length. Diatoms have a length of 200 µm or more. The single-celled alga, *Acetabularia* which consists of a stalk and a cap is exceptionally large-sized and measures up to 10 cm in height.

The size of the cells of multicellular organisms ranges between 20 to 30 µm. Among animals, the smallest cells have a diameter of 4 µm (*e.g.*, polocytes); human erythrocytes being 7 to 8 µm in diameter. Largest animal cell is the egg of ostrich, having a diameter of 18 cm (its yolk or deutoplasm is about 5 cm in diameter); though, some nerve cells of human beings have a meter long "tails" or axons. Among the multicellular plants, the largest cell is the ovule of *Cycas*. The fibre cells (*i.e.*, sclerenchyma cells) of Manila hemp are over 100 cm in length.

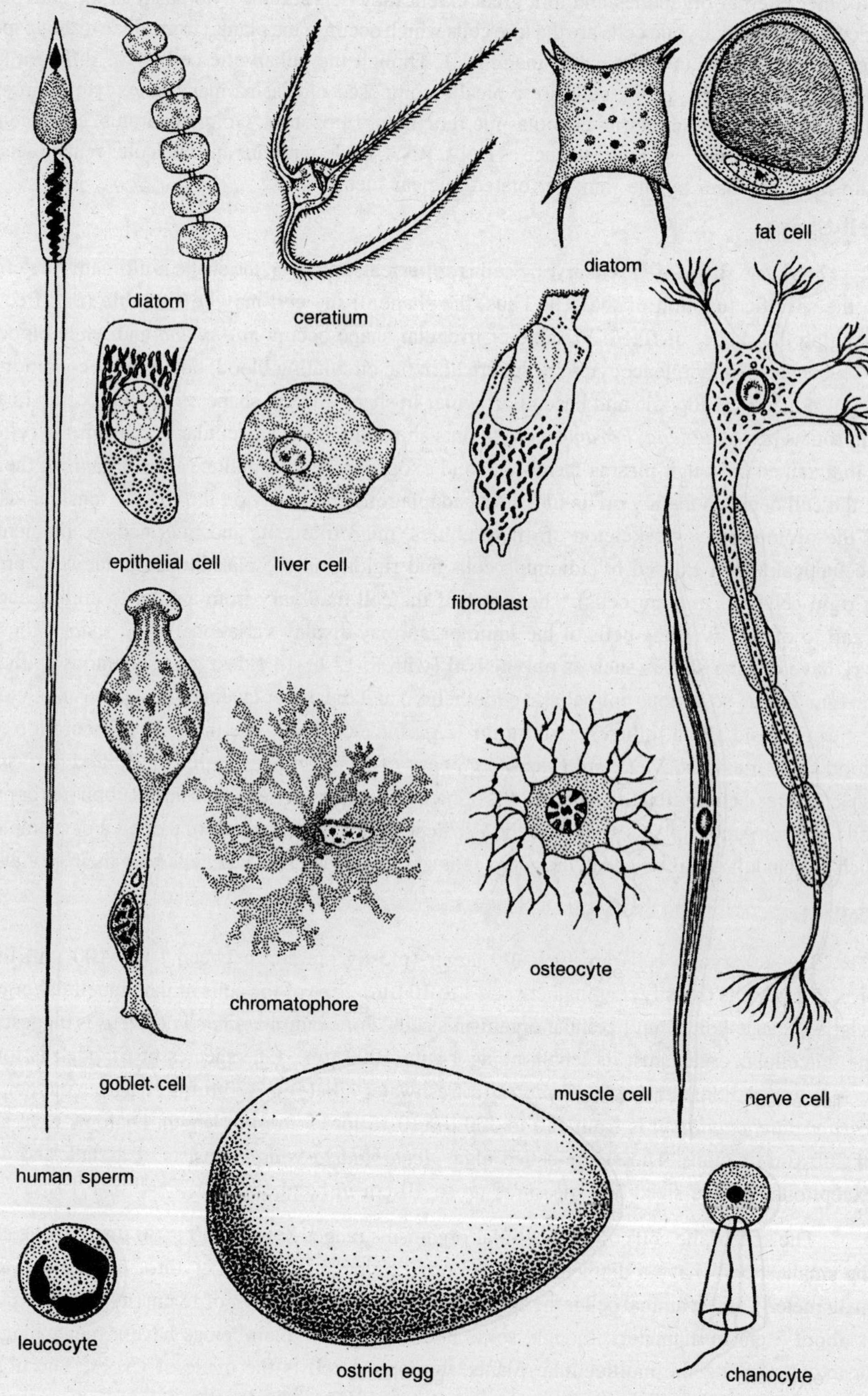

Fig. 55.10. Various types of eukaryotic cells showing different shapes.

Cell Volume

The volume of a cell is fairly constant for a particular cell type and is independent of the size of the organism. (This is called the **law of constant volume**). For example, kidney or liver cells are about the same size in the bull, horse and mouse. The difference in the total mass of the organ or organism depends on the number, not on the volume of the cells. Thus, the cells of an elephant are not necessarily larger than those of other tiny animals or plants. The large size of the elephant is due to the larger number of cells present in its body.

If a cell is to be efficient, the ratio of volume to surface should be within a limited range. An increase in cell volume is accompanied by a much smaller expansion in the surface area of the cell (In fact, volume increases as cube of radius, while surface area increases as square of radius). In other words, a large cell has a proportionately smaller surface and a higher volume: surface ratio than a smaller cell. Further, a large cell volume has to accommodate many organelles simultaneously limiting the exchange of information and materials through the surface. This problem is partially overcome by developing a cylindrical shape or by forming numerous extensions (*e.g.*, microvilli) of the plasma membrane. It is also for this reason that metabolically active cells, tend to be smaller in size.

Cell Number

The number of cells present in an organism varies from a single cell in a **unicellular organism** (Protists such as Protozoa and Protophyta) to many cells in multicellular organisms (most plants, fungi and animals). The number of cells in the multicellular organisms usually remains correlated with size of the organisms and, therefore, small-sized organism has less number of cells in comparison to large-sized organisms. For example, a human being weighing about 80 kg may contain about 60 thousand billion cells in his body. This number would be more in certain other multicellular organisms.

Further, the number of cells in most multicellular organisms is indefinite, but the number of cells may be fixed in some multicellular organisms (*e.g.*, nematodes, rotifers). For example, in rotifers, number of nuclei in the various organs are found to be constant in any given species. This phenomenon of cells or nuclear constancy is called **eutely**. In one species of rotifer, **Martini** (1912) always found 183 nuclei in the brain, 39 in the stomach, 172 in the cornea epithelium, and so on. Among plants, colonial green algae exhibit cell constancy. For example, the green alga, *Pandorina* has a colony consisting of 8, 16 or 32 cells. Likewise, another green alga, *Eudornia*, has 16, 32 or 64 cells in its colony.

Animal Cell

A typical animal cell, which is an eukaryotic cell, consists of following three main parts:

1. Plasma membrane
2. Cytoplasm
3. Nucleus

1. Cell membrane or plasma membrane. Every kind of animal cell is bounded by a living, extremely thin and delicate membrane called **plasmalemma, cell membrane** or **plasma membrane**. In plant cells, plasma membrane occurs just inner to cell wall, bounding the cytoplasm. The plasma membrane exhibits a trilaminar (*i.e.*, three-layered) structure with a translucent layer sandwiched between two dark layers. At molecular level, it consists of a continuous bilayer of lipid molecule (*i.e.*, phospholipids and cholesterol) with protein molecules embedded in it or adherent to its both surfaces. Some carbohydrate molecules may also be attached to the external surface of the plasma membrane, they remain attached either to protein molecules to form **glycoproteins** or to lipids to form **glycolipids**. Such a fluid mosaic plasma membrane is a **selectively permeable membrane;**

its main function is to control selectively the entrance and exit of materials. This allows the cell to maintain a constant internal environment **(homeostasis)**. Transport of small molecules such as water, oxygen, carbon dioxide, ethanol, ions; glucose, etc., across the plasma membrane takes place by various means such as osmosis, diffusion and active transport. The process of active transport is performed by special type of protein molecules of plasma membrane called **transport proteins** or **pumps,** consuming energy in the form of ATP molecules. For bulk transport of large-sized molecules, plasma membrane performs **endocytosis** (*i.e.,* endocytosis, pinocytosis, receptor-mediated endocytosis and phagocytosis) and **exocytosis** both of these processes also utilise energy in the form of ATP molecules.

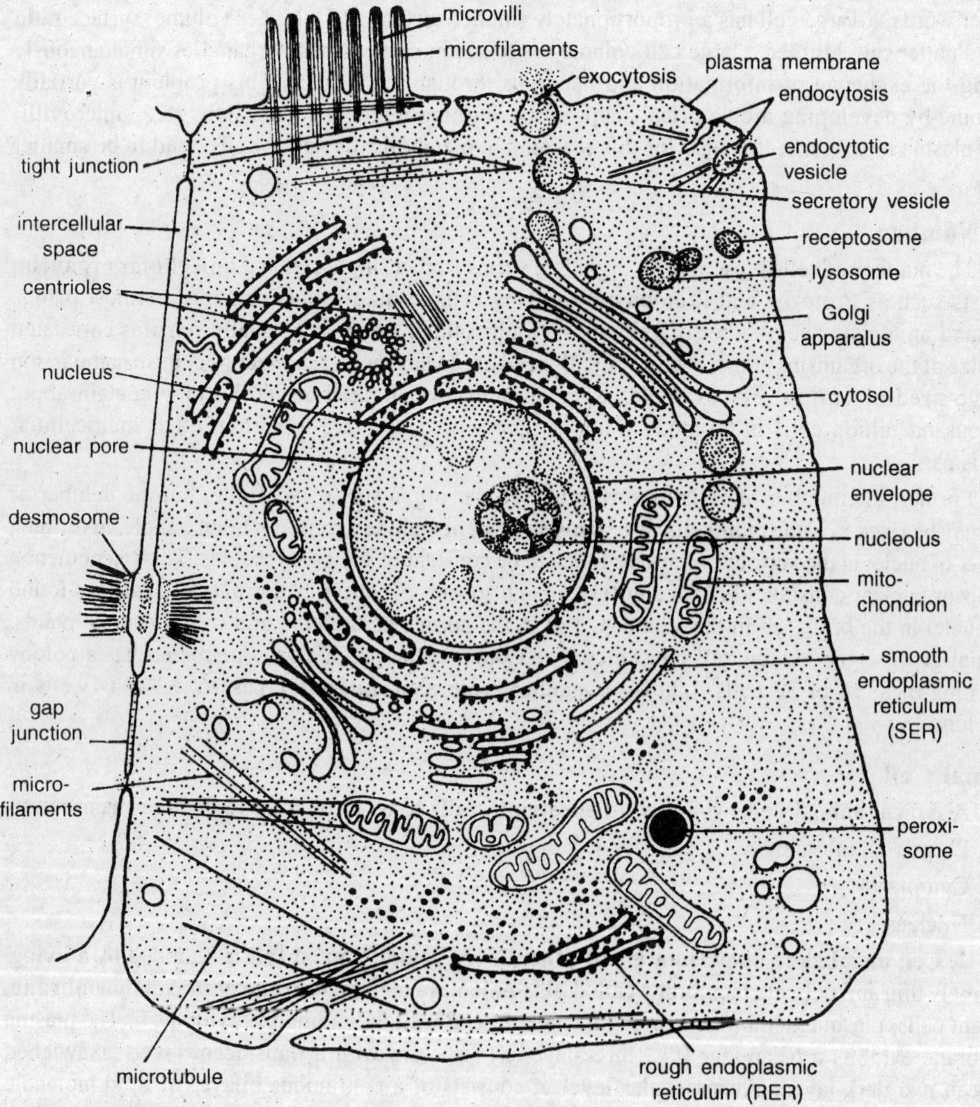

Fig. 55.11. Ultrastructure of a typical animai cell as seen in the electron microscope.

Unit membrane. Various cell organelles such as chloroplasts, mitochondria, endoplasmic reticulum and lysosomes are also bounded by membranes similar to the plasma membrane. All the cellular membranes have a basic trilaminar fluid mosaic **unit membrane** construction. However, their structure and extent of activity are mainly depended on the relative proportion of their constituent protein and lipid molecules. Thus, membranes which are metabolically highly active, *e.g.*, those of mitochondria and chloroplasts have a greater proportion of proteins and more granular appearance than those membranes which are relatively less active, *e.g.*, myelin sheath of certain nerve fibres.

2. Cytoplasm. The plasma membrane is followed by the cytoplasm which is distinguished into following structures:

A. Cytosol. The plasma membrane is followed by the colloidal organic fluid called **matrix** or **cytosol**. The cytosol is the aqueous portion of the **cytoplasm** (the extra nuclear protoplasm) and of the **nucleoplasm** (the nuclear protoplasm). It fills all the spaces of the cell and constitutes its true **internal milieu**. Cytosol is particularly rich in differentiating cells and many fundamental properties of cell are because of this part of the cytoplasm. The cytosol serves to dissolve or suspend the great variety of small molecules concerned with cellular metabolism, *e.g.*, glucose, amino acids, nucleotides, vitamins, minerals, oxygen and ions. In all type of cells, cytosol contains the soluble proteins and enzymes which form 20 to 25 per cent of the total protein content of the cell. Among the important soluble enzymes present in the matrix are those involved in glycolysis and in the activation of amino acids for the protein synthesis. In many types of cells, the cytosol is differentiated into following two parts: (*i*) **Ectoplasm** or **cell cortex** is the peripheral layer of cytosol which is relatively non-granular, viscous, clear and rigid. (*ii*) **Endoplasm** is the inner portion of cytosol which is granular and less viscous.

Cytoskeleton and microtrabecular lattice. The cytosol of cells also contains **fibres** that help to maintain cell shape and mobility and that provide anchoring points for the other cellular structures. Collectively, these fibres are termed as the **cytoskeleton.** At least three general classes of such fibres have been identified. 1. The thickest are the **microtubules** (20 nm in diameter) which consists primarily of the **tubulin** protein. The function of microtubules is the transportation of water, ions or small molecules, cytoplasmic streaming (cyclosis), and the formation of fibres or asters of the mitotic or meitotic spindle during cell division. Moreover, they form the structural units of the centrioles, basal granules, cilia and flagella. 2. The thinnest are the **microfilaments** (7 nm in diameter) which are solid and are principally formed of **actin** protein. They maintain the shape of cell and form contractile component of cells, mainly of the muscle cells. 3. The fibres of middle order are called the **intermediate filaments (IFs)** having a diameter of 10 nm. They have been classified according to their constituent protein such as **desmin filaments, keratin filaments, neurofilaments, vimentin** and **glial filaments**.

Recently, cytoplasm has been found to be filled with a three-dimensional network of interlinked filaments of cytoskeletal fibres, called **microtrabecular lattice.** Various cellular organelles such as ribosomes, lysosomes, etc., are found anchored to this lattice. The microtrabecular lattice being flexible, changes its shape and results in the change of cell shape during cell movement.

B. Cytoplasmic structures. In the cytoplasmic matrix certain non-living and living structures remain suspended. The non-living structures are called **paraplasm** or **inclusions,** while the living structures are membrane bounded and are called **organoids** or **organelles.** Both kinds of cytoplasmic structures can be studied under the following headings:

(*a*) Cytoplasmic inclusions. The stored food and secretory substances of the cell remain suspended in the cytoplasmic matrix in the form of refractile granules forming the cytoplasmic inclusions. The cytoplasmic inclusions include oil drops, triacylglycerols (*e.g.*, fat cells of adipose tissue), yolk granules (or **deutoplasm,** *e.g.*, egg cells), secretory granules, glycogen granules (*e.g.*, muscle cells and hepatocytes of liver) and starch grains (in plant cells).

(*b*) Cytoplasmic organelles. Besides the separate fibrous systems, cytoplasm is coursed by a multitude of internal membranous structures, the organelles (literally the word organelle means a tiny organ). Membranes close off at specific regions of the eukaryotic cells performing specialized tasks: oxidative phosphorylation and generation of energy in the form of ATP molecules in mitochondria; formation and storage of carbohydrates in plastids; protein synthesis in rough endoplasmic reticulum; lipid (and hormone) synthesis in smooth endoplasmic reticulum; secretion by Golgi apparatus; degradation of macromolecules in the lysosomes; regulation of all cellular activities by nucleus; organization of spindle apparatus by centrosomes and so forth. Membrane-bound enzymes catalyze reactions that would have occurred with difficulty in an aqueous environment. The structure and function of some important organelles are as follows:

1. Endoplasmic reticulum (ER). Within the cytoplasm of most animal cells is an extensive network (reticulum) of membrane-limited channels, collectively called **endoplasmic reticulum** (or **ER**). Some portion of ER membranes remains continuous with the plasma membrane and the nuclear envelope. The outer surface of **rough ER** has attached ribosomes, whereas **smooth ER** do not have attached ribosomes. Functions of smooth ER include **lipid metabolism** (both catabolism and anabolism; they synthesize a variety of phospholipids, cholesterol and steroids); **glycogenolysis** (degradation of glycogen; glycogen being polymerized in the cytosol) and **drug detoxification** (by the help of the **cytochrome P-450**).

On their membranes, rough ER (RER) contain certain ribosome-specific, transmembrane glycoproteins, called **ribophorins I and II**, to which are attached the ribosomes while engaged in polypeptide synthesis. As a growing secretory polypeptide emerges from ribosome, it passes through the RER membrane and gets accumulated in the lumen of RER. Here, these polypeptide chains undergo tailoring, maturation, and molecular folding to form functional secondary or tertiary protein molecules. RER pinches off certain tiny protein-filled vesicles which ultimately get fused to cis Golgi. RER also synthesize membrane proteins and glycoproteins which are cotranslationally inserted into the rough ER membranes. Thus, endoplasmic reticulum is the site of biogenesis of cellular membranes.

2. Golgi apparatus. It is a cup-shaped organelle which is located near the nucleus in many types of cells. Golgi apparatus consists of a set of smooth **cisternae** (*i.e.*, closed fluid-filled flattened membranous sacs or vesicles) which often are stacked together in parallel rows. It is surrounded by spherical membrane bound **vesicles** which transport proteins to and from it.

Golgi apparatus consists of at least three distinct classes of cisternae: **cis Golgi, median Golgi** and **trans Golgi**, each of which has distinct enzymatic activities., Synthesized proteins appear to move in the following direction: rough ER → cis Golgi → median Golgi → trans Golgi → secretory vesicles/cortical granules of egg/lysosomes or peroxisomes. Thus, the size and number of Golgi apparatus in a cell indicate the active metabolic, mainly synthetic, state of that cell. Plant cells contain many freely distributed sub-units of Golgi apparatus, called **dictyosomes**, secreting cellulose and pectin for cell wall formation during the cell division.

Generally, Golgi apparatus performs the following important functions:

1. The packaging of secretory materials (*e.g.*, enzymes, mucin, lactoprotein of milk, melanin pigment, etc.) that are to be discharged from the cell.
2. The **processing** of proteins, *i.e.*, glycosylation, phosphorylation, sulphation and selective proteolysis.
3. The synthesis of certain polysaccharides and glycolipids.
4. The sorting of proteins destined for various locations (*e.g.*, lysosomes, peroxisomes, etc.) in the cell.
5. The proliferation of membranous element for the plasma membrane.
6. Formation of the acrosome of the spermatozoa and cortical granules of an ovum.

3. Lysosomes. The cytoplasm of animal cells contains many tiny, spheroid or irregular-shaped, membrane-bounded vesicles known as **lysosomes**. The lysosomes are originated from Golgi apparatus and contain numerous (about 50) hydrolytic enzymes (*e.g.*, **acid phosphatase** that is cytochemically identified) for intracellular and extracellular digestion. They digest the material taken in by endocytosis (such as phagocytosis, endocytosis and pinocytosis), parts of the cell (by autophagy) and extracellular substances. Lysosomes have a high acidic medium (pH 5) and this acidification depends on ATP-dependent **proton pumps** which are present in the membrane of lysosomes and which accumulate protons (H^+) inside the lysosomes. Lysosomes exhibit great **polymorphism,** *i.e.*, there are following four types of lysosomes: primary lysosomes (storage granules), secondary lysosomes (digestive vacuoles), residual bodies and autophagic vacuoles. The lysosomes of plant cells are membrane-bounded storage granules containing hydrolytic digestive enzymes, *e.g.*, large **vacuoles** of parenchymatous cells of corn seedlings, **protein** or **aleurone bodies** and **starch granules** of cereal and other seeds.

4. Cytoplasmic vacuoles. The cytoplasm of some animal cells (*e.g.*, ciliate protozoans) contains numerous small or large-sized, hollow, liquid-filled structures, the **vacuoles.** These vacuoles are supposed to be greatly expanded endoplasmic reticulum or Golgi apparatus. The vacuoles of animal cells are bounded by a lipoproteinous membrane and their function is the storage, transmission of the materials and the maintenance of internal pressure of the cell.

5. Peroxisomes. These are tiny circular membrane-bound organelles containing a crystal-core of enzymes (such as urate oxidase, peroxidase, D-amino oxidase and catalase) of certain cells (*e.g.*, liver cells and kidney cells). These enzymes are required by peroxisomes in **detoxification** activity, *i.e.*, in the metabolism or production and decomposition, of hydrogen peroxide or H_2O_2 molecules which are produced during neutralization of certain **superoxides** the end products of mitochondrial or cytosolic reactions. Peroxisomes are also related with β-oxidation of fatty acids and thermogenesis like the mitochondria and also in degradation of the amino acids. In green leaves of plants, peroxisomes carry out the process of **photorespiration**.

6. Mitochondria. Mitochondria are oxygen-consuming ribbon-shaped cellular organelles of immense importance. Each mitochondrion is bounded by two unit membranes. The outer mitochondrial membrane resembles more with the plasma membrane in structure and chemical composition. It contains **porins,** proteins that render the membrane permeable to molecules having molecular weight as high as 10,000. Inner mitochondrial membrane is rich in many enzymes, coenzymes and other components of electron transport chain. It also contains **proton pumps** and many **permease** proteins for the transport of various molecules such as citrates, ADP, phosphate and ATP. Inner mitochondrial membrane gives out finger-like outgrowths **(cristae)** towards the lumen of mitochondrion and contains tennis-racket shaped **F_1 particles** which contain ATPase enzyme for ATP synthesis.

Mitochondrial matrix which is the liquid (colloidal) area encircled by the inner membrane, contains the soluble enzymes of Krebs cycle which completely oxidize the **acetyl-CoA** (an end product of cytosolic glycolysis and mitochondrial oxidative decarboxylation) to produce CO_2, H_2O and hydrogen ions. Hydrogen ions reduce the molecules of NAD and FAD, both of which pass on hydrogen ions to respiratory or electron transport chain where oxidative phosphorylation takes place to generate energy-rich ATP molecules. Since mitochondria act as the 'power-houses' of cells, they are abundantly found on those sites where energy is earnestly required such as sperm tail, muscle cell, liver cell (up to 1600 mitochondria), microvilli, oocyte (more than 300,000 mitochondria), etc. Mitochondria also contain in their matrix single or double circular and double stranded DNA molecules, called **mtDNA** and also the 55S ribosomes, called **mitoribosomes**. Since mitochondria can synthesize 10 per cent of their proteins in their own protein-synthetic machinery, they are considered as **semi-autonomous organelles.** Mitochondria may also produce heat (brown fat), and accumulate iron-containing pigments (Heme, ferritin), ions of Ca^{2+} and HPO_4^{2-} (or phosphate; *e.g.*, osteoblasts of bones or yolk proteins).

7. Ribosomes. Ribosomes are tiny spheroidal dense particles (of 150 to 200 A° diameter) that contain approximately equal amounts of RNA and proteins. They are primarily found in all cells and serve as a scaffold for the ordered interaction of the numerous molecules involved in protein synthesis. Ribosome granules may exist either in the **free state** in the cytosol (*e.g.,* basal epidermal cells) or **attached** to RER (*e.g.,* pancreatic acinar cells, plasma cells or antibodies-secreting lymphocytes, osteoblasts, etc.). Ribosomes have a sedimentation coefficient of about **80S** and are composed of two subunits namely **40S** and **60S**. The smaller 40S ribosomal subunit is prolate ellipsoid in shape and consists of one molecule of 18S ribosomal RNA (or rRNA) and 30 types of proteins (named as S_1, S_2, S_3, and so on). The larger 60S ribosomal subunit is round in shape and contains a **channel** through which growing polypeptide chain makes its exit. It consists of three types of rRNA molecules, *i.e.,* 28S rRNA, 5.8 rRNA and 5S rRNA, and 40 types of proteins (named as L_1, L_2, L_3 and so on).

8. Microtubules and microtubular organelles. With rare exceptions, such as human erythrocyte, microtubules are found in the cytoplasm of all types of eukaryotic cells. They are long fibres (of indefinite length) about 24 nm in diameter. In cross section each microtubule appears to have a dense wall of 6 nm thickness and a light or hollow centre. In cross section, the wall of a microtubule is made up of 13 globular subunits, called **protofilaments,** about 4 to 5 nm in diameter. Chemically, microtubules are composed of two kinds of protein subunits: **α-tubulin (tubulin A)** and **β-tubulin (tubulin B)**, each of M.W. 55,000 daltons. The wall of a microtubule is made up of a helical array of repeating α and β tubulin subunits. Assembly studies have indicated that the structural unit is an **αβ dimmer** of 8 nm length. Thus, in each microtubule, there are 13 protofilaments, each composed of αβ dimers that run parallel to the long axis of the tubule: The repeating unit is an αβ heterodimer which is arranged 'head to tail' within the microtubule, that is αβ → αβ → αβ. Thus, all microtubules have a defined **polarity:** their two ends are not structurally equivalent.

Microtubules undergo reversible assembly-disassembly (*i.e.,* polymerization and depolymerization), depending on the need of the cell or organelles. Their polymerization is regulated by certain **MAPs** or **microtubule-associated proteins** (*e.g.,* Tau protein). The assembly of microtubules involves preferential addition of subunits (αβ dimers) to one end of tubule, called **A end** (or **net assembly end**); the other end of the tubule is called **D end** (or **net disassembly end**). Such an assembly involves the hydrolysis of GTP to GDP. Thus, assembly of tubulin in the formation of microtubules is a specifically oriented and programmed process. Centrioles, basal bodies and centromeres of chromosomes are the sites of orientation for this assembly. Calcium and **calmodulin** (an acidic protein having four Ca^{2+} binding sites) are some other regulating factors in the *in vivo* polymerization of tubulin. Certain drugs such as **colchicine** and **vinblastin,** are found to block the polymerization of tubulin.

The following cell organelles are derived from special assemblies of microtubules:

1. Cilia and flagella. Ciliary and flagellar cell motility is adapted to liquid media and is executed by minute, specially differentiated appendices, called **cilia** and **flagella.** Both of these organelles have very similar structure; they differ mainly in size and number (*i.e.,* flagella are longer and fewer in number, while cilia are short and numerous). Cilia are used for locomotion in isolated cells, such as certain protozoans (*e.g., Paramecium*) or to move particles in the medium, as in air passages and oviduct. Flagella are generally used for locomotion of cells, such as the spermatozoon and *Euglena* (protozoan). All cilia and flagella are built on a common fundamental plan: a bundle of microtubules called the **axoneme** (1 to 2 nm in length and 0.2 μm in diameter) is surrounded by a membrane that is part of the plasma membrane. The axoneme is connected with the basal body which is an intracellular granule lying in the cell cortex and which originates from the centrioles. Each axoneme is filled with **ciliary matrix,** in which are embedded two central **singlet** microtubules, each with the 13 protofilaments and nine outer pairs of microtubules, called **doublets**. This recurring motif is known as the 9 + 2 array. Each doublet contains one complete microtubule, called the **A subfibre**, containing all the 13 protofilaments. Attached to each A subfibre is a **B subfibre** with 10 protofilaments. Subfibre A has two **dynein arms** which are oriented in a clockwise direction. Doublets are linked together by **nexin links.** Each subfibre A is also connected to the central microtubules by **radial spokes** terminating in fork-like structures, called **spoke knobs** or **heads.**

Propulsion by both cilia and flagella is caused by bending at their base. Cilia move by a whip-like **power stroke** fueled by hydrolysis of ATP, followed by **a recovery stroke.** Flagellar movement is also powered by ATP hydrolysis. In contrast to cilia, flagella generally move by waves that emanate from the base and spread outward toward the tip.

2. Basal bodies and centrioles. Basal bodies and centrioles are similar in structure and function; both act as nucleating centres from which microtubules grow. **Centrioles** are cylinders that measure 0.2 μm × 0.5 μm. This cylinder is open on both ends, unless it carries a cilium or flagellum (then it is called **basal body** or **kinetosome**). The wall of a centriole has nine groups of microtubules arranged in a circle. Each group, called **blade** is a **triplet** formed of three tubules—*A*, *B*, and *C* that are skewed toward the centre. Tubule *A* has 13 protofilaments, while tubules *B* and *C* have only 10 protofilaments. each. There are no central microtubules in the centrioles and no dynein arms like the cilia; however, triplets are linked by connectives. The **procentriole** (or daughter centriole) is formed at right angles to the centriole and is located near the proximal end of the centriole. Both centrioles are found in a specially differentiated region the **centrosome, cell centre** or **centrosphere.** The centrosome is juxtanuclear (L., juxta = near) and firmly attached to the nuclear envelope. At the time of cell division two pairs of centrioles are formed and make up the spindle of microtubules which help in the separation and movement of chromosomes during concluding stages of cell divisions.

3. Nucleus. The nucleus is centrally located and spherical cellular component which controls all the vital activities of the cytoplasm and carries the hereditary material the DNA in it. The nucleus consists of the following three structures:

1. Chromatin. Nucleus being the heart of every type of eukaryotic cell, contains the **genes,** the hereditary units. Genes are located on the **chromosomes** which exist as **chromatin network** in the non-dividing cell, *i.e.,* during interphase. The chromatin has two forms:

(*i*) Euchromatin is the well-dispersed form of chromatin which takes lighter DNA-stain and is genetically active, *i.e.,* it is involved in gene duplication, gene transcription (DNA-dependent RNA synthesis) and **phenogenesis** or phenotypic expression of a gene through some type of protein synthesis.

(*ii*) Heterochromatin is the highly condensed form of chromatin which takes dark DNA-stain and is genetically inert. Such type of chromatin exists both in the region of centromere (called **constitutive heterochromatin**) and in the sex chromatin (called **facultative heterochromatin**) and is late-replicating one.

Chemically, the chromatin contains a single DNA molecule, equal amount of histone proteins. (which are of five basic types), some RNA molecules and variable amount of different types of acidic proteins. In fact, the chromatin has its unit structures in the form of **nucleosomes**. The chromatin binds strongly to the inner part of **nuclear lamina,** a 50 to 80 nm thick fibrous lamina lining the inner side of the nuclear envelope. Nuclear lamina is made up of three types of proteins, namely **lamin *A*, *B*** and ***C*.** Lamin proteins are homologous in structure to IF proteins and serve the following functions:

(*i*) They anchor parts of interphase chromatin to the nuclear membrane. They tend to interfere with chromatin condensation during interphase of cell cycle.

(*ii*) Lamins may play a crucial role in the assembly of interphase nuclei after each mitosis.

2. Nuclear envelope and nucleoplasm. Nuclear envelope comprises two nuclear membranes— an **inner nuclear membrane** which is lined by nuclear lamina and an **outer nuclear membrane** which is continuous with rough ER. At certain points the nuclear envelope is interrupted by structures called **pores** or **nucleopores**. Nuclear pores contain octagonal **pore complexes** which regulate exchange between the nucleus and cytoplasm. The number of nucleopores is found to be correlated with the transcriptional activity of the cell. For example, in the frog *Xenopus laevis* oocytes (which are very active in transcription) have 60 pores/μm^2 (and upto 30 million pore complexes per nucleus), whereas frog's mature erythrocytes (inactive in transcription) have only about 3 pores/μm^2 (and a total of only 150 to 300 pores per nucleus).

The nuclear envelope binds the **nucleoplasm** which is rich in those molecules which are needed for DNA replication, transcription, regulation of gene actions and processing of various types of newly transcribed RNA molecules (*i.e.,* tRNA, mRNA and other types of RNA).

3. Nucleolus. Nucleus contains in its nucleoplasm a conspicuous, darkly stained, circular sub-organelle, called **nucleolus.** Nucleolus lacks any limiting membrane and is formed during interphase by the ribosomal DNA (rDNA) of **nucleolar organizer (NO)**. Nucleolus is the site where ribosomes are manufactured. It is here where ribosomal DNA transcribes most of rRNA molecules and these molecules undergo processing before their step-wise addition to 70 types of ribosomal proteins to form the ribosomal subunits.

Plant Cell

All the cytoplasmic and nuclear components found in the animal cells are found in the plant cells (Fig. 55.12). In addition to these cell components, plant cells contain certain structures not found in animal cells. These are **cell wall, plasmodesmata, plastids, large vacuoles** and **glyoxysomes**.

1. Cell wall. The outermost structure of most plant cells is a dead and rigid layer called **cell wall**. It is mainly composed of carbohydrates such as cellulose, pectin, hemicellulose and lignin. and certain fatty substances such as waxes. Ultrastructurally cell wall is found to consist of a microfibrilar network lying in a gel-like matrix. The microfibrils are mostly made up of cellulose. There is a pectin-rich cementing substance between the walls of adjacent cells which is called **middle lamella**. The cell wall which is formed immediately after the division of cell, constitutes the **primary cell wall.** Many kinds of plant cells have only primary cell wall around them. Primary cell wall is composed of pectin, hemicellulose and loose network of cellulose micro-fibrils. In certain types of cells such as phloem and xylem, an additional layer is added to the inner surface of the primary cell wall at a later stage. This layer is called **secondary cell wall** and it consists mainly of cellulose, hemicellulose and lignin. In many plant cells, there are tunnels running through the cell wall called **plasmodesmata** which allow cominunication with the other cells in a tissue.

The cell wall constitutes a kind of exoskeleton that provides protection and mechanical support to the plant cell. It determines the shape of plant cell and prevents it from desiccation.

2. Plastids. Plastids occur only in the plant cells. They contain pigments and may synthesize and accumulate various substances. Plastids are of the following types:

***(i)* Leucoplasts** are colourless plastids of embryonic and germ cells lacking thylakoids and ribosomes.

***(ii)* Amyloplasts** produce starch.

***(iii)* Proteinoplasts** accumulate protein.

***(iv)* Oleosomes** or **elaioplasts** store fats and essential oils.

(*v*) **Chromoplasts** contain pigment molecules and are coloured organelles. Chromoplasts impart a variety of colours to plant cells, such as red colour in tomatoes, red chillies and carrots, various colours to petals of flowers and green colour to many plant cells. The green coloured chromoplasts are called **chloroplasts**. They have chlorophyll pigment and are involved in the photosynthesis of food and so act like the kitchens of the cell.

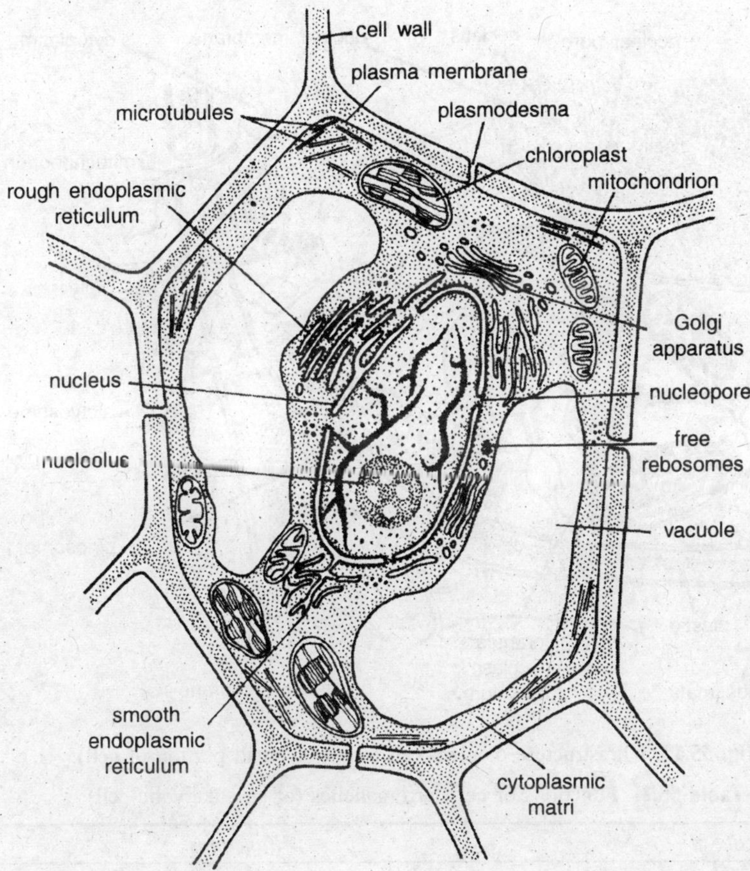

Fig. 55.12. Ultrastructure of a typical eukaryotic cell of a higher plant cell.

Chloroplasts have diverse shapes in green algae but are round, oval or discoid in shape in higher plants. Like mitochondria, each chloroplast is bounded by two membranous envelopes, both of which have no chlorophyll pigment. However, unlike mitochondria there occurs third system of membranes within the boundary of inner membrane, called **grana**. The grana form the main functional units of chloroplast and are bathed in the homogeneous matrix, called the **stroma**. Stroma contains a variety of photosynthetic enzymes and starch grains. Grana are stacks of membrane-bounded, flattened discoid sacs, arranged like neat piles of coins. A chloroplast contains many such interconnected grana on which are located various photosynthetic enzymes and the molecules of green pigment chlorophyll and other photosynthetic pigments to trap the light energy. They contain DNA, ribosomes and complete protein synthetic machinery.

3. Vacuoles. Often most of the volume (up to 90 per cent) of a mature cell is occupied by one or more large vacuoles. The vacuoles of the plant cells are bounded by a single, semipermeable membrane known as **tonoplast**. These vacuoles contain water, phenol, flavonols, anthocyanins (blue and red pigment), alkaloids and storage products, such as sugars and proteins.

4. Glyoxysomes. These organelles develop in a germinating plant seed (*e.g.*, castor bean or *Ricinus*) to utilize stored fat of the seed (*i.e.*, to metabolise the triglycerides). Glyoxysomes consist of an amorphous protein matrix surrounded by a limiting membrane. The membrane of glyoxysomes originates from the ER and their enzymes are synthesized in the free ribosomes in the cytosol. Enzymes of glyoxysomes are used to transform the fat stores of the seed into carbohydrates by way of **glyoxylate cycle**.

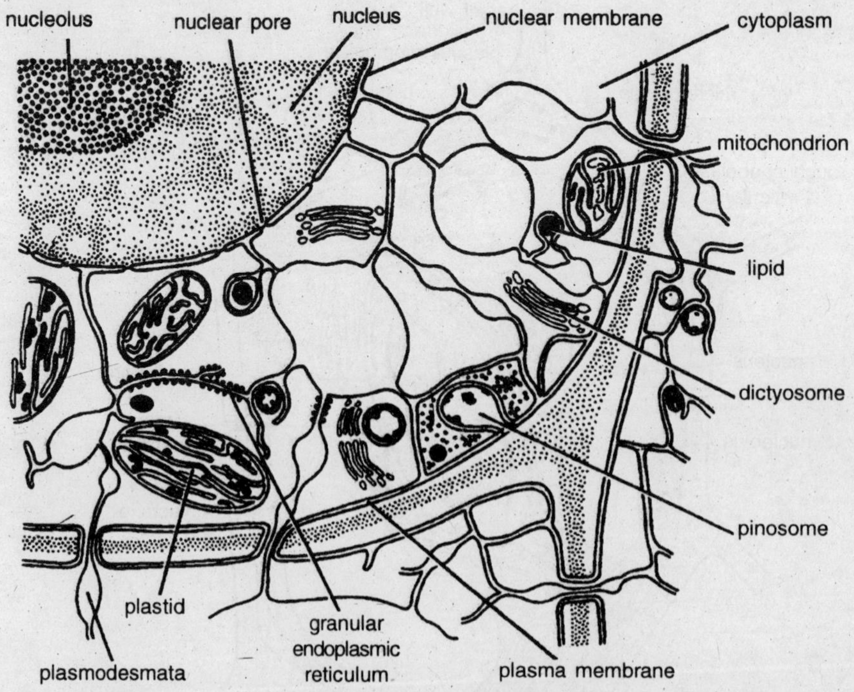

Fig. 55.13. Ultrastructure of plant cell (showing small portion of cell).

Table 55.1 Functions of cellular organelles (of an eukaryotic cell)

Cell organelle	*Functions*
Plasma membrane	It is differentially permeable membrane through which extracellular substances may be selectively sampled and cell products may be liberated.
Cell wall (plant cells)	It forms a thick cellulose wall around the plasma membrane giving strength and rigidity to the cell.
Cytoplasm	It contains machinery for carrying out the instructions sent from the nucleus.
Hyaloplasm or matrix	It contains enzymes for glycolysis and structural materials such as sugars, fatty acids, amino acids, water, vitamins, nucleotides, etc.
Microtubules	They transport water, ions and small molecules of various substances, form structural unit of centrioles, basal granules, cilia, flagella and mitotic spindle and play active role in the cyclosis (amoeboid movement) of cytoplasm and in movement of chromosomes during the cell divisions.
Microfilaments	They maintain the structure of cell and form contractile components of muscle cell.
Centrioles	They form the spindle and have significant role in the movement of chromosomes during cell divisions.
Basal granules	They originate the cilia or flagella and have their importance in ciliary or flagellar movement.

Cell organelle	*Functions*
Cilia and flagella	These are extracellular projections of the cell which provide movement to the cells and serve various other roles of cells.
Endoplasmic reticulum	It provides greatly expanded surface area for biochemical reactions which normally occur at or across membrane surfaces and provides skeletal support to the colloidal complex of hyaloplasm. It also has transporting business of extracellular and intracellular chemical molecules.
Ribosomes	These are the sites for synthesis of protein molecules.
Golgi apparatus	It stores the synthetic proteins and enzymes of cells and forms secretory vesicles and lysosomes.
Lysosomes (animals only)	It contains intracellular digestive enzymes which aid in disposal of bacteria and other foreign bodies.
Peroxisomes (Microbodies)	They contain enzymes for hydrogen peroxide metabolism, purine metabolism, gluconeogenesis and photorespiration.
Sphaerosomes	They store fats.
Vacuoles	They are storage depots for excess water, waste products, soluble pigments, etc.
Mitochondria	They are the organelles for the oxidation of food by Krebs cycle, production of energy by electron transport chain and beta-oxidation of fatty acids.
Plastids (Plants only)	They are the structures for storage of starch, pigments, and other cellular products. Photosynthesis occurs in chloroplasts.
Nucleus	It regulates growth and reproduction of cell.
Nuclear membrane	It provides selective continuity between nuclear and cytoplasmic materials.
Nucleoplasm	It contains materials for building DNA and messenger molecules (different kinds of RNA molecules, *viz.*, rRNA, mRNA and tRNA) which act as intermediates between nucleus and cytoplasm.
Chromosomes	They are the bearers of hereditary instructions and regulations of cellular processes.
Nucleolus	It disappears during cellular replication and synthesizes the ribosomes.

Differences between Prokaryotic and Eukaryotic Cells

Main differences between prokaryotic cells (of bacteria, blue green algae and mycoplasma) and eukaryotic cells (of protozoa, fungi, plants and animals) have been tabulated in Table 55.2.

Table 55.2 Differences between prokaryotic and eukaryotic cells.

Feature	*Prokaryotic cell*	*Eukaryotic cell*
1. Size	Mostly 1–10 mm	Mostly 10–100 μm
2. Multicellular forms	Rare	Common, with extensive tissue formation
3. Cell wall	Present in most but not in all cells	Present in plant and fungal cells only
4. Plasma membrane	Present	Present
5. Nucleus	Absent	Present
6. Nuclear membranes	Absent	Present
7. Chromatin with histone	Absent	Present
8. Genetic material	Circular or linear, double-stranded DNA: genes are not interrupted by intron*	Linear double-stranded DNA: genes frequently interrupted by intron sequences, especially in higher eukaryotes

*Intron is an intervening sequence of nucleotides in DNA, located within a gene that is not included in the mature mRNA.

Feature	Prokaryotic cell	Eukaryotic cell
9. Nucleoli and mitotic apparatus	Absent	Present
10. Plasmids	Commonly present	Rare
11. Cellular organelles:		
(*i*) Mitochondria	Absent	Present
(*ii*) Endoplasmic reticulum	Absent	Present
(*iii*) Vacuoles	Absent	Present
(*iv*) Lysosomes	Absent	Present
(*v*) Chloroplasts	Absent	Present (only in plants)
(*vi*) Centrioles	Absent	Present (absent in higher plants)
(*vii*) Ribosomes	Present (70S)	Present (80S)
(*viii*) Microtubules	Absent	Present
(*ix*) Flagellae	Simple structure composed of the protein flagellin	Complex 9 + 2 structure of tubulin and other proteins
12. Respiration	Many strict anaerobes (oxygen fatal)	All aerobic, but some facultative anaerobes by secondary modifications
13. Metabolic patterns	Great variations	All share cytochrome electron transport chains, Krebs cycle oxidation, Embden-Meyerhof glucose metabolism or glysolysis
14. Photosynthetic enzymes	Bound to plasma membrane as composite chromatophores	Enzymes packaged in plastids bound by membrane
15. Sexual system	Rare: if present one way (and usually partial); transfer of DNA from donor to recipient cell occurs	Both sexes involved in sexual participation and entire genomes transferred; alternation of haploid and diploid generations is also evident

Differences between Animal and Plant Cells

Differences between typical animal and plant cells have been tabulated in Table 55.3.

Table 55.3 Differences between animal and plant cells.

Animal cell	Plant cell
1. Animal cell are generally small in size.	1. Plant cells are larger than animal cells.
2. Cell wall is absent.	2. The plasma membrane of plant cells is surrounded by a rigid cell wall of cellulose.
3. Except the protozoan *Euglena* no animal cell possesses plastids.	3. Plastids are present.
4. Vacuoles in animal cells are many and small.	4. Most mature plant cells have a large central sap vacuole.
5. Animal cells have a single highly complex and prominent Golgi apparatus.	5. Plant cells have many simpler units of Golgi apparatus, called dictyosomes.
6. Animal cells have centrosome and centrioles.	6. Plant cells lack centrosome and centrioles.

QUESTIONS

Long Answer Question

1. Describe the structure or a prokaryotic cell.
2. Describe generalized structure of bacterial cell. *(CCS 1996)*
3. Describe the ultrastructure of *E.coli*.
4. What are the chief characteristics of a cyanobacteria? *(CCS 1996)*
5. Prokaryotes are considered to be ancestors of eukaryotes. Discuss. *(B. U. Coimbatore 1997)*
6. Define cell. Give electron microscopic structure of PPLO cell. Give a well labeled diagram of an eukaryotic animal cell. *(CCS 1991)*
7. Define the cell. What is an eukaryotic cell? Describe the general shape and size, number of eukaryotic cells in the living organism.
8. Give an account of the structure of an animal cell and state the functions which are attributed to its component parts.*(Delhi 1989; Lucknow 1989, 1991, 1992; CCS 1989, 1999; Vikram 1989, 1992)*
9. Give an illustrated account of the functional anatomy of animal cell. *(Kanpur 1990, 1993)*
10. Enumerate the various organelles found in a typical animal cell and briefly describe the functions of Golgi body and ribosomes. *(Gorakhpur 1998)*
11. Give an illustrated account of ultrastructure of animal cell. *(Garhwal 1989; Utkal 1989; Rohelkhand 1989; Bundelkhand 1991, 1992; Jiwaji 1992; Kanpur 1996; CCS 1997)*
12. Give a detailed labelled diagram of electron microscopic structure of animal cell along with important cell organelles. *(Punjab (G) 1990; Kanpur 1994; Lucknow 1994; Gorakhpur 1996)*
13. Give distinguishing characteristics of an eukaryotic plant cell.
14. Describe structure of a typical animal cell and mention the differences between a plant and animal cell. *(Kanpur 1998)*

Short Answer Questions

1. Summarize important diagnostic features of prokaryotes.
2. Write a short note on PPLOs. *(CCS 1991)*
3. Write a short note on 'Prokaryotic cell'. *(Jiwaji 1992: Guwahati (G) 1996)*
4. Mention the functions of the following cellular components:
   (*a*) Cell wall
   (*b*) Cell membrane
   (*c*) Golgi apparatus
   (*d*) Lysosomes
   (*e*) Ribosomes
   (*f*) Mitochondria
   (*g*) Centrioles
   (*h*) Plastids
   (*i*) Chromatin
5. Discuss the factors which control the shape of the cells.
6. What are the conditions which set upper limit in cell size?
7. What are the differences between prokaryotic and eukaryotic cells? *(Nagarjuna 1994; Guwahati (G) 1994; CCS 1995, 1997; NEHU (Major) 1996; Kanpur 1996)*
8. Draw a neat and labeled diagram of animal cell. Explain differences between prokaryotic and eukaryotic cells. *(CCS 1992)*
9. Differentiate between 'Animal cell and Plant cell'. *(Lucknow 1991)*
10. Draw neat and labeled diagram of the following:
    1. Prokaryotic cell *(CCS 1999)*
11. Give the structure of Eukaryotic cell. *(H.N.B., Garwal 2008)*

Very Short Answer Questions

1. Name the smallest known cell. Give its size.

Ans. Smallest cells are bacterium *Dialister pneumogintes* and *Mycoplasma gallisepticum.*

2. Enumerate cell organelles that fall under following categories:

(*i*) Cytomembranes

Ans. Endoplasmic reticulum, Goigi and nuclear envelope.

(*ii*) Membrane bound

Ans. Lysosomes, mitochondria and chloroplasts.

(*iii*) Membraneless

Ans. Ribosomes and centrioles.

3. What will happen if all the mitochondria of a cell are destroyed?

Ans. There will be no metabolic energy for metabolic activities and cell will die.

4. How can you change an animal cell into a plant cell?

Ans. By introducing plastids, vacuoles and cell wall.

5. Name the organelle where photosynthesis occurs in the cell.

Ans. Chloroplast.

6. A cell wall is a part of all plant cells. Is it living or nonliving?

Ans. Nonliving.

7. Apart from nucleus, which two other cell organelles have their independent DNA?

Ans. Mitochondria and chloroplasts.

Fill in the Blanks

1. The longest animal cell is..............

2. Vacuole is generally present in .... cells but absent in .... cells.

3. .... are the fundamental units of life.

4. .... are not surrounded by cell membrane.

5. The nuclear region of prokaryotic cells is called ....

Matching of Columns

Match appropriately the items in column A with those of column B

Column A	Column B
1. Mitochondria	(*a*) Animal cell
2. Centriole	(*b*) Bacteria
3. Nucleus	(*c*) Eukaryotic cell
4. Chromosome without histone	(*d*) Protoplasm
5. The nerve cell of human beings	(*e*) Hyaloplasm
6. The ground substance or cytoplasm	(*f*) Control room of cell
7. DNA	(*g*) 90 cm
8. Colloidal substance	(*h*) Hereditary material

Multiplc Choice Questions

1. Which of the following structures is present only in animal cell?

(*a*) cell membrane (*b*) lysosomes
(*c*) centrioles (*d*) ribosomes

2. Which of the following structures is associated with the formation of aster during nuclear division?

(*a*) endoplasmic reticulum (*b*) centrioles
(*c*) sphaerosome (*d*) ribosome

3. Which of the following statement is not correct?
(*a*) ribosomes are power house of cell (*b*) lysosomes are called suicide bags
(*c*) viruses lack their own enzyme system (*d*) bacteria lack a definite cell wall

4. Plant cells differ from animal cells in the absence of
(*a*) endoplasmic reticulum (*b*) ribosomes
(*c*) mitochondria (*d*) centrioles

5. A well organized nucleus with a distinct nuclear membrane is absent in
(*a*) bacterial cells (*b*) eukaryotic cells
(*c*) protozoan cells (*d*) fungal cells

6. Eukaryotic cells differ from prokaryotic cells in having
(*a*) endoplasmic reticulum (*b*) ribosomes
(*c*) true nucleus (*d*) cytoplasm

7. Which of the cell organelle is considered to be rich in catabolic enzymes?
(*a*) lysosomes (*b*) Golgi apparatus
(*c*) mitochondria (*d*) endoplasmic reticulum

8. Lysosomes are called suicide bags because of
(*a*) hydrolytic enzymes (*b*) phagocytic activity
(*c*) proteolytic activity (*d*) respiratory enzymes

9. Cellular organelle containing hydrolytic enzymes are called
(*a*) microsomes (*b*) lysosomes
(*c*) oxysomes (*d*) ribosomes

10. Single envelope system is characteristic of *(NECU (Major) 1996)*
(*a*) prokaryotic cell (*b*) eukaryotic cell
(*c*) none (*d*) both

11. Cell boundary of animal cell is *(NEHU (Major) 1996)*
(*a*) cell wall (*b*) plasma membrane
(*c*) both (*d*) none of these

12. The granular structures found attached with endoplasmic reticulum are *(NEHU (Major) 1996)*
(*a*) ribosomes (*b*) lysosome
(*c*) desmosome (*d*) kinetosome

13. The power house of cell is *(NEHU (Major) 1996)*
(*a*) Golgi complex (*b*) histone
(*c*) mitochondria (*d*) mitochondria

14. Prokaryates and eukaryotes have the common *(B.U. Coimbatore 1997)*
(*a*) mitotic apparatus (*b*) histone
(*c*) genetic code (*d*) mitochondria

ANSWERS

Fill in the Blanks

1. Nerve cell; **2.** Plant, animal; **3.** Cells; **4.** Ribosomes;
5. Nucleoid.

Matching of Columns

1. (*c*) **2.** (*a*) **3.** (*f*) **4.** (*b*)
5. (*g*) **6.** (*e*) **7.** (*h*) **8.** (*d*).

Multiple Choice Questions

1. (*c*) **2.** (*b*) **3.** (*a*) **4.** (*d*) **5.** (*a*) **6.** (*c*)
7. (*a*) **8.** (*b*) **9.** (*b*) **10.** (*a*) **11.** (*b*) **12.** (*a*)
13. (*c*) **14.** (*c*).

56

Viruses

Viruses (L., venoum or poisonous fluid) are very small submicroscopic biological entities which though lack cellular organization (*viz.,* plasma membrane and metabolic machinery) possess their own genetic material, genetically determined macromolecular organization and characteristic mode of inheritance. For their multiplication, they essentially require the presence of some host cell, *i.e.,* they are obligate cellular parasites of either bacteria, plants or animals.

56.1 STRUCTURE

Viruses are quite a varied group (Fig. 56.1). They range in between 30 to 300 nm or 300 to 3000 Å in size, so they can be observed only by electron microscopy and X-ray crystallography. They have a regular geometrical and macromolecular organization. Basically an infectious virus particle (called **virion**) is composed of a **core** of only one type of nucleic acid (DNA or RNA) which is wrapped in a protective coat of protein, called **capsid**. The capsid consists of numerous **capsomeres**, each having a few **monomers** or **structural units**. Each structural unit is made up of one or more polypeptide chains. The capsomerers are of different shapes such as hollow prism, hexagonal, pentagonal, lobular or any other shape. The specific arrangement of capsomeres in the capsid determines the shape of a virion. Viruses have the following three different types of symmetry:

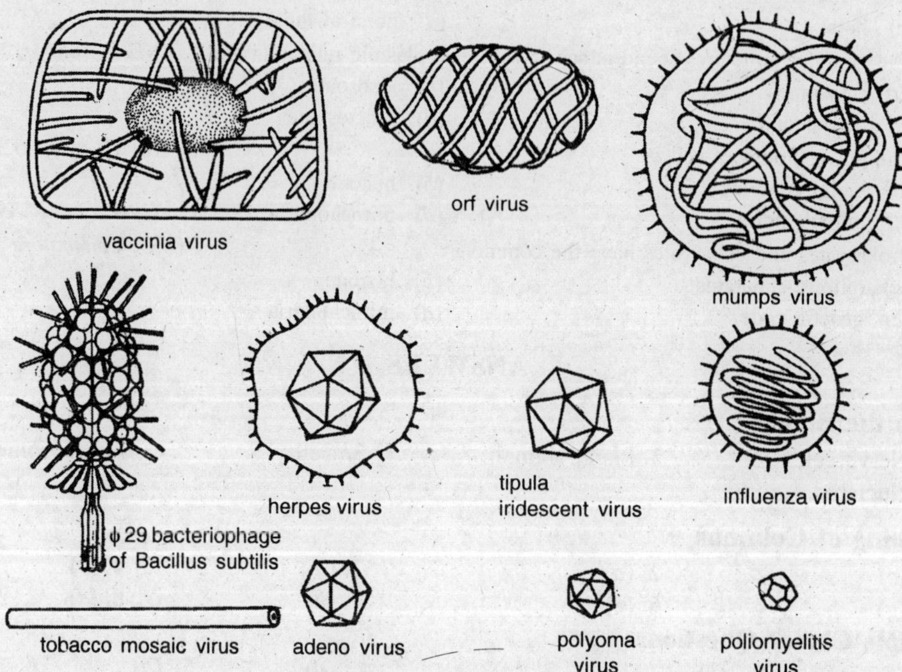

Fig. 56.1. Different kinds of viruses.

1. Icosahedral symmetry. Many viruses have spherical, cubical or polygonal shape which is basically **icosahedral** or 20-sided. Icosahedral symmetry depends on the fact that the assembly of the capsomeres causes the capsid of the virus to be at a state of minimum energy (**Caspar** and **Klug**, 1962). An icosahedral capsid comprises both **pentameres** (*i.e.*, capsomeres containing 5 structural units) and **hexameres** (*i.e.*, capsomeres having 6 structural units). In an icosahedral virus the minimum number of capsomeres is 12 or its multiple such as 32, 42, 72, 92, 162, 252, 362, 492, 642 and 812.

For example, the total number of capsomeres of different icosahedral viruses are:

(*i*) Bacteriophage ϕ (phi) × 174 = 12 pentameres;

(*ii*) Turnip yellow mosaic virus or TYMV = 32 capsomeres;

(*iii*) Poliovirus = 32 capsomeres;

(*iv*) Polyoma virus and papilloma virus = 72 capsomeres;

(*v*) Reovirus = 92 capsomeres;

(*vi*) Herpes virus = 162 capsomeres;

(*vii*) Adenovirus = 252 capsomeres; and

(*viii*) Tipula iridescent virus = 812 capsomeres.

In all of these icosahedral viruses, only 12 capsomeres are pentameres, occupying 12 corners of five-fold symmetry, while the rest are hexameres (Fig. 56.2). Since a polyhedron of 20-sided icosahedron basically has triangular faces, it is also known as **deltahedron.**

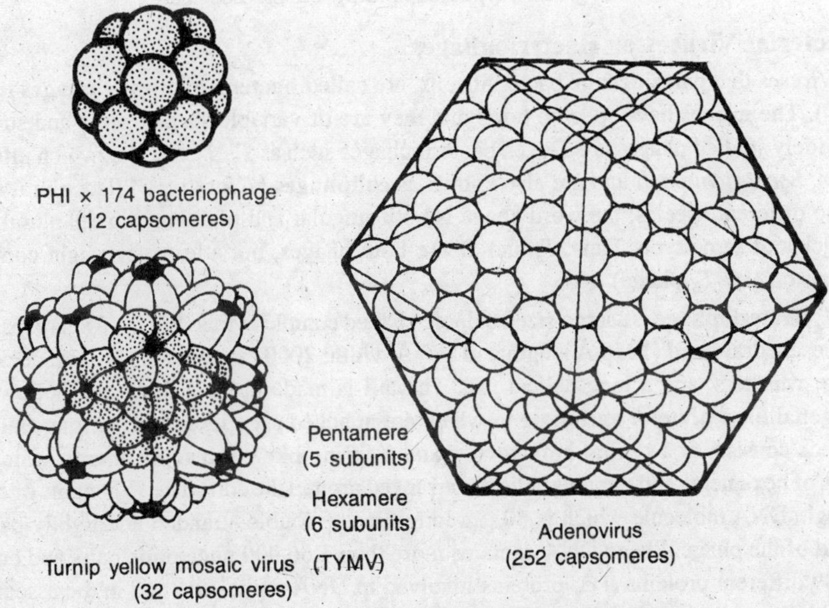

Fig. 56.2. Certain polyhedral (icosahedral) viruses.

2. Helical or cylindrical symmetry. The rod-shaped helical capsid of viruses such as tobacco mosaic virus (TMV), bacteriophage M13 and influenza virus, consists of numerous identical capsomeres arranged into a helix because they are thicker at one end than the other.

3. Complex symmetry. Viruses with complex shaped capsids are of two shapes; those without identifiable capsids (*e.g.*, pox viruses such as vaccinia, cowpox, extromelia and orf viruses) and those with tadpole-shaped structures in which each part has different sort of symmetry (*e.g.*, T-even phages of *E.coli;* T_2 phage has an icosahedral head, helical tail sheath, hexagonal end plate and rod-shaped tail fibres). Some viruses such as rabies virus are bullet-shaped.

Some viruses such as herpes virus, influenza virus, mumps virus and Semliki forest virus are surrounded by a 100–150 Å thick spiked membrane. This membrane contains lipid bilayer of plasma membrane from which projects the virus-specific protein molecules or spikes. It is not made by or specified by the virus itself but is derived from the plasma membrane of the host cell (*i.e.,* animal cell).

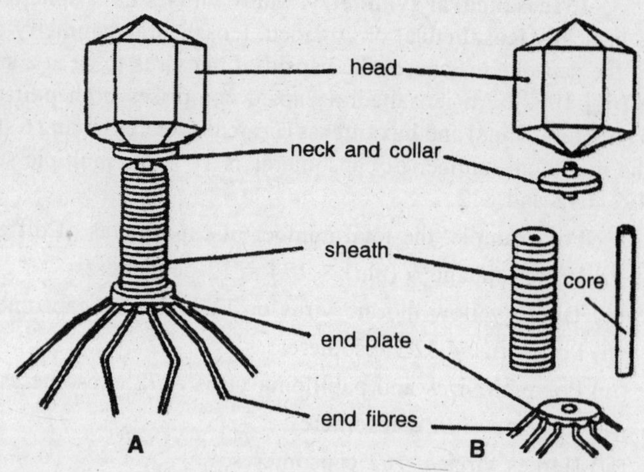

Fig. 56.3. A T_4 bacteriophage. A–External structure; B–Parts.

56.2 TYPES OF VIRUSES

Viruses are of following three types depending on the host cell.

A. Bacterial Viruses or Bacteriophages

Viruses that parasitize the bacterial cells, are called **bacteriophages** or **phages** (phage means 'to eat'). The phages have specific hosts and they are of variable shapes, sizes and structures. The most widely studied phages are *T*-even bacteriophages such as T_2, T_4, T_6, etc., which infect the colon bacillus. *Escherichia coli* and are also known as **coliphages** (*T* for "type". The plural word phages refers to different species; the word phage is both singular and plural and in the plural sense refers to particles of same type. Thus, T_4 and T_7 are both phages, but a test tube might contain either 1 T_4 phage or 100 T_4 phage).

T_4 bacteriophage is a large-sized tadpole-shaped complex virus (Fig. 56.3). Its capsid comprises of an icosahedral **head** (1250 Å length and 850 Å width; 2000 capsomeres), a short neck with **collar** bearing 'whiskers' and a long helical **tail.** The tail is made up of a thick and hollow **mid-piece,** a hexagonal **base plate** or **end plate** to which are attached six **spikes** and six long **tail fibres.** The mid-piece consists of a central hollow **core** and a spring-like **contractile sheath** which comprises 24 rings of hexameres and remains helically arranged around the core. The T_4 genome or chromosome is a single DNA molecule which is 60 μm long, linear, double-stranded and tightly-packed within the head of the phage. Phage DNA contains more than 1,66,000 nucleotide pairs and encodes more than 200 different proteins (*i.e.,* proteins involved in DNA replication and in the assembly of head and tail). For example, T_4 phage DNA codes for at least 30 different enzymes (*e.g.,* helicases, topoisomerases, DNA polymerases, DNA ligases, etc.) all of which ensure rapid replication of phage chromosome in preference to DNA of *E.coli* (host cell). Further, during DNA replication, an unusual nitrogen base, called **5-hydroxymethylcytosine** is incorporated in place of cytosine in the phage DNA. This unusual base makes phage DNA recognisable from that of host DNA and selectively protects it from the, nuclease enzymes. The nucleases are encoded in T_4 phage genome to degrade only the DNA of host cell. Some other phage proteins alter host cell's RNA polymerase enzymes, so that they transcribe different sets of T_4 genes at different stages of viral infection according to the phage's needs.

B. Plant Viruses

The plant viruses parasitize the plant cells and disturb their metabolism and cause severe diseases in them. All plant viruses consist of ribonucleoproteins in their organization. The important plant viruses are tobacco rattle virus (TRV), tobacco mosaic virus (TMV), potato virus, beet yellow virus (BYV), southern bean mosaic virus (SBMV) and turnip yellow viruses (TYV). Among plants, few hundred viral diseases are caused, e.g., mosaic diseases of tobacco, cabbage, cauliflower, groundnut and mustard; black-ring spot of cabbage; leaf roll of tomato; leaf curl of papaya, cotton, bean and soyabean; yellow diseases of carrot, peach; little-leaf of brinjal. These deseases are spread mainly by insects such as aphids, leaf hoppers and beetles.

Tabacco mosaid virus (TMV). TMV is the most extensively studied plant virus. It was discovered by **Iwanowski** (1892) and obtained in a pure state (*i.e.,* in paracrystalline form) by **Stanley** (1935). **Bawden** and **Pirie** (1937) extensively purified TMV and showed it to be a nucleoprotein containing RNA. **H. Fraenkel-Conrat** experimentally demonstrated that RNA is the genetic substance of TMV.

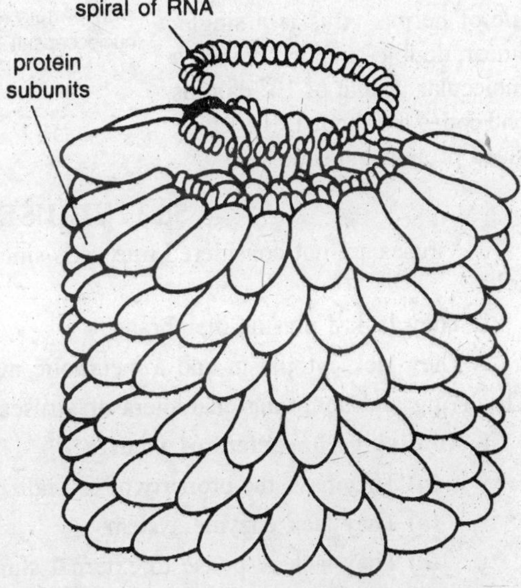

Fig. 56.4. Molecular organization of the tobacco mosaic virus (TMV).

TMV is a rod-shaped, helically symmetrical RNA virus (Fig. 56.4). Each virus particle is elongated, cigarette-like in shape having the length of 3000 Å (300 nm) and diameter of 160 Å (16 nm). In each rod of TMV, there are about 2130 identical elliptical protein subunits or capsomeres. The capsomeres are closely packed and arranged in a helical manner around the RNA helix, forming a hollow cylinder. Thus, there is a hollow core (axial hole) of about 40 Å (4nm) diameter which runs the entire length of the rod and contains the RNA molecule. The RNA molecule does not occupy the hole but is deeply embedded in the capsomeres. RNA of TMV is a single-stranded molecule consisting of 6500 nucleotides and is in the form of a long helix extending the whole length of viral particle. Lastly, there are about 16 capsomeres in each helical turn. Each capsomere contains about 158 amino acids and has a molecular weight of 18000 daltons. The whole TMV capsid has all amino acids found in other plant proteins.

C. Animal viruses. The animal viruses infect the animal cells and cause different fatal diseases in animals including humans. Generally, they have a polyhedron or spherical shape and genetic material in the form of DNA or RNA. The protein coat or capsid of animal viruses is surrounded by an envelope.

Certain common viral infections of human beings are: common cold, influenza, mumps, measles, rubella (German measles), chicken pox, small pox, polio, viral hepatitis, herpes simplex, viral encephalitis, fever blisters, warts and some types of cancer. Among livestock and fowl, viruses cause encephalitis, foot and mouth disease, fowl plague, Newcastle disease, pseudorabis, hog cholera and a variety of warts and other tumors. A virus usually displays some specificity for a particular animal group.

Poliomyelitis is a most extensively studied RNA-containing animal virus. The polio virus has comparatively very simple organization. It consists of a protein shell built up of 60 structurally equivalent asymmetric protein subunits of approximately 60 Å diameter, packed together in such a way that they form a spherical shell of about 300 Å in diameter. The shell or capsid encloses a single stranded RNA molecule of 5,200 nucleotides.

Herpes virus is another most extensively studied animal virus. It possesses a DNA containing core embedded in a regular icosahedral **capsid** (162 capso-meres) and an outer **envelope** of lipids, proteins and carbohy-drates. The DNA molecule of herpes virus is a single, linear, double-stranded having a molecular weight of 10^8 daltons and codes for about 100 average sized protein molecules.

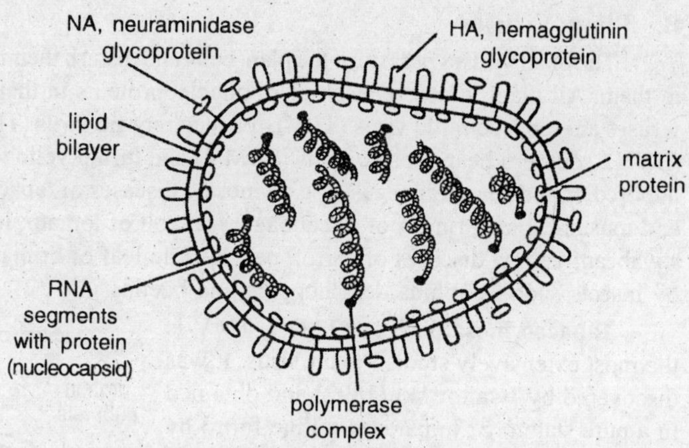

Fig. 56.5. The influenza viron.

56.3 VIRUSES ARE LIVING!

Viruses are not considered true cells since they lack two very important characteristics of a cell:

1. They lack a plasma membrane;
2. They lack cytoplasm and a metabolic machinery to sustain life activities.

The living and nonliving characters of viruses are summarized below:

1. Nonliving characters of viruses

(*a*) They have the property of crystallization like other inorganic and organic compounds.

(*b*) They lack enzyme system.

(*c*) They lack response to external stimuli.

(*d*) They lack growth.

2. Living characters of viruses

(*a*) Viruses have the capacity to multiply (inside a cellular host).

(*b*) They have the property of undergoing mutations.

(*c*) They have the property of recombination and inheritability.

56.4 LIFE CYCLE OF THE BACTERIOPHAGE

Bacteriophages may have the following two types of life cycles:

1. **Lytic cycles**, in which viral infection is followed by **lysis** (bursting and death) of the host cell and release of new infective phages, *e.g.*, virulent phages such as T_4 and all other *T*-even coliphages.
2. **Lysogenic cycles,** in which infection rarely causes lysis, *e.g.*, temperate phages such as P_1 and lambda (λ) phages.

1. Lytic Cycle of a Virulent Phage

Life cycle of a T_4 bacteriophage (Fig. 56.6) involves the following steps:

(*a*) **Attachment** or **adsorption** of phage to bacterial (host) cell.

(*b*) **Injection** or **penetration** of viral genetic material (DNA) into the host cell.

(*c*) **Eclipse period,** during which synthesis of new phage DNA and protein coats takes place.

(*d*) **Assembly** of phage DNA into protein coats.

(*e*) **Lysis** of host cell and release of the infective progeny phages. Such a phage is called **virulent or lytic phage** since it has infectiousness and it causes death of host cell by lysis.

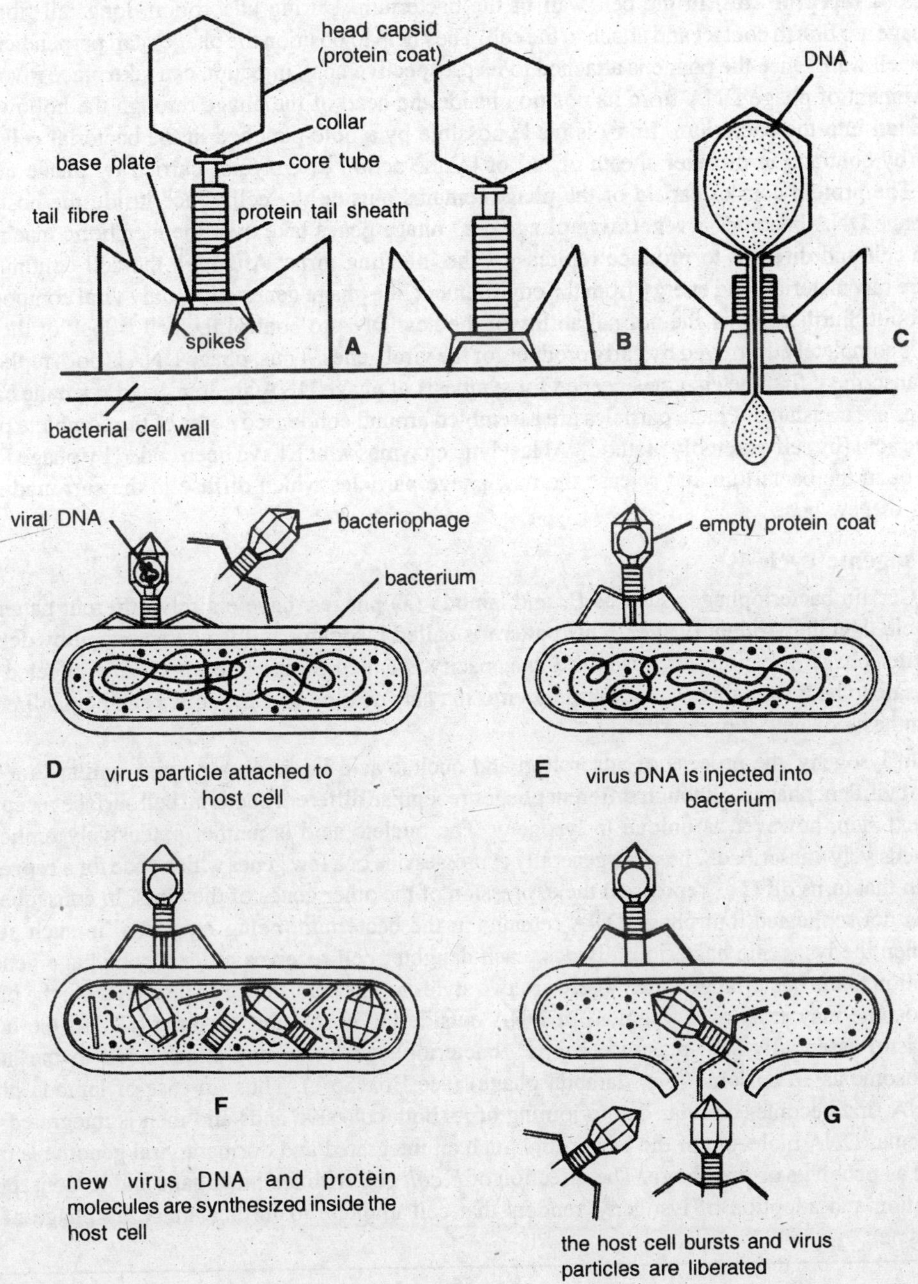

Fig. 56.6. A, B, C—Mode of attachment of a *T*-even phage (bacteriophage) on a bacterial cell wall and injection of DNA into the bacterium (host cell); D, E, F and G—Steps of viral reproduction inside the host cell.

The adsorption of the phage to its host is made possible by a reaction of chemical groups on the two during a random collision. Reactive groups (called **adsorption protein** or **pilot protein; Kornberg,** 1974) at the end of the tail of the phage can join with a complementary set of chemical groups (a **receptor site**) in the cell wall of the bacterium. During adsorption, long tail fibres of the phage are first to contact and attach to the cell. They help to position the phage's tail perpendicularly to the cell wall. Once the phage is attached to its prospective host, injection can take place involving a movement of phage DNA from its position inside the head of the phage through the hollow core of the tail into the bacterium. Entry is made possible by a hole punched in the bacterial cell wall, either by contraction of outer sheath of tail or by the action of enzymes carried by phage tail, or both. The protein coat or capsid of the phage remains outside the cell. Once inside the host cell, the phage DNA becomes a **vegetative phage,** *i.e.*, phage genes take over the metabolic machinery of the cell and direct it to produce replicas of the infecting virus. Although the cell continues to procure raw materials and energy from the environment, the phage genes allow only viral components to be built. Further, either the normal ability of the host DNA to control the cell is lost, or the host DNA is completely destroyed by early products of the viral genes. Thus, phage DNA is both replicated and transcribed; first the enzymes needed for synthesis of phage DNA are translated, then the capsid proteins are translated. Phage particles are assembled around condensed cores of the complete phage nucleic acid (by self-assembly method). At last lytic enzymes which have been coded by phage DNA, break open the bacterium and release the new phage particles which diffuse in the surrounding in search of new host.

2. Lysogenic cycle

Certain bacteriophages such as P_1 and lambda (λ) phages, have entirely different pattern of life cycle than the virulent phages. This pattern is called **lysogeny** and is characterized by delayed lysis after phage infection. A virus with this capacity is called **temperate virus.** The infected host cell is said to be **lysogenic** because dormant virus may at any time become active and begin directing the synthesis of new virus particles.

In lysogeny, the process of adsorption and nucleic acid injection are quite similar to a lytic cycle of virulent phages, although different phages recognise different bacterial cell surface receptors. The next step, however, is unique to lysogeny. The nucleic acid is neither extensively replicated nor extensively transcribed. The virus generally expresses one or a few genes which code for a **repressor protein** that turns off (*i.e.*, represses) the expression of the other genes of the virus. In consequence, virus is not replicated, but phage DNA remains in the bacterium, being replicated in such a way that when the lysogenic bacterium divides, each daughter cell receives at least one phage genome in addition to the bacterial genome. There are two styles to this persistence of phage DNA: the phage chromosome may exist as a fragment of DNA outside the host's chromosome (*i.e.*, in the host's cytoplasm) essentially as a plasmid (*e.g.*, P_1 bacteriophage) or it may attach itself to the host's chromosome as an episome (*e.g.*, lambda phage) (see Box 56.1). Thus, in case of lambda phage the DNA first becomes circular due to joining of its both cohesive ends and then is integrated into the circular DNA molecule of the bacterium. Such an integrated and dormant viral genome is often termed as **provirus** or **prophage.** The infection of *E.coli* cell with lambda phage and its consequent integration and adoption of lysogeny, renders that cell immune to further attack by phage of the same type.

Box 56.1.

Episome is an extra-chromosomal, circular, transposable, closed DNA molecule which can exist either integrated into the bacterial chromosome or separately and autonomously in the cytoplasm. **Plasmid** is that bit of autonomous genetic material of bacteria (*i.e.*, circular DNA) that exists only extra-chromosomally and cannot be integrated into the bacterial chromosome (DNA). Now only the term plasmid is used for all kinds of extra-chromosomal autonomous transposable circular DNA fragments.

The calm lysogenic period is ended by some type of shock (*e.g.*, temperature changes, UV irradiation or conjugation of a lysogenic bacterium with a non-lysogenic bacterium) to the lysogenic culture. A shock evidently inactivates the repressor of the phage so that all the phage genes can be expressed. Then the lysogenic bacterium is ruined, for the phage DNA that the host bacterium is harbouring enters the lytic phase. It replicates, transcribes, translates, assembles viron particles and lyses the bacterium (Fig. 56.7).

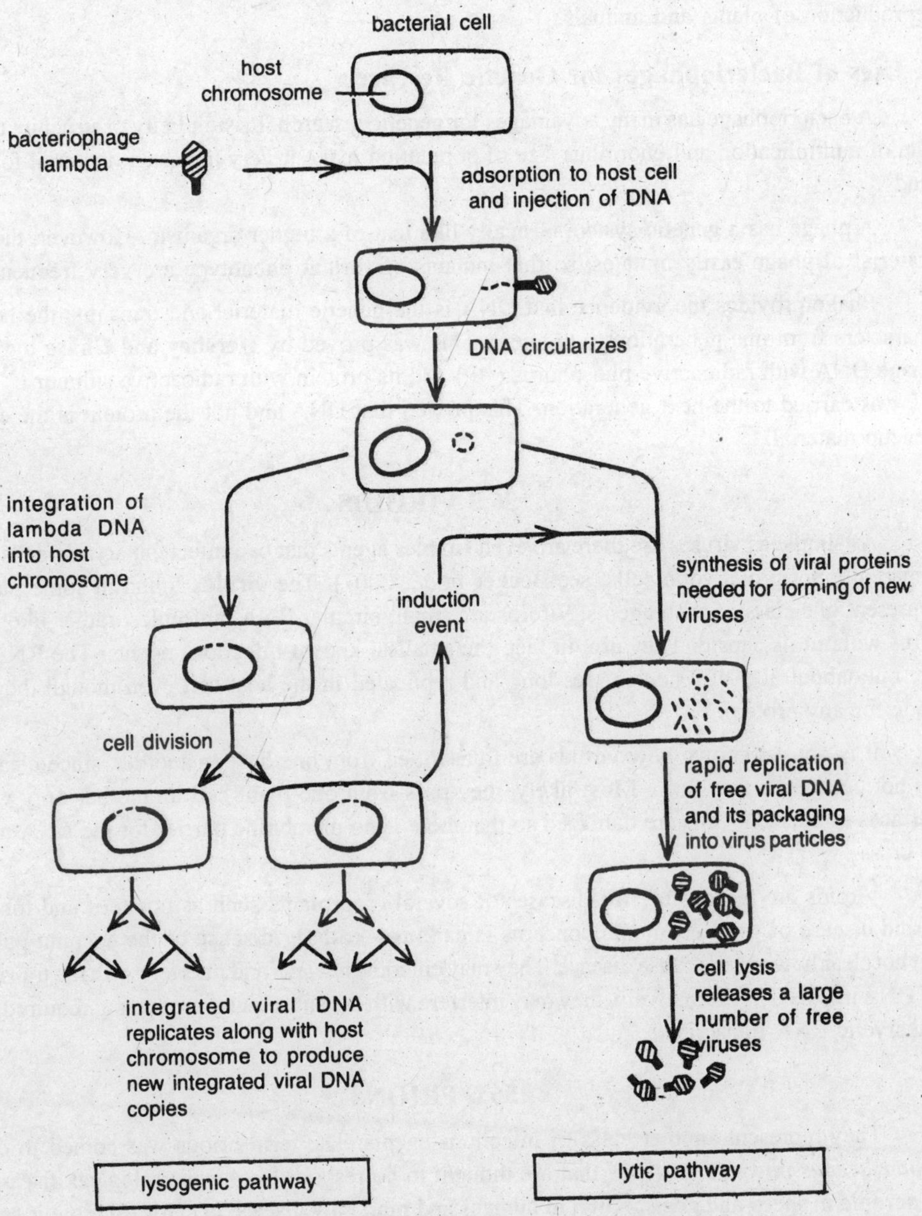

Fig. 56.7. The life cycle of bacteriophage lambda. Its double stranded DNA can exist in both linear and circular forms. The lambda phage can multiply by either a lytic or a lysogenic pathway in the *E.coli*.

3. Recombination in Phages

Recombination is any process generating new combinations of pre-existing genetic material. Recombination in phages was described by **Max** and **Mary Delbruck** (1894). When two different types of phages T_2 and T_4 are placed in a culture of bacteria which is sensitive to both, the new generation of phages consists of the two parental types. But some of them were found to have combination of T_2 and T_4 genes. The recombination in bacteriophages can be equated to the sexual reproduction of plants and animals.

4. Uses of Bacteriophages for Genetic Research

A bacteriophage has many advantages for genetic research. Its simplicity of structure and rapid rate of multiplication and enormous size of population make it very important material for genetic study.

A phage has a genetic system basically like that of a higher organism. However, the genetic material of phage easily mutates, so that mutants of variant phenotype are very frequent.

Phage provides the evidence that DNA is the genetic material and transmits the hereditary characters from one generation to the next. This was proved by **Hershey** and **Chase** by labelling phage DNA with radioactive phosphorus (^{32}P) and its protein with radioactive sulphur (^{53}S). Only ^{32}P was carried to the-next generation. This proved that DNA and not the protein is the carrier of genetic material.

56.5 VIROIDS

As simple as viruses are there are even simpler agents that can infect enkaryotic cells (though apparently not prokaryotic cells; see **Becker** *et al.*, 2007). The **viroids** found in some plant cells represent one class of such agents. Viroids are small, circular RNA molecules, rather like an RNA virus without its capsid. They are, in fact, the smallest known infections agents. The RNA circles are only about 300–400 nucleotides long and replicated in the host cell even though they do not code for any protein.

It is not yet known how viroids are transmitted from one host to another, since clearly they do not occur in a free form. Most likely, they pass from one plant cell to another only when the surfaces of adjacent cells are damaged so that there is no membrane barrier for the RNA molecule to cross.

Viroids are responsible for diseases of several crop plants such as potatoes and tobacco. A viroid disease of great economic concerns is **cadang—cadang disease** of the coconut palm. It is yet not clear how viroids cause disease. They may enter the nucleus and interfere with the transcription of DNA into RNA. Alternatively, they may interfere with the subsequent processing required of most enkaryotic RNA transcripts.

56.6 PRIONS

They represent another class of infections agents. The term prions was coined to describe *proteinaceous infective particles* that are thought to be responsible for neurological diseases such as **scrapie** in sheep and goats, **Kuru** in humans and **mad cow** disease in cattle. Scrapie is so named because infected animals rub incessantly against trees or other objects, scraping off most of their wool in the process. Kuru is a degenerative disease of the central nervous system originally rcported among native peoples in New Guinea patients with this or other prion-based disease suffer initially from mild physical weakness and dementia, but the effects slowly become more severe, and the diseases are eventually fatal. Scrapie like brain-wasting disorder occurs in case of mad cow disease. **Stanley Prusiner**, who received a Nobel prize in 1997 for his pioneering work in this field, has proposed that these diseases are transmitted by infectious protein-containing particles called **prions.** Because prions do not appear to contain DNA or RNA, **Prusiner** has formulated a unique theory to explain how prions might transmit disease by causing the infections spread of abnormal **protein folding**. According to this theory, a prion protein (designated Pr Psc) is simply a misfolded version of a normal cellular protein (designated Pr Pc). When the misfolded Pr Psc encounters a normal Pr Pc peptide chain in the process of folding, it causes the normal polypeptide to hold improperly. The resulting, abnormally folded protein triggers extensive nerve cell damage in the brain, leading to uncontrolled muscle movements and eventual death. Both the normal and variant forms of prion protein are found on the surfaces of neurons, suggesting that the protein may somehow affect the receptors that detect nerve signals.

QUESTIONS

Long Answer Questions

1. Describe general organisation of viruses. Why are viruses considered both living as well as non-living?

Short Answer Questions

1. Why viruses are not true cells? *(CCS 1995, 1996)*
2. Write short notes on the following:
   (*i*) Bacteriophage
   (*ii*) Reproduction in bacteriophage
3. Draw a neat and labelled diagram of the following:
   (*i*) T_4 bacteriophage

Very Short Answer Questions

1. Define the virus.

   Ans. Virus is a subcellular parasite composed of a protein coat and DNA or RNA incapable of independent existence; it invades and infects cells and redirects the host cell's synthetic machinery toward the production of more virus.

2. What is viroid?

   Ans. Viroid is a small, circular RNA molecule that can infect and replicate in host cells even though it does not code for any protein (see **Becker** *et al.*, 2007).

Multiple Choice Questions

1. AIDS causing factors are associated with

   (*a*) protozoa (*b*) DNA virus
   (*c*) bacteria (*d*) RNA virus

2. Viruses are living because they

   (*a*) multiply in the host cells (*b*) carry anaerobic respiration
   (*c*) carry metabolic activity (*d*) cause infection

3. Virus parasite in bacteria is

(*a*) bacteriophage (*b*) viroid

(*c*) cyanophage (*d*) mycoplasma

4. Number of capsomeres in TMV is

(*a*) 231 (*b*) 1230

(*c*) 2130 (*d*) 3120

5. Polio virus has

(*a*) ss DNA (*b*) ds DNA

(*c*) ss RNA (*d*) ds RNA

6. T_2 has

(*a*) ds DNA (*b*) ss RNA

(*c*) ds RNA (*d*) ss DNA

7. When a viral DNA integrates with a host DNA, then the process is referred to as

(*a*) lysis (*b*) prophase

(*c*) lysogeny (*d*) none of these

ANSWERS

Multiple Choice Questions

1. (*d*) **2.** (*a*) **3.** (*a*) **4.** (*c*)

5. (*c*) **6.** (*a*) **7.** (*c*).

57

The Macromolecules of the Cell (Cytosol and Biomolecules)

Within the cells of any organism, the living substance, or **protoplasm,** is itself comprised of a multitude of non-living constituents: proteins, nucleic acids, fats (lipids), carbohydrates, vitamins, minerals, waste metabolites, crystalline aggregates, pigments, and many others, all of which are composed of molecules and their constituent atoms. The *protoplasm is alive because of the highly complex organization of these non-living substances and the way they interact with one another.*

57.1 PHYSICAL NATURE OF CYTOSOL

The fluid and soluble portion of the cytoplasm that exists outside the organelles is called **cytoplasmic matrix** or **cytosol.** The cytosol (cytoplasmic matrix) is a colourless or grayish, translucent, viscid, gelatinous or jelly-like **colloidal** substance. It is heavier than water and capable of flowing. In past, there has been a lot of controversy about the physical nature of the matrix. Different workers advanced different theories about the physical characteristics of the matrix. Their theories can be represented as follows:

1. Reticular theory suggests that the matrix is composed of reticulum of fibres or particles in the ground substances (Fig. 57.1A)

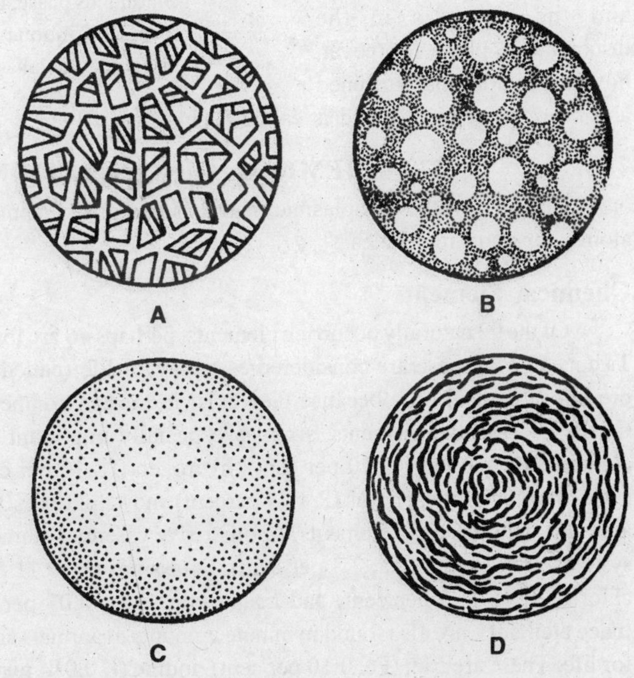

Fig. 57.1. Physical appearance of protoplasm. A—Reticular, B—Alveolar, C—Granular, and D—Fibrillar.

2. Alveolar theory was proposed by **Butschili** in 1892 and according to it, the matrix consists of many suspended droplets or alveoli or minute bubbles resembling the foams of emulsion (Fig. 57.1B).

3. Granular theory was propounded by **Altmann** in 1893. This theory supports the view that the matrix contains many granules of smaller and larger size arranged differently. These granules were known as bioplasts (Fig. 57.1C).

4. Fibrillar theory was proposed by **Fleming** and it holds that the matrix is fibrillar in nature (Fig. 57.1D).

5. Colloidal theory has been forwarded recently after the electron microscopical observations of the matrix. According to this concept, the matrix is partly a true **solution**, partly a **colloidal system.**

Phase Reversal

Cytosol (cyto-plasmic matrix) like many colloidal systems, shows the property of **phase reversal**. For example, gelatin particles (discontinuous phase) are dispersed through water (continuous phase) in a thin consistency that is freely shakable (Fig. 57.2 A). Such a condition is called a **sol**. When the solution cools, gelatin now becomes the con-tinuous phase and the water is in the discontinuous phase. Moreover, now the solution has stiffened and becomes semisolid and is called a **gel**.

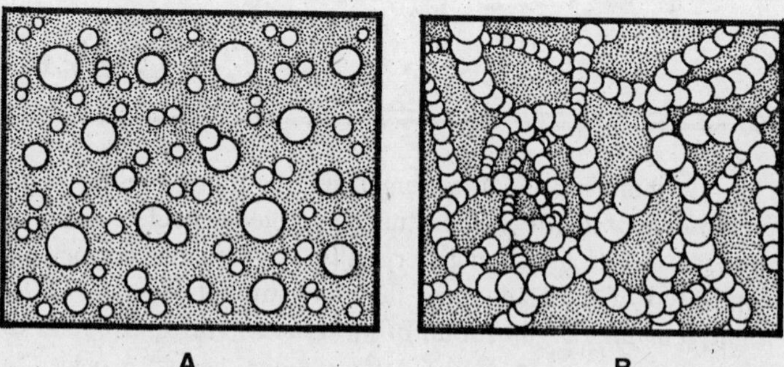

Fig. 57.2. Sol and gel state of the cytosol. A—Sol condition in which gelatin particles are the discontinuous phase, water the continuous phase; B—Gel condition in which gelatin particles form continuous phase (net work), enclosing water as discontinuous phase.

57.2 CHEMICAL ORGANIZATION OF CYTOSOL

Chemically, the cytoplasmic matrix is composed of many chemical elements in the form of atoms, ions and molecules.

Chemical Elements

Of the 92 naturally occurring elements, perhaps 46 are found in the cytosol (cytoplasmic matrix). Twenty four of these are considered essential for life (called **essential elements**), while others are present in cytosol only because they exist in the environment with which the organism interacts. Of the 24 essential elements, six play especially important roles in living systems. These **major elements** are carbon (C, 20 per cent), hydrogen (H, 10 per cent), nitrogen (N, 3 per cent), oxygen (O, 62 per cent), phosphorus (P, 1.14 percent) and sulphur (S, 0.14 per cent). Most organic moleceules are built with these six elements. Another five essential elements found in less abundance in living systems are calcium (Ca, 2.5 per cent), potassium (K, 0.11 per cent), sodium (Na, 0.10 per cent), chlorine (C1, 0.16 per cent) and magnesium (Mg, 0.07 per cent). Several other elements, called **trace elements,** are also found in minute amounts in animals and plants, but are nevertheless essential for life. These are iron (Fe, 0.10 per cent), iodine (I, 0.014 per cent), molybdenum (Mo), manganese (Mn), Cobalt (Co), zinc (Zn), selenium (Se), copper (Cu), chromium (Cr), tin (Sn), vanadium (V), silicon (Si), nickel (Ni), fluorine (F) and boron (B).

Ions

The cytoplasmic matrix consists of various kinds of ions. The ions are important in maintaining osmotic pressure and acid-base balance in the cells. Retention of ions in the matrix produces an increase in osmotic pressure and, thus, the entrance of water in the cell. The concentration of various ions in the intracellular fluid (matrix) differs from that in the interstitial fluid. For example, in the cell K^+ and Mg^{++} can be high, and Na^+ and Cl^- high outside the cell. In muscle and nerve cells a high order of difference exists between intracellular K^+ and extracellular Na^+. Free calcium ions (Ca^{++}) may occur in cells or circulating blood. Silicon ions occur in the epithelium cells of grasses. The free ions of phosphate (primary, $H_2PO_4^-$ and secondary, HPO_4^-) occur in the matrix and blood. These ions act as a buffering system and tend to stabilize pH of blood and cellular fluids. The ions of different cells also include sulphate (SO_4^-), carbonate (CO_3^-), bicarbonate (HCO_3^-), magnesium (Mg^{++}) and amino acids. Cellular functions of certain ions have been tabulated in Table 57.1.

Table 57.1 Cellular functions of certain ions.

Element	*Ionic form present*	*Functions*
1. Molybdenum	MoO_4^{2-}	Cofactor or activator of certain enzymes (*e.g.*, nitrogen fixation, nucleic acid metabolism, aldehyde oxidation).
2. Cobalt	Co^{2+}	Constituent of vitamin B_{12}.
3. Copper	Cu^+, Cu^{2+}	Constituent of plastocyanin and cofactor of respiratory enzymes.
4. Iodine (Heaviest trace element)	I^-	Constituent of thyroxin, triiodothyronine and other thyroid hormones
5. Boron	BO_3^{3-}, $B_4O_7^{2-}$	Activates arabinose isomerase.
6. Zinc	Zn^{2+}	Cofactor of certain enzymes (*e.g.*, carbonic anhydrase, carboxypeptidase).
7. Manganese	Mn^{2+}	Cofactor of certain enzymes (*e.g.*, several kinases, isocitric decarboxylase).
8. Iron	Fe^{2+}. Fe^{3+}	Constituent of haemoglobin, myoglobin and cytochromes.
9. Magnesium	Mg^{2+}	Constituent of chlorophyll; activates ATPase enzyme.
10. Sulphur	SO_4^{2-}	Constituent of coenzyme A, biotin, thiamine, proteins.
11. Phosphorus	PO_4^{3-}, $H_2PO^-_4$	Constituent of lipids, proteins, nucleic acids, sugar phosphates, nucleoside phosphates.
12. Calcium	Ca^{2+}	Constituent of plant cell walls; matrix component of bone tissue; cofactor of coagulation enzymes.
13. Potassium	K^+	Cofactor for pyruvate kinase and K^+– stimulated ATPase.

Types of Compounds of Cytosol

Chemical compounds are conventionally divided into two groups: **organic** and **inorganic**. Organic compounds form 30 per cent of a typical cell, rest are the inorganic substances such as water and other substances.

Table 57.2 The approximate percentage composition of the human body.

Substance	*Percentage*
1. Water	65
2. Protein	18
3. Fat	10
4. Carbohydrate	5
5. Other organic	1
6. Inorganic	1

I. Inorganic Compounds

The inorganic compounds are those compounds which normally found in the bulk of the physical, non-living universe, such as elements, metals, non-metals, and their compounds such as water, salts and variety of electrolytes and non-electrolytes. In the previous section, we have discussed a lot about the inorganic substances except the water which will be discussed in the following paragraph.

Water. The most abundant inorganic component of the cytosol is the water (the notable exceptions are seeds, bone and enamel). Water constitutes about 65 to 80 per cent of the matrix. In the matrix the water occurs in two forms, *viz.,* **free water** and **bound water**. The 95 per cent of the total cellular water is used by the matrix as the solvent for various inorganic substances and organic compounds and is known as **free water**. The remaining 5 per cent of the total cellular water remains loosely linked with protein molecules by hydrogen bonds or other forces and is known as **bound water**.

The water contents of the cellular matrix of an organism depend directly on the age, habitat and metabolic activities. For instance, the cells of the embryo have 90 to 95 per cent water which decreases progressively in the cells of the adult organism. The cells of lower aquatic animals contain comparative high percentage of the water than the cells of higher terrestrial animals. Further the percentage of water in the matrix also varies from cell to cell according to the rate of the metabolism.

Unique physical properties of water and their biological utility. There are several extraordinary properties of water that make it especially fit for its essential role in the protoplasmic systems (*i.e.,* cytosol or matrix). Some of the unique properties of water are the following:

1. Water as a solvent. Water is most stable yet versatile of all solvents. Water's properties as a solvent for inorganic substances as mineral ions, solids, etc., and organic compounds such as carbohydrates and proteins, depend on water's dipole nature. Because of this polarity, water can bind electrostatically to both positively and negatively charged groups in the protein. Thus, each amino group in a protein molecule is capable of binding 2.6 molecules of water. The solvency is of great biological importance because all the chemical reactions that take place in the cells do so in aqueous solution. The water also forms the good dispersion medium for the colloidal system of the matrix.

2. Water's thermal properties. Water is the only substance that occurs in nature in the three phases of solid, liquid and vapour within the ordinary range of earth's temperatures. Water has a **high specific heat:** it requires 1 calorie (4.185 joules) to elevate the temperature of 1 gram of water by 1°C (such as from 15 to 16°C). Such a high thermal capacity of water has a great moderating effect on environmental temperature changes and is a great protective agent for all life.

Water also has a **high heat of vaporization.** It requires more than 540 calories (2259 joules) to change 1 gram of liquid water into water vapour. Thus, water tends to have a remarkably high boiling point (100°C) for a substance of such low relative molecular mass. Were it not for this lucky accident, it is likely that liquid water would never have existed on earth and would have been lost to outer space. Further, for terrestrial plants and animals, cooling produced by the evaporation of water is an important means of getting rid of excess heat. Moreover, at the other temperature extreme, large amounts of energy (335 joules or 80 cal per gram) must be lost for water to be converted from the liquid to the solid state. This is called **heat of fusion.** Water's melting point being 0°C.

Another important property of water from a biological standpoint is its unique **density behaviour** during change of temperature. Most liquids become continually more dense with cooling. Water, however, reaches its maximum density at 4°C and then becomes lighter with further cooling. Therefore, ice floats rather than settling on the bottom of lakes and ponds. This protects the aquatic life from freezing.

3. Surface tension. Water has a **high surface tension**. This property, caused by the great cohesiveness of water molecules, is important in the maintenance of protoplasmic form and movement. Despite its high surface tension, water has **low viscosity,** a property that favours the movement of blood through minute capillaries and of cytoplasm inside cellular boundaries.

Molecules dissolved in water, lower its surface tension and tend to collect at the interface between its liquid phase and other phases. This may have been important in the development of the plasma membrane, and certainly plays an important role in the movement of molecules across it.

4. Transparency. The water is transparent to light, enabling the specialized photosynthetic organelles, the chloroplast, inside the plant cell to absorb the sunlight for the process of photosynthesis.

II. Organic Compounds (Micromolecules and Macromolecules)

The chemical substances which contain carbon (C) in combination with one or more other elements as hydrogen (H), nitrogen (N), sulphur (S), etc., are called **organic compounds.** The organic compounds usually contain large molecules which are formed by the similar or dissimilar unit structure known as the **monomers.** A monomer (Gr., *mono* = one, *meros* = part) is the simplest unit of the organic molecule which can exist freely. Some organic compounds such as carbohydrates occur in the matrix as the monomers. The monomers usually link with other monomers to form **oligomers** (Gr., *oligo* = few or little, *meros* = part) and **polymers** (Gr., *poly* = many, *meros* = part). The oligomers contain small number of monomers, while the polymers contain large number of monomers. The oligomers and polymers contain large-sized molecules or marcromolecules. When a polymer contains similar kinds of monomers in its macromolecule it is known as **homopolymer** and when the polymer is composed of different kinds of monomers it is known as the **heteropolymer**.

The main organic compounds of the matrix are the carbohydrates, lipids, proteins, vitamins, hormones and nucleotides.

Carbohydrates. The carbohydrates (L., *carbo* = carbon or coal, Gr., *hydro* = water) are the compounds of the carbon, hydrogen and oxygen. They form the main source of the energy of all living beings. Only green part of plants and certain microbes have the power of synthesizing the carbohydrates from the water and CO_2 in the presence of sunlight and chlorophyll by the process of photosynthesis. All the animals, non-green parts of the plants (*viz.*, stem, root, etc.), non-green plants (*e.g.*, fungi), bacteria and viruses depend on green parts of plants for the supply of carbohydrates.

Chemically the carbohydrates are polyhydroxy aldehydes or ketones and they are classified as follows:

A. Monosaccharides (Monomers),

B. Oligosaccharides (Oligomers), and

C. Polysaccharides (Polymers).

A. **Monosaccharides.** The monosaccharides are the simple sugars with the empirical formula $Cn\ (H_2O)n$. They are classified and named according to the number of carbon atoms in their molecules as follows:

(*i*) Trioses contain three carbon atoms in their molecules, *e.g.*, glyceraldehyde and dihydroxy acetone.

(*ii*) **Tetroses** contain four carbon atoms in their molecules, *e.g.*, erythrulose, and erythrose.

(*iii*) **Pentoses** contain five carbon atoms in their molecules, *e.g.*, ribose, ribulose, deoxyribose, arabinose and xylulose.

(*iv*) **Hexoses** contain six carbon atoms in their molecules, *e.g.*, glucose, mannose, fructose and galactose.

(*v*) **Heptoses** contain seven carbon atoms in their molecules, *e.g.*, sedoheptulose.

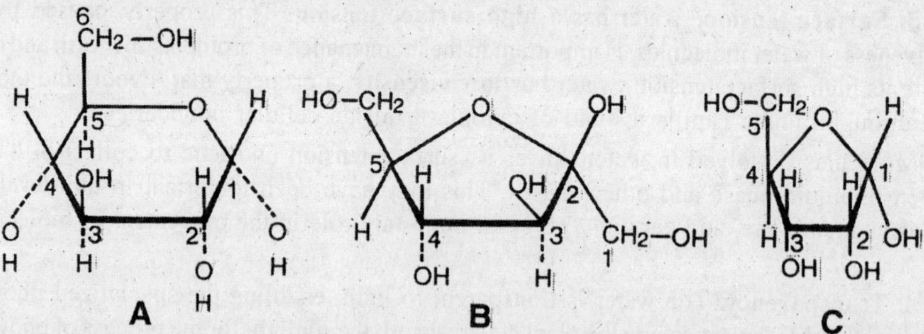

Fig. 57.3. Ring structures of monosaccharides A—Glucose; B—Fructose; C—Ribose.

The monosaccharides are the monomers and cannot split further or hydrolysed into the simpler compounds. The pentoses and hexoses are the most abundantly occurring monosaccharides of the matrix.

The pentose sugar, **ribose** is the important constituent molecule of the ribonucleic acid (RNA) and certain coenzymes as nicotinamide adenine dinucleotide (NAD), NAD phosphate (NADP), adenosine triphosphate (ATP) and coenzyme A (CoA). Another pentose sugar the **deoxyribose** is the important constituent of the deoxyribonucleic acid (DNA). The **ribulose** is a pentose sugar which is necessary for photosynthetic mechanism.

The **glucose**, a hexose sugar, is the primary source of the energy for the cell. The other important hexose sugars of the matrix are the fructose and galactose.

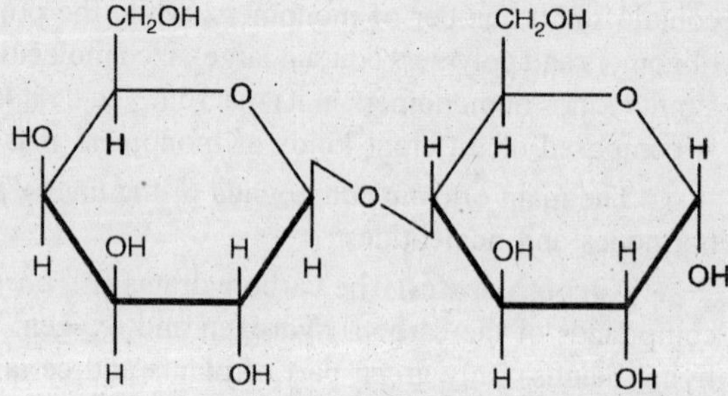

Fig. 57.4. Chemical formula of lactose.

B. Oligosaccharides. The oligosaccharides consist of 2 to 10 monosaccharides (monomers) in their molecules. The monomers remain linked with each other by the **glycosidic bonds** or **linkages**. Certain important oligosaccharides are as follows:

(*i*) **Disaccharides** contain two monomers, *e.g.*, sucrose, maltose, lactose, etc.

(*ii*) **Trisaccharides** contain three monomers, *e.g.*, raffinose, mannotriose, rabinose, rhaminose, gentianose and melezitose.

(*iii*) **Tetrasaccharides** contain four monomers, *e.g.*, stachyose, and scordose.

(*iv*) **Pentasaccharides** contain five monomers, *e.g.*, verbascose.

The most abundant oligosaccharides of the animal and plant cells are the disaccharides such as sucrose, maltose and lactose. The sucrose and maltose occur mainly in the matrix of plant cells, while the lactose occurs exclusively in the matrix of animal cells. The molecules of sucrose are composed of D-glucose and D-fructose. The molecules of maltose consist of two molecules of

D-glucose. The molecules of lactose are composed of two monomers, *viz.*, D-glucose and D-galactose. Like monosaccharides all disaccharides are sweet, soluble in water and crystallizable.

C. Polysaccharides. The polysaccharides are composed of ten to many thousands monosaccharides as the monomers in their macromolecules. Their empirical formula is $(C_6H_{10}O_6)n$. The molecules of the polysaccharides are of colloidal size having high molecular weights. The polysaccharides can be hydrolysed into simple sugars.

Polysaccharides can be divided into two main functional groups: the structural polysaccharides and the nutrient polysaccharides. The **structural polysaccharides** serve primarily as extracellular or intracellular supporting elements. Included in this group are **cellulose** (found in plant cell wall), **mannan** (a homopolymer of mannose found in yeast cell walls), **chitin** (in the exoskeleton of arthropods and the cell walls of most fungi and some green algae), **hyaluronic acid, keratin sulphate** and **chondroitin sulphate** (these three are found in cartilage and other connective tissues) and the **peptidoglycans** (in bacterial cell wall). The **nutrient polysaccharides** serve as reserves of monosaccharides and are in continuous metabolic turnover. Included in this group are **starch** (plant cells and bacteria), **glycogen** (animal cells), **inulin** (plants such as artichokes and dandelions) and **paramylum** (an unbranched nutrient and storage homopolymer of glucose found in certain protozoa, *e.g., Euglena*).

Molecules of some polysaccharides are **unbranched** (*i.e.,* linear) chains whose structure may be ribbon-like or helical (usually a left-handed spiral). Other polysaccharides are **branched** and, like many proteins, assume a globular form.

On the chemical basis, the polysaccharides can be divided into two broad classes: the homopolysaccharides and the heteropolysaccharides.

Homopolysaccharides. The homopolysaccharides contain similar kinds of monosaccharides in their molecules. The most important homopolysaccharides of the matrix are the starch, glycogen, paramylum and cellulose.

Heteropolysaccharides. The polysaccharides which are composed of different kinds of the monosaccharides and amino-nitrogen or sulphuric or phosphoric acids in their molecules are known as heteropolysaccharides. The most important heteropolysaccharides are as follows:

(*a*) Hyaluronic acid, keratin sulphate and chondroitin sulphate. Cartilage tissue contains the related acidic heteropolysaccharides such as hyaluronic acid, keratin sulphate and chondroitin sulphate. **Hyaluronic acid** is an unbranched heteropolysaccharide containing repeating disaccharides of **N-acetylglucosamine** (or D-glucosamine) and glucuronic acid. In addition to cartilage, hyaluronic acid is also found in other connective tissues, in the synovial fluid of joints, in the vitreous humor of the eyes, and also in the capsules that enclose bacteria.

Keratin sulphate, like hyaluronic acid, is a repeating disaccharide forming an unbranched chain. Each disaccharide unit of the polysaccharide consists of **D-galactose** and **sulphated N-acetylglucosamine.** It is found in cartilage and cornea.

Chondroitin sulphate is a repeating disaccharide consisting of alternating **glucuronic acid** and **sulphated N-acetyl galactosamine** residues. It is found in cartilage, bone, skin, notochord, aorta and umbilical cord.

(*b*) Heparin. Heparin is a blood anticoagulant and found in the skin, liver, lung, thymus, spleen and blood. Its molecule contains the repeated disaccharide units, each having **D-glucuronic acid** and **D-glucosamine.**

(*c*) Proteoglycans, glycoproteins and glycolipids. Polysaccharides also occur in covalent combination with proteins and lipids, to form the following three types of molecules:

(*i*) **Proteoglycans.** The molecules of proteoglycans consist of much longer portion of polysaccharide and a small portion of protein. They are also called **mucoproteins.** The proteoglycans are amorphous and form gels which are able to hold large amounts of water.

The **cartilage proteoglycan** is found extracellularly in cartilage and bone. In its molecule, strands of protein, called **core protein,** extend radially from a long, central hyaluronic acid molecule. In each core protein strand, three carbohydrate bearing regions may be identified. The first region contains numerous oligosaccharides, the second region contains keratin sulphate chains and the third region contains chondroitin sulphate chains. This arrangement gives cartilage its resilience and tensile strength.

(*ii*) **Glycoproteins** (or **glycosaminoglycans** or **mucopolysaccharides**). In these molecules, the carbohydrate portion consists of much shorter chains which are often branched. Glycoproteins serve diverse roles in cells and tissues and include certain enzymes, hormones, blood groups, saliva, gastric mucin, ovomucoids, serum, albumins, antibodies or immunoglobins.

(*iii*) **Glycolipids.** These molecules are covalent combinations of carbohydrate and lipid. The carbohydrate portion may be a single monosaccharide or a linear of branched chain. Glycolipids form the component of most cell membranes, *e.g.*, cerebrosides and gangliosides.

Lipids (Fats). The lipids (Gr., *lipos* = fats) are the organic compounds which are insoluble in the water but soluble in the nonpolar organic solvents such as acetone, benzene, chloroform and ether. The cause of this general property of lipids is the predominance of long chains of aliphatic hydrocarbons or benzene ring in their molecules. The lipids are non-polar and hydrophobic. The common examples of lipids are cooking oil, butter, ghee, waxes, natural rubber and cholesterol. Like the carbohydrates, lipids serve two major roles in cells and tissues:

```
      H
      |
  H — C — OH
      |
  H — C — OH
      |
  H — C — OH
      |
      H
```

Glycerol

1. They occur as constituents of certain structural components of cells such as membranous organelles; plant pigments such as **carotene** found in carrots and **lycopene** that occurs in tomatoes; vitamins such as *A*, *E* and *K*; menthol and eucalyptus oil; and

2. They may be stored within cells as reserve energy sources. Like the starch and glycogen, fat is compact and insoluble and provides a convenient form in which energy-yielding molecules (the fatty acids) can be stored for use when occasion arises.

Lipids are all made of carbon, hydrogen and sometimes oxygen. The number of oxygen atoms in a lipid molecule is always small compared to the number of carbon atoms. Sometimes small amounts of phosphorus, nitrogen and sulphur are also present. Natural fats and oils are compounds of **glycerol** (*i.e.,* glycerine or propane– 1, 2, 3 triol) and **fatty acids**. They are esters which are formed due to reaction of organic acids with alcohols. There is only one kind of glycerol: its molecular configuration shows no variation and it is exactly same in all lipids. The formula of glycerol is $C_3H_8O_3$ and following is its molecular structure:

Fatty acids. A fatty acid molecule is **amphipathic** and has two distinct regions or ends; a long **hydrocarbon chain,** which is **hydrophobic** (water insoluble) and not very reactive chemically, and a **carboxylic acid group** which is ionized in solution (COO^-), extremely **hydrophilic** (water soluble) and readily forms esters and amides. In neutral solutions, salts of the fatty acids form small spherical droplets or **micelles** in which the dissociated carboxyl groups occur at surface and the hydrophobic carbon chains project towards the centre. In cells, the fatty acids only sparingly occur freely; instead, they are esterified to other components and form the saponifiable lipids.

A fatty acid molecule may be either saturated or unsaturated. The **saturated fatty acids** consist of long hydrocarbon chains terminating in a carboxyl group and conform to the general formula:

$$CH_3 - (CH_2)_n - COOH$$

In nearly all naturally occurring fatty acids, *n* is an even number from 2 to 22. In the **saturated fatty acids,** most commonly found in animal tissues, *n* is either 12 (*i.e.,* **myristic acid**), 14 (*i.e.,* **palmitic acid**) or 16 (*i.e.,* **stearic acid**). In **unsatutated fatty acids,** at least two but usually no more than six of the carbon atoms of the hydrocarbon chain are linked together by double bonds (–C = C–), *e.g.,* **oleic acid, linoleic, linolenic, arachidonic** and **clupanadonic acids.** Double bonds are important because they increase the flexibility of the hydrocarbon chain, and thereby the fluidity of biological membranes. Unsaturated fatty acids predominate in lipids of higher plants and in animals that live at low temperatures. Lipids in the tissues of animals inhabiting warm climates contain larger quantities of saturated fatty acids.

Essential fatty acids. Some animals, especially mammals, are unable to synthesize certain fatty acids and, therefore, require them in their diet. They are called **essential fatty acids** and include linoleic acid, linolenic acid and arachidonic acid. Such essential fatty acids have to be obtained from plant material by the animal.

Types of lipids. The lipids are classified into three main types:

1. simple lipids,
2. compound lipids and
3. derived lipids.

1. Simple lipids. The simple lipids are alcohol esters of fatty acids:

$$\text{Triglyceride (Simple lipids)} \xrightarrow[H_2O]{\text{Lipase}} \text{Glycerol} + 3 \text{ Fatty acids}$$

Simple lipids are also of following two types:

(*a*) Neutral fats (Glycerides or triglycerides). They are triesters of fatty acids and glycerol. Neutral fats represent the major type of stored lipids and so accumulate in the cytoplasm.

(*b*) Waxes. Waxes have a higher melting point than neutral fats and are the esters of fatty acids of high molecular weight with the alcohol except the glycerol. The most important constituent alcohol of the molecules of waxes is the **cholesterol**, *e.g.,* bees wax.

2. Compound lipids. The compound lipids contain fatty acids, alcohols and other compounds as phosphorus, amino-nitrogen carbohydrates, etc., in their molecules. Some of the compound lipids are important structural components of the cell, in particular of cell membranes. The compound lipids of the cell are of the following types:

(*i*) Phospholipids (or Glycerophosphatides). Such type of lipids form the major constituent of cell membranes. In a molecule of phospholipid two of the —OH or hydroxyl groups in glycerol are linked to fatty acids, while the third —OH group is linked to phosphoric acid. The phosphate is further linked to a hydrophilic compound such as **etanolamine, choline, inositol** or **serine.** Each phospholipid molecule, therefore, has hydrophobic or water-insoluble tail which is composed of two fatty acid chains and a hydrophilic or water-soluble polar head group, where the phosphate is located. Thus, in effect the phospholipid molecules are detergents, *i.e.,* when a small amount of phospholipid is spread over the surface of water, there forms a monolayer film of phospholipid molecules; in this thin film, tail regions pack together very closely facing the air and their head groups are in contact with the water (Fig. 57.5). Two such films can combine tail to tail to make a phospholipid sandwich or self-sealing **lipid bilayer,** which is the structural basis of cell membranes.

Various membranes of cell contain the following four types of phospholipids:

1. **phosphatidyl choline** or **lecithin;**
2. **phosphatidyl ethanolamine** or **cephalin;**

3. phosphatidyl serine; and

4. phosphatidyl inositol.

The other important phospholipids of the matrix are the phosphoinositides (occur mostly in the cells of liver, brain, muscle and soyabean), plasmalogens and isositides.

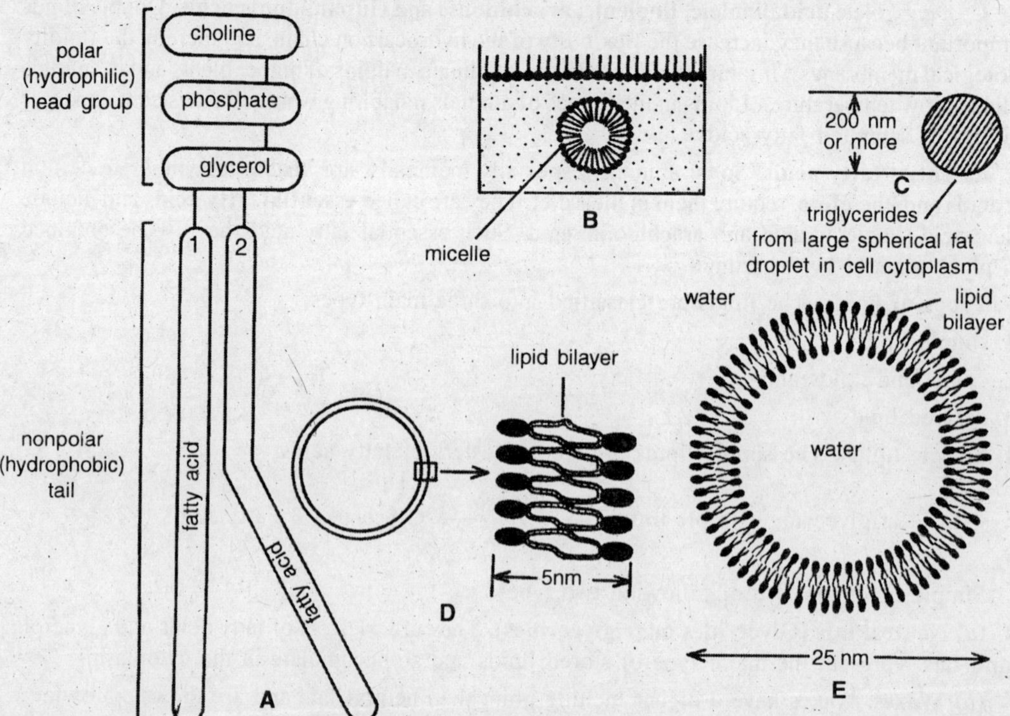

Fig. 57.5. Formation of various types of lipid aggregates. A—Schematic representation of a phospholipid molecule; B—Formation of micelle and monolayer film; C—Formation of a fat droplet by triglycerides; D—Formation of self-sealing lipid bilayer (*e.g.*, liposome); E—Cross-section of a liposome.

Plasmalogens are a special class of phospholipids which are especially abundant in the membranes of nerve and muscle cells and are also characteristic of cancer cells.

Liposomes. When aqueous suspensions of phospholipids are subjected to rapid agitation by using ultrasound (*i.e.*, insonation), the lipid disperses in the water and forms **liposomes** or **lipid vesicles.** Liposomes are small spherical bodies (25 nm to 1 μm in diameter) whose surface is formed by a bilayer of phospholipid molecules enclosing a small volume of the aqueous medium. They exhibit many of the permeability properties of natural membranes, *i.e.*, water soluble small molecules or ions can be enclosed by the liposomes and they can also traverse the lipid bilayer of latter. Recently, liposomes have been found to have great therapeutic promise, since, they can be used as vectors for the transfer of specific drugs, proteins, hormones, nucleic acids, ions or any other molecule into the specific types of animal cells. The contents of the liposomes can enter the target cells by two routes:

1. The liposomes can attach to the surface of target cells and may fuse with the plasma membrane, following which their contents are released into the cytosol or cytoplasmic matrix.
2. The entire liposomes may be endocytosed and degraded intracellularly.

***(ii)* Sphingolipids.** The sphingolipids occur mostly in the cells of the brain. Instead of the glycerol, they contain in their molecules amine alcohol (sphingol or sphingosine). For instance, the myelin sheaths of the nerve fibres contain a lipid known as **sphingomyelin** which contains sphingosine and phospholipids in its molecules.

***(iii)* Glycolipids.** The glycolipids contain in their molecules the carbohydrates and the lipids. The matrix of the animal cells contains two kinds of glycolipids, *viz.*, cerebrosides and gangliosides.

***(a)* Cerebrosides.** The cerebrosides contain in their molecules sphingosine, fatty acids and galactose or glucose. The cerebrosides are the important lipids of the white matter of the cells of brain and the myelin sheath of the nerve. The important cerebrosides are the kerasin, cerebron, nervon and oxynervons.

***(b)* Gangliosides.** The gangliosides have complex molecules which are composed of sphingosine, fatty acids and one or more molecules of glucose, lactose, galactosamine and neuraminic acid. The gangliosides occur in the grey matter of the brain, membrane of erythrocytes and cells of the spleen. Gangliosides act as **antigens**.

One type of ganglioside, called **GM_2**, may accumulate in the lysosomes of the brain cells because of a genetic deficiency that results in the failure of the cells to produce a lysosomal enzyme that degrades this ganglioside. This condition is called **Tay-Sachs disease** and leads to paralysis, blindness and retarded development of human beings.

3. Derived lipids (or Nonsaponifiable lipids). Some types of lipids do not contain fatty acids in their constituents and they are of following three types:

A. Terpenes. The terpenes include certain **fat-soluble vitamins** (*e.g.*, vitamins A, E and K) **carotenoids** (*e.g.*, photosynthetic pigments of plants), and certain **coenzymes** (such as coenzyme Q or ubiquinone). All the terpenes are synthesized from various numbers of a five-carbon building block, called **isoprene unit.** The isoprene units are bonded together in a head-to-tail organization. Two isoprene units form a **monoterpene**, four form a **diterpene**, six a **triterpene**, and so on. The monoterpenes are responsible for the characteristic odours and flavours of plants (*e.g.*, **geraniol** from geraniums, **menthol** from mint and **limoneme** from lemons). **Dolicol phosphate** is a **polyisoprenoid** (*i.e.*, long chain polymer of isoprene) and is used to carry activated sugars in the membrane-associated synthesis of glycoproteins and some polysaccharides.

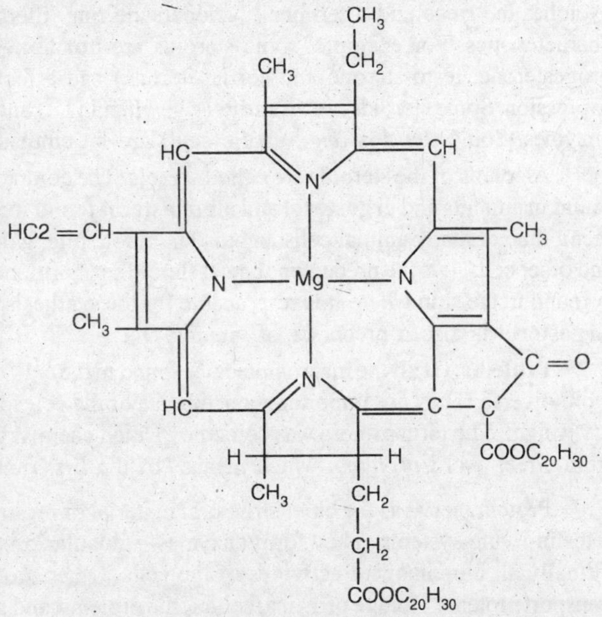

Fig. 57.6. Chemical formula of chlorophyll.

The **carotenoids** are the compound lipids and they form the pigments of the animal and plant cells. There are about 70 carotenoids occurring in both types of cells. The important carotenoids of cells are the α, β and γ carotenes, retinene, xanthophylls, lactoflavin in milk, riboflavin (vitamin B_2), xanthocyanins, coenzyme Q, anthocyanins, flavones, flavonols and flavonones, etc. Chemically all carotenoids are long-chain isoprenoids having an alternating series of double bonds. They are synthesized by plant tissues and are located in the chloroplast lamellae to help in light absorption during photosynthesis. In animal cells, carotenoids serve as precursors of vitamin A.

The **chlorophylls** are essential photosynthetic green pigments of the chloroplasts. A chlorophyll molecule (Fig. 57.6) consists of a **head** and a **tail**. The head consists of a **porphyrin ring** or **tetrapyrrole nucleus** from which extends a hydrophobic tail which is made up of a 20-carbon grouping, called the **phytol**. Phytol (C_{20} H_{39}) is a long straight-chain alcohol containing a single double bond. It may be regarded as a hydrogenated carotene (vitamin A). The porphyrins (Gr., *porphyra* = purple) are complex carbon-nitrogen molecules that usually surround a metal, *i.e.*, it is formed from four pyrrol rings linked together by methane bridges and metal atom (Mg or Fe) is linked to pyrrol rings. In chlorophyll molecule, the porphyrin surrounds a magnesium ion, while in haeme of **haemoglobin**, it surrounds an iron ion (Fig. 57.7). Many other pigments of animal cells such as **myoglobin** and **cytochromes** have porphyrin rings in their molecules.

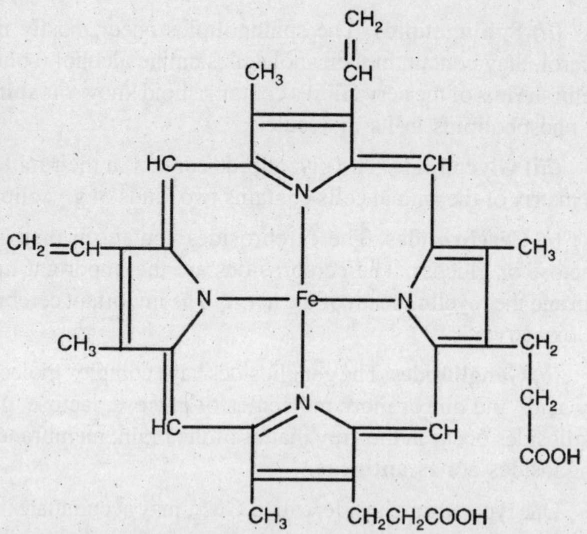

Fig. 57.7. Chemical formula of haeme portion of haemoglobin.

B. Steroids. The steroids consist of a system of fused cyclohexane and cyclopentane rings. All are derivative of **perhydro-cyclopentano-phenanthrene,** which consists of three fused cyclohexane rings and a terminal cyclopentane ring. Steroids have widely different physiological characteristics. For example, some steroids are **hormones** (*e.g.,* sex hormones such as estrogen, progesterone, testo-sterone and corticosterone) and affect cellular activities by influencing gene expression. Some steroids are **vitamins** (*e.g.,* vitamin D_2) and influence the activities of certain cellular enzymes. Some steroids (*e.g.,* cholic acid) are fat emulsifier found in the bile.

Alcohols of the steroids are called **sterols**. The common examples of the sterols are **cholesterol** found in animals and **ergosterol** and **stigmasterol** found in plants. Cholesterol is found in the plasma membrane of many animal cells and also in blood, bile, gallstone, brain, spinal cord, adrenal glands and other cells. It is the precursor of most steroid sex hormones and cortisones. **7-dehydro-cholesterol** is found in the skin where it is responsible for the synthesis of vitamin D in the presence of sunlight. **Ergosterol** is also a precursor of vitamin D.

Proteins. Of all the macromolecules found in the cell, the proteins are chemically and physically more diverse. They are important constituents of the cell forming more than 50 percent of the cell's dry weight. The term protein was coined by Dutch chemist **G.J. Mulder** (1802–1880) and is derived from Greek word *proteios*, which means "of the first rank".

Proteins serve as the chief structural material of protoplasm and play numerous other essential roles in living systems. They form enzymes—globular proteins specialized to serve as catalysts in virtually all bio-chemical activities of the cells. Other proteins are antibodies (immunoglobulins), transport proteins, storage proteins, contractile proteins, and some hormones. In every living organism, there are thousands of different proteins, each fitted to perform a specific functional or structural role. Indeed, a single human cell may contain more than 10,000 different types of protein molecules. Chemically, proteins are polymers of amino acids.

1. Amino acids. Nobel Laureate **Emil Fischer** (1902) discovered that all proteins consist of chains (linear sequence) of smaller units that he named **amino acids.** There are about 20 different amino acids which occur regularly as constituents of naturally occurring proteins (Table 57.3). An organic compound containing one or more amino groups ($-NH_2$) and one or more carboxyl groups (—COOH) is known as amino acid. The amino acids occur freely in the cytoplasmic matrix and constitute the so called **amino acid pool**. Of the 20 commonly occurring amino acids, 19 may be represented by the following general formula (Fig. 57.8).

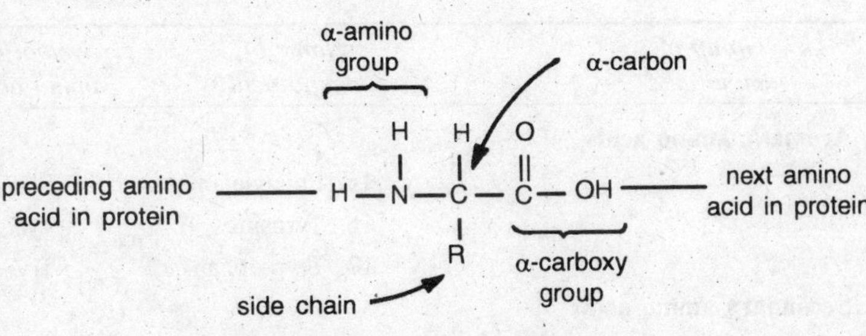

Fig. 57.8. Basic structure of an amino acid.

The sole exception is proline, where the amino group forms part of a ring structure. The central or **alpha** carbon atom of each amino acid is covalently bonded to four groups:

(*i*) A hydrogen atom,

(*ii*) an amino group (—NH_2),

(*iii*) an acid (or carboxyl) group, and

(*iv*) a side chain called an **R-group**. It is the particular chemical structure of the R-group that distinguishes one amino acid from another.

In certain amino acids R group is either a hydrogen atom (*e.g.*, glycine, the simplest amino acid) or a hydrophobic aliphatic (*e.g.*, leucine) or aromatic (*e.g.*, pheylalanine) hydrocarbon. In other cases, R group contains either an extra carboxyl group or an extra amino group or its equivalent. Glutamic acid and aspartic acid each have an extra carboxyl (—COOH) group. Lysine and arginine both contain an additional amino group or equivalent structure. Histidine also contains a N group. Other amino acids such as serine and tyrosine have hydroxyl groups in their side chains. Of a particular importance is the amino acid cysteine which possesses a thiol (SH) group.

Table 57.3 The 20 amino acids which occur naturally.

Group of amino acid	*Name of amino acid*	*Symbol of amino acid*	*Single letter code of amino acid*
A. Aliphatic amino acids			
1. Simple amino acids or Mono-amino-monocarboxylic amino acids	1. Glycine	Gly	G
	2. Alanine	Ala	A
	3. Valine	Val	V
	4. Leucine	Leu	L
	5. Isoleucine	Ile	I
2. Acidic amino acids or Monoamino-monocarboxylic acids	6. Aspartic acid	Asp	D
	7. Glutamic acid	Glm	E
3. Basic amino acids or Diamino-monocarboxylic amino acids	8. Lysine	Lys	K
	9. Arginine	Arg	R
	10. Histidine	His	H
4. Hydroxyl containing amino acids	11. Serine	Ser	S
	12. Threonine	Thr	T
5. Sulphur-containing amino acids	13. Cysteine	Cys	C
	14. Methionine	Met	M

Group of amino acid	*Name of amino acid*	*Symbol of amino acid*	*Single letter code of amino acid*
B. Aromatic amino acids			
	15. Phenylalanine	Phe	F
	16. Tyrosine	Tyr	Y
	17. Tryptophan	Try	W
C. Secondary amino acids			
	18. Proline	Pro	P
D. Amino acid amides			
	19. Aspargine	Asn	N
	20. Glutamine	Glu	Q

2. Formation of proteins. Because a molecule of the amino acid contains both basic or amino (—NH_2) and acidic or carboxyl (—COOH) group, it can behave as an acid and base at a time. The molecules of such organic compounds which contain both acidic and basic properties are known as **amphoteric molecules**. Due to amphoteric molecules, the amino acids unite with one another to form complex and large protein molecules. When two molecules of amino acids are combined then the basic group (—NH_2) of one amino acid molecule combines with the carboxylic (—COOH) group of other amino acid and the loss of a water molecule takes place. This sort of condensation of two amino acid molecules by —NH—CO linkage or bond is known as **peptide linkage** or **peptide bond**. A combination of two amino acids by the peptide bond is known as **dipeptide**. When three amino acids are united by two peptide bonds, they form **tripeptide**. Likewise, by condensation of few or many amino acids by the peptide bonds the **oligopeptides** and **polypeptides** are formed respectively. The various molecules of polypeptides unite to form the **peptones**, **proteases** and **proteins**. Thus, protein macromolecules are the polymers of many amino acid monomers. The size (molecular weight), shape, and function of proteins are determined by the number, type and distribution of the amino acids present in the molecule. Proteins occur in a wide spectrum of molecular sizes from small molecules such as the hormone ACTH (or adrenocorticotrophic hormone) which consists of only 39 amino acids and has a molecular weight of 4500, to extremely large proteins such as haemocyanin (an invertebrate blood pigment) which consists of 8200 amino acids and has a molecular weight greater than 900,000.

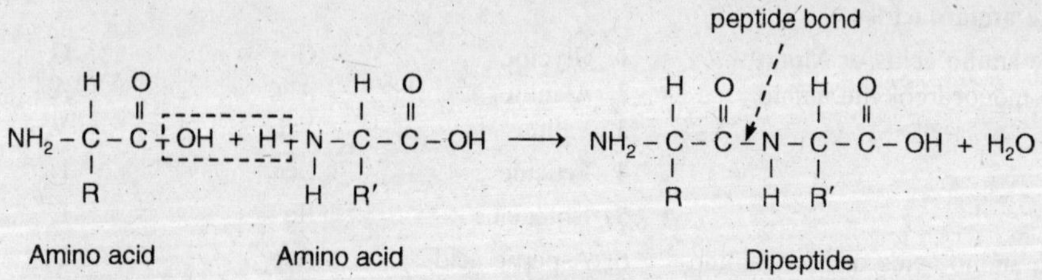

$$\underbrace{NH_2-\underset{R}{\overset{H}{C}}-\overset{O}{\overset{\|}{C}}-OH}_{\text{Amino acid}} + \underbrace{H-\underset{H}{N}-\underset{R'}{\overset{H}{C}}-\overset{O}{\overset{\|}{C}}-OH}_{\text{Amino acid}} \longrightarrow \underbrace{NH_2-\underset{R}{\overset{H}{C}}-\overset{O}{\overset{\|}{C}}-\underset{H}{N}-\underset{R'}{\overset{H}{C}}-\overset{O}{\overset{\|}{C}}-OH}_{\text{Dipeptide}} + H_2O$$

peptide bond

Fig. 57.9. A chemical reaction showing formation of a dipeptide.

3. Types of proteins. Many different methods have been used to classify proteins, no method of their classification being entirely satisfactory:

1. Classification based on biological functions. Based on function proteins fall into nine major classes (Table 57.4). Main proteins are **enzymes,** serving as catalysts that greatly increase the rates of the thousands of chemical reactions on which life depends. **Structural proteins**, on the other hand, provide support and shape to cells and organelles. **Motility proteins** play key roles in the contraction and movement of cells and intracellular structures. **Regulatory proteins** are responsible for control and coordination of cellular functions, ensuring that cellular activities are regulated to meet cellular needs. **Transport proteins** are involved in the movement of other substances into, out of and within the cell. **Hormonal proteins** mediate communication between cells in distant parts of an organism, and **receptor proteins** enable cells to respond to chemical stimuli from their environment. Finally, **defensive proteins** provide protection against disease, and **storage proteins** serve as reservoirs of amino acids.

Thus, with virtually everything that a cell is or does dependent on the proteins it contains (see **Becker** *et al.*, 2007). Most proteins are **monofunctional**, which simply means that a specific protein has a single function, whether catalytic, structural, motile, or regulatory. However, some proteins are **bifunctional**, which means that an individual protein plays two distinctly different roles.

Table 57.4 Protein functions (Source: **Becker** *et al.*, 2007).

Type of protein	*Function*	*Example*
1. Enzymes	Selective catalysis	Digestive enzymes catalyze the hydrolysis of the macromolecules in food.
2. Structural proteins	Support of cellular structures	Collagen and elastin provide fibrous framework in animal connective tissues; keratin is the protein of hair, horns, feathers and other skin appendages.
3. Motility proteins	Movement of cells and cell parts	Actin and myosin are responsible for the movement of muscles; tubulin is the main constituent of cilia and flagella.
4. Regulatory proteins	Regulation of cellular functions	DNA-binding proteins, called transcription factors regulate the read-out of genetic information.
5. Transport proteins	Transport of substances, into, out of, and within cells	Transport of glucose and other small molecules across membranes is mediated by specific transport proteins embeded in the membranes.
6. Hormonal proteins	Communication between distant parts of an organism	Insulin is secreted by the pancreas and regulate the concentration of glucose in the blood of vertebrates.
7. Receptor proteins	Response of cells to chemical stimuli	Receptors in the membrane of a nerve cell detect chemical signals released by other nerve cells.
8. Defensive proteins	Protection against disease	Antibodies in the circulatory system of vertebrates detect and destroy bacteria and viruses.
9. Storage proteins	Storage and release of amino acids	Storage proteins in seeds are broken down during germination to release amino acids.

2. Classification based on shape of proteins. According to the shape or conformation, two major types of proteins have been recognized:

(*a*) Fibrous proteins. Fibrous proteins are water-insoluble, thread-like proteins having greater length than their diameter. They contain secondary protein structure and occur in those cellular or extracellular structures, where strength, elasticity and rigidity are required, *e.g.*, collagen, elastin, keratin, fibrin (blood-clot proteins) and myosin (muscle contractile proteins).

(*b*) Globular proteins. Globular protein are water-soluble roughly spheroidal or ovoidal in shape. They readily go into colloidal suspension. They have tertiary protein structure and are usually functional proteins, *e.g.*, enzymes, hormones and immunoglobulins (antibodies). **Actin** of microfilaments and **tubulins** of microtubules are also globular proteins.

3. Classification based on solubility characteristics. According to this criterion proteins can be classified into two main types:

***(a)* Simple proteins.** These proteins contain only amino acids in their molecules and they are of following types:

***(i)* Albumins.** These are water soluble proteins found in all body cells and also in blood stream, *e.g.,* **lactalbumin**, found in milk and **serum albumin** found in blood.

***(ii)* Globulins.** These are insoluble in water but are soluble in dilute salt solutions of strong acids and bases, *e.g.,* **lactoglobulin** found in milk and **ovoglobulin**.

***(iii)* Glutelins.** These plant proteins are soluble in dilute acids and alkalis, *e.g.,* **glutenin** of wheat.

***(iv)* Prolamines.** These plant proteins are soluble in 70 to 80 per cent alcohol, *e.g.,* **gliadin** of wheat and **zein** of corn.

***(v)* Scleroproteins.** They are insoluble in all neutral solvents and in dilute alkalis and acids, *e.g.,* keratin and collagen.

***(vi)* Histones.** These are water soluble proteins which are rich in basic amino acids such as arginine and lysine. In eukaryotes histones are associated with DNA of chromosomes to form nucleoproteins.

***(vii)* Protamines.** These are water soluble, basic, light weight, **arginine** rich polypeptides. They are bound to DNA in spermatozoa of some fishes, *e.g.,* **salmine**, of salmon and **sturine** in sturgeons.

***(b)* Conjugated proteins.** These proteins consist of simple proteins in combination with some non-protein components, called **prosthetic groups**. The prosthetic groups are permanently associated with the molecule, usually through covalent and/or non-covalent linkages with the side chains of certain amino acids. Conjugated proteins are of following types:

***(i)* Chromoproteins.** Chromoproteins are a heterogeneous group of conjugated proteins which are in combination with a prosthetic group that is a pigment, *e.g.,* respiratory pigments such as **haemoglobin, myoglobin** and **haemocyanin; catalase, cytochromes, haemerythrins;** visual purple or **rhodopsin** of rods of retina of eye and yellow enzymes or **flavoproteins.**

***(ii)* Glycoproteins.** Glycoproteins are proteins that contain various amounts (1 to 85 per cent) of carbohydrates. Of the known 100 monosaccharides, only nine are found to occur as regular constituents of glycoproteins (*e.g.,* glucose, galactose, mannose, fucose, acetylglucosamine, acetylgalactosamine, acetylneuraminic acid, arabinose and xylose). Glycoproteins are of two main types: **1. Intracellular glycoproteins** which are present in cell membranes and have an important role in membrane interaction and recognition. They also serve as antigenic determinants and receptor sites. **2. Secretory glycoproteins** are **plasma glycoproteins** secreted by the liver; **thyroglobulin,** secreted by the thyroid gland; **immunoglobulins** secreted by the plasma cells; **ovoalbumins** secreted by the cells of oviduct of hen; **ribonucleases** and **deoxyribonucleases. Mucus** and **synovial fluid** are also glycoproteins with lubricative properties.

***(iii)* Lipoproteins.** Lipid containing proteins are called **lipoproteins.** Their lipid contents are 40 to 90 per cent of their molecular weight and this tends to affect the density of the molecule. There are four types of lipoproteins: 1. **High density lipoproteins (HDL)** or α-lipoproteins; 2. **Low density lipoproteins (LDL)** or β-lipoprotiens; 3. **Very low density lipoproteins (VLDL)** or pre-β-lipoproteins; and 4. **Chylomicrons.** Lipoproteins include some of the blood plasma proteins, various types of membrane proteins, lipovitellin of egg yolk and proteins of brain and nerve tissue.

***(iv)* Nucleoproteins.** Nucleoproteins are proteins in combination with nucleic acids (DNA and RNA). However, these proteins are not true conjugated proteins since the nucleic acid involved cannot be regarded as prosthetic groups. Nucleoproteins are of two types: 1. **Histones** which are quite similar in all plants and animals. Their highly basic nature accounts for the close associations histones form with the nucleic acids. Histones are involved in the tight packing of DNA molecules during the condensation of chromatin into chromosomes for the mitosis. 2. **Nonhistones** have great heterogeneous amino acid composition and are acidic in nature. They have selective combination with certain stretches of nuclear DNA and, thus, are involved in the regulation of gene expression.

(*v*) Metalloproteins. Metalloproteins are proteins conjugated to metal ions which are not part of the prosthetic group, *e.g.*, **carbonic anhydrase** enzyme contains zinc ions and amino acids in its molecule; **caeruloplasmin,** an oxidase enzyme containing copper; and **siderophilin** contains iron.

(*vi*) Phosphoproteins. Phosphoproteins are proteins in combination with a phosphate group, *e.g.*, casein of the milk and ovovitellin of eggs.

4. Structural levels of proteins. The protein as synthesized on the ribosome is a linear sequence of amino acids, polymerized by the elimination of water between successive amino acids to from the peptide bond, and existing as a randomly coiled chain without specific shape and possessing no biological (*i.e.*, catalytic) acitivity. Within seconds of synthesis being completed, the protein folds into a specific three-dimensional form, which is the same for all molecules of the same type of protein and which now is capable of doing catalysis. According to their mode of folding the following four levels of protein organization have been recognized (Fig. 57.10):

(*a*) **Primary structure of protein,** *e.g.*, insulin.

(*b*) **Secondary structure of protein,** *e.g.*, α-Keratin, collagen, globular proteins.

(*c*) **Tertiary structure of protein,** *e.g.*, myoglobin.

(*d*) **Quaternary structure of protein,** *e.g.*, haemoglobin.

(*a*) Primary protein structure. The primary protein structure is defined as the particular sequence of amino acids found in the protein. It is determined by the covalent peptide bondings

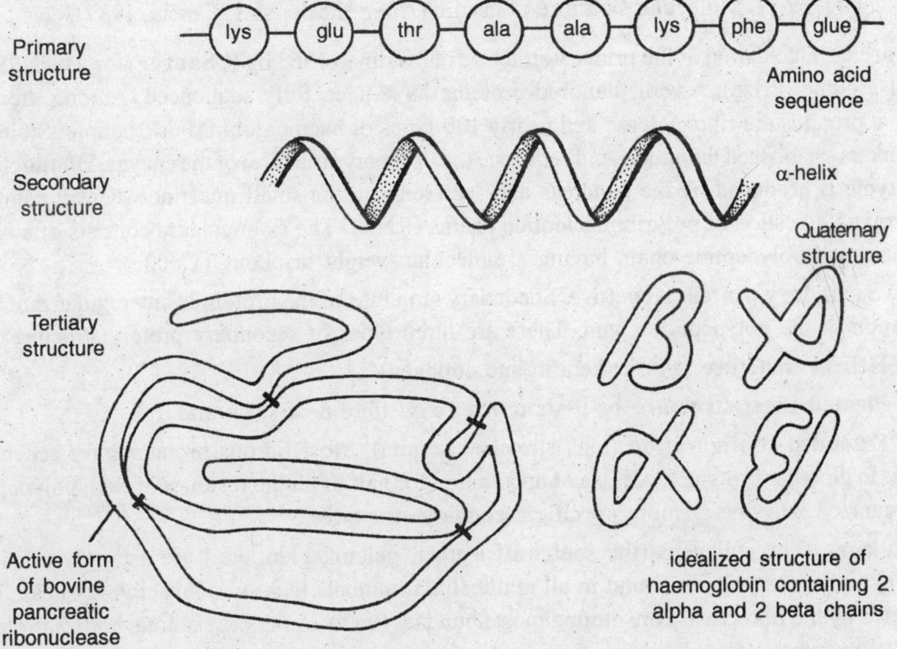

Fig. 57.10. The four levels of structural organization in protein molecules.

between amino acids. Primary structure also includes other covalent linkages in proteins, for example the linkages that may exist between sulphur atoms of cysteine amino acids located in the chain of the protein insulin. The first protein to have its primary structure determined was of insulin, the pancreatic hormone that regulates glucose metabolism in mammals. Insulin has a molecular weight of 5,800 daltons and contains 51 amino acids. Insulin consists of two polypeptide chains of 21 and 30 amino acid residues, called the A and B chains, respectively (Fig. 57.11). (An amino acid residue is that which is left when the elements of water are split out during polymerization).

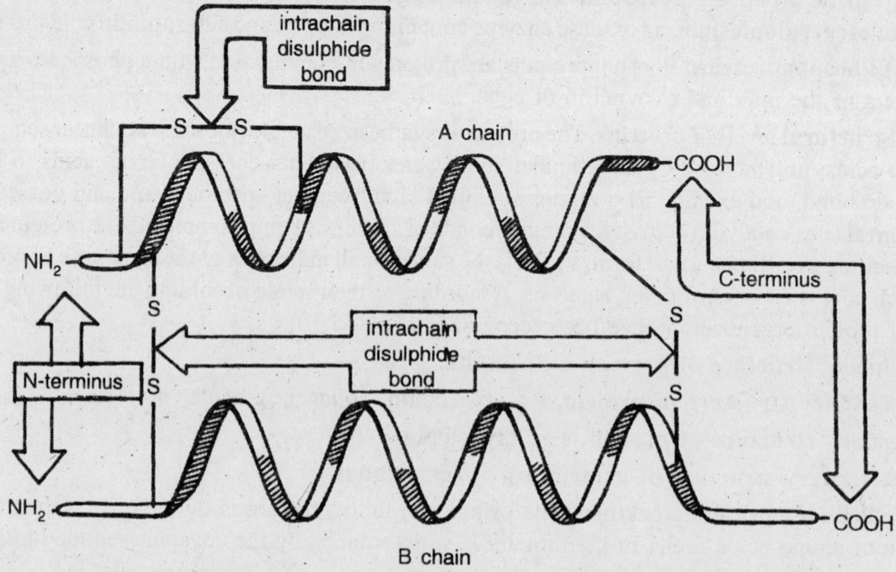

Fig. 57.11. Molecular structure of insulin (After Sheeler and Bianchi, 1987).

Since the elucidation of the primary structure of insulin in 1953 by **F. Sanger** (for which Sanger received a Nobel Prize), several hundred proteins have been fully sequenced. Among the fully sequenced proteins are ribonuclease and nearly 100 types of haemoglobin. For example; **Stein** and his coworkers established the amino acid sequence (*i.e.*, primary structure) of the enzyme **ribonuclease.** This enzyme is produced by the pancreas and secreted into the small intestine where it catalyzes the hydrolytic digestion of polyribonucleotide chains (RNA). The ribonuclease consists of a single 124 amino acid polypeptide chain having a molecular weight or about 12,000.

(*b*) Secondary protein structure. Secondary structure of the protein is any regular repeating organization of the polypeptide chain. There are three types of secondary protein structure:

1. **Helical structure** (*e.g.*, α-keratin and collagen);

2. **Pleated sheet structure or β-structure** (*e.g.*, fibroin of silk); and

3. **Extended configuration** (*e.g.*, stretched keratin). Most fibrous proteins have secondary structure. In globular protein, too, it is not uncommon for half of all the residues of each polypeptide to be organized into one or more specific secondary structures.

Collagen. The collagens (the source of leather, gelatin, glue, etc.) are a family of highly characteristic fibrous proteins found in all multicellular animals (*e.g.*, in connective tissues). They are secreted by the fibroblasts constituting most abundant (up to 25 per cent of total body's proteins) proteins of mammals. The characteristic feature of collagen (or **tropocollagen**) molecules is their stiff, triple-stranded helical structure (which was discovered by **Rich, Crick** and **Ramachandran**). Three collagen polypeptide chains are left-handed α-helices or alpha chains, each is about 1000 amino acid residues long. These chains are wound around one another in a regular superhelix to generate a rope-like collagen or tropocollagen molecule which is about 300 nm long and 1.5 nm is diameter (Fig. 57.12).

Collagens are exceptionally rich in **proline** (and **hydroxyproline;** both accounting for more than 20 percent of collagen's amino acids) and **glycine**. Other dominant amino acids of collagens are **lysine** and **alanine**.

So far, about 20 distinct collagen-chains have been identified, each encoded by a separate gene. About 10 types of triplet-stranded collagen molecules have been found to assemble from various combinations of 20 types of α-chains. The best defined are types I, II, III and IV. **Type I collagen** is present in the dermis, tendons, ligaments, bone, cornea, dentine of teeth and internal organs and accounts for 90 per cent of body's collagen. **Type II collagen** is present mainly in cartilage, intervertebral disc, embryonic notochord and vitreous humour of eye. **Type III collagen** occurs in skin, cardiovascular system, gastrointestinal tract and uterus. **Type IV collagen** is present in basal laminae or basement membranes of epithelia. Type I, II and III collagens are the fibrillar collagens showing typical striated fibres. Type IV collagen lacks a distinct fibrillar structure (Fig. 57.13).

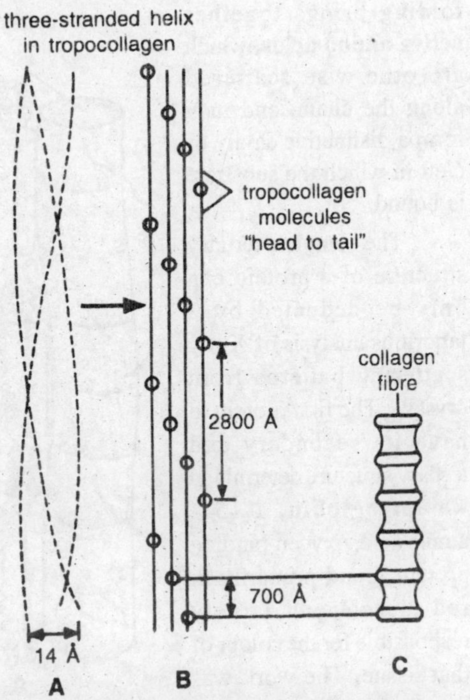

Fig. 57.12. A—A tropocollagen molecule (collagen or superhelix) with three intertwined left-handed α-helices; B—Staggered arrangement of tropocollagen molecules (super helices) in a collagen fibril; C—Striated appearance of a collagen fibril under the electron microscope (after Dyson, 1978).

The individual collagen polypeptide chains (α-chains) are synthesized on membrane bound ribosomes and injected in the lumen of ER as larger precursors, called pro-α-chains. These precursors have distinct polarity, containing terminal propeptides at their *N*-and *C*-terminus. Each pro-α-chain then combines with two others to form a hydrogen-bonded, triple stranded helical molecule, called **procollagen**. Such a procollagen of fibrillar collagens (I, II, and III) is serceted from the fibroblast in the extracellular space. Due to enzyme action its terminal propeptides are removed and the procollagen is converted into a **tropocollagen molecule**. Many tropocollagen molecules spontaneously assemble into the ordered arrays, called **collagen fibrils**. The collagen fibrils are thin (10 to 300 nm in diameter), cable-like structures, many micrometres long, exhibiting cross-striations every 67 nm and are clearly visible in the electron micrographs. The collagen fibrils often aggregate into larger bundles which can be seen in light microscope as **collagen fibres**. Type IV collagen molecules assemble to form a sheet-like **meshwork** that constitutes a major part of all basal laminae (**Martin** *et. al.,* 1985, **Burgeson**, 1988).

(*c*) Tertiary protein structure. Tertiary protein structure refers to a more compact structure in which the helical and non-helical regions of a polypeptide chain are folded back on themselves. This structure is typical of globular protein structure, in which it is the non-helical region that permits the folding. The folding of a polypeptide chain is not random but occurs in a specific fashion, thereby imparting certain steric (*i.e.,* three-dimensional) properties to the protein. For example, in enzymes folding brings together **active amino acids,** which are otherwise scattered along the chain, and may form a distinctive cavity or cleft in which the substrate is bound.

The complete tertiary structure of a protein can only be deducted by a laborious analysis of X-ray scattering patterns from crystals. The first protein to have its secondary and tertiary structure determined was **myoglobin,** a 153-amino acid, oxygen-binding protein found primarily in red muscle and largely responsible for the colour of that tissue. The work was done at Cambridge under the direction of **J.C. Kendrew** (1961). Although at some points the polypeptide chain does have secondary structure (alpha-helical structure), the chain is mainly characterized by seemingly random loops and folds.

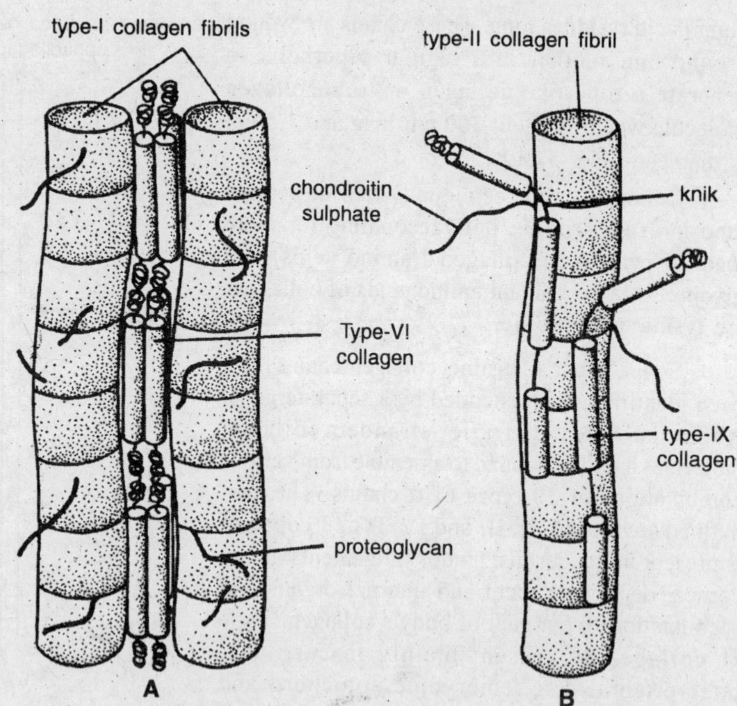

Fig. 57.13. Interactions of fibrous collagens with nonfibrons collagens. A—In tendons, type I collagen fibrils are associated with type-VI collagen. B—In cartilage, type II collagen fibrils are associated with type IX collagen (after Lodish *et. al.,* 2004).

In a tertiary protein the polypeptide chain is held in position by weak secondary bonds which are of different types such as **ionic bonds** (or electrostatic bonds or salt or salt bridges); **hydrogen bonds; hydrophobic bonds** and **disulphide bonds**.

***(d)* Quaternary protein structure.** In proteins that are composed of two or more polypeptide chains, the quaternary structure refers to the specific orientation of these chains with respect to one another and the nature of the interactions that stabilize this orientation. The individual polypeptide chains of the protein are called **sub-units** and the active protein itself is called **multimer**. While multimeric proteins containing up to 32 subunits have been described, the most common multimers are **dimers, trimers, tetramers, pentamers** (*e.g.,* RNA polymerase) and **decamers** (*e.g.,* DNA polymerase III) (Table 57.5). If the protein consists of identical sub-units, it is called **homopolymers** and is said to have **homogeneous quaternary structure,** *e.g.,* the isozymes H_4 and M_4 of lactic dehydrogenase (LDH), enzyme phosphorylase and L-arabinose isomerase. The enzyme β-galactosidase, consists of four identical polypeptide chains. Lastly, when the sub-units of the protein are different, the protein is called **heteropolymer** and is said to have a heterogeneous quaternary structure, *e.g.,* haemoglobin and immunoglobulins. Quaternary proteins are usually joined by hydrophobic forces. Hydrogen bonds, ionic bonds and possibly disulphide bonds may also participate in forming quaternary structures.

Table 57.5 Subunits and molecular weight of some multimer proteins (Source: **Sheeler** and **Bianchi**, 1987).

Protein	*Molecular weight*	*Subunits*		
		Number	*Designation*	*Molecular weight*
1. Haemoglobin A (human)	64,500	4	Alpha chains (2) Beta chains (2)	15,700 16,500
2. Lactate dehydrogenase	135,000	4	A chain (0 to 4) B chain (4 to 0)	33,600 33,600
3. Immunoglobulin G	150,000	4	Light chains (2) Heavy chains (2)	25,000 50,000
4. Tryptophan synthetase (*E.coli*)	150,000	4	Alpha chains (2) Beta chains (2)	29,500 45,000
5. Aspartate transcarbamylase	306,000	12	C chains (6) R chains (6)	34,000 17,000
6. L-arabinose isomerase (*E.coli*)	360,000	6	(identical)	60,000
7. Apoferritin (iron storage protein)	456,000	24	(identical)	19,000
8. Thyroglobulin	670,000	2	(identical)	3,35,000

Some Examples of Tertiary and Quaternary Proteins

***(i)* Ribonuclease. C.B. Anfinsen** initiated and confirmed the notion that, acting in concert, the specific primary structure of a polypeptide and the innate properties of the side chains of its amino acids cause the polypeptide to spontaneously assume its biologically active tertiary structure. In 1972, he got the Nobel Prize for this definitive work. **Anfinsen** identified four disulphide bridges in the ribonuclease protein, suggesting that the enzyme is highly folded. As is the case with almost all enzymes, the catalytic activity of ribonuclease depends on the maintenance of a particular three-dimensional shape. In concentrated solutions of β-mercaptoethanol and urea, the disulphide bridges of the enzyme are broken and the resultant unfolding of the polypeptide chain is accompanied by a loss of enzyme activity. The enzyme is said to be **denatured**. If the β-mercaptoethanol and urea are removed by dialysis and the denatured ribonuclease reacted with oxygen, the four disulphide bridges reform spontaneously, and essentially all the catalytic activity of the protein is restored. Similar observations have been made with other proteins, that is, they are capable of spontaneously reestablishing their biologically active tertiary (or even quaternary, *e.g.*, haemoglobin) structure after having undergone extensive molecular disorganization.

***(ii)* Haemoglobin.** Haemoglobin is one of the fully sequenced protein. Our present understanding of the structure and function of haemoglobin is the outcome of 50 years of research of **M.F. Perutz**. He got the Nobel Prize in 1962, along with **J.C. Kendrew**, for their studies of haemoglobin and myoglobin.

The haemoglobin is a conjugated globular protein, that is, it contains some non-protein part. In all but the lowest vertebrates, haemoglobin is a tetramer (a heteropolymer). In lampreys, however, haemoglobin is monomeric, that is, it contains a single globin chain like the myoglobin. In humans, most common type of haemoglobin is haemoglobin A (HbA), which consists of 574 amino acid residues and has a molecular weight of 64,500. Its secondary, tertiary and quaternary structure is typical of all higher vertebrate haemoglobins. The protein portion of the haemoglobin molecule, called **globin**, is composed, of four polypeptide chains, each of which is also globular in shape.

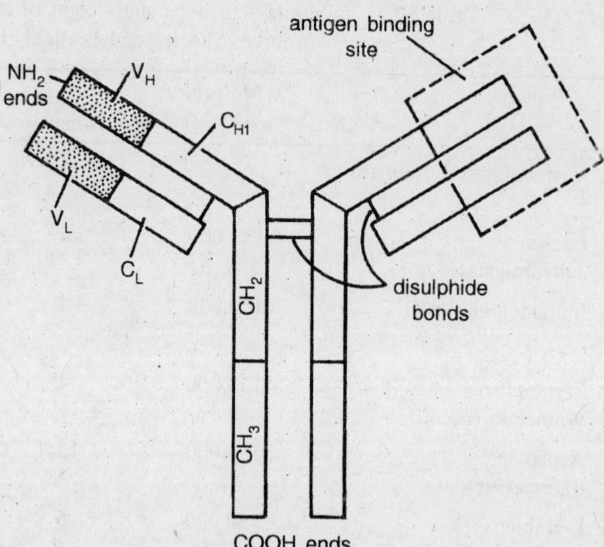

Fig. 57.14. Molecular structure of an immunoglobulin molecule (after Stansfield 1986).

The four globin chains consist of, two identical pairs: two **alpha chains** (141 amino acids each) and two **beta chains** (146 amino acids each). The non-protein portion of haemoglobin consists of four iron-containing haem groups, one associated with each of the four globin chains. Nineteen of the twenty biologically important amino acids are included in the globin of haemo-globin.

Haemoglobin molecule is highly symmetric; it can be divided into two identical halves, each consisting of an αβ-dimer. The com-plete tetramer is similar to a mildly flattened sphere having a maximum diameter of about 5.5 nm. The four polypeptide chains are arranged in such a way that unlike chains have numerous stabilizing interactions, whereas, like chains have few. A cavity about 2.5 nm long and varying in width from about 5 to 10 Å passes through the molecule along the axis. Each globin chain envelops its haem group in a deep cleft.

***(iii)* Immunoglobulins.** The ability of animals to resist infection by pathogens (viruses, bacteria and other unicellular parasites) and by multicellular endoparasites, is called **immunity**. Specific immune response function by recognizing particular chemical structures, known as **antigens**—on the surface of invading cells. An antigen can be a protein, lipid, carbohydrate or any other molecule. These antigens interact with protein molecules produced by the host, the **immunoglobulins**, which bind the antigen in much the same way as an enzyme binds its substrate. Specific immune responses involve many different types of cells. One type, the **B-lymphocyte** or **B-cell**, is capable of producing **free immunoglobulins**, called **antibodies**.

An immunoglobulin (Ig) molecule is a Y-shaped heteropolymer and is composed of two identical **H (heavy)** polypeptide chains and two smaller identical **L (Light)** polypeptide chains (Fig. 57.14). Heavy chains contain antigenic determinants in the "tail" (carboxyl) segments by which they can be classified as IgG, IgM, IgA, IgD or IgE. Light chains can likewise be typed as kappa or lambda. Within a H chain class or L chain type, these segments exhibit very little variation in primary structure from one individual to another and are called **constant regions (C)**. The amino (—NH_2) ends, however, are extremely diverse in primary structure, even within a class and are called **variable (V) regions.** The V_H and V_L regions together form two **antibody-combining sites** (called **antigen-binding sites**) for specific interactions with homologous antigen molecules. The C_H region consists of three or four similar segments, presumably derived evolutionarily by duplication of an ancestral gene and subsequent modification by mutations; the similar segments are called **domains** and are labelled CH_1, CH_2, CH_3, etc. A mature lymphocyte (plasma cell) produces antibodies with a single class H chain and a single type of L chain, hence, also a single antigen-binding specificity. The first antibodies produced by a developing plasma cells are usually of class IgM.

Enzymes. The cytoplasmic matrix and many cellular organelles contain very important organic compounds known as the **enzymes.** The word enzyme (Greek, "in yeast") had been proposed by **Kuhne** in 1878. The enzymes are the specialized proteins and they have the capacity to act as catalysts in chemical reaction. Like the other catalysts of chemical world, the enzymes are the catalysts of the biological world and they influence the rate of a chemical reaction, while themselves remain quite unchanged at the end of the reaction. The substance on which the enzymes act is known as **substrate**. The enzymes play a vital role in various metabolic and biosynthetic activities of the cell such as synthesis (anabolism) of DNA, RNA and protein molecules and catabolism of carbohydrates, lipids, fats and other chemical substances. The enzymes of the matrix and cellular organelles are classified as follows:

1. Oxireductases. The enzymes catalyzing the oxidation and reduction reaction of the cell are known as oxireductases. These enzymes transfer the electrons and hydrogen ions from the substrates, *e.g.*, hydrogenases or reductases, oxidases, oxygenases and peroxidases.

2. Transferases. The enzymes which transfer following groups from one molecule to other are known as transferases: one carbon, aldehydic or ketonic residues, acyl, glycosyl, alkyl, nitrogenous, phosphorus containing groups and sulphur containing groups.

3. Hydrolases. These enzymes hydrolyse a complex molecule into two compounds by adding the element of the water across the bond which is cleaved. These enzymes act on the following bonds–ester, glycosyl, ether, peptide, other C–N bonds, acid anhydride, C–C, halide and P–N bonds. Certian important hydrolase enzymes are the proteases, esterases, phosphatases, nucleases and phosphorylases.

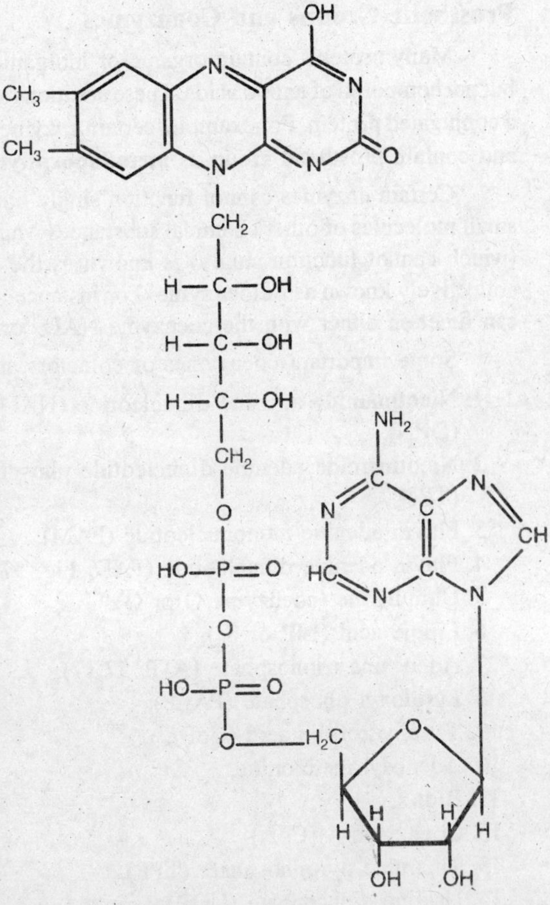

Fig. 57.15. Chemical formula of flavin adenine dinucleotide (FAD).

4. Lyases. The lyase enzymes add or remove group to or from the chemical compounds containing the double bonds. The lyases act on C–C, C–O, C–N, C–S and C–halide bonds.

5. Isomerases. These enzymes catalyse the reaction involving in the isomerization or intramolecular rearrangements in the substrates, *e.g.*, intramolecular oxidoreductases, intramolecular transferases, intramolecular lysases, cis-trans-isomerases, racemases and epimerases.

6. Ligases or synthetases. These enzymes catalyze the linkage of the molecules by splitting a phosphate bond. The synthetase enzymes form C–O, C–S, C–N and C–C bonds.

According to the chemical nature of the substrate the enzymes have also been classified as follows: 1. Carbohydrases, 2. Proteases (endopeptidases and exopeptidases), 3. Amylases, 4. Esterases, 5. Dehydrogenases, 6. Oxidases, 7. Decarboxylases, 8. Hydrases, 9. Transferases, and 10. Isomerases.

The enzymes are **specific** in action and many factors such as pH, temperature and concentration of the substrate affect the rate of the activity of enzymes. Certain enzymes occur in the inactive form (called **proenzymes** or **zymogens**) and these are activated by other enzymes known as **kinases** to perform catalytic activities. Likewise, the enzyme **trypsinogen** of the pancreatic cells is activated in the intestine by the enzyme **enterokinase** and the enzyme **pepsinogen** of the Chief cells of the stomach is activated by the **hydrochloric acid** which is secreted by parietal cells.

Prosthetic Groups and Coenzymes

Many proteins contain organic or inorganic units that are the integral part of the molecule, but not composed of amino acids. These additions are called **prosthetic groups;** their presence defines a conjugated protein. For example, certain enzymes such as **cytochromes** are the conjugated proteins and contain prosthetic group as **metalloporphyrins** complex in their molecules.

Certain enzymes cannot function singly but they can function only by the addition with the small molecules of other chemical substances which are known as **coenzymes**. The inactive enzyme (which cannot function singly) is known as the **apoenzyme**. The apoenzyme and coenzyme are collectively known as **holoenzyme**. For instance, the enzyme **hydrogenase** is an apoenzyme which can function either with the coenzyme NAD^+ or NADP.

Some important coenzymes or cofactors are as follows:

1. Nicotinamide adenine dinucleotide (NAD, Fig. 57.16) or Diphosphopyridine nucleotide (DPN),
2. Nicotinamide adenine dinucleotide phosphate (NADP) or Triphosphopyridine nucleotide (TPN),
3. Flavin adenine mononucleotide (FAM),
4. Flavin adenine dinucleotide (FAD, Fig. 57.15)
5. Ubiquinone (coenzyme Q or Q),
6. Lipoic acid (LIP or S_2),
7. Adenosine triphosphate (ATP, 57.17),
8. Pyridoxyl phosphate (PALP),
9. Tetrahydrofolic acid (CoF),
10. Adenosyl methionine,
11. Biotin,
12. Coenzyme A (CoA),
13. Thiamine pyrophosphate (TPP),
14. Uridine diphosphate (UDP).

Nucleic Acids

The nucleic acids are the complex macromolecular organic compounds of immense bio-logical importance. They control the important biosynthetic activities of the cell and carry hereditary informations from generation to generation. There occur two types of nucleic acids in living organisms, *viz.*, **ribonucleic acid (RNA)** and **deoxyribonucleic acid (DNA)**. Both types of nucleic acids are the polymers of the nucleotides. A nucleotide is composed of nucleoside and phosphoric acid. Even the nucleoside is composed of the pentose sugars (Ribose or Deoxyribose) and nitrogen bases (Purines or Pyrimidines). The **purines** are **adenine** and **guanine** and the **pyrimidines** are the **cytosine, thymine** and **uracil.** The cytoplasmic matrix contains only RNA, while DNA exclusively remains concentrated in the nucleus.

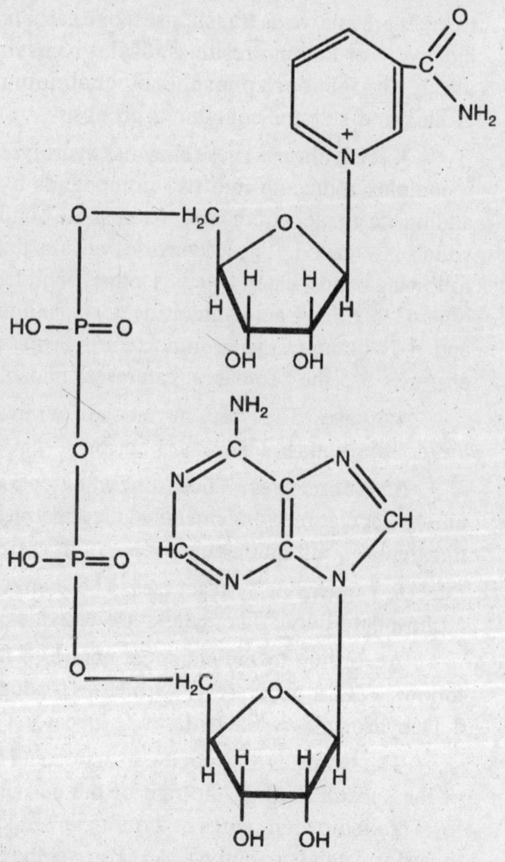

Fig. 57.16. Chemical formula of nicotinamide adenine dinucleotide (NAD^+).

The DNA and RNA have almost similar chemical compositions except a few differences. Both have been compared in Table 57.6.

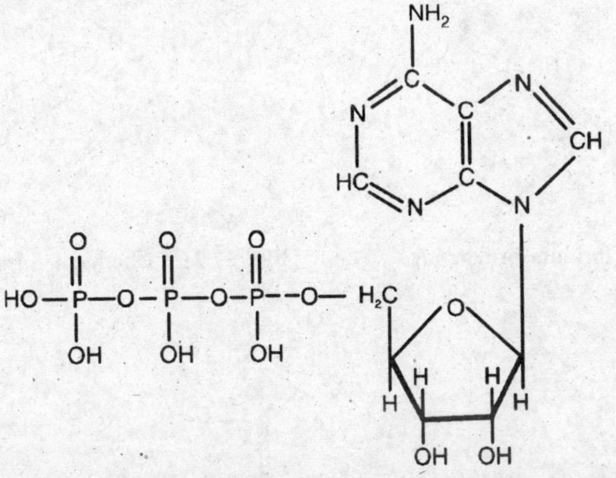

Fig. 57.17. Chemical formula of adenosine triphosphate (ATP).

Table 57.6 Comparison of DNA and RNA

DNA	*RNA*
1. It contains pentose sugar known as deoxyribose.	1. It contains pentose sugar called the ribose.
2. The molecule contains the phosphoric acid (phosphate) molecule which connects various sugars with one another.	2. The molecule contains the phosphoric acid (phosphate) molecule which connects various sugars with one another.
3. The nitrogen bases are: (*i*) **Purines**—adenine and guanine. (*ii*) **Pyrimidine**—cytosine and thymine.	3. The molecule contains following nitrogen bases in its molecule: (*i*) **Purines**—adenine and guanine. (*ii*) **Pyrimidines**—cytosine and uracil.
4. Molecules have four nucleotides as deoxyadenosine monophosphate, deoxyguanosine monophosphate, deoxycytidine monophosphate and thymidine monophosphate.	4. Molecules have four nucleotides as adenosine monophosphate, guanosine monophosphate, cytidine monophosphate and uridine monophosphate.
5. The molecule contains a double stranded helix structure in which many nucleotides remain arranged in pair.	5. The molecules consist of single chain of polynucleotides.
6. DNA is a genetic material and occurs in chromosomes, nucleoplasm and mitochondria, chloroplasts, etc.	6. RNA is a carrier of genetic informations and it plays very significant role in the mechanism of protein synthesis. It mostly occurs in nucleolus, nucleoplasm and cytoplasm.

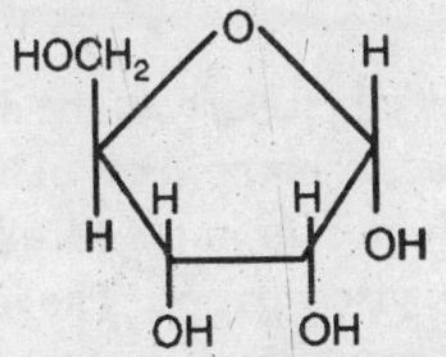

Fig. 57.18 Chemical Formula of ribose sugar.

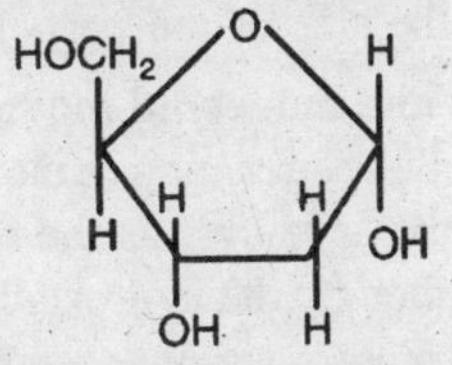

Fig. 57.19 Chemical formula of deoxyribose sugar.

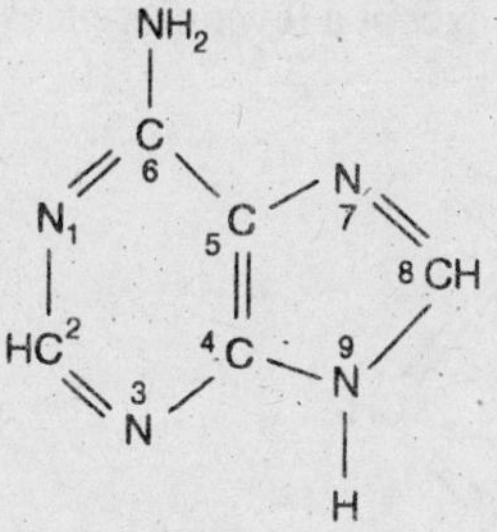

Fig. 57.20 Chemical formula of adenine.

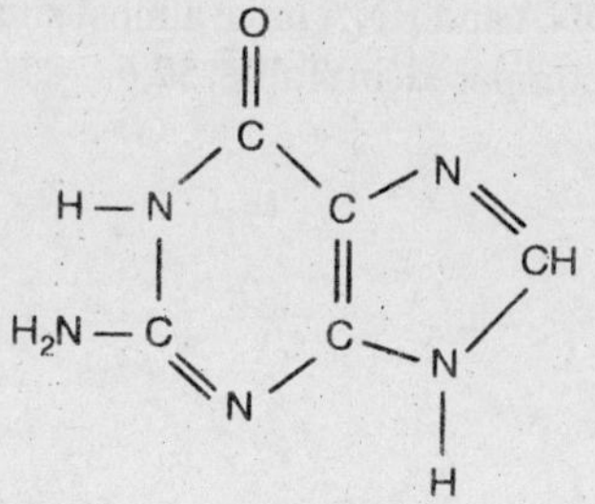

Fig. 57.21 Chemical formula of guanine.

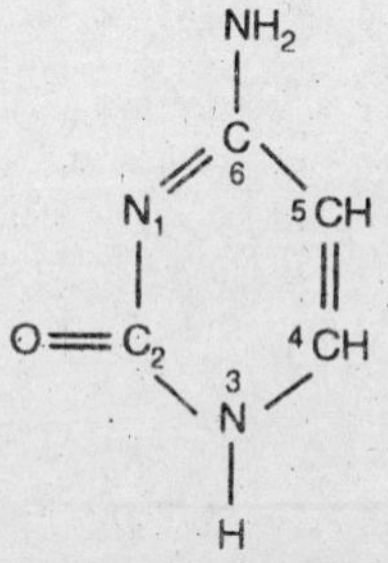

Fig. 57.22 Chemical formula of cytosine.

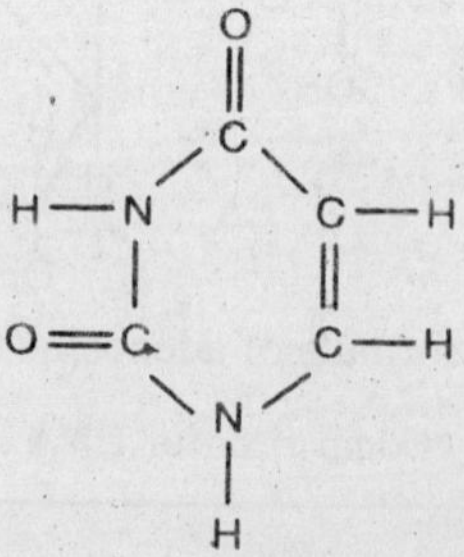

Fig. 57.23 Chemical formula of uracil.

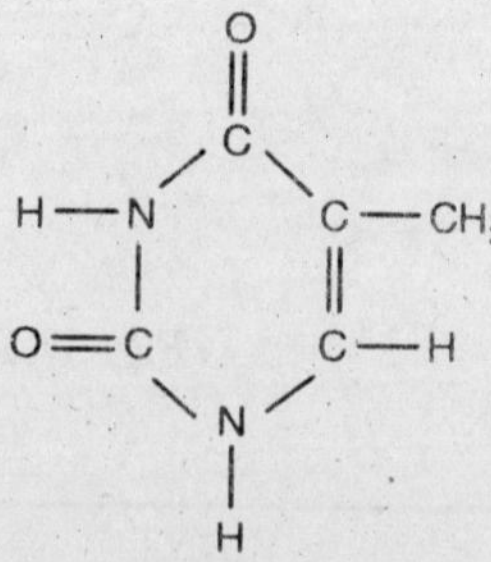

Fig. 57.24. Chemical of formula of thymine.

57.3 PHYSICAL AND BIOLOGICAL PROPERTIES OF CYTOPLASMIC MATRIX

The matrix is a living substance and it has following physical and biological properties:

Physical Properties

The most of the physical properties of the matrix are due to its colloidal nature and these are as follows:

1. Tyndall's effect. When a beam of strong light is passed through the colloidal system, of the matrix at right angles in the dark room, the small colloidal particles which remain suspended in the colloidal system reflect the light. The path of the light appears like a cone. This light cone is known as **Tyndall's cone** because, this phenomenon has been first of all reported by **Tyndall** (1820–1893) in colloids.

2. Brownian movement. The suspended colloidal particles of the matrix always move in zig-zag fashion. This movement of molecules is caused by moving water molecules which strike with the colloidal molecules to provide motion to them. This type of movement was first of all observed by Scottish botanist **Robert Brown** in 1827 in the colloidal solution. Therefore, such movements are known as **Brownian movement**. The Brownian movement is the peculiarity of all colloidal solutions and depends on the size of the particles and temperature.

3. Cyclosis and amoeboid movement. Due to the phase reversal property of the cytoplasmic matrix, the intracellular streaming or movement of the matrix takes place. This property of intracullular movement of matrix is known as the **cyclosis**. The cyclosis usually occurs in the sol-phase of the matrix and is effected by the hydrostatic pressure, temperature, pH, viscosity, etc. The intracellular movements of the pinosomes, phagosomes and various cytoplasmic organelles such as the lysosomes, mitochondria, chromosomes, centrioles, *etc.*, occur only due to cyclosis of the matrix. The cyclosis has been observed, in most animal and plant cells.

The amoeboid movement depends directly on the cyclosis. The amoeboid movement occurs in the protozoans, leucocytes, epithelia, mesenchymal and other cells. In the amoeboid movement the cell changes its shape actively and gives out cytoplasmic projections known as **pseudopodia**. Due to cyclosis matrix moves these pseudopodia and this causes forward motion of the cell.

4. Surface tension. The molecules in the interior of a homogeneous liquid are free to move and are attracted by surrounding molecules equally in all directions. At the surface of the liquid where it touches air or some other liquid, however, they are attracted downward and sideways or inward, more than upward; consequently they are subjected to unequal stress and are held together to form a membrane. The force by which the molecules are bound is called the **surface tension** of the liquid. The cytoplasmic matrix being a liquid possesses the property of surface tension. The proteins and lipids of matrix have less surface tension, therefore, occur at the surface and form the membrane, while the chemical substances such as NaCl have high surface tension, therefore, occur in deeper part of the matrix.

5. Adsorption. The increase in the concentration of a substance at the surface of a solution is known as adsorption (*L.*, *ad*=to, *sorbex*=to draw in). The phenomenon of adsorption helps the matrix to form protein boundaries.

6. Other mechanical or physical properties of matrix. Besides surface tension and adsorption, the matrix possesses other mechanical properties, *e.g.*, elasticity, contractility, rigidity and viscosity which provide to the matrix many physiological utilities.

7. Polarity of the egg. The colloidal system due to its stable phase determines the polarity of the cell matrix which cannot be altered by centrifugation of other mechanical means.

8. Buffers and pH. The matrix has a definite pH value and it does not tolerate significant variations in its pH balance. Yet various metabolic activities produce small amount of excess acids or bases. Therefore, to protect itself from such pH variation the matrix contains certain chemical compounds as carbonate-bicarbonate system known as **buffers** which maintain a constant state of pH in the matrix.

Biological Properties

The matrix is a living substance and it has following biological properties:

1. Irritability. The irritability is the fundamental and inherent property of the matrix. It possesses a sensitivity to stimulation, an ability to transmission of excitation and ability to react according to stimuli. The heat, light, chemical substances and other factors stimulate the cytoplasmic matrix to contract.

2. Conductivity. The conductivity is the process of conduction or transmission of excitation from the place of its origin to the region of its reaction. The matrix of nerve cells possesses the property of the conductivity.

3. Movement. The cytoplasmic matrix can perform movement due to cyclosis. The cyclosis depends on the age, water contents, heredity factors and composition of the cells.

4. Metabolism. The matrix is the seat of various chemical activities. These activities may be either constructive or destructive in nature. The constructive processes such as biosynthesis of proteins, lipids, carbohydrates and nucleic acids are known as **anabolic processes,** while the destructive processes such as oxidation of foodstuffs, *etc.*, are known as **catabolic processes.** The anabolic and catabolic processes are collectively known as **metabolic process**.

5. Growth. Due to the secretory or anabolic activities (*Gr., anabolism* = a throwing up) of the cell, new protoplasm continuously increases in its volume. The increase in the volume of the matrix causes into the growth of the cell which ultimately divides into daughter cells by the cell division.

6. Reproduction. The cytoplasm has the property of asexual and sexual reproduction.

QUESTIONS

Long Answer Questions

1. Write in brief the physical properties of the protoplasm.
2. Write in brief the chemical properties of the protoplasm.
(*Guwahati (G) 1993, 1994; Kanpur 1996*)
3. Describe biological properties of protoplasm.

Short Answer Questions

1. Describe in brief nine types of proteins according to function.
2. Differentiate between DNA and RNA. (*Guwahati (G), 1993*)

Very Short Answer Questions

1. In what forms carbohydrates are stored in plants and animals?

Ans. In the form of starch in plants and in the form of glycogen in animals.

2. Why amino acids are called amphoteric?

Ans. Since amino acids contain at least one amino group and at least one carboxyl group, they are called amphoteric.

3. Name two animal hormones that are peptides (or proteins) and steroids.

Ans. Peptide (or protein) hormones: Insulin and growth hormone. Steroid hormone: Estrogen and Testosterone.

4. What is cholesterol?

Ans. Cholesterol is a lipid constituent of animal cell plasma membrane; serves as a precursor to the steroid hormones.

5. Explain the antibody.

Ans. Antibody is a class of proteins produced by lymphocytes that bind with extraordinary specificity to substances, referred to as antigens, that provoke an immune response.

6. Define antigen.

Ans. Antigen is a foreign or abnormal substance that can trigger an immune response.

7. What is adenine?

Ans. Adenine (A) is a nitrogen-containing aromatic base, chemically designated as a purine, that serve as an informational monomeric unit when present in nucleic acids with other bases in a specific sequence; forms a complementary base pair with thymine (T) or uracil (V) by hydrogen bonding.

8. Define cytosine.

Ans. Cytosine (C) is a nitrogen-containing aromatic base, chemically designated as a purimidine, that serves as an informational monomeric unit when present in nucleic acids with other bases in a specific sequence; forms a complementary base pair with guanine (G) by hydrogen bonding.

9. What is guanine?

Ans. Guanine (G) is a nitrogen-containing aromatic base, chemically designated as a purine, which serves as an informational monomeric unit when present in nucleic acids with other bases in a specific sequence; forms a complementary base pair with cytosine (C) by hydrogen bonding.

10. Define thymine.

Ans. Thymine (T) is a nitrogen-containing aromatic base, chemically designated as a purimidine, which serves as an informational monomeric unit when present in DNA with other bases in a specific sequence; forms a complementary base pair with adenine (A) by hydrogen bonding.

11. What is uracil?

Ans. Uracil (U) is a nitrogen-containing aromatic base, chemically designated as a pyrimidine, that serves as an information monomeric unit when present in RNA with other bases in a specific sequence; forms a complementary base pair with adenine (A) by hydrogen bonding.

Multiple Choice Questions

1. The four elements that form the 99% of the protoplasm are

(*a*) N, O, P, S (*b*) C, H, N, O
(*c*) H, N, Ca, Cl (*d*) C, O, Ca, S

2. Which of the following contain iodine?

(*a*) insulin (*b*) thyroxine
(*c*) ocytocin (*d*) adrenalin

3. Sucrose is composed of

(*a*) fructose + galactose (*b*) glucose + fructose
(*c*) glucose + glucose (*d*) glucose + galactose

4. Which of the following lipids are found in largest number in cell membrane?

(*a*) glycolipids (*b*) phospholipids
(*c*) sphingolipids (*d*) mucolipids

5. Name the gland which acts both as an endocrine as well as exocrine gland

(*a*) pancreas (*b*) pituitary
(*c*) thyroid (*d*) adrenal

6. The "repeating unit" of glycogen is

(*a*) fructose (*b*) mannose
(*c*) glucose (*d*) galactose

7. Which of the following is basic amino acid?

(*a*) threonine (*b*) glycine
(*c*) lysine (*d*) cysteine

8. Collagen is

(*a*) phosphoprotein (*b*) globulin
(*c*) derived protein (*d*) scleroprotein

ANSWERS

Multiple Choice Questions

1. (*b*) **2.** (*b*) **3.** (*b*) **4.** (*b*)
5. (*a*) **6.** (*c*) **7.** (*c*) **8.** (*d*).

58

Cellular Energy Transactions (Role of Mitochondria and Chloroplasts)

A living cell bustles with activity. Macromolecules of all types are assembled from raw materials, waste products are produced and excreted, genetic instructions flow from the nucleus to the cytoplasm, vesicles are moved from the Golgi apparatus to the plasma membrane, ions are pumped across cell membranes, and so forth. To maintain such a high level of activity, a cell must acquire and expend energy. The study of various types of energy transformations that occur in living organisms is referred to as **bioenergetics**.

58.1 ENERGY TRANSDUCTIONS

Energy is defined as the capacity to do work, that is, the capacity to change or move some thing. Energy exists in two alternate states: potential and kinetic. A rock perched on the edge of a cliff possesses **potential energy** because it has the potential to perform work. Rock has this potential because it exists in a field of force, in this case, a gravitational field. If the rock is pushed over the edge, then gravity can act on it and cause it to fall, while falling, the rock has **kinetic energy** and can accomplish work, for example, by lifting another object (*i.e.*, fish in Fig. 58.1). Similarly, a resting nerve cell, has potential energy when it maintains a high concentration of sodium ions on the outer side of its plasma membrane and a low concentration of sodium ions on the inner side. Like the flow of water through a dam, the opening of specific channels in the plasma membrane allows sodium ions to flow across the membrane into the cell. The directed movement of sodium ions into the cell is a form of kinetic energy that can be used to accomplish work, as occurs when a nerve impulse passes along the membrane to the nerve cell.

In both of these cases just described–the falling rock and the directed movement of sodium ions–as in all cases of measurement of energy at work, two factors must be considered: a potential factor and a capacity factor (Table 58.1). The **potential factor** is proportional to the intensity of the field of force, whereas the **capacity factor** provides some measure of the "size" of the subject being considered. In the case of falling rock, the **potential factor** is the distance it will fall, and the **capacity factor** is the rock's mass. For the movement of charged ions, the potential factor is the voltage, and the capacity factor is the combined charged of the particles. The work, or energy released during these events, is a multiple of these two factors; as either one increases, so too does the amount of energy.

Rock has potential energy

Falling rock has kinetic energy

Rock has accomplished work

Fig. 58.1. Accomplishing work. The rock perched on top of the cliff has potential energy. The energy available for work is proportional to the mass of the rock and the distance it can fall. Once it leaves the edge of the cliff, the rock has kinetic energy that can be used to accomplish work, in this by lifting both water and fish (after Karp, 2002).

Table 58.1 Potential and capacity factors in measuring energy **(source: Karp 2002)**

Type of energy	*Potential factor*	*Capacity factor*
1. Mechanical		
Falling of rock	Height	Mass
Compression	Pressure	Volume
Stretching	Tension	Length
2. Electrical		
Movement of ions	Electric potential (voltage)	Charge
3. Osmotic		
Movement of water into a cell	Solute concentration gradient	Mass
4. Energy available to do work	Temperature	Entropy

The Laws of Thermodynamics and the Concept of Entropy

Thermodynamics is the study of the changes in energy that accompany events in the universe.

The First Law of Thermodynamics. The first law of thermodynamics is the law of conservation of energy. It states that **energy** can neither be created nor destroyed. Energy can, however, be converted **(transduced)** from one form to another. Electric energy is converted to mechanical energy when we plug in a fan, and chemical energy is converted to thermal energy when fuel is burned in an oil heater. Cells are also capable of energy transduction. The chemical energy present in certain biological molecules, such as ATP, is converted to mechanical energy when organelles are moved from place to place in a cell, to thermal energy when heat is released during muscle cell contraction, or to electrical energy when ions flow across a membrane. *The most important energy transduction in the biological world is the conversion of sunlight into chemical energy–the process of photosynthesis*–which provides the fuel that directly or indirectly powers the activities of nearly all forms of life. A number of animals including fireflies and luminous fish are able to convert chemical energy back into light. Regardless of the transduction process, however, the total amount of energy in the universe remains constant.

To discuss energy transformations involving matter, we need to divide the universe into two parts: the **system** under study and the remainder of the universe, which we will refer to as the

surroundings. A system can be defined in various ways: it may be a certain space in the universe or a certain amount of matter. For example, the system may be a living cell. The changes in a system's energy that occur during an event are manifested in two ways—as a change in heat content of a system and in the performance of work. Even though the system may gain or loss, energy, the first law of thermodynamics indicates that the loss or gain must be balanced by a corresponding gain or loss in the surroundings, so that the amount in the universe as a whole remains constant. The energy of the system is termed as **internal energy (E)**, and its change during a transformation in **ΔE**. One way to describe the first law of thermodynamics is that $\Delta E = Q - W$, where Q is the heat energy and W is the work energy.

Depending on the process, the internal energy of the system at the end can be greater than, equal to, or less than its internal energy at the start, depending on its relationship to its surroundings (Fig. 58.2). In other words ΔE (*i.e.*, change during the transformation) can be positive, zero, or negative.

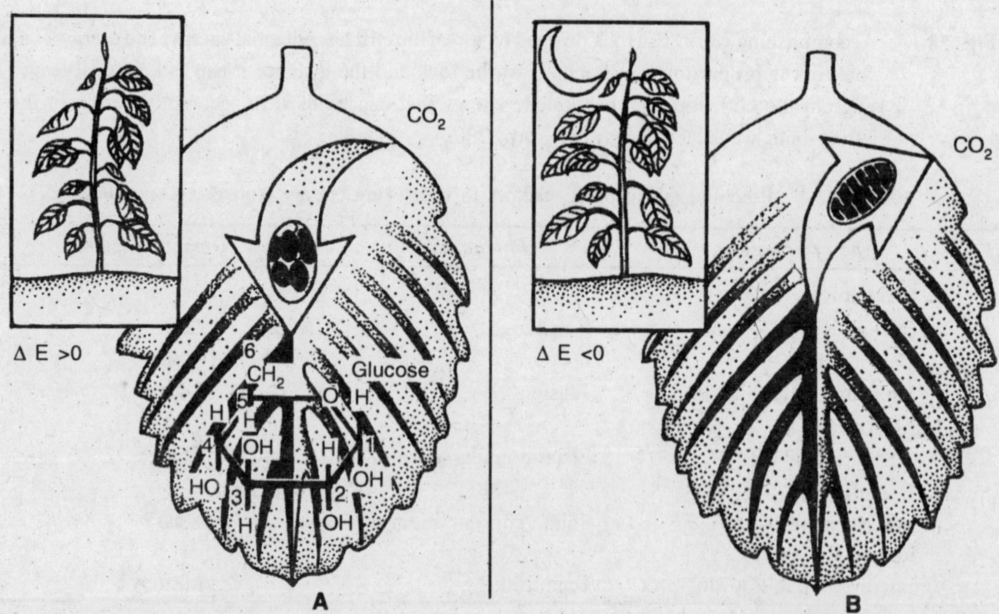

Fig. 58.2. A change in system's internal energy. In this example, the system will be defined as a particular leaf of a plant. A–During the day, sunlight is absorbed by photosynthetic pigments in the leaf's chloroplasts and used to convert CO_2 into carbohydrates, such as glucose molecule shown in the drawing (glucose is subsequently incorporated into sucrose or starch). As the cell continues to absorb light, its internal energy increases; the energy present in the remainder of the universe has to decrease. B–At night, the relationship between the cell and its surroundings is reversed as the carbohydrate produced during the day are oxidized to CO_2 in the mitochondria and the energy is used to run the cell's nocturnal activities (after Karp, 2002).

W e can consider a system to be the contents of a reaction vessel. As long as there is no change in pressure or volume of the contents, there is no work being done by the system on its surroundings, or vice versa. In that case, the internal energy at the end of the transformation will be greater than at the beginning if heat is absorbed and less of heat is released. Reactions that lose heat are termed **exothermic**, and ones that gain heat are **endothermic**, and there are many reactions of both types.

The Second Law of Thermodynamics. The second law of thermodynamics expresses the concept that events in the universe have direction; they tend to proceed "downhill" from a state of higher energy to a state of lower energy. Thus, in any energy transformation, there is a decreasing availability of energy for doing additional work. Rocks fall off cliffs to the ground below, and once at the bottom, their ability to do additional work is reduced; it is very unlikely that they will lift themselves back to the top of the cliff. Similarly, opposite charges normally move together, not apart, and heat flows from a warmer to a cooler body not the reverse. Such events are said to be spontaneous, a term that indicates that they are thermodynamically favourable and *can occur without the input of external energy.*

The concept of the second law of thermodynamics was formulated originally for engines, and the law carried with the idea that it is thermodynamically impossible to construct a perpetual-motion machine. In other words, it is impossible for a machine to be 100 percent efficient, which would be required if it were to continue functioning without the input of external energy. Some of the energy is inevitably lost as the machine carries out its activity. A similar relationship holds true for living organisms. For example, when a giraffe browses on the leaves of a tree or a lion preys on the giraffe, most of the chemical energy in the food never becomes available to the animal having the meal. The energy that is unavailable for doing additional work after an event has occurred has a potential factor and a capacity factor (Table 58.1). The potential factor is temperature (in degrees) and the capacity factor is **entropy (S)**, which has the dimensions of energy per degree (*e.g.*, calories per degree). The unavailable energy is equal to $T\Delta S$, where ΔS is the change in entropy between the initial and final states.

The loss of available energy during a process is a result of a tendency for the randomness, or disorder, of the universe to increase every time there is a transfer of energy. Entropy provides a measure of this disorder; it is also associated with the random movements of the particles of matter, which, because they are random, cannot be made to accomplish a *directed* work process. According to the second law of thermodynamics, every event is accompanied by an increase in the entropy of the universe. When a sugar cube is dropped into a cup of hot water, for example, there is a spontaneous shift of the molecules from an ordered state in the crystal to a much more disordered condition when the sugar molecules are spread throughout the solution (Fig. 58.3A). As the molecules of the sugar cube dissolve into solution, their freedom of movement increases, as does the entropy of the system. The change from the concentrated to the dispersed state results from the random movements of the molecules. The sugar molecules eventually spread themselves equally through the available volume because the state of uniform distribution is the most probable state.

For example, the release of heat from the oxidation of glucose within a cell or from the friction generated as blood flows through a vessel, is another example of an increase in entropy. The release of thermal energy by living organisms increases the rate of random movements of atoms and molecules; it cannot be redirected to accomplish additional work. Since the energy of molecular and atomic movements increases with temperature, so too does the entropy. It is only at absolute zero (0°K), when all movements ceases, that the entropy is zero.

As with other spontaneous events, one must distinguish between the system and its surroundings. The second law of thermodynamics indicate only that the total entropy in the universe must increase; the disorder within one part of the universe (the system) can decrease at the greater expense of its surroundings. The dissolved sugar can decrease in entropy it can be recrystallized by evaporating the water (Fig. 58.3B). The consequence of this process, however, is an increase in the entropy of the surroundings. The increased freedom of movement of the water molecules in the gaseous phase more than balances the decrease in freedom of the molecules of the sugar crystals.

Life operates on in a similar principle. Living organisms are able to decrease their own entropy by increasing the entropy of the environment. Entropy is decreased in an organism when relatively simple molecules, such as amino acids, are ordered into more complex molecules, such as the protein myoglobin in a muscle cell. For this to occur, however, the entropy of the environment must increase, which is accomplished as complex, ordered molecules such as glycogen stored in liver or muscle tissue are converted into heat and smaller less ordered compounds (such as CO_2 and H_2O) are released to the environment. It is this feature of metabolism that allows living organisms to maintain such a highly ordered and improbable state.

Another measure of the energy state of a living organism is provided by the information content of its macromolecules. **Information** is a subject that is difficult to define but easy to recognise. Information can be measured in terms of the ordered arrangement of a structure's subunits. For example, proteins and nucleic acids, in which the specific linear sequence of the subunits is highly ordered, are low in entropy and high in information content. Maintaining a state of high information content (low entropy) requires the input of energy. We can consider here just one molecule of DNA located in one cell of our liver. That cell has dozens of different proteins whose sole job is to patrol the DNA, looking for damage and repairing it. Nucleotide damage in an active cell can be so great that, without this expenditure of energy, the information content of DNA would rapidly deteriorate (See **Karp** 2002).

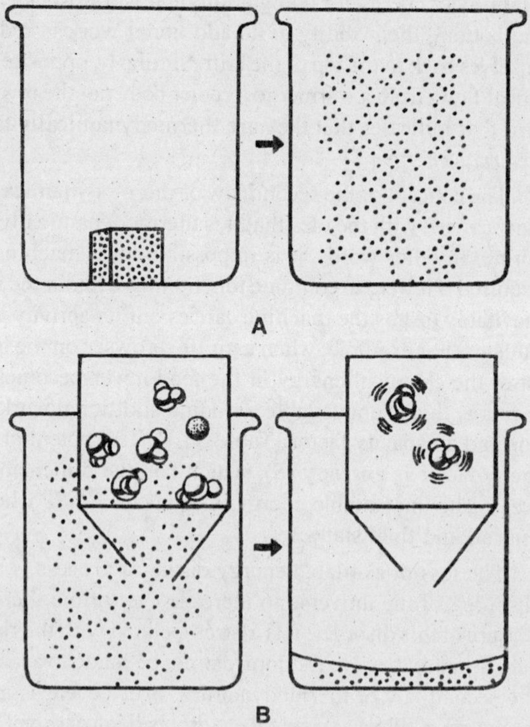

A

B

Fig. 58.3. Events are accompanied by an increase in the entropy of the universe. A–A sugar cube contains sucrose molecules in highly ordered arrangement in which the freedom of movement of the sucrose molecules is greatly increased, and their random movement causes them to become equally distributed throughout the available space. Once this occurs, there will be no further tendency for redistribution, and to entropy of the system is at a maximum. B–sugar molecules spread randomly through a solution can be returned to a ordered state, but only if the entropy of the surroundings is increased, as occur when the more ordered water molecules of the liquid phase become disordered following evaporation (after Karp, 2002).

Free Energy. Together, the first and second laws of thermodynamics indicate that the energy of the universe is constant, but the entropy continues to increase toward a maximum. The concepts inherent in first two laws were put together by the American chemist, **Josiah Willard Gibbs** in 1878 into the expression $\Delta H = \Delta G + T\Delta S$, where ΔG is the change in **free energy**, that is, the change during a process in the energy available to do work; ΔH is the change in **enthalpy** or total energy content of the system (equivalent to ΔE), T is the absolute temperature (°K = °C + 273), and ΔS is the change in the entropy of the system. The equation states that the total energy change is equal to the sum of the changes in useful energy (ΔG) and energy that is unavailable to do further work ($T\Delta S$).

Processes that can occur spontaneously, that is, processes that are thermodynamically favoured (have a $-\ \Delta G$), are described as **exergonic**. In contrast, if the ΔG for a given process is positive, then it cannot occur spontaneously. Such processes are thermodynamically unfavourable and are described as **endergonic**.

58.2 METABOLISM

Metabolism is the collection of biochemical reactions that occur within a cell, which include a tremendous diversity of molecular conversions. Most of these reactions can be grouped into **metabolic pathways** containing a sequence of chemical reactions in which each reaction is catalyzed by a specific enzyme, and the product of one reaction is the substrate of the next. The enzymes constituting a metabolic pathway are usually confined to a specific region of the cell, such as mitochondria or the cytosol. Increasing evidence suggests that the enzymes of a metabolic pathway are often physically linked to one another, a feature that allows the product of one enzyme to be delivered directly as a substrate to the active sites of the next enzyme in the reaction sequence.

The compound formed each step along the pathway are **metabolic intermediate** (or **metabolites**) that lead ultimately to the formation of an **end product**. End products are the molecules with a particular role in the cell, such as an amino acid that can be incorporated into a polypeptide, or a sugar that can be consumed for its energy content. The metabolic pathways of a cell are interconnected at various points so that a compound generated by one pathway may be shuttled in a number of directions depending on the requirements of the cell at the time.

Metabolic pathways can be divided into two broad types. **Catabolic pathways** lead to the disassembly of complex molecules to form simpler products. Catabolic pathways serve two functions: they make available the raw materials from which other molecules can be synthesized, and they provide chemical energy. The energy released by catabolic pathways is stored temporarily in two forms: a high energy phosphate (primarily ATP) and high-energy electrons (primarily in NADPH). **Anabolic pathways** lead to the synthesis of more complex compounds from simple starting materials. Anabolic pathways are energy requiring and utilize chemical energy released by the exergonic catabolic pathways.

Oxidation and Reduction

Both catabolic and anabolic pathways include key reactions in which electrons are transferred from one reactant to another. Reactions that involve a change in the electronic state of the reactants are called **oxidation-reduction** (or **redox) reactions**. Changes of this type involve the gain or loss of electrons, *e.g.*, the conversion of metallic iron (Fe^0) to the ferrous state (Fe^{2+}), by which iron atom loses a pair of electrons, thereby attaining a more positive state. When an atom loses one or more electrons, it is said to be **oxidized**. The reaction is reversible, meaning ferrous ions can be converted to metallic iron, a more negative state, by the acquisition of a pair of electrons. When an atom gains one or more electrons, it is said to be **reduced**.

For metallic iron to be oxidized, there must be some substance to accept the electrons that are released. Conversely, for ferrous ions to be reduced, there must be some substance to donate the necessary electrons. In other words, the oxidation of one reactant must be accompanied by the simultaneous reduction of some other reactant, and vice versa. One possible reaction involving iron might be

$$Fe^0 + Cu^{2+} \rightleftharpoons Fe^{2+} + Cu^0$$

The substance that is oxidized during an oxidation-reduction reaction, that is, the one that loses electrons, is called a **reducing agent**, and the one that is reduced, that is, the one that gains electrons, is called an **oxidizing agent**.

The oxidation or reduction of metals, such as iron or copper, involves the complete loss or gain of electrons. The same cannot occur with most organic compounds for the following reason; the oxidation and reduction of organic substrates during cellular metabolism involve carbon atoms that are covalently bonded to other atoms. When a pair of electrons are attracted more strongly to one of the two atoms of the polarized bond. In a C – H bond, the carbon atom has the strongest pull on the electrons; thus, it can be said that carbon atom is in a reduced state. In contrast, if a carbon atom is bonded to a more electronegative atom, as in a C – O or C – N bond, the electrons are pulled away from the carbon atom, which is thus is an oxidised state, since carbon has four outer-shell electrons it can share with other atoms, it can exist in a variety of oxidation states. This is illustrated by the carbon atom in a series of one carbon molecules (Fig. 58.4) ranging from the fully reduced state in methane (CH_2) to the fully oxidized state in carbon dioxide (CO_2). The relative oxidation state of an organic molecule can be roughly determined by counting the number of hydrogen versus oxygen and nitrogen atoms per carbon atom. The oxidation state of the carbon atoms in an organic molecule provides a measure of the molecules free-energy content.

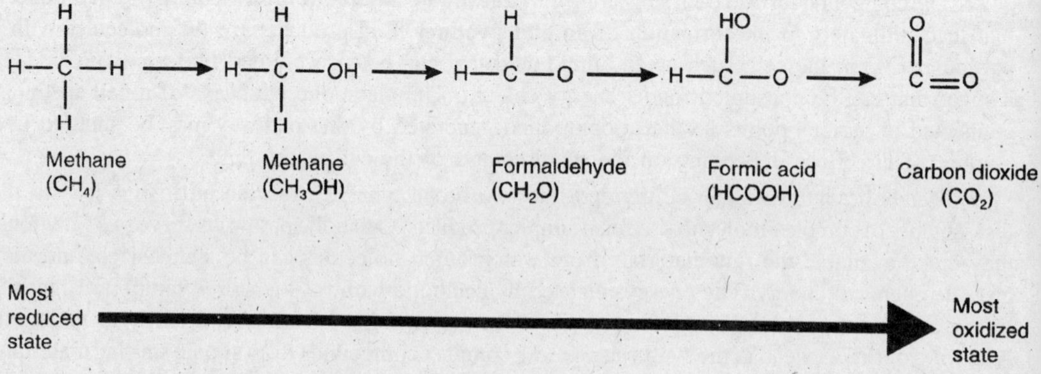

Methane (CH_4) → Methane (CH_3OH) → Formaldehyde (CH_2O) → Formic acid (HCOOH) → Carbon dioxide (CO_2)

Most reduced state → Most oxidized state

Fig. 58.4. The oxidation state of a carbon atom depends on the other atoms to which it is bonded. Each carbon atom can form a maximum of four bonds with other atoms. This series of simple, one-carbon molecules illustrates the various oxidation states in which the carbon atom can exist. In its most reduced state, the carbon is bonded to four hydrogens (forming methane); in its most oxidized state, the carbon atom is bonded to two oxygen (forming carbon dioxide) (after Karp 2002).

The capture and utilization of Energy

The compounds we use as chemical fuels to run automobiles are highly reduced organic compounds, such as natural gas (CH_4) and petroleum derivatives. Energy is released when these molecules are burned in the presence of oxygen, converting the carbon to more oxidized state, as in carbon dioxide and carbon monoxide gases. The degree of reduction of a compound is also a measure of its ability to perform chemical work within the cell. The more hydrogen atoms that can be stripped from a "fuel" molecule, the more ATP that ultimately can be produced. Carbohydrates are rich in chemical energy because they contain strings of (H–C–OH) units. Fats contain even greater energy per unit weight because they contain strings of more reduced (H–C–H) units.

As the sole building block of both starch and glycogen, glucose is a key molecule in the energy metabolism of plants and animals. The free energy released by the complete oxidation of glucose is very large:

$$C_6H_{12}O_6 + 6O_2 \longrightarrow 6CO_2 + 6H_2O$$
$$\Delta G^{0'} = -686 \text{ kcal/mol}$$

By comparison, the free energy required to convert ADP to ATP is relatively small:

$$ADP + Pi \longrightarrow ATP + H_2O$$
$$\Delta G^{0'} = 7.3 \text{ kcal/mol}$$

It is evident from these numbers that the complete oxidation of a molecule of glucose to CO_2 and H_2O can release enough energy to produce a large number of ATPs. As we will see in this chapter later on, upto 36 molecules of ATP are formed per molecule of glucose oxidized under conditions that exist in most cells. For this many ATP to be produced, the sugar molecule is disassembled in many small steps. Those steps in which the free-energy difference between the reactants and products is relatively large can be coupled to reactions that lead to the formation of ATP.

There are basically two stages in the catabolism of glucose, and they are virtually identical in all aerobic organisms. The first stage, **glycolysis**, occurs in the soluble phase of the cytoplasm (the cytosol) and leads to the formation of pyruvate. The second stage is the **tricarboxylic acid (TCA) cycle** or **Krebs cycle**, which occurs within the mitochondria of eukaryotic cells and the cytosol of prokaryotic cells and leads to the final oxidation of the carbon atoms to carbon dioxide. Most of the chemical energy of glucose is stored in the form of high-energy electrons, which are removed as substrate molecules are oxidized during both glycolysis and the TCA cycle. It is the energy of these electrons that is ultimately used to synthesize ATP.

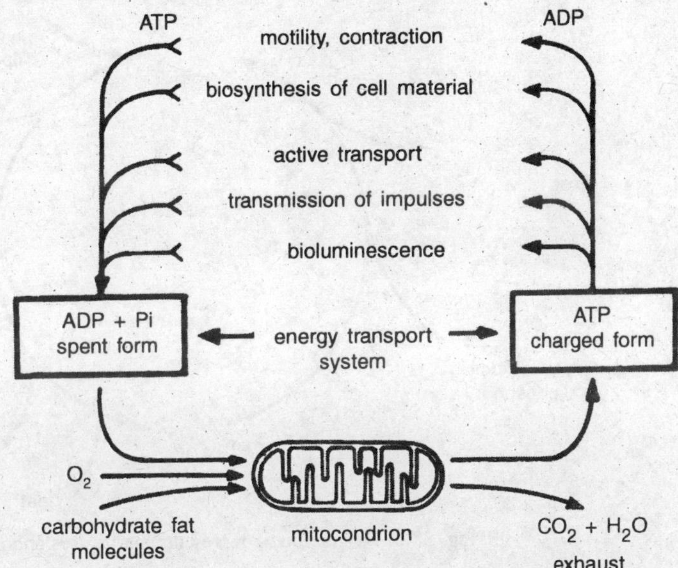

Fig. 58.5. Diagrammatic representation of the energy liberation in mitochondrion and its utilization in various cellular functions.

Adenosine Triphosphate or ATP

The ATP consists of a purine base **adenine,** a pentose sugar **ribose** and three molecules of the **phosphoric acids.** The adenine and ribose sugar collectively constitute the nucleoside **adenosine** which by having one, two or three phosphate groups forms the adenosine monophosphate (AMP), adenosine diphosphate (ADP) and adenosine triphosphate (ATP) respectively. In ATP the last phosphate group is linked with ADP by a special bond known as **"energy rich bond"** because when the last phosphate group of the ATP is released the large amount of energy is released as shown by the following reaction:

A-P~P~P =	A–P~P +	Pi +	7300 calories
ATP	ADP	phosphate group	Energy

In the above reaction, we have seen that by the breaking of the energy rich bond about 7300 calories of energy are released, while the common chemical bond releases only 300 calories of energy. The chemical reactions which synthesize the energy rich bond or ~P bond require great amount of energy which is supplied by the oxidation of the food-stuffs in the mitochondria. The utility of energy rich phosphate bond (~P) of the ATP is that great amount of energy is kept stored in the ready state in a very limited space of the cell. The stored chemical energy is disposed of very quickly at the time of the need in various cellular functions such as respiratory cycle, protein and nucleic acid synthesis, nervous transmission, cell division, transportation and bioluminescence, etc. (Fig. 58.6).

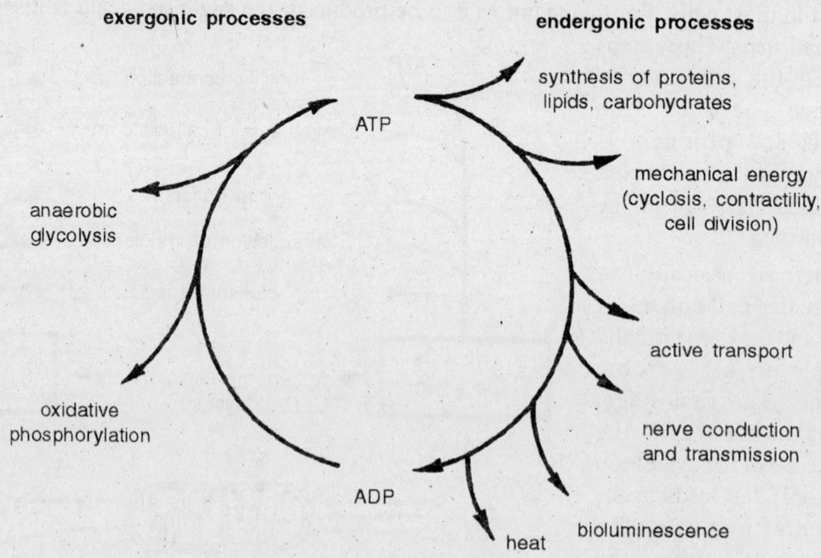

Fig. 58.6. Diagrammatic representation of uses and synthesis of ATP.

Because the terminal phosphate linkage in ATP is easily cleaved with release of free energy, ATP acts as an efficient phosphate donor in a large number of different phosphorylation reactions. In this way, ATP acts as a carrier molecule like the acetyl CoA and as coenzyme like the CoA or NAD.

Later on, besides ATP, certain other energy rich chemical compounds have been found to be active in the cellular metabolism. These are **cytosine triphosphate (CTP), uridine triphosphate (UTP)** and **guanosine triphosphate (GTP)**. These compounds, however, derive the energy from the ATP by nucleoside diphosphokinases. The energy for the production of ATP or other energy rich molecules is produced during the breakdown of food molecules including carbohydrates, fats and proteins **(catabolic** and **exergonic activities).**

Energy transducers. The cell organelles, such as chloroplasts and mitochondria, which converts one form of energy to another are known as **energy transducers**. The plants and animals have not only evolved highly effective energy transducers but have also developed very efficient control systems to regulate the energy transformations and to enable the cells to adjust to variations in environmental conditions.

58.3 MITOCHONDRIA

The mitochondria (Gr, *mito* = thread, *chondrion* = granule) are filamentous or granular cytoplasmic organelles of all aerobic cells of higher animals, plants and also of certain microorganisms including algae, Protozoa and Fungi. These are absent in bacterial cells and red blood cells of mammals. The mitochondria have lipoprotein framework which contains many enzymes and coenzymes required for energy metabolism. They also contain a specific DNA for the cytoplasmic inheritance and ribosomes for the protein synthesis.

Box 58.1 Historical

1. The mitochondria were first observed by **Kolliker** in 1850 as granular structures in the striated muscles. In 1888, he isolated them from insect muscles and named them **sarcosomes**.
2. **Richard Altmann** (1890) developed a specific stain that had useful specificity for the mitochondria He named this organelle, the **bioblast**. He suggested that bioblasts were independent organelles.
3. The present name mitochondria was assigned by **Benda** (1897–98) to them. He stained mitochondria with alizarin and crystal violet.
4. **Sir Hans Adolph Krebs** (German biochemist in England; born 1900), in 1937 worked out various reactions of the **citric acid cycle** (or tricarboxylic acid or TCA cycle). Krebs received the Nobel Prize in 1953 along with **Lipmann** for his discovery of the citric acid cycle.
5. **Hogeboom** and coworkers (1948) first demonstrated that cellular respiration occurs in mitochondria.
6. **Kennedy** and **Lehninger** (1948–1950) showed that the citric acid cycle (Krebs cycle), oxidative phosphorylation and fatty acid oxidation took place in the mitochondria. In 1951, **Lehninger** proved that oxidative phosphorylation requires electron transport.
7. **Palade** (1954) described the ultra structure of mitochondrial cristae.
8. In 1961, **Mitchell** proposed the highly acclaimed **"chemiosmotic-coupling hypothesis"** for the ATP-production in mitochondria. He got the Nobel Prize in 1978 for the development of this model.
9. In 1963, **Nass** and **Nass** demonstrated the presence of **DNA fibres** in the matrix of mitochondria of embryonic cells.
10. **Attardi, Attardi** and **Aloni** (1971) reported the 70S type ribosomes inside the mitochondria.

Mitochondria are called **"power houses"** or **'atom bombs'** of the cell, since they produce and provide chemical energy (ATP) to the cell.

Distribution or Localization

The mitochondria move autonomously in the cytoplasm, so they generally have uniform distribution in the cytoplasm, but in many cells their distribution is very restricted. The distribution and number of mitochondria (and also of mitochondrial cristae) are often correlated with type of function the cell performs. Typically mitochondria with many cristae are associated with mechanical and osmotic work situations, where there are sustained demands for ATP and where space is at a premium, *e.g.*, between muscle fibres, in the basal infoldings of kidney tubule cells, and in a portion of inner segment of rod and cone cells of retina.

Morphology

Number. The number of mitochondria in a cell depends on the type and functional state of the cell. It varies from cell to cell and from species to species. Certain cells contain exceptionally large number of the mitochondria, *e.g.*, the *Amoeba, Chaos chaos* contain 50,000; eggs of sea urchin contain 140,000 to 150,000 and oocytes of amphibians contain 300,000 mitochondria. Certain cells, *viz.*, liver cells of rat contain only 500 to 1600 mitochondria. The cells of green plants contain less number of mitochondria in comparison to animal cells because in plant cells the function of mitochondria is taken over by the chloroplasts. Some algal cells may contain only one mitochondrion.

Shape. The mitochondria may be filamentous or granular in shape and may change from one form to another depending upon the physiological conditions of the cells. Thus, they may be of club, racket, vesicular, ring or round-shape. Mitochondria are granular in primary spermatocyte of rat, or club-shaped in liver cells (Fig. 58.7).

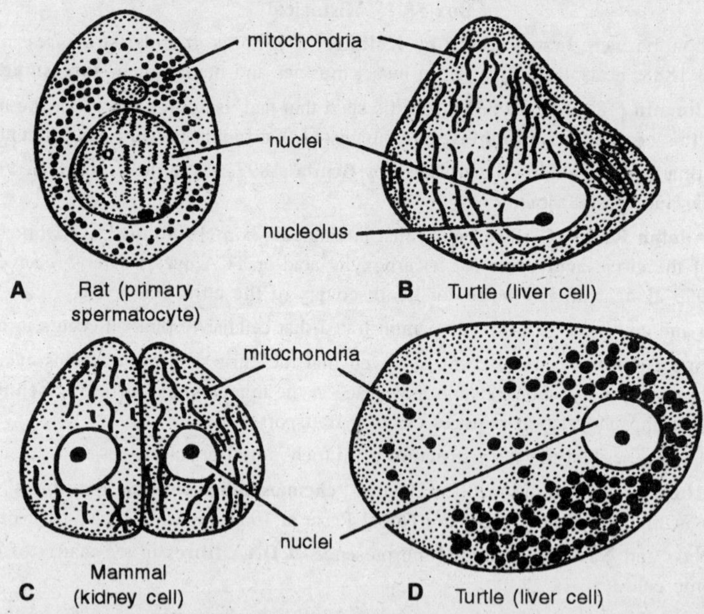

Fig. 58.7. Mitochondria of different type of animal cells.

Size. Normally mitochondria vary in size from 0.5 μm to 2.0 μm and, therefore, are not distinctly visible under the light microscope. Sometimes their length may reach up to 7 μm.

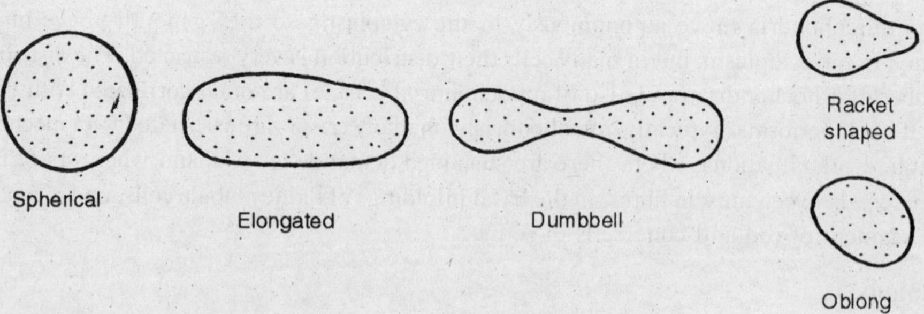

Fig. 58.8. Different shapes of mitochondria.

Structure. Each mitochondrion is bound by two highly specialized membranes that play a crucial part in its activities. Each of the mitochondrial membrane is 6 nm in thickness and fluid-mosaic in ultra-structure. The **outer membrane** is quite smooth and has many copies of a transport protein called **porin** which forms large aqueous channels through the lipid bilayer. This membrane, thus, resembles a sieve that is permeable to all molecules of 10,000 daltons or less, including small proteins. Inside and separated from the outer membrane by a 6–8 nm wide space is present the **inner membrane** (Fig. 58.9). The inner membrane is not smooth but is impermeable and highly convoluted, forming a series of infoldings, known as **cristae**, in the matrix space.

Thus, mitochondria are double membrane envelopes in which the inner membrane divides the mitochondrial space into two distinct chambers:

1. The **outer compartment, peri-mitochondrial space** or the **inter-membrane space** between outer membrane and inner membrane. This space is continuous into the core of the crests or cristae.

2. The **inner compartment, inner chamber** or **matrix space**, which is filled with a dense, homogeneous, gel-like proteinaceous material, called **mitochondrial matrix.** The mitochondrial matrix contains lipids, proteins, circular DNA molecules, 55S ribosomes and certain **granules** which are related to the ability of mitochondria to accumulate ions. Granules are prominent in the mitochondria of cells concerned with the transport of ions and water, including kidney tubule cells, epithelial cells of the small intestine, and the osteoblasts of bone-forming cells. Further, the inner membrane has an outer **cytosol** or **C face** toward the perimitochondrial space and an inner **matrix** or **M face** toward matrix.

In general, the cristae of plant mitochondria are tubular, while those of animal mitochondria are lamellar or plate-like.

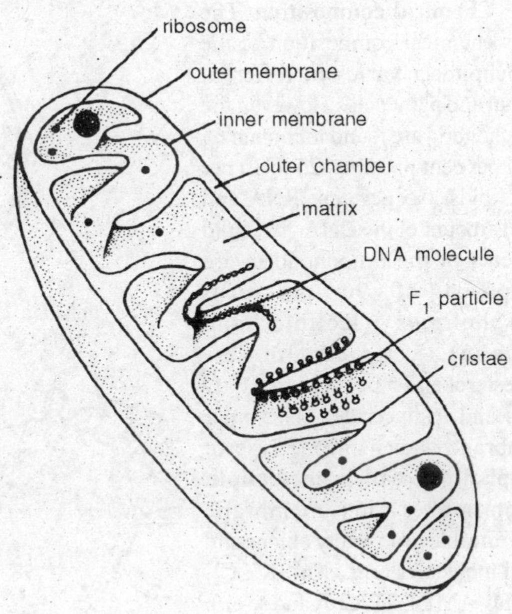

Fig. 58.9. A longitudinally cut mitochondrion showing its internal structure.

Attached to M face of inner mitochondrial membrane are repeated units of stalked particles, called **elementary particles, inner membrane subunits** or **oxysomes** (Fig. 58.10). They are also identified as $\mathbf{F_1}$ **particles** or $\mathbf{F_0 - F_1}$ **particles** and are meant for ATP synthesis (phosphorylation) and also for ATP oxidation (*i.e.*, acting as ATP synthetase and ATPase). $F_0 - F_1$ particles are regularly spaced at intervals of 10nm on the inner surface of inner mitochondrial membrane. According to some estimates, there are 10^4 to 10^5 elementary particles per mitochondrion.

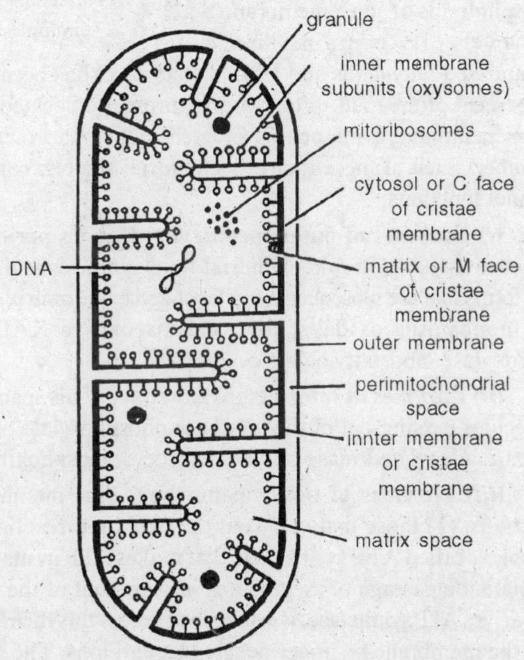

Fig. 58.10 A mitochondrion in sectional view to show its numerous oxysomes.

Mitoplast. When outer mitochondrial membrane is removed with digitonin (a detergent), the so-called **mitoplast** is formed. Mitoplast includes inner mitochondrial membrane with unfolded cristae and matrix. Mitoplast is found to carry out oxidative phosphory-lation. The isolated outer membrane is revealed by negative staining and shows a "folded-bag" appearance (Fig. 58.11). Such isolation of two membranes and compartments has enabled localization of various enzyme systems of mitochondria.

Chemical composition. The gross chemical composition of the mitochondria varies in different animal and plant cells. However, the mitochondria are found to contain 65 to 70 per cent proteins, 25 to 30 per cent lipids, 0.5 per cent RNA and small amount of the DNA. The lipid contents of the mitochondria are composed of 90 per cent phospholipids (lecithin and cephalin), 5 per cent or less cholesterol and 5 per cent free fatty acids and triglycerides. The inner membrane is rich in one type of phospholipid, called **cardiolipin** which makes this membrane impermeable to a variety of ions and small molecules (*e.g.*, Na^+, K^+, Cl^-, NAD^+, AMP, GTP, CoA and so on).

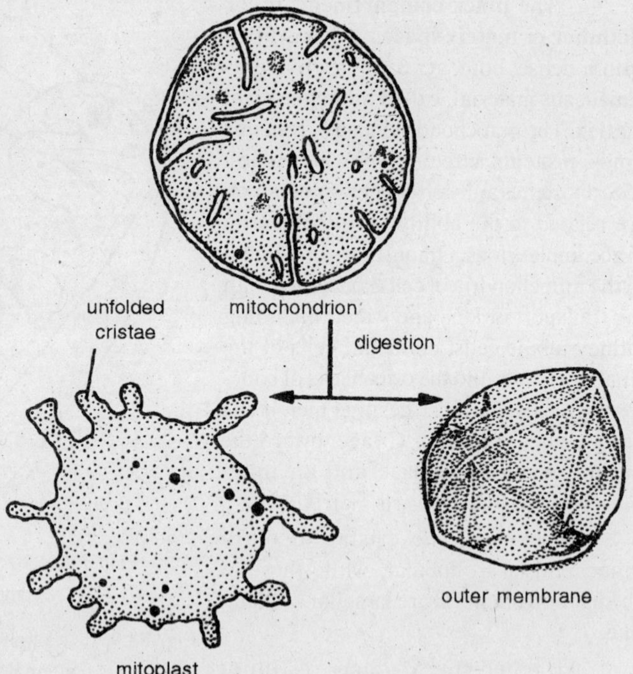

Fig. 58.11. Action of digitonin on liver mitochondrion to produce mitoplast and "folded-bag" of outer membrane.

The outer mitochondrial membrane has typical ratio of 50 per cent proteins and 50 per cent phospholipids of 'unit membrane'. However, it contains more unsaturated fatty acids and less cholesterol. It has been estimated that in the mitochondria of liver 67 percent of the total mitochondrial protein is located in the matrix, 21 per cent is located in the inner membrane, 6 per cent is situated in the outer membrane and 6 per cent is found in the outer chamber. Each of these four mitochondrial regions contains a special set of proteins that mediate distinct functions:

***(i)* Enzymes of outer membrane.** Besides porin, other proteins of this membrane include enzymes involved in mitochondrial lipid synthesis and those enzymes that convert lipid substrates into forms that are subsequently metabolized in the matrix. Certain important enzymes of this membrane are monoamine oxidase, rotenone-insensitive NADH-cytochrome-C-reductase, kynurenine hydroxylase, and fatty acid CoA ligase.

***(ii)* Enzymes of intermembrane space.** This space contains several enzymes that use the ATP molecules passing out of the matrix to phosphorylate other nucleotides. The main enzymes of this part are adenylate kinase and nucleoside diphosphokinase.

***(iii)* Enzymes of inner membrane.** This membrane contains proteins with three types of functions: (1) those that carry out the oxidation reactions of the respiratory chain; (2) an enzyme complex, called **ATP synthetase** that makes ATP in matrix; and (3) specific transport proteins that regulate the passage of metabolites into and out of the matrix. Since an electrochemical gradient, that drives ATP synthetase, is established across this membrane by the respiratory chain, it is important that the membrane be impermeable to small ions. The significant enzymes of inner membrane are enzymes of electron transport pathways, *viz.*, nicotinamide adenine dinucleotide (NAD), flavin adenine dinucleotide (FAD), diphosphopyridine nucleotide (DPN), four cytochromes (Cyt. *b*, Cyt. *c*, Cyt. c_1, Cyt.a and Cyt.a_3), ubiquinone or coenzyme Q_{10}, non-heme copper and iron, ATP synthetase, succinate dehydrogenase; β-hydroxy butyrate dehydrogenase; carnitine fatty acid acyl transferase (Fig. 58.12).

***(iv)* Enzymes of mitochondrial matrix.** The mitochondrial matrix contains a highly concentrated mixture of hundreds of enzymes, including those required for the oxidation of pyruvate and fatty acids and for the citric acid cycle or Krebs cycle. The matrix also contains several identical copies of the mitochondrial DNA, special 55S mitochondrial ribosomes, tRNAs and various enzymes required for the expression of mitochondrial genes. Thus, the mitochondrial matrix contains the following enzymes: malate dehydrogenase, isocitrate dehydrogenase, fumarase, aconitase, citrate synthetase, α-keto acid dehydrogenases, β-oxidation enzymes. Moreover, the mitochondrial matrix contains different nucleotides, nucleotide coenzymes and inorganic electrolytes—K^+, HPO_4^-, Mg^{++}, Cl^- and SO_4^-.

Box 58.2 Coenzymes

Some coenzymes have a central role in mitochondrial function. **Coenzyme A** (CoA) is part of a group (Table 58.2) that is derived from a nucleoside (adenine-D-ribose) and contains pantothenic acid (a vitamin of B complex) linked to the ribose by pyrophosphoric acid. CoA can be easily transformed into an ester at the thiol end (–SH) by acetyl group making acetyl-CoA. Acetyl-CoA is a carrier molecule in which acetyl group is linked by reactive bonds so that they can be transferred efficiently to other molecules. The same carrier molecule will often participate in many different biosynthetic reactions in which its group (*i.e.,* acetyl group) is needed, *e.g.,* growing fatty acid.

Other mitochondrial coenzymes are **nicotinamide adenine dinucleotide (NAD^+)** which contains the vitamin nicotinic acid of B complex and **flavin mononucleotide (FMN)** and **flavin adenine dinucleotide (FAD)**, both of which contain riboflavin or vitamin B_2. NAD^+, FMN and FAD are important coenzymes not only in mitochondria but also in chloroplasts.

Table 58.2 Some important coenzymes which act as carrier molecules in transfer of a group.

Coenzyme	*Group transferred*
1. ATP	phosphate
2. NADH, NADPH	hydrogen and electron (hydrogen ions)
3. Coenzyme A	acetyl
4. Biotin	carboxyl
5. 5′–Adenosyl-methionine	methyl

Functions

The mitochondria perform most important functions of the cell such as oxidation, dehydrogenation, oxidative phosphorylation and electron transport through the respiratory chain. Their structure and enzymatic system are fully adapted for their different functions. They are the actual respiratory organs of the cells where the foodstuffs *i.e.,* carbohydrates and fats are completely oxidised into CO_2 and H_2O. During the biological oxidation of the carbohydrates and fats large amount of energy is released which is utilized by the mitochondria for synthesis of the energy rich compound known as **adenosine triphosphate** or **ATP**. Because mitochondria synthesize energy rich compound ATP, they are also known as "**power houses**" of the cell. In animal cells mitochondria produce 95 per cent of ATP molecules, remaining 5 per cent is being produced during anaerobic respiration outside the mitochondria. In plant cells, ATP is also produced by the chloroplasts.

Besides the ATP production, mitochondria serve the following important functions in animals:

1. Heat production or thermiogenesis. Only 45 per cent of the energy released during the oxidation of glucose is captured in the form of ATP, the rest 55 per cent is either lost as heat or used to regulate body temperature of warm-blooded animals. In some mammals, especially young animals and hibernating species, there is a specialized tissue called **brown fat**. This tissue, typically located between the shoulder blades, is especially important in temperature regulation; it produces large quantities of body heat necessary for arousal from hibernation. The colour of brown fat comes from its high concentration of mitochondria, which are sparse in ordinary fat cells. The mitochondria of brown fat catalyze electron transport in the usual way but are much less efficient at producing ATP. Hence, a higher than usual fraction of the oxidatively released energy is converted directly to heat (called **non-shivering thermiogenesis**).

2. Biosynthetic or anabolic activities. Mitochondria also perform certain biosynthetic or anabolic functions. Mitochondria contain DNA and the machinery needed for protein synthesis. Therefore, they can make less than a dozen different proteins. The proteins so far identified are subunits of the ATPase, portions of the reductase responsible for transfer of electrons from CoQ to the iron of Cyt c, and three of the seven subunits in cytochrome oxidase. Altogether, no more than 5–10 per cent of mitochondrial components can be attributed to mitochondrial genes.

Some biosynthetic functions of mitochondria are of primary benefit to the rest of the cell. For example, the synthesis of **haeme** (needed for cytochromes, myoglobin and haemoglobin) begins with a mitochondrial reaction catalyzed by the enzyme, delta or δ-aminolevulinic acid synthetase. Likewise, some of the early steps in the conversion of cholesterol to steroid hormones in the adrenal cortex are also catalyzed by mitochondrial enzymes.

3. Accumulation of Ca^{2+} and phosphate. In the mitochondria of **osteoblasts** present in tissues undergoing calcification large amount of Ca^{2+} and phosphate (PO_4^-) tend to accumulate. In them microcrystalline, electrone-dense deposits may become visible. Sometimes, the mitochondria assume storage function, *e.g.,* the mitochondria of amphibian ovum store large amounts of yolk proteins and are transformed into **yolk platelets**.

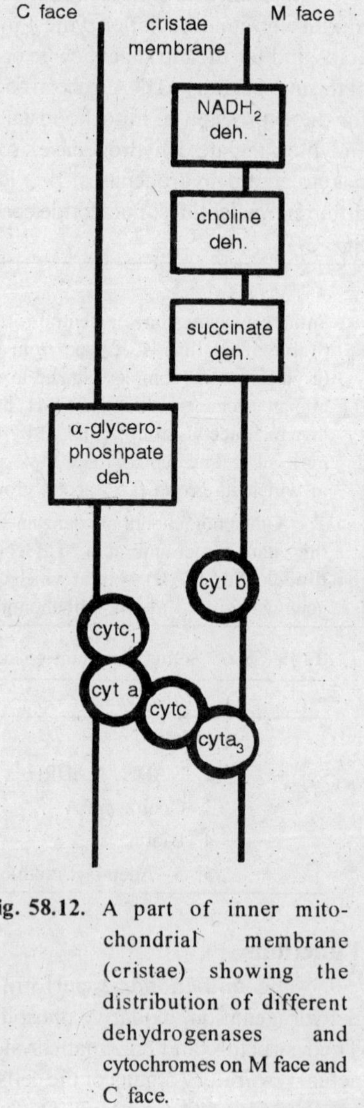

C face | cristae membrane | M face

NADH$_2$ deh.

choline deh.

succinate deh.

α-glycerophoshpate deh.

cyt b

cytc$_1$

cyt a

cytc

cyta$_3$

Fig. 58.12. A part of inner mitochondrial membrane (cristae) showing the distribution of different dehydrogenases and cytochromes on M face and C face.

58.4 MODE OF ENERGY PRODUCTION

The mitochondrion is the centre of oxidative metabolism in the cell, converting the products from carbohydrate, fat and protein catabolism into chemical energy stored in ATP.

Oxidation of Carbohydrates

The carbohydrates enter in the cell in the form of monosaccharides such as glucose. The hexose sugars are first broken down into 3-carbon compound (pyruvic acid) by a series of chemical reactions by many known enzymes. The pyruvic acid enters in the mitochondria for its complete oxidation into CO_2 and water. The reactions which involve in the oxidation of glucose in to CO_2 and water are known to form the **metabolic pathways** and they can be grouped under the following heads:

1. Glycolysis or Embden-Meyerhof pathways (EMP) or Embden-Meyerhof Parnas pathways (EMPP);
2. Oxidative decarboxylation;
3. Krebs cycle; citric acid cycle or tricarboxylic acid cycle;
4. Respiratory chain and oxidative phosphorylation.

1. Glycolysis. Under anaerobic conditions (*i.e.*, in the absence of oxygen) glucose is degraded into lactic acid or lactate by a process called **glycolysis** (*i.e.*, lysis or splitting of glucose), *e.g.*, it commonly occurs in vertebrate muscles when the energy demand in heavy exercise exceeds the available oxygen.

If glycolysis is carried out under aerobic conditions the final products are pyruvic acid and coenzyme NADH. Glycolysis is achieved by a series of 10 enzymes all of which are located in the cytosol (cytoplasmic matrix). As shown in (Fig. 58.13), in this chain of reactions, the product of one enzyme serves as a substrate for the next reaction. To facilitate its analysis, the sequence of glycolysis can be subdivided into following three main steps: (*i*) activation (stage I); (*ii*) cleavage (stage II); and (*iii*) oxidation (stage III).

(*i*) Activation. In reactions 1 to 3 the glucose molecules is converted into **fructose-1-6-diphosphate**. This step uses two molecules of ATP and involves the following enzymes: hexokinase, phosphoglucose isomerase (or phosphohexoisomerase) and phosphofructokinase.

(*ii*) Cleavage. In reactions 4 and 5, fructose-1-6-diphosphate splits into two (3-carbon) end products, **glyceraldehyde-3-phosphate** molecules. During the step of cleavage only two enzymes are used: aldolase (fructoaldolase) and triose isomerase (triosephosphate isomerase).

(*iii*) Oxidation. In reactions 6 to 10, two molecules of glyceraldehyde-3-phosphate are oxidized and ultimately converted into two molecules of **pyruvic acid**. This step produces four molecules of ATP by **substrate-level-phosphorylation** and involves the following enzymes: phosphoglyceric dehydrogenase (glyceraldehyde phosphate dehydrogenase), phosphoglyceric kinase, phosphoglyceromutase, enolase and pyruvic kinase.

The net energy yield of chain reactions of glycolysis is the production of two ATP molecules from one molecule of glucose. Under aerobic conditions, the end products of glycolysis are pyruvic acid and reduced coenzyme NAD (*i.e.*, NADH). NADH carries two electrons, taken from glyceraldehyde-3-phosphate and contains little energy. However, under anaerobic conditions, pyruvic acid remains in the cytosol (cytoplasmic matrix) and is used as a hydrogen acceptor and converted into lactic acid:

$$\text{Pyruvic acid} \xrightarrow[+\text{NADH}]{\text{Lactic dehydrogenase}} \text{Lactic acid} + \text{NAD} + \text{Energy}$$

In above case, the following equation represents the overall reaction of glycolysis:

$$\underset{\text{Glucose}}{C_6H_{12}O_6} + \text{Pi} + 2\text{ADP} \longrightarrow \underset{\text{Lactic acid}}{2C_3H_6O_3} + 2\text{ATP} + H_2O$$

2. Oxidative decarboxylation. In aerobic organisms, since pyruvic acid still contains a larger amount of energy, it must undergo further degradation, but this time inside the mitochondria. This is done in three consecutive steps: **oxidative decarboxylation** (removal of carboxyl or—COOH group), **Krebs cycle** and **oxidative phosphorylation**. Pyruvic acid directly enters the mitochondrial matrix and is converted into **acetyl-CoA** by the help of a huge enzyme, called **pyruvic acid dehydrogenase** (Fig. 58.14). The two NADH molecules (which are generated during glycolysis) cannot penetrate directly into the mitochondria, so their electrons are transferred to **dihydroxyacetone phosphate**, which shuttles them into the mitochondria. This process utilizes, one ATP molecule for each NADH; in all two ATP molecules are consumed for two NADH molecules. When both of these NADH pass through ETS, they tend to generate 6 ATP molecules.

Pyruvic acid dehydrogenase and its action. Sometimes two enzymes that catalyze sequential reactions form an **enzyme complex** and the product of the first enzyme does not have to diffuse through the cytoplasm to encounter the second enzyme. The second reaction begins as soon as the

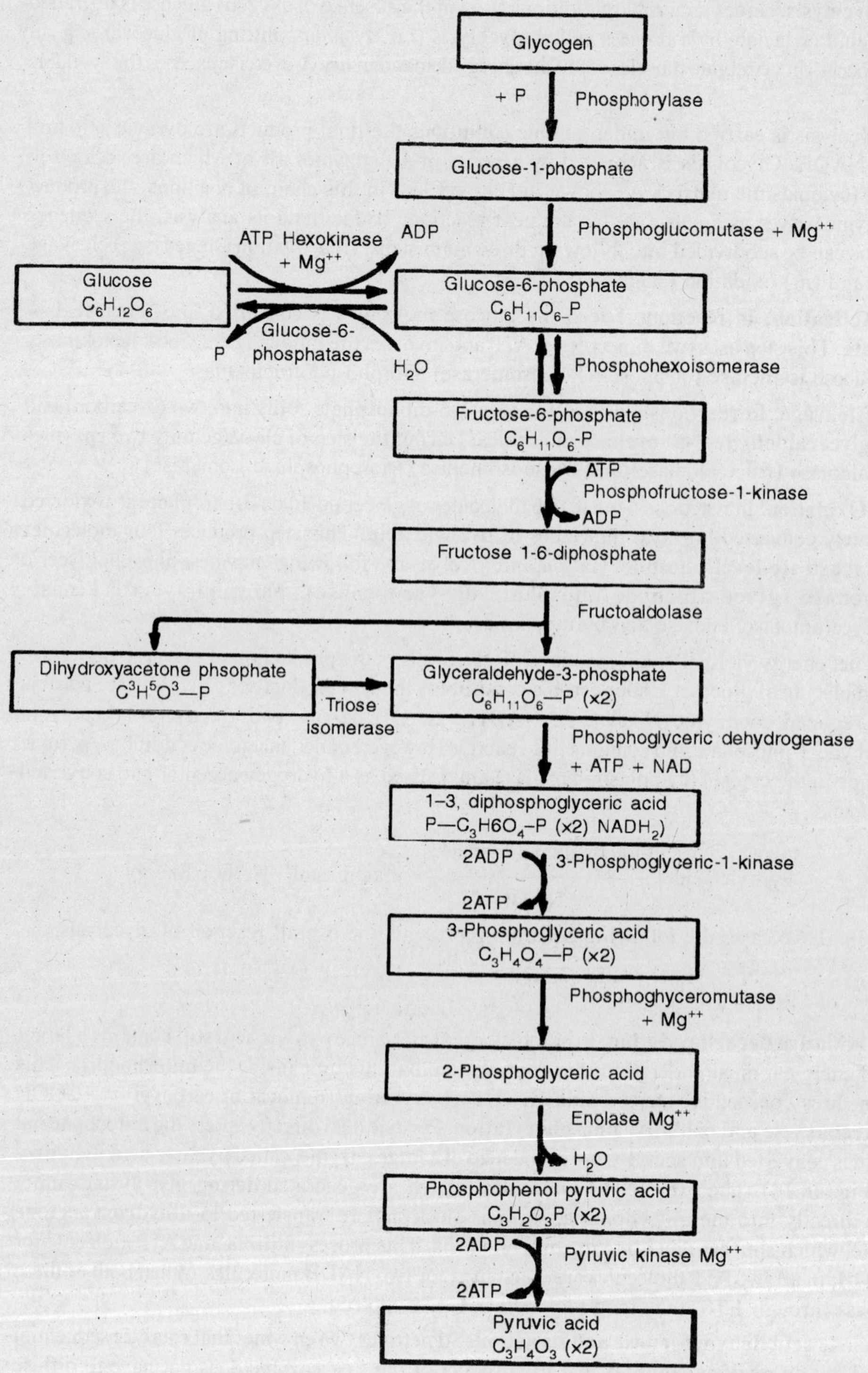

Fig. 58.13. Glycolysis or Embden-Meyerhof pathway showing the step-wise degradation of glucose to pyruvic acid.

first is over. Some large enzyme (multienzyme) aggregates carry out whole series of reactions without losing contact with the substrate. For example, the conversion of pyruvic acid to acetyl CoA proceeds in *three* chemical steps, all of which take place on the same large multienzyme complex (*i.e.*, pyruvate dehydrogenase). Pyruvic acid dehydrogenase occurs in the mitochondrial matrix and is larger than a ribosome in size. It contains multiple copies of **three** enzymes namely **pyruvic acid dehydrogenase, dihydrolipoyl transacetylase** and **dihydrolipoyl dehydrogenase.** It also contains five coenzymes (*e.g.*, NAD, coenzyme A, etc.) and **two regulatory proteins** (*e.g.*, protein kinase and protein phosphatase; both regulating the activity of pyruvic acid dehydrogenase, turning it off whenever ATP levels are high).

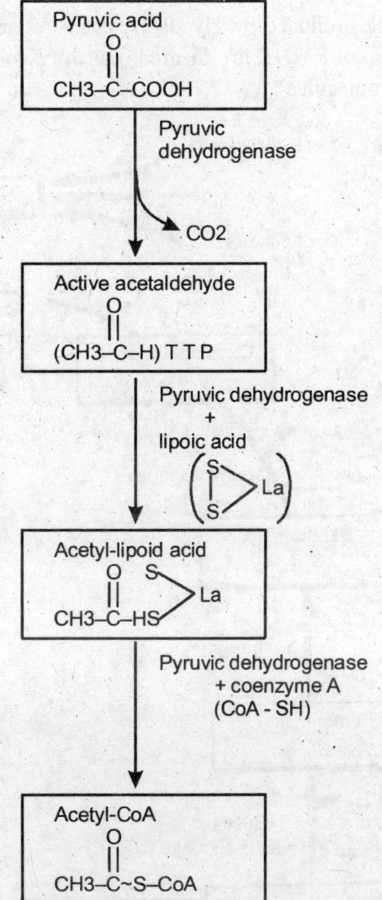

Fig. 58.14. The oxidative decarboxylation of pyrurvic acid to acetyl coenzyme A.

Thus, during oxidative decarboxylation of one molecule of pyruvic acid, one mole of CO_2 is produced and one NAD is reduced to NADH. The end product of this reaction is a 2-carbon compound, the **acetyl group** which is attached to coenzyme A to produce the carrier molecule, called **acetyl CoA.**

3. Krebs cycle. Two acetyl CoA molecules, produced above by oxidation of one molecule of glucose pass through a series of reactions of Krebs cycle to produce CO_2, H_2O and electrons. Enzymes and coenzymes of Krebs cycle are located in the mitochondrial matrix (some of them such as succinic dehydrogenase, are attached to M face of inner mitochondrial membrane, Fig. 58.12). As illustrated in (Fig. 58.15), the Krebs cycle involves the condensation of the acetyl group with oxaloacetic acid to make **citric acid** (6-carbon compound). This step is directed by the enzyme **citrate synthase.** From citric acid, H_2O is released twice by **aconitase** enzyme to produce **isocitric acid**. This is followed by a decarboxylation (loss of CO_2) by **isocitric dehydrogenase**, producing 5 carbon **α-ketoglutaric acid**. CO_2 is released by **α-ketoglutarate dehydrogenase** in the presence of CoA to produce **succinyl CoA** which changes by **succinyl kinase** enzyme (also called succinyl CoA synthetase) into a 4-carbon compound, the **succinic acid** (at this stage one GTP is generated by substrate level phosphorylation). The next enzyme of Krebs cycle, the **succinic dehydrogenase** converts succinic acid into **fumaric acid** and then **fumarase** enzyme produces **malic acid.** The mediation of **malate dehydrogenase** enzyme produces **oxaloacetic acid**, and thereby closes the Krebs cycle.

At each turn of the Krebs cycle, four pairs of hydrogen atoms are removed from the substrate intermediates by enzymatic dehydrogenation and two CO_2 molecules are released. These hydrogen atoms (or equivalent pairs of electrons) enter the respiratory chain, being accepted by either NAD^+ or FAD. Three pairs of hydrogen molecules are accepted by NAD^+, reducing it into NADH, and one pair by FAD, reducing it into $FADH_2$ (this pair of electrons come directly from the succinic dehydrogenase reaction). Since it takes two turns of the cycle to metabolize the two acetyl groups that are produced by glycolysis from one molecule of glucose, a total of six molecules of NADH and two of $FADH_2$ are formed. During Krebs cycle are also produced two ATP molecules (*i.e.*, via GTP molecules).

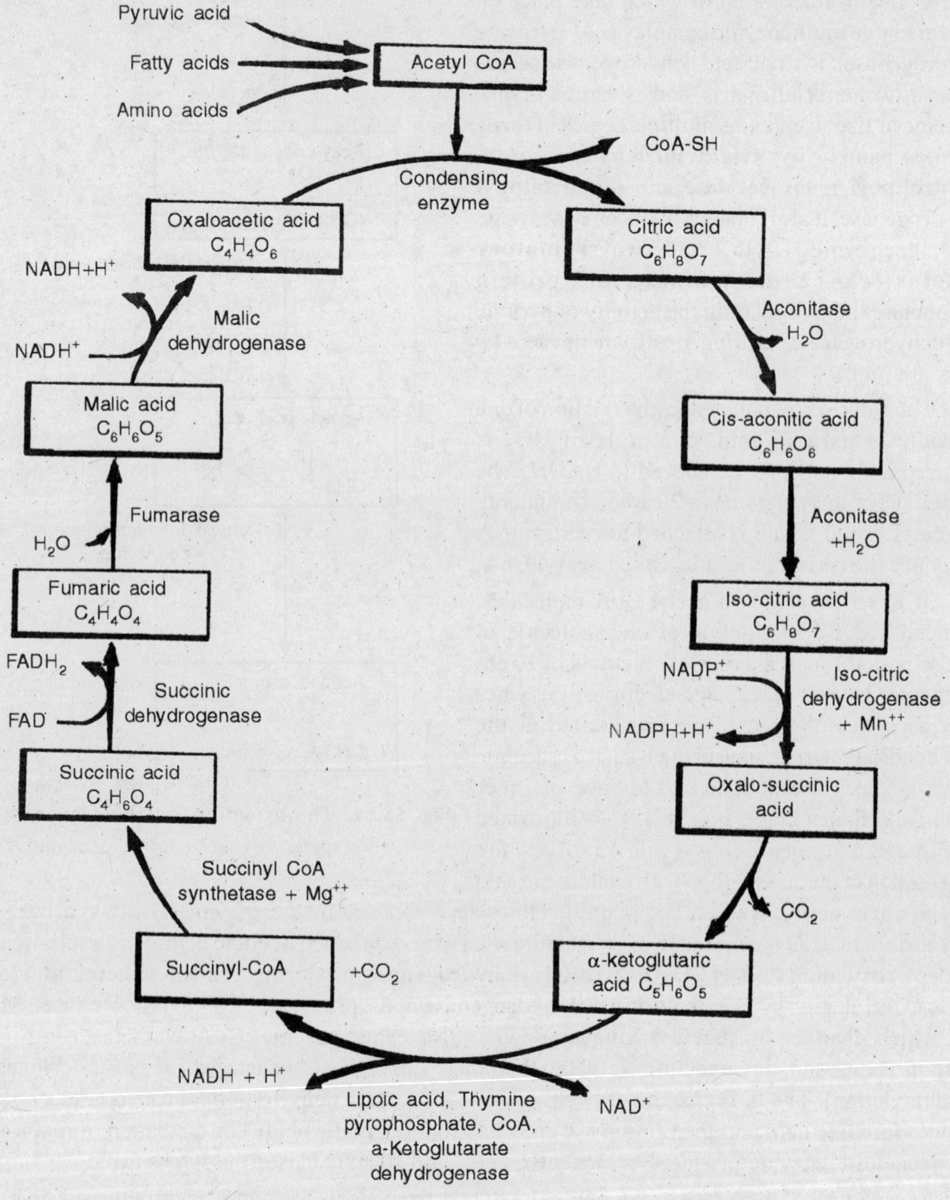

Fig. 58.15. Krebs cycle or tricarboxylic acid cycle.

The Krebs cycle itself releases no energy, but as the glucose breaks up, it also frees the hydrogen atoms attached to the carbons. Each hydrogen atom contains one proton (H^+) and one electron (e^-) The electrons which are released during complete oxidation of glucose (*i.e.*, glycolysis, oxidative decarboxylation and Krebs cycle) carry most of the energy of the glucose. In the final stage of cell respiration, the electron transport system, these electrons will at last release their energy to the cell.

4. Respiratory Chain and Oxidative Phosphorylation. Two molecules of $FADH_2$ and six molecules of NADH produced in Krebs cycle (from two molecules of acetyl-CoA) are oxidized by molecular O_2 in a **respiratory chain** or **electron transport system** or **ETS** involving a series of enzymes and coenzymes.

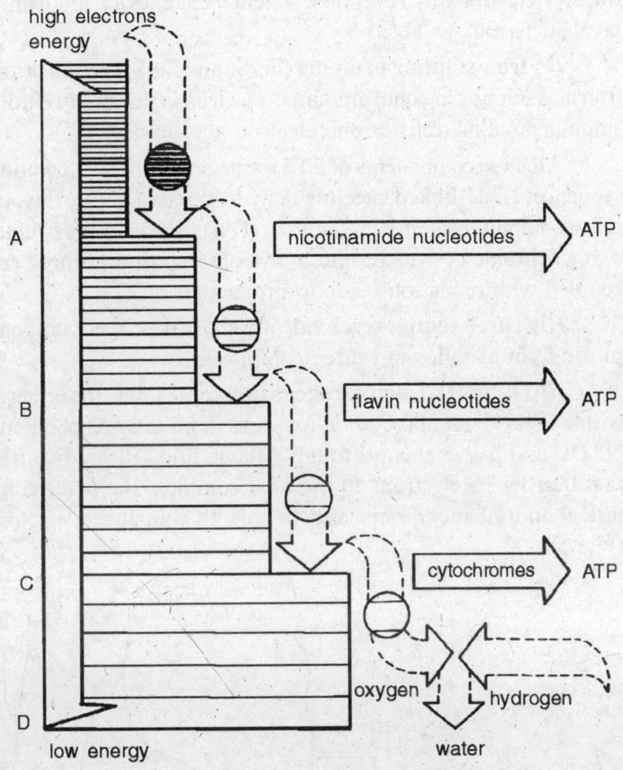

Fig. 58.16. Electron transfer in respiratory cascade or electron transport system.

In the electron transport system, the successive electron acceptors are at lower and lower energy levels. With each transfer to a lower energy level, the electrons release some of their potential energy. That is why this series is called an **electron cascade**, like a cascade of falling water. At each stage, the released energy is used to form ATP. Since electron transport involves oxidation as well as phosphorylation (*i.e.*, ADP + P = ATP) this process by which cell system traps chemical energy is called **oxida-tive phosphorylation** (Fig. 58.16). The passage of electrons from NAD to oxygen generates 3 ATP molecules, whereas the passage of electrons from FAD to oxygen generates only 2 ATP molecules.

(*A*) Compounds that occur in ETS. Following five types of compounds are associated with electron transport system of inner mitochondrial membrane:

(*i*) Pyridine-linked dehydrogenases require as their coenzyme either NAD^+ or $NADP^+$ both of which can accept two electrons at a time. There are about 200 dehydrogenases for which NAD^+–linked compounds such as pyruvic acid dehydrogenase, are involved in ETS.

(*ii*) Flavin-linked dehydrogenases (often called **flavoproteins** or **FPs**) require either FAD or FMN. Both are prosthetic groups whose isoalloxazine ring can accept two hydrogen atoms. Flavin-linked enzymes are commonly involved in a number of enzyme systems such as fatty acid oxidation, amino acid oxidation and Krebs cycle activity (*e.g.*, succinic dehydrogenase or SDH).

(*iii*) Ubiquinones were so named because of their occurrence in so many different organisms and their chemical resemblance to quinone. They are found in several different forms including the **plastoquinones** of chloroplasts. The form of ubiquinones present in mitochondria is often called **coenzyme Q_{10}** (CoQ_{10} or Q). CoQ_{10} is a lipid soluble and accepts two hydrogen atoms (or two protons and two electrons) at a time.

(*iv*) Cytochromes are proteins containing iron-porphyrin (haem) groups. There are a large number of cytochrome in cells; most are found in mitochondria, although some also function in the *ER* and in chloroplasts. Mitochondria have five types of cytochromes which are arranged in the following order in an inner membrane: cyt. *b*, cyt. c_1, cyt. *c*, cyt. *a* and cyt. a_3. All of them transfer electrons by reversible valence changes of the iron atom (trivalent ferric or $Fe^{3+} \rightleftharpoons$ bivalent ferrous or Fe^{2+}).

***(v)* Iron-sulphur proteins** (Fe_2S_2 and Fe_4S_4) are electron carriers of mitochondria containing iron and sulphur in equal amounts. The iron is reversibly oxidized during the electron transfer. Iron-sulphur proteins transfer one electron at a time.

All these components of ETS are arranged in the inner mitochondrial membrane in the following sequence: NAD-linked succinic dehydrogenase (SDH), flavoprotein (FAD), non-haem iron protein or iron-sulphur protein, flavoprotein (FAD), cytochrome *b*, ubiquinone or coenzyme Q_{10}, cytochrome c_1, cytochrome *c*, cytochrome *a*, cytochrome a_3 and three coupling sites, where phosphorylation coupled with oxidation leads to production of ATP.

***(B)* Three complexes.** Evidently above described components of ETS occur in the mitochondria in the form of following three complexes:

***(i)* The NADH-dehydrogenase complex.** It is the largest of the respiratory enzyme complexes, with a mass about 800,000 daltons and more than 22 polypeptide chains. It accepts electrons from NADH and passes them through a flavin and at least five iron-sulphur centres to ubiquinone (Q) that transfer its electrons to the next complex, the b–c_1, complex. This complex spans the inner mitochondrial membrane and is able to translocate protons across it from M side to C side (Fig. 58.17).

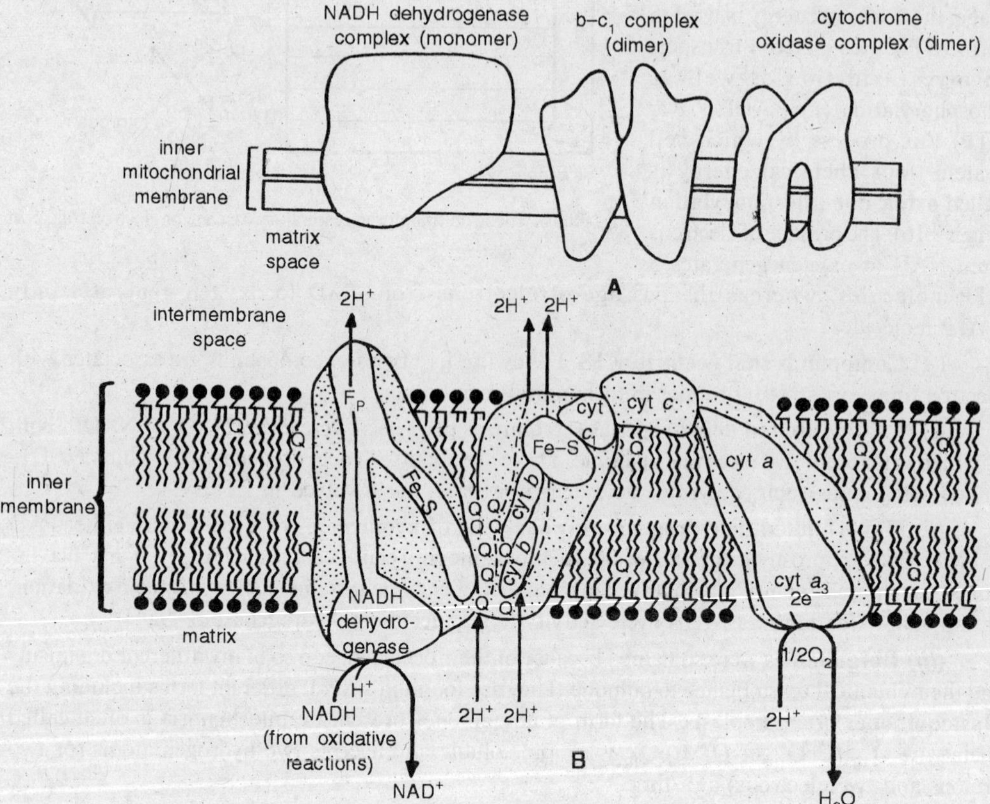

Fig. 58.17. Respiratory enzyme complexes. A—The relative sizes and shapes of the three respiratory enzyme complexes; B—Mode of transfer of H^+ through the three respiratory enzyme complexes of inner mitochondrial membrane.

***(ii)* The b–c_1 complex.** It contains at least 8 different polypeptide chains and is thought to function as a dimer of about 500,000 daltons. Each monomer contains three haemes bound to cytochromes and iron-sulphur protein. This complex accepts electrons from ubiquinone (Q) and passes them to cytochrome, a small peripheral membrane protein that carries its electrons to the cytochrome oxidase complex. In the topology of this complex the Q-site may be in the middle of the membrane in the hydrophobic area and the cytochrome c-site on the C side.

***(iii)* The cytochrome oxidase complex.** It comprises at least eight different polypeptide chains and is isolated as a dimer of about 300,000 daltons; each monomer contains two cytochromes (a, a_3) and two copper atoms. This complex accepts electrons from cytochrome c and passes them to oxygen and is thought to traverse the mitochondrial membrane, protruding on both surfaces. Such a trans-membrane orientation is associated with the vectorial transport of protons across the membrane.

The cytochrome oxidase reaction is estimated to account for 90 per cent of the total oxygen uptake in most cells. The toxicity of the poisons such as cyanide and azide is due to their ability to bind tightly to this complex and thereby block all electron transport.

***(C)* F_0–F_1 complex or coupling factors.** One of the main proteins in the inner mitochondrial membrane is the multisubunit **coupling factor** (Fig. 58.18), the enzyme that actually synthesizes ATP and simultaneously acting as a **proton pump.** A quite similar enzyme complex is located in the thylakoid membranes of chloroplasts and in the plasma membrane of bacterial cell. The coupling factor has two principal components:

***(a)* F_0-complex.** It is an integral membrane complex, composed of very hydro-phobic proteins—3 or 4 distinct polypeptides and one proteolipid—which together span the mitochondrial membrane. F_0–complex possesses the proton translocating mechanism. F_0-complex can be extracted only with strong detergents.

***(b)* F_1-particle.** Attached to the F_0 complex is F_1 particle, a complex of five distinct polypeptides: alpha (α), beta (β), gamma (γ), delta (δ) and epsilon (ε), with the probable composition of $\alpha_3\ \beta_3\gamma\delta\varepsilon$. F_1 forms the knob or 'tadpole' that protrudes on the matrix side of the inner mitochondrial membrane. F_1 particle can be detached from the membrane by mechanical agitation and is water soluble. When physically separated from the membrane, F_1 particle is capable only of catalysing the hydrolysis of ATP into ADP and phosphate. Hence, it is often called the **F_1-ATPase**. However, its natural function is the synthesis of ATP.

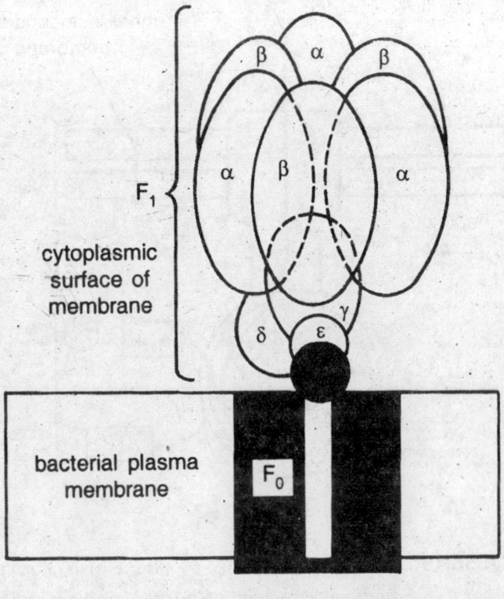

Fig. 58.18. Model of molecular structure of F_0 and F_1 particles in the bacterial plasma membrane. The core of F_1 particle is an $\alpha_3\ \beta_3$ complex. The δ polypeptide of F_1 and probably the γ are involved in binding F_1 to the membrane-embedded F_0-particle. The role of ε subunit of F_1 complex is not known.

***(D)* Redox reactions and redox couples.** The movements of electrons between cellular reductants and oxidants represent a form of energy transfer in cells. A reductant (or **reducing agent**) is a substance that loses or donates electrons to another substance; the latter substance is the oxidant (or **oxidizing agent**). Conversely, an oxidant is a substance that accepts electrons from another substance, the latter being the reductant. Reactions that involve the movement of electrons between reductants and oxidants are called **redox reactions**.

***(E)* ATP synthesis.** The potentials drop in three large steps, one across each major respiratory enzyme complex. The change in redox potential between any two electron carriers is directly proportional to the free energy released by an electron transfer between them. Each complex acts as an **energy-conversion-device** to harness this free-energy change, pumping H^+ across the inner membrane to create an electrochemical proton gradient as electrons pass through. The energy conversion mechanism underlying oxidative phosphorylation requires that each protein complex be inserted across the inner mitochondrial membrane in a fixed orientation, so that all protons are pumped in the same direction out of the matrix space. Such a **vectorial organization** of membrane proteins has been experimentally proved.

Just as a flow of water from a higher to a lower level can be utilized to turn a water wheel or a hydroelectric turbine, the energy released by the flow of the protons down the gradient is utilized in the synthesis of ATP. Similarly, resultant electrochemical proton gradient is harnessed to make ATP by F_0–F_1 complex (acting as ATP synthetase and proton pump), through which the protons flow back into the matrix (Fig. 58.19). The ATP synthetase is a reversible coupling device that normally converts a back-flow of protons into ATP phosphate-bond energy, but it can also hydrolyze ATP (into ADP and phosphate) to pump protons in the opposite direction, if the electrochemical proton gradient is reduced.

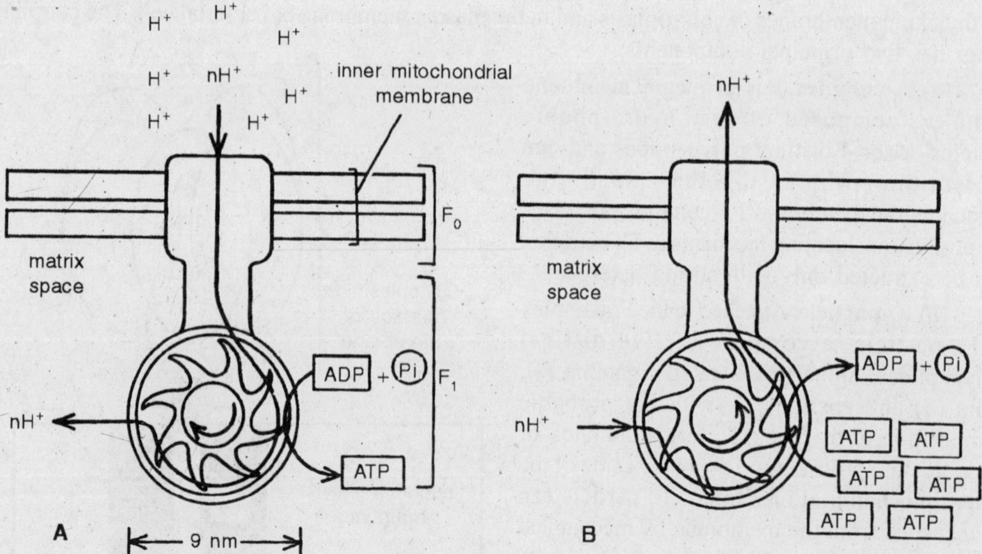

Fig. 58.19. ATP synthetase (F_0–F_1 ATPase) is a reversible coupling device that interconverts the energies of the electrochemical proton gradient and chemical bonds. The ATP synthetase can either synthesize ATP by harnessing the proton motive force (A) or pump protons against their electrochemical gradient by hydrolyzing ATP (B). The direction of operation at any given instant depends on the net free-energy change for the coupled processes of proton translocation across the membrane and the synthesis of ATP from ADP and Pi.

Chemiosmotic mechanism. The controlled movement of protons back across the membrane through ATP-synthesizing enzyme provides the energy required to phosphorylate ADP to ATP. The importance of proton movements on the formation of ATP was first proposed in 1961 by **Peter Mitchell** of the University of Edinburgh. This proposed mechanism is called **chemiosmotic mechanism.** For this contribution, Mitchell was awarded Nobel Prize in Chemistry in 1978.

According to the chemiosmotic model, the main energetic driving force for the synthesis of ATP is an ion gradient across a selectively permeable biological membrane; in chloroplasts and mitochondria, this gradient is known to be a **proton gradient** that is generated as a consequence of electron transfer. This gradient establishes a concentration difference of protons and can also result in difference in electric potential of the membrane. These sources of potential energy can be used for the phosphorylation of ADP by ATP synthase, an enzyme that energetically favourable proton flow to the synthesis of ATP (**Buchanan** *et al.*, 2007).

***(F)* Energetics of glucose oxidation.** Of the 686,000 calories contained in a mole of glucose, less than 10 per cent (*i.e.*, 58,000 calories) can be released by anaerobic glycolysis. The cell is able to store only 45 per cent of the chemical energy liberated by the combustion of glucose in the form of ATP (*i.e.*, only 36 ATP molecules). The rest of the energy is dissipated as heat or used for other cell functions.

At this stage, let us do stocktaking of ATP generation during aerobic respiration of one mole of glucose. We have seen that glycolysis and Krebs cycle can each generate 2 molecules of ATP per molecule of glucose by substrate level phosphorylation (total 4 ATP molecules). In addition 10 NADH (*i.e.*, 2NADH in glycolysis, 2NADH in oxidative phosphorylation and 6NADH in Krebs cycle) and $2FADH_2$ are produced which are equivalent to 34 ATP molecules. Thus, a total of 38 ATP molecules are produced per glucose molecule oxidized. However, in most eukaryotic cells 2 molecules of ATP are used in the transportation of 2 mole of NADH produced during glycolysis into the mitochondrion (*via* the malate shuttle) for their further oxidation (*via* ETS). Hence, the net gain of ATP is 38 – 2 = 36 molecules; since one high energy phosphate bond is equal to 36.8 kJ; so 36ATP = 1325 kJ or 360,000 calories :

$$\underset{\text{Glucose}}{C_6H_{12}O_6} + \underset{\substack{\text{Inogranic}\\ \text{phosphate}}}{36Pi} + 36ADP + 6O_2 + 6H_2O \longrightarrow$$

$$6CO2 + 36\ ATP\ (360{,}000\ \text{caloiries or } 1325\ kJ) + 12H_2O$$

Mitochondria as Semi-autonomous Organelles

Recently the study of mitochondrial and chloroplast biogenesis became of great interest because it was demonstrated that these organelles contain DNA as well as ribosomes and are able to synthesize proteins. The term **semiautonomous organelles** was applied to the two structures in the recognition of these findings. This term also indicated that the biogenesis was highly dependent on the nuclear genome and the biosynthetic activity of the ground cytoplasm. It is well established now that the mitochondrial mass grows by the integrated activity of both genetic systems, which cooperate in time and space to synthesize the main components. The mitochondrial DNA codes for the mitochondrial, ribosomal and transfer RNA and for a few proteins of the inner membrane. Most of the proteins of the mitochondrion, however, result from the activity of the nuclear genes and are synthesized on ribosomes of the cytosol (cytoplasmic matrix). The cooperation of two genomes has been greatly clarified by studies on the molecular assembly of cytochrome oxidase. This cytochrome, as studied in *Saccharomyces cerevisiae* is made up of seven polypeptide subunits for a combined molecular weight of 139,000 daltons. Three of the polypeptides are coded by mtDNA and assembled on mitochondrial ribosomes. They are very hydrophobic and high in molecular weight (23,000–40,000 daltons). The remaining four subunits are coded by nuclear DNA and made on cytoplasmic ribosomes. These are hydrophilic polypeptides of lower molecular weight (4500–14,000 daltons).

Mitochondrial DNA. Mitochondrial DNA (mtDNA) molecule is relatively small, simple, double-stranded and except for the DNA of some algae and protozoans, it is circular. The size of mitochondrial genome is very much large in plants than in animals. Thus, mtDNA varies in length from about 5 μm in most animal species to 30 μm or so in higher plants. The mtDNA is localized in the matrix and is probably attached to the inner membrane at the point where DNA duplication starts. This duplication is under nuclear control and the enzymes used (*i.e.*, polymerases) are imported from the cytosol.

Mitochondrial ribosomes. Mitochondria contain ribosomes (called **mitoribosomes**) and polyribosomes. In yeast and *Neurospora*, ribosomes have been ascribed to a 70S class similar to that of bacteria; in mammalian cells, however, mitoribosomes are smaller and have a total sedimentation coefficient of 55S, with subunits of 35S, and 25S. In mitochondria, ribosomes appear to be tightly associated with the inner membrane.

Mitochondrial protein synthesis. As already described, mitochondira can synthesize about 12 different proteins, which are incorporated into the inner mitochondrial membrane. These proteins are very hydrophobic (*i.e.*, they are proteolipids). Thus, on the mitoribosomes are made the following proteins: three largest subunits of cytochrome oxidase, one protein subunit of the cytochrome b–c_1 complex, four subunits of ATPase and a few hydrophobic proteins. One of the best known differences between the two mechanisms of protein synthesis (*i.e.*, in the cytosol and in the mitochondrial matrix) is in the effect of some inhibitors. The mitochondrial protein synthesis is inhibited by **chloramphenicol**, while synthesis in the cytosol (cytoplasmic matrix) is not affected by this drug. In contrast, **cycloheximide** has the reverse effect.

Import mechanism of mitochondrial proteins. Most mitochondrial proteins are coded by nuclear genes and are synthesized on free ribosomes in the cytosol (cytoplasmic matrix). The import of these polypeptides involves similar mechanism both in mitochondria, and chloroplasts. The transport processes involved have been most extensively studied in mitochondria, especially in yeasts.

Mitochondrial lipid biosynthesis. The biogenesis of new mitochondria and chloroplasts requires lipids in addition to nucleic acids and proteins. Chloroplasts tend to make the lipids they require. For example, in spinach leaves, all cellular fatty acid synthesis takes place in the chloroplast. The major glycolipids of the chloroplast are also synthesized locally.

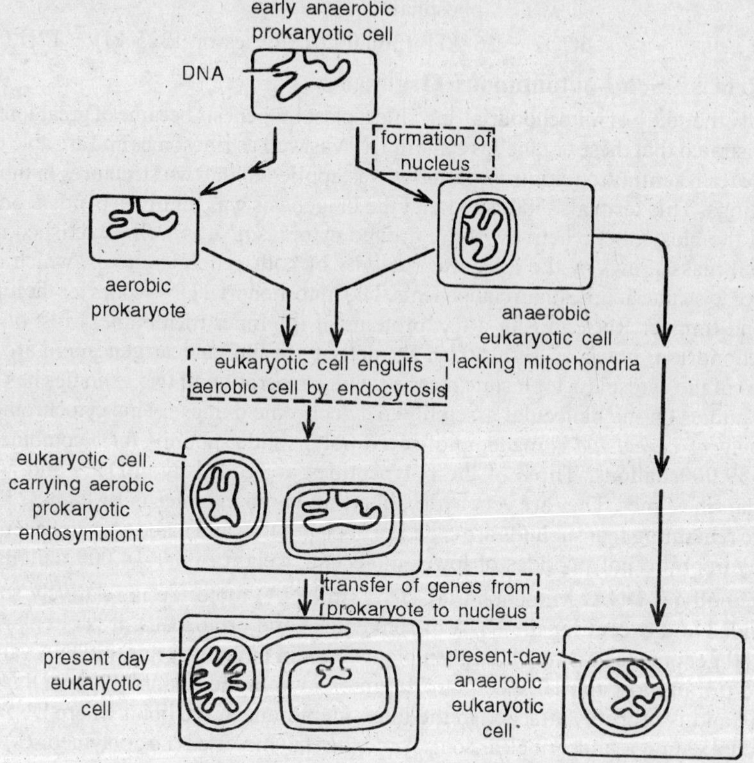

Fig. 58.20. Symbiotic origin of mitochondria and chloroplast.

Mitochondria, on the other hand, import most of their lipids. In animal cells the phospholipids—**phosphatidyl-choline** and **phosphatidyl-serine**—are synthesized in the ER and then transferred to the outer membrane of mitochondria.

Symbiont Hypothesis or Prokaryotic Origin of mitochondria. Early cytologists such as **Altmann** and **Schimper** (1890) have suggested the possibility of origin of the mitochondria from the prokaryotic cells. According to their hypothesis, the mitochondria and chloroplasts may be considered as intra-cellular parasites of the cells which have entered in the cytoplasm of eukaryotic cells in early evolutionary days, and have maintained the symbiotic relations with the eukaryotic cells. The mitochondira are supposed to be derived from the bacterial cells (purple bacteria) while chloroplasts are supposed to be originated from the blue green algae (**Margulis**, 1981). Due to these reasons **Altmann** suggested the name **"bioblasts"** to the mitochondria and he also hinted about their self-duplicating nature.

Recent cytological findings have also suggested many homologies between the mitochondria and the bacterial cells. The similarities between the two can be summarised as follows:

1. Similarity in inner mitochondrial membrane and bacterial plasma membrane.

(*i*) In the mitochondria the enzymes of the respiratory chain are localized on the inner mitochondrial membrane like the bacteria in which they remain localized in the plasma membrane. The bacterial plasma membrane resembles with the inner mitochondrial membrane in certain respects.

(*ii*) The plasma membrane of certain bacterial cells gives out finger-like projections in the cytoplasm known as mesosomes. The mesosomes can be compared with mitochondrial crests. **Salton** (1962) has reported respiratory chain enzymes in the mesosomes.

(*iii*) Because the outer mitochondrial membrane resembles with the plasma membrane, therefore, it may be assumed that the mitochondrial matrix and the inner mitochondrial membrane represent the symbiont which might be enclosed by the membrane of the cellular origin (outer mitochondrial membrane).

2. Similarity in DNA molecule. The DNA molecule of the mitochondria is circular like the DNA molecule of the bacterial cells. Further the replication process of the mitochondrial DNA is also similar to bacterial DNA.

3. Similarity in ribosomes. The mitochondrial ribosomes are small in size and resemble the ribosomes of the bacteria.

4. Similarity in the process of protein synthesis. The process of protein synthesis of both mitochondria and bacteria is fundamentally same because in both, the process of protein synthesis can be inhibited by same inhibitor known as chloramphenicol.

Further, the mitochondria for the process of protein synthesis depend partially on the mitochondrial matrix and DNA and partially on the nucleus and cytoplasm of the eukaryotic cells. This shows the symbiotic nature of the mitochondria.

Due to the above-mentioned similarities between the bacteria and mitochondria, the symbiont hypothesis postulated that the host cell (eukaryotic cell) represented an anaerobic organism which derives the required energy from the oxidations of food by the process of glycolysis. While the mitochondria represent the symbionts which respire **aerobically** and contain the enzymes of **Krebs cycle** and **respiratory chain.** The symbionts seem to be capable to get the energy by **oxidative phosphorylation** from the partially oxidised food (pyruvic acid) of the host cell.

58.5 PLASTIDS

Plastids are present in all living plant cells and in *Euglena* (a protozoan). They are small bodies found free in the cytoplasm. Plastids are often more or less spherical or disc-shaped (1 μm to

1 mm in diameter), but may be elongated or lobed or show amoeboid characteristics. Their other identifying features are their double bounding membranes, the possession of **plastoglobuli** (spherical lipid droplets; store of lipids surplus to current requirements) and an internal membrane fretwork of many discrete internal vesicles. All plastids in a particular plant species contain multiple copies of same relatively small genome (DNA) and 70S-type ribosomes. They are self-replicating organelles containing a protein-synthesizing capacity comparable to that of mitochondria. They perform most important biological activities as the synthesis of food and storage of carbohydrates, lipids and proteins. Plastids are absent in the cells of fungi, bacteria, animals and male sperm cells of certain higher plants.

Types of Plastids. The term 'plastid' is derived from the Greek word "*plastikas*" (= formed or moulded) and was used by **A.F.W. Schimper** in 1885. **Schimper** classified the plastids into following types according to their structure, pigments and the functions:

1. Leucoplasts. The leucoplasts (Gr., *leuco* = white; *plast* = living) are the colourless plastids which are found in embryonic and germ cells. They are also found in meristematic cells and in those regions of the plant which are not receiving light. Plastids located in the cotyledons and the primordium of the stem are colourless (leucoplasts) but eventually become filled with chorophyll and transform into chloroplasts. **True leucoplasts** occur in fully differentiated cells such as epidermal and internal plant tissues. They never become green and photosynthetic. True leucoplasts do not contain thylakoids and even ribosomes. They store the food materials as carbohydrates, lipids and proteins and accordingly are of following types:

(*i*) Amyloplasts. The amyloplasts (L., *amyl* = starch; Gr., *plast* = living) are those leucoplasts which synthesize and store the starch. The amyloplasts occur in those cells which store the starch. The outer membrane of the amyloplast encloses the stroma and contains one to eight starch granules. In some plant tissues such as potato tuber, amyloplasts can grow to be as large as an average animal cell. In them starch granules may become so large that they rupture the encasing membrane. Starch granules of amyloplasts are typically composed of concentric layers of starch.

(*ii*) Elaioplasts. The elaioplasts store the lipids (oils) and occur in seeds of monocotyledons and dicotyledons. They also include sterol-rich **sterinochloroplast**.

(*iii*) Proteinoplasts. The proteinoplasts are the protein storing plastids which mostly occur in seeds and contain few thylakoids.

2. Chromoplasts. The chromoplasts (Gr., *chroma* = colour; *plast* = living) are the coloured plastids containing **carotenoids** and other pigments. They impart colour (*e.g.*, yellow, orange and red) to certain portions of plants such as flower petals (*e.g.*, daffodils, rose), fruits (*e.g.*, tomatoes) and some roots (*e.g.*, carrots). Chromoplast structure is quite diverse; they may be round, ellipsoidal, or even needle-shaped, and the carotenoids that they contain, may be localized in droplets or in crystalline structures. The function of chromoplasts is not clear but in many cases (*e.g.*, flowers and fruits) the colour they produce probably plays a role in attracting insects and other animals for pollination or seed dispersal.

In general, chromoplasts have a reduced chlorophyll content and are, thus, less active photosynthetically. The red colour of ripe tomatoes is the result of chromoplasts that contain the red pigment **lycopene** which is a member of carotenoid family. Chromoplasts of blue-green algae or cyanobacteria contain various pigments such as **phycoerythrin, phycocyanin, chlorophyll a** and **carotenoids**. Chromoplasts are of following two types:

(*i*) Phaeoplast. The phaeoplast (Gr., *phaeo* = dark or brown; *plast* = living) contains the pigment **fucoxanthin** which absorbs the light. The phaeo-plasts occur in the diatoms, dinoflagellates and brown algae.

(*ii*) **Rhodoplast.** The rhodoplast (Gr., *rhode* = red; *plast* = living) contains the pigment **phaeoerythrin** which absorbs the light. The rhodoplasts occur in the red algae.

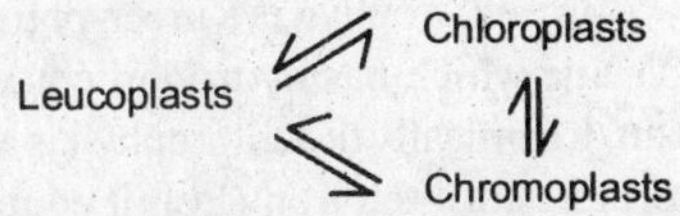

Fig. 58.21. Interconversion of three kinds of plastids.

3. Chloroplasts. The chloroplast (Gr., *chlor* = green; *plast* = living) is most widely occurring chromoplast of the plants. It occurs mostly in the green algae and higher plants. The chloroplast contains the pigment chlorophyll *a* and chlorophyll *b* and DNA and RNA.

According to **Schimper**, different kinds of plastids can transform into one another, as shown in Fig. 58.21.

58.6 CHLOROPLASTS

Chloroplasts are **chlorophyll**-containing specialized organelles of plant cells which are involved in the photosynthesis of food (*i.e.*, carbohydrates). So chloroplasts act as **"kitchens of cells".** They contain their own DNA, ribosomes and protein synthetic machinery to synthesize certain proteins of their use; that is why, they are termed as **semi-autonomous organelles**.

Distribution. The chloroplasts remain distributed homogeneously in the cytoplasm of plant cells. But in certain cells, the chloroplasts become concentrated around the nucleus or just beneath the plasma membrane. The chloroplasts have a definite orientation in the cell cytoplasm. Since chloroplasts are motile organelles, they show passive and active movements.

Morphology

Shape. Higher plant chloroplasts are generally biconvex or plano-convex. However, in different plant cells, chloroplasts may have various shapes, *viz.,* filamentous, saucer-shaped, spheroid, ovoid, discoid or club-shaped. They are vesicular and have a colourless centre.

Size. The size of the chloroplasts varies from species to species. The chloroplasts generally measure 2–3 μm in thickness and 5–10 μm in diameter (*e.g.*, *Chlamydomonas*). The chloroplasts of polyploid plant cells are comparatively larger than the chloroplasts of the diploid plant cells. Generally, chloroplast of plants grown in the shade are larger and contain more chlorophyll than those of plants grown in sunlight.

Number. The number of the chloroplasts varies from cell to cell and from species to species and is related with the physiological state of the cell, but it usually remains constant for a particular plant cell. The algae usually have a single huge chloroplast. The cells of the higher plants have 20 to 40 chloroplasts. According to a calculation, the leaf of *Ricinus communis* contains about 400,000 chloroplasts per square millimeter of surface area. When the number of chloroplasts is inadequate, it is increased by division; when excessive, it is reduced by degeneration.

Chemical Composition

The chloroplasts are composed of the carbohydrates, lipids, proteins, chlorophyll, carotenoids (carotene and xanthophylls), DNA, RNA and certain enzymes and coenzymes. The chloroplasts also contain some metallic atoms as Fe, Cu, Mn and Zn.

The carbohydrates occur in very low percentage in the chloroplasts. The most common carbohydrates of the chloroplasts are the starch and sugar phosphates.

The chloroplasts contain 20–30 per cent lipids of its dry weight. The lipids are composed of 50 per cent fats, 20 per cent sterols, 16 per cent waxes and 7 to 20 per cent phospholipids. The most common alcohols of the lipids are the choline, inositol, glycerol, ethanolamine.

The proteins constitute 35 to 55 per cent of the chloroplast. About 80 per cent proteins are insoluble and forming the unit membranes of the chloroplasts along with the lipids. The 20 per cent proteins are soluble and occur in the form of the enzymes.

The chlorophyll is a green pigment of the chloroplasts. The chlorophyll contains an asymmetrical molecule which has hydrophilic head of four rings of the pyrols and hydrophobic tail of the phytol chain. Chemically the chlorophyll is a porphyrin like the animal pigment haemoglobin and cytochromes except besides the iron (Fe), it contains Mg atom in between the rings of the pyrols which remain connected with each other by the methyl groups. The chlorophyll consists of 75 per cent **chlorophyll *a*** and 25 per cent **chlorophyll *b***.

The carotenoids are carotenes and xanthophylls, both of which are related to vitamin A. The carotenes have hydrophobic chains of unsaturated hydrocarbons in their molecules. The xanthophylls contain many hydroxy groups in their molecules.

DNA of chloroplast of *Chlamydomonas* represents non-chromosomal genetic system and has been found to be related with cytoplasmic heredity. **Ruth Sager**, who is pioneer on nonchromosomal genes, was able to prepare genetic map of *Chlamydomonas* chloroplast. She has shown that the genome of the chloroplast of *Chlamydomonas* is circular like that of bacteria.

Ultrastructure

A chloroplast comprises the following three main components (Fig. 58.22):

1. Envelope. The entire chloroplast is bounded by an **envelope** which is made up of double unit membranes. Across this double membrane envelope occurs exchange of molecules between chloroplast and cytosol (cytoplasmic matrix). Isolated membranes of envelope of chloroplast lack chlorophyll pigment and cytochromes but have a yellow colour due to the presence of small amounts of carotenoids. They contain only 1 to 2 per cent of the total protein of the chloroplast.

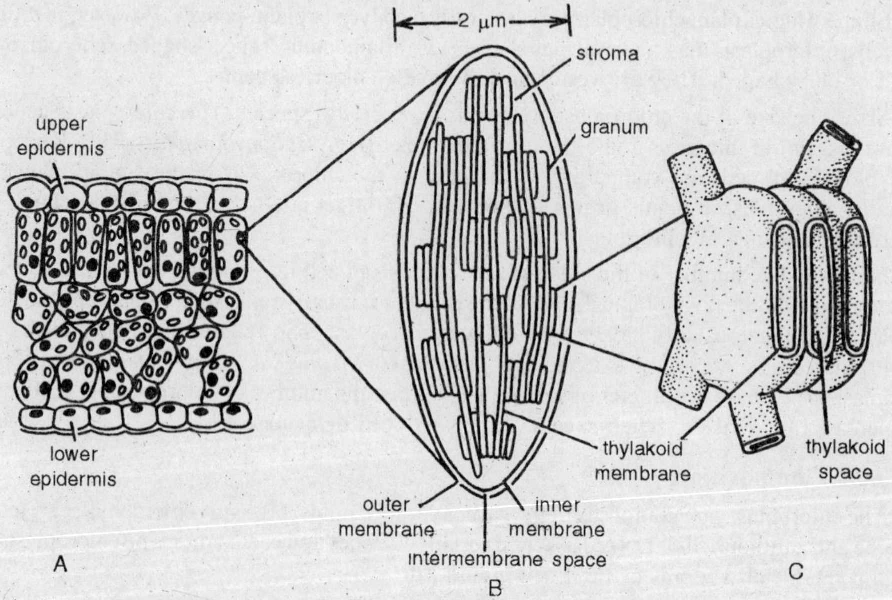

Fig. 58.22. A—Distribution of chloroplasts in mesophyll cells of a leaf; B—Ultrastructure of a chloroplast; C—Details of a granum.

2. Stroma. The matrix or stroma fills most of the volume of the chloroplasts and is a kind of gel-fluid phase that surrounds the thylakoids (grana). It contains about 50 per cent of the proteins of the chloroplast, most of which are soluble type. The stroma also contains ribosomes and DNA molecules (*i.e.*, 80 DNA molecules per chloroplast per cell of *Chlamydomonas*; 20 to 40 DNA molecules per chloroplast per cell of leaf of maize), both of which are involved in the synthesis of some of the structural proteins of the chloroplast. The stroma is the place where CO_2 fixation occurs and where the synthesis of sugars, starch, fatty acids and some proteins takes place.

3. Thylakoids. The thylakoid (thylakoid = sac-like) consists of flattened and closed vesicles arranged as a membranous network. The outer surface of the thylakoid is in contact with the stroma, and its inner surface encloses an **intrathylakoid space** (the third compartment). Thylakoids may be stacked like a neat pile of coins, forming **grana** or they may be unstacked, **intergranal**, or **stromal thylakoids**, forming a system of anastomosing tubules that are joined to the **grana thylakoids**. There may be 40 to 80 grana in the matrix of a chloroplast. The number of thylakoids per granum may vary from 1 to 50 or more. For example, there may be single thylakoid (*e.g.*, red alga), paired thylakoids (*e.g.*, Chrysophyta), triple thylakoids and multiple thylakoids (*e.g.*, green algae and higher plants).

Molecular organization of thylakoids. Molecular organization of the membrane of thylakoids is based on the fluid-mosaic model of the membrane which represents following main characteristics: **fluidity, asymmetry** and **economy** (*i.e.*, lack of movement in the third dimension). Lipids represent about 50 per cent of the thylakoid membrane; these include those directly involved in photosynthesis (called **functional lipids**) such as **chlorophylls, carotenoids** and **plastoquinones. Structural lipids** of thylakoids include glycolipids, sulpholipids and a few phospholipids. Most of these structural lipids are highly unsaturated which confer to the membrane of thylakoids a high degree of fluidity.

The protein components of thylakoid membrane are represented by 30 to 50 polypeptides which are disposed in the following five major supramolecular complexes (Fig. 58.23), which can be isolated with mild detergent:

1. Photosystem I (PS I). This complex contains a reactive centre composed of P700 (Type of pigment which is bleached at the wavelength of 700 nm), several polypeptides, a lower chlorophyll *a*/*b* ratio and β-carotene. It acts as a light trap and is present in unstacked thylakoid membranes. In it light induced reduction of $NADP^+$ takes place.

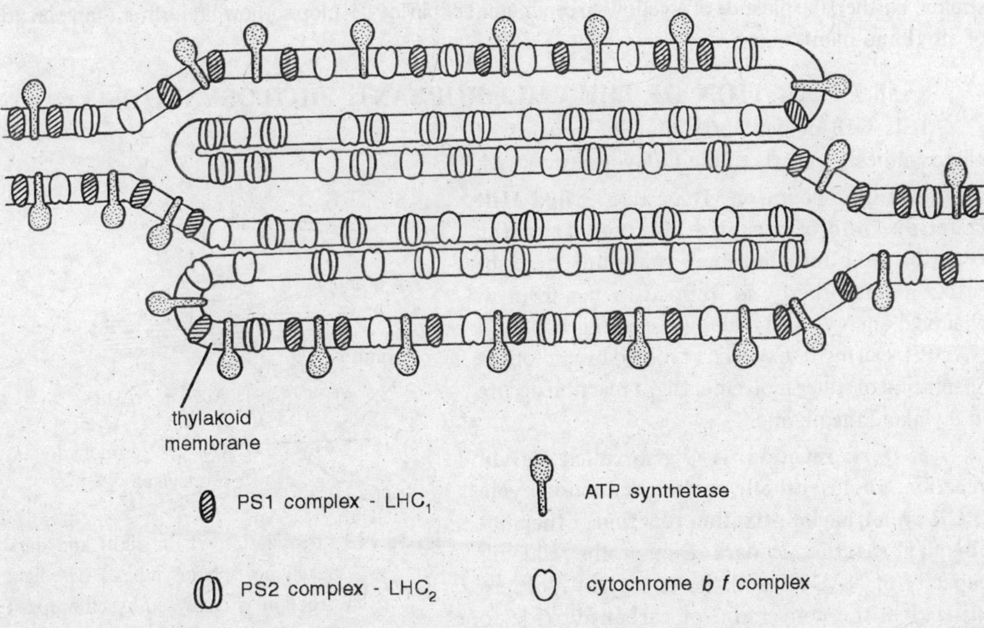

Fig. 58.23. Diagram showing the distribution of the main complexes within the thylakoid membranes both in the granal or stacked and stromal or unstacked regions.

2. Photosystem II (PS II). This complex comprises two intrinsic proteins that bind to the reaction centre of chlorophyll P680 (The pigment that bleaches when absorbing light at 680 nm). It contains a high ratio of chlorophyll *a*/*b* and β-carotene. Frequently, the PS IIs are associated with the light-harvesting complex and are involved in light induced release of O_2 from H_2O (*i.e.*, photolysis of water). PS II works as a light trap in photosynthesis and is mainly present in the stacked thylakoid membranes of grana.

3. Cytochrome b/f. This complex contains one cytochrome *F*, two cytochromes of *b*563, one FeS centre and a polypeptide. It is uniformly distributed in the grana and acts as the electron carrier.

These three complexes are related to the electron transport and are linked by **mobile electron carriers** (*i.e.*, plastoquinone, plastocyanin and ferredoxin). Electron transport through PS II and PS I finally results in the reduction of the coenzyme $NADP^+$. Simultaneously, the transfer of protons from the outside to the inside of the thylakoid membrane occurs.

4. ATP synthetase. As in mitochondria, this complex consists of a $\mathbf{CF_0}$ hydrophobic portion, a proteolipid that makes a proton channel, and a $\mathbf{CF_1}$ (or coupling factor one) that synthesizes ATP from ADP and Pi, using the proton gradient provided by the electron transport. ATP synthetase complexes are located in stacked membrane (grana).

5. Light harvesting complex (LHC). The main function of LH complex is to capture solar energy. It contains two main polypeptides and both cholorophyll *a* and *b*. LH complex is mainly associated with PS II, but may also be associated with PS I. LHC is localized in stacked membranes and lacks photochemical activity.

Mutation and chloroplast structure. The organization of chloroplasts and other plastids is often modified due to mutation. **D. Von Wettstein** (1956) reported that the plastids of normal barley plant have a well organized system of grana and stroma. But the plastids of an albino mutant of barley, fail to develop beyond a particular stage and there occurs no differentiation of grana and stroma. Further, the plastids of a yellow-green mutant of barley develop somewhat further than plastids of an albino plant.

58.7 FUNCTION OF THE CHLOROPLAST: PHOTOSYNTHESIS

It is well evident now that the process of photosynthesis consists of the following two steps:

1. Light reaction. It is also called **Hill reaction, photosynthetic electron transfer reaction** or **photochemical reaction.** In light reaction solar energy is trapped in the form of chemical energy of ATP and as reducing power in NADPH. During it, oxygen is evolved by photolysis or splitting of water molecule. Light reaction occurs in thylakoid membranes.

2. Dark reaction. It is also called **Calvin reaction, photosynthetic carbon reduction cycle (PCR cycle), carbon-fixation reaction** or **thermo-chemical reaction.** In dark reaction, the reducing capacity of NADPH and the energy of ATP are utilized in the conversion of carbon dioxide to carbohydrate. Such a process of **"carbon fixation"** or **"CO_2-fixation"** occurs in the stroma of chloroplast (Fig. 58.24).

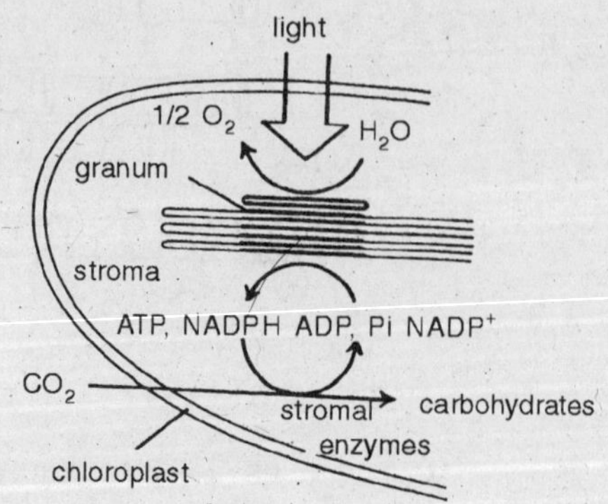

Fig. 58.24. Localization of the light and dark reaction of photosynthesis. The light reaction is catalyzed by chloroplast lamellae, espiecially in the grana. The dark reaction is catalyzed by enzymes of the stroma.

1. Light Reaction

The most important step of light reaction is harvesting of the maximum amount of solar energy for conversion into chemical energy. When the P680 of photosystem II acquires a sufficient quantum of energy, it emits a pair of electrons. These electrons with high potential energy move down an electron transport chain (of thylakoid membrane) and during this process ATP molecules are formed (Fig. 58.25). Two electrons are passed through an electron acceptor Q (which is a quinone) to an electron transport chain involving four electron carriers (**plastoquinone, cytochrome-559, cytochrome-553** or **cytochrome *f*** and **plastocyanin**), before being passed on to PS I. The electrons are passed through four carriers successively at lower energy levels, so that at each step energy is released, which is harvested in the production of two ATP molecules (from ADP + Pi). The electron lost by P 680 is ultimately accepted by P700 of photosystem I. P700 (PS I) also captures light, and for absorbing each photon, it ejects an electron. This electron is replaced by an electron from PS II and this flow of electron continues as long as the light is available. The electrons liberated from P700 are passed through acceptor X, to an electron transport chain **(ferredoxin, ferredoxin NADP reductase)** at successive lower energy levels. Finally these electrons reach NADP coenzyme, each molecule of which receives an electron, enabling it to pick up a H^+ ion (proton), thus, producing two molecules of NADPH from one molecule of H_2O used in PS II:

$$2e^- + 2H^+ + 2NADP^+ \xrightarrow{2hv} 2\,NADPH$$

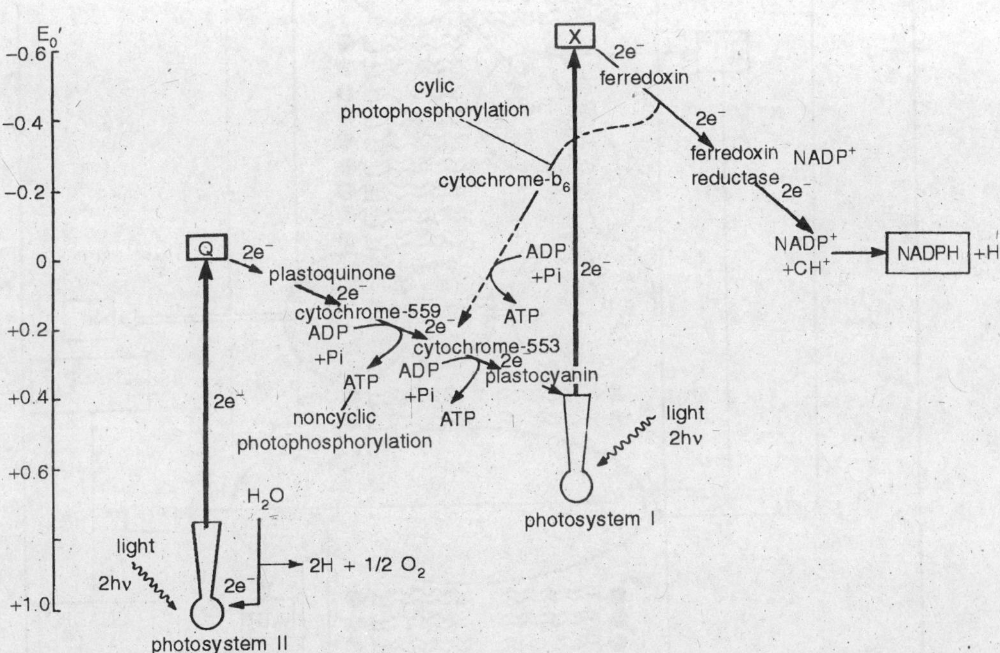

Fig. 58.25. Photosynthetic light reaction showing major events occurring in photosystem I and photosystem II of chloroplasts.

The oxidized P680 regains its electrons by the photolysis of water into $2H^+$, $2e^-$ and oxygen:

$$H_2O \xrightarrow{2 \text{ quanta or } 2hv} \tfrac{1}{2}O_2 + 2H^+ + 2e^-$$

Oxygen is given out by photosynthesizing plants. The protons (H^+) accumulate inside the thylakoid membrane resulting in a **proton gradient.** The energy released by the protons when they diffuse across the thylakoid membrane into the stroma (along H^+ concentration gradient) is used to produce ATP molecules, by **CF_0 – CF_1 ATP synthetase** in the membrane. As synthesis of ATP occurs in light and the process is not cyclic (*i.e.*, it needs a constant supply of water molecules to be oxidized and NADP to be reduced), the process is called **non-cyclic photophosphorylation**.

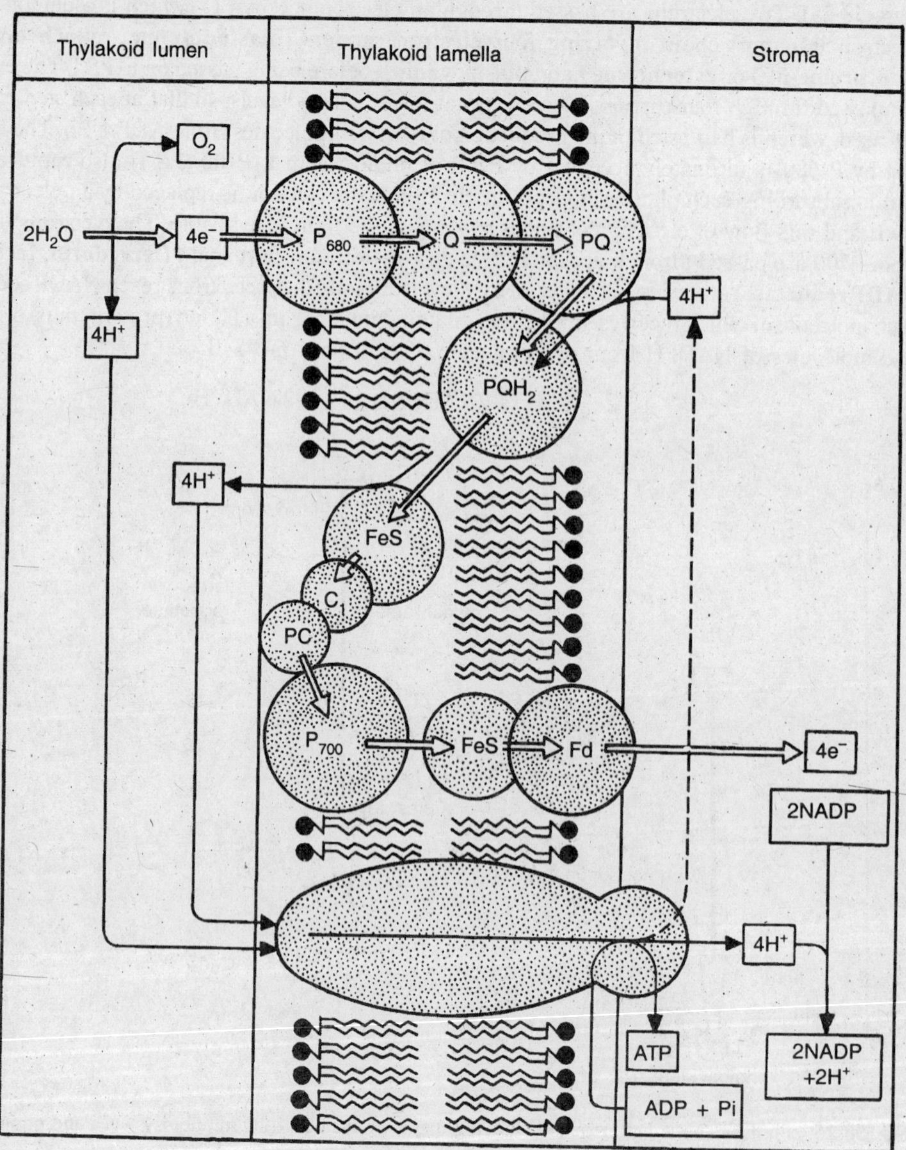

Fig. 58.26. Model for chemiosmotic coupling. Locations of electron transfer intermediates in the thylakoid lamella (membrane) and enzymes involved in ATP synthesis and proton (H^+) transport have been tentatively depicted.

ATP production by **cyclic photophosphorylation** also occurs during the light reaction. This process involves another electron transfer mechanism involving **cytochrome b_6** and starting with P700; the ultimate acceptor of the de-energized electron is also P700 of photosystem I.

All the molecules of ATP and NADP generated in the light reaction of photosynthesis are used by soluble enzymes of stroma of chloroplast during the dark reaction. Total light reaction can be summarized as follows:

$$H_2O + 2\,NADP + 2\,ADP + 2\,Pi \xrightarrow{4h\nu} ½\,O_2 + 2\,NADPH + 2\,ATP$$

2. Dark Reaction

The dark reaction is completed by passing through following three main phases:

(*i*) Phase 1: Carboxylation. During this phase of dark reaction, three molecules of carbon dioxide (3C) are attached to three molecules of **ribulose 1, 5, biphosphate (RuBP;** this pentose sugar was previously termed RuDP or ribulose diphosphate; 15C) to produce short-lived six-carbon intermediates. This process is called **carboxylation** and is catalyzed by the enzyme **RuBP carboxylase, carboxydismutase** or **"Rubisco"** (which is widely acclaimed as one of the most abundant proteins present on the planet Earth). Rubisco is a large protein molecule (500,000 dalton MW) comprising 16 per cent of chloroplast protein. The six carbon intermediates are immediately broken down into six molecules of **PGA** or **3-phosphoglyceric acid** (*i.e.*, a *C*–3 or three carbon compound, 6C × 3C = 18C).

(*ii*) Phase 2: Glycolytic reversal. By utilizing six ATP molecules, these six molecules of PGA are transformed into six molecules of **1, 3, diphosphoglyceric acid** (*i.e.*, 6C × 3C = 18C; 3C from $3CO_2$ and 15C from 3 RuBP). These in turn get converted into six molecules of **glyceraldehyde-3-phosphate, 3-phosphoglyceraldehyde (PGAL)** or **3-phosphoglyceric acid** (triose) by utilizing six NADPH molecules.

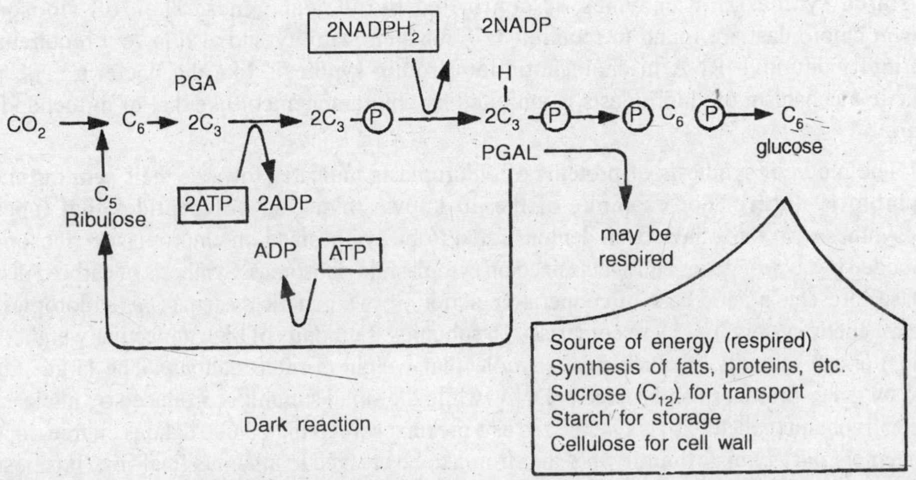

Fig. 58.27. Synthesis of carbohydrate during dark reaction.

(*iii*) Phase 3: Regeneration of RuBP. For the continuous running of Calvin cycle, there must be a regular supply of ATP and NADPH and also sufficient amount of RuBP. Three molecules of RuBP (3C × 5C = 15C) are regenerated by a complex series of reactions which utilize three ATP molecules and five molecules of 3-phosphoglyceric acid (5C × 3C = 15C). This leaves one molecule of PGAL as the net gain of one Calvin cycle. Two turns of Calvin cycle result in the production of one molecule of **glucose**. This glucose is used by the plant to form a large variety of organic compounds required for its structure and function (*e.g.*, starch, cellulose, lipids, amino acids and proteins, etc.). The dark reaction may be summed up as follows:

$$6\,RuBP + 6CO_2 + 18\,ATP + 12\,NADPH \longrightarrow 6\,RuBP + C_6H_{12}O_6 + 18\,ADP + 18\,Pi + 12\,NADP$$

Chloroplast as Semiautonomous Organelle

Like the mitochondria, the chloroplasts have their own DNA, RNAs and protein synthetic machinery and are semiautonomous in nature.

1. DNA of chloroplast. The chloroplasts of the algae and higher plants are found to contain DNA molecules. First of all **Ris** and **Plant** (1962) have reported DNA molecule in the chloroplast of the *Chlamydomonas*. Later on DNA molecule has been reported from the chloroplasts of other algae and higher plants. In general, chloroplasts have a double helical DNA circle with an average length of 45 μm (about 135,000 base pairs). The replication of chloroplast DNA has been followed with ^3H– thymidine. Maps of the location of genes (genetic maps) have been made in several chloroplast DNAs by the help of restriction enzymes. The gene for the large subunit of carboxydismutase enzyme has been fully sequenced and is found to contain 1425 nucleotides.

2. Ribosomes of chloroplasts. The chloroplasts contain the ribosomes which are smaller than the cytoplasmic ribosomes. The ribosomes of the chloroplast are of 70S type and resemble with the bacterial ribosomes. The ribosomes of the chloroplasts consist of two ribosomal RNAs, 23S rRNA and 16S rRNA. **Lyttleton** (1962) has separated polyribosomes or polysomes from the chloroplast. The chloroplasts also contain aminoacyl-tRNAs, aminoacyl-tRNA synthetases, methionyl-tRNA.

3. Protein synthesis. The DNA of chloroplast codes for chloroplast mRNA, rRNA, tRNA, and ribosomal proteins. It also codes for certain structural proteins of thylakoid membranes. The synthesis of other chloroplast components as chlorophyll, carotenoids, lipids and photosynthetic and starch synthesizing enzymes, is controlled by nuclear genes. The 70S ribosomes of *Euglena* chloroplast are found to require Mg^{++} for their stability and also have a requirement for N-formyl methionyl-tRNA in chain initiation protein synthesis like the bacteria. The protein synthetic mechanism of chloroplasts is inhibited by chloramphenicol like that of mitochondria and bacteria.

The mode of synthesis of proteins of chloroplasts indicates towards their **semiautonomous** or **symbiotic nature**. For example, of the 30 known thylakoid polypeptides that function in photosynthesis, so far 9 have been demonstrated to be synthesized on chloroplastic ribosomes and 9 are coded by nuclear genes and synthesized on cytoplasmic ribosomes. Synthesis of carboxydismutase (C Dase) presents a good case of cooperative action of two genetic systems (*i.e.*, chloroplastic and nuclear genetic systems). C Dase comprises 16 subunits: 8 subunits of high molecular weight (55,000 daltons) and 8 subunits of much smaller molecular weight (14,000 daltons). The large subunit is coded by genes present in chloroplastic DNA, while the small subunit is produced by nuclear genes. The small subunit (called P20) is synthesized as a precursor weighing 20,000 daltons on free ribosomes; it then enters **post-translationally** into the stroma to be cleaved to attain its final size. It is postulated that the chloroplastic envelope has receptor sites that recognize the proteins that are to be incorporated into the organelle. The extra sequence (acting as the **signal**) that is present in P20 is composed of acidic amino acids, in contrast to the hydrophobic ones in the signal sequence of secretory proteins. After entering the chloroplast the signal sequences are removed by a protease enzyme, which is present in the envelope of chloroplast, and the small subunit of C Dase is released into the stroma (**Ellis**, 1981). Thus, chloroplast proteins may be synthesized by three avenucs:

1. by an exclusive chloroplastic mechanism,
2. by a mechanism involving nuclear genes and chloroplastic ribosomes, and
3. by nuclear genes and cytoplasmic ribosomes.

Protein transport into chloroplasts resembles transport into mitochondria in many respects: both occur post-translationally, both require energy, and both utilize hydrophilic amino-terminal signal peptides that are removed after use. However, there is at least one important difference that while mitochondria exploit the electrochemical gradient across their inner membrane to help drive the transport, chloroplasts (which have an electrochemical gradient across their thylakoid but not their inner membrane) appear to employ only ATP hydrolysis to import across their double-membrane outer envelope.

Translocation of proteins into the thylakoid space of chloroplasts requires two signal peptides and two translocation events. The precursor polypeptide contains an amino-terminal chloroplast signal peptide followed immediately by a thylakoid signal peptide. The **chloroplast signal peptide** initiates translocation into the stroma through a membrane contact site by a mechanism similar to that used for translocation into mitochondrial matrix. The signal peptide is then cleaved off, unmasking the thylakoid signal peptide, which initiates translocation across the thylakoid membrane (Fig. 58.28).

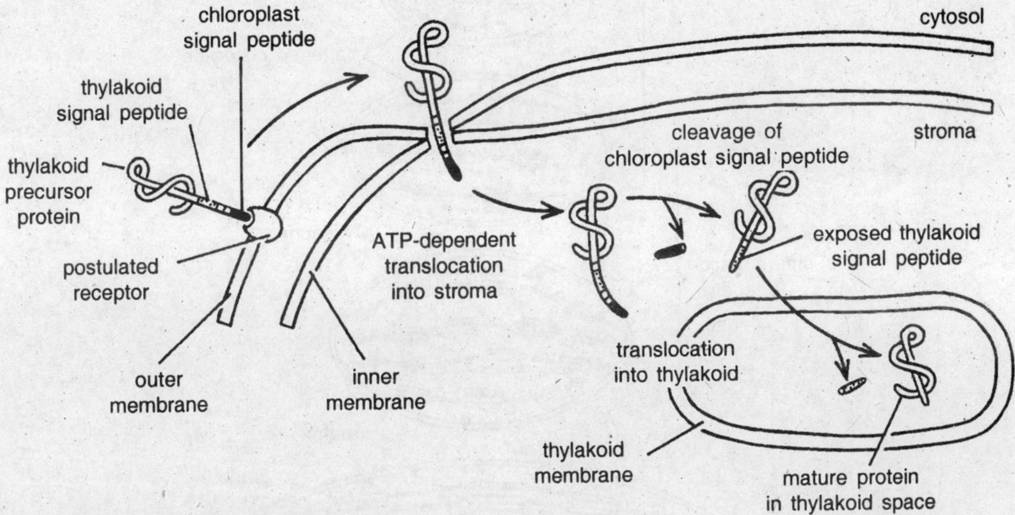

Fig. 58.28. Mode of translocation of proteins into stroma and thylakoid of chloroplast by the help of signal peptides.

Biogenesis of Chloroplast

The chloroplasts never originated *de novo*. Since the classic work of **Schimper** and **Meyer** (1883) it has been accepted that chloroplasts multiply by fission, a process that implies growth of the daughter organelles. This is easily observed in the alga *Nitella,* which contains a single huge chloroplast. In *Nitella* a division cycle of 18 hours has been cinematographically recorded for the chloroplast.

During the development of the chloroplast, the first structure to appear is the so-called **proplastid,** which has a double membrane. Development of proplastid into chloroplast takes place in the following steps:

1. In the presence of light, the inner membrane grows and gives off vesicles into the matrix that are transformed into discs (Fig. 58.29). These intrachloroplastic membranes are the thylakoids which, in certain regions, pile closely to form the grana. In the mature chloroplast the thylakoids are no longer connected to the inner membrane, but the grana remain united by inter-granal thylakoids.
2. In the absence of light, a reverse sequence of changes takes place. This is the process of **etiolation,** in which the leaves lose their green pigment and the chloroplast membranes become disorganized. The chloroplasts are transformed into **etioplasts**, in which there is a paracrystalline

arrangement of tubules forming the so-called **prolamellar body**. Attached to these bodies are young thylakoid membranes that lack photosynthetic activity.

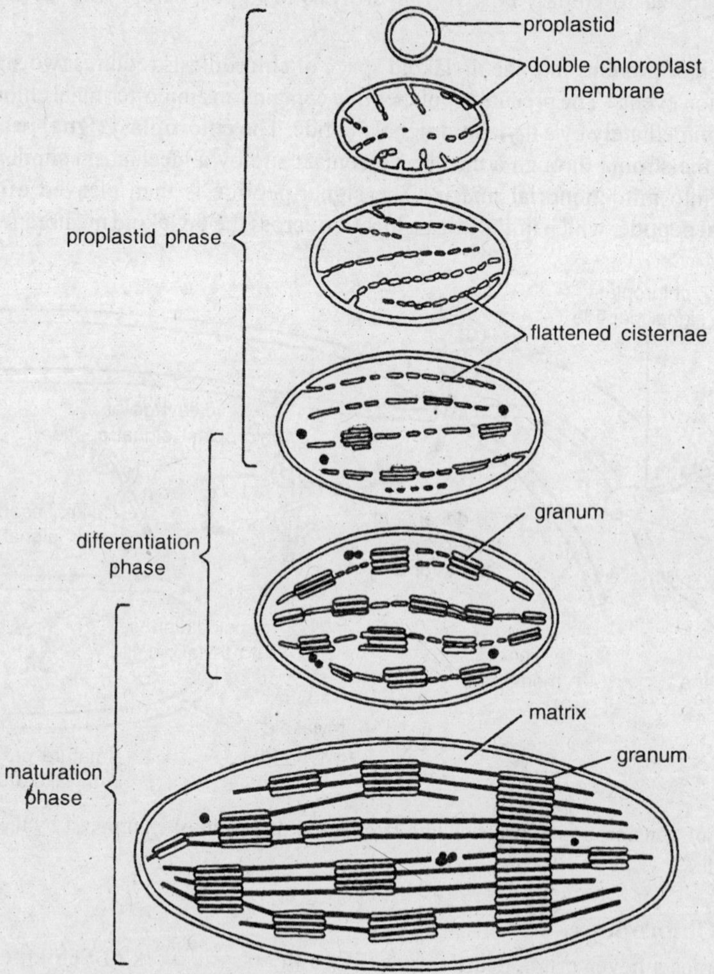

Fig. 58.29. Development of a chloroplast from a submicroscopic proplastid in the presence of light.

The regular crystal lattice of two prolamellar bodies surrounded by young thylakoid membranes is observed by **Osumi** *et al.*, (1984). If etiolated plants are re-exposed to light, thylakoids are reformed and the prolamellar material is used for assembly.

The symbiotic origin of the chloroplast. In certain characteristics, the chloroplasts are comparable with that of a semiautonomous or symbiotic organism living within the plant cells. They divide, grow and differentiate; they contain circular DNA, ribosomal RNA, messenger RNA and are able to conduct protein synthesis. By visualizing these similarities between chloroplast and micro-organism; it has been suggested that chloroplast might have resulted from a symbiotic relationship between an autotrophic microorganism, one which is able to transform radiant energy from sunlight and heterotrophic host cell. The symbiotic origin of the chloroplast appears very justified but **Kirk** (1966) has shown that certain important enzymes which are necessary for the development of the chlorophyll and for the photosynthetic mechanism are synthesized according to the codes of the nuclear DNA.

58.8 COMPARISON OF CHLOROPLASTS AND MITOCHONDRIA

Chloroplasts carry out their energy inter-conversions by chemiosmotic mechanisms in much the same way that mitochondria do and they are organized on the same principles. Thus, each chloroplast contains three distinct membranes which define three separate internal compartments—the intermembrane space, the stroma and the thylakoid space. The thylakoid membrane contains all the energy generating systems of chloroplasts.

Like the mitochondria, chloroplasts have a highly permeable outer membrane; a much less permeable inner membrane, in which special carrier or transport proteins are embedded; and a narrow intermembrane space in between. The inner membrane surrounds a large space called the **stroma**, which is analogous to the mitochondrial matrix and contains various enzymes, ribosomes, RNAs and DNA.

However, there is an important difference between the two: the inner membrane of the chloroplast is not folded into cristae and does not contain an electron-transport chain. Instead, the photosynthetic light-absorbing system, the electorn-transport chain and an ATP synthetase are all contained in a third distinct membrane that forms a set of flattened disc-like sacs, the **thylakoids** (Fig. 58.31).

In a general way, one might view the chloroplast as a greatly enlarged mitochondria in which the cristae are converted into a series of interconnected submitochondrial particles in the matrix space. The knobbed ends of the chloroplast ATP synthetases ($F_0 - F_1$ coupling factors), where ATP is made, protrude from the thylakoid membrane into the stroma, just as they protrude into the matrix from the membrane of each mitochondrial crista.

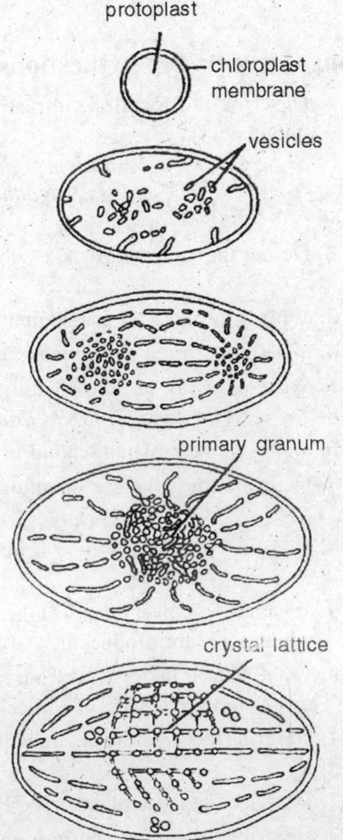

Fig. 58.30. Development of a chloroplast from submicroscopic proplastid in dark.

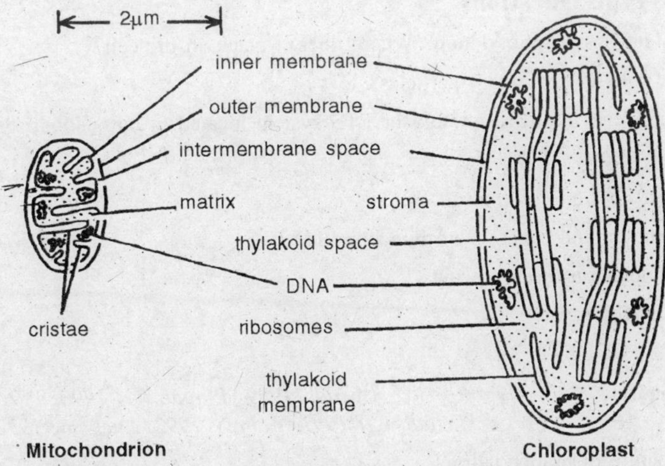

Fig. 58.31. Comparison of a mitochondrion and a chloroplast.

QUESTIONS

Long Answer Type Questions

1. Describe the location, ultrastructure, biochemistry (or chemical composition) and functions of mitochondria. *(Shree Venkateshwar 1989; Garhwal 1989; Delhi 1989, 1991; Kanpur 1989, 1994, 1995; CCS 1989, 1993, 1997; Bihar 1990; Awadh 1990; Magadh (Subs.) 1990, 1994; Purvanchal 1991; Sagar 1991; Vikram 1992; Allahabad 1993; Lucknow 1992, 1996, 1998; Guwahati (G) 1996)*
2. Define the chloroplast. Describe its structure, chemical composition and main functions. *(CCS 1991, 1992, 1994, 1996)*
3. Why chloroplast is considered as a semiautonomous organelle. Discuss it.
4. What are stroma and grana? Discuss their role in the process of photosynthesis.
5. Write an essay on mitochondria. *(Berhampur 1989; Kanpur 1989, 1993; Bhopal 1990; North Bengal 1991; Mithila 1992)*
6. Write the role of mitochondria in cellular activities. *(Punjab (G) 1992; NEHU 1996)*
7. Describe the structure of mitochondria. Why is it known as power house of the cell? *(Allahabad 1994)*
8. How is pyruvate (pyruvic acid) metabolized via Krebs cycle? Describe the sequence of reactions. *(Bihar 1989)*
9. What is the 'power house' of the cell? Describe the structure and chemical composition of mitochondria where ATP are produced. *(CCS 1999)*
10. What do you understand from 'power house' and 'atom bomb' of the cell? How energy is produced in the 'power house' of the cell?
11. Describe the structure and function of mitochondria with special reference to electron transport system. *(Guwahati 1992)*
12. Describe steps involved in ATP synthesis in mitochondrion. *(Allahabad 1996)*
13. Describe the ultrastructure of mitochondrion with the help of neat and labeled diagrams. How does it provides energy to the cell? *(Kanpur 1997)*
14. Describe energetics of cellular respiration. *(Kumaun 1990)*
15. Explain Krebs cycle. *(Rajasthan 1990; Agra 1992)*
16. Mitochondria are "Power House" of the cell. Justify this statement with cellular respiration. *(CCSU Meerut 2005)*

Short Answer Type Questions

1. What is oxidative phosphorylation? Where does it occur in the cell?
2. What is the role of elementary particles?
3. Where are the enzyme of respiratory or ETS system located in mitochondria?
4. Give a brief account of the plastids.
5. Write a short note on chloroplast. *(Purvanchal 1992; CCS 1997)*
6. Give the diagram and function of mitochondria. *(HNB., Garwal 2008)*
7. Write short notes on the following:

   (*i*) FAD or FMN *(Lucknow 1998)*

   (*ii*) ATP

   (*iii*) Mitochondria *(Awadh 1989; Punjab (G) 1990, 1992; CCS 1990, 1997; Magadh (G) 1991, Kanpur 1992; Lucknow 1992; Nagarjuna 1994)*

   (*iv*) Ultrastructure of mitochondria *(Gorakhpur 1997)*

   (*v*) Functions of mitochondria *(Guwahati (G) 1989; Jiwaji 1991; Magadh (G) 1995)* (*vi*) Mitochondrial enzymes (*Punjab (G) 1991*)

(*vii*) Oxysomes (*Lucknow 1998*)

(*viii*) Glycolysis (*Kanpur 1993; Lucknow 93*)

(*ix*) Krebs cycle (*Kanpur 1991, 92*)

(*x*) Electron transport chain

(*xi*) Chloroplasts (*HNB Garwal 2005*)

8. Compare the chloroplasts with mitochondria.
9. Differentiate between leucoplasts and chromoplasts.
10. Draw a well labelled diagram of ultrastructure of mitochondria. (*Gorakhpur 1995; Lucknow 1997*)
11. 'Mitochondrion as a semi-autonomous organelle'. (*B.H.U. (Hons.) 1990*)
12. Describe the origin or biogenesis of chloroplasts.

Very Short Answer Questions

1. Who coined the term mitochondria?
   Ans. C. Benda in 1897.
2. Name the cell type in which mitochondria are absent.
   Ans. Mammalian red blood corpuscles (RBCs).
3. What will happen if all the mitochondria of a cell are destroyed?
   Ans. Cell will not get energy for performing various activities.
4. What will happen if elementary particles of mitochondria are removed?
   Ans. ATP molecules will not be produced.
5. Name the organism which possess only one mitochondria.
   Ans. *Micromonas.*
6. Which cells possess highest number of mitochondria?
   Ans. Muscle cells.
7. For which structure the name bioblast was used by Altman?
   Ans. Mitochondria.
8. What is metabolism?
   Ans. All chemical reactions occurring within a cell.
9. Where are enzymes of ETS located in mitochondria?
   Ans. Inner membrane of mitochondria.
10. What is the location in enzymes involved in oxidation of pyruvic acid in the mitochondria?
    Ans. Mitochondrial matrix.
11. Where is the site of oxidative phosphorylation?
    Ans. F_1 particles (of mitochondria).
12. Who gave the chemiosmotic hypothesis?
    Ans. Peter Mitchell.
13. State the significance of thylakoids of chloroplasts.
    Ans. Thylakoids have light absorbing system, electron transport system and ATP synthetase enzyme.
14. Name the type of chlorophyll found in all green plant.
    Ans. Chlorophyll–*a*.

15. Name the site of occurrence of photochemical or light reaction.

Ans. Thylakoids of the chloroplasts.

16. What is the use of sunlight in light reaction?

Ans. Radiant energy present in sunlight is used by chloroplasts to form ATP and to oxidize water molecules.

17. Where does dark or synthetic reaction occur in chloroplasts?

Ans. Stroma of chloroplasts.

18. What is antenna complex?

Ans. Clusters of hundreds of chlorophyll molecules which trap sunlight energy is called **antenna complex**.

19. Define the chlorophyll.

Ans. Chlorophyll is a light-absorbing molecule that donates photoenergized electrons to organic molecules, initiating photo chemical events that lead to the generation of the NADPH and ATP required for the Calvin cycle; because of its absorption properties, chlorophyll gives plants their characteristic green colour.

20. What is Calvin cycle?

Ans. It is a cyclic series of reactions used by photosynthetic organisms for the fixation of carbon dioxide and its reduction to form carbohydrates.

21. What is the biological significance of carbon atom?

Ans. Carbon is most important atom in biological molecules, capable of forming upto four covalent bonds.

22. Define the chloroplast.

Ans. It is a double membrane-enclosed cytoplasmic organelle of plants and algae that contains chlorophyll and enzymes necessary to carry out photosynthesis.

23. What is cellular respiration?

Ans. It is oxidation-driven flow of electrons from reduced coenzymes to an electron acceptor, usually accompanied by the generation of ATP.

24. What is bioenergetics?

Ans. It is an area of science that deals with the application of thermodynamic principles to reactions and processes in the biological world.

25. Define the metabolic pathway.

Ans. It is a series of cellular enzymatic reaction that convert one molecule to another via a series of intermediates.

26. What is free energy?

Ans. Free energy (G) represents thermodynamic function that measures the extractable energy content of a molecule under conditions of constant temperature and pressure, the change in free energy is a measure of the ability of the system to do work.

27. Define exergonic.

Ans. It is pertaining to an energy-releasing reaction characterized by a negative free energy change ($\Delta G < O$).

28. What is endergonic?

Ans. It is pertaining to an energy-requiring reaction characterized by a positive free energy change ($\Delta G > O$).

29. What is ATP?

Ans. ATP or adenosine triphosphate is a adenosine with three phosphates linked to each other by phosphoanhydride bonds and to 5′ carbon of the ribose by a phosphoester bond; it is a principal energy storage compound of most cells, with energy stored in the high-energy phosphoanhydride bonds.

30. Define coenzyme?

Ans. It is a small organic molecule that functions along with an enzyme by serving as a carrier of electrons or functional groups.

31. What is coenzyme A (CoA)?

Ans. It is a organic molecule that serves as a carrier of acyl groups by forming a high-energy thioester bond with an organic acid.

32. Define coenzyme Q.

Ans. Coenzyme Q (CoQ) is a nonprotein (quinone) component of the mitochondrial electron transport system that serves as a collection point for electrons from both FMN and FAD-linked dehydrogenases; also called **ubiquinone.**

33. What is electron transport system (ETS)?

Ans. It is a group of membrane-bound electron carriers that transfer electrons from the coenzymes NADH and $FADH_2$ to oxygen.

34. What are cytochromes?

Ans. These are heme-containing proteins of the ETS involved in the transfer of electrons from coenzyme Q to oxygen by the oxidation and reduction of the central iron atom of the heme group.

35. Define FAD.

Ans. FAD or flavin adenine dinucleotide is a coenzyme that accepts two electrons and two photons from an oxidizable organic molecule to generate the reduced form, $FADH_2$; it is an important electron carrier in energy metabolism.

36. Define F_0–F_1 complex.

Ans. It is a protein complex in the mitocnondrial inner membrane and the bacterial plasma membrane that consists of the F_1 complex bound to F_0 complex; the flow of proton through the F_0 component leads to the synthesis of ATP by the F_1 component.

37. What is chemiosmotic coupling model?

Ans. It is a model postulating that electron transport pathways establish proton gradients across membranes and that the energy stored in such gradients can then be used to drive ATP synthesis.

True and False Statements

Indicate true or false statements by a tick mark:

1. ATP synthesis takes place in ribosome. *(NEHU (M) 1996)*
(*a*) True (*b*) False

2. In a biological system an important source of energy is glucose. *(NEHU (M) 1996)*
(*a*) True (*b*) False

3. Krebs cycle takes place in mitochondrial matrix *(NEHU (M) 1996)*
(*a*) True (*b*) False

Yes or No Questions

Write Yes or No for the following statements:

1. Mitochondrion is made up of double membrane. *(Guwahati (G) 1993, 1994)*

2. Cytochrome oxidase enzyme is found in mitochondria. *(Guwahati (G) 1991, 1996)*

3. Citric acid cycle was first described by Bateson. *(Guwahati (G) 1995)*

Fill in the Blanks

1. Metabolic reactions in the body are catalysed by .... *(NEHU (M) 1996)*

2. Citric acid cycle was worked out by .... *(NEHU (M) 1996)*

Multiple Choice Questions

Choose the correct answer from the four alternatives given:

1. One of the following is a coenzyme (*B.U. Coimbatore 1997*)
(*a*) glucose (*b*) protease
(*c*) ATP (*d*) NAD

2. Unit of carbohydrate is (*NEHU (M) 1996*)
(*a*) monosaccharide (*b*) amino acid
(*c*) fatty acid (*d*) all

3. Cellular respiration, *i.e.*, transformation of chemical energy into usable bioenergy takes place in
(*a*) lysosomes (*b*) ribosomes
(*c*) mitochondria (*d*) endoplasmic reticulum

4. Mitochondria are not found in
(*a*) liver cells (*b*) nerve cells
(*c*) maturing RBCs (*d*) mature RBCs

5. In mitochondria, cristae act as sites for
(*a*) protein synthesis (*b*) oxidation-reduction reactions
(*c*) breakdown of macromolecules (*d*) phosphorylation of flavoproteins

6. Cristae helps in
(*a*) respiration (*b*) transpiration
(*c*) photosynthesis (*d*) photo-oxidation

7. Cytochrome oxidases are found
(*a*) on cristae of mitochondria (*b*) on outer wall of mitochondria
(*c*) in the matrix of mitochondria (*d*) in the lysosomes

8. If the outer membrane is removed from a mitochondrion, the remaining structure is called
(*a*) mitoribosome (*b*) mitoplast
(*c*) microsome (*d*) peroxisome

9. The two types of cellular organelles that transform energy are:
(*a*) chromoplasts and leucoplasts (*b*) mitochondria and chloroplasts
(*b*) mitochondria and chromoplasts (*d*) chloroplasts and leucoplasts

10. Carbon dioxide is fixed in
(*a*) light reaction (*b*) dark reaction
(*c*) aerobic respiration (*d*) anaerobic respiration

11. Calvin cycle is
(*a*) inhibited by light (*b*) dependet upon light
(*c*) independent of light (*d*) supported by light

12. Which of the following occurs in dark reaction of photosynthesis
(*a*) formation of ATP (*b*) release of O_2
(*c*) release of H_2 (*d*) synthesis of PGAL

13. End product of glycolysis is
(*a*) acetyl Co-A (*b*) 1 molecule of pyruvic acid
(*c*) 2 molecules of pyruvic acid (*d*) glucose 1-phosphate

14. At the end of the Krebs cycle, most of the energy removed from the glucose molecule has been transferred to the
   (*a*) NADH and $FADH_2$ (*b*) ATP
   (*c*) oxaloacetic acid (*d*) citric acid
15. When a molecule is reduced it always
   (*a*) loses electrons (*b*) loses both proteins and electrons
   (*c*) loses neutrons (*d*) gains electrons
16. Last electron acceptor in ETS in
   (*a*) O_2 (*b*) water
   (*c*) cytochrome *C* (*d*) cytochrome a_3
17. Enzymes of glycolysis are present in
   (*a*) mitochondrial matrix (*b*) cristae
   (*c*) cytosol (*d*) all of the above
18. Oxidative phosphorylation means
   (*a*) photo-oxidation during photosynthesis (*b*) photolysis of water
   (*c*) ATP production in respiration (*d*) anaerobic respiration

ANSWERS

True and False Statements

1. (*b*) **2.** (*a*) **3.** (*a*)

Yes or No Questions

1. Yes **2.** Yes **3.** No

Fill in the Blanks

1. Enzymes **2.** Hans A. Krebs.

Multiple Choice Questions

1. (*d*) **2.** (*a*) **3.** (*c*) **4.** (*d*)
5. (*b*) **6.** (*a*) **7.** (*a*) **8.** (*b*)
9. (*b*) **10.** (*b*) **11.** (*d*) **12.** (*d*)
13. (*c*) **14.** (*a*) **15.** (*d*) **16.** (*a*)
17. (*c*) **18.** (*c*).

59

Plasma Membrane and Cell Wall

59.1 PLASMA MEMBRANE

A plasma membrane encloses every type of cell, both prokaryotic and eukaryotic cells. It physically separates the cytoplasm from the surrounding cellular environment. Plasma membrane is a ultrathin, elastic, living, dynamic and selective transport-barrier. It is a fluid-mosaic assembly of molecules of lipids (phospholipids), proteins and carbohydrates. Plasma membrane controls the entry of nutrientes and exit of waste products, and generates differences in ion concentration between the interior and exterior of the cell. It also acts as a sensor of external signals (for example, hormonal, immunological, etc.) and allows the cell to react or change in response to environmental signals. The cells of bacteria and plants have the plasma membrane between the cell wall and the cytoplasm. For cells without cell walls (*e.g.*, mycoplasma and animal cells), plasma membrane forms the cell surface.

The plasma membrane is also called **cytoplasmic membrane, cell membrane,** or **plasmalemma.** The term cell membrane was coined by **C. Nageli** and **C. Cramer** in 1855 and the term plasmalemma has been given by **J.Q. Plowe** in 1931.

Isolation and Analysis

The plasma membrane is so thin that it cannot be observed by the light microscope. Structure of the plasma membrane of various cells has been studied by their isolation from the living systems and also by their artificial synthesis by using their constituent molecules (*e.g.*, liposome, see Chapter 57). The pure and isolated membranes are then studied by biochemical and biophysical methods. The purity of isolated membranes is controlled by **electron microscopy, enzyme analysis** and the **study of surface antigens**. A variety of cells such as mammalian red blood cell (erythrocytes), medullated nerve fibres, Ehrlich mouse ascites tumor cells, liver cells, striated muscle, *Amoeba proteus*, sea urchin eggs and bacteria, have been used in studying the ultrastructure of the plasma membrane. The mammalian erythrocytes and the myelin sheath of the nerve fibre, however, have provided the bulk of information regarding the structure and properties of the plasma membrane. For such experiments, human red blood cells or erythrocytes have been selected by **E. Gorter** and **F. Grendel** (1925) for following advantages: these cells are easy to obtain and are known to be extremely simple. Since these cells contain no intracellular organclles or membrane, so the only membrane structure to be considered is almost entirely that of the cell surface. Lastly, the plasma membrane of erythrocytes is relatively tough and does not readily fragment (See **Lucy**, 1975).

Plasma membranes are more easily isolated from erythrocytes subjected to haemolysis. The cells are treated with hypotonic solutions (to be discussed elsewhere in the chapter) that due to endosmosis produce swelling and then loss of the heamoglobin content (*i.e.*, haemolysis). The resulting membrane is called a **red cell ghost**. If haemolysis is mild, permeability functions of the membrane can be restored by certain treatment, such a ghost is called **resealed ghost**. But if haemolysis is more drastic (*i.e.*, there is complete removal of the haemoglobin) and there is no chance of its resealing, the resulting membrane is called **white ghost**. While the resealed ghosts can be used for the study of physiological as well as biochemical properties, white ghosts can only be used for the study of biochemical properties.

The cell wall of yeast, *Saccharomyces cerevisiae*, can be enzymatically removed by the help of a snail gut enzyme, and the resultant protoplast serves as a source of plasma membrane in a manner similar to that of mammalian erythrocytes.

Chemical Composition

Chemically, plasma membrane and other membranes of different organelles are found to contain proteins, lipids and carbohydrates, but in different ratios (Table 59.1). For example, in the plasma membrane of human red blood cells proteins represent 52 per cent, lipids 40 per cent and carbohydrates 8 per cent.

Table 59.1 Chemical composition of some purified membranes (in percentages).

Membrane	*Protein*	*Lipid*	*Carbohydrate*
1. Myelin (Nerve cell)	18	79	3
2. Plasma membrane:			
(*i*) Mouse liver	44	52	4
(*ii*) *Amoeba*	54	42	4
(*iii*) Human erythrocyte	52	40	8
3. Spinach chloroplast lamellae	70	30	0
4. Mitochondrial inner membrane	76	24	0

1. Lipids. Four major classes of lipids are commonly present in the plasma membrane and other membranes: **phospholipids** (most abundant), **sphingolipids, glycolipids** and **sterols** (*e.g.*, **cholesterol**). All of them are amphipathic molecules, possessing both hydrophilic and hydrophobic domains. The relative proportions of these lipids vary in different membranes. Phospholipids may be **acidic phospholipids** (20 per cent) such as **sphingomyelin** or **neutral phospholipids** (80 per cent) such as **phosphatidyl choline, phosphatidyl serine,** etc. Many membranes contain cholesterol. Cholesterol is especially abundant in the plasma membrane of mammalian cells and absent from prokaryotic cells. **Cardiolipin** (diphosphatidyl glycerol) is restricted to the inner mitochondrial membrane.

2. Proteins. The amount and types of proteins in the membranes are highly variable: in the myelin membranes which serve mainly to insulate nerve cell axons, less than 25 per cent of the membrane mass is protein, whereas, in the membranes involved in energy transduction (such as internal membranes of mitochondria and chloroplasts), approximately 75 per cent is protein. Plasma membrane contains about 50 per cent protein.

According to their position in the plasma membrane, the proteins fall into two main types: **integral** or **intrinsic proteins** and **peripheral** or **extrinsic proteins**, both of which may be either **ectoproteins**, lying or exposing to external or extracytoplasmic surface of the plasma membrane or **endoproteins**, lying or sticking out at the inner or cytoplasmic surface of the plasma membrane. The intrinsic proteins tend to associate firmly with the membrane, while the extrinsic proteins have a weaker association and are bound to lipids of membrane by electrostatic interaction. On the basis of their functions, proteins of plasma membrane can also be classified into three main types: structural proteins, enzymes and transport proteins (permeases or carriers). Some of them may act as **antigens**,

receptor molecules (*e.g.*, insulin-binding sites of liver plasma membrane), **regulatory molecules** and so on. **Structural proteins** are extremely lipophilic and form the main bulk (*i.e.*, backbone) of the plasma membrane. **Enzymes** of plasma membrane are either **ectoenzymes** or **endoenzymes** and are of about 30 types (Table 59.2). **Transport proteins** transport specific substances across the plasma membrane and other cellular membranes.

Table 59.2 Some important enzymes present in the plasma membrane.

1. Acetyl phosphatase	10. Aminopeptidase
2. Acetyl cholinesterase (Ectoenzyme of erythrocyte)	11. Cholesterol esterase
	12. Guanylase cyclase
3. Acid phosphatase	13. Monoglyceride lipase
4. Adenosine triphosphatase	14. NAD-ase (Ectoenzyme of erythrocyte)
5. Mg^{2+} ATPase (Endoenzyme of erythrocyte)	15. Protein kinase (Endoenzyme of erythrocyte)
6. Na^+-K^+ ATPase (Ectoenzyme of erythrocyte)	16. Phospholipase A
	17. Lactase
7. Adenylate cyclase	18. Maltase
8. RNAase	19. Sialidase
9. Alkaline phosphatase	20. UDP glycosidase

3. Carbohydrates. Carbohydrates are present only in the plasma membrane. They are present as short, unbranched or branched chains of sugars (**oligosaccharides**) attached either to exterior ectoproteins (forming **glycoproteins**) or to the polar ends of phospholipids at the external surface of the plasma membrane (forming **glycolipids**). No carbohydrate is located at the cytoplasmic or inner surface of the plasma membrane. All types of oligosaccharides of the plasma membrane are formed by various combinations of six principal sugars (all of which are glucose-derivatives): **D-galactose, D-mannose, L-fucose, N-acetylneuraminic acid** (also called **sialic acid**), **N-acetyl-D-glucosamine** and **N-acetyl-D-galactosamine.**

Structure of Plasma Membrane

1. Evolution of Fluid Mosaic Model of Membrane

The existence of the plasma membrane of the cell was difficult to prove by direct examination before 1930's (when electron microscopy was invented) because of technological limitations. The membrane is beyond the resolution of the light microscope, rendering a morphological approach of its study quite unfeasible with this instrument. Thus, most of the experimental approaches have been provided by only indirect evidences of the existence of such a membrane around the cells. Let us narrate in brief the saga of evolution of presently well accepted fluid-mosaic model of structure of the plasma membrane:

1. The plasmolysis of plant cells in hypertonic solutions suggests the existence of the plasma membrane in the plants.
2. The very fact that a cell, especially an animal cell which has no cell wall, can exist as a physically defined entity suggests that it must have some sort of boundary around it.
3. The presence of plasma membrane can be inferred because protoplasm leaks out of animal cells when cell surface is punctured.
4. After performing some 10,000 experiments with more than 500 different chemicals, in 1899, **Overton** concluded that the peculiar osmotic properties of living protoplasts are due to a **selective solubility mechanism**. Hydrophobic compounds entered cells more rapidly than hydrophilic ones. **Overton** believed this was because of an outer lipoid layer in which

hydrophobic compounds were more soluble. He correctly speculated that this layer might contain cholesterol, lecithin and fatty oils.

5. **Hober** (1910) and **Fricke** (1925) found that the intact cell had low electrical conductivity, indicating the presence of a lipid layer around it.
6. If a lipid containing **hydrophilic groups** (such as the carboxyl groups of fatty acids or the phosphate groups of phospholipids) is dissolved in a highly volatile solvent (*e.g.*, benzene) and several drops of it are then carefully applied to the surface of the water, the lipid spreads out to form a thin, one-molecule-thick or **monomolecular film.** In this film, it is found that the hydrophilic parts of each molecule project into the water surface and the hydrophobic parts are directed up, away from the water.
7. In 1917, **Langmuir** (Nobel Laureate of 1932 in chemistry) fabricated a **trough** or **film balance** (Fig. 59.1) for measuring the specific minimum surface area occupied by a monomolecular film of **lipid** and the force necessary to compress all the lipid molecules into this area. Langmuir trough consists of a shallow trough filled with water on which lipid substance can be spread to make a monomolecular film. A barrier can be pushed across the trough to compress the film.

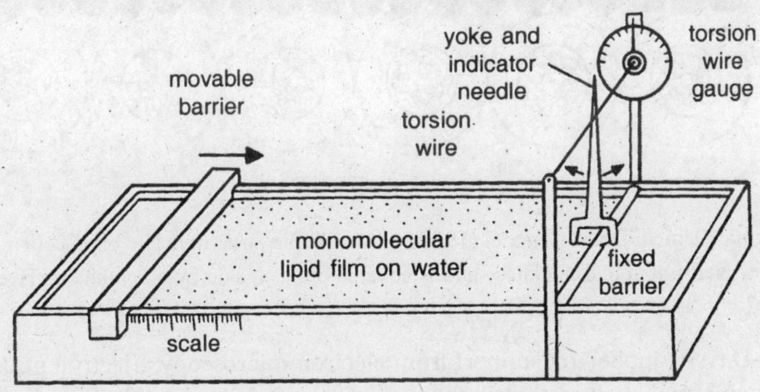

Fig. 59.1. Langmuir trough.

8. In 1925, **Gorter** and **Grendel** extracted the lipids from erythrocyte ghosts of a variety of mammals (such as dogs, sheep, rabbits, guinea pigs, goats and humans) and spread them out on monolayers in the Langmuir trough. These investigators discovered that the area covered by the lipid monomolecular layer film was twice than what was needed to cover the surface of the cells from which the lipid was extracted. Consequently, they safely concluded that *erythrocytes were covered by a layer of lipids two molecules thick* (**lipid bilayer or bimolecular lipid layer**) oriented with polar groups toward the inside and outside of the cell.
9. By studying the surface tension of cells **Harvey** (**Harvey** and **Cole**, 1931, **Danielle** and **Harvey**, 1935) suggested the presence of proteins in the plasma membrane, in addition to the lipids.
10. In 1935, **Danielli** and **Davson**, proposed a model, called **sandwich model**, for membrane structure in which a lipid bilayer was coated on its either side with hydrated proteins (globular proteins). Mutual attraction between the hydrocarbon chains of the lipids and electrostatic forces between the protein and the "head" of the lipid molecules, were thought to maintain the stability of the membrane. From the speed at which various molecules penetrate the membrane, they predicted the lipid bilayer to be about 6.0 nm in thickness, and each of the protein layer of about 1.0 nm thickness, giving a total thickness of about 8.0 nm.

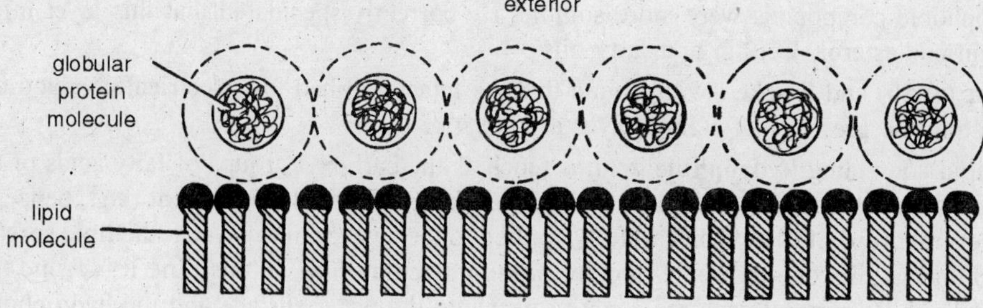

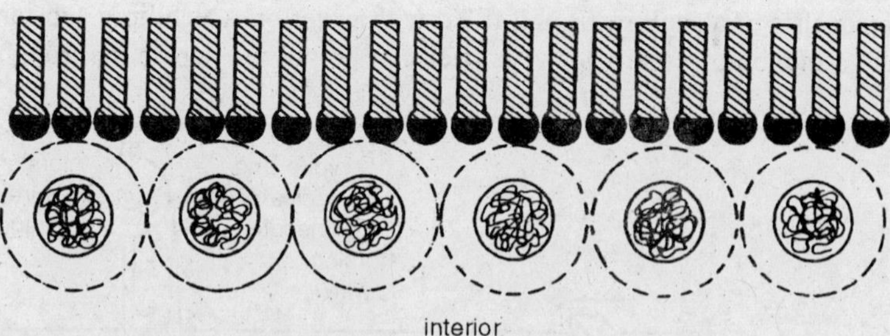

Fig. 59.2. The original Danielli-Davson model (1935) of membrane structure. The bimolecular layer of lipid molecules is of undefined thickness and is covered on each side by a continuous layer of globular proteins.

The Danielli-Davson model got support from electron microscopy. Electron micrographs of the plasma membrane showed that it consists of two dark layers (electron dense granular protein layers), both separated by a lighter area in between (the central clear area of lipid bilayer). The total thickness of the membranes is also turned out to be about 7.5 nm.

11. **Myelin as a model for unit membrane model.** To understand the relationship between the trilaminar appearance of plasma membrane and its underlying molecular structure, **Robertson** chose myelin as the model system for study. **Myelin** surrounds the axons of certain nerve cells and is composed of multiple layers of plasma membrane successively wound around one another. This stack of membrane is derived from the specialised **Schwann cells** that are intimately associated with the axon (Fig. 59.3). During the early stages of myelin formation, axons become surrounded by Schwann cell. The plasma membrane of Schwann cell start wrapping around the axon in a spiral manner. At the later stage of development, the plasma membrane of the Schwann cell become closely stacked upon each other and the cytoplasmic content of the cell becomes disorganised from the space between the membranes. This stack of closely packed membranes around the axon is called **myelin**.

Robertson used myelin, rather than a typical membrane as a model for study because, the repcating layers of membranes present in myelin creates a quasicrystalline structure agreeable to analysis by X-ray diffraction. The X-ray diffraction analyses, carried out by **F.O. Schmitt** and later on **J.B. Fineam** indicated that myelin consists of lipid bilayer alternating with layers of protein. These studies revealed that the general pattern of plasma membrane had one of three possible arrangements of protein and lipid (Fig. 59.4).

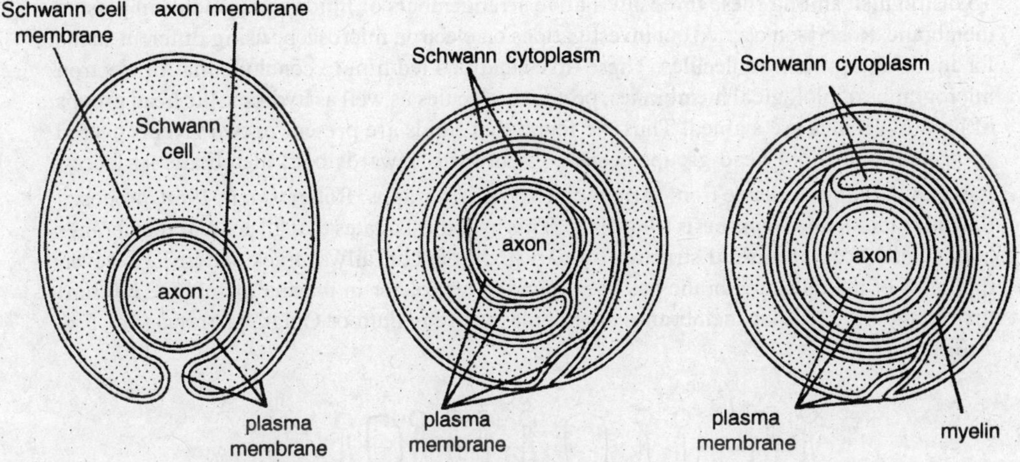

Fig. 59.3. Myelin formation around axon of neuron by Schwann cell.

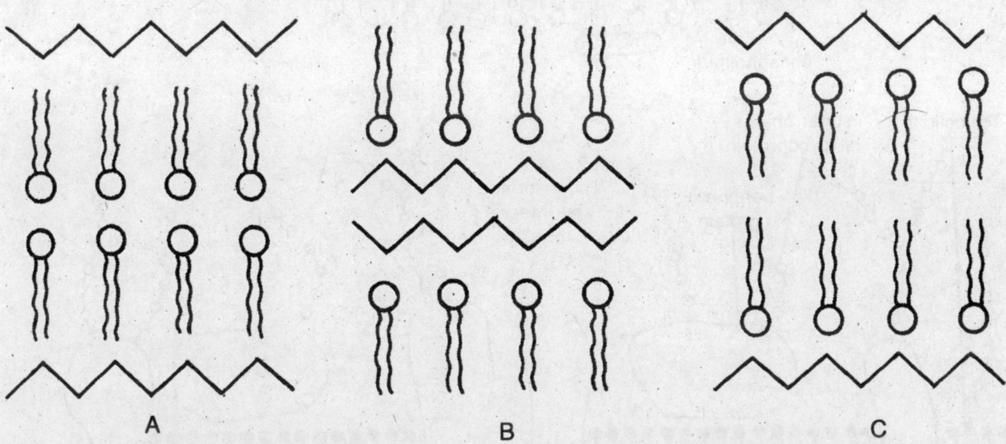

Fig. 59.4. Three possible arrangements of protein and lipid in plasma membrane.

Fig. 59.5. Schematic diagram of the Robertson model of membrane structure. The lipid layer is defined as bimolecular, and the protein is extended but different on the two faces of the membrane.

To distinguish among these three alternative arrangements of lipids and proteins in plasma membrane, Robertson carried out investigations on electron microscope using different stains for lipids and protein molecules. These investigations led him to conclude that in electron micrographs of biological membranes, protein molecules as well as hydrophillic head groups of lipid molecules are stained. Thus, he found that lipids are present in two layers covered by protein with lipid head groups projecting outwards towards both membrane surfaces. Later on by using evidence from various electron micrographs, **Robertson** in 1960, proposed the **unit membrane hypothesis** (Fig. 59.5). This hypothesis states that all cellular membranes have an identical **trilaminar** structure (or dark-light-dark or railway track pattern). However, thickness of the unit membrane has been found to be greater in plasma membrane (10 nm) than in the intracellular membranes of endoplasmic reticulum or Golgi apparatus (*i.e.*, 5 to 7 nm).

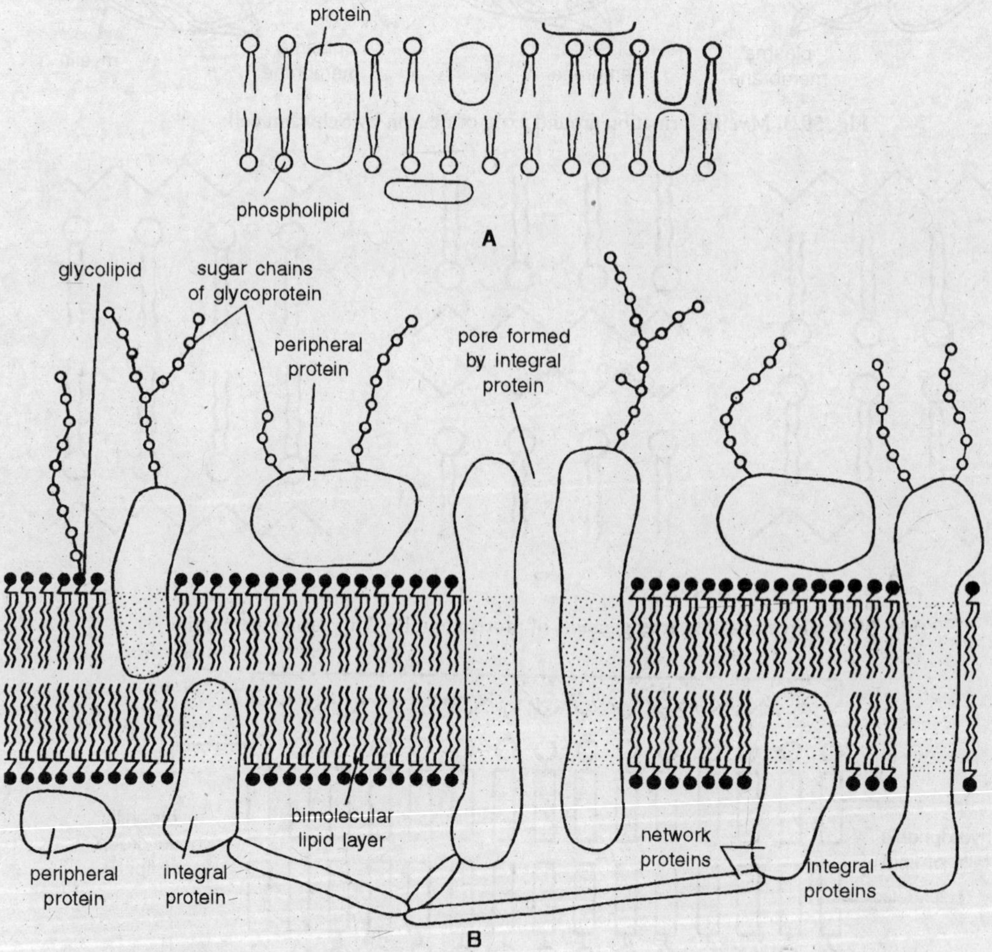

Fig. 59.6. Fluid mosaic model of the plasma membrane. A—Simplistic view; B—Complex view.

12. **S.J. Singer** and **G. L. Nicolson** (1972) suggested the widely accepted **fluid mosaic model** of biological membranes. According to this model (Fig. 59.6), the plasma membrane contains a bimolecular lipid layer, both surfaces of which are interrupted by protein molecules. Proteins occur in the form of globular molecules and they are dotted about here and there in a mosaic

pattern. Some proteins are attached at the polar surface of the lipid (*i.e.*, the extrinsic proteins); while others (*i.e.*, integral proteins) either partially penetrate the bilayer or span the membrane entirely to stick out on both sides (called **transmembrane proteins**). Further, the peripheral proteins and those parts of the integral proteins that stick on the outer surface (*i.e.*, ectoproteins) frequently contain chains of sugar or oligosaccharides (*i.e.*, they are glycoproteins). Likewise, some lipids of outer surface are glycolipids.

The fluid-mosaic membrane is thought to be a far less rigid than was originally supposed. In fact, experiments on its viscosity suggest that it is of a fluid consistency rather like the oil, and that there is a considerable sideways movement of the lipid and protein molecules within it. On account of its fluidity and the mosaic arrangement of protein molecules, this model of membrane structure is known as the "fluid mosaic model" (*i.e.*, it describes both properties and organization of the membrane). The fluid mosaic model is found to be applied to all biological membranes in general, and it is seen as a dynamic, ever-changing structure. The proteins are present not to give it strength, but to serve as enzymes catalysing chemical reactions within the membrane and as pumps moving things across it.

Experimental Evidence in Support of Fluid Mosaic Model of Plasma Membrane

There is a good deal of evidence to support the fluid mosaic model of the plasma membrane:

A. Evidence in support of mosaic arrangement of proteins. Freeze-fracture electron microscopy of the plasma membrane by **Branton** (1968) revealed the presence of bumps and depressions (7 to 8 nm in diameter) which are randomly distributed. These were later shown to be transmembrane integral protein particles (Fig. 59.7). (For details of freeze-fracture technique, see Chapter 54).

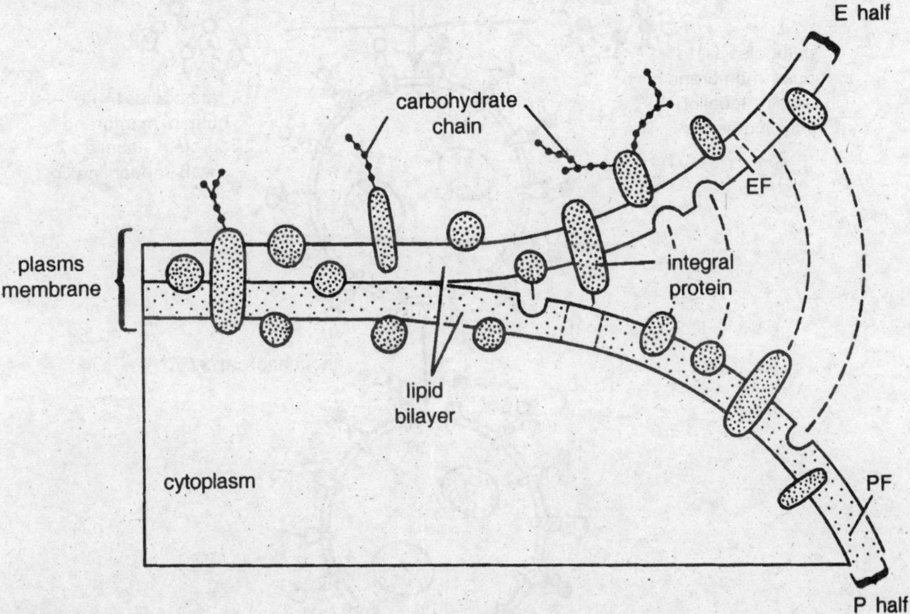

Fig. 59.7. Freeze fracturing of the plasma membrane. The fracture plane occurs at the centre of the lipid bilayer and passes over (or under) the integral membrane proteins. E = exterior, P = protoplasmic (cytosolic) side, F = fracture face.

B. Evidence in support of fluid property of lipid bilayer. Mobility of membrane proteins due to fluid property of lipid bilayer was demonstrated by a classical experiment of **D. Frye and**

M. Edidin (1970). They fused two different types of cultured cells having different surface antigens (proteins). The **cell fusion** is achieved by the use of some fusogen such as an inactivated parainfluenza virus, called Sendai virus (named after a city of Japan). A **fusogen** is a membrane fusion promoting factor such as Sendai virus, lysophosphatides, oleic acid or an electric field. Sendai virus facilitates

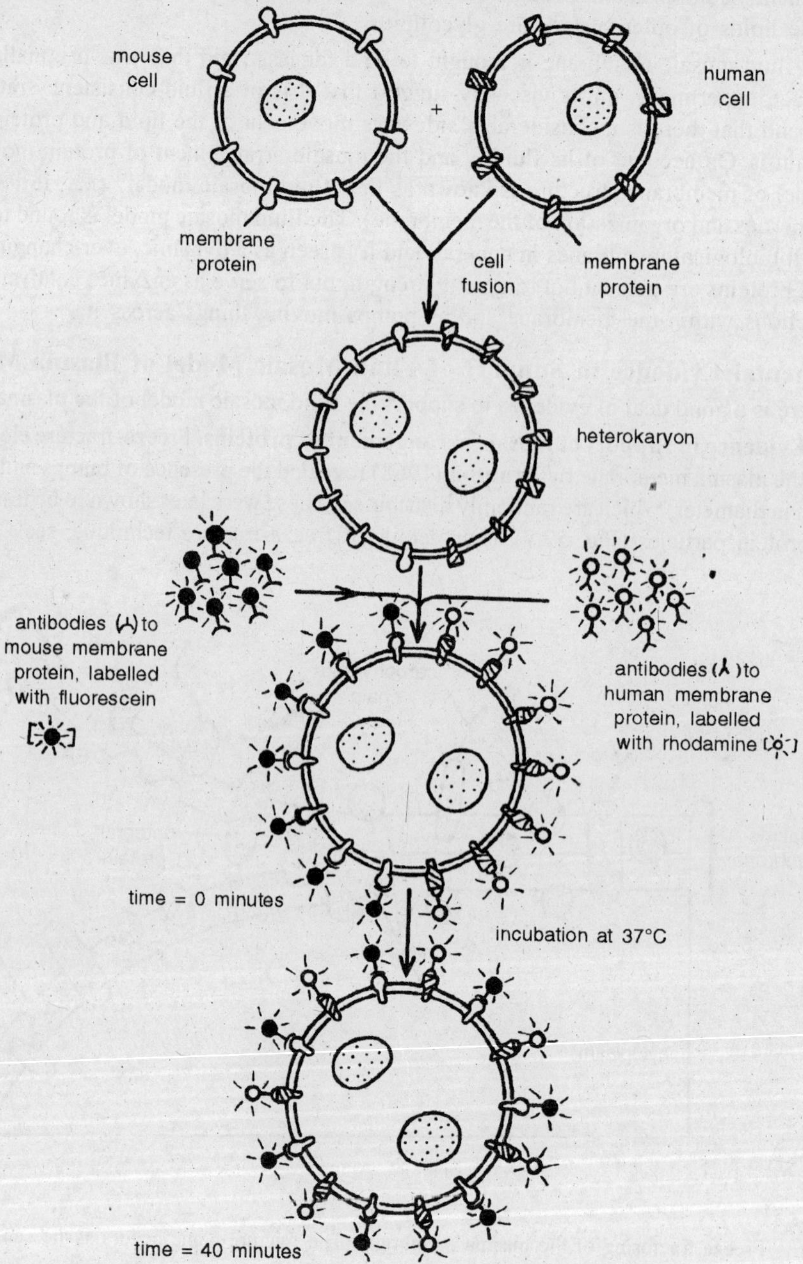

Fig. 59.8. Frye and Edidin's experiment demonstrating the mixing of plasma membrane proteins on mouse-human hybrid cells. In the plasma membrane of newly formed heterokaryon, the mouse and human proteins intially confined to their own halves, but they intermixed with time.

fusion of the plasma membranes and cytoplasms of both cells to produce a **hybrid cell** or **heterokaryon** with two types of nuclei. If the two cells are originally labelled with fluorescent antibodies of different colours, such as fluorescein (green) and rhodamine (red), it is possible at the onset of fusion to recognise the parts of the plasma membrane corresponding to each cell. However intermixing occurs as the antigens are dispersed and the two colours become less and less detectable. After 40 minutes (at 37° C) the intermixing of two colours is complete and the two antigens can no longer be distinguished (Fig. 59.8).

Role of Lipid Molecules in Maintaining Fluid Property of Membrane

(*i*) Types of movements of lipid molecules. Lipid molecules very rarely migrate from one lipid monolayer to other monolayer of lipid bimolecular layer. Such a type of movement is called **flip-flop** or **transbilayer movement** and occurs once a month for any individual lipid molecule. However, in membranes where lipids are actively synthesized, such as smooth ER, there is a rapid flip-flop of specific lipid molecules across the bilayer and there are present certain membrane-bound enzymes, called **phospholipid translocators** (*e.g.*, flippase) to catalyze this activity. On the other hand, lipid molecules readily exchange places with their neighbours within a monolayer (~ 10^7 times a second). This results in their rapid **lateral diffusion**. Individual lipid molecules **rotate** very rapidly about their long axes and their hydrocarbon chains are flexible, the greatest degree of **flexion** occurring near the centre of the bilayer and the smallest adjacent to the polar head groups (Fig. 59.9).

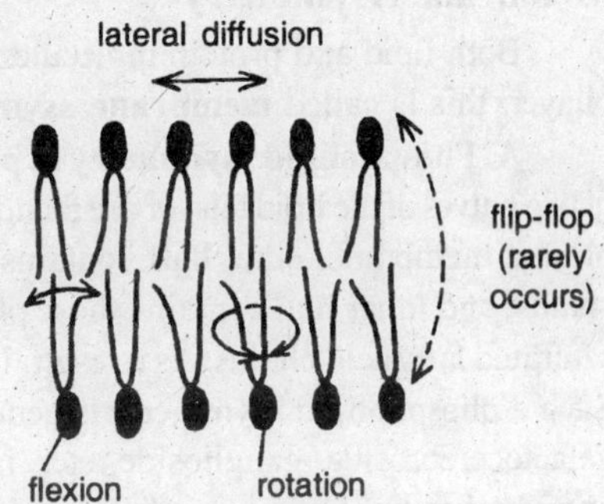

Fig. 59.9. The types of movement possible for phospholipid molecules in a lipid bilayer.

(*ii*) Role of unsaturated fats in increasing membrane fluidity. A synthetic bilayer made from a single type of phospholipid changes from a liquid state to a rigid crystalline or gel (viscous) state at a characteristic freezing point. This change of state is called a **phase transition** and the temperature at which it occurs becomes lower if the hydrocarbon chains are short or have double bonds. Double bonds in unsaturated hydrocarbon chains tend to increase the fluidity of a phospholipid bilayer by making it more difficult to pack the chains together. Thus, to maintain fluidity of the membrane, cells of organisms living at low temperatures have high proportions of unsaturated fatty acids in their membranes, than do cells at higher temperatures.

In fact, certain membrane **transport processes** and **enzyme activities** are found to cease when the lipid bilayer's viscosity increases beyond a threshold level. In contrast, if lipid bilayer's fluidity is increased, the membrane's receptors for the hormone are withdrawn from the cell surface, thereby hampering hormone action.

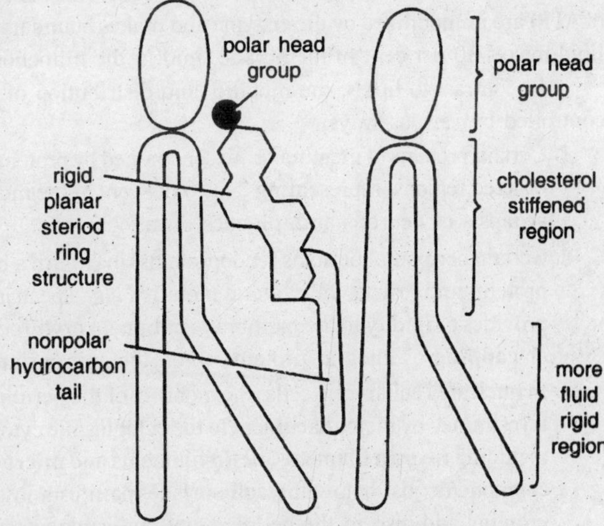

Fig. 59.10. Cholesterol molecule (schematic) interacting with two phospholipid molecules in a monolayer.

***(iii)* Role of cholesterol in maintaining fluidity of membrane.** Eukaryotic plasma membranes (in animals) are found to contain a large amount of cholesterol; up to one molecule for every phospholipid molecule. Cholesterol molecules orient themselves in the lipid bilayer in such a way that their hydroxyl groups remain close to polar head groups of the phospholipids, their rigid plate-like steroid rings interact with and partly immobilise those regions of hydrocarbon chains that are closest to the polar head groups, leaving the rest of the chain flexible (Fig. 59.10). Cholesterol inhibits phase transition by preventing hydrocarbon chains from coming together and crystallizing. Cholesterol also tends to decrease the permeability of lipid bilayers to small water-soluble molecules and is thought to enhance both the flexibility and the mechanical stability of the bilayer.

Membrane Asymmetry

Both lipid and protein molecules have irregular distribution in both monolayers of the lipid bilayer, this is called **membrane asymmetry**.

A. Phospholipid asymmetry in plasma membrane. The lipid composition and state of fluidity of two halves of the lipid bilayer are found to be strikingly different. For example, in human erythrocyte's plasma membrane, outer half contains those phospholipids which have more saturated fatty acid chains, and inner half contains those phospholipids which contain terminal amino groups and less saturated fatty acid chains. As a result, inner monolayer is more fluid than the outer lipid monolayer. Such a phospholipid asymmetry is generated in smooth ER. The asymmetry of glycolipids such as galactocerebroside, ganglioside, etc., in myelin sheath of nerves (*i.e.,* they are found only in the outer half of lipid bilayer) is found to be originated in lumen of Golgi apparatus. The specific role of lipid asymmetry of the membrane is still not clear.

B. Protein asymmetry in plasma membrane. The outer and inner sides of the plasma membrane and other membranes do not contain either the same types or equal amounts of the various peripheral and integral proteins, *e.g.,* erythrocyte's plasma membrane.

Constraints on the Mobility of Membrane Molecules

In the fluid mosaic plasma membrane, there is not complete and independent freedom of movement for its different component molecules. The mobility of some part of lipid molecules is constrained since that remains tightly bound to some of the integral membrane proteins. For example, the mobility of lipid molecules surrounding **cytochrome oxidase** (an enzyme involved in the synthesis of ATP) are immobilized by the enzyme and makes boundary lipid layer. The immobilized boundary lipid makes 30 per cent of membrane lipid in the mitochondrial membrane.

In contrast to lipids, the mobility and distribution of protein molecules in the membrane is controlled by various ways:

1. Certain proteins of membrane are constrained by protein-protein interactions to form specialized ordered regions, representing 2 to 20 per cent of the membrane of a system, *e.g.,* gap junctions, synapsis of neurons and, plaques of halobacteria.
2. Certain peripheral proteins (endoproteins) may form a bridge-like lattice work between integral proteins and restrict their lateral mobility, *e.g.,* spectrin-ankyrin-actin cytoskeletal meshwork provides a rigidity to the membrane of human erythrocytes and does not permit the clustering or **capping** of integral proteins when the appropriate antibodies or lectins are added.
3. In nucleated eukaryotic cells, the mobility of the peripheral endoproteins and integral proteins is restrained by their attachment to the ectoplasmic cytoskeleton. The cytoskeleton is extensive, including **myosin** filaments, **actin** filaments and **microtubules.** Rearrangement of cytoskeletal components just below the cell surface manifests in the distribution of integral membrane proteins and also in the cellular motions, endocytosis and exocytosis.

Functions of Plasma Membrane

Plasma membrane performs the following vital functions for the cell:

1. Regulation of passage of materials. The plasma membrane regulates the passage of materials into and out of the cell. It is **selectively permeable**. It facilitates the entrance of required nutrients into the cells and permits the exit of digestive and nutritive wastes. It checks the exit of useful substances from the cell.

Passage of materials across the plasma membrane occurs by **passive transport** (such as osmosis, simple diffusion), **facilitated diffusion** (*e.g.*, ion transport through charged pores, Ca^{2+}-channels and D-hexose and anion exchange permeases of RBCs), **active transport** (by $Na^+ - K^+$ ATPase, calcium ATPase, proton pump) and **bulk transport** (by exocytosis, phagocytosis, pinocytosis and receptor mediated endocytosis).

2. Maintenance of differential distribution of ions. Plasma membrane regulates and maintains the differential distribution of ions inside and outside the cell. For example, potassium ions are concentrated inside a living cell, whereas sodium and chloride ions are distributed outside. Such an unequal distribution of ions leads to a potential difference. This electric potential energy in all living cells is known as **membrane potential**.

3. Response to environment. The plasma membrane responds to changes in its environment with the help of **receptor proteins**. These receive chemical messages from other cells and thus help in cell recognition. The receptor proteins respond to **hormones, growth factors** and **neurotransmitters.**

4. Contact with neighbouring cells. The plasma membrane maintains structural and chemical relationships with the neighbouring cells by the help of certain glycoproteins.

5. Protection. Plasma membrane protects the cell and may be involved in cell movement, cell secretion and transmitting impulses.

Origin of Plasma Membrane

There is hardly any cell structure more important to the immediate health of the cell than the plasma membrane. If it is weakened or injured, the cell loses its ability to maintain gradients, to carry out the selective transport of nutrients, and to contain the pool of enzymes and organelles essential for the homeostasis. In consequence, new membranes may be added to existing membranes without altering the functions as a barrier and selective transporter. Also for maintaining the characteristic membrane asymmetry, the membrane must be assembled with precisely the correct molecular topography.

Thus, all cellular membranes grow from pre-existing membranes which act as **templates** for the addition of new precursors. All cells divide, daughter cells receive a full complement of membrane systems which undergo growth until the next division, to be passed on to subsequent progeny. Meanwhile the molecules within the membrane undergo continuous replacement.

The protein molecules of the plasma membrane are synthesized on both attached and free ribosomes. Proteins synthesized by free ribosomes may be inserted into the plasma membrane following their completion and release from the ribosomes. Proteins of plasma membrane synthesized on attached ribosomes of rough ER are **inserted** first into the membrane of RER and then **transferred** to the Golgi apparatus, **processed** there (*e.g.*, glycosylation) and ultimately are dispatched to the plasma membrane via the secretory vesicles. Likewise, the synthesis of phospholipid molecules of the plasma membrane takes place by the smooth ER (SER). Like the proteins, newly synthesized lipids are inserted into SER membranes, then they are passed to Golgi apparatus for the processing and ultimately are dispatched to the plasma membrane via small secretory vesicles. The cytosol also contains a number of **phospholipid transport proteins** that function to transfer phospholipid molecules from one cellular membrane to another (*e.g.*, from ER membranes to plasma membrane).

In fact, the process of glycosylation (or glycosidation, *i.e.*, addition of oligosaccharides containing the sugars such as galactose, fucose and/or sialic acid, to the molecules of proteins and phospholipids of the plasma membrane) is completed at the level of Golgi apparatus. However, some sugars are added to the proteins in the lumen of RER.

59.2 MEMBRANE TRANSPORT

The transport of substances into the cell is performed by the following methods:

A. Transport of Water

1. Osmosis

B. Transport of Ions and Small Molecules

2. Passive transport
    (*i*) Simple diffusion
    (*ii*) Facilitated diffusion
3. Active transport

C. Transport of Large Molecules (Bulk Transport)

4. Phagocytosis
5. Endocytosis
    (*i*) Pinocytosis
    (*ii*) Receptor mediated endocytosis

The plasma membrane acts as a semipermeable barrier between the cell and the extracellular environment. This permeability must be highly **selective** if it is to ensure that essential molecules such as glucose, amino acids and lipids can readily enter the cell, that these molecules and metabolic intermediates remain in the cell, and that waste compounds leave the cell. In short, the **selective permeability** of the plasma membrane allows the cell to maintain a constant internal environment (**homeostasis**). Transport across the membrane may be passive or active. It may occur via the phospholipid bilayer or by the help of specific integral membrane proteins, called **permeases** or **transport proteins**.

A. Passive transport. It is a type of **diffusion** in which an ion or molecule crossing a membrane moves down its electrochemical or concentration gradient. *No metabolic energy is consumed in passive transport*. Passive transport is of following three types:

1. Osmosis

The plasma membrane is permeable to water molecules. The to and fro movement of water molecules through the plasma membrane occurs due to the differences in the concentration of the solute on its either sides. The process by which the water molecules pass through a membrane from a region of higher water concentration to the region of lower water concentration is known as **osmosis** (*Gr.*, *osmos* = pushing). The process in which the water molecules enter into the cell is known as **endosmosis**, while the reverse process which involves the exit of the water molecules from the cell is known as **exosmosis**. In plant cells due to excessive exosmosis the cytoplasm along with the plasma membrane shrinks away from the cell wall. This process is known as **plasmolysis** (*Gr.*, *plasma* = molded, *lysis* = loosing) (Fig. 59.11).

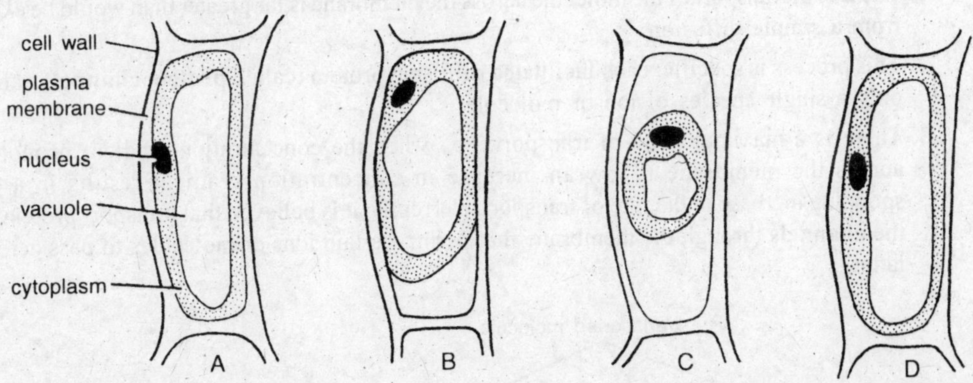

Fig. 59.11 Plasmolysis and deplasmolysis in plant cells. Plasmolysis occurs when a normal plant cell (A) is placed in a hypertonic solution. Water leaves the cell and the plasma membrane shrinks away from the cell wall (B, C). If solutes can penetrate the plasma membrane, the cell will eventually regain water—a process termed deplasmolysis (D).

Box 59.1 Plasma Membrane and Isotonic, Hypotonic and Hypertonic Solutions

A cell contains a variety of solutes in it, for instance, the mammalian erythrocytes contain the ions of potassium (K^+), calcium (Ca^+), phosphate (PO^-_4), dissolved haemoglobin and many other substances. If the erythrocyte is placed in a 0.9% solution of sodium chloride (NaCl), then it neither shrinks nor swells. In such case, because the intra-cellular and extra-celluar fluids contain same concentration, so no osmosis takes place. This type of extra-cellular solution or fluid is known as **isotonic solution** or **fluid**. If the concentration of NaCl solution is increased above 0.9% then the erythrocytes are shrinked due to excessive exosmosis. The solutions which have higher concentrations of solutes than the intracellular fluids are known as **hypertonic solutions**. Further, if the concentration of NaCl solution decreases below 0.9% the erythrocytes will swell up due to endosmosis. The extra-cellular solutions having less concentration of the solutes than the cytoplasm are known as **hypotonic solutions**.

Due to endosmosis or exosmosis the water molecules come in or go out of the cell. The amount of the water inside the cell causes a pressure known as **hydrostatic pressure**. The hydrostatic pressure which is caused by the osmosis is known as **osmotic pressure**. The plasma membrane maintains a balance between the osmotic pressure of the intra-cellular and inter-cellular fluids.

2. Simple Diffusion

In simple diffusion (Fig. 59.12), transport across the membrane takes place unaided, *i.e.*, molecules of gases such as oxygen and carbon dioxide and small molecules (*e.g.*, ethanol) enter the cell by crossing the plasma membrane without the help of any permease. During simple diffusion, a small molecule in aqueous solution dissolves into the phospholipid bilayer, crosses it and then dissolves into the aqueous solution on the opposite side. There is little specificity to the process. The relative rate of diffusion of the molecule across the phospholipid bilayer will be proportional to the concentration gradient across the membrane.

3. Facilitated Diffusion

This is a special type of passive transport, in which ions or molecules cross the membrane rapidly because specific permeases in the membrane facilitate their crossing. Like the simple diffusion, facilitated diffusion does not require the metabolic energy and it occurs only in the direction of a concentration gradient. Facilitated diffusion is characterized by the following special features:

1. The rate of transport of the molecule across the membrane is far greater than would be expected from a simple diffusion.
2. This process is specific; each facilitated diffusion protein (called **protein channel**) transports only a single species of ion or molecule.
3. There is a maximum rate of transport, *i.e.,* when the concentration gradient of molecules across the membrane is low an increase in concentration gradient results in a corresponding increase in the rate of transport. Currently, it is believed that transport proteins form the **channels** through the membrane that permit certain ions or molecules to pass across the latter.

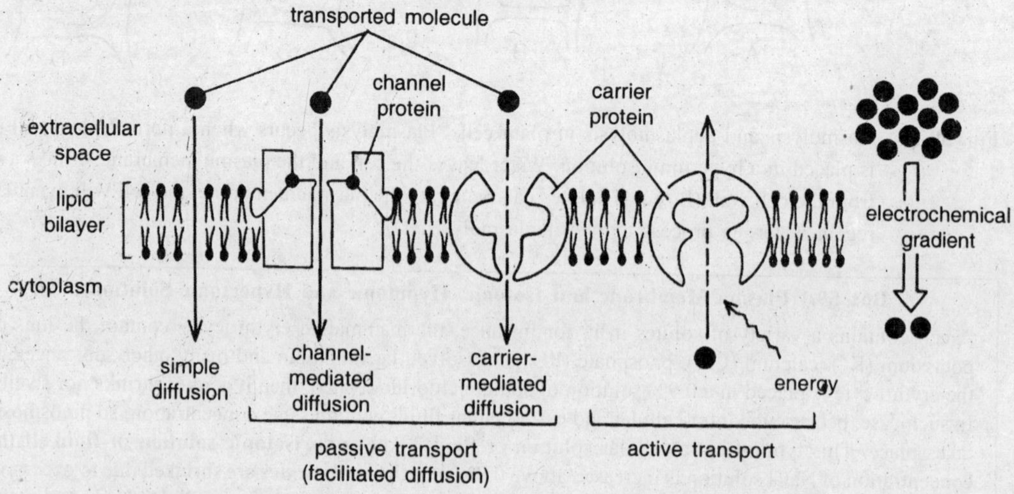

Fig. 59.12. Schematic diagram showing various types of transports across the membrane: simple diffusion, passive transport (down an electrochemical gradient) and active transport (against an electrochemical gradient).

Examples of Facilitated Diffusion

(*i*) Ionic transport through charged pores. Nerve conduction is propagated along the axonal membrane by **action potential** which regulates opening and closing of two main types of **ion channels** (*i.e.,* channel proteins with water filled pores): **Na^+ channels** (or **voltage-gated Na^+ channels**) and **K^+ channels** (or **K^+ leak channels**). At the point of stimulation there is a sudden and several hundred fold increase in permeability to Na^+, which reaches its peak in 0.1 millisecond (*i.e.,* the membrane potential may depolarise from –90 mV and overshoot to +50 mV). At the end of the period, the membrane again becomes essentially impermeable to Na^+, but the K^+ permeability increases and this ion leaks out of the cell, repolarising the nerve fibre. In other words, during the rising phase of the spike, Na^+ enter through the Na^+ channels, and in descending phase K^+ is extruded through the K^+ channels.

Such ion channels also occur in other types of cells such as muscle, sperm and unfertilized ovum. They are not coupled to an energy source (ATP), so the transport they mediate is always passive ("down hill"), allowing specific ions mainly Na^+, K^+, Ca^{2+} and Cl^- to diffuse down their elecurochemical gradient across the lipid bilayer.

Calcium ion channels (Ca^{2+} channels) occur in axonal membranes and other membranes for the entrance of Ca^{2+} ions in the cell. Ca^{2+} ions have a fundamental role in many cellular activities such as exocytosis, endocytosis, secretion, cell motility, cell growth, fertilization and cell division.

In the neuronal membrane, there are a number of Ca^{2+} channels that are driven by the membrane potential and are essential in the release of neurotransmitters (acetylcholine).

***(ii)* D-hexose permease of erythrocyte.** The plasma membrane of mammalian erythrocytes (Fig. 59.13) and other body cells, contain specific channel proteins for the facilitated diffusion of glucose into the cells. They are called **glucose transporter, glucose permease** or **D-hexose permease**. After the glucose is transported into the erythrocyte, it is rapidly phosphorylated (by **hexokinase** enzyme and ATP) to form **glucose-6-phosphate**. Once phosphorylated, the glucose no longer leaves the cell; moreover, the concentration of the simple glucose in the cell is lowered. As a result, the concentration gradient of glucose across the membrane is increased, allowing the facilitated diffusion to continue to import glucose. Since no cellular membrane (except the mitochondrial membranes) contains any permease for facilitated diffusion of phosphorylated compounds, so a cell can retain any type of molecule by phosphorylating them, *e.g.*, ATP and phosphorylated nucleosides are never released from the cells containing a normal intact plasma membrane. However, permeases for ATP and ADP do exist in a mitochondrial membrane to allow these molecules to move across it.

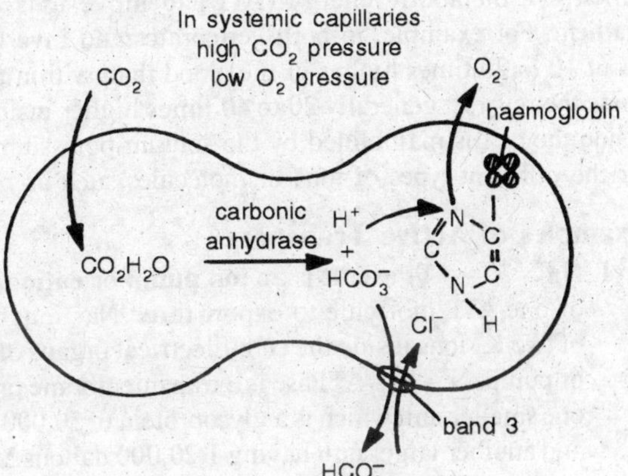

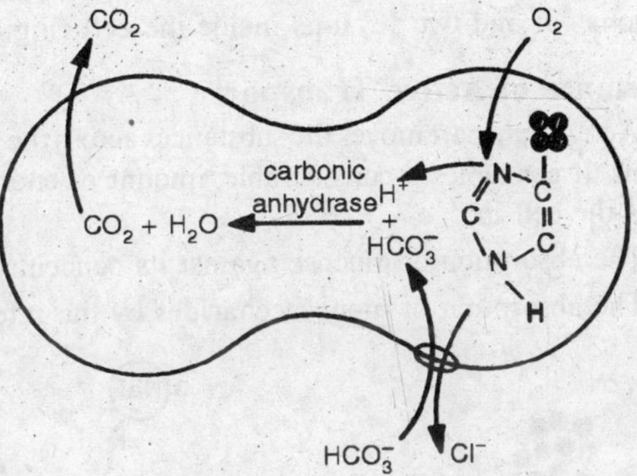

Fig. 59.13. Schematic drawing showing anion transport through the erythrocyte membrane in the capillaries and in the lungs. Band 3 protein (= anion exchange permease) catalyzes the exchange of the anions: Cl^- and HCO_3^- across the erythrocyte membrane.

***(iii)* Anion exchange permease of erythrocyte.** Band 3 polypeptide of plasma membrane of the erythrocytes and other cells is an **ion exchange permease protein** which catalyzes an one-for-one exchange of anions such as chloride (Cl^-) and bicarbonate (HCO_3^-) across the membrane (called **chloride shift;** erythrocyte has 100,000 times more permeability of Cl^- than other cells). The rapid flux of anions in the erythrocyte facilitates the transport in the blood of CO_2 from the tissues to the lungs. Waste CO_2 that is released from cell into the capillary blood, diffuses across the membrane of erythrocyte. In its gaseous form, CO_2 dissolves poorly in aqueous solutions such as blood plasma, but inside the erythrocyte the potent enzyme **carbonic anhydrase** converts it into a bicarbonate anion:

$$CO_2 + H_2O \underset{\text{anhydrase}}{\overset{\text{carbonic}}{\rightleftharpoons}} \underset{\text{proton}}{H^+} + \underset{\text{bicarbonate}}{HCO_3^-}$$

This process occurs while the haemoglobin in the erythrocyte is releasing its oxygen into the blood plasma. The removal of oxygen from haemoglobin induces a change in its conformation that enables a globin histidine (amino acid) side chain to bind to the proton produced by carbonic anhydrase enzyme. The bicarbonate anion formed by carbonic anhydrase is transported out of the erythrocyte in exchange for a chloride (Cl^-) anions:

$$\underset{(in)}{HCO_3} + \underset{(out)}{Cl} \rightleftharpoons \underset{(out)}{HCO_3} + \underset{(in)}{Cl}$$

B. Active transport. Active transport uses specific transport proteins, called **pumps,** which use metabolic energy (ATP) to move ions or molecules against their concentration gradient. For example, in both vertebrates and invertebrates, the concentration of sodium ion is about 10 to 20 times higher in the blood than within the cell. The concentration of the potassium ion is the reverse, generally 20 to 40 times higher inside the cell. Such a low sodium concentration inside the cell is maintained by the sodium-potassium pump. There are different types of pumps for the different types of ions or molecules such as calcium pump, proton pump, etc.

Examples of Active Transport

1. **Na^+– K^+– ATPase.** It is an **ion pump** or **cation exchange pump** which is driven by energy of one ATP molecule to export three Na^+ ions outside the cell in exchange of the import of two K^+ ions inside the cell. Electrical organs of eels are found to be very rich in this enzyme or pump. N^+– K^+– ATPase is a transmembrane protein which is a dimer having two subunits: one smaller unit which is a glycoprotein of 50,000 daltons M.W., having an unknown function; and another larger unit having 1,20,000 daltons M.W. The larger subunit of Na^+– K^+ – ATPase performs the actual function of cation transport. It has three sites on its extra cytoplasmic surface: two sites for K^+ ions and one site for the inhibitor **ouabain**. On its cytosolic side, the larger subunit contains three sites for three Na^+ ions and also has one catalytic site for a ATP molecule. It is believed that the hydrolysis of one ATP molecule somehow drives conformational changes in the Na^+– K^+– ATPase that allows the pump to transport three Na^+ ions out and two K^+ ions inside the cell (Fig. 59.14).

Importance of Active Transport

Active transport moves the substances across the plasma membrane against their concentration gradient. It consumes a considerable amount of energy and is very essential for the cell since it enables the cell in:

1. The absorption of glucose against its concentration gradient in the uriniferous tubules.
2. The absorption of monosaccharides by the intestinal cells.

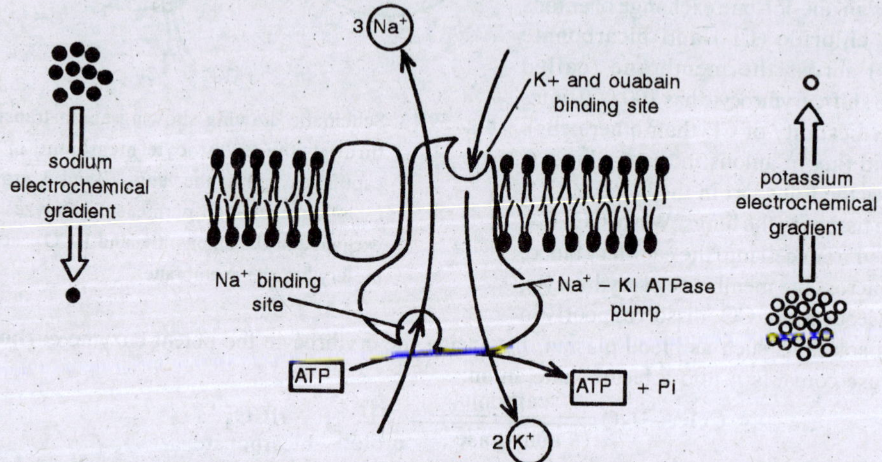

Fig. 59.14. The Na^+ – K^+ ATPase in the plasma membrane actively pumps Na^+ out and K^+ into a cell against their electrochemical gradients. For every molecule of ATP hydrolyzed inside the cell, 3 Na^+ ions are pumped out and $2K^+$ ions are pumped in.

3. Maintenance of $Na^+ - K^+$ pump for establishing the unequal distribution of ions between inside and outside the plasma membrane. It helps in the conduction of nerve impulse in nerve cells and in maintaining a specific level of Na^+ in the blood.

Functions of Proteins of Membrane

Biological membranes (*e.g.*, plasma membrane and unit membrane) contain the following five different classes of proteins that carry out different functions:

1. Structural proteins. These form backbone of plasma membrane. These proteins provide elasticity and mechanical stability to plasma membrane. These have little catalytic activity

2. Enzymes. These are catalytic proteins. They carry out different biological reactions in different parts of the cell and also in various types of cells.

3. Membrane transport proteins. These serve for the transport of polar molecules of various substances such as ions, monosaccharides, amino acids, nucleotides and certain metabolites across the plasma membrane in and out of the cell. These occur in many forms. Each transport protein is designed to transport a particular class of molecules. These are basically **multipass transmembrane proteins**, projecting on both the sides of lipid bilayer and traversing it many times. Membrane transport proteins are of following types:

(*i*) **Carrier proteins** bind the specific solute to be transported by active transport, *i.e.*, these undergo conformational change in order to transfer the solute across the membrane. These carrier proteins are of following three types:

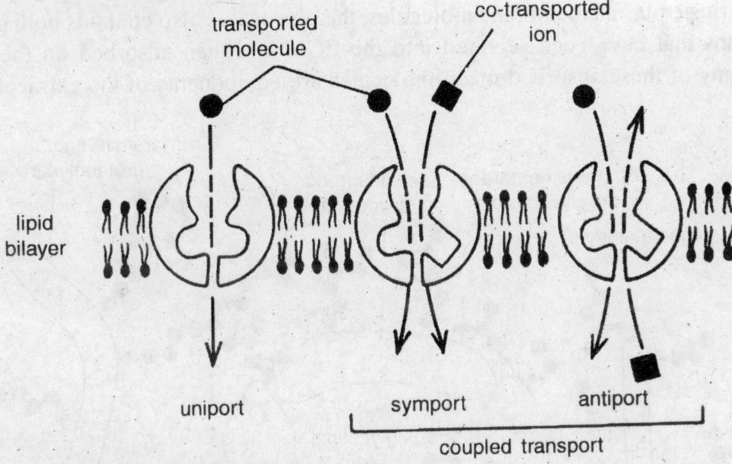

Fig. 59.15. Carrier proteins of membrane functioning as uniports, symports and antiports.

Uniport, symport and antiport. Those carrier proteins which simply transport a single solute from one side of the membrane to the other are called **uniports**. Others function as **coupled transporters**, in which the transfer of one solute depends on the simultaneous transfer of a second solute, either in the same direction (**symport**) or in the opposite direction (**antiport**). Both symport and antiport collectively form the **cotransport**. Most animal cells, for example, must take up glucose from the extracellular fluid, where the concentration of the sugar is relatively high, by passive transport through the glucose carriers (such as **D-hexose permease**) that operate as the uniports. By contrast, intestinal and kidney cells must take up glucose from the lumen of the intestine and kidney tubules, respectively, where the concentration of the sugar is low. These cells actively transport glucose by symport with Na^+ ions whose extracellular concentration is very high. The anion exchange permease of human erythrocytes operates as an antiport to the exchange of Cl^- for HCO_3^-.

(*ii*) **Channel proteins.** These are water filled pores that extend across the lipid bilayer. When these pores are open, they allow specific solutes to pass through them across the membrane.

4. Glycoproteins act as **cell receptors** (*e.g.*, **integrins**) and **cell antigens**.
5. Lipoproteins are **drug receptor proteins**.

Cell Coat or Glycocalyx

Plasma membrane proteins, as a rule, do not protrude naked from the exterior of the cell. They are usually decorated by carbohydrates, which coat the surface of all eukaryotic cells. These carbohydrates occur as **oligosaccharide chains** covalently bound to membrane proteins (glycoproteins) and lipids (glycolipids). They also occur as the polysaccharide chains of integral membrane proteoglycan molecules. Proteoglycans, which consist of long polysaccharide chains linked covalently to a protein core, are found mainly outside the cell as part of the extra-cellular matrix or ECM (discussed in Chapter 60). But for some proteoglycans, the protein core either extends across the lipid bilayer or is attached to the bilayer by a GPI (= glycosyl phosphotidyl inositol) anchor.

The term **cell coat** or **glycocalyx** is often used to describe the carbohydrate-rich zone of the cell surface. This zone can be visualized by a variety of stains, such as **ruthenium red**, as well as by its affinity for carbohydrate-binding proteins called **lectins**, which can be labelled with a fluorescent dye or some other visible marker. The cell coat is a 10 to 20 nm thick layer and is in direct contact with the outer leaflet of the plasma membrane. In *Amoeba*, the cell coat is formed by fine filaments—5 to 8 nm thick and 100 to 200 nm long. Although most of the carbohydrate is attached to intrinsic plasma membrane molecules, the glycocalyx also contains both glycoproteins and proteoglycans that have been secreted into the ECM and then adsorbed on the cell surface (Fig. 59.16). Many of these absorbed macromolecules are components of the extracellular matrix,

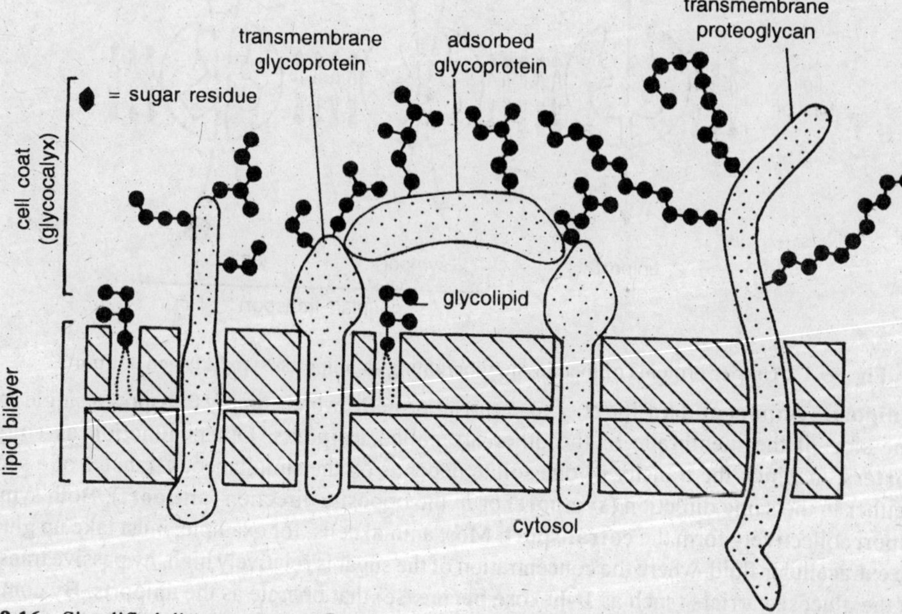

Fig. 59.16. Simplified diagram of the cell coat (glycocalyx). The cell coat is made up of the oligosaccharide side chains of glycolipids and integral membrane glycoproteins and the polysaccharide chains on integral membrane proteoglycans. In addition, absorbed glycoproteins and absorbed proteoglycans (not shown) contribute to the glycocalyx in many cells. Note that all of the carbohydrate is on the noncytosolic surface of the membrane (after Alberts *et al.*, 2002).

so that where the plasma membrane ends and the ECM begins is largely a matter of semanitics. One of the possible functions of cell coat is to protect cells against mechanical anu chemical damage and to keep foreign objects and other cells at a distance, preventing undersirable protein-protein interaction. The cell coat of the intestinal epithelium is quite strong it resists vigorous mechanical and chemical attacks; in other cells the coat is labile and may be depleted by washing or enzyme exposure.

Cell Surface Differentiations

In addition to specific molecular components plasma membrane consists of various cell surface dif-ferentiations such as microvilli, invaginations, etc.

1. Invaginations. The bases (inner ends) of certain cells, such as the cells of the kidney, perform active transpor-tation and contain many **invaginations** or **infold-ings** of the plasma membrane. At the base of these folds, there deve-lops a septa and, thus, narrow compartments of basal cytoplasm are formed. These infoldings contain many mitochondria. These mitochondria along with the enzymes of plasma membrane possibly provide energy rich compound, *viz.*, ATP to the plasma membrane for the active transportation of solutes.

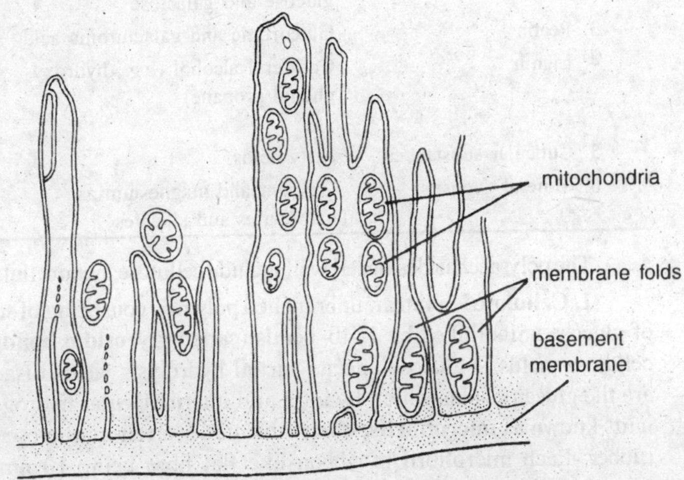

Fig. 59.17. Infoldings of plasma membrane.

2. Microvilli. Microvilli are finger-like, slender projections of plasma membrane which are found in mesothelial cells, hepatic cells, epithelial cells of intestine (**striated border**), uriniferous tubules (**brush border**), gall bladder, uterus, growing oocyte and yolk sac. Microvilli increase the effective surface of absorption. For example, a single epithelial cell of intestine may have as many as 3000 microvilli and in a square millimetre of intestine there may be 200,000,000. These microvilli are 0.6 to 0.8 μm long and 0.1 μm in diameter. The narrow spaces between the microvilli form a kind of sieve through which substances may pass during the process of absorption. Within the cytoplasmic core of a microvillus fine microfilaments are observed which in the underlying cytoplasm form a **terminal web.** The microfilaments contain actin and are attached to the tips of the microvilli by α-actinin; their function is to produce contraction of microvilli.

59.3 CELL WALL

The plant cell is always surrounded by a **cell wall** and this feature distinguishes them from animal cells. The cell wall is a non-living structure which is formed by the living protoplast (A plant cell without its cell wall is called a **protoplast**). In most of the plant cells, the cell wall is made up of cellulose, hemicellulose, pectin and protein. In many fungi, the cell wall is formed of chitin and in bacteria, the cell wall contains protein-lipid-polysaccharide complexes. Thus, the cell wall is a rigid and protective layer around the plasma membrane which provides the mechanical support to the cell. The cell wall also determines the shape of plant cells. Due to the shape of cell walls many types of plant cell as the parenchymatous, collenchymatous, etc., have been recognised.

Chemical Composition of Cell Wall

Chemically speaking, the plant cell wall is composed of a variety of polysaccharides (carbohydrates), lipids, proteins and mineral deposits, all exhibiting distinct staining reactions (Table 59.3).

Table 59.3 Chemical nature and staining reaction of various components of plant cell wall.

Substance	*Chemical unit*	*Staining reaction*
1. Cellulose	Glucose	Chlorzinc iodide (stains violet)
2. Hemicellulose	Arabinose, xylose, mannose, glucose and galactose	No specific stain
3. Pectin	Glucuronic and galacturonic acid	Ruthenium red
4. Lignin	Coniferyl alcohol (*e.g.*, hydroxy-phenyl propane)	Phloroglucinol hydrochloride (stains rose); chlorzinc iodide (stains yellow)
5. Cuticular substances	Fatty acids	Sudan III (stains orange)
6. Mineral deposits	Calcium and magnesium as carbonates and silicates	

The polysaccharides of cell wall include cellulose, hemicelluloses, pectin compounds and lignins.

1. Cellulose is a linear, unbranched polymer, consisting of straight polysaccharide chains made of glucose units linked by 1–4b–bonds (called **glycosidic bonds**). (**Note.** Complete hydrolysis of cellulose yields D-glucose and its partial hydrolysis yields disaccharide units, **cellobiose**). These are the glucan chains which by intra and intermolecular hydrogen bonding produce the structural units known as **microfibrils,** observable under electron microscopy and having toughness like the rubber. Each microfibril is ribbon-like flat fibre being 10 nm wide and 3 nm thick (or 25 to 30 nm in diameter) and is composed of about 2000 glucan chains in it. According to a classical estimate, each cellulose microfibril comprises three **micelles** or **elementary fibrils:** each elementary fibril contains about 100 cellulose molecules and each cellulose molecule is made up of 40 to 70 glucan chains (*i.e.*, one microfibril = 3 × 100 × 70 = 21000 glucan chains). Often numerous microfibrils get associated to form the **macrofibrils** having up to 0.5 μm diameter and observable under the light microscopy. Cellulose is synthesized by a wide variety of cells that include bacteria (*e.g.*, acetobacter, agrobacter and rhizobium), algae, fungi, cryptogams and seed plants.

2. Hemicelluloses are short but branched heteropolymers of various kinds of monosaccharides such as arabinose, xylose, mannose, galactose, glucose and uronic acid. Some of the common hemicelluloses go under the names xylans, arabinoxylans, glucomannans, galactomannans and xyloglucans.

3. Pectins are water soluble, heterogeneous branched polysaccharides that contain many negatively charged D-galacturonic acid residues along with D-glucuronic acid residues. Because of their negative charge, pectins are highly hydrated and intensely bind cations. When Ca^{2+} is added to a solution of pectin molecules, it cross-links them to produce a semirigid gel. Such Ca^{2+} cross-links are thought to help hold cell-wall component together.

4. Mannan is a homopolysaccharide of mannose and is found in the cell wall of yeast, fungi and bacteria.

5. Agar is a polysaccharide, found in the cell wall of sea weeds and containing D and L-galactose residues.

6. Lignin is a biological plastic and non-fibrous material. It occurs only in mature cell walls and is made of an insoluble hydrophobic aromatic polymer of phenolic alcohols (*e.g.*, hydroxyphenyl propane).

7. The **chitin** is a polymer of glucosamine.

Glycoproteins (present up to 10 per cent in primary cell wall) are hydroxyproline-rich proteins (like the collagen). In them, many short oligosaccharide side chains are attached to hydroxyproline and serine side chains. Thus, more than half the weight of glycoprotein is carbohydrate. These glycoproteins are known to act like the glue to increase the strength of the cell wall.

Cutin is also a biological plastic and is made of fatty acids (waxes). **Suberin** is a water-resistant substance, comprising of fatty acids and found in the cork and cell wall of many plants. **Sporopollenin** is a lipoidal polymer forming tough wall (with species-specific patterns) of pollen grains.

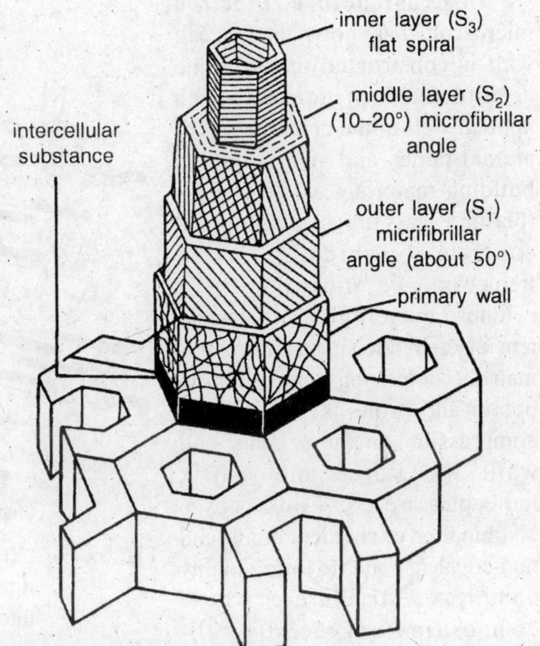

Fig. 59.18. Structure of a cell wall showing middle lamella, primary cell wall and three regions (S_1, S_2 and S_3) of the secondary cell wall.

Mineral deposits occur in cuticle in the form of calcium and magnesium carbonates and silicates. Deposits of calcium compounds are found in the cell wall of cruciferous and cucurbitaceous plants. Silicate deposits are common in the cell wall of Graminae family.

Structure. The cell wall is complex in nature and is differentiated in the following layers (Fig. 59.18):

(*i*) Primary cell wall;

(*ii*) Secondary cell wall;

(*iii*) Tertiary cell wall.

(*i*) Primary cell wall. The first formed cell wall is known as primary cell wall. It is the outermost layer of the cell and in the immature meristematic and parenchymatous cells it forms the only cell wall. The primary cell wall is comparatively thin and permeable. Certain epidermal cells of the leaf and the stem also possess the cutin and cutin waxes which make the primary cell wall impermeable. The primary cell wall of the yeast and the fungi is composed of the chitin.

(*ii*) Secondary cell wall. The primary cell wall is followed by secondary cell wall. The secondary cell wall is thick, permeable and lies near the plasma membrane of the tertiary cell wall, if the latter occurs. It is composed of three concentric layers (S_1, S_2 and S_3) which occur one after another. Chemically the secondary cell wall is composed of compactly arranged macrofibrils of the cellulose, in between which sometimes occurs lignin as a inter-fibrillar material.

(*iii*) Tertiary cell wall. In certain plant cells, there occurs another cell wall beneath the secondary cell wall which is known as tertiary cell wall. The tertiary cell wall differs from the primary and secondary cell wall in its morphology, chemistry and staining properties. Besides the cellulose, the tertiary cell wall consists of another chemical substance known as the xylan.

Middle lamella. The cells of plant tissues generally remain cemented together by an intercellular matrix known as the **middle lamella.** The middle lamella is mainly composed of the pectin, lignin and some proteins.

Ultrastructure. Electron microscopy has shown that the cell wall is constructed on the same architectural principle which applied well in the construction of animal bones and such common building materials as fibre glass (plastic + glass fibres) or reinforced concrete (concrete + metal framework), *i.e.*, strong fibres (*e.g.*, cellulose microfibrils) resistant to tension embedded in an amorphous matrix (comprising hemicellulose, pectin and proteins) resistant to compression. In the primary cell wall, the fibres and matrix molecules are cross-linked by a combination of covalent bonds and non-covalent bonds to form a highly complex structure whose composition is generally cell-specific (Fig. 59.19). In fact, hemicellulose molecules (*e.g.*, **xyloglucans**) are linked by hydrogen bonds to the surface of the cellulose microfibrils. Some of these hemicellulose molecules are cross-linked in turn to acidic pectin molecules (*e.g.*, rhamno-galacturonans) through short neutral pectin molecules (*e.g.*, arabinogalactans). Cell wall glycoproteins are tightly woven into the texture of the wall to complete the structure of matrix.

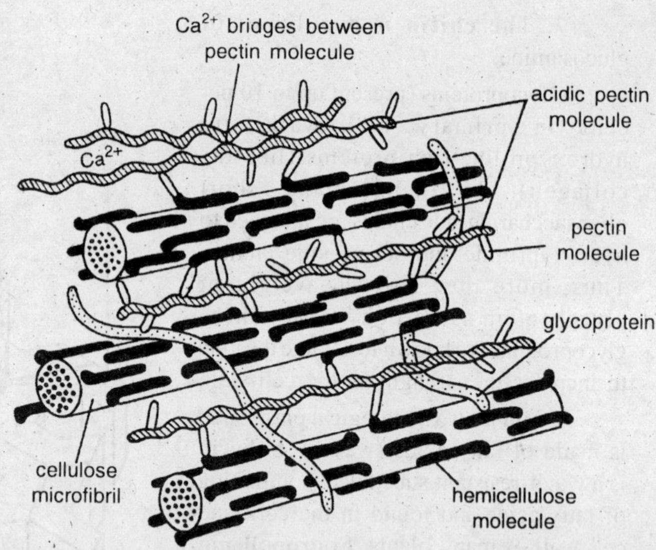

Fig. 59.19. Ultrastructure of primary cell wall showing interconnections between the two major components of the primary cell wall: the cellulose microfibrils and the matrix.

In the multilamellar secondary cell wall, cellulose microfibrils are laid down in layers, the microfibrils of each layer running roughly parallel with each other but at an angle to those in other layers (Fig. 59.18).

Plasmodesmata. Every living cell in a higher plant is connected to its living neighbours by fine cytoplasmic channels, each of which is called a **plasmodesma** (Gr., *desmos* = ribbon, ligament; plural, **plasmodesmata**) which pass through the intervening cell walls. The plasma membrane of one cell is continuous with that of its neighbour at each plasmodesma. A plasmodesma is a roughly cylindrical, membrane-lined channel with a diameter of 20 to 40 nm. Running from cell to cell through the centre of most plasmodesmata is a narrower cylindrical structure, the **desmotubule**, which remains, continuous with elements of the SER membranes of each of the connected cells. Between the outside of the desmotubule and the inner face of the cylindrical plasma membrane is an **annulus of cytosol**, which often appears to be constricted at each end of the plasmodesmata. These constrictions may regulate the flux of molecules through the annulus that joins the two cytosols.

Plasmodesmata are formed around the elements of smooth endoplasmic reticulum that become trapped during cytokinesis (of mitotic cell division) within the new cell wall that will bisect the parental cell. Plasmodesmata function in intercellular communication, *i.e.*, they allow molecules to pass directly from cell to cell. For example, plasmodesmata are especially common and abundant in the walls of columns of cells that lead toward sites of intense secretion, such as in nectar-secreting glands (trichomes of *Abutilon* nectaries). In such cells there may be 15 or more plasmodesmata per square micrometer of wall surface, whereas there is often less than 1 per square micrometer in other cell wall.

In fact, experimental evidence has suggested that the plasmodesmata mediate transport between adjacent plant cells, much as gap junctions of animal cells. They allow the passage of molecules with molecular weights of less then 800 daltons. Transport through the plasmodesmata is also found under complex regulations which may involve Ca^{2+} and protein phosphorylation. Thus, no dye movement is observed between the cells of the root cap and the root apex or between epidermal and cortical cells in either roots and shoots. However, certain plant viruses such as TMV can enlarge plasmodesmata in order to use this route to pass from cell to cell. Tobacco mosaic virus is known to synthesize a protein, called **P30** (30,000 dalton M.W.) that nullifies the normal regulatory mechanisms of plasmodesmata.

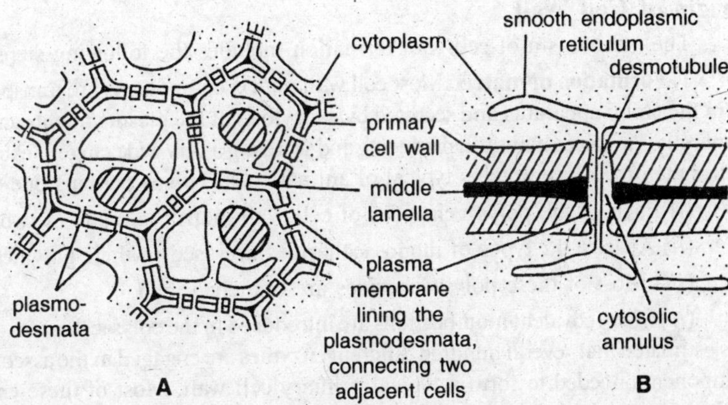

Fig. 59.20. A—Plasmodesmata pierce the cell wall and connect all cells in the plant together to form symplast; B—Details of the structure of a plasmodesma.

Lignification. The structure of cell wall is stabilized by the deposition of lignin in the cell wall matrix. Such a process of lignification was required in connection with the transition from aquatic to the terrestrial plant life during organic evolution of plants. A lignified cell wall is composed of microfibrils of cellulose embedded in the matrix containing large amount of lignin. Usually the primary cell wall becomes more lignified than secondary cell wall.

Functions of Cell Wall

The chief function of cell wall in plant cells is that it provides mechanical strength to the latter. Like the exoskeleton or endoskeleton of animals, cell wall acts like a skeletal framework of plants. Particularly in vascular plants, the cell walls provide the main supporting framework.

Despite its strength, the plant cell wall is fully **permeable** to water and solutes. This is because the matrix of cell wall is perforated by minute water-filled channels through which free diffusion of water and water soluble substances such as gases, salts, sugars, hormones and like can take place. Moreover, the molecules of the matrix are strongly hydrophilic ("water-loving") with the result that in normal circumstances the cell wall is saturated with water like a sponge (*e.g.*, primary cell wall is 60 per cent water by weight): The cross-linked structure of the cell wall is, however, found to slightly block the diffusion of small molecules such as water, sucrose and K^+. The average diameter of the spaces between the cross-linked macromolecules in most cell wall is about 5 nm, this is small enough to make the movement of any globular macromolecules with a M.W. much above 20,000 daltons extremely slow. Therefore, plants must survive on molecules of low molecular weight, and any intercellular signaling molecules that have to pass through the cell wall must also be small and water soluble. In fact, most of the known plant signaling molecules, such as growth regulating substances—auxins, cytokinins and gibberellins—have molecular weights of less than 500 daltons.

During lignification, lignin is deposited in spaces between the cellulose molecules, making the cell wall much more rigid, and making it **impermeable**. Once lignification is complete the protoplasm can no longer absorb materials from outside the cell, which, therefore, dies. Hence, lignified tissue is always dead. Thus, a lignified tissue becomes well adapted for two types of functions:

1. It provides the mechanical strength due to its ligno-cellulose composition.
2. It transports water and salts, since, lignification involves loss of the protoplasm resulting in the formation of a hollow waterproof tube.

Origin of Cell Wall

The mechanism of cell wall formation includes the following steps:

1. Formation of matrix. Most cell wall matrix components are transported via vesicles derived from the Golgi apparatus and secreted by exocytosis at the plasma membrane. Golgi apparatus of the plant cells is involved mainly in producing and secreting a very wide range of extracellular polysaccharides, rather than the glycoproteins typical of animal cells. This is perhaps due to following facts:

(*i*) Each of the polysaccharides of cell wall matrix is made of two or more sugars;

(*ii*) At least 12 types of monosaccharides are used in their polymerization;

(*iii*) Most of these polysaccharides are branched; and

(*iv*) Many covalent modifications are introduced in the polysaccharides after they are synthesized. It is estimated that several hundred different enzymes are engaged in the assembly of the polysaccharide components needed to form a typical primary cell wall. Most of these enzymes are found in the endoplasmic reticulum and Golgi apparatus. Some enzymes, which are concerned with later covalent modifications of the polysaccharides, are found in the cell wall itself.

Since, the cell wall varies in composition and morphology at different locations around the cell, the Golgi-derived vesicles are directed at specific regions of the plasma membrane by the help of cytoskeleton (*i.e.*, microtubules and microfilaments). For example, we can consider the case of formation of new primary cell wall that separates two daughter cells after the karyokinesis of mitosis. At the end of telophase, a barrel-shaped or disc-like region, called **phragmoplast** (Gr., *phragma* = hedge, enclosure; *plasso* = to form) forms in the plane of former spindle equator. The phragmoplast comprises a double-ring of short microtubules on either side of, and terminating at, the division plane,

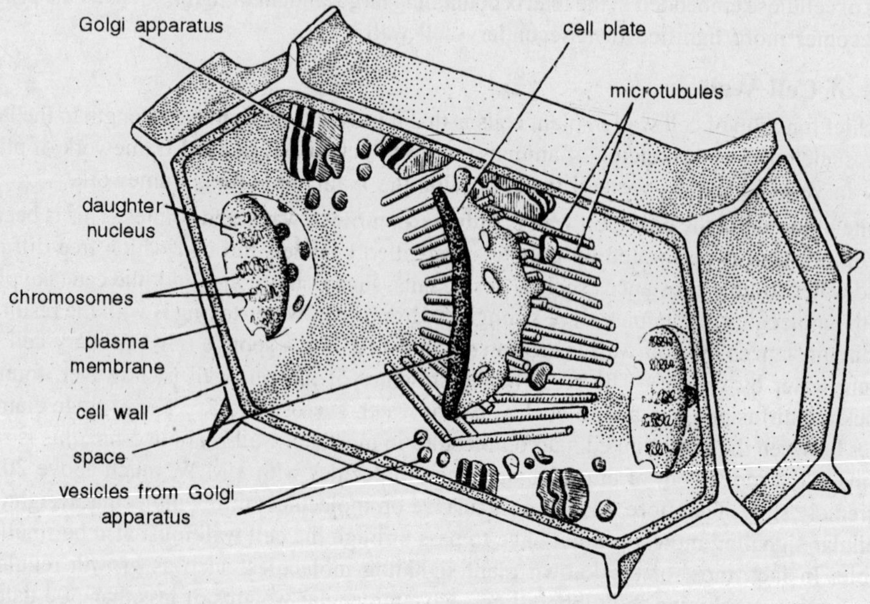

Fig. 59.21. Schematic representation of a cell of a higher plant as seen at telophase in mitosis. In the phragmoplast region, a region of membranes and microtubules, a cell plate forms and grows until it separates the cytoplasm into two daughter cells. The cell plate develops as a membrane-delimited structure enclosing a space in which new cell wall will form. The Golgi apparatus contributes many vesicles to the phragmoplast membrane; the vesicle membranes apparently are incorporated into the membrane of the cell plate and the vesicle contents enter the forming cell wall.

and a set of microfilaments are coaligned with the microtubules. Golgi derived vesicles containing cell wall precursors, especially **pectin**, are guided inward along these oriented microtubules until they reach the phragmoplast, where they fuse with one another to form the **cell plate** (Fig. 59.21). Cell plate, at this stage, comprises following structures: (1) central middle lamella; (2) primary cell walls on both sides of middle lamella; and (3) plasma membrane lining cytoplasm of each daughter cell. Ultimately, the microtubular ring moves centrifugally outward as Golgi vesicles continue to add precursors to the growing cell plate. The cell plate fuses with the mother cell wall to create two separate daughter cells. It is not clear which component of the phragmoplast—the microtubules or the actin filaments (or both)—are responsible for the movement and guidance of the Golgi derived vesicles.

2. Synthesis and orientation of cellulose microfibrils. In most plants, cellulose is synthesized on the external surface of the cell by a plasma membrane bound enzyme complex, called **cellulose synthetase** which uses a sugar nucleotide precursor supplied from the cytosol, probably UDP- glucose. As they are being synthesized, the nascent cellulose chains spontaneously assemble into **microfibrils** that form a layer on the surface of the plasma membrane (a lamella) in which all the microfibrils have more or less the same alignment (**Note.** Each cellulose molecule has a polarity, having a 1′ and a 4′ end). Cellulose synthetase complexes are thought to be associated with the ends of growing microfibrils and the sugars present in the extracellular matrix are polymerized into cellulose at these "terminal complexes". Extension of a cellulose microfibril is presumably achieved by lateral movement of the enzyme complex in the fluid phase of plasma membrane, with the microfibril "spun out" on the outer surface of the membrane behind the moving enzyme complex (Fig. 59.22). The di:ection in which the complex moves and the orientation of the microfibril depend on some interactions between the membrane complex and the underlying cytoplasmic microtubules (*i.e.,* microtubules of cell, cortex). Because the cellulose is synthesized at the plasma membrane, each new wall lamella forms internally to the last formed lamella. The cell wall, therefore, consists of concentrically arranged lamellae, with the oldest on the outside.

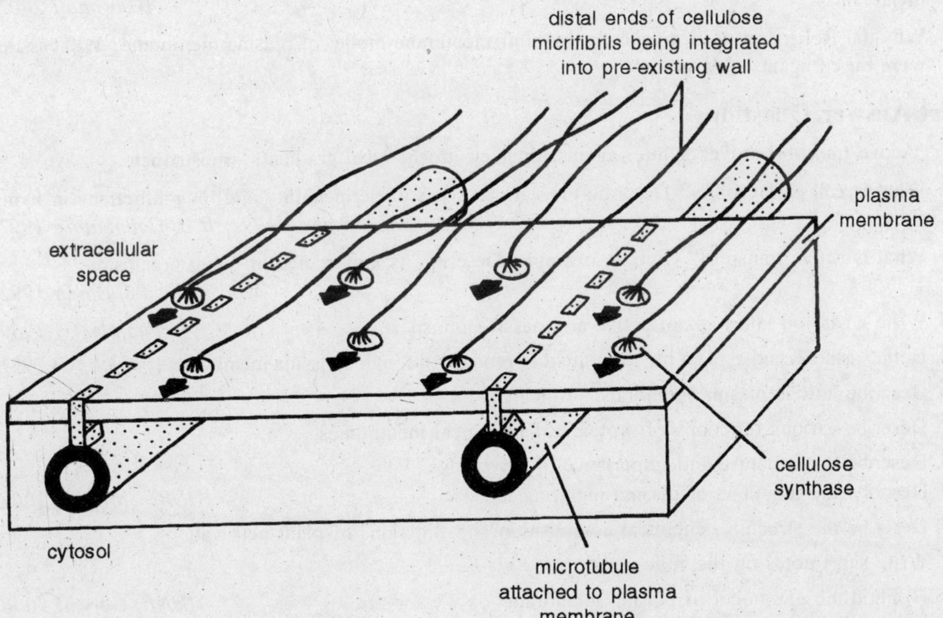

Fig. 59.22. A model explaining the mode of orientation of newly deposited cellulose microfibrils according to orientation of cortical microtubules.

QUESTIONS

Long Answer Questions

1. Describe the structure, nature and functions of plasma membrane.
(*Allahabad 1989, 93; Bundelkhand 1992; Agra 1991, 1992; CCS 1990, 1992; Guwahati (G) 1992, 1993, 1995*)
2. Describe the ultrastructure and functions of plasma membrane.
(*Magadh (Subs.) 1989, 1990, 1993; Punjab (G) 1989; BHU (Hons.) 1990*)
3. Describe the structure of plasma membrane in the light of recent findings.
(*Magadh (G) 1992, 1995*)
4. How many models of plasma membrane do you know? Explain which of the model of it is more dynamic and why? Describe the mechanism of active transport. (*CCS 1991, 1995, 1999*)
5. Describe the accepted model of the structure and chemical composition of cell membrane. State briefly the functional significance of cell membrane. (*Guwahati (M) 1996*)
6. Describe briefly the ultrastructure of cell membrane based upon various theories. (*Lucknow 1992*)
7. Give an account of the fluid mosaic model of plasma membrane. Briefly describe the important functions of this organelle. (*CCS 1992; Gorakhpur 1995*)
8. What are the characteristic features of a cell membrane? Describe the permeability and chemical composition of a cell membrane. (*Guwahati (G) 1991, 1995*)
9. Describe fluid mosaic model of structure of plasma membrane. Discuss various experimental evidences in support of this model.
10. Describe the fluidity of plasma membrane. (*Allahabad 1996*)
11. Discuss various types of diffusion that occur across the cell membrane.
12. What is passive transport? Describe the mechanism of transport of small molecules through the cell membrane. (*Guwahati 2007*)
13. Why did Robertson used myelin for his unit membrane model of plasma membrane? What issues were raised against this model?

Short Answer Questions

1. 'Active transport moves solutes against their electrochemical gradients'–substantiate.
2. What is cell permeability? Describe the various types of permeability and their functions in living cells. (*Guwahati 1992; B.U. Coimbatore 1997*)
3. What is active transport? Give an example. Describe its importance in living organisms.
(*Delhi 1989; Rajasthan 1992*)
4. Write what you know about active and passive transport. (*Punjab (G) 1989*)
5. Define osmosis and explain the movement of products through a plasma membrane. (*Punjab (G) 1991*)
6. Describe how is plasma membrane originated.
7. Describe various functions of proteins of biological membranes.
8. Describe the structure and importance of glycocalyx.
9. Describe the functions of plasma membrane. (*HNB, Garhwal, 2009*)
10. Describe the structure, chemical composition and functions of plant cell wall.
11. Write short notes on the following:
   (*i*) Fluid mosaic model of plasma membrane (*HNB, Garwal 2008*)
   (*ii*) Channel proteins
   (*iii*) Unit membrane (*Jiwaji 1992; Gorakhpur 1998*)
   (*iv*) Permeability of cell membrane (*Guwahati 1993, 1996*)

(*v*) Active transport (*Sagar 1989; Kanpur 1990; Bundelkhand 1989, 1992*)

(*vi*) Osmosis

(*vii*) Role of semipermeability of plasma membrane

(*viii*) Passive transport.

12. Differentiate between the following terms:

(*i*) Simple diffusion and facilitated diffusion. (*Guwahati (M) 1995*)

(*ii*) Active transport and passive transport.

(*iii*) Plasma membrane and endoplasmic reticulum. (*CCS 1997*)

(*iv*) Cell wall and karyotheca. (*CCS 1999*)

(**Hint.** Cell wall is nonliving, porous, cellulose-containing part of cell that occurs outside plasma membrane of the cell of plants. Karyotheca is nuclear envelope which consists of two unit membranes).

13. Differentiate between uniport transport and passive diffusion. (*M.S.U., Tamil Nadu 2006*)

14. Draw labelled diagrams of the following (No description is required):

(*i*) Fluid-mosaic model of plasma membrane. (*Lucknow 1994*)

(*ii*) Cell wall (*Lucknow 1998*)

Very Short Answer Questions

1. Who gave the concept of unit membrane?

Ans. Robertson.

2. Name the scientists who proposed bimolecular lipid layer model of plasma membrane.

Ans. Gorter and Grendel.

3. Who proposed fluid mosaic model of plasma membrane?

Ans. Singer and Nicholson.

4. Define passive transport.

Ans. Transport of substances through a membrane along their concentration gradients is called passive transport.

5. Define diffusion.

Ans. Movement of molecules and ions along an electrochemical gradient is diffusion.

6. What is flip-flop movement?

Ans. Movement of lipid molecules from one layer to another of plasma membrane.

7. Name three substances passing the tunnels in the integral proteins of plasma membrane.

Ans. Water, oxygen, carbon dioxide, urea and glycerol.

8. What cause diffusion of ions across the plasma membrane?

Ans. Electrical difference between the inside and outside of a cell.

9. What is the name given to the proteins separated from plasma membranes of RBC?

Ans. Tektins.

10. Name any three sugars found in the glycocalyx.

Ans. D–galactose, D–mannose, L–fucose, sialic acid, N–acetyl–D–glucosamine. (Note. Students may choose any three of them).

11. Is glycocalyx found in prokaryotic plasma membrane.

Ans. No.

12. How do large molecules of sugars and amino acids diffuse through plasma membrane quickly along the concentration gradient?

Ans. Transportation of these substances across the plasma membrane takes place by specific carrier protein molecules of the plasma membrane.

13. Name the integral components acting as a sodium-potassium pump.

Ans. Na^+, K^+–ATPase.

14. Name the processes responsible for bulk transport of substances across the plasma membrane.

Ans. Pinocytosis, phagocytosis, endocytosis, receptor-mediated endocytosis and exocytosis.

15. Name any two substances taken in by the plasma membrane through receptor-mediated endocytosis

Ans. Cholesterol, steroid hormones, antibodies, some proteins and phagocytosis.

16. Name the organelle present in both prokaryotic and enkaryotic cells.

Ans. Ribosome or Plasma membrane.

Mark the Statements True of False

(*i*) Cell membrane is composed of liporotein. *(NEHU (M) 1996)*

(*ii*) Cellulose of cell wall is a protein.

(*iii*) Plasma membrane is present in all cells.

(*iv*) Cell wall of plant cell is a living structure.

Yes or No Questions

(*i*) Thickness of plasma membrane ranges from 100 Å to 200 Å. *(Guwahati (G) 1994)*

(*ii*) The concept of 'Unit membrane' was put forward by Robertson. *(Guwahati (G) 1995)*

Multiple Choice Questions

1. Plasma membrane is made up of

(*a*) glycoproteins and lipids (*b*) phospholipids and glycolipids

(*c*) phospholipids and proteins (*d*) phosphoproteins and phospholipids

2. Phospholipid molecules are

(*a*) acidic (*b*) alkaline

(*c*) amphoteric (*d*) amphipathic

3. In lipid bilayer, the phospholipid molecules are arranged in such a fashion that:

(*a*) their hydrophobic ends are directed outwards

(*b*) their hydrophobic ends are directed inwards

(*c*) the hydrophobic ends of molecules of one layer make contact with the hydrophilic ends of th molecules of other layer

(*d*) their hydrophobic ends are scattered indiscriminately

4. The hydrophobic ends of phospholipid molecules are

(*a*) polar (*b*) nonpolar

(*c*) neutral (*d*) bipolar

5. Spectrin, cytochrome-C and enzymes present on the plasma membrane are examples of

(*a*) integral or intrinsic proteins (*b*) peripheral or extrinsic proteins

(*c*) glycoproteins (*d*) lipoproteins

6. The membrane proteins that extend through both sides of lipid bilayer

(*a*) acidic proteins (*b*) glycoproteins

(*c*) intrinsic proteins (*d*) glycocolic acid

7. According to fluid mosaic model, the plasma membrane presents

(*a*) a quasifluid structure (*b*) a micellar structure

(*c*) a polar structure (*d*) a unit membrane structure

8. Fluidity of plasma membrane at body temperature is due to
(*a*) fluidity of lipid bilayer
(*b*) fluidity of glycolipids
(*c*) presence of numerous unsaturated fatty acids in the lipid bilayer
(*d*) presence of integral proteins in the form of glycophorins

9. Fluid mosaic model to explain structure of plasma membrane was proposed by
(*a*) Hoffman (*b*) Singer and Nicolson
(*c*) Robertson (*d*) Davson and Danielli

10. Unit membrane model of plasma membrane was proposed by
(*a*) Nicolson (*b*) Danielli and Davson
(*c*) Robertson (*d*) Mitchel

11. Egestion of residual substances through plasma membrane of cell is called
(*a*) egestion (*b*) ephagy
(*c*) exocytosis (*d*) emeiocytosis

12. Active transport is
(*a*) movement of molecules against concentration gradient
(*b*) movement of molecules along the concentration gradient
(*c*) movement of molecules to and fro of plasma membrane
(*d*) mone of the above

13. Plasmodesmata allow direct exchange of materials
(*a*) between two distant cells (*b*) between neighbouring plant cells
(*c*) within the cell (*d*) all these.

14. Element necessary for the middle lamella
(*a*) Ca (*b*) Zn
(*c*) K (*d*) Cu

ANSWERS

True and False Statements

1. True **2.** False **3.** True **4.** False.

Yes or No Questions

1. No **2.** Yes.

Multiple Choice Questions

1. (*c*) **2.** (*d*) **3.** (*b*) **4.** (*b*)
5. (*b*) **6.** (*c*) **7.** (*a*) **8.** (*c*)
9. (*b*) **10.** (*c*) **11.** (*d*) **12.** (*a*)
13. (*b*) **14.** (*a*).

60

Cell Junctions, Cell Adhesion and the Extracellular Matrix (ECM)

The building technologies of multicellular animals and plants are different, and each type of organism is formed of many types of **tissues**, in which the cells are assembled and bound together in different ways. In both animals and plants, however, an essential part is played in most tissues by the **extracellular matrix (ECM)**. This complex network of secreted extracellular macromolecules has many functions, but the first and salient is that it forms a supporting framework. ECM helps hold cells and tissues together, and in animals, it provides an organized environment within which migratory cells can move and interact with one another in orderly ways. The extracellular matrix, however, is only half the story. In animals especially, the cells of most tissues are bound directly to one another by **cell-cell junctions**. These junctions too are of many types, serving many purposes in addition to mechanical attachment; but without them, bodies of animals would disintegrate.

In vertebrates, the major tissue types are nerve, muscle, blood, lymphoid, epithelium, and connective tissues. Connective tissues and epithelial tissue represent two extremes of organization. In *connective tissue*, the extracellular matrix is plentiful, and cells are sparsely distributed within it. The matrix is rich in fibrous polymers, especially **collagen**, and it is the matrix-rather than the cells—that bears most of the mechanical stress to which the tissues is subjected. Direct attachments between one cell and another are relatively rare.

In **epithelial tissue**, by contrast, cells are tightly bound together into sheets called **epithelia**. The extracellular matrix is scanty, consisting mainly of a tiny mat called the **basal lamina**, which underlies the epithelium. The cells are attached to each other by cell-cell adhesions, which bear most of the mechanical stresses. For this purpose, strong intracellular protein filaments (component of cytoskeleton) cross the cytoplasm of each epithelial cell and attach to specialized junctions in the plasma membrane. The junctions, in turn, tie the surfaces of adjacent cells either to each other or to the underlying basal lamina.

Epithelial cell sheets line all the cavities and free surfaces of the animal body. The specialized junctions between the cells enable epithelia to form barriers that inhibit the movement of water, solutes and cells from one body compartment to another. As shown in (Fig. 60.1), epithelia almost always rest on a supporting bed of connective tissue. This supporting bed may in turn attach them to other tissues, such as the muscle shown in figure. In this way, tissues join together in various combinations to larger functional units called **organs**.

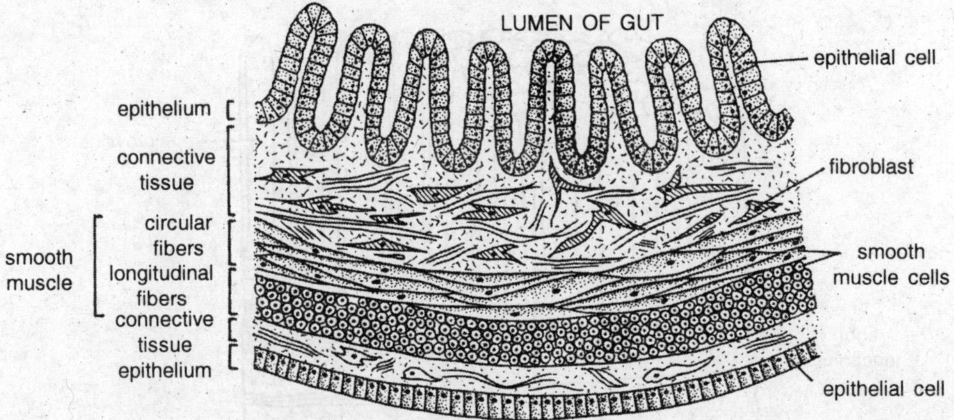

Fig. 60.1. A cross-sectional view of part of the wall of the intestine. This long, tube-like organ is constructed from epithelial tissue, connective tissue and muscle tissue. Each tissue is an organized assembly of cells held together by cell-cell adhesions, extracellular matrix or both (after Alberts *et al.*, 2002).

60.1 CELL JUNCTIONS

Specialized **cell junctions** occur at points of cell-cell and cell-matrix contact in all tissues, and they are particularly abundant in epithelia. Cell junctions are best visualised using either conventional or freeze-fracture electron microscopy, which reveals that the interacting plasma membranes (and often the underlying cytoplasm and the intervening intercellular space as well) are highly specialized in these regions.

Cell junctions (Fig. 60.2) are often divided into following three functional groups:

1. Occluding junctions. They seal cells together in an epithelium in a way that prevents even small molecules from leaking from one side of the sheet to the other. Indeed they form a selectively permeability barrier across epithelial cell sheet.

2. Anchoring junctions. They mechanically attach cells (and their cytoskeleton) to their neighbors or to the extracellular matrix. Thus, they connect the cytoskeleton of a cell to the cytoskeleton of it neighbors or to the ECM.

3. Communicating junctions. They mediate the passage of chemical or electrical signals from one interacting cell to its partner.

The major kinds of intercellular junctior within each group are listed in Table 60.1.

Table 60.1 The functional classificatıoıı of cell junctions (after **Alberts** *et al.,* 2002).

A. Occluding Junctions
1. Tight junctions (vertebrates only)
2. Septate junctions (invertebrates only)

B. Anchoring Junctions

I. Actin Filament Attachment Sites
1. Cell-cell junctions (adherens junctions)
2. Cell-matrix junctions (focal adhesions)

II. Intermediate Filament Attachment Sites
1. Cell-cell junctions (desmosomes)
2. Cell-matrix junctions (hemidesmosomes)

C. Communicating Junctions
1. Gap junctions
2. Chemical synapses
3. Plasmodesmata (plants only)

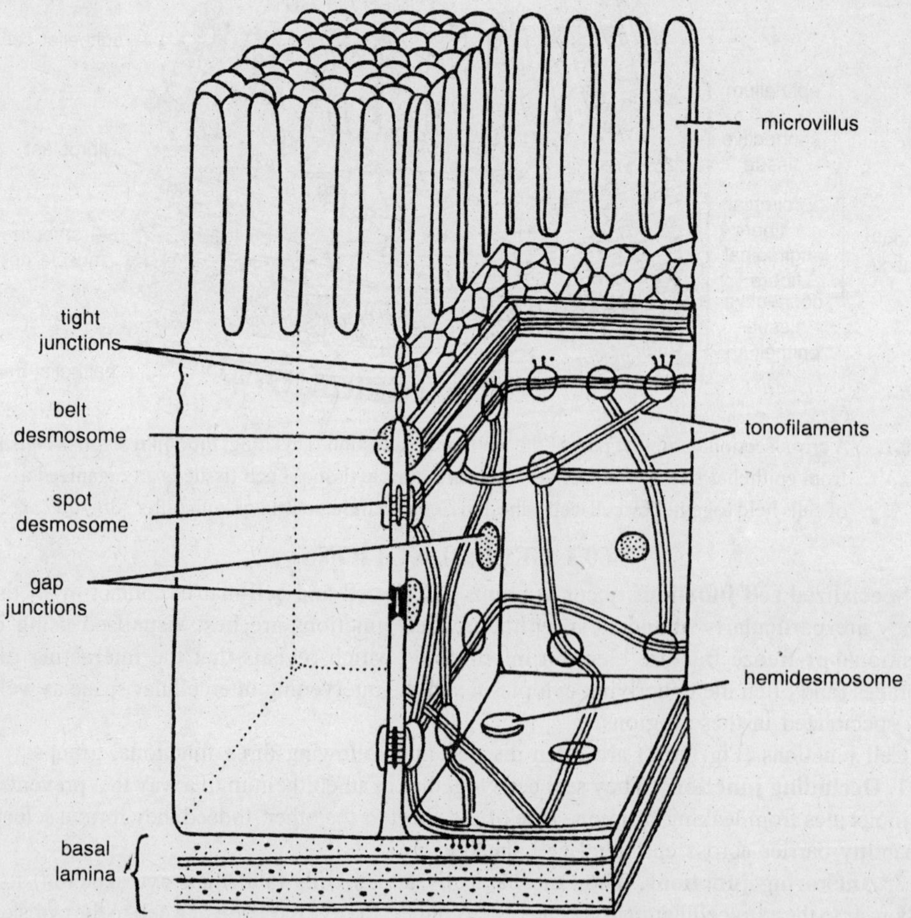

Fig. 60.2. Schematic diagram of the principal types of cell junctions, as found in the intestinal epithelial cell.

A. Occluding Junctions.

All epithelia have at least one important function in common: they serve as selective permeability barriers, separating fluids on either side that have a different chemical composition. This function requires that the adjacent cells be sealed together by occluding junctions.

Occluding junctions are of following two types:

1. Tight junctions. They occur in vertebrates in the organs such as mammalian small intestine, or gut. Tight junctions are also abundant in the ducts and cavities of glands, such as liver and pancreas, that connect with the digestive tract, as well as in the urinary bladder, where they ensure that urine stored in the bladder does not seep out between cells.

The epithelial cells lining the small intestine form a barrier that keeps the gut contents in the gut cavity, the **lumen**. At the same time, however, the cells must transport selected nutrients across the epithelium from the lumen into the extracellular fluid that permeates the connective tissue on the other side (Fig. 60.1). From there these nutrients diffuse into small blood vessels to provide nourishment to the organism. This is called **transcellular transport** and it depends on two sets of membrane-bound **transport proteins**. One set is confined to the apical surface of the epithelial cell (*i.e.*, the surface facing th^ lumen) and *actively transports* selected molecules into the cell from the gut (Fig. 60.3). The other set is confined to the *basolateral* (basal and lateral) *surfaces* of the cell, and it allows the same molecules to leave the cell by *facilitated diffusion* into the extra-cellular fluid on the other side of the epithelium. To maintain this *directional transport,* the apical set of transport proteins must not be allowed to migrate to the baso-lateral surface of the cell, and the basolateral set must not be allowed to migrate to the apical surface. Furthermore,

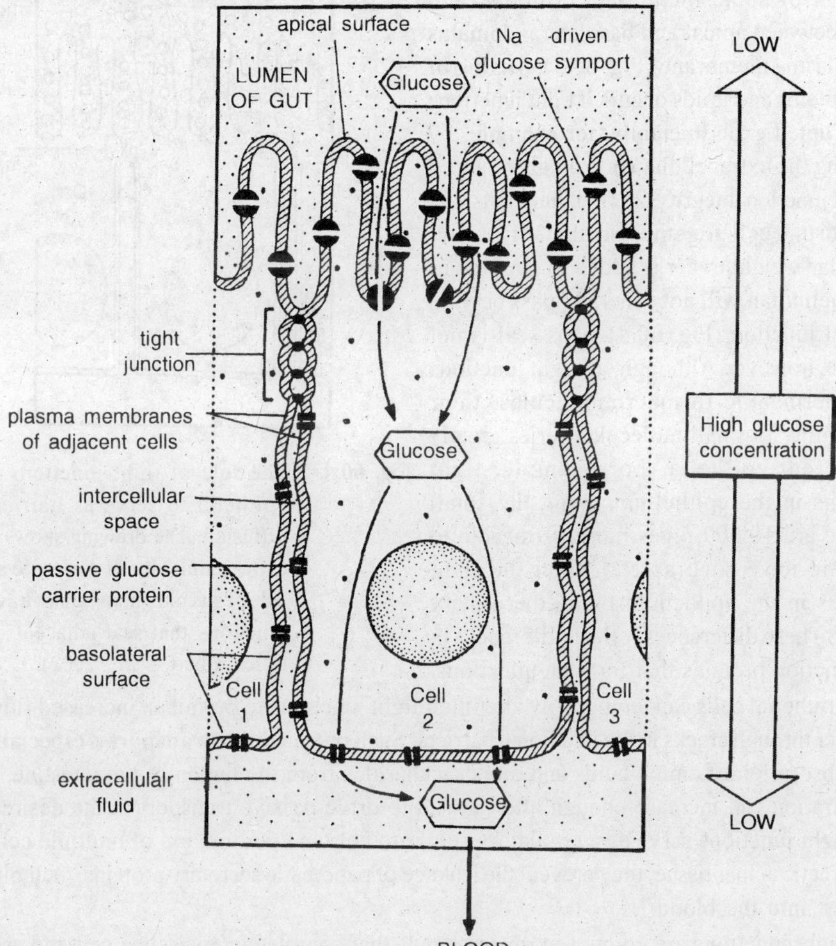

Fig. 60.3. The role of tight junctions in transcellular transport. The transport proteins are confined to different regions of the plasma membrane in epithelial cells of the small intestine. This separation permits a vectorial* transfer of nutrient across the epithelium from the gut lumen to the blood. In the example shown, glucose is actively transported into the cell by Na^+–driven glucose symports at the apical surface, and it diffuses out of the cell by facilitated diffusion mediated by glucose carriers in the basolateral membrane. Tight junctions are thought to confine transport proteins to their appropriate membrane domains by acting as diffusion barriers within the lipid bilayer of the plasma membrane; these junctions also block the backflow of glucose from the basal side of the epithelium into the gut lumen (after Alberts *et al.*, 2002).

***Vectorial**. Vector-like; vector represents a quantity that has magnitude and direction and that is usually represented by part of a straight line with the given direction and with a length representing a magnitude.

the space between the epithelial cell must be tightly sealed, so that the transported molecules cannot diffuse back into the gut lumen through these spaces.

The tight junctions between epithelial cells are thought to have following two types of functions: 1. They function as barriers to the diffusion of some membrane proteins (and lipids) between apical and basolateral domains of the plasma membrane (Fig. 60.3). Mixing of such proteins and lipids occurs if tight junctions are disrupted experimentally, for example, by removing the extracellular Ca^{2+} that is required for tight junction integrity. 2. Tight junctions seal neighboring cell together so that, if a low-molecular-weight tracer is added to one side of an epithelium, it will not generally pass beyond the tight junction (Fig. 60.4). This seal is not absolute, however. Although, all tight junctions are impermeable to macromolecules, their permeability to small molecules varies greatly in different epithelia. For example, tight junctions in the epithelium lining the small intestine are 10,000 times more permeable to inorganic ions, such as Na^+, than the tight junctions in the epithelium lining the urinary bladder. These differences reflect differences in tight junction proteins that form the junctions.

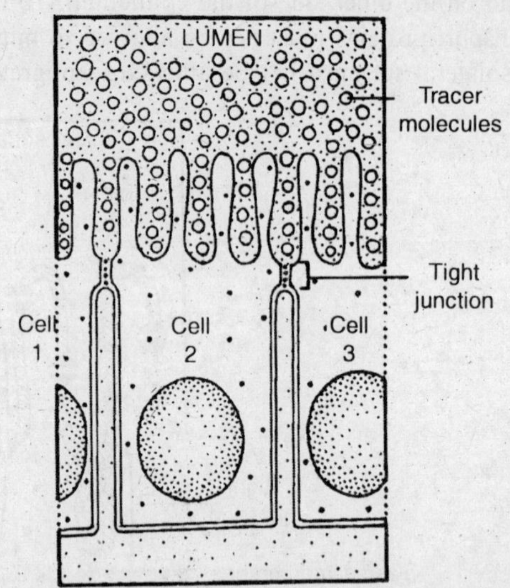

Fig. 60.4. The role of tight junctions in allowing epithelia to serve as barriers to solute diffusion. The drawing shows how a small extra-cellular tracer molecule added on one side of an epithelium cannot traverse the tight junctions that seal adjacent cell together (after Albert *et al.*, 2002).

Epithelial cells can temporarily alter their tight junctions to permit an increased flow of solutes and water through cracks in the junctional barriers. Such **paracellular transport** is especially important in the absorption of amino acids and monosaccharides from the lumen of the intestine, where their concentration can increase enough after a meal to drive passive transport in the desired direction. Thus, tight junctions serve as a regulatory center to help in coordination of multiple cell processes. In pancreatic acinar tissue, they prevent the leakage of pancreatic secretary proteins, including digestive enzymes, into the blood.

Tight junctions are composed of thin bands that completely encircle a cell and are in contact with thin bands of adjacent cells. In a thin section, in the tight junction two adjacent plasma membranes appear to be fused at a series of points. However, in three-dimensional structure, revealed by freeze-fracture technique the tight junctions appear as a network of ridges on the cytoplasmic half of the membrane, with complementary grooves in the outer half. The ridges appear to be composed of two rows of protein particles, as in zipper, each one belonging to the adjacent cells. The lines of these particles produce the sealing and for this reason have been named **sealing strands.** Often, sealing strands form a series of interconnected and anastamosing lines, like a row of stitches in a quilted surface.

Each tight junction sealing strand is composed of long rows of transmembrane adhesion proteins embedded in each of the two interacting plasma membranes. The extracellular domains of these proteins joint directly to one another to occlude the intercellular space (Fig. 60.5). The major transmembrane proteins in a tight junction are the **claudins**, which are essential for tight junction formation and function and differ in different tight junctions. A specific claudin found in kidney epithelial cells, for example, is required for Mg^{2+} to be resorbed from the urine into the blood. A mutation in the gene encoding this claudin results in excessive loss of Mg^{2+} in the urine. A second major transmembrane protein in tight junctions is **occludin**, the function of which is uncertain. Claudins and occludins associate with intracellular peripheral membrane proteins called **ZO proteins** S (Z = zonula, O = occludens, another name of tight junctions) which anchor the strands to the actin cytoskeleton.

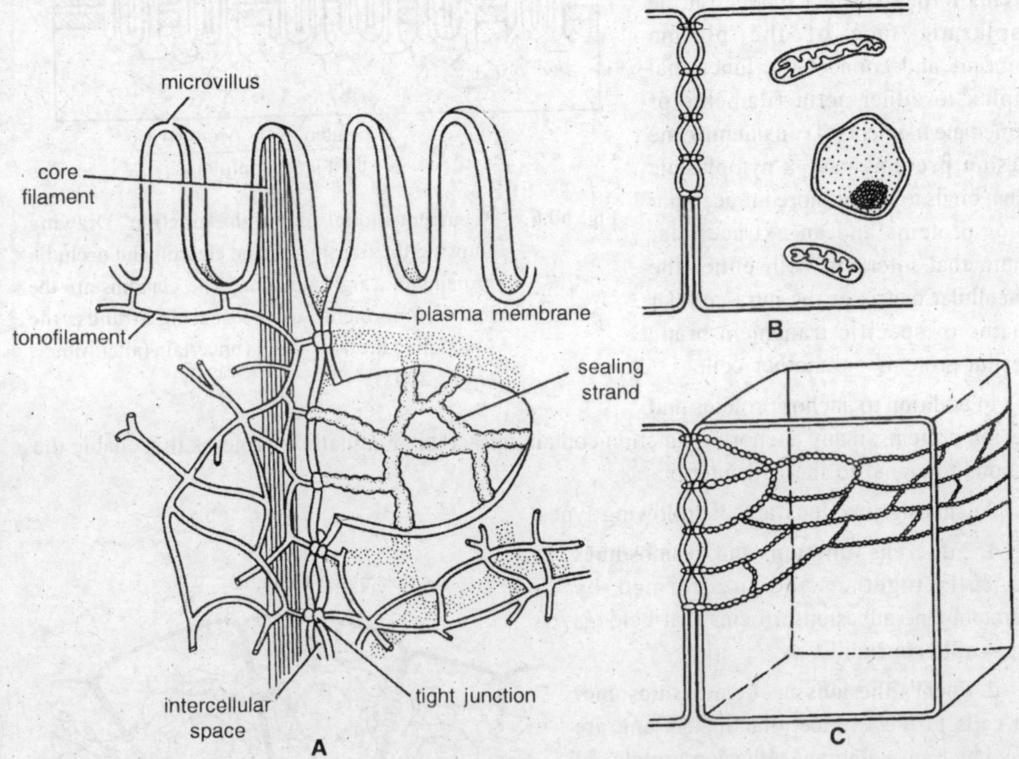

Fig. 60.5. Tight junctions: A—The adjacent plasma membranes are held firmly at the sealing strand which is composed of two rows of particles, as in zipper; B, C—Three-dimensional representations.

2. Septate junctions. These occluding junctions occur in invertebrates. Septate junctions are more regular in structure than the tight junctions and they form a continuous band around each epithelial cell. But their morphology is distinct because the interacting plasma membranes are joined by proteins that are arranged in parallel rows with a regular periodicity. A protein called **Discslarge**, which is required for the formation of septate junction in *Drosophila*, is structurally related to the ZO proteins found in vertebrate tight junction. Mutant flies that are deficient in this protein not only lack septate junctions but also develop epithelial tumors.

B. Anchoring Junctions.

The lipid bilayer is flimsy and cannot by itself transmit large forces from cell to cell or from cell to extracellular matrix. Anchoring junctions solve the problem by forming a strong membrane-spanning structure that is tethered (*i.e.*, fastened) inside the cell to the tension-bearing filaments of the cytoskeleton (Fig. 60.7).

Anchoring junctions are widely distributed in animal tissues and are most abundant in tissues that are subjected to severe mechanical stress, such as heart, muscle and epidermis. They are composed of two main classes of proteins (Fig. 60.8). **Intracellular anchor proteins** form a distinct plaque on the cytoplasmic face of the plasma membrane and connect the junctional complex to either actin filaments or intermediate filaments. **Transmembrane adhesion proteins** have a cytoplasmic tail that binds to one or more intracellular anchor proteins and an extracellular domain that interacts with either the extracellular matrix or the intra-cellular domains of specific transmem-brane adhesion proteins on another cell.

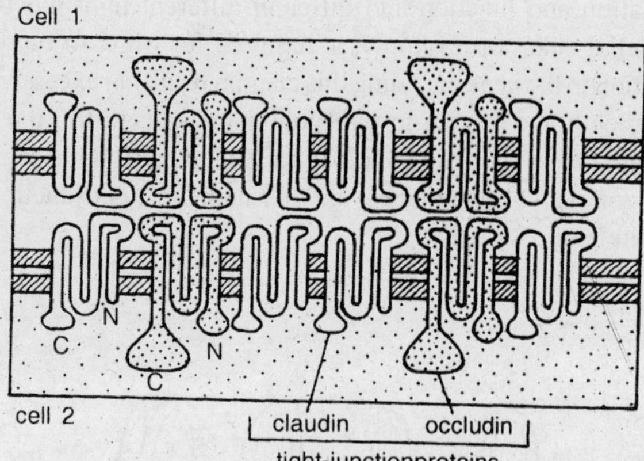

Fig. 60.6. A current model of a tight junction. Drawing showing the transmembrane claudin and occludin proteins in a tight junction. The claudins are the main components of the sealing strands; the function of the occludins is uncertain (after Alberts *et al.*, 2002).

In addition to anchor proteins and adhesion proteins, many anchoring junction contain intracellular signalling proteins that enable the junctions to signal to the cell interior.

Anchoring junctions are of following types:

1. Adherens junctions and **desmosomes** hold cells together and are formed by transmembrane adhesion proteins that belongs to the **cadherin** family.

2. Focal adhesions and **hemidesmosomes** bind cells to the extracellular matrix and are formed by transmembrane adhesion proteins of the **integrin**.

On the intracellular side of the membrane, adherens junctions and focal adhesions serve as connection sites for actin filaments while desmosomes and hemidesmosomes serve as connection sites for intermediate filaments.

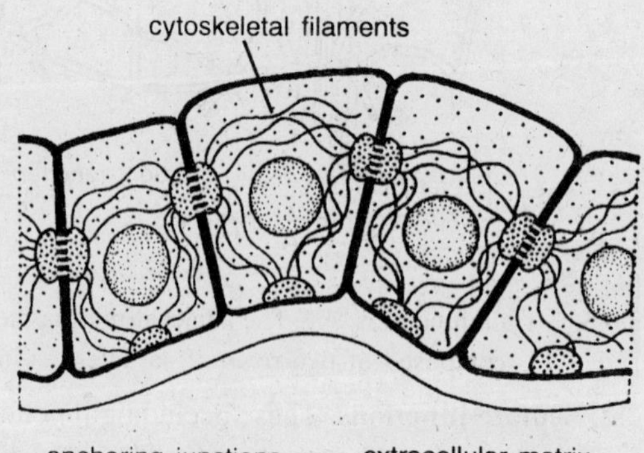

Fig. 60.7. Anchoring junctions in an epithelium. This drawing shows in a general way, how anchoring junctions join cytoskelet filaments from cell to cell and from cell to the extracellular matrix (after Alberts *et al.*, 2002).

I. Actin Filament Attachment Sites

1. Adherens Junctions (Zonula adherens). Adherens junctions commonly occur in epithelia where they often form a continuous **adhesion belt** or **zonula adherens** just below the tight junctions, encircling each of the interacting cell in the sheet. The adhesion belt are directly apposed in adjacent epithelial cells, with the interacting plasma membranes held together by the **cadherins** that serve as transmembrane adhesion proteins.

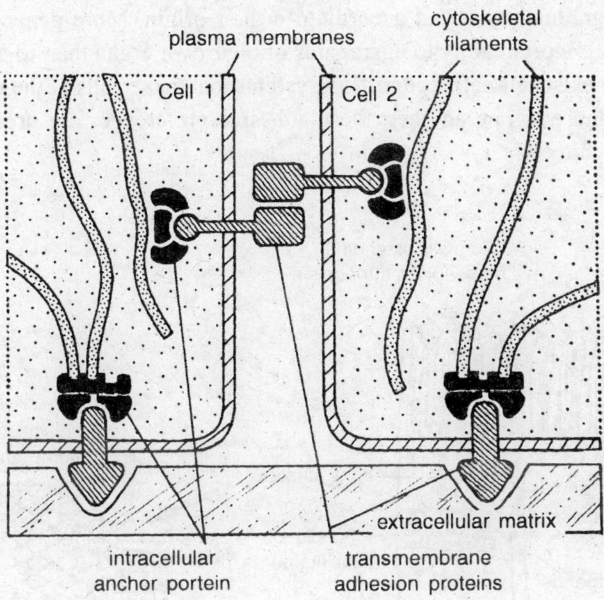

Fig. 60.8. The construction of an anchoring junction from two classes of proteins. This diagram shows how intracellular anchor proteins and transmembrane adhesion proteins form anchoring junctions (after Alberts *et al.*, 2002).

Within each cell, a contractile bundle of actin filaments lies adjacent to the adhesion belt, oriented parallel to the plasma membrane. The actin is attached to this membrane through a set of intracellular anchor proteins, including **catenins, vinculin** and **a-actinin.** The actin bundles are thus linked, via the cadherins and anchor proteins, into an extensive transcellular network (Fig. 60.9). The network can contract with the help of **myosin** motor proteins and it is thought to help in mediating a fundamental process in animal morphogene-sis—the folding of epithelial cell sheets into tubes and other related structure (Fig. 60.10).

The assembly of tight junctions between epithelial cells seems to require the prior formation of adherens junctions. Anti-cadherin antibodies that block the formation of adherens junctions, for example, also block the formation of tight junctions.

Selective Cell-Cell Adhesion and CAMs

Unlike adult vertebrate tissues, which are difficult to dissociate, embryonic vertebrate tissues are easily dissociated. This is usually done by treating the tissue with low concentrations of a proteolytic enzyme such as trypsin. Sometimes this technique is combined with the removal of extracellular Ca^{2+} and Mg^{2+} with a divalent-cation chelator (such as EDTA). These reagents disrupt the protein-protein interactions (many of which are divalent-cation-dependent) that hold cells together. Remarkably, the dissociated cell often reassemble *in vitro* into structures that resemble the original tissue. Findings of this nature reveal that tissue structure is not just a product of history (*i.e.*, origin and development); it is actively maintained and stabilized by the system of affinities that cells have for one another and for the extracellular matrix (ECM).

A salient example of this phenomenon is seen when dissociated cells from two embryonic vertebrate organs, such as liver and the retina, are mixed together and artificially formed into a pellet: the mixed aggregates gradually sort out according to their origin. More generally, disaggregated cells are found to adhere more readily to aggregates of their own organ than to aggregates of other organs. Evidently there exist **cell-cell recognition systems** that make cells of the same differentiated the mixed aggregates gradually sort out according to their origin. More generally, disaggregated cells are found to adhere more readily to aggregates of their own organ than to aggregates of other organs. Evidently there exist **cell-cell recognition systems** that make cells of the same differentiated tissue preferentially adhere to one another; these adhesive preferences are presumably important in stabilizing tissue architecture.

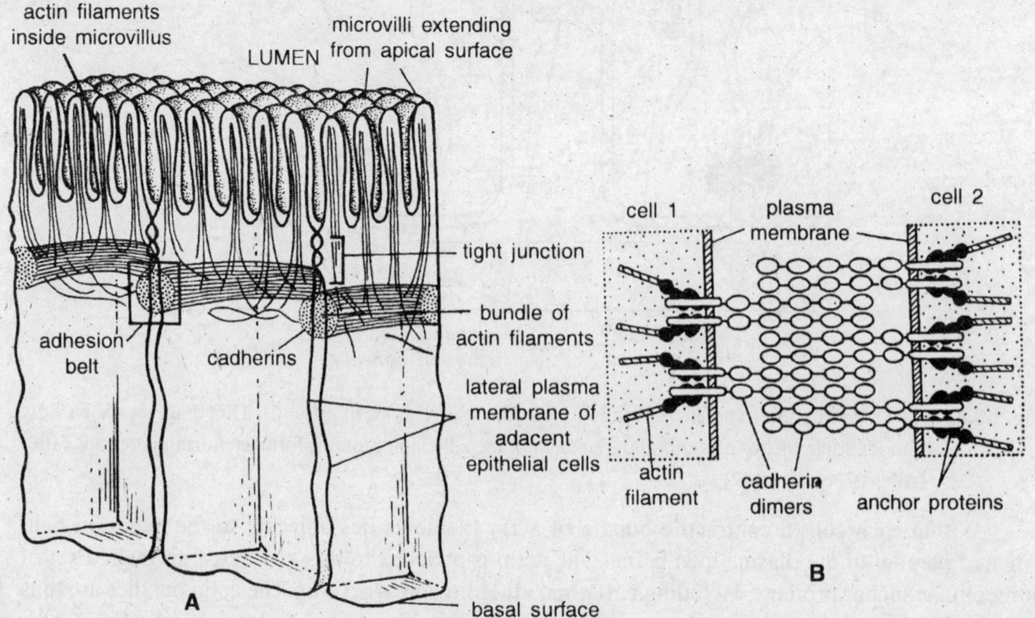

Fig. 60.9 Adherens junctions. A—Adherens junctions in the form of adhesion belts between epithelial cells in the small intestine. The belt-like junction encircles each of the interacting cells. Its most obvious feature is a contractile bundle of actin filaments running along the cytoplasmic surface of the junctional plasma membrane. B—The actin filaments are joined from cell to cell by transmembrane adhesion proteins called **cadherins.** The cadherins form homodimers in the plasma membrane of each interacting cell. The extracellular domain of one cadherin dimer binds to the extracellular domain of an identical cadherin dimer on the adjacent cell. The intracellular tails of the cadherins bind to anchor proteins that tie them to actin filaments. These anchor proteins include α-catenin, β-catenin, γ-catenin (also called plakoglobin), α-actinin and vinculin (after Alberts *et al.*, 2002).

Cells adhere to each other and to the extracellular matrix through cell surface proteins called **cell adhesion molecules (CAMs)**—a category that includes the transmembrane adhesion proteins such as cadherins and integrins. CAMs can be **cell-cell adhesion molecules** or **cell matrix adhesion molecules.** Some CAMs are Ca^{2+}-dependent (*e.g.*, cadherins), whereas others are Ca^{2+}-independent (*e.g.*, immunoglobulins). The Ca^{2+}-dependent CAMs seem to be primarily responsible for the tissue-specific cell-cell adhesion seen in early vertebrate embryos, explaining why these cells can be disintegrated with Ca^{2+}-chelating agents.

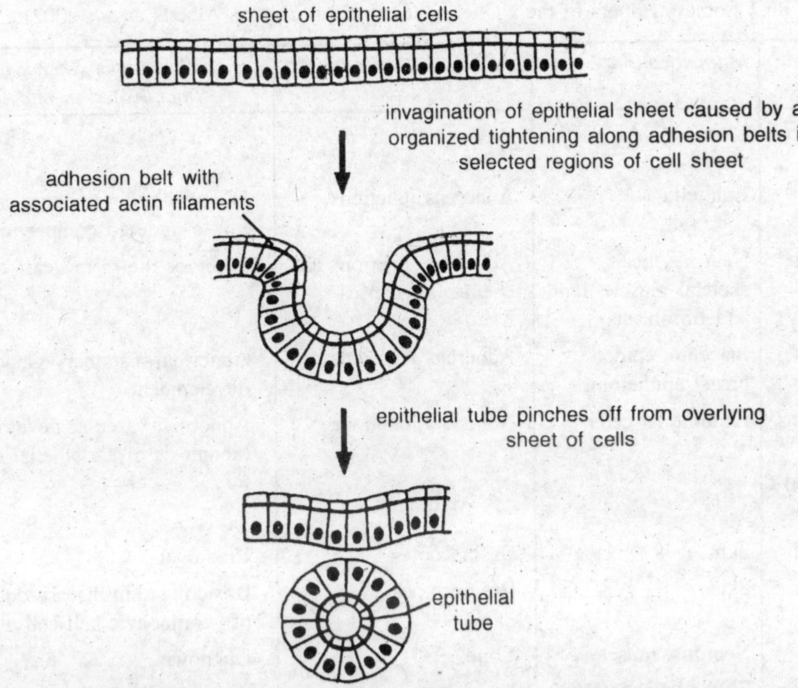

Fig. 60.10. The folding of an epithelial sheet due to adhesion belt to form an epithelial tube. The oriented contraction of the bundles of actin filaments running along adhesion belts causes the epithelial cells to narrow at their apex and helps the epithelial sheet to roll up into a tube. An example is the formation of the neural tube in early vertebrate development (during neurula stage). However, rearrangements of the cells within the epithelial sheet are also thought to have an important role in the process (after Alberts *et al.*, 2002).

Examples of CAMs

1. Cadherins.

The cadherins are the major CAMs which are responsible for Ca^{2+}-dependent cell-cell adhesion in vertebrate tissues. They are of following two main types (Table 60.2).

***(i)* Classical cadherins.** The first three catherins that were discovered were named according to the main tissues in which they were found: **E-cadherin** is present on many types of epithelial cells; **N-cadherin** on nerve, muscle and lens cells; and **P-cadherin** on cells in the placenta and epidermis. All are also found in various other tissues. For example, N-cadherin is expressed in fibroblasts and E-cadherin is expressed in parts of the brain. All these cadherins are **classical cadherins** and they are related in sequence throughout their extracellular and intracellular domains.

***(ii)* Nonclassical cadherins.** They are of many types; more than 50 nonclassical cadherins are expressed in the brain alone. The nonclassical cadherins include proteins with known adhesive function, such as the **desmosomal cadherins** and diverse **proto-caderins** found in the brain. They also include proteins that appear to have nonadhesive functions, such as **T-cadherins**, which lack a transmembrane domain and is attached to the plasma membrane of nerve or muscle cells by a glycosulphosphatidylinositol (GPI), anchor, and the **Fat protein**, which was first identified as the product of a tumor-suppressor gene in *Drosophila*. Together, the classical and nonclassical cadherin proteins form the **cadherin superfamily**.

Table 60.2 Some members of the cadherin superfamily (Source. Alberts *et al.*, 2002).

Name	*Main location*	*Junction association*	*Phenotype when inactivated in mice*
A. Classical cadherins			
1. E-cadherins	Epithelia	Adherens junctions	Die at blastocyst stage; embryos fail to undergo compaction
2. N-cadherins	Neurons, heart, skeletal muscle, lens and fibroblasts	Adherens junctions and chemical synapses	Embryos die from heart defects
3. P-cadherins	Placenta, epidermis, breast epithelium	Adherens junctions	Abnormal mammary gland development
4. VE-cadherins	Endothelial cells	Adherens junctions	Abnormal vascular development (apoptosis of endothelial cells)
B. Nonclassical cadherins			
1. Desmocollin	Skin	Desmosomes	Unknown
2. Desmoglein	Skin	Desmosomes	Blistering skin disease due to loss of keratinocyte cell-cell adhesion
3. Cadherin	Neurons, muscle	None	Unknown
4. Fat (in *Drosophila*)	Epithelial and CNS	None	Enlarged imaginal discs and tumors
5. Protocadherins	Neurons	Chemical synapses	Unknown

Occurence. Cadherins are found in both invertebrates and vertebrates. Virtually all vertebrates seem to contain one or more cadherins, according to the cell type. They are the main adhesion molecules holding cells together in early embryonic tissues. In culture, the removal of extracellular Ca^{2+} or treatment with anti-cadherin antibodies disrupts embryonic tissues. Mutations that inactivate the function of E-cadherin cause mouse embryos to fall apart and die early in development.

Molecular structure. Most cadherins are single-pass transmembrane glycoproteins about 700–750 amino acids long. Structural studies suggest that they associate in the plasma membrane to form **dimers** or larger **oligomers**. The large extracellular part of the polypeptide chain is usually folded into five or six **cadherin repeats**, which are structurally related to immunoglobulin (Ig) domains (Fig. 60.11A). The crystal structures of E-cadherin and N-cadherin have helped to explain the importance of Ca^{2+} binding for cadherin function. The Ca^{2+} ions positioned between each pair of cadherin repeats, locking the repeats together to form a stiff, rodlike structure: the more Ca^{2+} ions that are bound, the more rigid the structure is. If Ca^{2+} is removed, the extracellular part of the protein becomes floppy and is rapidly degraded by proteolytic enzymes (Fig. 60.11).

Functions of Cadherins

(*i*) Role of cadherins in development. E-cadherin is the best characterized cadherin. It is usually concentrated in adherens junctions in mature epithelial cells, where it helps connect the cortical actin cytoskeletons of the cell it holds together (Fig. 60.9). E-cadherin is also the first cadherin expressed during mammalian development. It helps cause **compaction**, an important morphological change that occurs at the eight cell stage of mouse embryo development. During compaction, the loosely attached cells, called **blastomeres**, become tightly packed together and joined by intercellular junctions.

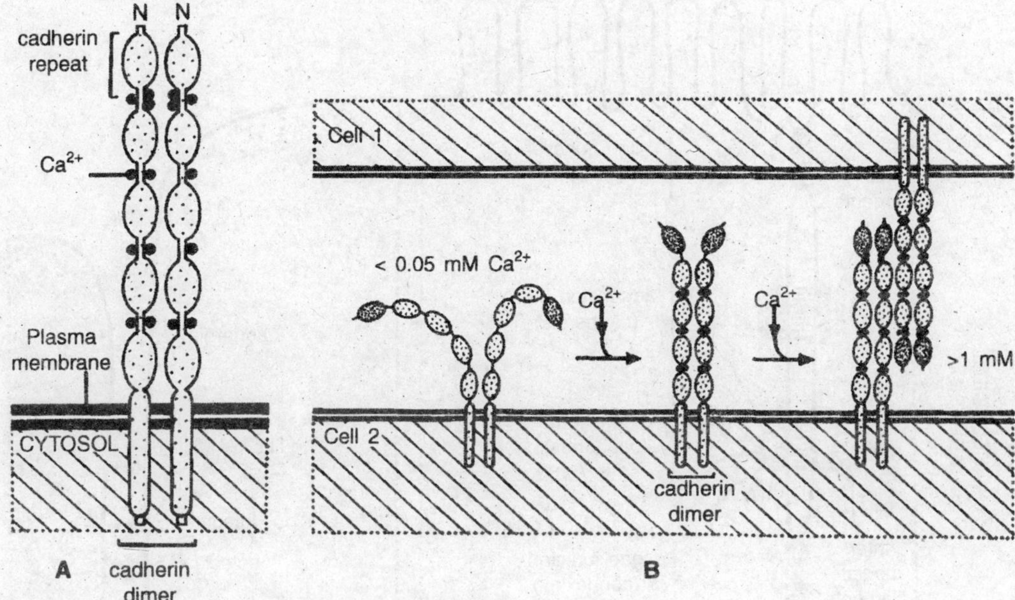

Fig. 60.11. The structure and function of cadherins. A—A classical cadherin molecule. The protein is a homodimer, with the extracellular part of each polypeptide folded into five cadherin repeats. There are Ca^{2+}-binding sites between each pair of repeats. B—The influence of extracellular Ca^{2+}. As the amount of Ca^{2+} increases, the extracellular parts of the cadherin chains become more rigid. When enough Ca^{2+} is bound, the cadherin dimer extends from the surface, where it can bind to a cadherin dimer on a neighboring cell. If Ca^{2+} is removed, the extracellular part of the protein becomes floppy and is degraded by proteolytic enzymes (after Alberts *et al.*, 2002).

Further, as the neural tube forms during the tabulation embryonic stage and pinches off from the overlying ectoderm, neural tube lose E-cadherin and acquire other cadherins, including N-cadherin, while the cells in the overlying ectoderm continue to express E-cadherin.

***(ii)* Role of cadherins in cell-cell adhesion.** How do cell-cell adhesion molecules such as the cadherins bind cells together? There can occur following three mechanisms (Fig. 60.13): 1. In **homophilic binding**, molecules on one cell bind to other molecules of the same kind on adjacent cells; 2. In **heterophilic binding**, the molecules on one cell bind to molecules of a different kind on adjacent cells; 3. In **linker-dependent binding**, cell surface receptors on adjacent cells are linked to one another by secreted multivalent linker molecules. Although all three mechanisms have been found to operate in animals, cadherins usually link cells by the homophilic mechanism.

***(iii)* Role of cadherins in their linkage to actin filaments.** Most cadherins, including all classical and some nonclassical ones, function as transmembrane adhesion proteins that indirectly link the actin cytoskeletons of the cell they join together. This arrangement occurs in adherens junctions (Fig. 60.9B). The highly conserved cytoplasmic tail of these cadherins interacts indirectly with actin filaments by means of a group of intracellular anchor proteins called **catenins** (Fig. 60.14). This interaction is essential for efficient cell-cell adhesion, as classical cadherins that lack their cytoplasmic domain cannot hold cells strongly together.

Some embryonic cells (*e.g.*, neural tube formation) are known to regulate the adhesive activity of their cadherins. This is done by phosphorylation of anchor protein.

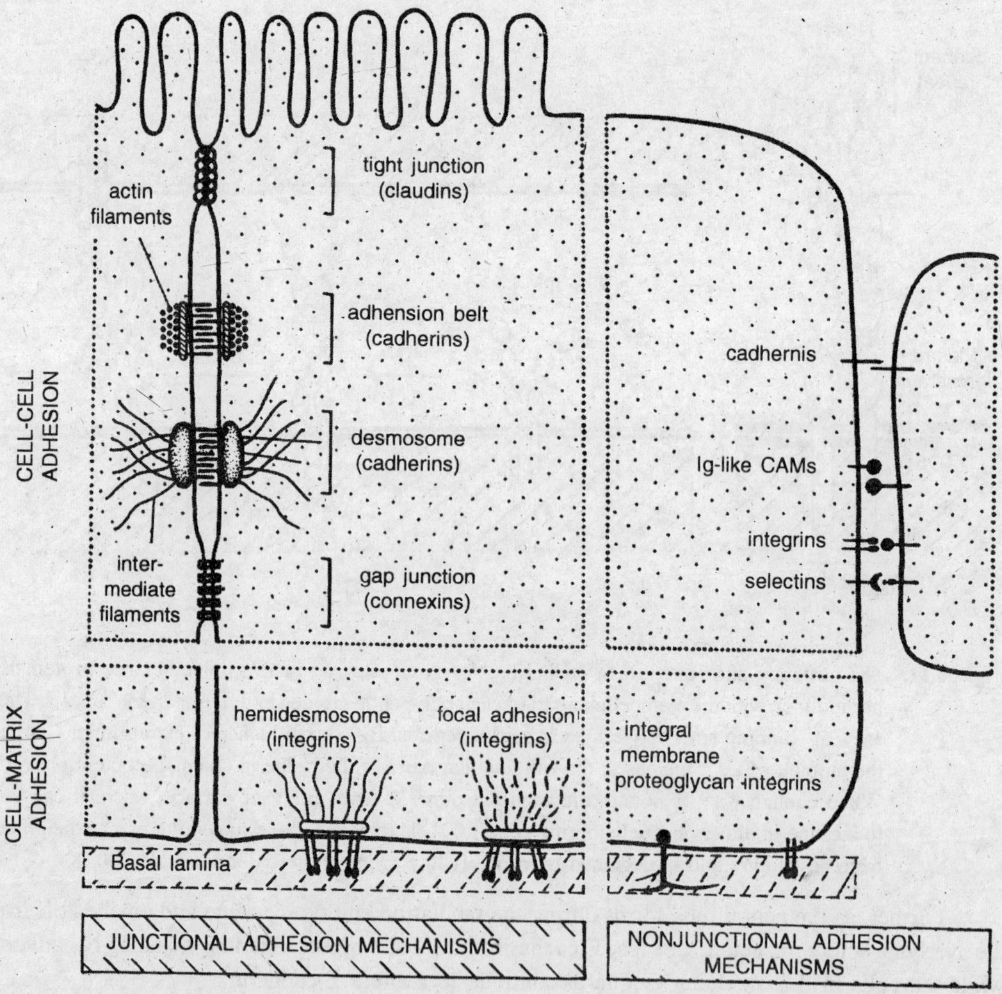

Fig. 60.12. A summary of the junctional and nonjunctional adhesive mechanisms used by animal cells in binding to one another and to the extracellular matrix (after Alberts *et al.*, 2002).

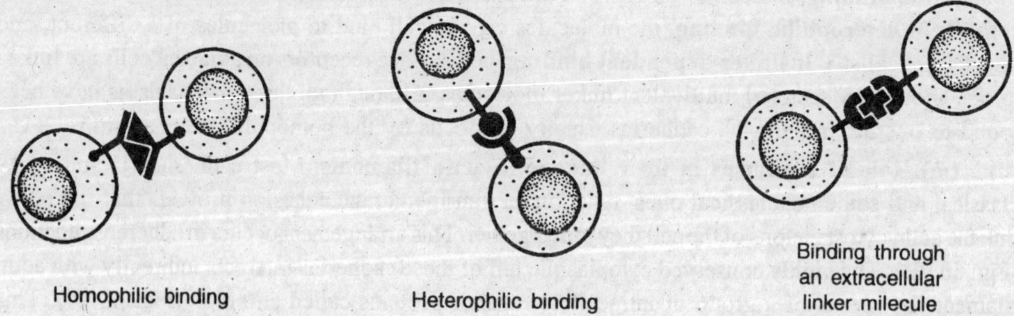

Fig. 60.13. Three possible mechanisms by which cell-surface molecules can mediate cell-cell adhesion (after Alberts *et al.*, 2002).

***(iv)* Transient cell-cell adhesions.** Leucocytes or white blood cells lead a nomadic life, moving to and fro between the blood-stream and the tissues, and this necessitates special adhesive properties. These properties depend on **selectins.** Selectins are cell surface carbohydrate-binding proteins (**lectins**) that mediate a variety of transient, Ca^{2+}-dependent, cell-cell adhesion interactions in the blood stream. These are of three types: 1. **L-selectin** on white blood cells, 2. **P-selectin** on blood platelets and on endothelial cells that have been locally activated by an inflammatory response, and 3. **E-selectin** on activated endothelial cells. Each selectin is a transmembrane protein with a highly conserved lectin domain that binds to a specific oligosaccharide on another cell (Fig. 60.15).

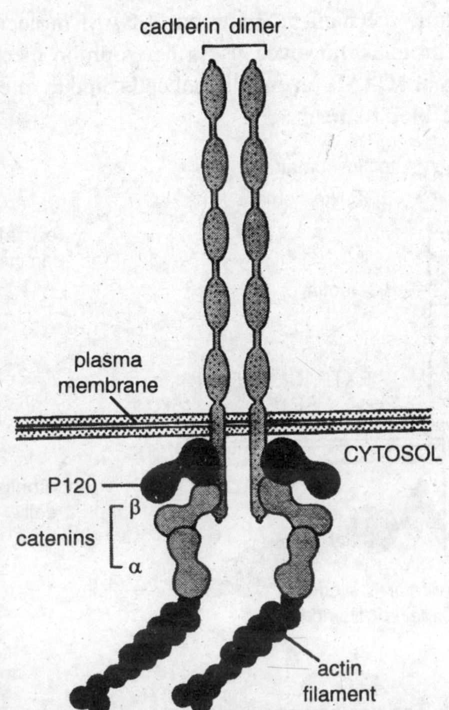

Fig. 60.14. The linkage of classical cadherins to actin filaments. The cadherins are coupled indirecty to actin filaments by the anchor protein α-catenin and β-catenin. A third intracellular protein, called p120, also binds to the cadherin cytoplasmic tail and regulates cadherin function (after Albert *et al.,* 2002).

Selectins have an important role in binding white blood cells to endothelial cells lining blood vessels, thereby enabling the blood cells to migrate out of the bloodstream into a tissue. In a lymphoid organ, the endothelial cells express oligosaccharides that are recognized by L-selectin on lymphocytes, causing the lymphocytes to hang about and become trapped. Conversely, at sites of inflammation, the endothelial cells switch on expression of selectins, flagging the cells down to help deal with the local emergency. Selectins do not act alone, however; they colloborate with *integrins,* which strengthen the binding of the blood cells to the endothelium. The cell-cell adhesions mediated by both selectin and integrins are heterophilic (Fig. 60.13): selectins bind to specific oligosaccharides on glycoproteins and glycolipids, while integrin bind to specific proteins.

Selectins and integrins act in sequence to let white blood cells leave the bloodstream and enter tissues. The selectins mediate a weak adhesion because the binding of the lectin domain of the selectin to its carbohydrate ligand is of low affinity. This allows the white blood cell to adhere weakly and reversibly to the endothelium, rolling along the surface of the blood vessel propelled by flow of blood. The rolling continues until WBC activates its integrins (Fig. 60.16), now causing the cell to bind strongly to the endothelial cell surface and to crawl out of the blood vessel between adjacent endothelial cells (Fig. 60.15).

2. Immunoglobin Superfamily.

Cadherins, selectins and integrins all depend on extra-cellular Ca^{2+} (or Mg^{2+} for some integrins) to function in cell adhesion. The molecules responsible for **Ca^{2+}-independent cell-cell adhesion** belong mainly to the large and ancient **immunoglobulin (Ig) superfamily** of proteins. These proteins contain one or more Ig-like domains that are characteristics of antibody molecules. One of the best-studied examples is the **neural cell adhesion molecule (N-CAM),** which is expressed by a variety of cell types, including most nerve cell. N-cam is the most common of the Ca^{2+}-independent cell-cell adhesion molecules in vertebrates, and, like cadherins, it is thought to bind cells together by a homophilic mechanism (between N-CAM molecules on adjacent cells). Some Ig-like cell-cell adhesion proteins, however, use a heterophilic mechanism. For example, **intercellular adhesion molecules** or **ICAMs** on endothelial cells bind to integrins on white blood cells when WBCs migrate out of the blood stream.

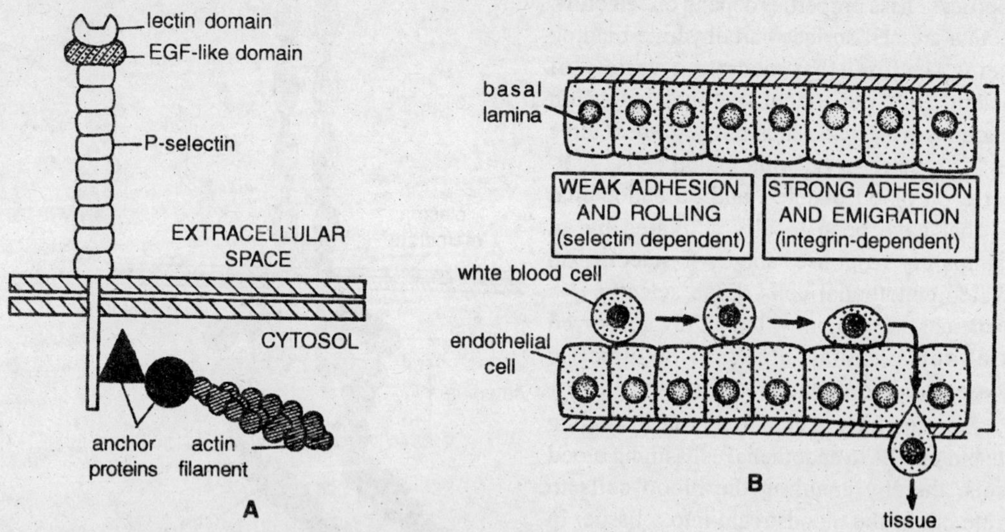

Fig. 60.15. The structure and function of selectins. A—The structure of P-selectin. The selectin attaches to the actin cytoskeleton through anchor proteins that are still poorly characterized. B—How selectins and integrins mediate the cell-cell adhesions required for a white blood cell to migrate out of the blood stream into a tissue (after Alberts *et al.*, 2002).

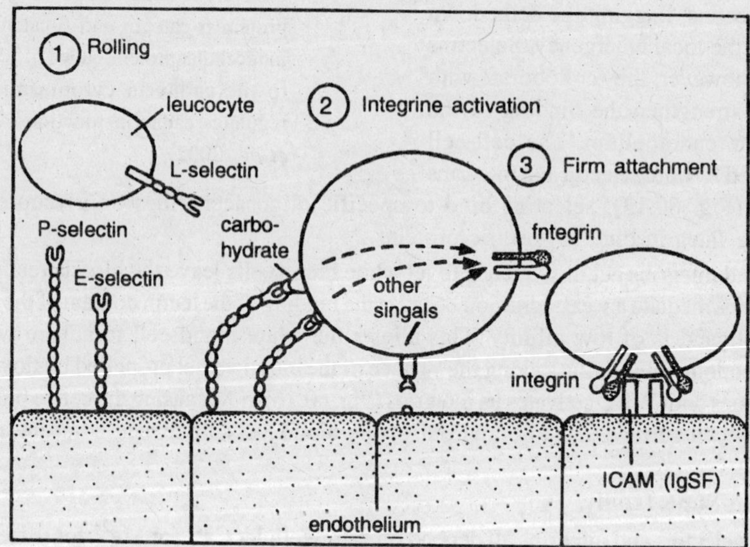

Fig. 60.16. Leucocyte adhesion and selectin. The initial attachment of leucocytes to endothelial cells that line blood vessels is mediated by selectins. (1) Attached leucocytes or white blood cells roll along the blood vessel wall, where they sense activating factors that are deposited on the surface of endothelial cells. (2) This leads to activation of leucocyte integrins, which (3) bind to IgSF proteins (such as ICAMs), allowing leucocytes to adhere firmly to endothelial cells. Such adhesion allows leucocytes to stop and to pass through blood vessels to sites of inflammation (after Becker *et al.*, 2007).

Types of N-CAM. These are about 20 forms of N-CAM, all generated by alternative splicing of an RNA transcript produced from a single gene. In all forms, the large extracellular part of the polypeptide chain is folded into five Ig-like domains (Fig. 60.17). Some forms of N-CAM carry on unusually large quantity of sialic acid. By virtue of their negative charge, these long polysialic acid chains prevent cell adhesion, and there is increasing evidence that N-CAM heavily loaded with sialic acid serves to prevent adhesion, rather than cause it.

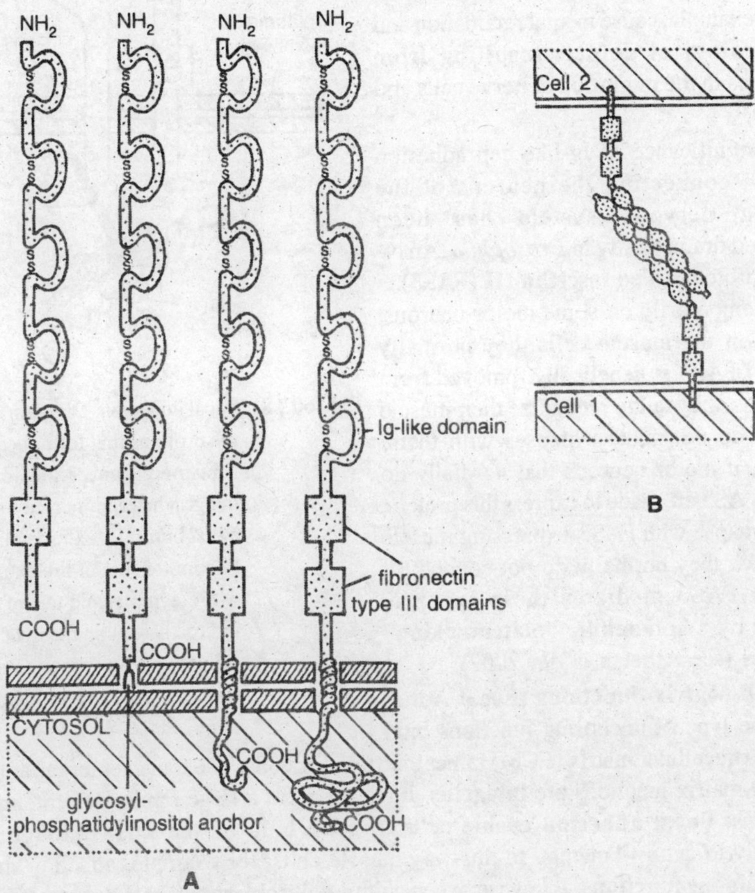

Fig. 60.17. The cell adhesion protein N-CAM. A—Four forms of N-CAM. The extracellular part of the polypeptide chain in each case is folded into five Ig-like domains (and one or two other domains called fibronectin type III repeats). Disulphide bonds (S S) connect the ends of each loop that forms an Ig-like domain. B—A model for the homophilic interaction that allows N-CAM to mediate cell-cell adhesion (after Alberts *et al.*, 2002).

While cadherins and Ig family members are frequently expressed on the same cells, the adhesions mediated by cadherins are much stronger, and they are largely responsible for holding cell together, segregating cell collectives into discrete tissues, and maintaining tissue integrity. N-CAM and other members of the Ig family seem to contribute more to the **fine-tuning** of these adhesive interactions during development and regeneration. For example, in the developing rodent pancreas, the formation of the **islets of Langerhans** requires cell aggregation, followed by cell sorting. Whereas inhibition of cadherin function prevent cell aggregation and islet formation, loss of N-CAM only impairs the cell sorting process, so that disorganized islets forms.

Likewise, whereas mutant mice that lack N-cadherin die early in development, mutant mice that lack N-CAM develop normally, although they do have some defects in neural development. Mutations in other genes that encode Ig-like cell adhesion proteins, however, can cause more severe neural defects. *L1* gene mutations in human, for example, cause mental retardation and other neurological defects resulting from abnormalities in the migration of nerve cells and their axons.

The significance of Ig-like cell adhesion proteins in connecting the neurons of the developing nervous system has been demonstrated dramatically in *Drosophila*. An N-CAM-like protein called **fasciclin III (FAS3)** is expressed temporarily on some motor neurons, as well as on the muscle cells they normally innervate. If FAS3 is genetically removed from these neurons, they fail to recognize their muscle targets and do not make synapses with them. Conversely, if motor neurons that normally do not express FAS3 are made to express this protein, they now synapse with FAS3-expressing muscle cells to which they normally do not connect. It seems that FAS3 mediates these synaptic connections by a **homophilic "matchmaking" mechanisms** (see **Alberts** *et al.*, 2002).

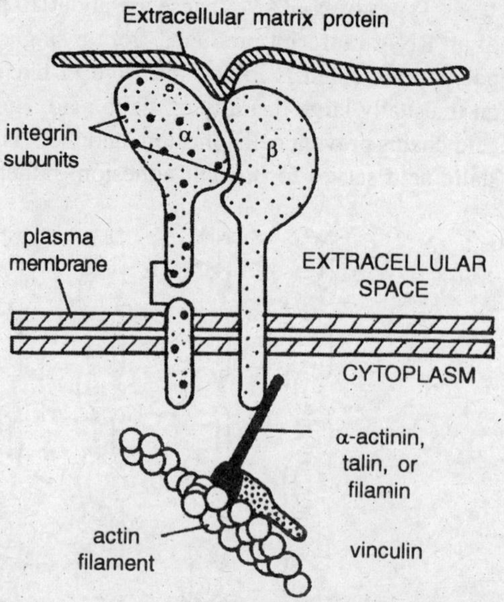

Fig. 60.18. Focal adhesion. Diagram showing some of the proteins that form focal adhesions. The transmembrane adhesion protein is an integrin heterodimer, composed of an α and a β subunit. Its extracellular domains bind to components of the extracellular matrix, while cytoplasmic tail of β subunit binds indirectly to actin filament via several intra-cellular anchor proteins such as vinculin and α-actinin, talin or filamin (after Alberts, *el al.*, 2002).

2. Cell-Matrix Junctions (Focal Adhesions). Some type of anchoring junctions bind cells to the extracellular matrix (ECM) rather than to other cells. The transmembrane adhesion proteins in these cell-matrix junctions are **integrins**. Integrins form a large family of proteins distinct from the cadherins. **Focal adhesion** enable cells to get a hold on the ECM through integrins that intracellularly to actin filaments. In this way muscle cells, for example, attach to their tendons at the **myotendinous junctions**. Likewise, when cultured fibroblasts migrate on an artificial substratum coated with extracellular matrix (ECM) molecules, they also grip the substratum as focal adhesions, where bundles of actin filaments terminate. At all such adhesions, the extra-cellular domains of transmembrane integrin protein bind to a protein component of the extracellular matrix, while their intracellular domains bind indirectly to bundles of actin filaments via the intracellular anchor protein **talin, α-actinin, filamin** and **vinculin** (Fig. 60.18).

II. Intermediate Filament Attachment Sites of Anchoring Junctions

1. Desmosomes (or Maculae adherens). Desmosomes are disc-shaped adhesive junctions approximately 1 μm in diameter having a narrow (30 nm) extracellular gap. They are particularly numerous in tissues that are subjected to mechanical stress, such as cardiac muscle and the epithelial layers of the skin and uterine cervix (see **Karp**, 2002). In a way, desmosomes are buttonlike points of intercellular contact that rivet cells together (Fig. 60.19). Inside the cell, desmosomes serve as anchoring sites for ropelike **intermediate filaments (IFs)**, which form a structural framework of great tensile strength (Fig. 60.19A). Through desmosomes, the intermediate filaments of adjacent cells are linked into a net that extends throughout the many cells of a tissue. The particular type of intermediate filaments attached to the desmosomes depends on the cell type: for example, they are **keratin filaments** in most epithelial cells and **desmin filaments** in heart muscle cells.

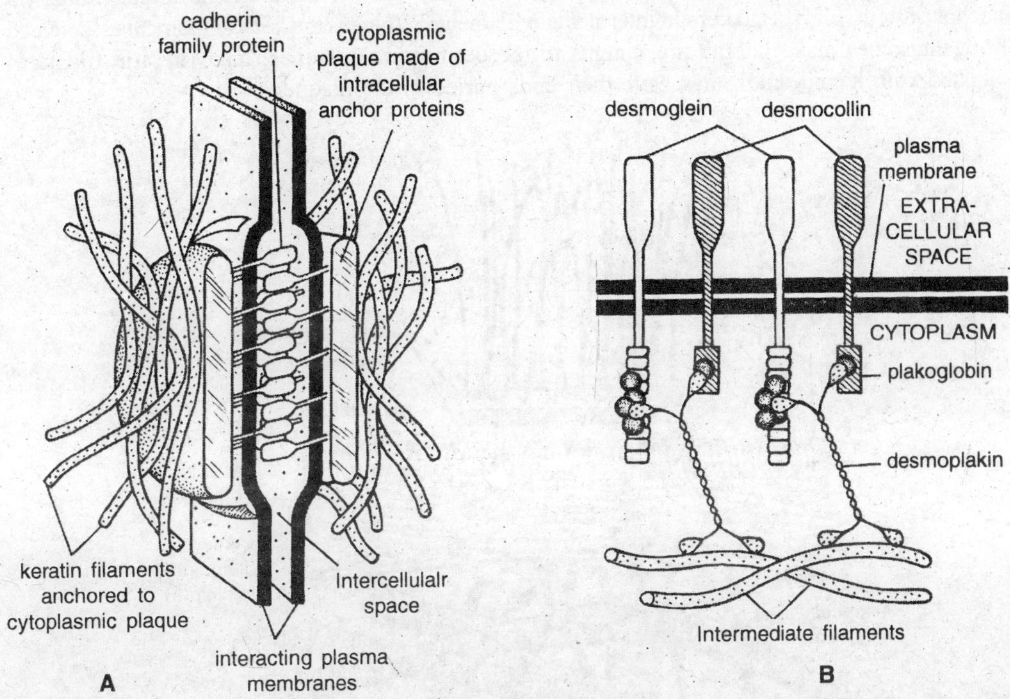

Fig. 60.19. Desmosomes. A—The structural component of a desmosome. On the cytoplasmic surface of each interacting plasma membrane is a dense plaque composed of a mixture of intracellular anchor proteins. A bundle of keratin/desmin intermediate filaments is attached to the surface of each plaque. Transmembrane adhesion proteins of the cadherin family bind to the plaques and interact through their extracellular domains to hold the adjacent membranes together by a Ca^{2+}-dependent mechanism. B—Some of the molecular components of a desmosome. Desmoglein and desmocollin are members of the cadherin family. Their cytoplasmic tails bind plakoglobin, which in turn binds to desmoplakin. Desmoplakin also binds to the sides of intermediate filaments, thereby tying the desmosome to these filaments (after Alberts *et al.*, 2002).

The desmosome has a dense cytoplasmic plaque composed of a complex of intracellular anchor proteins (**plakoglobin** and **desmoplakin**) that are responsible for connecting the cytoskeleton to the transmembrane adhesion proteins (Fig. 60.19B). These adhesion proteins (**desmoglein** and **desmocollin**) like those at an adherens junction belong to the cadherin family. They interact through their extracellular domains to hold the adjacent plasma membrane together.

The importance of desmosome junctions is demonstrated by some forms of the potentially fatal autoimmune skin disease called **pemphigus vulgaris** of human beings. Affected individuals make antibodies against one of their own desmosomal cadherin protein (desmogleins). These antibodies bind to and disrupt the desmosomes that hold their skin epithelial cells (keratinocytes) together. This results in a severe blistering of the skin, with leakage of body fluids into the loosened epithelium.

2. Hemidesmosomes. Hemidesmosomes or half-desmosomes resemble desmosomes morphologically and in connecting to intermediate filaments and, like desmosomes, they act as rivets to distribute tensile and shearing forces through an epithelium (Box 60.1). Instead of joining adjacent epithelial cells, however, hemidesmosomes connect the basal surface of an epithelial cell to the underlying basal lamina (Fig. 60.20). The extracellular domains of the $\alpha_6\beta_4$ **integrins** that mediate the adhesion bind to a **laminin** protein in the basal lamina, while an intracellular domain binds via an anchor protein (**plectin**) to keratin intermediate filaments. Whereas the keratin filaments associated with desmosomes make lateral attachments to the desmosomal plaques, many keratin filaments associated with hemidesmosomes have their ends buried in the plaque.

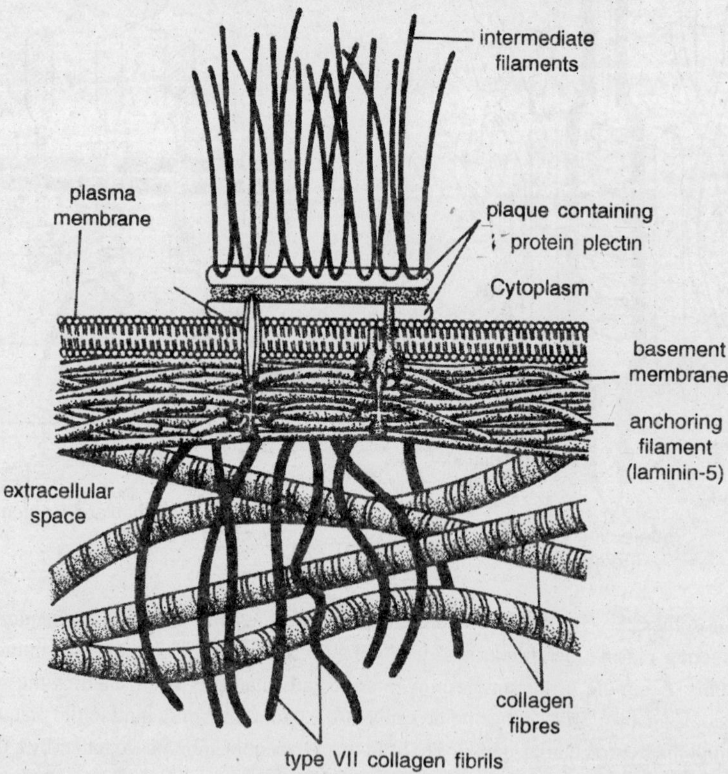

Fig. 60.20. Hemidesmosomes are differentiated sites at the basal surfaces of epithelial cells where the cells are attached to the underlying basement membrane. Schematic diagram showing the major components of a hemidesmosome connecting the epidermis to the underlying dermis. The $\alpha_6\beta_4$ integrin molecules of the epidermal cells are linked to cytoplasmic keratin intermediate filaments by a protein called plectin that is present in plaque and to the basement membrane by anchoring filaments of a particular type of laminin. A second transmembrane protein (BP180) is also present in hemidesmosomes. The collagen fibers are part of the underlying dermis (after Karp, 2003).

Box 60.1

Within body where occur **tightest attachment** between a cell and its extracellular matrix is seen at the basal surface of epithelial cells where the cells are anchored to the underlying basement membrane by a specialised adhesive structure called the **hemidesmosome**.

In addition, another transmembrane protein called BPAG2* and its associated plakin, called BPAG1 can serve as bridge between keratin and laminin (Fig. 60.21). (* BPAG stands for *b*ullous *p*emphigoid *a*ntigen).

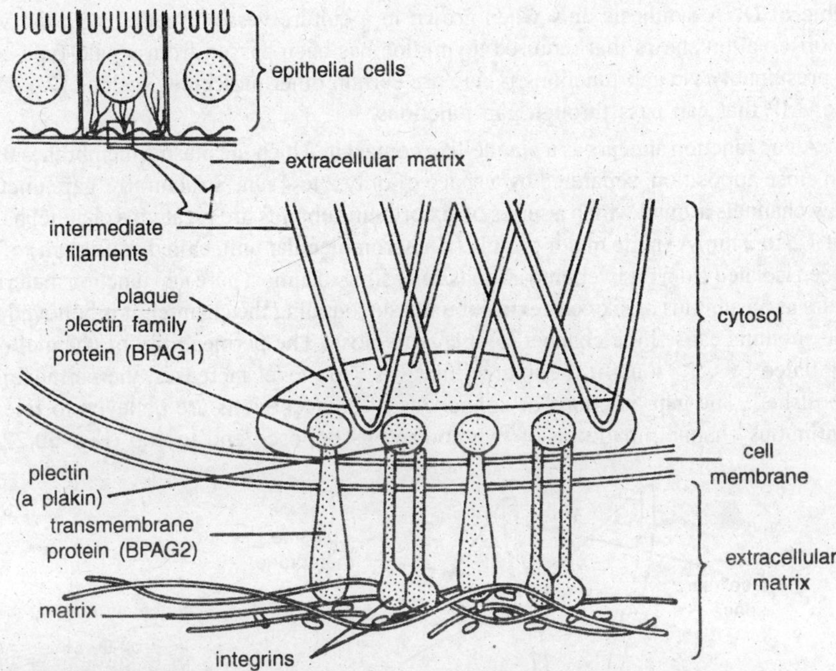

Fig. 60.21. Ultrastructure of a hemidesmosome (after Becker *et al.*, 2007).

The importance of hemidesmosomes is revealed by a rare disease of elderly people, called **bullous pemphigoid**, in which individuals produce antibodies that bind to proteins (the bullous pemphigoid antigens) present in these adhesive structures. Disease is caused by production of antibodies directed against one's own tissues (*i.e.*, autoantibodies), are called **autoimmune disorders** and are responsible for a wide variety of conditions. The presence of autoantibodies causes the lower layer of the epidermis to lose attachment to the underlying basement membrane (and thus to the underlying connective tissue layer of the dermis). The leakage of fluid into the space beneath the epidermis results in severe blistering of the skin (**Karp**, 2002). A similar inherited blistering disease, called **junctional epidermolysis bullosa**, can occur in patients with mutation in β_4 subunit of integrin of the hemidesmosome (see **Becker** *et al.*, 2007).

C. Communicating Junctions

1. Gap Junctions. Many cells of the tissues of higher animals are coupled together by interconnecting **gap junctions, nexus** or **communicating junctions**. The presence of gap junctions explains the ionic or electronic connections between adjacent cells, *i.e.*, there are some cells which are **electrically coupled** and have regions of low resistance in the membrane through which there is a rather free flow of electrical current carried by ions. Such electrical coupling is found extensively in embryonic cells. In adult tissues it is usually found in epithelia, cardiac cells and liver cells. Skeletal muscles and most neurons do not show electrical coupling.

Gap junctions are found to permit molecules such as inorganic ions, sugars, amino acids, nucleotides and vitamins to pass with comparative freedom between one cell and another within a tissue, but they prevent larger molecules, such as proteins, nucleic acids and polysaccharides from being transferred. This observation also explains the phenomenon of **metabolic cooperation** or **metabolic coupling** between cells, *i.e.*, cells can transfer to neighbouring cells, the molecules which cannot be synthesized by the recipient cells. For example, in the tissue-culture experiments, the mutant cells which are deficient in the enzyme **thymidine kinase** can be shown by autoradiography to be capable of DNA synthesis only when grown in a culture vessel together with the wild type cells. This observation shows that required thymidine has been passed from a wild-type cell to a mutant cell, presumably via gap junctions. There are certain other molecules such as IP_3, AMP, ADP, ATP and cAMP that can pass through gap junctions.

A gap junction appears as a plaque-like contact in which the plasma membranes of adjacent cells are in close apposition, separated by a space of only 2 to 4 nm. Structurally, gap junctions consist of hollow channels around which a series of six protein subunits are located; a channel has a diameter of about 1.5 to 2 nm. A single major protein (a macromolecular unit, called **connexon**) of 27000 daltons has been isolated from rat liver preparations consisting of almost pure gap junction material. A connexon appears as an annulus of six **connexin** subunits surrounding the channel. It is believed that the sliding of the subunits caused the channel to open and close. The permeability of channel of gap junction is regulated by Ca^{2+} ions; if the intercellular Ca^{2+} ion level increases, the permeability is reduced or abolished. The gap junctions or connexons of adjacent cells are believed to line up to provide a continuous channel, made up of two connexons opposed end to end (Fig. 60.22).

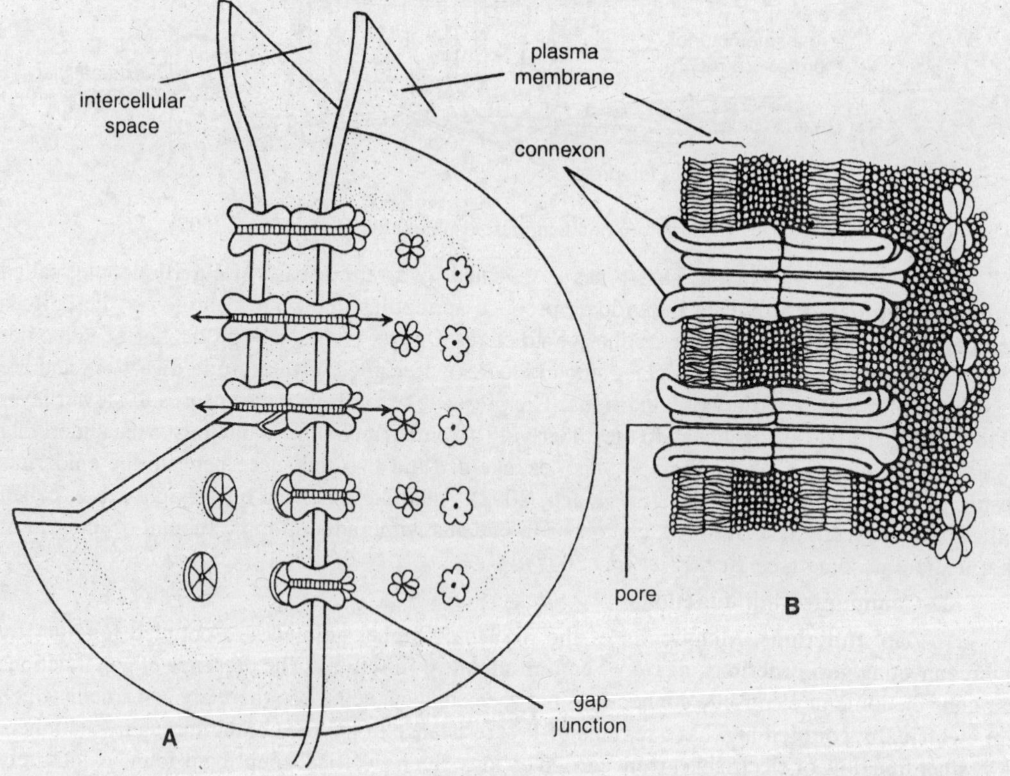

Fig. 60.22. Gap junctions. A—Location of gap junctions between two cells. The channels are made by particles in each membrane that traverse the intercellular space. The flow of fluid between the cells is indicated by arrows. B—Finer structure of unit structure or connexion of gap junctions. The channel has a pore about 2 nm and is formed by two hexamers (six subunits) traversing the lipid bilayers of two plasma membranes.

Gap junctions in different tissues can have different properties. The permeability of their individual channels can vary, reflecting differences in the connexins that form the junctions. In humans, for example, there are 14 distinct connexins, each encoded by a separate gene and each having a distinctive, but sometimes over-lapping, tissue distribution. Most cell types express more than one type of connexin, and two different connexin protein can assemble into a **heteromeric connexon,** the properties of which differ from those of a **homomeric connexon** (Fig. 60.23) constructed from a single type of connexin.

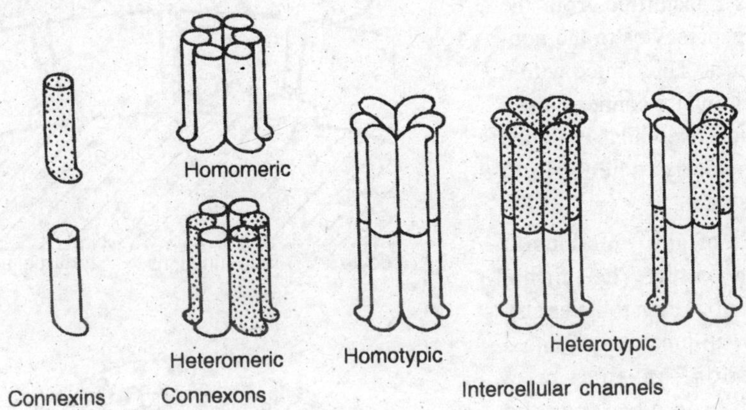

Fig. 60.23 The organization of connexins into connexons and connexons into inter-cellular channels. The connexons can be homomeric or heteromeric, and the intercellular channels can be homotypic or heterotypic (after Albert *et al.*, 2002).

Innexins. When connexin mRNA is injected into frog oocyte or gap-junctions-deficient cells, channels with the properties expected of gap-junction channels can be demonstrated electrophysiologically where pairs of injected cells make contact. The mRNA injection approach has been useful for identifying new gap-junction proteins. Genetic studies in the fruit fly *Drosophila* identified the gene *shaking B*, when mutated, resulted in flies that failed to jump in response to a visual stimulus. Although these flies had defective gap junctions, the sequence of shaking B protein did not resemble a connexin, and the function of the protein was unclear. An injection of the shaking B mRNA into frog oocytes, however, led to the formation of functional gap-junctional channels, just-like those formed by connexins. Shaking B thus became first member of a new family of invertebrate gap-junction proteins called **innexins.** There are more than 15 innexin genes in *Drosophila* and 25 in the nematode *C.elegans* (see **Alberts** *et al.*, 2002).

Functions of gap junctions. With the exception of a few terminally differentiated cells such as skeletal muscle cells and blood cells, most cells in animal tissues are in communication with their neighbour via gap junctions. Gap junctions occur in most vertebrate and invertebrate cell types. They are especially abundant in tissues such as muscle and nerve where extremely rapid communication between cells is required. In heart tissue, for example, gap junctions facilitate the flow of electrical current that causes the heart to beat; in the brain; they they are concentrated in the cerebellum, which is involved in coordinating rapid nuclear (*i.e.,* gray matter related) activities. These roles are confirmed by analyzing mutations in connexins and in nexins. For example, mice lacking in one type of connexin have defects in conducting electrical impulses in the heart.

Role of gap junctions in the liver. Gap junctions also occur in many tissues that do not contain electrically excitable cells. In principle, the sharing of small metabolites and ions provides a mechanism for coordinating the activities of individual cells in such tissues and for smoothing out random fluctuations in small molecule concentrations in different cells. For example, in the liver, the release of noradrenaline from sympathetic nerve endings in response to a fall in blood glucose levels stimulates hepatocytes to increase glycogen break-down and release of glucose into blood.

Not all liver cells (hepatocytes) are innervated by sympathetic nerves, however. By means of the gap junctions that connect hepatocytes, the signal is transmitted from the inner-vated hepatocytes to the non-innervated ones. Thus, mice with a mutation in the major connexin gene expressed in the liver fail to mobilize glucose normally when blood glucose levels fall.

2. Role of gap junctions in mammalian ovary. The normal development of ovarian follicles also depends on gap-junction-mediated communication—in this case, between the oocyte and the surrounding granulosa cell. A mutation in the gene that encodes the connexin that normally couples these two cell types causes **infertility** (Fig. 60.24).

3. Role of gap junctions in embryogenesis. Cell coupling via gap junctions also seem to be important in embryogenesis. In early vertebrate embryos, beginning with the late eight call stage in mouse embryos, most cells are electrically coupled to one another. As specific groups of cells in the embryo develop their distinct identities and begin to *differtiate*, they commonly uncouple from surrounding tissue. As the neural plate folds up and pinches off to form the neural tube (Fig. 60.10) for example, its cells uncouple from the overlying epidermal ectoderm. Meanwhile, the cells within each group remain coupled with one another and therefore tend to behave a cooperative assembly all following a similar developmental pathway in a coordinated fashion.

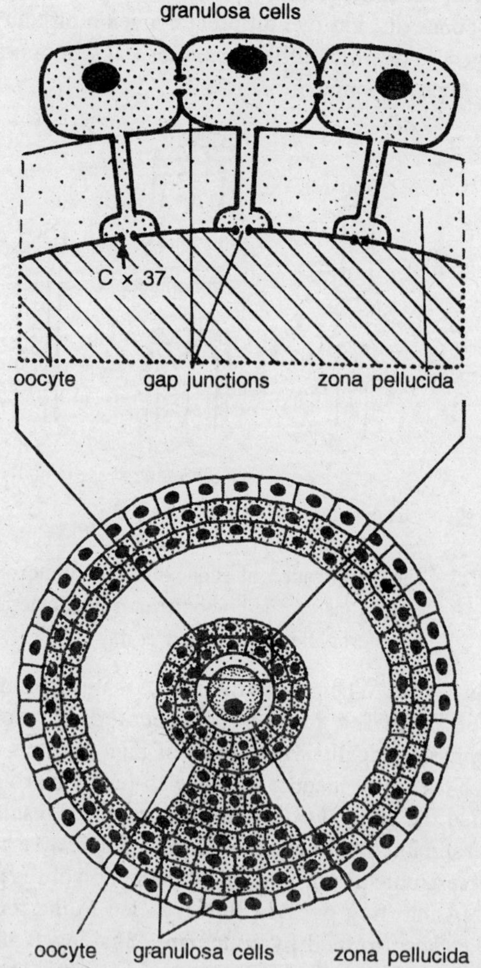

Fig. 60.24. Gap junction coupling in the ovarian follicle. The mammalian ovum is surrounding by a thick layer of extracellular or matrix called the zona pellucida. The surrounding granulose cells are coupled to each other by gap junctions formed by protein connexin 43 (*Cx* 43). In addition, the granulosa cells extend processes through the zona pellucida and make gap junctions with the oocyte. These gap junctions contain a different connexin protein, called Cx 37. Mulation in the genc cncoding Cx 37 causes infertility by disrupting the development of both the grannlosa cells and the oocyte (after Alberts *et. al.*, 2002).

Distinctions between Gap junctions, Tight junctions and Desmosomes

Table 60.3 Differences between gap junctions, tight junctions and desmosomes.

Gap junctions	*Tight junctions*	*Desmosomes*
1. Have channels or pores through the two plasma membranes of adjacent cells and across the inter-cellular space.	1. In them there occurs a fusion of two adjacent cell membranes.	1. They contain buttonlike electron-dense regions in each of the opposing cell membranes.
2. Tonofibrils are absent.	2. Tonofibrils are absent.	2. Tonofibrils are present.
3. Inter-celluar space is narrowed to 5 nm.	3. Inter-cellular space is absent as the two adjacent membranes fuse.	3. Inter-cellular space 30 to 35 nm.
4. Provide electrical communication between cells and for flow of ions.	4. Form a complete barrier between the cavities.	4. Hold cells tightly together.

60.2 EXTRACELLULAR MATRIX (ECM) OF ANIMAL CELLS

Tissues are not made up exclusively of cell. An ample part of their volume is **extracellular space,** which is mostly filled by a complicated network of macromolecules forming the **extracellular matrix (ECM)**. This matrix is composed of a variety of proteins and polysaccharides that are secreted locally and assembled into an organized meshwork in close association with the surface of the cell that produced them.

Whereas the cell junctions have been discussed mainly in the context of epithelial tissues, this account of the extracellular matrix concentrates on connective tissues (Fig. 60.25). The extracellular matrix in connective tissue is commonly more plentiful than the cell it surrounds, and it determines the tissue's physical properties. In fact, connective tissues form the framework of the vertebrate body, but the amounts found in different organs vary greatly-from cartilage to bone, in which they are the major component, to brain and spinal cord, in which they are minor component.

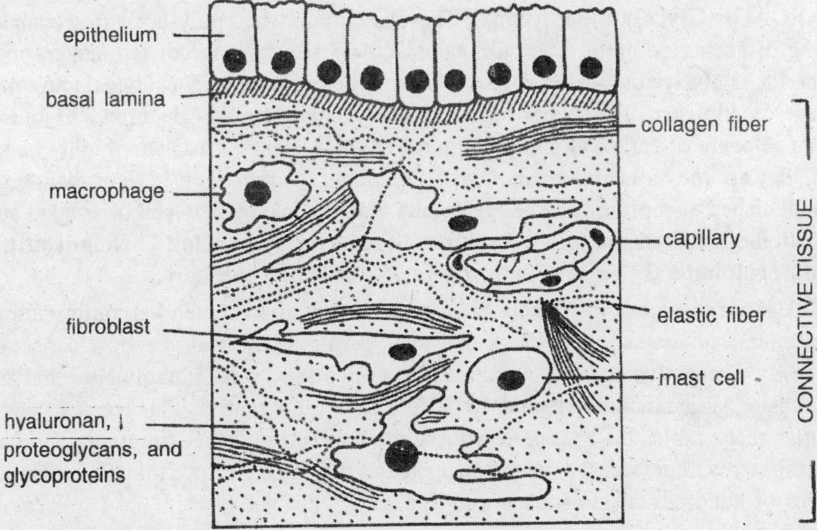

Fig. 60.25. The connective tissue underlying an epithelium. This tissue contains a variety of cells and ECM components. The predominant cell type is the fibroblasts, which secretes abundant extracellular matrix (after Alberts *et al.*, 2002).

Variations in the relative amounts of the different types of matrix macromolecules and the way in which they are organised in the ECM give rise to a surprising diversity of forms, each adapted to the functional requirements of the particular tissue. For example, matrix can become calcified to form the rock-hard structures of bone or teeth, or it can form the transparent matrix of the cornea, or it can adopt the ropelike organization that gives tendons their massive tensile strength. At the interface between an epithelium and connective tissue, the matrix forms a **basal lamina,** which is important in controlling cell behaviour.

Once it was believed that ECM of vertebrates serve mainly as a relatively inert scaffold to stabilize the physical structure of tissue. But now, it is well established that the matrix has a far more active and complex role in regulating the behaviour of the cells that contact it, influencing their survival, development, migration, proliferation, shape and function.

The origins of the ECM are very ancient and virtually all multicellular organisms make it; examples include the cuticles of worms and insects, the shell of mollusks and the cell walls of plants.

1. Cells and Molecules of ECM. The macromolecules that constitute the ECM are mainly produced locally by cells in the matrix. In most connective tissues, for example the matrix macromolecules are secreted largely by cells called **fibroblasts.** In certain specialized types of connective tissues, such as cartilage and bone, however, they are secreted by cells of the fibroblast family that have more specific names: **chondroblasts** form cartilage and **osteoblasts** form bone.

Two main classes of extracellular macromolecules make up the matrix: 1. polysaccharide chains of the class called **glycosaminoglycans (GAGs)**, which are usually found covalently linked to protein in the form of **proteoglycans;** and 2. **fibrous proteins** including **collagen, elastin, fibronectin** and **laminin**, which have both structural and adhesive functions.

The proteoglycan molecules in connective tissue form a highly hydrated, gel-like **"ground substance"** in which fibrous proteins are embedded. The polysaccharide gel resists compressive forces on the matrix while permitting the rapid diffusion of nutrients, metabolites and hormones between the blood and the tissue cells. The collagen fibres both strengthen and help organize the matrix, and rubber-like elastic fibers give it resilience. Finally, many matrix proteins help cells attach in the appropriate locations.

***(i)* GAGs (Glycosaminoglycans).** These are unbranched polysaccharide chains composed of repeating disaccharide units. They are called GAGs because one of two sugars in the repeating disaccharide is always on **amino sugar** (*N*-acetylglucosamine or *N*-acetylgalactosamine), which in most cases is sulphated. The second sugar is usually a **uronic acid** (glucuronic or iduronic). Because there are sulphate or carboxyl groups on most of their sugars, GAGs are highly negative charged. Indeed, they are *the most anionic molecules produced by animal cells.* Four main groups of GAGs are distinguished according to their sugars, the type of linkage between the sugars and the number and location of sulphate groups: 1. **hyaluronan** (or hyaluronic acid); 2. **chondroitin sulphate** and **dermatan sulphate**; 3. **heparan sulphate** and 4. **keratin sulphate.**

GAGs are strongly hydrophilic. They tend to adopt highly extended conformations that occupy a huge volume relative to their mass, and they form **gels** even at very low concentrations. Their high density of negative charges attracts a cloud of cations, most notably Na^+, that are osmotically active causing large amounts of water to be sucked into the matrix. This creates a swelling pressure or **turgor**, that enables the matrix to withstand compressive forces (in contrast to collagen fibrils, which resist stretching forces). The cartilage, matrix that lines the knee joint, for example, can support pressures of hundreds of atmospheres in this way.

The GAGs in connective tissue usually constitute less than 10% of the weight of the fibrous proteins. But, because they form porous hydrated gels, the GAG chains fill most of the extracellular space, providing mechanical support to the tissue. In some rare human genetic disease, there is a severe deficiency in the synthesis of the **dermatan sulphate** disaccharide (Fig. 60.26). The affected individuals have a short stature, permanently aged appearance and generalized defects in their skin. joints, muscles and bones.

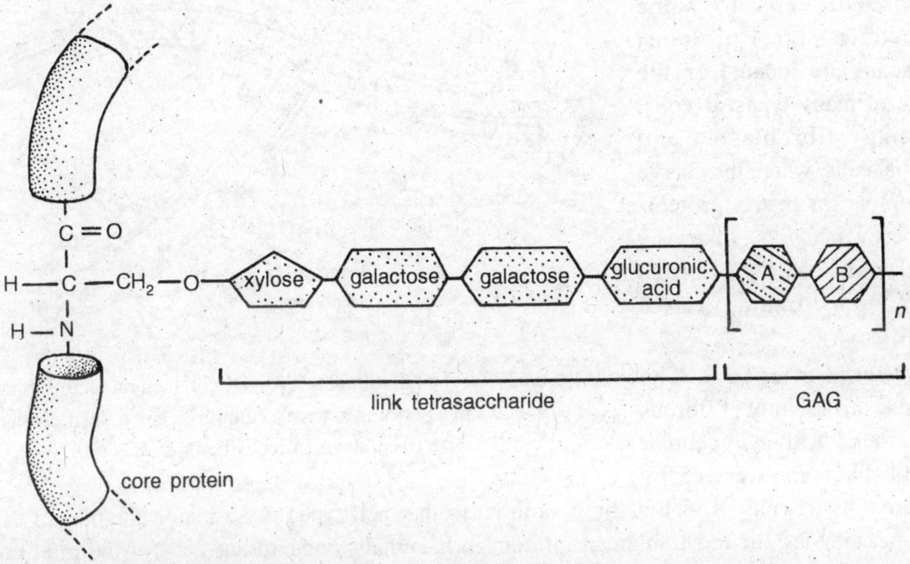

Fig. 60.26. The linkage between a GAG chain and its core protein in a proteoglycan molecule. A specific link tetrasaccharide is first assembled on a serine side chain. The rest of the GAG chain, consisting of mainly of a repeating disaccharide unit is then synthesized, with one sugar being added at a time (after Alberts *et al.*, 2002).

***(ii)* Proteoglycans.** Except for hyaluronan, all GAGs are found covalently attached to proteins in the form of **proteoglycans**, which are made by most animal cells. The polypeptide chain, called **core protein**, of a proteoglycan is made on membrane-bound ribosome and threaded in the lumen of the endoplasmic reticulum. The polysaccharide chains are mainly assembled on this core protein in the Golgi apparatus. First a special link **tetrasaccharide** is attached to a serine side chain on the core protein to serve as a primer for polysaccharide growth; then one sugar at a time is added by specific glycosyl transferase (enzyme) (Fig. 60.27). While still in the Golgi apparatus, many of the polymerized sugars are covalently modified by a sequential and coordinated series of reactions.

The GAG chains of proteoglycans can form gels of varying pore size and charge density; so, they serve as **selective sieves** to regulate the traffic of molecules and cells according to their size, charge, or both. Evidence suggests that a heparan sulphate proteoglycan called **perlecan** has this role in the basal lamina of the kidney glomerulus, which filters molecules passing into the urine from the blood stream.

Proteoglycans are thought to have a major role in chemical signalling between cell. They bind various secreted signal molecules, such as certain protein growth factors and can enhance on inhibit their signaling activity.

Proteoglycans as coreceptors. Not all proteoglycans are secreted components of the ECM. Some (*e.g.*, syndecans, betaglycans, etc.) are integral components of plasma membranes and have their core protein either inserted across the lipid bilayer or attached to the lipid bilayer by a GPI (*i.e.*, glycosylphosphatidylinositol) anchor. Some of these plasma membrane proteoglycans act as

coreceptors that collaborate with conventional cell-surface receptor proteins, in both binding cells to the ECM and initiating the response of cells to some extracellular signal proteins. **Syndecans** are located on the surface of many types of cells, including fibroblasts and epithelial cells, where they serve as receptors for matrix proteins (fibronectin, integrins). Likewise, **betaglycan** binds TGF-β (transforming growth factor-β).

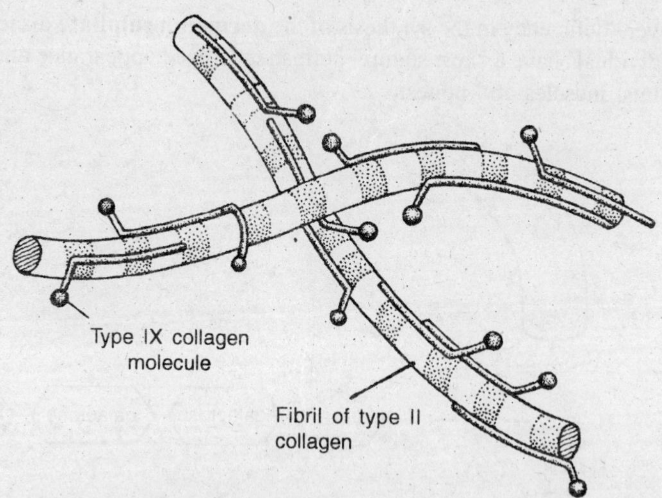

Fig. 60.27. Type IX collagen. Type IX collagen molecules binding in a periodic pattern to the surface of a fibril containing type II collagen (after Alberts *et al.*, 2002).

(iii) **Collagen.** The **collagens** are a family of fibrous protein found in all multicellular animals. They are secreted by connective tissue cells as well as by a variety of other cell types. As a major component of skin and bone, they are the most abundant protein in mammals, constituting 25% of the total protein mass in these animals (see Chapter 57 for more details).

The **fibrillar collagens** (type I, II, III, V and XI collagens) are ropelike, triple-stranded helical molecules that aggregate into long fibrils in the extracellular space. The fibrils in turn can assemble into a variety of highly ordered arrays. **Fibril-associated collagen** molecules, such as type IX and XII, decorate the surface of collagen fibrils and influence the interactions of the fibrils with one another and with other matrix components (Fig. 60.28). In contrast, **type IV collagen** molecule assemble into a sheetlike meshwork that is a crucial component of all mature basal laminae.

(*iv*) Elastin. Many vertebrate tissues, such as skin, blood vessels and lungs, need to be both strong and elastic in order to function. A network of **elastic fibers** in the extra-cellular matrix of these tissues gives them the required resilience so that they can recoil after transient stretch. Elastic fibers are at least five times more extensible than a rubber band of the same cross-sectional area. Long, inelastic collagen fibrils are interwoven with the elastic fibers to limit the extent of stretching and prevent the tissues from tearing.

The main component of elastic fibers is **elastin,** a highly hydrobhobic protein (about 750 amino acid long), which like collagen, is unusually rich in proline and glycine, but unlike collagen, is not glycosylated and contain some hydroxy-proline but no hydroxylysine. The biosynthetic precursor of elastin is soluble **tropoelastin** which is secreted into the extracellular space and assembled into elastic fibers close to the plasma membrane, generally in cell-surface infoldings. After secretion, the tropoelastin molecules, generating an extensive network of elastin fibers and sheets. The cross-links are formed between lysines.

The elastin protein is composed mostly of two types of short segments that alternate along the polypeptide chain: hydrophobic segments, which are responsible for the elastic properties of the molecule; and alanine-and lysine-rich α-helical segments, which form cross-links between adjacent molecules. Each segment is encoded by a separate exon of a split gene. According to one popular model, the elastin polypeptide chain, like the polymer chains in ordinary rubber, adopts a loose "random coil" conformation, and it is the random coil structure of the component molecules cross-linked into the elastic fiber network that allows the network to stretch and recoil like a rubber band (Fig. 60.28).

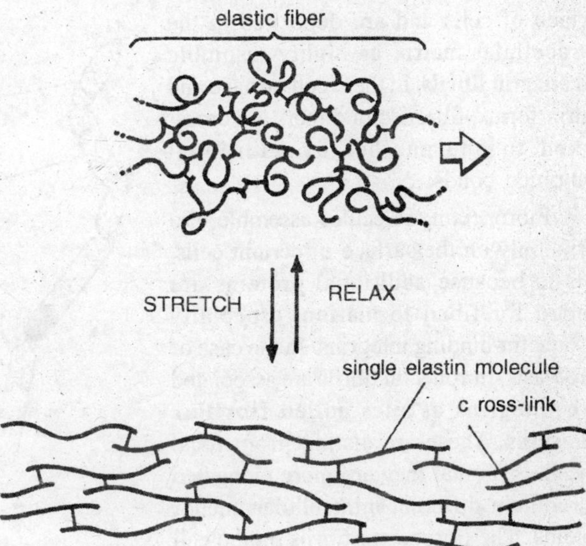

Fig. 60.28. Stretching a network of elastin molecule. The molecules are joined together by covalent b onds to generate a cross-linked network. Eac. elastin molecule in the network can expand and contract as a random coil.

(v) Fibronectin. Fibronectin and laminin are examples of large, multidomain matrix glycoproteins. By means of their multiple binding domains, such proteins help organize the matrix and help cells adhere to it.

Fibronectin protein is found in all vertebrates. It is a dimmer composed of two very large subunits joined by disulphide bonds at one end. Each subunit is folded into a series of functionally distinct domains separated by regions of flexible polypeptide chain (Fig. 60.29). The domains in turn consist of smaller modules, each of which is serially repeated and usually uncoded by a separate exon, suggesting that the fibronectin gene, like the collagen gene, evolved by multiple exon duplications. All forms of fibronectin are encoded by a single large gene that contains about 50 exons of similar size. Transcription produces a single large RNA molecule that can be alternatively spliced to produce the various isoforms of fibronectin. The main type of module, called the **type III fibronectin repeat,** binds to integrins. It is about 90 amino acid long and occurs at least 15 times in each subunit. The type III fibronectin repeat is the most common of all protein domains.

The RGD sequence. One way to analyze a complex multifunctional protein molecule such as fibronectin is to chop it into pieces and determine the function of its individual domains. When fibronectin is treated with a low concentration of a proteolytic enzyme, the polypeptide chain is cut in the connecting regions between the domains, leaving the domains themselves intact. One can then show that one of its domain binds to collagen, another to heparin, another to specific receptors on the surface of various types of cells and so on (Fig. 60.29) Synthetic peptide corresponding to different segments of the cell-binding domain have been used to identify a specific **tripeptide sequence-Arg-Gly-Asp** or **RGD,** which is found in one of the type III repeats as a central feature of the binding site. Even very short peptides containing this **RGD sesequence** can compete with fibronectin for the binding site on cells, thereby inhibiting the attachment of the cells to a fibronectin matrix. If these peptides are coupled to a solid surface, they cause cell to adhere to it.

The RGD sequence is not confined to fibronectin. It is found in a number of extracellular protein, including, for example, the blood-clotting factor *fibrinogen*. Fibrinogen peptides containing this RGD sequence have been useful in the development of anti-clotting drugs that mimic these peptides. Snakes use a similar strategy to cause their victims to bleed: they secrete RGD-containing anti-clotting proteins called **disintegrins** into their venom.

Forms and functions of fibronectin. There are multiple isoforms of fibronectin. One, called **plasma fibronectin,** is soluble and circulates in the blood and other body fluids, where it is thought to enhance blood clotting, wound healing and phagocytosis. All of the other forms assemble on the surface of cells and are deposited in the extracellular matrix as highly insoluble **fibronectin fibrils.** In these cell-surface and matrix forms, fibronectin dimers are cross-linked to one another by additional disulphide bonds.

Fibronectin molecules assemble into fibrils only on the surface of certain cells. This is because additional proteins are needed for fibril formation, especially fibronectin-binding integrins. In the case of fibroblasts, fibronectin fibrils are associated with integrins at sites called **fibrillar adhesions.** These are distinct from focal adhesions, in that they are more elongated and contain different intracellular anchor proteins. The fibronectin fibrils on the cell surface are highly stretched and under tension. The tension is exerted by the cell and is essential for fibril formation. Some secreated proteins function to prevent fibronectin assembly in inappropriate places. **Uteroglobin,** for example, binds to fibronectin and prevent it form forming fibrils in the kidney. Mice that have a mutation in the uteroglobin gene accumulate insoluble fibronectin fibrils in their kidneys.

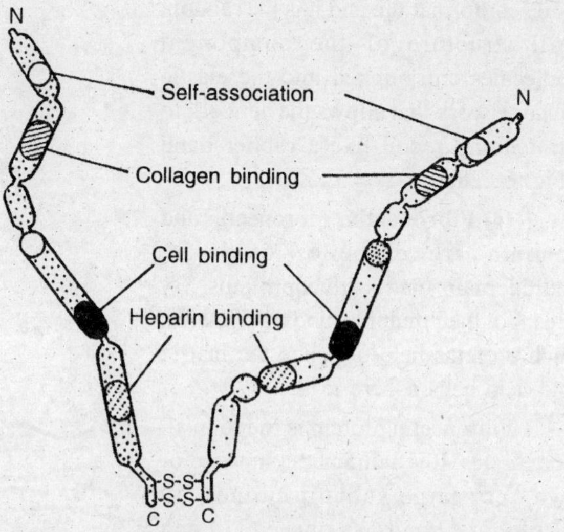

Fig. 60.29. The structure of a fibronectin dimmer. The two polypeptide chains are similar but generally not identical (being made from the same gene but from different spliced mRNAs). They are joined by two disulphide bonds near the C-termini. Each chain is almost 2500 amino acids long and is folded into five or six domains connected by flexible polypeptide segments. Individual domains are specialized for binding to a particular molecule or to a cell as indicated for five of the domains (after Alberts *et al.*, 2002).

Fibronectins are also found to have a role in embryogenesis. Mutant mice that are unable to make fibronectin die early in embryogenesis because their endothelial cells fail to form proper blood vessels.

Fibronectin is important not only for cell adhesion to the matrix but also for guiding cell migrations in vertebrate embryos. Large amounts of fibronectin, for example, are found along the pathway followed by migrating prospective mesodermal cells during amphibian gastrulation (*i.e.*, involution of pharyngeal endoderm and chordamesoderm in frog's gastrulation).

Many matrix proteins are believed to have a role in guiding cell movements during development. The **tenascins** and **thrombospondins,** for example, are composed of several types of short amino acid sequences that are repeated many times and form functionally distinct domains. They can either promote or inhibit cell adhesion, depending on the cell type. Indeed, anti-adhesive interactions are as important as adhesive ones including cell migration (see **Alberts** *et al.*, 2002).

2. Basal Laminae and Laminin. Basal laminae are flexible, thin (40–120 nm thick) mats of specialized extracellular matrix that underlie all epithelial cell sheets and tubes. They also surround individual muscle cells, fat cells and Schwann cells. The basal lamina thus separates these cells and epithelia from the underlying or surrounding connective tissue. In other locations such as the kidney glomerulus, a basal lamina lies between two cell sheets and functions as a highly selective filter (Fig. 60.30). Basal laminae have more than simple structural and filtering functions, however. They are able to determine cell polarity, influence cell metabolism, organize the proteins in adjacent plasma membranes, promote cell survival, proliferation, or differentiation, and serve as specific highways for cell migration.

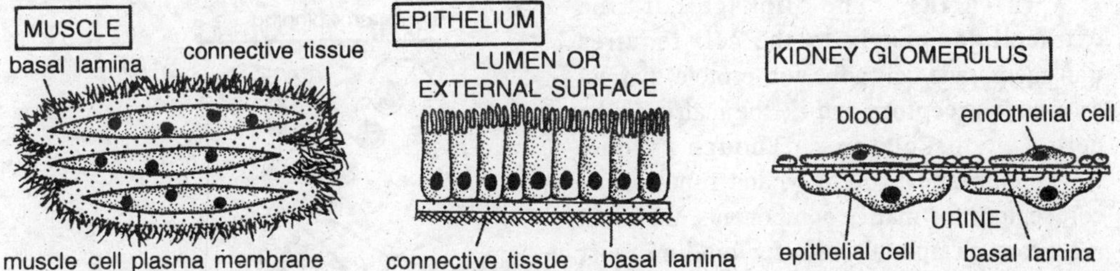

Fig. 60.30. Three ways in which basal laminae are organized. Basal laminae surround certain cells (such as skeletal muscle cells), underlie epithelia and are interposed between two cell sheets (as in the kidney glomerulus). Note that in the kidney glomerulus both cell sheets have gap in them, so that the basal lamina serves as the permeability barrier determining which molecules will pass into the urine from the blood (after Alberts *et al.*, 2002).

The basal lamina is synthesized largely by the cells that rest on it. In some multilayered epithelia, such as the stratified squamous epithelium that forms the epidermis of the skin, the basal lamina is tethered to the underlying connective tissue by specialized anchoring fibrils made of type VII collagen molecules. The term **basement membrane** is often used to describe the composite of the basal lamina and this layer of collagen fibrils (see **Alberts** *et al.*, 2002).

Although precise composition of the basal lamina varies from tissue to tissue and even from region to region, most mature basal laminae contain type IV collagen, the large heparan sulphate proteoglycan perlecan (*e.g.*, agrin) and the glycoproteins laminin and nidogen (or entactin).

Laminin. Early in development, basal laminae contain little or no type IV collagen and consist mainly of laminin molecules. **Laminin-1** (classical laminin) is a large, flexible protein composed of three vary long polypeptide chain (α, β, and γ) arranged in the shape of an asymmetric cross and held together by disulphide bonds (Fig. 60.31). Several isoforms of each type of chain can associate in different combinations to form a large family of laminins. The laminin γ-1 chain is a component of most laminin heterotrimers, and mice lacking it die during embryogenesis because they are unable to make a basal lamina. Like many other proteins in the ECM, the laminin in basement membranes consists of several functional domains: one binds to *perlecan*, one to *nidogen*, and two or more laminin receptor proteins on the surface of cells.

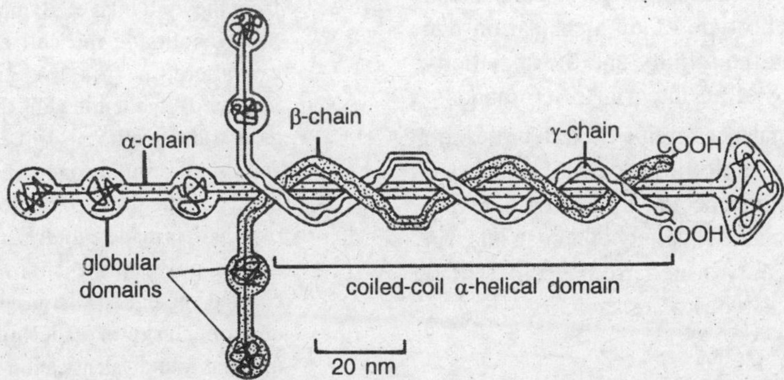

Fig. 60.31. The structure of laminin. The subunits of a laminin-I molecule. This multidomain glycoprotein is composed of three polypeptides (α, β, and γ) that are disulphide bonded into an asymmetric cross-like structure. Each of the polypeptide chains, three types of β chains and three types of γ-chains are known; in principle, they can assemble to form 45 (5 × 3 × 3) laminin isoforms. Several such isoforms have been found each with a characteristic tissue distribution (after Alberts *et al.*, 2002).

Integrins. The linkage of the extracellular matrix to the cell requires transmembrane cell adhesion protein that act as **matrix receptors** and tie the matrix to the cells cytoskeleton. Though some transmembrane proteoglycans function as coreceptors for matrix components, the main receptors on animal cells for binding most extracellular matrix proteins–including collagen, fibronectin and laminins–are the **integrins**. Integrins form a large family of homologous transmembrane, cell-matrix adhesion receptors. In blood cells, integrins also serve as cell-cell adhesion molecules, helping the cells bind to other cells, as well as the ECM.

Integrin are crucially important because they are the main receptor proteins that cells use to both bind to and respond to the extracellular matrix. An integrin molecule is composed of two noncovalently associated transmembrane glycoprotein subunits called α and β (Fig. 60.32). Because the same integrin molecule in different cell types can have different ligand-binding specificities, it seems that additional cell-type-specific factors can interact with integrins to modulate their binding activity.

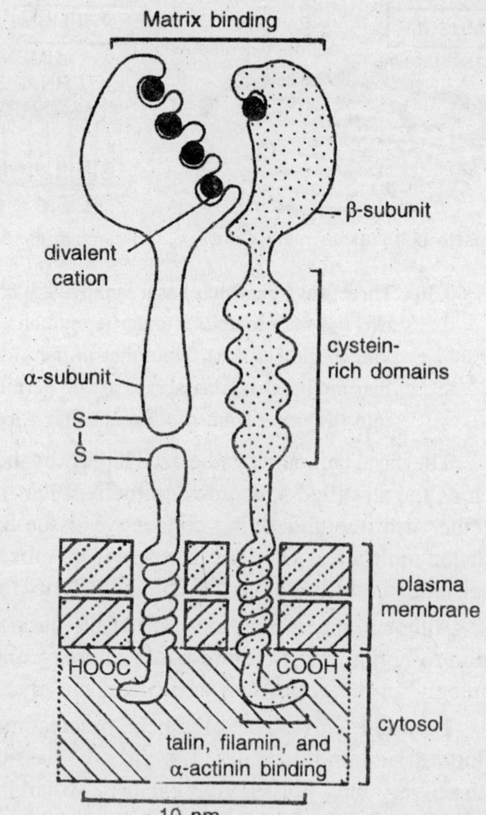

Fig. 60.32. The subunit structure of an integrin cell-surface matrix receptor. Electron micrographs of isolated receptors suggest that the molecule has approximately the shape shown here, with the globular head projecting more than 20 nm from the lipid bilayer. By binding to a matrix protein outside the cell and to the actin cytoskeleton (via the anchor proteins indicated) inside the cell, the protein serves as a transmembrane linker. The α and β subunits are held together by noncovalent bonds. In the fibronectin receptor shown, the α subunit is made initially as a single 140,000 dalton polypeptide chain, which is then cleaved into one small transmembrane domain and one large extracellular domain that contains four divalent-cation-binding sites; the two domains held together by a disulphide (S–S) bond. The extracellular part of the β subunit contains a single divalent-cation-binding site, as well as repeating cysteine-rich region where intrachain disulphide bonding occurs (after Alberts *et al.*, 2002).

The binding of integrins to their ligands depend on extracellular divalent cations (Ca^{2+} or Mg^{2+}), depending on the integrin, reflecting the presence of divalent-cation-binding domains in the extracellular part of the α and β subunits. The type of divalent cation can influence both the affinity and the specificity of the binding of an integrin to its ligands.

Many matrix proteins in vertebrates are recognized by multiple integrins. At least 8 integrins bind fibronectin, for example, and at least 5 bind laminin. A variety of human integrin heterodimers are formed from 9 types of β subunit and 24 types of α subunits.

Function of ECM

In addition to the protection of the plasma membrane, the cell coat performs the following important functions:

***(i)* Filtration.** The extraneous coats sometimes act as filters. For instance, the extraneous coats surrounding the blood capillaries of most vertebrates, especially the kidney glomerulus act as filter and regulate the passage of molecules through it. The extracellular coats of connective tissues contain the chemical compound **hyaluronate** which controls the diffusion.

***(ii)* Maintenance of the micro-environment of the cell.** The extraneous coats of animal cell can affect the concentrations of different substances at the surface of the cell. For example, a muscle cell with its excitable plasma membrane which is surrounded by a glycocalyx is found to maintain the micro-environment of muscle cell by trapping the sodium ions.

***(iii)* Enzymes.** The cell coat of intestinal microvilli are found to contain a variety of enzymes which are involved in the terminal digestion of carbohydrates and proteins. For example it contains the enzyme alkaline phosphatase.

***(iv)* Immunological properties of the extraneous coats.** Some substances of extraneous coats provide immunological properties to the cell. As for instance the plasma membrane of mammalian erythrocytes is found to contain some specific, genetically determined substances (carbohydrates and proteins) corresponding to the A, B and O blood groups. The major sialic acid containing glycoproteins of the red blood cell membrane carry the M and N antigens that appear infrequently in humans. The cell coat also contains the receptor sites for the influenza virus and for various lectins.

(*v*) Histocompatibility. The cell coats of some cells contain some antigens which provide **histocompatibility**, *i.e.*, they permit the recognition of the cells of one organism and rejection of other cells that are foreign to it (*e.g.*, the rejection of grafts from another organism).

60.3 FUNCTIONS OF CELL

The cell is the fundamental unit of life. It has all the chemical and physical factors necessary for its growth and maintenance so *it is a complete unit of metabolism*. Living organisms have various types of cells which have lot of differences in their shape and size. However, there exist some basic similarities in them at the functional level. Cells are found to contain the following four similar basic functions.

1. Functions as barrier. All cells maintain a barrier that protects the contents of cell from the external environment. Both prokaryotic and enkaryotic cells are enclosed by a thin membrane called **plasma membrane**. This membrane as a barrier maintains the concentration of the solutes in the cell by regulating the transport of materials from in and out of the cell. According to the requirement of the cell sometimes the plasma membrane is folded or invaginated to increase the surface area for more exchange of materials. The plasma membrane has the following roles in the cell:

(*i*) Membranes are selectively permeable allowing certain molecules to pass in and out of the cell.

(*ii*) Membranes can selectively allow certain specific molecules against a concentration gradient (active transport).

(*iii*) The cell receives substances from its environment by the process of **endocytosis.**

(*iv*) Undesired and excess amounts of substances present in the cell are thrown out with the formation of vesicles. These vesicles are pinched off and released from the cell. This type of outward movement is called **exocytosis.**

(*v*) The cell perform certain specific metabolic functions by the help of enzymes and proteins present on plasma membrane.

2. Functions of genetic material. Inheritance and transmission of genetic material from one generation to another are performed in the cell through cell division (mitosis and meiosis). In eukaryotic cells the genetic material is present in the nucleus which is enclosed by nuclear envelope. The nuclear envelope regulates the transport of materials (chemical substances) from the nucleus to the cytoplasm and vice versa. It also serves as an anchoring point of chromosomes/choromatin fibers during interphase. In prokaryotic cells, organised nucleus is absent. The genetic material (DNA) is folded into a compact structure known as **nucleoid.**

The genetic material (DNA) both in prokaryotes and enkaryotes control the synthesis of protein on small cytoplasmic structures called **ribosomes**. Three type of RNA molecules (*i.e.*, rRNA, tRNA and mRNA) are transcribed from DNA to form specific proteins in adequate amounts in correct form. The structure and function of ribosomes, in both prokaryotes and eukaryotes are similar.

3. Functions in cellular metabolism. All cells carry out series of chemical reactions for the synthesis of macromolecules, tapping energy, degradation of unusued molecules, converting food substances into sugars, amino acids, etc. Most of the reactions are performed in the cytoplasm of the cell. The initial reactions of the cell start through the degradation of reserve foods (*e.g.*, glycogen) in the fluid phase of the cytoplasm (cytosol). The main biochemical reactions, which release energy such as ATP occur in the cytoplasmic organelle called **mitochondria**. The desired proteins are synthesized on the ribosomes according to genetic information of genes.

In photosynthetic eukaryotes, the energy from sunlight is used by converting the light energy into chemical energy. This conversion is done by some specialized organelles called **chloroplasts**. In prokaryotes such as photosynthetic bacteria and blue green algae (cyanobacteria) similar trapping of energy from sunlight is done by folding of the plasma membrane.

4. Functions in motility. Cells perform different types of motility, *i.e.*, from locomotion to the movement of some components of the cell. The movements may be either the movement of cell from one place to another (*e.g.*, amoeba, paramecium, white blood cells, sperm, etc.) or the movement of liquid over the surface of a cell by cilia and flagella (*e.g.*, ciliated surfaces of trachea, oviduct, etc.).

In eukaryotic cells, microtubules (Mts) are present which consist of a special protein called **tubulin**. The microtubules play an important role in forming spindle apparatus that helps in the movement of the chromosomes during the cell division. Microtubules also provide a rigid set of tracks for the transport of a variety of membrane-enclosed organelles and vesicles. The mechanical work needed for movement depends on **microtubule-associated motor proteins (motor MAPs),** which attach to vesicles or organellels and then "walk" along the microtubule, using ATP to provide the needed energy. **Motor protein** or **mechanoenzyme** is a protein that uses energy derived from ATP to change shape in a way that exerts force and causes attached structures to move; includes three families of proteins (myosin, dynein, and kinesin) that interacts with cytoskeletal elements (microtubules and microfilaments) to produce movement (see **Becker** *et al.*, 2007).

In eukaryotic cells, other smaller filaments are found. These are made up of actin proteins and are called **microfilaments**. These microfilaments help in muscle contraction and cytoplasmic streaming (cyclosis).

In this way, the cells are capable of performing all the necessary metabolic activities essential for continuity of life.

QUESTIONS

Long Answer Questions

1. Describe the structure and functions of animal cell junctions. What is extracellular matrix?

2. Describe the main functions of cell.

3. What is cell adhesion? Describe the various components of extracellular matrix which are responsible for this process.

4. Describe various types of cell adhesion molecules which provide differential affinity.

Short Answer Questions

1. What are glycocalyx?

2. What are desmosomes? How many layers of desmosomes are found? Describe their function.

3. Write a short note on extracellular matrix.

4. How does adhesion takes place?

5. Differentiate between gap junctions, tight junctions and desmosomes.

6. Write a short note on tight junctions or zona occludens.

Very Short Answer Questions

1. What is full form of ECM?

   Ans. The extracellular matrix.

2. What are cadherins?

   Ans. Cadherinrs are Ca^{2+} dependent cell adhesion glycoproteins.

3. Write the names of main cell adhesion proteins.

   Ans. Adhesion proteins such as fibronectin, laminin and tenacin.

4. What is fibronectin?

   Ans. It is adhesive glycoprotein which is found in the extracellular matrix and loosely associated with the surface; it binds cells to the ECM and is important in determining the cell shape and guiding cell migration.

5. What is laminin?

   Ans. It is a glycoprotein of the extracellular matrix localized predominantly in the basal lamina of epithelial cells.

6. Define the basal lamina?

   Ans. It is a thin sheet of specialized extra-cellular matrix material that separates epithelial cells from underlying connective tissue.

7. What is collagen?

   Ans. Collagen is a family of closely related proteins that form high-strength fibers found in high concentration in the ECM of animals.

8. Define proteoglycans.

   Ans. They are complex between proteins and glycosaminoglycans found in the extracellular matrix.

9. Define the plaque.

   Ans. It is a dense layer of fibrous material located on the cytoplasmic side of adhesive junctions such as desmosomes, hemidesmosomes, and adherens junctions; composed of intracellular attachment proteins that link the junction to the appropriate type of cytoskeletal filament.

10. Define integrins.

    Ans. Integrins are any of several plasma membrane receptors that bind to extracellular matrix components at the outer membrane surface and interact with cytoskeletal components at the inner membrane surface; includes receptors for fibronectin, laminin and collagen.

11. In which tissue the gap junctions are mostly found?

    Ans. Embryonic tissue.

12. What prevents leakage of pancreatic secretary proteins in blood?

    Ans. Zona occludens.

13. Name the structure that provide mechanical adhesion between adjacent cells.

Ans. Desmosomes.

Multiple Choice Questions

1. The cell junctions called tight, adhering and gap junctions are found in

(*a*) connective tissue

(*b*) epithelial tissue

(*c*) neural tissue

(*d*) muscular tissue

2. Desmosomes are concerned with

(*a*) cell division (*b*) cellular excretion

(*c*) cytolysis (*d*) cell adherance

3. Cellular recognition and adhesion are facilitated by components of plasma membranes. These components are generally

(*a*) lipid molecules

(*b*) protein molecules

(*c*) both lipid and protein molecules

(*d*) glycolipids and glycoproteins

4. Tonofibrils are anchored at this cell junction

(*a*) desmosome (*b*) nexus

(*c*) zonula occludens (*d*) zonula adherens

5. Collagen contains

(*a*) hydroxyproline (*b*) tryptophan

(*c*) arginine (*d*) peptone.

ANSWERS

Multiple Choice Questions

1. (*b*) **2.** (*d*) **3.** (*d*) **4.** (*a*) **5.** (*a*).

61

Cell Signalling (Messengers, Receptors and Signal Transduction)

No cell lives in isolation. In eukaryotic microorganisms such as yeast, slime molds and protozoans, secreted molecules called **pheromones** coordinate the aggregation of free-living cells for sexual mating or differentiation under certain environmental conditions (see **Lodish** *et al.*, 2004). The amoeboid cells of the slime mold *Dictyostelium* secrete a compound called cyclic AMP (cAMP) at one stage in its life cycle. The compound binds to receptors on the surfaces of neighbouring cells, triggering a process whereby thousands of individual amoeboid cells aggregate and differentiate into a multicellular organism (see **Becker** *et al.*, 2007).

Factors determining the mating-types of yeast are a well understood example of pheromone-mediated cell-to-cell signalling. More important in plants and animals are extracellular **signalling molecules** that function *within* an organism to control metabolic processes within cells, the growth and differentiation of tissues, the synthesis and secretion of proteins, and the composition of intracellular and extracellular fluids. Adjacent cells often communicate by direct cell-cell contact. For example, gap junctions in the plasma membrane of adjacent cells permit them to exchange small molecules and to coordinate metabolic responses.

Extracellular signalling molecules are synthesized and released by **signalling cells** and produce a specific response only in **target cells** that have receptors for the signalling molecules. In multicellular organisms, an enormous variety of chemicals including small molecules (*e.g.,* amino acid or lipid derivatives, acetylcholine, peptides) and proteins are used in this type of cell-to-cell communication. Some signalling molecules, especially hydrophobic molecules, such as steroids, retinoids, and thyroxine, spontaneously diffuse through the plasma membrane and bind to intracellular receptors.

In this chapter we focus on signalling from a diverse group of receptor proteins located in the plasma membrane (*e.g.,* G protein-linked receptors, cytokine receptors, receptor for tyrosine kinases, TGFB receptors, etc.). The signalling molecule acts as a **ligand**, which binds to a structurally complementary site on the extracellular or membrane-spanning domains of the receptor. Binding of a ligand to its receptor causes a conformational change in the cytosolic domain (or domains) of the receptor that ultimately induces specific cellular responses. The overall process of converting signals into cellular responses, is termed **signal transduction**. As we will see, signal transduction pathways may involve relatively few or many components.

Communication by extracellular signals usually involves the following 7 step: 1. synthesis and 2. release of the signalling molecule by the signalling cell; 3. transport of the signal to the target cell; 4. binding of the signal by a specific receptor protein leading to its activation; 5. initiation of one or more intracellular signal transduction pathways by the activated receptor; 6. specific changes in cellular function, metabolism or development; and 7. removal of the signal, which often terminates the cellular response.

The vast majority of receptors are activated by binding of secreted or membrane bound molecules (*e.g.*, hormones, growth factors, neurotransmitters and pheromones). Some receptors, however, are activated by changes in the concentration of a metabolite (*e.g.*, oxygen or nutrients) or by physical stimuli (*e.g.*, light, touch, heat). In bacterium *E. coli*, for example, receptors in the cell surface membrane trigger signalling pathways that help the cell respond to changes in the external level of phosphate and other nutrients.

Signalling Pathways in *E. coli*

Bacteria (*e.g.*, *E. coli*) have two-component regulatory systems, one protein acts as a **sensor**, monitoring the level of nutrients or other components in the environment. The other protein acts as **response regulator** which in phosphorylated state binds to DNA regulatory sequences, thereby stimulating or repressing transcription of specific genes (Fig. 61.1).

For example, in *E. coli* two membrane proteins **PhoR** and **PhoB** are found to regulate transcription in response to concentration of free phosphate. PhoR is a transmembrane protein (sensor), located in the inner (*i.e.*, plasma) membrane, whose periplasmic domain binds phosphate with moderate affinity and whose cytosolic domain has protein kinase activity; PhoB (*i.e.*, response regulator) is a cytosolic protein.

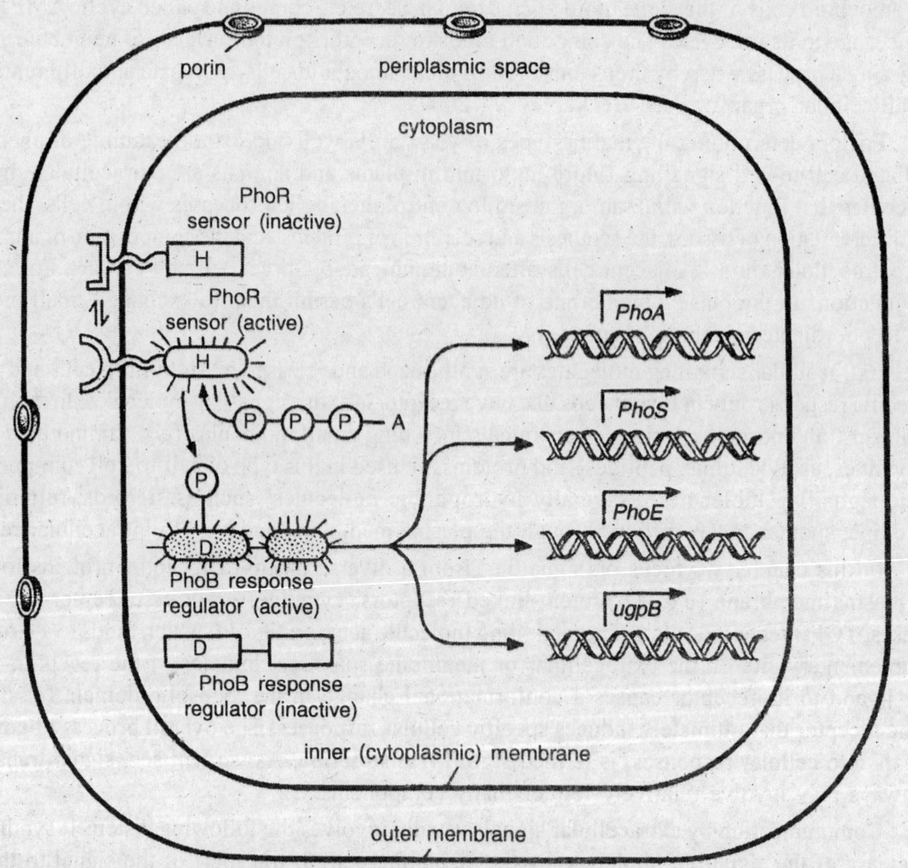

Fig. 61.1. The PhoR/PhoB two-component regulatory system in *E. coli* (see text for explanation of figure, after Lodish *et. al.*, 2004).

Large protein pores in the *E. coli* outer membrane allow ions to diffuse freely between the external environment and the periplasmic space. Consequently, when the phosphate concentration in the environment falls, it also falls in the periplasmic space, causing phosphate to dissociate from the PhoR periplasmic domain (Fig. 61.1). This causes a conformational change into the PhoR cytoplasmic domain that activates its protein kinase activity. The activated PhoR initially transfers a γ-phosphate from ATP to a histidine side chain in the PhoR kinase domain itself. The same phosphate in then transferred to a specific aspartic acid side chain in PhoB; converting it from an inactive to an active transcriptional activator. Phosphorylated, active PhoB then induces transcription from several genes (*e.g., phoA, phoS, phoE* and *ugpB*) that help the cell cope with low phosphate condition.

61.1 TYPES OF CHEMICAL SIGNALS

A variety of compounds can function as chemical messengers that relay signals between cells (Fig. 61.2). Signalling molecules are often classified based on the distance between their site of production and the target tissue(s) upon which they act.

1. Endocrine signals. These messengers are called **hormones** and they act as **endocrine signals** (endocrine = from Greek words that mean to secrete into). They are produced at great distances from their target tissues, and are carried by the circulatory system to various sites of the body.

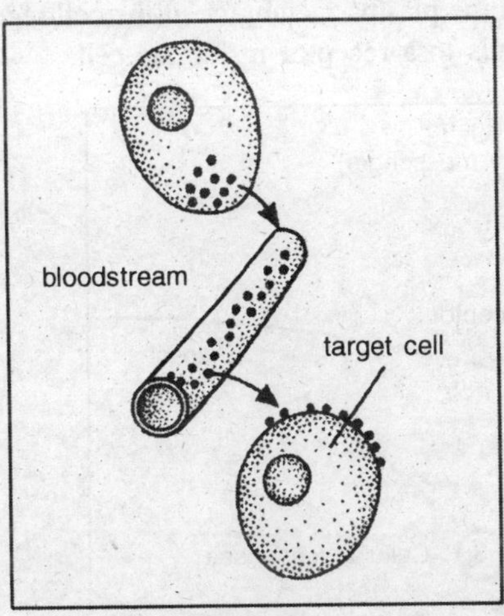

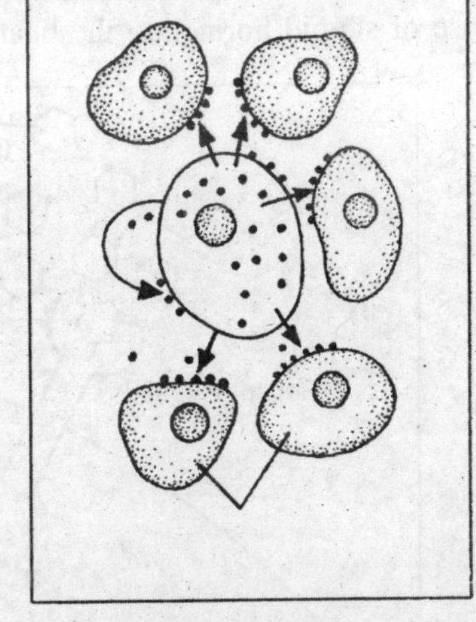

Fig. 61.2. Cell-to-cell signalling by hormones and local mediations. The main distinction between classes of signalling molecules is the distance the molecule travels before encountering its target cell or tissue. Hormones act as endocrine signals, and are carried by the blood stream. Local mediators, such as growth factors, can act on nearby cells (paracrine signals), or on the cell that produces them (autocrine signals). Although many signalling molecules act on the cell surface, certain hydrophobic molecules enter their target cells (after Becker *et al.*, 2007).

2. Paracrine signals. Some signals, such as growth factors, are released locally, where they diffuse to act at short range on nearby tissues. Such signals are referred to as **paracrine signals** (paracrine = form the Greek *para*, which means "beside"). The conduction by a neurotransmitter of a signal from one nerve cell to another or from a nerve cell to a muscle cell (inducing or inhibiting muscle contraction) occurs via paracrine signalling (Chapter 68). Many growth factors regulating development in multicellular organisms also act at short range. Some of these molecules bind tightly to the extra-cellular matrix, unable to signal, but subsequently can be released in an active form. Many develop-mentally important signals diffuse away from the signalling cell, forming a concentration gradient and inducing various cellular responses depending on their concentration at a particular target cell.

3. Autocrine signals. Some local mediators act on the same cell that produces them; such signals are called **autocrine signals**. Some growth factors act in this fashion. Cultured cells often secrete growth factors that stimulate their own growth and proliferation. Autocrine signalling is particularly common in tumor cells, many of which overproduce and release growth factors that stimulate inappropriate, unregulated proliferation of themselves as well as adjacent nontumor (normal) cells; this process may lead to formation of a tumor mass.

Some signalling molecules can act at both short range and long range. **Epinephrine**, for example, functions as a neurotransmitter (paracrine signalling) and as a systemic hormone (endocrine signalling). Another example is epidermal growth factor (EGF), which is synthesized as an integral plasma membrane protein. Membrane bound EGF can bind to and signal an adjacent cell by direct contact. Cleavage by an extracellular protease releases a soluble form of EGF, which can signal in either an autocrine or a paracrine manner.

Primary and Secondary Messengers

Ligands are the **first** or **primary messengers**. Once a messenger reaches its target tissue, it binds to receptors on the surface of the target cells, initiating the signalling process (Fig. 61.3). A molecule coming from either a long or a short distance functions as a ligand by binding to a receptor. A ligand often binds to a receptor embedded within the plasma membrane of the cell receiving the signal. In case of steroid hormones, the ligand binds to a receptor inside the cell.

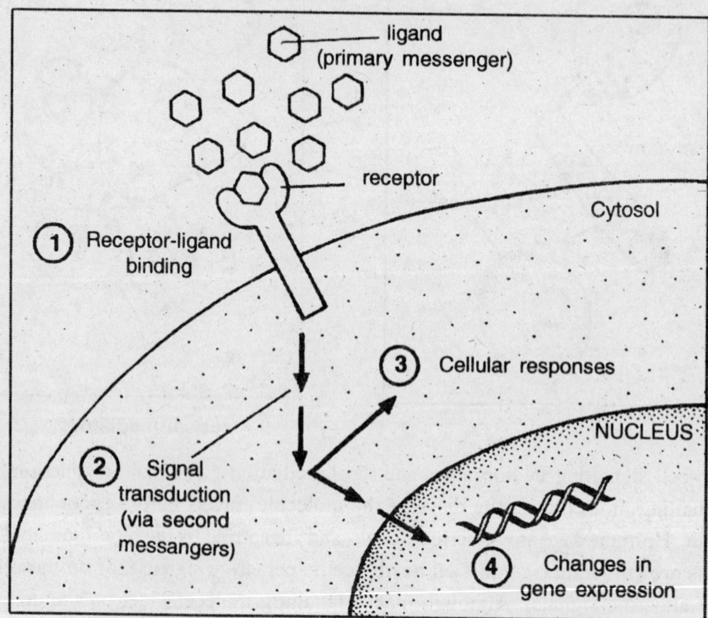

Fig. 61.3. The overall flow of information during cell signalling. Binding of ligand, the primary messenger, by a receptor activates a series of events known as signal transduction, which relays the signal to the interior of the cell, resulting in specific cellular responses and/or changes in gene expression (after Becker *et al.*, 2007).

The binding of ligand to receptor often results in a short-lived increase (or decrease) in the concentration of certain low-molecular weight intracellular signalling molecules termed **second** or **secondary messengers**. These molecules include cAMP, cGMP, DAG and IP_3 (Fig. 61.4). Other important second messengers are Ca^{2+} and various inositol phospholipids, also called **phosphoinositides**, which are embedded in cellular membranes. Second messengers relay the signals from one location in the cell, such as the plasma membrane, to the interior of the cell, initiating a cascade of changes within the receiving cell. Often these events affect the expression of specific genes within the receiving cell. The ability of a cell to translate a receptor-ligand interaction to changes in its behaviour or gene expression is the **signal transduction** pathways.

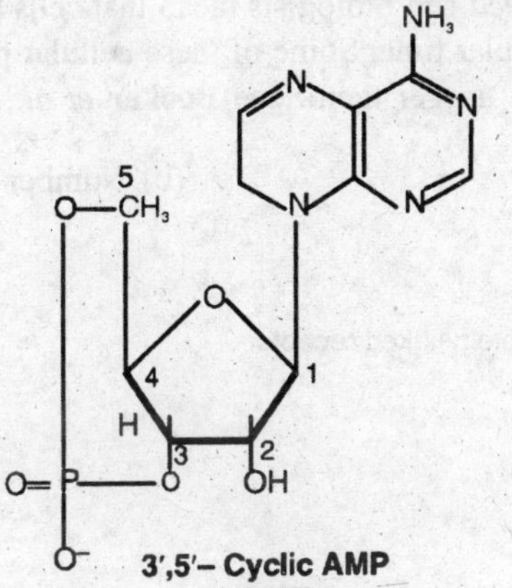

3′,5′– Cyclic AMP (cAMP)

Activates protein kinase A (PKA)

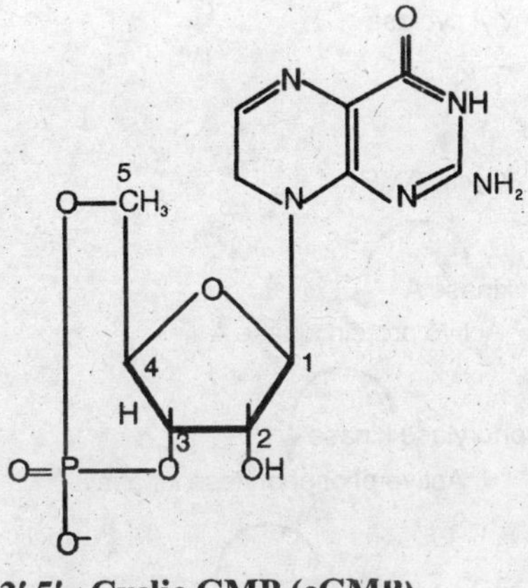

3′,5′– Cyclic GMP (cGMP)

Activates protein kinase G (PKG) and opens cation channels in rod cells

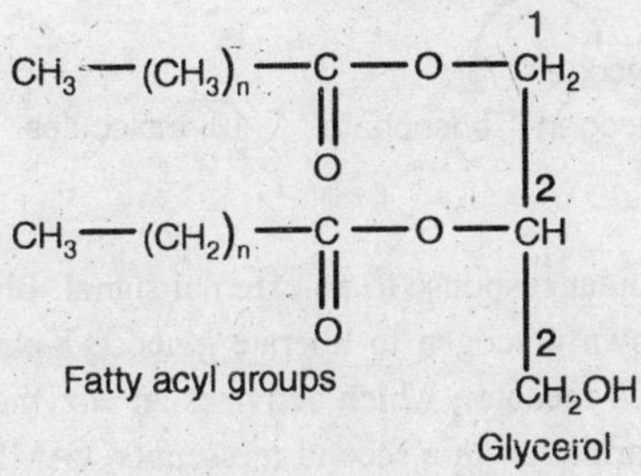

1,2-Diacylglycerol (DAG)

Activates protein kinase C (PKC)

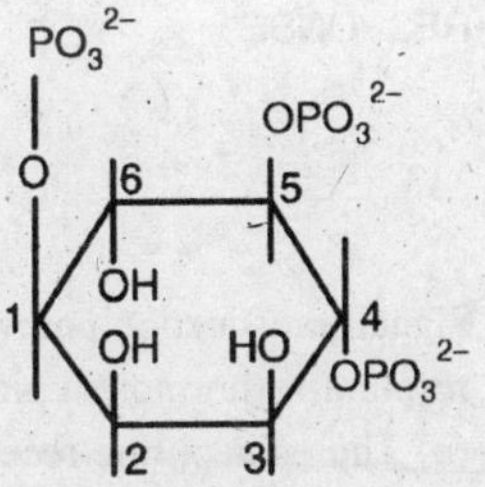

Inositol 1,4,5-trisphosphate (IP_3)

Opens Ca2+ channels in the endoplasmic reticulum

Fig. 61.4. Four common intracellular second messengers. The major direct effect or effects of each compound are indicated below its structural formula (after Lodish *et al.*, 2004).

Signal transduction pathways. When a ligand (first messenger) binds to its specific (or cognate) receptor, the receptor is altered in a way that causes changes in cellular activities. In general, the binding of a ligand either induces a change in receptor conformation or causes receptors to cluster together. Once one of these changes takes place, the receptor initiates a pre-programmed sequence of events inside the cell. By *preprogammed* cell biologists mean that cells have a greater repertoire of functions than are in use at any particular time. Some of these cellular processes remain unused until particular signals are received that trigger them (see **Becker** *et al.*, 2007).

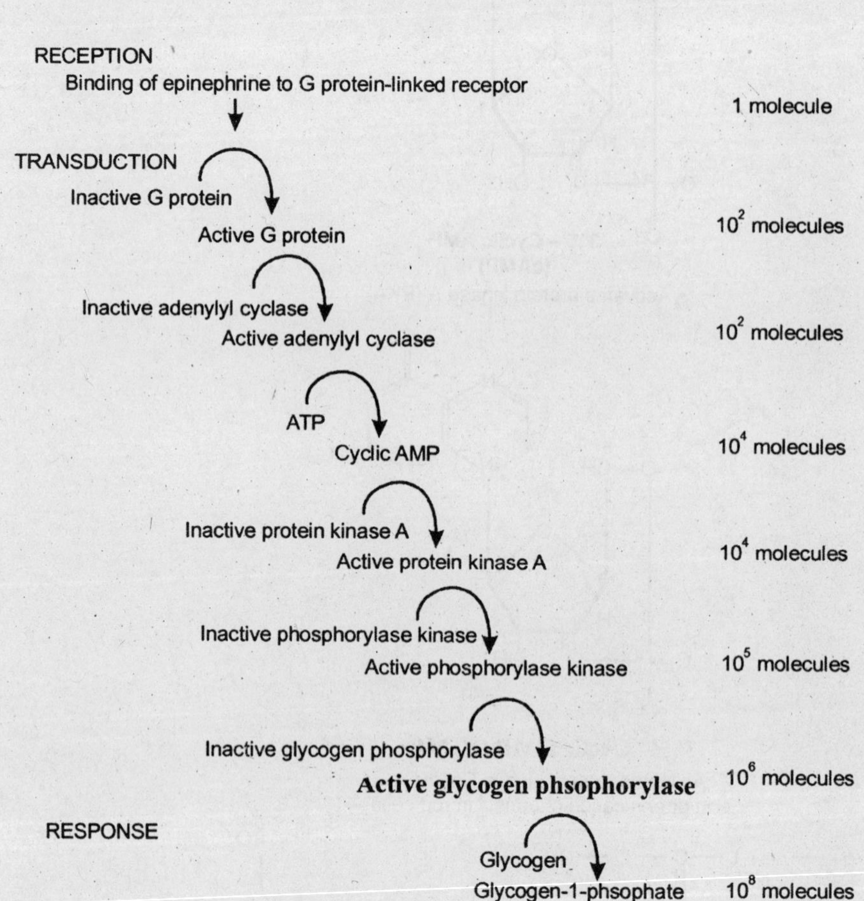

Fig. 61.5. Signal transduction pathways can amplify the cellular response to an external signal. Liver cells respond to hormone epinephrine by breaking down glycogen to liberate glucose-1-phosphate. (*a*) The epinephrine receptor is a G protein-linked receptor, which activates an enzyme known as adenylyl cyclase. Adenylyl cyclase catalyses formation of a second messenger, cAMP, which activates protein kinase (protein kinase A), which in turn activates another kinase (phosphorylase kinase). Ultimately, the enzyme glycogen phosphorylase is activated, which catalyses the breakdown of glycogen. (*b*) The approximate number of molecules produced at each step is shown on the right. One epinephrine molecule is capable of triggering the production of hundreds of millions of glucose-1-phosphate molecules (after Becker *et al.*, 2007).

Signal amplification. Signal transduction pathways allow another important aspect of a cell's response to an external signal; it is called **signal amplification**. Often exceedingly small quantities of a ligand are sufficient to elicit a response from a target cell, yet the responding cell however responds in dramatic ways. Often the strong response of the target cell results from a signalling cascade with the responding cell. At each step in the cascade, a signalling intermediate persists long enough to stimulate the production of many molecules required for the next step in the cascade, thereby multiplying the effects of a single receptor/ligand interaction on the cell surface. A well-known example of such signal amplification involves the **breakdown of glycogen** in liver cells in response to the hormone *epinephrine* (Fig. 61.5). This process involves a G protein-linked receptor on the cell surface, which in turn initiates a signalling cascade within the liver cell. As a result of the cascade, a single epinephrine ligand can elicit the production of hundreds of millions of glucose molecules.

61.2 G PROTEIN-LINKED RECEPTORS

G protein-linked receptor family is so named because ligand binding causes a change in receptor conformation that activates a particular **G protein** (an abbreviation for **guanine-nucleotide binding protein**). The activated G protein in turn binds to a target protein such as an enzyme or a channel protein, thereby altering the target's activity. All G protein-linked receptors initiate signal transduction inside the cell in this way. Examples of G protein-linked receptors include olfactory receptors (responsible for human's sense of smell), norepinephrine receptors, and hormone receptors such as those for thyroid-stimulating hormone and follicle-stimulating hormone.

Structure of G protein-linked Receptor

The G protein-linked receptors are remarkable in that they all have a similar structure yet differ significantly in their amino acid sequences. In each case, the receptor protein forms seven transmembrane α helices connected by alternating cytosolic or extracellular loops (Fig. 61.6). The N-terminus of the protein is exposed to the extracellular fluid while the C-terminus resides in the cytosol (Fig. 61.7). The extra-cellular portion of each G protein-linked receptor has a unique messenger binding site, and a cytosolic loop connecting the fifth and sixth transmembrane α helices is a specific for a particular G protein. G protein-linked receptors therefore provide a versatile method for linking different messengers to different signal transduction pathways.

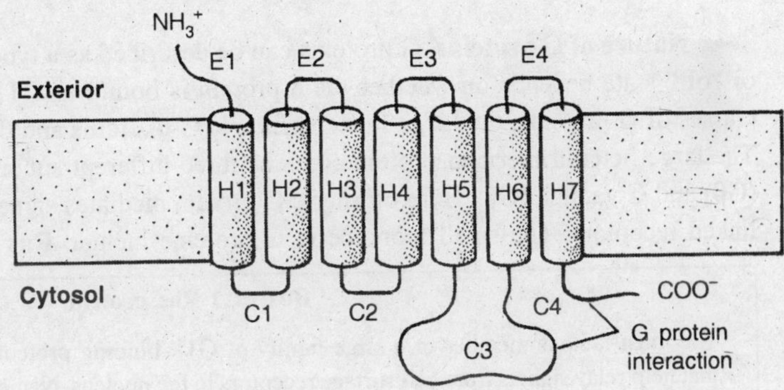

Fig. 61.6. Schematic diagram of the general structure of G protein-linked receptors. All receptors of this type have the same orientation in the membrane and contain seven transmembrane α-helical regions (H1 to H7), four extracellular segments (E1-E4), and four cytosolic segments (C1-C4). The carboxyl-terminal segment (C4), the C3 loop, and, in some receptors, also the C2 loop are involved in interactions with a linked/coupled trimeric G protein (after Lodish *et al.*, 2004).

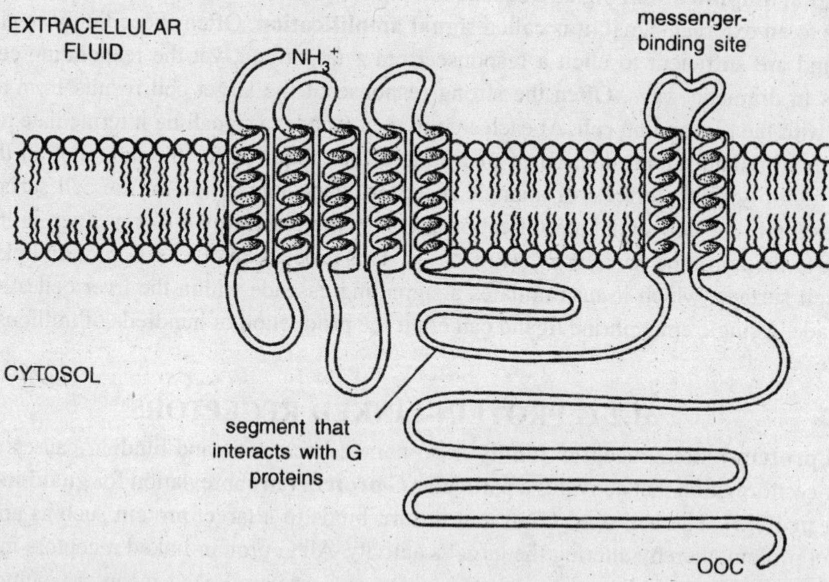

Fig. 61.7. The structure of G protein-linked receptor. Each G protein-linked receptor has seven transmembrane α helices. The primary messenger binds to the extracellular portion of the receptor. This binding causes an intracellular portion of the receptor to activate an adjacent G protein (after Becker *et al.*, 2007).

Nature of G proteins. G proteins can be described as a type of molecular switch whose "on" or "off" state depends on whether the a protein is bound to GTP or GDP. There are two distinct classes of G proteins: the **large heterotrimeric G proteins** and the **small monomeric G proteins**. The large heterotrimeric G protein contains three different subunits, called G alpha (Gα), G beta (Gβ) and G gamma (Gγ). Heterotrimeric G protein mediates signal transduction through G protein-linked receptors. The small monomeric G proteins include **Ras** protein (Box 61.1).

Box 61.1 Ras protein

The most famous member of a large family of GTP-binding proteins (called **monomeric GTPases**) that help relay signals from cell surface receptors to the nucleus. Named for the *ras* gene, first identified in viruses that cause rat sarcomas (**Alberts** *et al.*, 2002).

All G proteins have the same basic structure and mode of activation. Of the three subunits in the Gαβγ heterotrimer, Gα, the largest, binds to a guanine nucleotide (GDP or GTP). When Gα binds to GTP, it also detaches from the Gβγ complex. The Gβ and Gγ subunits, on the other hand, are permanently bound together. Some G proteins, such as **Gs,** act as stimulators of signal transduction (hence *s*, for stimulatory); others, such as G*i*, act to inhibit signal transduction (hence *i*, for inhibitory).

When a messenger (ligand) binds to a G protein-linked receptor on the surface of the cell, the change in conformation of the receptor causes the G protein to associate with the receptor, which in turn causes the *G*α subunit to release its bound GDP, acquire a GTP, and then detach from the complex (Fig. 61.8). Depending on the G-protein and the cell type, either the free GTP-Gγ subunit or the Gβγ complex can then initiate signal transduction events in the cell. Each portion of the G protein exerts its effect by binding to a particular enzyme or other protein in the cell. In some cases, both the GTP-Gγ and GTP-Gβγ, subunits simultaneously regulate different processes in the cytosol. However, the activity of the G protein persists only as long as the Gα is bound to GTP and the subunits remain separated. Because the Gα subunit hydrolyzes GTP, it will only remain active for a short time before reverting to the GDP-bound state and reassociated with Gβγ. This characteristic allows the signal transduction pathway to shut down rapidly when the messenger is removed.

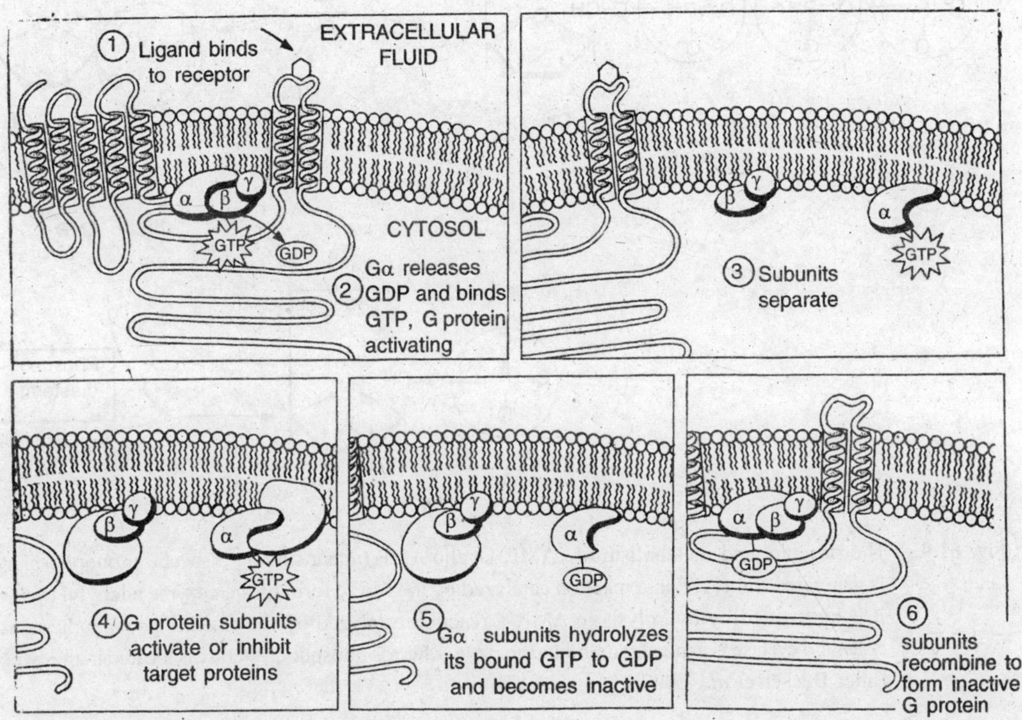

Fig. 61.8. The G protein activation and inactivation. 1. When the ligand binds, 2. The receptor activates a G protein by causing the Gα subunit to release GDP and acquire GTP, 3. The Gα and Gβγ subunits then separate and 4. Initiate signal transduction events. 5. The GTP-Gα subunit eventually hydrolyzes its bound GTP, converting the subunit back to its inactive GTP-Gα form. 6. The inactive GTP-Gα subunit then recombines with Gβγ to form the inactive heterotrimeric G protein (after Becker *et al.*, 2007).

Role of Cyclic AMP in Signal Transduction by G protein.

The second messenger, the cyclic AMP (cAMP) is formed from cytosolic ATP by the enzyme **adenylyl cyclase** (Fig. 61.9) anchored in the plasma membrane with its catalytic portion protruding into cytosol. Normally, the enzyme is inactive until it binds to the activated α subunit of a specific G protein, Gs. When the G protein-linked receptor is coupled to Gs, the binding of ligand stimulates the Gsα subunit to release GDP and acquires GTP (Fig. 61.10). This in turn causes GTP-Gsα to detach from the Gsβγ subunits and bind to adenylyl cyclase, the enzyme becomes active and converts ATP to cAMP.

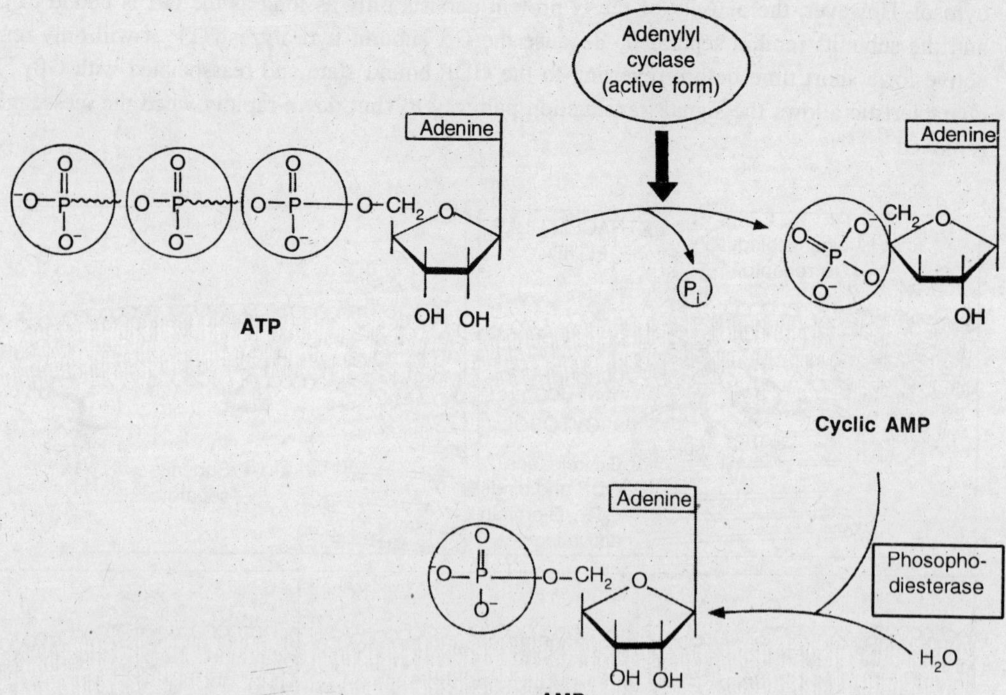

Fig. 61.9. The structure and metabolism of cAMP. Cyclic AMP (adenosine-3′, 5′-cyclic monophosphate) is generated from ATP in a reaction catalyzed by the active form of the enzyme adenylyl cyclase; it is inactivated by hydrolysis to AMP, a reaction catalyzed by the enzyme phosphodiesterase. Adenylyl cyclase is a membrane-bound enzyme, whereas phosphodiesterase is located in the cytosol (after Becker *et al.*, 2007).

The G proteins respond quickly to changes in ligand concentration because they remain active for only a short period of time before the Gα subunit hydrolyzes its bound GTP and converts to the inactive state. Once the G protein becomes inactive, the adenylyl cyclase cease to make cAMP. However, cAMP levels would still remain elevated in the cell if not for the enzyme **phosphodiesterase**, which degrades cAMP. This process further ensures that the signal transduction pathway will shut down promptly when the concentration of the ligand outside the cell falls.

Cyclic AMP has one main intracellular target, an enzyme known as cAMP-dependent kinase, or **protein kinase A (PKA)**. PKA enzyme phosphorylates a wide variety of cellular proteins by transferring phosphate from ATP to a serine or threonine found within the target protein. cAMP tends to regulate the activity of PKA.

An increase in cAMP concentration can produce many different effects in different cell types. 1. When cAMP is elevated in skeletal and liver cells, the breakdown of glycogen is stimulated. 2. In cardiac muscles, the elevation of cAMP strengthens heart contraction; 3. whereas in smooth muscle contraction is inhibited. 4. In blood platelets, the elevation of cAMP inhibits their mobilization during blood clotting, and 5. In intestinal epithelial cells, it causes the secretion of salts and water into the lumen of the gut. Each of these reactions is an example of the preprogrammed response discussed earlier.

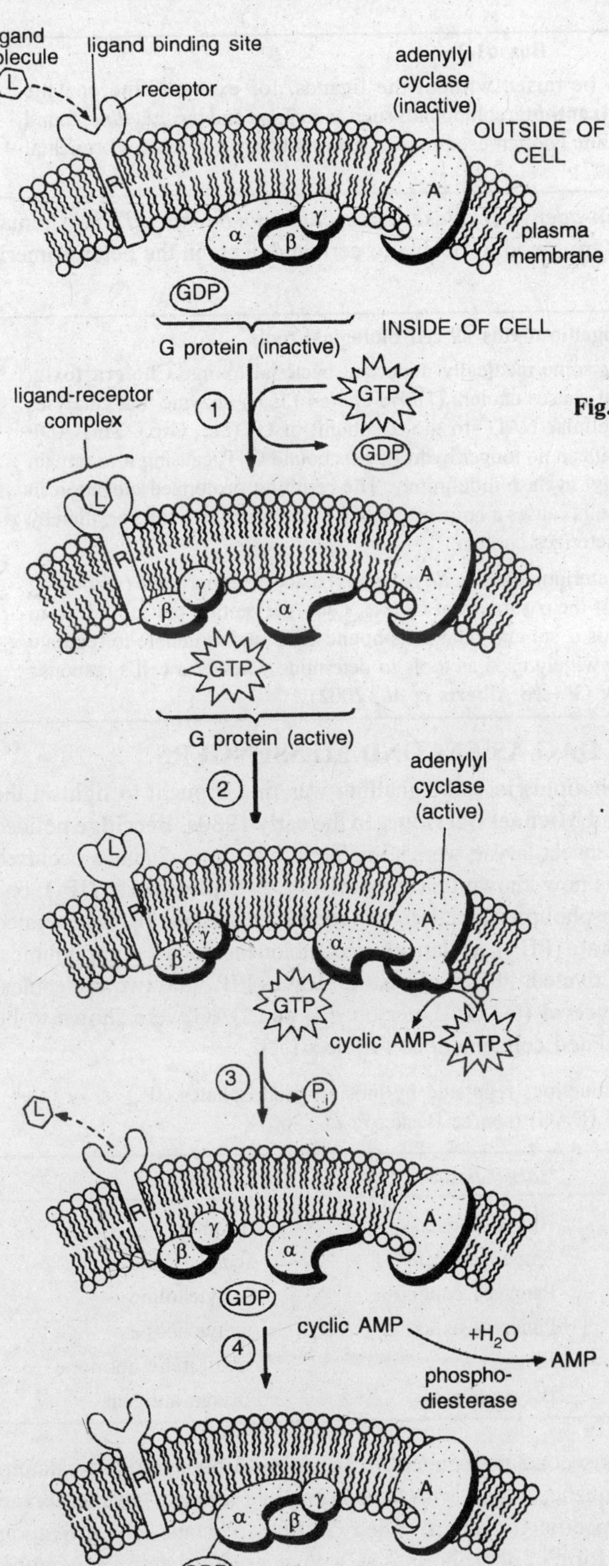

Fig. 61.10. The roles of G proteins and cyclic AMP in signal transduction. G proteins mediate signal transduction through G protein-linked receptors. When a messenger binds to its receptor, the Gs protein is activated. In the inactive state, the α, β and γ subunits of the Gs protein are present as a complex, with GDP bound to the α subunit. 1. When a receptor is activated by binding of its specific ligand on the outer surface of the plasma membrane, the receptor-messenger complex associates with Gs protein, causing the displacement of GDP by GTP and the dissociation of Gsα-GTP complex. 2. The GTP-Gsα complex then binds tightly to a molecule of membrane-bound adenylyl cyclase, activating it for synthesis of cAMP. 3. Activation ends when the ligand leaves the receptor, the GTP is hydrolyzed to GDP by the GTPase activity of the Gsα subunit, and the Gsα dissociates from the adenylyl cyclase. 4. Adenylyl cyclase then reverts to the inactive form, the Gsα reassociates with the Gsβγ complex, and cAMP molecules in the cytosol are hydrolyzed to AMP by enzyme phosphodiesterase (after Becker *et al.*, 2007).

Box 61.2

The intracellular level of cAMP can be raised without the ligands, for example, the enzyme phosphodiesterase is inhibited by **methylxanthine**, compounds such as *caffeine* and *theophylline*, found in coffee, tea and soft drinks. Theophylline is often used to treat **asthma** because it relaxes bronchial smooth muscles (see **Becker** *et al.*, 2007).

What would happen if the G protein-adenylyl cyclase system could not be shut off? Two human diseases caused by pathogenic bacteria are found to be due to certain defects in the heterotrimeric G protein (Box 61.3).

Box 61.3 Pathogenic toxins as cell biological tools

Both G*s* and G*i* proteins are targets for some medically important bacterial toxins. **Cholera toxin,** which is produced by the bacterium that causes cholera (*Vibrio cholere*) is an enzyme that catalyses the transfer of ADP ribose from intracellular NAD^+ to the α subunit of G*s* (*i.e.*, G*s*α). This ADP ribosylation alters the α subunit so than it can no longer hydrolyse its bound GTP, causing it to remain in an active state that stimulates adenylyl cyclase indefinitely. The resulting prolonged elevation in cAMP level within intestinal epithelial cells causes a large efflux of Cl^- and water into the gut, thereby causing the severe diarrhoea that characterizes cholera.

Pertussis toxin, which is made by the bacterium (*Bordetella pertussis*) that causes pertussis (whooping cough), catalyses the ADP ribosylation of the α subunit of G*i* (*i.e.*, G*i*α), preventing the subunit from interacting with receptors; as a result, this α subunit retains its bound GDP and is unable to regulate its target proteins. These two toxins are widely used as tools to determine whether a cell's response to a signal is mediated by the G*s* or by G*i* (see **Alberts** *et al.*, 2002).

ROLE OF IP_3 AND DAG AS SECOND MESSENGERS

The importance of inositol phospholipids in cell signalling was first brought to light in the pioneering studies of **Robert Michell** and **Michael Berridge**. In the early 1980s, **Berridge** noticed that when the salivary glands of certain insect larvae were stimulated to secrete, changes occurred in membrane inositol phospholipids. It is now known that **inositol 1, 4, 5-triphosphate (IP_3)**, one of the breakdown products of inositol phospholipids functions as a second messenger. IP_3 is generated from **phosphatidylinositol-4, 5-biphosphate (PIP_2)**, a relatively uncommon membrane phospholipids, when the enzyme **phospholipase C** is activated. Phospholipase C cleaves PIP_2 into two molecules: inositol triphosphate (IP_3) and **diacylglycerol (DAG)**. Later on, IP_3 and DAG were shown to be *second messengers* in a variety of regulated cell functions (Table 61.1).

Table 61.1 Examples of cell functions regulated by inositol triphosphates (IP_3) and diacylglycerol (DAG) (source Becker *et al.*, 2007).

Regulated function	*Target tissue*	*Messenger*
1. Platelet activation	Blood platelets	Thrombin
2. Muscle contraction	Smooth muscle	Acetylcholine
3. Insulin secretion	Pancreas endocrine	Acetylcholine
4. Amylase secretion	Pancreas exocrine	Acetylcholine
5. Glycogen degradation	Liver	Antidiuretic hormone
6. Antibody production	B lymphocytes	Foreign antigens

The functions of IP_3 and DAG as second messengers in the inositol-phospholipid-calcium pathway are shown in Fig. 61.11. The sequence begins with the binding of a ligand to its membrane receptor, leading to the activation of a specific G protein, called Gp (the "p" stands for activation of phospholipase C). Gp then activates a form of phospholipase C known as Gβ, thereby generating both IP_3 and DAG.

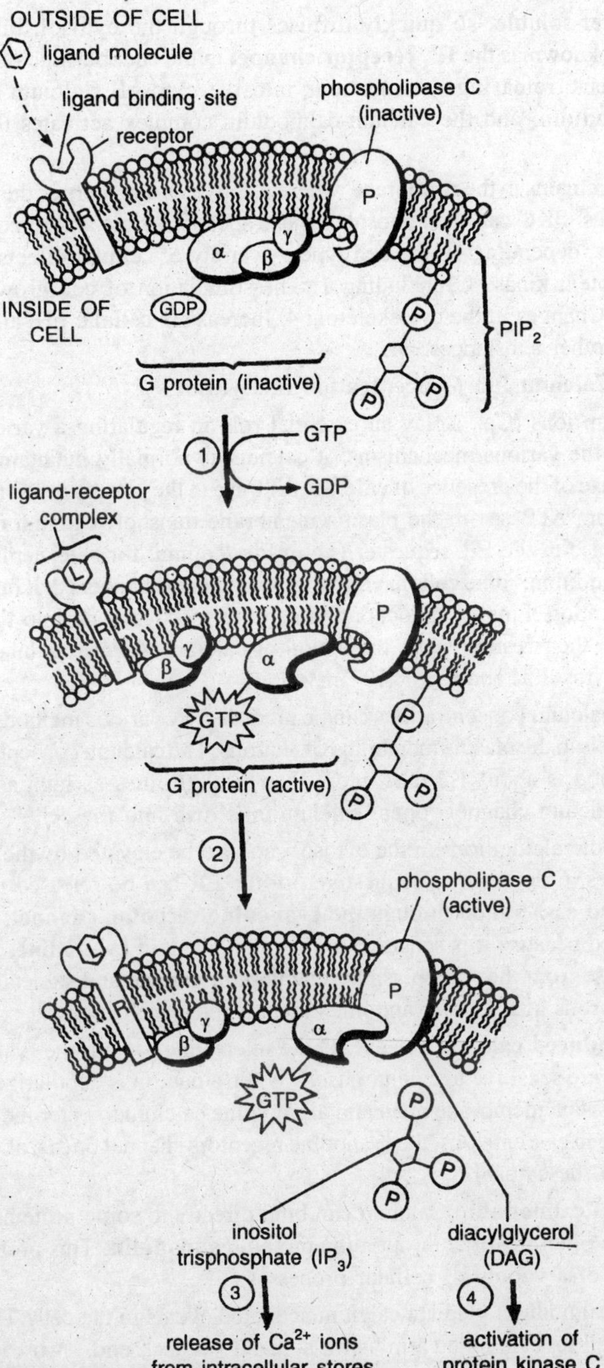

Fig. 61.11. The role of IP_3 and DAG in signal transduction. 1. When a receptor (R) is activated by the binding of its ligand (L) on the outer surface of the plasma membrane, the receptor-ligand complex associates with G protein Gp, causing the displacement of GDP by GTP and the dissociation of the GTP-Gα complex. 2. The GTP-Gα complex then binds to phospholipase (P), activating it and causing cleavage of PIP_2 into one molecule each of IP_3 and DAG 3. The IP_3 is released into the cytosol, where it triggers the release of calcium. 4. The DAG remains in the membrane, where it activates protein kinase C (after Becker *et al.*, 2007).

IP_3 is water-soluble, so quickly diffuses through the cytosol, binding to a ligand-gated calcium channel known as the **IP_3 receptor channel** in the endoplasmic reticulum. When IP_3 binds, the channel opens, releasing calcium ions into the cytosol. Calcium then binds to a protein known as **calmodulin**, and the calcium-calmodulin complex activates the desired physiological process.

The DAG remains in the membrane, where it activates members of the **protein kinase C (PKC)** family of enzymes. PKC can then phosphorylates specific serine and threonine groups on a variety of target proteins, depending on the cell type. A variety of cellular effects have been linked to the activation of protein kinase C, including 1. The stimulation of cell growth; 2. The regulation of ion channels; 3. Changes in the cytoskeleton; 4. Increase in cellular pH, and 5. Effects on secretion of proteins and other substances.

Regulation of Calcium Ion Concentration in Cytosol

The calcium ions (Ca^{2+}) play an essential role in regulating a variety of cellular functions. An overview of the various mechanisms of calcium is normally maintained at very low levels in the cytosol because of the presence of **calcium ATPase** in the plasma membrane and the endoplasmic reticulum. Calcium ATPases in the plasma membrane transport calcium out of the cell, where as the calcium ATPase in the ER sequester (Latin for "commit for safekeeping") calcium ions in the lumen of ER. In addition, some cells have sodium-calcium exchangers that further reduce the cytosolic calcium concentration. Finally, mitochondria can transport calcium into the mitochondrial matrix. For most cells in their resting state, the action of calcium ATPase maintains the concentration of the calcium in cytosol at about 1×10^{-4} m*M*.

Cytosolic calcium concentrations can be increased by various methods. One way is by opening of calcium channels in the plasma membrane of neurons. The calcium concentration in the extracellular fluid and the blood is about 1.2 mM, more than 10,000 times as high as that of the cytosol. As a result, when calcium channels open, calcium ions rush into the cell.

The levels of calcium ions in the cytosol can also be elevated by the release of calcium from intracellular stores. Calcium ions sequestered in the ER can be released through the IP_3 receptor channel, discussed earlier, and through the **ryanodine receptor channel**. The ryanodine receptor channel, so named because it is sensitive to the plant alkaloid **ryanodine**, is particularly important for calcium release from the sarcoplasmic reticulum of cardiac and skeletal muscle, but nonmuscle cells such as neurons also have ryanodine calcium release channels.

Calcium induced calcium release. Quite interestingly, both the ryanodine receptor channel and IP_3 receptors are sensitive to calcium itself. When a neuron is depolarized, for example, calcium channels in the plasma membrane open and allow some calcium to enter the cytosol. Upon exposure to a rapid increase in calcium ions, the ryanodine receptors channel opens, allowing calcium to escape from the ER into the cytosol.

Function of calmodulin. Calcium can bind directly to some proteins and thereby alter their activity, but more often it works through the protein **calmodulin**. This protein has pivotal function in the regulation of a variety of cellular processes.

How does calmodulin mediate calcium activated events in the cell? The calmodulin molecule has been compared to a flexible "arm" with a "hand" at each end. Two calcium ions bind at each of two "hand" regions, causing the calmodulin to undergo a change in shape that form the active **calcium-calmodulin complex** (Fig. 61.12). When a protein is present that contains a calmodulin-binding site, the hands and arm bind to it by wrapping around the binding site (Fig. 61.12).

Calmodulin protein is found to have a remarkable affinity for calcium: calmodulin binds to calcium when the cytosolic calcium concentration increases to about 10^{-3} m*M*, but it releases calcium when cytosolic calcium levels decline back to the resting level of 10^{-4} mM. Thus calmodulin protein is uniquely suited to operate within the typical range of cytosolic calcium concentrations.

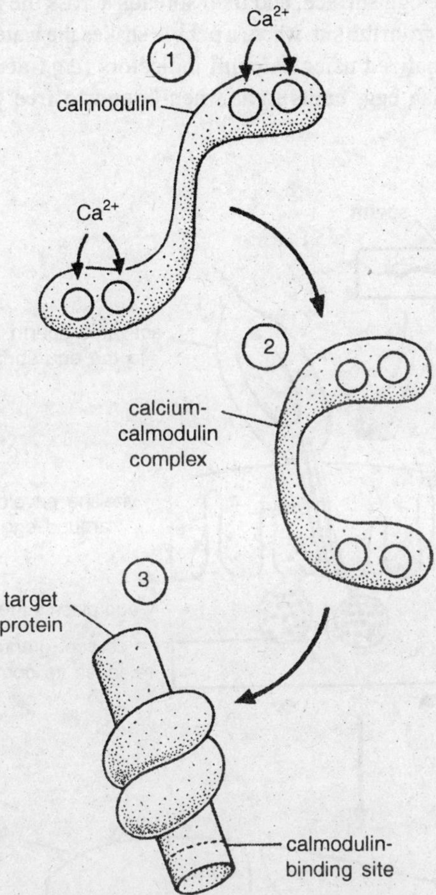

Fig. 61.12. The structure and function of the calcium-calmodulin complex. A—Calmodulin is a cytosolic calcium-binding protein. This model of molecular structure of calmodulin is based on data from X-ray crystallography. The molecule consists of two globular ends ("hands") joined by a helical region ("arm"). Each end has two calcium binding sites. B—To form a functional calcium-calmodulin complex, 1. Calmodulin binds four calcium ions. 2. Calmodulin changes conformation, resulting in an active complex. 3. The two globular "hands" of the complex wrap around a calmodulin binding site on a target protein (after Becker *et al.*, 2007).

Most calmodulin binding proteins are enzymes such as *protein kinases* and *protein phosphatases*. The response of a target cell to an increase in calcium concentration depends on the particular calmodulin-binding proteins that are present in the cell. *This means that the same change in calcium concentration can produce markedly different effects in two target cells if each possesses different calmodulin-sensitive enzyme system.*

Function of calcium in slow block to polyspermy. The process of fertilization of animal eggs (*i.e.*, ova) is an outstanding example of the importance of calcium-mediated signal transduction following a receptor-ligand interaction. In many animals, when sperm that have undergone prior steps of activation bind to the surface of mature eggs and unite with them at fertilization, a striking sequence of events arises. One of the early responses of the egg-within 30 seconds to several minutes after fertilization is the release of calcium from internal stores. Calcium release occurs initially at the site where the sperm penetrates the egg surface, and then spreads across the egg, much as a ripple on the surface of a pond spreads away from the site where a pebble strikes the water. The wavelike propagation of calcium release can be visualized using calcium indicators (*e.g.*, **aequorin**, which is a dye that when injected in fertilizing fish egg, emits light when bound to free Ca^{2+} ions in the cytosol).

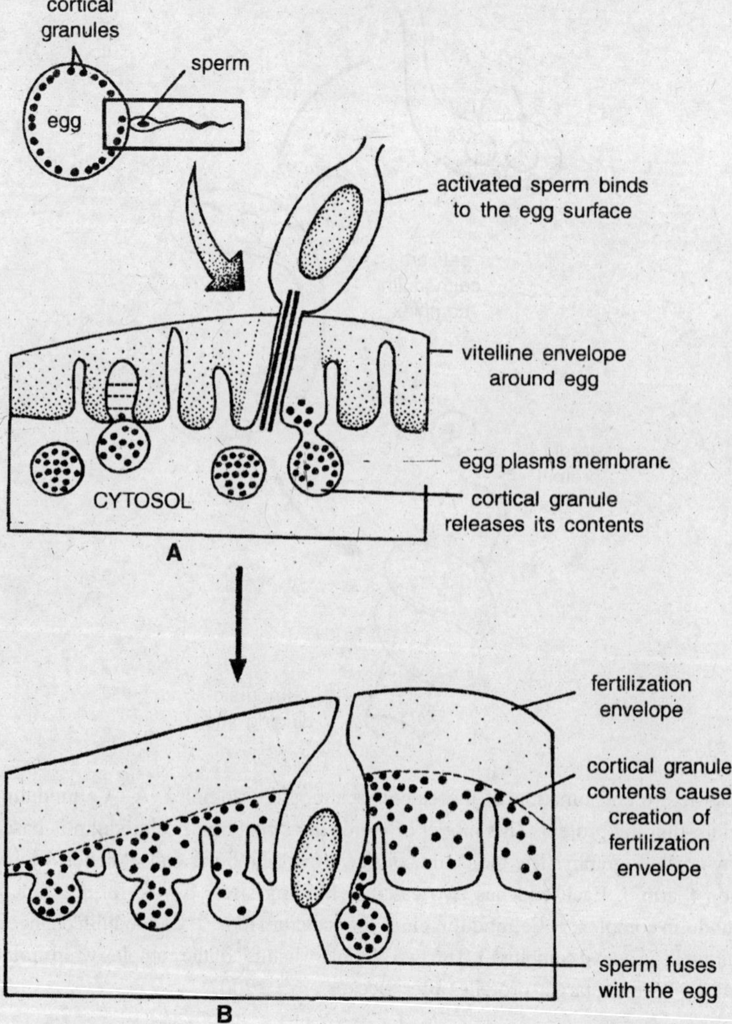

Fig. 61.13. The role of calcium in the slow block to polyspermy in sea urchins. A—A sperm cell that has been activated binds to the egg surface, resulting in local calcium release and exocytosis of cortical granule. B—The result is the creation of the fertilization envelope, which prevents additional sperm from penetrating the egg (after Becker *et al.*, 2007).

Release of calcium is necessary for two crucial events. First, it stimulates the fusion of vesicles, known as **cortical granules**, with the egg plasma membrane, resulting in the release of granule contents outside the cell (exocytosis; Fig. 61.13). Cortical granules contain several proteins and enzymes, the release of which results in alterations of the protein coat surrounding most eggs (typically known as **vitelline envelope**). These changes make the egg unable to bind additional sperm thereby preventing more than one sperm from fertilizing the egg. This process is known as **slow block to polyspermy** (An earlier process called **fast block to polyspermy** involves a transient depolarization of the egg plasma membrane).

The second major function of calcium is **egg activation**. Egg activation involves the restart of many metabolic processes, the reorganization of the internal contents of the egg, and other events that initiate the process of embryonic development. Many features of the slow block to polyspermy and egg activation can be initiated bv treating unfertilized eggs with calcium ionophore in the sperm, demonstrating the key role elevated calcium levels within the egg play in its activation.

Oscillation of Ca^{2+} concentration. The oscillation of calcium concentration over time is found to elicit a cellular response in neurons and fertilized mammalian eggs. This contributes to stable changes in the state of these cells. Calcium oscillations are also important in regulating the opening and closing of stomata in plants.

FUNCTION OF NITRIC OXIDE IN G PROTEIN-LINKED RECEPTOR

An important signalling molecule in the cardiovascular system is **nitric oxide (NO)**, a toxic short-lived (half life about 5–10 seconds) gas molecule produced by the enzyme **NO synthetase**, which converts the amino acid arginine to NO and citrulline. Nitric oxide's mode of action on blood vessels involves G protein. NO is hydrophobic and small enough to pass readily across target cell plasma membrane.

It has been known for many years that *acetylcholine,* which is released by autonomic nerves, dilates blood vessels by causing their smooth muscles to relax. In 1980, **Robert Furchgott** demonstrated that acetylcholine dilated blood vessels only if the endothelium (the inner lining of the blood vessel) was intact. He concluded that blood vessels are dilated because the endothelial cells produce a **signal molecule** (or vasodilator) that makes vascular smooth muscles relax. In 1986, work by **Furchgott** and parallel work by **Louis Ignarro** identified NO as the signal released by endothelial cells that causes relaxion of the vascular smooth muscles (Fig. 61.14).

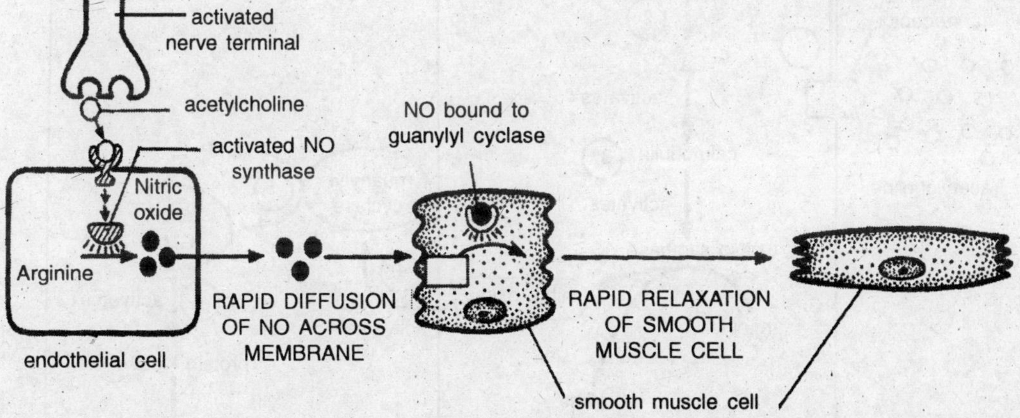

Fig. 61.14. The role of nitric oxide (NO) in smooth muscle relaxation in a blood vessel wall. Acetylcholine released by nerve terminals in the blood vessel wall activates enzyme NO synthetase in endothelial cells, lining the blood vessel, causing the endothelial cells to produce NO. The NO gas diffuses out of the endothelial cells and into the underlying smooth muscle cells, where it binds to and activates guanylyl cyclase to produce cyclic GMP. The cyclic GMP triggers a response that causes the smooth muscle cells to relax, enhancing blood flow through the blood vessel (after Alberts *et al.*, 2002).

Mode of action of NO on smooth muscle cells. In Fig. 61.15, it has been shown that how the binding of acetylcholine to the surface of vascular endothelial cells result in release of NO. This process involves the following six steps: 1. Acetylcholine binds to G protein-linked receptors that activate the phosphoinositide signalling pathway, causing IP_3 to be produced by the endothelial cells. 2. IP_3 causes the release of calcium from the endoplasmic reticulum. 3. The calcium ions bind to calmodulin, forming a complex that stimulates NO synthetase enzyme to produce nitric oxide. 4. Nitric oxide is a gas that readily diffuses through plasma membranes, allowing it to pass from the endothelial cell into the adjacent smooth muscle cell. 5. Once inside the smooth muscle cell, NO activates the enzyme **guanylyl cyclase**, which catalyzes the formation of cyclic GMP (cGMP). Cyclic GMP is derived from GTP in a manner similar to the production of cAMP from ATP, and like cAMP, cGMP concentration activates a protein known as **protein kinase G**, which induces muscle relaxation by catalyzing the phosphorylation of the appropriate muscle proteins.

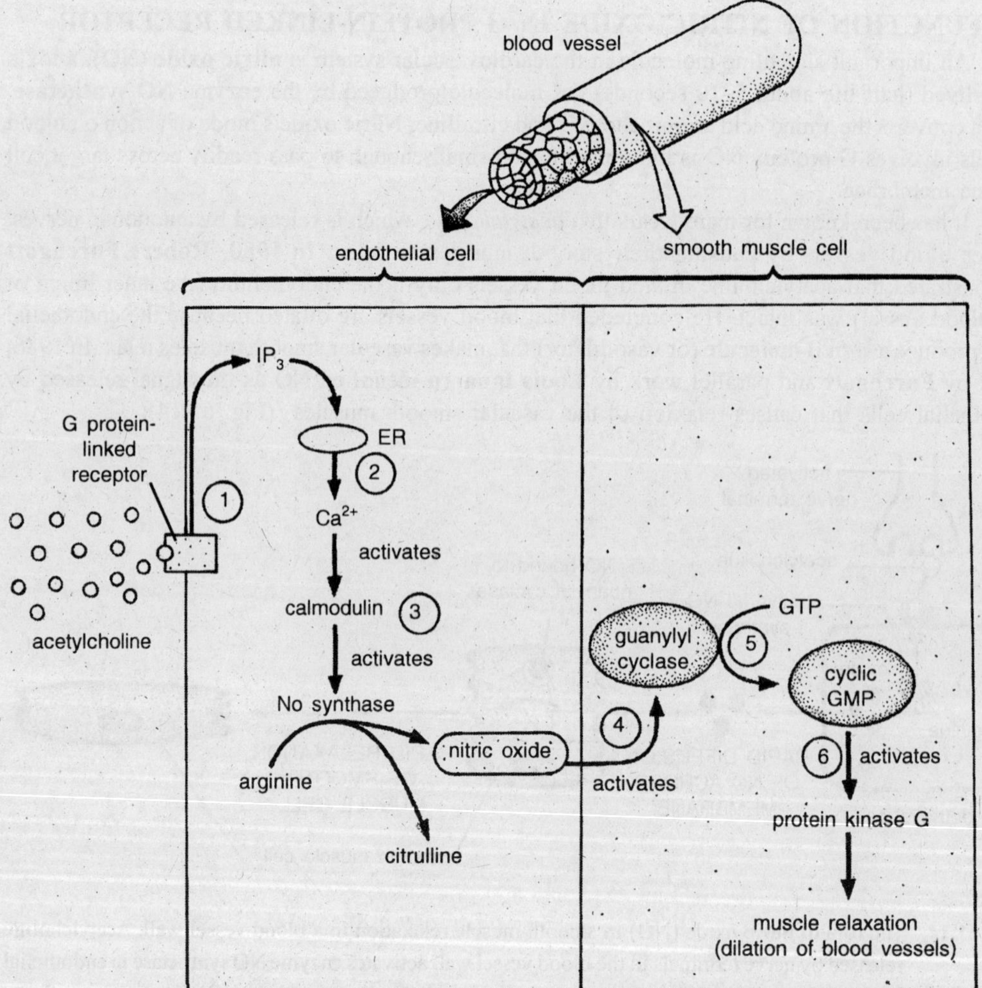

Fig. 61.15. The action of nitric oxide on blood vessels. The binding of acetylcholine to endothelial cells triggers the production of nitric oxide (NO), which diffuses into the adjacent smooth muscle cells and stimulates guanylyl cyclase, thereby leading to muscle relaxation (after Becker *et al.*, 2007).

Box 61.4

Gases such as NO (nitric oxide) and CO (carbon monoxide) are paracrine signaling molecules. They are used as intercellular signals in mammals and certain plants. For example, activated macrophages and neutrophils use NO in killing invading microorganisms. In plants, NO is involved in defensive responses to injury or infection.

Use of NO in treatment of some human ailments.

1. **Role of NO in treatment of angina.** The mechanism by which acetylcholine stimulation of the endothelial cells leads to smooth muscle relaxation also explains the mechanism of action of **nitroglycerin**. Nitroglycerin is often taken by patients with **angina** (chest pain due to inadequate blood flow to the heart) to relieve constriction of coronary arteries. In 1977, **Ferid Murad** found that nitroglycerin and similar vasodilators elicit release of nitric oxide (NO), which relaxes arterial smooth muscle cells. In 1998, **Furchogott, Ignarro** and **Murad** received a Nobel Prize for their explanation of NO's effects on the cardiovascular system.

2. **Role of NO in penile erection.** Nitric oxide (NO) is also used by neurons to signal nearby cells. For example, NO released by the neurons in the penis results in the blood vessel dilation responsible for penile erection. The drug *slidenafil,* sold under the trade name of **viagra**, is an inhibitor of a phosphodiesterase that normally catalyzes the breakdown of cyclic GMP. By maintaining elevated levels of cyclic GMP in erectile tissue, this pathway is stimulated for a longer time period following NO release.

61.3 SIGNALLING BY PROTEIN KINASE-ASSOCIATED RECEPTORS

Some proteins not only function as receptors, but are also serve as protein kinases. Thus, when they bind to the appropriate ligand, their kinase activity is stimulated, and they transmit signals through a cascade of phosphorylation events within the cell. Kinases are enzymes and tend to add phosphate groups to particular amino acids within substrate proteins. Receptor kinases are of following two types: 1. Those that phosphorylate a tyrosine residue (**tyrosine kinases**) and 2. Those that phosphorylate a serine or threonine residue (serine/threonine kinases). The action of tyrosine kinases are well understood, so are described here.

Receptor Tyrosine Kinases.

Many **receptor tyrosine kinases (RTKs)** trigger a chain of signal transduction events inside the cell that ultimately leads to cell growth, proliferation or in the specialization of cells in a process known as **differentiation**. These processes are competently controlled, so that only specific cells respond when the proper ligand is available. Examples of RTKs include the insulin receptor, the nerve growth factor receptor, and epidermal growth factor (EGF) receptor.

Structure of RTKs. These receptors often consist of a single polypeptide chain with only one transmembrane segment. Within this polypeptide chain are several distinct domains (Fig. 61.16). The extracellular portion of the RTK contains the **ligand-binding domain**. The other end of the peptide protrudes through the plasma membrane into the cytosol. On the cytosolic side, a portion of the receptor forms the tyrosine kinase while the remainder constitutes a cytosolic tail. The cytosolic portion of the receptor (RTK contains tyrosine residues that are in fact substrates or targets for the tyrosine kinase portion of the receptor).

The Activation of RTKs. Signal transduction is initiated when a ligand binds, causing the receptor tyrosine kinases (RTKs) to aggregate (Fig. 61.16B). Often two receptor molecules cluster together within the plasma membrane when they bind ligand. Once the receptors cluster in this way, the tyrosine kinase associated with each receptor phosphorylates the tyrosines of neighbouring receptors (**cross-phosphorylation**). Since the receptors phosphorylate other receptors of the same type, this process is referred to as **autophosphorylation**.

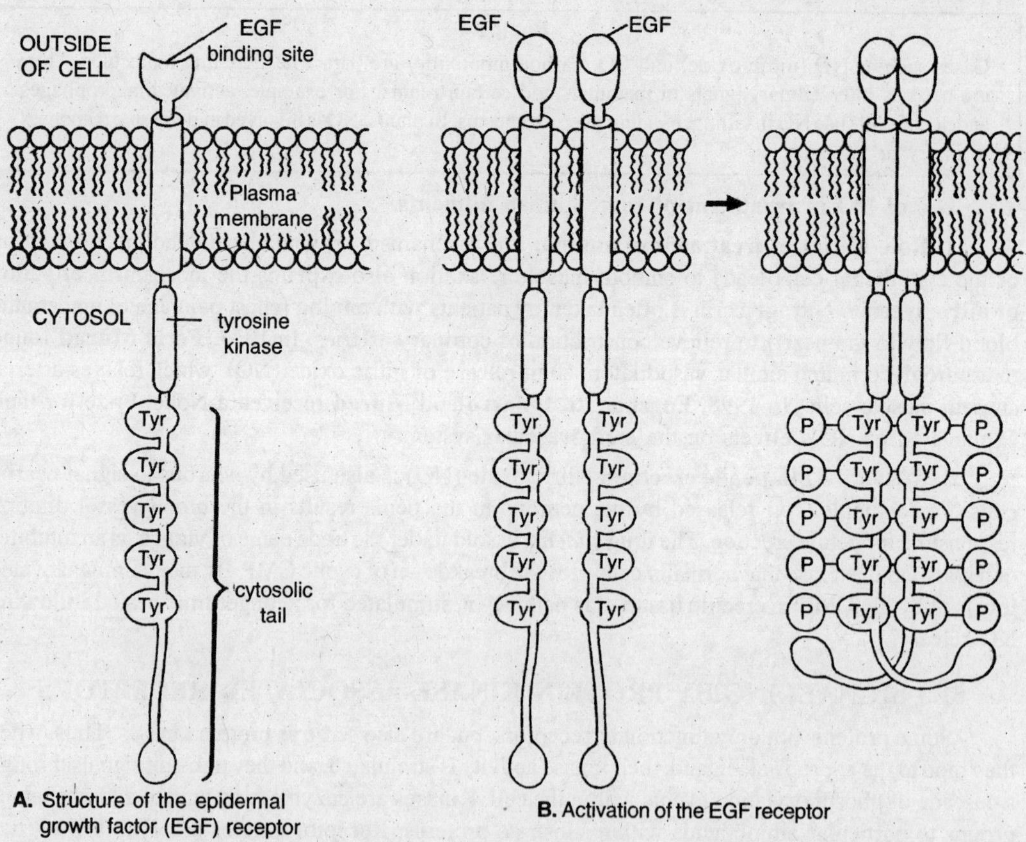

Fig. 61.16. The structure and activation of a receptor tyrosine kinase (RTK). A—The receptor for epidermal growth factor (EGF), shown here, is typical of many RTKs. These receptors often have only one transmembrane segment. The extracellular portion of the receptor binds to the ligand (EGF in this case). Inside the cell, a portion of the receptor has tyrosine kinase activity. The remainder of the receptor contains a series of tyrosine residues that are substrates for the tyrosine kinase. B—The activation of a receptor tyrosine kinase starts with the binding of a messenger (EGF), causing receptor aggregation or clustering. Once the receptors aggregate, they cross-phosphorylate each other at a number of tyrosine amino acid residues. The formation of tyrosine phosphate (Tyr-P) residues on the receptor creates binding sites for cytosolic proteins that contain SH_2 domains (after Becker *et al.*, 2007).

Signal transduction through RTKs. Once autophosphorylation of tyrosine residues on the cytosolic portion of the receptor occurs in response to ligand binding [(1) and (2) in Fig. 61.17], the recruits a number of cytosolic proteins to interact with itself. Each of these proteins binds to the receptor at a phosphorylated tyrosine residue. To bind to the receptor, each cytosolic protein must contain a stretch of amino acids that recognizes the phosphotyrosine and a few neighbouring amino acids on the receptor. The portion of a protein that recognizes one of these phosphorylated tyrosines is called an **SH_2 domain** [SH_2=Src homology $(\text{domain})^2$].

Enrolment of different SH_2 domain-containing proteins activates different signal transduction pathways. As a result, receptor tyrosine kinases can activate several different signal transduction pathways at the same time. These include the inositol-phospholipid-calcium second messenger pathway, which we have already discussed, and the **Ras pathway**, which ultimately activates the expression of genes involved in growth or development.

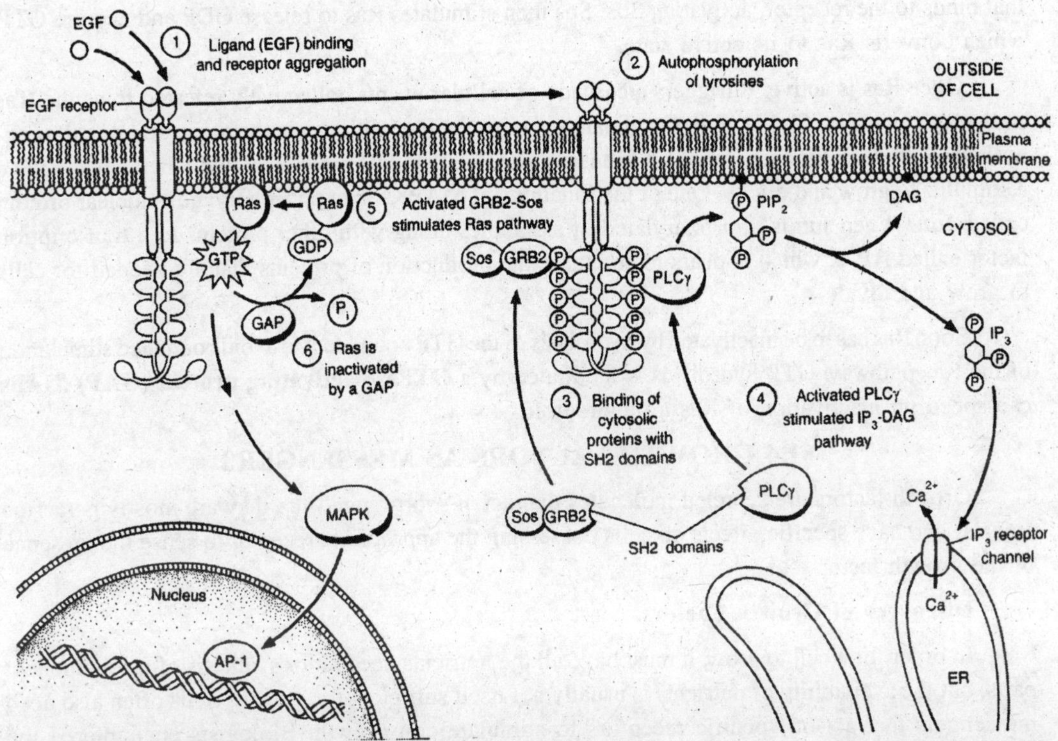

Fig. 61.17. Signal transduction through receptor tyrosine kinases. 1. Upon ligand binding, receptor tyrosine kinases, such as that shown here for the epidermal growth factor (EGF). aggregate and 2. Undergo autophosphorylation. Once a receptor is phosphorylated at residues in its cytosolic tail, 3. Proteins with SH_2 domains such as phospholipase C (PLC) and GRB2 bind to the receptor. 4. The binding of phospholipase C results in its activation and the cleavage of PIP_2 into IP_3 and DAG 5. The binding of GRB2 causes the activation of Sos, to which it is bound. Sos (a GEF) then causes the activation of the Ras protein by helping it release GDP and acquire GTP. Activated Ras initiates a cascade of events that ultimately results in the formation of AP-1, a transcription factor in the nucleus that stimulates the expression of genes needed for growth of cell. 6. Ras is inactivated by hydrolysis of its bound GTP, a step facilitated by a GAP (after Becker *et al.*, 2007).

Characteristics of Ras pathway. Ras is important in regulating the growth of cells. It comprises a single subunit. Ras and other small monomeric G proteins are important signalling molecules. Like other types of G proteins, Ras can be bound to either GDP or GTP, but is active only when bound to GTP. In the absence of receptor stimulation, Ras is normally in the GDP-bound state. For Ras to become active, it must release GDP and acquire a molecule of GTP. For this to take place, Ras requires the help of another type of protein called a **guanine-nucleotide exchange factor (GEF)**.

The GEF that activates Ras is a Sos (so called because it was originally identified from a genetic mutation in fruit flies called *Son of sevenless* that results in the failure of cells in the compound eye to develop properly). For Sos to become active, it must bind indirectly to the receptor tyrosine kinase through another protein called **GRB2** (3. in Fig. 16.17), which contains a SH_2 domain. Thus, to activate Ras, the receptor becomes tyrosine phosphorylated, and GRB2 and Sos form a complex that binds to the receptor, activating Sos. Sos then stimulates Ras to release GDP and acquire GTP, which converts Ras to its active state.

Once Ras is active, it triggers a cascade of cellular events collectively referred to as the **Ras pathway** (5. in Fig. 61.17). One significant event in the Ras pathway is the activation of **mitogen-activated protein kinases**, or **MAP kinases (MAPKs)**. MAPKs are activated when cells receive a stimulus to grow and divide. One of the functions of MAPKs is to phosphorylate a nuclear protein called **Jun**. When Jun is phosphorylated, it assembles along with other proteins in a transcription factor called **AP-1**, which appears to stimulate the production of proteins that are needed for cells to grow and divide.

Soon Ras has to be inactivated by hydrolysis of the GTP bound to it to avoid continued stimulation of the Ras pathway. GTP hydrolysis is facilitated by a **GTPase activating protein (GAP)**. GAPs can speed up inactivation of Ras a hundredfold.

61.4 GROWTH FACTORS AS MESSENGERS

Growth factors are secreted molecules that act at short range (*i.e.*, they are mostly paracrine signals) and have specific effects on cells possessing the appropriate receptor to sense the presence of the growth factor.

Discovery of Growth Factors.

In order for a cell to grow, it must have all the nutrients needed for synthesis of its component parts, but the availability of nutrients is usually not itself sufficient for growth. Cells often also need messengers that act on specific receptors to stimulate cell growth. Biologists encountered the requirements for cell growth when they first tried to culture cells in nutrients, including the presence of blood plasma, cells would not grow. A turning point came when blood serum was used instead of plasma: *Serum was able to support the growth of cells, whereas plasma would not*. Many of the messengers present within the serum have now been purified, and they are members of various classes of proteins known as **growth factors**.

The difference between blood serum and blood plasma held an important clue about growth factors. *Plasma* is whole blood, including unreacted platelets (which contain clotting components) but without the red and white cells. *Serum* is the clear fluid remaining after blood has clotted. During clotting, platelets secrete growth factors into the blood that stimulate the growth of cells called **fibroblasts**, which form the new connective tissue that makes up a scar. After clotting, the resulting serum is full of **platelet-derived growth factor (PDGF)**. Plasma does not contain this factor because the clotting reaction has not taken place (see **Becker** *et al.*, 2007).

Examples of Growth Factors. Receptor for PDGF is found to be a receptor for tyrosine kinase. In fact, several growth factors act by stimulating receptor tyrosine kinases, including **insulin, insulin-like growth factor-I, fibroblast growth factor, epidermal growth factor**, and **nerve growth factor**. Many other types of growth factors have been isolated (Table 61.2).

Although collectively known as growth factors, however, these proteins have roles in many diverse events. These include not only growth and cell division, but also crucial events during the development of embryos, responses to tissues injury, and many other activities.

Table 61.2 Examples of growth factor families/superfamilies (source: Becker *et al.*, 2007).

Growth factor	*Target cells*	*Type of receptor complex*
1. Epidermal growth factor (EGF)	Wide variety of epithelial and mesenchymal cells	Tyrosine kinase
2. Transforming growth factor-α (TGFα)	Same as EGF	Tyrosine kinase
3. Platelet-derived growth factor (PDGF)	Mesenchyme, smooth muscle, trophoblast	Tyrosine kinase
4. Transforming growth factor-β (TGFβ)	Fibroblastic cells	Serine-threonine kinase
5. Fibroblast growth factor (FGF)	Mesenchyme, fibroblasts, many other cell types	Tyrosine kinase
6. Interleukin-2 (IL-2)	Cytotoxic T lymphocytes	Complex of three units
7. Colony stimulating factor-I (CSF-1)	Macrophage precursors	Tyrosine kinase
8. Wnts	Many types of embryonic cells	Seven-pass protein

1. Disruption of growth factor signalling through receptor tyrosine kinases. Many receptors for growth factors are tyrosine kinases, falling into families based on which growth factor (s) they can bind. One well-studied class of growth factors are the **fibroblast growth factors (FGFs)** and their receptor tyrosine kinases, the **fibroblast growth factor receptors (FGFRs)**. FGFs and FGFRs are used in signalling events in both adult animals and embryos.

For example, FGFRs have been shown to play an important role in the development of cells derived from the embryonic cell layer or germinal layer of early embryo, known as **mesoderm**. The mesoderm forms many cell types including muscle, cartilage, bone and blood cells, as well as the forerunner of the vertebral column (the notochord). When specific FGFRs fail to function properly, the development of particular mesodermal tissue is affected. In one class of FGFR defects, a **mutation** in the receptor results in *dominant effects* on the developing embryo. In other words, even though the embryo makes a substantial quantity of normal functional receptor, the presence of the mutant receptor within the embryo's cells prevents the normal receptors from functioning properly. Normal function is **inhibited** because FGFRs must act together as dimers to bind FGFs. If a normal receptor dimerizes with a mutant receptor, then the phosphorylation event that normally occur within the tyrosine kinase portion of the receptor fail to occur, blocking signal transduction (Fig. 61.18). Such a mutation that cancels the function of the normal receptor is called a **dominant negative mutation.**

Examples of dominant negative mutations. Dominant negative mutations are found to have dramatic effects on cells in developing embryos. For example, when genetically engineered dominant negative FGFRs are expressed in frog eggs and the eggs are fertilized, the embryos fail to develop mesodermal tissues in the trunk and tail, resulting in tadpoles with heads, but no bodies. In humans, dominant mutations in the transmembrane portion of the FGFR-3 gene result in the most common form of dwarfism, called **achondroplasia**. Heterozygous individuals have abnormal bone growth, in which the long bones suffer from abnormal ossification (*i.e.*, the process by which cartilage is converted to bone during childhood). A related condition known as **thanatophoric dysplasia** (a congenital dwarfism) often results from a single amino acid change in the cytosolic portion of FGFR-3 protein. In this case, more severe bone abnormalities result, and affected individuals die soon after birth.

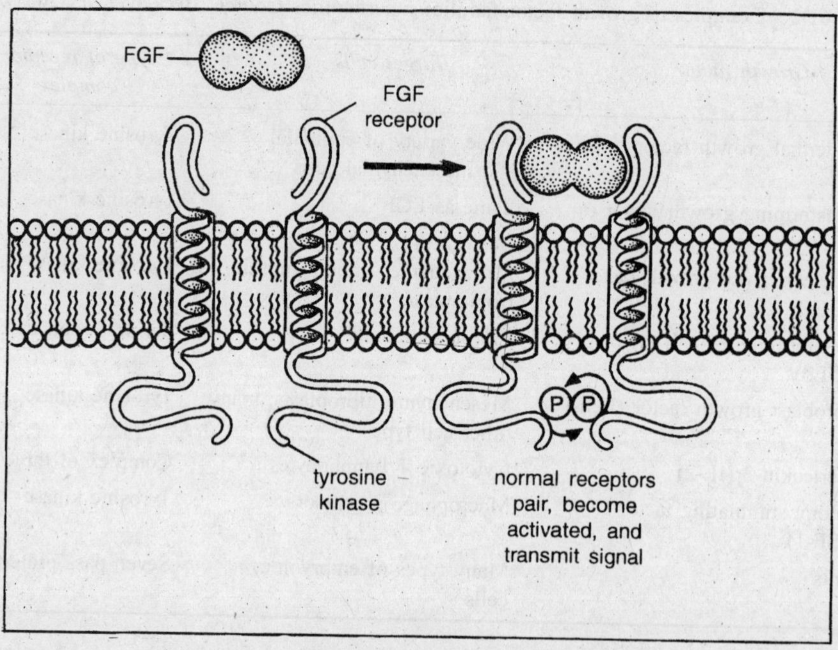

A. Normal FGFR: FGF binds; FGFRs form a dimer

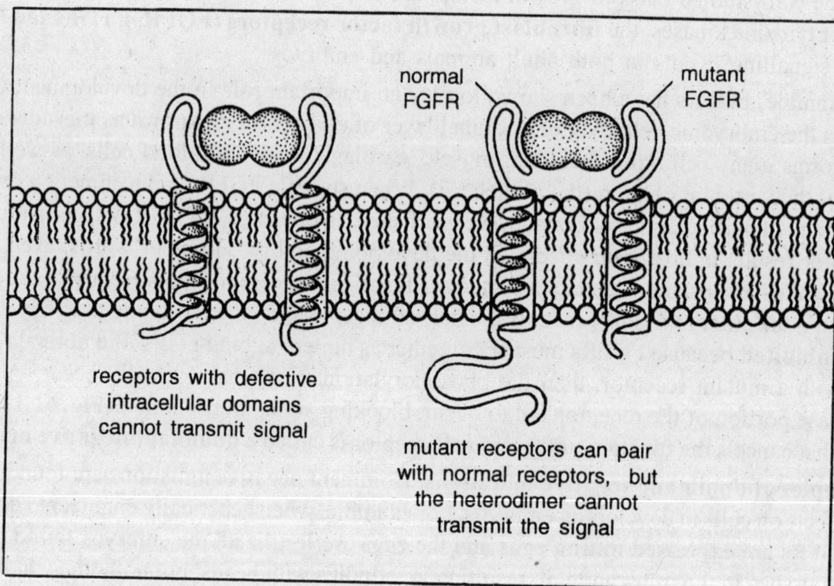

B. Dominant negative mutation in FGFR

Fig. 61.18. Dominant negative disruption (due to mutation) of FGF receptor (FGFR) function. A—normal receptors form dimers after binding FGF, and transmit the appropriate signal via Ras and MAPK. B—When a cell makes mutant FGF receptors, normal receptors can dimerize as in A, or the defective receptors can bind to FGF and dimerize with normal receptors. In this case, no signal is transmitted (after Becker *et al.*, 2007).

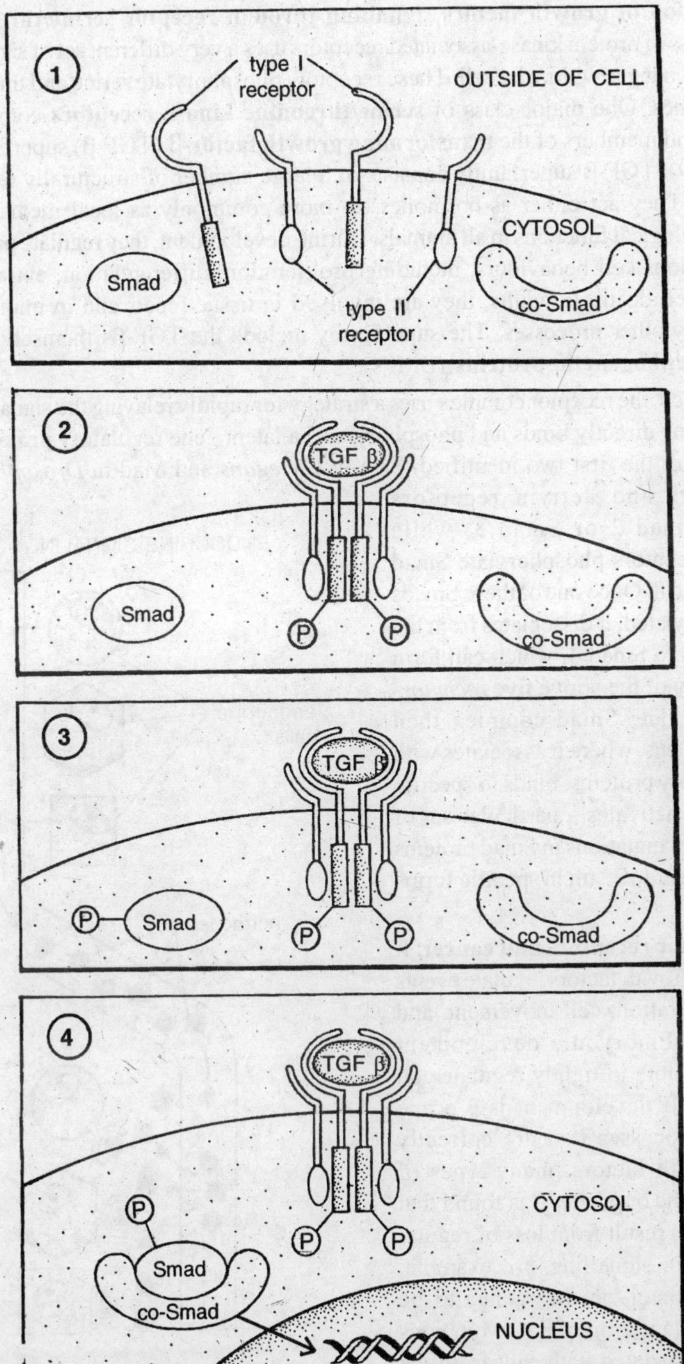

Fig. 61.19. Signal transduction by TGFβ receptor family proteins. 1. Type I and Type II receptors for TGFβ in a cell prior to binding of the growth factor. 2. Binding of growth factor results in clustering of type I and type II receptors. 3. The activated type I receptors then phosphorylate particular receptor-mediated smads. 4. These smads then bind to other smads, and together they enter the nucleus (after Becker *et al.*, 2007).

2. Disruption of growth factors signalling through receptor serine/threonine kinases. Another major class of protein kinase-associated receptors uses a very different set of signal transduction pathways to elicit changes within the cell. These receptors phosphorylate serine and threonine residues rather than tyrosines. One major class of **serine/threonine kinase receptors** comprises a family of proteins that bind members of the **transforming growth factor-β** (TGF-β) superfamily of growth factors (Fig. 61.19). TGF-β superfamily consists of a large number of structurally related, secreted, dimeric proteins. They act either as hormones or, more commonly as local mediator, to regulate a wide range of biological functions in all animals. During development, they regulate pattern formation and influence various cell behaviours, including proliferation, differentiation, extracellular matrix production, and cell death. In adults, they are involved in tissue repair and in immune regulation, as well as in many other processes. The superfamily include the TGF-βs themselves, the *activins* and the **bone morphogenetic proteins** (BMPs).

Once activated, the receptor complex uses a strategy for rapidly relaying the signal to the nucleus. The **type I receptor** directly binds and phosphorylates a latent gene regulatory protein of the **Smad family** (named after the first two identified, Sma in *C.elegans* and Mad in *Drosophila*). Activated TGF-β receptors and activin receptors phosphorylate Smad 2 or Smad 3, while activated BMP receptors phosphorylate Smad 1, Smad 5 and Smad 8. Once one of these Smads has been phosphorylated, it dissociates from the receptor and binds to Smad 4, which can form a complex with any of the above five *receptor-activated Smads*. The Smad complex then moves into the nucleus, where it associates with other gene regulatory proteins, binds to specific sites in DNA, and activates a particular set of target genes. Several mutations in Smad proteins in human cells appear to result in specific forms of cancer.

Growth factor receptors and cancer. It is well known that growth factors regulate events such as cell proliferation, cell movement, and gene expression. Embryonic development requires growth factors to tightly regulate cell biological events. If development is a prime example of the processes that are correctly regulated by growth factors, many types of cancer are exactly the opposite. It is found that a number of cancers result from loss of regulation of growth factor signalling. For example, one of the first cancer-causing genes to be identified was a mutant form of Ras. Mutations in Ras are often associated with cancer. In the case of receptor tyrosine kinases, for example, mutations in the epidermal growth factor receptors can result in **breast cancer, glioblastoma** (a cancer of glial cells in the brain), and **fibrosarcoma** (a type of cancer of long bones). Similar discoveries have been made

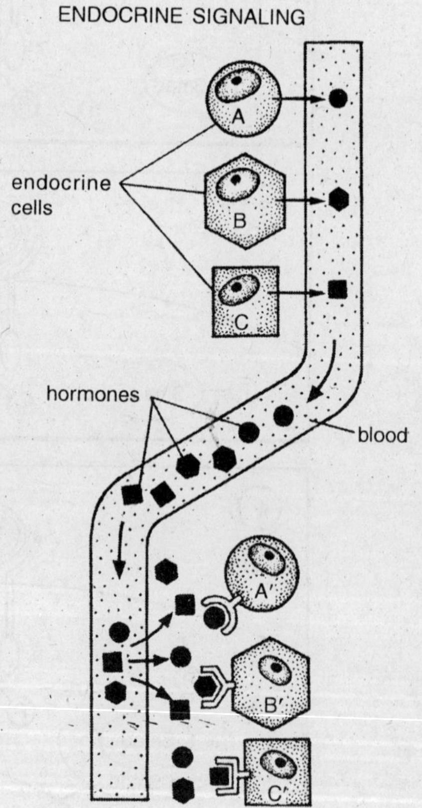

Fig. 61.20. Endocrine cells secrete hormones into the blood, which signal only the specific target cells that recognize them. These target cells have receptors for binding a specific hormone, which the cells "pull" from the extracellular fluid (after Albert *et al.*, 2002).

for receptor serine threonine kinases. Mutations in TGF-β type I receptor occur in one-third of **ovarian cancers**, and mutations in the type II receptor occur in many **colorectal cancers**. Mutations in one of the Smads (Smad 4) occur in one-half of all **pancreatic cancers**.

61.5 SIGNALLING BY HORMONES

Hormones are the chemical signals which are used by the animals to regulate the function of various cells and tissues. They are chemical messengers secreted by one tissue that regulate the function of other cells or tissues in the same organisms. In contrast to growth factors, hormones are often act over large distances. In animals, hormones are often transported via the vascuiature. Hundreds of different hormones regulate a wide variety of functions, many critical for maintaining the physiological steady state of an organism.

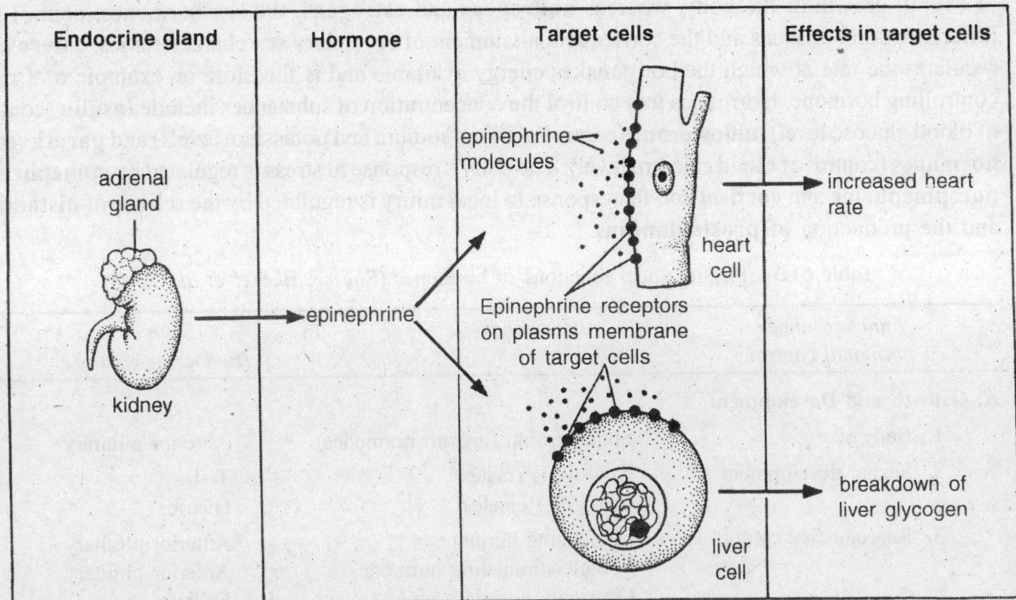

Fig. 61.21. Target tissues for endocrine hormones. Cells in a target tissue have hormone-specific receptors embedded in their plasma membrane (or in case of the steroid hormones, present in nucleus and cytosol). Heart and liver cells can respond to epinephrine synthesized by the adrenal glands because these cells have epinephrine specific receptors on their outer surfaces. A specific hormone may elicit different responses in different target cells. Epinephrine causes an increase in heart rate but stimulates glycogen breakdown in the liver (after Becker *et al.*, 2007).

Although hormones can be considered as a group based on their regulatory functions, however, they differ in many ways. Some hormones are steroids or other hydrophobic molecules that are targeted to intracellular receptors. Other hormones, such as adrenergic hormones are targeted to a wide variety of different G protein-linked receptors. Still others, such as insulin, are ligands, for receptor tyrosine kinase.

1. Classification of Hormonal Signals

Every animal hormone can be placed in one of two categories depending on the distance over which it operates. An **endocrine hormone** travels by means of the circulatory system from the cells where it is released to other cells (target cells), where it regulates one or more specific functions. A **paracrine hormone** is a more local signal that is taken up, destroyed, or immobilized so rapidly that it can act only on cells in the immediate environment.

Endocrine hormones are synthesized by the **endocrine tissue** of the body and are secreted directly into the bloodstream. Once secreted into the circulatory system endocrine hormones have a limited life span, ranging from a few seconds for *epinephrine* (a product of the adrenal gland) to many hours for *insulin* (a product of a beta (β) cells of islets of Langerhans of endocrine pancreas). As they circulate in the bloodstream, hormones come into contact with receptors in tissues throughout the body. A tissue that is specifically affected by a particular hormone is called a **target tissue** for that hormone (Fig. 61.21). For example, the heart and the liver are target tissues for epinephrine, whereas the liver and skeletal muscles are targets for insulin.

2. Functions of Hormones. Hormones regulate a wide range of physiological functions, including growth and development, rate of body processes (metabolism), concentration of substances and responses to stress and injury (Table 61.3). For example, **somatotropin** is involved in the regulation of overall growth of the body, whereas **androgens** and **estrogens**, the sex hormones, control the differentiation of tissues and the consequent attainment of secondary sex characteristics. **Thyroxine** regulates the rate at which the body makes energy available and is therefore an example of a rate-controlling hormone. Hormones that control the concentration of substances include **insulin** (control of blood glucose level), **aldosterone** (control of blood sodium and potassium levels) and **parathyroid hormones** (control of blood calcium level). The body's response to stress is regulated by **epinephrine, norepinephrine** and **cortisol** and its response to local injury is regulated by the release of **histamine** and the production of **prostaglandins.**

Table 61.3 Physiological functions of hormones (Source: Becker *et al.*, 2007).

Function under hormonal control	*Hormones*	*Source (Endocrine tissue)*
A. Growth and Development		
1. Body size	Somatotropin (growth hormones)	Anterior pituitary
2. Sexual development	Androgens (males) Estrogen (females)	Testes Ovaries
3. Reproductive cycle	Luteinizing hormone Follicle-stimulating hormone Chorionic gonadotropin	Anterior pituitary Anterior pituitary Follicle
B. Rates of body processes		
1. Hormone secretion	Tropic hormones	Anterior pituitary
2. Basal metabolism	Thyroxine	Thyroid
3. Glucose uptake	Insulin	Pancreas
4. Kidney filteration	Antidiuretic hormone (vasopressin)	Posterior pituitary
5. Uterine contraction	Oxytocin	Posterior pituitary
C. Concentration of substances		
1. Blood glucose	Glucagon, insulin	Pancreas
2. Mineral balance	Corticosteroids	Adrenal cortex
3. Blood calcium	Parathyroid hormone	Parathyroid
D. Responses to stress and injury		
1. Heart rate	Epinephrine	Adrenal medulla
2. Blood pressure	Epinephrine	Adrenal medulla
3. Inflammation	Histidine Prostaglandins Corticosteroids (cortisol)	Mast cells All tissues Adrenal cortex

3. Chemical Type of Hormones

Chemically, the endocrine hormones are of four types: amino acid derivatives, peptides, proteins and lipid-like hormones such as steroids. An example of amino acid derivative is epinephrine, derived from tyrosine. **Antidiuretic hormone** (also called **vasopressin**) is an example of peptide hormone, whereas **insulin** is an example of protein hormone. **Testosterone** is an example of a steroid hormone. The steroid hormones are derivatives of cholesterol that are synthesized either in the gonads (the **sex hormones**) or in the adrenal cortex (the **cortisteroids**).

Examples of paracrine hormones are histamine and the prostaglandins. **Histamine** is produced by decarboxylation of the amino acid histidine and is responsible for local inflammatory responses. The **prostaglandins**, so named because they were first discovered in human semen as a secretion of prostate gland, are derived from arachidonic acid and are important in smooth muscle function.

Table 61.4 Chemical classification and functions of hormones (Source: Becker *et al.*, 2007).

Chemical classification	*Examples*	*Regulated functions*
A. Endocrine Hormones		
1. Amino acid derivatives	1. Epinephrine (adrenaline) and norepinephrine (both derived from tyrosine)	Stress response: regulation of heart rate and blood pressure; release of glucose and fatty acids from storage sites
	2. Thyroxine (derived from tyrosine)	Regulation of metabolic rate
2. Peptides	1. Antidiuretic hormones (vasopressin)	Regulation of body water and blood pressure
	2. Hypothalamic hormones (releasing factors)	Regulation of tropic hormone release from pituitary gland
3. Proteins	1. Anterior pituitary hormones	Regulation of other endocrine systems
	2. Sex hormones (androgens and estrogens)	Development and control of reproductive capacity
	3. Corticosteroids	Stress responses; control of blood electrolytes
B. Paracrine Hormones		
1. Amino acid derivatives	Histamine	Local responses to stress and injury
2. Arachidonic acid derivatives	Prostaglandins	Local responses to stress and injury

4. Modes of Endocrine Regulation of Target Cells.

Since hormones modulate the function of particular target tissues, an important aspect of studying hormones is understanding the specific functions of these target tissues. To understand how endocrine hormones act, we will look more closely at the water-soluble (nonlipid) hormones such as **adrenergic hormones** epinephrine and norepinephrine. (Epinephrine is also called adrenaline, the two words are of Greek and Latin derivation, respectively, and mean "above or near, the kidney", referring to the location in the body of the adrenal glands, which synthesize this hormone). When secreted into the blood-stream, epinephrine and norepinephrine stimulate changes in many different tissues or organs, all aimed at preparing the body for dangerous or stressful situations (the so-called "fight-or-flight response"). On the whole, the adrenergic hormones trigger increased cardiac output, shunting blood from the visceral organs to the muscles and the heart, as well as dilation of arterioles to facilitate oxygenation of the blood. In addition, these hormones stimulate the breakdown of glycogen to supply glucose to the muscles.

Adrenergic hormones bind to a family of G protein-linked receptors known as **adrenergic receptors**. The individual members of this family differ mainly in their preference for epinephrine or norepinephrine and in which G protein is linked to the receptor. They are broadly classified into α- and **β-adrenergic receptors**. The α-adrenergic receptors bind both epinephrine and norepinephrine. These receptors are located on the smooth muscles that regulate blood flow to visceral organs. The β-adrenergic receptors bind epinephrine much better than norepinephrine. These receptors are found on smooth muscles associated with arterioles that feed the heart, smooth muscles of the bronchioles in the lungs, and skeletal muscles.

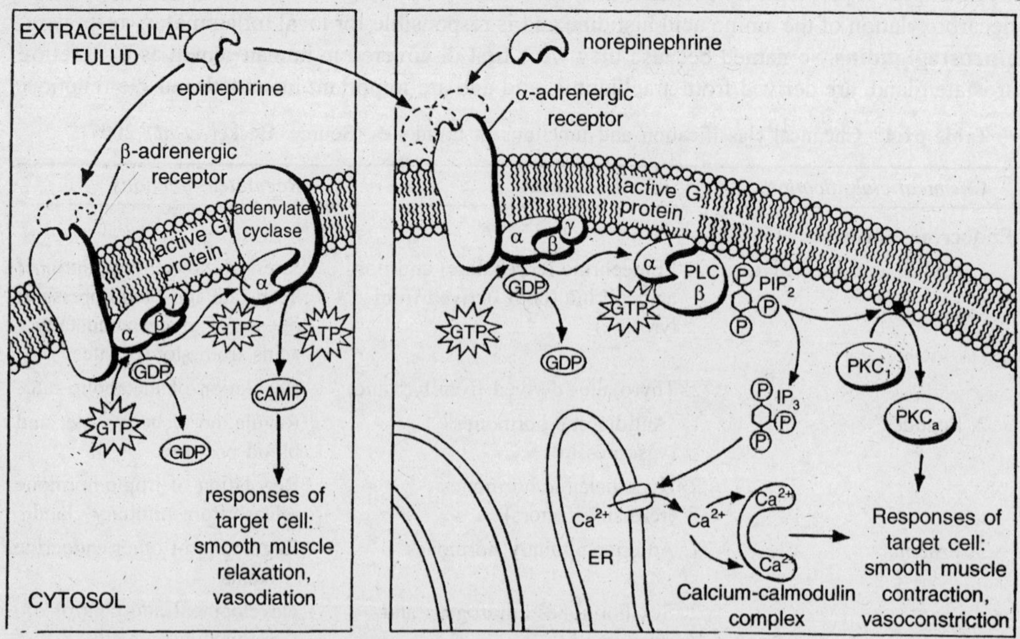

Fig. 61.22. The stimulation of G protein-linked signal transduction pathways by α- and β-adrenergic receptors (after Becker *et al.*, 2007).

The α- and β-adrenergic receptors stimulate different signal transduction pathways, because they are linked to different G proteins (Fig. 61.22). For example, the G proteins activated by one type of α-adrenergic receptor, the *α_1-adrenergic receptors*, are Gp proteins, whereas the β-adrenergic receptors activate Gs. As we discussed earlier, activation of G*s* stimulates the *cAMP signal transduction pathway*, leading to relaxation of certain smooth muscles. Activation of G*p* stimulates phospholipase C, leading to the production of IP_3 and DAG, which in turn elevates intracellular calcium levels. Table 61.5 summarises some of the main physiological functions that are regulated by cAMP. As a specific example of cAMP-mediated regulation, here will be considered the control of glycogen degradation by epinephrine in liver or muscle cells.

Table 61.5. Examples of cell functions regulated by cAMP (Source: Becker *et al.*, 2007).

Regulated function	*Target tissue*	*Hormone*
1. Glycogen degradation	Muscle, liver	Epinephrine
2. Fatty acid production	Adipose	Epinephrine
3. Heart rate, blood pressure	Cardiovascular	Epinephrine
4. Water reabsorption	Kidney	Antidiuretic hormone
5. Bone resorption	Bone	Parathyroid hormone

***(i)* Control of glycogen degradation by cAMP pathway.** The breakdown of glycogen to provide muscle cells with an adequate supply of glucose is stimulated by the adrenergic hormones. The breakdown of glycogen is catalysed by the enzyme **glycogen phosphorylase**, which cleaves glucose units from glycogen as glucose-1-phosphate by the addition of inorganic phosphate (Pi). The glycogen phosphorylase system was the first cAMP-mediated regulatory sequence to be explained. The original work was published in 1956 by **Earl Sutherland** who received a Nobel Prize in 1971 for this discovery.

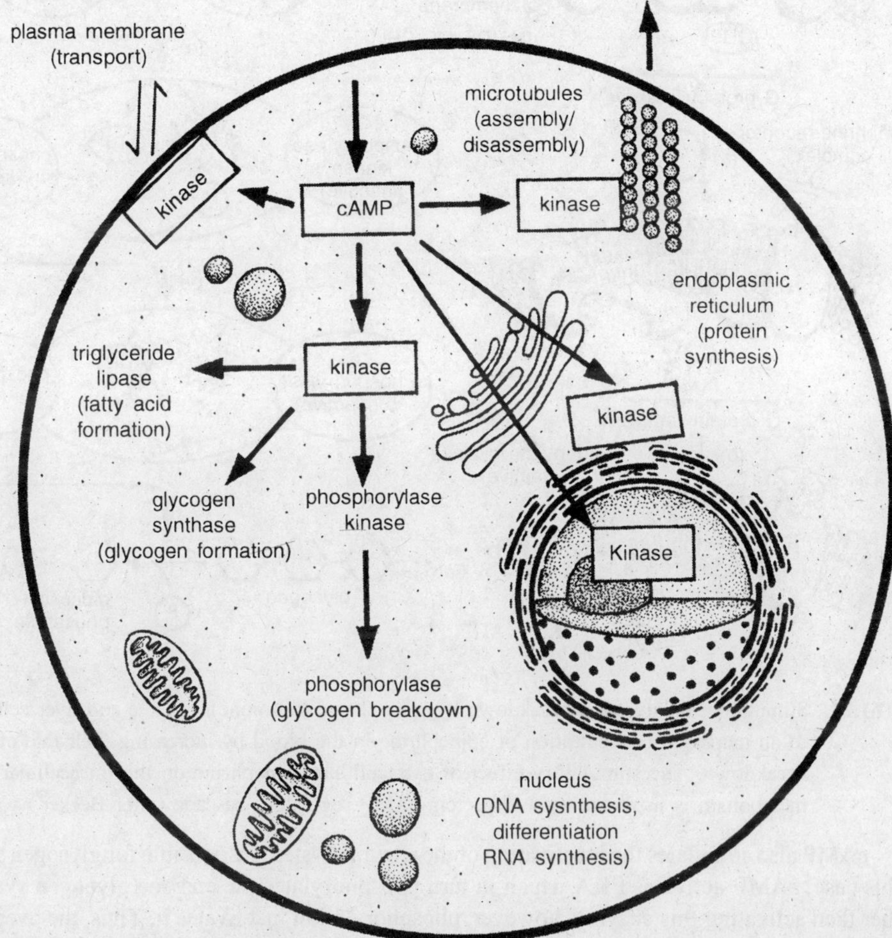

Fig. 61.23. Schematic illustration of the variety of processes that can be affected by changes in cAMP concentration. All of these effects are thought to be mediated by activation of the same basic enzyme, protein kinase A (after Karp, 2002).

Increased glycogen metabolism begins when an epinephrine (hormone) molecule binds to a β-adrenergic receptor on the plasma membrane of a liver or muscle cell. As described earlier, the receptor activates a neighboring Gs protein, and the Gs protein in turn stimulates adenylyl cyclase, the membrane-bound enzyme that generates cAMP from ATP (Fig. 61.24). The resulting transient increase in the concentration of cAMP in the cytosol activates protein kinase A (or PKA). PKA then activates another cascade of events that begins with the phosphorylation of the enzyme

phosphorylase kinase. This leads to the conversion of **glycogen phosphorylase** from phosphorylase b form and thus to an increased rate of glycogen breakdown (Fig. 61.23).

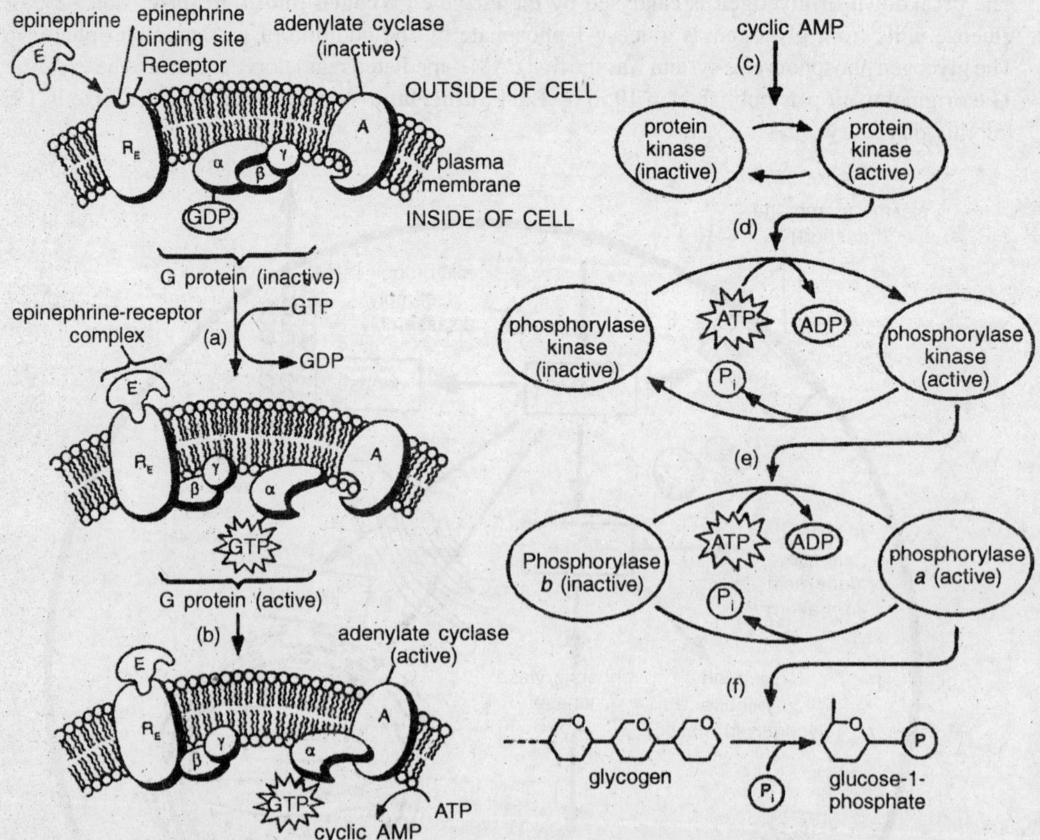

Fig. 61.24. Stimulation of glycogen breakdown by epinephrine (hormone). Muscle and liver cells respond to an increased concentration of epinephrine in the blood by increasing their rate of glycogen breakdown. The stimulatory effect of extracellular epinephrine on the intracellular glycogen metabolism is mediated by a G protein-cAMP regulatory cascade (after Becker *et al.*, 2007).

cAMP also stimulates the inactivation of the enzyme system responsible for glycogen synthesis. In this case, cAMP activates PKA which in turn phosphorylates the enzyme **glycogen synthetase**. Rather than activating this enzyme, however, phosphorylation inactivates it. Thus, the overall effect of cAMP involves both an increase in glycogen breakdown and a decrease in its synthesis.

***(ii)* α-Andrenergic receptors and IP_3-DAG-Ca^{2+} pathway.** The α_1-andrenergic receptors are found mainly on smooth muscles in blood vessels, including those controlling blood flow to the intestines. When α_1-andrenergic receptors are stimulated, the formation of IP_3 causes an increase in the intercellular calcium concentration. The elevated level of calcium causes smooth muscle contraction, resulting in constriction of the blood vessels and diminished blood flow. Thus, the activation of α_1-adrenergic receptors affects smooth muscle cells in a manner opposite to that of β-adrenergic activation, which causes smooth muscles to relax.

Example of Paracrine Regulation by Prostaglandins

Typically, the prostaglandins act on G protein-linked receptors to stimulate either the cAMP or IP_3-DAG-calcium (second messenger) pathway. Prostaglandins have a variety of effects, many of which involve smooth muscle. For example, prostaglandins contained in semen stimulate uterine smooth muscle contraction, which help transport sperm to egg. Prostaglandins also help to initiate smooth muscle contraction during labour and can be used to induce labour clinically. Some prostaglandins cause smooth muscle relaxation and can, for example, cause bronchiole dilation or lowering of blood pressure. Prostaglandins or also important in the activation of blood platelets, essential components of the blood-clotting mechanism that plug sites where blood vessels are ruptured.

Box 61.5

Prostaglandins were first discovered in 1935 by **Dr. Von Euler**. For this landmark discovery he got Nobel Prize in 1970.

61.6 CELL SIGNALS FOR APOPTOSIS

Cell signalling also regulates a kind of **programmed cell death (PCD)** or **apoptosis**. Apoptosis is a key event in many biological processes. In embryos, apoptosis occurs in a variety of circumstances. Examples include removal of the webbing between the digits (fingers and toes) during the development of hand and feet, the resorption of the tail of tadpoles when they undergo metamorphosis (Fig. 61.25) and the "pruning" of neurons that occur in human infants during first few months of life as connections mature within the developing brains. In adult humans, apoptosis occurs continually; when cells become infected by pathogens or when white blood cells reach the end of their life span, they are eliminated through apoptosis. As a result, millions of cells die every minute in the human body. When cells that should die via apoptosis do not, the consequences can be critical.

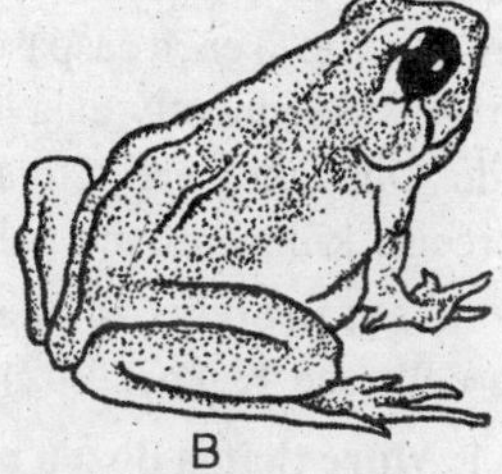

Fig. 61.25. Apoptosis during the metamorphosis of a tadpole into a frog. All the changes that occur during metamorphosis, including the induction of apoptosis of tail, are stimulated by an increase in thyroid hormone in the blood (after Albert *et al.*, 2002).

It is currently known that mutations in some of the proteins that participate in apoptosis can lead to cancer; melanoma frequently results from a mutations in Apaf-1, a protein that participates in apoptosis.

Apoptosis is very different from another type of cell death, known as **necrosis**, which sometimes follows massive tissue injury. Whereas necrosis involves a specific series of events that leads to the dismantling of the internal contents of the cell (Fig. 61.26). During the early phases of apoptosis, the cell's DNA segregates near the periphery of the nucleus and volume of the cytoplasm decreases. Next, the cell begins to produce small bubble-like cytoplasmic extensions ("blebs"), and the nucleus and organelles begin to fragment. The cell's DNA is cleaved by an apoptosis-specific **DNA endonuclease** or **DNase** (an enzyme that digests DNA), at regular intervals along the DNA. As a result, the DNA fragments, which are multiples of 200 base pairs in length, form a diagnostic "ladder" of fragments. Eventually the cell is demolished into small pieces called **apoptotic bodies**. Ultimately the remnants of the affected cell are engulfed by other nearby cells (typically macrophages) via phagocytosis. The macrophages act as scavengers to remove the resulting cellular debris.

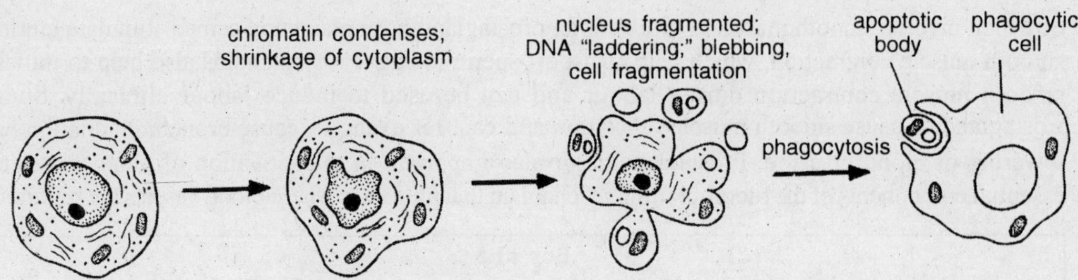

Fig. 61.26 Major steps in apoptosis. As a cell begins to undergo apoptosis, its chromatin condenses and the cytoplasm shrinks. Eventually the nucleus becomes fragmented, its DNA is digested at regular intervals ("laddering"), the cytoplasm becomes fragmented, and the cell extends numerous blebs. Ultimately the remnants of the dead cells (apoptotic bodies) are ingested by phagocytic cells (after Becker *et al.*, 2007).

Discovery of Apoptosis

The cells have a "death program" was first conclusively demonstrated in the nematode, *Caenorhabditis elegans*, where key genes that control apoptosis were first identified (Box 61.5). Subsequently research showed that many other organisms, including mammals, use similar proteins during apoptosis. As a result of these investigations many of the molecular events that underlie apoptosis are now known. For example, a key event in apoptosis of mammals is the activation of a series of enzymes called **caspases** (Caspases get their name because they contain a *c*ysteine at their active site, and they cleave proteins at sites that contain *aspar*tic acid residue followed by four amino acids that are specific to each caspase). Caspases are produced as inactive precursors known as **procaspases,** which are subsequently cleaved to create active enzymes, often by other caspases, in a proteolytic cascade. Once they are activated, caspases cleave other proteins within cells, resulting in efficient and precise killing of the cell in which they are activated.

Target proteins for caspases. Caspases are known to cleave the following target proteins of the cell (see **Karp**, 2002):

1. **More than a dozen protein kinases, including focal adhesion kinases (FAK), PKB, PKC and Raf 1**. Inactivation of FAK, for example, is presumed to disrupt cell adhesion, leading to detachment of the apoptotic cell from its neighbours.

Box 61.6

Horvitz, Sulston and **Brenner** shared the Nobel Prize in Physiology or Medicine in 2002. They worked out different aspects of apoptosis in the nematode *Caenorhabditis elegans*. **John Sulstan** performed initial cell lineage studies on *C.elegans*. He discovered occurrence of apoptosis during development of this nematode **H.Robert Horvitz** *et al.*, provided key insight in to apoptosis and cell signalling in the formation of vulva of this nematode. **Sydney Brenner** suggested genetic basis of apoptosis in *C.elegans*.

In *C.elegans*, development relies on programmed cell death. The number of cells that die during the worm's development is always the same: 131 cells of total 1090 cells in the hermaphrodites and 147 cells of total 1178 cells in males. In addition, the point in development at which a given cell dies and the identity of the cells that die are always the same.

Genetic analysis of mutants indicates that although apoptosis occurs in cells of different developmental origins, all the cells die by using the same genetic pathway. In *C. elegans*, 15 genes are involved in cell death (apoptosis). These genes form four groups: 1. decision makers; 2. execution of the decision; 3. engulfment of dying cells, and 4. degradation of cell debris in the engulfing cells. For example, expression of *Ced-3* and *Ced-4* (= *ce*ll *d*eath defective) are necessary for execution of the cell death programme; mutations that inactivates either of these genes result in survival of cells that normally die. Expression of *Ced-3* and *Ced-4* is controlled by *Ced-9*. Gain-of-function mutations that cause constitutive expression or overexpression of *Ced-9* prevent cell death. Conversely, loss-of-function mutations that inactivate *Ced-9* cause embryonic lethality. Based on these observations, it has been concluded that *Ced-9* gene works by preventing expression of *Ced-3* and *Ced-4* genes in cells that survive. This means that Ced-9 is a **binary switch gene** for programmed cell death. Cells that express *Ced-9* survive and that do not die.

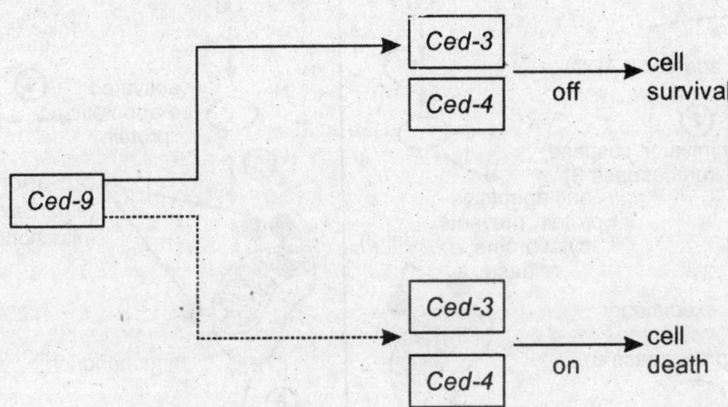

Fig. 61.27 The genetic pathway controlling cell death, the gene *Ced-9* act as a binary switch. If *Ced-9* is active, it repressed the expression of *Ced-3* and *Ced-4*, and the cell lives. If *Ced-9* is inactive, *Ced-3* and *Ced-4* are expressed and the cell die (after Klug and Cummings, 2001).

2. **Lamins**, which makes up the inner lining of the nuclear envelope. Cleavage of lamins leads to the disassembly of the nuclear lamina and shrinkage of the nucleus.

3. **Proteins required for cell structure**, such as those of intermediate filaments, actin and gelsolin. Cleavage and consequent inactivation of these proteins lead to changes in cell shape.

4. **An endonuclease called capsin activated DNase (CAD)**, which is activated following caspase cleavage of an inhibitory protein. Once activated, CAD translocates from the cytoplasm to the nucleus where it attacks DNA, serving it into fragments.

5. **Enzymes involved in DNA repair**, which are inactivated by caspase cleavage. DNA repair is a homeostatic activity that is inappropriate in an apoptotic cell in which the DNA is being dismantled.

Triggering of Apoptosis. There are two main routes by which cells can activate caspases and enter the apoptic pathway (Fig. 61.28).

1. **Cell death signals**. When cells in the human body are infected by certain viruses, a population of **killer lymphocytes** are activated and induce the infected cells to initiate apoptosis. How do lymphocytes induce cells to commit suicide via apoptosis? Typically, such activation is triggered when cells receive **cell death signals**. Two well known death signals are *tumor necrosis factor* and *CD95/Fas*. For example, **CD95** is a protein which is present on the surface of infected cells. Lymphocytes have a protein on their surfaces that binds to CD95, causing the CD95 within the infected cell aggregate (Fig. 61.28). CD95 aggregation results in the attachment of adaptor proteins to the clustered CD95, which in turn recruits a procaspase (**procaspase-8**) to sites of receptor clustering, when the procaspase is activated, it acts as an initiator of the caspase cascade. A key action of such **initiator caspases**, known as **caspase-3**. Active caspase-3 is important for activating many steps in apoptosis.

2. **Survival Factors**. In other cases, apoptosis is triggered indirectly. One of the best-studied cases of this type of apoptosis involves **survival factors**. Survival factors mediate one of the many events associated with growth factors: cell survival. When such factors are withdrawn, the cell enters

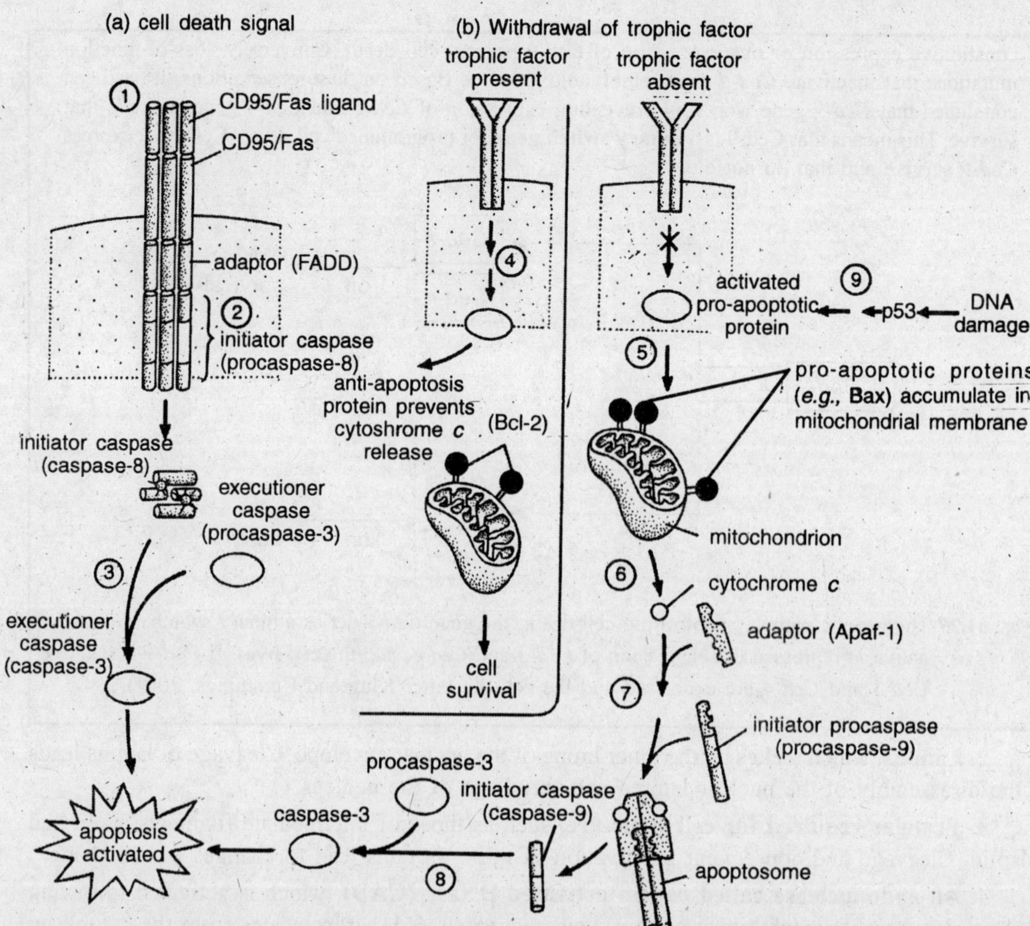

Fig. 61.28. Induction of apoptosis by cell death signals or by withdrawal of survival factors. A—Cell death signals, such as CD95/Fas ligand on the surface of a killer lymphocyte, can lead to apoptosis. 1. CD95/Fas ligand binds to CD95/Fas protein on the surface of a target cell. Binding causing clustering of receptors and recruitment of adoptor proteins in the target cell, resulting in clustering of interior procaspase (procaspase-8) protein. 2. Initiator caspases then become activated. After they are activated, 3. The initiator capases in turn activate the executioner caspase, caspase-3, a key initiator of apoptosis. B—4. When trophic factors are present, cell signalling results in inactivation of pro-apoptotic proteins and the cell does not enter apoptotis. 5. When trophic factors are withdrawn, pro-apoptotic proteins are activated, and some accumulate in the outer membrane of the mitochondrion. 6. The balance between pro-apoptotic and anti-apoptotic proteins (such as Bcl-2) at the mitochondrial outer membrane determines whether the mitochondrion releases cytochrome *C*. 7. Cytochrome *C* forms a complex with other proteins, resulting in activation of an initiator caspase (caspase-9).8. The initiator caspase in turn activates the executioner caspase, caspase-3, triggering apoptosis. 9. DNA damage can also lead to apoptosis through the activity of the P53 protein (after Becker *et al.*, 2007).

apoptosis. Surprisingly, a key site of action of this second pathway is the mitochondrion. The connection between mitochondria and cell death may be surprising, but it is that in addition to their role in energy production, mitochondria are important in apoptosis. If withdrawal of trophic factors is the sentence of execution, then the executioners are the mitochondria.

How does mitochondrion speed up cell death? In a healthy cell that is not committed to apoptosis, there are several **anti-apoptotic proteins** in the outer mitochondrial membrane that prevent apoptosis, but only as long as a cell continues to be exposed to trophic factors. These proteins are structurally related to a protein known as **Bcl-2**, the best understood of these anti-apoptotic proteins. Bcl-2 and other anti-apoptotic proteins exert their effects by counteracting other proteins that are also structurally similar to Bcl-2. These proteins, however, *promote* apoptosis, and so they are collectively called **pro-apoptotic proteins**. These proteins (such as proteins called **Bax** and **Bad**), can insert into the mitochondrial outer membrane, but they only do so at high levels when trophic factors are no longer present. Thus, pro- and anti-apoptotic proteins carry on an ongoing battle in many cells; when the balance shifts toward pro-apoptotic proteins, a cell is more likely to undergo apoptosis.

Mitochondria trigger apoptosis by releasing **cytochrome c** into the cytosol. As we already know that cytochrome c is normally involved in electron transport, it also has other important functions in triggering apoptosis in at least two ways. 1. Cytochrome c stimulates calcium release from adjacent mitochondria and from the endoplasmic reticulum, where it binds IP_3 receptors. 2. Cytochrome can activate an initiator procaspase associated with mitochondria, known as **procaspase-9**. It does this by recruiting a cytosolic adaptor protein (known as **Apaf-1**) that assembles procaspase-9 into a complex sometimes called an **apoptosome**; the apoptosome promotes the production of active caspase-9. Like other initiator caspases, caspase-9 activates the executioner caspase, caspase-3. Thus, in the end, both cell death mechanism lead to activation of a common caspase that sets apoptosis in motion.

There is another situation that can trigger the mitochondrial pathway to apoptosis. When a cell suffers so much damage that it is unable to repair itself, it may trigger its own demise. In particular, when a cells DNA is damaged (for example, by radiation or ultraviolet light), it can enter apoptosis via the activity of a protein known as **P53** (P53 is an important regulator of the cell cycle). Like the withdrawal of trophic factors, the P53 pathway can also activate pro-apoptotic proteins to trigger apoptosis (see **Becker** *et al.*, 2007).

QUESTIONS

Long Answer Questions

1. What is cell signalling? Describe various type of receptors responsible for the function of cell signalling.
2. What are cell surface receptors? Discuss their types and roles in cell signalling.
3. Discuss the role of second messengers in cell signalling.
4. 'Apoptosis in mediated by an intracellular proteolytic cascade'-substantiate.

Short Answer Questions

1. Explain the mechanism of cell signalling. (*V.B.S. Purvanchal, 2004*)
2. What is ligand? What is its function?
3. What are second messengers? How do they participate in cell signal transduction?
4. Explain the role of mitochondria in apoptosis.
5. Write short notes on the following:
   (*i*) Signal transduction;
   (*ii*) Apoptosis.

Very Short Answer Questions

1. What is a ligand?
   Ans. Ligand is any molecule that binds to a specific site on a protein or other molecule (From Latin *ligare*, to bind).
2. What is the function of ligand?
   Ans. Binding of ligand changes the property of the surface receptors and activate changes.
3. What is the function of cell surface receptors?
   Ans. Communication with neighbouring cells, to bind matrix components with extracellular matrix and cell adhesion.
4. What is the chemical nature of receptors?
   Ans. The receptors are protein molecules. These are proteoglycans.

5. What is receptor?
Ans. A protein that contains a binding site for a specific signalling molecule.
6. Define cAMP.
Ans. cAMP is an intracellular molecule which acts as second messenger in enkaryotes and mediates the effects of various signalling molecules by activating protein.
7. What is full form of IP_3?
Ans. Inositol-1, 4, 5-triphosphate.
8. What is the function of IP_3?
Ans. IP_3 functions as a second messenger by triggering the release of calcium ions from storage sites within the endoplasmic reticulum.
9. Define calmodulin.
Ans. Calmodulin is a calcium-binding protein involved in mediating many of the intracellular effects of calcium ions in eukaryotic cell.
10. What is a paracrine signal?
Ans. It is a growth factor or other secreted molecule that acts on nearby cells.
11. Define the endocrine signal.
Ans. It is a hormone or other molecule that is released into the circulatory system so that it can act on distant target cell.
12. What is a autocrine signal?
Ans. It is a growth factor or other molecule secreted by a cell that acts on the cell that produces it.
13. Define signal transduction.
Ans. Signal transduction is detection of specific signals at the cell surface and the mechanisms by which such signals are transmitted into the cell's interior, resulting in changes in cell behaviour and/or gene expression.
14. What is G protein?
Ans. Any of numerous GTP-binding regulatory proteins located in the plasma membrane that mediate signal transduction pathways, usually by activating a specific target protein such as an enzyme or channel protein.

Multiple Choice Questions

1. One of the following is a calcium-binding protein of the cell
(a) actin (b) tubulin
(c) cal modulin (d) dynein
2. A molecule acting as 'second messenger' in biological system is
(a) cDNA (b) cAMP
(c) tRNA (d) hn RNA
3. Epinephrine and nor-epinephrine together are known as
(a) steroids (b) proteins
(c) catecholamine (d) none of these
4. Substances which are formed at one place and express their effect at a distant place are called as
(a) enzymes (b) hormones
(c) pheromones (d) WBC
5. Insulin facilitates glucose uptake in
(a) kidney tubules (b) red blood cells
(c) brain (d) skeletal muscle
6. Cortical granules found in the egg cortex of many animals consist of
(a) lipids (b) proteins
(c) mucopoly saccharides (d) none of the above

ANSWERS

1. (c) 2. (b) 3. (c) 4. (b)
5. (d) 6. (c).

62

Cytoskeletal Systems and Cellular Movement

(Microtubules, Microfilaments, Intermediate filaments, Centrioles, Motor MPA, Cilia and Flagella)

Advances in microscopy and other investigative techniques have revealed that the interior of an eukaryotic cell is highly structured. Part of this structure is provided by the **cytoskeleton:** a complex network of interconnected filaments and tubules that extends throughout the cytosol, from the nucleus to the inner surface of the plasma membrane (Box 62.1).

The three major elements of the cytoskeleton are **microtubules (MTs), microfilaments (MFs)** and **intermediate filaments (IFs)**, each of which is unique to eukaryotic cells. However, prokaryotes, such as rod-shaped bacteria, have recently been shown to have proteins that function in a manner very similar to microfilaments (proteins from the *MreB* family), microtubules (the *FtsZ* protein) and intermediate filaments (one type is called *crescentin*) (see **Becker** *et al.*, 2007).

Box 62.1

The existence of an organized fibrous array or cytoskeleton in the structure of the protoplasm was postulated in 1928 by **Koltzoff**. He conceived of a cytoskeleton that determines both the shape of the cell and the changes in its form.

The main proteins that are present in the cytoskeleton are **tubulin** (in the microtubules), **actin**, **myosin**, **tropomyosin** and other (in the microfilaments) and **keratins**, **vimentin**, **desmin**, **lamin** and others (in intermediate filaments). Tubulin and actin are globular proteins, while subunits of intermediate filaments are fibrous proteins.

Cellular architecture depends on the unique properties of the different cytoskeletal components working together. Microtubules generally thought to resist bending when a cell is compressed, while microfilaments serve as contractile elements that generate tension. Intermediate filaments are elastic and can withstand tensile force. The mechanical integration of IFs, MFs and MTs is made possible by specific linker proteins that connect them. **Plectin**, for example, is a versatile linker protein that is found at sites where IFs are connected to MFs and MTs.

Functions of Cytoskeleton

The cytoskeleton plays important roles in cell movement and cell division, and its position and activity moves membrane-bounded organelles within the cytosol. It also plays a similar role for messenger RNA and other cellular components. Many enzymes in the cytosol are in fact probably not soluble at all, but are physically clustered and attached to the cytoskeleton in close proximity to other enzymes involved in the same pathways, thereby facilitating the channeling of intermediates within each pathway. The cytoskeleton is also involved in many forms of cell movement and is intimately related to other processes such as cell signalling and cell-cell adhesion. The cytoskeleton is altered by events at the cell surface and, at the same time, appears to participate in and modulate these events.

62.1 MICROTUBULES

Microtubules were first of all observed in the axoplasm of the myelinated nerve fibres by **Robertis** and **Franchi** (1953). They called them **neurotubules**. The exact nature of microtubules was brought into light when **Sabatini**, **Bensch** and **Barnett** (1963) made use of the glutarldehyde fixative in the electron microscopy. Microtubules of plant cells were first described in detail by **Ledbetter** and **Porter** (1963).

Occurrence

With rare exceptions such as the human erythrocytes, microtubules are found in all eukaryotic cells, either free in the cytoplasm or forming part of centrioles, cilia and flagella. The most abundant source of microtubules for the biochemical studies is vertebrate brain—high densities of microtubules exist in axons and dendrites of nerve cells. In the cytoplasm of animal and plant cells, microtubules occur at following seven sites: 1. Cilia and flagella, 2. Centrioles and basal bodies, 3. Nerve processes, 4. The mitotic apparatus, 5. The cortex of meristematic plant cells, 6. Elongating cells such as during the formation of the lens or during spermatogenesis of certain insects. 7. Selected structures in Protozoa such as the axostyle of parasitic flagellates, the axoneme of *Echinosphaerium*, the fibre systems of *Stentor*, and the cytopharyngeal basket of *Nassula*.

Types of Microtubules

Eukaryotes have following two types of microtubules:

1. Axonemal microtubules. These include the highly organized, stable microtubules found in specific subcellular structures associated with cellular movement, including cilia, flagella, basal bodies and centrioles to which these appendages are attached. The central shaft, or **axoneme**, of a cilium or flagellum consists of a highly ordered bundle of axonemal MTs and associated proteins.

2. Cytoplasmic microtubules. They include loosely organized, dynamic microtubules which pervade the cytosol of most eukaryotic cells. They were detected by the help of fluorescence microscopy around early 1960's.

Structure

Microtubules constitute a class of morphologically and chemically related filamentous rods which are common to both plant and animal cells. A microtubule consists of a long, unbranched, hollow tubules 24–25 nm in diameter, several micrometers long and with 6 nm thick wall having 13 subunits or **protofilaments**. Thus, the wall of the microtubule consists of 13 individual linear or spiralling filamentous structures about 5 nm in diameter, which in turn, are composed of tubulin. These protofilaments have a centre-to-centre spacing of 4.5 nm. Application of negative staining techniques has shown that microtubules have a lumen 14 nm wide and a protofilament or subunit structure in the wall (Fig. 62.1).

Chemical Composition

Biochemically, a protofilament of microtubule is made of a protein called **tubulin**. Tubulin is an acidic protein with a molecular weight of 55,000 and a sedimentation constant of 6S. It occurs in two different forms, called **α-tubulin** and **β-tubulin**, each containing about 450 amino acids. Both of these proteins have a distinct, though closely related, amino acid sequences and are thought to have evolved from a single ancestral protein. The two proteins show very little divergence from the lowest to the highest eukaryotes; for example, the β-tubulins of sea urchin flagella and chick brain cells differ only in one amino acid. Similarities such as this suggest that most mutations disrupt the functions of microtubules and are thus lethal and are eliminated by selection (see **King**, 1986).

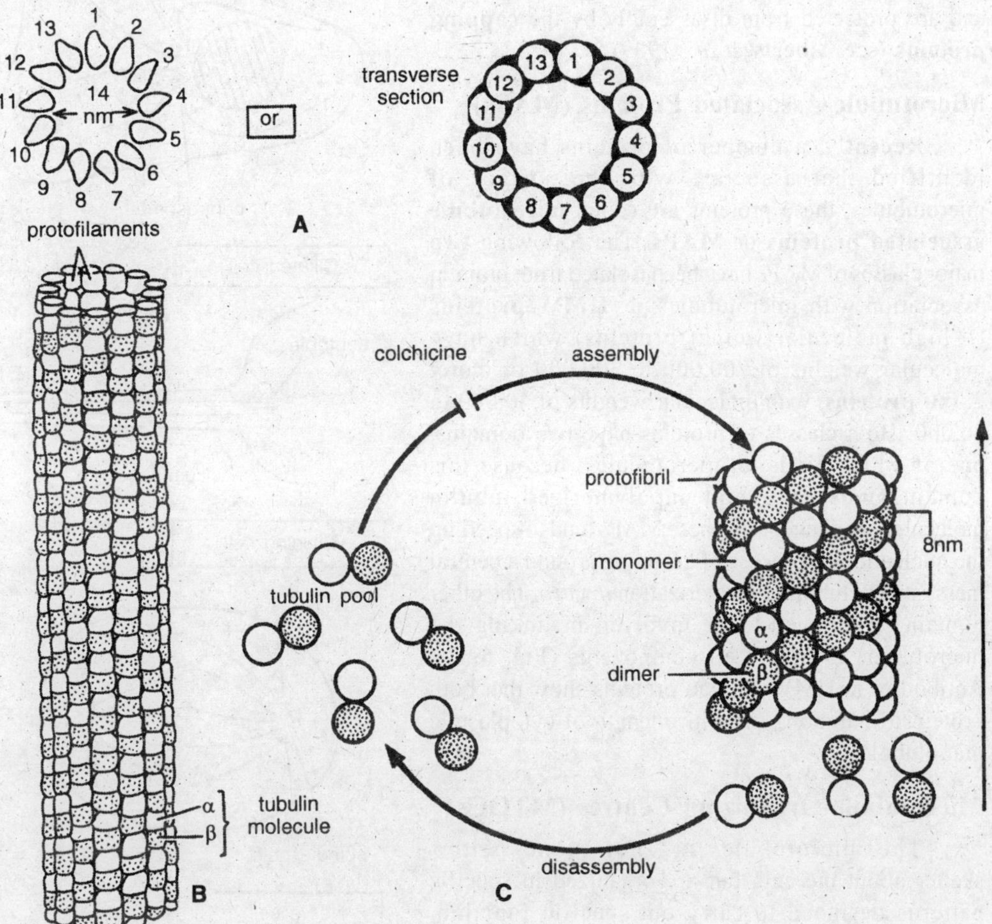

Fig. 62.1. Schematic diagrams of a microtubule, showing how the tubulin molecules pack together to form the cylindrical wall. A—13 tubulin molecules (subunits of protofilaments) in cross section; B—Side view of a short section of a microtubule, with the tubulin molecules aligned into rows, or protofilaments. Each of the 13 protofilaments is composed of a series of tubulin molecules, each with an α/β heterodimer; C—Assembly and disassembly of the microtubule. The microtubule is being disassembled at the bottom while being simultaneously assembled at the top. Colchicine, by blocking the assembly process, produces depolymerization of the microtubules (after De Robertis and De Robertis, Jr., 1987; Alberts *et al.*, 1989).

Tubulin in the form of dimers (rather heterodimers of α- and β-tubulins; each with 115,000 MW, see **Berns**, 1983) polymerizes into the microtubules. Thus, heterodimers of tubulins assemble to form linear "protofilaments" with the β-tubulin of one dimer in contact with the α-tubulin of the next. Since all the 13 protofilaments are aligned parallely with the same polarity, the microtubules are the **polar** structures having a **plus** or **fast growing end** and **minus** or **slow growing end**. The minus ends of cytoplasmic microtubules in cells are bound tightly to **microtubule organizing centres (MTOCs)** from which their assembly or polymerization starts. MTOCs also protect the minus ends of the microtubules from the disassembly. Generally, the plus ends of microtubules terminate near cell margins (Fig. 62.2) and are protected from disassembly by the **capping proteins** (see **Alberts** *et al.*, 1989).

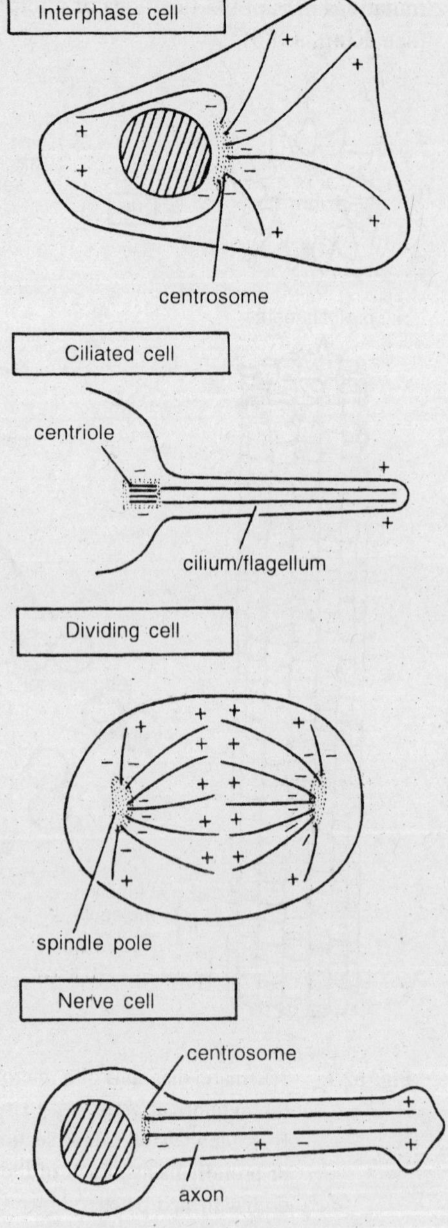

Fig. 62.2. The minus ends of microtubules in cells are generally embedded in a microtubule-organizing centre; while the plus ends are often near the plasma membrane (after Alberts *et al.*, 1989).

Microtubule-Associated Proteins (MAPs)

Recently, a number of proteins have been identified that associate with the surface of microtubules; these proteins are called **microtubule associated proteins** or **MAPs**. The following two major classes of MAPs have been isolated from brain in association with microtubules: 1. **HMW proteins** (= high molecular weight proteins) which have molecular weights of 200,000 to 300,000 or more; 2. **tau proteins**, with molecular weights of 40,000 to 60,000. Both classes of proteins have two domains, one of which binds to microtubules; because this domain binds to several unpolymerized tubulin molecules simultaneously, these MAPs tend to speed up the nucleation (= process of grouping around a central mass) step of tubulin polymerization *in vitro*. The other domain is believed to be involved in linking the microtubule to other cell components (Fig. 62.3). Antibodies to HMW and tau proteins show that both proteins; bind along the entire length of cytoplasmic microtubules.

Microtubule Organizing Centres (MTOCs)

The microtubules are not found helter-skelter about the cell, but are organized in specific patterns designed to carry out specific function. Spontaneous nucleation, as seen *in vitro* (Fig. 62.1C), probably does not occur *in vivo*. Rather, initiation of assembly occurs at **microtubule organizing centres (MTOCs)**. Thus, MTOCs are nucleating centres that serve as templates for the polymerization of tubulin (see **Thorpe**, 1983). MTOCs exist in basal bodies (*e.g. Chlamydomonas*); in centrioles (*e.g.*, most animal cells); at the poles of mitotic spindles in dividing cells that do not have centrioles (*e.g.*, most plant cells); on chromosomes (*i.e.*, **kinetochore**); in membranes and probably many other places as well. Later studies have revealed that most cytoplasmic microtubules do not arise directly from the centrioles, but from a densely staining **pericentriolar material** that surrounds the centriole (see **King**, 1986).

Turning on and off of these organizing centres for microtubule assembly at different times in the cell's life are probably regulated by one or all the following factors: changes in nucleation centres, changes in Ca^{2+} concentration and involvement of MAPs.

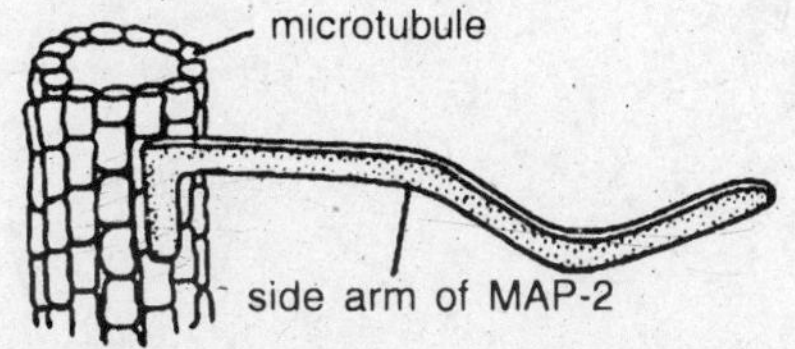

Fig. 62.3. A microtubule associated protein (known as MAP-2) showing its two domains (after Alberts *et al.*, 1989).

Assembly and Disassembly of Microtubules

Cytoplasmic microtubules are highly dynamic structures, constantly forming and disappearing depending on cell activities. They, like the microfilaments, grow by the reversible addition of subunits, accompanied by nucleotide (GTP) hydrolysis and conformational change. The process of polymerization (assembly) and depolymerization (disassembly) of the microtubules appears to be a form of **self-assembly**. The assembly of microtubules from the tubulin dimers is a specifically oriented and programmed process. In the cell, the sites of orientation are MTOCs from which the polymerization is directed. The quantity of polymerized tubulin is high at interphase (cytoplasmic microtubules) and metaphase (spindle microtubules), but low at prophase and anaphase.

Within the cell, microtubules are in equilibrium with free tubulin. Phosphorylation of the tubulin monomers by a cyclic AMP-dependent kinase favours the polymerization. A definite relationship has been found between cell shape, the number and direction of microtubules and cAMP. The assembly and disassembly of tubulin constitute a polarized phenomenon. In a microtubule, the assembly of tubulin dimers takes place at one end, while disassembly is common at the other end (Fig. 62.1C). If a cell is treated with certain drugs such as **colchicine, vincristine** or **vinblastine**, the assembly of the microtubules is inhibited, while the disassembly continues, leading to the disorganization of the microtubule. Further, the assembly is accompanied by the hydrolysis of guanosine triphosphate (GTP) to guanosine diphosphate (GDP) and lack of GTP stops the assembly. *In vivo* control of assembly and disassembly of tubulin involves Ca^{2+} and the calcium-binding protein **calmodulin**. The addition of Ca^{2+} inhibits polymerization of tubulin; this effect is also enhanced by the addition of calmodulin.

The *in vivo* mechanism involved in self-assembly of the microtubules is still little understood, however, *in vitro* studies have revealed various interesting facts about it. Thus, in a classical study using isolated bovine brain tubulin, **Weingarten** *et al.*, (1975) demonstrated that tubulin alone was not sufficient to bring about *in vitro* assembly into microtubules. Under normal conditions, if brain microtubules are isolated and caused to depolymerize into tubulin subunits, the tubulin molecules will reassemble into microtubules if Mg^{2+} and GTP (an energy source) are added to the mixture. However, according to **King** (1986), *in vitro* assembly of microtubules can occur in the presence of low calcium concentration, MAPs, GTP, and a level of free tubulin monomers above a threshold concentration.

In vitro polymerization evidently involves two distinct phases, one of **initiation** and the other of **elongation**. The initiation event seems to involve the formation of some multimeric "**nucleating**" centre, following which the addition of more subunits proceeds rapidly during elongation. Thus, during *in vitro* polymerization of microtubules, α- and β-tubulins combine to form heterodimers (Fig. 62.4). The heterodimers associate to form multimeric **rings**, **spirals** and other intermediate structures which eventually open up to form strands or protofilaments. Side-by-side assembly of the protofilaments creates sheet-like structures that curl to form a tube. Elongation of this short cylinder occurs by direct addition of new heterodimers at one end of the tubule (*i.e.*, the plus end of tubule). It is believed that during anaphase, addition of dimers to one end of a microtubule is accompanied by the loss of dimers from the other end.

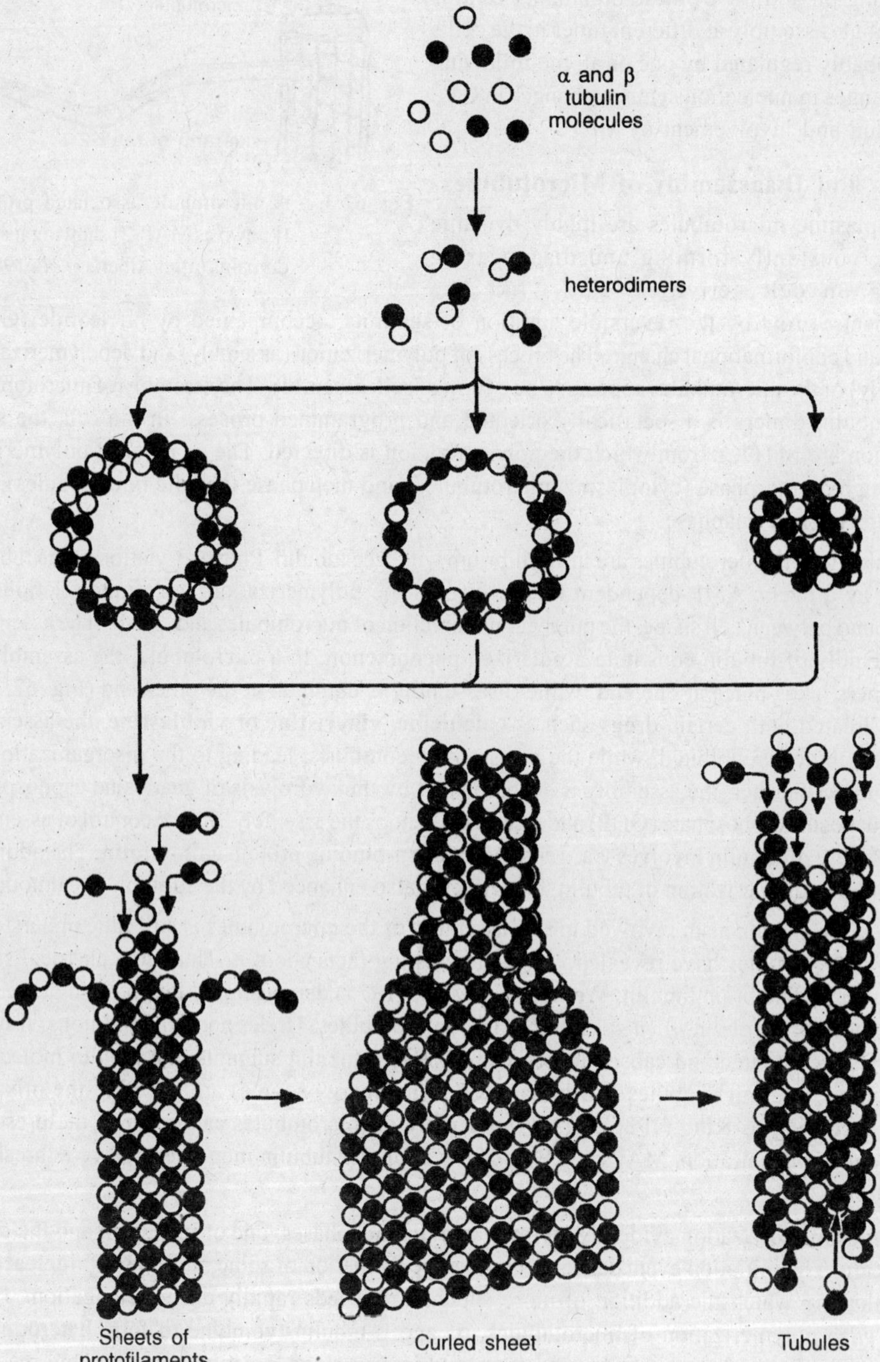

Fig. 62.4. Various steps *in vitro* polymerization or assembly of the microtubules (After Thorpe, 1984).

General Functions.

Cytoplasmic MTs are responsible for a variety of functions. For example, in animal cells they are required to maintain axons, nerve cell extensions. Some migrating animal cells require cytoplasmic

MTs to maintain their polarized shape. In plant cells, cytoplasmic MTs are thought to govern the orientation with which cellulose microfibrils are deposited during the growth of cell wall. Significantly, cytoplasmic MTs form the mitotic and meiotic spindles that are essential for the movement of chromosomes during mitosis and meiosis (see Chapter 71).

Specific Functions of Cytoplasmic Microtubules

Microtubules (MTs) have several functions in the eukaryotic cells such as follows:

1. Mechanical function. The shape of the cell (*e.g.*, red blood cells of non-mammalian vertebrates) and some cell processes or protuberances such as axons and dendrites of neurons, microvilli, etc., have been correlated to the orientation and distribution of microtubules.

Cytoplasmic microtubules also contribute to the spatial deposition and directional movement of vesicles and other organelles by providing an organized system of fibre to guide their movement. For example, cytoplasmic MTs help to govern the location of organelles, such as the Golgi apparatus and the endoplasmic reticulum, and are involved in active movement of vesicles.

2. Morphogenesis. During cell differentiation, the mechanical function of microtubules is used to determine the shape of the developing cells. For example, the enormous elongation in the nucleus of the spermatid during spermiogenesis is accompanied by the production of an orderly array of microtubules that are wrapped around the nucleus in a double helical arrangement. Likewise, the elongation of the cells during induction of the lens placode in the eye is also accompanied by the appearance of numerous microtubules.

3. Cellular polarity and motility. The determination of the intrinsic polarity of certain cells is also related to the microtubules. Directional gliding of cultured cells is found to depend on the microtubules.

4. Contraction. Microtubules play a role in the contraction of the spindle and movement of chromosomes and centrioles as well as in ciliary and flagellar motion.

5. Circulation and transport. Microtubules are involved in the transport of macromolecules, granules and vesicles within the cell. **Examples:** 1. The protozoan *Actinosphaerium* (Heliozoa) sends out long, thin pseudopodia within which cytoplasmic particles migrate back and forth. These pseudopodia contain as many as 500 microtubules disposed in a helical configuration. 2. In the protozoan *Nassula*, microtubules drive the food in the gullet. 3. In melarocytes, melanin granules move centrifugally and centripetally with different stimuli. These granules have been observed moving between channels created by the microtubules in the cytoplasmic matrix. 4. In the erythrophores found in fish scales the pigment granules may move at a speed of 25 to 30 μm per second between the microtubules. 5. They have a role in axoplasmic transport of proteins, glycoproteins and enzymes.

62.2 MICROFILAMENTS

Thin, solid **microfilaments** of actin protein, ranging between 5 to 7 nm in diameter and indeterminate length; represent the active or motile part of the cytoskeleton. They appear to play major role in cyclosis and amoeboid motion. With high voltage electron microscopy a three-dimensional view of microfilaments has been obtained (*i.e.*, an image of **microtrabecular lattice**). These microfilaments are sensitive to **cytochalasin-B,** an alkaloid that also impairs many cell activities such as beat of heart cell, cell migration, cytokinesis, endocytosis and exocytosis. It is generally assumed that the cytochalasin-B-sensitive microfilaments are the contractile machinery of non-muscle cells.

Distribution

Microfilaments are generally distributed in the cortical regions of the cell just beneath the plasma membrane. In contrast, intermediate filaments and microtubules are found in subcortical and deeper regions of the cell. Microfilaments also extend into cell processes, especially where there is movement. Thus, they are found in the microvilli of the brush border of intestinal epitheliun and in cell types where amoeboid movement and cytoplasmic streaming are prominent.

Chemical Composition

Actin is the main structural protein of microfilaments. Actin is an extremely abundant protein in virtually all eukaryotic cells, including those of plants, algae, and fungi. Actin is synthesized as a single polypeptide consisting of 375 amino acids, with a molecular weight of about 42 kDa. Once synthesized, it folds into a roughly U-shaped molecule, with a central cavity that binds ATP or ADP. Individual actin molecules are referred to as **G-actin (globular actin)**. Under the right conditions, G-actin molecule polymerize to form **microfilaments**; in this form the actin is referred to as **F-actin (filamentous actin)**. Actin in the G or F form also binds to a wide variety of other proteins. These actin-binding proteins either regulate and modify the function of actin, or are themselves regulated or organized by their association with actin.

Of the three types of cytoskeletal proteins, actin is the most highly conserved. In functional assays, all actins appear to be identical and actins from diverse organisms will copolymerize into filaments. Despite this high degree of sequence similarity, actins do differ among different organisms. Based on sequence similarity, actins can be broadly divided into two major groups: 1. The **muscle-specific actins** (*e.g.*, **α-actin**) and 2. **Non-muscle actins** (*e.g.*, **β-actin** and **γ-actin**).

In addition to the various types of actin, another class of protein known as **actin-related proteins (Arps)** show recognizable sequence similarity to actin Arp 2 and Arp 3 are found to be involved in nucleating assembly of new microfilaments in migrating cells.

Polymerization of G actins. Like tubulin dimers, monomers of G-actin polymerize into long filaments of F-actin with a diameter of about 7 nm (Fig. 62.5). In F-actin, a full turn of the helix occurs every 36–37 nm, with about 13.5 monomers required for a full turn. The addition of each G-actin monomer is usually followed by hydrolysis of ATP molecules so that it is tightly bound to the monomer, although the energy of ATP hydrolysis is not required to drive the polymerization reaction.

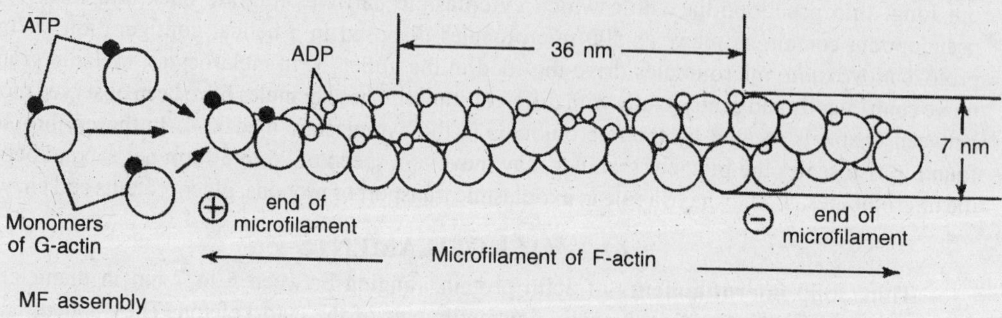

Fig. 62.5. A model for microfilament assembly in vitro (after Becker *et al.*, 2007)

Within a microfilament, all the actin monomers are oriented in the same direction, so that an MF, like a microtubule, has an inherent polarity, with one end differing chemically and structurally from the other end. This polarity can be readily demonstrated by incubating MFs with **heavy myosin (HMM)** or **myosin subfragment (S1)**, a proteolytic fragment of myosin (Fig. 62.6). HMM/S1 fragments bind to, or "decorate" the actin MFs to give a distinctive arrowhead pattern, the term **pointed end** and **barbed end** are commonly used to identify the **minus** and **plus ends** of a MF, respectively. The polarity of the microfilament is important, because it allows for independent regulation of actin assembly or disassembly of each end of the filament. Normally, plus end grows faster than the minus end.

Capping of MFs. Whether microfilament ends are available for further growth depends on whether or not the filament end is **capped**. Capping occurs when a **capping protein** binds the end of a filament and prevents further addition or loss of subunits, thereby stabilize it. One such protein that acts as a cap for the plus end of microfilament is appropriately named **Cap Z** (see **Becker** *et al.*, 2007).

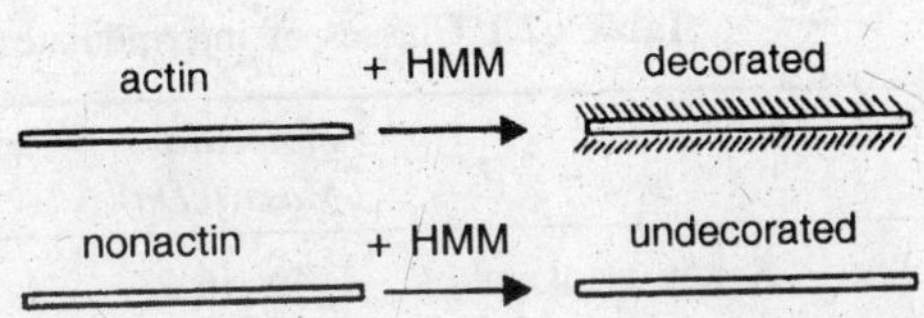

Fig. 62.6. Mode of "decoration" of actin filaments by heavy myosin or HMM. Intermediate filaments are not decorated by HMM. (after Thorpe, 1984).

Function

Microfilaments are found to be involved in movement associated with furrow formation in cell division, cytoplasmic streaming in plant cells (*e.g.*, *Nitella* and *Chara*) and cell migration during embryonic development.

62.3 INTERMEDIATE FILAMENTS

Intermediate filaments (IFs) are tough and durable protein fibres in the cytoplasm of most higher eukaryotic cells. Constructed like woven ropes, they are typically between 8 nm to 10 nm in diameter, which is "intermediate" between the thin and thick filaments in muscle cells, where they were first described; their diameter is also between microfilaments (actin filaments) and microtubules. IFs are found resistant to colchicine and cytochalasin B and are sensitive to proteolysis.

In most animal cells IFs form a "basket" around the nucleus and extend out in gentle curving arrays to the cell periphery. IFs are particularly prominent where cells are subjected to mechanical stress, such as in epithelia, where they are linked from cell to cell at desmosomal junctions, along the length of axons, and throughout the cytoplasm of smooth muscle cells. Various names have been attached to the intermediate filaments that have a basis in the cell type in which they are observed. Thus, IFs in epidermal cells are called **tonofilaments**, in nerve cells they are referred to as **neurofilaments** and in neuroglial cells they are designated as **glial filaments**.

In cross-section, intermediate filaments have a tubular appearance. Each tubule appears to be made up of 4 or 5 protofilaments arranged in parallel fashion (**Thorpe**, 1984). IFs are composed of polypeptides of a surprisingly wide range of sizes (from about 40,000 to 240,000 daltons).

Types of Intermediate Filaments

Intermediate filaments are only found in multicellular organisms, in contrast to MTs and MFs, differ markedly in amino acid composition from tissue to tissue. Based on cell type in which they are found, IFs and their proteins can be grouped into six classes (Table 62.1).

Class I and II IFs comprise the **keratins**, proteins that make up the **tonofilaments** found in the epithelial cells that cover the body surfaces and line its cavities (the IFs visible beneath the terminal web in the intestinal mucosa cell consists of keratin). Class I keratins are **acidic keratins**, whereas class II are **basic** or **neutral keratins**.

Class III IFs include vimentin, desmin, synemin and glial fibrillarly acidic protein. **Vimentin** is present in connective tissue and other cells derived from non-epithelial cells. Vimentin-containing filaments are often prominent features, in cultured fibroblast cells. **Desmin** is found in muscle cells. **Synemin** is a protein of 230,000 daltons, which is also present in the intermediate filaments of muscle, together with desmin and vimentin. Vimentin and synemin containing IFs can be observed in the chicken erythrocytes. **Glial fibrillary acidic (GFA) protein** is characteristic of the glial cells that surround and insulate nerve cells.

Table 62.1 Classes of intermediate filaments (*Source:* Becker *et al.*, 2007)

Class	*IF protein*	*Molecular Mass (kDa)*	*Tissue*	*Function*
I	Acidic cytokeratins	40–56.5	Epithelial cells	Mechanical strength
II	Basic cytokeratins	53–67	Epithelial cells	Mechanical strength
III	Vimentin	54	Fibroblasts; cells of mesenchymal origin; lens of eye	Maintenance of cell shape
III	Desmin	53–54	Muscle cells, especially smooth muscle	Structure support for contractile machinery
III	GFA protein	50	Glial cells and astrocytes	Maintenance of cell shape
IV	Neurofilament protein		Central and peripheral nerves	Axon strength; determines axon size
	NF-L (major)	62		
	NF-M (minor)	102		
	NF-H (minor)	110		
V	Nuclear lamins		All cell types	Form a nuclear scaffold to give shape to nucleus
	Lamin A	70		
	Lamin B	67		
	Lamin C	60		
VI	Nestin	240	Neuronal stem cell	Unknown

Class IV IFs are the **neurofilaments** (NF) **proteins** found in the neurofilaments of nerve cells.

Class V IFs are the **nuclear lamins** A, B and C, which form a filamentous scaffold along the inner surface of the nuclear membrane of virtually all eukaryotic cells.

Neurofilaments found in cells in the embryonic nervous system are made of **nestin** protein, which constitute class VI IFs.

Each of these IF proteins tends to assemble spontaneously *in vitro* to form homopolymers and will also co-assemble with the other Types III IF proteins to form **co-polymers** and **hetero-polymers**. In fact co-polymers of vimentin and desmin, or of vimentin and glial fibrillary acidic protein, are found in some type of cells. For example, desmin remains concentrated in the Z-lines and T-tubule system of striated or skeletal system, together with vimentin, synemin and α-actinin. Since desmin links actin to plasma membrane, from this fact the name of desmin has been derived by **Lazarides** and coworkers in 1976 (in Greek desmin means link or bond).

General Structure of IFs

All six types of intermediate filaments (IFs) contain proteins which have a central rod like domain consisting of four **helical segments**. The central domain is highly conserved in size, secondary structure, and sequence, though sequence homologies are confined to the helical regions. In types I-IV, the helical segments contain a total of 276 amino acids and the linker segments are non-helical (Fig. 62.7). In type V, the helical segments contain 318 amino acids and their linker segments are also helical. The N- and C-terminal domains that flank the central section are non-helical and are much more variable in size and sequence.

Assembly of IFs

A popular model of assembly of a intermediate filament includes the following steps: 1. Two identical **monomers** pair to form a **dimer** in which the conserved helical central regions are aligned

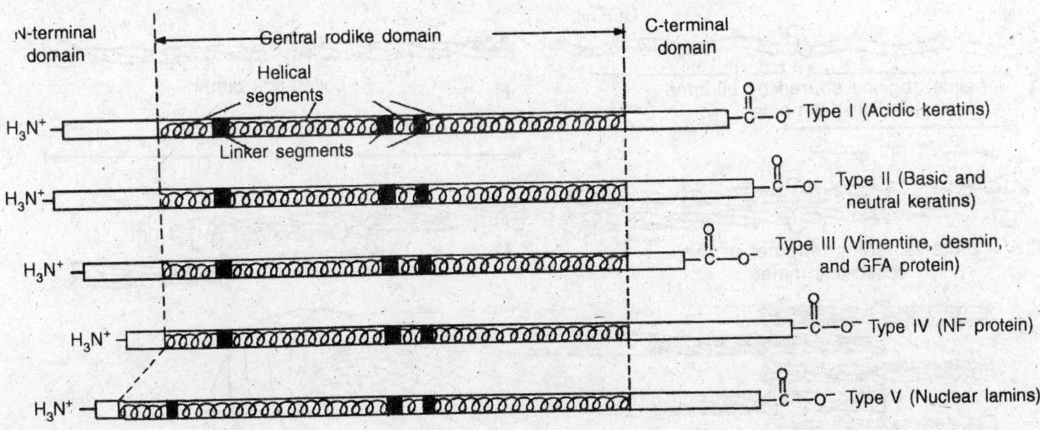

Fig. 62.7. Structural similarities of intermediate filament protein. Except VI or nestin, all the five IFs are shown (after Becker *et al.*, 2007).

in parallel and are wound together into a coiled coil. 2. Two dimers then line up side-by-side to form a 48 nm by 3 nm **protofilament** containing four polypeptide chains. 3. These protofilaments then associate in a staggered manner to form successively larger structures. 4. The final 10 nm diameter of the intermediate filament is thought to be composed of 8-protofilaments (*i.e.*, 32 polypeptide chains) joined end on end to neighbours by staggered overlap to form the long rope-like filaments (Fig. 62.8). It is still not known whether IFs are polar structures (like actin and tubulin) or non-polar (like the DNA double helix).

IFs During Mitosis

Mitosis of cultured epithelial cells shows striking changes in intermediate filaments of cytokeratin and vimentin. During prophase the 10 nm filaments unwind into threads of 2 to 4 nm and into spheroidal aggregates containing both types of proteins. At metaphase and anaphase most vimentin and cytokeratin appear as spheroid bodies, while at telophase the filamentous cytoskeleton become gradually reestablished. From these experimental studies, **Franke** (1982) has concluded that the living cells contain factors that promote the reversible disintegration and restoration of intermediate filaments during mitosis.

Functions of IFs

The main function of most intermediate filaments is to provide mechanical support to the cell and its nucleus. IFs in epithelia form a transcellular network that seems designed to resist external forces. The neurofilaments in the nerve cell axons probably resist stresses caused by the motion of the animal, which would otherwise break these long, thin cylinders of cytoplasm. Desmin filaments provide mechanical support for the sarcomeres in muscle cells, and vimentin filaments surround and probably support the large fat droplets in the fat cells.

Comparison of Microtubules, Intermediate Filaments and Microfilaments

The three components of the cytoskeleton, namely microtubules, intermediate filaments and microfilaments have been compared in Table 62.2.

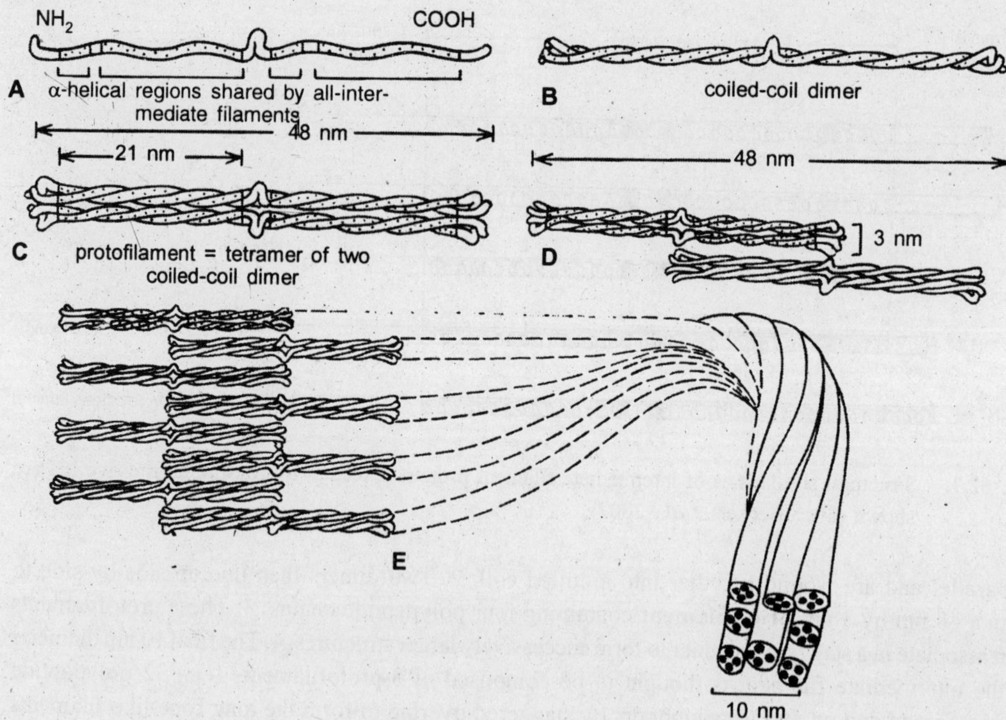

Fig. 62.8. A model to explain the mode of assembly of a 10-μm thick intermediate filament (after Albert *et al.*, 1989).

Table 62.2. Comparison of some properties of microtubules, intermediate filaments and microfilaments (*Source:* Thorpe, 1984).

Property	*Microtubules*	*Intermediate filaments*	*Microfilaments*
1. Structure	Hollow with walls made up of 13 protofilaments	Hollow with walls made up of 4 to 5 protofilaments	Solid made up of polymerized actin (F-actin)
2. Diameter (nm)	24–25	10	7–9
3. Monomer units	α- and β-tubulin	Six types of protein defining six major classes	G-actin
4. ATPase activity	Present in dynein arms	None	None
5. Functions	1. Motility of eukaryotes	1. Integrate contractile units in muscle	1. Muscle contraction
	2. Chromosome movement	2. Cytoskeletal structural function in cytoplasm	2. Cell shape changes
	3. Movements of intracellular materials		3. Protoplasmic streaming
	4. Contribute toward maintaining cell shape		4. Cytokinesis

62.4 CENTRIOLES AND BASAL BODIES

Cytoplasm of some eukaryotic cells contains two cylindrical, rod-shaped, microtubular structures, called **centrioles**, near the nucleus. Centrioles lack limiting membrane and DNA or RNA and form a spindle of microtubules, the mitotic apparatus during mitosis or meiosis and sometimes get arranged just beneath the plasma membrane to form and bear flagella or cilia in flagellated or ciliated cells (**Fulton**, 1971). When a centriole bears a flagellum or cilium, it is called **basal body**. There are many synonyms for basal body, including kinetosome, blepharoplast, basal granule, basal corpuscle, and proximal centriole.

Due to the classic works of **Henneguy** and **Lenhossek** (1897), it has been proposed that basal bodies of cilia and flagella are homologous/with the centrioles found in mitotic spindle.

Occurrence

Centrioles occur in most algal cells (notable exception being red algae), moss cells, some fern cells and most animal cells. They are absent in prokaryotes, red algae, yeast, cone-bearing and flowering plants (conifers and angiosperms) and some non-flagellated or non-ciliated protozoans (such as amoebae). Some species of amoebae have a flagellated stage as well as an amoeboid stage; a centriole develops during the flagellated stage but disappears during the amoeboid stage.

Structure

Centrioles and basal bodies are cylindrical structures which are 0.15–0.25 μm in diameter usually 0.3–0.7 μm in length, though, some are as short as 0.16 μm and others are as long as 8 μm (see **Fulton**, 1971). Both have following ultrastructural components:

1. Cylinder wall. The most striking and regular ultrastructural feature of centrioles and basal bodies is the array of nine triplet microtubules equally spaced around the perimeter of an imaginary cylinder (Fig. 62.9). The space between and immediately around the triplet is filled with an amorphous, electron-dense material. In transverse section the triplets are arranged like vanes or blades of pinwheel or turbine. Each triplet or blade is tilted inward to the central axis at an angle of about 450 to the circumference; within each blade the tubules twist from one end to the other or have a helical course. Since centrioles have no outer membrane, the triplets are considered to form the wall of the cylinder, and arbitrarily define the inside and outside of the centriole.

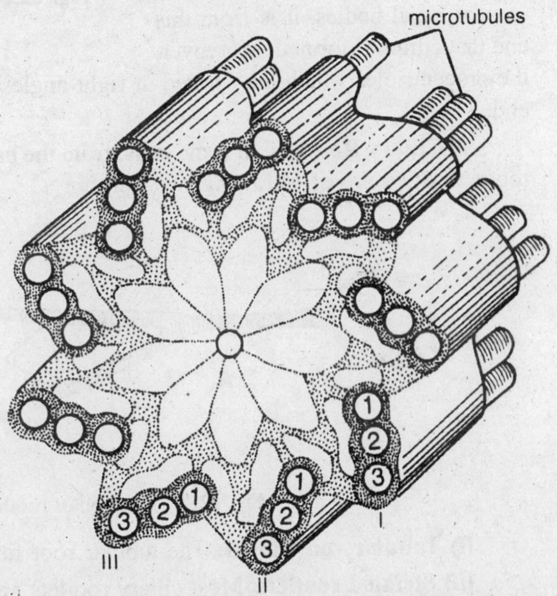

Fig. 62.9 Cross-sectional structure of a centriole. There are nine groups of three microtubules (after De Witt, 1977).

2. Triplets. The nine triplets that make up the wall are basically similar in centrioles and basal bodies. The three subunit microtubules have been designated A, B, and C, with the innermost tubule being A. Individual tubules are 200–260 Å in diameter. Only the A tubule is round; the others are incomplete, C-shaped and share their wall with the preceding tubule. At both ends the C tubule often terminates before the A and B tubules.

The substructure of A, B and C tubules, is similar to the structure of other microtubules (**Stephens**, 1 970). The A tubule has 13, 40–45 Å globular subunits around its perimeter. Three or four of these subunits are shared with the B tubules, which in turn share several of its subunits with the C tubules.

Often the triplets are thought to run parallel to one another and to the long axis of the cylinder, but this is not always the case. In the basal bodies of some organisms, the triplets get closer toward, the proximal end, so the diameter of the cylinder gets smaller. In some centrioles the triplets are parallel to one another but turn in a long-pitched helix with respect to the cylinder axis (**Fulton**, 1971).

3. Linkers. The A tubule of each triplet is linked with C tubule of neighbouring triplet by protein **linkers** at intervals along their entire length. These linkers hold the cylindrical array of the microtubules and maintain the typical radial tilt of the triplets.

4. Cartwheel. There are no central microtubules in the centrioles and no special arms. However, often faint protein spokes are radiate out to each triplet from a central core, forming a pattern like a cartwheel. Such a cartwheel configuration determines the **proximal end** of a centriole and, thus, provides a structural and functional **polarity** to it. The growth of the centriole takes place from the distal end, and in the case of basal bodies, it is from this end that cilium is formed. Moreover, the procentrioles which are formed at right angles to the centriole, are located near the proximal end.

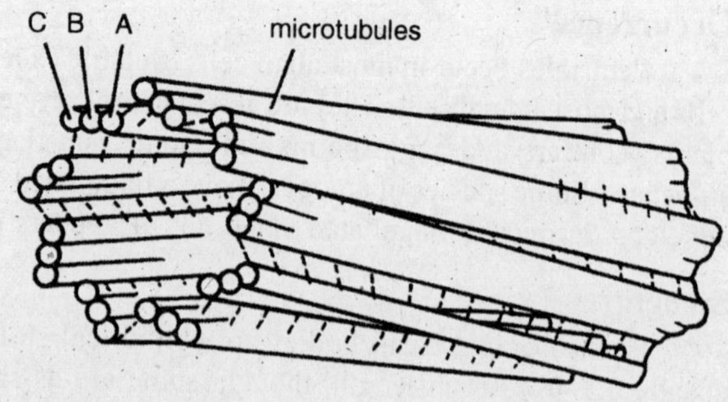

Fig. 62.10. Ultrastructure of a centriole (after Alberts *et al.*, 1989).

5. Ciliary Rootlets. In some cells, from the basal ends of the basal bodies originate the ciliary rootlets which are of following two types:

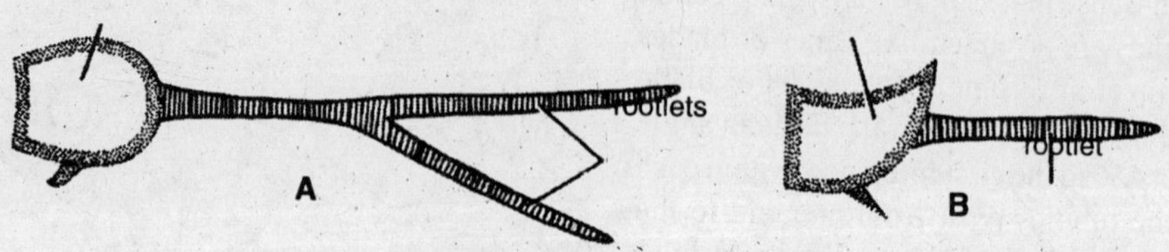

Fig. 62.11. Relation of rootlets with the basal body.

***(i)* Tubular root fibrils.** The tubular root fibrils have the diameter of 200Å.

***(ii)* Striated rootlets.** Most ciliary rootlets are striated, having a regular cross-banding with a repeating period of 55 to 70 nm. The striated fibres of rootlets are composed of parallel microfilaments; 3 to 7 nm in diameter, which in turn are formed of globular subunits. These fibres and filaments may have a structural role such as anchoring the basal body. Due to microfilaments, ciliary rootlets also have a contractile role. The rootlet may be double (*e.g.*, molluscs) or single (*e.g.*, the frog *Rana*).

6. Basal feet and satellites. The **basal feet** are dense processes that are arranged perpendicularly to the basal body. These processes impose a structural asymmetry on the basal body that is related with direction of the ciliary beat. A basal foot is composed of microfilaments that terminate in a dense bar. It may act as a focal point for the convergence of microtubules.

Satellites or **pericentriolar bodies** are electron-dense structures lying near the centriole that are probably nucleating sites for the microtubules.

Chemical Composition

The microtubules of centrioles and basal bodies contain the structural protein, **tubulin**, along with lipid molecules (**Fulton**, 1971). The centrioles and basal bodies contain a high concentration of ATPase enzyme. There exists a controversy that whether centrioles and basal bodies have DNA and RNA. **Fulton** (1971) has doubted the presence of nucleic acids in these organelles.

Origin of Centrioles and Basal Bodies

The idea prevalent years ago that new centrioles arise by the division of existing centrioles is no longer accepted. Rather it appears that new centrioles are either produced *de novo* or are synthesized using an existing centriole as some form of template (semi-autonomous replication).

1. Origin of centrioles by duplication of pre-existing centrioles. In cultured fibroblasts, centriole doubling begins at around the time that DNA synthesis begins (interphase). First the two members of a pair of centrioles separate; then a daughter centriole, called **procentriole**, is formed perpendicular to each original centriole, the two organelles being separated from each other by a distance of 50 to 100 nm. An immature centriole contains a ninefold symmetric array of single micro tubules; each microtubule then presumably acts as a template for the assembly of the triplet microtubules of mature centrioles. Each daughter centriole grows to mature size in late prophase maintaining their close proximity to and orientation at right angles to the mother centriole. As a result, when the interphase nuclei reform at the end of nuclear division, a centrosome containing two centrioles exists beside each nucleus.

Development of centriole (or basal body) has been studied in the ciliates *Paramecium* and *Tetrahymena* and in tracheal epithelium of *Xenopus* and chicks. The stages of development are virtually the same in all of them. Development of a basal body begins with the formation of a single microtubule in an amorphous mass. Microtubules are added one at a time until there is an equally spaced ring of nine. As the microtubules appear, the amorphous mass is lost, as though it were being consumed in the production of the microtubules. There is some evidence that connectives exist between the microtubules, which could act to set the distance between them. Thus, a ring of nine complete microtubules (*i.e.*, A tubules) is formed. The C-shaped B microtubule develop next and finally the C microtubules are added. The hub and cartwheel are added in the centre. The A – C links are not established until the end of development (see **Sheeler** and **Bianchi**, 1987).

2. Origin of basal bodies. In a ciliated vertebrate cell, which may contain hundreds of cilia, the centrioles of the precursor cell give rise to the many basal bodies required to nucleate the cilia in the mature cell. For example, during the differentiation of the ciliated epithelial cells that line the oviduct and the trachea, the centriole pair migrates from its normal location near the nucleus to the apical region of the cell where the cilia will form. There, instead forming a single daughter centriole in the typical manner, each centriole in the pair forms numerous electron-dense fibrogranular satellites. Many basal bodies then arise from these satellites and migrate to the membrane to initiate the formation of cilia.

3. The de novo origin of centrioles and basal bodies. There are certain cases where centrioles seem to arise *de novo*. For example, unfertilized eggs of many animals lack functional centrioles and use the sperm centriole for the first mitotic division (for cleavage), however, under certain experimental conditions—such as extreme ionic imbalance or electrical stimulation—the unfertilized egg can produce a variable number of centrioles. Each of these centrioles nucleates the formation of a small aster, one of which can be used by the egg for cleavage division, so that a haploid organism develops by a process called **parthenogenesis**. In fact, centriole precursors are stored in the cytoplasm of unfertilized eggs and can be activated to form a new centriole under special situation.

Like the centrioles, the basal bodies are found to possess some capacity for self-assembly and they appear suddenly in *Naegleria* as it changes from its amoeboid form to a typical ciliate (see **Reid** and **Leech**, 1980).

The unusual mode of duplication of centrioles and their continuity over many generations led to the earlier suggestion that centrioles might be fully **autonomous, self-replicating** organelles. Although it is now known that this is not the case and that under certain situation centrioles can arise *de novo* in the cytoplasm. So, it is possible that some information necessary for centriole formation is usually carried in the centriole itself (just as the replication of mitochondria and chloroplasts depends on extrachromosomal genes carried in the organelles). For example, in *Chlamydomonas* a set of genes that encode proteins involved in basal body structure and flagellar assembly is carried on a separate genetic element that segregate independently of the major chromosomes. The nature and location of this genetic element have still to be investigated (**Ramanis** and **Luck**, 1986).

Functions

1. Formation of basal bodies and ultimately the cilia is the specialized function of the centrioles in the cell.
2. The normal function of a pair of centrioles in most animal cells is to act as a focal point for the centrosome. The centrosome (also called the **cell centre**) organizes the array of cytoplasmic microtubules during interphase and duplicates at mitosis to nucleate the two poles of the **mitotic spindle**.
3. Sometimes centrioles can serve first one function and then another in turn: for example, prior to each division in *Chlamydomonas*, the two flagella resorb and the basal bodies leave their position to act as mitotic poles.
4. In spermatozoon one centriole give rise to the tail fibre or flagellum.
5. Centrioles and basal bodies are also found to be involved in ciliary and flagellar beat.
6. Centrioles and basal bodies have a role in the reception of optical, acoustic and olfactory signals.
7. Recently, it has been suggested that centrioles could serve as devices for locating the directions of signal sources. Such a role for them has been conceived by comparing the geometric design of centrioles (with their disposition in pairs at the right angles and their ninefold symmetry) with human-made devices such as radar scanner, that detect directional signals (**Albrecht-Buehler**, 1981).

62.5 CELLULAR MOVEMENT

Cellular movement involves motility and contractility. **Motility** occurs at the tissue, cellular and subcellular levels. The most conspicuous examples of motility, particularly in the animal world, take place at **the tissue level**. The muscle tissues common to most animals consists of cells specifically adapted to for *contraction*, and the movements produced are often obvious, whether manifested as the bending of a limb, the beating of a heart, or a uterine contraction during childbirth.

Equally important is the movement of intracellular components, which might be regarded as **motility at the subcellular level**. For example, highly ordered microtubules of the mitotic spindle play a key role in the separation of chromosomes during cell division (see Chapter 71). In addition, some cells display the phenomenon of cytoplasmic streaming, in which the cytoplasm undergoes rhythmic patterns of flow (*e.g.*, cyclosis in *Paramecium*). Other examples of mechanical work at the subcellular level include the characteristic movements of molecular structures that occur during cell growth and differentiation. An example of such a process is the transport of cellulose to the growing wall of a dividing or differentiating plant cell.

The microfilaments and microtubules of the cytoskeleton provide a basic scaffolding for specialized **motor proteins**, which interact with the cytoskeleton to produce motion at the molecular level (Table 62.3). The combined effects of these molecular motions produce movement at cellular level. In cases such as muscle contraction, the combined effects of many cells moving simultaneously produce motion at the tissue level.

Table 62.3 Selected motor proteins (or mechanoenzymes) of eukaryotic cells; they produce motion at the molecular level (Source: Becker *et al.*, 2007).

Molecules	*Typical function*
A. Microtubule associated proteins	
1. Cytoplasmic matrix	Motion toward minus end of microtubule
2. Axonemal dynein	Activation of sliding in flagellar microtubule
3. Kinesins	Motion toward plus end of microtubules
B. Microfilament-associated (actin-binding) proteins	
1. Myosin I, monomer	Motion along actin filament
2. Myosin II, filament	Slides actin filaments in sarcomere of muscle

In eukaryotes, there are two motility systems. The first involves interactions between specialized motor molecules and microtubules. Such **microtubule-based movement** occurs in **fast axonal transport**, one of the processes by which a nerve cell transport materials between the central part of the cell and outlying regions. The second type of eukaryotic motility is based on interactions between the actin microfilaments and members of the myosin family of motor molecules. A familiar example of **microfilament-based movement** in muscle contraction.

Intracellular Microtubule-Based Movement

Microtubules (MTs) provide a rigid set of tracks for the transport of a variety of membrane-enclosed organelles and vesicles. The centrosome (including centrioles) provides organization and orientation to MTs, because the minus ends of most microtubules are embedded in the centrosome. The centrosome is generally located near the center of cell, so traffic towards the minus ends of MTs might be considered "inbound" traffic. Traffic directed toward the plus ends might likewise be considered "outbound", meaning that it is directed toward the periphery of the cell.

While microtubules provide an organized set of tracks along which organelles can move, they do not directly generate the force necessary for movement. The mechanical work needed for movement depend on **microtubule-associated motor proteins (motor MAPs)** which attach to vesicles or organelles and then "walk" along the microtubule using ATP to provide the needed energy. These motor MAPs recognize the polarity of the MT, with each motor MAP having a preferred direction of movement. At present, only two major families of motor MAPs namely **kinesins** and **dyneins** are known.

depend on **microtubule-associated motor proteins (motor MAPs)** which attach to vesicles or organelles and then "walk" along the microtubule using ATP to provide the needed energy. These motor MAPs recognize the polarity of the MT, with each motor MAP having a preferred direction of movement. At present, only two major families of motor MAPs namely **kinesins** and **dyneins** are known.

Role of Kinesins and Dyneins in Movement of Organelles

A historically important cell type for studying microtubule-dependent intracellular movement is the neuron. In particular, the squid (*Loligo*, a molluscan animal) giant axon has been extremely useful for the biochemical purification of components that interact with MTs. A receptor protein or neurotransmitter synthesized in cell body of the neuron must be transported over distances upto a meter between the cell body and the nerve ending. The need for such transport arises because ribosomes are present only in the cell body, so no protein synthesis occurs in the axons or synaptic knobs. Instead proteins and membranous vesicles are synthesized in the cell body and transported along the axons to the synaptic knobs. Some form of energy-dependent transport is clearly required mechanism. The process is called **fast axonal transport** and appears to involve the movement of protein containing vesicles and other organelles.

The role of microtubules in axonal transport was initially suggested because the process is inhibited by **colchicine** and other drugs that impair MT function but is insensitive to drugs such as **cytochalasins** that affect microfilaments. Since then, MTs have been visualized along the axon by the deep-etch technique and shown to be prominent feature of the axonal cyto-skeleton. Moreover, axonal MTs have small membranous vesicles and mitochondria associated with them.

Further extensive investigations have revealed that in a nerve cell, **kinesin** protein mediates transport from the cell body down the axon to the nerve ending; it is called **anterograde axonal transport**. Likewise, **cytoplasmic dynein** mediates transport in the opposite direction, towards the minus ends of the MTs; it is called **retrograde axonal transport**. These two motors transport materials in opposite direction within the cytoplasm (Fig. 62.12).

Structure and Function of Kinesins

The first kinesins were originally identified in the cytoplasm of squid giant axons. Such "classic" kinesins consists of three parts (Fig. 62.13): 1. A globular head region that attaches to the microtubules and is involved in hydrolysis of ATP; 2. A coiled helical region; and 3. A light-chain region that is involved in attaching the kinesin to other proteins or organelles.

Classic kinesins move along microtubules in 8 nm steps: one of the two globular heads moves forward to make an attachment to a new β-tubulin subunit, followed by detachment of the trailing globular head which can now make an attachment to a new region of the MT (Fig. 62.13B). This movement occurs in a "hand-over-hand" fashion, as the two heads take turns serving as the front "foot". This movement is coupled to the exchange of ATP and ADP at specific sites within the heads. The result is that kinesin moves toward the plus end of a MT in an ATP-dependent fashion. A single kinesin molecule exhibits **processivity**. It can cover long distances before detaching from a MT. A single kinesin molecule can move as far as 1 μm which is a great distance relative to its size. Lastly, as a molecular motor, kinesin appears to be quite efficient, estimates of its efficiency in converting the energy of ATP hydrolysis to useful work are on the order of 60–70% (**Endow**, 2003).

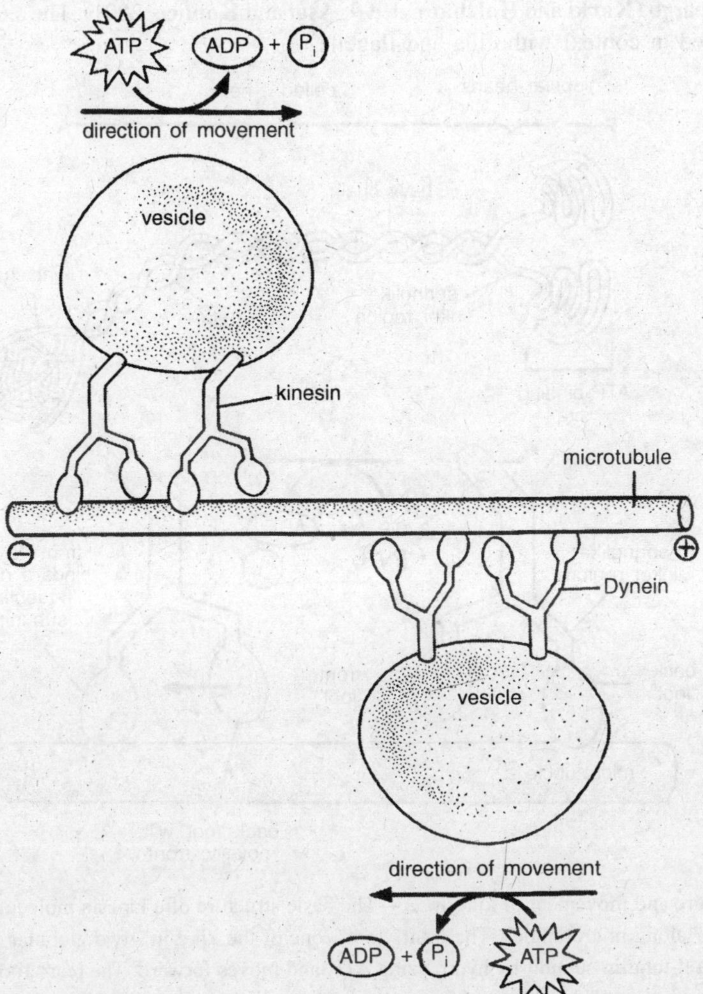

Fig. 62.12. Microtubule-based motility. Kinesins and dyneins are families of molecules that use the energy of ATP hydrolysis to "walk" along microtubules. In the process, they move intracellular structures (vesicles and organelles) along MTs. In general, members of the kinesin family move vesicles or organelles toward the plus ends of MTs-that is, from the center of the cell to the periphery. Dynein move in opposite direction toward the minus ends of MTs, and therefore toward the center of cell where the MTOC (microtubule-organizing center) is located (after Becker *et al.*, 2007).

Structure and Functions of Cytoplasmic Dynein

The dynein family of motor MAPs consists of two basic types: 1. Cytoplasmic dynein and 2. Axonemal dynein. **Cytoplasmic dynein** contains two heavy chains that interacts with microtubules, three intermediate chains and four light chains. In contrast to most kinesins, cytoplasmic dynein moves toward the minus ends of MTs. Also, unlike kinesins, cytoplasmic dynein cannot efficiently bind to cytoplasmic organelles on its own. Instead, dynein is associated with the protein complex known as **dynactin** (Fig. 62.14). The dynactin complex helps to link cytoplasmic dynein to the cargo (such as membraneous vesicle) it transports along MTs, by binding to proteins such as spectrin attached to the membrane cargo (**Karki** and **Holzbaur**, 1999; **Asai** and **Koonce**, 2001). The axonemal dynein has been described in context with cilia and flagella.

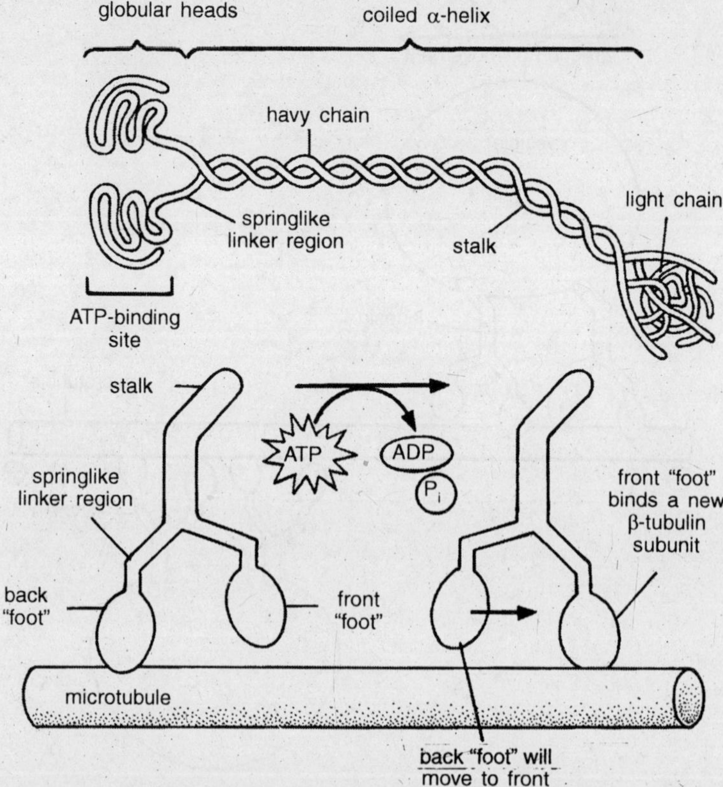

Fig. 62.13. Structure and movement of kinesin. A—The basic structure of a kinesin molecule. B—Kinesins "walk" along microtubules. The front "foot", one of the kinesin's two globular heads, detaches from a β-tubulin subunit by hydrolyzing ATP, and moves forward. The rear "foot" then releases by hydrolyzing ATP and springs forward, moving to the front in a "hand-over-hand fashion" (after Becker *et al.*, 2007).

Figure 62.15 has shown how motor MAPs are involved in transport of intracellular vesicles (**Allan, Thompson** and **McNiven**, 2002).

Microtubule-Based Motility

Microtubules are crucial not only for movement within cells, but also for the movement of **flagella** and **cilia**, the motile appendages of eukaryotic cells (cilia and flagella have been dealt in quite detail in the next section of this chapter).

Actin-Based Cell Movements: The Myosins

Movement of molecules and other cellular components also occurs along another major filament system in the cell, the actin cytoskeleton. As with microtubules, mechanoenzymes act as ATP-dependent motors that exert force on actin microfilaments within cells. These mechanoenzymes are all members of a large superfamily of proteins known as **myosins**. Currently there are at least 18 known classes of myosins. All myosins have at least one polypeptide chain, called the **heavy chain**, with a globular head group at one end attached to a tail of varying length (Fig. 62.16). The globular head binds to actin and uses the energy of ATP hydrolysis to move along an actin filament.

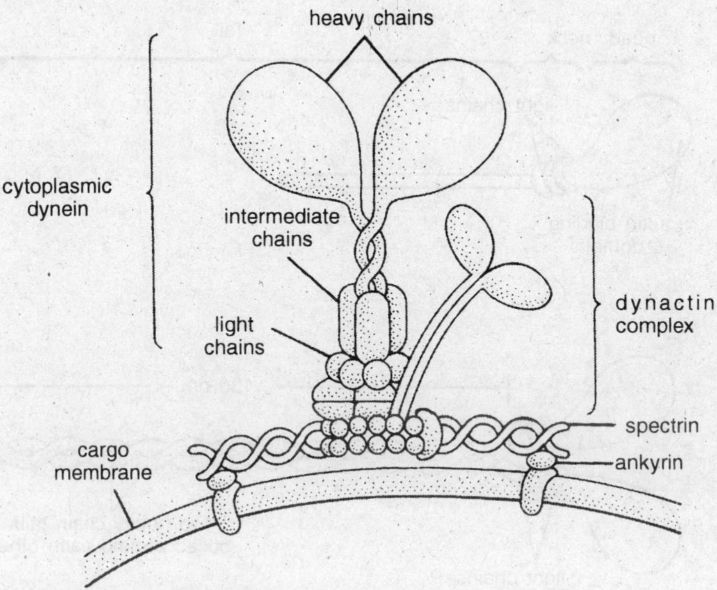

Fig. 62.14. Schematic representation of the cytoplasmic dynein/dynactin complex. Cytoplasmic dynein is linked with cargo membrane indirectly through the dynactin multiprotein complex. The dynactin complex binds to spectrin in the membrane of the cargo vesicle (after Becker *et al.*, 2007).

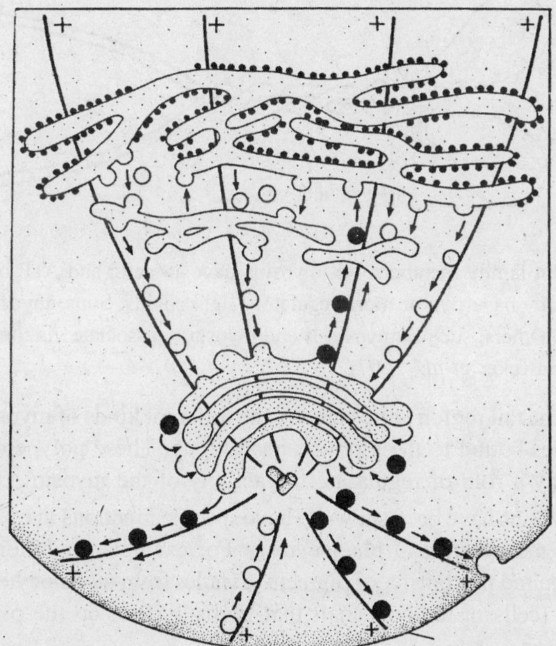

Fig. 62.15. Microtubules, Motor MAPs and the Golgi apparatus. According to this model, vesicles going to and from the Golgi apparatus are attached to microtubules and are thought to be carried by motor MAPs similar to dynein and kinesin. Dynein is a minus-end directed motor MAP, whereas kinesin is plus-end directed MAP. Thus, vesicles derived from either the ER or the plasma membrane are carried toward the Golgi apparatus and MTOC by dynein, where as vesicle derived from the Golgi apparatus are carried toward either the ER or the cell periphery by kinesins (after Becker *et al.*, 2007).

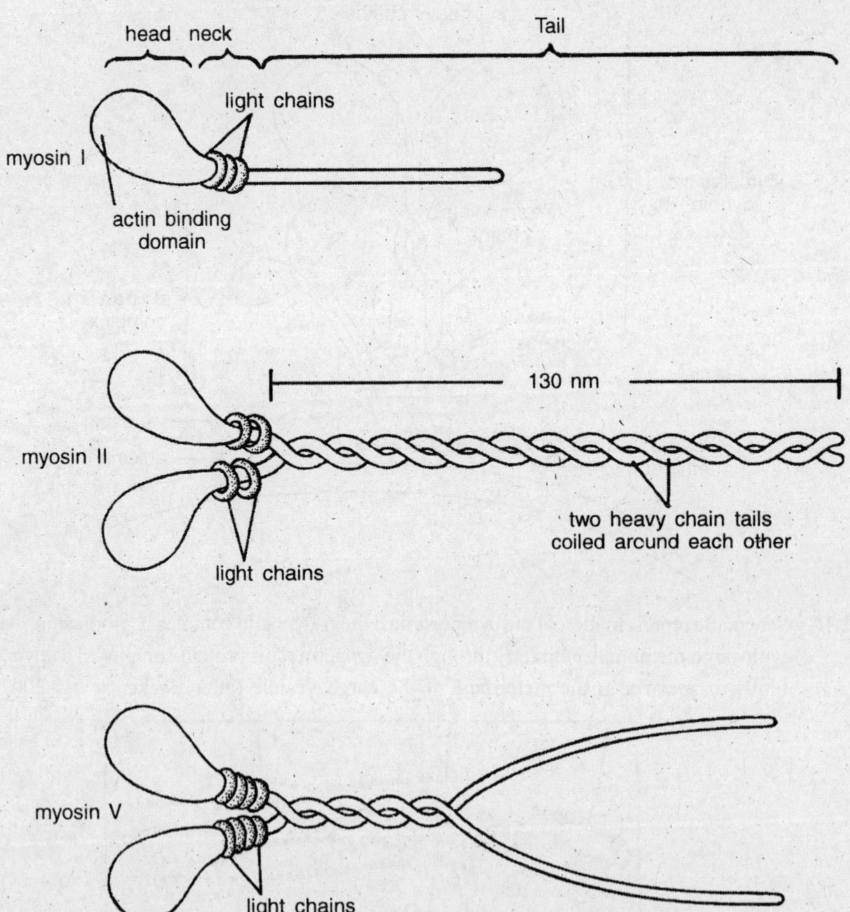

Fig. 62.16. Myosin family members. All myosins have an actin-and ATP-binding heavy chain "head" and typically have two or more regulatory light chains. Some myosins, such as myosin I, have one head. Others, such as myosin II and myosin V, associate via their tails into two headed proteins (after Becker *et al.*, 2007).

The structure of the tail region varies among the different kinds of myosin. Myosins typically contain small polypeptides bound to the globular head group. These polypeptides, referred to as the **light chains**, often play a role of regulating the activity of the myosin ATPase.

Several myosins have been shown to have specific functions in events as wide-ranging as muscle contraction, cell movement and phagocytosis. For example, mutations in the mouse indicate that a myosin V is required for transfer of pigment granules from melanocytes (cells that produce pigment) to keratinocytes (cells in the hair shaft that normally take up the pigment).

62.6 CILIA AND FLAGELLA

The cilia (*L.*, *cili* = eye lash) and flagella (*L.*, little whip) are microscopic, contractile and filamentous processes of the cytoplasm which create food currents, act as sensory organs and perform many mechanical functions of the cell. Morphologically and physiologically, the cilia and flagella are identical structures but even then both can be distinguished from each other by their number, size and functions. Their distinguishing features are as follows:

1. The flagella are less (1 or 2) in number than the cilia which may be numerous (3000 to 14000 or more) in number.
2. The flagella occur at one end of the cell, while the cilia may occur throughout the surface of the cell.
3. The flagella are longer (up to 150 μm) processes, while the cilia are short (5 to 10 μm) appendages of the cytoplasm.
4. The flagella usually beat independently, while the cilia tend to beat in a coordinated rhythm.
5. The flagella exhibit undulatory motion, while the cilia move in a sweeping or pendular stroke.

Stereocilia and Kinocilia

The cell sometimes gives out immobile cytoplasmic extensions known as **stereocilia**. The stereocilia differ from the true cilia which are known as **kinocilia**. The stereocilia occur in most epithelial cells of the epididymis and macula and crista of the internal ear. Stereocilia of the hair cells of the inner ear are responsible for the transduction of sound. These and other stereocilia do not contain microtubules. They contain, however, about 3000 microfilaments which are disposed longitudinally but have a definite polarity and a helical symmetry, with cross-bridges around the filaments (**De Rosier**, 1980).

Distribution of the Cilia and Flagella

The flagella occur in the protozoans of the class Flagellata, choanocyte cells of the sponges, spermatozoa of the Metazoa and among plants in the algae and gamete cells. The cilia occur in the protozoans of the class Ciliata and members of other classes and ciliated epithelium of the Metazoa. The cilia may occur on external body surface and may help in the locomotion of such animals as the larvae of certain Platyhelminthes, Nemertines, Echinodermata, Mollusca and Annelida. The cilia may line the internal cavities or passages of the metazoan bodies as air passage of the respiratory system and reproductive tracts. The nematode worms and arthropods have no cilia.

Except for sperm, the cilia in mammalian systems are not organelles of locomotion. But their effect is the same, that is, to move the environment with respect to the cell surface.

Structure of the Cilia and Flagella

The ciliary apparatus is composed of following basic components: 1. The **shaft** or **cilium**, which is the slender cylindroid process that projects from the free surface of the cell; 2. The **basal body** or granule, a centriole like cellular organelle from which the cilium originates; and 3. In some cells fine fibrils—called **ciliary rootlets**. Basal body remains separated from cilium by a **ciliary** or **basal plate** which has two functions: termination of the C Tubule of each triplet of basal body; and beginning of two central microtubules. The cilia and flagella are extremely delicate, permanently formed, thread-like extension of cytoplasm and their thickness is often at the limit of the resolving power of light microscope.

Isolation and Chemical Composition of Cilia and Flagella

The first detailed chemical analysis of the protein components of the cilia of *Tetrahymena pyriformis* was conducted by **I. R. Gibbons** (1963). Ciliary movements can be analyzed easily by scraping the pharyngeal epithelium of a frog or toad with a spatula and placing the scrapings in a drop of physiological salt solution between a slide and a coverglass. By certain recent techniques, a flagellum can be severed from a cell by a laser beam and ciliary membrane can be peeled off by detergent treatments.

Axoneme of cilia has a variety of proteins such as A and B **tubulins** in the microtubules, **dynein** (the microtubule ATPase), **nexin** and others.

Ultrastructure of the Cilia and Flagella

An eukaryotic cilium or flagellum is composed of three major parts: a central axoneme or shaft, the surrounding plasma membrane and the interposed cytoplasmic matrix (Fig. 62.17).

1. Ciliary membrane. Though the ciliary membrane (9.5 nm thick) is physically continuous with plasma membrane of the cell, but it contains far less amount of proteins than the latter (*i.e.*, it is a typically protein poor; **Satir**, 1977). Further, some of the proteins present in the ciliary membrane are specific to it and have a role as the barrier against the loss of ATP and certain essential ion that are required at appropriate concentrations to provide the energy for the ciliary movement.

Ciliary necklace. An unusual feature of the membranes of all somatic cilia is the presence of multiple strands (2 to 6 and up to 11) of particles, called **ciliary necklace**, at the base of the organelle. These particles can be seen in the electron micrographs of freeze-fractured cilium (see **Fumi Suzuki**). The ciliary necklace is found at a region in the cilium where microtubules and basal bodies make contact with the membrane. According to **Thorpe** (1984), ciliary necklace may have following two functions: 1. It may position the underlying basal body from which the cilium is originated. 2. It may help in the differentiation of ciliary membrane; *i.e.,* the rings of particles may retain proteins that would otherwise diffuse out and be incorporated into ciliary membrane.

2. Matrix. The bounded space of the cilium contains a watery substance known as **matrix**. In the ciliary matrix are embedded eleven microtubules of axoneme and other interconnecting proteins.

3. Axoneme. The axial basic microtubular structure of cilia and flagella is called **axoneme**. It is the essential motile element of these organelles. The axoneme is about 0.2 to 10 μm in diameter and may range from a few microns to 1 to 2 mm in length. The cilia may be thicker at the base and may become thinner gradually along the length.

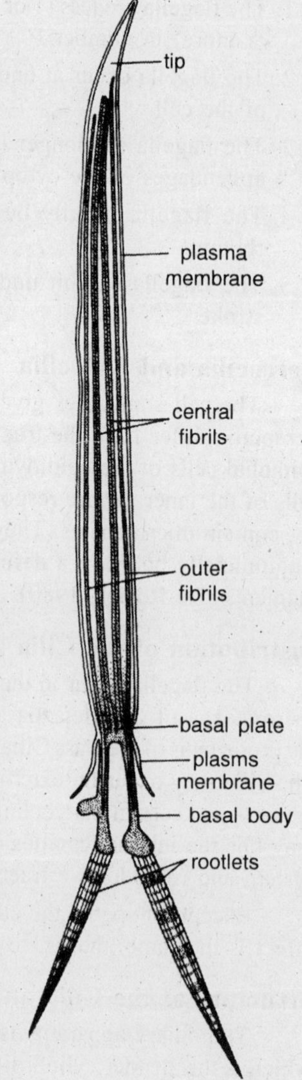

Fig. 62.17. A ciliary apparatus in L.S. showing fundamental structure of cilium or shaft, basal bodies and ciliary rootlets.

The axonemal elements of nearly all cilia and flagella (as well as the tails of sperm cells) contain the same 9 + 2 arrangement of microtubules. In the centre of the axoneme are two **singlet microtubules** or fibrils that run length of the cilium. Each of the central microtubules (25 nm in diameter) is composed of 13 protofilaments. The central fibrils, each has a wall of 6 nm thick and are located 35 nm away from each other. Both central fibrils are connected by a bridge and are enclosed in a common central **sheath**. A plane perpendicular to line joining the two central tubules divides the axoneme into a right and a left symmetrical half. It is generally accepted that the plane of the ciliary beat is perpendicular to this plane of symmetry.

Table 62.4 Major protein structures of the axoneme of the cilia and flagella (Source: Alberts *et al.*, 1989).

Axoneme component (periodicity along axoneme)	*Function*
1. Tubulin (8 nm)	Principal component of microtubules.
2. Dynein (24 nm)	Project from microtubule doublets and interact with adjacent doublets to produce bending.
3. Nexin link (86 nm)	Hold adjacent microtubule doublets together.
4. Radial spokes (29 nm)	Extend from each of the nine outer doublets inward to the central pair.
5. Sheath projections (14 nm)	Project as a series of side arms from the central pair of microtubules; together with the radial spokes these regulate the form of the ciliary beat.

Nine **doublet microtubules** (each 36 nm in diameter) surround the central sheath; they remain separated from each other by a distance of 20 nm and from the ciliary membrane by a distance of 25 nm. Each peripheral doublet consists of two microtubules or subfibres (18 to 25 nm in diameter), one is smaller (A) and complete, having 13 protofilaments of tubulin and lying closer to the axis; the other subfibre (B) is larger and incomplete, having only 11 protofilaments. The B subfibre lacks the wall adjacent to A subfibre and is skewed at about 10° away from the axis. Other associated structures of the doublets are the following:

1. Dynein arms. Extending from each A subfibre are two **dynein arms**—an **outer arm** and an **inner arm**, that are oriented in the same direction in all microtubules (*i.e.*, peripheral doublets). This orientation is clockwise when the axoneme is viewed from base to tip. The arms contain **dynein**, which is large protein complex (nearly 2 million daltons) composed of 9 to 10 polypeptide chains, the largest of which are about 450,000 daltons. Dynein is a Mg^{2+} and Ca^{2+}-activated ATPase enzyme which after solubilization can recombine at the same position on the A microtubule. Dynein contains two or three elliptical or globular **heads** (depending on the source) linked to a common **root**, **foot** or **base** by the thin flexible **strands** or **stalks** (Fig. 62.19). Thus, the base of the dynein molecule attaches only to A subfibre, leaving the heads free to make contact with the adjacent B tubules of neighbouring doublet. It indicates that B tubule has different structure than the A tubule, so that, the base of dynein cannot attach to it. The resulting asymmetry is required to prevent a fruitless tug-of-war between the neighbouring microtubules, which presumably explains why each of nine outer microtubules is an A – B doublet (see **Alberts** *et al.*, 1989).

2. Nexin links. Adjacent doublets are joined or linked by **peripheral**, **interdoublet** or **nexin links**; the nexin links have a periodicity of 86 nm. Nexin links extend from A tubule of one doublet to B tubule of adjacent doublet. Nexin protein has a molecular weight of about 150,000 to 160,000 daltons. Nexin links are highly elastic: their normal length is 30 nm, but they can be stretched to 250 nm without breaking (see **Darnell**, 1986). They are thought to function like the rubber bands to resist the sliding between adjacent double microtubules (*i.e.*, they maintain integrity of the axoneme during the sliding motion.

3. Radial spokes. There are 36 nm long radial bridges or links between the A subfibre and the sheath containing the central microtubules. These spokes terminate in a dense **knob** or **head**, which may have a fork-like structure. Earlier observation of **Warner** and **Satir** (1974) that the spokes are attached perpendicularly to the ciliary axis where it is straight and that they are relatively detached in bent or tilted regions of the axis has led to the hypothesis that *they may be active in the conversion of active sliding between outer doublets into local axial bending.*

Lastly, the structures of a cilium at the base and tip are slightly different from features described above (Fig. 62.18).

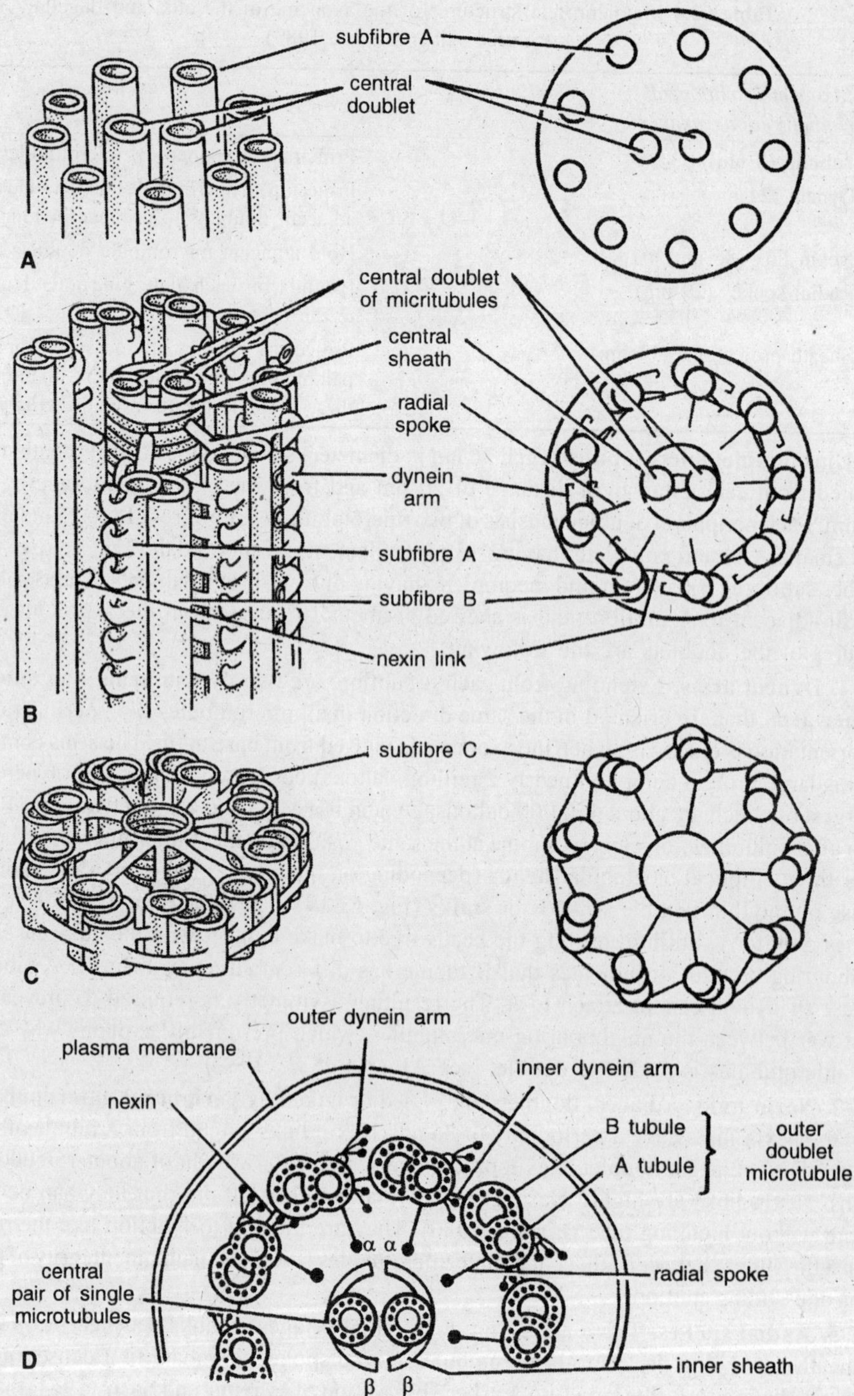

Fig. 62.18. Schematic diagram of a flagellar axoneme, showing relationships between its components at the flageller tip (A), in the middle of the flagellum (B), and at the basal body (C). More detailed picture of flagellum in the cross section (after Berns, 1983; Albert *et al.*, 1989).

Physiology of Ciliary Movement

The cilia and flagella serve many purposes and their movements propel the organism. The cilia are contractile structures and in them two types of rhythms known as **metachronic** and **isochronic** or **synchronous rhythms** produce the wave of contractions in the cilia. In the **metachronic** type of rhythm the cilia of a row beat one after the other, while in the **synchronous** or isochronic rhythm, all the cilia of a row beat simultaneously. In contrast to the cilia, the flagella exhibit undulant motion and beat independently.

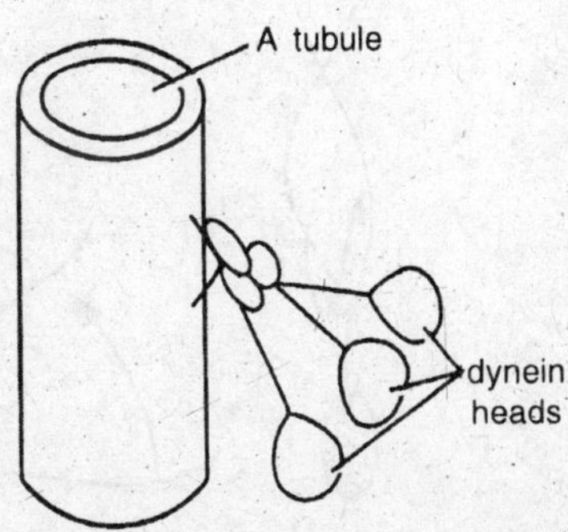

Fig. 62.19 The base of the dynein molecule binds tightly to an A tubule in an ATP-independent manner, while the large globular heads have an ATP-dependent binding site for a second microtubule (B tubule) (after Alberts *et al.*, 1989).

The beating of cilia or flagella is caused by the intraciliary excitation which is followed by the interciliary conduction. **Hayashi** (1961) has reported that the two inner filaments of a cilium transmit excitation and the nine outer filaments are the seat of ATP splitting. The movement of cilia may be under nervous or cytoplasmic control. In a few invertebrate embryos the cilia are probably under nervous control since their movement may be stopped upon stimulation of the embryo. In ciliates they are thought to be coordinated by a neuromotor centre near the mouth since destruction of the fibres connecting the centre to the cilia results in uncoordinated movements (**Taylor**, 1920). However, **Okajima** (1966) reported coordinated movements in *Euplotes* even after complete dissection of the neuromotor fibres. Recent studies have shown that cytoplasm is necessary for the ciliary movements. The ATP provides necessary amount of energy for the motion of the cilia and flagella. Four types of ciliary movements have been recognized which are as follows:

1. The pendulus ciliary movement. The pendulus type of ciliary movement is carried out in a single plane. It occurs in the ciliated protozoans which have rigid cilia. In such cases the movement of the cilia is carried out by a flexion at its base.

2. The unciform ciliary movement. The unciform (hook-like) ciliary movement occurs commonly in the metazoan cells. In such type of movement, when the cilia contract it becomes double and acquires a hook-like shape.

3. The infundibuliform ciliary movement. The infundibuliform ciliary movement occurs due to the rotary movement of the cilium and flagellum. In this case, the cilium or flagellum is passed through three mutually perpendicular planes in the space and makes conical or funnel-shaped shape.

4. The undulant movement. The undulant movement is the characteristic of the flagellum. In undulant movement the waves of the contraction proceed from the site of implantation and pass to the border.

Each beat of cilium or flagellum involves the same pattern of microtubule movement. Each cilium moves with a whip-like motion and its beat may be divided into two phases: 1. The fast **effective stroke** (or forward active stroke or power stroke) in which the cilium is fully extended and beating against the surrounding liquid (*i.e.*, it is like the action of an oar in a rowboat; Fig. 62.20B). 2. The slow **recovery stroke**, in which the cilium returns to its original position with an unrolling movement that minimizes viscous drag (Fig. 62.20C). The cycles of adjacent cilia are almost but not quite in synchrony, creating a wave-like pattern.

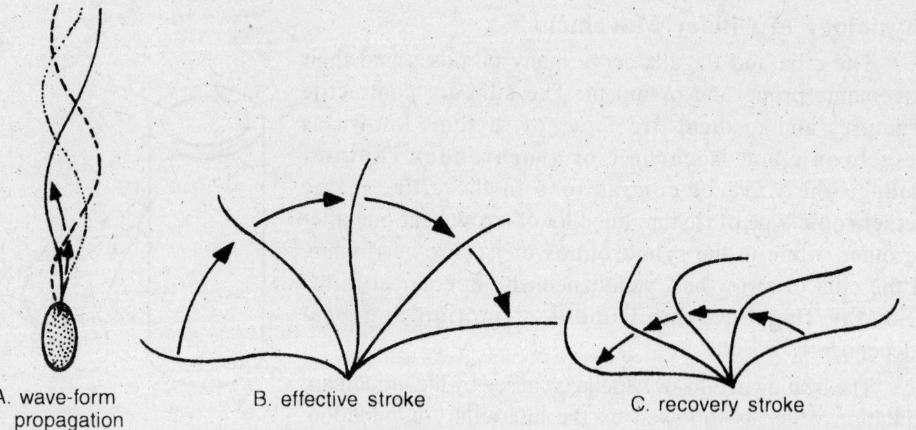

Fig. 62.20. Movement of cilia and flagella. A—Typical wave-form propagation of a flagellum, such as that found in sperm; B—Effective stroke common with cilia of all types; C—Recovery of cilium prior to the next effective stroke (after Dyson, 1978).

The flagellum instead of making whip-like movements, propagates quasi-sinusoidal waves (Fig. 62.20A), *i.e.*, successive waves move toward the tip of the flagellum, propelling the cell (*e.g.*, sperm) in the opposite direction.

The mechanism of force and movement (bending) by the flagellum has recently been studied extensively. It is well established now that the ciliary movement is generated by the microtubules and the associated structures of the flagellum. It was shown that the cell free flagella can be caused to move by adding an energy source such as ATP. Even broken pieces of cilia or isolated axoneme itself continue to beat, suggesting the role of microtubules in the movement. The contractile **axostyle** of some microorganisms such as *Metamonadida* (a dinoflagellate that lives in the gut of termite; Fig. 62.21) is another example of microtubule mediated motile process (see **Berns**, 1983). In fact, bending force is produced by the sliding of microtubules. This has been shown by exposing isolated axonemes to proteolytic enzymes, which disrupts both the nexin links and the radial spokes but leaving the dynein arms and the microtubules themselves intact. If such a partially digested structure is exposed to as little as 10 μM ATP, the axoneme elongates until it is up to nine times its original length, the component microtubules in the axoneme telescoping out of the loosened structure. It seems that the adjacent peripheral doublets can actively slide against each other once they have been freed of their lateral cross-links (such as those made of nexin). Apparently, in intact structure this sliding movement is converted to bending. Further, since the adjacent outer doublets actively slide against each other, a force must be generated between them. This force is apparently generated by dynein arms which 'walk' along the doublets, as has been suggested by **Peter Satir's** (1968) sliding filament hypothesis.

Sliding Filament Hypothesis

Recent experimental work on ciliary motion has shown notable similarities with the sliding mechanism involved in the interaction of actin and myosin in muscle. The dynein arms attached to subfibre A have been compared with the cross bridges of myosin and it has been postulated that they form intermittent attachments, by which one doublet (N_1) is able to push the adjacent one ($N_1 + 1$) toward the tip of the axoneme (Fig. 62.22). Under normal conditions, the attachment of subfibre A of N to subfibre B of N + 1 by dynein arms is not observed in an intact cilium. Only when the ciliary membrane is extracted with a detergent, the axoneme enters in a state of **rigor**

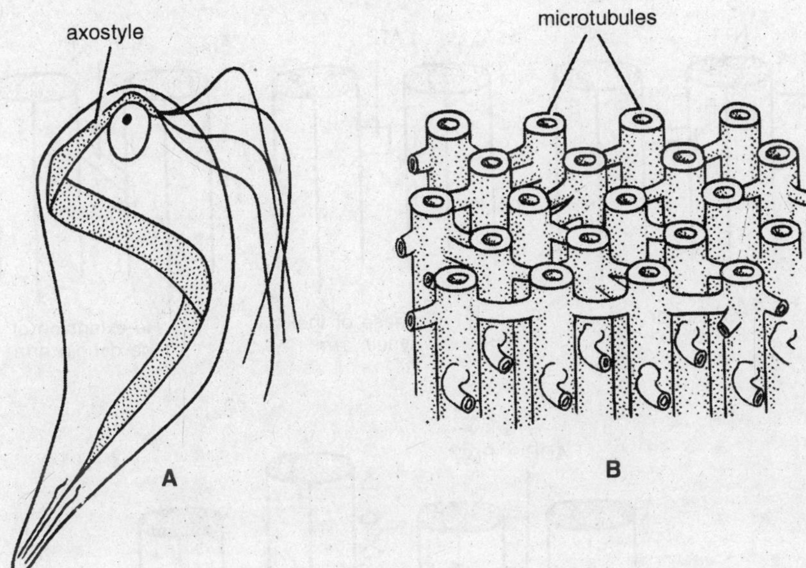

Fig. 62.21 Axostyle of *Metamonadida* flagellate. A—The relationship of the axostyle with the cell body; B—Schematic representation of the arrangement of microtubules into parallel sheets with protein cross bridges between them as seen in a T.S. of axostyle (after Berns, 1983).

in which the attachment is produced (Fig. 62.22A). Addition of ATP to axonemes in the state of rigor restores motility and causes release of the dynein arm (Fig. 62.22B). In this mechano-chemical cycle, the next step would be **re-extension** of the dynein arm (Fig. 62.22C) and its **rebinding** at an angle, with a new, more proximal site on subfibre B (Fig. 66.22C). This step involves the hydrolysis of ATP to ADP + Pi. In the last step, the arm returns to the rigor position and displacement of the doublets results (Fig. 62.22E).

Force is generated when dynein arms move. The movement of sliding is converted to bending by virture of radial spokes that bridge each other doublet to the inner pair of microtubules (**Warner** and **Satir**, 1974; **Huang** *et al.,* 1981). The wave that is generated by sliding is propagated down the organelle from base to tip, with the cell generally moving in a direction opposite from that of wave propagation.

Immotile Cilia Syndrome (Kartagenre's Syndrome)

Ciliary motion can be affected by many deficiencies in the protein composition of the organelle. For example, in immotile cilia syndrome, a condition characterized by severe respiratory difficulty (chronic bronchitis and sinusitis) and male sterility, the underlying genetic defect is the absence of inner and outer dynein arms on the peripheral doublets of both cilia and flagella. The symptoms of this syndrome result from the immobility of cilia in the respiratory tract and of the flagella in the sperm.

In *Chlamydomonas* several mutational defects have been studied in the axoneme of flagellum which may lead to paralysis of the flagellar function (**Luck**, 1984).

Other Functions of the Cilia and Flagella

1. The ciliary or flagellar movement provides the locomotion to the cell or organism.
2. The cilia create food currents in lower aquatic animals.
3. In the respiratory tract, the ciliary movements help in the elimination of the solid particles from it.

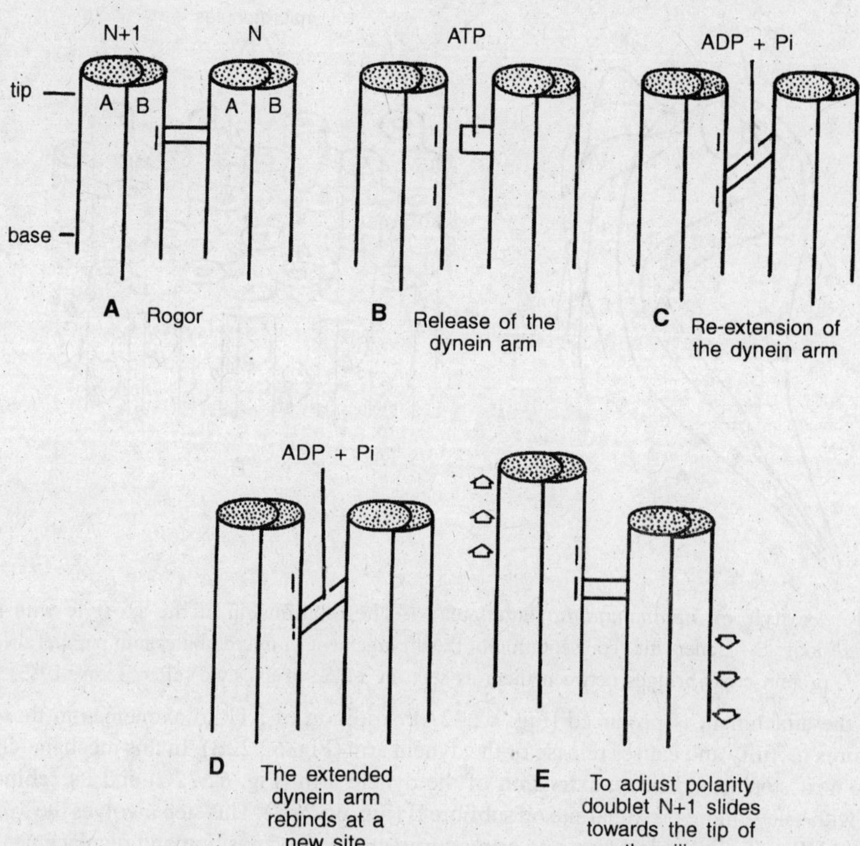

Fig. 62.22. Schematic representation of the mechano-chemical cycle involved in sliding of filament in ciliary movement (after De Robertis and De Robertis, Jr., 1987).

4. The eggs of amphibians and mammals are driven out from the oviduct by the aid of vibratile cilia of the latter.

Thus, the cilia and flagella serve many physiological processes of the cell, such as locomotion, alimentation, circulation, respiration, excretion and perception of sense.

Origin of Cilia

The newly formed basal bodies become aligned in rows beneath the apical plasma membrane and each basal body may then produce satellites from the side, a root from its base and a cilium from its apex.

The formation of the cilia from the basal bodies is started by the formation of a vesicle-like structure of the cytoplasm towards the distal end of the basal bodies. The walls of the vesicle are invaginated due to rapid growth of ciliary shaft. The walls of the vesicle are temporary and are replaced by the new and permanent ciliary sheaths.

Derivatives of Cilia

The cilia are modified into a variety of structures such as the rods and cones of the retina, crown cell of saccus vasculosus of third ventricle of fishes, primitive sensory cells of the pineal eye and cnidocil of the nematocysts of the coelenterates.

QUESTIONS

Long Answer Questions

1. What are microtubules? Describe their structure, assembly, disassembly and functions.
2. What are the intermediate filaments? Describe their types, structure and cellular functions.
3. Define the term microfilament. Describe the structure and function of microfilament in the cell.
4. Describe the ultrastructure of basal body and centrioles.
5. Differentiate between cilia and flagella. Describe the structure of the axoneme.

Short Answer Questions

1. What is cytoskeleton? Write a short note on the cytoskeleton.
2. How are the centrioles and basal bodies (kinetosomes) are originated in the cell?
3. What are the main functions of the basal bodies and the centrioles?
4. Explain sliding microtubule hypothesis of ciliary movement.
5. Flagella and cilia, though identical in structure, commonly exhibit a quite different pattern of movement. Describe the two patterns and conditions under which one or the other would be more appropriate.
6. Make a comparison of three main components of the cytoskeleton: microtubules, intermediate filaments and microfilaments.
7. Write short notes on the following:
   (*a*) axoneme; (*b*) sterocilia; (*c*) ciliary necklace; (*d*) axostyle; (*e*) sliding filament hypothesis·
   (*f*) Kartagenre's syndrome.
8. Describe structure and function of kinesins.
9. Give an account of structure and function cytoplasmic dyneins.

Very Short Answer Questions

1. Define the terms basal bodies and centrioles.

   Ans. *Basal body* is a microtubule containing structure located at the base of a eukaryotic flagellum or cilium that consists of nine sets of triplet microtubules; identical in appearance to a centriole. *Centriole* is microtubule containing structure embedded within the centrosome of animal cells, where centrioles lie at right angles to each other; identical in structure to the basal body of eukaryotic cilia and flagella.

2. Write the names of main constituents of cytoskeleton.

   Ans. Microtubules, microfilaments and intermediate filaments.

3. How many subunits are found in a microtubule?

   Ans. 13 subunits.

4. What is microtubule?

   Ans. Microtubule (MT) is a polymer of the protein tubulin, with a diameter of about 25 nm, that is an integral part of cytoskeleton and that contributes to the support, shape and motility of eukaryotic cells.

5. Define microfilament.

   Ans. Microfilament (MF) is a polymer of actin, with a diameter of about 7 nm, that is an integral part of the cytoskeleton, contributing to the support, shape and mobility of eukaryotic cells.

6. What is intermediate filament?

   Ans. Intermediate filament (IF) is a group of protein filaments that are the most stable components of the cytoskeleton of eukaryotic cells, exhibit a diameter of 8 to 12 nm, which is intermediate between the diameter of actin microfilaments and microtubules.

7. Define the MTOC.

   Ans. MTOC stands for microtubule-organizing center. It is structure that initiates the assembly of microtubules, the primary example being the centrosome.

8. What is cilium?

   Ans. Cilium (plural cilia) is a membrane-bounded appendage on the surface of a eukaryotic cell, composed of a specific arrangement of microtubules and responsible for motility of the cell. It is shorter and more numerous than closely related organelles called flagella.

9. What is axoneme?
   Ans. Axoneme is a group of interconnected microtubules that form the backbone of a eukaryotic cilium or flagellum, usually arranged as nine outer doublet microtubules surrounding a pair of central microtubules.
10. What is microtubule-based movement?
    Ans. It is motility based on microtubules; it includes motility involving cilia, flagella and sperm tails, as well as chromosomal movements mediated by spindle microtubule.
11. What is microfilament-based movement?
    Ans. It is motility based on microfilaments composed of actin and their interaction with myosin; includes muscle contraction, amoeboid movement, cytoplasmic streaming, and cytokinesis in animal cells.
12. Define motor MAP.
    Ans. MAP stands for microtubule-associated motor protein; it includes proteins, such as kinesin or dynein, that uses energy derived from ATP to drive the transport of vesicles and organelles along microtubules or to generate sliding forces between microtubules.
13. What is kinesin?
    Ans. It is family of motor proteins that generate movement along microtubules using energy derived from ATP hydrolysis.
14. Define the cytoplasmic dynein.
    Ans. Cytoplasmic dynein is a cytoplasmic motor protein that moves along the surface of microtubules in the plus-to-minus direction driven by energy derived from ATP hydrolysis; it is associated with dynactin which links cytoplasmic dynein to cargo vesicle.
15. Define myosin.
    Ans. Myosin is a family of motor proteins that create movements by exerting force on actin microfilaments using energy derived from ATP hydrolysis; makes up thick filaments that move the actin thin filament during muscle contraction.
16. What is found in the core of cilia? *(H.N.B. Garhwal 2009)*
    Ans. Matrix with axoneme having 9 + 2 arrangement of microtubules

Multiple Choice Questions

1. Tubulin protein is absent in: *(H.N.B., Garhwal 2008)*
   (*a*) cilia (*b*) flagella
   (*c*) plasma membrane (*d*) microtubules
2. ATPase activity is associated with structural component of cilia called
   (*a*) A tubule (*b*) B tubule
   (*c*) radial spoke (*d*) dynein
3. Cytoskeleton is made up of
   (*a*) callose deposits (*b*) cellulosic microfibrils
   (*c*) proteinaceous filaments (*d*) calcium carbonate granules
4. Organelle important in spindle formation during nuclear division is
   (*a*) centriole (*b*) Golgi apparatus
   (*c*) chloroplast (*d*) mitochondrion
5. Stereocilia can be found in
   (*a*) trachea (*b*) epididymis
   (*c*) sperm (*d*) buccal cavity
6. Intermediate filaments are composed of
   (*a*) tubulin (*b*) actin
   (*c*) flagellin (*d*) keratin, vimentin, desmin or lamin
7. Microtubular structure of cilium is called
   (*a*) basal body (*b*) ciliary rootlet
   (*c*) axonem (*d*) kinetosome

ANSWERS

1. (*c*) **2.** (*d*) **3.** (*c*) **4.** (*a*) **5.** (*b*) **6.** (*d*) **7.** (*c*).

63

Endoplasmic Reticulum (ER)

The cytoplasmic matrix is traversed by a complex network of inter-connecting membrane bound vacuoles or cavities. These vacuoles or cavities often remain concentrated in the endoplasmic portion of the cytoplasm; therefore, known as **endoplasmic reticulum**, a name derived from the fact that in the light microscope it looks like a "net in the cytoplasm." (Eighteenth-century European ladies carried purses of netting called **reticules**).

The name "endoplasmic reticulum" was coined in 1953 by **Porter**, who in 1945 had observed it in electron micrographs of liver cells.

63.1 OCCURRENCE

The occurrence of the endoplasmic reticulum varies from cell to cell. The erythrocytes (RBC), egg and embryonic cells lack in endoplasmic reticulum. [**Note**. In the reticulocytes (immature red blood cells) which produce only proteins to be retained in the cytoplasmic matrix (cytosol) (*e.g.*, haemoglobin, the ER is poorly developed or non-existent, although the cell may contain many ribosomes]. The spermatocytes have poorly developed endoplasmic reticulum. The adipose tissues, brown fat cells and adrenocortical cells, interstitial cells of testes and cells of corpus luteum of ovaries, sebaceous cells and retinal pigment cells contain only **smooth endoplasmic reticulum (SER)**. The cells of those organs which are actively engaged in the synthesis of proteins such as acinar cells of pancreas, plasma cells, goblet cells and cells of some endocrine glands are found to contain **rough endoplasmic reticulum (RER)** which is highly developed. The presence of both SER and RER in the hepatocytes (liver cells) is reflective of the variety of the roles played by the liver in metabolism.

63.2 ER AND ENDOMEMBRANE SYSTEM

The endoplasmic reticulum is the main component of the **endomembrane system**, also called the **cytoplasmic vacuolar system** or **cytocavity network**. This system comprises following structures: 1. The **nuclear envelope**, consisting of two non-identical membranes, one opposed to the nuclear chromatin and other separated from the first membrane by a perinuclear space (both forming a cisternae), the two membranes being in contact at the nuclear pores; 2. The **endoplasmic reticulum**; and 3. The **Golgi apparatus**, which is mainly related to some of the terminal processes of cell secretion. **GERL** (or Golgi, ER and lysosome) refers to a special region of endomembrane system, which is more related to the Golgi apparatus and is involved in the formation of lysosomes.

The entire endomembrane system represents a barrier separating cytoplasmic compartments. The membrane of each component of this system has two faces: (*i*) The **cytoplasmic** or **protoplasmic face** and (*ii*) The **luminal face** (Fig. 63.1). The luminal face borders the perinuclear cisternae, the cavities of ER and SER, and the Golgi elements. It also corresponds to the interior of the secretory granules, the lysosomes and peroxisomes and also to faces of mitochondrial membranes confronting to outer mitochondrial chamber.

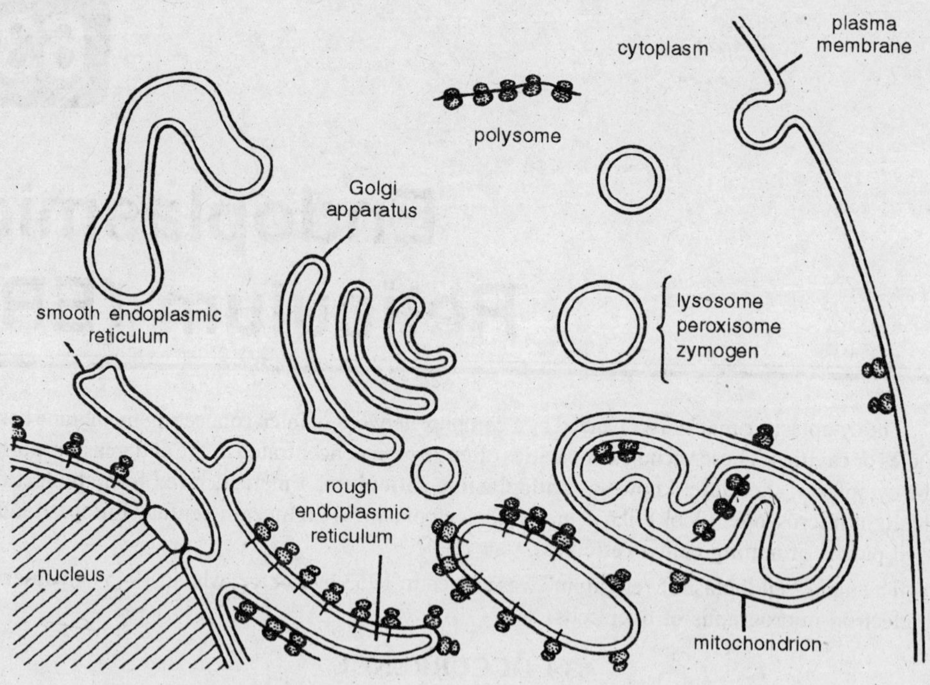

Fig. 63.1. Two faces of membranes of endomembrane system. In each membrane the luminal faces are shown in thick lines, while the cytoplasmic faces are depicted by thin lines. Ribosomes are always located on the cytoplasmic or matrix side.

63.3 MORPHOLOGY (STRUCTURE)

Morphologically, the endoplasmic reticulum may occur in the following three forms: 1. Lamellar form or cisternae (A closed, fluid-filled sac, vesicle or cavity is called **cisternae**); 2. Vesicular form or vesicle and 3. Tubular form or tubules.

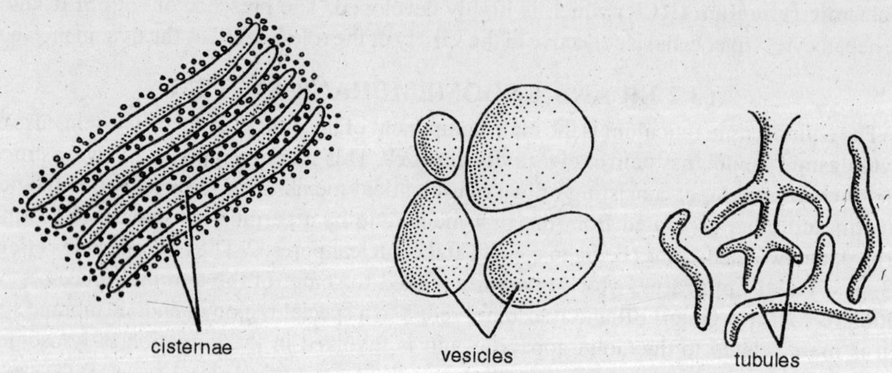

Fig. 63.2. Various components of the endoplasmic reticulum.

1. Cisternae. The cisternae are long, flattened, sac-like, unbranched tubules having the diameter of 40 to 50 μm. They remain arranged parallely in bundles or stakes. RER usually exists as cisternae which occur in those cells which have synthetic roles as the cells of pancreas, notochord and brain.

2. Vesicles. The vesicles are oval, membrane-bound vacuolar structures having the diameter of 25 to 500 μm. They often remain isolated in the cytoplasm and occur in most cells but especially abundant in the SER.

3. Tubules. The tubules are branched structures forming the reticular system along with the cisternae and vesicles. They usually have the diameter from 50 to 190 μm and occur almost in all the cells. Tubular form of ER is often found in SER and is dynamic in nature, *i.e.*, it is associated with membrane movements, fission and fusion between membranes of cytocavity network.

Ultrastructure

The cavities of cisternae, vesicles and tubules of the endoplasmic reticulum are bounded by a thin membrane of 50 to 60 Å thickness. The membrane of endoplasmic reticulum is fluid-mosaic like the unit membrane of the plasma membrane, nucleus, Golgi apparatus, etc. The membrane, thus, is composed of a bimolecular layer of phospholipids in which 'float' proteins of various sorts. The membrane of endoplasmic reticulum remains continuous with the membranes of plasma membrane, nuclear membrane and Golgi apparatus. The cavity of the endoplasmic reticulum is well developed and acts as a passage for the secretory products. **Palade** (1956) has observed secretory granules in the cavity of endoplasmic reticulum.

Sometimes, the cavity of RER is very narrow with two membranes closely apposed and is much distended in certain cells which are actively engaged in protein synthesis (*e.g.*, acinar cells, plasma cells and goblet cells). **Weibel** *et al.*, 1969, have calculated that the total surface of ER contained in 1 ml of liver tissue is about 11 square metres, two-third of which is of rough type (*i.e.*, RER).

63.4 TYPES OF ENDOPLASMIC RETICULUM

Two types of endoplasmic reticulum have been observed in same or different types of cells which are as follows:

1. Agranular or Smooth Endoplasmic Reticulum

This type of endoplasmic reticulum possesses smooth walls because the ribosomes are not attached with its membranes. The smooth type of endoplasmic reticulum occurs mostly in those cells, which are involved in the metabolism of lipids (including steroids) and glycogen. The smooth endoplasmic reticulum is generally found in adipose cells, interstitial cells, glycogen storing cells of the liver, conduction fibres of heart, spermatocytes and leucocytes. The muscle cells are also rich in smooth type of endoplasmic reticulum and here it is known as **sarcoplasmic reticulum.** In the pigmented retinal cells it exists in the form of tightly packed vesicles and tubes known as **myeloid bodies**.

2. Granular or Rough Endoplasmic Reticulum

The granular or rough type of endoplasmic reticulum possesses rough walls because the ribosomes remain attached with its membranes. Ribosomes play a vital role in the process of protein synthesis. The granular or rough type of endoplasmic reticulum is found abundantly in those cells which are active in protein synthesis such as pancreatic cells, plasma cells, goblet cells, and liver cells. The granular type of endoplasmic reticulum takes basiophilic stain due to its RNA contents of ribosomes. The region of the matrix containing granular type of endoplasmic reticulum takes basiophilic stain and is named as **ergastoplasm, basiophilic bodies, chromophilic substances** or **Nissl bodies** by early cytologists.

In RER, ribosomes are often present as polysomes held together by mRNA and are arranged in typical "rosettes" or spirals. RER contains two transmembrane glycoproteins (called **ribophorins I and II** of 65,000 and 64,000 dalton MW, respectively), to which are attached the ribosomes by their 60S subunits.

3. Annulate Lamellae

Usually the endoplasmic reticulum has no pores or annuli in it but in certain cases the pores or annuli have been reported, *e.g.*, ER of invertebrates, oocytes and spermatocytes of the vertebrates. These annuli resemble with the pores or annuli of the nuclear membranes. Like the annuli of nuclear membranes it contains a diaphragm across it and possesses an octagonal symmetry. The annulate lamellae (pores) of the ER arise by the evagination from the nuclear envelope and have their association with the ribosomes.

4. Glycosomes

Although the SER forms a continuous system with RER, it has different morphology. For example, in liver cells it consists of a tubular network that pervades major portion of the cytoplasmic matrix. These fine tubules are present in regions rich in glycogen and can be observed as dense particles, called **glycosomes**, in the matrix. Glycosomes measure 50 to 200 nm in diameter and contain glycogen along with enzymes involved in the synthesis of glycogen. Many glycosomes attached to the membranes of SER have been observed by electron microscopy in the liver and conduction fibre of heart.

63.5 ISOLATION

The membranes of the endoplasmic reticulum can be isolated by subjecting homogenized tissues to differential centrifugation. Electron microscopy of such ER preparations reveals that the membranes disrupt to form closed vesicles (~100 nm diameter) of either a rough or a smooth form. These membranous entities were coined the term "**microsomes**" by **Claude** in 1940, and the relationship between microsomes and the elements of endoplasmic reticulum in the intact cell was established

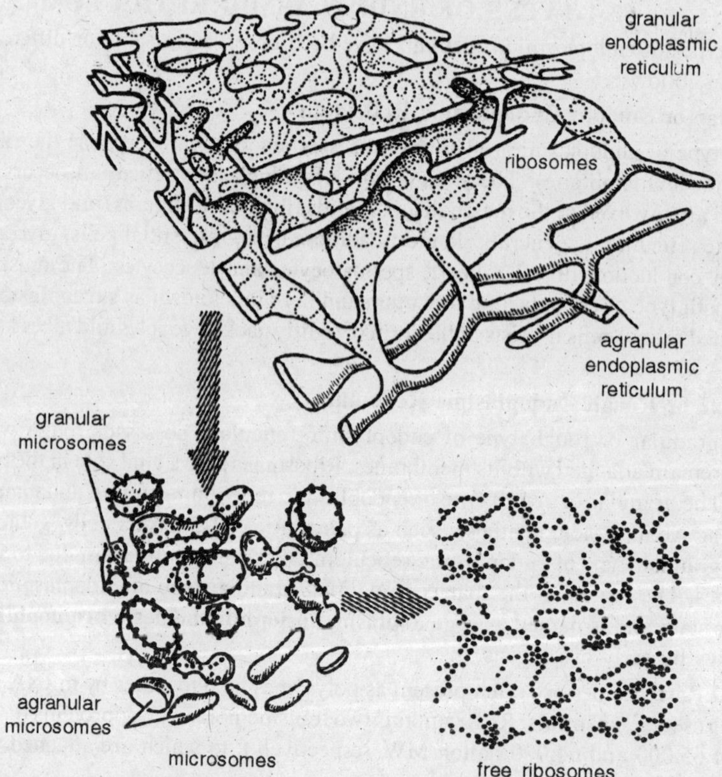

Fig. 63.3. Three-dimensional structure of endoplasmic reticulum showing microsomes and ribosomes.

by **Palade** and **Siekevitz** in 1956. Microsomes are not found in intact cells, *i.e.*, they are not natural cell organelles.

Microsomes derived from rough ER are studded with ribosomes and are called **rough** or **granular microsomes**. The ribosomes are always found on the outside surface, the interior being biochemically equivalent to the luminal space of the ER. Homogenate also contains **smooth** or **agranular microsomes** which lack attached ribosomes. They may be derived in part from smooth portion of the ER and in part from fragments of plasma membrane, Golgi apparatus, endosomes and mitochondria. Thus, while rough microsomes can be equated with rough portions of ER, the origin of smooth microsomes cannot be so easily assigned. However, since the hepatocytes of liver contain exceedingly large quantities of smooth ER, therefore, most of the smooth microsomes in liver homogenates are derived from smooth ER.

As rough microsomes can be readily purified in functional form, they are especially useful for studying many biochemical processes carried out by the ER, *e.g.*, protein synthesis, glycosylation and lipid synthesis.

In rat liver, the membranes of microsomes are 60 to 70 per cent protein and 30 to 40 per cent phospholipid by weight.

63.6 CHEMICAL COMPOSITION

ER membranes contain more proteins, both in amount and kind (having about 33 types of polypeptides) than the plasma membrane. They are also richer in phosphotidyl-choline and poorer in sphingomyelin.

The membranes of the endoplasmic reticulum are found to contain many kinds of enzymes which are needed for various important synthetic activities. Some of the most common enzymes are found to have different transverse distribution in the ER membranes (Table 63.1). The most important enzymes are the stearases, NADH-cytochrome C reductase, NADH diaphorase, glucose-6-phosphotase and Mg++ activated ATPase. Certain enzymes of the endoplasmic reticulum such as nucleotide diphosphate are involved in the biosynthesis of phospholipid, ascorbic acid, glucuronide, steroids and hexose metabolism. The enzymes of the endoplasmic reticulum perform the following important functions:

1. Synthesis of glycerides, *e.g.*, triglycerides, phospholipids, glycolipids and plasmalogens.

Table 63.1 Transverse distribution of various enzymes in the membranes of endoplasmic reticulum.

Enzymes	*Surface localization*
1. Cytochrome b_5 (involved in synthesis of unsaturated fatty acids)	Cytoplasmic face
2. NADH-cytochrome b_5 reductase	Cytoplasmic face
3. NADP-cytochrome c reductase	Cytoplasmic face
4. Cytochrome P-450 (most abundant)	Both on cytoplasmic and luminal face
5. ATPase	Cytoplasmic face
6. 5-nucleotidase	Cytoplasmic face
7. Nucleoside pyrophosphatase	Cytoplasmic face
8. GDP-mannosyl transferase	Cytoplasmic face
9. Nucleoside diphosphatase	Luminal face
10. Glucose-6-phosphatase (histochemical marker enzyme)	Luminal face
11. Acetanilide-hydrolysing esterase	Luminal face
12. β-glucuronidase	Luminal face

2. Metabolism of plasmalogens.
3. Synthesis of fatty acids.
4. Biosynthesis of the steroids, *e.g.*, cholesterol biosynthesis, steroid hydrogenation of unsaturated bonds.
5. $NADPH_2 + O_2$—requiring steroid transformations: Aromatization and hydroxylation, side-chain oxidation, deamination, thio-ether oxidations, desulphuration.
6. L-ascorbic acid synthesis.
7. UDP-uronic acid metabolism.
8. UDP-glucose dephosphorylation.
9. Aryl-and steroid sulphatase.

63.7 ORIGIN OF ENDOPLASMIC RETICULUM

The exact process of the origin of endoplasmic reticulum is still unknown. But because membranes of ER resemble with the nuclear membrane and plasma membrane and also at the telophase stage the ER membranes are found to form the nuclear envelope. Therefore, it is normally assumed that the ER has originated by evagination of the nuclear membranes. **Seikevitz** and **Palade** (1960) have reported that the granular type of ER has originated first and later it synthesizes the agranular or smooth type of endoplasmic reticulum. According to **Leskes** (1971) and **Eytan** and **Ohad** (1972), the cell receives ER from its parent cell. Later on, ER grows by expansion.

The synthesis of membranes of ER is found to proceed in the following direction: RER → SER. In fact, membrane biogenesis is a multi-step process involving , first, the synthesis of a basic membrane of lipid and intrinsic proteins and thereafter the addition of other constituents such as enzymes, specific sugars, or lipids. The process by which a membrane is modified chemically and structurally is called **membrane differentiation**. The ER (especially SER) is the organelle containing the main phospholipid synthesizing and translocating enzymes (*i.e.*, there occurs an intense flip-flop of lipid components). The insertion of proteins into ER membranes occurs at the level of RER. Most of these proteins are formed on membrane-bound ribosomes. However, some of these are synthesized by free ribosomes in the cytosol (cytoplasmic matrix) and then are inserted into the membrane. For example, the enzyme **NAD-cytochrome-b_5-reductase** is synthesized in the cytosol (cytoplasmic matrix) and then becomes incorporated in various parts of the endomembrane system (*i.e.*, RER, SER and Golgi apparatus) and in the outer mitochondrial membrane.

63.8 FUNCTIONS OF ENDOPLASMIC RETICULUM

The endoplasmic reticulum acts as secretory, storage, circulatory and nervous system for the cell. It performs following important functions:

A. Common Functions of Granular and Agranular Endoplasmic Reticulum

1. The endoplasmic reticulum provides an ultrastructural skeletal framework to the cell and gives mechanical support to the colloidal cytoplasmic martix.
2. The exchange of molecules by the process of osmosis, diffusion and active transport occurs through the membranes of endoplasmic reticulum. Like plasma membrane, the ER membrane has permeases and carriers.
3. The endoplasmic membranes contain many enzymes which perform various synthetic and metabolic activities. Further the endoplasmic reticulum provides increased surface for various enzymatic reactions.
4. The endoplasmic reticulum acts as an intracellular circulatory or transporting system. Various secretory products of granular endoplasmic reticulum are transported to various organelles as follows: Granular ER → agranular ER → Golgi membrane → lysosomes, transport vesicles

or secretory granules. Membrane flow may also be an important mechanism for carrying particles, molecules and ions into and out of the cells. Export of RNA and nucleoproteins from nucleus to cytoplasm may also occur by this type of flow.

5. The ER membranes are found to conduct intra-cellular impulses. For example, the sarcoplasmic reticulum transmits impulses from the surface membrane into the deep region of the muscle fibres.
6. The ER membranes form the new nuclear envelope after each nuclear division.
7. The sarcoplasmic reticulum plays a role in releasing calcium when the muscle is stimulated and actively transporting calcium back into the sarcoplasmic reticulum when the stimulation stops and the muscle must be relaxed.

B. Functions of Smooth Endoplasmic Reticulum.

Smooth ER performs the following functions of the cell:

1. Synthesis of lipids. SER performs synthesis of lipids (*e.g.*, phospholipids, cholesterol, etc.) and lipoproteins. Studies with radioactive precursors have indicated that the newly synthesized phospholipids are rapidly transferred to other cellular membranes by the help of specific cytosolic enzymes, called **phospholipid exchange proteins**.

2. Glycogenolysis and blood glucose homeostasis. The process of glycogen synthesis (glycogenesis) occurs in the cytosol (in glycosomes). The enzyme **UDPG-glycogen transferase**, which is directly involved in the synthesis of glycogen by addition of **uridine diphosphate glucose (UDPG)** to primer glycogen is bound to the glycogen particles or glycosomes.

SER is found related to **glycogenolysis** or breakdown of glycogen. An enzyme, called **glucose-6-phosphatase** (a marker enzyme) exists as an integral protein of the membrane of SER (*e.g.*, liver cell). Generally, this enzyme acts as a glucogenic phosphohydrolase that catalyzes the release of free glucose molecule in the lumen of SER from its phosphorylated form in liver (Fig. 63.4). Thus, this process operates to maintain homeostatic levels of glucose in the blood for the maintenance of functions of red blood cells and nerve tissues.

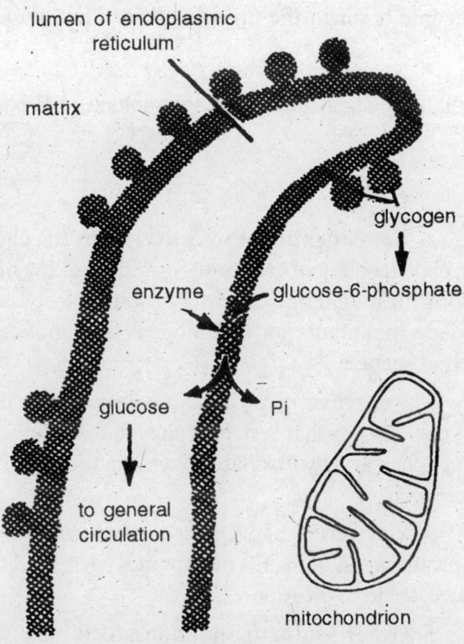

Fig. 63.4. Diagram of the intervention of the smooth endoplasmic reticulum in glycogenolysis with the consequent release of glucose. The enzyme glucose-6-phosphatase, is present in the membrane and has a vectorial deposition by which it receives the glucose-6-phophate from the matrix surface. The product glucose penetrates the lumen of the endoplasmic reticulum.

3. Sterol metabolism. The SER contains several key enzymes that catalyze the synthesis of **cholesterol** which is also a precursor substance for the biosynthesis of two types of compounds—the steroid hormones and bile acids:

(*i*) **Cholesterol biosynthesis.** The cholesterol is synthesized from the acetate and its entire biosynthetic pathway involves about 20 steps, each step catalyzed by an enzyme. Out of these twenty enzymes, eleven enzymes are bounded to SER membranes, rest nine enzymes are the soluble enzymes located in the cytosol and mitochondria. Examples of SER-bound enzyme include **HMG-Co A reductase** and **squalene synthetase**.

(*ii*) **Bile acid synthesis.** The biosynthesis of the bile acids represents a very complex pattern of enzymes and products. Enzymes involved in the biosynthetic pathway of bile acids are hydroxylases, mono-oxygenases, dehydrogenases, isomerases and reductases. For example, by the help of the enzyme **cholesterol 7α-hydroxylase**, the cholesterol is first converted into 7α-hydroxyl cholesterol, which is then converted into bile acids by the help of hydroxylase enzymes. The latter reaction requires NADPH and molecular oxygen and depends on the enzymes of electron transport chains of SER such as **cytochrome P-450** and **NADPH-cytochrome-c-reductase.**

(*iii*) **Steroid hormone biosynthesis.** Steroid hormones are synthesized in the cells of various organs such as the cortex of adrenal gland, the ovaries, the testes and the placenta. For example, cholesterol is the precursor for both types of sex hormones—estrogen and testosterone—made in the reproductive tissues, and the adrenocorticoids (*e.g.*, corticosterone, aldosterone and cortisol) formed in the adrenal glands. Many enzymes (*e.g.*, dehydrogenases, isomerases and hydroxylases) are involved in the biosynthetic pathway of steroid hormones, some of which are located in SER membranes and some occur in the mitochondria. This biosynthetic pathway has the following steps:

Acetate (Cytosol) → Cholesterol (SER) → Pregnenolone (Mitochondria) → Progesterone (SER) → 11-Deoxycortisol (SER) → Aldosterone (Mitochondria)

Progesterone (SER) → 11-Deoxycortisterone (SER) → Cortisol (Mitochondria)

4. Detoxification. Protectively, the ER chemically modifies **xenobiotics** (toxic materials of both endogenous and exogenous origin), making them more hydrophilic, hence, more readily excreted. Among these materials are drugs, aspirin (acetyl-salicylic-acid), insecticides, anaesthetics, petroleum products, pollutants and carcinogens (*i.e.*, inducers of cancer; *e.g.*, **3-4-benzopyrene** and **3-methyl cholanthrene**).

The enzymes involved in the detoxification of aromatic hydrocarbons are **aryl hydroxylases**. It is now known that benzopyrene (found in charcoal-broiled meat) is not carcinogenic, but under the action of aryl hydroxylase enzyme in the liver, it is converted into **5, 6-epoxide**, which is a powerful carcinogen.

A wide variety of drugs (*e.g.*, phenobarbital), when administrated to animals, they bring about the proliferation of the ER membranes (first RER and then SER) and/or enhanced activity of enzymes related to detoxification.

5. Other synthetic functions. SER plays a role in the synthesis of triglycerides in intestinal absorptive cells and of visual pigments from vitamin A by pigmented epithelial cell of retina. In plant cells, SER forms the surface where cellulose cell walls are being formed.

C. Functions of Rough Endoplasmic Reticulum

1. Protein synthesis. The major function of the rough ER is the synthesis of protein. It has long been assumed that proteins destined for secretion (*i.e.*, export) from the cell or proteins to be used in the synthesis of cellular membranes are synthesized on rough ER-bound ribosomes, while cytoplasmic proteins are translated for the most part on free ribosomes. In fact, the array of the rough endoplasmic reticulum provides extensive surface area for the association of metabolically active enzymes, amino acids and ribosomes. There is more efficient functioning of these materials to synthesize proteins when oriented on a membrane surface than when they are simply in solution, mainly because chemical combinations between molecules can be accomplished in specific geometric patterns.

The membrane-bound ribosomes are attached with **specific binding sites** or **receptors** of rough ER membrane by their large 60S subunit, with small or 40S subunit sitting on top like a cap. These receptors are membrane proteins which extend well into and possibly through the lipid bilayer. The receptor proteins with bound ribosomes can float laterally like other membrane proteins and may facilitate formation of the polysome and probably translation which requires that mRNA and ribosome move with respect to each other.

Further, the secretory proteins, instead of passing into the cytoplasm, appear to pass instead into the cisternae of the rough ER and are, thus, protected from protease enzymes of cytoplasm. It is calculated that about 40 amino acid residues long segment at the—COOH end of the nascent protein remains protected inside the tunnel of 'free' or 'bound' ribosomes and rest of the chain, with—NH_2 end, is protected by the lumen of RER. The passage of nascent polypeptide chain into the ER cisterna takes place during translation leaving only a small segment exposed to the cytoplasm at any one time.

How the polypeptide chain gets through the lipid bilayer is not so clear, but it is quite reasonable to propose that the membrane proteins serving as ribosomal receptors also has a very fine channel through its core that opens into the cisterna of the rough ER. The chain may have great flexibility, permitting the amino acids to snake their way single file through the proposed pore. As soon as growing polypeptide chain reaches the cisterna, it folds into its secondary and tertiary structures and thus trapped in the cisterna of the rough ER.

2. Protein glycosylation. The covalent addition of sugars to the secretory proteins (*i.e.*, glycosylation) is one of the major biosynthetic functions of rough ER. Most of the proteins that are isolated in the lumen of RER before being transported to the Golgi apparatus, lysosomes, plasma membrane or extracellular space, become glycoproteins (a notable exception is albumin). In contrast, very few proteins in the cytosol (cytoplasmic matrix) are glycosylated and those that carry them have a different sugar modification.

The process of **protein glycosylation** in RER lumen is one of the most well understood cell biological phenomena. During this process, a single species of **oligosaccharide** (which comprises N-acetyl-glucosamine, mannose and glucose, containing a total of 14 sugar residues) is transferred to proteins in the ER. Because it is always transferred to the NH_2 group on the side chain of an asparagine residue of the protein, this oligosaccharide is said to be **N-linked** or **asparagine-linked** (Fig. 63.5A). The transfer is catalyzed by a membrane-bound enzyme (*i.e.*, **glycosyl transferase**) with its active site exposed on the luminal surface of the ER membrane. The preformed precursor oligosaccharide is transferred *en bloc* to the target asparagine residue in a single enzymatic step almost as soon as that residue emerges in the lumen of ER during protein translocation (Fig. 63.5B). Since most proteins are co-translationally imported into the ER, N-linked oligosaccharides are almost always added during protein synthesis, ensuring maximum access to the target asparagine residues, which are present in the sequences—*Asn-X-Ser* or *Asn-X-Thr* (where *X* is amino acid except proline). *These two sequences, thus, function as signals for N-linked glycosylation.*

The precursor oligosaccharide is held in the ER membrane by a special lipid molecule, **dolicol** (the carrier). The oligosaccharide is linked to the dolicol by a high-energy **pyrophosphate bond** which activates the oligosaccharide for its transfer from the lipid to an asparagine side chain (*i.e.*, it provides activation energy for the glycosylation reaction). The oligosaccharide is built up sugar by sugar on the membrane-bound dolicol (towards the cytosolic side) prior to its transfer to a protein. Sugars are first activated in the cytosol (cytoplasmic matrix) by the formation of **nucleotide-sugar**

intermediates (*e.g.*, UDP-glucose, UDP-N-acetylglucosamine, and GDP-mannose), which then donates their sugar (directly or indirectly) to the lipid in an orderly sequence. At some step of this process, the lipid-linked oligosaccharide is flipped from the cytosolic to the luminal side of the ER membranes. Dolicol is long and very hydrophobic: its 22 five-carbon units can span the thickness of lipid bilayer more than three times, so that the attached oligosaccharide is firmly anchored to the membrane.

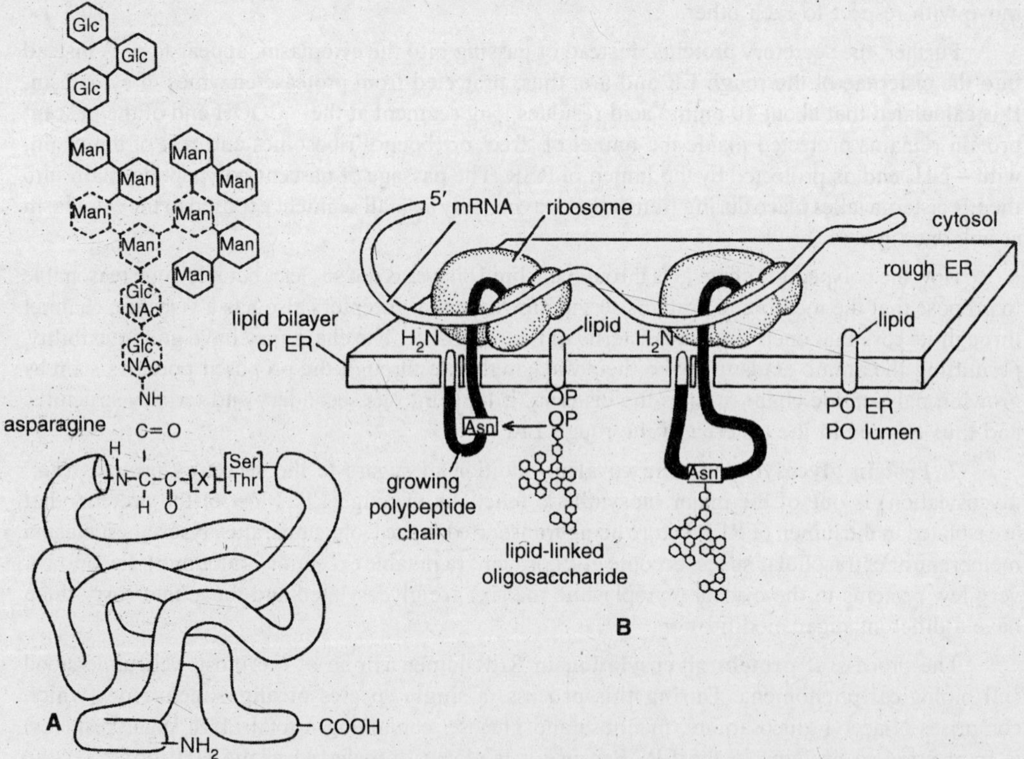

Fig. 63.5. N-linked glycosylation of protein in RER. A—The structure of asparagine-linked oligosaccharide. The sugars shown in shaded-form the 'core-region' of this oligosaccharide. For many glycoproteins, only the core sugars survive the extensive oligosaccharide trimming process in the Golgi apparatus; B—Mode of transfer of the oligosaccharide to the asparagine residues of the nascent protein inside RER lumen.

While still in RER lumen, three glucose residues and one mannose residue are quickly removed from the oligosaccharides of most glycoproteins. Such oligosaccharide "trimming" or "processing" continues in the Golgi apparatus. If a glycoprotein is to contain a terminal glucose, fucose or sialic acid, then those sugars are added in the Golgi apparatus where the appropriate sugar transferase enzymes are localized.

3. Transportation of secretory proteins. From RER secretory proteins pass to Golgi apparatus, secretory vesicles and then to the outside of the cell (exocytosis).

63.9 SIGNAL HYPOTHESIS

The proteins for the secretion, the lysosomes and the membrane formation, are synthesized on the membrane bound ribosomes. The free and bound ribosomes were found to be continuously interchanging and show no differences between them. The **signal hypothesis** was proposed by **Blobel** and **Sabatini** (1971; see Box 63.1) to explain how the ribosomes which are meant for the biosynthesis of secretory type proteins get specifically attached to RER membranes. According to this hypothesis, the mRNA is able to recognize free or bound ribosomes. It is postulated that the mRNA for secretory proteins contain a set of **special signal codons** localized after the initial codon AUG. Once the ribosome "recognizes" the signal the ribosome becomes attached to the membrane of ER and the polypeptide penetrates. It is also postulated that at the luminal surface there is a **signal peptidase enzyme** that removes the signal peptide. Thus, the mRNA produces a **preprotein** of larger molecular weight than the final protein. This signal peptide has between 15 to 30 amino acids which are generally hydrophobic. Such a signal peptide probably establishes the initial association of the ribosome with the membrane, but some protein factors are involved. A **signal recognition protein** (SRP) complex binds to the nascent signal peptide and stops the translation until it reaches the ER membrane (Fig. 63.6). It is suggested that a **SRP receptor** or **docking protein** which is a pore-containing integral membrane protein of ER, removes the SRP block, allowing for the translocation of the polypeptide into lumen of RER.

Box 63.1

Guenter Blobel of Rockefeller University, New York won the Nobel Prize for Medicine on 11.10.1999 for signal hypothesis. His research has shed new light on human disease including cystis fibrosis and early development of kidney stones.

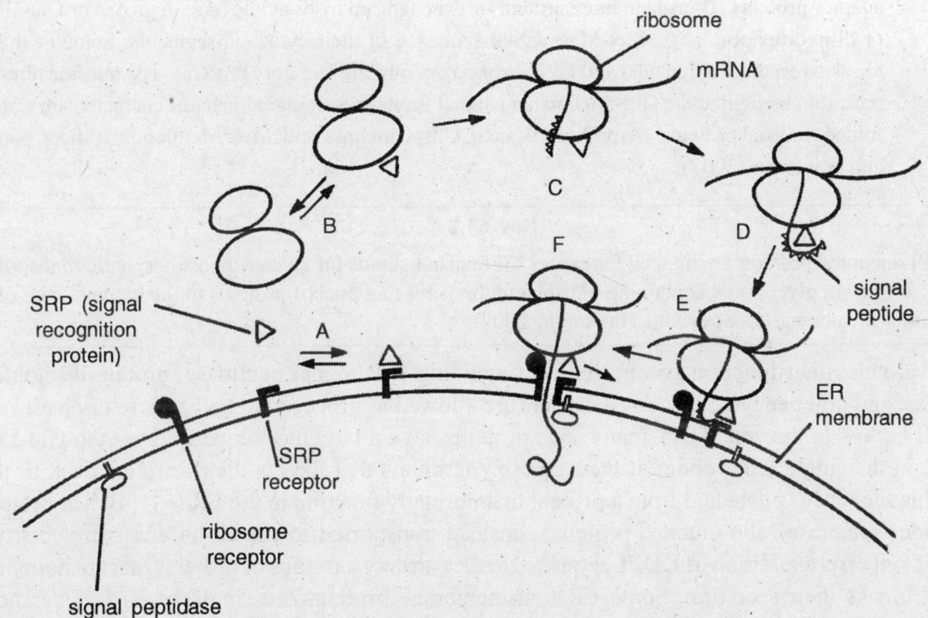

Fig. 63.6. Diagram explaining the signal hypothesis.

63.10 ER EXPORT SIGNALS

Most proteins that enter the transitional ER move through the ER-Golgi intermediate compartment (ERGIC) and on the Golgi. These proteins are marked by sequences that signal either their export from or retention within the ER (Fig. 63.7). Many transmembrane proteins possess di-acidic or di-hydrophobic amino acid sequences in their cytosolic domains that function as ER export signals. Both GPI-anchored proteins and lumenal secretory proteins appear to be recognised and sequestered by these transmembrane receptor proteins (Box 63.2). Very few ER export signals have been detected on lumenal secretory proteins and their recognition may dpend on the shape of the correctly folded proteins. It is also possible that there is a default pathway where otherwise unmarked proteins, in the ER lumen move to the Golgi and beyond.

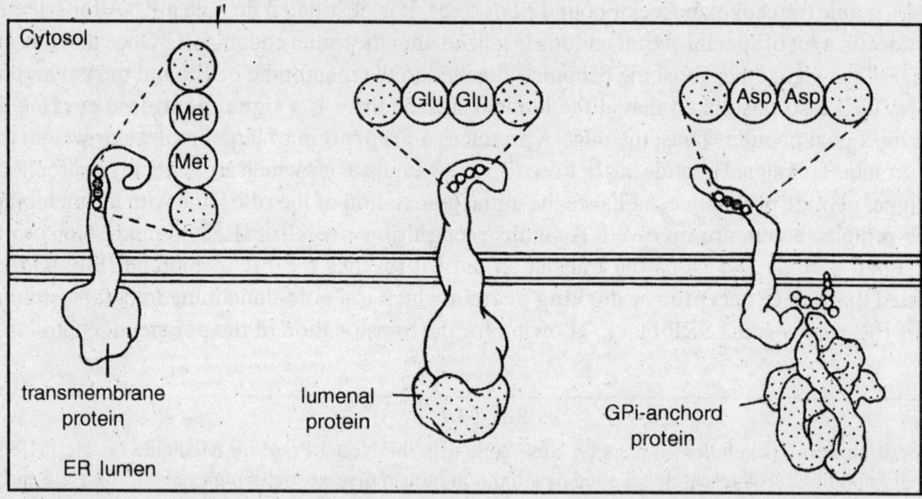

Fig. 63.7. ER export signals. Three types of secretory proteins are recruited into budding vesicles by cytosolic adaptor proteins. Transmembrane proteins are recognised by di-acidic (*e.g.*, Asp-Asp or Glu-Glu) or di-hydrophobic (*e.g.*, Met-Met) signal sequence in their cytosolic segments. Some of these are also receptor for lumenal and GPI-anchored membrane secretory proteins. The transmembrane receptors recognize the GPI-anchors and signal sequences or signal patches characteristic of the folded-lumenal proteins. (Asp=Aspartic acid; Glu=Glutamic acid; Met=Methionine) (after Cooper and Hausman, 2007).

Box 63.2

The GPI-anchored proteins are marked for export. GPI anchor stands for glycosylphosphosphatidylinositol anchor. These are glycolipids containing phosphatidylinositol that anchor proteins to the external face of the plasma membrane (**Cooper** and **Hausman**, 2007).

If proteins that function within the ER (including BiP, signal peptidase, protein disulphide isomerase, and other enzyme discussed earlier) are allowed to proceed along the secretory pathway, they will be lost to the cell. Thus, many such proteins have a targeting sequence Lys-Asp-Glu-Leu (KDEL, in the single letter code) at their carboxy terminus that directs their retrieval back to the ER. If this sequence is detected from a protein that normally function in the ER (*e.g.*, BiP or protein disulphide isomerase), the mutated protein is instead transported to the Golgi and secreted from the cell. Conversely, addition of KDEL sequence to the carboxy terminus of proteins that are normally secreted blocks their secretion. Some ER transmembrane proteins are similarly marked by short C-terminal sequences that contain two lysine residues (KKXX sequences).

Curiously, the KDEL and KKXX signals do not prevent ER proteins from being packaged into vesicles and carried to the Golgi. Instead, these signals cause these ER resident proteins to be selectively retrieved from the ER-Golgi intermediate compartment or the Golgi-apparatus and returned to the ER via a recycling pathway (Fig. 63.8). Proteins bearing the KDEL and KKXX sequences bind to specific recycling receptors in the membranes of these compartments and are then selectively transported back to the ER. The KDEL and KKXX sequences are the best characterised retention/retrieval signals but there may be others. Other proteins are retrieved because they specifically bind to KDEL-bearing proteins such as BiP. Thus continued movement along the secretory pathway or retrieved back from the Golgi to the ER is the first branch point encountered by proteins being sorted to their correct destinations in the secretory pathway. Similar branch points arise at each subsequent stage of transport, such as retention in the Golgi versus export to lysosomes or the plasma membrane. In each case specific localization signals target proteins to their correct intracellular destinations.

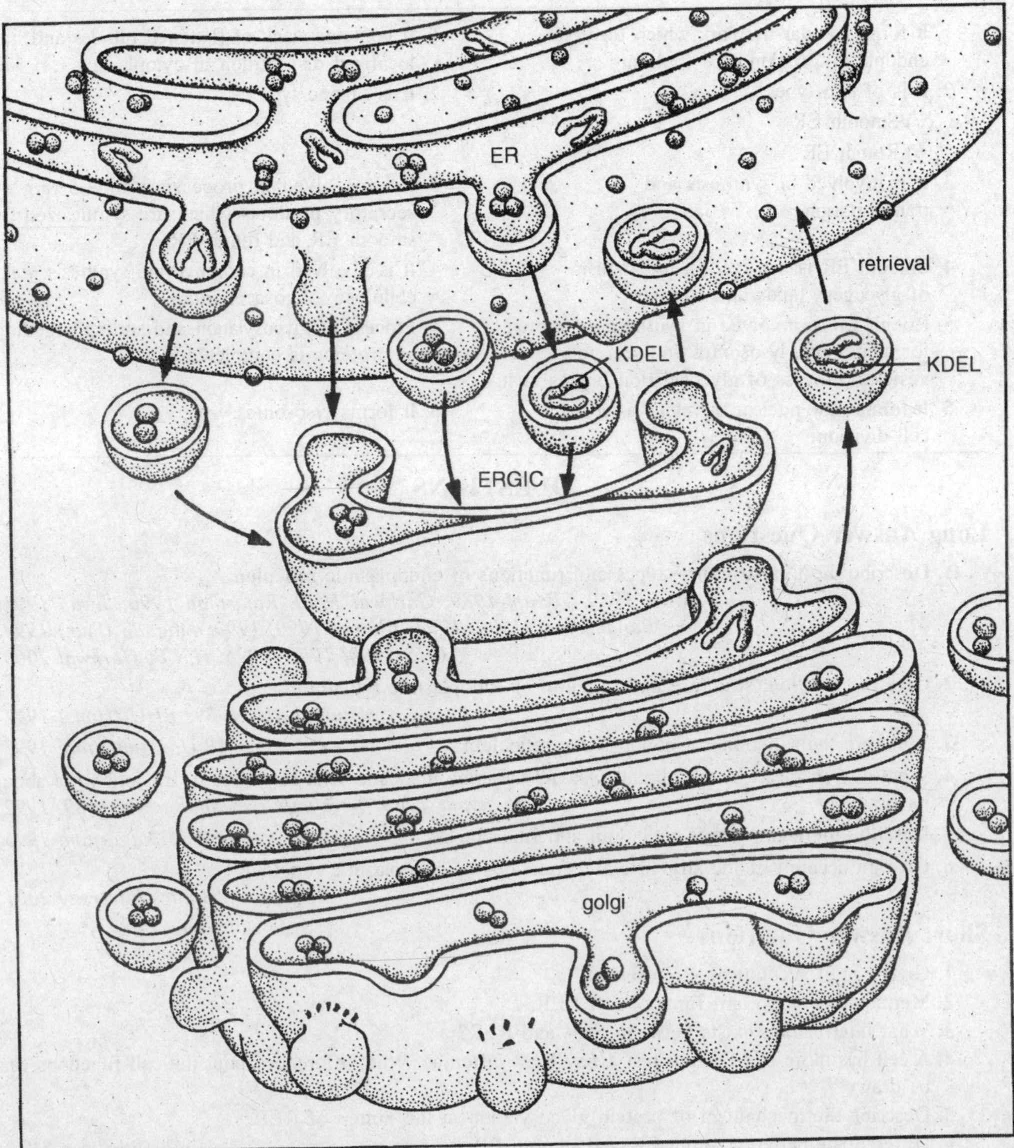

Fig. 63.8. Retrieval of resident ER proteins. Proteins destined to remain in the lumen of the ER are marked by the sequence Lys-Asp-Glu-Leu (KDEL) at their carboxy terminus. These proteins are exported from the ER to the Golgi, but they are recognised by a receptor in the ERGIC (ER Golgi intermediate compartment) or the Golgi apparatus and selectively returned to the ER (after Cooper and Hausman, 2007).

Differences in Endoplasmic Reticulum and Golgi Body/Apparatus

Table 63.2 Differences in endoplasmic reticulum and Golgi apparatus.

Endoplasmic reticulum	*Golgi apparatus*
1. It forms tubular network which traverses endoplasmic region of cytoplasm.	1. It contains stack of flattened tubules and has localized distribution in cytoplasm.
2. It is of two types: (*a*) Smooth ER (*b*) Rough ER	2. It is of one type.
3. It is involved in synthesis and transportation.	3. It is involved in processing and sorting of secretory products which are synthesized by smooth ER and rough ER.
4. Smooth ER is involved in metabolism of glycogen, lipids and steroids.	4. It is involved in carbohydrate synthesis, *e.g.*, cellulose, oligosaccharides.
5. Rough ER is involved in synthesis of protein for membrane, lysosomes and secretory vesicles. Process of glycosylation begins in it.	5. Process of glycosylation and sorting/targetting of proteins is completed in it.
6. It forms new nuclear envelope after each cell division.	6. It forms lysosomes.

QUESTIONS

Long Answer Questions

1. Describe the ultrastructure, types and functions of endoplasmic reticulum.
(*A.P. 1989; Vikram 1989; Garhwal 1990; Rajasthan 1990; Jiwaji 1991; Bundelkhand 1990, 1992; Bhopal 1990, 1993; Magadh (Sub.) 1994; CCS 1990, 1995, 1996; H.N.B. Garhwal 2009*)

2. Give the location, structure and function of endoplasmic reticulum.
(*Punjab (G) 1989; BHU (Hons) 1991*)

3. What are the functions of endoplasmic reticulum? (*B.U. Coimbatore 1997*)

4. What do you mean by membrane system? Describe the same with reference to ER and write about its functions. (*Jadavpur 1989; Berhampur 1990; Baroda 1989, 1992*)

5. Describe the endoplasmic reticulum and ribosomes. (*Nagarjuna 1995*)

6. Give an account of the structure and functions of endoplasmic reticulum.
(*Bihar University 2001*)

Short Answer Questions

1. Give a brief account of microsomes.

2. Mention two important functions of RER.

3. What interrelation exists between SER and RER?

4. A cell has more of SER but just a few RER cisternae. What inference about the cell functions can be drawn?

5. Describe the mechanism of protein glycosylation in the lumen of RER.

6. Differentiate between rough ER and smooth ER. (*Burdwan 1990*)

7. Distinguish between endoplasmic reticulum and Golgi body.
(*C.C.S.U. Meerut 2009*)

8. Write short notes on the following:

(*i*) Endoplasmic reticulum or transporting system
(*Garhwal 1989; Magadh (G) 1990, 1992, 1993, 1995; Sagar 1991; Awadh 1991; Purvanchal 1991; Kanpur 1992; Punjab (G) 1992; Guwahati 1992, 1994, 1996; Lucknow 1994; Gorakhpur 1998; Bihar 1999*)

(*ii*) Functions of endoplasmic reticulum (*Kanpur 1993*)

(*iii*) Functions of smooth ER.

Very Short Answer Questions

1. Who coined the term endoplasmic reticulum?
   Ans. Porter.
2. Which eukaryotic cell is devoid of ER?
   Ans. RBC or erythrocyte of mammals.
3. Name the type of ER associated with glycogen metabolism.
   Ans. Smooth endoplasmic reticulum.
4. What do you understand by sarcoplasmic reticulum?
   Ans. Smooth E.R. of muscle cells is called sarcoplasmic reticulum.
5. Which cell organelle is responsible for the transport of secretory products to various other cell organelles?
   Ans. Endoplasmic reticulum.

Fill in the Blanks

(*i*) The type of ER is associated with glycogen metabolism is ....

(*ii*) It was possible to observe elements of ER with the help of ....

(*iii*) .... discovered endomembrane system.

(*iv*) .... coined the term endoplasmic reticulum.

(*v*) The eukaryotic cell which is without ER is ....

(*vi*) SER stands for ....

(*vii*) .... membranes develops from ER during cell division.

(*viii*) .... ER is associated with glycoprotein synthesis.

(*ix*) .... ER is developed in cell that secretes mucus.

(*x*) The appearance of ER is dependent on the .... of the cell.

Multiple Choice Questions

1. What part of the cell forms nuclear envelope during telophase? (*B.U. Coimbatore 1997*)
   (*a*) cytoskeleton (*b*) centriole
   (*c*) golgi complex (*d*) endoplasmic reticulum
2. Internal membranes or cytomembranes are extensions of cell membrane.
   (*a*) the membranes have unit membrane structure
   (*b*) all the organelles are bounded by the same membrane
   (*c*) have same thickness as that of cell membrane
   (*d*) the membrane of ER are connected at certain points with the cell membrane
3. The enzyme glucose-6-phosphatase is present in
   (*a*) endoplasmic reticulum (*b*) lysosome
   (*c*) mitochondria (*d*) nucleus
4. RER is well developed in cells engaged in synthesis of (*Bihar University, 2004*)
   (*a*) nucleotides (*b*) proteins
   (*c*) lipids (*d*) secretory products
5. The cells lacking the endoplasmic reticulum are (*Bihar University 2001*)
   (*a*) amphibian monocytes (*b*) mature erythrocytes of mammals
   (*c*) mature leucocytes (*d*) mammalian monocyte

ANSWERS

Fill in the Blanks

(*i*) SER (*ii*) Electron microscope
(*iii*) Porter (*iv*) Porter
(*v*) Red blood cells of mammal (*vi*) Smooth endoplasmic reticulum
(*vii*) Nuclear (*viii*) Rough
(*ix*) Rough (*x*) Secretion.

Multiple Choice Questions

1. (*d*) **2.** (*a*) **3.** (*a*) **4.** (*b*) **5.** (*b*).

64

Golgi Apparatus

For the performance of certain important cellular functions such as biosynthesis of polysaccharides, packaging (compartmentalizing) of cellular synthetic products (proteins), production of exocytotic (secretory) vesicles and differentiation of cellular membranes, there occurs a complex organelle called **Golgi complex** or **Golgi apparatus** in the cytoplasm of animal and plant cells. The Golgi apparatus, like the endoplasmic reticulum, is a canalicular system with sacs, but unlike the endoplasmic reticulum it has parallely arranged, flattened, membrane-bounded vesicles which lack ribosomes and stainable by osmium tetraoxide and silver salts.

64.1 HISTORICAL

An Italian neurologist (*i.e.*, physician) **Camillo Golgi** in 1873 discovered and developed the **silver chromate method** (termed *la reazione nera*) for studying histological details of nerve cells. He, thus, opened a new field of scientific inquiry, called **neuromorphology**. In 1898, Golgi found that Purkinje cells (*i.e.*, nerve cells of cerebral cortex of brain) of barn owl contained an internal reticular network which stains black with the silver stain. He called this structure *apparato reticolare interno* (= internal reticular apparatus).

Due to their presumed high lipid contents, Golgi apparatuses were called **lipochondria** (**Baker**, 1951, 1953). Since originally these were known to be networks, they were also called **"dictyosomes"** (Gr., *dictyes*=net). Currently, the term **Golgi apparatus** is more prevalent one, than many other names such as **Golgi complex, Golgiosome, Golgi bodies, Golgi material, Golgi membrane**, etc. The Golgi apparatus of the cells of plants and lower invertebrates is usually referred to as **Golgi body** or **dictyosome**.

64.2 OCCURRENCE

The Golgi apparatus occurs in all cells except the prokaryotic cells (*viz.*, mycoplasmas, bacteria and blue green algae) and eukaryotic cells of certain fungi, sperm cells of bryophytes and pteridiophytes, cells of mature sieve tubes of plants and mature sperm and red blood cells of animals. Their number per plant cell can vary from several hundred as in tissues of corn root and algal rhizoids (*i.e.*, more than 25,000 in algal rhizoids), to a single organelle in some algae. Certain algal cells such as *Pinularia* and *Microsterias*, contain largest and most complicated Golgi apparatuses. In higher plants, Golgi apparatuses are particularly common in secretory cells and in young rapidly growing cells.

In animal cells, there usually occurs a single Golgi apparatus, however, its number may vary from animal to animal and from cell to cell. Thus, *Paramoeba* species has two Golgi apparatuses and nerve cells, liver cells and chordate oocytes have multiple Golgi apparatuses, there being about 50 of them in the liver cells.

64.3 DISTRIBUTION

In the cells of higher plants, the Golgi bodies or dictyosomes are usually found scattered throughout the cytoplasm and their distribution does not seem to be ordered or localized in any particular manner. However, in animal cells the Golgi apparatus is a localized organelle. For example, in the cells of ectodermal or endodermal origin, the Golgi apparatus remains polar and occurs in between the nucleus and the periphery (*e.g.*, thyroid cells, exocrine pancreatic cells and mucus-producing goblet cells of intestinal epithelium) and in the nerve cells it occupies a circum-nuclear position.

64.4 ULTRASTRUCTURE (ELECTRON MICROSCOPIC STRUCTURE)

The Golgi apparatus is morphologically very similar in both plant and animal cells. However, it is extremely **pleomorphic**: in some cell types it appears compact and limited, in others spread out and reticular (net-like). Its shape and form may vary depending on cell type. Typically, however, Golgi apparatus appears as a complex array of interconnecting *tubules*, *vesicles* and *cisternae*. There has been much debate concerning the terminology of the Golgi's parts. The classification given by **D.J. Morre** (1977) is most widely used. In this scheme, the simplest unit of the Golgi apparatus is the **cisterna**. This is a membrane-bound space in which various materials and secretions may accumulate. Numerous cisternae are associated with each other and appear in a stack-like (lamellar) aggregation. A group of these cisternae is called the **dictyosome**, and a group of dictyosomes makes up the cell's Golgi apparatus. All dictyosomes of a cell have a common function.

The detailed structure of three basic components of the Golgi apparatus can be studied as follows (Fig. 64.1 and Fig. 64.2):

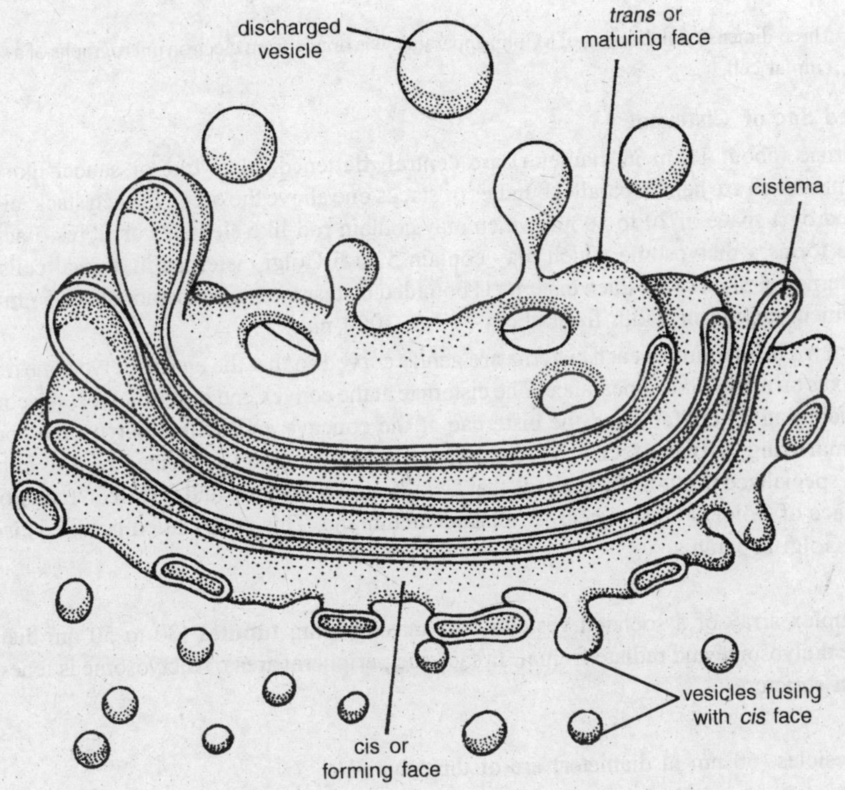

Fig. 64.1. Structure of Golgi apparatus.

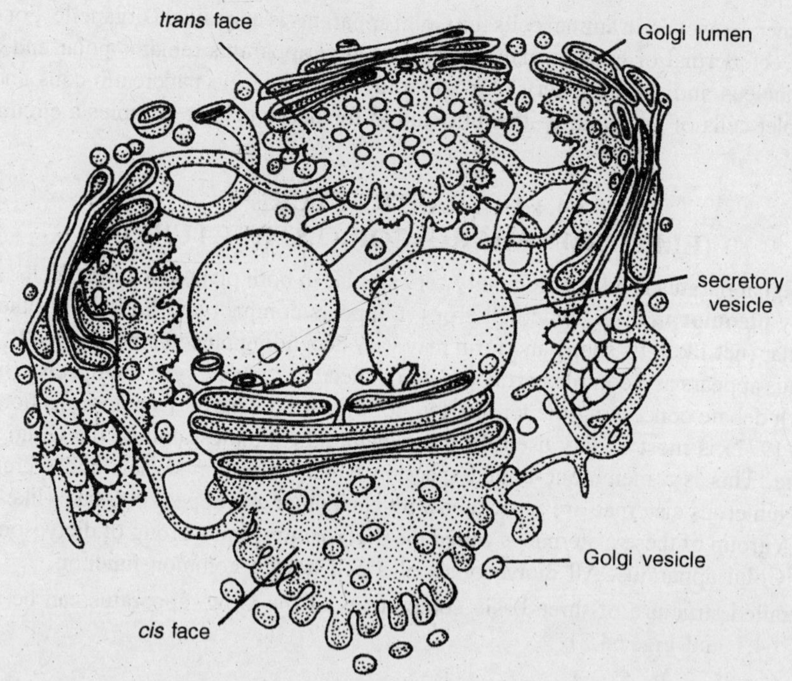

Fig. 64.2. Three-dimensional drawing of a Golgi apparatus; it is drawn from electron micrographs of a secretory animal cell.

1. Flattened Sac or Cisternae

Cisternae (about 1 μm in diameter) are central, flattened, plate-like or saucer-like closed compartments which are held in parallel bundles or stacks one above the other. In each stack, cisternae are separated by a space of 20 to 30 nm which may contain rod-like elements or fibres. Each stack of cisternae forms a dictyosome which may contain 5 to 6 Golgi cisternae in animal cells or 20 or more cisternae in plant cells. Each cisterna is bounded by a smooth unit membrane (7.5 nm thick), having a lumen varying in width from about 500 to 1000 nm.

Polarity. The margins of each cisterna are gently curved so that the entire dictyosome of Golgi apparatus takes on a bow-like appearance. The cisternae at the convex end of the dictyosome comprise **proximal, forming** or **cis-face** and the cisternae at the concave end of the dictyosome comprise the **distal, maturing** or **trans-face**. The forming or cis face of Golgi is located next to either the nucleus or a specialized portion of rough ER that lacks bound ribosomes and is called **"transitional" ER**. Trans face of Golgi is located near the plasma membrane. This polarization is called **cis-trans axis** of the Golgi apparatus.

2. Tubules

A complex array of associated **vesicles** and anastomosing **tubules** (30 to 50 nm diameter) surround the dictyosome and radiate from it. In fact, the peripheral area of dictyosome is fenestrated (lace-like) in structure.

3. Vesicles

The vesicles (60 nm in diameter) are of three types:

***(i)* Transitional vesicles** are small membrane limited vesicles which are thought to form as blebs from the transitional ER to migrate and converge to cis face of Golgi, where they coalasce to form new cisternae.

(ii) **Secretory vesicles** are varied-sized membrane-limited vesicles which discharge from margins of cisternae of Golgi. They, often, occur between the maturing face of Golgi and the plasma membrane.

(iii) **Clathrin-coated vesicles** are spherical protuberances, about 50 μm in diameter and with a rough surface. They are found at the periphery of the organelle, usually at the ends of single tubules, and are morphologically quite distinct from the secretory vesicles. The clathrin-coated vesicles are known to play a role in intra-cellular traffic of membranes and of secretory products, *i.e.*, between ER and Golgi, as well as, between GELR region and the endosomal and lysosomal compartments.

The GERL Region

Golgi apparatus is a differentiated portion of the endomembrane system found in both animal and plant cells. This membranous component is spatially and temporally related to the endoplasmic reticulum (ER) on one side and by way of secretory vesicles, may fuse with specific portions of the plasma membrane. To the trans face of Golgi is associated the **trans-reticular Golgi, TGN** (=trans-Golgi-network) or **GERL** (=Golgi + smooth ER + lysosomal), in which acid phosphatase enzyme (a characteristic lysosomal enzyme) makes its first appearance. GERL is found to be involved in the origin of **primary lysosomes** and of **melanin granules**; in the processing, condensing and packaging of secretory material in endocrine and exocrine cells; and in lipid metabolism. GERL is also a region of sorting of cellular secretory proteins.

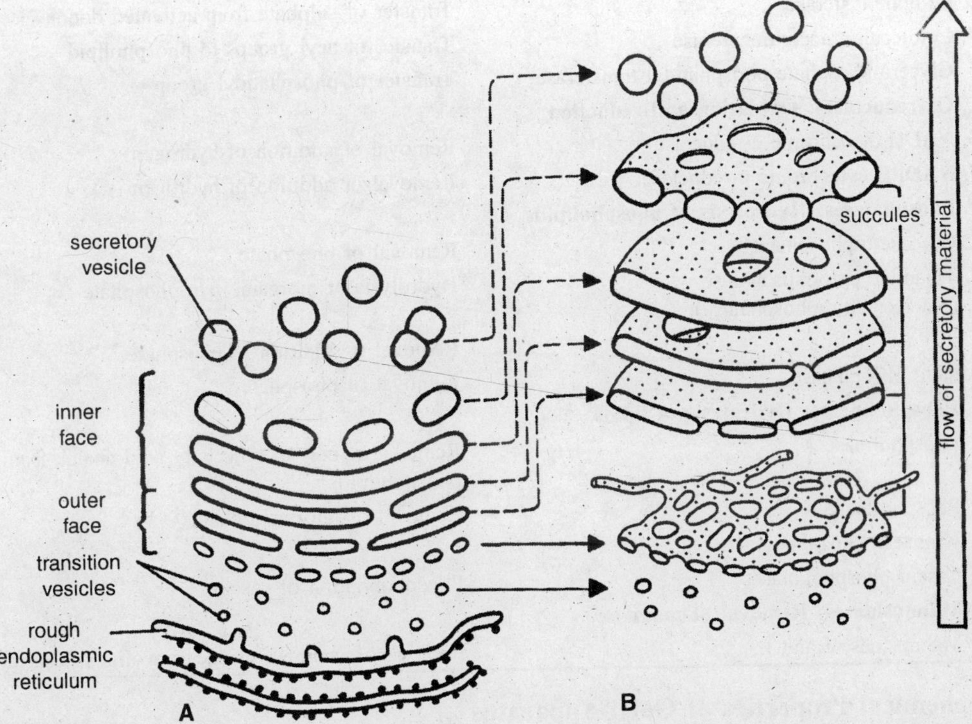

Fig. 64.3. The position and orientation of the Golgi apparatus in the secretory pathway. A—The components as they might be seen by electron microscopy of thin sections; B—A three-dimensional reconstruction of Golgi apparatus.

Zones of Exclusion

A Golgi body or Golgi apparatus is surrounded by a differentiated region of cytoplasm where ribosomes, glycogen, and organelles such as mitochondria and chloroplasts are scarce or absent.

This is called **zone of exclusion** (**Morre** *et al.*, 1971). Endoplasmic reticulum within the zone of exclusion has a smooth surface (lacking ribosomes), and coated vesicles of the Golgi apparatus are restricted to this region. Similar zones of exclusion are associated with microtubules, centrioles, and regions of centriole formation.

64.5 CHEMICAL COMPOSITION

Chemically, Golgi apparatus of rat liver contains about 60 per cent lipid material. The Golgi apparatus of animal cells contains phospholipids in the form of **phosphatidyl choline**, whereas, that of plant cells contains **phosphatidic acid** and **phosphatidyl glycerol**. The Golgi apparatus also contains a variety of enzyme (Table 64.1), some of which have been used as cytochemical markers.

Table 64.1 Some important enzymes of the Golgi apparatus of animal cells.

Enzymes: class and types	*Function*
A. Glycosyl transferases: Glycoprotein biosynthesis	
1. Sialyl transferases	Transfers sialic acid from CMP-sialic acid
2. Galactosyl transferases	Transfer galactose to lipids or proteins
B. Sulpho- and glyco-transferases: Glycolipid biosynthesis	
3. Sulphotransferase	Transfer of sulphate from activated donor
4. Lysolecithin acetyltransferase	Transfer of acyl groups to phospholipid
5. Glycero-phosphate phosphatidyl transferase	Transfer of phosphatidyl group
C. Oxireductases: Oxidation and reduction	
6. NADH-cytochrome c-reductase	Removal or addition of hydrogen
7. NADPH-cytochrome c-reductase	Removal or addition of hydrogen
D. Phosphatases: Hydrolysis of phospholipids	
8. Glucose-6-phosphatase	Removal of phosphate
9. Thiamine pyrophosphatase (Nucleoside diphosphatase)	Hydrolysis of inorganic pyrophosphate
10. ATPase	Removal or addition of phosphate
11. Acid phosphatase	Removal of phosphate
E. Phospholipases: Hydrolysis of lipids	
12. Phospholipase A_1	Removal of non-specific fatty acid chains from phospholipids
13. Phospholipase A_2	Removal of fatty acid chains
F. Kinases: Phosphorylation	
14. Casein phosphokinases	Phosphorylation of casein
G. Mannosidases: Removal of mannose	
15. Mannosidase I and II	Removal of mannose residue from oligosaccharide

Cytochemical Properties of Golgi Apparatus

Different parts of Golgi apparatus have been histochemically identified by specific staining properties.

1. Osmium tetroxide (OsO_4) selectively impregnates the outer face (cis face) of the Golgi apparatus. This stain adheres well to lipids, especially phospholipids and unsaturated fats.

2. Phosphotungstic acid (H_3PO_4. 12 WO_3. 24 H_2O) selectively stain the maturing or trans face of Golgi stack. This stain is an anionic stain having special affinity for polysaccharides and proteins.

3. Glycosyl transferase and **thiamine pyrophosphatase** can be localized cytochemically in the trans cisternae of Golgi apparatus. Transferase enzymes are found to be located in the membranes of Golgi, not in the lumen of cisternae.

4. Acid phosphatase enzyme is cytochemically marked in the GERL region.

64.6 ORIGIN

Origin of Golgi apparatus involves the formation of new cisternae and there is great variation in shape, number and size of cisternae in each stack (dictyosome). The process of formation of new cisternae may be performed by any of the following methods: 1. Individual stacks of cisternae may arise from the pre-existing stacks by division or fragmentation. 2. The alternative method of origin of Golgi is based on *de novo* formation. In fact, various cytological and biochemical evidences have established that the membranes of the Golgi apparatus are originated from the membranes of the smooth ER which in turn have originated from the rough ER. The proximal Golgi saccules are formed by fusion of ER-derived vesicles, while distal saccules "give their all" to vesicle formation and disappear. Thus, Golgi saccules are constantly and rapidly renewed.

64.7 FUNCTIONS

Golgi vesicles are often, referred to as the **"traffic police"** of the cell (**Darnell** *et al.*, 1986). They play a key role in **sorting** many of cell's proteins and membrane constituents, and in **directing** them to their proper destinations. To perform this function, the Golgi vesicles contain different sets of enzymes in different types of vesicles—**cis, middle** and **trans cisternae**—that react with and modify secretory proteins passing through the Golgi lumen or membrane proteins and glycoproteins that are transiently in the Golgi membranes as they are *en route* to their final destinations (Fig. 64.4). For example, a Golgi enzyme may add a "signal" or "tag" such as a carbohydrate or phosphate residues to certain proteins to direct them to their proper sites in the cell. Or, a proteolytic Golgi enzyme may cut a secretory or membrane protein into two or more specific segments (*e.g.*, molecular processing involved in the formation of pancreatic hormone insulin: preproinsulin → proinsulin → insulin).

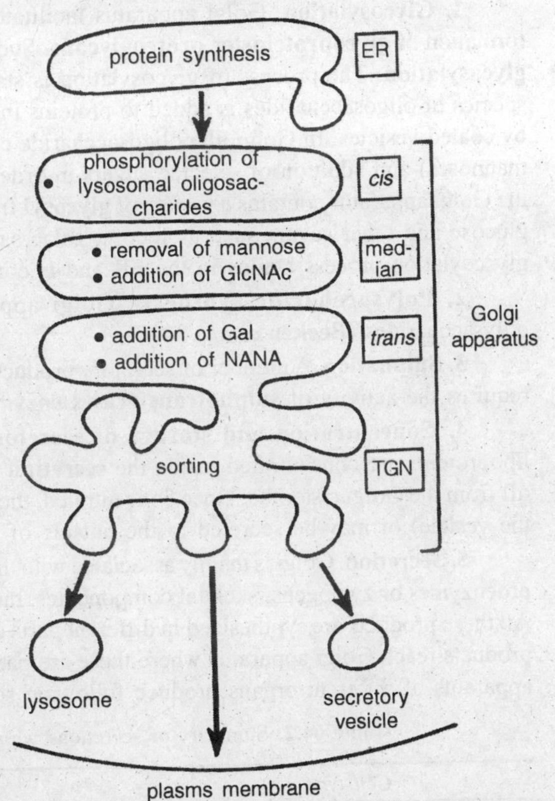

Fig. 64.4. The compartmentalization of the Golgi apparatus. GlcNAc = N-actyl-glucosamine galactose; NANA = N-acetyl neuraminic acid (sialic acid).

Recently, in the function of Golgi apparatus, subcompartmentalization with a division of labour has been proposed between the *cis* region (in which proteins of RER are sorted and some of them are returned back possibly by coated vesicles), and the *trans* region in which the most refined proteins are further separated for their delivery to the various cell compartments (*e.g.*, plasma membrane, secretory granules and lysosomes) (See Box 64.1).

Box 64.1

Following two contrasting models have been forwarded regarding the movement of materials through the Golgi complex.

1. Maturation model. Until the mid-1980s, it was generally accepted that Golgi cisternae were transient structures; it was supposed that they formed at the *cis* face of the stack by fusion of vesicles from the ER and that each cisterna physically moved from the *cis* to *trans* end of the stack, changing in composition as it progressed. This is known as the **maturation model,** since according to it, each cisterna "matures" into a trans cisterna.

2. Rothman and Orci model. According to **J.E. Rothman** and **L. Orci** (1990, 92), the cisternae of a Golgi stack remain in place as stable entities held together by a protein scaffold. In this model, materials moves through the Golgi stack, from the *cis* face to the TGN, by vesicles that bud from one membrane compartment and fuse with a neighbouring compartment farther along the stack.

The Golgi apparatus is primarily a **processing plant** of the cell. It is a centre of *reception, finishing, packaging*, and *dispatch* for a variety of materials in animal and plant cells. Thus, Golgi apparatus performs the following functions in the eukaryotic cells.

1. Glycosylation. Golgi apparatus facilitates linking of carbohydrates and proteins in the formation of **glycoproteins** or **proteoglycans**. Such addition of carbohydrates to proteins is called **glycosylation**. The process of glycosylation is started in the endoplasmic reticulum. Mainly one species of oligosaccharides is added to proteins in the ER. These proteins are transferred to Golgi by coated vesicles. In Golgi, the oligosaccharide chains of proteins are modified by removal (*e.g.*, mannoses) and addition of selected sugars in order to produce the final product. For this purpose the Golgi apparatus contains a variety of **glycosyl transferase enzymes** and sugar nucleotides. More glucose and other sugars such as fucose, galactose and sialic acid are added at the Golgi. Thus, glycosylation process begins in the ER and is completed in the Golgi apparatus.

2. Polysaccharide synthesis. Golgi apparatus is involved in synthesis of complex polysaccharides (**Becker** *et al.*, 2007).

3. Sulphation. A number of secretory products also contain sulphates. The sulphation process requires the activity of **sulphotransferases** enzymes which are present in Golgi apparatus.

4. Concentration and storage of secretory products. The proteins, glycoproteins and lipoproteins are concentrated inside the **secretion vesicles** or in **condensing vacuoles** which bud off from the Golgi cisternae. Once concentrated, the products may be stored in the cytoplasm (inside the vesicle) or may be secreted to the outside of the cell through exocytosis.

5. Secretion. Golgi is mainly associated with the secretory activity of the cell. Various hormones, proenzymes or zymogens, skeletal componentes, mucus, etc., are secreted through Golgi apparatus. All these products are synthesized in different parts of the cell. Through endoplasmic reticulum these products reach Golgi apparatus where these are packed and concentrated for final secretion. Golgi apparatus of different organs produce following secretion (Table 64.2).

Table 64.2. Summary of secretions which are produced in Golgi apparatus.

Cell type	*Golgi function*
1. Exocrine cells of pancreas	Secretion of zymogens (proenzymes of digestive enzymes such as proteases, lipases, carbohydrases and nucleases)
2. Parotid (salivary) gland cells	Secretion of mucus and zymogen
3. Goblet cells of intestinal epithelium	Secretion of mucus and zymogen
4. Goblet cells of respiratory epithelium	Secretion of mucus (*i.e.*, mucin, a glycoprotein)
5. Paneth cells of intestine (*i.e.*, of crypts of Lieberkuhn)	Secretion of digestive enzymes (*e.g.*, chitinase)

Cell type	Golgi function
6. Brunner's gland cells of intestine	Secretion of enzymes, mucus and hormones
7. Follicle cells of thyroid	Prethyroglobulin
8. Plasma cells of blood	Immunoglobulins
9. Hepatic cells of liver	Transformation and secretion of lipids
10. Cells of alveolar epithelium of mammary gland	Secretion of milk proteins (casein)
11. Nervous tissue (myelocytes, sympathetic ganglia, Schwann cells)	Sulphation reaction
12. Endothelial cells of blood vessels	Sulphation reaction
13. Cornea of birds	Secretion of tropocollagen and collagen
14. Most plant cells	Secretion of pectin and cellulose
15. Melanocytes	Secretion of melanin
16. Primary oocyte	Secretion of yolk and vitelline membrane
17. Root cap cells	Mucus

Types of Secretion

Molecular transport and diffusion across membranes account for the transfer of some materials from cytoplasm to extra cellular space. Cellular secretion may be of the following three types: holocrine, apocrine and merocrine.

***(i)* Holocrine secretion.** In this kind of cellular secretion, the entire cell fills with secretory product and then is broken up and sacrificed. An example is the sebaceous gland, which secretes a fatty discharge onto adjacent hair and skin. It is this gland that, when overstimulated, gets plugged to form blackheads and ultimately the pimples of adolescent acne.

***(ii)* Apocrine secretion.** In this kind of secretion, the secretion vesicle buds through the plasma membrane carrying with it a portion of that membrane and hence a thin layer of cytoplasm. The secretion vesicle is thus covered by two layers of membrane, its own plus the plasma membrane. Milk fat droplets (MFD) of the cells of the lactating mammary gland are released as apocrine secretion (Fig. 64.5).

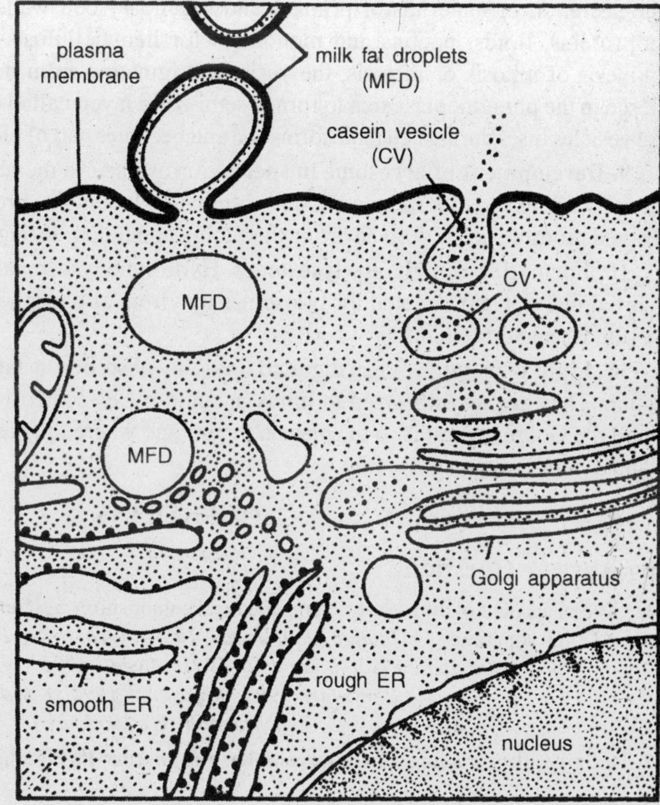

Fig. 64.5. Secretion in lactating mammary gland. Milk fat droplets are released by the apocrine mechanism, whereas casein vesicles normally release their protein by merocrine secretion. The former removes plasma membrane, the latter adds to it.

(*iii*) **Merocrine or eccrine secretion.** It is a most common type of cellular secretion. This process is also called **exocytosis** because it is the function opposite of endocytosis. Thus, it involves the fusion of secretory vesicles with the plasma membrane and their opening to the exterior. It is the usual mode of secretion of most glands and of most other cells, both animal and plant. It is, for example, the mode of secretion of the **true sweat glands**, which are distributed over most of the human skin. In cells of lactating mammary gland, casein vesicles (CV) release their protein by merocrine secretion (Fig. 64.5).

Pancreas which includes exocrine as well as endocrine cells has provided a best example of merocrine secretion.

6. Recycling of plasma membrane components. There occurs a continuous exchange of membrane between plasma membrane and Golgi apparatus due to exocytosis and endocytosis. The membrane material is reprocessed in the Golgi for reuse.

7. Intracellular sorting of proteins. Proteins are exported from the Golgi apparatus to different intracellular and extracellular destinations. The accurate sorting of proteins for selective export is one of the major functions of Golgi. This sorting is done by **coated vesicles** which surround the cisternae. These contain specific receptors which recognize specific markers on the proteins to be exported.

8. Formation of cell plate and cell wall in plants. In plants, Golgi apparatus is mainly involved in the secretion of materials of primary and secondary cell walls (*e.g.*, formation and export of glycoproteins, lipids, pectins and monomers for hemicellulose, cellulose, lignin, etc.). During cytokinesis of mitosis or meiosis, the vesicles originating from the periphery of Golgi apparatus, coalesce in the phragmoplast area to form a semi-solid layer, called **cell plate**. The unit membrane of Golgi vesicles fuses during cell plate formation and becomes part of plasma membrane of daughter cells.

9. Development of acrosome in sperm. Acrosome is a membrane-bounded cap-like structure at the forwarding end of a spermatozoon (in animals). It contains proteolytic enzymes. The acrosome is formed from vesicles derived from the Golgi apparatus during sperm maturation.

10. Formation of primary lysosomes. Hydrolytic lysosomal enzymes produced in rough ER are sorted in TGN and targeted for lysosomes by virtue of phosphorylated mannose residues within their oligosaccharide chains.

11. Lipid transport. When digested lipids are absorbed as fatty acids and glycerol in the small intestine. They are resynthesized to lipids in smooth ER, coated with protein and then transported through the Golgi apparatus to the plasma membrane where they leave the cell (exocytosis), mainly to enter lymph system.

QUESTIONS

Long Answer Questions

1. Give an account of the ultrastructure, chemical composition and functions of Golgi body, Golgi complex or Golgi apparatus. (*Punjab1989; Jiwaji 1989; Agra 1989; Bundelkhand 1989; Jabalpur 1989; Allahabad 1993; Bhagalpur 1991; Awadh 1991, 1992; CCS 1990, 1991, 1992, 1996; Bangalore 1992; Rajasthan 1992; Kanpur 1991, 1992, 1993, 1995, 1998; Magadh (G) 1993; 1995; Magadh (Sub.) 1996; Bihar 2004*)
2. Describe ultrastructure of Golgi complex and discuss its role in cell secretion. (*B.U. Coimbatore 1997*)
3. State the physical and chemical structure and functions of Golgi apparatus. (*Guwahati (G) 1989*)
4. Give an account of discovery, ultrastructure and functions of Golgi complex. (*Bihar 2002*)

Short Answer Questions

1. Describe main functions or role of Golgi complex. (*Punjab (G) 1991, 1992; CCS 1991; Kanpur 1997*)
2. What is Golgi? Mention its functions.

3. Write short note on Golgi complex. (*Punjab (G) 1990, 1995; Magadh (G) 1991; Magadh (Sub.) 1993; Nagarjuna 1994; Lucknow 1994; Kanpur 1994; Gorakhpur 1997*)

4. Write short notes on the following

(*i*) Golgi apparatus (*Avadh 1989; 1990; Purvanchal 1991*)

(*ii*) Functions of Golgi body (*Garhwal 1990, Jiwaji 1990, 1991*)

(*iii*) Acrosome formation (*Rohilkhand 1990*)

(*iv*) Zones of exclusion

Very Short Answer Questions

1. Who discovered the Golgi apparatus/Golgi complex?
   Ans. Camillo Golgi.
2. Which face of Golgi cisternae lie towards the plasma membrane?
   Ans. Trans face.
3. From which face the proteins enter in the Golgi apparatus?
   Ans. Cis face.
4. Which cell organelle is responsible for glycosylation and sulphation processes?
   Ans. Golgi apparatus.
5. Which cell organelle is responsible for concentration and storage of secretory products?
   Ans. Golgi apparatus.
6. Which part of mammalian sperm is formed by Golgi apparatus?
   Ans. Acrosome.
7. In which organelle synthesis of polysaccharides takes place? (*H.N.B. Garhwal 2009*)
   Ans. Golgi apparatus.

Yes or No Questions

(*i*) Golgi was first described by Bateson. (*Guwahati 1993*)

(*ii*) Acrosome is formed by Golgi body. (*Guwahati (G) 1992, 1994*)

(*iii*) Golgi apparatus was first described by Morgan. (*Guwahati (G) 1991*)

Multiple Choice Questions

1. Golgi apparatus is concerned with (*Bihar 2004*)
   (*a*) excretion (*b*) secretion
   (*c*) ATP synthesis (*d*) RNA synthesis
2. Golgi apparatus is absent in (*H.N.B. Garhwal 2008*)
   (*a*) yeast (*b*) liver cell
   (*c*) higher plant (*d*) blue green algae
3. Golgi apparatuses present in invertebrates and plant cells are called as
   (*a*) polysomes (*b*) golgisomes
   (*c*) dictyosomes (*d*) cisternae

ANSWERS

Yes or No Questions

1. No **2.** Yes **3.** No

Multiple Choice Questions

1. (*b*) **2.** (*d*) **3.** (*c*).

65

Lysosomes

The lysosomes (*Gr., lyso*=digestive + *soma*=body) are tiny membrane-bound vesicles involved in intracellular digestion. They contain a variety of hydrolytic enzymes that remain active under acidic conditions. The lysosomal lumen is maintained at an acidic pH (around 5) by an ATP-driven proton pump in the membrane. Thus, these remarkable organelles are primarily meant for the digestion of a variety of biological materials and secondarily cause aging and death of animal cells and also a variety of human diseases such as cancer, gout, Pompe's disease, silicosis and I-cell disease.

> **Box 65.1.**
>
> Lysosomes were observed and named by **Christian de Duve** in 1955 as cytoplasmic rounded dense bodies. They were termed **"suicide bags"** by **de Duve** in 1959.
>
> Lysosomes were investigated according to following two schools: 1. **C. de Duve** and his co-workers (1963, 1964, 1974) worked in Belgium and their approach was biochemical one. 2. **Alex Novikoff** and his research group (1962, 1964) worked in United States and their approach was morphological and cytochemical. For the discovery of lysosomes and a brilliant series of experiments on them, **de Duve** shared the 1974 Nobel Prize for physiology with **Palade** and **Claude**, both of them were pioneer cell biologists.

65.1 OCCURRENCE

The lysosomes occur in most animal and few plant cells. They are absent in bacteria and mature mammalian erythrocytes. Few lysosomes occur in muscle cells or in acinar cells of the pancreas. Leucocytes, especially granulocytes are a particularly rich source of lysosomes. Their lysosomes are so large-sized that they can be observed under the light microscope. Lysosomes are also numerous in epithelial cells of absorptive, secretory and excretory organs (*e.g.*, intestine, liver, kidney, etc.). They occur in abundance in the epithelial cells of lungs and uterus. Lastly, phagocytic cells and cells of reticuloendothelial system (*e.g.*, bone marrow, spleen and liver) are also rich in lysosomes.

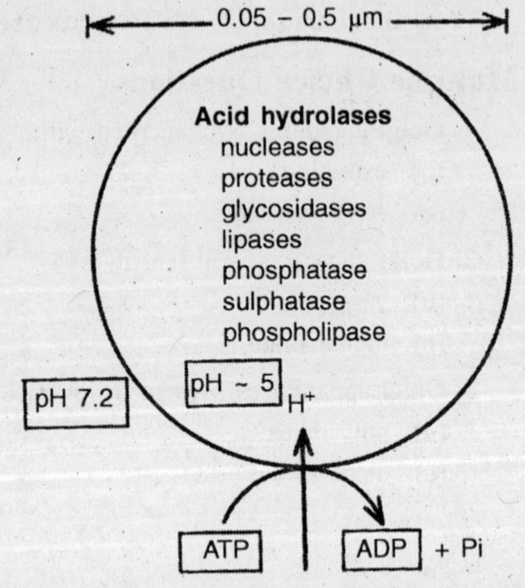

Fig. 65.1. A lysosome showing its various characteristics.

65.2 STRUCTURE

The lysosomes are round vacuolar structures which remain filled with dense material and are bounded by single unit membrane. Their shape and density vary greatly. Lysosomes are 0.2 to 0.5 µm in size. Since, size and shape of lysosomes vary from cell to cell and time to time (*i.e.*, they are polymorphic), their identification becomes difficult. However, on the basis of the following three criteria, a cellular entity can be identified as a lysosome: 1. It should be bound by a limiting membrane; 2. It should contain two or more acid hydrolases; and 3. It should demonstrate the property of enzyme latency when treated in a way that adversely affects organelle's membrane structure.

65.3 CHEMICAL COMPOSION

The **acid phosphatase** is the principal enzyme which is used as a marker for the lysosomes by the use of **Gomori' staining technique** (**Gomori**,1952). Specific stains are also used for other lysosomal enzymes such as **B-glucuronidase, aryl sulphatatase, N-acetyl-B-glucosaminidase** and **5-bromo-4-chloroïndolacetate esterase**.

Lysosomal Enzymes

According to a recent estimate, a lysosome may contain up to 50 types of hydrolytic enzymes. They include **proteases** (*e.g.*, cathepsin for protein digestion), **nucleases, glycosidases** (for digestion of polysaccharides and glycosides), **lipases, phospholipases, phosphatases** and **sulphatases** (Table 65.1). All lysosomal enzymes are acid hydrolases, optimally active at the pH5 maintained

Table 65.1 Some lysosomal enzymes and their substrates.

Enzyme	*Substrate*
A. Phosphatases	
1. Acid phosphatase	Phosphate monoesters
2. Acid pyrophosphatase	ATP, FAD
3. Acid phosphodiesterase	Oligonucleotides, phosphodiesters
B. Sulphatase	
4. Arylsulphatase	Aryl sulphates
C. Proteases and peptidases	
5. Cathepsin A, B, C, D, and E	Various proteins and peptides
6. Collagenase	Collagen
7. Peptidases	Peptides
D. Nucleases	
8. Acid ribonuclease	RNA
9. Acid deoxyribonuclease	DNA
E. Lipases	
10. Triglyceride lipase (Acid lipase)	Triacyglycerols
11. Phospholipase	Lecithin, phosphatidyl ethanolamine
12. Esterase	Fatty acid esters
13. Sphingomyelinase	Sphingomyelin
F. Glycosidases	
14. α-Glucosidase	Glycogen
15. β-Glucosidase	Glycoproteins (β-Glucosides)
16. β-Galactosidase	β-Galactosides
17. α-Mannosidase	Mannosyl oligosaccharides
18. α-Fucosidase	Glycoprotein
19. β-Xylosidase	Glycoprotein
20. β-Glucocerebrosidase	Glycolipids
21. α-N-Acetylhexosaminidase	Heparin
22. β-N-Acetylhexosaminidase	Glycoproteins, glycolipids
23. Sialidase	Sialic acid derivatives
24. Lysozyme	Mucopolysaccharides, bacterial cell wall
25. Hyaluronidase	Hyaluronic acid, chondroitin sulphates
26. β-Glucuronidase	Polysaccharides, mucopolysaccharides

within lysosomes. The membrane of the lysosome normally keeps the enzymes latent and out of the cytoplasmic matrix or cytosol (whose pH is about ~ 7.2), but *the acid dependency of lysosomal enzymes protects the contents of the cytosol (cytoplasmic matrix) against any damage even if leakage of lysosomal enzymes should occur.*

The so-called **latency** of the lysosomal enzymes is due to the presence of the membrane which is resistant to the enzymes that it encloses. Most probably this is due to the fact that most lysosomal hydrolases are membrane-bound, which may prevent the active centres of enzymes to gain access to susceptible groups in the membrane.

Lysosomal Membrane

The lysosomal membrane is slightly thicker than that of mitochondria. It contains substantial amounts of carbohydrate material, particularly **sialic acid**. In fact, lysosomal membranes contain two groups of acidic highly glycosylated integral proteins, called **lgp-A** and **lgp-B**, whose function may be to protect the membrane from attack of its enclosed enzymes.

The entire process of digestion is carried out within the lysosome. Most lysosomal enzymes act in an acid medium. Acidification of lysosomal contents depends on an ATP-dependent proton pump which is present in the membrane of the lysosome and accumulates H^+ inside the organelle. Lysosomal membrane also contains transport proteins that allow the final products of digestion of macromolecules to escape so that they can be either excreted or reutilized by the cell.

65.4 KIND OF LYSOSOMES (POLYMORPHISM IN LYSOSOMES)

Lysosomes are extremely dynamic organelles, exhibiting polymorphism in their morphology. Following four types of lysosomes have been recognized in different types of cells or at different times in the same cell. Of these, only the first is the **primary lysosome**, the other three have been grouped together as **secondary lysosomes**.

1. Primary Lysosomes

These are also called **storage granules, protolysosomes** or **virgin lysosomes**. Primary lysosomes are newly formed organelles bounded by a single membrane and typically having a diameter of 100 nm. They contain the degradative enzymes which have not participated in any digestive process. Each primary lysosome contains one type of enzyme or another and it is only in the secondary lysosome that the full complement of acid hydrolases is present.

2. Heterophagosomes

They are also called **heterophagic vacuoles**, **heterolysosomes** or **phagolysosomes**. Heterophagosomes are formed by the fusion of primary lysosomes with cytoplasmic vacuoles containing **extracellular substances** brought into the cell by any of a variety of endocytic processes (*e.g.*, pinocytosis, phagocytosis or receptor-mediated endocytosis). The digestion of engulfed substances takes place by the enzymatic activities of the hydrolytic enzymes of the secondary lysosomes. The digested material has low molecular weight and readily passess through the membrane of the lysosomes to become the part of the matrix (Fig. 65.2).

3. Autophagosomes

They are also called **autophagic vacuole**, **cytolysosomes** or **autolysosomes**. Primary lysosomes are able to digest **intracellular structures** including mitochondria, ribosomes, peroxisomes and glycogen granules. Such autodigestion (called **autophagy**) of cellular organelles is a normal event during cell growth and repair and is especially prevalent in differentiating and dedifferentiating tissues (*e.g.*, cells undergoing programmed death during metamorphosis or regeneration) and tissue under stress. Autophagy takes several forms. In some cases the lysosome appears to flow around the cell structure and fuse, enclosing it in a double membrane sac, the lysosomal enzymes being initially confined between the membranes. The inner membrane then breaks down and the enzymes are able to penetrate to the enclosed organelle.In other cases, the organelle to be digested is first encased by smooth ER, forming a vesicle that fuses with a primary lysosome (Fig. 65.4). Lysosomes also regularly engulf bits of cytosol (cytoplasmic matrix) which is degraded by a process, called **microautophagy**.

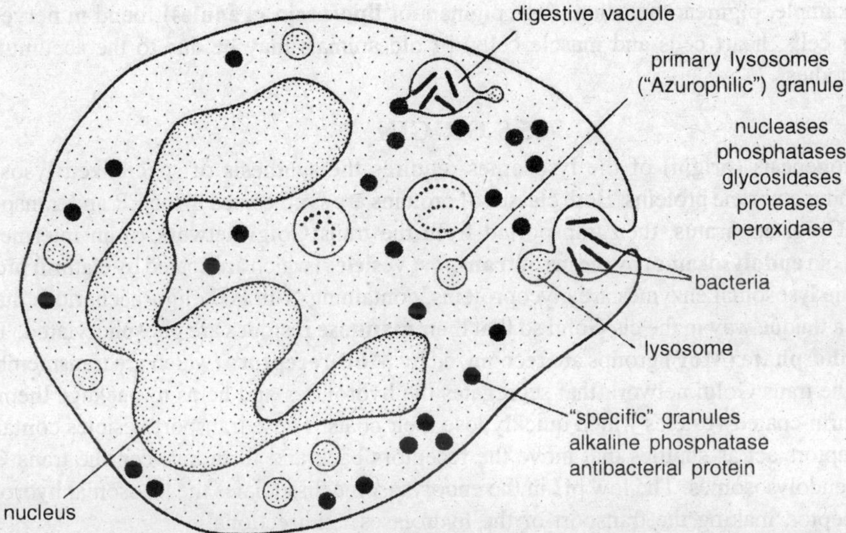

Fig. 65.2. Diagram of a white blood cell (neutrophil) ingesting bacteria. Two types of granules fuse with the phagocytotic vacuoles and contribute digestive enzymes and other components.

As digestion proceeds, it becomes increasingly difficult to identify the nature of the original secondary lysosome (*i.e.*, heterophagosome or autophagosome) and the more general term **digestive vacuole** is used to describe the organelle at this stage.

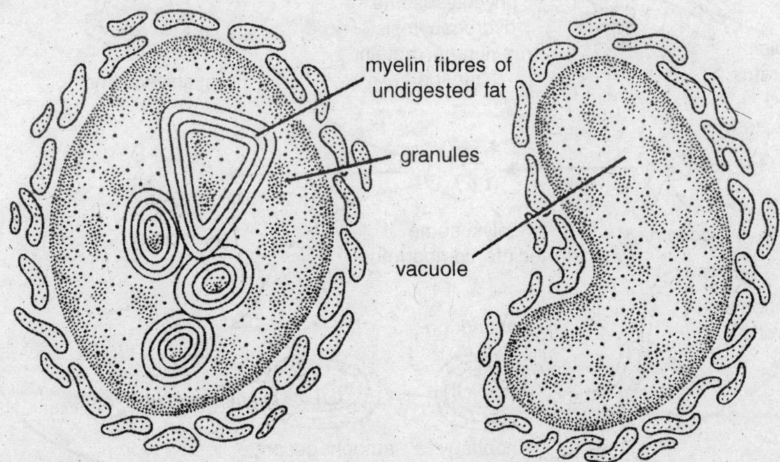

Fig. 65.3. Lysosomes of the kidney cells of rat, showing the presence of residues.

4. Residual Bodies

They are also called **telolysosomes** or **dense bodies**. Residual bodies are formed if the digestion inside the food vacuole is incomplete. Incomplete digestion may be due to absence of some lysosomal enzymes. The undigested food is present in the digestive vacuole as the residues and may take the form of whorls of membranes, grains, amorphous masses, ferritin-like or myelin fibres (Fig. 65.3).

Residual bodies are large, irregular in shape and are usually quite electron-dense. In some cells, such as *Amoeba* and other potozoa, these residual bodies are eliminated by **defecation**. In other cells, residual bodies may remain for a long time and may load the cells to result in their aging. For example, pigment inclusions (age pigment or **lipofuscin granules**) found in nerve cells (also in liver cells, heart cells and muscle cells) of old animals may be due to the accumulation of residual bodies.

65.5 ORIGIN

The biogenesis (origin) of the lysosomes requires the synthesis of specialized lysosomal hydrolases and membrane proteins. Both classes of proteins are synthesized in the ER and transported through the Golgi apparatus, then transported from the trans Golgi network to an intermediate compartment (an **endolysosome**) by means of **transport vesicles** (which are coated by clathrin protein; Fig. 65.4). The lysosomal enzymes are glycoproteins, containing N-linked oligosaccharides that are processed in a unique way in the cis Golgi so that their **mannose** residues are phosphorylated. These **mannose 6-phosphate (M6P)** groups are recognized by **M6P-receptors** (which are transmembrane proteins) in the trans Golgi network that segregates the hydrolases and helps to package them into budding clathrin-coated vesicles which quickly lose their coats. These transport vesicles containing the M6P-receptors act as shuttles that move the receptors back and forth between the trans Golgi network and endolysosomes. The low pH in the endolysosome dissociates the lysosomal hydrolases from this receptor, making the transport of the hydrolases unidirectional.

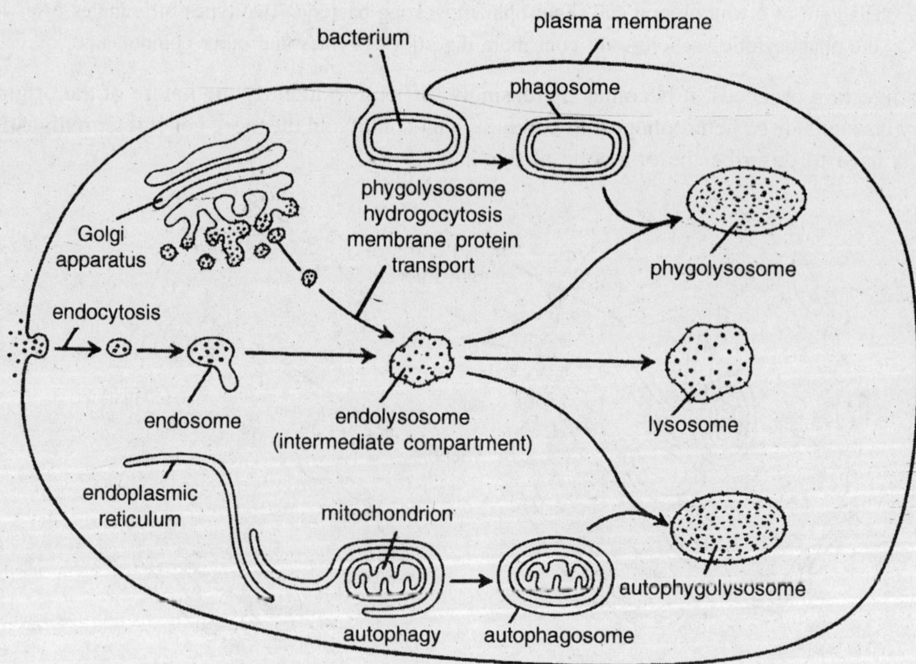

Fig. 65.4. Recently understood mechanism of the origin of three types of lysosomes : phagolysosome, lysosome (the classical secondary lysosome) and autophagolysosome. Transport vesicles (the classical primary lysosomes) originate from trans Golgi network to fuse with endolysosome which contains already endocytosed materials for digestion.

65.6 FUNCTIONS OF LYSOSOMES

The important functions of lysosomes are as follows:

1. Digestion of large extracellular particles. The lysosomes digest the food contents of the phagosomes or pinosomes. The lysosomes of leucocytes enable the latter to devour the foreign proteins, bacteria and viruses.

2. Digestion of intracellular substances. During the starvation, the lysosomes digest the stored food contents, *viz.*, proteins, lipids and carbohydrates (glycogen) of the cytoplasm and supply to the cell necessary amount of energy.

3. Autolysis. In certain pathological conditions the lysosomes start to digest the various organelles of the cells and this process is known as **autolysis** or **cellular autophagy**. When a cell dies, the lysosome membrane ruptures and enzymes are liberated. These enzymes digest the dead cells. In the process of metamorphosis of amphibians and tunicates many embryonic tissues, *e.g.*, gills, fins, tail, etc., are digested by the lysosomes and utilized by the other cells.

4. Extracellular digestion. The lysosomes of certain cells such as sperms discharge their enzymes outside the cell during the process of fertilization. The lysosomal enzymes digest the limiting membranes of the ovum and form penetration path in ovum for the sperms. Acid hydrolases are released from **osteoclasts** and break down bone for the reabsorption; these cells also secrete lactic acid which makes the local pH enough for optimal enzyme activity. Likewise, preceding ossification (bone formation), **fibroblasts** release cathepsin D enzyme to break down the connective tissue.

65.7 LYSOSOMAL STORAGE DISORDERS (LYSOSOMES AND DISEASE)

Malfunctioning of lysosomes often results in various pathological disorders affecting the life of the cell or an individual. Some of these are **inborn diseases**, caused by gene mutation (*e.g.*, I-cell disease, gout, Pompe's disease, Tay-Sach's disease, etc.) and others are induced by some environmental pollutants (*e.g.*, silicosis). Typically, the accumulated materials (*e.g.*, low-molecular weight materials, drugs, dyes, etc.) may cause **malignant** transformation of cells by bringing about leakage of lysosomal enzymes that attack the genetic material in the DNA.

Defects in normal functioning of lysosomes cause the following diseases (about 30 diseases):

1. Silicosis and asbestosis. Silicosis is a miner's disease which results from the uptake of silica fibres by macrophages (phagocytic cells) in the lungs. The fibres become enclosed within lysosomes but cannot be digested; instead they cause the lysosomal membrane to leak, spilling the contents of digestive enzymes into the cell and damaging the tissue of the lungs.

A similar result occurs when asbestos fibres are taken up by scavenging cells, resulting in the disease **asbestosis**. Both of these conditions, can be exhausting and may even be fatal.

2. Rheumatoid arthritis. This inflammatory disease results, in part, from the release of lysosomal enzymes from the immune cells into the extracellular space, causing damage to materials in the joints.

3. I-Cell disease. Patients of this inheritable disease has a deficiency of a lysosomal enzyme, called **N-acetylglucosmine phosphotransferase** (a lysosomal enzyme) which is required for phosphorylation of **mannose**. This lysosomal enzyme lacks mannose phosphate residues and due to this defect is not included in the lysosomes of fibroblasts of patients of I-cell disease. As a result, such lysosomes have bloated appearance due to undigested materials.

4. Pompe disease. In the absence of another lysosomal enzyme, called **α-glucosidase**, undigested glycogen is accumulated in the lysosomes. This results in swelling of these organelles and irreversible damage to the cells and tissues.

5. Tay-Sachs disease. This disease is caused from the deficiency of the enzyme **β-N-hexosaminidase A**, an enzyme required to degrade the ganglioside G_{M2}, a common component of the membranes of brain cells. In its severest form, which strikes during infancy, the disease is characterized by progressive mental and motor retardation, as well as skeletal, cardiac and respiratory abnormalities. Tay-Sachs disease results in mental retardation, blindness and death by the age of 3.

65.8 LYSOSOMES IN PLANTS

Plants contain several hydrolases, but they are not always as neatly compartmentalized as they are in animal cells. Many of these hydrolases are found bound to and functioning within the vicinity of the cell wall and are not necessarily contained in membrane-bound vacuoles at these sites. Many types of vacuoles and storage granules of plants are found to contain certain digestive enzymes and these granules are considered as lysosomes of plant cell. According to **Matile** (1969) the plant lysosomes can be defined as membrane-bound cell compartments containing hydrolytic digestive enzymes. **Matile** (1975) has divided vacuoles of plants into following three types:

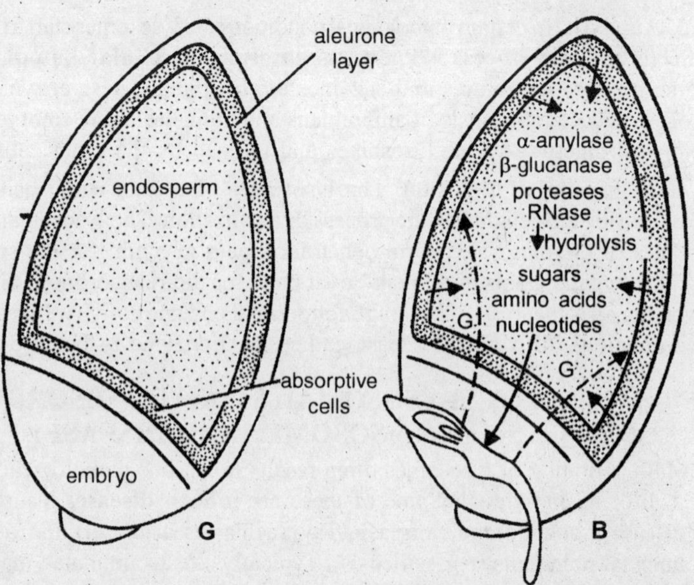

Fig. 65.5. Role of hormone and hydrolytic enzymes in seed germination. A—Ultrastructure of seed before germination; B–During germination, gibberellins (G) pass from the embryo to the aleurone layer, where the *de novo* synthesis of hydrolytic enzymes is induced. These enzymes break down the macromolecular stored reserves and the low molecular weight products are transported to the embryo, where they function as nutrients.

1. Vacuoles. The vacuole of a mature plant cell is formed from the enlargement and fusion of smaller vacuoles present in meristomatic cells; these **provacuoles**, which are believed to be derived from the ER and possibly the Golgi and contain acid hydrolases. These lysosomal enzymes are associated with the tonoplast of large vacuole of differentiating cells. Sometimes, mitochondria and plastids are observed inside the vacuole suggesting autophagy in plants.

2. Spherosomes. The spherosomes are membrane-bounded, spherical particles of 0.5 to 2.5 μm diameter, occurring in most plant cells. They have a fine granular structure internally which is rich in lipids and proteins. They originate from the endoplasmic reticulum (ER). Oil accumulates at the end of a strand of ER and a small vesicle is then cut off by constriction to form particles, called **prospherosomes**. The prospherosomes grow in size to form spherosomes. Basically, the spherosomes are involved in lipid synthesis and storage. But, the spherosomes of maize root tips and spherosomes of tobacco endosperm tissue have been found rich in hydrolytic digestive enzymes and so have been considered as lysosomes. Like lysosomes they are not only responsible for the accumulation and mobilization of reserve lipids, but also for the digestion of other cytoplasmic components incorporated by phagocytosis.

3. Aleurone grain. The aleurone grains or protein bodies are spherical membrane-bounded storage particles occurring in the cells of endosperm and cotyledons of seeds. They are formed during the later stages of seed ripening and disappear in the early stages of germination. They store protein (*e.g.*, globulins) and phosphate in the form of phytin. **Matile** (1968) has demonstrated that aleurone grains from pea seed contain a wide range of hydrolytic enzymes including protease and phosphatase which are required for the mobilization of stored protein and phosphate, although the presence of other enzymes such as β-amylase and RNAase suggest that other cell constituents may also be digested. Thus, like spherosomes, aleurone grains store reserve materials, mobilize them during germination and in addition form a compartment for the digestion of other cell components. The aleurone grains are derived from the strands of the endoplasmic reticulum.

During germination of barley seed, the activity of hydrolases is found to be controlled by hormones such as **gibberellic acid**. Gibberellic acid, a plant growth hormone, is released by the embryo to the aleurone layer where, in turn, the hydrolases are released to the endosperm. This hormone operates by derepressing appropriate genes in the aleurone cells, which then begin to crank out (*i.e.,* produce regularly and routinely) new hydrolytic proteins.

Extra-cellular digestion by plants. Plant cells are generally unable to engulf large particles, presumably because of the restrictions imposed on the cell by cell wall. The secretion of hydrolases to carry out extracellular digestion, therefore, becomes an important process. Hydrolases are commonly secreted by fungi, enabling the organism to degrade and grow on macromolecules it cannot transport into the cell. Higher plants also secrete hydrolases, a notable example being the insectivorous pitcher plants, which produce a proteinase-containing liquid in which victims are trapped and digested.

QUESTIONS

Long Answer Questions

1. Describe in detail the structure and functions of lysosomes. (*Punjab 1989; Jiwaji 1991; Garhwal 1990; CCS 1990, 1992; Nagpur 1992; Delhi 1992;Lucknow 1993; Magadh (G) 1995; Gorakhpur 1995, 1997*)
2. What are lysosomes? Give the ultrastructure of lysosomes and their functional significance. (*CCS 1989, 1991; Punjab (G) 1990; Allahabad 1992*)
3. "Lysosomes are said to be the suicidal bags in the cells". Explain it. Give structure, chemical composition and functions of lysosomes. (*CCS 1993, 1995*)
4. Describe lysosomes in plants.
5. Write an essay on lysosome. (*Bihar 2003*)

Short Answer Questions

1. Give an account of lysosomal polymorphism. (*Punjab (G) 1991*)
2. Explain the involvement of lysosomes in human disease. (*B.U. Coimbatore 1997*)
3. What are suicide bags of cell? Describe their functions. (*C.C.S.U. Meerut 2007*)
4. Write short notes on the following:
   (*i*) Lysosome. (*Allahabad 1993; Bihar 2004; H.N.B. Garhwal, 2008*)
   (*ii*) Lysosomes or Suicide bags (*Garhwal 1989; Rajasthan 1989; Sagar 1990; Indore 1990; Rohilkhand 1990; Kanpur 1990, 1993, 1998; Purvanchal 1992; Magadh (G) 1989, 1992, 1994; Magadh (Subs.) 1994; Guwahati (G) 1993, 1996; Gorakhpur 1996; Lucknow 1997, 1998*)
   (*iii*) Polymorphism in lysosome (*Gorakhpur 1994*)
   (*iv*) Enzymes of lysosomes (*Punjab 1991*)
5. Lysosomes are suicide bags of the cell. Explain. (*CCS 1990*)
6. Write a brief account of polymorphism in lysosomes.
7. Give a brief account of the functions of lysosomes.
8. How and where lysosomes are formed in the cell?
9. Give significance of autophagic vacuole.
10. Lysosomal enzymes are inactive as long as they remain inside the lysosomes but become active as soon as they come in contact with some other substance. Why?
11. Differentiate between the following:

(*i*) Primary and secondary lysosomes
(*ii*) Autophagic vacuole and digestive vacuole
(*iii*) Peroxisomes and lysosomes
(*iv*) Lysosomes and zymogen granules.

Very Short Answer Questions

1. What is the role of acrosomal granule in spearm penetration?
Ans. Lysis.

2. Name the animal cells which do not contain lysosomes.
Ans. Red blood corpuscles.

3. Where does lysosome formed in the cell?
Ans. Golgi apparatus.

4. Which secondary lysosome stores undigested wastes of cellular digestion?
Ans. Residual bodies.

5. What is autophagy?
Ans. Digestion of portions of a cell's own organelles by lysosomes.

6. What is pH of lumen of lysosomes?
Ans. 5.

7. Give the name of marker enzyme of the lysosome.
Ans. Acid phosphatase.

Fill in the Blanks

1. Lysosomal function is related to ....of substance.
2. Lysosomes with phagosomes form ....
3. Microbodies containing oxidative enzymes are called...........
4. The microbodies with specific lipid contents in their matrix are known as.............

Yes or No Questions

Write Yes or No as answers of the following:

1. Lysosomes help in protein synthesis. (*Guwahati (G) 1991, 1995*)
2. Vaculoes of a cell are known as "suicide bags". (*Guwahati (G) 1994*)

Multiple Choice Questions

1. Lysosomes are absent in (*Bihar 2004*)
(*a*) erythrocytes (*b*) plasma cells (*c*) nerve cells (*d*) muscle cells

2. Which one of the following is called as the suicidal sac of the cell? (*Bihar 2001, 2003*)
(*a*) centrosome (*b*) lysosomes (*c*) microsomes (*d*) desmosomes

3. At which pH, enzymes of lysosomes are usually active
(*a*) pH5 (*b*) pH7 (*c*) pH8 (*d*) at any pH

4. Hydrolytic enzymes are stored in
(*a*) Golgi bodies (*b*) lysosomes (*c*) endoplasmic reticulum (*d*) mito chondria

ANSWERS

Fill in the Blanks

1. Hydrolysis (*i.e.,* intracellular digestion)
2. Heterolysosomes (or food vacuoles)
3. Peroxisomes
4. Sphareosomes.

Yes or No Questions:

1. No **2.** No.

Multiple Choice Questions

1. (*a*) **2.** (*b*) **3.** (*c*) **4.** (*b*)

66

Peroxisomes

Peroxisomes are small, simple, single-membrane enclosed organelles with a diameter of 0.1 to 1.0 μm. They often contain a dense, crystalline core of oxidative enzymes. Peroxisomes are multifunctional organelles, so contain enzymes involved in a variety of metabolic reactions, including several aspects of energy metabolism. They are found in all eukaryotic cells; most human cells contain about 500 peroxisomes. Peroxisomes do not have their own genomes and ribosomes and all their proteins, called **proxins** (Pex 1, Pex 2, etc.) are synthesized from the nuclear genome. Most proxins are synthesized on free ribosomes and then imported into peroxisomes as completed polypeptide chains.

Comparison of Peroxisomes and Mitochondria.

Peroxisomes share several properties with mitochondria:

1. Both mitochondria and peroxisomes form by splitting from preexisting organelles. (However, unlike mitochondria, peroxisomes can also be regenerated even if entirely lost to the cell).
2. Both types of organelles import preformed proteins
3. Both types of organelles engage in similar types of oxidative metabolism. In fact, at least one enzyme, alanine/glyoxylate aminotransferase, is found in the mitochondria of some mammals (*e.g.*, cats and dogs) and the preoxisomes of other mammals (*e.g.*, rabbit and humans; see **Cooper** and **Hausman**, 2007).
4. Peroxisomes use molecular oxygen like mitochondria, but instead of having cytochromes and capacity of ATP synthesis like them, they contain flavin-linked oxides and catalases for the hydrogen peroxide metabolism and also enzymes for fatty acid metabolism.

   Lastly, while mitochondrial (and plastid) proteins resemble those of prokaryotes, reflecting their endosymbiotic origin, the peroxins resemble typical eukaryotic proteins.

Peroxisomes differ from mitochondria and chloroplasts in many ways. Most notably, these organelles are surrounded only by a single membrane, and they do not contain DNA (genome) or ribosomes. However, they resemble ER in being self-replicating membrane bound organelle.

66.1 HISTORICAL

Since the mid-1950s electron microscopists have observed small structures or bodies in cells that on morphological grounds have been aptly termed **microbodies. C.De Duve** and **P. Baudbuin** (1966) coined the term **peroxisome** for the microbodies of mammalian systems and studied their structure and function. **Leaf peroxisomes** were first isolated from spinach leaf homogenate (*i.e.*, from mesophyll cells) by **Tolbert's** group in Michigen in 1968. Glyoxylate cycle containing peroxisomes, called **glyoxysomes**, were discovered in 1969 by **Beevers** in the endosperm cells of germinating castor bean (*Ricinus*). Biogenesis of peroxisomes has been investigated by following cell biologists: **Purdue** and **Lazarow** (2001); **Mullen, Flynn** and **Trelease** (2001); and **Subramani** (2001) and **Lazarow** (2003).

66.2 OCCURRENCE

Peroxisome is found in all eukaryotic cells but is especially prominent in mammalian kidney and liver cells, in algae and photosynthetic cells of plants, and in germinating seedlings of plant species that store fat in their seeds.

66.3 STRUCTURE

Peroxisomes are variable in size and shape, but usually appear circular in cross section having diameter between 0.2 and 1.5μm (0.15 to 0.25 μm) diameter in most mammalian tissues; 0.5 μm in rat liver cells). They have a single limiting unit membrane of lipid and protein molecules, which encloses their granular matrix. In some cases (*e.g.*, in the festuciod grasses) the matrix contains numerous threads or fibrils, while in others they are observed to contain either an amorphous nucleoid or a dense inner core which in many species shows a regular crystalloid structure (*e.g.*, tobacco leaf cell, **Newcomb** and **Frederick**, 1971). Little is known about the function of the core, except that it is the site of the enzyme urate oxidase in rat liver peroxisomes and much of the catalase in some plants (see **Hall** *et.al.*, 1974).

66.4 ISOLATION AND CHEMICAL COMPOSITION

The technique of **isolation** of microbodies from animal and plant tissues includes the following steps: 1. Tissues are ground very carefully to save microbodies from disruption. 2. The homogenate is treated with differential centrifugation to obtain a fraction of the cell homogenate which is rich in microbodies. 3. The enriched fraction is subjected to isopycnic ultra-centrifugation on discontinuous or continuous sucrose density gradient.

Animal peroxisomes often contain a distinct **crystalline core**, which is usually consists of a crystalline form of *urate oxidase.* Crystalline cores are also often present in the peroxisomes of plant leaves, but these usually consist of *catalase* instead. When such cores are present, it is easy to identify microbodies as peroxisomes, since urate oxidase and catalase are two of the enzymes by which peroxisomes are defined. Indeed, these organelles were named "peroxisomes" because they are the site of synthesis and degradation of hydrogen peroxide.

In the absence of crystalline core, it is not always easy to spot peroxisomes ultrastructurally. A useful technique in such cases is a **cytochemical test** for catalase called the **diaminobenzidine (DAB) reaction**. This assay depends on the ability of catalase to oxidize DAB to a polymeric form that causes deposition of electron dense osmium atoms when the tissue is treated with osmium tetroxide ($Os\ O_4$). The resulting electron-dense deposits can be readily seen in cells from stained tissue. In animal peroxisomes, the entire internal space often stains intensely with DAB, indicating that catalase exists as a soluble enzyme uniformly distributed throughout the matrix of the organelle. In plant leaf cells, DAB treatment preferentially stains the crystalline cores of the peroxisomes, thereby definitively identifying the cores as crystalline catalase. Because catalase is the single enzyme present in all peroxisomes and does not routinely occur in any other organelle, the DAB reaction is a very reliable and highly specific means of identifying organelles unambiguously as peroxisomes.

Peroxisomes contain at least 50 different enzymes, which are involved in a variety of biochemical pathways in different types of cells. However, peroxisomes are diverse organelles and even in various cell types of a single organism they may contain different sets of enzymes. They can also adapt remarkably to changing conditions. Yeast cells grown on sugar, for example, have small peroxisomes. But when some yeasts are grown on methanol, they develop large peroxisomes that oxidise methanol; and when grown on fatty acids, they develop large peroxisomes that break down fatty acids to acetyl CoA by β–oxidation (see **Alberts** *et al.,* 2002).

66.5 FUNCTIONS

There are at least **five** general categories of peroxisomal functions:

1. Hydrogen peroxide metabolism;
2. Detoxification of harmful substances;
3. Oxidation of fatty acids;
4. Metabolism of nitrogen-containing compounds, and
5. Catabolism of unusual substances.

1. Hydrogen peroxide metabolism. The most obvious role of peroxisomes in eukaryotic cells is the detoxification of H_2O_2, which is accomplished by the *coexistence* in the same organelle of catalase and the oxidases that generate H_2O_2. The oxidases in peroxisomes vary considerably in the specific reactions they catalyze but they all share the property of **transferring** electrons from their respective substrates directly to molecular oxygen (O_2), forming H_2O_2. Using RH_2 to represent an oxidizable organic substrate, the general reaction catalyzed by oxidases can be written as follows:

$$RH_2 + O_2 \longrightarrow R + H_2O_2 \quad \text{...(Equation 66.1)}$$

The hydrogen peoxide formed in this manner is broken down by catalase in one of two ways. Usually catalase functions in what is called its **catalytic mode**, in which one molecule of H_2O_2 is oxidized to oxygen and a second is reduced to water:

$$2H_2O_2 \longrightarrow O_2 + 2H_2O \quad \text{...(Equation 66.2)}$$

Dividing reaction 66.2 and adding it to reaction 66.1 yields a summary reaction for the two-step process:

$$RH_2 + \frac{1}{2}O_2 \longrightarrow R + H_2O \quad \text{...(Equation 66.3)}$$

Alternatively, catalase can function in its **peroxidatic mode**, in which electrons derived from an organic donor are used to reduce H_2O_2 to water:

$$R'H_2 + H_2O_2 \longrightarrow R' + 2H_2O \quad \text{...(Equation 66.4)}$$

(The prime on R group (*i.e.*, R′) simply indicates that this substrate is likely to be different from the substrate in reaction 66.1) The corresponding summary reaction in this case is

$$RH_2 + R'H_2 + O_2 \longrightarrow R + R' + 2H_2O \quad \text{...(Equation 66.5)}$$

The result is the same in either case: Hydrogen peroxide is degraded without ever leaving the peroxisome. Given the toxicity of hydrogen peroxide (which is the main active ingredient in a variety of disinfectants), it makes good sense for the enzymes responsible for peroxide generation to be compartmentalized together with the catalase that catalyzes its degradation. Indeed, catalase is the most abundant protein in most peroxisomes, representing upto 15% of the total protein content of this organelle. In this way, virtually every molecule of H_2O_2 generated by oxidases will encounter a molecule of catalase almost immediately be promptly degraded.

H_2O_2 and aging. Most cytosolic H_2O_2 is produced by mitochondria and membranes of ER, although there are also H_2O_2 producing enzymes localized in the cytoplasmic matrix. Catalase acts as a **"safety valve"** for dealing with the large amounts of H_2O_2 generated by peroxisomes; however, other enzymes such as **glutathione peroxidase;** are capable of metabolizing organic hydroperoxides and also H_2O_2, in the cytosol (cytoplasmic matrix) and mitochondria. The production of superoxide anion (O_2–) in mitochondria and cytosol (cytoplasmic matrix) is regulated mainly by the enzyme

superoxide dismutase. All of these protective enzymes are present in high levels in aerobic tissues.

Recently, a possible relationship has been stressed between peroxides and free radicals (such as superoxide anion–O_2^-) with the process of aging. These radicals may act on DNA molecule to produce **mutations** altering the transcription into mRNA and the translation into proteins. In addition, free radicals and peroxides can affect the membranes by causing peroxidation of lipids and proteins. For these reasons reducing compounds such as vitamin E or enzymes such as superoxide dismutase could play a role in keeping the healthy state of a cell.

2. Detoxification of harmful compounds. In its peroxidatic mode (reaction 66.4), catalase can use a variety of substances as electron donors, including methanol, ethanol, formic acid, formaldehyde, nitrites and phenols. Because all of these compounds are harmful to cells, *their oxidative detoxification by catalase* may be a vital peroxisomal function. The prominent peroxisomes of liver and kidney cells are thought to be important in such detoxification reactions.

3. Oxidation of fatty acids. Peroxisomes found in animal, plant and fungal cells contain enzymes necessary for oxidising fatty acids. This process, called **β-oxidation**, also occurs in the mitochondrion as was discussed in chapter 58. About 25 to 50% of fatty acid oxidation in animal tissues occurs in peroxisomes, with the remainder localized in mitochondria. In plant and yeast cells, on the other hand, all β-oxidation is confined to peroxisomes.

In animal cells, peroxisomal β-oxidation appears to be especially important for the catabolism of long-chain (16 to 20 carbons), very long chain (24–26 carbons), and branched fatty acids. The primary product of β-oxidation, acetyl-CoA, is then exported to the cytosol and enters biosynthetic pathways. Once fatty acids are shortened to fewer than 16 carbon, further oxidation usually occurs in the mitochondria. Thus, in animal cells, the peroxisome is important for shortening fatty acids in preparation for subsequent metabolism in the mitochondrion rather than completely breaking them down to acetyl-CoA. In plants and yeast, on the other hand, peroxisomes are essential for the complete catabolism of fatty acids to acetyl-CoA.

4. Metabolism of nitrogen containing compounds. Except for primates, most animals require urate oxidase (also called uricase) to oxidize urate, a purine that is formed during the catabolism of nucleic acids and some proteins. Like other oxidases, urate oxidase catalyzes the direct transfer of electrons from the substrate to molecular oxygen, generating H_2O_2:

$$\text{Urate} + O_2 \longrightarrow \text{Allantion} + H_2O_2 \qquad \text{...(Equation 66.6)}$$

As noted earlier, the H_2O_2 is immediately degraded in the peroxisome by catalase. The allantoin is further metabolized and excreted by the organism, either as allantoic acid, or in case of crustaceans, fish, and amphibians, as urea.

Additional peroxisomal enzymes involved in nitrogen metabolism include aminotransferase. Members of this collection of enzymes catalyze the transfer of amino acids to α-keto acids.

$$\underset{\text{Amino acid}}{R-\overset{H_3N^+}{\overset{|}{C}}-\overset{O}{\overset{||}{C}}-O^-} + \underset{\alpha\text{-keto acid}}{R'-\overset{O}{\overset{||}{C}}-\overset{O}{\overset{||}{C}}=O^-} \rightleftharpoons \underset{\alpha\text{-keto acid}}{R-\overset{O}{\overset{||}{C}}-\overset{O}{\overset{||}{C}}-O^-} + \underset{\text{Amino acid}}{R'-\overset{H_3N^+}{\overset{|}{C}}-\overset{O}{\overset{||}{C}}=O^-}$$

...(Equation 66.7)

Such enzymes play important roles in the biosynthesis and degradation of amino acids by moving amino groups from one molecule to another.

5. Catabolism of unusual substances (e.g., xenobiotics). Some of the substrates for peroxisomal oxidases are rare compounds for which the cells has no other degradive pathways. Such compounds include D-amino acids, which are not recognized by enzymes capable of degrading the L-amino acids that constitute polypeptides. In some fungi, the peroxisomes also contain enzymes that break down unusual substances called **xenobiotics,** chemical compounds foreign to biological organisms. This category include **alkanes**, short-chain hydrocarbon compounds found in oil and other petroleum products. Fungi containing enzymes capable metabolizing such xenobiotics may turn out to be useful for cleaning up oil spills that would otherwise contaminate the environment (see **Becker** *et.al.*, 2007).

6. Synthetic activities. In addition to providing a compartment for oxidation reactions, peroxisomes are involved in biosynthesis of lipids and the amino acid, lysine. In animal cells, cholesterol and dolicol are synthesized in peroxisomes as well as in ER. In the liver, peroxisomes are also involved in the synthesis of bile acids, which are derived from cholesterol. In addition, peroxisomes contain enzymes required for the synthesis of **plasmalogens** a family of phospholipids in which one of the hydrocarbon chains is joined to glycerol by an **ether bond** rather than by ester bond (Fig. 66.1). Plasmalogens are important membrane components in some tissues. Particularly heart and brain, although they are absent in others. Deficiency of plasmalogens causes profound abnormalities in the **myelination** (formation of myelin-sheath) of nerves, which is one reason why many peroxisomal disorders lead to neurological disease (See **Alberts** *et.al.*, 2002) (Box 66.1).

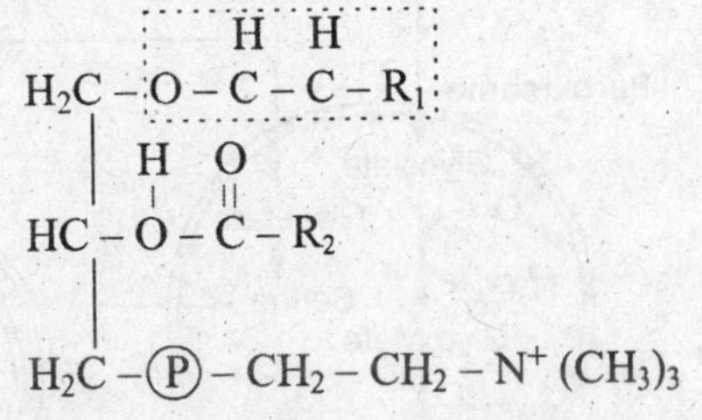

$$
\begin{array}{l}
H_2C - O - \overset{H}{C} - \overset{H}{C} - R_1 \\
\;\;| \\
HC - \overset{H}{O} - \overset{O}{\overset{\|}{C}} - R_2 \\
\;\;| \\
H_2C - (P) - CH_2 - CH_2 - N^{+}(CH_3)_3
\end{array}
$$

Fig. 66.1. Structure of a plasmalogen.

Box 66.1

The most common peroximal disorder is **X-linked adrenoleukodystrophy.** The defective protein causing this disorder is an integral membrane protein that may be responsible for transporting very long-chain fatty acids into the peroxisomes for β-oxidation. Accumulation of these long-chain fatty acids in human body fluids destroys myelin sheath in nervous system (See **Becker** *et. al.*, 2007).

Roles of Plant Specific Peroxisomes

1. Leaf peroxisomes. In green leaves, there are peroxisomes that carry out a process called **photorespiration** which is a light-stimulated production of CO_2 that is different from the generation of CO_2 by mitochondria in the dark. In photorespiration, **glycolic acid (glycolate)**, a two-carbon product of photosynthesis is released from chloroplasts and oxidized into **glyoxylate** and $\mathbf{H_2O_2}$ by a peroxisomal enzyme called **glycolic acid oxidase**. Later on, glyoxylate is oxidized into CO_2 and **formate**:

$$CH_2OH.\ COOH + O_2 \longrightarrow CHO - COOH + H_2O_2$$

$$CHO - COOH + H_2O_2 \longrightarrow HCOOH + CO_2 + H_2O$$

Photorespiration is so-called because light induces the synthesis of glycolic acid in chloroplasts. The entire process involves intervention of two basic organelles : chloroplasts and peroxisomes.

Lastly, photorespiration is driven by atmospheric conditions in which the O_2 tension is high and the CO_2 tension low. Apparently O_2 competes with CO_2 for the enzyme **ribulose diphosphate carboxylase** which normally is the key enzyme in CO_2 fixation during photosynthesis. When O_2 is used by the enzyme, an unstable intermediate is formed which breaks down into **3-phosphoglycerate** and **phosphoglycolate**. The latter tends to increase the glycolate concentration by removal of its phosphate group and, therefore, more glycolate is available for additional oxidation and CO_2 release.

Photorespiration is a wasteful process for the plant cell, since, it significantly reduces the efficiency of the process of photosynthesis (*i.e.*, it returns a portion of fixed CO_2 to the atmosphere).

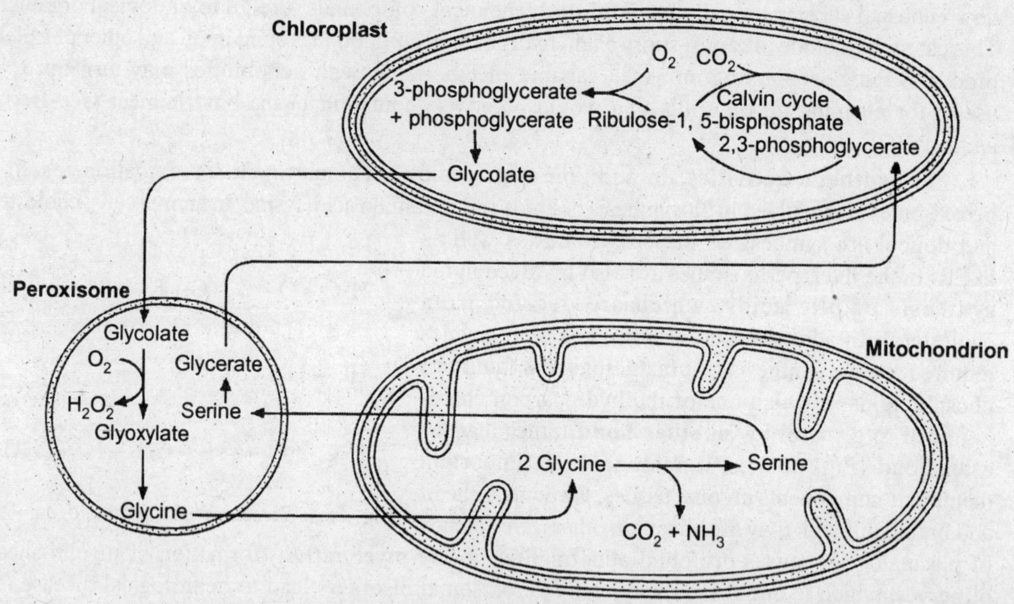

Fig. 66.2. Role of peroxisomes in photorespiration. During photosynthesis CO_2 is converted to carbohydrates by the Calvin cycle, which initiates with the addition of CO_2 to the five-carbon sugar ribulose–1, 5–diphosphate. However, the enzyme involved sometimes catalyzes the addition of O_2 instead, resulting in production of the two-carbon compound phosphoglycolate. Phosphoglycolate is converted to glycolate, which is then transferred to peroxisomes, where it is oxidized and converted to glycine. Glycine is then transferred to mitochondria and converted to serine. The serine is returned to peroxisomes and converted to glycerate, which is transferred back to chloroplasts (after Cooper and Hausman, 2007).

Glycolate cycle. Peroxisomes of plant leaves contain catalaze together with the enzymes of **glycolate pathway**, as glycolate oxidase, glutamate glyoxylate, serine-glyoxylate and asparate-α-ketoglutarate aminotransferases, hydroxy pyruvate reductase and malic dehydrogenase. They also contain FAD, NAD and NADP coenzymes. The glycolate cycle is thought to bring about the formation of the amino acids-glycine and serine-from the non-phosphorylated intermediates of photosynthetic carbon reduction cycle, *i.e.*, glycerate to serine, or glycolate to glycine and serine in a sequence of reactions which involve chloroplasts, peroxisomes, mitochondria and cytosol (**Tolbert,** 1971). The glycolate pathway also generates C_1 compounds and serves as the generator of precursors for nucleic acid biosynthesis.

2. Glyoxysomes. It occurs transiently in seedlings of plant species that store carbon and energy reserves in the seed of fat (primarily triacylglycerols). In such species, stored triglycerols are mobilized and converted to sucrose during early postgerminative development by a sequence of events that includes β-oxidation of fatty acids as well as a pathway known as the glyoxylate (= glycolate) cycle. All of the enzymes needed for these processes are localized to specialized peroxisomes called **glyoxysomes**. Glyoxysomes are found only in the tissues in which the fat is stored (endosperm or cotyledons, depending on the species) and are present only for the relatively short period of time required for the seedling to deplete its supply of stored fat. *Once they fulfill their role in the seedlings, the glyoxysomes are converted to peroxisomes*. Glyoxysomes have been reported to appear again in the senescing (aging) tissues of some plant species, presumably to degrade lipids derived from the membranes of the senescent cells.

3. Other kinds of plant peroxisomes. A specialized kind of peroxisomes is found to occur in **nodules**, the structures on plant roots in which plant cells and certain bacteria cooperate in the fixation of atmospheric nitrogen (that is, the conversion of nitrogen (N_2) into organic form). The peroxisomes in these cells are involved in the processing of fixed nitrogen.

66.6 BIOGENESIS OF PEROXISOMES

Although most peroxins are synthesized on free cytosolic ribosomes and then imported to peroxisomes. Recent experiments indicate that peroxin assembly begins on the rough ER, where two peroxins, Pex 3 and Pex 19, initially localise (Fig. 66.3). Pex 3 is an integral transmembrane protein while Pex 19 is a fernesylated protein found largely in the cytosol (Box 66.2). Pex 3 recruits Pex 19 to the ER membrane, where their interaction causes Pex 3/Pex 19-containing vesicles to bud off the ER. These vesicles may then fuse either with pre-existing peroxisomes or with one another to form new peroxisomes.

Box 66.2 Farnesylated protein

Lipid-anchored proteins are located on one of the surfaces of lipid bilayer but are covalently bond to lipid molecules embedded within the bilayer. Farnesylated proteins are the *prenylated membrane proteins* which are synthesized as soluble cytosol proteins before being modified by addition of a prenyl group, usually a 15 carbon **farnesyl** group. After attachment, the farnesyl group is inserted into the lipid bilayer of the membrane (see **Becker** *et al.*, 2007).

Pex 3, Pex 19, and other peroxisomal membrane proteins then act as receptors for import of the other peroxins, which are translated on free cytosolic ribosomes and then transported into peroxisomes as completed and folded polypeptides. They are targeted to the interior of peroxisomes by at least two pathways, which are conserved from yeasts to human.

Most peroxins are targeted by the simple amino acid sequence Ser-Lys-Leu at their carboxyl terminus (**peroxisome targeting signal 1 or PTS1**). A small number of peroxins are targeted by a sequence of nine amino acids at their amino terminus (**peroxisome targeting signal 2** or **PTS2**). PTS1 and PTS2 are recognized by distinct cytosolic receptors and then passed through a poorly understood channel in the peroxisomal membrane into the matrix. The receptors are subsequently retrieved from the peroxisome and recycled. Unlike the translocation of polypeptide chains across the membranes of the endoplasmic reticulum, mitochondria and chloroplasts, the targeting signals are usually not cleaved during the import of proteins into peroxisomes and the mechanism of translocation is not known (see **Cooper** and **Hausman** 2007).

Protein import together with the continuing addition of lipids from the rough ER, results in peroxisome growth, and new peroxisomes can be formed by division of old ones. In addition, peroxisomes undergo a complex **maturation process** that involves the import of different classes of proteins from the cytosol at different times. As a result, the enzyme content, and the metabolic activities of peroxisomes may change as they mature.

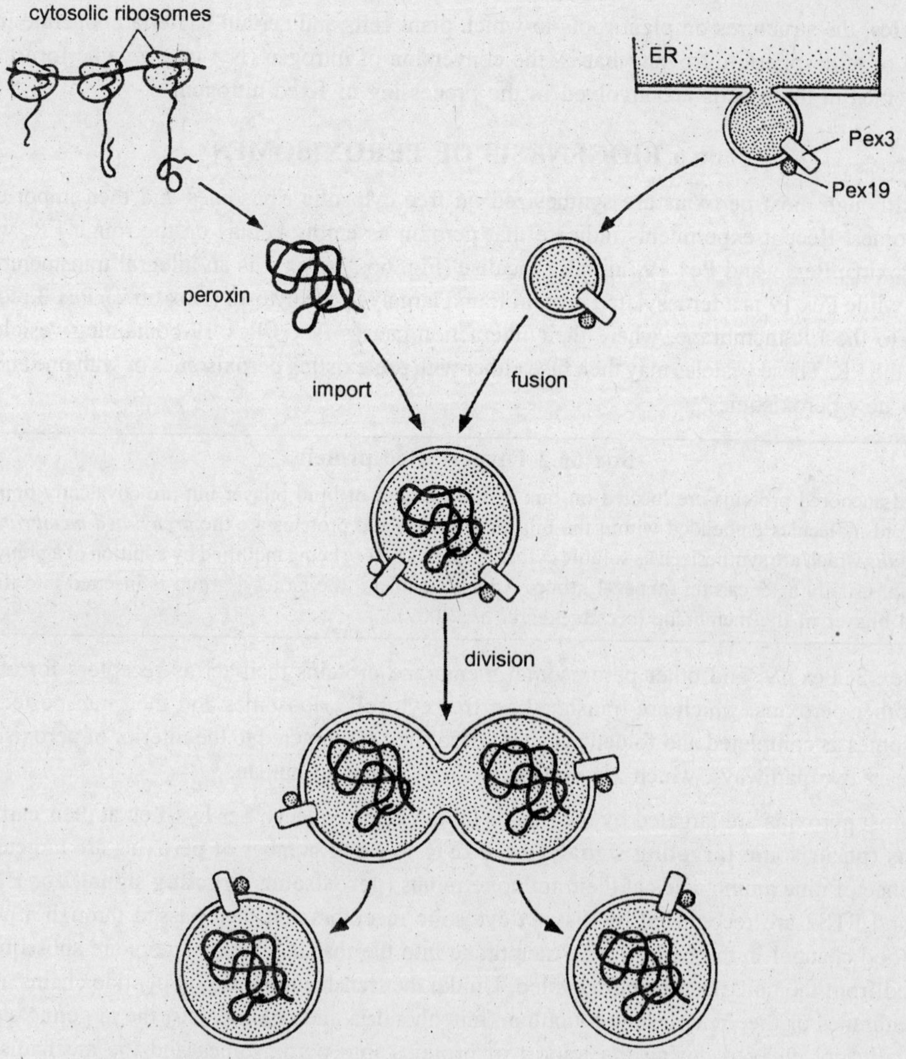

Fig. 66.3. Assembly of peroxisomes. Initiation of peroxisome assembly begins in the rough ER when the transmembrane protein peroxin (Pex 3) recruits the soluble farnesylated protein peroxin (Pex 19) and initiate budding of a nascent peroxisome. The nascent peroxisomes fuse either with each other or with existing peroxisomes. Additional peroxins are synthesized on free cytosolic ribosomes and imported as completed polypeptide chains to form functional peroxisomes, which grow larger and divide (after Cooper and Hausman, 2007).

QUESTIONS

Long Answer Questions

1. Write an account of structure, synthesis, function and histochemical localization of the catalase enzyme in eukaryotic cells.
2. Describe the mode of biogenesis and various functions of peroxisomes in animal cells.

Short Answer Questions

1. What would be the most likely role of peroxisome in cells lacking mitochondria?
2. What is the function of the glyoxylate cycle?
3. What is photorespiration and how do peroxisomes contribute?
4. Compare peroxisomes with mitochondria.
5. Give an account of DAB reaction.
6. Give an account of functions of plant's peroxisomes.

Very Short Answer Questions

1. What are microbodies?

   Ans. Microbody is an early term for a peroxisome based on its appearance in electron micrographs.

2. Define the terms peroxisome and glyoxysome.

   Ans. *Peroxisome.* It is a sungle membrane-bounded organelle that contains catalase and one or more hydroxide-generating oxidases and is therefore involved in the metabolism of hydrogen peroxide.

   Glyoxysome. This is a specialized type of plant peroxisome that contains some of the enzymes responsible for conversion of stored fat to carbohydrate in germinating seeds.

3. Which enzyme is found in peroxisomes for breakdown of H_2O_2? *(H.N.B., Garhwal 2009)*

   Ans. Catalase

Multiple Choice Questions

1. Peroxisomes are rich in
   (*a*) catalytic enzymes (*b*) DNA
   (*c*) RNA (*d*) oxidative enzymes
2. The substrate of photorespiration is
   (*a*) glycolate (*b*) glucose
   (*c*) pyruvic acid (*d*) acetyl coA
3. The function of glyoxysomes is
   (*a*) protein metabolism (*b*) carbohydrate metabolism
   (*c*) fat metabolism (*d*) protein synthesis
4. Peroxisomes contains
   (*a*) synthetase (*b*) catalase
   (*c*) hydrolase (*d*) lyase.

ANSWERS

Multiple Choice Questions

1. (*d*) **2.** (*a*) **3.** (*c*) **4.** (*b*).

67

Ribosomes

The ribosomes are small, dense, rounded and granular particles of the ribonucleoprotein. They occur either freely in the matrix of mitochondria, chloroplast and cytoplasm (*i.e.*, cytoplasmic matrix) or remain attached with the membranes of the endoplasmic reticulum and nucleus. They occur in most prokaryotic and eukaryotic cells and are known to provide a scaffold for the ordered interaction of all the molecules involved in protein synthesis.

Box 67.1 Some Historical Facts.

1. **George E. Palade**, in 1952, described the ribosomes.
2. **Philip Siekevitz** and **Paul Zamecnik**, in 1952, showed that radioactive amino acids were incorporated into proteins on ribosomes.
3. In 1953, ribosomes were observed in plant cells (roots of bean) by **Robinson** and **Brown**.
4. Presence of ribosomes in both free and membrane attached form was confirmed in the animal cells by **Palade** and **Siekevitz** by electron microscopy.
5. In 1956, **Palade** detected RNA in ribosomes.
6. In 1958, the term **ribosome** was coined by **R.B. Roberts.**
7. **Tissieres** and **Watson** (1958) isolated *E.coli* ribosomes and showed that they consist of two subunits, 50S and 30S.
8. The first definitive evidence linking the nucleolus with eukaryotic ribosome formation was provided in 1960s by **Robert Perry.**
9. **James A. Lake**, in 1981, suggested the well accepted **asymmetrical model** for 70S ribosome.

67.1 OCCURRENCE

The ribosomes occur in cells, both prokaryotic and eukaryotic cells. In prokaryotic cells the ribosomes often occur freely in the cytoplasm. In eukaryotic cells the ribosomes either occur freely in the cytoplasm or remain attached to the outer surface of the membrane of endoplasmic reticulum. The yeast cells, reticulocytes or lymphocytes, meristamatic plant tissues, embryonic nerve cells and cancerous cells contain large number of ribosomes which often occur freely in the cytoplasmic matrix. The cells in which active protein synthesis takes place, the ribosomes remain attached with the membranes of the endoplasmic reticulum. Such cells are the pancreatic cells, plasma cells, hepatic parenchymal cells, nissl bodies, osteoblasts, serous cells, or the submaxillary gland, chief cells of thc glandular stomach, thyroid cells and mammary gland cells. The cells which synthesize specific proteins for the intracellular utilization and storage often contain large number of free ribosomes. Such cells are the erythroblasts, developing muscle cells, skin and hair.

67.2 TYPES OF RIBOSOMES

Recently according to the size and the sedimentation coefficient (S) two types of ribosomes have been recognised (Fig. 67.1).

1. 70S ribosomes. The 70S ribosomes are comparatively smaller in size and have sedimentation coefficient 70S and the molecular weight 2.7 × 106 daltons. (**Dalton** is the unit of molecular weight (**MW**); one dalton equals the weight of hydrogen atom. For example, a water molecule weighs 18 daltons). They occur in the prokaryotic cells of the blue green algae and bacteria and also in mitochondria and chloroplasts of eukaryotic cells.

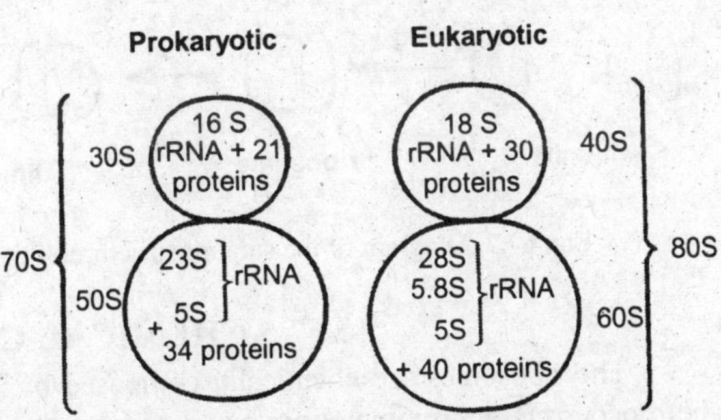

Fig. 67.1. Various components of prokaryotic (70S) and eukaryotic (80S) ribosomal subunits.

2. 80S ribosomes. The 80S ribosomes have the sedimentation coefficient of 80S and the molecular weight 40 × 10^6 daltons. The 80S ribosomes occur in eukaryotic cells of the plants and animals.

The ribosomes of mitochondria and chloroplasts are always smaller than 80S cytoplasmic ribosomes and are comparable to prokaryotic ribosomes in both size and sensitivity to antibiotics, although their sedimentation values vary in different phyla, *e.g.*, 77S in mitochondria of fungi, 77S in mitochondria of mammals and 60S in mitochondria of animals in general. The ribosomes of chloroplasts are 70S type.

67.3 NUMBER OF RIBOSOMES

An *E. coli* cell contains 10,000 ribosomes, forming 25 per cent of the total mass of the bacterial cell. In contrast, mammalian cultured cells contain 10 million ribosomes per cell, each of which is about twice as large as a prokaryotic ribosome.

67.4 STRUCTURE OF RIBOSOMES

The ribosomes are oblate spheroid structures of 150 to 250 Å in diameter. Each ribosome is porous, hydrated and composed of two subunits. One ribosomal subunit is large in size and has a dome-like shape, while the other ribosomal subunit is smaller in size and occurring above the larger subunit and forming a cap-like structure.

The 70S ribosome consists of two subunits, *viz.*, 50S and 30S. The 50S ribosomal subunit is larger in size and has the size of 160 Å to 180 Å. The 30S ribosomal subunit is smaller in size and occurs above the 50S subunit like a cap.

The 80S ribosome also consists of two subunits, *viz.*, 60S and 40S. The 60S ribosomal subunit is dome-shaped and larger in size. In the ribosomes which remain attached with the membranes of endoplasmic reticulum and nucleus, etc., the 60S subunit remains attached with the membranes. The 40S ribosomal subunit is smaller in size and occurs above the 60S subunit forming a cap-like structure. Both the subunits remain separated by a narrow cleft (Fig. 67.1).

The two ribosomal subunits remain united with each other due to high concentration of the Mg^{++} (.001M) ions. When the concentration of Mg^{++} ions reduces in the matrix, both ribosomal subunits get separated. Actually in bacterial cells the two subunits are found to occur freely in the cytoplasm and they unite only during the process of protein synthesis. At high concentration of Mg^{++} ions in the matrix, the two ribosomes (called **monosomes**) become associated with each other and known as the **dimer**. Further, during protein synthesis many ribosomes are aggregated due to common messenger RNA and form the **polyribosomes** or **polysomes** (Fig. 67.2).

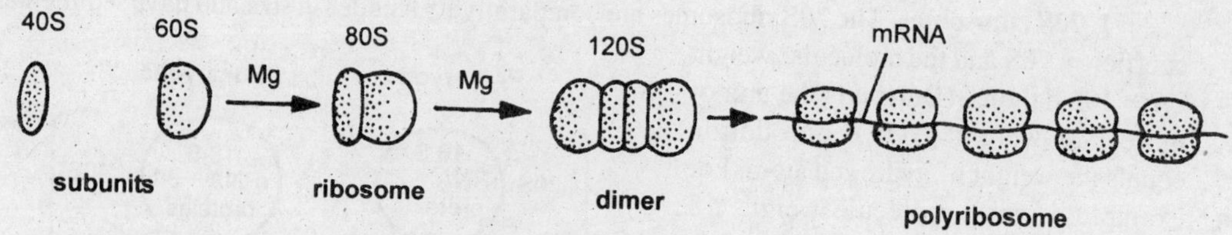

Fig. 67.2. Diagram of the subunit structure of the ribosome and the influence of Mg^{++} ions.

67.5 CHEMICAL COMPOSITION

The ribosomes are chemically composed of RNA and proteins as their major constituents; both occurring approximately in equal proportions in smaller as well as larger subunit. However, the 70S ribosomes contain more RNA (60 to 40%) than the proteins (36 to 37%), *e.g.*, the ribosomes of *E. coli* contain 63% rRNA and 37% protein. While the 80S ribosomes contain less RNA (40 to 44%) than the proteins (60 to 56%), *e.g.*, yeast ribosomes have 40 to 44% RNA and 60 to 56% proteins; ribosomes of pea seedling contain 40% RNA and 60% proteins. There is no lipid content in ribosomes.

1. Ribosomal RNAs

The 70S ribosomes contain three types of rRNA, *viz.*, **23S rRNA, 16S rRNA, 5S rRNA**. The 23S and 5S rRNA occur in the larger 50S ribosomal subunit, while the 16S rRNA occurs in the smaller 30S ribosomal subunit. Assuming an average molecular weight for one nucleotide to be 330 daltons, one can calculate the total number of each type of rRNA. Thus, the 23S rRNA consists of 3300 nucleotides, 16S rRNA contains 1650 nucleotides and 5S rRNA includes 120 nucleotides in it.

The 80S ribosomes contain four types of rRNA, *viz.*, **28S rRNA** (or **25-26 rRNA** in plants, fungi and protozoa), **18S rRNA, 5S rRNA** and **5.8S rRNA**. The 28S, 5S and 5.8S rRNAs occur in the larger 60S ribosomal subunit, while the 18S rRNA occurs in the smaller 40S ribosomal subunit. About 60 per cent of the rRNA is helical (*i.e.*, double stranded) and contains paired bases. These double stranded regions are due to hairpin loops between complimentary regions of the linear molecule.

The 28S rRNA has the molecular weight 1.6×10^6 daltons and its molecule is double stranded and having nitrogen bases in pairs. The 18S rRNA has the molecular weight 0.6×10^6 daltons and consists of 2100 nucleotides. The 18S and 28S ribosomal RNA contain a characteristic number of methyl groups, mostly as 2′-O-methyl ribose. The molecule of 5S rRNA has a clover leaf shape and a length equal to 120 nucleotides. The 5.8S rRNA is intimately associated with the 28S rRNA molecule and has, therefore, been referred to as **28S-associated ribosomal RNA (28S A rRNA).**

The 55S ribosomes of mammalian mitochondria lack 5S rRNA but contain **21S** and **12S rRNAs**. The 21S rRNA occurs in larger or 35S ribosomal subunits, while 12S rRNA occur in smaller or 25S ribosomal subunit.

It is thought that each ribosomal subunit contains a highly folded ribonucleic-acid filament to which the various proteins adhere. But as the ribosomes easily bind the basic dyes so it is concluded that RNA is exposed at the surface of the ribosomal subunits, and the protein is assumed to be in the interior in relation to non-helical part of the RNA.

2. Ribosomal Proteins

Each 70S ribosome of *E. coli* is composed of about 55 **ribosomal proteins**. Out of these 55 proteins, about 21 different molecules have been isolated from the 30S ribosomal subunit, and some 32 to 34 proteins from the 50S ribosomal subunit. The primary structure of several of these proteins has been elucidated. Most of the recent knowledge about the structure of ribosomal proteins has been achieved by dissociation of ribosomal subunits into their component rRNA and protein molecules. When both 50S and 30S ribosomal subunits are dissociated by centrifuging both of them in a gradient of 5M cesium chloride, then there are two inactive core particles (40S and 23S, respectively) which contain the RNA and some proteins called **core proteins** (**CP**) at the same time several other proteins—the so-called **split proteins** (**SP**) are released from each particle (Fig. 67.3). There are SP50 and SP30 proteins which may reconstitute the functional ribosomal subunit when added to their corresponding core. Some of the split proteins are apparently specific for each ribosomal subunit. The split proteins have been further fractionated and divided into acidic (A) and basic (B) proteins. **Nomura** *et al.*, (1968) fractionated at least six different groups of proteins in the ribosome (Fig. 67.3).

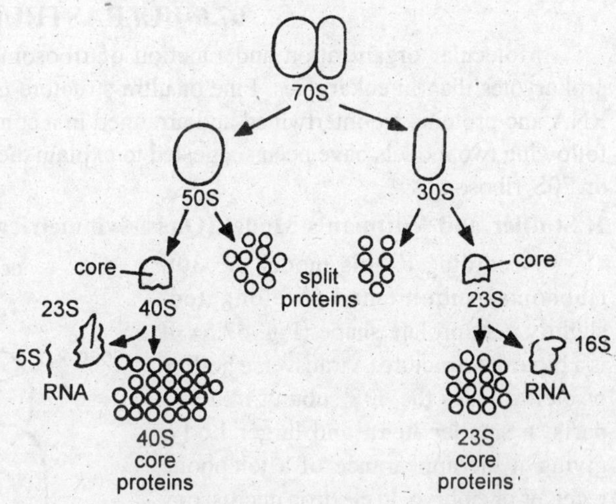

Fig. 67.3. Diagram showing the stepwise dismantling of the two subunits of 70S ribosome. Note that the proteins may be separated into split and core proteins. The 50S subunit contains 23S and 5S RNAs, and the 30S subunit has 16S RNA.

In all, 21 types of proteins have been isolated in smaller subunit (30S) of ribosome of *E. coli.* These are designated as S1 to S21. Similarly, in larger subunit (50S) 34 different proteins designated as L1 to L34, have been isolated. Thus, the 70S ribosome was thought to consist of 55 different proteins. However it was later shown that protein S20 is identical to L6, thus, the correct number of S proteins is 20. Likewise, L8 was shown to be an aggregate of proteins L7, L12 and L10; thus, the correct number of L proteins is 33. Thus, the prokaryotic 70S ribosome consists of 53 different proteins (20S + 33L = 53 proteins). Similar organization of ribosomal proteins and RNA is found in 80S ribosomes (Fig. 67.4). Small 40S subunit comprises 33 proteins and large 60S subunit contains more than 45 proteins (*i.e.,* 50 proteins; **Twyman** 1998).

Different rRNA molecules evidently play a central role in the catalytic activities of ribosomes in the process of protein synthesis. Various ribosomal proteins have been found to mainly enhance the catalytic function of the rRNA in the ribosomes.

3. Metallic Ions. The most important low molecular weight components of ribosomes are the divalent metallic ions such as Mg^{++}, Ca^{++} and Mn^{++}.

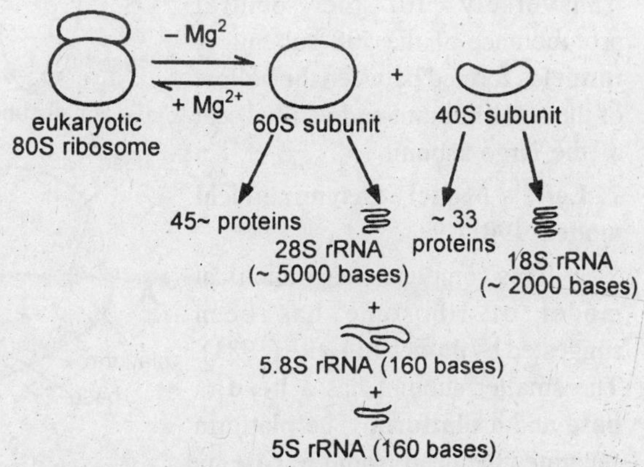

Fig. 67.4. Different RNA and protein components of eukaryotic ribosomes.

67.6 ULTRASTRUCTURE

Molecular organization and function of ribosomes have been studied more intensively in prokaryotes than in eukaryotes. Fine or ultra-structure of 70S ribosome is very complex. In it, the RNA and proteins are intertwined and arranged in a complex manner in the two subunits. Recently following two models have been suggested to explain the three-dimensional structure of prokaryotic or 70S ribosomes:

1. Stoffler and Wittman's Model (Quasi-symmetrical model, 1977)

According to this model the 30S ribosomal subunit has an elongated, slightly bent prolate shape (Fig. 67.5). It is a bipartite structure. A transverse hollow or cleft divides the 30S subunit into two parts, a smaller **head** and larger **body**, giving it the appearance of a telephone receiver or embryo. In electron microscopy 50S ribosomal subunit showed various shapes depending on structure seen in different views such as frontal-maple leaf, lateral-kidney shaped or rear view-rounded. In a frontal view, the 50S subunit appears bilaterally symmetrical and shows three protuberances arising from a rounded base (maple leaf structure). The **central protuberance** being the most prominent.

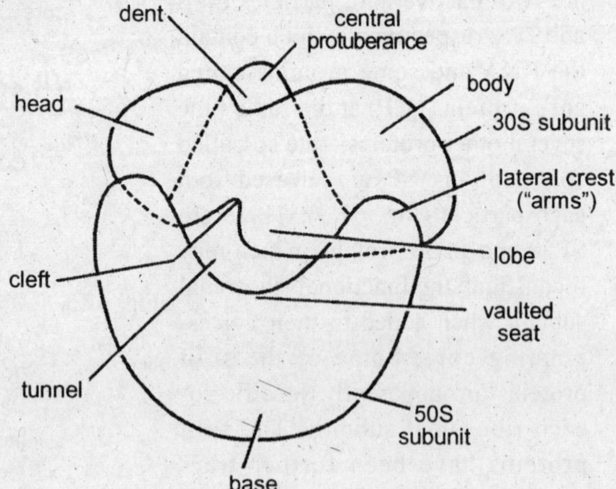

Fig. 67.5. Stoffler and Wittman's model of 70S ribosomes.

The 50S subunit is often compared with an armchair, with the rounded base forming a **vaulted seat**, the central protuberance forming the **back** and the lateral protuberances the **arms** of chair.

When 30S and the 50S subunits become associated to form the 70S ribosome, the frontal face of the 30S subunit with its hollow faces the vaulted seat of the 50S subunit. The long axis of 30S subunit is oriented transversely to the central protuberance of the 50S subunit. A **tunnel** is formed between the hollow of the small subunit and vaulted seat of the large subunit.

2. Lake's Model (Asymmetrical model, 1981)

This completely asymmetrical model of ribosome has been suggested by **James A. Lake** (1981). The smaller subunit has a **head**, a **base** and a **platform**. The platform separates the head from the base by the help of a **cleft**. This cleft is an important functional region; it is

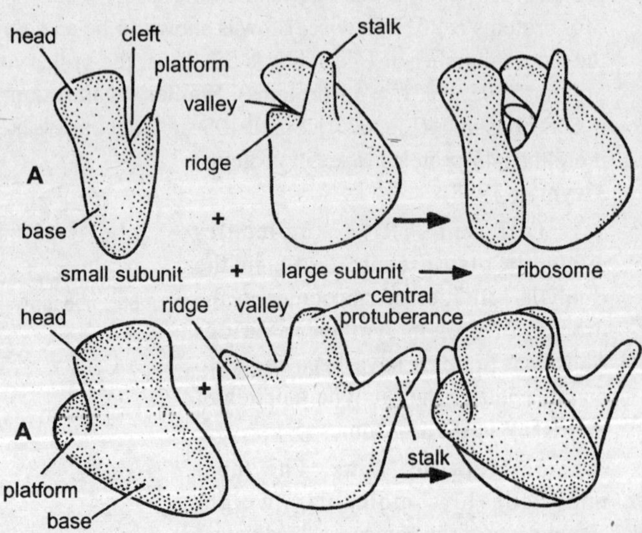

Fig. 67.6. Lake's model of the 70S ribosome. Note the three dimensional structure of the ribosome in two different orientations.

suggested to be the site of codon-anticode interaction and as a part of binding site for initiation factors of protein synthesis.

The large subunit consists of a **ridge**, a **central protuberance** and a **stalk**. The ridge and the central protuberance are separated with the help of a valley (Fig. 67.6).

67.7 THREE DIMENSIONAL MODEL OF 80S RIBOSOME

In spite of the difference in overall sizes (as manifested in the greater molecular weights, sedimentation constants, sizes and numbers of rRNAs and proteins), the cytoplasmic ribosomes of eukaryotes (80S) are remarkably similar in morphology to those of prokaryotes. As in 30S subunits of prokaryote ribosomes, the 40S ribosomal subunit of eukaryotes is divided into **head** and **base** segments by a transverse groove (Fig. 67.7). The 60S ribosomal subunit is generally rounder in shape than the small subunit, although its one side is flattened; this is the side that becomes confluent with the small subunit during the formation of the monomer or monosome (*i.e.*, functional 80S ribosome).

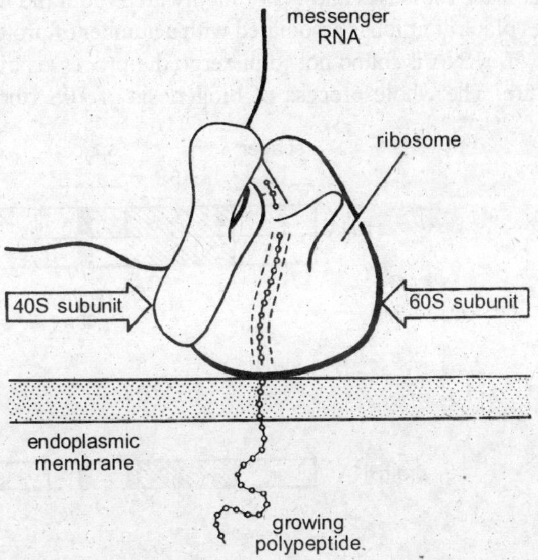

Fig. 67.7. A three dimensional model of eukaryotic ribosome.

Several **domains** of the ribosomes have particularly important functions during protein synthesis. The small subunit contains a binding site for tRNA. The **A-site (amino-acyl-tRNA site)** binds incoming charged tRNAs during elongation, the **P-site (peptidyl-tRNAs site)** binds the tRNA carrying the nascent polypeptide chain. Bacterial 70S ribosomes possess a third **E-site (exit site)** to which spent tRNA are dispatched prior to ejection. The large subunits possess a **peptidyl-transferase domain**, which provides the catalytic activity for peptide bond formation and a GTPase domain whose activities are required for translocation of the ribosome.

67.8 COMPARISON OF 70S AND 80S RIBOSOMES

Eukaryotic 80S ribosomes differ from prokaryotic 70S ribosomes in the following respects: 1. they are considerably larger; 2. they contain a large number of proteins (70–80 types of proteins instead of 53); 3. they have four types of RNA molecules instead of three types; 4. their proteins and nucleic acids are large-sized; 5. the RNA-protein ratio is near to 1 : 1 instead of 2 : 1 and 6. several antibiotics, such as **chloramphenicol**, inhibits bacterial but not eukaryotic ribosomes (this is the basis of the use of many antibiotics in medical treatment). Protein synthesis by eukaryotic ribosomes is inhibited by **cycloheximide**.

However, eukaryotic ribosomes do not differ functionally from those in prokaryotes in a fundamental way; they perform the same functions, by the same set of chemical reactions. The genetic code is the same for all living organisms, and eukaryotic ribosomes are able to translate bacterial mRNAs efficiently, provided that a "cap" is added enzymatically.

67.9 BIOGENESIS OF RIBOSOMES

Ribosomes are not self-replicating particles. Synthesis of various component of ribosomes such as rRNAs and proteins, are under genetic control, *i.e.*, rRNAs and mRNAs (for various ribosomal proteins) are transcribed by genes (DNA). Since the mechanism of biogenesis of 70S and 80S ribosomes differ greatly, so can be studied separately as follows:

1. Biogenesis of 70S Ribosomes

In bacteria a single gene transcript containing the sequences of 16S, 23S and 5S rRNAs, is synthesized by a rRNA operon and this larger molecule is thought to undergo both tailoring and chemical modifications before each rRNA molecule assumes its mature form. During tailoring of larger rRNA molecule, 16S rRNA sequence is first of all cleaved off and is separated from the 23S and 5S sequences. The fragment containing 16S information is still larger than the mature 16S rRNA by at least 100 bases and is not methylated: both the methylation and the tailoring of this molecule takes place after it has associated with a number of proteins to form the precursor ribosomal subunits. The 5S rRNA is found not to undergo the processes of tailoring and methylation before it becomes mature. The whole process of biogenesis of 70S ribosomes takes place in cytoplasm.

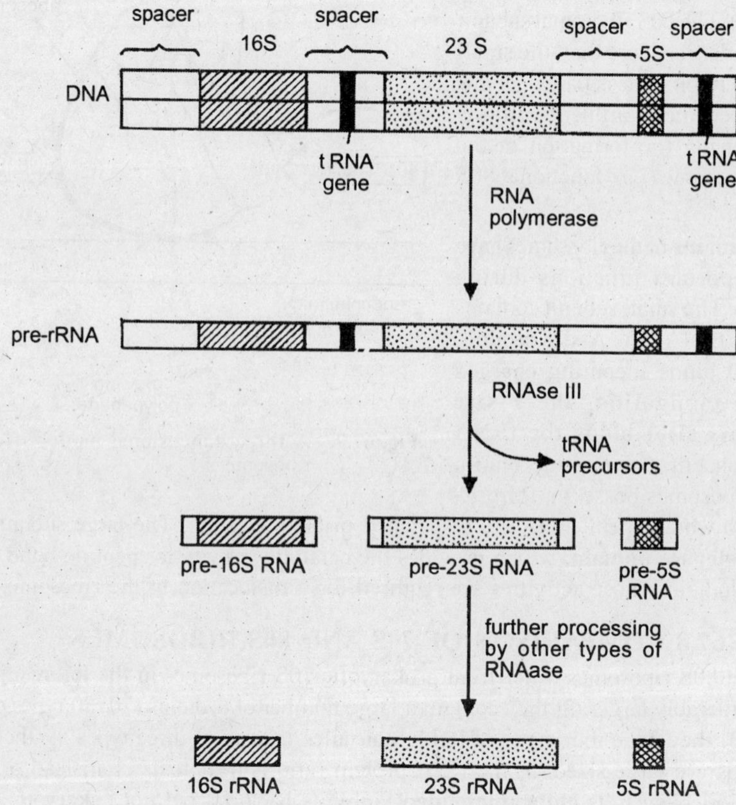

Fig. 67.8. Ribosomal RNA processing in the bacterium *E.coli*.

2. Biogenesis of 80S Ribosomes.

In eukaryotes, the biogenesis of ribosomes is much more complex and involves a long-lasting process in which several regions of cell are involved. The 5.8S, 18S and 28S rRNAs are transcribed as a much larger molecule in the **nucleolar organizer (NO)** which contains many copies 5.8S, 18S and 28S rRNA genes or **ribosomal DNA** (*i.e.*, there is gene redundancy or amplification). The DNA coding for the 5S rRNA is also highly repetitive, but the molecule is synthesized outside the nucleolus. It is in the nucleolus that newly synthesized rRNA accumulates and becomes associated, presumably by a self-assembly process, with 50 or more ribosomal proteins that have been synthesized in the cytoplasm by usual mechanism of protein synthesis and then migrate to the cytoplasm of cell, in the form of ribosomal subunits.

A. Ribosomal RNA synthesis inside nucleolus. All organisms have multiple rRNA genes. In case of *Xenopus*, each nucleolar organizer contains 450 rRNA genes. These genes are **tandemly repeated** or **reiterated** along the DNA molecule (in a head to tail arrangement) and are separated from each other by stretches of **spacer DNA,** which is not transcribed. These rRNA genes are being actively transcribed and the nascent RNA chains are spread perpendicularly to the DNA axis. Each gene is transcribed into a long RNA molecule (which varies in size from 40S to 45S according to species) which will eventually be processed to give rise to 18S, 28S and 5.8S rRNA. Because each rRNA gene has a fixed initiation site (promoter) and a fixed termination site, the transcripts adopt the characteristic "christmas tree" or "fern leaf" configuration (Fig. 67.8). Nucleolar rRNA genes are transcribed by **RNA polymerase I** (about 100 enzymes per gene). RNA polymerase I molecules are found to remain bound to the nucleolar organisers during mitotic metaphase and anaphase. During this period there is no RNA synthesis and so the enzyme molecules must remain in an inactive state.

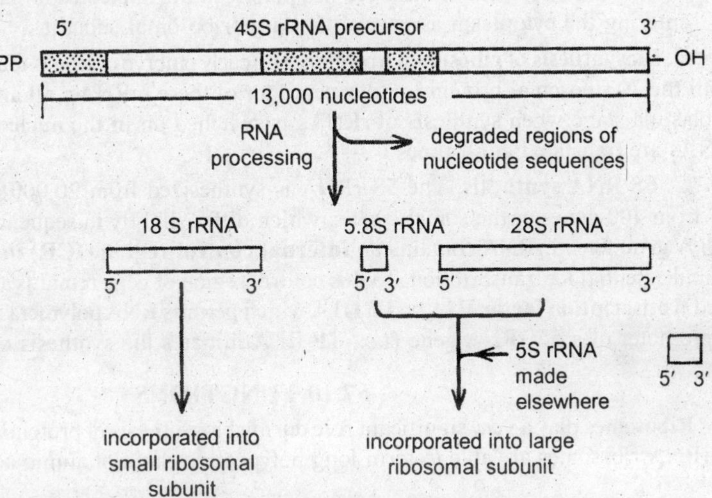

Fig. 67.9. The pattern of processing of 45S rRNA precursor molecule into three separate ribosomal RNAs.

Processing of rRNAs inside nucleolus. As already described, rRNA genes are transcribed into a long precursor RNA (which is 40S in *Xenopus* and 45S in HeLa and other human cells); this precursor must be cleaved into 18S, 28S and 5.8 rRNA. In the cleavage process about 50 per cent of the precursor RNA is degraded. In HeLa cell, the processing of rRNA involves the following steps:

(*i*) The first ribosomal RNA in HeLa cells is a large 45S molecule of 14,000 nucleotides. Within this precursor molecule the rRNAs are separated by stretches of spacer RNA and the order of transcription is: 5′ end—18S—5.8S—28S—3′ end. On a fully active gene about 100 RNA polymerase I enzymes (along with transcription factor I or TFI) are transcribing simultaneously on the rRNA gene.

(*ii*) In nucleolus, 45S RNA is rapidly **methylated**, even before transcription is completed. Methylations occur mostly on the ribose moiety (producing 2′-O- methyribose) and occur only in the 18S (46 methylations) and 28S (71 methylations) sequences that have to be conserved. Those segments of 45S which have to be degraded are not methylated.

(*iii*) 45S RNA has a lifetime of about 15 minutes and is then cleaved into smaller components as follows:

$$45S \longrightarrow 41S \begin{cases} \nearrow 32S \longrightarrow 28S + 5.8S \\ \searrow 20S \longrightarrow 18S \end{cases}$$

(*iv*) 20S RNA is rapidly processed in 18S rRNA and probably due to this reason the small ribosomal subunits appear in the cytoplasm earlier than the large ribosomal subunits. The large ribosomal subunits have a slower RNA processing.

(*v*) 32S RNA remains in the nucleolus for about 40 minutes and is then cleaved into 28S rRNA and 5.8S rRNA. Both of these rRNAs persist in the nucleolus for another 30 minutes before entering the cytoplasm as part of the large ribosomal subunit.

B. Biosynthesis of ribosomal proteins. The early embryo of frog is found to contain the mRNAs for all the 70 ribosomal proteins, and except four of these mRNAs, all are not translated until the midblastula stage when synthesis of rRNAs is switched on in the nucleolus. The remaining four mRNAs are translated at all times.

C. 5S RNA synthesis. The 5S rRNA is synthesized from 20,000 genes in the oocytes but only from 400 genes in the somatic cells, which differ slightly in sequence (six nucleotides out of 120). A gene for 5S rRNA contains an **internal control region (ICR)** in its middle region which is found essential for transcription. To this control region of gene remains attached a special protein, called **transcription factor IIIA** or **TF III A** which permits RNA polymerase III enzyme to recognize the promoter of a 5S rRNA gene (*i.e.*, TF III A initiates the synthesis of 5S rRNA).

67.10 FUNCTIONS

Ribosomes play a very significant role during biosynthesis of proteins. With the help of mRNA and rRNA ribosomes are able to form long polypeptide chain of amino acids from the cytoplasm.

67.11 DISTINCTION BETWEEN LYSOSOMES AND RIBOSOMES

Ribosomes are the granules of protein and RNA present in the cytoplasm. They lack any limiting membrane and are often attached to endoplasmic reticulum. They are involved in protein synthesis.

Lysosomes are cytoplasmic organelles (they are limited by a unit membrane) containing about 50 digestive enzymes for intracellular digestion of bacteria and foreign bodies which enter the cells by the process of phagocytosis or pinocytosis. They also remove worn out cellular organelles from the cells. Lysosomes are not found in prokaryotic cells.

QUESTIONS

Long Answer Questions

1. Describe the structure, biogenesis and function of ribosomes.
(*Bengal 1989, 1992; Garhwal 1990; Gorakhpur 1991, 1994; Guwahati (G) 1993, 1996*)
2. Define ribosome. Give an account of the ultrastructure, chemistry and functioning of ribosomes. Explain their role in protein synthesis.
(*Bihar 1989, 1992; Kanpur 1992; Burdwan 1992; CCS 1991, 1994, 1997*)
3. Describe the types and roles of ribosomes. (*Punjab (G) 1989, 1996*)
4. Give an account of structure and functions of eukaryotic ribosomes (*Allahabad 1989*)
5. Describe the ultrastructure of ribosomes and discuss the function of nucleolus as a ribosome factory.
(*B.U. Coimbatore 1997*)
6. Write an essay on ribosomes. (*Bihar 2000*)

Short Answer Questions

1. Write short notes on the following:
(*i*) Ribosomes. (*Magadh (G) 1991; Purvanchal 1992; Guwahati (Major) 1992, 1996; Gorakhpur 1993, 1996; Guwahati (G) 1994; Magadh (Subs.) 1994; Kanpur 1990, 1994, 1995, 1997; Lucknow 1994, 1995, 1997, 1998; CCS 1995, 1996*)

(*ii*) Polysomes. (*Lucknow 1996*)

(*iii*) Ultrastructure of ribosome. (*CCS 1990, 1993*)

(*iv*) Ribosome biogenesis. (*Guwahati (G) 1992*)

(*v*) Types of ribosomes. (*Punjab (G) 1991, 1992*)

(*vi*) Different types of RNA. (*Punjab 1992*)

(*vii*) 80S subunit of ribosome. (*Jiwaji 1992*)

2. Distinguish between ribosome and lysosome. (*Madras 1989, 1992; Guwahati (M) 1994, 1996*)
3. Define ribosome. Describe Lake's asymmetrical model of ribosome. (*CCSU Meerut 2002*)
4. Prepare a well-laballed diagram of J.A. Lakes model of 70S ribosome. (*Gorakhpur 1995*)

Very Short Answer Question

1. Define the ribosome.

   Ans. Ribosome is a small particle composed of rRNA and protein that function as the site of protein synthesis in the cytoplasm of prokaryotes and in the cytoplasm, mitochondria and chloroplasts of eukaryotes, it is composed of large and small subunits.

2. What is polyribosome?

   Ans. Polyribosome is a cluster of two or more ribosomes simultaneously translating a single mRNA molecule.

3. What is rRNA?

   Ans. The rRNA is ribosomal RNA which is any of several types of RNA molecules used in the construction of ribosome.

4. What is the main function of ribosome?

   Ans. Protein synthesis.

5. What are the main chemical constituents of ribosome?

   Ans. RNA and proteins.

6. Who gave the asymmetrical model of the ribosome?

   Ans. J.A. Lake in 1981.

7. Name the three parts of smaller subunit of ribosome.

   Ans. Head, base and platform.

Yes or No Answer

(*i*) RNA plays an important role in protein synthesis. (*Guwahati (G) 1993*)

Ans. Yes.

Fill in the Blanks

(*i*) Amino acid are linked together to form polypeptide chain by ....

(*NEHU (Major) 1996*)

(*ii*) Ribosome is the site of .... (*NEHU (Major) 1996*)

(*iii*) Ribosome were first observed by ....

(*iv*) The term ribosome was used by ....

(*v*) The principal site of ribosomal RNA in the cell is the ....

(*vi*) During protein synthesis mRNA gets attached to .... subunit of ribosome.

(*vii*) Ribosome are formed in .... as .... and .... subunits.

(*viii*) The bacterial ribosomes are formed of .... and .... subunits.

(*ix*) Ribosomal RNA of large subunit is synthesized as .... RNA.

Multiple Choice Questions

Pick the correct option:

1. Ribosomes help in

(*a*) lipogenesis (*b*) cellular digestion

(*c*) protein synthesis (*d*) photosynthesis

2. The sedimentation constant of eukaryotic ribosome is generally 80S. What is the sedimentation constant of its subunits?

(*a*) 40S and 30S (*b*) 50S and 30S

(*c*) 60S and 30S (*d*) 60S and 40S

3. Proteins are synthesized by

(*a*) ribosomes (*b*) mitochondria

(*c*) Golgi apparatus (*d*) mesosomes

4. Ribosomes on hydrolysis produce

(*a*) acidic proteins and rRNA (*b*) basic proteins and rRNA

(*c*) split proteins and rRNA (*d*) proteins, rRNA and DNA

5. If all the ribosomes of a cell are destroyed

(*a*) respiration will not take place (*b*) protein synthesis will not take place

(*c*) photosynthesis will stop (*d*) it will have no effect

6. Cell organelle devoid of DNA is (*Bihar 2004*)

(*a*) ribosomes (*b*) nucleus

(*c*) plastids (*d*) mitochondria

7. Prokaryotic ribosomes are (*Bihar 2004*)

(*a*) 30S (*b*) 50S

(*c*) 70S (*d*) 80S

8. Which of the following are non-membranous organelles?

(*a*) microbodies (*b*) vacuoles

(*c*) ribosomes (*d*) chloroplasts

9. Ribosomes are attached to ER through

(*a*) ribophorins (*b*) rRNA

(*c*) tRNA (*d*) hydrophobic interaction

ANSWERS

Fill in the Blanks

(*i*) Peptide bond (*ii*) Protein synthesis (*iii*) Palade

(*iv*) R.B. Roberts (*v*) Ribosome (*vi*) Small

(*vii*) Nucleolus, ribonucleoproteins, ribosomal (*viii*) 30S, 50S (*ix*) 45S.

Multiple Choice Questions

1. (*c*) **2.** (*d*) **3.** (*a*) **4.** (*c*)

5. (*b*) **6.** (*a*) **7.** (*c*) **8.** (*c*)

9. (*a*).

68

Vesicular Traffic in the Secretory and Endocytic Pathway

Transport vesicles play a central role in the **traffic** or **transport** of molecules between different membrane-enclosed compartments of the secretory pathway. Vesicles are similarly involved in the transport of materials taken up at the cell surface (*i.e.,* at the level of plasma membrane through endocytosis). **Vesicular traffic (= transport)** is thus a major cellular activity, responsible for molecular traffic between a variety of specific membrane-enclosed compartments. The selectivity of such transport is therefore key in maintaining the functional organization of the cell. For example, lysosomal enzymes must be transported specifically from the Golgi apparatus to lysosomes not to the plasma membrane or to the ER. There exist some signals that target proteins to specific organelles, such as lysosomes. These proteins are transported within vesicles, so the specificity of transport is based on the selective packaging of the intended cargo into vesicles that recognize and fuse only with the appropriate target membrane. Because of the central importance of vesicular transport to the organization of eukaryotic cells, understanding the molecular mechanisms that control vesicle packaging, budding and fusion is a major area of research in cell biology.

Experimental Approaches for Understanding Vesicular Transport.

Progress toward explaining the molecular mechanisms of vesicular transport has been made by three distinct experimental approaches: (*i*) Isolation of yeast mutants that are defective in protein transport and sorting; (*ii*) Reconstitution of vesicular transport in cell-free systems; and (*iii*) Biochemical analysis of synaptic vesicles, which are responsible for the regulated secretion of neurotransmitters by neurons (see **Cooper** and **Hausman**, 2007).

1. In 1980, **Randy Schekman** and his colleagues have pioneered the isolation of yeast mutants defective in vesicular transport. These include mutants that are defective at various stages of protein secretion (*sec* mutants), mutants that are unable to transport proteins to the vacuole, and mutants that are unable to retain resident ER proteins. The isolation of such mutants in yeasts led directly to the molecular cloning and analysis of corresponding genes, thereby identifying a number of proteins involved in various steps of the secretory pathway. For example, the protein Sec 61 forms a major component of the protein acting as translocation channel in RER.
2. Biochemical studies of vesicular transport using reconstituted systems have complemented these genetic studies and have enabled the direct isolation of proteins from mammalian cells. The first cell-free transport system was developed in 1984 by **James Rothman** and his colleagues, who analysed protein transport between compartments of the Golgi apparatus. Similar reconstituted systems have been developed to analyse transport between other compartments, including transport from ER to Golgi apparatus and transport from the Golgi

apparatus to secretory vesicles, vacuoles and the plasma membrane. The development of these *in vitro* systems has enabled biochemical studies of the transport process and functional analysis of proteins identified by mutations in years, as well as direct isolation of some of the proteins involved in vesicles budding and fusion.

3. Insights into the molecular mechanisms of vesicular support have come from studies of synaptic transmission in neurons, which represents a highly specialised form of regulated secretion. A **synapse** is the junction of a neuron with another cell, which may be either another neuron or an effector, such as muscle cell. Information is transmitted across the synapse by chemical neurotransmitters to be released into the synapse and stimulating the post-synaptic neuron of effector cell. Synaptic vesicles are extremely abundant in the brain, allowing them to be purified in large amounts for biochemical analysis. Some of the proteins isolated from synaptic vesicles are closely related to proteins that have been shown to play critical roles in vesicular transport by yeast genetics and reconstitution experiments, so biochemical analysis of these proteins has provided important insights into the molecular mechanisms of vesicle fusion.

Lastly, recent studies using GFP fusion proteins (Box 68.1) have allowed transport vesicles carrying specific proteins to be visualized by immunofluorescence as they move through the secretory pathway. In these experiments, cells are transfected with cDNA constructs encoding secretory proteins tagged with green fluorescent protein (GFP). The progress of the GFP-labelled proteins through the secretory pathway can then be followed in living cells, allowing characterisation of many aspects of the dynamics and molecular interactions involved in vesicular transport.

Box 68.1

GFP or green fluorescent protein is derived from the pacific jellyfish, *Aequoria victoria.*

68.1 SECRETORY PATHWAYS

Integral to the vesicular traffic shown in Figure 68.1 are **secretory pathways** by which proteins move from the ER through the Golgi apparatus to **secretory vesicles** and **secretory granules**, which then discharge their contents to the exterior of the cell. The collective roles of the ER and the Golgi apparatus in secretion were first demonstrated by **Georg Palade** and his colleagues, who used autoradiography to trace the fate of radioactively labelled protein through cells in the salivary gland.

In eukaryotic cells, mode of secretion may be of two types: 1. Constitutive secretion; and 2. Regulated secretion.

1. Constitutive secretion.

After budding from the TGN (trans Golgi network), some secretory vesicles move directly to the cell surface where they immediately fuse with the plasma membrane and release their contents. This process, which is continuous and independent of specific extracellular signals, occurs in most eukaryotic cells and is called **constitutive secretion**. Examples include the continuous release of mucus by cells that line human intestine and the secretion of the glycoproteins of the extracellular matrix.

Default pathway model. Constitutive secretion was once assumed to be a **default pathway** for proteins synthesized by ribosomes attached to the rough ER (see **Becker** *et al.*, 2006). According to this model, all proteins destined for transport from the ER to the Golgi apparatus must have a tag that diverts them from constitutive secretion. Unless an amino acid sequence of oligosaccharide chain identifies a protein for retention in, or transport to, a specific organelle, the protein moves through the endomembrane system and is released outside the cell by default. Support for this model came from studies in which the **retrieval tags** on ER-specific proteins were removed and the fates of the proteins were traced. Removal of the KDEL sequence from resident protein of the ER generally led to secretion of the protein. More recent evidence, however, points to a variety of short amino acid sequence, that identify proteins for constitutive secretion by one of several distinct pathways for the process.

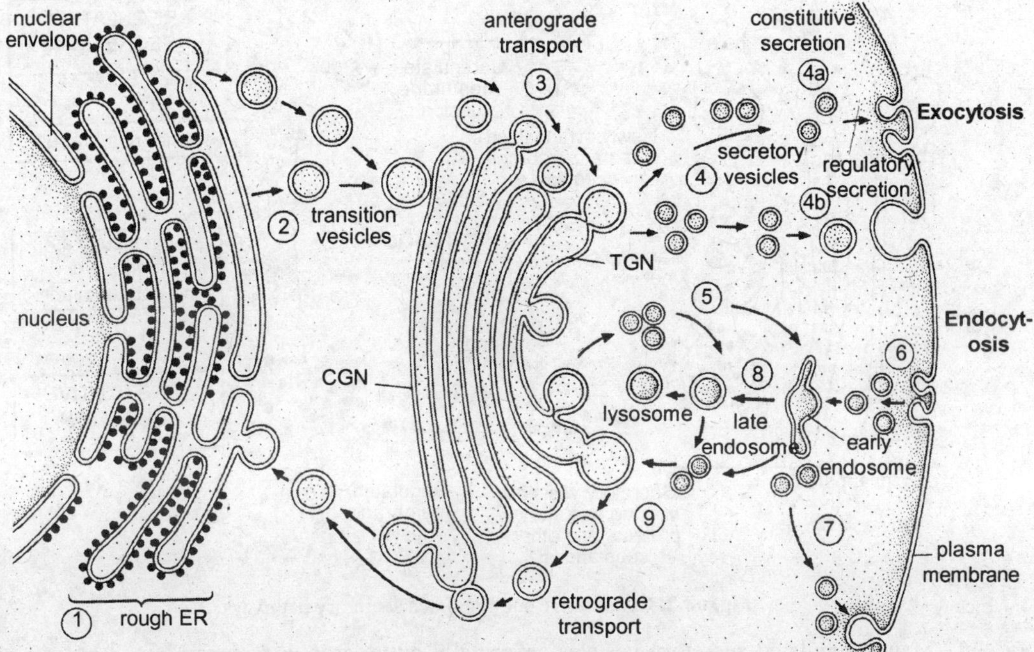

Fig. 68.1. Trafficking through the endomembrane system. Vesicles carry lipids and proteins along several routes from the ER through Golgi apparatus to various destinations, including secretory vesicles, endosomes, and lysosomes. 1. Proteins are synthesized by ribosome attached to the cytosolic surface of the rough ER. Initial glycosylation steps occur within the ER lumen. 2. Transition vesicles carry newly synthesized lipids and glycosylated proteins to the CGN. 3. Lipids and proteins move through the cisternae of the Golgi stack via shuttle-vesicles or as cisternae mature. At the TGN, vesicles bud off to form secretory vesicles (4) or endosomes (5), depending on their protein content. Secretory vesicles move to the plasma membrane, where they release their contents by exocytosis, either constitutively (4a) or in response to an appropriate signal (4b). 6. Protein and other materials are taken into the cell by endocytosis, forming endocytic vesicles that fuse with early endosomes. 7. Cellular components not destined for digestion following endocytosis are recycled to the plasma membrane. 8. Early endosomes containing material for digestion mature to form late endosomes and then lysosomes. 9. Retrograde traffic returns compartment-specific proteins to earlier compartments (after Becker *et al.*, 2007).

2. Regulated Secretion

While vesicles containing constitutively secreted proteins move continuously and directly from the TGN to the plasma membrane, secretory vesicles involved in **regulated secretion** accumulate in the cell and then fuse with the plasma membrane only in response to specific extracellular signals. An important example of regulated secretion is the release of neurotransmitters. Two additional examples of regulated secretion are the release of insulin from the β cells of the pancreatic islets of Langerhans and the release of digestive enzymes from acinar cells of the pancreas.

Regulated secretory vesicles form by budding from the TGN as immature secretory vesicles, which lose their clathrin coats and undergo a maturation process. Maturation involves concentration of the proteins-referred to as **condensation** and frequently also the proteolytic processing of secretory proteins. The mature secretory vesicles then move close to the site of secretion and remain near the plasma membrane, until receiving the signal that triggers release of their contents by exocytosis.

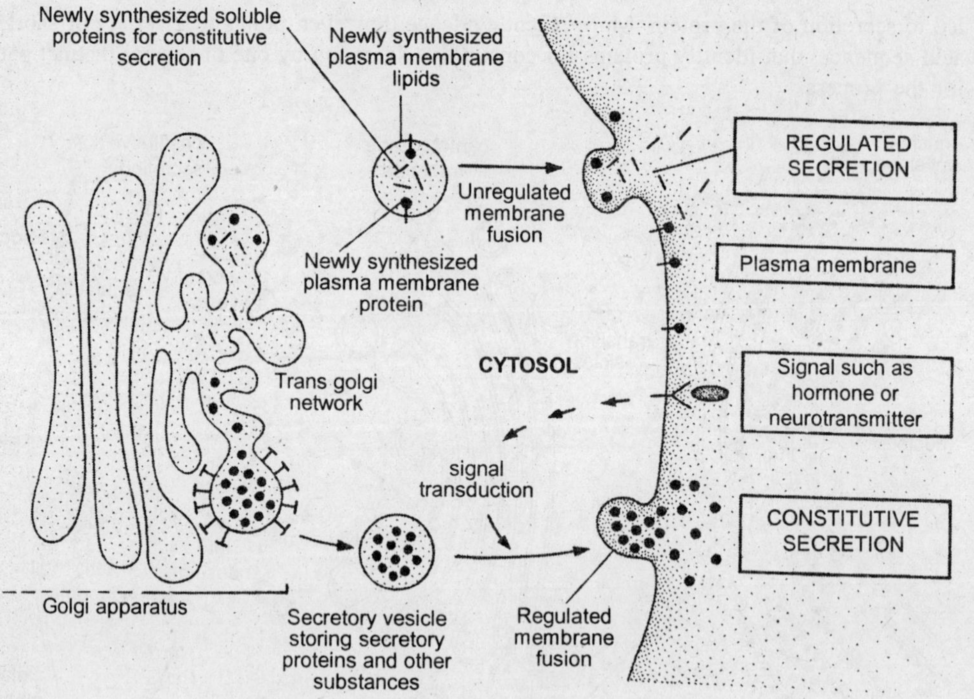

Fig. 68.2. Constitutive and regulated secretory pathways.

Mature regulated secretory vesicles are usually quite large and contain much more highly concentrated proteins than the constitutive secretory vesicles. Such large dense vesicles are often called **secretory** or **zymogen granules** to distinguish them from other secretory vesicles. For example, in an acinar (secretory) cell of exocrine pancreas of the rat zymogen granules (Z, G) are concentrated in the region of the cell between the Golgi stacks from which they arise and the portion of the plasma membrane bordering the lumen into which contents of the granules are eventually discharged.

The information needed to direct a protein to a regulated secretory vesicle is presumably inherent in the amino acid sequence of the protein, though the precise signals and mechanisms for **sorting proteins** to secretory vesicles are not clear. Current evidence suggests that high concentrations of secretory proteins in secretory granules promote the formation of large **protein aggregates** that exclude non secretory proteins. This could occur in the TGN, where only aggregates would be packaged in vesicles destined for secretory granules, or it could occur in the secretory granules itself. The pH of the TGN and the secretory granule lumens may serve as a trigger favouring aggregation as material leaves the TGN. The soluble proteins that do not become part of an aggregate in the TGN or a secretory granules would be carried by transport vesicle to other locations.

68.2 TRANSPORTING MATERIAL ACROSS THE PLASMA MEMBRANE (BULK TRANSPORT BY THE PLASMA MEMBRANE)

Cells routinely import and export large molecules across the plasma membrane. Macromolecules are secreted out from the cell by **exocytosis** and are ingested into the cell from outside through **phagocytosis** and **endocytosis**.

1. Exocytosis.

It is also called **emeiocytosis** and **cell vomiting**. In all eukaryotic cells, **secretory vesicles** are continually carrying new plasma membrane and cellular secretions such as proteins, lipids and carbohydrates (*e.g.*, cellulose) from the Golgi apparatus to the plasma membrane or to cell exterior by the process of exocytosis. The protein to be secreted are synthesized on the rough endoplasmic reticulum (RER). They pass into the lumen of the ER, glycosidated and are transported to the Golgi apparatus by ER-derived **transport vesicles.** In the Golgi apparatus the proteins are modified, concentrated, further glycosidated, sorted and finally packaged into vesicles that pinch off from trans Golgi tubules and migrate to plasma membrane to fuse with it and release the secretion to cell's exterior. In contrast, small molecules to be secreted (*e.g.*, histamine by the mast cells) are actively transported from the cytosol (where they are synthesized on the free ribosomes) into preformed vesicles, where they are complexed to specific macro-molecules (*e.g.*, a network of proteoglycans, in case of histamine), so that, they can be stored at high concentration without generating an excessive osmotic gradient.

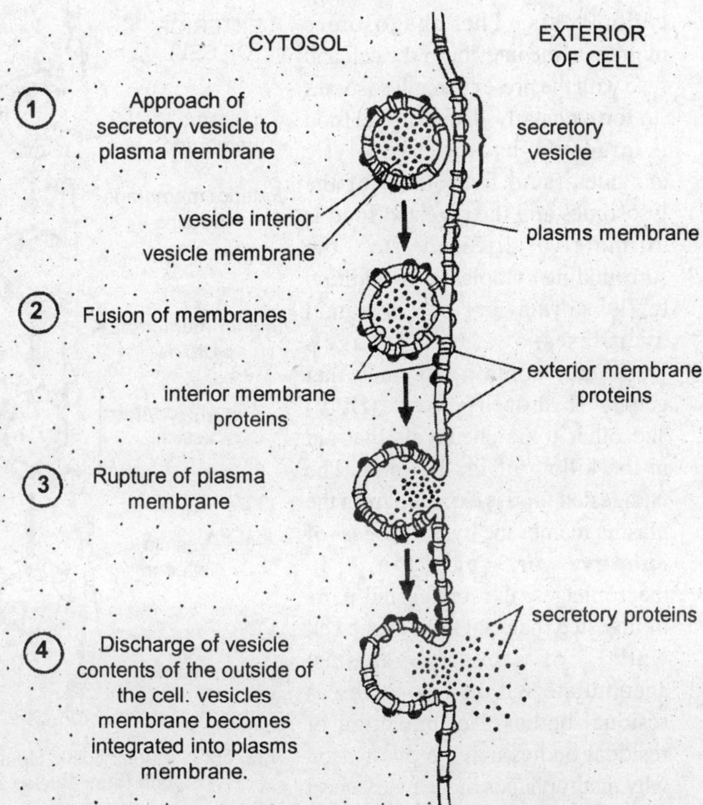

Fig. 68.3. Exocytosis. Release of the contents of a secretory vesicle or granule to the exterior of the cell (after Becker *et al.*, 2007).

During exocytosis the vesicle membrane is incorporated into the plasma membrane. The amount of secretory vesicle membrane that is temporarily added to the plasma membrane can be enormous: in a pancreatic acinar cell discharging digestive enzymes, about 900 μm^2 of vesicle membrane is inserted into the apical plasma membrane (whose area is only 30 μm^3) when the cell is stimulated to secrete.

2. Phagocytosis

Sometimes the large-sized solid food or foreign particles are taken in by the cell through the plasma membrane. The process of ingestion of large-sized solid substances (*e.g.*, bacteria and parts of broken cells) by the cell is known as **phagocytosis** (Gr., *phagein*=to eat, *kytos*=cell or hollow vessel).

Occurrence of phagocytosis. The process of phagocytosis occurs in most protozoans and certain cells of multicellular organisms. In multicellular organisms such as mammals, the phagocytosis occurs very actively in granular leucocytes and in the cells of mesoblastic origin. The cells of the mesoblastic origin are collectively known as the cells of **macrophagic** or **reticuloendothelial system**. They can ingest bacteria, protozoa, cell debris or even colloidal particles by the process of phagocytosis.

Process of phagocytosis. In phagocytosis, first the target particle is bound, to the specific receptors on the cell's surface (process is called **adsorption**), then the plasma membrane expands along the surface of the particle and eventually engulfs it. Vesicle formed by phagocytosis is called **phagosome** and it is typically 1 to 2 μm or larger in diameter, much larger than those formed during pinocytosis and receptor-mediated endocytosis. The phagosomes migrate to the interior of the cell and fuse with the pre-existing lysosomes (to form phagolysosome). The food is digested by the hydrolytic enzymes (acid hydrolase) of the lysosomes and the digested food is ultimately diffused to the surrounding cytoplasm. In addition to the normal set of lysosomal hydrolases, macrophage's lysosomes contain enzymes that generate hydrogen peroxide (H_2O_2) and other toxic chemicals that aid in the killing of the bacteria. The undigested food is expelled from the plasma membrane by the process of **ephagy** or **egestion**. In macrophages, the undigested parts of ingested material such as the cell walls of microorganisms, accumulate within lysosomes as residual bodies. Accumulation of residual bodies may be one reason why macrophages have a very short life time (*i.e.*, less than a few days).

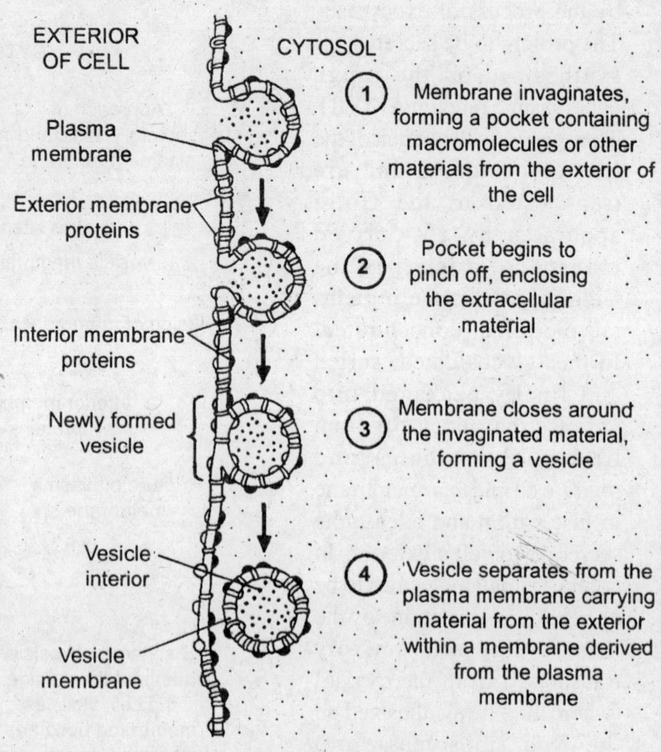

Fig. 68.4. Endocytosis. Uptake of materials from the exterior of the cell (after Becker *et al.*, 2007).

3. Endocytosis

In endocytosis, small regions of the plasma membrane fold inwards or **invaginate** until it has formed new intracellular membrane limited vesicles. In eukaryotes, the following two types of endocytosis can occur: pinocytosis and receptor-mediated endocytosis.

***(i)* Pinocytosis.** Pinocytosis (Gr., *pinein*=to drink; 'cell drinking') is the non-specific uptake of small droplets of extracellular fluid by **endocytic vesicles** or **pinosomes**, having diameter of about 0.1 μm to 0.2 μm. Any material dissolved in the extracellular fluid is internalized in proportion to its concentration in the fluid.

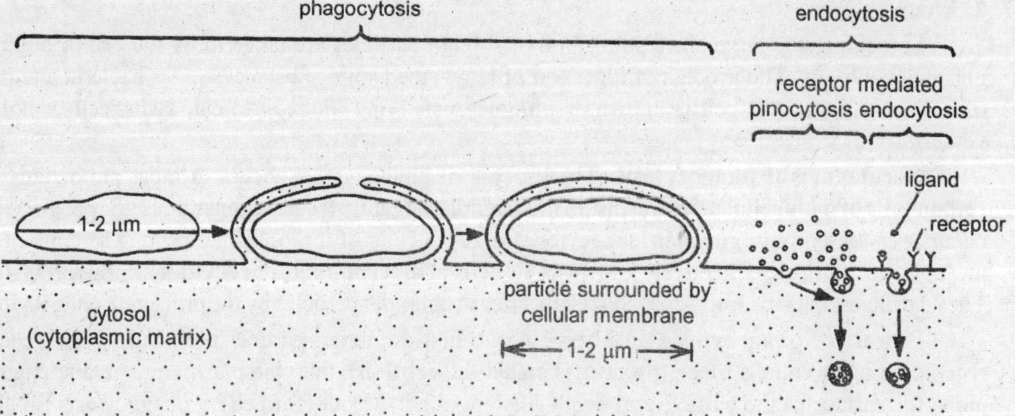

Fig. 68.5. The process of phagocytosis and endocytosis.

The light microscopy has shown that in *Amoeba* tiny **pinocytic channels** are continually being formed at the cell surface by invagination of the plasma membrane. From the inner end of each channel small vacuoles or pinosomes are pinched off, and these move towards the centre of the cell; where they fuse with primary lysosomes, form **food vacuoles**. Ultimately, ingested contents are digested, small breakdown products such as sugars and amino acids diffuse to cytosol.

***(ii)* Receptor-mediated endocytosis.** In this type of endocytosis, a specific receptor on the surface of the plasma membrane "recognizes" an extracellular macromolecule and binds with it. The substance bound with the receptor is called the **ligand.** Examples of ligands may include viruses, small proteins (*e.g.*, insulin, vitellogenin, immunoglobin, transferrin, etc.), vitamin B_{12}, cholesterol containing LDL or low density lipoprotein, oligosaccharide, etc. The region of plasma membrane containing the receptor-ligand complex undergoes endocytosis. The whole process of receptor-mediated endocytosis, includes the following events:

1. Interaction of ligands and cell surface receptors. The macromolecules (ligands) bind to complementary cell-surface receptors. There are more than 25 different types of receptors which are involved in receptor-mediated endocytosis of different types of molecules. Such a receptor is a transmembrane protein which contains two specific binding sites: (1) **ligand-binding site** at the external surface of plasma membrane; and (2) **coated-pit binding site** at the inner or cytosolic face of the plasma membrane.

2. Formation of coated-pits and coated-vesicles. The endocytic cycle begins at specialized regions of the plasma membrane, called **coated-pits.** Coated-pits are depressions of plasma membrane having a coat of bristle-like structure towards their cytosolic side. The ligand-loaded receptors diffuse into these coated-pits. They invaginate into the cell and pinches off to form the **coated-vesicles.** The coat at coated

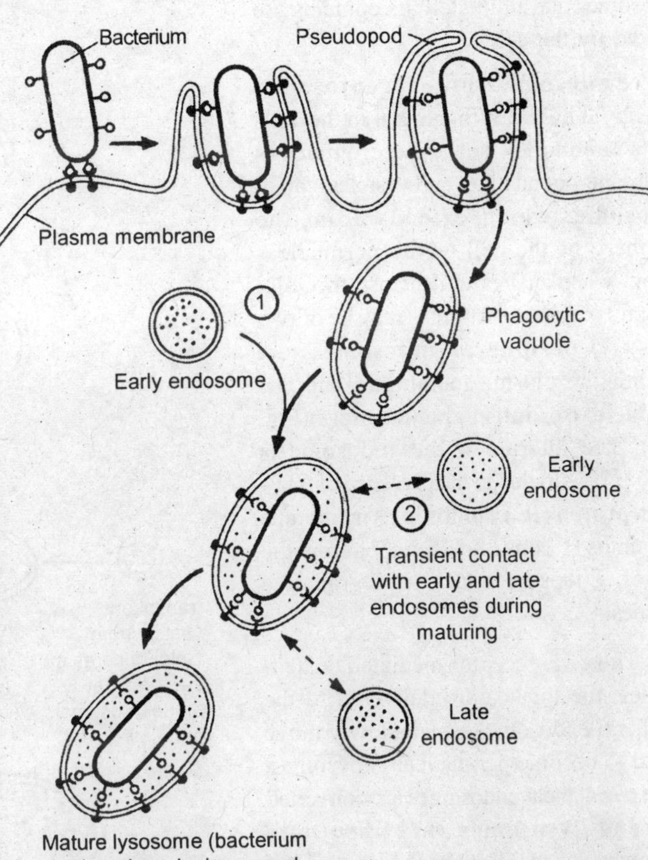

Fig. 68.6. Phagocytosis. Particle or microorganisms bind to receptors on the cell surface, triggering the onset of phagocytosis. In a process involving polymerization of actin, folds of membrane called pseudopods gradually surround the particle. Eventually, the pseudopods meet and engulf the particle, forming a phagocytic vacuole. The vacuole then (1) Fuses with an early endosome or (2) Forms transient connections (indicated by dashed lines) with early and late endosomes and matures into a lysosome, in which digestion of the internalised material occurs. (after Becker *et al.*, 2007).

vesicles is made up of protein, called **clathrin**.

3. Fusion of endocytic vesicle and endosome. Once a coated vesicle is formed, the clathrin and associated proteins dissociate from the vesicle membrane and return to the plasma membrane to form a new coated-pit. The resultant **endocytic vesicle** gets fused with pre-existing endosomes and ultimately its contents are utilized by the cell.

Endosome or receptosome. Recently it has been found that in the cells exists a complex set of heterogeneous membrane-bound tubes and vesicles, called **endosome**, which extends from the periphery of the cell to the perinuclear region, where it lies quite close to Golgi apparatus. Thus, endosomes may be of two types: (*i*) **peripheral endosomes** just beneath the plasma membrane and (*ii*) **perinuclear** or **internal endosomes** (Fig. 68.7). The interior of the endosome is acidic (pH 5–6) due to the presence of ATP-driven **proton (H^+) pumps** in its membrane that pumps H^+ ions into the lumen from the cytosol. Endosomes lack in degradative enzymes.

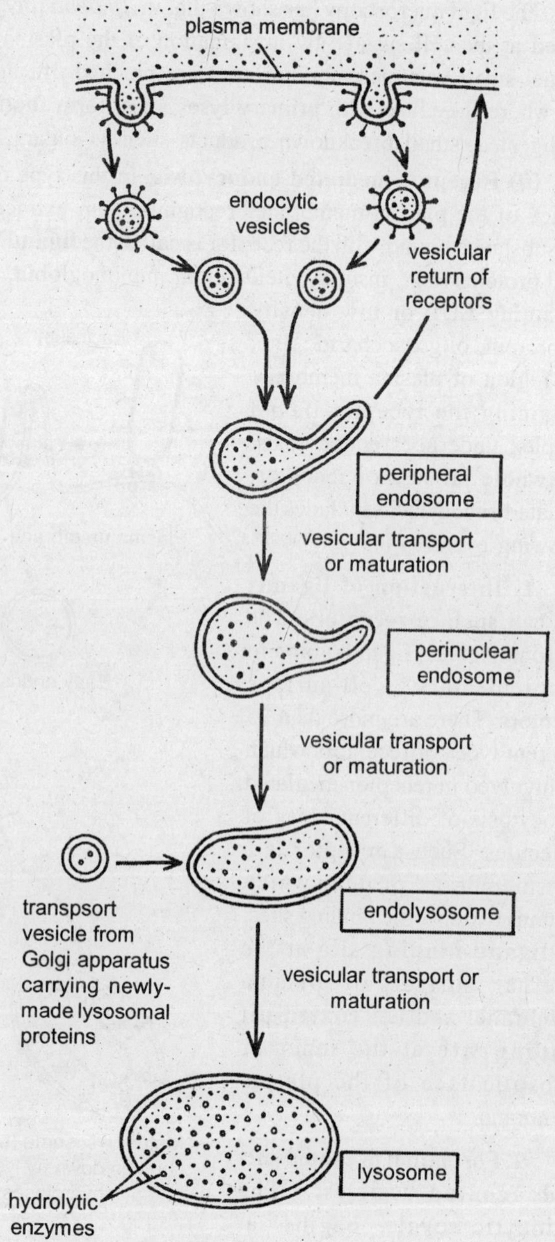

Fig. 68.7. An endocytic pathway: ligands which are endocytosed via coated pits, are delivered to the peripheral endosomes and then move sequentially to perinuclear endosomes, endolysosomes and lysosomes.

Thus, via receptor-mediated coated-vesicles, the ligands are delivered to the peripheral endosomes which slowly move inward to become perinuclear endosomes. These perinuclear endosomes are converted into **endolysosomes** and then into **lysosomes** due to following three activities: 1. The fusion of transport vesicles from the Golgi apparatus. (**Note. Transport vesicles** capture a cargo of molecules. *e.g.*, proteins, from the lumen of one compartment as they pinch off from its membrane and then discharge that cargo into another compartment as they fuse with it. Thus, in such vesicular transport, the transported proteins do not cross any membrane and they are transferred from lumen to lumen). 2. Continuous membrane retrieval, and 3. Increased acidification. The endosomal compartment also acts as the **main sorting station** in the endocytic pathway. The acidic environment of the endosome causes dissociation of ligands from their receptors. Such ligands are destined for destruction in the lysosomes

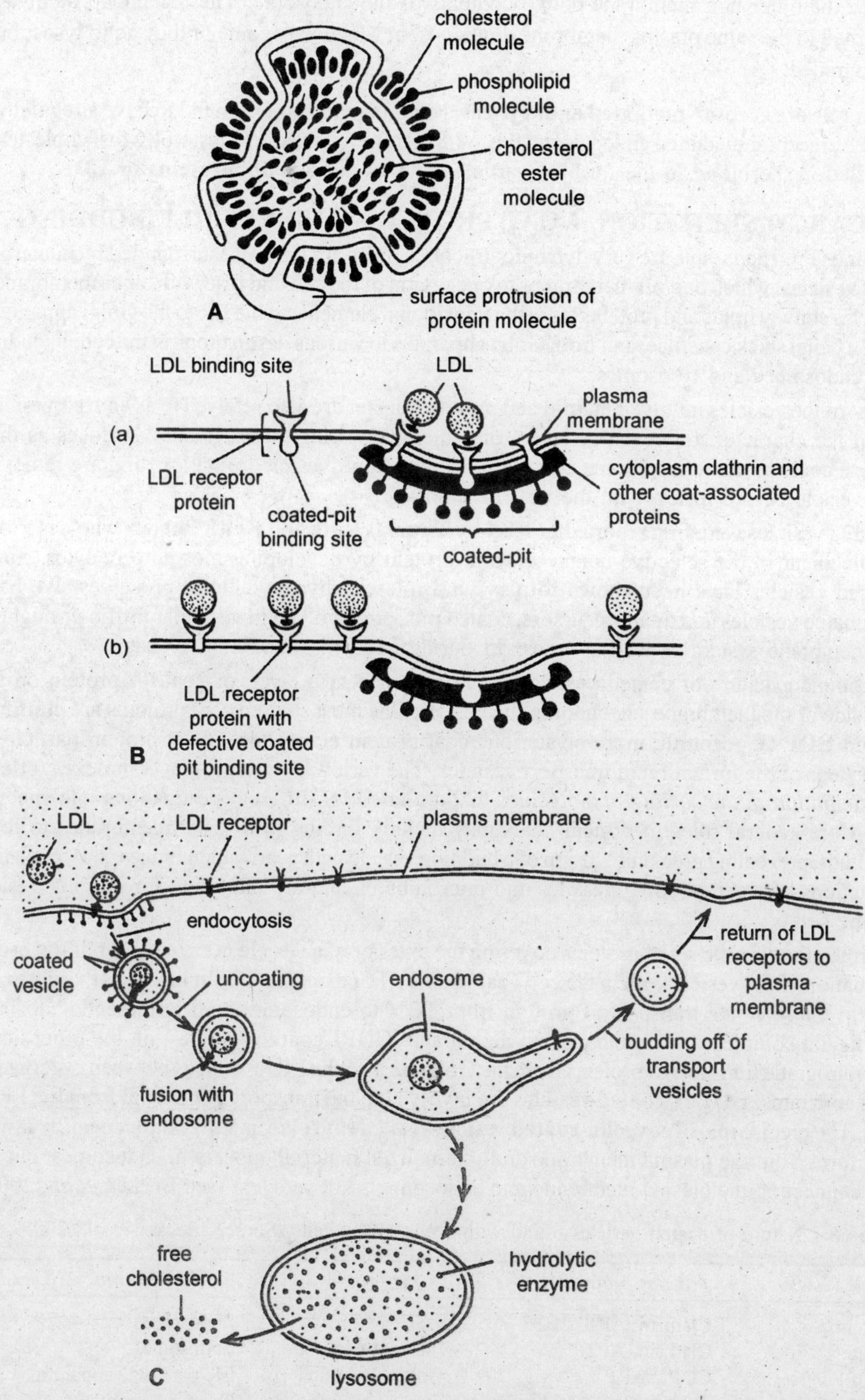

Fig. 68.8. Receptor-mediated endocytosis. A—A LDL particle in cross section; B—LDL receptor proteins binding to a coated-pit in the plasma membrane; C—Mechanism of receptor-mediated endocytosis of LDL particles.

along with the other non membrane-bound contents of the endosome: The receptor-proteins are either returned to the same plasma membrane domain from which they come or they go to lysosomes and are degraded.

Example of receptor-mediated endocytosis. Most animal cells are found to have a regulatory pathway (*i.e.*, receptor mediated endocytosis; Fig. 68.8) for the uptake of cholesterol. Most cholesterol is transported in the blood in the form of particles of **low-density lipoproteins** or **LDL**.

68.3 CARGO SELECTION, COAT PROTEINS AND VESICLE BUDDING

ER and Golgi apparatus are very dynamic structures. Both are typically surrounded by numerous **transport vesicles**, which bud off membranes in one region of the cell and fuse with other membranes. Such vesicles convey lipids and proteins from the transitional elements of the ER to the Golgi apparatus, between the Golgi stack cisternae, and from Golgi apparatus to various destinations in the cell, including granules, endosomes and lysosomes.

Most of the vesicles involved in lipid and protein transfer are also referred to as **coated vesicles** because of the characteristic coats, or layers of proteins covering their cytosolic surfaces as they form. These coats promote membrane curvature, which allows vesicle formation, and are removed from the vesicle before fusion with the target membrane.

Coated vesicles were first reported in 1964 by **Thomas Roth** and **Keith Porter**, who described their involvement in the selective uptake of yolk protein by developing mosquito oocytes. Since then, coated vesicles have been shown to play vital roles in diverse cellular processes. We have seen that coated vesicles and their precursors, coated pits, are involved in vesicular traffic throughout the endomembrane system as well as transport during exocytosis and endocytosis.

A common feature of coated vesicles is the presence of a layer, or coat, or protein on the cytosolic side of the membrane surrounding the vesicle. The most studied coat proteins are **clathrin, COP I** and **COP II**. A fourth, more mysterious coat protein is **caveolin**. Coat protein participate in several steps of the formation of transport vesicles. The variety of coat protein complexes reflects *their participation in the sorting of molecules* that are fated for different destinations into specific vesicles. More general roles of protein coats may include forcing nearly flat membranes to form spherical buds, preventing premature fusion of a budding vesicle with nearby membranes, and regulating the interactions between budding vesicles and microtubules that are important for moving vesicle through the cell.

Further, the specific set of proteins covering the exterior of a vesicle is an indicator of the origin and destination of the vesicle with in the cell (Table 68.1). For example, **clathrin-coated vesicles** are involved in the selective transport of protein from TGN to endosomes, and in the endocytosis of receptor-ligand complexes from the plasma membrane. **COPI-coated vesicles**, on the other hand, facilitate retrograde transport of proteins from the Golgi back to the ER, as well as between cisternae of the Golgi apparatus. **COP II-coated vesicles** are involved in the transport of material from the ER to the Golgi. The precise role of **caveolin-coated vesicles,** called **caveolae**, is still controversial. Transient caveolae form from the plasma membrane and internalized material appears in endosomes, but the pathway connecting the plasma membrane and endosome is not yet clear (see **Becker** *et at.*, 2007).

Table 68.1 Nature of coated vesicles found within eukaryotic cells (Source: Becker *et al.*, 2007).

Coated vesicle	*Coat protein**	*Origin*	*Destination*
1. Clathrin	Clathrin, AP1, ARF	TGN	Endosomes
2. Clathrin	Clathrin, AP2	Plasma membrane	Endosomes
3. COPI	COPI, ARF	Golgi apparatus	ER or Golgi apparatus
4. COP II	COP II (Sec 13/31 and sec 23/24), Sar 1	ER	Golgi apparatus
5. Caveolin	Caveolin	Plasma membrane	?

* *ARF designates ADP ribosylation factor 1; AP1 and AP2 designate different adaptor protein complexes (also called assembly protein complexes).*

1. Clathrin-coated vesicles. Clathrin-coated vesicles are surrounded by coats composed of two multimeric proteins, **clathrin** and **adaptor protein (AP)**. The term *clathrin* comes from *clathratus* the Latin word for a "lattice" as clathrin and AP assemble to form polygonal protein lattices. *Flat clathrin lattices* are composed entirely of hexagons and pentagons.

Clathrin-coated vesicles readily dissociate into soluble clathrin complexes, adaptor protein complexes, and uncoated vesicles. These complexes and vesicles, in turn, can spontaneously reassemble under appropriate conditions. In a slightly acidic solution containing calcium ions, clathrin complexes will even assemble independent of adaptor protein and membrane-bounded vesicles, resulting in empty shells called **clathrin cages.**

(*i*) Components of clathrin lattices. In 1981 **Ernst Ungewickell** and **Daniel Branton** visualised the basic structural units of clathrin lattices, three-legged structures called **triskelions** (Fig. 68.9A). Each triskelion is a multimeric protein composed of three large polypeptides and three small polypeptides radiating from a central vertex. The large polypeptides, each with a molecular weight of 192,000 are **clathrin heavy chains**; they form the legs of triskelion. Each leg is slightly curved near the middle, or knee, and has a globular domain at its outer tip. The small polypeptides, with a molecular weights in the range of 30,000 – 36,000, are **clathrin light chains**. The light chains are found to be associated with the inner half of each leg.

By combining information collected from electron microscopy and X–ray cyrstallography, researchers have assembled a model for the organization of triskelions into the characteristic hexagons and pentagons of clathrin-coated pits and vesicles (Fig. 68.9B). According to one popular model, one clathrin triskelion is located at each vertex of the polygonal lattice, with the knee of the clathrin heavy chains are located at adjacent vertices, each edge of the clathrin lattice is composed of overlapping legs from three triskelions. This arrangement of triskelions into overlapping networks ensures extensive longitudinal contact between clathrin polypeptides. Such contact may impart the mechanical strength needed when a coated vesicle forms from a membrane. The knees, however, do not interact with other polypeptides. This may provide flexibility for forming both hexagons and pentagons and also may accomodate vesicles of different sizes.

(*ii*) Adaptor protein. The second major component of clathrin coats-the **adaptor protein (AP)**– was originally identified simply by its ability to promote the assembly of clathrin coats around vesicles; thus, AP is also called **assembly proteins**. Eukaryotic cells contain at least-four types of AP complexes, each composed of four polypeptides—two adaptin subunits, one medium chain, and one small chain. The four polypeptides, which are highly different in each type of AP complex, bind to different transmembrane receptor proteins and *impart specificity during vesicle budding and targeting*. In addition to ensuring that appropriate macromolecules will be concentrated in coated pits, AP complexes also mediate the attachment of clathrin to the plasma membrane. AP complexes are found to be sites for regulation of clathrin assembly and disassembly. For example, the ability of AP complexes to bind to clathrin is affected by pH, phosphorylation and dephosphorylation.

(*iii*) The assembly of clathrin coats drives the formation of vesicles from the plasma membrane and TGN. The binding of AP complexes to the plasma membrane and the concentration of receptors or receptor-ligand complexes in coated pits require ATP and GTP—though perhaps only for regulation of the process. However, the assembly of clathrin coat around budding vesicles can occur without additional expenditure of ATP and GTP. An exception is the assembly of clathrin coats around vesicles forming from the TGN, which depends on the hydrolysis of nucleotide triphosphates.

The accumulation of clathrin triskelions and the assembly of a clathrin coat on the cytosolic side of a membrane appear to provide part of the **driving force** for formation of a vesicle at the site. In the case of receptor-mediated endocytosis, the assembly of the clathrin coat on the inner side of the plasma membrane causes the membrane to fold inward. Initially, all clathrin units are hexagonal and form a planar, two-dimensional structure. As more clathrin triskelions are incorporated into the growing lattice, a combination of hexagonal and pentagonal units allows the new clathrin coat to curve around the budding vesicle.

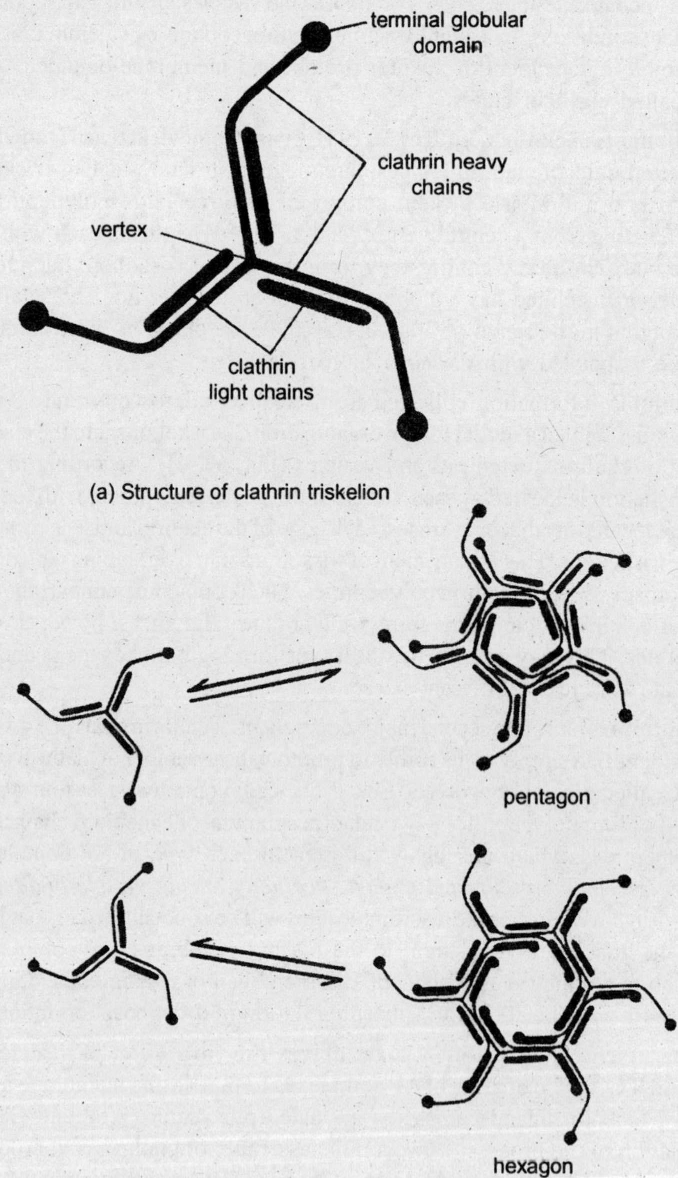

Fig. 68.9. Clathrin triskelion. A—Each triskelion consists of three clathrin heavy chains radiating from a central vertex, with a terminal globular domain at the tip of each triskelion leg and a clathrin light chain bound to the inner half of each leg. B—Under appropriate conditions, clathrin triskelions assemble into the pentagonal and hexagonal structures characteristic of coated pits and vesicles (after Becker *et al.*, 2007).

***(iv)* Sealing of the coated-vesicles.** As clathrin accumulates around the budding vesicle, at least one more protein-**dynamin**–participates in the process. Dynamin is a cytosolic GTPase (enzyme) required for coated pit *constriction and closing of the budding vesicle*. This essential protein was first identified in *Drosophila*. Flies expressing a temperature-sensitive form of dynamin were instantly paralyzed after a temperature shift. Further investigation revealed an accumulation of coated pits in the presynaptic membranes of neuromuscular junctions in affected flies. Binding of GTP probably allows dynamin to form helical rings around the neck of the coated pit. As GTP is hydrolyzed, the dynamin rings tighten and separate the fully sealed endocytic vesicle from the plasma membrane.

***(v)* Uncoating of coated-vesicles.** Uncoating is done in a regulated manner because, in most cases, the clathrin coat remains intact as long as the membrane is part of a coated pit or budding vesicle but dissociate rapidly once the vesicle is fully formed. Like assembly, dissociation of the clathrin coat is an energy-consuming process, accompanied by the hydrolysis of about three ATP molecules per triskelion. At least one protein, an **uncoating ATPase** is essential for this process, though the uncoating ATPase releases only the clathrin triskelion from the APs; the factors responsible for releasing APs from the membrane have not yet been identified.

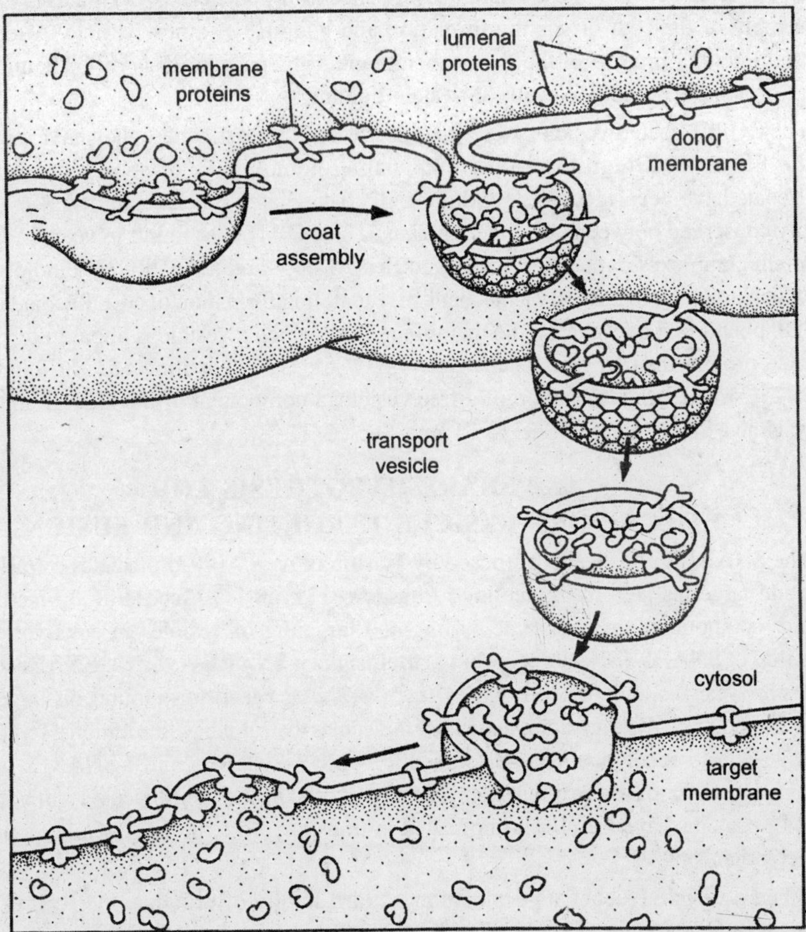

Fig. 68.10. Formation and fusion of a transport vesicle. Membrane and lumenal secretory proteins are collected into selected regions of a donor membrane where the formation of a cytosolic coat results in the budding off of a transport vesicle. During transport the coat is disassembled, and the transport vesicles docks and fuses with the target membrane (after Cooper and Hausman, 2007).

2. Functions of COP I-coated and COP II-coated vesicles. COPI-coated vesicles (COP indicates coat protein) have been found in all eukaryotic cells examined, including mammalian, insect and yeast cells. Such vesicles are surrounded by coats composed of COPI and **ADP ribosylation factor (ARF)**. Viewed by electron microscopy, COPI-coated vesicles do not display polyhedral lattices like those surrounding clathrin-coated vesicles. Instead, they have dense "fuzzy" coats. The major component the coat, COPI, is a protein multimer composed of seven subunits.

COPI-coated vesicles are likely involved in transport from the Golgi apparatus back to the ER as well as between Golgi apparatus cisternae, though their precise role is controversial. Early evidence strongly suggested that COPI-coated vesicle, do not bud from the ER, lending support to the view that COPI-coated vesicles are involved in retrograde transport.

Assembly of a COPI coat is mediated by ARF, which is a small GTP-binding protein. In the cytosol, ARF occurs as part of an ARF-GTP complex. When ARF encounters a specific *guanine nucleotide exchange factor* associated with the membrane from which a new coated vesicle is about to form, however, the GDP is exchanged for GTP. ARF is then able to bind to the membrane by inserting a hydrophobic tail into the lipid bilayer. Once firmly anchored, ARF binds to COPI multimers, and assembly of the coat drives the formation and budding of a new vesicle. After the formation of a free vesicle, a protein in the donor membrane triggers hydrolysis of GTP, and ARF releases the coat proteins for another cycle of vesicle budding.

The COP II-coated vesicles were first discovered in yeast, where they have a role in transport from the ER to the Golgi apparatus. Mammalian homologues of some of the components of COP II coats have been identified, and the COP II-mediated mechanism of ER export appears to be highly conserved between yeast and humans. The COP II coat found in yeast is assembled from two protein complexes-called **Sec 13/31** and **Sec 23/24** – and a small GTP-binding protein called **Sar I**, which is similar to ARF. By a mechanism resembling formation of a COPI coat, a Sar I molecule with GDP bound to it approaches the membrane from which a vesicle is about to form. A peripheral membrane protein then triggers exchange of GTP for GDP, enabling Sar I to bind to Sec 13/31 and Sec 23/24. After the formation of a free vesicle, a component of the COP II coat triggers GTP hydrolysis and Sar I releases Sec 13/31 and Sec 23/24.

68.4 SNARE HYPOTHESIS FOR TRANSPORT VESICLE TARGETING AND FUSION

The **SNARE hypothesis** (proposed by **Rothman** *et al.*, 1993) provides a working model for sorting and targeting step in intracellular transport (Fig. 68.11). According to this hypothesis, the molecular components that facilitate sorting and targeting of vesicles in eukaryotic cells include two families of **SNARE (SNAP receptor) proteins**: the ***v*-SNAREs (vesicle-SNAP receptors)** found on transport vesicles and the ***t*-SNAREs (target-SNAP receptors)** found on target membranes. *V*-SNAREs and *t*-SNAREs are complementary molecules that, along with additional **tethering proteins**, allow a vesicle to recognise a target organelle. Both *v*-SNAREs and *t*-SNAREs were originally investigated because of their role in neuronal exocytosis. Since their discovery in brain tissue, both families of proteins have also been implicated in transport from the ER to Golgi apparatus in yeast and other organisms.

When a vesicle reaches its destination, a third family of proteins, the **Rab GTPases**, come into play. Rab GTPase are also specific (Table 68.2); vesicles fated for different destinations have distinct members of the Rab family associated with them. As shown in Fig. 68.11, the affinity of complementary *v*-SNAREs and *t*-SNAREs for one another forms a four-helix bundle. This ensures that when they collide they will remain in contact long enough for a Rab protein associated with the vesicle to lock the complementary *t*- and *v*-SNARE, together.

Following the vesicle fusion, another set of proteins, including **N-ethylmaleimide-sensitive factor (NSF)** and a group of **SNAPs (soluble NSF attachment proteins)**, mediates release of the *v*- and *t*-SNAREs of the donor and target membranes possibly accompanied by the hydrolysis of

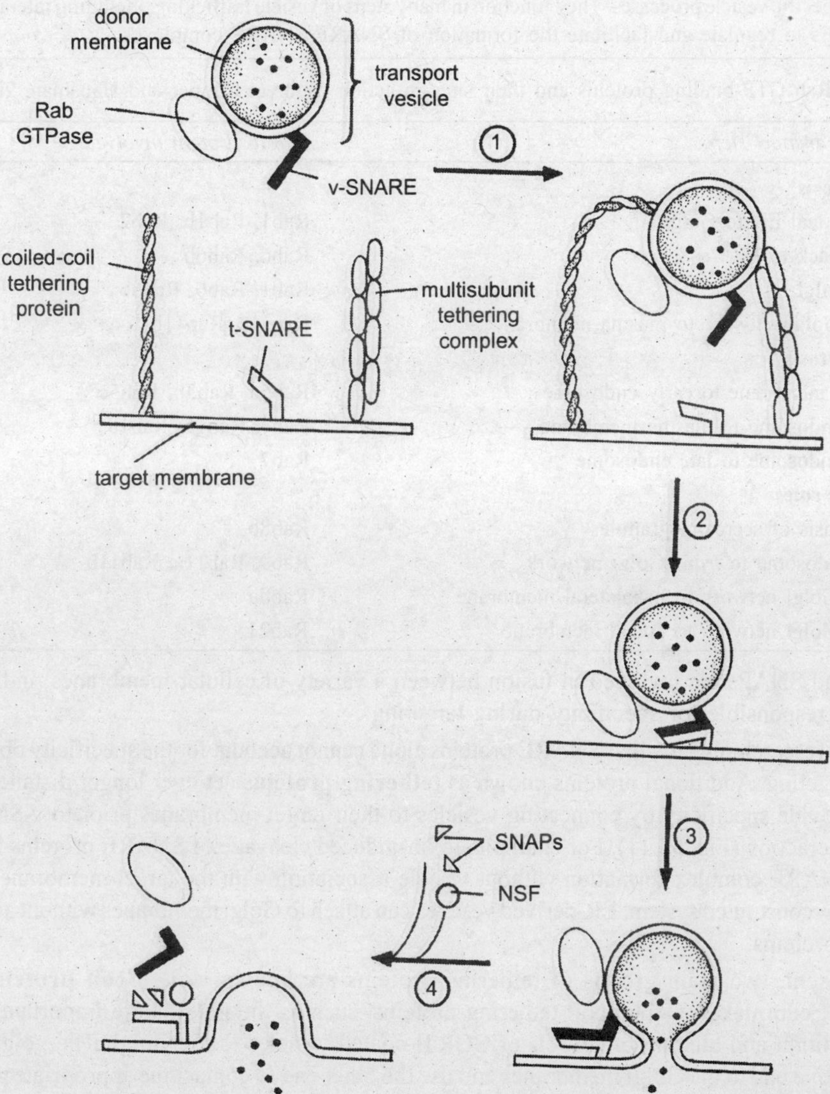

Fig. 68.11. The SNARE hypothesis for transport vesicle targeting and fusion. The basic molecular components that mediate sorting and targeting of vesicles in eukaryotic cells include tethering proteins, *v*-SNAREs on the transport vesicles, t-SNAREs on target membranes, Rab GTPase, NSF, and several SNAPs. 1. The proper vesicle is recognised and bound by particular membrane-anchored tethering proteins: a coiled coil protein and/or a multisubunit tethering complex. 2. A Rab GTPase bound to the incoming vesicle stimulates formation of a stable four-helix bundle composed of one *v*-SNARE helix and three *t*-SNARE helices (the tethering proteins are no longer shown). 3. Fusion of the vesicle membrane with the target membrane is promoted by the *v*-SNARE/*t*-SNARE interaction. 4. Binding of NSF and SNAPs promotes dissociation of SNARE complexes. The exact timing of GTP or ATP hydrolysis is still unclear, but most likely occurs after vesicle fusion (after Becker *et al.*, 2007).

Box 68.2 Rab Proteins

Rab proteins, like the ARF family, participate in many of the vesicle budding and fusion reactions during vesicular transport. More than 60 different mammalian Rab proteins have been identified and shown to function in specific vesicle processes. They function in many steps of vesicle trafficking, including interacting with SNAREs to regulate and facilitate the formation of SNARE/SNARE complexes.

Table 68.2 Rab GTP-binding proteins and their sites in action (Source: Cooper and Hausman, 2007).

Transport step	*Rab protein involved*
I. Exocytosis	
1. Transitional ER to Golgi	Rab1, Rab1b, Rab2
2. Golgi back to ER	Rab6, Rab6b
3. Intra-Golgi	Rab1, Rab6, Rab6b
4. *Trans* Golgi network to plasma membrane	Rab11a, Rab11b
II. Endocytosis	
1. Plasma membrane to early endosome	Rab5a, Rab5b, Rab5c
2. Early endosome to plasma membrane	Rab4, Rab15, Rab18
3. Early endosome to late endosome	Rab7
III. Special roles	
1. Exocytosis of secretory granules	Rab8b
2. Late endosome to *trans* Golgi network	Rab9, Rab11a, Rab11b
3. *Trans* Golgi network to basolateral membrane	Rab8a
4. *Trans* Golgi network to apical membrane	Rab21

ATP. NSF and SNAPs are involved in fusion between a variety of cellular membranes, indicating they are not responsible for specificity during targeting.

Recent research suggests that SNARE proteins alone cannot account for the specificity observed in vesicle targeting. Additional proteins known as **tethering proteins** act over longer distances and appear to provide specificity by connecting vesicles to their target membranes prior to *v*-SNARE/*t*-SNARE interaction (Fig. 68.11). For example, toxin-induced cleavage of SNARE proteins in vivo can block SNARE complex formation without vesicle association with the target membrane. Also, in an in vitro reconstituted system, ER-derived vesicles can attach to Golgi membranes without addition of SNARE proteins.

At present, two main groups of tethering proteins are known: **coiled coil proteins** and **multisubunit complexes**. Coiled-coil tethering proteins, such as the **golgins** are important in the initial recognition and binding of COP I- or COP II-coated vesicles to the Golgi. The golgins are anchored by one end to the Golgi membranes and use the other end to contact the appropriate passing vesicle. These proteins are also important in connecting Golgi cisternae to each other. Antibodies to certain golgian block the action of the golgins and disrupt the structure of the Golgi medial cisternae.

A second class of tethering proteins consists of several families of multisubunit protein complexes containing four to eight or more individual polypeptides. For example, the **exocyst** complex of yeast and mammals is important for protein secretion, binding both to the plasma membrane and to vesicles from the TGN whose contents are destined for export. Other types of multisubunit tethering complexes such as **COG (conserved oligomeric Golgi)** complex, the **GARP (Golgi-associated retrograde protein)** complex and the **TRAPP (transport protein particle)** complex are implicated in the initial recognition and specificity of vesicle-target membrane interaction. Most of these proteins in these complexes are will conserved among organisms, as different as yeast and humans (see **Becker** *et al.*, 2007).

68.5 TRANSPORT OF NEUROTRANSMITTERS ACROSS NERVE SYNAPSES

Nerve cells (neurons) communicate with one another and with glands and muscles at **synapses**, which are of following two types:

1. Electrical synapse. Here, an axon of one neuron, called the **presynaptic neuron**, is connected to the dendrite (s) of another cell, the **postsynaptic neuron**, by gap junctions. As ions move back and forth between the two cells, the depolarisation in one cell spreads passively to the connected cell. Electrical synapses provide for transmission with virtually no delay and occur in places in the nervous system where speed of transmission is critical.

2. Chemical synapse. In this case, the presynaptic and postsynaptic neurons are close to each other but not directly connected (Fig. 68.12). Typically, the presynaptic membrane is separated from the postsynaptic membrane by a gap of about 20–25 nm, known as the **synaptic cleft**. A nerve signal arriving at the terminals of the presynaptic neuron cannot bridge the synaptic cleft as an electrical

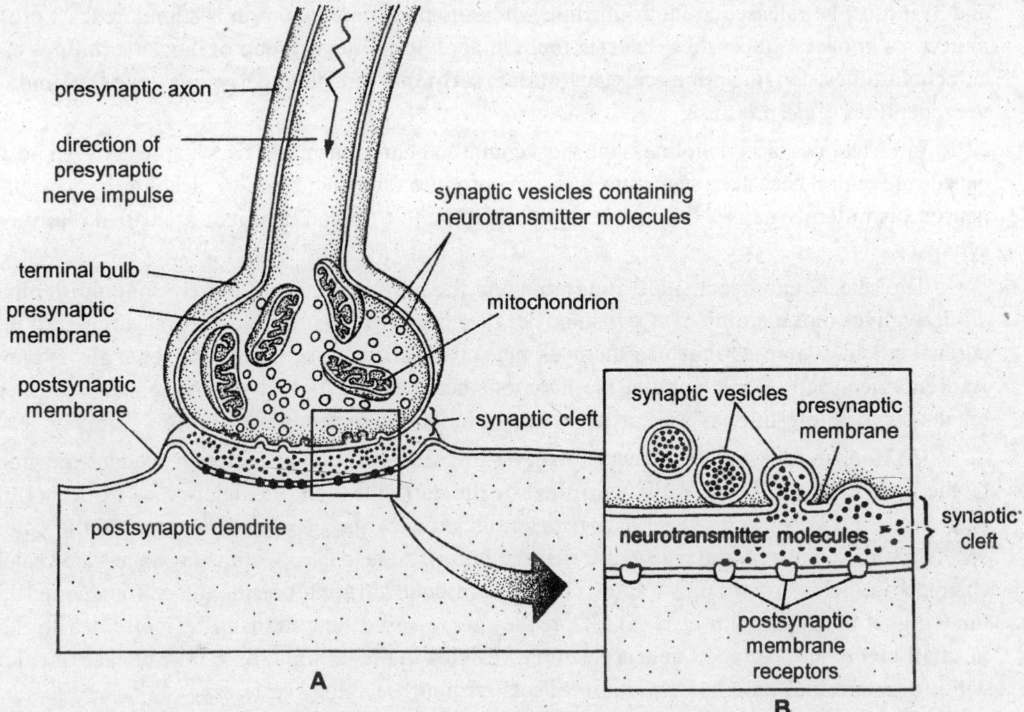

Fig. 68.12. Structure of a chemical synapse. A—When a nerve impulse from the presynaptic axon arrives at the synapse, it causes synaptic vesicle containing neurotransmitters in the terminal bulb to fuse with the presynaptic membrane, releasing their contents into the synaptic cleft. B—Neurotransmitter molecules diffuse across the cleft from the presynaptic (axonal) membrane to the postsynaptic (dendritic) membrane, where they bind to specific membrane receptors and change the polarisation of the membrane, either exciting or inhibiting the postsynaptic cell (after Becker *et al.*. 2007).

impulse. For synaptic transmission to take place, the electrical signal must be converted at the presynaptic neuron to a chemical signal carried by a neurotransmitter. Neurotransmitter molecules are stored in the **terminal bulbs** (also called **synaptic knobs**) of the presynaptic neuron. An action potential arriving at the terminal bulb causes the neurotransmitter to be secreted into and diffuse across the synaptic cleft. The neurotransmitter molecules then bind to specific proteins (called **receptors**) embedded within the plasma membrane of the postsynaptic neuron and are converted back to electrical signals, setting in motion a sequence of events that either stimulates or inhibits the production of an action potential in the postsynaptic neuron, depending on the kind of synapse.

(*i*) Neurotransmitters. A **neurotransmitter** is a small molecule whose function is to bind to a cell surface protein called a **receptor** within the membrane of a postsynaptic neuron. When a neurotransmitter molecule binds to its specific receptor, the properties of receptor are altered and the postsynaptic neurons responds accordingly. An **excitatory neurotransmitter** causes depolarization of the postsynaptic neuron, whereas an **inhibitory neurotransmitter** typically causes the postsynaptic cell to hyperpolarise.

A neurotransmitter should have the following three properties: 1. It must elicit the appropriate response when introduced into the synaptic cleft; 2. it must occur naturally in the presynaptic neuron; and 3. it must be released at the right time when the presynaptic neuron is stimulated. At present, molecules known to meet these criteria include **acetylcholine**; a group of biogenic amines called **catecholamines**; certain amino acids (**histamine**, **serotonin**) and their derivatives (**GABA**) and some neuropeptides (**eukephalins**).

In vertebrates, acetylcholine is the most common neurotransmitter for synapse between neurons outside the central nervous system as well as for neuromuscular junctions. Acetylcholine is an excitatory neurotransmitter. Synapses that use acetylcholine as their neurotransmitter are called **cholinergic synapses**.

The catecholamines include **dopamine** and the hormone **norepinephrine** and **epinephrine**, all derivatives of the amino acid tyrosine. Because the catecholamines are also synthesized in the adrenal gland, synapses that use them as neurotransmitters are termed **adrenergic synapses**. Andrenergic synapses are found at the junctions between nerves and smooth muscles in internal organs such as intestine, as well as nerve-nerve junctions in the brain.

(*ii*) Role of calcium ions in secretion of neurotransmitters. The secretion of neurotransmitters by the presynaptic cell is directly controlled by the concentration of calcium ions in the terminal bulb (Fig. 68.13). Each time an action potential arrives, the depolarization causes the calcium concentration in the terminal bulb to increase temporarily due to the opening of voltage-gated calcium channels in the terminal bulbs (**Note.** The calcium concentration outside the cell is about 10,000 times higher than that of the cytosol). Before they are released, neurotransmitter molecules are stored in small membrane-bounded **neurosecretory vesicles** in the terminal bulb. The release of calcium within the terminal bulb has two main effects on neurosecretory vesicles.

1. Vesicles held in storage are mobilised for rapid release. For any given action potential, only a tiny fraction of the total number of vesicles stored in the terminal release their contents. *Neurons hold vesicles in reserve by linking them to the cytoskeleton so that they cannot move close to the synaptic membrane for secretion.*
2. Calcium ions induced sensitive effect on the neurotransmitter release is the rapid docking and fusion of neurosecretory vesicles with the plasma membrane in the terminal bulb region. During this process, the membrane of a vesicle moves into close contact with the plasma membrane of the axon terminal and then fuses with it to release the content of the vesicle.

(*iii*) Mechanism of docking and exocytosis. For the neurotransmitter to act on a postsynaptic cell, it must be secreted by the process of **exocytosis**. The key event in secretion is the fusion of neurosecretory vesicles with the plasma membrane, which discharges the vesicle contents into the synaptic cleft. The process requires ATP and proceeds through several steps, one of which is calcium dependent.

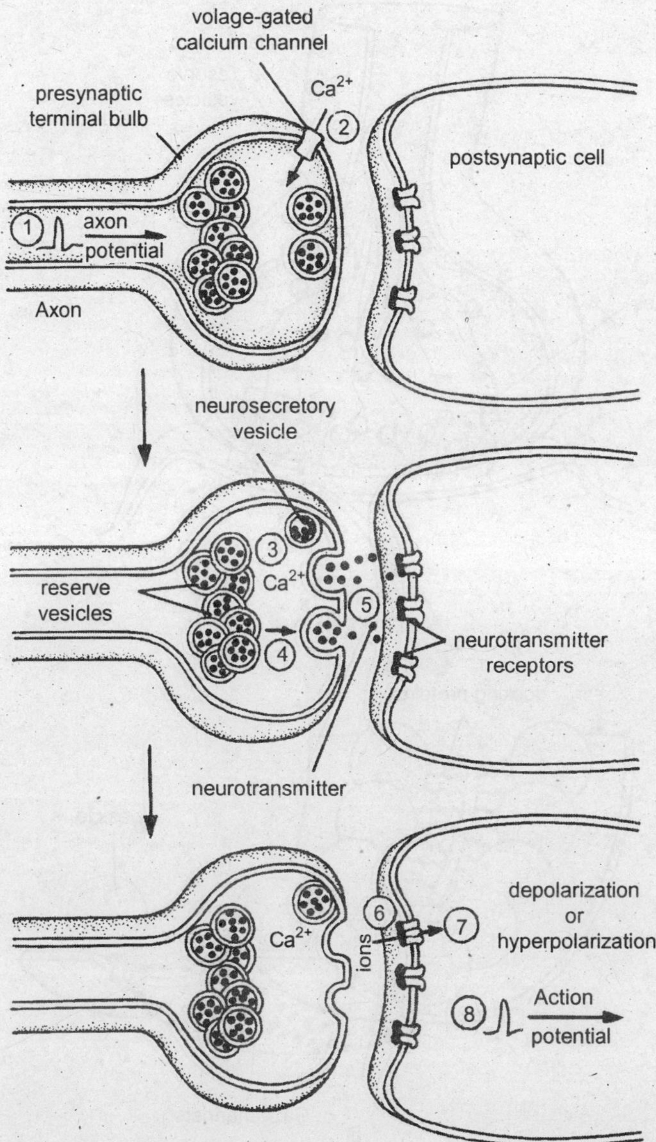

Fig. 68.13. The transmission of a signal across a synapse. 1. An action potential arrives at the presynaptic terminal bulb, resulting in a transient depolarization. 2. Depolarization opens voltage-gated calcium channels, allowing calcium ions to rush into the terminal. 3. This increase in the calcium concentration in the terminal bulb induces the secretion of a fraction of the neurosecretory vesicles. 4. Calcium also causes reserve vesicles to be released from the actin cytoskeleton so that they are ready for secretion. 5. Secreted neurotransmitter molecules diffuse across the synaptic clefts to receptors on the postsynaptic cell. 6. Binding of neurotransmitter to the receptor alters the receptor properties. 7. For receptors that are ligand-gated channels, the channel opening leads to either depolarization or hyperpolarization of the postsynaptic plasma membrane. 8. If depolarization results, a sufficient amount of excitatory neurotransmitter will result in an action potential in the postsynaptic cell (after Becker *et al.*, 2007).

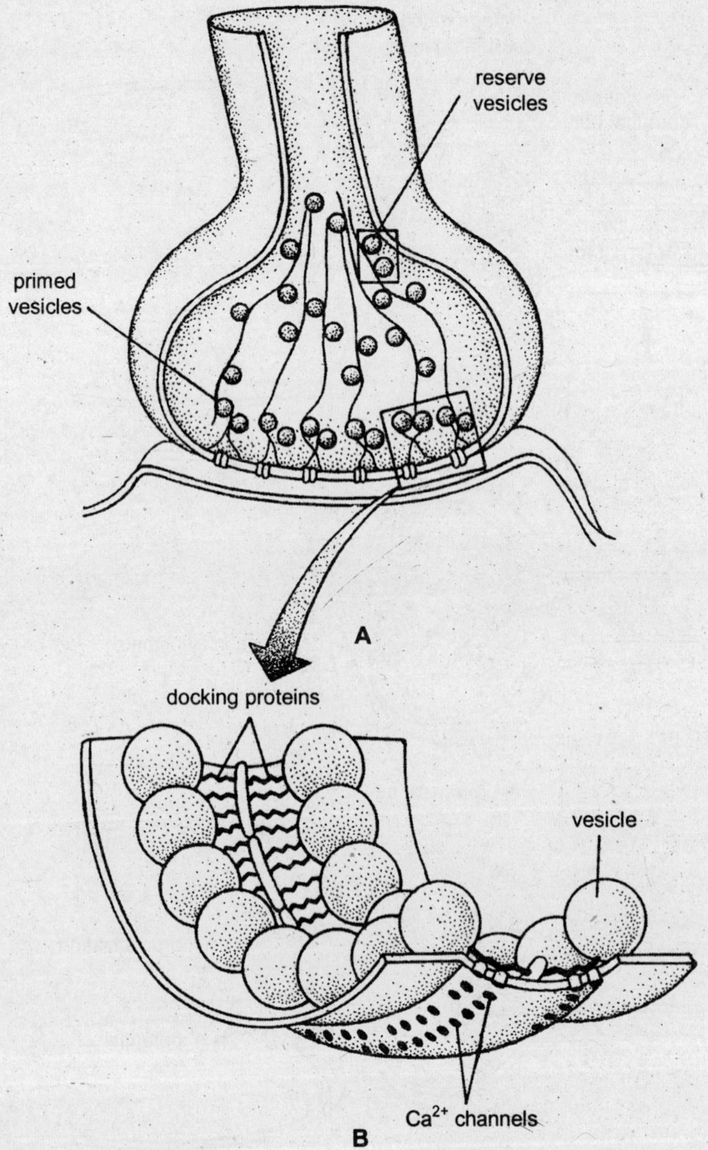

Fig. 68.14. Docking of synaptic vesicles with the plasma membrane of the presynaptic neuron. A—In response to local elevation of calcium in the presynaptic neuron, synaptic vesicles are released from the cytoskeleton, allowing them to fuse with the plasma membrane of the terminal bulb within the active zone. Vesicles bound to the cytoskeleton via synapsin form a reserve pool of vesicles. Once released from the cytoskeleton, vesicles move near the plasma membrane and become "primed" (*i.e.*, charged or prepared). Some vesicles become tightly associated with the plasma membrane, or "docked". When nearby calcium channels open, such docked vesicles fuse with the plasma membrane, releasing their contents. B—A drawing based on an actual TEM reconstruction of the active zone of a motor neuron from a frog. Docked secretory vesicles are arranged in rows and connected by a complex of proteins associated with docking. Calcium channels lie immediately beneath the vesicles (after Becker *et al.*, 2007).

When the action potential arrives at an axon terminal and triggers the opening of voltage-gated calcium channels, calcium enters the terminal bulb. For neurotransmitter vesicles to become available for secretion, they must become disengaged from the cytoskeleton. This process is stimulated by calcium release, and the addition of phosphate groups to **synapsin**, an integral membrane protein found in the membrane of neurosecretory vesicles. When phosphorylated synapsin no longer binds to the cytoskeleton, freeing the vesicles so that they can proceed to the next step in neurotransmitter release.

Once synaptic vesicles are released from the cytoskeleton, they are not quite ready to fuse with the plasma membrane. Such vesicles cluster near, but are not in contact with, the plasma membrane (Fig. 68.14). In order to be ready to fuse with the plasma membrane, such released vesicles must be "primed". Once such priming has occurred, the freed vesicles are now capable of docking and fusing with the plasma membrane of the presynaptic neuron in response to elevated calcium.

Docking takes place at a specialized site, called **active zone**, within the membrane of the presynaptic neuron. The active zone is a highly organised structure. Synaptic vesicle, the proteins to which the vesicles attach and the calcium channels that elicit their release are all clustered in ordered rows poised for secretion (Fig. 68.14B). The close proximity of calcium channels to the vesicles helps to explain the extremely rapid fusion of the primed, releasable population of vesicles with the presynaptic neuron's plasma membrane when that neuron is stimulated.

The processes of docking and fusion of neurosecretory vesicle are mediated by docking proteins within the vesicle and within the plasma membrane of the active zone. Within the vesicle membrane exist **synaptotagmin**, **synaptobrevin** and other proteins which dock with proteins, such as **syntaxin** at the plasma membrane of the active zone (**Koh** and **Bellen**, 2003). Once the initial docking occurs, additional protein from the cytosol stabilise the attachment, forming a multiprotein docking complex. Fusion depends on the successful docking of vesicles at the active zone and the formation of a complete docking complex.

Human diseases due to defects in docking events. Two common and prospective fatal illnesses of humans result from interference with the docking events. Both **tetanus** and **botulism** result from interference by **neurotoxins** with vesicle docking and release. **Tetanus toxin** prevents the release of neurotransmitter from inhibitory neurons in the spinal cord, resulting in uncontrolled muscle contraction (which is why tetanus has been familiarly referred to as "lockjaw"). **Botulinum toxin** prevents release of neurotransmitter from motor neurons, resulting in muscle weakness and paralysis (see **Becker** *et al.*, 2007).

(*iv*) The acetylcholine receptor. Acetylcholine binds to a ligand-gated sodium channel called **acetylcholine receptor**. When two molecules of acetylcholine bind with receptor, the channel opens and lets sodium ions rush into the postsynaptic neuron causing a depolarization (Fig. 68.15).

The purified acetylcholine receptor has molecular weight of about 300,000 and consists of four kinds of subunits-α, β, γ and δ-each containing about 500 amino acids (Fig. 68.15). Acetylcholine receptors play a key role in the transmission of nerve impulses to muscle. In some cases, human patients develop a autoimmune response to their own acetylcholine receptors (in other words, their immune system attacks their own receptors as if they were foreign invaders). When this happens, the patient can develop a condition known as **myasthenia gravis**, in which degenerative muscle weakness occurs.

Whether excitatory or inhibitory, once the neurotransmitter has been secreted, it must be rapidly removed from the synaptic cleft. If it were not, stimulation or inhibition of a postsynaptic neuron would be abnormally prolonged even in the absence of further signals from presynaptic neurons. In fact, the persistence of an excitatory neurotransmitter such as acetylcholine renders muscles unable to relax, ultimately leading to death.

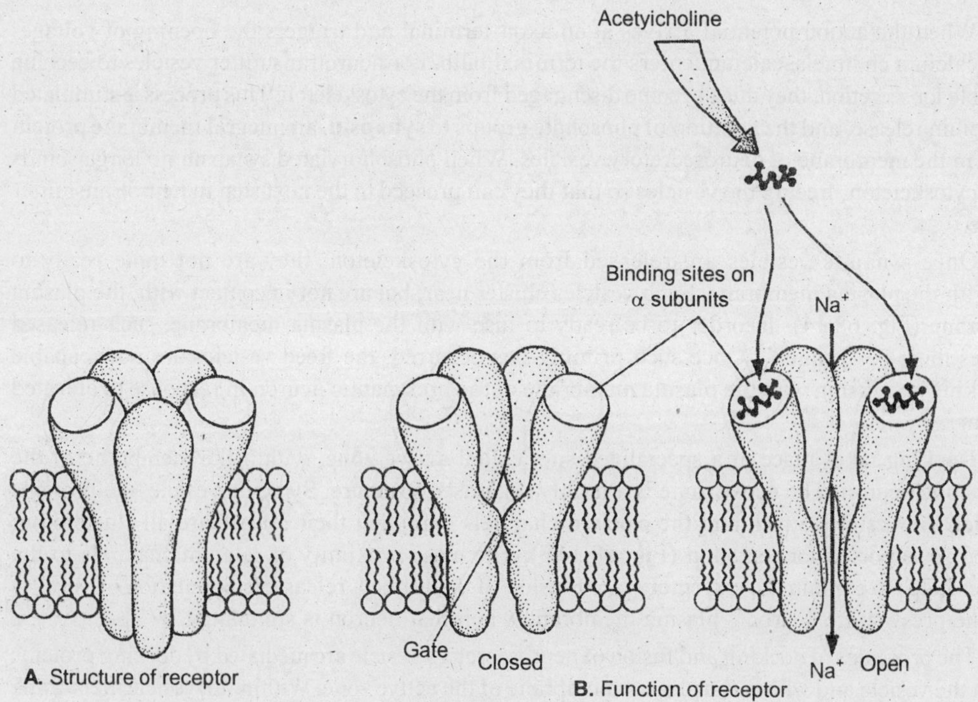

Fig. 68.15 The acetylcholine receptor which is a primary excitatory receptor of central nervous system. A—This receptor contains five subunits, including two α subunits with binding sites for acetylcholine and one each of β, γ and δ. The subunits aggregate in the lipid bilayer in such a way that the transmembrane portion form a channel. B—The channel (shown here with β subunit removed) is normally closed, but when acetylcholine binds to the two sites on the α subunits, the subunits are altered in such a way that the channel opens to allow sodium ions to cross (after Becker *et al.*, 2007).

Neurotransmitters are removed from the synaptic cleft by two specific mechanisms: degradation into inactive molecules or reuptake into the presynaptic terminals (*e.g.*, Serotonin). An example of the first mechanism is found in the case of the acetylcholine into acetic acid (or acetate ions) and choline, neither of which stimulatory to the acetylcholine receptor.

QUESTIONS

Long Answer Questions

1. What is bulk transport? Describe the process of endocytosis and exocytosis in brief.

2. What is vesicular transport? Give an outline of biosynthetic (secretory) pathway.

Short Answer Questions

1. What is the fate of receptor proteins that take part in receptor mediated endocytosis?

2. Write short notes on the following:

(*i*) Phagocytosis

(*ii*) Receptor-mediated-endocytosis

(*iii*) Coated vesicles

(*iv*) Exocytosis

(*v*) Pinocytosis

3. Describe the mechanism of transport of neurotransmitters in nerve cells.

4. Differentiate between the following terms:

(i) Phagocytosis and pinocytosis. *(Lucknow 1992; Guwahati (G), 1993)*

Very Short Answer Questions

1. What is early endosome?

Ans. Vesicles budding off the trans-Golgi network that are sites for the sorting and recycling of extracellular material brought into the cell by endocytosis.

2. What is late endosome?

Ans. Late endosome is a vesicle containing newly synthesized acid hydrolases plus material fated for digestion; activated either by lowering the pH of the late endosome or transferring its material to an existing lysosome.

3. Define secretory vesicle.

Ans. Secretory vesicle is a membrane bounded compartment of a eukaryotic cell that carries secretory proteins from the Golgi apparatus to the plasma membrane for exocytosis and that may serve as a storage compartment for such proteins before they are released.

4. What is secretory granule?

Ans. It is a large, dense secretory vesicle.

5. Define secretory pathway.

Ans. Secretory pathway is a pathway in which newly synthesized proteins move from ER through the Golgi apparatus to secretory vesicles and secretory granules, which then discharge their contents to the exterior of the cell.

6. What is exocytosis?

Ans. Exocytosis is fusion of vesicle membranes with the plasma membrane so that contents of the vesicle can be expelled or secreted to the extracellular environment.

7. Define endocytosis.

Ans. Endocytosis is uptake of extracellular materials by infolding of the plasma membrane, followed by pinching off of a membrane-bound vesicle containing extracellular fluid and materials.

8. What is endocytic vesicle?

Ans. It is a membrane vesicle formed by pinching off of a small segment of plasma membrane during the process of endocytosis.

9. What is SNARE hypothesis?

Ans. It is a model explaining how membrane vesicles fuse with the proper target membrane; based on specific interactions between *v*-SNAREs (vesicle-SNAP receptor) and *t*-SNAREs (target-SNAP receptors) (SNAP=Soluble NSF attachment proteins).

10. Define transition vesicles.

Ans. It is a membrane vesicle that shuttles lipids and proteins from the endoplasmic reticulum to Golgi apparatus.

11. What is transport vesicle?

Ans. Vesicle that buds off from a membrane in one region of the cell and fuses with other membranes; includes vesicles that convey lipids and proteins from the ER to the Golgi apparatus, between the Golgi stack cisternae, and from the Golgi apparatus to various destinations in the cell, including secretory vesicles, endosomes and lysosomes.

12. What is coated vesicle?

Ans. Any of several types of membrane vesicles involved in vesicular traffic within the endomembrane system; surrounded by a coat protein such as clathrin, COP I, COP II or caveolin.

13. Define phagocytosis.

Ans. It is a type of endocytosis in which particulate matter or even an entire cell is taken up from the environment and incorporated into vesicles for digestion.

14. What is receptor-mediated endocytosis?

Ans. It is a type of endocytosis initiated at coated pits and resulted in coated vesicles; believed to be a major mechanism for selective uptake of macromolecules and peptide hormones.

15. Define synapse.

Ans. Synapse is a tiny gap between a neuron and another cell (neuron, muscle fibrc, or gland cell), across which nerve impulse is transferred by direct electrical connections or by chemicals called neurotransmitters.

16. What is neurosecretory vesicle?

Ans. It is a small vesicle containing neurotransmitter molecules; located in the terminal bulb of an axon.

17. Write the names of two types of cellular secretions?

Ans. Constitutive secretion and regulated secretion.

18. Name of the cells in mammals that are engaged in phagocytosis.

Ans. Macrophages and neutrophils.

Multiple Choice Questions

1. Unidirectional transmission of a nerve impulse through nerve fibre is due to the fact that

(*a*) nerve fibrc has nodes of Ranvier

(*b*) nerve fibrc is insulated by a medullary sheet

(*c*) neurotransmitters are released by dendrites and not by axon endings

(*d*) neurotransmitters are released by the axon endings and not by dendrites

2. Among which one of the following groups of chemicals, all are neurotransmitters?

(*a*) somatostatin, serotonin, acetylcholine (*b*) nor-adrenaline, somatostatin, threonine

(*c*) glycine, dopamine, melatonin (*d*) acetylcholine, nor-adrenaline, dopamine

3. The process by which a cell secretes macromolecules through fusion of secretory vesicles to the plasma membrane is called

(*a*) pinocytosis (*b*) endocytosis

(*c*) phagocytosis (*d*) exocytosis

4. Membrane receptors are

(*a*) carbohydrates (*b*) nucleic acids

(*c*) proteins (*d*) lipids

5. Pinocytosis means endocytosis of

(*a*) droplet of extracellular fluid (*b*) sub-microscopic particles

(*c*) solid particles (*d*) extracellular fluid with particles

6. The neurotransmitter produced at electric synapse is

(*a*) GABA (*b*) acetylcholine

(*c*) nor-epinephrine (*d*) none of these

7. A person with Parkinson's disease shows defective action of which neurotransmitter in brain

(*a*) dopamine (*b*) serotonin

(*c*) noradrenaline (*d*) enkalphia

ANSWER

Multiple Choice Questions

1. (*d*) **2.** (*d*) **3.** (*d*) **4.** (*c*)

5. (*d*) **6.** (*b*) **7.** (*a*).

69

Nucleus, Nucleopore and Nucleolus

The nucleus (*L.*, *nux* = nut) is the heart of the cell. It is here that almost all of the cell's DNA is confined, replicated and transcribed. The nucleus, thus, controls different metabolic as well as hereditary activities of the cell. A synonymous term for this organelle is the Greek word **karyon**. Nucleus serves as the main distinguishing feature of eukaryotic cells, *i.e.*, this is the true nucleus as opposed to the nuclear region, prokaryon or nucleoid of the prokaryotic cells.

Nuclei were first discovered and named by **Robert Brown** in 1833 in the plant cells and were quickly recognized as a constant feature of all animal and plant cells. Nucleoli were described by **M.J. Schleiden** in 1838, although first noted by **Fontana** (1781). The term nucleolus was coined by **Bowman** in 1840. In 1879, **W. Flemming** coined the term chromatin for chromosomal meshwork.

69.1 OCCURRENCE AND POSITION

The nucleus is found in all the eukaryotic cells of the plants and animals. However, certain eukaryotic cells such as the mature sieve tubes of higher plants and mammalian erythrocytes contain no nucleus. In such cells nuclei are present during the early stages of development. Since mature mammalian red blood cells are without any nuclei, they are called red blood "corpuscles" rather than cells (*L. corpus* = body, especially dead body or corpse).

Usually the nucleus remains located in the centre. But its position may change from time to time according to the metabolic states of the cell. For example, in the embryonic cells the nucleus generally occupies the geometric centre of the cell but as the cells start to differentiate and the rate of the metabolic activities increases, the displacement in the position of the nucleus takes place. In certain cells such as the glandular cells the nucleus remains located in the basal portion of the cell.

69.2 MORPHOLOGY

Number. Usually the cells contain single nucleus but the number of the nucleus may vary from cell to cell. According to the number of the nuclei following types of cells have been recognised:

1. Mononucleate cells. Most plant and animal cells contain single nucleus, such cells are known as **mononucleate cells.**

2. Binucleate cells. The cells which contain two nuclei are known as **binucleate cells**. Such cells occur in certain protozoans such as *Paramecium* and cells of cartilage and liver.

3. Polynucleate cells. The cells which contain many (from 3 to 100) nuclei are known as **polynucleate cells**. The polynucleate cells of the animals are termed as **syncytial cells**, while the polynucleate cells of the plants are known as **coenocytes**. The most common example of the syncytial cells are the osteoblast (polykaryocytes of the bone marrow) which contain about 100 nuclei per cell and striated muscle fibres each of which contains many hundred nuclei. The siphonal algae *Vaucheria* contains hundreds of nuclei and certain fungi are the best example of the coenocytic plant cells.

Shape. The shape of the nucleus normally remains related with the shape of the cell, but certain nuclei are almost irregular in shape. The spheroid, cuboid or polyhedral cells (isodiametrical cells) contain the **spheroid** nuclei. The nuclei of the cylindrical, prismatic or fusiform cells are **ellipsoid** in shape. The cells of the squamous epithelium contain the **discoidal** nuclei. The leukocytes, certain infusoria, glandular cells of some insects and spermatozoa contain the irregular shaped nuclei. Nuclei of cells of silk glands of silk worm have finger-like extensions that greatly increase their surface area (Fig. 69.1).

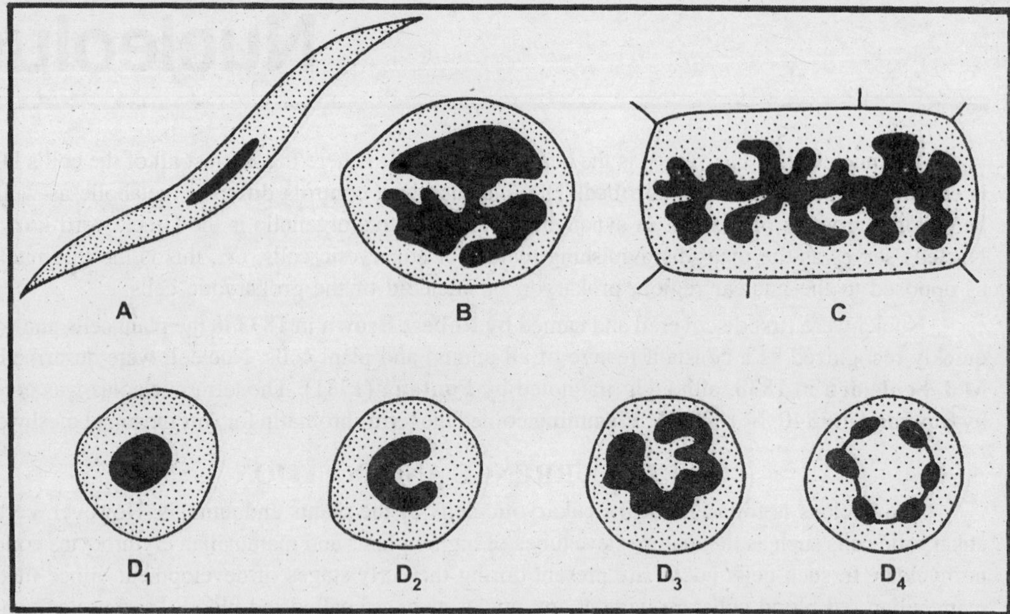

Fig. 69.1. Different shapes of the nucleus in animal cells. A—Elongated in muscle cell: B—Lobed in a human neutrophil cell: C—Branched in a silk spinning cell of an insect larva: D1 to D4—Variable shapes in leucocytes.

Size. Generally nucleus occupies about 10 per cent of the total cell volume. Nuclei vary in size from about 3 μm to 25 μm in diameter, depending on cell type and contain diploid set of chromosomes. The size of the nucleus is directly proportional to that of the cytoplasm. **R. Hertwig** has given the following formula for the deduction of the size of the nucleus of a particular cell.

$$NP = \frac{V_n}{V_c - V_n}$$

where NP is the nucleoprotein index, V_n is the volume of the nucleus and V_c is the volume of the cell.

Moreover, the size of the nucleus is related with the number of the chromosomes or ploidy. The haploid cells contain small sized nuclei in comparison to the nuclei of the diploid cells. Likewise the polyploid cells contain larger nuclei than the diploid cells. Thus, the size of the nucleus of a cell depends on the volume of the cell, amount of the DNA and proteins and metabolic phase of the cell.

69.3 ULTRASTRUCTURE

The nucleus is composed of following structures: 1. The nuclear membrane or karyotheca or nuclear envelope; 2. The nuclear sap or nucleoplasm; 3. The chromatin fibres; and 4. The nucleolus

1. Nuclear Envelope

The nuclear envelope (or perinuclear cisterna) encloses the DNA and defines the nuclear compartment of interphase and prophase nuclei. It is formed from two concentric unit membranes, each 5–10 nm thick. The spherical **inner nuclear membrane** contains specific proteins that act as binding sites for the supporting fibrous sheath of intermediate filaments (IF), called **nuclear lamina**. Nuclear lamina has contact with the chromatin (or chromosomes) and nuclear RNAs. The inner nuclear membrane is surrounded by the **outer nuclear membrane**, which closely resembles the membrane of the endoplasmic reticulum, that is continuous with it. It is also surrounded by less organized intermediate filaments (Fig. 69.2). Like the membrane of the rough ER, the outer surface of outer nuclear membrane is generally studded with ribosomes engaged in protein synthesis. The proteins made on these ribosomes are transported into space between the inner and outer nuclear membrane, called **perinuclear space**. The perinuclear space is a 10 to 50 nm wide fluid-filled compartment which is continuous with the ER lumen and may contain fibres, crystalline deposits, lipid droplets or electron-dense material.

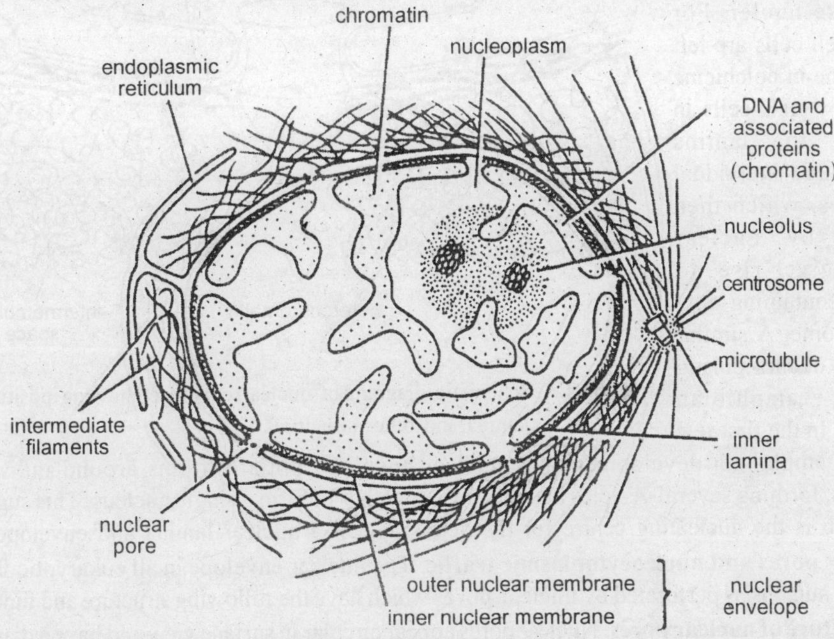

Fig. 69.2. Cross-section of a typical cell nucleus showing ultrastructure.

Nuclear lamina. It is also called **fibrous lamina, zonula nucleum limitans, internal dense lamella, nuclear cortex** and **lamina densa**.

The nuclear lamina is a protein meshwork which is 10 nm thick. It lines the inside surface of the inner nuclear membrane, except the areas of nucleopores, and consists of a square lattice of intermediate filaments. In mammals, these intermediate filaments are of three types: **lamins A, B** and **C** having M.W. 74,000, 72,000 and 62,000 daltons, respectively. The lamins form dimers that have a rod-like domain and two globular heads at one end. Under appropriate conditions of pH and ionic strength, the dimers spontaneously associate into filaments that have a diameter and repeating structure similar to those of cytoplasmic filaments.

The nuclear lamina is a very dynamic structure. In mammalian cells undergoing mitosis, the transient phosphory-lation of several serine residues on the lamins causes the lamina to reversibly disassemble into tetramers of hypo-phosphorylated lamin A and lamin C and mem-brane associated lamin B. As a result, lamin A and C become entirely soluble during mitosis, and at telophase they become dephosphory-lated again and polyme-rize around chromatin. Lamin B seems to remain associated with mem-brane vesicles during mitosis, and these vesi-cles in turn remain as a distinct subset of mem-brane components from which nuclear envelope is reassembled at telophase. Inside an interphase nucleus, chromatin binds strongly to the inner part of the nuclear lamina which is believed to interfere with chromosome condensation. In fact, during meiotic chromosome condensation, the nuclear lamina completely disappears by the pachytene stage of prophase and reappears later during diplotene in oocytes, but does not reappear at all in spermatocytes.

The lamins may play a crucial role in the assembly of interphase nuclei. For example, when cells are left for a long time in colchicine (drug which arrests cells in metaphase), the lamins assemble around individual chromosomes, which then surrounded by nuclear envelopes give rise to micronuclei containing only one chromosome. A similar phenomenon occurs during normal amphibian development. In the first few cleavages of amphibian development, the nuclear envelope initially forms around individual chromosomes, forming several vesicles that then fuse together to form a single nucleus. This suggests that chromatin is the nucleating centre for the deposition of a nuclear lamina and envelope.

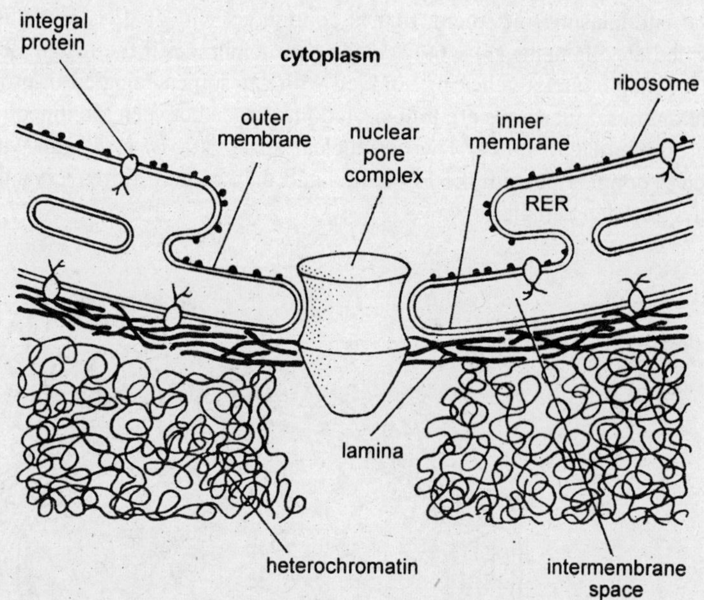

Fig. 69.3. Schematic diagram of nuclear envelope showing position of lamina and other structures.

Nuclear pores and nucleocytoplasmic traffic. The nuclear envelope in all eukaryotic forms, from yeasts to humans, is perforated by **nuclear pores** which have the following structure and function:

1. Structure of nuclear pores. Nuclear pores appear circular in surface view and have a diameter between 10 nm to 100 nm. Previously it was believed that a diaphragm made of amorphous to fibrillar material extends across each pore limiting free transfer of material. Such a diaphragm called **annulus** has been observed in animal cells, but lacking in plant cells. Recent electron microscopic studies have found that a nuclear pore has far more complex structure, so it is called **nuclear pore complex**. Each pore complex has an estimated molecular weight of 50 to 100 million daltons. Negative staining techniques have demonstrated that pore complexes have an eight-fold or octagonal symmetry. More recent computerized image-processing techniques of **Unwin** and **Mulligan** (1982) have shown that the pore complex consists of two **"rings"** (**R** or annuli) at its periphery with an inside diameter of 80 nm, a large particle that forms a **central plug (C)** and radial '**spokes**' (S) that extends from the plug to the rings (Figs. 69.4B and C). **Particles (P)** are anchored to cytoplasmic ring and are thought to be inactive ribosomes. The '**hole**' in the centre of the pore complex is an aqueous channel through which water-soluble molecules shuttle between the nucleus and the cytoplasm. This hole often appears to be plugged by a large central granule (**central plug**) which is believed to consist of newly made ribosomes or other particles caught in transit.

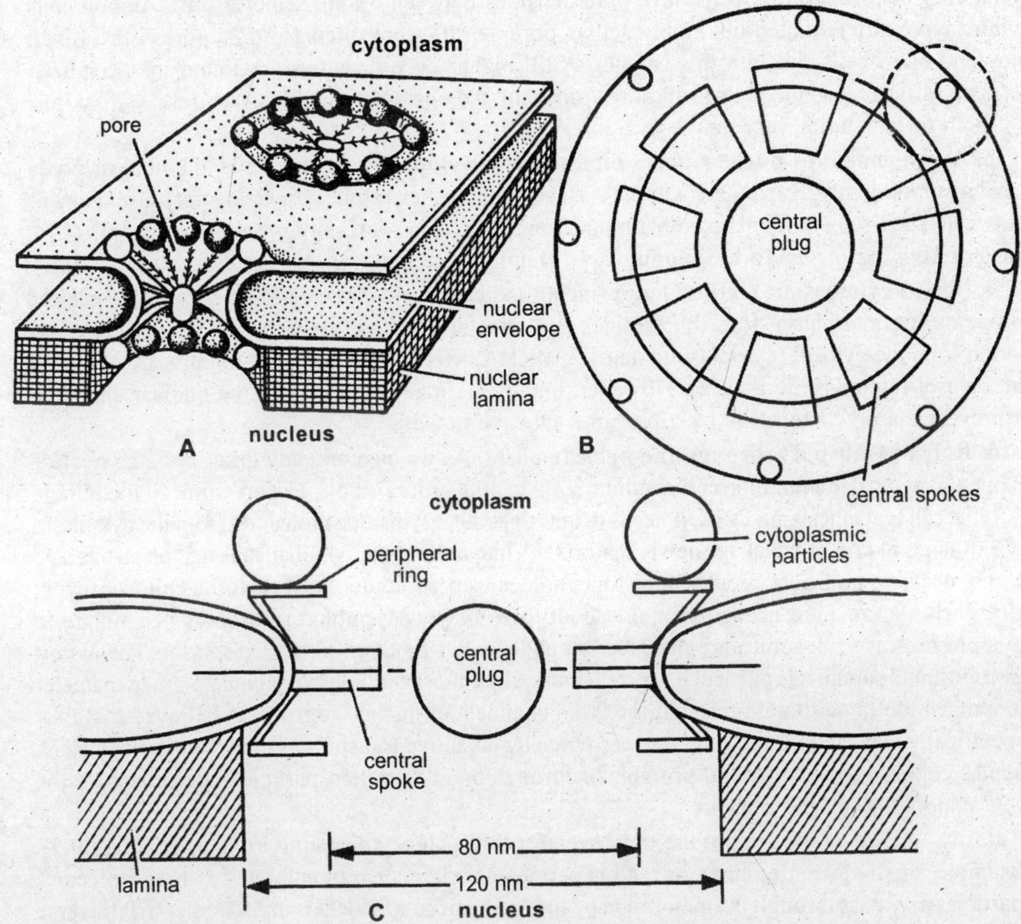

Fig. 69.4. Nuclear pore. A—A part of nuclear envelope with 80 nm pores occupied by pore complexes of octagonal radial symmetry; B—Nuclear pore in top view; C–A nuclear pore in cross section.

The pore complex perforates the nuclear envelope bringing the lipid bilayers of the inner and outer nuclear membrane together around the margins of each pore. Despite this continuity, which would seem to provide a pathway for the diffusion of membrane components between the inner and outer membranes, the two membranes remain chemically distinct.

Quite recently, following two proteins have been found to be associated to the nuclear pores : one is an integral membrane protein, a glycoprotein of 120,000 daltons that may anchor the annuli to the lipid bilayer. The second protein is a 63,000 dalton protein (that has covalently bound acetyl neuraminic acid) located on the cytoplasmic side of the electron-dense material that occludes the nuclear pores. This protein may be involved in the transport of materials through the nuclear pores.

2. Number of nuclear pores (Pore density). In nuclei of mammals it has been calculated that nuclear pores account for 5 to 15 per cent of the surface area of the nuclear membrane. In amphibian oocytes, certain plant cells and protozoa, the surface occupied by the nuclear pores may be as high as 20 to 36 per cent. The number of pores in the nuclear envelope or **pore density** seems to correlate with the transcriptional activity of the cell. Thus, pore densities as low as ~3 pores/ μm^2 are seen in nucleated red blood cells and lymphocytes (which are inactive in transcription). These cells are highly differentiated but metabolically inactive and they are non- proliferating cells.

The majority of proliferating cells have pore densities between 7 and 12 pores/μm^2. Among cells of a third type, differentiated but highly active, pore densities are often 15 to 20 pores/μm^2. Liver, kidney and brain cells fall into this category. Still higher pore densities are found in specialized cells, such as salivary gland cells (~40 pores/μm^2) and the oocytes from *Xenopus laevis* (~50 pores/μm^2), both of which are very active in transcription.

3. Arrangement of nuclear pores on nuclear envelope. In somatic cells, the nuclear pores are evenly or randomly distributed over the surface of nuclear envelope. However, pore arrangement in other cell types is not random but rather range from **rows** (*e.g.,* spores of *Eqisetum*) to **clusters** (*e.g.,* oocytes of *Xenopus laevis*) to **hexagonal** (*e.g.,* Malpighian tubules of leaf hoppers) packing order.

4. Nucleo-cytoplasmic traffic. Quite evidently there is considerable trafficking across the nuclear envelope during interphase. Ions, nucleotides and structural, catalytic and regulatory proteins are imported from the cytosol (cytoplasmic matrix); mRNA, tRNA and ribosome subunits are exported to the cytosol (cytoplasmic matrix). However, one of the main functions of the nuclear envelope is to prevent the entrance of active ribosomes into the nucleus.

5. Rate of transport through the nuclear pores. As we have already described, the nuclear envelope of a typical mammalian cell contains 3000 to 4000 pores (about 11 pores/μm^2 of membrane area). If the cell is synthesizing DNA, it needs to import about 10^6 histone molecules from the cytoplasm every 3 minutes in order to package newly made DNA into chromatin, which means that on an average each pore needs to transport about 100 histone molecules per minute. Further, if the cell is growing rapidly, each nuclear pore needs to export about three newly assembled ribosomes per minute to the cytoplasm, since ribosomes are produced in nucleus but function in the cytoplasm. The export of new ribosomal subunits is particularly problematic since these particles are about 15 nm in diameter and are much too large to pass through the 9 nm channels of nuclear pores, it is believed that they are specifically exported through the nuclear pores by an active transport system. Similarly, mRNA molecules complexed with special proteins to form ribonucleoprotein particles, are thought to be actively exported from the nucleus.

Lastly, nuclear pores are not the only avenues for nucleocytoplasmic exchanges (Fig. 69.5). For example, small molecules and ions readily permeate both nuclear membranes. Larger molecules and particles may pass through the membrane by formation of small pockets and vesicles that traverse the envelope and empty on the other side.

2. Nucleoplasm.

The space between the nuclear envelope and the nucleolus is filled by a transparent, semi-solid, granular and slightly acidophilic ground substance or the matrix known as the **nuclear sap** or **nucleoplasm** or **karyolymph**. The nuclear components such as the chromatin threads and the nucleolus remain suspended in the nucleoplasm.

The nucleoplasm has a complex chemical composition. It is composed of mainly the nucleoproteins but it also contains other inorganic and organic substances, *viz.*, nucleic acids, proteins, enzymes and minerals.

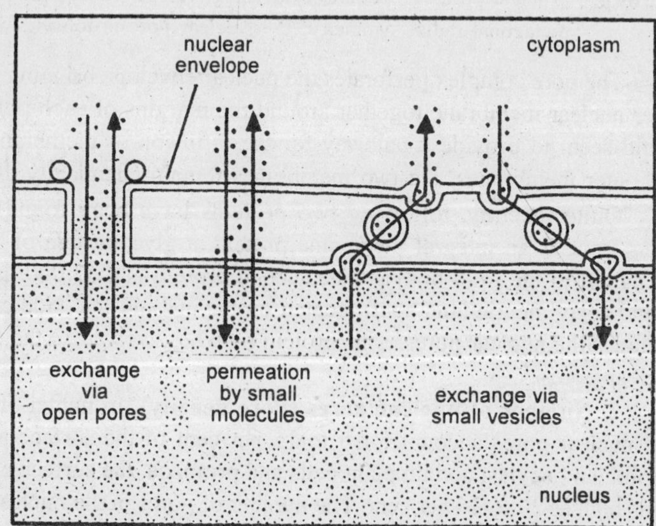

Fig. 69.5. Various avenues for transport of material from the nucleus to the cytosol (cytoplasmic matrix).

1. Nucleic acids. The most common nucleic acids of the nucleoplasm are the DNA and RNA. Both may occur in the macromolecular state or in the form of their monomer nucleotides.

2. Proteins. The nucleo-plasm contains many types of complex proteins. The nucleoproteins can be categorized into following two types:

***(i)* Basic proteins.** The proteins which take basic stain are known as the basic proteins. The most important basic proteins of the nucleus are **nucleoprotamines** and the **nucleohistones**.

The nucleoprotamines are simple and basic proteins having very low molecular weight (about 4000 daltons). The most abundant amino acid of these proteins is **arginine** (pH 10 to 11). The protamines usually remain bounded with the DNA molecules by the salt linkage. The protamines occur in the spermatozoa of the certain fishes. The nucleohistones have high molecular weight, *e.g.*, 10,000 to 18,000 daltons. The histones are composed of basic amino acids such as **arginine, lysine** and **histidine**. The histone proteins remain associated with the DNA by the ionic bonds and they occur in the nuclei of most organisms. According to the composition of the amino acids following types of histone proteins have been recognised, *e.g.*, histones rich in lysine, histones with arginine and histones with poor amount of the lysine.

***(ii)* Non-histone or Acidic proteins.** The acidic proteins either occur in the nucleoplasm or in the chromatin. The most abundant acidic proteins of the euchromatin (a type of chromatin) are the **phosphoproteins** (Box 69.1).

Box 69.1

Nucleoproteins are the main and the most important component of nucleus. These are compounds of nucleic acids and proteins. The proteins present in the nucleus are specific and are of following two types:

1.Histones. Histones (which are of five basic types such as H2A, H2B, H3, H4 and H1 or H5) are highly basic proteins rich in positively charged **lysine** and **arginine** residues. Histones H3 and H4 are most strongly conserved, while linker histones (H1 or H5) show most diversity.

All the histones except H4 exist as multiple isoforms (**isohistones**) whose relative predominance in chromatin varies in a cell-type-specific manner. Histones represent the bulk of protein in chromatin and are relatively homogeneous in nature.

2.Non-histone proteins. Non-histone proteins are acidic in nature and are rich in **tryptophan** and **tyrosine**. Non-histone proteins represent small but extremely heterogenous fraction. The non-histone proteins include **enzymes** involved in DNA and histone metabolism, replication, recombination, and transcriptional regulation. They also include **scaffold proteins** which organize higher order chromatin structure and **high-mobility group proteins** (*HMG proteins*) which are highly charged proteins with various functions in gene regulation and structural organization. A further class of non histone proteins, termed **protamines,** facilitate the packaging of DNA into sperm head (**Twyman** 1998).

3. Enzymes. The nucleoplasm contains many enzymes which are necessary for the synthesis of the DNA and RNA. Most of the nuclear enzymes are composed of non-histone (acidic) proteins. The most important nuclear enzymes are the **DNA polymerase, RNA polymerase, NAD synthetase, nucleoside triphosphatase, adenosine diaminase, nucleoside phosphorylase, guanase, aldolase, enolase, 3-phosphoglyceraldehyde dehydrogenase** and **pyruvate kinase**. The nucleoplasm also contains certain cofactors and coenzymes such as **ATP** and **acetyl CoA**.

4. Lipids. The nucleoplasm contains small lipid content.

5. Minerals. The nucleoplasm also contains several inorganic compounds such as phosphorus, potassium, sodium, calcium and magnesium. The chromatin comparatively contains large amount of these minerals than the nucleoplasm.

3. Chromatin Fibres.

The nucleoplasm contains many thread-like, coiled and much elongated structures which take readily the basic stains such as the basic fuchsin. These thread-like structures are known as the **chromatin** (*Gr., chrome*=colour) **substance** or **chromatin fibres**. Such chromatin fibres are observed only in the interphase nucleus. During the cell division (mitosis and meiosis) chromatin fibres become thick ribbon-like structures which are known as the **chromosomes**.

Chemically, chromatin consists of DNA and proteins. Small quantity of RNA may also be present but the RNA rarely accounts for more than about 5 per cent of the total chromatin present. Most of the protein of chromatin is histone, but "non-histone" proteins are also present. The protein: DNA weight ratio averages about 1:1. Histones are constituents of the chromatin of all eukaryotes except fungi, which, therefore, resemble prokaryotes in this respect.

The fibres of the chromatin are twisted, finely anastomosed and uniformly distributed in the nucleoplasm. Two types of chromatin material have been recognised, *e.g.*, heterochromatin and euchromatin.

A. Heterochromatin. The darkly stained, condensed region of the chromatin is known as heterochromatin. The condensed portions of the nucleus are known as **chromocenters** or **karyosomes** or **false nucleoli** (Box 69.2). The heterochromatin occurs around the nucleolus and at the periphery. It is supposed to be metabolically and genetically inert because it contains comparatively small amount of the DNA and large amount of the RNA.

Box 69.2

Chromocenters are darkly stained heterochromatic regions of chromatin of interphase nucleus. They may represent a heterochromatic region of several chromosomes or of all the chromosomes of the nucleus. Chromocenters are well marked in the nuclei of salivary gland cells of *Drosophila.* In this case, pericentromeric heterochromatin of all polytene chromosomes coalesces in a chromocenter.

B. Euchromatin. The light stained and diffused region of the chromatin is known as the euchromatin. The euchromatin contains comparatively large amount of DNA.

Heteropyknosis

Differential condensation of certain chromosomes, such as sex chromosomes, or chromosome parts is called **heteropyknosis** (*Gr.*, differential staining). During interphase and prophase stages of cell division most of the chromosomal substance is uncoiled and is in the form of light stained fine threads. But some chromosomes or their parts remain condensed and can be recognized by their dark stain. Thus, heteropyknosis can be defined as *the property of certain chromosomes or their parts to remain more condensed and stain more intensely than other chromosomes or their parts during the nuclear cycle.*

Heteropyknosis is the characteristic of sex chromosomes of many species, but may also be observed in other chromosomes as well. For example, one of two X chromosomes in females of all the mammals undergoes heteropyknosis. The heteropyknosis may involve entire chromosome, may be intercalated or localized at the extremities.

When the heteropyknotic chromosomes (*e.g.* sex chromosomes) are darkly stained, it is called **positive heteropyknosis,** however, in the toad, *Bufo arenarus,* the chromosomes exhibit **negative heteropyknosis** (under-condensation).

4. Nucleolus.

Most cells contain in their nuclei one or more prominent spherical colloidal acidophilic bodies, called **nucleoli**. However, cells of bacteria and yeast lack nucleolus. The size of the nucleolus is found to be related with the synthetic activity of the cell. Therefore, the cells with little or no synthetic activities, *e.g.*, sperm cells, blastomeres, muscle cell, etc., are found to contain smaller or no nucleoli, while the oocytes, neurons and secretory cells which synthesize the proteins or other substances contain comparatively large-sized nucleoli. The number of the nucleoli in the nucleus depends on the species and the number of the chromosomes. The number of the nucleoli in the cells may be one, two or four. The position of the nucleolus in the nucleus is eccentric.

A nucleolus is often associated with the **nucleolar organizer (NO)** which represents the secondary constriction of the nucleolar organizing chromosomes, and are 10 in number in human beings (Fig. 69.6). In corn, *Zea mays* chromosome 9 and 6 contain 'darkly staining knobs' or nucleolar organizers. Nucleolar organizer consists of the genes for 18S, 5.8S and 28S rRNAs. The genes for fourth type of rRNA, *i.e.*, 5S rRNA occur outside the nucleolar organizer.

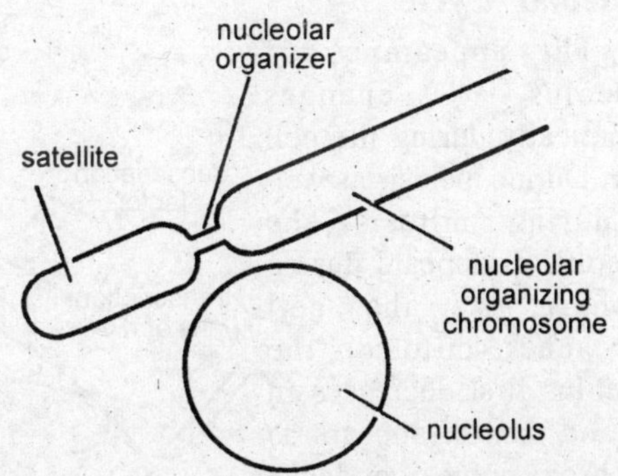

Fig. 69.6. A satellited chromosome and an attached nucleolus.

Chemical composition of nucleolus. Nucleo-lus is not bounded by any limiting membrane; calcium ions are supposed to maintain its intact organization. Chemically, nucleolus contains DNA of nucleolar organizer, four types of rRNAs, 70 types of ribosomal proteins, RNA binding proteins (*e.g.*, nucleolin) and RNA splicing nucleoproteins (U_1, U_2......U_{12}). It also contains phospholipids, orthophosphates and Ca^{2+} ions. Nucleolus also contains some enzymes such as acid phosphatase, nucleoside phosphorylase and NAD^+synthesizing enzymes for the synthesis of some coenzymes, nucleotides and ribosomal RNA. **RNA methylase** enzyme which transfers methyl groups to the nitrogen bases, occurs in the nucleolus of some cells.

Ultrastructure and function of nucleolus. Nucleolus are the sites where biogenesis of ribosomal subunits (*i.e.*, 40S and 60S) takes place. In it three types of rRNAs, namely 18S, 5.8S and 28S r RNAs, are transcribed as parts of a much longer precursor molecule (45S transcript) which undergoes processing (RNA splicing, for example) by the help of two types of proteins such as nucleolin and U3 snRNP (U3 is a 250 nucleotide containing RNA, snRNP represents small nuclear ribonucleoprotein). The 5S rRNA is transcribed on the chromosome existing outside the nucleolus and the 70 types of ribosomal proteins are synthesized in the cytoplasm. All of these components of the ribosomes migrate to the nucleolus, where they are assembled into two types of ribosomal subunits which are transported back to the cytoplasm. The smaller (40S) ribosomal subunits are formed and migrate to the cytoplasm much earlier than larger (60S) ribosomal subunits; therefore, nucleolus contains many more incomplete 60S ribosomal subunits than the 40S ribosomal subunits. Such a time lag in the migration of 60S and 40S ribosomal subunits, prevents functional ribosomes from gaining access to the incompletely processed heterogeneous RNA (hn RNA; the precursor of mRNA) molecule inside the nucleus.

Different stages of formation of ribosomes are completed in three distinct regions of the nucleolus. Thus, their **initiation, production** and **maturation** seem to progress from centre to periphery. Following three regions have been identified in the nucleolus (Fig. 69.7):

***(i)* Fibrillar centre.** This pale-staining part represents the innermost region of nucleolus. The RNA genes of nucleolar organizer of chromosomes are located in this region. The transcription (*i.e.*, ribosomal RNA synthesis) of these genes is also initiated in this region.

***(ii)* Dense fibrillar component.** This region surrounds the fibrillar centre and RNA synthesis progresses in this region. The 70 ribosomal proteins (rps) also bind to the transcripts in this region.

***(iii)* Cortical granular components.** This is the outermost region of the nucleus where processing and maturation of pre-ribosomal particles occur.

Nucleolar Cycle

The appearance of nucleolus changes dramatically during the cell cycle. During meiosis as well as during mitosis the nucleolus disappears during prophase. As the cell approaches mitosis, the nucleolus first decreases in size and then disappears as the chromosomes condense and all RNA synthesis stops, so that generally there is no nucleolus in a metaphase cell. When ribosomal RNA synthesis restarts at the end of mitosis (in telophase), tiny nucleoli reappear at the chromosomal locations of the ribosomal RNA genes (NOs).

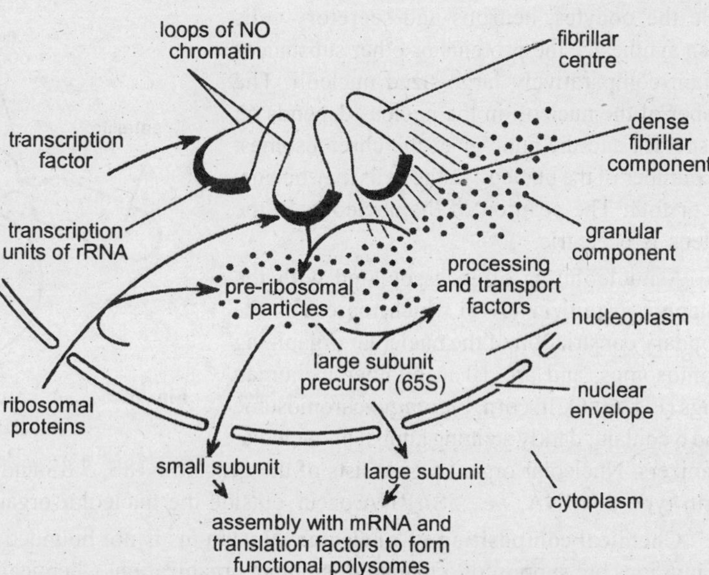

Fig. 69.7. Three different regions of the nucleolus and their involvement in ribosome assembly.

For example, in humans the rRNA genes are located near the tips of each of the 5 different chromosomes (*i.e.*, paired autosomes 13, 14, 15, 21 and 22). Accordingly, 10 small nucleoli are formed after mitosis in a human diploid cell, although they are rarely seen as separate entities because they quickly grow and fuse to form the single large nucleolus typical of many interphase cells.

Now let us see what happens to the RNA and protein components of the disintegrated nucleolus during mitosis? It seems that at least some of them become distributed over the surface of all of the metaphase chromosomes and are carried as cargo to each of the two daughter cell nuclei. As the chromosomes decondense at telophase, these "old" nucleolar components help re-establish the newly emerging nucleoli.

Functions of Nucleus

Nucleus is the controlling centre of the cell.

(*i*) It controls all the metabolic activities of the cell by controlling the synthesis of enzymes required there in.

(*ii*) The nucleus controls the inheritance of characters from parents to offsprings.

(*iii*) It is responsible for the development of characters. Our phenotype is determined by the genes we have inherited from our parents. These genes are present in the chromosomes of nucleus.

Differences in Nucleus and Nucleolus

Table 69.1 Differences between nucleus and nucleolus.

Nucleus	*Nucleolus*
1. It is a large sized organelle.	1. It is a very small part of nucleus. There occurs one nucleolus for each haploid set of chromosomes.
2. It is bounded by a nuclear envelope of double unit membranes. Nuclear envelope bears nucleopores.	2. It is not bounded by any membrane.
3. It is site of replication, recombination and transcription of DNA.	3. It is site of transcription of rDNA of nucleolar region of chromosomes. Tailoring and modifications of various types of rRNAs take place in nucleolus and they are assembled with ribosomal proteins to form ribosomal subunits.

69.4 HAMMERLING'S EXPERIMENT

The evidences for nucleo-cytoplasmic communication as a factor in cell maintenance and development have been known before the rediscovery of Mendel's "genes". In the late nineteenth century, **Verworm, Balbiani** and others showed that following microsurgery, nucleated halves of various protozoans survived and grew, whereas the enucleated halves degenerated and died. Later, in the 1930s it was shown that insertion of nuclei into enucleated amoebae restored pseudopodial activity, a feeding behaviour and growth. It was also shown that the nucleus was essential for the growth and regeneration of the morphologically complicated ciliate *Stentor*. In a classical series of experiments spanning between 1934 and 1954, on the unicellular alga *Acetabularia,* **Hammerling** demonstrated by means of interspecific nuclear transplants, that morphological features, notably the shape of cap, were determined by the nucleus. He also showed that even after removal of the nucleus, the cell was able to continue morphogenesis for a time and proposed that the cytoplasm contained a store of morphogenetic material (later on recognized as mRNA molecules) that had been produced by the nucleus. Let us closely examine the Hammerling's classical nuclear transplantation experiments:

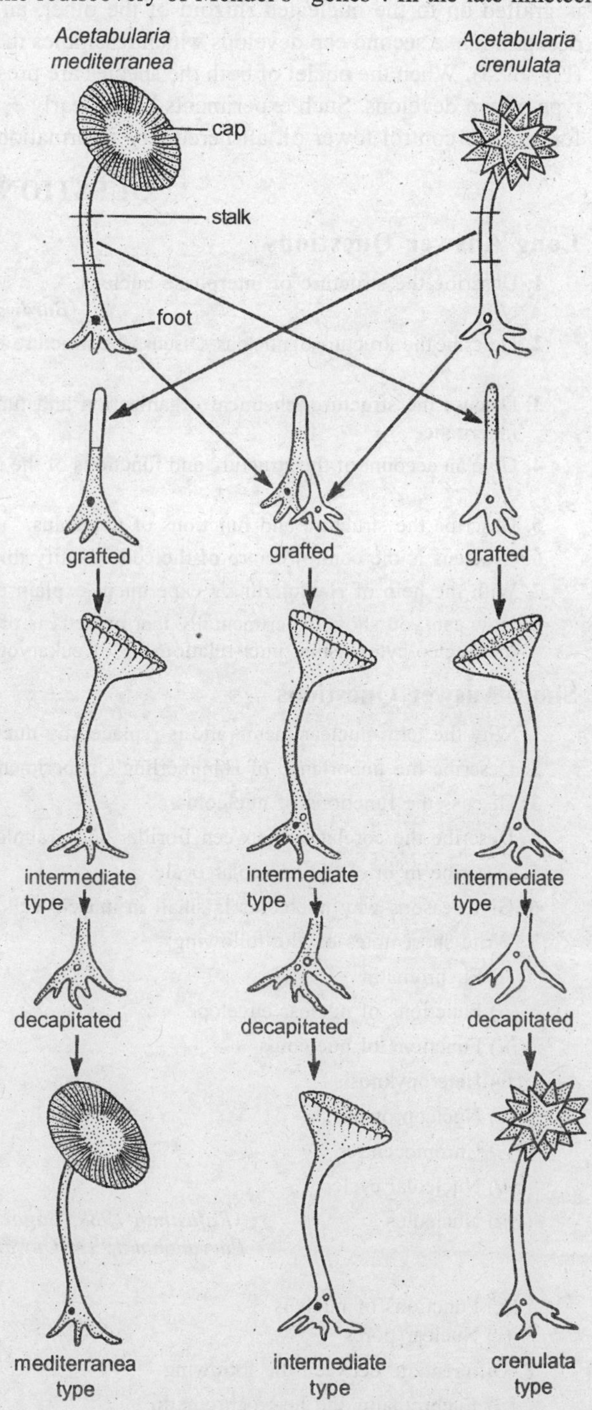

Fig. 69.8. Summary of the grafting experiments with *Acetabularia* to prove that the hereditary characters are determined by the nucleus and not by the cytoplasm.

Hammerling's experi-ment. The body of an alga *Acetabularia* is about six centimeters long and is differentiated into a foot, a stalk and a cap. The cap has a characteristic shape for each species and is easily regene-rated if removed. The single nucleus is situated in the rhizoid portion. *Acetabularia crenulata* has a cap, with about 31 rays, the tips of which are pointed, but *Acetabularia mediterranea* has about 81 rays with rounded tips. If the cap, stalk or even the nucleated portion of the rhizoid is removed, the remaining portion of the alga has the capacity to regenerate into a whole plant. The enucleated part loses the regeneration capacity after a few decapitations, but the nucleated portion always maintains this ability. When the stalk of one species is grafted on to the nucleated rhizoid of the other, an **intermediate type** of cap is formed. On decapitation, a second cap develops which resembles the cap of species which provides the nucleus (Fig. 69.8). When the nuclei of both the species are present in the same cytoplasm, an intermediate type of cap develops. Such experiments have clearly established that the nucleus is the storehouse for and the control tower of, all hereditary information.

QUESTIONS

Long Answer Questions

1. Describe the structure of interphase nucleus. (*Burdwan 1989; Guwahati 1989, 1991; Jiwaji 1991*)

2. Describe the structure of nucleus. Discuss the structure and functions of different components of nucleus. (*Kanpur 1998*)

3. Discuss the structure, chemical organization and functions of nucleus. Throw light on its role in inheritance. (*Kanpur 1992, 1996*)

4. Give an account of the structure and functions of the nucleus with special reference to nucleic acids. (*Guwahati (G) 1994*)

5. Describe the structure and functions of nucleolus. (*CCS 1993*)

6. 'Nucleus is the control centre of the cell'. Justify this statement with the help of two experiments.

7. With the help of Hammerling's experiment explain the importance of nucleus. (*CCS 1990, 1994*)

8. How can you show experimentally that nucleus is of central importance in cell heredity? Describe the nucleo-cytoplasmic inter-relationship of eukaryotic cell.

Short Answer Questions

1. Why the term nuclear membrane is replaced by nuclear envelope?

2. Describe the importance of Hammerling's experiment on *Acetabularia*. (*CCS 1991*)

3. Discuss the functions of nucleolus.

4. Describe the corelation between fibrillar and granular portions of nucleolus.

5. Describe in brief the nucleolar cycle.

6. Give reasons why nucleolus is small in muscle cell but large in liver cells.

7. Write short notes on the following:

(*i*) Euchromatin

(*ii*) Functions of nuclear envelope (*Gujarat 1989, 1990; Garhwal 1990*)

(*iii*) Functions of nucleolus (*Garhwal 1990; Jiwaji 1992*)

(*iv*) Heteropyknosis

(*v*) Nucleoproteins

(*vi*) Chromocentres

(*vii*) Nucleolar cycle

(*viii*) Nucleolus (*Rajasthan 1989; Sagar 1990; BHU (Hons.) 1990; Punjab (G) 1992, Purvanchal 1992; Guwahati (M) 1992, 1996; CCS 1991, 1992, 1994; Gorakhpur 1993; Lucknow 1994, 1998*)

(*ix*) Functions of nucleus (*Jiwaji 1990*)

(*x*) Nuclear pores (*CCS 1990*)

8. Differentiate between the following:

1. Euchromatin and heterochromatin (*NEHU (M) 1996*)
2. Nucleus and nucleolus.

9. Draw a neat and labelled diagram of nucleus. *(CCS 1999)*

Very Short Answer Questions

1. Why mammalian RBCs have a very short life span?
   Ans. Mammalian RBCs are without nucleus.
2. What will happen if nucleus in *Amoeba* is destroyed?
   Ans. Such *Amoeba* will die.
3. Who discovered the nucleus?
   Ans. Robert Brown in 1833.
4. What is the nucleus?
   Ans. Nucleus is a large, double membrane enclosed organelle that contains the chromosomal DNA of a eukaryotic cell.
5. Define the nucleolus.
   Ans. Nucleolus is a large, spherical structure present in the nucleus of an eukaryotic cell; it is site of ribosomal RNA synthesis and processing and of the assembly of ribosomal subunits.
6. What is nuclear envelope?
   Ans. Nuclear envelope is a double membrane around the nucleus that is interrupted by numerous small pores.
7. Define nuclear lamina.
   Ans. Nuclear lamina is a thin, dense, meshwork of fibre that lines the inner surface of the inner nuclear membrane and helps support the nuclear envelope.
8. What is the nuclear pore?
   Ans. It is a small opening in the nuclear envelope through which molecules enter and exist the nucleus; it is lined by an intricate protein structure called the nuclear pore complex (NPC).
9. Define nuclear pore complex.
   Ans. It is a intricate protein structure, composed of 100 or more different polypeptide subunits, that lines the nuclear pores through which molecules enter and exit the nucleus, both by simple diffusion of smaller molecules and active transport of large molecules.
10. What is chromatin?
    Ans. Chromatin is DNA-protein fibre that make up chromosomes. It is constructed from nucleosomes spaced regularly along a DNA chain.
11. How many types of chromatin occur in the nucleus?
    Ans. Euchromatin and heterochromatin.
12. What is heterochromatin?
    Ans. Heterochromatin is highly compacted form of chromatin present during interphase; it contains DNA that is not being transcribed.
13. Define euchromatin.
    Ans. Euchromatin is loosely packed, uncondensed form of chromatin present during interphase; it contains DNA that is being actively transcribed.
14. What is NOR?
    Ans. NOR stands for nucleolus organizer region which is a stretch of DNA in certain chromosomes where multiple copies of the genes for ribosomal RNA are located and where nucleoli form.
15. What is main function of nucleolus?
    Ans. Nucleolus is a ribosome producing organ.

Multiple Choice Questions

1. Site of formation of immature ribosomal subunits in a eukaryotic cell is:
   (*a*) cytoplasm (*b*) nucleolus
   (*c*) nucleus (*d*) nuclear pore complex
2. The role of nucleus in heredity was at first demonstrated by

3. Who coined the term nucleolus?
   (*a*) Fontana
   (*b*) Robert Brown
   (*c*) Camillo Golgi
   (*d*) Stransburger
4. Nucleolus is the site for the synthesis of
   (*a*) DNA
   (*b*) m-RNA
   (*c*) tRNA
   (*d*) ribosomes
5. An example for the enucleated living plant cell is
   (*a*) xylem parenchyma
   (*b*) RBCs
   (*c*) sieve tubes of phloem
   (*d*) companion cells of phloem

ANSWERS

1. (*b*) **2.** (*a*) **3.** (*a*) **4.** (*d*)
5. (*c*).

70

Chromosomes

The chromosomes (Gr., *chrom* = colour, *soma* = body) are the nuclear components of special organisation, individuality and function. They are capable of self-reproduction and play a vital role in heredity, mutation, variation and evolutionary development of the species.

Chromosomes are most distinct in metaphase stage of cell division. In an interphase nucleus these form an interwoven net of fine chromatin threads. The term **chromosome** was introduced by **Waldeyer** in 1888 because of their affinity towards dyes or stains.

Types of Chromosomes

1. Viral chromosomes. In viruses, a single, linear or circular molecule of DNA or RNA represent the chromosome. Nucleic acid of the viral chromosome may be single-stranded or double-stranded.

2. Prokaryotic chromosomes (or Bacterial chromosomes). In bacteria and cyanobacteria (blue green algae), the hereditary material (DNA) is organized into a single circular chromosome. Chromosomal DNA is double stranded and lacks histone proteins. Prokaryotic chromosomes are also found in certain cellular organelles such as mitochondria and chloroplasts.

3. Eukaryotic chromosomes. In eukaryotic cells (*e.g.*, protistans plants, fungi and animals), chromosomes are formed of DNA and proteins (histones and non-histones). DNA is linear and double stranded.

In the chapter the structure and function of eukaryotic chromosomes only will be described.

Box 70.1 Distinction between viral and bacterial chromosomes

In viruses (including bacteriophages) a single molecule of DNA or RNA forms the viral chromosome. It is about 20 Å wide corresponding to the width of DNA molecule. The viral chromosome may be linear or circular.

Bacterial chromosome is a single, large circular chromosome, formed of a circular DNA molecule.

Box 70.2 Distinction between prokaryotic and eukaryotic chromosomes

Prokaryotic or bacterial chromosome is a large circular DNA molecule. It is without histones but may contain some polyamines. It lacks centromere and nucleosomes, but is folded into a series of **looped domains.**

Euchromatic chromosome is much complicated structure having centromere and nucleosomes. Each chromosome is formed by the condensation of a long chromatin thread. The chromatin thread is formed of supercoiling of a series of nuclesosomes. Each nucleosome is formed of histones and DNA.

70.1 CHEMICAL COMPOSITION OF CHROMOSOMES

Chromatin which has been isolated from rat liver contains DNA, RNA and protein. The protein of chromatin is of two types : the histones and the non-histones. Rat liver chromatin has been used as a model for chromatin. It possesses a histone to DNA ratio near 1 : 1, a non-histone protein to DNA ratio of 0.6 : 1 and a RNA/DNA ratio of 0.1 : 1.

1. DNA.

DNA is the most important chemical component of chromatin, since it plays the central role of controlling heredity. The most convenient measurement of DNA is picogram (10^{-12} gm). DNA of chromatin represent the following two phenomena:

The C-value. The DNA in nuclei was stained using the **Feulgen reactions** and the amount of stain in single nuclei was measured using a special microscope (called **cytophotometer**). Both of these techniques demonstrated that nuclei contain a constant amount of DNA. Thus, all the cells in an organism contain the same DNA content (2C) provided that they are diploid. Gametes are haploid and, therefore, have half the DNA content (1C). Some tissues such as liver, contain occasional cells that are polyploid and their nuclei have a correspondingly higher DNA content (4C or 8C) (see Table 70.1).

Table 70.1 DNA content and chromosome component

	Cells	*Mean DNA-Feulgen content*	*Presumed chromosome set*
1.	Spermatid	1.68 (1C)	Haploid (n)
2.	Liver	3.16 (2C)	Diploid (2n)
3.	Liver	6.30 (4C)	Tetraploid (4n)
4.	Liver	12.80 (8C)	Octoploid (8n)

Thus, each species has a characteristic content of DNA which is **constant** in all the individuals of that species and has, thus, been called the C-value.

The C-value paradox. Eukaryotes vary greatly in DNA content but always contain much more DNA than prokaryotes. Lower eukaryotes in general have less DNA, such as nematode *Caenorhabditis elegans* which has only 20 times more DNA than *E.coli*, or the fruit fly *D. melanogaster* which has 40 times more DNA (*i.e*, 0.18 pg or picogram per haploid genome). Vertebrates have greater DNA content (about 3 pg), in general about 700 times more then *E. coli*. One of the highest DNA content is that of the salamander *Amphiuma* which has 84 pg of DNA. Each human being has about 3 pg of DNA per haploid genome, or 3×10^9 base pairs, *i.e.*, the human genome could accommodate about 3 million average sized proteins. If all the DNA were coding (or containing structural genes) and if this was true, salamanders would have 30 times more genes than human beings. This is called **C-value paradox** (**Gall**, 1981). It was detected quite early that there was little connection between the morphological complexity of eukaryotic organisms and their DNA content. For example, *E. coli* (containing 3,400,000 base pairs in its DNA) has about 3000 genes. Although it is difficult to estimate how many different genes exist in the human genome, there are probably not more than 20,000 to 30,000 genes (Note : According to a most recent estimate, there are 100,000 genes in human genome, see **Deviah**, 1994). There is no reason to believe that salamanders should have any more. From these facts, it can be easily concluded that most of the DNA in the eukaryotic genomes must be of a non-coding nature.

2. Histones.

Histones are very basic proteins, basic because they are enriched in the amino acids arginine and lysine to a level of about 24 mole present. Arginine and lysine at physiological pH are cationic and can interact electrostatically with anionic nucleic acids. Thus, being basic, histones bind tightly to DNA which is an acid. There are five types of histones in the eukaryotic chromosomes, namely H1, H2A, H2B, H3 and H4.

One of the important discoveries that has come from chemical studies is that the primary structures of histones have been highly conserved during evolutionary history. For example, histone H4 of calf and of garden pea contains only two amino acid differences in a protein of 102 residues (**DeLange**, 1969). Likewise, the sequence of histone H3 from rat differs only in two amino acids from that of peas, out of 102 total amino acid residues. These organisms are estimated to have an evolutionary history of at least 600 million years, during which they diverged structurally. This conservation of structure suggests that over the eras, histones have had a very similar and crucial role in maintaining the structural and functional integrity of chromatin. Such an evolutionary conservation suggests that the functions of these two histones involve nearly all of their amino acids so that a change in any position is deleterious to the cell.

Histone H1 is the least rigidly conserved histone protein. It contains 210 to 220 amino acids and may be represented by a variety of forms even within a single tissue. H1 is present only once per 200 base pairs of DNA (in contrast to rest of the four types of histones each of which is present twice) and is rather loosely associated with DNA. H1 histone is absent in yeast, *Saccharomyces cerevisiae*. Histones besides determining the structure of chromatin, play a regulatory role in the repression activity of genes.

3. Non-histones. In contrast to the modest population of histones in chromatin, non-histone proteins display more diversity. In various organisms, number of non-histones can vary from 12 to 20. Heterogeneity of these proteins is not conserved in evolution as the histones. These non-histones differ even between different tissues of the same organism suggesting that they regulate the activity of specific genes.

About 50 per cent non-histones of chromatin have been found to be structural proteins and include such proteins as **actin**, and **α-** and **β-tubulins** and **myosin**. Although for sometime these contractile proteins were thought to be contaminants, it is now believed that they are vital ingredients of the chromosome, functioning during chromosome condensation and in the movement of chromosomes during mitosis and meiosis (see **Thorpe**, 1984). Many of the remaining 50 per cent of non-histones include all the enzymes and factors that are involved in DNA replication, in transcription and in the regulation of transcription. These proteins are not as highly conserved among organisms, although they must carry out similar enzymatic activities. Apparently they are not as important as the histones in maintaining chromosome integrity.

70.2 STRUCTURE OF CHROMOSOMES

Shape

The shape of the chromosomes is changeable from phase to phase in the continuous process of the cell growth and cell division. In the resting phase or interphase of the cell, the chromosomes occur in the form of thin, coiled, elastic and contractile, thread-like stainable structures, the chromatin threads. In the metaphase and the anaphase, the chromosomes become thick and filamentous. Each chromosome contains a clear zone, known as **centromere** or **kinetocore**, along their length. The centromere divides the chromosomes into two parts, each part is called **chromosome arm**.

Depending upon the number of centromeres, an eukaryotic chromosome may be of following types:

1. Monocentric chromosome has one centromere. Most normal chromosomes are monocentric.

2. Holocentric chromosome has a diffuse centromere, so that spindle attachment occurs over the entire chromosome, *e.g.*, Lepidoptera.

3. Dicentric chromosome is with two centromeres.

4. Polycentric chromosome is with more than two centromeres.

5. Acentric chromosome is without any centromere. Such chromosomes represent freshly broken segments of chromosomes which do not survive for long.

In a monocentric chromosome, the position of centromere varies form chromosome to chromosome and it provides different shapes to the latter which are following (Fig. 70.1):

1. Telocentric (Telomictic). It is a rod-like monocentric chromosome with a terminal centromere. It is also known as **T-chromosome** or **monobrachial chromosome**. Such chromosomes are very rare in nature.

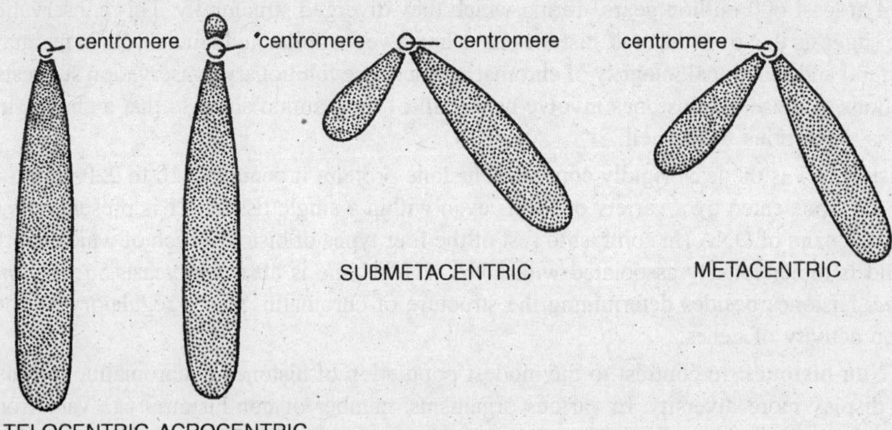

Fig. 70.1. The four morphological types of chromosomes according to the position of the centromere.

2. Atelocentric (Atelomictic). It is a monocentric chromosome with a nonterminal centromere. It is also called **dibrachial chromosome**. Atelocentric chromosomes are of following four types:

***(i)* Metacentric.** It is a monocentric V-shaped chromosome with a central or near central (Fig. 70.2) chromosome. If the centromere is exactly at the median point, it is classified as a **M-chromosome**, whilst if the centromere is not central but lies within the median region, it is classified as a **m-chromosome**. M-chromosomes are rare in nature.

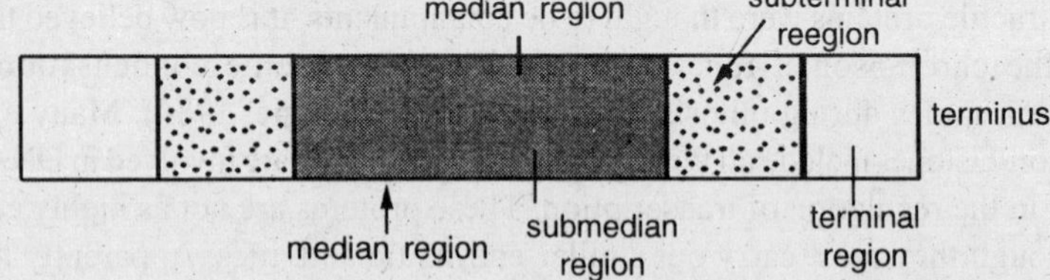

Fig. 70.2. Various regions of a chromosome used in classification of chromosomes.

***(ii)* Submetacentric.** It is a monocentric chromosome with a centromere in the **submedian region.** It is J-or L-shaped.

***(iii)* Subacrocentric.** It is a monocentric chromosome with a centromere in the **subterminal region** (Fig. 70.2).

***(iv)* Acrocentric.** It is a monocentric chromosome with its centromere in **terminal** region. It is also called **t-chromosome**, since most so-called telocentric chromosomes are actually acrocentric.

Structure (Parts of Chromosomes)

In the metaphase chromosomes following structures or parts have been recognized:

1. Chromatid. At mitotic metaphase each chromosome consists of two symmetrical structures, called **chromatids**. Each chromatid contains a single DNA molecule. Both chromatids are attached to each other only by the centromere and become separated at the beginning of anaphase, when the sister chromatids of a chromosome migrate to the opposite poles.

2. Chromonema (ta). During mito-tic prophase the chromosomal material becomes visible as very thin filaments, called **chromonemata** (a term coined by **Vejdovsky** in 1912). A chromonema represents a chromatid in the early stages of condensation. Therefore, 'chromatid' and 'chromonema' are two names for the same structure: a single linear DNA molecule with its associated proteins. The chromonemata form the gene-bearing portions of the chromosomes.

According to old view, a chromosome may have more than one chromone-mata which are embedded in the achro-matic and amorphous substance, called **matrix**. The matrix is enclosed in a sheath or **pellicle**. Both matrix and pellicle are non-genetic materials and appear only at metaphase when the nucleolus disapp-ears. It is believed that nucleolar material and matrix are interchangeable, *i.e.,* when chromosomal matrix disappears, the nucleolus appears and vice versa. Electron microscopic observations, however, have questioned the occurrence of pellicle and matrix in them.

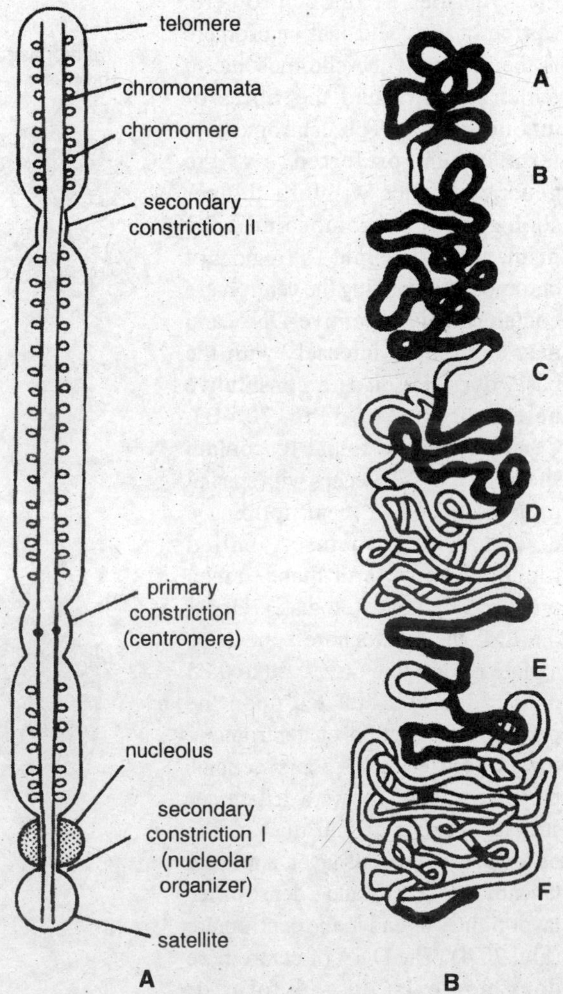

Fig. 70.3. A—Structure of a typical chromosome; B–Model of constitutive heterochromatin in a mammalian metaphase chromosome. [A—Constitutive heterochromatin; B—Secondary constriction; I or nucleolar organizer; C–Primary constriction or centromere; D—Euchromatin; E—Secondary constriction. II—possible site of 5S rRNA cistrons; F—Telomere].

3. Chromomeres. The chromomeres are bead-like accumulations of chromatin material that are sometimes visible along interphase chromosomes. The chromomere-bearing chromatin has an appearance of a necklace in which several beads occur on a string. Chromomeres become especially clear in the polytene chromosomes, where they become aligned side by side, constituting the chromosome beads (Fig. 70.3). At metaphase the chromosomes are tightly coiled and the chromomeres are no longer visible.

Chromomeres are regions of tightly folded DNA and have great interest for the cell biologists. They are believed to correspond to the units of genetic function in the chromosomes. In fact, for long time most geneticists considered these chromomeres as genes, *i.e.*, the units of heredity.

4. Centromere and kinetochore. Originally it was considered that the centromere consists of small granules or spherules. The centromere of the chromosome of the *Trillium* has the diameter of 3μm and the spherules have the diameter of 0.2 μm. The chromonema remains connected with the spherules of the centromere. Currently it is held that centromere is the region of the chromosome to which are attached the fibres of mitotic spindle. The centromere (a term much preferred by the geneticists) lies within a thinner segment of chromosome, the **primary constriction.** The regions of chromosome flanking the centromere contain highly repetitive DNA and may stain more intensely with the basic dyes (*i.e.,* it is a constitutive heterochromatin, Fig. 70.4B). Centromeres are found to contain specific DNA sequences with special proteins bound to them, forming a disc-shaped structure, called **kinetochore** (a term that is much preferred by the cytologists). Under the EM, the kinetochore appears as a plate-or cup-like disc, 0.20 to 0.25 nm, in diameter situated upon the primary constriction or centromere. In thin electron micro-scopic sections, the kinetochore shows a trilaminar structure, *i.e.,* a 10 nm thick dense **outer proteinaceous** layer, a **middle** layer of low density and a dense **inner** layer tightly bound to the centromere (Fig. 70.4). The DNA of centromere does not exist in the form of nucleosome.

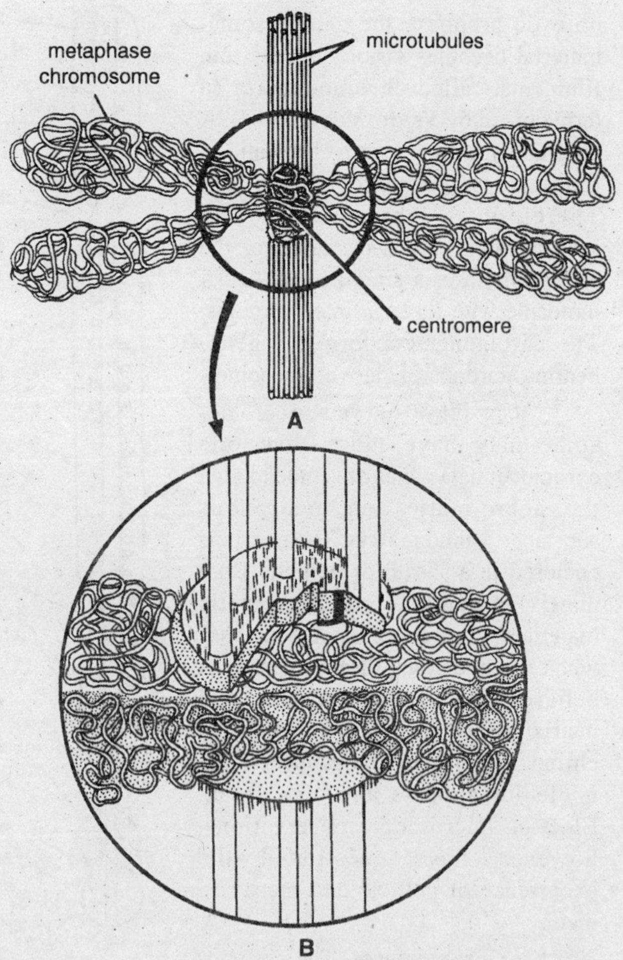

Fig. 70.4. A—A metaphase chromosome showing the folded fibre structure and the centromere with implanted microtubules; B—Inset: a higher magnification, showing the convex electron dense layer and the fibrillar material forming the "corona" of the centromere.

5. Telomere. (Gr., *telo*=for; *meros*=part). Each extremity of the chromosome has a polarity and therefore, it prevents other chromosomal segments to be fused with it. The chromosomal ends are known as the **telomeres.** If a chromosome breaks, the broken ends can fuse with each other due to lack of telomeres.

6. Secondary constriction. The chromosomes besides having the primary constriction or the centromere possess secondary constriction at any point of the chromosome. Constant in their position and extent, these constrictions are useful in identifying particular chromosomes in a set. Secondary constrictions can be distinguished from primary constriction or centromere, because chromosome bends (or exhibits angular deviation) only at the position of centromere during anaphase.

7. Nucleolar organizers. These areas are certain secondary constrictions that contain the genes coding for 5.8S, 18S and 28S ribosomal RNA and that induce the formation of nucleoli. The secondary constriction may arise because the rRNA genes are transcribed very actively and, thus, interfering with chromosomal condensation. In human beings, the nucleolar organizers are located in the secondary constrictions of chromosomes 13, 14, 15, 21 and 22, all of which are acrocentric and have satellites.

8. Satellite. Sometimes the chromosomes bear round elongated or knob-like appendages known as **satellites**. The satellite remains connected with the rest of the chromosome by a thin chromatin filament. The chromosomes with the satellite are designated as the **sat chromosomes.** The shape and size of the satellite remain constant.

Chromosome satellites are a morphological entity and should not be confused with satellite DNAs which are highly repeated DNA sequences.

Material of the Chromosomes

The material of the chromosomes is the chromatin. Depending on their staining properties, the following two types of chromatin may be distinguished in the interphase nucleus:

1. Euchromatin. Portions of chromosomes that stain lightly are only partially condensed; this chromatin is termed **euchromatin.** It represents most of the chromatin that disperse after mitosis has completed. Euchromatin contains structural genes which replicate and transcribe during G_1 and S phase of interphase. The euchromatin is considered genetically active chromatin, since it has a role in the phenotype expression of the genes. In euchromatin, DNA is found packed in 3 to 8 nm fibre.

2. Heterochromatin. In the dark-staining regions, the chromatin remains in the condensed state and is called **heterochromatin.** In 1928, **Heitz** defined heterochromatin as those regions of the chromosome that remain condensed during interphase and early prophase and form the so-called **chromocentre.** Heterochromatin is characterized by its especially high content of repititive DNA sequences and contains very few, if any, structural genes (*i.e.,* genes that encode proteins). It is **late replicating** (*i.e.,* it is replicated when the bulk of DNA has already been replicated) and is not transcribed. It is thought that in heterochromatin the DNA is tightly packed in the 30 nm fibre.

Types of heterochromatin. In an interphase nucleus, usually there is some condensed chromatin around the nucleolus, called **perinucleolar chromatin**, and some inside the nucleolus, called **intranucleolar chromatin.** Both types of this heterochromatin appear to be connected and together, they are referred to as **nucleolar chromatin.**

Dense clumps of deeply staining chromatin often occur in close contact with the inner membrane of the nuclear envelope (*i.e.,* with the nuclear lamina) and is called **condensed peripheral chromatin.** Between the peripheral heterochromatin and the nucleolar heterochromatin are regions of lightly staining chromatin, called **dispersed chromatin.** In the condensed chromosomes, the heterochromatic regions can be visualized as regions that stain more strongly or more weakly than the euchromatic regions, showing the so-called **positive** or **negative heteropyknosis** of the chromosomes (Gr., *hetero* = different + *pyknosis* = staining).

Heterochromatin has been further classified into the following types:

1. Constitutive heterochromatin. In such a heterochromatin the DNA is permanently inactive and remains in the condensed state throughout the cell cycle. This most common type of heterochromatin occurs around the centromere, in the telomeres and in the C-bands of the chromosomes. In *Drosophila virilis*, constitutive heterochromatin exists around the centromeres and such **pericentromeric heterochromatin** occupies 40 per cent of the chromosomes. In many species, entire chromosomes become heterochromatic and are called **B chromosome, satellite chromosomes** or **accessory chromosomes** and contain very minor biological roles. Such chromosomes comprising wholly constitutive heterochromatin occur in corn, many phytoparasitic insects and salamanders. In the fly *Sciara*, large metacentric heterochromatic chromosomes are found in the gonadal cells, but are absent in somatic cells. Entire Y chromosome of male *Drosophila* is heterochromatic, even though containing six gene loci which are necessary for male fertility.

Constitutive heterochromatin contains short repeated sequences of DNA, called **satellite DNA.** This DNA is called satellite DNA because upon ultracentrifugation, it separates from the main component of DNA. Satellite DNA may have a higher or lower G + C content than the main fraction. For example, the mouse satellite DNA is a 240 base pair sequence that is repeated about 1000,000

(10^6) times in the mouse genome, constituting 10 per cent of the total mouse DNA. The exact significance of constitutive heterochromatin is still unexplained.

2. Facultative heterochromatin. Such type of heterochromatin is not permanently maintained in the condensed state; instead it undergoes periodic dispersal and during these times is transcriptionally active. Frequently, in facultative heterochromatin one chromosome of the pair becomes either totally or partially heterochromatic. The best known case is that of the X-chromosomes in the mammalian female, one of which is active and remains euchromatic, whereas the other is inactive and forms at interphase, the **sex chromatin** or **Barr body** (Named after its discoverer, Canadian cytologist **Murray L. Barr**). Barr body contains DNA which is not transcribed and is not found in males. Indeed, the number of Barr bodies is always one less than the number of X chromosomes (*i.e.*, in humans, XXX female has two Barr bodies and XXXX female has three Barr bodies; **M.L. Barr**, 1959).

70.3 ULTRASTRUCTURE AND MOLECULAR ORGANIZATION

The structural unit of each chromosome is single DNA complexed with basic proteins (histones) in 1 : 1 ratio. This is called **single stranded hypothesis**.

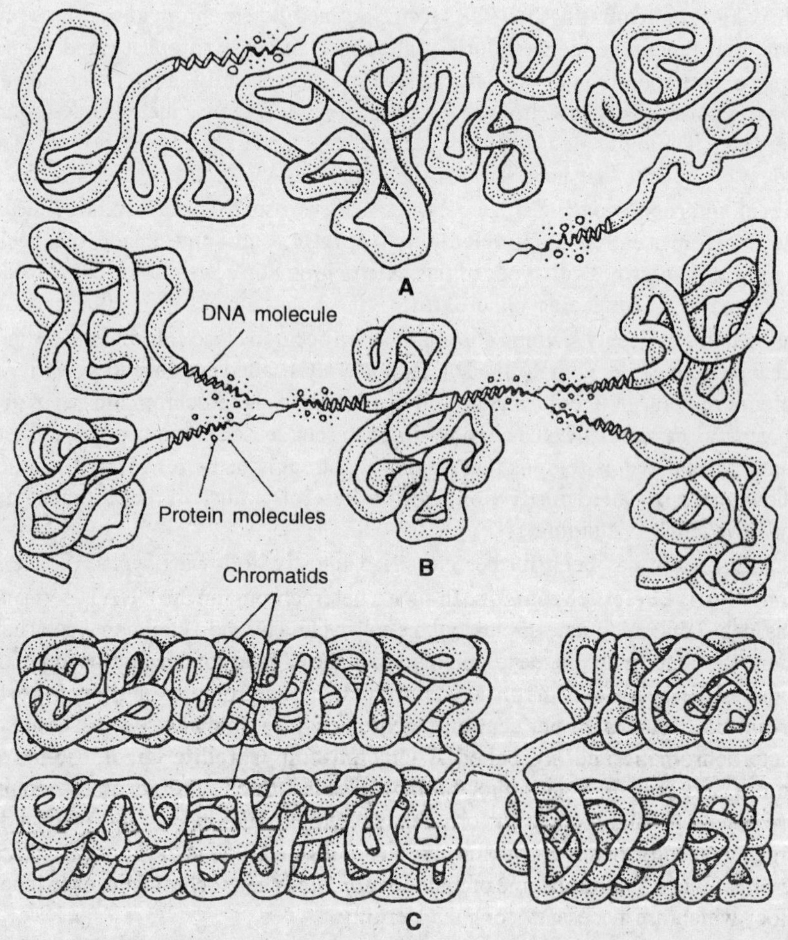

Fig. 70.5. Dupraw's folded fibre model of chromatin in interphase (A and B) and in metaphase (C).

The manner of coiling and folding of DNA was a matter of debate and dozens of models were available for this purpose; of them only two stand out and are important. A popular model is the

folded fibre model, proposed by **E.J. Dupraw** in 1965. According to it, the bulk of the chromosome is visualized to be composed of a tightly folded fibre which has a rather homogeneous diameter of 200 to 300 A°. This folded fibre is supposed to contain the DNA histone helix (of 30 A° diameter) in a supercoiled condition (Fig. 70.5). Another model is most significant and universally accepted one and is called **nucleosome model** which was proposed by **R.D. Kornberg** (1974) (Fig. 70.6) and confirmed and christened by **P. Oudet** *et al.*, (1975). Thus, while in the folded-fibre model, it was proposed that the histones were bound on the outside of the DNA coils (*i.e.*, histone shell around DNA), the nucleosome model has proposed the opposite (*i.e.*, histone particle with DNA round it). In other words, the earlier theory that basic chromatin fibre had DNA core surrounded by histones was incorrect. In fact, from a genetic perspective, a significant feature of packing mechanism through the nucleosomes lies in its topology: *at no point is the DNA buried;* instead, it is freely exposed along the entire surface of the "spool", available for genetic expression. Nucleosomes seem to be universal device for compacting the long DNA molecules of eukaryotic cells.

Nucleosomes and Solenoid Model of Chromatin

In eukaryotes, DNA is tightly bound to an equal mass of histones, which serve to form a repeating array of DNA-protein particles, called **nucleosomes**. If it was stretched out, the DNA double-helix in each human chromosome would span the cell nucleus thousands of time. His-tones play a crucial role in packing this very long DNA molecule in an orderly way (*i.e.*, nucleosome) into nucleus only a few micro-metres in diameter. Thus, nucleo-somes are the fundamental pack-ing unit particles of the chromatin and give chromatin a **"beads-on-a-string"** appearance in electron micorgraphs taken after treat-ments that unfold higher- order packing.

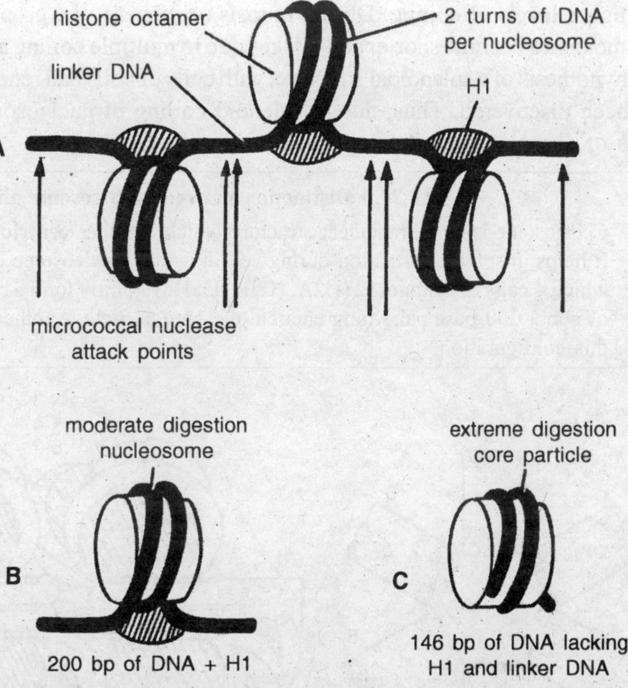

Fig. 70.6. Diagram showing the effect of nuclease enzymes on chromatin. A—Intact chromatin, note that the nucleosome is flat and that each histone octamer has two turns of DNA sealed off by histone H1. In living cells, the nucleosomes would be touching each other to form a 10 nm fibre and not stretched out (to form beads-on-a-string preparation); B—A nucleosome released by moderate digestion; C—Core particle of nucleosome obtained by extensive digestion.

The nucleosome 'beads' can be removed from long DNA "string" by digestion with enzy-mes that degrade DNA, such as bacterial enzyme, **micrococcal nuclease**. After digestion for a short period with micrococcal nuclease, only the DNA between the nucleosome beads is degraded (Fig. 70.6). The rest is protected from digestion and remains as double-stranded DNA fragments 146 nucleotide pairs long bound to a specific complex of 8 nucleosome histones (the **histone octamer**). The nucleosome beads obtained in this way have been crystallized and analyzed by X-ray diffraction

Each nucleosome is a disc-shaped particle with a diameter of about 11 nm and 5.7 nm in height containing 2 copies of each 4 nucleosome histones–H2A, H2B, H3 and H4. This histone octamer forms a protein core [(*i.e.*, a core of histone tetramer $(H3, H4)_2$ and the apolar regions of 2(H2A and H2B)] around which the double-stranded DNA helix is wound 1¾ time containing 146 base pairs. In undigested chromatin the DNA extends as a continuous thread from nucleosome to nucleosome. Each nucleosome bead is separated from the next by a region of **linker DNA** which is generally 54 base pair long and contains single H1 histone protein molecule. Generally, DNA makes two complete turns around the histone octamers and these two turns (200 bp long) are sealed off by H1 molecules. [**Note** : In some organisms nucleosome DNA may vary from 162 base pairs (*e.g.*, rabbit cortical neurons) to 242 base pairs (*e.g.*, sea urchin sperm)]. Thus, on an average, nucleosomes repeat at intervals of about 200 nucleotides or base pairs. For example, an eukaryotic gene of 10,000 nucleotide pairs will be associated with 50 nucleosomes and each human cell with 6×10^9 DNA nucleotide pairs contains 3×10^7 nucleosomes.

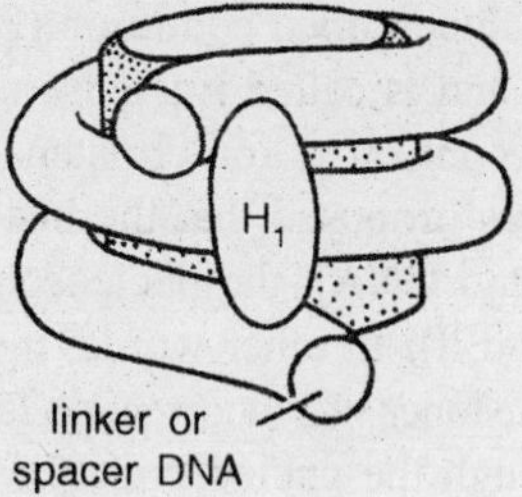

Fig. 70.7. A nucleosome.

Solenoid Models

H1 is reported to be phosphorylated just before mitotic and meiotic cell division to make possible the higher levels of coiling. During mitosis or meiosis, the prophase is the stage during which the chromosomes become shorter and thicker due to multiple coiling as proposed by **DuPraw** and others. The hypothesis of a solenoidal structure, with coils of coils had renewal since nucleosomal substructure has been discovered. Thus, due to solenoid coiling of nucleosome containing fibre, the following types of chromosomal structures can be observed during the cell cycle.

> **Box 70.3 Distinction between centrosome and nucleosome**
>
> Centrosome is an extranuclear structure which includes **centriole** with surrounding cytoplasm. It helps in spindle formation during cell division. Nucleosome is the unit of chromatin fibre in which 4 pairs of histones [2 (H2A, H2B, H3, H4)] jointly form a core particle and a DNA segment of about 200 base pairs long encircle the core particle. Another histone (H1 or H5) maintains this configuration.

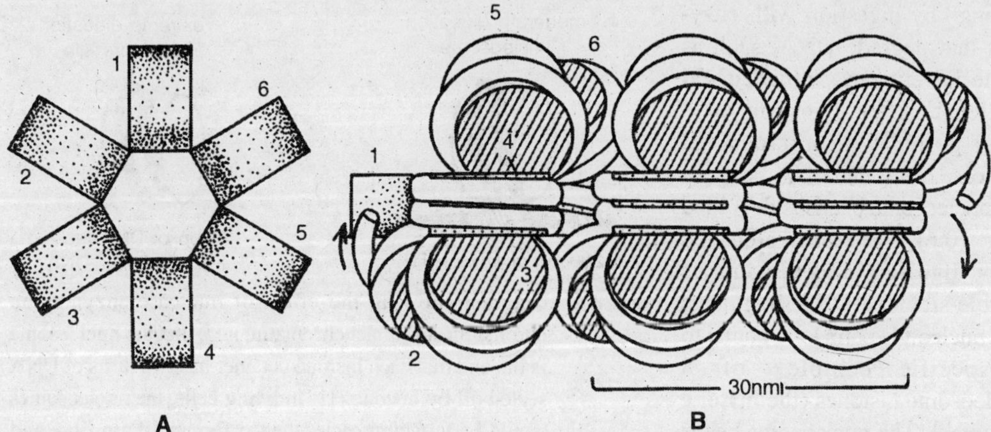

Fig. 70.8. A model suggested to explain how the "Beads-on-a-string" form of nucleosomes is packed to form the 30-nm fibre. A —Top view of 30-nm fibre; B—Side view of 30 nm fibre.

1. The 10-nm fibre. When nucleosomes are in close apposition, they form the **10-nm filaments**, in which packing of DNA is about five-to seven-fold, *i.e.*, five to seven times more compact than free DNA.

2. The 30-nm fibre. When nuclei are very gently lysed onto an electron microscopy grid, most of the chromatin is seen to be in the form of a fibre, with a diameter of about 30 nm. Such **30-nm fibres** can be observed in metaphase chromosomes and in interphase nuclei and it probably represents the natural conformation of transcriptionally inactive chromatin.

The 30-nm fibre consists of closely packed nucleosomes. It probably arises from the folding of the nucleosome chain into a **solenoid structure** having about six nucleosomes per turn. The DNA of 30-nm solenoid has a packing that is about 40-fold.

Histone H1 molecules are found responsible for packing nucleosomes into the 30-nm fibre. The H1 histone molecule has an evolutionarily conserved globular core or central region linked to extended amino-terminal and carboxyl-terminal "arms", whose amino acid sequence has evolved much more rapidly. Each H1 molecule binds through its globular portion to a unique site on a nucleosome and has arms that are thought to extend to contact with other sites on the histone cores of adjacent nucleosomes, so that the nucleosomes are pulled together into a regular repeating array. The binding of H1 molecule to chromatin tends to create a local polarity that the chromatin otherwise lacks.

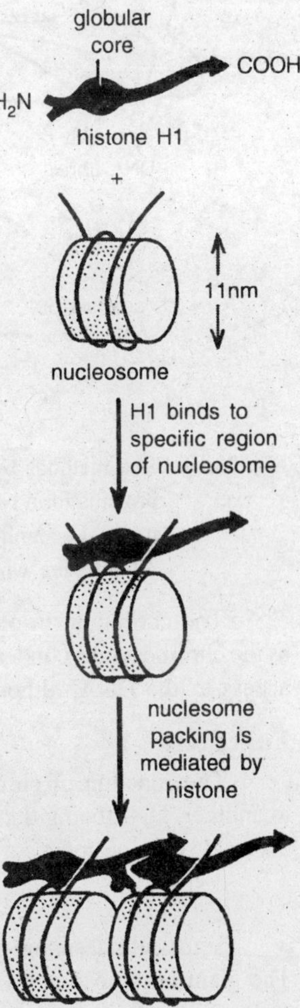

Fig. 70.9. The way histone H1 is thought to help to pack adjacent nucleosomes together.

3. Radial loops of 30-nm fibre and metaphase chromosome. The nucleus is typically about 5μm (5×10^{-4} cm) in diameter. The packaging of DNA into a 30-nm chromatin fibre leaves a human chromosome about 0.1 cm long, so there must be several higher orders of folding. The probable nature of one further level of folding was originally suggested by the appearance of specialized chromosomes—the lampbrush chromosomes and polytene chromosomes. These two types of chromosomes seem to contain a series of **looped domains**— loops of chromatin that extend at an angle from the main chromosome axis. Since such loops do occurs in *E.coli* chromosome, so the presence of loops may be a general feature of chromosomes.

In principle, looped domains in chromatin could be established and maintained by DNA binding proteins that clamp two regions of the 30- nm fibre together by recognizing specific DNA sequence that will form the neck of each loop. Alternatively, they could be formed by binding of DNA at the base of loop to a chromosome axis. Structural non-histone proteins could be involved in organizing the 30-nm fibres into loops. In an experiment, the histones are removed from the metaphase chromosome by adding the polyanion dextran sulphate. Histone-depleted chromosomes are found to have a central core of **scaffold** surrounded by a halo made of hoops of DNA (Fig. 70.10A). The scaffold is made of non-histone proteins and retains the general shape of the metaphase chromosome. Each chromosome has two scaffolds, one for each chromatid, and they are connected together at the centromere region. When the histones are removed, the DNA which has packed about 40-fold in the 30-nm chromatin fibre, becomes extended and produces loops with an average length of 25μm (75,000 base pairs). In each loop the DNA exits from the scaffold and returns to an adjacent point. On the basis of these observations a model of chromosome structure has been proposed by **Laemmli** and co-workers (1979, 1984). In Laemmli's radial loop model DNA is arranged in loops anchored to the non-histone scaffold. Because the lateral loops have 25μm DNA, after contracting 40-fold in the 30-nm fibre, they would be only about 0.6μm long, a length consistent with the diameter of metaphase chromosome (1μm). Figure 70.10B shows how the chromatin is arranged in loops which during metaphase, become arranged so that the base of the loops forms a scaffold in the centre of the chromatid. The base of the loop might be arranged on a helical coiled path (*e.g., Trillium* and *Tradescantia*).

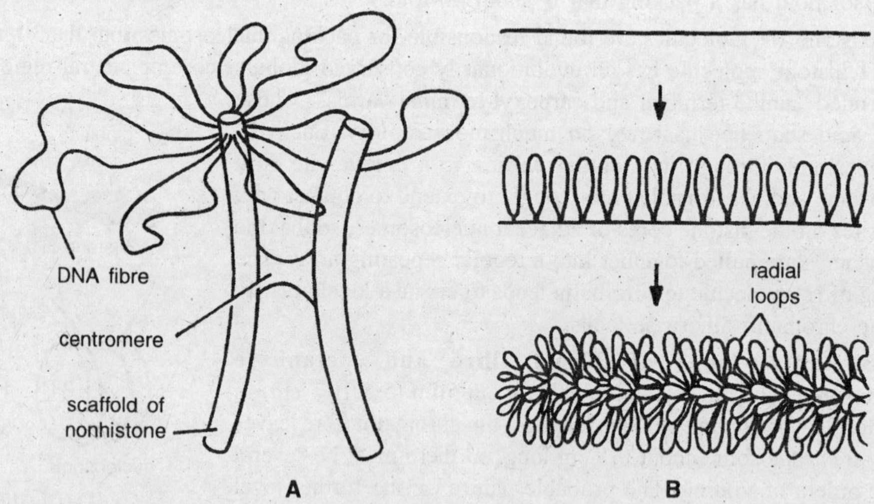

Fig. 70.10. Two methods by which DNA loops may form a metaphase chromosome. A—The non-histone proteins form two scaffolds, one per chromatid, while the naked DNA fibres form a halo around it; B—Laemmli's radial loop model of chromosome structure showing how 10-nm fibre form 30-nm fibre which further folds into radial loops.

The chromatin in mitotic chromosomes is transcriptionally inert: all RNA synthesis ceases as the chromosomes condense. Presumably the condensation prevents RNA polymerase from gaining access to the DNA, although other controlling factors might also be involved.

Functions

The function of chromosomes is to carry the genetic information from one cell generation to another. DNA being the only permanent component of chromosome structure, is the sole genetic material of eukaryotes.

70.4 GIANT CHROMOSOMES

Some cells at certain particular stages contain large nuclei with giant or large-sized chromosomes. The giant chromosomes are the **polytene** and **lampbrush** chromosomes.

1. Polytene Chromosome (Salivary Gland Chromosomes)

An Italian cytologist **E.G. Balbiani** (1881) had observed peculiar structures in the nuclei of certain secretory cells (*e.g.*, of salivary glands) of midge, *Chironomus* (Diptera). These structures were long and sausage-shaped and marked by swellings and cross striations (transverse bands). Unfortunately, he did not recognize them as chromosomes, and his report remained buried in the literature. It was not until 1933 that **Theophilus Painter**, **Ernst Heitz**, and **H. Bauer** rediscovered them in *Drosophila* and recognized them as the chromosomes. Since these chromosomes were discovered in the salivary gland cells, they were called **salivary gland chromosomes** (Fig. 70.11). The present name **polytene chromosomes** was suggested by **Kollar** due to the occurrence of many chromonemata (DNA) in them.

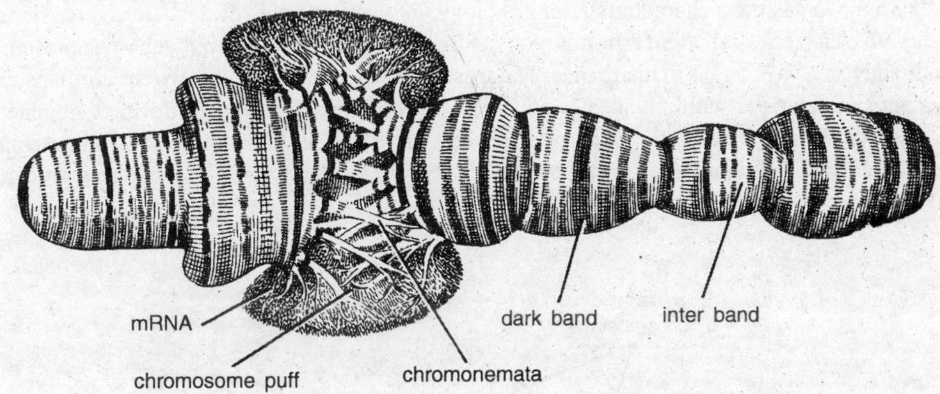

Fig. 70.11. Polytene chromosome of an insect, showing bands and interbands and a puff or Balbiani ring.

Thus, some cells of the larvae of the dipteran insects such as flies (*e.g., Drosophila*), mosquitoes and midges (*Chironomus*) become very large having high DNA content. These cells are unable to undergo mitosis and are destined to die during metamorphosis (Those cells of larva which are destined to produce the adult structures after metamorphosis, *i.e.,* imaginal discs remain diploid). Such polytenic cells are located most prominently in the salivary gland, but also occur in Malpighian tubules, rectum, gut, foot pads, fat bodies, ovarian nurse cells, etc. Polyteny of giant chromosomes is achieved by replication of the chromosomal DNA several times without nuclear division (endomitosis); and the resulting daughter chromatids do not separate but remain aligned side by side. In the process of endomitosis the nuclear envelope does not rupture and no spindle formation takes place. In fact, polyteny differs from polyploidy, in which there is also an excess DNA per nucleus, but in which the new chromosomes are separated from each other.

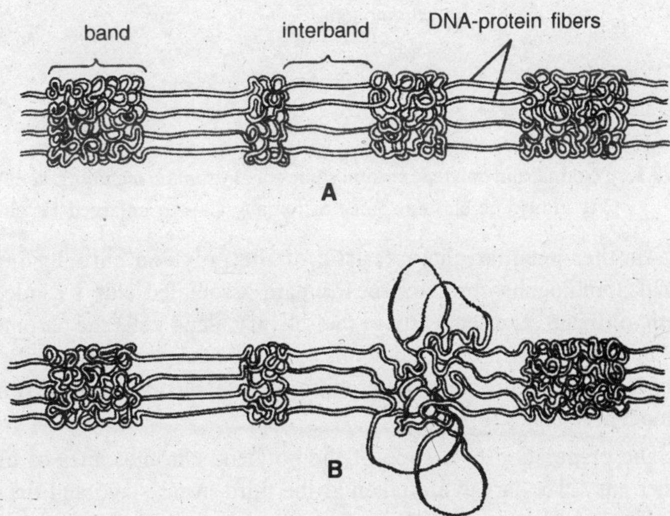

Fig. 70.12. A—Model of Dupraw and Rae showing band and interband structure in giant chromosome. B—The same model showing a possible method of "puffing".

A polytene chromosome of *Drosophila* salivary **gland** has about 1000 DNA molecules which are arranged side by side and which arise from 10 rounds of DNA replication ($2^{10} = 1024$). Other dipteran species have more

DNA, for example, *Chironomus* has 16000 DNA molecules in their each polytene chromosomes. Further, the polytene chromosomes are visible during interphase and prophase of mitosis. In them the chromomere (regions in which the chromatin is more tightly coiled) alternate with regions where the DNA fibres are folded more loosely (Fig. 70.12). The alignment of many chromomeres gives polytene chromosomes their characteristic morphology, in which a series of dark transverse **bands** alternates with clear zones called **interbands**. About 85 per cent of the DNA in polytene chromosomes is in bands and rest 15 per cent is in interbands. The crossbanding pattern of each polytene chromosome is a constant characteristic within a species and helps in chromosome mapping during cytogenetic studies. For example, in *Drosophila melanogaster* there are about 5000 bands and 5000 interbands per genome and each band and interband represent a set of 1024 identical DNA sequences arranged in file.

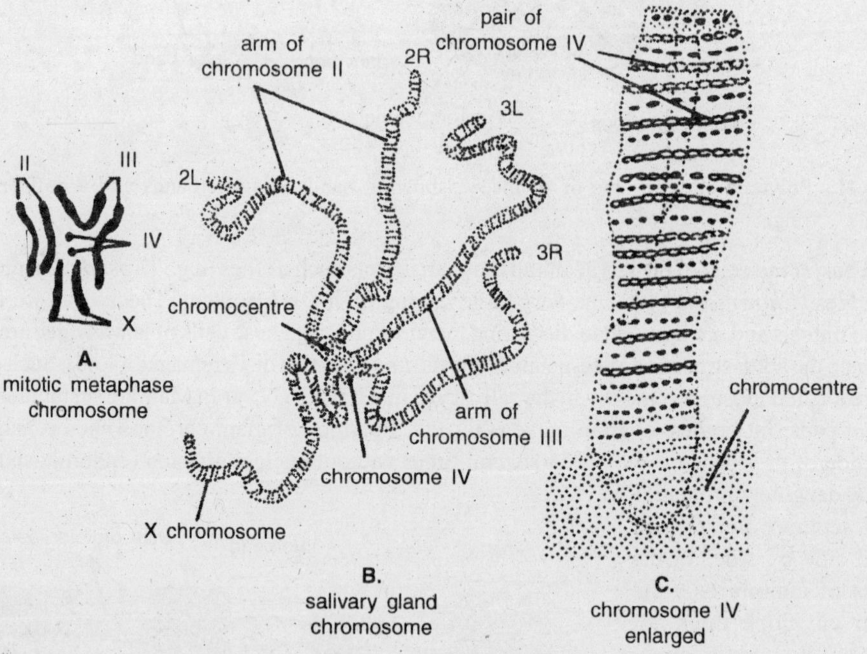

Fig. 70.13. Normal and polytene chromosomes of *Drosophila melanogaster.* A—Normal mitotic chromosomes; B—Polytene chromosomes of female; C—An enlarged IV chromosome.

Another peculiar characteristic of the polytene chromosomes is that the maternal and paternal homologous chromosomes remain associated side by side. This phenomenon is called **somatic pairing**. Consequently in the salivary gland cells the chromosome number always appear to be half of the normal somatic cells, *e.g., D. melogaster,* has only 4 polytene chromosomes. In *Drosophila*, pericentromeric heterochromatin of all polytene chromosomes also coalesces in a **chromocentre** (Fig. 70.13).

The preparation of a slide of the polytene chromosomes of dipterans for light microscopy is rather easy. The larvae are taken at the third instar stage and the salivary glands are dissected out and squashed in acetocarmine. In such preparations, these chromosomes in aggregate reach a length of as much as 2000 μm in *D. melanogaster*. In female *Drosophila,* the polytene chromosomes are found in the form of five long and one short strands radiating from a single more or less amorphous chromocentre (Fig. 70.13). One long strand corresponds to the X chromosome and remaining four long strands are the left and right arms of II and III chromosomes. The shortest strand represents the small dot-like IV chromosome. Each of these chromosomes contains maternal and paternal homologs in somatic pairing which lacks in the sex chromosomes of male fruit flies. Thus, in male *Drosophila,* X chromosome remains single and thin and Y chromosome exists indistinctly fused with the chromocentre.

One-gene, one-band hypothesis. The fixed pattern of bands and interbands in a *Drosophila* polytene chromosome suggested the early cytologists such as **Painter** (1933) and **Bridges** (1936) that each band might correspond to a single gene. Accordingly, they concluded that *Drosophila* might contain only 5000 essential genes. It was also believed that bands were the sites of genes (DNA) and interbands were relatively inert linker regions. Recent data, however, have contradicted this simple "one-band, one-gene" hypothesis, now it is held that bands as well as inter-bands contain active genes and a band may even contain more than one genes.

At this juncture a question arises, why is the single long strand of chromatin in each chromosome subdivided into so many distinct regions ? Exact explanation of this question is still not known. However, **Alberts** *et al.,* (1989) believed that this type of organization (*i.e.,* banding and interbanding of chromosomes in general) may help to: (1) Keep the DNA organized; (2) Isolate genes from their neighbours and thereby prevent biological "crosstalk", or (3) Regulate gene transcription for the cytodifferentiation, for example, constitutively expressed "**housekeeping**" **genes** could be located in interbands, whereas **cell-type-specific genes** could be confined to the bands.

Chromosome puffs or Balbiani rings. Chromosome puffs or **Balbiani rings** are the swellings of bands of the polytene chromosomes (Fig. 70.14) where DNA unfolds into open loops as a consequence of intense gene transcription (*e.g.*, mRNA formation). In 1954, **W. Beerman** compared the polytene chromosomes of different tissues of *Chironomus* larvae and showed that although the pattern of bands and interbands was similar in all tissues, the distribution of puffs differred from one tissue to another. These puffs represent regions where the tightly coiled chromosomal fibres open out to form many loops. In fact, puffing is a cyclic and reversible phenomenon: at definite times and in different tissues of the larvae, puffs may appear, grow and disappear. In salivary glands the appearance of some puffs has been corelated with the production of specific proteins which are secreted in large amounts in the larval saliva. The process of puffing involves several processes such as the accumulation of acidic proteins, despiralization of DNA, accumulation of **RNA polymerase II** (an enzyme involved in the transcription of mRNA molecules), synthesis of mRNA and release of newly synthesized mRNA in the cytoplasm.

2. Lampbrush chromosomes. The lampbrush chromosomes were first observed by **Flemming** in salamander (amphibian) oocytes in 1882. He coined the name because the chromosomes look like the brushes which were used for cleaning the glass chimneys of old-fashioned paraffin or kerosene lamps. They were described in detail in shark oocytes by **R. Ruckert** in 1892. **Thorpe** (1984) and **Burns** and **Bottino** (1989) preferred the term **test tube brush chromosomes** for them. However, due to recent investigations the structure of these exceptionally large-sized chromosomes has been interpreted in functional terms, *i.e.*, now they are merely visualized as means of *"turning on and turning off*" of the genes.

The lampbrush chromosomes occur at the diplotene stage of meiotic prophase in the primary oocytes of all animal species, both vertebrates and invertebrates. Thus, they have been described in *Sagitta* (Chaetognatha), *Sepia* (Mollusca), *Echinaster* (Echinodermata) and in several species of insects, shark, amphibians, reptiles, birds and mammals (humans). Lampbrush chromosomes are also found in spermatocytes of several species, giant nucleus of *Acetabularia* and even in plants. Generally, they are smaller and "hairy" in invertebrates than in vertebrates. Lamphrush chromosomes are best visualized in salamander oocytes because they have a high DNA content. For example, the largest chromosome having a length up to 1 mm have been observed in urodele amphibian. Thus, lampbrush chromosomes are much larger (longer) than the polytene chromosomes of insects.

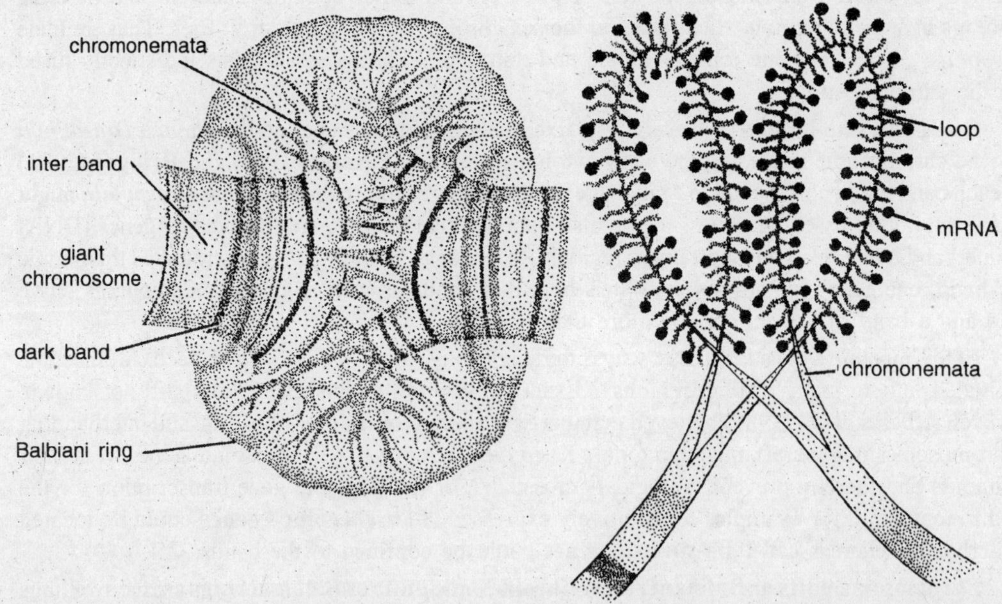

Fig. 70.14. A Balbiani ring of a polytene chromosome.

Fig. 70.15. Suggested form of the individual loops in chromosome puffs. Note the tiny fibrils of RNA and proteins attached to the loop.

Since the lampbrush chromosomes are found in the prolonged diplotene stage of meiotic prophase I, they are present in the form of **bivalents** in which the maternal and paternal chromosomes are held together by **chiasmata**, at those sites where crossing over has previously occurred (Fig. 70.16). The paired homologs are not condensed as usual chromosomes would be; instead, they are very long and stretched out. Each bivalent has four chromatids, two in each homolog. The axis of each homologs consists of a row of granules or **chromomeres** from which **lateral loops** extend. The loops are always symmetrical, each chromosome having two of them, one for each chromatid. The loops can be categorized by size, thickness and other morphological characteristics. Each loop appears at a constant position in the chromosome; this fact helps in the chromosome mapping. There are about 10,000 loops per chromosome set or haploid set (*e.g.*, oocytes of slamander *Triturus*). Each loop has an **axis** which is made of single DNA molecule that is unfolded from the chromosome for the intense RNA synthesis. Thus, about 5 to 10 per cent of the DNA exists in the lateral loops, the rest being tightly condensed in the chromomeres which are transcriptionally inactive. The centromeres of the chromosomes bear no loops.

Each loop of lampbrush chromosomes is found to perform intense transcription of **hnRNA** or **heterogeneous RNA** molecules. (*i.e.,* precursors of mRNA molecules for various ribosomal proteins and for five types of histone proteins). Electron microscopy of the loops has shown that **RNA polymerase** enzyme molecules are attached to the principal axis (DNA) of the loop from which RNA fibrils of increasing length extend. As transcription continues along the DNA strand of loop, the fibrils of RNA (*i.e.*, hnRNA) lengthen. Proteins get associated with these RNA fibrils as they are formed and ultimately ribonucleoprotein product is released.

Thus, each lateral loop is covered by a **matrix** that consists of RNA transcripts with hnRNA-binding proteins attached to them. Generally this matrix is asymmetrical, being thicker at one end of the loop than at the other. RNA synthesis starts at the thinner end and progresses toward the thicker end. Preparations spread for electron microscopy exhibit the typical '**christmas tree**" images with nascent ribonucleoprotein chains attached perpendicularly to the DNA axis. Many of the loops correspond to a single transcriptional unit (or single gene) and they are transcribed continuously from end to end; they form a continuous thin-thick matrix. However, other loops contain several units of transcription (or many genes); such loops include an extended section of chromatin that is not transcribed at all. For example, **Gall** *et al.*, (1981) found that in the American newt *Notophthalmus viridescens*, clusters of the five histone genes are tande-mly repeated in the genome but separated by about 50,000 base pairs of repeats of a 225-base pair satellite DNA.

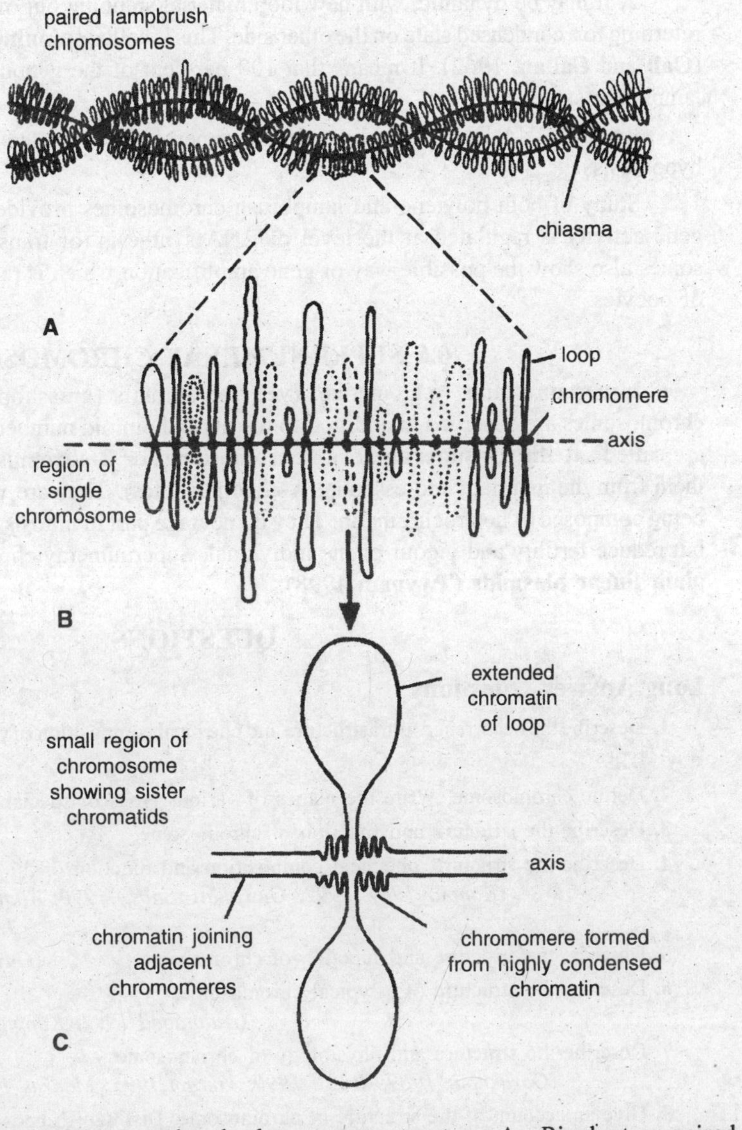

Further, the number of pairs of loops gradually increa-ses during meiosis till it reaches maximum in diplotene. Such a lampbrush stage may persist for months or years as the primary oocyte builds up a supply of mRNA molecules and other materials required for its ultimate de-velopment into a new individual. As meiosis proceeds further, num-ber of loops gradually decreases and the loops ultimately disappear either due to disintegration or by reabsorption back into the chromosome. For example, the addition of histone proteins to the lampbrush chromosomes stops the synthesis of RNA on the loops and causes the loops to retract into the chromosomes.

Fig. 70.16. Lampbrush chromosome structure. A—Bivalent or paired homologous chromosomes in pairing showing chiasmata, B—A part of one homologue (or homolog) showing paired loops given out by two chromatids; C—Single pair of loop.

Certain hypothesis regarding nature of loops. The loop of the lampbrush chromosomes can be viewed in the following two ways:

1. It may be static, unchanging in length and constantly exposing the same stretch of chromosome fibre.

2. It may be dynamic, with new loop material spinning out of one side of a chromomere and returning to a condensed state on the other side. This is called **spinning out** and **retraction hypothesis** (**Gall** and **Callan**, 1962). It means that 100 per cent of the genome can be expressed during the lampbrush stage.

Recent, DNA-RNA hybridization experiments have rejected the spinning out and retraction hypothesis.

Study of both polytene and lampbrush chromosomes provided the evidence that eukaryotic gene activity is regulated at the level of RNA synthesis (or transcription). Lampbrush chromosomes also show the possible way of gene amplification which is required during the growth phase of oocytes.

70.5 SUPERNUMERAY CHROMOSOMES

In certain plants (e.g., maize, rye) and animals (grasshopper), one or more additional chromosomes are observed in addition to the normal diploid number. Such accessory chromosomes are called **satellite** or **supernumeray chromosomes** or **B-chromosomes in plants** (to distinguish them from the invariant and essential **A-chromosomes**). They are very small and genetically inert being composed of heterochromatin. They do not take part in mitosis and do not affect the phenotype but reduce fertility and vigour of the individual. Supernumeray chromosomes may be regarded as **giant linear plasmids** (**Twyman** 1998).

QUESTIONS

Long Answer Questions

1. Describe the occurrence, ultrastructure and chemical composition of chromosome. What are its different types? *(CCS 1999)*
2. Define chromosome. Write the names of various types of special chromosomes.
3. Describe the structure and function of chromosome. *(CCS 1995)*
4. Describe the structure, chemical composition and function of the eukaryotic chromosomes. *(Kanpur 1989, 1992; Guwahati (Subs.) 1990; Allahabad 1991; Magadh (G) 1995; Lucknow 1995, 1997; CCS 1997)*
5. Describe the structure and function of chromosomes. *(Guwahati (M) 1994; Kanpur 1998)*
6. Describe the structure of a typical chromosome. *(Allahabad 1992; Kanpur 1994; Gorakhpur 1996, 1998)*
7. Describe the structure and physiology of chromosomes. *(Gorakhpur 1989; Jiwaji 1990; Vikram 1991; Madras 1992; Bhopal 1992; Bihar 1992)*
8. Give an account of the structure of chromosome. Distinguish between chromonema and chromatid. Write a short note on the different chromosomal shapes at anaphase. *(Gorakhpur 1994)*
9. Give a detailed account of chromatin. *(B.U. Coimbatore 1997)*
10. Describe the Oudet and other's nucleosome models of chromosomes and explain as to how this model is considered better than Dupraw's folded fibre model? Describe as to how many structure groups human chromosomes belong? *(CCS 1991, 1993)*
11. Give the biochemical components of the chromatin and discuss the structural organization of chromatin emphasizing the nucleosome organization. *(NEHU (Major) 1996)*
12. What are chromosomes? Describe the chemical composition and organization of metaphase chromosome. *(BHU (Hons.) 1990)*
13. Give an account of the chemical organization of chromosomes. What are nucleosomes? *(Allahabad 1989)*
14. What do you mean by giant chromosome? Describe the physico-chemical structure of polytene chromosome. *(Guwahati (Major) 1989, 1996)*

15. What is a giant chromosome? Describe the structure of a giant chromosome and its role in heredity. *(Magadh (G) 1991)*

16. Give a brief description of the significance of polytene chromosomes in understanding gene regulation. *(B.U. Coimbatore 1997)*

17. Describe the structure of polytene chromosome and state their significance. *(Delhi 1989, 1991; Madurai 1990; Kanpur 1997)*

18. What are lampbrush chromosomes and Balbiani rings? *(Bangalore 1989)*

19. Discuss the structure and functions of lampbrush chromosomes. *(Guwahati (G) 1989; Delhi 1990, 1992; B.U. Coimbatore 1997)*

20. State the difference between the lampbrush and the salivary gland chromosomes. Write the structural and functional significance of the salivary gland chromosomes. *(Guwahati (Major) 1994)*

Short Answer Questions

1. What is the role of basic proteins which are associated with the DNA molecule?

2. What are Feulgen positive regions?

3. Why secondary constriction is known as nucleolar organizer?

4. What are supernumerary chromosomes?

5. Describe the role of centromere.

6. What is the nucleolar organizer (NO) region in the chromosome? What is its role?

7. Describe unistranded folded fibre model of chromosome structure.

8. How protein molecules are associated with DNA molecule?

9. What is nucleosome concept?

10. Describe structure of a nucleosome.

11. What do you mean by telocentric and metacentric chromosome?

12. What do you mean by polyteny? Who introduced this concept?

13. Explain significance of puffing.

14. What do you mean by gene amplification? What is its need? Explain.

15. Write short note on the following:

(*i*) Chromosomes *(Nagarjuna 1995)*

(*ii*) Chromosome shapes *(Gorakhpur 1995)*

(*iii*) Types of chromosomes *(Garhwal 1989)*

(*iv*) Centromere *(Lucknow 1994, 1996; CCS 1995)*

(*v*) Autosomes *(Lucknow 1996, 1997)*

(*vi*) Barr bodies *(Lucknow 1997)*

(*vii*) Nucleosome *(Allahabad 1991; Lucknow 1993)*

(*viii*) Heterochromatin and euchromatin *(Kanpur 1995; 1996; CCS 1995)*

(*ix*) Fine structure of chromosome *(Punjab (G) 1989)*

(*x*) Polytene (or polytenic) chromosome *(BHU (Hons.) 1990; CCS 1990; Kanpur 1992, 1996; Allahabad 1993; Guwahati (G) 1993, 1995)*

(*xi*) Lampbrush chromosomes *(Kanpur 1989, Allahabad 1991; Bangalore 1992; Guwahati (G) 1994, 1996; Lucknow 1995, 1997)*

(*xii*) Balbiani chromosome *(CCS 1999)*

(*xiii*) Histones

16. Prepare a well labelled diagram of each of the following:

1. Lampbrush chromosomes *(Lucknow 1998)*

2. Fine structure of chromosome *(Punjab(G) 1992)*

17. Differentiate between the following:

1. Viral and bacterial chromosomes

2. Prokaryotic (bacterial) and eukaryotic chromosomes
3. Primary constriction and secondary constriction
4. Centrosome and nucleosome
5. Euchromatin and heterochromatin (*H.N.B., Garhwal 2009*)
6. Lampbrush chromosome and salivary gland chromosome (*Jiwaji 1990, 1991*)
7. Heterochromatin and Barr bodies (*Guwahati (Major) 1995*)
8. Centromere and chromocentres (*H.N.B., Garhwal 2009*)
9. Histones and non-histones (*H.N.B., Garhwal 2009*)

Very Short Answer Questions

1. Name the scientists.

(*i*) Who coined the term chromosome?

Ans. Waldeyer

(*ii*) Who described parallelism between chromosomes and transmission of hereditary factors?

Ans. S.W. Sutton

2. Name the part of a chromosome where its two chromonemata are joined together.

Ans. Centromere

3. Name the organism where RNA act as hereditary material.

Ans. Some viruses, *e.g.*, TMV.

4. What is nucleoid?

Ans. The nuclear regions in bacterial cells (= prokaryotic cells) where DNA molecule of bacterial chromosome lies in a coiled state. Bacterial cells lack a nucleus.

5. How many DNA double helical fibres are present in a chromosome?

Ans. One in each chromosome.

6. 'Beads in the string' appearance of chromatin represents what?

Ans. The 10 nm chromatin fibre (*i.e.*, nucleosomes in close apposition).

7. What is the importance of telomere in chromosome? (*H.N.B., Garhwal 2009*)

Ans. The chromosomal ends are called **telomeres**. Each telomere provides polarity to the chromosome and prevents other chromosomal segments to fuse with it.

8. Lampbrush chromosomes are found during which stage of cell division? (*H.N.B., Garhwal 2009*)

Ans. Diplotene of prophase I of meiosis.

True/False Questions

(*i*) Histone are basic proteins rich in lysine and arginine. (*H.N.B., Garhwal 2009*)

Ans. True.

Fill in the Blanks

1. Polytene chromosomes are also known as.................. (*Lucknow 1995*)

2. Number of chromosomes in human genome is .... (*Lucknow 1995*)

Multiple Choice Questions

1. Number of chromatids present in a metaphase chromosome is

(*a*) four (*b*) three
(*c*) two (*d*) one

2. Number of chromosomes in definite nucleus
(*a*) haploid (*b*) diploid
(*c*) triploid (*d*) polyploid

3. Diploid chromosome number in humans is
(*a*) 42 (*b*) 44
(*c*) 46 (*d*) 48

4. Two sister chromatids are attached with the
(*a*) spindle fibre (*b*) chromatid
(*c*) centromere (*d*) chromocentre

5. Autosomes in humans are
(*a*) 11 pairs (*b*) 22 pairs
(*c*) 23 pairs (*d*) 43 pairs

6. Eukaryotic chromosome is made of
(*a*) DNA (*b*) DNA + lipid
(*c*) DNA + proteins (*d*) RNA

7. The term 'nucleosome' was given by
(*a*) Dupraw (*b*) Oudet
(*c*) Flemming (*d*) Emil Heitz

8. The nucleosome
(*a*) surrounds nuclear pores
(*b*) has only DNA and non-histone proteins
(*c*) is fully responsible for packing DNA into chromosomes
(*d*) contains a core of histones with DNA wrapped around it

9. Balbiani rings occur in
(*a*) polytene chromosomes (*b*) lampbrush chromosomes
(*c*) polysomes (*d*) heterosomes

10. Chemical composition of chromomere is
(*a*) DNA and proteins (*b*) DNA and lipids
(*c*) DNA and carbohydrates (*d*) proteins and lipids

11. A chromosome in which the centromere is situated near one end is known as
(*a*) metacentric (*b*) submetacentric
(*c*) acrocentric (*d*) telocentric

12. The total number of genes present in a haploid cell is
(*a*) genome (*b*) genotype
(*c*) karyotype (*d*) gene pool

13. The centromere is found at the centre of the two arms in......... (*B.U. coimbatore 1997*)
(*a*) telocentric (*b*) acrocentric
(*c*) submetacentric (*d*) metacentric

14. Chromosomes can be stained with one of the following chemicals?
(*a*) acetocarmine (*b*) safranine
(*c*) light green (*d*) eosin

ANSWERS

Fill in the Blanks

1. Salivary gland chromosomes. **2.** 23

Multiple Choice Questions

1. (*c*) **2.** (*b*) **3.** (*c*) **4.** (*c*) **5.** (*b*) **6.** (*c*)
7. (*b*) **8.** (*d*) **9.** (*a*) **10.** (*a*) **11.** (*c*) **12.** (*a*)
13. (*d*) **14.** (*a*).

71

Cell Cycle and Mechanics of Cell Division

71.1 CELL CYCLE

All cells are produced by divisions of pre-existing cell. Continuity of life depends on cell division. A cell born after a division, proceeds to grow by macromolecular synthesis, reaches a species-determined division size and divides. This cycle acts as a unit of biological time and defines life history of a cell. **Cell cycle** can be defined as the entire sequence of events happening from the end of one nuclear division to the beginning of the next. The cell cycle involves the following three cycles.

1. Chromosome cycle. In it **DNA synthesis** alternates with **mitosis** (or karyokinesis or nuclear division). During DNA synthesis, each double-helical DNA molecule is replicated into two identical daughter DNA molecules and during mitosis the duplicated copies of the genome are ultimately separated.

2. Cytoplasmic cycle. In it **cell growth** alternates with **cytokinesis** (or cytoplasmic division). During cell growth many other components of the cell (RNA, proteins and membranes) become double in quantity and during cytokinesis cell as a whole divides into two. Usually the karyokinesis is followed by the cytokinesis but sometimes the cytokinesis does not follow the karyokinesis and results into the multinucleate cell, *e.g.*, cleavage of egg in *Drosophila*.

3. Centrosome cycle. Both of the above cycles require that the **centrosome** be inherited reliably and duplicated precisely in order to form the two poles of the mitotic spindle; thus, centrosome cycle forms the third component of cell cycle.

Howard and **Pelc** (1953) have divided cell cycle into four phases or stages: G_1, S, G_2 and M phase. The G_1 phase, S phase and G_2 phase are combined to from the classical **interphase.**

1. G_1 Phase. After the M phase of previous cell cycle, the daughter cells begin G_1 of interphase of new cell cycle. G_1 is a resting phase. It is called **first gap phase**, since no DNA synthesis takes place during this stage; currently, G_1 is also called **first growth phase**, since it involves synthesis of RNA, proteins and membranes which leads to the growth of nucleus and cytoplasm of each daughter cell towards their mature size.

During G_1 phase, chromatin is fully extended and not distinguishable as discrete chromosomes with the light microscope. This is a time of resumption of normal cell metabolism which has slowed down during the previous cell division. Thus, G_1 involves transcription of three types of RNAs, namely rRNA, tRNA and mRNA; rRNA synthesis is indicated by the appearance of nucleolus in the interphase (G_1 phase) nucleus. Proteins synthesized during G_1 phase are (1) Regulatory proteins which control various events of mitosis; (2) Enzymes (*e.g.*, DNA polymerase) necessary for DNA synthesis of the next stage; and (3) Tubulin and other mitotic apparatus proteins.

G_1 phase is most variable as to duration (Table 71.1); it either occupies 30 to 50 per cent of the total time of the cell cycle or lacks entirely in rapidly dividing cells (*e.g.*, blastomeres of early embryo of frog and mammals). Terminally differentiated somatic cells (*i.e.*, end cells such as neurons and striated muscle cells) that no longer divide, are arrested usually in the G_1 stage; such a type of G_1 phase is called **G_0 phase**.

Table 71.1 Different stages of a mitotic cell cycle and their duration in hours.

Parts of cell cycle	*Phases*	*Description of phases*	*Duration in hours*		
			Vicia faba	*Mouse L cells*	*Human HeLa cells*
Interphase	G_1	Pre-DNA-synthesis phase	12	12	12
	S	DNA-synthesis phase	6	6–8	10
	G_2	Post-DNA synthesis phase	12	3–4	3
Mitosis	M	Mitotic phase	1	1	1

2. S phase. During the S phase or **synthetic phase** of interphase, replication of DNA and synthesis of histone proteins occur. New histones are required in massive amounts immediately at the beginning of the S period of DNA synthesis to provide the new DNA with nucleosomes. Thus, at the end of S phase, each chromosome has two DNA molecules and a duplicate set of genes. S phase occupies roughly 35 to 45 per cent of cell cycle.

3. G_2 phase. This is a **second gap** or **growth phase** or resting phase of interphase. During G_2 phase, synthesis of RNA and proteins continues which is required for cell growth. It may occupy 10 to 20 per cent time of cell cycle. As the G_2 phase draws to a close, the cell enters the M phase.

General events of interphase. The interphase is characterized by the following features:

The nuclear envelope remains intact. The chromosomes occur in the form of diffused, long, coiled and indistinctly visible chromatin fibres. The DNA amount becomes double. Due to accumulation of ribosomal RNA (rRNA) and ribosomal proteins in the nucleolus, the size of the latter is greatly increased. In animal cells, a daughter pair of centrioles originates near the already existing centrioles and, thus, an interphase cell has two pairs of centrioles.

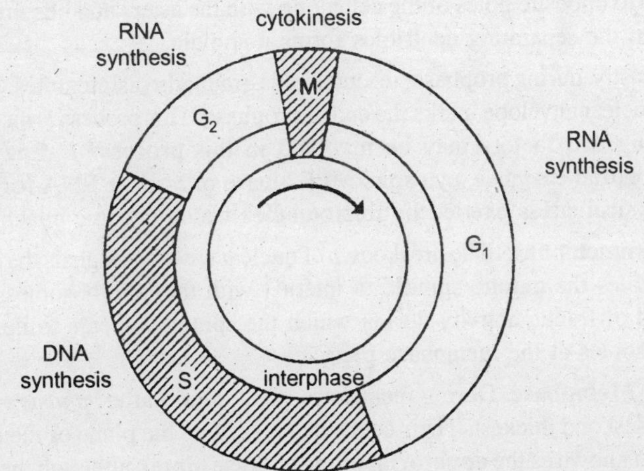

Fig. 71.1. The cell cycle or mitotic cycle, showing relative duration of phases (*e.g.*, interphase and mitotic phase) in a growing cell. S, synthesis of DNA; G_1 the first gap or growth phase; G_2, the second gap or growth phase; and M, mitotic phase.

In animal cells, net membrane biosynthesis increases just before cell division (mitosis). This extra membrane seems to be stored as **blebs** on the surface of the cells about to divide.

4. M phase or mitotic phase. The mitosis (Gr., *mitos* = thread) occurs in the somatic cells and it is meant for the multiplication of cell number during embryogenesis and blastogenesis of plants and animals. Fundamentally, it remains related with the growth of an individual from zygote to adult stage. Mitosis starts at the culmination point of interphase (*i.e.*, G_2 phase). It is a short period of chromosome condensation, segregation and cytoplasmic division. Mitosis is important for replacement of cells lost to natural friction (**attrition**), wear and tear and for wound healing.

71.2 MITOSIS

Mitosis is the process that distributes the two sets of duplicated chromosomes into two daughter nuclei. As a process, mitosis is remarkably similar in all animals and plants. It is a smoothly continuous process and is divided arbitrarily into following stages or phases for convenient reference (Fig. 71.2 and Fig. 71.3):

1. Prophase. The appearance of thin-thread like condensing chromosomes marks the first phase of mitosis, called **prophase** (Gr., *pro* = before; *phasis* = appearance). The cell becomes spheroid, more refractile and viscous.

Each prophase chromosome is composed of two coiled filaments, the **chromatids**, which are the result of the replication of DNA during the S phase. As prophase progresses, the chromatids become shorter and thicker and two sister chromatids of each chromosome are held together by a special DNA-containing region, called the **centromere** or **primary constriction**. During prophase, proteins of the trilaminar **kinetochores** (one for each chromatid) start depositing or organizing on the centromere of each chromosome. Further, during early prophase, the chromosomes are evenly distributed in the nuclear cavity; as prophase progresses, the chromosomes approach the nuclear envelope, causing the central space of the nucleus to become empty.

In the cytoplasm, the most conspicuous change is the formation of the spindle or **mitotic apparatus**. In the early prophase, there are two pairs of centrioles, each one surrounded by the so-called **aster** which is composed of microtubules radiating in all directions. The two pairs of centrioles migrate to opposite poles of the cell along with the asters and become situated in antipodal positions. Between the separating centrioles forms a spindle.

Lastly, during prophase, the nucleolus gradually disintegrates. Degeneration and disappearance of the nuclear envelope marks the end of prophase. This process is incompletely understood. However, following two factors may be involved in this process: 1. **Enzymatic action** either by some mitochondrial enzymes, cytosolic MPF kinase or nuclear RNA (or ribozyme). 2. **Physical action**, *i.e.*, physical stress exerted by microtubules which become attached to the nuclear envelope.

Prometaphase. The breakdown of nuclear envelope signals the commencement of prometaphase and enables the mitotic spindle to interact with the chromosomes. This stage is characterized by a period of frantic activity during which the spindle appears to be trying to contain and align the chromosomes at the metaphase plate.

2. Metaphase. During metaphase (Gr., *meta* = after; *phasis* = appearance) the chromosomes are shortest and thickest. Their centromeres occupy the plane of the equator of the mitotic apparatus (a region known as the **equatorial** or **metaphase plate**), although the chromosomal arms may extend in any direction. At this stage the sister chromatids are still held together by centromere and the kinetochores of the two sister chromatids face opposite poles; this would permit their proper separation in the next phase (anaphase).

Metaphase occupies a large portion of the mitotic phase (see Table 71.2), as if the cell pause until all their chromosomes are lined up correctly on the metaphase plate. During metaphase, subunits (tubulin dimers) are added to the plus end of a microtubule at the kinetochore and are removed from the minus end at the spindle pole. Thus, a poleward flux of tubulin subunits occurs, with the microtubules remaining stationary and under tension.

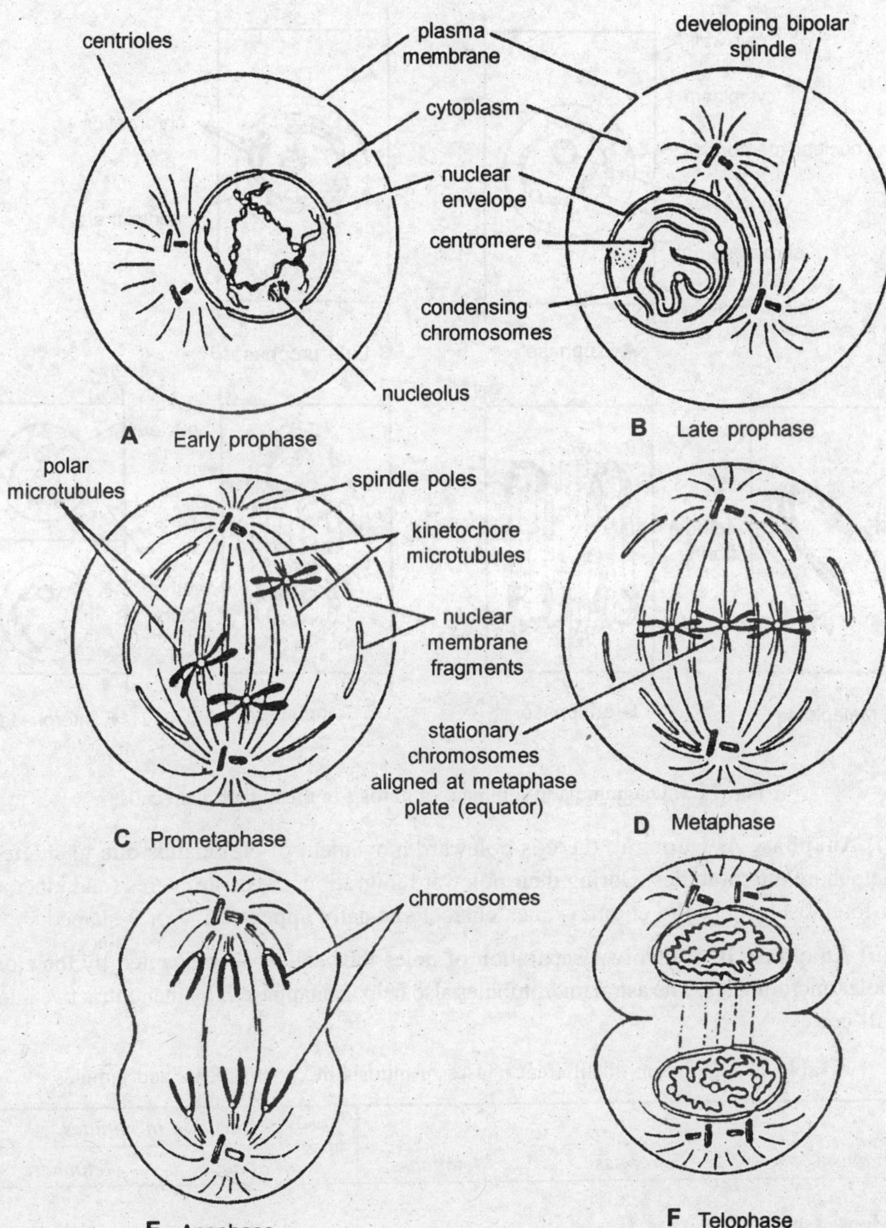

Fig. 71.2. Diagrammatic summary of mitosis in the animal cell.

3. Anaphase. The anaphase (Gr., *ana* = up; *phasis*=appearance) begins abruptly with the synchronous splitting of each chromosome into its sister chromatids, called **daughter chromosomes**, each with one kinetochore. Synchronous splitting of each centromere during prophase is evidently caused by an increase in cytosolic Ca^{2+}. In fact, Ca^{2+-}containing membrane vesicles accumulate at spindle poles and release calcium ions to initiate anaphase. Anaphase involves the following two steps:

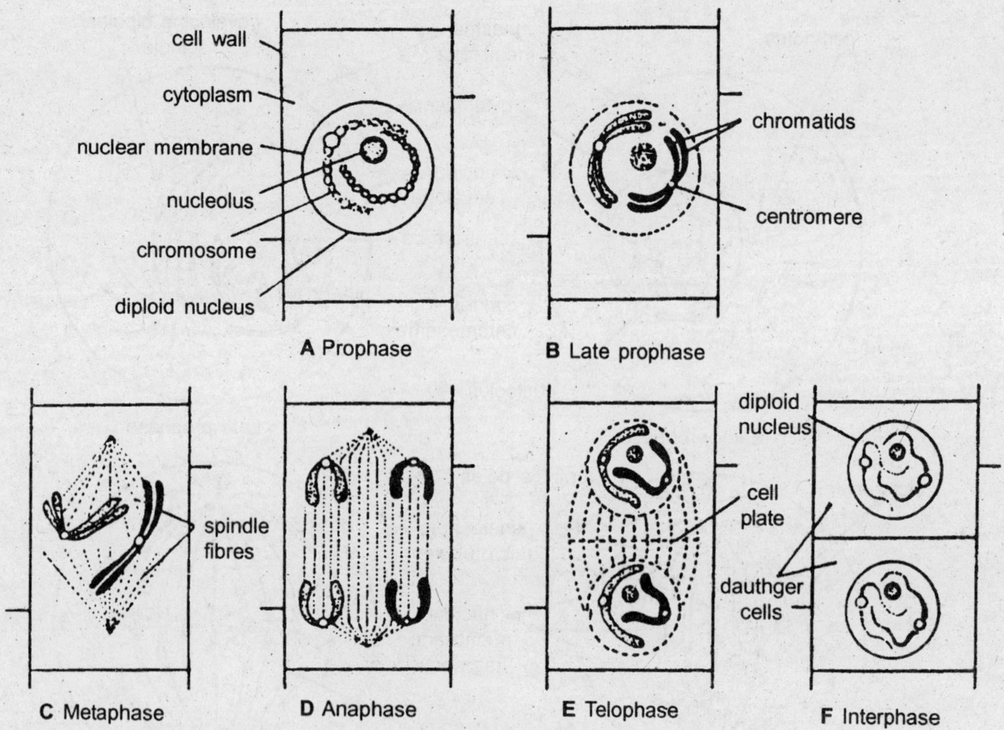

Fig. 71.3. Diagrammatic summary of mitosis in the higher plant cells.

***(i)* Anaphase A**. During it, there is poleward movement of chromatids due to shortening of the kinetochore microtubules. During their poleward migration, the centromeres (and kinetochores) remain foremost so that the chromosomes characteristically appear U, V or J-shaped.

***(ii)* Anaphase B.** It involves separation of poles themselves accompanied by the elongation of the polar microtubules. The astral microtubules also help in anaphase B by their attractive interaction with cell cortex.

Table 71.2 Duration of different phases of mitosis in certain plants and animals.

Organism	*Prophase*	*Metaphase*	*Duration in minutes*	
			Anaphase	*Telophase*
1. Mouse (spleen)	21	13	5	4
2. Grasshopper (neuroblasts)	102	13	9	57
3. Pea (root tip)	78	14.4	4.2	13.2
4. Onion (root tip)	71	6.5	2.4	3.8

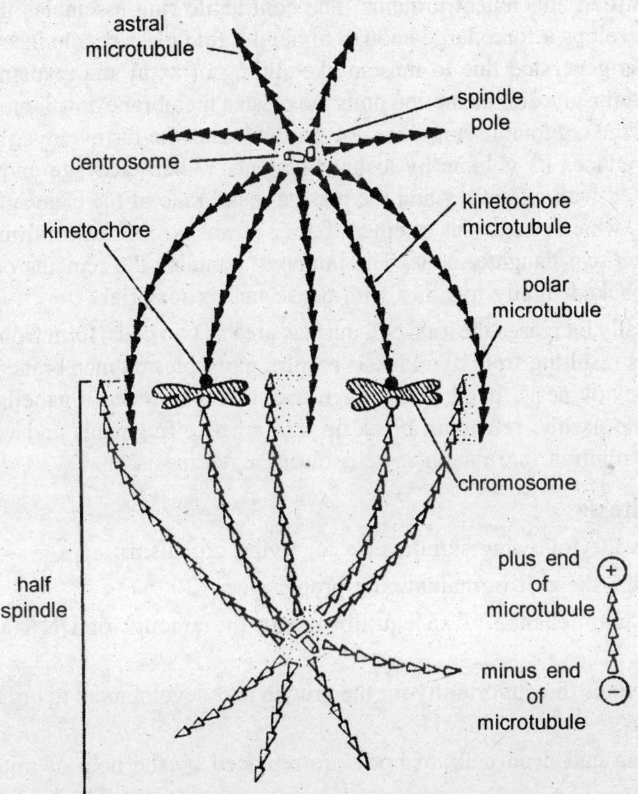

Fig. 71.4. The mitotic spindle at metaphase. The spindle is constructed from two half spindles, each composed of kinetochore, polar and astral microtubules. The polarity of the microtubules is indicated by the arrowheads.

4. Telophase. The end of the polar migration of the daughter chromosomes marks the beginning of the telophase; which in turn is terminated by the reorganization of two new nuclei and their entry into the G_1 phase of interphase. In general terms, the events of prophase occur in reverse sequence during this phase. A nuclear envelope reassembles around each group of chromosomes to form two daughter nuclei. The mitotic apparatus except the centrioles disappears; high viscosity of the cytoplasm decreases; the chromosomes resume their long, slender, extended form as their coils relax; and RNA-synthesis restarts causing the nucleolus to reappear.

Cytokinesis

Both DNA synthesis and mitosis are coupled to cytoplasmic division, or cytokinesis—the constriction of cytoplasm into two separate cells. During cytokinesis, the cytoplasm divides by a process, called **cleavage**. The mitotic spindle plays an important role in determining where and when cleavage occurs. Cytokinesis usually begins in anaphase and continues through telophase and into interphase. The first sign of cleavage in animal cells is **puckering** and **furrowing** of the plasma membrane during anaphase. The furrowing invariably occurs in the plane of the metaphase plate, at right angles to the long axis of the mitotic spindle. A cleavage furrow tends to form midway between asters originating from two centrosomes in fertilized sand dollar eggs.

Cleavage is accomplished by the contraction of a ring composed mainly of actin filaments. This bundle of filaments, called **contractile ring**, is bound to the cytoplasmic face of the plasma membrane by unidentified attachment proteins. The contractile ring assembles in early anaphase, once assembled, it develops a force large enough to bend a fine glass needle inserted into the cell. Evidently this force is generated due to muscle-like sliding of actin and myosin filaments in the contractile ring. The actin-myosin interaction pulls the plasma membrane down into a furrow. During a normal cytokinesis, the contractile ring does not get thicker as the furrow invaginates, suggesting that it continuously reduces its volume by losing filaments. When cleavage ends, the contractile ring is finally dispensed with altogether and the plasma membrane of the cleavage furrow narrows to form the **midbody**, which remains as a tether (Tether means a rope for confining a beast within certain limits) between two daughter cells. The midbody contains the remains of the two sets of polar microtubules, packed tightly together with dense matrix material.

Cytokinesis greatly increases the total cell-surface area as two cells form from one. Therefore, the two daughter cells resulting from cytokinesis require more plasma membrane than in the plant cell. Lastly, prior to cytokinesis, in M phase large membrane-bounded organelles such as Golgi apparatus and the endoplasmic reticulum break up into smaller fragments and vesicles; this may ensure their even distribution into daughter cells during cytokinesis.

Significance of Mitosis

The mitosis has the following significance for living organisms:

1. The mitosis helps the cell in maintaining proper size.
2. It helps in the maintenance of an equilibrium in the amount of DNA and RNA in the cell.
3. The mitosis provides the opportunity for the growth and development to organs and the body of the organisms.
4. The old decaying and dead cells of body are replaced by the help of mitosis.
5. In certain organisms, the mitosis is involved in asexual reproduction.
6. The gonads and the sex cells depend on the mitosis for the increase in their number.
7. The cleavage of egg during embryogenesis and division of blastema during blastogenesis, both involve mitosis.

71.3 MECHANISM OF MITOTIC APPARATUS

The microtubules of the spindle are arranged like two cones base to base, broad at the centre or equator of the cell and narrowing to a point at either end or pole. Mitotic spindle contains three main types of fibres (Fig. 71.5): (1) **Polar fibres**, which extend from the two poles of the spindle toward the equator; (2) **Kinetochore fibres**, which attach to the kinetochores of centromeres of each mitotic chromosomes and extend toward the poles; and (3) **Astral fibres**, which radiate outward from the poles toward the periphery or cortex of cell. In cells of most higher plants, however, spindle forms without the aid of centrioles and lacks asters.

Working of mitotic spindle during anaphase. During anaphase A, a surprisingly large force acts on a chromosome as it moves from the metaphase plate to the spindle pole. By hydrodynamic analysis it has been calculated that to move a chromosome, a force about 10^{-11} dynes is needed, and that the entire displacement—from equator to the pole of a chromosome—may require the use of about 30 ATP molecules. As each chromosome moves polewards, its kinetochore microtubules disassemble, so that they have nearly disappeared at telophase. The site of subunit loss can be determined by injecting labelled tubulin into cells during metaphase. The labelled subunits are found to be added to the kinetochore end of kinetochore microtubules and then lost as anaphase A proceeds, indicating that kinetochore "eats" its way poleward along its microtubules at anaphase (Fig. 71.5). However, typically microtubule disassemble at kinetochores, poles or at both sites is probably necessary for equator-to-pole movement.

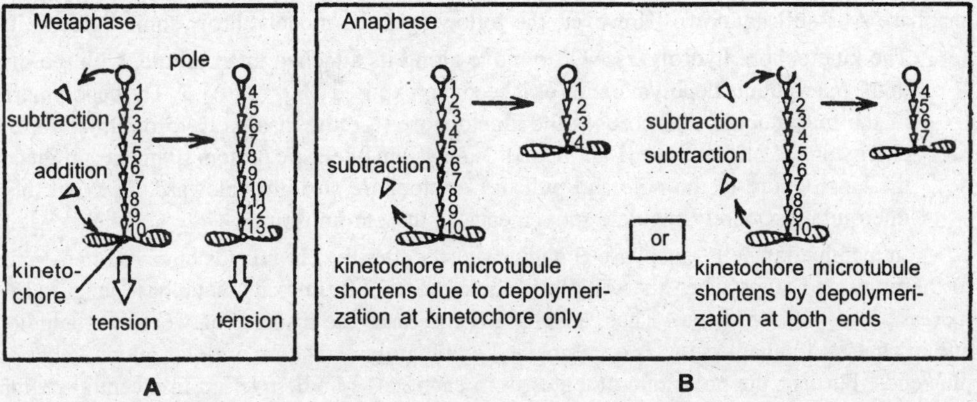

Fig. 71.5. Comparison of behaviour of kinetochore microtubules during metaphase (A) and anaphase (B). A—At metaphase, subunit are added to the plus end of a microtubule at the kinetochore and are removed from the minus end at the spindle pole. Thus, a constant poleward flux of tubulin subunits occurs, with the microtubules remaining stationary and under tension; B—At the anaphase the tension is released, and the kinetochore moves rapidly up the microtubule, removing subunits from its plus end as it goes (B I). Its attached chromatid is thereby carried to a spindle pole. In some organisms, part of the chromatid movement is due to the simultaneous shortening of the microtubules (B II).

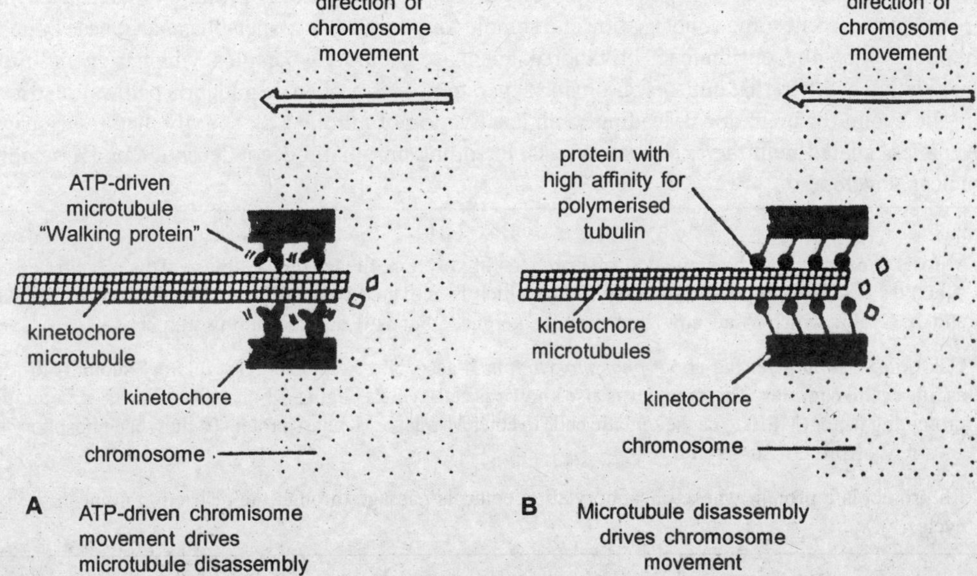

Fig. 71.6. Two alternative models to explain the generation of poleward force by the kinetochore for its chromosome during anaphase. A—Microtubule-walking proteins that resemble dynein or kinesin are part of the kinetochore, and they use the energy of ATP hydrolysis to pull the chromosome along its bound microtubules; B—Chromosome movement is driven by microtubule disassembly; as tuɒulin subunits dissociate, the kinetochore tends to slide poleward in order to restore its binding to the walls of the microtubule.

Further, the mechanism by which kinetochore of the chromosome moves up the spindle during anaphase A is still unknown. However, the following three models throw some light on it:

1. The kinetochore hydrolyzes ATP to move along its attached microtubule, with the plus end of the microtubule depolymerizing as it becomes exposed (Fig. 71.6A). 2. The depolymerization of the microtubule itself causes the kinetochore to move passively to optimize its binding energy on the microtubule (Fig. 71.6B). 3. A system of elastic protein filaments might connect the kinetochore to the pole and pull the kinetochore steadily poleward. Thus, in this case, microtubules merely regulate movements of the chromosomes.

In mammalian cells, anaphase B begins shortly after the chromatids have begun their voyage to the poles and stops when the spindle is about 1.5 to 2.0 times its metaphase length (15 times increase in certain protozoa). Thus, anaphase B increases the distance between two spindle poles and in contrast to anaphase A, is accompanied by the polymerization of polar microtubules at their plus ends. Further, the polar microtubules from each half-spindle overlap in a central region near the spindle equator (*e.g.*, diatoms). During anaphase B these two sets of antiparallel polar microtubules appear to slide away from each other in the region of overlap. Dynein-like force generating protein may be involved in such a directed movement of chromosomes.

71.4 REGULATION OF CELL CYCLE

Progression through the eukaryotic cell cycle is regulated by cyclin-dependent kinases (cdks) which bind to cyclins to form Cdk-cyclin complexes (Box 71.1). At the restriction point (in yeast this point is called start), a **G1-Cdk-cyclin** catalyzes the phosphorylation of the *Rb protein* to trigger passage into S phase (Fig. 71.7). At the G_2–M boundary, a mitotic Cdk-cyclin triggers entry into mitosis by catalyzing the phosphorylation of various proteins, thereby promoting nuclear envelope breakdown, chromosome condensation and spindle formation. And at metaphase-anaphase boundary, mitotic-Cdk-cyclin contribute to activation of anaphase-promoting complex, which triggers a protein degradation pathway that initiates chromatid separation. This protein degradation pathway also targets mitotic cyclin for breakdown, leading to an inactivation of mitotic Cdk activity that in turn triggers events associated with the exit from mitosis, including chromatin decondensation and assembly of nuclear envelope.

Box 71.1

Mitosis depends upon the activation of a pre-existing protein, the **M-phase kinase**. This protein has two subunits: 1. **Cyclin dependent kinase (Cdk)** which is activated by modification at the start of M-phase and 2. **Cyclin** which is accumulated during interphase but is destroyed during mitosis.

Cdk (kinase subunit) bring about *phosphorylation*. It associates with cyclin (regulatory subunit) to form a **Cdk-cyclin complex**. This complex is also known as **cell-cycle's engine**. M-phase kinase acts as maturation promoting factor (MPF) to cause somatic cells to enter M-phase. M-phase protein (dimer) can phosphorylate a variety of protein substrates.

Rb protein is a protein whose phosphorylation controls passage through the restriction point in the cell cycle.

Lastly, cell utilises a series of **checkpoint mechanism** to monitor conditions within the cell and transiently halt the cell cycle if conditions are not suitable for continuing. The DNA replication checkpoint monitors the state of DNA replication to ensure that DNA synthesis is completed prior to permitting the cell to exit from G_2 and begin mitosis. The **p53 protein** plays a central role in a series of DNA damage check-points that halt the cell cycle at various points of DNA damage. Finally, the *anaphase promoting complex* is involved in the spindle checkpoint, which prevents anaphase chromosome movements from the beginning before the chromosomes are attached to the spindle.

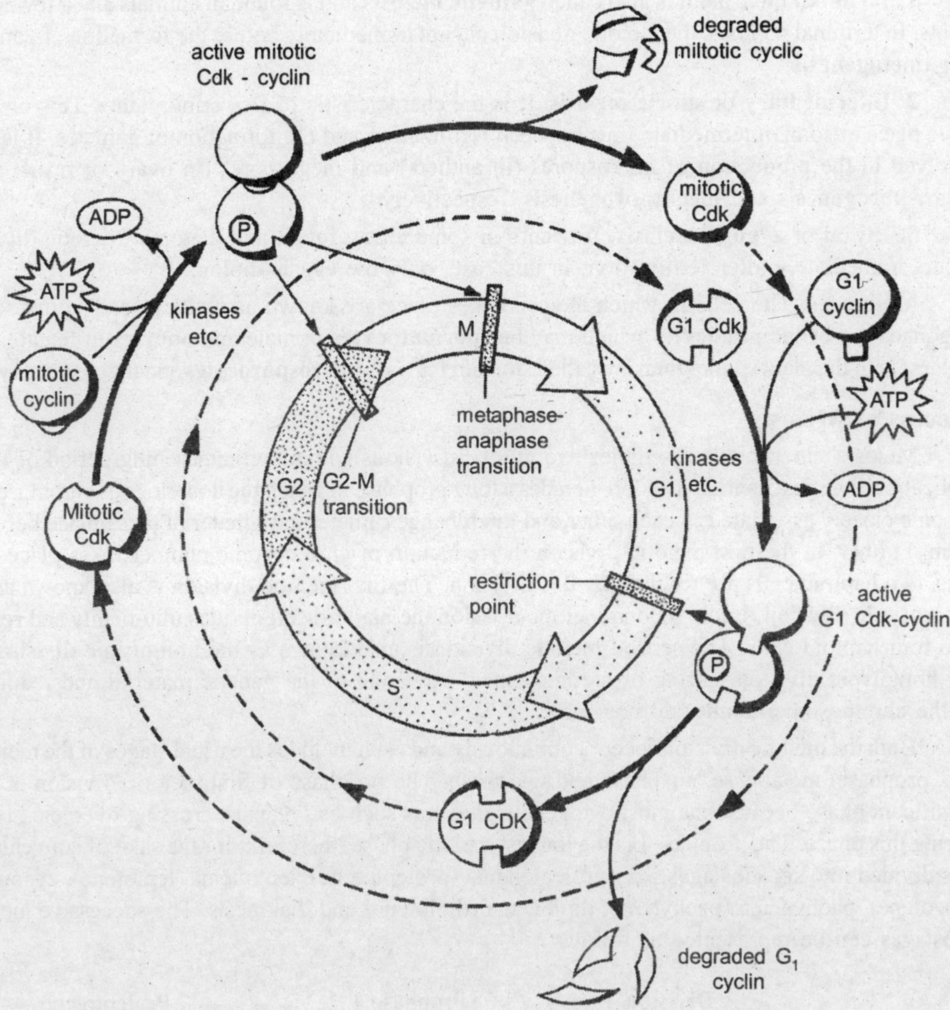

Fig. 71.7. A general model for cell cycle regulation. According to this model, passage through the three main transition points in the cell cycle is triggered by protein complexes made of cyclin and Cdk, whose phosphorylation of other proteins induces progression through the cell cycle (after Becker *et al.*, 2007).

71.5 MEIOSIS AND REPRODUCTIVE CYCLE

The term meiosis (Gr., *meioum*=to reduce or to diminish) was coined by **J.B. Farmer** in 1905. Meiosis produces a total of four haploid cells from each original diploid cell. These haploid cells either become or give rise to gametes, which through union (fertilization) support sexual reproduction and a new generation of diploid organisms. Thus, meiosis is required to run the **reproductive cycle** of eukaryotes such as microorganisms *Chlamydomonas, Neurospora*; bryophytes; plants and animals.

Kinds of Meiosis

Meiosis occurs in the germ cells of sexually reproducing organisms. In both plants and animals, germ cells are localized in the gonads. The time at which meiosis takes place varies among different organisms, and on this basis the process can be classified into: **terminal, intermediate** or **initial**.

1. Terminal meiosis. It is also called **gametic meiosis** and is found in animals and a few lower plants. In terminal meiosis, the meiotic division occurs immediately before the formation of gametes or **gametogenesis.**

2. Intermediary or **sporic meiosis.** It is the characteristic of flowering plants. This meiosis takes place at some intermediate time between fertilization and the formation of gametes. It is also involved in the production of microspores (in anthers) and megaspores (in ovary or pistil) or in **microsporogenesis** and **megasporogenesis**, respectively.

3. Initial or **zygotic meiosis.** It occurs in some algae, fungi, and diatoms. Meiotic division occurs immediately after fertilization; in this case, only the egg is diploid.

Meiocytes. The cells in which meiosis takes place are known as **meiocytes**. The meiocytes of gonads are called **gonocytes** which may be **spermatocytes** in male and **oocytes** in female. The meiocytes of the plant sporangium are called **sporocytes** (*i.e.*, **microsporocytes** and **megasporocytes**).

Process of Meiosis

Meiosis superficially resembles two mitotic divisions without an intervening period of DNA replication. The first meiotic division includes a long prophase in which the homologous chromosomes become closely associated to each other and interchange of hereditary material takes place between them. Further, in the first meiotic division the reduction of chromosome number takes place and, thus, two haploid cells are resulted by this division. The first meiotic division is also known as the **heterotypic division**. In the second meiotic division the haploid cell divides mitotically and results into four haploid cells. The second meiotic division is also known as the **homotypic division**. In the homotypic division pairing of chromosomes, exchange of the genetic material and reduction of the chromosome number do not occur.

Both the meiotic divisions occur continuously and each includes the usual stages of the meiosis, *viz.*, prophase, metaphase, anaphase and telophase. The prophase of first meiotic division is very significant phase because the most cytogenetical events such as synapsis, crossing over, etc., occur during this phase. The prophase I is the longest meiotic phase, therefore, for the sake of convenience it is divided into six substages, *viz.*, proleptonema (proleptotene), leptonema (leptotene), zygonema (zygotene), pachynema (pachytene), diplonema (diplotene) and diakinesis. The successive meiotic substages can be represented as follows:

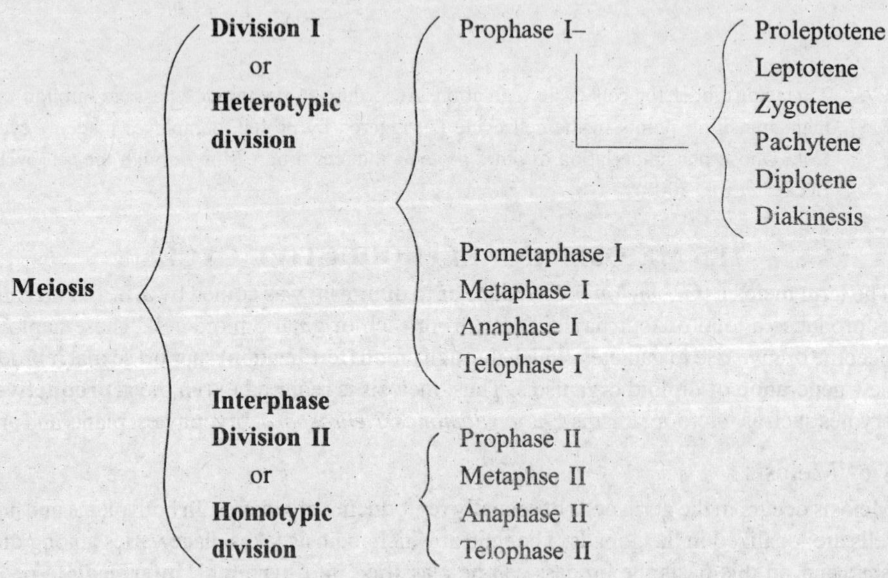

Heterotypic Division or First Meiotic Division

Meiosis starts after an **interphase** which is not very different from that of an intermitotic interphase. During the premeiotic interphase DNA duplication has occurred at the S phase. In the G_2 phase of interphase apparently there is a decisive change that directs the cell toward meiosis, instead of toward mitosis (**Stern** and **Hotta**, 1969). Further, in the beginning of the first meiotic division the nucleus of the meiocyte starts to swell up by absorbing the water from the cytoplasm and the nuclear volume increases about three folds. After these changes the cell passes to the first stage of first meiotic division which is known as prophase.

Prophase I

The first prophase is the longest stage of the meiotic division (Fig. 71.9). It includes following substages:

1. Proleptotene or **Proleptonema.** (Gr., *pro* = before; *leptas* = thin; *nema* = thread). The proleptotene stage closely resembles with the early mitotic prophase. In this stage the chromosomes are extremely thin, long, uncoiled, longitudinally single and slender thread-like structures.

2. Leptotene or **Leptonema.** In the leptotene stage the chromosomes become more uncoiled and assume a long thread-like shape. The chromosomes at this stage take up a specific orientation inside the nucleus; the ends of the chromosomes converge toward one side of the nucleus, that side where the centrosome lies (the **bouquet stage**). The centriole duplicates and each daughter centriole migrates towards the opposite poles of the cell. On reaching at the poles, each centriole duplicates and, thus, each pole of cell possesses two centrioles of a single diplosome.

3. Zygotene or **Zygonema.** (Gr., *zygon* = adjoining). In the zygotene stage, the pairing of homologous chromosomes takes place. The homologous chromosomes which come from the mother (by ova) and father (by sperm) are attracted towards each other and their pairing takes place. The pairing of the homologous chromosomes is known as **synapsis** (Gr., *synapsis* = union). The synapsis begins at one or more points along the length of the homologous chromosomes. Three types of synapsis have been recognised.

***(i)* Proterminal synapsis.** In proterminal type of synapsis the pairing in homologous chromosomes starts from the end and continues towards their centromeres.

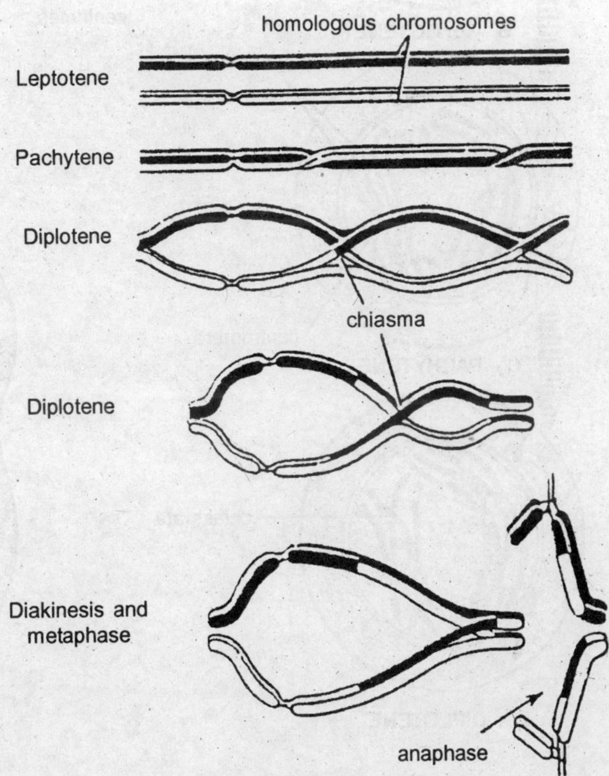

Fig. 71.8. Diagram of chromosomal exchange in the four-strand stage and of terminalization during first meiotic division.

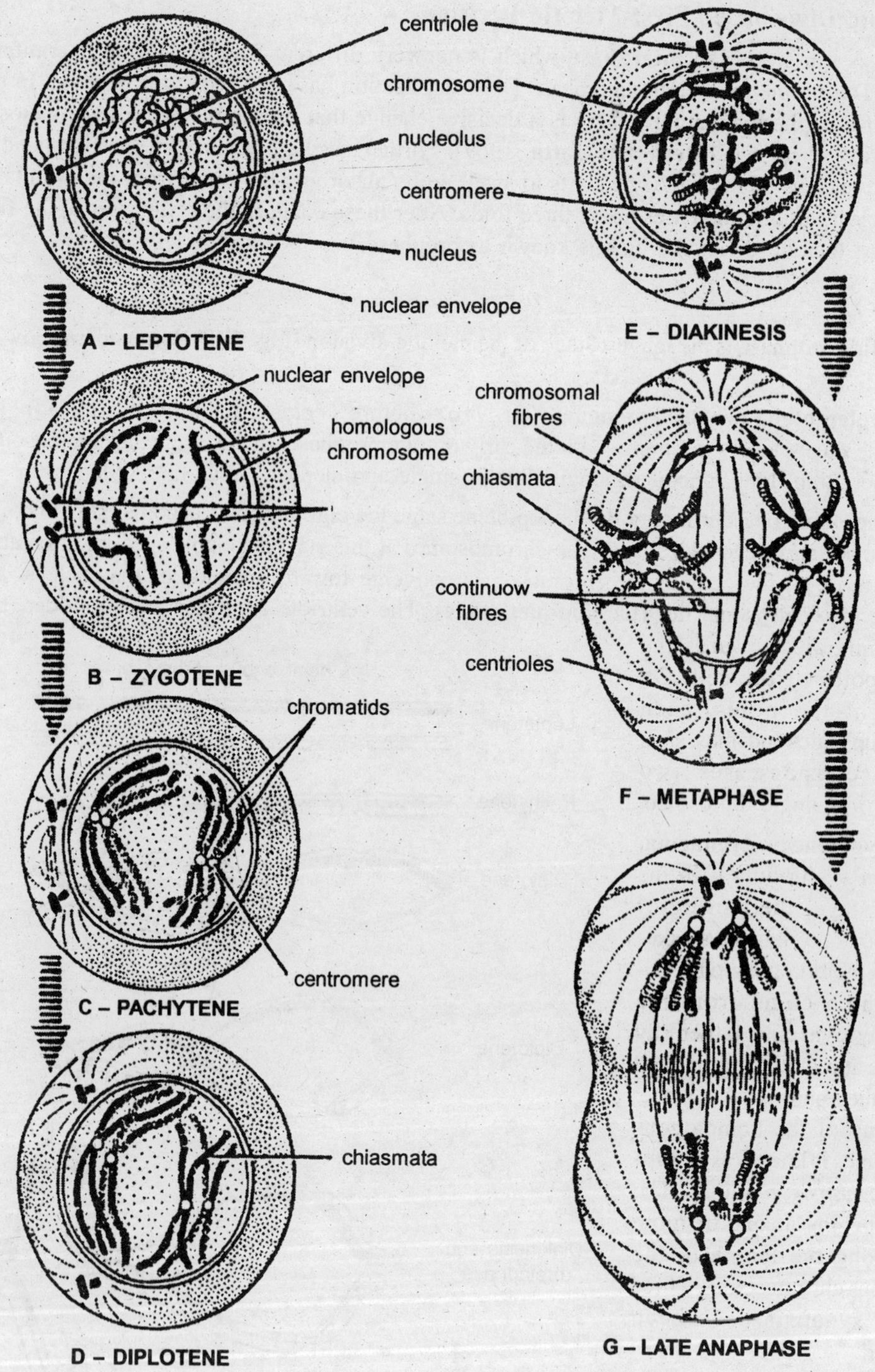

Fig. 71.9. Different stages of first meiotic division.

***(ii)* Procentric synapsis.** In procentric synapsis the homologous chromosomes start pairing from their centromeres and the pairing progresses towards the ends of the homologs.

***(iii)* Localized pairing or random synapsis.** The random type of synapsis occurs at various points of the homologous chromosomes.

The pairing of the homologous chromosome is very exact and specific (*i.e.*, alignment of chromosomes is exactly gene-for-gene). The paired homologous chromosomes are joined by a roughly 0.2 μm thick, protein-containing framework called a **synaptonemal complex (SC)**. This complex extends along the whole length of the paired chromosomes and is usually anchored at either end to the nuclear envelope. SC helps to stabilize the pairing of homologous chromosomes and to facilitate the cytogenetical activity, called **recombination** or **crossing over** (occurring during pachynema). SC is not found in those organisms in which crossing over does not occur (*e.g.*, the male fruitfly, *Drosophila melanogaster*).

4. Pachytene or Pachynema. (Gr., *pachus* = thick). In the pachynema stage the pair of chromosomes become twisted spirally around each other and cannot be distinguished separately. In the middle of the pachynema stage each homologous chromosome spilts lengthwise to form two chromatids. Actually, the doubling of the DNA molecule strands which is necessary for the subsequent duplication of chromosomes occurs earlier, before the beginning of meiotic prophase. Through the earlier part of the meiotic prophase, however, the DNA molecule in each chromosome behaves as a single body. In the pachynema stage, this is now changed, the two chromatids of each chromosome containing half of the DNA present in the chromosome at start, become partially independent of one another, although they still continue to be linked together by their common centromere. Each synaptonemal pair at this point is commonly referred to as **bivalents** or **dyads** because it consists of two visible chromosomes, or as a **quadrivalent** or **tetrad** because of the four visible chromatids.

During pachynema stage an important genetic phenomenon called **crossing over** takes place. The crossing over involves reshuffling, redistribution and mutual exchange of hereditary material of two parents between two homologous chromosomes. According to recent views, one chromatid of each homologous chromosome of a bivalent may divide transversely by the help of an enzyme the **endonuclease** which is reported to increase in the nucleus during this stage by **Stern** and **Hotta** (1969). After the division of chromatids, the interchange of chromatid segments takes place between the non-sister chromatids of the homologous chromosomes. The broken chromatid segments are united with the chromatids due to the presence of an enzyme, **ligase** (**Stern** and **Hotta**, 1969). The process of interchange of chromatin material between one non-sister chromatid of each homologous chromosome is known as the **crossing over** which is accompanied by the **chiasmata formation.**

Stern and **Hotta** (1969) have reported that during the pachytene and zygotene stage, synthesis of small amount of DNA takes place. This DNA amount is utilized in the repairing of broken DNA molecule of the chromatids during the chiasmata formation and crossing over.

The nucleolus remains prominent up to this stage and it is found to be associated with the nucleolar organizer region of the chromosome.

5. Diplotene or Diplonema. In diplonema, unpairing or **desynapsis** of homologous chromosomes is started and chiasmata are first seen. At this phase the chromatids of each tetrad are usually clearly visible, but the synaptonemal complex appears to be dissolved, leaving participating chromatids of the paired homologous chromosome physically joined at one or more discrete points called **chiasmata** (singular, **chiasma**; Gr., *chiasma* = cross piece). These points are where crossing over took place. Often there is some unfolding of the chromatids at this stage, allowing for RNA synthesis and cellular growth.

6. Diakinesis. In the diakinesis stage the bivalent chromosomes become more condensed and evenly distributed in the nucleus. The nucleolus detaches from the nucleolar organizer and ultimately disappears. The nuclear envelope breaks down. During diakinesis the chiasma moves from the centromere towards the end of the chromosomes and the intermediate chiasmata diminish. This type of movement of the chiasmata is known as **terminalization**. The chromatids still remain connected by the terminal chiasmata and these exist up to the metaphase.

Prometaphase

In the prometaphase the nuclear envelope disintegrates and the microtubules get arranged in the form of spindle in between the two centrioles which occupy the position of two opposite poles of the cell. The chromosomes become greatly coiled in the spiral manner and get arranged on the equator of the spindle.

Metaphase I

Metaphase I consists of spindle fibre attachment to chromosomes and chromosomal alignment at the equator. During metaphase I, the microtubules of the spindle are attached with the centromeres of the homologous chromosomes of each tetrad. The centromere of each chromosome is directed towards the opposite poles. The repulsive forces between the homologous chromosomes increase greatly and the chromosomes become ready to separate.

Anaphase I

At anaphase I homologous are freed from each other and due to the shortening of chromosomal fibres or microtubules each homologous chromosome with its two chromatids and undivided centromere move towards the opposite poles of the cell. The chromosomes with single or few terminal chiasma usually separate more frequently than the longer chromosomes containing many chiasmata.

The actual reduction and **disjunction** occurs at this stage. Here it should be carefully noted that the homologous chromosomes which move towards the opposite poles are the chromosomes of either paternal or maternal origin. Moreover, because during the chiasma formation out of two chromatids of a chromosome, one has changed its counterpart, therefore, the two chromatids of a chromosome do not resemble with each other in the genetical terms.

Telophase I

The arrival of a haploid set of chromosomes at each pole defines the onset of telophase I, during which nuclei are reassembled. The endoplasmic reticulum forms the nuclear envelope around the chromosomes and the chromosomes become uncoil. The nucleolus reappears and, thus, two daughter chromosomes are formed. After the karyokinesis, cytokinesis occurs and two haploid cells are formed.

Both cells pass through a short resting phase of **interkinesis**. During this phase, no DNA replication occurs, so that chromosomes at the second prophase are the same double-stranded structures that disappeared at the first telophase. In case of *Trillium* telophase I and interphase I do not occur and the anaphase I is followed by prophase II directly.

Homotypic or Second Meiotic Division

The homotypic or second meiotic division is actually the mitotic division which divides each haploid meiotic cell into two haploid cells. The second meiotic division includes following four stages.

Prophase II

In the prophase second, each centriole divides into two and, thus, two pairs of centrioles are formed. Each pair of centrioles migrates to the opposite pole. The microtubules get arranged in the form of spindle at the right angle of the spindle of first meiosis. The nuclear membrane and the nucleolus disappear. The chromosomes with two chromatids become short and thick.

Metaphase II

During metaphase II, the chromosomes get arranged on the equator of the spindle. The centromere divides into two and, thus, each chromosome produces two monads or daughter chromosomes. The microtubules of the spindle are attached with the centromere of the chromosomes.

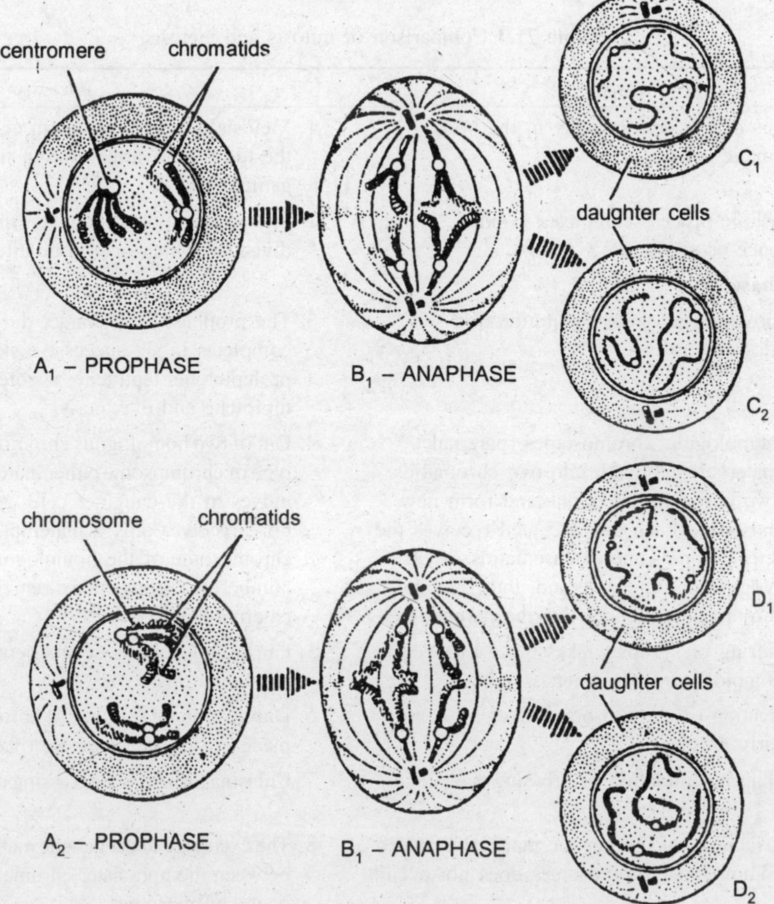

Fig. 71.10. Different stages of second meiotic division.

Anaphase II

The daughter chromosomes move towards the opposite poles due to the shortening of chromosomal microtubules and stretching of interzonal (polar) microtubules of the spindle.

Telophase II

The chromatids migrate to the opposite poles and now known as chromosomes. The endoplasmic reticulum forms the nuclear envelope around the chromosomes and the nucleolus reappears due to synthesis of ribosomal RNA (rRNA) by rDNA and also due to accumulation of ribosomal proteins.

After the karyokinesis, in each haploid meiotic cell, the cytokinesis occurs and, thus, four haploid cells are resulted. These cells have different types of chromosomes due to the crossing over in the prophase I.

Significance of Meiosis

The meiosis has the greatest significance for the biological world because of its following uses:

1. The meiosis maintains a definite and constant number of the chromosomes in the organisms.
2. By crossing over, the meiosis provides an opportunity for the exchange of the genes and, thus, causes the genetical variations among the species. The variations are the raw materials of the evolutionary process.

Thus the meiosis is a peculiar taxonomic, genetical and evolutionary process.

71.6 COMPARISON BETWEEN MITOSIS AND MEIOSIS

Table 71.3 Comparison of mitosis and meiosis.

Mitosis	*Meiosis*
1. Mitosis occurs continuously in the body or somatic cells.	1. Meiosis occurs in the germ cells (the cells of the testes or ovaries) during the process of gametogensis.
2. The whole process completes in one sequence or phase.	2. The whole process completes in two successive divisions which occur one after the other.
Prophase	
3. The prophase is of short duration and includes no substage.	3. The prophase is of longer duration and it completes in six successive stages, *viz.*, proleptotene, leptotene, zygotene, pachytene, diplotene and diakinesis.
4. The homologous chromosomes (paternal and maternal) duplicate into two chromatids. The two chromatids separate and form new chromosomes. Each daughter cell receives the daughter chromosome or chromatids of each homologous chromosome and, thus, having the chromosome number like the parental cells.	4. Out of two homologous chromosomes only one type of chromosome either maternal or paternal moves to the daughter cells. A daughter cell, thus, receives only a maternal or paternal chromosome of the homologous pair and the number of chromosomes remain half than the paternal cells.
5. No pairing or synapsis takes place between the homologous chromosomes.	5. Pairing or synapsis occurs between the homologous chromosomes.
6. Duplication of chromosomes takes place in the early prophase.	6. Duplication or splitting of chromosomes takes place in the late prophase (pachytene stage).
7. No chiasma formation or crossing over takes place.	7. Chiasma formation or crossing over takes place.
8. The exchange of the genetic material between the homologous chromosomes does not occur.	8. The exchange of the genetic material takes place between the non-sister chromatids of homologous chromosomes.
Metaphase	
9. The chromatids occur in the form of dyads.	9. The chromatids of two homologous chromosomes occurs as the tetrads.
10. The centromeres of the chromosomes remain directed towards the equator and the arms of the chromosomes remain directed towards the poles.	10. The centromeres of the chromosomes remain directed towards the poles and the chromosomal arms remain directed towards the equator.
Anaphase	
11. The chromosomes are the monads, *i.e.*, having single chromatid.	11. The chromosomes are the diads, *i.e.*, having two chromatids and single centromere.
12. The chromosomes are long and thin.	12. The chromosomes are short and thick.
Telophase	
13. The telophase always occurs.	13. The first telophase is sometimes omitted.
Significance	
14. The chromosome number in each daughter cell remains the same like the parent cell.	14. In meiotic division the chromosome number is reduced to half in the daughter cells than the parental cells.
15. A diploid cell produces two diploid cells by a mitotic division.	15. A diploid cell produces four haploid cells by a meiotic division.

QUESTIONS

Long Answer Questions

1. What do you understand by cell cycle? Give an account of the salient features of various phases of cell cycle. *(NEHU (Major) 1996)*
2. Give an illustrated account of mitosis. *(Allahabad 1996, Lucknow 1997)*
3. Give a detailed account of mitotic cell division and add a note on the significance of mitosis. *(Kanpur 1993, 1995)*
4. Describe the meiosis. *(Punjab (G) 1990; Kanpur 1993, 1995; Nagarjuna 1995; B.U. Coimbatore 1997)*
5. Describe meiosis in detail and explain its significance. *(Lucknow 1995)*
6. Describe meiotic cell division and state the significance of crossing over. *(Lucknow 1992; Kanpur 1994)*
7. Describe the prophase of meiosis. Give its significance. *(Punjab (G) 1989, 1992)*
8. What do you mean by cell division? Describe the different stages of meiotic cell division and its significance. *(C.C.S.U., Meerut 2007)*

Short Answer Questions

1. Describe the molecular organization and functions of mitotic apparatus. *(B.U. Coimbatore 1997)*
2. What are the hypotheses about the mechanism of anaphase movement of chromosomes? *(Allahabad 1996; B.U. Coimbatore 1997)*
3. Compare mitosis with meiosis.
4. Compare mitosis and first meiotic division with special reference to prophase. *(Gorakhpur 1998)*
5. Explain the difference between metaphase of mitosis and meiosis. *(Punjab (G) 1992)*
6. Differentiate between prophase of mitosis and meiosis. What will happen if there is no meiosis in the germinal cells? *(Punjab (G) 1991)*
7. Describe 'mitosis' and compare it with meiosis. *(Lucknow 1994, 1996)*
8. Write short note on the 'significance of meiosis'. *(Gorakhpur 1995)*
9. What are the main differences between mitosis and meiosis? *(Punjab (G) 1989, 1990)*
10. Differentiate between karyokinesis and cytokinesis. *(CCS 1999)*
11. Explain the meiotic cell division and its significance. *(C.C.S.U. Meerut 2009)*
12. Give the meiosis and its importance with diagram. *(H.N.B., Garhwal 2008)*
13. Differentiate between mitosis and meiosis cell division. *(C.C.S.U., Meerut 2008; H.N.B., Garhwal 2009)*
14. Write short notes on the following:
    (*i*) Mitotic metaphase *(Gorakhpur 1993, 1995)*
    (*ii*) Metaphase *(Lucknow 1998)*
    (*iii*) S-phase of cell cycle. *(BHU (Hons.) 1990; Lucknow 1998)*
    (*iv*) Leptotene *(Lucknow 1993)*
    (*v*) Zygotene *(Lucknow 1993)*
    (*vi*) Pachytene *(Lucknow 1993)*
    (*vii*) Diplotene *(Lucknow 1993)*
    (*viii*) Diakinesis *(Lucknow 1993)*
    (*ix*) Anaphase of meiosis *(Allahabad 1992)*
15. Draw neat and labelled diagram of the various stages of mitotic cell division (No description is required). *(Kanpur 1997)*

16. Draw neat and labelled diagram of the various stages of meiotic cell division (No description is required). (*Gorakhpur 1994, 1997; Kanpur 1996*)

Very Short Answer Questions

1. When does synapsis occur during meiosis? (*HNB, Garhwal 2008*)
   Ans. Zygonema.
2. In which phase of meiosis chromosome number reduces to half? (*H.N.B., Garhwal 2009*)
   Ans. Anaphase I.
3. In which type of cell mitosis occurs?
   Ans. Somatic cell.
4. In which type of cell meiosis occurs?
   Ans. Germ cells.
5. In which phase the karyotype can be studied?
   Ans. Metaphase.
6. What is cyclin?
   Ans. Cyclin is any of proteins that activate the cyclin dependent kinases (Cdks) involved in regulating progression through the eukaryotic cell cycle.
7. Define Cdk.
   Ans. Cdk stands for cyclin-dependent kinase, it is any of several protein kinases that are activated by different cyclins and that control progression through the eukaryotic cell cycle by phosphorylating various proteins.

Fill in the Blanks

(*i*) The division of cytoplasm during cell division is called .... (*Lucknow 1996*)

(*ii*) In mitotic prophase the chromatids are plectonemic but at the same phase of meiosis they are .... (*Lucknow 1995*)

(*iii*) Give one word or few words for following sentences:
Mode of cell division involved in the formation of sperm cells in haploid male bee .... (*Lucknow 1995*)

(*iv*) Polytene chromosomes were discovered by .... (*H.N.B. Garhwal 2009*)

True and False Statements

(*i*) Polytene chromosomes are in permanent interphase state. (*B.H.U. (Hons.) 1991*)

(*ii*) Visualization of condensed chromosome during G_1 or G_2 phase of the cell cycle is possible.

(*iii*) Each chromatid is composed of single DNA duplex.

(*iv*) Crossing-over always takes place after duplication of chromonema.

(*v*) Anaphase movement is diphasic.

Multiple Choice Questions

1. Which of the following cells do not divide once it is differentiated?
   (*a*) intestine cells (*b*) nerve cells
   (*c*) blood cells (*d*) glial cells
2. Diploid somatic cells divide by
   (*a*) meiosis (*b*) mitosis only
   (*c*) both meiosis and mitosis (*d*) none of these
3. Correct sequence of stages in cell cycle is
   (*a*) G_1, S, G_2, M (*b*) G_1, G_2, S, M

(*c*) M_1, S, G_1, G_2 (*d*) G_2, G_1, M, S

4. Condensation of chromosome with visible centromere occurs during

(*a*) G_1 phase (*b*) G_2 phase

(*c*) S phase (*d*) M phase

5. DNA-replication takes place during

(*a*) G_1 phase (*b*) G_2 phase

(*c*) S phase (*d*) prophase

6. Histone proteins are synthesized in

(*a*) S phase (*b*) G_1 phase

(*c*) prophase (*d*) G_2 phase

7. The term 'Meiosis' was coined by

(*a*) Knoll and Ruska (*b*) Farmer and Moore

(*c*) Beadle and Tatum (*d*) A. Flemming

8. Meiosis occurs in

(*a*) liver (*b*) kidneys

(*c*) gonads (*d*) brain

9. Pairing of homologous chromosomes occur during

(*a*) leptotene (*b*) zygotene

(*c*) pachytene (*d*) diplotene

10. Crossing-over occurs during

(*a*) leptotene (*b*) pachytene

(*c*) diakinesis (*d*) diplotene

11. Interphase is the resting stage between (*B.U. Coimbatore 1997*)

(*a*) two mitotic divisions (*b*) two meiotic divisions

(*c*) mitotic and meiotic divisions

12. The structural basis for pairing of homologous chromosome during prophase I. (*B.U. Coimbatore 1997*)

(*a*) kinetochore (*b*) synaptonemal complex

(*c*) chromomere (*d*) centromere

13. The spindle microtubules attach to the (*B.U. Coimbatore 1997*)

(*a*) telomere (*b*) nucleosome

(*c*) kinetochore (*d*) none

14. Synapsis between homologous chromosome occurs in (*NEHU (M) 1996*)

(*a*) leptotene stage (*b*) pachytene

(*c*) zygotene (*d*) diplotene

15. Chromosomes become tetravalent during (*NEHU (M) 1996*)

(*a*) leptotene (*b*) zygotene

(*c*) pachytene (*d*) diplotene

16. The phase of cell cycle which lasts for longer duration (*NEHU (M) 1996*)

(*a*) G_1 (*b*) G_2

(*c*) S (*d*) M

17. Cells in Go phase of cell cycle

(*a*) exit cell cycle (*b*) enter cell cycle

(*c*) suspend cell cycle (*d*) terminate cell cycle

18. In which phase of meiosis homologous chromosome separate?

(*a*) metaphase I (*b*) metaphase II

(*c*) anaphase I (*d*) anaphase II

19. The cellular structure which always disappears during mitosis is

(*a*) mitochondria (*b*) nuclear membrane

(*c*) plastids (*d*) plasma membrane

20. Interkinesis is followed by

(*a*) prophase II (*b*) prophase I

(*c*) interphase (*d*) telophase II

ANSWERS

Fill in the Blanks

(*i*) Cytokinesis (*ii*) Paranemic (*iii*) Mitosis (*iv*) E.G. Balbiani

True and False Statements

(*i*) True (*ii*) False (*iii*) True (*iv*) True

(*v*) True

Multiple Choice Questions

1. (*b*)	**2.** (*b*)	**3.** (*a*)	**4.** (*d*)	**5.** (*c*)	**6.** (*b*)
7. (*b*)	**8.** (*c*	**9.** (*b*)	**10.** (*b*)	**11.** (*a*)	**12.** (*b*)
13. (*c*)	**14.** (*c*)	**15.** (*c*)	**16.** (*a*)	**17.** (*c*)	**18.** (*c*)
19. (*b*)	**20.** (*a*)				

72

Cell Transformation and Cancer (Biology of Cancer)

The proliferation, differentiation and survival of individual cells in multicellular organisms are carefully regulated to meet the needs of the organism as a whole. This regulation is lost in cancer cells, which grow and divide in an uncontrolled manner, ultimately spreading throughout the body and interfering with the function of normal tissues and organs (**Cooper** and **Hausman**, 2007).

Cells which undergo rapid, abnormal and uncontrolled growth at the cost of remaining cells are called **neoplastic cells**. The growths resulting from the division of such cells are called **neoplastic growths** or **tumours**. Thus, any abnormal proliferation of cells is the tumour. Almost all type of differentiated cells of animals can become neoplastic or cancerous. The process of cell change in which a cell loses its ability to control its rate of division, and thus becomes a tumour cell, is called **cell transformation**.

Cancer is a genetic disease because it can be traced to alterations within specific genes, but in most cases, it is not an inherited disease. In an inherited disease, the genetic defect is present in the chromosomes of a parent and is transmitted to the zygote. In contrast, the genetic alterations that lead to most cancers arise in the DNA of a somatic cell during the lifetime of the affected individual. Because of these genetic changes, cancer cells proliferate uncontrollably, producing malignant tumours that invade surrounding healthy tissue (see **Karp**, 2002).

72.1 TYPES OF CANCER

Cancer can result from abnormal proliferation of any of the different kinds of cells in the body, so there are more than a hundred distinct types of cancer. These cancers can vary substantially in their behaviour and response to treatment. Based on differences in their growth patterns, tumours are classified as either benign or malignant. A **benign tumour** or **primary tumour**, such as a common skin wart, remains confined to its original location. A **malignant tumour** or **secondary tumour**, however, is capable of both invading surrounding normal tissue and spreading throughout the body via the circulatory or lymphatic systems (**metastasis**). Only malignant tumours are properly referred to as cancers and it is their ability to invade and metastasize that makes cancer so dangerous. The term **cancer** refers to any malignant tumour—that is, any tumour capable of spreading from its original location to other sites (Box 72.1). Whereas benign tumours can usually be removed surgically, the spread of malignant tumours to distant body sites frequently makes them resistant to such localized treatment.

Both benign and malignant tumours are classified according to the type of cell from which they arise. About two hundred distinct types of cancer have been recognized. These can be grouped into four main types: carcinomas, sarcomas, lymphomas and leukemias.

Box 72.1

The earliest recorded description of cancer is found in an Egyptian papyrus dating back to around 3000 B.C. (Papyrus refers to a manuscript written on paper prepared from papyrus plant *Cyperus papyrus*). The term cancer, which means "crab" in Latin, was coined by Hippocrates in the fifth century B.C., to describe diseases in which tissues grow and spread unrestrained throughout the body, eventually choking off life (**Becker** *et al.*, 2007); Hippocrates used the words *carcinus, carcinoma* and *cancer*, which refer to crabs in Latin, to describe the tumours (**Cooper** and **Hausman**, 2007).

1. Carcinomas. These cancers account for 90 per cent of all human cancers. Carcinomas arise from the epithelial cells that cover external and internal body surfaces. These epithelial cells may be of ectodermal or endodermal origin. Carcinomas include cervical, breast, skin, brain, lung, colon, prostate, kidney, bladder cancers. Carcinomas affect cells of ectodermal or endodermal origin.

2. Sarcomas. These cancers account for 4 per cent of all human cancers. Sarcomas are solid tumours of connective tissues such as cancers of muscle, bone, cartilage, fat cells and fibrous tissue. Sarcomas affects the cells of mesodermal origin.

3. Lymphomas. These cancers constitute about 4 per cent of human cancers. In lymphomas, there is an excessive production of lymphocytes by the lymph nodes and spleen. Hodgkin's disease is an example of lymphoma (Hodgkin's disease is a neoplastic disease that is characterized by progressive enlargement of lymph nodes, spleen, and liver and by progressive anaemia).

4. Leukemias. These cancers form about 4 per cent of human cancers. Leukemias are neoplastic growth of leucocytes (W.B.C.) and are characterized by excessive production of white blood cells (Box 72.2). Leukemias affect cells of mesodermal origin and are a class of sarcomas (**Lodish** *et al.* 2004). In addition to the four types of cancer mentioned above there may be **mixed malignant** tumours arising from ectodermal and mesodermal tissues, *e.g.*, tumours arising from ectodermal and mesodermal tissues.

Box 72.2

The leukemias grow as individual cells in the blood. The name **leukemia** is derived from the Latin for "white blood": the massive proliferation of leukemic cells can cause a patients blood to appear milky.

Box 72.3

Prior to 1900, most human deaths were due to **infectious diseases** such as pneumonia and tuberculosis, and life expectancy was less than 50 years. Cancer was rare disease that accounted for only a small percentage of deaths.

According to a recent survey conducted by American cancer society in 2005, more than 5 lac Americans die of cancer each year. The four most common cancers accounting for more than half of all cancer cases are those of the prostate, breast, lung, and colon/rectum. Lung cancer, by far the most lethal, is responsible for nearly 30% of all cancer death.

72.2 GROWTH PROPERTIES OF NORMAL AND CANCEROUS CELLS

1. Growth of cancer cells produce multilayered clumps. Information on the behaviour of human cancer cells has come largely from studies on cells grown in vitro. At the cellular level, the most important characteristic of a cancer cell—whether residing in the body or on a culture dish—...s loss of growth control. The **capacity** for growth and division is not drastically different between ...r cell and most normal cells. When normal cells are grown in tissue culture, under conditions ...ote cell proliferation, they grow and divide at a rate similar to that of their malignant ...wever, when the normal cells proliferate to the point where they cover the bottom ...dish, their growth rate decreases markedly, and they tend to remain as a single layer (monolayer) of cells (Fig. 72.1A). Growth rates drop as normal cells respond to inhibitory influences from their environment. Growth-inhibiting influences may arise as the result of depletion of growth factors in the culture medium or from contact with surrounding cells on the dish (called **contact inhibition**). In contrast, when malignant cells are cultured under the same conditions, they continue to grow, piling on top of one another to form clumps (Fig. 72.1B). It is evident that malignant cells are not responsive to the types of regulatory signals that causes their normal counterparts to cease growth and division.

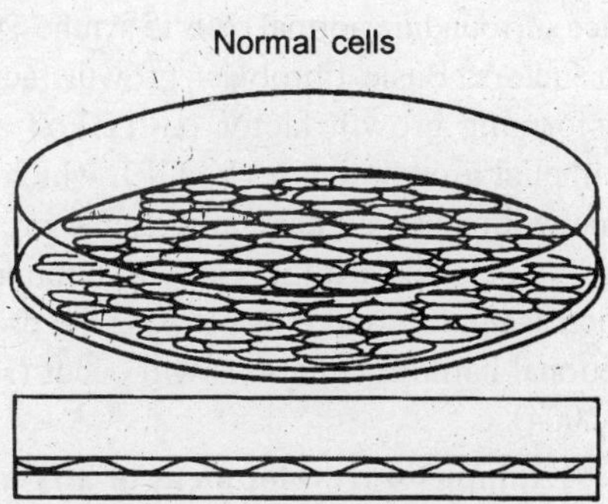

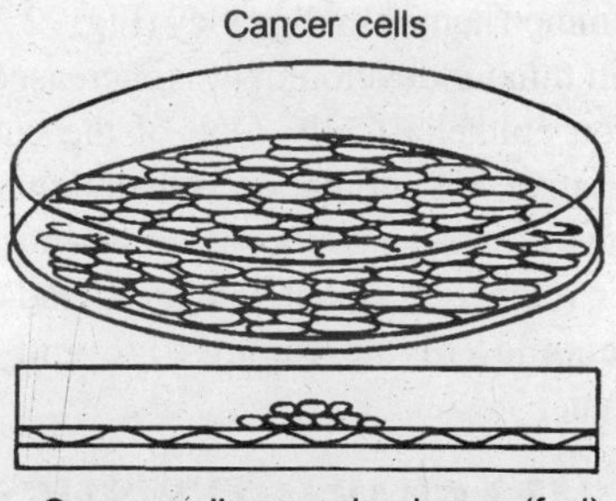

Fig. 72.1. Growth properties of normal and cancerous cells. A—Normal cells typically grow in a culture dish until they cover the surface as a monolayer. B—In contrast, cells that have been transformed by viruses or carcinogenic chemicals (or malignant cells that have been cultured from tumours) typically grow in multilayered clumps or foci (after Karp, 2002).

Stages of Tumour Development.

At the cellular level, the development of cancer is viewed as multistep process involving **mutation** and **selection** for cells with progressively increasing capacity for proliferation, survival, invasion, and metastasis (Fig. 72.2).

1. Initiation. The step of tumour initiation is thought to be the result of a genetic alteration due to mutation leading to abnormal proliferation of a single cell. Cell proliferation then leads to the outgrowth of population of clonally derived tumour cells.

2. Tumour progression. This process continues as additional mutations occur within cells of the tumour population. Some of these mutations confer a selective advantage to the cell, such as more rapid growth, and the descendants of a cell bearing such mutations will consequently become dominant within the tumour population. The process is called **clonal selection**, since a new clone of tumour cells has evolved on the basis of its increased growth rate or other properties (such as survival, invasion or metastasis) that confer a selective advantage. Clonal selection continues throughout tumour development, so tumours continuously become more rapid-growing and increasingly malignant.

Angiogenesis.

Both primary and secondary tumours require *angiogenesis*, the formation of new blood vessels, in order to grow to a large mass. Most tumours induce the formation of new blood vessels that invade the tumour and nourish it, a process called **angiogenesis**. This complex process requires several distinct steps:

1. Degradation of the basal lamina that surrounds a nearby capillary,
2. Migration of endothelial cells lining the capillary into the tumour,
3. Division of these endothelial cells and
4. Formation of a new basement membrane around the newly elongated capillary.

Many tumours produce growth factors that stimulate angiogenesis; other tumours somehow induce surrounding normal cells to synthesize and secrete such factors. Basic fibroblast growth factors (bFGF), transforming growth factor α (TGFα) and vascular endothelial growth factor (VEGF), which are secreted by many tumours, all have angiogenic properties. New blood vessels nourish the growing tumour, allowing it to increase in size and thus increase the probability that additional harmful mutations will occur (see **Lodish** *et al.* 2004).

Example of Development of a Human cancer: Colon carcinomas

Studies of colon carcinoma have provided a clear example of tumour progression during the development of a common human malignancy (Fig. 72.3). The earliest stage in tumour development is increased proliferation of colon epithelial cells. One of the cells within this proliferative cell population is then thought to give rise to a small benign neoplasm (an **adenoma** or **polyp**). Further rounds of clonal selection lead to the growth of adenomas of increasing size and proliferative potential.

72.3 CHARACTERISTICS OF CANCER CELLS

The cancerous cell generally retains the structural and functional characteristics of the normal cell type from which it is derived. For example, the cancerous cells of thyroid gland continue to secrete thyroxin. Neoplastic cells, however, differ from their normal counterparts in several ways.

1. Immortalization. Normal cell cultures do not survive indefinitely. For example, human cell cultures die after about 50 generations, and chicken cell cultures have a much shorter life expectancy. In contrast, transformed cell cultures are immortal and can grow indefinitely. A striking example is provided by HeLA cells, which were obtained in 1953 from a uterine cancer diagnosed in a women named **Henreitta Lacks** (hence the term "HeLa" cells). After the tumour was removed by surgeons, some of its cells were placed in culture. The cultured cells quickly began to proliferate and have continued to do so for more than 50 years, dividing more than 18,000 times with no sign of stopping (see **Becker** *et al.*, 2007). Similarly, cell cultures infected with mouse sarcoma virus can be maintained as long as nutrition is provided and overcrowding avoided.

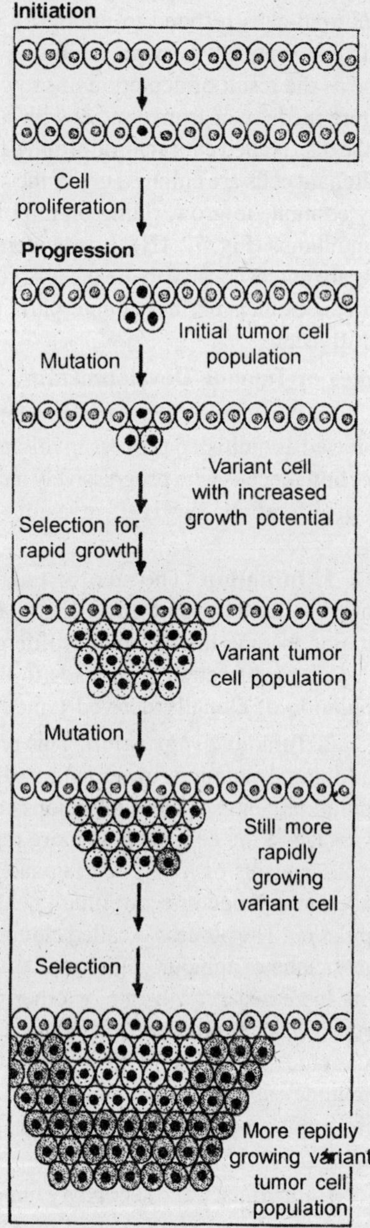

Fig. 72.2. Stages of tumour development. The development of cancer initiates when a single mutated cell begins to proliferate abnormally. Additional mutations followed by the selection for more rapid growing cells within the population then result in progression of the tumour to increasingly rapid growth and malignancy (after Cooper and Hausman, 2007).

Box 72.4

Cancer cells, like germ cells and stem cells but unlike most differentiated cells, produce **telomerase** enzyme, which prevent shortening of chromosomes during DNA replication and may contribute to their **immortalization**. The absence of telomerase is associated with resistance to generation of certain types of human tumours (*e.g.*, papillomas and colon tumours of mice) (see **Lodish** *et al.*, 2004).

2. Loss of contact inhibition. Normal cells in a culture stop growing when their plasma membranes come into contact with one another. This inhibition of growth after contact is called **contact inhibition**. Transformed cells are unable to go into a quiescent stage after division. They will grow (and divide) continuously until they kill themselves. Transformed cells apparently undergo a change in the property of their plasma membranes, which become less adhesive. This change enables the cells to dissociate from neighbouring cells and to infiltrate other organs, where they form metastatic tumours. Cancer cells thus, lack proper recognition and communication.

3. Reduced cellular adhesion. When normal cells become cancerous there is change in the **stickiness** or **adhesiveness** of their plasma membranes. If grown in a nutrient medium kept in a glass vessel, the normal cells stick to the glass rather than float in the medium. Transformed cells show a **decreased adhesiveness** and if grown in solid media, they stick to each other less than do normal cells.

Adhesiveness shows considerable specificity. For example, liver cells tend to stick to other liver cells and not to other cell types, *e.g.*, kidney cells. If the cells of the liver and the kidney are separated by the enzyme trypsin and incubated together, they aggregate to form small pieces of liver tissue and kidney tissue. Thus, kidney cells stick to kidney cells and liver cells to liver cells. Cancerous cells do show this property (*i.e.*, specificity of adhesiveness). Thus, if malignant skin cancer cells are mixed with normal kidney cells then aggregates formed contain both kidney and skin cells mixed together. This probably explains why malignant cell can invade several normal organs.

4. Invasiveness. One of the most significant characteristics of transformed cells is their invasiveness, *i.e.*, the ability of cancer cells, to invade other tissues. In contrast to normal cells, transformed cells can penetrate the chorioallantoic membrane of the hen's egg. This invasiveness could be due to changes in the plasma membrane and/or proteases released by the cancerous cells.

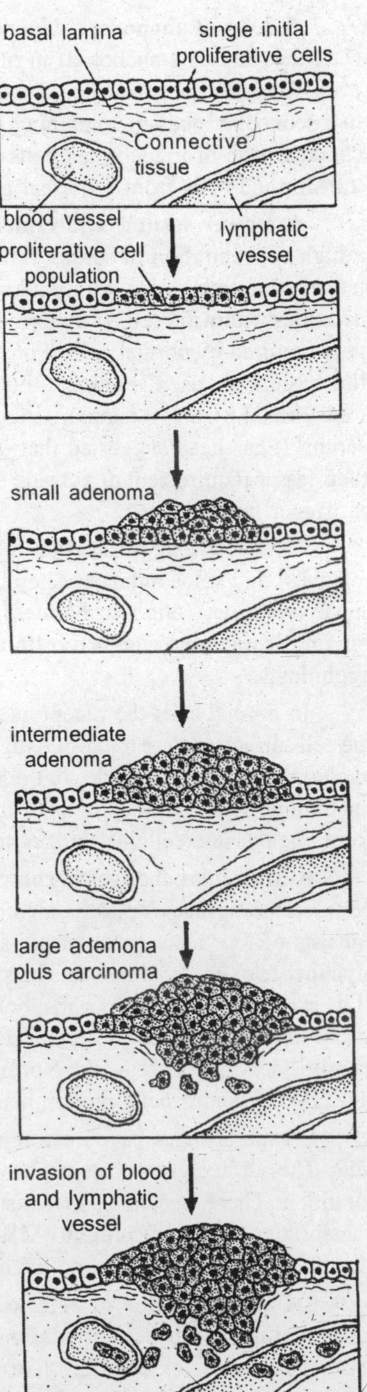

Fig. 72.3. Development of colon carcinomas. A single initially altered cell gives rise to a proliferative cell population, which progresses first to bening adenomas of increasing size and then to malignant carcinoma. The cancer cells invade the underlying connective tissue and penetrate blood and lymphatic vessels, thereby spreading throughout the body (after Cooper and Hausman, 2007).

5. Loss of anchorage dependence. Most normal cells must be attached to a rigid substratum (*i.e.*, they must be anchored) in order to grow in a culture medium. Transformed cells can grow even when they are not anchored to the substratum. For example, when transformed cells are suspended in a semisolid medium containing agar or methyl cellulose. This loss of anchorage is most striking characteristic of transformed cells which form malignant tumours. This property is used to select transformed cells from a normal cell population.

6. Lower serum requirements. Growth of normal cells in tissue culture medium requires a high concentration of serum. Some serum growth factors (*e.g.*, somatomedins) resemble insulin hormone in interacting with external receptors on the plasma membrane to regulate biochemical activities within the cell. Transformed cells can grow in a culture medium containing much less serum than required by normal cells. For example, normal 3T3 cells (established fibroblasts of the mouse line commonly used is tissue culture) grow optimally in 10 per cent foetal calf-serum, while cells transformed by SV40 (simian virus number 40) can grow equally well in 1 per cent or 10 per cent serum. It has been suggested that the lower serum requirement of transformed cells is because of their lesser requirement of out side substances to lower their intracellular cAMP (cyclic AMP) level to trigger mitosis.

7. Selective agglutination by lectins. Lectins are proteins widely distributed in plants (*e.g.*, legumes) and some animals. They have the ability to bind to receptors, which are branched chain sugar molecules (oligosaccharides), on the surface of the plasma membrane. As a result of such binding lectin cause **agglutination** or **clumping** of cells. Due to this property, lectins work as agglutinins.

In normal cells the receptors or agglutin binding sites for lectins lie in a diffuse manner on the cell surface and are immobile. In such cases, lectins make few intercellular bridges, and therefore agglutination is not possible. In transformed cells, the receptors are more mobile within the plasma membrane. Local regions of high binding site concentration are formed. Lectins are thus able to form enough intercellular bridges to result in agglutination in the transformed cells.

8. Molecular changes in components of the plasma membrane. There are several differences between the surface of plasma membranes of normal and transformed cells. The plasma membrane consists of four main types of **phospholipids**, which form the lipid bilayer, with **glycolipids** and **glycoproteins** inserted into this bilayer. Cancerous cells apparently do not differ from normal cells in their relative amounts of phospholipids. However, **gangliosides** (glycolipids which contain sialic acids) become reduced in certain mouse cancer cells. Enzymes involved in biosynthesis of these gangliosides are also reduced. Normal cells possess four types of gangliosides, GM1a, GM1, GM2 and GM3. Tumour cells predominantly contain the simplest type, GM3.

Certain changes have been reported in the **glycoproteins** of plasma membrane of cancerous cells. The surface glycoprotein of MW 46,000 disappears early in transformation to the cancerous condition. There is also a slow disappearance of a major protein called **LETS** (large, external, transforming sensitive) protein (MW 240,000). Probably the most important protein to disappear after transformation is the one having MW 200,000.

9. Disorganisation of cytoskeleton. Normal cells have a well organised cytoskeleton which consists of **microtubules** (of tubulins), **microfilaments** (of actins) and **intermediate filaments** (of collagen for example). The fibre-like proteins have a regular arrangement and bring about coordinated cell movement. In transformed or cancer cells, the fibres are much fewer in number and usually much thinner. In transformed cells the cytoskeleton is found to undergo **depolymerisation**. The microtubules disaggregate. The microfilaments undergo depolymerisation and disappear, but diffuse actin persists. The myosin-like filaments also disappear. Thus, in transformed cells the cytoskeleton proteins become less organised than in normal cells. It has been suggested that due to this disorganisation of the cytoskeleton, there is an increased mobility of plasma membrane proteins.

The disorganisation of the cytoskeleton also affects the cell surface in another way. When cancer cells are touched, there is a constant and uncoordinated throwing out and retraction of **blebs, microvilli** and **ruffles** from the cell surface. Tumour cells have a more ruffled surface than normal cells, with many more surface processes.

10. Increase in negative surface charge. Comparisons of surface membrane charge by microelectrophoresis have been made between normal and malignant cells. In malignant cells anodic mobility is usually higher, indicating increase in negative surface charge.

11. Defective electrical communication. Electrical connections normally occur between individual normal cells. In some cancer cells, however, it has been reported that such electrical connections are defective.

12. Increased sugar transport. Tumour cells consume much more glucose than normal cells because they have to grow and multiply. There is a great increase in the rate of sugar transport across the surface membrane (plasma membrane) after transformation of the cell. This tends to increase in sugar intake by malignant cells.

13. Increased rate of glycolysis. In the 1920's **Warburg** pointed out that oxidative (aerobic) respiration is depressed in tumour cells and that glycolysis (anaerobic respiration) increases. This has been demonstrated by an increase in lactic acid production in cells of solid tumour.

14. Appearance of virus-specific transplantation rejection antigens. Plasma membranes of most transformed cells contain **antigens** which are not present in normal cells. Thus in cells transformed by adenoviruses and papovaviruses the **T-antigen** is always present. Similarly all cells transformed by the Epstein-Barr virus (EB virus) contain an antigen called the **EB nuclear antigen (EBNA)**. Tumour antigens can bring about an **immunity response** against themselves in genetically similar hosts. This is in contrast to normal histocompatibility antigens. The immunity response brought about by the antigens results in recognition and destruction of newly formed cancer cells and their descendants. Such a defence mechanism is called **immunological surveillance** and can lead to the elimination of transformed or cancer cells under favourable conditions. It has been suggested that only in the rare cases when this defence mechanism fails that tumours are formed.

15. Increased secretion of proteolytic enzymes. Large amount of proteolytic enzymes are secreted by all types of cancer cells, except those of blood forming tissues. The cancer cell secretes a protease called the **cell factor** (MW ~40,000). The cell factor acts on an inert serum protein **plasminogen** (MW 85,000) to form a **plasmin**, a **proteolytic enzyme** (MW 76,000). It has been suggested that plasmin removes many proteins projecting from the cell surface by enzymatic digestion and signals the cell into division.

If normal cells are treated with proteases they show many of the characteristics of transformed cells. It has been speculated that viral proteins cause the release of extracellular protease. However, there is no direct evidence for this.

16. Aldolases. In most mammalian tissues the enzyme **aldolase** exists in the form of three isozymes A, B and C. Isozymes A and C predominate in embryonic tissues while in adult differentiated tissues the B isozyme is predominant. In some tumours, especially in poorly differentiated and rapidly growing cancers such as **hepatomas**, isozyme B is replaced by isozyme A, the embryonic form.

17. Most cancers develop later in life. Because the multiple mutations that lead to formation of a tumour may require many years to accumulate, most cancers develop later in life. The occurrence of cancer after the age of reproduction may be one reason that evolutionary restraints have not done more to supress cancer. The requirement for multiple mutations also lowers the frequency of cancer compared with what it would be if tumerogenesis were triggered by a single mutation. However, huge numbers of cells are, in essence, mutagenized and tested for altered growth during our lifetime, a sort of evolutionary selection for cells that proliferate. *Fortunately the tumour itself is not inherited* (see **Lodish** *et al.*, 2004).

72.4 WHAT CAUSES CANCER ?

Cancers are caused mainly by environmental agents and lifestyle factors, most of which act by triggering DNA mutations.

1. Epidemiological data. The first indication that a particular agent may cause cancer is usually provided by an approach called **epidemiology,** the branch of medical science that investigates the frequency and distribution of diseases in human populations. Epidemiological studies have revealed that cancers arise with differing frequencies in different parts of the world. For example, stomach cancer is frequent in Japan, breast cancer is prominent in the United States and liver cancer is common in Africa and Southeast Asia. To determine whether differences in heredity or environment are responsible for such differences, scientists have examined cancer rates in people who move from one country to another. For example, in Japan the incidence of stomach cancer is greater and the incidence of colon cancer is lower than in United States. When Japanese families move to the United States, their cancer rates come to resemble the rates in the United States, indicating that cancer rates are determined more by environment and lifestyle factors than by heredity.

Epidemiological data have played an important role in identifying environmental factors that can cause cancer. The most forcible statics involve lung cancer, a disease that has increased over tenfold in frequency in the United states since 1900. When the possible causes for this epidemic of lung cancer were investigated, it was discovered that virtually all lung cancer patients share one trait: a history of smoking cigarettes. As might be expected if cigarettes were responsible, heavy smokers develop lung cancer more frequently than light smokers, and long-term smokers develop lung cancer more frequently than do short-term smokers.

While the association between cigarette smoking and cancer was first detected through epidemiological studies, definitive proof of a cause-and-effect relationship requires direct experimental evidence. In the case of cigarettes and lung cancer, such evidence has come from studies showing that cigarette smoke contains several dozen chemicals that cause cancer when administered to animals. Such substances that cause cancer are called **carcinogens**.

2. Carcinogens. The idea that certain chemicals, such as those found in tobacco smoke, can cause cancer was first proposed more than 200 year's ago. In 1761 a London doctor, **John Hill**, reported that people who routinely use snuff (a powdered form of tobacco that is inhaled) experience an abnormally high incidence of nasal cancer, suggesting the presence of carcinogen (cancer-causing) chemicals in tobacco. A few years later another British physician, **Percival Pott**, observed an elevated incidence of scrotum cancer among men who had served as chimney sweepers in their youth. It was common practice at the time to employ young boys to clean chimney flues because they fit into narrow spaces more readily than adults. Pott speculated that the chimney soot became dissolved in the natural oils of the scrotum, irritated the skin, and eventually triggered the development of cancer. This theory led to the discovery that scrotum cancer could be prevented among chimney sweepers through the use of protective clothing and regular bathing practices.

In forthcoming years, the list of known and suspected carcinogens has grown to include hundreds of different chemicals. Chemicals are usually labelled as carcinogens because humans or animals develop cancer when exposed to them. This does not mean, however, that each of these substances causes cancer through its own direct action. For example, consider the behaviour of **2-naphthylamine**, a potent carcinogen that causes bladder cancer in industrial workers and is present in tobacco smoke. As might be expected, feeding 2-naphthylamine to laboratory animals induces a high incidence of **bladder cancer**. But if 2-naphthylamine is implanted directly into an animal's bladder, cancer rarely develops. The explanation for the apparent difference is that

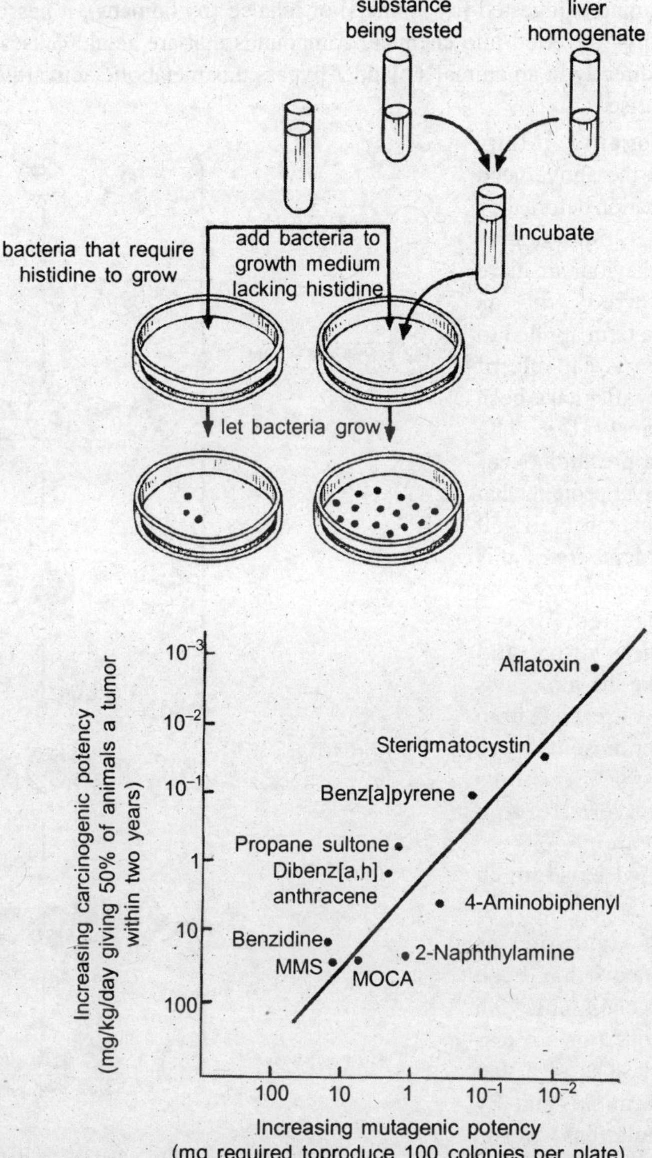

Fig. 72.4. Procedure of Ames test. Ames test is based on the grounds that most carcinogens are mutagens. A—The ability to induce mutations is measured in bacteria that lacks the ability to synthesize the amino acid histidine. When placed in a growth medium that lacks histidine, the only bacteria that can grow are those that have acquired a mutation that allows them to make histidine. The number of bacterial colonies that grow is therefore related to the mutagenic potency of the substance being tested. Chemicals being investigated with the Ames test are first incubated with a liver homogenate because many of the chemicals to which humans are exposed only become carcinogenic after they have undergone biochemical modification in the liver. B—The data in the graph reveal that substances that exhibit strong mutagenic activity in the Ames test also tend to be strong carcinogens. Here, aflatoxin is the most potent mutagen and the most potent carcinogen. (*Abbreviations:* MOCA = 4–4′ methylene-bis 2-chloroanaline, MMS = methyl methane sulphonate; after Becker *et. al.,* 2007).

2-naphthylamine is ingested (by animals) or inhaled (by humans), it passes through the liver and is metabolically converted into chemical compounds that are actual causes of cancer. Placing 2-naphthylamine directly in an animal's bladder bypass this metabolic activation and consequently cancer does not arise.

Precarcinogens. Many carcinogens share the above need for metabolic activation before they can cause cancer. Substances exhibiting such behaviour are more accurately referred to as **precarcinogens**, a term applied to any chemical that is capable of causing cancer only after it has been **metabolically activated**. The activation of most precarcinogens is carried out by liver proteins that are members of cytchrome P450 enzyme family. Members of this enzyme family catalyze the oxidation of ingested foreign chemicals, such as drugs and pollutants, to make the molecules less toxic and easier to excrete from the body. However, in some cases these oxidation reactions accidentally convert foreign chemicals into carcinogens—a phenomenon called **carcinogen activation**.

Mode of action of carcinogens. Once it has been determined that chemicals can cause cancer, the question arose as to how they work. The idea that carcinogenic chemicals act by triggering **DNA mutations** was first proposed around 1950, but there was little supporting evidence at that time because nobody had systemically compared the mutagenic potency of different chemicals with their ability to cause cancer.

Ames Test. The need for identifying potential carcinogens, inspired **Bruce Ames** (1973) to develop a simple rapid laboratory test for measuring a chemical's

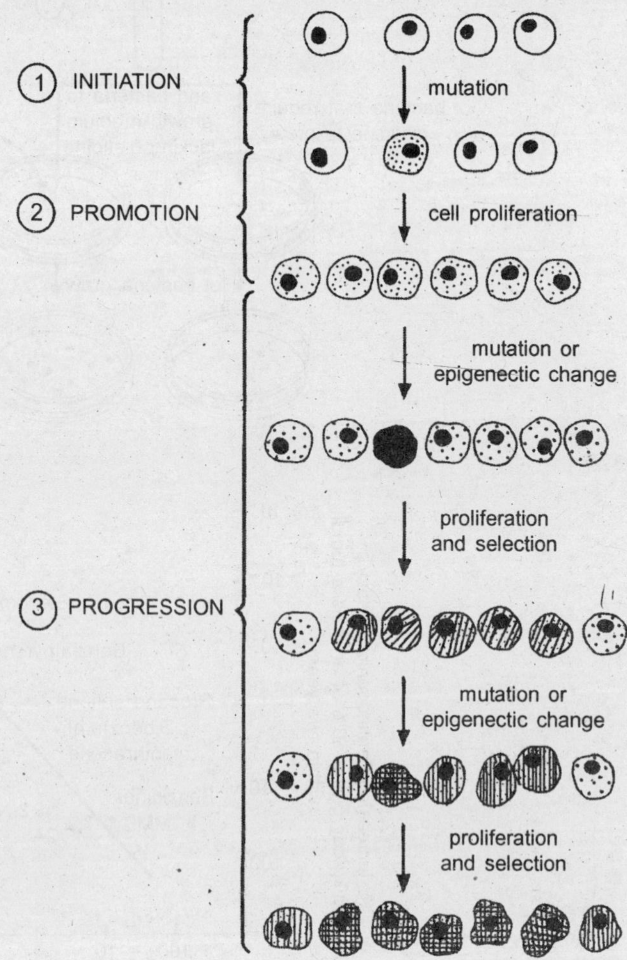

Fig. 72.5. Main stages in the development of cancer arises by a complex process involving three main stages. 1. Initiation. It is based on DNA mutation. 2. Promotion. During this stage the initiated cell is stimulated to proliferate. 3. Progression. During tumour progression, further mutations and changes in gene expression create variant cells exhibiting enhanced growth rates or other aggressive properties that give certain cells a selective advantage. Such cells tend to outgrow their companions and become the predominant cell population in the tumour. Repeated cycles of this **selection process** create a population of cells whose properties gradually change over time (after Becker *et al.*, 2007).

mutagenic activity. The procedure he developed, called the **Ames test**, utilise bacteria as a test organisms because they can be quickly grown in enormous numbers in culture. The bacteria used for the Ames test are a special strain that lack the ability to synthesize the amino acid **histidine**. As shown in Fig. 72.4, the bacteria are placed in a culture dish containing a growth medium without histidine, along with the chemicals being tested for mutagenic activity. Normally, the bacteria would not grow in the absence of histidine. However, it will trigger random mutations, some of which might restore the ability to synthesize histidine. Each bacterium acqui-ring such a mutation will grow into a visible colony, so that the total number of colonies is a measure of the **mutagenic potency** of the substance being investigated.

Since many chemicals that cause cancer only become car-cinogenic after they have been modified by liver enzymes, the Ames test includes a step in which the chemical being tested is first incubated with an extract of liver cells to mimic the reactions that normally occur in the liver. The resulting chemcial mixture is then tested for its ability to cause bacterial mutations. When the Ames test is performed in this way, a strong correlation is observed between a chemical's ability to cause mutations and its ability to cause cancer.

Cancer arise through a multistep process. Chemicals induce the development of cancer through a multistep process involving **initiation, promotion**, and **tumour progression**. **Initiation** is based on DNA mutation, whereas **promotion** involves proliferation of the initiated cells for a prolonged period of time, a accompanied by a gradual **selection** of cells exhibiting enhanced growth properties. During **tumour progression**, cells acquire additional mutations and undergo changes in gene expression that produce cells with increasingly aberrant traits.

3. Role of ionizing and ultraviolet radiations in carcinogenesis. Shortly after the discovery of **X-rays** by **Wilhelm Roentgen** in 1895, it was noticed that people working with this type of radiation developed cancer at abnormally high rates. Animal studies subsequently confirmed that X-rays create DNA mutations and cause cancer in direct proportion to the dose administered.

X-rays type of radiations is emitted by many **radioactive elements.** An early example of the carcinogenic hazards posed by radioactivity occurred in 1920's in a New Jersey factory that produced glow-in-the-dark watch dials. A luminescent paint containing the radioactive element **radium** was used for painting the dials, and this paint was applied with a fine tipped brush that the workers frequently wetted with their tongues. As a result, tiny

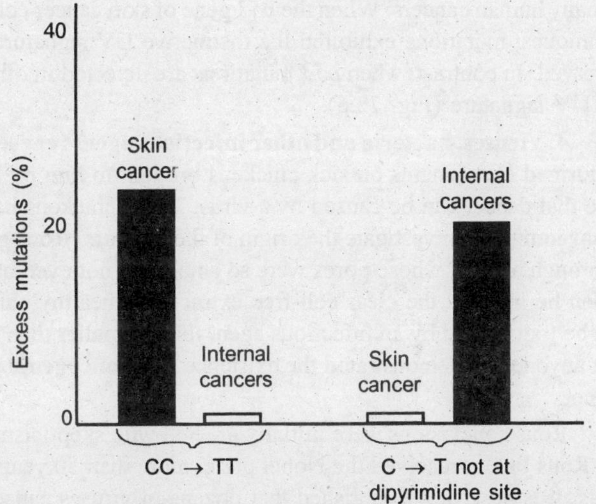

Fig. 72.6. Incidence of two types of *p53* mutations in skin cancer and internal cancers. The two bars on the left represent the frequency of CC → TT mutations, which are triggered by UV radiation. The two bars on the right represent the frequency of C → T mutations not located at dypyrimidine sites, which are not caused by UV radiation. Note that the UV-triggered type of mutation is found in the *p53* gene of skin cancers (squamous cell carcinomas), but not in cancers of internal organs. Mutations trequencies are plotted relative to what would be expected to occur randomly (after Becker *et al.*, 2007).

quantities of radium were carelessly ingested and became concentrated in their bones, leading to the development of **bone cancer**.

High rates of cancer caused by exposure to radioactivity have also been observed in people exposed to radioactive fallout from nuclear explosions. The most dramatic incidents occurred in the Japanese cities of Hiroshima and Nagasaki after atomic bombs were dropped there in 1945. Another incident occurred in the area surrounding the Chernobyl nuclear power plant in the former Soviet Union (now Ukraine), which exploded in 1986.

X-rays and related forms of radiation emitted by radioactive elements, are called **ionizing radiation** because they remove electrons from molecules, thereby generating highly reactive ions that create DNA damage. **Ultraviolet radiation (UV)** is another type of radiation that causes cancer by damaging DNA. The ability of the UV radiation in sunlight to cause cancer was first deduced from the observation that skin cancer is most prevalent in people who spend long hours in the sun, especially in tropical regions where the sunlight is very intense. UV radiation is absorbed mainly by the skin, where it imparts enough energy to trigger **pyrimidine dimer** formation—that is, the formation of covalent bonds between adjacent pyrimidine bases in DNA. If the damage is not repaired, distortion of the double helix causes improper base-pairing during DNA replication that leads to distinctive mutation patterns. For example, a CC $\rightarrow$ TT mutation (conversion of two adjacent cytosines to thymines) is a unique product of UV exposure and can therefore be used as a distinctive "signature" to identify mutations caused by sunlight.

The existence of such **signature mutations** provided a way to prove that UV-induced mutations cause skin cancer. One of the first genes studied was the *p53* gene, which is known to be mutated in many human cancers. When the *p53* gene of skin cancer cells, is examined using DNA sequencing techniques, mutations exhibited the distinctive UV signature (such as CC $\rightarrow$ TT) are frequently observed. In contrast, when *p53* mutations are detected in other types of cancer, they do not exhibit the UV signature (Fig. 72.6).

4. Viruses, bacteria and other infectious agents cause some cancers. Peyton Rous, in 1911, performed experiments on sick chickens brought to him by local farmers that showed for the first time that cancer can be caused by a virus. These chickens had cancers of connective tissue origin, or **sarcomas**. To investigate the origin of the tumours, **Rous** ground up the tumour tissue and passed it through a filter, whose pores were so small that not even bacterial cells could pass through them. When he injected the clear cell-free extract into healthy chickens, Rous concluded that sarcomas can be transmitted by an infectious agent that is smaller than a bacterial cell. This was the first time that anyone had demonstrated the existence of an **oncogenic viruses**—that is, a virus that can cause cancer.

Rous conclusions were initially greeted with skepticism. It was not until 1966 that an 87-year old Rous finally received the Nobel prize, more than 50 years after his discovery of the first cancer virus. It is now well established that dozens of viruses cause cancer in animals and that a smaller number cause cancer in humans. The first human example was discovered by **Danis Burkitt**, a British surgeon working in Africa in the late 1950's. At certain times of the year, Burkitt noted that a large number of his young patients developed massive **lymphocytic cancers** of the neck and jaw. Because this kind of cancer, now known as **Burkitt's lymphoma** occurred in periodic epidemics localized to specific geographical regions, Burkitt proposed that it was transmitted by an infections agent.

Burkitt's ideas soon attracted the attention of two virologists, named **Epstein** and **Barr**, whose electronic microscopic studies revealed virus particles in Burkitt's lymphoma cells, the virus is now called the **Epstein-Barr virus** or **EBV** in recognition of these virologists (Box 72.5).

Box 72.5 EBV and Carcinogenesis

Unlike viruses that cause cancer in animals, it is difficult to prove that a virus such as EBV causes cancers in humans because ethical considerations prevent directly testing the hypothesis by injecting the virus into healthy individuals. Nonetheless, some indirect evidence supports the conclusion that EBV causes Burkitt's lymphomas: 1. DNA sequences and proteins encoded by EBV are detected in tumour cells obtained from patients with Burkitt's lymphoma, but not in normal cells from the same individuals; 2. Adding purified EBV to normal human lymphocyte cultures cause the cells to acquire the properties of cancer cells; and 3. Injecting EBV into monkeys induces the formation of lymphomas.

Following the discovery of EBV, several additional viruses have been identified as causes of human cancer. For example, liver cancer is caused by **hepatitis B** and **hepatitis C** viruses and cervical cancer is caused by **human papillomavirus (HPV)**. In addition to viruses, the bacterium *Helicobacter pylori* (*H. pylori*) causes stomach cancer, and parasitic flatworm infections have been linked to a small number of bladder and bile duct cancers.

Infectious agents trigger the development of cancer in two fundamentally different ways. One involves those agents such as the hepatitis B and C viruses, *H. pylori*, and parasitic flatworms that cause tissue destruction and chronic inflammation. Such destruction creates chronic inflammatory conditions in which cells of the immune system infiltrate the tissue and attempt to kill the infectious agent and repair the tissue damage. Unfortunately, the mechanisms used by immune cells to fight infections often produce mutagenic chemicals, such as **oxygen free radicles** (highly reactive forms of oxygen containing an unpaired electrons). This means that proliferation of replacement cells for the injured tissue takes place under conditions in which DNA damage is likely, thereby increasing the likelihood that cancer-causing mutations will arise.

The other way in which infectious agents cause cancer is related to the ability of certain viruses to directly stimulate the proliferation of infected cells. The mechanism by which this is accomplished varies among viruses. In some cases, it involves cancer-causing viral genes, whereas in other cases a virus alters the behaviour of host cell genes. The types of genes involved in such events play a role not just in viral cancers, but in cancers caused by chemicals and radiation as well.

72.5 GENES INVOLVED IN CANCERS

Oncogenes are genes whose presence can cause cancer. While they are sometimes introduced into cells by viruses, more often they arise from normal cellular genes called **proto-oncogenes** by *point mutation, gene amplification, chromosomal translocation, local DNA rearrangements* or *insertional mutagenesis.* Most proteins produced by oncogenes are signalling pathway components, such as growth factors, receptors, plasma membrane GTP-binding proteins, non-receptor protein kinases, transcription factors, and cell-cycle or cell-death regulators. Oncogenes code for abnormal forms or excessive quantities of such proteins, thereby leading to excessive stimulation of cell proliferation.

Tumour suppressor genes are genes whose loss or inactivation can lead to cancer. Susceptibility to developing cancer is increased in people who inherit defective tumour suppressor genes. Three important tumour suppressor genes are:

1. The *RB* gene, which produces a protein that restrains passage from G1 into S phase.
2. The *p53* gene, which produces a protein that prevents cells with damaged DNA from proliferating.
3. The *APC* gene, which produces a protein that inhibits the Wnt pathway. The **Wnt pathway** plays a prominent role in controlling cell proliferation and differentiation during embryonic development. The central component of this wnt pathway is a protein called **β-catenin**.

72.6 DIAGNOSIS, SCREENING AND TREATMENT OF CANCER

Much progress has been made in recent years in explaining the genetic and biochemical abnormalities that underlie cancer development. One of the hopes for such research is that our growing understanding of the molecular alterations exhibited by cancer cells will eventually lead to improved strategies for cancer diagnosis and treatment.

1. Microscopic Analysis of Cancer

Because cancer can arise in almost any tissue, few regular generalizations are possible regarding disease symptoms. A definitive diagnosis typically requires a **biopsy,** which involves surgical removal of a tiny tissue sample for microscopical examination. Under the microscope, cancer cells show a number of features that together indicate the presence of cancer (Table 72.1). For example, cancer cells often have large, irregularly shaped nuclei, prominent nucleoli, and a high ratio of nuclear-to-cytoplasmic volume. Cancers also tend to exhibit significant variability in cell size and shape, as well as a loss of normal tissue organisation. To varying extents, cancer cells lose the specialized structural and biochemical properties of the cells normally residing in the tissue of origin. Cancers usually have more dividing cells than normal, which means that the mitotic index (Box 72.6) will be elevated.

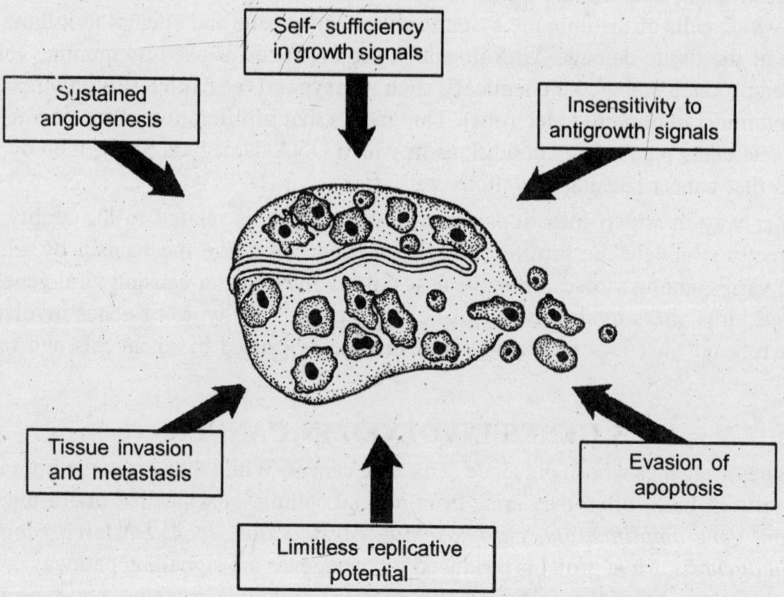

Fig. 72.7. Overview of changes in cells that cause cancer. During carcinogenesis, six fundamental cellular properties are altered, as shown here, to give rise to the complete, most destructive cancer phenotype Less dangerous tumours arise when only some of these changes occur (after Lodish *et. al.*, 2004).

Table 72.1 Some differences in the microscopic appearance of benign and malignant tumours (source: Becker *et al.*, 2006)

Trait	*Benign*	*Malignant*
1. Nuclear size	Small	Large
2. N/C ratio (Ratio of nuclear to cytoplasmic volume)	Low	High
3. Nuclear shape	Regular	Pleomorphic (irregular shape)

Trait	Benign	Malignant
4. Mitotic index	Low	High
5. Tissue organisation	Normal	Disorganised
6. Differentiation	Well-differentiated	Poorly differentiated (anaplastic)
7. Tumour boundry	Well-defined	Poorly-defined

Box 72.6 Mitotic index

Percentage of cells in a population that are in any stage of mitosis at a certain point in time; used to estimate the relative length of the M phase of cell cycle. For example, the mitotic index for cultured mammalian cells is often about 3–5 per cent, which means that M phase lasts less than an hour (usually 30–45 minutes).

Tumour grading. If a sufficient number of these seven traits (Table 72.1) are observed upon microscopic examination of sample, it can be concluded that cancer is present. In other words, the presence of these traits indicates a tumour that, if left untreated will eventually spread by invasion and metastasis. The severity of these observed microscopic, abnormalities varies significantly among cancers, even when they arise from the same cell type and in the same organ. These variability form the basis for **tumour grading**, which is the assignment of numerical grades to tumours based on differences in their microscopic appearance.

Lower numerical grades (*e.g.*, grade 1) are assigned to tumours whose cells exhibit normal differentiated features, divide slowly, and display only modest abnormalities in the traits listed in Table 72.1. Higher numbers (*e.g.*, grade 4) are assigned to tumours containing rapidly dividing, poorly differentiated cells that bear less resemblance to normal cells and exhibit severe abnormalities in the trait, listed in Table 72.1. The highest grade cancers contain cells that are **anaplastic**, which means that they are so poorly differentiated and abnormal in appearance and organization that they bear no resemblance to the cells of the tissue in which the tumour arose. Such anaplastic, high-grade cancers tend to grow and spread more aggressively and be less responsive to therapy than lower grade cancers (see **Becker** *et. al.*, 2007).

2. Screening Techniques of Cancer

When cancer is detected before it has spread, cure rates tend to be very high. Therefore, a great need exists for screening techniques that can routinely detect cancers at an early stage. Some of these are following:

***(i)* Pap smear.** It is one of the most successful screening procedures that was developed in the early 1930s by **George Papanicolaou** (for whom it is named). The logic underlying this procedure is that the microscopic appearance of cancer cells is so distinctive that it is possible to detect the likely presence of cancer by simply examining a few isolated cells. A Pap smear is performed by taking a tiny sample of a woman's vaginal secretions and examining it with a microscope. If the cells in the fluid exhibit unusual features, such as large irregular nuclei or prominent variations in cell size and shape, it is a sign that cancer may be present and further tests need to be done. Because a Pap smear allows **cervical cancer** to be detected in its early stages before metastasis has occurred, this procedure has prevented hundreds of thousands of cancer deaths.

***(ii)* Mammography and other techniques.** The success of the Pap smear has led to development of screening techniques for other cancers. For example, **mammography** utilizes a special X-ray technique to look for early signs of **breast cancer** and **colonoscopy** uses a slender fiber-optic instrument to examine the colon (*i.e.*, the part of the large intestine that extends from the cecum to the rectum) for early signs of **colon cancer**. The ideal screening test would allow doctors to detect cancers anywhere in the body with one simple procedure, such as a blood test. **Prostate cancer** is an example of cancer that can sometimes be detected this way. Men over the age of 50 are often advised to get a **PSA test,** which measures how much **prostate-specific antigen (PSA)** is present in the bloodstream. PSA, which is a protein produced by cells of the prostate gland, normally appears in only tiny concentrations in the blood. If a PSA test reveals a high concentration of PSA, it indicates the existence of a prostate problem and further tests are performed to determine whether or not cancer is actually present.

***(iii)* Proteomic analysis.** Other cancers also release small amounts of specific proteins into the bloodstream, where there presence might be used to signal the existence of early disease. To investigate such tiny changes in blood proteins, scientists are experimenting with a general approach called **proteomic analysis** to analyse proteins present in the blood (the term proteomic refers to the complete set of proteins produced by an organism's genome). The key to most proteomic techniques is **mass spectrometry**, a high speed, extremely sensitive method for identifying proteins based on differences in mass and electrical charge. Because a blood sample contains thousands of different proteins, the data generated by proteomic analysis can be extremely complex. To deal with this problem, artificial intelligence software programs are used to compare the complex protein patterns seen in blood samples from individuals with or without cancer.

One of the first cancer to be investigated by proteomic analysis was **ovarian cancer**. When ovarian cancer is detected before it spreads, the five year survival rate exceeds 95 per cent. Using proteomic analysis, scientists have recently identified a pattern of five proteins in the blood of women with ovarian cancer that is not seen in the blood of other women.

3. Treatments of Cancer

People diagnosed with cancer have various treatment options that depend both on the type of cancer involved and how far it has spread.

***(i)* Surgery.** The most common treatment of cancer involves surgery to remove the primary tumour followed (if necessary) by radiation therapy and/or chemotherapy to destroy any remaining cancer cells.

***(ii)* Radiation therapy.** This approach employs high energy X-rays or other forms of ionizing radiation to kill cancer cells. We already know that DNA damage created by ionizing radiation can cause cancer, but paradoxically, the same type of radiation is also used in higher doses to destroy cancer cells in people who already have the disease. Ionizing radiation kills cells in two different ways. First, DNA damage caused by radiation activates the p53 signaling pathway, which then triggers cell death. However, many cancers have mutations that disable the p53 pathway, so p53-induced apoptosis plays only a modest role in the response of most cancers to radiation treatment. In the second mechanism, radiation kills cells by causing chromosomal damage that is so severe that it prevents cells from progressing through mitosis, and the cells therefore die while trying to divide.

***(iii)* Chemotherapy**. Most forms of chemotherapy use drugs that, like radiation, are intended to kill dividing cells. Such drugs can be subdividing into four major categories:

***(a)* Antimetabolites.** They inhibit metabolic pathways required for DNA synthesis by acting as competitive inhibitors that bind to enzyme active sites in place of normal substrate molecules. Examples of such antimetabolites include methotrexate, fluorouracil and mercaptopurine.

***(b)* Alkylating agents.** They inhibit DNA function by chemically cross-linking the DNA double helix, Examples of such drugs include cyclophosphamide, chlorambucil and cisplatin.

(c) Antibiotics. These are substances made by microorganisms that inhibit DNA function by either binding to DNA or inhibiting topoisomerases required for DNA replication. Examples of such antibiotics include bleomycin and doxorubicin.

(d) Plant-derived drugs. These drugs either inhibit topoisomerases or disrupt the microtubules of the mitotic spindle. Examples of such drugs include the topoisomerase-inhibitor **etoposide** and the microtubule disrupting drug, **taxol**.

One problem with such drugs (and radiation therapy) is that they are toxic to normal dividing cells as well as to cancer cells. When cancer arises in a tissue whose growth requires a specific hormone, it may be treated in a less toxic manner using drugs that block the action of that particular hormone. For example, many breast cancers require estrogen for their growth. Estrogens tend to exert their effects by binding to nuclear receptor proteins that activate the expression of specific genes. The drug **tamoxifen**, a common treatment for breast cancer, binds to estrogen receptors in place of estrogen and prevents the receptors from being activated.

Newer treatment approaches include **immunotherapies** that exploit the ability of the immune system to attack cancer cells, **molecular** targeting drugs aimed at proteins that are critical to the cancer cells, and **antiangiogenic agents** that attacks a blood supply or the tumour.

QUESTIONS

Long Answer Questions

1. Define the cancer. Describe different type of cancer. Explain various characteristics of cancer cells.
2. Explain various causes of cancer.
3. Describe diagnosis, screening and treatment of cancer.
4. Write an essay on cancer.

Short Answer Questions

1. What are preventive and control measures of cancer? *(Bangalore 2005)*
2. What is cancer? Mention any four carcinogens. *(Bangalore 2006)*
3. Write a short note on cancer. *(H.N.B, Garhwal, 2008)*
4. Explain cancer and cell transformation. *(H.N.B. Garhwal 2009)*

Very Short Answer Questions

1. Define cancer.

   Ans. Cancer is uncontrolled, growing mass of cells that is capable of invading neighboring tissues and spreading via body fluids, especially the bloodstream, to other parts of body. Cancer is also called **malignant tumour**.

2. What is carcinogen?

   Ans. Carcinogen is any cancer-causing agent.

3. What is Ames test?

   Ans. It is screening test for potential carcinogens that assesses whether a substance causes mutations in bacteria.

4. What is a carcinoma?

   Ans. It is a malignant tumour (*i.e.*, cancer) arising from the epithelial cells that cover external and internal body surfaces.

5. Define the leukemia.

   Ans. It is cancer of blood or lymphatic origin in which the cancer cells proliferate and reside mainly in the blood stream rather than growing as a solid masses of tissue.

6. What is lymphoma?

   Ans. It is cancer of lymphatic origin in which the cancer cells grow as solid masses of tissue.

7. Define the sarcoma.

   Ans. Sarcoma is any cancer arising from a supporting tissue, such as bone, cartilage, fat, connective tissue and muscle.

Fill in the Blanks

1. One of the first cancers to be investigated by proteomic analysis was ....
(**Ans.** Ovarian cancer)

Multiple Choice Questions

1. Carcinoma refers to
(*a*) malignant tumours of the connective tissue
(*b*) malignant tumours of the skin or mucous membrane
(*c*) benign tumours of the connective tissue
(*d*) none of the above
2. Cancer cells are more easily damaged by radiation than normal cells because they are
(*a*) non-dividing (*b*) starved of food
(*c*) undergoing rapid division (*d*) different in structure
3. Which one of the following gives rise to cancer? (*Bihar 2003*)
(*a*) unrestricted and fast multiplication of cells by meiosis
(*b*) breakdown of mechanism regulating meiosis
(*c*) unlimited enlargement of cells by failure of formation of cleavage furrow after telophase to separate the daughter cells
(*d*) the two daughter cells after mitosis carrying different amount of DNA
4. Cancer can be defined as
(*a*) hyperplastic growth, infiltration and destruction of tissues
(*b*) non-malignant epithelial growth
(*c*) repair of damaged parts
(*d*) metabolic disorder
5. Blood cancer is called
(*a*) leukaemia (*b*) haemophilia
(*c*) thrombosis (*d*) haemolysis
6. The cells affected by leukaemia are
(*a*) plasma cells (*b*) erythrocytes
(*c*) thrombocytes (*d*) leucocytes
7. One of the faster and less expensive procedures for preliminary screening of potential carcinogens include
(*a*) Biuret (*b*) Dick
(*c*) Ames (*d*) ELISA
8. Sarcoma is the cancer of
(*a*) epithelial tissue (*b*) connective tissue
(*c*) blood (*d*) endocrine tissue
9. Cancer treatment includes
(*a*) surgery (*b*) radiotherapy
(*c*) treatment with anticancer drug (*d*) all of these
10. Which type of cancer is found in lymph nodes, gland and spleen?
(*a*) sarcoma (*b*) carcinoma
(*c*) adenoma (*d*) leukaemia
11. The spread of cancerous cells to distant sites is termed
(*a*) metamorphosis (*b*) metagenesis
(*c*) metastasis (*d*) metachrosis

ANSWERS

Multiple Choice Questions

1. (*b*)	**2.** (*c*)	**3.** (*d*)	**4.** (*a*)	**5.** (*a*)	**6.** (*d*)
7. (*c*)	**8.** (*b*)	**9.** (*d*)	**10.** (*c*)	**11.** (*c*).	

UNIT – III
GENETICS

UNIT–III
GENETICS

73

Introduction

Genetics is the study of inheritance in all of its manifestations, from the distribution of human traits in family pedigree to the biochemistry of the genetic material in our DNA (**Tamarin** 2002). The fundamental concept of genetics is that inherited traits are determined by the elements of heredity, that are transmitted from parents to offspring in reproduction; these elements of heredity are the genes.

73.1 HISTORY OF GENETICS

According to **Tamarin** (2002), the history or growth of genetics can be studied under the following four periods: 1. Before 1860; 2. 1860 to 1900; 3. 1900–1944, and 4. 1944 to the present.

1. Growth of Genetics before 1860

1665 **Robert Hooke** coined the term *cell* in his studies of cork. He saw, in fact, empty cells which were observed at a magnification of thirty power.

1674–1683 **Anton van Leeuwenhoek** discovered living organisms (protozoa and bacteria) in rainwater. Leeuwenhoek was a master lens maker and produced magnifications of several hundred power from single lenses.

1833 **Robert Brown** discovered the nuclei of cells.

1835–1839 **Hugo von Mohl** described mitosis in nuclei.

1859 **Rudolf Virchow** summed up the concept of the cell theory with his Latin aphorism *omnis cellula e cellula*—all cells come from pre-existing cells.

1859 The publication of **Charles Darwin's** *The Origin of Species* took place.

Thus, by 1959, biologists had an understanding of the continuity of cells and knew of the cell's nucleus.

2. Growth of Genetics in period from 1860 to 1900

1865 **Gregor Mendel** read his paper in the British Society for Natural History.

1866 Mendel's paper published in the proceeding of the British Society for Natural History.

1875 **O. Hertwig** described the fusion of sperm and egg to form zygote.

1879–1885 With the aid of new staining technique, **W. Flemming** described the chromosomes which were noticed by **C. Von Nageli** in 1842. Flemming studied the way chromosomes split during division and the separation of sister chromatids and their movement of opposite poles of the dividing cell during mitosis.

1880's **Theodor Boveri**, as well as **K. Rabi** and **E. van Benden**, hypothesized that chromosomes are individual structures with continuity from one generation to the next despite their "disappearance" between cell divisions.

1885 **August Weismann** stated that inheritance is based exclusively in the nucleus. He is famous for his **germplasm theory** (1883).

1887 **Weismann** predicted the occurrence of a reductional division, which we now call meiosis.

1888 **W. Waldeyer** first used the term **chromosome**.

1890 **O. Hertwig** and **T. Boveri** had described the process of meiosis in detail.

Thus, this period (*i.e.*, 1860–1900) includes the discoveries of chromosomes and their behaviour insights that shed new light on Mendel's research.

3. Growth of Genetics in period from 1900 to 1944

1900 Three biologists working independently–**Hugo de Vries, Carl Correns** and **Erich von Tschermak**– re-discovered Mendel's landmark work on the rules of inheritance.

1902 **Boveri** and **Sutton** demonstrated the presence of paired chromosomes (homologs) in diploid species.

1903 **Walter Sutton** hypothesized that the behaviour of chromosomes during meiosis explained Mendel's rules of inheritance, thus leading to the discovery that genes are located on chromosomes.

1905 **William Bateson** named the science **genetics**.

Fig. 73.1. **Hugo de Vries** (1848–1935). Dutch botanist who discovered Mendel's principles (or rules or laws) in 1900. He was a pioneer in plant genetics (after Gardner *et al.*, 2006).

1908 **Hardy** and **Weinberg** formulated the "Hardy-Weinberg law" relating genotypic frequencies to gene frequencies in randomly mating populations.

1909 **Johannsen** introduced the term **gene**.

1909 **Garrod's** book *Inborn Errors of Metabolism* published.

1910 **T.H. Morgan** (Nobel Prize 1933) established the *sex-linked inheritance* of white eyes in *Drosophila melanogaster* (fruit fly).

1911 **Morgan** postulated the chromosomal basis of linkage.

1913 **Alfred Sturtevant** created the first genetic map using fruit fly. He showed that genes existed in a linear order on chromosomes.

Fig. 73.2. **William Bateson** (1861–1926), English biologists who repeated Mendel's experiments and supported his principles and introduced Mendelian genetics to the English-speaking world (after Gardner *et al.*, 2006).

1927 **I. Stadler** and **H.J. Muller** showed that genes can be mutated artificially by X-rays.

1927 **Muller** (Nobel Prize 1946) reported the ClB technique to demonstrate that X-rays are mutagenic.

1928 **Griffith** discovered the phenomenon of **transformation** in *Diplococcus pneumoniae.*

1930–1932 **R.A. Fisher, S. Wright** and **J.B.S. Haldane** developed the algebraic foundations for our understanding of the process of evolution.

1931 **Creighton** and **McClintock's** (maize) and **Stern's** (*Drosophila*) papers appeared, demonstrating that *genetic recombination* is correlated with the *exchange of morphological marker on chromosomes.*

1940 **Oliver** demonstrated recombination within the lozenge functional unit in *Drosophila.*

1941 **Beadle** and **Tatum's** work (Nobel Prize 1958) on *Neurospora* was published, establishing *one gene-one enzyme concept.*

1943 **S. Luria** and **M. Delbruck** demonstrated that bacteria have normal genetic systems and thus could serve as models for studying genetic processes.

Thus, during this period (*i.e.*, 1900–1944), modern genetics flourished with the development of the chromosome theory, which showed that chromosomes are linear arrays of genes. In addition, the foundations of modern evolutionary and molecular genetics were derived.

4. Growth of Genetics in period from 1944 to present

1944 **Avery, MacLeod and McCarty** demonstrated that the pneumococcal "transforming principle" is DNA.

1946 **Lederberg** (Nobel Prize 1958) and **Tatum** discovered **conjugation** in bacteria.

1950 **Barbara McClintok** (Nobel Prize 1983) published her first paper on "**transposable elements**" in maize.

1952 **Hershey** (Nobel Prize 1969) and **Chase** demonstrated that the *genetic material of bacteriophage T2 is DNA.*

1952 **Zinder** and **Lederberg** discovered phage mediated **transduction** in bacteria.

1953 **Watson** and **Crick** (Nobel Prize 1962) worked out the double-helix structure of DNA using the X-ray diffraction data of **Wilkins** (Nobel Prize 1962) and base composition data of **Chargaff**.

1955 **Benzer's** first paper was published on *fine structure of the phage T4 rII* locus.

1956 **Tjio** and **Levan** established that the normal *diploid chromosome number in humans is 46.*

1957 **Fraenkal-Conrat** and **Singer** demonstrated that *the genetic information of tobacco mosaic virus is stored in RNA.*

1958 **Meselson** and **Stahl** demonstrated that *DNA replication is semiconservative.*

1958 **Kornberg** (Nobel Prize 1959) *isolated DNA polymerase I from E.coli.*

1959 **Ochoa** (Nobel Prize 1959) discovered the first RNA polymerase.

1961 **Jacob** and **Monod** (Nobel Prize 1965) proposed "*operon model*" for regulating gene expression.

1964 Colinearity between genes and polypeptide products established by the work of **Yanofsky** and colleagues and by **Brenner** and colleagues.

1964 **Temin** (Nobel Prize 1975) proposed the *DNA provirus* form of RNA tumour viruses.

1965 **Holley** (Nobel Prize 1968) worked out the first complete nucleotide sequence of a tRNA, a yeast alanine tRNA.

1966 The complete genetic code was established by the work of **Nirenberg** and **Khorana** (Nobel Prize 1968) and coworkers.

1968–1973 **W. Arber, H. Smith** and **D. Nathans**, along with their colleagues discovered and described **restriction endonucleases**, the enzymes that opened up our ability to manipulate DNA through recombinant DNA technology.

1970 **Nathans** and **Smith** (Nobel Prize 1978) isolated the first *restriction endonucleases.*

1970 *Reverse transcriptase* of RNA tumour viruses identified by **Baltimore** (Nobel Prize 1975).

1972 First *recombinant DNA* produced *in vitro* in laboratory of **Paul Berg** (Nobel Prize 1980).

1976 **Bishop** and **Varmus** (Nobel Prize 1989) demonstrated the *protooncogene to oncogene.*

1976 **Hozumi** and **Tonegawa** (Nobel Prize 1987) demonstrated *somatic rearrangements of genes encoding antibodies.*

1977 Demonstration of **introns** in eukaryotic genes by **Breathnach, Mandel** and **Chambon** and by **Jeffreys** and **Flavell**.

1977 Publication of *the DNA sequencing techniques* of **Maxam** and **Gilbert** and of **Sanger, Nicklen** and **Coulson** (Sanger and Gilbert, Nobel Prize 1980).

1977 **Sanger** and colleagues published *complete 5387 nucleotide sequence of phage $\phi \times 174$.*

1978 Discovery of "*splicing*" of adenovirus RNAs in three different laboratories.

1982 **Sanger** and colleagues published the complete 48,502-nucleotide-pair sequence of phage lambda.

1983 **Cech** and **Altman** (Nobel Prize 1989) established the existence of catalytic RNAs (*i.e.*, ribozymes).

1988 **Watson** became co-ordinator of the "*human genome project*".

1989 **Tsui, Collins** and colleagues cloned the "*cystic fibrosis gene*", the gene whose mutant alleles account for majority of the cases of this dreaded disease that afflicts about one out of every 2000 children in United States.

1997 The first mammal was cloned, a sheep named Dolly.

2000 The sequence of the entire human genome was determined.

2003 Completion of draft of the human genome.

2009 World's first cloned camel (named Inaz) in Dubai was the outcome of efforts of Indian Scientist **Nisar Ahmad Wani** and his colleagues.

Thus, this period (*i.e.*, 1944-present) is the era of molecular genetics, beginning with the demonstration that DNA is the genetic material and culminating with our current explosion of knowledge due to recombinant DNA technology. Today, genetics is a mature, but still vigorously growing science.

73.2 THREE GENERAL BRANCHES OF GENETICS

Historically, geneticists have worked in three different areas, each with its own particular problems, terminology, tools and organisms. These areas (or branches) are classical genetics, molecular genetics and evolutionary genetics.

1. Classical genetics. In this branch of genetics, geneticists are concerned with the chromosomal theory of inheritance, that is, the concept that genes are located in a linear fashion or chromosomes and that the relative positions of genes can be determined by their frequency in offspring.

2. Molecular genetics. It is the study of the genetic material: its structure, replication and expression, as well as the information revolution emanating from the discoveries of recombinant DNA techniques (genetic engineering, including the human genome project).

3. Evolutionary genetics. It is the study of the mechanisms of evolutionary changes, or changes in gene frequencies in populations. **Darwin's** concept of evolution by natural selection finds a firm genetic footing in this area of the study of inheritance (Table 73.1).

Table 73.1 The three major areas of genetics—Classical, Molecular and Evolutionary—and the topics they cover (Source: Tamarin 2006).

Classical Genetics	*Molecular Genetics*	*Evolutionary Genetics*
1. Mendel's principles	1. Structure of DNA	1. Quantitative genetics
2. Meiosis and mitosis	2. Chemistry of DNA	2. Hardy-Weinberg equilibrium
3. Sex determination	3. Transcription	3. Assumptions of equilibrium
4. Sex linkage	4. Translation	4. Evolution
5. Linkage	5. DNA cloning and genomics	5. Speciation
6. Crossing over and chromosome mapping	6. Control of gene expression	
7. Cytogenetics (chromosomal changes)	7. DNA mutation and repair	
	8. Extrachromosomal inheritance	

QUESTIONS

Long Answer Questions

1. Give an account of growth of genetics during year 1900 to 1944.

2. Describe the growth of genetics in period 1944 to 2009.

Short Answer Questions

1. Mention topics covered by three main branches of genetics.

2. Write short notes on the following:

(*a*) Molecular genetics; (*b*) Classical genetics.

Very Short Answer Questions

1. Define genetics.

Ans. Genetics is a biological science of heredity and variation.

2. What is heredity?

Ans. Transmission of traits from parents to offspring.

3. Define variation.

Ans. The occurrence of differences among individuals.

Fill in the Blanks

1. The term genetics was coined by ....

2. Mendel began his scientific investigations on inheritance in ....

3. Punnett square was devised by ....

4. The term allelomorph was coined by ....

5. Operon model was proposed by ....

Multiple Choice Questions

1. Genetics is the branch which deals with the study of

(*a*) relation between plants and environment (*b*) inheritance and variation
(*c*) cell structure (*d*) thermal structure

2. The term 'genetics' was given by

(*a*) Morgan (*b*) Mendel
(*c*) Johannsen (*d*) Bateson

3. Mendel was a

(*a*) physiologist (*b*) plant breeder
(*c*) cytologist (*d*) taxonomist

4. Hybridisation means

(*a*) removal of stamens (*b*) mixing parental characters
(*c*) production of large sized grains (*d*) study of pollination habit

5. Mendel carried out his investigations on garden pea from

(*a*) 1851–1859 (*b*) 1856–1864
(*c*) 1865–1834 (*d*) 1870–1875

ANSWERS

Fill in the Blanks

1. Bateson **2.** 1856 **3.** R.C. Punnett **4.** Bateson
5. Jacob and Monod

Multiple Choice Questions

1. (*b*) **2.** (*d*) **3.** (*b*) **4.** (*b*) **5.** (*b*).

74

Laws of Heredity (Mendelian Laws of Inheritance)

Gregor Johann Mendel was born in Moravia in 1822 to peasant parents. In 1843 he joined an Augustinian monastery at Brno, in Czechoslovakia where he took Holy Orders. From there he went to the University of Vienna where he spent two years studying natural history and mathematics before returning to monastery in 1853, where he lived and worked till his death in 1884.

Whilst in Vienna, Mendel had become interested in the process of hybridization in plants, the different forms in which hybrid (see Box 74.1) appear and statistical relationship between them. At Brno, his breeding experiments on the garden pea (*Pisum sativum*) were designed and conducted over a period of ten years (1856–1865). He appears to be the first scientist to appreciate the probabilistic nature of inheritance and use probability reasoning in biological research. He published a paper "*Experiments on plant hybridization*" in 1866 which laid the foundations for the present-day science of genetics. He demonstrated that *characteristics do not blend but pass from parents to offspring as discrete units* (which were named *genes* by Danish botanist **Johannsen**). *These units, which appear in the offspring in pairs, remain discrete and are passed on to subsequent generations by the male and female gametes which each contain a single unit.*

Partly because his ideas were ahead of his time, and partly because he published his work in an obscure journal (*i.e.*, Proceedings of the Brunn Natural Science Society), the significance of his monumental work went unnoticed till 1900. In that year, three eminent biologists, **Hugo de Vries, Carl Correns** and **Erich von Tschermak** independently 're-discovered' Mendel's work while studying inheritance in plant hybrids. These biologists had obtained results similar to those of Mendel, only to find that Mendel had published similar findings 34 years earlier. Mendel's original paper was republished in Flora, 89, 361 (1901). **Bateson** confirmed Mendel's work by a series of hybridization experiments.

74.1 REASONS OF MENDEL'S SUCCESS

The success of Mendel's hybridization experiments depended on the following factors: 1. Luck; 2. Foresight; 3. Mathematical background, and 4. Scientific aptitude. The following factors led to Mendel's success.

I. Method of Working

1. Mendel first analysed the reasons of failure of his predecessors and then chalked out the programme of his hybridization experiments.
2. Preliminary investigations were carried out to obtain familiarity with the experimental organism.

Box 74.1

1. Allele or allelomorph. A variant form of a gene specifying a particular trait is called **allele.** It is one of possible mutational states of a gene, distinguished from other alleles by phenotypic effects. **Gene mutations** convert one allelic form of a gene into another. For many gene loci, there is **wild-type allele** which predominates in the population because it confers great **fitness** (*i.e.*, ability to survive and reproduce). This generally encodes the **normal, functional** product associated with the gene, and the **wild-type phenotype** reflects this normal gene activity. Other, rare alleles are designated **mutant alleles**. The quantity and/or structural properties of the encoded product of the mutant alleles may differ greatly, generating distinct **mutant phenotype**. Most genes are **diallelic** having only two alleles, but some are **multiple allelic** containing more than two alleles of a gene.

2. Phenotypes. They are observable properties of an organism that are genetically controlled.

3. Hybrid. An offspring of two animals or plants of different varieties.

4. Transmission genetics. It includes the study of mode of gene transmission from generation to generation. The kind of studies that Mendel performed are now included in the discipline of transmission genetics.

5. Variation. Divergence in one or more characteristics of an organism or biotype from those typical of or usual for its group. Variation can be discontinuous or continuous.

6. Segregation is the separation of alleles into different gametes and then into different offsprings, of any pair of alleles possessed by an individual. The two alleles having previously been brought together in the individual, one from each of its parents.

7. Genotype. Genotype is the specific allelic or genetic constitution of an organism; often the allelic composition of one or a limited number of genes under investigation.

8. Phenocopy. Phenocopy is an environmentally induced phenotype which is nonheritable and closely resembles the phenotype produced by a known gene. Epilepsy disease of human beings, for example, may result from either genetic or environmental causes. As a result, the definitive diagnosis of an inherited abnormality requires family pedigree data or other information indicating that the condition is inherited, in addition to the appropriate phenotypic manifestations in an individual.

9. Genome. Genome is the array of genes carried by an individual. It is the genetic constitution or the sum total of hereditary determinants or genes of an organism which it has received from its parents. For example, the genes for tallness in a tall pea plant is one of the allelic pair of its total genotype. Genome will include total number of gene pairs present in the chromosomes. For example, *Drosophila* has total eight chromosomes and its genotype is the total number of genes present on these chromosomes.

10. Proteome. Proteome represents the entire collection of proteins which are encoded by the genome. The proteome is a complex system, even though it represents only a fraction of the genome (*i.e.*, the coding sequences). This is because there are numerous ways to use single genes to generate multiple products, so the proteome contains many overlapping, structurally similar products.

11. Phenome. When proteome interacts with the environment, it generates the **phenome** which is the representation of the total sum of characters (phenotypes) displayed by an organism.

3. Mendel studied the inheritance of one character at a time unlike his predecessors who had considered the inheritance of an organism as a whole.
4. Mendel carried out his hybridization experiments to F_2 and F_3 generations.
5. Mendel scrupulously maintained the statistical records of all the experiments and analysed them carefully. Sufficient data were obtained to have statistical significance.
6. He experimentally produced and maintained genetically pure parent plants (*i.e.*, pure breed lines). He tested the purity by self-crossing the progeny for several generations.

7. Mendel made crossing between those genetically pure parents which have sharply visible contrasting characters.
8. He grew different pure lines in separate garden plots, preventing chances of cross pollination.

II. Selection of the Experimental Plant.

For his hybridization experiments Mendel had certain consideration in his mind about the choice of a suitable material. Mendel's considerations about the material were as follows:

1. Variation. The organisms which are to be chosen for the genetical experiments, should have a number of detectable differences and at a time only single detectable character should be considered.

2. Reproduction. The chosen organisms should be sexually reproducing (*i.e.*, by fusion of male and female sex cells) because only then the offsprings will be able to receive different characters from both the male and female parents.

3. Controlled mating. The chosen organisms should be able to mate in controlled or well-planned conditions. Because in genetical experiments sometimes we have to rear genetically pure parents by methods of controlled mating. One should maintain careful records of the offsprings of many generation.

4. Short life cycle. The chosen organisms should have very short life cycles.

5. Large number of offsprings. The organisms which have been chosen for the genetical experiments should produce large number of offsprings after each successive mating because it will help in deducing the correct conclusions.

6. Convenience in handling. The experimental species should be of a type that can be raised and maintained conveniently and inexpensively in the laboratory. For instance, the elephant will prove entirely useless material for genetical experiments than the *Drosophila*, pea plants, tomato, rats, guinea pigs, etc., which have been generally used and are still used in hybridization experiments.

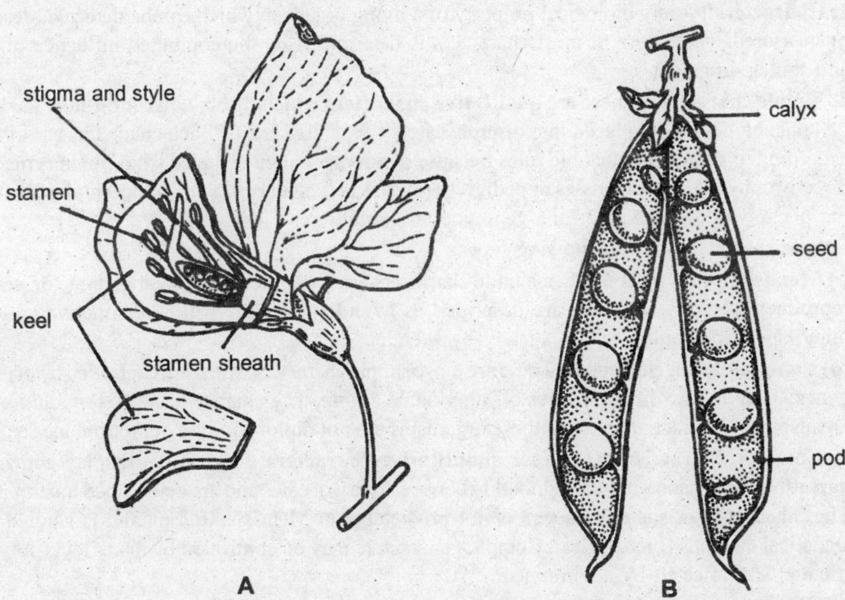

Fig. 74.1. A self-fertilizing flower of *Pisum sativum*. A—Portion of the keel which encloses the reproductive organs has been cut open to show the stamen and stigma; B—The mature fruit (pod) containing seeds that develop from the fertilized pea flower.

Mendel selected garden pea (*Pisum sativum*) as his experimental material because it has the following advantages:

1. Garden pea plant is an annual plant. Its short life cycle make it possible to study several generations within a short period.
2. It has perfect bisexual flowers containing both male and female parts (Fig. 74.1). The flowers are predominantly self-pollinating.
3. Because of self-fertilization plants remain genetically pure. It is, therefore, easy to get pure lines for several generations.
4. Experimental cross pollination can be done easily in pea plant.

III. Selection of Traits

Mendel selected seven pairs of contrasting characters (see Box 74.2) in pea plant for his hybridization experiments (Table 74.1).

Table 74.1 Seven pairs of contrasting characters in pea plant and their dominance-recessive relationship as worked out by Mendel.

Characters	*Dominant trait*	*Recessive trait*
1. Stem length	Tall	Dwarf
2. Flower position	Axial	Terminal
3. Pod shape	Inflated	Constricted
4. Pod colour	Green	Yellow
5. Seed shape	Round	Wrinkled
6. Cotyledon colour	Green	Yellow
7. Seed coat colour	Coloured (Grey)	Colourless (White)

Box 74.2 Characters and Traits

1. Character. It is any biological property of a living organism which can be detected, described or measured. The nature of most characters is determined by the combined influence of genes and the environment.

2. Simple characters. These are **qualitative characters** which display **discontinuous variation,** *i.e.*, phenotypes can be placed into discrete categories, called **traits**. Such characters are inherited according to simple, predictable rules because **genotype** can be deduced from **phenotype**, either directly or by analysis of crosses or pedigrees (in case of human beings). For the simplest character, the phenotype depends upon the genotype at a single gene locus.

Simple characters are of following types:

(*i*) Mendelian characters. When simple characters are controlled by nuclear genome of sexually reproducing eukaryotes, they are described as **Mendelian**. They follow distinctive patters of inheritance first studied systematically by Mendel.

(*ii*) Non-Mendelian characters. These are those simple characters which are controlled by **organelle genes**. They follow different rules of inheritance. (**Note.** The characters of bacteria and viruses are also Non-Mendelian because these organisms are not diploid and do not reproduce sexually).

3. Complex characters. These are **quantitative characters** which often display **continuous variation**, *i.e.*, phenotypes vary smoothly between two extremes and are determined quantitatively. The inheritance of such characters is not predictable in Mendelian terms and is studied using statistical methods (**biometrics**). Complex characters may be controlled by many loci (**polygenic theory**) and affected by environment.

4. Character mode. A general type of character, eye colour in human beings.

5. Trait. A specific type of character, *e.g.*, blue eye colour.

IV. Mendel's Crossing Techniques

The flowers of pea plants are so constructed that the pollens of a flower normally fall on the stigma of the same flower and, thus, affects self-pollination or self-fertilization. For the required cross-pollination, the anthers have to be removed from the flower in bud stage (*i.e.*, before their maturity). This operation of removal of anthers is called **emasculation**. The stigma is protected against any foreign pollen with the help of its covering by a bag. The pollen, then at the dehiscence stage is brought from the plant to be used as a male parent and by the help of a brush is dusted on the feathery stigma of the emasculated flower. At the time of such **cross pollination**, the pollen should be mature and stigma should be receptive.

For each of seven pairs of characters listed in Table 74.1, plants with one alternative trait were used as female, and those with the other alternative as male. Reciprocal crosses were also made, *i.e.*, each of the crosses was made in two ways, depending on which phenotype is used as male or female. For example the following two crosses are reciprocal crosses:

phenotype A (*e.g.*, Tall) C × phenotype B (*e.g.*, Dwarf) F

phenotype B (*e.g.*, Dwarf) C × phenotype A (*e.g.*, Tall) F

The population obtained as a result of crossing plants showing contrasting characters is called **F_1 generation**. The progeny of F_1 plants was then obtained by self-fertilization and it forms the **F_2 generation**. Similarly, F_3, F_4, etc., generations can also be obtained.

Further, for getting the exact results in the breeding experiments, it was necessary for Mendel to rear genetically pure variety of pea plants for a single character. Mendel adopted self-fertilization technique for it. For instance, to get pure character for tallness, he self fertilized a tall pea plant for many generations till the resulted offsprings always produced only tall plants. Likewise, he got genetically pure variety for dwarf pea plants. Mendel cross pollinated these two varieties of pea plants which were differing in a pair of contrasting characters, viz, tallness and dwarfness of the stem. When he made observations on the offsprings of first generation he found only tall plants. He allowed the self pollination in the offsprings of first generation and made further observations on the offsprings of second generation. He was astonished to note both tall and dwarf offsprings in the second generation. This showed him that the character of dwarfness disappeared in first generation but again reappeared in second generation. Further, the tall and dwarf plants of second generation were always in the ratio of 3:1 [3 (75%) tall: 1 (25%) dwarf). He self-pollinated the dwarf offsprings of second generation and found only dwarf plants in third generation. But when he self-pollinated the tall plants of second generation then he found that one-third (1/3) tall plants yield only tall plants in third generation, while rest two-third (2/3) tall plants yield tall and dwarf plants in the ratio of 3:1.

V. Mendel's Principles of Inheritance

On the basis of the results of Mendel's experiment the following seven principles have been recognised:

1. The heredity and variation of characters are controlled by factors, now called **genes**, which occur in pairs.
2. Contrasting traits are specified by different forms of each gene (different **alleles**).
3. When two dissimilar alleles are present in the same individual (*i.e.*, in a **hybrid** or **heterozygote**), one trait displays **dominance** over the other: the phenotype (see box 74.3) associated with one allele (the **dominant allele**) is expressed at the expense of that of the other (the **recessive allele**). This forms the Mendel's **phenomenon of dominance.**
4. Genes do not blend but remain discrete (**particulate**) as they are transmitted.
5. During meiosis, pairs of alleles **segregate** equally so that equivalent numbers of gametes carrying each allele are formed. This is called **principle of segregation** or **Mendel's first law**.
6. The segregation of each pair of alleles is independent from that of any other pair. This is called principle of **independent assortment** or **Mendel's second law.**

Actually Mendel himself did not postulated any genetical principle or laws, he simply gave conclusive theoretical and statistical explanations for his hybridization experiments in his research paper. However, it was **Correns**, the discoverer of Mendel's work, who thought that Mendel's discovery could be represented by these fundamental laws of heredity.

Box 74.3 Some Common Terms Used in Transmission Genetics

Allele. Broadly, a variant form of a gene specifying a particular trait. At the molecular level, a sequence variant of a gene.

Gene. Broadly a hereditary factor controlling or contributing to the control of a particular character. At the molecular level, a segment of DNA (or RNA in some viruses) which is expressed, *i.e.*, used to synthesize one or more products with particular functions in the cell.

(Gene) locus. The position of a gene on a chromosome or physical or genetic map.

Genotype. The genetic nature of an individual, often used to refer to the particular combination of alleles at a given locus.

Hemizygous. Containing one allele in a diploid cell, often used to refer to sex-linked genes.

Heredity. Passed from parent to offspring.

Heterozygous. Containing different alleles at a particular locus.

Phenotype. The outward nature of an individual, often used to refer to the nature of particular characters.

Pleiotropic gene. Affecting more than one character simultaneously.

Zygosity. The nature of alleles at a locus-homozygous, heterozygous or hemizygous.

74.2 MENDEL'S TWO LAWS

1. Law of Segregation

Mendel's 'first law—the law of segregation is also known as **law of purity of gametes**. The law states that *the hybrids or heterozygote's of F_1 generation have two contrasting characters or allelomorphs of dominant and recessive nature. These alleles though remain together for long time but do not contaminate or mix with each other and separate or segregate at the time of gametogenesis, so that each gamete receives only one allele of a character either dominant or recessive.*

For the proper understanding of the Mendel's law of segregation it will be helpful to study one of the Mendel's monohybrid cross (Fig. 74.2). Mendel crossed a homozygous red flowered pea plant with a homozygous white flowered pea plant. The F_1 heterozygotes or hybrids were found to be pink or purple flowered, thus, showing the incomplete dominance of red colour over white colour. When the F_1 hybrids were allowed to be self-fertilized, they produce both coloured (red or purple) and white flowered pea plants in F_2 generation in the ratio of three and one (3 : 1). The re-appearance of white colour in F_2 generation indicates towards the process of segregation.

Mechanism of segregation. The mechanism of segregation in above mentioned monohybrid cross between red and white flowered pea plants can be understood by assuming that the homozygous red flowered pea plant has the allele *RR* for redness and white flowered pea plant has the alleles rr for whiteness. The pea plant with *RR* alleles produces the gametes with single allele *R* and pea plant with *rr* alleles produces the gametes with the allele *r*. The gametes of both pea plants are united to form a hybrid or heterozygote having the alleles *Rr* both for redness and whiteness. Due to the phenomenon of incomplete dominance the allele *R* for red colour partially expresses itself in hybrids of F_1, while the allele *r* for white colour remains latent or recessive. Both the allele *R* and *r* remain together for long time but they do not affect each other. Neither they mix nor they contaminate each other. Because a gamete can contain only one chromosome of a homologous pair, therefore, each gamete can carry single allele *R* or *r*. At the time of gametogenesis two types of gametes are produced by F_1 hybrids in equal numbers. Half of the gametes carry the allele *R* and other half carry the allele *r*. These gametes during the process of fertilization can unite in three possible combinations, *viz.*, *RR*, *Rr* and *rr*, to produce three types of individuals in F_2 generation. Thus, in F_2 75% individuals have coloured flowers and 25% white flowers. The appearance of white colour

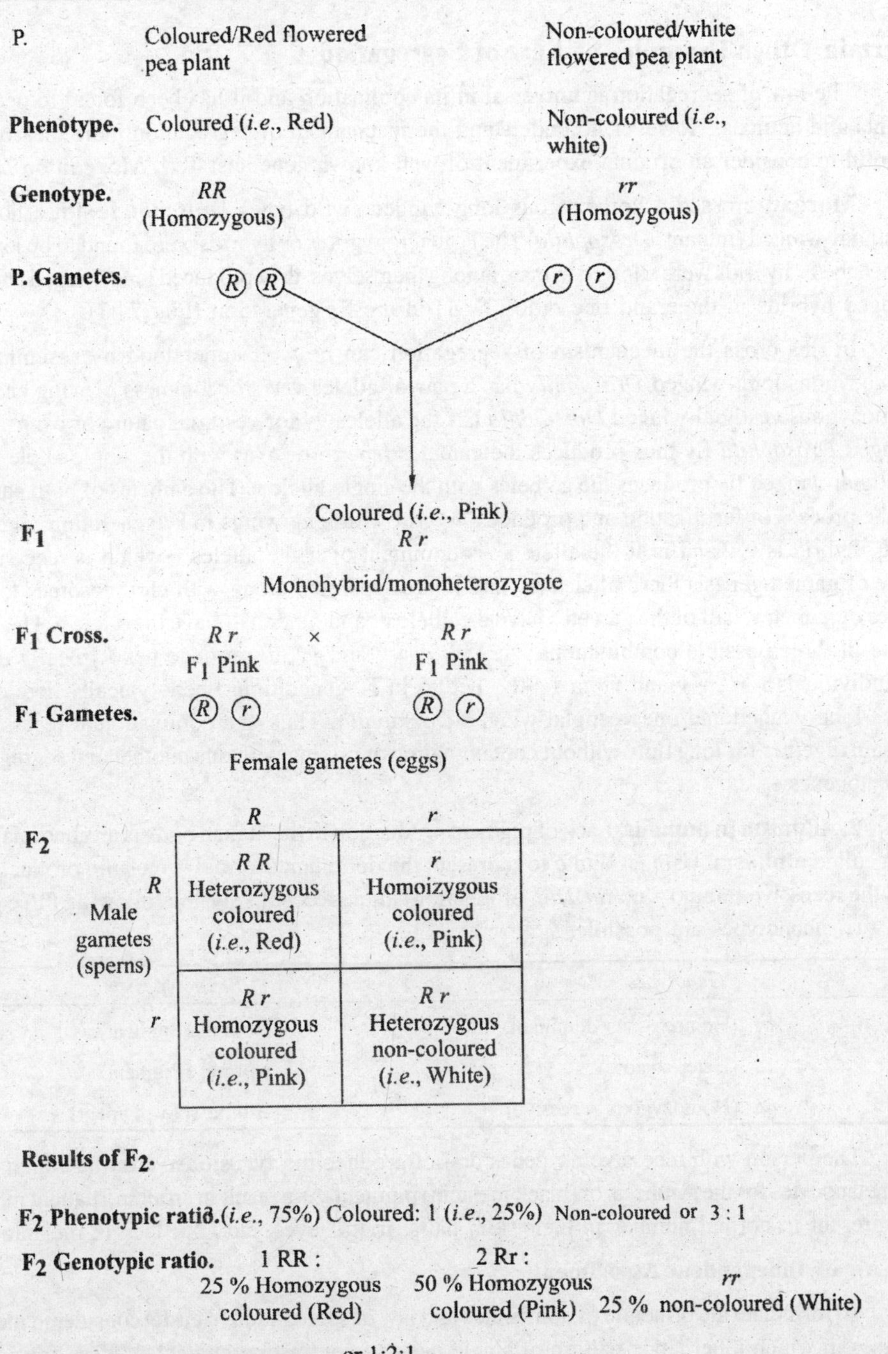

Fig. 74.2. Diagrammatic representation of a monohybrid cross of coloured (Red) and non-coloured (White) pea plants.

in F_2 generation indicates that in the hybrid the allele (*r*) for white colour remains along with allele (*R*) for red colour but does not mix with it or contaminated by it and it separates or segregates during gametogenesis.

Certain Other Examples of Law of Segregation

The law of segregation is universal in its application and it has been found to occur in both plants and animals. However, to understand the mechanism of segregation more clearly it will be helpful to consider an original experiment of well known geneticist **T.H. Morgan** on *Drosophila*.

Morgan crossed a homozygous long-winged (wild type) *Drosophila* with a homozygous vestigial-winged (mutant) *Drosophila*. The F_1 heterozygotes or hybrids were found to be long winged. When the F_1 hybrids were allowed to mate among themselves, they produced long-winged and vestigial-winged hybrids in three and one ratio (3 : 1) in the F_2 generation (Fig. 74.3).

In this cross the mechanism of segregation can be well understood by assuming that the homozygous long-winged *Drosophila* has a pair of alleles v^+v^+ for longness of wing and similarly homozygous vestigial-winged *Drosophila* has the alleles vv for vestigial nature of wings. The long-winged *Drosophila* fly thus produces the gametes (sperm or ova) with the single allele v^+ and the vestigial-winged fly produces the gametes with the single allele v. The gametes of both parents unite in the process of fertilization and produce a hybrid with long wings in F_1 generation. The genotype of F_1 hybrid is v^+v and in it the allele v^+ is dominant over the alleles v which is recessive. At the time of gametogenesis these alleles (v^+ and v) are separated along with chromosomes to form two types of gametes, half of the gametes have the allele v^+ and other half have the allele v. These gametes unite in three possible combinations, viz., v^+v^+, v^+v and vv, to produce genotypically three types of individuals v^+v^+, v^+v and vv in 1 : 2 : 1 ratio in F_2 generation. Phenotypically here occur only three long-winged and one vestigial-winged *Drosophila*. Thus, the dominant and recessive alleles remain together for long time without contaminating or mixing with one another and segregate during gametogenesis.

2. Albinism in humans. Lack of pigment production in the human body is an abnormal recessive trait called **albinism**. Using *A* and *a* to represent the dominant (normal = melanin production) allele and the recessive (albino = no melanin production) allele respectively, the following three genotypes and two phenotypes are possible:

Genotypes	*Phenotypes*
AA (Homozygous dominant)	Normal (Pigment)
Aa (Heterozygote)	Normal (Pigment)
aa (Homozygous recessive)	Albino (No pigment)

The person with recessive *aa* genes do not produce the **tyrosinase** enzyme which is needed by melanocytes for the synthesis of (black melanin) pigment. As a result in an albino patient melanocytes are present in normal number in their skin, hairs, iris of eyes, etc., but lack in melanin pigment.

2. Law of Independent Assortment

To formulate the principle of dominance and law of segregation **Mendel** considered monohybrid crosses in which inheritance pattern of single pairs of contrasting characters were considered at a time. But he tried to find out how different characters would behave in relation to each other in their inheritance from generation to generation. For this purpose Mendel crossed two varieties of pea plants which were differing in two pairs of contrasting characters. Because such crosses yielded **dihybrids** and at a time two pairs of contrasting characters had been considered in them, therefore, these crosses were known as **dihybrid crosses**.

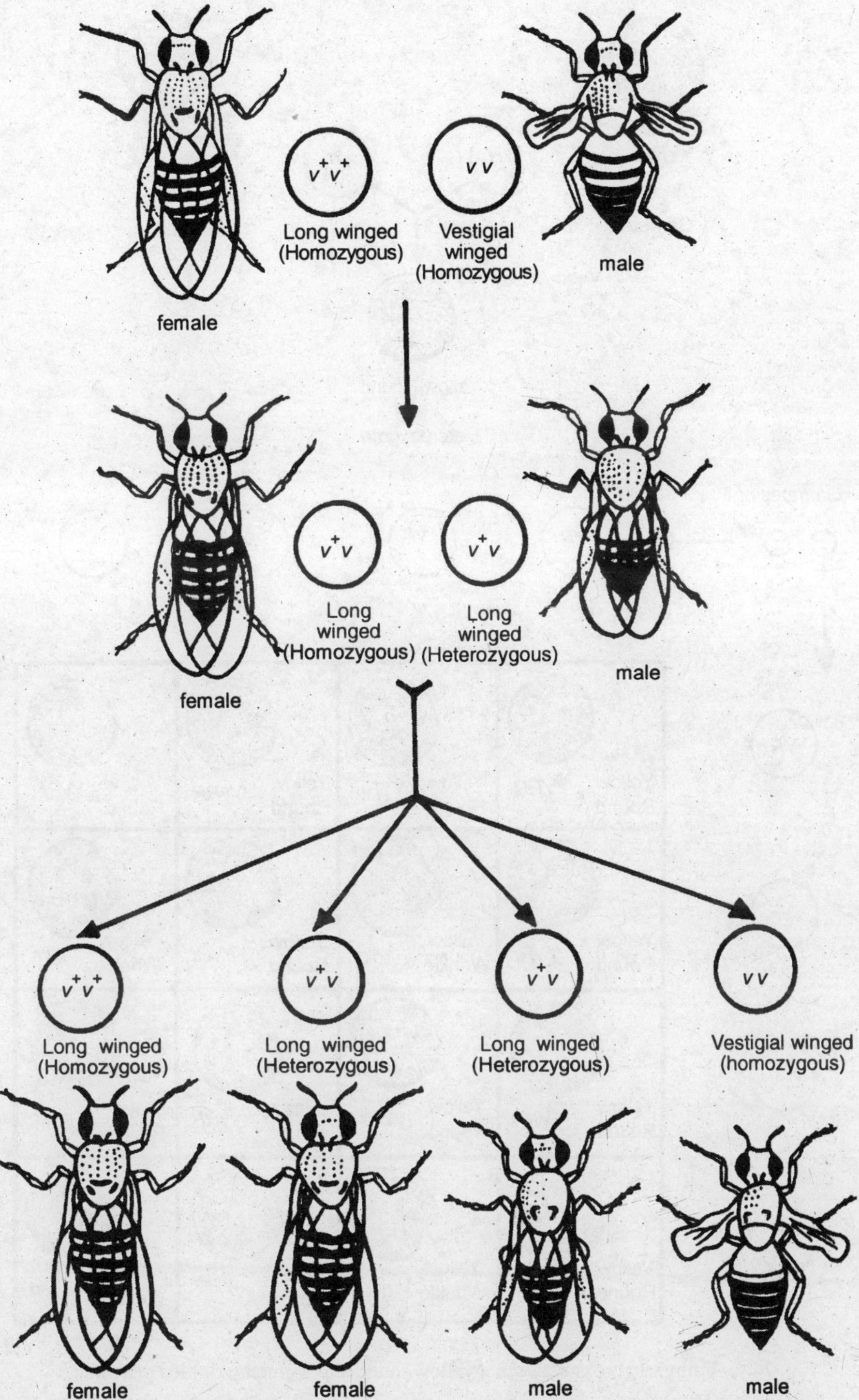

Fig. 74.3. Cross between long-winged and vestigial-winged *Drosophila*.

Mendel's dihybrid cross. In one of his hybridization experiment Mendel crossed a homozygous pea plant having **yellow round seeds** with the homozygous pea plant having **green wrinkled seeds**

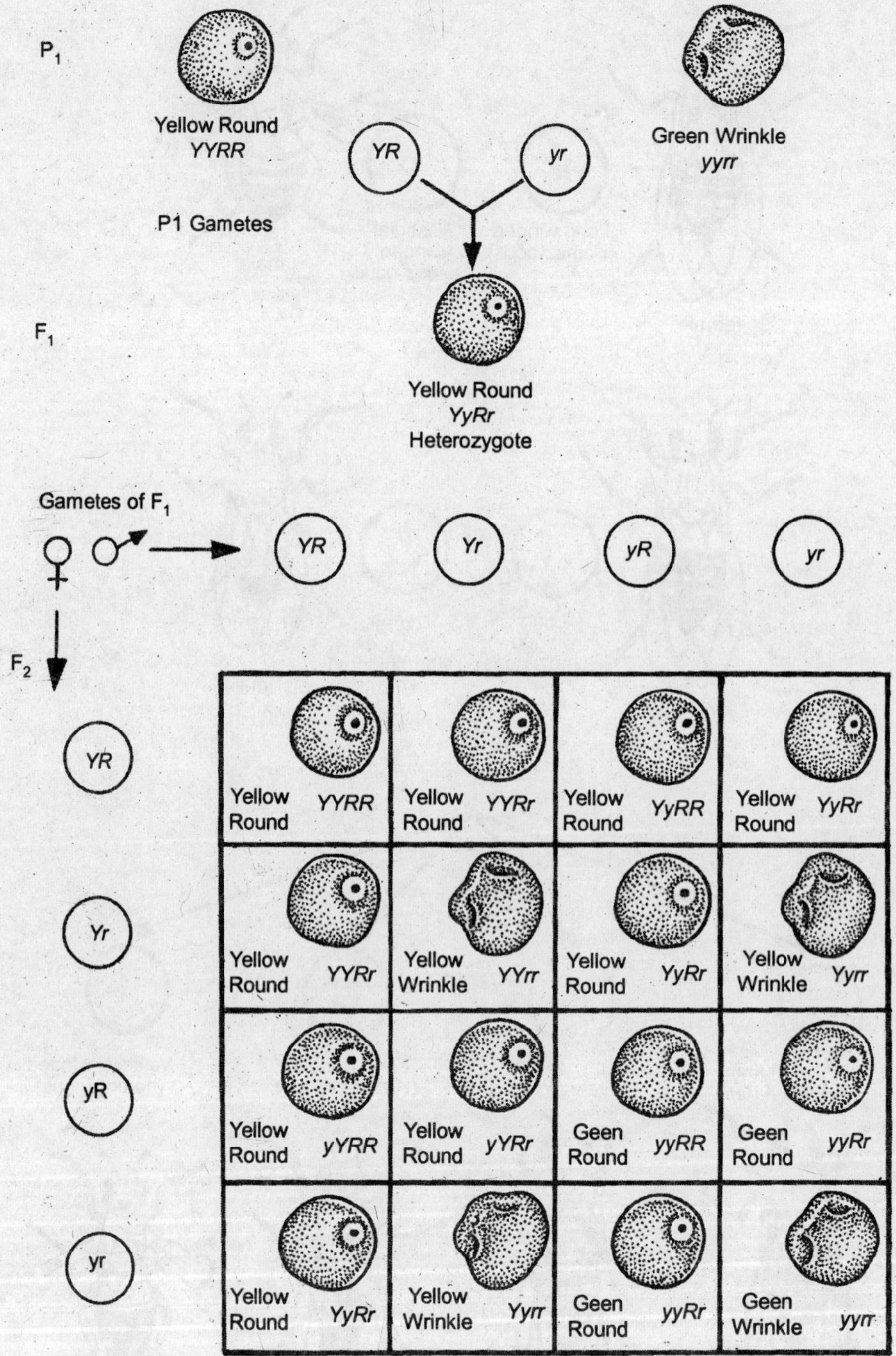

Fig. 74.4. A dihybrid cross between a yellow round and a green wrinkled pea plant.

(Fig. 74.4). The F_1 hybrids were found to have yellow round seeds. When the F_1 hybrids were allowed to cross among themselves they produced four types of seeds in the ratio of 9 : 3 : 3 : 1 given as follows:

1. Yellow Round – 9
2. Yellow Wrinkled – 3
3. Green Round – 3
4. Green Wrinkled – 1

Beside getting the ratio of 3 : 1 of the monohybrid crosses Mendel got the ratio of 9 : 3 : 3 : 1. This irregularity in the ratio of F_2 offspring was explained by Mendel stating that "*when the parents differ from each other in two or more pairs of contrasting characters or factors then the inheritance of one pair of factors is independent to that of the other pair of factors*". This is the Mendel's law of independent assortment.

Mechanism of independent assortment. The mechanism of independent assortment can be understood easily by assuming that the homozygous pea plant with yellow round seeds has the alleles *YY* and *RR* for the yellow colour and roundness of the seed, respectively. Similarly the homozygous pea plant with green wrinkled seeds contains the alleles *yy* and rr for the green colour and wrinkledness of seeds. The gametes which are produced by *YYRR* and *yyrr* plants are *YR* and *yr* types respectively. When both parents are crossed the union of both types of gametes takes place to give the F_1 hybrid (*Yy Rr*). The F_1 hybrid have been found to contain yellow round seeds showing the dominance of allele *Y* for yellow colour over the recessive allele *y* for green colour and the dominance of allele *R* for roundness over the recessive allele *r* for wrinkledness of seed. Now the F_1 hybrids have four types of alleles, viz., *Y* for yellow colour, *y* for green colour, *R* for round shape and *r* for rinkledness of seed.

During gametogenesis these four alleles, may combine in following four combination:

1. The allele *Y* may be associated with the allele *R* to give rise *YR* combination.
2. The allele *Y* may be associated with the allele *r* to give rise *Yr* combination.
3. The allele *y* may be associated with the allele *R* to give rise *yR* combination.
4. The allele *y* may be associated with the allele *r* to give rise *yr* combination.

Thus, four types of alleles are assorted independently to produce four types of gametes, viz., *YR, Yr, yR* and *yr.* These four types of gametes (pollens or ovules) of F_1 hybrid unite at random in the process of fertilization and produce sixteen types of individuals in F_2 generation which are given in the Table 74.2.

Table 74.2 The phenotype and genotype of F_2 generation of a dihybrid cross between pea plants with yellow round and green wrinkled seeds.

Number of individuals	*Genotypes class*	*Phenotype class*
1	*YY RR*	Homozygous Yellow Round
2	*YY Rr, YY Rr*	
2	*Yy RR, Yy RR*	Heterozygous Yellow Round
4	*Yy Rr, Yy Rr, Yy Rr, Yy Rr*	
		Yellow Round = 9
1	*YY rr*	Homozygous Yellow Wrinkled
2	*Yy rr, Yy rr*	Heterozygous Yellow Wrinkled
		Yellow Wrinkled = 3
1	*yy RR*	Homozygous Green Round
2	*yy Rr, yy rR*	Heterozygous Green Round
		Green Round = 3
1	*yy rr*	Homozygous Green Wrinkled
		Green Wrinkled = 1
16	– –	9 : 3 : 3 : 1

Thus, the sixteen F_2 individuals have the ratio of 9 yellow round: 3 yellow wrinkled: 3 green round: 1 green wrinkled. These results have proved the law of independent assortment and showed that each pair of contrasting characters behaves independently and bears no permanent association or relation with a particular character. The allele *Y* was associated with allele *R* in parent but it does not always remain associated with it and it is also associated with the allele *r*.

Dihybrid cross in *Drosophila*. In a dihybrid cross homozygous long-winged and black-bodied *Drosophila* fly is crossed with a vestigial-winged, grey-bodied fly, they produced in F_1 long-winged grey-bodied *Drosophila*. The hybrids of F_1 generation after mating with each other produced in F_2 generation 9 long-winged grey-bodied, 3 long-winged black-bodied, 3 vestigial-winged grey-bodied and 1 vestigial-winged black-bodied *Drosophila* (Total = 16). Therefore, the 16 individuals of F_2 have the ratio of 9 : 3 : 3 : 1 (Fig. 74.5).

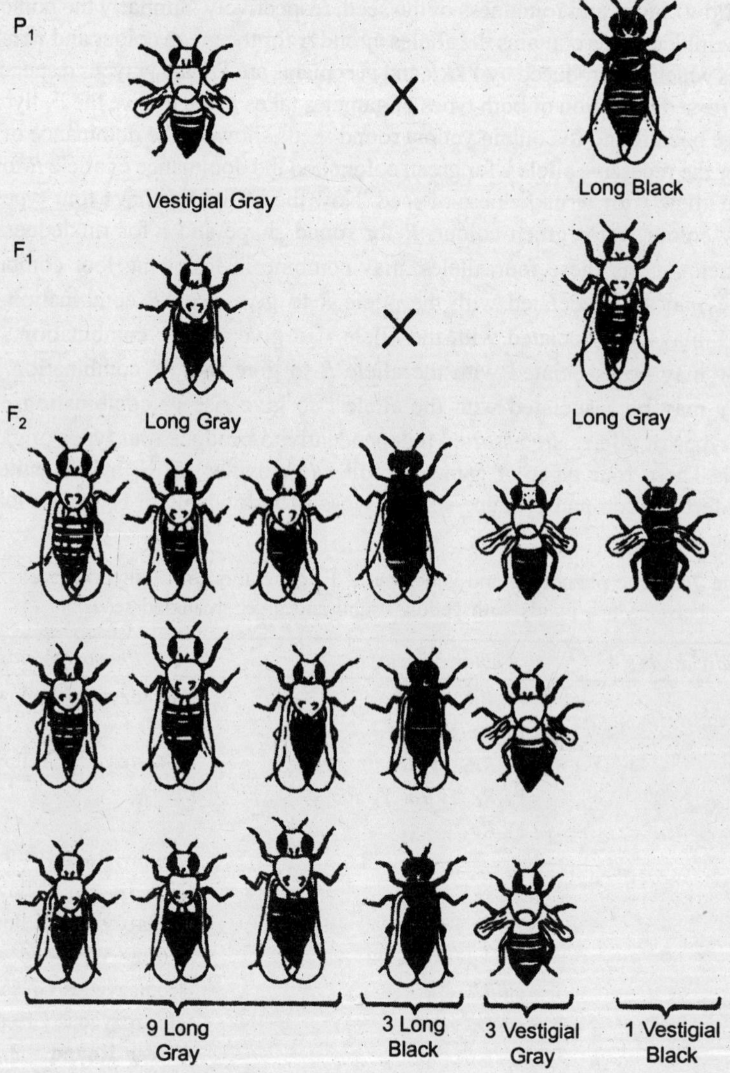

Fig. 74.5. A dihybrid cross between the vestigial-grey and long-black *Drosophila*.

Practical Application of Mendelism

Since the time of their re-discovery in 1900, Mendel's laws are being applied to various branches of breeding. In poultry, these are used for improving the varieties of fowls and their eggs. Various improved breeds of horses, dogs, sheep and cows are also obtained. A number of rust-resisting, disease-resistant and high-yielding varieties of wheat, rice, maize, etc., have been produced by cross-breeding experiments. Eugenics is mainly based on Mendelism.

74.3 CHROMOSOME THEORY OF HEREDITY (CYTOLOGICAL BASIS OF HEREDITY)

American cytologist **W. Sutton** in early 20th century noticed striking similarities between the behaviour of chromosomes during gamete formation and fertilization, and the transmission of Mendel's hereditary factors. These similarities have been summarised in Table 74.3.

Table 74.3 A summary of the similarities between events occurring during meiosis and fertilization and Mendel's hypothesis.

Meiosis and fertilization	*Mendel's hypothesis*
1. Deploid cells contain *pairs* of chromosomes (homologous chromosomes).	1. Characteristics are controlled by *pairs* of factors.
2. Homologous chromosomes *separate* during meiosis.	2. Pairs of factors *separate* during gamete formation.
3. *One* homologous chromosome passes into each gamete cell.	3. Each gamete receives *one* factor.
4. Only the *nucleus* of the male gamete fuses with the egg cell nucleus.	4. Factors are transmitted from generation to generation as *discrete* units.
5. Homologous pairs of chromosomes are restored at fertilization, each gamete (male and female) contributing one homologous chromosome.	5. Each organism inherits one factor from each parent.

On the basis of the evidence suggested above, **Sutton** and **Boveri** in 1902 proposed that chromosomes were the carriers of Mendel's factors, *i.e.*, they formulated the so called **chromosome theory of heredity**. According to this theory *each pair of factors is carried by a pair of homologous chromosomes, with each chromosome carrying one of the factors*. Since the number of characteristics of any organism vastly outnumbers the chromosomes, as revealed by microscopy, each chromosome must carry many factors.

In 1909, term **factor** was replaced by the term **gene** by **Johannsen**. Whilst term gene is used to describe the unit of heredity, its variants are called alleles which influence phenotypic expression. Alleles occupy corresponding positions or **loci** on the homologous chromosomes.

Chromosomal Explanation of Segregation

Mendel's law of segregation of factors could now be explained in terms of the separation of homologous chromosomes which occurs during anaphase I of meiosis and the random distribution of alleles into gamete cells. These events have been summarised in Fig. 74.6.

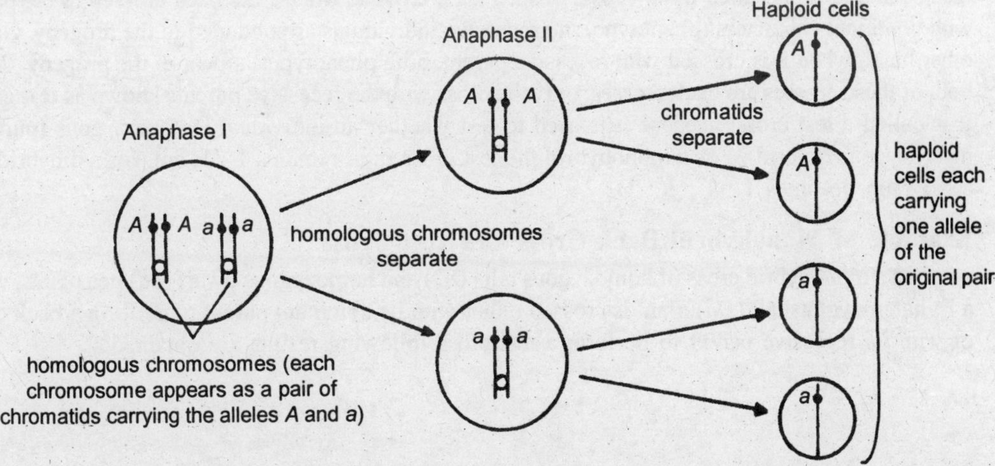

Fig. 74.6. Mendel's principle of segregation of factors (alleles) *A* and *a* described in terms of the separation of homologous chromosomes which occurs during meiosis.

Chromosomal Explanation of Independent Assortment

Mendel's law of independent assortment may also be explained in terms of the movement of chromosomes during meiosis. During gamete formation, the distribution of each allele from a pair of homologous chromosomes is entirely independent of the distribution of alleles of other pairs. This situation is described in Fig. 74.7. It is the random alignment or assortment of homologous chromosomes on the equatorial spindle during metaphase I of meiosis, and other subsequent separation during metaphase I and anaphase I that leads to the variety of alleles recombination in the gamete cells.

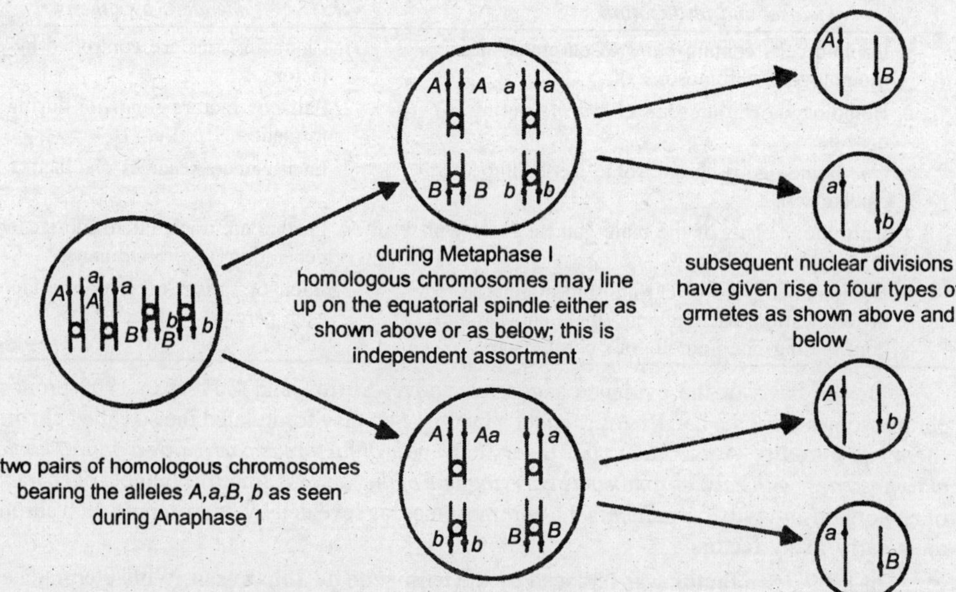

Fig. 74.7. Mendel's principle of independent assortment of factors (or alleles) *A*, *a*, *B*, *b* described in terms of the separation of homologous chromosomes which occurs during meiosis.

74.4 BACK CROSS AND TEST CROSS

When F_1 individuals are crossed with one of the two parents from which they were derived, then such a cross is called **back cross**. In such back crosses, when F_1 is back crossed to the parent with dominant characters (phenotype), no recessive individuals are obtained in the progeny. On the other hand, when it is crossed with recessive parent, both phenotypes appear in the progeny. While both of these crosses are back crosses, only the cross with the recessive parents known as **test cross.** It is called a test cross, because it is used to test whether an individual is homozygous (pure) or heterozygous (hybrid). For a monohybrid the test cross ratio remains 1 : 1, but, for a dihybrid test cross ratio becomes 1 : 1 : 1 : 1.

Example of Monohybrid Back Cross and Test Cross

In a monohybrid cross of homozygous tall (*DD*) and homozygous dwarf (*dd*) pea plants, when a F_1 heterozygous tall (*Dd*) plant is crossed either with its dominant parent to perform a back cross or with its recessive parent to perform a test cross following results are obtained:

P_1:	Homozygous tall	×	Homozygous dwarf
	DD	↓	*dd*
		Dd	
F_1:		Heterozygous tall	
A. Back cross:	F_1 Tall	×	P_1 Tall
	Dd	↓	*DD*
Back cross progeny:	½ *DD* Homozygous tall	:	½ *Dd* Heterozygous tall

(Back cross produces all tall)

B. Test Cross:	F_1 Tall	×	P_1 Dwarf
	Dd	↓	*dd*
Test cross progeny:	½ *Dd* Heterozygous tall	:	½ *dd* Homozygous

(Test cross produces a ratio of 1 : 1 of dwarf tall and dwarf)

Example of Dihybrid Test Cross

The test cross of a heterozygous yellow round seeded pea plant with a double recessive parent (green wrinkled, yyrr), yields the test cross genotypic and phenotypic ratio of 1 : 1 : 1 : 1 as follows:

Parents:	F_1 Heterozygous Yellow Round	×	F_1 Homozygous Green Wrinkled
	Yy Rr		*yy rr*
	↓		↓
Gametes:	*YR, Yr, yR, yr*		*yr*

Test cross progeny: ¼ *Yy Rr* : ¼ *Yy rr* : ¼ *yy Rr* : ¼ *yy rr* ¼ Green Round : ¼ Green Wrinkled or ¼ Yellow Round : ¼ Yellow Wrinkled: or 1 : 1 : 1 : 1.

74.5 DISTINCTIONS

1. Phenotype and Genotype

Table 74.4 Differences between phenotype and genotype.

Phenotype	*Genotype*
1. Phenotype is the expression of the character of an individual. Character may be morphological (such as form, sex, colour, behaviour) or biochemical (ABO blood group).	1. Genotype is the genetic make up of the individual which determines the characters.
2. Phenotype can be read out from the individuals by direct observation.	2. Genotypes can be ascertained from the progeny of the individuals.
3. Individuals having identical phenotypes may or may not have the same genotype.	3. Individuals having identical genotypes usually have the same phenotype provided they are present in the same environment.

2. Homozygous and Heterozygous

Table 74.5 Differences between homozygous and heterozygous individuals.

Homozygous	*Heterozygous*
1. These individuals possess identical alleles of a gene such as either dominant alleles *DD* for for tallness or recessive alleles *dd* for dwarfness.	1. These individuals possess different alleles of a gene such as a dominant allele *D* and a recessive allele *d*.
2. They produce only one type of gametes, e.g., only *D* by *DD* individuals and *d* by *dd* individuals.	2. They produce two types of gametes in case of monoheterozygote (monohybrid), e.g., *D* and *d*.
3. They breed true to the particular character, *i.e.*, when a dominant parent (or a recessive parent is self-pollinated or self-fertilized, its all the offsprings possess the same genotype and phenotype).	3. When self-pollinated or self-fertilized, heterozygote of a monohybrid cross produces three different genotypes (*i.e.*, *DD*, *Dd* or *dd*) which may have two or three different phenotypes.

3. Dominant and Recessive

The allele which gains phenotypic expression in F_1/F_2 hybrid (heterozygote) is known as **dominant allele**, while the allele which is unable to express itself in the presence of the dominant allele is called the **recessive allele**. Though recessive allele fails in expressing itself in hybrids, it is transmitted to the next generation without change. It expresses only in the homozygous state.

For example, a heterozygous pea plant has one allele for tallness (*D*) and one allele for dwarfness (*d*). The allele for tallness (*D*) is dominant in nature and has it phenotypic expression both in homozygous and heterozygous states. The allele for dwarfness (*d*) is recessive and is expressed only in homozygous condition.

QUESTIONS

Long Answer Questions

1. Why did Mendel select *Pisum sativum* for his experiments?

2. Describe Mendel's method of experimentation. How was the Mendel's work re-discovered? Briefly describe the factors that led to Mendel's success. (*Gorakhpur 1989*)

3. State and discuss Mendel's laws. Why was Mendel successful in enunciating these laws. (*Punjab (G) 1991*)

4. Describe Mendel's laws of inheritance with the help of suitable examples. (*Agra 1988; Avadh 1991, 1992; Gorakhpur 1989; Garhwal 1989; Nagpur 1991; C.C.S.U. Meerut, 1997, 2009; Kanpur 1992; Lucknow 1992; Avadh 1997*)

5. Write an essay on Mendelism and its practical applications. (*Bundelkhand 1988; Avadh 1996*)

6. Give an account of the Mendel's laws of inheritance and discuss their importance in genetics. (*Allahabad 1992; C.C.S.U. Meerut, 2007*)

7. Write an essay on Mendelian inheritance. (*Gorakhpur 1993, 1996; Kanpur 1995, 1998*)

8. Describe Mendel's two laws of heredity and explain their mechanism. (*Rohilkhand 1989*)

9. Explain the Mendelian laws of segregation and independent assortment. (*Gorakhpur 1988; Rewa 1989; Lucknow 1989; Purvanchal 1992*)

10. Explain Mendel's laws of segregation with the help of two examples. (*Bangalore 1994; Kanpur 1994, 1995, 1996; Avadh 1998*)

11. With suitable examples explain Mendel's law of independent assortment. (*Gorakhpur 1993, 1996; Kanpur 1995, 1998*)

12. Write Mendel's law of 'independent assortment of factors'. Draw a checker board to justify. *(Kanpur 1989)*

13. What do you understand by dihybrid cross? State and explain Mendel's second law by an experiment. *(N. Bengal 1989)*

14. Discuss the meiotic and genetic basis of the Mendel's laws of inheritance. *(BHU (Hons.) 1990)*

15. Explain briefly the second law of Mendel with a suitable example. *(Bangalore 1992)*

16. Describe in detail the Mendel's law of independent assortment giving suitable example and also describe the seven characters chosen by Mendel to perform his cross-breeding experiments. *(Kanpur 1994)*

17. Give a critical appreciation of Mendel's laws of Genetics. Describe the Law of Independent Assortment and state reasons why this law is not universally applicable. *(Avadh 1999)*

18. State Mendel's law of independent assortment in man. Brown eye colour (*B*) is dominant over blue eye colour (*b*). A brown eyed man, whose mother was blue eyed, married a brown eyed woman whose father was blue eyed. What are genotypes of man and woman. *(Bangalore 1992)*
(**Hint.** Male *Bb*; Female *bb*).

19. State and explain Mendel's law of independent assortment and explain the following terms: 1. Recessive characters; 2. Heterozygous; 3. Genotype; 4. Test cross; 5. Monohybrid cross. *(Punjab (G) 1992)*

Short Answer Questions

1. Give a brief life sketch of Gregor Johann Mendel.
2. Write briefly the contribution to biology by Mendel. *(Lucknow 1992)*
3. State briefly what is meant by 'dominant' and 'recessive' as applied to inherited character. *(Guwahati (M) 1995)*
4. Describe the chromosomal basis of Mendel's crosses.
5. Explain giving examples Mendel's law of segregation. Interpret it in the light of the chromosome theory of heredity.
6. Explain "what is the heredity bridge between parents and offspring". *(Guwahati (Major) 1995)*
7. Differentiate between the following:
   (*i*) Phenotype and Genotype.
   (*ii*) Genotype and Phenotype *(Garhwal 1989, 1990; Gorakhpur 1989; Bundelkhand 1989; Rohilkhand 1990; Delhi 1992; Lucknow 1992, 1993, 1994; Guwahati (G) 1992, 1993, 1995; Allahabad 1995; CCS 1997)*
   (*iii*) Testcross and Backcross *(C.C.S.U. Meerut 2007, 2008; Lucknow 1997; Allahabad 1995)*
   (*iv*) Homozygous and Heterozygous *(Bundelkhand 1989; Gorakhpur 1989; Garhwal 1989; Guwahati (G) 1994, 1996; CCS 1999)*
   (*v*) Dominant and Recessive *(Kanpur 1989)*
   (*vi*) Dominance and Segregation
   (*vii*) Monohybrid and Dihybrid crosses
   (*viii*) Gene and Allele *(Allahabad 1995)*
8. Write short notes on the following:
   (*i*) Mendelism *(Kanpur 1990; Allahabad 1993)*
   (*ii*) Monohybrid cross *(Bundelkhand 1991, 1992; Avadh 1995)*
   (*iii*) Dihybrid cross *(Rohilkhand 1990; Avadh 1990, 1996; Kanpur 1992, 1993)*
   (*iv*) Mendel's first law or Law of segregation *(Allahabad 1996; Gorakhpur 1998)*
   (*v*) Law of independent assortment or Independent assortment.

(*Lucknow 1989; Punjab (G) 1990; CCS 1993; Kanpur 1993*)

(*vi*) Test cross (*Guwahati (G) 1992, 1993, 1995*)

(*vii*) Allele or Allelomorph (*Kanpur 1989; Guwahati (G) 1992, 1995; Lucknow 1997*)

(*viii*) Phenotype

(*ix*) Genotype (*Magadh 1990*)

(*x*) Phenocopies (*Bangalore 1993*)

(*xi*) Genome (*Lucknow 1997*)

(*xii*) Proteome

(*xiii*) Phenome

(*xiv*) Second law of Mendel (*H.N.B., Garhwal, 2008*)

(*xv*) Mendel's law of independent assortment (*H.N.B., Garhwal 2009*)

(*xvi*) Difference between test cross and back cross and their significance (*H.N.B., Garhwal 2009*)

(*xvii*) Back cross.

9. What is test cross and back cross? What is its significance?
(*Kumaon 1989, 1991, 1992; Rohilkhand 1990; Purvanchal 1991*)

10. What is test cross? How it helps in interpreting of the results of a monohybrid cross?
(*Berhampur 1989; Kanpur 1990*)

11. What is the principle of independent assortment? How did Mendel test this hypothesis with the help of test cross? (*NEHU 1996*)

Very Short Answer Questions

1. What do the following genetic symbols mean? *AA*, *Aa*, *aa*.

Ans. *AA*—homozygous dominant; *Aa*—heterozygous dominant; *aa*—homozygous recessive.

2. Name the scientist who re-discovered Mendel's law.

Ans. Hugo De Vries, Carl Correns and Eric Von Tschermak.

3. What is pure line?

Ans. Homozygous population.

4. Who is considered as the 'Father of Genetics'?

Ans. Gregor Johann Mendel.

5. When did the results of Mendel's experiments published?

Ans. 1866.

6. What do the two forms of genes called?

Ans. Alleles.

7. What is hybridization?

Ans. Crossing of two genetically different individuals is called hybridization.

8. Test cross and Back cross:

Back cross. The cross of an F_1 hybrid to one of the parental types. The offspring of such a cross are referred to as back cross generation or back cross progeny.

Test cross. Back cross to the recessive parental type, or a cross between genetically unknown individuals with a fully recessive tester to determine whether an individual in question is heterozygous or homozygous for a certain allele. Test cross is also used as a test cross for linkage.

Example. Test crosses in pea plant:

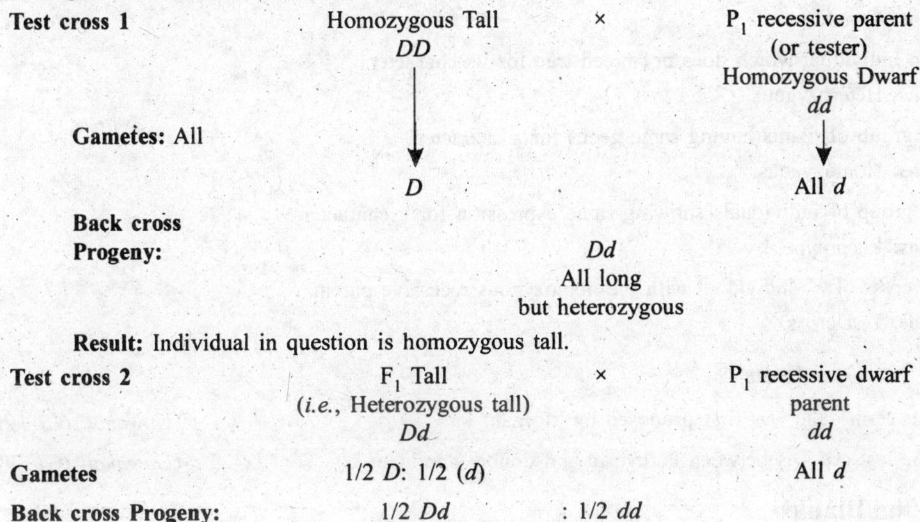

Test cross 1 Homozygous Tall × P_1 recessive parent (or tester)
DD Homozygous Dwarf dd

Gametes: All D All d

Back cross Progeny: Dd
All long but heterozygous

Result: Individual in question is homozygous tall.

Test cross 2 F_1 Tall (*i.e.*, Heterozygous tall) Dd × P_1 recessive dwarf parent dd

Gametes 1/2 *D*: 1/2 (*d*) All *d*

Back cross Progeny: 1/2 *Dd* Heterozygous tall : 1/2 *dd* Homozygous dwarf

Result: 1:1 phentypic ratio indicates that individual in question is a monohybrid or single heterozygote for dominant allele *D* for tallness and recessive allele *d* for dwarfness.

9. What is meant by heterozygote?

Ans. A heterozygote (or hybrid) possesses two contrasting alleles of a gene, but only one allele of the two is able to express itself, while the other remains hidden. The allele which gains phenotypic expression in the F_1 heterozygote is called **dominant allele**, while the other allele which fails to express itself in the presence of the dominant allele is the **recessive allele**. Recessive allele, though, is unable to express itself in F_1, is transmitted to the next generation without change. For example, a heterozygous pea plant has one allele for tallness (*D*) and one allele for dwarfness (*d*). The allele for tallness (*D*) is dominant and the allele for dwarfness (*d*) is recessive.

10. Define the allele.

Ans. Alleles are variants of a gene controlling the same characteristic (*e.g.*, length of pea stem) but producing different effects (*e.g.*, tall or dwarf) and occupying corresponding positions (loci of gene) on homologous chromosomes. An allele may be a dominant allele, recessive or null allele, wild type allele or mutant allele.

The genes for the characters described by Mendel occurred in two forms, *i.e.*, they were **diallelic**. For many genes, however, a greater degree of **allelic variation** is apparent, *i.e.*, they are **multiple allelic**. The human ABO blood group locus, for instance, has three physiologically distinct alleles, *i.e.*, I^A, I^B, or *i*.

11. Define F_2 generation.

Ans. It stands for **second filial generation** which is resulted by interbreeding or selfing of F_1 offspring (which are heterozygotes). F_2 phenotypic ratio of a monohybrid cross is 3 : 1 and of a dihybrid cross is 9 : 3 : 3 : 1.

12. What is dihybrid cross? *(H.N.B. Garhwal 2009)*

Ans. A cross between homozygous parents differing in two respects. Progeny of such a cross is called dihybrid; it is heterozygous for two pairs of alleles.

Coin one word for the following:

1. A group of individuals which produce offsprings for several generations having same specific character.

Ans. Pure line.

2. A study of resemblances and differences between the parents and their offsprings.
Ans. Genetics.

3. An individual which does not breed true for its character.
Ans. Heterozygous.

4. A group of plants having same genes for a character.
Ans. Homozygous.

5. A group of individuals showing same expression for a character.
Ans. Phenotype.

6. A cross of F_1 individual with the homozygous recessive parent.
Ans. Test cross.

Yes or No Questions

1. The term gene was first proposed by Morgan. *(Guwahati (G) 1995)*

2. The test cross is between F_1 hybrid and double dominant. *(Guwahati (G) 1995)*

Fill in the Blanks

1. The expression of both alleles in a heterozygote is called .... *(NEHU 1996)*

Multiple Choice Questions

1. The various forms of a given gene are called *(NEHU 1996)*
(*a*) genotype (*b*) phenotype
(*c*) gamete (*d*) alleles

2. The expression of more than one character by a single gene is called. *(NEHU 1996)*
(*a*) pleiotropism (*b*) codominance
(*c*) epistasis (*d*) none

3. Mendel has chosen pea plant for his experiments to get successful results because *(M.S.U., Tamil Nadu, 2006)*
(*a*) it had a short life (*b*) it had several pairs of contrasting characters
(*c*) it was easy to breed (*d*) all the above

4. One of the Mendel's pure strains of pea plants has green peas. How many different kinds of ovum could such a plant produce with regard to pea colour? *(M.S.U., Tamil Nadu, 2006)*
(*a*) 1 (*b*) 2
(*c*) 4 (*d*) 8

5. Emasculation is achieved by
(*a*) removal of anthers (*b*) removal of style
(*c*) removal of petals (*d*) removal of sepals

6. Organisms phenotypically similar but genotypically different are said to be
(*a*) heterozygous (*b*) monozygous
(*c*) multizygous (*d*) homozygous

7. When a gene exists in more than one form, the different forms are called
(*a*) heterozygous (*b*) complementary genes
(*c*) genotypes (*d*) alleles

8. Pea plants
(*a*) are cross-pollinated (*b*) can be selfed
(*c*) are selfed and can be cross-pollinated (*d*) can be selfed and are cross-pollinated

9. Phenomenon of dominance and recessiveness is the result of

(*a*) back cross (*b*) incomplete dominance

(*c*) dihybrid cross (*d*) monohybrid cross

10. Law of independent assortment can be explained with the help of

(*a*) test cross (*b*) back cross

(*c*) monohybrid cross (*d*) dihybrid cross

11. A cross made between F_1 hybrid and any of its parent is

(*a*) back cross (*b*) monohybrid cross

(*c*) dihybrid cross (*d*) reciprocal cross

12. The cross between homozygous recessive and its hybrid is called

(*a*) back cross (*b*) test cross

(*c*) monohybrid cross (*d*) dihybrid cross

ANSWERS

Yes or No Questions

1. No **2.** No.

Fill in the Blanks

1. Codominance

Multiple Choice Questions

1. (*d*) **2.** (*a*) **3.** (*d*) **4.** (*a*)

5. (*a*) **6.** (*a*) **7.** (*d*) **8.** (*c*)

9. (*d*) **10.** (*d*) **11.** (*a*) **12.** (*b*).

75

Dominance, Incomplete Dominance and Codominance

75.1 PHENOMENON OF DOMINANCE

The cross between the pea plants differing in single pair of contrasting characters is known as **monohybrid cross.** As we have already noticed that when Mendel made a monohybrid cross (Fig. 75.1) between tall and dwarf pea plants then only tall pea plants appeared in the first filial generation (F_1). But when the F_1 progeny were allowed to be self-fertilized, both tall and dwarf characters appeared in the second filial generation or F_2. This shows that in F_1 hybrid the character of tallness dominates or conceals the character of dwarfness and so the character of dwarfness could not express itself in F_1 generation. The character which expresses itself (*i.e.,* tall) in F_1 generation is called by Mendel as **dominant character;** while the character which remained unexpressed or latent had been called **recessive**. According to these results Mendel described the phenomenon of dominance in following way: *in crossing between pure (homozygous) organisms for contrasting characters of a pair, only one character of the pair appears in the first filial generation.*

In pea plant Mendel found following characters to be dominant or recessive in various pairs of contrasting characters (See Table 75.1). In each of Mendel's crosses, the trait associated with one allele was fully dominant over the other, so that the phenotype of the heterozygote was identical to that of the dominant homozygote. At the biochemical level, such **complete dominance** often reflects the presence of a (recessive) **null allele** which is totally compensated by the presence of a (dominant) functional allele.

Certain Examples of Phenomenon of Dominance

After Mendel several geneticists tested the validity of the phenomenon of dominance on several plants and animals. They found its wide application in various plants and animals. A few important examples are described as follows:

1. Phenomenon of Dominance in Plants. Besides pea plant the phenomenon of dominance has also been observed in the following plants (See Table 75.2).

2. Application of Phenomenon of Dominance in Animals

A. The phenomenon of dominance is also applicable well to the animals. For instance, when homozygous black guinea pig is crossed with a homozygous white guinea pig (Fig. 75.2) then all hybrids of first filial generation (F_1) are found to be black. The black hybrids of F_1 when mated among themselves they produced black and white offsprings in 3:1 ratio. This shows that black coat colour dominates over white coat colour.

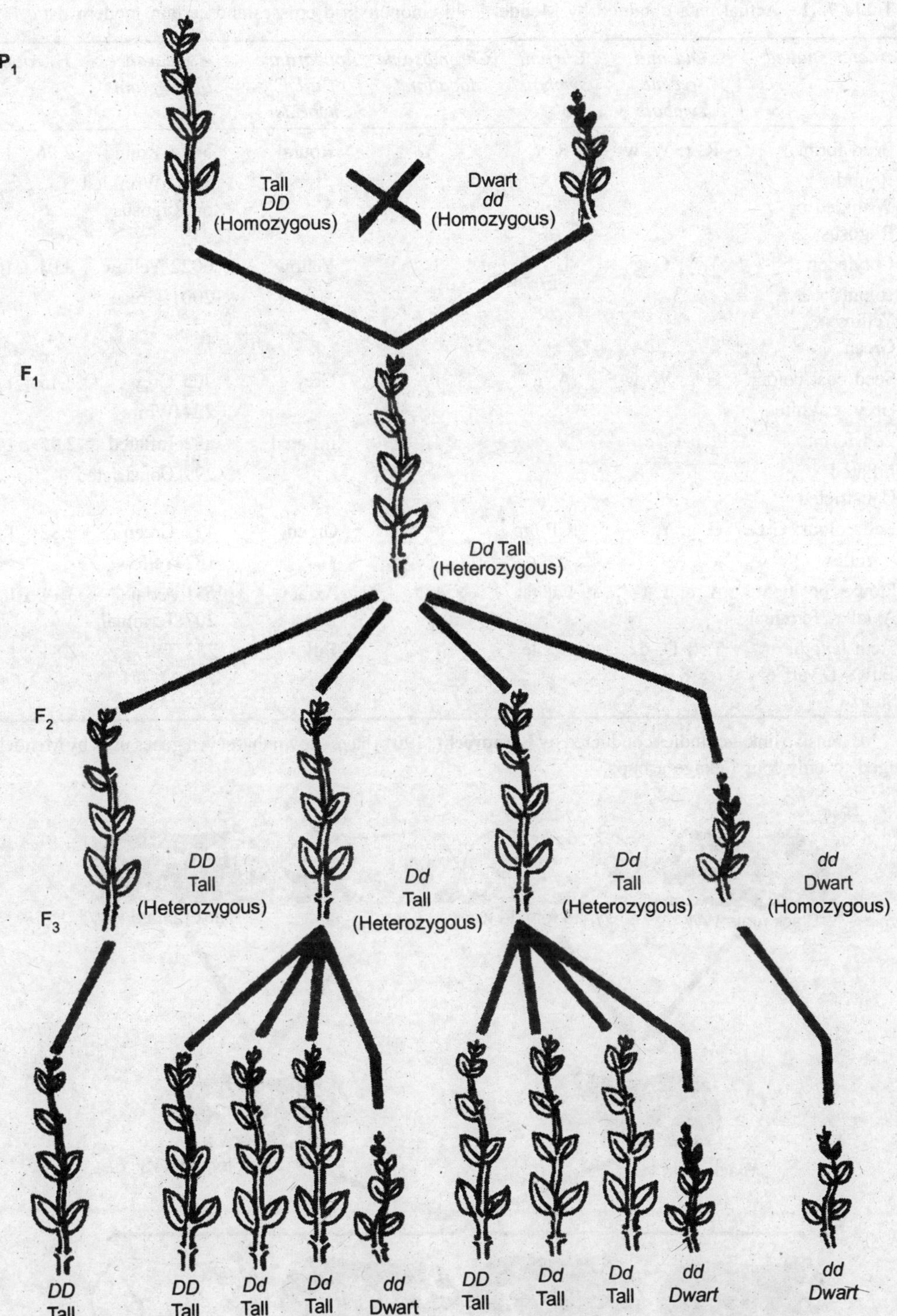

Fig. 75.1. A cross between a tall (*DD*) and a dwarf (*dd*) pea plant and their offsprings of F_1 and F_2 generations.

Table 75.1 Actual data obtained by Mendel in his monohybrid crosses and certain modern data.

Character studied	*Old and recent symbols*	*Current symbols*	*Chromosome location**	*Appearance of all F_1 hybrids*	*Appearance of F_2 plants*	*F_2 ratio*
1. Seed form: Round × Wrinkled or Rugosus	R, r; W, w	R, r	7	Round	5474 Round 1850 Wrinkled or Rugosus	2.96 : 1
2. Cotyledon colour: Yellow × Green	Y, y; G, g	I, i	1	Yellow	6022 Yellow 2001 Green	3.01 : 1
3. Seed coat colour: Grey × White	G, g; W, w	A, a	1	Grey	705 Grey 224 White	3.15 : 1
4. Pod form: Inflated × Constricted	I, i; C, c	V, v	4	Inflated	882 Inflated 299 Constricted	2.95 : 1
5. Pod colour: Green × Yellow	G, g; Y, y	GP, gp	5	Green	428 Green 152 Yellow	2.82 : 1
6. Flower position: Axial × Terminal	A, a; T, t	Fa, fa	4	Axial	651 Axial 207 Terminal	3.14 : 1
7. Stem length: Tall × Dwarf or Short	T, t; D, d	Le, le	4	Tall	787 Tall 277 Dwarf	2.84 : 1

* Extensive linkage studies conducted by **Lamprecht** (1961) have shown that seven genes used by Mendel belonged to only four linkage groups.

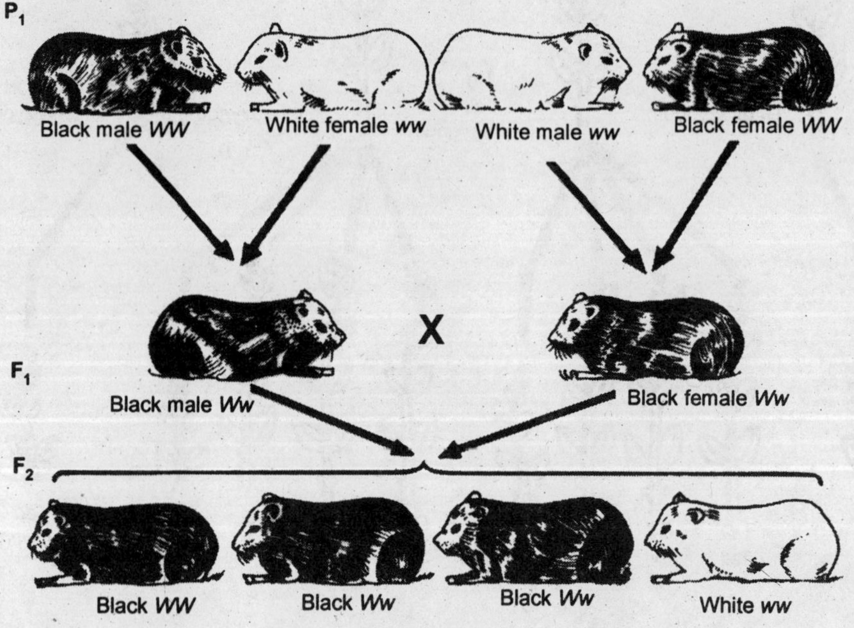

Fig. 75.2. A cross between black and white guinea pigs.

Table 75.2 The dominant and recessive characters in plants.

Name of the plant	*Dominant*	*Recessive*
1. Nettle	Serrated leaves	Smooth margined leaves
2. Sunflower	Branched habit	Unbranched habit
3. Cotton	Coloured lint	White lint
4. Maize	Round starchy kernel	Wrinkled sugary kernel
5. Snapdragon	Red flower	Non-red flower
6. Barley	Beardlessness	Beardness
7. Wheat	Susceptibility to rust	Immunity to rust
8. Tomato	Two celled fruit	Many celled fruit

B. Certain other examples of phenomenon of dominance in animals. The dominant and recessive traits or characters of some animals can be tabulated in following table:

Table 75.3 The dominant and recessive characters in animals.

Name of animal	*Body character*	*Dominant*	*Recessive*
1. Cat	Skin colour	Tabby or Stripped	Black or blue
	Length of hair	Short hairs	Long hairs (Angora)
2. Dog	Skin colour	Grey	Black
	Tail	Stumpy	Normal tail
3. Cattle	Colour of face	White	Coloured
	Horn	Polled or hornless	Horned
4. Horse	Skin colour	Black	Red
	Movement	Trotting	Pacing
5. Sheep	Hair or wool or fleece	White	Black
6. Swine or Pig	Skin colour	Black	Red
	Hoof	Uncleft	Normal
7. Salamander	Body colour	Dark	Light
8. Fruit fly	Eye colour	Red	White
(*Drosophila*)	Wings	Flat and yellow	Curled and white
	Body colour	Grey	Black
9. Land snail	Shape of shell	Unbanded shell	Banded shell

C. Dominant and recessive characters in human beings. In human beings the following traits normally behave as the dominant and recessive traits (Table 75.4).

Table 75.4 Some normal dominant and recessive traits in human beings.

Body part	*Character*	*Dominant*	*Recessive*
1. Hair	Form	Curly	Straight
	Colour	Dark	Light
2. Skin	Colour	Dark	Light
	Pigment	Normal (with melanin)	None (albinism, without melanin pigment)
3. Eyes or Iris	Colour	Brown	Blue

Mechanism of dominance. Mendel carefully studied the results of his experiments and it became evident to him that there is a clear cut difference in between the actual visible character and that something which caused its production. Because the character cannot be present as such in sex cells (sperm and ova) which form the only link between a new individual and the parents.

Therefore, there must be something which represents the characters and is responsible for their production. This something of Mendel is now called the **factor** or **gene**. The gene, thus, can be considered as the unit of inheritance that is transmitted in a gamete and determines or controls the development of a character by interaction with the other genes, the cytoplasm and the enviornment.

The cytological investigations have now established that the genes are the units of deoxyribonucleic acid (DNA) which along with ribonucleic acid (RNA) and nucleoproteins constitute the thread like stainable structures the **chromosomes.** The chromosomes are specific in number, shape and size to a particular species.

A diploid cell has two sets of chromosomes which come from two different parents (male and female) via gametes (sperm and ova). The chromosomes of similar size and nature often form pairs during meiotic cell division and such identical chromosomes are known as **homologous chromosomes**. Each character of a pair of contrasting characters is represented by an **allele**. (When a gene for a unit character contains two or more alternative forms, they are called **allelomorphs** or **alleles**. All alleles of a gene are produced due to mutation of a wild gene (or normal gene). Thus, homozygous tall pea plant has two identical alleles *TT* on both gene loci of the homologous chromosomes; likewise, homozygous dwarf pea plant contains *tt* alleles.

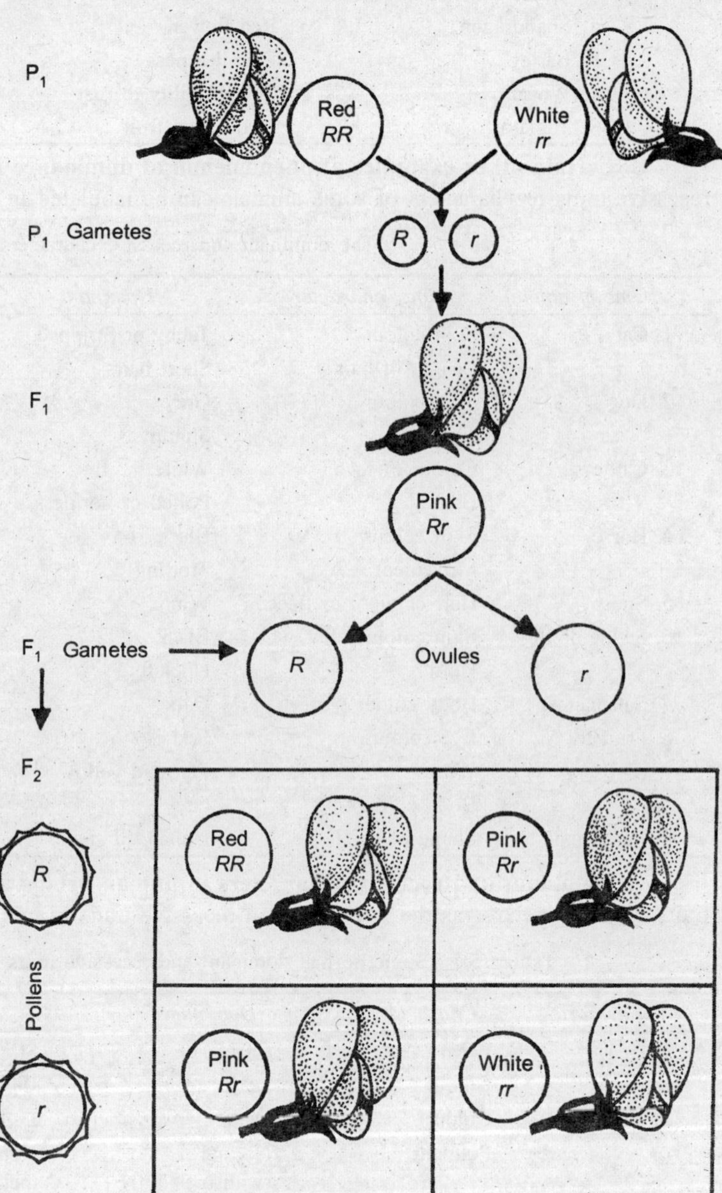

Fig. 75.3. A cross between a red flowered and a white flowered pea plant showing incomplete dominance.

During the gametogenesis, the homologous chromosomes with *TT* or *tt* genes are separated and each chromosome with *T* or *t* gene is passed to the gamete. The gametes of both parents unite during the process of fertilization and produce a new individual containing both tall (*T*) and dwarf (*t*) characters. This new individual of first generation (F_1) contains two different genes (*i.e.,* alleles *T* and *t*) of a contrasting pair of characters, therefore, it is known as **heterozygote** or **hybrid**. Because the hybrids of F_1 have tall stems so the character of tallness (*T*) is considered as **dominant** and because the character for dwarfness could not express itself in F_1 generation, therefore, it is considered as **recessive**.

Variation in Dominance Relations

Mendel reported full dominance and recessiveness for all the seven gene pairs (or allele pairs) he studied (see Table 75.1). He might have been selective in his choice of pea characters to study, because variations on the basic theme appear quite often in analysis. For instance, in some cases the phenotypes of heterozygotes are found to be different from either of the homozygotes (see Box 75.1).

Box 75.1 Types of Dominance

1. **Complete dominance.** The dominant allele completely masks the effect of the recessive allele. The phenotype of the heterozygote is identical to that of one of the homozygote. The monohybrid ratio is 3: 1. This is classical dominance, effect described by Mendel, and often occurs where the recessive allele is null.
2. **No dominance and partial dominance (Incomplete dominance).** Neither allele is fully dominant over the other. The phenotype of the heterozygote is somewhere in between those of the homozygotes and the monohybrid ratio is 1 : 2 : 1.
   If the heterozygote phenotype is exactly intermediate between the two homozygotes, **there is no dominance.** If the phenotype is closer to one homozygote than the other, there is **partial dominance.**
3. **Over dominance and underdominance.** The phenotype of the heterozygote lies outside the range characterized by those of the homozygote. If the heterozygous phenotype is greater than either homozygous phenotype, the locus shows **overdominance**; if less, it shows **underdominance**. The monohybrid ratio is 1 : 2 : 1.
4. **Codominance.** The phenotype associated with each allele is expressed independently of that of the other. Codominance occurs when there is no competition between alleles. The monohybrid ratio is 1 : 2 : 1.
5. **Pseudodominance.** An allele appears dominant because the locus is hemizygous. This is applicable to sex-linked loci in the heterogametic sex, *e.g.*, in male mammals (sex-linkage) and to individuals with chromosome deletion or chromosome loss.

75.2 INCOMPLETE DOMINANCE

Sometimes in a heterozygote the dominant allele does not completely mask the phenotypic expression of the recessive allele and there occurs an intermediate phenotype in the heterozygote. This is called **incomplete dominance**.

Examples. 1. When a red flowered pea plant (*RR*) is crossed with white flowered pea plant (*rr*) tnen the F_1 hybrid pea plants are found to have pink flowers (*Rr*). It shows that allele for red colour could not completely dominate the allele for white colour as shown in the Fig. 75.3. In such a case, F_2 phenotypic ratio and genotypic ratio are the same, (Fig. 74.2) as follows:

F_2 phenotypic ratio = 1 Red: 2 Pink : 1 White

F_2 genotypic ratio = 1 *RR*: 2 *Rr* : 1 *rr*

2. In four-o'clock plants (*Mirabilis jalapa*) or snapdragons (*Antirrhinum majus*), when a pure line or homozygous with red petals (C_1C_1) is crossed to a pure line with white petals (C_2C_2), the F_1 progeny has no red petals but pink petals (C_1 C_2). It shows incomplete dominance. If a F_2 is produced, its progeny exhibits the following results:

¼ Red petals –1C_1C_1 ; ½ Pink petals – 2C_1C_2; ¼ White petals –1C_2 C_2

3. When a homozygous Andalusian fowl with black feathers is crossed with a homozygous fowl with splashed white feathers the F_1 hybrids are found to contain blue feathers (Fig. 75.4).

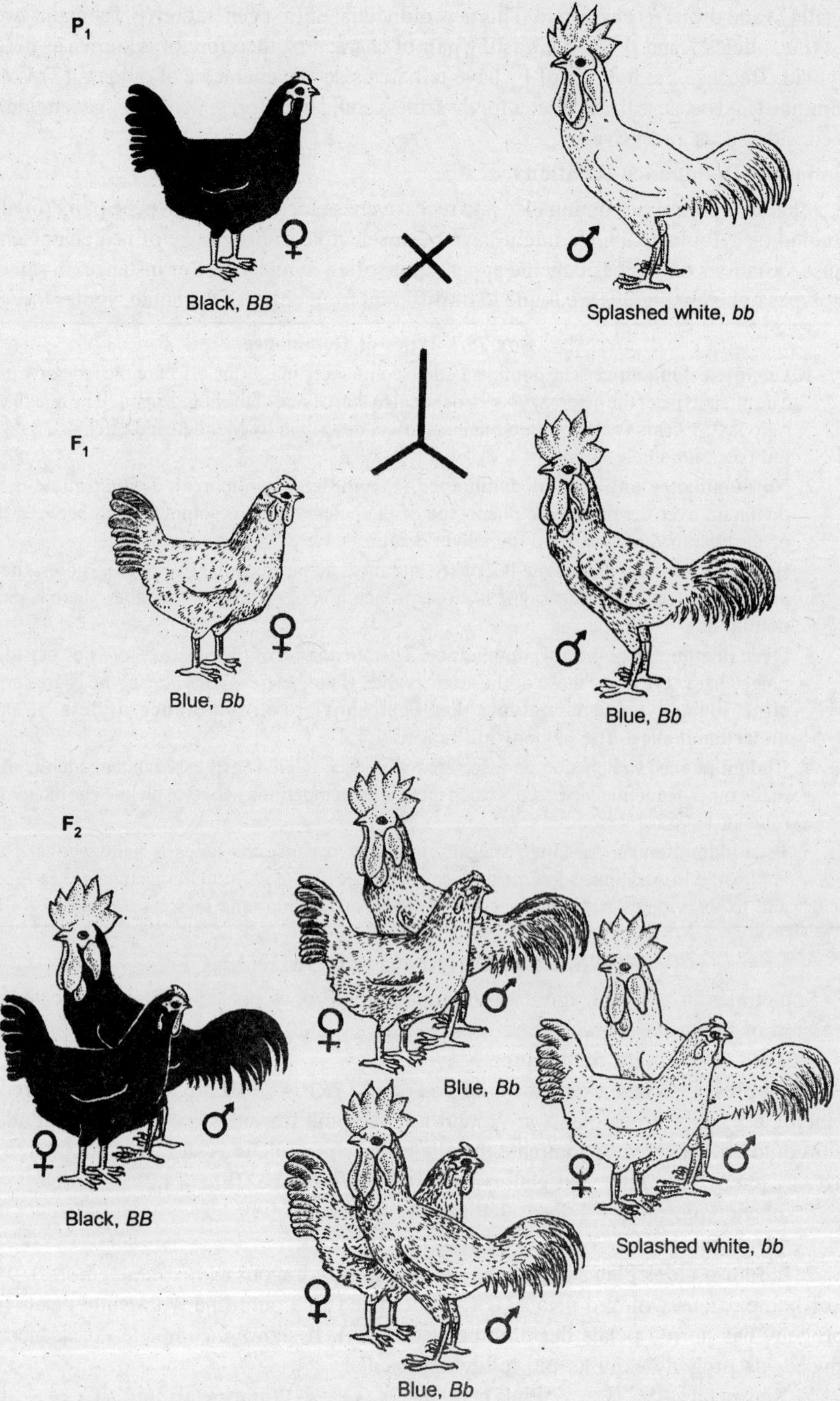

Fig. 75.4. A monohybrid cross between black and splashed white Andalusian fowl showing the incomplete dominance.

75.3 CODOMINANCE

Sometimes both alleles of a gene in a heterozygote have an equal phenotypic expression. In a sense, codominance is no dominance at all, the heterozygote showing the phenotypes of both homozygotes. Hence, heterozygote genotype gives rise to a phenotype distinctly different from either of the homozygous genotypes.

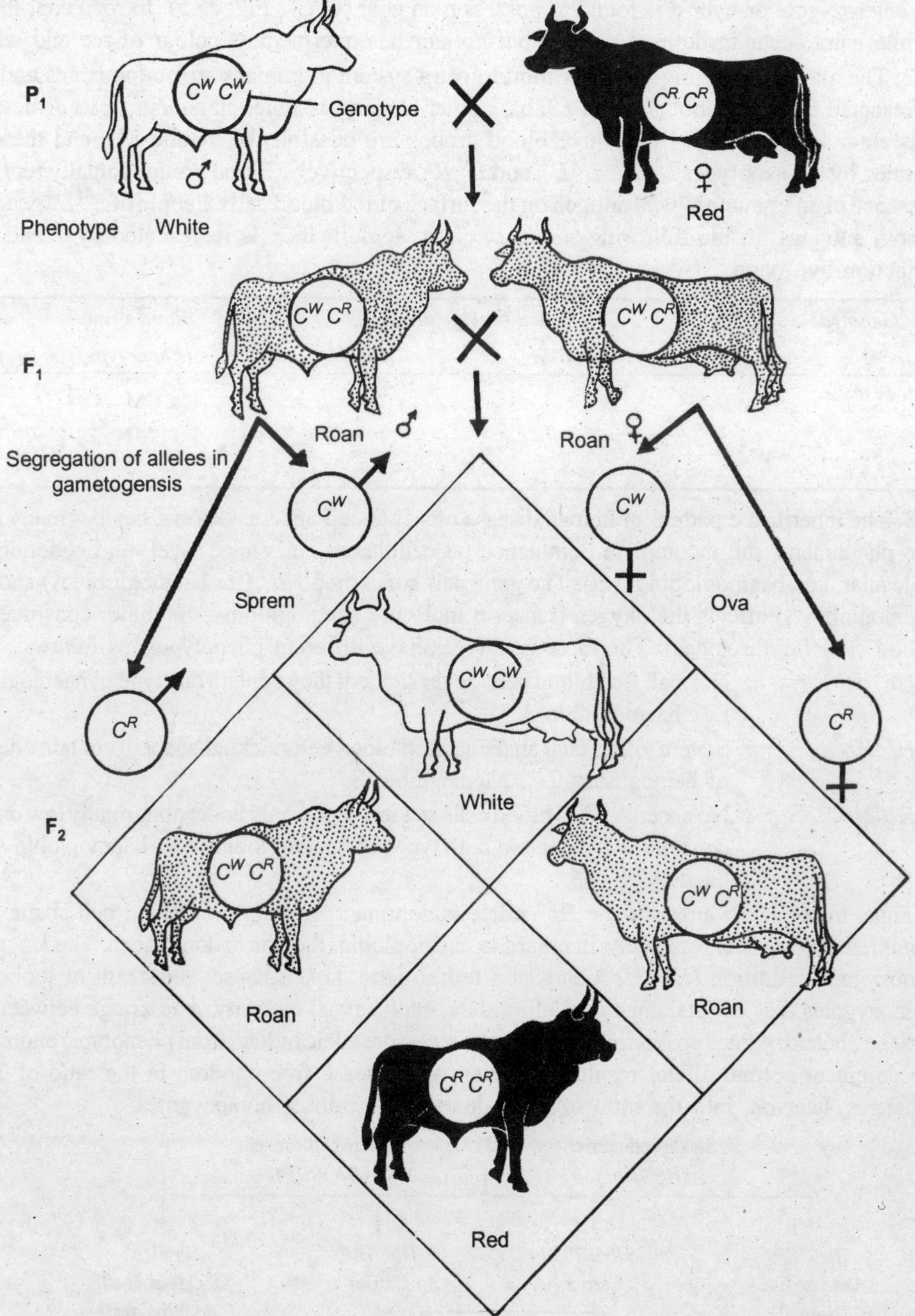

Fig. 75.5. A monohybrid cross in a white-coated and a red-coated cattle showing codominance.

Symbolism for codominant alleles. For codominant alleles, all upper case base symbols with different superscripts are used. The upper case letters indicate that each allele can express itself to some degree even when in the presence of its alternative allele (heterozygous).

Example 1. The coat colour of the shorthorn breed of cattle represent a classical example of codominance. When a cattle of red coat ($C^R C^R$) is crossed with the cattle of white coat ($C^W C^W$), the F_1 heterozygote or hybrid is found to possess roan coat ($C^R C^w$; Fig. 75.5). In roan coat the red and white hairs occur in definite patches but no hair has intermediate colour of red and white.

2. The alleles governing the **M-N blood group** system in humans are codominants and may be represented by the symbols L^M and L^N, base letter L being assigned in honour of its discoverers (**Landsteiner** and **Levine**). Here, three blood groups are possible–M, N and MN-and these are determined by the genotypes $L^M L^M$, $L^N L^N$, and $L^M L^N$, respectively. Blood groups actually represent the presence of an immunological antigen on the surface of red blood cells. People of $L^M L^N$ genotype have both antigens. In the following summary chart, agglutination is represented by + and non-agglutination by – sign:

Genotype	*Reaction with antisera:*		*Blood group*
	Anti-M	*Anti-N*	*(Phenotype)*
1. $L^M L^M$	+	–	M
2. $L^M L^N$	+	+	MN
3. $L^N L^N$	–	+	N

3. The inheritance pattern of human disease of sickle-cell anaemia shows, besides many other genetic phenomena, the incomplete dominance (at cellular or cell shape level) and codominance (at molecular, *i.e.*, haemoglobin level). The gene pair concerned Hb^A (for haemoglobin A) and Hb^S for haemoglobin S) affects the oxygen transport molecule haemoglobin—the major constituent of red blood cells (erythrocytes). The three genotypes have different phenotypes, as follows:

$Hb^A Hb^A$: Normal. Red blood cells never sickled; they contain one type of haemoglobin, i.e., haemoglobin A.

$Hb^S Hb^S$: Severe, often fatal anaemia. Red blood cells sickled-shaped; contain one type of hemoglobin, *i.e.*, haemoglobin S.

$Hb^A Hb^S$: No anaemia. Red blood cells sickle-shaped only under abnormally low oxygen concentration; contain both types of haemoglobins, *i.e.*, haemoglobin A and haemoglobin S.

Thus, in regard to anaemia the Hb^A allele is dominant. In regard to blood cell shape there is incomplete dominance. And lastly, in regard to haemoglobin, there is codominance. The Hb^S allele in homozygous condition ($Hb^S Hb^S$) acts as a **lethal gene**, *i.e.*, it causes the death of its bearer; the homozygotes dies of fatal anaemia before they attain sexual maturity. A marriage between two carriers (*i.e.*, heterozygotes possessing a deleterious recessive allele hidden from phenotypic expression by the dominant normal allele) results in carriers and disease free children in the ratio of 3 : 1, that changes, later on, into the ratio of 2 : 1 due to the death of homozygotes:

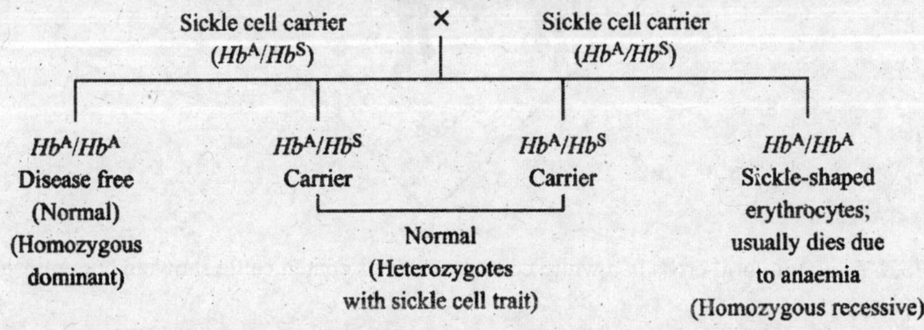

75.4 DEVIATION FROM MENDEL'S MONOHYBRID AND DIHYBRID PHENOTYPIC RATIOS

1. Deviation from Mendel's Monohybrid (3 : 1) Ratio

Mendel's monohybrid ratio of 3 : 1 becomes modified in case of incomplete dominance (*i.e.*, 1 : 2 : 1) and codominance (1 : 2 : 1) (see Sections 75.2 and 75.3).

2. Deviation from Mendelian Dihybrid (9 : 3 : 3 : 1) Ratio.

The Mendelian dihybrid phenotypic ratio of 9 : 3 : 3 : 1 is obtained only when the alleles at both gene loci display dominant and recessive relationship. If one or both gene loci have incompletely dominant alleles, or codominant alleles or lethal alleles, the dihybrid ratio becomes modified variously, such as follows:

1. 3 : 6 : 3 : 1 : 2 : 1 Ratio

When the dihybrid parent have dominant and recessive alleles at one gene locus and codominant alleles at second gene locus, the F_2 9 : 3 : 3 : 1 phenotypic ratio becomes 3 : 6 : 3 : 1 : 2 : 1.

Example. In cattles, hornless or polled (*P*) condition is dominant to horned (*p*) condition, and trait of white (*W*) coat is codominant to the trait for red (*w*) coat colour. The mating of homozygous white polled (*PP WW*) cattle with a homozygous red horned (*pp ww*) cattle produces F_1 heterozygotes (dihybrids) with the phenotype of hornless, roan and genotype of *Pp Ww*. These hornless roan F_1 dihybrids produce a F_2 progeny in the ratio of 3 : 6 : 3 : 1 : 2 : 1 as illustrated in following diagram:

P_1 : Phenotype	Hornless White	×	Horned Red
Genotype	*PP WW*		*pp ww*
	↓		↓
P_1 : Gametes:	*PW*		*pw*
F_1:		*PpWw* Hornless Roan	

F_2: 3 Hornless White: 6 Hornless Roan: 3 Hornless Red:
1 Horned White: 2 Horned Roan: 1 Horned Red (or 3 : 6 : 3 : 1 : 2 : 1).

2. 1 : 2 : 1 : 2 : 4 : 2 : 1 : 2 : 1 Ratio

When each parent of a dihybrid cross has incompletely dominant alleles at both gene loci, then in F_2 large number of phenotypic classes are produced.

Example. In snapdragons, red flower colour (*R*) is incompletely dominant to white flower colour (*r*) and trait for broadness of leaf (*B*) is incompletely dominant to the trait for narrowness of leaf (*b*). The dihybrid cross between red, broad plant (*RR BB*) and white, narrow plant (*rr bb*) produces heterozygotes (dihybrids) having pink flowers and leaves of intermediate width and genotype of *Rr Bb*. These F_1 dihybrids produce the F_2 progeny in 9 phenotypic classes as follows:

P_1 : Phenotype	Red Broad	×	White Narrow
Genotype	*RR BB*		*rr bb*
	↓		↓
P_1 : Gametes:	*RB*		*rb*
F_1 : Phenotype		Pink Intermediate *Rr Bb*	

F_2 : 1 Red Broad; 2. Red Intermediate: 1 Red Narrow: 2 Pink Broad: 4 Pink Intermediate: 2 Pink Narrow: 1 White Broad: 2 White Intermediate: 1 White Narrow (or 1 : 2 : 1 : 2 : 4 : 2 : 1 : 2 : 1).

3. 3 : 1 : 6 : 2 Ratio

When the F_1 dihybrids have dominant-recessive alleles at one gene locus and recessive lethal alleles at second gene locus, the F_2 offsprings manifest the phenotypic ratio of 3 : 1 : 6 : 2.

4. 1 : 2 : 1 : 3 : 4 : 2 Ratio

When the F_1 dihybrids contain codominant alleles at first gene locus and recessive-lethal alleles at second gene locus, then their F_2 progeny display the phenotypic ratio of 1 : 2 : 1 : 3 : 4 : 2.

5. 4 : 2 : 2 : 1 Ratio

This phenotypic ratio is obtained when at both gene loci of F_1 dihybrids occur the recessive lethal alleles.

QUESTIONS

Long Answer Questions

1. Explain giving example, what do you understand by complete and incomplete dominance. (*C.C.S. 1989; Avadh 1990*)
2. The classical Mendelian ratios, such as 3 : 1, 9 : 3 : 3 : 1, etc., do not by any means occur in all crosses. Explain. (*NEHU 1996*)
3. In 4 'O' clock plants red flower colour *R* is incompletely dominant over the white *r*. The heterozygous plants are with pink flowers. If a red coloured 4 'O' clock plant is crossed with a white one, what will be flower colour of F_1, of the F_2, of the offspring of the cross F_1 with its red parent and its white parent. (*Bangalore 1992*)

Short Answer Questions

1. Write short answer/notes on the following:
   (*i*) Dominance (*Bundelkhand 1988; Garhwal 1989*)
   (*ii*) Incomplete dominance (*Agra 1992: Bangalore 1993; Lucknow 1998*)
   (*iii*) Codominance (*Delhi 1989, 1990; BHU 1990; Lucknow 1996; Avadh 1999*)
   (*iv*) Back cross (*Guwahati (G)* 1991, 1996)
2. Differentiate between the following:
   (*i*) Complete and incomplete dominance (*Delhi 1989; Indore 1989; Madras 1993*)
   (*ii*) Codominance and incomplete dominance (*Lucknow 1994*)
   (*iii*) Dominance and codominance (*Lucknow 1995*)

Very Short Answer Questions

1. What is dominance?
   Ans. Condition in which one member of an allele pair is manifested to the exclusion of the other.
2. Define incomplete dominance?
   Ans. Expression of two alleles in a heterozygote that allows the heterozygote to be distinguished from either of its homozygous parents.
3. What is codominance?
   Ans. The condition in heterozygote where both members of an allelic pair contribute to phenotype, which is then a mixture of the phenotypic trait produced in either homozygous conditions.
4. What are codominant alleles?
   Ans. Alleles that produce independent effects when heterozygous.

Fill in the Blanks

1. A cross between homozygous recessive and heterozygous organism is called. (*H.N.B., Garhwal, 2008*)

True and False Type Questions

(*i*) *Mirabilis jalpa* hybrid or pink colour is an example of codominance. (*H.N.B., Garhwal 2008*)

Multiple Choice Questions

1. Recessive gene can express in

(*a*) homozygous condition (*b*) heterozygous condition
(*c*) both the above conditions (*c*) none of these conditions

2. A gene is said to be dominant if

(*a*) it expresses its effect both in homozygous condition
(*b*) it expresses its effect only in heterozygous condition
(*c*) it expresses its effect both in homozygous and heterozygous condition
(*d*) it never expresses its effect in any condition

3. Pink flower in the progeny of parents of *Antirrhinum majus*, having red and white flowers, are produced because of

(*a*) dominance (*b*) codominance
(*c*) epistasis (*d*) incomplete dominance

4. Alleles that produce independent effects in heterozygous condition are called

(*a*) codominant alleles (*b*) epistatic alleles
(*c*) complementary alleles (*d*) supplementary alleles

5. Mendel did not propose

(*a*) segregation (*b*) independent assortment
(*c*) dominance (*d*) incomplete dominance

6. A blue Andalusion fowl is the product of mating between black fowl and white fowl. A cross of blue fowl with another blue fowl results in

(*a*) 1 black : 2 white : 1 blue (*b*) 1 black : 2 blue : 1 white
(*c*) 2 black : 1 blue : 1 white (*d*) none of these

ANSWERS

Fill in the Blanks

1. Test cross.

True and False

(*i*) False.

Multiple Choice Questions

1. (*a*) **2.** (*c*) **3.** (*d*) **4.** (*a*)
5. (*d*) **6.** (*b*).

76

Linkage

The hereditary units or genes which determine the characters of an individual are carried in the chromosomes and an individual usually has many genes for the determination of various different characters. As there are more genes than the chromosomes, it can be expected that each chromosome contains more than one gene. The genes for different characters may be either situated in the same chromosome or in different chromosomes. When the genes are situated in different chromosomes, the characters they control appear in the next generation either together or apart, depending on the chance alone. They assort independently according to Mendel's law of independent assortment. But if the genes are situated in the same chromosome and are fairly close to each other, they tend to be inherited together. This type of coexistence of two or more genes in the same chromosome is known as **linkage**.

76.1 DIFFERENCE BETWEEN LINKAGE AND INDEPENDENT ASSORTMENT

The difference between independent assortment and linkage can be understood by the following two examples:

Example 1. Genes on different chromosomes assort independently giving a 1 : 1 : 1 : 1 test cross ratio which is as follows:

P_1:	*AA BB*	×	*aa bb*
P_1 gametes:	(*AB*)	↓	(*ab*)
F_1:		*Aa Bb*	
Test cross:	F_1 *Aa/Bb*	×	P_1 *aa/bb*
Gametes:	(*AB*) (*Ab*) (*aB*) (*ab*)		(*ab*)
F_2:	¼ *Aa Bb*: ¼ *Aa bb*: ¼ *aa Bb*: ¼ *aa bb* or 1 : 1 : 1 : 1.		

Example 2. Linked genes do not assort independently but tend to stay together in the same combination as they were in the parents. In the following figure, the genes to the left of the slash line (/) are on one chromosome and those on the right are on the homologous chromosome.

P_1:	*AB/AB*	×	*ab/ab*
P_1 gametes:	(*AB*)	↓	(*ab*)
F_1:		*AB/ab*	
Test cross:	F_1 *AB/ab*	×	P_1 *ab/ab*
Gametes:	(*AB*) (*ab*)		(*ab*)
F_2:	½ *AB/ab*	:	½ *ab/ab* or 1 : 1.

This type of large deviation of F_2 results of example 2 from a 1 : 1 : 1 : 1 ratio in the test cross progeny of dihybrid is an evidence for linkage.

Further, linked genes are separated from each other by the process of crossing over. In that case, linkage is termed as **incomplete linkage** and causes genetical variations due to recombination of genes.

76.2 THEORIES OF LINKAGE

Linkage can be defined as close physical association between two loci on a chromosome, preventing independent assortment (see Box 76.1).

Box 76.1 Linkage

The term **linkage** has three distinct meanings in genetics.

1. Firstly, linkage (or **linking**) describes the coiling of double-stranded DNA and is used to quantify topological parameters of duplex DNA.

2. Secondly, linkage indicates when a particular gene (and by extension, its associated phenotype) is found on a certain chromosome, *e.g.*, genes and the characters they control associated with the X-chromosome are described as **X-linked** (*i.e.*, **sex-linked inheritance**).

3. Finally, **genetic linkage** refers to genes which tend to be inherited as a unit rather than assorting independently. This reflects close **physical linkage**, *i.e.*, genes are physically joined together on the same chromosome.

Development of Concept of Linkage

The hypothesis that linked genes tend to remain in their original combinations because of their location in the same chromosome was advanced by **T.H. Morgan** in 1911. But prior to Morgan, **W. Sutton** and **T. Boveri** (1902), **Sutton** (1903) and **Bateson** and **Punnet** (1906) had given some hints about the phenomenon of linkage. It is strange that **Mendel** could not detect the phenomenon of linkage in pea plant, since extensive linkage studies of **Lamprecht** (1961) have demonstrated that seven genes used by Mendel belonged to only four linkage groups (chromosomes 1,4,5 and 7). The views of various geneticists about the phenomenon of linkage can be represented as follows:

1. Sutton-Boveri chromosome theory of heredity. Sutton and **Boveri** (1902) concluded that genes or Mendelian factors were located on chromosomes. This generalization is now known as **chromosome theory of inheritance (or of heredity)**. This theory provides a physical basis of heredity.

2. Sutton's view on linkage. Sutton (1903) predicted that the chromosomes are the bearers of the units of heredity or genes and since the number of hereditary units is much larger than the number of chromosomes, therefore, each chromosome must contain a number of genes. He further stated that since the chromosomes move as units during meiosis to the gametes, all the genes which are situated in the same chromosome will be linked together. As a result each species of animals and plants would have a specific number of groups of linked genes which would correspond with the number of chromosomes found in that species. Unfortunately **Sutton** could not prove his predictions experimentally.

3. Bateson and Punnet's coupling and repulsion hypothesis. Bateson and **Punnet** (1905–1908) formulated the 'hypothesis of coupling and repulsion' to explain the unexpected F_2 results of a dihybrid cross between a homozygous sweet pea (*Lathyrus odoratus*) having dominant alleles for blue or purple flowers (*RR*) and long pollen grains (*Ro Ro*) with another homozygous double recessive plant (*rr*, *roro*) with red flowers and round pollen grains. When they test crossed a heterozygous blue or purple long (*Rr*, *Ro ro*) plant with recessive parent (*rr*, *ro ro*), besides getting the 1 : 1 : 1 : 1 test cross ratio, they received phenotypic ratio of 7 : 1 : 1 : 7 as has been illustrated in following figure:

Parent:	Blue or purple long (*RR RoRo*)	×	Red Round (*rr roro*)
F_1:	All Blue or purple long (*Rr Roro*)		
Test cross:	F_1 Blue or purple long (*Rr Roro*)	×	Red Round (*rr roro*)

Test cross progeny:	Blue or purple long	=	192
	Red round	=	182
	Blue or purple round	=	23
	Red long	=	30
			427

Test cross ratio: 7 Blue or purple Long : 1 Blue or purple Round : 1 Red Long : 7 Red Round or 7 : 1 : 1 : 7.

The 7 : 1 : 1 : 7 test cross ratio clearly indicated that there was a tendency in the dominant alleles (*R Ro*) to pass together to the same gamete. Similar was the case with recessive alleles (*r ro*). This tendency of dominant or recessive alleles to inherit together was explained as "**gametic coupling**" by **Bateson** and **Punnett**.

Further, when they crossed blue or purple round (*RR ro ro*) with red long (*rr Ro Ro*), the F_1 hybrids were found to be heterozygous blue or purple long (*Rr Roro*). The F_1 hybrid when test crossed with recessive (*rr roro*) parent, the test cross ratio was 1 blue or purple long : 7 red long : 7 blue or purple round : 1 red round, as has been illustrated in following figure:

Parent:	Blue or purple Round (*RR roro*)	×	Red Long (*rr RoRo*)
F_1:		All Blue or purple Long (*Rr Roro*)	
Test cross:	F_1 Blue or purple Long (*Rr Roro*)	×	P_1 Red Round (*rr roro*)

Test cross progeny: 1/16 Blue or purple Long : 7/16 Blue or purple Round : 7/16 Red Long : 1/16 Red Round or 1 : 7 : 7 : 1.

Hence, the two dominant pairs of alleles repelled each other. The tendency of both dominant or both recessive alleles to repel each other, so that the gametes of genotypes of *R ro* and *r Ro* are formed more frequently, was termed **repulsion**.

Bateson and Punnett could not explain the exact reasons of coupling and repulsion, and it was **T.H. Morgan** who while performing experiments with *Drosophila*, in 1910, found that coupling or repulsion was not complete. He further suggested that the two genes are found in coupling phase or in repulsion phase, because they are present on the same chromosome (coupling) or on two different homologous chromosomes (repulsion). Such genes are then called **linked genes** and the phenomenon of inheritance of linked genes is called **linkage** by Morgan. The terms **cis** and **trans** were employed later (**Haldane**, 1942), the former replacing coupling and the latter, repulsion.

4. Morgan's views on linkage. Morgan stated that the pairs of genes of homozygous parent tend to enter in the same gametes and to remain together, whereas same genes from heterozygous parent tend to enter in the different gametes and remain apart from each other. He further stated that the tendency of linked genes remaining together in original combination is due to their location in the same chromosome. According to him the degree or strength of linkage depends upon the distance between the linked genes in the chromosome. Morgan's concept about the linkage developed the theory of linear arrangement of genes in the chromosomes which helped the cytogeneticists in the construction of genetic or linkage maps of chromosomes.

5. Chromosome theory of linkage. Morgan along with **Castle** formulated the chromosome theory of linkage which is as follows:

1. The genes which show the phenomenon of linkage are situated in the same chromosomes and these linked genes usually remain bounded by the chromosomal material so that they cannot be separated during the process of inheritance.

2. The distance between the linked genes determines the strength of linkage. The closely located genes show strong linkage than the widely located genes which show the weak linkage.
3. The genes are arranged in linear fashion in the chromosomes.

Why did Mendel failed to detect linkage?

Because genes for some of the traits Mendel followed, are on different chromosome pairs; linked traits with which he worked (seed colour/flower colour; pod colour/plant height, among others) are so far apart on their chromosomes that genes for these traits appear to segregate randomly. In one other case Mendel did not make the cross that would have revealed linkage.

76.3 KINDS OF LINKAGE

T.H. Morgan and his co-workers by their investigation on the *Drosophila* and other organisms have found two types of linkage, *viz.*, complete linkage and incomplete linkage.

1. Complete linkage. The complete linkage is the phenomenon in which parental combinations of characters appear together for two or more generations in a continuous and regular fashion. In this type of linkage the genes are closely associated and tend to transmit together.

Example. The genes for bent wings (*bt*) and shaven bristles (*svn*) of the fourth chromosome mutant of *Drosophila melanogaster* exhibit complete linkage.

Complete linkage in male *Drosophila*. In most of the organisms crossing over takes place both in males and females. But in male *Drosophila* and female silkworm, *Bombyx mori* (see **Swanson,** 1957) crossing-over takes place either very rarely or not at all. This becomes clear from Morgan's experimental results from *Drosophila*. In 1919, **T.H. Morgan** mated gray bodied and vestigial winged (b^+ *vg*/b^+ *vg*) fruit flies with flies having black bodies and normal wings (*bvg*$^+$/*bvg*$^+$). F_1 progeny had gray bodies and normal long wings (b^+*vg*/*bvg*$^+$), indicating thereby that these characters are dominant. When F_1 males (b^+ *vg*/*bvg*$^+$), were backcrossed (*i.e.*, test crossed) to double recessive females (*bvg*/*bvg* or black vestigial), only two types of progeny (one with gray bodies and vestigial wings, *b+vg*/*bvg* and the other with black bodies and normal wings (*bvg*$^+$/*bvg*) instead of four types of phenotypes were obtained.

Parents:	Gray, Vestigial b^+vg/b^+vg	×	Black, Long bvg^+/bvg^+
Gametes:	(b^+vg)		(bvg^+)
F_1:		All Gray, Long b^+vg/bvg^+	
Test cross:	F_1 male Gray, Long b^+vg/bvg^+	×	Female Black, Vestigial bvg/bvg
Gametes:	(b^+vg) (bvg^+) (only two types of gametes due to complete linkage and lack of crossing over in male *Drosophila*)		(bvg)
Test cross ratio:	½ Gray, Vestigial b^+vg/bvg	:	½ Black, Long or 1: 1. bvg^+/bvg

Here, the use of the testcross is very important. Because one parent (the tester) contributes gametes carrying only recessive alleles, the phenotypes of the offspring represent the gametic contribution of the other double heterozygote parent. So the genetical analyst can concentrate on one meiosis and forget the other. This is in contrast to the situation in an F_1 selfing where there are two sets of meiotic divisions to consider one for the F_1 male parental gametes and one for the F_1 female.

2. Incomplete linkage. The linked genes do not always stay together because homologous non-sister chromatids may exchange segments of varying length with one another during meiotic prophase. This sort of exchange of chromosomal segments in between homologous chromosomes is known as **crossing over**. The linked genes which are widely located in chromosomes and have chances of separation by crossing over are called **incompletely linked genes** and the phenomenon of their inheritance is called **incomplete linkage.**

Example. The incomplete linkage has been reported in female *Drosophila* and various other organisms such as tomato, maize, pea, mice, poultry and human beings, etc. Here, the examples of incomplete linkage have been considered only for *Drosophila* and maize.

Incomplete linkage in female *Drosophila*. When F_1 females of the Morgan's classical cross in *Drosophila* between gray, vestigial (b^+vg/b^+vg) and black, normal or long (bvg^+/bvg^+) were test-crossed to double-recessive (bvg/bvg) males, all four types of progeny were obtained in following ratio, showing occurrence of crossing over:

Parents:	Gray, Vestigial b^+vg/b^+vg	×	Black, Long bvg^+/bvg^+
Gametes:	(b^+vg)		(bvg^+)
F_1:		Gray, Long b^+vg/bvg^+	
Test cross:	F_1 Female Gray, Long b^+vg/bvg^+ ↓	×	Male Black, Vestigial bvg/bvg ↓
Gametes:	(b^+vg) (bvg^+) = Non-crossovers (b^+vg^+) (bvg) = Recombinants		(bvg)

Test cross ratio:	1. Gray, Vestigial; b^+vg/bvg	= 41.5%	83% parental combination showing linkage
	2. Black, Long; bvg^+/bvg	= 41.5%	
	3. Gray, Long; b^+vg^+/bvg	= 8.5%	17% recombinants due to crossing over
	4. Black, Vestigial; bvg/bvg	= 8.5%	

Similar different test cross ratios (showing complete linkage in males and incomplete linkage in females) were obtained for F_1 males and females of *Drosophila* by **Bridges**, one of the student of Morgan. He made a cross between fruit flies having wild dominant alleles for red eye colour and normal wings (pr^+vg^+/pr^+vg^+) and having mutant recessive alleles for purple eye colour and vestigial wings ($pr\ vg/pr\ vg$).

2. Incomplete linkage in maize. In *Zea mays* (maize) a case of incomplete linkage between the alleles for colour and shape of the seed has been observed by **Hutchinson**. When a maize plant with seeds having coloured and full endosperm (*CS/CS*) is crossed with another plant having recessive alleles for colourless, shrunken seeds (*cs/cs*), the F_1 heterozygotes are found with the phenotype of coloured full and genotype of *CS/cs*. When F_1 hybrid is test crossed with double recessive parent (*cs/cs*) four classes of descendants are obtained instead of two as shown in following figure:

Parents:	Coloured Full *CS/CS*	×	Colourless Shrunken *cs/cs*
F_1:		Coloured Full (*CS/cs*)	
Test cross:	F_1 Coloured Full *CS/cs*	×	Colourless Shrunken *cs/cs*

Test cross results:	Coloured : Full	Coloured : Shrunken	Colourless : Full	Colourless : Shrunken
	CS/cs	*Cs/cs*	*cS/cs*	*cs/cs*
	48%	2%	2%	48%

The test cross results are clearly showing that parental combination of alleles (*e.g.*, *CS/cs* and *cs/cs*) are those expected from complete linkage and appear in 96% cases, the other two are new combinations (*e.g.*, *Cs/cs* and *cS/cs*) and appear in 4% cases. Thus, in 4% cases crossing over has occurred between linked genes.

Arrangement of Linked Genes.

In an individual, which is heterozygous for two pairs of linked genes, the linkage can be either of the following two types:

1. Cis-arrangement. The dominant alleles of both linked genes are located in one member of the homologous pair of chromosome and their recessive alleles are located in the other chromosome of the pairs. This arrangement is known as **cis-arrangement** and the heterozygotes with such arrangement (*AB/ab* or *R Ro/r* ro in Fig. 76.1A) are known as **cis-heterozygotes.**

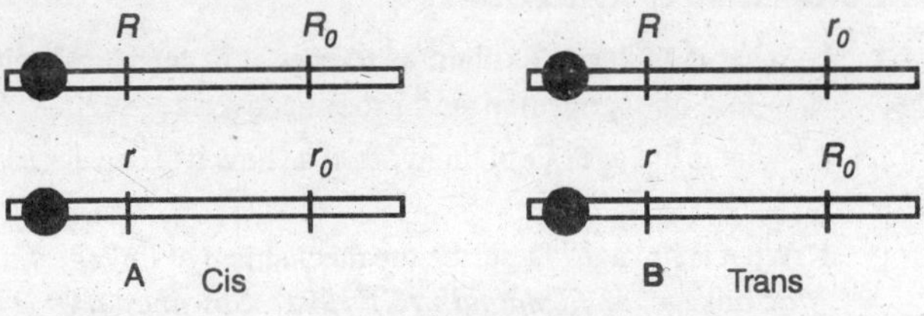

Fig. 76.1. Cis (A) and trans (B) arrangement of two pairs of linked genes of *Lathyrus odoratus*.

2. Trans-arrangement. The dominant allele of one linked gene and recessive allele of the other linked gene are located on one chromosome of the homologous pair and the recessive allele of first linked gene and dominant allele of the second linked gene are located in the second chromosome of the homologous pair (*Ab/aB* or *Rro/rRo* in Fig. 76.1B). Such an arrangement of a dominant and recessive alleles of two linked genes in the same chromosome is called **trans-arrangement** and heterozygotes with such arrangement are called **trans-heterozygotes.**

Significance of Linkage

The phenomenon of linkage has one of the great significance for the living organisms that it reduces the possibility of variability in gametes unless crossing over occurs.

76.4 LINKAGE GROUPS

All the linked genes of a chromosome form a **linkage group**. A linkage group is a collection of genes which demonstrate linkage and therefore map together. In physical terms, linkage group is equivalent to a chromosome. Because, all the genes of a chromosome have their identical genes (alleomorphs) on the homologous chromosome, therefore linkage groups of a homologous pair of chromosome is considered as one. The number of linkage group of a species, thus, corresponds with haploid chromosome number of that species.

Example.

1. *Drosophila* has 4 pairs of chromosomes and 4 linkage groups.
2. Each human being has 23 pairs of chromosomes and 23 linkage groups.
3. Corn (*Zea mays*) has 10 pairs of chromosomes and 10 linkage groups.

However, in organisms where female or male sex has dissimilar sex chromosomes (*e.g.*, human beings, *Drosophila*, fowl, etc.), one more linkage group occurs than the haploid number.

Example.

1. Female human beings: = 22 pairs of autosomes or non-sex chromosomes + 1 pair of homomorphic X chromosomes
= 22 autosomal linkage groups + 1 X chromosomal linkage group
= 23 linkage groups.
2. Male human beings = 22 pairs of autosomes + 2 heteromorphic sex chromosomes, *i.e.*, 1 X chromosome + 1Y chromosome
= 22 autosomal linkage group + 1X chromosomal linkage group + 1Y chromosomal linkage group
= 24 linkage groups.

QUESTIONS

Long Answer Questions

1. What is linkage? Explain as to how it is correlated with Mendel's law of independent assortment or to law of segregation and crossing over. (*CCS 1992; Kanpur 1993*)
2. What is linkage? Explain in detail as how it is correlated with Mendel's laws of independent assortment. (*CCS 1991*)
3. What is linkage? Describe the mechanism of linkage with the help of an example. State the significance of linkage. (*Magadh (G) 1991; Guwahati 1993; 1995; Bangalore 1994; Kanpur 1994, 1998*)
4. What is linkage? Describe in detail the complete and incomplete linkage. (*Burdwan 1989; Kumaun 1989; Osmania 1990; Jiwaji 1990; BHU (Hons.) 1990*)
5. Describe in detail the phenomenon of linkage. (*Allahabad 1990; Magadh (Subs.) 1996*)
6. Define linkage. Give an account of different theories of linkage with a note on its significance. (*CCS 1997*)
7. Describe the process of linkage and give its significance. (*Gorakhpur 1989*)
8. Give an account of linkage in *Drosophila melanogaster*. (*Magadh (Subs.) 1994*)
9. Describe the independent assortment law of Mendel. How it is affected by linkage? (*C.C.S.U. Meerut, 2008*)
10. Define linkage and write about the theories of linkage. (*M.S.U., Tamil Nadu, 2006*)

Short Answer Questions

1. Mendel studied seven pairs of contrasting characters in the garden pea. Why do you suppose, did he not discover the principle of linkage?
2. What is coupling and repulsion?
3. What are the advantages of test cross method for determining linkage relation?
4. Write short notes on the following:
   (*i*) Linkage group
   (*ii*) Linkage (*Magadh (Subs.) 1989; Garhwal 1989; CCS 1990, 1993; Allahabad 1992; Lucknow 1992; Kanpur 1995; Magadh (G) 1995; Avadh 1995, 1997; Gorakhpur 1998*)
   (*iii*) Complete linkage. (*Guwahati (G) 1992*)
5. Give differences between the following:
   (*i*) Linkage and law of independent assortment (*CCS 1995*)
   (*ii*) Complete linkage and incomplete linkage (*Guwahati 1991*)
   (*iii*) Linkage and crossing over
   (*iv*) Cis-arrangement and trans-arrangement.

Very Short Answer Questions

1. *Drosophila melanogaster* has 8 chromosomes and human beings have 46. How many linkage groups are there in each case?
Ans. *D. melanogaster* has 4 linkage groups; each human being has 23 linkage groups.
2. Define linkage.
Ans. Tendency of the genes located on the same chromosome to remain together in the process of inheritance is called linkage.
3. Name the genes that do not follow the Mendel's law of independent assortment.
Ans. Linked genes.
4. Write the name of an organism in which no crossing over is observed.
Ans. Male fruitfly (*Drosophila melanogaster*).

Fill in the Blanks

(*i*) All the genes which occur in a given chromosome are known as ....

(*ii*) If test cross of offsprings of a dihybrid cross show a ratio of 1 : 1 : 1 : 1 it means the genes are ....

(*iii*) Mendel's law of independent assortment fails in case of .... (*Lucknow 1995*)

(*iv*) Chromosome theory of linkage is formulated by ....

Yes or No Questions

(*i*) Muller propounded the chromosome theory of heredity. (*Gawahati (G) 1992*)

(*ii*) The term gene was first used by Morgan. (*Guwahati (G) 1992*)

(*iii*) Linkage reduces the frequency of hybrid. (*H.N.B. Garhwal, 2008*)

Multiple Choice Questions

1. Bateson and Punnet (1909) discovered linkage while working on
(*a*) *Pisum sativum* (*b*) *Lathyrus odoratus*
(*c*) *Drosophila melanogaster* (*d*) *Lychmis alba*
2. The genes controlling the seven pea characters studied by Mendel are now known to be located on how many different chromosomes?
(*a*) seven (*b*) six
(*c*) five (*d*) four
3. When two or more characters do assort independently in F_2 generation, it indicates that characters are not
(*a*) dominant (*b*) recessive
(*c*) linked (*d*) lethal
4. Linked genes may be separated by
(*a*) crossing over (*b*) polyploidy
(*c*) gene mutation (*d*) segregation
5. When a cluster of genes show linkage behaviour they
(*a*) do not show a chromosome map (*b*) show recombination during meiosis
(*c*) do not show independent assortment (*d*) induce cell division

ANSWERS

Fill in the Blanks

(*i*) Linked genes (*ii*) Not linked (*iii*) Linked genes
(*iv*) Morgan and Castle

Yes or No Questions

(*i*) No (*ii*) No (*iii*) Yes

Multiple Choice Questions

1. (*b*) 2. (*d*) 3. (*c*) 4. (*a*) 5. (*c*).

77

Crossing Over

In the chapter of linkage, we have stated that the genes located in the same chromosome show linkage. These linked genes may either remain together during the process of inheritance and, thus, showing complete linkage or they may be segregated or separated during gametogenesis and, thus, displaying the incomplete linkage. The incomplete linkage takes place due to the occurrence of new combinations or recombinations of linked genes. The recombination in its turn is accomplished through a process known as **crossing over** in which the non-sister chromatids of homologous chromosomes, exchange the chromosomal parts or segments. In another words "*the crossing over is a process that produces new combinations (recombinations) of genes by interchanging of corresponding segments between non-sister chromatids of homologous chromosomes*". The chromatins resulting from such interchanges of chromosomal parts are known as **cross overs**. Thus, a cross-over is a site of chromosome breakage and strand exchange; it is called **Holliday Junction** in molecular terms and leads to homologous recombination between the parental chromosomes and the production of recombinant combination of alleles. [**Note** that *crossing over* refers to the exchanges of segments of chromosomes (the process), whereas *recombination* implies the formation of new combination of genes (one possible outcomes of the process, see **Gardner** *et al.*, 2001)].

The term **crossing over** was coined by **Morgan**. No crossing over occurs in male *Drosophila* and female silk worm, *Bombyx mori*. Certain external agents such as heat shock, chemicals, radiations, etc., have profound effect on crossing over.

Characteristics of Crossing Over

1. Crossing over or recombination occurs at two levels (*i*) at gross chromosomal level, called **chromosomal crossing over** and (*ii*) at DNA level, called **genetic recombination**.
2. A reciprocal exchange of material between homologous chromosomes in heterozygotes is reflected in crossing over.
3. The crossing over results basically from an exchange of genetic material between non-sister chromatids by break-and-exchange following replication.
4. The frequency of crossing over appears to be closely related to physical distance between genes on chromosome and serves as a tool in constructing genetic maps of chromosomes.

77.1 TYPES OF CROSSING OVER

According to its occurrence in the somatic or germ cells following two types of crossing over have been recognized:

1. Somatic or mitotic crossing over. When the process of crossing over occurs in the chromosomes of body or somatic cells of an organism during the mitotic cell division it is known as **somatic** or **mitotic crossing over**. The somatic crossing over is rare in its occurrence and it has no genetical significance. The somatic or mitotic crossing over has been reported in the body or somatic cells of *Drosophila* by **Curt Stern** and in the fungus *Aspergillus nidulans by* **G. Pontecorvo** (Box 77.1).

Box 77.1 Parasexual Exchange

Any recombination occurring in the absence of meiosis (*i.e.*, in the absence of sexual reproduction) is termed **parasexual exchange**, and this is the form of genetic exchange predominantly seen in bacteria (*i.e.*, gene transfer in bacteria). Parasexual exchange in eukaryotes occurs between **organelle genome** and between nuclear chromosomes in somatic cells (**mitotic recombination**).

2. Germinal or meiotic crossing over. Usually the crossing over occurs in germinal cells during the gametogenesis in which the meiotic cell division takes place. This type of crossing over is known as **germinal** or **meiotic crossing over**. The meiotic crossing over is universal in its occurrence and is of great genetic significance.

77.2 MECHANISM OF MITOTIC RECOMBINATION

Mitotic recombination between homologous chromosomes lead to **mitotic segregation**, where heterozygous loci segregate into somatic daughter cells generating reciprocal homozygous clones identifiable by their distinct phenotypes. The first evidence for mitotic recombination came from *Drosophila*.

Example. Stern (1936) observed that *Drosophila* females which are heterozygous for sex-linked recessive genes, yellow body colour (*y*) and singed (=burned or scorched) hairs and bristles (*sn*), are phenotypically like wild types but occasionally show yellow spots on their body. Homozygous recessive individuals for these genes are yellow bodied and with bent blunted bristles. Normally, all the cells of a heterozygous should be normal and should not express either of these genes. Yellow skin spots can appear only if cells are homozygous for *y*, indicating thereby that whenever yellow spots appear in heterozygous flies, there is prior crossing over between *y* and *sn* alleles. Stern found that equal sections of singed and yellow tissues lie adjacent to each other or form **twin spots**. This indicate that such tissues arise as a result of mitotic crossing over, rather than due to mutation. Somatic or mitotic crossing over occurs at a four strand stage and during this process there is pairing of homologous chromosomes (Fig. 77.1).

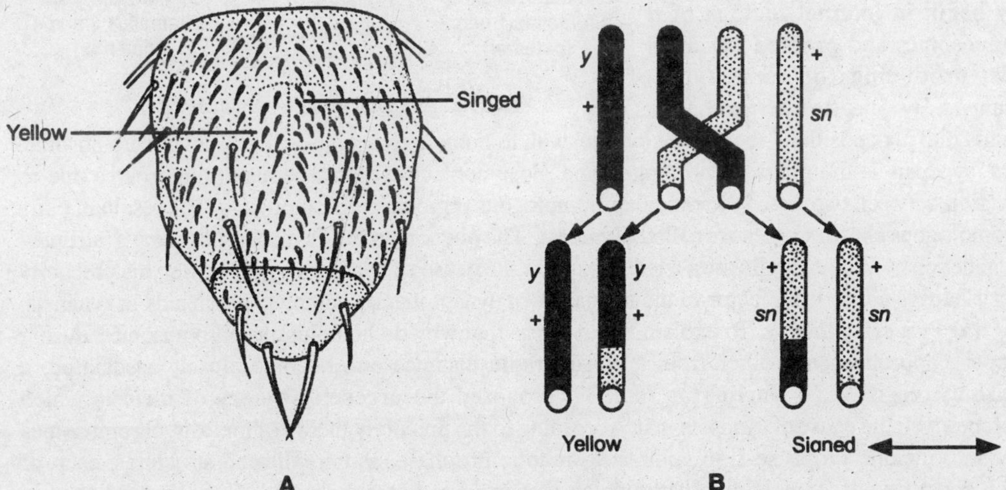

Fig. 77.1. Somatic or mitotic crossing over in *Drosophila*. A—Thorax of a fruit fly heterozygous for *y* (yellow body) and *sn* (singed hairs and bristles) with a twin spot; B—Mechanism of crossing over leading to homozygosity for distal genes.

Mitotic recombination can be stimulated by exposure to X-rays, which generate double-strand breaks in DNA. This suggests that the process may be a by-product of DNA-repair.

A second form of mitotic recombination occurs not between homologous chromosomes but between identical chromatids–**sister chromatid exchange (SCE)**. This is also stimulated by damage to DNA.

77.3 MECHANISM OF MEIOTIC CROSSING OVER

The process of crossing over includes following stages in it, *viz.*, synapsis, duplication of chromosomes, crossing over and terminalization. The chromosomes which tend to undergo recombination due to meiotic crossing over necessarily complete two functions: 1. 99.7 per cent replication of DNA and 75 per cent synthesis of histones, both of which take place prior to onset of prophase I and 2. attachment of each chromosome by its both ends (**telomeres**) to the nuclear envelope (*i.e.*, to nuclear lamina) *via* the specialized structure, called **attachment plaques** (see **Suzuki** *et al.*, 1986). This event occurs during the **leptotene** stage of prophase I and though each chromosome at this stage is visually long and thin thread, but contains material of two sister chromatids (*i.e.*, two DNA molecules plus almost duplicated amount of histones).

1. Synapsis.

Synapsis or intimate pairin between the two homologoı chromosomes (one maternal and anoth paternal) is initiated during **zygoter** stage of prophase I of meiosis I. Synaps often starts when the homologous enc of the two chromosomes are brougl together on the nuclear envelope and continues inward in a zipper-like mann from both ends, aligning the tw homologous chromosomes side by sic (*e.g.*, mammals). In other cases, synaps may begin in internal regions of tl chromosomes and proceed toward tl ends, producing the same type of alignment. By synapsis each gene is, thus, brought into juxtaposition (=being side by side) with its homologous gene on the opposite chromosome. Thus, synapsis is the phase of prolonged and close contact of homologous chromosomes due to attraction between two exactly identical or homologous regions or chromomeres. The resultant pairs of homologous chromosomes are called **bivalents**. The phenomenon of synapsis has always intrigued cytogeneticists. While **Darlington** tried to explain the cause of synapsis by proposing his precocity theory, **Moses** identified a factor in the formation of synaptonemal complex which aids in synapsis.

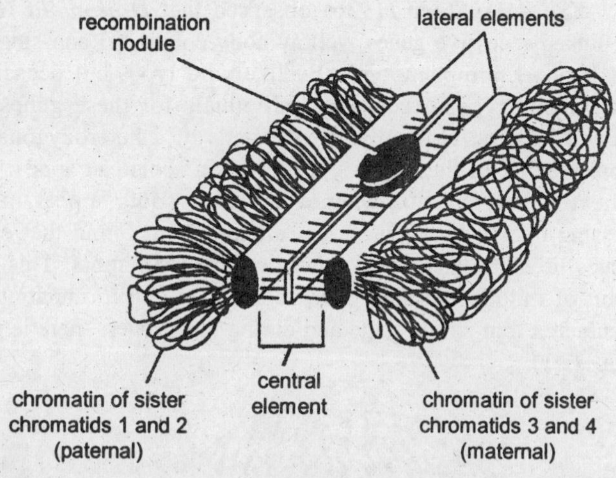

Fig. 77.2. A typical synaptonemal complex.

(a) Precocity theory. To explain the question that why do homologous chromosomes, during synapsis, approach each other from a considerable distance and become closely associated, a British cytologist **C.D. Darlington** in 1937, proposed the **precocity theory of meiosis** which embraces well the cause of synapsis in it. According to the precocity theory of meiosis, chromosomes enter into meiotic prophase I, in contrast to mitotic prophase, as unreplicated structures, each of which consisting of a single chromatid which he considered as unbalanced or unsaturated state in electrostatical relations. In order to become saturated or balanced the chromosomes must pair. Thus, the sequences of meiotic pairing or synapsis are determined precociously.

Precocity theory is now insupportable since it erroneously stressed upon the idea that DNA synthesis or chromosome duplication takes place later in pachytene or diplotene and then results in separation of homologous chromosomes.

(*b*) Synaptonemal complex. Montrose J. Moses (1956) has revealed a highly organized structure of filaments called **synaptonemal complex** in between the paired chromosomes of zygotene and pachytene stages in crayfish by electron microscopy. Synaptonemal complex has also been observed in a wide variety of species of plants and animals.

In electron-micrographs the synaptonemal complex appears as three parallel dense lines that lie equally spaced in a plane and are flanked by chromatin. The elements of two lateral lines usually appear densest, while the element of central line is of variable prominence. Some fine transverse strands also cross between lateral elements, connecting them with the central element. Though, the morphology of lateral and central elements may vary from species to species, but the basić structure and the spacing of the synaptonemal complex is constant within the species.

Cytochemical studies have shown that the lateral elements of synaptonemal complex are rich in DNA, RNA and proteins, but that the central element of it contains mainly RNA, protein and little DNA.

Functions of synaptonemal complex. The synaptonemal complex is found to contribute both chiasma formation and crossing over. For instance, the synaptonemal complex may serve crossing over by facilitating effective synapsis in one or more of the following ways—(*a*) to maintain pairing in fixed state for an extended period, (*b*) to provide a structural framework within which molecular recombination may occur, and (*c*) to segregate recombination DNA from the bulk of the chromosomal DNA.

Robert King (1970) suggested that the synaptonemal complex may orient the non-sister chromatids of homologs in a manner to facilitate enzymatically induced exchanges between their DNA molecules. In 1971, **D.A. Comings** and **T.A. Okada** have shown electron microscopically that synapsis occurs at two levels one at chromosomal level and the other at the molecular level. According to them, the synaptonemal complex pulls homologous chromosomes into approximate association with each other but plays no role in molecular pairing of DNA strand.

Later on each event of recombination due to crossing over is found to be performed by a large protein assembly of 90 nm diameter, called **recombination nodule**, which is placed on the synaptonemal complex (**Carpenter**, 1977, 1987). Each nodule marks the site of a large multienzyme 'recombination machine' which brings local regions of DNA on maternal and paternal chromatids together across the 100-nm wide synaptonemal complex.

2. Duplication of Chromosomes

The synapsis is followed by duplication of chromosomes (in **pachytene**). During this stage, each homologous chromosome of bivalents splits longitudinally and form two identical sister chromatids which remain held together by an unsplitted centromere. The longitudinal splitting of chromosomes is achieved by the separation of already duplicated DNA molecules along with certain chromosomal proteins. At this stage each bivalent contains four chromatids, so it is known as **tetrad**.

3. Crossing Over by Breakage and Union

It is well evident that crossing over occurs in the homologous chromosomes only during the four stranded or tetrad stage. Homologs continue to stay in synapsis for days during pachytene stage and chromosomal crossing over occurs due to exchange of chromosomal material between non-sister chromatids of each tetrad. In pachytene, the recombination nodules become visible between synapsed chromosomes.

During the process of crossing over, two non-sister chromatids first break at the corresponding points due to the activity of a nuclear enzyme known as **endonuclease**. Then a segment on one side of each break connects with a segment on the opposite side of the break, so that the two non-sister chromatids cross each other. At this stage 3 per cent synthesis of DNA (*i.e.*, P–DNA replication) occurs to fill the gap. The fusion of chromosomal segments with that of opposite one takes place due to the action of an enzyme known as **ligase**. The crossing of two chromatids is known as **chiasma** (Gr., *chiasma* = cross) formation. The crossing over, thus, includes the breaking of chromatid segments, their transposition and fusion.

Chiasma frequency or percentage of crossing over. The crossing over may take place at several points in one tetrad and may result in the formation of several chiasmata. The number of chiasmata depends on the length of the chromosomes because the longer the chromosome the greater the number of chiasmata. In a species each chromosome has a characteristic number of chiasmata. The frequency by which a chiasmata occurs between any two genetic loci has also a characteristic probability. The more apart two genes are located on a chromosome, the greater the opportunity for a chiasma to occur between them. The closer two genes are linked lesser the chances for a chiasma occurring between them.

Limits of recombination. If two gene loci are so far apart in the chromosome that the probability of chiasma forming between them is 100% then 50% of the gametes will be parental type (non-cross-over) and 50% recombinant (cross over) type. Such dihybrids give the test cross ratio of 1 : 1 : 1 : 1. Recombination between two linked genes cannot exceed 50% even when multiple cross overs occur between them.

Box 77.2 Recombination Proteins in Eukaryotes

RAD52 group of genes is involved in recombinational repair. **RND51** gene is inducible by DNA damage. **RNA50** gene encodes a DNA-binding protein which functions in the repair of double strand breaks in mitotic cells and induces and processes such breaks in meiotic cells. The product of **RAD52** gene is essential for both mitotic and meiotic crossing over, *i.e.*, in the repair of double-strand DNA breaks and mating-type switching in yeast cells. RAD52 interacts with RAD51. **RAD54** encodes **helicase** required for recombination and other forms of DNA repair.

4. Terminalisation.

After the occurrence of process of crossing over, the non-sister chromatids start to repel each other because the force of synapsis attraction between them decreases. During **diplotene**, desynapsis begins, synaptonemal complex dissolves and two homologous chromosomes in a bivalent are pulled away from each other. During **diakinesis**, chromosomes detaches from the nuclear envelope and each bivalent is clearly seen to contain four separate chromatids with each pair of sister chromatids linked at their centromeres, while non-sister chromatids that have crossed over are linked by chiasmata. The chromatids separate progressively from the centromere towards the chiasma and the chiasma itself moves in a zipper fashion towards the end of the tetrad (Fig. 77.3). The movement of chiasma is known as **terminalisation**. Due to the terminalisation the homologous chromosomes are separated completely.

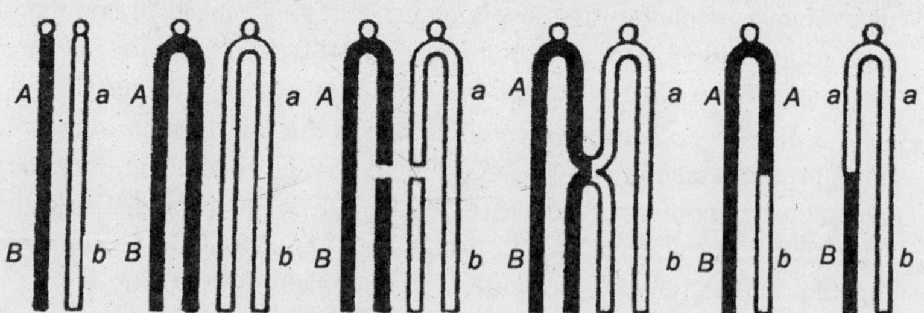

Fig. 77.3. Diagram showing the mechanism of crossing over.

77.4 KINDS OF CROSSING OVER

According to the number of chiasma following types of crossing over have been described.

1. Single crossing over. When the chiasma occurs only at one point of the chromosome pair then the crossing over is known as **single crossing over**. The single crossing over produces two cross over chromatids and two non-cross over chromatids.

2. Double crossing over. When the chiasmata occur at two points in the same chromosome, the phenomenon is known as **double crossing over**. In the double crossing over, the formation of each chiasma is independent of the other and in it four possible classes of recombination occur.

In the double crossing over following two types of chiasma may be formed:

***(i)* Reciprocal chiasma.** In the reciprocal chiasma the same two chromatids are involved in the second chiasma as in the first. Thus, the second chiasma restores the order which was changed by the first chiasma and it produces two non-cross over chromatids. The reciprocal chiasma occurs in two strand double crossing over in which out of four chromatids only two are involved in the double crossing over (Box 77.3).

***(ii)* Complimentary chiasma.** When both the chro-matids taking part in the second chiasma are different from those chromatids involved in the first chiasma, the chiasma is known as **complimentary chiasma.** The complimentary chiasma produces four single cross overs but no non-cross over. The complimentary chiasma occurs when three or four chromatids of the tetrad undergo the crossing over.

3. Multiple crossing over. When crossing over takes place at more than two places in the same chromosome pair then such crossing over is known as **multiple crossing over**. The multiple crossing over occurs rarely.

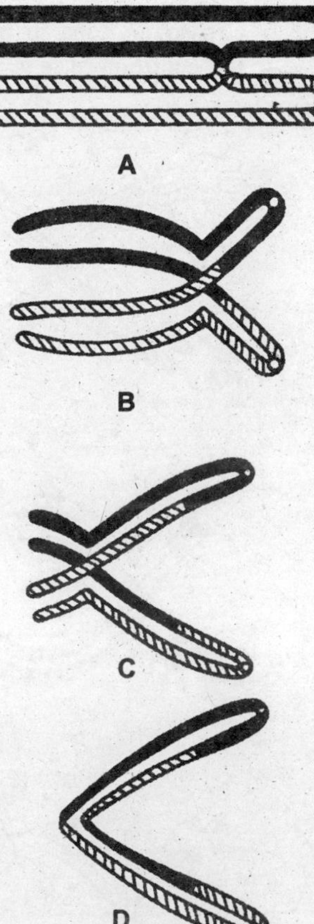

Fig. 77.4. Diagram showing the terminalisation.

Factors Affecting Crossing Over.

Many extrinsic and intrinsic factors have found to affect the rate of crossing over. Among these are the effects of age, sex, temperature, water content, ionizing radiation, certain chemicals, heterochromatic regions, chromosomal aberrations such as inversions, etc. Some examples of factors affecting the process of crossing over are as follows:

1. The frequency of crossing over decreases with increasing age in female *Drosophila*.
2. Crossing over is often depressed in one sex of certain species, *e.g.*, male *Drosophila*, female silkworm. There is higher frequency of recombination in females.
3. Low water contents can lead to reduced chiasma fre-quencies, in extreme cases it can result in complete failure of pairing and exchange.

Box 77.3 Reciprocal and Non-Reciprocal Recombination

1. Reciprocal recombination. Reciprocal recombination is both symmetrical and conservative, *i.e.*, information is *exchanged* between duplexes and is neither lost nor gained. Thus, if recombination occurs between parental chromosomes carrying alleles *AB* and *ab*, the reciprocal recombinants are *Ab* and *aB*.

2. Non-reciprocal recombination. It involves a unidirectional transfer of information. An example of non-reciprocal recombinants arising from above cross would be *Ab* and *AB*, where information has been transferred-unidirectionally from *A* to *a* allele and *a* allele has been lost. This type of recombination is also called **gene conversion**, because it appears that one gene has been converted to the other. Non-reciprocal recombination can be observed in fungal tetrads (*i.e.*, tetrad analysis).

3. Ionizing radiation increases the rate of mitotic as well as meiotic crossing over.

4. Certain chemicals such as **colchicine** and **selenium** have a reducing effect on crossing over.

5. The distance between two linked genes is the main factor for the occurrence of crossing over. If the two genes are distant apart, they have better chances of separation than those which are closely set together on the chromosome.

4. Increase in temperature enhances the amount of crossing over.
5. Crossovers tend to be much more frequent in telomeric chromosome regions than around the centromere, *e.g.*, human chromosomes.

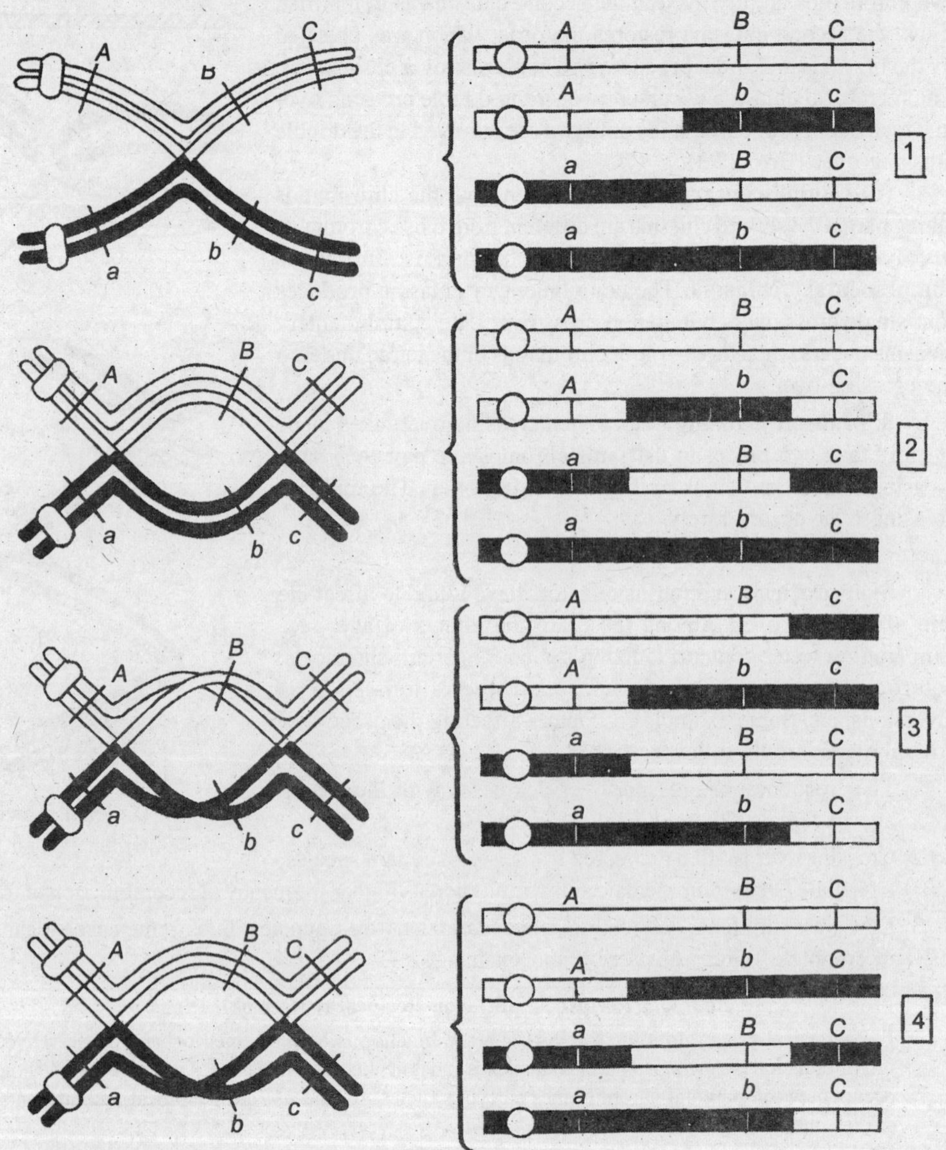

Fig. 77.5. Diagram showing the results of single and double crossing over. 1. Single chiasma. 2. Two strand double cross over. 3. Four strand double cross over. 4. Three strand double cross over.

77.5 THEORIES ABOUT THE MECHANISM OF CROSSING OVER

Following theories have been propounded to explain the mechanism of crossing over:

1. Duplication theory. This theory was proposed by **John Belling** (1928) while studying meiosis in some plant species. Belling believed that crossing over might occur during duplication of homologous chromosomes and might be brought about due to novel attachments formed between newly synthesized genes. He visualized genes as beads (described as **chromomeres**), connected by non-genic linking elements-the **interchromomeric regions**. During duplication of chromosomes, initially the chromomeres are duplicated and the newly formed chromomeres remain tightly juxtaposed to the old ones. When interchromomeric regions are synthesized to join these new genes or chromomeres, they may switch from a newly synthesized chromomere on one homologous chromosome to an adjacent chromomere of other homologue. This results in the formation of recombinants or crossovers in new sets of chromatids (Fig. 77.6).

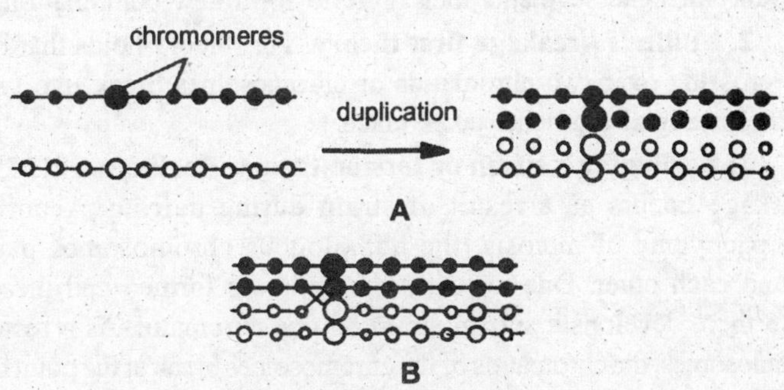

Fig. 77.6. Three steps of Belling's duplication theory.

2. Copy choice hypothesis or Switch model. To explain recombination in microorganisms, **J. Laderberg** (1955) proposed a modified version of Belling's hypothesis. This is called copy choice mechanism of recombination or crossing over. According to this model a daughter chromosome is formed by alternate use of recipient and donor chromosome material as a kind of model or template. The daughter chromosome is then like the recipient chromosome except for portions "copied" from the donor chromosome segment. At the end of process a recipient cell would contain (1) Original donor segment, (2) The originally recipient whole chromosome, and (3) A "hybrid" daughter chromosome. In succeeding divisions of the recipient cell, the donor segment is lost.

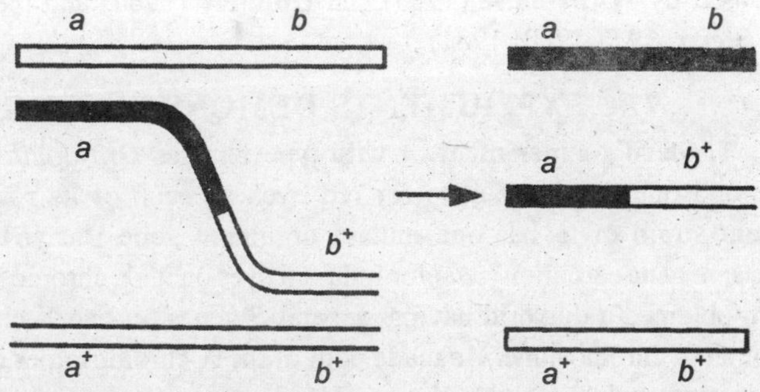

Fig. 77.7. Copy choice model for the mechanism of crossing over.

The copy choice model to explain the mechanism of recombination has been criticised for the following two reasons: (1) Copy choice model relies on conservative mode of DNA replication, but all experimental evidences suggest that DNA replication occurs in a semi-conservative manner. (2) The copy choice model proposed the involvement of only two chromatids in the process of crossing over, while cytological investigations have made it clear that crossing over occurs at four strand stage.

3. Break and exchange theory. The break and exchange theory is the most accepted theory to explain the process of crossing over. This theory states that in the crossing over breaks occur in the non-sister chromatids of the tetrad and the exchange of chromosomal segments occurs between the non-sister chromatids.

There exists a lot of controversy about the manner and time of the break of chromatids during the crossing over. Following theories have been advanced by different workers to explain the process of breaking of chromatids:

1. Serebrovsky's contact first theory. According to this theory two chromatids of a tetrad first touch and cross each other and then at the point of cross the breakage of chromatids occur. The chromosome segments then unite to form new combinations.

2. Muller's breakage first theory. This theory holds that before the occurrence of chiasma and crossing over two chromatids or chromosomes break into two fragments and then the union of chromosomal segments takes place.

3. Darlington's strain or torsion theory. Darlington (1935) propounded that chromosomal breakage occurs as a result of strain during pairing. According to this theory during the late pachytene of meiosis, the homologous chromosomes pair and become spirally twisted around each other. Due to twists the coils are formed and these are known as **relational coil.** Then there develops a sort of strain on one chromatid. As a result of the strain due to torsion of chromosomes, the chromatids of the chromosomes break at the point of contact and, thus, recombination results.

4. Hybrid DNA models. In the last three decades several hybrid DNA models have been proposed to explain the mechanism of recombination. In these models, only one strand in each of the two DNA duplexes belonging to non-sister homologous chromatids breaks. The single strands released from these breaks then pair crosswise with unbroken strands by complementary base pairing. This results in the formation of hybrid DNA segments. Some important hybrid DNA models have been proposed by **Whitehouse** (1963) and **Holliday** (1964) and these have been supported by (**Hotta** and **Stern,** 1977,1980, 1981).

77.6 EXPERIMENTS DEMONSTRATING CROSSING OVER

1. Stern's experiment. A wild type females *Drosophila* has one recessive gene for round eyes and one dominant gene for red eyes on each of its rod-shaped X chromosome, while, a mutant strain of it, has one mutant dominant gene Bar (*B*) for narrow eyes and one mutant recessive gene carnation (*car*) for light red eyes on its X chromosomes. By crossing these two strains, Stern obtained a dihybrid having *car* and *B* genes on one X chromosome and normal genes (+ +) on other X chromosome. He made both of the X chromosomes of this heterozygote female aberrant by treating such flies with X-rays. The X chromosome having the genes *car* and *B* was broken into two segments, one fragment having both of the genes, while, the other X chromosome having a fragment of the Y chromosome attached to it and contained normal alleles (+ +). Thus, both aberrant X chromosomes of heterozygote were cytologically detectable. A female heterozygote with the aberrant X chromosomes was mated with a normal male having X chromosome with car + alleles, four classes of eggs (*e.g.*, two types of eggs are cross overs and two types of eggs are non-cross overs) were produced which by fertilization produced following four kinds of females:

1. Carnation Bar females with broken X without any fragment of Y.
2. The red round females with unbroken X with the attached Y fragment.
3. The carnation round have the unbroken X without an attached Y fragment.
4. The red Bar have the broken X with the attached Y fragment.

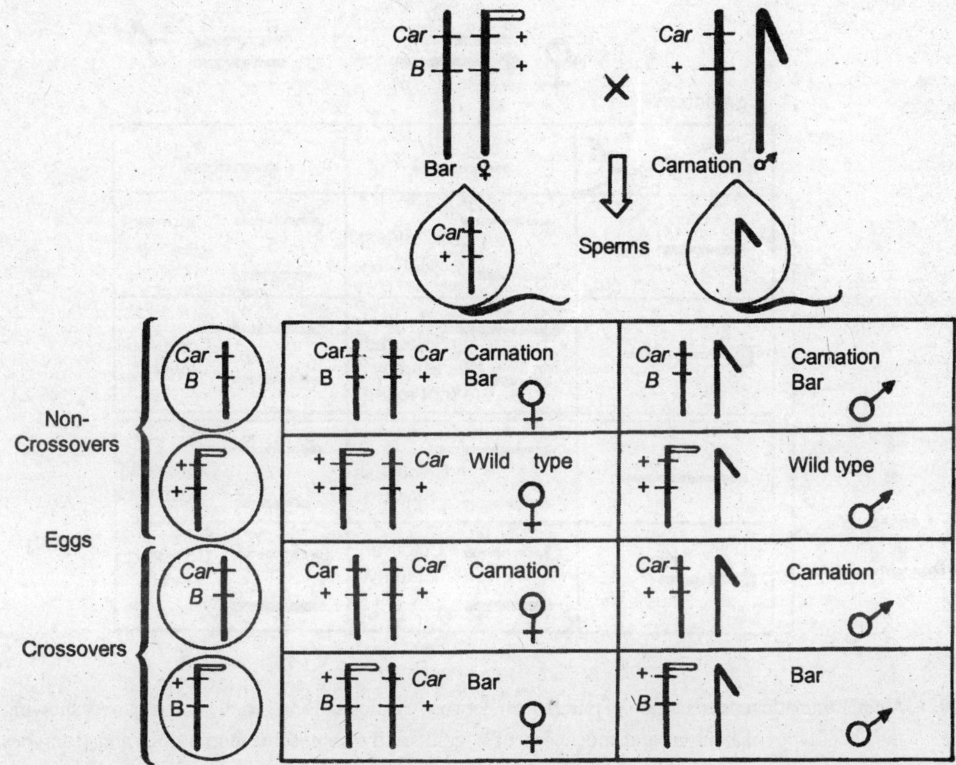

Fig. 77.8. Stern's experiment for the detection of crossing over in *Drosophila* cytologically.

Thus, flies in which crossing over was indicated phenotypically showed microscopic evidence of exchanges between homologous chromosomes. The physical or cytological basis of crossing over was thus established.

2. Creighton and McClintock's experiment. Creighton and **McClintock** like **Stern** made convincing correlation between cytological evidence and genetical results of crossing over in maize. They made the use of knob of 9th chromosome of maize which in two different strains might had one allele for either coloured aleurone (*C*) or colourless aleurone (*c*) and one allele for either starchy endosperm (*Wx*) or waxy endosperm (*wx*). The results of their experiment have been illustrated in Fig. 77.9.

77.7 SIGNIFICANCE OR UTILITY OF CROSSING OVER

The study of mechanism of crossing over is significant in the following ways:

1. Crossing over furnishes a proof for the linear arrangement of linked genes on the chromosome.
2. Crossing over helps in the construction of linkage maps or genetic maps of chromosomes. Linkage maps of *Drosophila* and maize chromosomes have been constructed in details.
3. As a result of crossing over, new combinations of genes are produced which change the frequency of genes in the genetic pool of a population and therefore, play an important role in the process of evolution (micro-evolution).

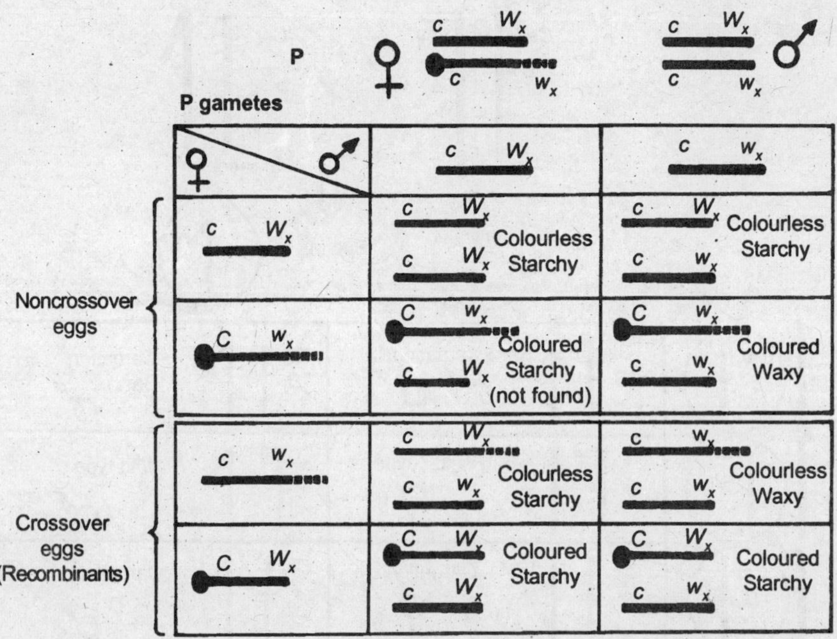

Fig. 77.9. A diagrammatic representation of parallelism between cytological and genetic crossing over showing the phenotypes of aleurone and endosperm of F_1 grains and chromosome morphology and genotypes for microsporocytes produced by F_1 plants.

QUESTIONS

Long Answer Questions

1. What do you understand by crossing over? At what stage does crossing over occur?
2. Discuss the mechanism and significance of crossing over. *(Avadh 1989, 1995; Bundelkhand 1989; Sagar 1990; Gorakhpur 1991; Bhagalpur 1991; Rohilkhand 1992; Purvanchal 1992; Guwahati (G) 1994)*
3. Define crossing over. Describe various types and theories regarding crossing over and explain its mechanism. *(Lucknow 1991; CCS 1993)*
4. Define crossing over. Explain the mechanism of crossing over and its significance. *(Allahabad 1995)*
5. Write an essay on crossing over? *(CCS 1995)*
6. What do you understand by crossing over? Describe the different types of crossing over.
7. Discuss the different factors influencing the crossing over. *(Guwahati (G) 1992)*
8. Discuss the phenomenon of recombination of genes by giving suitable examples. *(Allahabad 1991)*
9. Describe the crossing over, chiasma formation and mechanism of genetic recombination. *(Magadh (G) 1990)*
10. What is crossing over and how does it explain recombination between linked genes? *(Guwahati (M) 1996)*
11. What do you understand by crossing over and what part does it play in heredity? *(Kanpur 1991)*

12. Explain with example the mitotic recombination or crossing over.
13. Describe various theories to explain the mechanism of crossing over. *(CCS 1990; Purvanchal 1992)*
14. What do you understand by crossing over? Describe an experiment to show the process of crossing over. *(Allahabad 1989; CCS 1996)*
15. Give an account of the phenomenon of linkage and crossing over. Add a note on their significance. *(Punjab (G) 1991; Allahabad 1993; Lucknow 1996)*
16. What do you mean by linkage and crossing over? Write a note on various theories proposed for explaining the mechanism of crossing over. *(Punjab (G) 1990)*
17. Briefly describe linkage and crossing over. What is the significance of this phenomenon? *(Avadh 1998)*
18. Define crossing over and explain the mechanism with *Drosophila* as example. *(M.S.U. Tamil Nadu, 2006)*

Short Answer Questions

1. Write differences between chiasma and crossing over.
2. Write short notes on the following:
   (*i*) Crossing over *(Bundelkhand 1991; Rohilkhand 1989, 1990; Berhampur 1990; Sagar 1991; Purvanchal 1991; Magadh (G) 1993; Magadh (Subs.) 1993; Gorakhpur 1993, 1996, 1998; Avadh 1997; H.N.B., Garhwal 2008)*
   (*ii*) Biological significance of crossing over *(Magadh (M) 1994)*
   (*iii*) Cytological basis of crossing over *(Allahabad 1989)*
   (*iv*) Synapsis
   (*v*) Theories of crossing over
   (*vi*) Mechanism of crossing over *(Magadh (Subs.) 1990)*
   (*vii*) Precocity theory *(Avadh 1999)*

Very Short Answer Questions

1. What gemetes are produced by an individual whose genotype is *AB*/*ab*, if crossing over can occur between the genes? *(NEHU 1996)*

   Ans. *AB*, *Ab*, *aB* and *ab*.

2. Define crossing over.

   Ans. Mutual exchange of exactly equal and homologous parts of genetic material or genes between the non-sister chromatids of a pair of homologous chromosomes is called crossing over.

3. How can genes located on a single chromosome assort independently?

   Ans. Through crossing over.

4. Give one word for the following statement:

   "An exchange of segments of homologous chromosomes including linked genes at meiosis". *(Lucknow 1994)*

   Ans. Crossing over.

Yes or No Question

1. Crossing over occurs in the same chromosome. *(Guwahati (G) 1991, 93)*

   Ans. No.

Fill in the Blanks

(*i*) The chances of recombination between two gene loci are inversely proportional to the ....

(*ii*) Recombination of genes is caused as a result of .... between chromatids of homologous chromosomes.

(*iii*) Crossing over occurs at .... stage of meiosis.
(*iv*) Crossing over requires the formation of .... between homologous chromosomes.
(*v*) Crossing over percentage can never exceed .... per cent. (*H.N.B. Garhwal 2009*)
(*vi*) Complete linkage is due to absence of .... between genes on same chromosome.

Multiple Choice Questions

1. Exchange of chromosome segments between maternal and paternal chromatids at pachytene is called
(*a*) linkage (*b*) crossing over
(*c*) disjunction (*d*) segregation

2. Crossing over occurs at
(*a*) 1 strand stage (*b*) 2 strand stage
(*c*) 3 strand stage (*d*) 4 strand stage

3. Longitudinal splitting of chromosomes and crossing over occurs in
(*a*) leptotene (*b*) zygotene
(*c*) pachytene (*d*) diplotene

4. The pairing of homologous chromosomes during zygotene (of meiosis) is called
(*a*) synapse (*b*) terminalization
(*c*) synapsis (*d*) crossing over

5. Crossing over involves
(*a*) duplication of the chromosomes (*b*) deletion of the chromosomes
(*c*) exchange of genetic material (*d*) addition of chromosome

6. Synaptonemal complex helps in
(*a*) gamete formation
(*b*) recombination during cell division
(*c*) production of enzymes during cell division
(*d*) chromosomal movements towards poles

7. Recombination of parental characters are due to
(*a*) chiasma (*b*) synapsis
(*c*) linkage (*d*) crossing over

8. Crossing over occurs in
(*a*) leptotene stage (*b*) pachytene stage
(*c*) anaphase stage (*d*) diakinesis stage

9. Crossing over is advantageous because it brings about
(*a*) linkage (*b*) variation
(*c*) inbreeding (*d*) stability

ANSWERS

Fill in the Blanks

(*i*) Difference in distance (*ii*) Exchange (*iii*) Pachytene
(*iv*) Synaptonemal complex (*v*) 50% (*vi*) Crossing over.

Multiple Choice Questions

1. (*b*) **2.** (*d*) **3.** (*c*) **4.** (*c*)
5. (*c*) **6.** (*b*) **7.** (*d*) **8.** (*b*)
9. (*b*).

78

Linkage Maps

Gene mapping is the assignment of a gene to a locus on a chromosome. By mapping the relative positions of many genes and other markers it is possible to generate a **chromosome map** or a map of entire genome. Gene maps are of following three types:

1. Genetic or linkage maps which assign genes to **linkage groups**. Genetic maps are calibrated in arbitray units, reflecting the likelihood of marker separation by recombination (*e.g.*, in meiotic mapping) or by chromosome fragmentation (*e.g.*, in **radiation hybrid mapping** and **HAPPY mapping**, Box 78.1). The principle behind genetic mapping is that the chance fragment of DNA decreases with increasing distance between them.

Genetic maps have a resolution between cytogenetic and physical maps but those based on recombination frequencies are distorted by areas of relatively high or relatively low recombination (*i.e.*, recombination hot spots and recombination cold spots; Box 78.1). Additionally, the frequency of recombinations is not the same in all species, thus, genetic map units reflect different physical distances in different species.

Box 78.1

1. **Radiation hybrid mapping.** Where chromosome breaks are introduced by irradiating somatic cell hybrids. Radiation hybrid maps are calibrated in **CentiRays** (cR_x) where *x* is the dosage of X-rays in Rads; lcR being equivalent to 1% frequency of separation.
2. **HAPPY mapping**. Where genomic DNA is sheared by vortexing (mixing by means of a rapid whirling or circular motion) or sonication (disrupting by treatment with high-frequency sound waves).
3. **Recombination hotspots** are often endonuclease target sites because cleavage provides single-stranded DNA for the initiation of homologous recombination.
4. **Recombination coldspots** are often sites where homologous DNA fail to synapse effectively. This usually occurs where there is steric hindrance, *e.g.*, in an individual heterozygous for a chromosome inversion or translocation where synapsis at the break points is prevented.

2.Cytogenetic map. It is created by correlating phenotypes with observable chromosomal re-arrangements and deficiencies. This type of map has a low resolution and is applicable to the few species, *e.g.*, mammals and *Drosophila*, which display either natural or artificially inducible reproducible chromosome banding patterns. Cytogenetic maps of simple genomes (viruses, plasmids, etc.) are used to locate regions which differ in their physical properties (*e.g.*, denaturation mapping).

3. Physical or molecular map. It is created by ordering cloned fragments of genomic DNA and is calibrated in real units (base pairs, kilobase pairs, megabase pairs). The physical map has the highest resolution and is the ultimate aim of mapping projects.

78.1 LINKAGE MAPPING

Our discussion of linkage and crossing over has made clear so far, that 1. Because the number of genes usually exceeds the number of chromosomes in different species, therefore, many genes have to be located on same chromosomes. 2. The genes remain arranged linearly on the chromosomes and they have no option before them, except to behave according to the chromosomes during gametogenesis and inheritance, therefore, all the genes on a chromosome inherit together and are said to be linked with each other to form linkage groups. 3. The number of linkage groups corresponds to the number of homologous pairs of chromosomes or bivalents of the species. 4. The linked genes do not always remain linked, but, occasionally are departed from other members of their linkage groups by crossing over. 5. The closely linked genes have less chances of departure or frequency of crossing over than the widely located genes. 6. And each gene has definite order and location in a linkage group or chromosomes, as the crossing over frequency has been found constant for two given linked genes of a species. For example various genetic crosses have shown that in *Drosophila*, among three mutant sex-linked genes (*viz*., genes for white eyes, yellow body and cut wings), the genes for white and yellow are always found to have 1% crossing over and similarly, white and cut genes are found to have 20% crossing over. The percentage or frequency of crossing over, thus, appears to be closely related to physical distance between genes. When one knows all the genes, linkage groups and number of linkage groups of a species, it becomes possible for him that by adopting the crossing over as a tool, he may determine the relative distances between the genes in a linkage group and also their order and may give diagrammatic representation of chromosomes showing the genes as points separated by distances proportional to the amount of crossing over. Such a diagrammatic, graphical representation of relative distances between linked genes of a chromosome is called **linkage** or **genetic map**. Because, such a linkage or genetic map is the outcome of crossing over studies, is also called **cross over map**.

Construction of a Linkage Map or Genetic Mapping

The method of construction maps of different chromosomes is called **genetic mapping**. The genetic mapping includes following processes:

1. Determination of Linkage Groups

Before starting the genetic mapping of the chromosomes of a species, one has to know the exact number of chromosomes of that species and then, he has to determine the total number of genes of that species by undergoing hybridization experiments in between wild and mutant strains. By the same hybridization techniques, it can also be easily determined that how many phenotypic traits remain always together or linked and consequently their determiners or genes during the course of inheritance. And thus, the different linkage groups of a species can be worked out.

2. Determination of Map Distance

The intergene distance on the chromosomes cannot be measured in the customary units employed in light microscopy, geneticists use an arbitary unit to measure the **map unit**, to describe distances between linked genes. A map unit is equal to 1 per cent of crossovers (recombinants); that is, it represents the linear distance along the chromosome for which a recombination frequency of 1 per cent is observed. These distances can also be expressed in **morgan units;** one morgan unit represents 100 per cent crossing over. Thus, 1 per cent crossing over can also be expressed as **1 centimorgan (lcM),** 10 per cent crossing over as **l decimorgan** and so on. The Morgan unit is named in honour of **T.H. Morgan**; however, most geneticists prefer map units.

Quite interestingly, it is now possible to calculate the size of many genes, as well as distances separating them, and to photograph genes in the electron microscope (see **Burns and Bottino**, 1989).

Examples

1. If a F_1 hybrid having the genotype *Ab*/*aB* produces 8% of cross over gametes *AB* and *ab*, then the distance between *A* and *B* is estimated to be 16 map units or centimorgan.
2. If the map distance between the gene loci *B* and *C* is 12 centimorgan, then 12% of gametes of genotype *BC*/*bc* should be crossover types, *i.e.*, 6% *bC*.

Because, each chiasma produces 50% crossover products, 50 per cent crossing over is equivalent to 50 map units or centimorgans. If the mean number of chiasmata is known for a chromosome pair, the total length of the map for that linkage group may be predicted:

Total length = mean number of chiasmata × 50.

Two Point Test Cross.

The percentage of crossing over between two linked genes is calculated by test crosses in which a F_1 dihybrid is crossed with a double recessive parent. Such crosses because involved crossing over at two points, so called **two point test crosses.** For example, a dihybrid having the genotype *AC/ac* is test crossed with a double recessive parent (*ac/ac*), then among F_2 test cross hybrids we may get 37% dominant genes at both gene loci (*AC/ac*), 37% recessive genes at both gene loci (*ac/ac*), 13% dominant gene at first gene locus and recessive gene at the second gene locus (*Ac/ac*), and 13% recessive gene at first gene locus and dominant gene at second gene locus (*aC/ac*). The last two groups (*i.e.*, 13% *Ac/ac* and 13% *aC/ac*) were produced by crossover gametes (13 + 13) from the dihybrid parent. Thus, 26% of all gametes (13 + 13) were of cross over types and the distance between the loci *A* and *C* is estimated to be 26 centimorgans. Because, double crossovers usually do not occur between genes less than 5 centimorgans apart, so for genes further apart, the **three point test crosses** are used.

Three Point Test Cross.

A three point test cross or trihybrid test cross (involving three genes) gives us information regarding relative distances between these genes, and also shows us the linear order in which these genes should be present on chromosome. Such a three point test cross may be carried out if three points or gene loci on a chromosome pair can be identified by marker genes. If, in addition to genes *A* and *C* indicated above, a third marker gene *B* is located in fairly close proximity in the same linkage group, all three markers may be used together in conducting a more precise analysis of the map distance and the relative position of the three points.

Suppose that we test cross trihybrid individuals of genotype ABC/abc and find in the progeny the following:

36% *ABC/abc*	9% *Abc/abc*	4% *ABc/abc*	1% *AbC/abc*
36% *abc/abc*	9% *aBC/abc*	4% *abC/abc*	1% *aBc/abc*
72% Parental type	18% Single crossovers: between *A* and *B* (region I)	8% Single crossovers: between *B* and *C* (region II)	2% Double crossover

To find the distance *A–B* we must count all crossovers (both singles and doubles) that occurred in region I = 18% + 2% or = 20% or 20 map units between the loci *A* and *B*. To find the distance *B – C* we must again count all crossovers (both singles and doubles) that occurred in region II = 8%+ 2% = 10% or 10 map units between the loci *B* and *C*. The *A – C* distance is, therefore, 30 map units when double crossovers are detected in a three point linkage experiment and 26 map units when double crossovers are undetected in the two-point linkage experiment above.

Without the middle marker (*B*), double crossovers would appear as parental types and hence we underestimate the true map distance (crossover percentage). In this case the 2% double crossovers would appear with the 72% parental types, making a total of 74% parental types and 26% recombinant types. Therefore, for any three linked genes whose distances are known, the amount of detectable crossovers between the two outer markers *A* and *C* when the middle marker *B* is missing is; (*A – B* crossover percentage) plus (*B – C* crossovers percentage) minus (2 × double crossover percentage).

3. Determination of Gene Order

After determining the relative distances between the genes of a linkage group, it becomes easy to place genes in there proper linear order. For example, if the linear order of three genes *ABC* is to be determined, then these three genes may be in anyone of three different orders depending upon that which gene is in the middle. For the time being we may ignore left and right end alternatives. If double crossovers do not occur, map distances may be treated as completely additive units. Now, if we suppose that the distance between the genes *A – B* = 12, *B – C* = 7 *A – C* = 5, we can determine the order of genes correctly in the following manner:

Case I. Let us assume that gene A is in the middle (*e.g.*, *B – A – C*):

B 12 *A*

A 5 *C*

B 7 *C*

In this case because, the distances between *B – C* are not equitable, genes *A* cannot be in the middle.

Case II. Let us assume that gene *B* is in the middle (*e.g.*, *A – B – C*):

A 12 *B*

B 7 *C*

A 5 *C*

In this case, because the distance between *A – C* are not equitable, therefore, gene *B* cannot be in the middle.

Case III. Let us assume that gene *C* is in the middle (*e.g.*, *A – C – B*).

A 5 *C*	*C* 7 *B*
A 12 *B*	

In this case, because the distances between *A – B* are equitable, therefore, gene *C* must be in the middle.

Thus, the relative distances and ordering of genes in a linkage group are determined in separate segments by two point test crosses or three point crosses, as the case may be.

4. Combining Map Segments.

Finally, the different segments of maps of a complete chromosome are combined to form a complete genetic map of 100 centimorgans long for a chromosome.

Example. For example, suppose we have to combine following three map segments:

(*i*) | *a* 8 *b* 10 *c* |

(*ii*) | *c* 10 *b* 22 *d* |

(*iii*) | *c* 30 *e* 2 *d* |

We can superimpose each of these segments by aligning the genes shared in common.

(*i*) | *a* 8 b 10 *C* |

(*ii*) | *d* 22 *b* 10 *c* |

(*iii*) | *d* 2 *e* 30 *c* |

Then finally we may combine the three segments into one map:

The *a* to *d* distance = (*d* to *b*) – (*a* to *b*) = 22 – 8 = 14

The *a* to *e* distance = (*a* to *d*) – (*d* to *e*) = 14 – 2 = 12

d 2 *e* 12 *a* 8 *b* 10 *c*

Model Organism for Gene Mapping

Currently, there is an intense collaborative international efforts to generate genetic and physical maps of several model organisms, *e.g.*,

1. Bacteria (*E. coli, B. sublitis*);
2. Yeast (*Sacchromyces cerevisiae* and *S. pombe*);
3. Nematode (*Caenorhabditis elegans*);
4. Fruit fly (*Drosophila melanogaster*);
5. Puffer fish (*Fugu rubripes*);
6. Zebra fish (*Danio rerio*);
7. Mouse and rats;
8. Commercial and domestic animals (cattle, pigs, sheep, cats and dogs);
9. Human beings (*Homo sapiens*);
10. Angiosperm plant (*Arabidopsis thaliana*);
11. Plants of commercial value, *e.g.,* rice, wheat and maize.

The first linkage map has been constructed for two chromosomes of *Drosophila* by **Strutevant** in 1911. The linkage maps of other chromosomes of *Drosophila* have been constructed by **C.B. Bridges.** The linkage or genetic mapping in maize has been done by **McClintock** under the leadership of **R.A. Emerson**.

Significance of Chromosome Maps

The study of chromosome maps reveals following facts:

1. Shows the number of chromosomes and the number of linkage groups of an organism.
2. Shows exact location, arrangement and exact sequence of genes on a chromosome.
3. Shows the relative distance and chances of crossing over between the genes on each of the chromosome pair.

78.2 INTERFERENCE

In most higher organisms it has been found that one chiasma formation reduces the probability of another chiasma formation in an immediately adjacent region of the chromosome, probably because of physical inability of the chromatids to bend back upon themselves within certain minimum distances. The tendency of one crossover to interfere with the other crossover is called **interference**. The phenomenon of interference was discovered by **Muller** in 1911. It is found to be **inversely proportional** to the distance between the points of crossing over. It means interference is maximum over a short distance and decreases as the distance increases. In *Drosophila* interference prevents double crossing over almost together between genes ten units apart.

Types of Interference

1. Positive interference. Where one crossover inhibits another.

2. Negative interference. Where one crossover stimulates another.

In eukaryotes, it is generally observed that recombination shows **positive interference,** *i.e.,* one crossover inhibits the initiation of a second crossover nearby. Thus, when the double crossovers are altogether absent, we can say that interference is cent-per cent. If the frequency of double crossovers is equal to the expected value, it means that there is no interference. Recombination between very close markers shows evidence of **negative interference,** *i.e.,* several crossovers appear to be clustered.

Factors controlling interference. Interference varies from chromosome to chromosome and within a chromosome from one point to the other. Generally, it is maximum near the centromere and at the two ends of the chromosome (telomeres). Therefore, for accurate mapping interference in different parts of different chromosomes should be taken into consideration.

78.3 CROSSING OVER AMONG THREE LINKED GENES

A test-cross for three linked genes, *e.g.*, *C* (colourless aleurone), *sh* (shrunken grains) and *wx* (waxy endosperm) in maize/corn, yields eight types of gametes and phenotypes (Table 78.1). Remember that in a test-cross, the phenotype of a progeny is not affected by the test-cross parent as it contributes only the recessive alleles of the genes. Two of the eight types of progeny (*e.g.*, *C, Wx, Sh* and *c wx sh*) are the most frequent and represent the parent or non-recombinant types, while two others (*e.g.*, *C, Wx, sh* and *c wx Sh*) the least frequent and are **double crossovers**. The remaining four types (*e.g.*, *C wx sh, c Wx sh, cWx Sh* and *C wx sh*) are produced by single crossing over between the three linked genes. As a rule, the parental types are the most frequent, the double cross overs are the least frequent, while the single crossovers are intermediate between the two.

Table 78.1 The different types of progeny obtained from a test-cross for three genes *c*, *sh* and *wx* in maize.

Genotype	*Phenotype*	*Number*	*Per cent frequency**
1. *C Wx Sh/c wx sh*	Coloured, nonwaxy, full	2777	39.7
2. *c wx sh/ c wx sh*	Colourless, waxy, shrunken	2708	38.7
3. *C wx sh/c wx sh*	Coloured, waxy, shrunken	116	1.7
4. *c Wx Sh/c wx sh*	Colourless, nonwaxy, full	123	1.8
5. *C wx Sh/ c wx sh*	Coloured, waxy, full	643	9.2
6. *c Wx sh/c wx sh*	Colourless, nonwaxy, shrunken	626	8.9
7. *C Wx sh/c wx sh*	Coloured, nonwaxy,	4	0.06
8. *c wx Sh/c wx sh*	Colourless, waxy full	3	0.04

$$\text{*Per cent frequency} = \frac{\text{Number of progeny in a phenotypic class}}{\text{Total number of progeny}} \times 100$$

Coefficient of Coincidence

Double crossovers are produced by two simultaneous crossing overs one each on either side of the gene located between two genes (*e.g.* the gene *sh* located between *c* and *wx* [Fig. 78.1]). If the occurrence of crossing over in the two regions (*e.g.*, between the genes *c* and *sh*, and between *sh* and *wx*,) were independent of each other, the frequency of double crossovers will be product of the frequencies of crossing overs in two regions. The independence of crossing over in two regions means that the occurrence of crossing over in one regions does not affect the chance of its occurrence in the other region. Therefore, the expected frequency of double crossovers may be estimated from the data of Table 78.1 as follows:

The expected frequency of double crossovers may be estimated from the data of Table 78.1 as follows:

Frequency (%) of crossing over between *c* and *sh* (= α) = 3.6

Frequency (%) of crossing over between *sh* and *wx* (= β) = 18.2

The expected frequency (%) of double crossing overs = the product of the frequencies of crossing overs between *c* and *sh*, and between *sh* and *wx*.

$$= \alpha \times \beta/100$$
$$= (3.6 \times 18.2)/100$$
$$= 0.66\%$$

The product $\alpha\beta$ (= 3.6 × 18.2) is divided by 100 since these frequencies are already in per cent. The frequency observed double crossovers is only 0.1% (= sum of the frequencies of double cross over types *C Wx sh* and *c wx Sh* = 0.06 + 0.04 per cent). The observed frequency is only 15.1 per cent (= 0.1/0.66 × 100) of the expected frequency. The ratio between the observed and the expected frequencies of double crossovers is called **coefficient of coincidence**. Thus,

$$\text{Coefficient of coincidence} = \frac{\text{Observed frequency of double crossovers}}{\text{Expected frequency of double crossovers}}$$

$$= 0.10/0.66 = 0.151$$

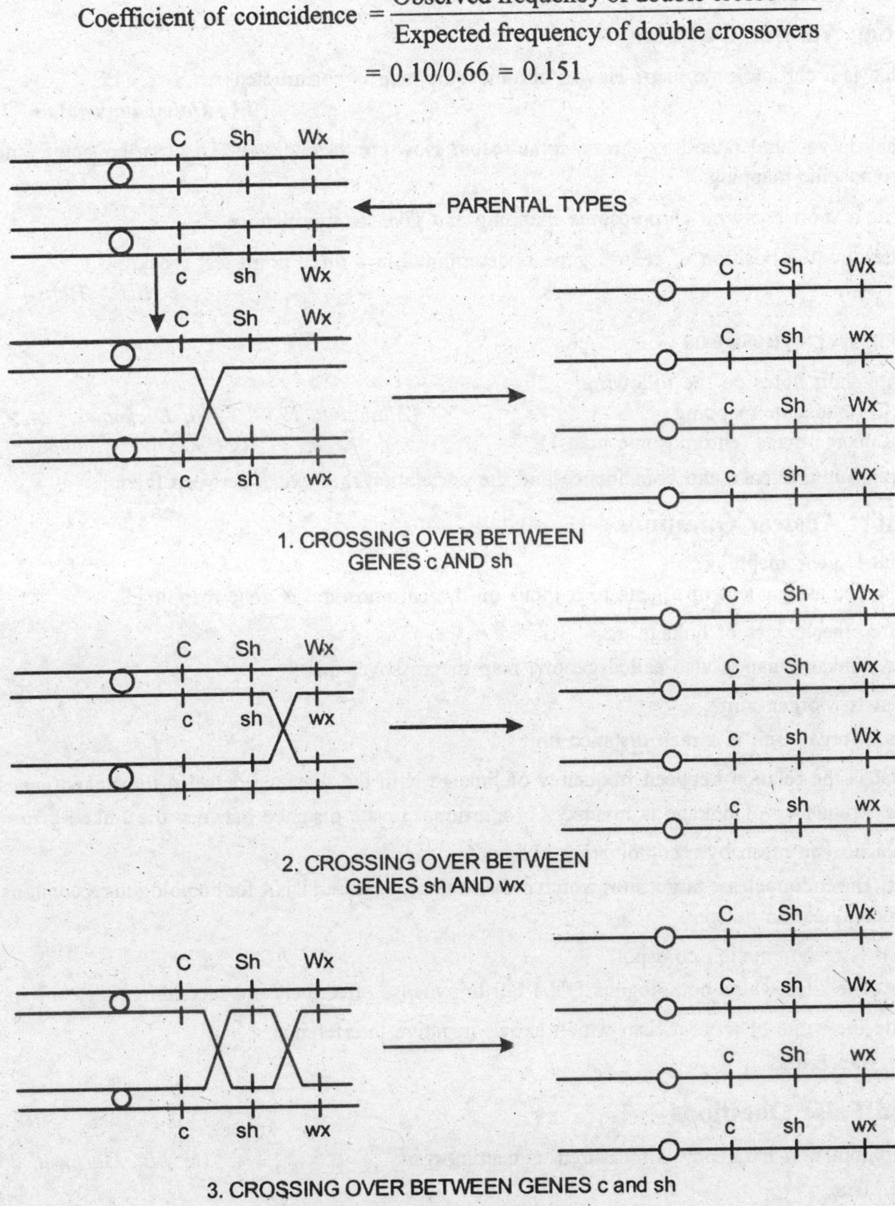

Fig. 78.1. Crossing over in the different regions between three linked genes *c*, *sh* and *wx*, and their consequences (1) crossing over between genes *c*, and *sh*, (2) crossing over between *sh* and *wx* and (3) simultaneous crossing over between genes *c*, and *sh*, and between *sh* and *wx* (double crossing over). Note that in double crossing over only the gene located in the middle is exchanged.

Coefficient of coincidence (%) = 0.151 × 100 = 15.1%.

The estimate of coefficient indicates the degree of agreement between the observed and the expected frequencies of double crossovers.

QUESTIONS

Very Long Answer Questions

1. What is a chromosome map? How a chromosome map is constructed? (*B.U. Coimbatore (Main) 1997*)
2. What do you understand by chromosome maps? How are these drawn? Discuss the significance of chromosome mapping.
3. Write a short essay on chromosome mapping and give its significance.
4. State how the position of central gene is determined in a three point test cross? (*BHU (Hons.) 1991*)

Short Answer Questions

1. Write short notes on the following:
   (*i*) Chromosome mapping (*Allahabad 1989; 1990; Lucknow 1994, 1996*)
   (*ii*) Linkage maps (chromosome maps) (*CCS 1989; Allahabad 1989*)
2. Explain interference and coincidence and the correlation that exists between them.

Very Short Answer Questions

1. What is gene mapping?
   Ans. The assignment of a gene to a locus on the chromosome is gene mapping.
2. Write other names of linkage map.
   Ans. Linkage map is also called genetic map or crossover map.
3. What is Morgan unit?
   Ans. Morgan unit is a map distance unit.
4. What is the relation between frequency of linkage with the distance between the linked genes?
   Ans. Frequency of linkage is inversely proportional to the distance between the linked genes.
5. What do you mean by recombination hotspot?
   Ans. The endonuclease target sites which provide single stranded DNA for homologous recombination is recombination hotspot.
6. What is recombination coldspot?
   Ans. The sites where homologous DNA fail to synapse effectively are recombination cold spots.
7. Write the name of an organism which shows negative interference.
   Ans. *Aspergillus*.

True and False Questions

1. Recombination frequency is measured as centimorgan. (*H.N.B., Garawal, 2008*)
   Ans. True.

Multiple Choice Questions

1. The unit used in indicating the distance between two genes is
   (*a*) centimeters (*b*) centimorgans
   (*c*) microns (*d*) angstrom
2. For the preparation of genetic maps, the recombination frequencies between genes are additive over short distances, but not over long distances due to
   (*a*) multiple cross overs (*b*) lethal mutations
   (*c*) epistasis (*d*) synaptonemal complex

3. Based on the relative proportion of parental and recombinant phenotypes in the test cross progeny involving two recessive genes *r* (round) *y* (yellow) controlling the seed shape and seed colour respectively, the observed percent recombination is 10. The distance between *r* and *y* is

(*a*) 10 map units (*b*) 20 map units

(*c*) 30 map units (*d*) 40 map units

4. Individuals homozygous for *cd* genes were crossed with wild type (++). The F_1 dihybrid thus produced was test crossed. It produced progeny in the following ratio:

+ +	900
c d	880
+ d	115
+ c	105

What is the distance between *c* and *d* genes?

(*a*) 5.75 units (*b*) 11 units

(*c*) 47 units (*d*) 88 units

ANSWERS

Multiple Choice Questions

1. (*b*) **2.** (*a*) **3.** (*a*) **4.** (*b*).

79

Varieties of Gene Expressions: I Interactions of Genes (Factor Hypothesis)

During the discussion of Mendel's monohybrid and dihybrid crosses, we encountered with the fact that for the determination of single phenotypic trait of an organism, two alleles or allelomorphs of a single gene interacted in various ways. Such as out of two allelomorphs of a single gene, one allelomorph might show simple (complete) dominance over the action of other which was recessive; or both allelomorphs might have partial or incomplete dominant relationship or both allelomorphs might have equal expression or codominant relationship. These kinds of genic or genetic interactions occur in between the two allelomorphs of a single type of gene and are usually referred to as **intra-alleic** or **allelic genetic interactions.** These kinds of genetic interactions give the classical ratios of 3 : 1 and 9 : 3 : 3 : 1. But, in addition to intra-allelic genetic interactions, **factor hypothesis, non-allelic** or **inter-allelic genetic (genic) interactions** also occur. In inter-allelic genetic interactions, the independent (non-homologous) genes located on the same or on different chromosomes interact with one another for the expression of single phenotypic trait of an organism. The discovery of the interallelic genetic interactions has been made after Mendel. Some different type of non-allelic interactions which generate modified dihybrid ratios have been described in Table 79.1.

79.1 SUPPLEMENTARY GENES OR COLLABORATOR FACTORS (9 : 3 : 1 : 1)

In certain cases, two pairs of genes determine a same phenotype but assorted independently, produce new phenotypes by mutual non-epistatic interactions and the F_2 phenotypic ratio 9 : 3 : 3 : 1 remains unaltered. Two pairs of genes which interact to affect size and shape of comb but are independently transmitted exist in chicken.

Example

Combs in fowl (9 : 3 : 3 : 1). The classical case of genetic interaction of two genes is discovered by **Bateson** and **Punnett** (1905–1908) in fowls. There are many different breeds of domestic chicken. Each breed possesses a characteristic type of comb. The Wyandotte breed has a comb called "rose", the Brahmas breed has a comb called "pea", and the Leghorns have a comb called 'single'. Each of these types breeds true.

A cross of chicken with a rose comb to one with a single comb produces 3/4 rose and 1/4 single, showing dominance of rose over single. Another cross between pea combed and single combed chickens produces pea and single combed chickens in the ratio of 3 : 1, showing dominance of pea over single. But, when a rose combed chicken crossed with that of pea combed, the F_1 progeny was found with a different type of comb known as 'walnut' (Malay breed). When the F_1 walnut combed chickens were breed together, in F_2 all four types of combs, *i.e.*, 9/16 walnut, 3/16 rose. 3/16 pea and 1/6 single appeared (Fig. 79.1).

Table 79.1 Some different types of non-allelic interactions.

Interaction	*Description*
1. None	Two segregating loci which are unlinked and control distinct characters. Dihybrid ratio is 9 : 3 : 3 : 1.
2. Additive effect	The segregating loci contribute to the same character in an additive fashion. Alleles are generally regarded as positive or neutral in effect and resulting phenotypes is determined by the net contribution of all addition loci. The dihybrid ratio is 9 : 3 : 3 : 1. (**Note.** For further details see chapter 80 of Quantitative Inheritance).
3. Complementary action of genes	A situation where specific alleles at two gene loci are required to generate a particular phenotype. The dihybrid ratio would be 9 : 7 if **dominant** alleles were required at both loci and 15 : 1 if **recessive** alleles were required at both loci.
4. Epistasis	The situation where an allele at one locus prevent or masks the effect of gene expression at a second locus, which is said to be **hypostatic.** Epistasis may be **dominant** or **recessive** and dihybrid ratio would be 12 : 3 : 1 and 9 : 3 : 4, respectively.
5. Suppression	An allele at one locus (the **suppressor allele**) prevents expression of a specific (and usually deleterious) allele at a second locus by compensating in some way for its effect. **The term is used to describe an interaction where a mutation compensates for a second mutation found at different locus and restore the wild-type phenotype.** Suppression may be **dominant** or **recessive** dependent upon whether one or two alleles are required at the *suppressor locus*. The dihybrid ratio would be 12 : 3 : 1 and 9 : 3 : 4, respectively.
	Epistasis differ from suppression in that the former is gene-specific but not allele-specific (*i.e., epistasis prevents the expression of all alleles at a particular locus*), whereas the latter is gene-specific and allele-specific (*i.e.*, suppression compensates for the effect of a particular allele at the second locus).
6. Synergism, Enhancement	Similar to suppression mechanism expect that the **enhancer allele** specifically increases the effect of second mutation instead of suppressing it. As with suppressor alleles, enhancer alleles may be dominant or recessive. Synergic enhancement differs from additive effects because the enhancer locus alone does not contribute to the phenotype associated with the target (enhanced) locus, *i.e.*, if the target allele is not present, the enhancer locus has no effect on the phenotype.

These peculiar results were interpreted by **Bateson** and **Punnett** as follows: The rose comb is caused by the combination of homozygous recessive genes "*pp*" and homozygous or heterozygous dominant genes '*RR*' or '*Rr*'. The pea comb is supposed to be produced by combination of a homozygous recessive condition (*rr*) and homozygous or heterozygous dominant condition (*PP* or *Pp*). While, the single type comb is produced by the double recessive, *rrpp*, genes. Thus, *R* gene determines the shape of rose comb and *P* gene determines the shape of pea comb, but when both genes happen to come together in single individual due to cross between rose and pea combed chickens, they interact to produce a walnut comb in F_1. In the cross of two walnut chickens, two genes interact variously and produces four types of offsprings in F_2.

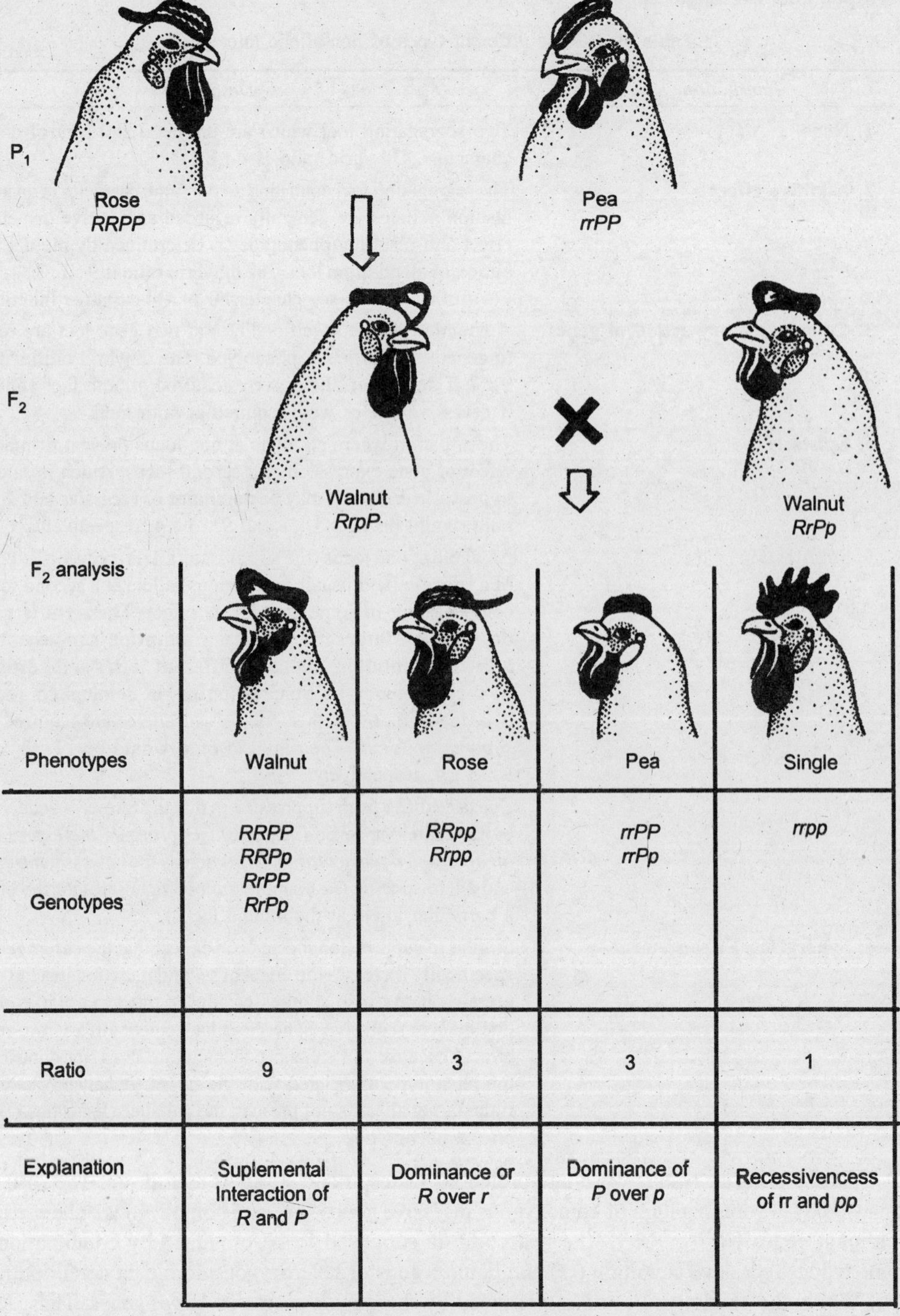

Fig. 79.1. Supplementary inheritance. Inheritance of the shape of the comb in chickens.

Thus, here two pairs of genes interact to produce comb size and shape in fowl. During the inheritance of combs in fowls, the genes themselves do not determine the development of a character (presence or absence of comb) and simply modify a character determined by a basic gene and, therefore, known as **supplementary** or **modifying genes.**

79. 2 DOMINANT COMPLEMENTARY GENES OR DUPLICATE RECESSIVE GENES (9 : 7)

If both gene loci have homozygous recessive alleles and both of them produce identical phenotypes, the F_2 ratio 9 : 3 : 3 : 1 would become 9 : 7. In such case, the genotypes *aa BB, aa Bb, AA bb, Aa bb,* and *aa bb* produce one phenotype (Table 79.2). Both dominant alleles when present together, complement each other and are called **dominant complementary genes** and produce a different phenotype. A case of such complemental inheritance, resulting from the combined action of complemental genes is known in sweet peas.

Table 79.2 Mode of action of complementary genes.

Epistatic allele	*Hypostatic allele*	*Phenotypic expression of alleles*
1. *aa*	*BB, Bb, bb*	No phenotype production
2. *AA, Aa, aa*	*bb*	
3. *AA, Aa*	*BB, Bb*	Phenotypie expression due to dominant alleles on both loci (complementation of gene action)

Example

When a pure line variety of white flowered sweet pea (*Lathyrus odoratus*) was crossed with another pure line variety of white flowered sweet pea, in F_1 purple or red flowered plants were produced (Table 79.3). The F_1 plants when self-pollinated or crossed among themselves, produced the F_2 generation with the phenotypic ratio of 9 coloured and 7 white flowered plants.

Table 79.3 Mode of action of complementary alleles in the production of coloured flowers in sweet pea.

Epistatic alleles	*Hypostatic alleles*	*Phenotypic expression of allele*	*F_2 Phenotypic ratio*
1. *c c*	*EE, Ee, ee*	Neither *c* nor *E* or *e*	White = 7
2. *CC, Cc*	*e e*	Neither *C* or *c* nor *e*	
3. *CC, Cc*	*EE, Ee*	Both *C* + *E* (complementation)*	Coloured = 9

* Chromogen production due to complementation of *C* and *E* (*i.e.*, dominant epistatic and hypostatic alleles).

These surprising results could be understood by analysing the mechanism of colour production in flowers. A given enzyme (genetically controlled as to absence or presence in a given individual) acts upon chromogen (a colourless colour base whose absence or presence is also genetically controlled) to produce the purple or red colour of flowers. The dominant allele or alleles (*CC* or *Cc*) of gene *C* are responsible for the presence of chromogen, while the homozygous recessive alleles (*cc*) of this gene are responsible for the absence of chromogen. Likewise, the dominant alleles of gene E in homozygous (*EE*) or heterozygous (*Ee*) conditions caused the production of an enzyme which is necessary for colour production from chromogen, while homozygous recessive (*ee*) condition does not produce any such enzyme.

The appearance of 9 : 7 ratio instead of 9 : 3 : 3 : 1 ratio from the cross of two white flowered sweet pea plants (*CC ee* × *cc EE*) can be illustrated in Fig. 79.2.

79. 3. RECESSIVE COMPLEMENTARY GENE OR DUPLICATE DOMINANT GENE (15 : 1).

If the dominant alleles of both gene loci produce the same phenotype without cumulative effect, the 9 : 3 : 3 : 1 ratio is modified into a 15 : 1 ratio (Table 79.4). They are also called **recessive complementary genes** since both *aa* and *bb* are necessary for the phenotype.

P:	White flowers	×	White flower
	CC ee		*cc EE*
	↓	↓	↓
P gametes:	(*Ce*)		(*cE*)
F_1:	Purple flower	×	Purple flower
	Ce Ee		*Cc Ee*

F_1 Female gemetes →

F_1 Male gemetes ↓

F_2:

	CE	*Ce*	*cE*	*ce*
CE	*CC EE* Purple	*CC Ee* Purple	*Cc EE* Purple	*Cc Ee* Purple
Ce	*CC Ee* Purple	*CC ee* White	*Cc Ee* Purple	*Cc ee* White
cE	*Cc EE* Purple	*Cc Ee* Purple	*cc EE* White	*cc Ee* White
ce	*Cc Ee* Purple	*Cc ee* White	*cc Ee* White	*cc ee* White

F_2 phenotypic ratio: 9/16 Purple : 7/16 White or 9 : 7.

Fig. 79.2. Complementary genes. Production of 9 : 7 phenotypic ratio in sweet peas due to complementary genes.

Table 79.4 Mode of action of duplicate dominant genes.

Epistatic allele	*Hypostatic allele*	*Phenotypic expression of alleles*
1. *AA, Aa*	*bb*	Same phenotype (*A, B, A + B*) (Presence of one dominant allele on both or either loci result in same phenotype)
2. *AA, Aa*	*BB, Bb*	
3. *a a*	*BB, Bb*	
4. *a a*	*bb*	Another phenotype (*a* or *b*) (Absence of dominant allele on both or either loci results in different pnenotype)

Example

The seed capsules of shepherd's purse (genus *Capsella*) occur in two different shapes, *i.e.*, triangular and top-shaped. When a plant with triangular seed capsule is crossed with one having top-shaped capsule, in F_1 only triangular, character appears. The F_1 offsprings by self crossing produce the F_2 generation with the triangular and top-shaped seed capsules in the ratio of 15 : 1. Two independently segregating dominant genes (*A* and *B*) have been found to influence the shape of the capsule in the same way (Table 79.5). All genotypes having dominant alleles of both of these genes (*A* and B) would produce plants with triangular-shaped capsules. Only those with the genotype *aa bb* would produce plants with top-shaped capsules. The results of this example has been shown in Fig. 79.3.

Table 79.5 Mode of action of duplicate genes in producing triangular seed capsules in *Capsella*.

Epistatic alleles	*Hypostatic alleles*	*Phenotypic expression of allele*	*Phenotypic ratio*
1. *AA, Aa*	*bb*	Phenotype of either or both dominant alleles, *i.e., A* or *B* or both	Triangular = 15
2. *AA, Aa*	*BB; Bb*		
3. *a a*	*BB, Bb*		
4. *a a*	*bb*	Recessive phenotype of *a* and *b* alleles	Top-shaped = 1

P:	Triangular *AA BB*	× ↓	Top-shaped *aa bb*
P gametes:	(*AB*)		(*ab*)
F_1:	Triangular *Aa Bb*	×	Triangular *Aa Bb*

F_1 Female gemetes → / **F_1 Male gemetes ↓**	*AB*	*Ab*	*aB*	*ab*
AB	*AA BB* Triangular	*AA Bb* Triangular	*Aa BB* Triangular	*Aa Bb* Triangular
Ab	*AA Bb* Triangular	*AA bb* Triangular	*Aa Bb* Triangular	*Aa bb* Triangular
F_2: *aB*	*Aa BB* Triangular	*Aa Bb* Triangular	*aa BB* Triangular	*aa Bb* Triangular
ab	*Aa Bb* Triangular	*Aa bb* Triangular	*aa Bb* Triangular	*aa bb* Top shaped

F_2 phenotypic ratio: 15/6 triangular : 1/16 Top-shaped or 15 : 1.

Fig. 79.3. Duplicate dominant genes: a cross between two strains of *Capsella* having triangular and top-shaped seed capsules to get 15 : 1 F_2 dihybrid ratio.

79.4 EPISTASIS

The gene is a chemical determiner. Whereas a phenotypic trait results from the combined action of many genes and their products constantly interacting with the environment. The environment includes not only ecological factors such as temperature and the amount or quality of light, but also internal factors such as hormones and enzymes. The enzymes are proteins and the specific molecular organization of protein is determined by genes. The enzymatic proteins perform catalytic function in various cellular chemical (metabolic) reactions and causing the splitting or union of various molecules. Each cellular chemical reaction involves stepwise conversion of one substance called **precursor** into another, called **end product**. Each step being mediated by a specific enzyme. All the subsequent steps of a chemical reaction constitute the **biosynthetic pathway**.

Thus, a simplest biosynthetic pathway includes various steps, each step is catalyzed by a specific enzymatic protein and each enzymatic protein in its turn depends on a specific gene for its production. For example, we may consider a simple biosynthetic pathway which transforms a precursor substance 'P' into the end product 'C' in following three subsequent steps:

$$\begin{array}{ccccccccc} & g_1^+ & & & g_2^+ & & & g_3^+ & \\ & \downarrow & & & \downarrow & & & \downarrow & \\ \textbf{P}\text{ (precursor)} — & e_1 & \longrightarrow & \textbf{A} — & e_2 & \longrightarrow & \textbf{B} — & e_3 & \longrightarrow \textbf{C}\text{ (end product)} \end{array}$$

In this biosynthetic pathway each metabolite (A, B, C) is produced by the catalytic action of different enzymes (e_1, e_2, e_3 ... e_x, specified by different wild type genes (g_1^+, g_2^+, g_3^+, ... g_x^+). When more than two or more genes become involved in the specification of enzymes for different steps of a common biosynthetic pathway, the phenomenon of genetic interaction occurs. If substance C is essential for the production of a normal phenotype and the recessive mutant alleles g_1, g_2, g_3 produce defective enzymes, then a mutant or abnormal phenotype would result from a genotype homozygous recessive at any of the three loci. If wild gene g_3^+ becomes mutant, the conversion of metabolite B to C does not occur and substance B tends to accumulate in excessive quantity; if g_2^+ becomes mutant, substance A will accumulate. Thus, the mutant genes caused "**metabolic blocks**" in synthetic pathway.

An organism with a mutation only in gene g_2^+ could produce a normal phenotype, if it was given either substance B or C, but an organism with a mutation in gene g_3^+ has a specific requirement for substance C for the production of normal phenotype. Thus, gene g_3^+ becomes dependent upon g_2^+ for its expression as a normal phenotype. If the genotype is homozygous for the recessive g_2 allele, then the biosynthetic pathway ends with substance A. Neither g_3^+ nor its recessive allele g_3 has any effect on the phenotype. Thus, genotype (g_2g_2) can hide or mask the phenotypic expression of alleles at the g_3^+ locus. Originally a gene or locus which suppressed or masked the action of a gene at another locus was termed **epistatic gene.** The gene or locus which was suppressed by a epistatic gene was called **hypostatic gene**. Later studies revealed the fact that both loci or genes (*i.e.*, epistatic and hypostatic) could be epistatic to one another. Presently, the term epistasis (Greek, standing upon) is used for almost any type of allelic genetic interaction.

Difference Between Dominance and Epistasis

The phenomenon of dominance involves intra-allelic gene suppression, or the masking effect which one allele has upon the expression of another allele at the same locus, while the phenomenon of epistasis involves inter-allelic gene suppression or the masking effect which one gene locus has upon the expression of another. The classical phenotypic ratio of 9 : 3 : 3 : 1 observed in the progeny of dihybrid parents becomes modified by epistasis into ratios which are various combinations of the 9 : 3 : 3 : 1 groupings.

Epistasis is of following two main types:

I. Dominant Epistasis (12 : 3 : 1)

When out of two genes, the dominant allele (*e.g.*, *A*) of one gene masked the activity of alleles of another gene (*e.g.*, *B*) and expressed itself phenotypically, then *A* gene locus is said to be epistatic to the *B* gene locus (Table 79.6). Because, the dominant allele *A* can express itself only in the presence of either *B* or *b* allele, therefore, such type of epistasis is termed as **dominant epistasis**. The alleles of hypostatic locus or gene *B* will be able to express themselves phenotypically only when gene locus *A* may contain two recessive alleles (*aa*). Thus, the genotype *AA BB* or *Aa Bb* and *AA bb* or *Aa bb* produce the same phenotype whereas the genotype *aa BB* or *aa Bb* and *aa bb* produce two additional phenotypes. The dominant epistasis modify the classical ratio of 9 : 3 : 3 : 1 into 12 : 3 : 1 ratio.

Table 79.6 Mode of dominant epistasis.

Epistatic allele	*Hypostatic allele*	*Phenotypic expression of alleles*
1. *AA, Aa*	*BB, Bb, bb*	*A*
2. *aa*	*BB, Bb*	*B*
3. *aa*	*bb*	*b*

Example

Dominant epistatis in dogs. Among dogs, the colours of coats depend upon the action of two genes. One gene locus has a dominant epistatic inhibitor allele (*I*) of coat colour pigment (see Table 79.7). The allele *I* prevents the expression of colour allele at another independently assorting, hypostatic gene locus (*B* or *b*) and produces white coat colour. The alleles of hypostatic gene locus (*BB, Bb*, or *bb*) express only when two recessive alleles (*II*) occur on the epistatic locus, *i.e., ii BB* or *ii Bb* produces black and *ii bb* produces brown individuals. When two such white coat colour dogs are crossed, in F_1 the white, black and brown coat colours appear in 12 : 3 : 1 ratio as shown in Fig. 79.4.

Table 79.7 Mechanism of dominant epistasis in dogs.

Epistatic alleles	*Hypostatic alleles*	*Phenotypic expression of allele*	F_2 *Phenotypic ratio*
1. *II, Ii*	*BB, Bb, bb*	*I* (no pigment)	White = 12
2. *ii*	*BB, Bb*	*B* (Black)	Black = 3
3. *ii*	*bb*	*b* (Brown)	Brown = 1

White (Male) × White (Female)

P: *Ii Bb* *Ii Bb*

P Female gemetes → / **P Male gemetes ↓**	*IB*	*Ib*	*iB*	*ib*
IB	*II BB* White	*II Bb* White	*Ii BB* White	*Ii Bb* White
Ib	*II Bb* White	*II bb* White	*Ii Bb* White	*Ii bb* White
iB	*Ii BB* White	*Ii Bb* White	*ii BB* Black	*ii Bb* Black
ib	*Ii Bb* White	*Ii bb* White	*ii Bb* Black	*ii bb* Brown

F_1:

F_1 **Phenotypic ratio:** 12/16 White : 3/16 Black : 1/16 Brown or 12 : 3 : 1.

Fig. 79.4. Checkerboard derived from a cross between two heterozygous white coated dogs showing 12 : 3 : 1 ratio due to dominant epistatic inhibitor genes.

2. Recessive Epistasis (9 : 3 : 4)

Sometimes the recessive alleles of one gene locus (*aa*) mask the action (phenotypic expression) of alleles of another gene locus (*BB, Bb* or *bb* alleles) (see Table 79.8). This type of epistasis is called **recessive epistasis.** The alleles of *B* locus express themselves only when epistatic locus

Table 79.8 Mode of action of recessive epistasis.

Epistatic allele	*Hypostatic allele*	*Phenotypic expression of alleles*
1. *a/a*	*BB, Bb, bb*	*a*
2. *AA, Aa*	*BB, Bb*	*B*
3. *AA, Aa*	*bb*	*b*

has dominant alleles (*e.g.*, *AA* or *Aa*). Due to recessive epistasis the phenotypic ratio 9 : 3 : 3 : 1 becomes modified into 9 : 3 : 4 ratio.

Example

Recessive epistasis in mice. In mice various types of epistatic genetic interactions have been reported. The most interesting case is of recessive epistasis in coat colours. The common house mouse occurs in a number of coat colours, *i.e.*, agouti, black and albino. The agouti colour pattern is commonly occurred one (wild type) and is characterized by colour banded hairs in which the part nearest the skin is gray, then a yellow band and finally the distal part is either black or brown. The albino mouse lacks totally in pigments and has white hairs and pink eyes.

Table 79.9 Mode of action of recessive epistasis in alleles for coat colour in mice.

Epistatic alleles	*Hypostatic alleles*	*Phenotypic expression of allele*	*F_2 Phenotypic ratio*
1. *cc*	*AA, Aa, aa*	*c*	Abino = 4
2. *CC, Cc*	*AA, Aa*	*A*	Agouti = 9 (due to supplementary genes)
3. *CC, Cc*	*aa*	*a*	Black = 3

When a homozygous black (*CC aa*) is crossed with a homozygous albino (*cc AA*) in F_1 all agouti (*Cc Aa*) offsprings appear. When, the F_1 agouti are crossed among themselves in F_2 agouti, black and albino offsprings appear in the ratio of 9 : 3 : 4 as shown in the Fig. 79.5.

P_1: Black × Albino
CCaa *cc AA*

P_1 gametes: (*Ca*) ↓ (*cA*)

F_1: Agouti
Cc Aa

F_1 Male gemetes → **F_1 Female gemetes ↓**	*CA*	*Ca*	*cA*	*ca*
CA	*CC AA* Agouti	*CC Aa* Agouti	*Cc AA* Agouti	*Cc Aa* Agouti
Ca	*CC Aa* Agouti	*CC aa* Black	*Cc Aa* Agouti	*Cc aa* Black
F_2: *cA*	*Cc AA* Agouti	*Cc Aa* Agouti	*cc AA* Albino	*cc Aa* Albino
ca	*Cc Aa* Agouti	*Cc aa* Black	*cc Aa* Albino	*cc aa* Albino

F_2 Phenotypic ratio: 9/16 Agouti : 3/16 Black : 4/16 Albino.

Fig. 79.5. Checkerboard derived from a cross between coloured (black) and albino mice showing 9 : 3 : 4 ratio due to recessive epistatic genes.

Supplementary genes. In the cross between black (*CC aa*) and albino (*cc AA*) mice, one thing becomes apparent that two independent pairs of genes (*i.e.*, $C - c$ and $A - a$) have interacted in the production of the phenotypic trait (*i.e.*, coat colour) in such a way that one dominant (*C*) produces its effect whether or not the second (*A*) is present, but the second (*A*) gene can produce its effect only in the presence of the first. These genes (*i.e.*, *C* and *A*) have been termed as **supplementary genes**.

79.5 DOMINANT AND RECESSIVE INTERACTIONS (13:3)

Sometimes, the dominant alleles of one gene locus (*A*) in homozygous (*AA*) and heterozygous (*Aa*) condition and the homozygous recessive alleles (*bb*) of another gene locus (*B*) produce the same phenotype, the F_2 phenotypic ratio becomes 13 : 3 instead of 9 : 3 : 3 : 1. In such case, the genotype *AA BB, AA Bb, AA bb, Aa BB, Aa Bb* and *Aa bb* produce same phenotype and the genotype *aa BB, aa Bb* and *aa bb* produce another but same phenotype (Table 79.10).

Table 79.10 Mode of interaction between dominant and recessive alleles.

Epistatic allele	*Hypostatic allele*	*Phenotypic expression of alleles*
1. *AA, Aa*	*BB, Bb, bb*	*A* inhibits *B* or *b* (No phenotype of *B* or *b*)
2. *a a*	*BB, Bb, bb*	*a* does not inhibit *B* or *b* (phenotype of *B* or *b*)

Example

In Leghorn type of fowl the white colour of feather is caused by the dominant genotype *CC II*, similarly the white colour of feathers of Plymouth Rock breed is caused by the recessive genotype *cc ii* (Table 79.11). When both white varieties of fowl are crossed, in F_1 white coloured hybrids appear. The F_1 hybrids in F_2 produce the white and coloured offsprings in the ratio of 13 : 3, as have been illustrated in Fig. 79.6.

Table 79.11 Mode of interaction between dominant and recessive genes.

Epistatic alleles	*Hypostatic alleles*	*Phenotypic expression of allele*	*F_2 Phenotypic ratio*
1. *II, Ii*	*CC, Cc, cc*	*I* (dominant inhibitor)	White = 12
2. *ii*	*CC, Cc*	*C* (due to recessive inhibitor *i*)	Coloured = 3
3. *ii*	*cc*	*c* (due to c and *i*)	White = 1

	White Leghorn		White Plymouth Rock
P:	*CC II*	×	*cc ii*
	↓	↓	↓
P gametes:	(*C I*)		(*c i*)
F_1:	White	×	White
	Cc Ii		*Cc Ii*

F_1 Female gemetes → **F_1 Male gemetes ↓**	*CI*	*Ci*	*cI*	*ci*
CI	*CC II* White	*CC Ii* White	*Cc II* White	*Cc Ii* White
Ci	*CC Ii* White	*CC ii* Coloured	*Cc Ii* White	*Cc ii* Coloured
F_2: *cI*	*Cc II* White	*Cc Ii* White	*cc II* White	*cc Ii* White
ci	*Cc Ii* White	*Cc ii* Coloured	*cc Ii* White	*cc ii* White

F_2 Phenotypic ratio: 13/16 White : 3/16 coloured or 13 : 3

Fig. 79.6. A cross between two white coloured breeds of fowls to get 13 : 3 F_2 dihybrid ratio.

QUESTIONS

Long Answer Questions

1. What is factor hypothesis? Write various types of interactions of genes. Explain with example any one type of interaction of gene. *(Jiwaji 1990; Bundelkhand 1990; Vikram 1991; Punjab 1992; CCS 1991, 1992, 1995)*
2. Write an essay on the interaction of genes. *(CCS 2009; Rajasthan 1991; Delhi 1991; Vikram 1991)*
3. What do you understand by gene interaction? Explain with suitable examples supplementary and complementary genes. *(Jiwaji 1990)*
4. What is Interaction of genes? Describe it with a suitable example. *(Avadh 1996)*
5. What do you understand by intra-allelic and interallelic interactions? Explain the interallelic interaction by taking suitable examples.
6. Explain the inheritance of complementary genes. *(Bangalore 1994)*
7. What does gene interaction means? How does epistasis differ from dominance?
8. What is epistasis of gene? Explain with suitable examples. *(Avadh 1991)*
9. Explain with the help of suitable examples the modification of Mendelian 9 : 3 : 3 : 1 ratio to (*i*) 9 : 7; (*ii*) 9 : 3 : 4; and (*iii*) 13.3. *(Bhagalpur 1988)*
10. What do you understand by 'Interaction of genes'? What is the difference between 'Dominance and Epistasis'? Describe with examples. *(Avadh 1999)*
11. Discuss the different type of gene interactions. *(H.N.B., Garhwal 2008, 2009)*
12. Describe the kinds of epistatic interaction with suitable examples. *(M.S.U. Tamil Nadu 2006)*

Short Answer Questions

1. Write short note on the following:
   (*i*) Epistasis *(CCS 1989, 1999; BHU (Hons.) 1990; Kanpur 1996; Lucknow 1997)*
   (*ii*) Interaction of genes *(Rohilkhand 1989)*
   (*iii*) Intra-allelic and inter-allelic interaction
   (*iv*) Complementary genes *(Lucknow 1992; CCS 1996)*
2. Write differences between following:
   Complementary genes and supplementary genes. *(Lucknow 1991, 1994, 1996; CCS 1994, 1996)*

Very Short Answer Questions

1. Who proposed factor hypothesis?
   Ans. Bateson.
2. Define epistasis.
   Ans. Epistasis is masking of the action of alleles of one gene by allelic combinations of another.
3. Define hypostatic gene.
   Ans. A gene whose expression is masked by an epistatic gene.
4. Name the type of genetic interaction in which action of one non-allelic gene is completed by the other.
   Ans. Complementary gene interaction.
5. How does a walnut comb fowl produced in a generation?
   Ans. A walnut comb of fowl is produced when the dominant alleles of rose (R) and pea (P) are present together ($R - P -$).

6. What phenotypic ratio is found in the F_2 generation in a cross between two varieties having duplicated genes for a single trait?

   Ans. 15 : 1.

7. What is the result of cross between two white flowered varieties (*CCpp* × *CcPp*) of *Lathyrus odoratus*?

   Ans. In F_1 generation purple flowered (*CcPp*) variety is produced.

8. What is complementation?

   Ans. Complementation means production of the wild-type phenotypes by a cell or an organism that contains two mutant genes. If complementation occurs, the mutants are almost certainly nonallelic.

Multiple Choice Questions

1. The phenomenon in which an allele of one gene suppresses the activity of an allele of another gene, is known as

   (*a*) segregation (*b*) dominance
   (*c*) epistasis (*d*) suppression

2. The ratio of 9 : 7 is due to

   (*a*) lethal genes (*b*) supplementary genes
   (*c*) complementary genes (*d*) epistatic gene

3. In the duplicate dominant gene for shape of fruit in shepherd's purse (*Capsella*), the ratio of progeny in F_2 generation is

   (*a*) 9 : 3 : 3 : 1 (*b*) 12 : 3 : 1
   (*c*) 13 : 1 (*d*) 15 : 1

ANSWERS

1. (*c*) **2.** (*c*) **3.** (*d*).

80

Varieties of Gene Expressions: II – Polygenic Inheritance, Lethal Genes and Pleiotropic Genes

80.1 POLYGENIC INHERITANCE

Qualitative and Quantitative Traits. The phenotypic traits of the different organisms may be of two kinds, *viz.*, qualitative and quantitative. The **qualitative traits** are the classical Mendelian traits of **kinds** such as form (*e.g.*, round or wrinkled seeds of pea); structure (*e.g.*, horned or hornless condition in cattles); pigments (*e.g.*, black or white coat of guinea pigs); and antigens and antibodies (*e.g.*, blood group types of human beings) and so on. Each qualitative trait may be under genetic control of two or many alleles of a single gene with little or no environmental modifications to obscure the gene effects. The organisms possessing qualitative traits have distinct (separate) phenotypic classes and are said to demonstrate **discontinuous variations**. The **quantitative traits**, however, are economically important measurable phenotypic traits of **degree** such as height, weight, shape, skin pigmentation, metabolic activity, reproductive rate, behaviour, eye-facet or bristle number in *Drosophila*, susceptibility to pathological diseases or intelligence in human beings; amount of flowers, fruits, seeds, milk, meat or egg produced by plants or animals, etc. Economically important traits such as body weight gains, mature plant heights, egg or milk production records, yield of grain per acre, etc., are also quantitative traits. The quantitative traits are also called **metric traits**. They do not show clear cut differences between individuals and forms a spectrum of phenotypes which blend imperceptively from one type to another to cause continuous **variations**.

In contrast to qualitative traits, the quantitative traits may be modified variously by the environmental conditions and are usually governed by many factors or genes (perhaps 10 or 100 or more), each contributing such a small amount of phenotype that their individual effects cannot be detected by Mendelian methods but by only statistical methods. Such genes which are non-allelic and affect the phenotype of a single quantitative trait, are called **polygenes** or **cumulative genes**. The inheritance of polygenes or quantitative traits is called **quantitative inheritance**, **multiple factor inheritance**, **multiple gene inheritance** or **polygenic inheritance** (Box 80.1).

Multiple Factor Hypothesis

Two or more different pairs of alleles, with presumed cumulative effects, govern the quantitative traits. Those alleles which contribute to the trait involved are called **contributing, effective** or **active alleles**; those alleles which do not appear to do so are referred to as **non-contributing, non-effective** or **null alleles**. A gene, individually exerting a slight effect on the phenotype but along with a few or many other genes, controls a quantitative trait is called a **polygene** (a term coined by **K. Mather**). Since there are usually many genes of this kind for one quantitative trait, they are named **multiple factors**, each factor having too small an effect to be traced. This **multiple factor hypothesis** (*i.e.*, large number of genes, each with a small effect, are segregating to produce quantitative variation) has long been the basic model of quantitative genetics.

Box 80.1 Some Definitions of Quantitative Inheritance

I. Types of complex characters. Many characters show continuous variation, *i.e.*, phenotypes are measured in quantitative terms and cannot be placed into discrete traits. The phenotype often show a normal distribution about a mean value. Such quantitative characters are inherited in a complex manner: genotype cannot be deduced from phenotype and no simple rules of heredity can be used to predict the outcome of a cross. The inheritance of such characters is studied using statistical methods (**biometrics**).

The discipline of quantitative inheritance embraces following three types of characters:

1. **Continuous characters.** They demonstrate true continuous phenotypic variation (*i.e.*, there exist no boundaries between different phenotypes).
2. **Meristic characters.** They vary in a similar manner to continuous characters but the intrinsic nature of the character itself demands that phenotypes are placed into discrete categories, usually because the value of the phenotype is determined by counting hence such characters may also be termed **countable characters.**
3. **Threshold or dichotomous characters.** They have two phenotypes—a certain condition can be either present or absent. Such characters thus appear very much like Mendelian diallelic traits but in this case an underlying quantitative mechanism controls liability to display the phenotype, which is manifested once some triggering level has been exceeded.

II. Factorial hypothesis. Since quantitative characters are unsuitable for Mendelian analysis, they were originally considered to be controlled by **factors** *which were fundamentally different from the genes that control Mendelian characters.*

III. Major and minor genes. The relative contribution of each gene to the quantitative phenotype would be different-some loci would have strong effects and others only weak effects. These are termed major genes and minor genes, respectively.

Differences between Quantitative and Qualitative Genetics

The genetical studies of qualitative and quantitative traits are called **qualitative genetics** and **quantitative genetics**, respectively. The major differences between the two are following:

Table 80.1 Differences between qualitative and quantitative traits or genetics.

Qualitative genetics	*Quantitative genetics*
1. Characters of kind.	1. Characters of degree.
2. Discontinuous variation; distinct phenotypic classes.	2. Continuous variations; phenotypic measurements form a spectrum.
3 Single gene effects.	3. Polygenic control; effects of single genes too slight to be detected.
4. Concerned with individual matings and their progeny.	4. Concerned with population of organisms consisting of all possible kinds of matings.
5. Analyzed by making counts and ratios.	5. Statistical analysis give estimates of population parameters such as the mean and standard deviation.

Characteristics of Multiple Genes

Multiple genes for quantitative traits have following characteristics:

1. Each contributing allele in the series of multiple genes produces an equal effect.
2. Effects of each contributing allele are cumulative or additive.
3. There is no dominance, rather, there exist pairs of contributing and non-contributing alleles.
4. There is no epistasis (masking of the phenotypes) among genes at different loci.
5. There is no linkage involved.
6. The environmental conditions have considerable effect on the phenotypic expression of polygenes for the quantitative traits. The genotype determines the **range** an individual will occupy with regard to a given quantitative character; environment determines the **point** within the genetically determined range at which an individual's measurement will fall.

Examples of Quantitative Inheritance

1. Kernel colour in wheat. A whole grain or seed of a cereal plant such as corn, wheat, barley, etc., is called **kernel**. Kernel colour in wheat is a quantitative trait and its inheritance was studied by Swedish geneticist **H. Nilsson–Ehle** for the first time in 1908. When he crossed a certain red strain to a white strain, he observed that the F_1 was all light red and that approximately 1/16 of the F_2 was as extreme as the parents, *i.e.,* 1/16 was white and 1/16 was red. He interpreted these results in terms of two genes, each with a pair of alleles exhibiting cumulative effect (Fig. 80.1).

P : Red kernel × White kernel

$R_1\ R_1\ R_2\ R_2$ $\qquad$ $r_1\ r_1\ r_2\ r_2$

F_1 : Light red

$R_1\ r_1\ R_2\ r_2$

F_2 : Summary of checker board derived results, *i.e.,* F_2 genotypic and phenotypic ratios:

Genotype	*Genotypic ratio*	*Number of contributing alleles*	*Phenotype*	*Phenotypic ratio*	
$R_1\ R_1\ R_2\ R_2$	1	4	Red	1	Coloured (15/16)
$R_1\ R_1 R_2\ r_2$ $R_1\ r_1 R_2\ R_2$	2 2	3	Medium red	4	
$R_1\ r_1\ R_2\ r_2$ $R_1\ R_1\ r_2\ r_2$ $r_1\ r_1\ R_2\ R_2$	4 1 1	2	Light red	6	
$R_1 r_1 r_2\ r_2$ $r_1 r_1 R_2 r_2$	2 2	1	Very light red	4	
$r_1\ r_1\ r_2\ r_2$		0	White	1	Colourless (1/16)

Fig. 80.1. Results of a cross between two varieties of wheat having red kernel and white kernel showing cumulative effect of alleles.

Each of the contributing alleles R_1 or R_2 adds some red to the phenotype of kernel colour, so that the genotypes of whites contain neither of these alleles however a red genotype contains only R_1 and R_2 alleles. These results are plotted as histograms in Fig. 80.2. Here five phenotypic classes are obtained in F_2; each 'dose' of a contributing allele for pigment production increases depth of colour. At this stage one point should be clear that in case there were two genes involved, there would be obtained 15 : 1 ratio (15 coloured: 1; see Fig. 80.1).

2. Skin colour in human beings. Another classical example of polygenic inheritance was given by **Davenport** (1913) in Jamaica. He found that two pairs of genes, *A-a* and *B-b* cause the difference in skin pigmentation between Negro and Caucasian people. These genes were found to affect the character in additive fashion. Thus, a true Negro has four dominant genes, *AABB*, and a white has four recessive genes *aabb*. The F_1 offspring of mating of *aabb* with *AABB*, are all *AaBb* and have an intermediate skin colour termed **mulatto**. A mating of two such mulattoes produced a wide variety of skin colour in the offspring, ranging from skins as dark as the original black (negro) parent to as white as the original white parent. The result of this cross have been shown in Fig. 80.3. The F_2 phenotypic and genotypic ratios have been tabulated in Table 80.2.

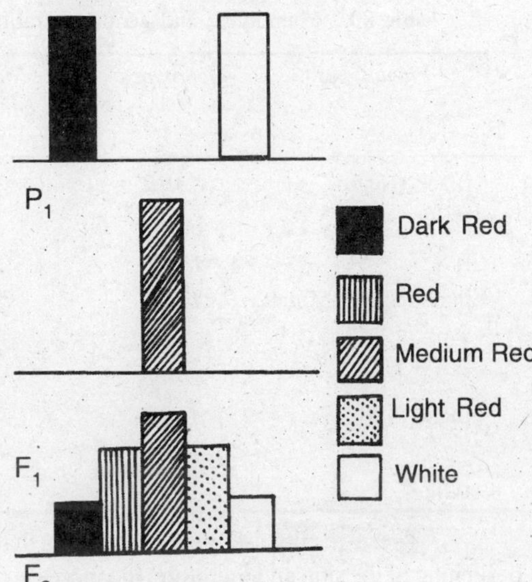

Fig. 80.2. Wheat colour as an example of a quantitative trait.

P_1:	*AABB* Negro ↓	×	*abb* White ↓
P_1 gametes:	*AB*	↓	*ab*
F_1:	○	*AaBb* Mulattoes Intermediate skin colour	○
Intercross:	*Aa Bb* Mulattoes	×	*AaBb* Mulattoes

♀/♂	*AB*	*Ab*	*aB*	*ab*
AB	*AABB* Like negro	*AABb* Darker than mulattoes	*AaBB* Darker than mulattoes	*AaBb* Like mulattoes
Ab	*AABb* Darker than mulattoes	*AAbb* Like mulattoes	*AaBb* Like mulattoes	*Aabb* Lighter than mulattoes
aB	*AaBB* Darker than mulattoes	*AaBb* Like mulattoes	*aaBB* Like mulattoes	*aaBb* Lighter than mulattoes
ab	*AaBb* Like mulattoes	*Aabb* Lighter than mulattoes	*aaBb* Lighter than mulattoes	*aabb* Like white

F_2:	Negro colour	Colour between mulattoes and negro	Colour of mulattoes	Colour between mulattoes and white	White skin colour
	(1/16)	(4/16)	(6/16)	(4/16)	(1/16)

Fig. 80.3. Checkerboard exhibiting segregation of skin colour in F_2 generation from a marriage between a negro and a white person.

Table 80.2 Phenotypic and genotypic ratios of F_2 generation of cross shown in Fig. 80.3.

Phenotypes	*Genotypes*	*Number of contributing alleles*	*Genotypic frequency*	*Ratio*
Black (Negro)	*AABB*	4	1	1
Dark	*Aa BB,*	3	2	4
	AA Bb	3	2	
Intermediate (Mulatto)	*Aa Bb*	2	4	6
	aa BB	2	1	
	AA bb	2	1	
Light	*Aa bb*	1	2	4
	aa Bb	1	2	
White	*aa bb*	0	1	1

These results are clearly showing that A and B genes produce about the same amount of darkening of the skin and therefore, the increase or decrease of *A* and *B* genes cause variable phenotypes in F_2 in the ratio of 1 negro: 4 dark: 6 intermediate: 4 light : 1 white.

Significance of Quantitative Genetics

Quantitative genetics has great agricultural importance and has helped in the increase of yield of verious economically important crops.

Environmental Influence and the Norm of Reaction

In fact, a large number of gene loci is not necessary for continuous variation because all biological characters are influenced to some degree by the **environment** as well as by genotype. The environment in which an organism lives will interact with the genotype to produce the phenotype. Thus, if a single genotype is exposed to a range of environments, a range of phenotypes is observed which is described as the **norm of reaction**. This explains much of the phenotypic variance in the **isogenic populations** (*i.e.*, populations where each individual has the same genotype) as all individuals are not exposed to identical environments. The degree to which a phenotype can be shaped by the environment is described as its **phenotypic plasticity**.

Simple characters (*i.e.*, monogenic Mendelian trait) thus exist because the differences between the mean phenotypic values of each genotype are larger than the norm of reaction for each genotype (In other words, the variance between genotypes is greater than the variance within genotypes). For continuous characters the opposite is true: the differences between the mean phenotypic value of each genotype is smaller than the norm of reaction of each genotype, so that the latter overlap. Such an overalp means that the genotype cannot be predicted from phenotype and Mendelian analysis is impossible – the character is quantitative. By controlling the environment, so that, the norms of reaction are small, continuous character controlled by few loci can be resolved into discrete traits and their transmission can be dissected in terms of Mendelian inheritance.

80.2 LETHAL GENES

Lethal genes are mutant genes and result in the death of the individual which carries them. Death of the individual occurs either in the prenatal or postnatal period prior to sexual maturity. A **fully** (completely) **dominant lethal allele** kills both in homozygous and heterozygous states. Individuals with a dominant lethal allele die before they can leave progeny. Therefore, the mutant dominant lethal is removed from the population in the same generation in which it arose. **Recessive lethal genes** kill only when they are in a homozygous state and they may be of two kinds: 1. One which has no obvious phenotypic effect in heterozygotes and 2. One which exhibits a distinctive phenotype when heterozygous.

The **completely lethal genes** usually cause death of the zygote, later in the embryonic development or even after birth or hatching. Complete lethality, thus, is the case where no individuals of a certain genotype attain the age of reproduction. However in many cases lethal genes become operative at the time the individuals become sexually mature. Such lethal genes which handicap but do not destroy their possessor are called **subvital, sublethal** or **semilethal** genes. The lethal alleles modify the 3 : 1 phenotypic ratio into 2 : 1.

Examples of Lethal Alleles

A. Lethal alleles in plants. In plants, recessive lethal alleles are known which produce **albinism**, where absence of chlorophyll is lethal (fatal) to them. Following two examples illustrate this fact:

1. In snapdragons (*Antirrhinum majus*) three types of plants occur: 1. Green plants with chlorophyll; 2. Yellowish green plants with carotenoids, usually are referred to as pale green, **golden** or **auria** plants and 3. White plants without any chlorophyll. The homozygous green plants have the genotype *CC* and the homozygous white plant has the genotype cc. The auria plants have the genotype *Cc* because they are heterozygotes of green and white plants. When two such auria plants are crossed, the F_1 progeny has identical phenotypic and genotypic ratio of 1 : 2 : 1 (*viz*., 1 green (*CC*): 2 auria (*Cc*) : 1 white (*cc*). But the white plants because lack chlorophyll pigment, therefore, die to modify the ratio of 1 : 2 : 1 into 1 : 2 or 2 : 1. In this case the homozygous recessive genotype (*cc*) is lethal.

F_1 heterozygote:		Auria	×	Auria		
		Cc		*Cc*		
F_2:		1 *Cc*	:	2*Cc*	:	l*cc*
		Green		Auria		White (lethal)
	or	1*CC*	:	2 *Cc*	or	1 : 2

Thus, *c* allele exhibits a lethal effect when homozygous and a distinctive phenotypic effect (*e.g*., auria) when heterozygous.

2. In maize (*Zea mays*) the amount of chlorophyll is controlled by a recessive allele (*g*) which exhibits a lethal effect in homozygous (*gg*) and in heterozygous condition (*Gg*) has phenotype similar to homozygous condition for dominant gene *GG*. It modifies 3 : 1 phenotypic ratio into 2 : 1.

F_1 heterozygote:		Green	×	Green		
		Gg		*Gg*		
F_2:		1 *GG*	:	2 *Gg*	:	l*gg*
		Green		Green		White (lethal)
	or	1*GG*	:	2 *Gg*	or	1 : 2 or 2 : 1.

B. Lethal alleles in animals. Among animals, the following three examples exhibit the role of recessive lethal alleles:

1. The inheritance of mouse body colour was studied by the French geneticist, **L. Cuenot** in 1905. The coat colour of mice is governed by a multiple allelic series (see Chapter 81) in which *A* allele determines **agouti** or mousy-coloured coat, A^Y allele determines **yellow** coat and *a* allele forms **black** coat. The dominance hierarchy is as follows: $A^Y > A > a$. The A^Y allele also acts as a **recessive lethal**, since in the homozygous state (A^YA^Y), it kills the individual in early embryonic stage (*i.e.*, during gastrulation). Thus, when two yellow coated heterozygotes

($A^Y A$) are crossed, they produce a progeny showing a ratio of 2 : 1 since homozygous yellow ($A^Y A^Y$) individuals are never borned due to lethal effect of A^Y gene:

Parents: Yellow $A^Y A$ (Hybrid of yellow and agouti) × Yellow $A^Y a$ (Hybrid of yellow and black)

↓

Progeny: $1A^YA^Y$: $1AY A$: $1AYa$: $1\ Aa$

1 Homozygous yellow (die in uterus) ; 2. Heterozygous Yellow ; 1 Agouti

or 2 Yellow: 1 Agouti or 2 : 1.

2. In the chicken an incompletely dominant gene, creeper (*cp*) in heterozygous condition (cp/+) cause "creeper" condition. The creeper birds have much shortened and deformed legs and wings, giving them a squatty appearance and creeping gait. A cross of two creeper birds yields viable offsprings in the ratio of 2 creepers: 1 normal. The homozygous creepers having such a gross deformities that they die during incubation.

F_1: *cp/+* Creeper (Heterozygous) × *cp/+* Creeper (Heterozygous)

↓

F_2: 1 *cp/cp* Creeper (Homozygous; dies) : 2*cp/+* Creeper (Heterozygous) : 1+/+ Normal (Homozygous)

3. In cattle, a recessive lethal gene in homozygous condition (*aa*) causes calves to born "amputated" which die soon after birth. The cross between two carriers (*a*+) produces the following result:

Parents: Normal *a*+ × Normal *a*+

P gametes: Ⓐ⊕ Ⓐ⊕

Progeny: 1*aa* : 2*a* + : 1 ++

Amputated (die) ; Normal

4. In *Drosophila* various sublethal and sex-linked lethal genes have been reported. The genes for vestigial wings and genes for white eyes are best examples of sublethal genes of *Drosophila*. Both of these genes reduce the viability of flies up to great extent. In *Drosophila*, certain recessive lethal genes such as curly wings (*Cy*), plum eyes (*Pm*) and stubbles (*Sb*) influence the viability of the flies when present in homozygous condition.

C. Lethal alleles in human beings. In humans several hereditary diseases have lethal effects. Few important lethal genes of human beings are following:

1. Congenital ichthyosis. Onc of the most typical cases of a recessive lethal gene in human beings is expressed in congenital ichthyosis. At birth children afflicted with this disease have a crusted leathery skin with deep fissures down to the subcutaneous tissue; the fissures lead to bleeding, infection and death. Congenital ichthyosis occurs only when there occurs homozygous condition for its recessive lethal genes.

2. Infantile amaurotic idiocy. A recessive allele in homozygous condition causes a fatal disease called **infantile amaurotic idiocy** in juvenile stage. Bearers of this genotype begin to lose their eye sight between the age of four to seven years. The complete blindness is followed by mental degeneration and finally death before adolescence.

3. Thalassemia. Thalassemia or **Cooley's anaemia** is a haemoglobin disease somewhat similar to sickle cell anaemia. It occurs mostly in children (in India and other countries such as Italy, Greece and Syria) and is nearly 100 per cent fatal (lethal). Thalassemia is controlled by a single gene *c* which in homozygous condition (*cc*), produces the severe Cooley's anaemia or **thalassemia major** and causes death of the patient. The heterozygous condition of this lethal gene (*Cc*) results in a mild form of the disease called **thalassemia minor** or **microcythemia.**

80.3 PENETRANCE

The ability of a given gene or gene combination to be expressed phenotypically to any degree is called **penetrance**. It is of following two kinds:

1. Complete penetrance. Most dominant and recessive genes in homozygous conditions and many completely dominant genes even in heterozygous conditions give their complete phenotypic expressions. Such genes are called to have **complete penetrance.**

Examples of Complete Penetrance

(*i*) In pea the alleles (*RR*) for red flowers and the alleles (*rr*) for white flowers have complete penetrance in homozygous conditions.

(*ii*) In *Drosophila* the recessive alleles (*vv*) for vestigial wings in homozygous conditions have complete penetrance.

(*iii*) In guinea pigs the dominant allele '*B*' for black coat has complete penetrance both in homozygous and heterozygous conditions.

2. Incomplete penetrance. Some genes in homozygous as well as in heterozygous conditions fail to provide complete (cent per cent) phenotypic expression of them. Such genes are called to have **incomplete penetrance.**

Examples of Incomplete Penetrance

(*i*) Polydactyly in human beings is thought to be produced by a dominant gene *P*. The normal condition with five digits on each limb is produced by the recessive genotype (*pp*). Some heterozygous individuals (*Pp*) are not polydactylus and, therefore, has a penetrance of less than 70%.

(*ii*) In human beings, the tendency to develop diabetes mellitus (a condition in which there is an excess of sugar in the blood) is controlled by certain genes. However, not everyone carrying the genes for diabetes actually develops the condition, for these genes have incomplete penetrance.

Effects of environment on penetrance. The environmental factors and genetical background have some definite effect on the degree of penetrance of a gene. For example, when various twins which carry genes for diabetes mellitus are studied, it is found that the disease appears only in those cases which ate more carbohydrate foods (starch and sugars).

Expressivity.

A trait though penetrant, may be quite variable in its phenotypic expressions. The degree of effect produced by a penetrant genotype is called **expressivity.**

Example of expressivity. In human beings the polydactylous condition may be penetrant in the left hand (6 fingers) and not in the right (5 fingers); or it may be penetrant in the feet and not in the hands.

Effects of environment on expressivity. The expressivity of a given gene is often influenced by environmental conditions. For instance, the expressivity of completely penetrant gene for vestigial wings in *Drosophila* is influenced by the temperature at which the fly develops, with the effect being most obvious at lower temperature.

Other examples of environmental effects on the expressivity of a gene includes such cases as the differences in the severity of symptoms of an inheritable allergy, or the differences in height of identical twins who have been raised in different home (with different diets), or who have had different medical histories (one with a serious childhood disease, the other escaping this disease).

80.4 PLEIOTROPISM

The Mendelian concept of a gene controlling a single character has also expanded to take into account genes which affect several characters simultaneously (**pleiotropy**) and genes which cooperate in groups to control individual characters (*i.e.,* quantitative inheritance).

A single gene often influences more than one phenotypic trait. However, it may be that one gene may cause evidently well marked expression of some phenotypic trait (**major effect**) than the others with less evident phenotype (**secondary effect**). Most genes have their multiple effects and are called **pleiotropic genes**. The phenomenon of multiple effect (multiple phenotypic expressions) of a single gene is called **pleiotropism**.

Examples of Pleiotropism

(*i*) In *Drosophila* the recessive gene for vestigial wings cause vestigial wings in homozygous condition. However, careful observations show that other traits as well are affected: (*i*) The tiny wing-like balancer behind the wings; (*ii*) Certain bristles; (*iii*) The structure of the reproductive organs; (*iv*) Egg production is lowered, and, (*v*) Longevity is reduced.

(*ii*) In human, the gene for disease **phenylketonuria** has pleiotropic effect and produces various abnormal phenotypic traits, collectively called **syndrome**. For example, the affected individuals secrete excessive quantity of amino acid phenylalanine in their urine, cerebrospinal fluid and blood. They become short stature, mentally deficient, with widely spaced incisors, with pigmented patches on skin, with excessive sweating, and with non-pigmented hairs and eyes.

(*iii*) The A^Y allele for yellow coat in mice is also a good example of pleiotropic gene. It affects two characters: coat colour and survival. It is most probable that both effects of the A^Y allele are the result of same basic cause which promotes the yellowness of coat in a single dose and death in double dose. Genetic analysis has revealed that lethal pleiotropic A^Y allele basically affects the cartilage of mice and cause death.

Sickle cell anaemia. The inheritance pattern of the trait of sickle-cell anaemia in humans presents an example of various genetical phenomena such as codominance, incomplete dominance, lethal effect, pleiotropism and polymorphism. When an allele results in a number of related changes, then such changes are collectively called **syndrome**. The syndrome, called sickle cell anaemia, is caused by an autosomal mutant allele, Hb^s, that in homozygous condition causes the production of an abnormal haemoglobin, called **haemoglobin S**; this indicates **primary effect** of mutant gene. In 1949, **Pauling** and his co-workers reported that the formation of an abnormal haemoglobin (*i.e., Hb^S*) which differed in its electrophoretic mobility from normal haemoglobin (Hb^A or adult haemoglobin) was the cause of this hereditary disease.

Further, due to **subsidiary effect** of abnormal haemoglobin (*i.e., Hb^S*), red blood cell of some persons suffered from this fatal form of haemolytic anaemia undergo a reversible alteration in shape when oxygen tension of the plasma falls slightly and they assume elongate, filamentous and sickle cell forms. Such sickle-shaped red blood cells show a lower life span, since they clump together and often causes vascular obstruction and are rapidly destroyed. This results in severe anaemia and death of the patient due to damaged heart, kidney, spleen and brain as a result of their clogged blood vesseles.

Biochemical investigations have shown that globin protein of haemoglobin is a **heteropolymer** having two identical α polypeptide chains, each with 141 amino acid residues and two identical β polypeptide chains, each with 146 amino acid residues. In 1957 **Vernon Ingram** showed that β chain amino acid of 6th position is occupied by **glutamic acid** in haemoglobin A, when this amino acid is substituted by another amino acid, the **valine**, it forms the haemoglobin S:

	1		2		3		4		5		6		7		8	
Hb^A =	val	-	his	-	leu	-	thr	-	pro	-	glu	-	glu	-	lys	-
HbS =	val	-	his	-	leu	-	thr	-	pro	-	val	-	glu	-	lys	-

At the molecular level, it is found that the genetic code of DNA undergoes a change from GAG to GTC.

Pleiotropy of sickle cell allele. The mutant allele Hb^s causes the production of abnormal haemoglobin. As a consequence the shape of the RBC containing it becomes distorted and sickle-shaped. Homozygous individuals (Hb^S Hb^S) normally die early in life due to severe anaemia caused by premature destruction of the sickled red blood cells. But heterozygotes who have both normal (haemoglobin A) and abnormal (haemoglobin S) haemoglobins and who also have mild anaemia, are naturally protected against contracting malaria as the parasite cannot live in these distorted cells. Heterozygotes may, therefore, survive better in region where malaria is endemic. Such population have both normal individuals and individuals heterozygous for the gene. The latter act as carriers of the gene from generation to generation as shown below:

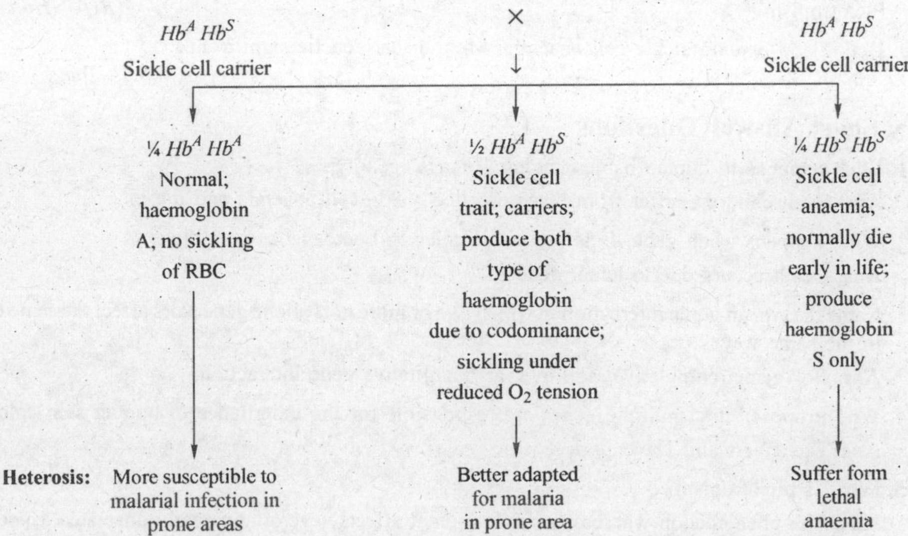

Polymorphism of sickle cell trait. The existence of two or more genetically different classes in the same interbreeding population at the same time, in the same habitat, is known as **polymorphism**. These classes are controlled by two competing alleles in the same locus. In **balanced polymorphism,** there establishes an intermediate equilibrium between the frequencies of these alleles and two classes are maintained together in the same population. For example, in case of sickle-cell anaemia superiority retains the two types at equilibrium. Here, heterosis or superiority of heterozygotes (Hb^A Hb^s which is resistant to malaria) over the normal homozygotes (Hb^A Hb^A) maintains the two alleles of haemoglobin (*i.e.*, normal and sickling) in an equilibrium.

QUESTIONS

Long Answer Questions

1. What are polygenes? Explain the inheritance of skin colour in man. (*Bangalore 1995*)

2. Explain multiple factors with suitable examples. (*Bangalore 1992*)

3. What are the differences between the inheritance of quantitative and qualitative characters? Give an example of each. *(Bangalore 1992)*

4. Define polygenic inheritance. Describe skin colour in man. *(M.S.U. Tamil Nadu 2006)*

Short Answer Questions

1. What is interaction of gene? Describe the quantitative interaction of gene. *(C.C.S.U. Meerut 2008)*

2. With suitable examples discuss the influence of environment over gene action. *(Lucknow 1992; Bangalore 1994)*

3. Make a brief account of pleiotropism. *(M.S.U. Tamil Nadu 2006)*

4. Write short note on the polygenic inheritance. *(CCS 1999)*

5. Write short notes on the following:

(*i*) Lethal genes *(Avadh 1995; Kanpur 1996)*

(*ii*) Sickle cell anaemia. *(Allahabad 1990; BHU (Hons.) 1991; Bangalore 1993)*

(*iii*) Sickle cell anaemia and sickle cell gene. *(Lucknow 1997)*

(*iv*) Pleiotropism *(BHU (Hons.) 1990)*

6. Define and explain sickle cell anaemia. Mention its genetic significance. *(M.S.U. Tamil Nadu 2006)*

Very Short Answer Questions

1. What happens to human babies having $Hb^S\ Hb^s$ genotypes?

Ans. Such children suffer from disease called sickle cell anaemia and die.

2. What happens when gene A^Y is present in mice in homozygous condition?

Ans. Death occurs due to lethal gene A^Y.

3. Name the type of genic interaction in which two or more non-allelic gene pairs affect the same character in the same way?

Ans. Polygenic, cumulative, additive, or quantitative gene interaction.

4. Who proposed that multiple genes are responsible for the inheritance of human skin colour?

Ans. Davenport and Davenport.

5. What is pleiotropism?

Ans. The phenomenon whereby a single mutant affects several apparently unrelated aspects of the phenotype.

6. What is quantitative inheritance?

Ans. The mechanism of genetic control of traits showing continuous variation.

7. Define Quantitative trait loci.

Ans. Chromosomal regions contributing to inheritance of a quantitative trait. These regions may contain one or more polygenes that contribute to the phenotype.

True or False Questions

1. Sickle cell anaemia as an example of lethal gene. *(H.N.B. Garhwal 2008)*

Ans. True.

Multiple Choice Questions

1. A gene which kills the bearer is termed as: *(M.S.U. Tamil Nadu, 2006)*

(*a*) polethal (*b*) lethal

(*c*) important (*d*) protective

2. Pleiotropic genes are those which have:
   (*a*) single phenotype
   (*b*) multiple phenotype
   (*c*) multiple genotype
   (*d*) single genotype

3. The polygenic genes show
   (*a*) different karyotypes
   (*b*) different genotypes
   (*c*) different phenotypes
   (*d*) none of these

4. Sickle cell anaemia has not been eliminated from the African population because
   (*a*) it is controlled by dominant genes
   (*b*) it is controlled by recessive genes
   (*c*) it is not a fatal disease
   (*d*) it provides immunity against malaria

5. Which one of the following is an example of polygenic inheritance?
   (*a*) skin colour in humans
   (*b*) flower colour in *Mirabilis Jalapa*
   (*c*) production of male honeybee
   (*d*) pod shape in garden pea

6. Sickle cell anaemia is caused due to:
   (*a*) valine at 6th position at β chain of Hb
   (*b*) glutamic acid at 6th position in β chain of Hb by valine
   (*c*) glutamic acid at 6th position in α chain of Hb
   (*d*) valine at the 6th position in α chain of Hb by glutamic acid

7. In which of the following disorders, blood has a defective haemoglobin?
   (*a*) haemophilia
   (*b*) haematuria
   (*c*) haematoma
   (*d*) sickle cell anaemia.

8. The inheritance of blood group in human beings shows
   (*a*) quantitative characters
   (*b*) qualitative characters
   (*c*) multiple characters
   (*d*) polygenic characters

9. Kernel colour in wheat shows an F_2 ratio of 1 : 4 : 6 : 4 : 1. It is due to
   (*a*) different number of dominant genes
   (*b*) different number of recessive genes
   (*c*) two polygenes
   (*d*) supplementary genes

ANSWERS

Multiple Choice Questions

1. (*b*)	**2.** (*b*)	**3.** (*c*)	**4.** (*d*)
5. (*a*)	**6.** (*a*)	**7.** (*d*)	**8.** (*b*)
9. (*c*).			

81

Varieties of Gene Expressions: III – Multiple Alleles

A given phenotypic trait (character) of an individual depends on a single pair of genes, each of which occupies a specific position called the gene **locus**, on a homologous chromosome. Moreover, a particular gene has been found to occur in two alternative forms or **allelomorphs**, one being dominant and other recessive; one being wild form and the other mutant form. For example, a gene (L) for length wings of *Drosophila* may occur in two alternative forms : a allele or gene (L^+) for normal development of wings and another allele or gene (L^{vg}) for vestigial wings. Because, most flies have normally developed wings, so, it can be. easily concluded that gene L^+ is the original form of gene or allele from which the other form of gene or allele (L^{vg}) might have originated by certain mutational event at sometime in past. The gene L^+ for normal development of wing is called the normal or **wild type allele** of the gene L and usually symbolized as L^+, while the mutated gene L^{vg} for vestigial wing is called reduced type or **mutant allele** of the gene L. A fly with normal wings, thus, has two wild type alleles (L^+L^+) and the vestigial winged fly has two mutant alleles (L^{vg} L^{vg}). Both of these allelic forms (L^+ and L^{vg}) of gene L occur at corresponding positions on genetically identical (homologous) chromosomes of same or different individual.

If the mutant allele has developed from the wild form of allele due to mutation, one may expect that the wild form of allele can mutate in more than one way. The mutant form of allele too can mutate once again to give rise to another mutant form of allele. Therefore, it is possible to have more than two allelic forms, *i.e.*, **multiple alleles**, of one kind of gene. Although only two actual alleles of a gene can exist in a diploid cell (and only one in a haploid cell), the total number of possible different allelic forms that might exist in a population of individuals is often quite large. This situation is termed as **multiple allelism**, and the set of alleles itself is called a **multiple allelic series** (Box 81.1).

81.1 CHARACTERS OF MULTIPLE ALLELES

The most important and distinguishing features of multiple alleles are summarized below:

1. Multiple alleles of a series always occupy the same locus in the chromosome.
2. Because, all the alleles of multiple series occupy same locus in chromosome, therefore, no crossing-over occurs within the alleles of a same multiple allele series.
3. Multiple alleles always influence the same character.
4. The wild type allele is nearly always dominant, while the other mutant alleles in the series may show dominance or there may be an intermediate phenotypic effect.
5. When any two of the mutant multiple alleles are crossed, the phenotype is mutant type and not the wild type.

Box 81.1 Allelic Variation

The character described by Mendel occurred in two forms, *i.e.*, they were **diallelic**. For many characters, however, a greater degree of **allelic variation** is evident. The human ABO blood group locus, for example, has three physiologically distinct alleles, and in the extreme example of the self-incompatibility loci of clover and tobacco, over 200 different alleles may be detected in a given population. The observed allelic variation also depends upon the level at which the phenotype is determined. *At the molecular level, there is often more diversity than is evident at the morphological level* because many of the alleles identified as sequence variant or **protein polymorphism** are neutral with respect to their effect on the morphological phenotype; these are termed **isoalleles.** No matter how many alleles can be distinguished for a particular locus in a population, only two are present in the same diploid individual at anyone time. Morphologically distinct alleles can be often arranged in order of dominance, as so called **allelic series.**

Symbolism for Multiple Alleles

The dominance hierarchy is defined at the beginning of each problem involving multiple alleles. A capital letter is commonly used to designate the allele which is dominant to all other alleles in the senes.

The corresponding lower case letter designates the allele which is recessive to all others in the series. Other alleles which are intermediate in their degree of dominance between these two extremes, are usually assigned the lower case letter with some suitable superscript.

81.2 EXAMPLES OF MULTIPLE ALLELES

The best examples of multiple allelic system have been observed in coat colour of rabbits, wings of *Drosophila* and blood groups in human beings.

1. The *C* gene in Rabbit (Coat Colour)

The coat of rabbit may have following different colours.

(i) **Full colour.** The coat of the ordinaryr (wild type) rabbit is reffered to as "**agouti**" or **full colour**, in which individuals have banded hairs, the portion nearest the skin being gray, succeeded by a yellow band, and finally a black or brown tip. The genes for full colour may be represented by capital letter C or c^+.

(ii) **Chinchilla.** In some individuals, the coat lacking the yellow pigment and due to the optical effect of black and gray hairs, have the appearance of silver-grey. The gene for chinchilla is represented as c^{ch}.

(iii) **Himalayan.** The Himalayan type coat is white except for black extremities (nose, ears, feet and tail). The eyes are pigmented. The gene for Himalayan coat is represented by c^h.

(iv) **Albino.** The albino coat totally lacks in pigmentation and the eyes of an albino also remain pink due to lack of pigment in iris of eye. The gene for albino is represented by c.

Crosses of homozygous agouti (c^+c^+) and albino (cc) individuals produce a uniform agouti F_1; interbreeding of the F_1 produces an F_2 ratio of 3 agouti : 1 albino (Fig. 81.2). Two third of F_2 agouti are found to be heterozygous by testcrosses. Thus, it is a case of monohybrid inheritance, with agouti completely dominant to albino. Likewise, crosses between chinchilla and agouti (Fig. 81.3) produce all agouti individuals in the F_1 and a 3 agouti : 1 chinchilla ratio in the F_2. Such complete dominance of agouti also occurs on Himalayan (Fig. 81.4). Further crosses, reveal that c^{ch} allele for chinchilla though is recessive to c^+ allele for agouti coat or skin, is incompletely dominant over Himalayan (c^h) and albino (c) alleles (Fig. 81.5 and Fig. 81.6). Likewise, c^h allele for Himalayan coat is recessive to c^+ (agouti) and c^{ch} (chinchilla), but dominant over albino (c, Fig. 81.7). The results of all these crosses exhibit that c^+ agonti c^h (Himalayan) and c (albino) are allelic to each other and the alleles of this multiple allelic series have following dominance hierarchy:

$$c^+ > c^{ch} > c^h > c$$

Fig. 81.1. Different coat colours in rabbits.

The possible phenotypes and their associated genotypes of this multiple allelic series can be summarized in Table 81.1.

P$_1$: Agouti × Albino
c^+c^+ ↓ cc
Agouti
F$_1$: c^+c
F$_2$: $1c^+c^+ : 2c^+c : 1cc$
3 Agouti : 1 Albino

Fig. 81.2. A monohybrid cross between agouti and albino rabbits.

P$_1$: Agouti × Chinchilla
c^+c^+ ↓ $c^{ch}c^{ch}$
Agouti
F$_1$: c^+c^{ch}
F$_2$: $1c^+c^+ : 2c^+c^{ch} : 1c^{ch}c^{ch}$
3 Agouti : 1 Chinchilla

Fig. 81.3. A monohybrid cross between agouti and chinchilla rabbits.

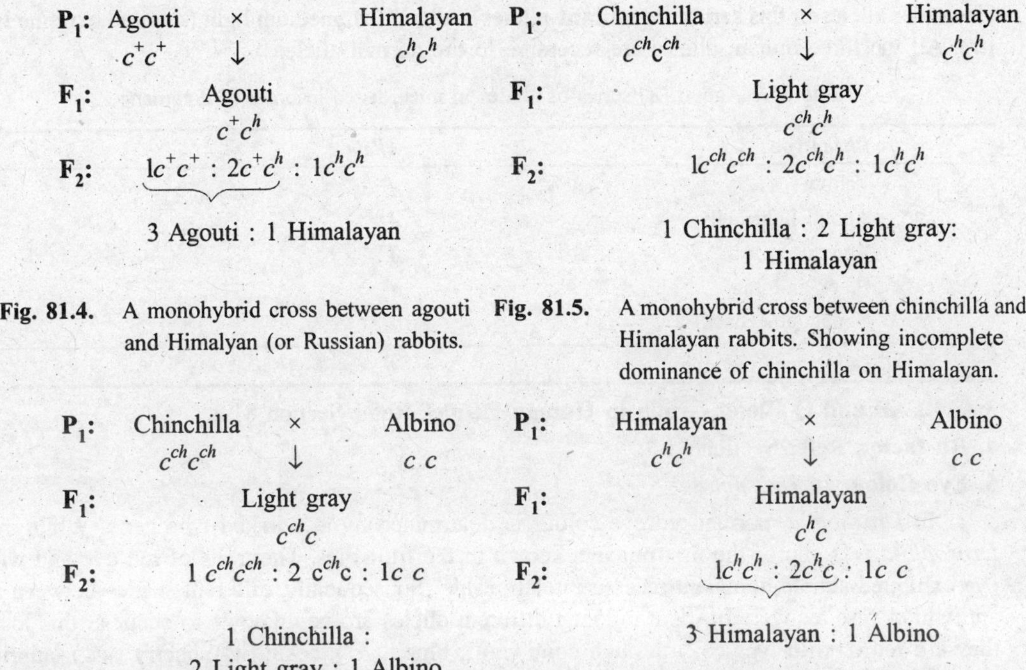

Fig. 81.4. A monohybrid cross between agouti and Himalyan (or Russian) rabbits.

Fig. 81.5. A monohybrid cross between chinchilla and Himalayan rabbits. Showing incomplete dominance of chinchilla on Himalayan.

Fig. 81.6. A monohybrid cross between chinchilla and albino rabbits showing incomplete dominance of chinchilla over albino.

Fig. 81.7. A monohybrid cross between Himalayan and albino rabbits.

Table 81.1 The phenotypes and genotypes of multiple allelic series for coat colour in rabbit.

Phenotypes	*Genotypes*
Full colour (Agouti)	c^+c^+, c^+c^{ch}, c^+c^h, c^+c
Chinchilla	$c^{ch}c^{ch}$
Light gray	$c^{ch}c^h$, $c^{ch}c$
Himalayan	c^hc^h, c^hc
Albino	cc

2. Skin Colour in Rodents (Mice)

In mice, the skin or coat colour is determined by two series of multiple alleles; both allelic series occur at different chromosomes. Series number 1 is at the black locus.

In mice, the skin or coat colour is governed by a multiple allelic series. The gene *A* controlling the agouti pattern in mice has several alleles. Among the members of the series a gradation exists in dominance, similar to that of the *c* series in rabbits, but in this case two members, A^y (yellow) and A^l (agouti light belly) are dominant over the wild-type allele A^+ (agouti). The alleles of agouti (*A*) series in mice can be listed according to dominance in Table 81.2.

In this multiple allelic series, the allele A^y, when homozygous, (*i.e.*, $A^y A^y$) is lethal early in embryonic development but produces yellow colour when in heterozygous condition with other alleles.

The second series of multiple alleles for skin colour is located at the albino locus and comprises of four alleles. The normal allele in this series is for **gray coat colour** and is represented by a^+.

The other alleles in this series are **mutant alleles**—albino (a), medium light (a^m) and extreme light (a^l). All the three mutant alleles are recessive to the normal allele.

Table 81.2 Agouti (A) series of alleles in mice, listed in order of dominance.

Phenotype	*Allele*
Yellow	A^y
Agouti light belly	A^1
Agouti	A^+
Black and Tan	a^t
Nonagouti (Black)	a
Dominance hierarchy:	$A^y > A^1 > A^+ > a^t > a$

3.A, B, AB and O Blood Group in Human Beings. Refer Section 81.3.

4. Rh factor. Refer Section 81.3.

5. Eye Colour in *Drosophila*

In *Drosophila*, normal red eye colour is determined by a X-linked type gene. White eyed *Drosophila* was one of the first mutants known in the fruit flies. The traits of red eye and white eye exhibited simple dominant recessive relationship. Subsequently, different shades between red and white were recovered. About a dozen different alleles are now known to occur at this locus; they are red or wild type (w^+) through coral (w^{co}), blood (b^{bl}), eosin (w^e), cherry (w^{ch}), appricot (w^a), honey (w^h), buff (w^{bf}), tinged (w^t), pearl (w^p) and ivory (w^i), to white (w). All of these were considered, on the basis of F_2 ratios, to form a multiple series, wild being dominant to all others and white recessive to all-$w^+ > w^{co} > w^{bl} > w^e > w^{ch} > w^a > w^{bf} > w^t > w^p > w^i >$ w. When any two recessive alleles were brought together, intermediate types, called **compound**, is obtained. However, some of the members of this multiple allelic series have been found to be pseudoalleles by **Lewis** (1951) (see Chapter 82: Fine structure of Gene).

6. Self-sterility Alleles.

A series of self-sterility alleles insures cross pollination in many plants, it is well known for long that some plants just will not self-pollinate in contrast to selfing in pea plant by Mendel. Thus, a single plant may produce both male and female gametes but pollen grains of this plant fail to fertilize the ovules of the same plant; so as a result no seed will ever be produced. The same plants, however, will cross with certain other plants, so evidently they are not sterile. This phenomenon is called **self-incompatibility** or **self-sterility.**

Kolreuter (1764) described self-sterility in tobacco plant (*Nicotiana*). Later on, the phenomenon of self-sterility is found to be common in many other dicot and monacot plants such as sweet cherries, petunias and evening primroses.

81.3 INHERITANCE OF HUMAN BLOOD GROUP

1. A, B, AB and O Blood Group in Humans.

Landsteiner in 1900 and 1902 discovered two kinds of **agglutinogens** or **antigens**, called A and B **antigens** from the surface of red blood cells of human blood. He found that out of A and B antigens, a person may contain either one (*i.e.*, A or B antigen) or neither of them. Accordingly, he recognised three kinds of

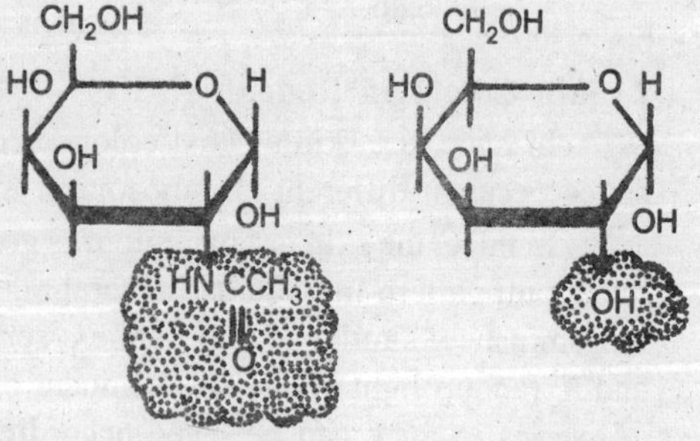

Fig. 81.8. Sugars associated with the surface of erythrocytes in blood type A and type B persons.

blood types or **blood groups: type A, type B,** and **type O.** The fourth and most rare, the **AB blood group** or type, was discovered in 1902 by two of **Landsteiner's** students, **Von Decastello** and **Sturli**. For A and B antigens, there occur two **agglutinins** or **antibodies:** anti-A (or α) and anti-B (or β). Recent, chemical investigations have shown that A and B antigens are not proteins but, are mucopolysaccharides (sugars + aminoacids) of 300,000 molecular weights.

The four blood types have different agglutinizing properties. To determine the blood group types of different persons, an **agglutination test** is performed. On a glass slide is placed a drop of type A serum (containing anti B antibodies and a separate drop of type B-serum (containing anti-A antibodies). When a drop of type O blood is added to each drop there is no agglutination or clumping of red blood cells in either drop takes place. This shows that O blood group has neither A nor B antigen. If a drop of type B blood is added agglutination occurs with type A serum; type A red blood cells are agglutinated by type B serum and type A B red blood cells are agglutinated by both sera. The agglutination tests for four types of human blood has been illustrated in Fig. 81.9.

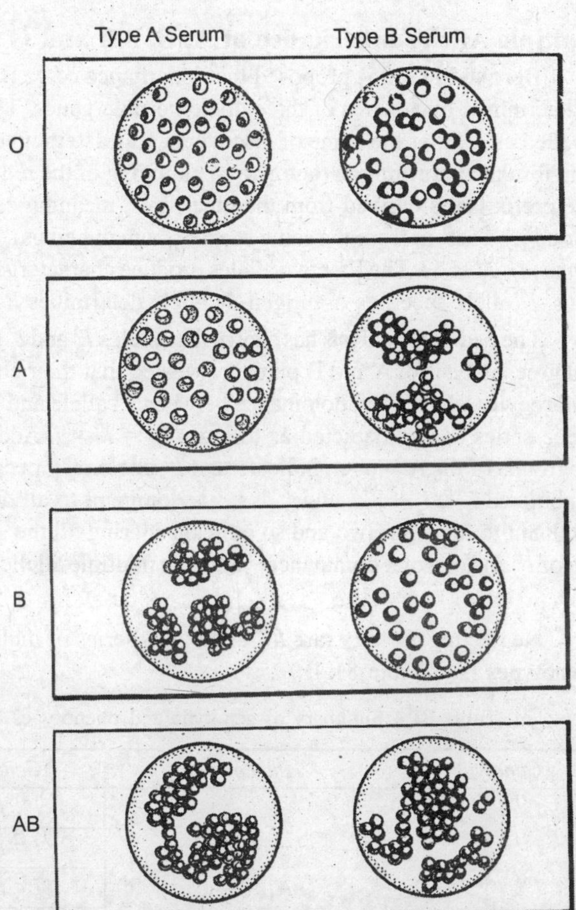

Fig. 81.9. Agglutination tests for A, B, AB and O human blood groups.

Each person, therefore, can use the blood of its own blood group in emergency, otherwise, clumping of red blood cells may take place, if blood group of different type is transfused in him. The clumped red blood cells occlude capillaries and, thus, deprive vital organs of normal blood supply and may lead to death. The characteristics of blood groups and the types of transfusions can be summarized in following Table 81.3.

Table 81.3 Human blood groups.

Blood groups (phenotype)	*Antigen in red blood cells*	*Antibodies in plasma*	*Can give blood to groups*	*Can receive blood from group*	*Genotype*
O	None	Anti-A Anti-B	O, A, B, AB	O	*ii*
A	A	Anti-B	A, AB	O, A	I^AI^A or I^Ai
B	B	Anti-A	B, AB	O, B	I^BI^B or I^Bi
AB	A and B	None	AB	O, A, B, AB	I^AI^B

Multiple Allelic Inheritance of A, B, AB and O Blood Types

Bernstein (1925) proposed that inheritance of A, B, AB and O blood types of human beings is determined by a series of three allelomorphic genes. The gene controlling blood types has been labelled as ***L*** (after the name of discoverer, **Landsteiner**) or ***I*** [from **isoagglutination**, the technical term for the **agglutinogen** (antigen) or clumping of the red blood cells by an agglutinin or antibody]. The prefix **iso** is derived from the greek *isos*, meaning equal and indicates that the agglutinations caused by a serum from the same species, human beings). The *I* gene exists in three different allelic forms : I^A, I^B and *i*. The first two alleles produce characteristic antigens on the surface of erythrocytes. Thus, I^A allele specifies A antigen, I^B allele determines *B* antigen and *i* allele specifies no antigen.

The pedigree analysis has shown that alleles I^A and I^B have dominance over *i* allele. Likewise the pedigree analysis of A and B parents revealed that their children have both A and B antigens on the erythrocytes, showing codominance between I^A allele and I^B allele. The dominance hierarchy of this allelic series can be depicted as follows: $I^A = I^B > i$. Additional studies that take into account the subgroups of the A antigen indicate that I^A allele may occur in at least four allelic forms. These are symbolized I^{A1}, I^{A2}, I^{A3}, and I^{A4}. I^{A1} is dominant to all other I^A alleles, I^{A2} is recessive to I^{A1} but dominant to the other two and so on. Considering all the six alleles of gene *I*, *i.e.*, four forms of I^A, one of I^B and one of *i*, dominance within the multiple allelic series can be shown in the following way:

$$[(I^{A1} > I^{A2} > I^{A3} > I^{A4}) = I^B] > i$$

Neglecting the very rare I^{A4} allele, this series of multiple alleles produces 15 genotypes and 8 phenotypes (see Table 81.4).

Table 81.4 Summary of genotype and phenotypes of A, B, AB and O blood group.

Genotype	*Phenotype*	*Genotype*	*Phenotype*
$I^{A1}\,I^{A1}$		$I^{A1}\,I^B$	A_1B
$I^{A1}\,I^{A2}$		$I^{A2}\,I^B$	A_2B
$I^{A1}\,I^{A3}$	A_1	$I^{A3}\,I^B$	A_3B
$I^A\,i$		$I^B\,I^B$	
$I^{A2}\,I^{A2}$		$I^B i$	B
$I^{A2}\,I^{A3}$	A_2	*ii*	O
$I^{A2}\,i$			
$I^{A3}\,I^{A3}$			
$I^{A3}\,i$	A_3		

2. The H Antigen and Bombay Phenotype

Antigens A and B of A, B and O blood phenotype are synthesized from a precursor mucopolysaccharide in the presence of the dominant allele of another pair designated as *H* and *h*. With genotypes *HH* or *Hh* the precursor is converted to an **H antigen** which, in turn, in the presence of I^A and/or I^B allele is partly converted to antigen A and/or antigen B. Gene *h* is termed **amorph** because it is producing no demonstrable product. So long as persons are of genotype *H*—(*i.e.*, *HH* or *Hh*), persons of group A produce antigens A and H, group B persons produce antigens B and H, and group AB persons produce antigens A, B and H. However, group O persons produce only antigen H if they are of the genotype *ii H*— (Fig. 81.10). On the other hand, blood of person of genotype -*hh* does not react with anti-A, anti-B, or anti-H. This is very rare **Bombay phenotype** (*i.e.*, one case in 13,000 persons; it is so named because it was first described in a family from the Bombay metropolitan. The allele *h* is found to be **epistatic** to the multiple alleles at the A-B-O locus. Erythrocytes of person having- -*hh* genotype give no reaction with anti-A or anti-B sera (even though they possess I^A or I^B genes); in fact, they contain no antigen of this multiple allelic series.

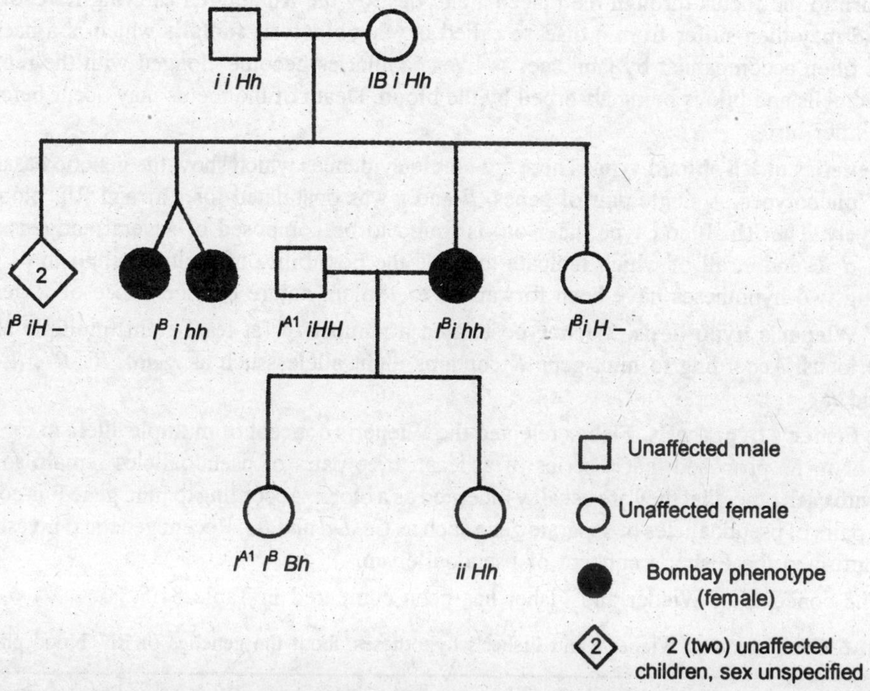

Fig. 81.10. Pathways leading to production of antigens on the red blood cells or erythrocytes showing how the Bombay phenotype is caused.

3. Rh Factor

The surface of erythrocytes (RBC) of some individuals contain one more type of antigen called **Rh** factor besides the A and B antigens. It is named after the *Macaca rhesus* monkey in which Rh factor was first discovered by **Landsteiner** and **Wiener** in 1940. Human beings are found to contain eight different types of Rh antigens.

The production of Rh antigen (Rh blood phenotype) depends on three closely set autosomal genes (pseudoallels). If any one of them is dominant, a Rh antigen is produced, but if all of them are recessive, no Rh antigen is formed. The individuals possessing the Rh antigen are called **Rh-positive** (Rh^+) and those lacking it are **Rh-negative** (Rh^-). Both of these types of persons are normal and none has natural anti-Rh antibodies in their blood plasma. However, a Rh-negative person can develop these antibodies on receiving Rh antigens through transfusion of Rh-positive blood. Such a blood transfusion will be safe only when the recipient had never been exposed to Rh-positive blood earlier. If already exposed, the previously developed anti-Rh antibodies will agglutinate the donor's RBC. In fact, "the degree of RBC agglutination depends upon the amount of anti-Rh antibodies present. The high concentration of anti-Rh causes severe agglutination of RBC which sometimes proves fatal.

Erythroblastosis foetalis. The incompatibility (mismatch) of Rh-positive and Rh-negative bloods may also be noticed in death of the child before or soon after birth. If a Rh-negative women marries a Rh-positive man and bears a Rh-positive foetus, sometime due to some placental defect, some of the foetal RBC carrying Rh antigens may pass into her own blood stream and cause the production of anti-Rh antibodies. The concentration of anti-Rh antibodies is gradually built up in the mother and she, thus, becomes **sensitized** only at or just before birth of the first Rh-positive child. In a second or subsequent pregnancy involving a Rh-positive child; these anti-Rh antibodies may return to the foetus through the placenta and destroy the Rh antigen carrying RBC of foetus. The child may then suffer from a disease called **erythroblastosis foetalis** which is a haemolytic anaemia often accompanied by jaundice, as liver capillaries become clogged with the remains of red blood cells and bile is being absorbed by the blood. Death of the foetus may occur before birth or soon after birth.

Genetics of Rh^+ blood type. There are ample evidences which show the genetic basis of Rh^+ and Rh^- phenotypes. A single pair of genes, *R* and *r* was postulated for Rh^+ and Rh^- blood types respectively. The Rh^+ blood type, later on, is found to be composed of several antigens such as *C, c, D, d, E* and *e*, all of which indicate towards the possibility of multiple allelism of gene *R*. Following two hypotheses have been forwarded to explain nature of inheritance of *R* gene:

1. Wiener's hypothesis. Wiener postulated a number of (at least eight) multiple alleles at a single locus. According to him, gene *R* contains eight alleles such as *r, Ro, R′, R″*, R_1, R_2, R_x or R_z and R_y.

2. Fisher's hypothesis. Fisher rejected the Wiener's concept of multiple allelism for R gene, instead of it, he proposed that a series of at least three pairs of pseudoalleles remain so closely linked with each other that they are usuall y inherited as a block. According to him, gene R is composed of three pairs of pseudoalleles or separate gene such as *Cc, Dd* and *Ee*. Recent genetical investigations have confirmed the Fisher's concept of pseudoallelism.

The concepts of Wiener and Fisher has been compared in Table 81.5.

Table 81.5 Comparison of Wiener's and Fisher's hypotheses about the genetics of Rh^+ blood phenotype.

	Gene symbol	*Antigen produced*	*Phenotypes*
Wiener	*Fisher*		
r	*cde*	none	Rh^-
Ro	*cDe*	Ro	Rh^+
R′	*Cde*	R′	Rh^+
R″	*cdE*	R″	Rh^+
R_1	*CDe*	Ro and R″	Rh^+
R_2	*cDE*	Ro and R″	Rh^+
Rx or *Rz*	*CDE*	R′ and R″	Rh^+
Ry	*CdE*	R′ and R″	Rh^+

QUESTIONS

Long Answer Questions

1. What are multiple alleles? Describe in brief the inheritance of anyone character determined by multiple alleles of *Drosophila*. *(Kanpur 1997)*
2. Explain the inheritance of multiple alleles with a suitable example. *(Bangalore 1992)*
3. Write an essay on multiple allelism. *(Lucknow 1991; Kanpur 1995)*
4. Discuss the inheritance of skin colour in rodents. *(Bangalore 1994)*
5. Write an essay on the genetics of ABO type of blood group in man. *(Kanpur 1996)*
6. Write an essay on the various human blood groups. *(Kanpur 1992, 1993)*
7. Explain the inheritance of blood groups of man. *(Bangalore 1992)*
8. Describe different blood groups in man and discuss their genetic basis. *(Allahabad 1993)*
9. Describe the immunological criterion in blood grouping.

10. What do you understand by blood groups? Discuss them in detail. Discuss their significance. (*Osmania 1988; Bombay 1989*)

11. Explain Rh factor. How can a marriage between homozygous rhesus-negative female and Rh^+ male cause erythroblastosis foetalis?

Short Answer Questions

1. Give a brief account of multiple allelism. (*M.S.U. Tamil Nadu, 2006*)
2. Write short notes on the following:
   (*i*) Multiple alleles (*Lucknow 1994; CCS 1997*)
   (*ii*) Multiple allelism (*Kanpur 1994*)
   (*iii*) Blood groups (*Agra 1989; Guwahati (G) 1993*)
   (*iv*) Effect of transmission of blood (*Allahabad 1988*)
   (*v*) Inheritance of blood groups (*Allahabad 1990, 1991; Lucknow 1995*)
   (*vi*) Rh^- blood groups (*Magadh (G) 1994; Kanpur 1995, 1998*)
   (*vii*) Erythroblastosis foetalis (*Bangalore 1994*)

Very Short Answer Questions

1. What are multiple alleles?

   Ans. It is a condition in which a particular gene occurs in three on more allelic forms in a population of organisms.

2. Define Rh factor.

   Ans. Rh factor is an antigen in the red blood corpuscles of certain people. A pregnant Rh-negative woman carrying an Rh^- positive child may develop antibodies, causing the child to develop a hemolytic disease.

3. Can a Rh^- person be given the blood of Rh^+ person?

   Ans. No.

4. Predict the possible blood group or groups of children whose parents have blood group 'O' and 'A'.

   Ans. O and A.

5. Does multiple alleles of a gene undergo crossing over?

   Ans. No.

Multiple Choice Questions

1. Inherited Rh gene is found in
   (*a*) Rh^+ individual (*b*) Rh^- individual
   (*c*) AB blood group individuals (*d*) O blood group individuals
2. Marriage between one of the following pairs may cause child death
   (*a*) Rh^- woman and Rh^+ man (*b*) Rh^- woman and Rh^- man
   (*c*) Rh^+ woman and Rh^- man (*d*) Rh^+ woman and Rh^+ man
3. A, B, O blood group system is due to
   (*a*) epistasis (*b*) multiple factor inheritance
   (*c*) incomplete dominance (*d*) multiple allelism

ANSWERS

Multiple Choice Questions

1. (*a*) 2. (*a*) 3. (*d*)

82

The Gene (Fine Structure of Gene)

82.1 EVOLUTION OF GENE CONCEPT

The hereditary units which are transmitted from one generation to the next generation are called **genes**. A gene is the fundamental biologic unit, like the atom which is the fundamental physical unit. **Mendel** while explaining the results of his monohybrid and dihybrid crosses, first of all conceived of the genes as particulate units and referred them by various names such as **hereditary factors** or **hereditary elements** (or **Merkmal**). The term **gene** was coined by **Wilhelm Johansen** in 1909 to describe a heritable factor responsible for the transmission and expression of a given biological character but without reference to any particular theory of inheritance. Thus, in its original context, the gene had no specific material basis and could be treated purely in abstract terms, as it is in the study of transmission genetics (Mendelian inheritance).

A more precise idea of the physical and functional basis of the gene arose from many sources in the first half of the twentieth century. In 1909, **Archibald Garrod** showed that the human metabolic disorder alkaptonurea resulted from the failure of some specific enzyme and could be transmitted in an autosomal recessive fashion. He called this disorder as an **'inborn error of metabolism'**. Garrod was unaware about Mendelian inheritance and the significance of his discovery was not appreciated for 30 years, until **George Beadle** and **Edward Tatum** (1940) found that X-ray induced mutations in fungi (*e.g., Neurospora*) often caused specific biochemical defects. This led to the **one gene one enzyme** model of gene function. In 1911, **Thomas Hunt Morgan** showed that genes were located on chromosomes and were physically linked together (chromosome theory of inheritance). Thus, genes were considered as beads and chromosomes as strings of beads. Mutation was supposed to alter the bead structure and recombination (or crossing over) was regarded to involve a breakage between two beads followed by their exchange between paired chromosomes. Each bead was thought of to control one character by controlling some biochemical step. Thus, a gene was considered to be a unit of mutation, recombination and function **(Benzer)**.

Muller (1932) have compared the mutant genes with wild type genes (or standard genes) and classified them into the following types: **(1) Hypomorphs.** These are the genes which have the same effect as the standard gene but their phenotypic expression is less effective. **(2) Amorphs.** These are genes which have very less effect than the standard gene, *i.e.*, they do not cause any significant change in phenotype. *e.g.*, h antigen of Bombay phenotype. **(3) Hypermorphs.** These are genes which have greater effect than the standard gene, *i.e.*, they produce much phenotypic change. **(4) Antimorphs.** These are genes which result in opposite phenotypic effect than the standard gene, *i.e.*, they produce abnormal phenotype. **(5) Neomorphs.** These are genes which produce a phenotypic effect that is qualitatively different from that of standard gene, *e.g.*, Bar locus of *Drosophila*.

In 1944, **Oswald Avery** and colleagues showed by elimination that DNA was the genetic material. Thus, evolved a simple picture of gene: *a length of DNA in a chromosome which encoded the information for an enzyme.*

This definition of gene has to be expanded to include what is currently known about gene function: **not all genes encode enzymes** (some encode polypeptides with different functions; and some encode functional RNA molecules such as rRNA and tRNA); also in viruses, genes may be RNA not DNA. Furthermore, the information in the gene may be used selectively to generate more than one product. The Mendelian concept of a gene controlling a single character has also been expanded to include genes which affect several characters simultaneously (**pleiotropy,** *e.g.*, sickle cell anaemia) and genes which cooperate in groups to control individual characters (**quantitative inheritance**). With the advent of molecular biology, it is now possible to define the gene exactly in terms of its structure and function.

82.2 UNIT OF GENETIC STRUCTURE AND FUNCTION

(*i*) Indivisible units of genetic structure. After having established that genes were carried on chromosomes and could be mapped on them by recombination (**genetic mapping** or **linkage mapping**), individual genes came to be regarded as **fundamental** and **indivisible units** of genetic information, in terms of both structure and function, linked together in a linear fashion (the **beads on a string theory**). This view was challenged by recombination studies in T_4 bacteriophage which showed that the gene could be subdivided into smaller units. **Seymour Benzer** (1950) introduced the terms **muton, recon** and **cistron** to define indivisible units of mutation, recombination and function, respectively. By crossing independently derived mutants (*i.e.*, rII) of the same gene in a phage infection, it was shown that wild type phage (r +) could be produced. This could happen in **intragenic recombination** (recombination within a gene) if distinct sub-elements of the gene were mutated. This showed that the gene was divided into smaller units. The muton and recon are equivalent to a single nucleotide pair (of DNA molecule).

(*ii*) The gene as an indivisible unit of genetic function. Benzer's term cistron means an indivisible unit of genetic function. This can be determined by **complementation analysis** where the gene (or its product) is tested for its ability to compensate for a mutation in a homologous gene in the same cell. Successful complementation restores the wild type phenotypes.

Cis-trans test. The basis of complementation analysis is the cis-trans test (from which the term cistron arises), where pairs of independently derived mutations are considered in the *cis* and *trans* configuration. The *cis* test is a control, because of it both mutations (mutant alleles) are present in one genome, the other should be wild type at both loci and produce the normal gene product(s), thus granting the wild type phenotype upon the cell. The *trans* test is the complementation test and defines the boundaries of the unit of function. If both mutations are within the same gene, when they are present in the *trans*-configuration, each genome carries a mutant copy of the gene and no functional product would be generated in the cell—there would be no complementation (**negative complementation**). If the mutations lie in different genes, when they are present in the *trans*-configuration, each genome can supply the gene product the other lack. With all necessary gene products, the cell is wild type-there is **positive complementation** (Fig. 82.1).

Complementation analysis in bacteria and yeast established that the gene is a cistron, *i.e.*, the gene could be defined as a unit of function.

82.3 GENE-CISTRON RELATIONSHIP IN PROKARYOTES AND EUKARYOTES

(*i*) Equivalence of gene and cistron in simple genomes. In prokaryotes and lower eukaryotes, there is usually a simple relationship between the gene and its product. In most cases, there is one gene-one product correspondence and the gene is usually colinear with its product. In these organisms, therefore, *the gene and cistron are equivalent, i.e.*, the gene is a unit of genetic function; a unit of expressed genetic information. In bacteria, the gene is synonymous with the coding region (open reading frame), whereas in eukaryotes it is synonymous with the transcription unit. This is because bacterial genes are often arranged as an **operon**, so that many products are translated from a common **polycistronic mRNA**. In eukaryotes, conversely, most genes are transcribed as **monocistronic mRNA**.

(*ii*) Nonequivalence of gene and cistron in complex genomes. In higher eukaryotic genomes, there is often a complex relationship between the gene and its product. Most higher eukaryotic genes contain **introns**, which are intervening sequences not represented in the final product and are not therefore part of its function. While the gene represents the entire transcription unit, the cistron would be interrupted by the introns. The *cistron is therefore* equivalent to the **exons** of the eukaryotic gene.

82.4 GENE SIZE

Bacterial genes are characteristically small (average 1 kbp) and show little diversity in size, whereas genes of higher eukaryotes are large (average 16 kbp) and show great size diversity. Although higher eukaryotic genes are generally much longer than bacterial genes, the same is not true of the mRNAs derived from them. Such a difference is caused by **introns** which are intervening sequences which interrupt the transcription unit and must be spliced out at the RNA level (*i.e.*, during RNA processing). The remaining parts of the transcription unit, which become spliced together and expressed, are termed **exons**. *Gene size is inversely proportional to the percentage of exon material in the gene*. Bacterial genes generally lack introns and are therefore 100 per cent exon material. Introns are also rare in many microbial eukaryotes, such as yeast *Saccharomyces cerevisiae*, whose average, size, 1–2 kbp, is similar to that of bacteria. In humans, the smallest genes have the fewest and smallest sized introns (*e.g.*, the 500 bp histone H4 gene has no introns). Conversely, the largest human gene, the **dystrophin gene** which is 2500 kbp in length has > 95 per cent intron material. The dystrophin gene has 75 introns with an average size of 30 kbp; only 0.5 per cent of the gene is exon material.

82.5 GENE NOMENCLATURE

Genes are named according to species convention (Table 82.1). It is normal to present the names of genes, alleles, genotypes, and where necessary, the mRNA transcribed from a gene in italic and the protein products and phenotypes in Roman. In mot species, genes are designated by a symbol comprising several letters and numbers. Some species conventions (*e.g.*, *Drosophila*, *E. coli*) dictate the use of lower case letters for genes identified as recessive mutations and initial capital letters for those identified as dominant mutations. In other species, including humans, genes are designated by all capital letters.

Table 82.1 A brief summary of conventional genetic nomenclature (Source : Twyman 1998).

Species	*Convention*
1. *E. coli and other bacteria*	Three lower case letters to designate an operon followed by upper case letters to denote different loci, *e.g.*, *lac* operon; lac Z, lac Y, lac A. Proteins Lac Z, Lac Y, Lac A.
2. Yeast	Three letters identify gene function followed by a number to specify different loci, *e.g.*, *Saccharomyces cerevisiae* gene GAL 4, CDC 28; proteins GAL4, CDC28.
3. *Caenorhabditis elegans*	Three lower case letters indicative of the mutant phenotype hyphenated to a number if more than one locus involved, *e.g.*, genes unc-86, ced-9; proteins UNC-86, CED-9.
4. Plants	Although no convention covers all plants, most named with 1–3 letter lower case symbols. *Arabidopsis* genes named in style of *Drosophila* but using capital letters, *e.g.*, gene *AGAMOUS* (*AG*), protein AGAMOUS.

Species	*Convention*
5. *Drosophila*	Name descriptive of mutant phenotype, may be represented by symbol of 1–4 letters, *e.g.*, genes *white* (*w*), *tailless* (*tll*), *hedgehog* (*hh*); proteins white, Tailless, Hedgehog.
6. Vertebrates	Generally a 1–4 letter lower case symbol with letters and numbers, representative of gene function, *e.g.*, genes *sey, myc,* proteins Sey, Myc.
7. Humans	As above but in capitals, *e.g.*, genes *MYC*, *ENO1*, proteins MYC, ENO1.

82.6 PSEUDOALLELES OR LEWIS EFFECT

Apricot eye colour in *Drosophila*. In 1951, **E.W. Lewis** reported results of one of his experiments on *Drosophila* where frcm a cross of apricot (w^{apr}) and white (*w*) eye he obtained F_1 having intermediate eye colour. In F_1 generation he expected segregation for apricot and white, but to his surprise, he recovered from F_2 wild type (red eyes) also, showing that the two presumed alleles were non-allelic, *i.e.*, both occupying different sites, **Lewis,** therefore, preferred to called them **pseudoalleles** and the phenomenon as **pseudoallelism**. Thus, pseudoalleles are those genes which are located almost at the same place on the linkage map but show recombination, indicating thereby that although they are functionally allelic, structurally they are non-allelic. **Pontecorvo** later preferred to use the term **Lewis effect** for the pseudoalleles. With accumulated cross over data, the distance between the two sites was placed at about 0.01.

Pseudoalleles show complementation, thereby, indicating their functional individuality. Thus, if *a* and *b* are non-allelic the parents should be represented as *aa* ++ and ++ *bb* and in F_2, wild type (++) would be recovered as expected. Thus, the appearance of wild type in F_2 generation could be explained. However, it was difficult to explain why F_1 progeny *a*+/+*b* was not wild type. For explaining this, another criterion, called **complementation**, was employed. Complementation means that in allelic forms there will be lack of complementation due to which F_1 individuals will not be wild type. But since intragenic interalleic crossing over (recombination) is possible, in the F_2 generation wild type individual appear. In F_1 individual *a*+/+*b*, the **position effect** does not allow the wild type to be expressed. As a result of crossing over, ++ and *ab* gametes are produced from the F_1 individuals and in the F_2 generation ++/++ or ++/*ab* will express the wild phenotype. Thus, it has been demonstrated that within the same gene ++/*ab* and *a*+/+*b* exhibit different phenotypes. This is known as the **cis-trans effect** (Fig. 82.1).

$$\frac{++}{ab} \text{ or } \frac{apr/w}{+/+} \qquad\qquad \frac{a+}{+b} \text{ or } \frac{apr/+}{+/w}$$

cis (wild) red — trans (mutant) apricot

Fig. 82.1. Cis and trans arrangements of alleles a and b or apr and w alleles of *Drosophila* in heterozygous condition.

Why do the *w* and *apr* genes behave differently in different positions? The production of red pigment is a two-step chemical process. One gene may control one step, while the other gene in close proximity may control another step in the same reaction. When the two genes w^+ and apr^+ (++) occur next to each other on the same chromosome (cis position) they produce the normal red eye colour. When they are opposite to each other; however, chemical blocks occur that stop the reaction at one or more points. So cis-trans effect is a type of position effect of the genes with respect to each other rather than their presence or absence determines the end result.

82.7 COMPLEX GENE LOCI

The Russian geneticist **N.P. Dubinin** in 1939 recognized **step allelism** (that is, the graded effect of different alleles) in *Drosophila*. This discovery foreshadowed the idea that "genes" could be subdivided, *i.e.*, certain genes are **compound** or **complex genes**. The complex loci were discovered when techniques became available in particular organisms for exploring the fine structure of the gene. The first evidence came from *Drosophila* studies in which presumed, alleles were found to be pseudoalleles, that is, *closely linked genetic units that behaved ordinarily as if they were alleles but were not in exactly corresponding position*. They could be separated by crossing over. Virus studies then showed that many sites were present in a particular locus (*e.g.*, *rII* locus in T_4 phase). Around 1970's, studies of gene's fine structure were extended to bacteria (*Salmonella*), yeast, mold (*Neurospora*), wasp (*Mormoniella*), flies, cotton and maize. It appears that complex gene structure is basic to all organisms.

Examples of complex loci. The lozenge gene of *Drosophila* provides a good example of complex locus. In *Drosophila* several other complex loci have been demonstrated which include the following loci: *bx* (bithorax), *sb* (stubble bristles), *N* (notch wing), *g* (garnet eye colour), *sn* (singed bristles), *dp* (dumpy wing) and s (star).

Complementary interaction between alleles indicates complex loci even when recombination between them has not been observed. By this technique, the *Td* region in *Neurospora crassa* which controls the final step in tryptophan synthesis, has been shown to be a complex of several sites in linear order.

In maize, complex loci have been reported at the *A* locus, which controls pigment production; the *R* locus, which controls pigmentation; and the *wx* locus, which influence the nutrient content of the seed.

QUESTIONS

Long Answer Type Questions

1. Write an essay on gene. (*Lucknow 1998*)
2. Explain the fine structure of gene according to Benzer. (*H.N.B. Garhwal 2009*)

Short Answer Questions

1. Describe the fine structure of gene. (*H.N.B. Garwal 2008*)
2. Write short notes on the following:
   (*a*) Modern concept of gene (*Lucknow 1993*)
   (*b*) Structure of genes (*Punjab (G) 1991*)
   (*c*) Cistron (*Kanpur 1997*)
   (*d*) Gene (*Magadh (Subs.) 1993*)
   (*e*) Pseudoalleles
   (*f*) Lewis effect
   (*g*) Complex gene loci

Very Short Answer Questions

1. Define the gene.

   Ans. Gene is a hereditary determinant of a specific biological function; a unit of inheritance (DNA) located in a fixed place on the chromosome.

2. What is cistron?

   Ans. 1. Cistron is a unit of function in DNA. One DNA cistron specifies one polypeptide chain in protein synthesis. 2. Cistron is a term coined by **Benzer** for the smallest genetic unit that exhibits the *cis-trans* position effect; this term is synonymous with gene.

3. What is *trans*?

Ans. *Trans* means "across" and refers usually to the geometric configuration of mutant alleles across from each other on a homologous pair of chromosomes.

4. Define *cis*.

Ans. *cis* mean "on the near side of", refers to geometric configurations of atoms or mutants on the same chromosome.

5. What is position effect?

Ans. An alteration of phenotype caused by a change in the relative arrangement of the genetic material.

6. What is *cis-trans* complementation test?

Ans. It is a mating test to determine whether two mutants on opposite chromosomes complement each other; a test for allelism.

Multiple Choice Questions

1. The carriers of hereditary material are

(*a*) chromosomes (*b*) genes
(*c*) gametes (*d*) gametocytes

2. The terms cistron, muton and recon were coined by

(*a*) Jenner (*b*) Benzer
(*c*) Carlson (*d*) Bateson

3. A gene that specifies the amino acid sequence of a polypeptide chain is termed as

(*a*) structural gene (*b*) regulator gene
(*c*) operator gene (*d*) split gene

4. Smallest part of DNA which takes part in crossing over is

(*a*) recon (*b*) muton
(*c*) replicon (*d*) exon

5. The smallest segment of genetic material affected by mutation is

(*a*) recon (*b*) cistron
(*c*) muton (*d*) exon

6. A broad definition of gene would be that it is

(*a*) a unit of heredity (*b*) a part of DNA
(*c*) a unit of mutation (*d*) all the three

ANSWERS

1. (*b*) **2.** (*b*) **3.** (*a*) **4.** (*a*)
5. (*c*) **6.** (*d*).

83

Sex Chromosomes, Sex Determination and Sex Differentiation

83.1 SEX DETERMINATION

Members of almost all species are often divided into two sections according to the kind of gamete or sex cell produced by them, *i.e.*, male sex and female sex. The word sex has been derived from Latin word *sexus* meaning section or separation. However, some of the lowest forms of plant and animal life are found to have several sexes. For example, in one variety of the ciliated protozoan *Paramecium bursaria* there are eight sexes or "mating types" all morphologically identical. Each mating type is physiologically incapable of conjugating with its own type, but may exchange genetic material with any of the seven other types within the same variety. Further, in organisms in which the number of sex reduced to just two, sexes may reside in different individuals or within the same individual. An animal possessing both male and female reproductive organs is usually referred to as a **hermaphrodite**. The sex cells and reproductive organs form the primary sexual characters of male and female sexes. Besides these **primary sexual characters**, the male and female sexes differ from each other in many somatic characters known as **secondary sexual characters**. The phenomenon of molecular, morphological, physiological or behavioral differentiation between male and female sexes is called **sexual dimorphism.**

The phenomenon of sexual dimorphism has been a biological riddle for the thinkers and biologists of all time. People always tried to know those factors which determine the male and female sexes of a species. Literally hundreds of mistaken hypotheses and wild guesses were proposed before 1900 in vain attempts to find out a solution to the problem of determination of sex. Modern geneticists have reported many different mechanisms of determination of sex in living organisms. Some important and common mechanisms of sex determination are following:

83.2 GENETICALLY CONTROLLED SEX DETERMINING MECHANISMS

Most of the mechanisms of the determination of the sex are under genetic control and they are of following types:

1. Sex chromosome mechanism or Heterogamesis;
2. Genic balance mechanism;
3. Male haploidy or haplodiploidy mechanism;
4. Single gene effects.

A. Sex Chromosomal Mechanisms (Heterogamesis)

Sex chromosomal mechanism or chromosome theory of sex determination was proposed by **McClung** (1902). The chromosomes of an individual are of two types:

***(i)* Autosomes.** The chromosomes which have no relation with the sex and contain the genes which determine the somatic characters of the individuals are known as **autosomes (A).**

***(ii)* Sex chromosomes.** The chromosomes which are responsible for the determination of sex are known as **sex chromosomes**, *e.g.,* X and Y chromosomes.

Structure of sex chromosomes. The X and Y sex chromosomes exhibit structural differences. The cytological studies have shown that the X chromosomes of most organisms are straight, rod-like and comparatively larger than Y chromosomes. The Y chromosome is smaller in size with one end slightly curved or bent to one side in Drosophila (Fig 83.1); in human beings and *Melandrium* no such curvature of Y chromosome occurs. The X chromosomes have large amount of euchromatin and small amount of heterochromatin. The euchromatin has large amount of DNA material, hence, much genetic information. The Y chromosome contains small amount of euchromatin and large amount of heterochromatin. The Y chromosome has little genetic information, therefore, sometimes it is referred to as genetically **inert** or **inactive**.

Types of Sex Chromosomal Mechanism of Sex Determination.

In dioecious diploidic organisms following two systems of sex chromosomal determination of sex have been recognized:

(*a*) Heterogametic males; (*b*) Heterogametic females.

Heterogametic males. In this type of sex chromosomal determination of sex, the female sex has two X chromosomes, while the male sex has only one X chromosome. Because, male lacks a X chromosome, therefore, during gametogenesis produces two types of gametes, 50 per cent gametes carry the X chromosomes, while the rest 50 per cent gametes lack in X chromosomes. Such a sex which produces two different types of gametes in terms of sex chromosomes, is called **heterogametic sex**. The female sex, because, produces similar type of gametes, is called, **homogametic sex**. The heterogametic males may be of following two types:

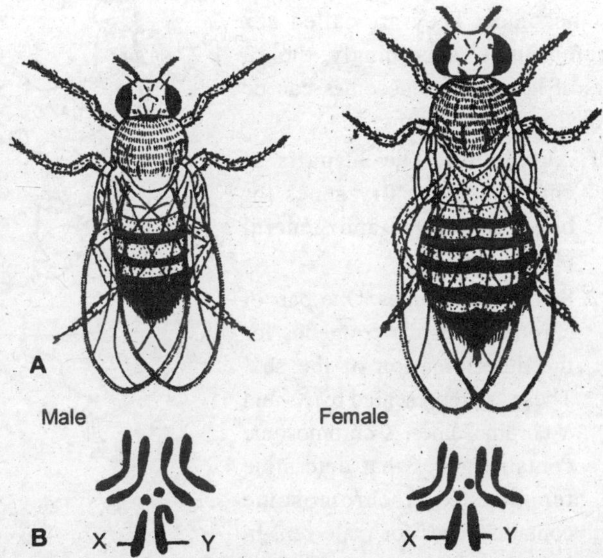

Fig. 83.1. Male and female *Drosphila melanogaster* and their chromosomes.

***(i)* XX-XO type.** In certain plants (*e.g., Vallisneria spiralis, Dioscorea sinuata, etc.,*) and insects specially those of the orders Hemiptera (true bugs) and Orthoptera (grasshoppers and roaches), the female having two X chromosomes (hence, referred to as XX) and are, thus, homogametic, while the male having only one X chromosome (hence, referred to as XO). The presence of an unpaired X chromosome determines the masculine sex. The male lacking in one X chromosome produces two types of sperms: half with X chromosome and half without X chromosome. The sex of the offspring depends upon the type of sperm that fertilizes the egg (each of which carries a single X chromosome).

***(ii)* XX-XY type.** In human beings, other mammals, certain insects including *Drosophila* and *Lygaeus turicicus* and in certain angiospermic plants such as *Melandrium album* (*Lychis*), *Humulus lupulus, Coccinia indica*, the female possesses two homomorphic X chromosomes in their body cells (hence, referred to as XX) and they being homogametic, produce one kind of eggs, each with one X chromosome. The males of these organisms possess one X chromosome and one Y chromosome (hence, referred to as XY). The males having two heteromorphic sex chromosomes produce two kinds of sperms: half with X chromosome and half with Y chromosome. The sex of embryo depends on the kind of sperm. An egg fertilized by a X-bearing sperm, produces a female, but, if fertilized by a Y-bearing sperm, a male is produced.

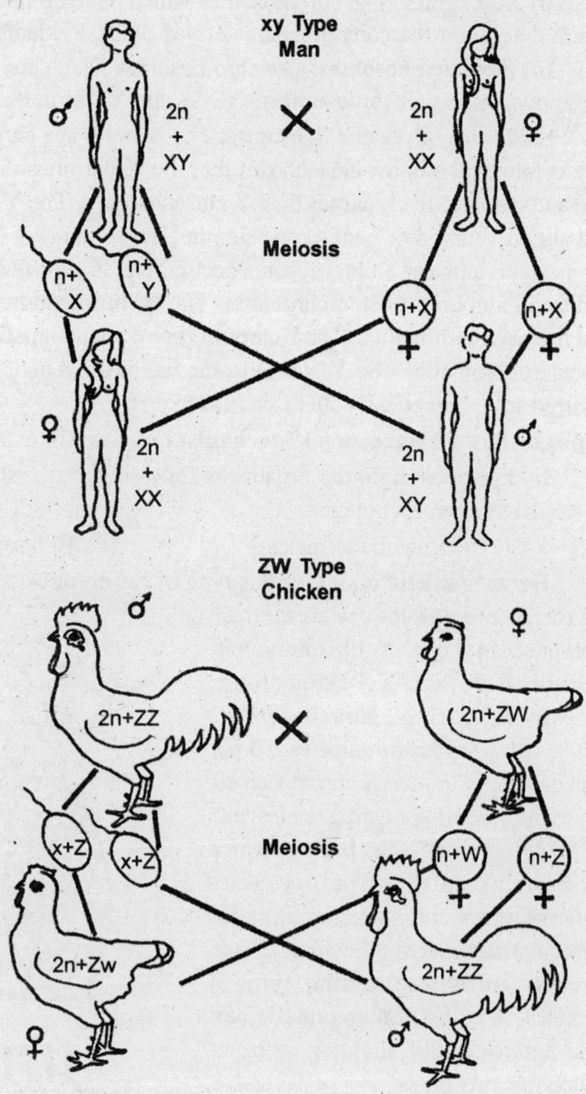

Fig. 83.2. The XX-XY and ZW-ZZ type determination of sex in human beings and fowl.

Sex determination in human beings. Sex has the genetic basis. Genes determining the sex in human beings are located on a special pair of chromosomes. They are called **sex chromosomes**. Accordingly, the 23 pairs of human chromosomes can be divided into two groups:

1. **Autosomes.** The 22 pairs of chromosomes carry genes for body characters and general physiology.
2. **Sex chromosomes.** One pair of chromosomes carries genes for the determination of the sex. These are represented by X- and Y-chromosomes. X-chromosome contains genes that determine female sex; Y-chromosome contains genes for male sex. In humans, sex determination is of XX-XY type (Fig. 83.2). Female is homogametic producing only one type of gametes (ova), each containing one X chromosome; male is heterogametic, producing two types of sperms, half carry X and other half carry Y chromosome.

	Man (Heterogametic)		Woman (Homogametic)
	44A + XY		44A+XX
Sperms:	(22 A + X)	**Ova:**	(22A + X)
	(22 A + Y)		(22 A + X)

Sperms	*Ova*	
	22 A + X	22 A + X
22A + X	44 A + XX Girl	44A + XX Girl
22A+ Y	44 A + XY Boy	44 A + XY Boy

In human beings like *Drosophila* XX-XY type sex determining mechanism occurs but here the Y chromosome contains potent male sex-determining genes which can almost completely overcome the feminizing action of the rest of the genotype. The conclusive evidences that Y chromosome is a determiner of fertility and sex of male individual came from certain abnormal conditions (called **syndromes**) which contained aneuploidic sex-chromosomal abnormalities (for details see Chapter 87, Human Genetics). For instance, **Turner's syndromes** (XO) are sterile female individuals having certain abnormalities such as short stature, congenital malformations, shield chest, pronounced webbing of the neck, short fourth metacarpel, colour blindness, etc. Similarly, **Klinefelter's syndromes (XX Y)** are males, despite the presence of two X chromosomes. A person with extra one X and Y chromosome display true **hermaphroditism** having both ovarian and testicular tissues and variable degrees of intersexual development of the genitalia.

Heterogametic females. In this type of sex chromosomal determination of sex, the male sex possesses two homomorphic X chromosomes, therefore, is homogametic and produces single type of sperms, each carries a single X chromosome. The female sex either consists of single X chromosome or one X chromosome and one Y chromosome. The female sex is, thus, heterogametic and produces two types of eggs, half with a X chromosome and half without a X chromosome (with or without a Y chromosome). To avoid confusion with that of XX-XO and XX-XY types of sex determining mechanisms, instead of the X and Y alphabets, Z and W alphabets are generally used respectively.

The heterogametic females may be of following two types:

***(i)* ZO-ZZ system**. This system of sex determination is found in certain moths and butterflies. In this case, the female possesses single Z chromosome in its body cells (hence, is referred to as ZO) and is heterogametic, producing two kinds of ova, half with a Z chromosome and half without any Z chromosome. The male possesses two Z chromosomes (hence, referred to as ZZ) and is homogametic, producing single type of sperms, each of which carries a single Z chromosome. The sex of the offspring depends on the kind of ovum as shown below:

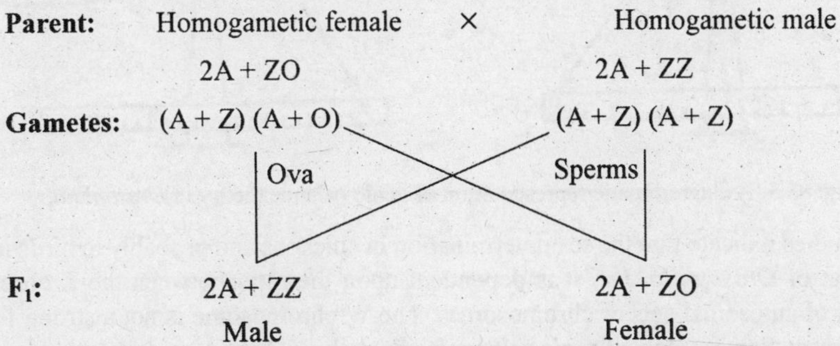

***(ii)* ZW-ZZ system.** This system of sex determination occurs in certain insects (gypsy moth) and vertebrates such as fishes, reptiles and birds and plants such as *Fragaris elatior*. Here the female sex has one Z chromosome and one W chromosome. It is heterogametic and produces two types of ova, 50 per cent ova carry the Z chromosomes, while rest 50 per cent ova carry W chromosomes. The male sex has two homomorphic Z chromosomes and is homogametic producing single type of sperms, each carries a Z chromosome. The sex of the offspring depends on the kind of egg, the Z bearing eggs produces males but the W bearing eggs produce females (Fig. 83.2).

B. Genic Balance Mechanism

By studying sex chromosomal mechanism of sex determination, it may appear at first glance that some genes carried by the sex chromosomes (X and Y) were entirely responsible for sex. But this is not the case. Extensive experimentation of different workers (**Wilson**, 1909 ; **Bridges**, 1921 and **Goldschmidt**, 1934) on different organisms have revealed the fact that most organisms generally have inherent potentialities for both sexes and each individual is found to be more or less intermediate between male and female sexes (Hence may be referred to as **intersex**). There seems to exist a delicate **balance** of masculine and feminine tendencies in the hereditary compliment of an individual and mechanisms like the XY, ordinarily serve to trip the balance in one direction or another. Such **genic balance** mechanism of determination of sex was first of all studied in *Drosophila* by **C.B. Bridges** in 1921.

Sex determination in *Drosophila*. In *Drosophila*, the presence of Y chromosome has been found essential for the fertility of male sex but that has nothing to do with the determination of male sex. In this fly, the sex is determined **polygenically**. The sex determining genes were so distributed that the net effect results in the autosomes determining maleness and the X chromosomes femaleness. The sex of an individual then depends upon the ratio of X chromosomes to autosomes. If each haploid set of autosomes carries factors with a male-determining value equal to one (1), then each X chromosome carries factors with a female determining value of one and half (1½). Let A represent a haploid set of autosomes. In a normal male (AAXY) the male and female determinants are in the ratio of 2 : 1½ and, therefore, the genic balance is in the favour of maleness. A normal female (AAXX) has a male and female ratio of 2 : 3 and, therefore, the balance is in the favour of femaleness (See Fig. 83.3).

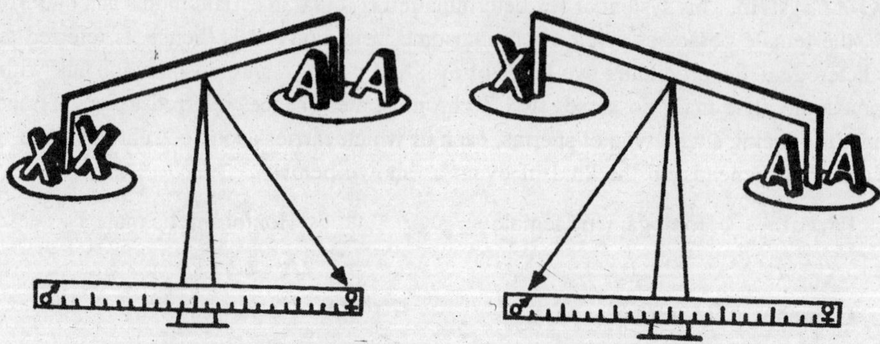

Fig. 83.3. A diagrammatic representation of genic balance theory in *Drosophila.*

Recent studies indicate that the sex determination in chickens and probably in birds in general, is similar to that of *Drosophila*, *i.e.*, it is dependent upon the ratio between the Z chromosomes and the number of autosomal sets of chromosomes. The W chromosome is not a strong female sex determining element. Same system of genic balance in which the X/A ratio is critical, has been reported for the flowering plant (angiosperm) *Rumex acetosa*.

The genic balance theory of Bridges has been supported by the following findings:

1. Polyploid flies. Bridges crossed the experimentally produced triploid (3n individual having three whole sets of chromosomes) female *Drosophila* (3A : 3X) to diploid males (2A : XY). The results obtained from such a cross are shown in Fig. 83.4. From this cross he obtained normal diploid

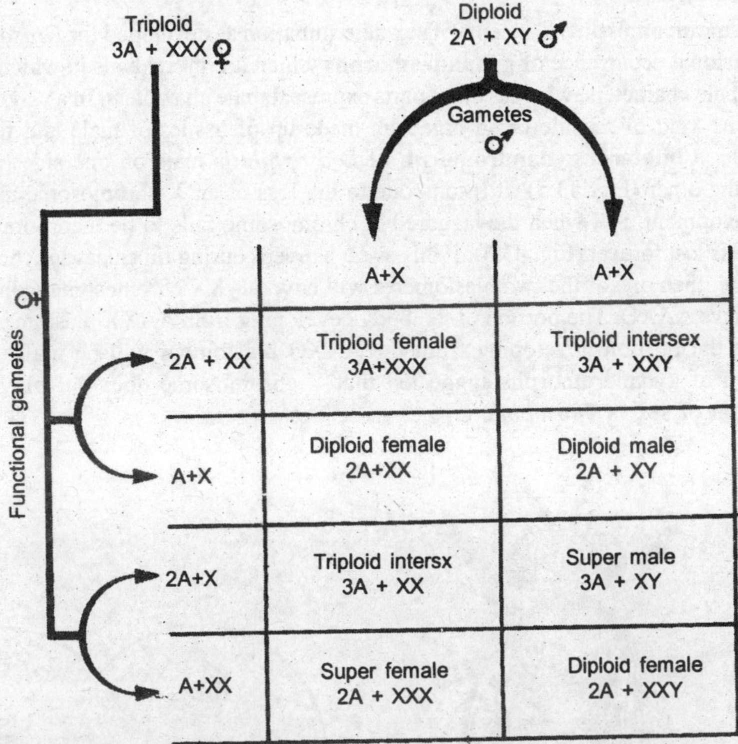

Fig. 83.4. Results obtained from a Bridge's classical cross of a triploid (3A + XXX) female fly and a diploid (2A + XY) male fly (*Drosophila*).

males triploid females, intersexes, super males and super females. The occurrence of triploid intersexes from such a cross, clearly established that the autosomes also carry genes for sex determination. These intersexes were sterile individuals and had phenotype in between male and female sexes. The occurrence of such intersexes, super males and super females were explained by him by genic balance mechanism. Different combination of X chromosomes and autosomes and corresponding sex expression in *Drosophila* have been summarized in Table 83.1.

Table 83.1 Different doses of X-chromosomes and autosome sets and their effect on sex determinations in *Drosophila*.

Phenotypes		*Number of chromosomes*	*Number of autosomes (A sets)*	*Sex index No. X's/No. A sets*
Super female		3	2	1.5
Normal female	tetraploid	4	4	1.0
	triploid	3	3	1.0
	diploid	2	2	1.0
	haploid	1	1	1.0
Intersex		2	3	0.67
Normal male		1	2	0.50
Super male		1	3	0.33

As shown in the table, when the X/A ratio is 1.0, the individual will be female and if it is 0.50, it would be male. When this balance is disturbed, the sex of individual deviate from normal male or normal female. For example, when the X/A ratio falls between 1.0 and 0.50, it would be **intersex**; when it is below 0.50 it would be **supermale** and when above 1.0, it would be **metafemale** or **super female**.

2. Gynandromorphs. Concepts of sex determination as developed for *Drosophila* are verified by the occasional occurrence of **gynandromorphs** which are individuals in which part of the body expressed male characters, whereas other parts express female characters. In a way, gynandromorphs represent one kind of **mosaic** or an organism made up of tissues of male and female genotypes. For example, a **bilateral gynandromorph** of *Drosophila* is male on one side (right or left) and female on the other (Fig. 83.5). It results due to the loss of an X-chromosome in a particular cell during development, *i.e.*, when the laggered X chromosome fails to be incorporated in a daughter nucleus and is lost forever (Fig. 83.6). If this event happens during first cleavage (or mitotic division) of the zygote, then one of the two blastomeres will have AAXX chromosomal complement and the other will have AAXO. The portion of the body developing from AAXX blastomere will be normal female and the portion developing from the AAXO blastomere will be male. The cytological examination of gynandromorphs suggested that Y chromosome does not play any role in the determination of sex in *Drosophila*.

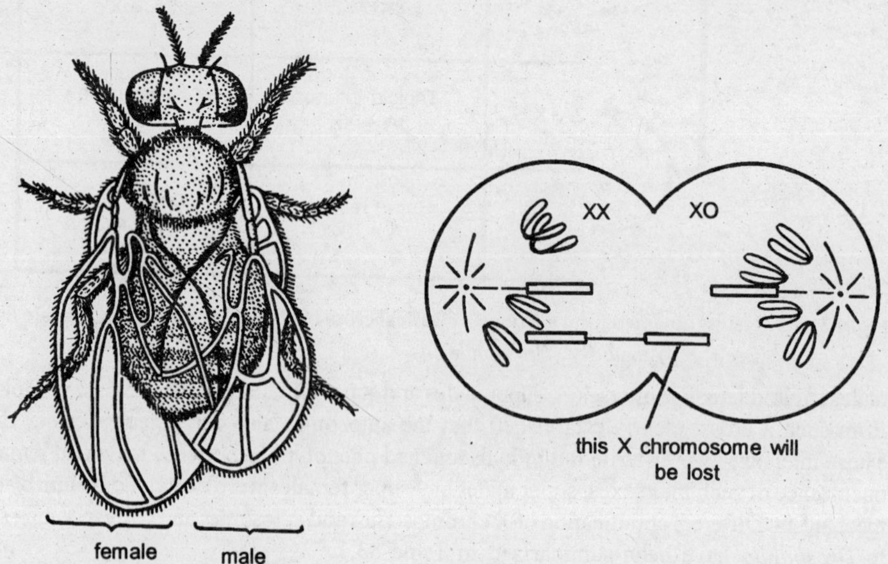

Fig. 83.5. A bilateral gynandromorph of *Drosophila*.

Fig. 83.6. The loss of an X chromosome during mitosis in a 2A + XX cell and formation of two types of cells—XX and XO.

C. Male Haploidy or Haplodiploidy Mechanism

Male haploidy or haplodiploidy or arrhenotokous parthenogenesis is particularly common in the hymenopterous insects such as ants, bees, sawflies and wasps (*e.g.*, *Bracon hebetor*). In these insects, since, fertilized eggs develop into diploid females and unfertilized ones into haploid males; so arrhenotoky is both a form of reproduction and a means of sex determination. Meiosis is normal in females, but crossing over and reduction division in chromosome number fail to occur during spermatogenesis in males due to their haploidy.

For example, a honeybee queen (whose diploid number is 32) can lay two types of eggs. By controlling the sphincter of her sperm receptacle (which holds sperms previously obtained in matings with males during nupital flight), she produces a fertilized egg (a diploid zygote having 32 chromosomes and developing into a female) or an unfertilized egg (a haploid zygote having 16 chromosomes and developing into a male). The diploid female zygotes can differentiate into either workers (sterile) or queens (fertile) depending on the diet they consume during their development.

D. Single Gene Control of Sex

In certain organisms, for example *Chlamydomonas*, *Neurospora*, yeast, *Asparagus*, maize, *Drosophila*, etc., individual single genes are found to be responsible for the determination or expression of sex. Following cases exemplified the single gene control of sex:

(*a*) Monogenic sex determination in *Drosophila*. In *Drosophila*, a **transformer gene (*tra*)** has been recognized which when present in homozygous condition (*tra/tra*) transforms a female fly into a sterile male, but, it does not act upon normal male individuals. Thus, a XX female with *tra/tra* genotype will be a sterile male, but, a XY male with *tra/tra* genotype will still be a normal male fly.

(*b*) Sex reversal gene (Sxr) in mammals. Recently a **sex reversal gene (*Sxr*)** has been discovered in human beings, so that in the presence of this gene XX female individuals may become male. Such cases of sex reversal are also reported in goat and mice. Mice also contain two other genes *Tdy* and *Tda*-1 which interact to cause sex reversal in XY male individuals to transform them into females.

83.3 METABOLICALLY CONTROLLED SEX DETERMINING MECHANISM

Certain workers have seen the possibility of sex determination in the phenomenon of metabolism. **Crew** suggested that sex is a physiological equitable division between anabolic and catabolic individuals. **A.F. Shull** and **D.D. Whitney** have shown that by increasing metabolic rate in rotifers the occurrence of male individuals increases than females. Likewise, **Riddle** found that metabolism had some definite role in the determination of sex in pigeons and doves, because, increased rate of metabolism developed the male potency, while decreased rate of metabolism caused femaleness.

83.4 HORMONALLY CONTROLLED SEX DETERMINING MECHANISM

In many cases it has been observed that sexual differentiation is controlled by hormones. Following classical examples would make clear that how hormones control the sex of an individual:

(*a*) Sex in *Bonellia*. An excellent example of environmental determination of sexual phenotype is provided by *Bonellia viridis*, a marine echiuroid worm studied extensively by **F. Baltzer** (1935). The adult female is about an inch (2.53 cm) long and has fairly complex anatomical organization. Male is of the size of large Protozoa and has rudimentary organs (Fig. 83.7). The males live as parasites in the uterus of the females. All larvae of *Bonellia* are genetically and cytologically similar. If a particular larva settles on the proboscis of an adult female, it becomes a male individual. On the other hand, if a larva develops in isolation (*i.e.,* in water) it develops into a female. Further, if a incompletely developed male is detached from the proboscis of female, it becomes an intersex. Thus, it becomes evident that proboscis of adult female secretes some hormone-like substance which suppresses femaleness and initiating maleness in the larvae which are attached with it.

(*b*) The Crew's hen. In birds only one gonad of a normal female develops into a functional ovary. The other gonad remains rudimentary. If the functional ovary of a hen is destroyed, the rudimentary gonad develops into a testis (**Crew**, 1923). Thus, the female sex is reversed into male sex due to the phenomenon called **sex reversal**. Such a sex reversed hen can even father chicks which will be expected to show a sex ratio of two females to one male, since in birds the female is the heterogametic sex. The case of sex reversal of hen can be interpreted as follows:

During embryonic development, the XY (ZW) genotype stimulates the pituitary gland to produce female hormones that cause the gonad of the hen to develop into an ovary. After the development of ovary, the pituitary ceases to produce female hormones, due to inhibition of the pituitary by hormones produced by the ovary, thus, acting as **developmental feed-back system**. The high level of female hormones secreted sequentially by the pituitary and the ovary is sufficient to suppress the action of male hormone producing cells of the body such as the steroid producing cells of the adrenal glands. When the ovary of a hen is removed, the steroid cells of adrenal become active and provoked the development of rudimentary gonad into testis which itself is an endocrine gland. Both endocrine glands (*e.g*, adrenal and testis) produce large amount of male hormones which sufficiently suppress the action of the female hormone cells of pituitary. Such cases of sex reversals have also been observed in amphibians (*e.g.*, in salamander by **E. Witschi**).

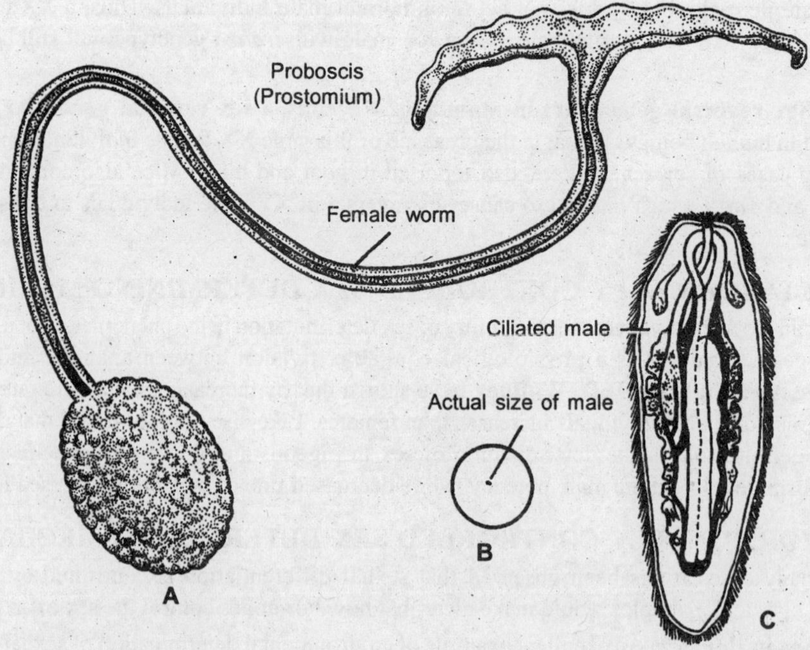

Fig. 83.7. The female (A) and male (B, C) of the echiuroid *Bonellia viridis*.

***(c)* Hormones and freemartin.** Another classical example of hormonal control of sex determination has been found in cattles. In cattle, when twin calves of different sexes occur, the female member is usually a sterile intersex called a **freemartin**. The freemartin has external female genitalia but internal sex organs are more or less like those of male. The male twin is usually normal. **F.R. Lillie** (1917) has suggested that the formation of a freemartin was due to a fusion of the foetal membranes of the twin calves, while they were in uterus of the mother. The fusion of the foetal membranes permitted the blood of each twin to circulate in the blood vessel of the other. The male hormones produced by the male twin are presumed to suppress the differentiation of the female internal sex organs of the co-twin.

83.5 ENVIRONMENTALLY CONTROLLED SEX DETERMINING MECHANISM

In some organisms, the environment determines the sexual phenotype of the individuals. Thus, some environmental factors as size of parent body or of egg, age of parent and temperature are found to determine the sex in many cases.

Coe (1943) demonstrated in mollusc genus *Crepidula* that its every zygote contains all the genes which are necessary for the development of male and female reproductive systems. But, in this species it has been found that young *Crepidula* remains functional male and produce only sperms but no ova. As it matures the male phase is changed by transitional phase to a functional female phase. The female phase continues for the remainder of the animal's life. It is suggested that environment somehow regulates the activity of certain genes to produce only male phenotype during young stages, while inhibits them during adult stages and simultaneously initiating other genes to produce female phenotype.

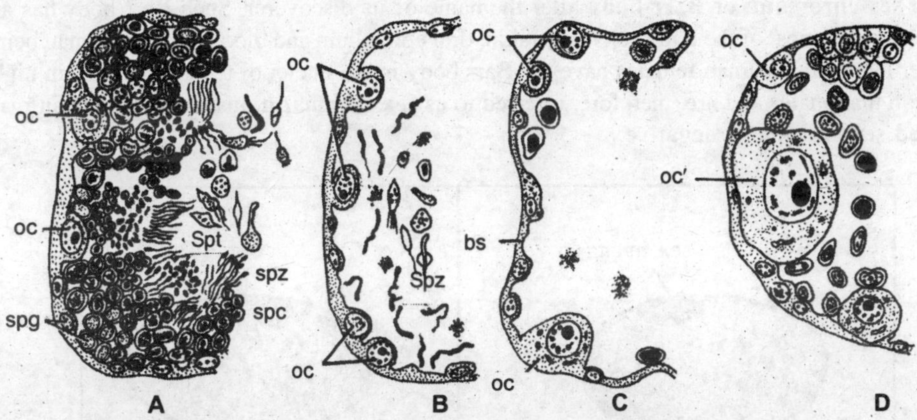

Fig. 83.8. Change in the gonad of the mollusc *Crepidula* experiencing consecutive sexuality. A–Male phase; B, C–Transitional phases: D–Female phase. oc–Immature occyte, oc′–Mature occyte, spg–Spermatogonia, spt–Spermatids, spz–Spermatozoa, spc–Spermatocytes.

83.6 SEX DIFFERENTIATION

Although the primary determination of sex is made at fertilization, but the embryo acquires its definite sex characteristics by a more complex mechanism called **sex differentiation**. Thus, sex determination is not a single, once-and-for all decision. But rather it is a gradual *awakening* that spreads from the ovary to testis to the reproductive tract, to the general body tissues and even to the brain itself; finally almost the whole body is overtaken by the process of sexual differentiation. In human beings and other animals sex differentiation occurs in the following steps:

1. Genetic Sex

Normal females ordinarily have two X chromosomes; normal males have one X and one Y. The genes on these sex chromosomes determine femaleness or maleness. Thus, one can speak of females having the genetic sex designation XX and males having the genetic sex designation XY.

Further, since the X-chromosome carries much more genetic information in striking contrast to Y chromosome, one might wonder how it is that the female can carry a double dose of many vital X-linked genes, whereas the male has only a single dose of these X-linked genes. Such inequality in fact cannot be tolerated and so female mammals and male *Drosophila* seem to have developed their own types of dosage compensation mechanisms.

Dosage Compensation of Genes

The term dosage compensation was coined by **Muller** (1932) and this phenomenon was originally discovered in *Drosophila*. Dosage compensation of the genes is done either by **hypoproduction due** to inactivation of one X-chromosome in homogametic female sex, as observed in mammals, or by the **hyperproduction** due to hyperactivity of X-chromosome in the heterogametic male sex as observed in *Drosophila*.

(*i*) X-chromosome inactivation in mammals. It has been demonstrated that in homogametic XX female individuals, one X-chromosome gets characteristically condensed and inactivated. Such chromatin material is called **facultative heterochromatin**, since it becomes inactive in certain part of the life cycle and resumes activity before entering the germ line. The phenomenon of inactivation of chromosome was confirmed by the observation of the Barr body (Box 83.1).

Barr body. Barr and **Bertram** (1949) reported a deeply stained chromatin body (*i.e.*, a chromocentre) in the nerve cells of female cat which was absent in the male. This chromatin body is called **sex chromatin** or **Barr body** after the name of its discoverer. Such Barr body has also been observed in most of the body cells (*e.g.*, skin, oral epithelium and blood cells) of human beings and other mammals. Human females have the Barr body in the nuclei of their body cells in higher proportion than males and are, therefore, referred to as **sex chromatin positive**. The human males are called **sex chromatin negative**.

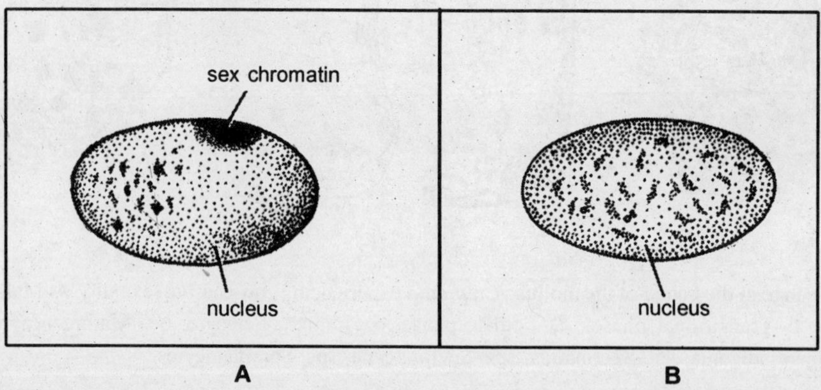

Fig. 83.9. Sex chromatin in the nucleus of female cat (A). Male cat (B) has no sex chromatin in its nucleus.

The sex chromatin appears in the interphase nucleus as a small chromocentre, heavily stained with basic dyes (Fig. 83.9). It gives a positive Feulgen reaction and has a relatively constant position in each tissue and species. It can be found in four positions: (*i*) attached to the nucleus as in nerve cells of certain species; (*ii*) attached to the nuclear membrane as in cells of epidermis or of the oral mucosa; (*iii*) free in the nucleoplasm as in neurons after electric stimulation, (Fig. 83.10) and (*iv*) as a nuclear expansion (Fig. 83.11). The best known example of nuclear expansion is that of the neutrophil leukocyte of female in which the sex chromatin (Barr body) appears as a small rod called the **drumstick**.

Box 83.1 Molecular Basis of X-chromosome Inactivation

In the somatic cells of female mammals, **dosage compensation** for the X-chromosome is promoted by **X-chromosome inactivation**. One of the X-chromosomes become condensed into **heterochromatin** and so become transcriptionally inert. The **inactive X** (called Xi) remains condensed throughout the cell cycle as a densely staining **Barr body**, while the **active X** (called **Xa**) acts normally.

In the female zygote, both X-chromosomes are active. The inactivation process initiates at the late blastocyst stage in the extraembryonic lineage, and slightly later in the epiblast lineage which gives rise to the somatic cells of the embryo. In embryonic tissues of placental mammals, the choice of active versus inactive X chromosome is random. Once choice is made, the process is irreversible: the active and inactive chromosomes are clonally propagated, so that an individual heterozygous for a visible X-linked marker demonstrates patchy variegation for the marker phenotype (*e.g.*, tortoise shell coat colours in cats). In extraembryonic tissue of many mammals (but not humans), the parental X-chromosome is preferentially inactivated (it is called **parental imprinting**; Box 83.2) and paternal preference occurs in both embryonic and extraembryonic tissue of marsupials.

The molecular basis of X-inactivation can be divided into four stages: **counting, initiation, propagation (spreading)** and **maintenance**. A X-linked *cis*-acting site, apparently involved in all four stages, is the **X-inactivation centre** (*Xic*), which has been mapped by studying deletions and translocations of the X-chromosome resulting in the disruption of inactivation. If the region containing *Xic* is deleted, the mutant X chromosome remains active; if the X chromosome is involved in a translocation, the fragment containing *Xic* is inactivated (as is part of the autosome to which it is attached) while other fragment remains active.

Micromapping of *Xic* originally revealed a single locus, called *X-ist*, which is expressed specifically on Xi (= inactive X chromosome). *X-ist* encodes a large **untranslated RNA** which associates with Xi in the nucleus and is thought to induce heterochromatinization by coating the X chromosome. The overall regulation of X-inactivation exhibits the balance between X-chromosome and autosomes. **Polysomy X** is tolerated because supernumerary X-chromosomes are inactivated, generating multiple Barr bodies, whereas **polyploidy** causes more X-chromosomes remain active. The nature of X chromosome counting mechanism is unknown.

During gametogenesis the promoter of *Xist* allele is subjected to sex-specific **methylation**, causing a maternal **imprint** (Box 83.2) in the zygote. This results in preferential expression of the paternal *Xist* allele and hence preferential inactivation of the paternal X-chromosome. Such a pattern of methylation is maintained in the mouse extraembryonic lineage (and the paternal X-chromosome remains inactive), while in the epiblast lineage the methylation pattern is *reset* by a **global demethylase** enzyme and *de novo* methylation of the *X-ist* promoter occurs at random (for methylation see Box 83.2). Thus, the choice of active versus inactive X chromosome lies ultimately with the *de novo* **methylase,** and it is still not clear how the choice is made. In humans, random X-inactivation occurs in both extra-embryonic and epiblast lineages. In some extra-embryonic cells both X-chromosomes remain active.

Box 83.2 Methylation of DNA

1. Methylation involves covalent modification of DNA bases but it does not alter base-pairing specificity, so the information carried in DNA is preserved. Methylation can influence DNA function, *e.g.*, by changing the way it interacts with transcription factors and other proteins. The state of DNA methylation may thus regulate gene expression and has a direct connection with the phenotype of the organism.

2. Enzymes which add methyl group to DNA, **DNA methyltransferases** (or **DNA methylases**) uses S–adenosyl methionine as methyl donor. In eukaryotes, the only modified base commonly found in DNA is 5-methyl cytosine (5-mec).

   Two types of enzymes can be recognized: ***de novo* methylases** add methyl groups to unmethylated DNA at specific sites and can therefore initiate a pattern of methylation, whereas **maintenance methylases** add methyl groups to DNA which is already methylated on one strand **(hemimethylated DNA)** and thus perpetuates patterns of methylation through successive rounds of replication.

3. DNA carries two forms of information: **genetic information** in its nucleotide sequence and epigenetic information in its structure. *Epigenetic information is any heritable property of DNA which influences its activity* (this ultimately contributes to the phenotype of the organism) but which lies outside the nucleotide sequence itself). **DNA methylation is one source of epigenetic information: it is heritable** (due to maintenance methylation) **and able to control gene expression**. A methylation state is thus termed as an **epigenotype** and a change in methylation state as an **epimutation.**

4. **Parental imprinting.** An **imprint** is a form of epigenetic information, and **parental imprinting** is thus an epigenetic, mechanism by which chromosome region can retain an epigenetic memory of its parental origin. In somatic cells, most genes are expressed on both maternal and paternal chromosomes but **imprinted genes** are expressed in a parent chromosome-specific manner. Conventionally, an allele is said to be imprinted if it is repressed (*e.g.*, at maternally imprinted loci, only the paternal allele is expressed). At the molecular level, *the imprint is DNA methylation* but in some loci methylation corresponds to gene activity and in others it corresponds to inactivity.

The relationship between sex chromatin and sex chromosomes has been elucidated by **Ohno, Kaplan** and **Kinosita**, 1959. It has been demonstrated that the sex chromatin is derived only from one of the two X-chromosomes. The other X chromosome behaves like a autosome and is not heteropycnotic at interphase. Later, **Lyon** (1972) confirmed the existence of Barr body in normal females, metafemales or superfemales (XXX) and in Klinefelter males (XXY).

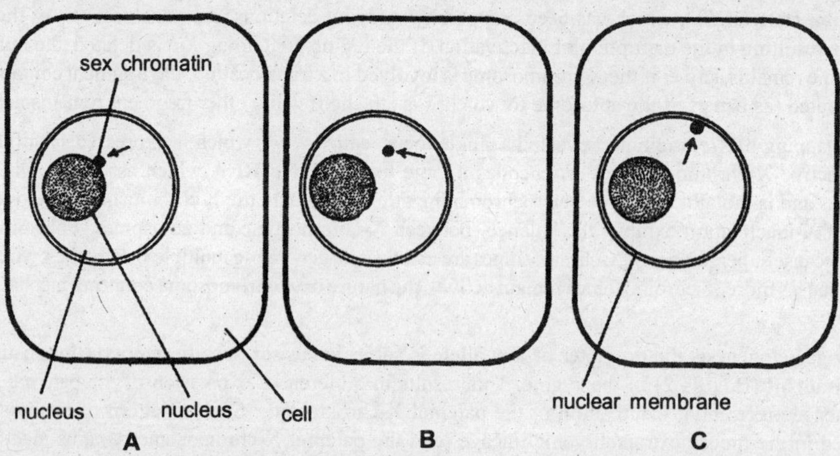

Fig. 83.10. A–Position of sex chromatin near nucleolus; B–In nucleoplasm and C–Attached to the nuclear membrane.

Lyon's hypothesis. It was demonstrated by **M.F. Lyon** of U.K. that whenever the number of X chromosomes was two or more than two, the number of Barr bodies was one less than the number of X chromosomes (nX-1; *i.e.*, one Barr body in XX females and XXY males; two Barr bodies in XXXY males and XXX metafemales). Thus, in normal female only one active X chromosome is present.

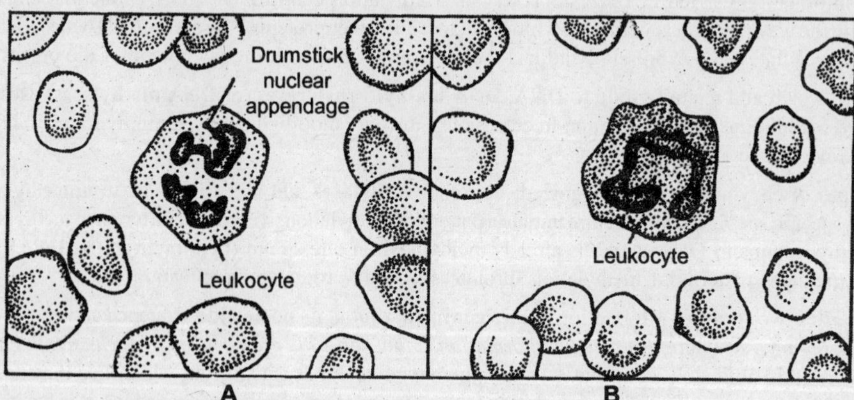

Fig. 83.11. A-Leukocyte of human female with drumstick-like sex chromatin; B-Leukocyte of human male without any drumstick.

Which of the two X chromosomes remains active in female individuals, is determined at the early stages of development. It was observed by Lyon in 1961 that each of the paternal (P) and maternal (M) X chromosomes has a chance to become inactive (*i.e.*, Barr body). In other words, the inactivation of X chromosome is a random phenomenon. This fact has been demonstrated in human diseases linked to X chromosome. The **Lesch-Nyhan syndrome,** in which a deficiency of one enzyme of the purine metabolism (*i.e.*, hypoxanthineguanine phosphoribosyl transferase) produces mental retardation and increased uric acid levels results, from a recessive mutation in the X chromosome. This is shown as follows: if fibroblasts of these patients are cultured *in vitro*, two types of cell clones are obtained. Half the clones contain the enzyme, whereas the other half (in which the X carrying the normal gene is condensed) lack the enzyme (Box 83.1 and Box 83.2).

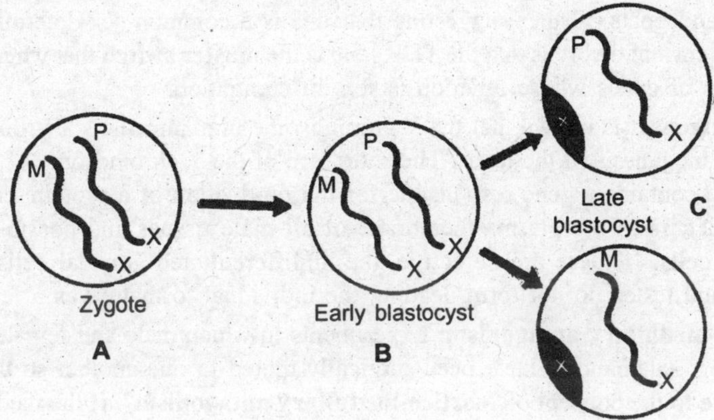

Fig. 83.12. Diagram of the evolution of XX chromosome. A–In the zygote both paternal (P) and maternal (M) X-chromosomes are euchromatic; B–In the early blastocyst the same is true as in A; C–In late blastocyst 50% of the cells have a maternal heterochromatic X chromosome and the other 50% have the paternal heterochromatic X chromosome. In diagram C formation of sex chromatin by one of the X chromosome has been shown.

In the human embryo X chromosome inactivation starts in the late blastocyst about the 16th day of life. Once the inactivation is established, it is irreversibly maintained in somatic cells, however, in germ cell line reactivation occurs at a specific stage of germ cell development (*i.e.*, entry into meiotic prophase; **Martin**, 1982).

The good illustration of X chromosome inactivation is seen with tortoise-shell cats, where the coat is a mosaic patchwork of black and yellow hair. Black hair is produced by the dominant allele *B*, and yellow by its recessive allele *b*. This gene is X-linked, so if one X chromosome contains the dominant allele *B* and other X chromosome the recessive allele *b*, random inactivation will allow both coat colours to be expressed. Male tortoise-shell cats are understandably rare, since it has only one X chromosome.

(*ii*) Hyperactivity of X chromosome in male *Drosophila*. The phenomenon of dosage compensation in *Drosophila* has been shown to be due to hyperactivity of one X chromosome in male *Drosophila*. This fact was discovered by **Dr. A.S. Mukherjee** of Calcutta University. In mosaic individuals with XX and XO cells, it could be shown that X chromosome in XO cells was always hyperactive. Due to this fact, mutant and wild type flies showed same intensity of eye colour in male and female flies. Similarly, enzyme levels for several enzymes including **6PGD** (6 phosphogluconate dehydrogenase), **G6PD** (glucose-6 phosphate dehydrogenase), **tryptophan pyrrolase** and **fumarase** were found to be similar in female and male flies.

2. Gonadal Sex

In the human embryos until the six weeks the gonads and primordia of the urinogenital tract are identical in males and females. At this stage (time) the gonad has already been invaded by the primary XX or XY cells. At this point, a gene or set of genes, called **testis-determining factor** or **TDF**, present in the Y chromosome causes the undifferentiated gonad to differentiate into a testis and the absence of this gene allows the gonad to become an ovary. The development into a testis starts as soon as the gonocytes (*i.e.*, primordial germ cells) from the yolk sac have finished their migration into the gonadal ridge. Gonocytes of the male (XY) migrate deeper into the gonadal blastema forming the medulla and female gonocytes (XX) remain at the periphery, forming a thick cortical layer. Hence, the XX genetic sex is ordinarily associated with **ovarian gonadal sex**, and XY is associate with **testicular gonadal sex**.

Further, identical TDF region of Y chromosome has been identified in gorillas, monkeys, dogs, cattle, horses and goats, further suggesting that this is a common sex-determining mechanism in mammals. The current theory is that the TDF gene is the **master switch** that when turned on, activate an entire series of genes whose function is sex differentiation.

H-Y antigen. It is known that the Y chromosome in mammals is a strong male inducer and directs the organogenesis of the testis. The short arm of the Y chromosome of human beings, near the centromere contains a gene responsible for the production of a protein, called **H-Y antigen**. This antigen occurs in the plasma membrane of all male tissues and possibly tends to regulate H-Y receptor cells. If these are defective, the undifferentiated gonadal cells fail to respond to H-Y antigen and testes do not form, leading the individual to mature as a female.

Cortico-medullary antagonism. Experiments in which male and female amphibians larvae (*e.g.*, toad, frog, salamander) have been surgically united to one another so that their circulation fused, gave rise to the concept of "**cortico-medullary antagonism**". If the medulla was stimulated to develop, it was thought to produce a substance that could act locally or via the circulation to inhibit the ingrowth of ovarian cortex and vice versa. For example, in case of **freemartin** which occurs in cows, sheep, pigs and goats, a female foetus that is twin to a male and shares a conjoined placental circulation undergoes a partial sex reversal of her ovaries, which interferes with the subsequent sexual differentiation of the reproductive tract.

3. Hormonal or Genital Sex

The embryonic gonads produce hormones that, in turn, determine the morphology of the external genitalia and the genital ducts. XX embryos normally develop ovaries, female external genitalia and Mullerian ducts. XY embryos, on the other hand, ordinarily develop testes, male external genitalia, and Wolffian ducts. In XX embryos the Wolffian ducts are suppressed; in XY embryos the Mullerian duct remain under-developed. Thus, there is a distinction between male and female genital sex.

For example, the human gonad differentiates into a definite testicle at the seventh week, whereas the female gonad begins to develop into an ovary between the eight and nine weeks of development. At the time of differentiation an important epigenetic factor is the production of androgens (testosterone) by somatic cells in the embryonic male gonad. In mammals, administration of testosterone hormone to the mother produces in the foetus a shift in the differentiation of XX genitalia to a male type, producing **masculine pseudohermaphroditism**. This hormone locally accelerates the development of the testis, whereas in the female the absence of the hormone permits the slower development characteristic of the ovary.

4. Somatic Sex

Production of gonadal hormone continues to increase until secondary sex characters appear at puberty. These include amount and distribution of hair (*e.g.*, facial, body, axillary, pubic), pelvis dimensions, general body proportions, subcutaneous fats over hips and thighs and breast development in human females, as well as, increased larynx size and deepening of the voice in the human male.

5. Socio-psychological Sex

In most individuals, genetic sex, gonadal sex, genital sex and somatic sex are consistent; XX persons, for example, develop ovaries, female genitalia and female secondary sexual characteristics. Ordinarily these persons are raised as females and adopt the faminine gender role under whatever culture pattern has been established in the society of which they are members. A similar consistency from genetic sex to socio-psychological sex is seen for XY individuals. Intersexuality occurs when the chromosomal contribution is inconsistent with the gonadal or other secondary sexual characters.

Sex Mosaics or Gynandromorphs in Humans

***(i)* Sex mosaics.** These are those persons whose body contains two or more populations of cells, each with a different sex chromosome complement. For example, **Murray Barr** found a girl in which cells of buccal and vaginal smears showed two Barr bodies thus indicating XXX chromosomes but her blood cells showed XO chromosomal complement. The girl showed Turner syndrome indicating XO condition. Such sex mosaics are produced due to the phenomenon of non-disjunction during mitosis.

***(ii)* Pseudohermaphroditism.** This is an intersexual condition in human beings. In pseudohermaphroditism, an individual exhibits characters of both sexes in one body. It is either male or female but owing to deficiency of sex hormones, he fails to show all the characters of his sex in full.

***(iii)* Sex reversal in human beings.** In human beings the sex is determined by XX-XY mechanism, however, sex is differentiated by the hormones. If testes of male is removed before puberty, the male secondary sexual characters fail to develop. In him female secondary sexual characters, such as body form, voice, hair pattern, *etc*., appear. If such sex reversed male is injected with male hormones, the male sex characters reappear.

In fact, in human embryo, the gonads are quite similar regardless of the chromosomal sex of the embryo. These gonads are called **ovotestes**. Each gonad is formed of inner **testicular tissue**, and outer **ovarian tissue**. In the presence of Y chromosome, the testicular tissue develops first and secretes male hormone. These male hormones have following functions. 1. Limit the development of ovarian tissue of the gonads. 2. Accelerate development of testicular tissue. 3. Control differentiation of male external genitalia. These changes result in the formation of male foetus.

QUESTIONS

Long Answer Questions

1. What is sex determination? Name various mechanisms of sex-determination. Give an account of genetically controlled sex-determination mechanism. *(Kanpur 1990)*
2. Describe the mechanism of sex-determination in animals. *(Bundelkhand 1989, 1990; Bangalore 1993; Kanpur 1998)*
3. Write an essay on 'sex determination'. *(Sagar 1990; Kanpur 1992; Lucknow 1996)*
4. Write an account of modern theories of sex determination. *(Kurukshetra 1988; Burdwan 1989)*
5. Describe sex determination in *Drosophila*. Explain how environmental factors and internal secretions influence sex differentiation in *Bonellia* and cattle. *(Lucknow 1991)*
6. Describe sex-determination in *Drosophila*, bugs, birds, mammals. Explain how environmental factors and internal secretions influence sex determination in *Bonellia* and cattle.
7. Write an essay on chromosomal sex determination. *(Punjab (G) 1990)*
8. Give an account of chromosomal basis of sex determination in *Drosophila*. *(Jiwaji 1991; Guwahati (M) 1994; Allahabad 1994)*
9. What is sex chromosome? Describe the phenomenon of sex determination with special reference to balance theory. *(Guwahati (M) 1994)*
10. Explain genetic (genic) balance theory of Bridge with an example. *(Madras 1989; Guwahati 1989; West Bengal 1989; Bangalore 1992, 1995; Guwahati (M) 1996)*
11. Describe the mechanism of sex-determination in *Drosophila*. *(Allahabad 1991; Kanpur 1996)*

12. Give an account of the chromosomal mechanism of sex determination in man. How does it differ from the avian type of sex determination. *(Lucknow 1994; Guwahati (M) 1995)*

13. Discuss the mechanism of sex determination in human beings. Does the Y-chromosome bear any gene for sex determination. *(B.H.U. (Hons) 1991)*

14. Give an account of sex differentiation.

15. What do you understand by sex determination? How does it differ from sex-differentiation? Describe the chromosome theory of sex determination. *(CCS 1994)*

16. What do you understand by sex-determination? How does it differ from sex-differentiation? Describe in detail the genic balance theory of sex determination and various theories of sex differentiation in different animals. *(CCS 1991, 1993)*

17. Explain the chromosomal mechanism of sex determination. *(M.S.U., Tamil Nadu, 2006)*

18. Discuss the role of X and Y chromosomes in the determination of sex in man.

19. What do you understand by sex chromosomes? Explain chromosomal theory of sex determination in animals. *(H.N.B. Garhwal 2009)*

Short Answer Questions

1. Given an account of Lyon's hypothesis. *(Allahabad 1990)*

2. Give an account of the chromosomal basis of sex determination in man. What are Barr bodies? *(Allahabad 1982, 1992)*

3. What is Barr body? Discuss the relationship between the Barr body and X-chromosome.

4. Describe the phenomenon of sex determination in animals. *(H.N.B. Garwal, 2008)*

5. Write short notes on the following:

(*i*) Sex chromosomes *(Delhi 1988; Sardar Patel 1989; Kanpur 1995; Guwahati (G) 1995)*

(*ii*) Y Chromosome

(*iii*) Sex determination *(Punjab (G) 1991; Purvanchal 1991; Allahabad 1993)*

(*iv*) Sex determination in *Drosophila*. *(Indore 1990)*

(*v*) Environmental sex determination *(Bundelhand 1989)*

(*vi*) XY female-XX male type sex determination *(Bundelkhand 1989)*

(*vii*) Dosage compensation *(Allahabad 1991)*

(*viii*) Barr bodies *(CCS 1995; Kanpur 1997)*

(*ix*) Lyon's hypothesis *(Allahabad 1994, 1996)*

(*x*) Gyandromorph *(CCS 1992)*

(*xi*) Sex mosaics

(*xii*) Pseudohermaphroditism

6. Write differences between the following:

(*i*) Sex determination in *Drosophila* and *Apis*. *(Lucknow 1995)*

(*ii*) Sex chromosomes and autosomes *(Guwahati (G) 1992, 1994, 1996)*

(*iii*) Sex determination in birds and human beings.

(*iv*) Autosome and allosome *(Rohilkhand 1990)*

Very Short Answer Questions

1. How many Barr bodies will be found in the cells of XXXY human beings? *(W.B. 1989)*

Ans. Two Barr bodies.

2. Explain how sex-chromatin helps in the identification of sex?

Ans. The sex chromatin or Barr body is condensed heterochromatic X-chromosome which is found in the nuclei of somatic cells of female mammals.

3. The nucleus from buccal epithelium cells of a person has a Barr body but he is a male. How do you account for this?

   Ans. A man with Barr body will have XXY chromosomes. It means he is suffering from Kleinefelter syndrome.

4. Sex determination. (*C.C.S.U., Meerut, 2009*)

   Ans. Sex is a hereditary character and is determined by chromosomes and genes in various organisms.

5. What are the sex chromosomes?

   Ans. Sex chromosomes are heteromorphic chromosomes whose distribution in a zygote determines the sex of the organism.

6. Define sex-determining Y (SRY).

   Ans. The sex switch, or testes-determining factor (TDF), in human being, located on the Y chromosome (It is celled *Sry* in mice).

Give one word for the following sentence:

1. A special condition is *Dropsophila*, where one half on the body exhibits male sex characters and the other half female sex-character.

   Ans. Gyandromorph. (*Lucknow 1994*)

Multiple Choice Questions

1. In *Drosophila*, sex is determined by the genic balance between

   (*a*) X and Y (*b*) X and sets of autosomes

   (*c*) Y and sets of autosomes (*d*) none of the above

2. In sex determination of human beings the key factor is

   (*a*) Y chromosome (*b*) X chromosome

   (*c*) both (a) and (b) (*d*) none of these

3. Barr body is

   (*a*) nucleolus (*b*) facultative heterochromatin

   (*c*) constitutive heterochromatin (*d*) euchromatin

4. Foetal sex can be determined by examining cells from the amniotic fluid by looking for:

   (*a*) chiasmata (*b*) kinetochore

   (*c*) Barr bodies (*d*) autosomes

5. Human gene determining sex at seventh week of pregnancy called TDF is present

   (*a*) X chromosome (*b*) Y chromosome

   (*c*) 21st pair of autosomes (*d*) none

6. In *Melandrium* the sex determination type is

   (*a*) XX–XY type (*b*) XX–XO type

   (*c*) ZZ–ZW type (*d*) ZZ–ZO type

7. Where are Barr bodies found?

   (*a*) sperm (*b*) ova

   (*c*) body / somatic cells of human male (*d*) body/somatic cell of human female

8. AMH (= anti-Mullerian duct hormone) is secreted by

   (*a*) Sertoli cells (*b*) Leydig cells

   (*c*) by both types of cells (*d*) by none of these cells

9. Who discovered sex chromosomes

   (*a*) Robert Brown (*b*) M.J.D. White

   (*c*) Nettle Stevens (*d*) G.J. Mendel

10. Genetic identity of a human male is determined by

(*a*) sex chromosomes (*b*) cell organelles
(*c*) nucleolus (*d*) autosomes

11. Chromosome theory of sex determination was propounded by

(*a*) McClung (*b*) Balbiani
(*c*) Goldschmidt (*d*) Mendel

12. Lyon's hypothesis deals with

(*a*) centromere position (*b*) genetic compatibility
(*c*) genetic incompatibility (*d*) number of Barr bodies

13. Genetic balance theory was proposed by

(*a*) Bridges (*b*) Morgan
(*c*) Boveri (*d*) Lillie

14. A sex mosaic is known as

(*a*) hemizygous (*b*) hermaphrodite
(*c*) gynandromorph (*d*) heterozygous

15. A medical technician while observing a human blood smear under the microscope notes the presence of Barr body close to the nuclear membrane in the WBC. This indicates that person under investigation is

(*a*) colour blind (*b*) haemophilic
(*c*) normal female (*d*) normal male

16. Webbed neck in a human being is characteristic of

(*a*) XXY male (*b*) YY male
(*c*) XO female (*d*) XXX female

ANSWERS

Multiple Choice Questions

1. (*b*)	**2.** (*a*)	**3.** (*b*)	**4.** (*c*)
5. (*b*)	**6.** (*a*)	**7.** (*d*)	**8.** (*a*)
9. (*c*)	**10.** (*a*)	**11.** (*a*)	**12.** (*d*)
13. (*a*)	**14.** (*c*)	**15.** (*c*)	**16.** (*c*).

84

Sex Linkage (Sex-Linked Inheritance)

In Mendel's crosses we have already observed that the progeny of a cross between two individuals of pure lines is the same regardless of which individual is female or which individual male or we can say that sex makes no difference in Mendel's crosses. But the Mendel's laws are not applicable on those genes which are exclusively located either in X or Y chromosome. It has been observed that the genes occurring only in the X chromosomes are represented twice in female (because female contains 2X chromosomes) and once in male (because male has only one X chromosome). Moreover, if the recessive type of genes occur in X chromosomes of males, they express themselves phenotypically. Because in such case Y chromosome contains no dominant allelomorph or gene to overcome the recessive gene of X chromosome. The genes which occur exclusively on the X chromosome (mammals, *Drosophila*, *Melandrium*, etc.) or on the analogous Z chromosome (in birds and other species with ZO or ZW mechanism of sex determination) are called **X- or Z -linked genes**. The genes which exclusively occur in Y chromosome are called **holandric genes**. The inheritance of X- or Z-linked and holandric genes is called **sex-linked inheritance**. In XX-XY type organisms, sex-linked genes (see Fig. 84.1) can be classified into following three types:

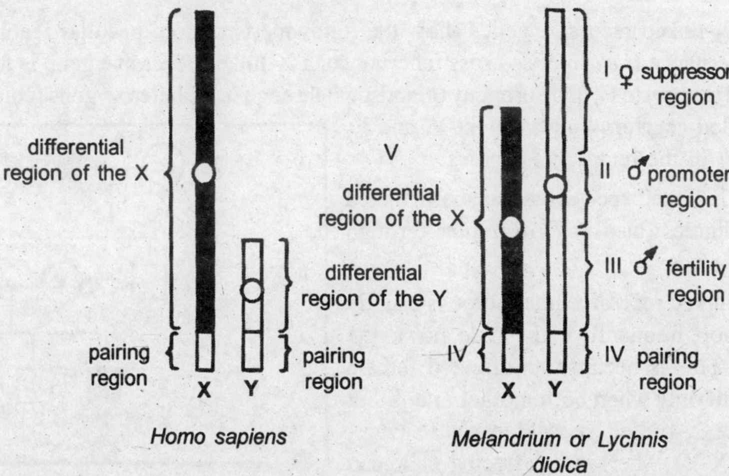

Fig. 84.1. Differential and pairing regions of sex chromosomes of humans and of the plant *Melandrium album.*

A. X-linked. The X-linked type sex-linked inheritance is performed by those genes which are localized in the non-homologous sections of X-chromosome, and that have no corresponding allele in Y chromosome. The X-linked genes are commonly known as **sex-linked genes**.

B. Y-linked. The Y-linked type sex-linked inheritance is performed by those genes which are localized in the non-homologous section of Y chromosome, and that have no alleles in X-chromosome. The Y-linked genes are commonly known as **holandric genes** (Greek, *holos* = whole, and *andros* = man).

C. XY-linked. The XY-linked type sex-linked inheritance is performed by those genes which are localized in homologous sections of X and Y chromosomes.

Box 84.1

The **sex chromosomes** control **sex determination** in animals and are asymmetrically distributed between the sexes. There is a **homogametic sex** which possesses a pair of identical sex chromosomes and thus produces one type of gamete, and **heterogametic sex,** which possesses either a pair of non-identical sex chromosomes or a single unpaired sex-chromosome and produces two types of gametes.

In mammals, females are homogametic: they carry two copies of the X-chromosome where as males carry one X-chromosome and one Y-chromosome and are therefore heterogametic. There, are two short regions of homology between the X-and Y-chromosomes, the **major** and **minor pseudoautosomal regions,** which facilitate pairing during meiosis. The major pseudoautosomal region is the site of an obligatory cross-over and genes located there are inherited in normal autosomal fashion (pseudoautosomal inheritance). Other genes are described as **sex-linked** because their expression depends upon how the sex chromosomes are distributed.

84.1 INHERITANCE OF X-LINKED (SEX-LINKED) GENES

Basic Characteristics of Inherence of X-linked Recessive Genes

The X-linked genes exhibit following characteristic patterns of inheritance:

1. The differential region of each chromosome (*i.e.*, X) contains genes that have no counterparts on the other kind of sex chromosome. These genes, whether dominant or recessive, show their effects in the male phenotype. Genes in the differential regions are called **hemizygous** ("half-zygous") in the males.

2. The X-linked recessive genes show the following two more peculiar features: **criss-cross** pattern of inheritance (*i.e.*, in criss-cross inheritance, a X-linked recessive gene is transmitted from P_1 male parent (father) to F_2 male progeny (grandsons) through its F_1 heterozygous females (daughters), which are called **carriers** and different F_1 and F_2 results (ratios) in the reciprocal crosses.

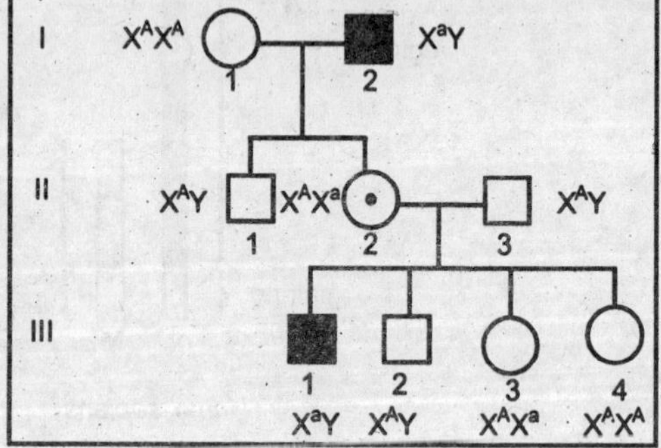

Fig. 84.2. Pedigree showing how X-linked recessive genes are expressed in males, then carried unexpressed by females in the next generation, to be expressed in their sons. III.3 and III.4 heterozygous or carrier females are not distinguished phenotypically.

3. The X-linked recessives can be detected in human pedigrees (also in *Drosophila*) through the following clues:

(*i*) The X-linked recessive phenotype is usually found more frequently in the male than in the female. This is because an affected female can result only when both mother and father bear the X-linked recessive allele (*e.g.*, $X^AX^a \times X^aY$), whereas an affected male can result when only the mother carries the gene. Further, if the recessive X-linked gene is very rare, almost all observed cases will occur in males.

(*ii*) Usually none of the offspring of an affected male will be affected, but all his daughters will carry the gene in masked heterozygous condition, so one half of their sons (*i.e.*, grandsons of F_1 father) will be affected (Fig. 84.2).

(*iii*) None of the sons of an affected male will inherit the X-linked recessive gene, so not only will they be free of the defective phenotype, but they will not pass the gene along to their offspring.

4. Dominant X-linked genes can be detected in human pedigrees (also in *Drosophila*) through the following clues:

(*i*) It is more frequently found in the female than in the male of the species.

(*ii*) The affected males pass the condition on to all of their daughters but to none of their sons (Fig. 84.3).

(*iii*) Females usually pass the condition (defective phenotype) on to one-half of their sons and daughters (Fig. 84.4).

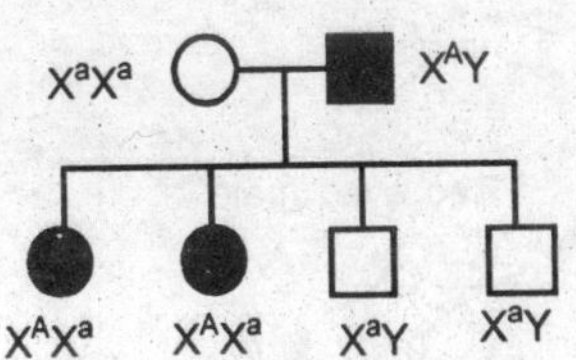

Fig. 84.3. Pedigree chart showing how X-linked dominants are expressed in all the daughters of affected male.

X^AX^a X^aY

X^aX^a X^AX^a X^aY X^AY

Fig. 84.4. Pedigree chart showing that females affected by a X-linked dominant condition usually are heterozygous and pass the condition to one-half of their progeny.

1. Inheritance of X-Linked Gene for Eye Colour in *Drosophila*

In *Drosophila*, the gene for white eye colour is X-linked and recessive to another X-linked, dominant gene for red-eye colour. It was discovered by **Morgan** in 1910. Following crosses between white eyed and red eyed *Drosophila* will make clear the characteristic criss-cross inheritance of gene for white eyed colour in it:

(*a*) Red eyed female × White eyed male. If a wild red eyed female *Drosophila* is crossed with a mutant white eyed male *Drosophila*, all the F_1 individuals irrespective of their sex have red eyes (Fig. 84.5). When the red eyed male and red eyed female individuals of F_1 are intercrossed, the F_2 progeny is found to include an exclusively red eyed female population and a male population with 50 per cent red eyed individuals and 50 per cent white eyed individuals. Thus, F_2 generation includes red eyed and white eyed individuals in the ratio of 3: 1.

(*b*) White eyed female × Red eyed male. When a white eyed female *Drosophila* is crossed with a red eyed male *Drosophila*, all the female individuals in the F_1 generation are red eyed (Fig. 84.6). When these red eyed female individuals and white eyed male individuals of F_1 are intercrossed, the female population of F_2 generation is found to include 50 per cent red eyed and 50 per cent white eyed flies. Similarly, the male population of F_2 includes 50 per cent red eyed and 50 per cent white eyed flies.

The inheritance of X-linked recessive gene for white eyes can be understood more properly by considering the behaviour of X and Y chromosome, such as follows:

In the cross between red-eyed female and white-eyed male *Drosophila*, the red-eyed female contains the gene ++ for red colour of eye. The white-eyed male contains a single gene *w* for white colour of eye which remains located in X chromosome. This one allelic condition of male is termed **hemizygous** condition in contrast to the homozygous or heterozygous possibilities in the female. The female being homogametic produces only one type of gametes or eggs each with the gene +

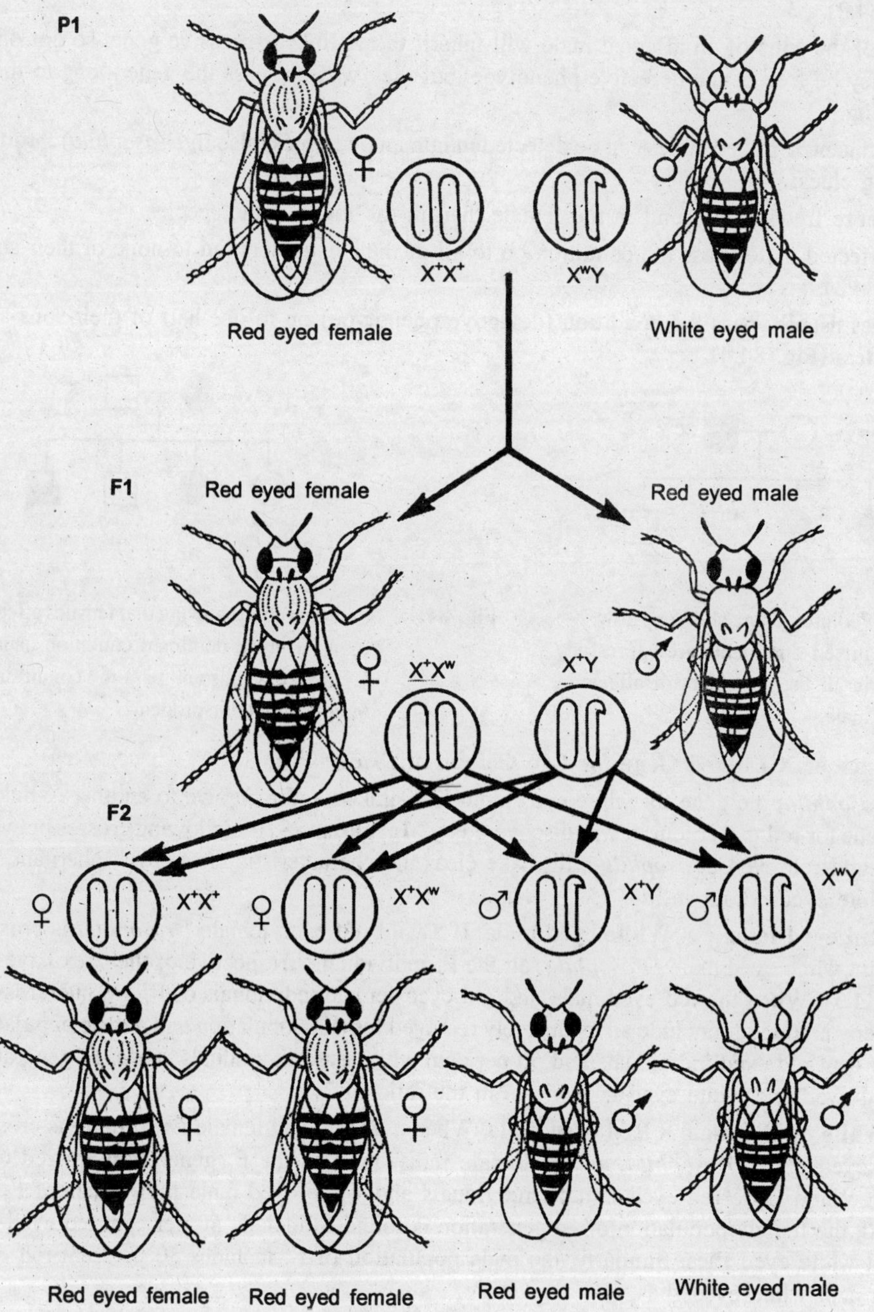

Fig. 84.5. A cross between red-eyed female and white-eyed male showing sex-linked inheritance in *Drosophila*.

for red coloured eyes. The male being heterogametic produces two types of gametes, 50 per cent sperms with X chromosome containing *w* gene, 50 per cent sperms with 'Y' chromosome without *w* or + gene. The gametes of both parents unite in fertilization to produce F_1 progeny. The F_1 hybrids which receive a X-chromosome with + gene from the female and a X-chromsome with *w* gene from the male becomes red-eyed female because gene '+' is dominant over gene '*w*'. Further the hybrids which happen to receive a gene '+' from mother and Y chromosome from father produce red eyeu males.

The F_1 red-eyed female with the gene '+ *w*' when crossed with F_1 red-eyed male having the gene '+' The female hybrids produce two types of eggs, 50 per cent eggs carry the gene '+' and the remaining 50 per cent carry the gene '*w*'. The males produce two types of sperms, half carry the '+' and half carry no such gene on Y chromosomes. The union of sperms and ova of F_1 offsprings may produce four possible types of F_2 individuals:

1. The eggs with '+' genes if fertilized by sperms with '+' genes produce homozygous red-eyed females.
2. The eggs with '+' gene if fertilized by the sperms with 'Y' chromosomes produce the hemizygous red-eyed males.
3. The eggs with the gene '*w*' when fertilized by the sperms having the gene '+' produce heterozygous red-eyed females.
4. The eggs with the gene '*w*' when fertilized by the sperms having the Y chromosome hemizygous white-eyed males are produced.

Likewise in other experiment in which a white-eyed female is crossed with red-eyed male, similar X-linked inheritance of recessive gene for white eye colour is revealed (Fig. 84.6). The white-eyed female contains the gene '*ww*' located on both X chromosomes. The red-eyed male contains the gene '+' located on single X chromosome. The female being homogametic produces single type of eggs with single '*w*' gene for whiteness, while the male being heterogametic produce two types of sperms, 50 per cent sperms carry the gene '+' and remaining 50 per cent sperms carry no such genes on the Y chromosomes. In the mating of both parents by the union of these eggs and sperms, two kinds of F_1 hybrids are produced. The eggs with '*w*' genes by union with the sperms having '+' genes produce heterogametic red-eyed female and the eggs with '*w*' gene by union with the sperms having Y chromosome produce white-eyed male. When the F_1 brothers and sisters with the genes '*w*' and '+*w*' breed together they produce four types of individuals in F_2 such as white-eyed female (ww), red-eyed female (+*w*), white-eyed male (*w*) and red-eyed male (+).

The results of these experiments, thus, are clearly indicating that the trait located on a sex chromosome alternates the sex from one generation to the next generation, *i.e.*, the trait of white eyes transfers from P_1 father to F_1 daughter and from F_1 daughter to F_2 son. Such X-linked genes are also called **digenic genes**; they exhibit a distinct criss-cross pattern of inheritance.

The trait of barred eye in *Drosophila* is another sex-linked trait which is dominant over the normal eye shape but display inheritance pattern like eye-colour.

Box 84.2

In human beings more than 150 confirmed or highly probable X-linked traits are known; most of these are recessives. Certain well known examples of X-linked recessive genes in humans are those for red-green colour blindness or **daltonism**, haemophilia and Duchenne's muscular dystrophy. Some other examples of X-linked recessive traits include (1) deficiency of enzyme glucose-6 phosphate dehydrogenase (G6PD deficiency) in erythrocytes causing haemolytic anaemia during allergy reaction of persons for the drugs such as sulphonamides or for the broad bean (*Vicia faba*), called **favism;** (2) two forms of diabetes insipidus; (3) one form of anhidrotic ectodermal dysplasia (absence of sweat glands and teeth); (4) absence of central incisors; (5) certain forms of deafness; (6) spastic parapelagia (*i.e.* tetanoid or partial paralysis of lower extremities with increased irritability and spasmodic contraction of the muscles); (7) uncontrollable rolling of the eye balls (nystagmus); (8) a form of cataract; (9) night blindness; (10) optic atrophy; (11) juvenile glaucoma (hardening of eye ball); (12) juvenile muscular dystrophy and (13) white frontal patch of hair.

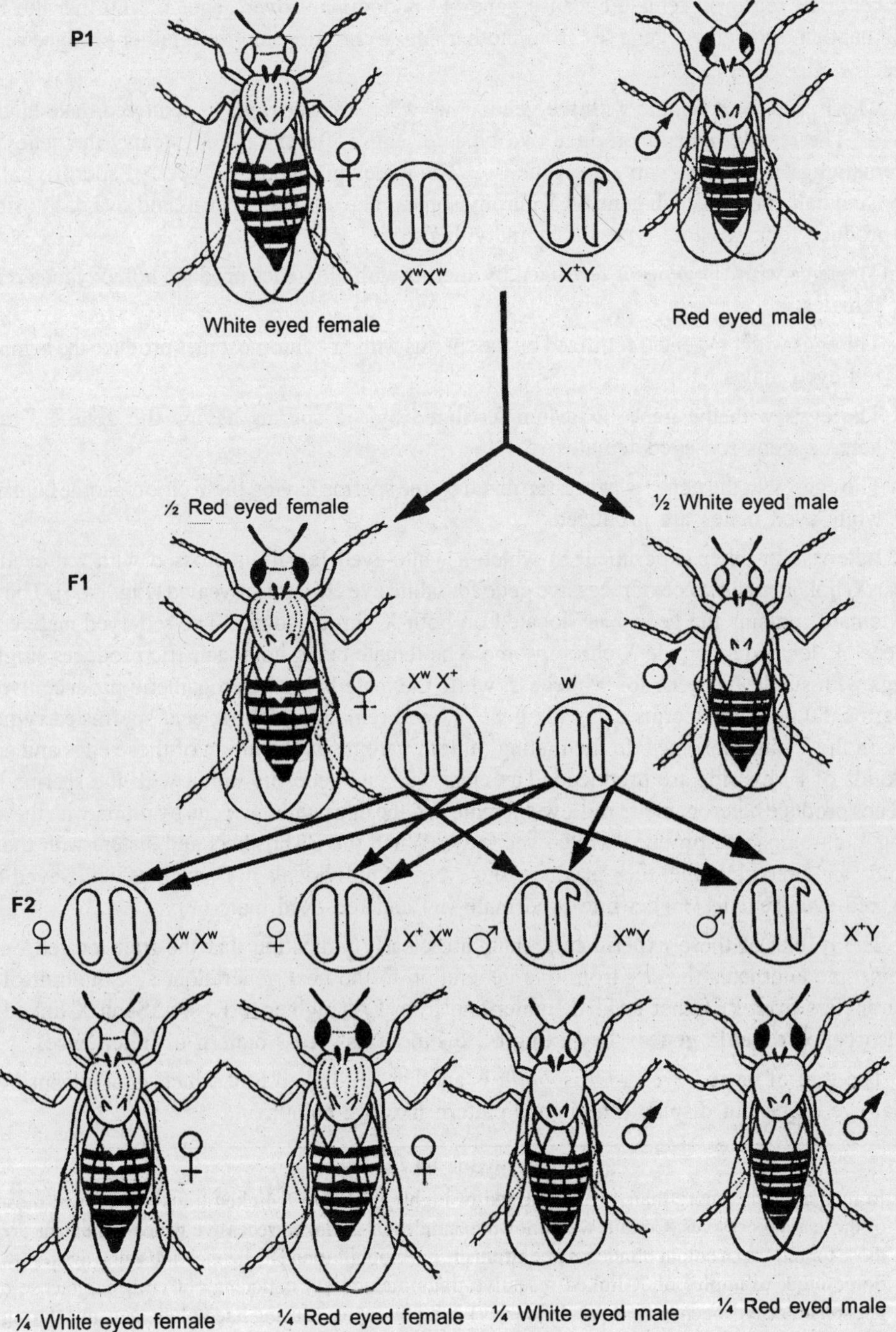

Fig. 84.6. A cross between white-eyed female and red-eyed male showing X-linked inheritance in *Drosophila.*

2. Inheritance of X-Linked Recessive Genes in Humans

1. Colour blindness. In human beings, a dominant X-linked gene is necessary for the formation of the colour sensitive cells, the **cones**, in the retina of eye. According to **trichromatic theory** of colour vision, there are three different types of cones, each with its characteristic **pigment** that react most **strongly** to red, green and violet light. The recessive form of this gene (*i.e.*, presence of recessive X-linked allele for colour blindness) is incapable of producing the colour sensitive cones and the homozygous recessive females ($X^C X^C$) and hemizygous recessive males (X^CY) are unable to distinguish between these two colours (**Wilson,** 1911). The frequency of colour blind women is much less than colour blind man (Box 84.2 and Box 84.3).

Box 84.3

Detailed studies indicate that there are two closely linked gene loci each with several multiple alleles controlling the colour blind trait. Lack of the **colorable pigment** in the retinal cones results in an inability to discriminate green colours. This defect is known as **deuteranopia** or **deuton colour blindness.** It affects about 8 per cent of human males but only about 0.7 per cent of females. The deuton or green colour blindness is first to be described in literature and is most commonly encountered sex-linked trait in human beings. Likewise, lack of **erythrolable pigment** which is necessary for discrimination in red end of the spectrum results in **protanopia** or **protan colour-blindness.** This abnormality is much less common than the deutan type, occurring in only about 2 per cent of males and in only 4 women out of 10,000. It is also caused by an X-linked recessive gene present quite close to the deuteranopia locus. Still other forms of colour blindness, some X-linked and some autosomal, are also known in humans.

The inheritance of colour blindness can be studied in the following two types of marriages:

***(i)* Marriage between colour-blind man and normal visioned woman.** When a colour-blind man marries with a normal visioned woman, then they will produce normal visioned male and female individuals in F_1. The marriage between a F_1 normal visioned woman and normal visioned male will produce in F_2 two normal visioned female, one normal visioned male and one colour-blind male (*i.e.,* 3 Normal visioned and 1 colour-blind progeny; Fig. 84.7).

Parents:	$X^+ X^+$ Normal female	×	$X^C Y$ Colour-blind male
Gametes:	(X^+) (X^+)		(X^C) (Y)
F_1:	½ $X^+ X^C$ Normal but carrier (heterozygous) female		½ $X^+ Y$ Normal but hemizygous male

(Marriage between a carrier female and a normal male produces the progeny as follows—)

F_2:	¼ $X^+ X^+$	= Normal female	3 Normal: 1 Colour-blind or 3 : 1
	¼ $X^+ X^C$	= Normal but carrier female	
	¼ $X^+ Y$	= Normal male	
	¼ $X^C Y$	= Colour-blind male	

Fig. 84.7. A marriage between a normal visioned female and a colour-blind man and result of the marriage of F_1 carrier female with a normal male.

***(ii)* Marriage between normal visioned male and colour-blind female.** If a woman (XX) is colour blind and she happens to marry a normal visioned male (XY), then all F_1 sons will be colour blind and daughters will be normal visioned. Because male receives one X-linked recessive gene for colour-blindness from colour-blind mother. The daughter receives one X-linked dominant gene for normal vision from father and one X-linked recessive gene for colour-blindness from the mother. The F_1 brother and sisters, if inbred or married they will produce in F_2 a colour-blind homozygous daughter, a normal visioned heterozygous daughter, a normal visioned hemizygous son and a hemizygous colour-blind son (*i.e.,* 1:1 ratio of normal and colour-blind; Fig. 84.8).

2. Haemophilia. Haemophilia is the most serious and notorious disease which is more common in men than women. This is also known as bleeder's disease. The person which contains the recessive

Parents:	$X^C X^C$ Colour-blind female	×	$X^+ Y$ Normal male
Gametes:	(X^C) (X^C)		(X^+) (Y)
F_1:	½ $X^+ X^C$ Normal but carrier (heterozygous) female		½ $X^C Y$ Colour-blind but hemizygous male

(Marriage of F_1 carrier female with a colour blind male produces the progeny as follows–)

F_2:	¼ $X^+ X^C$	= Normal but carrier female	2 Normal: 2 Colour-blind or 1 : 1
	¼ $X^C X^C$	= Colour-blind female	
	¼ $X^+ Y$	= Normal male	
	¼ $X^C Y$	= Colour-blind male	

Fig. 84.8. A marriage between a colour-blind female and a normal male and result of the marriage of F_1 carrier female with a colour-blind male.

gene for haemophilia lacks in normal clotting substance (**thromboplastin**) in blood so minor injuries cause continuous bleeding and ultimate death of the person due to haemorrhages. This hereditary disease was reported by **John Cotto** of Philadelphia in 1803 in man. Recently, two types of X-linked haemophilia have been recognized:

***(a)* Haemophilia A.** It is characterized by lack of antihaemophilic globulin (Factor VIII). About four fifths of the cases of haemophilic are of this type.

***(b)* Haemophilia B.** It is also called "**christmas disease**" after the family in which it was first described in detail. Haemophilia B results from a defect in plasma thromboplastic component (factor IX). This is milder form of haemophilia.

Woodliff and **Jackson** in 1966 have found the occurrence of both types of haemophilia in an unusual family, both segregating independently; therefore, they concluded that the two loci (*i.e.*, haemophilia A and haemophilia B) were far apart on the X chromosome (*i.e.*, having a distance of more than 40 map units between them).

Haemophilia is well known in the royal families of Europe, where it is traceable to Queen Victoria of England, who must have been heterozygous (carrier). No haemophiliac is known in her ancestry; hence, it is concluded that her haemophilia allele arose from a mutant gamete. Since it is caused by recessive X-linked gene, a lady may carry the disease (*i.e.*, she is carrier but nonsufferer and would transmit it to 50 per cent of her sons, even if the father is normal (Fig. 84.9).

Parents:	$X^+ X^h$ Normal (Carrier) mother	×	$X^+ Y$ Normal father
Gametes:	(X^+) (X^h)		(X^+) (Y)
Progeny:	¼ $X^+ X^+$	= Normal daughter	3 Normal: 1 Haemophilic or 3 : 1
	¼ $X^+ X^h$	= Normal but carrier daughter	
	¼ $X^+ Y$	= Normal son	
	¼ $X^h Y$	= Haemophilic son	

Fig. 84.9. Inheritance of haemophilia disease in human beings.

84.2 HOLANDRIC GENES

Genes in the non-homologous region of the Y chromosome pass directly from male to male (Fig. 84.10). In man, the Y-linked or holandric genes such as **ichthyosis hystrix gravis hypertrichosis** (excessive development of hairs on pinna of ear) are transmitted directly from father to son. Recently, certain other holandric genes have been reported in humans, *e.g.*, genes for H-Y antigen, histocom-patibility antigen, spermatogenesis, height (stature) and slower maturation of individual. The Y-linked gene of H-Y antigen is located on short arm of Y chromosome, while genes controlling spermatogenesis occur on the long arm of Y chromosome. Y chromosome of *Drosophila* too contain Y-linked gene for male fertility.

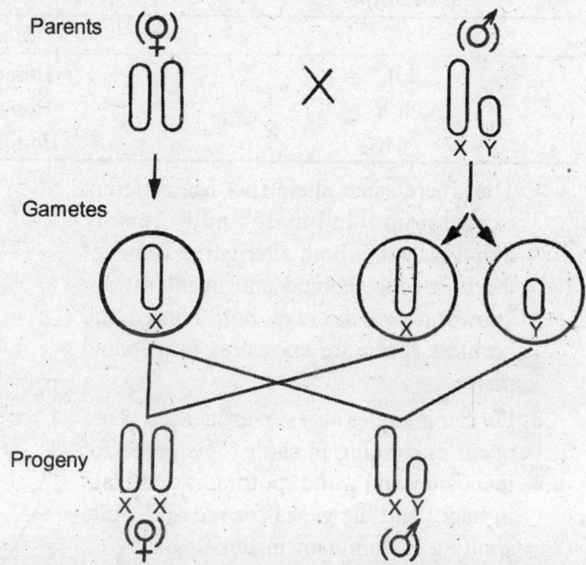

Fig. 84.10. Mode of inheritance of Y-linked genes from father to son.

84.3 SEX INFLUENCED GENES

Sex influenced genes are those whose dominance is influenced by the sex of the bearer. Thus, male and female individuals may be similar for a particular trait but give different phenotypic expressions of the same trait.

Table 84.1 In heterozygotic human females (Bb) dominant allele B for baldness fail to make its phenotypic expression.

Genotype	*Phenotypes*	
	Men	*Women*
BB	Bald	Bald
Bb	Bald	Non-bald
bb	Non-bald	Non-bald

Examples.

1. In man the baldness may occur due to disease, radiation or thyroid defects but in some families baldness is found to be inherited trait. In such inherited baldness the hairs gradually become thin on head top, leaving ultimately a fringe of hair low on the head and commonly known as **pattern baldness**. The gene B for baldness is found to be dominant in males and recessive in females (Table 84.1). In heterozygous condition it expresses itself only in the presence of male hormones (in male sex).
2. In sheep, the genes for the development of horns in dominant in males and recessive in female (Table 84.2).

Table 84.2 Phenotypic expression of dominant sex-influenced gene for horned trait occurs only in male sheep.

Genotype	*Phenotypes*	
	Males	*Females*
h^+h^+	Horned	Horned
h^+h	Horned	Hornless
h h	Hornless	Hornless

Thus, here same allele (h^+) has different expressions in male and female individuals, but both alternative forms of the trait (*e.g.*, homed and hornless) are known in each sex, *i.e.*, both horned and hornless sheep are known as female and male.

3. Further in cattles also sex-influenced genes occur as spotting in cattle. The genes for mahogany and white spotting is dominant in males and the genes for red and white spotting is dominant in females.
4. **Length of index finger in human.** A pair of sex-influenced genes control the length of index finger in humans. Index finger is mostly long in females and short in males in comparison to ring finger (Fig. 84.11).

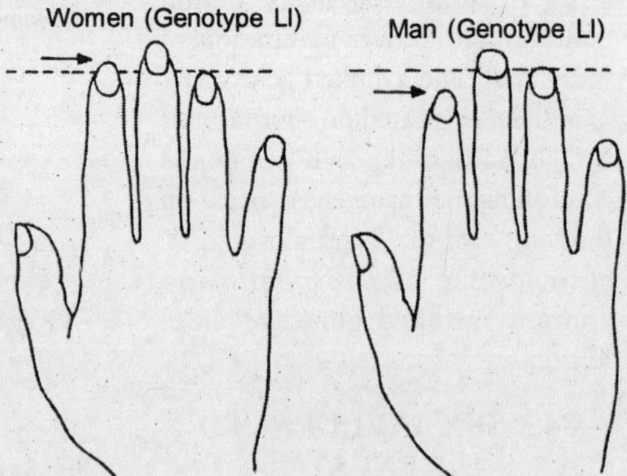

Fig. 84.11. Length of index finger in human female and human male.

84.4 SEX-LIMITED GENES

Sex-limited genes are autosomal genes whose phenotypic expression is determined by the presence or absence of one of the sex hormones. Their phenotypic effect is limited to one sex or other. In other words, the penetrance of a sex-limited gene in one sex remain zero. The sex-limited genes are mainly responsible for secondary sex characters in cattle, humans and fowl.

Examples

1. The bulls have genes for milk production which they transmit to their daughters, but they or their sons are unable to express this trait. The production of milk is, therefore, limited to variable expression only in the female sex.
2. Beard development in human beings is a sex limited trait as men normally have beards, whereas women normally do not. Likewise, the genes for male voice, body hair and physique are autosomal in human beings, but they are expressed only in the presence of androgens which are absent in females.
3. In chicken the recessive gene (*h*) for cock feathering is male sex-limited (*i.e.*, it is penetrant only in male environment).

Genotype	*Phenotypes*	
	Males	*Females*
HH	Hen-feathering	Hen-feathering
Hh	Hen-feathering	Hen-feathering
hh	Cock-feathering	Hen-feathering

84.5 NON-DISJUCTION

The failure of homologs (at meiosis) or sister chromatids (at mitosis) to separate properly to opposite poles is called **non-disjunction**. The sex chromosomes pass through the phenomenon of non-disjunction and present various interesting situations. The non-disjunction may be of following types:

1. Primary Non-disjunction.

The phenomenon of primary non-disjunction was discovered by **C.B. Bridges** (1916) during his classical matings of *Drosophila melanogaster*. Let us consider one of his classical crosses as follows:

The gene for wild type red eyes (+) is carried by the X chromosome; a recessive allele (*v*) produces vermilion eyes in homozygous females (X^VX^V) and in all males (X^VY). Normally vermilion-eyed females mated to red-eyed males produce only red-eyed daughters and vermilion eyed sons (Fig. 84.12).

Parents:		$X^V\ X^V$ Vermilion female (or mother)	×	$X^+\ Y$ Red male (or father)
Gametes:	**Eggs:**	(X^V)	(X^V)	
	Sperms:	(X^+)	(Y)	
F_1:	½ $X^+\ X^V$ Red female (or daughter)	:	½ $X^V\ Y$ Vermilion male (or son)	

Fig. 84.12. Results of a normal cross between vermilion-eyed female and red-eyed male *Drosophila*.

Crosses of this type, however, in rare cases produce unexpected vermilion-eyed daughters and red-eyed sons with a frequency of one per 2,000 to 3,000 offspring. **Bridges** speculated that these unusual progeny are due to a failure of the X chromosomes in an XX female to disjoin during meiosis I of oogenesis. This phenomenon was termed **primary disjunction** by Bridges and this event tends to produce three types of eggs: the majority of eggs with normal single X chromosome and a small number of eggs with their two X chromosomes or no X at all. These eggs on normal fertilization produce the following results:

Parents:		A $AX^V\ X^V$ Vermilion female	×	$AX^+\ Y$ Red male
Gametes:	**Eggs:**	(A X^V) (numerous)	+ ($AX^V\ X^V$) + (rare due to non-disjunction)	(AO) (rare called **nullo**)
	Sperms:	(A X^+)	+	(AY)

F_1:

1. AA X^+X^v = Red female (numerous; normal)
2. AA $X^+X^VX^V$ = Red "metafemales" (rare; die)
3. AA X^+O = Red male (rare; sterile)
4. AA X^VY = Vermilion male (numerous; normal)
5. AA X^VX^VY = Vermilion female (rare: fertile)
6. AAOY = Die (rare)

Fig. 84.13. Inheritance of eye colour in *Drosophila* due to primary non-disjuction of sex chromosomes.

The **metafemales** (AA XXX) are weak and seldom live beyond the pupal stage; the AAOY individuals die in the egg stage. Note that the AA XXY female indicate that the presence of a Y chromosome does not determine maleness itself, though males without it (AAXO) are sterile. From these results three facts became established: 1. In *Drosophila* sex determination occurs according to genetic balance theory of Bridges. 2. In *Drosophila*, Y chromosome contains genes for the spermatogenesis. 3. It provides strong support to chromosome theory of inheritance (*i.e.*, Mendelian factors or genes are carried on the chromosomes).

2. Secondary Non-disjunction. In next cross, **Bridges** mated the exceptional vermilion-eyed females (AAX^VX^VY) that arose as a result of primary non-disjunction to normal red-eyed males (AAX^+Y). The progeny of this cross included 96 per cent female with red eyes and only 4 per cent females had vermilion eyes indicating that there was 4 per cent secondary non-disjunction (Fig. 84.14).

Thus, primary non-disjunction may occur in either XX females or XY males. In the former it leads to the production of XX and O eggs. Occurrence in the first meiotic division of male produces XY and O sperms results. Secondary non-disjunction, on the other hand, occurs in XXY females, where it gives rise to XX, XY, X and Y eggs (Fig. 84.15). As the term non-disjunction implies, these aberrant gametes are produced only as a result of failure of the sex chromosomes to disjoin after synapsis; they are not physically attached.

Similar results were obtained by **Bridges** (1916) from the cross of red eyed male and white eyed female *Drosophila*.

Parents: $AAX^+X^VX^VY$ Vermillion females × $AA\,X^+Y$ Red males

Gametes:

Egg: $(AX^V) + (AX^VY) + (AX^VX^V)$ [Due to normal disjunction] $+ (AY) + (AX^V) + (AX^VY)$ [Due to secondary non-disjunction]

Sperms: $(AX^+) + (AY)$

F_2:

			Sperms		
			AX^+(50%)	AY (50%)	
		AXVXV (4%)	$AAX^+X^VX^V$ Metafemale (die) (2%)	$AAX^v\,X^vY$ Vermilion female (2%)	4% secondary exceptional progeny phenotype
	X-Y	AY (4%)	AAX^+Y Red male (2%)	AAYY (die) (2%)	
	pairing (16%)	AX^VY (4%)	AAX^+X^VY Red female (2%)	$AA\,X^VY\,Y$ Vermilion male (2%)	95% regular (i.e., expected) progeny phenotypes
Eggs		AX^V (4%)	AAX^+X^VY Red female (2%)	AAX^VY 96% Vermilion male (2%)	
		AX^V (42%)	AAX^+X^V Red female (21%)	AAX^VY Vermilion male (21%)	
	X-X pairing (84%)	AX^VY (42%)	AAX^+X^VY Red female (21%)	AAX^VYY Vermilion male (21%)	

Fig. 84.14. Result obtained from secondary non-disjuction.

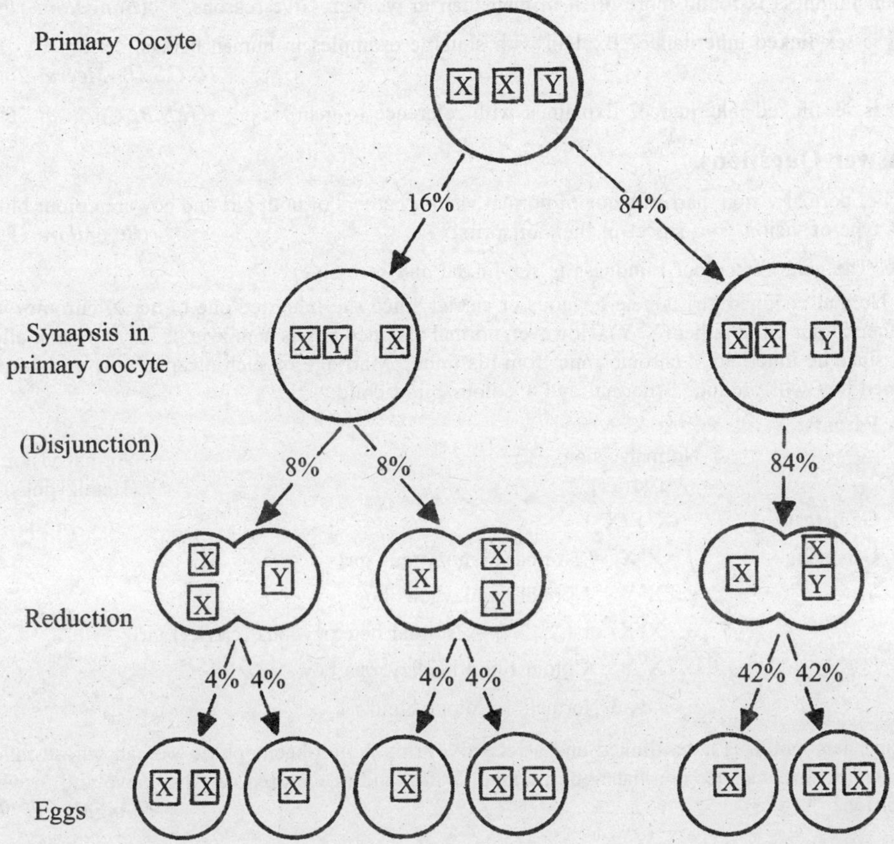

Fig. 84.15. Secondary non-disjunction in XXY female *Drosophila*, resulting in 46 per cent X, 46 per cent XY, 4 per cent XX, and 4 per cent Y eggs.

QUESTIONS

Long Answer Questions

1. Write an essay on sex-linked inheritance. (*Agra 1989; CCS 1989; Gorakhpur 1989; Sagar 1992; Kanpur 1994; Lucknow 1997*)

2. What is sex-linked inheritance? Illustrate sex-linked inheritance with suitable examples. (*Guwahati (G) 1995*)

3. What is sex-linked inheritance? Explain this phenomenon with reference to man or *Drosophila* giving suitable examples. (*Bundelkhand 1989; CCS 1990; Kanpur 1990*)

4. Describe sex-linked inheritance in the fruitfly, *Drosophila.* (*Allahabad 1996*)

5. Give an account of the chromosomal basis of sex-linked inheritance in *Drosophila* and man. (*CCS 1996*)

6. What is sex-linked inheritance? Explain the inheritance pattern of haemophilia/red green blindness in man. (*CCS 1992, 1994; Kanpur 1993, 1995, 1997*)

7. What is criss-cross inheritance? Describe mechanism of inheritance of haemophilia in man. (*Bundelkhand 1988*)

8. What do you mean by sex-linked inheritance? Describe the role of genes with reference of haemophilia and colour blindness. (*Lucknow 1993*)

9. Colour blindness is found more often in man than in women. Give reasons. *(Bangalore 1993)*

10. What is sex linked inheritance? Explain with suitable examples in human beings. *(C.C.S.U., Meerut 2008)*

11. What is sex linked inheritance? Explain it with reference to man. *(H.N.B. Garhwal 2009)*

Short Answer Questions

1. A girl of normal vision marries a boy of normal vision. Fathers of both girl and boy are colour blind. What type of vision you expect in their offspring? *(Bangalore 1994)*

(**Note.** The gene for colour blindness is sex-linked and recessive).

Ans. Normal visioned girl is heterozygous or carrier since she inherited one of her X chromosome from her colour blind father (X^CY). However, normal visioned boy is hemizygous for dominant allele (X^+), since he inherited Y chromosome from his father. Marriage of such a carrier girl and normal visioned boy will produce 3 normal and 1 colour blind child:

Parents:	♀ $X^+ X^C$ Normal vision (Carrier)	×	♂ X^+Y Normal vision (Hemizygous)
Gametes:	(X^+) (X^C)		(X^+) (Y)

Offspring
1. X^+X^+ = Normal homozygous girl
2. X^+Y = Normal hemizygous boy
3. X^CX^+ or (X^+ X^C) = Normal heterozygous (carrier) girl
4. X^CY = Colour blind hemizygous boy

i.e., 3 Normal: 1 Colour blind.

2. In man haemophilia is a sex linked and a recessive trait. A non-haemophilic woman whose father was haemophilic, marries non-haemophilic man. What kind of children can they have and in what proportion? *(Bangalore 1996)*

Ans.	♀ $X^+ X^h$ Non-haemophilic (Carrier)	×	♂ X^+Y Non-haemophilic (Hemizygous)
Gametes:	(X^+) (X^h)	×	(X^+) (Y)

Offspring:
1. X^+X^+ = Normal or non-haemophilic homozygous girl
2. X^+Y = Normal or non-haemophilic hemizygous boy
3. X^+X^h = Normal or non-haemophilic carrier (heterozygous) girl
4. X^hY = Haemophilic hemizygous boy

or 3 Non-haemophilic : 1 Haemophilic.

3. Can a normal son have a colour-blind mother? Explain in only five sentence. *(NEHU 1996)*

Ans. A normal son (X^+Y) gets his X-linked dominant allele for normal vision from his mother. A colour blind mother would have two homozygous recessive alleles for the colour blindness (*i.e.*, X^CX^C), Therefore mother of a normal visioned son is always normal visioned (*i.e.*, X^+X^+ or X^+X^C) never colourblind (X^CX^C).

4. Describe basic characteristics of inheritance in sex-linked characters.

5. Discuss the inheritance of Y-linked genes.

6. What are holandric genes? How does their inheritance differ from the inheritance of digenic genes? Describe with suitable examples, the inheritance of digenic genes in man. *(Delhi 1988)*

7. Write a short note on non-disjunction. *(Bangalore 1993)*

8. What is non-disjunction? Explain this phenomenon with suitable examples. *(B.U. Coimbatore (Main) 1997)*

9. Give an account of sex-influenced characters. (*Bangalore 1992*)

10. Give a brief account of sex-limited genes.

11. How colour blindness is inherited in man? (*M.S.U. Tamil Nadu 2006*)

12. Describe the sex linked inheritance with suitable examples. (*C.C.S.U., Meerut 2009*)

13. Describe the sex-linked inheritance in *Drosophila.* (*C.C.S.U., Meerut, 2007*)

14. Describe the sex-linked inheritance and genetic disorder in human. (*H.N.B., Garwal, 2008*)

15. Write short notes on the following:

(*i*) Colour blindness (*Jiwaji 1992; CCS 1995*)

(*ii*) Haemophilia (*Bangalore 1993, 1994; Lucknow 1996; H.N.B. Garwal, 2008*)

(*iii*) Sex-linked genes (*Bangalore 1993*)

or

Sex-linked inheritance (*Allahabad 1993, 1994; Guwahati (G) 1993*)

or

Sex-linked characters (*Lucknow 1991*)

(*iv*) Carrier

(*v*) Sex-linked inheritance in man (*Bundelkhand 1989; Kanpur 1990*)

(*vi*) Sex-linked and sex-limited genes (*Delhi 1992*)

16. Differentiate between X-linked and sex-limited characters. (*Lucknow 1995*)

17. Show diagrammatically the inheritance of eye colour in *Drosophila* (*CCS 1999*)

Very Short Questions

1. What do you mean by sex-chromosome?

Ans. Chromosomes which differ in two sexes (male and female sexes).

2. What do you mean by sex-linked genes?

Ans. Genes located on sex-chromosomes, X and Y are called sex-linked genes. These genes are located on non homologous parts of two chromosomes.

3. Define holandric inheritance.

Ans. Characters passes from father to son directly (*e.g.*, Y-linked genes in human male).

4. Define hologenic inheritance.

Ans. Character goes directly from mother to daughter (female to female, *e.g.*, W-linked genes in birds).

5. What is digenic sex linked character?

Ans. When a character passes from male to male through female (*e.g.*, X-linked recessive gene such as allele for haemophilia in humans).

6. What do you mean by diandric sex-linked inheritance?

Ans. When a character passes from female to female through male (*e.g.*, Barred plumage in poultry).

7. A son cannot receive X-linked gene directly from his father, why?

Ans. Son inherits only Y-chromosome from his father.

8. What is sex-limited trait?

Ans. In case of sex-limited trait expression of a trait is only in one sex. Examples–milk production in mammals; horns in Rambouiltet sheep, egg production in chicken. Sex-limited traits may be controlled by sex-linked or autosomal loci.

9. Define sex-influenced traits.

Ans. Sex-limited traits are also called **sex-conditioned traits**. These are traits that appear more often in one sex than in another.

Yes or No Question

(*i*) Muller propounded the chromosome theory of heredity. (*Guwahati (G) 1996*)

Ans. No.

Fill in the Blanks

(*i*) Haemophilia is an example of a human trait with .... pattern of inheritance.

Ans. Criss-cross.

Multiple Choice Questions

1. A gene present only in one dose on the X-chromosome in heterogametic males is called (*NEHU 1996*)

(*a*) heterozygous (*b*) hemizygous
(*c*) homozygous (*d*) none of the above

2. The number of Barr body in XXY men is (*M.S.U., Tamil Nadu 2006*)

(*a*) one (*b*) two
(*c*) none (*d*) not known

3. Which of the following is sex linked?

(*a*) sickle-cell anaemia (*b*) albinism
(*c*) haemophilia (*d*) phenylketonuria

4. In protanopia, a person cannot distinguish

(*a*) green colour (*b*) red colour
(*c*) blue colour (*d*) blue and green colour

5. According to **Gardner** *et al.,* (1981), in humans, only two Y-linked genes have been discovered till date. These are

(*a*) *H-Y* and *SRY* (*b*) *H-Y* and hairy ear pinna
(*c*) *SRY* and hairy ear pinna (*d*) *H-Y, SRY* and hairy ear pinna

6. Sex linked characters are always

(*a*) lethal (*b*) recessive
(*c*) dominant (*d*) pleiotropic

7. Colour blindness, in which all colours are perceived as gray, is termed as

(*a*) chromasia (*b*) dichromasia
(*c*) monochromasia (*d*) all of these

8. Moustaches and beared in human males are example of

(*a*) sex-linked traits (*b*) sex-limited traits
(*c*) sex-differentiating traits (*d*) sex-determining traits

9. Deuteronopia is a disease when there is no perception of ________ colour

(*a*) red (*b*) blue
(*c*) green (*d*) white

10. One of the following is not true of haemophilia

(*a*) royal disease (*b*) bleeder's disease
(*c*) X-linked disease (*d*) more common in males than in females
(*e*) Y-linked disease

ANSWERS

Multiple Choice Questions

1. (*b*) **2.** (*a*) **3.** (*c*) **4.** (*b*)
5. (*a*) **6.** (*b*) **7.** (*c*) **8.** (*b*)
9. (*c*) **10.** (*e*).

85

Non-Chromosomal Inheritance (Cytoplasmic Inheritance)

In preceding and forthcoming chapters, we have discussed different roles of genes of the nuclear chromosomes in inheritance, cellular metabolism, development and mutation of the organisms in which they occur. Though, the genes of nuclear chromosomes have a significant and key role in the inheritance of almost all traits from generations to generations, but they altogether cannot be considered as the sole vehicles of inheritance, because certain experimental evidences suggest the occurrence of certain extranuclear genes or DNA molecules in the cytoplasm of many prokaryotic and eukaryotic cells. For example, bacterial cells such as *E. coli* possesses a single main chromosome in the nucloid and often extra DNA elements called **plasmids** in the cytoplasm; the eukaryotic cells possess a main complement of chromosomes in the nucleus and extra-DNA molecules or chromosomes in their mitochondria and chloroplasts (plastids). Qualifying also as extra hereditary elements are certain viruses, bacteria, and algae (which live as endosymbionts inside the eukaryotic cells). These cytoplasmic extra-nuclear genes or DNA molecules of plasmids, mitochondria, chloroplasts, endosymbionts and cellular surfaces have a characteristic pattern of inheritance which does not resemble with that of genes of nuclear chromosomes and is known by different terms such as **non-Mendelian, non-chromosomal, uniparental, maternal, extra-chromosomal, cytoplasmic** and **extra-nuclear inheritance.**

85.1 EVIDENCES FOR CYTOPLASMIC FACTORS (GENES)

Traits with extranuclear basis are identified by the accumulated evidence from a number of diagnostic criteria such as follows:

1. Because the female gamete contributes almost all of the cytoplasm to the zygote and male gamete (sperm or pollen) contributes only a nucleus, an inheritance pattern that differs between reciprocal crosses suggests a cytoplasmic involvement. This is clearly the basis for **uniparental** or **maternal inheritance** where the progeny always resemble one parent, most commonly the female parent (*e.g.*, shell coiling in *Limnaea peregra*).
2. Differences in reciprocal crosses which cannot be attributed to sex-linkage or some other chromosomal basis tend to associate extranuclear factors (*e.g.*, chloroplast inheritance in *Mirabilis jalapa*).
3. The uniparental inheritance of a trait which cannot be attributed to unequal cytoplasmic contributions from parental gametes may, however, involve cytoplasmic factors (*e.g.*, streptomycin resistance in *Chlamydomonas*).

4. Whenever trait fails to demonstrate the classical segregation patterns and deviates from standard ratios, the conclusion is again a cytoplasm based type of inheritance (*e.g.*, mitochondrial inheritance in yeast).
5. When the trait fails to show linkage to any known nuclear linkage groups and assort independently from nuclear genes, a cytoplasmic mode of inheritance is suggested.
6. Many types of mutants that fit the above criteria will show segregation during mitotic division. This is very common in variegated plants that carry more than one type of plastid (chloroplast) per cell.

The cytoplasmic inheritance, therefore, will be understood to be based on cytoplasmically located, independent, self-replicating nucleic acids, which differ from chromosomal genes by their location within the cell, and have their own nucleotide sequences. The smallest heritable extra chromosomal unit is called a **plasmagene** or **cytoplasmic gene**. All of the plasmagenes of a cell constitute the **plasmon**.

85.2 NON-CHROMOSOMAL INHERITANCE IN EUKARYOTES

Many geneticists have studied various cases of extra-nuclear inheritance in different eukaryotes. Certain most important examples of extra-nuclear inheritance in eukaryotes are the following:

1. Maternal Inheritance

In certain cases, it has been observed that certain characteristic phenotypic traits of F_1, F_2 or F_3 progeny are not the expression of their own genes, but rather those of the maternal parents. Such phenotypic expressions of maternal genes (genotype) may be short-lived or may persist throughout the life-span of the individual. The substances which produce the maternal effects in the progeny are found to be transcriptional products (*i.e.*, mRNA, rRNA and tRNA) of maternal genes which have been manufactured during oogenesis and which exist in the ooplasm of unfertilized eggs in the form of inactive protein coated and late translating mRNA molecules (**informosomes; masked messenger** hypothesis of **Spirin**) or inactivated rRNA and tRNA. These transcriptional products of maternal genes produce their phenotypic effects during early cleavage and blastulation when there occur little or no transcription since, maternal and paternal genes of zygote remain engaged in mitotic replication or duplication of DNA. There may be other reasons of maternal effect which are still little understood.

The maternal inheritance has been studied in some of the following cases:

(*a*) Shell coiling in *Limnaea*. In the snails (gastropods), the shell is spirally coiled. In most cases the direction of coiling of the shell is clockwise, if viewed from apex of the shell. This type of coiling is called **dextral**. However, in some snails the coiling of shell may be counter clockwise or **sinistral**. Both types of coilings are produced by two different types of genetically controlled cleavages, one being **dextral cleavage**, another being **sinistral cleavage** (Fig. 85.1).

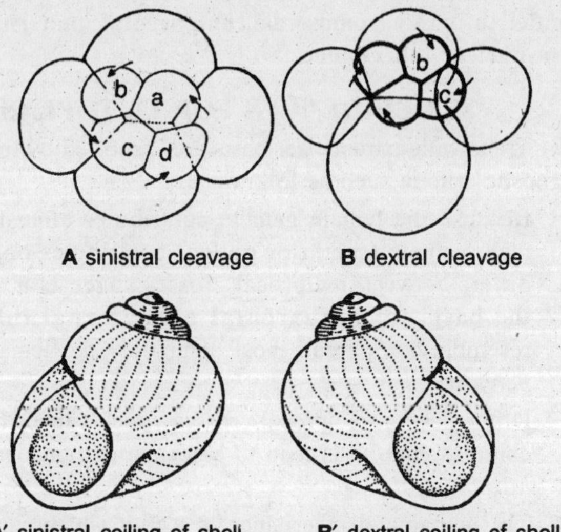

Fig. 85.1. Correlation of sinistral (A) and dextral (B) cleavage with sinistral (A′) and dextral (B′) coiling of the shell in gastropods.

There are some species of gastropods in which all the individuals are sinistral but the main interest attaches to a species in which sinistral individuals occur as a mutation among a population of normal dextral animals. Such a mutant was discovered in the freshwater snail *Limnaea peregra* (**A. Sturtevant,** 1923).

Breeding and cross breeding of dextral and sinistral snails showed that the difference between the two forms is dependent on a pair of allelomorphic genes, the gene for sinistrality being recessive (S), and the gene for the normal dextral coiling being dominant (S^+). The two genes are inherited according to Mendelian laws, but the action of any genic combination is visible only in the next generation after the one in which a given genotype is found.

The eggs of a homozygous sinistral individual (SS) are fertilized by the sperm of a dextral individual (S^+S^+), the eggs cleave sinistrally and all the snails of this F_1 generation show a sinistral coiling of the shell. Thus, the gene of sperm (S^+) do not manifest themselve, although the genotype of the F_1 generation is S^+S. If a second generation (F_2) is bred from such F_1 sinistral individuals, it is all dextral, instead of showing segregation as would be expected in normal Mendelian inheritance. In fact, segregation does take place in the F_2 generation so far as the genes are concerned, but the new genic combinations fails to manifest themselves, since the coiling is determined by the genotype of the mother. The genotype of F_1 mother being S^+S, the gene for dextrality dominates and is responsible for the exclusively dextral coiling of the second generation. Only in the F_3 generation does segregation in the ratio of 3 : 1 becomes apparent, since the individuals of the F_2 generation had the genotypes —1 S^+S^+; 2 S^+S, 1SS, 1/4 of them, on the average, produce eggs developing into sinistral individuals (Fig. 85.2).

It is easy to understand that the results of a reciprocal cross that is, of the fertilization of the eggs of a homozygous dextral individual (S^+S^+) by the sperm of a sinistral individual (SS)— will lead to a somewhat different type of pedigree: the F_1 generation will be dextral (with genotype S^+S) and the F_2 generation again all dextral (with genotypic ratio of $1S^+S^+ : 2S^+S : 1SS$). The F_3 generation will show segregation among broods, just as observed in the first cross.

The whole case becomes clear if it is realized that the type of cleavage (sinistral or dextral) depends on the organization of the egg which is established before the maturation division of the oocyte nucleus. The type of cleavage is, therefore, under the influence of the genotype of the maternal parent. The sperm enters the egg after this organization is already established. Lastly, the direction of coiling of shell depends upon the orientation of the mitotic spindle of first cleavage of the zygote. If the spindle is tipped toward the left of the median line of the egg cell, the sinistral pattern will develop; conversely if the mitotic spindle is tipped toward the right of the median line of the cell, the dextral pattern will develop. The spindle orientation is, thus, controlled by the organization of ooplasm which becomes established during oogenesis and before fertilization.

2. Non-Chromosomal Inheritance by Cellular Organelles

Chloroplasts and mitochondria are organelles that contain their own DNA and protein-synthesizing apparatus. A widely held theory concerning their origin proposes that they were once infectious endosymbiotic prokaryotes that evolved such a dependence on the gene products of the host that they are no longer able to function autonomously.

This theory has been supported by the fact that the genetic components of these organelles are often similar to those found in prokaryotes. For example, the chloroplasts of certain algae and *Euglena* contain 70S type small ribosomes and "naked" chromosomes or DNA which is circular. Their protein synthesis begins with the amino acid N-formylmethionine, as does prokaryotic protein synthesis, and their DNA-dependent RNA polymerase is sensitive to the inhibitor **rifampicin**. The genetic materials of chloroplasts and mitochondria will be transmitted to offspring almost exclusively via the egg. Maternal inheritance due to chloroplast and mitochondria is well illustrated by the following examples:

(*a*) Chloroplast inheritance in variegated four-o' clock plant. The cytoplasmic or extranuclear inheritance of colour in plant by plastids was first of all described by **C. Correns** in 1908 in the

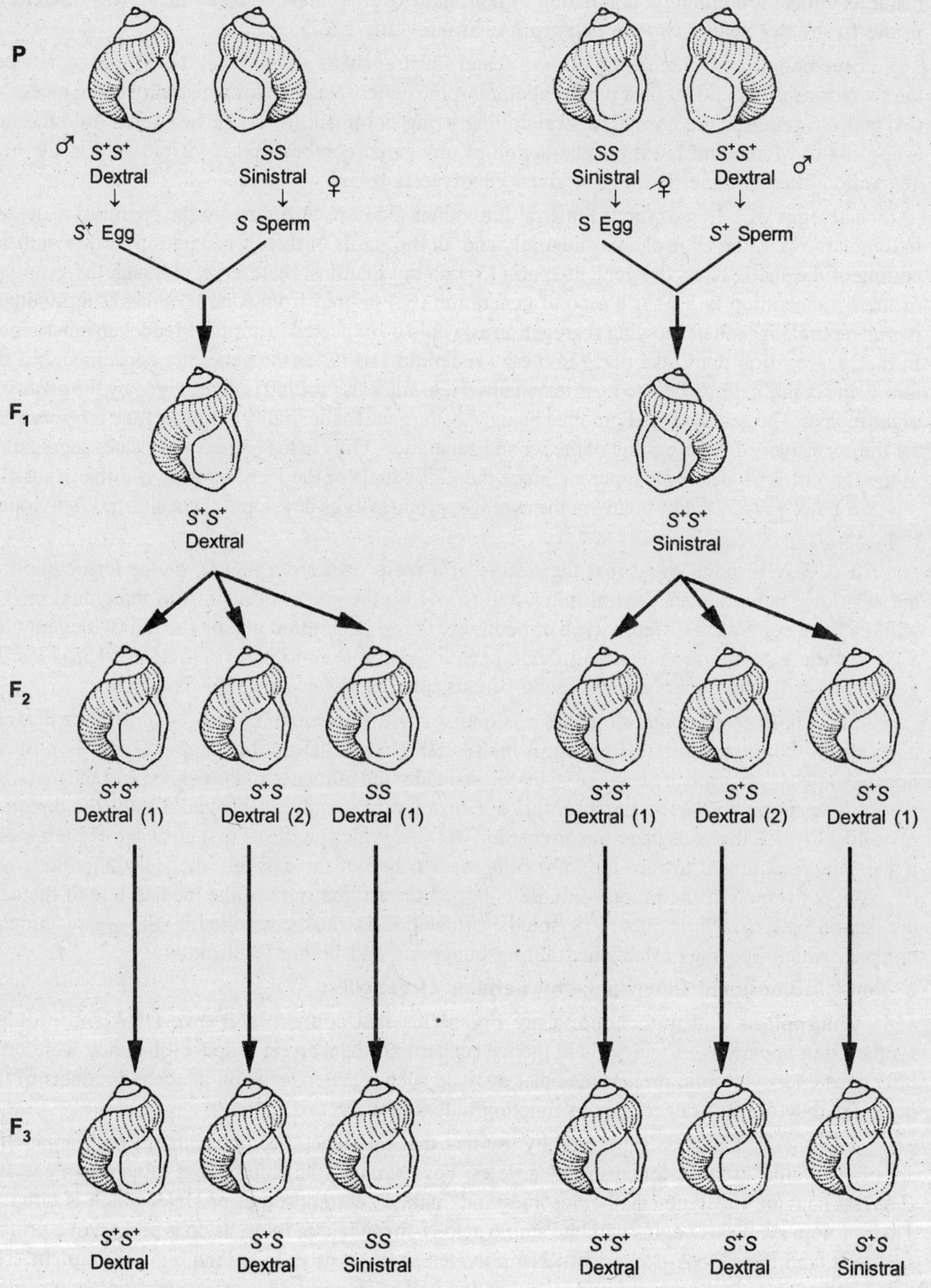

Fig. 85.2. Maternal effect in direction of coiling of the shell in *Limnaea.*

four-o' clock plant, *Mirabilis jalapa*. In contrast to other higher plants, *Mirabilis* contains three types of leaves and parts: (1) Full green leaves or branches having chloroplast, (2) White (pale) leaves and branches having no chloroplast, (3) Variegated branches having leucoplast in white (pale) areas and chloroplast in green patches (Fig. 85.3). Because, the chlorophyll pigment of chloroplast is related with photosynthesis of food and leucoplasts are incapable to perform photosynthesis, so the white or pale parts of plant survive by receiving nourishment from green parts. **Correns** reported that flowers on green branches produced only green offsprings, regardless of the genotype and phenotype of pollen parent and likewise, flowers from the white or pale branches produced only white or pale seedings regardless of genotype and phenotype of pollen parent. The plants developing from the white or pale seedings die because they lack chlorophyll and cannot carry on photosynthesis. **Correns** further reported that flowers from the variegated branches yielded mixed progeny of green, white (pale) and variegated plants in widely varying ratios (Fig. 85.4). These results are summarized in Table 85.1.

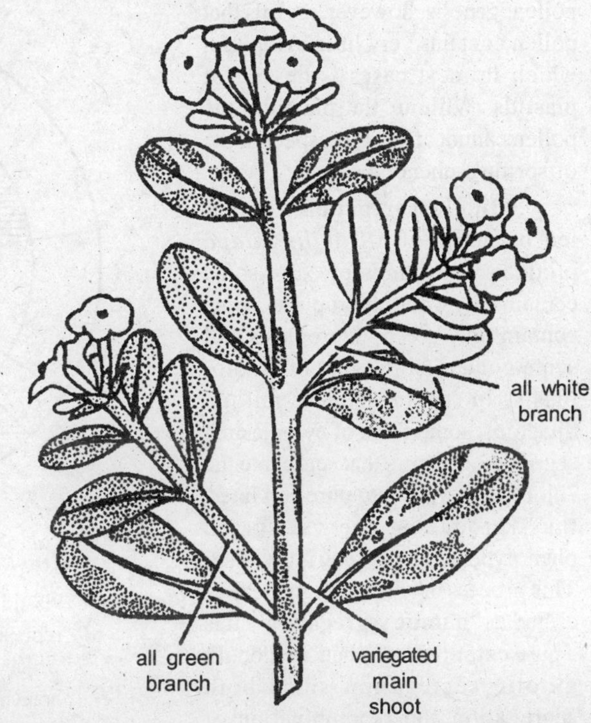

Fig. 85.3. Leaf variegation in *Mirabilis jalapa*, the four-o' clock plant. Flowers may form on any branch (variegated, green, or white), and these flowers may be used in crosses.

Table 85.1 Chloroplast inheritance in variegated four-o'clock plants.

Branch of origin of the male parent	*Branch of origin of the female parent*	*Progeny*
Green	Green	Green
	Pale or white	Pale or white
	Variegated	Green, pale or white, variegated
Pale or white	Green	Green
	Pale or white	Pale or white
	Variegated	Green, pale or white, variegated
Variegated	Green	Green
	Pale or white	Green, pale or white,
	Variegated	Variegated

The irregularity of transmission from variegated branches could be understood by considering cytoplasmic genes (plasmagenes) of plastids. A study of the egg during oogenesis in *Mirabilis* reveals that the ooplasm contains plastids like cytoplasm of other plant cells. If the egg cell is derived from green plant tissues, its ooplasm will contain coloured plastids; if derived from white plant tissues, its ooplasm will contain white plastids; if derived from variegated tissues, its cytoplasm may contain coloured plastids only, white plastids only or a mixture of coloured and white plastids. A study of the pollenogenesis, however, reveals that pollen contains very little cytoplasm which in most cases is devoid of plastids. Without the plastids, the pollen cannot affect this aspect of the offspring's phenotype.

Mitotic segregation. Variegated branches of *Mirabilis jalapa* produce three kinds of eggs: some contain only white chloroplasts, some contain only green chloroplasts and some contain both types of chloroplasts. In the subsequent mitotic divisions, some form of cytoplasmic segregation occurs that segregate the chloroplast types into pure cell lines, thus, producing the variegated phenotype in the progeny individual. This process of sorting might be described as "**mitotic segregation**". It is a pure extra-nuclear phenomenon. In mitotic segregation since both segregation and recombination of organelle genotype takes place, so it is called **cytoplasmic segregation and recombination** (its acronym is **CSAR**).

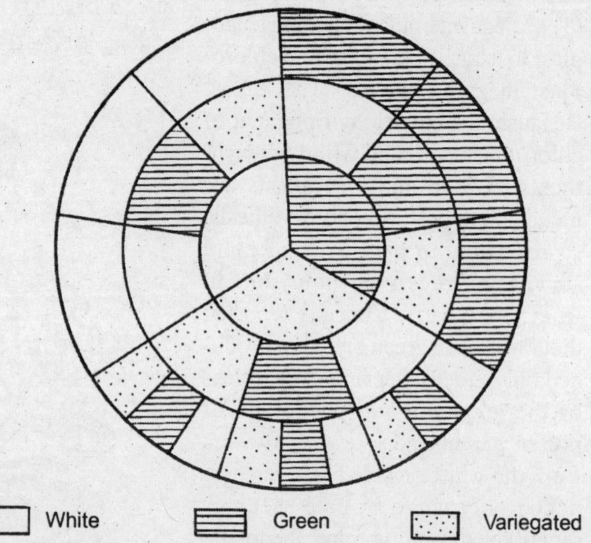

Fig. 85.4. Plastid inheritance of *Mirabilis jalapa*. the central circle represents the type of branch that produces flowers which are pollinated. Intermediate circle represents branch from which pollen is used and outer circle shows the progeny.

(*b*) Non-chromosomal inheritance by mitochondria. The most important work on the genetics of mitochondria done in yeast which was initiated by the discovery of petite mutants by **B. Ephrussi** (1953). Subsequently mtDNA was studied in several organisms including plants and animals.

1. Poky strain of *Neurospora*. In fungi, *Neurospora crassa* a number of mutations of mitochondria are inherited via the female parent. The best studied of these is the **poky strain** of *N. crassa*, first isolated by **Mitchell** and **Mitchell** (1952). A poky mutant differs from wild type strain of *Neurospora* in the following aspects: (1) it is slow growing; (2) it shows maternal inheritance, and (3) it has abnormal cytochromes. Of the three cytochromes—cyt a, b and c found in wild type, cyt a and cyt b are absent, and cyt c is in excess in poky mutant.

In reciprocal crosses, poky character shows maternal inheritance:

poky (female) × wild type (male) → all poky

wild type (female) × poky (male) → all wild type

However, there are other marker nuclear genes (ad^+/ad^-) which show. 1 : 1 mendelian segregation. The following evidences suggested that poky trait may be located in mitochondrial DNA: (*i*) slow growth may be due to lack of ATP energy and source of this energy is mitochondria; (*ii*) cytochromes in poky strain differ from those in wild type in quality and quantity and these cytochromes are found in mitochondria.

2. Non-chromosomal inheritance through mitochondria in human beings. In 1988, **Douglas Wallace** and his colleagues showed that **Leber's Hereditary Optic Neuropathy (LHON)** is a cytoplasmically inherited disease in human beings. The disease causes blindness due to degeneration of the optic nerve and it occurs at the age of fifteen to thirty five years. It is certain that the LHON disease is transmitted only maternally. The disease is caused by the mutation in mitochondrial DNA [*e.g.*, mutation at base pair 11778 affects one of the subunits of complex I of the electron transport chain (NADH hydrogenase), resulting in the substitution of a histidine for an arginine; see **Cooper** and **Hausman**, 2007].

3. Non-Chromosomal Inheritance by Endosymbionts

Certain intracellular parasites such as bacteria and virus particles maintain symbiotic relationship with host cells. They are self-reproducing and look like the cytoplasmic inclusions. Sometimes they exhibit an infection like transmission with a hereditary continuity of their own. Generally such symbionts are coined by letters of the Greek alphabets (sigma, kappa, mμ, etc.). The various types of infective symbionts are as follows:

(*i*) Sigma virus in *Drosophila*. A certain strain of *Drosophila melanogaster* shows a high degree of sensitivity to carbon dioxide, whereas the wild type strain can be exposed for long periods to pure CO_2 without permanent damage, the sensitive strain quickly becomes uncoordinated in even brief exposure to low concentrations of CO_2. This trait (extra-sensitivity) is transmitted primarily, but not exclusively, through the maternal parent. Tests have disclosed that CO_2 sensitivity is dependent upon an infectious DNA virus called **sigma**, found in the cytoplasm of **CO_2 sensitives** *Drosophila*. These infective particles are transmitted normally via the egg's larger amount of cytoplasm but occasionally through the sperms as well. Carbon dioxide sensitivity may even he induced in normal flies by injections of cell free extracts of sigma particles from CO_2 sensitive flies.

(*ii*) Spirochaetes and maternal sex ratio in *Drosophila*. Females of many *Drosophila* species can harbour a population of **spirochaete bacteria** known generally as **SR**. When SR spirochaetes infect the eggs of the host and when these eggs are fertilized, virtually all XY zygotes are killed early in embroyonic development and XX zygotes survive. Thus, the spirochaete can be considered as an endosymbiont of female but not of male *Drosophila*, and its presence in the female gives rise to the condition called **maternal sex ratio**, in which the progeny are exclusively or almost entirely female.

The SR spirochaete is infectious, for when isolated from the haemolymph of female carriers and introduced into normal females the latter become carriers. Why the female genotype permits their retention, and conversely, why XY cells are sensitive to their presence is not yet known. **K. Oishi** and **D. Poulson** (1970) have reported DNA-containing viruses in these endosymbiont spirochaetes of female *Drosophila*.

(*iii*) Kappa particles. In 1938, **T.M. Sonneborn** reported that some races (known as "**killers**" or **killer strain**) of the common ciliate protozoan, *Paramecium aurelia* produce a poisonous substance, called **paramecin** which is lethal to other individuals called "**sensitives**". The paramecin is water soluble, diffusible and depends for its production upon cytoplasmically located particles called **kappa**. Electron microscopic observations have shown that kappa particles are about 0.4μ long symbiotic bacteria, *Caedobacter taeniospiralis*; 20 per cent of kappa bacteria of the killer strain contain a refractile protein containing "**R body**" and are called "**brights**". They are infected with a virus that controls the synthesis of toxic viral protein, the paramecin. A killer *Paramecium* may contain hundreds (*e.g.*, 400) of kappa particles. The presence of kappa particles in the killer *Paramecium* is dependent for their maintenance and replication on the chromosomal dominant gene *K*. Paramecia with nuclear genotype *kk* are unable to harbour kappa particles.

When a *Paramecium* of killer strain having the genotype *KK* or ($K^+ K^+$) conjugates with the *Paramecium* of non-killer strain having the genotype *kk*, the exconjugants are all heterozygous for Kk genes (Fig. 85.5). The *Kk* genotype suggests that both exconjugant should be killers. But this is not the case. If conjugation is normal, *i.e.*, lasts only for a short time, and no exchange of cytoplasm takes place between the two, both killers and non-killers (sensitive) are produced. However, rare or prolonged conjugation (*i.e.*, lasting for long time) permits mixing of cytoplasm of both conjugants and results killers only. The killer trait is stable only in killer strain with *KK* genotype and is unstable in sensitive strain with *kk* genotype.

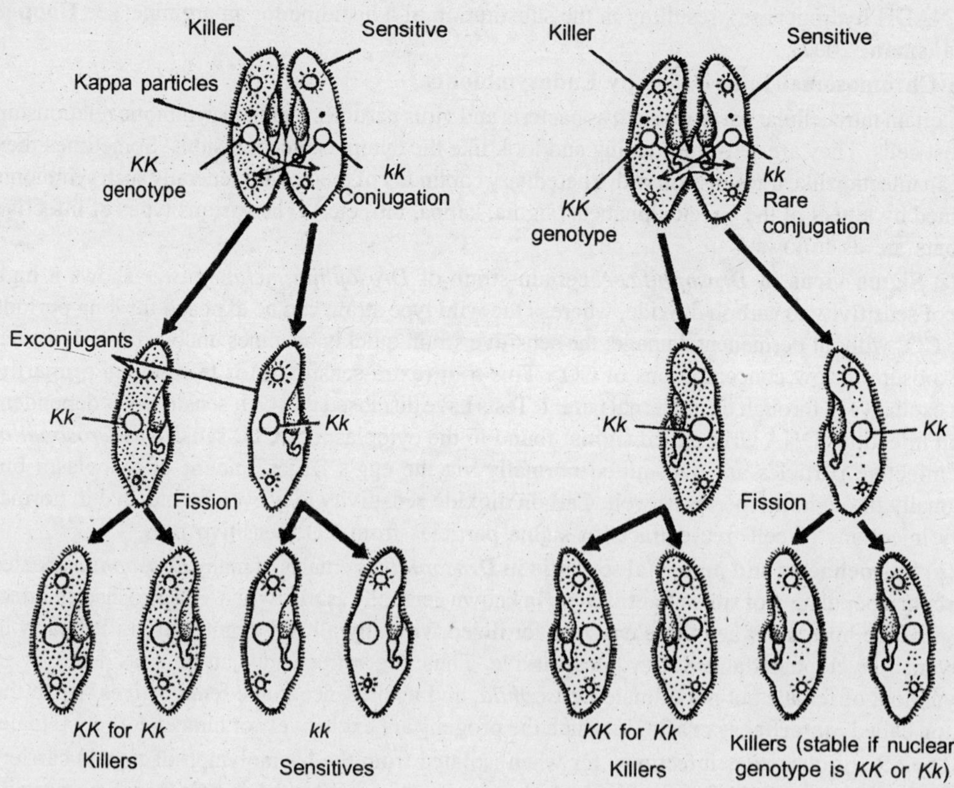

Fig. 85.5. Conjugation in *Paramecium* and the killer trait. Progeny of sensitives are killers only in rare situations where conjugation persists for a longer period so that kappa-containing cytoplasm is introduced into the conjugating sensitive. Kappa particles, however, are maintained only in the presence of *K*—nuclear genotype.

(*iv*) mμ particle. Another type of killer trait known as **mate killer** has been reported in *Paramecium* by **R.W. Siegel** in 1952. The mate killer trait is contributed by a cytoplasmic mμ particle and a *Paramecium* with a mμ particle is called **mate killer** because when it conjugates with a *Paramecium* without any mμ particle (called **mate sensitive**) then it kills the latter. The mμ particles exist only in those cells whose micronucleus contains at least one dominant gene of either of two pairs of unlinked chromosomal genes (M_1 and M_2). The mμ particles are composed of DNA, RNA and other substances and are symbionts.

(*v*) Milk factor or Breast cancer in mice. Bittner found that females of certain lines of mice are highly susceptible to mammary cancer and this trait was found to be maternally transmitted trait. Results of reciprocal crosses between these and animals of low-cancer-incidence strain depend on the characteristic of the female parent. When the young mice of a low-incidence strain are allowed to be nursed by susceptible foster mothers produces a high rate of cancer in them. Apparently this is a case of infective agent transmitted in the milk. This so called milk factor resembles in many respects with a virus and has been discovered to be transmissible also by saliva and semen. The presence of milk factor also depends on nuclear genes.

QUESTIONS

Long Answer Questions

1. What is meant by cytoplasmic inheritance? Explain with suitable examples. (*CCS 1992*)

2. Give an account of cytoplasmic inheritance.
(*Delhi 1991, 1992; Madras 1989, 1992; CCS 1991; Jiwaji 1992; Shri Venkat 1992*)

3. Write an account of non-chromosomal inheritance. (*CCS 1989; Bihar 1990*)

4. How will you distinguish between maternal effect and cytoplasmic inheritance? Illustrate with suitable examples. (*BHU (Hons.) 1991*)

5. How will you determine whether a particular trait is inherited via nucleus or as a cytoplasmic particle? (*Delhi 1989*)

6. Describe the non-chromosomal inheritance. Justify with suitable examples. (*CCSU Meerut 2005*)

Short Answer Questions

1. Write short notes on the following:

(*i*) Cytoplasmic genes (*CCS 1999*)

(*ii*) Plasma genes (*CCS 1989, 1990*)

(*iii*) Plastid inheritance (*CCS 1989*)

(*iv*) Cytoplasmic inheritance (*CCS 1989*)

2. Describe cytoplasmic inheritance in *Paramoecium*. (*C.C.S.U., Meerut, 2009*)

3. Differentiate between the following:

(*i*) Nuclear inheritance and cytoplasmic inheritance (*CCS 1995*)

4. What is cytoplasmic inheritance? Explain with suitable examples? (*C.C.S.U., Meerut 2008*)

Very Short Answer Questions

1. Dextral and sinistral coiling in snail. (*C.C.S.U., Meerut 2008*)

Ans. When the shells of the snails coil to right, it is called **dextral coiling**. In contrast, when the shells of the snails coil to left, it is called **sinistral coiling.**

2. What is cytoplasmic inheritance?

Ans. Extrachromosomal inheritance controlled by nonnuclear genomes.

3. Define non-Mendelian genes.

Ans. The non-Mendelian genes are plasma genes which are located in cytoplasm: DNA of mitochondria and plastids, mRNAs of unfertilized eggs, endosymbionts, etc.

4. How much cytoplasm is contributed to the zygote by the male gamete?

Ans. Negligible.

5. What are the kappa particles? (*C.C.S.U. Meerut, 2006*)

Ans. Kappa particles are DNA-containing, self-reproducing cytoplasmic particles in certain strains of *Paramecium aurelia*. They control a toxic substance, *Paramecin*, that is released into the culture medium and kills sensitive paramecia. Nuclear gene *K* is required for maintenance of kappa in the cytoplasm of killer strain.

6. Inheritance of plasma genes found in the cell?

Ans. Cytoplasm of egg.

7. Where are plasma genes found in the cell.

Ans. In mitochondria (DNA), plastids (DNA), mRNA of ooplasm (which are transcribed by maternal genome during oogenesis), endosymbionts (*e.g.*, kappa particles).

8. Who studied inheritance of plasma genes found in the cell?

Ans. L. Heritier and **Teissier.**

9. Who investigated inheritance of kappa particles in *Paramecium*?

Ans. T. Sonneborn.

Multiple Choice Questions

1. Cytoplasmic inheritance differ from nuclear inheritance in the absence of
   (*a*) similarity of reciprocal crosses (*b*) biparental contribution
   (*c*) effect of backcrossing (*d*) uniparental contribution.
2. The extrachromosomal autosomes unit within bacteria is
   (*a*) phasmid (*b*) plasmid
   (*c*) pyramid (*d*) nucleosome
3. The cytoplasmic units of inheritance are called
   (*a*) genome (*b*) lethal gene
   (*c*) plasmagene (*d*) all the above
4. Extra chromosomal genes can be traced in
   (*a*) colour blind person (*b*) *Paramecium*
   (*c*) sex-linked diseased person (*d*) sex-influenced person
5. Extranuclear genes are present in
   (*a*) mitochondria (*b*) mitochondria and chloroplasts
   (*c*) mitochondria, plastids and cytoplasm (*d*) ER and cytoplasm
6. Cytoplasmic male sterility is passed down
   (*a*) biparentally (*b*) maternally
   (*c*) paternally (*d*) through bacteriophage
7. Genes for cytoplasmic male sterility in plants are located in
   (*a*) nuclear genome (*b*) cytosol
   (*c*) chloroplast genome (*d*) mitochondrial genome
8. A plasmid is
   (*a*) genetic material of a virus (*b*) extrachromosomal DNA in a bacterial cell
   (*c*) smallest bacterium (*d*) starch granule
9. When a certain character is inherited only through the female parent, it probably represents case of
   (*a*) incomplete dominance (*b*) cytoplasmic inheritance
   (*c*) medelian nuclear inheritance (*d*) multiple plastid inheritance
10. When an albino plant of maize female is crossed with normal green male plant, all plants in the progeny are albino because
    (*a*) albinism is dominant over green characters
    (*b*) plastids are inherited through maternal plants
    (*c*) green plastids of male parent becomes maternal
    (*d*) crossing result in structural change in green plants

ANSWERS

Multiple Choice Questions

1. (*a*)	**2.** (*b*)	**3.** (*c*)	**4.** (*b*)
5. (*c*)	**6.** (*b*)	**7.** (*d*)	**8.** (*b*)
9. (*b*)	**10.** (*b*).		

86

Chromosomal Mutations

Genetics makes extensive use of deviations from the norms, and the study of chromosomes is no exception. The chromosomes of each species has a characteristic morphology (structure) and number. But, sometimes due to certain accidents or irregularities at the time of cell division, crossing over or fertilization some alterations in the morphology and number of chromosomes take place. The slightest variation in the organisation of chromosomes is manifested phenotypically and is of great genetical interest. The changes in the genome involving chromosome parts, whole chromosomes, or whole chromosome sets are called **chromosome aberrations** or **chromosome mutations**. Chromosome mutations have proved to be of great significance in applied biology—agriculture (including horticulture), animal husbandry and medicine.

Chromosome mutations are inherited once they occur and are of the following types (Box 86.1):

A. Structural changes in chromosomes:

1. Changes in number of genes
   (*a*) Loss: deletion
   (*b*) Addition: Duplication
2. Changes in gene arrangement:
   (*a*) Rotation of a group of genes 180° within one chromosome: inversion
   (*b*) Exchange of parts between chromosomes of different pairs: translocation.

B. Changes in number of chromosomes:

1. Loss, or gain, of the part of the chromosome set (aneuploidy)
2. Loss, or gain, of whole chromosome set (euploidy)
   (*a*) Loss of an entire set of chromosomes (haploidy)
   (*b*) Addition of one or more sets of chromosomes (polyploidy).

Both types of changes (structural and numerical) in chromosomes can be detected not only with a microscope (*i.e.,* cytologically) but also by standard genetic analysis. This gave birth to a hybrid science, called **cytogenetics** which attempts to correlate cellular events, especially those of chromosomes, with genetic phenomena.

86.1 STRUCTURAL CHANGES IN CHROMOSOMES

Structural changes in chromosome may be of the following types (Fig 86.1): 1. **deficiency** or **deletion** which involves loss of a broken part of a chromosome; 2. **duplication** involves addition of a part of chromosome (*i.e.*, broken segment becomes attached to a homolog which, thus, bears one block of genes in duplicate); 3. **inversion** in which broken segment is reattached to original chromosome in reverse order, and 4. **translocation** in which the broken segment becomes attached to a nonhomologous chromosome resulting in new linkage relations.

Further, structural, abnormalities can occur in both homologous chromosomes of a pair or in only one of them: When both homologous chromosomes: are involved, these are called **structural**

homozygotes; *e.g.*, deletion homozygate, duplication homozygote, etc. When only one homologous chromosome is involved, it is called **structural heterozygote**. In Fig 86.1 have been shown translocation heterozygote and translocation homozygote.

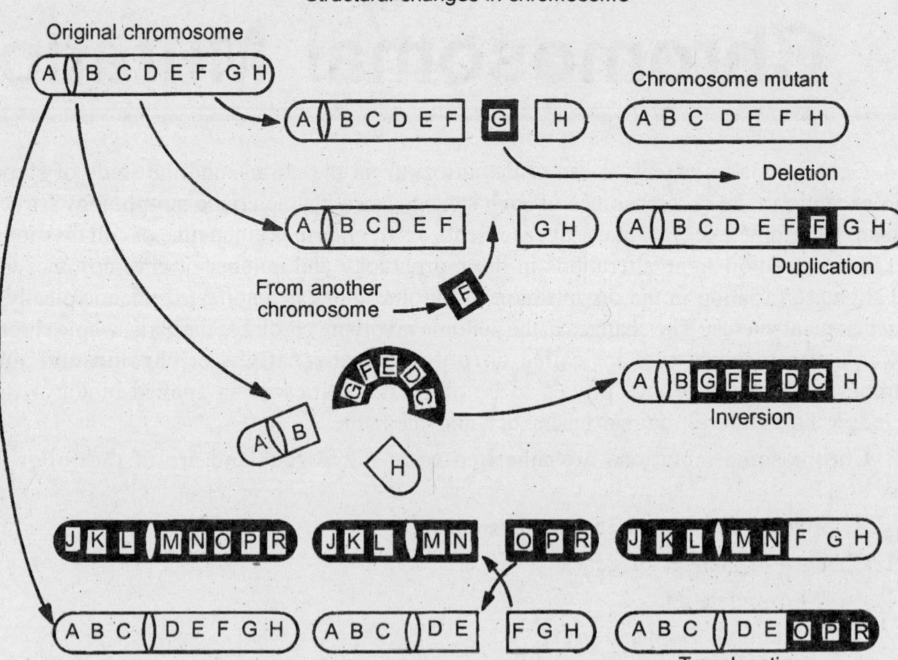

Fig. 86.1. Structural changes in chromosome.

Box 86.1

1. A **chromosome mutation** (or **chromosome aberration**) is a mutation involving a large segment of the genome. Such a mutation usually affects many genes and is often observable at the cytogenetic level, *i.e.*, it can be seen with a light microscope.
2. Chromosome mutuations are **numerical** if they involve a deviation from normal chromosome number and **structural** if they involve breakage and rearrangement of chromosome segments. Structural mutations are **balanced** if DNA is rearranged but there is no loss or gain of material, or **unbalanced** if DNA is lost or gained. All numerical mutations are unbalanced.
3. There are four consequences of chromosome mutations: disruption, fusion, position and dosage effects.
   (*i*) **Disruption and fusion effects.** These occur in structural mutations and reflect the nature of the chromosome breakpoints before and after mutation, *e.g.*, a breakpoint can interrupt a gene or separate a gene from its promoter, resulting in loss of gene function, fusion can join two genes, allowing a composite product to be synthesized perhaps with noval functions.
   (*ii*) **Position effects.** These also occur in structural mutations and reflect global influences on gene expression conferred by chromatic structure, *e.g.*, a translocation may bring a normally active gene adjacent to a region of heterochromatin causing transcriptional repression.
   (*iii*) **Dosage effect.** They occur in both structural and numerical mutations and concern the number of copies of each gene in the cell.

> **4. Constitution chromosome mutation.** These arise in the germline or in the zygote and thus affects every cell in the body. In humans, these give rise to specific clinical syndromes (*e.g.*, Down's syndrome, Kleinefelter syndrome).

1. Deletion (or Deficiency).

The simplest result of breakage is the loss of a part of a chromosome. Portions of chromosomes without a centromere (called **acentric fragments**) lag in anaphase movement and are lost from reorganizing nuclei or digested by nucleases. Such loss of a portion of a chromosome (and of some genes) is called **deletion**. The chromosomes with deletions can never revert to a normal condition. If gametes arise from the cells having a deleted chromosome, this deletion is transmitted to the next generation. Further, a deletion can be terminal or intercalary (interstitial). In terminal deletion a terminal section of a chromosome is absent and it is resulted by only one break (Fig. 86.3). While in the **intercalary deletion**, an intermediate section or portion of chromosome is lost and it is caused by two breaks–one on either end of the deleted region (Fig. 86.3). Thus, in the latter case, the chromosome is broken into three pieces, the middle one of which is lost and the remaining two pieces get joined again.

Two pairs of chromosomes	1 2 3 4 5 6 1 2 3 4 5 6	7 8 9 10 11 12 7 8 9 10 11 12
Translocation heterozygote	7 8 3 4 5 6 1 2 3 4 5 6	1 2 9 10 11 12 7 8 9 10 11 12
Translocation homozygoyte	7 8 3 4 5 6 7 8 3 4 5 6	1 2 9 10 11 12 1 2 9 10 11 12

Fig. 86.2. Chromosome constitution of a translocation heterozygote and a translocation homozygote.

In general, if a homozygous deletion is made, it is lethal. Even individuals heterozygous for deletion (deletion in one of the homologous chromosomes) may not survive. If meiotic chromosomes in such heterozygotes are examined, the region of deletion can be detected by the failure of the corresponding segment on the normal chromosome to pair properly; so a "**deletion-loop**" results (Fig. 86.4). Deletion loops are also detected in polytene chromosomes of *Drosophila,* where the homologs exist in permanent state of pairing. The cytological studies of pairing between normal and deleted chromosomes have helped a lot in finding out the relative positions of genes in chromosomes.

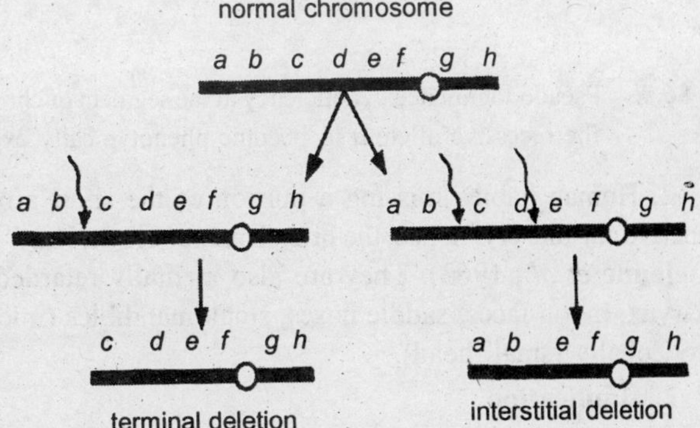

Fig. 86.3. Production of terminal and intersitial deletion. Chromosomes can be broken when struck by ionizing radiation (arrows).

Genetical effects of deletion. Deletion of some chromosome regions produce their own unique phenotypes. A good example of this is a dominant **notch-**

wing mutation in *Drosophila*. In fact, this is a small deletion and acts as a recessive lethal in this regard. Further, in the presence of a deletion, a recessive allele of the normal homologous chromosome will behave like a dominant allele, *i.e.*, it will be phenotypically expressed, this phenomenon is called **pseudodominance** (Fig. 86.5).

The phenomenon of pseudodominance exhibited by deficiency heterozygotes has been utilized for the location of genes on specific chromosomes and in preparing **cytological maps** in *Drosophila*, maize, bacteriophage and other organisms. Such cytological maps are often used to verify the **genetic maps** (based on linkage analysis) of these organisms.

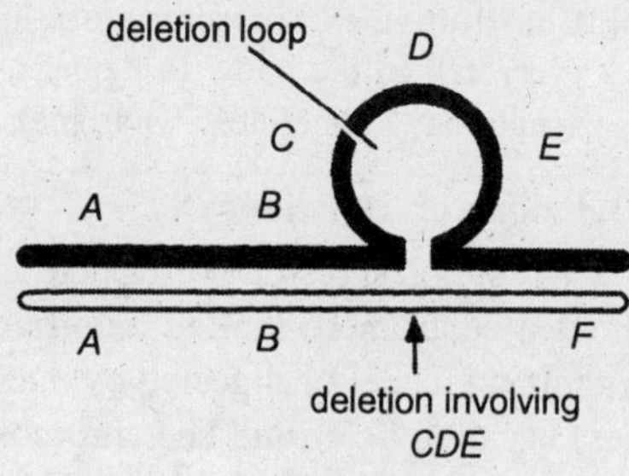

Fig. 86.4. Formation of a deletion loop during synapsis in a deletion heterozygote.

Examples of pseudodominance (deletion). 1. **Gates** (1921) demonstrated the effects of pseudodominance. Certain mice carrying the recessive allele (*v*) in homozygous condition move about erratically until exhausted. Such mice are called **'waltzing mice'**. When homozygous waltzing mice (*vv*) are mated with normal mice carrying the dominant allele (*VV*), all the affecting (*Vv*) are normal. However, in a cross, between homozygous normal female and a waltzer male, one of the seven offspring was a waltzer. A deletion had resulted in the elimination of the dominant allele (*V*); so that the recessive allele (*v*) for waltzing had expressed itself.

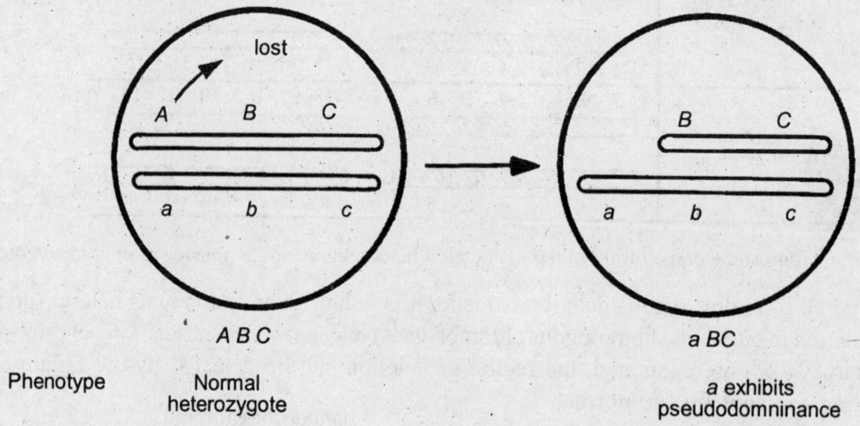

Fig. 86.5. Pseudodominance. A deficiency in the segment of chromosome bearing the dominant gene *A* allows the recessive allele *a* to become phenotypically expressed.

2. Human babies missing a portion of the short arm of chromosome 5 (autosome) have a distinctive cat-like cry; hence, the French name **"cri du chat"** (cry of the cat) **syndrome** (first described by **Lejeune** *et al.*, 1963). They are also mentally retarded (IQ below 20), have malformation in the larynx, moon faces, saddle noses, small mandibles (micrognathia), malformed low-set ears and microcephally (small head).

2. Duplication

The presence of a part of a chromosome in excess of the normal complement is known as **duplication**. Thus, due to duplication some genes are present in a cell in more than two doses. If duplication is present only on one of two homologous chromosomes, at meiosis the chromosome bearing the duplicated segment forms a **loop** to maximize the juxtaposi-tion (during pairing) of homologous regions (Fig. 86.6).

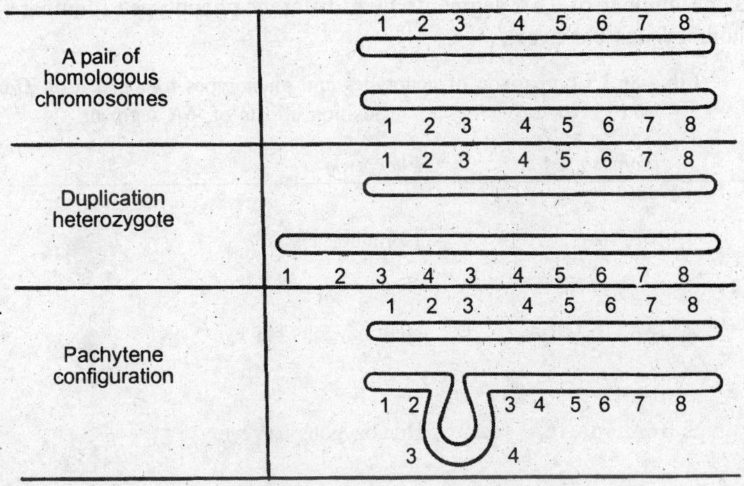

Extra segments in a chromosome may arise in a variety of ways such as follows:

1. Tandem duplication. In this case the duplicated region is situated just by the side of the normal corresponding section of the chromosome and the sequences of genes are the same in normal and duplicated region. For example, if the sequence of genes in a chromosome is ABC. DEFGH (The full stop depicts the centromere) and if the chromosomal segment containing the genes DEF is duplicated, the sequence of genes in tandem duplication will be ABC. DEFDEFGH.

A pair of homologous chromosomes
Duplication heterozygote
Pachytene configuration

Fig. 86.6. Loop formation during chromosome pairing in a duplication heterozygote.

2. Reverse tandem duplication. Here, the sequence of genes in the duplicated region of a chromosome is just the reverse of a normal sequence. In the above mentioned example, therefore, the sequence of genes due to reverse tandem duplication will be ABC. DEFFEDGH.

3. Displaced duplication. In this case the duplicated region is not situated adjacent to the normal section. Depending on whether the duplicated portion is on the same side of the centromere as the original section or on the other side, the displaced duplication can be termed either **homobranchial** or **heterobranchial**.

Example. Homobranchial duplication = ABC.DEFG DEFH

Heterobranchial duplication = ADEFB C. DEFGH

4. Transposed duplication. Here, the duplicated portion of chromosome becomes attached to a non-homologous chromosome. For example, if ABC.DEFGH and LMNOPQ. RST represent the gene sequences of two nonhomologous chromosomes, a transposed duplication will result into chromosomes with gene sequence ABC.GH and LMN DEF.OPQ. RST. Such a transposed duplication may be either interstitial (*e.g.*, LMNDEFOPQ. RST) or **terminal** (*i.e.*, LMN OPQ.RSTDEF).

5. Extra-chromosomal duplication. In the presence of centromere the duplicated part of a chromosome act as independent chromosome.

Genetical effects of duplication. Due to duplication, there occur unequal crossing over which results in deletion and reduplication which produce distinct phenotypes as shown by the following examples:

1. Bar eye in *Drosophila*. The Bar phenotype of *Drosophila* is characterized by narrower, oblong, bar-shaped eye with few facets. It is determined by a X-linked recessive allele *B*. The classical studies of **Bridges** (1936) showed that the bar trait of *Drosophila* is associated with the duplication of a segment of the X-chromosome, called **section 16A**, as observed in salivary gland chromosomes. Each added section 16A intensifies the bar phenotype (*i.e.,* duplication behaves genetically as a dominant factor). However, the narrowing effect is greater if the duplicated segments are on the same chromosome (called **position effect**) (Table 86.1). Thus, *cis* and *trans* arrangements of the same number of 16A segments give different phenotypes (compare heterozygous ultrabar and homozygous bar eyes).

Table 86.1 Comparison of genotypes and phenotypes for bar eye in *Drosophila* females showing position effects of 16A segment.

X chromosomes	*Phenotype*	*Mean number of facets*
1. 16A/16A	Normal	779
2. 16A, 16A/16A	Heterozygous bar eye	358
3. 16A, 16A/16A, 16A	Homozygous bar eye	68
4. 16A, 16A, 16A/16A	Heterozygous ultrabar (= double bar)	45
5. 16A, 16A, 16A/16A, 16A, 16A	Homozygous ultrabar	25

3. Inversion

Inversion involves a rotation of a part of a chromosome or a set of genes by 180° on its own axis. It essentially involves occurrence of *breakage* and *reunion*. The net result of inversion is neither a gain nor a loss in the genetic material but simply a rearrangement of the gene sequence. An inversion can occur in the following way: suppose that the normal order of segments within a chromosome is 1-2-3-4-5-6; breaks occur in regions 2-3 and 5-6 and broken piece is reinserted in reverse order. This results in an inverted chromosome having segments 1-2-5-4-3-6.

An **inversion heterozygote** has one chromosome in the inverted order and its homologue (homolog) in the normal order. The location of the inverted segment can be detected cytologically in the meiotic nuclei of such heterozygotes by the presence of an **inversion loop** in the paired homologs. The location of the centromere relative to inverted segment determines the genetic behaviour of the chromosomes. If the centromere is not included in the inversion it is called

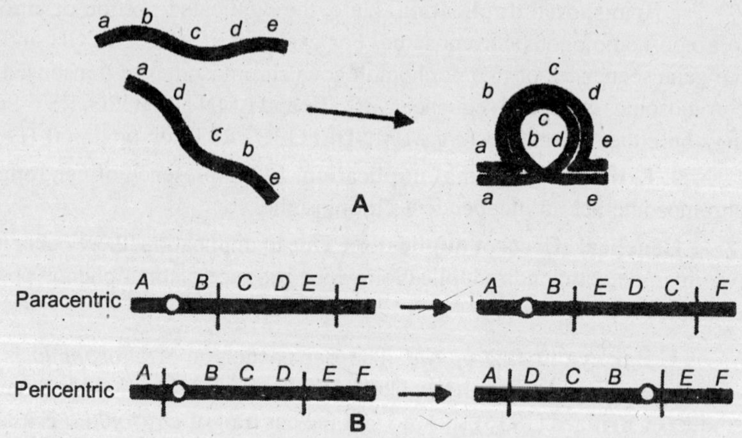

Fig. 86.7. A—An inversion loop in paired homologs of an inversion heterozygote; B—The location of the centromere relative to the inverted segment.

paracentric inversion and when inversion includes the centromere it is called **pericentric inversion** (Fig. 86.7). Homologous chromosomes; with identical inversions in each member, pair and undergo normal distribution in meiosis. However, crossing over in inversion heterozygotes produce, deletions, duplications and other curious configurations.

Advantages of inversions. Fertility of inversion homozygotes and sterility of inversion heterozygotes lead to establishment of two groups (or varieties) which are mutually fertile but do not breed well with the rest of the species. Both varities evolve in different directions and later become reproductively isolated species. There is plenty of cytological evidence to prove that such evolutionary mechanisms have and are operating in *Drosophila* and a number of other organisms.

4. Translocation

The shifting or transfer of a part of a chromosome or a set of genes to a non-homologous one, is called **translocation**. There is no addition or loss of genes during translocations, only a rearrangement (*i.e.,* change in the sequence and position of a gene). Translocations may be of following three types (Fig. 86.8).

1. Simple translocations. They involve a single break in a chromosome. The broken piece gets attached to one end of a nonhomologous chromosome.

2. Shift translocations. In this type of translocation, the broken segment of one chromosome gets inserted interstitially in a nonhomologous chromosome.

3. Reciprocal translocations. In this case, a segment from one chromosome is exchanged with a segment from another nonhomologous one, so that in reality two translocation chromosomes are simultaneously achieved.

Outcomes of reciprocal translocation. The exchange of chromosome parts between nonhomologous chromosomes creates new linkage relationships. Such translocations also drastically change the size of a chromosome as well as the position of its centromere. For example, a large metacentric chromosome is shortened by one-half in length to an acrocentric one, whereas the small chromosome becomes a large one (Fig. 86.9). Two types of translocations have been recognized: homozygous and heterozygous (Fig 86.9). The **translocation homozygotes** may have normal meiosis and in fact, are difficult to detect cytologically unless morphologically dissimilar chromosomes are involved, or banding patterns differ markedly. The **translocation heterozygotes** produce both translocated and normal chromosomes and exhibit characteristic cytological and genetical effects.

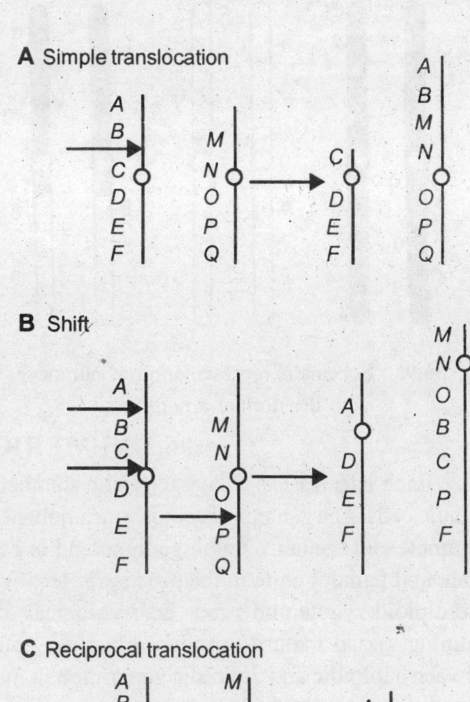

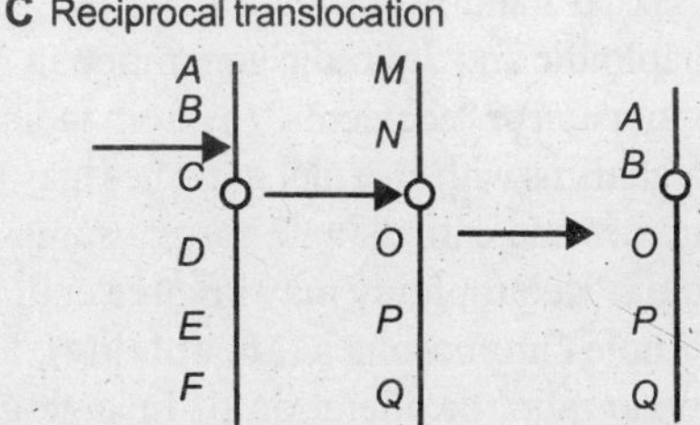

Fig. 86.8. Three types of translocations which occur in nonhomologus chromosomes. Arrows indicate the points of breaks.

The first case of translocation was studied in the evening primrose (*Oenothera*) which was originally described as a mutation by **de Vries**. In *Oenothera*, *Tradescalltia* and *Rhoeo* translocations in heterozygous condition are frequently found in nature. In many other crop plants translocations have been artificially induced by X-rays. They are well evident in many animals including humans. In *Drosophila* and other animals, translocations have also been induced by X-rays. For example, in humans, patients with **chronic myelocytic (myologenous) leukemia** (a kind of cancer) display an interesting chromosomal abnormality. In the bone marrow and in cells derived from it, is present a short chromosome, called the **Philadelphia (Ph^1) chromosome** (so named because it was discovered in that city). Detailed cytological study disclose Ph^1 to be a number 22 chromosome that has lost most of the distal part of its longer arm (22 *q*–). The deleted part of autosome 22 is translocated to one of the larger autosomes (most frequently to the distal end of chromosome 9). This translocation exhibits position effect and it is not transmitted to offspring of persons having Philadelphia chromosome (Ph^1 does not appear in gametes of the patients).

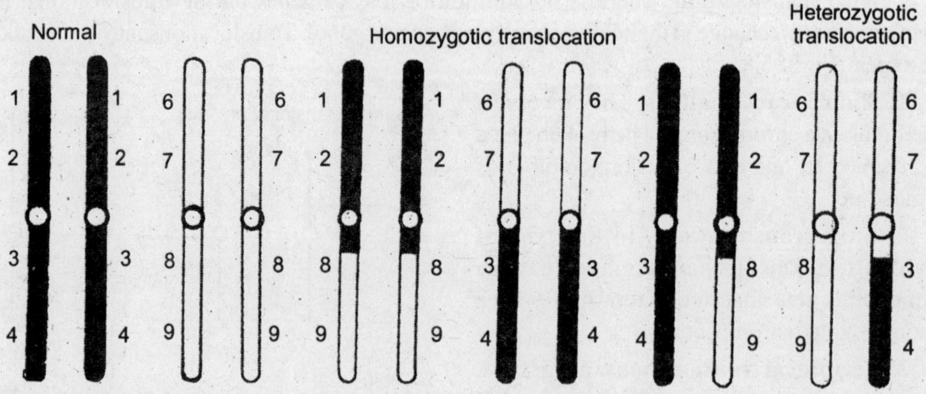

Fig. 86.9. Schematic representation of homozygotic and heterozygotic reciprocal translocations compared with the normal arrangement.

86.2 NUMERICAL MUTATIONS

Each species has a characteristic number of chromosomes in the nuclei of its gametes and somatic cells. The gametic chromosome number constitutes a basic set of chromosomes called **genome**. A gamete cell contains single genome and is called **haploid**. When haploid gametes of both sexes (male and female) unite in the process of fertilization a **diploid** zygote with two genomes is formed. The diploid zygote undergoes embryological development and forms an adult animal which upon attaining sexual maturity produces haploid gametes. And this alternation of generation continues between haploidic and diploidic generation in most species. However, sometimes irregularities occur in nuclear division or "accidents" (as from radiations) may befall interphase chromosomes so that cells or entire organisms with aberrant genomes may be formed. Such chromosomal aberrations may include whole genomes and entire single chromosomes. Changes in number of whole chromosomes is called **heteroploidy**. Heteroploidy may involve entire sets of chromosomes (**euploidy**), of loss or addition of single whole chromosomes (**aneuploidy**). Each may produce phenotypic changes, modifications of phenotypic ratio, or alteration of linkage groups. Many are of some evolutionary significance.

A. EUPLOIDY

The term **euploidy** (Gr., *eu* = even or true; *ploid* = unit) designates genomes containing chromosomes that are multiples of some basic number (x). The euploids are those organisms which contain balanced set or sets of chromosomes in any number. The number of chromosomes in a basic set is called the **monoploid number**, x. Those euploid types whose number of sets is greater than two are called polyploid. Thus, $1x$ is **monoploid**, $2x$ is **diploid**; and the polyploid types are $3x$ (**triploid**), $4x$ (**tetraploid**), $5x$ (**pentaploid**), $6x$ (**hexaploid**) and so on: The **haploid** (**n**) refers; strictly to the number of chromosomes in gametes. In most animals and many plants the haploid number and monoploid number are the same. Hence n or x (or $2n$ or $2x$) can be used interchangeably. However, in case of polyploids the usage of n may create confusions. For example; in modern wheat x *and* n are different. Wheat has 42 chromosomes, but careful study reveals that in hexaploid there are: six rather similar but not identical sets of seven chromosomes. So, $6x = 42$, and $x = 7$. However, gametes of wheat contain 21 chromosomes, hence, $2n = 42$ and $n = 21$. A triploid hybrid of wheat, from which a hexaploid *Triticum spelta* has been obtained due to colchicine treatment; contains $2n = 3x = 21$. Such haploids, since are obtained from the polyploids (*i.e.*, cross of tetraploid emmer wheat and diploid goat grass), they are called **polyhaploids**; just to differentiate them from the normal monoploids.

The lower organisms such as bacteria and viruses are called **haploids** because they have a single set of genetic elements. However, since they do not form gametes comparable to those of higher organisms, the term monoploid would seem to be more appropriate.

1. Monoploidy

Monoploids have a single basic set of chromosomes, *e.g.*, 7 in barley and 10 in corn. Monoploidy is common in plants and rare in animals.

Monoploids in some cases are found normally and are produced due to parthenogenesis, as in male (drone) hymenopteran insects such as bees, wasps and ants. In these insects, queen and workers are diploid females. In angiosperms (flowering plants) monoploids may also originate spontaneously due to parthenogenetic development of egg. Such rare monoploids have been obtained in tomatoes and cotton under cultivation. Rarely monoploid plants may originate from the pollen tube, synergids and antipodals of the embryo sac and are called **androgenic monoploids** (or androgenic haploids).

Monoploids can be produced by artificial means by the following methods: (1) X -ray treatments, (2) delayed pollination, (3) temperature shock (cold treatment), (4) colchicine treatment. (5) distant (interspecific or intergeneric) hybridization), (6) anther or pollen culture. Among these techniques, the most important ones are distant hybridization and anther culture.

Monoploid plants have reduced size of all vegetative and floral parts. Monoploid organisms are highly sterile. For instance, a monoploid in maize ($2n = 20$) will have 10 chromosomes and the number of chromosomes in a gamete can range from 0–10. Consequently, considerable sterility is found in such monoploid maize plants.

In contrast, in monoploid male honey bees during spermatogenesis the meiosis is bypassed by mitosis. As a result, their sperms are haploid and viable.

Uses of Monoploids. In a monoploid, since there is only one copy of each chromosome and only one allele of each gene, so, in it each gene is expressed whether it is dominant or recessive. This facilitates genetic experiments and this is the reason why microorganisms have been able helpful in genetic studies. For the same reason, scientists are trying hard to develop haploid strains of the flowering plants. Success has been achieved in developing monoploid strains of *Nicotiana, Datura* and *Triticum*. From these monoploid strains have also been developed pure breeding strains which are resistant for the insecticides and also for toxic compounds normally produced by the parasites of these plants.

2. Polyploidy

Any organism with more than two genomes ($2x$) is called a **polyploid**. Many plant genera include species whose chromosome numbers constitute a euploid series. For example, the rose genus *Rosa* includes species with the somatic numbers 14, 21, 28, 35, 42 and 56. These numbers are the multiples of 7. Therefore, this is a euploid series of the basic monoploid number 7, which gives diploid, triploid, tetraploid, pentaploid, hexaploid and octaploid species. Over 10 copies of the genome are represented by numbers, *e.g.*, 12-ploid. Except diploids, rest of these belong to polyploid category. Ploidy levels higher than tetraploid are not commonly encountered in natural populations, but our most important crops and ornamental flowers are polyploids, *e.g.*, wheat (hexaploid, 6*x*), strawberries (octaploid, 8*x*), many commercial fruits and ornamental plants. Generally, polyploidy is common in plants (more common in monocots) but rare in animals.

Types of polyploidy. There are following three different kinds of polyploids: (*i*) autopolyploids, (*ii*) allopolyploids and (*iii*) autoallopolyploids. Suppose that there are four different haploid sets of chromosomes A, B_1, B_2 and C, in which B_1 and B_2 genomes are related. By using these genomes, all three types of polyploids can be derived as have been shown in Fig. 86.10.

(*a*) Autopolyploids. The autopolyploids are those polyploids, which consist of same basic set of chromosomes multiplied. For example, if a diploid species has two similar sets of chromosomes or genomes (AA), an autotriploid will have three similar genomes (AAA), and an autotetraploid, will have four such genomes (AAAA).

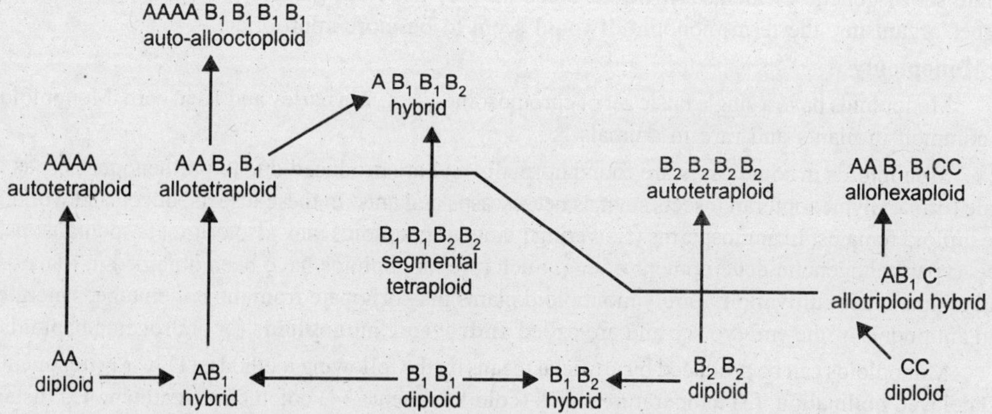

Fig. 86.10. Mode of formation of different kinds of polyploid.

(*i*) Origin and production of autopolyploids. The autopolyploids may occur in nature or may be produced artificially. When they are found in nature, their autopolyploidy nature is deduced by their meiotic behaviour. One of the common example of natural autopolyploidy is 'doob' grass (*Cynodon dactylon*) which is quite commonly cultivated in U.P. and Bihar. *Cynodon* is quite successful in the cultivation mainly due to its efficient way of vegetative propagation (since being triploid, it is sterile and setting no seed). Polyploids may arise naturally by following means: (*i*) in natural populations polyploidy may arise as a result of interference with cytokinesis, once chromosome replication has occurred; (*ii*) it may occur either in somatic tissues which give rise to tetraploid branches or during meiosis which produces unreduced gametes. All these natural inductions of polyploidy may occur due to chilling.

Some of common examples of autotriploid crop plants, which are mainly produced by artificial methods, are seedless varieties of watermelons, sugar beet, tomato, grapes and banana. Similarly, many important crop plants include **autotetraploids** such as rye (*Secale cereale*), corn (*Zea mays*), red clover (*Trifolium pratense*), berseem (*Trifolium alexandrium*), marigolds (*Tagetes*), snapdragons (*Antirrhinum*), *Phlox*, grapes, apples, *Oenothera lamarkiana* (which was recognized as mutation by **Hugo de Vries**).

Induced autopolyploidy. The autopolyploidy have been induced in many plant and animal cells by artificial means such as chemical (*e.g.*, chloral hydrate, colchicine, sulphanil amide, mercury chloride, hexachlorcyclohexane, *etc.*), radioactive substances, *e.g.*, radium and X-ray and temperature shocks. These inducers usually disturb the mitotic or meiotic spindle and cause non-segregation of already duplicated chromosomes, during cell divisions.

(*b*) Allopolyploids (Amphipolyploids). When the polyploidy results due the doubling of chromosome number in a F_1 hybrid which is derived from two distinctly different species, then, it is called **allopolyploidy** and the resultant species is called an **allopolyploid**. Let A represent a set of chromosomes (genome) in species X, and let B represent another genome in a species Y. The F_1 hybrids of these species then would have one A genome and another B genome. The doubling of chromosomes in the F_1 hybrids will give rise to allotetraploids with two A and two B genomes.

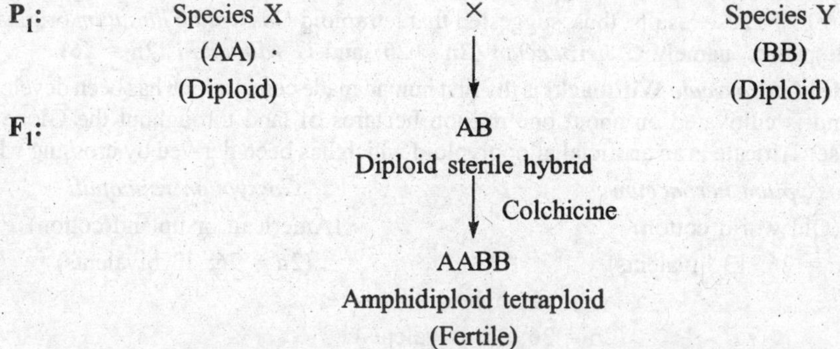

Fig. 86.11. Formation of an amphidiploid tetraploid.

Raphanobrassica is a classical example of allopolyploidy or amphipolyploidy. In 1927, a Russian geneticist, **G.D. Karpechenko** performed a cross between radish (*Raphanus sativum*, $2n = 18$) and cabbage (*Brassica oleracea*, $2n = 18$) and in F_1 got sterile (diploid) hybrids. Among these sterile F_1 hybrids, he found certain fertile plants which were found to contain 36 chromosomes. These fertile tetraploids were called *Raphanobrassica*.

Synthesized allopolyploids. Wheat, tobacco and cotton are naturally occurring allopolyploid crops. To find out the origin of naturally occurring allopolyploids some cytogeneticists produced certain allopolyploids in laboratory by employing artificial means. Common hexaploid wheat *Triticum spelta* and tetraploid cotton *Gossypium hirsutum* furnish two such examples.

(*i*) **Triticum spelta** is a hexaploid wheat which was artificially synthesized in 1946 by **E.S. McFadden** and **E.R. Sears** and also by **H. Kihara**. They crossed an emmer wheat, *Triticum dicoccoides*, (tetraploid: 2n = 28) with goat grass, *Aegilops squarrosa* (diploid; 2n = 14) and doubled the chromosome number in the F_1 hybrid (Fig. 86.12). This artificially synthesized hexaploid wheat was found to be similar to the primitive wheat *T. spelta*. When the synthesized hexaploid wheat was crossed with naturally occurrng *T. spelta*, the F_1 hybrid was completely fertile. This suggested that hexaploid wheat must have originated in the past due to natural hybridization between tetraploid wheat and goat grass followed by subsequent chromosome doubling.

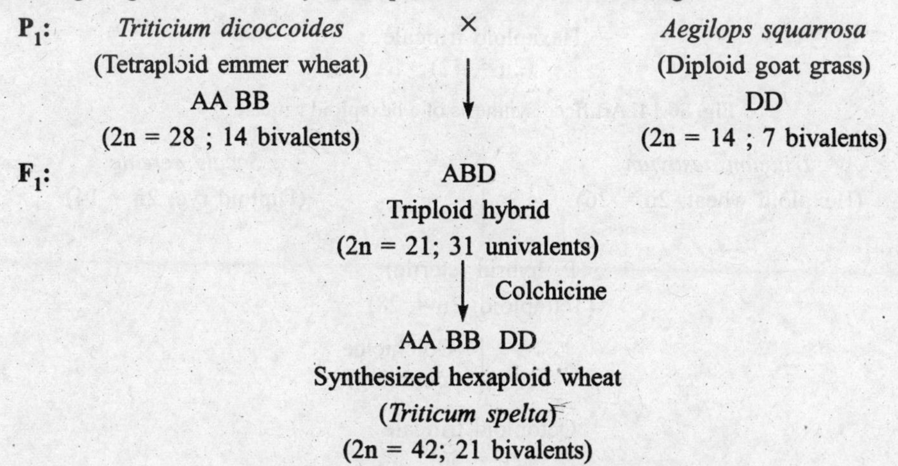

Fig. 86.12. Artificial synthesis of hexaploid wheat.

***(ii)* Gossypium hirsutum**, the New world cotton plant, is another interesting example of allopolyploidy. Old world cotton, *Gossypium herbaceum*, has 13 pairs of chromosomes, while, American or "upland cotton" also contains 13 pairs of chromosomes. **J.O. Beasley** crossed the old world and American cottons and doubled the chromosome number in the F_1 hybrids. The allopolyploids, thus, produced resembled the cultivated New world cotton and when crossed with it gave fertile F_1 hybrids (Fig. 86.13). These results, thus, suggested that tetraploid *Gossypium hirsutum* originated from two diploid species, namely *G. herbaceum* (2n = 26) and *G. raimondii* (2n = 26).

***(iii)* Triticale (*Triticosecale* Wittmack)** is the first human made cereal which has been developed in recent years and is cultivated on about one million hectares of land throughout the Globe for the commercial use. Triticale is an artificial allopolyploid which has been derived by crossing wheat

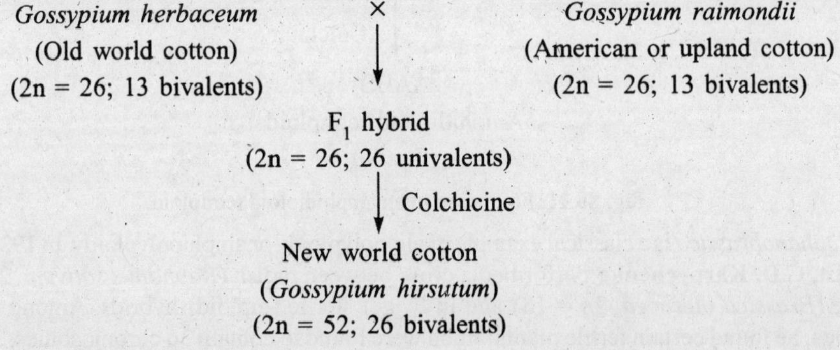

Fig. 86.13. Artificial synthesis of new world cotton.

(*Triticum*) and rye (*Secale*). Depending upon whether *Triticum* is a tetraploid (2n = 4*x* = 28) or hexaploid (2n = 6*x* = 42), one would get hexaploid triticale (2n = 6*x* = 42; Fig. 86.14) or octaploid triticale (2n = 8*x* = 56; Fig. 86.15), respectively. In each case, only diploid rye (2n = 4*x* = 14) was used.

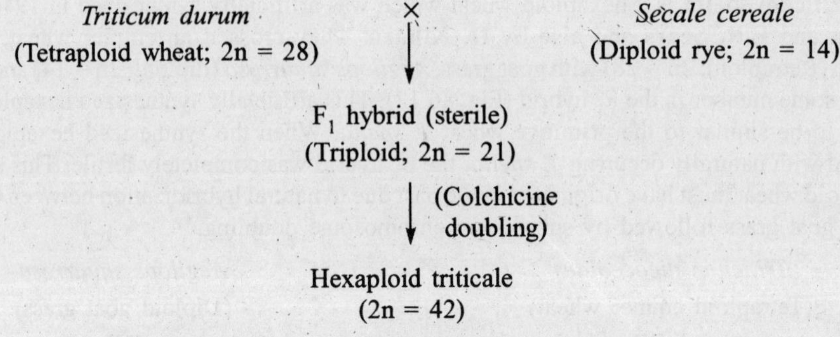

Fig. 86.14. Artificial synthesis of a hexaploid triticale.

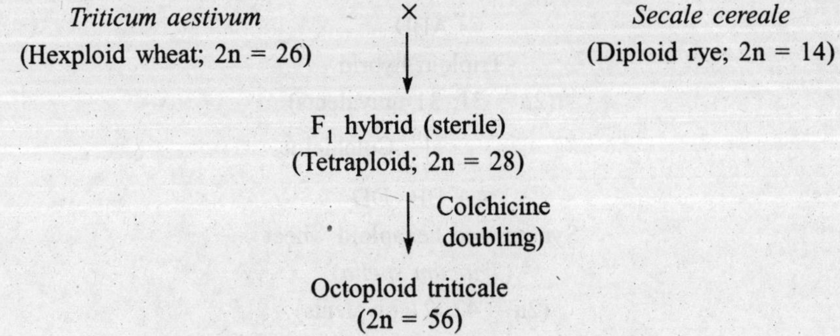

Fig. 86.15. Artificial synthesis of a octoploid triticale.

(*c*) **Segmental allopolyploids.** Different genomes of some allopolyploids are not quite different from each other. Consequently in these polyploids chromosomes belonging to different genomes do pair together to some extent. This indicates that segments of chromosomes and not the whole chromosomes are homologous. Therefore, such allopolyploids are called **segmental allopolyploids** (**Stebbins**, 1943, 1950). The segmental allopolyploids are intermediate between autopolyploids and allopolyploids and can be identified by their peculiar meiotic behaviour.

It is generally believed that most naturally occurring polyploids are segmental allopolyploids. Our common hexaploid bread wheat too is found to be a segmental hexaploid.

Box 86.2. Green Revolution in India

1. One of the most significant development of modern agriculture has been the production of **semi-dwarf** varieties of cereal plants (such as wheat, rice, maize, etc.).
2. **N.E. Borlaugh** was an agriculture scientist who was awarded the 1970 Nobel Peace Prize for his part in the development of improved varieties of crereal crops, important in green revolution of world.
3. **Borlaugh** and his assocaites at CIMMYT (International Centre for Wheat and Maize Improvement, Mexico) developed semi-dwarf wheat using *Norin-10* (a Japanese variety) as the source of dwarfing genes (see **M.U. Charaya**, 2008).
4. In 1963, ICAR (Indian Council for Agricultural Research) introduced many dwarf selections from CIMMYT. **Kalan Sona** and **Sonalika** were selected from these materials. For many years, these were most popular varieties of wheat in India.
5. A great majority of wheat varieties cultivated in India are semi-dwarf as these are lodging-resistant, fertilizer-responsive and high-yielding.
6. **Dr. M.S. Swaminathan** (b., 7th August 1925) is pioneer in mutation breeding (*i.e.*, induced mutation breeding) of cereal in India. He is called **"Founder of Radiation Genetics"** also in addition to **"Father of Green Revolution in India"**. While working at IARI, N. Delhi, during 1954–1972, Dr. Swaminathan developed a new variety of wheat called **Sharbati Sonora**.
7. Sharbati Sonora was produced from Sonora 64 through a mutation breeding using gamma rays.

Polyploidy in animals. Polyploidy is rare in animals but occur in flatworms, leeches and brine shrimp. In mice, also, 40 per cent liver cells are tetraploids, and about 5 per cent are octoploids. Polyploidy in humans have been found in liver cells and cancer cells. In them polyploidy is either complete or mosaic and it leads to gross abnormalities and death.

Phenotypic Effects of Polyploidy

The increase in the genome's size beyond the diploid level is often caused following detectable phenotypic characteristics in a polyploid organism:

(*i*) **Morphological effect of polyploidy.** The polyploidy is invariably related with, **gigantism**. The polyploid plants have been found to contain large-sized pollen grains, cells, leaves, stomata, xylem, etc. The polyploid plants are more vigorous than diploids.

(*ii*) **Physiological effect of polyploidy.** The ascorbic acid content has been reported to be higher in tetraploid cabbages and tomatoes than in corresponding diploids. Likewise corn meal of a tetraploid maize seed contain 40 per cent more vitamin A than corn meal from a diploid plant.

(*iii*) **Effect on fertility of polyploidy.** The most important effect of polyploidy is that it reduces the fertility of polyploid plants in variable degrees.

(*iv*) **Evolution through polyploidy.** Interspecific hybridization combined with polyploidy offers a mechanism whereby new species may arise suddenly in natural populations.

B. ANEUPLOIDY

Changes that involve parts of a chromosome set results in individuals, called **aneuploids** (Gr. *aneu* = uneven; *ploid* = unit). Aneuploidy can be either due to the loss of one or more chromosomes (**hypoploidy**) or due to addition of one or more chromosomes to the complete chromosome set (**hyperploidy**). Hypoploidy is mainly due to the substraction (or, loss) of a single chromosome, called **monosomy** (2*n*–l) or due to the loss of one pair of chromosome called **nullisomy** (2*n*–2; two, lost chromosomes are **homologs**). Likewise, hyperploidy may involve addition of either a single chromosome, called **trisomy** (2*n* + 1) or a pair of chromosomes, called **tetrasomy** (2*n* + 2). In the monoploid organisms, addition of single chromosome produces **disomy** (*n* + 1). All of these aneuploids are probably produced by non-disjunction during mitosis or meiosis.

1. Monosomy

Diploid organisms which are missing one chromosome of a single pair are monosomic with the genomic formula 2*n*–1. A monosomic individual forms gametes of two types, (*n*) and (*n*–1). The *n*–1 gametes do not survive in plants, but, in animals that may cause genetic imbalance, which is manifested by high mortality or reduced fertility of resulted organism.

Monosomy in diploids is not tolerated, since it creates imbalance due to loss of one complete chromosome. However, in polyploids, monosomy has no apparent effect, since they have several chromosomes of same type and loss of one chromosome can be easily tolerated. The number of possible monosomics in an organism will be equal to the haploid chromosome number. For example, in common wheat, since 21 pairs of chromosomes are present, 21 possible monosomics are known. These 21 monosomics in wheat were produced by **E.R. Sears** in variety cal-led "**chinese spring**" and being used for genetic studies all over the world. Monosomics were also reported in cotton (2*n* = 52) by **J. Endrizzi** and his coworkers and in tobacco

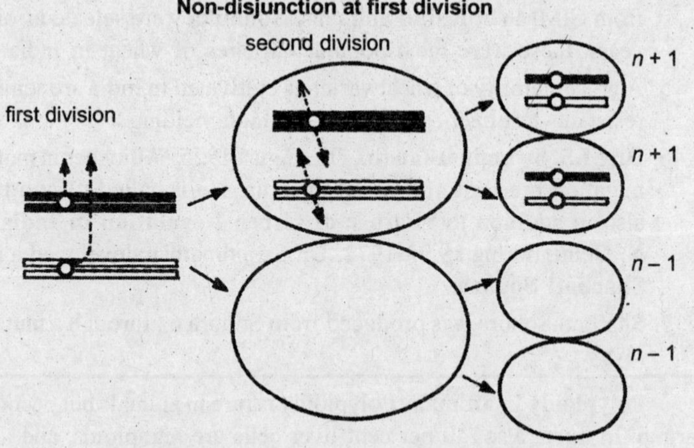

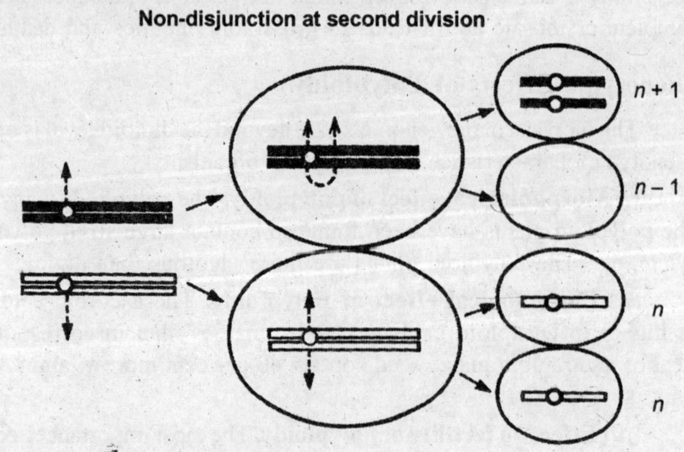

Fig. 86.16. The mode of origin of aneuploid gametes by non-disjunction at either the first or second meiotic division.

($2n$ = 48) by **E.R. Clausen** and **D.R. Cameron**. Monsomics have also been produced in maize and tomato ($2n$ = 24) despite their being diploids. Double mono-somics ($2n$–1–1) or triple monosomics ($2n$–1–1–1) could also be produced in polyploids such as wheat. Double monoso-mic means that the chro-mosome number is $2n$–2 like that in a nullisomic, but the missing chromo-somes are non-homologous. The same explana-tion is applied to the triple monosomics.

2. Nullisomy

An organism which has lost a chromosome pair is a nullisomic. The nullisomic organism has the genomic formula ($2n$–2). A nullosomic diploid often does not survive, however, a nullosomic polyploid (*e.g.*, hexaploid wheat, $6x$–2) may survive but exhibit reduced vigour and fertility.

3. Trisomy

Trisomics are those diploid organisms which have an extra chromosome ($2n$ + 1). Since the extra chromosome may belong to anyone of different chromosomes of a haploid complement, the number of possible trisomics will be equal to the haploid chromosome number. For example, haploid chromosome number of barley is 7, consequently in it seven trisomics are possible. Further, when the extra chromosome is identical to its homologs, such a trisomic is called **primary trisomic**. There are also secondary and tertiary trisomics. While the **secondary trisomic** means that the extra chromosome should be an isochromosome (*i.e.*, both chromosome arms genetically similar), a **tertiary trisomic** would mean that the extra chromosome should be the product of translocation (Fig. 86.17). Trisomics were obtained for the first time in jimson weed (*Datura stramonium*) by **A.F. Blakeslee** and **J. Belling** (1924). Since the haploid chromo-some number of this species is n = 12, so here, 12 primary trisomics, 24 secondary trisomics and a large number of tertiary trisomics are possible. Most of the trisomics were identified by the size, shape and other morphological features of the fruit of jimson weed. In barley, such a trisomic series is produced and extensively studied by **T. Tsuchiya.**

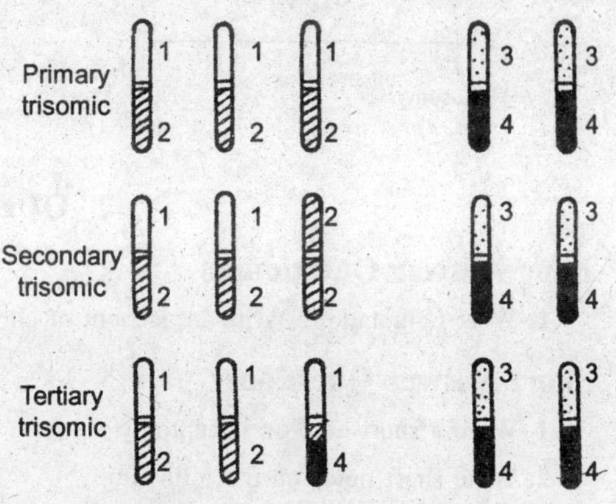

Fig. 86.17. Three kinds of trisomics.

Trisomy in humans. In human beings, the following three syndromes due to trisomy have been studied:

A. **Down's syndrome (DS) or trisomy-21**

B. **Edward's syndrome or trisomy-18**

C. **Patau's syndrome or trisomy-13**

(For details of these syndromes refer Chapter 87)

Trisomy in non-humans. Trisomy-22 has been reported in chimpanzees; this shows Down syndrome-like phenotypic features. **Trisomy-21** has been reported in the gorilla.

Cytology of trisomics. The trisomics have an extra chromosome which is homologous to one chromosome of the diploid complement. Therefore, it forms **trivalent** which may take a variety of shapes in primary and secondary trisomics. In a tertiary trisomic a characteristic pentavalent is observed.

4. Double Trisomy

In a diploid organism when two different chromosomes are represented in triplicate, the double trisomic is resulted. The double trisomic causes great genetic imbalance and has the genomic formula $2n + 1 + 1$.

5. Tetrasomy

The diploid organisms having two extra chromosomes are known as tetrasomic. They have the genomic formula $2n + 2$. All the 21 possible tetrasomics are available in wheat.

Different types of ploidies can be summarized in the following chart:

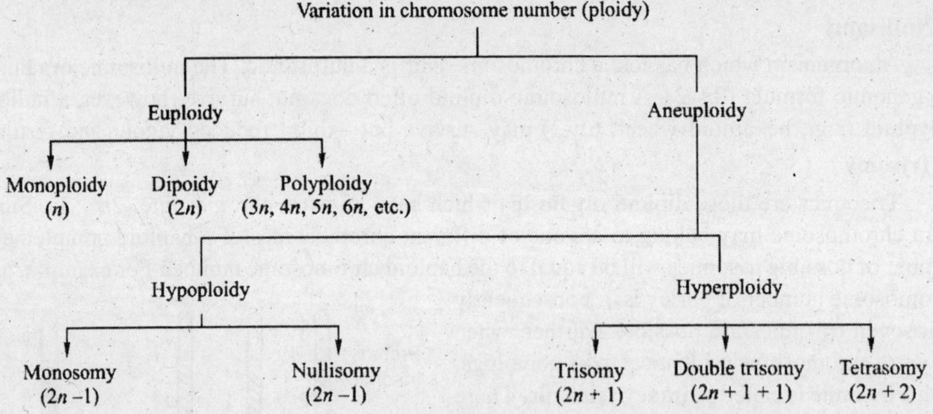

QUESTIONS

Long Answer Questions

1. What is mutation? Write an account of chromosomal mutations. (*Guwahati (G) 1996*)

Short Answer Questions

1. Write a short note on aneuploidy. (*CCS 1997*)
2. Write short notes on the following:
   (*i*) Polyploidy
   (*ii*) Trisomy
   (*iii*) Translocation.
3. Give an account of 'Frame-shift Mutation'. (*C.C.S.U. Meerut 2007*)
4. Describe the chromosomal aberration. What is their significance in organic evolution? (*C.C.C.S.U. Meerut 2008*)

Very Short Answer Questions

1. Define mutation. (*H.N.B. Garhwal 2009*)

   Ans. A change in the DNA at a particular locus in an organism. The term is used loosely to include point mutations involving a single gene change as well as a chromosomal change; the end result of this process.

Multiple Choice Questions

1. Polyploidy leads to rapid formation of new species through
   (*a*) genetic recombination (*b*) isolation
   (*c*) mutation (*c*) development of multiple sets of chromosomes
2. A syndrome in human in which an individuals somatic cells contain XXX set chromosomes is called (*M.S.U. Tamil Nadu 2006*)
   (*a*) Klinefelter's (*b*) Turner's
   (*c*) down's (*d*) superfemale

3. Genetic composition of Turner's syndrome is.....? (*H.N.B. Garhwal University 2008*)
   (*a*) 44 + XXY (*b*) 44 + XO
   (*c*) 44 + XXX (*d*) 44 + XYY
4. The bread wheat *Triticum aestivum* is a
   (*a*) hexaploid (*b*) diploid
   (*c*) haploid (*d*) heteropolyploid
5. The chromosome doubling in making polyploid plants is carried out by using
   (*a*) phosphoethylene glycol (PEG) (*b*) colchicine treatment
   (*c*) ethyl-methyl-sulphonate (EMS) (*d*) nicotinamide acedic acid (NAA)
6. The loss of one single chromosome creates a condition called
   (*a*) trisomy (*b*) nullisomy
   (*c*) monosomy (*d*) haploid
7. An autotetraploid plant bearing a genotype *AAAa* is called as
   (*a*) monosomic (*b*) trisomic
   (*c*) triploid (*d*) triplex
8. *Rhaphanobrassica* is a/an
   (*a*) autopolyploid (*b*) allotetraploid
   (*c*) hexaploid (*d*) none of these
9. Polyploidy means
   (*a*) occurrence of haploid set of chromosomes
   (*b*) occurrence of diploid st of chromosomes
   (*c*) occurrence of three or more sets of chromosomes
   (*d*) all of these
10. A polyploid derived from F_1 hybrid between two species is
   (*a*) amphidiploid (*b*) hexaploid
   (*c*) autodiploid (*d*) autopolyploid
11. Down's syndrome (DS) is the result of
   (*a*) an extra X chromosome (*b*) an extra Y chromosome
   (*c*) an additional twenty-first chromosome (*d*) an extra eighteenth chromosome
12. Colchicine interferes in
   (*a*) chromosome condensation
   (*b*) DNA replication
   (*c*) organisation and orientation of spindle elements
   (*d*) none of these
13. Who among the following merged cytology and genetics into cytogenetics?
   (*a*) H.J. Muller (*b*) O.T. Avery
   (*c*) G.W. Beadle (*d*) T.H. Morgan
14. Which is the human made cereal
   (*a*) wheat (*b*) rice
   (*c*) *Triticale* (*d*) barley
15. *Triticale* is obtained by crossing wheat with
   (*a*) oat (*b*) barley
   (*c*) maize (*d*) rye

16. Father of green revolution in India is
(*a*) T.S. Venkataraman (*b*) Boshi Sen
(*c*) K. Ramiah (*d*) M.S. Swaminathan

17. *Norin-10* is gene for
(*a*) dwarfness in wheat (*b*) dwarfness in rice
(*c*) flavour in rice (*d*) high yield in rice

18. Dr. M.S. Swaminathan could produce an amber-seeded high yielding variety called "Sharbati Sonora" from Mexican dwarf variety by using
(*a*) chemical mutagens (*b*) radiations of UV and gamma rays
(*c*) cross breeding techniques (*d*) tissue culture

19. The Indian dwarf wheat is
(*a*) *Triticum dicoccum* (*b*) *Triticum turgidum*
(*c*) *Triticum sphaerococcum* (*d*) *Triticum aestivum*

20. Most of our crop plants are of
(*a*) allopolyploidal origin (*b*) autopolyploidal origin
(*c*) mixed genotypic origin (*d*) structurally heterozygous origin

ANSWERS

Multiple Choice Questions

1. (*d*)	**2.** (*d*)	**3.** (*b*)	**4.** (*a*)
5. (*b*)	**6.** (*c*)	**7.** (*d*)	**8.** (*b*)
9. (*c*)	**10.** (*a*)	**11.** (*c*)	**12.** (*c*)
13. (*a*)	**14.** (*c*)	**15.** (*d*)	**16.** (*d*)
17. (*b*)	**18.** (*b*)	**19.** (*b*)	**20.** (*a*).

87

Human Genetics

Are humans not suitable subjects for genetic studies?

For the study of inheritance in any species geneticists prefer (1) to have **isogenic strains,** standard stocks whose members are genetically identical, (2) to mate members of different isogenic strains or inbred lines and (3) to raise the offspring under carefully controlled conditions. Judged by these criteria, a human being is not a very favourable subject for studies of inheritance, because (*i*) members of the human race are genetically diverse, *viz.*, they are heterozygous for many genes and there are wide variations in their physical, biological and social environments; (*ii*) in him controlled matings under standardized environment are impossible and non-ethical; (*iii*) he has small individual progenies, so unfavourable to the use of certain standard research techniques of the genetics, and (*iv*) he has long time between successive generations. Despite these difficulties, a great deal has been learned about human inheritance and the field is progressing rapidly. To study the genetics of various human traits several techniques such as pedigree analysis, amniocentesis (and cell culture), CVS, twin's study, etc., have been utilized.

87.1 PEDIGREE ANALYSIS

Since, in human beings controlled crosses cannot be made, so human geneticists have to resort to a scrutiny of established matings in the hope that informative matings have been made by chance. The scrutiny of established matings is called **pedigree analysis**. A member of a family who first comes to the attention of a geneticist is called the **propositus**. Usually the phenotype of the propositus is exceptional in some way—for example, a dwarf. The investigator then traces the history of the character shown to be interesting in the propositus back through the history of family, and a family tree or **pedigree** or **pedigree chart** is drawn up using certain standard symbols (see Fig. 87.1).

A family pedigree chart conventionally has circles for female individuals and squares for male individuals. A marriage is indicated by a horizontal bar connecting a circle and square and the symbols for offspring are shown suspended from a line drawn perpendicular to the marriage bar. Individuals on the same line are from the same generation, and individuals in each generation are numbered in sequence (such as II 1, II 2, II 3, and so on). The unshaded circles or squares designate people who are normal for the character being studied, while solid black squares or circles depict "affected" individual. Heterozygotes customarily are designated by colouring half of the symbol blacks (Fig. 87.1). Carrier of a sex-linked recessive gene is designated by a black dot in the middle of the symbol. Occasionally, an arrow pointing at a particular affected individual is added to indicate that this is the person who brought the trait to the geneticists attention.

Example. A case of polydactyly, the occurrence of extra fingers, in human beings can be considered as an example for pedigree analysis. A family pedigree for polydactyly has been illustrated in Fig. 87.2.

Here, we have a marriage between a polydactylous man and a normal woman (generation I). They reproduce three children, a polydactylous daughter, a polydactylous son, and a normal son (generation II). The first and third individuals of II generation, each marry normal (pentadactylous) persons, their children are shown in generation III. From the analysis of pedigree of polydactyly, one fact becomes clear that polydactylous trait is dominant on the normal pentadactylous trait which is recessive.

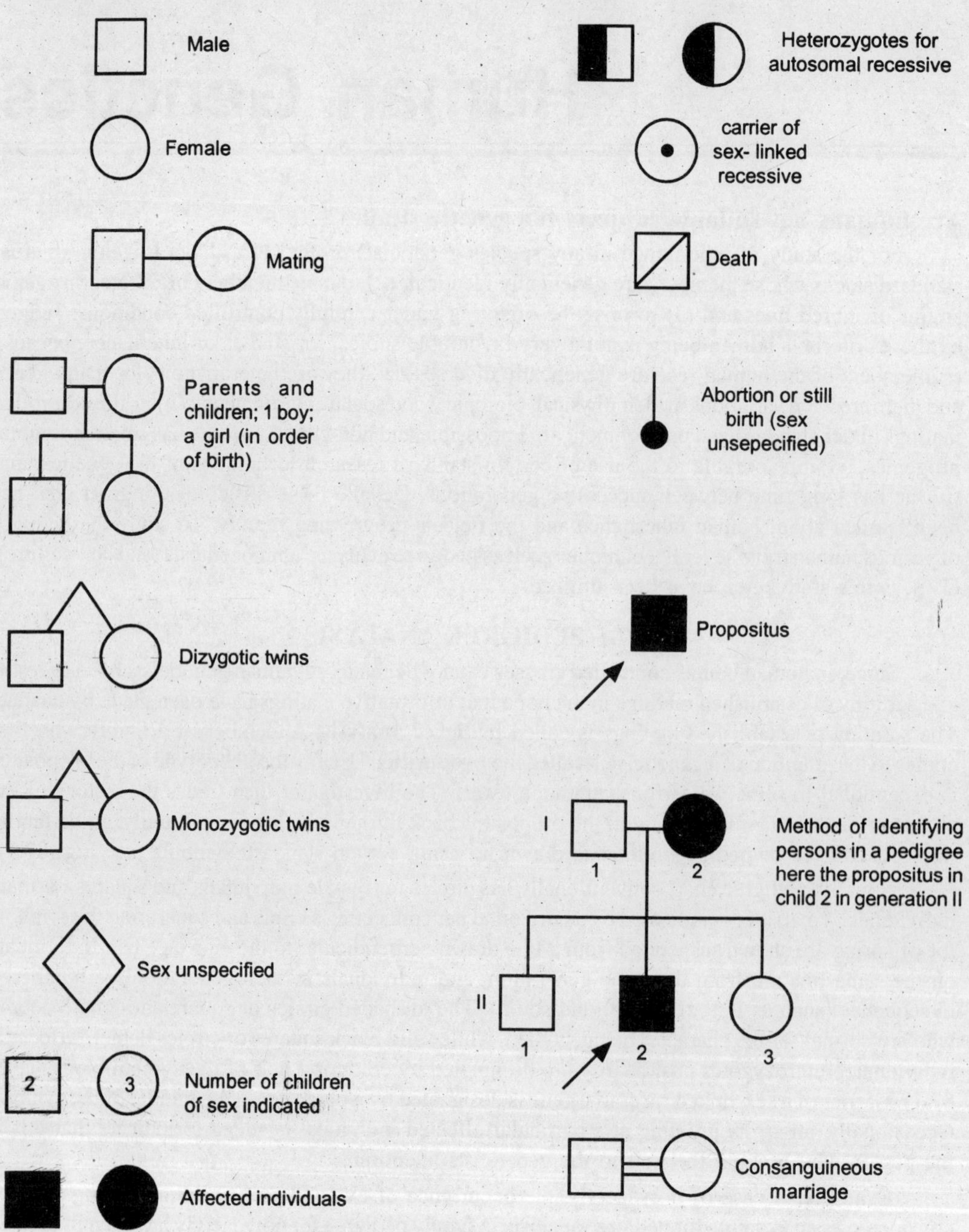

Fig. 87.1. Symbols commonly used in pedigree charts.

How pedigree charts are read? Many human diseases and other exceptional conditions are determined by simple mendelian recessive alleles. There are certain clues in the pedigree that must be sought. Characteristically the diseased condition appears in progeny of unaffected parents. Further more, two affected individuals cannot have an unaffected child. Quite often such recessive alleles are revealed by consanguineous matings-for example, cousin marriages. This is particularly true of rare conditions where chance matings of heterozygotes are expected to be extremely rare, *e.g.*, albinism. Tay-Sachs disease, cystic fibrosis and phenylketonuria (PKU).

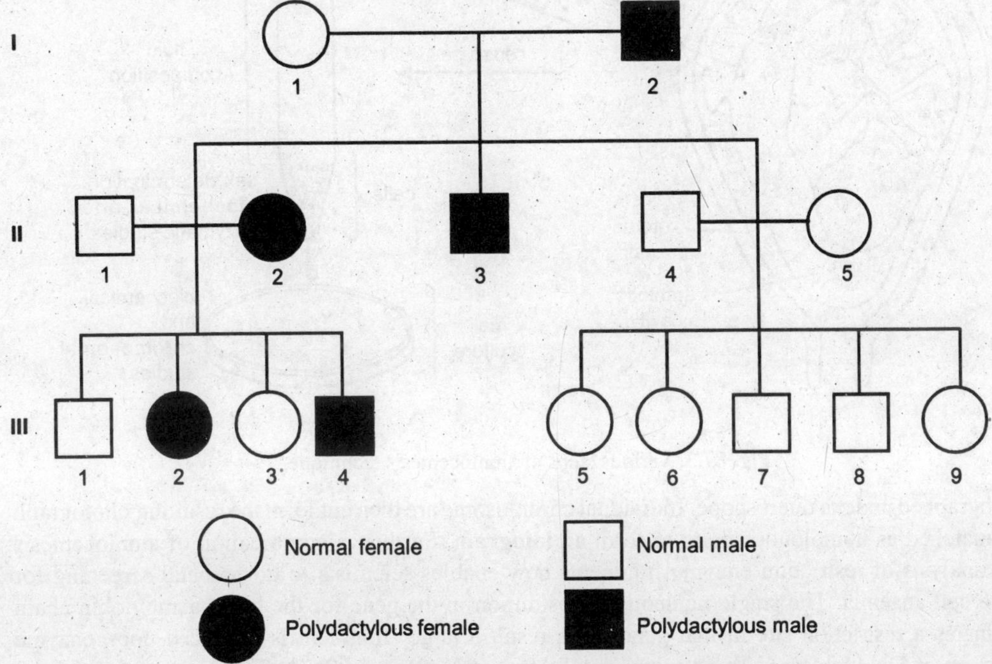

Fig. 87.2. A family pedigree for polydactyly in human beings.

There are also examples of exceptional conditions caused by dominant alleles. Once again, there are some simple rules to follow to ascertain from pedigrees a condition caused by a dominant allele: the condition typically occurs in every generation; unaffected individuals never transmit the condition to their offspring; two affected parents may have unaffected children; and the condition is passed, on average, to one-half of the children of an affected individual. Some examples of abnormal conditions caused by dominant alleles in humans are achondroplasia (a kind of dwarfism), Huntington's corea and brachydactyly (very short fingers).

2. Amniocentesis

Prenatal (=occurring before birth) screening of babies for gross chromosomal aberrations (such as polyploidy, aneuploidy, deletions, translocations, etc.), as well as, sex prediction, is now possible by the technique of amniocentesis.

Amniocentesis is usually performed from the fifteenth to the seventeenth week of pregnancy. Amniocentesis involves **removing** amniotic fluid by the help of a hypodermic needle from the amnion ("the bag of water") surrounding the foetus in uterus; **culturing** the foetal cells contained within the fluid; and **screening** the cultured cells for karyotype, enzyme production and restriction site pattern analysis of its DNA (Fig. 87.3). At least 35 genetic diseases have been detected in human beings by this technique; some of them have been listed in Table 87.1. If a disease is detected by this method, abortion of such a foetus can then be recommended.

For **karyotype** preparation, foetal cells found in amniotic fluid can be cultured *in vitro* in a highly nutritive solution, treated with colchicine to stop mitosis at metaphase, subjected to a hypotonic solution to cause the cells to swell and scatter the chromosome, placed on a slide, stained and photo-

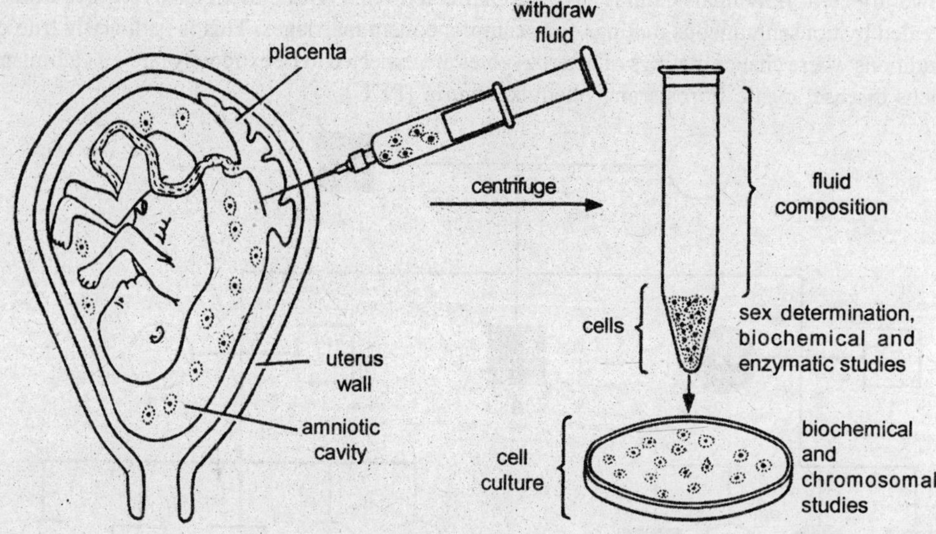

Fig. 87.3. Various steps of amniocentesis technique.

photographed under a microscope. Individual chromosome are then cut form the resulting photograph and matched as homologous pairs to form an **idiogram.** Further, a combination of amniocentesis with analysis of restriction enzyme fragments now enables scientists to do prenatal screening for sickle cell anaemia. The single nucleotide substitution in the gene for the beta haemoglobin chain eliminates a restriction site in that gene. As a result a larger fragment is produced upon enzyme digestion of DNA obtained from cultured cells from the amniotic fluid. This diagnosis could not be made until several months after birth, as the gene is inactive before that time

Table 87.1 Some common genetic diseases of humans which can be detected by amniocentesis.

Disease	*Possible cause*
I. Inbron errors of metabolism	
1. Cystic fibrosis	Autosomal recessive gene on chromosome 7
2. Duchenne muscular dystrophy	X-linked recessive
3. Gaucher's disease	Defective glucocerebrosidase enzyme
4. Tay-Sachs disease	Defective hexosaminidase A
5. Essential pentosauria	A benign condition
6. Classic haemophilia	Defective clotting factor VIII
7. Phenylketonuria	Defective phenylalanine hydroxylase enzyme
8. Cystinuria	Mutated gene unknown
9. Metachromatic leuko dystrophy	Defective arylsulphatase A
10. Galactosemia	Defective galactose 1-phosphate uridyl transferase
II. Haemoglobinopathies	
11. Sickle cell anaemia	Defective β-globin chain
12. β-thalassemia	Defective β-globin chain

3. Chorionic Villi Sampling (CVS).

CVS is usually performed from the fifth to the twelfth week of pregnancy. The doctor inserts a long, thin tube through the vagina into the uterus. With the help of ultrasound, which gives a picture of the uterus contents, the tube is placed between the lining of the uterus and the chorion (an extra-embryonic membrane, forming a part of the placenta). Then a sampling of the chorionic villi cells is obtained by suction (Fig. 87.4). Chromosome analysis and biochemical tests for genetic defects can be done immediately on these cells (see **Mader,** 1998).

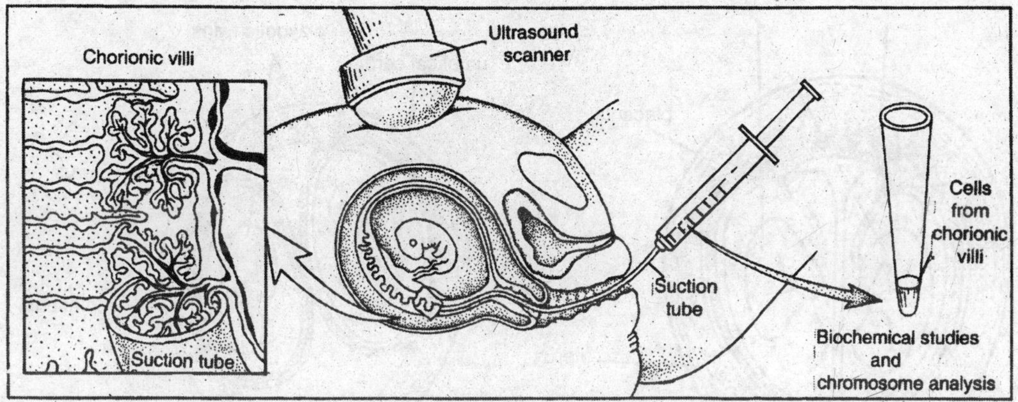

Fig. 87.4. Technique of chorionic villi sampling (CVS).

4. Twin Studies

Normally, a woman gives birth to only one young at a time, but sometimes more than one child are born to a woman at the same time. The delivery of more than one baby by a mother is called **multiple births**. Most commonly in the case of multiple births, the number of births is two and the individuals which are born in such a way are called **twins**. In case of multiple births, the number of infants may be three (**triplets**), four (**quadruplets**), five (**quintuplets**) or even more (*e.g.*, eleven). Twins are of two basic types:

1. Identical or monozygotic twins (MZ twins). Twins having no variability in their traits are called **identical twins**. They are produced from a single fertilized ovum (Fig. 87.5). Sometimes, the two blastomeres resulting from the first cleavage (mitosis) of the fertilized egg or zygote separate from each other and develop into independent embryos (hence they are also called **monozygotic twins**). Since, the two embryos have arisen by mitosis, they have the same chromosome sets in their body cells. Thus, identical twins possess identical genes and they are like one person walking around in two bodies. These twins belong to the same sex. The situation is the same for identical triplets and other multiplets, but they are quite rare in human beings.

Sometimes, identical twins fail to separate completely from each other, they are called **siamese twins.**

2. Fraternal or dizygotic twins (DZ twins). Non-identical or fraternal twins are produced by simultaneous fertilization of two separate ova by two separate sperms. Two foetuses so obtained from two zygotes are called **dizygotic twins** (Fig. 87.6). They contain different sets of genes and are like any other brothers and sisters, but just happen to grow simultaneously in the same uterus. So fraternal twins differ from each other in their traits, development and often even sex.

Twins are useful in detecting the relative effectiveness of heredity and environment upon the expression of a disease or trait. The hereditary trait is most likely shown by both members of the identical twins. If it is shown by one member of the identical twins, then the environmental factor is playing the major role since both possess the same genes. Very recently, studies of identical twins separated since birth revealed that genes (heredity) play a much greater role in determining behaviour than previously thought. Although these twins were not raised together, but even then they exhibited many similar behavioural traits. In any behavioural characteristic studied genes are found to exert at least 50 per cent influence on the trait. The conclusion is that genes play a major role in shaping almost any type of behaviour including **alcoholism, criminality, intelligence (I.Q.), political attitudes, schizophrenia** and **sociability.**

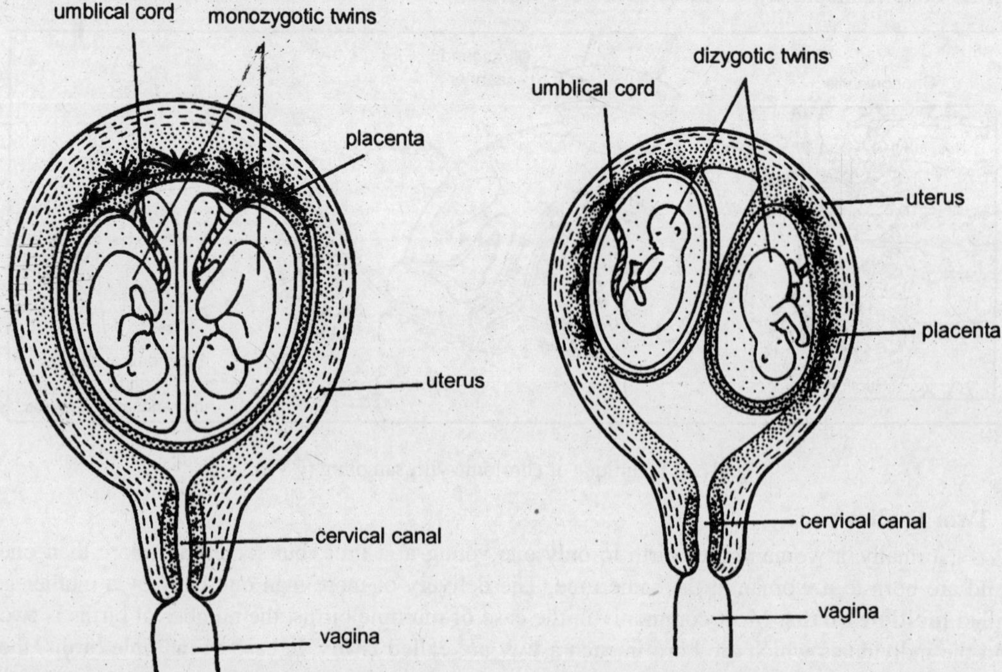

Fig. 87.5 Developing foetuses of monozygotic twins.

Fig. 87.6 Developing foetuses of dizygotic twins.

Genetic Significance of Study of Twins

Twin study has helped in understanding the effect of heredity and environment on the development of complex characters such as intelligence, talent, longevity, general health and personality and also behavioral characters. The idea for twin study in human genetics was introduced by **Galton. Muller, Freeman** and **Holzinger** reared the two members of monozygotic twins separately under different environmental and social conditions. These MZ twins showed variation in their intelligence test. It shows that environment plays an important role in the development of intelligence. However, the MZ twins reared under the same environmental and social conditions show almost no difference in their intelligence. This proves that inheritance of intelligence has a genetic basis on which environment plays its role.

Table 87.2 Comparison of monozygotic and dizygotic twins.

Feature	*Monozygotic twins*	*Dizygotic twins*
1. Development	Result from a single fertilized ovum	Result from two fertilized ova
2. Heredity	They have the same heredity in similar environments	They have different heredity in similar environments
3. Reaction to grafts	Skin grafts between the monozygotic twins are not rejected	Skin grafts between the dizygotic twins are usually rejected

Concordance and discordance

A. Concordance. In a pair of twins, if a character is either present or is absent in both members, they are called **concordants**. Their phenotype will be +/+ or –/–. Identical twins are always concordants.

B. Discordance. In a pair of twins, if a character is present in one but is absent in the other, the twins are called discordants (+/–). The fraternal twins may be discordant or concordant with +/– or +/+ or –/–.

A number of human traits such as diabetes mellitus, feeble mindedness, tumours, etc., have been studied with respect to discordance or concordance in identical and fraternal twins.

87.2 HUMAN TRAITS

Heredity affects a variety of human traits. They may involve physical appearance such as height, facial features, eye colour, skin colour, baldness, hair colour and form. Heredity also determines

Table 87.3 Some common phenotypic traits of humans.

Dominant	*Recessive*
1. Curly hair	Straight hair
2. Dark brown hair	All other colours
3. Coarse body hair	Fine body hair
4. Pattern baldness (dominant in males)	Baldness (recessive in females)
5. Normal skin pigmentation	Albinism
6. Brown eyes	Blue or grey eyes
7. Near or far sightedness	Normal vision
8. Normal hearing	Deafness
9. Normal colour vision	Colour blindness
10. Broad lips	Thin lips
11. Large eyes	Small eyes
12. Polydactylism (extra digits)	Normal digits
13. Brachydactylism (short digits)	Normal digits
14. Syndactylism (webbed digits)	Normal digits
15. Hypertension	Normal blood pressure
16. Diabetes insipidus	Normal excretion
17. Huntigton's chorea	Normal nervous system
18. Normal mentality	Schizophrenia
19. Migraine headache	Normal
20. Normal resistance to disease	Susceptibility to disease
21. Enlarged spleen	Normal spleen
22. Enlarged colon	Normal colon
23. A or B blood group (antigen)	O blood group
24. Rh blood factor	No Rh blood factor

biochemical conditions such as genetic diseases, *e.g.*, brachydactyly, Huntington's chorea, porphyria, phenylketonuria, sickle-cell anaemia, and diabetes. According to **McKusick** (1987), in human beings there are approximately 100,000 genes, however, definite positions of few genes on the chromosomes have been worked out. Thus, 1300 genes are found to be located on the autosomes, 107 on X-chromosomes and only one on the Y-chromosome. More than 500 human genetic diseases are known to be caused by **single-gene defects**. Certain well known dominant and recessive human genetic traits have been enlisted in Table 87.3.

87.3 CERTAIN GENETIC DISORDERS OF HUMAN BEINGS

1. Recessive Inherited Disorders

(*i*) Albinism. Albino (L., *albus* = white) humans are devoid of melanin pigment in the skin, hair and eyes. The skin is very light and hair whitish yellow and eyes appear pinkish. Albinos have poor vision, are sensitive to sun light and prone to skin cancer. Albinism is rare in its occurrence *i.e.*, one out of 20,000 individuals are albino.

Albinism is caused by a recessive, mutant allele in homozygous condition (*aa*).

(*ii*) Inborn errors of metabolism. In 1908, an English physician, **Archibald E. Garrod** published a book called *Inborn Errors of Metabolism*. He proposed that the genes produce enzymes whereas the mutant forms of these genes do not produce them, hence, resulting in physiological abnormalities. He studied the metabolism of phenylalanine (an essential amino acid) and showed that the metabolism of this amino acid proceeds in chains of enzyme-mediated reactions and a change or absence of an enzyme results in an abnormality. There are three important diseases associated with metabolic breakdown of the phenylalanine:

(*i*) Phenylketonuria (PKU). Persons having autosomal homozygous recessive alleles *pp* fail to produce an enzyme called **phenylalanine hydroxylase.** This enzyme is required by the body to convert phenylalanine into tyrosine. As a result, the concentration of phenylalanine increases in blood. This amino acid is then partly converted into phenylpyruvate, phenylacetate, phenyllactate and other derivatives, all of which (including phenylalanine) are excreted in the urine. The patients of PKU have light body pigmentation and are physically and mentally retarded. They are called **phenyl pyruvate idiots** having mean IQ of 65. The idiocy of PKU is thought to be due to damage of the brain tissues by the phenylpyruvate in the cerebrospinal fluid. Here, the abnormal gene is suppressed in heterozygous individual, it is called **recessive inheritance.**

(*ii*) Alkaptonuria. The persons with autosomal homozygous recessive genes *hh* fail to produce the enzyme **homogentisic acid oxidase** which catalyzes the oxidation of alkapton (*i.e.*, 2–5–dihydrophenylacetic acid or homogentisic acid). Therefore, in them, normal oxidation of alkapton into acetoacetic acid and ultimately into water and carbon dioxide does not take place and large amount of alkapton is accumulated in the blood and is excreted in the urine, which turns black upon exposure to the air. In such alkaptonuric persons, the disease is manifested by the darkenings of cartilaginous regions (such as ear pinna) and a proneness to arthritis (a condition called **ochronosis**).

(*iii*) Albinism. The persons with recessive *aa* genes do not produce the **tyrosinase** enzyme which is needed by melanocytes for converting DOPA (3, 4- dihydroxy- phenyl- alanine) into melanin (a dark brown pigment). As a result in an albino patient melanocytes are present in normal number in their skin, hair, iris of eyes, etc., but lack in melanin pigment. This trait is inherited like a typical Mendelian monohybrid cross.

(*iv*) Tyrosinosis. The recessive gene, *t* in its homozygous condition (*tt*), blocks the conversion of p-hydroxyphenylpyruvate into 2, 5-dihydroxyphenylpyruvate. This leads to the accumulation of tyrosine, excesses of which are excreted via the urine. This condition is called **tyrosinosis.** It is reported in only one human and cause no harmful effect.

(*v*) Goitrous cretinism. The persons with cc genotype fail to produce the enzyme which is required for the conversion of tyrosine into thyroxine, and triiodothyronine ,hormones in their thyroid glands. This condition is called **goitrous cretinism** which is accompanied by a considerable degree of physical and mental retardation and hypertrophy of thyroid gland

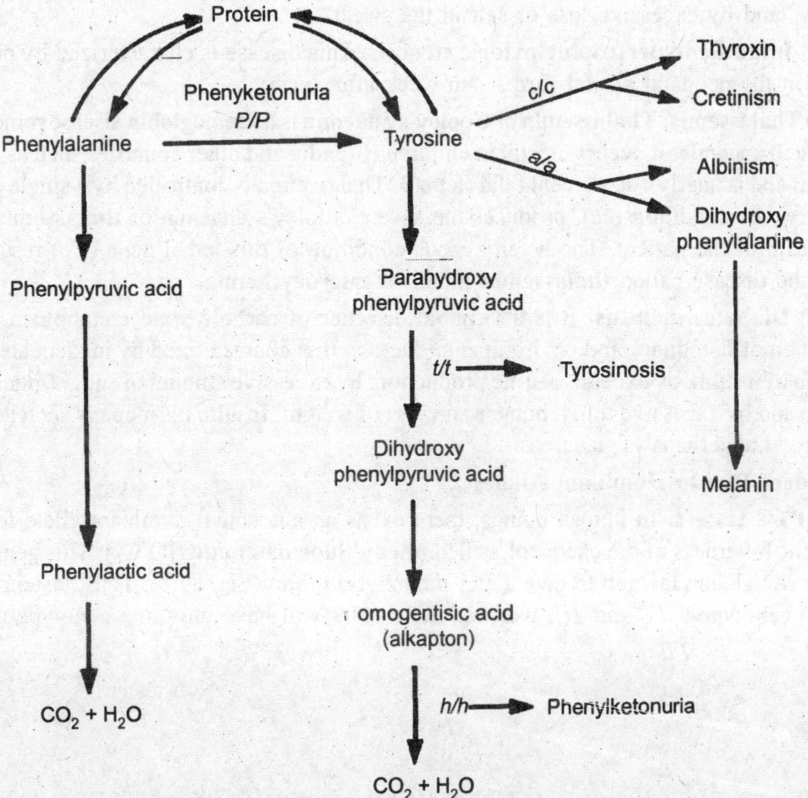

Fig. 87.7. Metabolism of phenylalanine in human body and various inborn errors of metabolism..

Since above discussed abnormal genes influence the human metabolism, these abnormalitie: are called **inborn metabolic disorders.**

***(iii)* Sickle cell anaemia.** Refer Chapter 80

2. Lethal Recessive Inheritance

Sometimes, a recessive gene has such a drastic effect that the infant dies before or after birth. Some common examples of such lethal recessive genes are the following:

***(i)* Congenital ichthyosis.** It is caused by recessive alleles and is characterized by the cracking and separation of skin into large scaly sections. The fissures lead to bleeding, infection and death. Affected infants are either aborted or are born premature.

***(ii)* Infantile amaurotic idiocy (Tay-Sach's disease).** This disease is common in persons of jewish origin. It is caused by homozygous condition of an autosomal recessive allele. Affected children appear normal and healthy at birth but within six months the nerves of the brain and spinal cord show remarkable degeneration. By the age of one year, child lies helplessly in his crib, later loses his sight and finally becomes paralyzed. The death occurs by the age of 3 or 4 years.

Due to deficiency of a lysosomal enzyme of lipid metabolism, the **hexosaminidase A,** lipids accumulate in the nervous tissue causing **macrocephaly** and mental retardation.

***(iii)* Cystic fibrosis.** It is a common disease of caucasians that appears in early childhood. It is caused by recessive alleles and involves functional disorder of the exocrine glands. The pancreas secrete abnormally thick or viscid mucus material (lacking pancreatic enzymes) which collects in intestine and blocks digestion. Patient suffers with difficult breathing due to mucus accumulation in airways and by excessive loss of salt in the sweat.

***(iv)* Infantile hypertrophic pyloric stenosis.** This disease is characterized by obstruction of intestine in the neonatal period (3rd – 7th week after birth).

(v) Thalassemia. Thalassemia or **Cooley's anaemia** is a haemoglobin disease somewhat similar to sickle cell anaemia. It occurs mostly in children (In India and other countries such as Italy, Greece and Syria) and is nearly 100 percent fatal (lethal). Thalassemia is controlled by a single gene *c* which in homozygous condition (*cc*), produces the severe Cooley's anaemia or **thalassemia major** and causes death of the patient. The heterozygous condition of this lethal gene (*Cc*) results in a mild form of the disease called **thalassemia** minor or **microcythemia.**

***(vi)* Diabetes mellitus.** It is a variable disorder of carbohydrate metabolism caused by a combination of hereditary and environmental factors. It is characterized by inadequate secretion or utilization of insulin, by excessive urine production, by excessive amount of sugar (glucose) in blood and urine and by excessive thirst, hunger and loss of weight. Insulin hormone is secreted by β- cells of Islets of Langerhans of pancreas.

3. Disorders Due to Dominant Alleles

1. PTC tasters. In human beings, there exists an autosomal dominant allele for the ability to taste [the bitterness of the chemical, called **phenylthiocarbamide (PTC)**]. This gene is inherited in simple Mendelian fashion to give a 3: 1 monohybrid ratio (Fig. 87.8). Thus, tasters of PTC may have two genotypes (*TT* and *Tt*), whereas non-tasters will have only one genotype (*tt*).

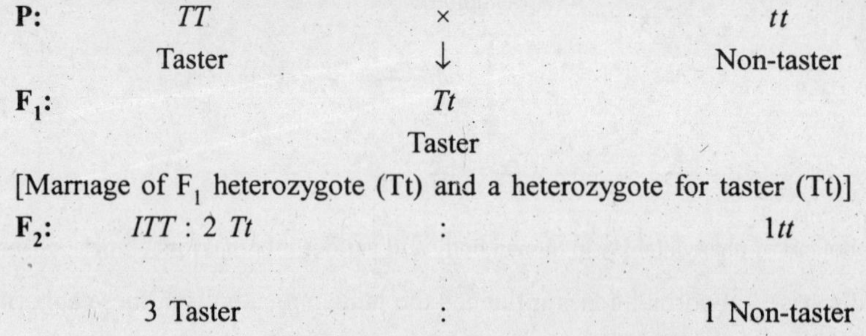

Fig. 87.8. Inheritance pattern of PTC taster trait.

2. Brachydactyly. In this disorder, the middle joints of the digits (phalanges) in hands or feet remain rudimentary and fuse with an adjacent joint (Fig. 87.9). Consequently, the fingers and toes become abnormally short. The defect of brachydactyly is caused by a dominant gene existing on an autosome. When an abnormal gene dominates the normal gene, this is called **dominant inheritance.**

3. Huntington's chorea. It is a fatal disease of human beings. It is characterized by uncontrolled jerking of body (due to involuntary twitching of voluntary muscles) and a progressive degeneration of the central nervous system (*i.e.*, brain and spinal cord) accompanied by gradual mental and physical deterioration. The mean age of onset of these symptoms is between 35 and 40. This disease is caused by an autosomal dominant gene.

4. Tongue rolling. The trait of **ability to roll one's tongue** is also controlled by a dominant allele.

5. Achondroplasia. It is a genetic disorder which is caused by an autosomal dominant allele. It affects normal growth of cartilage resulting in a form of dwarfism. The affected individuals are dwarf, having normal torso but abnormally short arms and legs.

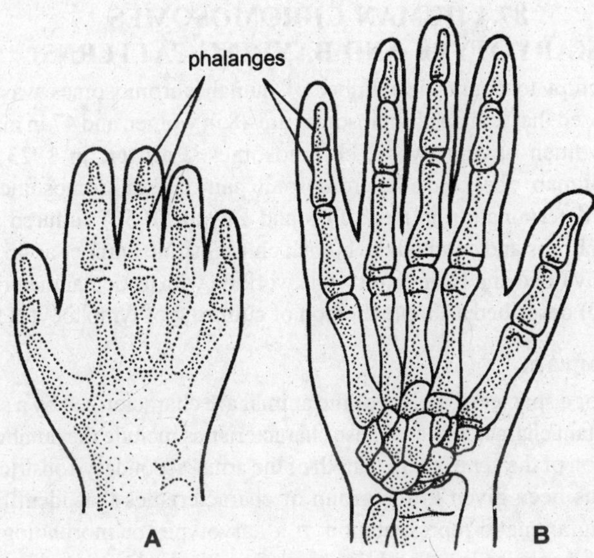

Fig. 87.9. Comparison of a brachydactyl hand and a normal hand. A—A brachydactyl hand showing two phalanges in each finger; B—A normal hand with three phalanges in each finger.

6. Tylosis. The persons with tylosis have excessive thick and hard skin on their palms and soles.

7. Anonychia. This congenital disease is characterized by the absence of some or all nails of the finger and toes.

8. Dentinogenesis imperfecta. It is an disorder of tooth development and is inherited as an autosomal dominant. It is characterized by relatively soft enamel in both the primary and permanent teeth that makes the teeth vulnerable to fracture, abrasion and wear.

9. Polydactyly. Presence of extra fingers or toes.

4. Disorders due to sex-linked genes.

***(a)* Fabry syndrome.** This syndrome is a lysosomal storage disease; it is a sex-linked inherited disorder. The affected persons lack an enzyme, **α-galactosidase-A,** needed for break down of lipids in the membranes of most cells of body (but mainly in the kidney and lining a blood vessels). The disease causes intense burning pain in the fingers and toes. A large number of tiny dark red skin lesions (pimple like swellings of the capillaries) develop along the lips, thighs and genital regions (*i.e.*, a rash in the inguinal, scrotal and umbilical regions) and corneal defects. The most serious symptoms is the accumulation of lipids is the kidney causing renal dysfunction. Lipids also accumulate in the heart and blood vessels and thus cause heart attacks. Death occurs between the age to 40 to 50 years. However, kidney transplantation can prolong life. This syndrome was discovered in 1898 by German dermatologist **Johannes Fabry.**

***(b)* Lesch-Nyhan syndrome.** This syndrome is caused by X-linked recessive allele and found only in males. The Lesch-Nyhan babies show an exceptional type of self destructive behaviour. Though irritation is the only sign a few months after birth, by the age of 2 years a serious nervous (=mental) disorder appears. The child grinds teeth and bites and chews his own lips, tongue and fingers. Kidney and brain damage cause a deficiency of an enzyme, *hypoxanthine-guanine-phosphoribosyl-transferase* or HGPRT, which is required in purine metabolism. Affected children excrete heavy amounts of uric acid in the urine (called **hyper uricemia**). This syndrome was discovered by an American pediatricians, **Michael Lesch** and his student **William Leo Nyhan** in 1964. The ultimate cause of this fatal syndrome, the deficiency of HGPRT enzyme was discovered in 1967.

87.4 HUMAN CHROMOSOMES (KARYOTYPE AND BANDING PATTERNS)

A successful attempt to count the number of human chromosomes was made in 1912 by **Winiwarter** who proposed that human chromosomes are 48 in women and 47 in man; men having one X chromosome and women having two X chromosomes. **Painter,** in 1923, while examining the testicular material of man, observed a heteromorphic pair of sex chromosomes and proposed the XY mechanism of sex determination in man. **Tjio** and **Levan** (1956) cultured somatic cells from fibroblasts of human embryos and counted the human chromosome number as 46. This chromosome number was confirmed by **Ford** and **Hamerton** while working with testicular material in the same year. **Moorhead** *et al.*, (1960) described a simple method of culturing of lymphocytes from human blood.

Karyotype and Idiogram

All the members of a species of a plant or the animal are characterized by a set of chromosomes which have certain constant characteristics. These characteristics include the number of chromosomes, their relative size, position of the centromere, length of the arms, secondary constrictions and satellites. The term **karyotype** has been given to the group of characteristics that identifies a particular set of chromosomes. A diagrammatic representation of a karyotype (or morphological characteristics of the chromosomes) of a species is called **idiogram** (Gr., *idios* =distinctive; *gramma* = something written). Generally, in an idiogram, the chromosomes of a haploid set of an organism are ordered in a series of decreasing size. Sometimes an idiogram is prepared for the diploid set of chromosomes, in which the pairs of homologues are ordered in a series of decreasing size.

Karyotyping human chromosomes. For karyotyping of human chromosomes venous blood is taken and blood leucocytes are stimulated to divide (by mitosis) *in vitro* by the addition of **phytohaemagglutinin.** Colchicine is added to arrest cell division at metaphase stage. It is further treated with hypotonic saline solution which results in swelling of cells and dispersal and better clarity of chromosomes for counting and morphological study. Thereafter, the material is stained (*e.g.*, with Giemsa technique) to demonstrate the banding patterns of chromosomes. Finally, a suitable metaphase spread is photographed through a high power microscope. The individual chromosomes are cut out form the photograph. The chromosomes are then arranged in an orderly fashion in homologous paris, to produce a standard arrangement, the karyotype.

To characterize a chromosome in the karyoype, the following parameters are used:

1. Shape of chromosome
2. Length of chromosome
3. Centromeric index; *i.e.*, this index is expressed in the form of ratio of the short-term length to the total chromosome length:

$$\text{Centromeric index} = \frac{\text{Short arm length } (p)}{\text{Total chrmosome length } (p+q)}$$

(q = length of long arm of chromosome)

For example, centrome-ric index in a metacentric chromosome is 0.5.

4. Proportion of the arms, *i.e.*, it is ratio between the long arm and short arm of the chromosome. This ratio is 1 : 1 in a typical metacentric chromosome.

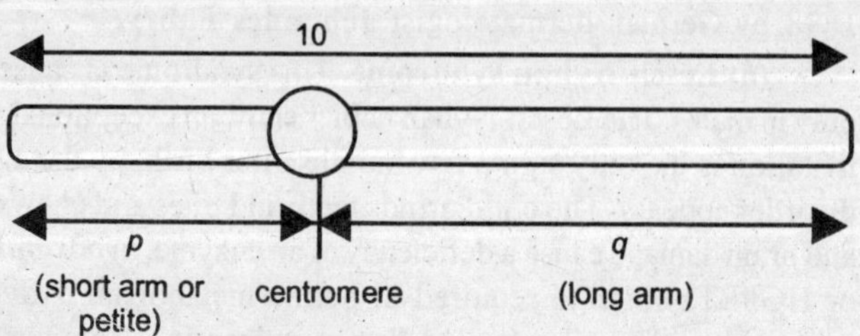

Fig. 87.10 Measurements used to define the morphological parameters of a chromosome.

Classification. The human metaphase chromosomes were first of all classified by a conference of cytogeneticists at Denver, Colorado in 1960 and is known as the **Denver classification.** To follow this classification, each of the 22 pairs of autosomes has been numbered from 1 to 22 according to their decreasing size. **Patau** (1960) divided the human chromosomes into the following seven groups designated A to G:

1. A group : 1 to 3 pairs — Metacentric
2. B group : 4 to 5 pairs — Submetacentric
3. C group : 6 to 12 pairs — Submetacentric
4. D group : 13 to 15 pairs — Acrocentric
5. E group : 16 to 18 pairs — Submetacentric (16 is metacentric)
6. F group : 19 to 20 pairs — Metacentric
7. G group : 21 to 22 pairs — Acrocentric

Group A consists of longest metacentric chromosomes.

Group G consists of the shortest acrocentric chromosomes. These chromosomes have satellites that correspond to nucleolar organizers. Chromosomes of group D also contains satellites. In males, group G includes a variable Y chromosome which lacks the satellites.

The X chromosome is the member of group C and can be identified by special banding or staining methods.

Banding Techniques

For identifying individual human chromosomes, different regions of individual chromosomes or changes in their structure a ingenious technique is being practiced since 1969. This technique involves staining the chromosomes with **fluorescent stains** after certain treatments. Such staining provides different patterns of **bands** and **interbands** (*i.e.*, dark and light bands or stained and, unstained regions) along the length of chromosomes (Fig. 87.11). The banding pattern of a particular chromosome remain constant for a particular treatment like finger prints. In human and other chromosomes, the following types of bands have been recognized:

1. Q banding. It uses fluorescent dyes (such as quinacrine mustard) and identifies the so-called **Q bands.**

2. G banding. It uses Giemsa stain and identifies the G bands. With G banding three major types of chromatin can be recognized—euchromatin, centromeric and intercalary heterochromatin. The Q and G bands are generally similar and correspond to intercalary heterochromatin.

3. C banding. It stains specifically centromeric constitutive heterochromatin.

4. R banding. It gives a pattern that is the reverse of that of Q and G banding.

5. T banding. It stains telomers of chromosomes.

Other banding techniques uses the Feulgen stain (**F bands**) and one selectively stains the nucleolar organizers (**N bands**) which are localized in the satellite of chromosomes 13, 14, 15, 21 and 22. G banding has become important tool in the analysis of mammalian, avain, reptilian and amphibian chromosomes; distinct G bands have not been found in plant chromosomes.

These staining techniques helped **D. Coming** in identifying **euchromatin, constitutive heterochromatin** and **intercalary heterochromatin.** In banding, the nature of DNA and the availability of its active sites are major factors.

Significance of Banding

1. Banding pattern is specific for each chromosome and for each fluorescent dye. It, therefore, helps in the identification of individual chromosomes in the karyotype of an individual.
2. It helps in the identification of different parts of various chromosomes and in localizing specific changes in the structure of chromosome caused by chromosomal aberrations or mutations,

such as loss of a very small part, insertion of an additional segment and addition of whole chromosome.

3. Banding has helped in the identification of association between various **syndromes** in human beings and the chromosome number or chromosome part. For example:
   (*i*) Trisomy of G-21 causes Down's syndrome.
   (*ii*) Trisomy of C-8 is recognized by banding pattern to be associated with mental retardation and skeleton abnormalities.
   (*iii*) Trisomy for partial extra chromosomes has been identified in certain cases through two or more banding patterns.
   (*iv*) cat-cry syndrome due to loss of small part of chromosome 5.

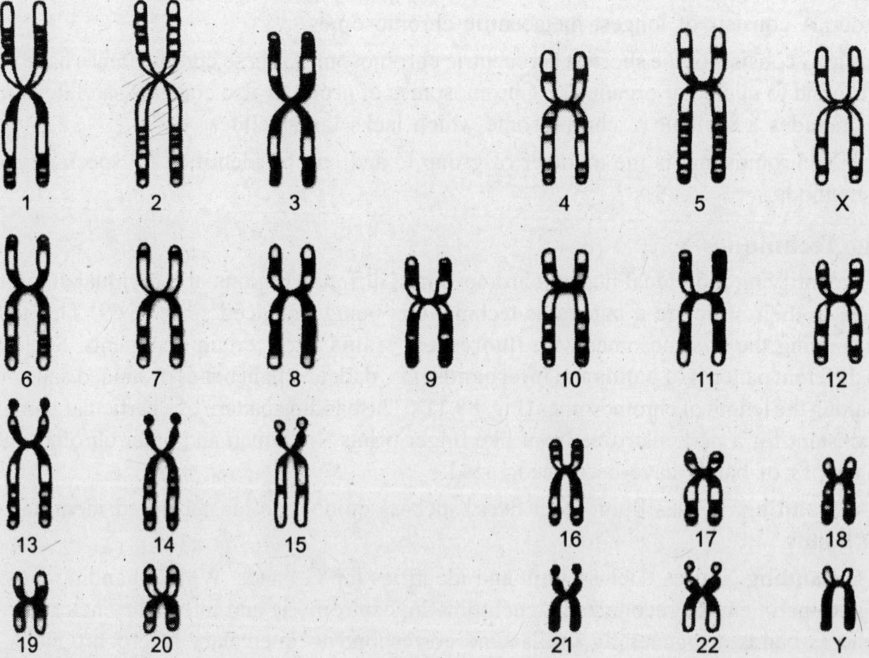

Fig. 87.11. Diagram showing the banding pattern of human chromosomes revealed with Giemsa stain. Since the banding is specific for each chromosome pair, this method, as well, as that involving fluorescent stains, is of great help in preparing a more refined analysis of normal and abnormal chromosomes.

Role of Sex Chromosomes in Sex Determination

In human beings, the presence of Y chromosome determines maleness and its absence determines femaleness. So, males are XY and females are XX in human beings. However, in 1986, certain peculiar cases have been reported which were found to be males with XX chromosomes and females with XY chromosomes. These can be due to any one of the following two reasons: (*i*) A sex reversal gene SRY located on the Y chromosome leads to XX males and XY females. (*ii*) Translocation of a small segment of the Y chromosome to an X chromosome in XX males and its deletion from the Y chromosome results in the XY females.

87.5 SYNDROMES

A **syndrome** is a group of signs and symptoms that occur together and characterize a particular abnormality. In humans, following syndromes are related with various chromosomal aberrations.

A. Autosomal Trisomy and Syndromes

1. Down's syndrome (DS) or Trisomy-21. Down's syndrome is named after the physician **J.Langdon Down** who first described this genetic defect in 1866 and it was formally called **mongolism** or **mongolian idiocy**. It is usually associated with a trisomic condition for one of the smallest human autosomes (*i.e.*, chromosome 21). It is the most common chromosomal abnormality in live births (1/650 births). There are about 50 physical characteristics shown by DS infants soon after birth. These include mild or moderate mental retardation; eyes that slant up and out with internal epicanthal folds; a tongue that is large, swollen and protruding; small and under developed ears; a single palmar crease; short stature; stubby fingers; an enlarged liver and spleen. Women over 45 years of age are about twenty times more likely to give birth to a child with DS than women aged 20. Nondisjunction of chromosome pair 21 during oogenesis is the main cause of occurrence of trisomy-21. This event is found to be affected either by senescence of oocytes, virus infection, radiation damage, etc. (*e.g.*, mothers who have had infectious hepatitis prior to pregnancy may have three times more chances to give birth to DS infants). Nondisjunction of chromosome pair 21 during spermatogenesis can also produce child with DS, but paternal age does not seem to be associated with its incidence.

Lastly, in about 2 to 5 percent cases, the normal chromosome number is present (2n = 46), but the extra chromosome 21 is attached (translocated) to one of the larger autosomes (usually chromosome 14).

2. Edward's syndrome or Trisomy-18. First described in 1960 by **John H. Edwards** and his colleagues, **trisomy-18** is found to contain an incidence of about 0.3 per 1000 births. It is characterized by multiple malformations, primarily low-set ears; small receding lower jaw; flexed and clenched fingers; cardiac malformations; and various deformaties of skull, face and feet. Harelip and cleft palate often occurs. Death takes place around 3 to 4 months of age. Trisomy-18 children show evidence of severe mental retardation, which is more pronounced in females (the reason is still not clear). Like the Down's syndrome, occurrence of Edward's syndrome is too related with maternal age (*i.e.*, 35 to 45 year old mothers have more chance of giving birth to trisomy-18 infant).

3. Patau's syndrome or Trisomy-13. This syndrome was described in 1960 by **Klaus Patau** and coworkers. Its incidence is about 0.2 per 1000 births. Individuals with Patau syndrome appear to be markedly mentally retarded; have sloping forehead, harelip and cleft palate. Polydactyly (of both hands and feet) is almost always present; the hands and feet are deformed. Cardiac and various internal defects (of kidney, colon, small intestine) are common. Death usually occurs within hours or days, but the foetus may abort spontaneously.

B. Sex Chromosomal Syndromes

1. Turner's syndrome (XO females). A female with 44 autosomes and only with one X chromosome in her body cells exhibits symptoms of Turner's syndrome. Such females are sterile and have short stature, webbed neck, a low hairline on the nape of the neck, broad shield-shaped chest, low intelligence, under developed breasts, poorly developed ovaries, sparse pubic hairs and no axillary hair.

2. Poly-X females (XXX females). Such females are called **superfemales** because they possess an extra X chromosome (44 autosomes+3X chromosomes). Some females may have 4 or 5X chromosomes besides the normal autosomes. All such poly-X females are mentally retarded and sterile showing abnormal sexual development.

3. Klinefelter's syndrome (XXY males). When an abnormal egg with XX chromosomes is fertilized by a sperm carrying Y chromosome, a zygote having three sex chromosomes (XXY chromosomes) is formed. The resulting young one is an abnormal sterile male showing the following features: small testicles, mental retardation, longer arms, feeble breasts, higher pitched voice and sparse body hairs.

4. XYY males. Presence of an extra Y chromosome in males (XYY) results in their unusual height, mental retardation, severe facial acne during adolescence and criminal bent of mind. Their genitals are affected by developmental abnormalities.

C. Autosomal Deletions or Translocations

1. Cri-du-chat syndrome. Human babies missing a portion of the short arm of chromosome 5 (autosome) have a distinctive cat-like cry; hence, the French name **"cri du chat"** (cry of the cat) **syndrome** (first described by **Lejeune** *et al.*, 1963). They are also mentally retarded (IQ below 20), have malformation in the larynx, moon faces, saddle noses, small mandibles (micrognathia), malformed low-set ears and microcephally (small head).

2. **Philadelphia chromosome.** In humans, patients with **chronic myelocytic (myologenous) leukemia** (a kind of cancer) display an interesting chromosomal abnormality. In the bone marrow and in cells derived from it, is present a short chromosome, called the **Philadelphia (Ph^1) chromosome** (so named because it was discovered in that city). Detailed cytological studies disclose Ph^1 to be a number 22 chromosome that has lost most of the distal part of its longer arm. The deleted part of autosome 22 is translocated to one of the larger autosomes (most frequently to the distal end of chromosome 9). This translocation exhibits position effect and it is not transmitted to offspring of persons having Philadelphia chromosome (Ph^1 does not appear in gametes of the patients).

Testicular feminization

Testicular feminization is a X-linked recessive disorder. It is also called **congenital insensitivity to androgen syndrome (CIAS).** In this case XY zygote develops into a sterile female, called **CIAS-female.** The CIAS-female is basically male. Her X chromosome has a recessive gene that inhibits the reactivity of body tissues to male sex hormones such as testosterones and MDI or Mullerian duct inhibitor. A CIAS – female has the following characteristics:

1. She has XY – sex chromosomes.
2. She has intra-abdominal testes (*i.e.*, testes fail to descend down or are partially descended).
3. She possess epididymis and sperm ducts.
4. She has normal level of testosterone.
5. She has no external male genitals.
6. She has short vagina but no ovaries, oviducts and uterus. Estrogen hormones are also lacking in her.
7. She has breast but little pubic or axillary hair.

Some XY human zygotes develop into CIAS female because her X-chromosome possesses a mutant sex-linked gene which inhibits the **differentiation** of external ginitalia and normal function of testes.

87.6 GENETIC COUNSELLING

A counsellor is a person trained to give advice on personal or psychological problems. **Genetic counselling** is the giving of information and advice about the risks of genetic diseases and their outcomes. Local doctors or other health professional can refer the patients to genetic counsellors. (Box 87.1). Genetic counselling is often carried out in special clinics in hospitals.

Box 87.1

Genetic counselling means medical education of affected individuals and the general public concerning inherited disorders that includes discussion of the probability of producing offspring with a disorder, diagnostic tests, and available treatment.

In fact, counsellors are doctors or paediatricians (child doctors). They must have a good knowledge of medical genetics as well as they must be trained in sympathetic counselling techniques.

During counselling, a counsellor must not impose his/her own views on the person being counselled. By giving information and encouraging discussion patient must be helped to make his/her own decisions.

The genetic counsellor studies the backgrounds of the couple and tries to determine if any immediate ancestor may have had a genetic disorder. A pedigree chart may be constructed. Then the counsellor studies the couple. As much as possible, laboratory tests (Box 87.2) are performed on all persons involved.

Box 87.2

Tests are now available for a large number of genetic diseases. For example, **chromosomal tests** are available for cystic fibrosis, neurofibrosis, and Huntington disease. **Blood tests** can identify carriers of thalassemia and sickle-cell diseases. By measuring enzyme levels in blood, tear or skin cells, carriers of **enzyme defects** can also be identified for certain inborn metabolic errors, such as Tay-Sachs disease. From this information, the counsellor can sometimes predict the chances of a child having the disorder.

Whenever the woman is pregnant, chorionic villi sampling (CVS) can be done early, and amniocentesis can be done later in the pregnancy. These procedures allow the testing of embryonic and fetal cells, respectively, to determine if the unborn child has a genetic disorder. If so, treatment may be available even before birth, or parents may decide whether or not to end the pregnancy.

A genetic counsellor is expected to discuss the following issues while giving counselling

1. Making a diagnosis. An accurate diagnosis, involving medical examinations and laboratory tests, is essential. For example, a woman aged 20 who has not begun to menstruate and whose breasts have not developed could be suffering from Turner's syndrome (a genetic disorder), a hormone deficiency or simply delayed puberty. Chromosome analysis (Karyotype study) of the patient could confirm whether or not Turner's syndrome is the cause.

2. Family history. Investigation of family history for previous cases of genetic diseases by drawing the family tree of the person being counselled is generally done.

3. Calculated risk. The genetic counsellor often has to calculate and explain the risks of having affected children. For example, if both members of a couple are carrier for a recessive gene, it can be explained that there is a 1 to 4 chance of any child being affected.

4. Explaining causes. The counsellor normally try to explain the cause of any clinical problem. For example, a lethal human disease such as sickle cell anaemia is caused by sickling of erythrocytes (RBCs) and it can be explained that this lead to anaemia and blockage of small blood vessels. These in turn cause many other symptoms.

5. Quality of life. A genetic counsellor has to discuss about the quality of life and expected length of life of the affected child/person. Availability of treatment, support group and financial help need to be discussed. It can also be discussed whether an affected child needs time consuming and expensive care and is likely effect on other family members.

6. Options. If a couple decides the risks are unacceptable, possible options must be explained by the genetic counsellor. These include contraception or sterilization adoption, artificial insemination to avoid the husband's genes or in vitro fertilization (IVF) if the problem lies with the mother. If the couple are willing to consider abortion of an affected foetus, CVS, amniocentesis and an abortion itself must be discussed.

7. Genetic screening. Part of the genetic counsellor's job is to discuss the results of genetic screening with the person who has been tested. Genetic screening includes prenatal diagnosis, carrier diagnosis, and predictive diagnosis. After a positive diagnosis genetic counselling of the affected person is essential. Genetic counsellor has to prepare a person mentally for the possibility of receiving a bad news from a test and helping him in making decision, which have long term consequences, which is very difficult when is feeling shocked and emotionally distressed. In this way, genetic counselling after genetic screening of a patient is one of the most challenging task for a counsellor.

8. Responsibility. If a person is responsible for transmission of a genetic defect, as in the case of dominant genes or sex linked conditions (such as colour blindness, haemophilia, etc.), guilt and blaming are common responses. The counsellor should try to emphasise sharing of responsibility and can point to the fact that we all carry some harmful genes, and in most cases we are lucky enough not to find out which ones.

Thus, the genetic counselling is most challenging task. It will become increasingly important as more genetic tests become available.

QUESTIONS

Long Answer Questions

1. Describe some common hereditary traits in humans and the pattern of their genetic transmission. (*Lucknow 1995; CCS 1997, 1998*)

2. What do you understand by inborn errors of metabolism in man? Name any two such metabolic disorders and explain one of them.

3. Write a short note on inborn errors of metabolism, in man. (*Allahabad 1989, 1992, 1994*)

4. What do you understand by Klinefelter's syndrome and Down's syndrome? Describe in detail their characteristics and the mechanism by which they are caused. (*CCS 1991, 1993*)

5. What are syndromes? Describe defects associated with Down syndrome and explain how does Down syndrome develop? (*Rohilkhand 1991*)

6. Describe different types and causes of chromosomal diseases in man. (*Rohilklzand 1990*)

7. How Karyotyping is done of human chromosomes? Also write regarding the banding methods.

8. Describe briefly simply Mendelian traits in man. (*M.S.U. Tamil Nadu 2006*)

9. What is genetic counselling. Describe the main points to be discussed by a genetic counsellor.

Short Answer Questions

1. Explain what is testicular feminization? Give reasons why sometimes XY-zygote develops into a female?

2. Briefly write the genetic basis of following disorders in man.

(*i*) Down's syndrome (*B.H.U. (hons) 1991*)

(*ii*) Phenylketonuria

(*iii*) Sickle cell anaemia.

3. What are the applications of the pedigree analysis in the study of human genetics. (*MSU Tamil Nadu 2006*)

4. Write short notes on the following: (*CCS 1991*)

(*i*) Pedigree analysis

or

Pedigree chart (*Bangalore 1993; H.N.B. Garwal 2008*)

or

Pedigree chart and its application. (*H.N.B. Garhwal 2009*)

(*ii*) Twins (*CCS 1991, 1992, 1995*)

or

Twin studies and their use (*CCS 1991*)

(*iii*) Albinism (*Lucknow 1992, 1998; Bangalore 1994*)

(*iv*) Phenylketonuria (*Allahabad 1991; Bangalore 1996*)

(*v*) Down's syndrome (*Jiwaji 1992, Bangalore 1996*)

(*vi*) Kleinefelter's syndrome (*Lucknow 1995*)
(*vii*) Turner's syndrome (*BHU (Hons.) 1990; Allahabad 1996*)
(*viii*) Fabry syndrome
(*ix*) Lesch-Nyhan syndrome

5. Differentiate between the following
(*i*) Kleinefelter's syndrome and Turner's syndrome (*CCS 1994, 1996*)

Very Short Answer Questions

1. Define pedigree (*H.N.B. Garwal 2008*)
Ans. Pedigree is a representation of the ancestry of an individual or family. Pedigree is also called **family tree.**

2. Name the syndromes caused by trisomy.
Ans. Down's syndrome, Edward's syndrome and Patau's syndrome.

3. Name the amino acid which cannot be metabolised by persons suffering from phenyketonuria (PKU).
Ans. Phenylalanine.

4. Addition of few drops of 5% solution of ferric chloride in the urine of certain individuals temporarily develops bluish-green colour. What does it indicate?
Ans. Presence of phenylpyruvic acid in the urine, or the disease of PKU (=phenylketoneuria).

5. What causes Down's syndrome?
Ans. Trisomy-21.

6. Edward's syndrome arises due to trisomy of which chromosome.
Ans. Chromosome number 18.

7. What is most important reason of origin of syndromes in humans?
Ans. Aneuploidy.

8. What causes Turner's syndrome in humans.
Ans. Aneuploidy of sex chromosome (*e.g.*, XO).

9. Name the human's syndrome in which accumulation of lipids occur in kidney.
Ans. Fabry syndrome.

10. What happens when conversion of DOPA (=3, 4–Dihydroxy phenylalanine) into melanin fail?
Ans. This causes albinism in humans.

11. What causes deficiency of HGPRT in humans?
Ans. HGPRT stands for hypoxanthine-guanine-phosphoribosyl transferase. Deficiency of HGPRT enzyme causes Lesch-Nyhan syndrome in humans.

12. Which scientists confirmed exact human chromosome number?
Ans. Tjio and **Levan** in 1956.

13. Define Genetic counselling. (*C.C.S.U. Meerut 2007*)
Ans. It is giving of information and advice about risks of genetic diseases and their outcomes.

Fill in the with the suitable words Blanks:

1. Genotype of Kleinefelter's syndrome is .... (*Lucknow 1995*)

2. Genotype of Turner's syndrome is .... (*Lucknow 1996*)

Multiple Choice Questions

1. Cri-due chat (cry of the cat) syndrome occurs in human babies due to
(*a*) deletion (*b*) duplication
(*c*) inversion (*d*) simple translocation

2. The X/O syndrome is called
   (*a*) Turner's (*b*) Klinefelter's
   (*c*) Down's (*d*) Cushing's
3. Polydactyly in humans is due to
   (*a*) autosomal recessive allele (*b*) autosomal dominant allele
   (*c*) sex-linked recessive allele (*d*) sex-linked dominant allele
4. Father of human genetics is
   (*a*) Cuvier (*b*) Bateson
   (*c*) Mendel (*d*) Garrod
5. A symbol of empty circles used in pedigree analysis represents
   (*a*) normal females (*b*) normal males
   (*c*) Edward's syndrome (*d*) affected females
6. A monosomic (2N – 1) abnormality in human's
   (*a*) Klinefelter syndrome (*b*) Turner's syndrome
   (*c*) Edward's syndrome (*d*) Down's syndrome
7. Klinefelter's syndrome has...............number of chromosomes
   (*a*) 45 (*b*) 47
   (*c*) 46 (*d*) 44
8. Edward's syndrome is an abnormality leading to mental deficiency caused by trisomy of chromosome number
   (*a*) 5 (*b*) 9
   (*c*) 15 (*d*) 18
9. The genetic disease cystic fibrosis includes symptom such as
   (*a*) failure of chloride ions transport (*b*) mucus clogging of lungs
   (*c*) defective function of pancreas and liver (*d*) all of these
10. Phenyketonuria is
   (*a*) sex-linked dominant trait (*b*) sex-linked recessive trait
   (*c*) autosomal dominant trait (*d*) autosomal recessive trait

ANSWERS

Fill in the Blanks

1. XXY **2.** XO.

Multiple Choice Questions

1. (*a*)	**2.** (*a*)	**3.** (*b*)	**4.** (*d*)
5. (*a*)	**6.** (*b*)	**7.** (*b*)	**8.** (*d*)
9. (*d*)	**10.** (*d*).		

88

Population Genetics and Evolutionary Genetics

All the individuals of a species constitute a **population**. The genetical studies for the inheritance of phenotypic traits in a given population is called **population genetics**. The population genetics is a quantitative science. To calculate the results of the mode of inheritance of genes in a given population various statistical and mathematical models are employed in it. Certain fundamental aspects of population genetics are the following:

88.1 MENDELIAN POPULATION

A population of a particular species includes many inbreeding groups. The inbreeding groups may form a community within defined geographical boundaries and are called **"Mendelian population"**. A Mendelian population, thus, is a group of sexually reproducing organisms with a relatively close degree of genetic relationship (such as species, subspecies, breed, variety, strain, etc.) residing within defined geographical boundaries where interbreeding occurs.

Gene Pool and Gene Frequency

To get a $F_2$3:1 phenotypic ratio of a monohybrid cross, we began with two homozygous parental strains, such as *AA* and *aa* : that is, we introduced the alleles *A* and *a* in **equal frequency**. (A **frequency** is the ratio of the actual number of individuals falling in a single class to the total number of individuals; see **Gardner**, 1972). But in a Mendelian population, frequencies of alleles may vary considerably. For example, in Mendelian population of humans, the gene for polydactyly is dominant, yet the polydactylous phenotype is fairly infrequent among infants. This indicates that frequency of the dominant allele in population is lower than that of its recessive allele and that both alleles do not exist in population in the 1 : 1 ratio like the individuals of monohybrid cross.

Further, if all the gametes produced by a Mendelian population are considered as a hypothetical mixture of genetic units from which the next generation will arise, we have the concept of a **gene** or **gamete pool**.

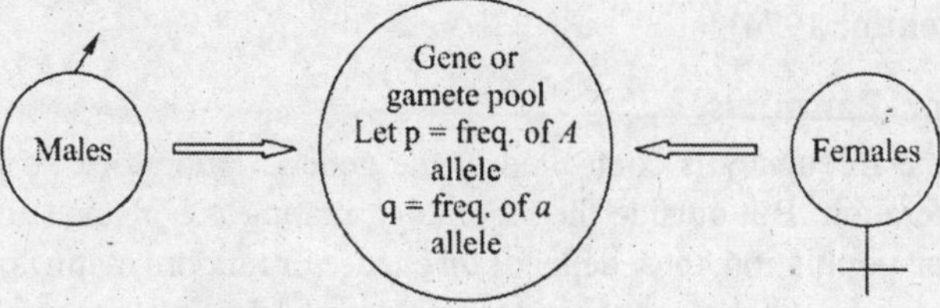

Fig. 88.1. The concept of gene pool.

The percentages of gametes in the gene pool for a pair of alleles (*A* and *a*) depend upon the genotypic frequencies of the parental generation whose gametes form the pool. Thus, if a population is of dominant genotype *AA*, then the frequency of dominant alleles in the gene pool will be relatively high and the percentage of gametes bearing the recessive (*a*) allele will be correspondingly low.

Two Models of Gene-Pool Structure

Regarding the genetic structure of the population, the following two hypothesis have been proposed:

1. Classical hypothesis. It was developed by **T.H. Morgan** (1932) and supported by **H.J. Muller** and **Kaplan** (1966). The classical hypothesis proposes that the gene pool of a population consists at each gene locus of a wild-type allele with a frequency approaching one. Mutant alleles in very low frequencies may also exist at each locus. A typical individual would be homozygous for the wild-type allele at most gene loci; at a very small proportion of its loci the individual would be heterozygous for a wild and a mutant allele. Except in the progenies of consanguineous matings, individuals homozygous for a mutant allele would be extremely rare. The "normal" ideal genotype would be an individual homozygous for the wild-type allele.

According to classical hypothesis, mutant alleles are continuously introduced in the population by mutation pressure, but are generally deleterious and, thus, are more or less gradually removed from the population by natural selection. Periodically a beneficial mutant allele might arise, conferring higher fitness upon its carriers than the pre-existing wild-type allele. This beneficial allele would gradually increase in frequency by natural selection to become the new wild type allele, while the former wild type allele would be eliminated. Evolution, thus, consists of the replacement at an occasional locus of the pre-existing wild-type allele by a new wild-type allele.

2. Balance hypothesis. This hypothesis was proposed by **Dobzhansky** (1970) and **E.B. Ford** (1971). This hypothesis was derived by direct study of natural populations. According to the balance model, there is generally no single wild-type or 'normal' allele. Rather, the gene pool of a population is envisioned as consisting at most loci of an array of alleles in moderate frequencies. A typical individual is heterozygous at a large proportion of its gene loci. There is no 'normal' or ideal genotype, only an adaptive norm consisting of an array of genotypes that yield a satisfactory fitness in most environments encountered by the population.

The proponents of the balance hypothesis argue that the common allelic polymorphisms are maintained in populations by various forms of balancing natural selection. The fitness granted on its carriers by an allele depends on what other alleles exist in the genotype at that and other gene loci. It also depends, of course, on the environment. Gene pools are co-adapted systems; the sets of alleles favoured at one locus depend on the sets of alleles that exist at other loci.

Evolution occurs by gradual change in the frequencies and kinds of alleles at many gene loci. As the configuration of the set of alleles changes at one locus, it also changes at many other loci.

The balance model of genetic structure of populations has now become definitely established, although some controversy remains regarding the process maintaining the common polymorphisms (**Ayala**, 1974; **Lewontin**, 1974).

Chance Mating or Panmixis

Just as the gene frequency is controlled in the genetic laboratory, so is the mating pattern, generation after generation. But outside the laboratory, mating is a chance or random affair. Every member of Mendelian population, thus, depends on chance or random matings. In other words, every male gamete in the gene pool has an equal opportunity of uniting with every female gamete. Zygotic frequencies expected in the next generation from such random gametic unions may be predicted from a knowledge of the gene (allelic) frequencies in the gene pool of the parental population. For example, the expected zygotic frequencies for allele *A* and *a* of a gene pool can be determined by chance mating or panmixis. If p stands for percentage of *A* alleles in the gene pool and q stands for the percentage of *a* alleles, the checker board of both alleles may predict possible chance combinations of *A* and *a* gametes as follow:

Male gametes→ / Female gametes ↓	p (*A*)	q (*a*)
p (*A*)	(*AA*) p^2	(*Aa*) pq
q (*a*)	(*Aa*) pq	(*aa*) q^2

or $(p + q)^2 = p^2 + 2pq + q^2$

Thus, p^2 is the fraction of the next generation expected to be homozygous (*AA*), 2pq is the fraction expected to be heterozygous (*Aa*) and q^2 is the fraction expected to be recessive (*aa*).

The zygotic combinations predicted in a randomly mating population may be represented by $p^2 : 2pq : q^2$, where p^2 represents the *AA* genotype, 2pq the *Aa*, and q^2 the *aa* genotype; or, in the form of equation, $p^2 + 2pq + q^2 = 1$. When only two alleles are involved, and, therefore, p and q represent the frequencies of all of the alleles concerned $p + q = 1$. Since $p + q = 1$, $p = 1 - q$. Now if $1 - q$ is substituted for p, all relations in the formula can be represented in terms of q as follows:

$$(1 - q)^2 + 2q(1 - q) + q^2 = 1.$$

88.2 HARDY-WEINBERG LAW

The formula $(p + q)^2 = p^2 + 2pq + q^2$ is expressing the genotypic expectations of progeny in terms of gametic or allelic frequencies of the parental gene pool and is originally formulated by a British mathematician **Hardy** and a German physician **Weinberg** (1908) independently. Both forwarded the idea, called **Hardy-Weinberg law equilibrium** after their names, that *both gene frequencies and genotype frequencies will remain constant from generation to generation in an infinitely large interbreeding population in which mating is at random and no selection, migration or mutation occur*. Should a population initially be in disequilibrium, one generation of random mating is sufficient to bring it into genetic equilibrium and thereafter the population will remain in **equilibrium** (unchanged in gametic and zygotic frequencies) as long as Hardy-Weinberg condition persists. Hardy-Weinberg law depends on the following kinds of genetic equilibriums for its full attainment:

1. The population is infinitely large and mate at random.
2. No selection is operative.
3. The population is closed, *i.e.*, no immigration or emigration occurs.
4. No mutation is operative in alleles.
5. Meiosis is normal so that chance is the only factor operative in gametogenesis.

The significance of Hardy-Weinberg equilibrium was not immediately appreciated. A rebirth of biometrical genetics was later brought about with the classical papers of **R.A. Fisher**, beginning in 1918 and those of **Sewall Wright**, beginning in 1920. Under the leadership of these mathematicians, emphasis was placed on the population rather than on the individual or family group, which had previously occupied the attention of most Mendelian geneticists. In about 1935, **T. Dobzhansky** and others started to interpret and to popularize the mathematical approach for studies of genetics and evolution.

Genetic Equilibrium

As shown by **Hardy** and **Weinberg,** alleles segregating in a population tend to establish an equilibrium with reference to each other. Thus, if two alleles should occur in equal proportion in a large, isolated breeding population and neither had a selective or mutational advantage over the other, they would be expected to remain in equal proportion generation after generation. This would be a special case because alleles in natural populations seldom if ever occur in equal frequency. They may, however, be expected to maintain their relative frequency, whatever it is, subject only to such factors as chance, natural selection, differential mutation rates or mutation pressure, meiotic drive and migration pressure, all of which alter the level of the allele frequencies. A genetic equilibrium is maintained through random mating.

88.3 APPLICATION OF HARDY-WEINBERG LAW IN CALCULATING GENE FREQUENCIES IN A POPULATION

The gene frequencies for the autosomal and sex-chromosomal allele can be determined by the help of Hardy-Weinberg law by the following method:

Calculation of Gene Frequencies of Autosomal Genes

An autosomal gene locus may have codominant alleles, dominant and recessive alleles or multiple alleles. If one desires to determine the gene frequencies for each of these kinds of autosomal alleles in a given population, he has to adopt the different methods.

Calculation of gene frequencies for codominant alleles. When codominant alleles are present in a two-allele system, each genotype has a characteristic phenotype. The numbers of each allele in both homozygous and heterozygous conditions may be counted in a sample of individuals from the population and expressed as a percentage of total number of alleles in a sample. If the sample is representative of the entire population (containing proportionately same numbers as found in the entire population) then we can obtain an estimate of the allelic frequencies in the gene pool. If in a given sample of N individuals of which D are homozygous for one allele ($A^1 A^1$), H are heterozygous (A^1A^2), and R are homozygous for the allele (A^2A^2), then N = D + H + R. Since each of the N individuals are diploid at this locus, there are 2N alleles represented in the sample. Each $A^1 A^1$ genotype has two A^1 alleles. Heterozygotes have only one A^1 allele. Letting p represents the frequency of the A^1 allele and q the frequency of the A^2 allele, we have

$$p = \frac{2D + H}{2N} = \frac{D + 1/2H}{N}, q = \frac{H + 2R}{2N} = \frac{1/2H + R}{N}$$

Example

The M-N blood type furnishes a useful example of a series of phenotypes due to a pair of codominant alleles. None of three possible phenotypes, M, MN and N, appears to have any selection value. The frequencies of the two alleles (*viz.*, L^M and L^N) for a sample from a group of white Americans living in New York City, Boston, and Columbus, Ohio, can be calculated by the following methods:

The sample of 6,129 Caucasian people includes the following three groups according to phenotypes and genotypes on M-N system:

Table 88.1

Phenotype	*Genotype*	*Number*
1. M	$L^M L^M$	1,787
2. M N	$L^M L^N$	3,039
3. N	$L^N L^N$	1,303
		Total = 6, 129

To calculate frequencies of the two codominant alleles, L^M and L^N, it should be kept in mind that these 6,129 persons possess a total of 6,129 × 2 = 12,258 genes. The number of L^M alleles, for example, is 1,787 + 1,787 + 3,039. Thus, calculation of the frequency of L^M and L^N alleles is worked out in this way.

$$(i)\ L^M = \frac{1,787 + 1,787 + 3,039}{12,258} = \frac{6,613}{12,258} = 0.5395$$

$$(ii)\ L^N = \frac{1,303 + 1,303 + 3,039}{12,258} = \frac{5,645}{12,258} = 0.4605$$

Thus, the frequencies of the two codominant alleles in this sample are almost equal, and this is reflected in the close approximation to a 1 : 2 : 1 ratio, which is a simple monohybrid ratio for codominant alleles in Mendelian genetics.

Gene frequencies expressed as decimals may be used directly to state probabilities (a *probability* is a function that represents the likelihood of occurrence of any particular form of an event). If we can assume this sample to be representative of the population, then there is a probability of 0.5395 that of the chromosomes bearing this pair of alleles, any one selected randomly will bear gene L^M, and 0.4605 that it will bear L^N. Let p represents genotypic frequency of L^M allele and q represents frequency of L^N allele, then the frequencies of three phenotypes to be expected in the population are as follows:

Table 88.2 Expansion of Hardy-Weinberg formula $(p + q)^2 = p^2 + 2pq + q^2$, according to L^M and L^N gene frequencies in a population sample.

Genotype	*Phenotype*	*Genotype frequencies*
L^ML^M	M	p^2 = 0.5395 × 0.5395 = 0.2911
L^ML^N, L^NL^M	MN	2pq = 2 (0.5395 × 0.4605) = 0.4968
L^NL^N	N	q^2 = 0.4605 × 0.4605 = 0.2121
		1.0000

88.4 FACTORS INFLUENCING ALLELE FREQUENCY OR DEVIATIONS FROM HARDY-WEINBERG EQUILIBRIUM

The Hardy-Weinberg explanation of equilibrium in the allele frequency pattern of a population required three assumptions– (1) individuals with each genotype must be as reproductively fit as those of any other genotype in the population; (2) the population must consist of a large number of individuals; and (3) random mating must occur throughout the population. The Hardy-Weinberg theorem with its assumptions does not account for any change in allele frequency within populations. That is just what Hardy and Weinberg intended because their formula described the statics of a Mendelian population. Something more was required to formulate a mathematical explanation of change or dynamics in terms of allele frequencies. This need was filled by **Fisher, Wright** and **J.B.S. Haldane,** who provided additional theoretical models and superimposed the mechanisms for change in allele frequencies upon the Hardy-Weinberg equilibrium. Population statics was, thus, extended to become population dynamics.

The shifts or changes in gene frequencies can be produced by a reduction in population size, selection, mutation, chance (genetic drift), meiotic drive and migration.

A. Selection

1. Evolution. Evolution has been described by **Lewontin** (1967) as a process that converts variation within a population into variation between populations, both in space (race formation and speciation) and in time (the evolution of phyla). The most prominent theory regarding the driving force behind evolution is of course the Darwin-Wallace theory of **natural selection** proposed in

1858 by **C. Darwin** and **A. Wallace**. The natural selection theory holds that as genetic variants arise within a population the fittest will be at a selective advantage and will be more likely to produce offspring than the rest. As the fit continue to enjoy greater survival, reproductively new species will eventually evolve.

2. Natural selection. Natural selection is generally believed to be the prominent agent for determining the relative frequency of alleles in a population. Natural selection differentiates between phenotypes in a population with respect to their ability to produce offspring. One phenotype may better survive endemic onslaughts of parasites or predators than another, one may penetrate new habitats more effectively than another; one may mate more efficiently than other; one may even prey on the other. The important point is that some natural situation or environmental feature selectively allows one organism to develop and to propagate more efficiently, and one genotype is thereby afforded greater representation in the populations gene pool. If this selective process continues over many generations, allelic frequencies will change significantly and the potential will arise for evolutionary change. Further, the natural selection process, while acting on the total phenotype, will, in fact, influence only the heritable portion of the phenotype. If a trait has a high heritability, selection can rapidly affect its frequency within a population, whereas selection will take a great deal longer to have any effect on a trait with a low heritability.

Natural selection includes the following three parameters—**survival rate, relative fitness** and **selection coefficient**. To calculate values for these parameters and, best understand their meaning, we can consider the data from a particular population. For simplicity we shall consider only a single gene locus in the population defined by the *A* and *a* alleles and we shall assume incomplete dominance, so that *A*/*a* heterozygote can be distinguished phenotypically from *A*/*A*, and *a*/*a* homozygotes. We then count the number of *A*/*A*, *A*/*a*, and *a*/*a* individuals in a given generation immediately before and immediately after some selective event—the introduction of a parasite, a change in temperature—has occurred. The representative numbers have been given in Table 88.3. From these we can calculate a survival rate, relative fitness and selection coefficient, as shown in Table 88.3.

The genotype with the largest survival rate is defined as the fittest, and it is used as the standard for the **relative fitness (W)** of all other genotypes. Specifically, in our example, *A*/*A* has a survival rate of 0.95 and is, thus, fitter than *A*/*a* or *a*/*a* (Table 88.3). In determining values for *W*, therefore, all survival rates are divided by 0.95 so that the relative fitness of *A*/*A* becomes 0.95/0.95=1, the relative fitness of *A*/*a* becomes 0.80/0.95=0.84, and that of *a*/*a* becomes 0.55/0.95=0.58.

The **selection coefficient** (s) is simply 1 – W. Just as W reflects the chances of an organism's reproductive success, so does s reflect the chances of its reproductive failure due to selection. They are simply two sides of the same coin.

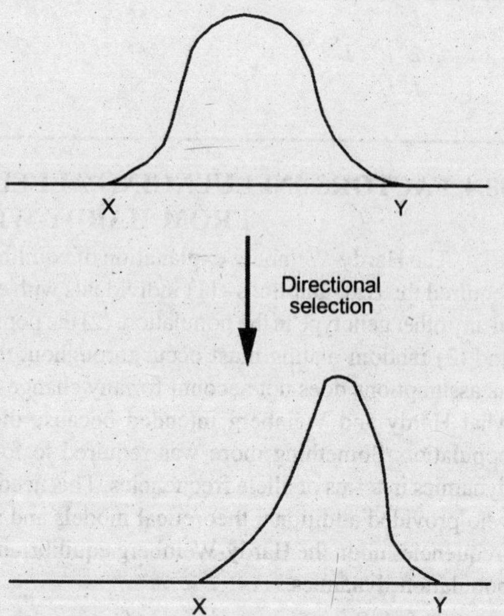

Fig. 88.2 The effects of directional selection on a population whose phenotypes are normally distributed between two extremes, X and Y. Directional selection acting against X and in favour of Y changes the distribution of the phenotypic variation.

3. Directional selection. A principal pattern followed by selection in natural populations is that of **directional selection**. Its effect is to eliminate or to reduce the reproductive potential of particular phenotypes in a population. Specifically, if the variance in a particular trait is normally distributed in a population before directional selection begins to operate, it might become markedly skewed in one direction afterwards (Fig. 88.2).

Table 88.3 Relative fitness and selection coefficients calculated from data taken in the same generation of population.

Data: Number of individuals in the population according to the genotype			
	A/A	*A/a*	*a/a*
Before selection	*4100*	*5000*	*2200*
After selection (same generation)	*3900*	*4000*	*1200*

Calculations:

(a) Survival rate

$$A/A = \frac{3900}{4100} = 0.95$$

$$A/a = \frac{4000}{5000} = 0.80$$

$$a/a = \frac{1200}{2200} = 0.55$$

(*b*) Relative fitness W

(compared with A/A's maximum survival rate)

$$W_{AA} = \frac{0.95}{0.95} = 1.00$$

$$W_{Aa} = \frac{0.80}{0.95} = 0.84$$

$$W_{aa} = \frac{0.55}{0.95} = 0.58$$

(*c*) Selection coefficients (s)

$$S_{AA} = 1 - W_{AA} = 0$$

$$S_{Aa} = 1 - W_{Aa} = 0.16$$

$$S_{aa} = 1 - W_{aa} = 0.42$$

Directional selection in its extreme form eliminates systematically the recessive homozygotes (*a*/*a*) from the population. In such cases the relative fitness of *a*/*a* is set at 0 and its selection coefficient, therefore, becomes 1 – 0 = 1. If we assume, for simplicity, that *A*/*A* and *A*/*a* have identical fitnesses and that no other alleles at the locus are involved, then by definition *A*/*A* and *A*/*a* each have fitness of 1 and selection coefficients of 0. Further, when directional selection remains in its incomplete form against homozygous recessives, then selection coefficients is less than 1 and the relative fitness W of the homozygous recessives is 1 – s, a number greater than zero.

4. Artificial selection. The artificial selection is simply a selection programme devised by humans rather than managed by natural situations. The process of enriching a population for a given trait normally occurs far more rapidly under the artificial situation, since all but the desired phenotypes can be prevented from reproducing. Countless experiments have shown that it is possible to select for just about any trait present in a population, a major limitation being the heritability of the trait. Strains with particular shapes, sizes, behaviour patterns, temperature optima, sexual preferences, and so on have been selected. **S. Spiegelman** and colleagues have even subjected isolated RNA molecules from RNA phage ϕ β to artificial selection and they reported that a rigorous selection for one trait can lead to the drastic exclusion of many other traits.

5. Significance of the heterozygote. When the frequency of an abnormal recessive gene becomes very low, most affected offspring (*aa*) will come from matings of two heterozygous carriers (*Aa*).

For example, in the human population, the vast majority of newly arising albino individuals (*aa*) in a given generation (more than 99 per cent of them) will come from normally pigmented heterozygous parents.

Detrimental recessive genes in a population are obviously harboured mostly in the heterozygous state. The frequency of heterozygous carriers is many times greater than the frequency of homozygous individuals afflicted with the trait. Thus, an extremely rare disorder such as **alkaptonuria** (blackening of urine) occurs in 1 in one million persons. This detrimental gene, however, is carried in the hidden state by 1 out of 500 persons. There are two thousand as many genetic carriers of alkaptonuria. For another recessive trait, **cystic fibrosis**, 1 out of 1,000 individuals is affected with this homozygous trait. One out of 16 persons is a carrier of cystic fibrosis (see Table 88.4).

Table 88.4 Frequencies of recessive homozygotes and heterozygous carriers.

Frequency of Homozygotes (aa)	*Frequency of Heterozygous Carriers (Aa)*	*Ratio of Carriers to Homozygotes*
1. 1 in 500 (Sickle cell anaemia)	1 in 10	50 : 1
2. 1 in 1,000 (Cystic fibrosis)	1 in 16	60 : 1
3. 1 in 6,000 (Tay-Sachs disease)	1 in 40	150 : 1
4. 1 in 20,000 (Albinism)	1 in 70	285 : 1
5. 1 in 25,000 (Phenylketonuria)	1 in 80	310 : 1
6. 1 in 50,000 (Acatalasia)	1 in 110	460 : 1
7. 1 in 1,000,000 (Alkaptonuria)	1 in 500	2,000 : 1

Genetic load (concealed variability) and price of evolution. Genetic load can be defined as the relative deficiency in viability or fecundity or both viability and fecundity as compared to that which would occur if the population were all of the most fit phenotype. Thus, genetic load is the difference between the actual fitness and an optimum fitness, this difference being caused by the presence of suboptimal genetic variants, concealed or not, within the population.

In fact, mutants arise in any species and many or most of them are deleterious. Some of these mutants may persist in the populations, *i.e.*, one or several generations intervene between the origin and the populations of any species, including the best adapted and most successful ones, carry burdens or loads of genetic defects. One can distinguish between **expressed** (overt) or **concealed genetic loads.**

The former comprises individuals with diseases, malformations, or constitutional weaknesses that are manifestly genetic. Concealed genetic burdens are due to recessive genes or gene complexes carried in heterozygous conditions in individuals who are themselves healthy or "normal". When homozygous, however, these genes too become lethal.

Types of genetic loads. According to **Fransworth** (1988), the genetic loads are of the following three types:

1. Mutational load. It refers to the deleterious effects on a population caused by recurrent mutation.

2. Segregational load. It arises when the **heterozygote** is the favoured genotype. The formation of heterozygotes is necessarily accompanied by the generation of the less well-adapted homozygous genotypes whose lowered viability and fecundity cause the segregational load. Segregational load is, thus, the loss in average fitness of a population caused by the segregation of the less fit homozygous genotypes. Because selection acts to preserve the heterozygous genotype, neither allele, even if one or both are lethal, is eliminated from the population. As an example, in malarial regions of Africa the loss or lowered fecundity of *HbS* and *HbA* homozygotes as compared to heterozygotes (*HbA HbS*) constitute the segregational load for that locus.

3. Substitutional or transient load. It occurs when a change in the environment results in directional selection in favour of a previously neutral or unfavourable allele whose effects in the altered environment now confer greater fitness. The substitutional load, thus, consists of the relative number of deaths and lack of birth which the population must allow in order to progress from the original new state of fitness. Substitutional load is transient in that it occurs only during the replacement of one allele by another. For example, in the United States where falciparum malaria is absent, *HbA* homozygotes possess the best adapted phenotype, and directional selection in favour of the normal allele is occurring. As a result, the lowered fecundity and viability of *HbS*/*HbA* heterozygotes and HbS homozygotes, as compared to normal individuals contribute a transient substitutional load to the black population of the United States.

The substitutional genetic load has been called the **price of evolution** and **J.B.S. Haldane** (1957) has made some theoretical calculations of the extent of this price in terms of reduced fitness or loss of individuals. He has estimated that a selection intensity of 0.1, approximately 300 generations will be required to replace a gene at a cost in members lost by death or sterility of 30 times the average number of individuals present in a single generation. Thus, according to these calculations, approximately 10 per cent of a population would be lost per generation if gene replacement occurred over a time span of 300 generations.

Because of the enormous amount of protein, and, therefore, genetic polymorphism recently revealed at the molecular level by immunological and electrophoretic methods, these calculations pose a dilemma. Evolution requires the concurrent replacement of many genes, and to the substitutional load borne by a population must be added to the mutational load as well as segregational load contributed by stabilizing selection. It would appear that no population could withstand such a drain on its members, especially if the effects of individual subvital genes were independent and, therefore, multiplicative. The problem, then, is how are these numerous genic polymorphisms maintained without increasing a genetic load so heavy as to cause the extinction of a population? The following two hypothesis have been proposed to resolve this problem:

1. Selection hypothesis of genetic load. This hypothesis proposes that observed polymorphisms are each due to heterozygote advantage and are maintained in a population by various types of stabilizing or balancing selection. Examples include balanced polymorphism, such as sickle cell heterozygotes, as well as cases where one or another allele or chromosomal inversion favoured in different seasons, regions or stages of life cycle or in different sexes. Frequency-dependent selection also serves to maintain polymorphism, and at equilibrium the alleles involved are selectively neutral, contributing no genetic load.

Supporters of the selection theory suggest that the primary cause of the extensive heterozygosity observed in nature is heterosis at hundreds or even thousands of gene loci, the cumulative effects conferring fitness. In 1967, various workers such as **Sved, King, Milkman** and others have proposed that the genetic load should not be estimated on the assumption that each locus contributes to this load independently and that the individual effects are multiplicative. In view of the known interaction between genes they suggest that the values of the sums of these effects exhibit a normal distribution in a population. *Since selection scans the entire organism and not single gene loci, only those individuals that fall below a certain threshold of fitness, that is, those with fewer heterotic loci than a minimum number required for average fitness, are collected from the population.* As a result, members of a population are divided into two classes, those above and those below a given threshold and under these circumstances it is estimated that perhaps 1,000 selectively maintained polymorphisms, each conferring an advantage to heterozygotes of around 1 per cent, could persist in a population with only minimal effects as genetic load.

2. Neutralist theory of genetic load. This theory was proposed and developed by **J.F. Crow, K. Kimura, T. Ohta,** and **T.H. Jukes** during 1968 and 1983. These workers suggest that the majority of thousands of genic polymorphisms discovered at the biochemical level have no significant effect on fitness and, therefore, selectively neutral, making no contribution to the genetic load. They propose that adaptation does occur through natural selection, but that selection acts primarily to remove deleterious mutations from a population. Proponents of the neutralist theory indicate that while mutations which result in the substitution in a protein of a compatible amino acid are detectable by sophisticated methodology, they may not be detected by natural selection because the function of the protein will remain essentially unchanged. Under these circumstances the frequencies of polymorphic genes conferring no advantage or disadvantage will be determined primarily by random genetic drift rather than by selective processes.

Genetic load in human populations. Estimates of the genetic load in the human population have been based principally on the incidence of defective offspring from marriages of close relatives (consanguineous marriages). It can be safely stated that every human individual contains at least one newly mutated gene. It can also be accepted that any crop of gametes contains, in addition to one or more mutations of recent origin, at least 10 mutant genes that arose in the individuals of preceding generations and which have accumulated in the population. The average person is said to harbour four concealed lethal genes, each of which, if homozygous, is capable of causing death between birth and maturity. The most conservative estimates place the incidence of deformities to detrimental mutant genes in the vicinity of 2 per 1,000 births.

B. Mutation

Mutation is an evolutionary agent and mutability is a required property of the genetic material if evolution is to occur. One can visualize several ways that mutation might bring about evolutionary changes. Mutation may be highly directed at a particular locus such that allele a_1 is selectively driven to the a_2 form. Alternatively, the mutation process might be random but with time the a_2 form would come to predominate over a_1. Finally, mutation might simply provide a population with new alleles (new mutational "currency") on which any and all evolutionary agents (natural selection, for example) can act.

C. Meiotic Drive and Migration Pressure

Besides directional selection (natural and artificial) and mutation pressure, meiotic drive and migration pressure are two agents that can shift gene frequencies in a population out of a Hardy-Weinberg equilibrium.

1. Meiotic drive. The Hardy-Weinberg concept of an equilibrium population assumes that alleles will segregate in a 1 : 1 fashion at meiosis and that all gametes in the pool have an equal probability of fertilizing one another. Clearly, if gametes of genotype a_1 have greater success in fertilization than gametes of genotype a_2 the frequency of a_1 in the population should increase as a_2 decreases. Such a mode of bringing about potential evolutionary change is in many respects a form of natural selection but it is called **meiotic drive**.

Normal segregation ratios can be biased by a number of factors. The most extensively studied example is the **segregation distorter (*SD*)** locus in *D. melanogaster* : a +/*SD* heterozygous male transmits to his offspring many *SD* alleles than + alleles. A high frequency of *SD* might, therefore, be expected in natural populations, but the allele is, in fact, uncommon, its frequency being less than 0.1. Some additional selective pressures, thus, appear to counteract meiotic drive at this particular locus presumably by agents that act against *SD*-carrying flies. The existence of meiotic drive is not easy to demonstrate experimentally. However, its specific role in evolution is still unassessed.

2. Migration pressure. Perhaps the most obvious way that gene frequencies in a population can be changed is the introduction of new individuals with new genotypes, a process known as **migration pressure**. The effect of migration pressure can be expressed mathematically as follows. In a population we can call X a particular gene is present with a frequency q_x (Fig. 88.3). A certain number of individuals from this population now migrate and join population Y in which the gene in question is present at frequency q_y. *The immigrant, thus, comes to represent some fraction m of the total number of individuals present in the expanded version of population Y and a new gene frequency, q'_Y, becomes established. The value of q'_Y, will be equal to the contribution made by the immigrants ($q_x m$) plus the contribution made by the original population* [q_Y (1 – m)], where 1 – m represents the proportion of non-migrants. Therefore,

$$q'_y = q_x m + q_y (1 - m)$$

and the change in q after one generation in such a population becomes

$$\begin{aligned} \Delta q &= q'_y - q_y \\ &= q_y - mq_y + mq_x - q_y \\ &= -m (q_y - q_x) \end{aligned}$$

when numerical values are substituted into this equation, it becomes clear that significant changes in the value of q can result if these two populations differ only slightly in the frequency of a given gene and if only a moderate degree of migration occurs between them.

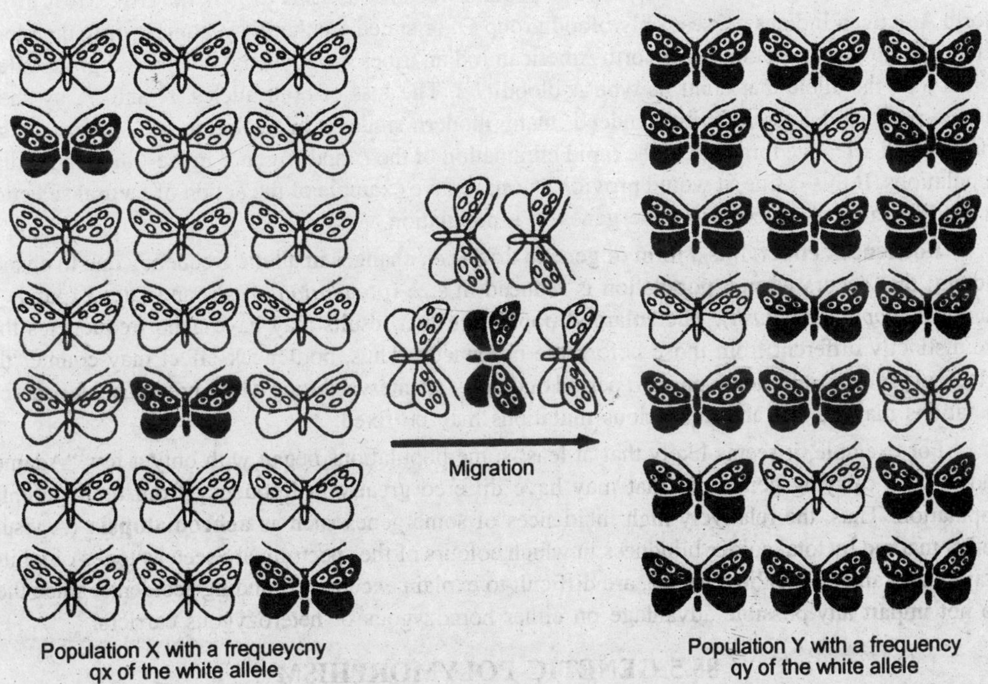

Fig. 88.3 Migration pressure as an evolutionary agent. The Y population of moths receives some immigrants from the X population, an event that changes its frequency of the white allele.

D. Random Genetic Drift

Random fluctuation in allele frequencies, called **genetic drift**, also occurs in breeding populations. The effect of genetic drift is negligible in large populations but in small breeding, populations all the limited number of progeny might be of the same type with respect to certain gene pairs because of chance alone. Should this happen, fixation or homozygosity will have occurred at the locus concerned. Fixation is defined as gene frequency reaching p = 1.00 or q = 1.00. Chance fluctuations may or may not lead to fixation.

E. Founder Principle

When a few individuals or a small group migrate from a main population, only a limited portion of the parental gene pool is carried away. In the small migrant group, some genes may be absent or occur in such low frequency that they may be easily lost. The unique frequency of genes that arise in population derived from small bands of colonizers or "founders", has been called the **founder effect** or **founder principle**. This principle was proposed in 1956 by Harvard evolutionist, **Ernst Mayr**. The founder principle essentially emphasizes the conditions or circumstances that support the operation of Sewall Wright's genetic drift. For example, North American Indian tribes, for the most part, lack the gene I^B that governs type B blood. However, in Asia, the ancestral home of the American Indians, the I^B gene is widespread. The ancestral population of Mongoloids that migrated across the Bering Strait to North America might have been very small. Accordingly, the possibility exists that none of the prehistoric immigrants happened to be of blood group B. It is also likely that a few individuals of the migrant band did carry the I^B gene but they failed to leave descendants.

Evolutionary geneticists interpret this peculiar feature in terms of genetic drift. Most of the North American Indians possess only blood group O, or stated another way, contain only the blood allele *i*. With few exceptions, the North American Indian tribes have lost not only blood group allele I^B but also the allele that controls type A blood (I^A). The loss of both alleles, I^A and I^B, by sheer chance perhaps defies credibility. Indeed, many modern students of evolution are convinced that some strong selective force led to the rapid elimination of the I^A and I^B genes in the American Indian populations. If this is true, it would provide an impressive example of the action of natural selection in modifying the frequencies of the genes in a population.

Bottleneck effect. It is a form of genetic drift (*i.e.*, changes in allelic frequency due to chance factors) that occurs when a population is reduced in size (*population crash*) and later expands in numbers (*population flush*). The enlarged population that results may have gene frequencies that are distinctly different from those before the bottleneck. Thus, bottleneck effect may counter the effects of previous selection for a short period of time — an interval during which previously favourable mutations may be lost and deleterious mutations may be fixed.

For example, it seems likely that at least some populations began with only a few "Adams" and "Eves" carrying genotypes that may have differed greatly in frequency from their parental population. Thus, the relatively high incidences of some genes such as **achromatopsia** (A visual defect marked by total colour blindness in which colours of the spectrum are seen as tones of white-gray-black) among the Pingelapese, are difficult to explain except as founding accidents, since they do not impart any possible advantage on either homozygous or heterozygous carriers.

88.5 GENETIC POLYMORPHISM

One of the most characteristic features of any natural population is its diversity. This diversity is obvious when we consider the human species, for we are attuned to sensing differences in human appearance, personality, sexuality, and so on. In the populations of flies or dandelions such well marked diversity does not occur but it exits nonetheless. In genetic terminology natural populations are said to be **polymorphic**.

Polymorphism is most apparent when it affects a visible or behavioural phenotype, but is not at all restricted to such traits. **R. Lewontin** and **J. Hubby,** in 1966, undertook the first extensive analysis of protein polymorphisms in natural population of *Drosophila pseudoobscura* by subjecting extracts of individual flies to get electrophoresis and observing the rates of migration of various proteins, which represented 18 gene loci. They found, quite unexpectedly, that many of the proteins existed in the population in the form of **isoelectric variants**, meaning that for a given type of protein some individuals possessed a fast-migrating species and others a slow-migrating species. Numerous subsequent studies of such diverse species as barley, wild oat, horse-shoe crab, mouse and humans have all produced the same result: an abundance of **protein polymorphism** is found wherever it is sought. Protein polymorphism signals the existence of **allelism**, and it has been estimated that 20 to 50 per cent of all structural gene loci in a given species exist in two or more allelic forms in any given population.

The polymorphism may arise in a population by the following three basic ways—**transient polymorphism, balanced polymorphism** and **random fixations of natural mutations.**

1. Transient polymorphism. Transient polymorphism is a by - product of directional natural selection. If we imagine that allele a_1 has a selective advantage over a_2, then with time a_1 should proceed toward fixation at p = 1, and a_2 should proceed toward elimination at q = 0. While this process is occurring, both a_1 and a_2 will be present in the gene pool and a_1/a_2 heterozygotes will be present in the population. As the name implies, transient polymorphism represents a temporary situation. For example, during the course of industrial melanization, both dark and light moths would be expected to cohabit the Manchester trees for the meantime, but the proportion of light moths would be seen to diminish with time as dark months gradually predominated.

2. Balanced polymorphism. Balanced polymorphism is also relatively permanent kind of equilibrium in which alleles a_1 and a_2 are present in the population at some steady-state frequencies. Balanced polymorphism is originated by **disruptive** or **diversifying selection** and **heterosis.**

3. Random fixation of natural mutations. The random fixation of natural mutations method of origin of polymorphism is also called **neutral mutation-random genetic drift hypothesis** or **"Non-Darwinian Evolution"** and this idea has been developed by **S. Wright** and **Kimura**. This hypothesis is based on the following two assumptions: (*i*) The first assumption states that selectively neutral mutations can occur in genes that code for proteins. This will clearly be true in the case of "synonymous" mutations in which one codon is replaced by another codon dictating the same amino acids, but it is also proposed to be true in the case of mutations that lead to amino acid substitutions. The idea is that an acidic amino acid might occur in an "unimportant" region of the protein, with the result that the emergent mutant protein is identical to the original in all functional aspects. (*ii*) The second assumption states that neutral alleles, being neither selectively advantageous nor disadvantageous, simply drift in the gene pool. Thus, if a neutral mutation arises in woman's germ cell and this germ cell gives rise to a female child, the probability is about 0.5 that the mutant allele will be transmitted to a grandchild and 0.5 that it will not. If it is not, then q becomes equal to zero and the allele is lost. Polymorphism of cytochrome c of all eukaryotes and haemoglobin protein of all vertebrates strongly support the Wright-Kimura hypothesis.

88.6 POPULATION GENETICS AND EVOLUTION

Population genetics has provided great support to the idea of organic evolution. Various aspects of evolution can be reviewed in terms of population genetics as follows:

The process of formation of new species is called **speciation**. Two genetically divergent populations can form new species only when they become geographically isolated from each other. If a large population is fragmented into two or more units which are geographically isolated from one another, each independent unit follows different evolutionary paths for the following reasons:

1. Each isolated unit of a population may have its own type of mutation which provides raw materials for organic diversity.
2. The mutations and gene combinations which appear in different isolated population units will have different adaptive values in the new environments.
3. The organisms which originally colonize a certain geographical area and form an isolated population may not be representative of the group from which they came so that different gene frequencies exist from the beginning.
4. The size of the new population may become quite small at various time so that a genetic "bottleneck" is formed, from which all subsequent organisms will arise.
5. During the period of small population size, the gene frequencies will fluctuate in unpredictable directions. The fluctuation in gene frequency is called **genetic drift**.

QUESTIONS

Long Answer Questions

1. Evaluate the origin, spread and fixation of genetic diversity in a population for understanding evolution. *(Calcutta (H) 1999)*
2. (*a*) What is Hardy-Weinberg equilibrium?
   (*b*) What are the conditions necessary for the maintenance of this equilibrium in any population? Discuss.
   (*c*) What evidence is required before concluding that an allele pair is in *Hardy Weinberg equilibrium*?
   (*d*) What kinds of sexually reproducing species are likely to show this type of equilibrium?
   (*e*) Clearly distinguish between allelic, genotypic and Hardy-Weinberg equilibrium, using a specific example to illustrate your answer.
3. Explain Hardy-Weinberg law. *(Calcutta (H) 1999; Kanpur 2001)*
4. Define Hardy-Weinberg law with particular reference to the synthetic theory of evolution. *(Calcutta (H) 1997)*
5. Explain the "genetic basis of evolution". *(H.N.B. Garhwal 2009)*

Short Answer Questions

1. Describe the significance of heterozygotes in the process of evolution.
2. What is genetic load? Discuss it in relation to evolution.
3. Explain the phenomenon of genetic polymorphism.
4. Write short note on panmixis.
5. Write an explanatory note on genetic drift. *(Allahabad 1994, 1996, 1997; Calcutta (H) 1995)*
6. Evaluate the bearing of genetic drift in evolution. *(Calcutta (H) 1998)*
7. Explain the genetic drift. *(Kerala 2001)*
8. Write explanatory note on the role of population genetics in evolution. *(Bombay 1996)*
9. Factors responsible for evolution. *(H.N.B., Garhwal 2008)*
10. Write a note on Hardy-Weinberg's law. *(H.N.B., Garhwal 2009)*

Very Short Answer Questions

1. What is a gene?
   Ans. Inherited determinant of the phenotype.
2. Define the allele.
   Ans. Allele is a alternative form of a gene.

3. What is allele frequency.

Ans. The proportion of one allele relative to all alleles at a locus in a population.

4. What is gene flow?

Ans. The movement of genes from one population to another by interbreeding between individuals in the two populations.

5. Define the gene pool.

Ans. All of the alleles available among the reproductive members of a population from which gametes can be drawn.

6. What is bottleneck?

Ans. A brief reduction in size of a population, which usually leads to random genetic drift.

7. Define the genetic load.

Ans. The relative decrease in the mean fitness of a population due to the presence of genotypes that have less than the highest fitness.

8. What is fitness?

Ans. Fitness (W) is the relative reproductive success of a genotype as measured by survival, fecundity, or other life-history parameters.

9. What is founder effect?

Ans. Genetic drift observed in a population founded by a small, non-reproductive sample of a larger population.

10. Define the population.

Ans. Population is a group of organisms of the same species relatively isolated from other groups of the same species.

11. What is a deme?

Ans. Deme is a local population of organism.

12. Define the meiotic drive.

Ans. Any mechanism that causes alleles to be recovered unequally in the gametes of a heterozygote.

13. What is natural selection?

Ans. Natural selection is a process in nature whereby one genotype leaves more offspring than another genotype because of superior life history attributes such as survival or fecundity.

14. Define neutral gene hypothesis.

Ans. The hypothesis that most genetic variation in natural populations is not maintained by natural selection.

15. What is genetic drift?

Ans. Changes in allelic frequency due to sampling error.

16. Define random mating.

Ans. The mating of individuals in a population such that the union of individuals with the trait under study occurs according to the product rule of probability.

17. What is panmictic?

Ans. Referring to unstructured (random mating) population.

18. Define polymorphism.

Ans. Two or more kinds of individuals maintained in a breeding population.

19. What is population genetics?

Ans. Population genetics is a branch of genetics that deals with frequencies of allele and genotypes in breeding populations.

20. Write the formula of Hardy-Weinberg law. (*H.N.B., Garhwal, 2008*)
   Ans. $(p + q)^2 = p^2 + 2pq + q^2$.

21. Heritable variations are "raw material" for evolution. (True/False) (*H.N.B., Garhwal, 2009*)
   Ans. True.

Fill in the Blanks

1. In natural selection .... are selected for or against and .... evolve.
2. A group of individuals with the potential to interbreed is a ....
3. A group of individuals with a high probability of interbreeding is a ....
4. A gene pool consists of all .... at all gene loci in all .... of the population.
5. When all members of a population are homozygous for the same allele, that allele is said to be .... in the gene pool.
6. Mendel described the frequency of .... for the offspring of a single .... while Hardy and Weinberg described the frequency of .... for an entire ....
7. Of all the causes of gene pool changes, only .... adapts a population to its environment.
8. A change in the gene pool of a small population due to chance is called .... Effects of this general phenomenon are of two types: .... effects and .... effects.
9. The northern population of elephant seals was reduced by hunting to only 20 animals, it then returns to a large population over 30,000. In the 24 gene loci examined by scientists, only a single allele was found at each of the loci. This is an example of allelic .... caused by a population ....
10. Genetic drift in a new colony is known as the .... effect.
11. The loss or gain of alleles in a population due to emigration or immigration is known as .... ; it decreases genetic differences among populations.
12. Non-random matings of individuals within a population is called .... and can cause deviations from the Hardy-Weinberg prediction.
13. When the environment causes a change in the gene pool by controlling reproductive success according to genotype; it is called ....
14. Diversity of genotypes at a single location is called .... variation; whereas diversity of genotypes over a broad region is called .... variation.
15. When two or more distinct forms of individuals, such as brown grasshoppers and green grasshoppers are found in the same population, such a population is known as a .... population.
16. In a polymorphic population when the frequency of each form does not change through the generations, this is called a ....
17. When and where a mutation occurs depends on ....
18. Heterozygosity protects .... from natural selection.
19. The diversity of human fingerprints is an example of .... variation, since it seems to confer no selective advantages.
20. In British land snails (*Capaea nemoralis*), there are several forms of shell colour and banding pattern. Each form is well-camouflaged from predatory birds within a particular microenvironment. This is an example of .... polymorphism; it is presented by ....
21. Selection acts on all traits that affect .... and ....
22. Darwinian fitness is the .... contribution of an individual to the gene pool of the next generation. Its values range from .... to ....
23. An individual who survives to old age but leaves no offspring has a fitness of ....

24. Abnormally large and abnormally small newborns have higher mortality rates than newborns of average weight. This is an example of .... selection.

25. When one population evolves in response to another, and vice versa, this is called .... An example is the evolution of flowers and their ....

26. Stabilizing selection tends to .... variation.

27. A population of mice, with coat colours that range from white to black, including all intermediate shades, placed in an environment of white sand and black lava and exposed to an aerial predator would experience .... selection.

28. The steady increase in body size that occurred during evolution of the horse is an example of .... selection.

Multiple Choice Questions

1. The unit of evolution is the

(*a*) species (*b*) individual
(*b*) population (*d*) social group

2. The total collection of genes, at anyone time, in a population is called the

(*a*) demotype (*b*) genotype
(*c*) multiple-allelic group (*d*) gene pool

3. The allelic frequency (*i.e.*, a change in the relative abundance of an allele) within a population, over a succession of generations, is called

(*a*) microevolution, or adaptive evolution (*b*) macroevolution, or speciation
(*c*) coevolution (*d*) phylogenetic evolution

4. Suppose there is a population of 100 snails in which shell colour is controlled by two alleles — *B* (black) and *b* (yellow). What is the total number of loci for the gene for the shell colour in this population?

(*a*) 3 (*b*) 60
(*c*) 100 (*d*) 200

5. In a snail population, 20 of the snails are yellow (*b*/*b*) and 80 are black (B/B or B/b). Of the black snails, 30 are homozygous and 50 are heterozygous. How many of the loci for this gene are occupied by the B allele?

(*a*) 80 (*b*) 90
(*c*) 100 (*d*) 110

6. In the snail population mentioned in MCQ 4, what is allelic frequency of B?

(*a*) 0.45 (*b*) 0.55
(*c*) 0.75 (*d*) 0.80

7. What will be frequency of the *b* allele in the next generation of the snail described in MCQ 4?

(*a*) 0.20 (*b*) 0.30
(*c*) 0.45 (*d*) 0.55

8. Genetic drift occurs in populations that are very

(*a*) isolated (*b*) small in number
(*c*) mobile (*d*) closely adapted to local environments

9. An isolated population of humans, with approximately equal numbers of blue-eyed and brown-eyed individuals was killed by an earthquake. Only a few brown-eyed people remained to form the next generation. This kind of change in the gene pool is called a

(*a*) hardy-Weinberg equilibrium (*b*) blocked gene flow
(*c*) bottleneck effect (*d*) founder effect

10. A potential danger to a population that has been greatly reduced in number is the
(*a*) loss of genetic; variability
(*b*) tendency toward assortive mating
(*c*) reduced gene flow
(*d*) Hardy-Weinberg disequilibrium

11. Genetic drift occurs when a few individuals of a species colonize an island. This particular phenomenon is known as
(*a*) the bottleneck effect
(*b*) the founder effect
(*c*) assortive mating
(*d*) random mating

12. The new mutations spread from one population to another by means of
(*a*) removed bottleneck
(*b*) emigrants and immigrants
(*c*) mutation pressures
(*d*) crossovers

13. If a allele suddenly becomes very abundant in a population, most-likely it is
(*a*) mutating rapidly
(*b*) flowing with emigrants
(*c*) strongly selected for
(*d*) a product of assortive mating

14. A plant population that reproduces by self-pollination is an extreme example of
(*a*) the bottleneck effect
(*b*) the founder effect
(*c*) rapid gene flow
(*d*) assortive mating

15. Natural selection is best defined as occurring when the environment causes
(*a*) differential success in reproduction
(*b*) differential mortality
(*c*) assortive mating
(*d*) a reduced gene pool

16. Microevolution can be measured by comparing observed allelic frequencies with those predicted by
(*a*) chance
(*b*) the Hardy-Weinberg equation
(*c*) mendelian ratios
(*d*) all known environmental factors

17. Human blood groups— A, B, AB and O — are an example of a
(*a*) gradualism
(*b*) cline
(*c*) gradient of diploidy
(*d*) polymorphism

18. All alleles originate from
(*a*) mutations
(*b*) crossovers
(*c*) gene flow
(*d*) non-disjunction

19. In diploid organisms, a great deal of genetic variation is conceald (*i.e.*, is not subject to natural selection) because it is present in the form of
(*a*) somatic mutations
(*b*) asexual differences
(*c*) recessive alleles
(*d*) gametic alleles

20. People who carry an allele for normal haemoglobin and one allele for sickle cell are resistant to malaria. They are examples of
(*a*) heterozygote advantage
(*b*) extreme diploidy
(*c*) outbreeding
(*d*) recessive superiority

21. Hybrid vigour is often seen in the offspring of two pure breed lines of crop plants (*e.g.*, corn). Such vigour is an example of
(*a*) heterozygote advantage
(*b*) extreme diploidy
(*c*) hermaphroditism
(*d*) recessive superiority

22. A lethal gene has a selection co-efficient of
(*a*) 0
(*b*) 0.50
(*c*) 1.00
(*d*) 100

23. A beneficial allele increases more rapidly in frequency if it is
(*a*) recently mutated (*b*) rare
(*c*) dominant (*d*) recessive

24. A deleterious allele decreases more rapidly in frequency if it is
(*a*) recently mutated (*b*) rare
(*c*) dominant (*d*) recessive

25. Interdependent genes, with related functions, form
(*a*) a coadapted gene complex (*b*) an inversion
(*c*) a fitness set (*d*) a super mutation

26. A single genotype can give rise to a range of phenotypes, called the genotype's
(*a*) norm of reaction (*b*) fitness set
(*c*) adaptive complex (*d*) supergene

27. Stabilizing selection favours
(*a*) both extreme form of a trait
(*b*) intermediate form of a trait
(*c*) environmental differences
(*d*) one extreme form over the other extreme form and over intermediate forms of a trait

28. Directional selection favours
(*a*) both extreme forms of a trait
(*b*) intermediate form of a trait
(*c*) environmental differences
(*d*) one extreme form over the other extreme form and over intermediate forms of a trait

29. Disruption selection favours
(*a*) both extreme forms of a trait
(*b*) intermediate forms of a trait
(*c*) environmental differences
(*d*) one extreme form over the other extreme form and over intermediate forms of a trait

30. In some birds, such as peacock and pheasant, the males are more colourful than the females. The selective agent producing the evolution of such conspicuous features is
(*a*) females (*b*) predators
(*c*) humans (*d*) climate

31. Quick changes in phenotypes in a small band of colonies is called
(*a*) founder effect (*b*) genetic bottleneck
(*c*) genetic drift (*d*) gene flow

32. In which condition gene ratio remains constant in a species?
(*a*) gene flow (*b*) mutation
(*c*) random mating (*d*) sexual selection

33. Genetic drift is found in
(*a*) small population with or without mutant genes
(*b*) large population with random mating
(*c*) animal population
(*d*) plant population

34. Hardy-Weinberg equilibrium is known to be affected by gene flow, genetic drift mutation, genetic recombination and ....
(*a*) evolution (*b*) overproduction
(*c*) saltation (*d*) natural selection

ANSWERS

Fill in the Blanks

1. Individuals, populations
2. Species
3. Population
4. Alleles, individuals
5. Fixed
6. Genotype, mated pair, alleles, population
7. Natural selection
8. Genetic drift, bottleneck, founder
9. Fixation, bottleneck
10. Founder
11. Gene flow
12. Assortive mating
13. Natural selection
14. Individual, geographic
15. Polymorphic
16. Balanced polymorphism
17. Chance
18. Recessive alleles
19. Neutral
20. Balanced, natural selection
21. Survival, reproduction
22. Relative, zero, one
23. Zero
24. Stabilizing
25. Co-evolution, pollinators
26. Reduce
27. Disruptive
28. Directional.

Multiple Choice Questions

1. (*c*)
2. (*d*)
3. (*a*)
4. (*d*)
5. (*d*, *i.e.*, (30 × 2) + 50 = 110)
6. (*b*, 110 loci occupied by *B* + 200 total loci = 0.55)
7. (*c*)
8. (*b*)
9. (*c*)
10. (*a*)
11. (*b*)
12. (*b*)
13. (*c*)
14. (*d*)
15. (*a*)
16. (*b*)
17. (*d*)
18. (*a*)
19. (*c*)
20. (*a*)
21. (*a*)
22. (*c*)
23. (*c*)
24. (*d*)
25. (*a*)
26. (*a*)
27. (*b*)
28. (*a*)
29. (*a*)
30. (*a*)
31. (*a*)
32. (*c*)
33. (*a*)
34. (*d*).

89

Immunogenetics and Genetics of Cancer

IMMUNOGENETICS

In 1882, Russian Zoologist, **Elie Metchnikoff** became the first to recognize that invertebrate animals possess immune defenses. On a beach in Sicily, he collected the tiny transparent larva of a common sea star. Carefully he pierced it with a rose thorn. When he looked at the larva the next morning, he saw a host of tiny cells covering the surface of the thorn as if trying to engulf it. The cells were attempting to defend the larva by ingesting the invader by phagocytosis. For this discovery of what came to be known as the **cellular immune response, Metchnikoff** was awarded the 1908 Nobel Prize, along with **Paul Ehrlich** for his work on the other major part of the immune defense, the antibody or **humoral immune response**. The invertebrate immune response shares several elements with the vertebrate immune response (see **Johnson**, 2000).

89.1 ANTIGEN AND ANTIBODIES

Vertebrates have evolved the ability to protect themselves against invading bacteria, viruses, and parasites (protists, fungi and animals) and against their own cancer cells by creating an enormous amount of immune diversity with relatively few genes. In this chapter, our attention will be concentrated on the genetic control of **immunity**, *the ability of an animal to resist infection.* The foreign substance from the bacterium virus, parasite, or cancer cell that evokes an immune response is called an **antigen**. The immune response itself is a complex interaction of various cell types, signalling pathways and other components. The immune system of a mammal can destroy millions of different antigens without harming its own cells.

The two major components of immune system are the B-and T-lymphocytes (Fig. 89.1), white blood cells that originate in bone marrow (B-cells) or the thymus gland (T-cells). The B-cells are responsible for producing very specific proteins called **antibodies** or **immunoglobulins (Igs)**, which protect the organism from antigens in three general ways: 1. Immunoglobulins can coat antigen so that they are more readily engulfed by phagocytes (white blood cells that engulf foreign material); 2. Immunoglobulins can combine with the antigens — for example, by covering the membrane recognition sites, of a virus and thereby directly prevent their ability to function; or 3. In combination with **complement**, a blood component, immunoglobulins can cause the cell to die if the antigen is from an intact cell. B-cells are the major component of **humoral immunity**, immunity controlled by antibodies in the serum and lymph; T-cells are the major components of **cellular immunity**, immunity against the infected cells.

While the B cells produce immunoglobulins, one type of T-cell is concerned with locating and destroying infected cells to prevent invading organisms from escaping detection within those infected cells. The **cytotoxic T lymphocytes** attack host cells infected by a virus, bacterium or parasite. Thus, infected cells are destroyed before new viruses, bacteria, or parasites can be produced, helping

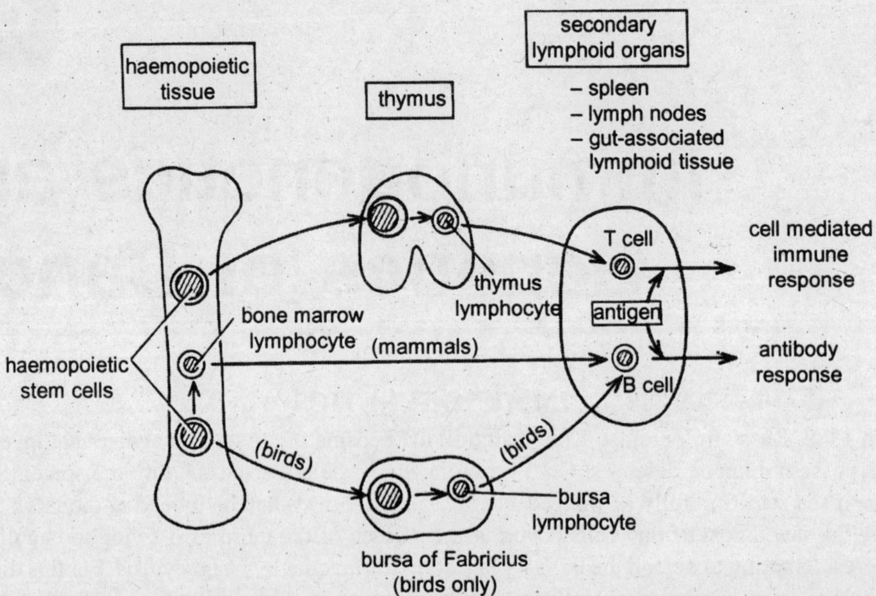

Fig. 89.1. Mode of development of T and B lymphocytes (after Alberts *et al.*, 1989).

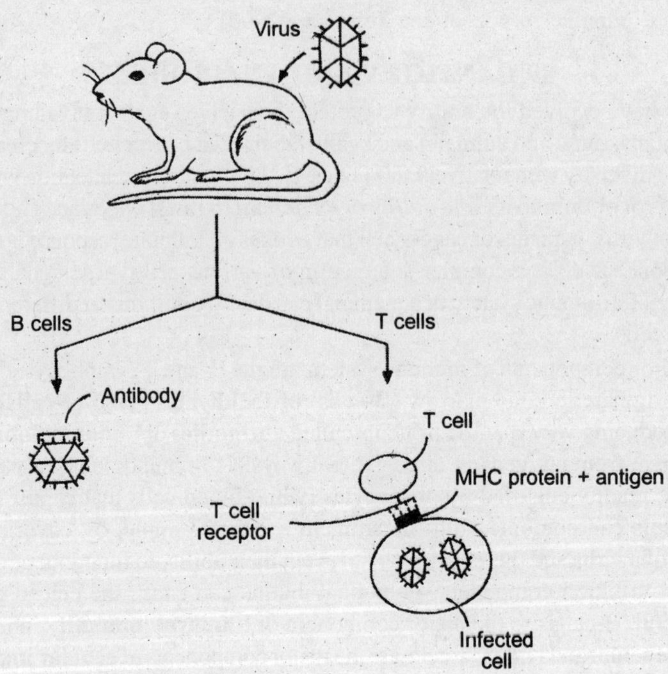

Fig. 89.2 When a mammal (*e.g.*, mouse) is infected by a virus, part of the viral coat is recognized as an antigen, triggering an immune response. B cells produce antibodies that specifically attach to the viral antigen (humoral immunity). Infected cells "present" the antigenic part of the viral coat to the outside at the major histocompatibility complex protein on the cell surface. T-cell receptors recognize this MHC-antigen complex and then trigger the destruction of the infected cell (cellular immunity) (after Tamarin 2002).

to terminate the infection. Cytoxic T-lymphocytes recognize the infected host cells by surface receptors called **T-cell receptors**. These receptors recognize an infected host cell by two aspects of the infected cell's surface **major histocompatibility complex (MHC)** gene products, and antigens. All host cells have MHC components on their surfaces; an infected cell has the ability to cause part of the antigen to appear on its surface with the MHC protein, as if the MHC protein were "presenting" the antigen to the T-cell receptor (Fig. 89.2).

The combined attack by B- and T-cells has three main components of genetic interest: 1. Antibodies (immunoglobulins); 2. T-cell receptors, and 3. Products of the major histocompatibility complex. These three protein families are evolutionarily related to each other, and each provides a diversity of protein products.

89.2 IMMUNOGLOBULINS AND THEIR GENETICS

Immunoglobulins, produced by the B-cells, are large protein molecules composed of four polypeptide chains: two identical **light chains** (about 214 amino acids) and two identical **heavy chains** (about 440 amino acids), held together by sulphydryl bonds (Fig. 89.3). Each polypeptide chains has a variable (V) and constant (C) region of amino acid sequences. The variable regions recognize the antigens and thereby give specificity to the immunoglobulins (Fig. 89.4). There are

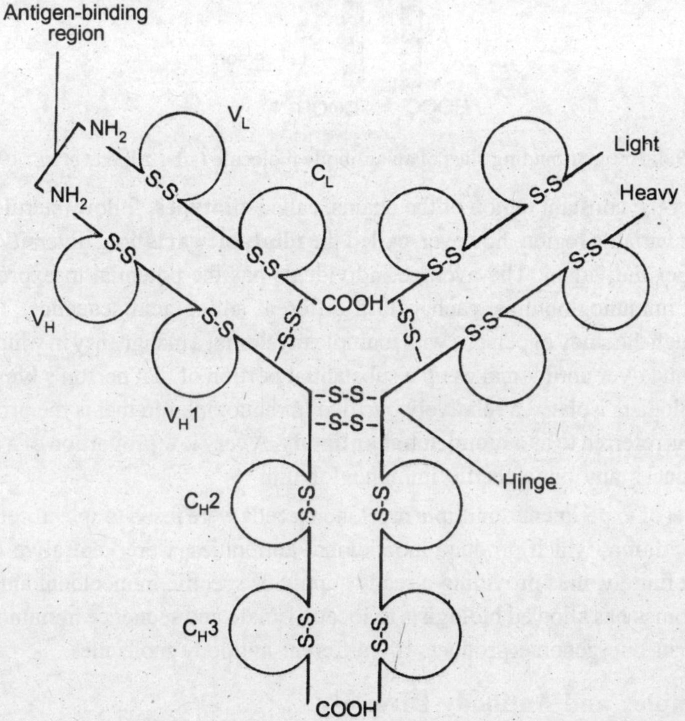

Fig. 89.3 Schematic view of an immunoglobulin protein (IgG). V = variable region, C = constant region; L = light chain; H = heavy chain. The S-S bonds are sulphydryl bridges across two cysteines (amimo acids). The NH_2 ends of the molecular form antigen-recognition parts. The internal sulphydryl bonds roughly mark areas called domains, two each on light chains and four each on the heavy chains. The heavy chain also has a hinge domain. Similar domains are found in the T-cell receptors and the MHC proteins. These domains indicate the evolutionary relatedness of these three types or molecules (after Tamarin 2002).

five major types of heavy chains (γ or gamma, α or alpha, μ or mu, δ or delta and ε or epsilon), giving rise to five types of immunoglobulins: IgG, IgA, IgM, IgD and IgE. Each has slightly, different properties; for example, only IgG (antibody) can cross the placenta, giving immunity to fetus. In addition, every immunoglobulin has one of two types of light chains, *K* or kappa and λ or lambda.

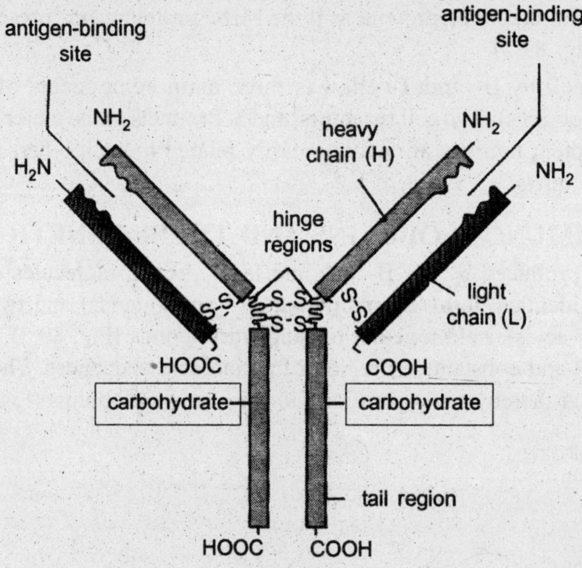

Fig. 89.4. Antigen-binding sites of an antibody molecule (after Alberts *et al.*, 1989).

Mutations of the constant region of the chains, called **allotypes,** follow the rules of Mendelian inheritance. In the variable region, however, called the **idiotypic variation**, diversity is much greater than two alleles per individual. The average individuals has the potential to express between 10^6 and 10^9 different immunoglobulins, each with a different amino acid sequence. The lower limit, 10^6 is arrived through the study of persons with multiple **myeloma,** a malignancy in which one lymphatic cell divides over and over until it makes up a substantial portion of that person's lymphocytes. From these persons, biologists isolated a relatively purified immunoglobulin that is the product of a single clone of cells and is referred to as a **monoclonal antibody**. A very low proportion of a normal person's lymphocytes produces any one specific immunoglobulin.

In another set of experiments, multiple myelonema cells were fused to spleen cells. The resulting cells, called **hybridoma**, which produce monoclonal antibodies, were kept alive (Maintained) in tissue culture indefinitely, thus providing a ready supply of specific monoclonal antibodies. Recent work with hybridomas has allowed biologists to locate, isolate and sequence immunoglobulin genes. Let us see how can one genome produce 10^9 different antibody molecules.

Multigene Complex and Antibody Diversity

Since the mammalian genome does not have 10^9 genes (10^6 genes is well accepted estimate, see **Tamarin**, 2002), different models were proposed to explain antibody diversity. In 1965, **W.J. Dreyer** and **J.C. Bennet** proposed that a given chain of an immunoglobulin was not the result of one gene, but of a combination of genes, one for the constant region and one for the variable region. In addition, they proposed that a particular organism, in its haploid genome had only one *constant gene* but several hundred or thousand *variable genes*. The final product would be the result of the action of a combination of one of the variable genes and the constant genes. Modern recombinant DNA technology has verified the meaning of the model.

In fact, several genes contribute to form the variable regions of the heavy and light chains, given that geneticists are using the word gene for DNA segment that code for a part of the final heavy or light chain of the immunoglobulin (antibody). Genes for the k or kappa, λ or lambda, a heavy chains are located on chromo-some's 2, 22 and 14, respectively, in human beings. Each is a **multigene complex**. Here we can examine the kappa (κ) light-chain gene complex as an example of how the DNA must be modi-fied to produce the final protein product (Fig. 89.5).

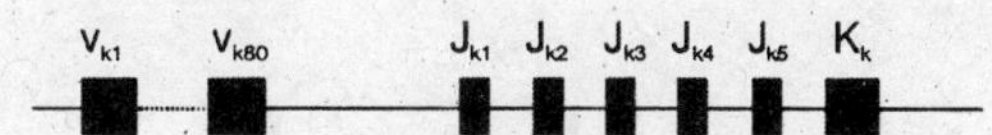

Fig. 89.5 The multigene complex for the human kappa light chain is composed of about eighty variable genes ($V_{k1} - V_{k80}$), five joining genes ($J_{k1} - J_{k5}$) and one constant gene, C_k, in the undifferentiated cell (germ line). The final kappa light chain will be composed of the product of one variable gene, one joining gene, and the constant gene (after Tamarin, 2002).

The first step in DNA rearrange-ment is the joining of a V (variable) and a J (joining) gene in a B cell (Fig. 89.5), a process called **V-J joining**. Since any one of eighty V genes can combine with any one of five J genes, four hundred different combinations are possible (80 × 5). We expect this to be an example of site specific recombination, recombinational signal (RS) sequences must be flanking all genes so that any two can be moved next to each other. Through DNA sequencing, these signals, called **recombination signal sequence**, have been determined to be a heptamer (seven bases) and a nonamer (nine bases) separated by twelve bases on one side and twenty-there bases on the other (called **the 12-23 rule**; Fig. 89.6).

5'GGTTTTTGT 3'CCAAAAACA Nonamer	12bp	CACTGTG GTG[illegible]AC Heptamer
5'CACAGTG 3'GTGTCAC Heptamer	23bp	ACAAAAACC TGTTTTTGG Nonamer

Fig. 89.6 In order for V-J joining to take place (site specific recombination), there must be signals at the V side and at the J side. One signal is a heptamer (seven base pairs) and a nonamer (nine base pairs) separated by twenty three base pairs, and the other signal is the same heptamer and nonamer separated by twelve base pairs, in reverse orientation (after Tamarin, 2002).

Two proteins, **RAG1** and **RAG2** (names are derived from *r*combination *a*ctivating *g*enes 1 and 2), form a recombinase enzyme capable of recognizing 12–23 signals and producing double strand breaks in DNA at the junction of coding (V and J), regions and signal sequences (12–23 regions; Fig. 89.7; Box 89.1). Recognition of the 12–23 signals is done with the help of **HMG1** or **HMG2** (*h*igh *m*obility *g*roup proteins 1 and 2), proteins that bind DNA and bend it, enhancing the activity of RAG recombinase. The recombinase brings together one of the variable regions and one of the joining regions and links them in a process that also generates a circle of DNA with the intervening segments, which is then eliminated (Fig. 89.7).

Box 89.1

Mutations in the genes for RAG1 and RAG2 can lead to disease, called **severe combined immunodeficiency (SCID)**. These patients are born without a functional immune system and develop lethal infections if left untreated. Treatments include living in a germ free environment (in plastic "bubbles"), transplantation with stem cells giving rise to a functional immune system, and gene therapy (**Cooper** and **Hausman**, 2007).

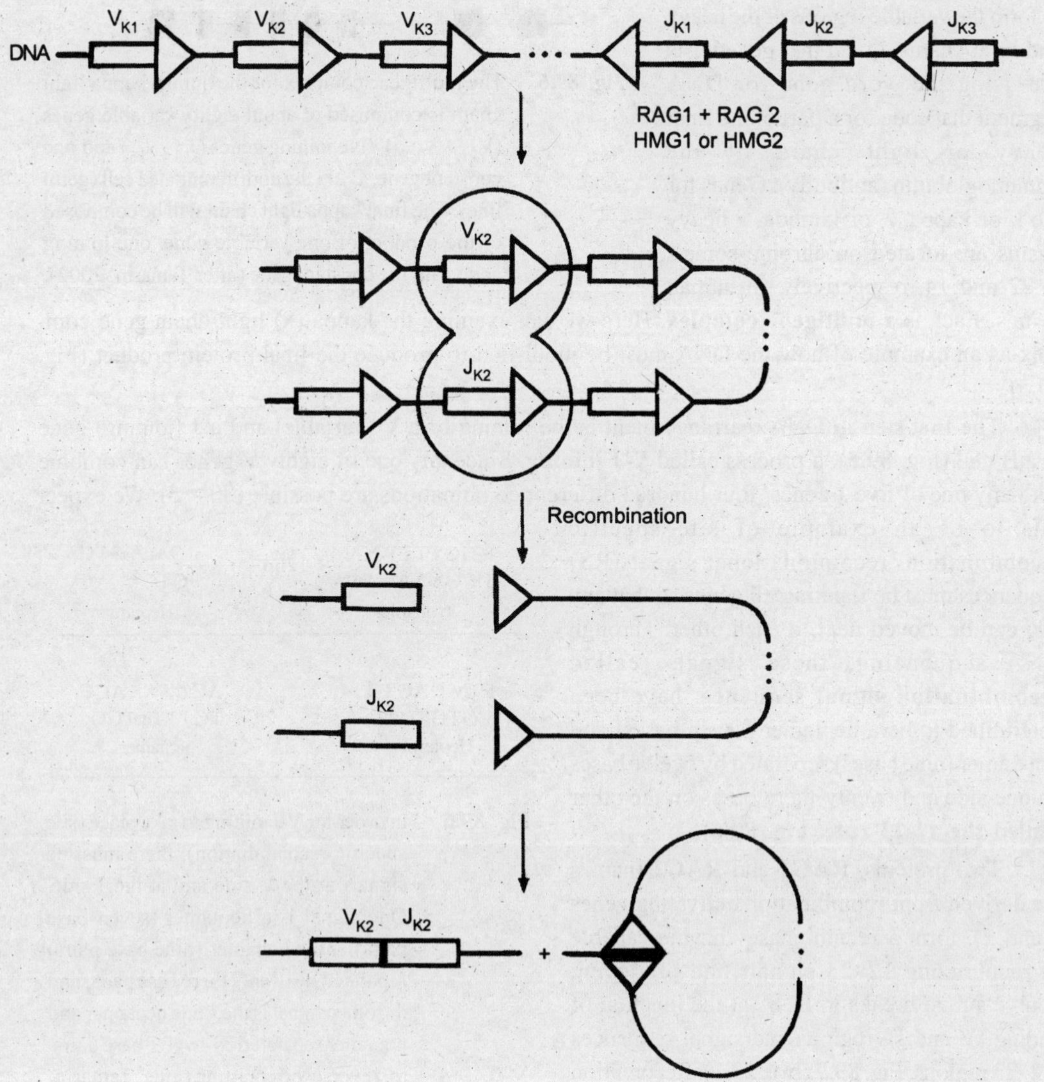

Fig. 89.7 V-J joining. The variable genes (V_{k1} to V_{k3}) are shown as boxes (left) with arrowheads as the recombination signal sequences (the 12 spacer signal); the joining genes (J_{k1} to J_{k3}) with their 23 spacer signals (right). With the RAG1-RAG2 recombinase and either HMG1 or HMG2, a variable and a joining gene are brought together. Recombination results in double-strand breaks that are then repaired so that a variable and a joining gene are spliced together and the intermediate material is released (after Tamarin 2002).

The RAG recombinase nicks (cuts) each DNA strand at its recombination signal sequence. Then by a **transesterification** (*i.e.*, by shifting a phosphodiester bond), the recombinase forms a **hairpin** and a blunt end, a double-strand break of the DNA (Fig. 89.8). The remainder of the DNA processing is done with double-strand break repair enzymes that repair radiation-induced DNA damage (Fig. 89.8). The result is an eliminated circle of intervening genes and a chromosome with a V region adjacent to a new J region.

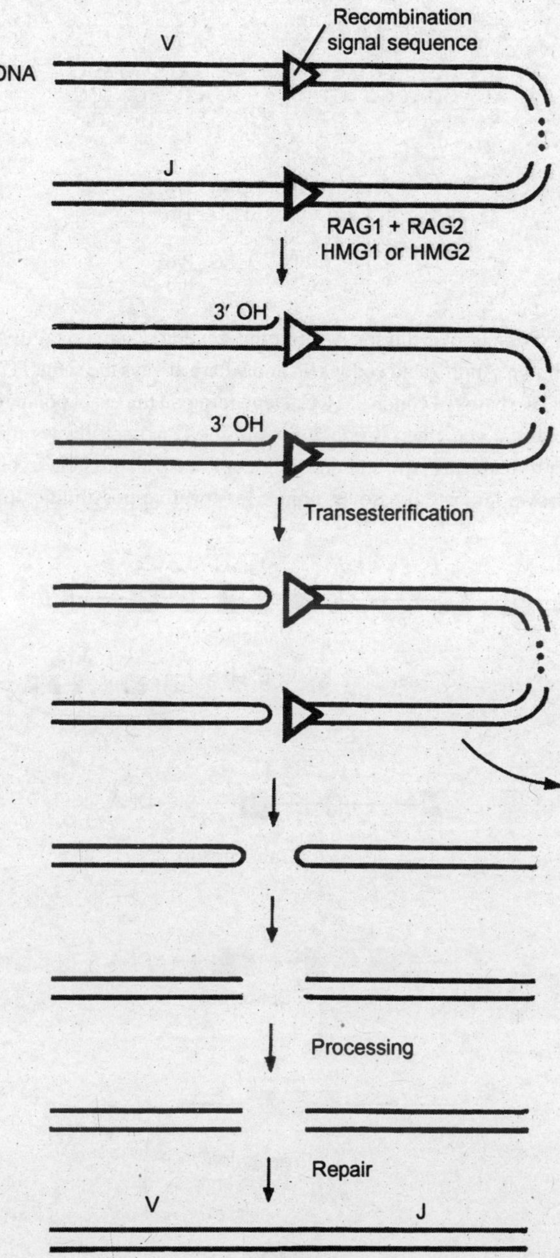

Fig. 89.8 The mechanism of site-specific recombination between a variable and joining gene. The RAG recombinase, with the help of HMG1 or HMG2, recognizes the recombination signal sequence (thin and thick arrow heads). A nick (cut) at each signal is made, producing 3′OH ends: transesterification then forms hairpin loops and recombinational signals with double-strand breaks. The hair pin loops are brought together, opened and repaired, with some processing taking place. This creates junctional diversity, including crossover point variability and N segments. The enzymes responsible for the processing are repair enzymes (after Tamarin 2002).

$V_{\kappa 41}$ $J_{\kappa 5}$

1 2 3 4 5 1 2 3 4 5

↓ ↓ ↓ ↓ ↓ ↓ ↓ ↓ ↓ ↓

... TCT CCT CCC ACA ... ▸ ... ◂ ... TGG TGG ACG ...

AGA GGA GGG TGT ACC ACC TGC

Ser Pro Trp Thr

94 95 96 97

	95		96	
1. Pro	CCG	–	TGG	Trp
2. Pro	CCT	–	TGG	Trp
3. Pro	CCT	–	CGG	Arg
4. Pro	CCT	–	CCG	Pro
5. Pro	CCT	–	CCC	Pro

Fig. 89.9 Variability in crossing over during V-J joining generates junctional diversity. In this case, V_{k41} and V_{k5} are shown. Amino acid codons 94 and 97 are always the same TCT and ACG, respectively, as are the first two bases of codon 95, CC. Depending on the exact point of crossover, five different codon pairs can be generated. Codons for proline (Pro) are always the first in each pair (95), but codons for tryptophan (Trp), arginine (Arg), or Pro are all possible second codons (96). Matching numbered arrows indicate crossover points for the five possibilities (after Tamarin, 2002).

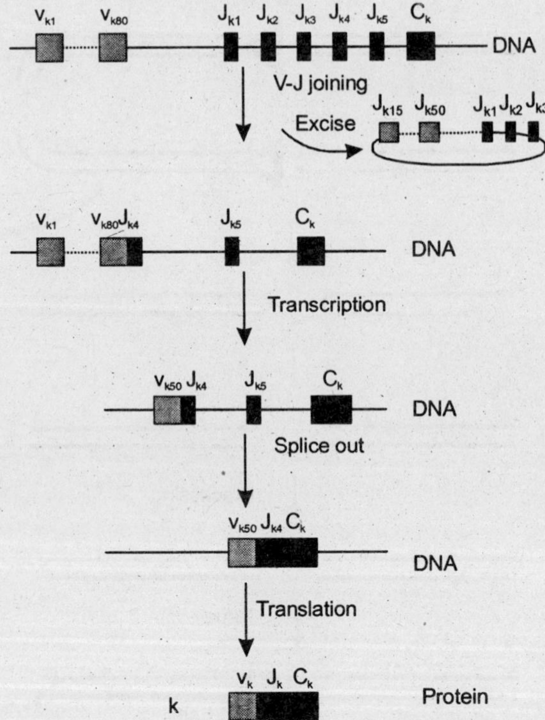

Fig. 89.10 V-J joining in the human κ region. In this example, $V_{\kappa 50}$ is joined to $J_{\kappa 4}$ and then to C_κ. First, V-J joining takes place using the heptamer-nonamer signals shown in figure 89.6. Then the region from the $V_{\kappa 50}$ to the C_κ genes is transcribed. Splicing the RNA removes the region containing the extra J genes, $J_{\kappa 5}$. The final RNA, containing $J_{\kappa 50}$ – $J_{\kappa 41}$ – C_κ, is then translated into the kappa (κ) light chain (after Tamarin, 2002).

The point of crossover at the V-J junction is itself variable, generating **junctional diversity.** Not only are any two V and J capable of coming together, but also the sequence at the junction of two genes can vary. For example, one can see in Fig. 89.9 that the junction in the protein at amino acids 95 and 96 can be Pro-Trp, Pro-Arg, or Pro-Pro, depending on exactly where the crossover occurred.

In Figure 89.10, we see the DNA after J_{k50} and J_{k4} join. This gene is now transcribed. The region between J_{k4} and C_k (the constant gene) is then removed by RNA splicing, leaving the final messenger RNA product, which is then translated into a kappa (κ) light chain. In this cell, the homologous k region is repressed as well as both λ regions; this phenomenon is called **allelic exclusion.** Thus, this cell produces only one light chain, the $V_{k50} - J_{k4} - C_k$ protein.

Similar types of events takes place in the heavy-chain gene and the lambda (λ) chain gene if it has been activated. There are some differences, however (Fig. 89.11). The lambda complex in human beings has only two variable genes, with four J genes and one C gene. The heavy chain complex has about one hundred to three hundred V genes, nine J genes, and five C genes of the five major types (g, a, m, d, e). In addition, heavy chain regions have another set of genes, called **diversity (D) genes**. At least five such genes are in the heavy chain complex, and they add still another variable region to the final protein. In the heavy chain, D–J joining first take place, then V-DJ joining; lastly, splicing creates the final heavy chain product (Fig. 89.12).

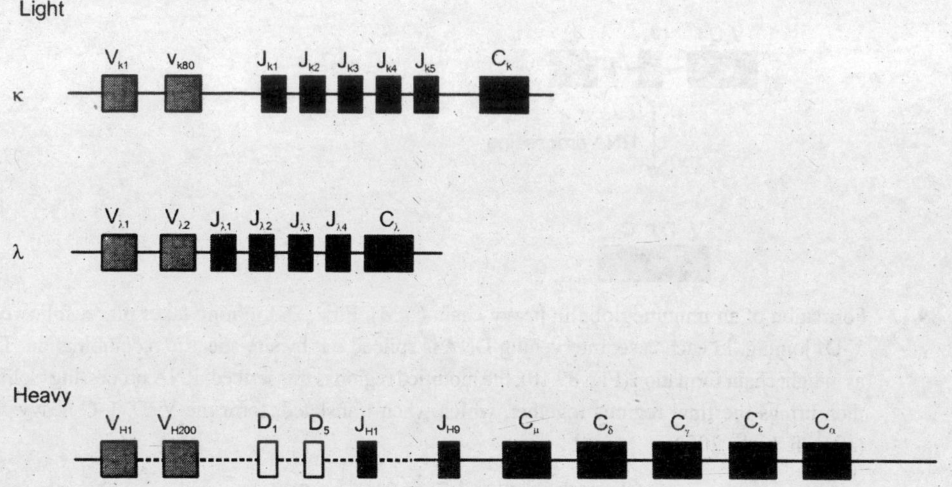

Fig. 89.11 Arrangement of the genes in the light and heavy chains of the human immunoglobulin complexes (after Tamarin, 2002).

As have been indicated earlier, the form of the heavy-chain protein in human beings has five regions or domains — CH_3, CH_2, hinge, CH_1, and variable region (Fig. 89.3). Each of the constant regions, as well as the hinge region, comes from its own exon (Fig. 89.13).The variable region, indeed comes from the extensive recombination just described: Fig. 89.12. Heavy-chain structure is based on exon shuffling (called **intron early model**).

The V – J, D – J, and V– DJ joining collectively called **V (D) J joining**, are the only examples of **site-specific recombination** in vertebrates. The genes responsible are active only in pre-B and pre-T cells.

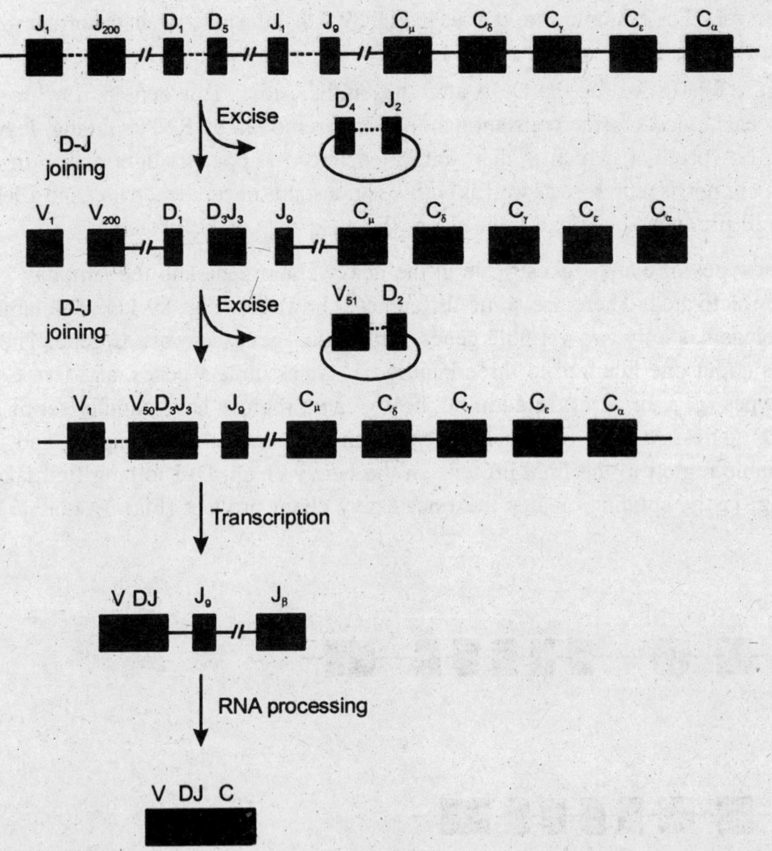

Fig. 89.12 Formation of an immunoglobulin heavy chain (IgM). First, D-J joining takes place; followed by V-DJ joining. In each case, intervening DNA is spliced out by site-specific recombination. Then, as in light chain formation (Fig. 89.10), the modified region is transcribed; RNA processing (splicing) then brings the final regions together, which when translated, form the V–D, J–C heavy chain (after Tamarin 2002).

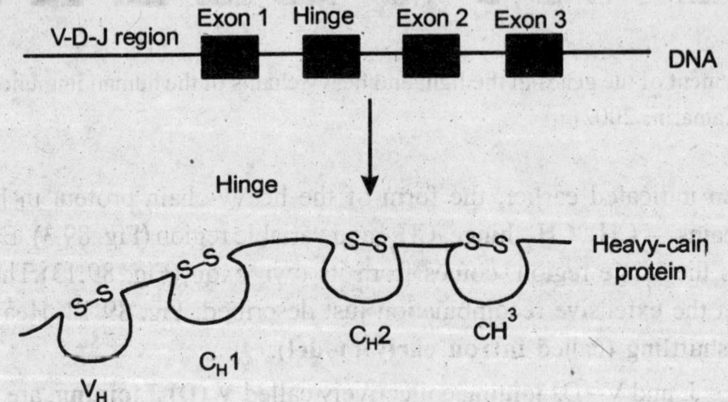

Fig. 89.13 The constant portion of heavy chain genes is made of four domains, each transcribed from its own exon (after Tamarin, 2002).

In addition to V (D) J joining, junctional diversity is also added during heavy chain recombination by the addition of nucleotides in a template-free fashion. In other words, added nucleotides, called **N segments,** appear at the joining junctions; they are not specified in the DNA.

For example, in one case, the sequence GTG, GGG, GCC (three codon long) was found at a D-J junction, but not seen in the undifferentiated (germ-line) genome. Mice that lack the gene for terminal deoxynucleotide transferase lack these N-segments; associating that gene in the process of N-segment formation. The enzyme adds nucleotides at 3′ ends of the DNA strands; these free ends are created during V (D) J joining. The enzyme is found in high levels in immature lymphocytes.

There is final way in which variability is generated. Sequencing studies indicate that mutation occurs in variable regions after recombination has taken place. The mechanism of this specific mutagenesis is called **somatic hypermutation**. Given the number of variable, constant, joining and diversity genes, as well as the variation at the joining junctions, it is easy to see how 10^9 different immunoglobulin combinations could be generated (Table 89.1).

Table 89.1 Three hundred immunoglobulin genes can generate 1.8 billion different antibodies (source: Tamarin, 2002).

Source	*Factor*
I. Light chains	
1. V genes	40 ×
2. J genes	5 ×
3. V–J recombination*	10 × = 2,000 ×
II. Heavy chains	
1. V genes	200 ×
2. D genes	5 ×
3. J genes	9 ×
4. V–D, D–J recombination	100 × = 900,000X
Total	2,000 × 900,000 = 1.8 billion

* Junctional diversity, N–segment formation and hypermutability.

Let us review the following three aspects of generation of diversity in immunoglobulins:

1. V (D) Jre combination and mutation;
2. Class switch recombination, and
3. Somatic hypermutation

1. V (D) J Recombination and Mutation

V (D) J recombination is mediated by a complex of two proteins, called **RAG1** and **RAG2,** which are specifically expressed in lymphocytes. The RAG proteins recognize **recombination signal (RS)** sequences adjacent to the coding sequences of each gene segment, and initiate DNA rearrangements by introducing a double strand break between the RS sequences and the coding sequence (Fig. 89.14). The coding ends of the gene segments are then joined to yield a rearranged immunoglobulin gene. Since these double strand breaks are joined by a **non-homologous end joining process,** the joining reaction is accompanied by the frequent loss of nucleotides. In addition, lymphocytes contain a specialized enzyme called **terminal deoxynucleotide transferase** that adds nucleotides at random to the ends of DNA molecule, so **mutations** corresponding to both the loss and the gain of nucleotides are introduced during the joining reaction. These mutations contribute substantially to the diversity of immunoglobulins.

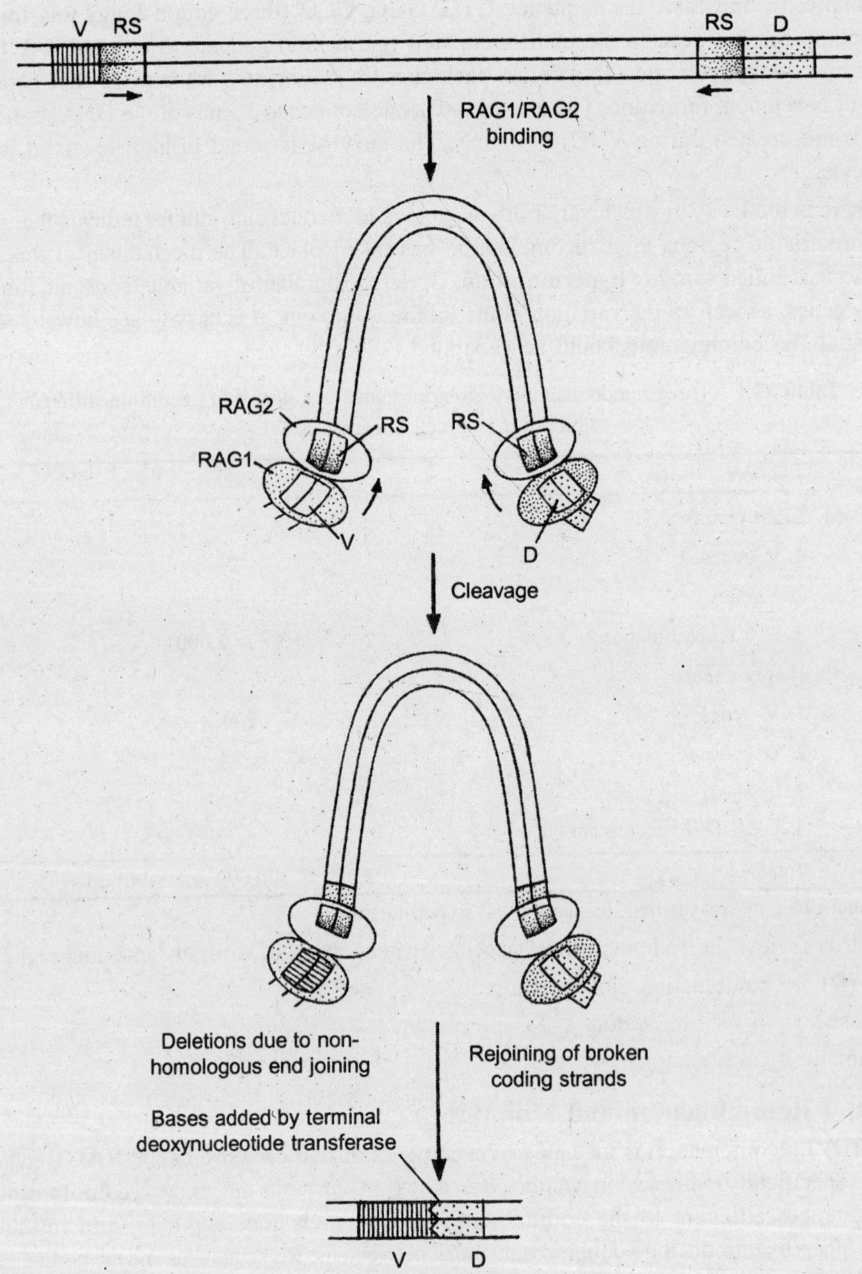

Fig. 89.14 V (D) J recombination. The coding segments of immunoglobulin and T-cell receptor genes (*e.g.*, a V and D segment) are flanked by short recombination signal (RS) sequences, which are opposite orientations at the $5'$ and $3'$ ends of the coding sequences. The RS sequences are recognized by a complex of the lymphocyte-specific recombination proteins RAG1 and RAG2, which cleave the DNA between the coding sequence and the RS sequence. The broken coding strands are re-joined to yield a rearranged gene segment. Mutations result from the loss of bases at the ends during non-homologous end joining, as well as from in addition of bases (N–segments) by terminal deoxynucleotide transferase (after Cooper and Hausman, 2007).

2. Class Switch Recombination

This process results in the reassociation of rearranged V (D) J regions with different heavy chain constant regions, leading to production of antibodies with distinct functional roles in immune response. Mammals produce four different classes of immunoglobulins — IgM, IgG, IgE, and IgA — with heavy chains encoded by a variable regions joined to the Cμ, and C_{γ}, C_{ε} and C_{α} constant regions, respectively. In addition, therefore subclasses of IgG, which are encoded by different C_{γ} regions. The different classes of immunoglobulins are specialized to remove antigens in different ways. IgM activates complement (a group of serum proteins that destroy invading cells or viruses), so **IgM antibodies** are an effective *first line defense* against bacterial or viral infections. **IgG antibodies**, the most abundant immunoglobulins in serum, not only activate complement but also bind receptors on phagocytic cells. In addition IgG antibodies can cross placenta from the maternal circulation, providing immune protection to the fetus. **IgA antibodies** are secreted into a variety of bodily fluids, such as nasal mucus, and saliva, where they can bind and eliminate invading bacteria or viruses to prevent infection. IgA antibodies are also secreted into the milk of nursing mothers, so they provide immune protection to newborns. **IgE antibodies** are effective in protecting against parasitic infections, and are also the class of antibodies responsible for allergies.

The initial V (D) J rearrangement produces a variable region joined to C_{μ}, resulting in the production of IgM antibodies. Class switch recombination then transfers a rearranged variable region to a new downstream constant region, with deletion of the intervening DNA (Fig. 89.15). Recombination occurs between highly repetitive sequences in **switch (S) regions** that are located immediately upstream of C region. The switch regions are 2–10 kb in length and recombination can take place anywhere within these regions, so class switching is more properly referred to as **region specific** rather than site specific recombination event. Because the switch regions are within introns, the precise sites at which class switch recombination takes place does not affect the immunoglobulin coding sequence.

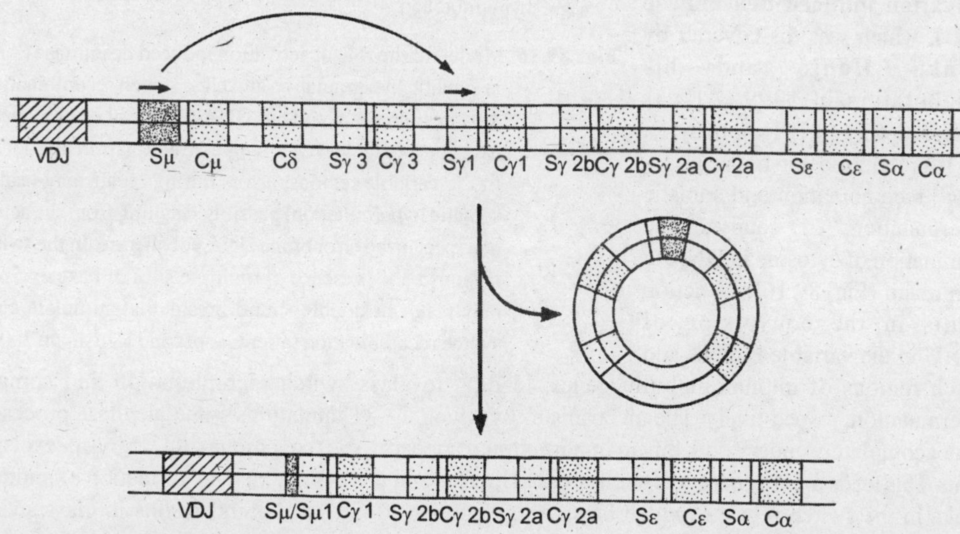

Fig. 89.15 Class switch recombination. Class switching takes place by recombination between repetitive switch (S) regions upstream of a series of constant (C) regions in the heavy chain locus (the mouse locus is shown). In the example shown, a V(D) J region is transferred from C_{μ} to $C_{\gamma 1}$ by recombination between the S_{μ} and $S_{\gamma 1}$ switch regions. The intervening DNA is excised as a circular molecule (after Cooper and Hausman, 2007).

3. Somatic Hypermutation

This process increases the diversity of immunoglobulins by producing multiple mutations within rearranged variable regions of both heavy and light chains. These mutations, principally **sin-gle base substitutions**, occur with frequencies as 10^{-3}, approxima-tely, a million times higher than normal rates of spontaneous muta-tion. They lead to the production of immunoglobulins with a substa-ntially increased affinity for anti-gens, and therefore are an impor-tant contributor to an effective immune response.

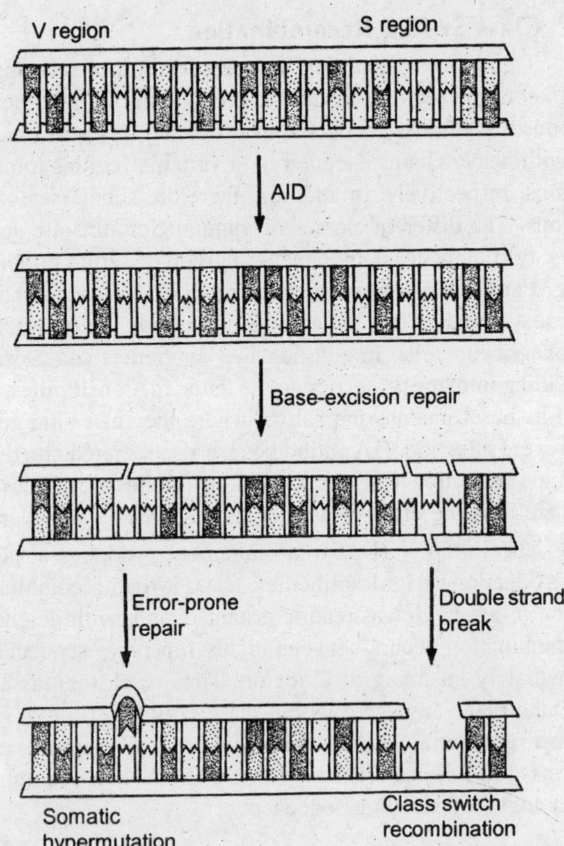

Fig. 89.16 Model for the role of activation induced deaminase (AID) in somatic hypermutation and class switch recombination. AID deaminates C to U in DNA. Removal of U by base-excision repair leaves a single-strand gap in the DNA. In the variable regions, errors during repair may lead to somatic hypermutation, possibly resulting from the action of a specialized error prone DNA polymerase. In the switch response, the presence of multiple sites of base excision may result in double strand breaks that stimulate class switch recombination (after Cooper and Hausman, 2007).

The AID Enzyme. Class switch recombination and somatic hypermutation are novel types of programmed genetic alterations, and the molecular mechanisms involved are very active areas of investigations. A key player in both processes is an enzyme called, **activation induced dea-minase (AID)**, which was dis-covered by **Tasuku Honjo** and his collaborators in 1999. AID is expressed only in B lymphocytes, and it is required for both class – switch recombination and somatic hypermutation. AID catalyzes the deamination of cytosine in DNA to form uracil (Fig. 89.16). Its action results in the conversion of C $\rightarrow$ U in the variable regions and switch regions of immunoglobulin genes, leading to class switch recombination and somatic hypermutation, respec-tively. The mechanisms by which C $\rightarrow$ U mutations stimulate these processes are not completely understood, but an important step appears to be the removal of U by base–excision repair. This leads to information of single–strand breaks, and it is thought that the formation of multiple breaks in the switch regions (which have a high content of GC base pairs) results in class switch recombination. In the variable regions, somatic hypermutation is thought to result from a high frequency of errors during repair of the C $\rightarrow$ U mutations, possibly resulting from repair by specialized error-prone DNA polymerases. Although the details of these processes remain to be explained, it is clear that AID enzyme is an extremely interesting with the novel role of introducing mutations into DNA at a specific stage of development (see **Cooper** and **Hausman,** 2007).

89.3 T-CELL RECEPTORS AND MHC PROTEINS

As has been mentioned earlier, genetic diversity also exists in the T-cell receptors and the major histocompatibility complex (MHC). From its function (recognizing both the antigen and the MHC "self" gene product), it seems evident that the T-cell receptor genes are very similar to the immunoglobulin genes. T-cell receptors are composed of α and β subunits; there are V, J and C components of the α subunit and V, J, D, and C components of the β subunits (Fig. 89.17). In a sampling of T-cell receptors from one individual, approximately one million different β chains and 25 different α chains were found, yielding approximately 25 million (1 million × 25) different T-cell receptors.

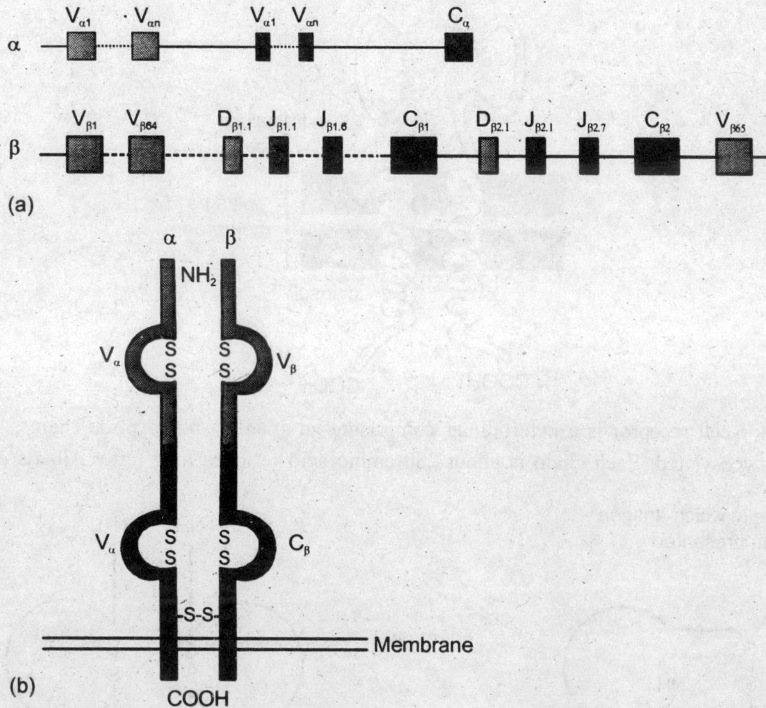

Fig. 89.17 The T-cell receptor is made up of two proteins, α and β, anchored in the cell membrane. Each protein is similar to an immunoglobulin chain and is created in the same way, with V-J types of joining taking place. A–The α gene complex is composed of numerous V and J genes and one constant gene. The β gene complex has V, D, J and C genes. B – Each protein chain has two domains, similar to the domains of the immunoglobulins, indicating a common evolutionary ancestry (after Tamarin, 2002).

The major histocompatibility complex (MHC) region (also known as the **human leukocyte antigen** or **HLA** region in people) comprises a region of 3.6 million base pairs with 224 identified gene loci. The genes are generally referred to as class I, II, and III genes. **Class III genes** code for proteins in the complement system, which is involved in the destruction of foreign cells. **Class I and II genes** code, in part, for proteins that present antigens to T_i cells. That is, class I and II proteins form structures with grooves on their surfaces that are shaped to hold small polypeptides. These polypeptides can be normal breakdown products of cellular metabolism in healthy cells ("self" proteins) or parts of foreign invaders or their gene products in infected cells. Although similar, the two types MHC proteins are found in different places and serve somewhat different functions.

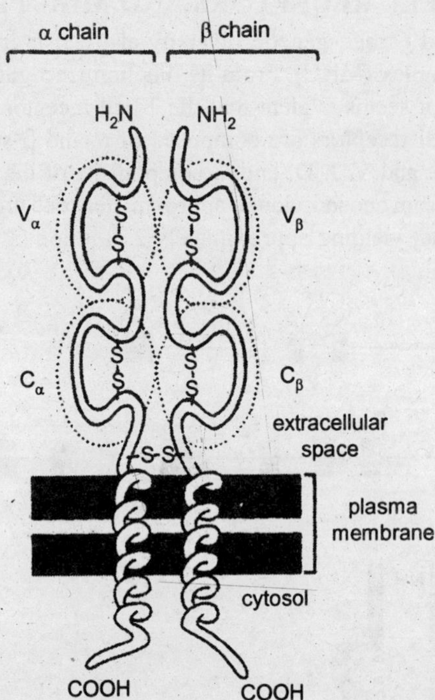

Fig. 89.18 A T-cell receptor is a heterodimer comprising an α and β polypeptide chains, both of which glycosylated. Each chain is about 280 amino acid residues long (after Alberts *et al.* 1989).

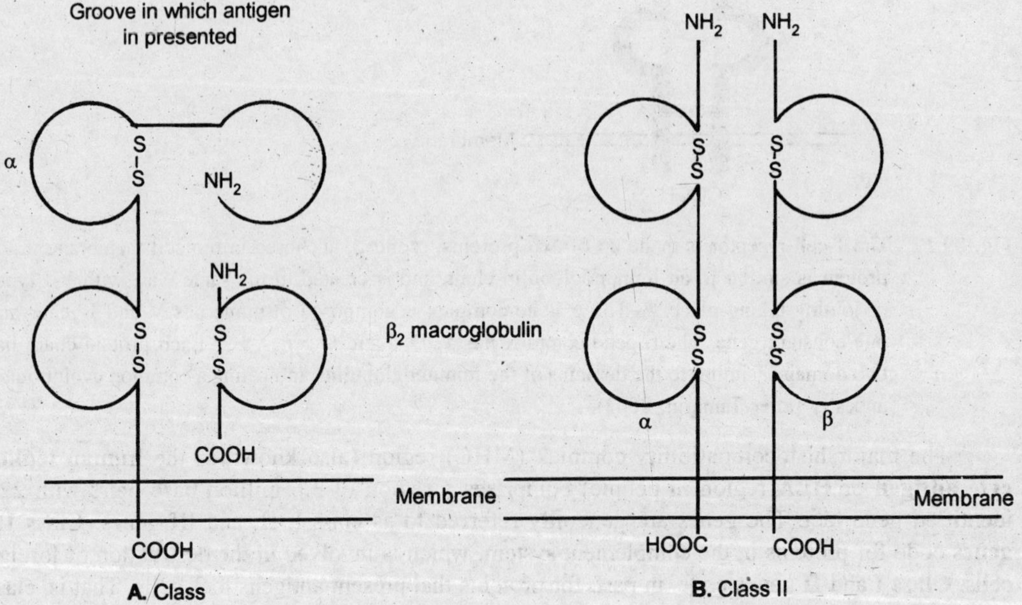

Fig. 89.19 The major histocompatibility complex (MHC) class I protein (A) is composed of two protein in chains. The α chain is composed of three domains similar to the immunoglobulins and T-cell receptors. The second chain is β_2 macroglobulin. The MHC II protein (B) is composed of an α and a β chain. The MHC proteins present antigens to the T-cell receptors to signal that a foreign agent has invaded the cell (after Tamarin 2002).

Class I MHC proteins consist of a membrane-bound α chain and a second chain called β_2 macroglobulin (Fig. 89.19A). **Class II protein consists of** α and β chains (Fig. 89.19B). These two proteins present antigens somewhat differently. In the class I molecules, groove is bound on both sides so that the presented polypeptide is small and defined. In the class II protein, the groove is unbound, allowing for a longer polypeptide.

1. Class I MHC proteins. These proteins are found on almost all cells. The polypeptides that are infected cell presents with the MHC I proteins come from the breakdown of proteins within the cytoplasm of the cell. Peptides are targeted for breakdown when a ubiquitin molecule binds to the protein, sending a cellular signal that the protein is to be degraded. (*Ubiquitin* is a small polypeptide of 76 amino acid residues highly conserved in eukaryotes). The ubiquitin-tagged protein is unfolded, in a ATP-dependent process and then fed into a **proteosome**, a barrel-shaped cellular organelle for protein breakdown (Fig. 89.20). Then the peptide fragments associate with two proteins, together called **TAP** (Transporter for Antigen Processing) that prevent further degradation of the peptide as well as transport the peptide into the endoplasmic reticulum, where the peptide binds

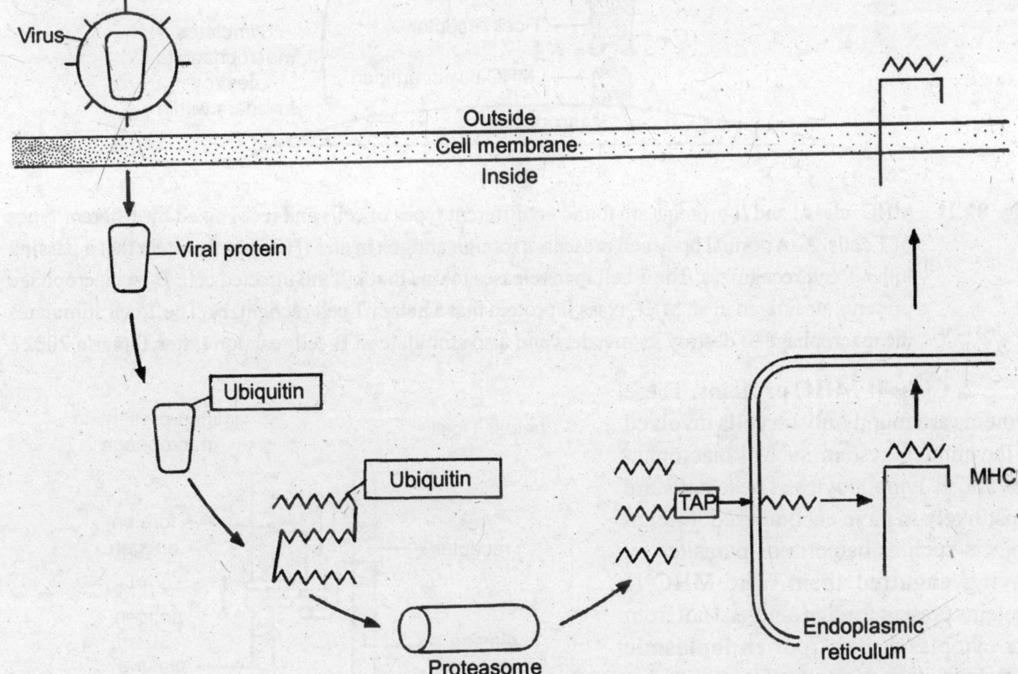

Fig. 89.20 Diagram showing how MHC I protein obtains foreign peptide to display at the cell surface. In this example, a virus attacks a cell. The viral protein is recognized as foreign and is tagged with ubiquitin. The tagged protein is then unfolded and fed into a proteosome with the aid of TAP, a piece of degraded protein enters the endoplasmic reticulum, where it combines with the MHC I protein, which is then transported to the cell surface (after Tamarin, 2002).

to the class I MHC proteins. The MHC I proteins with antigen are then transported to cell surface. Passing T-cells, called **killer T-cells** or **CD8T cells** because of their CD8 receptor protein, recognize the foreign antigen presented by the MHC I protein and release substances that kill the infected cell (Fig. 89.21). (White blood cells are classified by their surface antigens; the CD designation comes from cluster of differentiation antigens used for this purpose).

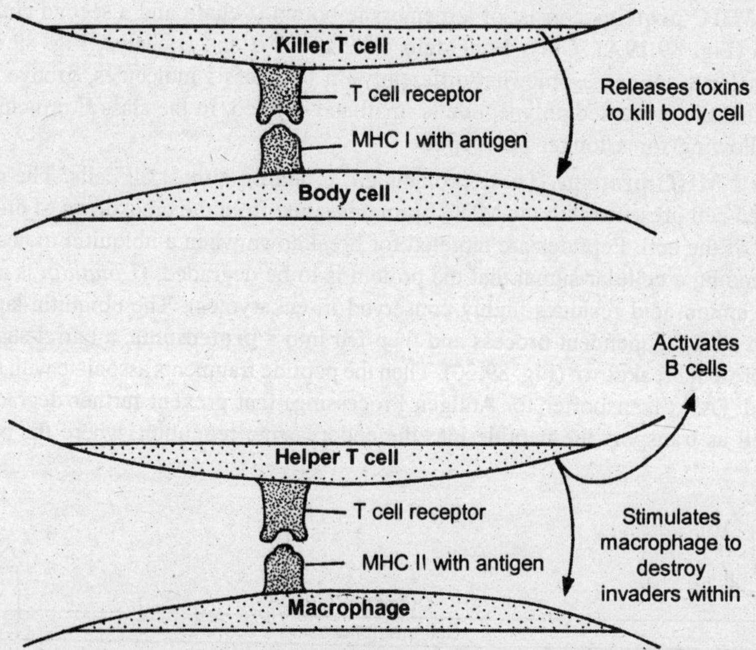

Fig. 89.21 MHC class I and II proteins are found on different types of cells and recognized by different types of T cells. A–A normal body cell presents a foreign antigen in an MHC class I protein that a passing killer T cell recognizes. The T cell then releases toxins that kill the infected cell. B–A macrophage presents an antigen in an MHC class II protein that a helper T cell recognizes. The T cell stimulates the macrophage to destroy its invaders and also stimulates a B cell reaction (after Tamarin 2002).

2. Class II MHC proteins. These proteins are found only on cells involved in the immune system, such as macrophages and B lymphocytes. These cells are most likely to have encountered foreign objects such as bacteria or parasites by having engulfed them. The MHC II proteins present foreign antigens not from the cytoplasm, but from **endoplasmic vesicles** within the cells. These vesicles form by budding from the cell surface and often contain foreign proteins and protease enzymes. The MHC II proteins migrate into the vesicles, where they pick up foreign polypeptides and then migrate to the cell surface. The response of passing T-cells to the presentation by MHC II proteins is different from the response to MHC I proteins. The MHC II proteins with antigens are recognized by helper T-cells, also called CD4 cells because of their surface receptor protein. Rather than kill cells such as infected macrophages that are useful in the immune system, the helper T-cells stimulate the macrophages to destroy the foreign bacteria in their endosomic vesicles. The helper T-cells are the prime targets of the HIV virus, making individuals and other immune problems.

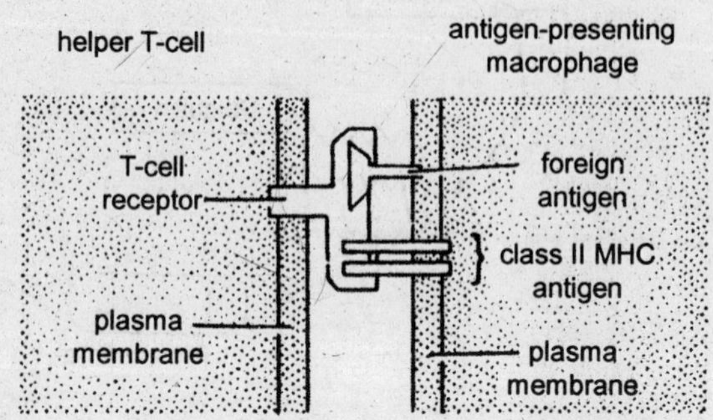

Fig. 89.22 Activation of helper T-cells by antigen-presenting macrophages involves an interaction between T-cell receptor and both foreign antigens and class II MHC antigens.

Lastly, the loci for MHC proteins do not have V(D) J joining to produce the high levels of variability found in B- and T-cells. The MHC loci are, however, very variable, with many alleles. (This variability is one reason why organ transplants are usually rejected without immunorepressive therapy). Each individual can have only two alleles at each locus, but hundreds of alleles exist in any population. Presumably, the different alleles allow for somewhat different affinities for different antibodies. They may have been selected over evolutionary time to give certain individuals in a population more chances to be able to identify and eliminate foreign substances (see **Tamarin**, 2002).

GENETICS OF CANCER

All cancers are **genetic** in nature. They originate due to alterations of genes that control cell growth. Most evidence indicates that cancers are **clonal**, *i.e.*, they arise from a single aberrant cell that then proliferates. Therefore, analysing the causes of cancer involves the understanding how one cell is changed or **transformed**, from a normal cell to a cancerous one. According to **clonal evolution theory of cancer**, most cancers come about from a series of genetic changes, progressively cancerous one (see **Tamarin** 2002).

Most evidently, cancers are caused by either **mutation** or **viruses.** Viruses are known to bring cancer causing genes (oncogenes) into cells, where their mutated forms or improper locations can lead to cancer. Thus, both the mutational and viral views of cancer are ultimately concerned with mutation. In other words, it can be said that *cancers result from those genes which were changed by mutation or were imported or activated by viruses.*

89.4 MUTATIONAL NATURE OF CANCER

Both forms of mutations, whether *point mutations* or *chromosomal mutations* have been investigated to be carcinogenic (cancer – causing; see Table 89.2). For example, the disease **xeroderma pigmentosum** (a human genetic condition inherited as a recessive autosomal trait that is caused by a defect in mechanisms that repair DNA mutations – as those caused by ultraviolet light) in human beings is caused by mutations in any of seven loci ($X_{pA} - X_{pG}$) that inactivate the mutation repair system that corrects UV-light damage; exposure to the sun then result in skin lesions that often become malignant. A related disease, **ataxia – telangiectasia**, is caused by a defect in the double-strand break repair mechanism, often the result of X-ray induced damage. (*Note: Ataxia* refers to difficulty in balance; *telangiectasia* refers to dilated blood vessels in the eye membranes). By binding to the ends of DNA, as would happen when a double-strand break occurs, ATM (the protein product of *ataxia – telangiectasia* locus, *atm*) begins a signalling pathway that tells the cell that there are broken ends of DNA. Persons with this defect are at risk for acute and chronic leukemia and lymphomas; women with it are also at risk for ovarian cancers.

Table 89.2 A small sample of chromosomal re-arrangements associated with specific cancers (Source: Tamarin, 2002).

Disease	*Chromosomal re-arrangement*	*Gene affected*
1. Burkitt's lymphoma	t (8; 14)*	C-MYC
2. Non-Hodgkin's lymphoma	t (3; 4)	Laz3, BCL-6
3. B-Cell chronic lymphocytic leukemia	t (11; 14)	BCL-l, PRAD-l
4. Follicular lymphoma	t (14; 18)	BCL-2
5. T/B-cell lymphoma	Inversion, chromosome 14	TCR-a
6. Chronic myelogenous leukemia/acute lymphocytic leukemia	t (9; 22)	CABL
7. Ewing's sarcoma	t (11; 22)	FL11, EWS
8. Melanoma of soft parts	t (12; 22)	ATF1, EWS
9. Liposarcoma	t (12; 16)	CHOP, FUS

* the notation of the form t (8; 14) indicates a translocation between chromosome 8 and 14.

Most types of cancer are associated with chromosomal defects; improved chromosomal banding techniques have shown that a specific chromosomal defect is often associated with a specific cancer (Table 89.2). The conclusion is that when a gene is in a new location (because of translocation or deletion of intervening material) that gene may fall under the control of more powerful promoters and promoters outside the range of that gene's normal control. Genes that are known to be able to transform cells (**oncogenes**) are often the ones that are relocated into regions of new control. These oncogenes then become more active, and transformation follows.

Cancer-Family Syndrome

In some cases, a proneness for malignancies (cancer) is inherited. When four thousand clinic registrants were interviewed, almost half reported virtually no family history of cancer, whereas 7% reported that many family members had cancer. This 7% segment was considered cancer prone because three or more close relatives of the interviewed person had cancer. The conclusion of this study is that some families are prone to the cancer, but most are not. This conclusion cancelled the earlier idea that everyone in the population has a uniform and low probability of developing cancer. This conclusion also supports the **cancer-family syndromes**, in which family members seem to inherit a normal proneness toward cancers of various types. According to **Tamarin** (2002), about twenty cancer family syndromes are reported. In Fig. 89.23 a pedigree for a cancer family syndrome in

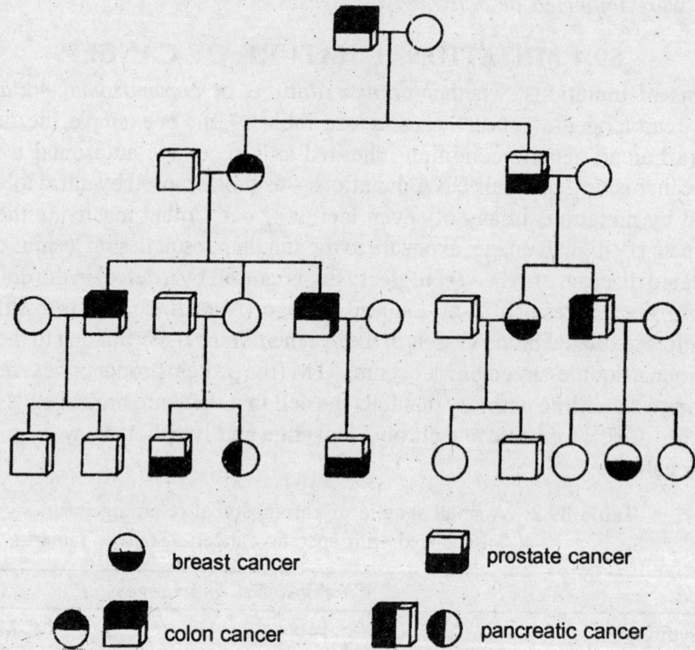

Fig. 89.23 Pedigree of a type I cancer-family syndrome. This is interpreted as the inheritance of the susceptibility for cancer rather than the inheritance of any specific type of cancer (after Tamarin, 2002).

which proneness for several different types of cancer seems to be inherited has been depicted. The female members of these families suffer from breast cancer, colon cancer and pancreatic cancer, where as male members are ailed by cancer of colon, prostate and pancreas.

Tumour-Suppressor Genes (Anti-oncogenes)

There exist certain genes which act by suppressing malignant growth. Such genes are known as **tumour-suppressor genes** or **anti-oncogenes.** When these genes are mutated in their recessive and homozygous state then they fail to suppress the cancer in the individual. More than twenty tumour-suppressor genes have been reported so far. Here, we will consider only following three such genes.

1. Retinoblastoma. This is a tumour of the retinoblast cells which are precursors of cone cells of retina of the eye. Retinoblastoma is a disease which affects young children, because after the retinoblast cells differentiate, they no longer divide and clearly no longer form tumours. This disease occurs both in a **hereditary** and a **sporadic** form. Both forms are suspected to be due to recessive homozygous state of locus. In the *hereditary form*, individuals inherit one mutant allele; a second mutation results in the disease. In the *sporadic form*, both alleles have apparently mutated spontaneously in the somatic tissue of retina. The retinoblastoma gene has also been implicated in certain other types of cancer such as sarcomas and carcinomas of the lung, bladder and breast.

Let us see that how does one know that retinoblastoma results from the loss of suppression rather than simply the activity of an oncogene? **J. Yunis** (1983), who examined cells from several retinoblastoma patients, found a frequently deleted part of chromosome 13, specifically band q14. **Yunis** noticed that the exact points of deletion varied from individual to individual, indicating that the phenomenon was due to loss of gene action rather than enhancement of gene activity due to the new placement of genes previously separated by the deleted material.

Molecular biology of retinoblastoma. Normally the retinoblastoma protein, **RB**, inhibits the cell cycle from advancing. If the appropriate check point is passed, RB is phosphorylated by cyclin-dependent kinase and cyclin complexes and the cell cycle progresses. The protein product of a homozygous recessive mutant, RB, no longer inhibits the cell cycle advance, even if the check point has not been cleared. Thus, DNA-damage and cancerous cells are allowed to continue to grow.

The retinoblastoma gene has been isolated and cloned. The gene specifies a 105-kilodalton protein (p105) found in the nucleus and acts as a suppressor of DNA transcription. It binds with at least three oncogenic proteins: (1) The **E1A protein** of adenovirus; (2) The **SV–40** (a simian virus) large T antigen; and (3) the **16E7 protein** of human papilloma virus, a virus associated with 50% of cervical carcinoma. From this, it can be concluded that these viruses may use a similar mechanism in transformation and this mechanism involves inactivation of the retinoblastoma p105 protein.

2. Wilm's tumour. This is a kidney cancer of children and is also called **nephroblastoma** and described by **Max Wilms** in 1899. Its genetics and was studies by **E. Stanbridge** and his colleagues in 1990. Wilm's tumour is also believed to be caused by loss of action in a tumour suppressor gene. This tumour is associated with the loss of band p13 on chromosome 11.

3. The p53 gene. The third tumour suppressor gene is the p53 gene, named for its 53-kilodalton protein product and located on chromosome 17. This gene is the most common mutation in cancers and is found in more than 50% of human tumours. The p53 protein molecule had gained the status of *Science* magzine's 1993 "**Molecule of the year**". Since the p53 protein is found in so many cases, it is clear that its role as a tumour suppressor was of great importance in the normal activity of cells. Normally, p53 protein is highly unstable: the MDM2 (or Mdm2; it is named for the cell culture in which it was cloned – *m*urine *d*ouble *m*inute chromosome clone number 2; see **Tamarin** 2002) protein binds its amino-terminal end and ubiquinates it, leading to the rapid degradation of p53 in the proteasome within several minutes. However, p53 is stabilized when it is phosphorylated by cell cycle checkpoint kinases. For example, ATM (a protein kinase) bind to double-stranded DNA breaks. Bound this way it activates the protein CHK2, a checkpoint kinase that then phosphorylate p53 protein. In the active state, p53 is a transcription factor that induces atleast 34 different genes, genes involved either in stopping the cell cycle, inducing apoptosis (*i.e.*, programmed cell death), or regulating itself.

Thus, p53 protein stops the cell cycle to give the cell a chance to repair its DNA. Cell growth is asserted by the induction (also called **upregulation**) of cyclin-dependent kinase inhibitors (proteins such as p21, WAF1 and CIPI). This action stops the cell cycle. In fact, if DNA repair does not take place, cells can be forced to persist permanently in G1 phase. Alternatively, the p53 protein can induce cell death by upregulating the *bax* gene. Its protein is involved in the pathway to induce **caspases**, proteinase enzymes, that destroy the cell (**Note:** Caspases get their name from the fact that they are *c*ysteine requiring *asp*artic acid proteinases. The *bax* gene's name comes from *b*cl – 2 *a*ssociated *x* gene and name *bcl – 2* is from *B* – cell *l*eukemia/*l*ymphoma – *2*).

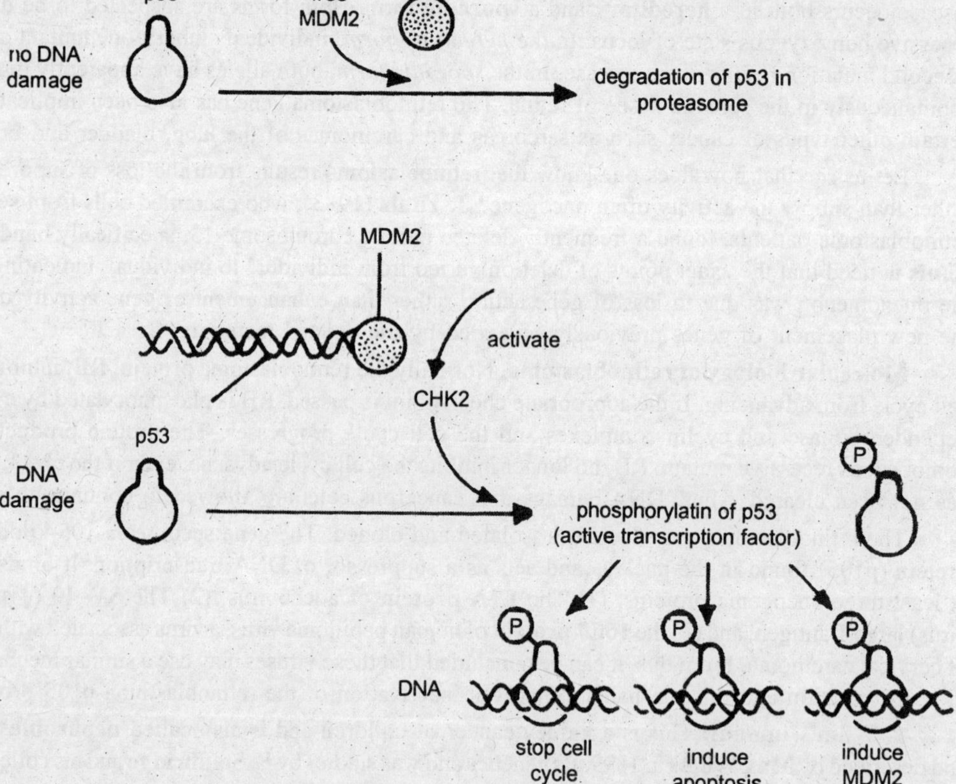

Fig. 89.24 Normally, MDM2 protein ubiquinates the p53 protein, which leads to its degradation in the proteosome. If there is DNA damage that produces broken DNA ends (double-strand breaks), MDM2 binds to these ends and activates the checkpoint kinase CHK2. The CHK2 phosphorylates p53 protein, making it stable and an active transcription factor. Next, the active *p*53 protein binds to the promoters of synexpression group genes for stopping the cell cycle and apoptosis. In addition, p53 protein induces the *MDM2* gene, which can then causes p53 degradation if the broken DNA has been repaired (after Tamarin, 2002).

Lastly, the p53 protein is a transcription factor for the MDM2, the protein that regulate p53 protein. Thus, the p53 protein has narrow window in which to stop the cell cycle or induce apoptosis, giving the cell a chance to repair its DNA damage or commit suicide. After this, p53 protein is itself repressed (Fig. 89.24).

Loss of p53 protein activity allows DNA damage to build up in a cell. This is why more than 50% of cancers involve loss of p53 activity.

89.5 VIRAL NATURE OF CANCER

Geneticists gained most of the information regarding oncogenes from studies of **RNA tumour viruses** or **retroviruses**. The name retrovirus is derived from the fact that these viruses store their genetic information in a single-stranded RNA genome and then convert it into a homologous double-stranded DNA form after infecting a host cell. Thus, they use a "backward" ("retro") flow of genetic information. Usually genetic information flows from DNA to RNA during transcription. The retroviruses code for a special enzyme, called **reverse transcriptase**, which catalyses the synthesis of homologous DNA sequences using RNA molecules as templates.

Life Cycle of Rous Sarcoma Virus

The best known of the retroviruses is **Rous Sarcoma Virus** ("Rous" for the scientist who discovered it; "Sarcoma" for the type of cancer induced; Box 89.2), which induces cancers in chicken cells. The life cycle of Rous sarcoma virus is diagrammed in figure 89.25. Shortly after Rous sarcoma virus infects a cell, its RNA genome replicated its DNA form by reverse transcription. The viral DNA integrated into chromosomal DNA of the host cell. In this integrated state, it is replicated and transcribed by the host cell's metabolic machinery, just like the normal genes of the host cell.

Box 89.2

In 1910, **Franchis Peyton Rous** (1879 – 1970), U.S. Physician who much later (in 1966) won the Nobel Prize for his work, discovered that a sarcoma (tumor) in chickens could be induced by a cell-free extract from a tumor in another chickens. The transmitted agent was later found to be a retrovirus, named **Rous Sarcoma Virus.**

The genome of Rous sarcoma virus contains just four genes:

1. *gag*, which codes for capsid protein of the virion.
2. *pol*, which codes for reverse transcriptase enzyme;
3. *env*, which codes for the protein spikes of the viral envelope;
4. oncogene *src* (name is derived from sarcoma), which codes for a membrane-bound protein kinase.

The viral genome also carries its own strong **promoter**, thus, the four viral gene products are synthesized in large amounts. The *src* gene is entirely responsible for the ability of Rous sarcoma virus to cause cancer; the deletion of this one gene yields a virus that infects and replicates just like the *src* – containing virus, but one that is totally non-cancerous.

When the virus enters the host, it is in the form of a plus (+) RNA strand (which is capable of acting as a messenger RNA; the minus strand is the complement to the plus strand). At either end of this RNA, is a repeated sequence (R) located outside two unique sequences (U_3 and U_5). Through reverse transcription, using the reverse transcriptase the virus brings in, the viral RNA is converted to a double-stranded DNA. During that process, the ends of the DNA take on the configuration of long terminal repeats (LTRs), repetitions of $U_3 - R - U_5$. The linear DNA then circularises and integrates into the host genome just as a transposon does, generating short direct repeats at either ends.

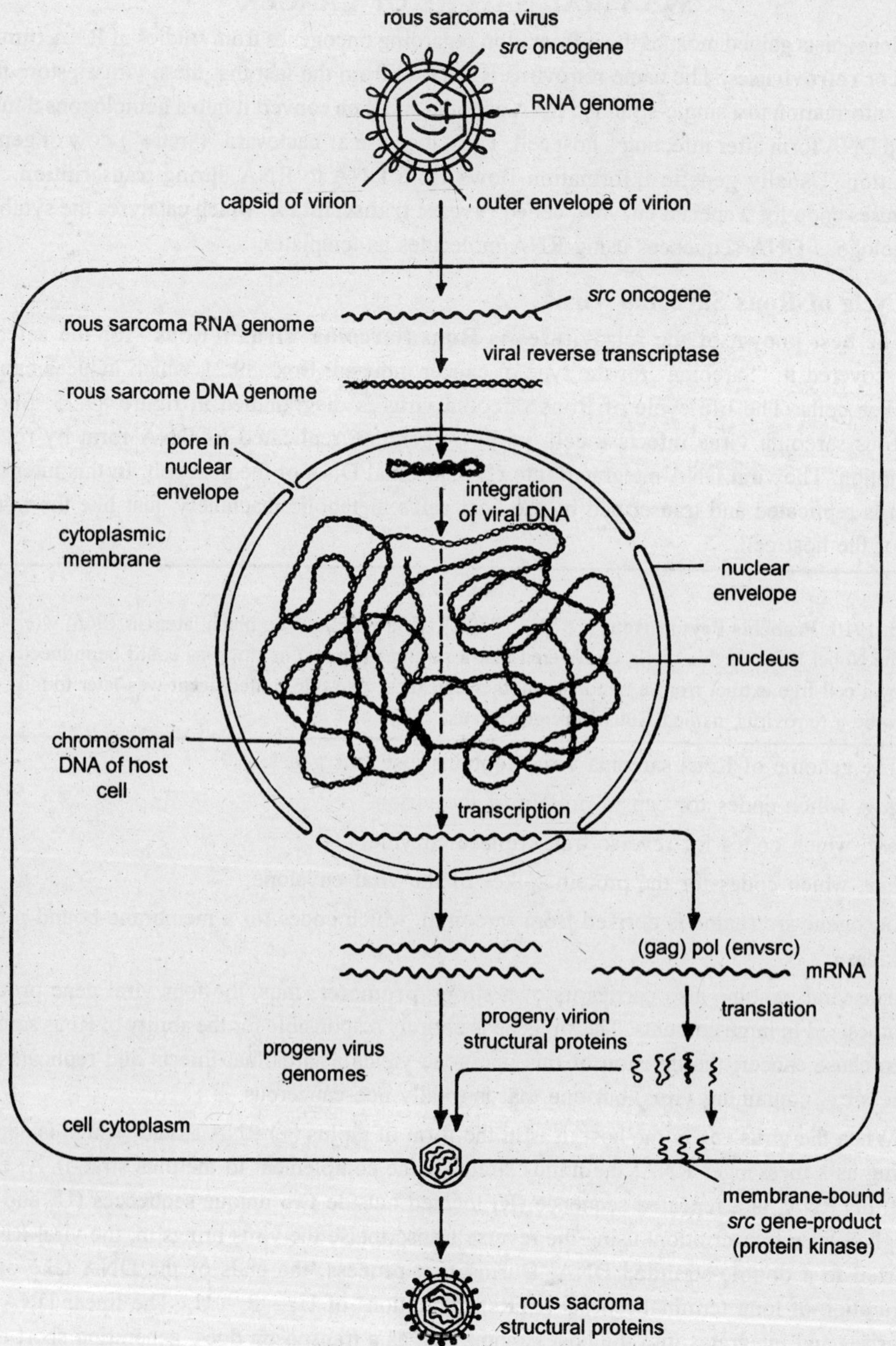

Fig. 89.25 Life cycle of Rous sarcoma RNA tumor virus. The ability of this virus to cause cancer resides in the *src* oncogene (after Gardner *et al.*, 2001).

As mentioned earlier, retroviruses can cause cellular transformation through direct integration or from the oncogenes they carry. Transformation from integration comes about because the integrated

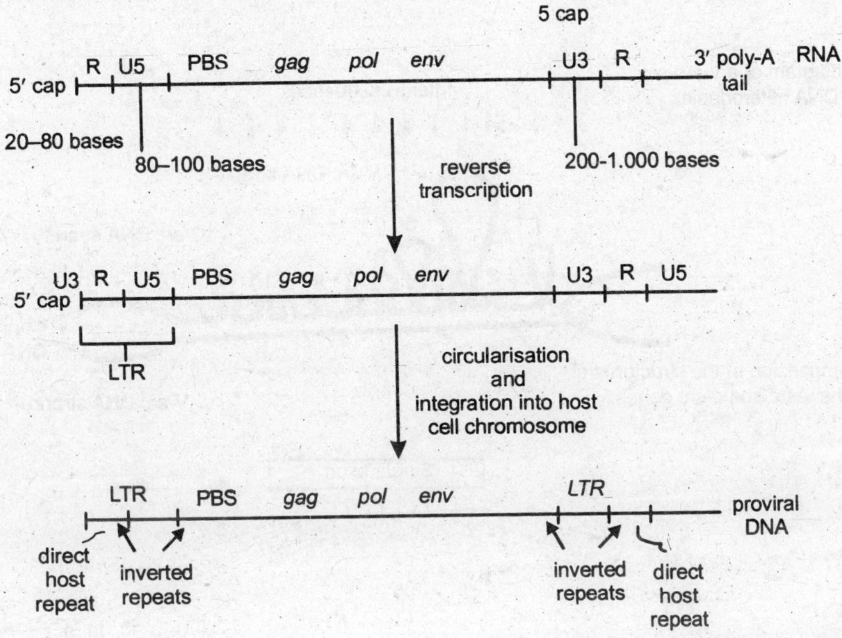

Fig. 89.26 A retroviral RNA genome. R is a repeated sequence; U_3 and U_5 are unique sequences; PBS is the primer binding site, LTR is the long terminal repeat; *gag, pol* and *env* are viral genes. During the process of reverse transcription, the LTRs are created at the ends of the DNA. Direct host repeats are created when the viral DNA integrates into the host chromosome (after Tamarin, 2002).

provirus either inactivates a tumor-suppressor gene or activates an oncogene in a process called **insertion mutagenesis.** The U_3 region of the retrovirus contains both a promoter and an enhancer. Since a long terminal repeat lies at either end of the provirus, cellular genes can be turned on when the virus integrates.

Table 89.3 Some oncogenes, their origin and their protein products (Source: Gardner *et al.*, 2001; Tamarin, 2002).

Oncogene	*Virus*	*Species of Origin*	*Gene function*
1. *abl*	Abelson murine leukemia virus	Mouse	Tyrosine kinase
2. *src*	Rous sarcoma virus	Chicken	Tyrosine kinase
3. *erb B*	Avian erythroblastosis virus	Chicken	Tyrosine kinase
4. *fms*	McDonough feline sarcoma virus	Cat	Growth factor
5. *mos*	Avian myeloblastosis virus	Chicken	Protein kinase
6. *sis*	Simian sarcoma virus	Woolly monkey	Growth
7. *Ha – ras*	Harvey murine sarcoma virus	Rat	GTP – binding protein
8. *Ki – ras*	Kristen murine sarcoma virus	Rat	GTP – binding protein
9. *fos*	FBJ (or FJB) osteosarcoma virus	Mouse	Binds DNA
10. *myb*	Avian erythroblastosis virus	Chicken	Binds DNA
11. *erb A*	Avian erythroblastosis virus	Chicken	Binds DNA
12. *rel*	Reticuloendotheliosis virus	Turkey	Binds DNA
13. *fun*	Avian sarcoma virus 17	Chicken	Binds DNA
14. *fes*	ST feline sarcoma virus	Cat	Tyrosine kinase
15. *ets*	E 26 virus	Chicken	Nuclear protein
16. *fgr*	Gardner – Rasheed feline sarcoma virus	Cat	Tyrosine kinase

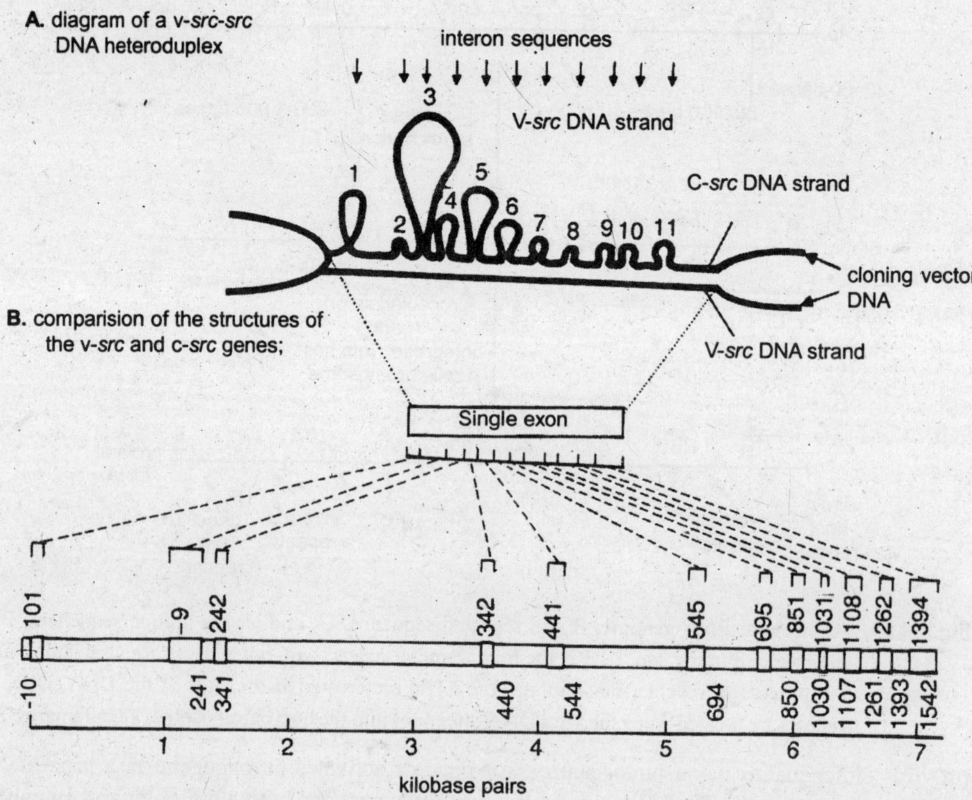

Fig. 89.27 Structure of the v – *src* oncogene and the c – *src* protooncogene of the chicken. A — Diagram of the predicted DNA heteroduplex formed by hybridisation of one strand carrying the c – *src* gene (top strand) with a (partially) complementary strand carrying the v – *src* gene (bottom strand) as observed by electron microscopy. The intron segments of c – *src* strand form single-stranded loops (numbered 1 – 11). (The c genes of top strand have been cloned in plasmid/vector DNA). B — Schematic comparison of the organisation of the coding sequences (exons) of the two genes. The v – *src* gene has a single, uninterrupted coding sequence. The c – *src* gene contains 12 exons separated by 11 introns. The starting and ending nucleotide positions of the exons are shown above and below, respectively, the map of the c – *src* gene; position 1 corresponds to the first nucleotide of the coding sequence. The first exon (position – 101 to – 10) is in the 5′ leader region of the transcript, not in the coding sequence. The dashed lines indicate the approximate corelations between the coding sequences of the two genes (after Gardner *et al.*, 2001).

Oncogenes

Genetic analysis and recombinant DNA studies have shown that Rous sarcoma virus transform cells through the action of a single gene. This gene, called *src*, was the first viral oncogene discovered. Since then about fifty viral oncogenes have been discovered, and each has been given a three letter designation (Table 89.3). Unlike tumor suppressors, which lead to cancer when in homozygous mutant (= recessive) condition, oncogenes act in dominant fashion, *i.e.*, only one copy of the activated gene is sufficient to cause transformation.

Researchers have exploited viral oncogene in creating probes for the identification of oncogenes within the DNA of the host (cell). To the utter surprise of everyone, these oncogenes were found in untransformed cells. Geneticists generally believe that oncogenes originated in the host and picked up, presumably as mRNAs, by the retroviruses. This belief rest on following two findings: 1. Transforming viruses can function quite well as viruses without their oncogenes and 2. Cellular oncogenes have introns and viral oncogenes do not.

It is believed that retroviruses pick up cellular genes by transcription read-through, transcribing beyond the end of the integrated virus and producing a mRNA that is then incorporated into a viral particle after removal of the intron. In this way, retroviruses can pick up genes which are adjacent to their point of integration.

The oncogenes of viruses and hosts are distinguished with each other by putting prefix to the name of a viral oncogene, such as *src*, with a *v* (v – *src*), and cellular oncogene with a c (c–*src*). Cellular oncogenes within a non-transformed cell are called **proto-oncogenes.**

Induction of Proto-Oncogenes

How are proto-oncogenes induced to become oncogenes, and what do proto-oncogenes normally do in the cell? Proto-oncogenes are found to be induced by following three ways and transform its host cell:

1. A *ras* proto-oncogene was converted (mutated) to an oncogene when one codon, *GGG* (glycine) was converted to *GTC* (valine).
2. A proto-oncogene can be activated if it is moved to a region with a strong promoter or enhancer. Burkitt's lymphoma, for example, is associated with a **translocation** involving the proto-oncogene *c-myc*, which is normally located on chromosome 8. When translocated to chromosome 14, *c-myc* is placed contiguous with the immunoglobulin IgM constant gene. This gene is very active in lymphocytes. Hence *c-myc* is now transcribed at a much higher rate than normal, resulting in cellular transformation. The *c-myc* gene normally occurs near a **fragile site**, a region of a chromosome that has tendency to break. Many proto-oncogenes occur near fragile sites on chromosomes.

   The simple capture of a gene by a retrovirus might be enough for transformation, since the gene is brought under the influence of viral transcriptional control. However, not all genes captured this way are oncogenes.
3. A proto-oncogene can be activated if it is amplified. Several cases are known in which amplified genes (*e.g.*, c-*ras* and c-*abl*) or genes on trisomic chromosomes are related to transformation.

Oncogene Products as Regulators of Cell Division

Considering that oncogenes induce uncontrolled cell growth resulting in tumors, one might well anticipate that the products of these genes would act by stimulating cell division in some manner. Indeed, it is now clear that the products of these oncogenes play various roles in regulating cell division in one or more cell types.

Essentially, proto-oncogenes are important to the cell because they have been conserved evolutionarily. For example, c –*src* is found in the fruit flies (*Drosophila*) as well as in vertebrates; c-*ras* is found in yeasts and in human beings. They are all genes that can promote cell growth; they are specific transcription factors or other components of growth stimulating factors. The known protein products of oncogenes are of following four types:

1. Growth factors;
2. Tyrosine kinase;
3. GTP-binding proteins; and
4. DNA-binding proteins

As proto-oncogenes, they normally function at low levels; in transformed cells, they (oncogenes) function at high level.

1. Growth factors. The product of the v-*sis* oncogene of simian sarcoma virus is closely related to a polypeptide growth hormone called **platelet-derived growth factor (PDGF).** PDGF is produced by platelet cells as the name indicates; it promotes wound healing by stimulating the growth of cells at the wound site. Simian sarcoma virus carrying the v-*sis* gene induce sarcomas when injected into woolly monkeys; they also transform fibroblasts growing in culture to a neoplastic or tumorous state. Presumably, this cellular transformation to the cancerous state occurs by a mechanism that is related to the effects of normal PDGF on cells at a wound site.

Other oncogenes encode products that are similar to growth factor and hormone receptors. For example, oncogenes *erb B* and *fms* encode proteins that are closely related to the receptors for **epidermal growth factor** and **CSF -1 growth factor**, respectively. CSF -1 is a growth-factor that stimulates growth and differentiation of macrophages. Both of these growth - factor receptors are transmembrane proteins with growth- factor binding domains on the inside of the cell. These receptors are key components in transmembrane signalling pathways. Finally, the *erb A* gene product is an analog of the nuclear receptor for the thyroid hormone T3. Thus, all of these gene products are undoubtedly involved in the **intercellular communication circuitry** that regulates cell division during the growth and development of highly differentiated tissues and organs in multicellular animals.

2. Tyrosine kinases. The enzyme tyrosine kinase (products of several oncogenes such as *abl, src, erbB, fes, fgr;* see Table 89.3) add a phosphate group to tyrosine residues in proteins, other kinases phosphorylate serine and threonine. Basically, these three amino acids have OH groups available for phosphorylation. Proteins that are phosphorylated at their tyrosine residues are involved in signal pathways, are of cytoskeleton shape (transformed cells are shaped differently than normal cells) and undergo glycolysis (cancer cells follow the anaerobic glycolytic pathway). Overactivation of the cellular oncogene can result in abnormal kinase activity, thereby changing many of the cellular activities and leading to cancer.

3. GTP-binding protein. These proteins are products of v-*ras* oncogene and they play a role in transmitting endocrine signals across membrane. Increased quantities of GTP-binding proteins can send continuous or amplified signals to certain cells and thus enhance growth.

4. DNA-binding proteins. Some oncogenes such as *jun, fos, erbA* and *myc* (*i.e.*, v–*myc*) encode nuclear transcription factors or DNA binding proteins that activate the expression of specific genes. Undoubtedly, some of the genes that they activate will prove to encode products that function as positive regulators of cell division.

In brief, oncogene products are simply proteins that play central roles in stimulating cell division in one or more cell types. In some cases, these oncogene products are probably altered or "mutant" proteins that trigger the division of cells that should normally not divide under the existing conditions. In other cases, the oncogene products (proteins) stimulate abnormal cell division by being overproduced (see **Gardner** *et al.*, 2001).

Steps of Growth of Full-Blown Cancer

From the initial lesion in a gene to full blown cancer normally takes many steps (Fig. 89.28). For example, in the **colorectal cancer** familial adenomatous polyposis (FAP), at least seven genetic changes are needed. Through these steps, a normal mass of cells passes through ***hyperplasia***, an increased growth without obvious changes in cells; to **dysplasia**, in which overgrowth continues

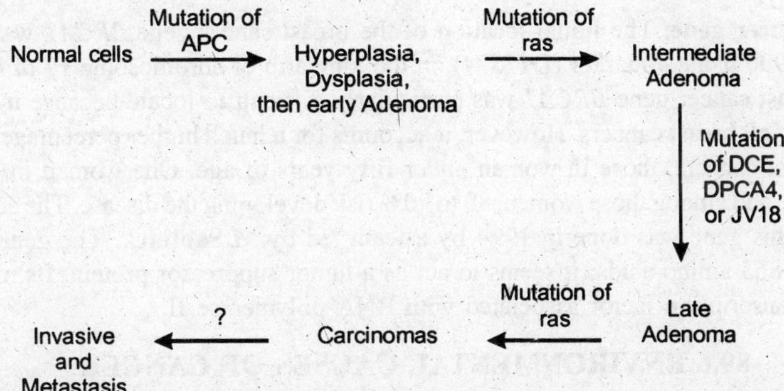

Fig. 89.28 Clonal evolution. Some of the known steps in converting a normal colon cell into a cancerous one. At least four known genes are involved, two oncogenes (*APC* and *ras*) and two tumor – suppressor genes (*DCC* – or *DPC4* or *JV18* – and p53) (after Tamarin, 2002).

with changes in cell and nuclear structures (polyp formation); to the cancerous state, with invasion of surrounding tissues and metastasis. **B.Vogelstein** and colleagues (1996) have discovered many of the genetic changes involved in the formation of this cancer. First, the *APC* gene (*a*denomatous *p*olyposis *c*oli) mutates, leading through hyperplasia to dysplasia, a condition known as **aberrant cryptic foci.** Though exact role of *APC* protein is still not known, it is found to bind with *B-catenin*, which is involved in cell adhesion and activates the *cyclin D1 promoter*, exerting a direct effect on cellular proliferation. Thus, mutation of *APC* results in an accumulation of B-catenin, which then affects progression of cell cycle and cell adhesion.

The next genetic change causes an early adenoma (a benign growth). Muta-tion of *ras* oncogene leads to intermediate adenoma, due to the autonomous growth signals set by the *Ras* GTP-binding protein. This is followed by late adenoma caused by the mutation of a gene in region q21 of chromosome 18, a gene called *deleted in colorectal cancer* (*DCC*). This gene codes for a transmembrane protein involved in the adhesion of cells to each other.

At this point, mutations resulting in the loss of p53 protein result in full-blown cancer. Throughout this series of events, it is a cell from the previous state that mutates into the next state, accordant with our concept of clonal evolution (Box 89.3).

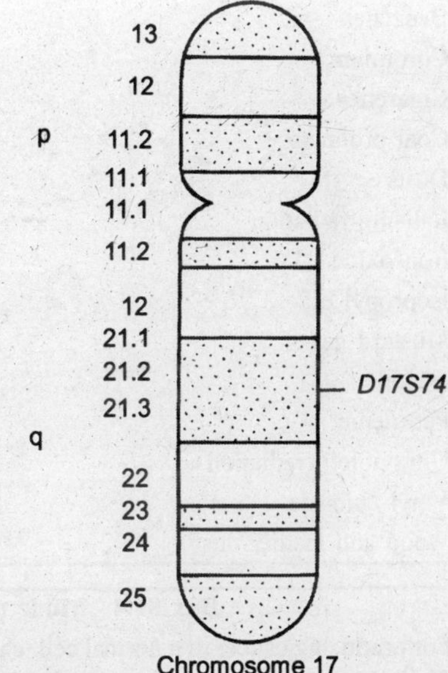

Fig. 89.29 A Giemsa-banded chromosome 17, showing the numbering of the regions and the location in region 21q in which marker *D17S74* is located. The terminology of the marker is that of section 74 of chromosome 17. This marker corelated to the position of *BRCA1* gene (after Tamarin, 2002).

Box 89.3

Although it may seem odd that so many mutations appear consequently in the same cells, remember that mutations in some genes, such as *p53*, result in overall higher mutation rate within the cells. In one study, when cancer cells were compared with non-cancerous progenitor cells, the cancer cells showed eleven thousand genetic changes (see **Tamarin**, 2002).

Breast cancer gene. The initial location of the breast cancer gene *BRCA1* was determined by **M.King** in 1990 using a marker (*D17S74*) on the long arm of chromosome 17 of humans (Fig. 89.29). The breast cancer gene *BRCA1* was particularly difficult to locate because it accounts for only about 5% of all breast cancers. However, it accounts for a much higher percentage of inherited, early onset breast cancers, those in women under fifty years of age. One woman in two hundred inherits this gene, and among those women, 80 to 90% risk developing the disease. The actual locating and cloning of this gene was done in 1994 by a team led by **M.Skolnick.** The gene codes for a protein having 1863 amino acids; it seems to act as a tumor suppressor protein. Its mechanism of action is as a transcription factor associated with RNA polymerase II.

89.6 ENVIRONMENTAL CAUSES OF CANCER

A **carcinogen** is a substance capable of causing cancer. Environment plays a major role in carcinogenesis and many carcinogens have been detected in the environment (Table 89.4 Box 89.4).

Table 89.4 Carcinogenic substances in the environment (Source: Tamarin, 2002).

Carcinogen	*Cancer site(s)*
1. Aromatic amines	Bladder
2. Arsenic	Liver, lungs, skin
3. Asbestos	Lung
4. Benzene	Bone marrow
5. Chromium	Lung, nose, nasopharynx sinuses
6. Cigarettes	Lung
7. Coal products	Bladder, lung
8. Dusts	Lung
9. Ionising radiation	Bone, bone marrow, lung
10. Iron oxide	Lung
11. Isopropyl oil	Nasopharynx sinuses, nose
12. Mustard gas	Lung
13. Nickel	Lung, nasopharynx, nose
14. Petroleum	Lung
15. Ultraviolet irradiation	Skin
16. Vinyl chloride	Liver
17. Wood and leather dust	Nasopharynx sinuses, nose

Box 89.4 Mode of action of chemical carcinogen

For producing cancer in a normal cell, chemical carcinogens act either as *initiators* or *promoters*. The **initiators** cause permanent heritable damage to a cell's DNA. These do not produce tumors on their own but makes the first move towards cancer development. For tumors to develop, initiation must be followed by promotion. Some carcinogens acting as initiators are nitrogen mustard, benzoyl chloride, polycyclic hydrocarbons and many metals.

The **promoters** enhance cell proliferation so that the genetic damage is passed on to future generations before the cell has time to repair its damaged DNA. One of the best studied promoter is tetradecanoyl phorbol acetate (TPA). It activates enzymes involved in transmitting signals that trigger cells division. Some hormones also act as promoters. For example, prolonged use of hormone diethystilbestrol (DES) which is used as a potent estrogen, causes endometrial carcinoma and vaginal cancer in women.

Proto-oncogenes and tumor suppressor genes are two critical targets of chemical carcinogens.

Many of these agents (see Table 89.4) are also mutagens. Avoidable substances in the environment and the diet are estimated to cause 80 to 90% of all cancer, although the exact mechanism by which these agents induce transformation are generally unknown. Perhaps, the most effective cancer prevention strategy would be to avoid as many carcinogens from the environment as possible.

89.7 THE SIX HALLMARK TRAITS OF CANCER

Because so many combinations of mutations involving tumor suppressor genes and oncogenes can lead to cancer, the question arises as to whether there are any common principles that would help simplify the picture. As a unifying concept, in 2000, **Douglas Hanahan** and **Robert Weinberg** have proposed that a series of six acquired traits are common, but each can be acquired through a variety of different genetic and epigenetic mechanisms. These six "hallmarks of cancer" are as follows:

1. Self-sufficiency in growth signals. Cells do not normally proliferate unless they are stimulated by an appropriate growth factor. Cancer cells escape this requirement through the action of oncogenes that produce excessive quantities or mutant versions of proteins involved in growth stimulation. One such pathway commonly activated in cancer cells is the *Ras pathway*. About 25–30% of all human cancers have mutant Ras proteins that provide an ongoing stimulus for the cell to proliferate independent of growth factors. Mutations affecting other components of the Ras pathway are common as well.

2. Insensitivity to antigrowth signals. Normal tissues are protected from excessive cell proliferation by a variety of growth-inhibiting mechanisms. Cancer cells must evade such anti- growth signals if they are to continue proliferating. Most antigrowth signals act during late phase of cell cycle and exert their effects through the Rb protein, whose phosphorylation regulates passage through the restriction point and into S phase. For example, **TGFβ** normally inhibits proliferation by triggering the TGF β- smad pathway, which produces Cdk inhibitors that block Rb phosphorylation and thereby prevent passage from G_1 into S phase. In cancer cells, the TGFβ-smad pathway is disrupted by a variety of different mechanisms, including mutations, epigenetic changes, and interactions with viral proteins. Mutations in *RB* gene also make cells insensitive to the antigrowth effects of TGFβ or any other growth inhibitor that exerts its effects through the Rb protein.

3. Evasion of apoptosis. Evading apoptosis, which otherwise destroy genetically damaged cells, is crucial to the survival of cancer cells. This is frequently done by loss-of-function mutation, in the p53 tumor suppressor gene, which disrupt the main pathway by which DNA damage would otherwise trigger apoptosis. Several oncogenes, such as *BCL2*, also promote cancer cell survival by producing proteins that interfere with apoptosis.

4. Limitless replicative potential. The overall effect of the preceding three traits is to uncouple cancer cells from the mechanisms that normally balance cell proliferation with an organisms need for new cells. However, this would not ensure unlimited proliferation in the absence of a mechanism for replenishing the **telomerase** sequences that are lost from the ends of each chromosome during DNA replication. Telomere maintenance is usually accomplished by activating the gene coding for telomerase, but a few cancer cells activate an alternative mechanism for maintaining telomeres that involves the exchange of sequence information between chromosomes. In either case, cancer cells maintain telomere length above a critical threshold and thereby retain the ability to divide indefinitely.

5. Sustained angiogenesis. In the absence of a blood supply, tumors will not grow beyond a few millimeters in size. Thus, at some point during early tumor development, cancer cells must trigger angiogenesis. A common strategy involves the activation a genes coding for angiogenesis stimulators combined with inhibition of genes coding for angiogenesis inhibitors. The mechanisms that underlie such changes in gene expression are not well understood, but in some cases they have been linked to the activities of known tumor suppressor genes or oncogenes. For example, the p53 protein activates the genes coding for the angiogenesis inhibitor thrombospondin; hence the loss of p53 function, which occurs in human cancers, can cause thrombospondin levels to fall. Conversely, *RAS* oncogenes trigger increased expression of the gene coding for the angiogenesis activator VEGF.

6. Tissue invasion and metastasis. The ability to invade surrounding tissues and metastasize to distant sites is the defining trait that distinguishes a cancer from a benign tumor. Three properties exhibited by cancer cells play a crucial role in these events: decreased cell-cell adhesion, increased motility, and the production of proteases (enzymes) that degrade the extracellular matrix and basal lamina. Decreased adhesiveness is caused by changes in E-cadherin, which is lost in the majority of epithelial cancers by either mutation, decreased gene expression, or destruction of E-cadherin itself. Changes in other molecules involved in cancer cell adhesion, motility, and protease production also play a role in invasion and metastasis. However, the mechanisms underlying these molecular changes are not completely understood and they appear to differ among tumor types and tissue environments.

Genetic instability. To acquire the preceding six traits, cancer cells need to accumulate more mutations than would be generated by normal mutation rates. Cells must therefore become genetically unstable before enough mutations can accumulate to cause cancer. **Genetic instability** arises most commonly from mutations that disrupt the ability of the p53 pathway (see **Becker** *et al.* 2007; Box 89.5) to trigger the destruction of genetically damaged cells. However, defects in genes coding for proteins involved in DNA repair and chromosome sorting also play a role. Genetic instability is placed in a category separate from the six "hall mark" traits, which are directly involved in cancer cell proliferation and spread, because genetic instability simply enables involving population of cancer cells to acquire the six hallmark traits.

Box 89.5 Role of the p53 protein in responding to DNA damage

Damaged DNA activates the ATM protein kinase, leading to phosphorylation of the p53 protein (also called "**guardian protein**"). Phosphorylation stabilizes p53 by blocking its interaction with Mdm2, a protein that otherwise mark p53 for degradation (The degradation mechanism involves Mdm2-catalyzed attachment of p53 to ubiquitin, which target molecules to the cell's main protein destruction machine, the proteosome). When the interaction between p53 and Mdm2 is blocked by p53 phosphorylation, the phosphorylated p53 protein accumulates and triggers two events, 1. The p53 protein binds to DNA and functions as a **transcription factor** that activates transcription of the gene coding for the p21 protein, a Cdk inhibitor. The resulting inhibition of Cdk-cyclin prevents phosphorylation of the Rb protein, leading to cell cycle arrest at the restriction/check point. 2. When the DNA damage cannot be repaired, p53 then activates genes coding for a group of proteins that are involved in triggering cell death by **apoptosis**. A key protein is **puma** ("p53 *u*pregulated *m*odulator of *a*poptosis"), which promotes apoptosis by binding to, and blocking the action, the apoptosis inhibitor, Bcl-2.

Rb protein. Protein phosphorylation controls passage through the restriction point of the cell cycle.

Proteosome. Multiprotein complex that catalyzes the ATP-dependent degradation of proteins linked to ubiquitin.

Ubiquitin. Small protein that is linked to other proteins as a way of marking the targeted protein for degradation by proteosome.

QUESTIONS

Long Answer Questions

1. Describe molecular structure of immunoglobulins.
2. Describe in detail multigene complex and antibody diversity.
3. Describe T-cell receptors and MHC proteins of mammals.
4. Describe the process of V(D)J joining.

5. What are three different sources of antibody diversity?
6. How can you settle the viral and mutational nature of cancer?
7. What chromosomal abnormality is associated with retinoblastoma? With Wilm's tumor?
8. What gross chromosomal abnormalities are associated with cancers?
9. What are the differences among *v-src*, *c-src*, and proto-*src* genes?
10. What is the evidence that the *c-src* gene came before the *v-src* gene?
11. What is the proposed mechanism of action of the retinoblastoma gene? What evidence supports this mechanism? Why is it called an anti-oncogene?
12. What evidence is there that the proto-oncogenes found in chromosomes of normal cells and normal animals are not simply integrated retroviral oncogenes?
13. The nucleotide-pair sequences of cellular proto-oncogenes are highly conserved in a large number of species in the animal kingdom. Does this conservation of proto-oncogene structure tells us anything about the functions of proto-oncogenes.

Short Answer Questions

1. Differentiate between the following:
   (*i*) Cellular immunity and humoral immunity
   (*ii*) B lymphocytes and T lymphocytes
   (*iii*) MHC class I protein and MHC class II proteins
2. Write short notes on the following:
   (*i*) AID enzyme
   (*ii*) Somatic hypermutations
   (*iii*) Class switch recombination
   (*iv*) T-cell receptors
   (*v*) Monoclonal antibody
   (*vi*) Immunogenetics (*H.N.B. Garhwal 2008*)
   (*vii*) Cancer genetics (*H.N.B. Garhwal 2009*)

Very Short Answer Questions

1. What is exon?

   Ans. In a gene that has intervening sequences (called **introns**), **exon** is a region that is actually exported from the nucleus to be expressed or become part of a transfer RNA (tRNA) or ribosomal RNA (rRNA).

2. Define exon shuffling.

   Ans. The hypothesis put forward by **Walter Gilbert** that exons code for the functional units of a protein, and that the evolution of new genes proceeds by recombination or the exclusion of exons.

3. What are intervening sequences or introns?

   Ans. Introns are DNA sequences within a gene that are transcribed but removed prior to translation.

4. Define spliceosome.

   Ans. Spliceosome is a protein-RNA complex that removes introns in eukaryotic nuclear RNAs.

5. What is transcription?

   Ans. The process whereby RNA is synthesized from a DNA template.

6. What is ubiquitin?

   Ans. Ubiquitin is a peptide of twenty-six amino acid residues that enzymes attach to proteins that the proteasome will degrade.

7. Define proteasome.

Ans. Proteasome is a barrel-shaped cellular organelle for protein breakdown involving the ubiquitin pathway.

8. What is V(D)J joining?

Ans. The process of joining variable, diversity and joining gene segments (V–J, V–D or D–J joining) in the formation of a functional immunoglobulin gene.

9. What are oncogenes? (*H.N.B., Garhwal, 2008*)

Ans. Oncogenes are genes capable of transforming a cell. They are found in the *active state* in retroviruses and transforming cells, in the *inactive state* in non-transformed cells, in which they are called **proto-oncogenes.**

10. Define immunity.

Ans. Immunity is the ability of an organism to resist infection.

11. What are immunoglobulins?

Ans. Immunoglobulins (Igs) are specific proteins produced by derivatives of B lymphocytes that protect an organism from antigens.

12. Define cellular immunity.

Ans. Immunity controlled by killer and helper T cells, which recognize infected body cells and either cause the infected cells to destroy internal invaders or destroy the infected cells directly.

13. What is humoral immunity?

Ans. Immunity due to antibodies in the serum and lymph.

14. What is antibody?

Ans. Antibody is a protein produced by B lymphocytes that protects the organism against antigens.

15. Define antigen.

Ans. Antigen is a foreign substance which is capable of triggering an immune response in an organism.

16. What is anti-oncogene?

Ans. A anti-oncogene is a gene that represents malignant growth and whose absence results in malignancy (*e.g.*, retinoblastoma).

17. What is proto-oncogene? (*H.N.B., Garhwal 2009*)

Ans. A cellular oncogene in an untransformed cell.

Multiple Choice Questions

1. The major phagocytic cells are

(*a*) lymphocytes (*b*) must cells
(*c*) plasma cells (*d*) macrophages

2. Which of the following immunoglobulins (Igs) is present in milk?

(*a*) IgA (*b*) IgG
(*c*) IgM (*d*) IgE

3. Lymphocytes that cause the formation of holes in plasma membranes are

(*a*) B cells (*b*) killer or cytotoxic T cells
(*c*) suppressor T cells (*d*) helper T cells

4. Hodgkins's disease is

(*a*) cancer of lymphoid tissue (*b*) cancer of mammary gland
(*c*) cancer of WBC (*d*) cancer of liver

5. Which one of the following genes is involved in the conversion of proto-oncogenes into oncogenes causing cancer?
(*a*) tumour suppressor gene
(*b*) jumping genes (transposons)
(*c*) angiogenesis genes
(*d*) metastasis genes

6. Migration of cancerous cells from the site of origin to other part of the body forming secondary tumours is called
(*a*) diapedesis
(*b*) proliferation
(*c*) metastasis
(*d*) none of these

7. Cancer is caused by
(*a*) uncontrolled meiosis
(*b*) loss of immunity of the cells
(*c*) uncontrolled mitosis
(*d*) rupturing of cells

8. Cancer of β-lymphocytes is called
(*a*) sarcoma
(*b*) melanoma
(*c*) myeloma
(*d*) carcinoma

9. The major form of cancer in Indian women is
(*a*) breast cancer
(*b*) throat and mouth cancer
(*c*) uterine-cervical cancer
(*d*) stomach cancer

10. Oncogenes are the cancer causing genes in the cells but they do not express usually. This is because of the presence of
(*a*) proto-oncogenes
(*b*) tumour promoter
(*c*) transposons
(*d*) tumour suppressor genes

11. Which of the following is an example of carcinoma?
(*a*) cancer of muscle
(*b*) cancer of blood
(*c*) melanoma cancer of skin
(*d*) cancer of lymph glands

12. Cancer cells can easily be destroyed by radiations due to
(*a*) rapid cell division
(*b*) lack of nutrition
(*c*) fast mutation
(*d*) lack of oxygen

13. In Rous Sarcoma Virus information flow is from
(*a*) DNA → protein → RNA
(*b*) RNA → DNA → protein
(*c*) DNA → RNA → protein
(*d*) RNA → DNA → RNA → protein

14. Burkitt's lymphoma is a cancer of
(*a*) T-lymphocytes
(*b*) B-lymphocytes
(*c*) lymphoid tissues
(*d*) leucocytes

15. An allergic reaction is initiated by antibodies of the
(*a*) IgG group
(*b*) IgM group
(*c*) IgA group
(*d*) IgE group

16. The major histocompatibility complex is a
(*a*) series of complement-enhanced reactions to antigens
(*b*) group of antigens, coded by a family of genes, on the surface of body cell
(*c*) form of autoimmunity
(*d*) the major cause of cancer

17. Prespiration, saliva and tears contains an enzyme, lysozyme that kills
(*a*) virus-infected bacteria
(*b*) protozoa
(*c*) bacteria
(*d*) viruses

18. T-cells of the cell-mediated immune system make specific proteins, similar to antibodies that become
    (*a*) embedded in interstistial fluid
    (*b*) embedded within membranes of lymph nodes
    (*c*) receptors on the plasma membrane of the T cell
    (*d*) linked to antibodies by disulphide bridges

ANSWERS

Multiple Choice Questions

1. (*d*)	**2.** (*a*)	**3.** (*b*)	**4.** (*d*)
5. (*b*)	**6.** (*c*)	**7.** (*c*)	**8.** (*c*)
9. (*c*)	**10.** (*d*)	**11.** (*c*)	**12.** (*a*)
13. (*d*)	**14.** (*b*)	**15.** (*d*)	**16.** (*b*)
17. (*c*)	**18.** (c).		

90

Eugenics, Euthenics and Euphenics

90.1 EUGENICS

Eugenics deals with the application of the laws of genetics to the improvement of human race. The term eugenics (Gr. *eugenes* = well born) was coined by an English scientist **Francis Galton** in 1885. The science of eugenics can be defined as a science of well born, improving the inborn qualities of race and obtaining the better heritage by judicious breeding.

Eugenics and Euthenics

The betterment of human society can be achieved by following two inter-related methods:

1. By one of the method we can deal with the already existing human beings. The improvement of already existing human beings can be achieved by improving the environmental conditions, *e.g.*, by subjecting them to better nutrition, better unpolluted ecological conditions, better education and sufficient amount of medical facilities. This type of method of improving the human race is known as **euthenics**.
2. By another method we can improve the future generations by improving the germplasm of existing individuals. The type of method is known as **eugenics**. Eugenics believes in artificial selection of physically and mentally sound individuals and discouragement of defective individuals for the inheritance of their defective germplasm to the future generations. In other words, eugenics seeks the measures to preserve the best type of germplasm and to eliminate defective germplasm from the human society by applying the laws of inheritance to human beings.

History

The primary aim of many of the ancient systems of eugenics was to produce a race of physically perfect human beings. The Greeks had definite ideas regarding eugenics. In Sparta a physically perfect manhood was the chief aim, where as the Atheniens carried more for the intellectual achievements.

Following the doctrine of Greeks and until the 19th century there was little interest in the eugenics. Of the particular importance was the eugenics movement in England in the last part of the nineteenth century. The movement, spearheaded by persons of outstanding intellect such as **Francis Galton** and **Karl Pearson,** had its objective in the application of biologically sound principles to human populations. Since the biological basis of heredity was unknown, the first objective was to establish the nature of heredity.

Galton and **Pearson** choose to work with human beings and with what they considered important human traits such as intelligence, stature and special abilities. We now know that these characters are very complex traits and are under the control of many genes (polygenes) interacting with environmental variations. So, quite naturally the early investigators made little progress. However, **Francis Galton** should be credited with being the real founder of the modern movements of eugenics. He defined the eugenics as the study of all the agencies under social control which may improve or impair the inborn qualities of future generations of human beings either physically or mentally. **Darwin** also attached great importance to the eugenics and he compared it to a signpost with three directions. One of these indicates the influence of heredity on the fate of nations. Another point to the rules that an individual should try to carry out in regard to parenthood based on the law of human heredity. The third arm indicates the regulations to be adopted by the society to encourage racial progress.

Eugenics and Human Betterment

For better type of development both good heredity and suitable environment are necessary. When we consider the future welfare of the human race then the following two factors alarm us greatly: (*i*) the declining birth rate among the normal and superior people (those having best germplasm) (*ii*) a relative rapid increase of the abnormal and defective individuals (those having defective germplasm). For the betterment of future generation, it is necessary to increase the population of outstanding people and to decrease the population of abnormal and defective people by applying the principle of eugenics. The eugenics can be applicable by adopting following two methods:

(A) By encouraging the marriages between desirable persons (constructive method or **positive eugenics**).

(B) By discouraging the marriages between undesirable persons (restrictive method or **negative eugenics**).

A. Positive Eugenics

The positive eugenics attempts to increase consistantly better or desirable germplasm and, thus, to preserve best germplasm of the society. The percentage of desirable traits can be increased by adopting following measures:

1. Early marriage of those having desirable traits. It is most commonly observed fact that the highly placed persons of the society often have great ambitions for the future life. In achieving their ambitious goals, they often devote the best part of their youth and they are able to marry in their mature age (*e.g.*, 30 to 35 years). The biological and psychological investigation have revealed that the aged persons often lack in necessary amount of emotional warmth for the sexual activities and moreover, their germplasm also lost its vigour. Therefore, some laws should be formulated to prevent the late marriages of highly endowed persons by applying high taxation on them and at the same time the young persons having best hereditary traits should be encouraged for early marriage.

2. Subsidizing the fit. Because the highly endowed persons lead a well-planned life and to avoid unnecessary difficulties in nursing the children they often prefer to have small number of children. Therefore, the selected young men and women of best eugenic value should be encouraged to increase their birth rate.

3. Eutelegenesis or germinal choice. H.J. Muller has suggested that eugenically sound persons not only should increase their family size but through artificial insemination the outstanding man can serve as father to many more children than would be otherwise possible. The artificial insemination is already widely practiced to permit those women whose husbands are sterile or have some serious hereditary diseases to bear children. The sperms and eggs of outstanding persons can be stored for future use by quick freezing and storing them in deep freeze. These germ cells, thus, can be stored for 100 or more years.

Very recently the scientists felt the urgent need of establishment of **sperm** and **egg banks** to protect these precious germ cells from the radiation. The germ cells could be collected during early adulthood and stored in lead lined containers in the deep freeze. In this state the germ cells would not be subjected to radiation exposure which might affect the donors. In a modern technique, a woman's ovum is taken out and artificially fertilized in a test-tube. The zygote is, then implanted into woman's uterus for the embryonic development. The resulting baby is called **test-tube baby**.

4. Education. For the eugenically oriented reforms in the society, the people should be educated about the basic principles of human biology, human genetics, eugenics and sex. The children should be instructed about basic laws of health and they should be encouraged to develop a physically and mentally healthy body. Moreover, the sex should be free from the wide spread confusion, narrow-minded concepts and religious and ethical bindings because that is a natural biological instinct. The children ignorant about the facts of sex may do more harm to society than otherwise.

5. By avoiding germinal waste. The wastage of best type of germplasm can be avoided by adopting following measures:

(*i*) The selection of marriage partners should be made with intelligence.

(*ii*) The social hinderance which do not allow the religious gurus, nuns and priests to get married, must be removed. By adopting such measures the wastage of best type of germplasm due to lack of opportunities can be prevented.

(*iii*) The wars must be avoided because in wars the best germplasm of the society is wasted.

6. Genetic counselling. Genetic counselling can do great benefit to human society. The role of the genetic counsellor is to inform concerned individuals of the nature of the mutant condition that concern them. If it is inherited in a Mendelian fashion, then probability of producing affected offspring can be calculated. The final decision for taking a risk, is entirely the responsibility of the individual involved, it cannot be made by the counsellor (consult chapter 87 for more details).

7. Improvement of environmental conditions. Both heredity and environment have inter-related role in the development of eugenically better persons. Therefore, every person should get better food, living conditions, education and medical guidance, etc., so his or her hereditary traits can do their best development.

8. Promotion of genetic research. Our knowledge about the genetics is not sufficient enough because we still have little informations about various human diseases and metabolic disorders which are generally related with the genes. Therefore, the research in the field of genetics should be increased so that we can learn more and more about the human beings.

9. Genetic engineering. During the late 1970s, the science of genetics entered a new era dominated by the use of **recombinant DNA technology** or **genetic engineering** (or **biotechnology**) to produce novel life forms not found in nature. Through this technology, it has been possible to transfer genes from mammals into bacteria, causing the microbes to become tiny factories for making (in relatively large quantities) proteins of great economic importance such as hormones (including growth hormone) and interferons (lymphocyte proteins that prevent replication of a wide variety of viruses). These proteins are produced in such small quantities in humans that the cost of their extraction and purification from tissues have been very expensive, thus, limiting their medical use in prevention and treatment of disease. By genetic engineering, it has become possible to produce various blood clotting factors, complement proteins (part of immune system) and other substances for the correction of genetic deficiency diseases (euphenics).

Recently, experiments have been conducted in which human cells deficient in the synthesis of purines have been obtained from the patients with **Lesch-Nyhan syndrome** and grown in culture; these cells have been converted to normal cells by transformation with recombinant DNA. The exciting potential of this technique lies in the possibility of correcting genetic defects—for example, restoring the ability of a diabetic individual to make insulin or correcting immunological deficiencies. This technique is called **gene therapy**.

B. Negative Eugenics

The negative eugenics attempts to eliminate the defective germplasm of the society by adopting following measures:

1. Sexual separation of the defective. The defective persons may have various sex-linked diseases such as night blindness, haemophilia, colour blindness, etc., and various other defective traits which may be regulated by dominant or recessive genes. The increase of germplasm of the persons having such defective traits in the population can be checked by keeping them away and separated from the society. Different states have wisely adopted the restricted measures in segregation the mental defectives from the society and to place them in mental hospitals.

2. Sterilization. The sterilization is the best means to deny an individual from his power of reproduction without interfering with any of his normal functions. The sterilization method is based on surgical operation of sperm duct or vas deferens in males and oviducts or fallopian tubes in females. The former is known as **vasectomy** and the later is **tubectomy** or **salpingectomy**. The family planning movement in India has adopted the sterilization as the tool for controlling the rate of vastly increasing population and in that case the sterilization is euthenical in its application than the eugenical.

3. Control of immigration. Through immigration there are enough chances that undesirable or defective genes of different races and nationalities may intermingle with the normal germplasm of the population. Therefore, the immigration rules must be strict and the persons with undesirable hereditary traits must not be allowed to migrate from one place to another.

4. Regulation of marriage. Presently most human societies are money-minded and for the marriage relationship the wealthy or highly placed persons which, however, may contain several defective genetical traits, are more preferred than those which have eugenically sound hereditary traits but having no money and which fail to reach to highest status of the society due to lack of opportunities. Some rules must be enacted to encourage the marriages among desirable mates.

5. Birth control. People possessing more of undesired, dysgenic traits should be encouraged to have small families if at all allowed to reproduce. Their foetuses possessing dysgenic traits may be destroyed by abortion. (*dysgenic* = causing the biological impairment or deterioration of a race, especially humans. It is opposite of *eugenic*).

6. Control of consanguineous marriages. Most of the genetic diseases of humans are carried by mutant recessive genes. So to prevent the spreading of such genetic diseases, marriage between close relatives should be banned, since like inbreeding, this increases homozygosity and frequency of recessive traits.

90.2 EUTHENICS

Euthenics is a branch of applied genetics under which are studied the methods of providing, the best environmental conditions or 'nurture' to ensure complete expression of all eugenic traits in the developing children. Euthenics recommends the following methods to ensure maximum expression of genes of superior traits in the children.

1. Selection of children with superior traits. The primary requirement in euthenics is to select children having genes of superior traits. An effective way to do this is to put the children on intelligence tests through questions based upon the capabilities of thinking, memory, logic and debate, self-control, etc. Following two methods have been employed in psychobiology to test the intelligence of an individual.

(*a*) Stanford-Binet's classical method. In this method, the child is given progressively more difficult problems until he is finally unable to solve them. Thus, a wide range of mental ability from complete incompetence to excellent intelligence, has been reported for each age group. The mental ability of particular age group is called **mental age**. The mental age of a child when, is divided by his actual chronologic age and resultant quotient is then multiplied by 100 to give the **intelligence**

quotient or **I.Q.** of his. When I.Q. of different children of different mental ages are determined by this method, a continuous series of children from idiot to genius was obtained. The I.Q. of different children has been tabulated in Table 90.1.

Table 90.1 Classification of intelligence.

I.Q.	*Designation*	
140 or over	"Gifter"	
120–140	Very superior	
110–120	Superior	
90–110	Normal	
80–90	Dull	
70–80	Borderline	
50–70	Moron	"Feeble minded"
25–50	Imbecile	
0–25	Idiot	

If, the I.Q. values of different children are placed on a graph, one gets a normal distribution curve, a characteristic of multiple gene inheritance. The intelligence trait in an actual genetical experiments has been found to be determined by more than one pair of genes (polygenic).

***(b)* Modern methods.** More recent intelligence tests have provided measures of primary abilities such as the ability to reason inductively, the ability to memorize and the ability to visualize objects in three dimensions. Special abilities–musical, artistic, mechanical and mathematical have a hereditary basis and their inheritance is separate from that of general intelligence. From a study of inheritance of musical ability, **Philiptschenko** concluded that four pairs of genes are involved in it.

2. Promotion of superior students. After selecting superior children (and following genetic basis of their superior traits), provisions for their best education and 'nurture', must be made by adopting the following methods:

***(i)* Opening special schools.** Special schools with best facilities of education should be opened for superior students where their full mental development may occur.

***(ii)* Special curricula and methods.** Special curricula must be framed for these students and superior teachers and sophisticated equipments (*e.g.*, computer education) and techniques should be made available to them.

***(iii)* Extracurricular activities.** Besides prescribed curricula, facilities of extra-curricular activities such as games and sports, debates, quizes, scouting, military training, cultural activities, etc., should also be provided to these students for their multifaceted development.

***(iv)* Habitat and nutrition.** Best possible conditions of habitat and nutrition must be provided to superior children. Their daily routine should be well-regulated and co-ordinated.

***(v)* Tourism.** Tourism significantly helps in development of intelligence and experience. Therefore, the facilities of tourism should also be made available to superior children.

After subsidization as described above, it is also absolutely necessary to fix these well-endowed students to make maximum use of their hereditary superiority.

90.3 EUPHENICS

The symptomatic treatment of genetic diseases of human beings is called **euphenics**. The euphenics deals with the control of several inherited human diseases, especially inborn errors of metabolism in which the missing or defective enzyme has been identified. One example of this is the condition known as **phenylketonuria** or **PKU**, determined by an autosomal recessive gene.

Babies with this defect are unable to properly metabolize an amino acid, phenylalanine, the resulting chemical imbalance causes severe mental retardation. Now it is possible to distinguish PKU homozygotes from normal individuals by testing the urine of all newborn babies with ferric chloride. In affected children, the metabolic imbalance caused by the mutation will turn the urine green. Once such a child is detected, a diet free of phenylalanine is prescribed and the child can develop normally.

Although a number of inherited diseases can be treated in a similar euphenic manner, but these constitute only a small fraction of known inherited disease. For the most part, biochemical geneticists could not identify the biochemical errors of many genetic diseases. In other cases, such as **albinism**, even though the metabolic block leading to an abnormality is known, but, it is not possible to correct it. However, in future following euphenic measures can get rid of human beings from certain fatal genetic diseases:

1. Intake of missing enzyme. One possible euphenic measure for the future would be to supply the known missing enzyme to individuals that would allow their cells to complete the required biochemical reaction. Some attempts to do this have been made without much success. Immunological difficulties are encountered, since the enzymes being supplied are antigenic and the body produces antibodies against them.

2. Cure for inherited anaemia. Scientists studying the two anaemias of human beings, **Cooley's** and **Lepore anaemia**, resulting from abnormally low haemoglobin production in individuals homozygous for mutations at the beta-chain locus, hope that they may someday discover the factors regulating the beta gene activity and thereby increase the amount of beta chain synthesis. Furthermore, if the mechanism regulating haemoglobin synthesis in the foetus and the adult can be detected, they might be able to cure lethal conditions, such as **thalassemia major** and **sickle cell anaemia**, by suppressing synthesis of the abnormal beta chains and allowing foetal or gamma chains to be produced instead. In fact, individuals have been discovered whose β chain locus for some unknown reason never becomes active. Such individuals function normally with foetal haemoglobin even as adults.

3. Increasing role of genetics to medicine. The increasing number of human diseases that are being discovered to have a genetic basis lend great importance to the development of such euphenic measures. Three per cent of all humans have hereditary diseases which are transmitted in a Mendelian fashion. Two diseases which account for the deaths of hundreds of thousands a year, **cancer** and **heart disease**, are thought to have some heritable component. The future works of immunogenetics can suggest the ways by which an individual having genes for cancer may develop resistance for this disease.

A study has shown that some forms of heart diseases may be inherited as an autosomal dominant trait (**L.K. Altman**, 1972). An understanding of the genetic basis of such heart diseases would alert persons from families with an incidence of the disease to the possibility of incurring heart conditions and perhaps cause them to alter their diets and life habits accordingly. For example, those who may inherit genes for heart trouble would then quit smoking and avoid high fat diets.

Harmful evolutionary effect of euphenics. To restore affected individuals to normalcy by euphenic measures is the only compassionate goal for scientists to maintain but to do so is to counter the forces of natural selection that are the basis for the evolutionary strength of a species. PKU homozygotes, for example, would normally not reproduce and transmit the harmful mutations to future generations. Selection against them would be total; however, when they do develop normally and produce progeny, every member of the next generation would be a carrier of the mutation, assuming that the wife of such person being normal. This has to add to our genetic load and to weaken the human species from the evolutionary point of view.

Gene Targeting and Gene Therapy

Gene therapy is a technique of gene targeting which is an important tool of **genetic engineering** or **recombinant DNA technology**.

Gene targeting is a form of *in vivo* site-directed mutagenesis involving homologous recombination between a targeting **vector** containing one allele and an endogenous gene represented by a different allele. Following two types of targeting vectors are used:

1. Integration vectors (end-in vectors). Where cleavage within the homology domain stimulates a single crossover resulting in integration of the entire vector.

2. Transplacement vectors (end-out vectors). Where linearization occurs outside the homology domain or a double cross-over or gene-conversion event within the homology domain replaces part of the genome with the homologous region of vector.

The most efficient gene transfer vectors are based on mammalian viruses, specifically **retroviruses, adenoviruses** and the **adeno-associated viruses** (which integrate into the genome and may therefore mediate permanent correction of genetic defects) and the **herpes viruses** (which are neurotrophic and remain episomal). More direct approaches of gene transfer include **injection** (into muscles) and **microballistic techniques** (in which tungsten or gold particles coated with DNA are fired into cells at high velocity using a **gene gun**). A transfer procedure based on the packaging of DNA in **liposomes** is also popular in **cancer gene therapy**. **Aerosols** are used to introduce recombinant viral vectors into the lungs (they are internalized by receptor-mediated endocytosis) and are used for the treatment of **cystic fibrosis**.

Applications of Gene Targeting

Gene targeting has the following applications in genetic engineering:

1. Gene knockout or targeted disruption. It can be achieved by inserting a **cassette** (*i.e.*, a segment of exchangeable genetic information) anywhere in the integration vector or within the homology domain of transplacement vector. This cassette is usually a dominant selectable marker, such as bacterial *neo* gene, which allows selection of targeted cells.

2. Allele replacement. One allele is replaced by another *e.g.*, to investigate the effects of a implied mutation.

3. Gene knock-in. It is a novel application where one gene is replaced by another (non-allelic) gene.

4. Gene therapy. In this case, a mutant non-functional allele is replaced by a normal allele.

Scope of Gene Therapy

Gene therapy can involve genetic modification of cells in a living patient (*in vivo* **gene therapy**) or the genetic modification of cultured cells which are then returned to the patient (*ex vivo* **gene therapy**). *In vivo* gene therapy involves both genetic modification of target cells (**somatic transgenesis**) and therapeutic use of DNA as an epigenetic treatment (without changing the nucleotide sequence). The therapy can be used to improve or correct conditions caused by human gene mutuation or to prevent infectious diseases (*e.g.*, by interfering with the life cycle of a virus).

1. Gene augmentation therapy (GAT). In this technique DNA is added to the genome to replace a lost function. Its aim is to correct a loss of function effect, *e.g.*, caused by a **deletion**, by adding functional alleles. Transferred genes may be stably integrated into the genome (in which case there is the potential to permanently correct the defect, especially if stem cells are transformed) or may be maintained **episomally** (in which case there is an inevitable decay in the maintenance of gene expression and treatment may need to be repeated). Several GAT clinical trials are currently underway including **cystis fibrosis, adenosine deaminase deficiency** and **familial hypercholesterolemia**.

2. Gene inhibition therapy at the nucleic acid level. It involves treatment of disease by **targeted correction** (*i.e.*, allele replacement) or **gene knock-out** to remove the mutant allele. A novel approach which may play a role in gene therapy in the future is targeted correction at the RNA level by using **ribozymes** (Box 90.1) or RNA editing enzymes to correct pathogenic mRNA.

Use of antisense RNA. A alternative approach, which is presently undergoing clincial trials for the treatment of several types of cancer, is the use of nucleic acids to inhibit gene expression. The introduction of antisense genes allows the stable and permanent expression of **antisense RNA** (see Box 90.1), which binds to (mutant) mRNA and prevents translation (it may also target the mutant mRNA for degradation). Further more, increasing use is being made of **antisense constructs** containing ribozymes which degrade the mRNAs for **epigenetic gene therapy** (*i.e.*, therapy which does not involve changes in the genome).

3. Gene inhibition therapy at the protein level. Targeted inhibition of gene expression can also take effect at the protein level, by expression within a cell of genetically engineered antibodies (intrabodies) which bind to and inactivate mutant proteins. A novel approach is to use degenerate oligonucleotides to identify specific oligonucleotide sequences which interact with proteins. These oligonucleotides are called **aptamers** and can be used to inactivate specific mutant proteins.

Box 90.1

1. **Antisense RNA** (or **micRNA = mRNA-interfering complementary RNA**)	Antisense RNA is complementary to mRNA and can form a duplex with it to block protein synthesis. Naturally occurring antisense RNA is found in many systems but predominantly in bacteria, and is termed as mRNA-interfering complementary RNA.
2. **Ribozyme**	RNA molecule which can catalyze chemical reactions (RNA enzymes).

QUESTIONS

Long Answer Questions

1. What is eugenics? How does it differ from euthenics? Describe different aspects of positive and negative eugenics. (*CCS 1994, 1999*)
2. What is eugenics? Write an essay on future of mankind. (*Guwahati (M) 1994*)
3. Write an essay on eugenics. (*Kanpur 1993, 1998*)
4. What do you understand by eugenics? In what manner can eugenics improve the human race. (*Kanpur 1996*)
5. What is eugenics? Explain its importance. (*Bangalore 1993*)
6. Give an account of negative eugenics. (*Bangalore 1994*)

Short Answer Questions

1. Give a brief account of targeted gene modification and gene therapy.
2. Write short notes on the following:
   (*i*) Genetic counselling (*C.C.S.U.Meerut 2007*)
   (*ii*) Eugenics (*Punjab 1990, 1991; Lucknow 1994; Kanpur 1995; H.N.B. Garhwal 2008*)
   (*iii*) Euthenics (*Kanpur 1995, 1997, 1998*)
   (*iv*) Euphenics (*Kanpur 1994*)
3. Differentiate between the following:
   (*i*) Eugenics and Euthenics (*Lucknow 1998*)
   (*ii*) Genetic engineering and Genetic counselling (*C.C.S.U. Meerut 2008*)